SEAS AT THE MILLENNIUM:
AN ENVIRONMENTAL EVALUATION

SEAS AT THE MILLENNIUM: AN ENVIRONMENTAL EVALUATION

Edited by

Charles R.C. Sheppard
*Department of Biological Sciences,
University of Warwick,
Coventry, U.K.*

**Volume II
Regional Chapters:
The Indian Ocean to The Pacific**

2000
PERGAMON
An imprint of Elsevier Science

AMSTERDAM – LAUSANNE – NEW YORK – OXFORD – SHANNON – SINGAPORE – TOKYO

ELSEVIER SCIENCE Ltd
The Boulevard, Langford Lane
Kidlington, Oxford OX5 1GB, UK

© 2000 Elsevier Science Ltd. All rights reserved.

This work is protected under copyright by Elsevier Science, and the following terms and conditions apply to its use:

Photocopying
Single photocopies of single chapters may be made for personal use as allowed by national copyright laws. Permission of the Publisher and payment of a fee is required for all other photocopying, including multiple or systematic copying, copying for advertising or promotional purposes, resale, and all forms of document delivery. Special rates are available for educational institutions that wish to make photocopies for non-profit educational classroom use.

Permissions may be sought directly from Elsevier Science Global Rights Department, PO Box 800, Oxford OX5 1DX, UK; phone: (+44) 1865 843830, fax: (+44) 1865 853333, e-mail: permissions@elsevier.co.uk. You may also contact Global Rights directly through Elsevier's home page (http://www.elsevier.nl), by selecting 'Obtaining Permissions'.

In the USA, users may clear permissions and make payments through the Copyright Clearance Center, Inc., 222 Rosewood Drive, Danvers, MA 01923, USA; phone: (978) 7508400, fax: (978) 7504744, and in the UK through the Copyright Licensing Agency Rapid Clearance Service (CLARCS), 90 Tottenham Court Road, London W1P 0LP, UK; phone: (+44) 20 7631 5555; fax: (+44) 20 7631 5500. Other countries may have a local reprographic rights agency for payments.

Derivative Works
Tables of contents may be reproduced for internal circulation, but permission of Elsevier Science is required for external resale or distribution of such material.
Permission of the Publisher is required for all other derivative works, including compilations and translations.

Electronic Storage or Usage
Permission of the Publisher is required to store or use electronically any material contained in this work, including any chapter or part of a chapter.

Except as outlined above, no part of this work may be reproduced, stored in a retrieval system or transmitted in any form or by any means, electronic, mechanical, photocopying, recording or otherwise, without prior written permission of the Publisher.
Address permissions requests to: Elsevier Science Global Rights Department, at the mail, fax and e-mail addresses noted above.

Notice
No responsibility is assumed by the Publisher for any injury and/or damage to persons or property as a matter of products liability, negligence or otherwise, or from any use or operation of any methods, products, instructions or ideas contained in the material herein. Because of rapid advances in the medical sciences, in particular, independent verification of diagnoses and drug dosages should be made.

First edition 2000

Library of Congress Cataloging in Publication Data
Seas at the millennium: an environmental evaluation / edited by Charles Sheppard. -- 1st ed.
  p. cm.
  Includes bibliographical references.
  ISBN 0-08-043207-7 (alk. paper)
  1. Marine ecology. I. Sheppard, Charles (Charles R.C.)
QH541.5.S3 S35 2000
577.7--dc21
00-034738

British Library Cataloguing in Publication Data
Seas at the millennium: an environmental evaluation
1. Marine ecology 2. Coastal ecology 3. Marine resources conservation
I. Sheppard, Charles R.C.
577.7

ISBN: 0-08-043207-7

∞ The paper used in this publication meets the requirements of ANSI/NISO Z39.48-1992 (Permanence of Paper).

Printed in The Netherlands.

Contents

List of Authors xi

Acknowledgements xxi

Volume I. Regional Chapters: Europe, The Americas and West Africa

INTRODUCTION. 1
    Charles Sheppard

Chapter 1. THE SEAS AROUND GREENLAND 5
    Frank Riget, Poul Johansen, Henning Dahlgaard, Anders Mosbech, Rune Dietz and Gert Asmund

Chapter 2. THE NORWEGIAN COAST 17
    Jens Skei, Torgeir Bakke and Jarle Molvaer

Chapter 3. THE FAROE ISLANDS 31
    Maria Dam, Grethe Bruntse, Andrias Reinert and Jacob Pauli Joensen

Chapter 4. THE NORTH SEA 43
    Jean-Paul Ducrotoy, Mike Elliott and Victor N. de Jonge

Chapter 5. THE ENGLISH CHANNEL 65
    Alan D. Tappin and P. Chris Reid

Chapter 6. THE IRISH SEA 83
    Richard G. Hartnoll

Chapter 7. THE BALTIC SEA, ESPECIALLY SOUTHERN AND EASTERN REGIONS 99
    Jerzy Falandysz, Anna Trzosinska, Piotr Szefer, Jan Warzocha and Bohdan Draganik

Chapter 8. THE BALTIC SEA, INCLUDING BOTHNIAN SEA AND BOTHNIAN BAY. 121
    Lena Kautsky and Nils Kautsky

Chapter 9. THE NORTH COAST OF SPAIN 135
    Isabel Díez, Antonio Secilla, Alberto Santolaria and José María Gorostiaga

Chapter 10. SOUTHERN PORTUGAL: THE TAGUS AND SADO ESTUARIES 151
    Graça Cabeçadas, Maria José Brogueira and Leonor Cabeçadas

Chapter 11. THE ATLANTIC COAST OF SOUTHERN SPAIN 167
    Carlos J. Luque, Jesús M. Castillo and M. Enrique Figueroa

Chapter 12. THE CANARY ISLANDS 185
    Francisco García Montelongo, Carlos Díaz Romero and Ricardo Corbella Tena

Chapter 13. THE AZORES 201
    Brian Morton and Joseph C. Britton

Chapter 14. THE SARGASSO SEA AND BERMUDA 221
    Anthony H. Knap, Douglas P. Connelly and James N. Butler

Chapter 15. THE AEGEAN SEA 233
    Manos Dassenakis, Kostas Kapiris and Alexandra Pavlidou

Chapter 16. THE COAST OF ISRAEL, SOUTHEAST MEDITERRANEAN 253
    Barak Herut and Bella Galil

Chapter 17. THE ADRIATIC SEA AND THE TYRRHENIAN SEA 267
    Giuseppe Cognetti, Claudio Lardicci, Marco Abbiati and Alberto Castelli

Chapter 18. THE BLACK SEA 285
    Gülfem Bakan and Hanife Büyükgüngör

Chapter 19. THE GULF OF MAINE AND GEORGES BANK. 307
    Jack B. Pearce

Chapter 20. THE NEW YORK BIGHT 321
    Jack B. Pearce

Chapter 21. CHESAPEAKE BAY: THE UNITED STATES' LARGEST ESTUARINE SYSTEM 335
   Kent Mountford

Chapter 22. NORTH AND SOUTH CAROLINA COASTS 351
   Michael A. Mallin, JoAnn M. Burkholder, Lawrence B. Cahoon and Martin H. Posey

Chapter 23. THE GULF OF ALASKA. 373
   Bruce A. Wright, Jeffrey W. Short, Tom J. Weingartner and Paul J. Anderson

Chapter 24. SOUTHERN CALIFORNIA 385
   Kenneth C. Schiff, M. James Allen, Eddy Y. Zeng and Steven M. Bay

Chapter 25. FLORIDA KEYS. 405
   Phillip Dustan

Chapter 26. THE BAHAMAS. 415
   Kenneth C. Buchan

Chapter 27. THE NORTHERN GULF OF MEXICO 435
   Mark E. Pattillo and David M. Nelson

Chapter 28. COASTAL MANAGEMENT IN LATIN AMERICA. 457
   Alejandro Yáñez-Arancibia

Chapter 29. SOUTHERN GULF OF MEXICO 467
   Felipe Vázquez, Ricardo Rangel, Arturo Mendoza Quintero-Marmol, Jorge Fernández, Eduardo Aguayo, E.A. Palacio and Virender K. Sharma

Chapter 30. THE PACIFIC COAST OF MEXICO. 483
   Alfonso V. Botello, Alejandro O. Toledo, Guadalupe de la Lanza-Espino and Susana Villanueva-Fragoso

Chapter 31. BELIZE 501
   Alastair R. Harborne, Melanie D. McField and E. Kate Delaney

Chapter 32. NICARAGUA: CARIBBEAN COAST. 517
   Stephen C. Jameson, Lamarr B. Trott, Michael J. Marshall and Michael J. Childress

Chapter 33. NICARAGUA: PACIFIC COAST. 531
   Stephen C. Jameson, Vincent F. Gallucci and José A. Robleto

Chapter 34. EL SALVADOR 545
   Linos Cotsapas, Scott A. Zengel and Enrique J. Barraza

Chapter 35. JAMAICA 559
   Marjo Vierros

Chapter 36. PUERTO RICO 575
   Jack Morelock, Jorge Capella, Jorge Garcia and Maritza Barreto

Chapter 37. THE TURKS AND CAICOS ISLANDS 587
   Gudrun Gaudian and Paul Medley

Chapter 38. THE DUTCH ANTILLES 595
   Adolphe O. Debrot and Jeffrey Sybesma

Chapter 39. UK OVERSEAS TERRITORIES IN THE NORTHEAST CARIBBEAN: ANGUILLA, BRITISH VIRGIN ISLANDS, MONTSERRAT 615
   Fiona Gell and Maggie Watson

Chapter 40. THE LESSER ANTILLES, TRINIDAD AND TOBAGO 627
   John B.R. Agard and Judith F. Gobin

Chapter 41. VENEZUELA 643
   Pablo E. Penchaszadeh, César A. Leon, Haymara Alvarez, David Bone, P. Castellano, María M. Castillo, Yusbelly Diaz, María P. Garcia, Mairin Lemus, Freddy Losada, Alberto Martin, Patricia Miloslavich, Claudio Paredes, Daisy Perez, Miradys Sebastiani, Dennise Stecconi, Victoriano Roa and Alicia Villamizar

Chapter 42. THE CARIBBEAN COAST OF COLOMBIA 663
   Leonor Botero and Ricardo Alvarez-León

Chapter 43. THE PACIFIC COAST OF COLOMBIA 677
   Alonso J. Marrugo-González, Roberto Fernández-Maestre and Anders A. Alm

Chapter 44. PERU 687
   Guadalupe Sanchez

Chapter 45. THE CHILEAN COAST 699
   Ramón B. Ahumada, Luis A. Pinto and Patricio A. Camus

Chapter 46. TROPICAL COAST OF BRAZIL 719
    Zelinda M.A.N. Leão and José M.L. Dominguez

Chapter 47. SOUTHERN BRAZIL 731
    Eliete Zanardi Lamardo, Márcia Caruso Bícego, Belmiro Mendes de Castro Filho, Luiz Bruner de Miranda and Valéria Aparecida Prósperi

Chapter 48. THE ARGENTINE SEA: THE SOUTHEAST SOUTH AMERICAN SHELF MARINE ECOSYSTEM . 749
    José L. Esteves, Nestor F. Ciocco, Juan C. Colombo, Hugo Freije, Guillermo Harris, Oscar Iribarne, Ignacio Isla, Paulina Nabel, Marcela S. Pascual, Pablo E. Penchaszadeh, Andrés L. Rivas and Norma Santinelli

Chapter 49. THE GULF OF GUINEA LARGE MARINE ECOSYSTEM 773
    Nicholas J. Hardman-Mountford, Kwame A. Koranteng and Andrew R.G. Price

Chapter 50. GUINEA 797
    Ibrahima Cisse, Idrissa Lamine Bamy, Amadou Bah, Sékou Balta Camara and Mamba Kourouma

Chapter 51. CÔTE D'IVOIRE 805
    Ama Antoinette Adingra, Robert Arfi and Aka Marcel Kouassi

Chapter 52. SOUTHWESTERN AFRICA: NORTHERN BENGUELA CURRENT REGION 821
    David Boyer, James Cole and Christopher Bartholomae

Index 841

Volume II. Regional Chapters: The Indian Ocean to The Pacific

Chapter 53. THE ARABIAN GULF 1
    D.V. Subba Rao and Faiza Al-Yamani

Chapter 54. NORTHWEST ARABIAN SEA AND GULF OF OMAN 17
    Simon C. Wilson

Chapter 55. THE RED SEA 35
    Charles R.C. Sheppard

Chapter 56. THE GULF OF ADEN 47
    Simon C. Wilson and Rebecca Klaus

Chapter 57. THE INDIAN OCEAN COAST OF SOMALIA 63
    Federico Carbone and Giovanni Accordi

Chapter 58. TANZANIA 83
    Martin Guard, Aviti J. Mmochi and Chris Horrill

Chapter 59. MOZAMBIQUE 99
    Michael Myers and Mark Whittington

Chapter 60. MADAGASCAR 113
    Andrew Cooke, Onésime Ratomahenina, Eulalie Ranaivoson and Haja Razafindrainibe

Chapter 61. SOUTH AFRICA 133
    Michael H. Schleyer, Lynnath E. Beckley, Sean T. Fennessy, Peter J. Fielding, Anesh Govender, Bruce Q. Mann, Wendy D. Robertson, Bruce J. Tomalin and Rudolph P. van der Elst

Chapter 62. THE NORTHWEST COAST OF THE BAY OF BENGAL AND DELTAIC SUNDARBANS 145
    Abhijit Mitra

Chapter 63. SOUTHEAST INDIA 161
    Sundararajan Ramachandran

Chapter 64. SRI LANKA 175
    Arjan Rajasuriya and Anil Premaratne

Chapter 65. THE ANDAMAN, NICOBAR AND LAKSHADWEEP ISLANDS 189
    Sundararajan Ramachandran

Chapter 66. THE MALDIVES 199
    Andrew R.G. Price and Susan Clark

Chapter 67. THE CHAGOS ARCHIPELAGO, CENTRAL INDIAN OCEAN 221
    Charles R.C. Sheppard

Chapter 68. THE SEYCHELLES 233
    Miles Gabriel, Suzanne Marshall and Simon Jennings

Chapter 69. THE COMOROS ARCHIPELAGO.. 243
    Jean-Pascal Quod, Odile Naim and Fouad Abdourazi
Chapter 70. THE MASCARENE REGION 253
    John Turner, Colin Jago, Deolall Daby and Rebecca Klaus
Chapter 71. THE BAY OF BENGAL 269
    D.V. Subba Rao
Chapter 72. BANGLADESH 285
    Dihider Shahriar Kabir, Syed Mazharul Islam, Md. Giasuddin Khan, Md. Ekram Ullah and Dulal C. Halder
Chapter 73. THE GULF OF THAILAND 297
    Manuwadi Hungspreugs, Wilaiwan Utoomprurkporn and Charoen Nitithamyong
Chapter 74. THE MALACCA STRAITS 309
    Chua Thia-Eng, Ingrid R.L. Gorre, S. Adrian Ross, Stella Regina Bernad, Bresilda Gervacio and Ma. Corazon Ebarvia
Chapter 75. MALACCA STRAIT INCLUDING SINGAPORE AND JOHORE STRAITS 331
    Poh Poh Wong
Chapter 76. EAST COAST OF PENINSULAR MALAYSIA 345
    Zelina Z. Ibrahim, Aziz Arshad, Lee Say Chong, Japar Sidik Bujang, Law Ah Theem, Nik Mustapha Raja Abdullah and Maged Mahmoud Marghany
Chapter 77. BORNEO 361
    Steve Oakley, Nicolas Pilcher and Elizabeth Wood
Chapter 78. CONTINENTAL SEAS OF WESTERN INDONESIA 381
    Evan Edinger and David R. Browne
Chapter 79. THE PHILIPPINES 405
    Gil S. Jacinto, Porfirio M. Aliño, Cesar L. Villanoy, Liana Talaue-McManus and Edgardo D. Gomez
Chapter 80. THE CORAL, SOLOMON AND BISMARCK SEAS REGION 425
    Michael E. Huber and Graham B.K. Baines
Chapter 81. ASIAN DEVELOPING REGIONS: PERSISTENT ORGANIC POLLUTANTS IN THE SEAS 447
    Shinsuke Tanabe
Chapter 82. SEA OF OKHOTSK 463
    Victor V. Lapko and Vladimir I. Radchenko
Chapter 83. SEA OF JAPAN 473
    Anatoly N. Kachur and Alexander V. Tkalin
Chapter 84. THE YELLOW SEA 487
    Suam Kim and Sung-Hyun Kahng
Chapter 85. TAIWAN STRAIT 499
    Woei-Lih Jeng, Chang-Feng Dai and Kuang-Lung Fan
Chapter 86. XIAMEN REGION, CHINA 513
    Chua Thia-Eng and Ingrid Rosalie L. Gorre
Chapter 87. HONG KONG 535
    Bruce J. Richardson, Paul K.S. Lam and Rudolf S.S. Wu
Chapter 88. SOUTHERN CHINA, VIETNAM TO HONG KONG 549
    Zhang Gan, Zou Shicun and Yan Wen
Chapter 89. VIETNAM AND ADJACENT BIEN DONG (SOUTH CHINA SEA) 561
    Dang Duc Nhan, Nguyen Xuan Duc, Do Hoai Duong, Nguyen The Tiep and Bui Cong Que
Chapter 90. CAMBODIAN SEA 569
    Touch Seang Tana
Chapter 91. THE AUSTRALIAN REGION: AN OVERVIEW 579
    Leon P. Zann
Chapter 92. TORRES STRAIT AND THE GULF OF PAPUA 593
    Michael E. Huber
Chapter 93. NORTHEASTERN AUSTRALIA: THE GREAT BARRIER REEF REGION ... 611
    Leon P. Zann
Chapter 94. THE EASTERN AUSTRALIAN REGION: A DYNAMIC TROPICAL/TEMPERATE BIOTONE 629
    Leon P. Zann

Chapter 95. THE TASMANIAN REGION 647
    Christine M. Crawford, Graham J. Edgar and George Cresswell
Chapter 96. VICTORIA PROVINCE, AUSTRALIA 661
    Tim D. O'Hara
Chapter 97. THE GREAT AUSTRALIAN BIGHT 673
    Karen Edyvane
Chapter 98. THE WESTERN AUSTRALIAN REGION 691
    Diana I. Walker
Chapter 99. THE SOUTH WESTERN PACIFIC ISLANDS REGION 705
    Leon P. Zann and Veikila Vuki
Chapter 100. NEW CALEDONIA 723
    Pierre Labrosse, Renaud Fichez, Richard Farman and Tim Adams
Chapter 101. VANUATU 737
    Veikila C. Vuki, Subashni Appana, Milika R. Naqasima and Maika Vuki
Chapter 102. THE FIJI ISLANDS 751
    Veikila C. Vuki, Leon P. Zann, Milika Naqasima and Maika Vuki
Chapter 103. THE CENTRAL SOUTH PACIFIC OCEAN (AMERICAN SAMOA) 765
    Peter Craig, Suesan Saucerman and Sheila Wiegman
Chapter 104. THE MARSHALL ISLANDS 773
    Andrew R.G. Price and James E. Maragos
Chapter 105. HAWAIIAN ISLANDS (U.S.A.) 791
    James E. Maragos
Chapter 106. FRENCH POLYNESIA 813
    Pat Hutchings and Bernard Salvat
Index 827

Volume III. Global Issues and Processes

Chapter 107. GLOBAL STATUS OF SEAGRASSES 1
    Ronald C. Phillips and Michael J. Durako
Chapter 108. MANGROVES 17
    Colin D. Field
Chapter 109. CORAL REEFS: ENDANGERED, BIODIVERSE, GENETIC RESOURCES 33
    Walter H. Adey, Ted A. McConnaughey, Allegra M. Small and Don M. Spoon
Chapter 110. WORLD-WIDE CORAL REEF BLEACHING AND MORTALITY DURING 1998:
    A GLOBAL CLIMATE CHANGE WARNING FOR THE NEW MILLENNIUM? 43
    Clive R. Wilkinson
Chapter 111. SEA TURTLES 59
    Jeanne A. Mortimer, Marydele Donnelly and Pamela T. Plotkin
Chapter 112. WHALES AND WHALING 73
    Sidney Holt
Chapter 113. SMALL CETACEANS: SMALL WHALES, DOLPHINS AND PORPOISES 89
    Kieran Mulvaney and Bruce McKay
Chapter 114. SEABIRDS 105
    W.R.P. Bourne and C.J. Camphuysen
Chapter 115. FISHERIES EFFECTS ON ECOSYSTEMS 117
    Raquel Goñi
Chapter 116. BY-CATCH: PROBLEMS AND SOLUTIONS 135
    Martin A. Hall, Dayton L. Alverson and Kaija I. Metuzals
Chapter 117. FISHERIES MANAGEMENT AS A SOCIAL PROBLEM 153
    Douglas C. Wilson
Chapter 118. FARMING OF AQUATIC ORGANISMS, PARTICULARLY THE CHINESE AND THAI
    EXPERIENCE 165
    Krishen J. Rana and Anton J. Immink

Chapter 119. CLIMATIC CHANGES: GULF OF ALASKA 179
    Howard Freeland and Frank Whitney
Chapter 120. EFFECTS OF CLIMATE CHANGE AND SEA LEVEL ON COASTAL SYSTEMS 187
    Shiao-Kung Liu
Chapter 121. PARTICLE DRY DEPOSITION TO WATER SURFACES: PROCESSES AND CONSEQUENCES . . 197
    Sara C. Pryor and Rebecca J. Barthelmie
Chapter 122. MARINE ECOSYSTEM HEALTH AS AN EXPRESSION OF MORBIDITY, MORTALITY
    AND DISEASE EVENTS 211
    Benjamin H. Sherman
Chapter 123. EFFECT OF MINE TAILINGS ON THE BIODIVERSITY OF THE SEABED:
    EXAMPLE OF THE ISLAND COPPER MINE, CANADA 235
    Derek V. Ellis
Chapter 124. MARINE ANTIFOULANTS 247
    Stewart M. Evans
Chapter 125. EUTROPHICATION OF MARINE WATERS: EFFECTS ON BENTHIC MICROBIAL
    COMMUNITIES 257
    Lutz-Arend Meyer-Reil and Marion Köster
Chapter 126. PERSISTENCE OF SPILLED OIL ON SHORES AND ITS EFFECTS ON BIOTA 267
    Gail V. Irvine
Chapter 127. REMOTE SENSING OF TROPICAL COASTAL RESOURCES: PROGRESS AND
    FRESH CHALLENGES FOR THE NEW MILLENNIUM 283
    Peter J. Mumby
Chapter 128. SATELLITE REMOTE SENSING OF THE COASTAL OCEAN: WATER QUALITY
    AND ALGAE BLOOMS 293
    Bertil Håkansson
Chapter 129. ENERGY FROM THE OCEANS: WIND, WAVE AND TIDAL 303
    Rebecca J. Barthelmie, Ian Bryden, Jan P. Coelingh and Sara C. Pryor
Chapter 130. MULTINATIONAL TRAINING PROGRAMMES IN MARINE ENVIRONMENTAL SCIENCE . . . 323
    G. Robin South
Chapter 131. GLOBAL LEGAL INSTRUMENTS ON THE MARINE ENVIRONMENT AT THE YEAR 2000 . . . 331
    Milen F. Dyoulgerov
Chapter 132. COASTAL MANAGEMENT IN THE FUTURE 349
    Derek J. McGlashan
Chapter 133. SUSTAINABILITY OF HUMAN ACTIVITIES ON MARINE ECOSYSTEMS 359
    Paul Johnston, David Santillo, Julie Ashton and Ruth Stringer
Chapter 134. MARINE RESERVES AND RESOURCE MANAGEMENT 375
    Michael J. Fogarty, James A. Bohnsack and Paul K. Dayton
Chapter 135. THE ECOLOGICAL, ECONOMIC, AND SOCIAL IMPORTANCE OF THE OCEANS 393
    Robert Costanza

Index 405

Contributing Authors

Marco Abbiati
Dipartimento di Biologia Evoluzionistica, Università di Bologna, Via Tombesi dell'Ova 55, I-48100 Ravenna, Italy

Fouad Abdourazi
AIDE, Minizi, Mavouna, Moroni, Comoros

Nik Mustapha Raja Abdullah
Universiti Putra Malaysia, 43400 UPM Serdang, Malaysia

Giovanni Accordi
Centro di Studio per il Quaternario e l'Evoluzione Ambientale, C.N.R., Dipartimento di Scienze della Terra, Università degli Studi "La Sapienza", P. Aldo Moro, 5, 00185 Roma, Italy

Tim Adams
Secretariat of the Pacific Community (SPC), B.P. D5, 98848 Nouméa Cedex, New Caledonia

Walter H. Adey
Marine Systems Laboratory, Smithsonian Institution, NHB E-117, MRC 164, Washington, DC 20560, U.S.A.

Ama Antoinette Adingra
Centre de Recherches Océanologiques, BP V18, Abidjan, Côte d'Ivoire

John B.R. Agard
Department of Life Sciences, The University of the West Indies, St. Augustine, Trinidad and Tobago

Eduardo Aguayo
Instituto de Ciencias del Mar y Limnología, UNAM, Cd. Universitaria, A.P. 70-305, Mexico City, C.P. 04510 Mexico

Ramón B. Ahumada
Facultad de Ciencias, Universidad Católica de la Santísima Concepción, Campus San Andrés, Paicaví 3000, Casilla 297, Concepción, Chile

Faiza Al-Yamani
Kuwait Institute of Scientific Research, P.O. Box 1638, 22017 Salmiya, Kuwait

Porfirio M. Aliño
Marine Science Institute, University of the Philippines, 1101 Diliman, Quezon City, Philippines

M. James Allen
Southern California Coastal Water Research Project, 7171 Fenwick Lane, Westminster, CA 92683, U.S.A.

Anders A. Alm
Universidad de Cartagena, Facultad de Ciencias Químicas y Farmacéuticas, Zaragocilla, AA 1661, Cartagena, Colombia

Ricardo Alvarez-León
Universidad de la Sabana, Depto. Ciencias de la Vida, Campus Universitario Puente del Común, Edif. E-2, of. 232, Chía, Cundi., Colombia

Haymara Alvarez
Universidad Simon Bolivar, Apdo 89000, Caracas 1080-A, Venezuela

Dayton L. Alverson
Natural Resources Consultants, 1900 West Nickerson St., Suite 207, Seattle, WA 98119, U.S.A.

Paul J. Anderson
Kodiak Laboratory, National Marine Fisheries Service, National Oceanic and Atmospheric Administration, P.O. Box 1638, Kodiak, AK 99615, U.S.A.

Subashni Appana
Marine Studies Programme, University of the South Pacific, P.O. Box 1168, Suva, Fiji

Robert Arfi
Centre de Recherches Océanologiques, BP V18, Abidjan, Côte d'Ivoire

Aziz Arshad
Universiti Putra Malaysia, 43400 UPM Serdang, Malaysia

Julie Ashton
Greenpeace Research Laboratories, University of Exeter, Exeter EX4 4PS, U.K.

Gert Asmund
National Environmental Research Institute, Department of Arctic Environment, P.O. Box 358, DK-4000 Roskilde, Denmark

Amadou Bah
The National Center of Halieutic Sciences of Boussoura, B.P. 3060, Conakry, Republic of Guinea

Graham B.K. Baines
Environment Pacific, 3 Pindari St., The Gap, Brisbane, QLD 4061, Australia

Gülfem Bakan
Ondokuz Mayis University, Faculty of Engineering, Department of Environmental Engineering, 55139 Kurupelit, Samsun, Turkey

Torgeir Bakke
Norwegian Institute for Water Research (NIVA), P.O. Box 173 Kjelsaas, 0411 Oslo, Norway

Idrissa Lamine Bamy
The National Center of Halieutic Sciences of Boussoura, B.P. 3060, Conakry, Republic of Guinea

Enrique J. Barraza
Ministry of the Environment, San Salvador, El Salvador

Maritza Barreto
University of Puerto Rico Rio Piedras, Geography Department, Rio Piedras, Puerto Rico

Rebecca J. Barthelmie
Department of Wind Energy and Atmospheric Physics, Risø National Laboratory, DK-4000 Roskilde, Denmark

Christopher Bartholomae
National Marine Information and Research Centre, P.O. Box 912, Swakopmund, Namibia

Steven M. Bay
Southern California Coastal Water Research Project, 7171 Fenwick Lane, Westminster, CA 92683, U.S.A.

Lynnath E. Beckley
SAAMBR, P.O. Box 10712, Marine Parade, KwaZulu-Natal 4056, South Africa

Stella Regina Bernad
GEF/UNDP/IMO Regional Programme on Partnerships in Environmental Management for the Seas of East Asia (PEMSEA), P.O. Box 2502, Quezon City, 1165 Metro Manila, Philippines

Márcia Caruso Bícego
Universidade de São Paulo, Dept. Oceanografia Física do Instituto Oceanográfico, Pca do Oceanográfico, 191, Cidade Universitária, SP, 05508-900, Brazil

James A. Bohnsack
National Marine Fisheries Service, Southeast Fisheries Science Center, 75 Virginia Beach Drive, Miami, FL 33149, U.S.A.

David Bone
Universidad Simon Bolivar, Apartado 89000, Caracas 1080-A, Venezuela

Alfonso V. Botello
Institute for Marine and Limnology Sciences, National Autonomous University of Mexico, Marine Pollution Laboratory, P.O. Box 70305, México City 04510 D.F., Mexico

Leonor Botero
COLCIENCIAS, Trans. 9A # 133-28, Santafé de Bogotá, Colombia

W.R.P. Bourne
Department of Zoology, Aberdeen University, Tillydrone Avenue, Aberdeen AB24 2TZ, Scotland

David Boyer
National Marine Information and Research Centre, P.O. Box 912, Swakopmund, Namibia

Joseph C. Britton
Department of Biology, Texas Christian University, Fort Worth, Texas 76129, U.S.A.

Maria José Brogueira
Instituto de Investigação das Pescas e do Mar (IPIMAR), DAA, Av. Brasilia, 1400 Lisboa, Portugal.

David R. Browne
Department of Biology, McGill University, 1205 Docteur Penfield Avenue, Montreal, PQ, H3A 1B1 Canada

Grethe Bruntse
Kaldbak Marine Biological Laboratory, FO-180 Kaldbak, Faroe Islands

Ian Bryden
The Robert Gordon University, School of Mechanical and Offshore Engineering, Aberdeen, AB10 1FR, Scotland

Kenneth C. Buchan
Bahamian Field Station, San Salvador, Bahamas

Japar Sidik Bujang
Universiti Putra Malaysia, 43400 UPM Serdang, Malaysia

JoAnn M. Burkholder
Department of Botany, North Carolina State University, Raleigh, NC 27695-7612, U.S.A.

James N. Butler
Harvard University, Cambridge, Massachussetts, U.S.A.

Hanife Büyükgüngör
Ondokuz Mayis University, Faculty of Engineering, Department of Environmental Engineering, 55139 Kurupelit, Samsun, Turkey

Graça Cabeçadas
Instituto de Investigação das Pescas e do Mar (IPIMAR), DAA, Av. Brasilia, 1400 Lisboa, Portugal.

Leonor Cabeçadas
Direcção Geral do Ambiente (D.G.A.), R. da Murgueira, Zambujal, 2720 Amadora, Portugal

Lawrence B. Cahoon
Department of Biological Sciences, University of North Carolina-Wilmington, Wilmington, NC 28403, U.S.A.

Sékou Balta Camara
The National Center of Halieutic Sciences of Boussoura, B.P. 3060, Conakry, Republic of Guinea

C.J. Camphuysen
CSR Consultancy, Ankerstraat 20, 1794 BJ Oosterend, Texel, The Netherlands

Patricio A. Camus
Facultad de Ciencias, Universidad Católica de la Santísima Concepción, Campus San Andrés, Paicaví 3000, Casilla 297, Concepción, Chile

Jorge Capella
University of Puerto Rico R.U.M., Department of Marine Sciences, Mayagüez, Puerto Rico

Federico Carbone
Centro di Studio per il Quaternario e l'Evoluzione Ambientale, C.N.R., Dipartimento di Scienze della Terra, Università degli Studi "La Sapienza", P. Aldo Moro, 5, 00185 Roma, Italy

P. Castellano
Centro de Procesamiento de Imagenes Digitales (CPDI), Sartenejas, Caracas, Venezuela

Alberto Castelli
Dipartimento di Zoologia ed Antropologia, Corso Margherita di Savoia 15, I-17100 Sassari, Italy

Jesús M. Castillo
Departamento de Biología Vegetal y Ecología, Facultad de Biología, Universidad de Sevilla, Apdo 1095, 41080 Sevilla, Spain

María M. Castillo
Universidad Simon Bolivar, Apdo 89000, Caracas 1080-A, Venezuela

Michael J. Childress
Department of Biological Sciences, Idaho State University, Pocatello, ID 83209-8007, U.S.A.

Lee Say Chong
National Hydraulic Research Institute of Malaysia, Km 7 Jalan Ampang, 68000 Ampang, Malaysia

Nestor F. Ciocco
CENPAT-CONICET, Bv. Brown 3000, (9120) Puerto Madryn, Chubut, Argentina

Ibrahima Cisse
The National Center of Halieutic Sciences of Boussoura, B.P. 3060, Conakry, Republic of Guinea

Susan Clark
Department of Marine Sciences and Coastal Management, University of Newcastle upon Tyne, Newcastle upon Tyne, U.K.

Jan P. Coelingh
ECOFYS Energy and Environment, P.O. Box 8408, NL-3503 RK Utrecht, The Netherlands.

Giuseppe Cognetti
Dipartimento di Scienze dell'Uomo e dell'Ambiente, Università di Pisa, Via Volta 6, I-56124 Pisa, Italy

James Cole
2 Dolphin Cottage, 31 Penny St., Portsmouth PO1 2NH, U.K.

Juan C. Colombo
Química Ambiental y Bioquímica, Facultad de Ciencias Naturales y Museo, Universidad Nacional de La Plata, Paseo del Bosque s/n, (1900) La Plata, Argentina

Douglas P. Connelly
Bermuda Biological Station for Research, 17 Biological Station Lane, St. Georges, Bermuda, GE 01

Andrew Cooke
Cellule Environnement Marin et Côtier, Office National pour l'Environnement, B.P. 822, Antananarivo, Madagascar

Robert Costanza
Center for Environmental Science and Biology Department, and Institute for Ecological Economics, University of Maryland, Box 38, Solomons, MD 20688-0038, U.S.A.

Linos Cotsapas
Research Planning, Inc., 1121 Park St., Columbia, SC 29201, U.S.A.

Peter Craig
National Park of American Samoa, Pago Pago, American Samoa 96799, U.S.A.

Christine M. Crawford
Tasmanian Aquaculture and Fisheries Institute, University of Tasmania, Nubeena Crescent, Taroona, Tasmania 7053, Australia

George Cresswell
CSIRO Marine Research, Castray Esplanade, Hobart, Tasmania 7000, Australia

John Croxall
British Antarctic Survey, Natural Environment Research Council, High Cross, Madingley Road, Cambridge CB3 OET, U.K.

Deolall Daby
Faculty of Science, University of Mauritius, Reduit, Mauritius

Henning Dahlgaard
Risø National Laboratory, DK-4000 Roskilde, Denmark

Chang-Feng Dai
Institute of Oceanography, National Taiwan University, Taipei, Taiwan, Republic of China

Maria Dam
Food and Environmental Agency, Debesartrøð, FO-100 Tórshavn, Faroe Islands

Manos Dassenakis
University of Athens, Department of Chemistry, Division III, Inorganic and Environmental Chemistry, Panepistimiopolis, Kouponia, Athens 15771, Greece

Paul K. Dayton
Scripps Institution of Oceanography, 9500 Gilman Dr., La Jolla, CA 92093, U.S.A.

Adolpe O. Debrot
Carmabi Foundation, Piscaderabaai, P.O. Box 2090, Curaçao, Netherlands Antilles

E. Kate Delaney
Department of Geography, University of Southampton, Southampton, SO17 1BJ, U.K.

Tom Dahmer
Hyder Consulting Ltd, Hong Kong; Ecosystems Ltd, Hong Kong

Yusbelly Diaz
Universidad Simon Bolivar, Apdo. 89000, Caracas 1080-A, Venezuela

Rune Dietz
National Environmental Research Institute, Department of Arctic Environment, P.O. Box 358, DK-4000 Roskilde, Denmark

Isabel Díez
Departamento de Biología Vegetal y Ecología, Facultad de Ciencias, Universidad del País Vasco, Apdo. 644, Bilbao 48080, Spain

José M.L. Dominguez
Laboratório de Estudos Costeiros, Centro de Pesquisa em Geofísica e Geologia, Universidade Federal da Bahia, Rua Caetano Moura 123, Federação, Salvador, 40210-340, Bahia, Brazil

Marydele Donnelly
IUCN/SSC Marine Turtle Specialist Group, 1725 DeSales St. NW #600, Washington, DC 20036, U.S.A.

Bohdan Draganik
Sea Fisheries Institute, 1 Kollataja Str., PL 81-332 Gdynia, Poland

Nguyen Xuan Duc
Institute of Ecology and Biological Resources, Nghia Do, Hanoi, Vietnam

Jean-Paul Ducrotoy
University College Scarborough, CERCI, Scarborough YO11 3AZ, U.K.

Do Hoai Duong
Institute of Hydrometeology, Lang Thuong, Dong Da, Hanoi, Vietnam

Michael J. Durako
Center for Marine Science, The University of North Carolina at Wilmington, Wilmington, NC 28403, U.S.A.

Phillip Dustan
Department of Biology, University of Charleston, Charleston, SC 29424, U.S.A.

Milen F. Dyoulgerov
International Program Office, National Ocean Service, National Oceanic and Atmospheric Administration, 1305 East West Highway, Silver Spring, MD 20910, U.S.A.

Ma. Corazon Ebarvia
GEF/UNDP/IMO Regional Programme on Partnerships in Environmental Management for the Seas of East Asia (PEMSEA), P.O. Box 2502, Quezon City, 1165 Metro Manila, Philippines

Graham J. Edgar
Zoology Department, Tasmanian Aquaculture and Fisheries Institute, University of Tasmania, GPO Box 252-05, Hobart, Tasmania 7011, Australia

Evan Edinger
Department of Geology, St. Francis Xavier University, P.O. Box 5000, Antigonish, Nova Scotia B2G 2W5, Canada. Present address: Department of Earth Sciences, Laurentian University, Ramsey Lake Road, Sudbury, Ontario, P3E 2C6, Canada

Karen Edyvane
SA Research and Development Institute, P.O. Box 120, Henley Beach, South Australia 5022, Australia

Mike Elliott
IECS, University of Hull, Hull HU6 7RX, U.K.

Derek V. Ellis
Biology Department, University of Victoria, P.O. Box 3020, Victoria, B.C., V8W 3N5, Canada

Rudolph P. van der Elst
SAAMBR, P.O. Box 10712, Marine Parade, KwaZulu-Natal 4056, South Africa

Paul R. Epstein
Center for Health and the Global Environment, Harvard Medical School, Boston MA 02115, U.S.A

Mark V. Erdmann
Dept. of Integrative Biology, University of California, Berkeley, Berkeley, CA 94720, U.S.A.

José L. Esteves
CENPAT-CONICET, Bv. Brown 3000, (9120) Puerto Madryn, Chubut, Argentina

S.M. Evans
Dove Marine Laboratory (Department of Marine Sciences and Coastal Management, Newcastle University), Cullercoats, Tyne & Wear, NE30 4PZ, UK

Jerzy Falandysz
University of Gdansk, 18 Sobieskiego Str., PL 80-952 Gdansk, Poland

Kuang-Lung Fan
Institute of Oceanography, National Taiwan University, Taipei, Taiwan, Republic of China

Richard Farman
Southern Province, Department of Natural Resources, B.P. 3718, 98846 Nouméa Cedex, New Caledonia

Sean T. Fennessy
SAAMBR, P.O. Box 10712, Marine Parade, KwaZulu-Natal 4056, South Africa

Roberto Fernández-Maestre
Universidad de Cartagena, Facultad de Ciencias Químicas y Farmacéuticas, Zaragocilla, AA 1661, Cartagena, Colombia

Jorge Fernández
PEMEX-Exploración-Producción-Región Marina Suroeste, Calle 33 S/N, Edif. Cantarell, Cd. del Carmen, Campeche, C.P. 24170 Mexico

Renaud Fichez
IRD (Institute of Research for Development), B.P. A5, 98848 Nouméa Cedex, New Caledonia

Colin D. Field
Faculty of Science (Gore Hill), University of Technology, Sydney, P.O. Box 123, Broadway NSW 2007, Australia

Peter J. Fielding
SAAMBR, P.O. Box 10712, Marine Parade, KwaZulu-Natal 4056, South Africa

M. Enrique Figueroa
Departamento de Biología Vegetal y Ecología, Facultad de Biología, Universidad de Sevilla, Apdo. 1095, 41080 Sevilla, Spain

Belmiro Mendes de Castro Filho
Universidade de São Paulo, Dept. Oceanografia Física do Instituto Oceanográfico, Pca do Oceanográfico, 191, Cidade Universitária, SP, 05508-900, Brazil

William S. Fisher
U.S. Environmental Protection Agency, National Health and Environmental Effects Laboratory, Gulf Ecology Division, One Sabine Island Drive, Gulf Breeze, FL 32561, U.S.A

Michael J. Fogarty
University of Maryland Center for Environmental Science, Chesapeake Biological Lab., Solomons, MD, U.S.A. Present address: National Marine Fisheries Service, Northeast Fisheries Science Center, 166 Water St., Woods Hole, MA 02543, U.S.A.

Mark S. Fonseca
NOAA/National Ocean Service, Center for Coastal Fisheries and Habitat Research, 101 Pivers Island Road, Beaufort, NC 28516-9722, U.S.A.

Howard Freeland
Institute of Ocean Sciences, P.O. Box 6000, Sidney, B.C., V8L 4B2, Canada

Hugo Freije
Universidad Nacional del Sur, Química Ambiental, Av. Alem 1253, (8000) Bahía Blanca, Argentina

Miles Gabriel
Inter-consult Namibia (Pty) Ltd., P.O. Box 20690, Windhoek, Namibia

Bella Galil
Israel Oceanographic and Limnological Research, National Institute of Oceanography, P.O.Box 8030, Haifa 31080, Israel

Vincent F. Gallucci
University of Washington, School of Fisheries, Seattle, WA 98195, U.S.A.

Zhang Gan
Guangzhou Institute of Geochemistry, Chinese Academy of Sciences, Guangzhou 510640, People's Republic of China

María P. Garcia
Universidad Simon Bolivar, Apdo. 89000, Caracas 1080-A, Venezuela

Jorge Garcia
University of Puerto Rico R.U.M., Department of Marine Sciences, Mayagüez, Puerto Rico

Gudrun Gaudian
Sunny View, Main Street, Alne, N. Yorks, YO61 1RT, U.K.

Fiona Gell
ICLARM Caribbean and Eastern Pacific Office, PMB 158, Inland Messenger Service, Road Town, Tortola, British Virgin Islands

Bresilda Gervacio
GEF/UNDP/IMO Regional Programme on Partnerships in Environmental Management for the Seas of East Asia (PEMSEA), P.O. Box 2502, Quezon City, 1165 Metro Manila, Philippines

Ed Gmitrowicz
Hyder Consulting Ltd, Hong Kong; Ecosystems Ltd, Hong Kong

Judith F. Gobin
Institute of Marine Affairs, Hilltop Lane, Chaguaramas, Port of Spain, Trinidad and Tobago

Edgardo D. Gomez
Marine Science Institute, University of the Philippines, 1101 Diliman, Quezon City, Philippines

Raquel Goñi
Centro Oceanografico de Baleares Muelle de Poniente s/n, Apdo. 291, 07080 Palma de Mallorca, Spain

José María Gorostiaga
Departamento de Biología Vegetal y Ecología, Facultad de Ciencias, Universidad del País Vasco, Apdo. 644, Bilbao 48080, Spain

Ingrid Rosalie L. Gorre
GEF/UNDP/IMO Regional Programme on Partnerships in Environmental Management for Seas of East Asia (PEMSEA), P.O. Box 2502, Quezon City, 1165 Metro Manila, Philippines

Anesh Govender
SAAMBR, P.O. Box 10712, Marine Parade, KwaZulu-Natal 4056, South Africa

Stephen L. Granger
University of Rhode Island, Graduate School of Oceanography, Narragansett, Rhode Island 02882, USA

Martin Guard
Zoology Department, University of Aberdeen, Tillydrone Road, Aberdeen AB24 2TZ, U.K. and Department of Zoology and Marine Biology, University of Dar es Salaam, P.O. Box 35091, Dar es Salaam, Tanzania

Bertil Håkansson
Swedish Meteorological and Hydrological Institute, S-60176 Norrköping, Sweden

Dulal C. Halder
Independent University, Bangladesh (IUB), Plot 3 & 8, Road 10, Baridhara, Dhaka-1212, Bangladesh

Martin A. Hall
Inter-American Tropical Tuna Commission, 8604 La Jolla Shores Dr., La Jolla, CA 92037, U.S.A.

Alastair R. Harborne
Coral Cay Conservation, 154 Clapham Park Road, London, SW4 7DE, U.K.

Nicholas J. Hardman-Mountford
Centre for Coastal and Marine Sciences, Plymouth Marine Laboratory, Plymouth, U.K.

Guillermo Harris
Fundación Patagonia Natural, Marcos A. Zar 760, (9120) Puerto Madryn, Chubut, Argentina

Richard G. Hartnoll
Port Erin Marine Laboratory, University of Liverpool, Port Erin, Isle of Man IM9 6JA, British Isles

Frank Hawkins
Projet ZICOMA, BirdLife International, B.P. 1074, Antananarivo, Madagascar

Barak Herut
Israel Oceanographic and Limnological Research, National Institute of Oceanography, P.O.Box 8030, Haifa 31080, Israel

Sidney Holt
Hornbeam House, 4 Upper House Farm, Crickhowell, Powys, NP8 1BP, U.K.

Chris Horrill
Tanga Coastal Zone Conservation and Development Project, P.O. Box 5036, Tanga, Tanzania

Vicki Howe
Department of Maritime Studies and International Transport, University of Cardiff, Cardiff, Wales, U.K.

Michael E. Huber
Global Coastal Strategies, P.O. Box 606, Wynnum, QLD 4178, Australia

Manuwadi Hungspreugs
Department of Marine Science, Chulalongkorn University, Bangkok 10330, Thailand

George L. Hunt Jr.
Department of Ecology and Evolutionary Biology, University of California, Irvine, Irvine, CA 92697, U.S.A.

Pat Hutchings
The Australian Museum, Sydney, NSW 2010, Australia

Zelina Z. Ibrahim
National Hydraulic Research Institute of Malaysia, Km 7 Jalan Ampang, 68000 Ampang, Malaysia

Anton J. Immink
Fisheries Department, FAO, Rome, Italy

Oscar Iribarne
Universidad Nacional de Mar del Plata, Biologia, CC 573, Correo Central, (7600) Mar del Plata, Argentina

Gail V. Irvine
U.S. Geological Survey, Alaska Biological Science Center, 1011 E. Tudor Rd., Anchorage, AK 99503, U.S.A.

Ignacio Isla
Universidad Nacional de Mar del Plata, Centro de Geología de Costas

Syed Mazharul Islam
School of Liberal Arts and Science, Independent University, Bangladesh (IUB), Plot # 3 & 8, Road 10, Baridhara, Dhaka-1212, Bangladesh

Gil S. Jacinto
Marine Science Institute, University of the Philippines, 1101 Diliman, Quezon City, Philippines

Colin Jago
School of Ocean Sciences, University of Wales Bangor, LL59 5EY, U.K.

Stephen C. Jameson
Coral Seas Inc. – Integrated Coastal Zone Management, 4254 Hungry Run Road, The Plains, VA 20198-1715, U.S.A.

Woei-Lih Jeng
Institute of Oceanography, National Taiwan University, Taipei, Taiwan, Republic of China

Simon Jennings
CEFAS, Fisheries Laboratory, Lowestoft, NR33 OHT, U.K.

Jacob Pauli Joensen
Food and Environmental Agency, Debesartrøð, FO-100 Tórshavn, Faroe Islands

Poul Johansen
National Environmental Research Institute, Department of Arctic Environment, P.O. Box 358, DK-4000 Roskilde, Denmark

Paul Johnston
Greenpeace Research Laboratories, University of Exeter, Exeter EX4 4PS, U.K.

Victor N. de Jonge
National Institute for Coastal and Marine Management, Rijkswaterstaat, Ministry of Transport, Public Works and Water Management, P.O. Box 207, 9750 AE Haren, The Netherlands

Dihider Shahriar Kabir
School of Environmental Science and Management, Independent University, Bangladesh (IUB), Plot # 3 & 8, Road 10, Baridhara, Dhaka-1212, Bangladesh

Anatoly N. Kachur
Pacific Geographical Institute, Far East Branch, Russian Academy of Sciences, Vladivostok 690022, Russia

Sung-Hyun Kahng
Korea Ocean Research and Development Institute, Ansan P.O. Box 29, Seoul, 425-600, Korea

Kostas Kapiris
University of Athens, Department of Biology, Division of Zoology and Marine Biology, Panepistimiopolis, Kouponia, Athens 15784, Greece

Lena Kautsky
Department of Botany, Stockholm University, S-106 91 Stockholm, Sweden

Nils Kautsky
Department of Systems Ecology, Stockholm University, S-106 91 Stockholm, Sweden

W. Judson Kenworthy
NOAA/National Ocean Service, Center for Coastal Fisheries and Habitat Research, 101 Pivers Island Road, Beaufort, NC 28516-9722, U.S.A.

Oleg Khalimonov
International Maritime Organization, London, UK

Md. Giasuddin Khan
Centre for Environment and Geographical Information Systems Support, House 49, Road 27, Banani, Dhaka-1212, Bangladesh

Ruy K.P. Kikuchi
Departamento de Ciências Exatas, Universidade Estadual de Feira de Santana, BR-116, Campus Universitário, Feira de Santana, 44031-160, Bahia, Brazil

Suam Kim
Korea Ocean Research and Development Institute, Seoul, Korea. Present address: Dept. of Marine Biology, Pukyong National University, 599-1 Daeyeon 3-Dong, Nam-Gu, Pusan, 608-737, Korea

Rebecca Klaus
Department of Biological Sciences, University of Warwick, Coventry CV4 7RU, U.K.

Anthony H. Knap
Bermuda Biological Station for Research, 17 Biological Station Lane, St. Georges, Bermuda

Kwame A. Koranteng
Marine Fisheries Research Division, Ministry of Food and Agriculture, Ghana

Marion Köster
Institut für Ökologie der Ernst-Moritz-Arndt-Universität Greifswald, Schwedenhagen 6, D-18565 Kloster/Hiddensee, Germany

Aka Marcel Kouassi
Centre de Recherches Océanologiques, BP V18, Abidjan, Côte d'Ivoire

Mamba Kourouma
The National Center of Halieutic Sciences of Boussoura, B.P. 3060, Conakry, Republic of Guinea

Andreas Kunzmann
ZMT Bremen, Germany

Pierre Labrosse
Secretariat of the Pacific Community (SPC), B.P. D5, 98848 Nouméa Cedex, New Caledonia

Paul K.S. Lam
Department of Biology and Chemistry, City University of Hong Kong, 83 Tat Chee Avenue, Kowloon, Hong Kong

Eliete Zanardi Lamardo
University of Miami – RSMAS, Dept. Marine and Atmospheric Chemistry, 4600 Rickenbacker Causeway, Miami, FL 33149, U.S.A.

Guadalupe de la Lanza-Espino
Institute for Biological Sciences, National Autonomous University of Mexico, Marine Ecology Laboratory, P.O. Box 70233, México City 04515 D.F., México

Victor V. Lapko
Pacific Fisheries Research Centre, TINRO Centre, Vladivostok, Russia

Claudio Lardicci
Dipartimento di Scienze dell'Uomo e dell'Ambiente, Università di Pisa, Via Volta 6, I-56124 Pisa, Italy

Zelinda M.A.N. Leão
Laboratório de Estudos Costeiros, Centro de Pesquisa em Geofísica e Geologia, Universidade Federal da Bahia, Rua Caetano Moura 123, Federação, Salvador, 40210-340, Bahia, Brazil

Mairin Lemus
Instituto Oceanografico de Venezuela, Universidad de Oriente, Cumana, Venezuela

César A. Leon
Universidad Simon Bolivar, Apdo. 89000, Caracas 1080-A, Venezuela

Shiao-Kung Liu
Systems Research Institute, 3706 Ocean Hill Way, Malibu, CA, 90265, U.S.A.

Freddy Losada
Universidad Simon Bolivar, Apdo. 89000, Caracas 1080-A, Venezuela

Carlos J. Luque
Departamento de Biología Vegetal y Ecología, Facultad de Biología, Universidad de Sevilla, Apdo. 1095, 41080 Sevilla, Spain

Anmarie J. Mah
Vancouver Aquarium, P.O. Box 3232, Vancouver, British Columbia, Canada V6B 3X8

Michael A. Mallin
Center for Marine Science Research, University of North Carolina-Wilmington, Wilmington, NC 28403, U.S.A.

Bruce Q. Mann
SAAMBR, P.O. Box 10712, Marine Parade, KwaZulu-Natal 4056, South Africa

James E. Maragos
U.S. Fish and Wildlife Service, Pacific Islands Ecoregion, 300 Ala Moana Blvd., Box 50167, Honolulu, HI 96850, U.S.A.

Maged Mahmoud Marghany
Universiti Putra Malaysia, 43400 UPM Serdang, Malaysia

Alonso J. Marrugo-González
Universidad de Cartagena, Facultad de Ciencias Químicas y Farmacéuticas, Zaragocilla, AA 1661, Cartagena, Colombia

Michael J. Marshall
Coastal Seas Consortium, 5503 40th Avenue East, Bradenton, FL 34208, U.S.A.

Suzanne Marshall
Environment Agency, Kings Meadow Road, Reading, RG1 8DQ, U.K.

Alberto Martin
Universidad Simon Bolivar, Apdo. 89000, Caracas 1080-A, Venezuela

Ted A. McConnaughey
Marine Systems Laboratory, Smithsonian Institution, NHB E-117, MRC 164, Washington, DC 20560, U.S.A.

Melanie D. McField
Department of Marine Science, University of South Florida, 140 Seventh Ave South, St. Petersburg, FL 33701, U.S.A. and P.O. Box 512, Belize City, Belize

Derek J. McGlashan
Graduate School of Environmental Studies, Wolfson Centre, 106 Rottenrow East, University of Strathclyde, Glasgow, G4 0NW, U.K.

Bruce McKay
4058 Rue Dorion, Montreal, PQ H2K 4B9, Canada

Paul Medley
Sunny View, Main Street, Alne, N. Yorks, YO61 1RT, U.K.

Kaija I. Metuzals
Biological Sciences, University of Warwick, Coventry, U.K.

Lutz-Arend Meyer-Reil
Institut für Ökologie der Ernst-Moritz-Arndt-Universität Greifswald, Schwedenhagen 6, D-18565 Kloster/Hiddensee, Germany

Patricia Miloslavich
Universidad Simon Bolivar, Apartado 89000, Caracas 1080-A, Venezuela

Luiz Bruner de Miranda
Universidade de São Paulo, Dept. Oceanografia Física do Instituto Oceanográfico, Pca do Oceanográfico, 191, Cidade Universitária, SP, 05508-900, Brazil

Abhijit Mitra
Department of Marine Science, University of Calcutta 35, B.C Road, Calcutta 700 019, West Bengal, India.

Aviti J. Mmochi
Marine Environmental Chemistry, Institute of Marine Sciences, P.O. Box 668, Zanzibar, Tanzania

Jarle Molvaer
Norwegian Institute for Water Research (NIVA), P.O. Box 173 Kjelsaas, 0411 Oslo, Norway

Francisco García Montelongo
Department of Analytical Chemistry, Nutrition and Food Sciences, University of La Laguna, 38071 La Laguna, Spain

Jack Morelock
University of Puerto Rico R.U.M., Department of Marine Sciences, P.O. Box 3200, Lajas, Puerto Rico 00667

Jeanne A. Mortimer
Department of Zoology, University of Florida, Gainesville, FL 32611-8525, U.S.A. and Marine Conservation Society of Seychelles, P.O. Box 445, Victoria, Mahe, Seychelles

Brian Morton
The Swire Institute of Marine Science and Department of Ecology and Biodiversity, The University of Hong Kong, Hong Kong

Anders Mosbech
National Environmental Research Institute, Department of Arctic Environment, P.O. Box 358, DK-4000 Roskilde, Denmark

Kent Mountford
US Environmental Protection Agency, Chesapeake Bay Program, 410 Severn Ave., Suite 109, Annapolis, MD 21403, U.S.A.

Kieran Mulvaney
1219 W. 6th Avenue, Anchorage, AK 99501, U.S.A.

Peter J. Mumby
Centre for Tropical Coastal Management Studies, Department of Marine Sciences & Coastal Management, Ridley Building, The University, Newcastle upon Tyne, NE1 7RU, U.K.

Michael Myers
TCMC, University of Newcastle upon Tyne, Newcastle upon Tyne NE1 7RU, U.K.

Paulina Nabel
Museo Argentino de Ciencias Naturales-CONICET, Av. A. Gallardo 470, (1405) Buenos Aires, Argentina

Odile Naim
Laboratoire d'Ecologie Marine, Université de la Réunion, BP 9151, Saint-Denis messag 9, Réunion, France

Milika Naqasima
Marine Studies Programme, University of the South Pacific, P.O. Box 1168, Suva, Fiji

Milika Naqasima
Marine Studies Programme, University of the South Pacific, P.O. Box 1168, Suva, Fiji

David M. Nelson
National Ocean Service, 1305 East-West Highway, Silver Spring, MD 20910, U.S.A

Dang Duc Nhan
Institute of Nuclear Sciences and Techniques, P.O. Box 5T-160, Hoang Quoc Viet, Hanoi, Vietnam

Charoen Nitithamyong
Department of Marine Science, Chulalongkorn University, Bangkok 10330, Thailand

Scott W. Nixon
University of Rhode Island, Graduate School of Oceanography, Narragansett, Rhode Island 02882, USA

Steve Oakley
Institute of Biodiversity and Environmental Conservation, University of Malaysia, Kota Samarahan 93400, Sarawak, Malaysia

Tim D. O'Hara
Zoology Department, University of Melbourne, Parkville, Vic., Australia

E.A. Palacio
Instituto de Ingeniería, UNAM, Cd. Universitaria, Mexico City, C.P. 04510 México

Claudio Paredes
Universidad Simon Bolivar, Apartado 89000, Caracas 1080-A, Venezuela

Marcela S. Pascual
Instituto de Biología Marina y Pesquera "A. Storni", (8520) San Antonio Oeste, Río Negro, Argentina

Mark E. Pattillo
U.S. Army Corps of Engineers, Galveston District, P.O. Box 1229, Galveston, TX 77551-1229, U.S.A.

Alexandra Pavlidou
University of Athens, Department of Chemistry, Division III, Inorganic and Environmental Chemistry, Panepistimiopolis, Kouponia, Athens 15771, Greece

Jack B. Pearce
NMFS/NOAA, NE Fisheries Center, Woods Hole, MA 02543, U.S.A.

Pablo E. Penchaszadeh
Universidad Simon Bolivar, Apartado 89000, Caracas 1080-A, Venezuela

Daisy Perez
Universidad Simon Bolivar, Apartado 89000, Caracas 1080-A, Venezuela

Ronald C. Phillips
Commission of Environmental Research, Emirates Heritage Club, Abu Dhabi, United Arab Emirates. Correspondence: 1597 Meadow View Drive, Hermiston, OR 97838, U.S.A.

Niphon Phongsuwan
Phuket Marine Biological Center, P.O. Box 60, Phuket, 83000, Thailand

Nicolas Pilcher
Institute of Biodiversity and Environmental Conservation, University of Malaysia, Kota Samarahan 93400, Sarawak, Malaysia

Luis A. Pinto
Facultad de Ciencias, Universidad Católica de la Santísima Concepción, Campus San Andrés, Paicaví 3000, Casilla 297, Concepción, Chile

Pamela T. Plotkin
Center for Marine Conservation, 1725 DeSales St. NW #600, Washington, DC 20036, U.S.A.

Martin H. Posey
Center for Marine Science Research, University of North Carolina-Wilmington Wilmington, NC 28403, U.S.A.

Anil Premaratne
Coast Conservation Department, Sri Lanka

Andrew R.G. Price
Ecology and Epidemiology Group, Department of Biological Sciences, University of Warwick, Coventry, U.K.

Valéria Aparecida Prósperi
CETESB – Companhia de Tecnologia de Saneamento Ambiental, Setor de Ictiologia e Bioensaios com organismos aquáticos, Av. Prof. Frederico Hermann Jr., 345, Alto de Pinheiros, SP 05489-900, Brazil

Sara C. Pryor
Atmospheric Science Program, Department of Geography, Indiana University, Bloomington, IN 47405, U.S.A.

Bui Cong Que
Institute of Oceanology, Hoang Quoc Viet, Nghia Do, Hanoi, Vietnam

Arturo Mendoza Quintero-Marmol
PEMEX-Exploracíon-Produccion-Región Marina Noreste, Calle 31, Esq. Periferica, Cd. del Carmen, Campeche, C.P. 24170 Mexico

Jean-Pascal Quod
ARVAM, 14, Rue du stade de l'Est, 97490 Réunion, France

Vladimir I. Radchenko
Pacific Fisheries Research Centre, TINRO Centre, Vladivostok, Russia

Arjan Rajasuriya
National Aquatic Resources Research and Development Agency, Colombo, Sri Lanka

Sundararajan Ramachandran
Institute for Ocean Management, Anna University, Chennai 600 025, India

Krishen J. Rana
Fisheries Department, FAO, Rome, Italy

Eulalie Ranaivoson
Institut Halieutique et des Sciences Marines, B.P. 141, Toliara, Madagascar

Bemahafaly J. de D. Randriamanantsoa
Cellule des Océanographes de l'Université de Toliara (COUT), IHSM, B.P. 141, Toliara, Madagascar

Ricardo Rangel
Instituto de Ciencias del Mar y Limnología, UNAM, Cd. Universitaria, A.P. 70-305, Mexico City, C.P. 04510 Mexico

Onésime Ratomahenina
Ministère de la Recherche Scientifique, Direction Générale de la Recherche, B.P. 4258, Antananarivo, Madagascar

Haja Razafindrainibe
Cellule Environnement Marin et Côtier, Office National pour l'Environnement, B.P. 822, Antananarivo, Madagascar

P. Chris Reid
Sir Alister Hardy Foundation for Ocean Science, 1 Walker Terrace, The Hoe, Plymouth, PL1 3BN, U.K.

Andrias Reinert
Aquaculture Research Station of the Faroes, við Áir, FO-430 Hvalvík, Faroe Islands

Louise Richards
Hyder Consulting Ltd, Hong Kong; Ecosystems Ltd, Hong Kong

Bruce J. Richardson
Department of Biology and Chemistry, City University of Hong Kong, 83 Tat Chee Avenue, Kowloon, Hong Kong

Frank Riget
National Environmental Research Institute, Department of Arctic Environment, P.O. Box 358, DK-4000 Roskilde, Denmark

Andrés L. Rivas
CENPAT-CONICET, Bv. Brown 3000, (9120) Puerto Madryn, Chubut, Argentina

Victoriano Roa
Universidad Simon Bolivar, Apartado 89000 Caracas 1080-A, Venezuela

Wendy D. Robertson
SAAMBR, P.O. Box 10712, Marine Parade, KwaZulu-Natal 4056, South Africa

José A. Robleto
University of Mobile, Latin American Campus, San Marcos, Carazo, Nicaragua

Carlos Díaz Romero
Department of Analytical Chemistry, Nutrition and Food Sciences, University of La Laguna, 38071 La Laguna, Spain

S. Adrian Ross
GEF/UNDP/IMO Regional Programme on Partnerships in Environmental Management for the Seas of East Asia (PEMSEA), P.O. Box 2502, Quezon City, 1165 Metro Manila, Philippines

Bernard Salvat
Ecole Pratique des Hautes Etudes, URA CNRS 1453, Université de Perpignan, France, and Centre de Recherches Insulaires et Observatoire de l'Environnement, BP 1013, Moorea, Polynésia Française

Guadalupe Sanchez
Instituto del Mar del Peru, Callao, Peru

David Santillo
Greenpeace Research Laboratories, University of Exeter, Exeter EX4 4PS, U.K.

Norma Santinelli
Universidad Nacional de la Patagonia, Belgrano 504, (9100) Trelew, Chubut, Argentina

Alberto Santolaria
Departamento de Biología Vegetal y Ecología, Facultad de Ciencias, Universidad del País Vasco, Apdo. 644, Bilbao 48080, Spain

Suesan Saucerman
Environmental Protection Agency, EPA Region IX – WTR-5, 75 Hawthorne St., San Francisco, CA 94105-3901, U.S.A.

Kenneth C. Schiff
Southern California Coastal Water Research Project, 7171 Fenwick Lane, Westminster, CA 92683, U.S.A.

Michael H. Schleyer
SAAMBR, P.O. Box 10712, Marine Parade, KwaZulu-Natal 4056, South Africa

E.A. Schreiber
National Museum of Natural History, Smithsonian Institution, NHB MRC 116, Washington D.C. 20560, U.S.A.

Miradys Sebastiani
Universidad Simon Bolivar, Apdo. 89000, Caracas 1080-A, Venezuela

Antonio Secilla
Departamento de Biología Vegetal y Ecología, Facultad de Ciencias, Universidad del País Vasco, Apdo. 644, Bilbao 48080, Spain

Virender K. Sharma
Chemistry Department, Florida Tech., 150 West University Blvd., Melbourne, FL 32901-6975, U.S.A.

Charles R.C. Sheppard
Department of Biological Sciences, University of Warwick, Coventry CV4 7AL, U.K.

Benjamin H. Sherman
Climate Change Research Center, Institute for the Study of Earth Oceans and Space, OSP HEED MMED program, 206 Nesmith Hall, University of New Hampshire, Durham, NH 03824, U.S.A.

Zou Shicun
School of Chemistry and Chemical Engineering, Zhangshan University, Guangzhou 510301, People's Republic of China

Frederick T. Short
Jackson Estuarine Laboratory, University of New Hampshire, 85 Adams Point Road, Durham, NH 03824, U.S.A.

Jeffrey W. Short
Auke Bay Laboratory, National Marine Fisheries Service, National Oceanic and Atmospheric Administration, 11305 Glacier Highway, Juneau, AK 99801, U.S.A.

Jens Skei
Norwegian Institute for Water Research (NIVA), P.O. Box 173 Kjelsaas, 0411 Oslo, Norway

Allegra M. Small
Marine Systems Laboratory, Smithsonian Institution, NHB E-117, MRC 164, Washington, DC 20560, U.S.A.

G. Robin South
International Ocean Institute – Pacific Islands, The University of the South Pacific, P.O. Box 1168, Suva, Republic of the Fiji Islands

Don M. Spoon
Marine Systems Laboratory, Smithsonian Institution, NHB E-117, MRC 164, Washington, DC 20560, U.S.A.

Dennise Stecconi
Centro de Procesamiento de Imagenes Digitales (CPDI), Sartenejas, Caracas. Venezuela

Ruth Stringer
Greenpeace Research Laboratories, University of Exeter, Exeter EX4 4PS, U.K.

D.V. Subba Rao
Mariculture and Fisheries Department, Kuwait Institute For Scientific Research, P.O. Box 1638, Salmiya 22017, Kuwait

Jeffrey Sybesma
University of the Netherlands Antilles, Jan Noorduynweg 111, P.O. Box 3059, Curaçao, Netherlands Antilles

Piotr Szefer
Medical University of Gdansk, 107 Gen. Hallera Ave., 80-416 Gdansk, Poland

Liana Talaue-McManus
Marine Science Institute, University of the Philippines, 1101 Diliman, Quezon City, Philippines

Touch Seang Tana
Department of Fisheries, 186 Norodom Blvd., P.O. Box 582, Phnom Penh, Cambodia

Shinsuke Tanabe
Center for Marine Environmental Studies, Ehime University, Tarumi 3-5-7, Matsuyama 790-8566, Japan

Alan D. Tappin
Centre for Coastal and Marine Science, Plymouth Marine Laboratory, Prospect Place, Plymouth PL1 3DH, U.K.

Ricardo Corbella Tena
Department of Analytical Chemistry, Nutrition and Food Sciences, University of La Laguna, 38071 La Laguna, Spain

Gordon W. Thayer
NOAA/National Ocean Service, Center for Coastal Fisheries and Habitat Research, 101 Pivers Island Road, Beaufort, NC 28516-9722, U.S.A.

Law Ah Theem
Universiti Kolej Terengganu, Universiti Putra Malaysia, 21030 Kuala Terengganu, Malaysia

Chua Thia-Eng
GEF/UNDP/IMO Regional Programme on Partnerships in Environmental Management for the Seas of East Asia (PEMSEA), P.O. Box 2502, Quezon City, 1165 Metro Manila, Philippines

Nguyen The Tiep
Institute of Oceanology, Hoang Quoc Viet, Nghia Do, Hanoi, Vietnam

Alexander V. Tkalin
Far Eastern Regional Hydrometeorological Research Institute (FERHRI), Russian Academy of Sciences, 24 Fontannaya Street, Vladivostok 690600, Russia

Alejandro O. Toledo
Institute for Marine and Limnology Sciences, National Autonomous University of Mexico, Marine Pollution Laboratory, P.O. Box 70305, México City 04510 D.F., México

Bruce J. Tomalin
SAAMBR, P.O. Box 10712, Marine Parade, KwaZulu-Natal 4056, South Africa

Tomas Tomascik
Parks Canada – WCSC, 300-300 West Georgia St., Vancouver, British Columbia, Canada V6B 6B4

Michael S. Traber
University of Rhode Island, Graduate School of Oceanography, Narragansett, RI 02882, USA

Lamarr B. Trott
National Oceanic and Atmospheric Administration, National Marine Fisheries Service, 1315 East West Highway, Silver Spring, MD 20910, U.S.A.

Anna Trzosinska
Institute of Meteorology and Water Management, 42 Waszyngtona Str., PL 81-342 Gdynia, Poland

Caroline Turnbull
Department of Biological Sciences, University of Warwick, Coventry, UK

John Turner
School of Ocean Sciences, University of Wales Bangor, Marine Science Laboratories, Anglesey, Gwynedd LL59 5EY, U.K.

Md. Ekram Ullah
Environment Section, Water Resources and Planning Organisation (WARPO), House 4 A, Road 22, Gulshan-1, Dhaka-1212, Bangladesh

Wilaiwan Utoomprurkporn
Department of Marine Science, Chulalongkorn University, Bangkok 10330, Thailand

Marieke M. van Katwijk
University of Nijmegen, The Netherlands

Felipe Vázquez
Instituto de Ciencias del Mar y Limnología, UNAM, Cd. Universitaria, A.P. 70-305, Mexico City, C.P. 04510 Mexico

Marjo Vierros
UNEP-CAR/RCU, 14–20 Port Royal St., Kingston, Jamaica, Correspondence: Rosentiel School of Marine and Atmospheric Science, University of Miami, Dept. of Marine Geology & Geophysics, 4600 Rickenbacker Causeway, Miami, FL 33149-1098, U.S.A.

Alicia Villamizar
Universidad Simon Bolivar, Apartado 89000 Caracas 1080-A, Venezuela

Cesar L. Villanoy
Marine Science Institute, University of the Philippines, 1101 Diliman, Quezon City, Philippines

Susana Villanueva-Fragoso
Institute for Marine and Limnology Sciences, National Autonomous University of Mexico, Marine Pollution Laboratory, P.O. Box 70305, México City 04510 D.F., México

Maika Vuki
Chemistry Department, University of the South Pacific, P.O. Box 1168, Suva, Fiji

Veikila C. Vuki
Marine Studies Programme, University of the South Pacific, P.O. Box 1168, Suva, Fiji

Greg Wagner
Department of Zoology and Marine Biology, University of Dar es Salaam, Dar es Salaam, Tanzania

Diana I. Walker
Department of Botany, The University of Western Australia, Perth, WA 6907, Australia

Jan Warzocha
Sea Fisheries Institute, 1 Kollataja Str., PL 81-332 Gdynia, Poland

Maggie Watson
ICLARM Caribbean and Eastern Pacific Office, PMB 158, Inland Messenger Service, Road Town, Tortola, British Virgin Islands

Tom S. Weingartner
University of Alaska Fairbanks, Institute of Marine Science, School of Fisheries and Ocean Sciences, Fairbanks, AK 99775-7220, U.S.A.

Yan Wen
South China Sea Institute of Oceanology, Chinese Academy of Sciences, Guangzhou 510301, People's Republic of China

Frank Whitney
Institute of Ocean Sciences, P.O. Box 6000, Sidney, B.C., V8L 4B2, Canada

Mark Whittington
Frontier International, Leonard St., London EC2A 4QS, U.K.

Sheila Wiegman
American Samoa Environmental Protection Agency, Pago Pago, American Samoa 96799, U.S.A.

Clive R. Wilkinson
Australian Institute of Marine Science, PMB No. 3, Townsville MC 4810, Australia

Simon C. Wilson
Department of Biological Sciences, Warwick University, Coventry, CV4 7RU, U.K. and P.O. Box 2531, CPO 111, Seeb, Oman

Douglas C. Wilson
Institute for Fisheries Management and Coastal Community Development, P.O. Box 104, DK-9850 Hirtshals, Denmark

Poh-Poh Wong
Department of Geography, National University of Singapore, Singapore 119260

Elizabeth Wood
Marine Conservation Society, 9 Gloucester Road, Ross on Wye HR9 5BU, U.K.

Bruce A. Wright
Alaska Region, National Marine Fisheries Service, National Oceanic and Atmospheric Administration, 11305 Glacier Highway, Juneau, AK 99801, U.S.A.

Rudolf S.S. Wu
Department of Biology and Chemistry, City University of Hong Kong, 83 Tat Chee Avenue, Kowloon, Hong Kong

Sandy Wyllie-Echeverria
School of Marine Affairs, University of Washington, Seattle, WA 98105, U.S.A.

Alejandro Yáñez-Arancibia
Department of Coastal Resources, Institute of Ecology A.C., Km 2.5 Antigua Carretera Coatepec, P.O. Box 63, Xalapa 91000, Veracruz, México

Leon P. Zann
School of Resource Science and Management, Southern Cross University, P.O. Box 57, Lismore, NSW 2480, Australia

Eddy Y. Zeng
Southern California Coastal Water Research Project, 7171 Fenwick Lane, Westminster, CA 92683, U.S.A.

Scott A. Zengel
Research Planning, Inc., 1121 Park St., Columbia, SC 29201, U.S.A.

ACKNOWLEDGEMENTS

Several people greatly facilitated the logistical and editorial work of this series of 136 chapters. I am very grateful to Professor Leon Zann, in Australia, and Dr Jack Pearce in the USA who greatly assisted in the process of identifying sensible and manageable regions in their own respective parts of the world, and who helped to identify excellent people or groups of people to write about them. Both of them also contributed more than one excellent chapter themselves. Jack Pearce was also the co-editor of a more-or-less random collection of 16 of these chapters for a special issue of *Marine Pollution Bulletin* published simultaneously in 2000.

Bathymetry in the figures for these volumes was taken from 'GEBCO-97: The 1997 Edition of the GEBCO Digital Atlas'. This excellent product is published on behalf of the Intergovernmental Oceanographic Commission (of UNESCO) and the International Hydrographic Organisation as part of the General Bathymetric Chart of the Oceans (GEBCO) by the British Oceanographic Data Centre, Birkenhead. Further details are available at www.bodc.ac.uk. I wish to thank those who produced this invaluable digital data set for granting permission to use it. Coastlines, political boundaries and other cartographic details including some of the place names were taken from Europa Technologies 'Map elements: International & global map data components'. These GIS products were used in most 'Figure 1' sketch maps, and in many others. Early guidelines to authors bravely said that we would prepare 'standard' style maps from rough materials supplied by authors. The phrase 'rough materials' was taken rather literally by many, and for preparing the maps I am especially grateful to Anne Sheppard, whose interpretations of alleged draft maps often required prolonged diligence and clairvoyance. I thank Rebecca Klaus, who set up the GIS system, and I am also grateful to Olivia Langmead and Sheila O'Sullivan who provided editorial help with several draft chapters by converging numerous and variable dialects of the English language towards a common format. In the whole production process, Justinia Seaman in Elsevier provided masterful co-ordination of the huge project, and finally, when I thought that the editing process was complete, I had cause to be most grateful to Pam Birtles, whose production and copy editing skills added far more than just a final polish and a checking of references.

The final product is, of course, a credit to more than 350 authors. My emphasis throughout this series, particularly in the first two volumes, was on the lesser developed countries, precisely those areas with the least available information and commonly with the most pressing environmental difficulties. The breadth of material asked for in each chapter is considerable, and many large areas have very few resident marine scientists, who are in any case very over-stretched. Some had the data but insufficient time or resources to easily compile a review for a 'foreign language book'. Several had difficulty persuading their governments to allow them to do so, or to present data which might be embarrassing to their own employers. Several chapters have a long and interesting tale behind their gestation. That so many people did write is gratifying: an 'ordinary' chapter from some areas is anything but an ordinary achievement, when a range of obstacles conspired to prevent it. From personal experience, I know that in several countries these obstacles can include considerable censure and risk when describing, for example, environmental problems; which is understandable when it is realised that continuing aid may depend on the government pretending to comply with imposed 'sustainable use' measures. We should all be grateful to these authors, and I am also grateful for their subsequent acceptance of my sometimes drastic editorial changes done in the interests of brevity (mostly), format (usually) and language (sometimes). I hope that one of the benefits of this series will lie in its provision of information, so that one place may learn from the problems of another before repeating the same mistakes, and thus avoid impoverishing yet another bit of the world's coast and its people and, hopefully, reversing many of the problems.

Charles Sheppard

Chapter 53

THE ARABIAN GULF

D.V. Subba Rao and Faiza Al-Yamani

The Arabian Gulf, surrounded by arid land, contains 46% of the world's proven oil and about 16% of its natural gas reserves. Its main source of fresh water is the Shatt Al-Arab in the north which receives the waters of the Tigris, Euphrates and Karun rivers. It has a narrow connection with the Indian Ocean through the Straits of Hormuz. The basin is shallow, with a mean depth of less than 30 m, reaching as much as 170 m only towards its entrance with the Indian Ocean.

The Gulf's waters are more saline than the Indian Ocean, which limits the biodiversity, but the region supports a variety of biotopes including the world's most northerly coral reefs. Nutrient gradients exist, with maximum levels in the north and lower in the south, and gradients of phytoplankton biomass and production are also related to the hydrography. Activities associated with urbanization and petrochemical industries have resulted in loss of shallow marine habitat, general eutrophication, increasing levels of pollution, introduction of exotic species, and declines in fish stocks. River diversion, drainage of the northwestern marshes, and the construction of 22 dams of the SE Anatolia Project across the rivers to the north will drastically impact the hydrobiology of the Gulf. The area also suffers from high levels of oil contamination, and effects from large recent spillages have been severe, though in several cases not as long-term as was first feared.

The need for the Gulf countries to realize that their development and quality of life are closely connected to the functioning of the Gulf ecosystems is clear. Scientific research in the region is developing, but during the new millennium, long-range ocean science programmes integrated into the economic and social development of the Gulf region are recommended.

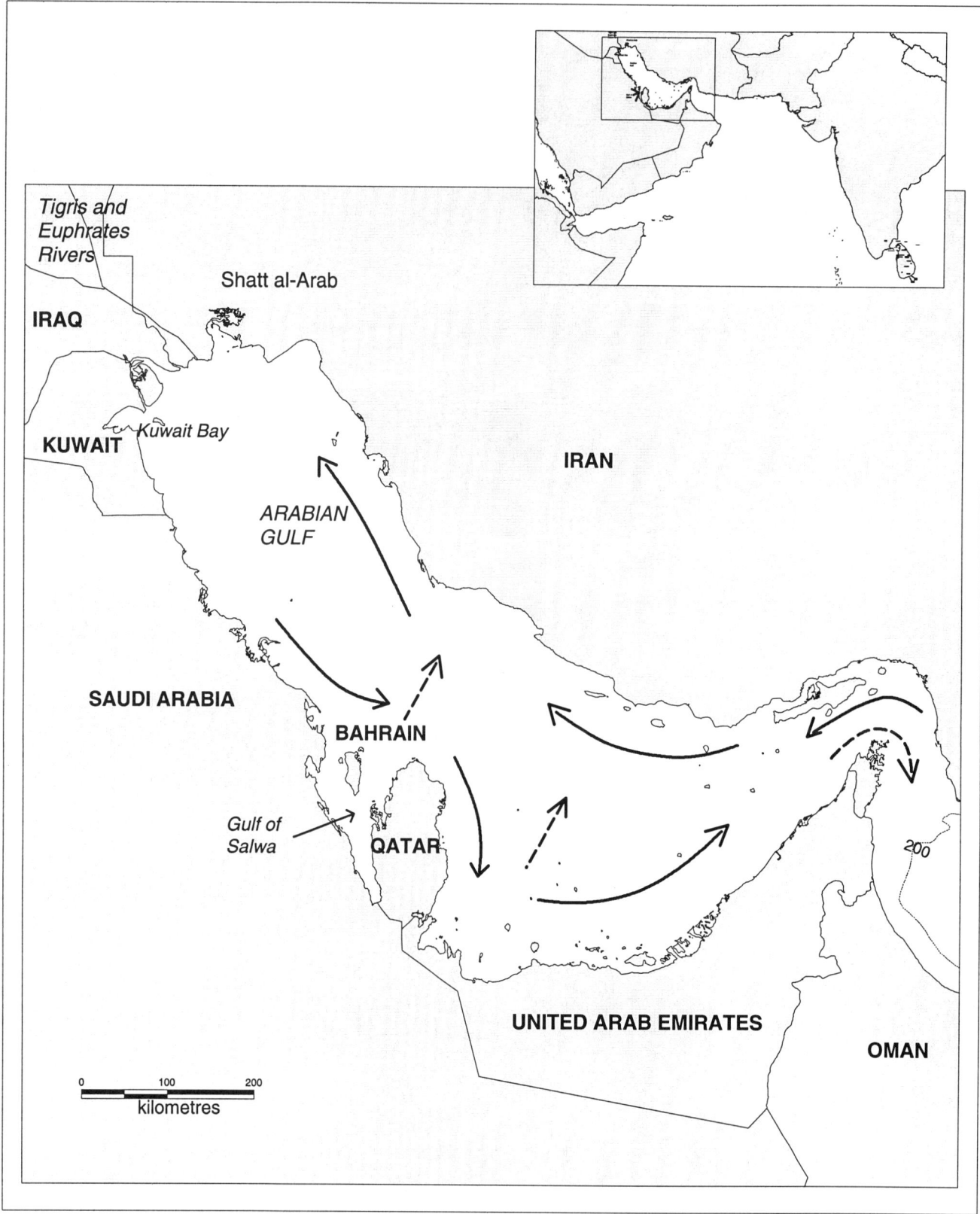

Fig. 1. The Arabian Gulf region and circulation pattern. Solid arrows and broken arrows correspond to surface currents and subsurface currents.

THE DEFINED REGION

The Arabian Gulf, also known as the Persian Gulf, is in biological terms part of the northwestern Indian Ocean. It is a shallow, sub-tropical sedimentary basin bound by the Iranian Coast on the east and the Arabian Coast on the west (Fig. 1). Its basin lies entirely upon the continental shelf and slopes into the Straits of Hormuz (Reynolds, 1993). It is about 1000 km long, and has a width between about 56 to 338 km. Its volume is about 6000 km^3 over an area of 226,000 km^2. In the south, the narrow Straits of Hormuz lead to the Arabian Sea and the Indian Ocean.

The average depth of the Gulf is 35 m and four zones are recognized: (a) Depth of <20 m. This is narrow on the Iranian coast and widest near the Bahrain Islands. This zone is usually dominated by rocks and coral reefs; (b) Shallows between 15 and 55 m depth, widest along the Arabian coast and mostly rough sea bed; (c) Transition zone with depths between 45 and 60 m; (d) Deep sea zone >50 m depth, parallel to the Iran coast and covered by river deposits, reaching 70–110 m near the entrance. Generally the sea bed is rough, covered with dolomite intrusion and coral reefs. Depths of up to 170 m are found at the Straits of Hormuz. There is no sill; the trough simply deepens to >100 m and drops to 2000 m within 200 km outside the Straits (Purser, 1973; Reynolds, 1993). The semi-enclosed nature of the Gulf makes it a "trap" for pollutants.

The Arabian Gulf is a Large Marine Ecosystem (Sherman, 1993) and contains distinct biotopes. A variety of environments exist in the Gulf, most notably hypersaline biotopes, estuarine habitats, intertidal flats, seagrass beds, mangroves and coral reefs around islands. Also there are artificial offshore oil structures which, as artificial reefs, provide habitat for a complex community of organisms.

Evaporation in Excess of Precipitation

Average annual rainfall is less than 5 cm (Evans, 1970). The only freshwater source in the north is the Shatt al-Arab, carrying waters from the Tigris, Euphrates, and Karun Rivers with an estimated annual flow of 5 to 100 km^3 (Grasshoff, 1976). The Tigris and Euphrates each discharge annually 45.3×10^6 m^3 water, and carry 57.6×10^6 and 4.8×10^6 t sediment respectively (Reynolds, 1993). The freshwater inflow is only about 1.3 to 28% of the water lost by evaporation (Al-Abdul-Razzak, 1984).

Oil

Discovery of oil in 1901 shaped events in the Gulf (Fig. 2). Today, it is one of the busiest waterways in the world. The Gulf's 461.7 billion barrels account for 46% of the world's proven oil reserves and supply over half the world's oil requirement. Estimates of 616×10^{12} standard cubic feet of natural gas reserves represent 15.4% of total world reserves (Dureja and Prasad, 1991). The Gulf's entry into the technological and industrial world has resulted in several environmental perturbations, partly from the annual oil spillage of about 160×10^6 t (Jacob and Al-Muzaini, 1995), or 47 times greater than average estimates for other similar surface areas (Linden et al., 1990).

Fig. 2. Oil export activity dominates the coastal industry of the Arabian Gulf. Shuwaikh Port, Kuwait.

Marine Studies

There is a growing interest in the marine environment in the Arabian Gulf region, as is evident from a survey of literature (Table 1). Areas of major emphasis are pollution, waste water management, impact of the petroleum and petrochemical industries, desalination, fisheries and aquaculture. In particular, the impacts of the Gulf War and its associated oil spills have been studied in great detail.

Oceanographic studies in the Gulf are hampered by non-availability of an oceanographic research vessel and by the absence of a regional program. However the Kuwait Institute for Scientific Research (KISR), with some programs sponsored by the Regional Organization for the Protection of the Marine Environment (ROPME) and several national universities are engaged in projects mostly related to pollution caused by petrochemical industries.

Table 1

Trends in number of publications on the Arabian Gulf since 1978

Category of Publications	1978–87	1988–96	1997–98
Arabian Gulf	115	385	65
Oceanography	27	30	12
Physical	4	5	1
Chemical	2	1	1
Biological	12	8	2
Geological	16	5	2
Pollution	18	27	32
Fisheries and Aquaculture	14	34	1
Oil	19	5	1
Desalination	12	9	1
Sewage	5	4	2
Gulf War		74	17

These focus on fauna and flora, coral reefs, short-term seasonal variations of plankton, sedimentation rates, trace metal levels and variations in sediments, on fisheries, mariculture, and on system response to and recovery from the Gulf war. Various special volumes have arisen from these studies (Basson et al., 1977; Al-Abdul-Razzak, 1984; Halwagy et al., 1986; Abuzinada and Krupp, 1994; Sheppard et al., 1992; Price and Robinson, 1993; ROPME, 1997; Al-Muzaini and Beg, 1998; Al-Muzaini and Hamoda 1998). Additionally a good number of technical reports, documents and a few theses with a regional emphasis are issued by academic and consultant organizations. Some of these are in Arabic, and some are classified and are not readily accessible.

NATURAL ENVIRONMENTAL VARIABLES

Climate

The Gulf climate is transitional between tropical and subtropical. The vast surrounding desert, passing winter cyclones formed above the Atlantic Ocean and the monsoon winds influence the Gulf's climate. The influence of the winter monsoon continues into March and April with southern monsoon winds occurring about 60% of the time (Al-Abdul-Razzak, 1984). The summer monsoon (April to October) is more settled and comes from the northwest (Al-Abdul-Razzak, 1984).

The Gulf region marks the boundary between tropical circulations and the synoptic weather systems of mid-latitudes (Reynolds, 1993). Descending dry air produces clear skies and arid conditions. The Taurus and Pontic mountains of Turkey, the Caucasus mountains of Iran, and the Hejaz mountains of the Arabian Peninsula, together with the Tigris–Euphrates Valley, form a northwest–southeast axis that strongly influences the tracks of extra-tropical storms which commonly travel in a southeast direction (Reynolds, 1993).

Shamal, meaning north in Arabic, signifies northwesterly winds which persist year round (Perrone, 1981). Unique to this region, the winter Shamal, which brings the strongest winds and high seas, sets in first in the north with great abruptness and a force of 10 m s^{-1} (<5% frequency) and spreads south (Reynolds, 1993). On the eastern seaboard the strongest southerly winds occur. In contrast, the summer Shamal is continuous from early June through July and is associated with the relative strengths of the Indian and Arabian thermal lows (Reynolds, 1993).

The temperature difference between the land and water surfaces is intense, which adds a landward component to all winds, i.e. from 10–15 m s^{-1}. Because of this, surface pollutants (e.g., oil) are beached relatively quickly (Reynolds, 1993). Occasionally, particularly in the west, Shamal winds that cause northwest storms change to "Kaus"—southerly winds and southeast storms. Consequently the net direction changes to northward, and with it transportation of surface pollutants (John, 1992).

Marine Circulation

Our knowledge of the oceanography in the Gulf is due to oceanographic cruises of the German Meteor (1966), Japanese Umitaka Maru (1974), U.S.S.R Akademician Kurchatov (1979), U.S.A. Atlantis II (1978), Oloum I (1986), Qatar Mukhtabar Al-Bihar (1987), U.S.A. Mt. Michell (1993) and Japanese Umitaka Maru (1993).

Seasonal winds, fresh water run-off from the Shatt Al-Arab and the excessive evaporation control hydrographical conditions. Due to excessive evaporation, the Gulf acts as a 'negative estuary' (Banse, 1997), with a circulation similar to that of the Mediterranean (Reynolds, 1993). During winter under the northwest winds a counter-clockwise circulation sets in. Surface water flows into the Gulf via the Straits of Hormuz while a deep outflow of saline water takes place (Fig. 3). Based on annual mean values of sensible, latent and infrared radiation fluxes, an annual heat flux surplus of 28 W m^{-2} was calculated (Sultan and Ahmad, 1993) which heats the surface water. There are two cyclonic circulation gyres, one in the southern Gulf, i.e. Straits of Hormuz, primarily driven by the evaporation from the Straits of Hormuz, and the other maintained by run-off from the Shatt Al-Arab in the northwest (Chao et al., 1992). This model also predicts a southward coastal jet east of Qatar, primarily wind-driven. In the near-coastal waters off Kuwait, the excessive evaporation creates an anomalous situation similar to an inverted estuarine circulation (Hunter, 1982). The resulting dense water overlies the lighter water, causing vertical mixing.

Flow velocities of residual currents vary from 4 cm s^{-1} in the northern Gulf and northern coast of Iran to 10–12 cm s^{-1} in the Straits of Hormuz (Lardner et al., 1993). Recent studies showed velocities (6 cm s^{-1}) at the surface and bottom (Abdelrahman and Ahmad 1995). The surface flow into the Gulf along the Iranian coast results in an anticlockwise gyre in the southern half of the Gulf during winter but this extends northwards in summer. Due to winter cooling and evaporation, dense water is formed which sinks to the bottom in the central Gulf and eventually flows out through the Straits of Hormuz in a thin layer into the Gulf of Oman (Grasshoff, 1976; Halim, 1984). This results in oxygenation down to the bottom (Siebold, 1973). Residence time, a measure of the mean amount of time a parcel of water will remain in the Gulf, is estimated at about 2–5 years (Reynolds, 1993), comparable to an earlier estimate of 3 years (Koske, 1972).

Temperature

Sea temperature follows the seasons, with the lowest mean of 12.5°C during February and the highest mean of 32°C in August. In the Straits of Hormuz values are 21 and 32°C. The water column is well mixed and therefore thermal stratification is usually absent except during summer between Qatar and Iran, and between Oman and Iran

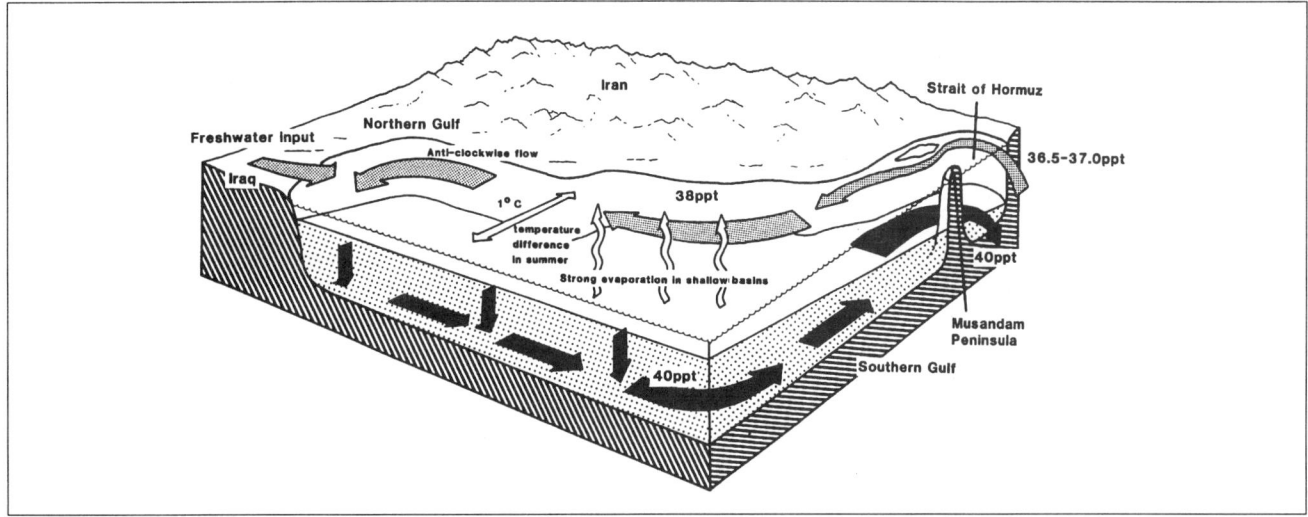

Fig. 3. Water flow through the Straits of Hormuz (from Sheppard et al., 1992).

Table 2
Comparison of ranges of nutrients (μmol) in the Arabian Gulf

Locality	PO_4-P	SiO_2-Si	NO_3-N	Reference
Shatt Al-Arab	1.55–6.01	135.6–306.9	21.5–52.7	Saad (1985)
Iraq-Basrah estuary	~0.1–0.77	~1.78–4.13	~0.2–1.21	Hadi et al. (1989)
Kuwait (South)	0.01–0.02	0.37–0.66	–	Halim (1984)
Kuwait	0–14.66	0–118.8	0–103.5	Al-Yamani et al. (1999)
Kuwait	0.97–2.26	4.27–47.35	12.14–59.25	Jacob et al. (1982)
Iran-Bushehr	<0.1–3.6	<1–13	<0.1–3.9	Hulburt et al. (1981)
Off Iran north of 27°N	0.002–0.02	0–0.002	–	Halim (1984)
Off Iran south of 27°N	0.01–0.014	0–0.002	–	Halim (1984)
Qatar	0.004–0.02	0.04–0.11	–	Halim (1984)
Qatar	0.03–1.23	0.66–5.12	0.12–0.90	Dorgham and Mofta (1989)
UAE	0–0.56	0.57–4.85	0.13–0.40	Dorgham and Mofta (1989)
Straits of Hormuz	0.23–0.49	0.39–0.99	0.15–0.23	Dorgham and Mofta (1989)
Gulf of Oman	0.19–0.79	1.62–5.48	0.12–0.59	Dorgham and Mofta (1989)
Transect north east to Oman	0.10–1.25	0–10	–	Grasshoff (1976)

(Reynolds, 1993). In such areas, maximum surface temperature of about 27–28°C decreased to <18°C at 60 m in the former site and to 20°C in the latter (Reynolds, 1993). Temperatures are low in mid-Gulf and increase towards the coast. Results of the Meteor Expedition show waters of <22°C rising from 90 m to 20 m in Hormuz Straits. These cold, dense waters are oxygen-depleted and enriched with phosphates (Hartmann et al., 1971).

Salinity

Salinity is elevated over ocean values, and ranges from 37–38 psu in the Straits of Hormuz to 38–41 psu near the northwestern end (Reynolds, 1993). Seasonal variations are low. Exceptions are some coastal waters that are diluted to low salinities of 23 psu while in isolated lagoons and coastal embayments such as the Gulf of Salwa, salinity increases to 60–70 psu (Basson et al., 1977). Many areas in the Gulf have a salinity exceeding 40 psu year round. The extent of these high-salinity areas is lowest during spring and summer but increases during winter (Al-Abdul Razzak, 1984). There is usually no salinity stratification.

Dissolved Oxygen and Nutrients

Oxygen distribution is fairly uniform due to the well-mixed nature of these shallow waters, and differences off Kuwait, for example, are insignificant (Al-Abdul-Razzak, 1984). In the Shatt Al-Arab estuary and the canals around Basrah city, oxygen saturation may be low (15.3 and 72.5%) due to BOD inputs (Hadi et al., 1989).

A north–south gradient seems to exist for phosphate (P), silicate (Si) and nitrate (N) (Table 2). These nutrients are most abundant in northern Gulf waters influenced by the

Shatt Al-Arab flow and may also be attributed to the discharge of sewage and industrial wastes. In canal waters of the Shatt Al-Arab, highest levels of nitrate and silicates are in September (Saad, 1984), and lowest nitrates are in June, when river run-off is usually lowest; for silicates, lowest values are in March, probably due to uptake by diatoms. Phosphates were exhausted in April and the highest values were in October (Saad, 1985). In these shallow waters absence of any vertical gradient or a diurnal rhythm in the nutrients is attributed to their well-mixed nature (Schiewer et al., 1982). Levels of nutrients in the Shatt Al-Arab were in general higher than many tropical and sub tropical waters, but are lower than many European estuaries (Schiewer et al., 1982).

In Shatt Al-Arab, average nitrate:phosphate ratios ranged from 0.2:1 in the upper reaches to 13.6:1 in lower reaches (Saad, 1985). The relative increase in nitrate is attributed to sewage, industrial pollution and land run-off (Saad, 1985). The higher nitrate levels suggest that it may not be a limiting nutrient. Off Kuwait, higher levels of nitrate, phosphate and silicate were associated with stations in the north impacted by the flow of Khor Al-Zubair leading to Khor Al-Sabiyah, which in turn is influenced by the man-made 'Third River' (Al-Yamani et al., 1999) (Fig. 4). Qatari waters have higher levels of phosphate and nitrate than the UAE, while silicates are higher in UAE water (Dorgham and Moftah, 1989). In the Straits of Hormuz, a tongue of water with >1.25 μmol PO_4-P originating from layers below the thermocline in the Gulf of Oman, penetrates at about 30 m (Grasshoff, 1976). In shallow coastal waters, turnover rates of dissolved and particulate nutrients can be rapid (Benitez-Nelson and Buesseler, 1999).

Brewer and Dryssen (1985) suggested that inflowing surface water through the Straits also enriches Arabian Gulf waters. A gradient in the phosphate from south to north along the Iranian coast is evident from "Meteor" data (Grasshoff, 1976). Although there is a similar gradient of silicate, Grasshoff (1976) suggested its low levels (<6 μmol SiO_2-Si) are due to slow mineralisation.

Redfield ratios (P:N:Si) range between 1:0.6 and 2.2:1.9–77.7 for the surface and 1:2 to 7.5:9.9–43.3 for bottom waters (El-Samra and El-Deeb, 1988) and thus show a wide variation in the Gulf. Ratios tend to be high in the northern waters and decrease to the south. The high P:Si ratios in the north are probably due to the contribution of silicates by run-off from the three rivers Tigris, Euphrates and Karun. Outside the Straits of Hormuz, P:Si ratios were below the Redfield ratio of 1:15. The P:N ratios in the whole Gulf region were less than the Redfield ratio of 1:16 which shows a substantial denitrification in these waters.

MAJOR COASTAL HABITATS AND BIODIVERSITY

The western Arabian Gulf is fairly well known (Basson et al., 1977) but the Iranian coast is not (Price and Robinson, 1993). The Gulf offers a variety of benthic habitats, from estuarine biotopes in the north to hypersaline lagoons, including exposed coastal beaches, tidal flats, a few mangroves, seagrass beds, coral reefs and islands, and artificial structures. Details on the fauna and flora of these biotopes can be found in Basson et al. (1977), Halwagy et al. (1986) and Sheppard et al. (1992).

Pelagic Biota

From the Gulf 527 algal species are reported (Al-Saadi and Hadi, 1987), with a north to south gradient in phytoplankton distribution (Subba Rao and Al-Yamani, 1998). The Shatt Al-Arab waters in the north have low species diversity (<116 species), high biomass (94 μg Chl a l^{-1}) and high primary production (3181 μg C l^{-1} h^{-1}). The northern Gulf waters off Kuwait have higher species diversity (<148 species), lower biomass (14 μg Chl a l^{-1}) and lower production (867 μg C l^{-1} h^{-1}). However, a few hot spots with a high biomass (55.4–262.7 μg Chl a l^{-1} and production 507.9–571.2 l^{-1} μgC l^{-1} h^{-1}) exit in these waters (Subba Rao et al., 1999). Further south the species diversity becomes highest (527 species) with a lower biomass (1.18 μg Chl a l^{-1}).

Tropical species constitute the bulk of phytoplankton in the Gulf. Based on the phytohydrographic associations of the floral elements, four broad groups are recognised (Table 3). There are 22 tychopelagic diatom taxa (Maulood and Hinton, 1979) compared with 205 fouling diatom taxa (Hendy, 1970) most of which are of benthic origin. The tidal exchange that is accelerated by the currents, eddies and

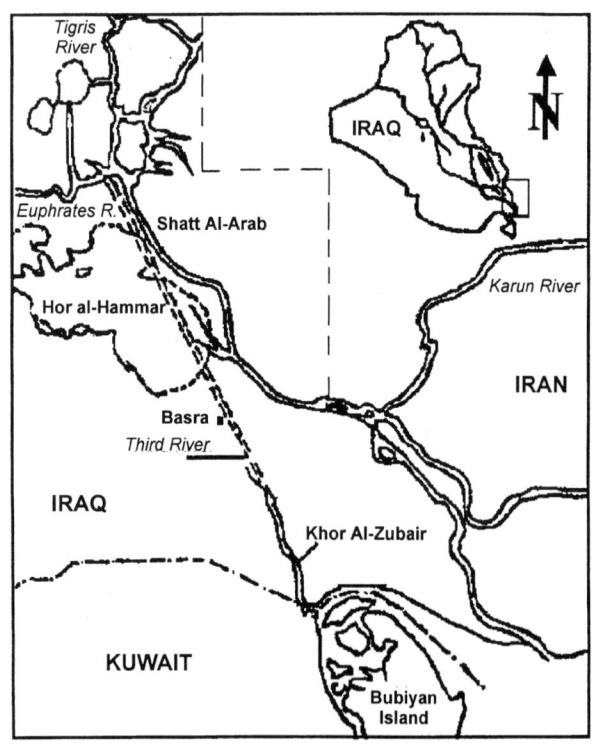

Fig. 4. Route of the 'Third River' in Iraq.

Table 3
Phytohydrographic associations of phytoplankton in the Gulf.

Tropical		Temperate	Ptychopelagic
Oceanic	Coastal		
Bacteriastrum delicatulum	Bellorochea malleus	Chaetoceros affine	Achnanthes
Chaetoceros peruvianum	Chaetoceros diversum	Chaetoceros compressum	Amphiprora
Chaetoceros coarctatum	Chaetoceros messanensis	Chaetoceros didymum	Amphora
Climacodium frauenfeldianm	Ditylum sol	Chaetoceros lasciniosum	Caloneis
Coscinodiscus centralis	Hemiaulus cuneiformis	Chaetoceros lorenzianum	Campylodiscus
Planktoniella sol	Leptocylindrus danicus	Hemiaulus sinensis	Diploneis
Proboscia alata f. indica (=Rhizosolenia alata f. indica)	Skeletonema costatum	Odontella sinensis (=Biddulphia sinensis)	Fragilaria
Proboscia alata f.gracillima (=Rhizosolenia alata f. gracillima)	Stephanopyxis palmeriana	Tropidoneis lepidoptera	Grammatophora
Rhizosolenia. castracanei	Thalassionema nitzschioides		Licmophora
Rhizosolenia styliformis	Triceratium favus		
Thalassiosira leptopus (=Coscinodiscus lineatus)			
Thalassiosira subtilis			
Thalassiothrix frauenfeldii			

bottom topography (Siebold et al., 1969) facilitates their redistribution. Extensive discoloration of intertidal sands particularly during winter is common. These algae seem to thrive well and probably play an important role in the trophodynamics of the coastal waters.

Western Indian Ocean components are marked and decrease northwards (Thorrington-Smith, 1971; Subba Rao and Al-Yamani, 1998). The number of taxa associated with equatorial sub-surface water which are common in the Gulf decreases from 79 in the south to 37 in the waters off Kuwait. Further north off Bushehr, Iran, probably due to less exchange with the open ocean, the number of species is limited to about 10 (Hulburt et al., 1981). A high species diversity of zooplankton species exists off Kuwait (Michel et al., 1986), with copepods constituting 83% of total counts. Sporadic appearance of typically offshore species such as *Cavolinia longirostris, Sagitta bedfordii, Oncaea conifera, Sagitta regularis* and *Krohnitta subtilis* in Kuwait Bay illustrate the influence of the offshore counterclockwise circulation (Michel et al., 1986).

Seagrasses and Macroalgae

Seagrasses play an important role, particularly for commercially important crustaceans. Penaeid shrimp is a major fishery, and are reported to complete their life cycles in macro algal and seagrass habitat. Migratory and postlarval stages of this shrimp take shelter in seagrass beds in Arabian waters (Basson et al., 1977). Many areas have exposed limestone, which offers substrate for algal attachment, but in many very sheltered areas, mixtures of algae and seagrasses co-exist. The Gulf has four species of seagrass: *Halodule uninervis, Halophila stipulacea, H. ovalis* and *Syringodium isoetifolium*, while the dominant macroalgae are *Hormophysa* and *Sargassum* (Basson et al., 1977).

Production estimates of seagrass blades, excluding the roots and rhizomes, are about 100 g C m^{-2} y^{-1} (Basson et al., 1977). The annual energy equivalent of seagrass beds is about 1.4×100^{10} kCal y^{-1} which is equivalent to 95,000 barrels of oil (Basson et al., 1977).

Mangroves

The Gulf has only about 125–130 km^2 mangroves (Sheppard et al., 1992), most of which is on the shores of Iran (90 km^2). Ten km^2 occur along the Gulf Coast of Saudi Arabia and Bahrain, with the rest along the UAE (Price et al., 1993). *Avicennia marina* is the only species, due to low winter temperatures and high salinities. Total gross production estimates from Red Sea are about 1690 kg O$_2$ d^{-1} and account for 86% of the relative productivity in mangal (Sheppard et al., 1992). The mangrove species *Rhizophora mucronata*, may have occurred earlier, but is now extinct due to over exploitation. In Kuwait Bay shoreline modifications due to anthropogenic activities have resulted in the slow erosion of the halophyte *Juncus arbicus* community.

Soft Substrates and Coral Reefs

There are few soft substrate benthic faunal studies of the Arabian Gulf. On the western side of the Arabian Gulf between Kuwait and Qatar, 80 species have been recorded, molluscs being dominant, followed by crustacea, polychaeta and echinodermata (Mohammed, 1995).

Corals and their distribution have been studied in considerable detail. In the northwestern half, Kuwait and Saudi Arabia have several coral islands and large patch reefs (Fig. 5). Probably because of salinity and temperature extremes, fewer than about 50 scleractinian coral species occur, 35 in Kuwait waters and 28 in Bahrain, and south of

Fig. 5. The Kuwaiti coral island of Kubbar, with fringing reef.

the latter in the Gulf of Salwa corals become very rare (Sheppard et al., 1992). Along much of the Arabian shore *Porites lutea, Acropora eurystoma* and *A. valida* are the reef builders, similar to the situation in the Red Sea (Carpenter et al., 1997). In Qatar, *Acropora* is most dominant. There are several areas with corals off the UAE, but near-shore, high salinity (>40 psu) restricts coral growth, and colonies that do occur tend to veneer older limestone and do not form reefs. Little information exists for Iran, but in the Straits of Hormuz, in Musandam, corals become abundant and diverse. In the Gulf, 200 fish species exist compared to 2000 in the Red Sea. The number of species and individuals vary within the nearshore and offshore locations, usually being greater in the reefs which grow in deeper water. Populations increase during summer–fall and decrease in late winter–early spring. Off Kuwait, the coral reef fish fauna comprise 124 species, representing 54 families. In the western Arabian Gulf the overall coral reef fish diversity seems to be limited because of both geographic isolation and environmental stress (Coles and Tarr, 1990). The green turtle *Chelonia mydas*, and the hawksbill turtle *Eretremochelys imbricata* are known to nest on at least two, and possibly three, of Kuwait's open water islands, as well as on Saudi Arabian islands. Four species of terns and dolphins live in the vicinity of coral reefs (Carpenter et al., 1997).

Sabkha

Sabkha is a low-lying, inter- and supratidal habitat, only seasonally inundated. They are very saline, often with crusts of white concentrated salts. Salinity in the pool reaches as high as 113 psu (Meshal, 1987). Sabkha development along the Arabian shores is the most developed in the world: euryhaline diatoms, mats of nitrogen-fixing cyanophytes and chaemosynthetic bacteria make up the bulk of the mats, which are inhabited by a few copepod species, flatworms and nematodes (Sheppard et al. 1992). Other sabkha fauna is of an impoverished Indo-Pacific nature, consisting of the imperforate foraminifer *Quinqueloculina, Triloculina, Cerithium* and cyprideid ostracods and some echinoderms (Evans et al., 1964). Production of the cyanophyta-dominated algal mats is likely to be higher than that of the mangroves (Price et al., 1993).

Rocky Shores

Rocky shore zonation of the algal species and animals conforms in general to worldwide patterns (Jones, 1986). However the eulittoral zone and sub-littoral fringe populations are characterized by warm temperate elements similar to those present at the southern end of the Indian Ocean. Compared to southern Gulf waters, on Kuwaiti shores there is an impoverishment of faunal elements attributed to low water temperatures (Jones, 1986). Only eurythermal species that can survive passage through tropical waters can occur in Kuwait (Jones, 1986).

Fisheries

A total of 465 marine fish species, representing 101 families in 24 orders, mostly tropical, were recorded from the Gulf (Kuronuma and Abe, 1986). In contrast with the Red Sea where 10–18% of fishes are endemic, in the Arabian Gulf only 1.5% are endemic, the rest being more widely spread in the Indian Ocean. Based on hydrological conditions, two main ichthyological areas are recognized: the northwest off Kuwait and the southeast off UAE and Qatar. A poorer benthic fauna characterizes the former, with sand and muddy bottom deposits, while the latter has a richer benthic fauna with rocks, gravel, coarse sand and broken shell fragments. The total number of species that are exclusive to the northwest is double (116) that of the southeast (60) (Kuronuma and Abe 1986). However trawl collections yielded 113 species from the southeast and 76 from the northwest, which suggests the former supported more benthic fishes than those from surface or mid layers. Of commercial fishes, Sideek et al. (1999) has recorded 350 fish species, 8 shrimp, 3 lobsters, and one species each of cuttle fish, crab and abalone. The demersals account for 40% of the total marine landings (475,000–552,000 tonnes).

Of shrimp species, four *Penaeus semisulcatus, Metapenaeus affinis, Parapenaeopsis stylifera* and *Metapenaeus stebbengi* contribute most to shrimp landings. In the Straits of Hormuz *Penaeus merguiensis* is important.

Mariculture

In the Gulf, demand for fish has increased more than its production. For example Kuwait's fish imports in 1982 were 7%, and this rose to 24.5% in 1989. In 1994 the imports were about 4500 tons compared to 1300 tons in 1983. It has been estimated that by the year 2008 Kuwait will need 17,560 tons of fresh fish, 4430 tons of frozen fish and 2670 tonnes of processed fish (KISR, 1989) and it is very unlikely that Kuwait will be self-sufficient. With a view to augmenting fish supply, mariculture has been actively pursued in Kuwait and progress has been made, particularly in development of grow-out technologies for sobaity (*Spaidentex hasta*), cage culture of gilthead sea bream and integration of *Tilapia* production with crop production. Projected estimates for mariculture and demands by the year 2008 are 9722 tonnes of zobaidy, hamoor, hamrah, nakoor, sheim, maid and sheem. This is about 40% more than that of 1987, worth almost U$ 140 million (KISR, 1989).

DEVELOPMENT ISSUES

Drainage and Irrigation

The discovery and exploration of oil in the Gulf region in 1901 changed its fortune, and the coastal landscape as well. Along the 9456 km coastline the urban population of 5.7×10^6 in 1980 and 13.38×10^6 in 2000 use the sea in several ways (Batanouny, 1999). A rapid urbanization necessitated coastal installations, development of ports, construction of power and desalination plants and other activities attendant with oil wealth. All continue to impact the Gulf environment. Today, the Gulf is one of the busiest waterways in the world. Despite its unique features and severe environmental perturbations, the need to study the Arabian Gulf remains acute. Extended baseline studies are absent as most of the studies are related to short-term projects and not to long-term institutionalized regional programs.

The Shatt Al-Arab watershed is the main source of freshwater entering the Gulf and therefore influences its productivity. Iraq has launched and completed in 1993 two major projects that may have detrimental effects on the functioning of the Gulf. The first is a 565 km Main Outfall Drainage (MOD), referred to as the Third River, running from northwest to southeast, situated between the Tigris and Euphrates (see Fig. 4). The associated farmland experiences chronic salt problems resulting in low crop production. The purpose of the Third River is to divert the flow of the Euphrates into a canal, collect the drainage water from salt-encrusted fields, as well as the wastewater. The discharge is about 210 m^3 s^{-1} into the Khor Al-Zubair and down to Khor Al-Sabiyah, flowing directly into the northwestern Arabian Gulf. This may prove beneficial to Iraq because it alleviates the chronic salt problem and produces a 50% increase in Iraq's crop production (Pearce, 1993). A second issue is that of draining the 15,540 km^2 marsh area in southern Iraq, which has resulted in the loss of about 60% of the marshes and associated wet land (Ryan, 1994) and probable alteration of the water quality in the Shatt Al-Arab.

After the Third River (1994–98) project, Kuwait experienced several hydrographic changes (Table 4), particularly during the spring peak river flow. These changes include a significant decrease in salinity, accompanied by higher nutrient levels and higher phytoplankton biomass (Al-Yamani et al., in press), probably due to altered flow

Table 4

Chemical constituents. Range, mean, standard error and number of observations during March–May for pre-1993 and post-1994 in Kuwait. Site located at 48°10'E and 29°20'N at a depth of 22 m. Mean, SE, *n* are given in parentheses.

	1985–1993	1994–1998
Salinity range (Psu)	36.00–41.79 (38.19; 1.54; 22)	30.60–37.48 (35.04; 1.96; 18)
Nitrate (μmol)	0.00–1.15 (0.30; 0.30; 20)	0.00–8.40 (1.61; 2.25; 18)
Phosphate (μmol)	0.00–0.30 (0.14; 0.09; 19)	0.00–2.1 (0.45; 0.52; 18)
Silicates μmol	0.11–11.60 (3.98; 3.50; 19)	0.84–17.77 (8.18; 5.20; 18)
Chlorophyll *a* (μg l^{-1})	0.52–2.55 (1.30; 0.48; 21)	0.15–23.66 (4.08; 5.64; 18)

regimes. This may have serious impacts on the functioning of the Gulf ecosystem. For example, the commercially important shrimp *Metapenaeus affinis*, the pomfret *Pampus argenteus*, and suboor *Tenualosa ilisha* seem to directly depend on the freshwater inflow for spawning or as a nursery habitat (Al-Yamani et al., 1997). Possible shifts in the dominance, phytoplankton community composition and changes in trophodynamics are to be expected.

The Southeast Anatolia Project in Turkey encompasses an area of 27,500 km^2, involving the construction of 22 dams on the upper Tigris and Euphrates Rivers to irrigate 1.7 million ha of agricultural lands and to generate 27.3 billion kilowatt-hours of electricity. As a consequence, a significant decrease in river flow would follow, with significant, far-reaching, and perhaps, irreversible environmental changes. Very likely, discharge of the Shatt Al-Arab will be drastically reduced, depriving the northern Gulf of nutrients and the lower salinities responsible for Kuwait's high productivity of commercially important species.

Landfill

Rapid urbanization has resulted in extensive landfill. Examples are the infilling of 40% of the Saudi coast, and dredging of >3000 ha of reclaimed area around the island of Bahrain (Price, 1993). In Kuwait, 64% of the most productive intertidal area of the South Kuwait Bay was eliminated by construction of 871 coastal structures including 64 km of sea walls, of which 93% are illegal. Because of infilling, longshore movements of sediment caused erosion on the down-current side. About 940 ha of intertidal area and another 405 ha are dredged in Kuwait. About 4.66×10^6 m^3 silt is dredged annually from shipping channels and the silt is disposed to make land and embayments, harbours and marinas. Being a depositional basin, dredging activity in Kuwait may resuspend the adsorbed pollutants that cause environmental problems, and disposal of spoil in Kuwait alone produces a turbidity envelope of 32 km^2 (Ghobrial and Kassim, 1986) which will seriously limit illumination for algal growth. Decline in habitats of the *Enteromorpha*, *Cladophora*, *Ulva* and *Sargassum* which shelter larval and juvenile shrimp could be reflected in a decline of the shrimp catches. The increased turbidity and excessive sedimentation results also in bleaching of corals, leading to mass mortality and elimination of this productive ecosystem.

Desalination and Power Plants

Desalination of sea water is the main source of potable water in most Gulf countries. For more than 20 years desalination plants have become a vital industry and played a significant role in the economy and welfare of this region. Often these are dual-purpose plants (DPPs) producing electricity as well. In Kuwait there are six power plants that produce 6789 mW which is 191 times higher than that three decades ago (Al-Mutaz, 1991). They also produce 1.173 million m^{-3} d^{-1} distilled water but discharge into the coastal zone 33.6 km^3 d^{-1} chlorinated cooling water and about 17 metric tonnes of residual oxidants. Saudi Arabia has 21 desalination plants with a total capacity of 5 Gigalitres per day, i.e. 50% of the global capacity of this form of water (Al-Sofi, 1994). The desalination plant at Bushehr, Iran produces 0.2 million m^{-3} d^{-1} distilled water. Similarly, more than 20 desalination plants are in operation in UAE in both the mainland and offshore islands. More than 80% of the 7×10^6 m^{-3} d^{-1} water produced in the Arabian Gulf is by multistage flash evaporation (MSF) plants (Ali-El-Saie, 1993), powered by fuels rich in sulphur and which operate usually at maximum brine temperature of 90°C and in some cases 110°C which upon discharge causes thermal pollution. By contrast, in the Red Sea effluents from desalination plants are normally +5°C and +3 PSU above the ambient (Linden et al., 1988).

The by-products of these plants include sulphur oxides, non-ferrous metals from tubeplates and sheets and polyphosphates used as anti-scalant agents in the evaporators. To prevent biofouling, 2.4 mg l^{-1} chlorine is added continuously with shock doses of 8 mg l^{-1} every 8 h for 20 min. The desalination blow-down contains dissolved minerals such as magnesium, calcium, potassium, chlorine, bromine, brominated methanes (Ali and Riley, 1986) and more specifically bromoform (BF), di-chloromethane (DBCM), dichlorobromomethane (DCBM) and chloroform (CF) (Shames-el-Din et al., 1991). In the vicinity of power and desalination plants, concentration of total haloforms could be as high as 90 μg l^{-1} and volatile liquid hydrocarbons 307 to 7882 ng l^{-1} (Ali and Riley 1986; Saeed et al., 1998). The annual average benzenoids level varies with the locality of the plant (Saeed et al., 1998); thus in Kuwait Bay it is low (677 ng l^{-1}) while at the intake near a power plant located near the oil-loading terminal and refineries, it may be higher (3006 ng l^{-1}). In Kuwait waters too, trihalomethanes constituted the bulk of the halogenated volatile liquid hydrocarbons (HVLHs) discharged (Saeed et al., 1998). From the brine, carbon dioxide is released between the evaporation stages and appreciable quantities of CO_3, $CaCO_3$ are formed (El-Din and Mohammed, 1989). Nearly 98% of haloforms are discharged into the atmosphere; the level of trihalomethanes in the distillate is small (Saeed et al., 1998). Although the levels of some of these chemicals are not alarming (Saeed et al., 1998), the impact of thermal and chemical pollution on the biota needs to be investigated. Use of zero-discharge direct-contact freezing/solar evaporation desalination plant complexes (Madani, 1992) merits serious consideration as an efficient system to reduce the environmental impacts.

Wastewater Pollution

The Gulf is considered as a sink for various liquid wastes, and so wastewater pollution is an acute problem. For example in Kuwait the Shuaiba industrial complex consists of 12 major industries including two petrochemical companies, three refineries, two power plants and an industrial gas corporation. This complex discharges directly into the Gulf 0.023×10^6 km^3 d^{-1} wastewater in addition to 0.003×10^6 km^3 d^{-1} sanitary wastewater (Al-Muzaini and Hamoda, 1998). Besides pathogens, these wastes contribute several chemicals including heavy metals. The input of raw sewage into the Gulf from three Saudi Arabian towns, Al-Khobar, Al-Hasa and Al-Qatif, is 40 km^3 d^{-1}, from 175,000 persons (IUCN, 1987). In Kuwait, with more than 90% of its urban population concentrated along the coastal towns Ardiya, Jahra and Reqqa, sewage is 0.282×10^6 km^3 d^{-1} (Al-Muzaini et al., 1991). Amongst the Gulf countries this sewage has the richest organic content (ROPME, 1986). Besides oil, these discharges have heavy metals, polychlorinated biphenyls and chlorinated pesticides at levels comparable to those in the UK and USA (Al-Muzaini et al., 1991). Chemical contamination due to sewage may be considerable. Levels of copper in the sewage may be 40–90 μg l^{-1}, zinc 380–700 μg l^{-1} and lead 80 μg l^{-1} (Al-Muzaini et al., 1991). BOD and COD in the presence of sewage is high and in Shuwaikh Port, Kuwait, corresponded to 209 and 390 mg l^{-1}.

Other measured contaminants in sewage include Hg (0.97 μg l^{-1}), Pb (16 μg l^{-1}), Cu (88 μg l^{-1}), Ni (24 μg l^{-1}), phosphates (2.4 mg l^{-1}), suspended solids (33 mg l^{-1}) and H_2S (6.1 mg l^{-1}), nitrate (1.8 mg l^{-1}), ammonia (5.6 mg l^{-1}) and low dissolved oxygen (<1 mg l^{-1}) (Al-Muzaini et al., 1991). In Shuwaikh Port, due to low flushing rates and shallow waters, one of the consequences is the anoxic and septic conditions (Al-Muzaini et al., 1991), confirmed by microbiological examination (Al-Muzaini and Hamoda, 1998). Besides these, effluents from slaughterhouses and dairy plants, for which no estimates exist, compound these problems.

Potentially Harmful Algal Blooms

Numerous algae and members of cyanophyceae may occur in bloom proportions in the Gulf (Subba Rao and Al-

Yamani, 1998), associated with inorganic nutrient enrichment caused by an increased disposal of untreated sewage (Al-Hasan et al., 1990).

There are 38 potentially harmful algal species in the Gulf (Subba Rao and Al-Yamani, 1998). Off Qatar, Al-Muftah (unpublished) recorded 46 potentially harmful taxa (17 diatoms, 25 dinoflagellates and four blue-green algae). In Kuwait Bay, numerous species of algae implicated elsewhere as being harmful do occur. During September–October 1999 several red water patches of algae comprising *Gymnodinium* spp., *Nitzschia longissima* and *Mesodinium rubrum* occurred off Kuwait. Concomitantly, 30 tonnes of wild mullets and 100 tonnes of cultured seabream died, but the linkage between the blooms and fish mortality could not be established.

Dwindling Fish Stocks

Changes in the finfish stocks, particularly zobaidy, hamoor and suboor seem to be related to events in the north. In general zobaidy and hamoor are declining. Between 1992 and 1995, the Kuwait zobaidy catch was 1108 t/yr on average, but decreased to 876 t in 1996. This fish is known to migrate during March–May to spawn in estuaries at the head of the Gulf. Similarly Hamoor catches declined from 532 t in 1992, to 286 t in 1996. However, the suboor catches increased from 518 t in 1992, to 1147 t in 1996, which is probably caused by high recruitment (Al Hossaini, unpublished).

The Gulf shrimp fishery is very important and constitutes about 15–20% of total fishery landings. Total shrimp landings for the Gulf were 13,600 tonnes, compared to 33,500 tonnes of shellfish and finfish for 1986. In the northwest Gulf, industrial and artisanal fleets are engaged in the shrimp fishery, and shrimp landings were 7 to 9 times larger than those from the Red Sea (Venema, 1984). Off Kuwait catches were around $4-5 \times 10^3$ t before the 1991 invasion and valued at US$10 million, the second largest export earner next to oil (Pauly and Mathews, 1986). Post invasion landings were reduced to an average of 2×10^3 t (Mohammed and Xucai, 1996; Mohammed et al., 1998). This postwar decrease of the shrimp stock is attributed to decreased water temperature resulting from oil-fire smoke in 1991–92, or to overfishing or other environmental conditions (Mohammed et al., 1998).

The life history of the shrimp fishery is closely related to hydrological events. They spawn at night during early February–March on the Saudi coast and a little later in Kuwait waters. A female lays about 200,000 eggs which hatch within 24 h. The pelagic stage is about 2–3 weeks and the juveniles settle amongst the macroalgae *Sargassum* and *Hormophysa*, at the period of maximum algal growth. Then they move to seagrass beds and subsequently migrate offshore where they are fished. Over-exploitation of the shrimp stocks resulted in a collapse of this fishery in the early 1960s. It was therefore necessary for the management to force seasonal (March to September) closure of this fishery (Ye, 1998).

Pearl fishing, once considered important, is on the decline. Seagrass beds and the *Sargassam* harbour juveniles of the pearl oysters *Pinctada radiata* and *P. margaritifera* while the adults inhabit rocky surfaces. *P. radiata* and *P. margaritifera* are probably commercially threatened. In the Gulf, over-exploitation, pollution and degradation of nursery areas has affected the demersal fisheries (Siddeek et al., 1999).

OIL SPILLS AND GULF WAR

Hydrocarbon input into the Gulf is 47 times more than the world average (Golob and Bruss, 1984) or 10 times greater than that in the North Sea (Readman et al., 1992). Much of it can be accounted for by the 1983 Nowruz oil spill during the Iran–Iraq war and the 1991 oil spill during the Gulf War. The six-week Gulf war has devastated both Kuwait and Iraq, and the oil pollution due to the Gulf War has been an enormous environmental disaster. The 600 oil-well fires had an estimated emission of 500,00 tonnes per month of smoke, mainly carbon as soot. The dimension of the oil slick was 90 km long and 15 km wide, 200 mm thick containing about 6–8 million barrels ($0.95-1.27 \times 10^6$ m³) of crude oil (Patel and Tiwari, 1991). Also, there are about 250 sunken and leaking ships, particularly the "Amouriya" loaded with 100,000 tonnes of Iraqi crude, a potential environmental threat. The smoke reduced solar radiation to 79% that of 1990, mean seawater temperature was lower by 2.5°C than the overall daily mean for 1986–90 (McCain et al., 1993). Initially, the intertidal habitats of Kuwait, the northern half of Saudi Arabia (about 640 km) and some parts of the Iranian coast were affected. These oil slicks also affected offshore Saudi Arabian coral islands.

Effects due to an elevation of trace metals and total hydrocarbons following oil spills seem to be limited. For example, halophytes (Price et al., 1993), seagrasses (Kenworthy et al., 1993), coral reefs (Downing and Roberts, 1993; Mohammed and Al-Sadh, 1996), and zooplankton (Al-Yamani et al., 1993) were not affected. The effects were immediate and confined to mangroves, fish eggs and crustacean larvae in the surface microlayer (Tafiq and Olsen, 1993), bird life (Evans et al., 1993), and prawn stocks (Kedidi, 1995). The postwar recovery of the environment appears to be faster than predicted (Price, 1998) considering its magnitude, and large spatial and temporal scales.

Hydrocarbons, Heavy Metals and Other Contaminants

Earlier data (1966–90) showed that in the Gulf the total organic carbon was 0.5 to 1.51% and exceeded the natural background level of 0.5% (Al-Ghadban et al., 1994). More recent data demonstrated an increase to 2.8% (Al-Ghadban et al., 1994), attributed to increased oil tanker traffic and spills. Petroleum hydrocarbons were highest in the northwest and decreased in a southwesterly direction, a pattern closely related to prevailing currents and localized sources (Al-Lihaibi and Al-Omran, 1996). Off Bahrain, land based oil inputs amount to 31 tonnes y^{-1}. Marked spatial and temporal variations existed in aromatic petroleum hydrocarbons; the highest overall mean value of 88.5 $\mu g\,l^{-1}$ was in industrial areas and the lowest 16.1 $\mu g\,l^{-1}$ at remote locations (Madany et al., 1994). Along Saudi Arabian Gulf coastline, petroleum hydrocarbons in sediments were relatively low; interestingly an increase in the benthic animals was associated with sediment petroleum hydrocarbons (Coles and McCain, 1990).

Concentrations of the metals Pb, Ni, and V in the sediments decreased in the order Kuwait > Saudi Arabia > Bahrain > Qatar > United Arab Emirates (Bu-Olayan et al., 1996), in other words, from west to east. In coastal surface waters off Bahrain, United Arab Emirates and the Sultanate of Oman Cu, Zn, Cd, Hg and Pb levels were low, but "hot spots" with 3–4 times higher Cd and Zn exist near more populated and industrialized areas (Fowler et al., 1984). Muddy sediments off Kuwait are rich in Cr, Cu, Co, Ni, V, Zn and Mn. These too decreased southwards, with an increase in sand fraction (Basaham and Lihaibi, 1993). Anthropogenic enrichment by these elements is minimal (Basaham and Lihaibi, 1993) and following a heavy spill of oil during the 1991 Gulf War, toxic metal enrichment was not severe (Al-Muzaini and Jacob, 1996). Concentrations of V, Ni, Cr and Pb in the marine sediments off Kuwait are not very high. Mean concentrations of numerous metals in seawater and particulate matter have been determined (Bu-Olayan et al., 1998); their distribution is associated more with the origin of the samples than with the season.

Recent PAH concentrations in seawater range between 21.1 and 321 $\mu g\,l^{-1}$, comparable to pre-Gulf War values (El-Samra et al., 1986). In habitats exposed to wave action, tidal flushing and sunshine, contamination due to the spill was limited. Saeed et al. (1996) showed high contamination of PAHs (486 $\mu g\,kg^{-1}$) in sediments soon after the Gulf War but these decreased subsequently to 52 $\mu g\,kg^{-1}$. Surprisingly, petroleum hydrocarbon contamination was relatively low in the sediments and in bivalve molluscs from Bahrain in June 1991, following Gulf War spill, even in the most heavily contaminated sites. These levels are comparable to those from coastal regions of the Baltic Sea, northeastern USA and UK, and are attributed to decreased oil tanker traffic and associated deballasting (Readman et al., 1992). However, two years after the war, in sheltered muddy areas along the Saudi Arabian coast which had low weathering, oil remained trapped under hardened near-surface crust (Sauer et al., 1998).

Oil degrading microbes belonging to the genera *Rhodococcus*, *Pseudomonas*, *Bacillus*, *Arthrobacter* and *Streptomyces* and the fungi *Aspergillus* and *Penicillium* seem to be immobilized within cyanobacterial mats and these can biodegrade 50% of the oil within 10–20 weeks (Sorokhoh et al., 1995). Cyanobacterial mats recovered from the Gulf War oil spills faster than other biota. Similarly, recovery of coastal vegetation on Qatari coastal sites polluted with tar seems to be rapid (Hegazy, 1995).

Mercury contamination in sediments was low (0.01–0.37 $\mu g\,g^{-1}$ dry weight), comparable to those in other regions (Al-Madfa et al., 1994). Near shore sediments off Qatar had slightly elevated levels (0.14 to 1.75 $\mu g\,g^{-1}$ dry weight) probably due to leaching from solid waste disposal sites (Al-Madfa et al., 1994). Mercury levels in fish do not seem to pose a health problem. In the commercially important fish *Cybium comersonii*, *C. guttatum*, *Lutjanus coccineus* and *Psettodes erimei* mean values of mercury corresponded to 0.17, 0.14, 0.19 and 0.15 mg kg^{-1} (Parvaneh 1979). Samples of canned tuna (*Euthynnus offinis*) yielded 0.22–0.44 mg kg^{-1} (Parvaneh, 1979).

Other contaminants include tributyltin (TBT) in coastal sediments originating from ship repair and painting. Nearshore TBT was 352–1330 ng g^{-1} (Hasan and Juma, 1992) and in offshore fish was about 25 ng g^{-1} (Watanabe et al., 1996) which is above acceptable levels in the USA (0.5 $\mu g\,g^{-1}$ wet weight). Arsenic (73.7 $\mu g\,g^{-1}$) in black-banded bream, *Acanthopagrus bifaciatus*, was nearly 150 times more than acceptable levels (Attar et al., 1992).

Alarming levels of waste plastic pellets were reported along Gulf beaches (Khordagui and Abu-Hilal, 1994). Besides the aesthetic considerations, waste plastic poses a hazard particularly to the turtles and coastal bird life. Recreational activities are causing damage to coral reefs; anchoring, discarded ghost nets, spear fishing, and dumping of litter and engine oil are major sources of pollution. Three tonnes of ghost nets and 20 tonnes of debris were removed recently from the coral reefs off Kuwait by volunteer teams, which is a measure of the acute litter problem.

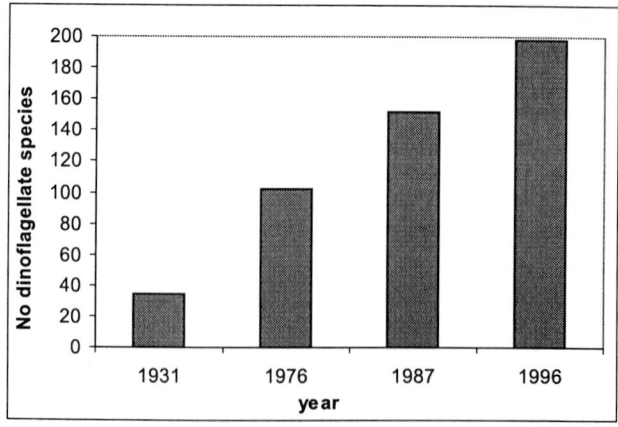

Fig. 6. Increase in numbers of dinoflagellate species in the Gulf.

Shipping Impacts

Annually about 20,000–30,000 tankers pass through the Straits of Hormuz (Linden et al., 1990). These may act as vectors for introducing exotic biota via ballast water. During the past 60 years there has been a steady increase in the number of dinoflagellate species in the Gulf (Fig. 6) (Subba Rao and Al-Yamani, 1998). This increase coincides with general organic enrichment (Dorgham et al., 1987), and some of the introduced tolerant species could survive and trigger extensive ecological changes in the phytoplankton community structure, leading to potential economic loss to commercial fisheries (Subba Rao et al., 1994). The mysidacean *Rhopalothalymus tattersalle*, the decapod *Exopalaeomon styliferus* from the Indian Ocean and the fish *Petrorhinus marmoratus* from Japan, are all known to have been introduced (Carleton and Geller, 1993).

PROTECTIVE MEASURES

When shrimp reserves were close to depletion, the Public Authority for Agricultural Affairs and Fish Resources (PAAFR) imposed a shrimp ban, which prohibits shrimping during the breeding season. This ban has had a positive effect on the shrimp stocks. For example in Kuwait territorial waters there were 750 tonnes in 1995 which had increased to 2552 tonnes by 1998. Each Gulf country has its Environment Public Authority (EPA) responsible for drafting marine Environment Protection Law. Recently the EPA of Kuwait recommended 42 Articles. When approved by the National Assembly, pollution of Kuwait's territorial waters will result in fines of up to US $120,000 and a maximum of one year imprisonment for violations.

FUTURE MARINE STUDIES

Several serious gaps in knowledge still exist for this region. Protection and sustainable development is essential for the eight Gulf countries as their survival depends on multiple-use development, and a deeper appreciation of this precious, beautiful environment and a vision to cultivate it and to protect it should be promoted through a firm commitment. A step in the right direction would be environmental education of the public, and at various levels of schooling as well. The relatively small size and shallow nature of the Arabian Gulf means that it is easily accessible and amenable for concerted marine studies.

In the new millennium, stresses on the Gulf are bound to increase. Consequently severe changes, particularly occurrence of harmful algal blooms and associated massive fish kills in the Gulf ecosystem may be anticipated. The opening of the Third River, draining of the marshes and damming of the rivers in the north, and the projected construction of 22 dams in Turkey, will drastically reduce the flow from the Tigris and Euphrates, which in turn will drastically affect the northern Gulf, and will possibly affect water quality and productivity of a wider area. Good baseline data on the spatial and temporal variations of the physical, chemical and biological characteristics of the Gulf are required. Food-web dynamics of filter feeders and pelagic feeders, and their dependence on environmental conditions in the Gulf need to be elucidated, as do effects of toxicants in sediments (Al-Ajimi et al., 1998). Protracted systematic studies remain necessary to assess the impacts likely to become apparent in the future. "Snapshot studies" do not provide a sufficient index for the functioning of the ecosystem (Price, 1998). Further, Price warned about the shortcomings and misjudgments about the health of the ecosystem resulting from incomplete time-series data and failure to dissociate synergism and antagonism due to oil pollution, war-related effects, background impacts and natural stresses. This is particularly so in view of the 'memory effect', i.e. reworking and return of oil from sediments over several years.

The main thrust needed in this area should be aimed at developing a regional, multinational, multi-disciplinary marine science program over several years. Regional cooperation, exchange of data, remote sensing technology and international expeditions are all desirable. It is crucial that the Gulf nations realize that their development is connected with the knowledge of the Gulf, so that they can support long-term ocean science programs which are integrated into the economic and social development of their region.

REFERENCES

Abuzinada, S.A. and Krupp, H. (eds.) (1994) The status of coastal and marine habitats two years after the Gulf War oil spill. *Courier Forsch. Inst. Senkenberg* **166**, 80 pp.

Al-Abdul-Razzak, F.H.Y. (1984) Marine Resources of Kuwait, Their Role in the Development of Non-Oil Resources. The University of Kuwait. Kuwait. 243 pp.

Al-Ajimi, D., Al-Muzaini, S. and Al-Sawari, M.A. (1998) Editorial recommendations: The long-term environmental effects of the Gulf War. *Environmental International* **24**, 1–4.

Al-Ghadban, A.N., Jacob, P.G. and Abdali, F. (1994) Total organic carbon in the sediments of the Arabian Gulf and need for biological productivity investigations. *Marine Pollution Bulletin* **28**, 356–362.

Al-Hasan, R.H., Sorokhoh, N.A., Al-Bader, D. and Radwan, S.S. (1994) Utilization of hydrocarbons by cyanobacteria from microbial mats on oily coasts of the Gulf. *Applied Microbial Biotechnology* **41**, 615–619.

Al-Lihaibi, S.S. and Al-Omran, L. (1996) Petroleum hydrocarbons in offshore sediments from the Gulf. *Marine Pollution Bulletin* **32**, 65–69.

Al-Madfa, H., Dahab, O.A. and Holail, H. (1994) Mercury pollution in Doha (Qatar) coastal environment. *Environmental and Toxicol. Chemistry* **13**, 725–735.

Al-Mutaz, I.S. (1991) Operation of dual purpose MSF plants at waterpower peak demand. *Proc. 12th Int. Symposium on Desalination and Water Re-use. Vol. 4. Evaporative Processes Operations, Electrodialysis, Material Selection and Corrosion* **84**, 105.

Al-Muzaini, S. and Beg, M.U. (eds.) (1998) The long-term environmental effects of the Gulf War. *Environmental International* **24**, 248 pp.

Al-Muzaini, S. and Hamoda, M.F. (eds.) (1998) Proceedings of The Third Middle-East Conference on Marine Pollution and Effluent Management. Kuwait, 305 pp.

Al-Muzaini, S. and Jacob, P.G. (1996) An assessment of toxic metals content in the marine sediments of the Shuaiba industrial area, Kuwait, after the oil spill during the Gulf War. *Water Science and Technology* **34**, 203–210.

Al-Muzaini, S., Samhanand, O. and Hamoda, M.F. (1991) Sewage-related impact on Kuwait's marine environment—a case study. *Water Science and Technology* **23**, 181–189.

Al-Saadi, H.A. and Hadi, R.A.M. (1987) Ecological and taxonomical studies on phytoplankton in Arab Gulf. Arabic section. *Journal of Biological Science Research* **18**, 7–31.

Al-Sofi, M.A.K. (1994) Water scarcity—the challenge of the future. Proc. IDA and WRPC World Conference on Desalination and Water Treatment, ed. M. Balaban, 98, 425–435.

Al-Yamani, F., Bishop, J., Ismaeil, W., Al-Rifaie, K., Al-Yaqoot, A., Kwarteng, A., Al-Ghadban, A., Al-Omran, L. and Sheppard, C. (1997) Assessment of the effects of the Shatt Al-Arab's altered discharge regimes on the ecology of the northern Arabian Gulf. Final report, FM006K. Kuwait Institute for Scientific Research. Kuwait, 247 pp.

Al-Yamani, F.Y., Al-Rifaie, K. and Ismail, W. (1993) Post-spill zooplankton distribution in the NW Arabian Gulf. *Marine Pollution Bulletin* **27**, 239–243.

Al-Yamani, F.Y., Subba Rao, D.V., Al-Omran, L., Al-Rifaei, K., Al-Said, T., Al-Yaqout, A., Bishop, J., Ismail, W., Al-Dashti, J. and Lennox, A. (in press) Impact of river diversion on the hydrobiology of the Northern Arabian Gulf. *Kuwait Journal of Science and Engineering*, in press.

Ali, M.Y. and Riley, J.P. (1986) The distribution of halomethanes in the coastal waters of Kuwait. *Marine Pollution Bulletin* **17**, 409–414.

Ali-El-Saie, M.H. (1993) The MSF desalination process and its prospects for the future. *Proc. Desal 92, Arabian Gulf Regional Water Desalination Symposium* **2**, 43–56.

Attar, K.M., El-Faer, M.Z., Rawdah, T.N. and Tawabini, B.S. (1992) Levels of arsenic in fish from the Arabian Gulf. *Marine Pollution Bulletin* **24**, 94–97.

Banse, K. (1997) Irregular flow of Persian (Arabian) Gulf water to the Arabian Sea. *Journal of Marine Research* **55**, 1049–1067.

Basaham, A.S. and Al-Lihaibi, S.S. (1993) Trace elements in sediments of the western Gulf. *Marine Pollution Bulletin* **27**, 103–107.

Basson, P.W., Buchard, Jr., J.E., Hardy, J.T. and Price, A.R.G. (1977) *Biotopes of the Western Arabian, Marine Life and Environments of Saudi Arabia*. Aramco Department of loss prevention and Environmental Affairs, Dahran, 284 pp.

Batanouny, K.H. (1999) The Mediterranean coastal dunes in Egypt: an endangered landscape. *Estuarine, Coastal and Shelf Science* **49**, 3–9.

Benitez-Nelson, C. and Buesseler, K.O. (1999) Variability of inorganic and organic phosphorus turnover rates in the coastal ocean. *Nature*, **398**, 502–505.

Brewer, P.G. and Dryssen, D. (1985) Chemical oceanography in the Persian Gulf. *Progress in Oceanography* **14**, 188–197.

Bu-Olayan, A.H., Al-Omran, L.A. and Subrahmanyam, M.N.V. (1996) Trends in the dissolution of metals from sediments collected during the Umitaka-Maru cruises using microwave-acid digestion technique. *Environment International* **22**, 711–716.

Bu-Olayan, A.H., Subrahmanyam, M.N.V., Al-Sarawi, M. and Thomas, B.V. (1998) Effects of the Gulf War oil spill in relation to trace metals in water, particulate matter, and PAHs from Kuwait coast. *Environment International* **24**, 789–797.

Carleton, J.T. and Geller, J.B. (1993) Ecological roulette: The global transport of nonindigenous marine organisms. *Science* **261**, 78–82.

Carpenter, K.E., Harrison, P.L., Hodgson, A., Al-Saffar, A.H. and Alhazeem, S.H. (1977) The corals and coral reef fishes of Kuwait. Kuwait Inst. Sci. Res. Environment Public Authority, 166 pp.

Chao, S.Y., Kao, T.W. and Al-Hajri, K.R. (1992) A numerical investigation of circulation in the Arabian Gulf. *Journal of Geophysical Research* **97** (C7), 11219–11236.

Coles, S.L. and Tarr, A.B. (1990) Reef fish assemblages in the western Arabian Gulf: a geographically isolated population in an extreme environment. *Bulletin of Marine Science* **47**, 696–720.

Coles, S.L. and McCain, J.C. (1990) Environmental factors affecting benthic infaunal communities of the western Arabian Gulf. *Marine Environment Research* **29**, 289–315.

Dorgham, M.M. and Moftah, A. (1989) Environmental conditions and phytoplankton distribution in the Arabian Gulf and Gulf of Oman, September 1986. *Journal of the Marine Biologists Association of India* **31**, 36–53.

Dorgham, M.M., Muftah, A. and El-Deeb, K.Z. (1987) Plankton studies in the Arabian Gulf. II. The autumn phytoplankton in the Northwestern Area. *Arab Gulf Journal of Scient. Agric. Biol. Science* **B5**, 215–235.

Downing, N. and Roberts, C. (1993) Has the Gulf War affected coral reefs of the northwestern Gulf? In: The 1991 Gulf War: coastal and marine environmental consequences, eds. A.R.G. Price and J.H. Robinson. *Marine Pollution Bulletin* **27**, 149–156.

Dureja, P. and Prasad, D. (1991) Gulf War and its effects on marine life. In *Gulf War and Environmental Problems*, ed. K.S. Ramachandran, pp. 201–205. Asia Publishing House, New Delhi.

El-Din, A.M.S. and Mohammed, R.A. (1989) The problem of alkaline scale formation from a study on Arabian Gulf water. *Desalination* **71**, 313–324.

El-Samra, M.I and El-Deeb, K.Z. (1988) Horizontal and vertical distribution of oil pollution in the Arabian Gulf and the Gulf of Oman. *Marine Pollution Bulletin* **19**, 14–18.

El-Samra, M.I., Emara, H.I. and Shunbo, F. (1986) Dissolved petroleum hydrocarbons in northwestern Arabian Gulf. *Marine Pollution Bulletin* **17**, 65–68.

Evans, G. (1970) Coastal and nearshore sedimentation: A comparison of clastic and carbonate deposition. *Proceedings Geological Association London* **81**, 493–508.

Evans, G., Kendall, C.G. and Skipwith, P.A. (1964) Origin of the coastal flats. The Sabka of the Trucial Coast. *Nature* **202**, 759–761.

Evans, M.I., Symens, P. and Pilcher, C.W.T. (1993) Short-term damage to coastal bird population in Saudi Arabia and Kuwait following the 1991 Gulf war marine pollution. *Marine Pollution Bulletin* **27**, 157–161.

Fowler, S.W., Huynh-Ngoc, L. and Fukai, R. (1984) Dissolved and particulate trace metals in coastal waters of the Gulf and western Arabian Sea. In M.V. Angel (ed.), Marine Science of the Northwest Indian Ocean and adjacent waters. *Deep Sea Research* **31**, 719–729.

Ghobrial, F. and Kassim, P. (1986) Environmental and hydraulic feasibility studies of the proposed navigation channel between Shuwaikh Port and Umm Al-Namel Island. EES-72 Final report, Kuwait Institute for Scientific Research, Report No. KISR2023, Kuwait.

Golob, R. and Bruss, E. (1984) Statistical analysis of oil pollution in the Kuwait Action Plan Region and the implications of selected worldwide oil spills to the region. Combatting oil pollution in the Kuwait Action Plan Region. IMO, London, 44, pp. 7–34.

Grasshoff, K. (1976) Review of hydrographic and productivity conditions in the Gulf Region. In Marine Sciences in the Gulf Area. *UNESCO Technical Papers in Marine Science* **26**, 39–62.

Hadi, R.A.M., Al-Mousawi, A.H. and Al-Zubaidy, A.J.M. (1989) A study on the primary productivity in the Shatt Al-Arab estuary at Basrah, Iraq. *Journal of Biological Sci. Research* **20**, 593–606.

Halim, Y. (1984) Plankton of the Red Sea and the Arabian Gulf. *Deep Sea Research* **34** (A,6–8), 969–982.

Halwagy, R., Clayton, D. and Behbehani, M. (eds.) (1986) *Marine Environment and Pollution*. Kuwait University, Kuwait, 348 pp.

Hartmann, M., Lange, E., Siebold, E. and Walger, E. 1971. Obser-

flaschen sedimente im Persischen Golf and Golf von Oman. I. Geologische Hydrologischer Rahmen und erste Sedimentologische Ergebnisse Meteor Forschungsergebnisse. Series C (4): 1-76.

Hasan, M.A. and Juma, H.A. (1992) Assessment of tributyltin in the marine environment of Bahrain. *Marine Pollution Bulletin* **24**, 408–410.

Hegazy, A.K. (1995) Phytoplankton monitoring and management of tar piles on the Qatar coastal marshes, Arabian Gulf. *Environmental Pollution* **90**, 187–190.

Hendy, N.I. (1970) Some littoral diatoms of Kuwait. *Nova Hedwigia* **31**, 101–167.

Hulburt, E.M., Mohmoodian, F., Russell, M., Stalcup, F., Lalezary, S. and Amirhor, P. (1981) Attributes of the plankton flora at Bushehr, Iran. *Hydrobiologia* **79**, 51–63.

Hunter, J.R. (1982) The physical oceanography of the Arabian Gulf: A review and theoretical interpretation of previous observations. Proceedings, *First Gulf Conference on Environment and Pollution*, Kuwait University, February 7–9, pp. 1–23.

IUCN (1987) Saudi Arabia: An assessment of biotopes and coastal zone management requirements for Arabian Gulf Coast. MEPA Coastal and Marine Management Series, Report 5. IUCN, Gland.

Jacob, P.G. and Al-Muzaini, S.L. (1995) Marine plants of the Arabian Gulf and effect of oil pollution. *Mahasagar* **28**, 83–101.

Jacob, P.G., Zarba, M.A. and Mohammad, O.S. (1982) Water quality characteristics of selected beaches of Kuwait. *Indian Journal of Marine Science* **11**, 233–238.

John, V.C. (1992) Circulation and mixing processes and their effect on pollution distribution in the western Arabian Gulf. *Applied Ocean. Research* **14**, 59–64.

Jones, D.A. (1986) A field guide to the sea shores of Kuwait and the Arabian Gulf. University of Kuwait, Kuwait.

Kedidi, S.M. (1995) Review of the Saudi Arabian Gulf shrimp fishery with a summarized assessment of the 1991 Gulf War effects. Papers presented at the Technical consultation on shrimp management in the Arabian Gulf, FAO, pp. 1–9.

Kenworthy, W.J., Durako, M.J., Fatemy, S.M.R., Valavi, H. and Thayer, G.W. (1993) Ecology of sea grasses in northern Saudi Arabia a year after the Gulf War oil spill. *Marine Pollution Bulletin* **27**, 213–222.

Khordagui, H.K. and Abu-Hilal, A.H. (1994) Industrial plastic on the southern beaches of the Arabian Gulf and western beaches of the Gulf of Oman. *Environmental Pollution* **84**, 325–327.

KISR (1989) Evaluation of fisheries by-catch and agricultural by-products for fish feed, Technical Appendix 4, Final Report; Strategic framework and master plan for fisheries development in Kuwait, report No 2907, Kuwait Inst. Scientific Res. 59 pp.

Koske, P. (1972) Hydrographische verhaltnisse im Persischen Golf Grund von beobachtungen von F.S. 'Meteor' im Fruhjahr 1965. 'Meteor' Forsch. *Ergebn., Gebruder Borntraeger, Berlin* **A11**, 58–73.

Kuronoma, K. and Abe, Y. (1986) *Fishes of Kuwait*. Kuwait Institute for Scientific Research, Kuwait, and the Dai Nippon Print Company, Tokyo, Japan.

Lardner, R.W., Al-Rabeh, A.H., Gunay, N., Hossain, M., Reynolds, R.M. and Lehr, W.J. (1993) Computation of the residual flow in the Gulf using the *Mt. Mitchell* data and the KFUPM/RI hydrodynamical models. *Marine Pollution Bulletin* **27**, 61–70.

Linden, O., Larsson, U. and Al-Alawi, Z.S. (1988) Effects of oil pollution in a shallow sub-tropical marine environment. *Oil Chem. Pollution* **5**, 65–79.

Linden, O., Abdulraheem, M.Y., Gerges, M.A., Alam, I., Behbehani, M., Borhan, M.A. and Al-Khassab, L.F. (1990) State of the marine environment in the ROPME Sea Area. UNEP Regional Seas Reports and Studies No.112. Rev. 1. UNEP, Nairobi, Kenya.

Madani, A.A. (1992) Zero-discharge direct-contact freezing/solar evaporation desalination complex. *Desalination* **85**, 179–195.

Madany, I.M., Al-Haddad, A. Jaffar, A. and Al-Shirbini, E.S. (1994) Spatial and temporal distributions of aromatic petroleum hydrocarbons in the coastal waters of Bahrain. *Archives of Environmental Contamin. Toxicology* **26**, 185–190.

Maulood, B.K. and Hinton, G.C.F. (1979) Tychoplanktonic diatoms from a stenothermal spring in Kurdistan. *British Phycology Journal* **14**, 175–183.

McCain, K., Beard, D.W. and Fadlallah, Y.H. (1993) Influence of the Kuwait oil well fires on seawater temperature in the Western Gulf. *Marine Pollution Bulletin* **27**, 79–83.

Meshal, A.H. (1987) Hydrography of hypersaline coastal lagoon in the Red Sea. *Estuarine and Coastal Shelf Science* **24**, 167–175.

Michel, H.B., Bhebehani, M., Herring, D., Arar, M., Shoushani, M. and Brakoniecki, T. (1986) Zooplankton diversity, distribution and abundance in Kuwait waters. *Kuwait Bulletin of Marine Science* **8**, 37–105.

Mohammed, S.Z. (1995) Observations on the benthic macrofauna of the soft sediment on western side of the Arabian Gulf (ROPME sea area) with respect to 1991 Gulf War spill. *Indian Journal of Marine Science* **24**, 147–152.

Mohammed, S.Z. and Al-Saadh, S. (1996) Coral reef grounds and its associated biota in the western side of Arabian Gulf (ROPME Sea Area) with respect to 1991 Gulf War oil spill. *Indian Journal of Marine Science* **25**, 35–40.

Mohammed, H.M.A. and Xucai, X. (1996) Variations in catch rates, species composition and size frequency of shrimp landings by dhow boats in Kuwait prior to and following the Gulf War. *Pakistan Journal of Zoology* **28**, 97–108.

Mohammed, H.M.A., Bishop, J.M. and Ye, Y. (1998) Kuwait's post Gulf-war shrimp fishery and stock status from 1991/92 through 1995/96. *Reviews in Fisheries Science* **6**, 253–280.

Parvaneh, V. (1979) An investigation on the mercury contamination of Persian Gulf fish. *Bulletin of Environmental and Contaminant Toxicology* **23**, 357–359.

Patel, M.K. and Tiwari, T.N. (1991) In *Oil Slicks: Their Fate at Sea and Methods of Control*, ed. K.S. Ramachandran. Ashish Publishing House, New Delhi, pp. 55–66.

Pauly, D. and Mathews, C.P. (1986) Kuwait's finfish catch three times more shrimp than its trawlers. *Naga ICLARM Quarterly, Philippines* **9**, 11–12.

Perrone, T.J. (1981) Winter Shamal in the Persian Gulf. Naval Environmental Prediction Research Facility, Monterey, CA, I.R.-79-06.

Price, A.R.G. and Robinson, J.H. (eds.) (1993) The 1991 Gulf War: coastal and marine environmental consequences. *Marine Pollution Bulletin*, **27**, 380 pp.

Price, A.R.G. (1998) Impact of the 1991 Gulf war on the coastal environment and ecosystems: current status and future prospects. *Environment International* **24**, 91–96.

Price, A.R.G., Sheppard, C.R.C. and Roberts, C.M. (1993) The Gulf: Its biological setting. *Marine Pollution Bulletin* **27**, 9–15.

Purser, B.H. (ed.) (1973) *The Persian Gulf*. Berlin, Springer-Verlag, Berlin.

Readman, J.W., Fowler, S.W., Villeneuve, J.P., Cattini, C., Oregioni, B. and Mee, L.D. (1992) Oil and combustion-product contamination of the Gulf marine environment following the war. *Nature* **358**, 662–665.

Reynolds, M. (1993) Physical oceanography of the Gulf, Straits of Hormuz, and the Gulf of Oman: Results from the *Mt. Mitchell* Expedition. *Marine Pollution Bulletin* **27**, 35–59.

ROPME (1997) Oceanography and pollution studies in the ROPME sea area. A bibliography. ROPME/GC-9/001, Kuwait.

Ryan, S. (1994) Saddam kills off wildlife in eco-disaster. *Arab Times* (April 19), 2.

Saad, M.A.H. (1984) Effect of pollution on three highly eutrophicated canals of the Shatt-Al-Arab Estuary at Basrah (Iraq). 2. Nutrient salts. *Rev. Int. Oceanogr. Med.* **73–74**, 61–73.

Saad, M.A.H. (1985) Distribution of nutrients in Shatt-Al-Arab Estuary. *Bulletin Institute Oceanogr. Fish. Cairo* **11**, 27–42.

Saeed, T., Al-Muzaini, S. and Al-Bloushi, A. (1996) Post-Gulf War assessment of the levels of PAHs in the sediments from Shuiaba Industrial Area, Kuwait. *Proc. 18th Bienniel Conf. International Assoc. Water Quality.* (Eds. Bally, D., Asano, T., Bhamidimarri, R., Chin, K.K., Brabow, W.O.K., Hall, E.R., Ohgaki, S., Orhon, D., Milburn, A., Purdon, C.D., Nagle, P.T.) 34: pp. 195–201.

Saeed, T., Khordagui, A. and Al-Bloushi, A. (1998) Distribution of volatile liquid hydrocarbons in the vicinity of power/desalination plants in Kuwait. The Third Middle East Conference on Marine Pollution and Effluent Management, State of Kuwait, pp. 122-131.

Sauer, T.C., Michel, J., Hayes, M.O. and Aurand, D.V. (1998) Hydrocarbon characterization and weathering of oiled intertidal sediments along Saudi Arabian coast two years after the Gulf War oil spill. *Environment International* **24**, 43–60.

Schiewer, U., Al-Saadi, H.A. and Hameed, H.A. (1982) On the diel rhythm of phytoplankton productivity in Shatt-Al-Arab at Basrah, Iraq. *Archives Hydrobiology* 93, 158–172.

Shams-el-Din, A.M., Arain, R.A. and Hammoud, A.A. (1991) A contribution to the problem of trihalomethane formation from the Arabian Gulf water. *Desalination* **85**, 13–32.

Sheppard, C., Price, A. and Roberts, C. (eds.) (1992) *Marine Ecology of the Arabian Region.* Academic Press, 359 pp.

Sherman, K. (1993) Large marine ecosystems as global units for marine resources management—an ecological perspective. In *Large Marine Ecosystems. Stress, Mitigation, and Sustainability,* eds. K. Sherman, L.M. Alexander and B. D. Gold, pp. 3–14. American Association for the Advancement of Science, Washington, DC.

Siddeek, M.S.M., Fouda, M.M. and Hermosa, Jr. G.V. (1999) Demersal fisheries of the Arabian Sea, the Gulf of Oman and the Arabian Gulf. *Estuarine, Coastal and Shelf Science* **49**, 87–97.

Siebold, E. (1973) Biogenic sedimentation of the Persian Gulf. In *The Biology of the Indian Ocean,* ed. B. Zeitzschel, pp. 103–114. Springer Verlag, Berlin.

Siebold, E., Ulrich, J. and Wollerecht, K. (1969) Bodengestaldes des Persischen Golfs. 'Meteor' *Forsch. Ergebn.,* Gebruder Borntraeger, Berlin **C2**, 29–56.

Sorokhoh, N.A., Al-Hasan, R.H., Khanafer, M. and Radwan, S.S. (1995) Establishment of oil-degrading bacteria associated with cyanobacteria in oil-polluted soil. *Journal of Applied Bacteriology* **78**, 194–199.

Subba Rao, D.V. and Al-Yamani, F. (1998) Phytoplankton ecology in the waters between Shatt al-Arab and Straits of Hormuz–Arabian Gulf. *Plankton Biology Ecology* **45**, 101–116.

Subba Rao, D.V., Al-Yamani, F., Lennox, A., Pan, Y. and Al-Said, T.F.O. (1999) Biomass and production characteristics of the first red tide noticed in Kuwait Bay, Arabian Gulf. *Journal of Plankton Research* **21**, 805–810.

Sultan, S.A.R. and Ahmad, F. (1993) Surface and oceanic heat fluxes in the Gulf of Oman. *Continental Shelf Research* **13**, 1103–1110.

Tafiq, N.I. and Olsen, D.A. (1993) Saudi Arabia's response to the 1991 Gulf oil spill. *Marine Pollution Bulletin* **27**, 333–345.

Thorrington-Smith, M. (1971) West Indian Ocean phytoplankton: a numerical investigation of phytohydrographic regions and their characteristic phytoplankton associations. *Marine Biology* **9**, 115–137.

Venema, S.C. (1984) Fishery resources in the north Arabian Sea and adjacent waters. *Deep Sea Research* **31**, 1001–1018.

Watanabe, M., Hashimoto, S., Fujita, K. and Otsuki, A. (1996) Organotin compounds in fish in ROPME area. Abstract. Report of ROPME/UNEP/IOC/(UNESCO) international symposium on the status of marine environment in the ROPME sea area after 1990-91 environmental crisis with special emphasis to UmitakaMaru cruises. Regional Organization for Protection of the Marine Environment, Kuwait. (ROPME/GC-8/1). ROPME, 13124 Safat, Kuwait.

Ye, Y. (1998) Assessing effects of closed seasons in tropical and subtropical penaeid shrimp fisheries using a length-based yield-per-recruit model. *ICES Journal of Marine Science* **55**, 1112–1124.

THE AUTHORS

D.V. Subba Rao
Kuwait Institute of Scientific Research,
P.O. Box 1638,
22017 Salmiya, Kuwait

Faiza Al-Yamani
Kuwait Institute of Scientific Research,
P.O. Box 1638,
22017 Salmiya, Kuwait

Chapter 54

NORTHWEST ARABIAN SEA AND GULF OF OMAN

Simon C. Wilson

The four countries bordering this region are Yemen, Oman, United Arab Emirates and Iran, which together comprise 2800 km of coastline. In the Arabian Sea, the Indian monsoon is the single greatest factor shaping the marine environment. The Southwest Monsoon blows for four summer months, inducing coastal and oceanic upwelling that lowers seawater temperatures by 10°C and greatly increases marine productivity. A northeast wind then blows during winter months, which reverses surface circulation. The resulting variable and productive marine ecosystem is semi-isolated by oligotrophic waters more typical of tropical areas.

The most diverse coastal habitats include coral communities, mangroves, seagrass beds, coastal lagoons, and seaweed beds, the latter only developing on the Arabian Sea coasts during the Southwest Monsoon. The region also contains the world's largest concentration of Loggerhead Turtles and large numbers of Green Turtles. However, despite these valuable resources, there are few protected areas, and management is only partially effective.

The human population remains low but is growing at 3.5% per year. While national economies have benefited from oil reserves, most coastal and rural areas still depend on agriculture and fishing, and these sectors, particularly fishing, will absorb much of the new labour force in coming years.

Current impacts to the environment from rural activities are greater than industrial or urban impacts. Serious consequences arise from the widespread use of gillnets which are causing long-term damage to turtle and cetacean populations and coral areas. Overgrazing in some areas and overfishing of some high-value fish stocks are growing issues, and industrial fishing has led to conflicts with artisanal fishers and environmental damage.

Coastal erosion is of growing concern, particularly in the Gulf of Oman, and large areas of low-lying coastal sabkha in central Oman are at risk from rising sea levels. Chronic contamination of beaches results from routine oil tanker operations and continues to worsen, while the impact from industry, though still relatively small, is increasing as nations diversify their economies. International and national environmental legislation are in place across much of the region but resources and capability for their enforcement is generally lacking.

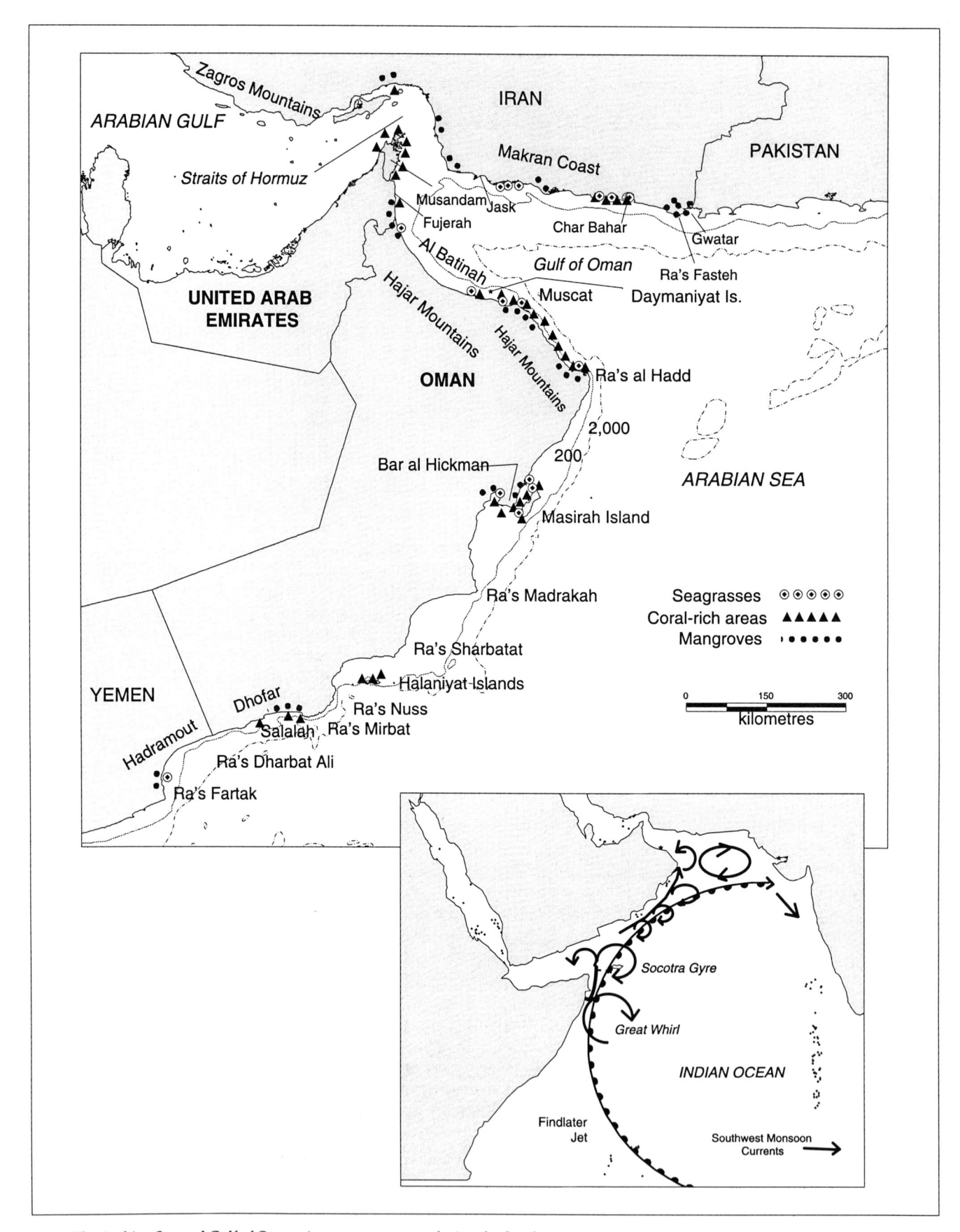

Fig. 1. The Arabian Sea and Gulf of Oman. Arrows are currents during the Southwest monsoon. Associated with the larger headlands are large eddies that transport productive coastal waters offshore 150 km or more. The arrow (marked *) is the mean flow of the Findlater Jet over the Arabian Sea in July. During this time, the Great Whirl is fully developed and spins out from the coast of Somalia to the south of Socotra.

THE DEFINED REGION

The northwestern Arabian Sea and Gulf of Oman extends in the southwest to the boundary of the Gulf of Aden, which runs between the Horn of Africa, and Ra's Fartak on the Arabian Peninsula (IMO, 1997). This chapter excludes archipelagoes within it which are covered elsewhere (e.g. the Indian Lakshadweep archipelago and Yemen's Socotra Archipelago). To the north, the Gulf of Oman is defined by the rhumb line between Ra's Al Hadd in Oman and Ra's Al Fasteh in Iran (IMO, 1997), and it extends to the straits of Hormuz. The eastern boundary is taken as a line south from the Iran–Pakistan border (i.e. from approximately 25°12′N, 061°36′E) (Fig. 1).

Four countries border these waters: Yemen, Oman, UAE and Iran. The southernmost 180 km, along Yemen's coast, is dominated by high limestone cliffs. The adjacent 300 km of southern Oman (Dhofar) are similar but are broken by the sandy coastal plain surrounding the city of Salalah. Stretching northward for 1000 km is a generally low, sandy desert coastline to Ra's Al Hadd, the easternmost point in Arabia. In the 200 km Between Ra's Al Hadd and Muscat lies a more rocky shore of raised fossil reefs and outcrops of dark ophiolite, which leads onto the sweeping beach of Al Batinah, some 320 km long. From here lies the short (80 km) stretch of mostly rocky coast of Fujeirah, an Emirate of UAE, which borders with the steep fjord-like coastline of Musandam, which is part of Oman. Across the Straits of Hormuz lies the Makran coast of southern Iran. Its desert coastal plain is 50–100 km wide in most places and stretches for over 800 km to the Iran–Pakistan border at Bandar Gwatar. In places this coast is steep with rugged outcrops, dominated by short, sandy beaches backing on to high, near vertical cliffs, sandy plains or dunes.

The seabed has a varied topography. The Straits of Hormuz, which is only 56 km wide at its narrowest point, is a simple trough that drops quickly from 100 m depth to more than 3000 m in the Gulf of Oman. The floor of the Arabian Sea lies at 3000–4000 m depth. From Ra's Al Hadd southwards, the continental shelf generally broadens to a maximum width of about 150 km before closing in again; along the Mirbat peninsula, water over 2000 m deep lies within 20 km of the shore.

Tectonically, the Arabian plate is moving northeastwards and colliding with the Eurasian plate at a rate of 3–4 cm/yr, which has given rise to the folded Zagros and Makran mountain ranges, marked by intensive earthquake and mud volcano activity (Ambrayseys and Melville, 1985). The oceanic part of the Arabian plate is subducting along the Makran Trench.

Although tropical, this region exhibits the greatest range of seasonal variability found in any ocean (Burkill, 1999). While its central waters are almost devoid of nutrients and exhibit low biological production, coastal regions are nutrient-rich and highly productive due to the Arabian upwelling. In biogeographical terms, Sheppard et al. (1992) consider the Arabian region to be partially separated by a barrier created across the Arabian Sea by the upwelling.

SEASONALITY, CURRENTS AND NATURAL ENVIRONMENTAL STRESSES

The over-riding influence on climate is the Indian monsoon system. The winds generate one of the five largest upwelling areas of the world, and its effect on both terrestrial and marine habitats is dramatic.

During summer months, low pressure over northern India and high pressure over the southern Indian Ocean gives rise to a strong south to north air flow which blows as a steady southwest surface wind of about 15 m/s over the Arabian Sea. Moisture falling in India releases latent heat which maintains and strengthens the pressure gradient. Over the Arabian Sea, these winds are focused into a narrow, strong, low-level stream known as the Findlater Jet that blows parallel with the coast of southern Arabia about 300 km offshore. The winds drive a strong surface Somali current across the Arabian Sea and induce both open ocean upwelling and coastal upwelling. Open-ocean upwelling occurs in an area along the Arabian Sea coast of Oman up to 300–400 km offshore and influences hydrography to a depth of about 250 m. Coastal upwelling is driven by the offshore deflection of surface waters by Ekman transport and is much more intense, affecting hydrography to a depth of about 400 m (Brock et al., 1992). This upwelling extends for about 150 km offshore along the full 1000 km length of the Arabian Sea coast from Ra's Fartak to Ra's Al Hadd, and its effect on coastal habitats is greatest where the continental shelf is narrowest.

Pulses of upwelling water are cool and rich in nutrients, triggering massive blooms of phytoplankton. However, some of the upwelled water comes from the Arabian Sea's oxygen minimum layer, so that if upwelling water is not mixed sufficiently with well-oxygenated surface water, it can result in widespread fish kills, particularly of demersal fish (Piechura and Sobajh, 1986).

The effect of the Southwest Monsoon on surface waters is remarkable. Coastal seawater temperatures fall from around 30–20°C, and nitrate and phosphate concentrations increase five fold to an average of 2 μg-at PO_4–Pl^{-1} and 20 μg-at NO_3–Nl^{-1} respectively (Savidge et al., 1990). The effect on productivity is still more dramatic, rising ten fold from less than 0.1 g C m^{-2} d^{-1} in the period between monsoons to above 1.1 g C m^{-2} d^{-1} during the Southwest Monsoon (Brock and McClain, 1992).

Upwellings associated with the monsoons vary considerably in strength between years depending on a number of factors including Indian Ocean sea surface temperatures, soil moisture content and even the El Niño phenomenon (Brock and McClain, 1992). For example, the weak monsoon of 1982 produced a phytoplankton bloom that was only 30% of that generated in 1980, a year with a particularly strong upwelling.

During winter, the pattern of differential heating reverses: the northern Indian landmass becomes cooler than the surrounding ocean, reversing the pressure gradient. Consequently a cool, dry northeast wind develops over the Arabian Sea from about November to April and skies are generally clear. These Northeast Monsoon winds are weaker, less than 5 m/s, and more variable than Southwest Monsoon winds.

The Northeast Monsoon also has a marked influence on productivity. Winds reverse the circulation pattern and biological production is elevated compared to the inter-monsoon months (Burkill, 1999). Cool northeast winds chill the warm surface waters to about 23–24°C which then sink and set in motion replacement by nutrient-rich water from around 100 m depth, generating 'entrainment blooms'. The effect of such spring upwelling is clearly seen in coastal seaweed communities: flushes of new algal growth have been observed in the Gulf of Aden (Ormond and Banaimoon, 1994), Socotra (Kemp, 1998) and Bar al Hickman. During inter-monsoon months winds are weak and variable, and when the supply of nutrients is exhausted, primary productivity falls dramatically.

Current flow generally mirrors the seasonal wind direction. During the Southwest Monsoon, a clockwise gyre is set up across the Arabian sea which persists until October when the transition to the Northeast Monsoon circulation begins (Brock and McClain, 1992).

During summer the Eastern Arabian Current runs parallel with the coast with a maximum speed of >1 m s^{-1} (Fig. 1). Its flow is broken by a series of large eddies that reach 300 km out to sea. Eddies of entrained coastal water are transported offshore beyond the continental shelf as plumes or filaments. Eddies tend to be associated with headlands, of which there are several in the region (Manghnani et al., 1998).

Where the Eastern Arabian Current reaches the Gulf of Oman, it divides into two eddies, cyclonic to the north and anti-cyclonic to the south, each about 200 km in diameter (Böhm et al., 1999). The northernmost of these eddies brings cool water from southern upwelling into the Gulf of Oman, strongly affecting the eastern Iranian coast. Reynolds (1993) suggests that upwelling is induced in winter months during the Northeast monsoon along the Iranian coast of the Gulf of Oman.

The diurnal tide has a mean spring range of 1 m in Yemen to almost 2 m in the northern Gulf of Oman. Tidal streams close to the coast are usually less than one knot, except through narrow straits; in the Straits of Hormuz, tidal streams reach 3 knots (National Hydrographic Office, 1998).

The prevailing climate is a hot steppe type, with humid coastal areas. This is strongly influenced by the monsoons, particularly south of 22°N where the summer monsoon and the cold waters modify the climate to a distance of up to 100 km inland. Mean 24 hour temperatures are lowest in January (20°C) and rise to a peak of 35–43°C in summer (Fig. 2). Maximum daytime summer temperatures may reach 50°C,

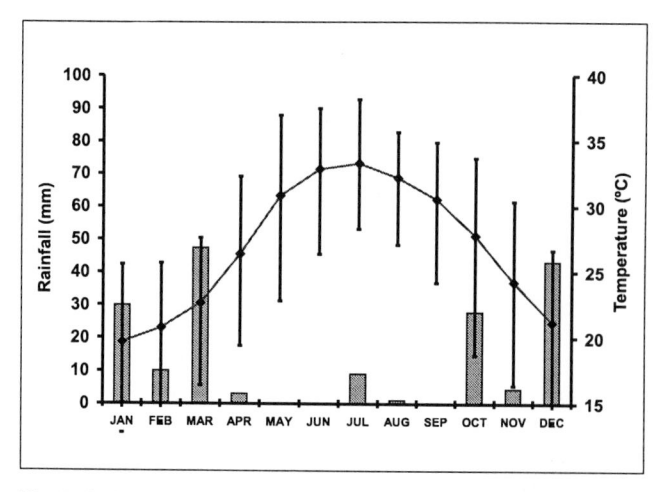

Fig. 2. Air Temperature and Rainfall, Sohar, 1993–1997. This graph illustrates the seasonal climate that prevails in the Gulf of Oman. The majority of rain falls in winter and is concentrated into a few heavy storms. Note that temperatures are 24-hour averages, and so do not reflect maximum day time temperatures. (Data from Directorate General of Civil Aviation and Meteorology, Ministry of Communications, Government of Oman).

while minimum night-time temperatures in winter can drop to 6°C. This may be strongly influenced by the winds that blow over cool seas which bring persistent clouds, fog and rain to coastal areas, in which case, temperatures during mid-summer are moderated by 10°C or more.

Over 90% of rainfall falls in winter, with much of the remainder falling during July and August. Most falls in heavy showers and storms, and in some years up to 55% of the years' rain—in total around 100 mm per year—may fall in a 24-hour period. Coastal areas adjacent to the upwelling have a different pattern, with most rain, totalling 120 mm per year, falling during the summer monsoon. As a result of the hot, dry conditions, evaporation exceeds precipitation by 1000 to 1500 mm per year.

The Arabian Sea is a cyclone-generating region, with tropical depressions developing most frequently around the Lakshadweep Islands (Murty and El-Sabh, 1984). Once formed, these generally move towards the Arabian Peninsula, sometimes curving towards Pakistan or the Gulf of Aden. About one in three reach the Arabian coast; storms hit the Arabian Sea coast about once in every three years and, of these, only half are likely to be fully developed cyclones. In all, 36 cyclones and tropical storms have affected the coast of Oman in the past 100 years. The last, in 1993 damaged coral communities in the Gulf of Oman, and there has been only a slight recovery since.

The extreme terrestrial and marine climate has important controlling effects on the biological systems. In the summer, the hot, dry desert conditions play a large part in determining the distribution of plants and animals. The marine climate in the sea can also be extreme. Savidge et al. (1990) recorded a minimum sea temperature of 15.9°C in the Arabian Sea during the 1989 Southwest Monsoon, while Salm (1992) reported a maximum seawater temperature of 39°C in 1990 in the Gulf of Oman, also during summer months. Exacerbating the potential problem of heat stress, seawater temperatures in the Gulf of Oman routinely fluctuate by up to 8°C in 24 hours (Coles, 1997).

The very low oxygen concentrations (less than 0.1 mg l^{-1}) that are normally found in water below 100 m is also important. This limits biomass living at intermediate depths, though a handful of pelagic species can tolerate these suboxic conditions. However, when this layer reaches the surface following upwelling, widespread mortality may result, particularly in demersal fish species.

The sporadic rainfall also leads to extremes. The thin and sandy soils in many watersheds are unable to absorb rainfall quickly, resulting in flash floods whose high peak flow rates carry large volumes of sediment and freshwater to the sea. Although infrequent, these extreme events are very important controlling factors affecting khawrs (embayments, often brackish, see later), mangrove stands and other shallow marine environments.

MAJOR SHALLOW WATER MARINE AND COASTAL HABITATS

Of the four countries included here, Oman has been most extensively surveyed in terms of its coastal and marine habitats, principally by Salm (IUCN, 1986; 1988; 1989a,b; 1991) and these form the basis for Oman's coastal management plans. The distribution of coastal habitats in Yemen and Iran is much less known (Huntington and Wilson, 1995).

The Arabian Sea spans the Tropic of Cancer, and habitats which are typical of sub-tropical and tropical environments such as coral reefs, mangrove woodlands and seagrass can be found along its shores (see Fig. 1). Superimposed on this, the major influence of upwelling creates an enclave of temperate marine conditions lasting four months that is surrounded by tropical conditions. This peculiar mixture of marine climates has led to a greater variety of marine and coastal life than would otherwise be possible.

About 70% of the coast has high and medium energy sandy beaches (McLachlan et al., 1998) separated by rocky headlands. Low-energy environments occur mainly in the Gulf of Oman where the influence of the Southwest Monsoon is reduced, although there are a few significant sheltered areas along the Arabian Sea coast, particularly around Bar Al Hickman.

Much of the Gulf of Oman is backed by dry and barren mountain ranges. In Iran, the Makran range rises over 1000 m from 50–100 km from the coast. Along the southern coast of the Gulf of Oman the Hajar Mountains reach over 1300 m and run parallel with the shore. Since rainfall is generally low, little terrestrial run-off reaches the sea from the relatively small and steep watersheds.

Along about 300 km of the coast of southern Arabia from Ra's Fartak in Yemen to Ra's Nuss in Oman lies a range of limestone mountains covered in a vegetation unique in Arabia that has developed in response to precipitation during the Southwest Monsoon. Peaks in this range rise to 2100 m and consist of a rolling grassy plateau, dissected by deep wooded valleys. Inland, the mountains drop gradually away to gravel hills and desert in the rain shadow, but on their seaward sides they fall sharply, forming spectacular escarpments and cliffs. From June to September they are blanketed in wet, cool clouds resulting in the development of a rich cover of vegetation. The vegetation itself plays an important role in the local climatic conditions, enormously increasing precipitation as condensation on plant surfaces. Over 750 plant species, of which 50 are endemic, have been recorded here (Miller and Morris, 1988) but floristically this assemblage has stronger affinities with that of the island of Socotra and drier regions of Northeast Africa than with northern Oman.

On the flat and drier coastal fringes, xerophytes and halophytes generally colonize low fore-dunes. Ghazanfar (1999) identified four assemblages which are widespread and characterised by species of *Atriplex*, *Halopyrum*, *Limonium*, *Sporobolus*, *Suaeda* and *Zygophyllum*. *Zygophyllum qatarense* is particularly widespread, since it tolerates drought and is unpalatable to goats and camels.

Extensive areas of sabkha, defined by Sheppard et al. (1992) as 'low-lying, sometimes intertidal but usually only seasonally inundated areas', are common in coastal sections of central Oman. Of these, the largest continuous sabkha is found in Bar Al Hickman. Here, large areas flood during the Southwest Monsoon, and the resulting area of shallow intertidal mudflat attracts hundreds of thousands of migrating, wading and sea birds.

Sandy shores, as noted, dominate the region. Beaches play a particularly important role in artisanal fisheries, being used for launching and recovering small fishing boats, landing the majority of the catch, as a transport corridor for the distribution of fish to markets, and in the important beach seining fishery for sardines. Four species of turtle nest on them, and the largest loggerhead turtle rookery in the world is located here.

Despite the wide distribution and importance of beaches, little work has been done to investigate their ecological or socio-economic importance. McLachlan et al. (1998) surveyed 10 beaches composed of fine to medium sand throughout Oman. Their data show that compared to 70 other physically similar beaches, species richness was generally high with at least 19–25 species per beach, but biomass was moderate to low at 26–90 g m^{-1}. Some zonation is evident, with ocipode crabs and the isopod *Tylos* in the supralittoral, cirolanid isopods on the upper shore and a variety of species on the lower shore. The coast of Oman appears to constitute a single zoogeographic zone, but with some regional differentiation between the Gulf of Oman and the Arabian Sea due to varying physical conditions. Beaches in the Gulf of Oman are particularly species-rich, probably as a result of the more varied mineralogy and the generally lower energy wave regime.

Fig. 3. Coral communities of the Gulf of Oman. Corals grow well in the Gulf of Oman, but rarely does more than a rudimentary framework develop. Coral communities that fringe the Daymaniyat Islands are amongst the best developed in the region, with coral cover exceeding 70% in places.

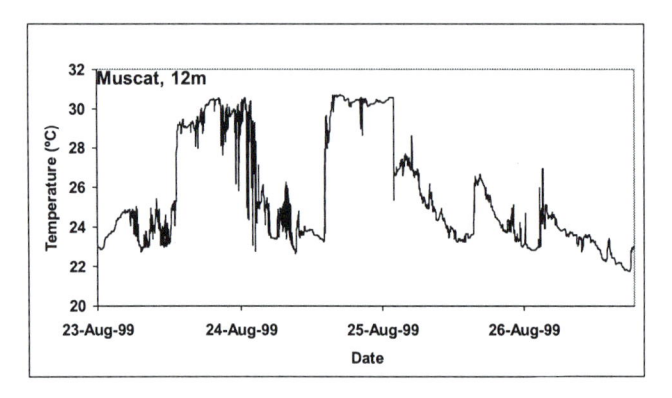

Fig. 4. Seawater temperatures recorded at 12 m depth in the Muscat, Oman. This graph shows near instantaneous fluctuations of over 7°C caused by the rise and fall of the sharp thermocline. Such rapid fluctuations do not appear to stress corals.

Corals, Reefs and Macroalgae

Corals are found on many of the rocky shores (Glynn, 1992; Salm, 1992a; Sheppard and Salm, 1988; Coles, 1995; Sheppard et al. 2000). Most often corals grow directly on exposed rock to form coral communities rather than forming true coral reefs (Fig. 3). However, a spectrum from isolated coral colonies to true coral reefs can be found in the region, making this an important area to study processes controlling coral reef formation. Other reef-associated organisms such as soft corals are also a conspicuous and important part of the fauna. Shores exposed to the Southwest Monsoon support communities with abundant sponges and ascidians.

A number of factors other than upwelling may limit the development of coral communities here, such as predation by Crown-of-Thorns Starfish (Glynn, 1992) and disease (Coles, 1994). However, low seawater temperatures, heavy swells and high nutrient concentrations during the Southwest Monsoon are probably the main factors on Arabian Sea coasts, and in the most affected areas, coral cover is typically <1–5%. Salm (1992a) reports that high seawater temperatures near Muscat reached 39°C in summer 1990, during a mass bleaching event. The typical maximum summer seawater temperature in this area is 32°C, but more significantly, temperatures routinely fluctuate up to 8°C in a 24-hour period (Coles, 1997) (Fig. 4).

The Arabian Sea and Arabian Gulf coral fauna are considered to be partly distinct from the Red Sea fauna (Sheppard 1998). The Arabian Sea assemblage comprises a large number of widespread species and a few endemic species. Of particular interest is the Arabian Sea endemic *Parasimplastrea simplicitexta*, a relict species known from fossils in Papua New Guinea, but not recorded elsewhere (Sheppard and Sheppard, 1991). Coral-rich areas have a discontinuous distribution due to the large stretches of sandy coast. In the Gulf of Oman, corals are relatively abundant, particularly around Muscat, the Daymaniyat Islands and in Musandam. Little is known of coral distribution in southern Iran, although abundant corals are reported to occur in Char Bahar Bay and Pizom Bay, and to fringe some of the islands in the Straits of Hormuz (Sheppard and Wells, 1988). In Yemen and southern Oman, much of the coast is either sandy or exposed to the full effect of the Southwest Monsoon and is therefore unsuitable for corals. In the most sheltered coves and on the more sheltered shores of the Hallaniyat Islands and Masirah, coral cover can reach 45%, but generally only a scattering of hard corals can be found. The only true coral reef described from this Arabian Sea coast is found to the south of Barr Al Hickman (Glynn, 1992; Sheppard et al., 2000), which is sheltered from high wave energy and from the most intense effects of the upwelling. This reef complex appears to be primarily composed of a single species, *Montipora foliosa*, and covers an area of several km^2.

Fig. 5. Sub-littoral seaweed community. In response to coastal upwelling along the southern Arabian Peninsula, a lush and diverse community of macroalgae develops in the intertidal and sub-littoral zones. *Sargassum* and another brown seaweed endemic to the Arabian Sea, *Nizamuddinia zanardinii*, are the most abundant species in the sub-littoral zone and together form a dense canopy below which a diverse assemblage dominated by red algae develops.

Along southern Arabia the coast takes on a temperate nature for part of each year. A luxuriant algal community develops at the beginning of the Southwest Monsoon and lasts for several months (Barratt et al., 1984, 1986). The main macrophytes are a thick, floating canopy of the brown algae *Sargassum* and *Nizzamudinia* (formerly *Sargassopsis*) with a diverse understorey of red, small brown and green algae and scattered coral colonies (Fig. 5). Remarkably, *Eklonia radiata*, a Southern Hemisphere kelp, is also present.

Macroalgal productivity during the peak monsoon period is high, over 60 g C m^2 d^{-1} for *Nizzamudinia* (Barratt et al., 1986). Over the entire season, macroalgal communities produce 0.5–1.4 kg (dry wt) of biomass per m^2. Much of the canopy is consumed by a diversity of grazers including large numbers of fish, particularly rabbitfish (*Siganus* spp.), while the understorey is grazed by parrotfish, abalone (*Haliotis mareae*), and abundant sea-urchins. A number of commercially important fish depend in some way on these algal beds.

Wadis, Khawrs and mangroves

No permanent river flows from Yemen, Oman, UAE or Iran. The only significant freshwater flow into the northern Arabian Sea comes from the Indus River in Pakistan. During rain showers water quickly percolates into groundwater through the sandy soils and coarse sediments of drainage channels ('wadis'). Flash floods are common after heavy downpours. Where wadis meet the sea, groundwater often rises to the surface forming a brackish pool separated from the sea by a sand bar. In the Arabian region, such brackish coastal wetlands ('khawrs') provide a unique habitat for a range of species, including brackish water plants, fish, and wading birds, as well as a focal point for human activity including fishing, livestock watering and grazing and recreation. The density of khawrs in an area depends on the relief of the hinterland, local geology and rainfall, but their contribution to the overall ecology and diversity in an area can be significant; there are 46 khawrs along the 450 km of coast in Dhofar, southern Oman.

Khawrs are naturally dynamic wetland environments. The salinity of their water varies according to the season, rainfall and the volume of freshwater seepage. Exceptionally heavy rainfall may flood them, breaching the sand bar and allowing marine conditions to dominate for several weeks until the sand bar is re-formed. This periodic opening of khawrs to the sea results in a temporal ecological succession from marine to brackish communities which are restored with the next heavy rains. Khawrs show distinct zones of lush green vegetation, with grasses (*Sporobolus virginicus* and *Paspalum vaginatum*) giving way to reed beds composed of *Typha* in the freshest khawrs, and *Phragmites* and *Schoenoplectus* in those with moderate or changing salinity (Ghazanfar, 1999). Submerged beds of water plants are also common: *Potamogeton pectinatus* is abundant in lower salinity pools, while *Ruppia maritima* and *Cladophora* dominate in more saline khawrs. Such lush vegetation attracts large numbers of livestock.

Mangrove development is limited. At a few sites, mangroves have colonized low-energy areas with saline or hyper-saline conditions. With so little in the way of freshwater inputs and such poor soil conditions, mangroves are mostly stunted or dwarf. The severe environmental stresses mean that only *Avicennia marina* exists, but experimental planting in Oman during the 1980s introduced *Rhizophora stylosa*, though only a handful survive today. On the Yemeni sector of this coast, only one small (1 ha) stand has been reported (Huntington and Wilson, 1996), a reflection of the exposed nature of that coast. In Oman, stands of mangroves are found at over 20 sites, totalling 1088 ha, with the largest stands found at Mahout Island (162 ha), Khawr Kalba on the Oman/UAE boarder (160 ha), and Khawr Jaramah (137 ha) (Fouda 1995). Little has been published of mangroves along the Iranian coast (see Sheppard et al., 1992).

As elsewhere, mangroves of the Arabian Sea provide habitat for a wide range of fish, birds, crustaceans, molluscs and other invertebrates. Fouda (1995) reports 86 species of fish, over 200 species of birds, 40 species of crustaceans, and 50 species of mollusc associated with mangroves in Oman. Of note is the endemic Kingfisher (*Todiramphus chloris*) whose range is limited to three mangrove stands in the northern Gulf of Oman. Considering the limited extent of mangroves, their contribution to the productivity of commercial fisheries is significant, particularly for the local shrimp (*Penaeus indicus* and *P. semisulcatus*) fisheries.

Seagrasses

Jupp et al. (1996) report four species of seagrasses in Oman (*Halodule uninervis*, *Halophola ovalis*, *Thalassodendron ciliatum* and *Syringodium isoetifolium*), the first two being the most widely distributed, probably due to their tolerance of high temperatures and silty conditions. Reports of two more species (*Cymodocea serrulata* and *Halodule wrightii*) by Weidleplan (1991), both from Ghubbat Hashish, require confirmation. For comparison, four species occur in the Arabian Gulf, 11 in the Red Sea (Sheppard et al., 1992) and 11 in southern India (Farshchi and Ansari, 1996). Biomass measurements (Jupp et al., 1996) indicate that the standing crop of *Halodule uninervis* at Masirah Island is low (77 g DW m^{-2}) compared to similar beds in the Red Sea (230–400 g DW m^{-2}; Lipkin, 1979) and Arabian Gulf (150–300 g DW m^{-2}; Kenworthy et al., 1993).

On the Yemen coast, seagrass has been reported growing at one site only (Huntington and Wilson, 1995). Further east, seagrass beds occur at Ghubbat Hashish ('Grass Bay' in Arabic) and Masirah Island (Jupp et al., 1996; Weidleplan, 1991). Many of these beds are well developed and play a crucial role in supporting diverse marine communities as well as commercial fisheries, particularly of the shrimp (*Penaeus semisulcatus*) (Siddeek et al., 1999), and support populations of Green turtles (Ross, 1985). Important seagrass beds occur in sheltered coastal lagoons near Jask and Char Bahar (Sheikholeslami, 1996). The generally lower energy conditions favoured by seagrasses are more common in the Gulf of Oman, where many of the sheltered embayments (e.g. Char Bahar, Ra's Sawadi) and creeks (Khawr Jaramah, Bandar Sur) support thin seagrass beds. Scattered seagrasses also occur in shallow water (5–15 m) along open stretches of coast, particularly Al Batinah (pers. obs.).

Turtles

Four species of turtle are common, the Green (*Chelonia mydas*), Hawksbill (*Eretmochelys imbricata*), Loggerhead (*Caretta caretta*) and the Olive Ridley (*Lepidochelys olivacea*). Leatherbacks (*Dermochelys coriacea*) are known to occur, but do not nest here. Green turtles nest throughout the Oman and Yemen coasts (Ross and Barwani, 1982; Salm, 1991; Huntington and Wilson, 1996) but the largest rookeries are clustered around a few beaches near Ra's Al Hadd where 6–13,000 females nest per year. Tagging studies have shown that Green turtles nesting here migrate to the more luxuriant seagrass beds in the Arabian Gulf and Southern Red Sea, presumably to feed, although a resident population of Green Turtles is present in Oman.

Hawksbill nesting beaches are more common in the Gulf of Oman; Salm (1991) estimates 600 nests are dug each year in Oman alone. The Loggerhead (*Caretta caretta*) nesting aggregation on Masirah Island is estimated to be at least 30,000 females each year, the largest known aggregation in the world (Ross and Barwani, 1982).

Cetaceans and Fishes

Five baleen whales and 16 toothed whales, dolphins and porpoises are known to occur off the Arabian Peninsula (Baldwin et al., 1998). A population of Humpback whales (*Megaptera novaeangliae*) is present in the Arabian Sea throughout the year.

Fish biodiversity in Oman has recently been reviewed (Fouda et al., 1998). A total of 1142 fish species from 520 genera and 164 families have been identified from the Gulf of Oman and Arabian Sea, almost double the number identified from the Arabian Gulf (Fouda et al., 1998). Of these, approximately 400 are demersal and 500 are associated with coral reef areas. There are four freshwater fish species that inhabit permanent streams and pools in desert and mountain areas.

While the distribution and composition of the most conspicuous habitats are known for much of the Arabian Sea region with the exception of Iran, relatively little work has been conducted on the ways in which habitats respond to the unusual environmental stresses here. Particularly important may be the link between the monsoon seasons and the timing of reproduction, recruitment and settlement of the inshore species.

OFFSHORE SYSTEMS

Two modes of wind-driven upwelling affect the northern Arabian Sea: coastal and open-ocean upwelling. Coastal upwelling is most intense where the continental shelf is narrow, and the resulting productive waters are transported up to 150 km offshore. Open-ocean upwelling is driven by Ekman pumping of strong positive wind stress to the northwest of the axis of the Findlater Jet (Brock et al., 1992).

Upwelling stimulates phytoplankton, whose blooms have chlorophyll concentrations of 2–5 mg m^{-3} in areas of upwelling (Brock and McClain, 1992). In terms of phytoplankton biomass, these values correspond to 3–9 g C m^{-2}. For comparison, oligotrophic waters in the central Arabian Sea contain 1–2 g C m^{-2} during the Southwest Monsoon season. Primary productivity in the Arabian Sea has been estimated to be 200–400 g C m^{-2} yr^{-1}, with daily rates during the Southwest Monsoon reaching 6 g C m^{-2} d^{-1} (Smith, 1995).

Diatoms (e.g. *Nitzschia delicatissima*) dominate the phytoplankton. This is heavily grazed: between 4 and 60% of daily primary productivity is removed by microzooplankton, especially by large copepod species *Calanoides carinatus* and *Eucalanus* sp. These exhibit very high growth rates and reach maturity in 15 days. Diapausing populations have been observed off the coast of Yemen and Somalia at depths of 1500 m (below the oxygen minimum layer), at which time they contain 50–70% lipid to support metabolism over the winter months. Such life-history characteristics make them extremely well adapted to upwelling areas (Smith, 1995).

The abundant detritus decomposes as it sinks. Bacterial decomposition in the well-oxygenated (4.2–5.2 mg/l) upper 100 m releases inorganic nutrients, causing nitrate to increase from about 5 mM in surface waters to about 20 mM at 100 m depth during pre-monsoon conditions. Concentrations of silicates follow a similar pattern, increasing from 5 mM on the surface to 15 mM at 100 m depth (Brock et al., 1992). However, at depths of about 100–1000 m, dissolved oxygen concentrations fall to about 10% of saturation or less (0.5–0.2 ml l^{-1}) as a result of aerobic decomposition of this detrital fallout. This creates the Oxygen Minimum Layer, in which bacteria use nitrate ions for oxidation of organic matter. This causes denitrification, and the Arabian Sea is one of three major water-column denitrification sites in the world (Morrison et al., 1999).

Most zooplankton are absent from the suboxic layer, and biomass is extremely low. However, some species, notably euphasiids and several species of myctophid fish, migrate between the surface and the suboxic layers.

The Arabian Sea has the largest concentration of mesopelagic fish found anywhere in the world's oceans, having a standing stock off Oman and Pakistan of 500,000 to 1,000,000 t (Gjøsaeter, 1984). These myctophids probably are the main prey of the larger, commercially important fish such as tuna, as well as some cetaceans. This abundance may be due to their daily migration from depths of 300–600 m (within the oxygen minimum layer where fish predators may be scarce) to 1–150 m, and to rapid growth. The dominant species, *Benthosema pterotum*, matures at 6–8 months, spawns and then probably dies. Extremely rapid growth rates of the dominant copepods and the dominant small fish of the region suggest that the food chain is well coupled with the fast turnover driven by the strong seasonal oscillations (Smith, 1995).

Large pelagic species, particularly Spanish mackerel (*Scomberomorus commerson*), Longtail tuna (*Thunnus tonggol*), Yellowfin tuna (*Thunnus albacares*), Kawakawa (*Euthynnus affinis*) and various pelagic sharks and rays support the valuable large-mesh gillnet, longline and troll fisheries. Of these, the Spanish mackerel is the most valued pelagic fish found in coastal waters of the Arabian Sea and is heavily targeted, accounting for 15–20% of the annual total catch in Oman (Al-Hosni and Siddeek, 1999). Highly migratory stocks, such as the Yellowfin tuna, are exploited by all countries bordering the Arabian Sea.

POPULATIONS

Throughout history the dichotomy between the ports of the coast and oases of the interior has shaped society, and this basic pattern can still be seen today, although modern communications are blurring this distinction. Constraints that used to keep the population low have now been released by modern development, and the region is undergoing a period of very rapid population growth.

Historically, the availability of water and the low productivity of land controlled population density. Most people were nomadic and made use of ephemeral desert grazing. Society was then firmly based on the tribal organisation and even today, Oman can be thought of as a confederacy of the many semi-autonomous tribes headed by the Sultan (Hoek, 1998).

Major trading routes have crossed the Arabian Sea for hundreds of years, driven by the predictable monsoon winds (Hoek, 1998). Regional trade allowed coastal settlements to develop into thriving towns. Later, these ports lay on important, longer distance trading routes between Indo-China and Europe that maintained thriving local economies until the advent of steam.

Oman expanded its trade network into East Africa, and in the 17th and 18th centuries the coast from Mogadishu to Zanzibar came under Omani control. Similar control was held over Baluchestan. Trading led to exchanges of goods, cultures and societies, leaving multi-cultural populations with many ethnic groupings, particularly in the trading ports.

In 1994, the population of Yemen was estimated to be 16 million, with 75% living in rural areas. It has been estimated that there are approximately 1200 fishermen operating from the Yemeni Governorate of Al Mahara (MEP, 1998). In 1993, the population of Oman was estimated at just over 2 million, of which 1.5 million were Omani and the remainder ex-patriate labourers. Much of Oman's coastal area is thinly populated. Sharqiyah, for example, with an area of approximately 40,000 km^2 and a population of 234,000 (Hoek, 1998) has an overall density of 5.8 people/km^2. The coastal strip is somewhat more densely populated as about 15% of Sharqiyah's population live in small fishing communities. The number of fishermen operating in the traditional fisheries sector was 16,000 in 1988, which indicates an average density of 8–9 fishermen per km of coast.

The two principal cities in Oman are Muscat, the capital, and Salalah in Dhofar. These are presently the main foci of industrial activity. However, Oman has embarked on two major industrial developments, one near Sur, where a liquid natural gas plant and associated industries have been built, and one near Sohar where a port and industrial estate are under construction. This move will help diversify the economy away from a reliance on oil, allowing direct economic benefits to filter through into the provinces.

Of the coastal towns of UAE, Fujeirah and its smaller neighbour, Khawr Fakkan, are the largest. Both are commercial ports strategically placed for handling trade between Gulf countries and India, Pakistan and Oman. There is a total of 1100 fishermen in 500 boats in Fujeirah (Anon, 1996).

The Iranian coastal provinces of Sistan and Baluchestan and Hormozgan have populations of about 1.5 million and 1 million respectively. The population density on the Gulf of Oman coast is the lowest of all coastal areas in Iran, with Sistan and Baluchestan having 7.7 people/km^2 and Hormozgan 13.8 people/km^2 (Ardalan, 1996). The two largest towns on the Iranian coast of the Gulf of Oman are Char Bahar, with a small commercial port, and Jask. No data are available for the proportion engaged in the fisheries sector in Iran.

Of greater relevance are the population growth rates, which are amongst the highest in the world. In Yemen, the population is growing at a rate of around 4% per year giving a population doubling time of 19 years. In Oman, the growth rate is 3.7% (Hoek, 1998). In both these countries, this is due to very high fertility rates (7.6 and 7.2 children per woman for Yemen and Oman respectively), combined with a marked decrease in child mortality. These trends are certain to have an impact on coasts through urban and rural development and increases in fishing pressure, particularly in Oman. In Oman, employment opportunities are limited in many rural coastal areas, which results in greater migration to cities to seek employment, and immigration of foreigners to undertake low-paid manual labour. One sector that will continue to absorb labour is fisheries, from which ex-patriates are excluded (Hoek, 1998).

RURAL FACTORS

Agriculture

Availability of water was the main factor influencing the development of both arable and pastoral farming in the region, since nowhere does sufficient rain fall to allow permanent agriculture. Around 500 BC, the Persians developed an ingenious system of irrigation channels and tunnels, called afalaj, that was introduced to Arabia a few hundred years later, since when afalaj and wells have irrigated small holdings and groves of date palms. This allowed small-scale cultivation. Today as in the past, the fertile coastal plains of Al Batinah and Salalah, as well as isolated river courses, are the centres of agricultural production in southeastern Arabia. However, over-abstraction from groundwater in some areas with more fertile soils, such as Al Batinah, has lowered the water table and encouraged intrusion of seawater into coastal aquifers.

Raising livestock is less dependent on a constant water supply, so is practised more widely than crop cultivation. Goats and camels are the more traditional stocks, but local breeds of cattle, sheep, chickens and donkeys are also raised. Traditionally, desert rangelands were grazed by camels and goats herded by bedouin while a semi-extensive system of grazing was adopted around more permanent settlements (Gaafar, 1995).

This pattern is now changing. Modern transport has facilitated the availability of water and supplementary feed, allowing stocking levels to exceed the natural carrying capacity of pastures. Overgrazing has been the inevitable consequence. The natural vegetation in upland areas of southern Yemen and Oman has been severely affected, and other areas such as Al Batinah and Masirah Island are also at risk. Excessive cutting of firewood has exacerbated the problem in Dhofar. The full significance of these changes to such forest and scrub areas has yet to be fully understood since the vegetation itself plays an important role in maintaining local conditions, enormously increasing precipitation as condensation on plant surfaces (Miller and Morris, 1988). Indicators of overgrazing such as abundance of the inedible *Calatropis procera* have become increasingly common in Dhofar, while planting of the invasive tree *Prosopis juliflora* has further perturbed the native floral communities.

Grazing and foraging livestock have also affected sensitive coastal habitats. Camels, goats, cattle and to a lesser extent sheep and donkeys graze sporadically on dune vegetation. Mangroves are cut for fodder, grazed, or are browsed (Fouda, 1995), and vegetation in and around khawrs has been overgrazed to the point where their functioning as wetland habitat has been impaired (Huntington and Wilson, 1996; TS-PCDEDG, 1993). In Oman, the government has erected fences around khawrs but this approach has not been effective as access is seen as a traditional right.

Fishing

Artisanal fishing accounts for 80–90% of national catches. In the Arabian Sea, fishing ceases almost completely during the Southwest Monsoon due to very rough sea conditions, but continues throughout the year in the Gulf of Oman (Fig. 6). Here, most effort is spent trolling for Longtail Tuna and beach seining for sardines. The density of fishermen working coastal waters is still relatively low in comparison to other parts of the world, and there is little or no evidence of overfishing except in a few high-value stocks, but again, the situation is changing fast.

The artisanal sector has seen a large increase in government investment. In the past 20 years, the availability of ice and the development of a modern road network have lead to an efficient fish distribution network supplying markets across the region. Traditionally, much of the catch was preserved by sun-drying and sold or exchanged for goods supplied from the interior, and today this can still be seen, particularly in the sardine and juvenile shark fisheries. Today, the majority of the catch is transported by traders using ice-boxes directly to public markets or fish processing factories in the main population centres. This efficient marketing network, developed hand in hand with mechanization of the artisanal fleet, has led to a major increase in fish production.

There has been a rapid increase in the numbers engaged in fishing. Between 1985 and 1992, the number of fishermen operating in Oman increased by 68% from 11,700 to 19,700 (Al-Requeshi, 1995) and increases in the artisanal fishery in Yemen doubled in five years between 1992 and 1997 (MEP, 1998). The type of boats and gears used has changed as investment has increased. Fibreglass skiffs with engines have replaced wooden traditional *huris* as the most popular craft, and fishing gears are now made of persistent modern materials. Finally, the distribution network can supply the necessary water and fuel to the most remote fishing camps as well as taking the fresh catch to market, which has facilitated the expansion of fishing into all areas. Further investment in the artisanal sector is being encouraged in Oman where the government is planning to introduce

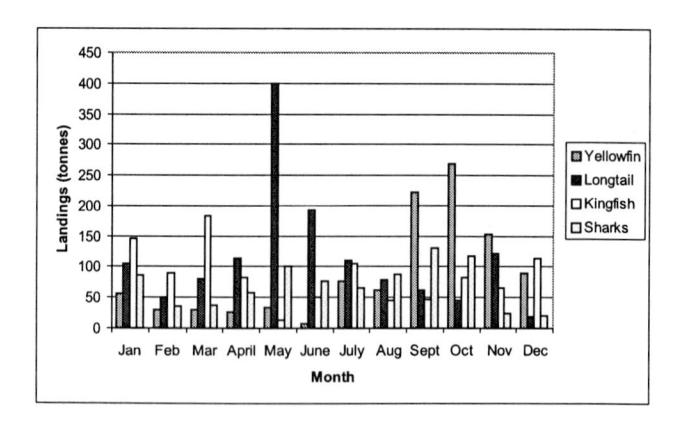

Fig. 6. Landings of large pelagic species in Al Batinah, Oman, 1995. Four species account for 95% of all large pelagic fish caught in Al Batinah: Longtail tuna (*Thunnus tonggol*, 1377 mt), Yellowfin Tuna (*Thunnus albacares*, 1053 mt), Kingfish (*Scomberomorus commerson*, 1022 mt) and Sharks (840 mt). Over 15,000 mt of sardines were landed here during 1995 making Al Batinah one of the most important fishing grounds in Oman. Note the clear seasonal peaks.

larger boats with crews of four to six that will allow the exploitation of stocks further offshore.

Many stocks (e.g. kingfish, tuna species, sardines, shark species) migrate throughout the region, but certain species, particularly demersal species, are found in greater concentrations in some areas only. The spiny lobster (*Panulirus* spp.) is most abundant on the rocky coasts of the Arabian Sea, particularly in Al Mahara, Dhofar and around Masirah Island. Abalone is only fished along a short stretch of coast in Dhofar, and there is a small shrimp fishery centred in Ghubbat Hashish with an annual production of about 200 mt. Along the sedimentary Al Batinah coast, large numbers of artificial reefs have been constructed from waste construction materials, old vehicles, date palm logs and mangrove branches which serve as nuclei for fish trapping.

A widely distributed and destructive fishing gear is the bottom-set gillnet which is used to target kingfish, jacks, and sharks. Environmental costs associated with this relatively non-specific gear has been the focus of a number of recent studies, particularly with regard to the by-catch of turtles and cetaceans, and its impact on coral communities. Fisheries-related mortality of Green turtles has been estimated at over 4000 per year in Oman (Siddeek and Baldwin, 1996) a level of mortality which is unsustainable. Cetaceans, especially dolphins and Humpback whales, are also routinely drowned (Salm, 1992b). A recent survey of coral areas in Oman revealed that fisheries activities had impacted 69% of sites (Al-Jufaili et al., 1999). Lost or abandoned gillnets were the single largest source of damage, impacting 49% of sites (Fig. 7).

So far, only a few high-value stocks have been shown to be over-exploited, and of these, the most valuable is the kingfish. The annual catch of kingfish in Oman fell from 27,700 mt in 1988 to 3200 mt in 1993 at a time when the number of boats in the artisanal fleet rose from 7400 to 11,750 (Al-Hosni and Siddek, 1999). The lobster fishery in Oman and Yemen, and the abalone fishery in Dhofar are also showing classic signs of overfishing. Contributing to overfishing of lobster in Yemen is the large number of traps lost each season, which then continue 'ghost-fishing'. Huntington and Wilson (1996) estimated that in Al Mahara, over the course of a three-month lobster season, 75% of all traps are lost, amounting to over 2000 lost traps per year on the Arabian Sea coast alone.

The underlying problem is that of maintaining adequate monitoring, surveillance and control which is notoriously difficult in these artisanal fisheries. While legislation and regulations are in place in all countries, enforcement is often lacking and management control is lost. Illegal fishing is widespread. The use of monofilament nets, locally known as *akrab*, Arabic for 'scorpion', is banned in Oman yet they are found in all regions of the country. Tangle nets, also banned, are commonly used to target lobster, and in the past ten years there has been increasing illegal use of *hyali*, a series of drifting gillnets linked together that can stretch for several km.

COASTAL EROSION AND LANDFILL

The nature of coastal dynamics, coastal erosion and landfill issues is different between the exposed Arabian Sea and more sheltered Gulf of Oman (James Dobbin, 1992). The most striking feature of the former is the phenomenal erosion–deposition annual cycle that occurs on some sandy beaches in response to the summer monsoon; for example, certain beaches in Dhofar erode 60–80 m during vigorous monsoons. Much of the beach material moved during the rough summer months slowly returns in the winter.

This natural cycle has important implications. Most directly affected are beach faunas and dune vegetation. Rough monsoon seas wash intertidal rock platforms free of sand, exposing surfaces that are colonised by seaweed (Ormond and Banaimoon, 1994; James Dobbin, 1992). Sandbars that separate khawrs from the sea erode, and once breached, the ecology of the lagoon changes significantly until a sandbar re-forms. Roads built between khawrs and the sea interfere with this process, and there are several examples where such development threatens important natural environments, e.g. Khawr Mugsayl, Sur.

The Al Batinah coast dominates the southern shore of the Gulf of Oman. Geologically it is a wadi plain flanked by long, mostly sandy beaches. Areas now eroding have intertidal rocky outcrops, and much of the coast experiences longshore drift northwards at a rate of 100,000 m^3 per year. Raised beaches and fossil marine terraces along the Iranian coast indicate tectonic emergence, but little is known about shoreline erosion or accretion rates (Rahimzadeh, 1996).

Development, particularly coastal construction and changes to inland drainage networks, affects natural coastal processes. Increased demand for fresh water and concurrent saltwater intrusion, particularly in Al Batinah, led to a plan to build low dams across wadis to allow more rainfall to percolate into groundwater. A consequence of this has been that large volumes of sediment are retained behind the dams, reducing the supply of material to the coast. While sand mining directly from beaches is a problem only in some areas, aggregate mining from wadi beds is a common practice across the region that also interferes with the supply of material along the coast.

The five major ports in the Arabian Sea and Gulf of Oman (Salalah, Muscat, Fujeirah, Khawr Fakkan, and Char Bahar) are located in naturally suitable sites that were enlarged or modified without causing major changes in the coastal sediment dynamics. However, in the late 1980s and early 1990s, Oman embarked on a program of constructing regional fishing harbours on sedimentary coastlines (e.g. Sharbithat, Lakbi, Quriyat, Al Khabura, Shinas) which in some cases led to increased rates of erosion downstream. Most recently, a new large industrial port under construction on a sedimentary coast near Sohar in northern Oman, has further disrupted longshore drift, putting a village and a mangrove area at risk.

Nowhere is the problem of coastal erosion more acute than along the Al Batinah coast where continuing natural re-alignment adds to the complex pattern of erosion. Here houses and other infrastructure have been built very close to the active beach to form an almost continuous strip along large sections of the shore.

Compared to other Arabian regions, there has been very little infilling in sensitive coastal areas. Where reclamation has impacted sensitive areas, these have been as a result of expanding urban infrastructure in the larger population centres. For example, a small area supporting a healthy coral community was reclaimed for use as a municipal park in Muscat, and the mangrove stand in the Qurm Nature Reserve, Muscat, is at risk from encroachment.

In the longer term, rising sea levels caused by global climate change are likely to change the nature of the coast in certain low-lying areas of the coast, particularly along central Oman. Coastal sabkha are particularly vulnerable since they lie only a few cm above sea level, and are naturally vulnerable to flooding following storm-surges.

URBAN AND INDUSTRIAL ACTIVITIES

Modern development has a history of about 40 years in this region. Before the discovery of oil in the late 1950s the mainstay of the national economies had been fisheries and limited agriculture, and in the more remote provinces this is still the case. Many of the centres of population developed where fresh water for agriculture could be found and where access to coastal fishing grounds was easy. This general pattern persists today.

All oil production in the region is from reserves found far inland, which has left the Iranian and Yemeni coastal provinces less developed than some inland centres (Ardalan, 1996). Even so, oil revenues have had a significant effect on coastal areas as governments have invested in infrastructure projects in an attempt to stimulate economic activities in areas without oil. In particular, the construction of roads, fishing harbours, and ports has greatly facilitated the development of the industrial and fisheries sectors particularly in Fujeirah, Al Sharqiyah (Hoek, 1998) and Al Batinah.

Although hydrocarbon production is entirely onshore at present, its transportation from this region is principally a marine activity. Also all shipping to and from the Arabian Gulf must pass through this area. Approaches to the Straits of Hormuz are amongst the busiest and most important shipping lanes in the world, with between 20,000 and 35,000 tankers passing through each year (Linden et al., 1990), carrying approximately 60% of the world's oil transported by sea. As in other areas, marine oil pollution originates from terrestrial discharges, tanker accidents and tanker operations. Terrestrial discharges arise from coastal refineries and loading terminals, of which there are two in this region: Fujeirah in UAE and Mina Al Fahal, near Muscat in Oman. Other ports (e.g. Salalah, Mina Qaboos near Muscat, Char Bahar) are a potential source of marine oil pollution.

Considering the vast volume of oil and the number of tankers passing through the region, there have been relatively few large accidental spills. The largest occurred in 1994, when the crude oil tanker 'Seki' spilled 16,000 tonnes of light crude off Fujeirah. Although dispersants were used to treat the slick offshore, about 20 km of coastline were contaminated (Sriadah, 1998). Subsequent monitoring over 18 months indicated that the impact to the coastal environment was relatively minor and short term (Böer, 1995).

Despite the Gulf of Oman's designation as a 'special area' in MARPOL 73/78 (IMO 1997), restricting the discharges of oil to sea from tankers, routine tanker operations are probably the major source of chronic oil pollution. There are a number of reasons why this may be so. Difficulties in detecting culprits and the lack of enforcement by the relevant authorities certainly contribute to the frequency of illegal discharges. However, in-bound, unladen tankers are under pressure to dump dirty ballast waters to prevent the need to discharge oily wastes in the Gulf. The problem is made worse by the lack of suitable waste reception facilities at oil loading terminals, and the high proportion of older tanker designs. Routine monitoring of beaches in Oman for oil contamination since the early 1980s (UNEP, 1982; Coles and Al-Riyami, 1996) has shown that beach tar concentrations have increased over the past decade, against the trend elsewhere. Chemical analysis indicates that the great majority of beach tar is crude oil from tank washings (M-Scan, 1997).

National economies are now diversifying away from dependence on oil revenues and are developing other industrial sectors, some of which will require access to the coast. While the region's coasts outside the main populated areas are still relatively untouched by development, economic diversification will take more industry into new sites, particularly those adjacent to existing ports and cities. As part of Oman's diversification, two new industrial developments are under construction. Faced with dwindling oil reserves, hydrocarbon production is shifting from oil to gas, and a new processing and exporting facility has recently been completed near Sur in the Gulf of Oman. A proportion of the production will be processed into ammonia and urea in an adjacent plant. These heavy industries form the core of a new industrial estate that will also contain a range of supporting manufacturing and light industry. Further north in the Gulf of Oman, at Sohar, a large industrial port designed to service a range of new heavy industries is under construction. Steel, aluminium and petro-chemical industries are planned.

Many coastal towns and cities are supplied with clean drinking water, and desalination plants and power stations discharge large volumes of heated water back to the sea. Little work has been done on assessing the impact of such discharges, particularly during summer months in the Gulf of Oman where natural seawater temperatures can reach up to 35°C. With fresh water being such a valuable commodity, several countries including Oman and UAE implement a policy of re-using waste-water for agricultural and municipal irrigation, and little treated sewage is discharged to sea from cities. However, most towns and villages do not have centralised sewage treatment facilities, relying on septic tanks and marine outfalls.

Industrial fishing has caused a disproportionate number of problems to the Arabian Sea. Mixed demersal stocks and cuttlefish are targeted in the industrial trawling fishery in Yemen which operates throughout the Gulf of Aden and into the Arabian Sea where the continental shelf is widest. In Oman, long-liners target tuna throughout the EEZ and trawlers have access to a concession block south of Masirah. In Iran the industrial sector comprises purse-seiners, long-liners, and trawlers.

Overfishing by industrial fishing methods is the greatest long-term threat. For example, trawling for cuttlefish in the Yemeni sector of the Gulf of Aden and Arabian Sea in the 1970s resulted in a crash in the stock, and production fell from 15,000 t in 1971 to 2000 t in the early 1980s; production has yet to recover. Enforcement of regulations is a problem for all countries, particularly for regulations designed to control foreign industrial fishing. In Oman, for example, Japanese and Korean trawlers contracted by Omani fishing companies to operate the artisanal fishing grounds during the cuttlefish season which results in conflict between the artisanal and commercial sectors. MEP (1998) also report that industrial trawling for cuttlefish within the excluded three-mile limit in Al Mahara, Yemen, results in damage to artisanal gear and has destroyed large areas of benthic habitat, thereby reducing long-term productivity.

Other industrial fishing techniques with the potential for environmental impacts are the long-lining and purse seine fisheries based in the Gulf of Oman that target surface-dwelling tuna. The Iranian commercial fleet is equipped for purse-seining, while long-lining is practised throughout the Gulf of Oman and the Arabian Sea. Very little published information is available on the impact of these fishing methods to wildlife or shallow coastal habitats, but significant by-catch of cetaceans and seabirds has been reported in similar fisheries elsewhere.

Although now part of the Indian Ocean Whale Sanctuary, the Arabian Sea was used as a hunting ground by Soviet whaling fleets in the late 1960s (Mikhalev, 1997). For example, in November 1966 a total of 238 humpback whales were killed off the coasts of Oman, Pakistan and northwestern India. Today, large cetaceans face the threat of incidental captures in gillnets used in the artisanal fishery and risk of disturbance from offshore seismic surveys.

PROTECTIVE MEASURES

In 1978, a regional conference of Gulf States, including Oman, Iran and UAE, was convened in Kuwait. Its purpose was to review and adopt an action plan for the 'protection and development of the marine environment and coastal areas', later known as the Kuwait Action Plan (KAP), and so the Regional Organisation for Protection of the Marine Environment (ROPME) was born. The area covered by ROPME is the Arabian Gulf and the Gulf of Oman, and its principal objective is the protection of the environment from pollution, and collaboration between member states on meteorological and environmental programmes. Today, ROPME remains the main regional organisation concerned with the marine and coastal environment.

In part, the establishment of ROPME had been precipitated by MARPOL in 1978. This encouraged promotion of technical co-operation for the prevention and mitigation of marine pollution. In MARPOL, the KAP Region was designated a 'Special Area' in which strict regulations exist governing the discharge of oil, ballast water, and other polluting discharges from ships. However, as noted, chronic oil pollution in the Gulf of Oman is worsening, despite its status. In accordance with MARPOL, investment in the provision of adequate oily waste reception facilities by ratifying countries would probably reduce pollution from routine tanker operations. Such a facility is currently under construction at the port of Fujeirah, UAE. However, of the four countries covered in this chapter, only Oman is a signatory to MARPOL, although other countries have alternative provisions in their national legislation controlling aspects of marine pollution. For example, Iran seeks to enforce its laws prohibiting dumping of oily wastes and other pollutants within its exclusive economic zone (IBRD, 1995).

All member countries of the Arabian Gulf Co-operative Council (AGCC) fund a Marine Emergency Mutual Aid Centre (MEMAC) and the region's oil industries have formed their own organisation, the Gulf Area Oil Companies Mutual Aid Organisation (GAOCMAO). Each country has a pool of equipment for combating oil pollution that can be used by MEMAC countries, and clean-up efforts are co-ordinated centrally.

The Arabian Sea region is also included in international declarations aimed at protecting migratory wildlife, particularly cetaceans and turtles. In 1979, the IWC declared the Indian Ocean, from 55°S northward, a whale sanctuary. More recently in 1995, the 'Sodwana Declaration' was drawn up by IUCN to recommend priority actions for the conservation of marine turtles in the Western Indian Ocean (IUCN, 1996), which includes the Arabian Sea region.

To date, there are no marine protected areas on the Yemeni coast of the Arabian Sea. The most significant habitat in this area lies in the coastal mountain areas which support monsoon woodlands. The regional and global significance of the monsoon forests around Hauf has been recognised by the Yemeni government and there are plans to protect 30,000 ha on the southern coastal escarpments. Other important habitats that are worthy of protection are turtle nesting beaches.

Fig 7. Damage to coral communities from lost or abandoned nets.

In Oman the situation is more complicated. Clark et al. (1986) proposed a network of conservation areas throughout the Sultanate of Oman, which formed the basis of the government's conservation strategy. The majority of the proposed sites are on the coast, and islands and khawrs are particularly well represented. Other sites are designed to conserve representative areas of habitat such as tidal inlets, coastal plains and cliffs, and yet others have been designed to conserve important archaeological resources. The latter category includes Khawr Rawri where the ancient trade in frankincense was centred, and reputed to be the site of the Palace of the Queen of Sheba. The system of conservation areas has three levels of protection: national nature reserve (NNR), national scenic reserve (NSR) and national resource reserve (NRR). NNR is the most strictly controlled, NSRs may have settlements within them, while NRR status is applied to areas of conservation interest but for which data are insufficient for planning NNR and NSR.

There are four national nature reserves (NNRs) with significant marine or coastal components in Oman, as well as a total of nine protected khawrs in Dhofar. Each receive differing levels of funding and consequently the effectiveness of their management varies considerably.

Daymaniyat Islands National Nature Reserve (Fig. 8), was established by Royal Decree in 1996. The islands are a nature conservation area of national and regional significance owing to their importance as nesting sites for sea birds, such as Sooty Falcons, Ospreys and Bridled Terns, as well as Green and Hawksbill turtles. The fringing reefs surrounding the islands are well developed, attracting artisanal fishermen and recreational visitors. Although management plans have been in place for over 10 years (Salm, 1986), enforcement of its regulations remains weak.

Situated in the heart of rapid development, the Qurm NNR protects 74 ha of mature mangrove woodland in the capital area. Within its boundaries lie the mangrove creeks, beaches, sabkha and archaeological sites of great local and national importance (Kogo, 1986). However, there are no rangers or wardens on hand for its day-to-day management and the area is now coming under increasing pressure from urban development, particularly road construction and recreational developments.

Fig. 8. The Daymaniyat Islands National Nature Reserve from the air. Formally protected by Royal Decree in 1996, these islands are located 15 km off Oman's Al Batinah coast in the Gulf of Oman. They are an important nesting site for a variety of seabirds and turtles, and have some of the best developed coral communities.

The Ra's Al Junayz NNR covers 40 km^2, including marine areas, and lies adjacent to the Ra's Al Hadd NSR. This park was set up to protect about 20 beaches between Ra's Al Hadd and Ra's Al Ruwais that are used as nesting sites by 6–13,000 female Green turtles. A permit is required to enter the park, where camping and turtle watching is controlled by park rangers. A turtle-tagging program has been designed to establish the population's migration and nesting behaviour and by 1995 over 32,000 female turtles had been tagged. The Ra's Al Hadd NSR encompasses an area of 80 km^2, including some shallow coastal marine habitats. The northern and western parts of the reserve cover mostly rugged limestone hills that are dissected by wadis and end as low sea cliffs. There are two large khawrs within the boundaries, both of which are feeding grounds for Green turtles for part of the year. This reserve serves as a buffer zone to the nearby NNR of Ra's Al Junayz and protects important turtle habitat.

Following the successful re-introduction of the Arabian oryx (*Oryx leucoryx*) back into its natural desert habitat in 1982, the Arabian Oryx Sanctuary was established by Royal Decree in 1994. Listed as a World Heritage Site by UNESCO, the sanctuary covers approximately 20,000 km^2 that includes a long stretch of coast. While the park rangers routinely patrol inland areas to protect the oryx, a degree of protection is afforded to the coastal khawrs and offshore islets within the sanctuary that represent important habitat for seabirds and other wildlife.

On the Gulf of Oman coast of UAE, only one site has been proposed to receive special protection: Khawr Kalba. The proposed 1230 ha site contains the best stand of mature mangrove woodland in the UAE, which offers nesting sites to the rare endemic kingfisher, *Todiramphus chloris* whose population was estimated to be 55 pairs in 1996 (Hornby, 1996). A provisional management plan for its protection has been prepared but as yet the site has not received formal protection.

Chiffings (1995) reports that there are no existing or proposed protected areas along the Iranian shores of the Gulf of Oman section. Harrington (1976, cited in Sheppard and Wells, 1988) recommended that Char Bahar and Pizom Bays were both suitable sites for some form of protection, although Char Bahar was not suitable for National parks status due to the relatively high degree of development existing at the time.

Key habitats of the Arabian Sea region are under-represented in the marine protected area network so far established. Indeed, there are only four national nature reserves, or equivalent, across the region, two in the Gulf of Oman and two on the Arabian Sea coast; all are located in Oman and some are under-resourced to the point where their effectiveness is questionable.

REFERENCES

Al Hosni, A.H.S. and Siddeek, S.M. (1999) Growth and mortality of the narrowbarred Spanish Mackerel, *Scomberomorus commerson*, (Lacepède), in Omani waters. *Fisheries Management and Ecology* **6**, 133–144.

Al-Jufaili, S., Al-Jabri, M., Al-Balushi, A., Baldwin, R.M., Wilson, S.C., West, F. and Matthews, A.D. (1999) Human Impacts on coral reefs in the Sultanate of Oman. *Estuarine, Coastal and Shelf Science* **49** (Suppl. A), 65–74.

Al-Requeshi, Y.A. (1995) Numbers of Traditional Fishermen and Traditional Fishing Boats. In: *Traditional Agriculture and Fishing in the Sultanate of Oman*. College of Agriculture, Sultan Qaboos University. pp 36–45.

Ambrayseys, N.N. and Melville, C.P. (1985) *A History of Persian Earthquakes*. Cambridge Earth Science Series, Cambridge University Press, Cambridge, UK.

Anon. (1996) Statistics. Ministry of Agriculture and Fisheries, Dubai, UAE.

Ardalan, A. (1996) Implementation of integrated coastal zone management in Iran: Difficulties and opportunities from a socio-economic point of view. In *Regional Workshop on Integrated Coastal Zone Management, Char Bahar, Iran. February 24–29, 1996*, ed. H. Zommorrodian.

Baldwin, R.M., Gallagher, M. and van Waerebeek, K. (1998) A Review of Small Cetaceans from Waters off the Arabian Peninsular. Document SC/50/SM6 Presented to 50th International Whaling Commission, 24 pp.

Barratt, L., Ormond, R.F.G. and Wrathall, T. (1986) Ecology and productivity of the sublittoral Algae *Ecklonia radiata* and *Sargassopsis zanardini*. Part 1. Ecological Studies of Southern Oman Kelp Communities. Council for the Conservation of the Environment and Water Resources, Muscat, Oman, and Regional Organisation for the Protection of the Marine Environment, Kuwait, pp. 1.1–1.22.

Barratt, L., Ormond, R.F.G., Campbell, A.C., Hiskock, S., Hogarth, P.J. and Taylor, J.D. (1984) An ecological study of the rocky shores on the south coast of Oman. Council for Conservation of the Environment and Water Resources, Muscat, pp. 99.

Böer, B. (1995) A follow up to the Fujeirah oil spill. *Tribulus* **5**, 20–23.

Böhm, E., Morrison, J.M., Manghnani, V., Kim, H-S. and Flagg, C.N. (1999) The Ras Al Hadd Jet: Remotely sensed and acoustic Doppler current profiler observations in 1994–1995. *Deep-Sea Research II* **46**, 1531–1549.

Brock, J.C. and McClain, C.R. (1992) Interannual variability in phytoplankton blooms observed in the northwestern Arabian Sea during the southwestern monsoon. *Journal of Geophysical Research* **97** (C1), 733–750.

Brock, J.C., McClain, C.R. and Hay, W.W. (1992) A Southwest Monsoon Hydrographic Climatology for the Northwestern Arabian Sea. *Journal of Geophysical Research* **97** (C6), 9455–9465.

Burkill, P.H. (1999) ARABESQUE: An overview. *Deep-Sea Research II* **46**, 529–547.

Chiffings, A.W. (1995) Arabian Seas. In *A Global Representative System of Marine Protected Areas. Volume III Central Indian Ocean, Arabian Seas, East Africa and East Asian Seas*, eds. G. Kelleher, C. Bleakley and S. Wells. GBRMPA, IUCN, World Bank, pp. 39–70.

Clarke, J.E., Al-Lumki F., Anderlini, V.C. and Sheppard, C.R.C. (1986) *Sultanate of Oman. Proposals for a System of Nature Conservation Areas*. IUCN, Gland, Switzerland, 477 pp.

Coles, S.L. and Al-Riyami, K.A. (1996) Beach tar concentrations on the Muscat Coastline, Gulf of Oman, Indian Ocean, 1993–1995. *Marine Pollution Bulletin* **32** (8/9), 609–614.

Coles, S.L. (1994) Extensive coral disease outbreak at Fahl Island, Gulf of Oman, Indian Ocean. *Coral Reefs* **13**, 242.

Coles, S.L. (1995) *Corals of Oman*. Richard Keech, Thorns, Hawes, North Yorkshire, DL8 3LJ, UK.

Coles, S.L. (1997) Reef corals occurring in a highly fluctuating temperature environment at Fahal Island, Gulf of Oman (Indian Ocean). *Coral Reefs* **16**, 269–272.

Farshchi, P. and Ansari, Z.A. (1996) Coastal Pollution and Sustainable Management of Living Resources, ed. H. Zommorrodian. In *Regional Workshop on Integrated Coastal Zone Management, Char Bahar, Iran. February 24–29, 1996*.

Fouda, M.M., Hermosa, G.V. and Al-Harthi, S.M. (1998) Status of fish biodiversity in the Sultanate of Oman. Unpublished manuscript.

Fouda, M.M. (1995) Status of Mangrove Resources in the Sultanate of Oman. *J. Fac. Sci., U.A.E. Univ.* **8** (2), 169–183.

Gaafar, O.M. (1995) Traditional Livestock Practices. In: *Traditional Agriculture and Fishing In the Sultanate of Oman*. College of Agriculture, Sultan Qaboos University, pp. 26–32.

Ghazanfar, S.A. (1999) Coastal Vegetation of Oman. *Estuarine, Coastal and Shelf Science* **49** (Suppl. A), 21–27.

Gjøsaeter, J. (1984) Mesopelagic fish, a large potential resource in the Arabian Sea. *Deep Sea Research* **31**, 1019–1035

Glynn, P.W. (1992) Monsoonal Upwelling and Episodic *Acanthaster* Predation as Probable Controls for Coral Reef Distribution and Community Structure in Oman, Indian Ocean. *Atoll Research Bulletin* **379**, 1–66.

Hoek, C.W. (1998) *Shifting Sands: Socio-economic Development in Al-Sharqiyah Region, Oman*. Nijmegen University Press, Netherlands.

Hornby, R.J. (1996) Khawr Kalba Proposed National Park: A Provisional Management Plan. Unpublished report prepared for the Arabian Leopard Trust, UAE.

Huntington, T.C. and Wilson, S.C. (1995) Coastal Habitats Survey of the Gulf of Aden Phase 1: Preliminary Habitat Classification and an Assessment of the Coast's Resources, Users and Impacts. Report to the Ministry of Fish Wealth, Government of the Republic of Yemen. MacAlister Elliott and Partners Ltd., UK, and Marine Sciences Resource Research Centre, Aden, Yemen.

IBRD (1995) Islamic Republic of Iran Environment Strategy Study. Report No. 12806-IRN.

IMO (1997) *MARPOL 73/78 Consolidated Edition 1997*. International Maritime Organisation, 4 Albert Embankment, London. UK. 419 pp.

IUCN (1986) Oman Coastal Zone Management Plan: Greater Capital Area. IUCN, Gland, Switzerland. 79 pp.

IUCN (1988) Oman Coastal Zone Management Plan: Quriyat to Ra's al Hadd. IUCN, Gland, Switzerland. 57 pp.

IUCN (1989a) Oman Coastal Zone Management Plan: Dhofar. Volume 1. IUCN, Gland, Switzerland.

IUCN (1989b) Oman Coastal Zone Management Plan: Dhofar. Volume 2. IUCN, Gland, Switzerland.

IUCN (1991) Oman Coastal Zone Management Plan: Musandam. IUCN, Gland, Switzerland. 70 pp.

IUCN (1996) A Marine Turtle Conservation Strategy and Action Plan for the Western Indian Ocean. Report prepared by IUCN East Africa Regional Office and IUCN/SSC Marine Turtle Specialist Group. 24 pp.

James Dobbin Associates Inc. (1992) Coastal Erosion in Oman—Draft Regulations for the Prevention of Erosion in the Sultanate of Oman. Report to the Ministry of Regional Municipalities and Environment, Government of Oman.

Jupp, B.P., Durako, M.J., Kenworthy, W.J., Thayer, G.W. and Schillak, L. (1996) Distribution, abundance, and species composition of seagrasses at several sites in Oman. *Aquatic Botany* **53**, 199–213.

Kemp, J.M. (1998) The occurrence of *Nizamuddinia zanardinii* (Schiffner) P.C. Silva (Phaeophyta: Fucales) at the Socotra Archipelago. *Botanica Marina* **41**, 345–348.

Kenworthy, W.J., Durako, M.J., Fatemy, S.M.R., Valavi, H. and Thayer, G.W. (1993) Ecology of seagrasses in northeastern Saudi Arabia one year after the Gulf War oil spill. *Marine Pollution Bulletin* **27**, 213–222.

Kogo, M. (1986) A Report of Mangrove Research in Sultanate of Oman. The Fourth Research on Mangroves in the Middle East. Japan Cooperation Centre for the Middle East.

Linden, O., Abdulraheem, M.Y., Gerges, M.A., Alam, I., Behbehani, M., Borham, M.A. and Al-Kassab, L.F. (1990) State of the Marine Environment in the ROPME Sea Area. UNEP Regional Seas Report and Studies No 112. 34 pp.

Lipkin, Y. (1979) Quantitative aspects of seagrass communities, particularly those dominated by *Halophila stipulacea*, in Sinai (northern Red Sea). *Aquatic Botany* **7**, 119–128.

MEP (MacAlister Elliott & Partners Ltd.) (1998) Republic of Yemen: Fisheries Sector Review. Report No. 501/R/108D. Produced under the Fourth Fisheries Development Project, Yemen. (Report commissioned and funded by the European Commission).

Manghnani, V., Morrison, J.M., Hopkins, T.S. and Böhm, E. (1998) Advection of upwelled waters in the form of plumes off Oman during the Southwest Monsoon. *Deep-Sea Research II* **45**, 2027–2052

McLachlan, A., Fisher, M., Al-Habsi, H.N., Al-Shukairi, S.S. and Al-Habsi, A.M. (1998) Ecology of Sandy Beaches in Oman. *Journal of Coastal Conservation* **4**, 181–190.

Mikhalev, Y.A. (1997) Humpback whales *Megaptera novaeangliae* in the Arabian Sea. *Marine Ecology Progress Series* **149**, 13–21.

Miller, A.G. and Morris, M. (1988) *Plants of Dhofar*. Office of the Advisor for Conservation and the Environment, Diwan of the Royal Court, Sultanate of Oman.

Morrison, J.M., Codispoti, L.A., Smith, S.L., Wishner, K., Flagg, C., Gardner, W.D., Gaurin, S., Navqi, S.W.A., Manghnani, V., Prosperie, L. and Gundersen, J.S. (1999) The oxygen minimum zone in the Arabian Sea during 1995. *Deep-Sea Research II* **46**, 1903–1931.

M-Scan (1997) Monitoring Pollutants in the Marine Environment: Report on the gas chromatographic analysis of tarballs collected from beaches in Oman. Unpublished report to Ministry of Regional Municipalities and Environment, Government of Oman.

Murty, T.S. and El-Sah, M.I. (1984) Cyclones and storm surges in the Arabian Sea: A brief review. *Deep Sea Research* **34**, 665–670.

National Hydrographic Office (1998) Sultanate of Oman Tide Tables 1999. National Hydrographic Office, Royal Navy of Oman, PO Box 113, Muscat, Oman.

Ormond, R.F.G. and Banaimoon, S.A. (1994) Ecology of intertidal macroalgal assemblages on the Hadramout coast of Southern Yemen, an area of seasonal upwelling. *Marine Ecology Progress Series* **105**, 105–120.

Piechura, J. and Sobajh, O.A.G. (1986) *Oceanographic Conditions of the Gulf of Aden. Scientific Investigations of the Gulf of Aden Series A: Oceanography No. 2*, pp. 1–15. Published by Marine Science and Resources Research Centre, Aden, Yemen.

Rahimzadeh, F. (1996) Geological setting, non-living resources and natural hazards of the Caspian Sea, Persian Gulf and the Gulf of Oman. In *Regional Workshop on Integrated Coastal Zone Management, Char Bahar, Iran, February 24–29, 1996*, ed. H. Zommorrodian.

Reynolds, R.M. (1993) Physical Oceanography of the Gulf, Straits of Hormuz, and the Gulf of Oman, Results from the Mt Mitchell Expedition. *Marine Pollution Bulletin* **27**, 35–59.

Ross, J.P. (1985) Biology of the Green Turtle, *Chelonia mydas*, on an Arabian feeding ground. *Journal of Herpetology* **19** (4), 459–468.

Ross, J.P. and Barwani, M.A. (1982) Review of sea turtles in the Arabian area. In *Biology and Conservation of Sea Turtles*, ed. K.A. Bjorndal, pp. 373–383. Smithsonian Institute Press, Washington, D.C.

Salm, R.V. (1986) The Proposed Daymaniyat Islands National Nature Reserve Management Plan. Report to the Ministry of Commerce and Industry, Government of Oman.

Salm, R.V. (1992a) Coral Reefs of the Sultanate of Oman. *Atoll Research Bulletin* **380**, 1–85.

Salm, R.V. (1992b) Impact of Fisheries on Environment and Wildlife, Sultanate of Oman. Scientific Results of the IUCN Coastal Zone Management Project CZMP4:F17. Report to the Ministry of Commerce and Industry, Government of Oman.

Salm, R.V. (1991) Turtles in Oman—Status, Threats and Management Options. Scientific Results of the IUCN Coastal Zone Management Project CZMP4:F11. Report to the Ministry of Commerce and Industry, Government of Oman.

Savidge, G., Lennon, J., Matthews, A.J. (1990) A shore-based survey of upwelling along the coast of Dhofar region, Southern Oman. *Continental Shelf Research* **10** (3), 259–275.

Sheikholeslami, M.R. (1996) Iranian Coastal Zone (Caspian Sea, Persian Gulf, and Sea of Oman): the Environment, Human impacts and the need for Integrated Coastal Zone Management. In *Regional Workshop on Integrated Coastal Zone Management, Char Bahar, Iran. February 24–29, 1996*, ed. H. Zommorrodian.

Sheppard, C.R.C. (1998) Biodiversity patterns in Indian Ocean corals, and effects of taxonomic error in data. *Biodiversity and Conservation* **7**, 847–868.

Sheppard, C.R.C. and Salm, R.V. (1988) Reef and coral communities of Oman, with a description of a new species (Order Scleractinia, genus *Acanthastrea*). *Journal of Natural History* **22**, 263–279.

Sheppard, C.R.C. and Sheppard, A.L.S. (1991) Corals and Coral Communities of Arabia. *Fauna of Saudi Arabia* **12**, 1–170.

Sheppard, C.R.C. and Wells, S. (1988) *Coral Reefs of the World. Volume*

2: *Indian Ocean, Red Sea and Gulf*. International Union for the Conservation of Nature, Gland, and United Nations Environment Programme, Nairobi. 389 pp.

Sheppard, C.R.C., Price, A.R.G. and Roberts, C.A. (1992) *Marine Ecology of the Arabian Region*. Academic Press, London.

Sheppard, C.R.C., Wilson, S.C., Salm, R.V. and Dixon, D. (2000) Reefs and Coral Communities of the Arabian Gulf and Arabian Sea. In *Coral Reefs of The Indian Ocean: Their Ecology and Conservation*, eds. T.R. McClanahan, C.R.C. Sheppard and D.O. Obura, Chapter 9. Oxford University Press.

Siddeek, M.S.M., Fouda, M.M. and Hermosa, G.V. Jr. (1999) Demersal Fisheries of the Arabian Sea, the Gulf of Oman and the Arabian Gulf. *Estuarine, Coastal and Shelf Science* **49** (Suppl. A), 87–97.

Siddeek, S.M. and Baldwin, R.M. (1996) Assessment of the Oman Green Turtle (*Chelonia mydas*) stock using a stage-class matrix model. *Herpetological Journal* **6**, 1–8.

Smith, S.L. (1995) The Arabian Sea: mesozooplankton response to seasonal climate in a tropical ocean. *ICES Journal of Marine Science* **52**, 427–438.

Sriadah, M.M.A. (1998) Impacts of an Oil Spill on the Marine Environment of the United Arab Emirates along the Gulf of Oman. *Marine Pollution Bulletin* **36** (11), 876–879.

TS-PCDEDG (1993) *Khawrs and Springs of the Dhofar Governorate – Land Use Proposals*. Report to the Technical Secretariat, Planning Committee for Development and Environment in the Governorate of Dhofar, Sultanate of Oman. P.O. Box 1781, Salalah, Oman.

UNEP (1982) Survey of tar, oil, chlorinated hydrocarbons and trace metal pollution in coastal waters of the Sultanate of Oman. *UNEP Regional Seas Reports and Studies* **5**, 1–48.

Weidleplan Consulting GmbH (1991) Study for Wildlife and Conservation Areas—Master Plan for the Coastal Areas of the Barr Al Hikman and Masirah Island. Report to Ministry of Regional Municipalities and Environment, Government of Oman.

THE AUTHOR

Simon C. Wilson
*Department of Biological Sciences,
Warwick University,
Coventry, CV4 7RU, U.K.
and
P.O. Box 2531, CPO 111,
Seeb, Oman.*

Chapter 55

THE RED SEA

Charles R.C. Sheppard

The Red Sea lies between about 13 and 30°N, straddling the Tropic of Cancer. It is a spreading centre, widening by a few cm each year; it is seismic, and the African and Arabian coasts mirror each other to a remarkable degree. In the north, the Gulf of Aqaba is a continuation of the Red Sea rift, but the Gulf of Suez is a shallow (<50 m) sedimentary basin. Widely known for its steep, clear-water coral reefs, the main reefal area is in fact focused in the northern half, while in the south, depth contours sweep well out from land, providing vast areas of sedimentary shallows much more suitable for mangroves and seagrasses than for reef growth. The total length of shoreline is probably more than doubled by small islands, which thus have a marked biological importance, especially in the provision of very sheltered habitat.

The Red Sea was hypersaline immediately before the Holocene sea-level rise about 17,000 years ago, so the present biota is therefore relatively recent and includes numerous endemic species. Most fringing reefs extend only a few tens of metres from the steep shores in the north, but where there are old alluvial fans, and further south, they commonly extend 1 km to seaward from alluvial plains 1 to 7 km wide. Offshore, extensive series of submerged limestone platforms form the foundations for a barrier reef. Further south, fringing reefs diminish and in many places are completely replaced by broad and thick stands of mangrove and extensive seagrass beds, by muddy flats and sand deposits. Calcareous red "algal reefs" exist in the absence of coral reefs, and a very conspicuous increase in brown algae, mainly *Sargassum*, occurs on shallow hard substrate.

Primary production is low in the north, but in the south there is considerable pumping of surface water into the Red Sea from the Indian Ocean, creating much richer conditions there. Intense evaporation removes an amount of water equivalent to a depth of 1 to 2 m per year, of which only a negligible part is replaced by fresh water, most coming from inflow through the southern entrance.

The Red Sea coastline for the most part has extremely low levels of population outside a few major ports and cities, which has greatly limited the degree of coastal shoreline alteration, pollution and resources abstraction. Petroleum hydrocarbon levels are relatively high in the Gulf of Suez, where there is substantial oil and tar on shores, and refuse and beach debris is also marked in localised areas. Lack of widespread sewage treatment facilities in most Red Sea countries has resulted in extensive localised damage to coral reefs and in some areas has caused eutrophication problems in ponded or lagoonal areas, or over extensive reef flats near towns.

Fishing is mainly artisanal, though purse seining and trawling activities occur in shallow areas, especially the Gulf of Suez. Fishing in most of the Red Sea has probably been mostly at sustainable levels to date, but such fisheries are very difficult to monitor and manage, and there is concern that regulations which do exist are commonly ignored. There is good potential for continued sustainable use, partly because of the existence of a regional programme and, equally importantly, because the overall population density of the coast remains very low.

Seas at The Millennium: An Environmental Evaluation (Edited by C. Sheppard)
© 2000 Elsevier Science Ltd. All rights reserved

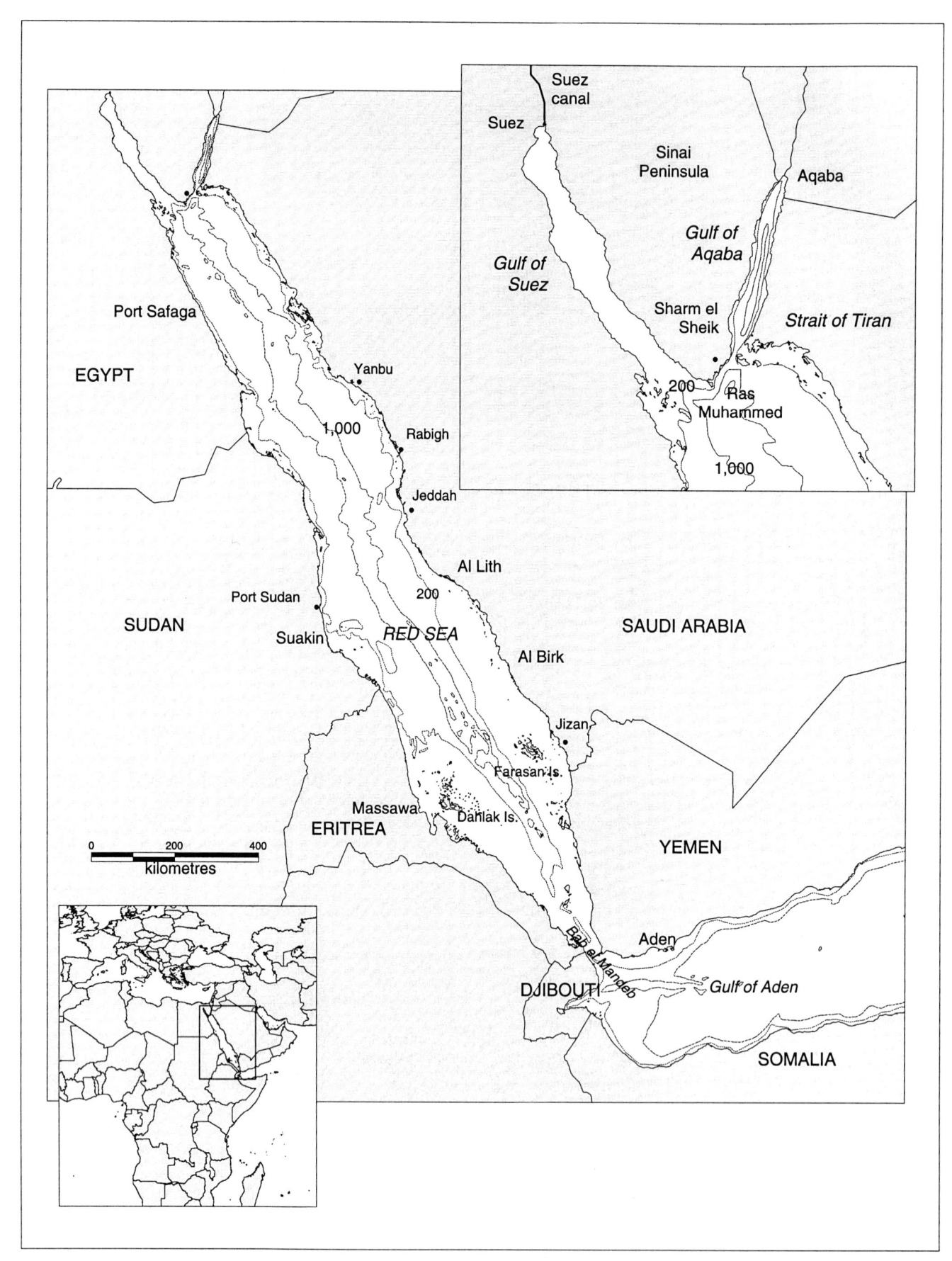

Fig. 1. The Red Sea. Inset: the northern Gulfs.

GEOGRAPHICAL AND HISTORICAL SETTING

The Red Sea extends for over 2200 km between the land masses of Arabia and Africa (Fig. 1). Until only 20 years ago it was assumed that the classical fringing reefs known from its northern shores extended throughout its length. Even recently the comment that "the longest continuous fringing reef... lies along the Red Sea coastline, over a total length of 4500 km" (Longhurst and Pauly, 1987) was understandable, though quite wrong. This is a measure of how recently access by scientists to this region has been possible. In fact, despite the fact that the Red Sea is a single 'Large Marine Ecosystem', habitats in the south are dominated by sedimentary environments, many of them very shallow and warm, by mangroves along the mainland coasts and by extensive macroalgal 'reefs'. The continental shelf sweeps well away from land in the south, creating vast sandy plains, and thus the coral reefs for which the Red Sea is well famed occur mainly in the northern half. In addition, the south is driven substantially by inputs of nutrients and plankton from the Gulf of Aden, the effects of which appear not to reach more than about half way northwards, and this compounds the difference between the north and south.

Sea passages by travellers in the Red Sea, many of them pilgrims from north Africa to Jeddah, traversed mainly the northern and central parts, and from their numerous accidents on reefs the view probably arose of a Red Sea filled with coral reef dangers. Only with more recent work have the marked differences in the southern half become clearer. Early science also focused heavily on northern parts. In Egypt, a series of patch reefs and fringing reefs around Hurghada, now an important tourist site, have been studied since Cyril Crossland founded a marine biological station there in 1930. Further south is Quseir, which was Carl Klunzinger's base during two periods between 1863 and 1875, whose studies are a landmark in tropical marine biology generally. Most importantly now perhaps, the substantial body of work carried out in the Sinai, mainly during its occupation by Israel, provided some of the most important fundamental reef research done anywhere, and the latter region is also the location today of most of the tourism in the region and of attempts at management.

Despite plans to develop marine science capacity in the region, most published work appears to be in English or German, and its quantity follows various expeditions or residencies by a few individuals in the region, rarely followed up to a significant degree. Two recent volumes have extensively summarised the marine science of the region: Edwards and Head (1987) and Sheppard et al. (1992), and parts of the latter have been updated by the present author in Medio et al. (2000). The following chapter attempts to avoid unnecessary repetition and so is relatively short, but retains the essential elements of the above within the theme of the present series, and reflects the fact that the Red Sea is a remarkable and very important area. It is especially important for three reasons: it contains many of the main tropical habitats commonly in close proximity, it has several natural stresses which enable us to understand the effects of stressors on marine systems, and most of it has a low human population density and so has not suffered from the ravages of high human pressures which have so severely affected many other otherwise similar tropical areas. To this may be added another recent factor; the high latitude in the north appears to have allowed that part to escape the massive reef and coral mortalities which affected so many parts of the tropics in 1998.

Geological, Geographical and Physical Background

The Arabian peninsula occupies the Arabian tectonic plate, whose southwestern boundary is marked by the line of sea floor spreading which passes down the Gulf of Aqaba and Red Sea. It is also the place where the earth's largest geological feature, the mid-ocean rift system, strikes a continental platform and splits it. The Red Sea is thus a spreading centre, and the Arabian plate is moving away from Africa. As Braithwaite (1987) puts it: "For the geologist, the Red Sea is an ocean. This disregard for the realities of scale is rooted in our understanding of the nature of the earth's crust and of what oceans are." It is now fairly well established that the Red Sea rift has been separating Arabia from Africa for about 70 million years (although the general Red Sea depression was flooded at times well before this).

The Gulf of Aqaba is a continuation of the Red Sea rift, which continues north almost to Syria. Its maximum depth is of the order of 2 km, but it is joined to the Red Sea via the Straits of Tiran at a relatively shallow 250 m deep sill. The Gulf of Suez is completely different. Its depths are mostly less than 50 m, it is sedimentary, with much less reef formation. It dips to 70 m at its southern end, and has no sill at its connection with the Red Sea.

The extent of past erosional processes can be seen by the major wadi systems which developed during the Pleistocene, and which were essentially cut by abrasion into the existing, older reefal, substrate. The depths of the wadis are generally 60 to 65 m deep, though some reach over 90 m below sea level. The excavation of these occurred during past greatly lowered sea levels when reefs and coasts constructed earlier were subjected to severe aerial erosion and scouring. Vast alluvial fans spread outwards from the wadis which, later and in turn, became the gradually consolidating foundations for subsequent periods of reef growth. The present reefs grow not only on older limestone reef deposits but also on the earlier alluvial fans, and some cores have shown that these different layers may alternate vertically as well as horizontally.

The Red Sea became hypersaline immediately before the Holocene sea level rise about 17,000 years ago (Braithwaite, 1987; Gvirtzman et al., 1977; Sheppard and Sheppard, 1991). Complete closure of the Red Sea would not have been necessary for highly saline conditions to have occurred, due to the sill at the Bab el Mandeb. Biologically, this is expected to have completely 'reset' the Red Sea biota.

The total length of shoreline is probably more than doubled by small islands, which thus have a marked biological importance, especially in the provision of very sheltered habitat. Volcanic remnants occur in a chain near the central axis, while on the mainland, volcanic cones occur in the south. Lava flows interrupt fringing reef development and are colonized by macroalgae more than by corals (Sheppard, 1985). The two largest archipelagoes are the Farasans and Dhalak islands in the south (Angelucci et al., 1981; 1985), derived from the relict of a large Pliocene–Pleistocene carbonate platform of a few hundred metres thick, which grew on evaporitic salt deposits up to 3 km thick of Miocene age. The southern part of the Red Sea is geologically active, experiencing frequent earthquakes.

The present, narrow entrance at the Bab el Mandeb, literally "Gate of Lamentations", is 29 km wide and only about 130 m deep. Somali folklore has a tradition that their ancestors crossed these straits from Arabia on a land bridge, and indeed an answer required of the Forskal expedition in the 1770s was whether a land bridge was possible (it was answered in the negative). This provides an important constriction into the Red Sea with marked biogeographical consequences.

Fossil Reefs

Series of raised fossil coral reefs higher than current sea level are one of the most striking features of the northern Red Sea coastline (Fig. 2). The elevation of the uplift also matches closely the ages of the reefs themselves. Coastal uplift occurs along both shores of the Red Sea and the mountain ranges lining its shores derive from faulting and uplift. Tectonic activity continues; in the Sinai the famous cracks containing a pan-tropical cave fauna appeared following an earthquake in 1968 (Holthuis, 1973).

BIOGEOGRAPHIC POSITION

There are numerous endemic species in the Red Sea. About 20 endemic species of zooxanthellate corals are reported (Sheppard and Sheppard, 1991), or 11.3% of the total. In reef fishes endemism as a whole is about 17%, the same order of magnitude as many other groups (Ormond and Edwards, 1987), though there are revealing differences between families. Pelagic, open-water fishes which have long larval dispersal phases have very low or no endemism, while small benthic or territorial fishes such as dottybacks and triplefins have levels greater than 90%, with their remaining species confined to the western Indian Ocean. Several broadly pan-tropical groups have endemicities of 30–50%. These include butterflyfishes, parrotfishes, blennies and pufferfishes. Few other groups have been assessed in this respect, but there is certainly enough evidence to suggest that the Red Sea is an unusual biogeographic region.

It is also richer than adjacent Indian Ocean sites, for two main reasons. Firstly, the Gulf of Aden experiences marked cold water upwelling which limits the development of much rich reef development. Secondly, climatic factors cause considerable pumping of surface water into the Red Sea from the Indian Ocean, thus ensuring that a rich supply of larvae continue to be pushed in. The more uniform conditions in the Red Sea may have helped ensure that many of them survived, developing a rich habitat.

In the north, an exchange of species with the Mediterranean followed the opening of the Suez Canal in 1869. This "Lessepian migration" (Por, 1978) has continued for less than 120 years, though two factors qualify this. First is the existence, from early Egyptian and Roman days, of sea level passages linking the Mediterranean and Red Sea via the River Nile, so some euryhaline species could have passed through long before the activities of the canal's architect M. de Lesseps. The killyfish *Alphanius dispar*, a sparid *Crenidens crenidens*, a herring *Etrumeus teres* and flyingfish *Parexocoetus mento* are likely candidates (Ormond and Edwards, 1987). Secondly, after the Suez Canal was completed, large salt deposits in the Bitter Lakes kept canal water at salinity levels of over 70 ppt until relatively recently, so that for most species a salt barrier existed until only 3–4 decades ago. Mean sea level at the south of the canal is about 0.5 m higher than in the Mediterranean, so water flows northwards. Many Red Sea fish have migrated through, and some are now well established in the western Mediterranean which had only one third of the number of fish species as the Red Sea. Several have now become important components of the fisheries there, extending west to at least the central Mediterranean. Currently about 12% of eastern Mediterranean coastal fishes are of Red Sea origin.

Fig. 2. Classical fringing reef of the northern Red Sea. Elevated fossil reefs 2–3 m high drop to the present reef flat. The latter is about 50–100 m wide, extending to the reef crest visible to the right. From there the reef slope drops at angles of 45° to vertical, to hundreds of metres depth. Yanbu, Saudi Arabia (photo C. Sheppard).

SEASONALITY, CURRENTS, NATURAL ENVIRONMENTAL VARIABLES

The extremely arid nature of the region, high temperatures and constant and intense sunshine gives the overall impression of a lack of seasonal variability. Overall, evaporation is intense, removing an amount of water variously estimated at equivalent to a depth of 1–2 m per year, of which only a negligible part is replaced by fresh water, most coming from inflow through the southern entrance.

The broad annual climate cycle is basically simple, and is driven mainly by the migration of the low pressure Inter-Tropical Convergence Zone (ITCZ). During the winter, the Red Sea has two air flows, both travelling along its axis: one from the northwest in the north, and one from the southeast in the south, which meet in a region of low prevailing winds in the middle. It is only during winter that most of Arabia is crossed by any potentially rain-bearing winds. Thus winter brings occasional heavy rain and sudden flash floods. In summer, dry winds flow down the full length of the Red Sea and continue through the Gulf of Aden.

Winds and Tides

Sea breezes may be extreme and, more than the prevailing seasonal winds, drive important high-energy wave conditions. Their effects are most noticeable in the alignment of coral reefs, whose general tendency to grow into prevailing waves makes them angle outward towards the waves wherever there is suitably shallow foundation for them to do so. It also has a marked effect on the distribution of mangrove stands within several embayments, both directly and via the soft substrate distributions. These afternoon winds in the summer are the strongest experienced in the Red Sea; in the central part they may induce a median wave height of nearly 0.6 m along unprotected outer edges of the barrier reef (Georeda, 1982).

Water broadly oscillates about a central node at 20–21°N which thus has almost no daily difference in tidal height. Tides at the northern and southern ends increase with distance from the central region, reaching about 0.6 m in the north and up to 0.9 m in the south. The Gulfs of Aqaba and Suez continue the trend, reaching maxima of 1.2 and 1.5 m respectively. Superimposed onto this is a seasonal tide of at least equal importance. In winter, mean sea level is over 0.5 m higher than in summer. In the central region this is the only 'tide'. The change from summer low to winter high occurs over less than a month in the spring and in early winter. The mechanism causing it is driven partly by the greater evaporation in the summer, but is mainly the result of wind-driven currents in the entrance to the Red Sea.

The biological importance of this annual tide is considerable. Added to it is a considerable water-level change which periodically occurs as a result of thermal winds. This may add an additional 0.5 m water height in the central Red Sea. Water forced shoreward commonly leads to reef flats being submerged in summer when they are usually dry, but the reverse may occur with opposite winds in winter when water is forced away from the shore to cause complete drying, leading to widespread mortality and community disturbance in shallow areas, as was first described by Loya (1976) for the Gulf of Aqaba.

Nutrient levels in the Red Sea are detailed in Weikert (1987). Water of its two northern Gulfs is poor in nutrients compared with the Indian Ocean, and there is evidence that the production potential of the Red Sea is low. The southern part of the Red Sea has higher levels due to the influx of Gulf of Aden water; there is a sudden drop in nutrients of surface waters around 19°N which corresponds to the limit of direct influence of the northward flow of Gulf of Aden water through the Bab el Mandeb.

Deep Environments

The Red Sea is unique amongst deep bodies of water for having an extremely stable, warm temperature throughout. Below about 250–300 m, temperature is a constant 21.5°C to the sea floor in all areas (except where heated brine pools exist). The mechanism which maintains this appears to be a density gradient which causes northerly drifting Red Sea water to become heavier and sink near the entrance to and in the Gulf of Suez, to turn under and to return south below the thermocline. Briefly, a northerly drift overall of water blown into the Red Sea continues to the northern end where, in or near the Gulf of Suez, it is cooled in winter to below 18°C. This and evaporation increases the salinity to 42.5 ppt or more in the north. There is thus a steep salinity gradient along the Gulf of Suez, so the dense, cool, saline water at the entrance to the Gulf of Suez then pours downwards into the deeper Red Sea, turns under the surface water and returns southward. (This does not also happen at the entrance to the Gulf of Aqaba where there is a sill.) The returning water has the constant temperature of 21.5°C found throughout the Red Sea below the thermocline at 250–300 m, and it has a constant salinity of 40.5 ppt, in other words close to the value of the water at the southern end of the Gulf of Suez.

Hot brine pools heated by tectonic activity occur below 2 km deep. They have elevated temperatures reaching values of over 40 and even 60°C in the case of the Atlantis, Chain and Discovery deeps (Karbe, 1987). Their chlorinity is four to eight times greater than that of average Red Sea water and is therefore dense, maintaining marked vertical stratification. They are thought to contribute only little to maintaining the temperature of the main body of water at 21.5°C.

Red Sea–Gulf of Aden water exchange

The flow through the Bab el Mandeb is important and not completely clear. Not only is there an inward flow to make up the Red Sea's large evaporation deficit but also an outward discharge of the more saline deeper layer. In the winter, a double layered contra-flow system operates. At the surface, water is driven into the Red Sea from the Gulf of Aden by the prevailing winds, beneath which is the deeper outward flow; this exists for about three quarters of each year (Edwards, 1987). For part of the summer, however, the upper layer divides vertically into two components, a wind-blown shallowest flow travelling south and out of the Red Sea over the main part which still flows inward. Beneath the latter, the outflow of denser water continues as before. The sill which lies at a depth of 130 m is thought to create considerable vertical mixing at the entrance and just within the Red Sea.

Estimates of turnover time of surface water (6 years) and of all Red Sea water (200 years) take this into account. However, an important point biologically is that a large amount of nutrients and plankton is forced into the Red Sea by this process. This has important consequences to local benthic and pelagic biota in the southern region and in the Red Sea as a whole.

MAJOR SHALLOW-WATER MARINE AND COASTAL HABITATS

Coral Reefs

The most northerly reefs of the Indian Ocean are those near Suez, and are small coral patches with elevations of 1–3 m, on calcareous sandy and silty substrate. They experience winter temperatures as low as 18°C and salinity as high as 41 ppt. Extensive reefs first appear in the southern Gulf of Suez, both on Sinai and around the Ashrafi islands on the African shore, at the point where the northerly surface flow of Red Sea water dips under to return southward at depth. At Ras Mohammed on the southern tip of the Sinai, the bathymetric profile changes abruptly into that of the main Red Sea system, with classical fringing reefs extending southward along both sides to 18–20°N.

Most fringing reefs are narrow, extending only a few tens of metres from the shore. Further south, reefs commonly extend 1 km to seaward from an alluvial plain 1–7 km wide, where old wadi systems add vast amounts of debris which form alluvial fans. Beaches here are well developed, and many fans are characterized by large back reef lagoons with shoals parallel to the shoreline. Reefs fringe the edges of these alluvial fans. Most fringing reefs in the north have steep sides, plummeting to depths of hundreds of metres very near land (Fig. 3).

Offshore in the northern Red Sea lies the Wedj Bank and its islands, with extensive seagrass, reefs and mangroves (Ormond et al., 1984). Also offshore, from Wedj to Jeddah, is an extensive series of submerged limestone platforms which form the foundation for a chain termed the "Little Barrier Reef". Coral communities on this are the most exposed, varied and vigorously growing known in the Arabian region.

Fig. 3. Fringing reef slope at Ras Mohammed peninsula, southern Sinai. 25 m depth at zone of peak diversity (photo C. Sheppard).

The central Red Sea is widest in the region of Jeddah and Sudan, and supports a correspondingly wide range of reef morphologies. The Arabian side is better studied. Reef flats begin to broaden in this central area (Fig. 4) and to develop more and larger lagoons, but mangroves begin to assume greater importance along the mainland shoreline. One atoll exists, Sanganeb Atoll, and is well studied, though several other similar, circular or annular reef structures which form part of the barrier reef system may not be dissimilar. Most work on the African coast has been done on the Towartit reefs and in the Suakin Archipelago, near and just south of Port Sudan.

On the Arabian side, fringing reefs become much reduced in size south of about Al Lith at 20°N. The continental shelf becomes broader, and the steady increase in muddy substrates and mangroves causes significant reef development to be pushed out further from shore, where there continue to be large expanses of limestone platform.

Fig. 4. Broadening reef flats in central Red Sea, with reef crest about 500 m or more from the shore. Shore with fairly rich halophytic vegetation, indicating the area is at the mouth of a wadi system. Reefs to seaward have developed on an old alluvial fan (photo C. Sheppard).

Fringing reefs diminish and have reduced coral diversity; in many places there is a complete replacement of fringing reef by broad and thick stands of mangrove. Sand shores and sandy sublittoral habitat with no coral or mangrove growth also increase. These conditions reach their maximum extent opposite the Farasan islands, and extend to the far south.

Algal Reef Constructions

Where wave energy is sufficiently high in the north, spur and groove systems develop. These reef-crest structures point into the prevailing waves, and are constructed from red calcareous algae. In the south there is a more unusual development of reefs constructed almost entirely by calcareous red algae on extensive, sublittoral sand. These develop an entirely different reef type in moderately and very sheltered conditions, and they appear to be formed from the same, or very similar, crustose coralline species. These "algal reefs" (Sheppard, 1985) exist in the absence of coral reefs. They arise from sandy substrata nearshore where water is 2–4 m deep, and develop steep sides which reach to the low tide level. They support very few and sparse corals, and are conspicuous from shore because they are covered in dense *Sargassum*, whose fronds float in thick mats at the water surface.

Coral Distributions

An analysis of coral data from nearly 200 sites along almost the full length of the Red Sea (Sheppard and Sheppard, 1991) showed the existence of 13 principal coral communities. Several could themselves be subdivided further into easily recognised units (see Medio et al., 2000). They are distributed along a clear north–south trend with most communities showing considerable localization. In addition, the general pattern of coral diversity with depth follows that of most Indo-Pacific reefs. On reef flats diversity is low, and increases sharply near the reef crest. In shallow water on the reef slope, diversity generally increases to a maximum between 5–30 m deep, below which it goes into a gradual, continuous decline as depth increases and light diminishes (e.g. Loya, 1972; Sheppard and Sheppard, 1985). Coral cover is usually less than 50%, but in sheltered areas, *Porites* may cover over 80% of the substrate, especially on offshore patch reefs. Several species of *Acropora* likewise occasionally show high cover, and there are many instances of particular, turbid areas where *Goniopora* or even *Galaxea* provide very high cover.

Mangroves, Seagrasses and Algae

In the Arabian region, *Avicennia marina* is by far the commonest mangrove, though *Bruguiera gymnorhiza* and *Ceriops tagal* also occur. The northernmost stands occur along the Sinai peninsula (Fig. 5), but throughout the north scattered mangroves occur in sheltered situations behind reef flats, in bays or creeks, and in the lee of offshore islands. At many sites the sediment is only a thin veneer overlaying fossil reef, leading to stunted mangrove bushes; mangroves growing in these conditions are usually termed 'hard-bottomed' or 'reef' mangals. Soft-bottomed mangals are characteristic of the southern Red Sea (Price et al., 1987), where stands may reach several hundred hectares, especially along mainland shores bordering broad shallow areas where the continental shelf lies well offshore. Although the southern Red Sea mangroves contain a rich fauna, diversity is limited in comparison with well developed mangals in other parts of the world. Mangroves in these southern regions are used as a fodder for camels (Fig. 6), though it seems likely that the generally low human population density has limited the overall impact of this to date.

Fig. 5. Some of the northernmost mangroves of the Red Sea, in sheltered embayment created by a cut in old fossil reefs. Trees are stunted 'hard substrate' mangroves, rooted in a relatively thin veneer of sediments. Ras Mohammed peninsula (photo C. Sheppard).

Fig. 6. Broad, shallow and muddy embayment, extensively fringed with mangroves, grazed by camels. Southern Saudi Arabia (photo C. Sheppard).

Fig. 7. *Sargassum* zone on reef crest of southern fringing reefs (photo C. Sheppard).

Eleven seagrass species occur, of seven genera. Several seagrass assemblages have been identified from cluster analysis (Price et al., 1988) revealing three groupings separated by latitude, again suggesting biogeographic trends. Seagrasses have colonised a range of unconsolidated sediments; in the Gulfs of Suez and Aqaba, fine to coarse sands are favoured, although the widespread *Halophila stipulacea* has also colonised fine-grained sediments and muddy 'lagoons'. *Thalassodendron ciliatum* and *Thalassia hemprichii* mostly inhabit coarser sediments, while *Enhalus acoroides* is characteristic of very soft mud substrata.

About 500 species of algae are known from the Red Sea and these also show a latitudinal distribution (Walker, 1987). Most algal turf species in northern and central areas are macroscopic, non-calcareous green, brown and red species, and may have standing stocks of approximately 25 g (dry wt.) m^{-2}. In the central Red Sea, turf algae generally provided around 4% cover below 5 m deep, but in the shallower algal lawn zone, this component may cover 50–80% of the substrate of many reefs.

On southern Red Sea sites, standing crop may be ten times higher than in the north. A feature which is characteristic of "marginal" reef conditions is increasing dominance of some of the Fucales, large brown algae which include the genera *Sargassum*, *Cystoseira* and *Hormophysa*. A very conspicuous trend which occurs on the crest of the fringing reefs in the Red Sea is the southward increase in these brown algae, mainly *Sargassum* (Fig. 7). The greatest increase is seen in the south where there is a corresponding decline in the extent of reefs and coral diversity. The *Sargassum* grows in thick stands, with numerous plants in each square metre, up to 2 m tall.

OFFSHORE SYSTEMS

Primary production in the Red Sea is low because the thermocline and halocline prevent recycling of nutrients from deeper water, and because there is almost no terrestrial run-off. The importance of the richer Gulf of Aden water which is forced into the southern Red Sea is also clear (Weikert, 1987): when in the north there were 180 cells m^{-3} and in the central Red Sea there were 300 cells m^{-3}, in the southern region there were 3000 cells m^{-3}. The latter were summer values while winter values were 14,000, 58,000 and 21,000 cells m^{-3} respectively. In the total euphotic zone of the Red Sea, productivity was tabulated at 0.21, 0.39 and 1.60 g C m^{-2} d^{-1} in the northern, central and southern regions. This demonstrates the relatively enormous input of nutrients and of plankton through the Bab el Mandeb, which Weikert (1987) calculated as being 96×10^{13} organisms, or 6×10^4 tonnes dry weight per month at peak times of influx. The increasingly harsh conditions of temperature and salinity mean that most die in the southern Red Sea and release nutrients in the process. The fishery of the southern section is the richest of the Red Sea and is supported by this.

Blooms of *Oscillatoria erythraeum* are common and are of enormous quantitative importance to pelagic productivity, in large part because of their ability to fix nitrogen in the nutrient-poor waters. The total carbon production is high: data reviewed in Weikert (1987) suggest that production in the photic zone is 250–500 mg C m^{-2} d^{-1} in the north, rising to over 1000 mg C m^{-2} d^{-1} in the south. In all areas there is also a trend of greater productivity near the shorelines than along the central axis.

Because of strong light intensities in the region, photoinhibition takes place in the upper 20 m of water. High production values and maximum chlorophyll *a* concentrations generally occur below this to as deep as 200 m, depending on the degree of vertical mixing and stratification. The most productive part is 50–80 m deep in offshore water, though this band is rather shallower nearer land.

Turtles and Dugong

Only the Green and Hawksbill are common turtles, and both nest throughout its length. The region is now very important for this group. There once was a much greater abundance in the Red Sea, and a significant reduction by humans has clearly taken place, perhaps from as early as Roman times; vast piles of turtle bones on several islands in the northern and southern Red Sea suggest a past, large exploitation. Today, exploitation is much lower than in many other regions, so that remaining breeding areas assume a correspondingly greater global importance. According to Miller (1989) there is still an artisanal turtle harvest and additional mortality from accidental trawling: in the Farasan Archipelago, about 2000–3000 are caught each year, of which about 25% die, while the remainder are released.

Rich seagrass beds around the Tiran Islands and Wedj Bank in the north, support moderate numbers of dugong.

They are recorded from the Gulf of Suez and the southern archipelagoes and mainland south of Al Lith (Preen, 1989), in extensive shallows of the Port Sudan and Suakin Archipelago, as well as in the Dhalak Archipelago (Frazier et al., 1987). Preen (1989) has estimated that the Red Sea population may be about 4000 individuals.

POPULATION, URBAN AND INDUSTRIAL ACTIVITIES

The Red Sea coastline for the most part has extremely low levels of population outside a few major ports and cities, which has greatly limited the degree of coastal shoreline alteration, pollution and resources abstraction. Petroleum hydrocarbon levels are relatively high in the Gulf of Suez, where there is substantial oil and tar on shores, in many cases due to accidents in that gulf's oil production industry. One report completed in 1979 concluded that 'coastal pollution by oil was found to be a very serious problem along major sections of the Gulf of Suez and Egyptian Red Sea coast. ... The degree of pollution is not uniform for the long coastline involved... However, a number of areas ... must rank with some of the worst polluted coastlines of the world' (Wennink and Nelson Smith, 1979). Oil contamination occurs also, but to a lesser extent, throughout the length of the Red Sea, not surprisingly being a greater problem nearer ports; along the coast of North Yemen beach tar was recorded at 58% of 131 coastal sites (Barratt et al., 1987). Much of the early work on the effects of oil pollution on coral reefs took place in the Red Sea and have been found to include colonisation failures by corals (Loya, 1975; Rinkevitch and Loya, 1979).

Refuse and beach debris is also marked in localised areas; such refuse was reported at 67% of 131 coastal sites in the southern Yemen sector, for example (Barratt et al., 1987). In the Gulf of Aqaba, inputs of phosphate, manganese and bauxite minerals are major pollutants (Walker and Ormond, 1982; Dicks, 1987), leading to death of corals. However, the overall impact from both industry and the very limited agriculture is much less in this region than in many other parts of the world.

Reclamation has been undertaken for residential developments, ports, bridges, causeways, corniche roads and other purposes. In the Red Sea, dredging has also been most prevalent in commercial and industrial areas, in particular Jeddah and Yanbu in Saudi Arabia. Pressures from recreation and tourism are higher in Egypt than other areas of the Arabian region, mainly due to tourism.

Lack of widespread sewage treatment facilities in most Red Sea countries has resulted in extensive localised damage to coral reefs, and in some areas has caused eutrophication problems in ponded or lagoonal areas, or over extensive reef flats near towns. Again, the lack of regionally serious effects appears to be a fortunate consequence of overall very low population densities, as even in the more wealthy countries it has not always been due to careful treatment of wastes.

Highly saline discharges from desalination plants are a particularly 'middle eastern' problem which are commonly cited as causing environmental problems. However, observations around large plants indicate that in the steep and deep Red Sea, discharge effects are limited to distances of less than 200–300 m from the release, a relatively small radius in view of the immense importance of the product. More important may be the addition of biocides. These have been known to kill important grazing gastropods which keep intertidal algae in check, and whose death reduces the diversity of mud flats and mangrove areas.

Several countries of this region have been torn by serious turmoil over the past few decades, but are still affected by some of the same environmental problems. In Yemen, pollution due to sewage and industrial development is acute in localised areas where it has led to visible degradation of the coastline. This is particularly relevant in the light of the rapidly expanding coastal population and the country's tourism aspirations.

Fishing

Artisanal methods catch about 70% of the total landings (52,700 t/year), though purse seining and trawling activities catch 22.0% and 8.8%, respectively, mostly from Egypt and the Gulf of Suez. Fishing in most of the Red Sea has probably been mostly at sustainable levels to date, especially where traditional techniques have been used, or where there has been limited efficiency and a lack of a proper market structure. One early exception to this trend was seen in the Gulf of Suez shrimp trawling industry which is thought to have shown signs of over fishing several decades ago. More recently, there have been increased levels of fishing using improved fishing methods to supply the ever-increasing populations. This has led to uncontrolled fishing or very selective fishing (Medio et al., 2000).

Heavily fished areas other than the Gulf of Suez include most areas adjacent to towns and cities, and several countries, such as Egypt and Yemen, have made great efforts to increase catches. In Yemen, extensive subsidies have made fishing much more profitable, and large numbers of Yemeni fishers now make trips throughout the southern Red Sea, including to Eritrean reefs where high-value reef-associated species are sought. Trawling is widespread on all shallow banks, and there is concern about overfishing several stocks. One concern also is that some of these fisheries are for providing more affluent parts of the region with luxury foods rather than sustaining local populations.

Reef fisheries in the south especially are very difficult to monitor and manage, and there is concern that regulations which do exist are commonly ignored; catch statistics, for example, show that trawling strays on to reef areas.

The increase in coastal populations and, to a lesser extent, tourist activities in the last few years have in some cases denuded large areas of reef organisms. Mother-of-pearl shell (*Trochus*) is very intensively collected in Sudan, and the few turtle colonies on the Saudi Arabian coast are typically used for food. Recently some Red Sea countries such as Jordan, Egypt and Israel have introduced legislation to control collection of reef organisms such as hard corals, shells and fish. Until recently the potentially high-earning aquarium industry faced practical limitations such as the lack of efficient airport infrastructure throughout the Red Sea, with the exception of Jeddah in Saudi Arabia. Recently, improved communication and airport infrastructure have led to an increase in aquarium trade and there is presently concern that the lucrative aquarium fish business may contribute to the deterioration of reef communities.

PROTECTIVE MEASURES

The main regional agreement for conservation and coastal management is ROPME (Regional Convention for the Conservation of the Red Sea and Gulf of Aden) and the associated action plan for the region of the UNEP Regional Seas Programme. Several countries are, at least nominally, fairly active in coastal matters, and have several measures in place to help ensure sustainable use of their coastal resources. Most activity perhaps is done by Egypt with its two major diving tourism centres at Hurghada on the African shore and its Sinai operations which focus on Sharm el Sheikh and the Ras Mohammed marine park area, by Saudi Arabia whose two main bodies with environmental interests (MEPA and NCWCD) are perhaps the most active, and Jordan. Work in this respect is also increasing in Yemen, via a GEF project (UNDP-GEF, 1992). Other countries have various regulations, mostly pertaining to fisheries and coastal development, though it is clear that in some cases these are nominal only. A detailed listing of all such measures has been prepared by the present author with A. Price (UNEP, in press).

REFERENCES

Angelucci, A. and 12 others (1985) L'arcipelago Delle Isole Dahlak Nel Mar Rosso Meridionale: Alcune Caratteristiche Geologiche. *Bollettino della Societa Geografica Italiana Series XI* **2**, 233–262.

Angelucci, A., Matteucci, R. and Praturlon, A. (1981) Outline of geology and sedimentary environments of the Dahlak Islands (Southern Red Sea). *Bollettino della Societa Geologica Italiana* **99**, 405–419.

Barratt, L., Dawson Shepherd, A.R., Ormond, R.F.G. and McDowall, R. (1987) Yemen Arab Republic marine conservation survey. Vol. I. Distribution of habitats and species along the YAR coastline. IUCN Red Sea and Gulf of Aden Environment Programme/TMRU York, UK. 110 pp.

Braithwaite, C.J.R. (1987) Geology and palaeogeography of the Red Sea region. In *Red Sea*, eds. A.J. Edwards and S.M. Head. Pergamon Press, Oxford. pp. 22–44.

Dicks, B. (1987) Pollution. In *Key Environments: The Red Sea*, eds. A.J. Edwards and S.M. Head. Pergamon Press, Oxford. pp. 383–404.

Edwards, A.J. and Head, S.M. (eds.) (1987) *Red Sea*. Pergamon Press, Oxford.

Edwards, F.J. 1987. Climate and oceanography. In *Red Sea*, eds. A.J. Edwards and S.M. Head. Pergamon Press, Oxford. pp. 45–69.

Ferry, R.E. and Kohler, C.C. (1987) Effects of trap fishing on fish populations inhabiting a fringing coral reef. *North American Journal of Fisheries Management* **7**, 580–588.

Frazier, J.G., Bertram, G.C. and Evans, P.G.H. (1987) Turtles and Marine Mammals. In *Red Sea*, eds. A.J. Edwards and S.M. Head. Pergamon Press. Oxford. pp. 288–314.

Georeda Ltd. (1982) Oceanography, final report. Royal Commission for Jubail and Yanbu Contract PID-0203 Kingdom of Saudi Arabia.

Gvirtzman, G., Buchbinder, B., Sneh, A., Nir, Y. and Friedman, G.M. (1977) Morphology of the Red Sea fringing reefs: a result of the erosional pattern of the last glacial low-stand sea level and the following Holocene recolonisation. *Mem. B.R.G.M.* **89**, 480–491.

Holthuis, L.B. (1973) Caridean shrimps found in land-locked saltwater pools of four Indopacific localities (Sinai peninsula, Funafuti Atoll, Maui and Hawaii Islands) with a description of one new genus and four new species. *Zoologische Verhandelingen* **128**, 1–48.

Karbe, L. (1987) Hot brines and the deep sea environment. In *Red Sea*, eds. A.J. Edwards and S.M. Head. Pergamon Press, Oxford. 441 pp.

Longhurst, A.R. and Pauly, D. (1987) *Ecology of Tropical Oceans*. Academic Press, London. 407 pp.

Loya, Y. (1972) Community structure and species diversity of hermatypic corals at Eilat, Red Sea. *Marine Biology* **13**, 100–123.

Loya, Y. (1975) Possible effects of water pollution on the community structure of Red Sea corals. *Marine Biology* **29**, 177–185.

Loya, Y. (1976) Recolonisation of Red Sea corals affected by natural catastrophes and man-made perturbations. *Ecology* **57**, 278–289.

Medio, D., Sheppard, C.R.C. and Gasgoine, J. (2000) The Red Sea. In *Reefs of the Western Indian Ocean: Ecology and Conservation*, eds. T. McClanahan, C.R.C. Sheppard and D. Obura. Oxford University Press.

Miller, J. (1989) Marine turtles. Vol 1. An assessment of the conservation status of marine turtles in Saudi Arabia. MEPA, Coastal and Marine Management Series Report No 9. MEPA, Jeddah. 209 pp.

Ormond, R.F.G., Dawson Shepherd, A.R., Price, A.R.G. and Pitts, J.R. (1984) Report on the distribution of habitats and species in the Saudi Arabian Red Sea. 1. IUCN/MEPA, Kingdom of Saudi Arabia No 1: 123 pp. No 2: 151 pp.

Ormond, R.F.G. and Edwards A.J. (1987) Red Sea fishes. In *Red Sea*, eds. A. Edwards and S.M. Head. Pergamon Press, Oxford, pp. 251–287.

Por, F.D. (1978) *Lessepsian Migration*. Ecological Studies 23. Springer Verlag, Berlin. 228 pp.

Preen, A. (1989) Dugongs. Vol. 1. The status and conservation of Dugongs in the Arabian Region. MEPA Coastal and Marine Management Series, Report No 10. MEPA, Jeddah, 200 pp.

Price, A.R.G., Crossland, C.J., Dawson Shepherd, A.R., McDowall, R.J., Medley, P.A.H., Ormond, R.F.G., Stafford Smith, M.G. and Wrathall, T.J. (1988) Aspects of seagrass ecology along the eastern coast of the Red Sea. *Botanica Marina* **31**, 83–92.

Price, A.R.G., Medley, P.A.H., McDowall, R.J., Dawson Shepherd, A.R., Hogarth, P.J. and Ormond, R.F.G. (1987b) Aspects of mangal ecology along the Red Sea coast of Saudi Arabia. *Journal of Natural History* **21**, 449–464.

Rinkevitch, B. and Loya, Y. (1979) Laboratory experiments on the effects of crude oil on the Red Sea coral *Stylophora pistillata*. *Marine Pollution Bulletin* **10**, 328–330.

Sheppard, C.R.C. and Sheppard, A.L.S. (1985) Reefs and coral assemblages of Saudi Arabia 1. The central Red Sea at Madinat al Sinaiyah. *Fauna of Saudi Arabia* **7**, 17–36.

Sheppard, C.R.C. and Sheppard, A.L.S. (1991) Corals and coral communities of Arabia. *Fauna of Saudi Arabia* **12**, 3–170.

Sheppard, C.R.C. (1985) Reefs and coral assemblages of Saudi Arabia. 2. Fringing reefs in the southern region, Jeddah to Jizan. *Fauna of Saudi Arabia* **7**, 37–58.

Sheppard, C.R.C., Price, A.R.G. and Roberts, C.J. (1992) *Marine Ecology of the Arabian Area: Patterns and Processes in Extreme Environments*. Academic Press, London.

UNEP (in press) The state of the marine environment in the Red Sea and Gulf of Aden. *UNEP Regional Seas Reports and Studies*.

UNDP-GEF (1992) Yemen: Protection of Marine Ecosystems of the Red Sea Coast. Project Document. United Nations Development Programme, Global Environment Facility, New York, No. 54.

Walker, D.I. and Ormond, R.F.G. (1982) Coral death from sewage and phosphate pollution at Aqaba, Red Sea. *Marine Pollution Bulletin* **13**, 21–25.

Walker, D.I. (1987) Benthic algae. In *Red Sea*, eds. A.J. Edwards and S.M. Head. Pergamon Press. Oxford. pp. 152–168.

Weikert, H. (1987) Plankton and the pelagic environment. In *Red Sea*, eds. A.J. Edwards and S.M. Head. Pergamon Press. Oxford. pp. 90–111.

Wennink, C.J. and Nelson Smith, A. (1979) *Coastal oil pollution study for the Gulf of Suez and Red Sea coast of the Republic of Egypt*. IMCO, London.

THE AUTHOR

Charles R.C. Sheppard,
*Department of Biological Sciences,
University of Warwick,
Coventry CV4 7AL, U.K.*

Chapter 56

THE GULF OF ADEN

Simon C. Wilson and Rebecca Klaus

The Gulf of Aden lies between southern Arabia and the Horn of Africa and connects with the Red Sea and Indian Ocean. The Socotra Archipelago lies at its entrance, off the Horn of Africa. The largest influence comes from the reversing monsoon system with strong and persistent winds that blow from the southwest in summer, and from the northeast in winter. These also cause a reversal in the direction of surface currents. Associated with the summer monsoon are upwelling areas along the eastern coast of Yemen, and one centred on the Somali coast southwest of Socotra. Both have a profound effect on coastal habitats and stimulate high marine productivity which supports a rich fishery.

Marine biodiversity is relatively high since the area is a transition zone between the Red Sea, Southern Arabia and East Africa. Terrestrial diversity, particularly in the flora of Socotra, is also elevated by high levels of endemism.

Coasts are mainly exposed sandy beaches separated by rocky headlands. Coral communities and reefs have developed most notably in Djibouti and offshore islands of Somalia. Seagrasses are relatively uncommon, and mangrove stands are most abundant to the west and southwest. A striking feature of rocky shores is the abundant macroalgae that appears following the onset of the Southwest Monsoon in particular. Green turtles nest in tens of thousands, and thousands of dolphins have also recently been observed.

Perhaps the most serious single threat to sustainable use of marine resources comes from overfishing, particularly by industrial fleets that operate with or without licences. Some stocks have collapsed or are showing signs of strain, including cuttlefish, shark and lobster. Wildlife species are also harvested and incidental mortality appears high. Levels of pollution are low except around larger towns where sewage and solid wastes are starting to affect resources. Chronic oil pollution originating from tankers is also cause for concern, but levels appear to be low.

Harsh environmental conditions and lack of infrastructure limits exploitation of coastal resources and traditional methods of limiting exploitation are still effective. Political instability and unrest, and lack of funding have hampered coastal management, though a strategic action plan for the conservation and protection of the marine environment has recently been prepared by PERSGA as a crucial first step. There are only two small marine parks in the region, both in Djibouti.

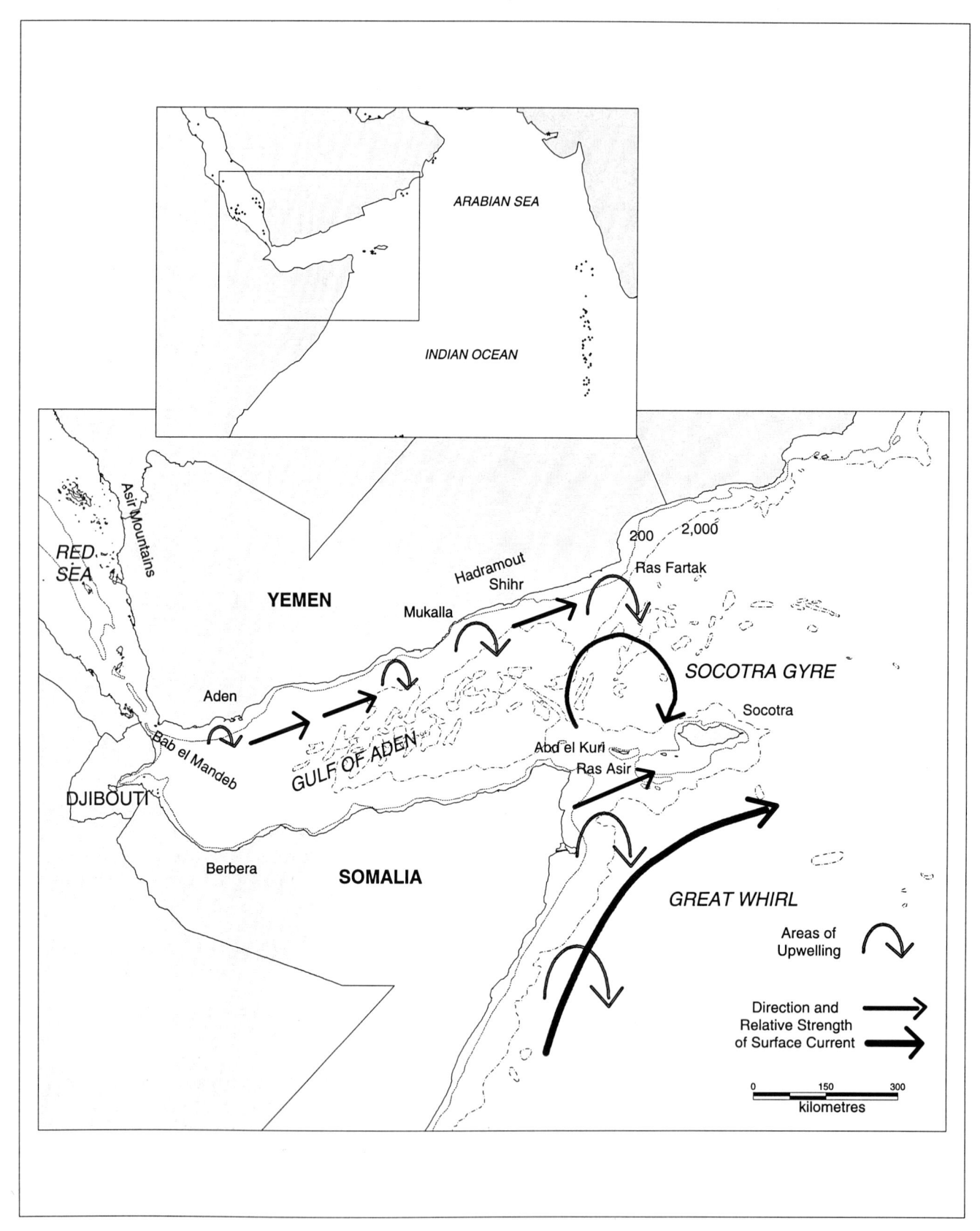

Fig. 1. The Gulf of Aden Area indicating the direction of surface current flows and the approximate size and position of upwelling areas during the Southwest Monsoon.

THE DEFINED REGION

The Gulf of Aden ('The Gulf') lies between the Horn of Africa and the Arabian peninsula, and connects the Red Sea to the Indian Ocean (Fig 1). Its boundary with the Red Sea is the narrow straits of Bab al Mandeb ('Gates of Lamentation' in Arabic). To the East the boundary is the rhumb line between Ra's Asir at the tip of the Horn of Africa and the 300 m high sea-cliffs of Ra's Fartak in Arabia. The Socotra Archipelago (Yemen), off Ra's Asir, is included in this chapter.

Three states border the Gulf: Yemen, Djibouti and Somalia. The coastline of Yemen is 1690 km; this includes both 1200 km of mainland coast which is relatively straight with few sheltered areas and the Socotra Archipelago of 490 km. The Socotra Archipelago consists of the main island of Socotra, the outer islands Samha and Darsa, otherwise known as "The Brothers", and Abd El Kuri to the southwest and two rock outcrops Sabunya and Kal Farun. The coast of Djibouti, approximately 340 km of which faces the Gulf, is dominated by the Golfe de Tadjora with its islands, sheltered bays and patch reefs. The coastline of Somalia is straight with few sheltered areas or offshore islands and has a total length of 1300 km.

The region's biological and oceanographic borders are less precise. Kemp (1998b) describes its major zoogeographic features and distinct coastal flora and fauna. Firstly, a Red Sea/Gulf of Aden barrier has been described by various authors, which probably lies between Bab al Mandeb and the north western corner of the Gulf of Aden (Ormond and Edwards, 1987; Sheppard and Sheppard, 1991; Kemp, 1998b). Secondly, to the south and east, cold upwelling caused by the Somali current in the Southwest Monsoon appears to present a partial barrier to larvae of tropical species from the more equatorial regions. The upwelling itself extends over 1000 km along the coast of southern Oman, and has a distinctive flora and fauna of its own.

In terms of marine biological diversity, the Gulf of Aden is a transition zone between three surrounding biogeographic sub-provinces: the Red Sea, Southern Arabia and East Africa. The total number of species found in the Gulf of Aden is relatively high due to the overlapping distributions of species from each. Recent surveys of Socotra have revealed very mixed and diverse marine communities and may prove to be an extremely important biogeographical stepping stone. (The underlying cause of high terrestrial diversity on Socotra is different, however. Species-richness is elevated by many endemic species; about a third of Socotra's 850 plant species are endemic, which has led WWF (cited in UNDP, 1996) to describe Socotra as an 'Indian Ocean version of the Galapagos'.)

Comparison of surface distributions of surface seawater temperature, salinity, oxygen, nutrients, transparency, and silicates all suggest that the Carlsberg Ridge, on which Socotra sits, forms a boundary between the East Somali coast and Gulf of Aden (Zeitzschel and Gerlach, 1973). During summer months, the northern arm of the Great Whirl arcs around from the coast of East Somalia to pass south of Socotra, carrying productive waters away from the Gulf of Aden and towards the central Arabian Sea. With it are carried the majority of larvae from tropical species in East Africa. Similarly, the coastal upwelling along the Yemeni coast of the northeastern Gulf of Aden is carried along the coast and into the Arabian Sea, forming part of a larger system to the east. The effect of this upwelling is moderate and the open waters of the central Gulf of Aden remain relatively oligotrophic during this period.

SEASONALITY, CURRENTS, NATURAL ENVIRONMENTAL VARIABLES

Oceanographic conditions in the Gulf of Aden are driven by the pattern of monsoon winds. The mechanism governing the monsoons is controlled by the annual oscillation across the equator of the Inter-Tropical Convergence Zone (ITCZ) that separates the trade winds of the Northern and Southern hemispheres. This rather diffuse belt migrates northward to reach the southern seaboard of Arabia in July, bringing the Gulf of Aden under the influence of strong southwesterly winds. During the four summer months (June–September), winds blowing down the Red Sea reinforce the clockwise airflow over the Arabian Sea causing sustained winds of over 15 m s^{-1} (Beaufort force 6–7) that run parallel to the southern coast of Arabia (Sheppard and

Fig 2. GIS rendering of Socotra island.

Dixon, 1998). Such strong and persistent wind with a fetch of many hundreds of kilometres generates severe swell with significant wave heights of 4 m or greater, more than 35% of the time in the waters surrounding Socotra. These heavy seas are an important factor in the distribution of coastal habitat types.

The southwest or summer monsoon commences in April/May and reaches its peak in July–August, and fades away again in September–October. It is replaced by the weaker northeast or winter monsoon which blows across the Indian Ocean during November to February (Currie et al., 1973). These influence the oceanographic cycling of waters in the Gulf of Aden. The Southwest Monsoon causes a dramatic reversal in the flow of the Somali current changing its southward flow along the East African coast to a northward flow. This seasonal reversal alters the circulation and the hydrographical conditions of the surface waters in the entire Indian Ocean as well.

The onset of the summer monsoon can vary by as much as a month from year to year (Piechura and Sobajh, 1986). In the Gulf of Aden, persistent southwesterly winds induce the northwards flow of the Somali Current between Abd El Kuri and the mainland Somalia and strengthens a weak eastward flow through the Gulf. By mid-May the East Arabian Current (EAC) (0.8 m s^{-1}) is fully established. The EAC and the increasing southwest winds blow warm surface water offshore and generate the Arabian Sea upwelling system along most of the coast of south Yemen and Oman during May to September. Displaced surface water is replaced by cold, nutrient-rich water drawn up from below 500 m depth, and its effect on coastal habitats is greatest where the continental shelf is narrow. This is intense and is the single most important phenomenon affecting coastal ecology here.

The Arabian Sea upwelling event is relatively short-lived (Bottero, 1969; Curie et al., 1973; Barrat et al., 1984, 1986). The fact that the upwelling is embedded in a western boundary current reduces its effectiveness for primary production—the swift current removes much of the additional biomass from the system before it can be utilised (Tomczak and Godfrey, 1994). The upwelling occurs particularly in the lee of headlands, being especially pronounced around Mukalla and Ra's Fartak (Stirn et al., 1985). Ormond and Banaimoon (1994) reported that in 1988–9, average seawater temperatures around Mukalla fell from 31°C in early June to 19–22°C in July, while phosphate–phosphorus concentrations increased fivefold from 0.5 μg-at l^{-1} to 2.78 μg-at l^{-1}. Smaller and more localised patches of upwelling are induced further west also. In August of 1992, much of the central Gulf of Aden was still oligotrophic, although an extensive phytoplankton bloom was encountered south of Bab al Mandeb that probably stemmed from upwelling along the Yemen coast (Baars, 1994).

South of Socotra, the flow of the Somali current northwards, together with increasing wind speeds, drives the current eastwards. At two sites, at about 3–5°N and 7–9°N, the current deviates offshore inducing wedges of upwelling to the north of these locations (Baars, 1994). The greater offshore current velocity (1–1.5 ms^{-1}) induces a more intense upwelling with surface temperatures below 20°C and a nitrate concentration of 18 μM. A two-gyre system develops known as the Great Whirl and the Socotran Gyre (Fisher et al., 1996). The clockwise gyres carry the cool, enriched productive waters eastwards south of the Socotra Archipelago back into the open Indian Ocean (Baars, 1994). This entire upwelling system migrates slowly northward over a period of about a month until it reaches a northern limit around Ra's Asir extending a short distance into the Gulf of Aden. Here, cool upwelled waters from around Ra's Asir meet warmer, nutrient-poor waters flowing east out of the Gulf of Aden to form a sharp boundary which often lies between the Somali mainland and Socotra (Baars, 1994).

The Somali coast upwelling is separated from the Arabian Sea upwelling by a stretch of warm water flowing east out of the Gulf of Aden, north of Socotra. The Somali upwelling is evidently a major influence on marine communities around Socotra (Turner et al., 1999).

The end of the summer monsoon is marked by the ITCZ migrating southwards in September. In winter months, the Gulf of Aden is dominated by the Northeast Monsoon with its gentler wind (Beaufort force 4–5) that generates a westerly current along the Yemen coast and more moderate sea conditions. This current drives Gulf of Aden water into the southern Red Sea before continuing to flow anti-clockwise along the Somali coast. Upwelling during this period has not been described in any detail but there is mounting evidence that elevated nutrient concentrations are common, which explains the well-defined spring bloom of macro-algae around Mukalla (Ormond and Banaimoon, 1994) and Socotra (Kemp, 1998a).

During periods of transition between the two monsoons, weaker and more localised diurnal winds generated by the differential heating and cooling of land and sea generally cause breezes to blow offshore in the mornings and onshore in the evenings. Stronger thermal winds are the 'belat', coming off coastal mountain ranges that blow offshore and can cause violent squalls (up to Beaufort force 7), particularly at night and at dawn. Other than these, tropical storms in the Gulf of Aden are rare, occurring approximately once every 50 years (Hydrographic Office, 1980). Similarly, surface water currents are weak and variable in the inter-monsoon seasons.

Upwelling extends northeast along the coast of Oman, and is considered one of the five largest upwelling areas of the world. Its effect on both terrestrial and marine habitats is dramatic. The cold seas affect local weather by lowering air temperatures and increasing cloud cover significantly, tempering the harsh summer conditions typical of other parts of Arabia. Rains fall on the coastal mountains of southern Yemen and trigger a seasonal flush of vegetation on which pasturalists depend for grazing, and which recharges groundwater. The upwelling creates an isolated and seasonal enclave of temperate sea conditions in an otherwise tropical environment that has a number of ecological implications. Little detail is available on the incidence of upwelling along the Somali coast of the Gulf of Aden, but dense schools of small pelagic fish from Berbera to 200 km east (World Bank, 1992) suggest that these waters are also highly productive.

The way in which this area of upwelling acts as a partial barrier to the flow of genes and the distribution of marine organisms has been discussed elsewhere (e.g. Sheppard et al., 1992), most recently by Kemp (1998b). 'Pseudo-temperate' conditions inhibit recruitment and settlement of larvae of tropical species. The seasonally reversing surface currents further complicate the temporal dimension of this barrier. Therefore, the timing of reproduction in marine species is a crucial factor determining reproductive success, since larvae can reach favourable settlement conditions only if released at precise times relative to the onset of the two monsoons. Sheppard et al. (1992) review evidence that the intensity of both the monsoons and upwelling during the Holocene was reduced compared to today, which potentially 'opened the gates' allowing tropical species to migrate across Southern Arabia.

Surface waters in the Gulf of Aden usually contain 4–5 ml l^{-1} dissolved oxygen. Deeper, at depths of 100–500 m, dissolved oxygen concentrations fall to about 10% saturation (0.5–0.2 ml l^{-1}), partly due to the decomposition of detrital fallout from upper layers and partly because this mass of water originates from deep water flowing out from the Red Sea. With the onset of upwelling, this body of water is drawn up into the otherwise productive photic zone where it can cause mass mortality, particularly of pelagic fish (Piechura and Sobajh, 1986). The temporal and spatial extent of these mass-mortality events is probably determined by the dynamics of small-scale mixing processes.

THE MAJOR SHALLOW WATER MARINE AND COASTAL HABITATS

The most immediate consequence of coastal upwelling in the Gulf is that a dense and diverse algal community flourishes along the Yemeni coast for approximately six months each year, particularly from Mukalla eastwards, which is almost completely absent for the rest of the year. In places, algal communities co-exist with well-developed coral communities, for example at Mukalla (Kemp and Benzoni, 1999). Mixed algal, hard and soft coral communities also occur on exposed southern coasts of the Socotra Archipelago (Turner et al., 1999).

In the past few years, several surveys have revealed the general character of shallow habitats of both the Yemeni (Huntington and Wilson, 1995; Watt, 1996; Huntington et al., 1996; Turner et al., 1999) and Somali (van der Elst and Salm, 1998; Schleyer and Baldwin, 1999) coasts. A few important groups have received detailed study including macroalgae (Ormond and Banaimoon, 1994; Kemp, 1998a)

Fig. 3. Sabellid colonies along coast of Al Hosu, Gulf of Aden (photo S. Wilson).

and reef-fish (Kemp, 1998b). Many groups remain poorly understood.

High-energy sandy beaches separated by low rocky headlands are the dominant coastal characteristics of the Gulf. In Yemen, high and medium energy beaches account for approximately 60% of the coastline's length (Huntington and Wilson, 1995). While sandy beaches generally support a low diversity of infauna, their importance to seabirds, nesting turtles and sardine fisheries is significant. Coarser grained beaches are often densely populated by ghost crabs *Ocypode* sp., and at some sites by high densities of gastropods such as Olive shells (*Oliva bulbosa*) and *Bullia* sp. In Yemen, large colonies of Sabellid worms (Fig. 3) develop in the intertidal zone on a number of the low promontories and provide a microhabitat for a diverse community of associated species in an otherwise unstable substrate.

In most cases, beaches are backed by low dunes colonised by xerophytic grasses (e.g. *Sporbolus spicatus*) and other halophytes (e.g. *Limonium* spp., *Sueda* spp. *Zygophyllum* spp., and *Atriplex* spp.) which lead to a dry coastal plain of alluvial sediments or desert dunes often backed by extensive mountain ranges. A variety of terrestrial plants are found in these habitats, particularly at higher elevations, and the flora of Socotra includes a spectacular level of endemism (277 of the 850 plant species are endemic). Among these plants are the Dragon Blood Tree *Dracaena cinnabari*, whose deep red resin has strong medicinal properties; the Cucumber Tree, *Dendrosicyos socotrana*, the only tree in the cucumber family; and *Punica protopunica*, the only wild relative of the commercially important pomegranate (Miller 1991).

On the mainland Yemen coast horizontal effusions of black basalt occasionally cross the coastal plain. These features are called *harik*, meaning 'burnt place' or 'hot one' since they support little vegetation. Where they meet the sea, they provide an ideal substrate for coral communities and, at a few sites, true coral reefs have developed (Sheppard et al., 2000).

Coral Communities

As with many coral communities in Southern Arabia, monospecific stands of some species, particularly *Montipora*, *Pocillopora* and *Porites*, are common along the Yemeni mainland (Kemp and Benzoni, 1999). The pattern in Somalia remains unclear, but relatively diverse coral communities grow on fossilised beach rock or fossil reefs in shallow water along the mainland coast (Schleyer and Baldwin, 1999), and appear to be more common than along the Yemeni coast. The most developed reefs described to date are found fringing the offshore islands and coast to the West of the Gulf, particularly in Djibouti (Sheppard and Wells, 1988) and around the Saardin Islands of Somalia (IUCN, 1997). Coral communities appear to be under pressure from natural stressors including predation by Crown-of-Thorns Starfish (Pichon and Jaubert, 1989) and bleaching (Schleyer and Baldwin, 1999; Turner et al., 1999; Kemp, pers. comm.).

Well-developed coral communities also occur around Socotra and surrounding islands, (Kemp, 1997; Turner et al., 1999; De Vantier, 1999). Although there has been little Holocene reef development, along the north and east coasts are areas which could be described as incipient reefs with some accretion. The south coasts exhibit mixed soft and hard coral communities dominated by macroalgae. Soft and hard coral cover is high and is composed of approximately 190 species (50 genera, 14 families) of scleractinian reef-building corals (DeVantier, 1999). The composition includes representative species from the Indo-Pacific (e.g. *Pocillopora damicornis*), Indo-west Pacific (e.g. *Acropora formosa*), Indian Ocean (e.g. *Siderastrea savignyana*), West Indian Ocean (e.g. *Anomastrea irregularis*), Arabian Sea, Gulf of Aden and Oman 'endemics' (e.g. *Acanthastrea maxima*, *Parasimplastrea simplicitexta*), and Red Sea 'endemics' (e.g. *Symphyllia erythraea*, *Stylophora wellsi*, *Echinopora forskaliana*) (De Vantier, 1999). The high diversity is consistent with the availability of a wide range of biotopes of different depth, light, temperature and productivity regimes and exposure. In 1998 there was significant coral bleaching and subsequent mortality during May–July. The worst affected were the table and staghorn *Acropora*, *Millepora* and soft corals along the north coast of the main island, whilst the outer islands appeared to have suffered lower mortality (Turner et al., 1999).

The biogeography of the reef-fish of Socotra has recently been described by Kemp (1998b) and demonstrates that conditions in the Gulf of Aden act as a partial barrier between north (Arabia) and south (Africa), and between east (Arabia) and west (the Red Sea).

Macroalgal Communities

Rock cliffs and platforms in the intertidal zone are invariably colonised by the rock oyster *Saccostrea cuccullata*, and barnacles (e.g. *Tetrachthalamus*), with limpets (e.g. *Cellana rota*) and mussels (e.g. *Perna indica*) and grapsid crabs often present. Further east on the northern shores, at sites such as Ra's Fartak, typically temperate communities of ascidians, sponges and urchins dominate sub-tidal rock surfaces for much of the year. However, following the onset of the southwest monsoon in June, a lush and diverse community of macroalgae develops which remains until November. At some sites (e.g. Mukalla), a smaller peak in algal growth is apparent in February and March (Ormond and Banaimoon, 1994) and may be triggered by upwelling associated with the Northeast Monsoon. These seasonal phenomena are responsible for very high intertidal productivity. Typical values for algal biomass from the lower shore in August range from 1–2 kg wet weight m^{-2}, and 163 species have been recorded. *Nizamuddinia zanardinii*, a large kelp endemic to the Arabian Sea area, is often a major component, and has now been observed in Somalia (Schleyer and Baldwin, 1999), Socotra (Kemp, 1998a), Yemen (Ormond and Banaimoon, 1994) as well as further east in Oman and Pakistan (Nizamuddin et al., 1993).

Wadis and *Khawrs*

No permanent rivers flow into the Gulf of Aden. Rain is absorbed into the porous soils or, if the volume is high enough, flows into a drainage network of dry river-beds (*wadis*). Much of the flow from *wadis* percolates into the groundwater, but after very heavy rainfalls the *wadis* flow to the sea. Thus, much of the coastal vegetation is concentrated in *wadis*, on which gazelle and goats browse. Wells are frequently dug for the same reason, and villages and towns have historically been situated near *wadis*, particularly in areas where date palms can be cultivated. Although they flow infrequently, *wadis* often cut shallow ravines in the coastal plain and are the main route by which sediment reaches the shores to replenish beaches.

Where *wadi* mouths meet the coast, groundwater can rise to the surface and form a brackish lagoon (*khawr*) (Fig. 4) behind the beach. Periodic floodwaters may breach the

Fig. 4. Coastal *khawr*, south Yemen (photo S. Wilson).

sand barrier until longshore drift and rough seas build up a sufficiently large sand barrier to block the flow of groundwater for a *khawr* to form again. Such *khawrs* are naturally ephemeral and rarely does a mature wetland ecosystem develop, but they are important habitats for a variety of migratory shorebirds and are commonly used for watering livestock.

Mangroves

Mangrove distribution has not been described in detail, but they generally occur in larger *khawrs*. One species of mangrove, *Avicennia marina*, dominates in the region due to its high tolerance of hypersaline conditions, but *Rhizophora mucronata* also occurs in isolated stands (IUCN, 1997). Huntington and Wilson (1995) report that no mangrove grows on the mainland Yemeni coast of the Gulf, although they are known from the southern Red Sea (Watt, 1996) and southern Oman (Fouda, 1995). A number of small *Avicennia* stands are associated with *khawrs* and tidal creeks along the Somali coast (van der Elst and Salm, 1998) including Khawr Shoora, 150 km east of Berbera (Schleyer and Baldwin, 1999).

Around the main island of the Socotra Archipelago, there are several mangrove forests (Fig. 5). Several have no direct sea access, the landward fringe being either salt flat sparsely vegetated with halophytes (Sabkhah), fossil reef cliff or a dune system that separates the mangrove from the sea (Turner et al., 1999). Similar mangrove habitats have also been found in Oman (Hywell-Davies, 1994). On the north coast of Socotra are the remains of a previously larger stand of *Avicennia marina*, the rest of which has been cut for firewood.

The greatest concentration of mangroves in the Gulf is to be found in sheltered parts of the Golfe de Tadjura in Djibouti, including Collines des Godoria and Les Iles de Moucha and Maskali (Burkett, pers. comm.).

In the high-energy environment of the Gulf, fine sediments blown offshore or carried down flooding *wadis* generally remain in suspension in coastal waters until they eventually settle in deeper water. Only in a few sheltered areas do extensive sand and mud flats develop. One of the best examples is found in and around the port of Aden, a natural anchorage that is sheltered from all but southerly swells by extinct volcanic craters. The 20 km^2 of intertidal flats and sewage-fed wetland support a wide range of waders and other birds including Lesser Flamingos that feed here in their thousands (Gascoigne, pers. comm.).

Seagrasses

Under sheltered conditions seagrasses often colonise fine sands sub-tidally, but the number of such sites is limited. Of particular significance is Khawr Umairah in Yemen (Hirth

Fig. 5. Mangrove area on the west coast of Socotra at Shuab (photo R. Klaus).

Fig. 6. Tatwah lagoon, northwest coast of Socotra (photo R. Klaus).

et al., 1973; Watt, 1996), since it is the most extensive area of mixed seagrass community (*Cymodocea serrulata, C. rotundata, Halophila ovalis, Halodule uninervis, Syringodium isoetifolium* and *Thalassia hemprichii*) so far described from the Yemen Gulf coast. More common are patches of seagrass with relatively sparse cover of *Halodule uninervis* and *Halophila ovalis* (Huntington and Wilson, 1995; IUCN, 1997; Schleyer and Baldwin, 1999).

Several seagrass beds have been identified around Socotra consisting of five species (*Halodule uninervis, Halophila ovalis, Thalassia hemprichii, Thalassodendron ciliatum, Cymodocea serrulata*) (Turner et al., 1999). The most extensive area of seagrass was found at Tatwah, a sandy lagoon on the northwest corner of Socotra (Fig. 6). Seagrasses elsewhere had relatively sparse cover and were mixed with macroalgae and in some areas coral communities. It is suspected that there are more extensive seagrass beds off the south coast, as piles of dead seagrass have been found washed onshore after the Southwest Monsoon. No seagrass beds were found around outer islands of the Socotra group (Turner et al., 1999).

Vertebrates

Five species of marine turtle are known to occur in the Gulf of Aden: the Green, Hawksbill, Olive Ridley, Loggerhead and Leatherback. Not only are significant numbers known to feed throughout the Gulf but several significant nesting beaches have been identified including Ra's Sharma (Shasar Beach) in Yemen where thousands of Green Turtles nest annually. In June 1996, the peak nesting season, Watt (1996) observed thousands of female Green Turtles nesting at Shasar Beach. An aerial survey along the Somali coast and offshore islands has also identified very high concentrations of feeding Greens and extensive nesting beaches, particularly between Berbera and Bossaso (van der Elst and Salm, 1998; IUCN, 1997). Green turtles are known to nest around Socotra between May and June (Sa'ad and Pilcher, 1999). Additionally, low concentrations of Greens and Hawksbill nests have been observed throughout the region. Green turtles nesting in Yemen and Oman migrate to the southern Red Sea and Arabian Gulf, probably to feed on the extensive seagrass meadows there. The abundant macroalgae forms a significant part of their diet during and following the Southeast Monsoon.

Cetacean distributions and abundances are poorly known (Baldwin et al., 1998). It is clear that large populations of dolphins feed and breed in coastal waters in northern Somalia (Schleyer and Baldwin, 1999). Further offshore in the Gulf, abundant whales have been reported, including the Indo-Pacific Beaked Whale *Indopacetus pacificus*, an exceedingly rare species known from only three skulls collected world wide (van der Elst and Salm, 1998). From skeletal remains, incidental sightings and recent surveys, the Yemeni mainland coast and Socotra also appear to be important for cetaceans, as well as for whale sharks and manta rays. Reports from fishermen suggest that dugong may occur in parts of the Somali coast (IUCN, 1997), and may represent the last fragments of a population that used to stretch from East Africa to the Red Sea.

Rich inshore productivity created by upwelling provides valuable feeding grounds for a variety of seabirds and waders. The southern seaboard of Arabia and the Red Sea are major flyways, giving an exceptional concentration and diversity of migratory bird species. Of the resident seabirds, the White-Eyed Gull *Larus leucophthalmus* is endemic, and others (e.g. Red-billed Tropic bird *Phaethon aethereus indicus*, Socotra Cormorant *Phalacrocorax nigrogulgaris*, Sooty Gull *Larus hemprichii*, White-cheeked tern *Sterna repressa*, Saunders Little Tern *Sterna albifrons saundersi*) have restricted ranges here. The beaches and mudflats of southern Yemen are particularly important for many species of gulls, terns, cormorants and waders. Globally and regionally significant populations of seabirds nest on the offshore islands and cliffs throughout the Gulf, including the Saardin Islands, Socotra and Ra's Fartak. Socotra is a particularly important seabird (e.g. Bulwer's Petrel *Bulweria fallax*, a Middle East endemic species) and raptor (e.g. Egyptian Vulture *Neophron percnopterus*,) nesting site and is home to six endemic terrestrial birds (see also Evans 1994; Cooper et al., 1984)

OFFSHORE SYSTEMS

During the Southwest Monsoon, primary productivity of nutrient-rich water that has very recently been drawn to the surface is relatively low (1 g C m^{-2} 24 h^{-1}) while phytoplankton stocks remain small. Typically the bloom takes 2–3 days to fully develop. With surface currents flowing eastward, upwelling water between Mukalla and Ra's Fartak is carried out of the Gulf of Aden, affecting a relatively narrow band of coastal water and triggering a significant macroalgal bloom along these rocky shores. Baars (1994) reports that the central Gulf of Aden remained relatively oligotrophic during the spring–summer of 1992, even though upwelling occurred along the Yemen coast near Oman and near Bab al Mandeb.

The high speed of the Great Whirl currents (1–1.5 m s^{-1}) carries recently upwelled water 175–250 km in two days, during which time the phytoplankton bloom develops, and diluting the water mass over 30,000 km^2, which is 20 times the original area of upwelling. Where small-scale eddies trap upwelled water close to shore, maximum productivity (2.7–3.2 g C m^{-2} 24 h^{-1}) occurs, while typical productivity of waters immediately south of Socotra may remain about 1 g C m^{-2} 24 h^{-1}. Winter rates are about half these summer values, yet biomass of primary consumers remains high throughout the year, switching to feed on secondary productivity of the decomposition food chain. In contrast to the Great Whirl, the slower speeds (0.8 m s^{-1}; Bauer et al., 1991) of surface currents associated with upwelling on the Yemeni coast carry the nutrient-rich waters into an upwelling system that continues for 1000 km along the coast of southern Oman.

This large biomass of zooplankton in winter sustains the development of the stocks of small and meso-pelagic fish that reach a maximum after autumn. Indian Mackerel *Rastrelliger kanagurta* and the Indian Oil Sardine *Sardinella longiceps* dominate small and meso-pelagic fish stocks. This stock increases from 12 tons km^{-2} in August to 34 tons km^{-2} in January (Venema, 1984). For comparison, trawl surveys indicate that the highest stock density of demersal fish was found in the Ra's Fartak area (6.4 tonnes/km^2) the lowest west of Mukalla (1.6 tonnes/km^2) and medium density in the Aden area (3.5 tonnes/km^2) (MEP, 1998).

The zooplankton community is dominated by the copepod *Calanoides carinatus*, a typical upwelling species also known from West Africa and Peru, that may consume half the total primary productivity. During the Northeast Monsoon this species is absent from the 0–300 m layer, and overwinters in the meso-pelagic (500–1500 m) zone throughout the Gulf. Another interesting feature is the great seasonal abundance of the pelagic crabs *Lupa pelagica* and *Charybdis smithii*. Juveniles are particularly abundant during the Southwest Monsoon upwelling, and develop into adults that spawn in winter months. A range of fish species including Yellowfin Tuna *Thunnus albacares*, Bigeye Tuna *T. obesus*, Dolphinfish *Coryphaena hippurus*, Wahoo *Acanthocybium solandri*, and Kingfish *Scomberomorus commerson* prey upon all life-cycle stages of these pelagic crabs and in turn they are targeted by the artisanal fishery both in mainland Yemen and Socotra. Yellowfin Tuna dominates artisanal landings in the Gulf of Aden and the industrial purse seine fishery associated with the Great Whirl.

Average primary productivity across the entire Gulf of Aden is greater in winter than in summer despite upwelling along the eastern coast of Yemen during the Southwest Monsoon. This is because easterly surface currents in summer export these rich waters into the Arabian Sea, and although weaker upwelling is also induced in the western sector of the Gulf, these events remain patchy and dilute quickly. Winter cooling assists winter productivity by increasing the depth of the mixed layer as well. Satellite data confirms that greater chlorophyll concentrations are present in the Gulf of Aden during winter than in summer, as illustrated in Sheppard and Dixon (1998). In winter, surface currents force relatively productive water into southern parts of the Red Sea which contributes significantly to carbon and nutrient cycles there.

POPULATIONS AFFECTING THE AREA

The aridity of the coastal zone has historically concentrated human settlements near available water supplies and created a traditionally heavy reliance on the marine environment as a source of food (PERSGA, 1998). The region has a long maritime history with the natural harbours on the Arabian coast being important trading settlements dating back to the pre-Islamic era (Sedov, 1994) (Table 1). Today, a major shipping lane passes through, and

Fig. 7. Mukalla, south Yemen (photo S. Wilson).

is serviced by the ports of Djibouti (estimated population in 1998 was 300,000) and Aden (estimated population in 1994 was 1.5 million). The other major city in the region is Mukalla (200,000) (Fig. 7).

Table 1

Ports in the Gulf of Aden area

	Position	Type of port	Comments
Yemen			
Aden	12°47'N 044°58'E	General cargo; oil export terminal	Undergoing expansion and revitalisation. Population 1.5 million
Mukalla	14°31'N 049°07'E	General cargo/bulk; fishing harbour	Population 200,000
Nishtun	15°49'N 052°14'E	Small commercial port and fishing harbour	Opened in 1984 but remains virtually unused since supporting road infrastructure is very poor.
Ash Shihr	14°41'N 042°31'E	Oil export terminal	One single point mooring (SPM) terminal
Djibouti			
Djibouti	11°36'N 043°08'E	Commercial pand oil import terminal	Population estimated at 300,000. Main port for trans-shipment of cargoes to and from Ethiopia. 13 berths.
Somalia			
Berbera	10°26'N 045°01'E	Commercial port and oil import terminal	Population of 15,000–30,000 depending on the season. Port formed by low spit that acts as a breakwater. 240,000 t cargo handled annually.

Table 2

National economic and social indicators (from PERSGA, 1998)

	Yemen	Djibouti	Somalia
Total Population (millions)	13	0.6	9.2
Projected population growth rate (1993–2000)	3.7	2.1	2.7
Urban population growth rate	6.6	7.6	4.3
GDP per capita	805	1270	712*
Literacy rate	41.1	45	24.9*
Human Development Index (HDI)	0.33	0.31	0.22*
HDI rank	148	162	172*

*Data from 1994.

The great majority of the population is rural, and poor (Table 2). About 75% of the estimated 16.5 million Yemenis are rural, and rates of population growth are high. In Yemen the average number of children in a family is 7.6, and as a result the population is young, with 47% of the population below the age of 14 years.

Approximately 20,000 men are involved in the artisanal fishery on the Yemeni mainland, and an additional 3000–5000 on Socotra, the variability of estimates being due to a number of part-time fishers (Huntington et al., 1996; Hariri and Yusif, 1999). Women participate in collecting shellfish, edible seaweed and driftwood from the shore, but generally their involvement with the marine environment is small, and is more focused on agricultural activities.

Relative wealth is spread unevenly, and in Yemen appears to be closely related to the abundance and value of fish stocks. In the extreme west where the catch contains relatively large portions of low-value fish, there are lower earnings of about US$40 per fisherman per month. The highest incomes are found in the rock lobster fishery in Al Mahra/Hadramout and in the Yellowfin fishery in Hadramout where seasonal incomes of US$317–793 per fisherman per month are reported (MEP, 1998). Some of the more affluent fishing families in the Mukalla-Shihr area own cars and houses with modern conveniences, and such families also benefit from relatively good social services and infrastructure. At the other end of the scale are fishing families living at subsistence level in isolated villages, where social facilities and services are lacking or absent.

Less is known about the other countries in the region, but with an average per capita income of less than US$200 pa, Somalia is among the poorest countries in the world. Development in rural Somalia is very low and only a very small proportion of the coastal community engages in fishing.

Environmental and political factors also play a significant role in the distribution and size of populations in the area. During the Southwest Monsoon, most fishing activity ceases and a proportion of the coastal communities migrates away from the coast to assist with agricultural activities such as date harvesting. This is common in the more traditional parts of the region such as Socotra and Somalia, and the population of some towns (e.g. Berbera) will vary by as much as 50% depending on the season. Tens of thousands of Somali refugees have started to return to their homelands from camps in Yemen and Djibouti, and the return of Yemeni ex-patriots from the Arabian Gulf countries in the early 1990s led to an increase in fishing pressure (MEP, 1998) and unplanned coastal development (Huntington and Wilson, 1995).

RURAL FACTORS

About 75% of the total population of the Gulf of Aden live in rural areas and the main income-generating activities are agriculture and fishing. In Yemen and Djibouti fishing is the more important but Somalia has a more pastoral population that herd livestock in the interior rangelands.

Fisheries

Information on fisheries in Yemen has been collected recently (MEP, 1998). Over 5500 small (7–9 m long) fibreglass skiffs and 150 traditional wooden *sambuqs* make up the artisanal fishing fleet on the mainland coast of Yemen. They use about 50 landing sites, most of them beaches, while the larger vessels work out of the few harbours, such as Mukalla. Approximately 50,000 tonnes of fish from the Gulf were landed in Yemen in 1995, worth US$ 39 million, and production from Socotra added a further 5% to this.

Several main categories of finfish and shell-fish are exploited. The most important are large pelagic fish, particularly Yellowfin Tuna *Thunnus albacares*, Kingfish *Scomberomorus commerson*, and Carangids, which are caught using baited hooks and gillnets. Yellowfin tuna dominate the catch in both mainland Yemen and Socotra, but shark is also targeted. Cast-nets are used to catch sardines for use as baitfish in the tuna fishery, but by far the largest yield of sardines comes from beach seining (Fig 8). In Yemen, for example, an estimated 12,000 tonnes of sardines are landed annually, the majority of which is sun-dried and fed to livestock, particularly camels, in the interior. There are high densities of demersal and pelagic fish in coastal waters of Somalia, but artisanal fishing remains at low subsistence levels (World Bank, 1992).

On Socotra there are few employment opportunities. There are approximately 40 fishing villages around the coast of the main island. Artisanal fishermen target reef fish, lobster and large pelagics, particularly shark. Until recently there have been very few facilities for processing or storage which has caused problems for marketing and has resulted in wastage. The only fresh fish trading takes place in the main town of Hadiboh (Fig. 9). With rock lobster, fishermen wait until there is a buyer in the area, negotiate a price, then fish to order. Despite this, the lobster fishery is already in decline (Hariri and Yusif 1999).

Fig. 8. Sardine drying, south Yemen (photo S. Wilson).

Fig. 9. Hawlaf, main port area, adjacent to the main town of Hadiboh (photo R. Klaus).

In the East of the Gulf of Aden, where the influence of the summer monsoon upwelling is greatest, there is an economically important lobster fishery. In 1996, over 300 tonnes, worth an estimated US$7 million, were caught by Yemeni fishers using wire traps, the majority of which was exported. Around Socotra, turtle eggs are collected for consumption (Sa'ad and Pilcher, 1999). Other minor fisheries in the region include sea-cucumber (e.g. Khawr Umeirah and Socotra), various snails (e.g. *Murex* sp.) whose operculae are used as an ingredient in incense and a range of other invertebrates gleaned from rocky shores (e.g. *Perna indicus, Octopus*) and beaches (e.g. *Tivela* sp.). Beaches are also routinely combed for ambergris, a waxy substance egested by Sperm whales and thought in Arabia to have powerful medicinal properties.

Dolphins are not regularly targeted, but some are harpooned as bait and for human consumption, despite local superstitions. It is estimated that 30–40 dolphins are caught annually on Socotra for shark bait (Huntington et al., 1996), and this practice is probably widespread, though relatively uncommon, throughout the region. The capture of Green turtles, *Chelonia mydas*, also occurs in the remoter parts of the Gulf including Socotra, although the numbers (hundreds) involved are thought to be lower than 10 years ago. In isolated locations, they are caught while nesting during the Southwest Monsoon, the meat and eggs being used to supplement diets when fishing is not possible due to rough seas. Elsewhere, for example at Ra's Sharma, turtles are killed for no particular reason.

Take of most targeted finfish appears to be below estimated sustainable limits. Fishing effort is limited by a number of factors. Since there are very few naturally sheltered landing sites, heavy seas associated with the Southwest Monsoon effectively prevent fishing using the small boats of the artisanal fleet. Otherwise, infrastructure, logistics and economics limit landings throughout the Gulf, particularly in Somalia: small boats are used for day fishing only and, since ice is not routinely used, the low-quality product can only find a limited market.

The main exception to these constraints is the shark fishery. *Sambuqs* operate in poor sea conditions and have a greater range than skiffs, so are preferred for shark fishing. Yemeni shark fishermen have started to move into new areas in the Gulf and now commonly operate in the waters of Socotra and Somalia. Sharks are caught using gill nets and long lines. Preliminary assessments estimated a total catch of shark from Socotra at 1350 tons by local fishermen during a 75-day season with a value of YR250 per kg, with an additional 1260 tons caught by Yemeni fishermen from the mainland (Hariri and Yusif, 1999). Although shark meat is taken and dried, often just the high-value fins are removed and the trunks discarded. There is also a fishery for juvenile sharks that are consumed in large amounts for their imagined improvement to sexual performance. As a result, sharks are heavily exploited and highly vulnerable to over-fishing.

Gillnetting, particularly in shallow inshore waters inevitably results in an incidental catch of cetaceans and marine turtles. Dolphin meat is considered prime bait for shark nets and longlines, so those caught in nets are generally taken even if they are still alive. Lost or abandoned static gear will continue to ghost fish, and in 1995 it was estimated that 75% of the wire traps set, amounting to some 4500, are lost during the three-month lobster season in southwestern Yemen (Huntington and Wilson, 1995).

Agriculture on the Coast

Agriculture is generally carried out at subsistence level, with the cultivation of date palms, small-scale market gardening, goat and camel herding. In the patriarchal societies, women and children most often engage in goat herding, while tending date groves and camels is traditionally a male activity. Status of a family is commonly measured in terms of the head of livestock; naturally, as human population increases so does the number of livestock. Goats are an introduced browsing species, and in parts of southern

Yemen, grazing pressure has destabilised coastal sand dunes and significantly reduced the vegetation cover in *wadis*. In areas with rapidly increasing human populations, such as Djibouti, mangroves are cut for fuel, construction and fodder for livestock. Other semi-domestic animals pose a significant threat to coastal wildlife: packs of feral dogs that roam on the fringe of settlements in Yemen cause significant damage to turtle nesting beaches (Huntington and Wilson, 1995).

The impact of agriculture on the limited water resources has the potential to affect coastal systems (PERSGA 1998). Over-extraction of water from wells in the coastal plain has led to saltwater intrusion. In an effort to increase the recharge rates of groundwater to prevent this intrusion and so maintain the quality of abstracted water, *wadis* are dammed, which in turn starves the coast of beach material. The situation is becoming significant and action is required to prevent further substantial damage (PERSGA, 1998).

COASTAL EROSION AND LANDFILL

Few man-made structures have been built on either the Yemeni or Somali coasts that might block longshore drift. Some do extend across beaches, for example the small harbour at the oil-loading facility at Ash Shihr, but there has been little erosion on either side, suggesting that here the seasonally reversing conditions result in no net sediment transport. Other beaches behave differently. Many beaches east of Berbera are very dynamic and undergo severe monsoon-induced erosion up to the backing dunes, followed by a period of strong accretion (Schleyer and Baldwin, 1999).

Some beaches in the Gulf of Aden are composed of deposits of ilmenite and rutile at concentrations that could be exploited commercially. Little mineral sand mining has occurred, except on beaches around Berbera, where this activity is likely to increase (van der Elst and Salm, 1998).

The most significant dredging works are carried out to maintain and improve the region's ports. Infilling has impacted mangroves around Djibouti (PERSGA, 1998) and further dredging and infilling are planned. The Port of Aden too has been deepened eight times in the past 100 years increasing the available draught from 8 to 12 m, and today Aden is under further improvement as a Free Trade Zone. The land concession for the new port developments covers approximately 2 km^2 of disused salt flat and contains the sewage fed-wetlands important for resident and migratory birds (Gascoigne et al., 1996).

The beaches of Socotra have a eulittoral berm of dead coral cobbles. Local villagers collect these for house construction, lime production and for export to the Emirates. At present the coral used is beach debris, and as such the impact is relatively minor, although the amount exported per annum is not known. With development increasing, the practice may lead to the mining of live coral.

EFFECTS FROM URBAN AND INDUSTRIAL ACTIVITIES

The level of urbanisation and industrialisation is low throughout the Gulf of Aden, particularly in Somalia. Environmental controls and regulations are weak and seldom enforced. Issues of sustainability arise only rarely and the level of public awareness or concern for pollution is similarly low. With sanitation and pollution control infrastructure being expensive, neither governments nor the private sector can afford such investments. Basic capital infrastructure is often funded externally and cannot be maintained or managed as intended. This situation is well illustrated by the standard of municipal services.

No coastal towns have adequate provision for waste water and large amounts are discharged directly to the sea or intertidal zone. The situation at Aden, the largest city on the Gulf coast, is particularly poor. Sewage enters Aden Harbour from three sources. Completed in 1986, the largest is the secondary treatment facility at Al-Shaab, that handles domestic and industrial sewage from the Mansura and Sheik Othmann quarters. It discharges 15,000–20,000 m^3 per day through leaking pipes, around which have grown artificial wetlands. A further 4000–5000 m^3 per day of untreated domestic waste water, and industrial waste water from a battery factory, a paint factory, and a textile factory are discharged into the harbour at Al Ma'alla. The outfall at Tawahi also discharges untreated domestic and industrial waste close to Ra's Marbut. However, with the exception of localised effects, a preliminary survey of water quality in 1996 showed that generally there were low concentrations of contaminants (Gascoigne et al., 1996) and the marine environment appears to be resilient to organic pollution. This apparent ability of the marine environment to absorb this level of additional nutrient loading may stem from its adaptation to upwelling.

The effect of sewage pollution on other aspects remains unclear. Untreated sewage discharges to the intertidal zone at Mukalla and Ash Shihr are known to cause throat infections in swimmers (Huntington and Wilson, 1995). Towns on the Somali coast including Berbera have no centralised waste-water collection system, and rely on septic tanks.

The management of solid wastes follows a similar pattern. The problem of waste disposal is particularly acute around the most densely populated areas, but still creates a problem in most rural villages. Waste is normally dumped and burnt in unprepared sites without containment, so spreads by wind or animals. In Yemen, 63% of coastal sites, particularly those close to towns and villages were contaminated with litter (Huntington and Wilson, 1995) from land-based sources. A similar survey in Somalia indicated very low levels of garbage (Schleyer and Baldwin, 1999), probably reflecting lower population density rather than better waste management practices. In Socotra there is currently no regulation on waste disposal, so solid waste is often dumped on the beach and in the *wadis* near the main town.

While not a serious ecological threat at the moment, litter is an increasing problem that can easily be controlled from source, with the development of a responsible solid waste disposal strategy and small changes in attitude.

Oil and gas production in Yemen currently takes place inland, and there are two crude oil loading terminals at Ash Shihr and Al Douqal and the oil refinery in Aden. Yemen has recently announced a new licensing round which includes coastal and offshore blocks (*Underwater News*, 6 August 1999), notably block 38 which covers the south coast of Socotra and the islands of Abd El Kuri, Samha and Darsa. The latter are of particular concern as they harbour a diverse marine community and were little affected by the coral-bleaching episode of 1998. In addition, the discovery of large quantities of natural gas in the interior has advanced plans to develop a Liquid Natural Gas (LNG) processing and export terminal on the Gulf of Aden shore at Belhaf, a site of considerable marine biological interest. In addition to the reef that would be affected, one of the very few in this part of Arabia, Belhaf has significant potential for tourism development.

Due to the major shipping route passing through the Gulf of Aden, tankers pass through the Gulf routinely. PERSGA (1998) describe the complex navigational hazards of the straits of Bab al Mandeb as causing 'extensive and routine risk of ship collisions and groundings' for the heavy maritime traffic. Despite this, oil contamination on Yemeni beaches is relatively low compared with the Arabian region as a whole, but low levels occur right along this coast. Studies of the Bir Ali to Ra's Fartak coast in the 1980s indicated that, although volatile fractions were seldom detected, heavy hydrocarbon residues were present in samples from all sites, suggesting that contamination has been taking place for many years. All beaches along the Yemen coast appear to be equally contaminated. A recent survey of the coast 150 km east of Berbera did not find evidence of significant oil pollution (Schleyer and Baldwin, 1999), although oil routinely leaks from storage tanks in Berbera port and spills occur (van der Elst and Salm, 1998). Contributing to chronic oil pollution are the lack of adequate reception facilities (e.g. Ash Shihr), for the discharge of ballast and washing water, and the lack of any monitoring system to detect tankers that discharge oily wastes. Of far greater concern than oil pollution are reports of deliberate dumping of toxic waste in the Somali waters of the Gulf of Aden (PERSGA, 1998), but details surrounding these reports remain scant.

Illegal and uncontrolled industrial fishing in the Gulf of Aden dates back more than 30 years and is still rife today. Severe overfishing of cuttlefish in Southern Yemen by Russian trawlers in the 1970s demonstrates the problem well. Twenty years ago more than 15,000 tonnes of cuttlefish per annum were being caught in Yemen waters until the stock crashed, and interest switched to the lobster fishery. Since then there has been little recovery and present catches are estimated at less than 2000 tonnes, and the lobster fishery is now in danger of collapse. Trawling in shallow waters destroyed cuttlefish breeding habitats and reduced the spawning biomass to dangerously low levels, and there are no means to control effort in the lobster fishery. Throughout the region destructive trawling continues, and foreign vessels, both licensed and unlicensed, have been observed fishing close to shore in depths of less than 20 m (Huntington et al., 1996; MEP, 1998). 'Back-storage', where small fish are dumped from storage to make way for larger fish, and the inherently large proportion of the catch that is discarded, further demonstrate poor fisheries practices common in the region.

PROTECTIVE MEASURES

Protective measures in the Gulf of Aden operate at a range of levels. At the international level (Table 3), the Gulf of Aden is classified as a 'special area' by IMO (1997). As such, the Gulf has particular restrictions aimed at controlling the discharge of oil, noxious liquids, garbage and sewage due to 'its oceanographic and ecological conditions and to the particular character of its [ship] traffic'.

The regional body responsible for coastal resource protection and conservation is the Regional Organisation for the Conservation of the Environment of the Red Sea and the Gulf of Aden (known as PERSGA) based in Jeddah, Saudi Arabia. This body has recently published a strategic action plan that aims 'to develop a regional framework for the protection of the environment and the sustainable development of marine and coastal resources' (PERSGA, 1998). Funding from national and international sources will be necessary to implement the action plan, and some projects are already underway.

Djibouti is the only country in the region to have a system of marine protected areas in place. The Parc Territorial de Mouscha was established in 1972 and contains the Réserve Intégrale de Maskali Sud established in 1980 (Sheppard and Wells, 1988). These areas are jointly managed by the Institute of Higher Study and Scientific Research and the Tourism Development Office. To the north, the islands of Les Sépt Frères and their fringing reefs have been proposed for protection, due to their potential for tourism and scientific research.

Table 3

Conventions signed by Gulf of Aden countries relating to the Marine Environment

	Yemen	Djibouti	Somalia
IMO Convention 48	X	X	X
IMO 73/78 (Annex I/II)		X	
CITES			X
Jeddah Convention 82	X	X	X
Biodiversity Convention	X		

Although currently there are no marine or coastal protected areas in Yemen, the Environment Protection Council is considering the establishment of a network of conservation areas. Amongst them is the Socotra Archipelago as a Biosphere Reserve, where the necessary scientific research is currently underway to underpin its new protected status. Other coastal sites in Yemen that have been recommended or proposed for protection are Nishtun, Perim Island and Ra's Abu Quizara (Chiffings, 1995), Bir Ali and Sikha Island (Kemp, 1998c). However, legislation to establish protected areas is only the very first step in a process leading to effective protection, and good management and enforcement are often lacking.

With its history of civil war, the national development priorities of Somalia are focused on peace, nation building and other socio-economic considerations, and environmental protection remains low on the political agenda. Despite this, IUCN under their Eastern African Programme has initiated a survey of coastal habitats and resources as part of a Somali Natural Resources Management Programme that will gather the necessary information to identify priority areas for management (van der Elst and Salm, 1998). Presently, there is no formal mechanism by which areas can be protected, but piracy and the general lawlessness in Somalia has tended to have a secondary protective effect on in-shore marine resources.

At the local level, examples of traditional fisheries management are rare, with the exception of Socotra. There, effort is controlled by an effective community management system. This is based on the island-wide village elders or *maqaddim*, who limit fishing effort through a number of rules that include seasons, gears and areas, as well as access to the fishery. There are several community enforced 'no fishing areas'. Policies are co-ordinated with other *maqaddim*, although rules may vary between villages. Infringements are dealt with severely, with either a ban on fishing (usually for seven days) or may even result in the destruction of nets. Strong peer pressure results in such infringements being rare (Huntington et al., 1996). More recently, the villages are beginning to set up fisheries co-operatives with the hope of receiving increased support from the government in terms of gear and boats. In addition, there are several military camps on the islands and these have been involved in the apprehension of illegal trawlers.

REFERENCES

Baars, M.A. (ed.) (1994) Monsoons and Pelagic Systems. Report on three cruises of RV *Tyro* in the Somali Current, the Gulf of Aden and the Red Sea during the Southwest monsoon of 1992 and the Northeast monsoon of 1993. Cruise Reports Netherlands Indian Ocean Programme, Vol. 1, National Museum of Natural History, Leiden, Holland. 143 pp.

Baldwin, R.M., van Waerebeek, K. and Gallagher, M. (1998) A Review of Small Cetaceans from Waters off the Arabian Peninsular. Document SC/50/SM5 presented to International Whaling Commission, Muscat, Oman, 28 pp.

Barrat, L., Ormond, R.F.G., Campbell, A.C., Hiscock, S., Hogarth, P.J. and Taylor, J.D. (1984) *An ecological study of the rocky shores on the South coast of Oman*. Council for Conservation of the Environment and Water Resources, Muscat, 54 pp.

Barrat, L., Ormond, R.F.G. and Wrathall, T.J. (1986) *Ecological studies of Southern Oman kelp communities*. Council for Conservation of the Environment and Water Resources, Muscat, 99 pp.

Bauer, S., Hitchcock, G.L. and Olson, D.B. (1991) Influence of monsoonally-forced Ekmann dynamics upon the surface layer depth and plankton biomass distribution in the Arabian Sea. *Deep Sea Research* 38, 531–553.

Bottero, J.S. (1969) An analysis of upwelling of the South East Arabian coast during the summer monsoon. Unpublished M.Sc. thesis, Oregon State University.

Chiffings, A.W. (1995) Arabian Seas. In *A Global Representative System of Marine Protected Areas. Volume III Central Indian Ocean, Arabian Seas, East Africa and East Asian Seas*, eds. G. Kelleher, C. Bleakley and S. Wells, pp. 39–70. GBRMPA, IUCN, World Bank.

Cooper, J., Williams, A.J. and Britton, P.L. (1984) Distribution, Population size and Conservation of Breeding Seabirds in the Afrotropical Region. In *Status and Conservation of the World's Seabirds*, eds. J.P. Croxall, P.G.H. Evans and R.W. Schreiber, pp. 403–419, ICBP Technical Publication No. 2, 778 pp.

Currie, R.I., Fisher, A.E. and Hargreaves, P.M. (1973) Arabian Sea upwelling. In: *The Biology of the Indian Ocean*, eds. B. Zeitshcel and S.A. Gerlach, pp. 37–52. Springer Verlag, New York.

DeVantier, L. (1999) Coral Communities of the Socotra Archipelago. Information for the Technical Meeting on Management and Zoning of Coastal and Marine Areas Mainz, 12–14 July 1999. UNDP/GEF Project YEM/96/G32 Conservation and Sustainable Use of Biodiversity of Socotra Archipelago.

Evans, M.I. (1994) *Important Bird Areas in the Middle East*. BirdLife Conservation Series No. 2, Cambridge, UK.

Fouda, M.M. (1995) Status of Mangrove Resources in the Sultanate of Oman. *J. Fac. Sci., UAE Univ.* 8 (2), 169–183.

Gascoigne, J., Rennis, D. and Brown, M. (1996) Port of Aden Preliminary Environmental Survey. Unpublished report to the Yemen Investment and Development Company. MacAlister Elliott and Partners Ltd., Lymington, UK.

Hariri, K.I. and Yusif, M.D. (1999) Fishing communities and status of the fisheries sector in the Socotra Archipelago. In Marine Habitat, Biodiversity and Fisheries Services. Report of Phase1. UNDP/GEF Project YEM/96/G32 Conservation and Sustainable Use of Biodiversity of Socotra Archipelago, pp. 161–180.

Hirth, H.F., Klikoff, L.G. and Harper, K.T. (1973) Seagrasses at Khor Umeirah, People's Democratic Republic of Yemen with reference to their role in the diet of the green turtle (*Chelonia mydas*). *Fishery Bulletin* 71, 1093–1097.

Huntington, T.C. and Wilson, S.C. (1995) Coastal Habitats Survey of the Gulf of Aden Phase 1: Preliminary Habitat Classification and an Assessment of the Coast's Resources, Users and Impacts. Report to the Ministry of Fish Wealth, Government of the Republic of Yemen. MacAlister Elliott and Partners Ltd., UK, and Marine Sciences Resource Research Centre, Aden, Yemen.

Huntington, T.C., Kemp, J.M., Watt, I. and Cheung, C. (1996) Mission Report (Marine Team) for the Conservation and Sustainable Use of the Biodiversity of Socotra. Report to UNDP/GEF. MacAlister Elliott and Partners Ltd., UK.

Hydrographic Office (1980) Admiralty Sailing Directions—Red Sea and Gulf of Aden. 12th Edition. Admiralty Charts and Publications (NP64), Hydrographic Department, Taunton, UK.

Hywell-Davies, A. (1994) A quantitative analysis of the horizontal and vertical zonation of brachiura and mollusca associated with the Qurm mangal, Muscat, Sultanate of Oman. M.Sc. thesis, University of Wales, Bangor, UK, 68 pp.

IMO (1997) MARPOL 73/78 Consolidated Edition 1997. International Maritime Organisation, 4 Albert Embankment, London. UK, 419 pp.

IUCN (1997) Preliminary Ecological Assessment of the Saardin Islands, Awdal Region. Somali Natural Resources Programme, IUCN Eastern Africa Programme, Nairobi, Kenya.

Kemp, J.M. (1997) Extensive Coral Communities of the Socotra Archipelago, Gulf of Aden. *Coral Reefs* **16**, 214.

Kemp, J.M. (1998a) The occurrence of *Nizamuddinia zanardinii* (Schiffner) P.C. Silva (Phaeophyta: Fucales) at the Socotra Archipelago. *Botanica Marina* **41**, 345–348.

Kemp, J.M. (1998b) Zoogeography of the coral reef fishes of the Socotra Archipelago. *Journal of Biogeography* **25**, 919–933.

Kemp, J.M. (1998c) Marine and Coastal Habitats and Species of the Bir Ali area of Shabwa Province, Republic of Yemen: Recommendations for Protection. Report to the Environment Protection Council of the Council of Ministers, Sana'a, Yemen. 10 pp.

Kemp, J.M. and Benzoni, F. (1999). Monospecific coral areas of the northern shore of the Gulf of Aden, Yemen. *Coral Reefs* **18**, 280.

MacAlister Elliott and Partners Ltd. (MEP). 1998. Republic of Yemen: Fisheries Sector Review. Report No. 501/R/108D. Produced under the Fourth Fisheries Development Project, Yemen (Report commissioned and funded by the European Commission).

Miller, A.G. (1991) Checklist of Socotra. Unpublished.

Nizamuddin, M.S., Hiscock, S., Barratt, L. and Ormond, R.F.G. (1993) The Occurrence and Morphology of *Sargassopsis* gen. Nov. (Phaeophyta, Fucales) in Southern Oman. *Botanica Marina* **36**, 109–121.

Ormond, R.F.G. and Banaimoon, S.A. (1994) Ecology of intertidal macroalgal assemblages on the Hadramout coast of Southern Yemen, an area of seasonal upwelling. *Marine Ecology Progress Series* **105**, 105–120.

Ormond, R.G.F. and Edwards, A.J. (1987) Red Sea Fishes. In *Red Sea*, eds. A.J. Edwards and S.M. Head, pp. 251–287. Pergamon Press, Oxford, UK.

PERSGA (1998) Strategic Action Programme for the Red Sea and Gulf of Aden. The World Bank, Washington, USA. 85 pp.

Pichon, M. and Jaubert, J. (1989) *Acanthaster planci*: Une menace pour les recif de Djibouti. Unpublished report to L'Institut Superieur D'Etudes et de Recherches Scientifiques et Techniques (Republique de Djibouti).

Piechura, J. and Sobajh, O.A.G. (1986) *Oceanographic Conditions of the Gulf of Aden. Scientific Investigations of the Gulf of Aden Series A: Oceanography No. 2*, pp. 1–15. Marine Science and Resources Research Centre, Aden, Yemen.

Sa'ad, M. and Picher, N. (1999) Marine Turtle Questionnaire. In Marine Habitat, Biodiversity and Fisheries Services. Report of Phase 1. UNDP/GEF Project YEM/96/G32 Conservation and Sustainable Use of Biodiversity of Socotra Archipelago. 191–194 pp.

Schleyer, M. and Baldwin, R.M. (1999) Biodiversity Assessment of the Northern Somali Coast East of Berbera. Somali Natural Resources Programme, IUCN Eastern Africa Programme, Nairobi, Kenya. pp. 46.

Sedov, A.V. (1994) Qana' (Yemen) and the Indian Ocean—The archaeological evidence. In *Proceedings of the International Seminar Techo-Archeological Perspectives of Seafaring in the Indian Ocean 4th Century BC–15th Century AD, New Delhi February 28–March 4, 1994*, eds. H. Prabba Ray and J-F. Salles.

Sheppard, C.R.C. and Dixon, D.J. (1998) Chapter 32: Seas of the Arabian Region Coastal Segment (29, S). In *The Sea, Volume 11*, eds. A. Robinson and K. Brink. John Wiley and Sons Inc.

Sheppard, C.R.C. and Sheppard, A.L.S. (1991) Corals and Coral Communities of Arabia. *Fauna of Saudi Arabia* **12**, 3–173.

Sheppard, C.R.C. and Wells, S.M. (eds.) (1988) *Coral Reefs of the World. Volume 2: Indian Ocean, Red Sea and Gulf*. UNEP Regional Seas Directories and Bibliographies. WCMC, Cambridge, UK, IUCN, Gland, Switzerland, and UNEP, Nairobi, Kenya. pp. 1–389.

Sheppard, C.R.C., Price, A.R.G. and Roberts, C.J. (1992) *Marine Ecology of the Arabian Region: Patterns and Processes in Extreme Tropical Environments*. Academic Press, London, UK. pp. 359.

Sheppard, C.R.C., Wilson, S.C., Salm, R.V. and Dixon, D. (2000) Reefs and Coral Communities of the Arabian Gulf and Arabian Sea. In *Coral Reefs of the Indian Ocean: Their Ecology and Conservation*, eds. T.R. McClanahan, C.R.C. Sheppard and D.O. Obura, Chapter 9. Oxford University Press.

Stirn, J., et al. (1985) Oceanographic Conditions, pelagic productivity and living resources in the Gulf of Aden. IOC/UNESCO Workshop on Regional Co-operation in Marine Science in the Central Indian Ocean and Adjacent Seas and Gulfs. Colombo, 8–13 July 1985. pp. 255–297.

Tomczak, M. and Godfrey, J.S. (1994) *Regional Oceanography*. Pergamon Press. 422 pp.

Turner, J., Klaus, R., Simoes, N. and Jones, D. (1999) Littoral and Sublittoral ground truthing survey of the Socotra Archipelago. In Marine Habitat, Biodiversity and Fisheries Services. Report of Phase 1. UNDP/GEF Project YEM/96/G32 Conservation and Sustainable Use of Biodiversity of Socotra Archipelago. pp. 33–140.

UNDP (1996) Conservation and Sustainable Use of Biodiversity of Socotra Archipelago Project Document. Unpublished document issued by UNDP/UNOPS/Government of Yemen.

van der Elst, R. and Salm, R.V. (1998) Overview of the Biodiversity of the Somali Coastal and Marine Environment. Somali Natural Resources Programme, IUCN Eastern Africa Programme, Nairobi, Kenya. 46 pp.

Venema, S.C. (1984) Fishery Resources in the North Arabian Sea and Adjacent Waters. *Deep Sea Research* **31**, 1001–1018.

Watt, I. (1996) Coastal Habitats Survey of the Gulf of Aden Phase 2: South Coast of Yemen. Report to the Ministry of Fish Wealth, Government of the Republic of Yemen. MacAlister Elliott and Partners Ltd., UK, and Marine Sciences Resource Research Centre, Aden, Yemen.

World Bank (1992) Project Completion Report. Somalia Fisheries Exploration/Pilot Project (Cr. 1465-SO). Agriculture Operations Division, Country Department II, Africa Regional Office. 8 pp.

Zeitzschel, B. and Gerlach, S.A. (eds.) (1973) The Biology of the Indian Ocean. *Ecological Studies* **3** (1/4).

THE AUTHORS

Simon C. Wilson
Department of Biological Sciences,
Warwick University,
Coventry, CV4 7RU, U.K.
and
P.O. Box 2531, CPO 111,
Seeb, Oman

Rebecca Klaus
Department of Biological Sciences,
Warwick University,
Coventry, CV4 7RU, U.K.

Chapter 57

THE INDIAN OCEAN COAST OF SOMALIA

Federico Carbone and Giovanni Accordi

Somalia has the longest national coastline (3025 km) in Africa with an estimated shelf area (depth 0–200 m) of 32,500 km^2. The country is divided into the northern coastal plain of Guban, which has a semi-arid terrain, the northern highlands with rugged mountain ranges containing the country's highest peak (2407 m), and the Ogaden region which descends to the south from the highlands and which consists of shallow plateau valleys, wadis and broken mountains; this region continues to the Mudug plain in central Somalia.

From Ras Caseyr to the Kenya border, the coast runs northeast to southwest, coinciding with the displacement caused by the Mesozoic marginal subsidence. This general structure is complicated by sedimentary troughs crossing the Horn of Africa, and by large sedimentary basins, cutting the coastline and extending inland into southern Somalia and northern Kenya (Juba-Lamu embayment, Mogadishu basin). Offshore, the western Somali Basin extends from Socotra to the Comores. The open shelf environments developed along the Somali coast are a consequence of an extensive marine transgression, connected to coastal subsidence or inland uplift.

The rocks along the southern coastal belt are Pliocene-Pleistocene, and are characterized by a sequence of both marine and continental deposits of skeletal sands, coral build-ups, eolian sands and paleosols. As well as eolian and biogenic sedimentary processes, sea-level fluctuations, Holocene climatic changes and neotectonic movements have combined to produce the modern coastline. A notable feature is an ancient dune ridge complex, known as the Merka red dune, which rims the coast extending beyond the Kenyan border and which separates the narrow coastal belt from the Uebi Shebeli alluvial plain. Two features of note are the Bajuni Archipelago which consists of islands, islets and skerries, forming a barrier island separated from the coast by a narrow marine sound, and a braided, channelized coastal area which originated from the drowning of a paleofluvial net.

The southern Somali coast, with that of Kenya and Tanzania, forms part of the Somali Current Large Marine Ecosystem, encompassing 700,000 km^2, and extending 800 km between Dar es Salaam and Ras Hafun. Abundant biomass develops here due to upwelling. The shelf area has a wide variety of coral reefs, mangroves, seagrass meadows, beaches and estuaries. In shallow water areas the abraded flats are colonized by scattered coral communities, with variable cover. A true fringing reef is achieved in places only in the Bajuni archipelago. All along the southern Somali coastal shelf there are spreading meadows of *Thalassodendron* seagrass, and benthic communities typical of mobile sandy substrates are limited to beach ridges and shoals developed along the coastline. Around the Bajuni barrier island and the channelized area there is more diversity. Mangroves grow on the tidal belts of the channels, and there are expanses of salt flats.

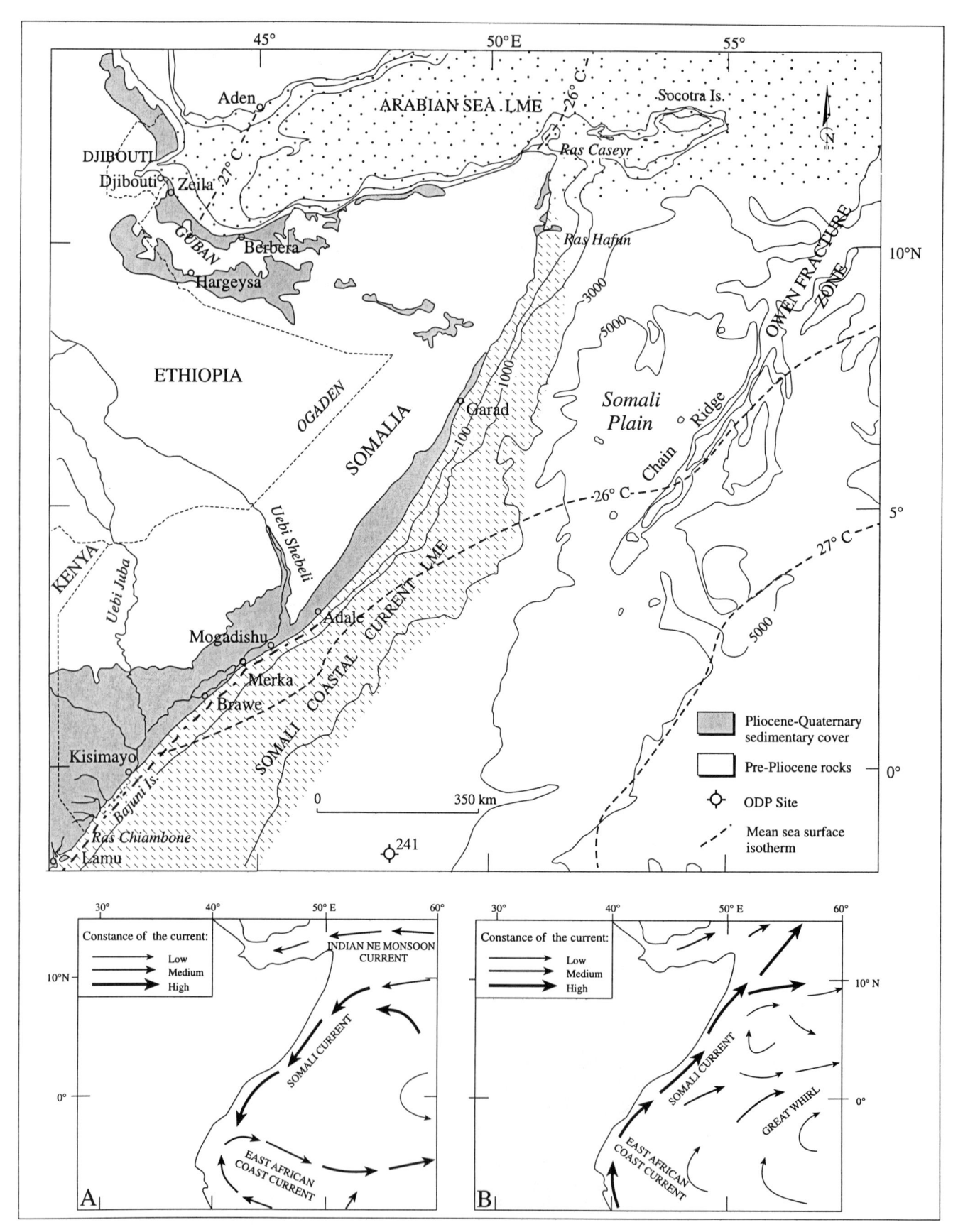

Fig. 1. Coast of Somalia and its main physiographic features. The map shows Pliocene–Quaternary sedimentary cover, the extension of two Large Marine Ecosystems of the Somali coast (cross-hatches) and Arabian Sea (dotted shading), and the annual mean surface isotherms. The dashed part along the 100 m isobath encloses the area colonized by coral reefs. Sketches A and B show the January and July Somali current.

Large-scale alteration produced by man on the Somali coast is relatively recent, but has accelerated in the last few decades, especially around major cities. This alteration affects especially backshore areas where the Pleistocene coral reefs are quarried. At present, the continental shelf is not adequately monitored or protected, so coastal habitats are being degraded, living marine resources are overexploited, and pollution levels are increasing, all of which impact on natural resources and biodiversity.

Somalia is one of the world's poorest and least developed countries, with few resources and devastated by civil war, but since 1993 it has been part of the Common Market for Eastern and Southern Africa (COMESA). This will affect fisheries and aquaculture in terms of the investment, production, trade and fish consumption of the member states. There are currently no Marine Protected Areas and no legislation concerning their establishment and management, although the WCMC (World Conservation Monitoring Centre) Protected Areas Database lists Busc Busc Game Reserve as an MPA. In 1992, The WCMC also listed the following coastal sites as proposed protected areas: Zeila (important sea bird colonies on offshore islets), Jowhar-Warshek, Awdhegle-Gandershe. The area from Kisimayo to Ras Chiambone is probably of highest priority, as it is important for coral reefs, marine turtles, and mangrove resources, although it is still poorly known.

INTRODUCTION

Somalia is located on the Horn of Africa, with Djibouti to the northwest, Ethiopia to the west, Kenya to the southwest, the Indian Ocean to the east and the Gulf of Aden to the north (Fig. 1). It extends over 637,600 km² and has the longest coastline (3025 km) in Africa with an estimated shelf area (depth 0–200 m) of 32,500 km². The country is divided into four geographical regions: the northern coastal plain of Guban, which has a semi-arid terrain; the northern highlands which are rugged mountain ranges that rise from the Guban region and contain the country's highest peak (2407 m); the Ogaden region which descends to the south from the highlands and consists of shallow plateau valleys, wadis and broken mountains; this region continues to the Mudug plain in central Somalia.

Little is known of the recent geological history of this huge coast. Some general geological and morphological information is available (Stefanini, 1930, 1933; Dainelli, 1943), as well as data on mineral composition (De Angelis, 1938). Clark (1954) provides numerous observations on the Quaternary raised coral reefs and associated sediments, and data on facies distribution are given by Carbone et al. (1984) and Carbone (1987).

Only few and fragmentary data exist on the northern part of the Somali Indian Ocean belt (Stefanini, 1930, 1933; Clark, 1954). The whole Somali coast of the Indian Ocean is part of the wider, shallow water ecosystem which encompasses the continental shelf of Somalia, Kenya and Tanzania, to form part of the Somali Current Large Marine Ecosystem (Alexander, 1998; Okemwa, 1998). The flat continental shelf supports a wide variety of coral reefs, mangroves, seagrass meadows, beaches and estuaries over an area of 700,000 km² between Dar es Salaam and just north of Ras Hafun. This continental shelf is mostly narrow and poorly surveyed, and in some places there appears to be no shelf along straight stretches of the coast, suggesting a fault origin. Indentations of the coast are generally accompanied by a widening of the shelf and the presence of strings of islands.

The existence of coral reefs developed along the coastline is formerly reported by Darwin (1842), who received from Capt. Owen and Lieut. Boteler information of "a coral-reef extending four or five miles along the shore" near Mogadishu, and of the "coast and islands formed of madrepore" from the Juba river to Lamu. North of Mogadishu, there are either no coral reefs whatsoever or, if there are some, they are poorly developed (Crossland, 1902, 1904). This is due to the presence of an ocean current which brings deep cold water to the surface. Fringing reefs stretch from between 500 and 1,500 metres offshore from Adale to the Kenyan border; the only major break in this reef is off Mogadishu, where corals are limited to patch reefs scattered within seagrass beds. More recent data on the distribution of coral reefs and other marine coastal habitats is summarised by Sheppard and Wells (1988), while Carbone et al. (1994) summarise the coast south of Mogadishu, where the Bajuni barrier island lies, flourishing with coral reefs, and where there is a drowned fluvial net made up of three wide braided channels, bordered by tidal flats and mangrove thickets (Fig. 9).

NATURAL ENVIRONMENTAL PARAMETERS

The climate in Somalia is tropical, from arid to semi-arid, with a bi-modal rainfall pattern influenced by monsoon winds. The country has an average annual rainfall of about 250 mm with droughts being common. Mean annual rainfall in the north is less than 250 mm; it is about 400 mm in the south and 700 mm in the southwest (FAO, 1995). Rainfall distribution is bimodal, falling in two seasons, the Gu from March to May and the Der from October to November. Occasionally the Gu season extends into June or July because of the Haggai rains which are produced by the onset of moist onshore winds. The Gu and Der rains are caused by the passage of the Inter Tropical Convergence Zone (ITCZ), where the surface winds of the northern and southern hemispheres meet and then rise in a low pressure zone of considerable atmospheric instability. This instability causes rain to fall in isolated storm cells, the result of which is an extremely irregular rainfall pattern. The ITCZ also controls wind direction. From May to September when the ITCZ is 15°S, the wind blows from the southwest and from December to February when the ITCZ is 15°N, the wind blows predominantly from the northeast. During the transitional periods (Tangambilis), the wind drops and becomes erratic in direction.

Mean daily temperature is very constant throughout the year, the hottest months, March and April, being only a few degrees warmer than the coolest months, July and August. At Afgoye, near Mogadishu, mean daily temperature for 1953–1976 ranges from 25.2°C to 28.8°C in March with an annual mean of 27°C. However diurnal temperature fluctuations are much greater and can range from 20°C to 35°C. The mean daily relative humidity pattern is the opposite of the temperature pattern, with high humidity corresponding to low temperatures. In Afgoye the relative humidity measured for 1953–1976 ranges from 66.3% in March to 76.9% in July with an annual mean of 71.8%.

The Somali Current runs parallel to and close to the coast. Though frequently strong, this current is narrow, and more than 100 miles offshore it is often weak. At about 2°S, along the east African coast, variations in coastal currents throughout the year are only slight. North of 2°S, the Somali Current reverses in direction during the year, as do the monsoon winds, though not necessarily at the same time. The southward flow of the current during the Northeast Monsoon is restricted to south of 10°N. It first occurs in early December near the equator, expanding rapidly northwards in January (Fig. 1A), with velocities of 0.7–1.0 m/s. The surface flow reverses in April, when the monsoon changes to southwest. The current develops into an intense jet with velocities of 2.0 m/s for mid-May and 3.5 m/s and higher for June, observed during the International Indian Ocean Expedition in 1964 (Schott, 1983; Rao and Griffiths, 1998). During the Southwest Monsoon (Fig. 1B), near the Somali continental margin, between 5° and 10°N, the onset of a two-gyre system is reported: the Great Whirl, with clockwise rotation, and a secondary one, the Socotra Eddy, further north (Bruce, 1973, 1979; Jensen, 1991; Tomczak and Godfrey, 1994).

The upwelling of cool subsurface water in the Arabian Sea produces an annual mean heat flux in the northern ocean; hence, there is a meridional circulation cell that carries warm surface water out of the region, and cool subsurface water into it (McCreary et al., 1993). This heat flux into the Arabian sea is associated with abrupt cooling during the Southwest Monsoon in summer, the time when, in other parts of the northern hemisphere, warming occurs (Wacongne and Pacanowski, 1996). The sea surface temperature is warmest in April and coldest in August, with a mean annual temperature of 26°C. and a minimum of 21°C at Ras Hafun in August. High variations from normal values are recorded in shallow water environments, especially in the southernmost part of the Somali coast, where the presence of wide floodable areas during the rainy seasons causes the drainage of fresh water towards the sea. Thus temporary mixed salinity and mixed temperature environments develop.

Mixed, semi-diurnal, and diurnal tides occur in the Indian Ocean with a mean range of 4.0 m. These cause strong localized currents around islands and reefs which are superimposed onto the overall longshore current. On the east African coast, the tide range increases from South Africa (1.5–2 m) to Mozambique (up to 5.5 m), then decreases to 3–3.5 m in Kenya, and to 1.5 m in Somalia, at 11°N. The tidal streams are generally weak along the Somali coast, except very close inshore. Their effect on the total water flow is negligible more than a few miles offshore. South of Kisimayo, in the vicinity of the outlets of the channels, the effect of the tidal streams becomes increasingly important.

The productivity of the Somali Current LME is similar to other upwelling areas in the region. The zooplankton biomass from this system is 4 g dry weight/m^2 and 75% of this biomass is composed of copepods, 25% of euphausids (Okemwa, 1998). Off Somalia, the dominant large copepods are *Calanoides carinatus* and *Eucalanus elongatus*, and there is no significant difference in total zooplankton biomass between the two monsoon seasons.

STRUCTURAL FRAMEWORK

From Ras Caseyr to the Kenya border, the Somali coast coincides with the displacement caused by the Mesozoic marginal subsidence, and there are sedimentary troughs crossing the Horn of Africa as well as large Meso-Cenozoic sedimentary basins (Fig. 2), cutting the coastline and extending inland into southern Somalia and northern Kenya (Juba-Lamu embayment, Mogadishu basin). These fault-controlled basins, with thick sedimentary piles (Angelucci et al., 1983; Piccoli et al., 1986; Bosellini, 1989), continue inland toward the southwest, to the southern part of the Rift Valley System. According to Kent (1982), both coastal faulting and inland rifting are related to the same tectonic control. Compressive phases which originated along northeast–southwest fracture zones are linked to oceanic systems of transform faults. These developed in connection with a change in the stress regime which occurred between the end of the Madagascar Rift about 120 Ma ago and the beginning of sea-floor spreading of the Mascarene Basin Province about 80 Ma ago. According to Boccaletti et al. (1988), asymmetric sedimentary basins with increasing thicknesses from east to west developed during the Cretaceous in connection with these tectonic movements.

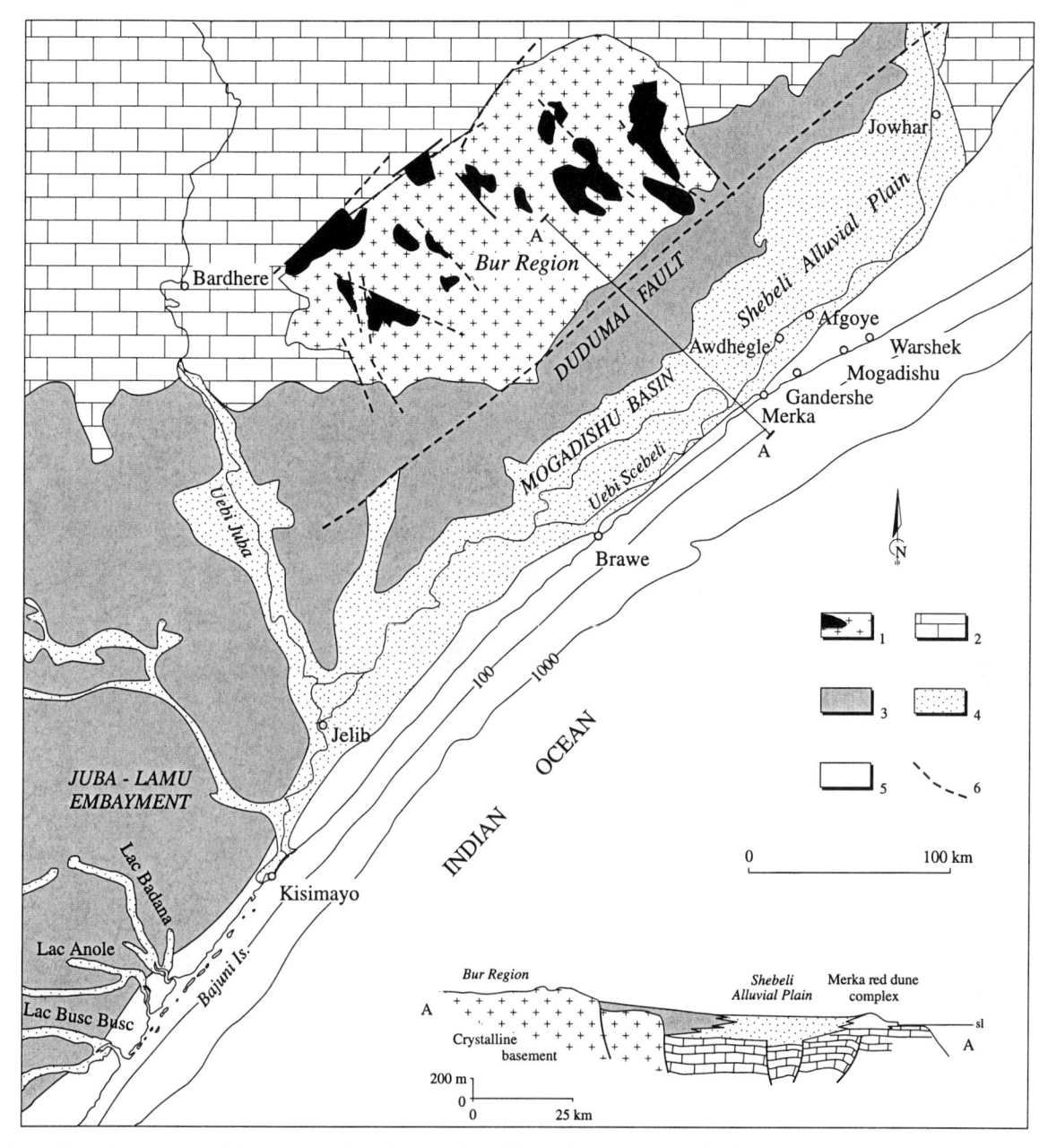

Fig. 2. Schematic geological map of the southern Somali coastal belt: (1) pre-Cambrian crystalline basement; the black areas indicate inselbergs (burs) formed of migmatitic granite and quartzite; (2) Meso-Cenozoic sedimentary sequence; (3) Pliocene–Quaternary eolian and eluvial sediment; (4) Quaternary alluvial sediment of Shebeli river and other floodable areas (gravel, sand, loam); (5) Merka red dune complex partially covering the Pliocene reef limestone along the narrow coastal belt; (6) Main faults.

Offshore, the Western Somali Basin extends from Socotra to the Comores and is bordered eastwards by the mid-ocean Owen Fracture Zone and the Chain Ridge (Fig. 1). The structural setting and the sedimentary features of the Somali Basin, described by Francis et al. (1966) and Bunce et al. (1967), have been more recently investigated by ODP oceanographic surveys. A site located 170 miles off the coast at the base of the continental rise in 4505 m water depth, recorded 470 m of Quaternary–Upper Oligocene deep water facies, overlying a 704 m section of Middle Eocene–Upper Cretaceous sediments, also of deep water environment. In Somalia and Kenya the main marine transgression took place during the Oligocene, whereas the widespread inland extension of marine water took place in the Lower and Middle Miocene. According to Kent (1974), the persistence of open shelf environments along the east African coast during the Neogene is a consequence of an extensive marine transgression which is partially worldwide, but can be connected as well to coastal subsidence or inland uplift. The Pliocene is not well documented in the Somali coastal area because of the absence of outcrops and of data deriving from hydrocarbon exploration. The development of late sedimentary basins between structural highs is documented in coastal Tanzania and Kenya (Kent et al., 1971; Nyagah, 1995) by means of seismic works and boreholes.

PRESENT-DAY GEOMORPHIC FEATURES OF THE COASTAL ZONE

The rocks outcropping along the coast from Mogadishu to the Kenya border represent the top of the Pliocene–Pleistocene sequence known (Piccoli et al., 1986) as the Merka formation (Fig. 2). These are deposits of skeletal sands, coral build-ups, eolian sands and paleosols, whose pattern is controlled by both eolian and biogenic sedimentary processes. These, with sea-level fluctuations, Holocene climatic changes and neotectonic movements combined to produce the modern coastline.

A notable feature of the area is an ancient dune ridge complex (Merka red dune) which rims the Somali coast from 6°N extending beyond the Kenyan border and separates the narrow coastal belt from the Uebi Shebeli alluvial plain. The wide coastal region extending from Kisimayo to the Kenyan border contains the Bajuni archipelago and three channels linked to the sea through three large outlets named, from north to south, Lac Badana, Lac Anole, and Lac Busc Busc (Fig. 9). Three villages, Istanbul, Kudai and Burgao are located on these channels. The Bajuni Archipelago consists of a series of islands, islets and skerries parallel to the coast at a distance of 2–4 km. Towards the coast, they separate a sound up to 10 m deep, where carbonate sedimentation takes place. The flat coastal belt shows a reefal substratum forming a marine terrace 4–6 m high, covered by reddish soils and dunes near the shoreline and by alluvial and marshy deposits inland.

Rivers and the Alluvial Plain

The country's only permanent rivers are the Juba and Shebeli, originating from the Ethiopian plateau, and draining into the Indian Ocean. There are large variations in discharge from year to year. The Shebeli river first flows to the coast, then, North of Mogadishu, changes direction, and near Mogadishu it forms a wide alluvial plain about 90 m above sea level which lowers southwards, forming a large marshy area.

According to Sommavilla et al. (1993) the outline of both the river and its valley have mostly resulted from recent tectonic movements, superimposed on the older faults (Fig. 2, cross section). From wells drilled for water supply on the alluvial plain southeast of Mogadishu (Faillace, 1964) the sediments deposited by the river are up to 350 m thick. More than 250 m of this alluvial cover lie below the present sea level.

The Merka Red Dune Complex

A wide dune ridge runs parallel to the coast from 6°N to 2°S, reaching its maximum elevation of 378 m a.s.l. in the Mudug region, with a maximum width of ca. 100 km at Adale. From Mogadishu to the Juba river mouth this runs very close to the coastline, and does not exceed 8–10 km in width, whereas it reaches an elevation of ca. 150 m south of Merka village (Figs. 3B and 5A).

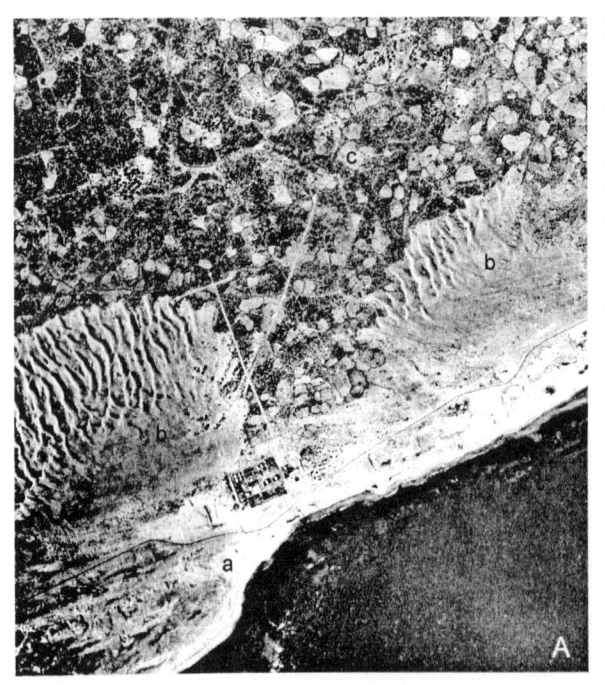

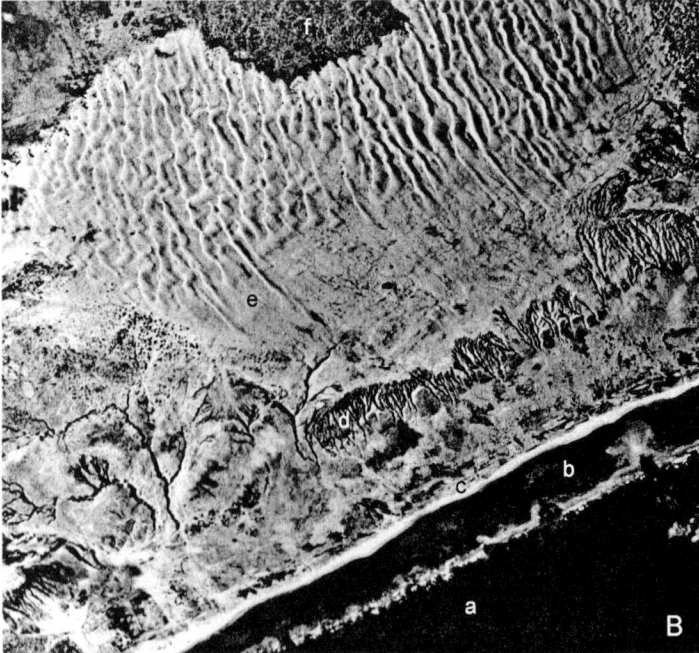

Fig. 3. Southern Somali coastal belt. (A) Effect of prolonged human use of the coastal belt just south of Mogadishu: (a) top of Pleistocene reef terrace affected by deep quarries; the oil-refinery is also clearly visible; (b) mobile dunes; (c) farmed land. (B) Coastal area near the village of Merka: (a) wave breaker line along a morphological step of the abraded shelf; (b) shallow shelf colonized by *Thalassodendron* seagrass and coral heads; (c) raised reef terrace covered by *Achatina* and *Georgia* eolianite and small mobile dunes; (d) gully erosion of the Merka red dune front; (e) wide mobile dune field; (f) Uebi Shebeli alluvial plain.

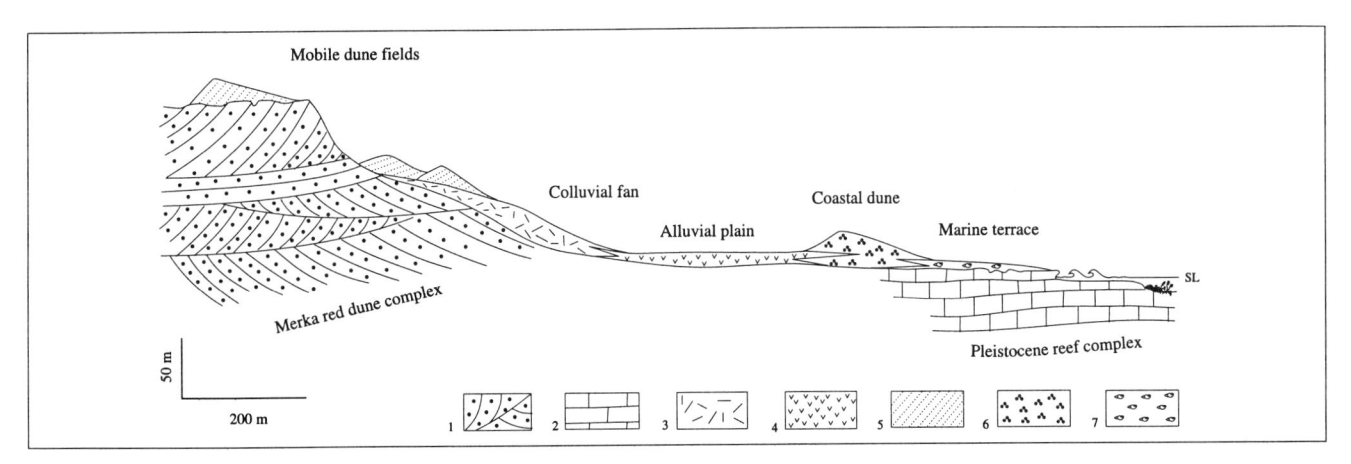

Fig 4. Schematic cross section from the red dune complex to the coastline near the village of Merka: (1) poorly cemented, cross bedded quartzose sand; (2) raised Pleistocene reef limestone interbedded with quartzose-carbonate beach sand; (3) reworked sand of the red dune complex, forming small accretionary alluvial fans developed at the foot of gullies; (4) reddish residual soil deposited in floodable, protected backshore areas; (5) mobile dune fields of quartzose sand; (6) coastal dunes built by accumulation of varying percentages of quartz and carbonate skeletal grains; (7) calichified eolianite with abundant *Achatina* and *Georgia* shells.

According to Clark (1954), the dune ridge is made up of an array of parallel dunes whose sand is similar to the recent deposits of the Juba river (Artini 1915, 1926; De Angelis, 1938; Clark, 1954; Angelucci et al., 1995). Seawards, the dune partially overlies a marine sequence (Fig. 4), rich in coral remains.

The Coastal Reef Terrace

The Pleistocene lithologic sequence outcropping along the coast, the channel cliffs, and in various quarries in the outskirts of Mogadishu (Fig. 6) and Kisimayo shows well diversified shallow marine environments. Near Mogadishu, cliffs and quarry walls, 4 to 6 m thick, show a sheltered facies of a well developed fringing reef (Carbone et al., 1984; Carbone and Matteucci, 1990), with coral colonies in growth position (massive *Porites, Lobophyllia* and *Galaxea* knobs and *Acropora* thickets) overlying sandy beach deposits containing small bivalves (*Donax, Atactodea, Gafrarium*) and gastropods (*Gibbula*). This marine sequence is topped by regressive rhodolith deposits and skeletal rubbles mainly consisting of coral fragments, generally encrusted by red algae, in places showing a spur and groove morphology. It is likely that reefal limestone extend tens of meters below the present sea level, and data from a well, ca. 50 m deep located close to the shoreline 15 km south of Mogadishu (Dal Prà and Salad, 1986) shows a sequence of sandy layers containing variable amounts of skeletal remains which are typical of shallow water coastal environments.

The reef complex also outcrops in the Kisimayo region, where reefal bodies are found in different places. Crusty–massive and massive types prevail in the coral community (*Favia, Favites, Gonyopora, Porites*). The colonies are loose in coarse skeletal sediment. The top of the reef terrace is generally altered by calichification processes, showing several cavities filled with reddish sandy sediments. At many sites this marine terrace is conformably overlain by strongly calichified eolianite, bearing a rich continental gastropod assemblage (*Achatina* and *Georgia*). In places, along the coast, the Pleistocene reefal limestone is overlain by a seaward dipping beachrock.

The Bajuni Barrier Island

The Bajuni archipelago shows many features of a barrier island complex (Hoyt, 1967; Schwartz, 1971; Purser and Evans, 1973). The islands consist of elongate ridges parallel to the coast and separated from it by a narrow and shallow marine sound (Fig. 9). The origin of this barrier island is linked to the migration of the coastal dune field towards the continental shelf edge during the last glacial sea-level lowering. Islands, islets and skerries, separated by inlets, have allowed a widespread coral colonization of the shelf. High carbonate sediment production from coral reefs and seagrass meadows causes the build-up of sandy bodies in the shape of bars and tails which emerge during the low tide (Fig. 7B). These bodies run from the islands towards the sound, forming wide protected intertidal flats, where fine sediment deposits, intensely bioturbated by infauna.

The islands reach an elevation of about 10 m a.s.l. Their indented seacliffs generally show a pronounced notch landward, where bioerosion is intense and wave action is weak. The boundary between eolianites and underlying skeletal limestone is barely detectable on the islands, and only in rare cases can it be found at the base of the seacliff, coinciding with the abraded flat.

The Braided Channelized Coast

Wide tidal flats along the coast south of Kisimayo seem to be connected to three braided channels flowing perpendicular to the coast in the Bajuni sound. These channels

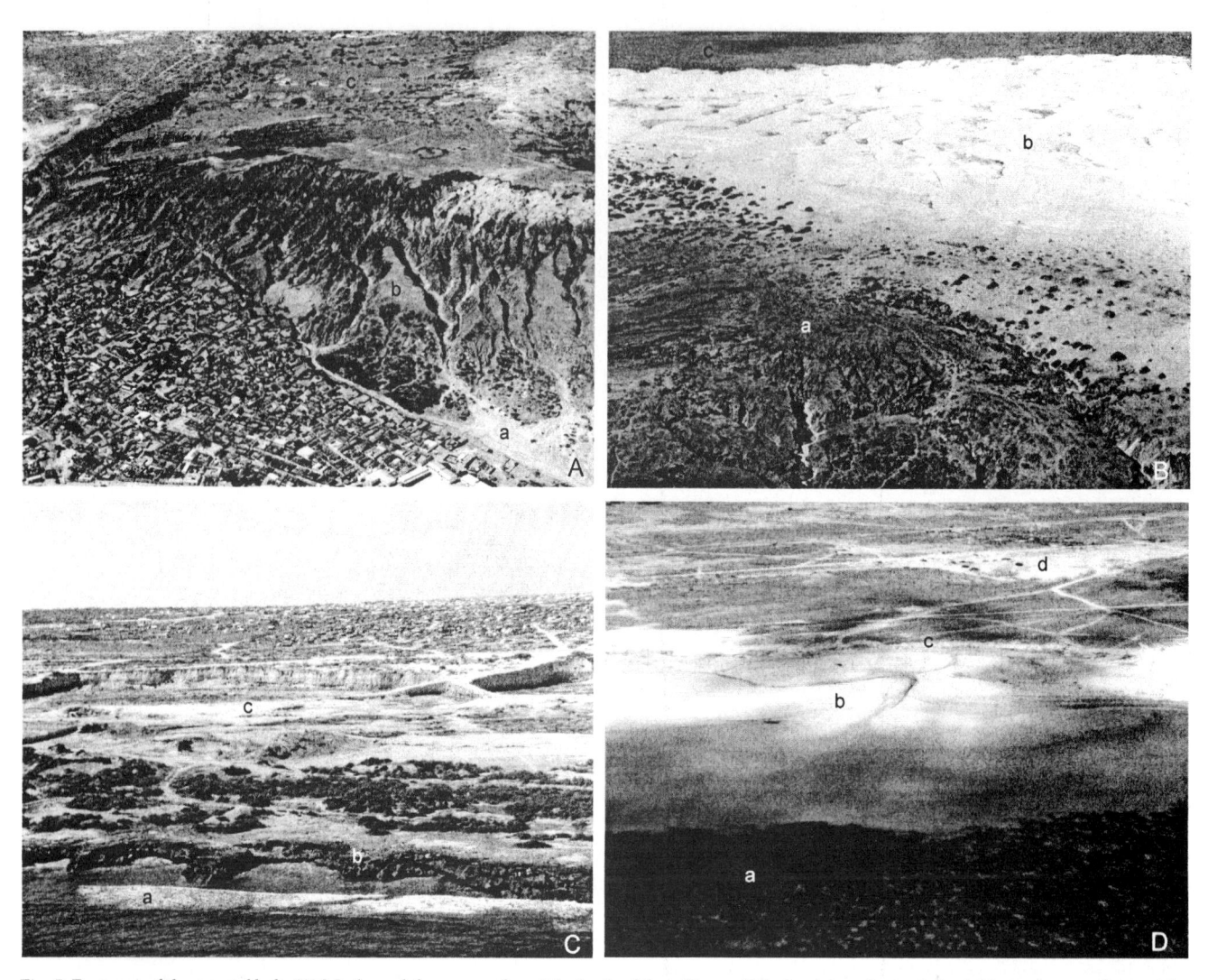

Fig. 5. Features of the coastal belt: (A) Merka red dune complex at the back of the village of Merka: (a) coalescent eluvial fans grown at base of the dune wall; (b) gully erosion of the steep dune wall; (c) top surface of the dune gently dipping towards the Uebi Shebeli alluvial plain. (B) Top surface of the Merka red dune covered by a mobile dune field: (a) poorly cemented sandstone showing a badland morphology; (b) mobile quartzose sand accumulated by monsoon winds; (c) Uebi Shebeli alluvial plain. (C) Man induced large-scale alteration of the coast near Mogadishu: (a) nearshore abraded flat; (b) wave notch in Pleistocene raised coral reef complex; (c) quarry surface often flooded during high tide. (D) Zonation of a flat coastal area south of Mogadishu: (a) *Thalassodendron* seagrass bed; (b) mobile sandy bottom; (c) coarse berm deposit; (d) flat backshore area, in places showing the effects of evaporative pumping of groundwater (white areas).

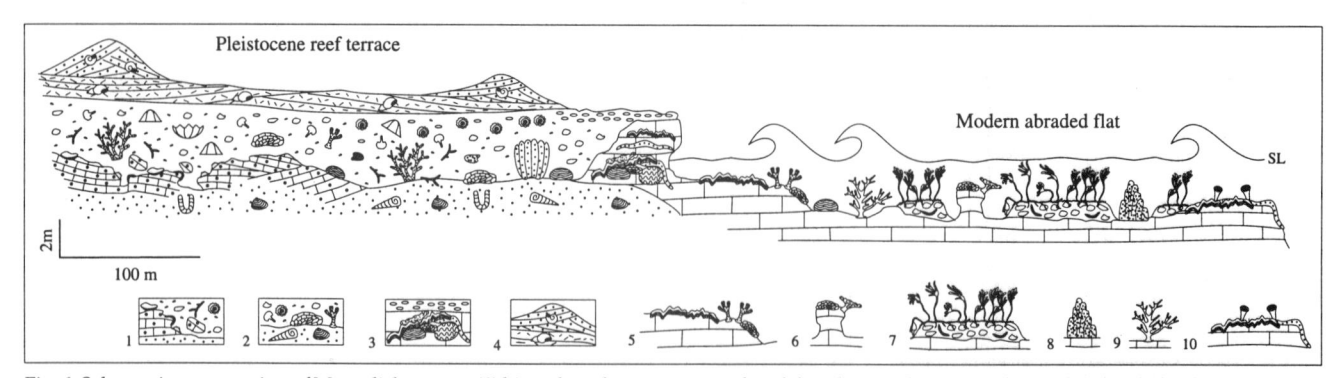

Fig. 6. Schematic cross section of Mogadishu coast: (1) bioturbated quartzose sand with bivalves and gastropods overlain by skeletal sand; (2) well sorted quartzose sand, with gastropod and bivalve infauna, covered by coarse skeletal sand showing several coral colonies in growth position, topped by rubble, mainly of coral remains often encrusted by red algae; (3) assemblage of corals and encrusting red corallinaceans, in places with spur and groove features, topped by coral rubble; (4) eolianites and mobile dunes, rich in *Achatina* and *Georgia* shells; (5) abraded flat encrusted by corallinaceans and bordered by small fringes of branching and massive *Porites*; (6) erosion remnant colonized by both massive and small branching corals; (7) *Thalassodendron* seagrass bed; (8) *Porites lutea* knob; (9) *Millepora* thicket; (10) shifting substrate encrusted by corals, corallinaceans and sponges.

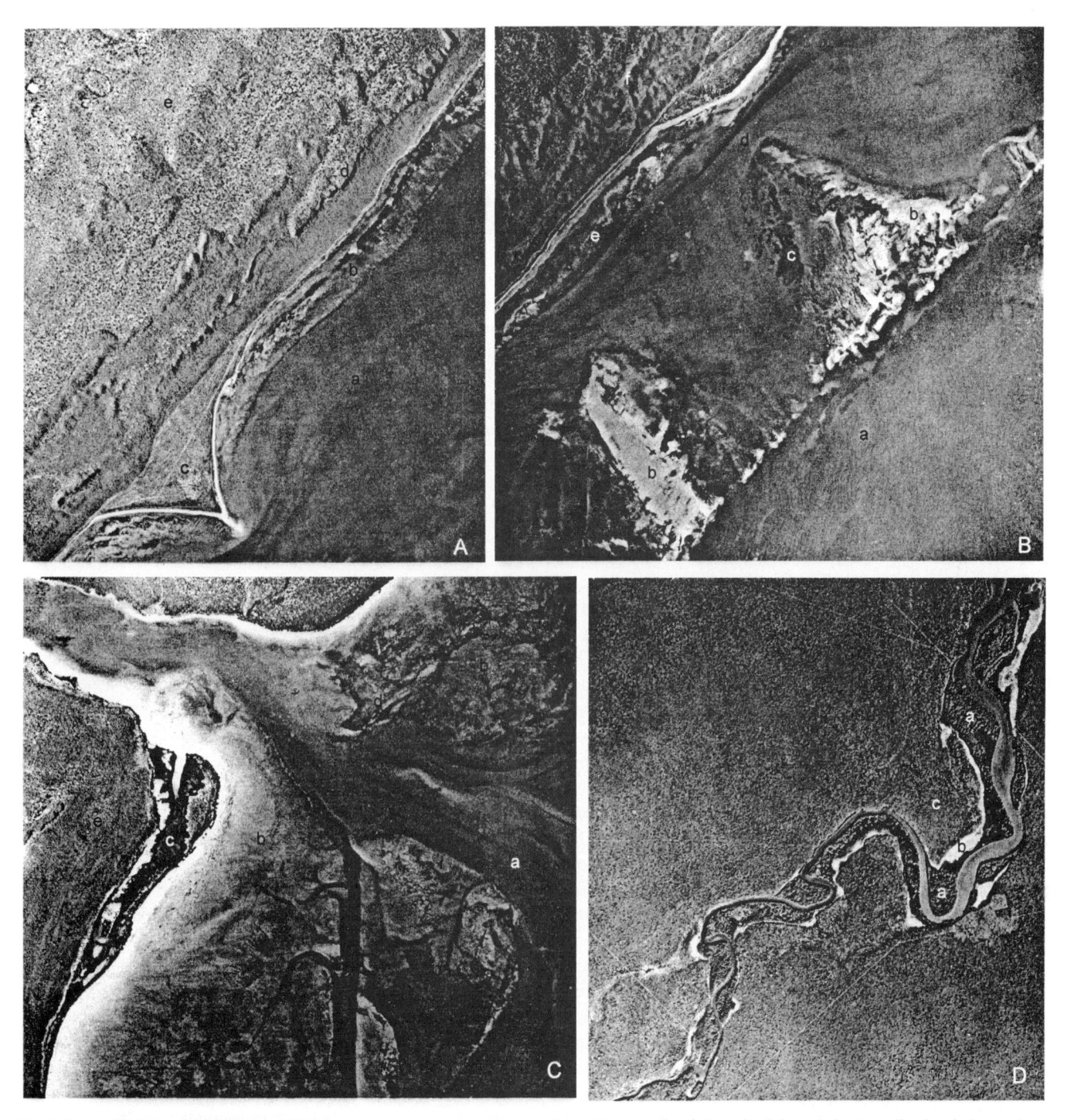

Fig. 7. Coastal belt south of Kisimayo. (A) Mainland coast protected by the Bajuni barrier island: (a) subtidal sandy bottom; (b) abraded nearshore shelf, colonized by seagrass and scattered coral heads; (c) sand spit prograding on the flat nearshore shelf; (d) old frontal dune ridge showing a fossil wave cut notch; (e) bedrock covered by eolian deposits stabilized by scrub. (B) Islets and skerries forming the Fuma island complex, showing sandy tails perpendicular to the coastline: (a) outer shelf largely colonized by *Thalassodendron* seagrass; (b) intertidal mounded and rippled sandy bottom, strongly bioturbated; (c) high subtidal bottom, mainly colonized by *Thalassodendron* seagrass; (d) subtidal mobile sandy bottom; (e) narrow shelf with seagrass and scattered coral heads; (f) old dune ridge covering the Pleistocene reef complex. (C) Lac Badana outlet: (a) low tide channel, cut in the bedrock, winnowed by tidal flows; (b) shelf where sand from the channel forms wave bars and shoals, in places stabilized by dense seagrass beds; (c) mangal bordering landward a salt flat; (d) abraded flat bordered by a small fringing reef; (e) old accretionary frontal beach and dune ridge. (D) Lac Badana channel upstream of Yamani village: (a) mangrove fringe; (b) sheltered floodable area with salt flat characteristics; (c) peneplaned area extensively covered by scrub.

originated during the Holocene from the drowning of a wide fluvial net extending from Kisimayo to Lamu in northeastern Kenya. Where there is low water energy, the intertidal zone is usually occupied by mangroves (Fig. 7D). They trap fine sediment to build mud flats, strongly bioturbated by crabs and gastropods. In places, salt flats develop at the back of this mangal flat in sheltered coastal areas. Relict quartzose sand migrates along the channels towards their mouths, mixing with carbonate skeletal sediment produced along the channels and transported to form sand drifts at the channel outlets and beach ridges along the shore (Fig. 7A, C).

During the rainy seasons, the fresh water flowing into the channels is generally insufficient to modify the flourishing marine biota of the lower reaches, whereas in the upper ones, it is probably responsible for the poor development of biota made up of small gastropods and thin-shell bivalves. The transition between the two environments is variable from one channel to another, depending on both their morphology and the extension of the tidal prism inside the channels.

LATE PLEISTOCENE TO PRESENT-DAY EVENT SEQUENCE

The modern setting of the Somali coast is the result of the interaction between tectonics of the passive east African margin and recent sea-level fluctuations. The presence of coral reef terraces, uplifted at different elevations, along the African coast of the Indian Ocean is well documented (Ase, 1978, 1981; Crame, 1980, 1981; Braithwaithe, 1984), though there is scarce and often contradictory age-dating. Braithwaithe (1984) established a complete description of depositional units outcropping along the Kenyan coast. He points out the existence of a complex of terraces ranging from 4.5 to 140 m. The dates range from about 240,000 years B.P. (^{230}Th/^{234}U dating recorded by Battistini, 1976) to about 25,000 to 27,000 years B.P. (radiocarbon dating by Hori, 1970 and Toyah et al. 1973), the last two being doubtful and coming from samples about 4–5 m a.s.l., near Mombasa and Malindi. More recent age-dating, ranging from 2,640±105 to 5,250±130 years B.P., are referred to as 'beachrocks' cropping out (0.9–0.2 m high) from Lamu to Mombasa (Hori, 1970; Toyah et al., 1973). In the southern Somali coast, some still unpublished ^{230}Th/^{234}U dating obtained by us on massive corals and *Tridacna* shells from the Mogadishu and Lamu areas places the older reef cycle cropping out along the coast between 105,000 and 131,000 years BP. Thus, a connection can be hypothesized between a basal transgressive episode and the depositional event of the last interglacial period of the isotope stage 5. In the area south of Kisimayo several evolutive phases have been identified, corresponding to eustatic sea-level variations starting from isotope stage 5e (Fig. 8).

MAJOR SHALLOW WATER MARINE AND COASTAL HABITATS

The habitat distribution along the southern Somali coast results not only from close interaction of biological, physical and chemical marine processes, but also from recent eustatic sea-level variations. In shallow water, the Pleistocene substratum is colonized by scattered coral communities, which are widely distributed but far from flourishing everywhere. The amount and kind of coral cover is very variable, giving rise to different coral facies. In many areas where the hard substratum is subject to stronger wave action, corals grow in the form of scattered small colonies. In more protected areas, several generations of corals succeed in forming knobs and patch reefs of varying sizes. A true fringing reef topography occurs only in the Bajuni archipelago.

All along the southern Somali coastal shelf, coral growth competes with *Thalassodendron* seagrass meadows, both inshore and offshore. Benthic communities typical of mobile sandy substrates are limited to beach ridges and shoals along the coast. A more diversified habitat is recognizable in the southernmost part of the Somali coast and in northern Kenya, because of the presence of the Bajuni barrier island and of the braided channelized belt. Mangal flats occur on the tidal areas of the channels.

Shelf and Fringing Reefs

The corals along the Somali coast of the Indian Ocean are poorly investigated. After general observations by Darwin (1842), Dana (1875) and Crossland (1904), only recently have some papers (Angelucci et al., 1982; Carbone et al., 1994) contributed to the knowledge of the coral associations and their distribution pattern. Various types of reef grow on both abraded rocky substrates and stabilized sandy bottoms. Coral colonization shows the same characteristics all along the Somali shelf; only south of the Juba river mouth, the widening of the shelf and the presence of the Bajuni barrier island cause a major diversification.

Coral carpets consist of dense assemblages mainly of massive and encrusting corals, on large flat areas near the coastline where mobile sandy cover and *Thalassodendron* meadow are lacking. Coral carpets are also found offshore, on the edge of the first shelf step, about 10 m deep, coinciding with the breaker zone (Fig. 3B and Fig. 6); here corals are scarce in high energy areas where sparse colonies of *Pocillopra eydouxi* are found together with a greater abundance of crustose coralline algae and soft caliculate sponges. In places, crustose coralline algae predominate and *Halimeda* tufts are common as well. Corals consist only of small colonies of *Psammocora contigua*, *Pavona decussata*, *Pocillopora damicornis*, *Acropora* spp., *Favites* spp., and scattered large encrusting colonies of soft corals.

In the Bajuni archipelago coral carpets are also developed on the protected shelf area. Massive corals generally colonize the edges of the abraded flat surrounding the islands, the most common being *Favia stelligera*, *F. pallida*, *Favites abdita*, *F. halicora*, *Goniastrea pectinata*, *G. retiformis*, *G. aspera*, *Goniopora lobata*, *Porites somaliensis*, *Platygyra lamellina*, *P. daedalea* and some colonies of branching *Pocillopora eydouxi*. Thickly branched corals, such as *Porites nigrescens* and *Stylophora pistillata*, together with some large crustose colonies of *Echinopora gemmacea* are also found.

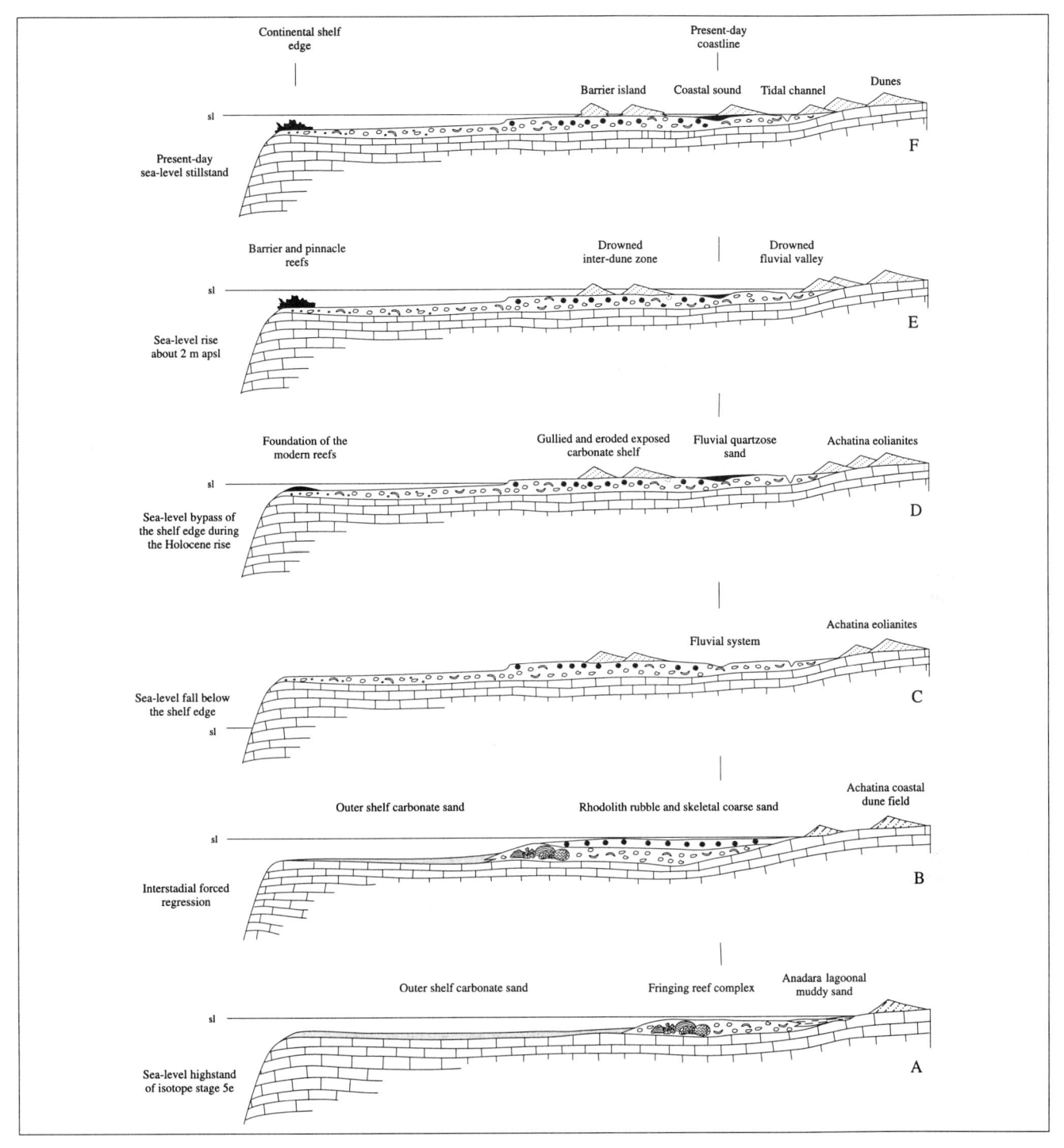

Fig. 8. Profiles showing the evolution of the shelf of the Bajuni Islands since sea-level highstand of isotope stage 5e, about 125,000 years BP. (A) *Sea-level highstand of isotope stage 5e*: sea-level rise with landward migration of a fringing reef complex forming the lower part of the marine carbonate sequence, overlying beach ridge facies along the coast from Kisimayo to Lamu. Protected lagoonal environments proved by muddy deposits rich in gastropod and bivalve communities. (B) *Interstadial forced regression*: sea-level drop with progressive seawards coastline migration. Deposition of coarse skeletal sand rich in coral remains and corallinaceans. Rhodolith rubble is present in places to form a regressive surface. (C) *Sea-level fall below the shelf edge*: long subaerial exposure of the continental shelf surface, with erosion and weathering. Development of a braided fluvial net, with its base level lower than the present one, and migration of a dune ridge towards the shelf edge. (D) *Sea-level bypass of the shelf edge during the Holocene rise*: rapid submergence of east African coast, beginning approximately 18,000 years BP (Colonna et al., 1996). A submerged ridge along the Bajuni shelf edge could signify the coral regrowth. The base of this ridge, located 20 m b.p.s.l., could indicate the starting point of the coral colonization 8000 years BP (Hopley 1994). (E) *Sea-level rise of about 2 m apsl*: inundation of the Pleistocene shelf with drowning of channels cut during the lowstand, and overflowing of the interdune areas. This caused formation of wide tidal flats. Deposition of beachrock takes place just above the present sea level. (F) *Present-day sea-level stillstand*: the present pattern, reached in very recent times, marked by beach ridges, longshore bars and wide tidal bars. Partially buried old notches, bordering the mangal flat, suggest a prograding tendency. Today, many areas inside the Bajuni sound appear to be shallowing, with increase of the mangrove colonization and decrease of the coral community in favour of *Thalassodendron* meadows.

Real fringing reefs grow only along the shelf edge of a few islands of the Bajuni archipelago. At Ilisi (Fig. 9A), the northernmost island of the archipelago, a 500–600 m wide shallow shelf is bordered by a steep reefal slope 8–12 m deep, connecting to a flat sandy bottom. The fore reef is covered by skeletal sand, locally stabilized by seagrass. Here coral pinnacles are found, at times reaching the sea surface and assuming the typical micro-atoll morphology. Among the corals, *Porites, Goniopora, Millepora* and *Faviids* prevail. Branching *Acropora* thickets are also found, with wide areas of dead coral, encrusted by corallinaceans and with dense *Fungia*. The reef front is very steep and densely covered by corals mostly in the middle and upper part. In the lower part of the wall, soft corals (*Sarcophyton*) are typical, whereas the remaining spaces are occupied by crusty and plate-like corals (*Montipora, Hydnophora, Echinopora*), small branching *Galaxea fascicularis* and Acroporids, and domal massive corals (*Favia, Favites, Porites*). The upper part of the reef wall shows a gradual upward increase of branching *Porites*. The reef crest, about 30 m wide, is characterized by an extensive cover of *Porites somaliensis*. Small *Acropora prolifera* thickets replace *Porites* colonies towards the reef flat. The reef flat is characterized by a landwards progressive decrease of the coral cover. Where coral cover decreases, the substrate is encrusted by *Melobesia* or covered by coarse bioclastic material originating from branching corals and mollusc shells. This material is encrusted by corallinaceans and polychaetes, and is densely bored by clionids and serpulids. The rough surface of the reef flat hosts algal mats and a rich population of ophiuroids and small gastropods. The back reef area is generally characterized by a dense *Thalassodendron* meadow. The meadow is frequently interrupted by irregularly shaped depressions, covered by sand which originates from the remains of molluscs, *Halimeda* and corallinaceans. The coral community is sparse and similar to that of the reef flat.

Smaller fringing reefs grow close to some islands where inlets allow water exchange between the sound and the open sea. At Cuvumbi island the reef front consists of a small steep wall generally showing a vertical zonation. The reef flat (Fig. 11A), very close to the island, is colonized by various species of branching *Acropora* and by *Tubipora musica*. Shoreward, the coral community consists of micro-atolls of *Favia stelligera* and small colonies of *Acropora abrotanoides, Favites flexuosa, F. abdita* and *Porites nigrescens*.

Along the barrier island, on sea floors particularly subject to wave and current action, corals are scarce and consist of small massive forms (*Porites somaliensis, Platygyra lamellina* and *Favites*), flat stout colonies of branching *Acropora* and encrusting forms of *Echinopora gemmacea* and *Hydnophora exesa*. Where maximum water energy is reached, corals are even less common, while encrusting coralline algae increase, locally forming rhodolith deposits.

In the Bajuni sound, coral knobs and patch reefs surrounded by sandy substrate rise up from the *Thalassodendron* beds. These build-ups may represent the colonization of successive generations of corals on erosion remnants of the hard substratum. These coral reefs show an ecological zonation: the central part of the top surface is covered by encrusting and articulate corallinaceans and tufts of *Halimeda*, whereas the elevated rims, exposed at low tide, are colonized by *Acropora, Pocillopora* and *Millepora*; the flanks often show *Porites* and faviids in the middle part and encrusting corals like *Echinopora gemmacea* and *Hydnophora exesa* in the lower part. The surfaces without living corals show intense bioerosion by bivalves and clionid sponges. Some of them resemble table reefs (James, 1983), some of which are 200 m wide and 5 m high. The distribution of the living corals along the wall-like margins is controlled by water energy: *Acropora* communities grow mostly on the landward side; leafy and plate-like communities grow on the seaward side. Huge single colonies of *Porites lutea* grow separated from the main patches.

Seagrass Meadows and Sandy Bottoms

Studies on seagrass beds and associated fauna have been carried out in various locations of the western Indian Ocean (Pichon, 1964; Mauge, 1967; Taylor and Lewis, 1968; Aleem, 1983), but little on the Somali coast. Few seagrass beds occur along the northern Somali coast, but they are extensive along the southern coast from Adale to Ras Chiamboni.

Thalassodendron meadows colonize wide areas of the shallow shelf (Fig. 5D and Fig. 6). The sea floor is generally covered by a thick layer of rubble, encrusted by corallinaceans, bound by *Thalassodendron* roots and characterized by whole bivalve shells (*Codakia*). The stalks and blades of the seagrass support a flourishing epiphyte community of both articulate and encrusting corallinaceans, together with bryozoans. The *Thalassodendron* meadows play an important role in the production of carbonate sediment, as well as in sediment trapping, including fine muddy particles (Scoffin, 1970; Land, 1970; Patriquin, 1972). In consequence, the substratum of the meadow is more elevated than the surrounding sea floor. Unvegetated areas, usually elongated parallel to the coast, are covered by rubble or coarse skeletal sediment, or show the hard substratum. Along the edges of these depressed areas the local increase of water energy causes lateral erosion of seagrass beds. *Thalassodendron* constantly competes with coral growth expanding laterally to cover areas colonized by corals, but also being replaced by corals when water energy increases cause its erosion.

Biota of mobile substrates are typical of inner coastal areas, where beach ridges and shoals develop. The coastline north of the Juba river shows cliffs (Fig. 5C) alternating with beaches. Locally, erosional remnants rising from the abraded flat are connected to the coast by tombolos. The beaches are normally colonized by suspension-feeding molluscs such as *Donax* and *Atactodea*. High-water beach and berm deposits often made of shell hash, are densely burrowed mainly by crabs. During the lowest low tides, the

The Indian Ocean Coast of Somalia

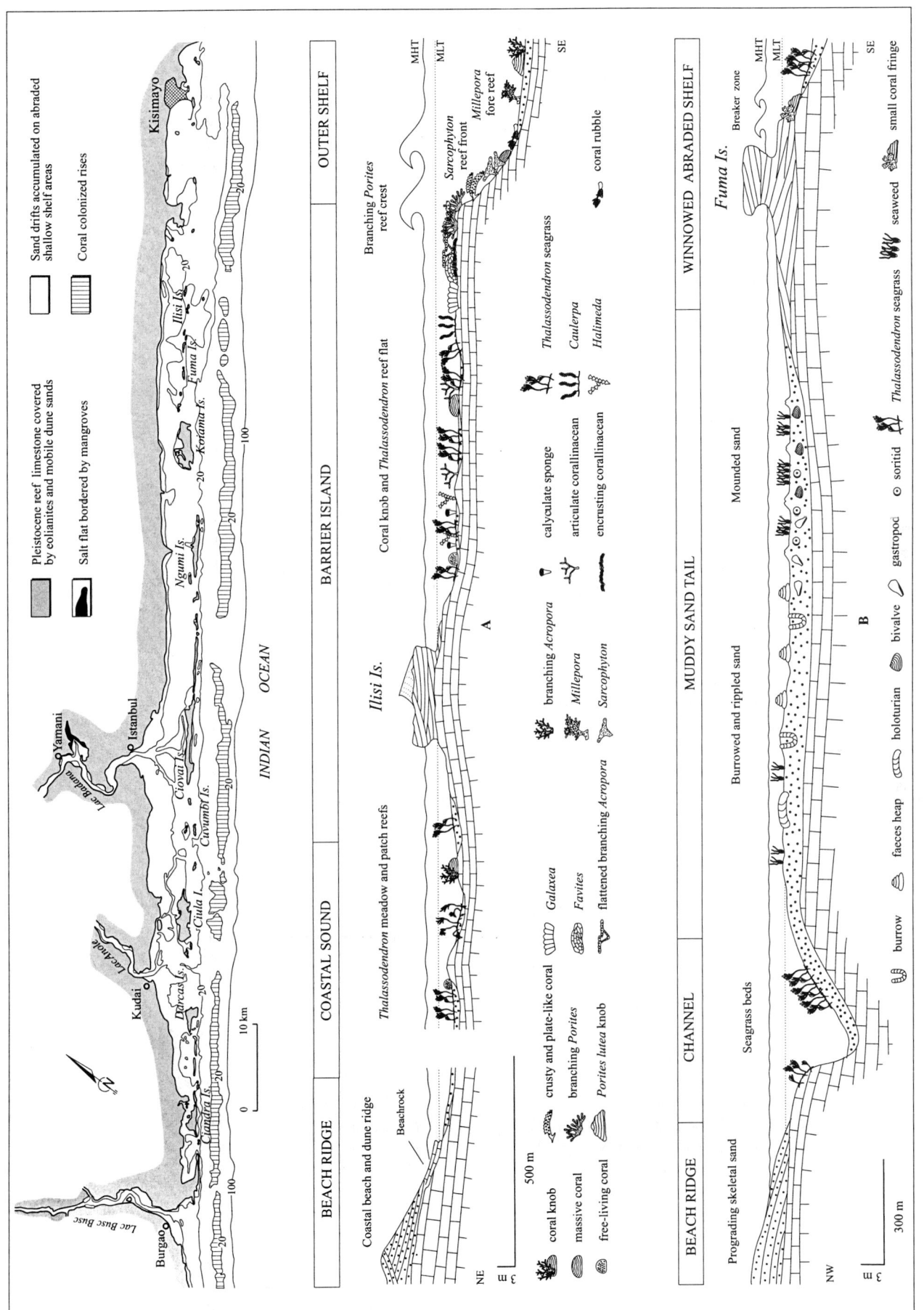

Fig. 9. The Bajuni barrier island. The benthic system is illustrated by two schematic ecological profiles through the Ilisi Island fringing reef (A) and the Fuma Island sandy tail (B).

mobile sandy bottoms are widely exposed and the sand is blown by the monsoon winds and accumulated to form backshore eolian deposits.

A mixture of habitats occurs in the Bajuni sound. The mixing and accumulation of carbonate and quartzose sand produces different shaped tidal bars at the channel outlets (Fig. 7C). The coastline has a frontal beach and dune complex, at the back of which older coastlines are present (Fig. 7A). Accretionary sandy tails and spits accumulate mainly on the landward side of the islands (Fig. 7A, B). Locally these tails develop perpendicular to the sound, forming wide intertidal areas, separated from the coast by a narrow channel no more than 10 m deep. At Fuma island, the intertidal area of the tail is characterized by various facies (Fig. 9B), mostly orientated perpendicular to the coast, with a distribution mainly controlled by the length of subaerial exposure and tidal energy.

Moving from the islands towards the centre of the sound, the high intertidal zone has beach deposits of poorly sorted skeletal sand, rimmed inshore by coarse beachrock formed by gastropod, bivalve and coral remains. They move on to a mounded zone where a fine sediment is densely bioturbated by crabs and worms (Fig. 11D). Here, there are seagrass carpets of small *Thalassodendron* and *Syringodium*, partly buried by fine sediment. Small gastropods (strombids and cypreids) are also present, and among foraminifers, soritids are common. Moving towards the axis of the channel, the bottom is flat and covered by asymmetric ripples orientated according to the tidal currents. The strong bioturbation is marked by the presence of abundant faecal pellets and faeces heaps, disgregated during floodtide. Holothurians are common. These accretionary tails end near the channel edge, where *Thalassodendron* colonization takes place.

Mangal and Braided Channels

Intertidal flats have been studied in different areas of the Indian Ocean, but again, little is known of the area between Kenya and Somalia (Carbone, 1987; Hughes and Hughes, 1992; Eisma, 1998). Some are colonized by mangroves (Fig. 7C, D). The mangal forms a low-energy intertidal environment, where black fine sediment forms the habitat for the development of a highly specialized oligotypic community.

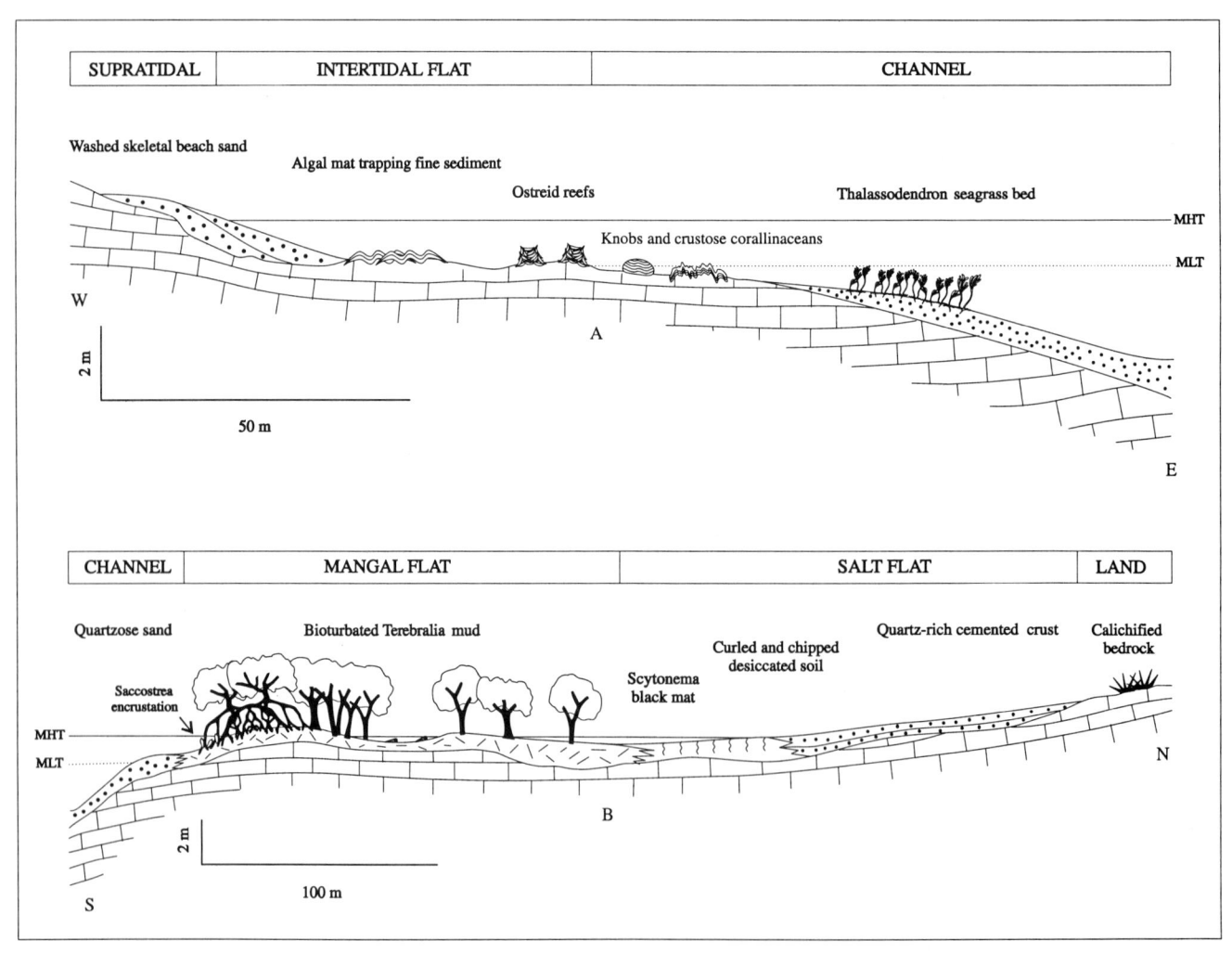

Fig. 10. Schematic ecological profiles of Lac Anole near the village of Kudai (A) and of Lac Badana near the village of Yamani (B).

The distinctive characters of mangal and its zonations are well known (Tomlinson, 1986), even if poor information is available for east Africa (Walter and Steiner, 1936). Most mangal in Somalia is along its southwestern coast, even if isolated stands of *Avicennia marina* are found behind sand spits along the northern coast (UNEP, 1987); mangrove habitats are also reported along the Gulf of Aden (Sheppard et al., 1992).

Biota in the channels is mainly controlled by the migration of quartzose sand along the channels and the increase of biogenic sediment towards the sea. In the upper reaches of the channels, where quartzose sediment predominates, foraminifers are rare and consist of small *Ammonia* and *Elphidium*. The foraminiferal assemblage gradually increases seawards, including miliolids, textulariids, *Amphistegina*, *Heterostegina*, *Ammonia* and the epiphytes *Rosalina* and *Planorbulina*. Coquina facies are in places found in the deepest channel areas, where coarse washed sand accumulates, consisting of shell hash of trochids, arcids, *Solen* and bullids and abundant bryozoan remains, mainly *Fenestella*. In high subtidal zones skeletal sediment prevails, with an abundant infauna of suspension- and detritus-feeder bivalves (*Codakia*, *Anodontia*, Pinnids, *Pitar*, Tellinids, *Modiolus*) and gastropods (Strombids, *Nassarius*, Conids, Cypreids, *Lambis*); in places this sediment is stabilized by small seagrass beds.

The channel levees show medium- to fine-grained sediment, highly burrowed by small crabs and with asymmetric ripples orientated according to the tidal flow (Fig. 11E). Linear tidal bars run parallel to the channel on its flat banks, forming tidal protected areas where small washover fans and semi-permanent pools originate. Towards the outlets, in areas of scarce sand accumulation, abraded bedrock forms the channel banks (Fig. 11B). This hard bottom is colonized (Fig. 10A) by encrusting red algae, thin algal mat trapping fine sediment, and small scattered coral colonies; the cliffs bordering the abraded flat are colonized by *Saccostrea cucullata*, barnacles and boring bivalves (*Lithophaga* and *Gastrochaena*). Where the channels flow into the Bajuni sound, the sediment consists of a mix of carbonate and quartzose sand, forming wave-generated bars alternating with areas colonized by seagrass. Seawards, typical swash bars and linear tidal shoals, often associated with sets of symmetrical megaripples, give rise to small tidal deltas (Fig. 7C).

The intertidal abraded flats facing the channels are widely colonized by red mangroves (*Avicennia* and *Rhizophora*) and inland are bordered by old wave notches topped by a wind-scoured surface where Acacia and Baobab grow. On mangrove roots emerging at low tide grow *Saccostrea cucullata*, barnacles, sedentary polychaetes, boring organisms (clionid sponges) and *Littorina* cf. *scabra* which lives on trunks and leaves, even when completely emerged (Fig. 11C). *Saccostrea* flourishes in the low and middle intertidal zone facing the channels and suddenly decreases towards inner areas. In sheltered muddy bottoms of mangrove thickets, where semi-permanent pools persist during low tides, the infauna is dominated by the fiddler crab *Uca*, while the gastropod *Nerita* occupies the peripheral part of the mangal flat. The inter-mangrove pools are highly populated by potamiid *Terebralia palustris* and *Cerithidae decollata* and by an infauna of polychaetes and arthropods (Fig. 10B). Scattered small build-ups, up to 1 m high, of successive generations of *Saccostrea cucullata* are also found where mangrove trees thin out, on a wide intertidal flat separating the channel from the land. This flat consists of Pleistocene bedrock and is locally covered by veneers of sandy sediment. The *Saccostrea* build-ups are often arranged in strings parallel to the levee on the channel bank (Fig. 10A); the shells adapt themselves to the substratum without a preferential orientation and are locally associated with *Nerita*, barnacles and polychaete worms.

Salt flats occur along the channels. In a few places, such as in Lac Badana channel near the village of Yamani, wide salt flats spread (Fig. 10B). These are affected by high evaporative processes which cause the development of a typical algal flat-sabkha plain. These gradually disappear landwards to be replaced by a sandy plain with small scattered dunes. In the higher sabkha areas, climatic conditions and water table fluctuations induced by tides are responsible for processes of alteration of the carbonate sediment and of adhesion of eolian quartz grains to the moist sabkha surface. In the lower areas, prism cracked surfaces and curled and chipped, desiccated algal mats are found, typical of the evaporative pumping of groundwater. During the highest tides the water flows from the channel into the lower part of the salt flat, forming semi-permanent pools, where algal marshes develop. They are characterized by black *Shitonema* algal mats with typical "pincushion" surfaces.

POPULATION AND NATURAL RESOURCES

The total population of Somalia is 9,077,000 of which 64% is rural; the annual demographic growth rate is approximately 3.1% and the average population density is about 14 inhabitants/km^2 (FAO, 1995). The coastal population is about 30% of the total and there is a strong tendency of migration to the major cities (Fig. 12). Somalia is mostly an agricultural and pastoral land with a cultivable area estimated at about 8 million ha in 1985, or 13% of the total area, and only about 980,000 ha cultivated with annual crops, i.e. 12% of the cultivable area. In 1993 about 18,000 ha consisted of permanent crops (FAO, 1995). Agriculture is the most important sector, while livestock accounts for about 40% of the gross domestic product and about 65% of the export earnings (CIA, 1999). About 60% of the indigenous population are nomads and semi-nomads and depend upon livestock for their livelihood. Crop production is only 10% of the gross domestic product and employs about 20% of the work force. Another export product is bananas, while

Fig. 11. The main shallow water marine and coastal habitats: (A) Cuvumbi Island reef flat, showing *Acropora* colonies during lowest low tide; (B) wide abraded flat bordering the Lac Badana outlet, emerged at low tide. The rugged surface is covered by a thin sandy sheet and scattered seagrass; (C) mangal flat at low tide near Yamani (Lac Badana). The roots are densely encrusted by ostreids (*Saccostrea cuccullata*); (D) Fuma Island sandy tail at low tide showing a densely bioturbated muddy bottom; (E) ebb-oriented ripples on the beach ridge at Lac Anole outlet.

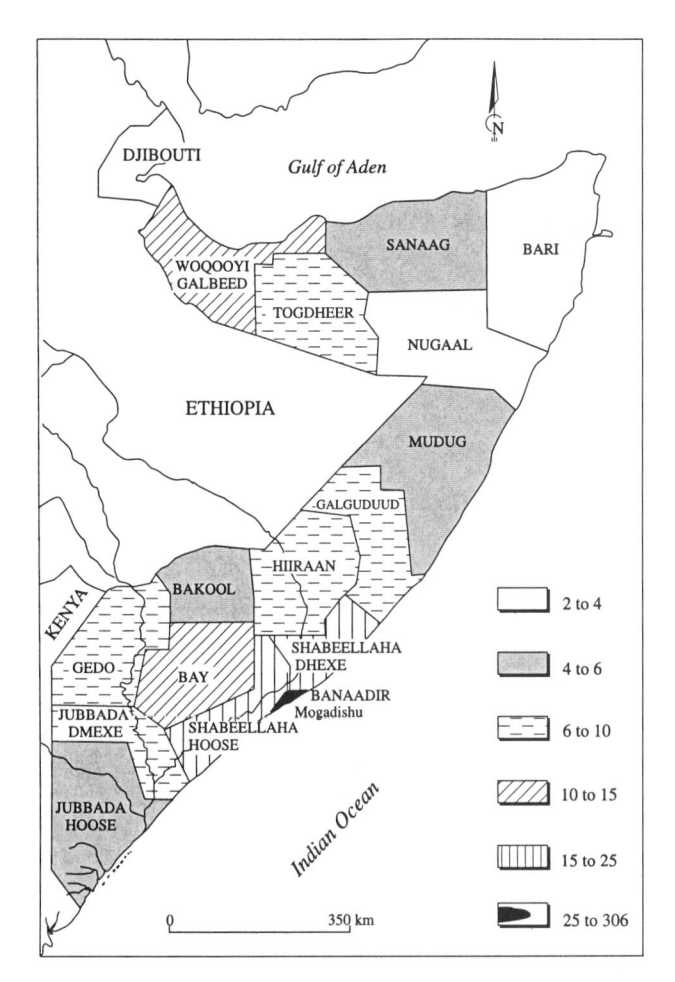

Fig. 12. Region boundaries and relative Somali population density indicated as inhabitants/km². Boundary representation is not necessarily authoritative.

various companies. The results are synthetized by Barnes (1976). Recently, the area of the Red Sea–Gulf of Aden was chosen by the World Bank (O'Connor, 1992) as a prototype of regional basin promotion for reasons both of geology and economic development. Bott et al. (1992) indicated good potential oil and gas deposits in northern Somalia, while, for the southern part of the coast, various foreign oil exploration plans were cancelled due to civil war after 1991.

Water resources in Somalia are mostly surface water. The total internally produced renewable water resources are estimated at 6 km³/year and the incoming surface water resources at 9.74 km³/year. In 1987, total water withdrawal was estimated at 0.81 km³. Agricultural water withdrawal is about 0.79 km³ or 97% of total withdrawal (FAO, 1989, 1995).

Along the Gulf of Aden, the mountainous zone is subject to torrential flows. The land slopes down towards the south and the south-flowing watercourses peter out in the desert sands. The rest of the country consists of a plateau, crossed by the two main rivers of Somalia, Uebi Shebeli and Uebi Juba. Over 90% of their discharge originates from the Ethiopian highlands and there is large variation from year to year. The discharge decreases rapidly through seepage, evaporation and overbank spillage due to low channel capacity and water abstraction. Often the rivers cease to flow in the lower reaches early in the year. Contribution to river flow inside Somalia occurs only during heavy rainfall, whereas that of other drainage basins to surface water is generally irrelevant. This normally consists of occasional runoff in seasonal watercourses. Groundwater potential is limited because of the scarce recharge potential.

Modern hydrogeological research in Somalia was first carried out by Wilson (1958), and by Hunt (1951) and Mac Fayden (1952) on British Somaliland. Faillace (1964) reported on the Uebi Shebeli valley, and Popov and Kidiwai (1972) on the Mudugh and Galgaduud regions. More recent papers deal with central Somalia (Pozzi et al., 1983), Uebi Shebeli valley (Sommavilla et al., 1993), the groundwater near Mogadishu (Dal Prà and Salad, 1986; Dal Prà et al., 1986). Sommavilla (1993) describes an ancient hydrological system of buried paleochannels, still visible in satellite images, when the Shebeli paleoriver probably flowed into the Indian Ocean just north of Mogadishu.

Irrigation and drainage development is very poor. In the main irrigated areas in the Juba and Shebeli valleys, there is no organized system of water allocation and management. There are no dams on the Shebeli river within Somalia, but off-stream storage exists in Jowhar (200 million m³). A second off-stream storage reservoir (130–200 million m³) is proposed for Duduble, upstream of Jowhar. Another proposed dam, primarily for hydropower, is in Bardhere on the Juba river. This should also provide maximum water control and storage for irrigation projects in the Juba valley. Total irrigation potential is estimated at 240,000 ha. In 1984, the total irrigated area was about 200,000 ha, only 50,000 ha of which had reasonably controlled irrigation (FAO, 1995).

sugar, sorghum and corn are produced for the domestic market.

Mineral resources are poorly exploited, even if they are believed to be potentially significant. Such deposits, mainly located in the north, include quartz and piezoquartz, uranium, gypsum anhydride and iron ore; meerschaum sepiolite is mined in the central region. In the south, several minerals are present in a narrow coastal belt, about 70 km long, at the mouth of Juba river. The occurrence of potentially valuable minerals in these placers was first reported by Artini (1915). More recently, unpublished data reported by Frizzo (1987) focused on the mineral characterization and economic evaluation of the placers. Even the Merka red dune complex, extending along 900 km of Somali coastline of the Indian Ocean, notable as a water reservoir in a region with generally arid climate, was recently evaluated (Angelucci et al., 1994, 1995) in the area between Mogadishu and Merka. The mineral composition (80 wt% of quartz combined with 15% wt of feldspar) and the grain size (between 420 and 88 micron) make this a suitable source of raw material for a potential national glass industry. Exploration for oil and natural gas was initiated by AGIP Mineraria in the 1950s and subsequently intensified by

The main official body in charge of water resources is the Ministry of Mineral and Water Resources (MMWR), and its Water Centre (NWC). The Water Development Agency (WDA) is responsible for operations exploiting groundwater resources for domestic water supply. A study carried out by the World Bank (1987) outlined a proposed strategy for the development of irrigation, drainage and water management systems.

EFFECTS FROM HUMAN ACTIVITIES AND PROTECTIVE MEASURES

Along the coast, modifications have occurred over countless centuries. Somalia has long been involved in trade in the western coast of the Indian Ocean. Chinese merchants have stopped here since the 10th century and Greek merchant ships and medieval Arab dhows also plied the Somali coast that formed the western part of the "bilad as Sudan" (the "Land of the Blacks"). More specifically, medieval Arabs referred to the Somalis and neighbouring peoples as the Berbers. By the 8th century, the Somalis essentially developed their present way of life, which is based on pastoral nomadism and Islamic faith. The large-scale alteration of the coastal belt, mainly near major human settlements, like Mogadishu and Kisimayo, began in the nineteenth century, during the colonial period (1891–1960), when the need to accommodate larger vessels induced harbour development. This led both to the destruction of wide coral fringes in nearshore shallow water areas and to direct alteration of the coastal belt on a massive scale, accelerated in the last few decades. In fact several quarries (Figs. 3A, 5C) have been cut into coral reef terraces bordering the coast near the major cities to provide building materials. The short thickness (4–6 m) of this uplifted reef terrace, which is buried by eolian sand a few tens of meters from the coast, has forced the inhabitants to exploit the narrow coastal belt extensively, digging quarries as far as the sea cliffs. As a consequence, the coastal environment of these areas has been deeply transformed. Inside the quarries, pools and marshes have developed and have given rise to the growth of oligotypic communities, where algal mats flourish.

An increase in the alteration of the coastal marine ecosystem occurred in the 1980s when a refinery was built in the outskirts of Mogadishu. The building of a slaughterhouse, which scattered its waste into the sea, has attracted large numbers of predatory fishes, mainly sharks, to the shallow waters.

Fisheries are another general problem for ecosystem protection. The Somalis are traditionally devoted to pastoral nomadism, thus only recently have fisheries been developed, particularly since the 1974–75 drought, when nomads were resettled in fishing co-operatives. Fish is caught for the domestic market, with a surplus for export; lobster fishing is rapidly increasing due to foreign market demand. In the late 1980s, about 17,000 metric tons of fish were caught annually. Tuna, now being processed, is a potential export product. Since 1993 Somalia has been part of the Common Market for Eastern and Southern Africa (COMESA), founded in 1993 in Kampala, Uganda. The COMESA treaty specifies objectives and activities regarding fisheries. This will have an impact on the fisheries and aquaculture in terms of investment, production, trade and fish consumption of the member states. Somali coasts that are subject to rapid and largely uncontrolled occupation by human activities are often the most at risk from environmental change.

Paradoxically, this coast, which is not subject to great industrial goals, is undergoing rapid degradation of the natural environment because of the absence of protective measures. At present, the coast is not adequately monitored. This has led to a situation in which coastal habitats such as mangroves and coral reefs are degraded, living marine resources are overexploited, and pollution levels are increased, while inadequate data is collected to characterize impact on natural resources and biodiversity.

Somalia is one of the world's poorest and least developed countries, having few resources, with much of its economy devastated by the civil war. Once this situation ends, this country will become a high priority. There is potential for the development of marine parks in the Bajuni archipelago and adjacent channelized coastal areas. The Lac Badana National Park (0°25' + 1°30'S; 42°30' + 43°30'E) could be extended to include part or all of the archipelago (UNEP 1987). This area, from Kisimayo to Ras Chiambone, is important for coral reefs, marine turtles and mangrove resources although little is known about it. There are currently no Marine Protected Areas and no legislation concerning their establishment and management although the WCMC (World Conservation Monitoring Centre) Protected Areas Database lists Busc Busc Game Reserve as an MPA. The WCMC in 1992 also listed as proposed Protected Areas the coastal sites of Zeila because of their important sea bird colonies on offshore islets, Jowhar-Warshek and Awdhegle-Gandershe.

REFERENCES

Aleem, A. (1983) Distribution and ecology of seagrass communities in the western Indian Ocean. In *Marine Science of the North-west Indian Ocean and Adjacent Waters. Part A. Oceanographic Rersearch Papers*, ed. M. Angel. Pergamon Press, Egypt, 31, Nos 6-8A, pp. 919–933.

Alexander, L. (1998) Somali Current Large Marine Ecosystems and related issues. In *Large Marine Ecosystems of the Indian Ocean: Assessment, Sustainability, and Management*, eds. K. Sherman, M. Ntiba and E. Okemwa. Blackwell Science, Oxford, pp. 327–333.

Angelucci, A., Barbieri, F., Cabdulqaadir, M.M., Faaduma, C.C., Franco, F., Carush, M.C. and Piccoli, G. (1983) The Jurassic stratigraphic series in Gedo and Bay regions (Southwestern Somalia). *Mem. Soc. Geol. Ital.*, **36**, 73–94.

Angelucci, A., Carbone, F. and Matteucci, R. (1982) La scogliera corallina di Ilisi nelle Isole dei Bajuni (Somalia meridionale). *Boll. Soc. Paleont. Ital.* **21** (2/3), 201–209.

Angelucci, A., De Gennaro, M., De Magistris, M. and Di Girolamo, P. (1994) Economic aspects of Red Sands from the Southern Coast of Somalia. *International Geology Review* **36**, 884–889.

Angelucci, A., De Gennaro, M., De Magistris, M. and Di Girolamo, P. (1995) Mineralogical, geochemical and sedimentological observations on recent and Quaternary sands of the littoral region between Mogadishu and Merka (Southern Somalia) and their economic implication. *Geologica Romana* **31**, 249–263.

Artini, E. (1915) Intorno alla composizione mineralogica di alcune sabbie ed arenarie. *Atti Società Italiana Scienze Naturali* **54**: 137–166.

Artini, E. (1926) Sulla composizione mineralogica di quattro campioni di sabbia provenienti dalle dune dei dintorni di Chisimaio nell'Oltre Giuba. *Agricoltura Coloniale* **40**, 101–102.

Ase, L. (1978) Preliminary report on studies of shore displacement at the southern coast of Kenya. *Geografiska Annaler* **60A** (3–4), 209–221.

Ase, L. (1981) Studies of shores and shore displacement on the southern coast of Kenya—especially in Kilifi district. *Geografiska Annaler* **63A** (3–4), 303–310.

Barnes, S. (1976) Geology and oil prospects of Somalia, East Africa. *Bulletin of the American Association of Petroleum Geologists* **60**, 389–413.

Battistini, R. (1976) Application des methotes Th^{230}-U^{234} à la datation des depots marins anciens de Madagascar et des iles voisines. Ass. Sénégal Etudes Quatern. *Afr. Bull. Liason Sénégal* **49**, 79–95.

Boccaletti, M. Dainelli, P., Angelucci, A., Arush, M.A., Cabdulqaadir, M.M., Nafissi, P., Piccoli, G. and Robba, E. (1988) Folding of the Mesozoic cover in SW Somalia: a compressional episode related to the early stages of the Indian Ocean evolution. *Journal of Petroleum Geology* **11** (2), 157–168.

Bosellini, A. (1989) The continental margins of Somalia: their structural evolution and sequence stratigraphy. *Mem. Sc. Geol. Univ. Padova* **41**, 373–458.

Bott, W., Smith, B., Oakes, G., Sikander, A. and Ibraham, A. (1992) The tectonic framework and regional hydrocarbon prospectivity of the Gulf of Aden. *Journal of Petroleum Geology* **15** (2): 211–243.

Braithwaite, C.J.R. (1984) Depositional history of late Pleistocene limestones of the Kenya coast. *Journal of the Geological Society of London* **141**, 685–699.

Bruce, J. (1973) Large scale variation of the Somali Current during the Southwest Monsoon. *Deep-Sea Research* **20**, 837–846.

Bruce, J. (1979) Eddies of the Somali coast during the Southwest Monsoon. *Journal of Geophysical Research* **84**, 7742–7748.

Bunce, E., Langseth, M., Chase, R. and Ewing, M. (1967) Structure of the Western Somali Basin. *Journal of Geophysical Research* **72**, 2547–2555.

Carbone, F. (1987) Modern and ancient coral reefs and coastal sediments along the southern Somali coast. In: F. Carbone (Ed.), Guidebook Excursion C, Geosom 87 International Meeting: Geology of Somalia and Surrounding Regions. Cotecno, Roma, Mogadishu, pp. 71.

Carbone, F. and Matteucci, R. (1990) Outline of Somali Quaternary coral reefs. *Reef Encounter* **7**, 12–14.

Carbone, F., Matteucci, R. and Arush, M.A. (1984) Schema geologico della costa del Benadir tra Gesira ed El Adde (Somalia centro-meridionale). *Boll. Soc. Geol. Ital.* **103**, 439–446.

Carbone, F., Matteucci, R., Rosen, B. and Russo, A. (1994) Recent coral facies of the Indian Ocean coast of Somalia, with an interim check list of corals. *Facies* **30**, 1–14.

CIA (1999) Somalia. In *The World Factbook*. US Government Printing Office.

Clark, J. (1954) *Prehistoric Cultures of the Horn of Africa*. Occasional publication of the Cambridge University, Museum of Archaeology and Ethnology, No. 255. Hoctagon Books, Farrar Division, Cambridge.

Colonna, M., Casanova, J., Dullo, W. and Camoin, G.F. (1996) Sea-level changes and $\delta^{18}O$ for the past 34,000 yr from Mayotte Reef, Indian Ocean. *Quaternary Research* **46**, 335–339.

Crame, J.A. (1980) Succession and diversity in the Pleistocene coral reefs of the Kenya coast. *Palaeontology* **23** (1), 1–37.

Crame, J.A. (1981) Ecological stratification in the Pleistocene coral reefs of the Kenya coast. *Palaeontology* **24** (3), 609–646.

Crossland, C. (1902) The coral reefs of Zanzibar. *Proc. Cambridge Philosoph. Soc. Mat. Phys. Sci.* **11**, 493–503.

Crossland, C. (1904) The coral reefs of the Pemba Island and of the east African mainland. *Proc. Cambridge Philosoph. Soc. Mat. Phys. Sci.* **12**, 36–43.

Dainelli, G. (1943) Geologia dell'Africa orientale, le successioni terziarie e i fenomeni del Quaternario. Reale Accademia Italiana, 3, 746 pp.

Dal Pra', A., De Florentiis, N., Hussen, S., Mumin, M.G., Omar S., Osman M., Sacchetto G.A. and Abukar M.A. (1986) Ricerche idrogeologiche sulla falda costiera della Somalia centrale tra Merka e Uarscek (Mogadiscio). *Mem. Sc. Geol. Univ. Padova* **38**, 91–110.

Dal Pra', A. and Salad, M.H. (1986) Ricerche sperimentali sui rapporti tra acque dolci di falda e acque salate di intrusione marina, lungo la costa della Somalia centrale nella zona di Jesira (Mogadiscio). *Mem. Sc. Geol. Univ. Padova* **38**, 169–186.

Dana, J. (1875) *Coral and Coral Islands*. Sompson Low, Marrstone, Low & Searle London, 348 pp.

Darwin, C. (1842) *The Structure and Distribution of Coral Reefs*. Smith Elder & Co., London.

De Angelis, A. (1938) Le rocce sedimentarie e le sabbie della Somalia italiana, Geologia della Somalia. Reale Società Geografica Italiana, pp. 69–121.

Eisma, D. (1998) *Intertidal Deposits: River Mouths, Tidal Flats, and Coastal Lagoons*. CRC Marine Science Series. CRC Press LLC, Boca Raton FL, 525 pp.

Faillace, C. (1964) Surface and underground water resources of Shebelle Valley, Minist. Publ. Works, Mogadishu, 98 pp.

FAO (1989) A brief description of major drainage basins affecting Somalia. National Water Centre, Mogadishu. Field document No. 14. FAO/SOM/85/008, prepared by D. Kammer, Rome.

FAO (1995) Irrigation in Africa in figures. Water Report No. 7, Rome.

Francis, T., Davies, D. and Hill, M. (1966) Crustal structure between Kenya and Seychelles. *Philosophical Transactions of the Royal Society of London, s. A* **259**, 240–261.

Frizzo, P. (1987) Mineral sand deposits at mouth of Juba River (Kisimayo, Southern Somalia). In: F. Carbone (Ed.), Guidebook Excursion C, Geosom 87, International Meeting: Geology of Somalia and surrounding regions, Mogadishu, pp. 59–64.

Hopley, D. (1994) Continental shelf reef systems. In *Coastal Evolution: Late Quaternary Shoreline Morphodynamics*, eds. R. Carter and C. Woodroffe. Cambridge University Press, Cambridge, pp. 303–340.

Hori, N. (1970) Raised coral reefs along the southeastern coast of Kenya. *Geogr. Rep. Tokyo Metrop. Univ.* **5**, 25–47.

Hoyt, J. (1967) Barrier island formation. *Geological Society of America Bulletin* **78**, 1125–1136.

Hughes, R. and Hughes, J. (1992) A directory of African wetlands, IUCN/UNEP/WCMC, 820 pp.

Hunt, J. (1951) A general survey of the Somaliland Protectorate (1944–1950), London.

James, N. (1983) Reef Environment. In *Carbonate Depositional Environments*, eds. P. Scholle, D. Bebout and C. Moore. Bull. Amer. Ass. Petrol. Geol., Tulsa, pp. 345–462.

Jensen, T. (1991) Modeling the seasonal undercurrents in the Somali Current System. *Journal of Geophysical Research* **96** (C12): 22151–22167.

Kent, P. (1974) Continental margin of East Africa—a region of vertical movements. In *The Geology of Continental Margins*, eds. C. Burk and C. Drake. Springer Verlag, New York, pp. 313–320.

Kent, P. (1982) The Somali Ocean Basin and the continental margin of East Africa. In *The Ocean Basins and Margins*, eds. A. Nairn and

F. Stehli. Plenum Press, New York, 185–204.

Kent, P., Hunt, J. and Johnstone, M. (1971) The geology and geophysics of coastal Tanzania. Geophysical Paper No. 6, Natural Environment Research Council, Institute of Geological Sciences, London.

Land, L. (1970) Carbonate mud production by epibiont growth on *Thalassia testudinum*. *Journal of Sedimentary Petrology* **40** (4), 1361–1363.

Macfadyen, W. (1952) Water supply and geology of parts of British Somaliland, Crown Agents, London, HMSO.

Mauge, L. (1967) Contribution preliminaire à l'inventaire ichtyologique de la région de Tuléar. *Récueil des Travaux de la Station Marine d'Endoume*, suppl. **7**, 101–132.

McCreary, J., Kundu, P. and Molinari, R. (1993) A numerical investigation of dynamics, thermodynamics and mixed-layer processes in the Indian Ocean. *Progress in Oceanography* **31** (3), 181–244.

Nyagah, K. (1995) Stratigraphy, depositional history and environments of deposition of Cretaceous through Tertiary strata in the Lamu Basin, Southeast Kenya and implications for reservoirs for hydrocarbon exploration. *Sedimentary Geology* **96**, 43–71.

O'Connor, T. (1992) The Red Sea–Gulf of Aden: hydrocarbon evaluation of multinational sedimentary basins. *Journal of Petroleum Geology* **15** (2), 121–126.

Okemwa, E. (1998) Application of the Large Marine Ecosystems concept to the Somali Current. In *Large Marine Ecosystems of the Indian Ocean: Assessment, Sustainability, and Management*, eds. K. Sherman, M. Ntiba and E. Okemwa. Blackwell Science, Oxford, pp. 73–99.

Patriquin, D. (1972) Carbonate mud production by epibionts on *Thalassia*: an estimate based on leaf growth rate data. *Journal of Sedimentary Petrology* **42** (3), 687–689.

Piccoli, G. Boccaletti, M., Angelucci, A., Robba, E., Arush, M.A. and Cabdulqadir, M.M. (1986) Geological history of central and southern Somalia since the Triassic. *Mem. Soc. Geol. Ital.* **31**, 415–425.

Pichon, M. (1964) Aperçu préliminaire des peuplements sur sables et sable vaseaux libres ou couvert par les herbiers de phanerogames de la région de Nossi-Be. *Cahiers Orstom* **2** (4), 5–15.

Popov, A. and Kidiwai, A. (1972) Groundwater in Somali Democratic Republic. Part I and Part II. Mineral and Groundwater Survey Project No. 141, UNDP, New York.

Pozzi, R., Benvenuti, G., Mohamed, C.X., and Shuurije, C.I. (1983) Groundwater resources in central Somalia. *Mem. Sc. Geol. Univ. Padova* **35**, 397–409.

Purser, B. and Evans, G. (1973) Regional sedimentation along the Trucial coast, SE Persian Gulf. In *The Persian Gulf*, ed. B. Purser. Springer Verlag, Heidelberg, pp. 211–231.

Rao, T. and Griffiths, R. (1998) *Understanding the Indian Ocean: Prospectives and Oceanography*. UNESCO Publishing, IOC Ocean Forum Series, Paris, 187 pp.

Schott, F. (1983) Monsoon response of the Somali Current and associated upwelling. *Progress in Oceanography* **12**, 357–382.

Schwartz, M. (1971) The multiple causality of barrier islands. *Journal of Geology* **79**, 91–94.

Scoffin, T. (1970) The trapping and binding of subtidal carbonate sediments by marine vegetation in Bimini lagoon, Bahamas. *Journal of Sedimentary Petrology* **40** (1), 249–273.

Sheppard, C., Price, A. and Roberts, C. (1992) *Marine Ecology of the Arabian Region*. Academic Press, London, 359 pp.

Sheppard, C. and Wells, S. (1988) *Coral Reefs of the World, II—Indian Ocean Region*. IUCN/UNEP, Nairobi, 389 pp.

Sommavilla, E., Sacdiya, C., Salad, M.H. and Farah, I. (1993) Neotectonic and geomorphological events in central Somalia. *Ist. Agron. Oltremare Firenze, Relaz. e Monogr.* **113**, 389–396.

Stefanini, G. (1930) I terrazzi fluviali dell'Africa italiana, Deuxieme Rapport Comm. Terrasses Pliocene Pleistocene. 17 pp., Firenze.

Stefanini, G. (1933) Paleontologia della Somalia. Fossili pliocenici e pleistocenici. Notizie sulle formazioni plioceniche e pleistoceniche. *Paleont. Ital.* **32**, suppl. 1, 55–66.

Taylor, J. and Lewis, M. (1968) The flora, fauna and sediments of the marine grass beds of Mahé, Seichelles. *Journal of Natural History* **4**, 199–220.

Tomczak, M. and Godfrey, J. (1994) *Regional Oceanography: An introduction*. Pergamon Press, Oxford. 422 pp.

Tomlinson, P. (1986) *The Botany of Mangroves*. Cambridge Tropical Biology Series. Cambridge University Press, Cambridge, 419 pp.

Toyah, H., Kadomura, H., Tamura, T. and Hori, N. (1973) Geomorphological studies in southeastern Kenya. *Geogr. Rep. Tokyo Metrop. Univ.* **8**:,51–137.

UNEP (1987) Coastal and marine environmental problems of Somalia. UNEP Regional Seas Reports and Studies No. 84, ESCWA/ FAO/ UNESCO/ IMO/ IAEA/ IUCN/ UNEP.

Wacongne, S. and Pacanowski, R. (1996) Seasonal heath transport in a primitive equations model of the tropical Indian Ocean. *Journal of Geophysical Research* **26** (12): 2666–2699.

Walter, H. and Steiner, M. (1936) Oekologie der Ost-Afrikanschen Mangroven. *Zeitschrift fuer Botanik* **30**, 65–93.

Wilson, G. (1958) Groundwater geology of Somalia, UNDP Report, 88 pp., New York.

World Bank (1987) Agricultural Sector Survey: Main report and strategy. Report No. 6131-SO. Washington DC.

THE AUTHORS

Federico Carbone

Centro di Studio per il Quaternario e l'Evoluzione Ambientale,
C.N.R., Dipartimento di Scienze della Terra,
Università degli Studi "La Sapienza"
P. Aldo Moro, 5
00185 Roma, Italy

Giovanni Accordi

Centro di Studio per il Quaternario e l'Evoluzione Ambientale,
C.N.R., Dipartimento di Scienze della Terra,
Università degli Studi "La Sapienza"
P. Aldo Moro, 5
00185 Roma, Italy

Chapter 58

TANZANIA

Martin Guard, Aviti J. Mmochi and Chris Horrill

The coastal zone of Tanzania contains a wealth of coastal and marine habitats that include coastal and mangrove forest, extensive coral reefs, seagrass, algae and sponge beds, large intertidal areas, coastal plains and sandy beaches. Sustaining a diverse array of plant and animal life, these habitats provide important sources of protein, livelihood, building materials and tourism potential for coastal and inland rural communities. Yet with increasing pressure from a rapidly rising population, continued over-exploitation, habitat degradation and urban pollution, these habitats are becoming degraded and the coastal human communities now face a serious threat to their food supplies. There is also a threat to the ability to alleviate poverty or to improve health or community development. Exacerbated by a lack of awareness, poor licensing systems, poor data collection and ineffective regulation, it is clear that a huge effort is required to successfully address and counter these problems. Institutions such as the University of Dar Es Salaam, the Institute of Marine Sciences, Zanzibar, NGOs and several overseas donor agencies are now working together in an attempt to develop and implement Integrated Coastal Zone Management (ICZM) in Tanzania. Recent initiatives to help build local and institutional capacity, develop local community marine management and prevent illegal and destructive extraction practices (such as dynamite fishing) have met with varying success but are nonetheless positive steps towards this goal.

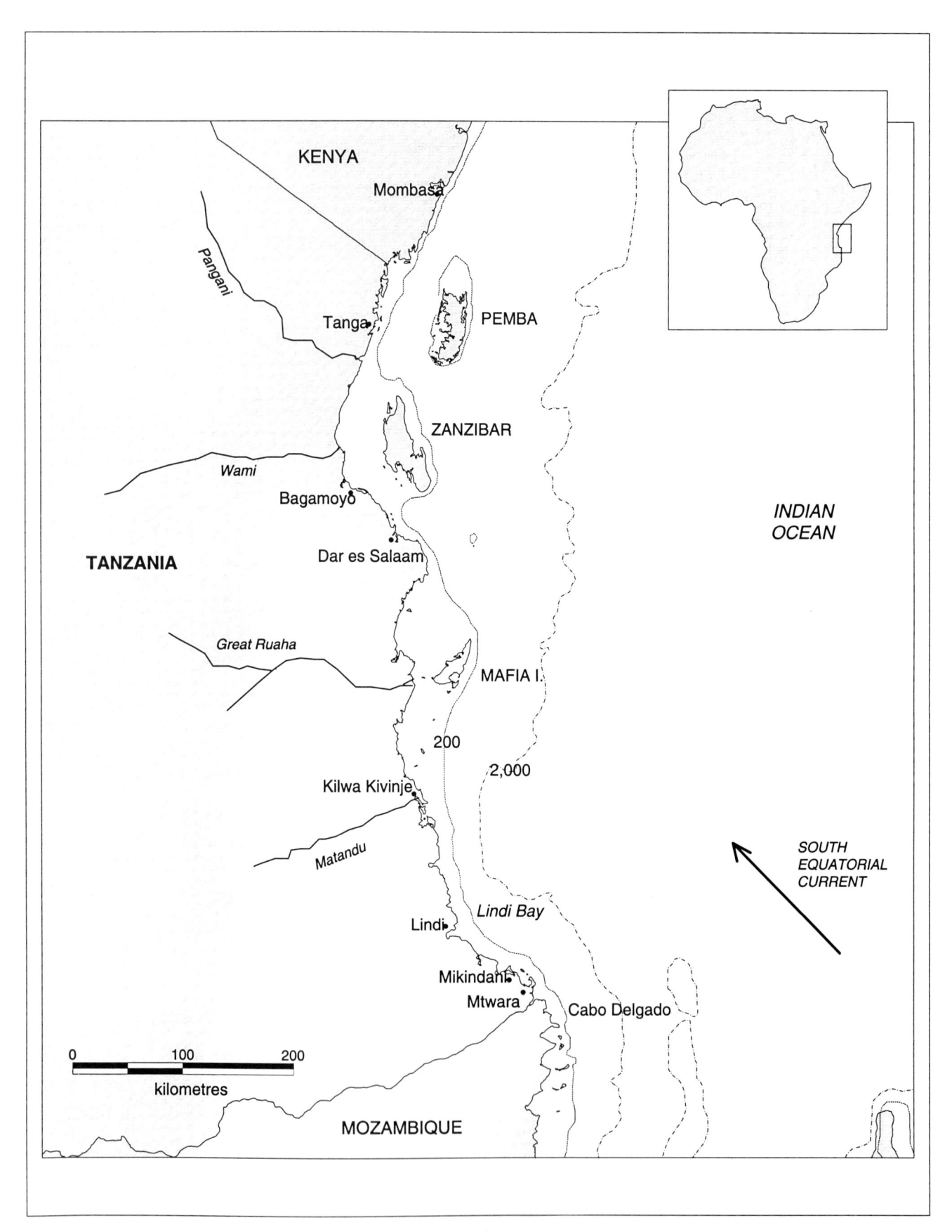

Fig. 1. Map of Tanzania.

THE DEFINED REGION

Tanzania is a large state (942,654 km²) with a coastline of 800 km along the Western Indian Ocean. Its coastal zone is approximately 30,000 km², and its continental shelf, which extends between 4 and 35 km offshore, adds a further 17,500 km² (Linden and Lundin, 1995). There are three large islands, the oceanic island of Pemba divided from the mainland by a 400 m deep channel, and two continental islands Unguja and Mafia. The shelf contains numerous coralline islets, patch reefs and fringing coral reefs broken by riverine basins (e.g. Rufiji Delta), deep-water channels, large embayments and small inlets (Fig. 1).

The area is part of the larger East African Marine Ecosystem which extends northwards through Kenya, even to Somalia with certain current patterns, and south to the border with Mozambique where its boundaries are demarcated by the Southern Equatorial Current.

By the year 2000, about a quarter of the total estimated population of 35 million will live at the coast. This is expected to place further pressure on an already threatened coastal zone and, if not controlled, may in the long term lead to a serious decline in food security.

SEASONALITY, CURRENTS, NATURAL ENVIRONMENTAL VARIABLES

The area is characterised by a tropical hot and humid climate dictated by two distinct seasonal monsoons winds. Air temperatures in the region range between 22 and 30°C and are hottest during the Northeast Monsoon (Swahili-'kaskazi') which blows from November to March. This is considered the more benign of the two seasons and is typified by lower wind speeds (Beaufort 1–4), and calmer seas. In contrast, the Southeast Monsoon (Swahili-'kusi') blows from May to October and is typified by stronger winds (average Beaufort force 2–5) reaching a maximum Beaufort force of 7–8, lower air temperatures and rougher seas. During inter-monsoon periods, wind direction can be variable but overall conditions tend to be calm.

The main rainy seasons coincide with the end of the Southeast Monsoon (March–May), with a shorter period of rainfall known as the 'short rains' during November and December. Annual rainfall in Tanzania averages 800 to 1000 mm per year with the highest and lowest levels recorded in Pemba Island (1500 mm/yr) and the Mtwara district (<500 mm/yr).

The monsoons influence the northward flowing East African Coastal Current (EACC). This originates as the north bifurcation of the Southern Equatorial Current at approximately 10°S, with the southern bifurcation becoming the south flowing Mozambique current. During the Northeast Monsoon the flow of the EACC is reduced to as little as 1 kt. Upon reaching the northern Kenyan border its path is redirected eastwards where it merges into the Equatorial Counter Current. During the Southeast Monsoon the stronger southerlies increase the flow of the EACC to 4–5 kt, which extends the EACC in a northerly direction.

The EACC is important for larval dispersal and downwelling along the East African coastline, but its main influence is offshore (McClanahan, 1988). Of greater influence on inshore bay areas, open fringing reefs and inlets are local semi-diurnal tidal currents. Tidal range is one of the highest of the Indian Ocean, averaging 3.3 m amplitude during spring tides and 1.1 m during neaps (Hamilton and Brakel, 1984). Tidal currents combined with inputs from many rivers but especially the Rufiji and Ruvuma, provide a major source of food and nutrients to adjacent inshore waters. Peak river outflow in the rainy season can markedly increase the amount of suspended sediments.

Water temperatures are highest between June to July and average between 25 and 29°C while maximum and minimum surface salinity occurs during November and May respectively (Newell, 1959). Thermocline depths range between 50 and 120 m depending on season, and oxygen concentrations are generally saturated throughout the upper surface waters throughout the year.

SHALLOW WATER MARINE AND COASTAL HABITATS OF TANZANIA

The three main coastal and marine habitats along the Tanzanian coast are coral reefs, mangrove forests and seagrass beds. Also present, to a lesser extent, are shallow lagoons, estuaries, mudflats, rocky shores, sandy beaches, dune systems and coastal forest. Containing diverse plant and animal communities, these habitats are important both in terms of national and global biodiversity and as a source of food and income for coastal and inland communities.

Coral Reefs

Tanzanian coral reefs are probably the most extensive and diverse in East Africa and cover an estimated 600 km of coastline (Table 1). The three main reef types described for Tanzania are outer fringing reef, inner fringing reef and patch reefs (Darwall, 1995b; Horrill and Ngoile, 1991; Guard et al., 1997; Muhando, 1998). Outer fringing reef is the most common and occurs along most of the coast and around outerlying islands. It is found close to shore (2–5 km) except around several islands where shallow water platforms extend up to 40 km offshore. Outer reef slopes may have 'spur and groove' structures (extending 90 m in Mtwara), or may have gentle gradients and steeper drop-offs in deeper water. Coral growth is generally extensive and diverse, often characterised by robust corals in response to wave action (Talbot, 1965). These are important wave breaks protecting the shoreline.

Inner fringing reef and patch reefs are generally found in bays or behind outer fringing reefs and are common at

Table 1
Outline of reef type and extent, reef condition and impacting activities of the main reef areas in Tanzania

	Tanga	Unguja Island	Pemba Island	Dar es Salaam	Mafia Island	Songo-Songo	Mtwara/Lindi
Reef type and extent	97 km outer fringing reef, 55 patch reefs	90 km^2 of coral reef is extensive fringing and patch reefs, numerous islands	1100 km of reef, 45% of total in Tanzania. Numerous islands and patch reefs	15 km outer fringing reef, 3 islands and numerous patch reefs	33 km outer fringing reef, numerous islands and patch reef	30 km outer fringing reef, 5 islands, 38 patch reefs	110 km outer fringing reef, scattered islands, 15 patch reefs
Reef condition	12% of reefs destroyed, 64% poor or moderate condition, 24% good condition	Coral cover 12–80%. Reefs in the south are more degraded. Chumbe Marine Park	15% coral cover on eastern reefs, 21–60% coral cover western reefs, 53–75% coral cover Misali Island	Coral cover 35-80%. The majority of reefs severely degraded	Outer reef sites relatively good condition. Shallow reefs and intertidal areas heavily fished and degraded	Shallow water reefs severely degraded. Outer reef relatively good condition.	Outer reef sites in pristine condition. Heavily fished sites severely degraded. Coral cover 30–70%
Impacting activities	Dynamite fishing, small mesh seine nets, over-fishing, coral bleaching, use of poisons	Dynamite fishing, 'Kigumi' fishing, tourism development, taking of under-sized crayfish, release of toxic compounds associated with seaweed farming, coral bleaching	Kigumi fishing technique, little dynamite fishing, unregulated tourism, coral bleaching	Urban pollution, over-exploitation and the use of dynamite, sedimentation	Over-exploitation of octopus, gillnets taking turtles, reef trampling, coral bleaching, sediment incursion from Rufiji delta	Dynamite fishing, over-exploitation by resident and migrant fishermen, reef trampling, collection of shells for curio trade, coral bleaching	Over fishing, use of dynamite, use of small mesh seine nets, over-collection of sea cucumbers, reef trampling

Tanga, Unguja, Pemba Mafia, Songo-Songo Archipelago and Mnazi Bay. Their size and coral community are highly variable depending on aspect, currents, turbidity, water depth and proximity to deep channels. Typically they display luxurious coral growth (Hamilton and Brakel, 1984; Talbot, 1965). Several coral reef studies of varying detail have been conducted along the mainland (e.g Tanga, Dar es Salaam, Mtwara) and outerlying islands (Unguja, Pemba, Mafia Island and the Songo-Songo Archipelago) since the 1960s (Anderson and Ngazi, 1995; Darwall et al., 1994, 1996; Franklin et al., 1999; Guard et al., 1997; Hanaphy and Muller, 1997; Horrill and Ngoile, 1991; Kamakuru, 1997). Results indicate a high diversity, but diversity and productivity have been shown to vary markedly depending on levels of local degradation and degree of fishing intensity.

For example, at Tanga, Horrill (1996) reported that 24% of 83 reef sites surveyed were in good condition, 64% remain in a good to moderate condition and that 12% were completely destroyed. The most degraded sites were characterised by low coral cover, with low densities of both commercially important fish and invertebrates due to the use of dynamite, beach seines, coral mining and damage from boat anchors. Degraded sites were generally located closer to shore and areas of high population density.

Similarly, in a study by Kamakuru (Kamakuru, 1997) at impacted reefs surrounding Mbudya, Bongoyo, and Pangavin Islands near Dar es Salaam, a direct correlation was indicated between hard coral cover and fish biomass. Although coral cover varied between 35 and 80% throughout the survey area, the more degraded reefs provided an average wet weight of fish of only 200 kg/ha, mostly of smaller less valuable species. The high sea urchin density observed (6680 kg/ha), further supported the conclusion of overfishing.

In the Songo-Songo Archipelago, marked differences in species richness were reported between reefs impacted by destructive fishing techniques and those where greater depth resulted in lower levels of degradation (Darwall et al., 1996). Reefs in good condition at Tanga, Misali Island, Mafia, the Songo-Songo Archipelago and Mtwara are mainly outer reef, deep water channel sites or offshore patch reefs (Horrill, 1996; Darwall et al., 1996; Horrill and Ngoile, 1991; Johnstone et al., 1999). Their distance from population centres and susceptibility to inclement weather conditions combined dramatically to reduce fishing.

Identified as one of the finest shallow-water complexes in East Africa, the southern part of Mafia Island was gazetted in April 1995, as Tanzania's first Multi User Marine Park. Covering an area of 400 km^2 the marine park includes four small islands and extensive coral reefs that have been the focus of many past and ongoing ecological, resource use, and reef restoration studies (Anderson and Ngazi, 1995; Choiseul and Darwall, 1996a; Darwall, 1995a; Dulvy et al., 1995; Horrill and Ngoile, 1991; Lindahl, 1999; Talbot, 1965).

The known inventory of coral reef species for Tanzania and the Western Indian Ocean is improving (Table 2). Over

Table 2

Diversity of flora and fauna for major study areas in southern Tanzania

Floral/Fauna Group	Mafia Island	Songo-Songo Archipelago	Mtwara/Lindi districts
Scleractinian coral	45 genera	49 genera	45 genera
Non-scleractinian	N/a	12 genera	11 genera
Algae	132	75	79
Seagrasses	12	9	9
Mangroves	8	2	8
Gastropods	89	93	N/a
Fish	396	270	402

55 genera of scleractinian corals are known (Darwall et al., 1996; Guard et al., 1997; Horrill and Ngoile, 1991; Johnstone et al., 1999), though reef fish inventories which number 244 species to just 400 species of fish identified depending on location (Darwall et al., 2000; Horrill et al., 2000) are probably well short of the true totals. Most invertebrate groups have still yet to be effectively studied, though gastropods, crustaceans, echinoderms and nudibranches are known to be diverse (Bathwondi and Mwaya, 1984; Brown, 1996; Choiseul, 1996; Kayombo, 1989; Newton et al., 1993; Richmond, 1998). Rare or less common species include the Bumphead parrotfish (*Bolbometopon muricatum*) and the Giant Grouper (*Epinephelus lanceolatus*), both susceptible to overfishing, and the coconut or 'robber' crab (*Birgus latro*) on Chumbe and Nyuni Islands. Three marine turtle species nest on Tanzanian beaches (Khatib, 1999; Cowper and Darwall, 1996; Frazier, 1976; Howell, 1993). Marine mammals have been little studied (Stensland et al., 1999), though several species of whales, dolphins and porpoises are reported along the coast (Ortland, 1997). Dugongs are now believed to be extinct in Tanzania (Linden and Lundin, 1995).

Mangrove and Other Habitats

Mangrove forests are prominent along river outlets and estuaries. The total area of mainland Tanzania and Mafia Island is 115,500 ha with a further 18,000 ha on Zanzibar islands; the Rufiji delta contains the largest single area of mangrove forest in East Africa (Semesi, 1999). Nine species of mangrove have been recorded in Tanzania with *Rhizophora mucronata*, *Avicennia marina* and *Ceriops tagal* the most common. Mangroves are used locally in a variety of ways (Table 3).

Seagrass and algae beds are extensive on sand and mud intertidal flats, sandy lagoons and at the base of shallow coral reefs. Twelve species of seagrass and 149 species of algae are known from Mafia Island Marine Park (Horrill and Ngoile, 1991) and over 300 species of algae are reported for Tanzania. Common seagrasses are *Thallasodendron*, *Thallasia*, *Halodule*, *Syringodium*, *Zostera* and *Cymodocea*.

Cessation of Dynamite Fishing: A Case Study of Mtwara

Martin Guard

University of Aberdeen/University of Dar es Salaam

Explosives have been used on coral reefs to catch fish in Tanzania since the 1960s. Many past attempts failed to eradicate this problem. Recently, successful enforcement operations have been facilitated by the Tanga Coastal Zone Conservation and Development Programme and the Mafia Island Marine Park. The main catalyst was outside donor involvement and the will of local communities and district government.

In March 1998 the extent of explosives fishing was made clear when a workshop 'to wage war against dynamite fishing' (convened by the Fisheries Department) included presentations from all areas along the Tanzanian coast. In the Mtwara district it was shown that dynamite use had virtually doubled over the previous six months. This was indicated by a near doubling in recorded dynamite blasts in 1997/98 compared with 1996/97. This led to discussion of the problems and possible solutions, culminating in the formulation of a national action plan. One of the first actions was that navy patrols would take place in the Mtwara district. Although these could be considered unconventional in that the navy arrived and beat up and harassed all fishermen, whether guilty of using dynamite or not, the desired effect was achieved. As reported by the patrol leader, the philosophy behind this easily criticised approach was that all fishermen could be considered guilty, if dynamite fishing was allowed to be practised in the area. The practice was already illegal but ignored, and thus was not acceptable and needed to be put across in the strongest terms. Nonetheless the next step was to be the turning point: the district commissioner for Mtwara called an amnesty for all users or dealers of dynamite to surrender their dynamite within a period of two weeks. Assisted by a list of names of users and dealers compiled over two years by local communities, the acceptance of this amnesty surprised everyone. At a series of meetings in 15 villages, a total of 228 surrendees came forward with a substantial amount of bombs and explosives (see Fig. 4). All culprits had to place their dynamite and swear in front of villagers, the district commissioner and the police that they would not use dynamite again, provide a set of fingerprints, sign a statement of intent and be photographed by police. They were then told that if they reneged they would face a 600,000 TSh fine ($700) and a minimum three years in jail.

As well as bringing about an end to dynamite fishing in the district, the amnesty was low cost and involved local communities in the initiative. Since the amnesty was initiated there has been no dynamite fishing in the Mtwara district for 18 months. Most local fishermen are now reporting an increase in fish densities and catches over reef areas.

Table 3

Uses of exploited mangrove timber

Mangrove species	Use of timber
Avicennia marina	Inferior firewood, used for fish smoking and lime production, dugout canoes and beehives, construction of beds and leaves for cattle fodder
Bruguiera gymnorrhiza	Good firewood and poles for building
Ceriops tagal	Good firewood, poles for building, timber for boat building, fishing stakes, fence posts
Rhizophora mucronata	Good firewood, poles for building and bark used for dyes
Sonneratia alba	Inferior firewood, preferred boat building timber.
Xylocarpus granatum	Good firewood, used for smoking fish, boat building, furniture, seeds and fruit used for medicinal purposes
Lumnitzera racemosa	Good firewood

Coastal forests of Tanzania are renowned for their high levels of diversity and endemism of both plants and animals (Burgess and Clarke, 1999; Hawthorne, 1993). Growing in a narrow strip along the coast and on Mafia Island, they are now one of the most threatened habitats in East Africa. At Mafia Island a small strip of coastal forest supports several roosts of the endangered fruit bat, the Seychelles flying fox.

Another important site is the oceanic Latham Island which is a very important seabird breeding site in East Africa.

OFFSHORE SYSTEMS

The 200-mile EEZ covers an estimated 223,000 km². Fish stocks in the offshore zone are not well known. The last estimated potential annual yield for commercial fisheries in 1984 was 38,000 tons/year for demersal species and 23,000 tons/year for pelagic resources such as sardines and small tuna. For Zanzibar, pelagic yields were estimated at 20,000 tons/year. Commercial exploitation is still relatively low, but factors limiting Tanzanian expansion of this fishery are the high capital costs and sophisticated equipment required.

POPULATION DEMOGRAPHY

By the year 2000 Tanzania will support an estimated total population of 35,000,000 which is expected to increase by 4% per annum. A quarter live at or near the coast; Dar es Salaam is one of the fastest-growing cities in the world (8% per annum). Government administration is conducted at both regional and district levels with a total of five coastal regions and 16 coastal districts. Outside of urban areas, coastal rural communities subsist through a combination of fishing, small-scale farming and local trading of coconuts, cashews, sea cucumbers and dried fish (Table 4).

RURAL FACTORS

Agriculture

Agriculture accounts for 80% of all export earnings and provides employment for over 80% of Tanzania's workforce (Linden and Lundin, 1995). Along the coast, agriculture is dominated by shifting or 'slash and burn' cultivation practised most intensively on fertile alluvial soils in floodplain areas and river valleys. Elsewhere, overlying soils are commonly highly alkaline sand and coralline deposits with poor water retention and low nutrient content. Thus small-scale farming is primarily restricted to periods of high rainfall. This shifting cultivation contributes markedly to local soil erosion and high suspended solid loads in rivers discharging into the sea.

Vermin infestation is also a common problem along the coast. During 1990 in the Mtwara and Lindi regions, rat infestations reached densities of 700–1000/ha while insect pest outbreaks (e.g locusts) commonly cause severe crop damage. To counteract such infestations, the known biocide use has tripled over 20 years (Fig. 2). Pesticide

Table 4

Population growth along the coast between 1978 to 2000

Regional and district towns	1978	1988	1998	2000 Est.	Yearly increase %
Dar es Salaam	757,346	1,360,850	2,449,530	2,667,266	8
Tanga	103,409	287,158	519,756	566,276	4.8
Lindi	27,308	41,587	63,212	67,537	4.0
Mtwara	48,510	76,362	121,079	130,022	4.1
Kilondoni (Mafia Island)	4,032	7,011	9,066	9,477	4.8
Total for mainland coast	N/a	4,160,000	7,113,600	7,704,320	4.8
Zanzibar Islands	N/a	600,000	900,000	960,000	4.0
Total for Tanzania	N/a	27,000,000	33,700,000	35,000,000	4.8

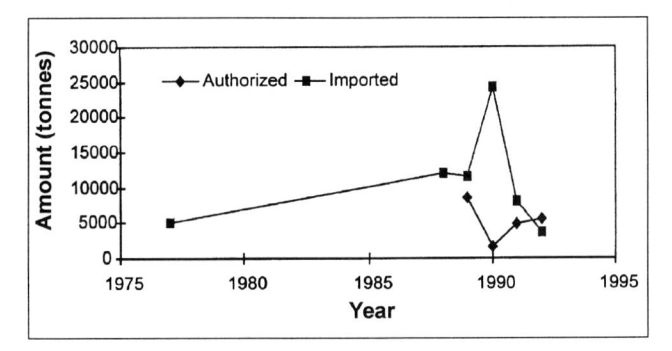

Fig. 2. Amounts of authorised and actually imported pesticides into Tanzania.

handling and storage in Tanzania is very poor, with large amounts disposed of carelessly. Almost half of Tanzania's drainage system empties into the Indian Ocean. Local consequences may be predictable, but are not monitored. Periodic fish mortality has been reported in the Mwanakombo and Zingwezingwe rivers in Zanzibar which drain the Mahonda sugar factory and sugar cane plantation (Mohammed, 1990). Similarly, there are reports by villagers of fish disappearance (Mmochi and Mberek, 1999) and mortality in Chwaka bay believed to result from the high herbicide use in adjoining government supported rain-fed rice farms. Leaching of pesticides and poisons have also been attributed for mortality of corals and other marine life along the coast.

Nitrogen and phosphate leaching is another increasing problem which accelerates local eutrophication, encouraging the growth of algae and reducing the oxygen content in the water body.

Deforestation and Mangrove Harvesting

The once extensive, ancient forests on the East African coastline are now reduced to a few fragmented isolated patches covering a total area of about 350 km^2 (Burgess and Clarke, 1999). Likewise, mangrove forests in Tanzania are also increasingly threatened. Clearance for salt production, agriculture, aquaculture and urban and industrial development, together with over-exploitation for firewood, pole and charcoal production, have led to declines in mangrove forest over the whole of Tanzania (Semesi, 1999). In Kunduchi over 60% of mangroves have been cleared (Banyikwa and Semesi, 1986) while at the Rufiji Delta, the largest mangrove area in East Africa, permission has been given to the African Fish Company (AFC) to clear 10,000 ha for shrimp aquaculture. This latter project is going ahead despite the government's own advisory body, the National Environment Management Council (NEMC), stating that the project is unsustainable and would lead to extensive environmental damage in the longer term. Local communities have also voiced objections and academics and several environmental NGOs have raised the issue in a debate that has continued for over two years. The proposed AFC shrimp farm has raised many issues relating to land ownership, structural adjustment policies and existing statutory legislation and highlights the ability and power of a few politicians to misdirect sustainable development policies in Tanzania (Semesi, 1999).

Legislation relating to both coastal and mangrove forest management in Tanzania is governed by the Forest Ordinance of 1957 and, for mangroves, the National Mangrove Management Plan (NMMP) prepared in 1991. Of the remaining coastal forest 70% is now contained within forest reserves while all mangrove forest was gazetted as forest reserves in 1928–32.

Restoration of Coral Reef and Mangrove Forest at Kunduchi and Mbweni Village, Dar es Salaam

Greg Wagner

Department of Zoology and Marine Biology, University of Dar es Salaam, Dar es Salaam, Tanzania

Habitats along the Dar es Salaam coast have been severely degraded in recent years through numerous destructive fishing practices, coral bleaching and pollution. Protection methods are insufficient and active restoration is required if the habitats are to be restored.

In an attempt to restore coral reef habitat local fishermen were trained to transplant coral fragments in cement-filled disposable plastic plates, onto fringing reef sites surrounding Mbudya Island, previously degraded by dynamite fishing. A total of 565 fragments of coral on 113 plates were prepared containing a mixture of *Galaxea*, *Acropora*, *Porites* and *Montipora* species. After three months a total of 342 coral fragments were relocated; average survival of transplanted corals was 77%. Survival of *Galaxea* spp. and *Porites* spp. was dependent upon site location and in some areas all *Galaxea* survived. A second survey eight months later showed similar survival rates, and for all species except those of *Acropora* there was significant growth.

Likewise, women were trained to begin restoring a highly degraded mangrove forest adjacent to their village by transplanting 3000 seedlings/propagules of *Rhizophora mucronata* with a few *Avicennia marina* over six months. About 50% survived. Triggered by the enthusiasm generated by this activity, the women involved formed a community-based organisation (CBO) called 'Kikundi cha Mazingira na Wanawake Mbweni' (Mbweni Environment and Women's Group) which is continuing with mangrove planting and is now developing plans for other environmental projects.

In another initiative, plans are being developed to promote and encourage ecotourism in Mbweni and Kunduchi whereby tourists will be taken by locals on guided tours to the mangrove forests and coral reefs where they will, at the same time, witness the on-going restoration activities. It is planned for the work to continue and that the mangrove forest at Kunduchi and coral reefs near Mbweni will be restored.

Despite forest reserve status and the development of the NMMP, management of both coastal and mangrove forest areas is not followed and community involvement is minimal. To counteract this failing, several collaborative pilot projects have been initiated in Tanzania to promote both coastal forest and mangrove protection, and the sustainable utilisation of mangroves for the benefit of local communities. Diverse activities have helped to raise awareness at a local level. For policy and decision makers a recent series of workshops organised by Sida/SAREC and the World Bank on Integrated Coastal Zone Management (ICZM) has highlighted the need for proper management.

Fisheries

As the population of Tanzania continues to rise, coastal fisheries are coming under increasing pressure. Unsustainable harvesting methods have led to reductions in productivity and degraded coastal habitats (Darwall, 1996a; Horrill, 1996; Guard et al., 1997). Poor licensing systems, unreliable data and a general lack of fisheries management exacerbate this problem.

Coastal fisheries provide the main source of income to impoverished coastal communities and account for over 60% of the protein consumed in the country. Over 95% of fishing is artisanal or traditional, using wooden canoes (mtumbwi), outrigger canoes (ngalawa), and wooden sailing vessels such as 'dhows' and 'mashua'. The main means of propulsion is by paddle, pole or sail, with engines being used in only a few boats due to limitations of cost and maintenance.

Commercial fishing is limited to prawn trawling adjacent to mangrove areas and small-scale exploitation of pelagic resources offshore (e.g. tuna). Although offshore pelagic resources are considered under-utilised as a result of the high investment costs and sophisticated equipment required, shallow-water near-shore fisheries are fished beyond sustainable yields (Linden and Lundin, 1995).

Government catch statistics from 1988 to 1995 (Table 5) for all marine fisheries indicate 39,000 to 56,000 tonnes taken in 1990 to 1993. No trends can be determined as government statistics are considered unreliable and rough estimates at best.

The main artisanal fishing gears used in Tanzania are large and small mesh entanglement nets, seine nets, fish traps, spears and handlines (Table 6). Entanglement nets ('jarife') are used as drift nets (kulambaza) or are bottom set. Sharks are the favoured target group, primarily for their liver and fins. The oil from the liver is used as a waterproof seal for the inside of wooden vessels and the fins have a high value (up to $50 per kilo) when sold to overseas buyers. Sharks are now overfished in many areas (Barnett, 1997).

The 'jarife' fishery also has a high turtle by-catch. Turtles are an important food for local communities, and it is estimated that over 1600 turtles are captured each year in southern Tanzania alone (Darwall et al., 2000).

Table 5

Marine fish production and values for Tanzania 1988–1998

Year	No. of fishermen	No. of vessels	Catches (tonnes) industrial	Catches (tonnes) artisanal	Total	Value (1000 TSh)
1988	13,855	4,390	2,190	47,193	49,383	2,568,859
1989	–		2,437	49,582	52,019	–
1990	16,178	4,354	2,015	54,527	56,542	4,663,859
1991	16,361	4,402	1,510	52,380	53,891	6,159,297
1992	15,027	3,514	1,119	42,183	43,302	6,014,473
1993			1,222	34,226	35,448	
1994			1,786	37,286	39,073	
1995	13,822	3,768	2,312	48,761	51,073	28,579,811

Table 6

Estimated catch yields for fin-fish and invertebrate fisheries in Tanzania

Gear type/ fishery	Mafia Island	Songo-Songo Archipelago	Mtwara/Lindi districts
Shark net	32	170	N/a
Seine net	291	300	64
Long line	12	0	0
Hand line	200	200	210
Box trap	50–60	31	93
Fence trap	19	n/a	86
Octopus	150	62	75
Sea cucumber	3.7 ind	8.1 ind	Fished out

Seine nets are used either from the beach or in open water reef areas. This method of fishing is highly destructive to reef areas and, due to the small mesh size, provides a catch mainly of immature fish (Benno, 1992; Darwall, 1996a; Horrill and Ngoile, 1991). It is however, one of the commonest methods used, especially on Zanzibar and Pemba Islands and in the Mtwara region. The 'kigumi' method is another highly damaging technique practised on Zanzibar and Pemba islands. Similar to the 'muro ami' method practised in the Philippines, a long net is placed around a large patch of coral reef and about a dozen fishermen then pound and break up the coral to scare and drive fish into the net. In some areas on the east coast of Pemba 'kigumi' fishing has transformed large areas of coral reef into rubble.

Women and children glean intertidal areas for invertebrates. Gastropods, bi-valves, octopus, shellfish and sea cucumbers are collected using a variety of techniques that include hand collection, the use of spears and hooks and natural and synthetic poisons. The high monetary value of, for example, sea cucumbers (up to $25 per kilo) and ease of collection has led to many areas being fished out.

Octopus resources are also coming under increasing pressure through unsustainable harvesting (Darwall, 1996a; Horrill and Ngoile, 1991) and at the Mafia Island Marine Park this is exacerbated by the competitive presence of two outside buyers. All harvested octopus are purchased, and no size restrictions exist, which now threatens future recruitment and productivity (Darwall, 1996a).

Lobster fishing is opportunistic as there is limited local demand, and problems with over-fishing are only evident where tourist demand is high, such as on Unguja Island. Also important locally is the catch of portunid crabs.

Collection of marine life for the curio trade is widespread in Tanzania. The most sought-after gastropods such as the

Table 7

Estimated scale and intensity of coral mining along the coast of Tanzania

	Dar es Salaam	Mafia Island	Songo-Songo Archipelago	Mikindani Town
No. people involved	N/a	475	1 full time; many part time	200
Est. tons/year	N/a	950	100	1000

Horned Helmet shell (*Cassis cornuta*), Triton Trumpet shell (*Charonias tritonis*), and the Mauritian Cowrie (*Cypraea mauritania*) are now very difficult to find (Linden and Lundin, 1995). In recent years there has also been a rise in the collection of seahorses for Chinese medicine (Fig. 3). Two species, *Hippocampus kuda* and *Hippocampus hystrix* fetch $30 per kg (approx. 200–300 seahorses). Although listed as endangered in CITES, no regulation exists in Tanzania (Mallela et al., 1998).

Coral Mining

Coral mining is conducted along the entire coastline. Mined coral is taken from living reefs or from ancient fossilised reefs on shore or inland. Both live and fossilised coral is used for building blocks and aggregate, but live coral is favoured to produce 'white lime' or 'chokaa' a cheap substitute for building cement and a whitewash for painting buildings. Extracted from the reef using iron bars, live coral is brought to shore, broken into smaller pieces and placed in kilns. Two to three tonnes of coral and one tonne of wood are used per kiln, resulting in a fine lime powder.

Coral mining is considered one of the major contributors to reef and forest degradation along the coast (Table 7). In Mikindani Town, Mtwara region, up to 30 kilns are burnt every month amounting to over 1000 tonnes of coral being mined per year. Consequences include increased erosion, declines in abundance of fish and invertebrates, changes in the composition of fish communities and reduced forest cover (Brown and Dunne, 1988; Dawson-Shepherd et al., 1992; Dulvy et al., 1995). Net financial benefits are much lower compared to the lost value of goods (e.g. marine resources) and services (e.g. tourism) provided from a properly managed reef (Berg et al., 1999). Alternatives are not in widespread use (Darwall et al., 1995), but as yet such schemes have to be encouraged in Tanzania and may only prove to be suitable in certain areas.

Mariculture

Mariculture in Tanzania is dominated by seaweed farming, with only a few attempts at shrimp, fish or mollusc farming (with the exception of the large shrimp farming project in the Rufiji delta as noted earlier). Today nearly 15,000

Fig. 3. Seahorses collected for sale.

Fig. 4. Dynamite, fuses and detonators surrendered at a recent amnesty initiated by the Mtwara district government.

people, 90% of which are women, harvest approximately 200 tons dry weight of algae per week (Linden and Lundin, 1995). The two main farmed species are *Euchema spinosum* and *E. cotonai*. Harvested every two weeks, an average plot (e.g. 50 ropes) provides about 13,000 Tanzanian shillings (US $22) to the farmer, a higher return than could be gained from activities such as handicrafts, intertidal gleaning or agriculture. Although clearly benefiting seaweed farmers, several negative impacts are associated with this practice. Increased trampling, disturbance and shading effects of algae may impact the abundance and diversity of the benthos, the food resource for other organisms is reduced, and farmed algae under stress may release potentially toxic chemicals such as bromides and hydrogen peroxide.

Tourism

The great variety of wildlife and scenic areas is a valuable attraction. Investment in tourist developments is encouraged by the Tanzanian Government and is important for generating much-needed foreign exchange. The Tanzanian tourist policy is, however, flawed in that it primarily focuses on income generation with little consideration given to potential environmental and social impacts. Tourist projects are being developed in protected areas, small islands and close to the shoreline with no requirement for the conducting of an Environmental Assessment first. In Zanzibar, and other areas where tourism is rapidly expanding, environmental concerns now evident include sewage pollution, environmental damage through poorly planned construction, damage to coastal and marine habitats through tourist activities, local resource over-extraction and increased erosion due to coral mining and the use of beach sand for construction (Linden and Lundin, 1995). In addition, tourism can lead to a sharp rise in local land value which can force out local communities. Large influxes of foreign tourists can also conflict with local values and attitudes.

There is a pressing need for the Government to recognise and address the issue and principles of sustainable tourism, defined as 'a process which allows development to take place without degrading or depleting the resources which make the development possible' (Linden and Lundin, 1995). To achieve this is difficult without appropriate guidelines, regulations, legislation and compliance.

Table 8

Comparison of seawater qualities in Dar es Salaam city and Zanzibar municipality coastal waters (Van Bruggen, 1990)

Parameter	Zanzibar	Dar Es Salaam
O_2 % saturation	119	60
Faecal Coliform per ml	589	287
Total Coliform per ml	1482	6748
BOD^5 mg/l	14.7	6
COD^5 mg/l	1235	602

There is also a need to initiate awareness raising programmes of the potential impacts of coastal tourism and to target these, not only for developers, but also for local communities and tourists themselves. The Government is now reportedly preparing a new tourism policy document that intends to address these issues. Preliminary recommendations will include the need to recognise and support the identity of local communities in the planning process and to make Environmental Assessments mandatory.

EFFECTS FROM URBAN AND INDUSTRIAL ACTIVITIES

Industrial Discharges

Most industries in Tanzania are situated in the major cities and towns. Dar es Salaam contains 80% of all industry in Tanzania (Mgana and Mahongo, 1997). Other coastal towns in order of importance are Zanzibar, Tanga, Mtwara, Lindi, Bagamoyo and Kilwa. The most polluting industries are agrochemicals, chemicals, breweries, metal, food and textiles. The total number of industries in Dar es Salaam has increased from 49 in 1982 to 74 in 1997, while at Tanga and Zanzibar the number of industries has decreased in recent years. About 68% of industries in Dar es Salaam discharge effluents through 12 main streams to the sea. Contaminants include oils, paint and metal wastes. The Msimbazi river in Dar es Salaam is practically devoid of life due to the anoxic conditions resulting from the pollution load of the brewing industry (Kimolo, 1989). Elevated concentrations of lead, cadmium, chromium and copper have been detected in shellfish at the mouth of the Msimbazi River (Mashauri and Mayo, 1989). In a study of seawater quality around Dar es Salaam and Zanzibar Town (Van Bruggen, 1990), the average coliform count (COD^5) for seawater samples collected were 602 mg/l and 1235 mg/l respectively (Table 8). The high frequency of communicable diseases in Zanzibar and the increasing frequency of red algae blooms in Zanzibar harbour reinforce the findings.

Most industrial wastes generated in Dar es Salaam are disposed of either on site, or at Vingunguti disposal area. Paper, sugar and cement industries attempt to recycle a proportion of wastes produced, but consumer industries have stimulated little recycling of used containers.

Although some laws in Tanzania have environmental implications no clear policy on the environment is evident. Industrial wastes are simply discharged into the city waterways and streams, many of which are used by local communities.

Coastal Urbanisation and Waste Disposal

Urban migration means that the rate of population rise in the major coastal towns is higher than the average population rise for Tanzania. This is of concern, as waste and

Table 9

Annual BOD pollutant load to the water environment and waste disposal data in selected cities and states of the region in 1982 and 1992

	BOD_5 (tons/year)		Solid waste/capita kg/year		% Population connected to sewage system		% Waste collected (tons/year)		Hazardous waste (tons/year)	
	1982	1992–1997	1982	1992–1997	1982	1992–1997	1982	1992–1997	1982	1992–1997
Tanga	200	274		150	10	13		24		278
Dar es Salaam	2250		15	150		13		23		65
Zanzibar		25		182	20	20		30		

sewerage facilities have not been upgraded to cope (Table 9). Only few urban dwellings are connected to a sewage system which is often pumped directly into rivers and coastal waters. This poses a threat to potable water supplies and to the clarity and quality of both freshwater and near shore coastal waters. Storm-water run-off in urban areas is also a significant source of land-based pollution to coastal waters. There are nine waste stabilisation ponds in Dar es Salaam, five for domestic wastes and the remainder for industrial wastes. At present, three ponds are out of order. Sewage from Dar es Salaam accounts for 55% (11,312 tonnes per year) of the BOD and 42% (18,418 tonnes per year) of the suspended solid load. For Tanga the same figures are 83% (2287 tonnes per year) and 99% (3258 tonnes per year). In Zanzibar, domestic sewage is considered the major source of pollution to nearshore coastal waters. Based on WHO guidelines (WHO, 1996) most drinking water around Zanzibar Town is unfit for human consumption due to coliform bacteria (Kivaisi and Van Bruggen, 1990). Furthermore, eutrophication from domestic sewage was identified as impacting coral reef building algae (Bjork et al., 1995). On Zanzibar the system was originally built in the 1920s, and while it has been recently renovated, nutrient concentrations have not improved (Mmochi and Francis, 1999).

Saline intrusion to ground water around Zanzibar is a natural phenomenon and perched aquifers are relied upon as the main source of potable water. During dry periods, the perched aquifer is separated from supply wells that intersect the deeper, contaminated aquifer. Increased water abstraction by the expanding population has led to reports by the Ministry of Water of increased salinity in water supplies. There have also been reports of domestic waste being dumped on mangroves around Zanzibar.

COASTAL EROSION

Coastal erosion is an increasing problem, with many beach fronts and islands now affected. Along the beaches of Kunduchi, north of Dar es Salaam, large areas have been lost, threatening the tourist industry (Linden and Lundin, 1995). Similarly in 1978, Maziwi island situated 50 km south of Tanga completely disappeared as the sea edge moved forward, most probably a result of vegetation clearance (Fay, 1992).

A number of factors are thought to be responsible for beach erosion but in reality little is known about the dynamic and often complex interactions which are involved. Wind and wave action, local sediment dynamics, sediment fluxes and tides all play a role in the erosion process which may simply be a result of natural change. Potential man-made causes are the extraction of sand and gravels from stream beds and beaches, vegetation clearance and coral reef and mangrove forest destruction, salt pans' creation and prawn farming. Demand for sand for building has increased dramatically since the 1980s; Griffiths (1997b) estimated a minimum of 100,000 m^3 of sand was extracted from four streams annually.

In order to control erosion rates, the Tanzanian Government have undertaken several measures. These include the placing of 54 groynes along Kunduchi beach and the building of seawalls in Dar es Salaam and adjacent to Zanzibar Harbour. However, in the case of the groynes, there has been no real change to the rate of erosion which, in some areas, has actually increased (Glasgow University, 1991).

The National Environment Management Council have recommended the use of a buffer zone in which the intensity of human activities would be reduced and vegetation removal prevented. At first it was suggested that the buffer be 200 m but, as a result of resistance, it has been reduced to just 60 m and, if shown to result in a loss of useful land, this may be reduced further to 30 m. Although the principle of buffer zones has been accepted they still have to be introduced effectively and monitored for compliance.

PROTECTED AREAS AND INTEGRATED COASTAL ZONE MANAGEMENT

Few areas along the Tanzanian coast are protected or actively managed. Past attempts to create a system of marine protected areas have all met with limited success due to a combination of poor co-ordination, lack of political will and a shortage of funds and trained personnel (Bryceson, 1981). In 1975, the first attempt to protect coastal resources was initiated when eight sites along the coast were designated as marine reserves. Yet, in the years that followed, none of the reserves were officially implemented and no management regulations were developed. In a second attempt to expand the number of protected areas in Tanzania and Zanzibar Islands (UNEP, 1989) 11 sites based

Table 10

Laws directly relating to the Marine Environment

Fisheries Act No. 6, 1970
The Fisheries Explosives, Poisons and Water Pollution Regulation, 1982 (Amending regulations 26,27 and 28 of the Fisheries General Regulation, 1973)
Territorial Sea and Exclusive Economic Zone Act No.3, 1989
Merchant Shipping Act, 1967
Inland Water Transport Ordinance, Cap. 172
The Petroleum (Exploration and Production) Act No. 27, 1980
Mining Act No.17, 1979
Fisheries Principal Regulations, 1979
Fisheries (Marine Reserves) Regulations, 1975
Marine Parks and Reserves Act, 1994

Table 11

Integrated Coastal Zone Management Projects in Tanzania

Project Name	Aims of Project
Tanga Coastal Zone and Development Programme	Community Management
Misali Island Conservation Project	Community Management
Menia Bay Conservation Area	Community Management
Chwaka Bay-Paje Integrated Coastal Area Management Project	Community Management
Kunduchi and Mbweni Village Coral Reef and Mangrove Restoration Project	Community Habitat Restoration
Mafia Island Marine Park	Protected Area, Community Management
Frontier Tanzania Marine Research Programme	Baseline data collection, research and community training
Rural Integrated Project Support/ Shirkisho Marine Environmental Protection Programme	Awareness Raising, Community Management

on their uniqueness of habitats, biological importance and commercial fisheries were recommended for marine park, marine reserve or management area status. The sites proposed are listed in Table 11.

Of these only at Mafia Island was there a committed effort towards the development of marine environmental management with the gazettment of Tanzania's first Multi User Marine Park in April, 1995. Regarded as a major achievement following an intensive period of lobbying, marine conservation and protection had finally been formalised and legitimised through the 'Marine Parks and Reserves Act' approved in 1994 (Table 10). Since gazettment, the Mafia Island Marine Park, covering 400 km^2, has essentially acted as a large pilot study aiming to achieve effective, well functioning and balanced participatory marine conservation and resource management. This, however, has not been an easy task and despite a general management plan being prepared prior to gazettment, active management of the park has suffered through institutional and key participant conflicts combined with an over-ambitious work programme (Darwall et al., 2000). Further, not until 1996 was the management plan translated into Swahili, and until 1997 were local communities shown the zoning plans. Such problems have led those responsible to re-think the overall direction and function of the marine park, appoint a new technical advisor and renew the drive towards a system of Integrated Coastal Zone Management throughout Tanzania.

Chumbe Island is the only other Marine Park in Tanzania, gazetted by the Zanzibar Government in 1994. Fundamentally different to the Mafia Island Marine Park in that it is privately owned and covers only 16 ha, Chumbe Island is situated off the south west side of Unguja Island and is fringed by pristine coral reefs of high biological importance. The ultimate goal of the project is to protect and conserve the surrounding coral reefs and the island's flora and fauna (Sterner and Andersson, 1999). To help raise local awareness, the project has initiated an environmental education programme and trained local fishermen to be employed as park rangers. Eco-tourism has been introduced to attract funding, with a design which is considered so revolutionary that the project has been chosen to represent Tanzania at the EXPO2000 in Germany (Carter, 1999). Although setting an example of how private ownership can contribute to habitat conservation, several problems were highlighted by Sterner and Andersson (Sterner and Andersson, 1999). For example, innovative design incurs high costs and there is often difficulty experienced in maintaining such technologies. The project has often been perceived by local authorities as being more eccentric than valuable and this has led to costly delays in gaining the necessary permission to develop. Finally, as the costs need to be recouped and the project maintained, the pricing has had to remain high which requires a sophisticated marketing strategy to ensure enough visitors come to stay.

A third marine park is proposed at Mnazi Bay, Mtwara. Funded by the GEF and co-ordinated by UNDP and IUCN, the park is still in the preliminary planning stages but will eventually cover an area of over 200 km^2. With lessons learnt at Mafia and with local communities already being involved, it is hoped that progress towards effective park management will be both quicker and more successful.

At local levels, several programmes are attempting to put ICZM into practice (Table 11), nearly all being based on a collaborative approach involving local communities, local and regional government and where necessary the private sector. At the Tanga Coastal Zone Conservation and Development Programme (TCZCDP), (funded by Irish Aid with technical support provided by IUCN) a collaborative approach has led to new local by-laws being developed and approved, and community action plans formulated to deal with the problems of overfishing, reef degradation, the development of alternative income strategies and mariculture. In the Kilwa, Lindi and Mtwara districts the identification, discussion and awareness-raising of marine issues has been co-ordinated through a community NGO, the Shirikisho, which successfully reaches a wider audience. Videos of community meetings discussing dynamite fishing were shown to the Prime Minister whose response was also filmed and shown back to the communities. Such films were integral to the cessation of dynamite fishing in southern Tanzania.

In another small conservation project for Misali Island, the need to co-operate with the private sector was recognised. After an extensive survey of tourism conducted through local tour operators, the project developed an ecotourist approach that not only meets project aims but also provides a mechanism for sustaining itself financially. These and many other similar projects in Tanzania are leading to widespread recognition of the problems of marine resource extraction and habitat degradation and are instrumental in changing attitudes both at local community and government levels.

At national level many government sectoral policy statements now recognise the need for an integrated and participatory resource management approach to resolve issues and take advantage of development opportunities. Directed by the Tanzanian Coastal Management Partnership (TCMP) (a joint initiative between the Government of Tanzania through the National Environment Management Council, USAID and the University of Rhode Island Coastal Resources Center) a national process has now been formulated to develop a comprehensive policy document for ICZM. The policy intends to guide and ensure cross-sectoral co-ordination at all levels. Areas covered include stakeholder participation, compliance with laws and regulations, support for research and training, and education and awareness. TCMP have also been involved with the collection and dissemination of information, and have facilitated workshops to formulate a scientific and technical advisory body responsible for the assessment and monitoring of Tanzania's marine resources.

At the international level Tanzania has actively supported ICZM principles and programs and has signed and ratified a number of international conventions that endorse the importance of ICZM. Towards this aim the Sida/SAREC Marine Science Programme has been instrumental in providing support for institutional development and the training of marine scientists. It has also played an important role in facilitating several ICZM workshops in the region and the development of research studies to counter the lack of scientific information for the coastal zone.

With the capacity of scientists in the region now improved, the Western Indian Ocean Marine Scientists Association (WIOMSA) was formed to promote the educational, scientific and technological development of all aspects of the marine sciences in the region. The main activities include the dissemination of scientific information and advice through quarterly newsletters, providing new updates on progress in the region. WIOMSA also provides an important international scientific forum for discussion of

Local Community Training and Education in Southern Tanzania

Vicki Howe

Department of Maritime Studies and International Transport, University of Cardiff, Cardiff, Wales, U.K.

Many local communities now find their livelihoods threatened, sometimes by causes lying outside their control. One NGO (Frontier-Tanzania) with support from local administration, initiated a marine education programme in the Mtwara district, southern Tanzania. The programme included training of fishery officers and local monitors to carry out simple marine survey and monitoring techniques, sustainable resource use and boat-handling skills, together with organising marine environmental education and awareness days for local primary schools. Liaison with the Shirikisho, a local community NGO, the Natural Resources Office, District and Town Education and Fisheries Offices, and the Rural Integrated Project Support (RIPS) assisted in defining the scope and specific aims and objectives of the programme. The programme was financed by the UK-based Society for Environmental Exploration, British High Commission and WWF.

Mtwara and its surrounding villages are located near Mnazi Bay, an area recently proposed as Tanzania's third marine park. Mnazi Bay is characterised by extensive fringing and patch coral reef, mangrove forest and seagrass beds with a high floral and faunal diversity. It was necessary to involve local users to collect basic information from which the health and status of local coral reefs could be determined, and to achieve this aim local representatives were selected from coastal villages and trained in various techniques to enable them to monitor the coral reefs and collect fisheries data. Methods taught included habitat assessment, fish diversity and abundance counts, fishery data collection and repetitive monitoring of a single area. Boat handling and maintenance skills were also taught. As a result, fishermen from 18 different villages who completed the training programme are now working as trainers to educate 102 surrounding villages.

In Tanzania the school curriculum places no emphasis on marine environmental issues, so another aim was to support local teachers through the provision of information (teachers handbook) and training in carrying out practical field projects. During 1999, 12 schools were involved. The majority were from poorer fishing villages within the Mtwara district. As a result, teachers now attempt to use more hands-on techniques to demonstrate various concepts concerning the marine environment.

Also, workshops held in 1998 on the destructiveness of dynamite fishing contributed to an amnesty against this fishing method. The need for continued education has been recognised by both authorities and by the local communities themselves, and this programme has been a first step towards establishing an effective management strategy in the Mtwara district and Mnazi Bay.

the main issues affecting the coastal zone, provides grants for coastal research, fosters inter-institutional linkages and co-ordinates seminars and meetings on marine science findings and applications.

Although all of the above developments are welcome steps, there are still many issues that need to be addressed and improved. Moffat et al. (1999) reviewed the main constraints facing the development and implementation of ICZM in the region:
- Most coastal projects still place emphasis more on biodiversity conservation with no real link to the development of local peoples or community involvement.
- Projects continue to be donor and NGO driven with little or no real national government ownership.
- Projects are rarely sustainable when donor funding ceases as financial mechanisms for their continuation are not included in project design.
- The work/research of many coastal projects does not fall in line with national policy strategies.
- Timeframes of projects need to be addressed, with donors better recognising advantages of longer term funding for objectives to be met.
- Time frames should incorporate flexibility so as to be responsive to particular project areas and situations.
- Employment and capacity building of local scientists should be afforded priority.
- There is a need for raising the awareness of policy and decision-makers on particular marine issues.
- Legislation and regulatory frameworks need to be improved to reflect the problems encountered today.
- More multi-disciplinary research needs to be conducted.
- Research projects should be designed to answer specific management questions rather than directed by academic goals.
- Monitoring activities should be built into projects.
- Project managers need to develop and refine communication strategies targeted to communities, the private sector, other ICZM programs, donors and different levels of government.
- Participation and local management are the key for successful ICZM.
- The private sector needs to fully integrate into ICZM activities.

The points mentioned here (Moffat et al., 1999) provide a clear and concise vision for how ICZM should progress in this region. To achieve this aim, however, much will depend on the commitment and attitudes of both project managers and local government to adapt to the concept of multi-disciplinary and collaborative approaches and ensure that local communities are integrated fully into the process. Building from the local level up will have the greatest impact on ICZM success in Tanzania. With valuable experience now being gained from collaboration, it seems there is no better time than at the start of a new millennium to bring this goal to fruition.

REFERENCES

Anderson, J.E.C. and Ngazi, Z. (1995) Marine resource use and the establishment of a marine park: Mafia Island, Tanzania. *Ambio* **24**, 475–48

Banyikwa, F.F. and Semesi, A.K. (1986) Endangered mangrove ecosystems. The case of the Kunduchi and Mbweni mangrove forests. In Proceedings of a Workshop on Save the Mangrove Ecosystems in Tanzania, 21–22 February 1986, Dar es Salaam, eds. J.R. Mainoya and P.J. Siegel. pp. 103–132.

Barnett, R. (1997) The shark trade in mainland Tanzania and Zanzibar. In: Shark Fisheries and Trade in the Western Indian and Southeast Atlantic Oceans. TRAFFIC Report.

Bathwondi, P.O.J. and Mwaya, G. (1984) The Fishery of Mollusc and Crustacea in Tanzania. In *Proc. of the NORAD Tanzania Seminar to Review the Marine Fish Stocks and Fisheries in Tanzania*, eds. A. Iverson and S. Myklevoll, pp. 19–26.

Benno, B.L (1992) Some Features of a Beach Seine Fishery along the coast of Dar es Salaam Coast, Tanzania. M.Sc Thesis. Applied Zoology. University of Kuopio. Finland. 68 pp.

Berg, H., Ohman, M.C., Troeng, S. and Linden, O (1999) Environmental Economics of Coral Reef Destruction. *Ambio* **27**, 627–634.

Bjork, M., Mohammed, M.S., Bjorkland, M. and Semesi, A. (1995) Coralline algae, important coral reef builders threatened by pollution. *Ambio* **24**, 502–505.

Brown, A. (1996) The prosobranch molluscs of Songo Songo, Tanzania: a species list and an investigation into the commercial shell trade and the methods used by the environmental organisation Frontier to assess it. Undergraduate Dissertation. Royal Holloway College, University of London.

Brown, B.E. and Dunne, R.P. (1988) The environmental impact of coral mining on coral reefs in the Maldives. *Environmental Conservation* **15**, 159–165.

Bryceson, I. (1981) A review of some of the problems of tropical marine conservation with particular reference to the Tanzanian coast. *Biological Conservation* **20**, 163–171.

Burgess, N.D. and Clarke, P. (1999) *Coastal Forests of Eastern Africa*. IUCN Forest Conservation Programme.

Carter, E. (1999) Sustainable management of a marine park resource using eco-tourism. Conference on Advances on Marine Sciences in Tanzania, University of Dar es Salaam, Institute of Marine Sciences, Tanzania. Conference Abstracts, p. 70.

Choiseul, V.M. (1996) The status of the marine gastropod fishery in the Mafia Island Marine Park, Tanzania. The Society for Environmental Exploration and The University of Dar es Salaam. Unpublished report.

Choiseul, V.M. and Darwall, W.R.T. (1996a) *Mafia Island Report No.5. Results of biological and resource use surveys of the western reefs of the Mafia Island Marine Park*. The Society for Environmental Exploration and The University of Dar es Salaam. ISSN 1369-8788

Cowper, D. and Darwall, W.R.T. (1996) The current status of marine turtles of the Songo Songo archipelago, Tanzania. *Miombo* **15**.

Darwall, W.R.T. (1995a) Mafia Island Report No. 4. The shark and ray fishery of the Mafia Island Marine Park: Current status and Management Recommendations. The Society for Environmental Exploration and The University of Dar es Salaam.

Darwall, W.R.T. (1995b) Marine biological and resource use surveys of the Songo Songo archipelago. Report No. 1: Simaya Island. A report submitted to the Royal Norwegian Embassy. The Society for Environmental Exploration and The University of Dar es Salaam. ISSN 1369-8788.

Darwall, W.R.T. (1996a) Marine biological and resource use surveys of the Songo Songo archipelago. Report No. 7: Marine resource use in the Songo Songo archipelago. A report submitted to the Royal Norwegian Embassy. The Society for Environmental Exploration and The University of Dar es Salaam. ISSN 1369-8788.

Darwall, W.R.T., Dulvy, N.K., Choiseul, V.M. (1994) Mafia Island Report No. 3: Results of biological and resource use surveys: Recommendations for management. The Society for Environmental Exploration and The University of Dar es Salaam.

Darwall, W.R.T., Edward, J. and Kabado, D. (1995) Results of a feasibility study for the production of sun-dried pressed mud bricks as an alternative to the use of mined live coral on Mafia Island. A report submitted to the Swedish Embassy, Dar es Salaam. The Society for Environmental Exploration and The University of Dar es Salaam.

Darwall, W.R.T., Guard, M., Choiseul, V.M. and Whittington, M. (1996) Marine biological and resource use surveys of the Songo Songo archipelago. Report No. 6: Survey of thirteen patch reefs (Vols. 1–4). A report submitted to the Royal Norwegian Embassy. The Society for Environmental Exploration and The University of Dar es Salaam. ISSN 1369-8788.

Darwall, W.R.T., Guard, M., and Andrews, G. (2000) Southern Tanzania In *Coral Reefs of the Western Indian Ocean*, eds. T.R. McClanahan, D.O. Obura and C.R.C. Sheppard, Oxford University Press.

Dawson-Shepherd, A.R., Warwick, R.M., Clarke, K.R. and Brown, B.E. (1992) An Analysis of Fish Community Response to Coral Mining in the Maldives. *Environ. Biol. Fishes.* **33**, 367–380.

Dulvy, N.K., Stanwell-Smith, D., Darwall, W.R.T., and Horrill, C.J. (1995) Coral mining at Mafia Island, Tanzania: A management dilemma. *Ambio* **24**, 358–365.

Fay, M.B. (1992) Maziwi Island off Pangani (Tanzania): History of its Destruction and Possible Cause. Regional Seas Reports and Studies No. 139. UNEP, 43 pp.

Franklin, H., Muhando, C.A. and Lindahl, U. (1999) Coral Culturing and Temporal Recruitment Patterns in Zanzibar, Tanzania. *Ambio* **27**, 651–655.

Frazier, J. (1976) Sea turtles in Tanzania. *Tanzania Notes and Records*. **77/78**, 11–14.

Glasgow University (1991) Study of Coastal Erosion along the Indian Ocean Coast in the Dar es Salaam Region. In: Glasgow University- Tanzania Expedition Report, Report IRA, Dar es Salaam

Griffiths, C.J. (1997b) The Impact of Sand Extraction from Seasonal Streams on Erosion of Kunduchi. In: Beach Erosion Along Kunduchi Beach, North of Dar es Salaam. Beach Erosion Monitoring Committee. Report for the National Environmental Management Council, pp. 36–47.

Guard, M., Muller, C. and Evans, D. 1997. Marine Biological and Marine Resource use surveys in the Mtwara district, Tanzania. Surveys of fringing and patch reef within and adjacent to Mnazi Bay Report No. 1 Vols. 1,2. The Society for Environmental Exploration and The University of Dar es Salaam.

Hawthorne, W.D. (1993) East African Coastal Forest Botany. In *Biogeography and Ecology of the Rain Forests of Eastern Africa*, eds. J.C. Lovert and S.K. Wasser, pp. 57–99. Cambridge University Press.

Hamilton, H.G.H. and Brakel W.H. (1984) Structure and coral fauna of East African reefs. *Bulletin of Marine Science* **34**, 248–266.

Hanaphy, F. and Muller, C. (1997) Report 8: Survey of six reefs: Amani; Mpovi; Mwanamkaya; Kiswani; Fungu Wango; and Rukyira, Kilwa district. Marine Biological and Resource Use Surveys in the Songo Songo Archipelago. Frontier-Tanzania Marine Research Programme. The Society for Environmental Exploration and the University of Dar es Salaam. ISSN 1369-8788.

Horrill, J.C. (1996) Coral reef survey: summary report. Tanga Coastal Zone Conservation and Development Programme, P.O. Box 5036, Tanga, Tanzania.

Horrill J.C., Mgaya Y.D. and Kamakuru, A.T. (2000) Northern Tanzania In *Coral Reefs of the Western Indian Ocean*, eds. T.R. McClanahan, D.O. Obura and C.R.C. Sheppard. Oxford University Press.

Horrill, J.C. and Ngoile, M.A.K. (1991) Mafia Island Report No. 2. Results of the Physical, Biological and Resource use Surveys: Rationale for the development of a management strategy. The Society for Environmental Exploration, London. pp. 46. ISSN 1 873070 02.

Howell, K.M. (1993) A review of the conservation status of sea turtles in Tanzania. A study by the Wildlife Conservation society of Tanzania.

Johnstone, R.W., Muhando, C.A. and Francis, J. 1999. The status of the coral reefs of Zanzibar: One example of a regional predicament. *Ambio* **27**, 700–707.

Kamakuru, A.T. (1997) Assessment of Biological Status of the Dar es Salaam Marine Reserve System off the Kunduchi Coast. Tanzania. Final Report to Western Indian Ocean Marine Science Association (WIOMSA).

Kayombo, N.A. (1989) Progress report: preliminary study of the ecology of the intertidal mollusks of Mafia Island, Tanzania. National Museums of Tanzania.

Khatib, A.A. (1999) Sea turtles nest recording program: Unguja Island. *Ambio* **27**, 763–764.

Kimolo, H.T. (1989) Contribution of brewing and malting to environmental pollution control and management. In *Environmental Pollution and its Management in East Africa*, eds. M.R. Khan and H.J. Gijzen). Faculty of Science, University of Dar es Salaam, Tanzania..

Kivaisi, A.K. and Van Bruggen, J.J.A. (1990) The quality of drinking water in Zanzibar Town. Prepared for the Zanzibar urban water supply development plan. Zanzibar, Tanzania.

Lindahl, U. (1999) Low-tech rehabilitation of degraded coral reefs through transplantation of Staghorn Corals. *Ambio* **27**, 645–650

Linden, O. and Lundin, C.G (1995) *Proceedings of the National Workshop and Integrated Coastal Zone Management in Tanzania*. The World Bank, Land, Water and Natural Habitats Division, Environmental Department and Sida, Marine Science Programme, Department of Research Co-operation, SAREC.

Mallela, J., Gallop, K. and Guard, M. (1998) The Seahorses of Southern Tanzania. *Miombo* **19**.

Mashauri, D.A. and Mayo, A. (1989) The environmental impact of industrial and domestic waste water in Dar Es Salaam, Tanzania. In *Environmental Pollution and its Management in East Africa*, eds. M.R. Khan and H.J. Gijzen. Faculty of Science, University of Dar Es Salaam, Tanzania.

McClanahan, T.R. (1988) Seasonality in East Africa's coastal waters. *Marine Ecology Progress Series* **44**, 191–199.

Mgana, S. and Mahongo, S. (1997) Land-based sources and activities affecting the quality and uses of the marine, coastal and associated fresh water environment. 47 pp.

Mmochi, A.J. and Francis, J. (1999) Water quality in Zanzibar town coastal waters, Makoba bay and Mapopwe creek of the Chwaka bay, Zanzibar, Tanzania. Report submitted to UNESCO/IOC for contract No. SC-298.273.7. 31 pp.

Mmochi, A.J. and Mberek, R.S. (1999) Trends in the types, amounts, and toxicity of pesticides used in Tanzania. *Ambio* **27**, 669–676

Moffat, D., Ngoile, M.N., Linden, O. and Francis, J (1999) The Reality of the Stomach: Coastal Management at the Local Level in Eastern Africa. *Ambio* **27**, 59–98

Mohammed, S.M. (1990) Pollution by industry and other users of chemicals. *Zanzibar Environmental Study Series* No. 2, 35 pp.

Muhando, C.A. (1998) The Status of Coral Reefs Around Zanzibar and the General Environmental Services they Provide. In *Coral Reefs: Values, Threats and Solutions. Proceedings of the National Conference on Coral Reefs. Zanzibar. Tanzania*, eds. R.W. Johnstone, J. Francis and C.A. Muhando. Institute of Marine Sciences, Zanzibar.

NEMC (1998) Chemical wastes. NEMC database.

Newell, B.S. (1959) The Hydrography of the British East African coastal waters. *Colon. Off. Fish. Publ. Lond.* **11**, 1–18.

Newton, L.C., Parkes, E.V.H. and Thompson, R.C. (1993) The effects of shell collecting on the abundance of gastropods on Tanzanian shores. *Biological Conservation* **63**, 241–245.

Ortland, N. (1997) Species Composition, Behaviour and Movement Patterns of Dolphins in Nungwi, Zanzibar. SIT, Marine Research Report. Institute of Marine Sciences, Zanzibar. University of Dar es Salaam.

Richmond, M.D. (1998) *A Guide to the Seashores of Eastern Africa and the Western Indian Ocean Island.* Pub: Sida/SAREC.

Semesi, A.K. (1999) Mangrove management and utilization in Eastern Africa. *Ambio* **27**, 620–626.

Stensland, E., Berggren, P., Johnstone, R. and Jiddawi, N. (1999) Marine mammals in Tanzanian waters: Urgent need for status assessment. *Ambio* **27**, 771–774.

Sterner, T. and Andersson, J. (1999) Private protection of the marine environment, Tanzania: A case study. *Ambio* **27**, 768–771.

Talbot, F.H. (1965) A description of the coral structure of Tutia Reef (Tanganyika territory, East Africa), and its fish fauna. *Proceedings of the Zoological Society, Lond.* **145**, 431–474.

UNEP (1989) *Coastal and marine environmental problems of the United Republic of Tanzania.* UNEP Regional Seas Reports and Studies No. 106.

Van Bruggen, J.J. (1990) Assessment of environmental pollution on Zanzibar. Consultancy report. Zanzibar environmental series. 33 pp.

WHO (1996) The WHO recommended classification of pesticides by hazard and guidelines to Classification, 1996–1997. International Programme for Chemical Safety. WHO/PCS/96.3. 64 pp.

THE AUTHORS

Martin Guard

Zoology Department, University of Aberdeen, Tillydrone Road, Aberdeen AB24 2TZ, U.K.
and
Department of Zoology and Marine Biology, University of Dar es Salaam, P.O. Box 35091, Dar es Salaam, Tanzania

Aviti J. Mmochi

Marine Environmental Chemistry, Institute of Marine Sciences, P.O. Box 668, Zanzibar, Tanzania

Chris Horrill

Tanga Coastal Zone Conservation and Development Project, P.O. Box 5036, Tanga, Tanzania

MOZAMBIQUE

Michael Myers and Mark Whittington

Mozambique, with its huge coastline, has an extensive territorial sea and an Exclusive Economic Zone (EEZ) with considerable economic value and significant biodiversity. The landmass of Madagascar shelters the coast from the Indian Ocean, so the marine environment is low-energy with high tidal ranges, warm water and a predominantly narrow continental shelf. Several island chains also run along the coast for some of its length, from the Quirimbas in the tropical north through the Bazaruto Islands in the central section, to Inhaca Island in the sub-tropical south. Fringing coral reefs are typical of this coast, while mangroves, seagrass beds and rocky and sandy shores are also common. Generally, all these habitats are in very good condition. Industrial activity is very limited but increasing; marine pollution is not yet a significant concern in Mozambique, but chronic pollution is probably increasing, partly as a result of the upstream activities of its neighbours. High volumes of tanker traffic move up and down this coast and there is consequently a high risk of accidental spillage of hydrocarbons. Lack of funds and technical expertise limit marine and coastal conservation initiatives.

Mozambique has recently put in place the legislative and institutional framework for good governance of the coast, but the structures are largely untested. There is capacity in some fish stocks to support greater fishing effort, while others are known to be over-fished. Prawn fishing in particular is an important source of national income. It is vital that government follows through on several promising measures initiated thus far, for conservation and for the sustainable development of coastal systems.

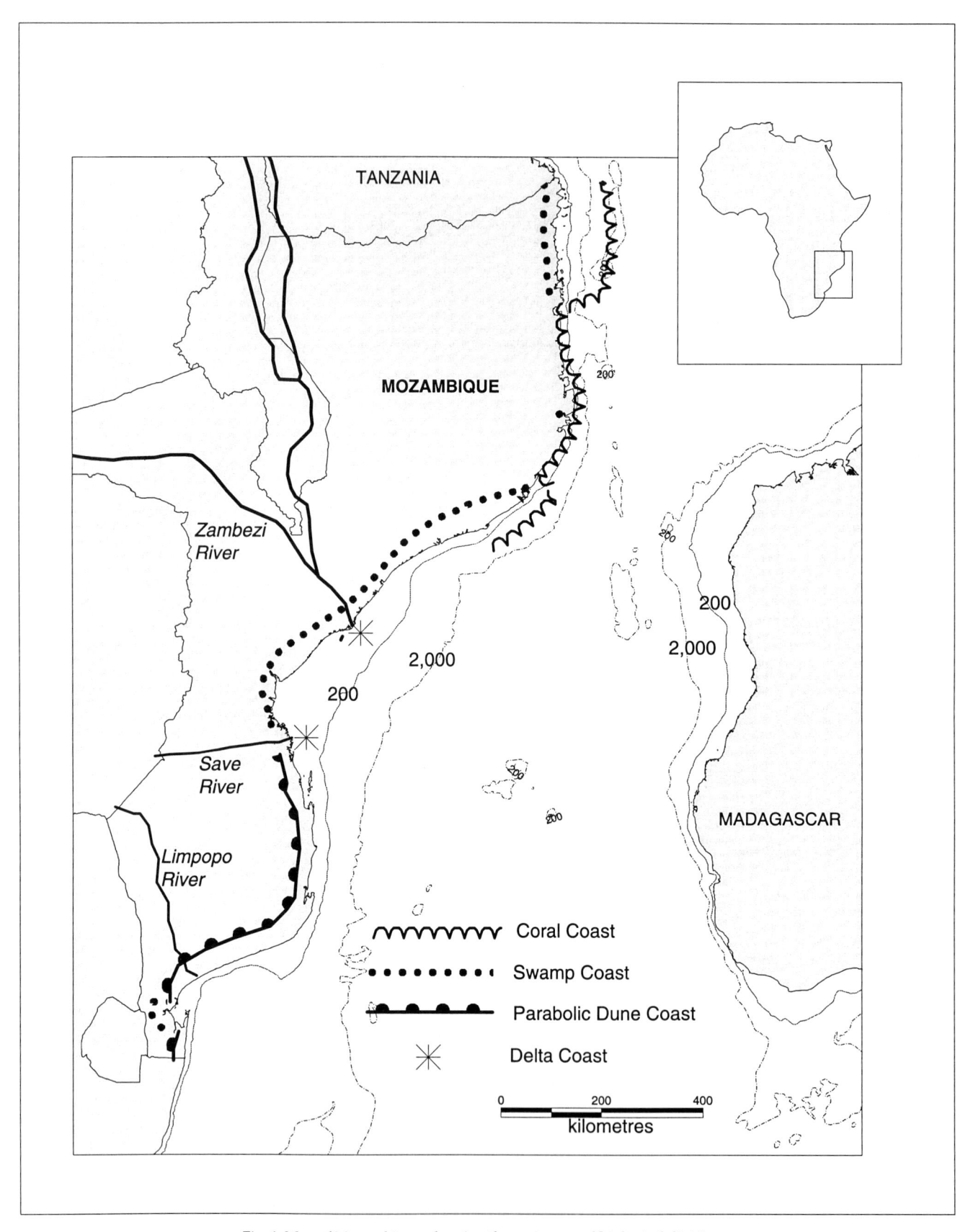

Fig. 1. Map of Mozambique, showing the main coastal/biological divisions.

THE DEFINED REGION

The sea off Mozambique—extending from 10°30'S to 26°49'S—is hydrodynamically distinct from other sections of the East African marine environment because of the sheltering presence of Madagascar several hundred kilometres to the East (Fig. 1). The influence of the Mozambique Current is important as it carries warm water from the Southern Equatorial Current down through the Mozambique Channel. These factors create a relatively sheltered warm water environment along much of the Mozambique coast. The coastline of the country is some 2770 km long (Massinga and Hatton, 1996) and the major habitats occurring there are mangrove, seagrass beds and coral reefs. Habitats immediately across the northern and southern borders in Tanzania and South Africa are not markedly different and the region discussed here is defined primarily by political demarcations.

The Mozambique coast may be divided into the following four sectors (following Hatton, 1996) (see Fig. 1).

(1) *Coralline coast*. From the Rovuma River at the Tanzanian border south as far as the Primeiras Archipelago at 17°20'S, an extent of 770 km. The reefs of this sector tend to be fringing reefs which line the exposed oceanic edges of island groups such as the Quirimbas and Segundas. They also fringe much of the continental coast as, for example, near Pemba, Mecúfi and Nacala. Recent studies on the reefs of the Segundos islands show them to be dominated by soft corals, while elsewhere hard corals are more significant (Whittington and Heasman, 1997).

(2) *Swamp coast*. The central section, some 980 km in length, is classed as swamp coast, with major estuaries, swamps and linear beaches opening onto a relatively wide continental shelf. Intense wave action contributes to high turbidity in this region. Extensive mangrove stands occupy the estuaries at the mouths of 24 rivers that discharge into the Mozambique Channel here. The littoral is characterised by low dunes running parallel to the coast.

(3) *Delta coast*. The Zambezi and Save river deltas, both in the central section, can be classified as delta coasts.

(4) *Parabolic dune coast*. From Bazaruto Island southwards for 850 km to the South African border (and continuing beyond it), the coast is characterised by high parabolic dunes and north-trending capes and barrier lakes. These vegetated dunes reach heights of 120 m. Within this zone scattered coral reefs do occur, usually offshore but also associated with the islands of Bazaruto and Inhaca (IUCN, 1988).

SEASONALITY, CURRENTS, NATURAL ENVIRONMENTAL VARIABLES

The Mozambique coast is sheltered from the Indian Ocean swell by the island of Madagascar 400 km to the East across the Mozambique Channel. The influence of the open ocean is felt only in the extreme North and extreme South of the country while elsewhere there is considerable deposition and longshore drift of river-derived sediments (IUCN, 1988). Cabo Delgado, the 'narrow cape', in the far north forms the dividing point for the South Equatorial Current. The southern-flowing branch is known as the Mozambique Current which takes warm water down the coast where it has a strong influence on the marine life and on the climate of southern Africa (Fig. 2). The current attains its greatest velocity during the Northeast monsoon in October and November, when it can exceed 6 km h^{-1} (Massinga and Hatton, 1996).

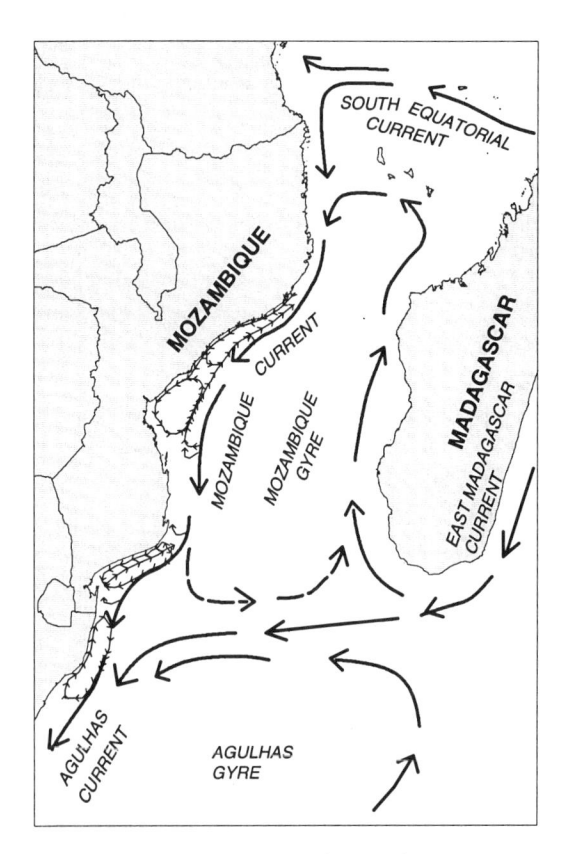

Fig. 2. Major currents of Mozambique.

Counter-currents occur in the Bights of Sofala and Maputo and these are responsible for the accretion of the northward trending peninsulas of Manchangulo, Inhambane and São Sebastião as well as the islands of the Bazaruto Archipelago and Inhaca (Massinga and Hatton, 1996).

Tidal ranges are generally high along this coast and Mean High Water Springs (MHWS) may attain 6.4 m at Beira where the continental shelf is at its widest. Tides are semi-diurnal for the most part; surface temperatures vary seasonally between 22° and 27°C (IUCN, 1988).

The climate along Mozambique's coast ranges from tropical humid to subhumid with the greatest rainfall experienced in the central sector around Beira and Pebane, where the Indian Ocean Sub-tropical Anticyclone System and the East African Monsoonal System compete for influence (IUCN/UNEP, 1988). Highest rainfall is in the summer (November to March). Along the coast, annual rainfall is typically from 800 to 900 mm; there is no monitoring of maritime weather. Four major rivers, including the Zambezi, and many smaller ones flow into the central section, and it is believed to be the low salinities and high turbidity associated with these that prevent scleractinian coral growth in this section as well as aiding the consolidation and expansion of the deltas.

Another factor that may be significant to marine and coastal habitats is the cyclones that occasionally affect the area (World Bank, undated), though generally the coast is not exposed to high energy levels despite the narrowness of the continental shelf.

THE MAJOR SHALLOW WATER MARINE AND COASTAL HABITATS

Beyond the general descriptions given above, there is a scarcity of more in-depth studies of the coastal and particularly of the sub-tidal habitats of this region, though quantitative estimates have recently been made for the northern parts of the coast. However, long-term and extensive study has occurred at the biological station at Inhaca Island (in the parabolic dune coast section) and to a lesser extent around the Bazaruto Archipelago (swamp coast section) and the Quirimba Archipelago (coralline coast section). Together these give a better insight into the condition of marine and coastal habitats in Mozambican waters.

The Quirimba Archipelago

The Quirimbas have been the subject of a two-year survey of marine and coastal habitats by the NGO Frontier-Moçambique (Heasman et al., 1998; Stanwell-Smith et al., 1998; Whittington et al., 1997, 1998; Whittington and Myers, 1997; Barnes et al., 1998; Gell, 1997).

Observations suggest that the oceanic waters near to the islands support a high diversity of pelagic fish, particularly Scombridae and Carangidae. Thus far there is no artisanal exploitation of the fish stocks in the offshore area due in part to the exposed nature of its coast and the low-technology fishing methods available to the local people.

The reef habitat consists primarily of exposed fringing reefs, with a well-developed hard coral cover. Eighteen scleractinian and five soft coral genera were identified during the surveys. In shallow waters *Acropora* spp. dominated whilst *Montipora* spp. and *Echinopora* spp. were prevalent on the lower reef. The main cause of damage to the reefs was judged to be wave action, and little anthropogenic impact was apparent at any site studied. Crown of Thorns starfish were widely noted, but not abundant.

On well-developed reefs with minimal fishing pressure, the diversity of reef-associated fish was considerable (Whittington and Myers, 1997); over 300 species were identified during reef surveys.

Gastropod diversity in surveyed taxa was high in both the subtidal and intertidal habitats although collection

Fig. 3. Extensive sandy beaches characterise much of the northern coast.

pressure for food and the curio trade, by islanders and itinerant fishermen, was substantial. Holothuria collection was another large-scale activity and sometimes involved the use of SCUBA diving as well as snorkelling gear. Most of the better equipped collection teams were made up of Tanzanian fishermen fishing illegally.

In the areas to the West of the Quirimbas, seagrass beds were the dominant habitat. These areas were unsuitable for much reef growth; they are flat and low lying and cover a vast expanse (Fig. 3). With 10 species of seagrass identified here diversity was relatively high, but the habitat may be classed as 'disturbed' due to intense and increasing fishing pressure. The socio-economic importance of the seagrass fishery to the islands is considerable, and the diversity of fish found in catch samples was 192 species. Both the biodiversity and the economic sustainability of the fishery appear to be at risk.

The mangrove stands of the area extend up to 1900 ha and are in good condition (Fig. 4). Intense cutting near to population centres seems to be causing or contributing to localised erosion. On the extensive intertidal flats around

Fig. 4. Dense mangroves line much of the coast which, with the seagrass beds, support most of the artisanal fishery of the region.

The Quirimba Archipelago: A Source of Productivity for South East Africa

Caroline Turnbull

Department of Biological Sciences, University of Warwick, Coventry, UK.

The Quirimba Archipelago lies along the northernmost section of coast in Mozambique, receiving warm water from the east and cold water from the south. The area is remote, with the capital lying far to the south, and has a few permanent communities on some of the larger islands. The inhabitants of the archipelago are highly dependent on the sea for their livelihood with a little commercial exploitation for local sale (Whittington and Myers, 1997). The area is largely unspoilt and is a rich mix of mangrove, seagrass and coral reefs common to the northern coast and beyond. For this reason it is a good representative site for both northern Mozambique and southern Tanzania from which productivity estimates can be estimated and extrapolated. Within it, coastal population is increasing, but there is currently no estimate of how much exploitation of the natural resources can be achieved sustainably.

Using GIS techniques, graphical representations of complex habitat distributions were made, along with quantitative areal estimates, in order to estimate gross area productivity (Hoare, 1998) (Fig. 1). The initial base layers of the GIS were made by digitising Russian nautical charts that were mapped in the 1980s. These delineated areas of coral reef and mangrove specifically, as well as lagoons and deeper water habitats where seagrass beds are found. Baseline data of the area was available from Frontier International and was integrated with the GIS to verify the charts. The areas covered by coral reef, seagrass and mangrove habitats were calculated, and net primary productivity (NPP) for these habitats was taken from literature and used to calculate the NPP over a year in the Quirimba Archipelago. Total NPP was calculated for the whole intertidal and subtidal area less than 20 m deep and secondary productivity was roughly estimated using the accepted ratio of 10:1.

It was assumed that habitats were uniform within the areas allocated to them and no attempt was made to incorporate algal habitats and sandy areas. Therefore, the estimates of NPP and secondary productivity are likely to be higher than reality but do give a good basic picture of what productivity is like in the area from which more detailed work can be done. Table 1 summarises the findings of the productivity calculations and compares them to accepted figures for major marine environments taken from Butcher et al. (1992).

The Archipelago was found to be very productive, comparing very favourably against the productivity per kilometre of major global habitats. The estimated figures for coral reef and mangrove are probably robust since these areas were confirmed by the ground truthing. The seagrass values are the most tentative since much less ground truthing was done

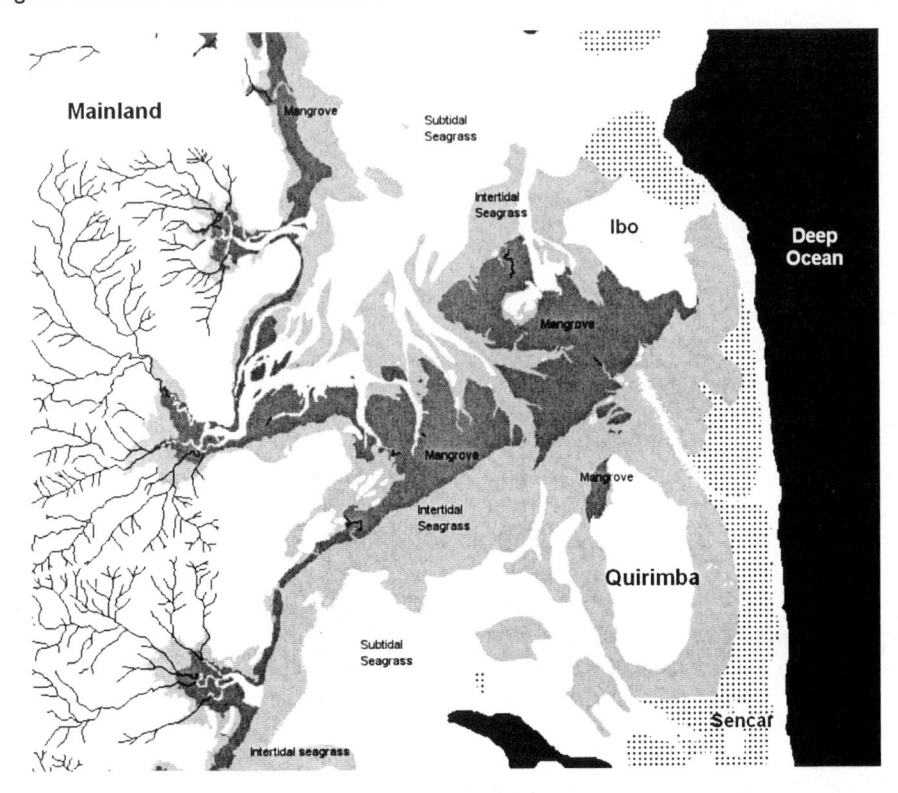

Fig. 1. A section of the Quirimba Archipelago. Dark green: Mangrove; turquoise: coral reef; light greens: intertidal areas designated as seagrass habitat; light blue: subtidal area less than 20 m deep designated as seagrass habitat; dark blue: open water greater than 20 m deep.

Table A

Estimated NPP values for the Quirimba Archipelago (values given at two decimal places as calculated) and for contrasting global habitat areas (values taken from Butcher et al., 1992)

Habitat in Quirimba Archipelago	Net primary productivity (kg C yr^{-1})	NPP per km^2 (kg C km^{-2} yr^{-1})
Coral reef	2.3×10^8	1.3×10^6
Seagrass	3.5×10^9	3.8×10^6
Mangrove	4.2×10^8	3.7×10^6
All	4.1×10^9	2.9×10^6
Global habitat estimates		
Algal beds, reefs and estuaries	1.6×10^{12}	8.0×10^5
Continental shelf	4.3×10^{12}	1.6×10^5
Upwelling	0.1×10^{12}	2.5×10^5
Open ocean	18.9×10^{12}	5.7×10^4

in these areas (but see panel entitled "The SeaGrass Fishery of Montepuez Bay"). Therefore, the present results may be an overestimation of the NPP for the area. These reservations aside, the Archipelago clearly can support high primary productivity.

Seagrass was seen to be the most abundant habitat accounting for 75% of the total area, and 84% of the total NPP could be attributed to this habitat. Perhaps 50% of dissolved organic carbon present in them may be exported to surrounding habitats (Sheppard, 1995). Mangroves likewise are a valuable source of organic carbon with 75% of carbon fixed in photosynthesis being exported from the system (Sheppard, 1995). Since 85% of the area covered by the three main habitats is accounted for by these two habitats, there is strong potential for a high level of export of carbon from the archipelago to the surrounding areas. Therefore, the Quirimba Archipelago is not only important locally as a source of carbon for fisheries and biological communities but may also be supporting productivity to areas both north and south, including to Tanzania where coastal areas are much more heavily exploited (Guard & Masaiganah, 1997).

The 'coralline coast' of Mozambique stretches for 770 km (Massinga and Hatton, 1996). Assuming that the width of the intertidal and subtidal is approximately the same as that within the archipelago, it is possible to roughly estimate an NPP of 3.3×10^{10} kg C yr^{-1}. When this is compared to the values in Table A it is clearly a significant quantity and provides a rich basis for a good resource base within northern Mozambican coastal waters. The same extrapolation can be applied to the coast of Tanzania, which is roughly 1000 km long (Shah et al., 1997), to give an annual NPP of 4.3×10^{10}.

The productivity of the Quirimba Archipelago can therefore support an important local fish community. An estimated figure of 4.1×10^5 t C yr^{-1} for secondary productivity was obtained for the area of the archipelago as mapped on the GIS, which provides a substantial quantity of biomass to be fished by local communities and be passed through the food web. Levels of exploitation are currently low using non-destructive techniques and exploitation would appear to be sustainable. Therefore, the Quirimba Archipelago is also an important source of carbon to areas further afield, and these estimates of primary and secondary productivity support the argument that the northern coast of Mozambique is a valuable resource for South East Africa.

REFERENCES

Butcher, S.S., Charlson, R.J., Orians, G.H. and Wolfe, G.V. (eds.) (1992) *Global Biogeochemical Cycles*. Academic Press, 379 pp.

Guard, M. and Masaiganah, M. (1997) Dynamite fishing in Southern Tanzania, geographical variation, intensity of use and possible solutions. *Marine Pollution Bulletin* **34**, 758–762.

Hoare, C.G. (1998) The determination of net primary productivity in the Quirimba Archipelago using a GIS and a relational database. MSc Thesis, University of Warwick.

Massinga, A. and Hatton, J. (1996) Country paper on the status of the coastal zone, Mozambique. Paper prepared for the ICZM workshop, Ilha da Inhaca, 5–10 May 1996, 28 pp.

Sheppard, C.R.C. (1995) Biological communities of tropical oceans. In *Environmental Biology, Vol 1*, ed. W.A. Nierenberg, pp. 277–289. Academic Press.

Whittington, M.W. and Myers, M. (1997) Technical Report 1: Introduction and Methods. Marine Biological and Resource Use Surveys of the Quirimba Archipelago, Mozambique. Society for Environmental Exploration, London, and the Ministry for the Coordination of Environmental Affairs, 47 pp.

---

the islands, there was a high diversity (195 taxa) of macroalgae and a similar diversity of invertebrates. The latter were collected intensively for subsistence use while the seaweeds apparently are not used.

Overall, this northern archipelago is a site of high habitat diversity with great species richness seen in each habitat type. Though still in generally unspoilt condition there are areas where human impacts are being felt and increasing resident and transient populations will accelerate this effect.

The Bazaruto Islands

At Bazaruto, as in the Quirimbas, there is a combination of shallow and shelf waters and thus an important range of marine habitats in a small area. Marine habitats include coral reef, sandy beach, rocky shore, seagrass beds and deep water. Additionally, the Bazaruto islands boast several fresh- and brackish-water lakes. The sand dunes play a central role in the hydrology of the islands and are also the habitat suffering the greatest impact from human activity.

In 1993, Guissamulo estimated a dugong (*Dugon dugon*) population of 300 individuals for the archipelago—one of the most significant populations on the entire East African coastline. However, preliminary indications from work in 1997 indicated a possible population collapse to less than 100 individuals. It is known that incidental capture of dugongs (and turtles and dolphins) takes place during shark netting in the area (MICOA, 1997). Other marine mammals found around Bazaruto are the common dolphin (*Delphinus delphis*), spinner dolphin (*Stenella longirostris*), bottlenose dolphin (*Tursiops truncatus*), humpback dolphin (*Sousa chinensis*) and the humpback whale (*Megaptera novamangliae*).

All five of Mozambique's marine turtles frequent the waters of the archipelago and nest on its beaches: Pacific green (*Chelonia mydas*), Loggerhead (*Caretta caretta gigas*), Hawksbill (*Eretmochelys imbricata*), Leatherback (*Dermochelys coriacea*) and Olive Ridley (*Lepidochelys olivacea*). Of the 180 bird species that have been identified in the area, many are marine and waterfowl and the islands are an important stopover site for several migratory birds.

More than 2000 species of fish have been identified so far and around 80% of the families known for the Indo-Pacific region are represented around Bazaruto (Magane, 1996).

Inhaca and Portuguese Islands Reserves

With an area of around 750 ha, these reserves were created to protect dune vegetation, turtle nesting beaches and coral reefs at this site adjacent to the capital at around 26°S. A biological research facility has existed on Inhaca for several decades and there has been a great deal of study of marine and terrestrial habitats (e.g. Macnae and Kalk, 1969; Campbell et al., 1988).

The main features of Inhaca Island, once a part of the Machangulo Peninsula, are the two east and west dune lines with subsidiary dune lines between them, along with associated shoals and fresh-water lagoons. The sandy soils of Inhaca are largely infertile and active natural erosion and deposition cycles sometimes threaten the integrity of the dune lines. Wind-blown sand also smothers the little terrestrial vegetation which does grow and is particularly detrimental to agricultural plots (MICOA, 1997).

Local marine fauna includes dugongs and dolphins (MICOA, 1998) and loggerhead and leatherback turtles. There are also mangroves, extensive seagrass beds in sheltered areas and well-developed coral reefs. Thirty-five hermatypic genera have been recorded at Inhaca; the fringing reef on the oceanic coast is composed of typically encrusting species adapted to the exposed conditions. This far south the mitigating influence of the landmass of Madagascar is reduced and the coastal environment is a high-energy one. Significant colonisation of soft substrata has occurred here because of the pioneering tendencies of *Porites* and *Goniopora* colonies, and total coral cover can reach 43% (MICOA, 1997).

The biodiversity within the coastal habitats of Mozambique can be seen to be substantial on the basis of the relatively few studies so far undertaken. The good condition of large areas of coral reef is important given the highly impacted state of this habitat type elsewhere in Eastern Africa. More studies are planned and these can be expected to reveal a still higher biological richness.

OFFSHORE SYSTEMS

Offshore systems of Mozambique have been little studied. The Mozambique Gyral current, along with the presence of Madagascar, mean warm sheltered water typifies this zone. Along most of the coastline deep water is encountered a short distance offshore, often as close as 2 km, though in the central section, between the southernmost of the Primeiras islands and Bazaruto island, the continental shelf is significantly wider, up to 180 km. Fishing for pelagic species such as tuna is the only economic activity in this zone. Migrating southern right whales and resident humpback and minke whales occur in deep waters as well as closer to shore. Surveys of whale populations in Mozambique waters are occasional, e.g. Best et al. (1991).

Mozambique is a signatory to the UN Convention on the Law of the Sea (Table 1) and claims an Exclusive Economic Zone of 562,000 km^2 (WCMC, 1997). Government fishery figures (SSF, 1994) put the level of exploitation of marine fish as very low (large pelagics) to low (small pelagics, shark, small demersals and bottom fish). Total catch from the industrial and semi-industrial fishery is given by the government as 19,700 tonnes for 1994, although FAO

Table 1
Participation in International Conventions (after MICOA, 1997)

Convention/Organisation	Status
Convention on Biological Diversity	Ratified 1996
Ramsar Convention	Steps being taken to accede
CITES	Acceded 1981
Convention on the Conservation of Migratory Species of Wild Animals	Not signed
Convention on Migratory Birds	Not signed
Convention for the Management, Protection and Development of the Marine and Coastal Environment of the Eastern African Region and related protocols	Ratified 1996
International Maritime Organisation	Member since 1991
Organisation on Indian Ocean Marine Affairs Cooperation (IOMAC)	Member since 1991
The Basel Convention (transboundary movement of toxic wastes and their disposal)	Ratified 1997
The Bamako Convention (ban on import and control of transboundary movement of Hazardous Wastes within Africa)	Ratified 1997
Protocol on Shared Watercourse Systems in the SADC Region	Signed 1995
The Zambezi River Basin Multilateral Agreement	Signed 1987
UNCLOS	Signed 1982; Ratified 1996
Convention on Climate Change	Ratified 1994
African Convention on the Conservation of Nature and Natural Resources	Ratified 1984

The Seagrass Fishery of Montepuez Bay, Northern Mozambique

Fiona Gell

Tropical Marine Research Unit, Department of Biology, University of York, York, YO10 5YW, UK.

On Quirimba Island coral reef fisheries play a small part in the local economy (Gell, 1999). Although the island has extensive and well developed reefs, most of them are exposed to strong winds and heavy wave action and lie a long distance from the village. They are therefore inaccessible to the local fishing fleet which uses traditional sailing boats and small outrigger canoes. Most of the island's fisheries are instead based on the extensive seagrass beds of the Montepuez Bay. The northern end of Montepuez Bay is shallow (1–10 m) and has extensive intertidal flats and banks. The intertidal seagrass beds are dominated by the seagrass *Thalassia hemprichii* and are used by local women who collect bivalves. Subtidally, the seagrass beds are dominated by the large seagrasses *Enhalus acoroides* and *Thalassodendron ciliatum*, both of which can grow to over one metre in height. These are the main fishing sites for local fishermen, who fish predominantly in *E. acoroides* beds using two main gears: basket traps, known locally as *marema*, and seine nets.

Marema are set by fishermen in outrigger canoes in shallow areas of *E. acoroides* at low tide and are hauled the following day at low tide. The mean daily catch for a fisherman setting 40 traps was nearly seven kilos, although catches could be as high as 27 kg per trip. The catch per unit effort (CPUE) for the trap fishery was 2.2 kg of fish per man hour spent fishing, or 1.7 kg of fish per man hour spent at sea. Trap catches were dominated by the seagrass parrotfish, *Leptoscarus vaigiensis* which accounted for nearly 75% of the fish caught by weight. Other important species included the parrotfish *Calotomus spinidens*, the rabbitfish *Siganus sutor*, the goatfish *Parupeneus barberinus*, the snapper *Lutjanus fulviflamma* and the wrasse *Pteragogus flagellifera*. A total of 61 species of fish were identified from 3500 fish sampled from the trap fishery, approximately 16 of them common in the fishery.

Seine nets are set from small sailing boats by teams of between five and twelve men in shallow areas of *E. acoroides*. The nets used are approximately 100 m in length with a mesh size of 4 cm stretch and a cod-end of 2 cm stretch or less. The mean duration of a fishing trip was nearly five hours and the mean catch per trip was 75 kg. Catch per unit effort was 3.6 kg of fish per man hour spent fishing or 2 kg of fish per man hour spent at sea. A total of 252 species were identified from the 46,600 fish sampled. Approximately 30 species were common in the catch. The most important species in the net fishery in terms of weight were: the rabbitfish *S. sutor*, the emperors *Lethrinus lentjan* and *L. variegatus*, the parrotfishes *Leptoscarus vaigiensis* and *Calotomus spinidens* and the mojarra *Gerres oyena*. The majority of fish caught in seine nets were less than 15 cm long and many were juveniles (see Table 2 in this chapter for catch composition and mean lengths of common species).

The seagrass fisheries of the Montepuez Bay support over 400 fishermen on Quirimba Island alone and many more in the mainland village of Quissanga and from other islands in the vicinity. The total fish catch from the 35 km^2 seagrass beds of the whole bay has been estimated at around 500 tonnes per year, or 14.3 tonnes per km^2 per year. This is high compared to many tropical reef and estuarine fisheries (Blaber, 1997; Dalzell, 1996). Locally this has an annual saleable value of approximately US $120,000. Roughly half the fish caught is consumed by the fishers and their families or exchanged for other goods. The other half is dried and traded on the mainland by the boat owners, or other traders.

CPUE and catch composition were compared with the habitat characteristics of the main fishing sites in the Montepuez Bay. There was a positive relationship between total seagrass cover and net fishing CPUE (Fig. 1), suggesting that seagrass cover influences fish biomass and fishery productivity. In experimental trap fishing, the preference of fishermen for areas of the seagrass *E. acoroides* was shown to be well-founded. The mean catch per trap for *E. acoroides* (220 g or 3.5 fish per trap) greatly exceeded that for other common seagrass species *Thalassodendron ciliatum* (84 g or 1.6 fish per trap) or *Cymodocea rotundata* (68 g or 0.9 fish per trap).

Ten species of seagrass are present in the seagrass meadows of the northern Montepuez Bay: *E. acoroides*, *T. ciliatum*, *C. rotundata*, *Cymodocea serrulata*, *Syringodium isoetifolium*, *Halodule uninervis*, *H. wrightii*, *Halophila ovalis*, *Halophila stipulacea* and *Thalassia hemprichii*. These species represent a variety of adaptations to the shallow water environment. It is therefore important that this variety of seagrass species is maintained in the seagrass beds. Although *E. acoroides* provides the highest yields in terms of fish catches, the other species undoubtedly play important ecological roles.

The importance of tropical seagrass beds in supporting productive and economically important fisheries is not widely acknowledged. Seagrass beds are most often thought of as nursery grounds for commercial fisheries species such as prawns or for their role in nutrient cycling or in coastal protection. The direct use value of seagrass beds is rarely considered and yet is important in coastal communities in developing countries from East Africa to the Philippines. The Montepuez Bay seagrass fishery is an excellent example of the importance of shallow seagrass habitats in supporting significant fisheries. It is also a highly diverse ecological system with species numbers in some taxa comparable with those found on coral reefs.

With so many people relying on the seagrass beds of the Montepuez Bay for their livelihoods, and with the paucity of alternative employment or sources of protein, it is vital that the conservation of the local seagrass beds and their sustainable use is promoted and monitored.

REFERENCES

Gell, F.R. (1999) Fish and fisheries in the seagrass beds of the Quirimba Archipelago, Northern Mozambique. DPhil Thesis, University of York., UK.

Blaber, S.J.M. (1997) *Fish and Fisheries of Tropical Estuaries*. Chapman and Hall, London.

Dalzell, P. (1996) Catch rates, selectivity and yields of reef fishing. In *Reef Fisheries*, eds. N.V.C. Polunin and C.M. Roberts. Chapman and Hall, London, pp. 161–192.

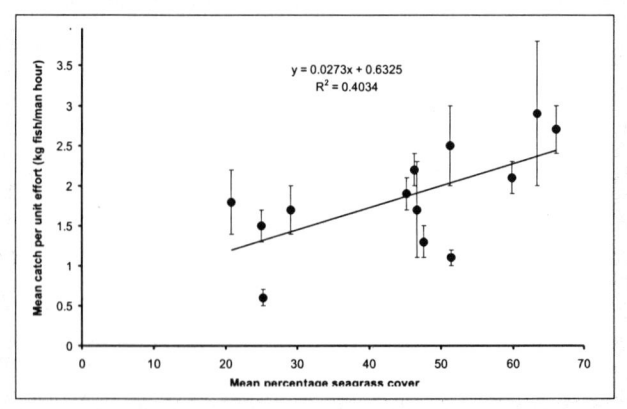

Fig. 1. Mean percentage cover of seagrass within fishing sites in the northern Montepuez Bay against mean fish catch per unit effort ($\pm$SE) for the sites. Each point represents an individual site.

estimates are closer to 33,500 tonnes (WCMC, 1997). The government fisheries department estimations suggest that the MSY is considerably higher than the current catch (SSF, 1994) and therefore there seems to be scope for expansion of the fish catch in Mozambique waters. Fishing remains vitally important to the economy though the productivity of Mozambican fishing grounds is low by comparison with the many upwelling areas on the West African coast (such as the Banc d'Arguin in Mauritania).

Impacts from the industrial fleet, including garbage dumping and torn net discard, (MICOA/UNEP, 1997) are covered later.

POPULATIONS AFFECTING THE AREA

The human population of Mozambique was assessed by census in 1997. The provisional total was 15.7 million of which more than 50% live in the coastal zone (defined as within 50 km of the shore). As the estimated population had been 18.5 million, growth since the previous census has been dramatically less than was assumed. A trend towards urbanisation is clear—the population of the capital, Maputo, doubled to 1.2 million between 1976 and 1996 (Massinga and Hatton, 1997) and other cities and towns have also seen greater than proportional growth. Three of the four largest cities are coastal; some of the move to the cities and to the coast is explained by the civil war, which drove people to areas of firmer government control and better security.

The arrival of internally displaced families has caused additional pressure on coastal resources. Not only were there more people fishing and cutting mangrove on the coast but these people had little understanding of the resources they were exploiting and little interest in the sustainability of their activities.

RURAL FACTORS

Subsistence farming on smallholdings, or *machambas*, using slash and burn techniques, combined with livestock rearing, is common on the nutrient-poor soils of the coast. Secondary or primary growth is burnt back and the plots are farmed for 3–4 years before being abandoned. Studies on Inhaca Island have shown that soil recovery from this activity is very slow, with no significant increase in nutrient status after 15 years (Serra King, 1995). The impacts of this activity are related mostly to the population density rather than technical differences in method. Soil exhaustion, diminishing fallow periods, especially in the south, pressure on forestry resources for firewood and construction, and poor irrigation and drainage leading to saltwater intrusion are the principal problems in this sector (MICOA, 1998). The capacity of coastal land to support agriculture thus deteriorates and has driven some farmers to take up fishing. At the same time coastal waters experience increasing turbidity.

Artisanal fishing is estimated to be worth 40 million dollars annually (MICOA, 1998) and is the primary source of protein and a significant employer in coastal communities (Table 2). Fishing technology is not advanced: outboard motors are a rarity and sailboats and canoes are generally not seaworthy enough to venture far offshore. Small vessels tend to be owner-operated while sailboat owners may employ a crew; payment is in the form of a share of the catch or of the money raised from its sale, and this system applies equally to another common technique which is beach seining.

Government figures show that on average artisanal fishers work 14 days per month, 10 months of the year. Landings in 1986 were put at 13.3 thousand tonnes and in 1994 at 3.3 thousand tonnes (MICOA, 1998). The decline appears to contradict the growth in the number of people involved in this activity (which has doubled to 80,000 since 1980). Although restricted to relatively localised areas, overfishing could certainly have contributed to the decline in catch, and it is likely that unreliable catch data is the major factor for the reported decline. Therefore all figures relating to the artisanal fishery need to be treated with caution.

In the Quirimbas, fishing techniques include box traps, fence traps, handlining, longlining, spearing and seine netting. Pressure on the near-shore and sheltered areas is

Table 2

Artisanal fishery production (tonnes) (from MICOA, 1998)

Type	1986	1987	1988	1989	1990	1991	1992	1993	1994
Prawn	143	96	14	135	832	469	237	375	105
Finfish	13195	10557	5093	5676	7436	4900	3447	3300	3205
Lobster	–	–	–	–	231	5	30	20	13
Crab	–	–	–	–	198	133	84	95	15
Squid	–	–	–	–	2	20	29	42	16
Holothuria	–	–	–	–	–	5	–	0	–
Bivalves	–	–	–	–	68	12	9	5	11
Octopus	–	–	–	–	–	–	–	2	–
Total	13338	10653	5108	5811	8767	5544	3835	3839	3362

great. Fishing on the seagrass beds in the area targets very small finfish which, given increasing pressure on the fishery, makes recruitment overfishing a probability. The reef fishery is generally under far less pressure; the reef areas which are more sheltered and accessible do have lower abundances of target fish, though no causality could be shown (Gell, 1997). Fishing on the reef sometimes involves trampling of the corals. Seagrass beds also suffer trample damage during invertebrate collection from the shallow subtidal; seine netting too damages seagrasses in this region, the impact being significant in heavily fished areas (Whittington et al., 1997).

Finfish exploitation is an overwhelmingly male activity, while collection of molluscs on the intertidal is primarily carried out by women and children at spring tides, and represents an important subsistence activity. Intense collection pressures suggest over-exploitation of these resources (Heasman et al., 1998; Stanwell-Smith et al., 1998; Whittington et al., 1998; Whittington and Myers, 1997).

COASTAL EROSION AND LANDFILL

Erosion is an issue of particular importance in the two largest cities. Both hard and soft coastal defences in Maputo are being swept away by wave and storm action and there is active erosion elsewhere along the southern coast. In Beira, the situation is more serious because the greater part of the city is below the current MHWS and in an area of active subsidence; many major parts of the infrastructure are at risk. The vital dune cordon in this area is protected by vegetation vulnerable to human disturbance, both by tourists in four-wheel drives and pastoralists grazing their herds. Felling of coastal *Casuarina* trees, sand mining for construction and bad siting of construction projects all exacerbate the situation (Hassan, 1997).

Another factor causing erosion is the damming of major rivers. The enormous Cahora Bassa dam on the Zambezi contributes to erosion in the Zambezi Delta, and the Kariba dam, further upstream, will also have an influence (MICOA, 1997).

At the local level, mangrove deforestation which is intense near population centres, will promote coastal erosion as it does on Quirimba Island in the far north (Whittington et al., 1997). Bad forestry practices, with their associated sediment runoff, cause turbidity and sedimentation that threaten reefs and seagrass meadows. Seagrass destruction will also increase coastal erosion rates in these areas.

Dredging is carried out often around Maputo (1,200,000 m^3 y^{-1}) and Beira (2,500,000 m^3 y^{-1}). These substantial dredged areas are needed to keep the ports functioning, and the scale of the activity is attributed to bad land use practices inland (MICOA/UNEP, 1997). The effect on the habitat of this combination of sedimentation and dredging has not been studied at these sites.

There is a rapid increase in the development of land for coastal tourism, for example in Pemba and near Xai-Xai. It is doubtful if these developments will be built to high environmental standards, and jurisdiction over these activities is confused between different government agencies and different levels of government. Plans are drawn up by district authorities in apparent contradiction of national policies, as has happened at Xai-Xai (Hatton et al., 1996). Without a clearer planning framework, impacts on coastal habitats are likely to be severe as development expands.

EFFECTS FROM URBAN AND INDUSTRIAL ACTIVITIES

Artisanal and Non-industrial Uses of the Coast

Shell collection and trade is a major activity along the coast (Whittington and Stanwell-Smith, in press; Barnes et al., 1998; Gessner, pers. comm.). Some of this activity is for a local market of Mozambicans and visitors, but greater pressure comes from artisanal fishers collecting for sale to traders for export, largely to Europe. Coral heads form a part of this curio trade. Fishing teams visit the Quirimbas from Nacala and other towns from Nampula province to the South, specifically to collect curio shells for trade. The degradation of this resource that is being seen now at the Quirimbas has probably already happened along the coast of Nampula province, hence the long journeys made by fishermen from that area. Indicative of over-exploitation is the switch from intertidal to subtidal collection, and this process has begun also in the southern Quirimbas (Whittington et al., 1998).

The story of sea cucumber exploitation is similar. Valuable resources are being intensively exploited and the benefits are being exported; in this instance a large part of the catch from Cabo Delgado and Nampula provinces is sold on via Tanzania. Indeed, many of the collectors are Tanzanians working in teams and using SCUBA gear to collect as deep as 50 metres. Holothuria are collected opportunistically by local people and sold to commercial buyers (Whittington et al., 1998); no size restrictions are known to be applied.

Coral mining is certainly carried out in Mozambique, for construction material, for lime production and for the curio trade. The extent of the activity is not known and it does not feature in the technical literature. Dynamite fishing, if it occurs at all, is at a very low level, in contrast to the situation across the border in southern Tanzania. The prevalence of dynamite fishing around Mtwara, the open nature of the border and the cultural continuity of communities either side of it suggest that the southward spread of dynamite fishing will continue into Mozambique if measures are not taken to prevent this.

Invertebrate collection in the intertidal zone, for consumption or barter, is a common activity for women and children of coastal and island communities throughout Mozambique. Collection pressure is greatest near to population centres. On the inhabited islands of the Quirimba chain it could be intense on spring low tides, to the extent that supra-tidal mollusc biodiversity is thought to be threatened.

Around Bazaruto fishing is the key economic activity, occupying 70% of the 3000 inhabitants; Dutton and Zolho (1991) estimated a catch of over 1300 tonnes per year, the main fishing methods being beach seining, hand lining and beach traps. This catch exceeds the requirement for subsistence and there is a surplus for sale to local tourist facilities and for export to the mainland.

The shrimp fishery is the major marine economic resource. On the Sofala Bank in central Mozambique the catch composition and overall tonnage are believed to have been affected by changes to the freshwater regime in the mangrove, reduced mangrove area and erosion of the delta. There is frequent disagreement between the artisanal shrimp fleet and the industrial vessels, which are said to come within 5 nm of the coast. The big trawlers operating so close to shore are said to cause environmental damage and also force the artisanal fleet into a smaller, near-shore area where their effort is concentrated in areas such as nursery grounds (MICOA, 1998).

Industrial Uses of the Coast

Coastal agriculture is increasing. Farming of marginal land and inappropriate farming techniques are causing problematic siltation and increased nutrient loadings in near-shore waters. Deforestation is continuing rapidly, for example in Cabo Delgado, and increased siltation in the coastal waters is to be expected. Intensive farming is on the increase in Mozambique but the greater problem is fertiliser and pesticide pollution caused by Mozambique's upstream riparian neighbours, such as Swaziland and Zimbabwe. Sugarcane farming in the Umbeluzi river valley in Swaziland is one particular example of a transboundary threat to Mozambique's coastal waters (Massinga and Hatton, 1997).

Shrimp aquaculture, promoted by government policy, is beginning to develop around the Beira area at the expense of mangrove stands. Mangrove is already felled for firewood and for the development of salt pans. MICOA (1997) quote an average annual mangrove deforestation rate, since 1972, of only 0.2% but with rates of up to 15% near population centres.

Industrial fishing is typically carried out on a joint venture basis, with nationals and expatriates in partnership. This applies to tuna fishing in deep waters as well as to shrimp fishing. With minimal surveillance systems in place, control of this fishery is very difficult and there is the additional problem of boats entering the EEZ from abroad to fish illegally, as the chances of apprehension are negligible. Industrial fishery production has declined from around 25,000 tonnes in 1986 to 17,000 tonnes in 1994 according to government figures (MICOA/UNEP, 1997).

Cities

Marine pollution in Mozambique is, so far as is known, a small-scale problem, confined to the coastal cities. The lack of work on the subject makes it impossible to give a comprehensive picture; what follows is based on Fernandez (1996).

The domestic sewage treatment facility in Maputo is the only one in the country and treats around 50% of the capital's waste. Some homes have septic tanks but outfalls are used for the remainder and for the other coastal cities. As coastal urbanisation continues, sewage disposal will increasingly affect the marine environment, along with garbage dumping from domestic and industrial sources. The impacts of sewage pollution around Maputo Bay include faecal coliform and streptococci contamination of marine waters and shellfish tissue, and *Vibrio* sp. contamination of clams. Total coliform counts in the bay are increasing and, in some areas, make swimming unsafe. In Beira and Nacala bays also, high levels of biological pollution have been recorded.

Industrial discharges are made into the rivers around Maputo/Matola and there is little if any treatment. Tyre, textiles, paper and beer production are large secondary industries, and heavy metals are building up in the river mouths and the bay. So far the level of industrial activity remains so low that the impacts of industrial pollution are localised.

Shipping and Offshore Accidents and Impacts

A huge volume of crude oil is transported down the Mozambique Channel—in the region of 450 million tonnes annually (MICOA, 1997). The risk of a catastrophic spill is high and increases with the age of the world tanker fleet. In typical weather any spill will be carried onto the coast, as was demonstrated by the 'Katina P' spill (16,000 tonnes of heavy fuel oil) in 1992 (Massinga and Hatton, 1997). There is believed to be no contingency plan in readiness for such an event (MICOA/UNEP, 1997).

Chronic hydrocarbon pollution occurs due to tanker discharges and bilge pumping while at sea. Mozambican ports have minimal facilities for handling bilge and there is also the likelihood of periodic spillage due to handling errors and poorly installed and maintained transfer pipes at the main petroleum ports of Beira and Maputo. Oil and gas exploration (and ultimately exploitation) in offshore areas, such as the Rovuma Basin near Tanzania, will contribute to the marine oil pollution hazard.

PROTECTIVE MEASURES

Mozambique lacks clear baseline data against which the trends in biodiversity and habitat condition can be measured, but in spite of this there are unarguable signs of

Table 3

Marine and Coastal Protected areas in Mozambique

Protected Area	Size (ha)	Features and Status
Marromeu Reserve	150,000	Composed of floodplains and mangrove. No active conservation in this reserve.
Maputo Reserve	70,000	Mangrove, dune forests, swamp forests, floodplains and grasslands. Possible home to turtle and dugong populations. Combined NGO/ministry management.
Pomene Reserve	20,000	Confusion over eastern boundary of this reserve means that the actual coastal strip may be outside the reserve. Possible turtle and dugong presence in the area. No active conservation currently.
Bazaruto National Park	8,000	Two islands within a small archipelago. Littoral waters contain significant dugong population and turtles
Inhaca and Portuguese Islands Biological Reserve	750	Turtle nesting, dune forests and coral reefs. Marine research station.

Table 4

Proposed coastal protected areas

Nacala–Mossuril	Turtle nesting, seabird nesting, seagrass beds and fringing coral reef. Thought to have tourist potential.
Primeiras and Segundas Islands	Fringing reefs, nesting turtles, seagrass beds, mangrove, dugongs and dune forest.
Quirimba Archipelago	Reefs, seagrass beds, extensive mangroves, turtles, tern colony, high biodiversity and habitat range.
San Sebastian Peninsula	No details.

decline in key marine habitats. Government has accepted the need for protection of marine mammals, mangrove forest, coral reefs and fish stocks (Table 3) and have recognised fundamental principles of biodiversity conservation, community participation, protected areas and research (MICOA, 1997).

The extent to which any of the terms of these conventions are reflected in national legislation has not been assessed.

Bazaruto was created as Mozambique's first marine national park in 1971 with the particular objective of protecting dugong and turtle populations. The protected area comprises three small islands and a 5-km zone around them. Work by WWF and DNFFB is under way to reassess the management plan for the park and also to perhaps extend the managed area, in response to a perception that the park is not yet fulfilling the role that it was intended to.

The current aims of the park authorities are to:
- conserve ecological integrity and biodiversity;
- provide environmental education and training to relevant individuals;
- promote the studies necessary for the correct management of the Park;
- permit greater community participation in the management and sustainable use of the resources of the park;
- permit the creation of recreation facilities for tourists within social and ecological constraints of the area;
- promote economic self-sufficiency of the park.

The Inhaca and Portuguese Islands Reserve is the only other existing protected area where marine conservation is of central importance.

As well as the reserves and parks mentioned in Table 3, there are four proposed coastal protected areas, listed in Table 4. The extents of these have not been defined.

Of the four, only the Quirimbas have been the subject of extensive study and investigation (see Whittington et al., 1998 for references) and the very fact that data are available may favour this northern archipelago as the site for the development of the next Mozambican MPA. It is worth noting that there has been no new MPA declared since 1981.

The setting for wise use of natural resources in Mozambique underwent a major change with the passing of the Framework Environmental Law in 1997. Sections of the law relate specifically to the conservation of biodiversity, protected areas, sustainable development and cross-sectoral integration. The net effect of the new legislation is to commit government to strong measures for environmental protection. Environmental Impact Assessments will be required for future projects 'likely to cause significant environmental impacts'. The creation of MICOA (Ministry for the Co-ordination of Environmental Affairs) is another important step as it will be the central government body responsible for the implementation of the National Environmental Management Plan, the document which expresses government policy and details the steps necessary to implement it.

Within MICOA is a Coastal Zone Management Unit with a mandate to develop an ICZM plan and to be the focal point for coastal zone issues.

The coastal zone is highlighted in the Environmental Law as one of four priority areas and within this the main issues are (i) fisheries, (ii) coastal and marine ecosystems management, (iii) coastal and marine protection, (iv) marine parks, and (v) tourism.

The development of good legislation and the ratification of relevant conventions are important steps but without actual implementation and enforcement of the policies, coastal and marine habitats will continue to be at risk. It is at this stage of the process that marine conservation in Mozambique faces the greatest challenge. For historical reasons, government institutions lack the experience and personnel needed to carry out research and to implement conservation programmes. There is a continuing dependence on donor finance and on foreign technical assistance for environmental projects. Capacity building is taken seriously but it is a long process.

Examples of governance/management

The Mecúfi Coastal Zone Management Project in Cabo Delgado province is an attempt to reduce pressure on natural resources in a single coastal district. Community structure strengthening and participatory approaches are central to the approach of this initiative and the lessons learned in this pilot-scale project will be applicable in rural development work elsewhere (Massinga and Hatton, 1997). The focus of the project is primarily terrestrial, however.

Another ongoing project is the installation at Xai-Xai of a Sustainable Development Centre for the Coastal Zone. This is conceived as a regional centre of excellence and of research into sustainable development in the coastal strip (MICOA, 1998). Xai-Xai will also be a pilot area for the testing of new coastal area management initiatives.

Measures taken to protect fisheries resources have focused on the prawn fishery which is the major export earner (typically 40% of total). Minimum mesh sizes have been increased from 37 mm to 60 mm since 1989 (de Sousa, pers. comm.) and there is also a two-month closed season, introduced in 1991, but stocks remain heavily exploited (DPAP).

To address the management of small-scale fisheries, the government's Fisheries Master Plan (SSF, 1994) has chosen the strategy of 'Co-management' and aims to formally devolve some power and responsibility for stock management to the fishing communities. Work so far is at the pilot level and is sponsored by ICLARM and the North Sea Centre in Denmark.

The Department for Wildlife and Parks (DNFFB) is responsible for activities within protected areas. However, the University of Eduardo Mondlane administers one of the two major marine reserves—the Inhaca and Portuguese Islands Biological Reserve. Beyond the boundaries of the few reserves, the marine environment falls under the jurisdiction of the Maritime Authority. It is doubtful that the majority of employees of this authority have any training in the field of ecological conservation and it is certain that they lack the means to undertake conservation initiatives; as an example, the Maritime Authority in Cabo Delgado Province, which sits on the border with Tanzania, has no vessel of any description with which it could carry out patrols.

Regarding sea turtle conservation, the legal framework was established in 1978 but the responsibilities are divided between different offices. Despite the protected status of turtles, turtle shell souvenirs can be found on sale at airports throughout the country. They are caught for food—either as by-catch or targeted—and nests are dug up for their eggs. Success has been reported in the protection of nesting sites on Inhaca island with a dramatic drop in the number and percentage of nests being raided for eggs by local people (Gove and Magane, 1996).

REFERENCES

Barnes, D.K.A., Corrie, A., Whittington, M., Carvalho, M.A. and Gell, F.R. (1998) Coastal shellfish resource use in the Quirimba Archipelago. *Mozambique Journal of Shellfish Research* **17**.

Best P.B., Findlay, K., Peddemors, V. and Gore, D. (1991) Whales of the Indian Ocean—survey of Humpback and Right Whale breeding grounds off Mozambique, 14 August–15 September 1991. Earthwatch, London.

Campbell, B.M., Attwell, C.A.M., Hatton, J.C., de Jager, P., Gambiza, J., Lynham, T., Mizutani, F. and Wynter, P. (1988) Secondary dune succession on Inhaca, Mozambique. *Vegetatio* **78**, 3–11.

Dutton, P. and Zolho (1991) Plano director de desenvolvimento sustentavel do arquipelago de Bazaruto. Relatorio submetido ao Ministerio de Agricultura e Pescas Notes: 109 pp.

Fernandez, A. (1996) Poluicao, factos e figuras In: Workshop on the Role of Research in Coastal Zone Management Departamento de Ciencias Biologicas, Universidade Eduardo Mondlane Conference, Maputo.

Gell, F.R. (1997)Technical Report 5· The seagrass fishery of Quirimba Island. Interim Report. Marine Biological and Resource Use Surveys of the Quirimba Archipelago, Mozambique Society for Environmental Exploration, London and Ministry for the Coordination of Environmental Affairs, Maputo. Technical Report: 5, 27 pp., ISSN 1360-0493.

Gove, D. and Magane, S. (1996) The status of sea turtle conservation and research in Mozambique In: Workshop on the Role of Research in Coastal Zone Management. Departamento de Ciencias Biologicas, Universidade Eduardo Mondlane, Maputo.

Hassan, T. (1997) Beira, Main environmental problems and potential solutions. In: *Workshop on Integrated Coastal Zone Management in Mozambique*, eds. O. Linden and C.G. Lundin. World Bank. Inhaca Island and Maputo, pp. 123–130. Proceedings published Upsaala 1997.

Hatton, J. (ed.), Macia, A., Guissamulo, A., Boane, C., Gaspar, A., Gove, D., Jose, V., Couto, M., Barbosa, F. and Wamusse, R. (1995) A status quo assessment of the coastal zone, Mozambique; Phase 1. Ponta do Ouro to Xai-Xai. MICOA and Universidade Eduardo Mondlane. 86 pp.

Hatton, J.C. (1996) Natural resource use, land tenure and conflicts in coastal Mozambique. Workshop on the Role of Research in Coastal Zone Management. Departamento de Ciencias Biologicas, Universidade Eduardo Mondlane, Maputo.

Hatton, J., Voabil, C. and Manjate, A. (1996) Tourism, Land Use and Conflicts, Bilene–Xai-Xai–Chongoene. *Workshop on Integrated Coastal Zone Management in Mozambique*, eds. O. Linden and C.G. Lundin. World Bank. Inhaca Island and Maputo. pp. 108–122.

Heasman, M., Antonio, C.M., Myers, M., Stanwell-Smith, D. and Whittington, M.W. (1998) Technical Report 4. Southern Islands Group—Mefunvo, Quisiava and Quipaco Islands. Marine Biological and Resource Use Surveys of the Quirimba Archipelago, Mozambique Institution: Society for Environmental Exploration, London and Ministry for the Coordination of Environmental Affairs, Maputo Technical Report No. 4. 168 pp., ISSN 1369-0493.

IUCN (1988) *Coral Reefs of the World.* Vols. 1–3. IUCN, Gland, Switzerland.

MacNae, W. and Kalk, M. (1969) *A Natural History of Inhaca Island, Mozambique.* Witwatersrand University Press, Johannesburg, 163 pp.

Magane, S. (1996) Environmental profile of the Bazaruto Archipelago. In *Workshop on Integrated Coastal Zone Management in Mozam-*

bique, eds. O. Linden and C.G. Lundin. World Bank. Inhaca Island and Maputo. pp 99–107. Proceedings published in Upsaala in 1997.

Massinga, A. and Hatton, J. (1997) Status of the coastal zone of Mozambique. In *Integrated Coastal Zone Management in Mozambique*, eds. O. Linden and C.G. Lundin. Proceedings of the National Workshop on Integrated Coastal Zone Management in Mozambique, Inhaca Island and Maputo, Mozambique, May 5–10, 1996. pp. 7–68.

MICOA (1997) First National Report on the Conservation of Biological Diversity in Mozambique. Impacto Lda, Maputo.

MICOA (1998) Macrodiagnostico da zona costeira de Mocambique; versao preliminar Institution: Ministerio da Coordenacao da Accao Ambiental. February 1998.

MICOA/UNEP (1997) Water-related environmental issues and problems of Mozambique and their potential regional and transboundary importance. Preparation of a transboundary diagnostic analysis and strategic action programme for the marine and coastal environment of the Western Indian Ocean.

SSF (State Secretariat of Fisheries) (1994). Master Plan, Republic of Mozambique. Republic of Mozambique, 39 pp.

Serra King, H.A. (1995) Estudo da Dinamica da Materia Organica do Solo Apos Corte Queimada, Ilha da Inhaca University: Universidade Eduardo Mondlane.

Stanwell-Smith, D., Antonio, C.M., Heasman, M., Myers, M. and Whittington, M.W. (1998) Technical Report 2. Northern Islands Group -Macaloe, Mogundula, Rolas and Matemo Islands. Marine Biological and Resource Use Surveys of the Quirimba Archipelago, Mozambique Institution: Society for Environmental Exploration, London and Ministry for the Coordination of Environmental Affairs, Maputo Technical Report No. 2, 211 pp., ISSN 1360-0493.

WCMC (1997) http://www.wcmc.org.uk/cgi-bin/mp_countryquery.p

Whittington, M.W., Antonio, C.M., Corrie, A. and Gell, F. (1997) Technical Report 3. Central Islands Group—Ibo, Quirimba, Sencar and Quilaluia Islands. Marine Biological and Resource Use Surveys of the Quirimba Archipelago, Mozambique Institution: Society for Environmental Exploration, London and Ministry for the Coordination of Environmental Affairs, Maputo. Technical Report No. 3, 277 pp., ISSN 1369-0493.

Whittington, M.W. and Myers, M. (1997) Technical Report 1. Introduction and Methods. Marine Biological and Resource Use Surveys of the Quirimba Archipelago, Mozambique Institution: Society for Environmental Exploration, London and Ministry for the Coordination of Environmental Affairs, Maputo. Technical Report no. 1, 47 pp., ISSN 1369-0493.

Whittington, M.W. and Heasman, M. (1997) A rapid assessment of the subtidal habitats and associated commercial fish populations of the Segundos Islands: Santo Antonio and Mafamede Islands. Institution: The Institute for the Development of Small-scale Fisheries (IDPPE), Mozambique.

Whittington, M.W. and Stanwell-Smith, D. In *Biodiversity and Conservation*. In press.

Whittington, M.W., Antonio, C.M., Heasman, M., Myers, M. and Stanwell-Smith, D. (1998) Technical Report 6. Results, Summary and Management Recommendations. Marine Biological and Resource Use Surveys of the Quirimba Archipelago, Mozambique Institution: Society for Environmental Exploration, London and Ministry for the Coordination of Environmental Affairs, Maputo. Technical Report no. 6, 49 pp., ISSN 1369-0493.

World Bank (undated). Africa: A Framework for Integrated Coastal Zone Management World Bank: Land, Water and Natural Habitats Division; Africa Environmentally Sustainable Development Division. 139 pp.

THE AUTHORS

Michael Myers
TCMC,
University of Newcastle upon Tyne,
Newcastle upon Tyne NE1 7RU,
U.K.

Mark Whittington
Frontier International,
Leonard St.,
London EC2A 4QS
U.K.

Chapter 60

MADAGASCAR

Andrew Cooke, Onésime Ratomahenina, Eulalie Ranaivoson and Haja Razafindrainibe

---

Madagascar extends from 10 to 25°S and supports a wide diversity of marine and coastal habitats, from equatorial reefs in the north to sub-tropical and temperate features in the south. The principal water bodies affecting the area are the oligotrophic waters of the South Equatorial Current, a gyre system in the northern Mozambique Channel and cooler, nutrient-rich waters in the south. Other controlling factors are the winds (southeast trade winds and the monsoon) and the influence of terrestrial processes, especially sedimentation. Cyclones have localised physical impacts on both the biota and the coastline, while elevated temperatures following the most recent El Niño event have had a substantial impact on coral communities.

The main environmental problems in the area are localised resource degradation (especially mangroves and coral reefs) through destructive or inefficient use practices, ecological impacts of the shrimp fishery and sedimentation from erosion of inland watersheds. Problems exist in the management of large-scale fisheries, and with the collection and use of information on the state of the marine environment and resources. Pollution is only a minor problem at present, although faecal pollution is an acute problem in coastal towns. Poisoning from the ingestion of marine organisms has increased in recent years. Problems also exist in relation to the co-ordination of development activities in the coastal zone, and conflict due to migration.

Projects established in the 1980s and early 1990s primarily addressed resource use practices, aquaculture and management of the industrial shrimp fishery and, to a lesser extent, the conduct of research. Recent projects have focused on management issues, including the introduction of environmental regulations, monitoring of coral reefs and ecotoxicology, pilot ICZM schemes and improved management of the shrimp and lobster fisheries. In 1997, a five-year national marine and coastal environment programme was initiated. This seeks to promote Integrated Coastal Zone Management, including protected areas, pollution prevention and monitoring as tools for ICZM.

The general prognosis is optimistic, in that Madagascar's marine ecosystems appear generally to be in healthy condition, with acute problems being limited to vulnerable species or to specific areas and issues. The essential challenge is to put in place programmes and solutions before existing problems get beyond a point where sustainable solutions are possible, due to population growth, overfishing, migration or other reasons.

Seas at The Millennium: An Environmental Evaluation (Edited by C. Sheppard)
© 2000 Elsevier Science Ltd. All rights reserved

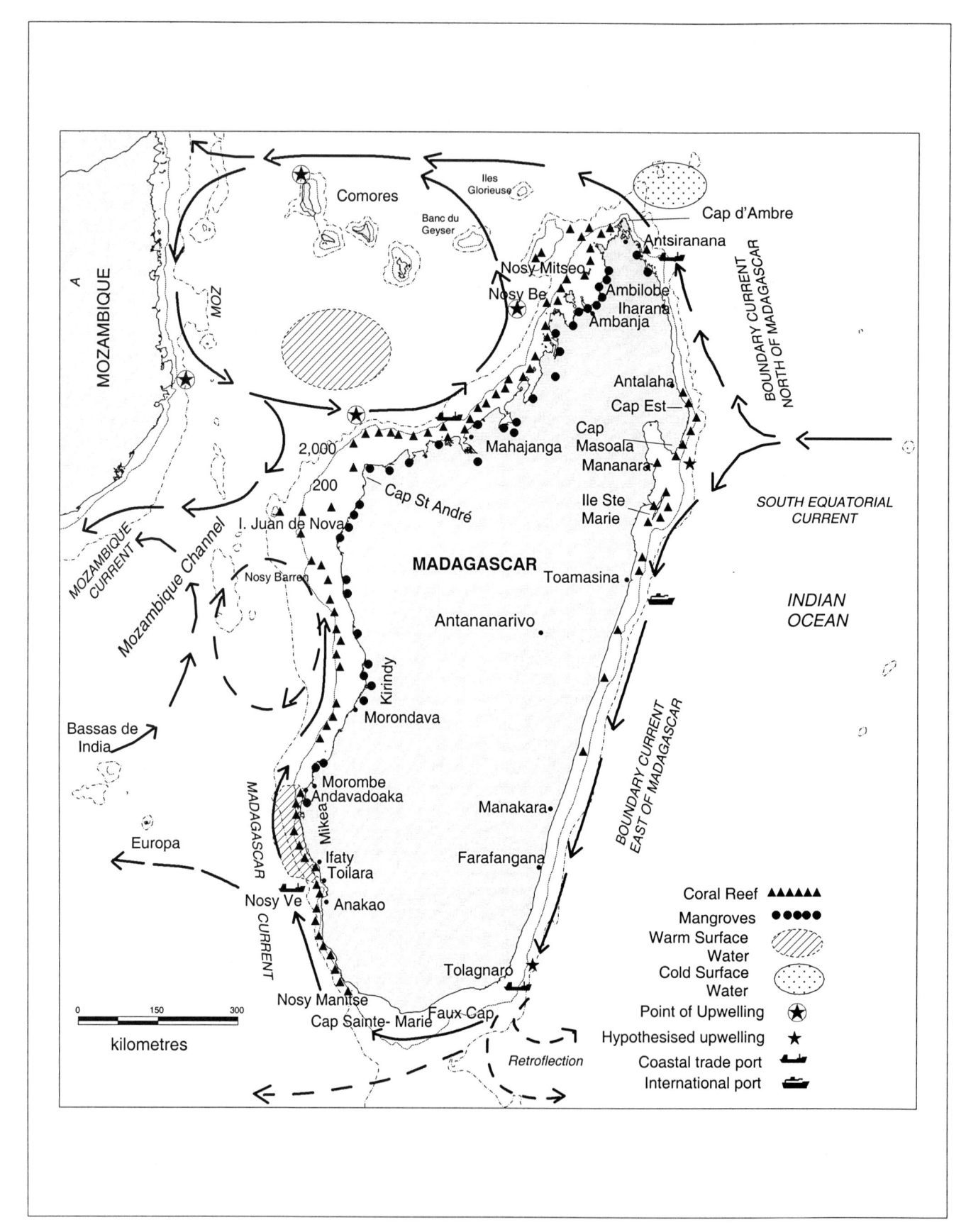

Fig. 1. Map of Madagascar.

THE DEFINED REGION

Madagascar is the world's fourth largest island. It is approximately 1600 km long, has a surface area of 590,000 km² and a coastline of about 4500 km. It includes about 270 small continental islets (Fig. 1).

Madagascar originally formed part of the Gondwana land mass, lying against what is now East Africa. Madagascar and India broke off from the African mainland by the late Middle Jurassic (~180 my) and themselves separated by the late Late Cretaceous (~80 my).

The continental shelf varies in width from a few km to over 100 km; and covers 117,000 km² (Ranaivoson, 1996). The land mass is tilted along its SSW–NNE axis to create an escarpment and a narrower shelf along the eastern edge. Local widening of the shelf is related to zones of continental uplifting or areas of vulcanism during the Miocene. The region of Toliara (southwest) is exceptional (Fig. 2): as a result of geological faulting, the 200 m isobath lies only 500 m from the shoreline; this is where the coelacanth (*Latimeria chalumnae*) has been reported in fishermen's catches (Heemstra et al., 1996). Madagascar's major rivers drain along the west coast.

The region may be regarded as a single marine ecoregion. It is dominated by a single current system derived from the South Equatorial Current (SEC) whose waters encircle most of Madagascar. The SEC divides to the east of Madagascar and sends branches around the north and south of the island and along the western coast (the Madagascar current). Madagascar is an island continent, separated by deep water (>2000 m) from its nearest neighbours. Its coastal and marine ecosystems are significantly influenced by terrestrial environmental processes, notably watershed erosion which results in a high discharge of sediments into coastal seas.

SEASONALITY, CURRENTS, NATURAL ENVIRONMENTAL VARIABLES

The marine biota are influenced by a variety of interacting seasonal environmental variables. Straddling almost 14° of latitude, mean annual open-water surface temperatures range from 22 to 28°C, with a maximum in February and a minimum in August. Variation from the seasonal mean increases to the south. Enclosed inshore waters experience greater extremes, with local shallow-water temperatures ranging from, for example, 19°C in Toliara lagoons during winter (COUT and Cooke, 1994) to 31–32°C at Morondava in summer (Razafindrainibe et al., 1996). A poorly defined thermocline occurs at about 150 m on the west coast which disappears completely between Maintirano (18°S) and Morondava (20°S). On the east coast, the thermocline is very marked and constant at 100 m (Ranaivoson, 1996).

Currents

The dominant surface current is the South Equatorial Current, which shifts within a band 6–23°S with peak flow in

Fig. 2. St. Augustin estuary, Toliara, showing the coastal escarpment (photo A. Cooke).

July–August and minimum flow in February–March (see Fig. 1). East of Madagascar it divides at 16–18°S into a northbound portion (average velocity 0.58 m s⁻¹ and depth 268 m) and the southbound Madagascar current (average velocity 0.33 m s⁻¹ and depth 166 m) (Schott et al., 1988). Upwelling occurs off the southeast coast north of Tolagnaro (Stequert et al., 1975) to the south of which there is a zone of retroflection centred around 25°S 46°E (Lutjerharms et al., 1981). The Madagascar current rounds the southern end of the island, where it splits to the south of Cap Ste. Marie into a northbound and a less distinct westbound component. The northbound portion sets north up the western Madagascar coast as far as Ile Juan de Nova/Banc de Pracel where it feeds into a quasi-permanent anticlockwise gyre to the north and into a less well defined gyre system to the south. The northern gyre is centred to the south of the Comoros and links waters of northern Madagascar with Mozambique and the Comoros archipelago. The gyre is bordered in the north by the westward-flowing south equatorial current as it rounds Cap d'Ambre, and encloses a zone of warmer surface water. South of Juan de Nova there is a quasi-stationary clockwise eddy (Saetra and Dasylva, 1984), and several authors report an intermittent upwelling along the western edge of the Banc de Pracel opposite Cap St. André (Ratomahenina, 1988).

Inshore currents over the continental shelf may run in the same or different directions. For example, the inshore current at Morondava is permanently southbound (Razafindrainibe et al., 1996) whereas offshore it is generally northbound. At Cap Est (northeast Madagascar) the offshore current is northbound, while the inshore current generally runs northwards in May to September and southwards in September to May (G. Schallkwyk, pers. comm.).

Nutrients and Salinity

Nutrient levels are low and oligotrophic off the east coast, as is typical of the central western Indian Ocean. Nutrient levels are slightly higher in the Mozambique channel, which may be related to river outflows from Mozambique and Madagascar, greater mixing with bottom waters and penetration of nutrient-rich water from the south. Satellite images indicate a "hot spot" off Cap St. André and other areas of enrichment associated with major rivers. Productivity over the continental shelf is also higher.

In general, isohalines between 15 and 40°S trend E/W, with salinity increasing towards the south and being higher in winter than summer. South equatorial surface waters which extend southwards to 21–22°S in the mid-Mozambique channel have salinities of 34.9–35.3 ppt. Subtropical surface waters, in the southern part of the channel, have salinities from 35.4–35.5 ppt (Magnier and Piton, 1973; Piton et al., 1981). River discharge rates are highly seasonal, especially along the west coast. At Antsabora, north of Morondava, salinities may be reduced by the Tsiribihina river to 24 ppt in the wet season from 32 ppt in the dry season (Razafindrainibe et al., 1996).

Winds and Tides

There are two principal wind systems—the Southeast Trade winds and the monsoon. The balance between these two systems is determined by shifts in position and intensity of the Mascarene Anticyclone (zone of high pressure, also known as the New Amsterdam anticyclone) over the Mascarenes, and a zone of low pressure in May–July over the Mozambique channel. The SE trade winds blow strongly with a median wind speed of 6 m s^{-1} for eight to nine months of the year. Accordingly, sea conditions along the east and southern coasts are rough. Northwestern and western coasts as far south as 17°S are directly affected by the monsoon and experience mostly lighter and more variable winds: during both the Northeast monsoon (November to March) and Southeast Monsoon (April to October) winds average 3.5 m s^{-1}. The Toliara region, while not always directly affected by strong winds, is noted for the almost constant presence of a long southwesterly swell which is considered partly responsible for the exceptional barrier reef in this area (Clausade et al., 1971). Daily onshore and offshore breezes are pronounced and influence the daily wind direction.

Madagascar is strongly affected by cyclones during the warm season (December to March). Between 1920 and 1972, 72 cyclones struck Madagascar, or an average of 1.3 yr^{-1}. Cyclones generally follow a parabolic path, starting to the northeast of Madagascar heading southwestwards, with the turning point lying over the Mozambique channel at about the latitude of Morondava, whereupon the cyclone moves southeastwards before blowing out.

Tidal range on the west coast is greatest between Toliara and Nosy Be, with a mean spring range of 3.8 m at Mahajanga, but up to 5 m in exceptional cases. Tides on the east coast are much smaller, typically about 0.5 m or less, and barely detectable at Tolagnaro.

Effects of Climate on Biota

The large water-temperature range is reflected in the different shallow-water faunal assemblages of the north and south. Coral reef communities of Nosy Be (Fig. 3) and Toliara are visibly different, although no systematic scientific comparisons have been made to date. The extreme south of Madagascar represents an ecotone between tropical and temperate systems, possessing reefs constructed of both coral and oyster shells (*Crassostrea* sp.) as well as algal reefs (Maharavo and Vicente, 1999). Important oyster fisheries occur around Tolagnaro in the southeast.

Seasonal temperature variation may be associated with migratory patterns of certain species. Tuna shoals move into the northern seas of Madagascar during February–April, following zones of planktonic productivity. Pelagic sharks migrate into shallower coastal waters to reproduce during austral spring (Cooke, 1997). Sea turtles mate and nest mainly between November and February, when temperatures are warmer (Rakotonirina and Cooke, 1994). Coastal fisheries are generally more productive during the warm season. Conversely, Humpback whales (*Megaloptera novaeangliae*) migrate northwards into waters around Madagascar during the austral winter (June to September) for calving and mating (Best et al., 1995; Rosenbaum et al., 1997).

Following the 1997 El Niño event, coastal water temperatures in April–May 1998 exceeded the seasonal average by about 2°C causing extensive bleaching of corals. Observations along the Masoala peninsula in the NE in May 1998 (McClanahan and Obura, 1998) indicated mortality of up to 80–90% on shallow water reefs around Cap Est, whereas well flushed, seaward reefs were less affected and more southerly reefs at Cap Masoala appeared unaffected. Shallow reefs at Cap Est were showing signs of recovery in December, 1999. Observations of offshore reefs at 15–20 m opposite Belo-sur-Mer in mid-1998 indicated mild bleaching (J. Maharavo, pers. comm.). Observations (unpublished) of lagoon reefs at Nosy Ve in January 2000 indicated extensive coral mortality since previous observations in 1995, with many signs of recovery in the form of young colonies.

Fig. 3. View of Nosy Tanikely, Nosy Be (photo: A. Cooke).

Table 1
Number of days with winds less than 10 km h^{-1} by coastal region (from Ralison, 1991, after Moal, 1974)

Region	Month												Total
	J	F	M	A	M	J	J	A	S	O	N	D	
N: Antsiranana	4	5	3	2	2	0	0	0	0	0	1	2	19
NE: Antalaha-Ste. Marie	8	5	7	6	6	3	4	4	4	4	4	4	59
E: Toamasina–Farafangana	3	3	3	1	0	0	0	2	2	2	2	3	21
S/SE: Tolagnaro-Cap Ste. Marie	4	4	6	6	5	5	7	8	8	8	8	7	73
SW/W: Toliara–Morondava	20	29	22	29	30	30	29	29	28	28	26	28	337
W/NW: Maintirano–Nosy Be	30	19	19	19	20	18	19	20	12	12	15	17	220

The dominance of the Southern Equatorial current means that seed stock for marine flora and fauna is mostly derived from further east. Thus, Madagascar shares many species of marine benthos with western Indian Ocean islands (Seychelles, La Réunion, Mauritius) and more distant Indo-Pacific areas. For example, Veron (1986) reports that about two thirds of the scleractinian coral genera occurring on the Great Barrier Reef of Australia are also found in Madagascar, but notes only two genera found in Madagascar which do not also occur in Australia. Generally, Madagascar appears to be part of a southwestern Indian Ocean grouping (Sheppard, 1998) characterised by a moderate group of regionally endemic species and genera.

A similar observation may be made for coral reef fishes. The most comprehensive published studies of coral reef fish communities to date are those of Harmelin-Vivien (1979) for the Grand Récif of Toliara, which refer to 552 species. Maugé (1967) indicates 700 species for the same region. Harmelin-Vivien (pers. comm. to Gabrié et al., 2000) estimates 1500 species for Toliara.

Nutrient levels influence the productivity, with the major coastal fisheries being located on the west coast. Shrimp fisheries in particular are more productive in the sediment-affected areas of the west coast. Salinity can also have a significant impact on fisheries. The combination of currents and winds in the Morondava region (where there are several large estuaries) results in a compression of the salinity gradient towards the coast, providing suitable conditions for all stages of the life cycle of the shrimp *Penaeus indicus* within a small distance of the coast. This drives industrial shrimp trawlers to operate within the official two-mile exclusion zone for traditional fishing, leading to conflicts between the two sectors (Razafindrainibe et al., 1996). The periodic presence of freshwater has prevented the development of coral reefs close to the shore along both coasts. The past influence of estuaries before sea levels rose about 6000 years ago is evident from gaps in the submerged barrier reef which runs along much of the western continental shelf.

The impacts of the various wind regimes around Madagascar have not been analysed. On the east coast, the prevailing southeasterly winds combined with the south equatorial current tend to drive nutrient-poor surface waters from the central Indian Ocean to Madagascar, thus limiting productivity in most eastern parts. Wave action resulting from these winds is also responsible for depositional processes such as beach formation and longshore drift which limit development of mangroves and reefs along this coast. The same winds generate large swells, which have been linked to the high degree of development of emergent barrier reefs along the southwest coast.

The northern and northwestern regions are influenced by monsoon winds which reverse direction in about March. Winds there are relatively light and variable, with wave and swell action low. As a result, massive reef structures are less well developed than further south, although coral reef communities are no less rich.

An important impact of wind has been to limit the development of coastal traditional fisheries in areas where high winds are frequent. Moal (1974), quoted in Ralison (1991), calculated the number of days with winds of less than 10 km h^{-1} around the Madagascar coast (Table 1), showing that the southwest and northwest coasts are clearly the most favourable for small-scale fisheries.

Cyclones, although brief in duration, have substantial impacts on the coastal zone, causing greatly accelerated erosion and extensive damage, for example at Toamasina (Guerra et al., 1998). It is likely that they play a role in limiting coral reef development along the east coast.

Tidal regimes influence the development of mangroves, coral reefs, estuaries and other ecosystems. Tidal ranges are maximal (up to 5 m at spring tides) at Cap St. André where the Mozambique channel is narrowest, and minimal along the east coast (no more than 1 m). As a consequence, intertidal and subtidal ecosystems are highly developed on the west coast, but occupy only a narrow band on the east.

THE MAJOR SHALLOW-WATER MARINE AND COASTAL HABITATS

The principal shallow-water marine and coastal habitats on Madagascar's 4500 km coast are coral reefs, mangroves, seagrass beds, estuarine mud flats, steeply shelving beach and

Fig. 4. Aquarium Nord, Nosy Ve Marine Reserve, anakao, Toliara (photo: A Cooke).

Fig. 5. Fringe of mangroves around the town of Mananara. Photo: A. Ebregt (courtesy UNESCO/MAP).

rocky shoreline. Coral reefs, mangroves, seagrass beds and mud flats are the dominant habitats along the shallow sloping west coast, whereas steeply shelving beach and rocky shoreline predominate on the east.

Coral Reefs

Coral reefs occur as emergent fringing (Figs. 4 and 5) and barrier reefs, patch reefs and submerged banks and shoals. Madagascar is particularly known for the spectacular barrier reefs of the southwest, but equally impressive is an ancient submerged barrier reef further north (Battistini, 1972). The ancient reef is manifested as a string of banks and shoals, sometimes broken by passes, at a general depth of 15 to 30 m with frequent peaks to 10 m or less and occasional coral islets. This formation was probably emergent along much of its length 6000 years ago. Today, in places the submerged reef no longer supports actively growing coral communities; for example, submerged reefs south of Anakao (Toliara) and opposite Andavadoaka at around 20–30 m depth lack living hard coral, and have been eroded into smooth, steeply sloping formations (COUT and Cooke, 1994), while submerged reefs south of Belo-sur-mer (10–15 km offshore and generally at 15–20 m depth) support rich hard-coral communities and appear to be actively growing (J. Maharavo, pers. comm.). Until more studies have been completed, it will be impossible to reconstruct the history of Madagascar's western reefs, but it would appear that, prior to sea-level rise, Madagascar possessed a barrier reef almost 2000 km in length.

The length of coral reefs in Madagascar has been estimated by several authors (e.g. Gabrié et al., 2000). A study of charts conducted for this chapter indicated 1130 km of fringing reef, 557 km of reef around islets and islands or patch reefs, 52 km of true barrier reef (all in the Toliara region) and 1711 km of submerged coral banks and shoals, or a total of 3540 km of coral reefs (Table 2).

The best described reefs are those of the Toliara region, with numerous publications during the 1960s and 1970s (e.g. Clausade et al., 1971; Thomassin, 1971), whereas subsequent publications tend to have concerned socio-economic issues such as fishing (Maharavo and Vicente, 1999).

Coral species diversity has been most thoroughly studied for the Grand Récif of Toliara, with 130 species of scleractinians and 71 alcyonarians (Pichon, 1978), 8 hydrocorals and 65 hydroids described. A total of 200 scleractinian species is estimated for Madagascar (Pichon, pers. comm. to Gabrié et al., 2000). Recent studies on Toliara's coral reefs (e.g. Vasseur et al., 1988; COUT and Cooke, 1994; IHSM, 1999) provide an update on the condition of shallow-water ecosystems. Coral reefs of Masoala (McClanahan and Obura, 1998) have at least 164 species (41 genera), - suggesting that coral diversity there is similar to that of reefs at Toliara, even if the reefs themselves are less well developed.

Mangroves

Madagascar possesses over 425,000 ha of tidal marshes, of which an estimated 327,000 are covered by mangroves (Kiener, 1972; Lebigre, 1990) (Fig. 5). About 70% of the mangroves are in large stands exceeding 500 ha, with the remainder in scattered small patches. Stands of over 35,000 ha occur in the estuaries of Mangoky, Tsiribihina, Betsiboka and Mahajamba. Mangroves occupy about 500 km of seaboard, or 10% of the coastline (Kiener, 1972). The vast majority (well over 95%) of mangroves are on the west coast, although narrow bands of mangrove occur frequently along the east coast and undoubtedly play an important role in coastal ecology and defence, despite their small size (Edmond and Grepin, 1999). Mangroves vary considerably in species composition, tree density and canopy height according to local environmental conditions. While many mangroves are subject to intensive human pressure for wood cutting, agriculture and other uses, mangroves on the west coast are also evolving in response to beach expansion and changes in local hydrology. Comparison of Landsat images and maps based on aerial

Table 2
Estimated lengths of coral reefs in Madagascar (based on an analysis of US Naval charts)

Region	Length of coastline (km)	Latitude range (N–S)	Fringing reef seaward edge (km)	Islets & patch reef edge (km)	Emergent barrier reef (length in km)	Submerged barrier reef (length in km)
Cap d'Ambre to Cap Est	435	11°57'–15°16'	153	56	0	19
Cap Est to Toamasina (inc. Ile Ste Marie)	560	15°16'–18°10'	187	24	0	108
Toamasina to Faroany	429	18°10'–21°48'	6	6	0	52
Faroany to Tolagnaro (Ft Dauphin)	408	21°48'–25°02'	38	0	0	0
Tolagnaro to Androka (via Cap Ste. Marie)	352	25°02'–25°35'	39	14	0	39
Androka to Morombe	412	25°02'–21°45'	363	49	46	75
Morombe to Cap Kimby	382	21°45'–18°52'	0	45	3	268
Cap Kimby to Cap St. André	374	18°52'–16°12'	0	72	3	383
Cap St. André to Mahajanga	275	16°12'–15°42'	43	12	0	297
Mahajanga–Pte. Andilana (Nosy Be)	619	15°42'–13°10'	149	127	0	318
Pte. Andilana to Cap d'Ambre	303	13°10'–11°47'	152	152	0	152
Total	4549	13°48'	1130	557	52	1711

photographs from the 1940s and 1950s indicates considerable change in mangrove cover over the last 40–50 years (P. Turcotte, pers. comm.).

Mangroves on the west coast are important as breeding areas for several commercial fish species, such as mullet (Mugilidae), sicklefish (Drepanidae) and ponyfish (Leignathidae) (Andrianjohany, 1992). Temporal patterns of fish assemblages in Toliara mangroves have been documented by Laroche et al. (1997). Large stands of mangrove have been associated with a high proportion of organic material in bottom sediments and high productivity of the shrimp *Penaeus indicus* (Razafindrainibe et al., 1996).

Species of mangrove trees for Madagascar include *Rhizophora mucronata*, *Bruguiera gymnorhiza*, *Ceriops tagal*, *Avicennia marina*, *Sonneratia alba*, *Lumnitzera racemosa*, *Xylocarpus granatum* and *Heritiera littoralis*, with *Ceriops boviniana* in the southwest (WWF, 1993; Razafindrainibe et al., 1996). Andrianarivelo et al. (1998) also report the existence of *A. marina*, *Carapa obovata*, *B. gymnorhiza*, *Ceriops candolleana*, *R. mucronata* and *S. alba* in the Masoala peninsula, making a total of 11 species for Madagascar. In larger stands, the Rhizophoraceae predominate. At the seaward fringe *Sonneratia alba* generally dominates, while *A. marina* is a pioneer, being the most common species in deltas where there are frequent shifts of the alluvial substrate.

Seagrass Beds

Seagrass beds are generally well developed in reefal lagoons, but have not been systematically studied. Andrianarivelo et al. (1998) report nine species from three sites on the Masoala peninsula: *Cymodocea serrulata*, *C. rotunda*, *Diplantera baudeti*, *Diplantera uninervis*, *Syringodium isoetifolium*, *Thassodendron ciliatum*: *Halophylla ovalis*, *Halophylla stipulacea*, *Thalassia hemprichii*. Chasse (1962) reports 10 species for Toliara.

Coastal Terrestrial Vegetation

Coastal terrestrial vegetation is highly varied in Madagascar due to marked variations in rainfall, wind and substrate. Along the east coast, where precipitation is generally high and soils generally acid, natural vegetation types include moist dune forest, moist littoral forest and, in a few places in the northeast, dense rainforest to the sea's edge. Moist dune forest is a particularly threatened forest type, remaining fragments being restricted to the southeast (Ste. Luce and Manombo Special Reserve). In the south, where cliffs predominate and precipitation reduces progressively towards the west, forest is absent and only coastal xerophytic scrub occurs. The southwest from Androka to Morombe, is dominated by a unique dry coastal bush growing on calcareous, rocky or sandy substrate, harbouring many endemic species. Substantial and exceptionally important areas of this bush remain between Manombo and Morombe, in the area known as Mikea (ANGAP, 1999). Further north along the west coast, where rainfall increases slightly, coastal lands are dominated by a degraded savannah. Some areas of original forest remain, such as at Kirindy to the north of Morondava.

Islets and Islands

Madagascar has about 270 small islands and islets. Of these, about 75% are located in the extreme north and northwest, and only 5% are found scattered along the eastern and southern coasts. Most of those in the north and west have coral formations. The total area of fringing reef associated with these islands is at least 137 km^2 (Cooke, 1996). Islands have considerable importance as refuges for terrestrial species, and as protected nesting sites for seabirds and sea turtles (Rakotonirina and Cooke, 1994). Islands also serve as seasonal bases for fishermen.

> ## Seabirds of Madagascar
>
> ### Frank Hawkins
>
> *Projet ZICOMA, BirdLife International, B.P. 1074, Antananarivo, Madagascar*
>
> Breeding seabirds in Malagasy waters are confined to two types of site that limit human predation of eggs and nestlings; steep seastacks (such as those of the Nosy Mitsio island group) and offshore islands protected by strong but variable winds in the breeding season (July–August). The greatest number of suitable sites occurs between Nosy Mitsio and Cap d'Ambre in the far north, where there are globally significant populations of Crested Tern *Sterna bergii* and Roseate Tern *Sterna dougallii* on offshore islets. Large populations of Brown Booby *Sula leucogaster* and White-tailed tropicbird *Phaeton lepturus*, as well as the only Malagasy populations of Frigatebirds *Fregata minor* and *F. ariel* nest on seastacks in the region. These species are specialist feeders on the relatively unproductive shallow tropical waters of the area. Further south, offshore winds are less frequent and strong, and there are no sea-stacks. The only breeding colonies are those of Roseate Terns on distant offshore islets, and south of Toliara, a breeding colony of Red-tailed Tropicbirds *Phaeton rubricauda* on Nosy Ve, where they are protected by a local taboo. Around the rest of the coast are only a few colonies of Roseate Terns and Wedge-tailed Shearwaters *Puffinus pacificus* on more distant and windswept islets.
>
> This is not to say that the situation was ever thus. Recent deep deposits of subfossil guano on the Barren Islands, west of central Madagascar, indicate former substantial breeding bird populations. Many species still breed, for example, on distant islands such as Europa. The presence of large flocks of non-breeding seabirds, especially in the upwelling area near Cap St André, indicate that significant food resources are still present. Equally, in the far south, the occasional observation of albatross from the headlands of Cap Ste Marie and Tolagnaro indicates the presence of cooler southern waters rich in prey species.

OFFSHORE SYSTEMS

The principal upwelling area of interest for offshore fisheries is opposite Cap St. André (Ralison, 1991) where chlorophyll *a* counts reach greater than 4 mg m^{-3} at 60 m (Citeau et al., 1973). It has also been theorised that an upwelling occurs in the vicinity of Cap Masoala, where FAO acoustic surveys indicated high abundance of plankton-eating fishes (Anon., 1983).

Madagascar offshore waters are important for fisheries of tuna, sharks, small pelagics, demersal fish, lobster (red and green) and deep-water shrimp (Andrianaivojaona et al., 1992) (Table 3). The shark fauna has been studied by Fourmanoir (1961) and the fishery by Cooke (1997) with a critical review by Smale (1998). While studies of the potential for commercial offshore and deep-water shark fisheries have recently been conducted by the Somapêche company, the results are not publicly available. A deep-water squaliform shark fishery was briefly operational near Ile Juan de Nova between 1990 and 1992 but subsequently abandoned, the reasons given being jurisdictional conflict with France (Cooke, 1997) and lack of profitability. The shark fauna has been studied by Fourmanoir (1961) and the fishery by Cooke (1997) with a critical review by Sinde (1998).

Small pelagics are actively fished in the bay of Toliara, with reported yields of 1.6 t km^{-2} yr^{-1} (Rakotoarinivo, 1998).

POPULATIONS AFFECTING THE AREA

Human settlement on Madagascar began only about 2000 years ago, with the last major wave of settlers arriving in about 1000 AD. The first European settlements date from

Table 3

Estimated potential of Madagascar offshore fisheries

Resource	Principal location	Estimated MSY (t yr^{-1})*	Considered status (in 1992)
Deep water shrimp	NW continental shelf	1,000	Unexploited
Red lobster	Continental shelf	340	Little exploited
Green lobster	Continental shelf	1,000	Under exploited; stocks little known
Small pelagics	Shelf waters	160,000	Mostly small non-commercial species
Demersal fish	Shelf waters	45,000	Under exploited
Tuna (albacore, dogtooth)	EEZ (principally NW, N, NE)	51,600	Under exploited (MSY estimated as a proportion of known stocks in the Seychelles–Madagascar–Comoros triangle)
Total		268,940	

Source: Andrianaivojaona et al. (1992).
*Yields of fish are estimated at 20% of standing stock biomass, within the recommended limit of 23% proposed by Beddington and Cooke (1983).

Table 4

Population increase in major coastal towns, 1975–1993. Sources: National censuses of 1975 and 1993

Coastal town	1975	1993	Increase
Mahajanga	86,859	135,700	+ 56%
Toliara	50,332	213,291	+324%
Toamasina	86,421	254,557	+195%
Antsiranana	87,972	99,936	+13.59%

1546. The origin of the earliest settlers (the *vazimba*) is obscure, while later settlers, a rice-growing culture which occupied the central highlands, were from the Malaysian region. The coastal zone, being the principal interface with the outside world, exhibits the greatest ethnic diversity, with elements of Arabian, African and European origin.

Madagascar's total population in 1993 was 13.7 million with an annual growth rate of about 3%. About 25% live within 50 km of the coast, and highest population densities are on the east coast.

Populations have been expanding especially fast in the towns (Table 4). Migration between rural areas is a factor influencing human pressures on the marine environment. For example, fishers of the Vezo ethnic group, traditionally confined to a band of about 80 km around Toliara in the south west (Fig. 6), have established new settlements on the southeast coast, bringing their specific style of fishing, gears and techniques, which are more efficient in some respects than local ones. This has led to conflicts in some areas (LEC, 1999), including the introduction of taboos on the use of nets at several places around Tolagnaro (unpublished observation). Along the southwest coast, ethnic groups such as the Mahafaly, Antandroy and Mikea, traditionally farmers or gatherers, have settled close to Vezo fishing villages in order to take up fishing, sometimes using destructive fishing methods, such as fine-meshed beach seines, discouraged by local custom.

Fig. 6. Traditional fishermen, Anakao, Toliara (photo A. Cooke).

The sea is used primarily for fishing and transportation. The number of artisanal fishing vessels remains relatively few and has changed little since 1982, when about 50 such vessels were reported for the entire country (Beurrier, 1982). Economic factors appear to favour traditional fishing, which has consistently increased from about 10,000 pirogues in 1980 (Beurrier, 1982) to about 22,000 in 1995 (ICRI, 1996).

Commercial Fisheries

Shrimp fisheries in Madagascar target mainly shallow-water species such as *Penaeus monodon* and *Penaeus indicus*, for which an industrial fishery has existed since 1967. Deep-water species are also targetted, but to a lesser extent. Shrimp is referred to as "pink gold" (*l'or rose*) in recognition of its economic importance. The current fleet of 67 trawlers is distributed mostly in the west and northwest, with outlying fishing operations as far south as Toliara in the southwest and Baie d'Antongil and Toamasina in the east. Yields have oscillated between 7000 and 9000 tonnes since 1987 despite an increase in the number of trawlers from 47 to 67. The national shrimp research project (PNRC) in collaboration with the IRD (formerly ORSTOM) of France has recently examined socio-economic aspects of the traditional fishery (in particular conflicts between traditional and industrial shrimp fisheries) and stock recruitment, but results are not yet publicly available. A preliminary finding has been that yields vary according to rainfall, wetter years giving higher yields (linked to the mangrove phase of the shrimps' life cycle). It is likely that the fishery is now operating at over-capacity, with an optimum of around 50 or fewer vessels (Z. Kasprzyk, pers. comm.).

Traditional fishermen complain that trawlers fish within the two-mile limit established to protect inshore fishing areas. Shrimp trawlers also take a large by-catch of fish, which is mostly jettisoned to maintain storage space for the shrimp. This wasted by-catch has been estimated at 30,000 tonnes annually (Jain, 1995). At the same time, traditional shrimp fishermen have introduced new techniques (nets) which considerably improve their efficiency, and a network of well-organised middlemen is developing, both of which threaten the industrial fishery. Active enforcement of a restricted zone for traditional fishing would further restrict industrial fishing, resulting in severe over-capacity of the industry. The industrial shrimp fishery around Nosy Be has been affected in some recent years by the proliferation of a sea urchin *Salmaciella erythracis*, the spines of which contaminate the catch. The cause remains unclear.

Shrimp trawlers in Madagascar are not fitted with turtle exclusion devices (TEDs) and take a substantial bycatch of sea turtles and sharks, especially in the west (Trawler captain, pers. comm.).

Industrial tuna fisheries have operated intermittently in Madagascar's EEZ since 1955 (FAO, 1995). Existing licensing arrangements date from 1984 (Rabeson, 1992). Standing stocks in the Madagascar EEZ have been estimated at 50,000 tonnes, with potential annual production at 10,000 tonnes i.e. 20% of standing stock biomass. Fisheries are of two types, a purse-seine fishery operating mainly in the north and west, and a long-line fishery operating in the south and east.

The purse-seine fishery is based on 80 m (2000 ton) vessels catching albacore (*Thunnus albacares*), big-eye tuna (*T. obesus*) and skipjack (*Katsuwonus pelamis*) and operates in the EEZs of the Seychelles, France (around Mayotte and Iles Glorieuses), Comoros and Madagascar, following the seasonal migration of target species. In 1996, under agreement with the EU, 42 French and Spanish purse-seiners operated in the northwest and west, descending as far south as Morombe. The purse-seine fishery is active in the Malagasy EEZ for four months per year (March/April–June/July).

Production from within Madagascar's EEZ is consistently reported officially at 10,000 tonnes yr^{-1} (Table 5) which is the previously estimated sustainable yield. However, transhipments in Madagascar (all at Antsiranana) greatly exceed this amount (72,420 tonnes in 1994; 60,000 tonnes in 1995) (PRE-COI, 1997). The Antsiranana cannery produced 22,000 tonnes of tinned tuna in 1995 (Kasprzyk and Andrianaijaona, 1997).

The long-line fishery is based on 49 m 350 ton vessels targeting swordfish (*Xiphias gladius*), sailfish (*Istiophorus platypterus*) and marlin. In 1995 licences were awarded for 25 long-line vessels from Asia, each capable of taking 500–700 tonnes annually, as well as 16 European vessels. The long-line fishery is seasonal, peaking in October–November.

One observer has recently reported that the purse seine fishery is responsible for a substantial but unquantified by-catch of sharks (Caira, pers. comm.). Limited data from the long-line fishery from the early 90s indicate that shark by-catch may be as much as 6–11% of the catch (F. Gilbert, in litt.).

Apart from the surveys of fish stocks by FAO, systematic surveys which might reliably be used for monitoring changes are restricted to a few areas, notably coral reefs of the Toliara region (e.g. COUT and Cooke, 1994; Laroche and Ramanarivo, 1995; IHSM, 1999), the Masoala region (McClanahan and Obura, 1998; Andrianarivelo et al., 1998) and sites which have been subject to coral reef monitoring under the PRE-COI programme (Ifaty (Toliara), Nosy Be, Mananara, Foulpointe). The numerous publications of the Marine Station of Toliara (Thomassin, 1971) may provide an indication of the former state of particular species or aspects of the environment.

Artisanal Fisheries

Artisanal fisheries target the full range of exploitable resources in shallow waters and deeper waters a little way offshore, including a wide range of fish species but also elasmobranchs, dolphins and dugongs, sea turtles, crustaceans and even seabirds and eggs.

Fishing activities are principally based on non-motorised pirogues (*lakana*), referred to as "traditional" fishing. The use of larger motorised vessels (referred to officially as "artisanal" fishing) is very little developed (there are perhaps as few as 100 operating motorised fishing boats in the entire country of which at least 37 are shrimp fishing boats (Ocean Consultant and Sepia, 1999). West coast pirogues

Table 5
Evolution of marine fisheries and aquaculture production in Madagascar (1990–1998) (tonnes)

	1990	1991	1992	1993	1994	1995	1996	1997	1998
Industrial	19,244	18,500	20,535	21,861	21,602	20,877	20,268	21,842	25,438
Shrimps	6,967	8,000	7,163	8,361	9,091	7,635	8,136	8,146	9,772
Finfish	2,277	500	3,372	3,500	2,511	3,242	2,132	3,696	3,666
Tuna*	10,000	10,000	10,000	10,000	10,000	10,000	10,000	10,000	12,000
Artisanal	–	–	–	–	441	686	557	809	623
Shrimps	–	–	–	–	188	284	334	609	446
Fish	–	–	–	–	253	402	223	200	177
Traditional	54,200	54,940	58,400	63,190	65,090	63,864	62,977	63,190	62,262
Shrimps	2,200	2,200	2,300	1,300	3,000	2,000	2,000	2,000	2,242
Crabs	1,200	960	850	1,100	1,300	1,300	1,000	1,000	1,500
Lobster	310	440	550	360	390	390	390	390	341
Bêche-de-mer	–	–	–	–	1,800	1,800	1,800	1,800	482
Seaweed	–	–	–	–	702	787	787	1,000	2,510
Finfish	50,000	50,000	50,000	57,500	50,200	50,200	50,000	50,000	50,000
Others	490	1,340	4,700	2,930	7,698	7,387	7,000	7,000	2,758
Aquaculture (shrimps)					408	1,535	2,425	2,477	2,429
TOTAL	73,444	73,440	78,935	85,051	87,541	86,962	86,227	88,318	88,323

*The figure given for industrial tuna fisheries production is the official estimate of potential yield—no production data are available.

generally possess an outrigger and a sail (*lakana fiara*), and may have a fishing range of 10 km or more. Pirogues on the east coast (except in a few areas where Vezo migrants are present) are simple dugouts with a minimal fishing range. A national census of 1995 indicated a total of 22,000 pirogues nation-wide of which about 75% were on the west coast, and that over 50% of these were in the province of Toliara alone (ICRI, 1996).

Quite detailed information exists on the types of fishing and target species for the Toliara area (e.g. FAO, 1998) and for Morondava (Razafindrainibe et al., 1996). There are also important reef-gleaning fisheries using spears, poison and traps. Beach seining is also carried out, particularly in the lagoons of the Toliara area (Vasseur et al., 1988). Dynamite fishing is virtually unknown at present.

Impacts of the traditional fishery are little known, though in Toliara the use of fine-meshed beach seines, reef-flat trampling and poisoning are widespread and have contributed to a pronounced degradation of shallow-water reef systems (Vasseur et al., 1988). Laroche and Ramanarivo (1995) report an annual finfish yield of 12.13 t km^{-1} yr^{-1} for the barrier reef of Toliara and note signs of over-exploitation. As Madagascar's most intensively fished reef, these may be considered maximal figures. McClanahan and Obura (1998) report standing fish biomass of about 350 kg ha^{-1} for the coral reefs of Masoala (north east) and compare these with figures of about 100 kg ha^{-1} for heavily fished Kenyan reefs and 1000 kg ha^{-1} for protected Kenyan reefs, indicating modest to intermediate fishing pressures for that region.

Sharp local declines have been observed for certain high value stocks, notably sea cucumbers (Remanevy et al., 1997), sharks (Cooke, 1997; Smale, 1998), sea turtles (Rakotonirina and Cooke, 1994), dugongs (Rakotonirina, 1999) and numerous gastropods (WWF, 1993). Formerly, shrimp and lobster stocks appeared resilient to pressures (Andrianaivojaona et al., 1992). Artisanal shrimp catches are increasing as a result of the introduction of new gears but at the expense of the industrial fishery. Recent statistics for lobster indicate a decline in annual catches (see Table 5).

The artisanal fishery has suffered from numerous events of intoxication, mainly associated with ingestion of shark, sardines and sea turtles (CNRE/CNRIT/IHSM, 2000). The most serious, fatal, poisonings have been associated with sea turtles (29 deaths at Salary in 1993: COUT and Cooke, 1994) and sharks, with several fatalities at Manakara in 1994 (Rabarison, 1994).

In the west and southwest, the traditional fishery has a considerable impact on the principal tree used in hull construction, the endemic *Givotia madagascariensis* ("farafatsy"), which has the texture and light weight of balsa. Large trees have become rare. The use of smaller trees requires extra jointing in the hull leading to reduced hull life, on average 2.5 seasons (Razafindrainibe et al., 1996), accelerating consumption of the smaller trees and exhaustion of the resource.

RURAL FACTORS

Agriculture

The principal agricultural activities in Madagascar are the production of food staples (mainly rice but also cassava and maize), cattle raising and cash crop farming (vanilla, coffee, cotton, butter beans (*pois de cap*), spices, oil palms, coconut palms and sugar cane).

In the central highlands and the west, rice is mainly grown in paddies along river valleys or flood plains, whereas in mountainous eastern and northern areas rice growing is based on forest cutting and burning (a process known as *tavy*) for planting hill rice, which yields only a single annual crop and leads to rapid erosion of slopes. Soil loss is further exacerbated by the annual burning throughout Madagascar of large areas of grassland to provide fresh grazing at the start of the rainy season for the sparse herds of cattle. As a result, vast quantities of topsoil are lost every year into rivers, mainly those draining along the west coast. This is probably the single greatest human impact on Madagascar's coastal ecosystems. Recent acceleration of sedimentation in the Fiherenana and Onilahy river basins near Toliara has been linked to a shift to *tavy* practice following increased incidence of crop pests in riverside paddies (CNRE/CNRIT/IHSM, 2000).

The growing of coffee and vanilla, often together with soft fruits and mainly on the east coast, causes less soil loss, but sugar cane farming, which occurs at Nosy Be and Ambilobe, necessitates the clearance of large areas and the use of fertilisers, causing both erosion and fertiliser run-off.

A comparison of maps based on aerial photographs taken in 1949–1951 with Landsat 7 images, indicates considerable changes in mangroves, with the disappearance of some "old growth" stands on fine sediments and the appearance of new, relatively sparse, stands primarily of *Avicennia marina* on coarse alluvial sediments. This has been correlated, for example, with the loss of suitable habitat for Bernier's teal (*Anas bernieri*) an endangered endemic duck which is confined to "old growth" mangrove forests with a high canopy (>20 m) (R. Lewis, pers. comm). Mangroves are also being lost at several locations near Toliara to encroachment by windblown terrigenous sands derived from dry river beds (WWF, 1993; Ramampiherika, 1998).

Sedimentation has also had a major impact on coral reefs. Vasseur et al. (1988) report a progressive inundation of the fringing coral reef of Songeritelo, north of the mouth of the Fiherenana river, by alluvial sands. Continuing inundation renders seaweed farming impossible in the same area (H. Rakotoarison, pers. comm.). Changes of the Grand Récif of Toliara between 1964 and 1996 include increased areas of eroded reef and seagrass bed and reduced areas of active reef construction (Randriamanatsoa, 1997). A boat-based survey in 1997, made mainly at inshore sites close to estuaries (Cockroft and Young, 1998), reports extreme levels of sedimentation for much of the west coast of

> ## Changes to the Grand Récif, Toliara, 1964–1996
>
> Bemahafaly J. de D. Randriamanantsoa
>
> *Cellule des Océanographes de l'Université de Toliara (COUT), IHSM, B.P. 141, Toliara, Madagascar*
>
> The great barrier reef of Toliara was the subject of intensive scientific study during the 1960s by French scientists of the University of Marseille, creating a unique legacy of data, which now permit an assessment of reef change over a time during which human activities have increased greatly. Comparative analysis of data on reef structure from 1964 and 1996 indicates a widening of the sediment-covered part of the reef and a corresponding narrowing of the zone of active reef growth by 150–300 m, or about 10%. The average time taken for an observer to cross the reef flat on foot dropped from 60 minutes in 1960 to 25 minutes in 1996, reflecting the loss of structural complexity. Analysis of sediments in 1996 indicated an overall increase of 1–1.5% in the proportion of terrigenous sediments since 1970, while Secchi disc data indicated an average reduction by 1 m of the depth of disc disappearance in the adjacent lagoon. Comparison of fish studies from 1969 and 1972 with data from 1987 (Vasseur et al., 1988) indicate a 37% reduction in the number of species encountered, extrapolated to 60% by 2002. Finally, observation of reef gleaning in 1996 indicated that fishers are capable of destroying all live coral over 22–36% of the reef flat in a single year. Collectively, these observations suggest that it is essential to place the Grand Récif of Toliara under special management (Randriamanantsoa, 1997).
>
> **REFERENCES**
>
> Randriamanantsoa, B. J de D. (1997) Modifications de la morphologie et des sédiments du Grand Récif de Toliara (comparaisons 1969/1970–1995/1996). Impacts sur la pêche et les ressources vivantes. Mémoire de DEA en Océanologie Appliquée, IHSM/Université de Toliara (7 avril 1997), 60 pp. + 5 annexes.
>
> Vasseur, P., Gabrié C. and Harmelin, V. (1988) Mission scientifique préparatoire pour la gestion rationnelle des récifs coralliens et mangroves dont les mises en réserves: Tuléar (S.W. Madagascar) Rapport définitif EPHE (3–30 avril 1997), 213 pp.
>
> Weydert, P. (1971) Etude sédimentologique et hydrodynamique d'une coupe de la partie médiane du Grand Récif de Tuléar. *Téthys* suppl. 1, 237–280.

Madagascar, noting only 10 out of 24 stations where coral reefs appeared in good condition. Most reefs between Nosy Ve, south of the Onilahy river, and Morombe, are visibly affected by sedimentation; only outer reef slopes in well flushed areas appear substantially free of fine sediments (Vasseur et al., 1988; WWF, 1993). Sedimentation is less apparent on the offshore submerged barrier reef. Maharavo (pers. comm.) reports pristine coral communities on reefs 20 km offshore from Belo-sur-Mer, south of Morondava.

Fertiliser run-off is more localised, and essentially comes from the sugar industry. Figures provided by SIRAMA, the nationally-owned sugar production company, indicate the use of about 800 tonnes of NPK fertilisers on 2500 ha of plantations on Nosy Be, where run-off occurs along a 10 km stretch of the west coast, rich in coral reefs, where sea water PO_4 concentrations of 0.06 mg l^{-1} have been reported in the bay adjacent to the refinery compared with a typical 0.03 mg l^{-1} for Nosy Be (CNRE/CNRO/CNRIT, 1999). Similar levels of use can be expected for plantations at other sites. Based on areas cultivated, total NPK fertiliser use by the sugar industry is at least about 8000 t yr^{-1}. In addition, effluents from each plant in the form of vinasse are rich in nutrients and organic compounds.

These point sources have visible impacts on coral reef structure (Rakotoarinjahary et al., 1994). In Nosy Be the situation has recently improved with the installation of a series of effluent treatment ponds. If such systems can be installed at SIRAMA's other refineries, there is hope that this pollution will be significantly reduced. Fertiliser run-off is likely to remain a problem, however. Sugar cane plantations also use substantial quantities of pesticides. Annual usage by SIRAMA, Nosy Be, is 6500 litres of concentrated insecticides, 10,000 litres of herbicides and 1000 litres of fungicides, for about 2000 ha of plantation (Rakotoarinjahary et al., 1994).

Cotton plantations to the north of Toliara of about 16,000 ha annually use 750 tonnes of N/P fertilisers and a wide range of pesticides, amounting to about 200,000 litres in the 1997/1998 anti-locust campaign, mostly of carbamades and organophosphates (CNRE/CNRIT/IHSM, 2000).

COASTAL EROSION AND LANDFILL

Coastal erosion is a problem where coastal construction has interfered with local hydrology, typically accentuated by natural disasters (cyclones) (Ratomahenina, 1999). No land filling has yet occurred, although it has been suggested for the lagoon at Morondava. Dredging is carried out regularly at Mahajanga, Morondava and Toamasina, in order to maintain shipping channels and harbours.

At Toamasina, on the east coast, extension of quays in 1930–1937 and 1957–1960, and extension of a breakwater narrowing the pass between the Grand Récif of Toamasina and Point Hastie (1972–1974) have altered local hydrology and interrupted the supply of sediment to beaches to the north of the port. This, coupled with four violent cyclones (Honorine, 1986; Deborah, 1987; Geralda, 1994; Bonita, 1996) led to a 160 m retreat of shoreline along a stretch of several kilometres since 1924, of which 125 m were lost in 1986–1996 alone (Guerra et al., 1998).

At Morondava, on the west coast, erosion of the sea front has been a chronic problem ever since records began in 1914, apparently due to a combination of anthropogenic (port construction, channel modification) and natural (shifting river channels, tides, cyclones, currents) factors (LCHF, 1977; Troadec and Delmas-Ferre, 1998). Morondava is situated on the north side of a small delta and the town is constructed on a south-pointing spit of land, with principal access by causeway across a shallow tidal lagoon. Losses of land, exacerbated by effects from cyclones, have led to establishment of a special local commission to identify solutions, one option being to re-focus development of the town inland, another to in-fill the lagoon behind the town and build on it. At Mahajanga, displacement of a sediment bank and the adjacent channel has resulted in the loss of a tourist village and other infrastructures, although this is probably due to natural processes associated with long-shore drift.

EFFECTS FROM URBAN AND INDUSTRIAL ACTIVITIES

Artisanal and Non-industrial Uses of the Coast

On the west coast, the sea provides a significant transport route in remote areas, with the result that numerous settlements have little or no access to the interior. Mangroves are an important, often the only, source of wood, and are intensively exploited for mangrove crab (*Scylla serata*) and gastropods such as the mangrove whelk (*Terebralia palustris*). Occasionally, the shells of such gastropods are fired using mangrove wood to produce chalk (Ramampiherika, 1998). Around Toliara, mangroves have come under intense pressure as a source of fodder for livestock, especially goats, and back mangrove has also been the focus of artisanal salt making, especially at Toliara and Antsiranana, where some loss of mangrove has resulted (WWF, 1993). Coral mining (mainly of *Porites* sp.) occurs to the north of Toliara for the production of porous blocks for septic tanks, and beach rock is mined for construction. The mining of coral to make building lime has occurred locally—e.g. at Foulpointe (Refeno, pers. comm.)—but is not reported to continue. Coral is still used locally in tomb and house construction by the *Vezo*. South of Toliara there is an artisanal brick-making area, using alluvial clays, where bricks are fired using wood, which has devastated the surrounding bush.

On the east coast, which supports some of the highest population densities, fishing plays a minor, subsistence rôle. In effect, the coast is a transitional settlement area, from which populations fan out inland to convert forest to agriculture. Terrestrial timber is still readily available in most areas, so mangroves suffer relatively little cutting. Locally, exploitation of corals, mainly *Porites* sp., occurs for making septic tank blocks, notably on Ile Ste. Marie (Razafindrainibe, pers. comm.). Corals are also harvested for the curio trade at Foulpointe, although this has recently been reduced as a result of local efforts in environmental education (Refeno, pers. comm.).

Industrial Uses of the Coast

Industrial uses of the coastal zone include agro-industry, mining, food industries, textile and leather industries and paper making, in addition to five ports, an oil refinery at Toamasina, a fish canning factory at Antsiranana, three substantial aquaculture installations (in the north west) and salt pans on an industrial scale (at Toliara and Antsiranana). In the absence of any pollution control system in Madagascar, data on discharges are not collected. Estimates have, however, been made of the total inputs of land-based pollutants into coastal waters (Rakotoarinajahary et al., 1994) for each major watershed. These results indicate generally low inputs of pollutants, in keeping with Madagascar's small population and non-industrialised economy, and that untreated domestic waste discharged directly into rivers constitutes the major source of marine pollution.

Cities

Five of the largest towns and cities lie on the coast (see Fig. 1). These are not large, but are growing rapidly (Table 4). At Toliara the population expansion poses the greatest threat to shallow marine and coastal ecosystems, due to the increased exploitation of resources. Marine ecosystems at Toamasina face the particular threat of oil pollution due to the presence of the country's only refinery, which already pollutes the Canal de Pangalan, an inland waterway which runs along much of Madagascar's east coast. All major coastal cities lack sewage treatment systems and only a small proportion of houses possess adequate sanitation. Thus, town beaches serve as a public latrine and faecal pollution of urban shoreline areas is an acute urban environmental health problem.

Shipping and Offshore Accidents

Madagascar has not so far suffered a major shipping accident, although the majority of world petroleum production passes Madagascar's coasts. Several minor oil spills have occurred in connection with the fleet of small tankers which regularly supply Madagascar's principal coastal towns. No serious case of ecosystem damage by oil spill has been reported to date.

Madagascar has, however, suffered a particular problem of shipwrecks. Most notable of these has been the *Wellborn*, a vessel carrying manganese ore, beached (apparently deliberately) at Tolagnaro in 1996. Studies have been carried out on the uptake of manganese by seaweeds, although the results await publication.

PROTECTIVE MEASURES

A considerable number of measures have been taken to promote wise or sustainable use of natural marine and coastal resources, ranging from local traditional management practices, through implementation of national environmental programmes to national legislation and adherence to international conventions on natural resource conservation, harvesting and trade.

Customary Management Practices

Customary resource management practices have received relatively little official recognition to date but are gaining importance. This may be attributed to several factors. Only a minority of the coastal people are orientated towards fishing and those that are, notably the Vezo, consider the sea as a commons (Koechlin, 1975). There is some evidence of customary tenure systems relating to intertidal fish barrages (*valakira*) in the northwest, but this appears to be on the decline in the face of immigration and arrival of new techniques of shrimp fishing. Taboos (*fady*) are frequently declared by indigenous populations as a means of controlling access to resources by immigrants, such as bans on small mesh shrimp nets (Goedefroit, 1999) and bans on nets in the Tolagnaro area. There are also a few examples of the use of traditional social conventions, known as *dina*, as a basis for local coastal management schemes, such as the creation of marine reserves. Dina now enjoy legal recognition under laws on decentralisation.

Fisheries Regulations, Management and Policies

Fisheries regulations date back to 1897, near the beginning of the colonial period. Comprehensive regulations were first issued in 1922, many of which are still in force (Beurrier, 1982). Their emphasis has been on the conservation of highly valuable species such as sea turtles (though this is mostly not enforced), lobster and *bêche-de-mer* rather than on species or habitat protection. Marine mammals (cetaceans and dugongs) have been legally protected since at least 1961. The most recent fisheries law, 93-002, provides a comprehensive code for fishing and aquaculture and states that ecological sustainability must be an integral part of fisheries policy. A key point in this is that fishing is no longer a free activity as it was previously (Beurrier, 1982), and may be carried out only in accordance with regulatory decree.

Fisheries laws also provide the basis for management measures, with the emphasis being on industrial fisheries for tuna and shrimp. Thus, the number of licences awarded to EC purse-seiners has not substantially increased since 1992 (Kasprzyk and Andrianaivojaona, 1997) and the shrimp fishery has become increasingly regulated, with the recent introduction of a zoning plan to supplement the existing licensing system.

As a result of fluctuations in supply and demand and excessive harvesting of undersize individuals, the sea cucumber export industry has established a national association (ONET) which aims to promote more rational harvesting of wild stock. Experimental hatcheries and growing basins (for permitting the culture of undersized catches) have been established by the Institut Halieutique et des Sciences Marines (IHSM) at Toliara.

The fisheries ministry has recently approved proposals for placing the management of certain fisheries under local control under a new law no. 96-025 known as GELOSE (Gestion locale sécurisée). The lobster fishery (which is exclusively artisanal or traditional) has been the subject of discussions for a revision of current management regulations relating to season and permitted catch sizes, and for the possible introduction of local management systems in the country's principal lobster fishery in the Tolagnaro area.

The development of fisheries policy in Madagascar has proceeded on the premise that, in general, current exploitation of fisheries is well below the sustainable yield, although recognising that over-exploitation of certain stocks and in certain areas is occurring. National fisheries policy in Madagascar is defined by the Plan Directeur for fisheries and aquaculture with the principal objectives being expanding fisheries, increasing efficiency and improving the living standards of traditional fishing communities. A feasibility study for a national strategy on ecologically sustainable fisheries was commissioned (Orgasys, 1997), although this has not yet been acted upon by MPRH. A new charter for responsible and sustainable shrimp aquaculture is in the course of preparation, which will regulate the industry and lay down a code of conduct.

Mangrove Harvesting Regulations

Mangroves are subject to forestry laws, and there have been examples of concessions being awarded for felling of mangroves for timber. In theory, the cutting of any mangrove tree requires a permit, but this is not enforced. Fisheries within mangroves are in theory subject to fisheries laws, although no regulations on fishing in mangroves are in place. Plans exist to place mangroves under local management through the GELOSE law in certain areas.

Environmental and Other Legislation

The Environment Charter of 1990 (law 90-033) lays the foundations for modern environmental law in Madagascar. While this is primarily concerned with terrestrial issues, the underlying principles of sustainable development apply to the marine environment. The law provides for obligatory environmental impact assessment (EIA) of investment projects liable to affect the environment, now regulated under Decree 95-377 concerning "la Mise En Compatibilité des Investissements avec l'Environnement" (MECIE). EIA is obligatory for investment projects in "sensitive zones" including coral reefs, small islands, mangroves, wetlands and zones subject to erosion, all of which are defined in specific regulations. Several coastal projects have already undergone the complete EIA process, resulting in negotiated mitigation and monitoring measures.

A new maritime code will constitute Madagascar's application of UNCLOS, and includes provisions on the base line, management and conservation of marine resources and protection of the marine environment.

Marine and Coastal Protected Areas

Marine and coastal protected areas in Madagascar may be split into two categories: (1) those forming part of the national protected areas network managed by the Association Nationale pour la Gestion des Aires Protégées (ANGAP); and (2) those outside the national network. Currently, there are two areas within the national network, both of which are in the northeast: (1) Mananara Nord, a biosphere reserve with terrestrial, intertidal and marine components which includes 1000 ha of mangrove and approximately 1000 ha of coral reefs and islets, notably the island of Nosy Atafana and surrounding reefs (Fig. 7); and (2) three marine reserves (Tampolo, Cap Masoala, Tanjona) which form part of the Masoala National Park, totalling about 10,000 ha. At this stage, only Nosy Atafana is effectively protected, with the reef being closed to all fishing every other day (L. Touber, pers. comm.). Of the three Masoala reserves, Tanjona and Cap Masoala are subject to negotiation with local users, on the basis of which management plans will be prepared (W. Rakotoarinivo, pers. comm.).

The only marine and coastal protected areas outside the national network are: (1) the waters surrounding Nosy Tanikely (Nosy Be) extending to 300 m from the shoreline according to a ministerial decree of 1968 and 500 m according to a local decree of 1995, covering an area of about 30 ha of coral reef; and (2) both land and waters of Nosy Ve, Toliara, under a local community convention or dina, which protects fully a small (~4 ha) area of lagoon reef and the entire island of about 21 ha, which harbours Madagascar's only colony of red-tailed tropic birds (*Phaeton rubricauda*) (Cooke and Randrianamindry, 1996). Protection of Nosy Tanikely has never been systematically assessed, whereas recent surveys confirm that the lagoon reef of Nosy Ve, Anakao, is genuinely protected, and several observers regularly diving on the site have confirmed markedly improved fish biomass in the protected zone (M. Vachoud and Ninés, pers. comms.).

Additionally, there may exist informally protected areas at: (1) Ifaty, Toliara (Jardin de Roses, Passe Sud, has been customarily avoided by fishermen for several years as a recognised tourist site (COUT and Cooke, 1994); and (2) Point Anjagnaharibe, Cap Masoala, a sacred site where no fishing is allowed and which has been reported to be well-conserved (Odendaal et al., 1995).

A specific action group on protected areas has recently been created through a partnership of ANGAP, the national marine and coastal environment programme (EMC) (see below) and interested organisations such as WWF, Conservation International, UNESCO, and the Wildlife Conservation Society.

Protected Species

Only marine mammals enjoy complete protection of the law. Sea turtles are not fully protected, although the export of turtle products is regulated (Rakotonirina and Cooke, 1994). Turtles are the subject of ongoing research in several areas and WWF has recently launched a sea turtle conservation programme in the north of Madagascar. The mandatory use of turtle exclusion devices (TEDs) by shrimp trawlers has been considered by the fisheries ministry but not so far implemented because destination countries for Malagasy shrimp (EU and Japan) do not insist upon it. Certain gastropods appear locally endangered and merit protective legislation (WWF, 1993).

International and Regional Conventions

Madagascar has either signed or ratified, or intends to sign or ratify, most of the major international conventions relevant to management and protection of marine and coastal resources (Table 6). It has also signed and participates in the UNEP Regional Seas Convention for East Africa 'Nairobi Convention'. A regional action plan and protocol on protected areas and wild flora and fauna were also agreed in 1985 and are soon to be revised to include provisions on integrated coastal zone management (ICZM).

As a member state of the Indian Ocean Commission (COI), Madagascar is a member of the Indian Ocean Tuna Association, whose office is based at Antananarivo. It also participates in the Western Indian Ocean Tuna Organisation and the Tuna Commission for the Indian Ocean (Kasprzyk and Andrianaivojaona, 1997). Negotiations have begun between the COI countries for a regional treaty on sustainable development, although they have been temporarily suspended for lack of funds.

Fig. 7. Parc National Marin de Nosy Atafana (Biosphere reserve). Photo: A. Ebregt (courtesy UNESCO-MAB).

Table 6

Status of relevant international conventions with respect to Madagascar

Name of Convention	Signed	Ratified
OILPOL (London 1954)	1954	1971
MARPOL	No	No
CITES	1973	1975
UNCLOS	25 Feb. 1983	Not yet
World Heritage Convention (UNESCO)	Yes	9 Dec. 1992
UNEP Regional Seas Convention for East Africa*	31 Dec. 1993	24 March 1998
Convention on Biological Diversity	1992	Yes
Ramsar** 1973	1998	1999

*Convention for the Protection, Management and Development of the Marine and Coastal Environment of the Eastern African Region.
**Convention on Wetlands of International Importance, especially as Waterfowl Habitat.

Finally, also as a member of the COI, Madagascar participates in a regional project to develop capacity to combat major oil spills, funded by the World Bank. The project will support legal training, the preparation of a national ecological sensitivity atlas, the formulation of pollution prevention plans and the purchase of appropriate equipment for dealing with spills.

Projects Concerning Marine and Coastal Resources

Projects established to date have mostly addressed the development of more efficient resource use practices, aquaculture and management of the industrial shrimp fishery and, to a lesser degree, the need for information for management of resources.

A project of FAO/UNDP aims to promote integrated development of fishing villages along the southwestern and southern coasts of Madagascar and sought to introduce new systems for monitoring marine resource exploitation, although it has limited funds. GTZ has supported a fisheries development project in Nosy Be which seeks to help fishermen develop more efficient offshore fishing methods, thus reducing pressures on shallow-water ecosystems, although concern has been raised about the sustainability of such techniques in so far as they specifically target sharks (Odendaal et al., 1995; Smale, 1998). The EU has funded projects to develop fish aggregation devices (FADs), shrimp aquaculture and seaweed culture. Significant, smaller, projects have included the culture of brine shrimp (*Artemia* sp.) in salt ponds at Ifaty north of Toliara and a project south of Toliara to develop captive growing and spawning of sea cucumbers.

The KEPEM project of USAID supported the National Environment Office in a variety of instruments and tools towards better management of resources, including environmental impact assessment. KEPEM also issued a baseline report on marine and coastal issues (Jain, 1995).

National Marine and Coastal Environment Programme

Madagascar is unusual in benefiting from a multi-donor national environmental action programme (NEAP), founded on the Charter for the Environment of 1990 (Law 90-033). The programme consists of three five-year phases, 1992–1996, 1997–2001 and 2002–2006.

Phase 2 includes for the first time a component for the marine and coastal environment (EMC), operated by the National Environment Office (ONE), under the aegis of the Ministry for Environment. The EMC programme is supported by funds from UNDP, the GEF and the World Bank. As described in the founding document (Gabrié, 1996) and subsequent work plans, the principal strategic objectives of the programme are: (1) to amass information on the coastal zone and to establish monitoring systems in order to help decision-making in integrated coastal zone management; (2) to establish and implement a national policy, legal/economic instruments and institutional framework for integrated coastal zone management; and (3) to develop integrated systems for sustainable management and development of specific coastal resources. The intention is that such regional and local structures should evolve into full programme partners, assuming responsibility for implementation, as their capacities develop.

The EMC programme includes pilot actions in the regions of Toliara and Nosy Be (with another pilot zone at Toamasina soon to be added), and places considerable emphasis on decentralised coastal governance, seeking to derive lessons for ICZM policy conception. The programme has established a national working group on coastal zone management and similar structures in the two pilot regions. It has also established numerous executive local committees in the regions. Parallel to the programme is an initiative by FAO and the fisheries ministry to improve management of the lobster fishery around Tolagnaro, using the GELOSE law and concertative processes of the type being piloted by EMC.

The EMC programme is subject to annual donor review, and a substantial mid-term review of NEAP will take place in mid-2000. The EMC programme has also been the subject of independent appraisal which has emphasised the need to adhere to an integrated approach, working at all levels in relation to a given region (Billé, 1999). In fact, the entire environment programme is evolving towards a regional approach, and the EMC has chosen to focus on the regions of Toliara and Nosy Be (already pilot areas), Toamasina (conveniently situated for access to a number of initiatives on the east coast), Menabe (to be transferred to EMC by the Regional Environment Programme of the Indian Ocean Commission, PRE-COI, which closes in March 2000) and Anosy (the region of Tolagnaro (Fort

Dauphin)) where major investments by a subsidiary of Rio Tinto Zinc in coastal titanium mining and by the World Bank in the form of a facilitation loan for infrastructure development provide a framework within which EMC will be able to promote ICZM objectives.

RECOMMENDATION AND PROGNOSIS

Experiences to date in the EMC programme indicate: (1) the necessity for a participatory approach in developing ICZM at all levels; (2) the need to place management of a resource at the level of resource users; (3) the importance of meeting local needs in ICZM initiatives; (4) the need to build local capacities for their implementation.

Our recommendation is that the EMC process should continue to emphasise the local and regional approaches for which opportunities exist, always working in partnership with other programmes which promote sustainable development, while regarding the national process primarily as a vehicle for dialogue on ICZM with concerned centrally represented institutions, rather than as necessarily leading to a finalised national policy in the near term. In general, the private sector and civil society should be encouraged to play a greater role.

The general prognosis is optimistic, in that the majority of Madagascar's marine ecosystems appear to be in reasonably healthy condition, with acute problems being limited to vulnerable species or to specific areas of intensive resource use and sedimentation. Coastal populations, at only 25% of the national total, are still relatively small, although unevenly dispersed. The essential challenge is to put in place sustainable management and development systems before any of the existing problems become too severe. At the same time, a shift to more enlightened, sustainable management systems and policies depends upon continued, balanced, national economic development. Probably the greatest single technical environmental management challenge facing Madagascar today, both for the coastal zone and upland areas, is the reduction of upland watershed erosion and its impact on coastal ecosystems.

ACKNOWLEDGEMENTS

The following provided essential unpublished information: Gilbert, François, Directeur des Ressources Halieutiques, DRH, Antananarivo; Kasprzyk, Zbigniew, Managing Director, Ocean Consultant, Antananarivo; Lewis, Richard, Scientific Co-ordinator, Durrell Wildlife Conservation Trust; Hawkins, Frank, Technical Adviser, Projet ZICOMA; Maharavo, Jean, Consultant, PRE-COI Programme, Antananarivo; Ninés, Diving instructor, Hotel Safari Vezo, Anakao, Toliara; Piso, Jean-Charles, Programme Manager, EMC/ONE, Nosy Be; Rakotoarinivo, William, Coordinator, Marine and Coastal Programme, Masoala Peninsula Integrated Conservation and Development Project, Antalaha; Rakotoarison, Hubert, Programme Manager, Technical Support Unit for the Marine and Coastal Environment programme, Toliara; Randriamiarana, Heritiana, Coordinateur National, Cellule EMC, ONE, Antananarivo; Razafimbelo, Honoré, Co-ordinateur National, Projet Sectoriel Pêche, FAO; Refeno, Germain, Co-ordinateur National, PRE-COI; Schallkwyk, George, Résidence du Cap, Cap Est, Masoala Peninsula; Touber, Luc, Technical Adviser, Mananara Nord Biosphere Reserve Project, MAB-UNESCO, Antananarivo; Turcotte, Paul-André, Land Resources Environmental Consultants, Antananarivo and Vachoud, Monique, Divemaster, Anakao, Toliara; Caira Janine, Dept. Ecology and Evolutionary Biology, University of Connecticut.

REFERENCES

Anon (1983) Cruise report *R/V Dr. Fridtof Nansen*, Resources Survey, Madagascar, 16–28 June 1983. Institute of Marine Research, Bergen. FAO.

Andrianaivojaona, C., Kasprzyk, Z.W. and Dasylva, G. (1992) Pêches et Aquaculture à Madagascar. FAO. 154 pp.

Andrianarivelo, N., Lope, J.C., Rakotonirina, J.S. and Randriamanantsoa, B.J-D. (1998) Suivi Ecologique des Trois Parcs Marins et d'un Site Touristique dans la Presqu'île de Masoala. Rapport Définitif. Cellule des Océanographes de l'Université de Toliara (C.O.U.T.) November–December, 1998.

Andrianjohany, S. (1992) Inventaire de Poissons des mangroves de Masoarivo. Memoire de Diplome d'Etudes Approfondies, Université d'Antananarivo, 1992. 60 pp.

ANGAP (1999) Plan Strategique du Réseau des Aires Protégées de Madagascar. Report of a scientific workshop, Antananarivo, 28–29 July 1999. Association Nationale pour la Gestion des Aires Protégées. 61 pp.

Battistini, R. (1972) Madagascar relief and main types of landscape. In *Biogeography and Ecology of Madagascar*, eds. G. Richard-Vindard and R. Battistini. Junk, The Hague.

Beddington, J. and Cooke, J.G. (1983) The Potential Yield of Fish Stocks. FAO Fisheries Technical Paper.

Beurrier, J.-P. (1982) Les Zones sous juridiction, la législation des pêcheries et l'organisation structurelle du secteur des pêches à Madagascar. Report prepared for the Government of Madagascar. FAO, Rome, 1982. 104 pp.

Billé, R. (1999) Financements Internationaux, engagements environnementaux: Une analyse organisationelle du dispositif de gestion intégrée du littoral à Madagascar. Mémoire du DEA, Ecole Nationale du Génie Rural, des Eaux et des Forêts (ENGREF), Université Paris X, Nanterre, France. 82 pp.

Chasse, C. (1962) Remarques sur la morphologie et la bionomie des herbiers de monocotylédones marines tropicales de la province de Tuléar (Madagascar). *Rec. Trav. Stat. Marine d'Endoume*, Fasc. RS, suppl. 1, 237–248.

Clausade, M., Gravier, N., Picard, J., Pichon, M., Roman, M-L., Thomassin, B., Vasseur, P., Vivien, M. and Weydert, P. (1971) Morphologie des récifs coralliens de la région de Tuléar (Madagascar): Eléments de terminologie récifale. Téthys, Suppl. 2. 74 pp.

Citeau, J., Piton, B. and Magnier, Y. (1973) Sur la circulation géostrophique dans l'ouest de l'Océan Indien Sud-Equatorial. Doc. Sci. du Centre ORSTOM de Nosy Be, no. 31, 20 p.

CNRE/CNRO/CNRIT (1999) Identification des caractéristiques des observatoires et establissement de plan local de prévention et réduction des pollutions et dégradations de la région de Nosy Be. Report prepared for EMC. 140 pp.

CNRE/CNRIT/IHSM (2000) Identification des caractéristiques des observatoires et establissement de plan local de prévention et réduction des pollutions et dégradations de la région de Toliara. Interim report to EMC. 102 pp.

Cockroft, V. and Young, D.D. (1998) An Investigation of the Status of Coastal Marine Resources along the West Coast of Madagascar. Centre for Dolphin Studies, Port Elizabeth, South Africa. 58 pp + tables and figures.

Cooke, A.J. (1996) Les îlôts. Définition et délimitation des zones sensibles et lignes directrices pour la préparation des études d'impact environnemental. Knowledge and Effective Policies for Environmental Management (KEPEM). Report No. 39d, November 1996. 41 pp, maps + island inventory.

Cooke, A.J. (1997) Survey of Elasmobranch Fisheries and Trade in Madagascar. In *The Trade in Sharks and Shark Products in the Western Indian and Southern Indian and South East Atlantic Oceans*, eds. N.T Marshall and R. Barnett, pp. 101–130. TRAFFIC.

Cooke, A.J. and Randriamanindry, J-J. (1996) Red-tailed tropicbird (*Phaeton rubricauda*) colony, Nosy Ve islet, Toliara. *Newsletter of the Working Group on Birds in the Madagascar Region* 6 (2).

COUT (Cellule des Océanographes de l'Université de Toliara) and Cooke, A.J. (1994) Toliara Coral Reef Expedition, Madagascar 1993. Final report. December 1994. 200 pp.

Edmond, R. and Grepin, G. (1999) Végétation Côtière. Report prepared for UNOPS Project MAG/96/G31 ("Inventaire des Ecosystèmes Marins et Côtiers et de leur Biodiversité, Profil Côtier National et Diagnostic Environnemental des Zones Marines et Côtières de Madagascar").

FAO (1995) Atlas of Industrial Tuna Fisheries Development and Management Programme. Compiled by J.D. Ardill. FAO, 1995.

FAO (1998) Aperçu sur les Zones de Pêche Traditionnelle Maritime dans la Region de Toliara. Programme Sectoriel Pêche PNUD/FAO/MAG/92/004-DT/10/98. January, 1998. 14 pp.

Fourmanoir, P. (1961) Requins de la Côte Ouest de Madagascar. *Mem. Inst. Sci. Madagascar (Sér. F Océanographie ORSTOM)* 4, 1–81.

Gabrié, C. (1996) Rapport d'Evaluation de la Composante "Environnement Marin et Côtier" du PEII du Plan D'Action pour l'Environnement (PAE). 45 pp.

Gabrié, C., Vasseur, P., Maharavo, J., Randriamiarana, H. and Mara, E. (2000). The Coral Reefs of Madagascar. In *Coral Reefs of the Western Indian Ocean*, eds. T.R. McClanahan, D.O. Obura and C.R.C. Sheppard, Oxford University Press.

Goedefroit, S., 1999. La Ruée Vers l'Or Rose - Stratégie d'accés et conflit d'usage de la ressourve crevettière chez les pêcheurs traditionnels à Madagascar. Programme National de Recherche Crevettière (PNRC), Madagascar, 1999.

Guerra, L. et al. (1998) Projet de Protection du Littoral Nord de Toamasina. 46p + Annexe. Published by Sauvons Toamasina NGO.

Harmelin-Vivien, M.-L. (1979) Ichtyofaune de récifs coralliens de Tuléar, Madagascar. Ecologie et rélations trophiques. Thèse Doctorat d'Etat-ès-Sciences, Univ. Aix-Marseille 2. 165 p + annexe.

Heemstra, P.C., Freeman, A.L.J., Yan H., Wong, Hensley, D.A. and Rabesandratana, H.D. (1996) First authentic capture of the coelacanth, *Latimeria chalumnae* (Pisces–Latimeridae) of Madagascar. *South African Journal of Science* 92, 160–171.

ICRI (1996) Atelier Regional de l'Océan Indien Occidental. Seychelles, 29 March–2 April, 1996. Report prepared by C. Gabrié and CAREX Environment. International Coral Reef Initiative. 4 pp.

IHSM (1999) Etude de la Biodiversité de l'Ile de Nosy Ve en vue de la création d'un Parc Marin. Institut Halieutique et des Sciences Marines, Université de Toliara. February, 1999.

Jain, M. (1995) Inventory of Marine and Coastal Activities—A Strategy Proposal for Madagascar. Knowledge and Effective Policies for Environmental Management (KEPEM). Report No. 20. 103 pp.

Kasprzyk, Z. and Andrianaivojaona, C. (1997) Situation et Perspectives de Développement de la Pêche et de l'Aquaculture à Madagascar. Volume 1: Bilan de la Pêche et de l'Aquaculture pour la Période 1990–1995. Antananarivo, April 1997.

Kiener, A.N. (1972) Ecologie, biologie et possibilités de mise en valeur des mangroves Malgaches. *Bulletin de Madagascar* 308, 49–84.

Koechlin, B. (1975) Les Vezo du Sud-Ouest de Madagascar—Contribution à l'étude de l'éco-système de semi-nomades marins. *Cahiers de l'Homme. Nouvelle Série XV*. Mouton & Co. & Ecole Pratique des Hautes Etudes, Paris, 1975. 247 pp.

Laroche, J. and Ramanarivo, N.V. (1995) A preliminary survey of the artisanal fishery on coral reefs of the Tuléar Region (Southwest Madagascar). *Coral Reefs* 14(4), 193–200.

Laroche, J., Baran, E. and Rasoandrasana, N.B. (1997) Temporal patterns in a fish assemblage of a semiarid mangrove zone in Madagascar. *Journal of Fish Biology*, 50.

Lebigre, J.M. (1990) Les marais maritimes du Gabon et de Madagascar. Thesis, University of Bordeaux III, Institut de Géographie. 470 pp.

LEC (1999) Addressing the Ecological and Social Impacts if Migrating Vezo People on Coastal Areas and Marine Ecosystems in Southern Madagascar. Proposal submitted to the John D. and Catherine T. McArthur Foundation. Libanona Ecology Centre, Tolagnaro, Madagascar. July, 1999. 7 pp + Annexes.

LCHF (1977) Laboratoire Centrale Hydraulique de France. Etude du Port de Morondava. PNUD-Min. de l'Aménagement du Territoire RM. Projet MAG/72/013 1A. Hydrographie, Océanographie, Météorologie. pp. 18–120.

Lutjeharms, J.R.E., Bang, N.D. and Duncan, C.P. (1981) Chracteristics of the currents east and south of Madagascar. *Deep Sea Research, Oceanographic Abstracts* 28, 879–899.

Magnier, Y. and Piton, B. (1973) Les masses de l'eau de l'Ocean Indien à l'Ouest et du nord de Madagascar au début de l'été austral. *Cahiers ORSTOM sér. Océanogr.* XI (1), 97–113.

Maharavo, J. and Vicente, V. (1999) Milieu Marin. Report prepared for UNOPS Project MAG/96/G31 ("Inventaire des Ecosystèmes Marins et Côtiers et de leur Biodiversité, Profil Côtier National et Diagnostic Environnemental des Zones Marines et Côtières de Madagascar").

Maugé, L.A. (1967) Contribution préliminaire à l'inventaire ichtyologique de la région de Tuléar. *Rec. Trav. Sta. Mar. Endoume, Marseille, Fasc. Hors sér.*, suppl. 7, 101–132.

McClanahan, T. and Obura, D. (1998) Monitoring, Training and Assessment of the Coral Reefs of the Masoala Peninsular. Wildlife Conservation Society, June 1998.

Ocean Consultant and Sepia (1999) Analyse des avantages et inconvenients du système actuel d'allocation des droits de pêche à Madagascar. Rapport Intermédiaire. November, 1999.

Odendaal, F., Kroese, M. and Jaomanana (1995) The Strategic Plan for the Management of the Coastal Zone of the Masoala Peninsular, Madagascar. Eco-Africa Consultants Madagascar Working Paper No. 4. Report prepared for CARE Madagascar.

Orgasys (1997) Etude pour une Politique de Développement de la Pêche et de l'Aquaculture Ecologiquement Durable. Description de la Situation (Vol 1). Antananarivo, June, 1997. Sections 1–4.5. 4 Annexes.

Pichon, M. (1978) Recherches sur les peuplements à dominance d'Anthozoaires dans les récifs coralliens de Tuléar (Madagascar). *Atoll Research Bulletin* 222, 1–447.

Piton, B., J.H. Pointeau and Ngoumbi, J.S. (1981) Atlas Hydrologique du Canal de Mozambique (Océan Indien). Travaux et doc. de l'ORSTOM. no. 132, 41 p.

PRE-COI (1997) Pré-Audit, Première Partie: La Gestion Intégrée de la Zone Côtière et le Programme Régional Environnement de la COI. Programme Régional Environnement, Commission de l'Océan Indien (PRE-COI).

Rabarison, A.G.A. (1994) Observations océanographiques sur

l'intoxication collective par ingestion de requin à Manakara (Madagascar). 5 pp. CNRO, Nosy Be, Madagascar.

Rabeson, C. (1992) La Pêche Thonnière a Madagascar: Impacts socio-economiques. Mémoire de DEA, 1992. IHSM, Toliara. 59 pp., 5 annexes.

Rakotoarinjanahary, H., Razakasoniaina, T.N., Kaderbay, B. and Ranaivoson, E. (1994) Inventaire des sources de pollution et quantités de polluants à Madagascar. CNRO/OMS/AFRO. 58 pp.

Rakotoarinivo, W. (1998) Les Petits Poissons Pélagiques de la Région de Toliara (Sud-Ouest de Madagascar): Biologie, Ecologie, Exploitation et Aménagement. Doctorate thesis in Applied Oceanography. IHSM, Toliara, Madagascar.

Rakotonirina, B.P. and Cooke, A.J. (1994) Sea turtles of Madagascar—their status, exploitation and conservation. *Oryx* **28** (1), 51–61.

Rakotonirina, B.P. (1999) Etude préliminaire des Tortues Marines dans la Presqu'île de Masoala—Rapport Final, Novembre–Décembre, 1998. Wildlife Conservation Society, Madagascar. 28 pp.

Ralison, A. (1991) Les Potentiels des Ressources Halieutiques Maritimes et leur Niveau d'Exploitation. In Rapport du Seminaire National sur les Politiques et la Planification du Développement des Pêches à Madagascar, Antananarivo, 15–19 October, 1990, eds. C. Andrianaijaona, Z. Kasprzyk and G. Dasylva, pp. 137–150. FAO, Madagascar, 1991.

Ramampiherika, K.D (ed.) (1998) Analyse Diagnostic de l'Etat de Santé de l'Environnement Côtier et Marin de la Région de Toliara. Toliara, November 1998. 91 pp.

Ranaivoson, E. (1996) Environnement Marin et Côtier. In: *Monographie sur la Biodiversité de Madagascar*. ONE.

Randriamanantsoa, B. J de D. (1997) Modifications de la morphologie et des sédiments du Grand Récif de Toliara (comparaisons 1969/1970–1995/1996). Impacts sur la pêche et les ressources vivantes. Mémoire de DEA en Océanologie Appliquée, IHSM/Université de Toliara (7 avril 1997), 60 pp. + 5 annexes.

Ratomahenina, O.R.J. (1988) Etude Bibliographique du Canal de Mozambique—Hydrologie et Circulation. Memoire de Diplôme d'Etudes Approfondies. Univ. Paris VI, 35 pp. + annexe (45 figs.).

Ratomahenina, O.R.J. (1999) Impacts et érosion. Report prepared for UNOPS Project MAG/96/G31 ("Inventaire des Ecosystèmes Marins et Côtiers et de leur Biodiversité, Profil Côtier National et Diagnostic Environnemental des Zones Marines et Côtières de Madagascar").

Razafindrainibe, H., Razakafoniaina, N.T., Rasolofo, V.M., Rakotoarinjanahary, H., Randriambololona, C. and Rasolonjatovo, H. (1996) La pêche tradtionnelle maritime dans le Menabe (côte ouest de Madagascar). Rapport des Recherches. Projet "Etudes biologiques et socioéconomiques des pêches artisanales traditionnelles à Madagascar". Phase II. Centre National de Recherches Océanographiques. August 1996. 85 pp.

Remanevy, M.E., Rabesandratana, H.D., Conand, C., Rakotoarinivo, W., Rasolofonirina, R., Rasoandrasana, N.B. and Ravelo, I. (1997) Etude de la Pêcherie aux Holothuries dans le Sud-Ouest de Madagascar et Propositions de Mésures d'Aménagement. IHSM, December 1997.

Rosenbaum, H.C., Walsh, P.D., Razafindrakoto, Y., Vely, M. and de Salle, R. (1997) Preliminary and final reports on humpback whale, *Megaptera novaeangliae*, research and conservation in Baie d'Antongil, Madagascar, July–September, 1996. 13 pp.

Saetre, R., and Dasylva, A.J. (1984) The Circulation of the Mozambique Channel. *Deep Sea Research* **31** (5), 485–508.

Schott, F., Fieux, M., Kindle, J., Swallow, J.C. and Zantopp, R. (1988) The boundary currents east and north of Madagascar, 2. Direct measurements and model comparisons. *Journal of Geophysical Research* **93**, 4963–4974.

Sheppard, C.R.C. (1998) Biodiversity patterns in Indian Ocean corals, and effects of taxonomic error in data. *Biodiversity and Conservation* **7**, 847–868

Smale, M.J. (1998) Evaluation of Shark Populations around the Masoala Peninsular, North East Madagascar. Final Report. Wildlife Conservation Society, Madagascar Country Programme. May, 1998. 40 pp.

Stequert, B., Marcille, J. and Piton, B. (1975) La pêche thonnière à Madagascar de mai 1973 à février 1975. Doc. Sci. du Centre ORSTOM de Nosy Be, no. 52, 56 pp.

Thomassin, B. (1971) Révue bibliographique des travaux de la Station Marine de Tuléar (Madagascar) 1961–1970. Téthys, Suppl. 1, 3–49.

Troadec, R., and Delmas-Ferre, M. (1998) Mission pilote d'évaluation de l'érosion côtière et les risques sur les aménagements de la ville de Morondava - Madagascar. PRE-COI/UE/GREEN. 81 p.

Vasseur, P., Gabrié C. and Harmelin, V. (1988) Mission scientifique préparatoire pour la gestion rationnelle des récifs coralliens et mangroves dont les mises en réserves: Tuléar (S.W. Madagascar) Rapport définitif EPHE (3–30 avril 1997), 213 pp.

Veron, J.E.N. (1986) *Corals of Australia and the Indo-Pacific.* University of Hawaii Press Edition (1993). 644 pp.

WWF (1993) Coral Reefs and Coastal Zone of Toliara—Conservation and Development through Ecotourism: Pre-project. Report edited by A. J. Cooke. 235 pp. + maps & photos. WWF Madagascar, January 1993.

ZICOMA (1999) Les Zones d'Importance pour la Conservation des Oiseaux à Madagascar. Projet ZICOMA, Antananarivo, Madagascar. 266 pp.

THE AUTHORS

Andrew Cooke
Cellule Environnement Marin et Côtier,
Office National pour l'Environnement, B.P. 822,
Antananarivo, Madagascar

Onésime Ratomahenina
Ministère de la Recherche Scientifique,
Direction Générale de la Recherche, B.P. 4258,
Antananarivo, Madagascar

Eulalie Ranaivoson
Institut Halieutique et des Sciences Marines,
B.P. 141, Toliara, Madagascar

Haja Razafindrainibe
Cellule Environnement Marin et Côtier,
Office National pour l'Environnement, B.P. 822,
Antananarivo, Madagascar

Chapter 61

SOUTH AFRICA

Michael H. Schleyer, Lynnath E. Beckley, Sean T. Fennessy,
Peter J. Fielding, Anesh Govender, Bruce Q. Mann,
Wendy D. Robertson, Bruce J. Tomalin and Rudolph P. van der Elst

---

The South African coastline stretches from the mouth of the Orange River in the northwest to Kosi Bay in the northeast, a distance of some 3000 km. It covers a wide range of habitats from kelp beds to mangrove forests and coral reefs. The Benguela ecosystem on the west coast is characterised by upwelling and extends from Cape Point northwards, constituting the Namaqua biogeographic province. The warm temperate Agulhas biogeographic province is found on the south coast from Cape Point to East London and the Natal biogeographic province along the sub-tropical east coast. Marine biodiversity in South Africa is particularly high as a result of this environmental variability. Commercial, recreational and subsistence fisheries are discussed, many of which are harvested sustainably because of conservation legislation instituted by management agencies. There is no room for complacency, however, as there has been a serious depletion in some instances. Demographic pressures on the estuarine and coastal environments are cause for concern as well as the popular misconception of the sea providing an endless supply of food and as a repository for unlimited waste.

Seas at The Millennium: An Environmental Evaluation (Edited by C. Sheppard)
© 2000 Elsevier Science Ltd. All rights reserved

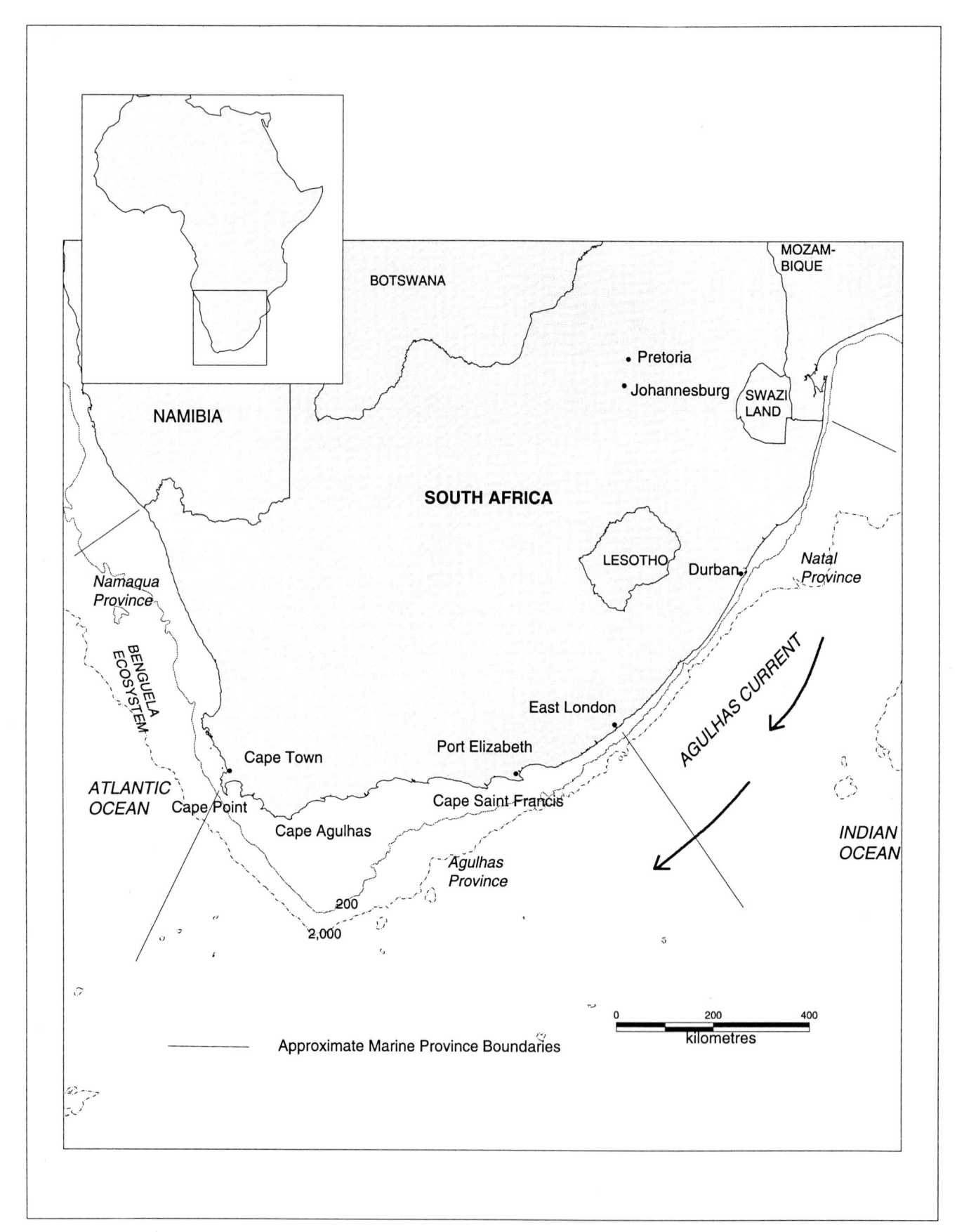

Fig. 1. Map of the coast of South Africa.

THE SOUTH AFRICAN COASTLINE

The South African coastline stretches from the mouth of the Orange River in the northwest to Kosi Bay in the northeast and totals some 3000 km (Fig. 1). It covers a wide range of habitats including rocky shores, sandy beaches, estuaries, embayments, mangrove forests, kelp beds and coral reefs. Overviews of its oceanography and biogeography are provided by Brown and Jarman (1978), Branch and Branch (1981), Smith and Heemstra (1986) and Payne et al. (1989).

The continental shelf is extremely narrow along the east coast (<5 km wide in areas) but expands to form the extensive, wedge-shaped Agulhas Bank which is up to 200 km wide south of the continent. Off the southwestern coast, the shelf is again relatively narrow, widening once more near the Namibian border.

The west coast or Benguela ecosystem is characterised by upwelling. This process is wind-driven, surface water being blown offshore and replaced by deeper, colder water that is rich in nutrients as it wells up to the surface along the coast. Phytoplankton bloom in this nutrient-rich water and form the basis of a rich food chain. The rocky coast is characterised by kelp beds, an ecosystem in which numerous harvestable resources such as mussels, abalone and rock lobsters are found.

The most conspicuous feature of the east coast of South Africa is the warm Agulhas Current, a major western boundary current which flows southwest along the edge of the continental shelf until it retroflects and heads eastward south of the Agulhas Bank. The warm water which it transports southwards extends the range of numerous tropical and sub-tropical Indo-Pacific species, including corals and fishes, into this region.

The shores of South Africa are particularly rich in biodiversity at the species level; close to 10,000 species of marine plants and animals have been recorded from the waters around the southern African subcontinent. This represents almost 15 per cent of all the coastal marine species known worldwide (Branch et al., 1994).

The fauna of these rocky shores can be separated into specific zoogeographic provinces on the basis of distribution records (Emanuel et al., 1992). The cool temperate southwest coast from southern Namibia to Cape Point constitutes the Namaqua province; the warm temperate south coast from Cape Point to East London, the Agulhas province; and the sub-tropical east coast from East London to southern Mozambique, the Natal province. Sandy shore fauna and seaweeds mirror similar trends, although the exact boundaries may differ slightly.

Species diversity increases from the west coast to the east coast as a result of the southward dispersal of Indo-Pacific species by the Agulhas Current. Endemism in southern African marine biota is relatively high at about 12%, with the south coast being particularly well-endowed with endemic species (e.g. the fish families *Sparidae* and *Clinidae* (Turpie et al., 1999)). Several cosmopolitan species and wide-ranging oceanic species such as tunas, billfishes and whales frequent these waters, as well as limited numbers of Atlantic and southern ocean species.

The wide variety of oceanographic and hydrological conditions found along the South African coast has also richly endowed this region with a high biodiversity at structural and functional levels. There are few countries in the world that can boast such diversity per kilometre of coast, a heritage of great value that merits wise and sustainable management.

ESTUARIES

Estuaries are an important ecosystem with great aesthetic and recreational value. This is clear from their popularity as venues for picnicking, swimming, boating, water-skiing, canoeing, bird-watching and fishing. They provide convenient sites for harbour construction and safe launch sites for skiboats. Their unique habitats are rich in biodiversity. They are highly productive systems as they receive an input of nutrients from both the sea and the freshwater environment.

Many of South Africa's estuaries have been severely affected by human activities. A stable mix of physical conditions within an estuary is largely dependent on a delicate balance between fresh- and seawater input into the system. Any human activity that interferes with this balance has a profound influence on the estuarine environment and consequently on its fauna and flora.

This is well illustrated in the St Lucia system in KwaZulu-Natal. During extended periods of low freshwater input, the upper reaches of the St Lucia system become hypersaline from the combined effects of evaporation and seawater flow into the lake. This naturally occurring event has been exacerbated by a reduction of freshwater input as a result of damming of the Hluhluwe River for irrigation, and afforestation in the Mkuzi River catchment. The reduced freshwater flow also allows sediments of marine origin to clog the mouth of the estuary and constant dredging is required to ensure that the vital connection with the sea is maintained for the exchange of nutrients and the migration of animals.

Badly designed bridges can obstruct water flow in estuaries and reduce the scouring effect of floods, contributing to estuarine siltation. This is particularly true when large embankments are built to reduce cost, as was done in the construction of the coastal bridges over the Bushmans and Kariega estuaries in the Eastern Cape. Interference with tidal exchange in estuaries can also have deleterious effects. A healthy mangrove forest in the Mgobezeleni estuary at Sodwana Bay, KwaZulu-Natal, was decimated by hyposaline water which was dammed above a causeway built over the estuary in the 1970s.

Poorly planned urban development adjoining estuaries has often created another set of problems. Nutrient enrichment has resulted from sewage seepage into a number of systems and the danger of properties being flooded when estuary mouths have been closed has led to their artificial and unseasonal breaching.

Poor farming practices, particularly amongst sugar cane growers, are largely responsible for the badly silted and turbid condition of most of the estuaries of KwaZulu-Natal. Natural vegetation has been cleared from river banks and replaced by crops whose roots do not have the same binding properties. Subsequent bank erosion has resulted in large quantities of silt being carried down into estuaries. Swamp lands have also been filled and canalised, resulting in faster water flow and an increase in silt transfer. The reed beds associated with such swamps would normally act as huge filters, trapping both sediment and excess nutrients from fertilizers that now end up in many estuaries. Removal of these reeds has thus often resulted in large algal blooms and a proliferation of water hyacinth.

A study in the early 1980s of 51 of KwaZulu-Natal's 73 estuaries revealed that, owing to the degraded condition of most of them, only six make a significant contribution to the recruitment of estuarine-dependent marine stocks (Begg, 1983). Stocks of many important angling fishes such as spotted grunter, white steenbras, garrick (leervis), tenpounder, Cape and Natal stumpnose and river bream, which are associated with estuaries for part of their life cycle, are thus being threatened by estuarine degradation.

Probably the greatest threat to South African estuaries in the future will be the chronic shortage of water in this country. Demands on meagre water supplies will continue to increase with the government's commitment to meet the basic needs of a growing urban population. More dams will be needed, but water budgets will have to be carefully compiled to ensure that the functioning of valuable estuarine systems is not jeopardised.

The need for an Environment Impact Assessment (EIA) by competent and unbiased ecologists prior to any development within or adjacent to an estuary or its catchment cannot be over-emphasized. Farmers must be educated on the far-reaching consequences of poor farming practices and fishermen need to understand the importance of sustainable utilisation of resources. Education of all sectors of society is essential if our fragile estuarine systems are to survive. If water use is going to be the greatest threat to estuaries in the future, then reduction of water consumption is surely an important start in combating that threat—an activity in which we can all participate.

RESOURCE USE

Estuarine Resources

Estuarine resources are harvested for recreational and subsistence or artisanal purposes. Substantial numbers of kob, grunter, stumpnose and river bream are caught by recreational fishermen, while swimming prawns, sand and mud prawns and mud crabs are other popular food and bait organisms collected in this environment. One of the best examples of subsistence harvesting in estuaries is found in the Kosi Bay system in northern KwaZulu-Natal. Fish are caught in traditional fish traps, by spearing and with hook and line, and women and girls collect crabs for food and reeds for weaving. More recently, controlled gill netting has been allowed for the capture of fish, such as pouter, tilapia and mullet, which are not heavily exploited by other methods of fishing. However, illegal gill netting and the commercialisation of this activity are fast becoming major threats to these fish stocks in several of the country's largest estuarine systems e.g. Kosi Bay, St Lucia and the Richard's Bay sanctuary.

Commercial Fisheries

Although not traditionally a nation which consumes much fish, South Africans nevertheless enjoy a considerable marine harvest. The total catch of marine organisms is about 500,000 tonnes per annum at present, and provides direct employment to about 20,000 people while contributing significantly to the protein needs of the country. Diminishing global returns from fisheries are a cause for concern (Parfit, 1995) and South Africa must be cautioned not to repeat the mistakes of the world in the management of its fisheries.

The entire coastline is exploited by one or more types of commercial fishermen, be they shore-based net fishermen (Fig. 2), divers, line fishermen or deep-sea trawlermen. In 1993, the major commercial fisheries and their main catches were (in order of total catch by weight): purse seining (anchovies and sardines); trawling (hakes and horse mackerel); linefishing (tunas, snoek and reef fish); jigging (squid); pot fishing (rock lobsters); seine and other small nets (mullet and sharks); seaweed gathering; abalone harvesting and crustacean trawling (prawns).

There are also several other small commercial fisheries with low catch volumes which are of local importance. These include oyster collecting, and crab and whelk harvesting. Whales have not been exploited since 1975 and there has been a moratorium on the harvesting of seals since 1990.

Commercial fisheries are those for which a permit or licence is issued, a catch levy is payable and the catch may be sold. The wholesale value of harvested marine resources in South Africa was R 1.6 billion in 1994, and exports earned foreign exchange to the value of R 898 million (Stuttaford, 1996). Licence fees, levies and taxes from the fishing industry also contribute significantly to state revenue. The industry provides considerable employment opportunities, both directly and also via associated concerns, such as fish processing plants, ship builders, maintenance personnel, etc.

Most of the South African commercial fishing enterprises are concentrated in the Cape provinces because of the extremely productive nature of the waters over their continental shelf. The pelagic purse seine fisheries which

Fig. 2. A seine net harvest of sardines caught on the KwaZulu-Natal South Coast during the winter sardine run in this province, a local phenomenon. (Photo: Rudy van der Elst).

target fish in surface waters are concentrated here, as well as a large part of the deep water hake fishery. The Agulhas Bank off the south coast is a very large area, much of which is accessible to demersal trawlers. Linefishing is spread along the entire South African coastline, with most catches again being made in Cape waters.

Most of the seine and other small net fisheries as well as the main rock lobster fisheries are found on the south and west coasts, while the squid jigging fishery is concentrated off the Eastern Cape coast. Seaweeds are harvested along most of the south coast while abalone are mainly harvested on the southern coast of the Western Cape. The trawl fishery for crustaceans (prawns, langoustines and crabs) is restricted to the KwaZulu-Natal coast.

The commercial fisheries are of national importance and are managed by the Chief Directorate: Marine and Coastal Management, of the Department of Environmental Affairs and Tourism. This agency essentially limits the amount that can be harvested annually by commercial operators by setting quotas for each resource. In some fisheries, such as the linefishery, it controls fishing effort by the closure of seasons or areas for fishing, or by limiting the size at which a species can be harvested. This legislation is based on research undertaken by several government departments and private organisations which are contracted to do the work. Staff of the fisheries inspectorate of Marine and Coastal Management, assisted by members of provincial conservation authorities, ensure that compliance with licence conditions is maintained.

The early 1960s and 1970s were boom years for the major South African fisheries, i.e. the purse seine and trawl fisheries. Catches of sardines reached a peak in 1963 and those of hake in 1973. These peaks were followed by a decline in the catch of these species, attributable to over-exploitation. Many foreign boats were operating in South African waters at the time, and this necessitated the declaration of a 200 nautical mile Exclusive Economic Zone (EEZ) in 1977.

This action, combined with conservative quota policies, has led to the gradual stabilisation of catches, particularly those of the valuable hakes. The collapse of the sardine fishery in the late 1960s was followed by vastly increased catches of anchovies by purse seiners, although the proportion of sardines in catches is increasing gradually again as the stock recovers. Catches of hake are improving slowly, but pressure on these stocks is likely to increase as previously disadvantaged communities are demanding access to the fishery.

Commercial linefish landings also peaked in the 1970s, although the composition of catches has changed considerably from those of the early 1900s. Reef fishes initially comprised the bulk of the catch, but as stocks of these declined, pelagic fishes, such as snoek and tuna, assumed more importance. The 'seventy-four', once the mainstay of the KwaZulu-Natal fishery, has now almost vanished from commercial catches.

This trend of switching targets as once common fish become less available, continues today. Improved fishing aids increase the fishing pressure on vulnerable species. The long-term prospects for our slow-growing reef fish are, therefore, generally bleak. As a result, commercial linefishermen will have to rely more and more on pelagic gamefish. This may lead to large interannual variations in catches, as the availability of these fish is inconsistent.

Catches of squid also vary considerably from year to year, largely as a result of fluctuations in recruitment, a factor influenced by oceanographic processes. Rock lobster catches on the west coast have declined dramatically in recent years, partly because of over-exploitation and partly owing to unfavourable environmental conditions which retard lobster growth. This fishery dates back to the early part of the 20th century, and the decline in the resource has had a serious socio-economic effect on local communities. South coast rock lobster catches have remained fairly stable since commencement of the fishery in the 1970s.

Catches of mullet and St Joseph sharks by beach seine and gill-net fisheries have also been fairly constant in recent years. The volume of seaweed harvested varies from year to year, depending on availability and demand. Commercial catches of abalone have been steady for several years as they are strictly controlled by quota. However, growth in the market price has led to a massive increase in poaching and the resource is thus under more pressure. Crustacean catches off the KwaZulu-Natal coast have been variable, partly because of environmental factors, but also because of varying fishing effort.

In summary, most of the current South African commercial fisheries are being maximally exploited. With the exception of the west coast rock lobster and several species of reef fish, catches should remain reasonably stable in the foreseeable future. Some resources, such as abalone and the hakes, are facing growing pressure and will have to be carefully managed.

There is potential for expansion of certain fisheries, either by exploiting species not currently utilised or developing new fishing grounds. New markets may have to be sought for the former and the latter may require the development of new fishing technology. Optimal utilisation, involving improved processing methods and the use of currently discarded by-catches, will be a priority. Although not a panacea, mariculture may relieve the pressure on some species; however, rugged local conditions and limited sites suitable for this practice restrict its application in South Africa.

Recreational Angling

Angling is popular in South Africa (Fig. 3) and, since the establishment of the first angling club (Umgeni in 1885), its popularity has grown to reach an estimated total of half a million participants.

Angling can be divided into two broad categories and is undertaken for recreation and sport. Recreational anglers fish primarily for fun and hopefully for some fresh fish for the table. Sport anglers, on the other hand, are strongly competitive, are affiliated to national and international bodies and use highly specialized equipment and technologies. Six different types of angling can be identified: rock and surf, skiboat, estuarine, flyfishing, spearfishing and partyboat angling.

Rock and surf angling is most popular and most accessible to the general population. Though catches per angler are on average low, many anglers do succeed in taking fresh fish home. The other forms of angling are less accessible and more expensive, requiring boats and specialized fishing equipment. Despite some of the costs, growth in all facets of recreational angling has been substantial, especially in the past three decades. The resources that sustain marine recreational angling in South Africa are diverse and include at least 250 species of fishes. Galjoen, hottentot, bronze bream, white steenbras, elf, kob, snoek, king mackerel and a range of other fishes are popular target species. However, many are either endemic or limited in abundance and their catches have diminished, significantly in many cases. It is not easy to convince a single angler that his activities, together with thousands of others, can deplete a resource and hence the conservation of certain stocks has been a problem.

Conflict between recreational and commercial interests has become a more serious problem. Although this is a worldwide phenomenon, it has special relevance in South Africa where wealthy anglers with sophisticated equipment fish for fun alongside commercial fishermen living close to poverty, with both targeting the same species. Although resource managers have attempted to resolve this problem by introducing legislation that allocates certain species either for commercial or recreational use, the issue is far from resolved.

Fig. 3. Tournament angling in the Eastern Cape. (Photo: Rudy van der Elst).

From Management to Co-management: The *Pomatomus saltatrix* Fishery

In KwaZulu-Natal, the elf (*Pomatomus saltatrix*) is the most heavily harvested and targeted of all near-shore linefish species. Research revealed a declining catch per unit effort (CPUE) for elf in the late 1970s and this led many to believe that the stock was biologically overexploited. A combination of a minimum size limit, a bag limit and a closed season were introduced and was stringently enforced, despite the fact that no quantitative stock assessment had been carried out on this species. Other regulatory measures included the prohibition of commercial harvesting and the sale of elf catches. These regulations have been in effect for over 20 years.

In more recent years, the KwaZulu-Natal Nature Conservation Service has been promoting co-management of the elf stock. The co-management of a fisheries resource presupposes that all participants in the management process are capable of choosing between disparate harvesting strategies on the basis of clearly defined objectives. They must also understand the risks and uncertainty involved in assessing the status of the stock. A modified yield-per-recruit analysis was developed to facilitate such decision-making and evaluate different fisheries regulations. The combined effects of different daily bag limits, closed seasons and minimum size limit options were examined in the analyses and compared with the current regulations (minimum size 30 cm, closed season from September to November and a daily bag limit of five elf per angler per day).

The best harvesting strategies were considered to be those that increased egg production and decreased fishing mortality rates. Different combinations of regulations were initially recommended by fisheries managers and by the fishermen themselves. By providing a decision table that weighed the harvesting strategies against each other, co-managers were finally able to agree on a harvesting strategy that met the above objectives. It was shown that lay persons can participate in stock assessment modelling and decision-making of this nature, provided that the science is pitched at a level that is understandable to them.

Demands by recreational anglers, who claim it is their democratic right to catch and sell what they can, are refuted by commercial fishermen who assert that their livelihood and a food source of the nation are at risk. These socio-economic issues will have to be resolved if the potential benefits of our marine fish resources are to be realised for all South Africans on a sustainable basis.

Angling is not equally popular among the different population groups in South Africa and is likely to increase in future. It is accessible to all, and provides quality leisure time, a valuable accompaniment to social upliftment. Angling also generates substantial revenue through support industries, tourism and, by implication, employment. The challenge will be to develop recreational angling wisely without jeopardising the legitimate claims of both commercial and subsistence fishermen.

Invertebrate Fisheries

The collection of marine invertebrates for bait or human consumption is popular in South Africa and a wide range of organisms is gathered. While recreational harvesting occurs in all the coastal provinces, licensing of harvesters and the collection of catch statistics for all species is most intensive in the province of KwaZulu-Natal. Information from this province thus provides a case study of the situation.

Total catches have been estimated since 1963 using the number of licences issued and data from the approximately 10 per cent of licence holders who render catch returns. Catches seem to have levelled off in recent years owing to a limit on the number of licences sold.

Approximately 126 t rock lobsters (worth R 5 million), 268 t mussels (R 1.6 million), 851,000 mole crabs (R 500,000), 14 t octopus (R 65,000), one million sandprawns (R 352,000), 5.6 tonnes mud crabs (R 116,000), 78,000 oysters (R 78,000) and 56,000 ghost crabs were landed annually between 1992 and 1994. None of these organisms may be sold by recreational licence holders, although there is a small commercial fishery for oysters (R 500,000 per year) and mud crabs, and a thriving black market for rock lobsters and sandprawns. Restrictions such as daily bag limits, closed seasons, and closed areas (Tomalin, 1995) are stringently enforced by a large contingent of KwaZulu-Natal Nature Conservation Service law enforcement officers. In stark contrast to the Cape provinces, KwaZulu-Natal may have one of the best patrolled coastlines in the world.

The recreational harvesting of abalone and rock lobster is also subject to licensing in the Western Cape. The catches amount to 550 tonnes (or 47% of the total catch) in the case of the former and 391 tonnes (or 15% of the total catch) in the case of the latter. There is thus conflict between commercial and recreational harvesters of these resources in this province.

Much debate has ensued as to the effects of these fisheries on the stocks themselves and their associated biological communities. Also debated are the issues of whether these levels of offtake are suboptimal, sustainable or too high.

Estimates of catch per unit effort (CPUE) are obtained from voluntary catch returns in KwaZulu-Natal. These are, essentially, rather crude, as variables such as the skill and motivation of the licence holder cannot be taken into account. They nevertheless provide a rough indication of stock density, as overfishing causes a reduction in the stock and decline in CPUE.

Examination of CPUE data for the recreational invertebrate fisheries of KwaZulu-Natal reveals that only the oyster fishery is undergoing a steady decline. However, the commercial oyster fishery has remained relatively stable and the drop in the recreational catch is probably attributable to reduced interest in an animal difficult to collect, rather than a reduction in the stock. Although the CPUE of rock lobsters showed a decline for a number of years, this has again rebounded to earlier levels.

The distribution of the catch is of consequence in KwaZulu-Natal. The largest harvest is collected within the proximity of the Durban metropolitan area with no effect on the catch per outing; the demand in this, the most urbanised area, does not appear to be affecting the CPUE. However, the percentage offtake of the total stock in this area is as high as 40% for mussels and 50% for oysters, and models have indicated that this is close to the maximum in the case of the mussels.

The overall conclusion is that the recreational invertebrate fisheries of KwaZulu-Natal are heavily fished but well managed. This is not necessarily the case in the other coastal provinces, as found during a recent survey of these resources on the Transkei coast (Fielding et al., 1994). Fisheries of this nature are not suitable for open or unlimited access and the management systems employed in KwaZulu-Natal will hopefully be introduced to the other coastal provinces. However, even this has inherent problems in view of South Africa's history of poor apportionment of its resources, an issue currently under review in various forums.

Artisanal and Subsistence Fishing

As recreational and commercial fishermen constitute the most conspicuous (and historically vociferous) of the extractive user groups on South Africa's coast, most research and data collection have been directed at them in the past. However, artisanal and subsistence fishermen form an important and often overlooked category of fishermen. They harvest both finfish and shellfish in order to make a living and generally have a traditional heritage of resource utilisation.

Historically, these fishermen have been marginalised in South Africa, in the sense that subsistence and artisanal fishermen have not been accommodated in fishery legislation and their activities are often in contravention of the law. Frequently their activities are seen as ecologically detrimental by other categories of fishermen and there has often been friction between artisanal fishermen and other sectors of the fishing community, e.g. False Bay trek netters, Transkei shellfish collectors, St Lucia gill-netters etc.

It is extremely difficult to categorise or define artisanal and subsistence fishing. Both groups are commonly referred to as small-scale fishermen. Artisanal fishermen are generally those who can demonstrate a historic fishing right and who fish exclusively with native or traditional methods; they may sell or barter their catch. Subsistence fishermen rely on fishing as a means of survival for themselves and their families; they may not have any other source of income and use their catch primarily as a source of protein for personal consumption. The problem is that there are very few fishermen that fit neatly into these categories.

On a global scale, small-scale fisheries are far more important than large-scale commercial fisheries in terms of manpower utilisation, although similar in terms of yield. It is thus important that they be included in fisheries policy development. Their importance in the South African economy in providing jobs and food for impoverished communities is now being recognised, and management practices and legislation are being tailored to meet their needs.

KwaZulu-Natal is probably ahead of the rest of the country in this respect. Subsistence and artisanal fishermen are now being allowed to pursue their previously illegal activities in a controlled manner. More importantly, the fishing communities are being included in the management of resources in places such as the Kosi Bay Nature Reserve (Kyle, 1995), Lake St Lucia (Mann, 1995) and parts of the northern KwaZulu-Natal coast.

Artisanal and subsistence collectors have operated along the coast of Transkei for centuries, harvesting finfish and shellfish for barter, sale and personal consumption. However, in the last decade, the economic value of these fish and shellfish resources has been recognised and much of the harvest is now sold to visitors along the coast (Fielding et al., 1994; Robertson and Fielding, 1997). These sales provide a limited cash income for coastal communities that have no access to the formal job market. Since Transkei became re-incorporated into South Africa, legislation governing harvesting practices and the rights of access to the resources has become confused and there is no clear directive as to how these resources should be managed. This has created tension between the various user groups and the law enforcement agencies. Legislation is currently being drafted that will legalise the activities of small-scale coastal fishermen, but the problems surrounding resource access rights, particularly to users with a commercial interest, remain unresolved.

Small-scale fishing communities are also found along the Western and Northern Cape coasts where they often compete with major commercial fisheries. Most commercial quota species such as Western Cape rock lobster and hake,

> ### Intertidal Harvesting in the Maputaland Marine Reserve
>
> Rural communities living adjacent to the Maputaland Marine Reserve in northern KwaZulu-Natal have harvested marine and estuarine resources for centuries. Mussels, red bait, limpets, barnacles, ghost crabs and mole crabs and various other invertebrate organisms are harvested from the intertidal zone and constitute an important component of the diet of these people.
>
> When the Maputaland Marine Reserve was proclaimed in 1986, traditional subsistence harvesting of the intertidal resources was allowed to continue. Visitors to the area were denied the right to harvest these organisms and this created some tension between local people and visitors as well as between the visitors and conservation authorities. In an attempt to resolve this conflict, temporary permits were issued to local harvesters allowing them access to limited resources while a project was undertaken to determine the sustainability of their harvesting practices. After considerable consultation with the community authorities and the resource users, an intensive monitoring programme was initiated. Monitors, selected from the community, were employed to record quantities of all organisms harvested from the seashore on a daily basis. After four years, the data showed no decline in catch per unit effort, and effort appeared to be declining. Harvesting was thus allowed to continue, as was the monitoring programme. Stock assessments of key organisms have since been undertaken and these, together with harvesting trends, indicate that most of the harvestable stocks are being used sustainably. However, the conservation authorities are concerned that harvesting practices may be changing the community structure of rocky shore organisms and reducing biodiversity in an important marine reserve. Co-management structures are presently being established with the communities who use the resources which will enable resource-users and conservation managers to work together to find solutions to these problems.

for example, are already fully subscribed in the sense that there is very little room for expansion in the harvest. An attempt was recently made to allocate some of the quota of commercial fishing operations to artisanal fishing communities who could demonstrate a historic right of access to the resource. This is frequently not a practical solution to the need for socio-economic upliftment in disadvantaged artisanal/subsistence communities; in many cases, the primitive fishing methods used by artisanal fishermen prevent them from competing in harvesting the resource and processing and marketing the catch. The artisanal subsistence quota is therefore often simply sold back to the large commercial concerns, with little or no long-term benefit to the artisanal community.

Artisanal and subsistence fisheries basically require the same fisheries management as the larger commercial fisheries. The fishermen desire high catches, a stable catch per unit effort and, when there is a premium on the size of individual fish, a large mean size is preferred. In the past, overfishing has generally not been a problem when the artisanal subsistence fishermen have been the sole users of a resource, because of their primitive traditional fishing methods. A notable exception is the over-exploitation of mussels in the Transkei. However, conflict arises when traditional fishermen and other users view each other's activities as being detrimental to their continued use of a resource. In South Africa, such conflict is presently exacerbated by an increased demand for most resources as a result of rapid population growth and a poor job market caused by a sluggish economy. Increased commercialisation of fisheries, even by the informal sector, has resulted in increased polarisation in the attitudes of fishing groups, management bodies and enforcement agencies.

There are no easy solutions to these problems, but clearly the rights of artisanal and subsistence fishermen must be acknowledged and have been incorporated into fisheries policy. More importantly, the issue of the right of access to resources needs to be addressed. The requirements of the different sectors using a resource need clarification as different fishing techniques often harvest completely different components of a particular resource.

However, the central, crucial problem of overfishing is unlikely to be resolved until the people involved in resource utilisation are given a clearly defined stake in the resource and allowed to participate in decisions regarding its management. South African fisheries managers are confronted with the challenge of defining the relationship of fisheries resources to human welfare, and promoting conservation of the resources for the use of future generations.

Non-consumptive Uses

The expressions 'ecotourism' and 'aquasports' provide the collective terms for non-consumptive use of the resources in question here. Local demand largely occupied these markets during the later apartheid years, particularly relative to the tremendous potential of this country to attract foreign ecotourists and watersport enthusiasts. This is likely to change in the post-apartheid era if South Africa earns an international reputation of peace and stability. The amenable climate and wide range of marine and estuarine environments in South Africa offer a full spectrum of opportunities for activities such as bathing, boating, waterskiing, SCUBA diving, game viewing and bird-watching.

An increase in these activities will not occur without problems, however. Expansion of the existing facilities will be needed to accommodate an increase in tourism, and it is important that this is planned so that development is not deleterious to the environment being exploited. An economic analysis must be included in the planning to ensure that capitalization is adequate and cost-effective for this purpose.

Reports in the media indicate that South Africans provide poor service to tourists, another matter that will need redress. Continued free access, as in the past, to an environment for ecotourism and water sport can also give rise to problems. An example of this is beach driving, undertaken

---

Resolution of Conflict in the Use of the Aliwal Shoal

Aliwal Shoal, south of Durban, provides the southern-most habitat for tropical coral communities on the African coast. Its proximity to a major metropolitan area has made it popular with anglers and sport-divers, giving rise to both over-exploitation and user conflict. It has potential for ecotourism but local communities derive little benefit from its use. A public workshop was held for interested parties to discuss its importance and to seek solutions to the user conflict. This was attended by 121 people representing the full spectrum of interests in the shoal. Despite approaching the situation from disparate positions, the participants discovered they had much in common.

It was agreed that, for the sustainable use of Aliwal Shoal, the desired objective can only be accomplished if its use is regulated in a management plan refined by research. The users wished to participate in its management and agreed to appropriate limitations on their activities. It was further agreed that an interim management plan should be implemented before completion of the research and that users would assist in funding the latter through levies.

Interested parties were thus given a mandate to form an Aliwal Shoal Forum with the purpose of formulating an interim management plan and liaising in matters concerning management. Members underwent a long and arduous process of negotiation in formulating the interim management plan which has the following salient features; it will be self-funded from user-levies, regulated by a management board representing all the interested and affected parties, policed by a Baywatch type of operation, and will make a substantial contribution to the Reconstruction and Development Programme in the area. Its implementation is now dependent on the imminent proclamation of the shoal as a protected area. The research needed for the refinement of the management plan has commenced, and includes rapid survey monitoring techniques using underwater videography and bathymetric mapping, both of which will be useful in zoning Aliwal Shoal.

A large local industry, Sappi Saiccor, discharges pulp effluent in the vicinity of the shoal. As a result, the company is frequently targeted in the media concerning the effects of its effluent on Aliwal Shoal, which are largely unknown. Such effects are also being investigated as part of the biological survey and, if any are detected, target organisms will be subjected to further analysis to validate the results.

as a recreational pursuit in its own right or as a means of access for other activities. Beach traffic was a rarity in the past, but has increased with the burgeoning off-road vehicle market, to the extent that lives are being endangered and dunes, with their associated fauna and flora, are being damaged. A government white paper leading to restrictive legislation on this activity is being implemented, an unpopular development in the present dispensation of a democratic right of free access!

Finally, the increased use of certain environments is giving rise to user conflict, a problem that is likely to increase. Examples of this are the reefs at Aliwal Shoal and Sodwana Bay that are visited by large numbers of anglers, divers and spearfishermen. Conflict resolution has been initiated in public workshops in both instances and is progressing. Agreement is being sought to zone use of these resources in space and time and the process has been documented in the case of Aliwal Shoal (Schleyer, 1995). The development of this site for ecotourism would be of great economic value to the area and participants in the process have made a clear commitment that all members of the local community will benefit from its use.

Non-Renewable Resources

Oil, diamonds and a range of minerals are regular, non-renewable harvests from the seabed. Although South Africa may not rank amongst the world leaders in the exploitation of non-renewable marine resources, many ventures do exist. Most lucrative of these resources are the rich diamond deposits along the west coast.

These originate from the alluvial deposits of the Orange and other rivers and are mined by divers or remotely operated devices using suction equipment in shallow water. Deposits of heavy metals in the form of rutile, ilmenite and titanium occur at many places on the east and west coasts, and intensive dune mining is underway in KwaZulu-Natal. Oil and gas deposits are also being tapped, although this remains at a modest and possibly sub-economic level. Beach sand and salt are another two products taken from the sea.

While these mining operations have undoubtedly provided the country with valuable economic and employment benefits, they have invariably been associated with public controversy. Mining at sea in South Africa is now being subjected to scrutiny by EIA. Dune mining, diamond mining and oil exploration have environmental consequences, preclude people from certain regions, and have at times questionable economic benefits.

The strongly autocratic approach in handling these resources in the past will have to be replaced with more 'transparency' in their management so that people can perceive the potential benefits to the nation. If this is achieved, then these non-renewable resources will undoubtedly play a significant role in underpinning the Reconstruction and Development Programme.

POLLUTION AND ENVIRONMENTAL QUALITY

There are two main sources of estuarine and marine pollution in South Africa: domestic and industrial effluent from urban and metropolitan areas along the coast, much of which is pumped out to sea through pipelines, and waste from informal settlements which is washed into estuaries and the sea by rivers, together with eroded soil and agricultural chemicals from farms. Much of the silt is retained in estuaries.

The major waste disposal pipelines are found on the east coast, where the volume and strength of flow of the Agulhas Current are employed for dispersal and dilution of the effluent. Despite a healthy public concern over the design and use of these pipelines, considerable effort is made to ensure that they are monitored and operated safely. In this way, the assimilative capacity of the sea for pollutants is sustainably managed.

Connell (1988) provides a comprehensive review of the pipelines in KwaZulu-Natal, relative to other forms of marine pollution in the province, and is satisfied that all but one are operating within the assimilative capacity of the area. The Sappi Saiccor pipeline is the one that emits effluent of poor quality, but the cellulose mill which feeds it is being upgraded to improve the situation. The total volume discharged per day in KwaZulu-Natal is 600,000 m^3.

A historical case of interest is that of the sewage pipelines off Durban. Livingstone (1990) monitored faecal contamination on Durban beaches prior to construction of the two pipelines, when sewage was released from the harbour on the outgoing tide and into the surf, as well as for some decades after the pipelines were commissioned. This provided a case study of the improvement of water quality in the surf zone with the installation of such pipelines and also the opportunity to develop a simple microbial indicator system for evaluating seawater quality.

Connell (1988) reviews considerable information on metal, pesticide and oil-derived pollution on the east and south coasts but expresses no concern that any of these are excessive. Moldan (1991) gives more detailed information on oil pollution in False Bay in the Western Cape, which has experienced pollution events from tanker spills and shipwrecks. Apart from episodic catastrophes caused by these events, it would appear that an increasing amount of oil in stormwater, resulting from rapid urbanization, is cause for greater concern.

Herein lies a major problem for the future. Informal settlements are growing rapidly along the coast, particularly in coastal catchments, and are generally devoid of waste or sewage disposal facilities. The densest developments are closest to the water supply, a river or stream, and, apart from the obvious health hazards, stormwater run-off ensures a heavy load of pollution during peak flow. Pollution of this nature is much publicised during canoe races, such as the annual event held in the Umsindusi River in KwaZulu-Natal.

Estuaries and the sea are the final recipient of this contamination and the most comprehensive studies on this phenomenon have again recently been undertaken in False Bay (Brown et al., 1991; van der Merwe et al., 1991). An increased load of suspended solids, nutrients and metals has been measured, apart from the oil pollution mentioned earlier. Imaginative solutions are urgently needed to address the root cause of the problem, viz. the need for adequately serviced housing in these areas.

In conclusion, first-world controls and solutions appear to be functioning adequately in first-world situations to which they are being applied. Hopefully, the South African economy will be able to sustain this approach which is essential for the maintenance of a healthy environment. The need for environmental auditing in the acquisition of foreign investment will probably play a role in this regard. However, the solution of third-world pollution problems will also have to be sought to prevent pollution of the marine environment.

MANAGEMENT STRUCTURES AND POLICY

South African marine resources are common-property resources. This type of collective ownership creates a major conservation dilemma. For instance, a commercial fisherman who catches too many fish to satisfy his individual short-term interests (a bigger profit) will over the long term undermine the collective, long-term interests of other users by over-exploitation of the resource. If sufficient individuals adopt such behaviour and their combined harvest rate is above that which is sustainable and optimal, they will collectively deplete the resource.

The principal aim of management of renewable marine resources is thus to curb or restrain the individual from pursuing selfish, short-term interests at the expense of the long-term sustainability of a resource.

Legislation

The Marine Living Resources Act (No. 18 of 1998; hereafter referred to as the Act) was promulgated on 27th May 1998. This is the first time that a formal policy for the rational harvesting of living marine resources has been incorporated in legislation in South Africa. Development of the marine fisheries policy was a long process. It was started on 27th October 1994 by the Minister of Environmental Affairs and Tourism at a public meeting. A Fisheries Policy Development Committee (FPDC) was formed which executed the work through a working committee and technical subcommittees, and which co-ordinated and integrated contributions from a range of relevant interest groups, institutions and individuals. The FPDC submitted its report to the Minister on 4th June 1996. The Act was finalized from this report and included input from the provincial and national environmental portfolio committees, the public and interested bodies.

Principles

The Act incorporates three fundamental principles; (a) the sustainable use of living marine resources, (b) the maintenance of stability in the fishing industry, and (c) equitable access to these resources by all users.

In terms of the Act, the management of South Africa's marine living resources is now a national function. Nevertheless, the management of marine living resources at the provincial, local authority or even community level is not ruled out by the Act.

The rights of access to marine living resources by recreational and subsistence fishers have been formally recognized for the first time. All fishers, including commercial, recreational and subsistence fishers, will have to purchase a permit to fish. Permit fees will be paid into a Marine Living Resources Fund which will be used to manage the resources and fund marine research.

Advisory Bodies

The Minister, in terms of the Act, has to appoint members to serve on the Consultative Advisory Forum for Marine Living Resources (CAF). The CAF advises the Minister on any matter referred to it by him or her, and deals in particular with:

- The management and development of the fishing industry.
- Any issues concerning total allowable catch (TAC) i.e. the maximum allowable harvest of a resource, usually established on an annual basis on scientific grounds.
- The management of marine living resources and the related legislation.
- The establishment and amendment of operational management procedures (OMP; a scientific process that determines the use of available data to establish a control measure such as a TAC) and management plans.
- Recommendations on research direction.
- The allocation of money from the Marine Living Resources Fund.

The Minister also appoints a Fisheries Transformation Council (FTC) whose main function is to facilitate the achievement of fair and equitable access to fishing rights. The Minister allocates fishing rights to the FTC which, in turn, leases them to persons from historically disadvantaged sectors of society and to small and medium size enterprises. The FTC determines the price of these leases.

A Bill accompanies the Act and contains all the regulations (drafted at the national level) pertaining to the marine living resources and their utilization. This incorporates and amalgamates all previous regulations (national regulations in terms of the Sea Fishery Act, 1988; provincial ordinances and the regulations of the former homelands such as Transkei and Ciskei) and is presently being amended.

REFERENCES

Begg, G.W. (1983) The comparative ecology of Natal's smaller estuaries, PhD thesis, University of Natal, Pietermaritzburg.

Brown, A.C. and Jarman, N.G. (1978) Coastal and marine habitats. In *Biogeography and Ecology of Southern Africa*, eds. M.J.A. Werger and A. C. van Bruggen, pp. 1239–1277. Dr. W. Junk, The Hague.

Brown A.C., Davies, J.A. and Gardiner, A.J.C. (1991) Chemical pollution loading of False Bay. *Transactions of the Royal Society of South Africa* **47** (4–5), 703–717.

Branch, G.M. and Branch, M.L. (1981) *The Living Shores of Southern Africa*, Struik, Cape Town.

Branch, G.M., Griffiths, C.L., Branch, M.L. and Beckley, L.E. (1994) *Two Oceans—A Guide to Marine Life of Southern Africa*. David Philip, Cape Town.

Connell, A.D. (1988) Pollution and effluent disposal off Natal. In *Coastal Ocean Studies off Natal, South Africa*, ed. E.H. Schumann, pp. 226–252. Springer Verlag, New York.

Emanuel, B.P., Bustamante, R.H., Branch, G.M., Eekhout, S. and Odendaal, F.J. (1992) A zoogeographic and functional approach to the selection of marine reserves on the west coast of South Africa. *South African Journal of Marine Science* **12**, 341–354.

Fielding, P.J., Robertson, W.D., Dye, A.H., Tomalin, B.J., van der Elst, R.P., Beckley, L.E., Mann, B.Q., Birnie, S., Schleyer, M.H. and Lasiak, T.A. (1994) Transkei coastal fisheries resources, Special Publication No. 3, Oceanographic Research Institute, Durban.

Kyle, R. (1995) The wise use of wetlands by rural indigenous people: the Kosi Bay Nature Reserve: A case study. In *Wetlands of South Africa*, ed. G.I. Cowan, pp. 273–291. Department of Environmental Affairs and Tourism, Pretoria.

Livingstone, D.J. (1990) Microbial studies on seawater quality off Durban: 1964–1988, CSIR Research Report No. 704, Council for Scientific and Industrial Research, Durban.

Mann, B.Q. (1995) Quantification of illicit fish harvesting in the Lake St. Lucia Game Reserve, South Africa. *Biological Conservation* **74** (2), 107–113.

Moldan, A. (1991) Petroleum hydrocarbon input into False Bay. *Transactions of the Royal Society of South Africa* **47** (4–5), 731–737.

Parfit, M. (1995) Diminishing returns. *National Geographic* **188**, 2–37.

Payne, A.I.L., Crawford, R.J.M. and van Dalsen, A. (1989) *Oceans of Life off Southern Africa*. Vlaeberg, Cape Town.

Robertson, W.D. and Fielding, P.J. (1997) Transkei Coastal Fisheries Resources Phase 2: Resource utilisation, development and tourism. Special Publication No. 4, Oceanographic Research Institute, Durban.

Schleyer, M.H. (1995) Proceedings of the Aliwal Shoal Workshop held at the Amanzimtoti Town Hall on 8th June 1995, unpublished report No 117, Oceanographic Research Institute, Durban.

Smith, M.M. and Heemstra, P.C. (1986) *Smith's Sea Fishes*. Macmillan, Johannesburg.

Stuttaford, M. (ed.) (1996) *Fishing Industry Handbook—South Africa, Namibia and Mozambique*, Marine Information C.C., Stellenbosch.

Tomalin, B.J. (1995) Invertebrates harvested in KwaZulu-Natal: their ecology, fishery and management. Information Booklet No. 1, Oceanographic Research Institute, Durban.

Turpie, J.K., Beckley, L.E. and Katua, S.M. (1999) Biogeography and the selection of priority areas for conservation of South African coastal fishes. *Biological Conservation* **92** (2000), 59–72.

van der Merwe, I.J., Vlok, A.C. and van der Merwe, J.H. (1991) Land use and population characteristics in the False Bay coastal frame. *Transactions of the Royal Society of South Africa* **47** (4–5), 693–703.

THE AUTHORS

Michael H. Schleyer
SAAMBR, P.O. Box 10712, Marine Parade, KwaZulu-Natal 4056, South Africa

Lynnath E. Beckley
SAAMBR, P.O. Box 10712, Marine Parade, KwaZulu-Natal 4056, South Africa

Sean T. Fennessy
SAAMBR, P.O. Box 10712, Marine Parade, KwaZulu-Natal 4056, South Africa

Peter J. Fielding
SAAMBR, P.O. Box 10712, Marine Parade, KwaZulu-Natal 4056, South Africa

Anesh Govender
SAAMBR, P.O. Box 10712, Marine Parade, KwaZulu-Natal 4056, South Africa

Bruce Q. Mann
SAAMBR, P.O. Box 10712, Marine Parade, KwaZulu-Natal 4056, South Africa

Wendy D. Robertson
SAAMBR, P.O. Box 10712, Marine Parade, KwaZulu-Natal 4056, South Africa

Bruce J. Tomalin
SAAMBR, P.O. Box 10712, Marine Parade, KwaZulu-Natal 4056, South Africa

Rudolph P. van der Elst
SAAMBR, P.O. Box 10712, Marine Parade, KwaZulu-Natal 4056, South Africa

Chapter 62

THE NORTHWEST COAST OF THE BAY OF BENGAL AND DELTAIC SUNDARBANS

Abhijit Mitra

The Indian Sub-continent lies entirely to the north of the equator with the Tropic of Cancer cutting the country roughly into two halves. About half of the Sub-continent in the northern portion lies outside the Tropics in the mid-latitudes or the Temperate zone. However, the whole of the country is still considered as a tropical country mainly because of its clearly defined isolation from the rest of the Asia by the Himalayan range and the prevalence of a tropical monsoon climate.

With a coastline of 7515 km, an exclusive economic zone of 2,014,900 km^2 and a shelf area of 452,100 km^2, India occupies a unique position with regard to coastal biodiversity. The eastern, western and southern coasts of peninsular India are replete with majestic rivers which have extensive and highly productive estuarine areas. West Bengal, a maritime state of the northeastern part of the country, adjacent to Bangladesh, is indented in the south by numerous river openings. The important rivers from east to west are the Harinbhanga, Gosaba, Matla, Thakuran, Saptamukhi, Muriganga and Hugli which ultimately terminate in the Bay of Bengal, but on the way encompass approximately 54 islands, criss-crossed and intersected by various creeks and delta distributaries. The deltaic complex at the apex of the Bay of Bengal is the Indian Sundarbans. With a total land and water area of approximately 1,000,000 ha, the entire Sundarbans ecosystem of India and Bangladesh supports the world's largest mangrove block, a well-known ecosystem of the Tropics. Almost 62% of the Sundarbans is situated in Bangladesh, while the remaining 38%—the western sector—lies within India.

The Indian Sundarbans, with rich mangrove floral and faunal diversity, swamps and backwaters, form a productive and protective margin for coastal West Bengal. These forest systems are dominated by the salt-tolerant halophytic seed plants that range in size from tall trees to shrubs. There are some similarities in general architecture (e.g. the presence of pneumatophores, cyrptoviviparous seeds or propagules, xerophyllous leaves etc.) and physiology (such as the presence of salt excretory glands or salt regulatory glands). A total of 69 floral species (included within 29 families and 50 genera) have been identified in this ecosystem, out of which 34 species are true mangrove types. These specialised vegetations play an important role in maintaining the economic structure of the coastal population of West Bengal state, as they are the reservoirs of various forestry products ranging from firewood, timber and construction materials for thatching houses, to honey, wax, alcohol, tannins and fisheries.

Seas at The Millennium: An Environmental Evaluation (Edited by C. Sheppard)
© 2000 Elsevier Science Ltd. All rights reserved

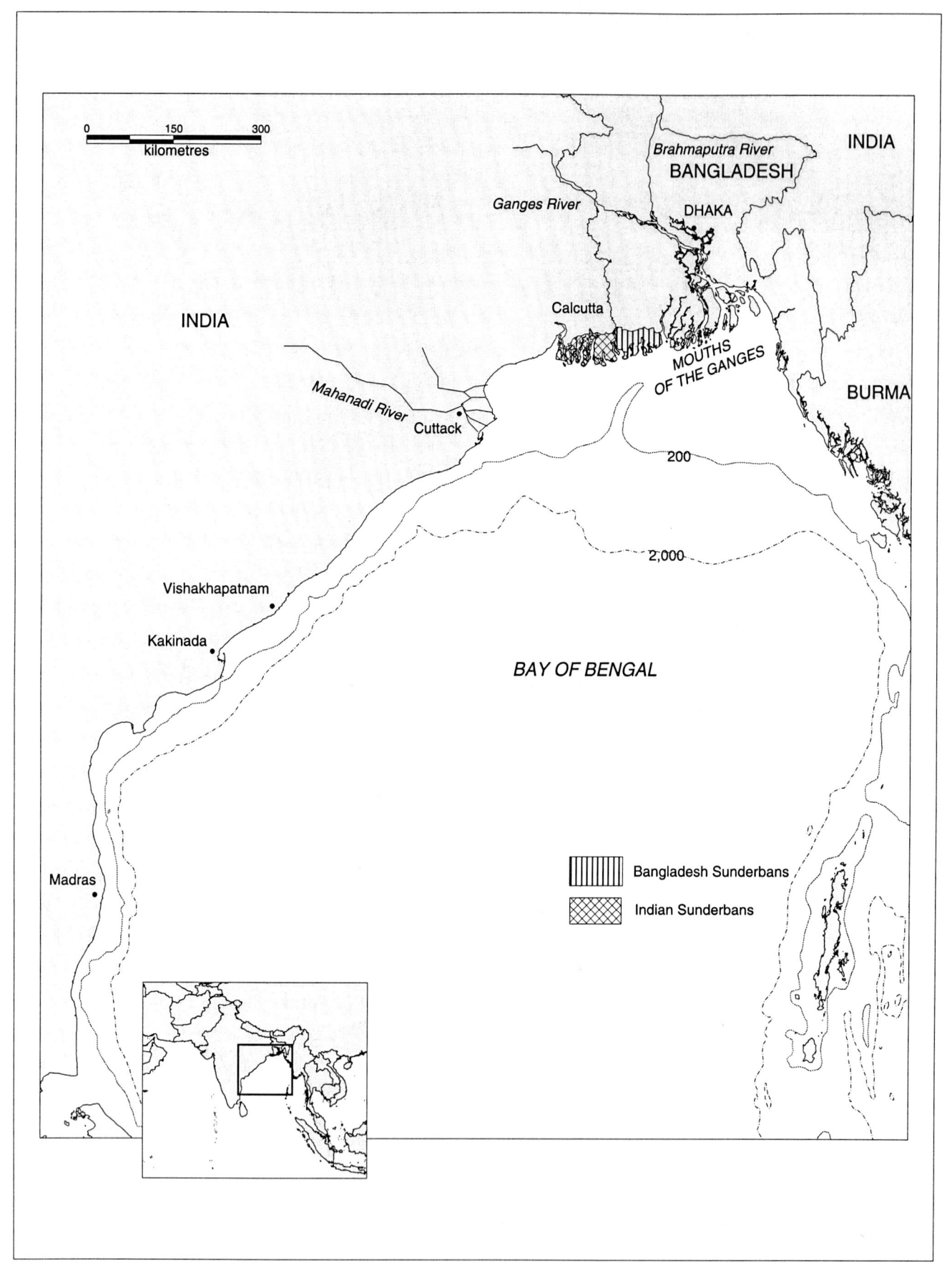

Fig 1. Sundarbans Mangrove ecosystem in India and Bangladesh at the apex of the Bay of Bengal.

The detritus supplied by this ecosystem to the aquatic phases of the Bay of Bengal and adjacent estuaries provides nutritional input; as a result, the coastal zone of the Bay of Bengal has become a nursery and breeding ground for a large variety of finfish and shellfish. Approximately 70 species of finfish juveniles and 25 species of shellfish juveniles have been recently recorded in the neritic zone of the Bay of Bengal, although a considerable portion of this ecological crop is wasted during the wild harvesting of prawn seeds. This operation is performed by employing nets of special mesh size to screen the seeds of tiger prawn (*Penaeus monodon*) from the coastal waters. It is needed to meet the growing demands of the large number of shrimp culture farms that have recently sprung up in this part of the Sub-continent.

The rapid industrialisation and urbanisation of the cities of Calcutta, Howrah and the newly developing Haldia industrial complex has posed another negative stress to this productive ecosystem: the factories and industries situated on the western bank of the Hugli estuary have significantly contributed to the degradation of this taxonomically diverse ecosystem. As a considerable number of small and large rivers find their way into the Bay of Bengal, a seasonal variation of aquatic salinity and pH is observed. This alters the speciation of compounds present in the marine and estuarine compartments; consequently, the concentrations of several conservative pollutants in the aquatic phase and biological samples oscillate markedly with seasons.

Against this background, this chapter discusses the macrofloral and macrofaunal diversity of the coastal region of the Bay of Bengal and the adjacent Sundarbans mangrove ecosystem, along with the various stresses operating in the area.

THE DEFINED REGION

The eastern side of peninsular India is bordered by the Bay of Bengal. The coastal region in this zone is not particularly rocky, with a predominance of sand, silt and clay in the substrate. The coastal waters are shallow as a consequence of a vast stretch of continental shelf. This results in the breaking of waves from a considerable distance from the shore; this is very common between the months of April to August, when the wind speed ranges from 15 to 40 km/h and the waves attain an average height of about 2 m.

The Indian Sundarbans at the apex of the Bay of Bengal (between 21°32' to 22°40'N latitude and 88°85' to 89°00'E longitude) are located on the southern fringe of the state of West Bengal, covering the major portions of the North and South 24 Paraganas districts (Figs. 1 and 2). The region is bordered by Bangladesh in the east, the Hugli river in the west, the 'Dampier and Hodges Line' in the north and the Bay of Bengal in the south.

With a considerable degree of marine characteristics in the major portion of the ecosystem, the important morphotypes of the deltaic Sundarbans are beaches, mudflats, coastal dunes, sand-flats, estuaries, creeks, inlets and mangrove swamps. The mangrove forests of the Indian Sundarbans have been variously estimated to cover 418,888 ha (Sidhu, 1963; UNEP, 1985), 200,000–300,000 ha (IUCN, 1989) and 426,300 ha by the West Bengal Forest Department. It has further been estimated that approximately 178,100 ha are comprised of water areas. These discrepancies in the figures for forest cover possibly arise from the fact that some denote areas designated as forest land which includes both the water bodies and degraded forests, while others represent an assessment of the area covered with vegetations only. The present investigation encompasses an area of about 9630 km^2 of the Indian part of Sundarbans which includes both forest and reclaimed areas as outlined by Sundarban Development Board. This area has recently been declared a "Sundarban Biosphere Reserve" and includes a Tiger Reserve, National Park, Wildlife Sanctuaries etc. (Table 1).

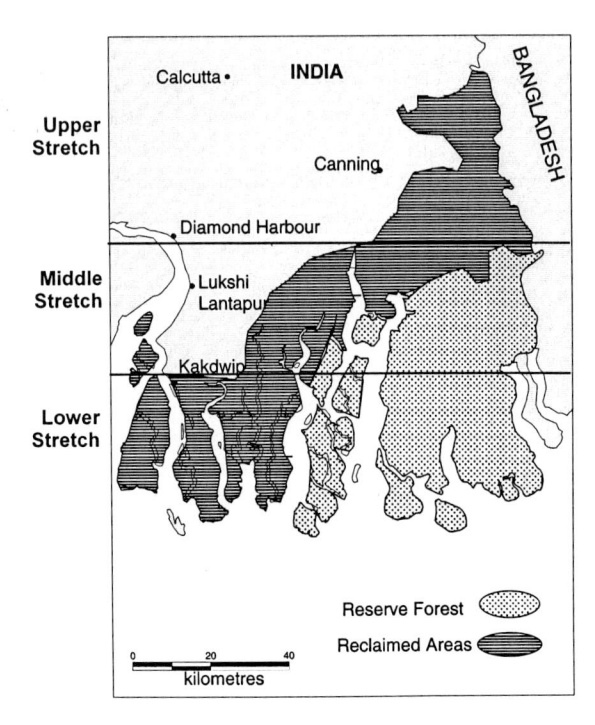

Fig 2. Deltaic Sundarbans of India showing three zones based on salinity, reserve forests and reclaimed areas.

Table 1

Demarcated areas under Sundarbans Biosphere Reserve (after Banerjee, 1964)

Zone	Area (km^2)
Tiger Reserve	2585
National Park	1330
Water area	1781
Sand char without forest	153
Actual forest coverage	2333

The Sundarbans has the distinction of being the first mangrove forest in the world to be brought under scientific management. A Forest Management Division was established in 1869, for the natural resource management of the deltaic Sundarbans. The first management plan for the forest was implemented in 1892. The concept of sustainable management was thus implemented in the Sundarbans long before the concept became popular elsewhere in the developing world. This has resulted in the retention of a wide spectrum of flora and fauna in spite of several negative stresses such as the semi-intensive method of shrimp culture in the coastal zone, urban development, construction of harbours and channels, liquid or solid waste disposal and spillage of oil and other hazardous chemicals. The presence of the highly urbanised cities of Calcutta (22°30'N latitude and 88°30'E longitude) and Howrah (22°30'N latitude and 88°25' E longitude) along the bank of the Hugli river have been of significant detriment to the coastal zone of the Bay of Bengal due to the discharge of various municipal and industrial wastes. The recent development of the Haldia industrial complex (22°03' N latitude and 88°05' E longitude) on the western bank of the Hugli river has had a further negative impact on the health of the marine and estuarine system. Although the three major cities (Calcutta, Howrah and Haldia) are located outside the Sundarban Biosphere reserve, their multivarious discharges are conveyed to the coastal zone of the Bay of Bengal mainly through the Hugli and Kulti estuaries. Prior to the early 1940s, waste from Calcutta and Howrah was discharged into the Bidyadhari river, but subsequently, after a drastic reduction was observed in upland freshwater supply and silting, Kulti—part of an offshoot of the Raimangal estuary (also known as Harinbhanga or Ichamati) which flows at the extreme eastern sector of Indian Sundarbans—was chosen for the disposal of such waste (Majumder, 1952; David, 1956).

SEASONALITY, CURRENTS AND NATURAL ENVIRONMENTAL VARIABLES

The regional climate of the North Eastern part of the Bay of Bengal is a function of various active factors like wind, wave, temperature and precipitation.

Surface waves in the Bay of Bengal region are due to the wind system, the direction and velocity of which are mainly controlled by the northeast and southwest monsoon. The wind from the north and northeast commences at the beginning of October, continuing until the end of March. The months of January and February are relatively calm, with an average wind speed around 3.5 km/h. Violent wind speed recommences from the southwest around the middle of March and continues until September. During this period, several low pressure systems occur in the region, a number of which take the form of depressions and cyclonic storms of varying intensity. The wind speed during this time rises above 100 km/h and is usually accompanied by huge tidal waves, which cause much loss of life, as well as damage to property and forests. Sea waves in this region rarely become destructive except during these cyclonic storms. When the cyclonic incidences coincide with the spring tides, wave height can rise over 5 m above the mean tide level. Ripple waves appear in the months of October, November and December when wind generated wave height varies approximately from 0.20 to 0.35 m. In the months of April to August, when the normal wind speed ranges from between 15 and 40 km/h, large wavelets are formed in the shelf region; these start breaking as they approach the coastal margin. Wave height rises by up to 2 m, which causes maximum scouring of land masses during this period.

The adjacent Sundarban ecosystem has a moderate climate as a result of its location at the apex of the Bay of Bengal and the regular tidal flushing in the estuaries. Wave actions, micro and macrotidal cycles and longshore currents are recorded in most of the islands in the ecosystem. Coastal processes are very dynamic and are accelerated by tropical cyclones—a process known locally as Kal Baisakhi (Nor' Wester). Field investigations have shown that the coastline is changing in the reclaimed areas due to abnormal cliff erosion.

Indeed, Sundarban is a biogenous coast where biological factors are the key players in the processes of coastal evolution. Mangrove vegetations of the intertidal environment are treated as the builder of new landmasses and the intertidal vegetations have an important role in swamp morphology.

The seasonal climate in the Northeast coast of the Bay of Bengal and the adjacent deltaic Sundarbans may be conveniently categorised into pre-monsoon (March–June), monsoon (July–October) and post-monsoon (November–February).

The pre-monsoon period in the deltaic Sundarbans is characterised by a mean surface water temperature of around 34°C and is occasionally accompanied by rains and thunderstorms. Surface salinity in the aquatic phase is maximum in this season (average 32.5‰) due to an extremely low dilution factor and a high rate of evaporation. The pH is also relatively high (mean 8.40) and the average amount of dissolved oxygen is about 5.5 mg/l. During this season, the

Table 2

Seasonal variations of physico-chemical variables in the Sundarbans ecosystem

Season	Dissolved oxygen (mg/l)	Surface water temp. (°C)	Dilution factor	Salinity (‰)	pH	Nitrate (μgm.at/l)	Phosphate (μgm.at/l)	Silicate (μgm.at/l)
Upper Stretch								
A	6.0	34.0	0.96	1.5	8.00	19.87	3.36	56.30
B	5.9	33.5	1.00	0.0	7.48	21.05	4.18	63.85
C	5.8	22.1	0.97	1.0	7.90	20.60	2.84	70.36
Middle Stretch								
A	6.3	33.5	0.70	10.5	8.30	17.80	1.37	79.84
B	6.2	33.0	1.00	0.0	7.89	18.55	2.08	91.28
C	5.9	25.1	0.90	3.3	8.20	20.29	2.37	101.20
Lower Stretch								
A	6.9	33.9	0.23	26.1	8.36	19.50	1.36	161.85
B	6.8	33.2	0.67	8.5	8.12	26.53	1.89	182.30
C	6.3	26.0	0.40	20.0	8.28	18.75	1.37	137.40

A = Premonsoon, B = Monsoon, C = Postmonsoon.

mean salinity and pH of the open sea are 35‰ and 8.45, respectively.

The monsoon period is characterised by considerable surface water temperature of around 32°C (mean), minimum salinity (average 10‰), relatively low pH (mean 8.10) and a maximum dilution factor. The southwest wind triggers the precipitation in the monsoon period with an average rainfall of about 165 cm on the deltaic region. The dissolved oxygen shows minimum seasonal fluctuation with an average value around 5.0 mg/l. The average salinity in the open sea during monsoon drops to 28‰ along with a decrease in the pH value (mean 8.25).

The post-monsoon period is characterised by cold weather with a mean surface water temperature around 23°C. Rainfall is negligible in this period and therefore the dilution factor around the deltaic islands increases, resulting in a moderately high surface water salinity (mean value 25‰). The average pH of the surface water is about 8.25 and the mean dissolved oxygen is about 4.8 mg/l. The open seawater exhibits an average salinity of 31‰ and a mean pH of 8.40 at the surface layer.

With the change in seasons, tidal interactions in the estuarine system also change (Pillay, 1958). In the monsoon season the effect of flood tide is more or less countered and nullified by freshets and there is a strong predominance of ebb-tide. The strength of flood tide over ebb-tide is at a minimum during the post-monsoon season. Conversely, during the pre-monsoon season the effect of flood tide is considerably stronger than that of the ebb-tide.

The unique physico-chemical characteristics of the Sundarban ecosystem and the adjacent coastal region, in addition to occasional cyclonic storms, result in a marked seasonal oscillation of dilution factor, salinity and to some extent pH of the surface waters; although other variables such as dissolved oxygen and nutrients do not show any sharp seasonal trend (Table 2). Due to a sharp decrease in salinity in the monsoon period, the growth of *Penaeus monodon* is often restricted, thus causing problems for the shrimp culture industries in the coastal zone. The lowering of salinity and pH also enhance the process of dissolution of metallic compounds from the sediment bed to the water column and hence high concentrations of dissolved heavy metals are normally recorded in the monsoon period both in the aquatic phases and biological samples.

THE MAJOR SHALLOW WATER MARINE AND COASTAL HABITATS

The coastal zone of the Bay of Bengal contains the luxuriant mangrove vegetations which support innumerable taxa of invertebrate and vertebrate fauna, illustrating the high productivity of the ecosystem.

A total of 69 floral species included within 29 families and 50 genera have been recognised in the mangrove forests of the Indian Sundarbans, out of which 34 species are true mangrove types. The total area covered by mangroves in the Indian Sub-Continent is approximately 656,000 ha. (Anon., 1987), 65% of which falls within the Sundarbans region. The ecosystem includes almost all the floral species available in most of the Indian mangroves and also in the coastal zones of Andamans, Burma and other South East Asian countries. The area of tidal flats in which mangroves occur can be divided into three distinct zones on the basis of physico-chemical variables, each sustaining characteristic vegetation (Table 3). The three compartments in this horizontal zonation are not particularly distinct, as there is a continuous admixture of fresh and salt water due

Table 3

Horizontal mangrove zonation of deltaic Sundarbans

Zone	Dominant vegetation
Lower stretch (true estuarine and coastal zone)	*Avicennia marina* *Aegialites rotundifolia* *Avicennia alba* *Bruguiera cylindrica* *Bruguiera parviflora* *Lumnitzera racemosa* *Sonneratia griffithi*
Middle stretch (estuarine zone)	*Bruguiera gymnorhiza* *Ceriops decandra* *Ceriops tagal* *Kandelia kandel* *Rhizophora apiculata* *Rhizophora mucronata* *Sonneratia apetala*
Upper stretch (riverine zone)	*Exocoecaria agallocha* *Heritiera fomes* *Sonneratia caseolaris*

to tidal actions: hence, overlapping of species is very common in this ecosystem. The composition of mangrove vegetation also undergoes changes with the gradual elevation of the shores from the sea level.

The brackish waters of the Sundarbans support huge masses of phytoplankton and macrobenthic algal species which constitute the dominant primary producers of the ecosystem, which in turn support a diverse array of zooplankton. The planktonic survey in the Hugli–Matla estuarine system and mangrove back waters between 1985 and 1994 documented a total of 35 phytoplankton species distributed over 18 genera. Among all the phytoplanktons, diatoms seem to dominate the ecosystem. The genus *chaetoceros* contributed the largest number of species and accounted for the bulk of the diatom population during the peak phase of phytoplankton bloom. Genera like *Coscinodiscus*, *Biddulphia*, and *Pleurosigma* are commonly found in the ecosystem throughout the year.

The zooplankton community comprises a heterogeneous assemblage of animals covering many taxonomic groups including copepods, mysids, lucifers, gammarid, amphipods, cladocerae, ostracods, cumacea, hydromedusae, ctenophores and chaetognaths among holoplankters and polychaete larvae, molluscan and echinoderm larvae, crustacean larvae and fish eggs and larvae among meroplankters. Among the zooplankton, calanoid copepods constitute the major bulk, followed by cyclopoids and harpacticoids. In general, the higher abundance of zooplankton is encountered during the pre-monsoon periods. The main cause of the high species diversity index of zooplankton during the pre-monsoon months may be attributed to the coincidence of the breeding period of most of the coastal and estuarine fishes in the months of March and April in this part of the Sub-continent.

Invertebrates are another important component of this ecosystem which provide valuable food resources for humans and other higher animals. The scat analysis of Royal Bengal Tiger (*Panthera tigris tigris*) revealed the presence of *Toxocara cati*, proving the direct dependency of higher vertebrates on the invertebrate species. In addition to the resident invertebrate fauna of the ecosystem, the area also invites two important marine creatures, *Carcinoscorpius rotundicauda* and *Tachypleus gigas*. These are present during the onset of pre-monsoon, which is their breeding season (Fig. 3). During this period of increasing salinity, the living fossils visit this ecosystem in pairs along with the expanding tidal waters. The macrobenthic invertebrates exhibit a distinct vertical zonation on hard substrata like boulders, lighthouses and mangrove tree trunks in the lower stretch of the Hugli estuary, which remains in close association with different types of macroalgal species.

The Sundarban mangrove ecosystem is rich in terrestrial, aquatic and avian faunal species. The forests and water bodies provide dwelling places, habitats, breeding sites and roosting grounds for a wide range of vertebrate species encompassing approximately 250 species of fish, 8 species of amphibians, 57 species of reptiles, 161 species of birds and 40 species of mammals (Table 4), many of which are endangered in other parts of the world (Choudhury and Choudhury, 1994). The mangrove forest of the Sundarbans is an excellent habitat for the largest tiger (*Panthera tigris tigris*) population in the world. *Heritiera* and *Phoenix* vegetations, which are rarely or periodically inundated, make ideal tiger habitats. During the period of extremely high tides, these habitats are submerged and under such circumstances the tigers are compelled to lead an amphibious life. Thus, the tiger population in this part of the world has the unusual adaptation of being able to drink saline water. Tigers in this ecosystem have taken a toll of about 1500 human lives over the last twenty years. The victims are mostly the peoples of the adjacent islands who are living well below the poverty line. These people venture into the deep mangrove forests for the purposes of collecting

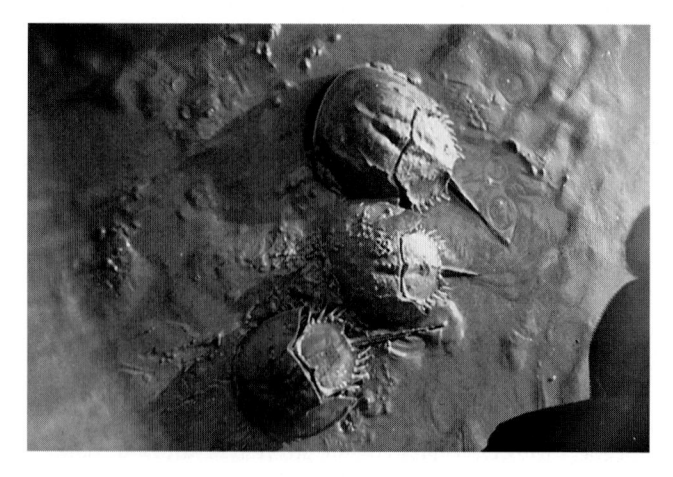

Fig 3. Horseshoe crab—a visitor to the northeast coast of the Bay of Bengal.

Table 4
Vertebrate faunal diversity in deltaic Sundarbans

Class	Genera/Species	
	Source: Chaudhuri & Choudhury (1994)	Source: Mandal & Nandi (1989)
Pisces	96/250	100/141
Amphibia	04/07	4/8
Reptilia	40/57	40/57
Aves	120/300	122/161
Mammalia	33/40	32/40

Table 5
Extinct species from Sundarbans mangrove ecosystem

Cairina scutulata	White-winged wood duck
Muntiacus muntjac	Barking deer
Cervus deruchea	Swamp deer
Bubalis bubalis	Wild buffalo
Rhinoceros sondaicus	Javan rhinoceros
Rhinoceros unicornis	Indian one-horned rhinoceros

Table 6
Endangered species of Sundarbans mangrove ecosystem

Mammals

Felis viveina	Fishing cat
Panthera tigris	Tiger
Platanista gangetica	Gangetic dolphin
Orcella brevirostris	Irrawady dolphin
Neophocaena phocaenoides	Little porpoise
Manis pentadactyla	Chinese pangolin

Birds

Ardea goliath	Giant heron
Leptoptilos dubius	Adjunct stork
Haliaeetus leucogaster	White bellied sea eagle
Pandion haliaetus	Osprey

Reptiles

Crocodilus porosus	Saltwater crocodile
Lepidochalys olivacea	Olive Ridley Turtle
Batagur baska	Batagur turtle
Kachuga teeta	Indian tent turtle
Lissemys punctata	Indian flap shelled turtle
Troonyx gangeticus	Indian soft shelled turtle
Troonyx hurun	Peacock marked soft shelled
Varanus bengalensis	Common indian monitor
Varanus flavesceus	Yellow monitor
Varanus salvator	Water monitor
Python molurus	Indian rolled python

honey, wood, fish, crabs, tiger prawn seeds and various miscellaneous mangrove resources; they pose stress to the virginity of the forest and its associated biodiversity.

The mangrove species of the Sundarbans ecosystem are currently under serious threat as a result of habitat destruction, unwanted human intrusion, conversion of mangrove patches into aquaculture, salinity fluctuation and the deterioration of water quality caused by wastes of industrial and anthropogenic origin. About half of the mangrove areas in West Bengal have been clear felled (Blasco, 1977). *Heritiera fomes* has become extremely rare in the Indian Sundarbans and may be considered endangered in this belt, although the species is abundant in Bangladesh (Chaudhuri and Choudhury, 1994). The species *Nypa fruticans* also has a very limited population in the Indian Sundarbans. The faunal species of the Sundarbans have undergone a compositional change due to the salinity fluctuations and habitat destruction. About six vertebrates have disappeared from this ecosystem in the last 200 years (Table 5) and about 20 species are on the list of endangered species (Table 6). A conservation approach has recently been adopted by declaring this area a 'National Biosphere Reserve', to ensure long-term sustainability of the diversity of species in this ecosystem.

SOCIAL HISTORY AND POPULATION PROFILE

Human settlements started in the Sundarbans region at a very early age. Anthropological evidences collected from the region indicate that early settlers were of Negroid stock (Das, 1981). They were followed by Proto-Australoid and Homo-Alpinus groups. Proto-Nordic or Aryans invaded the area much later. Most of the people of the Sundarbans belt have either a Negroid or Proto-Australoid type of morphological appearance.

Records of early Greek and Chinese travellers reveal the existence of a prosperous kingdom in the lower Bengal region. Hieun Tsang's visit to this region (629–641 A.D.) demonstrated that the coastal belt of the Bay of Bengal was a fertile land inhabited by an intelligent group of people. Historical records of the area focus on the interchange of rulers in the lower Bengal belt, along with religious beliefs and customs. The region was under the control of the Pal (760–1142 A.D.) and the Sen (1118–1199 A.D.) dynasties for several prosperous centuries. Buddhism was a strong influence during the time of the Pal, but, with the advent of the Sen dynasty, the region was soon dominated by Hinduism. The Muslims occupied Bengal in the 13th century and, in 1576, following an invasion by the Mughal emperor, Akbar, the region was secured by Muslims. Many Islamic landlords are still inhabiting various pockets of the deltaic ecosystem of the Sundarbans. In 1757, the British East India Company gained possession of the region from Mir-Jaffar, who was then the Nawab (Supreme Authority) of Bengal. It was from this period that large-scale reclamation of land and resettlement began in the coastal belt of Bengal, especially around the Matla and Bidyadhari basins.

The settlers had to battle against hostile environmental conditions such as floods, cyclonic storms, lack of drinking water and wild animals—in particular, tigers and crocodiles. Systematic leases of land in the Sundarbans region were granted to the settlers from 1831 to 1833. Since then the roots of colonisation gradually started to spread in the deltaic region which paved the way for the clearing of the mangrove forests and construction of embankments in the coastal zone to protect against tidal ingression. The early settlers were mainly migrants of tribal origin and a small number of indigenous people. Their principal occupations were wood cutting, honey gathering and fishing. The tribal people were mainly composed of several groups like Bhumij, Mundas, Oraon and Santhal. The co-existence of several groups in the same environmental area gradually led to the emergence of a new type of society with a heterogeneous culture (cultural convergence). Although there are both Muslims and Hindus in the region, the religious barrier has been mostly dissolved as a result of the unrelenting struggle against natural disasters, wild animals and diseases such as cholera. Folk rituals and rites of this region are closely associated with the seasons, agriculture, trees, rivers, plants, animals and crops. A large number of folk deities and Goddesses are worshipped by the people (Table 7) with the desire to rid the region of its natural hurdles and the wild animals of the forests, which often attack villagers.

The deltaic region of the Sundarbans ecosystem is rich in natural resources, but is inhabited by an extremely poor population. Although approximately 94.61% of the total population depends on agriculture, more than 54% of the people do not own any land (Chattapadhyay, 1983). These people are serving as landless labourers in the agricultural sector. About 5.39% of the population are engaged in fishery, forestry and handicrafts. The economic situation is very miserable; this is reflected by the average monthly income and expenditure per family (Table 8).

With the sudden rise of a semi-intensive prawn culture practice in the coastal zone of the Indian Sub-Continent during the early nineties, a new class with a distinct identity became dominant in the population profile of the deltaic region. This class—constituted mainly of the poor and landless people—is the major supplier of tiger prawn seeds or juveniles in the shrimp culture farms of the state, as the hatchery supply is negligible to the farms located in the Northeastern coast of the Bay of Bengal. Most of the prawn hatcheries are concentrated in the Southeastern sector of the country and their products are either consumed locally or sent long distance by air using the oxygen packing method. Since the mortality rate of prawn seeds greatly increases during such long distance transport, most of the farms of the state of West Bengal depend on the supply of local prawn seed collectors who screen the coastal and estuarine waters for the seeds of tiger prawn. It has been found that a significant part of the coastal population of the state of West Bengal, in the process of earning for this livelihood, constantly expose themselves, irrespective of sex or age, to sea waters for catching prawn seeds. They employ nets of particular mesh sizes to haul the drifting community from the coastal waters and throw away the major portion

Table 7
Principal folk deities and goddesses worshipped in deltaic Sundarbans

Dakshin Rai is believed to be the God of the Tiger and is worshipped below a banyan (*Ficus bengalensis*), peepul (*Ficus religiosa*) or neem tree (*Azadirachta indica*). The deity is dressed like a warrior with bow and arrow in his hands and is seated on a tiger. Wood cutters, boat men and honey collectors worship Dakshin Rai with the hope of overcoming all hurdles in the dense mangrove forests.

Bonbibi is believed to be the guardian deity of the forests and is worshipped by the people of both Hindu and Muslim religions before venturing into the forests in search of natural resources. She has been given several names like Bibima, Basuli, Bon Durga and Bon Kali and is placed either on a tiger or on a hen.

Manasa is regarded as the Goddess of snakes and is worshipped to protect the people from venomous snakes.

Jagatguru There is a common belief among the people of coastal Sundarbans that worship of Jagatguru can protect them from cobras.

Manik Pir is worshipped for the welfare of cattle, specially cows.

Olabibi is worshipped to get rid of cholera which is a common disease in some periods of the year.

Ganga. The Goddess Ganga is worshipped by the people of deltaic Sundarbans to protect the embankment that are the main defensive structures against tidal surges and waves.

Table 8
Average monthly income and expenditure of the people of deltaic Sundarbans

Economic group	Average monthly income per family (in Rupees)	Average monthly expenditure per family (in Rupees)
During 1980 to 1985 (before the initiation of Semi intensive method of shrimp culture)		
Marginal/small farmer	328.50	380.65
Medium farmer	450.75	440.70
Large farmer	855.10	625.00
Landless labour	260.35	298.80
Fisherman	690.25	582.00
During 1990 to 1995 (Peak period of semi-intensive method of shrimp culture)		
Marginal/small farmer	1085.70	1000.55
Medium farmer	1568.85	1200.80
Large farmer	2800.90	1575.00
Landless labour	875.00	710.50
Fisherman	2500.00	1280.25
Prawn seed collectors	2900.00	1300.85

People of almost all the economic tiers of Sundarbans region are indirectly linked with prawn seed collection either by providing nets, boats etc. or by way of dealership though satellite collectors.

Fig. 4. Dragging of nets along the coast in search of tiger prawn seeds.

of the haul (containing the juveniles of several species of finfish and shellfish) after sorting out the post larval stage (PL 20) or seeds of tiger prawn. This practice paves the way for soil erosion due to the constant dragging of nets along the coast (Fig. 4), uprooting the mangrove seedlings and saltmarsh grass. Furthermore, major portions of the juveniles of the fish population, which are both commercially and ecologically important, are also destroyed during this screening process, with a long-term impact on demersal fishery of the offshore area. The economic platform of a significant portion of the coastal population has risen, however, after adopting this alternative source of income (Table 8).

According to the recent census, there are more than 5.2 million people in the Sundarbans area with a considerable fraction (about 45% to 65%) of socially lower classes of people placed under the category of scheduled castes and scheduled tribes. These people rear pigs and fowls, and are engaged in a wide range of activities such as fishing in the offshore region, basket making, fishing in the mud flats, swamps, marshes and paddy fields, honey collection, timber collection from the deep forest areas; prawn seed collection from coastal zones and earth excavation work on daily wage basis. Many people of this class have also been engaged in recent times in betel and chilli cultivation. The non-scheduled category mainly comprises land owners, the majority of whom rear cattle, sheep and goats. Many people of this category are also engaged in giving loans to the poor at high interest rates for making boats, fishing trawlers and nets for prawn seed collection.

Industrial development is essentially very poor in the Sundarbans area. Except for a few rice mills, ice factories and vessel repairing units, organised industry is notably absent. There are few hand-loom, pottery and handicraft industries in the region. Manufacturing of salt from the coastal waters is practised in some pockets of the Sundarbans—which provides employment for a small number of people. A fish-drying industry has been recently initiated at Jambu island where marine and estuarine fishes like *Harpodon neherius, Trichiurus savala, Trichiurus gangeticus, Arius gagora, Ilisha filigera* etc. are dried. Two thousand five hundred metric tons of fish are caught and dried annually by approximately 15,000 fishermen. In the early nineties, the rapid development of semi-intensive shrimp culture industries in the area involved many local people in various jobs like earth digging, pond construction, embankment raising, water quality monitoring, water exchange, feed spraying in culture pond, water aerating by paddler, harvesting etc. Subsequently, however, these industries have been closed down after the Supreme Court of India imposed restrictions on all the semi-intensive shrimp farms within 500 m of the high tide level (HTL) throughout the entire Indian coast. Although most of the shrimp farms were closed, many shifted to a modified extensive method of culture in which the stocking density is very low and there is little chance of aquatic pollution from the aquaculture farms.

MAJOR RURAL ACTIVITIES IN DELTAIC SUNDARBANS AND ENVIRONMENTAL IMPACT

Agriculture is the main occupation of the people of Sundarbans, but due to the saline nature of the soil nitrogen, fertilisers do not have substantial positive effects in boosting crop production. Presently, only 4.45% of the area under cultivation has irrigation facilities. The average productivity of high yielding kharif paddy in some blocks is 2.5 MT/ha while the boro paddy has a yield of 3.0 MT/ha. A negligible area is under multiple cropping. Betel and chilli are grown in many pockets of the deltaic region, in response to demand from cities in the Calcutta and Haldia industrial belt. Deterioration of coastal water quality due to leaking of pesticides and fertilisers is practically nil in this part of the country as the people rarely use artificial fertilisers and pesticides to accelerate agricultural production.

Shrimp culture is an important activity practised in this region. During the early eighties, the traditional shrimp culture system was prevalent in the coastal parts of the Bay of Bengal. This was not particularly productive, yielding only 200–500 kg/ha. The method of culture was potentially sustainable, except that the wet feed prepared from locally available items like trash fishes, oyster flesh, cereal flour, and beef without any adhesive component, could well have led to the increase of BOD and COD values of the aquatic medium. As the stocking density was extremely low and no input of artificial feed and fertilisers were applied to the culture ponds, the chances of deterioration of the coastal waters from the traditional system was very low.

During the early nineties, a radical change in the area of shrimp culture took place, replacing the traditional method of culture by a semi-intensive method. Using this approach, about 15–25 tiger prawn seeds (PL 20) were stocked per m^2 and artificial feed, mainly imported from Taiwan, was supplied at different stages of the growth. To supply additional oxygen, aerators and manual paddlers were also used. A detailed scrutiny of the major impacts of shrimp culture using the semi-intensive method in the northeast coastal zone of the Bay of Bengal revealed that the gain was minimal in comparison to the loss incurred in various sectors. Although returns were high on at least a short-term basis, the shrimp culture in the mangrove ecosystem aggravated problems such as eutrophication, siltation, deterioration of water quality, growth of pathogenic microbes, ecological crop loss, acidification of soil, salinity intrusion and so forth. Of these, the deltaic Sundarbans experienced problems of ecological crop loss, deterioration of water quality and microbial attack on tiger prawns at the highest scale between 1992 and 1995.

The ecological crop loss occurred at random due to a meagre supply of tiger prawn seeds from the hatchery; almost nil in the state of West Bengal. Most of the farms that emerged in the coastal belt of the Bay of Bengal in the early nineties were semi-intensive in nature with high demands for prawn seeds. As the supply line, mainly originating from the hatchery, was far below the demand line, the gap was filled by supplying prawn seeds collected from coastal waters. A large percentage of the coastal population engaged themselves in collecting prawn seeds from the coastal water bodies. During this operation, varieties of finfish and shellfish juveniles, other than tiger prawn seeds (non-target species), were also caught in the net; but as the non-target species were non-remunerative, the major portion of the haul was thrown away and wasted (Banerjee and Singh, 1993). A study carried out on the impact of mass collection of prawn seeds in the mangrove ecosystem of the Sundarban Biosphere Reserve [Project No. J-22014/11/93-CSC (M)] on behalf of the Ministry of Environment and Forests (Government of India) from 1994 to 1996 highlighted the following: that due to wild harvest of tiger prawn seeds from the coastal water bodies, the mean annual weight of finfish and shellfish juveniles wasted per net were about 1.190 tonnes, 1.588 tonnes respectively around the Sagar Island area (21°56′ to 21°88′N latitude and 88°08′ to 88°16′E longitude) of the deltaic Sundarbans. About 22 species of finfish juveniles and nine species of shellfish juveniles were wasted in the screening process per net per day during 1994 (Mitra et al., 1998). The same process continues unrestricted today in the coastal belt of the Bay of Bengal. Many of these species are not commercially important, but their presence in bulk amount in each and every net employed for harvesting tiger prawn seeds indicates their importance in energy transference through the food webs spun in this unique mangrove ecosystem through a long evolutionary period of time. The process of wild harvesting of tiger prawn seeds may have an adverse effect on the delicate ecological balance of the system (Trivedi et al., 1994).

Deterioration of coastal water bodies is another major problem that has emerged as a result of the semi-intensive method of prawn culturing in the region. Due to the

Table 9

Mean values of physico-chemical variables during three different periods in the waters around Chemaguri aquaculture farms at Sagar Island

Physico-chemical variables	1989	1994	1997
Salinity (‰)	15.0	19.8	18.1
pH	8.21	8.00	8.25
D.O. (mg/l)	6.8	4.9	6.3
PO_4 (μg-at/l)	2.33	4.65	2.80
NO_3 (μg-at/l)	16.1	29.9	20.0

1989: before initiation of semi-intensive method of shrimp culture;
1994: peak period of semi-intensive method of shrimp culture;
1997: period of modified traditional method of shrimp culture.

increased inputs of fertilisers and artificial feeds, the potential for increasing the nutrients and organic load per unit area have increased (Mitra and Choudhury, 1995). The artificial feed supplied to the tiger prawns during the different stages of their development often settles at the pond bottom as residue leading to an increase of the H_2S level. In fact, H_2S is generated due to the reduction of sulphur and other sulphur compounds present in the extra feed which becomes waste, faecal matter, etc. by certain heterotrophic bacteria under anaerobic conditions. No systematic studies have been conducted in the deltaic region to monitor the impact of wastes released from the semi-intensive shrimp farm on the biological diversity of the area, but considerable amounts of nutrient enrichment were observed in the released water and adjacent coastal waters of certain farms (Table 9). As there were no scientific treatment plants in the majority of the coastal shrimp culture farms of the state of West Bengal, the waste waters were released into the Bay without any dilution, treatment, or biological filtering during the processes of water exchange, harvesting and pond preparation for a new crop. These operations caused much deterioration in the coastal waters of the Bay of Bengal and encouraged many pathogens and microbes of an unknown nature. Severe tiger prawn mortality throughout the entire coastal belt was witnessed during 1994 and 1995 due to various types of microbial diseases. The condition became so acute that on 11 December, 1996, the Supreme Court of India ordered the coastal states to demolish aquaculture farms, other than traditional and improved traditional ones, operating within 500 m of the HTL in the coastal zone by March 1997 (Sen, 1998).

EFFECTS FROM URBAN AND INDUSTRIAL ACTIVITIES

With the gradual increase of human population and industrialisation, the problem of environmental pollution has become critical, particularly in Calcutta, which

Table 10
Classification of the Calcutta people on the basis of living status

Class	Percentage
Elite	1
Upper middle class	8–9
Lower middle class	40
Poor and slum dwellers	50

effectively discharges its municipal wastes into the Bay of Bengal via the Kulti estuary. The city of Calcutta, with a metropolitan area of 1350 km², sustains a population of approximately 11.86 million population. Out of 100 municipal wards at the core of Calcutta there are four wards where the population is about 1.0 *lakh* per km² and 35 wards having a population density of about 70,000 per km². A large proportion of the people in the Calcutta city are the poor and middle classes (Table 10) who are the major contributors of the city's domestic wastes.

The industrial crowding in the city of Calcutta is also at a critical level. Within the city, there are 11,516 small and large factories and a considerable percentage of these factories are engaged in the manufacturing of toxic chemicals. In some pockets of the city, there are a number of lead factories producing lead ingots and lead alloys. Many tanneries are also located in the outskirts of the city that contribute appreciable amounts of chromium to the adjacent water bodies. These huge quantities of Calcuttan metropolitan wastes, arising from both domestic and industrial sources, are currently being discharged into the Kulti estuary, situated 35 km southeast of the eastern fringe of the city, via two long canals. A study conducted at the Kulti outfall region between 1996 and 1997 revealed significant correlations of lead and chromium with the Shannon Weaver index of the stress tolerant molluscan community (Table 11). A considerable amount of lead and chromium is also detectable in the prawn tissues (*Penaeus monodon* and *Penaeus indicus*) that are cultured in a nearby traditional farm with this contaminated water. About 20% of the waste water of the city of Calcutta that ultimately finds its way into the Bay of Bengal is comprised of industrial wastes. A large number of industries (including chemicals, fertilisers, pharmaceuticals, metallurgical, galvanising, electroplating, pulp, paper, textiles, safety matches, soaps and detergents, cosmetics, battery and tanneries) discharge their crude wastes, without any treatment, into the city's drainage system where they mix with domestic sewage that ultimately flows to the Kulti outfall region. The industrial effluents have changed the primarily organic nature of waste water into a more complex one by adding metals and other industrial chemicals.

The Hugli estuary, located at the extreme western sector of the deltaic Sundarbans, mainly carries the industrial discharges of the Haldia region—a newly emerging industrial complex on the western bank of the estuary. The high concentration of heavy metals in the estuarine water of this region may be attributed to the existence of a large number of coastal industries, whose principal activities include the refining of crude oil, and the manufacture of automotive, industrial and submarine storage batteries (Mitra et al., 1994). The presence of a large number of floating old, rusty and stranded barges in the Haldia port region is another cause of contamination in the ambient media (Mitra et al., 1995). These barges are the major sources of particulate iron which settles down and mixes with the bottom sediments. The precipitated iron in the form of hydroxide has a tendency to scavenge other metals like copper, zinc and lead as they pass through the water to the sediment compartment (Waldichuk, 1985).

The concentrations of six heavy metals recorded at six sampling stations in the estuarine and Bay region between 1994 and 1996 also exhibited a unique seasonal pattern (Mitra, 1998). The concentrations were usually high during monsoon months, the period characterised by maximum precipitation and dilution factor, and minimum salinity and pH. In the pre-monsoon season of extremely high salinity and pH, the concentrations of dissolved heavy metals were normally low. The high concentrations during monsoon may be due to the increased run off to coastal waters from the adjacent cities (Mitra and Choudhury, 1992). The second cause of higher levels of metal concentration in the aquatic phase may be explained in the light of their chemical *speciation*. During monsoon, the dilution factor of the sampling stations increased significantly due to heavy precipitation, especially during the months of August and September. This resulted in the lowering of salinity and pH which possibly facilitated the dissolution of precipitated forms of metal and increased the amount of metallic ions in solutions (Mitra and Choudhury, 1993). A significant negative correlation between salinity and dissolved copper was observed by Krishnamurty (1972) in Portonovo water in the southeast part of the Bay of Bengal. On the northeast coast, an interrelationship study performed concerning dissolved heavy metals and salinity (Mitra et al., 1994) also confirmed the dissolution processes from the sediment compartment to the aquatic phase due to low salinity and pH.

Table 11

Mean concentrations of Pb and Cr (in μg/g dry wt.) in the muscles of *Penaeus monodon* and *Penaeus indicus* analysed during March 1997 to February 1998

Prawn species	Premonsoon (Mar.–Jun.)		Monsoon (Jul.–Oct.)		Postmonsoon (Nov.–Dec.)	
	Pb	Cr	Pb	Cr	Pb	Cr
Penaeus monodon	3.86	2.99	4.02	3.09	2.09	2.67
Penaeus indicus	2.84	2.60	3.00	2.85	2.98	2.77

Table 12

Mean seasonal variations of salinity, pH and dilution factor (df) at different sampling stations in the Sagar Island on the same date and time

Station	Premonsoon Date: 4.5.96. Time: 10.45 a.m.			Monsoon Date: 11.9.96. Time: 2.45 p.m.			Postmonsoon Date: 20.12.96. Time: 11.30 a.m.		
	Salinity	pH	df	Salinity	pH	df	Salinity	pH	df
Sagar light house area (south-western tip of the Sagar Island)	28.2	8.34	0.17	16.8	8.30	0.42	24.1	8.31	0.24
Chema-guri (southeastern part of the Sagar Island)	23.7	8.21	0.30	10.1	8.20	0.65	19.6	8.22	0.39
Kachu-beria (northern-most tip of the Sagar Island)	12.8	8.01	0.62	2.5	8.00	0.91	8.8	8.10	0.73

The bioaccumulation of heavy metals by certain selected sedentary animals, mangrove plants and macroalgal species also exhibited a similar type of seasonal variation with high values during the monsoon and low values during the pre-monsoon (Mitra et al., 1993; Mitra and Choudhury, 1993; Mitra and Choudhury, 1994; Trivedi et al., 1995; Mitra et al., 1995; Mitra et al., 1996). In an attempt to find the indicator species of heavy metal pollution in the coastal waters, it was observed that bivalves cannot serve the purpose of indicator organisms, particularly in the deltaic Sundarbans. This is mainly because in the upper stretch of the estuarine system, the survival of bivalves like *Crassostrea cucullata*, *Crassostrea cuttackensis*, *Enigmonea aenigmatica*, *Macoma birmanica*, *Anadara granosa*, etc. is impossible. This is due to a reduction of salinity to below 5‰ in certain regions which completely eradicates the settlement of bivalves. In fact, the deltaic region of the Sundarbans mangrove ecosystem is a unique type of ecosystem where the riverine input of fresh water has a dominant role in maintaining the community structure of flora and fauna. With a different horizontal transect of the ecosystem, there is a drastic variation of different physico-chemical variables. The dilution factor, salinity and pH reveal high variations at the same time in different locations (Table 12). This alters the species composition and structure of the ecosystem as one approaches from the coastal region and lower estuarine stretch to the upper estuarine zone. However, a recent study (Trivedi et al., 1994) observed that the barnacles, *Balanus balanoides*, which have a wide range of tolerance to environmental variables, are readily available at all times of the year, easily accessible, and sufficiently tolerant to low salinities and high suspended solids. The species has also been recorded in the vicinity of the Haldia port-cum-industrial complex where the salinity value drops to 2‰ during the monsoon months. This suggests it has a wide range of osmotic adaptability. The magnitude of accumulation of heavy metals (Zn, Cu, Mn, Fe, Co, Ni, Pb and Cd) by *Balanus balanoides* at different sampling stations also exhibited unique positive correlations with the dissolved heavy metals of the ambient media, which suggests the suitability of the species to be used for the purpose of biological monitoring of some heavy metals in the north east coast of the Bay of Bengal and adjacent Sundarbans ecosystem (Mitra et al., 1995).

PROTECTIVE MEASURES

The Sundarbans mangrove ecosystem is situated very near the densely populated city of Calcutta. The wastes generated due to the various anthropogenic and industrial activities from this highly urbanised and industrialised city are discharged into the Bay of Bengal through the Sundarbans. The Haldia industrial complex releases all its wastes arising from industrial or port activities into the Bay of Bengal through the Hugli estuary at the extreme western sector of the Sundarbans. These stresses have a negative impact on the health of the ecosystem. Large numbers of faunal and floral species have vanished as a consequence (Chaudhuri and Choudhury, 1994). To improve this critical situation, several protective measures have been adopted which may be broadly classified into four categories: conservation of biological resources, conservation policies, ecological rehabilitation of the Sundarbans under the Integrated Wasteland Mangrove Project and captive breeding programmes.

Conservation of Biological Resources

Several sanctuaries and national parks have been created in the mangrove forest area with the aim of preserving floral and faunal diversity. The establishment of the Sundarban Biosphere Reserve in the southern sector of the Dampier Hodges line (in the deltaic region) over an area of 9630 km^2 has made a significant contribution to protective measures. It includes a Tiger reserve under the Project Tiger Programme. The core area of Project Tiger has been converted into a National Park with the aim of preserving the existing genetic diversity of the representative ecosystem. For the purpose of protecting resident and migratory birds, a sanctuary has been created at Sajnakhali. Two more

Table 13

Zonation of Sundarbans Biosphere Reserve

Zone	Area (km^2)	Dominant feature
Core zone	1700	Dense mangrove forests; includes the tiger reserve and the 1330 km^2 of the Sundarbans National Park
Manipulation zone	2400	Mangrove forests
Restoration zone	230	Degraded forests, saline banks and the degraded mudflats within the non-forest areas
Development zone	5330	Non-forest areas utilized for aquaculture, agriculture, apiculture etc.

sanctuaries have been established to the west outside of the Project Tiger area at Haliday Island and Lothian Island. In order to achieve a sustainable conservation programme, the entire biosphere reserve has been classified into four zones (Table 13). In the manipulation zone, the involvement of local people has been greatly increased to reduce the direct human pressure on forest resources.

The most significant among all the steps of conservation of biological resources in the deltaic Sundarbans region was the initiation of Project Tiger in 1973, in an area covering 2585 km^2. Keeping pace with the effort of the International Union for the Conservation of Nature (IUCN) to conserve the global tiger population, the killing and hunting of tigers and other wild animals was banned in India in 1970. A country wide tiger census in 1972 revealed that the tiger population had steeply declined to less than 2000 from a figure of 40,000 at the turn of the present century (Panwar, 1991). Against this background, Project Tiger was launched initially in nine reserves of the country covering a total area of 13,723 km^2, out of which 4242 km^2 were core areas. Presently there are eighteen tiger reserves in thirteen states across the country, covering a total area of 28,017 km^2 with 12,634 km^2 of core areas. The Sundarbans Tiger Reserve presently has a core area of 1330 km^2 and a buffer zone of 1255 km^2. As a result of these efforts, the population of tigers has increased in the Project Tiger area since 1979 (Table 14).

Table 14

Tiger population in deltaic Sundarbans

Year	Tiger population
1979	205
1984	264
1989	269
1993	251
1998	289

Conservation Policies

A number of Acts and policies have been adopted to protect the Indian ecosystem and its biotic and abiotic components, which are also applicable for the protection of the Sundarbans mangrove ecosystem. There are three main Acts for the protection of the environment. These are the Wildlife Protection Act (1972), the Forest Conservation Act (1980) and the Environment Protection Act (1986), with their various amendments (Government of India, 1992a).

The National Conservation Strategy and Policy Statement on Environment and Development (Government of India, 1992b) provides directives for the integration and internalisation of environmental issues in the policies and programmes of various sectors. The National Forest Policy (1988) complements the Forest Conservation Act (1980) by increasing the participation of local people in forest conservation activities.

The deterioration of coastal water bodies due to untreated discharges from the semi-intensive shrimp culture farms has also been restricted by the Supreme Court of India on the basis of the Environment Protection Act (1986). On 11 December 1996, the Supreme Court of India ordered the coastal states with aquaculture farms other than traditional and improved traditional ones, operating within 500 m of the HTL in the coastal zone to demolish them by March 1997. The Court also directed that no shrimp farms could be set up within 500 m of the HTL in the zone. In accordance with the principles of 'sustainable development' and 'polluter pays', the Court also directed the Government to set up a body under clause (3) of Section 3 of the Environment Protection Act 1986. The said body would issue permits to the farmers of traditional aquaculture to adopt improved traditional farming systems. The Court also ruled that the owners of the closed aquaculture units would be liable to pay six years wages as compensation to the workers as well as for eco-restoration of the affected areas by way of creating an Environment Protection Fund. There was an excellent response to this notification: now, only traditional and improved traditional shrimp farms exist in the north-east coast of the Bay of Bengal.

India has several legislative Acts that are relevant to various uses of the oceans and coastal zones under which the Coastal Regulation Zone (CRZ) was notified in 1991 through the Ministry of Environment and Forests. Under Section 3(1) and (2) of the Environment Protection Act (1986) and Rule 5(3) (D) of the Environment Protection Rules (1986), the CRZ covers the coastal stretches of seas, bays, estuaries, creeks, rivers and backwaters which are influenced by tidal actions in the landward side up to 500 m from the High Tide Line (HTL) and the land between the Low Tide Line (LTL) and the HTL. In general, the HTL is estimated to be the line which the highest tide reaches during the spring time. In relation to the state of West Bengal which supports the Indian Sundarbans at the apex of the Bay of Bengal, the CRZ has been classified into three

Table 15
Structure of Coastal Regulation Zone (CRZ) in relation to West Bengal coast

Category	Area	Important conservation oriented activities
CRZ -I	All the ecologically sensitive and important areas viz. the Sajnekhali, Lothian and Halliday sanctuaries; reserve forests, wildlife habitat and mangrove forests in the Sundarbans; New Moore and Sandhead islands; important breeding sites of finfishes, shellfishes and other species; important hotspots from the point of biodiversity; the lower long sand area for its aesthetic beauty and biological diversity; historical and heritage sites at Sagar Island that fall within the high tide level (HTL) and low tide level (LTL).	– some eco friendly communication facilities, construction of schools and hospitals will be allowed; – aquaculture will not be permitted in mangrove rich area; – seed hatchery will be encouraged where by 5 % will be released in the creek; – a marine park will be established in the southern part of the Sagar Island; – Culture of algae and edible oyster will be encouraged;
CRZ - II	Areas that have already been developed and close to the shore line. Such areas include those within the municipal limits (e.g., Haldia along the Hugli river) as and other legally designated areas which have drainage facilities, approach roads, water supply etc. The Haldia town and eastern part of Digha fall within this zone.	– shore protection measures will be adopted; – dredging activities will be regulated; – EIA will be compulsory for any projects; – Regular monitoring of pollution will be carried out;
CRZ - III	Areas that are yet to be developed and include Shankarpur and western Digha.	– aquaculture sites will be selected on the basis of recommendation by a special body of the State Government

categories with distinct gradation of activities for a more scientific management of the ecosystem (Table 15).

Ecological Rehabilitation of the Sundarbans under the Integrated Wasteland Project

The project envisages a total economic uplift of local communities presently living below the poverty line in remote areas of Sundarbans. The project is being implemented through the Sundarbans Biosphere Reserve and the scheme has been designed to combine a full spectrum of activities comprising afforestation, conservation of fragile areas, development of pasture, soil conservation, minor irrigation, cottage industries and other socio-economic and ecological components.

Captive Breeding Programmes

Crocodile

The estuarine crocodile (*Crocodylus porosus*) is one of the largest saline water reptiles which was indiscriminately killed for the purpose of making luxury goods from its skin. The level of poaching became so severe that the population subsequently declined, making the species endangered.

On 1st April 1975, the Government of India, with the assistance of FAO and UNDP, requisitioned the services of an FAO expert and initiated crocodile farming in India. The salt water crocodile scheme was initiated in the Sundarbans in March 1976 at Bhagabatpur on Lothian Island, and efforts were focused on reducing the high mortality rate at the egg and newly hatched stages. Today, the salt water crocodile scheme in the Sundarbans is one of the principal crocodile breeding centres within India. By 1990, more than 197 crocodiles had been released into the Sundarbans (Chaudhuri and Choudhury, 1994). This is a clear indicator of the programme's success.

Olive Ridley Turtles

Marine turtles are protected under schedule 1 of the Wildlife Protection Act. The Olive Ridley turtle (*Lepidochelys olivacea*), which extends its migration to the Sundarbans deltaic region, is an endangered species. Previously there was a period when millions of Olive Ridley eggs were harvested immediately after the breeding season. The poaching of mating pairs was a lucrative business during the breeding season (January and February) when thousands of animals were brought to the shore at Digha, West Bengal for selling in the Calcutta market. This reduced their population to a great extent and, therefore, attempts were made to rear and rehabilitate this species at the Bhagabatpur Crocodile Rearing Centre and at the Saptamukhi Hatchery by the Department of Forests, Government of West Bengal. The programme showed encouraging results when out of 117 hatchlings that emerged from artificial nests at Bhagabatpur, 99 healthy hatchlings were released into the ambient water and 18 were segregated for further research works (Banerjee, 1985).

Horseshoe Crabs

Horseshoe crabs, chelicerate arthropods belonging to the class merostomata, are considered to be the oldest 'living fossils'. There are only four species of the horseshoe crab reported in the world, among which *Carcinoscorpius rotundicauda* is plentiful in the deltaic Sundarbans region. Although *Tachypleus gigas* is also reported from the deltaic Sundarbans, the frequency of occurrence is very low (3–5%) among the total horseshoe crab population. The species is destroyed at random during the fish catch in the neritic zone and wild harvests of tiger prawn seeds. With the unfounded belief that they can cure arthritis, the extracts of different organs of the species are sold in the Calcutta markets. These activities caused a decline in the population of the species in recent years in the northeast coast of the Bay of Bengal. The species has been found to be a potential source of a bioactive substance, Limulus Amoebocyte Lysate, from its blue blood (Chatterjee and Abidi, 1993). This reagent is highly sensitive and useful for the rapid and accurate assay of gram negative bacteria, even if present in a very minute quantity, up to the level of 10^{-10} g. Hence the LAL has proved to be a valuable diagnostic reagent in the detection of endotoxins in several pharmaceutical products, especially injectibles. Given the immense biomedical value of the species, a project entitled "Application of biotechnology and molecular biology in the resource exploration and sustainable management of Sundarbans mangroves" was sanctioned to the S.D. Marine Biological Research Institute at Sagar Island by the Department of Biotechnology under the Ministry of Science and Technology, Government of India, on 26th December 1997. This aims to develop culture technology for *Carcinoscorpius rotundicauda* and *Tachypleus gigas* suited to the Sundarbans environment. The project has started to investigate the optimum range of various physico-chemical variables (such as water temperature, salinity, pH, nitrate, phosphate, silicate, dissolved oxygen, dissolved copper, organic carbon of sediment and acid-leachable copper of sediment) in relation to the fertilisation and condition value of the three different weight groups. The development and standardisation of culture technology of this species may pave the way for manufacturing Carcinoscorpius Amoebocyte Lysate (CAL) and Tachypleus Amoebocyte Lysate (TAL) from the amoebocyte of these species without damaging their natural population.

REFERENCES

Anon (1987) *Status of the Mangroves in India*. Ministry of Environment and Forests, New Delhi.

Banerjee, A.K. (1964) *Forests of Sundarban; Centenary Volume*. West Bengal Forest Department, Calcutta, pp. 166–185.

Banerjee, R. (1985) Captive rearing of Olive Ridley turtle. *Hamdryad*. 12–14.

Banerjee, B.K. and Singh, H. (1993) The shrimp fry by-catch in West Bengal, India. BOBP/WP/88.

Blasco, F. (1977) *Outlines of Ecology, Botany and Forestry of the Mangals of the Indian Subcontinent, West Coastal Ecosystems*. Elsevier Scientific Publishing Company, Oxford.

Chatterji, A. and Abidi, S.A.H. (1993) The Indian horseshoe crab—a living fossil. *Journal of Indian Ocean Studies* **1** (1), 43–48.

Chattopadhyay, K. (1983) A preliminary report of a Pilot enquiry on the social ecology of Sundarbans. Sociological Research Unit, Indian Statistical Institute, Calcutta (Mimeo).

Chaudhuri, A.B. and Choudhury, A. (1994) *Mangroves of the Sundarbans, India*. IUCN.

Das, A.K. (1981) *A Focus on Sundarbans*. Calcutta Editions, India.

David, A. (1956) Effect of Sewage upon fish of Kulti estuary. *Journal of the Asiatic Society* **1** (4), 339–363.

Government of India (1992a) *Conservation Of Wetlands in India*. Ministry of Environment and Forests, New Delhi.

Government of India (1992b) *National Conservation Strategy and Policy Statement on Environment and Development*. Ministry of Environment and Forests, New Delhi.

IUCN (1989) *Marine Protected Areas Needs in the South Asian Seas Region*, **2**, India.

Krishnamurty, T.M. (1972) Occurrence of copper in Portonovo waters. *Health Physics* **24** (3), 12–15.

Majumder, S.B. (1952), Rivers of Bengal Delta- River Problem in West Bengal and their solutions. *Journal of the Asiatic Society, Science* **18**, 103–120.

Mandal, A.K. and Nandi, N.C. (1989) *Fauna of Sundarban Mangrove Ecosystem, West Bengal, India*. Fauna of Conservation Areas, Zoological Survey of India.

Mitra, A. (1998) Status of coastal pollution in West Bengal with special reference to heavy metals. *Journal of Indian Ocean Studies, Society for Indian Ocean Studies* **5** (2), 135–138.

Mitra, A. and Choudhury, A. (1993) Seasonal variation in metal content in metal content in the gastropod *Nerita articulata* (Gould). *Indian Journal of Environmental Health*, NEERI, **35** (1), 31–35.

Mitra, A. and Choudhury, A. (1993) Trace metals in macrobenthic molluscs of the Hooghly Estuary, India. *Marine Pollution Bulletin* **26** (9), 521–522.

Mitra, A. and Choudhury, A. (1994). Dissolved trace metals in surface waters around Sagar Island, India. *Journal of Eco-Biology* **6** (2), 135–139.

Mitra, A., Mishra, A., Singh, B., Trivedi, S., Gupta, A., Choudhuri, A. and Choudhury, A. (1995). Biodiversity loss due to intense wild harvest of Prawn seed in mangrove environment of Sunderban, W.B. India. *National Symposium on Perspective in Biodiversity*, The Zoological Society, Calcutta, India.

Mitra, A. and Choudhury, A. (1995) Causes of water pollution in Prawn culture farms. *Journal of Indian Ocean Studies, Society for Indian Ocean Studies* **2** (3), 230–235.

Mitra, A., Choudhury, A. and Zamadar, Y.A. (1992) Seasonal variation in metal content in the gastropod *Cerithidea* (*Cerithideopsis*) *cingulata*. *Proceedings of the Zoological Society (Calcutta)* **45**, 497–500.

Mitra, A., Niyogi, S. Aich, A. and Choudhury, A. (1996) Seasonal variations of trace metals in the Mangrove associate algae inhabiting the Nayachar Island, W.B. *International Conference on Environmental Science, India*, pp. 125–130.

Mitra, A., Trivedi, S., Gupta, A., Choudhuri, A. and Choudhury, A. (1994) Effects of Heavy Metals on Oyster Population in Hooghly estuary. *National symposium on Aquaculture for 2000 A.D.*, Madurai Kamaraj University, Madurai-625021, India.

Mitra, A., Niyogi, S. and Choudhury, A. (1998) Impact of wild harvest of prawn seeds on the community structure of juvenile fish in the coastal part of West Bengal. Living Resources of India's Exclusive Economic Zone, Omega Scientific Publishers, New Delhi, pp. 86–96.

MOEN (1996) Final Report on Project. Impact of mass collection of Prawn seeds in mangrove ecosystem of Sundarbans biosphere Reserve. Sanction No. J-22014/11/93- CSC (M). S.D. Marine Biological Research Institute, Sagar Island, W.B. (India).

Panwar, H.S. (1991) Status on management of protected areas in India: problems and prospects. *Tigerpaper* **XVIII** (3), 17–25.

Pillay, T.V.R. (1958) Biology of *Hilsa ilisha*. *Indian Journal of Forestry* **5** (2), 201–257.

Sen, S. (1998) *Environment, law of the sea and coastal zone*. World Wide Fund for Nature, India, Eastern Region.

Sidhu, S.S. (1963) Studies on Mangrove. *Proceedings of the National Academy of Sciences, India* **33** (b), Part 1, 129–136.

Trivedi S., Mitra, A., Chaudhuri, A. Gupta, A., Singh, B. and Choudhury, A. (1994) A case study on the loss of biodiversity during prawn seed collection from the Hoogly Estuary, India. *Proceedings of the National Convention on Environment of India—Challenges for the 21st Century*. Institution of Public Health Engineers, India, T-V/16-T-V/24.

Trivedi, S., Mitra, A., Bag, M., Ghosh, I. and Choudhury, A. (1995) Heavy Metal Concentration in Mud skipper Boleopthalmus boddaerti of Nayachar Island. *Indian Journal of Environmental Health, NEERI,* **37** (2), 120–122.

UNEP (1985) *Environmental problems of the Marine and Coastal areas of India*. UNEP Regional Seas Reports and Studies, No. 59.

Waldichuk, M. (1985) Biological availability of metals to marine organisms. *Marine Pollution Bulletin* **16**, 7–11.

THE AUTHOR

Abhijit Mitra
*Department of Marine Science,
University of Calcutta
35, B.C Road, Calcutta 700 019
West Bengal, India.*

Chapter 63

SOUTHEAST INDIA

Sundararajan Ramachandran

The seas along the southeastern states of India, totalling approximately 1860 km, are described in this chapter. Covering the States of Tamilnadu and Andhrapradesh, the ecologically important and distinctive sites are the Gulf of Mannar, Palk Bay, Vedaranyam, Pichavaram, Pulicat Lake, the Godavari-Krishna delta and the Coringa mangroves. Economically important areas and centres of industry include Tuticorin, Nagapattinam, Cuddalore, Chennai, Nizampattinam, Kakinada and Vishaghapatnam. The coastal population density is 20 to 600/km^2, and more than 9 million people live along these coastal areas.

The major activities that cause degradation of coastal ecosystems in these areas are disposal of untreated domestic and industrial wastes, port and harbour activities, ocean traffic, exploration and exploitation of minerals, oil and gas, as well as natural hazards such as storm surges. Both coastal erosion and siltation of river mouths are major problems in some of these areas as well. Reduction of freshwater flow in the rivers due to the construction of dams for irrigation purposes has affected coastal ecosystems and the stability of coasts. The rapid development of activities and the increasing coastal population are threatening the health of seas and their resources. Fisheries have stagnated during the past five years and environmental degradation is evident through the reduction of biodiversity, loss of habitats, reduction in mangroves and by impacts on coral reefs. Community participation in integrated coastal zone management plans and strict implementation of those plans are required for the sustainable utilisation of marine resources and conservation of biodiversity.

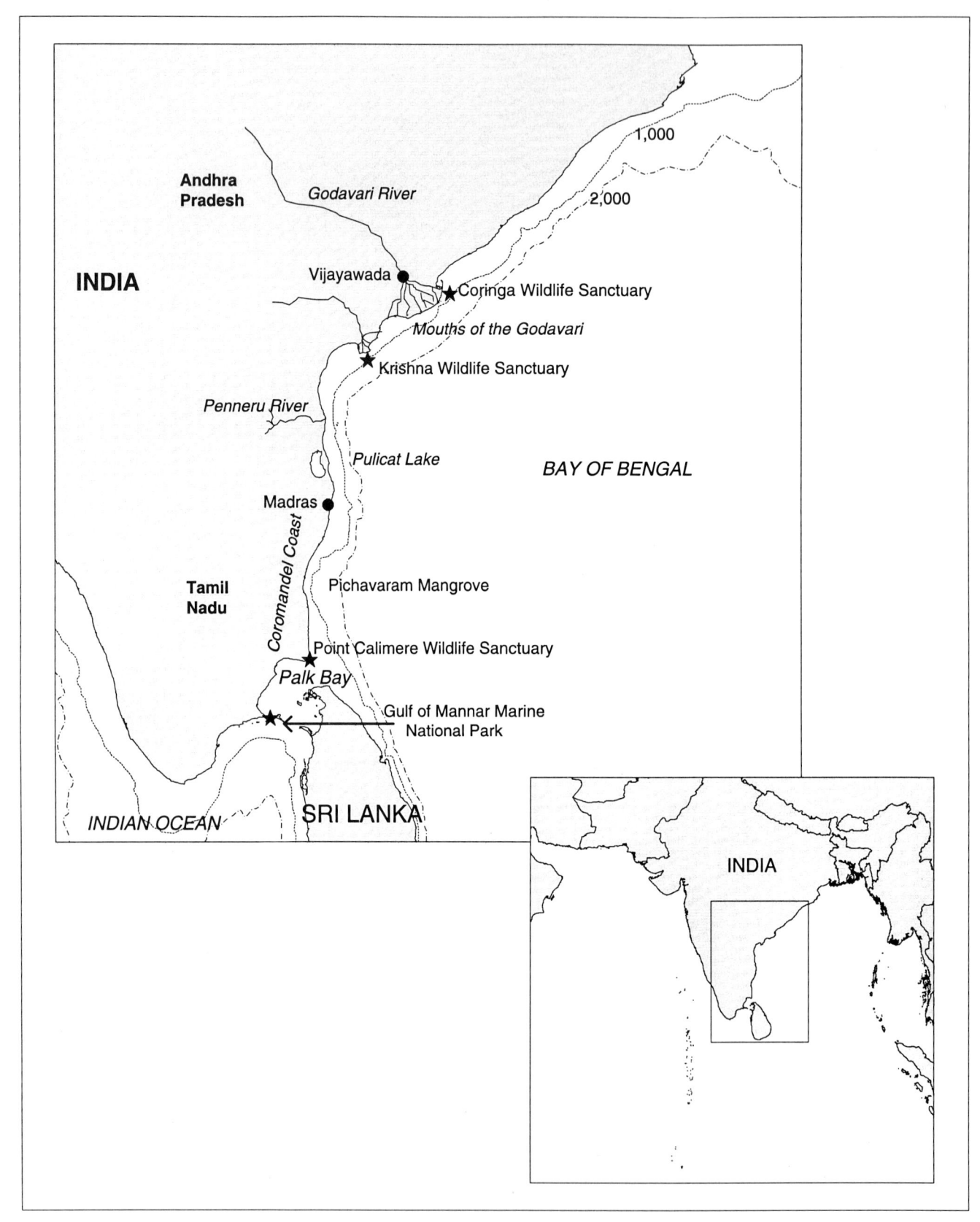

Fig. 1. Map of the Southeastern India region.

INTRODUCTION

Seas of South East India include the areas of Gulf of Mannar, Palk Bay, North of Palk Bay to Madras coast and Krishna-Godavari deltaic coast (Fig. 1). These areas are rich in biological diversity and have a long history of human settlement, use and exploitation. These tropical coastal marine areas contain most of the common diverse and productive ecosystems that are sensitive to human activities, impact and intervention, such as estuaries, salt marshes, coral reefs, seagrass beds and mangroves. Conservation and sustainable use of biological diversity is of critical importance for meeting the food, health and other needs of the growing population. Pressure and demand on the habitats, resources and ecosystems of this coast are growing intensely due to rapid development and population growth along the coast, and new models and mechanisms are required to move towards multiple use of the habitats, and towards a systems-oriented and community-based management approach, because, as is the case in many places, current sectoral approaches to the management of marine and coastal resources in this area are not capable of conserving marine biodiversity, nor can they ensure sustainable development.

The various sectors of this coast are first described in turn, northwards from near the southern tip of India.

GULF OF MANNAR

The Gulf of Mannar is situated between India and Sri Lanka, and is spread between 8°35′ to 9°15′N latitude and 78°08′ to 79°30′E longitude. Coral reefs are characteristic of this area and are distributed on the shelves of 21 islands, lying between Rameshvaram and Kanyakumari (Table 1). These islands are situated up to 8 km from the mainland and the Gulf of Mannar has been declared a National Marine Park and Marine Biosphere Reserve by the Government of India.

Climate and Coastal Hydrography

This area is affected by Southwest and Northeast monsoons. The Southwest monsoon contributes little towards the total annual rainfall of this area which is around 1000 mm/yr. Relative humidity ranges between 40 and 80%. The atmospheric temperature varies from 24 to 38°C. The currents are strong; during the Northeast monsoon, they run southwards, while during the southwest monsoon, the currents flow northwards. Between March and April and during October, the tidal stream is significant. The sea is rough between April and August and calm during September. Tidal elevation is around 1 m and the wind speed between 2 and 6 km/h, directed towards the northeast or northwest.

Several rivers, the Thamirabarani, Vaipar, Vaigai and Guridar all drain into the Gulf of Mannar which alters the physical parameters to a certain extent, including increasing the siltation. Values for this and other sites are given in Table 2.

The Main Marine Ecosystems

Coral Reefs

The reef of the Gulf of Mannar forms a barrier about 140 km long and 25 km wide, running between Pamban and Tuticorin. Different types of reef form within this complex, such as platform, patch, coral pinnacles and even atoll types (Fig. 2). The numerous islands have fringing reefs around them, mostly located at a distance of 50–100 m from the islands, and patch reefs lie between them, arising from depths of 2–9 m and extending 1–2 km in length, with a width as much as 50 m. The total area occupied by reef and

Table 1

Islands in the Gulf of Mannar

1.	Van island	12.	Talaiari island
2.	Koswari island	13.	Valai island
3.	Kariyashulli island	14.	Mulli island
4.	Vilanguchalli island	15.	Hare island
5.	Upputhanni island	16.	Manoli island
6.	Puluvinichalli island	17.	Manoli putti island
7.	Nallathanni island	18.	Poomarichan island
8.	Anaipar island	19.	Pullivasal island
9.	Valimunai island	20.	Krusudai island
10.	Appa island	21.	Shingle island
11.	Poovarasampatti island		

Table 2

Some physical and biological parameters of surface waters from southeast India (North to South) (Ramachandran, 1997)

Coast	Surface temp (°C)	Susp. solids (mg/l)	Surface salinity (ppt)	pH	Inorganic P (μg/l)	Ammonia N (μg/l)	Total N (μg/l)	Nitrate N (μg/l)	Nitrite N (μg/l)	BOD (mg/l)	Dissolved O_2 (mg/l)	Ch a (mg/m^3)
Godavari to Krishna		6–56			31.7–38.4	43.7–66.2	213–231				4.3–5.7	0.2–0.4
Madras–Palk Bay	25.5–32	5–30	30–35	7.3–8.3	10–177	20–127	18.9–823			0–1.2	2.2–6.5	0.12–2.45
Palk Bay	21.7–32	13–120	24.5–34.1	7.1–8.1				0.2–5.5	0.5–6.3	0–2.2	3.8	0.98–14.40
Gulf of Mannar		10–940	31.3–35	7.3–8.2	0.10–42.5	0.2–80.0	1.4–196	0.6–372		0–3.1	2.3–5.5	2–23

Fig. 2. Exposed patch reefs typical of the Gulf of Mannar.

its associated features is 94.3 km². Along with the reef size, waves also play a major role in the formation and development of the islands of the Gulf of Mannar. The wave-generated currents move the sediment from the reef edge towards the centre of the reef which eventually fills the lagoons with sediment. The beach rock formation on the islands was formed by the process of colonization by plants and rainwater percolation. The development of the coral islands themselves has depended on various geological factors such as reef types, size, shape, orientation and exposure to other reefs, and has led to a good diversity (Table 3).

Ninety-four Scleractinian coral species of 37 genera have been recorded in the Gulf of Mannar (Gopinadhapillai, 1973). The southern side of the islands is usually covered with dense corals, which are more diverse than the northern side. This may be due to the physical processes working upon the reef; the steep southern slopes have greater circulation with open oceanic water. Due to the severe anthropogenic pressure and natural climatic events, the coral health is very poor. A coral transplantation programme here found that *Acropora formosa* was the most successful and suitable species for transplantation in the Gulf of Mannar as well as in Palk Bay.

Mangroves

It is believed that the Gulf of Mannar once supported thick mangrove forest. Even now, their remnants and relics are evident. Krusadai Island has the maximum number of species (8) followed by Manoli Island (7 species). *Aegiceras corniculatum*, *Excoecaria agallocha*, *Brugiera cylindrica*, *Rhizophora apiculata*, *Luminitzera racemosa* and *Rhizopora mucronata* are all critically endangered here now. The *Avicennia* species distributed here are stunted in growth, which also makes them vulnerable. Mangroves here appear to require low salinity water in their early stage of growth. However, the mangroves which grow on the islands of the Gulf of Mannar do not have a freshwater supply from rivers, as is the case on the mainland shore, and consequently the plants have to depend heavily on the monsoon rainfall for germination of their seeds.

Algae

The Gulf of Mannar is very well known for its wealth of algal species—the standing crop of the macroalgae from a total area of 17,125 ha is 22,044 tonnes (wet weight), consisting of 1709 tonnes of agarophytes and 10,266 tonnes of alginophytes and 10,069 tonnes of other seaweeds. A total number of 147 species of algae in the islands of the Gulf of Mannar includes 42 species of green algae, 31 species of brown algae, 69 species of red algae and 5 species of blue-green algae (Kaliaperumal, 1998). As the demand for raw materials of agar-yielding seaweeds is increasing, their natural resource is dwindling, and the agarophytes *Gelidiella acerosa* and *Gracilaria edulis* especially are over-exploited. Thirteen species of seagrasses are recorded from the Gulf of Mannar area.

Pearl Oyster and Chank beds

The Gulf of Mannar is also known for its vast beds of pearl oysters and chanks. Six species of pearl oysters have been reported from the rocky beds which lie 10–20 m deep. *Pinctada fucata* is the most common species. The Chank fishery is a major industry in the Gulf of Mannar; out of the five subspecies of chanks, the dominant variety includes *Xancus phyrum* var. *acuta* and *Xancus phyrum* var. *obtusa*, but in recent years, due to pollution, the oyster and chank beds have disappeared from all but a few small areas.

Fish and Fisheries of Gulf of Mannar

There are about 75 fish-landing centres in this region, including two major fishing harbours, one at Tuticorin and

Table 3

Recorded biodiversity in the Gulf of Mannar

Name of group	No. of species
Phytoplankton	51
Sponges	108
Scleractinian Corals	100
Ahermatypic corals	10
Polychaetes	119
Molluscs	450
Crabs	106
Lobster	6
Prawn	12
Finfishes	450
Turtles	4
Dolphin	5
Whales	10
Dugong	1

the other at Chinnamuttom. Coastal fish production has almost stabilized or is showing only a slow rate of growth. Although there are as many as 450 known species in the region, only about 200 are of commercial value, either for direct human consumption or for yields of fish by-products e.g. fish oil, liver oil, isinglass, shark fin and processed fish skins. The oil sardine (*Sardinella longiceps*) which was previously caught in only limited numbers, is beginning to form a substantial industry (1419 t in 1996). Similarly the Indian Mackerel (*Rastrelliger kanagurta*) registered a substantial increase from 213 t in 1992 to 3711 t in 1996 (Dorairaj, 1998). The annual marine fish production is around 81,000 t.

Human Population and Environmental degradation

There are about 23,000 fishing households with a population of 115,000 in 90 villages in the Gulf of Mannar. The literacy rate is 31%, which is far less than the state average of about 64%. The major towns along the Gulf of Mannar are Kanyakumari, Thiruchendur, Tuticorin, Mandapam and Rameshwaram, whose total population is 700,000 (Census Data, 1991).

These local communities depend on coral reefs and reef resources. The Gulf of Mannar's coral reef ecosystem is now in poor condition due to both natural calamities and anthropogenic pressure. These impacts, especially coral quarrying (Fig. 3), pose a severe threat to the coral reef ecosystem of the Gulf of Mannar. In addition, due to the use of trawl nets over the seagrass beds, the seagrass dependent marine system is also becoming affected: notable examples include the seahorse, *Hippocampus* sp. which is reduced, and the sea cow, *Dugong* sp. which is now endangered. In areas around Rameswaram, the coastal waters are polluted due to tourism and domestic sewage disposal, while around Vallinokkam, industrial effluents and aquaculture discharges affect the coastal water quality. Apart from these, siltation on the shoreward sides of islands influences the growth of corals, a problem which is aggravated during the Northwest monsoon periods. Export-led and indiscriminate large-scale fishing of attractive marine living organisms such as gorgonians, sea cucumbers, ornamental echinoderms and molluscs are also now threatening the existence of these organisms.

Impact of Human Activities on the Ecosystem

There are several groups of activity or event which are causing particular problems in this area.

Deforestation is a problem affecting the coastal zone, and the large scale felling of coastal trees for firewood causes significant coastal erosion. Simultaneously, the high growth rate and urbanisation of the human population in the coastal areas, both due to migration and natural increase, are increasing the demands made on the coastal habitats. Much of the previous forest land along the coast of the Gulf of Mannar has changed into housing developments. Domestic sewage disposal into the coastal waters has increased the pathogenic bacterial population; more than 30,000 million litres per day of untreated domestic sewage are discharged into the Gulf of Mannar.

Upstream of the islands, Tuticorin new harbour has been constructed with a breakwater of 3 km in length. Construction of the breakwater has changed the current flow pattern and sediment movement along the Tuticorin coast. Periodic dredging operations in the entrance of the harbour have caused silt plumes and increased water turbidity. Quarrying of corals for the construction of houses is another form of destructive activity, which continues in spite of efforts to control it. A casual survey of old buildings in the Gulf of Mannar area, including the oldest building, a Dutch church dated about 1750, reveals that big coral blocks have long been used for construction in the place of bricks, as they were cheap and available locally.

The fish processing industries, located near to the shore, discharge enormous quantities of fresh and cold water to the coastal habitats and alter the habitat structure. Since the new Tuticorin port became operational, the coast of the Gulf of Mannar has experienced accelerated industrialization. The Thermal Power Station located on the coastline of Tuticorin presently has five units of 210 MW each and consumes 2800 tonnes of coal per day, producing huge quantities of fly ash, whose dumping causes significant pollution and which has significantly affected marine filter feeding organisms. More than 10 major marine chemical and fertiliser industries are located along the Gulf of Mannar coast. Their industrial effluent discharge has already polluted the coastal water, such that water quality data have revealed high concentrations of toxic metals and nutrients.

Indiscriminate use of nets having meshes less than 3 cm traps juvenile fish and causes damage to fishing of species with late maturity and low fecundity. Further, trawling within 30 m depth has caused extensive damage to the sensitive coastal habitat, which is faunastically rich and high in biodiversity. Dynamite fishing, though banned, is usually

Fig. 3. Quarried corals used for lime preparation. This is the livelihood of some traditional coastal populations of the Gulf of Mannar.

carried out to capture sailfishes since the meat of the animal is highly desired and fetches a good price. Activities such as the anchoring of fishing boats, trawl net operation, and ship anchoring on the coral beds have also all caused serious damage. Keelakkarai is the foremost place for catching *Dugong* species for consumption. Data are few, but during 1960, the capture of *Dugong* in the Gulf of Mannar and Palk Bay was around 250 individuals. Today at Keelakkari a special bottom set gill net is operated for capturing *Dugong*, and now this species has become nearly extinct and has been declared endangered.

Currently there is very little ecotourism in the Gulf of Mannar. Tiruchendur and Rameshwaram are coastal II type towns and are visited by thousands of pilgrims every year. Tourism development, construction of restaurants, hotels etc. have affected the coastal water quality. The key challenges relate to the need for community-supported improved sewage and garbage disposal, creation of a local environmental ethic, and the need for proper planning, management and enforcement.

There are 18 major aquaculture industries, using around 1500 million litres per day of seawater for culture operations and discharging the same amount of water into the coast. The aquaculture discharges contain high nutrient and organic loads. Hatchery production and farming of shrimps along the Gulf of Mannar coast have led to alteration of coastal morphology and to marine pollution. Other problems come from user conflicts due to traditional fishing, the use of sea frontage and beach, and loss of coastal habitats and benthic life.

PALK BAY

Geological Features

Palk Bay is situated between longitude 78°58′ and 79°55′N and latitude 9°55′ and 10°45′E. It is a very shallow flat basin and its depth does not exceed 9 m. The area has a distinctive and unusual marine environment and this is being affected by sea level variation. As the sea-level oscillation is being aided by sediments, its marine environment is constantly changing. Due to its prograding nature, microdeltas are being formed and there is a noticeable advance in the swampy area. Recent changes in Palk Bay are mainly due to fluvial activities and marine action (Loveson et al., 1990).

Climate and Coastal Hydrography

This area is affected by southwest and northeast monsoons. The mean annual rainfall varies from 820 to 1650 mm and Palk Bay is usually calm except during the Northeast monsoon when turbulent conditions prevail. The northeast wind velocity at Nagapattinam is about 8 to 10 km, and at Pamban the wind strength reduces to between 2 and 4 km. Table 2 shows the values of several water parameters. The salinity of the water decreases gradually along an axis in the southwest direction running from the strait, and high saline water forms pockets in the southwest corner of the bay, possibly due to the incursion of Gulf of Mannar water through Pamban Pass. Inshore waters of the Palk Bay become muddy due to the presence of suspended sediments and silt stirred up by wave action during the monsoon. Soil erosion is a major problem in this area; based on observations made by several researchers, the plume of suspended material from Vedaranyam moves towards Jaffna (in Sri Lanka) and towards Rameshwaram islands so that all of Palk Bay contains suspended sediment during both monsoon seasons, a considerable amount of which enters the Gulf of Mannar. The Cauvery tributaries Adappar, Mulliyur, Valavanar and Korayar drain into Palk Bay.

Coastal Ecosystems

The coastline of Palk Bay has coral reef, mangrove, lagoon and seagrass ecosystems (Table 4). The coral reef runs parallel to land (east to west direction) between latitude 10°15′N and longitude 79°55′E. It starts from Munakad as a wall-like formation (1–2 m broad) and runs east up to Tonithurai to a distance of about 5.5 km, where the width of the reef reaches more than 300 m. Gopinadhapillai (1969) listed 20 coral species in this area and classified the reefs of the Palk Bay into five zones, viz. shore, lagoon, shoreward slope, reef crest and seaward slope. More recent observations indicate the area has 25 species of corals.

The shore of the Palk Bay reef is mostly sandy, with dead pieces of corals, except at the extreme eastern side and near the Pamban Bridge where there are traces of sandstone. The width of the lagoon behind the reef varies from 200 to 600 m at different places with a depth of 1–2 m. The bottom is sandy with molluscan shells and pieces of disintegrating corals. There are few or no living corals in the lagoonal area due to the absence of any hard substratum (Gopinadhapillai, 1969), though sponges such as *Hercina fusca*, *Dysidea*

Table 4

Recorded biodiversity of Palk Bay

Group	No. of species	Remarks
Mangroves	8	*Avicennia* sp. is the dominant
Macroalgae	61	*Halimeda* sp. is the dominant
Boring sponges	20	*Cliona* sp. is the dominant
Scleractinian coral	25	*Acropora* sp. is the dominant
Molluscs	73	–
Boring Bivalves	17	*Lithopaga* sp. is the dominant
Crabs	89	–
Lobster	6	–
Prawn	12	–
Turtles	4	–
Dolphin	5	–
Birds	250	65 of them are migratory

fragilis, Spirastrella inconstans and *Calispongia diffusa* are fairly common on the bottom.

Thirty years ago (Gopinadhapillai, 1969), the shoreward slope of the reef supported encrusting and massive corals including types with comparatively large polyps such as *Favia pallida, F. favus, Favites virens, Goniastrea pectionata, G. retiformis, Platygyra lamellina, Hydnophora* sp., *Symphillia* sp. and *Goniopora* sp. Living colonies of *Porites* sp. were rare or small in size. *Galaxea fascicularis, Turbunaria peltata* and *Pavona varians* are the rarest species. This zone of the reef supported a good number of associated reef dwellers like encrusting sponges, bryozoans and calcareous algae. Among the soft corals, *Lobophytum* sp. and *Sarcophytum* sp. were occasionally present.

The reef crest is often exposed at low tides, with corals being rare at the exposed reef crest. On the seaward side of the reef slope, coral growth was comparatively richer, and the majority of the corals were ramose forms, viz. *Pocillopora* sp., *Acropora* sp. and *Montipora* sp.

Today, the coral reef ecosystem has completely changed in its pattern of distribution. Now the lagoon has large numbers of boulders occupied by various species of scleractinian corals. Six scleractinian coral species have been recorded from the lagoon of Palk Bay. Fishermen report that the sponge and soft coral populations have been decreasing during the past two decades. Boring sponge species are increasing (Asir Ramesh and Kannupandi, 1997) whereas macrosponges are declining. Table reefs are also found in the lagoons. These newly found boulders and table reefs in the lagoon are due to changes in substrate nature due to changed drift and wind processes. The native green algal community has increased compared to other algal groups. On the shoreward slope of the reef, the coral population has increased significantly in both distribution and diversity, and Palk Bay has 20 coral species with a density of around 50 colonies/10 m^2. Further, most of them are ramose corals. Sponge populations are higher on the shoreward slope than in the lagoon. Some coral species, *Platygyra lamellina, Hydnophora* sp. and *Galaxea fascicularis*, which were recorded by Gopinadhapillai (1969), have now disappeared from Palk Bay, but the total number of recorded species has increased to 10 (Asir Ramesh and Kannupandi, 1997) on the shoreward slope.

Vedaranyam Wildlife Sanctuary

Vedaranyam or Point Calimere is one of the six major wild life sanctuaries in the Tamilnadu part of Palk Bay, extending over an area of 17.29 km^2. It is one of the major wintering grounds for migrant birds from North India, Europe, Asia and Africa, and is found at the southern end of the Kaveri delta where the Paminiyar, Korayar, Kilathangi, Marakkakoriyar, Akini, and Valanar and various tributaries of the river Kaveri flow through. The wetland has been proposed for designation as wetland of international importance under the RAMSAR convention.

About 250 species of birds have been recorded here by the Bombay Natural History Society. These birds migrated here from "Arctic Siberia", Iran and the Himalayas. The bird migration commences in August and lasts until March.

Muthupet Lagoon

Muthupet coastal lagoon (latitude 10°25'N and longitude 79°39'E) covers an area of 20,000 ha and opens into Palk Strait. Eight mangrove species are recorded from this area. The forest extends to an area of approximately 6800 ha, of which only 77.20 ha (4%) is occupied by well developed mangrove and the remaining 96% of the area is covered by poorly growing trees. *Avicennia marina* is the dominant mangrove species. The southern portion of the lagoon is formed by a sand spit, devoid of mangroves. The mangrove zone of this forest is restricted to the edges of the brackish water lagoon where true mangrove plant species are distributed in varying degrees of abundance.

Fish and Fisheries of Palk Bay

Perna viridis, which is a very rare bivalve in the coral reef ecosystem, can be collected in good numbers and in huge sizes from Palk Bay. The fin-fishing season starts here in October and lasts till February every year. However, the peak season is only in December and January. Fishermen come here from many places with their vessels and stay until the season ends (October to February). Around 2000 mechanised boats and 5000 non-mechanised boats are involved in active fishing during the season, and the annual average fish production is around 85,000 t.

Human Population and Environmental Degradation

Of around 420,000 people living along Palk Bay coast, the fishing population is around 157,000 and increases during the fishing season. Sixty percent of the coastal population depend upon the coastal ecosystems.

Natural climatic events cause damage, sometimes annually, to the marine system. High velocity, cyclonic winds during the monsoon season cause mechanical damage to the corals and mangrove ecosystems in this area, and huge quantities of silt in the northeast monsoon have profound effects on the distribution and diversity of the coral reef and associated plants and animals. Sedimentary deposits progressively fill the lagoons. The area affected by salt also increases about 4.5 times due to drying up of mud flats and salt pans. Due to the expansion of saltpans and other human activities, mangroves have degraded in the Muthupet and Vedaranyam areas (Ramachandran et al., 1998)

Impact of Human Activities on the Ecosystem

More than 40% of the mangrove forests in this area have been cleared due to indiscriminate tree felling for firewood (from 1976 to 1998). The area occupied by salt pans has also increased by about 30% at the expense of the mangrove forest. Enrichment of nutrients (total nitrogen 152 µg/l) in the coral reef ecosystem has increased the algal and phytoplankton populations and reduced the amount of penetrating sunlight. Processing industries and nearby households discharge wastewater directly into the sea. Wherever these industries are located, changes are evident in the coral reef ecosystem.

Before the establishment of the salt industries at Vedaranyam, the area was accessible to boats, and fishing hamlets were located in the area up to 5 km. from the sea. Now, due to around 40 major salt industries, the entire swamp area is divided into many compartments by the construction of innumerable bunds, which has reduced the wetland area. As a result of damming sea water for long periods in order to obtain various grades of brine from the vast reservoirs, sea water percolates and seeps into the subsoil up to a distance of 3.5 km inland.

Other than the salt pans, there are no major industries polluting the coastal environment of Palk Bay except some processing industries located near the coasts.

The dried sea horse (*Hippocampus kuda*) is in great demand in Southeast-Asian countries, especially Singapore and China. Due to this demand, the animal has been overexploited from Palk Bay where it is now 'endangered'. Due to over-harvesting, *Dugong* are extinct in this area. Overexploitation and selective fishing of the herbivorous fishes reduces the grazing pressure on the algae (especially green algae); consequently, the algal growth is now competing with the corals and has reduced the amount of coral growth in Palk Bay.

Point Calimere bird sanctuary, the Manohara heritage site and Mandapam coral reef areas are all tourist spots. It is estimated that around 1,200,000 tourists visit these areas every year. These tourism activities generate a lot of solid waste, and the developing infrastructure, such as restaurants, hotels, lodges etc. contribute to coastal pollution. These have greatly affected the coastal water quality.

The mangrove area, which was about 21,000 ha, has been reduced to about 13,240 ha. This reduction is mainly due to the establishment of salt and aquaculture industries. It is estimated that around 5400 ha of mangrove forests of this area have been converted into aquaculture ponds. Sandy beaches along Palk Bay have also been converted into aquaculture ponds.

NORTH OF PALK BAY TO MADRAS COAST

Geological Features

The region between Palayar and Kovalam area, north of Palk Bay to the Madras coast, has a length of around 350 km. This area is rich in mineral resources such as copper, zinc, lead and lignite in addition to clay, iron ore and limestone. This coastline has around 20 major estuaries, 40 major backwater channels and is endowed with a variety of fauna and flora. The important areas are Pichavaram mangrove forest (11°26′N and 79°48′E), Pondicherry beach (11°53′N and 79°50′E), Mahabalipuram heritage area (12°37′N and 80°12′E), Kovalam beach (12°46′N and 80°15′E), Marina beach (12°59′N and 81°15′E), and Pulicat lake (13°24′ and 13°43′N and 80°03′ and 80°18′E). Each and every area of this coast is significant, with its distinct ecosystem characters.

Climate and Coastal Hydrography

The average annual rainfall of this coast is 1200 mm, with a maximum during October to December (Northeast monsoon). Wind speed ranges between 2 and 12 km/h directed towards the northeast or northwest. Relative humidity is between 42 and 86%. The average tidal levels vary from 0.73 to 1.85 m for spring tides and 1.49 to 1.02 m for neap tides, with an observed maximum tidal level in Chennai of about 2.3 m. The tide is semi-diurnal. High amplitude, short waves are very common in the region; observations indicate that waves of amplitude 0.5 to 1.5 m can occur with periods of 5–10 s. Littoral drift along the coast moves sand from south to north, resulting in the formation of sandbars across the mouths of rivers entering the Bay of Bengal. Water chemistry values are shown in Table 2.

Coastal Ecosystems

This coast is lined with productive mangrove forests, estuaries, sandy beaches and lagoons.

Mangrove Ecosystem

Mangroves are present in Pitchavaram, Madras, where they are transplanted in the Adyar, Cooum and Ennore estuaries, and Ariyanguppam, where Rhizoporaceae is transplanted. In the East Coast next to the Sunderbans, Pichavaram covers an area of about 1100 ha and contains a heterogeneous mixture of mangrove elements. This area includes 51 islets, waterways, channels, gullies and rivulets and accounts for 40% of the total area of the mangrove. Of this, 50% is forest and the remaining part consists of mudflats and sandy plains. The depth of the channels of the Pitchavaram mangrove ecosystem varies from 0.5 to 1.8 m, with an average of 1 m. The tidal amplitude within the creek varies from 0.3 to 0.9 m.

About 20 species of typical mangrove vegetation are reported from Pichavaram. *Rhizophora apiculata, R. stylosa* and *R. mucronata* form the fringe community, while *Avicennia marina* dominates at the back. The plants are distributed in varying degrees of abundance in the forest and *Avicennia marina* is the most common species. *Rhizophora apiculata, Rhizophora mucronata, Bruguiera cylindrica, Aegiceras corniculatum* follow in that order. *Avicennia marina* alone constitutes nearly 30.96% of the total population followed by *Bruguiera cylindrica* (16.77%) and *Avicennia officinalis* (16.12%). The population density of other species is poor and many of the species are on the verge of extinction here. The mangrove plants such as *Kandelia candel*, *Xylocarpus granatum*, *Rhizophora stylosa* and *Sonneratia apetala* are now fast disappearing from this area. At Madras city along Adyar estuary, mangroves such as *Avicennia marina*, *Excoecaria agallocha* and *Sonneratia apetala* have been replanted. In Adyar the population density of *Avicennia marina* is 105 per ha, *Excoecaria agallocha* is 36 per ha and *Sonneratia apetala* is 3 per ha. In Cooum estuary the population density of *Avicennia marina* is less, though in Ennore, it is 486 trees per ha.

Estuarine Ecosystems

The important estuaries along this part of the coast are the Edaiyur-Sadras estuarine complex (12°33'N and 80°10'E), Araniyar (13°25'N and 80°15'E), Ennore (13°15'N and 80°19'E), Cooum (13°10'N and 80°16'E), Adyar (13°01'N and 80°17'E), Uppanar (11°42'N and 79°46'E), and Vellar (11°29'N and 79°46'E). All the estuaries are perennial and during the summer season their estuarine mouths are closed. The estuaries at Adyar, Cooum and Ennore in Madras city are highly polluted, while all the other estuaries are relatively unpolluted.

Lagoon Ecosystem

Pulicat Lake is located about 40 km north of Madras City and is the second largest brackish water lake in India (after Chilka lake). The lake is about 60 km in length and 0.2–17.5 km in breadth, covering an area of 460 km^2 at high tide and 250 km^2 at low tide. Due to deltaic deposits, the lake is extensively shallow with an average depth of about 1.5 m. The main source of freshwater is land runoff through three small seasonal rivers that open into the lake. They are Arani at its northern end, Kalangi at its mid-western side and Swarnamukhi at its mid-western end. Due to damming, freshwater flow into this lake is restricted mostly to monsoon months. The organic productivity of the waters of Pulicat lake generally ranges from 250 to 312 mg/m^3/day and the production peaks coincides more or less with that of the phytoplankton peaks. Pulicat is the third most important wetland for migratory shorebirds in India, and is especially important during the spring and autumn migratory season.

Biodiversity

The increased nutrient input into the coast due to river discharges promoted a remarkable growth of the plant and animal communities. More than 205 species of fish, 41 of zooplankton, 33 phytoplankton and 35 mangrove species are recorded on this coast. Algae and seagrasses are abundant in the lower reaches of the mangroves, where there is a narrow range of salinity fluctuations.

Fish and Fisheries

More than 250,000 fishermen live along this coast, using around 6000 small and large mechanised boats. Fishermen regard the pelagic fishery (50,000 t/yr.) to be important.

Environmental Degradation

The reduction in the flow of freshwater into the rivers (Arani, Kalangi, Swarnamughi, Ennore, Cooum, Adyar, Uppanar, Vellar and Coleroon) is leading to environmental and ecosystem changes. The freshwater flow into the estuaries has reduced by 50% compared to 1950. As it is on an erosion site, the coastal heritage area of Mahabalipuram is facing severe problems, especially near the shore temple area. Madras Marina beach is an accretion site with a mean rate of accretion of 1.7 m/year. However, north of Madras harbour, there is severe erosion, currently at the rate of 6.6 m/yr. Near to Pulicat lagoon-fishing villages, the seaside is also subjected to beach erosion, the rate of which is estimated to be 3.2 m/yr. Due to a limited freshwater supply and tidal action (39 cm), the mouth of the lagoon is silted up and closed during the summer season.

A wide variety of pollutants enter these coastal waters. Of this, the most common are sewage, suspended solids, heavy metal, oil, pesticides and pathogens. Domestic and municipal wastes are discharged directly into the coastal waters. In Chennai city alone, there are more than 4 million people living near the coast and as a result, more than 90 million litres of sewage is discharged into the coastal waters.

Impacts of Human Activities on the Ecosystem

At Pichavaram the coastal community illicitly fells trees for fire wood and house construction materials. Around 3000 cattle utilize the mangrove forest as grazing land, where they affect *Avicennia* populations. The mangrove here had a cover of about 2000 ha, but it has now been reduced to an area of about 1000 ha. The grazing pressure is also heavy during the monsoon season, when the *Avicennia marina* produces seeds and seedlings. The loss of mangrove areas in Pichavaram between 1986 and 1993 is estimated to be 0.36 km^2 (Ramachandran et al, 1998).

Being one of the four metropolitan cities, Madras has a population of 5.36 million and has rapid development along the coast. Domestic sewage, along with industrial effluents, are discharged into the estuaries of Adyar, Cooum, Ennore and Buckingham canal. Residential, industrial, commercial, institutional and recreational wastes discharge degradable and non-degradable pollutants. Studies of the mass transport of pollutants revealed that the Adyar transported a maximum of 9.82 kg/d of ammonia;

61.62 kg/d of nitrate; 8.93 kg/d of phosphate, 0.045 kg/d of cadmium, 0.646 kg/d of lead and 8.79 kg/d of zinc to the coastal waters. Similarly, the Cooum river transported a maximum of 29.35 kg/d of ammonia, 14.62 kg/d of nitrate and 5.36 kg/d of phosphate, 0.01 kg/d of cadmium, 0.219 kg/d lead and 1.93 kg/d of zinc (Gowri and Ramachandran, 1999).

The Buckingham canal was built parallel to the Tamil Nadu and Andhrapradesh coasts as a salt water navigation canal stretching to about 315 km north and about 103 km south of Madras city. Since the canal is only 1 km inland from the coast, there are numerous points where the sea flows into the canal. Within the city, the canal links the mouth of the Cooum and Adyar over a stretch of about 8 km. The canal has become a public health hazard on account of the sewage and garbage dumped into it, and it is also heavily silted. Within the city, the embankments of the canal have been badly damaged and a slum population of about 50,000 now live on the riverbanks. Madras harbour is very near to the Marina beach. The harbour activities disturb and change the current pattern, which leads to shoreline changes. Recent developments in tourism with construction of cottages within the mangrove forest at Pichavaram and sandy beaches are damaging the natural ecosystem.

Over-harvesting of fish resources including berried fishes, prawns and crabs are reducing the stock availability of this area. Aquaculture industries have been established in many places along this coast. This has reduced the wetlands to 35% of the area they occupied in 1987. Saltwater intrusion to around 3 km inland is also reported in coastal towns, and this affects the agriculture.

There is a long stretch of sandy beach which attracts tourists and also serves as a recreation site for local people. The little towns of Pondicherry, Mahabalipuram (a World Heritage site) and Kovalam beaches are also important tourist areas. The status of the beaches is clean except for some sewage discharges and solid waste dumping.

GODAVARI AND KRISHNA DELTAIC COAST

Geological Features

The Godavari river originates in Nasik District of the Western Ghats in Maharastra, flowing eastward, covering 300,000-km² drainage along the East Coast of India (latitude 10°15′N and longitude 82°05′E). The maximum discharge of the river is 3500 m³/s. The upper Godavari river basin is occupied by the Deccan traps of late Cretaceous–Eocene age. The traps yield minerals such as magnetite, zircon, rutile, apatite and chlorite. The deltaic region mostly contains black cotton soil with deep extensions in the subsurfaces. The texture of the soil is "silty clay" with 50–60% clay. The huge sediment discharges generated by erosion and carried by the fluvial system eventually contribute to the build-up of the delta. The Godavari drainage basin is second largest in India, and discharges correspondingly heavy sediment loads into the Bay of Bengal.

Climate and Coastal Hydrography

Annual rainfall is around 1200 mm, and relative humidity is around 60%. Wind speed ranges between 1 and 7 km/h. During monsoon seasons, the near-shore currents flow towards the north with an estimated speed of 1 km/h. Temperature is between 22.4°C and 38.4°C, and water chemistry values are given in Table 2. The Krishna - Godavari basin has rich hydrocarbon resources.

The Mangrove Ecosystem

The influence of seawater in the Krishna–Godavari delta and estuary extends up to a distance of about 40 km from the sea, and the Godavari and Krishna basin annually transports 27.5×10 tonnes of solute load from the Indian sub continent. Thus the delta has rich estuarine and mangrove ecosystems. The total area covered by the mangrove complex is around 575 km² containing 32 species of mangroves and associated plants belonging to 26 genera and 18 families. Two species *Scyphipora hydrophyllacea* (Godavari delta) and *Aegialites rotundifolia* (Krishna delta) are rare mangrove species of the Indian mainland. The most common species are *Avicennia marina*, *Excoecaria agallocha* and *Sonneratia apetala*. High densities of mangroves have been observed in the sediments with high silt fraction. The total mangrove forest of the Godavari basin covering Coringa (Fig. 4), Balusutippa and Pandi zones extends to an area of 316.21 km² (Raman, 1995).

Biodiversity

There is a wide diversity of marine invertebrates in this region. Notably, 18 species of prawns belonging to 8 genera and 3 families occur here, 11 species of crabs belonging to 10 genera and 3 families, and 106 finfish species belonging to 78 genera and 52 families. Several species of birds are also seen in this mangrove ecosystem: *Podiceps ruficollis*, *Pelicanus philippensis*, *Anhinga rufa*, *Palacrocorax niger*, *Ardea cinerea*, *Egretta* sp., *Ardeola greylil*, *Ibis leucocephalus*, *Pseudibis papillosa*, *Anser indicus* and *Anas* sp. (Ramanamurthy and Kondala Rao, 1993; Raman, 1995).

Fish and Fisheries

Both the fin fish species and the shrimps support major fisheries. An estimated 1027 mechanised trawlers operate from Kakinada Fisheries harbour. Of these, approximately 180 to 200 boats land their catches locally, including daily fishing catches, depending on the resources available. Landings have shown that there is an increase in the fin fishery but the prawn catches show a decreasing trend.

Fig. 4. IRS LISS-III image showing part of the Coringa mangroves in the Godhavari delta. Healthy mangrove appears in different shades of red, coastal vegetation is dark brown, aquaculture areas are blue and occur in regular patterns. The discharge of sediment load by the Gauthami–Godhavari river into the Bay of Bengal can also be seen.

Table 5
Fish and fisheries of the southeast Indian region

Groups of fishes	Landings (tonnes)		
	1994–1995	1995–1996	1996–1997
Total fishery	7188.37	6702.03	7647.37
Marine fishes	5600.79	5145.96	6146.95
Prawns	1587.88	1556.07	1500

Table 5 shows fish catches for the entire coast of southeastern India.

The fishing population in the Godavari and Krishna delta is around 150,000, of which a total number of 42,255 individuals are engaged in the actual fishing operations. Some of them are full-time (26,117 persons) and the rest are part-time (16,138).

Environmental Degradation

In Nizampatnam, there is little industrial pollution. The water quality data support this. However, most of the pollution in this region is due to discharges from the brackish-water aquaculture farms, which have developed very recently. Between Machilipatnam and Godavari estuaries, many other industries have also been established. These are sugar, distillery and fertilizer manufacturing industries. However, the water quality data shows that this region is still relatively unpolluted.

The Godavari and Krishna mangrove ecosystems have, like other biotopes, been experiencing human impact in the form of resource utilization and alteration. Mangrove exploitation has been indiscriminate, with excessive deforestation, injudicious harvesting of juveniles and overseining of mangrove waters for finfish and shellfish seeds for aquaculture practices. Further, the current boom of aquaculture practice in brackish-water habitats has been leading to irrevocable alterations in these habitats. Also, natural calamities take a heavy toll on these systems. For example, the 1977 Diviseema cyclone cleared mangrove forests in several locations (e.g. Sorlagondi, Palakayatippa etc.) in the Krishna mangrove belt. The utilization of Godavari and Krishna mangrove resources (both forest- and fishery-based) and the use of these habitats for alternative activities (particularly for brackish-water culture) leads to conflict in the uses of these systems.

It is estimated that nearly 150 million seeds of *Penaeus monodon* per annum are seined each year from the Godavari and Krishna mangrove ecosystems. Further, the unwanted organisms in these prawn seed collections are discarded improperly by the seed collectors. This indiscriminate scooping and sieving of mangrove waters for the seeds of some economically important finfish and shellfish and improper disposal of uneconomical young of other organisms must be checked through proper legislation and public education. If not, the present fishery may lead to a failure because of low recruitment.

The most important natural process relates to the development of Kakinada sand spit, which has brought various changes to areas surrounding the Bay. The bathymetry of Kakinada bay, the coastline from Kakinada to Uppada and the lives of the people living there have been affected considerably because of the development of the spit, which has, in turn, brought about a number of other morphological changes including depositional and erosion features on the beach.

Impacts of Human Activities on the Ecosystem

Conversion of mangrove forests into aquaculture industries is the major deforestation activity along the coast. Thirty percent of present aquaculture ponds have been converted from mangroves. The wild seed collection activities along the wetlands of this coast are also damaging the benthic habitat. On this coast, there are around 163,622 households, plus 260 hotels, 52 lodges and 30 restaurants. These generate considerable domestic sewage. The Andhra Pradesh Housing Board is also implementing their construction programmes nearer to the coastline. Urbanisation activities along Kakinada Bay and the Vishagapatnam coast have changed the coastal water quality significantly. Several waterways (Coringa, Gaderu, Kakinada Bay, and Gautami Godavari) remain the prime mode of transport.

Kakinada is a major port which, within Andhrapradesh, occupies second place only to Vishagapatnam. In recent years, Kakinada port is gaining prominence owing to development of the area in vital sectors such as agriculture, aquaculture and industries. At present, the port can cater for vessels up to a capacity of 30 to 40 thousand tonnes, and the port was recently converted into a deepwater harbour with proposals to expand the facilities further.

There are about 15 major industries generating 350,000 m^3 of effluents per day which are discharged into the coastal waters. The wastes released by the industries are mostly organic in nature and degradable, but increasing nutrient level at certain areas may be a problem resulting in eutrophication.

Due to the increased demand for wild seed stock for prawns, which are abundantly available at the mangrove, the wild seed collection is a major fishing activity in mangrove waters. This has reduced the natural prawn population in the mangrove environment significantly. The development of aquaculture industries has reduced the wetland extent (30%) and increased the nutrient level in the surrounding water. Concentrations of nitrites (1.32 μg/l), nitrates (21.7 μg/l) and phosphates (8.4 μg/l) in creeks near aquaculture ponds have confirmed this.

There are about 11 locations of importance to tourism in the Godavari–Krishna delta. They are mostly religious places with ancient temples, attracting devotees from all parts of India. Around 4,000,000 tourists visit the places every year. Impacts of tourism activities are seen in the solid waste dumping and also changes in the water quality.

Suggestions and Recommendations

The primary factors driving coastal and ocean pollution and degradation are the increase in populations, increase in demands on natural resources, and the improper management of wastes which enter the oceans in unacceptable amounts. The increasing human population alone is responsible for any rise in the nutrient levels in the water. Nutrients and metals form two of the major pollutant groups having a lasting effect on the aquatic ecosystem. Both are of considerable concern and of a global nature.

Many coastal communities discharge their wastes directly to the coast. These wastes cause measurable changes in the abundance, diversity and health of plant and animal communities, biological processes such as growth and reproduction and ecosystem functions like food web, energy transfer, etc. Domestic sewage results in eutrophication through overloading the marine environments with nutrients, and introduces pathogens and toxic matter. This disturbs the balance between the different species, often causing dramatic increases in populations of nuisance species (algae) at the expense of prized species (lobster and prawns).

The anthropogenic and natural activities are also changing the landscape of the coastal habitats and its ecosystems. Due to inadequate treatment facilities, the domestic sewage and sullage are directly discharged into the coastal waters. Cities, bordering the coasts, release a considerable quantity of pollutants into the ecosystem, well above the permitted levels, which lead to the deterioration, decay and ultimate destruction of the coastal ecosystem. To avoid this, treatment facilities for the wastewater should be created and only the treated effluents should be allowed to flow into the coastal waters.

The degradation of coastal ecosystems has been further intensified by the local fishing communities for their basic livelihood. Regeneration of species in suitable areas is one approach for regenerating the ecosystem. The development of extensive mariculture systems for economically viable species as an alternative livelihood for the fishing community may help in reducing their dependence on the ecosystems.

There is a need for training and awareness programmes involving local communities. This will lead to better conservation, and could be developed through volunteers implementing protective measures from conservation organizations, NGOs, water quality monitoring groups and citizens' groups.

The processes governing inputs, mixing, dispersion, transport and removal of pollutants in coastal marine environment is not understood fully. If the resources that are available in the coastal and marine areas are to be utilised on a sustainable basis, due attention has to be given to the physico-chemical and biological characteristics of the coastal waters. Hence, there is an urgent need to monitor the coastal waters in order to evaluate the impact of developmental activities on the coastal water quality, and to take measures against pollution, to reduce the adverse impacts and coastal environmental degradation.

Many ancient civilisations and societies survived successfully as waste flowed into the rivers, as tides helped to flush out the sewage as well as other community wastes, while sustaining the economic activities of the population living in the coastal belt. Today, the population is growing beyond sustainable limits and, along with other activities, degradation of the coast is increasing out of proportion. Hence there is an urgent need for an holistic approach to integrated coastal and marine area management in this region to control marine pollution, prevent habitat destruction and sustain the economic development and ecological balance of the coastal ecosystem.

REFERENCES

Asir Ramesh, D. and Kannupandi, T. (1997) Recent changes around three decades of Palk—a comparative study of the past records and researches. Regional workshop on the Conservation and sustainable Management of Coral reefs. BOBP and MSSRF, Chennai, C-123.

Dorairaj, K. (1998) Conservation and sustainable management of the marine living resources of Gulf of Mannar Marine Biosphere Reserve. In *Proceedings of the Symp. on Biodiversity of Gulf of Mannar Marine Biosphere Reserve.* MSSRF, Chennai, pp. 22–32.

Gopinadhapillai, C.S. (1969) The distribution of corals on a reef at Mandapam (Palk Bay), South India. *Journal of the Marine Biological Association of India* 11 (1), 62–72.

Gopinadhapillai, C.S. (1973) Coral resources of India with special reference to PalkBay and the Gulf of Mannar. In *Proceedings of the Symposium on Living Resources of the Seas around India*, pp. 700–705.

Gowri, V.S. and Ramachandran, S. (1999). Coastal pollution by Adyar river discharges. In *Proceedings of the 8th National Symposium on Environment.* IGCAR, Kalpakkam, pp. 9–11.

Kaliaperumal, N. 1998. Seaweed resources and biodiversity values. *Proceedings of the Symposium on Biodiversity of Gulf of Mannar Marine Biosphere Reserve.* MSSRF, Chennai, pp. 92–101.

Loveson, V.J., Victor Rajamanickam, G. and Chandrsekar, N. (1990) Environmental impact of microdeltas and swamps along the coast of Tamilnadu, India. *Sealevel Variation and its Impact on the Coastal Environment*, ed. Victor Rajamanickam, Tamil University, Tanjore. pp. 159–178.

Ramachandran, S. (1997) Case Studies on Coastal Fragile areas. Part 1. East Coast of India. Report submitted to Central Pollution Control Board, New Delhi. 261 pp.

Ramachandran, S., Sundaramoorthy, S., Krishnamoorthy, R., Devasenapathy, J. and Thanikachalam, M. (1998) Application of remote sensing and GIS to coastal wetland ecology of Tamil Nadu and Andaman and Nicobar group of islands with special reference to mangroves. *Current Science* 75 (3), 236–244.

Raman, A.V. (1995) An assessment of the ecological importance of Mangroves in Kakinada area, Andhra Pradesh, India. Annual Report. EC Int. Sci. Cooperation Project.

Ramanamurthy, K.V. and Kondala Rao, B. (1993). Studies on Mangrove Ecosystems of Godavari and Krishna estuaries Andhra Pradesh, India. UNESCO Curriculum workshop. Andhra University.

THE AUTHOR

Sundararajan Ramachandran
*Institute for Ocean Management,
Anna University,
Chennai 600 025, India*

Chapter 64

SRI LANKA

Arjan Rajasuriya and Anil Premaratne

The Indian Ocean island of Sri Lanka has a coastline of about 1,585 km excluding embayments, offshore islands, inlets and lagoons. Within this are tens of thousands of hectares of mangroves, salt marshes, coastal dunes, barrier beaches and spits, lagoons and estuaries. The climate is mainly driven by two alternating monsoon seasons, which generate substantial longshore drift of sediments around the coast. Offshore and in lagoons there are large quantities of seagrasses, and about 2% of the coastline contains fringing coral reefs, in addition to which are numerous patch reefs, generally 15 to 20 km offshore.

In the early 1990s, marine fisheries provided full or part-time employment to nearly 100,000 individuals and contributed 1.9% to GDP. Nearly 65% of the animal protein and about 13% of the total protein consumed in Sri Lanka is provided by marine fisheries, which yield about 242,000 tons annually. Coastal fish production reached its highest level in 1994 but has fluctuated since because of both over-exploitation and reduced effort in some areas due to the conflict between separatists and government. Offshore fish production shows similar fluctuations. Although fishing is the only source of income for about 80% of the country's fishermen, doubts have been raised whether the industry can offer adequate opportunities to new entrants to the labour force by the year 2000, though some hitherto under-exploited resources have the potential for increased production. The marine ornamental fishery sector is valuable, as is the chank fishery in the north which supplies mainly Bangladesh and India. A sea cucumber fishery exploits 13 species for export to far eastern countries, but following their collection using scuba diving in some parts, the resource was wiped out after two years.

Garbage and waste disposal systems are poor; currently trends are of increasing concentrations of urban wastes. Reclamation of wetland habitats is also an increasing problem which has led to rapid storm-water discharges, with sediment, domestic and industrial waste. Coastal erosion is a major problem and has resulted in damage of and loss to houses, hotels and other coastal infrastructure. Non-point agricultural pollution is a major problem; some farming districts use over 120 kg of fertilizer per ha, and total pesticide use is 2800 tons annually of which about 25% ends up in the sea. Shrimp farming has converted substantial mangrove areas. However, a framework for coastal zone management exists, and in several areas improvements have been made.

Seas at The Millennium: An Environmental Evaluation (Edited by C. Sheppard)
© 2000 Elsevier Science Ltd. All rights reserved

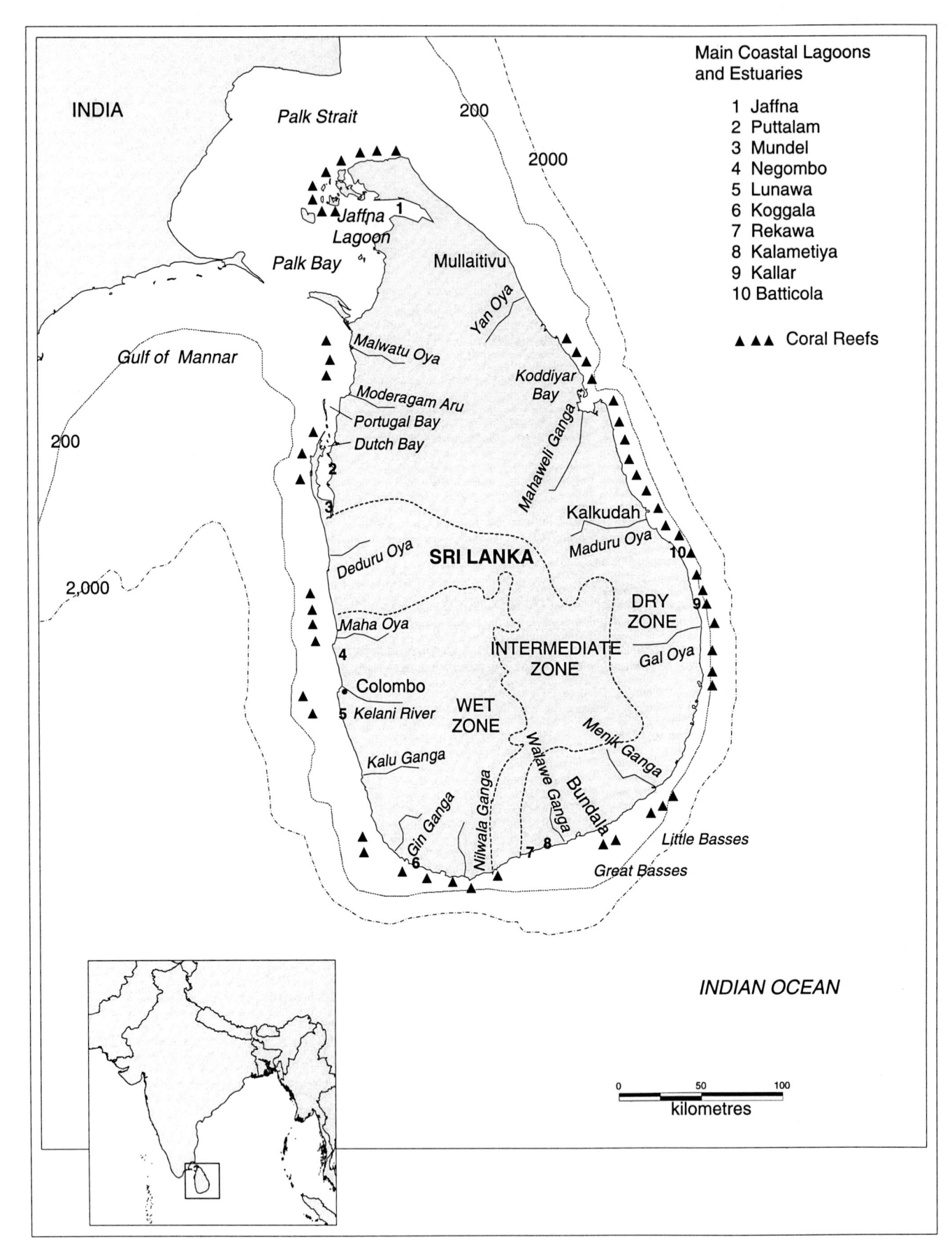

Fig. 1. Map of Sri Lanka.

THE GEOGRAPHICAL SETTING

Sri Lanka is located in the northern Indian Ocean between 5°54′ and 9°52′N, and 79°39′ and 81°53′E, and is separated from India by the Palk Straight which is about 40 km wide. Its land area is about 65,610 km² (Fig. 1). The island is part of the Indian shield and, geologically, nine-tenths of the country consists of crystalline, non-fossiliferous, Precambrian rocks. The rest, mainly the northwestern part, contains Mesozoic (Jurassic), Tertiary (Miocene) and Quaternary sedimentary formations (Cooray, 1984).

The country has a coastline of about 1585 km excluding embayments, offshore islands, inlets and lagoons (CCD, 1990). Much of it contains sandy beaches interspersed by rocky headlands and coastal lagoons and estuaries. Sri Lanka has a continental shelf area of about 28,000 km² which is narrower and deeper beginning in the southwest to Mullaitivu on the East Coast, but is wider and shallower (30 and 60 km) in the north and northwest, being continuous with the shelf of southern India (Swan, 1983). The EEZ of the country covers about 230,000 km² of ocean (Baldwin, 1991).

Sri Lanka has 12,189 ha of mangroves, 23,819 ha of salt marshes, 7606 ha of coastal dunes, 11,800 ha of barrier beaches and spits, 158,017 ha of lagoons and basin estuaries and 9754 ha of other marshes (CCD, 1986).

The true extent of coral reefs on the continental shelf has not been mapped, but Swan (1983) has estimated that approximately 2% of the coastline contains fringing coral habitats.

The country has a population of about 18 million, most of whom are in the west and south. Fifteen of the country's 25 principal urban areas are coastal and 34% of the total population are coastal dwellers occupying 24% of the total land area (Baldwin, 1991; Olsen et al., 1992).

SEASONALITY, CURRENTS, NATURAL ENVIRONMENTAL VARIABLES

Climate and Rainfall

The climate of Sri Lanka is greatly influenced by the surrounding ocean. Atmospheric conditions in the Bay of Bengal often affect climate and sometimes create gale-force winds and torrential rains mainly to the eastern and northeastern parts (Cooray, 1984). The climate is mainly driven by two monsoon seasons, namely the Southwest and Northeast Monsoons. Rainfall is variable and climatic zones have been categorized according to rainfall. These are the Dry zone, Wet zone and Intermediate Zone (Baldwin, 1991). The high mountains are included in the Wet zone. The highest point in the central hills is over 2500 m.

Climatologists have divided Sri Lanka's climate into five seasons. They are the Convectional–convergence period (March to mid April); Pre-monsoonal period (mid April to late May); Southwest monsoon period (late May to late September); Convectional cyclonic period (late September to late November) and the Northeast monsoon period (November to March) (Baldwin, 1991). Rainfall is variable; some places along the windward slopes of the central hills receive about 2500 mm annually, much of which falls during the southwest monsoon. Mean annual rainfall is about 2000 mm. The rest of the island receives about 1500 mm of rain annually. Sri Lanka receives a total of about 12 million cubic meters of rainwater annually (Baldwin, 1991).

The Southwest Monsoon brings much more rain and windy periods than the Northeast Monsoon. The latter commences with rain and strong gusts of wind in October–November but wanes in late January and the sea conditions improve after February along the eastern seaboard. In contrast, the Southwest Monsoon has several periods of intense rain and strong winds which continue from May to end of September.

Storms and cyclones are rare but may be experienced during the convectional cyclonic period from late September to late November. Storm surges are occasionally experienced along the southwest coast. A cyclone which affected the east coast in 1984 damaged the shallow coral reefs around Kalkudah and Passikudah near Batticoloa, and also caused coastal erosion problems.

Sri Lanka has 103 river basins that discharge rainwater through several major rivers and many smaller rivers and canals. Most are concentrated along the western and southern coasts within the wet zone. However, the Mahaweli ganga, which originates in the central highlands is one of the largest rivers that runs through the dry zone. It discharges about 11,016 million m³ of water annually at Trincomalee in the northeast. Next in size is Kalu ganga on the West Coast, which has an annual discharge of about 7862 million m³ (Table 1).

Table 1
Mean annual discharge of major rivers and streams

Coastal area	Name of river	Mean annual discharge (millions of m³)
NW	Malwatu oya	568
NW	Moderagam aru	161
NE	Yan oya	300
NE	Mahaweli ganga	11016
E	Maduru oya	8051
E	Gal oya	1250
SE	Menik ganga	486
SE	Walawe ganga	2165
S	Nilwala ganga	1104
SW	Gin ganga	1903
W	Kalu ganga	7862
W	Kelani ganga	5474
W	Maha oya	1608
W	Deduru oya	1608

Source: National Atlas of Sri Lanka in Baldwin (1991).

Currents and Tides

Regional currents driven by the monsoons in the northern Indian Ocean dominate the oceanic waters (Table 2). In addition, the equatorial oceanic zone south of Sri Lanka is characterized by the North Equatorial current, which flows to the west during the Northeast Monsoon and the Equatorial Counter current, which flows eastwards and intensifies during the Southwest Monsoon (Swan, 1983). The maximum intensity of currents around the southwestern and southern coastal and offshore areas occurs during the Southwest Monsoon whilst the east coast and offshore areas experience strong currents during the Northeast Monsoon. The currents along the east coast are stronger than those along the western coast; the strongest have been recorded along the southeastern coast with a velocity of about 1 m s^{-1} or more near the Little Basses Reef between late October to early January (Swan, 1983).

Stronger currents, however, have been reported from the Pamban Pass along the Indian Coast of the Adam's Bridge where velocities between 2.5 and 3 m s^{-1} are reported to be common during the full force of the monsoons (Swan, 1983). Coastal currents are subjected to various factors, which include seabed topography, tides, wind, and the shape of the coastline.

The predominant drift in the coastal waters is from south to north (Swan, 1983), except during the early phase of the Northeast Monsoon where the current may be reversed. During this period, winds are felt on the northwestern parts of the country and the adjacent waters as well, due to the absence of a blocking central massif. These currents cause substantial longshore drift, causing annual cycles of deposition and removal (Fig. 2).

Tidal range is low, between 0 and 1 m (Swan, 1983). Coastal lagoons are generally shallow with a mean depth range of 1–10 m (Wallberg and Johnstone, 1995). Salinity in the coastal waters is 35 ppt. but in lagoons it varies depending on the rainfall and on evaporation during the dry seasons. Studies carried out in Puttalam lagoon on the northwest coast reveal that the salinity can reach 60 ppt. in areas where there is very little freshwater input and high evaporation rates (Wallberg and Johnstone, 1995).

Fig. 2. Large, seasonal sand deposits on southeast coast (Yala) (photo C. Sheppard).

Temperature

Seawater temperature is usually 26–30°C, generally nearer the lower end of this range during the monsoon period, possibly due to oceanic waters being pushed onto the continental shelf or to minor upwelling caused by monsoon currents. No major upwelling currents have been recorded around Sri Lanka.

Sediment

Sediment input during the monsoons affects water clarity. The major river systems bring large amounts of suspended particles into the sea, particularly along the west and the southern coasts. On the east coast, water clarity is better due to lower rainfall. During non-monsoon periods, visibility varies from 5 to 30 m, depending on the location and the condition of the sea. Generally, visibility improves with increasing distance from shore.

COASTAL AND MARINE SHALLOW-WATER HABITATS

Lagoons and Basin Estuaries

In Sri Lanka most of the so-called lagoons are in fact estuaries as they are permanently connected to the sea (Fig. 3), and several of these grade inland changing to more freshwater conditions (Fig. 4). Some examples of basin estuaries are Puttalam Lagoon, Chilaw Lagoon, Negombo Lagoon, Batticoloa Lagoon and Jaffna Lagoon. Examples of true lagoons are Mundel Lake, Lunawa Lagoon in the West Coast, Koggala, Rekawa and Kalametiya in the south and Kallar Lagoon in the East Coast. The total extent of basin

Table 2

The average and maximum velocities of currents in the coastal waters around Sri Lanka

Location offshore	Average velocity (m/s)	Higher velocity (m/s)
Colombo	0.15–0.25	0.77
Galle	0.52	0.88
Weligama to Hambantota	0.25	0.50–0.90
Hambantota to Kubukkanoya outfall	0.25	0.50–1.00
Little Basses Reef	0.50	1.00–2.60
Trincomalee	0.25	1.50
Palk Bay	0.15	0.50
Pamban Pass	–	2.50–3.00

Source: Bay of Bengal & West Coast of India Pilots and Gerritsen. In Swan (1983).

Fig. 3. Flooded mangrove area south of Colombo (photo C. Sheppard).

Fig. 4. 'Lagoon' near its inland extremity, mainly fresh water (Muthurajawella, west coast) (photo C. Sheppard).

estuaries and lagoons has been estimated to be about 158,017 ha (Baldwin, 1991). Lagoons are important for coastal fisheries production, for birds including migratory species, and for biodiversity. Lagoons and Basin estuaries are utilized mainly for artisanal and small-scale fisheries. Some, such as Puttalam Lagoon have been subjected to a rapid increase in fisheries activities which may be unsustainable (Dayaratne et al., 1997). Coastal lagoons and estuaries have also been subjected to increasing levels of pollution from fishing boats and urban waste. A paper factory on the east coast has caused extensive water pollution problems in the Valachchenai Lagoon.

Mangroves

Mangroves are generally restricted to the banks of lagoons and estuaries. They are not common along coastal stretches exposed to strong wave action such as the west, south and eastern coasts, though they are very common in the lagoons and estuaries of these areas. Mangroves are found exposed to the open sea in the northwest and north where the continental shelf is wider, and where there is less wave impact on the coast. Although mangroves are common in coastal lagoons and estuaries, the largest mangrove stands are in the Puttalam Lagoon–Dutch Bay–Portugal Bay complex. The second largest is in Batticoloa Lagoon, followed by the Koddiyar Bay area in Trincomalee (Pinto, 1986).

The total extent of mangroves has been estimated to be about 12,189 hectares. Twenty-three species of mangroves and mangrove associates have been recorded for Sri Lanka (Baldwin, 1991). None of them are endemic, though the west coast mangroves are richer in species diversity than those in the east (Swan, 1983). Mangroves are important for their contribution to the production of coastal fisheries, protection of the banks of lagoons, estuaries and parts of the coastline, for trapping sediment and for avifauna. Numerous large expanses of mixed grasses, fringed with mangroves, extend several kilometres inland and are rich habitats for birds (Fig. 5).

Fig. 5. Expanse of grassy but saline land at sea level, fringed with mangroves. Bird reserve, near Tangalla, south coast (photo C. Sheppard).

This habitat is also important for aquaculture. Mangroves have been subjected to severe damage due to the expansion of shrimp farms. This has resulted in the loss of mangroves in the southern part of Puttalam Lagoon and Mundel Lake. Pollution resulting from shrimp farms and urban waste has contributed to declines in natural fisheries and productivity (Corea et al., 1995). Wetlands adjacent to mangrove areas have been severely damaged and have resulted in the loss of habitat for birds and other natural fauna (De Silva and Jakobsson, 1995). Mangroves are also utilized for firewood and for building temporary huts (Dayaratne et al., 1997).

Seagrass Beds

Seagrasses are commonly found in the shallow protected areas along the coast. They are, however, more common and extensive within the coastal lagoons in the dry zone areas. Large amounts occur in the Puttalam Lagoon–Dutch Bay–Portugal Bay system and northwards to the coastal lagoons in the Jaffna Peninsula. Large patches also cover parts of the Negombo and Batticoloa Lagoons. In the south coast small isolated patches occur in the leeward sides of fringing reefs. Seagrass beds are important nursery grounds for fish and other organisms and contribute to coastal fishery production. In the northwest and the north, large patches of sea grasses are the habitat of the endangered Dugong (*Dugong dugon*).

Salt Marshes

Salt marshes are mainly found in the coastal flats where there is hypersalinity. Extensive areas are found in the deltaic lands of the north and northwest and in sedimented lagoons in the southeast. These areas are in the dry zone, so receive comparatively little rainfall and have a very high rate of evaporation.

Five species of Halophytes in three genera that can tolerate salinity or exclude salt have been recorded. In addition, four species of grasses also occur in this habitat. In watery depressions, two species of rushes and two species of seagrasses have been recorded, particularly where there are strong tidal currents. Few specimens of true mangrove trees may occur in the lower areas of such habitats (Swan, 1983). Saltmarshes are important habitats and are also favoured by migrating and indigenous birds. Saltmarshes in the southeast are also the habitat of the marsh crocodiles, which are quite abundant in marshes near Hambantota and Bundala areas.

Reefs

Three types of reef habitats have been identified in Sri Lanka, namely true coral reef habitats, sandstone/beachrock habitats and crystalline rock habitats (Swan, 1983; Rajasuriya and De Silva, 1988; Rajasuriya et al., 1995; Rajasuriya and White, 1995; Rajasuriya et al., 1998). All three distinctly different habitats may be mixed together.

There are extensive areas of true coral habitats offshore in northwest Sri Lanka within the Gulf of Mannar region. Most other types occur as fringing reefs in the southwestern, eastern and northern coastal waters. Swan (1983) stated that about 2% of the coastline contains fringing coral reef. In addition there are patchy coral habitats offshore generally at a distance of about 15–20 km from the shore, at an average depth of 20 m to the west and east of the island. Sandstone/beachrock and rocky reef habitats are common as well. Whilst sandstone/beachrock habitats are widespread, rocky habitats occur mainly from the west coast beginning south of Colombo, extending through the southern areas to Trincomalee District in the northeast (Rajasuriya et al., 1995).

At present, 183 species of stony corals divided among 68 genera have been identified from Sri Lanka (Rajasuriya and De Silva, 1988; Rajasuriya, 1994). More than 300 reef-associated fish species from 51 families have been recorded, including 35 species of butterfly fish and three species of large angelfish. Six species of spiny lobsters have been recorded. The well-being of the demersal fisheries and coast protection are important services of reefs. Coastal reefs are also important for the tourist industry although the share of diving tourism in the tourism market is quite low compared to countries such as the Maldives that cater to large-scale diving tourists. Coral reefs were severely damaged as a result of a massive coral bleaching event in 1998. Recovery has been extremely poor among shallow water coral habitats (Rajasuriya et al., 1999; Wilkinson et al., 1999).

Sea Turtles

There are five species of sea turtles recorded for Sri Lanka. The common species are Greens (*Chelonia mydas*) and Olive Ridleys (*Lepidochelys olivacea*); the rarest is the Leather Back (*Dermochelys coriacea*). They are hunted for their meat and the eggs are taken for consumption. Gillnets and setnets cause mortality (De Bruin et al., 1994). More recently, with a greater sense of awareness, the taking of turtles for meat has been reduced, particularly in the west and southern coastal belt. The Turtle Conservation Project in the southern coast of Sri Lanka highlighted the tortoiseshell industry in Sri Lanka in 1994, which prompted the UK based Environmental Investigation Agency (E.I.A.) to produce a film that, in turn, prompted the Sri Lanka authorities to curtail this industry. However, there are still establishments that sell these products, although at a much lower level (Richardson, 1997).

Marine Mammals

There are 27 species of marine mammals recorded for Sri Lanka. These include the dugong, 10 species of dolphins, 10 species of small whales and six species of large whales including the Blue and Sperm whales (Leatherwood and Reeves, 1991). While whales and dolphins are found around the country, dugongs are rare and found in the shallow seagrass habitats in the Gulf of Mannar and Palk Bay areas. The Indian Ocean is a declared whale sanctuary which affords protection to large marine mammals only, whilst dolphins and dugongs remain at risk. Gill nets are a danger to both, and both are actively hunted in some parts of the country. Dayaratne and Joseph (1993) reported that about 5000 dolphins are landed annually in Sri Lanka. All marine mammals are protected by law through the Fauna and Flora Protection Ordinance and the Fisheries and Aquatic Resources Act of Sri Lanka, but actual protection is difficult due to the absence of policing and sea patrols.

NON-LIVING RESOURCES

The continental shelf around Sri Lanka is covered by detrital matter containing high proportions of biogenic materials such as shell and coral fragments, worm tubes, echinoderm body parts and foraminifera. Inorganic components include quartzose sand, coarse sand and bare rock (Swan, 1983). Sand is abundant but larger deposits are found towards the northern areas both off the west and east coasts. Sand accumulation is greater in the Gulf of Mannar, Palk Strait and Palk Bay due to detritus swept northwards along the west coast of Sri Lanka and southwards along the southeast coast of India.

Holocene sediments 3–5 m thick cover the shelf in the northwestern coast north of Trincomalee. Nearshore and seawards of the 30 m isobath, deposits are less than 1 m thick. Terrigenous silt and fine sand rich in minerals occur to a depth of about 8 m. Beyond 8 m to a depth of 15 m terrigenous sands occur with a strong biogenic component. Between 15 and 30 m isobath coarse sands are found, while beyond the 30 m isobath to a depth of 60 m mud and silty shell debris prevail (Meyer, 1979; Swan, 1983).

Mineral sands in the coastal areas are found mainly near estuaries, bays and raised beaches. These contain a high proportion of ilmenite, rutile, zircon, monazite and garnet. An estimated 4 million tons of raw sand containing nearly 70–80% of ilmenite, 8–10% rutile and 0.3% monazite and 1% sillimanite is found in the northeastern coastal belt (Fernando and Seneviratne, 1990). A smaller deposit of some 250 metric tonnes containing ilmenite has been identified about 70 km south of Battiicoloa. Sri Lanka exports ilmenite to Japan and Europe (Baldwin, 1991). In the mid 1980s NARA discovered a large deposit of monazite extending offshore about 60 km south of Colombo.

Peat is the only mineral which can be used as fuel. Deposits estimated to be about 50 million tons with an average thickness of 3–4.5 m occur in the swamps south of Negombo Lagoon (Baldwin, 1991).

Miocene limestone beds used for the production of cement are found in the Jaffna Peninsula and the northwestern coastal belt. The Jaffna deposit alone is estimated to be well over 10 million metric tonnes. Inland coral deposits are mainly in the south and southwestern coastal regions. Similar deposits occur in the Jaffna peninsula under the Miocene lime (Swan, 1983). Southwestern deposits are used mainly for the production of lime utilized in building construction.

Extensive shell deposits are found in the northwest and southeastern parts of the island. The shell deposit in the Hambantota District is estimated to be about 1 million metric tonnes and has a calcium carbonate content of nearly 98% Smaller deposits occur in isolated patches near some coastal lagoons (Baldwin, 1991).

MARINE RESOURCES USE AND POPULATIONS AFFECTING THE AREA

The civilization of Sri Lanka dates back more than 25 centuries. The ancient inhabitants of the country appear to have made little use of its coastal lands, so Sri Lanka's coastal zone was partially free of any stress conditions imposed by human activities. The early civilization was mostly inland, and the coastal zone was considered as waste lands, particularly the mangroves, saltmarshes and associated habitats.

With the advent of the Portuguese in 1505 followed by the Dutch in 1658 and the British in 1796, a profound change in man's influence on the island's coastal zone and attitudes towards it occurred (Wickremaratne, 1985). The rich natural resources attracted the attention of the invaders and many infrastructure facilities were established in the coastal zone to further their own economic necessities. With these developments the indigenous population began to migrate from the traditional inland areas to the coastal zone. Population centres were established along the coast with the development of more maritime activities, resulting in more intensive use of the resources within the coastal zone and adjacent waters.

After independence in 1948, more and more people were attracted to the coastal cities, so the pressure on the resources of the coastal areas began to increase proportionately, creating new stresses. Industrialization in the late 1950s, development of coastal fisheries and the initiation of the beach-based tourism of the late 1970s imposed further constraints on the coastal environment.

There are examples of fisheries harbours that have silted up, while structures designed to keep river outfalls open have sometimes had the opposite effect. There are instances of shore protection methods that have failed or merely transferred the problem to adjacent areas of the coast; flood protection schemes that have given rise to unintended salinity problems in paddy lands, and unplanned construction of tourist hotels that have led to changes in the coastline and which have had an adverse impact on the coastal fishing population.

Marine Fisheries

The ocean resources are used mainly by the fisheries industry (Table 4). These include offshore fisheries (pelagic fisheries) and nearshore fisheries (semi-pelagic and demersal). In the early 1990s the marine fisheries provided employment full-time to about 80,000 individuals, part-time to about 10,000 and indirectly to another 5000 individuals. Marine fisheries contributed 1.9% to GDP in 1988. Nearly 65% of the animal protein and about 13% of the total protein consumed in Sri Lanka is provided by the marine fisheries (Baldwin, 1991).

Sri Lanka has a fishing tradition dating back several centuries. However, the development of the fisheries industry has been a slow process until recently. This development took place mainly with government assistance, and there was a rapid growth from 1977 to 1983 (Joseph, 1993). Thereafter, fisheries production began to decline with increasing activity by separatist guerillas in the east and the northern parts of the country.

In 1997 Sri Lanka's fisheries production was estimated to be about 242,000 metric tonnes. The share of fisheries in

Table 3
Total fish landings in metric tonnes according to different sub-sectors, 1993 to 1997

	1993	1994	1995	1996	1997	96/95 Percentage	97/96 Difference
Coastal	169,900	174,500	157,500	149,300	152,750	* 5.0	2.3
Offshore/Oceanic	33,000	37,500	60,000	57,000	62,000	* 5.0	8.8
Inland	18,000	12,000	18,250	22,250	27,250	22.0	22.5
Total	220,900	224,000	235,750	228,550	242,000	* 3.0	5.9

* Decline in percentage. Source: Data Management Unit, Ministry of Fisheries and Aquatic Resources Development. In NARA (1998).

Table 4
Fisheries export earnings from tuna fish

Tuna	Japan		USA	
	Chilled US$/kg	Frozen US$/kg	Chilled US$/kg	Frozen US$/kg
Southern Blue fin	9.14–31.63	21.09–70.30		
Blue fin	10.54–15.46			
Big eye	2.81–28.12	7.03–33.74	6.6	1.2
Yellow fin	7.79–15.46	2.81–14.06		
Skip jack	1.26–1.40			

Source: Infofish Trade News, 3 August 1998. In NARA (1988).

GNP is 1.6% (NARA, 1998). Coastal fish production reached its highest level in 1994 (174,500 metric tonnes) and declined in 1995 and 1996, and then increased thereafter (Table 3) (NARA, 1998). These fluctuations have been attributed to several reasons, though one important factor is over-exploitation of coastal resources. Another reason is the decline in fishing activities in the north and the eastern sector of the country due to the ongoing conflict between separatists and the government. Offshore fish production also declined from 57,000 metric tonnes in 1996 from a total of 60,000 metric tonnes in 1995. The total offshore production had however, increased to 62,000 tonnes in 1997.

Coastal fisheries include a large number of different varieties such as carangids, scombrids, tuna-like species, sharks and skates, rock fish (reef fishes), species caught in beach seine nets, shrimps, spiny lobsters and slipper lobsters (NARA, 1998). Although the total catch of reef fishes in 1997 was 9100 metric tonnes (5.96% of coastal fish landings in 1997) many other species such as carangids, scombrids and lobsters are also caught from coastal and other shelf reefs. Production from the sub-sectors of coastal and offshore/oceanic fisheries in 1997 was 152,750 and 62,000 metric tonnes respectively (NARA, 1998).

Wild-caught shrimp and lobster catch increased from 1993 to 1994, but thereafter fisheries statistics have not recorded these groups separately. However, lobster catches have been affected mainly due to uncontrolled harvesting including that of berried females, undersized individuals and fishing techniques that are damaging to the habitat. In particular the use of bottom-set nets causes serious damage by destroying the corals, gorgonians and all other associated reef organisms that contribute to the reef structures. Furthermore bottom set nets entangle molluscs, echinoderms and crabs (Rajasuriya, 1991; Ohman et al., 1993; Rajasuriya et al., 1995).

Although fishing is the only source of income for about 80% of the country's fishermen, doubts have been raised whether the industry will be able to offer adequate opportunities of employment to new entrants to the labour force by the year 2000 (Joseph, 1993). Environmental concerns may also become crucial to the marine fisheries sector in the next decade. Sri Lanka depends heavily on the use of gill nets, while there is much concern about catching non-target species.

The coastal marine fishery is often described as heavily exploited (Joseph 1993), and certain resources have shown a decline, but it is also pointed out that hitherto untapped or under-exploited resources in the coastal sector have the potential for increased production.

The offshore fishery sector, which is based on the migratory pelagic resources such as tuna, billfish and sharks, is the fastest growing sector in the marine fishing industry and is actively promoted by the Ministry of Fisheries and Aquatic Resources Development.

Marine Ornamental Fisheries

The marine ornamental fishery sector, which has a history of about 40 years in Sri Lanka, is mainly based on snorkel and scuba diving. This industry provides employment for about 2000 marine ornamental fish collectors. At present about 200 species of marine fish from 40 families, and about 30 species of invertebrates are being exported (Wood, 1996). Of particular importance in the trade are ornamental shrimps such as the *Lysmata debelius* and *L. amboinensis*. In addition, the mainstays of the marine ornamental sector are butterfly fish, angelfish, groupers, damselfish, gobies, blennies, surgeon fish and wrasses.

The export value of ornamental species has been increasing steadily. In 1994 the value of all exported ornamental species was estimated at about US $ 4 million. However this amount includes the value of exported freshwater species. Although in 1985 about 80% of exports were attributed to the marine ornamental species, Wood (1996) stated that there could be a slight decline, possibly due to restrictions placed on collection and export and to the expansion of cultured freshwater fish and plants (Wood, 1996).

Currently, the list of protected marine fish species for Sri Lanka includes only ornamental species. They are listed in the Amendment (No. 49 of 1993) to the Fauna and Flora Protection Ordinance (FFPO) and the Fisheries and Aquatic Resources Act (FARA) No. 2 of 1996. The FFPO contains seven species of fish, 13 species of molluscs, three species of crustaceans, two species of echinoderms, gorgonians, cerianthus tube anemones, sabellid tube worms, soft corals and hard corals. FARA contains the seven species of ornamental fish and one reef shrimp (*Hymenocera elegans*) listed in the FFPO and contains an additional 16 species of marine ornamental fish listed under the 'Restricted Export' category. However exporting is permitted only through the Department of Fisheries and Aquatic Resources Development, and only for scientific purposes and to national zoological gardens.

Differences in the lists of species in FFPO and FARA pose problems for the exporters as well as to the Biodiversity Protection Unit of the Sri Lanka Customs Department. Sri Lanka also imports marine ornamental species from the Maldives, Red Sea, Arabian Gulf and the East African coast to be re-exported. Quarantine facilities are poor and lack the necessary infrastructure for proper management.

Chank and Sea Cucumber Fisheries

The chank (*Turbinella pyrum*) fishery has been carried out in Sri Lanka for more than 150 years. A special ordinance was passed in 1842 to manage the fishery, where a number of divers from India were involved. Subsequently this ordinance was amended to suit the needs of the changing patterns in the trade and the need to conserve the resource. Chanks are exported mainly to Bangladesh and India for producing bangles and other ornaments. In the past the resource was harvested mainly in the Gulf of Mannar and Palk Bay. More recently, with the use of scuba equipment, large-scale harvesting was carried out in coastal waters of the southeast and there is concern that this activity could be unsustainable, as regulations are not adhered to in the collection and export. Further, there is no monitoring of the catch rates in relation to the resource, thus there are only records of the weight of exported product.

The sea cucumber fishery was mainly restricted to the northwest where skin divers were employed (Fig. 6). There are 13 species of commercially important species and these are mainly exported to far eastern countries. More recently, scuba diving has been introduced as a method of collection and there is clear evidence that the resource in the east coast was wiped out after two years of intensive harvesting. There is no monitoring of catches or exports, although the

Fig. 6. Processing holothurians for export (Batticola, east coast) (photo Arjan Rajasuriya).

exporters are registered under the Ministry of Fisheries and Aquatic Resources Development where they submit the value and weight of the exported product.

No studies have been carried out to estimate the resources and to determine the sustainable yield. Therefore at present these two fisheries are not regulated or monitored.

RURAL AND URBAN FACTORS AFFECTING THE COASTAL ENVIRONMENT

Sri Lanka is one of the most densely populated countries in the world, second only to Bangladesh among the less developed countries. The population is unevenly distributed with a greater concentration in the wet zone. Densities estimated in 1988 indicated a range from a low of 42–46 individuals km^2 in the northeast to nearly 2900 km^2 in the Colombo District (Baldwin, 1991).

In recent times, squatters have increased along the beach front, particularly in and around Colombo. Waste generated from these households is discharged directly into the sea. Often the beaches in such areas are heavily polluted. During the rainy season the beaches around Colombo and in other cities such as Negombo and Galle receive large quantities of non-biodegradable matter such as polythene and plastics together with water hyacinth (*Eichhornia crassipes*) and salvinia (*Salvinia molesta*). Only the city of Colombo is provided with a sewage network but the sewage is pumped into the sea untreated, through two sea outfalls located about 1 km away from the shore.

Garbage and waste disposal systems are poor. Most of the garbage is dumped into designated open areas that are wetlands around urban centres. As a result, some of this matter ends up in the coastal environment, particularly during rainy periods. Currently, trends are of increasing concentrations of urban wastes, and inadequate management will extend into the 21st century (Deheragoda et al., 1992).

Reclamation of wetland habitats in the coastal region near urban centres for housing and industries is also an increasing problem. This has led to rapid storm-water discharge into the sea, with sediment together with domestic and industrial waste.

COASTAL EROSION AND LANDFILL

Coastal erosion is one of the major problems in Sri Lanka and results in damage to or loss of houses, hotels and other coastal structures; it undermines roads, contributes to the loss or degradation of valuable land and disrupts fishing, recreation and other activities (Hale and Kumin, 1992).

The net average erosion rate for Sri Lanka does not exceed 0.5 m/year. However, site-specific analysis of priority areas such as Negombo, Crow Island and Hikkaduwa have shown yearly net rates of erosion of more than 1 m/year (CCD, 1986). There are two causes. Natural causes are currents, current intensity and direction, wave height and direction, monsoon and cyclones. Man-made causes are sand mining from the rivers and beaches, coral mining, improperly constructed hotels, fishery harbours and other structures.

Erosion Control Measures

The coast protection measures generally adopted in Sri Lanka are revetments, groynes, offshore breakwaters and beach nourishment. The 685 km of the west, southwest and southern coastlines are the most prone to erosion, and the Coastal Erosion Management Plan for Sri Lanka has identified 340 km as being especially subject to erosion.

All structures for the protection of the coastline built up to 1985 total 49,000 m of revetment and 6500 m of groynes, providing protection to approximately 68 km of coastline in coastal sectors of the south, southwest and west (Kahawita, 1991). The proposed investment in 1985 for coastal protection has been approximately US$ 5 million, using foreign aid. This programme commenced in 1987 and is still ongoing. The most critical coastal stretch is being protected by the Coast Conservation Department.

MARINE POLLUTION

Land-based Sources

During the last two decades development, urbanization and industrialization have caused environmental degradation in these coastal waters. Most of the industrialization and development is concentrated in the west and the southern coastal belt, and although some of the industries are 8–10 km away from the shore, pollutants reach the sea through canals, storm-water drains and rivers. In some instances effluents are discharged directly into the sea.

Some 20,000 industrial units are concentrated around Colombo. Many have no facilities for separation of solid and liquid wastes. Often effluents are discharged directly into canals, rivers, wetlands and marshy areas. The most damaging industries identified are coconut and rubber-based industries, food-processing plants, paper mills and distilleries. Paper factories in the east and southeast continue to pollute rivers (De Silva, 1993) and several tanneries close to Colombo discharge chromium and other toxic residues into the Kelani River. Studies carried out in the Negombo Lagoon by Niwas and Guruge (1990) found elevated concentrations of zinc, chromium, cobalt, nitrogen, manganese and iron during rainy seasons. A higher concentration of metals was also detected in suspended sediments.

Agricultural wastes from non-point sources are a major problem. Some farming districts in Sri Lanka use more than 120 kg of fertilizer per ha. Pesticide use is widespread; it has been estimated that Sri Lanka uses 2800 tons annually of which about 25% end up in the sea.

Fish kills have also been recorded from lagoons, lakes and other inland water bodies. These are mainly diseases resulting from industrial and urban wastes, which lead to algal blooms and excessive loads of ammonia in combination with high pH values. A study carried out by the National Aquatic Resources Research and Development Agency (NARA) in 1984 found high levels of ammonia in the Kelani river originating from the Fertilizer Manufacture Corporation. Dassanayake et al. (1985) recorded heavy metals in sediments of the Kelani river, originating from industries. Fish kills in the sea, however have not been recorded although, the author (A.R.) has personally observed occasional mass mortalities of the blue triggerfish (*Odonus niger*), which is a common species on offshore reef habitats. This phenomenon, the cause of which is unknown, has been observed only in the west and southern waters of the island.

A study carried out by NARA in 1986 on edible molluscs from the coastal waters from Negombo to Beruwala revealed a significantly higher level of bacteria as compared to other areas, which clearly indicates the higher urban and industrial activities within this coastal belt (Dassanayake, 1994).

The aquaculture industry for shrimp farming has converted much of the mangrove areas around the Chilaw lagoon, Mundel Lake and Puttalam Lagoon and has been identified as a major source of pollution. Shrimp diseases due to intensive culture practices have also adversely affected the production of shrimp farms (Dayaratne et al., 1997; Corea et al., 1995, 1998).

Production of white and brown fibre from coconut husks is considered a traditional industry in Sri Lanka and is mainly located in the southwest and southern coasts. It is reported that 91,000 tons of brown fibre is produced annually (De Silva, 1993). The process of obtaining white coconut fibre involves soaking coconut husks in the sea or brackish water where the nutrients and suspended matter released has adversely affected coral habitats.

Tourism also contributes significantly to coastal pollution. This is most evident in locations such as Hikkaduwa in the southwest. Rapid and unplanned development of hotels, guest houses, restaurants and the town has polluted and damaged the beaches and the marine environment (Nakatani et al., 1994; HSAMMSCC, 1996; De Silva, 1997). No consideration has been given to waste management. Hoteliers along the waterfront are forced to discharge raw sewage, periodically, into the sea during the night. Although complaints have been lodged with relevant authorities, action has not been taken to prevent this situation.

Increasing levels of waste and pollutants degrade the coastal environment and threaten coastal tourism, which is Sri Lanka's second largest foreign exchange earner (Olsen et al., 1992).

SHIPPING ACTIVITIES AND FISHERY HARBOURS

A large volume of oil is transported close to the southern tip of India and Sri Lanka through the main tanker route between the Middle East and the Far East. It has been estimated that 3.6 million barrels of oil pass the southern shores of Sri Lanka each day. There is a traffic separation scheme in operation off the southern coast of Sri Lanka to avoid accidents, and the chances of ships colliding and spillage of oil is apparently minimal (Brown, 1997). However there is a more pervasive oil pollution of beaches from tar balls from petroleum hydrocarbons, mainly in the western and southern coastal areas (Wickremaratne and Pereira, 1986). Fishery harbours also pollute the marine environment with waste oil and oil discharged from boat traffic. This is a chronic problem, which has not been solved, although fishery harbours have been modernized in recent times.

Two small-scale spills occurred in the recent past in Colombo from ruptured pipelines from the offshore bunkering terminal of the Ceylon Petroleum Corporation. There was no lasting visible impact from these incidents, although beaches south of Colombo were polluted for a short period. More recently in September 1999, a cargo ship grounding in the southeast on the shores of the Bundala National Park resulted in spillage of bunker oil and fertilizer which has polluted the beaches there. Although there is a special government body (Marine Pollution Prevention Authority) to prevent marine pollution, Sri Lanka lacks the funds, basic equipment and manpower to handle even small-scale oil spills.

Toxic red tides have not been reported from Sri Lanka (Brown, 1997: Dassanayake, 1994) although deaths have been reported which are due to eating Hawksbill turtles (De Bruin et al., 1994).

COASTAL RESOURCES MANAGEMENT

The framework for Sri Lanka's coastal zone management programme is provided through the Coast Conservation Department (CCD), first set up in January 1978, although there are numerous other government bodies involved (Table 5), and several laws apply (Table 6). The legal mandate for the programme is provided by the Coast Conservation Act, No. 57 of 1981. The framework for management is provided through the Coastal Zone Management Plan (CZMP) adopted by the Parliament in 1990. The CZMP of 1990 was updated and published as the 'Revised Coastal Zone Management Plan' in 1997.

The CZMP provides the basic framework for management of activities within the coastal region and specifically within the coastal zone. The CCD is empowered to carry out Sri Lanka's EIA process for development projects in the coastal region, carry out coastal protection and protect coral reefs in nearshore areas to a distance of 2 km from the shore.

Table 5

Ministries and government departments with the responsibility of managing coastal waters. Direct responsibility over coral reef management (*).

Ministry of Fisheries and Aquatic Resources Development *
  Development and Management of fisheries, including licensing of fishermen, crafts and gear
Ministry of Environment and Forestry
  Responsible for conservation of terrestrial as well as aquatic ecosystems.
Ministry of Ports, Shipping and Rehabilitation
  Activities connected with shipping and commercial harbours.
Department of Wildlife Conservation *
  Conservation and management of protected areas and species.
National Aquatic Resources Research and Development Agency *
  Research, Development and their coordination on all aquatic living and non-living resources.
Coast Conservation Department *
  Regulating development activities within the coastal zone and conservation and management of coastal resources. Responsible for implementing the Coastal Zone Management Plan.
Central Environmental Authority *
  Responsible for the national environmental standards, coordination of environmental related matters including the EIA process for development.
Ceylon Fisheries Harbours Cooperation
  Development and maintenance of fishery harbours.
Sri Lanka Ports Authority
  Development and management of port services.
Urban Development Authority
  Responsible for the planning and development of towns, cities and their maintenance.
Ceylon Tourist Board
  Responsible for the planning, development and licensing of tourist facilities.
National Drainage and Water Resources Board
  Responsible for water supplies and sewerage facilities.

Sources: Adapted from Hale and Kumin (1992), Rajasuriya and White (1995) and Rajasuriya et al. (1995).

Table 6

Ordinances, Acts and Regulations, relevant to marine fisheries management in Sri Lanka

---

The Village Communities Ordinance (VCO) No. 24 of 1889. (The earliest enactment for artisanal fisheries management. Amended 13 times.)

The Local Board Ordinance (LBO) of 1887 (Chapter 196).

The Chank Fisheries Ordinance (CFO) No. 18 of 1842.

The Pearl Fisheries Ordinance (PFO) No. 2 of 1890.

Small Town Sanitary Ordinance (STSO) No. 18 of 1892 (Chapter 197).

Local Government Ordinance (LGO) of 1920 (Chapter 195) (Mainly fishing gear and fishing areas)

Whaling Ordinance (WO) No. 32 of 1928. (Concerned with all marine mammals, probably not actively operated (Sivasubramaniam 1997).

Crown/State Lands Ordinance and Amendments of 1947 & 1949. (Deals with the foreshore and inland waterways, allocation of land for fisheries activities)

The Fisheries Ordinance (FO) No. 24 of 1940 and later Amendments

The Chank Fisheries Act (CFA) Amendment No. 8 of 1953.

Fisheries (Regulations of Foreign Fishing Boats) Act (FFBA) No. 59 of 1979 and amendments, No. 59 of 1981 & No. 37 of 1982. (This Act was introduced before the Law of the Sea Convention and was amended subsequently to include fishing gear, storage, identification numbers for boats, joint venture arrangements, permits etc.)

Fauna and Flora Protection Ordinance. 1949, 1964, 1970 and 1993. (Includes protected marine species used in the ornamental fish trade).

Fisheries and Aquatic Resources Act (FARA) No. 2 of 1996. (The first comprehensive General Fisheries Act. This repealed the Fisheries Ordinance of 1940, all amendments including the Chank Fisheries Act, Pearl Fisheries Ordinance and the Whaling Ordinance. It also includes non-living aquatic resources.)

Maritime Zone Law (MZL) No. 22 of 1976. (Specifically to claim ownership of all aquatic resources within the waters belonging to the country including its exclusive economic zone.)

Source: Sivasubramaniam (1997).

Two Special Area Management Plans (SAMP) have been prepared, namely at Hikkaduwa and at Rekawa Lagoon in the south, and several management activities initiated. Based on the experience of these two sites, a further 23 sites have been identified for SAMP in the Revised Coastal Zone Management Plan of 1997. A few recommendations for management in the Hikkaduwa and Rekawa Lagoon SAM plans have been implemented. Due to lack of actual management at ground level even some actions taken to improve the Hikkaduwa Marine Sanctuary have not been fully successful.

There are two ongoing integrated coastal management (ICM) projects initiated in the southern coast funded by the Asian Development Bank (ADB) and GTZ (Technical Cooperation of the Federal Republic of Germany). An ADB project mainly involving coastal protection and development has been planned.

A major cause of coastal erosion in the southwestern coastal belt is coral extraction from the sea. Implementation of the ban on sea-coral mining and the removal of lime kilns from the coastal zone has reduced sea-coral mining in the recent past. However this has not been entirely successful due to lack of alternative employment for sea-coral miners. At present the government has decided to stop the use of lime from corals in the construction of government buildings, and alternative sources of lime have been investigated, particularly the dolomite deposits found in the central part of the country. This resource has not been very popular due to the lower quality of lime, as a result of excessive temperatures required in the processing, which makes the plaster brittle when applied to walls. The processing of dolomite has now been improved as a result of new kilns that have been developed by a special task force appointed by the Ministry of Science and Technology in order to reduce the dependency on corals mined from the sea.

REFERENCES

Baldwin, M.F. (ed.) (1991) *Natural Resources of Sri Lanka, Conditions and Trends*. Natural Resources, Energy and Science Authority of Sri Lanka and United States Agency for International Development (NARESA/USAID). 280 pp.

Brown, B.E. (1997) *Integrated Coastal Management: South Asia*. Department of Marine Sciences and Coastal Management, University of Newcastle, Newcastle upon Tyne, UK.

CCD (Coast Conservation Department) (1986) Master Plan, Coast Erosion Management. Coast Conservation Department, Colombo, Sri Lanka. Vol 1. 17 pp.

CCD (Coast Conservation Department) (1990) Coastal Zone Management Plan, Sri Lanka. Coast Conservation Department, Colombo, Sri Lanka. 81 pp.

CCD (Coast Conservation Department) (1997) Revised Coastal Zone Management Plan, Sri Lanka 1997. Coast Conservation Department of the Ministry of Fisheries and Aquatic Resources Development, Colombo, Sri Lanka, 121 p.

Cooray, P.G. (1984) *An Introduction to the Geology of Sri Lanka (Ceylon)*. National Museums of Sri Lanka. Colombo, 340 pp.

Corea, A.S.L.E., Jayasinghe, J.M.P.K., Ekaratne, S.U.K and Johnstone, R. (1995) Environmental impact of prawn farming on Dutch Canal: The main water source for prawn culture industry in Sri Lanka. *Ambio* 24 (7–8), 423–427.

Corea, A., Johnstone, R., Jayasinghe, J., Ekaratne, S and Jayawardene, K. (1998) Self-pollution: A major threat to the prawn farming industry in Sri Lanka. *Ambio* 27 (8), 662–668.

Da Silva, M.W.R.N. (1997) Trials and tribulations of Sri Lanka's First Marine Sanctuary—The Hikkadua Marine Sanctuary. In *Proceedings of the Regional Workshop on the Conservation and Sustainable Management of Coral Reefs, No. 22*, ed. V. Hoon. CRSARD, Madras, pp. 99–116.

Dassanayake, N.H. (1994) Sri Lanka. In *An Environmental Assessment of the Bay of Bengal Region*. BOBP/REP/67, Swedmar, BOBP, pp. 209–235.

Dassanayake, C.B., Weerasooriya, S.V.R. and Senaratne, A. (1985) The heavy metal pollution of the Kelani River in Sri Lanka. *Aqua*, No. 2. U.K.

Dayaratne, P. and Joseph, L. (1993) A Study on Dolphin Catches in Sri Lanka. BOBP/REP/56. Bay of Bengal Programme, Madras, India. 47 pp.

Dayaratne, P., Linden, O and De Silva, M.W.R.N. (eds.) (1997) The Puttalam/Mundel Estuarine System and Associated Coastal Wa-

ters: A report on environmental degradation, resource management issues and options for their solution. NARA, NARESA, Sida/SAREC Marine Science Programme, Colombo, 98 pp.
Deheragoda, C.K.M., Wanasinghe, Y.A.D.S. and Karunanayake, M.M. (1992) Urban growth in Sri Lanka: Implications for regional development. UNCRD Working Paper No. 92. 4. United Nations Center for Regional Development, Nagoya, Japan (unpublished) 53 p.
De Bruin, G.H.P., Russell, B.C. and Bogush, A. (1994) *The Marine Fishery Resources of Sri Lanka*. FAO Species Identification Field Guide for Fishery Purposes. Rome, FAO. 400 pp.
De Silva, R.I and Jakobsson, S. (1995) Mundel Lake (Sri Lanka): An important habitat for migratory shore birds. Wildlife and Nature Protection Society of Sri Lanka.
De Silva, S.H.C. (1993) Destruction of Natural Resources and Conservation Efforts. In *Proc. International and Interdisciplinary Symposium, Ecology and Landscape Management in Sri Lanka. Colombo, Sri Lanka, 12–26 March 1990*, eds. W. Erdelen, C. Preu, N. Ishwaran and Madduma Bandara, pp. 231–235.
De Silva, M.W.R.N. (1997) Trials and tribulations of Sri Lanka's first marine sanctuary–the Hikkaduwa Marine Sanctuary. In *Proceedings of the Regional Workshop on the Conservation and Sustainable Management of Coral Reefs, No. 22*, ed. V. Hoon. CRSARD, Madras, pp. 99–116.
Fernando, M.R.D. and Seneviratne, L.K. (1990) Mineral resources in the Coastal Area of Sri Lanka. Report prepared for the Coastal Resources management Project 1990 in the series: Conditions and Trends in the Use of Coastal Resources.
Hale, L.Z. and Kumin, E. (1992) Implementing a Coastal Resources Management Policy, The Case of Prohibiting Coral Mining in Sri Lanka. Coastal Resources Center, University of Rhode Island, 30 pp.
HSAMMSCC (Hikkaduwa Special Area Management and Marine Sanctuary Coordinating Committee) (1996) Special Area Management Plan for Hikkaduwa Marine Sanctuary and Environs, Sri Lanka. Coastal Resources Management Project, Coat Conservation Dept., National Aquatic Resources Agency, Colombo, Sri Lanka.
Joseph, L. (1993) Coastal Fisheries and Brackish water Aquaculture in Sri Lanka. Coastal Resources Management Project, Colombo, Sri Lanka, 46 pp.
Kahawita, B.S. (1991) Conceptual Framework of the Coast Erosion Management Plan. Proc. seminar on causes of coastal erosion in Sri Lanka. CCD/GTZ. Colombo. pp. 315–321.
Leatherwood, S. and Reeves, R.R. (eds.) (1991) Marine Mammal Research and Conservation in Sri Lanka, 1985–1986, 2nd edn. Nairobi, Kenya. United Nations Environmental Programme, Marine Mammal Technical Report No. 1. vi + 138 pp.
Meyer, K. (1979) *Placer Prospection off Pulmoddai, Sri Lanka*. Preussasg, A.G. Hannover.
Nakatani, K., Rajasuriya, A., Premaratne, A. and White, A.T. (eds.) (1994) *The Coastal Environmental Profile of Hikkaduwa Sri Lanka*. Coastal Resources Management Project, Colombo. 70 p.
NARA (National Aquatic Resources Research and Development Agency) (1998) *Sri Lanka Fisheries Year Book, 1998*. Colombo, 56 pp.
Niwas, J.M. and Guruge, K.S. (1990) Status of heavy metal abundance in the Negombo Lagoon. *Proc. of the 46th Ann. Sessions*, Sri Lanka Association for the Advancement of Science, Colombo. Sri Lanka.
Ohman, M.C., Rajasuriya, A. and Lindén, O. (1993) Human Disturbances on coral reefs in Sri Lanka: A case study. *Ambio* **22** (7), 474–480.
Olsen, S., Sadacharan, D., Samarakoon, J.I., White, A.T., Wickremaratne, H.J.M. and Wijeratne, M.S. (eds.) (1992) Coastal 2000: Recommendations for A Resource Management Strategy for Sri Lanka's Coastal Region, Vol. I & II. CRC Technical Report No. 2033. Coast Conservation Department, Coastal Resources Management Project, Sri Lanka and Coastal Resources Center, University of Rhode Island.
Pinto, L. (1986) *Mangroves of Sri Lanka*. Natural Resources, Energy and Science Authority of Sri Lanka, 54 pp.
Rajasuriya, A. and De Silva, M.W.R.N. (1988) Stony corals of the fringing reefs of the western, southwestern and southern coasts of Sri Lanka. *Proc. 6th Int. Coral Reef Symposium, Australia*, Vol. 3, pp. 287–296.
Rajasuriya, A. (1991) Location and Condition of Reefs along Sri Lanka's Coast. *Proc. Seminar on Causes of Coastal Erosion in Sri Lanka*, pp. 203–210. Coast Conservation Department, Sri Lanka.
Rajasuriya, A. (1994) Three Genera and Twelve Species of Stony Corals New to Sri Lanka (Abstr.). Paper presented at the Second Annual Scientific Sessions of the National Aquatic Resources Agency, Colombo, Sri Lanka.
Rajasuriya, A., De Silva, M.W.R.N and Öhman, M.C. (1995) Coral reefs of Sri Lanka; human disturbance and management issues. *Ambio* **24** (7–8), 428–437.
Rajasuriya, A. and White, A.T. (1995) Coral reefs of Sri Lanka: review of their extent, condition and management status. *Coastal Management* **23**, 77–90.
Rajasuriya, A., Öhman, M.C. and Svensson, S. (1998) Coral and Rock Reef Habitats in Southern Sri Lanka; Patterns in the Distribution of Coral Communities. *Ambio* **27** (8), 723–728.
Rajasuriya, A., Maniku, M.H., Subramanian, B.R. and Rubens, J. (1999) Coral Reef Ecosystems in South Asia. In Coral Reef Degradation in the Indian Ocean, Status Reports and Project Presentations, eds. O. Linden and N. Sporrong, pp. 11–24.
Richardson, P. (1997) Tortoiseshell Industry in Sri Lanka: A Survey Report. 1996. *Lyriocephalus*, Journal of the Amphibia and Reptile Research Organisation of Sri Lanka, pp. 6–24.
Swan, B. (1983) *An Introduction to the Coastal Geomorphology of Sri Lanka*. National Museums of Sri Lanka, Colombo, 182 pp.
Sivasubramaniam, K. (1997) *One Hundred Years of Fisheries Management in Sri Lanka: lessons for the future*. Department of Fisheries and Aquatic Resources Development, Colombo, Sri Lanka. 156 pp.
Wickremaratne, H.J.M. (1985) Environmental problems of the coastal zone in Sri Lanka. *Economic Review* (Peoples Bank Publication) **10** (2), 8–16.
Wickremaratne, S.W. and Pereira, R.P. (1986) Observations on the deposition of tarballs on beaches of Sri Lanka. *Proceedings of the 41st Annual Sessions of Sri Lanka Association for the Advancement of Science*. Colombo, Sri Lanka.
Wallberg, P. and Johnstone, R. (1995) Diurnal variations in pelagic bacterial and primary production during monsoon and inter-monsoon seasons: Puttalam Lagoon, Sri Lanka. *Ambio* **24** (7–8), 418–422.
Wood, E.M. (1996) *The Marine Ornamental Fishery in Sri Lanka: Current Status and Management Needs*. Marine Conservation Society, UK. 47 pp.
Wilkinson, C., Linden, O., Cesar, H., Hodgson, G., Rubens, J and Strong, A.E. (1999) Ecological and socioeconomic impacts of 1998 coral mortality in the Indian Ocean: An ENSO impact and a warning of future change? *Ambio* **28** (2) 188–196.

THE AUTHORS

Arjan Rajasuriya
*National Aquatic Resources Research and Development Agency, Colombo,
Sri Lanka*

Anil Premaratne
*Coast Conservation Department,
Sri Lanka*

Chapter 65

THE ANDAMAN, NICOBAR AND LAKSHADWEEP ISLANDS

Sundararajan Ramachandran

The Andaman and Nicobar islands comprise about 348 islands, located between 6° to 14°N latitude and 92° to 94°E longitude. Their total coastline is about 1926 km. The coastal area is richly fringed by mangroves, coral reefs, seagrass and seaweeds, and some atolls also occur. Being oceanic islands, the continental shelf area along these islands is limited in extent, but totals about 16,000 km². The continental slope is mostly absent. The coast is irregular and deeply indented by innumerable creeks, bays, inlets, estuaries and lagoons that have given rise to a rich biodiversity and high productivity. The population of these islands is about 0.29 million and it is growing rapidly. Fisheries and tourism play an important role in the economy of the islands, and both of these have great scope for further development. However, the coastal ecosystems are already degraded due to anthropogenic pressures, and an integrated developmental strategy is essential to ensure sustainable growth.

The Lakshadweep Islands are also a territory of India and are included here. They are a group of low-lying coral islands and atolls, located off southwest India, forming part of the long Lakshadweep–Chagos chain which includes the Maldive islands. Their marine life is diverse, and they create a very large EEZ, with an important fishery.

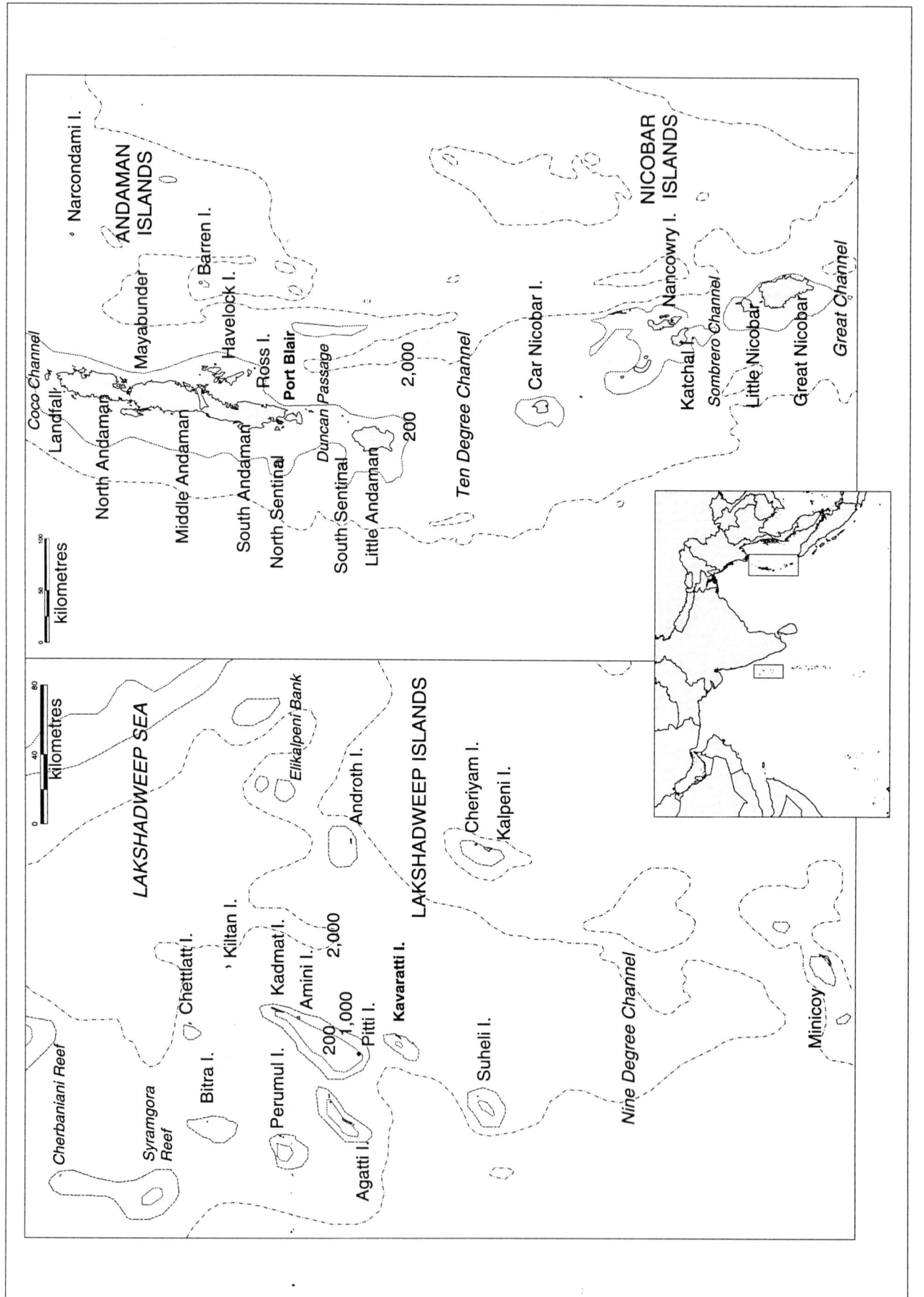

Fig. 1. Map of Andaman, Nicobar Islands and Lakshadweep Islands.

GEOGRAPHICAL FEATURES

The Andaman and Nicobar islands comprise about 340 islands of volcanic origin situated on a submerged mountain range, which arches down from Arallan Yoma in Burma in the north to Sumatra in the south. The islands themselves lie between 6° to 14°N latitude and 92° to 94°E longitude, occupying an area of about 8249 km² (Fig. 1). The area of the Andaman group of islands is the greater of the two, being 6408 km², while that of the Nicobar group is 1841 km², though otherwise both island groups have many similarities (Fig. 2). The 10° channel divides the two groups of islands, and their combined total coastline is about 1926 km long and is very irregular. Being oceanic islands, the continental shelf area along these islands is limited in extent to about 16,000 km². However, the continental slope is mostly absent due to the fact that the land drops steeply to great depths not far from the coastline. However, the Exclusive Economic Zone area lying around these islands accounts for about 0.6 million km². The limited continental shelf is compensated to a certain extent by the irregular coast, which is deeply indented by innumerable creeks, bays, inlets, estuaries and lagoons, which has led to a very rich biodiversity and high productivity (Silas, 1983).

CLIMATE AND COASTAL HYDROGRAPHY

The Andaman and Nicobar Islands have a humid, tropical monsoon climate with an average annual rainfall of 3180 mm. Since the Nicobars are very near the Equator, they are more humid and hot than the Andamans. Both receive precipitation from both the Southeast and Northwest Monsoons, which together last for 9–10 months each year. The dry months are February and March. Air temperature ranges from 16.7 to 32°C, and the mean relative humidity is 77% (Ramachandran, 1997a).

Fig. 2. Aerial view of Andaman Islands.

Studies on coastal water quality have not been carried out regularly in this region. Sea surface water temperature closely follows the trend of atmospheric temperature, ranging from 23 to 32°C, with the highest temperatures recorded during the period February to April (the dry months). The surface salinity ranges from 24.76 to 34.10 ppm in the open sea, while in creeks, salinity ranges from 9.98 to 33 ppt. Wide fluctuations in surface salinity are seen in most parts of the island seas during the months of May to September due to the heavy rainfall (Rangarajan and Marisamy, 1972). The dissolved oxygen was found to be in the range of 3.8 to 6.5 ml/l, and pH ranges from 7.5 to 8.5 (Marisamy, 1983).

Measurements of primary production in Andaman and Nicobar coastal waters yield values which vary greatly from 0.07 g C/m²/day to 3.64 g C/m²/day owing to the variety of coastal ecosystems present in these areas. The depth of the eutrophic zone ranges from 30 to 50 m in coastal waters, and from 75 to 90 m in open ocean areas (Ramachandran Nair and Gopinathan, 1983). Measurements of chlorophyll have produced values which range from 0.3 to 0.6 mg/m³.

The surface water of the Andaman Sea is influenced by large quantities of freshwater runoff from the perennial rivers of Myanmar, Thailand and Malaysia. The surface circulation is controlled by the monsoon system of the northern Indian Ocean. Circulation studies have indicated a flow of water from the Bay of Bengal and Andaman Sea through the Preparis channel.

COASTAL ECOSYSTEMS

All the islands of the Andaman and Nicobar area are richly fringed by mangroves, coral reefs, seagrasses and seaweed ecosystems.

Coral Reef Ecosystems

The coral formations of the Andaman and Nicobar Islands extend from 92°30' to 94°E longitude and 7° to 14°N latitude. The reefs are mostly of the fringing type (Fig. 3) though patch reefs also occur along the shores, especially in embayments. The west coast chain of reefs with interrupted banks extends over a length of 320 km and is similar to a barrier reef in its structure. The corals grow luxuriantly in protected bays and creeks. The Nicobar Islands have a richer coral growth than do the Andaman Islands due to a lesser degree of human interference (Pillai, 1983).

There may be a fairly high diversity of stony corals (Pillai, 1972, 1983; Bakus, 1994) though surveys have been few since the work of Scheer and Pillai (1974) and no surveys have been thorough in deeper water. Non-scleractinian corals are also abundant at certain locations and they include *Heliopora*, *Millipora* and *Tubipora*. Only 135 species representing 59 genera have so far been documented, though given the location of these islands in the

Fig. 3. Narrow sandy beaches and fringing coral reefs common in Andaman and Nicobar Islands.

eastern parts of the Indian Ocean, two or three times this species diversity might await discovery. Among those which have been recorded, 110 species in 45 of the genera are hermatypic, and 25 species of 14 genera are ahermatypic (Pillai, 1983). Based on remote sensing data, the total reef area coverage in the Andaman and Nicobar Islands combined is estimated as 1048.7 km^2 (SAC, 1992).

Mangrove and Algae

The mudflats with sand and coral rubble which are abundant along the creeks, bays and lagoons promote the growth of mangroves. The mangrove forests of the Andaman and Nicobar Islands (Fig. 4) occupy an area of 115,200 ha (Blasco, 1975). However, recent estimates by remote sensing satellite data showed a rather smaller areal extent of mangroves of only 77,000 ha, which is also the estimate of the Andaman & Nicobar Forest Department. The island's mangroves account for about 18% of the country's (India's) total mangrove area. About 27 species of mangroves are recorded from the Andamans, with 10 species recorded from the Nicobar group (Jagtap, 1992). The dominant mangrove species are *Rhizophora mucronata*, *R. stylosa*, *Bruguiera gymnorrhiza*, *Xylocarpus moluccansis*, *Avicennia marina*, *Sonnertia alba*, *Ceriops tagal*, *Nypa pulitans*, *Aegiceras corniculatum*, *Phoenix peludosa* and brackish-water marsh fern *Acrostichum aureum* (Gopinathan and Rajagopalan, 1983; Ramachandran, 1997a).

Gopinathan and Panigrahy (1983) have reported 55 species of seaweeds from all islands, of which 16 belong to the Chlorophyta, 17 to the Phaeophyta and 22 belong to the Rhodophyta. Alginophytes such as *Turbinaria*, *Sargasum* and *Padina* are the dominant species. Economically important agrophytes such as *Gracillaria* sp., *Gelidiella acerosa* and *Gelidium* sp. are poorly represented.

BIODIVERSITY

These islands are known for their rich biodiversity and high degree of endemism. The giant clam *Tridacna*, black lip, pearl oyster, some larger gastropods like *Trochus niloticus*, *Turbo marmarotus* and many echinoderms which are not found on the mainland coasts of India, occur in these islands.

The marine biodiversity of the Andaman and Nicobar Islands is rich, as has been reported by many authors. A comprehensive list is given in Table 1 (after Saldana, 1989).

FISH AND FISHERIES OF THE ANDAMANS

The islands did not have a large traditional fishing population and the limited fishing carried out there used to be done by the aborigines in the Andamans with bows and arrows, while the tribal Nicobarese used spears. Starting from the early 1950s fishermen from the mainland were employed, with incentives to settle down in the islands and carry out greater fishing activities. However, this has not led to any remarkable development in this industry. The remoteness of the island from the mainland, lack of infrastructure and the absence of any fish trade between the mainland and these islands are some of the factors responsible for the poor development of fisheries. Thus the fishery development in the Andaman and Nicobar Islands remained very meagre. Between the years 1990 and 1995, the growth in the fishery sector has been modest (Table 2).

There is an abundance of fish in the shallow as well as in the deep sea. A reliable estimate by the National Institute of Oceanography places the probable catch at 474,000 tons/annum. However, the present catch is between 30,000 to 40,000 only. This indicates that marine capture fisheries in the Andaman and Nicobar Islands have a vast scope, with a possibility of increasing production several fold (Saldana, 1989).

Fig. 4. Healthy mangroves and thick rain forest bordering the coast of Wandoor National Marine Park, Andamans.

Table 1
Wetland faunal diversity of Andaman and Nicobar islands

Name of group	No. of species/ sub-species	Endemic species sub-species	Remarks
Sponges	70	2	marine
Coelentrate	147	–	marine
Corals	179	–	marine
Polychaetes	161	–	marine
Pycnogonids	8	–	marine
Crustaceans	407	–	includes
Molluscs (fresh)	51	12	–
Molluscs (marine)	790	–	–
Meiofauna	324	193	marine
Fishes	750	–	Mostly marine
Reptiles	83	23	marine
Birds	242	95	No. of waterfowl not specified
Mammals	58	33	3 marine

Table 2
Some fisheries data for the Andaman Islands (Anon, 1994, 1995)

	1990	1995
No. of fishermen	3,500	3,650
Fishing crafts		
Country crafts	1,006	1,340
Mechanised ones	200	230
Fish landings		
Quantity	13,596 tons	25,080 tons
Value in Rupees	1.4 million	3.93 million

POPULATION

The native tribes of the Andamans are Negrito hunter–gatherers, while in the Nicobars the tribes are herders belonging to the Mongoloid race. At the end of the 18th century, their population was probably as low as 5000, while now their estimated population is around 26,000 (Anon, 1995). The British used the Andamans for a prison and a penal colony in the early 20th century. Until 1941, the population growth rate was only 0.5%. After the post-independence period, the population growth rate was phenomenal. Based on a 1961 census, the population was 63,548 and in 1991, the population was 280,661 (Anon, 1995).

ENVIRONMENTAL DEGRADATION

The tremendous increase in human population in both the Andaman and Nicobar Islands, and the increase in the influx of visitors, have brought about significant changes in these islands. The limited natural resources available were inadequate to meet the demands of the people. Several activities have increased which led to severe ecosystem degradation. Forest extraction increased from 49,000 m^3 per year in the 1950s to about 182,590 m^3 in 1994 (130,136 for timber and 52,456 for fuel wood). Similarly, the area cleared for agricultural plantations and horticulture has increased from 10,000 ha in the 1940s to about 48,000 ha in the 1990s. The increase in agriculture has created problems of contamination of nearshore waters due to runoff of various chemicals brought in by the high rainfall prevailing in these islands. Further, the farming is very ineffective and rapidly becomes uneconomical (Saldana, 1989).

IMPACTS OF HUMAN ACTIVITIES ON THE ECOSYSTEM

Under natural conditions, there is a regulated flow of water and nutrients from the forest ecosystem to the marine and mangrove ecosystems. The deforestation has caused soil erosion in the hills and undulating terrain of these islands. During heavy rainfall, the seas receive heavy loads of silt and this, together with the reduced salinity, affects the mangrove ecosystems. Terrigenous silts brought by the monsoon rains through newly constructed culverts at many places along the coast also cause siltation and turbidity which in turn reduces light penetration and productivity in the shallow coastal waters. Furthermore, the silt smothers the polyps of the corals, resulting in their death. A survey by INTACH (1989) indicated that such conditions also lead to bleaching. These changes in turbidity alter the biological processes in the coastal belt, but also affect the scenic value, tourism potential and many other benefits (Saldana, 1989). In the Little Andamans, about 75% of the area has recently been reported to have been deforested, but the effects of such massive deforestation on the estuary, mangrove and the coral ecosystem have not been investigated. Some 12,000 ha of mangroves have been removed mainly for firewood, with detrimental effects on offshore fisheries.

Before the immigration from the mainland, the natives of the Andaman and Nicobar Islands formed an integral part of the island ecosystem and it appears that they may have lived relatively harmoniously in their environment. With increased immigration and massive growth of population, urbanisation activities have also increased. As a result, construction, logging and several other industries have also increased. Construction of buildings, roads, wharves, jetties etc. have resulted in a rise in sand mining and quarrying. As a consequence, shorelines have started eroding in several locations which has also affected the mangrove and coral reef community.

Logging along the seashore, and the associated rafting and towing of logs over the reefs, damages the corals, especially when loose logs drift or sink to the bottom (INTACH,

Lakshadweep Islands

Lakshadweep is a group of islands located about 300 km to the west of the Kerala coast, occupying a vast expanse of the Arabian Sea between 8° and 12°30'N latitude and 71° and 74'E longitude. There are about 36 islands in 12 atolls, of which 10 are inhabited, the remainder being mainly submerged banks, open reefs and sand banks. The islands rise 3–9 m above the sea and are coralline, made up of fine coral sand and boulders which have been compacted into sandstone (Saldana, 1989; Ramachandran, 1997).

The group has an Exclusive Economic Zone of 400,000 km^2. The width of its intertidal areas range from 7 to 15 m, with gentle slopes and little wave action. Kavaratti is the capital. The average population density is 1258/km^2 (1981 Census) but is double this in Amini. Of the total population, 93% are Muslims, and the language in most of the islands is Malayalam except in Minicoy where people speak Mahl (Saldana, 1989).

CLIMATE AND COASTAL HYDROGRAPHY

The warm tropical climate gives a temperature of 17° to 38°C, with a relative humidity of over 70%. Beaches have varying exposure and generally low wave action. Tides are of mixed type with a strong semidiurnal influence. During the Southwest Monsoon, surface currents in the open ocean are eastwards and clockwise, and during the Northeast Monsoon, the general surface circulation is more or less reversed in the open ocean and is north-westward with counter clockwise circulation along the coast.

BIODIVERSITY

The atolls and seas are diverse (Ramachandran, 1997). The atolls, reefs and submerged banks total 4200 km^2. Almost all the atolls have a northeast–southwest orientation with islands on the east, broad, well-developed reefs on the west and a lagoon in between. The lagoons open into the sea by one or more channels (Saldana, 1989).

A total of 104 scleractinian corals of 37 genera are reported from this region. *Acropora* is the common genus, providing 25% of the total species known from Minicoy. There is a total absence of foliaceous forms, but varied massive forms such as *Porites solida*, *P. lutea* and *Diploastrea* sp. (Wafar, 1986).

The lagoons are well aerated, and have bottom sediments mainly of coral debris and calcareous sand. Production of particulate organic carbon by the reef community is about 20% of lagoonal gross production, the unit volume of production varying from 8 to 34 mg C/m^3/day, with a maximum rate at Minicoy. The annual net production from Minicoy lagoon is 3000 g C/m^2/day. The shallow lagoons support rich macrophyte growth; benthic algae and seagrasses are abundant, and the magnitude of their photosynthetic activity can be seen from the large diurnal changes in oxygen of 1.73 to 8.83 ml/l.

In Bingaram lagoon, seagrasses have been mapped and cover 5.57 km^2. Seagrass is also abundant and extensive in Kiltan, Minicoy, Agatti, Kalpeni, and Kavaratti. The major species are *Thalassia hemprichii* and *Cymodocea isoetifolia*, and *Cymodoceae serrulata*, the latter being rare in Kavaratti. *Thalassia* is a major food for adult sea turtles of the islands, and it has been observed that erosional processes are more severe where seagrass beds are absent.

Algae of substantial economic value are present in small and scattered patches on the inner reef margins of Cheriyapaniyam and Kiltan atolls. These algae are mainly *Ulva*, *Gracilaria*, *Gelidium*, *Hypnea*, *Laminari* and *Ceramium*.

FISHERIES

The Fishery Survey of India has conducted several surveys on the marine fishery potential of the islands' seas. Besides sail fishes, marlins, bill and sword fishes, vessels report an abundance of yellow fin tuna (76%) and pelagic sharks (24%). Pole-and-line seems to be the common method of tuna fishing. Annual fisheries yield 10–16 thousand tonnes/yr of which nearly 3000 tonnes are accounted for by tuna, mainly skipjack. Other fishes caught in this region are elasmobranchs, white bait, perches, red mullets, flying fish, half beaks, seer fishes and cephalopods. Ten fishing villages and 11 fish landing centres exist in the islands. There is a vast scope for improving commercial fishing activities as appropriate craft, gears, processing technologies and marketing are beyond the reach of the islanders. Since India is not exploiting these resources fully, other countries are poaching in these waters and depleting the resources (Ramachandran, 1997).

POPULATION AND TOURISM

Lakshadweep has a steadily increasing local population. It was about 13,000 in 1901 and had grown to more than 50,000 by 1995, in more than 8,000 households. The people are included among the 'Scheduled Tribes' (Saldana, 1989).

There are two types of tourists: day visitors who return to tourist ships at night, and the hotel tourists who stay in the huts and cottages in the islands. Since the islands constitute a tribal territory, tourist entry is by permit only. In one year about 18,900 tourists visit, mainly to the islands Kalpeni, Kavaratti, Kadmat, Bangaram and Minicoy. Boating, snorkelling, yachting, water and wind surfing are all being contemplated. Other future development activities planned are coral mining and guano collection. There are no other large-scale industries due to scarcity of raw materials and energy, though several small-scale industries like coconut processing or fish products exist. Two boat building yards and ten workshops also exist.

ENVIRONMENTAL DEGRADATION

The 200% rise in population in an area of 1615 km^2 has led to dramatic increases in the density of houses, passenger and cargo traffic. Erosion is a serious problem in these islands, which is aggravated by the removal of coral boulders for building material. Siltation in Minicoy lagoon and other islands has increased due to erosion, increased human activity, quarrying of corals from the shore and reefs, and from removal of surface soil. Further degradation of reefs has come from sponge infestation and *Acanthaster planci* (Wafar, 1986). The percentage of dead corals in Minicoy is 90%, Andrott 50%, Kalpeni 80% and Suheli Par 30% of the surface coverage.

> **CONSERVATION RECOMMENDATIONS**
>
> Special conservation measures are required (Saldana, 1989). Collection of shells at Pitti should be banned, and a marine park should be identified and constituted with the necessary protection and manpower for its proper management. In Lakshadweep, the natural and man-made ecosystems are closely linked together. Increasing population and developmental activities are causing depletion of land resources and degradation of marine resources. Damage to the lagoons and reefs especially by quarrying, over-exploitation and pollution has to be avoided. Proper sewage disposal should be introduced so that the sensitive coastal ecosystem is not harmed. On the other hand, deep-sea fishing to harvest the rich resources of the sea can receive priority, but this should be supported by a proper infrastructure.
>
> **REFERENCES**
>
> Ramachandran, S. (1997) Case studies on Coastal Fragile Area—Part 2 – West Coast of India. Project report submitted to Central Pollution Control Board, GOI. 261 pp.
>
> Saldana, C.J. (1989) *Andaman, Nicobar and Lakshadweep. An Environmental Impact Assessment.* Oxford & IBH Publishing Co. Pvt. Ltd., New Delhi. 114 pp.
>
> Wafar, M.V.M. (1986) Corals and coral reefs of India. *Proceedings of the Indian Academy of Science* (Suppl). **1**, 19–43.

1989). Dumping of sawdust has also destroyed some reefs. The effluents of industries and human settlements affect both mangroves and coral reefs through water quality degradation. Damage due to entanglement of nylon nets and fishing lines and hooks has damaged the corals. Furthermore, the collection of corals and shells by locals and visitors also has had an impact. Due to excessive fishing, including fishing in restricted areas, over-fishing has occurred in the nearshore waters. In offshore regions, however, it is still under-exploited. This over-fishing has resulted in depletion of some important invertebrate species such as *Trochus* and *Turbo*.

Tourism has not become a major activity in these islands due to restrictions imposed by the Government of India. Foreign visitors are limited to only a very few thousand in number (3000–4000) while permitted domestic visitors are in the range of 20 to 25,000 during the past five years. Hence, the direct impacts from this industry are thought to be small.

Coastal aquaculture has just been started in the islands. Conversion of coastal land for aquaculture purposes has caused destruction of coastal vegetation including seagrasses and seaweed beds.

NATIONAL PARKS AND WILDLIFE SANCTUARIES

Sixteen National Parks covering 36,157 ha and 94 wildlife sanctuaries covering 45,598 ha have been established for the protection and preservation of various unique or rare species of plants and animals in these islands. Table 3 lists Reserves with a coastal component. Many parts of these islands are suitable for protection, containing as they do some dramatic natural vegetation and geological features (Figs. 5 and 6).

CONSERVATION MEASURES

The Ministry of Environment and Forests of the Government of India has initiated activities toward conservation of mangroves which involve survey and identification of

Fig. 5. Natural coconut forest in Ross Island, Andamans.

Fig. 6. Volcanic eruption in Barren Islands, North Andamans.

Table 3

Reserves that include coastal habitat (Anon, 1995; Ramachandran, 1997a)

1. Marine National Park, Wandoor: 11° 22′–11°36′N, 92°30′–92°40′E; 28,150 ha in Andaman district; 0–85 m above sea level; includes coral reefs
2. North Button Island National Park: 12°18′N, 93°03′E; 44 ha; 0–49 m above sea level
3. Middle Button Island National Park: 64.74ha; 0–33 m above sea level
4. South Button Island National Park: 12°13′N, 93°01′E; 3 ha; 0–21 m above sea level
5. Saltwater (Lohabarruk) Crocodile sanctuary: 11°35′–11°40′N, 92°35′– 92°39′E; 2221 ha;
6. North Reef Island Sanctuary: 13°04′–13°05′N, 92°41′–92°43′E; 348 ha; 0–11 m above sea level
7. South Reef Island Sanctuary: 12°46′N, 92°39′E; 117 ha; 0–2 m above sea level
8. South Sentinel Island Sanctuary: 10°58′–10°59′N, 92°12′–92°13′E; 161 ha; 0–6 m above sea level
9. Dugong Sanctuary: 885 ha in Great Nicobar
10. Sir Hugh Ross Sanctuary: 106 ha to the southeast of Neil island

critical areas, development of strategies for eco-restoration of degraded areas, and ecological preservation and awareness programmes.

A field survey of seagrass habitats conducted by the Salim Ali Centre for Ornithology and Natural History proposes the creation of a Nancowrie Biosphere reserve for the protection of nine species of seagrasses and their high abundance of associated fauna, including dugong. Moreover, several sites in this island group, Camorta harbour, Camorta east and Nancowrie-Champion are highly disturbed sites requiring immediate conservation action. Considering the large area of forest range and richness of wildlife resources, wildlife units with wildlife camps at strategic points may be established. Islands such as Paget, Inglish-Henry Lawrence, Little Andamans, Camorta, Katchel and Little Nicobar may be considered as potential sites whose conservation would bring benefits.

RECOMMENDATIONS

The seas of the Andaman and Nicobar Islands are rich in biodiversity and fishery resources, but the present population of these islands exceeds the carrying capacity of some of the resources, and the land-based resources are insufficient to meet demands. However, there is a vast scope and potential for the development of inshore and deep-sea fishing as well as mariculture. Although deep-sea fishing is initially capital-intensive, requiring as it does proper processing and maintenance facilities, this could form the mainstay of the islands' economy. Mangrove fuelwood has been extracted in the past by wood-based industries as well as by private households. Given the ecological importance of the mangroves, the present ban on their exploitation should continue. Degraded mangrove should be identified, and wherever possible they should be restored. Some of the degraded areas could be earmarked as sites for mariculture, fisheries, jetties and other developmental needs. Rare and endangered marine fauna such as turtles should be identified, and the fishermen be informed of the need to protect them. The trade in seashells and corals should be banned. Corals should not be used as building material. Rapid population growth is likely to lead to resource deficiency, hence it is necessary to control immigration from the mainland. Town planning and measures for proper sewage and waste disposal should also be taken up. An integrated development strategy which permits environmental protection must be adopted in order to sustain development and preserve the uniqueness of the island ecosystem.

REFERENCES

Anon. (1994) Basic Statistics. Statistical Bureau, Andaman & Nicobar Administration, Port Blair, 249 pp.

Anon. (1995) Directorate of Information, Publicity & Tourism. Andaman & Nicobar Administration, Port Blair, 24 pp.

Ansari, Z.A., Ramani, P., Rivonkar, C.U. and Parulekar, A.H. (1990) Macro and meiofaunal abundance in six sandy beaches of Lakshadweep islands. *Indian Journal of Marine Science* 19, 159–164.

Ansari, Z.A., Rivonkar, C.U., Ramani, P. and Parulekar, A.H. (1991) Seagrass habitat complexity and macroinvertebrate abundance in Lakshadweep coral reef region, Arabian sea. *Coral Reefs* 10, 127–131.

Ananda Rao, T. and Chakraborti, S. (1987) Distributional Resume of coastal floristic elements in Andaman and Nicobar islands. *Current Science* 56 (20), 1045–1051.

Bakus, G.J. (1994) *Coral Reef Ecosystems*. Oxford & IBH Ltd., New Delhi, 232 pp.

Blasco, F. (1975) Mangroves of India. Inst. Francaise Ponich., *Trav. Sec. Sci. Tech.* 14(1), 154–156.

Gopinathan, C.P. and Panigrahy, R. (1983) Seaweed resources. In: Mariculture potential of Andaman & Nicobar islands—An indicative survey. *CMFRI Bulletin* 34, 47–51.

Gopinathan, C.P. and Rajagopalan, M.S. (1983) Mangrove resources. In: Mariculture potential of Andaman & Nicobar islands—An indicative survey. *CMFRI Bulletin* 34, 44–46.

INTACH (1989) An investigation into the effects of siltation, logging, blasting and human derived damage to corals in the Andaman & Nicobar islands. II Interim Report to NORAD.

Jagtap, T.G. (1992) Marine flora of Nicobar group of islands in Andaman sea. *Indian Journal of Marine Science* 21 (March), 56–58.

Jagtap, T.G. and Inamdar, S.N. (1991) Mapping of seagrass meadows from the Lakshadweep islands (India) using aerial photographs. *Photonirvachak* 19 (2), 77–81.

Marichamy, R. (1983) Zooplankton production in coastal waters. In: Mariculture potential of Andaman & Nicobar islands—An indicative survey. *CMFRI Bulletin* 34, 26–29.

Pillai, C.S.G. (1972) Stony coral of the seas around India. *Proceedings of the Symposium on Coral Reefs*. Marine Biological Association of India, pp. 191–216.

Pillai, C.S.G. (1983) Coral reefs and their environs, In: Mariculture potential of Andaman & Nicobar islands—An indicative survey. *CMFRI Bulletin* 34, 36–43.

Ramachandran Nair, P.V. and Gopinathan, C.P. (1983) Primary production in coastal waters. In: Mariculture potential of Andaman

& Nicobar islands—An indicative survey. *CMFRI Bulletin* **34**, 29–32.

Ramachandran, S. (1997a) Case studies on Coastal Fragile Area, Part 1: East Coast of India. Project report submitted to Central Pollution Control Board, GOI. 261 pp.

Ramachandran, S. (1997b) Case studies on Coastal Fragile Area, Part 2: West Coast of India. Project report submitted to Central Pollution Control Board, GOI. 261 pp.

Rangarajan, K. and Marichamy, R. (1972) Seasonal changes in the temperature, salinity and plankton volume at Port Blair, Andamans. *Indian J. Fish.* **19**, 60–69.

SAC (1992) Coastal environment. Space Application Centre (ISRO), Ahmedabad Report. Scientific note RSAM/SAC/COM/SN/ 11/92. 114 pp.

Saldana, C.J. (1989) *Andaman, Nicobar and Lakshadweep. An Environmental Impact Assessment.* Oxford & IBH Publishing Co. Pvt. Ltd., New Delhi. 114 pp.

Scheer, G. and Gopinadha Pillai, C.S. (1974) Report on the Scleractinia from the Nicobar islands. *Zoologica, Wein* **122**, 1–75.

Silas, E.G. (1983) Introduction: An indicative survey of the mariculture potential of Andaman and Nicobar islands. *CMFRI Bulletin* **34**, 1–9.

Wafar, M.V.M. (1986) Corals and coral reefs of India. *Proceedings of the Indian Academy of Science* (Suppl). **1**, 19–43.

THE AUTHOR

Sundararajan Ramachandran
*Institute for Ocean Management,
Anna University,
Chennai 600 025,
India*

Chapter 66

THE MALDIVES

Andrew R.G. Price and Susan Clark

The Maldives are a low-lying atoll nation comprising around 1120 islands within 26 geographic atolls. The country's continued physical and economic existence is dependent on the underlying coral reef platform. This provides a dynamic and unstable environment.

Coral species richness is high, with 187 known species and, based on cluster analysis, the coral fauna of the Maldives shows greatest affinities with Chagos and Seychelles to the west.

Both seagrasses and mangroves are uncommon, although mangrove diversity is remarkably high (10 species). Species of conservational significance include seabirds, which are used by Maldivian fishermen as a means of spotting tuna: 17 species are known to nest here. Endemic species include the fruit bat (*Pteropys giganteus ariel*); and there are several 'threatened' species listed in IUCN's Red Data Books, including five turtle species.

Since the 1970s the population has risen dramatically, particularly in the capital Malé. The population of Maldives was around 250,000 in 1995, with about 25% in Malé. Total population is projected to rise to c. 321,957 by 2005. Coastal uses and consequential environmental pressures include instances of heavy exploitation or degradation of coastal fisheries and are of much concern.

Other practical problems include unsustainable coral (and sand) mining. This has severely impaired the capacity of some reefs to act as natural sea defences, and undermined their biological role as areas for bait, other fish and as repositories of biodiversity. Problems of solid and liquid waste disposal can be acute, both in urban areas and on the tourist islands, eighty-seven of which are currently developed. Many aquifers are contaminated by faecal coliform bacteria, rendering the water unfit for direct consumption. Specific tourism-related concerns include impacts from infrastructures (i.e. resorts, jetties, vegetation clearance, turtle disturbance), and from diving and snorkelling. Other environmental concerns include occasional outbreaks of Crown-of-Thorns starfish, coral bleaching and global climate change. Threats posed by climatic changes and impending sea-level rise are taken very seriously, particularly since approximately 80% of Maldivian land area is less than 1 m above mean sea level.

A reasonable body of national and international environmental legislation exists, but implementation remains problematic. Environmental habitat restoration has included attempts to restore reefs degraded from coral mining but coral transplantation has not been cost-effective. Given the large number of unoccupied Maldivian islands (nearly 1000), prospects for tourism and other forms of development seem quite favourable, though more attention should be focused on carrying capacity. A critical challenge for the Maldives will be economic development that does not limit important future coastal uses, which could happen unexpectedly following incremental use and unrecognised environmental deterioration. If incidences of coral bleaching (as witnessed in 1998) are linked to global trends in temperature increase, then more serious problems are likely to arise. These are outside the scope of national reef resource managers and planners. However, healthy reefs are likely to better withstand adverse effects of global events than reefs degraded by the effects of local human activities. This may well become an increasingly important factor influencing the sustainability of reef uses on low-lying islands such as the Maldives.

Seas at The Millennium: An Environmental Evaluation (Edited by C. Sheppard)
© 2000 Elsevier Science Ltd. All rights reserved

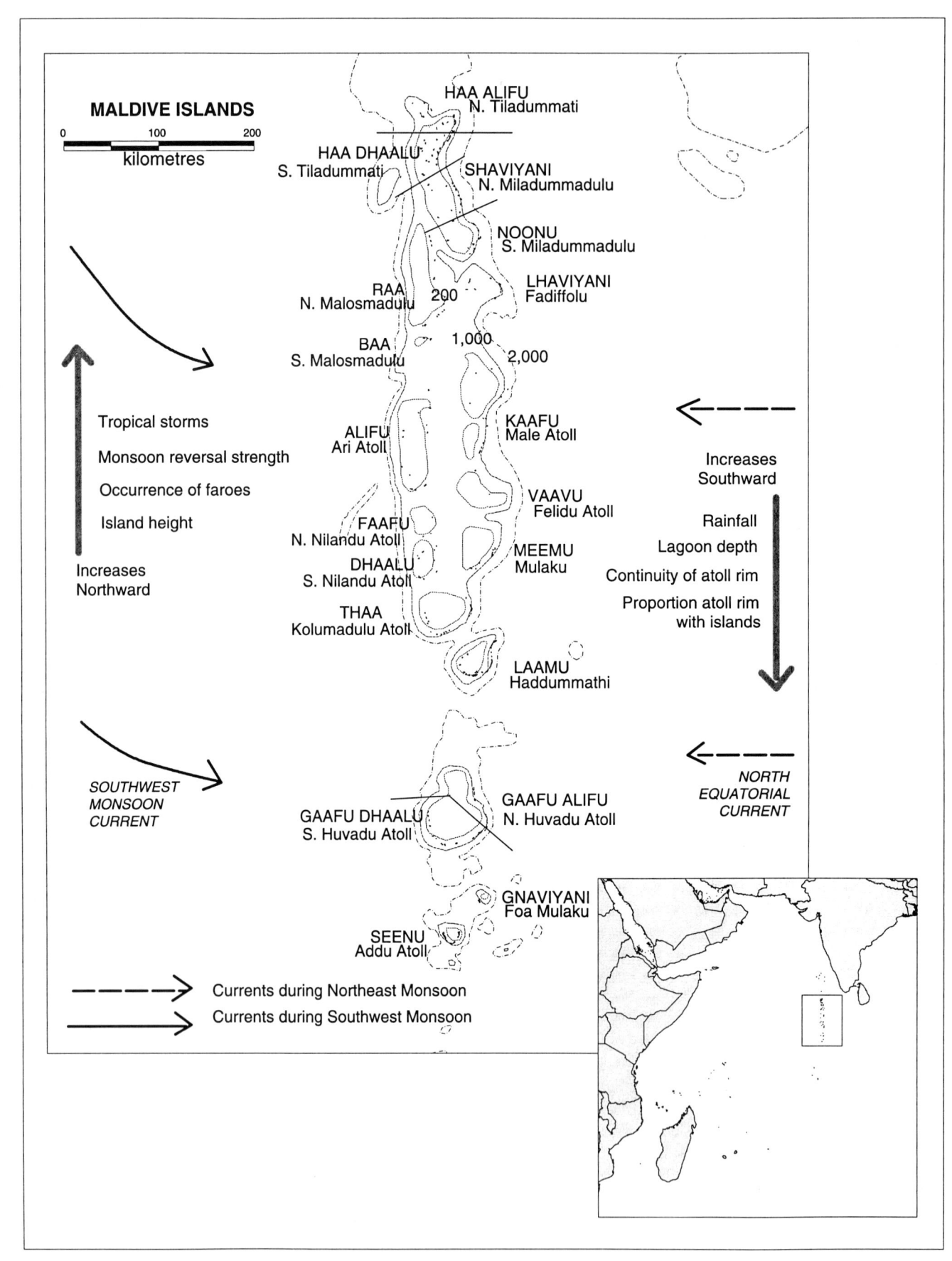

Fig. 1. Map of the Maldives showing the administrative atolls (upper case) and main geographic atolls (lower case; see also Pernetta, 1993).

THE DEFINED REGION

The Maldives form the central and largest part of the Lakshadweep (Laccadive)–Maldives–Chagos Ridge in the central Indian Ocean (Fig. 1). The archipelago comprises 26 natural atolls and some 1120 islands, the most southerly ones extending to south of the equator. The atolls form a single chain in the north and south, and a double chain in the central section. The islands are also grouped into 20 administrative atolls, with the capital island Malé (Fig. 2) forming a separate administration.

The islands are coralline in origin, made up of reef-derived rubble and sediments, providing a dynamic and unstable environment. Lagoons may contain patch reefs, knolls, micro-atolls and 'faros'. The latter, apparently unique to the Maldives, are special ring-shaped reefs, or 'mini-atolls', each with its own lagoon rising from the central lagoon floor. The diameter of faros ranges from 10s to 100s of metres. Together with the neighbouring Lakshadweeps, the chain of Maldivian atolls and faros is the largest and most extensive on earth (Risk and Sluka, 1999). Post-Eocene geological history, in particular sea-level changes during the Pleistocene, caused periodic exposure and flooding of the atolls and appreciable morphological changes (Bianchi et al., 1997; Risk and Sluka, 1999). Sea level is currently more than 130 m higher than some 20 000 years ago (Wilkinson and Buddemeir, 1994).

Of the 1120 islands only about 200 are permanently inhabited and 87 are used for tourist resorts. Despite extensive territorial waters (67,000 km^2) and a vast Exclusive Economic Zone (843,247 km^2), total land area is only 186 km^2. Further, 80% of Maldivian land area is less than 1 m above mean sea level and so threatened by impending climate change and sea-level rise.

Fig. 2. Aerial view of Malé, one of about 1120 islands which make up the Maldives. The continued physical and economic existence of the country depends on the underlying reef platforms. Extensive coastal infilling has occurred around Malé which is now almost square. This greatly impaired the natural protective capacity of the reefs, resulting in the need for an artificial breakwater on the south of the island (upper left of photo) for protection against flooding events (C.R.C. Sheppard).

SEASONALITY, CURRENTS, NATURAL ENVIRONMENTAL VARIABLES

Climate and Currents

The northern Maldivian islands are strongly influenced by monsoons, whereas Seenu (Addu) to the south is not. The northeast monsoon, the season characterised by relatively dry months and also popular for visiting tourists, lasts from December to March. The southwest monsoon extends from April to November, during which weather can be highly turbulent: severe storms, very strong winds and torrential rain. During the inter-monsoon period, weather is mixed and such conditions can also occur at the end of the dry season and in the early part of the wet/windy season.

Air temperatures are typically 28°C throughout the year, and average sea surface temperatures are similar (29°C) with a range of 27–30°C. Humidity seldom falls below 70% and, not surprisingly, the air is salty. Tides are mostly semi-diurnal, have a small range throughout, reaching a maximum of almost 1 m in the south (Wells and Sheppard, 1988).

Water circulation patterns vary seasonally. Ocean currents are strongly influenced by the trade winds and are of great strength. Early in the year northern atolls experience westerly surface currents created by the (NE) Monsoon, whereas the southern atolls have easterly currents due to the Equatorial Counter Current. In July, currents throughout most of the Maldives are easterly, as a consequence of the SW Monsoon, but these can be weak or even absent in the extreme south (Seenu Atoll).

In addition to seasonal influences, there are episodic and unpredictable events, such as cyclones, with devastating effects occurring about once every hundred years (Wells and Sheppard, 1988). One example was the severe storm which affected the Maldives in 1991. Widespread destruction resulted including damage to more than 3000 dwellings and the uprooting of or damage to more than 190,000 trees. It is possible that storms are becoming more intensive and unpredictable, particularly in the northern atolls (Maniku, unpublished observations). Small island states are particularly vulnerable to episodic events and to human-induced perturbations, including sea-level rise and other effects of global change.

Upper Water Column Characteristics

Maldivian waters are characterised by a seasonally fluctuating mixed layer of water from the Arabian Sea (ca. 36 ppt) and less saline water from the Bay of Bengal (ca. 34 ppt). Oceanographic surveys undertaken in the Kudhuvadhoo channel and Kaashidhoo channel in 1983 indicated that these areas were typical of the region (Pernetta, 1993).

The Veimandhoo channel between Thaa and Laamu atolls marks the southern boundary of the central plateau and is about 1800–2000 m deep. Local oceanographic conditions are not understood. However, atolls to the north experience greater seasonal upwelling than atolls to the south (Anderson et al., 1998a).

Primary productivity levels vary seasonally. Images from Coastal Zone Color Scanner (CZCS) reveal 'haloes' of enhanced production around oceanic atolls. These are common in Indian Ocean atolls, but poorly understood. They may represent low-level but sustained upwelling (Sheppard, 1999). Large plumes are associated with the Maldives, particularly during January to March and October to December, both of which match prevailing ocean circulation patterns (Sheppard, 1999). Zooplankton abundance in the upper water column (100 m) shows a similar spatial pattern to that of phytoplankton. Values in the Maldives and most of the northern Indian Ocean are relatively low (50–200 mg m^{-3}; Couper, 1983).

Gradients in Biophysical Features

Several features of Maldivian atolls, reefs and islands show geographic variation (Table 1). This is of ecological interest and has implications for tourism and other development since the biophysical environment is a major determinant of natural resources and also, directly or indirectly, of economic revenue. Environmental factors represent one set of criteria that can be used in development planning, to rank areas potentially suitable for future development Environmental criteria encapsulate two different concepts: risk to *development* (e.g. vulnerability of infrastructures to severe storms or other physically hazardous conditions); and risk to the *biological environment* (e.g. loss of biological diversity on a reef due to nearby dredging; see also Maxwell Stamp, 1995). Factors relating to atoll robustness and stability may be particularly important. In southern atolls there is a more continuous atoll rim than in the north, so providing better natural protection from wave action.

THE MAJOR SHALLOW WATER MARINE AND COASTAL HABITATS

Coral Reefs

Coral reefs have been studied in several atolls, although few resort island reefs have been studied, e.g. MPE, 1992a–d; Price and Firaq, 1996; Bianchi et al., 1997. Early studies (review by Wells and Sheppard, 1988) contributed significantly to the understanding of the reefs and corals of the Maldives. Latest assessments indicate 187 species of corals, with species and genus richness increasing from north to south down the Laccadive–Chagos Ridge (Sheppard, 1999). Of the 26 Indian Ocean sites assessed, only four contained more species than the Maldives: Chagos (220 species), Thailand, Mergui Archipelago (Myanmar) (214 species), NW Australia (311 species) and SW Australia (192 species). The coral fauna of the Maldives shows greatest affinities with Chagos and Seychelles to the west, based on cluster analysis (Sheppard, 1999).

The greatest coral growth and diversity generally occurs on the upper fore-reef slope at about 10 m depth, where coral cover averages 35–14% but can reach 100% (Risk and Sluka, 1999). Some 40 coral genera may be encountered even over a distance of 20 m, indicating a highly diverse coral fauna. One characteristic of Maldivian reefs is the high abundance and diversity of branching *Acropora* (Risk and Sluka, 1999). The rich coral fauna, together with the high diversity and abundance of reef fish are factors that attract tourists (Figs. 3 and 4). Overall reef condition is correlated significantly with reef fish species richness, coral species richness and coral cover, based on analysis of questionnaire data (Price and Firaq, 1996). More than 1200 species of coral fish are known for the Maldives, although only about 900 shallow-water species are reported to date (Randall and Anderson, 1993). Above depths of around 45 m reef building originates principally from zooxanthallate corals, but below this depth azooxanthallate corals (e.g. *Tubastrea micranthus*) can be important (Bianchi et al., 1997; Risk and Sluka, 1999).

Coral reef attributes on 43 resort islands have been assessed in terms of perceived present state (0–5) and also perceived magnitude of change in recent years (–4 to +4) (Table 2). Coral bleaching (loss of symbiotic algae, a sign of stress) was reported for 48% of the resort islands, and (not surprisingly) is associated with significantly reduced coral cover. Other analyses reveal a significant increase in algal

Table 1

Gradients in biophysical features in Maldives (from Wells and Sheppard, 1988; Anderson, 1992; Maxwell Stamp, 1995; quantitative data for many factors not available)

Environmental features	North–south pattern
Climatic factors	
Annual rainfall	increases to the south (N: 200 cm/yr; S: 230 cm/yr)
Strength of monsoon reversal	increases to the north
Incidence of severe tropical storms	increases to the north
Atoll features	
Depth of atoll lagoon	increases to the south (N: 50–60 m; S: 80–100 m)
Continuity of atoll rim	increases to the south
Proportion of atoll rim with islands	increases to the south
Occurrence of faroes (ring reefs)	increases to the north
Island height	increases to the north
Land availability (greatest number of uninhabited islands; largest total area of islands bigger than 150 ha)	greatest in south
Other features	
Proximity to oil spillages/pollution from tanker accidents	increase to the north
Biogeographic significance	increases to the south(?)

Fig. 3. The Maldives is renowned for its magnificent coral reefs and fish life which attract divers from around the world. More than 1200 species of coral fish are known for the Maldives, most of which (900 species) are found in shallow waters (A.R.G. Price).

Fig. 4. Table *Acropora* in Reethi Raa, North Malé Atoll. With 187 species of corals, the Maldives is the fourth richest area of the Indian Ocean in terms of coral species. This photograph was taken in 1994, prior to the major 'bleaching' event of 1998, which led to extensive mortality of coral reefs in the Maldives and other parts of the Indian Ocean (A.R.G. Price).

abundance, increase in the number of divers and decrease in overall reef condition. Changes in coral fish abundance and coral species were also evident.

Islands and Beaches

Maldivian islands, beaches and reef sediments are composed principally of coralline and other bioclastic skeletal material. Coral predominates in coarser fractions, but in sandy sediments calcareous algae (*Halimeda*) and foraminerans are common. Beach sands are fine and well sorted, while fine, poorly sorted calcareous sediments characterise lagoon floors (Pernetta, 1993). Sediments are composed of coral rubble at the approach of the reef edge. Outer reef slopes are steep and typically with luxuriant coral growth down to 15 m. Reef terraces, typically at depths of 3–6 m, 15–30 m and 50 m, are characteristic of the outer reef slope and represent past sea-level still stands (Risk and Sluka, 1999).

The unconsolidated nature of beaches and islands, together with seasonal opposing monsoon wind directions, create an environment that is naturally dynamic and unstable. Sands shift around the perimeter of islands, and, when lost to the lagoon or deep ocean floor, are gone from the island sub-system (Pernetta, 1993). Extreme episodic events, such as storms, cause more acute effects, which can even result in the loss of some islands and formation of others. Additionally, the building of tourist resorts and other physical infrastructures can interfere with natural sand movements and cause erosion. One example is causeways, which have caused severe erosion in Seenu atoll and led to beach undercutting and uprooting of coconut palms particularly on Feydhoo and Maradhoo.

Maldivian islands increase in height from the southern to the northern atolls (Table 1). This is probably of less

Table 2

Median (Mn) values for coral reef attributes and change in coral reef attributes for Maldivian resort islands using a ranked scale (0 to 5 for current conditions, –4 to +4 for change). Also shown is the significance of change based on two different tests (NS not significant; from Price and Firaq, 1996)

Coral reef attribute	Mn value of current state	Mn value of change in state	Significance of change in state	
			Wilcoxon's test	Mann–Whitney U test
Number of divers	3.0	2.0	$P<0.01$	$P<0.01$
Overall reef condition	4.0	–1.0	$P<0.01$	$P<0.01$
Live coral cover	3.5	0	NS	NS
Coral species	3.0	0	NS	$P<0.01$
Algal abundance	2.0	1.0	$P<0.01$	$P<0.01$
Reef fish abundance	4.0	0	$P<0.05$	NS
Reef fish species	4.0	0	NS	NS
Pelagic fish abundance	3.0	0	NS	NS
Urchin abundance	2.0	0	NS	NS
Coral bleaching	(48% of sites)			

importance as an indicator of island resilience, for example to sea level rise, than characteristics of the underlying reef platforms (Maxwell Stamp, 1995). Island configuration is also variable. Some islands (e.g. Gan on Gaaf Alifu/Gaaf Dhaalu) have a central depression and require infill before development. This represents an unknown, but possible future economic and environmental cost. Topographic features are described in detail for some islands (Stoddart et al., 1966; Woodroffe, 1989).

Mangroves

Mangrove ecosystems are uncommon in the Maldives, as is the case in Chagos to the south (Price, 1999), contrasting with the greater abundance of freshwater-dependent vegetation in both countries. Among the mangrove sites are Goidhoo (Goifurfehendu). While mangrove abundance is low, mangrove species richness is notably high (Spalding et al., 1997). Of the 14 species listed for the region comprising East Africa, the Middle East and Indian Ocean islands, ten are reported for the Maldives (Table 3). Mozambique is the only other country in the region with comparable species richness, although unlike the Maldives mangrove communities there have developed into extensive forests. Three of the species in the region are reported only for Maldives (*Bruguiera cylindrica, Exoecaria agallocha, Sonneratia caseolaris*), but are not actually endemic to the archipelago (Tomlinson, 1986). Other species, in particular *Avicennia marina and Rhizophora mucronata*, are more widespread (Table 3).

Knowledge of mangrove ecology in the Maldives is incomplete. Mangroves may be more prevalent in the north of Maldives, where there are also mangrove timber plantations (Maniku, unpublished data). Some northern islands (e.g. Kulhuddufhushi) may be suitable for mangrove replanting, offering potential for generating supplementary income (e.g. timber, honey production). Important mangrove areas in the southern Maldives include Hithadoo. The northeast of this island is of conservational importance, in view of its high biological productivity, important avifauna, and its role as a fishery/nursery area, for example for large shrimp (*Penaeus* sp.) and smaller shrimp species important as bait fish for tuna (Maxwell Stamp, 1997).

Table 3

Mangrove species in the Maldives and their relative occurrence across the East Africa, Middle East and Western/Central Indian Ocean region* (from Spalding et al., 1997)

Mangrove species	No. of countries in region (including Maldives) where species is recorded
Achrostichum aureum	6
Avicennia marina	19
Bruguiera cylindrica	1
Bruguiera gymnorrhiza	10 (excluding Yemen where species is extinct)
Ceriops tagal	9
Exoecaria agallocha	1
Lumnitzeria racemosa	9
Pemphis acidula	4
Rhizophora maculata	15
Sonneratia caseolaris	1

*Twenty two countries in the region: Bahrain, British Indian Ocean Territory (BIOT), Comoros, Djibouti, Egypt, Eritrea, Iran, Kenya, Madagascar, Maldives, Mauritius, Mozambique, Oman, Qatar, Saudi Arabia, Seychelles, Somalia, South Africa, Sudan, Tanzania, United Arab Emirates, Yemen.

Other Marine Floral Communities

Despite their documented ecological and bioeconomic significance elsewhere (e.g. Sheppard et al., 1992), seagrasses in the Maldives are poorly known. Extensive seagrass beds in natural environments are rare in the archipelago, as in Chagos. However, a large stand of *Thalassia hemprichii* and *Syringodium isoetifolium* is reported off Thuladhoo (Baa). *T. hemprichii* with smaller patches of *Thalassodendron ciliatum* are reported around Laamu (Pernetta, 1993). Seagrass beds may also be present in the vicinity of Funadhoo (Laamu Atoll) as the area is a feeding ground for green turtles (Frazier and Frazier, 1987). *Cymodocea* also occurs in the Maldives, although this seagrass and *Thalassia* were rare on the flats of nine atolls examined (Hackett, 1977).

A recent assessment of Indian Ocean benthic marine algae and Cyanobacteria indicate 237 taxa for the Maldives (Silva et al., 1996): Cyanophyta (27 taxa), Rhodophyceae (122 taxa), Chlorophycaea (69 taxa) and Phyophyceae (19 taxa). Earlier compilations indicate 285 species (Pernetta, 1993), the apparent reduction in total taxa suggesting that recent synonymizing ('lumping' species) has been greater than new records (additions) for the Maldives. Few if any detailed ecological studies have been undertaken on algal communities.

Proliferation of seagrass beds, and perhaps also algae, may be increasing on resort islands. Nutrient enrichment from sewage effluents leaching into lagoons is a likely contributing factor. Restricted water circulation due to sea walls, piers and other solid structures adds to the problem. Tourists consider seagrass beds irritating, since they dirty the bottom of lagoons and their detached leaves litter beaches. Much effort is therefore spent clearing the vegetation where it is seen as a problem, but without its environmental significance being properly determined. For example, seagrass might be useful by providing alternative energy pathways and absorbing 'stress' (e.g. from sewage). Leaf litter and natural terrestrial vegetation are similarly cleared from many resort islands to make them 'more attractive' (Wells and Sheppard, 1988).

Habitats, Sites and Species of Conservational Importance

Seabirds

Maldivian fishermen use seabirds as a means of spotting tuna. Hence, they are important socio-economically in addition to their conservational value. Of the 33 species of seabirds known to breed on oceanic islands or archipelagos of the northern Indian Ocean, 16 nest on Maldivian islands (Table 4). For comparison, 17 species are known to nest in Chagos and only six in the Lakshadweep (Laccadive) islands. The relatively low numbers recorded for the Lakshadweep may be partly or largely a reflection of the inaccessibility of these islands and limited observations.

Table 4

Seabird species breeding in the Maldives, Chagos and Lakshadweep (Laccadive) Islands (from Pernetta, 1993; Symens, 1996)

Seabird Species	Maldives	Chagos	Lakshadweeps
Puffinus pacificus		+	
Puffinus lherminieri	+	+	+
Phaeton lepturus	+	+	+
Sula dactylatra		+	
Sula sula		+	
Sula leucogaster	+	+	+
Fregata minor		+	
Fregata ariel	+	+	
Gelochidon nilotica	+		
Sternia bergii	+	+	+
Sterna bengalensis	+		
Sterna douagallii	+	+	
Sterna sumatrana	+	+	
Sterna albifrons	+	+	
Sterna saundersi	+		
Sterna anaethetus	+	+	
Sterna fuscata	+	+	+
Anous stolidus	+	+	+
Anous tenuirostris	+	+	
Gygis alba	+	+	
Total	16 species	17 species	6 species

Areas of known importance for seabirds in the Maldives include the following (Pernetta, 1993), although surveys have not been exhaustive:
- Filadu island, Thiladhunmathi atoll (tern rookery)
- Maalhosmadulu atoll (lesser frigate bird colony)
- Hithaadoo island, Gaaf Alifu (lesser frigate bird roosting area)
- Mulikolhu island, Seenu Atoll (white tern breeding and bird roosting in general)

Turtles

Five species of turtles are among species of conservational significance in the Maldives. All are listed as 'threatened' in the IUCN Red Data Books. An estimated 870–1300 green turtles (*Chelonia mydas*) nest annually on most uninhabited islands and some inhabited islands. Hawksbills (*Eretmochelys imbricata*) are also widespread and breed throughout the archipelago, but probably not in numbers exceeding 500 annually. Loggerhead turtles (*Caretta caretta*), the Olive Ridley turtle (*Lepidochelys olivacea*) and Leatherback turtle (*Dermochelys coriacea*) are less common.

Turtle breeding may be greater in the north than further south (Maxwell Stamp, 1995) based on consideration of two very limited datasets (not shown) which have been only partially analysed: (1) the number of nesting islands, which should be a reliable index of breeding, but the data are incomplete for north-central 'Region 2' and south-central 'Region 3' atolls (see Table 6 for atoll groupings); and (2) counts of the number of nesting females and of eggs, which may be a less reliable estimate, but the data are more complete. Despite the possibility of more suitable (less rocky) habitat in the south, the first dataset may give a better estimate of turtle nesting.

Marine Mammals

Those species which are of conservational importance are considered briefly below.

Endemic Species

Among the known endemics are five species of *Pandanus* trees/shrubs and the fruit bat, *Pteropys giganteus ariel*. Another fruit bat, *Pteropus hypomelanus maris*, is most common in Seenu and other southern atolls. However, knowledge for the Maldives is fragmentary and levels of endemism may be higher than previously supposed.

OFFSHORE SYSTEMS

Neither offshore pelagic nor benthic systems in the Maldives have been well studied, although useful regional information has been summarised (Couper, 1983; IUCN/UNEP, 1985; Pernetta, 1993). Surface temperatures decrease rapidly to below 20°C at about 90–100 m depth and at 500 m depth the minimum oxygen level was recorded (1.2 ml l^{-1}). Of the nekton, information on marine mammals is summarised below, while fish and fisheries are considered in greater detail in later sections. Knowledge of the benthos is also very limited, although some features are highlighted below.

Marine Mammals

Whales are uncommon in the Maldives, the Melon-headed whale (*Peponocephala electra*) being among the few species recorded (Pernetta, 1993). Other cetaceans known include at least five species of dolphins: the common dolphin (*Delphinus delphinus*), the spotted dolphin (*Stenella attuata*), the spinner dolphin (*Stennella longirostris*), the striped dolphin, (*Stenella coeruleoalba*) and Risso's dolphin (*Grampus griseus*). Records of the dugong (*Dugong dugon*) are uncertain. Additional species may occur in the Maldives, given that at least 25 species are known to exist in or around the South Asian Seas region and the actual number may exceed 30 species (IUCN/UNEP, 1985).

Benthos

Oceanic sediments in the Maldives are of mainly calcareous origin and oozes are predominant on the deep-sea floor. Benthic biomass in the Maldives region is around 1–10 g m^{-2}, which is the fourth category in an ascending scale of eight compiled for the world's oceans (Couper, 1983).

POPULATIONS AFFECTING THE AREA

This section summarises the main human population trends in the Maldives, as well as the main concerns arising from environmental use and other activities. Further details of particular resource uses and associated environmental pressures are given in subsequent sections.

Demographic Patterns

Until comparatively recently, the life styles of Maldivians had little direct impact on the environment, particularly since human populations were relatively small and stable. This situation is changing rapidly. Since the 1970s the population has risen dramatically, particularly in Malé, but also in some of the other atolls and as a result of increased tourism. The population of Maldives was around 244 600 in 1995 (Statistical Yearbook of Maldives, 1997) and is projected to increase to over 350,000 by 2005 (Maxwell Stamp, 1995).

The demographic situation is made more complex by migration, particularly since many rural atolls experience significant levels of out-migration. Estimates for 16 atolls predict a population increase of 73% within the next 25 years, assuming out-migration will stop, but not taking into account future in-migration which might result from proposed development (Maxwell Stamp, 1995). Population densities vary considerably between islands even in the same atoll. Higher density islands are more likely to justify infrastructure and perhaps private investment, due to economies of scale, which will clearly have environmental implications.

A programme is currently underway to develop regional economic centres with support from the Asian Development Bank. The project is financing port facilities and infrastructural development to promote economic activity in the north (Kulhudufushi in Haa Dhaal) and south (Hithadhoo in Seenu) of the country thereby reducing dependence on Malé.

Tourism Trends

The first tourist resorts in the Maldives opened in 1972, close to Hulhule Airport and the capital Malé (Fig. 5). This was followed by an exponential increase in cumulative bed space in resorts from 1972 to 1995 (Price et al., 1998). Until 1981 more than 91% of available bed space was provided by resorts on North and South Malé atolls (Kaafu Atoll). Thereafter development of new resorts in North and South Ari atolls and other atolls accounted for most of the increase in bed-space. Some additional bed space is provided by hotels, guest houses and registered vessels. Bed capacity of resorts has increased from 280 in 1972 to 11,444 in 1996, representing a steady annual increase.

The steady development of new and upgraded resorts in the Maldives (Fig. 6) has been mirrored by an equally

Fig. 5. Kurumba Village opened in the early 1970s and was the first resort island in the Maldives. Visitors come to the Maldives not only for diving and watersports but also to enjoy a tranquil holiday setting in the middle of the Indian Ocean. Tourism is now the principal source of national revenue (A.R.G. Price).

Fig. 6. Resort on an island in North Malé Atoll. The close proximity of Maldivian resorts to the Indian Ocean is a major attraction to visitors. Most of the Maldivian islands are less than 1 m above sea level, raising concern about the effects of impending climate change and sea-level rise (A.R.G. Price).

dramatic growth in tourist arrivals. A peak of 300,000 visitors occurred in 1995, spread over 74 resorts (Price et al., 1998). The number increased to 338,733 visitors in 1996 spread over 73 resorts. By the year 2000 tourist arrivals may approach 400,000. Clearly, demand has more than kept pace with supply. Some models predict a tourist demand of more than 1 million by the year 2001 (Hameed, 1993). Currently (Dec. 1998), 87 islands have resorts or resorts under construction (Min. of Tourism, pers. comm.).

Summary of Coastal Uses and Environmental Issues

A summary of the main environmental concerns in the Maldives arising directly or indirectly from human activities is given in Table 5. Most problems relate to practical problems, policy issues and cross-sectoral issues. Problems and issues relating more specifically to tourism are treated

Table 5

Summary of Main Environmental Concerns in the Maldives (from Price and Firaq, 1996; Brown, 1997).

Practical Problems
- coral and sand mining, which has severely impaired the capacity of some reefs to act as natural sea defences, and undermined their biological role as fishery areas and repositories of biodiversity
- marine resource exploitation
- waste disposal problems, in particular sewage and solid waste
- human population pressures, from high growth rates of the resident population in Malé and areas such as Seenu and Thulhaadoo, and from continued growth of tourism
- coastal erosion and sedimentation from coral and sand mining, dredging, coastal construction and reclamation, exacerbating impacts from natural events such as wave damage and flooding
- sea-level rise, in particular the potential threat to urban centres, populated atolls and tourist islands
- over-consumption and degradation of freshwater / land resources

Policy Issues
- policies leading to non-sustainable resource exploitation, and related to this
- further promotion of economic incentives and disincentives and other policies to conserve biological resources, following the example of the recent reduction of duty (30% to 15%) on imported construction materials
- implementation of EIA prior to new development projects, also recognising that human and financial resources are needed for this, and for monitoring compliance with EIA and related environmental guidelines/standards

Cross-Sectoral Issues
- limited understanding and data on cross-sectoral environment development issues
- the need for greater incorporation of environmental concerns into development planning
- conflicts between incompatible activities in the same area, such as fishing and diving
- the current land tenure system, which creates a climate of uncertainty and limits the commitment of lease holders to environmental management, and related to this
- the uncertain future status of islands which are currently unpopulated, and hence function as valuable fishery replenishment zones, turtle and bird breeding sites, i.e. as protected areas

Environmental Issues And Concerns Related To Tourism
- resort siting, construction and choice of construction materials
- resort operations (e.g. solid and liquid waste, other pollution, energy issues)
- tourist diving and souvenir or 'curio' species collection
- effects of cruise ships, boating, fishing and surfing
- indirect environmental effects of tourism
- environmental impact of other sectors on tourism

separately (see also Dawson Shepherd, 1995). A recent assessment ranked the following anthropogenic influences as being the most significant, listed in order of importance: (1) marine resource exploitation; (2) sedimentation (from dredging); (3) tourism; and (4) pollution (Brown, 1997).

In view of the strong role of the environment in virtually all aspects of Maldivian life, and the increasing demands placed upon it, the environment is a factor that is likely to assume increasing importance in development planning.

RURAL FACTORS

Agriculture

Details of agricultural and other terrestrial resource use are summarised in Table 6. These resources appear to be most available in southern atolls, based on a preliminary analysis. Regional differences in freshwater availability (e.g. aquifers) are not known, although some aquifers (e.g. Malé) have become heavily depleted. Evidently, freshwater resources are limited more by storage capacity than rainfall availability. Desalination is becoming more widespread at least on the resort islands, due to heavy tourism demand and decreasing capital and running costs of desalination plants.

Environmental effects of terrestrial resource use are not known in detail, although the following concerns have been identified (Maxwell Stamp, 1995).

- *loss of coconut trees and crops* (following projected urban expansion on Hithadoo and other islands of Seenu);
- *reduced soil fertility* (and hence vegetation growth from encroachment/degradation of productive land by urban development);
- *erosion* (from poorly planned construction projects (e.g. of causeways), which has caused severe erosion in Seenu and led to beach undercutting and uprooting of coconut palms particularly on Feydhoo and Maradhoo);
- *freshwater availability for agriculture* (and the possibility of stress to vegetation by increased salination of groundwater aquifers due to overdraw);
- *tidal flooding of cassava* (on Hulhumheedhoo and more widespread concern of possible saline intrusion of soils as a result of sea-level rise and associated events);
- *fire damage* (prevalent in the past and remains a threat in densely populated islands, can damage vegetation, food resources and housing);
- *potential impact of poisons for agricultural pests* (on the environment and water supplies, which are now being phased out by greater use of biological control and more environmentally acceptable agrochemicals).

Aquaculture

Despite the potential for aquaculture in the Maldives, it is practised very little. However, a pilot project involving pearl oyster culture is underway (Min. of Tourism, pers. comm).

Table 6

Regional differences in selected terrestrial natural resources on 16 atolls in the Maldives (from Maxwell Stamp, 1995; Jameel, pers. comm.)

Natural Resource	Region 1 (Haa Alifu, Haa Dhaalu)	Region 2 (Shaviyani, Noonu, Raa, Baa, Lhaviyani)	Region 3 (Meemu, Faafu, Dhaalu, Thaa, Laamu)	Region 4 (Gaafu Alifu, Gaafu Dhaalu, Gnaviyani, Seenu)
Agricultural crops and terrestrial resources (categories)*	++++	+	++++	++++
Agricultural crops and terrestrial resources (additional notes)	Agriculture crops include beetle leaf, pumpkins and other vegetables (Haa Dhaalu), root crops and bananas (Haa Alifu). In addition to coconuts there is a little timber production.	Agriculture crops include bananas and other fruits, and there is some timber production.	Many root crops, some bananas but otherwise few fruit trees. Some vegetables important and some timber and coconuts. Agriculture the main livelihood. Many islands are uninhabited.	Arable resources mainly root crops (e.g. cassawa, sweet potato, tara). These formed major component of the local diet up to about 30 years ago. Food preferences now changing with rice (imported) becoming more important. Fruits grown include mangoes and bananas. Timber production also important.
Coconut production	3 545 348	6 275 455	6 919 013	9 418 392

*Fewest no. of categories of agricultural crops = +, greatest no. of agricultural crops = ++++.

COASTAL EROSION AND LANDFILL

Coral and Sand Mining

For a nation built of coral reefs, coral traditionally has been a major building material (Sluker and Miller, 1998) (Figs. 7 and 8). Massive corals such as *Porites, Goniastrea* and faviids are preferred, although large *Acropora* is also used (Risk and Sluka, 1999). Sand for cement is commonly taken from lagoons and other shallow-water areas. Mining, particularly of coral, has a major impact on reefs and the practice cannot be sustained at the present rate (Brown and Dunne, 1988; Brown et al., 1990; Dawson Shepherd et al., 1992; Risk and Sluka, 1999). Surveys of mined reefs show few signs of recovery 20 years after mining has ceased (Risk and Sluka, 1999). Biological impacts include reduced biomass and altered community structure of bait fish (Fig. 9), which are used in the offshore tuna fishery. Weakening natural sea defence against waves provided by the reefs is an equally or more serious consequence.

Coral mining is prohibited on 'house reefs', the natural defence of resort islands, as well as on atoll rim reefs and important bait-fishing reefs. It is now legally permitted only at specified sites (Malé and Ari atolls). Nevertheless, the practice continues elsewhere, but due to the recent reduction of duty (to 10%) on imported construction materials it is on a smaller scale. The use of 'sacrificial' faros has also been proposed, but this possible solution is not without problems (Risk and Sluka, 1999).

It is not only sites mined but also areas where construction materials are used that become environmentally degraded (Fig. 10). Problems often arise from interference with natural (e.g. seasonal) erosion and deposition on beaches and islands. The interlinking of islands of Seenu Atoll to facilitate transportation led to unwanted erosion and other problems. However, re-opening up the causeway between Gan and Fehydhoo to re-establish water flow is reported to have increased the abundance of bait fish in the lagoon (Maxwell Stamp, 1997). Problems associated with sedimentation of reefs and other photosynthetic communities can be particularly severe.

Land Reclamation

The rapid increase in population and scarcity of land has resulted in many islands acquiring additional land through reclamation. This usually takes place at the reef edge and results in the loss of renewable resources. Coastal remodelling can affect beach accretion patterns and may also affect reefs downstream due to sediment loading. In addition it may have serious environmental consequences in terms of increased vulnerability to erosion, flooding and storm damage.

Various incidences of high waves and associated flooding have occurred in the Maldives. One such event occurred in April 1987 and caused significant damage to the capital island Malé, the airport and resort islands facing east on North and South Male atoll. On Malé, coral breakwaters built to protect the reclaim area were destroyed with subsequent loss of fill material. According to one estimate (JICA, 1987) 30–40% (360,000 m^3 tonnes) of the fill material was lost as a result of the high wave incident.

Malé's Artificial Breakwater

The economic costs of inadequate attention to the environment became all too apparent following extensive reclamation schemes in Malé. This resulted in the need for

Fig. 7. Houses, other buildings and small jetties in the Maldives have traditionally been built from coral. Massive corals such as *Porites*, *Goniastrea* and faviids are preferred, although large *Acropora* is also used. Coral mining cannot be sustained at the present rate. Coral mining is prohibited on 'house reefs', the natural defence of resort islands, as well as on atoll rim reefs and important bait-fishing reefs. It is now legally permitted only at specified sites including Malé and Ari atolls (C.R.C. Sheppard).

Fig. 10. Sand is mined for use in cement in the Maldives and is also dredged for use as infill. Harmful effects of sand and coral mining include direct habitat loss or degradation and increased sedimentation, which adversely affect coral reefs and other photosynthetic communities (A.R.G. Price).

Fig. 8. Coral piled up for use as a building material on an island in North Malé Atoll. Coral mining has had a major impact on Maldivian reefs. Surveys of mined reefs have shown few signs of recovery 20 years after mining has ceased (A.R.G. Price).

Fig. 11. Artificial breakwater along south of Malé to provide protection from flooding events following extensive coastal infilling and reclamation. This high technology solution, costing US $12 million, or US $8000 per linear metre of coastline, would not have been necessary had the possibility of adverse environmental impacts been considered (C.R.C. Sheppard).

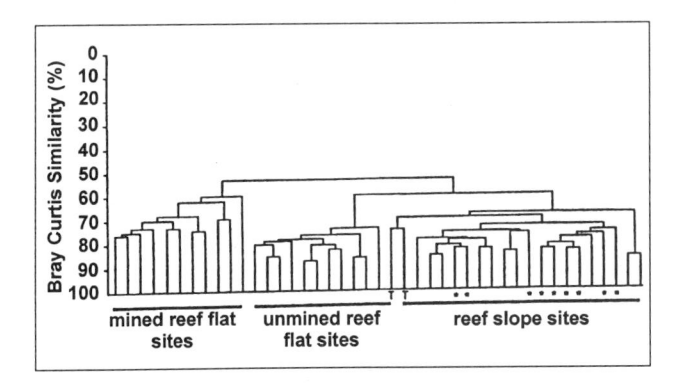

Fig. 9. Comparison of reef fish abundance in mined and unmined reef flat sites and at reef slope sites adjacent to unmined and mined reef locations, the latter shown by asterisk (from Brown et al., 1990).

an artificial breakwater on the south of the island for protection against flooding events (Figs. 2 and 11). This high technology solution, costing US$12 million, or US $8000 per linear metre of coastline, would not have been necessary had the possibility of adverse environmental impacts been considered. This provides a shadow price for the value of coral reefs in terms of sea defence function (Edwards, 1989).

Coastal Erosion from Pleasure Boats

Erosion and changes to the dynamics of water movement can be caused by speedboats. Environmental problems occur when these craft stop suddenly from high speed (Maniku, unpublished data). The effects are most

Fig. 12. Fishing boats in the capital, Malé. The Maldives is a nation with strong maritime traditions, and fishing remains a mainstay of the national economy (A.R.G. Price).

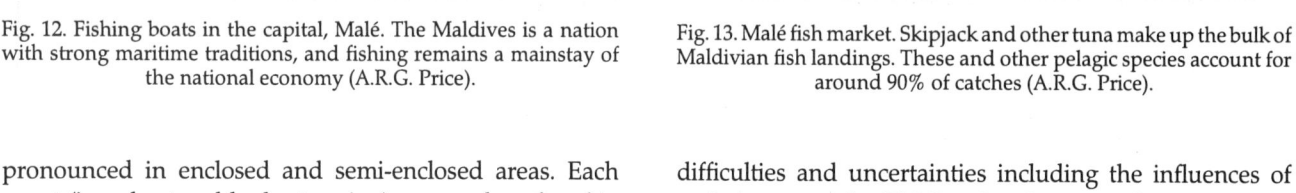

Fig. 13. Malé fish market. Skipjack and other tuna make up the bulk of Maldivian fish landings. These and other pelagic species account for around 90% of catches (A.R.G. Price).

pronounced in enclosed and semi-enclosed areas. Each event (i.e. a boat suddenly stopping) can produce four big and seven small waves, which increase shore erosion dramatically (Maniku, unpublished data). Hence a code of conduct for fast boats is needed, particularly since they can also pose a hazard to bathers and divers and can create unacceptable noise.

EFFECTS FROM URBAN AND INDUSTRIAL ACTIVITIES

Fisheries

Overview

Maldivian fisheries are based mainly on pelagic species (90% of catch), the remainder being demersal, mainly reef species. Tuna make up the bulk of catches, in particular skipjack (*Katsuwonus pelamis*) which represented 78% of catches in 1990 (Figs. 12 and 13). This is an open water, pole-and-line fishery dependent on reef fish bait from lagoons (see below). Skipjack are caught mainly to the east of atolls, yellowfin mainly to the west of them. Troll-fishing is also undertaken for other tuna-like fish such as *Neothunnus macropterus*, *Auzis thazard* and *Euthynnus affinis* (Pernetta, 1993).

Prior to tourism, fisheries were the major source of national revenue. The industry remains important and annual exports represent a multi-million dollar industry (Table 7). The tuna fishery still provides a major source of employment and food in the outer atolls. Dried shark are exported to Hong Kong and Japan. Shark liver is also in demand and the oil is extracted and exported. Preliminary stock assessments of tunas showed that Skipjack, Yellowfin, Frigate and Little tuna did not show signs of overfishing, whereas Bigeye tuna was heavily fished (Anderson, 1994). More recent stock assessments have been undertaken (Anderson et al., 1998a), which also highlight assessment difficulties and uncertainties including the influences of such factors as the El Niño Southern Oscillation (ENSO). Stock status is summarised below:

– Skipjack: some concerns about the stock, for example from stagnation of catches in recent years and a decline in catch per unit effort (Adam and Anderson, 1998);
– Yellowfin: concern about stock status, due to decline in catch rates in Maldivian west coast fishery and other signals (Anderson and Waheed, 1998);
– Frigate tuna: some data indicative of high exploitation rates, other data suggesting that the stocks are not being overexploited (Anderson et al., 1998b);
– Kawakawa: decline in catch and catch rates in 1997 (for reasons not fully understood), but no evidence to unsustainable fishing before then (Anderson et al., 1998c).

Regional variations in catch and catch rates are evident (Table 8) and are also reported for 23 fish species (Anderson, 1992). Large skipjack are more common in the north, although catch rates for this tuna may be higher in the south. Of the six species of reef fish assessed, all but one species (*Cephalapholis sonnerati*) appear to be more abundant in the south. The extent to which these availability/abundance trends reflect true stock abundance or other factors, such as proximity of a fishing area to ports, markets and fish catchability is not completely understood.

Lagoonal bait fisheries

Live bait is taken from reef areas and used for the offshore tuna fishery (above), a practice extending back to at least the 14th century. The twenty species regularly taken fall into three groups: (1) fusiliers (Caesionidae), (2) cardinalfish (Apogonidae) and (3) silver sprat (*Spratelloides gracilis*) or shorthead anchovy (*Encrasicholina heteroloba*) (Risk and Sluka, 1999). Formerly, collections were made from small sailing craft using small nets and harvesting was sustainable. Newer technologies have led to dramatically increased catches. While there is not yet documented

Table 7

Summary of fisheries data for Maldives (production and exports in metric tonnes $\times 10^3$; from MPHRE, 1994; Anderson et al., 1998)

	1987	1988	1989	1990	1991	1992	1993	1994	1995	1996
Fish Catch	57.0	71.5	71.2	76.4	80.7	82.0	89.9	104.0	104.0	105.4
Skipjack tuna	42.1	58.6	58.1	59.9	58.9	58.6	58.7	69.4	70.3	66.5
Yellow fin tuna	6.6	6.5	6.1	5.3	7.7	8.7	10.1	13.1	12.5	12.4
Other tuna	3.2	3.0	3.6	5.2	4.5	6.1	9.7	7.1	7.1	10.9
Reef and bottom fish	5.0	3.4	3.4	6.0	9.6	8.6	11.4	14.4	14.6	15.6
Exports of fresh/frozen fish	13.7	19.7	19.7	17.1	10.1	5.5	9.8	7.4	3.0	13.2
Island Enterprise	4.2	4.6	–	–	–	–				
State Trading Organisation	9.5	14.8	–	–	–	–				
Exports of other tuna fish*	25.8	17.3	32.1	41.5	48.9	45.2				
Dry skipjack tuna (Maldives fish)	6.1	6.1	9.9	12.1	16.4	15.5				
Dry salted skipjack tuna	8.5	1.3	3.7	6.3	6.9	4.0				
Dry salted reef tuna	4.3	1.7	1.9	2.3	4.0	3.3				
Canned fish	6.9	8.2	16.6	20.8	21.6	22.4				
Total fish exports	39.5	37.0	51.8	58.6	59.0	50.7				
(US$ million)	(17.3)	(24.2)	(30.9)	(31.1)	(39.9)	(31.6)				
Local consumption and changes in stock	17.4	34.5	19.4	17.8	21.7	31.3				
Fishing vessels										
Number of mechanised boats	1334	1338	1440	1449	1459	1533		1622	1628	1633

*Fresh fish equivalents.

Table 8

Regional variations in fish catch, effort and catch per effort for the Maldives (from Maxwell Stamp, 1995)

Region	Average Total Fish Catch (mt)		Mechanised Dhoni Data (av. 1991–94)		
	1985–90	1991–94	Fish catch (mt)	Fishing trips	Catch per effort (kg/dhoni trip)
Region 1 (Haa Alifu, Haa Dhaalu)	6,516	8,681	8,158	18,287	446
Region 2 (Shaviyani, Noonu, Raa, Baa, Lhaviyani)	20,441	25,546	24,331	68,472	355
Region 3 (Meemu, Faafu, Dhaalu, Thaa, Laamu)	13,410	17,936	17,629,	43,011	410
Region 4[1] (Gaafu Alifu, Gaafu Dhaalu, Gnaviyani, Seenu)	13,138	17,846	17,634	33,397	528
Maldives total[2]	66,201	89,184	86,628	212,193	408

[1] Data for Gaaf Alifu included.
[2] Includes data for other regions (i.e. Malé/central region).

evidence of stock decrease from this, fishers complain of local shortages. Of much concern are the adverse effects of coral and sand-mining on bait fish (above). It is possible that bait fish may become more important as a source of food on existing resort islands (e.g. Gan) or new ones (e.g. Vilingili).

Groupers and Other Coral Reef Fisheries

Coral reef fisheries target reef fish, molluscs, spiny lobsters and octopus. Although of lesser importance than the tuna fishery, subsistence fishing for groupers and other reef fish has always occurred. However, fishing has become more commercialised following increasing demands from tourists and overseas markets, and in some cases has led to spectacular stock crashes (Risk and Sluka, 1999).

At least 40 species of groupers are known in the Maldives, but only a few are targeted as fishery species. Since 1993 live grouper have been exported, via a Maldivian, usually to Hong Kong and Japan. Prices commanded for live groupers depend on species and size, *Plectropomus* spp., *Epinephelus fuscoguttatus* and *E. polyphekadion* attracting highest prices in Laamu Atoll (Risk and Sluka, 1999). Exports rose from 200 tons in 1994 to 1000 tons in 1995, and a 10-fold increase was projected for 1996 (Risk and Sluka, 1999). Capture of groupers by fishermen in favoured dive sites can conflict with tourism. However, night fishing and 'grouper watching' are also both important resort activities, which can lead to within-resort conflicts.

The long-term sustainability of this needs to be determined, as local stocks are showing signs of over-fishing. Protection of spawning aggregations is one conservation option. Resource-use conflicts and stock depletion are fairly new to the Maldives as fishermen have traditionally enjoyed open access to reef resources. Various recommendations have been developed to implement management measures for single species fishery whilst supporting the traditional multi-species fishery. These include open and closed periods in specific atolls, size restrictions, gear restrictions and aquaculture (Shakeel and Ahmed, 1997).

Holothurians

The fishery for holothurians, or sea cucumbers, began in 1985 and is directed at markets in China, Singapore and Hong Kong. Collecting is undertaken in shallow lagoons by snorkelling or scuba. Despite overfishing and severe stock reduction in less than a decade, prices obtained for quality specimens remain high at US $30 per kg dry weight (Risk and Sluka, 1999). Regulations for this fishery were developed but are not enforced.

Other Living Marine Resources

Giant clams (*Tridacna maxima* and *T. squamosa*) are not eaten locally (Pernetta, 1993). However, an export fishery for *Tridacna* adductor muscle began in 1990, when 125,000 clams were harvested. The fishery was considered unsustainable and licensing was discontinued in 1991 (Risk and Sluka, 1999). Small reef fishes and various invertebrates including large anemones (*Radianthus* and *Stoichactis*) and starfish (e.g. *Linckia* and *Fromia*), are also exported for the aquarium trade. Shells, coral, tortoiseshell (hawksbill turtle) are also exported as souvenir or curio species. Such species are on sale in resort shops and in Malé. The collection of such species (by licensed Maldivians) is apparently being phased out. Although the sale of such species still continues in resort shops and in Malé, it has been banned from export since December 1996. The money cowrie (*Cypraeea moneta*) was collected and exported in the past (Risk and Sluka, 1999).

Urban, Environmental and Health Problems

Conditions damaging to human health are often also unfavourable for the environment. The discharge of raw sewage into coastal waters can cause nutrient enrichment, algal blooms, deoxygenating and human health problems depending on the siting of the outfall. Effects include marine floral blooms, sometimes of undesirable species, and often a decrease in species diversity. The process can lead to deoxygenation and coral death. Sewage-related problems are of greater concern around densely populated islands (e.g. Kulhudhuffushi) and some tourist islands. Many aquifers are also contaminated by faecal coliform bacteria, rendering the water unfit for direct consumption.

Solid waste disposal includes burning, compaction of cans and dumping of refuse far out to sea, but the correct procedures are not always followed. It was seen to be a problem on Hithadoo, Gan (eutrophication, algal and seagrass blooms) and other islands in Seenu visited, as it is throughout much of the Maldives. Plastics are also burnt in Seenu and can emit toxic emissions. Much rubbish is evident on Kulhudhuffushi island and shark remains are dumped in the harbour. Urban expansion will stretch facilities which are already inadequate. Solid wastes encourage rats, which have health implications and impede seabird nesting. New planned solid waste sites should not conflict with uses that may depend more critically on the coastal strip (e.g. tourism/research infrastructures).

Lidkea (1993) investigated quantities of solid waste and found rural islands produce 0.1–0.15 kg/day/per capita whilst resort islands produce a much higher rate of 0.9–1.8 kg/day/per bed. Recommendations to improve existing solid waste management include recycling, composting and reduction effort.

In the capital island, solid waste management has been linked with a land fill development project. Waste from Malé is taken to a nearby reef (faro) by barge and dumped into excavated bunds and covered with sand and coral rubble dredged from the lagoon and reef flat area. In addition to direct physical damage, there are two major sources of pollutants that will cause damage to the coral reef environment: leachate from the waste material in the landfill and increased sedimentation and silt from dredging (Clark, pers. observ.).

Tourism Infrastructures and Activities

Impacts from Tourism Infrastructure

Tourism expansion carries environmental and social costs. Environmental impacts accrue both directly from use of reef resources and indirectly from construction and discharge of pollutants. Effects of tourist infrastructures are multiple and difficult to predict, since reefs and islands are naturally highly dynamic (see Pernetta and Sestini, 1989; Pernetta, 1993). Detection of change is compounded by limited baseline data, particularly in the Maldives (Price and Firaq, 1996). Resort siting, construction and operation are critical factors, since they affect the sustainability of tourism even before users get in the water. Environmental concerns associated with resort construction are summarised in Table 9 (see also Nethconsult/Transtec, 1995a; Price and Firaq, 1996; Price et al., 1998), including qualitative assessment of their severity.

The principal problems arise from lack of understanding of the fragility of tropical coastal environments and many are preventable. Following construction, environmental impacts from tourist operations continue and have caused concern in many parts of the world (Hawkins and Roberts, 1994), including the Maldives.

Small islands by their very nature have only a limited capacity to accommodate tourist waste. Tourist resorts tend to produce more solid waste per capita than is generated by locals. Waste production by resorts in the Maldives typically ranges from 40–200 t yr^{-1} (Hameed, 1993). Some of this ends up on the reefs themselves (e.g. plastics, fishing lines and building materials) but much increases pressure on local landfills or is improperly disposed of.

Table 9
Environmental impacts and issues associated with resort construction and their estimated severity (1: minor, 2: moderate, 3: high) in the Maldives (from Price et al., 1998)

Impact or issue	Severity
Limited understanding of environmental carrying capacity of an island and reefs prior to resort development.	2
Construction of solid jetties, piers, groynes and breakwaters, which restrict seawater circulation and can increase erosion, resulting in the need for costly beach replenishment. Locally severe disturbance to some fringing reefs has also occurred.	2–3
Creation of access routes by blasting of reefs for channel(s) and vegetation clearance, which can undermine the reef's protective capacity and increase soil erosion.	2
Clearance of peripheral vegetation and seagrass, without the latter's role being assessed, and disturbance of nesting turtle and seabirds.	2
Reclamation of more land for additional rooms and/or resort infrastructures.	1
Contamination of aquifer by pathogens during resort construction and from chemicals to clear/control vegetation; contaminants may also reach the marine environment.	1
Discharge of sediment onto reefs during construction	2
Improper disposal of construction materials	1
Non-sustainable use of coral and timber as construction materials	1

A recent survey in the Maldives revealed that at two thirds of tourist resorts sewage effluent is piped into septic tanks, and the untreated sludge is dealt with by natural processes and soil absorption (Hameed, 1993). The other third employed septic tanks with sea outfalls (Hameed, 1993). Problems are greatest in N and S Malé atolls, as a result of urban pressures in conjunction with resort operations. Nutrient enrichment encourages algal overgrowth of corals and can help transform actively growing reefs into eroding ones (Bell et al., 1989). Price and Firaq (1996) reported a significant decrease in overall condition of reefs around 43 resort islands in the Maldives. Declines in coral cover and number of species and increase in algal abundance may at least be partly linked to increasing nutrient levels. Risk et al. (1994) report lower coral cover and diversity and higher number of associates (commensals, epizoans, epiphytes and internal bioeroders) in sites at which sewage is the likely source of stress. Other differences at affected sites included higher coral growth rates and higher values of ^{15}N (used as a sewage tracer).

Impacts from Diving and Snorkelling

Reefs are living ecosystems easily damaged by scuba divers, snorkellers and bathers. The greater the use the more wear and tear there is (Hawkins and Roberts, in press). A study on Vihamanaafushi in North Malé atoll showed that on heavily impacted parts of the reef, breakage by snorkellers over a one-month period was 17% of susceptible coral cover and 7% of total coral (Allison, 1996). Snorkelling, trampling and scuba diving may also have local effects on reef fish since many species shelter in branching coral colonies (Brown et al., 1990).

Boating, Fishing and Other Effects of Tourism

In the Maldives cruises and boating are likely to expand in line with the national trend towards tourism diversification. Some charter and cruising vessels are apparently not allowed in lagoon and reef areas of certain resorts due to the anchor damage caused. The discharging of sewage from toilets in lagoons and near resorts is unacceptable in most areas. A holding tank is therefore required, and the contents pumped out only in deep water away from the resort (and inhabited) islands.

Speedboats and jet-skis also pose a hazard to bathers, divers and marine life such as turtles, and are horribly noisy. Environmental problems, including groundings and anchor use on seagrass beds, are most pronounced in enclosed and semi-enclosed areas, in addition to erosion from propwash. Sport fishing is undertaken from several resort islands. Favoured species are generally large pelagics not associated with the reef-fishery. Tourists also undertake limited fishing for groupers, snappers and barracudas, which are important fishery species. Brown et al. (1990) consider that reef fish landings probably make up no more than 2–3% of the total reef fishery.

Transport

The main environmental concerns associated with transport related to ports and shipping, airports and roads are as follows (Maxwell Stamp, 1997):

Harbours, Ports and Shipping

Most islands are surrounded by shallow reefs which make access to the islands difficult. Island development is often dependent upon the construction of harbours, jetties and boat channels. Around 100 of the 200 inhabited islands have already undergone dredging or blasting works for construction of harbours and boat channels. However, the negative environmental impacts must be weighed against the potential benefits which involve a reduction in boat damage, loss of goods and reduced risk of accidents and drowning. Often there are no or limited EIAs prior to port/harbour development or expansion projects (e.g. Maradhoo on Seenu atoll).

Related concerns include: (1) the potential for major shipping accidents which could release massive amounts of oil or other hazardous substances into the marine environment; and (2) no port facilities for waste oil and other substances, which are likely to become an environmental problem following port expansion in Seenu. Construction of ports, harbours/jetties and other infrastructures is a problem not only on Seenu but also on many other Maldivian islands.

Airports

The main environmental problems include accumulation of wastes, particularly during international flights, and the need to follow compliance with health and environmental regulations for disposal. There are also problems with EIAs, which are a requirement for construction of new airports and extension of existing ones, but resources for such work are insufficient.

Roads

Water run-off from roads onto the shore, leading to loss of a critical natural resource, erosion and sedimentation are the main environmental problems arising from roads. However, a cobbled road surface, for example in Malé, facilitates water penetration into the aquifer, which has helped to conserve rainwater.

Other Environmental Concerns

Crown-of-Thorns (CoT) starfish

Densities of CoT in the Maldives seem to be highly variable in time and space. Two outbreaks have been recorded in recent years: one in Ari Atoll during the 1970s and the other in Kaafu in the mid 1980s (Risk and Sluka, 1999). Current data indicate that outbreaks are not a problem at the present time, although there have been earlier reports of four tourist divers collecting 30,500 CoT from their dive sites in Kaafu (Risk and Sluka, 1999). CoT starfish are identified only as a potential environmental problem on resorts, and interestingly resorts with poor coral cover following CoT plagues maintain good occupancies (MPE, 1992b).

Coral Bleaching

Extensive coral bleaching is currently (1998/99) reported in the Maldives (D. Elder, pers. comm.) with tracts of dead coral reef common in many areas. The event is believed to be linked to the recent El-Niño which has been a particularly severe event. Scientific debate still continues over the proximate causes of the temperature anomaly but it is likely to involve a combination of physical, meteorological and climatic factors.

A warm water anomaly of +3°C occurred in the Maldives between late April and May 1998 resulting in extensive bleaching of reef invertebrates with symbiotic algae. In shallow reef-flat areas up to 100% coral bleaching was observed. However, widespread bleaching was also observed at depths greater than 30 m (Riyaz et al., 1998). The degree and extent of bleaching in 1998 was unprecedented in the Maldives and throughout the Indian Ocean (Wilkinson et al., 1999). In north Malé the percentage of living coral cover declined markedly immediately after the bleaching event, from around 30–60% pre-bleaching to 0–5% post-bleaching for shallow reef-flat areas (Zahir et al., 1998). Fast-growing, branching corals, particularly *Acropora* spp. were more susceptible to bleaching than slow-growing boulder corals, leading to a high mortality in the former. However, 10 months after the bleaching event, a rapid reef assessment found recolonisation of opportunistic species (fast growing branching growth forms), indicating that reef recovery processes were already underway (Clark et al., 1999).

The ultimate consequences of the 1998 bleaching event will not be fully understood for some time, possibly for decades. However, it is clear that reefs will be modified as a result of this event. In the short term (<5 years) reefs formerly dominated by branching species will be dominated by non-living substrate with a low cover of living corals dominated by massive species. The consequences of bleaching on the reef framework will depend largely on the transport and fate of $CaCO_3$ fragments. Where reef disturbance is severe, boring and grazing organisms may remove $CaCO_3$ faster than primary frame-builders can add to it. Such biogenic processes will determine whether the level of the reef structure will decrease or increase.

The consequences of the bleaching event on the socio-economic welfare of coastal communities are uncertain. Wilkinson et al. (1999) suggest that certain reef functions and services are likely to be affected in the short term, which will have an affect not only on coastal communities but also on national economies. These include a decrease in income from coastal tourism and diving, a reduction in fish composition and production and increases in bioerosion and loss of coastal protection. Estimates of the economic damage due to the 1998 Indian Ocean bleaching event were calculated as over 8 billion US$ over a 20-year period for the pessimistic scenario, compared to 0.7 billion US$ for the optimistic scenario. At present such estimates cannot be corroborated, as there are no long-term studies on socio-economic impacts associated with bleaching events. Understanding of key relationships between coral mortality and economic services such as coastal tourism or fish production is also limited. However, these issues highlight the importance of initiatives to reduce threats to reefs through adoption of appropriate management plans, especially where healthy reefs are crucial for the economic and social survival of the coastal communities.

Sea-level Rise

The Maldivian Government recognises that small islands' ecosystems are among the most vulnerable to both natural events and human impacts. Threats posed by climatic changes and impending sea-level rise are taken very seriously, particularly since approximately 80% of Maldivian land area is less than 1 m above mean sea level.

Best estimates for predicted global average sea-level rise is around 0.5 cm per year (see Houghton, 1997). Effects are exacerbated by coral mining, which effectively increases sea level over that area by 0.3–0.5 m (Brown, 1997). Tidal data have been recorded at Gan to determine possible annual changes in mean sea level. Unfortunately, gaps in the data have resulted in some incomplete time series. Data are sent to the Sea Level Center at the University of Hawaii under the TOGA project, which is expected to provide complete analysis of the data. In the absence of comprehensive estimates of sea-level rise, several points are significant and are highlighted below. Also, a current GEF Project on Climate Change (below) is expected to provide useful data.

- atolls and reefs are not static but living and have the capacity for vertical growth; sustained maximum rates of vertical accretion of 10 mm per year have been recorded (Wilkinson and Buddemeir, 1994);
- heavily degraded reefs have less capacity to respond to sea-level rise, associated events and other impacts than healthy ones, and linked to this;
- high phosphate levels (e.g. certain liquid effluents) in common with increased levels of other nutrients result in eutrophication, which often leads to algal increase and coral degradation; coral calcification rates are also impaired by high phosphate levels;
- atolls and reef systems in the south of the Maldives are probably inherently more robust in the south than in the north due to the greater atoll rim development (and hence protection), despite the generally lower island height in the south (Table 1).

PROTECTIVE MEASURES

Selected aspects of environmental management are covered in this chapter. More detailed accounts are available elsewhere (e.g. MPE, 1989, 1992a–d; MPHRE, 1994; Dawson Shepherd, 1995; Maxwell Stamp, 1997; Brown, 1997; Price et al., 1998). Enforcement, politics and institutional factors are especially important (Price et al., 1998).

Table 10

Some conservation requirements and environmental policy options identified in the Tourism Master Plan (Nethconsult/Trastec, 1995)

Integrating tourism with coastal zone management
Multidisciplinary EIA
Incorporation of new technologies and conservation measures (e.g. for waste disposal)
Further development of protected areas and zoning system
Development of contingency plans for oil spills and other Environmental hazards
Environmental public awareness
Combining tourism with marine research

Table 11

Important national environmental legislation and international agreements for the Maldives (from Bergesen and Parmann, 1996 and other sources)

	Notes
National	
National Environmental Action Plan (1989)	updated 1998/99
Environmental Protection and Preservation Act of Maldives (Law No. 4/93)	
The Law on Tourism (Law No. 15/79)	
Guidelines for Environmental Impact Assessment	
Regulations for coral mining (1992)	
FISHERIES LAW OF THE MALDIVES (LAW NO5/87)	
National Biodiversity Strategy	Project initiated late 1998
International	
Atmosphere	
Aircraft Engine Emissions (ICAO)	++
United Nations Framework Convention on Climate Change (FCCC)	+++
Ozone Layer Convention (1985)	+++
Montreal Protocol on Substances that Deplete the Ozone layer (including ratification of the 1989 London amendments)	+++
Hazardous Substances	
Basel Convention on Transboundary Movement of Hazardous Wastes	+++
Distribution and Use of Pesticides	++
Marine Environment	
Civil Liability for Oil Pollution	+++
Fund Convention	+++
UN Convention on the Law of the Sea (UNCLOS III)	+
Nature Conservation and Terrestrial Living Resources	
World Heritage Convention	+++
Convention on Biological Diversity (1992)	+++
Convention on International Trade in Endangered Species (CITES)	(under consideration)

+Signed, but not ratified, accepted or approved. ++Member States. +++Ratified, accepted, approved or acceded.

Many conservation measures and environmental legislation are in place in the Maldives, some broadscale, others targeted more specifically at tourism-related issues (Table 10). In the Maldives environmental impact assessment (EIA) is now required for major developments such as new resorts. Customary environmental management practices may be prevalent and are more likely to operate on 'inhabited' rather than tourist islands, and little documentation is available.

Environmental Legislation and Related Measures

Appreciable environmental and related legislation exists for the Maldives (Table 11). Also significant has been the Ministerial-level meeting hosted by the Maldives (Small States Conference on Sea Level Rise), which resulted in the Malé Declaration. This calls for greater international recognition of the unique and fragile nature of island environments and improved assistance from the UN system. Maldives also provided input for the Global Conference on the Sustainable Development of Small Island Developing States (SIDS) which has led to an agreed action plan. Due to shortages of physical and human resources, problems associated with implementation are often appreciable.

More recent measures include the National Biodiversity Strategy. The project was initiated in late 1998. A national coordinating committee is being set up, with workshops and finalisation of the strategy planned for 1999. This will call for gathering and collation of information on marine biodiversity as one component. Also significant is the current (1998/99) GEF Project on Climate Change. This involves strategic planning and other activities including measurements on coastal vulnerability, for example sea-level rise.

The law of the lease of islands has brought sharply divided opinion (Frontier, 1994). Leases on island resorts are now for a 21-year period in the first instance. Since there are no long-term tenure rights, it might be argued that incentives for conservation by resort operators are not particularly strong. Land tenure arrangements in the Maldives are complex and impinge on all areas of society.

National and Regional Development Plans

The environment is a major development parameter, since the continued physical and economic existence of the Maldives depends on healthy underlying reefs. These provide the basis for world-class tourism, fisheries production and livelihood. Reefs also give physical protection to urban communities and resort islands. The importance of the environment to development is highlighted in the Fourth National Development Plan (MPHRE, 1994), The Second National Environmental Action Plan (MHAHE, 1998) and the current Tourism Master Plan: 1996–2005 (Nethconsult/Transtec, 1995a,b,c). Increasing prominence is also reflected in the extensive national environmental legislation and international treaties to which the Maldives is party (Table 10). Hence environmental concerns are seen as a critical component of development planning.

In 1996 the Marine Research Centre/MOFA formulated an integrated reef resources management (IRM) framework to promote sustainable exploitation of reef resources. This framework is a result of project work conducted in four central atolls in consultation with resource users, community groups and various government departments. The project is currently installing interactive educational centres within each of the four atolls where data can be collated and retrieved. The IRM has prepared a set of guidelines and plans which will form an integral part of the National Environment Action Plan (NEAP) (Nickerson and Maniku, 1996).

The Maldives is a member of the South Asian Association for Regional Co-operation (SAARC) and the South Asia Co-operative Environment Programme (SACEP). In 1997 the Maldives adopted the SAARC environmental action plan to protect the environment in the SAARC region, and is committed to implementing the South Asian Seas Programme developed by UNEP.

Multidisciplinary Environmental Impact Assessment (EIA)

There is a strong need to incorporate marine ecological factors into the largely physical and terrestrial components of existing EIA procedures (Law No. 4/93), to provide a more multidisciplinary approach (Table 10). To address these issues a set of guidelines outlining the procedures for EIA were approved by Cabinet in December 1994, and an informational handbook for proponents and government agencies is being prepared. The National Commission for the Protection of the Environment (NCPE) appointed by the president was restructured in 1993 to broaden its mandate. NCPE advises government on all aspects of environmental assessment, planning and management.

While EIA is now a requirement for all new major development projects, shortage of physical and human resources makes this difficult. One partial solution might be for the prospective tourism developer or company to help fund EIAs as a precondition to the contract. MPHRE could provide a coordinating role, with assistance for EIA implementation from other ministries, NGOs and overseas as necessary, in collaboration with the developer.

Initiation of Protected Area System

Protected area initiatives during the mid-1990s included the identification of 15 key dive sites proposed for recommendation in five northern atolls: Lhaviyani, North Malé, South Malé, Ari and Vaavu Atoll. Legislation and management plans are under development for these areas, and should result in enhanced environmental protection. Two islands have also been given protected status: Hithaadhoo (In Huvadu Atoll) due to its important avian population, and Hurasdhoo (South Ari Atoll) due to its unique geological formation. The 27 Dive Sites commonly used by Ocean Reef Resort on Gan (Seenu Atoll), and an area to the northeast of Hithadhoo might augment the northerly candidate sites.

Plans are currently (1998/99) underway to develop a system of coastal protected areas to help safeguard marine biodiversity, in the interests of both the environment and the tourism industry. Planning of the system should be initiated in 1999, with assistance from AusAid. Field surveys will not be the main focus. Three sites will probably be chosen initially, which will require designation within three different IUCN protected area categories.

Environmental Restoration

A three-year nationwide programme is underway to plant 2 million trees. Special emphasis is given to the restoration of threatened habitats and species.

Studies have been undertaken in the Maldives to determine the feasibility of restoring shallow reef habitats degraded from coral mining. Artificial structures deployed included the following: (1) hollow concrete blocks; (2) Armorflex flexible concrete mattresses; (3) Armorflex plus transplanted corals; and (4) chain-like fencing anchored by paving slabs (Brown, 1997). Corals settled on the concrete surfaces within 6 months. Branching corals grew rapidly (up to 10 cm year^{-1}), while recruitment and growth of massive species was slower. Fish abundance and species diversity improved to levels equivalent or greater than on unmined reefs, except on the chain fencing.

Despite these findings, it was apparent that coral transplantation in this study was not cost-effective, due to high mortality of transplanted coral (51% survivorship after 28 months). It was concluded that where environmental conditions are suitable for coral growth (e.g. substrate, water quality), natural recruitment of corals can lead to significant restoration within 3–4 years (Clark and Edwards, 1995, 1999; Brown, 1997).

Carrying Capacity, Sustainability and Future Prospects

This concluding section assesses the overall environmental status and future prospects for the Maldives, with special reference to tourism. This is the nation's principal means of income-generation, but tourism activities and the marine environment are also intimately interconnected.

Pros and Cons of Tourism in the Maldives

Coral reefs in the Maldives in general are in good condition (Table 12). Impacts include some decline in reef quality and conflicts between tourism and other sectors (e.g. fisheries). To ensure the long-term sustainability of tourism and other human uses based around coral reefs more attention should be focused on carrying capacity. Estimates should not be fixed but will depend on local reef characteristics and environmental awareness and diving experience of visitors (Hameed, 1993). The cumulative effects of infrastructure development on the reef environment have scarcely been addressed to date and require concerted research effort.

Prospects in the Maldives for tourism (and other forms of development) seem quite favourable (Table 12). Carefully handled tourism can have many positive effects. For example, divers in the Maldives like to see giant clams and

Table 12

Summary of the status of the Maldivian coral reef environments and future prospects (from Price et al., 1998)

Feature or Issue	Key Feature(s) or Overall Status
Reef environment	26 geographic atolls and ca. 1200 coral islands of 186 km^2 land area
Diversity of reefs	High
Environmental status of reefs	Good nationally, but reef state declining towards Malé
Nature and extent of tourism	Diving, snorkelling, boating, fishing; 87 islands developed for tourism
Extent of conservation	National and international measures considerable, but constrained by physical and human resources
Short-term prospects for tourism	Favourable
Medium- and long-term prospects for tourism	Moderate to good, but increasing need to implement environmental measures in tourism planning and operations; the consequences of climate change and sea level rise of much concern.

protested when fishing began. Removal of clams is destructive to the reef. Divers' protests were accepted, and the time from the start to the end of the fishery was only one year (C. Anderson, pers. comm.). Another example is the reef fishery. Sharks are attractive to divers but are caught at night by fishermen. It is estimated that a single shark live is worth $3000 generated annually from divers, compared with $30 per shark dead for export (Anderson, 1997).

Tourism has also afforded protection to reefs from coral mining, which is prohibited on resort reefs in the Maldives. Hence tourist islands function as 'protected areas', safeguarding reefs from an activity that can be much more destructive than tourism pressures, provided that resorts are not constructed from coral mined from other islands (Price and Firaq, 1996). While it encourages economic growth of many kinds, coastal tourism can also deter more damaging forms of development, such as industrial and port facilities.

Challenges For the New Millennium

Many intense human activities, such as mass tourism, are incompatible with conserving the diversity, productivity and beauty of coral reefs. The Maldives have so far avoided the perils of mass tourism. To ensure that prospects remain good, it is vital that issues of carrying capacity and sustainability are tackled now and cautious long-term limits to growth determined. In this way both reefs and people will continue to coexist well into the future.

A critical challenge will be economic development that does not foreclose to later or even present generations important future resource-use options that may currently be unknown. This can happen unexpectedly following incremental and often unperceived environmental deterioration (e.g. progressive development of islands and gradual degradation of their natural resources). Healthy reefs are likely to better withstand adverse effects of global warming and sea-level rise than more heavily degraded reefs. This may well become an increasingly important factor influencing the sustainability of reef uses on low-lying islands such as the Maldives.

Individual developing countries, such as the Maldives and Marshall Islands in the Pacific, can probably do little to influence the western nations largely responsible for the increasing emissions of greenhouse gases, climate change and impending sea-level rise (Price, 1998). Regional initiatives help to voice national concerns at a regional level. Two points made earlier are emphasised here to demonstrate that national conservation efforts to combat effects of human-induced global climate change are also worthwhile:
- Atolls and reefs are not static but living and have the capacity for vertical growth;
- Healthy reefs have greater capacity to respond to sea level rise, associated events and other impacts than heavily degraded reefs.

These points are simple biological principles, and should serve as an environmental insurance policy: the premium for this is the cost of environmental management, while the benefits are minimisation of reef deterioration in the face of economic development.

REFERENCES

Adam, M.S. and Anderson R.C. (1998) The tuna fishery resources of the Maldives: Skipjack tuna (*Katsowonus pelamis*). *Maldives Marine Research Bulletin* 3, 47–77.

Allison, W.R. (1996) Snorkeler damage to reef coral in the Maldive Islands. *Coral Reefs* 15, 315–218.

Anderson, R.C. (1992) North-south variations in the distribution of fishes in the Maldives. *RASAIN* 12, 210–226.

Anderson, R.C. (1994) Executive review of the status of Maldivian fishery resources. Marine Research Section, Ministry of Fisheries and Agriculture, Maldives.

Anderson, R.C. (1997) Sharks mean business. *Scientific American*, 72–73.

Anderson R.C., Waheed, Z. and Adam, M.S. (1998a) The tuna fishery resources of the Maldives: 1. Introduction. *Mald. Marine Research Bulletin* 3, 5–45.

Anderson R.C. and Waheed, Z. (1998) The tuna fishery resources of the Maldives: 3. Yellowfin (*Thunnus albacares*). *Mald. Marine Research Bulletin* 3, 79–105.

Anderson R.C., Waheed, Z. and Nadheeh, I. (1998b) The tuna fishery resources of the Maldives: 4. Frigate tuna (*Auxis thazard*). *Mald. Marine Research Bulletin* 3, 107–126.

Anderson R.C., Waheed, Z. and Scholz, O. (1998c) The tuna fishery resources of the Maldives: 5. Kawakawa (*Euthynnus affinis*). *Mald. Marine Research Bulletin* 3, 127–144.

Bell, P.R.F., Greenfield, P.F., Hawker, D. and Connell, D. (1989) The impact of waste discharges on coral reef regions. *Water Science and Technology* 21, 121–130.

Bergesen, H.O. and Parmann, G. (eds) (1996) *Green Globe Yearbook of International Co-operation on Environment and Development.* Oxford University Press, Oxford, 368 pp.

Bianchi, C.N., Colantoni, P., Gesiter, J. and Morri, C. (1997) Reef geomorphology, sediments and ecological zonation at Felidu Atoll, Maldive Islands (Indian Ocean). *Proc. 8th International Coral Reef Symposium* 1, 431–436.

Brown, B.E. and Dunne, R. (1988) The environmental impact of coral mining on coral reefs in the Maldives. *Environmental Conservation* 152, 159–166.

Brown, B.E., Dawson Shepherd, A., Weir, I. and Edwards, A. (1990) Effects of degradation of the environment on local reef fisheries in the Maldives. Final report to Overseas Development Administration (ODA), London.

Brown, B.E. (1997) Integrated coastal management: South Asia. Dept. of Marine Sciences, and Coastal Management, University of Newcastle, Newcastle upon Tyne, UK.

Clark, S. and Edwards, A.J. (1995) Coral transplantation: an application to rehabilitate reef-flat areas degraded by coral mining in the Maldives. *Coral Reefs* 14, 201–213.

Clark, S. and Edwards, A.J. (1999) An evaluation of artificial reef structures as tools for marine habitat restoration in the Maldives. *Aquatic Conservation* 9, 5–21.

Clark, S., Akester, S. and Naeem, H. (1999) The status of coral reef communities in North Malé Atoll, Maldives: recovery following a bleaching event. Submitted to Ministry of Home Affairs, Housing And The Environment. February 1999, 12 pp.

Couper, A. (ed.) (1983) *The Times Atlas of the Oceans.* Times Books Ltd., London, 272 pp.

Dawson Shepherd, A.R., Brown, B.E., Warwick, R.M., Clarke, K.R. and Brown, B.E. (1992) An analysis of fish community responses to coral mining in the Maldives. *Environmental Fish. Biology* 33, 367–380.

Dawson Shepherd, A. (1995) Maldives: A framework for marine environmental management. *Coastal Management in Tropical Asia.* No. 4, 2–6.

Edwards, A.J. (1989) The implications of sea-level rise for the Republic of Maldives. Report to the Commonwealth Expert Group On Climate Change and Sea Level Rise. 68 pp. (and 41 pp. Annexes).

Frazier, J. and Frazier, S.S. (1987) Marine turtles in the Maldives Archipelago. Report to IUCN Conservation Monitoring Centre, Cambridge.

Frontier (1994) Trade Newsmagazine, Vol. 1, No. 1.

Hackett, H.E. (1977) Marine algae known from the Maldive Islands. *Atoll Research Bulletin* 210, 210.

Hameed, H. (1993) *Implications of tourism for the environment: A Maldives Case Study.* M.Phil. thesis, University of East Anglia, UK.

Hawkins, J.P. and C.M. Roberts (1994) The growth of coastal tourism in the Red Sea: present and future effects on coral reefs. *Ambio* 23, 503–508.

Hawkins, J.P. and Roberts, C.M. (1997) Estimating the carrying capacity of coral reefs for scuba diving. *Proc. 8th Int. Coral Reef Symp., Panama, June 1996.*

Houghton, J.T. (1997) *Global Warming: The Complete Briefing.* Cambridge University Press, Cambridge, 251 pp.

IUCN/UNEP (1985) Management and conservation of renewable marine resources in the South Asian Seas region. *UNEP Reg. Seas Rep. Stud.* No. 62, 60 pp.

Lidkea, T.R. (1993) Report on solid waste management. MPHRE (Maldives), 42 pp.

Maxwell Stamp (1995) Maldives Regional Development Planning Project. Phase 1 Report. ADB/Ministry of Planning, Human Resources and the Environment, Maldives.

Maxwell Stamp (1997) Maldives Regional Development Planning Project. Phase 2 Report. Vol. 1 The Northern Development Region; Vol. 2 The Southern Development Region. ADB/Ministry of Planning, Human Resources and the Environment, Maldives.

MHAHE (Ministry of Home Affairs, Housing and Environment) (1998) Second National Environmental Action Plan. Ministry of Home Affairs, Housing and Environment, Republic of Maldives.

Morri, C., Bianchi, C.N. and Aliani, S. (1995) Coral reefs at Gangehi

(North Ari Atoll, Maldives). *Publ. Serv. Geol.* Luxembourg 29: 3–12.
MPE (Ministry of Planning and Environment) (1989) National Environmental Action Plan. Ministry of Planning and Environment, Republic of Maldives.
MPE (Ministry of Planning and Environment) (1992a) Technical Co-operation Programme. Marine Environmental Management: The Maldives. Environmental Protocol (Tourism Sector). Section 2 of 4: Carrying Capacity.
MPE (Ministry of Planning and Environment) (1992b) Technical Co-operation Programme. Marine Environmental Management: The Maldives. Environmental Protocol (Tourism Sector). Section 3 of 4: Management Considerations.
MPE (Ministry of Planning and Environment) (1992c) Technical Co-operation Programme. Marine Environmental Management: The Maldives. Environmental Protocol (Tourism Sector). Section 3 of 4: Carrying Capacity.
MPE (Ministry of Planning and Environment) (1992d) Technical Co-operation Programme. Marine Environmental Management: The Maldives. Environmental Protocol (Tourism Sector). Section 4 of 4: The Protocol.
MPHRE (1994) *National Development Plan 1994–1996*. Ministry of Planning, Human Resources and Environment. Republic of Maldives, 249 pp.
MPHRE (1997) Statistical Yearbook of the Maldives. 352 pp.
Nethconsult/Transtec (in association with Borde Failte) (1995) *Maldives Tourism Master Plan: Assessment of Existing Conditions in the Tourism Sector*, EC Report to Report to Ministry of Tourism, Republic of Maldives.
Nethconsult/Transtec (in association with Borde Failte) (1995a) *Maldives Tourism Master Plan: Part I. Draft Final Report*, EC Report to Report to Ministry of Tourism, Republic of Maldives.
Nethconsult/Transtec (in association with Borde Failte) (1995b) *Maldives Tourism Master Plan: Part II. Guidelines for Implementation. Draft Final Report*, EC Report to Report to Ministry of Tourism, Republic of Maldives.
Nickerson, D.J. and Maniku, M.H. (1996) Workshop on Integrated Reef Resources Management in the Maldives. Bay of Bengal Programme. BOBP/Re/76, 312 pp.
Pernetta, J.C. and Sestini, G. (1989) Report of the mission to the Republic of Maldives. *UNEP Reg. Seas Rep. Stud.* **104**, 1–84.
Pernetta, J.C. (ed.) (1993) Marine Protected Area needs in the South Asian Seas region: Volume 3: Maldives. A Marine Conservation and Development Report. IUCN, Gland, Switzerland, 38 pp.
Price, A.R.G. and Firaq, I. (1996) The environmental status of reefs on Maldivian resort islands: a preliminary assessment for tourism planning. *Aquatic Conservation* **6** (2), 93–106.
Price, A.R.G. (1998) The environment and sustainable development in the Marshall Islands. Report to Maxwell Stamp plc/Asian Development Bank (ADB).
Price, A.R.G. (1999) Broadscale coastal environmental assessment of the Chagos Archipelago. *Linnaean Society Occasional Paper* No 2, 285–296.
Price, A.R.G., Roberts, C.M. and Hawkins, J.P. (1998) Recreational use of coral reefs in the Maldives and Caribbean. In *Biological Conservation and Sustainable Use*, eds. E.J. Milner Gulland and R. Mace. Blackwell Science, Oxford, pp. 242–260.
Randall, J.E. and Anderson, R.C. (1993) Annotated checklist of the epipelagic and shore fishes of the Maldives Islands. *Ichthyol. Bull. J.L.B. Smith Inst. Ichthyol.* **59**, 1–47.
Risk, M.J., Dunn, J.J., Allison, W.R. and Horrill, C. (1994) Reef monitoring in Maldives and Zanzibar: low-tech and high-tech science. In: *Proceedings of the Colloquium on Global Aspects of Coral Reefs: Health, Hazards and History*, ed. R.N. Ginsburg, pp. 66–72. Rosenstiel School of Marine and Atmospheric Sciences, University of Miami, Florida.
Risk, M.J. and Sluka, R. (1999) The Maldives: A nation of atolls. In *Coral Reefs of the Western Indian Ocean: Ecology and Conservation*, eds. T. McClanahan, C. Sheppard and D. Obura. Oxford University Press.
Riyaz, M., Shareef, M. and Elder, D. (1998) Coral bleaching event: Republic of Maldives, May 1998. Ministry Of Home Affairs, Housing and the Environment.
Shakeel, H. and Ahmed, H. (1997) Exploitation of reef resources: grouper and other food fishes. In *Report and Proceedings of the Maldives/FAO National Workshop on Integrated Reef Resources Management in the Maldives*, eds. D.J. Nickerson and M.H. Maniku, pp. 117–135. BOBP, Madras, Report 76.
Sheppard, C. (1998) Biodiversity patterns in Indian Ocean corals,m and effects of taxonomic error in data. *Biodiversity and Conservation* **7**, 847–848.
Sheppard, C. (1999) Corals of the Western Indian Ocean: An overview. In *Coral Reefs of the Western Indian Ocean: Ecology and Conservation*, eds. T. McClanahan, C. Sheppard and D. Obura. Oxford University Press.
Sheppard, C., Price, A. and Roberts, C. (1992) *Marine Ecology of the Arabian Region: Patterns and Processes in Extreme Tropical Environments*. Academic Press, London, 359 pp.
Silva, P.C., Basson, P.W. and Moe, R.L. (1996) Catalogue of the benthic marine algae of the Indian Ocean. *Univ. of Calif. Public. in Bot.* Vol. 79, Univ. of California Press, Berkeley, 1259 pp.
Sluka, R. and Miller, W. (1998) Coral mining in the Maldives. *Coral Reefs* **17**, 228.
Spalding, M., Blasco, F. and Field, C. (eds.) (1997) *World Mangrove Atlas*. The International Society for Mangrove Ecosystems. Okinawa, Japan, 178 pp.
Stoddart, D.R., Spencer-Davies, P. and Keith, A.C. (1966) Geomorphology of Addu Atoll. *Atoll Research Bulletin* **116**, 13–41.
Symens, P. (1996) Status and conservation of seabirds in the Chagos Archipelago, British Indian Ocean Territory. MSc Thesis, University of Warwick, UK, 90 pp.
Tomlinson, P.B. (1986) *The Botany of Mangroves*. Cambridge University Press, Cambridge, 409 pp.
Wells, S. and Sheppard, C. (1988) *Coral Reefs of the World. Vol. 2: Indian Ocean, Red Sea and Gulf*. IUCN/WCMC/UNEP.
Wilkinson, C.R. and Buddemeier, R.W. (1994) Global climate change and coral reefs: Implications for people and reefs. Report of the UNEP-IOC-ASPEI-IUCN Global Task Team on the implications of climate change on coral reefs. IUCN, Gland, Switzerland, 124 pp.
Wilkinson, C., Lindén, O., Cesar, H., Hodgson, Rubens, J. and Strong, A. (1999) Ecological and socio-economic impacts of 1998 coral mortality in the Indian Ocean: An ENSO impact and a warning of future change. *Ambio* **28** (2), 188–196.
Woodroffe, C. (1989) Maldives and sea-level rise: an environmental perspective. Report to Ministry of Planning and Environment, Republic of Maldives, 64 pp.
Zahir, H., Naeem, I., Rasheed, A. and Haleem, I. (1998) Reef Check Maldives: Reef Check 1997 and 1998. Marine Research Section, Ministry of Fisheries, Agriculture and Marine Resources, Republic of Maldives.

THE AUTHORS

Andrew R.G. Price
Ecology and Epidemiology Group,
Department of Biological Sciences,
University of Warwick, Coventry, U.K.

Susan Clark
Department of Marine Sciences and Coastal Management,
University of Newcastle upon Tyne,
Newcastle upon Tyne, U.K.

Chapter 67

THE CHAGOS ARCHIPELAGO, CENTRAL INDIAN OCEAN

Charles R.C. Sheppard

The Chagos Archipelago is a collection of atolls and banks covering an area about 250×400 km in the centre of the tropical Indian Ocean. The islands are few and small, and support large numbers of nesting seabirds on most of the islands which are not infested with rats. There are about 50 indigenous species of higher plants, which have since been joined by over 200 more species since the islands were occupied about 250 years ago. Some islands which were not inhabited or planted with coconut still support good stands of natural Indian Ocean hardwoods.

The marine habitat is mainly diverse coral reef with associated sandy areas; there are no mangroves and only patchy seagrass habitats. Even sandy habitat is very limited shallower than about 20 m deep and there are no muds within the photic zone. There is evidence from the high species richness that Chagos provides a significant stepping stone in the east–west oceanic passage of species, and is important in the maintenance of biodiversity in the Indian Ocean. The present uninhabited nature of most of these islands is the main reason for the richness and unimpacted nature of the marine habitats.

There is a military facility in the southernmost atoll of Diego Garcia, but the other four atolls are uninhabited. All atolls up to the 1970s were farmed for copra, but human impacts did not, as far as can be judged, impact the marine life of the region. While limited fishing activity takes place, there is no industry, no tourism and, apart from Diego Garcia, very little visitation of any kind to the region, which has recently been declared as a Ramsar site with a view to long-term protection. Chagos may be regarded as one of the very few locations with no major human impact. However, following the rise in Indian Ocean temperature in 1998, the coral populations have collapsed.

Much of this article is taken from the chapter: 'Chagos Archipelago' by Charles Sheppard, from Coral Reefs of the Western Indian Ocean: Their Ecology and Conservation, *edited by Tim R. McClanahan, Charles R.C. Sheppard and David Obura. Copyright © 2000 by Oxford University Press, Inc. used by permission of Oxford University Press, Inc.*

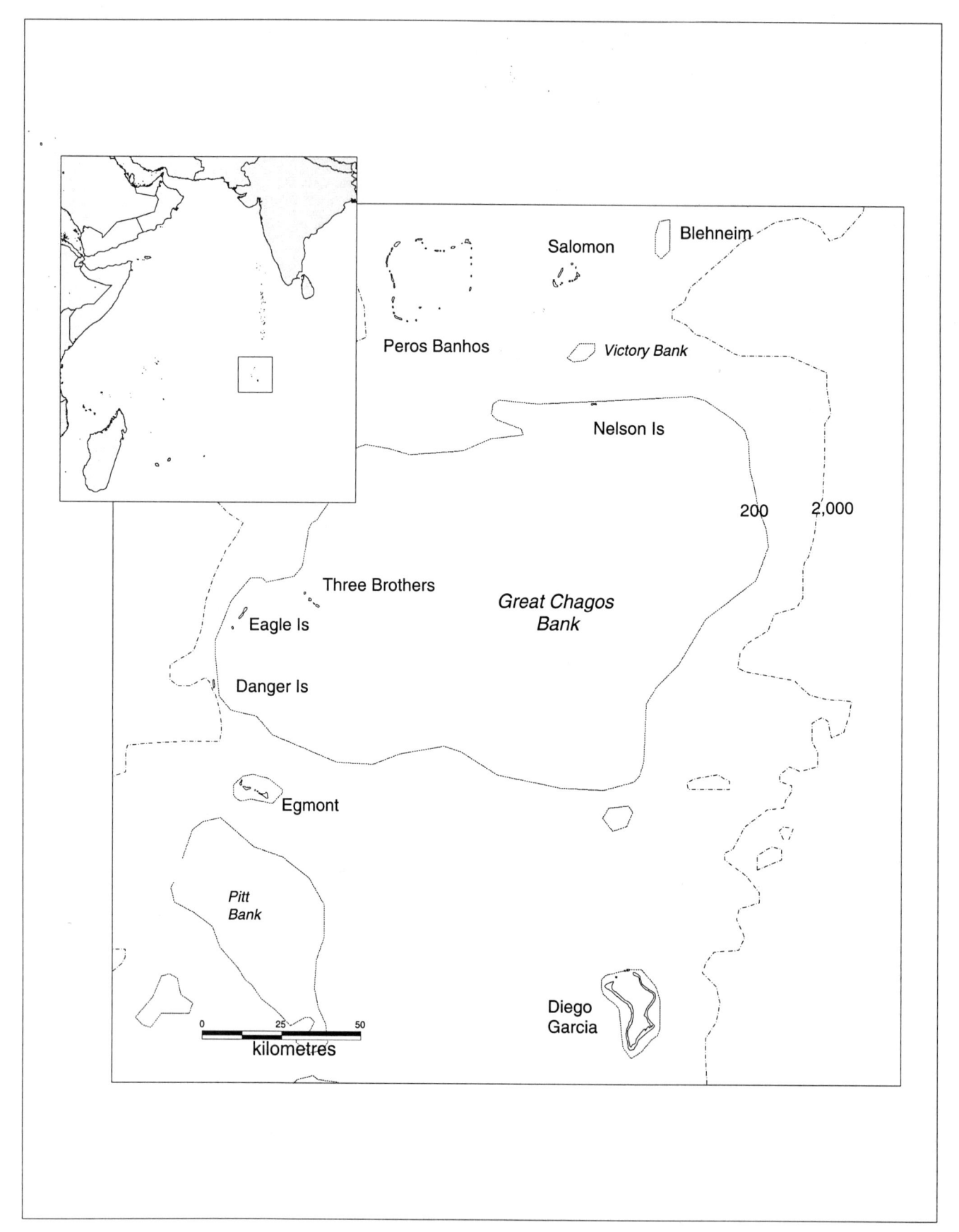

Fig. 1. Map of Chagos Archipelago.

GEOGRAPHICAL AND HISTORICAL SETTING

The Chagos Archipelago lies at the southern end of the Laccadive–Chagos ridge, in the centre of the tropical Indian Ocean. Politically it is a UK Dependent Territory, known as the British Indian Ocean Territory, administered from the Foreign and Commonwealth Office, London (Oldfield and Sheppard, 1997). The group has five islanded atolls, and several others which are awash or completely submerged (Fig. 1). Its central feature is the 200×100 km Great Chagos Bank, the world's largest atoll in terms of area. This is mostly submerged, but there are eight islands on its western and northern rim. This is surrounded by the smaller atolls: Peros Banhos and Salomon to the north, Egmont (on some charts the 'Six Islands') to the Southwest, and Diego Garcia to the south. Among these lie many submerged reefs, the whole complex covering about 250×400 km. The submerged structures are one of the most notable features of the central Indian Ocean.

It would be surprising if the widespread seafaring civilisations of 500 to 2000 years ago did not discover Chagos, but records of this are apparently absent. In this respect it is different from the relatively nearby Maldives. The islands' first documented discovery was by the Portuguese in the 16th century (Anon., 1979) but they were not inhabited for the next 200 years until, in the late 18th century, the larger ones were farmed for coconuts or copra. The supply of copra, which exceeded about 0.5 million litres annually in the 19th century (Moresby, 1884), caused the atolls to be known as the Oil Islands. In the 1970s when copra ceased to be an important global crop, the plantations were abandoned. Today all atolls except Diego Garcia are currently uninhabited, the latter supporting a communications, naval and air facility of a few thousand personnel.

The atolls and reefs attracted some distinguished early attention. All its reefs were mapped in detail in 1837 by Moresby (1884), which permitted Darwin (1842) to incorporate them extensively in his exposition of coral reef formation, though Darwin did not land there. Bourne (1888) visited its islands and interpreted its rock strata as evidence against reef formation by subsidence. Gardiner (1936) described several parts of Chagos which he had visited 30 years earlier. Then, few visits by scientists were undertaken until Stoddart and Taylor (1971) visited Diego Garcia and made collections of several groups of organisms from the island and reef flat. The first detailed, deeper reef ecology studies commenced in the 1970s when all the uninhabited atolls and several submerged banks were examined (Bellamy, 1979; Sheppard and Wells, 1988). No more scientific visits occurred until a visit in 1996 of about 12 island and reef ecologists (Sheppard and Seaward, 1999). This chapter summarises the above, together with results of a visit in 1999 following the severe coral mortality caused by the 1998 rise in Indian Ocean sea surface temperature.

Chagos contains the largest expanse of totally unexploited reef in the Indian Ocean as well as some of the richest. With the exception of part of Diego Garcia, Chagos' reefs have been almost completely undisturbed for at least 30 years and have never been subjected to major or direct human interference. A similar absence of human interference may be claimed for some other areas of the ocean such as the large swath of limestone banks running between the Seychelles and Mauritius, but the latter are heavily fished. Chagos is subjected to some fishing through a licenses issue by the UK authorities, and some illegal fishing occurs of shark, reef fish and sea-cucumbers, but this is probably minor compared with most other areas.

Geological, Geographical and Physical Background

Chagos is the southernmost and oldest part of the Chagos–Laccadive ridge, formed when the Indian tectonic plate migrated northwards towards Asia. The entire chain was created by hot-spot activity since the late Cretaceous, and the trace of this can be followed from the Deccan traps in western India and from there southwards down the chains of atolls to Chagos (Eisenhauer et al., 1999). The archipelago is a limestone cap a few hundred metres to a few kilometres thick, resting on volcanic rock (Francis and Shor, 1966).

Most of the present islands are those of typical atolls, located on atoll rims, with elevations of no more than two or three metres (Fig. 2). Two areas in Chagos have raised reefs; these are found in southern Peros Banhos and the adjacent, northwestern part of the Great Chagos Bank, both sites containing islands with small, uplifted and vertical cliffs up to six metres above high tide (Fig. 3). One of the main features of the Chagos reefs is the number of submerged banks and 'drowned' atolls (Table 1), the latter exhibiting typical atoll-like cross sections including 'lagoons' and passes cut through the submerged atoll rims. Most of the Great Chagos Bank itself lies between 3 and 20 m deep except in the west where shallow reef flats and islands exist. The atolls are classical islanded atolls, while Blenheim is an atoll of similar size but which is emerged at low tide only, while other structures are atoll-shaped but submerged to at least 5 m.

All reefs visited up to 1996 were found to have profusely growing corals, so the reason why some have islands, some are awash and others are drowned to 5 m or more is not known. Blenheim reef, for example, is a typical atoll in most respects except for a lack of islands, and indeed its wave resistant algal ridges are the best developed in the Archipelago and possibly in the entire ocean (see below). Blenheim is the same size as the islanded Salomon and Egmont atolls, so its small size alone has probably not precluded the emergence or development of islands. Indeed, some of the largest atoll-shaped structures in Chagos have no islands (Table 1).

Other reefs are crescent shaped (e.g. Benares, Colvocoresses) which may represent fragments of older atolls. On some large deeper reefs, such as Speakers Bank, corals and seagrasses are profuse, indicating high benthic

Fig. 2. Part of the atoll rim of Peros Banhos, showing typical structure of reef flat between two islands at low tide. Lagoonal reef is to right, seaward to left. The islands are no greater than 2 m elevation.

Fig. 3. Raised reef, part of Peros Banhos atoll, on southern side. The limestone cliffs are over 4 m high.

productivity. On the largest system of all, the Great Chagos Bank, complex patterns of reefs exist with several ring shaped structures within the main atoll rim, which attest to a complex past history of growth and erosion. Nowhere in Chagos are lagoonal faros seen, even though these structures are typical of the Maldives immediately to the north. Depths of over 1 km separate each atoll or reef.

BIOGEOGRAPHIC POSITION OF CHAGOS IN THE INDIAN OCEAN

Biogeographical work in Chagos is still in its early stages (Sheppard, 2000b), but several aspects are becoming clear. These are: that the central location of Chagos gives it a considerable importance in the Indian Ocean; the quantity of shallow reef habitat available is very large (Fig. 4); that the main east–west currents enable Chagos to function as a stepping stone in the Indian Ocean; and from the point of view of the corals at least, Chagos provides a strong east–west connecting link.

There are very few endemic species in Chagos. It is commonly assumed that for a site to possess many endemic species is a 'good thing' in conservation terms, which somehow raises its value. This is undoubtedly true of isolated systems or sections of water which are somehow separated from the rest, loss of which would cause a disproportionately high loss of biodiversity. However, a site located on a species thoroughfare, if indeed it is a thoroughfare, would be likely to have no or few endemic species, yet have a high diversity, with many species in common with sites to each side. Chagos has very few endemics but is one of the richest sites in the ocean, with many species in common with both east and west (in terms of coral) and is rich with stronger connections with the Maldives (in the case of fishes). This strongly supports the suggestion that Chagos is important in the oceanic movement of species, just as its geographical location would suggest.

Table 1

Physical characteristics of the major atolls and banks. N.B. Several small banks with atoll cross-section also exist, whose shallowest rim sections are <10 m, but which have not been studied.

Atoll	Atoll area (km^2)	Land area (ha)	Number of islands	Lagoon greatest depth (m)	Lagoon mean depth (m)	Shallowest depth (if no islands)	% Rim enclosure by islands or reef flats
Atolls							
Diego Garcia	<200	2733	4	31	10	–	95
Salomon	38	311	11	33	25	–	85
Peros Banhos	463	920	24	80	38	–	60
Great Chagos Bank	18,000	445	8	90	~30	–	<5
Egmont	40	~300	6	26	12	–	35
Submerged atolls							
Blenheim Reef	40	0	0	17	8	awash	60
Victory Bank	16	0	0	33	25	5	–
Speakers Bank	680	0	0	43	30	7	–
Pitt Bank	~1,200	0	0	44	35	<10	–

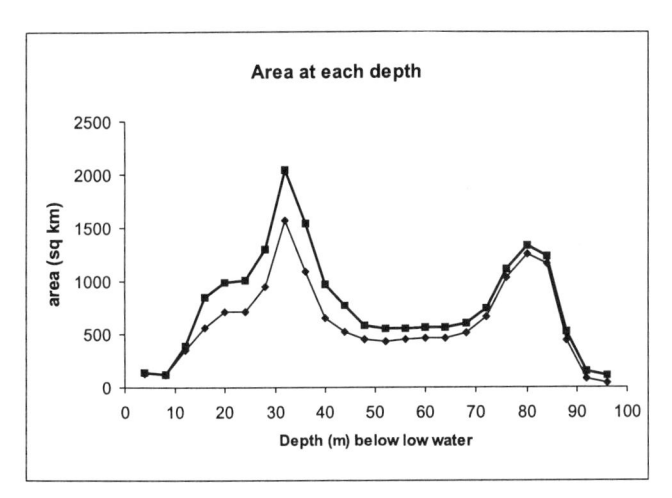

Fig. 4. Amount of reef substrate at different depths in Chagos. Method was a digital elevation model based on several thousand depth soundings taken from several different charts of the archipelago (Dumbraveanu and Sheppard, 1999).

Chagos is the most diverse coral site in the Indian Ocean, with 220 species in 58 genera of zooxanthellate corals, and when geographical distances between Chagos and 25 other Indian Ocean sites were replaced by biological distances with GIS techniques (Sheppard, 1999a), the map of the Indian Ocean is strongly 'compressed' in an east–west direction, suggesting that Chagos is part of a major east–west Indian Ocean 'species highway' which, if confirmed, adds enormously to the value of these reefs. The atolls of the Maldives and Laccadives to the north of Chagos, which are geographically closest, are less connected biologically, at least for corals, though this may not be true for fishes.

SEASONALITY, CURRENTS, NATURAL ENVIRONMENTAL VARIABLES

Climate, Tides and Nutrients

Being located between 5° and 7°S, Chagos is not regularly subject to cyclonic strength winds, although severe storms from the edges of cyclones sometimes occur, especially in Diego Garcia in the south. From October to April, winds are light or moderate and generally from the northwest. For the rest of the year, the Southeast trades blow strongly. Rainfall has a strong seasonal pattern (Stoddart, 1971), with most falling between October and April and much less falling during the Southeast trades. Total annual rainfall averages about 2500 mm or more in the southern atoll of Diego Garcia, but reaches 4000 mm in the northern atolls, making Chagos the wettest group of atolls in the Indian Ocean. December to February is the wettest time, when the Inter-Tropical Convergence Zone is over or just south of Chagos. Developing tropical cyclones are still weak at this low latitude. On average, about 11 tropical storms or cyclones develop each year in the region, building up to the east or southeast of Chagos but tracking away from the archipelago to the southwest.

Spring tidal ranges are about 1.2 m with neap ranges of about 0.2 m (Pugh and Rayner, 1981; Rayner, 1982). Several times each year, low water lasts until nearly noon, causing the reef flats to be emerged 30 cm during periods of strong sun (see Fig. 2). This is the supposed reason why the reef flats of Chagos are particularly poor in coral cover. Since other reefs in the Indian Ocean experience similar tidal conditions yet support rich reef flat corals and other fauna, the paucity of reef flat life on Chagos is probably attributable to greater absolute elevation of the atolls.

Tidal water exchange has a stabilising effect on the thermal characteristics of the lagoons. Even though there are marked differences in lagoon depths between atolls, temperature rise in lagoonal water is of the order of 2°C, varying with tidal height more than with lagoon depth. In both the open, large lagoon of Peros Banhos and the relatively small and enclosed Salomon lagoon, water is pumped by wave action on both the rising and falling tides over the reef flats into the lagoon. This water becomes increasingly oxygenated during transit over the flats, from photosynthesis as well as from aeration in the breaker zone. The water progressively accumulates nitrogen, presumably mainly from nitrogen-fixing and blue-green algae (Rayner and Drew, 1984). The implication of enhanced nitrogen in the lagoons, particularly Salomon, is that primary production should also be raised, and indeed carbon fixation was two to 10 times greater in the lagoon than is the case for the open ocean. Phosphate levels varied much less, suggesting fast recycling of this element. Dissolved phosphate was also found to rise after heavy rain with no parallel increase in nitrogen, suggesting that the run-off from many of the islands was rich in guano.

Climate Change and Warming

Temperatures are typically tropical, though a rise of about 1°C accompanied by a near halving of cloud cover, has taken place over the past 25 years (Sheppard, 1999b) (Fig. 5). Indirect measurements show that a sea surface high temperature anomaly which affected a very large part of the Indian Ocean in 1998 passed over Chagos. The effect of this on the corals is described below.

Seismology, Sea Level and Ages

The Chagos Bank is affected by frequent but generally low intensity seismic tremors, though one measuring 7.7 on the Richter scale was recorded in 1983 (Topp, 1997). Since measuring began in 1965, 123 tremors have been recorded, 70 being in one year, though tidal waves of no more than 0.5 m have been recorded. Historically, a notable tremor remarked on by Darwin (1842) concerned an island in Peros Banhos which was destroyed by an earthquake, the noteworthy consequence to the inhabitants being the loss of productive copra plantation. In 1996, that reef was seen to have been greatly fragmented to 30 m deep, lacking the

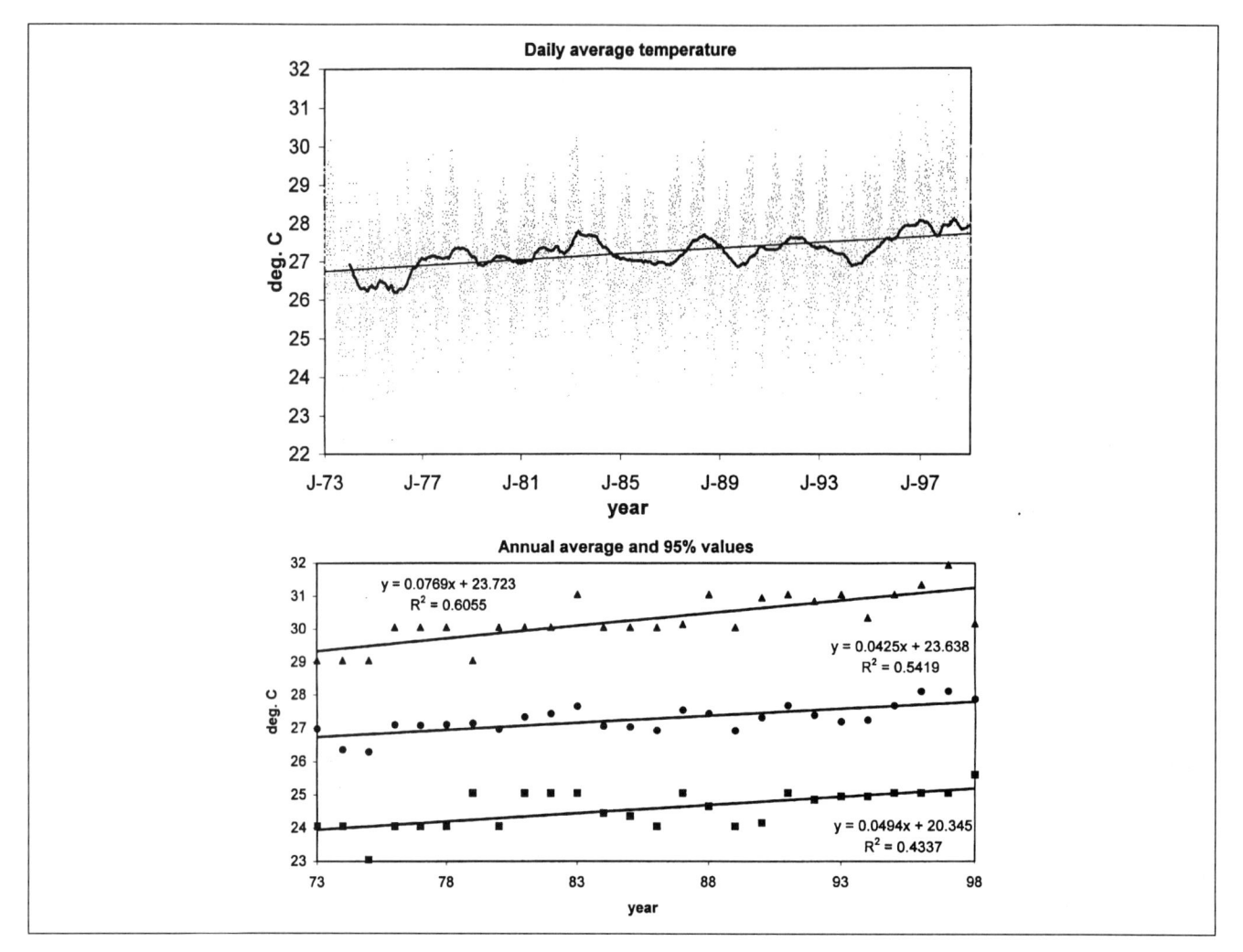

Fig. 5. Air temperature in Chagos between 1975 and 1998. (Top) daily averages. Lines are linear regression and a 365-day running mean. (Bottom) annual averages. Central line of best fit is mean, others are warmest and coolest 95% annual means.

classical profile of reef flat and reef slope. Evidence of the importance of seismic activity comes from the occasional raised reefs (Eisenhauer et al., 1999). Apart from seismic activity found on conventional plate boundaries, Chagos is possibly the most intense source of oceanic seismicity today, and it appears likely that there is a diffuse boundary between the Indo-Arabian and Australian plates extending from the Central Indian Ridge through the Chagos area to the Ninety East Ridge and Sumatra Trench.

The general sea-level profiles of the Pleistocene and Holocene have been confirmed by recent work in Chagos (Eisenhauer et al., 1999). In the lowest sea level stand of about 12,000 years BP, it is likely that about 13,000 km^2 more land was emerged than is the case today. It is also likely that sea level later on was about 1 to 2 m higher than at present. Corals collected from their positions of growth in reefs currently elevated on the islands have been dated to ages of 830 years, 2300 to 3000 years and 5200 years before present. This, and other results from fossil corals taken from emerged beach rock, suggests that the islands themselves are relatively recent.

MAJOR SHALLOW WATER MARINE AND COASTAL HABITATS

Reef Ecology

The distinction of reef scientists such as Darwin and Gardiner did not guarantee their accuracy when commenting on the richness of Chagos' reefs. Darwin reported that Chagos had 'scarcely any live coral', that the rim consisted of 'sand with a very little live coral' although some patches in the lagoons had 'luxuriantly growing coral'. Darwin never visited the archipelago but relied on conversations with Moresby, while Gardiner (1936) generally agreed with Darwin even after his own visit, quoting Darwin at length. These views were presumably based on observations on reef flats, which was the only part easily accessible to the early naturalists, and on the results of a little dredging and examination of what stuck to tallow coated weights. There are some very rich reef flats too, though these are uncommon, so the older views of coral paucity can only be explained by inadequate viewing and, of course, an inability to see the reef slopes at all. All atolls and submerged banks appear to be actively growing reefs. This was the case, at least, until the mass mortality of 1998.

Reef Flats, Coralline Algal Ridges, Spur and Groove Systems

Reef flats on Chagos dry and thus are depauperate compared to many others in the Indian Ocean. Areas with lowest biota lie nearest shore, to seaward of which is a boulder zone with storm tossed reef fragments colonised by surge resistant cup sponges and some limited small brown algae. Where the distance from seaward shores to reef slope is <100 m the entire reef flat has the appearance of a boulder zone. There is a marked contrast to this pattern in some parts of the lagoons where reef flats may be 1 m deep, and where corals may grow.

Algal ridges and associated spur and groove systems occur on the seaward edges of most Chagos seaward reef flats. These appear to be the largest and best developed in the Indian Ocean. Different stages of spur development occur, related to wave exposure (Sheppard, 1981). Most unusually, the lagoon of the large atoll of Peros Banhos is exposed enough to support rudimentary spurs along more than 1 km, each spur being blocks with sides 0.4 to 1.0 m, separated by gaps of equal size. They are completely emerged at low spring tides. Where exposure is greater, the rudimentary spurs become connected laterally at their landward ends. This can develop into a rudimentary algal ridge, connecting a dozen or more of the spurs.

On sheltered seaward reefs, spurs have a similar cross section but are longer, extending 20 m from shore. The most developed spurs are seen on exposed seaward reefs, where all are connected laterally by a ridge of *Porolithon* extending around the rim for several km. These ridges are up to 20 m wide and are elevated 0.5 m above low springs. The spurs extend for 20–75 m to a depth of 5 or 8 m and, like the ridges, are a rich pink as a result of the *Porolithon*.

One almost certain consequence of the spur and ridge system is a greatly increased resistance to erosion from waves. Oceanic swell is substantial in Chagos, even in periods of relative calm. No appreciable coral growth occurs on the shallowest parts where only *Porolithon* appears able to survive. Indeed, it thrives in, and may even require, such conditions.

Biological Patterns on the Reef Slope

On both seaward and lagoonal reef slopes, coral cover on hard substrate was measured as 50 to 80% (before 1998), at least to 40 m deep, with the stony corals providing most of this cover (Fig. 6, left). Below 40 m, coral cover dropped but was still 25% at 60 m deep, and soft corals provided a diminishing proportion of this cover (though gorgonians increased visibly). Encrusting red algal cover increased with increasing depth, covering up to 75% of the hard substrate at 50 to 60 m deep.

Coral diversity increases rapidly with increasing depth until 20 m deep, after which point it declines (Fig. 6, right). This is an unusually deep peak, contrasting with values of 5 to 10 m deep in many regions such as the Red Sea. Possible reasons for the peak being so deep include the exceptionally clear water and the oceanic swell.

Soft corals are still an under-sampled group in Chagos and the ocean as a whole, but in 1996 about 70 octocoral species were found, including two new genera and five new species (Reinicke and van Ofwegen, 1999). One transect on seaward reefs of SE Salomon atoll has been dominated by soft corals for at least 18 years.

Macromolluscs have also been examined to a depth of 40 m (Sheppard, 1984) and the pattern of this group matches those of the coelenterates, with a clear zonation with depth, and to a lesser extent with seaward or lagoonal position. As with corals, total mollusc diversity rises with depth to 10 to 20 m deep on all these reefs, and then declines to 40 m. Shallow water <2 m deep contained <1% of the total species present, while the region of the reef 'drop-off' at about 15 m deep contained up to 40% of the total. Sand-dwelling molluscs, which contained 28% of all species recorded, showed no zonation with depth; in sand, the nature of the substrate alone was important, though it should be noted that sandy habitat was only encountered in significant

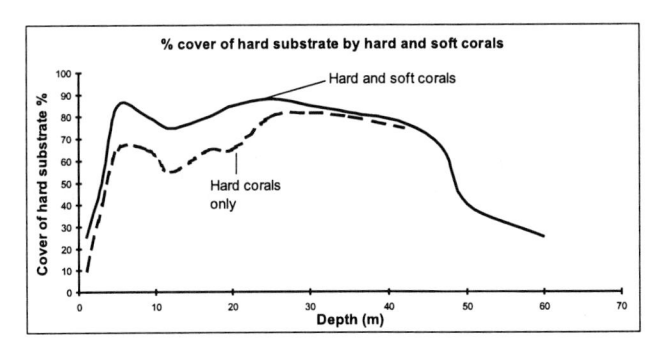

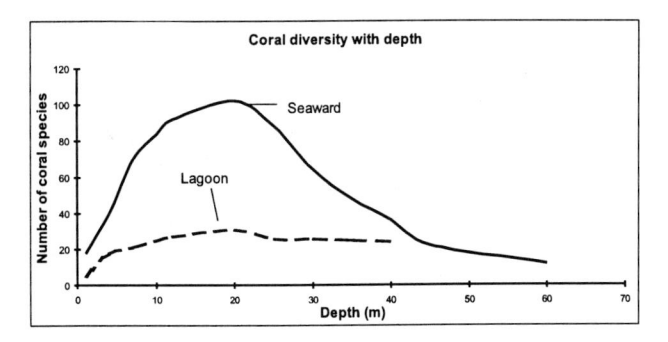

Fig. 6. *Left*: Percent cover of hard substrate (i.e. excluding sand) by stony corals (broken line) and by stony corals plus soft corals combined (solid line), with depth. All transects around two northern atolls of Peros Banhos and Salomon combined. *Right*: Coral diversity on reef slopes. Solid line is seaward reef slopes, broken line is lagoonal reef slopes. In excess of 20 transects on each were pooled for each series (from Sheppard, 1980).

amounts in the lagoons. Molluscan diversity in Chagos compares favourably with those of the Seychelles, Cocos Keeling, the Maldives and even the Red Sea.

The total number of fish species documented is presently nearly 800 (Winterbottom and Anderson, 1999), and their community structure as described through species accumulation and species abundance curves, is typical of many other complex and diverse ecosystems (Spalding, 1999). The reef fish community structures in Chagos appear to vary little across distances of up to 150 km. This apparent homogeneity can be linked to the oceanic location, the lack of terrigenous inputs and the very low impact of humans on these reefs. Reef fishes show greater affinities with the Maldives to the north than with other sites (in contrast to the corals).

Changes over Twenty Years

Chagos is one of very few Indian Ocean sites where comparisons may be made over a 20-year time span. This is a period which has seen increasingly severe and frequent sea-water warming episodes, the most severe being that of 1998 which resulted in mass mortality of reef life throughout the Indian Ocean. There is now substantial evidence of a massive coral mortality following the 1998 warming event, but there is evidence also that a decline began earlier.

Noticeable but not severe changes took place between the periods 1978 and 1996 (Sheppard, 1999c). Reef coral cover measured by tape transects showed that, despite visual impressions of totally unimpacted and vibrant reefs on both occasions, in 1996 total coral cover of all sites pooled had fallen from 59% to 36%, and sand, leafy and encrusting red algae, and soft corals, had all increased slightly. These changes occurred mainly in the shallowest zone to 10 m deep, and the main cause of the decline in cover was a loss of different groups of *Acropora*. Replacing these two species of corals was a near doubling of apparently unoccupied substrate, and to a lesser extent, by increased red algae. *A. palifera*, the most dominant of the two *Acropora* species in 1978/9, had particularly high cover in 1978 which declined markedly in 1996. No significant disease of the corals, or bleaching, were seen in any of the living *Acropora*, which were still far from being rare in 1996.

More recently, the warming of the Indian Ocean in 1998 had massive effects. A brief survey in 1999, in which snorkelling only was carried out, examined about 90 sites on 40 transects (Sheppard, 1999d). Initially it had been intended to measure cover of each reef component by line transects in the manner done at these sites previously but in fact not enough live coral existed in 1999 to justify them. It became a question in 1999 more of seeing what was still alive anywhere, rather than of obtaining a measure of cover which was accurate to 1% or so. Figure 7 shows cover values found on seaward reefs during all three periods of observation: 1978/9, 1996 and 1999. The moderate fall between the first two points (18 years) is clear, followed by a massive

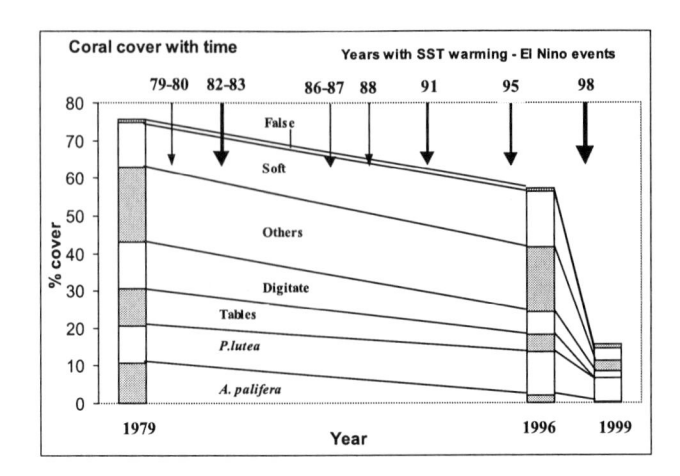

Fig. 7. Cover on seaward reefs by category in 1978/9, in 1996 and in 1999. Episodes of warming or El Niño activity is marked along the top (from Sheppard, 1999d).

reduction between the second and third. Typically, seaward reefs in Chagos which all previously had 50–70% living coral cover plus an additional 10–20% soft coral cover, in 1999 had at most a quarter of this amount, sometimes much less. Some sites appeared to have almost no remaining live coelenterates at all in 1999. There is a reduction in the amount of live calcareous algal cover too, which may be important given the role that red algae play in reef maintenance in shallow water. Almost no table corals remained alive, and on these seaward reefs, very few digitate corals survived either. *Acropora* appeared particularly badly affected though this may be because it is generally such a common genus, while the remaining live cover was largely as a result of the survival of the genus *Porites*. There was a mean fall in coral cover on seaward reefs of from about 65% to about 10%, and of soft corals from about 10% to almost none (Sheppard, 1999d).

Survival in lagoons was double that of seaward reefs (Sheppard, 1999d). Lagoons heat up more than seaward reefs, so might be expected to have suffered more, but the reverse was true. The simplest explanation is that lagoonal reefs experience warming more frequently, so quite simply the corals in the lagoons may be adapted to this.

Thus there was a fall in coral cover over the 18 years from 1979 to 1996, followed by a collapse in 1998. The latter severe event is attributable, or at least coincides with, the severe sea water warming. The earlier and smaller decline which occurred either at one time or gradually over a longer time, has unknown causes, but it is known that several more modest 'El Niño' events associated with coral bleaching also occurred during those years. The decline is not related to any change in local wind speed or direction, but there has been a gradual increase in air temperature, coupled with a fall in cloud and thus increase in insolation. The longer-term consequences of this are not yet known.

Another change between the two sampling periods was a major drop in numbers of sharks (Anderson et al., 1997). Two sets of diver records taken in 1975 and 1978/9 showed

Table 2
Summary of shark sightings by divers in Chagos, for 1975 to 1979 and 1996. Data from Anderson et al. (1997)

	1995+97	1996	Species seen in 1996 (number seen)
No. of dives	207	113	
No. of sharks seen	874	66	Nebrius ferrugineus (16) Carcharhinus albimarginatus (4) C. amblyrhynchos (38) C. melanopterus (7) Triaenodon obesus (1)
No. of sharks per dive	4.2±0.3	0.6±0.1	

high numbers of five species, averaging 4.2 sharks per dive (Table 2). These results also agree with anecdotal observations from an even earlier period in 1972. All three sets of observations in the 1970s were taken in the same months of February and March. In February and March 1996, however, shark numbers were greatly and significantly reduced ($p=0.05$), to about 14% of the earlier numbers.

This is probably a general Indian Ocean phenomenon, not limited to Chagos. Recorded shark fishing throughout the Indian Ocean shows a catch of over 50,000 tonnes per year for most of the past decade at least (FAO, 1992). Actual landings and mortality are likely to be much greater than recorded landings. While shark fishing is not permitted in Chagos waters, several arrested vessels filled with large quantities of shark attest to the activity and the continuing use of the area for shark fishing.

Island Ecology

The tiny size of the islands and their relatively recent emergence between about 1 to 3,000 years ago, makes them important for only a few groups. They are not of great botanical importance. Measures taken in the 1970s to encourage the survival of indigenous species of hardwood into areas where coconut trees were cleared appear to be succeeding, though this recovery may have taken place anyway, if more slowly. One island, Eagle on the Great Chagos Bank, is amongst very few tropical low islands to support a small peat deposit. Unfortunately its young age and small size has meant that its peat has not been able to preserve pollen—a useful tool for investigating pre-human vegetation. Of great importance is the fact that islands without rats and with a variety of vegetation types support globally important bird populations, which is also something which could be further encouraged, though it is notable that Red Foot Booby and other bird populations are increasing rapidly in Diego Garcia even where rats are present.

Vegetation on all the larger islands in the group was greatly disturbed by the copra industry and nearly all the larger islands are dominated by coconut trees, though most are still fringed by the shrubs *Argusia* and *Scaevola*. Some islands are dominated by *Casurina* instead. These have replaced the hardwood trees which were once abundant enough to "furnish good timber" (Moresby, 1884) and which included tatomacca and guiac. Other soft wood trees were also abundant. Many smaller islands have, however, remained unplanted and thus remain relatively undisturbed, supporting stands of the original hardwoods and other native vegetation. About 250 higher plants including ferns were recorded in 1997 (Topp and Sheppard, 1999), but none are endemic. The original native flora probably numbered about 45 species, the rest being introduced by humans, mainly during 30 years of military development in Diego Garcia, which is by far the largest atoll with about half the total land area. For these atolls, plant numbers bear a classical relationship with island size.

Earlier accounts of birds on Chagos showed a high diversity (Bellamy, 1979; Bourne, 1971). The main seabird nesting areas in Chagos are the islands of the Great Chagos Bank and the eastern islands of Peros Banhos atoll. In 1996, a total of 181,000 pairs of 16 breeding species were censused and, to date, a total of about 100 species have been sighted (Symens, 1999). Notable amongst the breeding birds are some such as the red-foot booby *Sula sula*, masked booby *Sula dactylatra* and lesser noddy *Anous tenuirostris* which are rapidly declining in other parts of the Indian Ocean.

The large numbers of birds transfer significant quantities of nutrients from the sea to the islands. Calculations show that they consume annually about 30,000 tons of food, of which 62% are fish and 38% are cephalopods. Through the production of guano, approximately 236 tons of free ammonia and 373 tons of phosphates are returned annually to the nearshore water and terrestrial habitats, providing a continuous replenishment of nutrients to underpin the rich vegetation. The presence of porous, limestone soils and high rainfall suggests a high degree of leaching from islands to the water. There are several indications that birds are steadily recovering from past impacts, but nevertheless, continuing infestation by rats on the majority of islands have continued to prevent full recovery of bird populations.

The islands are important breeding grounds for Green and Hawksbill turtles (Mortimer, 1999) and support the largest Hawksbills so far recorded (Dutton, 1980). Greens, *Chelonia mydas*, nest in at least four of the atolls, coming ashore in June–September during the Southeast trade winds. Hawksbills, *Eretmochelys imbricata*, also nest in the northern atolls, and this species tends to breed more during the Northwest Monsoon between December and March. Large numbers are seen in Diego Garcia lagoon.

OFFSHORE SYSTEMS

There is a licensed pelagic fishery with some inshore and reef fishery (Mees et al., 1999) though there are also known to be occasional illegal fisheries too, including that of reef fish and sea cucumbers. The area has not been included in any major oceanographic studies. Its waters are assumed to be very low in nutrients, as is typical for the central Indian Ocean, though examination of Coastal Zone Color Scanner images shows 'halos' of raised productivity which could be caused either by leakage of nutrients down current of the archipelago, or by a mild upwelling caused when the oceanic currents strike the 2 km high undersea mountains which underlie these atolls. Evidence that this is caused by upwelling rather than leakage comes from observations of the reefs of Diego Garcia whose south eastern tip is dominated by green algae, and probably has been for many years (Topp, personal communication). This site appears to be unique in Chagos. The condition exists in a limited area at the southeast. This part of the atoll, which is a steep 2 km high undersea mountain, receives the brunt of the Southeast trades and is probably one of the roughest parts of Chagos. Replacement of surface water by deeper nutrient-rich water in this part is at least a strong possibility. There is no corresponding, algal-dominated site off southeast Salomon or Peros Banhos.

POPULATION, URBAN AND INDUSTRIAL ACTIVITIES

Human Influences

There is currently no habitation on any atoll except Diego Garcia. Populations living there who were subsisting on a declining coconut industry were removed in the 1970s (Anon., 1979). In Diego Garcia there is a military base supporting a few thousand people, confined mostly to the western 'arm' of this horseshoe-shaped atoll. There is currently no agriculture.

Fishing

Fishing is licensed by the British authorities resulting in a number of vessels being permitted to catch tuna in territorial waters of Chagos (Mees et al., 1999). Undoubtedly shark and other large pelagics are caught also. Unlicensed vessels are confiscated when caught. A limited amount of fishing near reefs is permitted for some Mauritian vessels, but fishing inside the lagoons (even entry of the vessels) is not permitted. Recreational fishing in Diego Garcia removes over 100 metric tonnes per year (Wenban Smith, pers. comm.).

Gross Impacts, Trace Metal and Organic Contaminants

Broad-scale assessment of beaches shows that Chagos is not immune to the problem of trans-oceanic floating debris. Debris, mostly plastics but including bottles and metal containers were seen on all beaches, even piled up on some sand spits (Price, 1999). Oil contamination was detected by chemical means at 38% of sites, though mostly this was at barely discernible levels and no evidence of large slicks was seen. Analytical work confirmed the lack of oil inputs into water, sediments and tissues.

Organochlorine compounds were generally at limits of analytical detection, indicating a pristine environment, and the presence of more volatile poly-chlorinated biphenyls (PCBs) and lindane indicate atmospheric transport rather than marine transport to this site. Heavy metals in tissues and sediments were exceptionally low (Everaarts et al., 1999; Readman et al., 1999), an exception being nickel in some biota from Salomon atoll; nickel is a constituent of fungicides and may have originated from fungicide use in previous times of copra farming. Cadmium levels, even in bird livers and eggs where bio-concentration would be expected, were extremely low and similar to those generally detected in invertebrate fauna from remote open ocean areas. For PCBs and chlorinated biphenyl congeners, levels were generally below quantifiable levels in sediments, though some organochlorine pesticides could be detected in bird livers and eggs at levels of 12 to 300 ng/g lipid.

On the basis of chemical contamination and aside from the global problem of floating debris, Chagos can be considered to be among the least contaminated of all reefal sites in the Indian Ocean and even in the world (Everaarts et al., 1999). The copra farming ended in the early 1970s, before widespread use of modern pesticides and because of this, none of the usual pesticides commonly found in elevated levels around the world appear in the Chagos biota.

In Diego Garcia, human occupation has increased considerably since the 1970s due to the development of a military facility, in contrast to the trend of population and nature of activity in the rest of Chagos. Introductions of numerous plants have continued as already noted, together with several species of fauna. Construction has resulted in the loss of the protective ring of *Scaevola* in some places with the resultant inevitable erosion which is now countered by concrete. In the lagoon, brief work in the late 1970s showed little damage outside the immediate port area, despite dredging. No observations from the greater port construction work in the past 18 years have been published.

PROTECTIVE MEASURES

The UK/USA 1966 Treaty Agreement (3231 of 30 December 1966) specifically states that the whole of the Chagos Archipelago shall be available for defence purposes. In fact, the remoteness of all the atolls and banks, apart from the area on Diego Garcia which is used as a military base probably safeguards them in the most effective way. Nevertheless, several conservation provisions exist, and: "As far as possible the activities of the defence and its personnel shall not interfere with the flora and fauna of Diego Garcia" (BIOT, 1997). Controls are also applied to importation of plant species.

BIOT is included in the UK's ratification of CITES and the Convention on the Conservation of Migratory Species of Wild Animals (Bonn Convention). Several other conservation and pollution Conventions also apply. BIOT has asked the UK to arrange to have the provisions of the RAMSAR Convention applied to it, and will propose possible sites for listing under RAMSAR (BIOT, 1997). It intends also to designate Strict Nature Reserves as defined by the IUCN classification system, and to consider designating Marine Protected Areas. BIOT imposed a 200-mile Fisheries Conservation Zone around the Territory, and contracts fisheries consultants to manage the fishery including issue of licences for tuna and inshore fishing. No fishing is permitted inside lagoons. On Diego Garcia itself, access outside the area reserved for the military facility is carefully controlled, and the authorities are co-operating with the US on a Natural Resources Management Plan.

THE IMPORTANCE OF CHAGOS

This archipelago is one of the very few where most management concerns are absent, simply because people, industry, development and (for the last 20 years) agriculture are absent from most of it. A fortuitous environmental result of its present political status has been the survival of the most pristine set of reefs in the Indian Ocean. The islands are too small for economically viable development, 'eco-tourism' or diving tourism to take place without causing rapid and marked damage.

Much damage on reefs generally has come from gradual and sustained attrition of the natural resources by increased pollution and sewage discharges, coastal development, sedimentation and fishing. In addition to all these, there was in 1998 a massive die-off of corals from bleaching, attributable to warming sea water. As a result of recognition that much of this damage is now irreversible, calls have been made for the total protection of 20% of all marine areas. The reason is that recovery of damaged areas can only come about if there are sufficient unspoiled areas left, adjacent to the damaged ones, to act as species refuges and to provide sources of larvae. Total protection of carefully selected areas, rather than 'managed use', is increasingly being called for, simply because so many previous attempts at management have failed. While 'managed use' may be an ideal principle, commonly either the management is ignored and damage creeps along with the upward human pressure, or the management proposals (e.g. sustainable use estimates) were poor or miscalculated to begin with. Most 'multiple-use' areas have led inexorably to degradation.

Therefore, a network of totally protected habitats is increasingly thought to be essential for effectively sustaining habitats in a fully functioning form over a wide area. The responsibility for governance of Chagos is currently vested in a country which does not need to exploit it in any way for any urgent economic or survival reasons. This is unusual for the Indian Ocean. In other words, its government can afford a long term view. It may be argued that every ocean should have at least one set of reefs where no run-off, no dredging, no building, no fishing, no tourism, no nutrient enrichment and no pollutant and pesticide release whatsoever takes place. For the Indian Ocean, the reefs of Chagos are the prime candidate, and perhaps are the only sensible possibility. At the turn of the Millennium, there are now very few others which fit the bill.

REFERENCES

Anderson, C., Sheppard, C.R.C., Spalding, M., Crosby, R. (1997) Shortage of sharks at Chagos. *Shark News* 10, 1–3.

Anon. (1979) *Under Two Flags*. Ministry of Defence, London. 80 pp.

Bellamy, D.J. (1979) *Half of Paradise*. Cassell, London. 180 pp.

BIOT (1997) The British Indian Ocean Territory Conservation Policy. Foreign and Commonwealth Office, London. 8 pp.

Bourne, G.C. (1888) The atoll of Diego Garcia and the coral formations of the Indian Ocean. *Proceedings of Royal Society* 43, 440–461.

Bourne, W.R.P. (1971) The birds of the Chagos group, Indian Ocean. *Atoll Research Bulletin* 149, 175–207.

Darwin, C.R. (1842) *The Structure and Distribution of Coral Reefs*. Smith Elder and Co., London.

Dumbraveanu, D. and Sheppard, C.R.C. (1999) Areas of substrate at different depths in the Chagos Archipelago. In *Ecology of the Chagos Archipelago*, eds. C.R.C. Sheppard and M.R.D. Seaward. Special Issue of Linnean Society, London. pp. 35–44.

Dutton, R.A. (1980) The herpetology of the Chagos Archipelago. *British Journal of Herpetology* 6, 133–134.

Eisenhauer, A., Heiss, G., Sheppard, C.R.C. and Dullo, W.C. (1999) Reef and Island Formation and Late Holocene Sea-level Changes in the Chagos Islands. In *Ecology of the Chagos Archipelago*, eds. C.R.C. Sheppard and M.R.D. Seaward. Special Issue of Linnean Society, London. pp. 21–33.

Everaarts, J.M., Booij, K., Fischer, C.V., Maas, Y.E.M. and Nieuwenhuize, J. (1999) Assessment of the Environmental Health of the Chagos Archipelago (Indian Ocean). In *Ecology of the Chagos Archipelago*, eds. C.R.C. Sheppard and M.R.D. Seaward. Special Issue of Linnean Society, London. pp. 305–326.

FAO (1992) *Yearbook: Fishery Statistics, Catches and Landings*. Vol. 74. FAO, Rome, 673 pp.

Francis, T.J. and Shor, G.G. (1966) Seismic refraction measurements in the northwest Indian Ocean. *Journal of Geophysical Research* 71, 427–449.

Gardiner, J.S. (1936) The reefs of the western Indian Ocean. I. Chagos Archipelago. II The Mascarene Region. *Transactions of Linnean Society, London* (2) 19, 393–436.

Mees, C.C., Pilling, G.M. and Barry, C.J. (1999) Commercial fishing activity in the British Indian Ocean Territory. In *Ecology of the Chagos Archipelago*, eds. C.R.C. Sheppard and M.R.D. Seaward. Special Issue of Linnean Society, London. pp. 327–345.

Moresby, Commander Indian Navy (1884) Untitled. *Transactions of the Bombay Geographical Society* 1, 307–310.

Mortimer, Day, M.J. (1999) Sea turtle populations and habitats in the Chagos Archipelago. In *Ecology of the Chagos Archipelago*, eds. C.R.C. Sheppard and M.R.D. Seaward. Special Issue of Linnean Society, London. pp. 159–176.

Oldfield, S. and Sheppard, C.R.C. (1997) Conservation of biodiversity and research needs in the UK Dependent Territories. *Journal of Applied Ecology* 34, 1111–1121.

Price, A.R.G. (1999) Broadscale coastal environmental assessment of the Chagos Archipelago. In *Ecology of the Chagos Archipelago*, eds.

C.R.C. Sheppard and M.R.D. Seaward. Special Issue of Linnean Society, London. pp. 285–296.

Pugh, D.T. and Rayner, R.F. (1981) The tidal regimes of three Indian Ocean atolls and some ecological implications. *Estuarine Coastal and Shelf Science* **13**, 389–407.

Rayner, R.F. (1982) The circulation and water exchange properties of Salomon Atoll (Chagos Archipelago). Proceedings of 6th International Scientific Symposium, World Underwater Federation, London.

Rayner, R.F. and Drew, E.A. (1984) Nutrient concentrations and phytoplankton productivity in two Chagos Archipelago atolls. *Estuarine Coastal and Shelf Science* **18**, 121–132.

Readman, J.W., Tolosa, I., Bartocci, J., Cattini, C., Price, A.R.G., Jolliffe, A. (1999) Contaminant levels and the use of molecular organic markers to characterise the coastal environment of the Chagos Archipelago. In *Ecology of the Chagos Archipelago*, eds. C.R.C. Sheppard and M.R.D. Seaward. Special Issue of Linnean Society, London. pp. 297–304.

Reinicke, G.B. and Van Ofwegen, L.P. (1999) Soft corals (Alcyonacea: Octocorallia) from shallow water in the Chagos Archipelago: Species assemblages and their distribution. In *Ecology of the Chagos Archipelago*, eds. C.R.C. Sheppard and M.R.D. Seaward. Special Issue of Linnean Society, London. pp. 67–90

Sheppard, A.L.S. (1984) The molluscan fauna of Chagos (Indian Ocean) and an analysis of its broad distribution patterns. *Coral Reefs* **3**, 43–50.

Sheppard, C.R.C. (1980) Coral cover, zonation and diversity on reef slopes of Chagos atolls, and population structures of the major species. *Marine Ecology Progress Series* **2**, 193–205.

Sheppard, C.R.C. (1981) The groove and spur structures of Chagos atolls and their coral zonation. *Estuarine Coastal and Shelf Science* **12**, 549–560.

Sheppard, C.R.C. (1999a) Corals of Chagos, and the biogeographical role of Chagos in the Indian Ocean. In *Ecology of the Chagos Archipelago*, eds. C.R.C. Sheppard and M.R.D. Seaward. Special Issue of Linnean Society, London. pp. 53–66.

Sheppard, C.R.C. (1999b) Changes in some weather patterns in Chagos over 25 years. In *Ecology of the Chagos Archipelago*, eds. C.R.C. Sheppard and M.R.D. Seaward. Special Issue of Linnean Society, London. pp. 45–52.

Sheppard, C.R.C. (1999c) Changes in Coral Cover on Reefs of Chagos over 18 years. In *Ecology of the Chagos Archipelago*, eds. C.R.C. Sheppard and M.R.D. Seaward. Special Issue of Linnean Society, London. pp. 91–100.

Sheppard, C.R.C. (1999d) Coral Decline and Weather Patterns over 20 years in the Chagos Archipelago, central Indian Ocean. *Ambio* **28**, 472–478.

Sheppard, C.R.C. and Seaward, M.R.D. (eds.) (1999) *Ecology of the Chagos Archipelago*. Special Issue of Linnean Society, London. 350 pp.

Sheppard, C.R.C. and Wells, S.M. (1988) *Coral Reefs of the World. Volume 2: Indian Ocean, Red Sea and Gulf.* UNEP Regional Seas Directories and Bibliographies. IUCN, Gland, Switzerland and UNEP Nairobi. 389 pp.

Sheppard, C.R.C. (2000a) Coral reefs of the Indian Ocean. An Overview. In *Coral Reefs of the Indian Ocean: Ecology and Conservation*, eds. T. McClanahan, C.R.C. Sheppard and D. Obura. Oxford University Press. In press

Sheppard, C.R.C. (2000b) Chagos Archipelago. In *Coral Reefs of the Indian Ocean: Ecology and Conservation*, eds. T. McClanahan, C.R.C. Sheppard and D. Obura. Oxford University Press. In press.

Spalding, M. 1999. Biodiversity patterns in Chagos reef fishes. In *Ecology of the Chagos Archipelago*, eds. C.R.C. Sheppard and M.R.D. Seaward. Special Issue of Linnean Society, London. pp. 119–136.

Stoddart, D.R. (1971) Rainfall on Indian Ocean Coral Islands. *Atoll Research Bulletin* **147**, 1–21.

Stoddart, D.R. and Taylor, J. (1971) Geology and ecology of Diego Garcia atoll, Chagos Archipelago. *Atoll Research Bulletin* **149**, 1–237.

Symens, P. 1999. Breeding seabirds of the Chagos Archipelago. In *Ecology of the Chagos Archipelago*, eds. C.R.C. Sheppard and M.R.D. Seaward. Special Issue of Linnean Society, London. pp. 257–272.

Topp, J.M.W. (1997) Geography and climate of Chagos. Abstract: *Ecology of the Chagos Archipelago, British Indian Ocean Territory*. Linnean Society of London.

Topp, J.M.W. and Sheppard, C.R.C. (1999) Higher Plants of the Chagos Archipelago. In *Ecology of the Chagos Archipelago*, eds. C.R.C. Sheppard, and M.R.D. Seaward. Special Issue of Linnean Society, London. pp. 225–240.

Winterbottom, R. and Anderson, R.C. (1999) Fishes of the Chagos Archipelago. In: *Ecology of the Chagos Archipelago*, eds. C.R.C. Sheppard, and M.R.D. Seaward. Special Issue of Linnean Society, London. pp. 101–118.

THE AUTHOR

Charles R.C. Sheppard,
Department of Biological Sciences,
University of Warwick,
Coventry CV4 7AL,
U.K.

Chapter 68

THE SEYCHELLES

Miles Gabriel, Suzanne Marshall and Simon Jennings

The 115 islands of the Seychelles are scattered widely in the western Indian Ocean and occupy less than 0.1% of the area of the Seychelles' Exclusive Economic Zone. The climate is controlled by monsoonal wind shifts, the south Indian Ocean subtropical anticyclone, seasonal migrations of intertropical troughs and currents and sea surface temperatures in the South Indian Ocean. Climate has a profound impact on reef development. Reefs range from heavily exploited and intermittently polluted fringing reefs on the coasts of the large granitic islands to isolated atolls where fishing pressure and pollution levels are low. Reefs may sustain important artisanal fisheries, provide coastal protection or act as the focus of the dive tourism industry. Development and pollution threaten reefs around the large granitic islands, but the greatest impact in recent years was due to the 1998 coral bleaching event that has killed over 50% of corals in many localities. The offshore seas are largely unaffected by human activities other than fishing, although some oil exploration has taken place. Approximately 15% of the islands' reefs have protected status. The extensive reefs and EEZ of the Seychelles, coupled with a relatively small population, allow Seychellois to address conservation issues which may be insurmountable elsewhere.

Seas at The Millennium: An Environmental Evaluation (Edited by C. Sheppard)
© 2000 Elsevier Science Ltd. All rights reserved

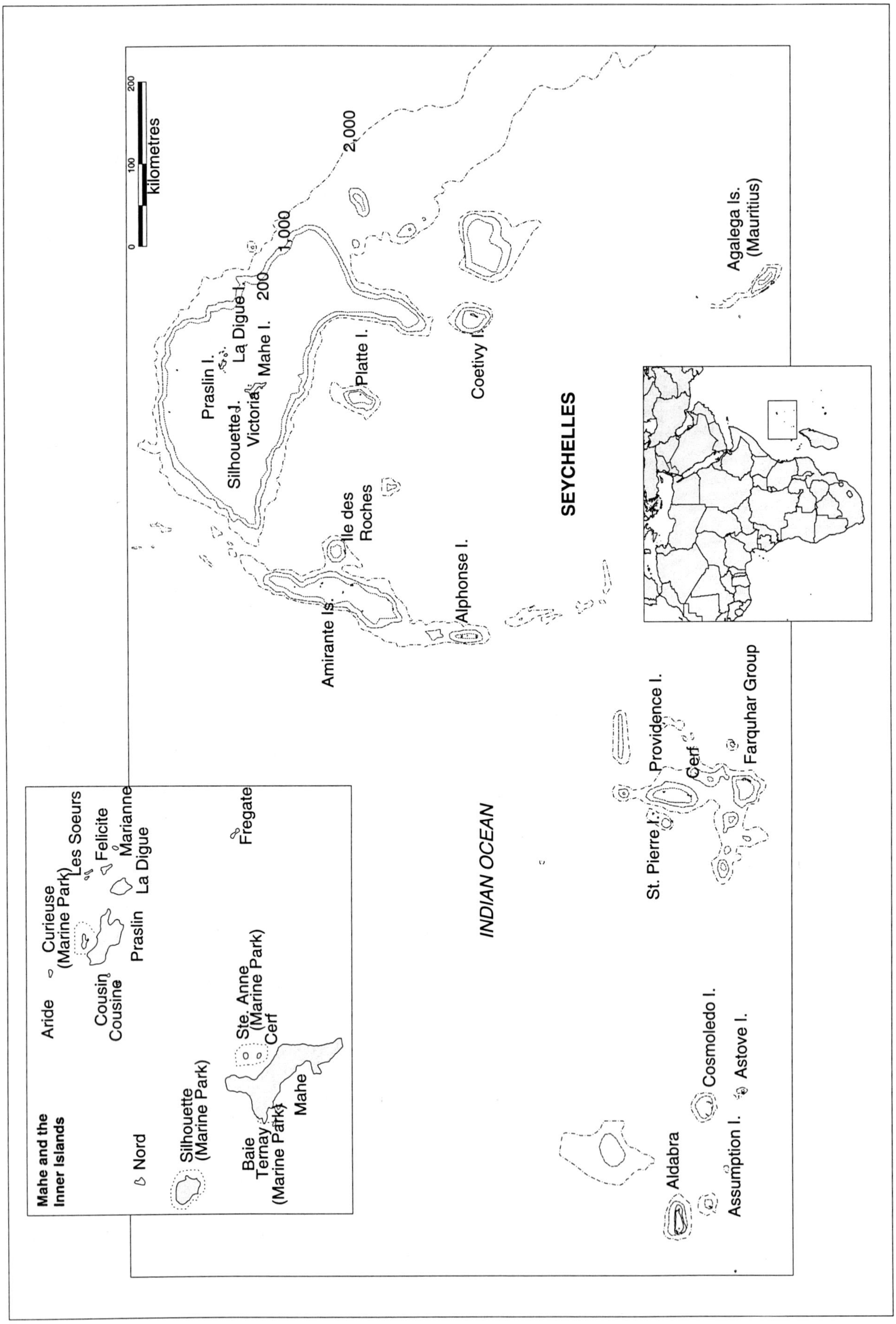
0
100
200
kilometres
2,000
1,000
200
Agalega Is.
(Mauritius)
Praslin I.
La Digue I.
Mahe I.
Silhouette I.
Victoria
Platte I.
Coetivy I.
SEYCHELLES
Ile des
Roches
Alphonse I.
Amirante Is.
INDIAN OCEAN
Providence I.
Farquhar Group
Cerf
St. Pierre I.
Cosmoledo I.
Astove I.
Aldabra
Assumption I.
Mahe and the
Inner Islands
Aride
Curieuse
(Marine Park)
Les Soeurs
Felicite
Marianne
La Digue
Cousin
Cousine
Praslin
Fregate
Nord
Silhouette
(Marine Park)
Ste. Anne
(Marine Park)
Cerf
Baie
Ternay
(Marine Park)
Mahe

THE REGION

The Republic of Seychelles is situated in the Western Indian Ocean between 3°41' and 10°17'S and 46°15' and 56°18'E. It is an archipelagic nation of 115 islands, with a total land area of 455 km² set within an exclusive economic zone of 1,374,000 km² (Fig. 1). The Seychelles are composed of 41 northern inner granitic islands, centred around Mahé and Praslin, and 74 outer coralline islands and atolls of the Amirantes, Farquhar and Aldabra groups to the south and west. The granitic group comprises a slightly larger landmass than the outer islands, but occupies a much smaller area of ocean. The area of Indian Ocean that surrounds the Seychelles is defined by the boundary of the EEZ and cannot be considered an isolated marine ecosystem.

The inner granitic islands of Seychelles are of Precambrian age, about 650 million years old, and constitute continental rocks, unlike other isolated oceanic islands which are composed of oceanic basalts or reef limestones (Braithewaite, 1984; Stoddart, 1984a,b). The islands were created as the ancient super-continent of Gondwanaland began to break up around 135 million years ago and stand on a shallow plateau known as the Seychelles Bank. Today, the Seychelles Bank is 43,000 km² in area, with mean water depths of between 44 and 65 m, and a rim that rises to within 20 m of the surface (Braithewaite, 1984). The granitic islands have a land area of 277 km² with the main island of Mahé accounting for 34% (148 km²) of the total area of the archipelago. The islands reach a maximum altitude of 914 m on Morne Seychellois on Mahé.

Within the western Indian Ocean, various other oceanic platforms exist with maximum depths of 33–90 m, most having a shallower rim around their margins up to depths of roughly 11 m. The Amirantes, Saya de Malha, and Nazareth Banks are examples. The banks are built on a base of oceanic basalt covered by a layer of granite up to 13 km thick, on top of which is a relatively thin (ca. 500 m) layer of sediment. During the Pleistocene period these banks would have been emergent, creating land areas exceeding 125,000 km² in the western Indian Ocean. At the present time, they are major areas of shallow-water carbonate sediment accumulation, of comparable importance to the better known Florida–Bahamas and Arabian Gulf provinces.

The outer coralline islands are classified as either low-lying coral or high limestone islands. The low-lying coral islands are basically reef platforms on which sand cays 2–3 m in height have formed through deposition; examples are found on the Seychelles and Amirantes Banks. Denis and Bird Island lie on the northern rim of the Seychelles Bank, rising from depths of about 55 m, while Coetivy rises from about 36 m on the southern rim of the bank. Other low-lying islands rise from the eastern margins of the Amirantes Bank to the west. The Amirantes Bank is approximately 6300 km² in area, with a central zone at depths of up to 70 m and a rim at depths of 11–27 m (Braithewaite, 1984).

The high limestone islands include the atolls of Aldabra, Assumption, Astove, Cosmoledo and St. Pierre to the southwest. These islands rise up to 8 m above sea level and are formed on top of basalt bases of volcanic origin rising from depths of over 2000 m. As a result, their outer reef slopes are steep and fall rapidly to depths of several hundred metres. The atolls show signs of successive sedimentations of calcium carbonate after the land has repeatedly emerged above and sunk below sea level. This phenomenon took place between one and two million years ago and indicates that the age of these islands is far greater than the low-lying coral islands (Braithewaite et al., 1973; Stoddart, 1984a).

SEASONALITY, CURRENTS, NATURAL ENVIRONMENTAL VARIABLES

The Seychelles experiences a humid tropical climate with annual rainfalls exceeding 700–800 mm and mean monthly temperatures always above 20°C. However, due to the geographic expanse of the Seychelles, over 1500 km from Denis (3°S, 56°E) in the northeast to Aldabra (9°S, 46°E) in the southwest, and differences in altitude of the high granitic islands, there are marked climatic differences within the islands.

The climate of the Seychelles is controlled by four factors: monsoonal wind shifts, the south Indian Ocean subtropical anticyclone, seasonal migrations of intertropical troughs, and currents and sea surface temperatures in the South Indian Ocean (Walsh, 1984). From May to October, southeast trade winds blow with great constancy and force (6–9 m s^{-1}) and the stable dry air results in dry weather and low rainfall. The southeast trades have a key effect on reef development around the islands and windward reefs are fundamentally different from those that develop on leeward shores. From December to March, the lighter and more variable northwest monsoon is dominant. During the transition months of April and November winds tend to be light and variable. The Northwest Monsoon carries moist air over the islands and tends to result in relatively wet weather (Walsh, 1984).

Mean annual lowland temperatures at Port Victoria on Mahé and at Aldabra are 26.6°C and 27.0°C respectively. Seasonal variations are minor. At Port Victoria, monthly means vary from 27.8°C in April to 25.7°C in August. The lowest temperature ever recorded at Port Victoria was 19.3°C and the highest 32.8°C. Rainfall throughout the Seychelles decreases towards the southwest with mean rainfalls in the northeastern atolls of Bird (1973 mm) and Denis (1730 mm) being twice as high as that of the southwestern atolls of Aldabra (966 mm) and Assumption (867 mm).

The inner granitic islands of Seychelles lie to the north of the tropical cyclone routes but cyclones have been recorded infrequently in the southwestern islands of Aldabra and Assumption. The passage of cyclones to the south of the granitic islands can however result in intensification of the intertropical convergence (and hence heavy rainfall) over the islands (Walsh, 1984).

The Seychelles lie in the south equatorial current, and sea-surface temperatures usually range from 26°C to 31°C and salinities from 34.5‰ to 35.5‰. Tidal ranges on the open ocean reefs are only 0.3 to 1 m, but they increase to 1 or 2 m around the granitic islands and 2 to 3 m in the Aldabra group. The oceanic water surrounding the Seychelles is characterised by low and variable rates of production. Atolls may, however, induce upwelling and this adds to downstream oceanic productivity (Dustan, 1992).

From March to May 1998, the sea temperatures in the Seychelles were the highest on record. They peaked at around 34°C, at least 3°C hotter than usual. The increase was associated with the warmest year in recorded history and a strong ENSO event (Wilkinson et al., 1999). The increase in sea temperature had a dramatic effect on coral reefs in the Seychelles, contributing to mass coral bleaching and subsequent mortality of many corals.

THE MAJOR SHALLOW-WATER MARINE AND COASTAL HABITATS

The shallow-water coastal habitats are dominated by carbonate reefs, sandy areas and corals growing on granite boulders. The reefs are among the most extensive in the western Indian Ocean. They range from heavily exploited and intermittently polluted fringing reefs on the coasts of the large granitic islands of Mahé, Praslin and La Digue to isolated atolls such as Aldabra where fishing pressure and pollution levels are low. Reefs may sustain important artisanal fisheries, provide coastal protection or act as the focus of the dive tourism industry. Reefs have been studied in most detail on the granitic islands and Aldabra. Reclamation, sedimentation and eutrophication threaten reefs, but the greatest recorded losses occurred in 1998 when sea temperatures rose dramatically due to an ENSO (Wilkinson et al., 1999). The reefs of Seychelles are now very different from those described by successive scientific studies since the 1960s (Lewis, 1968; Barnes et al., 1971; Braithwaite, 1971; Rosen, 1971; Stoddart, 1984; Hoeksema, 1994) with large areas of dead coral and coral rubble.

The shallow-water areas of the granitic islands consist of rocky shores or sandy beaches, rippled sand areas, marine grass beds, radial zone, algal ridge, reef edge, and outer slopes (Fig. 2). Seagrass and algae are abundant and diverse (Kalugina-Gutnik et al., 1992), and the *Sargassum* beds are some of the largest in the world. The marine fauna is typical of the Indo-Pacific coastal zone and includes two species of nesting turtles and some 900 species of marine fish, at least 70 of which are landed by local fishermen. There is a notable absence of dense bivalve mollusc settlements which may be due to the relatively low nutrient levels in Seychelles coastal waters, although nutrient upwellings sometimes spill over onto the Mahé bank. Fringing reefs have developed around the coasts of the granitic islands. Water depth and exposure govern the development of these reefs and they are most extensive around Mahé and Praslin where they occupy areas of 20 km² and 27 km² respectively. Fringing reefs on the other islands total 10 km². The vertical extent of fringing reefs is typically limited to 15 to 20 m by the shallow depths of the Seychelles Bank and their morphology is largely determined by their aspect (Braithwaite, 1984).

Fig. 2. Ste Anne National Park with Mahé in the background.

The southeast trade winds significantly affect coral growth and there are marked differences between reefs exposed to the southeasterly trades and those in sheltered locations. Thus reefs on the southeast coast of Mahé are exposed and tend to be continuous and unbroken with a width of 500 to 750 m whereas in northwest Mahé the reefs are sheltered and tend to be highly irregular, with many gullies and channels. On the west coast of Mahé, the granite cliffs between bays plunge steeply into the sea, and true fringing reefs have only developed in the sheltered bays such as Baie Ternay and Port Launay. In the exposed areas, corals grow directly on the granitic substrate. Fringing reefs around Praslin are extensive and may extend 3 km from the shore, but much of the coral is currently dead.

The two other types of reefs in Seychelles are the platform reefs and atolls. Platform reefs have developed in the Amirantes and Farquhar groups and total over 200 km² in area. Stoddart (1984) recognised two types of platform reef, those which have a considerable area of dry land following the accumulation of detrital deposits and those with shallow lagoons and reef and a very small land area as found at African Banks, Ile Plate and Providence. Assumption and St. Pierre are raised platform reefs reaching heights of 8 m above sea level. These raised reefs were used as nesting sites by seabird colonies and have been extensively mined for guano.

Farquhar, St. Joseph, St. Francois and Alphonse in the Amirantes are all atolls. The combined area of peripheral reef in these atolls is around 120 km². The remaining reefs in Seychelles are found around the raised atolls of Aldabra, Cosmoledo and Astove in the Aldabra group. The total area of peripheral reef in the raised atolls is approximately 200 km^2 and the lagoons are largely devoid of coral growth.

At the present time, much of the living veneer of coral tissue that covers the areas of reef accretion is dead, and the action of bioeroding species may be limiting any reef growth. This will reduce the coastal protection provided by reef structures and may lead to increased coastal erosion. Various reports from the islands (Wilkinson et al., 1999; Goreau, pers. comm.; Teleki, pers. comm.) suggest that the mortality of branching corals at many sites around the islands during the 1998 ENSO exceeded 50%. In some areas it approached 100%. From March to May 1998, when sea temperatures peaked, corals began to bleach. *Acropora* and other fast-growing branching species were most affected, though some massive *Porites* species also bleached. Some 40–50% of coral was bleached to a depth of 23 m around Aldabra, Providence and Alphonse from March to May 1998 (Wilkinson et al., 1999). By the time that sea temperatures had fallen again, most of the branching corals were dead and were colonised by algae, while many of the massive corals recovered from bleaching. At sites in the Inner Seychelles, where live coral cover was recorded prior to the ENSO in 1994 and several months after the ENSO in November 1998), total coral cover fell by at least 50%, and by 100% at one site near Cousin Island (Fig. 3a). Most of this fall was due to the loss of branching corals, which all died within 3 of the study sites (Fig. 3b).

While the main reason for coral bleaching is thought to be the stress imposed by abnormally high temperatures, other factors such as levels of sunlight, UV light, pollutants and local hydrography have an important influence (Brown, 1997). Interestingly, some of the reefs around Mahé, which were exposed to heavy sediment loads and considered to be in poor condition, are amongst the few areas where live coral cover remained. This is true, for example, in the muddy waters at the mouth of Victoria harbour. These reefs may have survived because the reduced water clarity limited the penetration of sunlight (Goreau, pers. comm.).

At present (1999), the reefs around the Seychelles have retained much of their structural complexity, and are still protecting coasts from wave action and providing suitable habitats for fish and invertebrates. However, now that the corals are dead, they may slowly be reduced to rubble, and the resulting habitat is likely to support a reduced abundance of fish and not provide such an effective barrier against wave action.

The principal terrestrial coastal zones on the granitic islands consist of 'plateaus' and marshy/mangrove areas. The plateaus are discontinuous elevated terraces that run along the shore in narrow bands up to 2 m above sea level and consist of calcareous reef material. Lowland coastal vegetation has been severely modified by human activity, with much of the well-drained plateau having been cleared of forests and planted with coconut which was previously evident only as a narrow belt on the littoral fringe (Shah, 1995).

OFFSHORE SYSTEMS

The Seychelles EEZ supports one of the most prolific tuna fisheries in the western Indian Ocean, and many large pelagic fishes are relatively abundant. Since the Seychelles Bank is quite shallow, similar species are found across much of this region and are also found inshore close to the islands at the base of reef slopes. For the atolls surrounded by deep water, local upwellings may fuel additional production and create rich fishing grounds. Apart from fisheries, and a little oil exploration that has not continued, the offshore systems are relatively unimpacted by human activities.

POPULATIONS AFFECTING THE AREA

In 1800, the population of Seychelles was just over 2000. This had risen to 10,000 by 1835 and 25,000 by 1930. The population has continued to grow steadily (Fig. 4) and is currently 77,000. About 99% of the population live on the three largest granitic islands of Mahé (88%), Praslin (8.2%)

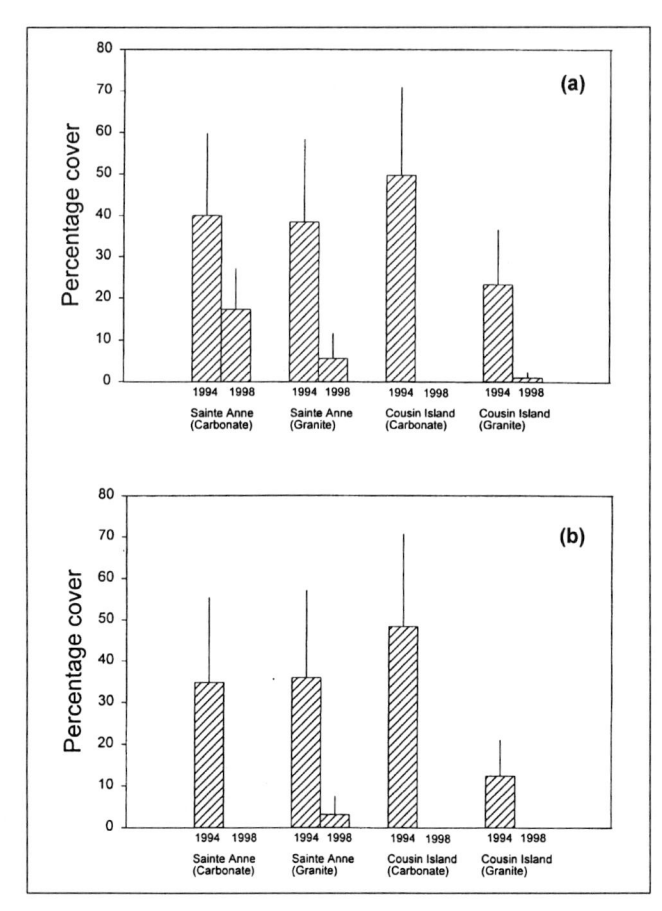

Fig. 3. (a) Total live coral cover and (b) live branching coral cover. Locations of sampling were near Ste Anne Island and Cousin Island. At both islands, measurements were made on a carbonate (true reef) and on a granitic substrate, and coral cover was recorded in 16 replicate areas of 150 m^2. (mean ± SD) in 1994 and 1998. (Jennings, Vogt and Callow, unpublished data).

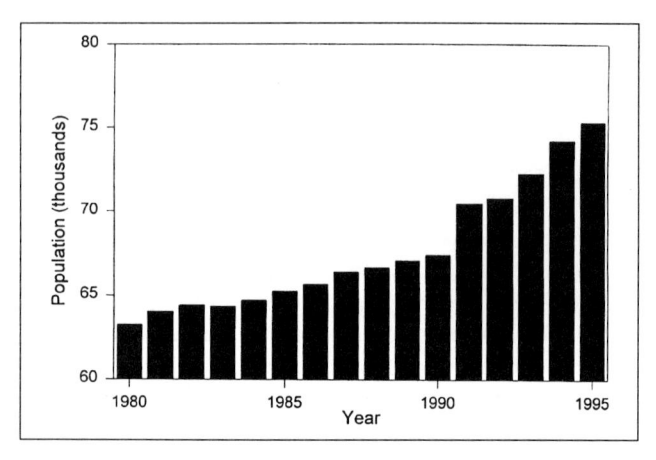

Fig. 4. Population trend in the Seychelles 1980–1995.

and La Digue (2.8%). The remaining population, roughly 300 persons, live on the outer coralline islands. The average family size is five and of the 20,000 wage earners a total of 2300 are employed in the agriculture, forestry and fishing sectors, 3300 are employed in mining, manufacture and construction, 2500 in tourism-related industries and 7500 in services (Government of Seychelles, 1991; Shah, 1995). The tourist industry leads to dramatic local increases in population density, visitor arrivals having grown steadily since an international airport opened on Mahé in 1972 (Fig. 5).

The Seychellois have, as an island state, always looked to the sea as the main source of food, with fish the main provider of protein and fish and rice the staple diet. Per capita annual consumption of fish is about 75 kg. The artisanal fishery is practised solely by Seychelles fishermen and is the only commercial fishery which uses coastal marine resources. Fishers operate on foot or by boat. Fishers on foot usually target octopus and catch 0.5% of the total estimated yield in the artisanal fishery. Four types of boat are used by artisanal fishers: pirogues, small open boats which are powered by oars or outboard engines of up to 15 hp (4–6 m); outboards, which have larger outboard engines and are usually a fibreglass design; the 'Mini-Mahé', which was introduced in 1971 (5–6 m); whalers, undecked wooden boats powered by inboard diesel engines (6.5–8 m), and

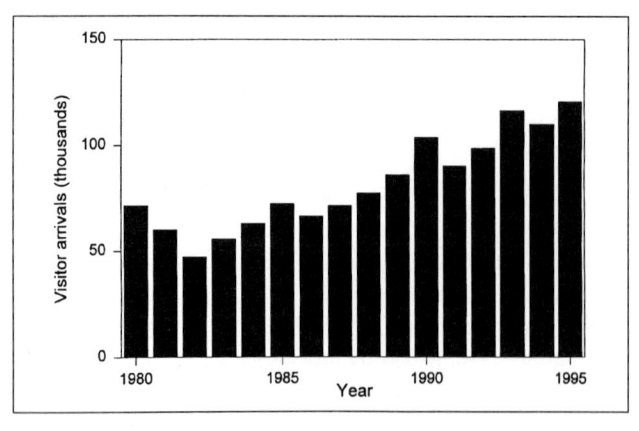

Fig. 5. Tourist arrivals to the Seychelles 1980–1995.

schooners, which are large fully decked fishing boats with inboard engines (10–25 m) (Anon., 1992, 1993). The schooners can travel considerable distances and are the only boats in the artisanal fleet which routinely fish on the Mahé and Amirantes plateaus. Fishing and catches are higher in the calmer inter-monsoonal periods. The number of schooners fishing in recent years has declined as their fishing activities have not been profitable (Mees, 1990a,b).

Small-boat fishing accounts for about 35% of the total catch, with whalers and schooners fishing further offshore. Although 70 or so species are caught, catches typically consist of coral-associated bottom fish, notably snappers (Lutjanidae), groupers (Serranidae) and emperors (Lethrinidae). Semi-pelagic jacks (Carangidae) also constitute a significant percentage of the catch as they feed around coral and sandy areas, near the coast. The artisanal fishing fleet comprises about 400 boats, with about 1000 men actively involved. The annual catch is around 4000–5000 tonnes: in 1994, the catch was 4428 tonnes, of which 370 tonnes valued at US$ 2 million were exported. In terms of trends, catches for the artisanal trap fishery in the inner reefs may have declined. Innovations in fishery techniques and gear, such as motherships/dories and electric handreels, have also led to rapid over-exploitation of demersal fish stocks even in remote areas.

Agriculture and traditional fishing used to be the major source of income but have now been replaced by tourism since the opening of the International Airport. Despite the large numbers of people working in the artisanal fisheries, the pelagic tuna fishery has greater economic importance. Foreign vessels are charged to fish in the Seychelles EEZ and tranship their catches in Mahé, where a large cannery also purchases a proportion of the catch. The fishing industry (mainly tuna) is now the second most important source of foreign revenue after tourism (Anon., 1995).

RURAL FACTORS

The Seychelles National Land Use Plan suggests that less than 6000 ha of land in the Seychelles has agricultural value, less than 0.1 ha per inhabitant. There are approximately 200 full-time and 700 part-time farmers, roughly 2% of the total population. Agricultural production in Seychelles has been of diminishing importance since the decline of the coconut based plantation economy of the 1970s. The sector has lost about 74% of its arable land to tourism and residential buildings, and now provides only a modest contribution to the GDP (<3% in 1992) (Government of Seychelles, 1988, 1991).

Early agricultural practices included the deforestation of coastal areas for the creation of coconut plantations. This practice resulted in significantly increased amounts of runoff from soils entering the river networks and ocean. Today these problems have been exacerbated by unwise development activities and poor agricultural practice, such as drainage of wetlands, and much of Mahé is susceptible to erosion even though regulations stipulate that cultivated land should be terraced. In instances, this red earth has been found on fringing reefs over a kilometre from the shore, causing fish deaths and settlement of heavy colloidal clay sediments.

The misuse of pesticides and fertilisers is also cause for concern, particularly when the regularity, intensity and widespread nature of run-off is considered. Often farmers are not aware of the detrimental effects these chemicals can have and proper storage and application levels are not ensured. Use of pesticides is of particular significance to coastal systems as high applications of chemicals such as Mataldehyde could lead to persistence in coastal environments and, in 1993, pesticide spillage led to fish deaths on the eastern coast of Mahé (Shah, 1995).

Other problematic issues within the agricultural sector include livestock rearing and new ventures, such as a hydroponics farm and orchid growing, which are being attempted on former coconut plantations. These activities are major sources of organic pollution in the coastal zone. Discharges from pig farms on Mahé need to be properly controlled and use should be made of the Seychelles Effluent Quality Standards to prosecute farms that continue to discharge raw sewage into rivers or mangroves. This practice could lead to severe problems within the important mangrove systems and also within the marine environment for both marine animals and bathers.

Unfortunately, it is difficult to assess the effects of agriculture (and other sectors) on the marine environment due to a lack of data concerning pollutant levels. Although Seychelles published the Seychelles Effluent Quality Standards, these are only useful if reliable pollution data exists within which to enforce the regulations. The Environmental Assessment and Pollution Control Section of the Division of Environment has monitored pollution levels for some time but the validity and reliability of these data was often inaccurate (Gabriel, 1996a,b). The data were found to contain discrepancies in the records of all pollutants tested. These discrepancies were due to excessive waiting periods between sampling and analysis, unsterile and unsealable containers, sample site discontinuity, analysis methods used, entry discrepancies, missing data etc. Recommendations have since been made and new monitoring programs with strict guidelines and procedures have been initiated. These should allow a better picture of the overall scale of the pollution problem and enable more precise evaluation of pollution effects and sources (Gabriel, 1996a,b).

COASTAL EROSION AND LANDFILL

Population growth and the developing tuna and tourist industries have led to increased requirements for housing, docks, factory units and hotels. Areas of flat land suitable for construction are very limited in Seychelles, and this has led to pressures to claim coastal land and reef areas for building. The largest such claim took place in the late 1960s and early 1970s, when the runway at the new International Airport was built over an area of shallow reef. There have also been many small losses of coastal habitat and reef to developers around Port Victoria. In 1998, builders also began to claim a large area of reef to the north of Port Victoria for development.

Apart from the direct loss of coastal habitat, many of these developments have led to increased sedimentation in shallow waters around the main population centres and, off Victoria, increased levels of sedimentation appear to have contributed to coral mortality in the Ste Anne Marine National Park, although the effects have been eclipsed by the effects of the recent ENSO.

EFFECTS FROM URBAN AND INDUSTRIAL ACTIVITIES

Polluting nutrient inputs have been recorded close to human population centres. At Beau Vallon Bay in northwest Mahé, the turn-over of water is very poor, leading to decreases in salinity when run-off is high, and the retention of nutrients and polluting substances. The release of polluted water from houses, hotels and farms may be acutely toxic to hard coral species and lead to a shift in the coral community until it is dominated by soft corals. Nutrient-rich effluents from pig farms have also been identified as a threat to reefs in Baie Ste Anne.

In the harbour at Port Victoria, there have been a number of spills of fuel oil and other chemicals that have proved difficult to trace. In addition, terrestrial run-off that follows heavy rain often exceeds the capacity of the sewage system and oils or other materials that have accumulated on the roads flow directly into the sea.

PROTECTIVE MEASURES

While the early history of Seychelles was characterised by unsustainable exploitation and environmental degradation, there are signs that public attitudes to the marine environment are beginning to change and that historic precedents can be nullified. The Seychelles government has expressed a desire to move away from the simple concept of 'nature conservation' to develop a broad-based strategy encompassing both environmental protection and the sustainable management of marine ecosystems (Marshall, 1994). One of the main reasons for this change is the recognition that the effective management of reef ecosystems can bring long-term socio-economic rewards to the islands (Jennings et al., 2000).

Despite the problems posed by population and industrial growth, and the need for foreign revenue, the Government of Seychelles has tried to adopt a broad-based nature conservation policy in the marine environment. The Government has ratified many treaties that control activities in the marine environment (Table 1) and has established a network of marine protected areas (Table 2).

The regulations which govern activities in these areas usually state that reef habitats must not be damaged and that fishes must not be captured. Some practical constraints, such as the lack of wardens and patrol boats, disregard for park regulations and a lack of visitor facilities, prevent the effective implementation of the regulations. Only Curieuse and Sainte Anne have sufficient staff to ensure that park regulations are largely enforced and the independently managed Cousin Island Special Reserve is also effectively patrolled (Jennings, 1998).

Seychelles is largely reliant on fish exports and tourism for foreign revenue and it is important that conflicts between conservation and exploitation can be resolved if sustainable development and the protection of natural resources are to be assured (Jennings and Marshall, 1995). Marine National Parks are viewed as a central component

Table 1

Key international conventions relating to the protection of the marine environment, as ratified by the Republic of Seychelles

Convention	Date ratified
Convention on International Trade in Endangered Species of Wild Fauna and Flora	8.2.1977
African Convention on the Conservation of Nature and Natural Resources.	14.11.1977
International Convention for the Regulation of Whaling	19.3.1979
Convention Concerning the Protection of the World Cultural and Natural Heritage	9.4.1980
Convention on the Prevention of Marine Pollution by Dumping Wastes and Other Matter.	29.10.1984
Treaty on the Prohibition of the Emplacement of Nuclear Weapons and Other Weapons of Mass Destruction on the Sea Bed and the Ocean Floor and in the Subsoil thereof	12.3.1985
Treaty Banning Nuclear Weapon tests in the Atmosphere, in Outer Space and Underwater	12.3. 1985
International Convention on Civil Liability for Oil Pollution Damage	11.7.1988
International Convention on the establishment of an International Fund for Compensation for Oil Pollution Damage	11.7.1988
UNEP Regional Seas Convention for the Protection, Management and Development of the Marine and Coastal Environment of the Eastern African Region	22.6.1990
Protocol Concerning Protected Areas and Wild Fauna and Flora of the East African Region	22.6.1990
Protocol Concerning Cooperation in Combating Marine Pollution in Cases of Emergency in the Eastern African Region	22.6.1990
Protocol of 1978 Relating to the International Convention for the Prevention of Pollution from Ships, 1973	8.2.1991
International Law of the Sea (UNCLOS)	16.9.1991
Convention on Biological Diversity	10. 6. 1992
United Nations Framework Convention on Climate Change	22.9.1992
Basel Convention on the Control of Transboundary Movements of Hazardous Wastes and their Disposal	11.5.1993

Table 2

The Marine Protected Areas of Seychelles. From Jennings et al. (2000).

Marine Protected Area	Year	Sea area (km^2)	Management responsibility	Management action
Ste Anne	1973	10	Marine Parks Authority	Managed & Policed[1]
Curieuse	1979	10.8	Marine Parks Authority	Managed & Policed
Baie Ternay	1979	0.8	Marine Parks Authority	Basic Management[2]
Port Launay	1979	1.6	Marine Parks Authority	Basic Management
Silhouette	1979	to 1 km	Marine Parks Authority	No Management
Cousin	1968	1.2[3]	BirdLife International	Managed & Policed
Aride	1973	to 200 m	Royal Society for Nature Conservation[4]	Managed & Policed
Aldabra	1981	350	Seychelles Islands Foundation[5]	Basic Management
Ile Cocos, Ilot Platte & Ile la Fouche	1997	0.03	Marine Parks Authority	Basic Management
African Banks	1987	0.3	Ministry of Defence[6]	No Management
Shell reserves	1981	–	None	No Management

1. A reserve which is managed and policed is delineated by boundary markers and regular anti-poaching patrols are mounted. See text for details of activities in specific MPAs.
2. Basic management implies that the boundary of the reserve is marked and the function of the reserve is advertised.
3. The reef is protected to 500 m from the low water mark.
4. Aride Island was also recognised as a Special Reserve in 1979 Government Legislation.
5. Aldabra was recognised as a Special Reserve in 1981 Government Legislation and as a UNESCO World Heritage Site in 1982.
6. African Banks is used as a naval firing range.

of the strategy to resolve these conflicts and to ensure the successful co-existence of activities which rely upon a shared resource base. When regulations are enforced, the abundance, biomass and diversity of fishes in the Marine National Parks is much higher than in areas where fishing occurs, and these fish are a major attraction for tourist visitors, many of whom come to dive and snorkel (Jennings et al., 1995, 1996). Sainte Anne MNP is one of the most popular tourist sites in Seychelles and more than 25% of all tourists arriving in Seychelles, pay to snorkel, dive or view the reef from glass-bottom boats (Anon., 1994). It remains to be seen whether the mass coral mortality that occurred in 1998 will discourage tourists from visiting the Seychelles or the Marine Parks.

ACKNOWLEDGEMENTS

We thank the Government of Seychelles, the Fisheries Management Science Programme of the then Overseas Development Administration (now DIFD) and the Shoals of Capricorn Programme for funding the collection of data referred to in this chapter. We thank BirdLife International for support while working on Cousin Island.

REFERENCES

Anon. (1992) Seychelles artisanal fisheries statistics for 1991. Seychelles Fishing Authority, Mahé.

Anon. (1993) Seychelles artisanal fisheries statistics for 1992. Seychelles Fishing Authority, Mahé.

Anon. (1994) Annual Report 1993: Conservation and National Parks. Government of Seychelles, Mahé.

Anon. (1995) Seychelles artisanal fisheries statistics for 1994. Seychelles Fishing Authority, Mahé.

Barnes, J., Bellamy, D.J., Jones, D.J., Whitton, B.A., Drew, E.A., Kenyon, L., Lythgoe, J.N., Rosen, B.R. (1971) Morphology and ecology of the reef front at Aldabra. *Symposia of the Zoological Society of London* **28**. 87–114.

Braithwaite, C.J.R. (1971) Seychelles reefs: structure and development. *Symposia of the Zoological Society of London* **28**, 39–63.

Braithwaite, C.J.R. (1984) Geology of the Seychelles. In *Biogeography and Ecology of the Seychelles Islands*, ed. D.R. Stoddart, pp. 17–38. Dr W Junk, The Hague.

Braithwaite, C.J.R., Taylor, J.D. and Kennedy, W.J. (1973) The evolution of an atoll: the depositional and erosional history of Aldabra. *Philosophical Transactions of the Royal Society* **B226**, 307–340.

Brown, B.E. (1997) Disturbances to reefs in recent times. In *Life and Death of Coral Reefs*.

Dustan, P. (1992) Estimates of Indian Ocean productivity using natural fluorescence. *Atoll Research Bulletin* **378**, 1–13.

Gabriel, M.E. (1996a) Marine Pollution Data, East Coast Monitoring Project, Mahé, Seychelles. 1995. A Critical Assessment. Report for the COI/FED Regional Environment Programme, Seychelles, September, 1996. 7.ACP RPR.068.

Gabriel, M.E. (1996b) Marine Pollution Data, Mahé, Seychelles, 1991–1995. A Critical Assessment. Report for the Indian Ocean Commission (COI/FED) Regional Environment Programme, Seychelles, August, 1996. 7.ACP RPR.068.

Government of Seychelles (1988) National Land Use Plan 1988. Ministry of Community Development, Republic of Seychelles.

Government of Seychelles (1991) National Land Use Plan 1991. Ministry of Community Development, Republic of Seychelles.

Hoeksema, B.W. (1994) Species diversity of stony corals and mushroom coral sizes. In *Oceanic Reefs of the Seychelles*, ed. J. Land, pp. 133–138. Netherlands Indian Ocean Programme, Leiden.

Jennings, S. (1998) Cousin Island, Seychelles: a small, effective and internationally managed marine reserve. *Coral Reefs* **17**, 190.

Jennings, S. and Marshall, S.S. (1995) Seeking sustainability in the Seychelles. *Biologist* **42**, 197–202.

Jennings, S., Grandcourt, E.M. and Polunin, N.V.C. (1995) The effects of fishing on the diversity, biomass and trophic structure of Seychelles' reef fish communities. *Coral Reefs* **14**, 225–235.

Jennings, S., Marshall, S.S. and Polunin, N.V.C. (1996) Seychelles' marine protected areas: comparative structure and status of reef fish communities. *Biological Conservation* **75**, 201–209.

Jennings, S., Marshall, S., Cuet, P. and Naim, O. (2000) The Seychelles. In: *Coral Reefs of the Western Indian Ocean: Their Ecology and Conservation*, eds. T.R. McClanahan, C.S. Sheppard and D.O. Obura, pp. 399–432. Oxford University Press, New York.

Kalugina-Gutnik, A.A., Perestenko, L.P. and Titlyanova, T.V. (1992) Species composition, distribution and abundance of algae and seagrasses of the Seychelles Islands. *Atoll Research Bulletin* **369**.

Lewis, M.S. (1968) The morphology of the fringing coral reefs along the east coast of Mahé. *Journal of Geology* **76**, 140–153.

Marshall, S.S. (1994) Proposed management strategy for Seychelles Marine National Parks. Division of Environment, Government of Seychelles, Mahé.

Mees, C.C. (1990a) Seychelles schooner fishery: an analysis of data collected during the period January 1985 to July 1990. Seychelles Fishing Authority, Mahé.

Mees, C.C. (1990b) The fishermen of Seychelles: results of a socioeconomic study of the Seychelles fishing community. Seychelles Fishing Authority, Mahé.

Rosen, B.R. (1971) Principal features of reef coral ecology in shallow water environments of Mahé, Seychelles. *Symposia of the Zoological Society of London* **28**, 163–183.

Shah, N.J. (1995) Managing coastal areas in the Seychelles. *Nature and Resources* **31**, 16–33.

Stoddart, D.R. (1984a) *Biogeography and Ecology of the Seychelles Islands*. Dr W. Junk, The Hague.

Stoddart, D.R. (1984b) Coral reefs of Seychelles and adjacent regions. In *Biogeography and Ecology of the Seychelles Islands*, ed. D.R. Stoddart, pp. 63–81. Dr W Junk, The Hague.

Walsh, R.P.D. (1984) Climate of the Seychelles. In *Biogeography and Ecology of the Seychelles Islands*, ed. D.R. Stoddart, pp. 39–62. Dr W. Junk, The Hague.

Wilkinson, C., Lindén, O., Cesar, H., Hodgson, G., Rubens, J. and Strong, A.E. (1999) Ecological and socioeconomic impacts of 1998 coral mortality in the Indian Ocean: an ENSO impact and a warning of future change. *Ambio* **28**, 188–196.

THE AUTHORS

Miles Gabriel
Inter-consult Nambia (Pty) Ltd.,
P.O. Box 20690,
Windhoek, Namibia

Suzanne Marshall
Environment Agency, Kings Meadow Road,
Reading, RG1 8DQ, U.K.

Simon Jennings
CEFAS, Fisheries Laboratory,
Lowestoft, NR33 OHT, U.K.

Chapter 69

THE COMOROS ARCHIPELAGO

Jean-Pascal Quod, Odile Naim and Fouad Abdourazi

The Comoros archipelago lies in the Indian Ocean east of Mozambique, and is composed of four volcanic islands and the off-shore bank of Geyser and Zélée. Its coastal and marine habitats range from rocky shores to pure white sandy beaches, coral reefs and mangroves.

Three islands belong to the Federal Islamic Republic of Comoros, and one is a French Collectivité. Grande Comore and Anjouan are steep-sided, while Moheli has a significant expanse of shallow water. These three are ringed by fringing reefs. Mayotte is a marine paradise with an immense and complicated barrier reef, fringing reefs and many patch reefs, and also has a very uncommon but well studied 'double-barrier' reef.

More than 80% of the Comorian population live in coastal areas and rely on agriculture and marine resources for their livelihood. Poverty and demographic trends (519,000 in Comoros, 133,000 in Mayotte), include a rate of population increase which is one of the highest in the world. This, and a shift of the economy from primary to secondary activities in Mayotte, create identified threats to the coastal environment. Human pressure on the marine and coastal resources is close to the limits of sustainability.

Available data show that anthropogenic threats include both over-harvesting of natural resources and habitat destruction. Marine ecosystems have suffered massively during the last decade from siltation due to poor land use practices and sewage pollution. Mayotte lagoonal communities, with their limited water exchange, are being affected by sediment and pollutant accumulation. Organic and toxic chemicals discharged into the lagoon from various sources over two decades will increase and become more serious over the next decade as well. This may affect, in the medium or long-term, neighbouring marine habitats as well as people living nearby.

Coastal and marine resources from the Comoros archipelago are of great importance in the maintenance of biodiversity in this part of the Indian Ocean, but loss of critical habitats such as coral reefs, mangroves and beaches will also be of general significant economic and social concern in the coming decade. Global phenomena such as the El Niño-related massive mortality in 1998 affected this region also. In addition, because of its location beside tanker routes, the Comoros islands remain at risk from oil pollution which, given a tidal range of 4 m, would affect all the coastal ecosystems.

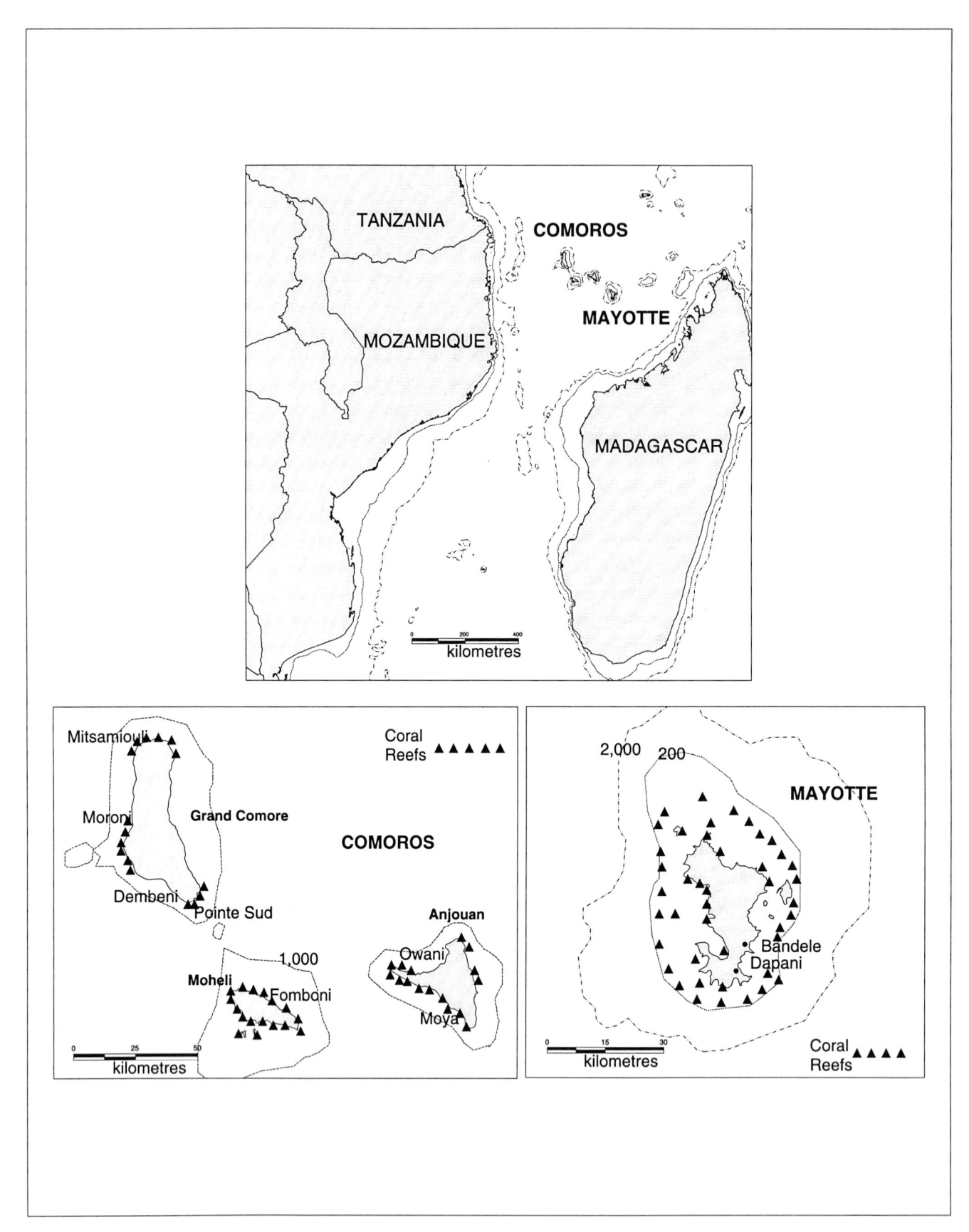

Fig. 1. The Comoros archipelago (inset is Mayotte).

THE DEFINED REGION

From West to East, the Comoros group has four main islands: the Grande Comore (Ngazidja), Moheli (Moili), Anjouan (Ndzouani) and Mayotte (Maore). They are located at the northern end of the Mozambique channel between 11°20' and 13°05' South, 43°05' and 46°59 (Fig. 1). These four islands are isolated seamounts of volcanic origin, and their age increases towards the East (Table 1). The youngest (and also the highest and largest) island, the Grande Comore, is an active volcano, with only scattered small fringing reefs; Moheli (the smallest and the lowest) and Anjouan, (two extinct volcanoes), have extensive fringing reefs, while Mayotte, the southernmost and the oldest, has a barrier reef which extends over a length of 76 nautical miles.

Grande Comore and Anjouan are steep-sided. Moheli is distinctive with a significant amount of shallow water. Mayotte is a marine paradise with a complicated barrier reef which is the largest in the Indian Ocean, fringing reefs, many patch reefs, and a rare double-barrier reef structure (Fig. 1). Eastward lies the Geyser and Zélée Bank which has the appearance of two incomplete atolls.

SEASONALITY, CURRENTS, NATURAL ENVIRONMENTAL VARIABLES

The southward-flowing Mozambique Current is derived from the South Equatorial Current (SEC), which splits at about 12°S, and flows past the northern part of Madagascar, mainly as the East Madagascar Current (Rao and Griffiths, 1998). Tides are semi-diurnal, and their amplitude exceeds 3.5 m during springs; the highest tides occur in Dzaoudzi. Sea surface temperature varies little from 24°C to 27°C, but lagoonal waters may experience higher temperatures, reaching 35°C in Mayotte. Swells predominantly come from the southeast and affect the southern and eastern coasts of all the islands.

There are three main seasons: *Kousi*, the dry southeast trade winds, bring cool air and good weather to the islands from April to September. *Niobene*, northern and variable winds, last from September to November, and *Kashkasi*, northwest monsoon winds, bring most of the annual rainfall (generally 100 to 250 cm) from December to March. The archipelago is not usually in the direct paths of cyclones (*"dambas"*), but some like Kamisy in April 1984 caused destruction both on land and in the sea. More usually, the islands are subjected to side effects of storms that pass by in the northern Mozambique Channel, and these can have important effects on the reefs and sand cays (Guilcher, 1965).

Islands of this size and altitude induce numerous local effects on the weather: frosts may occur on top of Karthala, but at sea level temperatures are equable. Rainfall varies tremendously according to altitude and is variable from place to place. In Grande Comore, the average annual fall can be 260 cm at Moroni while it can reach 800 cm on the western slopes of Karthala. Mayotte appears to be drier than the other islands; at Dzaoudzi the annual mean is 143.9 cm.

THE MAJOR SHALLOW WATER MARINE AND COASTAL HABITATS

The degree of coral reef development is directly related to the age of the islands, and the extent of the reef flats can vary from a few hundred metres to a few kilometres across (von Hentig, 1973; Prosperi, 1957; Frazier, 1985). The combined lengths of the coasts of the archipelago total 350 km (Mayotte alone is 185 km) and the coastal index (calculated by dividing the length of the shoreline by the land area) is 0.18.

Grande Comore (Ngazidja or Ngazidia)

The Grande Comore is the largest (64×31 km) and highest (Karthala is 2361 m) island of the archipelago. It is steep-sided. Lava fields and small volcanic cones are conspicuous over the island, notably in the northern half. The coast is rugged with 3 m high lava cliffs alternating with pockets of white, ochre or black beaches. The slopes of the volcano are cloaked in rain forest, and a crater lake (Niamaouy) in the extreme north is connected to the sea. The soil is poor and highly permeable, and consequently the majority of rainfall infiltrates deeply, leading to no permanent streams.

Mangroves are rare and beach ridges line most of the coast. The reefs are small (31 km^2), fringing and discontinuous. Nevertheless they line 60% of the 170 km perimeter of the island and are better developed in Mitsamiouli (northwest), Hantsindzi (northeast), Chindini (southeast) and Iconi (west). The outer slopes are steep (Guilcher, 1965), except in Mitsamiouli and Chindini, where an embryonic barrier reef can be recognized (Rouyer, 1988). Sea-grass beds are restricted to Choua-Chandroudé, which is the only off-shore island and which lies 300 m to the northeast (this 1 ha island is also called l'île aux tortues). To the West, the Vailheu bank, lying 10 miles from the coast, is reputed to be very rich in fish.

Table 1

Comparison of some features of the four main islands of the Comoros archipelago

	Maximum altitude (m)	Surface (km^2)	Reefs (km^2)
Grande Comore	2361	950	31
Anjouan	1595	378	?
Moheli	790	216	?
Mayotte	660	370	1500
Total		1914	

Anjouan (Ndzouani or Johanna)

Shaped like a triangle and 35×37 km, Anjouan is also steep-sided and geologically young with a rugged coastline, but this island has permanent streams, good soils and is often considered to be the most beautiful of the islands. The soil is completely vegetated but much of the island is devoted to coconut, ylang-ylang (*Cananga odorata*) of which Comoros is the world's largest supplier, vanilla, cloves and other crops. Even the steep upper slopes are cultivated. Some rain forest is left in the Forêt de Moya, toward the south. The fringing reefs are discontinuous but well-developed around the peninsulas of Jimiline (north), M'ramani (south) and particularly around Bimbini (west) where there is a complex fringing reef. Reefs cover 55 km (80%) of the coast of this island. No significant sea-grass beds are recorded. There are nearly a dozen offshore rocks, but only one significant island, l'île de la Selle, which occurs off the western point and which is about 0.3 km² in area and 101 m high.

Moheli (Moili or Mwali)

On this small (29×12 km) and populated island (210,000 inhabitants on a surface of 424 km²), there are several permanent streams and good soils. The soil is completely vegetated and much of the lower part is planted with coconut and ylang-ylang. The high ground has good forest cover, although this is secondary growth. A crater lake, Dziani Boundouni, occurs in the southeast. Small areas of mangroves occur. Unlike Grande-Comore and Anjouan, Moheli has an extensive area of shallow water reaching as much as 10 km wide off the south of the island. Fringing reefs are continuous and encircle the island, and are particularly developed and diverse around the southern offshore islets, as are seagrass beds.

Mayotte (Maore or Mahore)

Mayotte, the oldest island and so the most eroded, is made up of two parts: the main island of Grande Terre (40×20 km and 360 km²) where most of the population lives (Fig. 2), and the smaller Petite Terre which is 13 km² and which is embedded in the barrier reef. Mayotte is a well-watered island. Small forested areas remain, but most of the land is cultivated, especially with food crops which are grown over much of the island. Coconut, bananas (12,000 ha) and other export crops such as ylang-ylang and coffee occupy significant areas. Mayotte is almost entirely surrounded by a barrier reef 197 km long (Fig. 3), with a second, double-barrier in the southwest and the emerged Iris Bank in the north which has an area of 40 km². There are a dozen deep passes through the reefs, some of them being the beds of older rivers. The geological origin and structure have been well studied and described (Guilcher, 1965, 1971; Guilcher et al., 1965; Coudray et al., 1985). The inner enclosed lagoon is large (1500 km²) and averages 20 m deep, though in some

Fig. 2. Mamoudzou, the largest city of Mayotte. Sewage outputs from urban activities directly affect the lagoon communities. A dense mangrove stand is in the foreground. (Photo: J.P. Quod).

Fig. 3. Mayotte airport extension located on the barrier reef at Pamandzi island. (Photo: J.P. Quod).

places its depth is up to 80 m. The main island is surrounded by a fringing reef, which is discontinuous where there are river mouths. Numerous small islets of up to 2 km² are present in the lagoon, and these generally have fringing reefs also. A total of about 668 ha of mangroves occur in many places around the island (Gabrié, 1998), especially in protected bays. An extensive, healthy mangrove stand lies in Baie de Boueni.

OFFSHORE SYSTEMS

The Geyser and Zélée Bank (46°33'E, 12°28'E), is made up of two horseshoe-shaped reefs which are almost atolls, lying 12 km apart. The western Zélée reef is submerged at low tide to about 10 m depth, while the eastern Geyser rim is partly emergent. This bank is notoriously dangerous to shipping. Polunin and Frazier (1974) wrote that this reef had experienced virtually no human impact as a result of its isolation, and in 1996, Chabanet et al. (unpublished report) recorded 297 species of fish and a density varying from 300 to 700 fish/100 m². These banks are now regularly exploited by fishermen coming from Mayotte about 115 km away, and the snapper species *Lutjanus bohar* now comprises 48% of the fish caught on the bank. Between 1989 and 1992, a period of intensive fishing from one boat, the Yvalan, resulted in a rapid reduction of the number and size of the fish caught (Maggiorani et al., 1994, unpublished report).

POPULATIONS AFFECTING THE AREA

The early political history of Comoros is complex. The four islands came under French protection in 1886, and became a colony in 1912. A Unilateral Declaration of Independence (UDI) was made in 1975, creating the Federal and Islamic Republic of Comoros (FIRC), but Mayotte remained with France as a French "Collectivité territoriale" (Toussaint, 1966; Vienne, 1900). The Geyser and Zélée Bank lies in international waters.

In 1984, Madagascar, Mauritius and the Republic of Seychelles created COI (Commission de l'Océan Indien) that Reunion and Comoros joined two years later. COI includes 13 million inhabitants in 594,000 km² of lands, and collectively has 3 million km² of Exclusive Economic Zone (EEZ).

The populations of FIRC or Comoros (630,000 inhabitants in 1995) and of Mayotte (133,000 in 1997) are largely composed of homogeneous Islamic populations, predominantly rural (Table 2). Comoros falls into the category of countries with a low Human Development Ranking and is classified in the group of Least Developed Countries (LDCs). The small sizes of the three islands, and thus the small populations in absolute terms, place this country into the group of Small Island Developing States (SIDS).

THREATS TO THE ENVIRONMENT

Land-use Practices

The economy of the Comoros Archipelago is predominantly agricultural, and agricultural products contributed 39% of the GNP in 1995. One third of the total land is arable, and forests and woodlands cover less than one fifth of the land. Almost 80% of the labour force is engaged in agricultural activities (Bryceson et al., 1990; Shah et al., 1997). However, limited land and water severely constrain further agricultural and industrial development. In contrast, the people have access to vast areas of marine resources on the insular plateau of around 900 km², whose depths lie up to 400 m deep, and so villagers still have alternative sources of protein depending on the season. In economic terms, the annual growth rate of GNP is low in Comoros (2.7%), and is negative on a per capita basis (–0.7%) because of high population growth.

Soils are mostly relatively good, but cultivation increases the risk of soil erosion due to the mountainous topography, so soil conservation and watershed management require special attention. Unsustainable practices such as slash-and-burn shifting cultivation dominate in Comoros and Mayotte. With increasing pressure on land for subsistence farming, there is an increasing shifting cultivation and deforestation, soil erosion and decline in soil productivity, compared to fifty years ago.

Table 2
Population trends

	Date	Comoros	Mayotte
Population in millions	1970	0.3	0.04
	1995	0.6	0.13
	2015	1.1	0.24
Population growth rates	1970–1995	3.3	6.0
	1995–2015	2.9	
Population doubling date at 1995 growth rates (1)		2017	2018
Population densities per km²	reported	292 in 1995	349 in 1997
	projected	648 in 2020	650 in 2010

Sources: UNEP (1999).

The tropical moist forests which originally covered the Comoro Archipelago are greatly reduced. In the early 1990s, the average annual rate of deforestation was 5.6%, and less than 30% of natural forests is now left. In Grande Comore between 1983 and 1993, primary forest cover fell from 12,000 ha to less than 10,000 ha, generally between 200 m and 2000 m. In Anjouan and Moheli, the only pockets of primary forests occur at high altitude and inaccessible locations. Many indigenous tree species which existed fifty years ago, such as the tacamaca, ebony and baobab, have become almost extinct. In contrast, Mayotte appears to be more luxuriant now than it was in the 1950s when centuries of cultivation led to deforestation, exposing the lateritic soils. Silt running off the land has had an adverse effect on seawater quality in terms of both siltation and reduction of light.

While forest loss has been mainly due to land clearance for agriculture and urbanisation in all the islands, forest degradation is also linked to poverty and use of wood for domestic fuel. Illegal exploitation of forest trees for the distillation of ylang-ylang and furniture making is also compromising forest sustainability, and often creates conflicts with local villagers. The problems of deforestation and lack of access to clean water are linked in the archipelago, as deforestation is regarded as the primary cause of the drying up of water courses.

Coastal Development

In Mayotte, 21.5% of the population live in Mamoudzou. The development of new towns and villages in coastal areas and river plains has led to increased erosion problems of the sandy soils. Dredging of harbours and the removal and subsequent dumping of dredged material is a threat to nearby coral reefs due to excessive siltation. As an example, dredging of Longoni harbour on Mayotte has caused the nearby fringing reef to disappear under 0.3 m of mud.

The combination of poverty and negative economic growth on a per capita basis has compelled the poor and the unemployed to overfish the coral reefs, destroy primary forests and exploit beach sand, and to degrade the land. Most of these rural factors affect both coastal and marine ecosystems by siltation which has affected several square kilometres of reef. Exploitation of wildlife for commercial purposes is also a problem.

Exploitation of Sands, Corals, Pebbles, and Limestone

Beach sands are quarried in many parts of the Comoros, mainly for use in building. Beach and river pebbles are often taken near the building sites, especially in Anjouan, while limestone is also quarried, to be used for cement manufacture and for building material (Ahmed, 1988).

The practice of coral extraction is now disappearing, although this activity is still carried out by some people. The most sought-after corals belong to the genus *Porites*, *Acropora* and *Favia*, which are taken at low tide with metal bars. The collected corals are then loaded on a canoe and returned to the coast where they are put in heaps to be sold. The coral heaps are composed either of *Porites* alone, or are a mix of *Acropora* and *Favia*. They are also gathered directly from the reef flats in the form of dead colonies or rubble.

Demand for sand was estimated at 60,000 m^3 in 1997 (COI, 1999, unpublished report). Many people, especially women, are involved in this activity, and the sand from remote beaches may be transported by men across the mountains to the closest roads. One truck of coral or basaltic sand costs 25,000 to 30,000 FC, and by comparison, a truck of sand from the crushing of basaltic rocks (a centralised activity) costs 75,000 to 100,000 FC. Sand extraction is forbidden by law and villagers are aware of the problem, but there is no application of the law and no compensation for people if they were to stop the activity.

The consequences of sand and coral extraction can be severe. In Anjouan in front of the village of Muroey, for example, the reef front has been totally changed by quarrying, and consequently the shoreline is affected by swells. Numerous houses are now in danger, and only placement of artificial sea defences will be able to stop the erosive wave action. Beach loss has been reported in Comoros, and the assumption has been made that the loss has been entirely due to coral and sand extraction. Sea-level rise (by 15 to 95 cm by year 2100) may also contribute to these beach losses in the future, affecting probably turtle nesting too.

Degradation of Mangroves

Mangroves play an important role in stabilising shorelines and decreasing erosion, and their degradation worsens coastal erosion. Although some pockets of mangroves are expanding due to the influx of terrestrial sediment from the hillsides, large areas have been heavily cropped for poles, timber, firewood, charcoal, and the production and extraction of tannin. They can also be affected by land reclamation (Fig. 4). In Mayotte, a road will be built largely by reclamation of the mangroves of Passamainti and of the near-shore fringing reef.

Seagrass beds

Seagrass beds are particularly abundant in the lagoon of Mayotte but are of lesser abundance in Moheli. These meadows are of outstanding importance particularly for the green turtle and dugong, and provide a vital function in the breeding and feeding cycles of numerous species of the nearby coral habitats. The better studied seagrass assemblages apparently have not suffered yet from pollution and siltation.

Over-exploitation and Degradation of Corals and Associated Fauna

Fringing reefs are heavily threatened by the destruction of reef areas for mining for lime, collection of corals for trade, destructive fishing practices and siltation. In Mayotte, at the end of the 1980s, run-off and silt from construction of the littoral road were a major threat to fringing reefs all around the island, and led to progressive siltation in the lagoon. Coral reef degradation is highest near major towns and cities due to both land-based activities such as the road building, and from sewage discharge and marine species exploitation. Mayotte's live coral cover is currently less than 11%.

Mayotte lagoon is not homogenous from a hydrological point of view but is composed of a series of eight hydrologic

Fig. 4. Coastal development and land fill, causing disappearance of mangroves. (Photo: J.P. Quod).

basins, linked to the open ocean by passes which are sometimes the beds of submerged valleys. Each basin has its own particular hydrological characteristics. Currents are essentially tidal. Several important issues affect the health of the coral communities living in and outside the lagoon; on one hand, the thermal stratification of surface water during summer is possibly impairing recruitment and survival of coral and fish larvae, and juveniles; while on the other hand there is a low renewal rate of inner lagoonal waters.

Destructive fishing practices are still used in Comoros. Dynamite fishing, though now rare in Grande Comore and Mayotte, is still reported in Anjouan and Moheli. Generally, this type of degradation is never caused by local fishermen; poachers operate far from their own villages. In Mayotte, the utilisation of small mesh nets and the vegetable poison *Tephrosia*, locally named "*uruva*", are equally destructive, and during very low tides most of the population of villages collect everything that is edible on the reef flats.

Coral is used to make lime. In the lime ovens, the wood used for fuel is usually coconut, mango and kapok trees. *Porites* coral produces the best lime, which is used for traditional construction, notably for houses destined to be given to girls and houses of nobles with terraces. The lime from *Acropora* is particularly used as 'milk of lime' because it gives to walls of houses a vivid colour. *Porites* colonies were formerly also used as stone for building (copying an architectural tradition from the Red Sea). The stone for "sandalwood powder" also comes from *Porites* coral; one carves and decorates a sort of small table in the coral, on which is rubbed a piece of sandalwood; the powder thus obtained is used in manufacturing the "masks of beauty" that Comoro women use for their faces. In addition, precious black coral, the Antipatharian *Cirripathes* sp. is used to make jewellery in the archipelago. Its conservation status here is unknown.

Natural disturbances, such as the occurrence in large numbers of the coral-eating starfish *Acanthaster planci*, have been noticed in Grande Comore and Mayotte during the past two decades (Polunin, 1974). In 1983, there was a severe outbreak of coral bleaching on the Mayotte reefs (Faure et al., 1984), which may have been related to the abnormal El Niño that year. Severe episodes of coral bleaching were reported in the Indian Ocean reefs in 1988 and 1998, which are generally assumed to have resulted from higher than normal water temperatures. Mass bleaching was severe in Comoros in 1998, with 45% of corals now reported as dead in places like Mitsamiouli, Itsandra and Nioumachouoi. Only massive forms (*Porites*, *Montipora*) remained alive. In Mayotte, coral bleaching was detected between April and August 1998 with significant bleaching in both lagoonal waters and on the outer slope to depths greater than 25 m. Soft corals were also affected. On the outer slope, nearly 90% of the tabulate *Acropora* died and these are now covered by algae. Surveys conducted in March 1999 by the local authorities concluded that most of the outer slope colonies were still dead, but that in the inner reefs of the lagoon coral settlement was visible.

Fishing

Fishing remains artisanal. In Comoros, boats used vary from the "galawa", the traditional paddle boat, to the "japawa", an engine-powered boat of 9 m long. These last, provided by Japan, are glass fibre and have been introduced for the purpose of raising the fishing effort. In Comoros, some 8000 fishermen live along the coasts, grouped in 104 communities: 4500 in Grande Comore, 2,400 in Anjouan, and 1100 in Moheli. As the island shelf is wider in Moheli than in the two other islands, especially in the south where it extends for 10 nautical miles, this zone and its several islets is the most frequented, even by fishermen coming from Anjouan and Grande Comore. In Mayotte, there were 1666 fishermen in 1989, 2600 in 1992 and there are apparently 3600 in 1998.

In Mayotte, increased sedimentation in the lagoon has affected the fisheries. As early as the 1980s, Fourmanoir and Harmelin-Vivien (1984) recorded a change in fish population structure, with a lower species diversity and an increase of herbivorous species, presumably in response to the increasing amount of algae on the reef. In 1986 the coastal people were complaining about the death of corals on the fringing reefs which were being buried beneath sediments and consequently causing the disappearance of fish on these former rich fishing banks near their villages. They were forced to find boats to go fishing on the barrier reef further away.

There has been a marked increase of fish catch in Comoros. Local fish consumption has increased approximately from 4000 tons in 1986 to 8000 tons in 1991, and to 14,000 tons in 1995. In Mayotte, the total catch of reef-associated species is estimated to be 600 and 800 tons/year but this is continuously decreasing (1700 tons in 1989), and in markets and on the quays it has now become rare to see large groupers and other large reef species. These predator species have been overfished and in some places of the lagoon they have collapsed. Fish Aggregating Devices (FADs) which are floating materials anchored at sea for the purpose of attracting pelagic fishes such as tunas, have been developed, so that now there are 40 off Grande Comore, 22 off Anjouan, 14 off Moheli, and 9 in Mayotte. Some are close to the coast, accessible to galawas, others are deep and can be reached only by motor-boats.

Aquaculture is now being promoted as a solution, both for reducing pressure on the ecosystems and for providing alternative food sources. Also, the establishment of a marine reserve in Mayotte has had positive effects on commercial species as indicated by a higher biomass in them compared to that in non-reserve areas, and 239 species have been recorded (Letourneur, 1996).

Status and Exploitation of Protected Species

Species of special concern in Comoros are coelacanths, sea turtles and dugongs.

Coelacanths (*Latimeria chalumnea*) are locally named "Gombessa" or "M'tsambidoi". They are listed in Appendix I of CITES. Considered to be the last living fossil fish, they were known to reside only in Comoros until the recent discovery of two specimens in Toliara (Madagascar) in 1995 and 1997 and two others in North Sulawesi (Indonesia) in 1998. Their habitats are deep submarine caves located, in Comoros, mostly in southwest and west Grande Comore, and west Anjouan. The fish shelter in caves during the day and emerge at night, remaining at depths between 100 and 300 m. The population of coelacanths in Grande Comore was estimated to be 200 individuals in 1995, and 200–230 in 1991. From 1954 to the present, approximately 119 specimens have been fished, mainly in Itsandra and Vanambwani (COI-FED, 1977, unpublished report). Although their sale is forbidden, it seems that a relatively important trade exists to foreign countries. A feasibility study for a coelacanth park is being promoted by the Environment Programme/Indian Ocean Commission in the southwestern part of Grande Comore.

Two species of marine turtles, *Chelonia mydas* (Linnaeus), the green turtle, and *Eretmochelys imbricata* (Linnaeus), the hawksbill turtle, nest throughout the islands of the archipelago. A third species evidently occurs but is rarely reported and may be *Dermochelys coriacea* (Frazier, 1985). Moheli and Mayotte offer good pastures of sea grass for the green turtle, while the hawksbill turtle can feed all over the reefs of the four islands. Their reproduction in the islands is linked to the abundance of beaches available for nesting (Table 3), most of which occurs in May for the green turtle and between November and March for the hawksbill. The green turtle is eaten by many Comorians, while the hawksbill turtle is generally regarded as poisonous. The animals are nearly always captured while nesting, generally before laying. Of great concern is the fact that beaches frequented by turtles are generally not protected, and sand extraction puts in danger a number of sites essential for breeding. In the region as a whole, the populations of endangered turtles have declined greatly in recent years: in 1972, 185 turtles (13% of the population) were estimated to be killed (Frazier, 1985); in 1993, this number had increased to 432–802 turtles which is 35–65% of the population (COI-FED, 1997, unpublished report).

The Sirenian *Dugong dugon* which occurs in the lagoon of Mayotte, began to be captured with nets around the mid 1980s. The status of the population is largely unknown, and specimens can rarely be seen (one was observed in Mayotte in 1999).

Table 3

Number and length of beaches in the four islands (from Frazier, 1985) and evidence of turtle laying (pre-COI FED, 1997)

	Anjouan	Grande Comore	Moheli	Mayotte
Number of beaches	14	43	92	137
Length of beaches	6.5 km	10 km	26.3 km	27 km
Laying evidence	3 beaches: Bimbini, Moya, Mirontsy	4 beaches: Male, Ivoini, Maloudja, Mont de Mbachilé	40% of the beaches	Moya Saziley

Table 4

Number of endemic and non-endemic species of some taxonomic groups in Comoros (UNEP, 1999)

	Higher Plants	Reptiles	Birds	Mammals
Endemic	136	3	9	2
All species	935	22	99	12

Marine Reserve of Longogori, Mayotte

The creation of Marine Protected Areas (MPAs) is one of the recommended methods for conservation of two endangered species: the coelacanths and green turtle. MPAs should also contribute to long-term biodiversity in the region and should also foster ecotourism, especially that focused on coral reefs. There are no MPAs in the FIRC, but off Mohéli, the Nioumachoua islets have been proposed for protection since 1986 by UNEP. In the French island of Mayotte a marine reserve in Longogori Pass was created in 1990. This exceptional site, named locally 'S passage', is of great importance for tourism as it is a famous diving site located near the city of Mamoudzou. The original shape of the 'S passage' comes from a time when 60,000 years ago, the island's old Pleistocene coral reef barrier was eroded by the Kwale river and the hydrographic lagoonal basin was probably a coastal plain with mangroves. The passage was then the river mouth. Seismic activity associated with the up-surge of the Dziani volcano (Pamandzi islet) through the old barrier reef is supposed to be the origin of the S shape of the passage.

Spectacular dives down to the passage bottom (60 m) allow the observation of several sea-level steps, illustrating the seawater up-welling during the Holocene period.

The large variety of geomorphological and biological patterns make this site important. Activities have been restricted in the area and mooring buoys are developed for diving sites. The marine reserve was mapped in 1995 using airborne and sea-truthing techniques (Quod et al., unpublished) and a fish assessment was made (Letourneur, 1996). Following only three years of conservation, significant differences between the biomass of commercial species were seen in and outside the marine reserve.

Natural threats also affect the area. For example, the small sub-marine plateau on the south outer slope of the pass was healthy in 1995. The severe bleaching event which occurred in summer 1998 had dramatic effects here, killing more than 99% of corals and causing the disappearance of the fishes. Recovery processes may be longer now that reefs are increasingly affected by a wide range of impacts.

Development of Tourism

Comoros is trying to develop international tourism (23,700 in 1996 and 26,200 in 1997). However, this industry is very vulnerable to effects of sewage pollution which affects the condition of coastal and marine resources. About 80% of tourists go to the north of Grande Comore in Mitsamiouli (Galawa Beach and Maloudja hotels), while Mohéli and Anjouan are less visited. One consequence of uncontrolled and increasing tourism is the removal of shells and corals; the shell collectors use hammers and metal bars to smash coral bushes in their quest for rare species. Tourism is also increasing in Mayotte (9000 in 1998). Eco-tourism related to the "Grand Lagon" is recognised as a real potential especially linked to observation of reef habitats and popular species. Whale-watching may be developed: whales and their offspring can be observed during the hot season, particularly in Mayotte lagoon in which may be seen representatives of about a third of the world's whale species, and which is also a nursery for *Megaptera* (Gabrié, 1998). Diving tourism is also gaining a good reputation in these islands.

Loss of Biodiversity

All the islands of the southwest Indian Ocean are well-known centres of endemism. The Comoros (Table 4) have some of the least studied but most threatened biota of this Indian Ocean region. The major cause of species extinction is habitat loss and the introduction of alien species. Over-harvesting is also a major problem for marine turtles. Collecting of shells (*Charonia tritonis, Cassis* spp.) and crustaceans is also considered to be excessive. The coconut crab *Birgus latro* occurs in the islands (Jocque, 1984) but its status is unknown.

EFFECTS FROM URBAN AND INDUSTRIAL ACTIVITIES

Linked to population growth and economic development is increasing pressure on the limited land resources. The greatest concentrations of population and economic activity tend to be in the lowland areas. Therefore, land will be a critical and increasingly scarce resource, and pressures are likely to increase throughout the archipelago. Most of the impacts on the coastal zone result from land-based activities: domestic effluents, deforestation and poor agricultural practices, fertiliser run-off and mining, but oil spills can also be very damaging.

There is no serious industrial pollution, but garbage thrown onto the shoreline especially near the cities of Moroni, Foumbouni and Mitsamiouli where 75% of the population live, is a problem. A pilot operation of cleaning by villagers at Itsandra has been a success. Freshwater, the most critical resource, is considered to be one of the most pressing natural resource issues of the 21st century. The problem of sewage has to be closely considered as coastal population and development increases, and in Mayotte sewage treatment is being introduced in Mamoudzou to avoid polluting the lagoon. In Comoros, 53% of the population has access to safe water, but only 33% has access to sanitation services, the lack of which results in increased water-borne diseases such as diarrhoea and typhoid, as well as pollution of surface water and groundwater. There is a growing shortage of water on most of the islands: Comoros has been classified as becoming increasingly short of water, with freshwater resources predicted to fall from 1667 m^3 to 947 m^3 per capita by the year 2025 (on Grande Comore, there is an almost total absence of surface water due to high soil permeability). Losses from water distribution networks can also be an important factor, and can be as high as 50%.

In Mayotte, the lagoon has only a small water exchange with the deep sea, so questions remain concerning the cumulative concentration of land-based pollutants. Urban pollution is likely to increase in the coming years with increasing population densities. The natural biogeochemical processes of this area are not understood, and pollutants are not yet monitored, although bacterial contamination is now being examined by the Health Department.

The main transportation route of Middle-East oil to Europe and America traverses the western Indian Ocean. In the last 30 years, there were 30 major oil spills in the Mozambique channel, where between 20,000 and 90,000 tons of hydrocarbons were involved. Comoros and Mayotte have no contingency plans for major ballast discharges or accidents.

PROTECTIVE MEASURES

Both Comoros and Mayotte have regulations covering environmental protection and management. Mayotte comes under French legislation which prohibits destructive practices on land and sea. Mechanisms used to protect endangered species and ecosystems include protected areas, *in situ* conservation, species management, captive breeding programmes including the control of invasive species, and propagation of endemic tree species.

Comoros has only recently formulated its biodiversity strategy and has identified six areas for protection, but it requires financial support for implementation. One site is designated a Ramsar site. Projects are underway to establish a coelacanth park in Grande Comore. South Moheli and the Nioumachoua islets have been recommended as a marine protected area. Mayotte is undertaking ecosystems and associated resources monitoring programmes, and has strong and long-term help from the French scientific community (more than 450 publications are devoted to the lagoon; Thomassin et al., 1996, unpublished report). Other measures include:

1. Establishment of two marine protected areas: the Longogori Reserve (or S pass reserve, 400 ha) which is a strict fishing reserve and (ii) the Saziley territorial park (3100 ha of lagoon) with a management plan.

2. A ban on spearfishing and netting.
3. Nine monitoring stations contributing to the GCRMN (Global Coral Reef Monitoring Network) according to a regional methodology developed in 1997.
4. The SPEM (Service des Pêches et de l'Environnement Marin) from the Agriculture division which conducts activities for sustaining biodiversity and fisheries activities, including a lagoon patrol, but with limited means to apply policies and regulations.

Efforts are starting in respect of monitoring, but institutions in charge of environmental management and conservation lack human and financial capacities and there is a shortage of scientific local knowledge.

On the island watersheds, improvements in the sustainability of agriculture, including alternative farming methods and marketing, diversification into non-traditional export crops, and reviewing land tenure are all necessary in order to avoid soil erosion and improve the economy. Otherwise, in Comoros, the installation of supplementary Fish Aggregating Devices will be indispensable for increasing the food security of the populations. Today, the rich and dense forests of the past that provided habitats for thousand of plants, insects, birds and other animals have given way to settlements, road infrastructure, farming, open woodlands and pastures. It is thus a priority to protect the last pockets and areas of primary forests. No detailed assessment of endemic flora has yet been done and it is not even known which species are threatened. Maintaining the integrity of water resources must be recognised as a priority. Plans exist to improve water management, storage and availability of safe water through the construction of dams, reservoirs, sewage and treatment works, but these are major long-term investments and some projects lack finance.

The islands of the Comoros Archipelago have valuable and unique endangered resources, often unprotected and under-studied. Problems are considerable. Although agriculture is the main economic sector in Comoros and Mayotte, most of the food is imported. Thus, food security is a serious problem in these islands, and long-term food security measures must be devised to counteract increasing food imports, using the concept of regional complementarity and intra-regional trade. Also, biodiversity strategies and action plans to manage the valuable resources of the archipelago remain urgent.

REFERENCES

Ahmed, M. (1988) Extraction et utilisation des matériaux coralliens aux Comores. *J. Nat. (Univ. Réunion)* **1** (1), 24–26

Bryceson, I., De Souza, T.F., Jehangeer, I., Ngoile, M.A.K. and Wynter, P. (1990) State of the marine environment in the Eastern African Region. *UNEP Regional Seas Reports and Studies* **113**, 46 p.

Coudray, J., Thomassin, B.A. and Vasseur, P. (1985) Comparative geomorphology of New Caledonia and Mayotte Barrier reefs (Indo-Pacific Province). *Proc. 5th Int. Coral Reef Congr., Tahiti* **6**, 427–432

Faure, G., Guillaume, M., Payri, C., Thomassin, B.A., Van Praat, M. and Vasseur, P. (1984) Sur un phénomène remarquable de blanchiment et de mortalité massive des Madréporaires dans le complexe récifal de l'île Mayotte (SO Ocean Indien). *C. R. Acad. Sci. Paris* **299**, série 3 (15), 637–642

Fourmanoir, P. and Harmelin-Vivien, M.L. (1984) Evolution du peuplement ichtyologique de l'île Mayotte (Océan Indien) entre 1977 and 1983. Unpubl. report

Frazier, J. (1985) *Marine Turtles in the Comoro Archipelago*. North-Holland Publishing Company, Amsterdam, Oxford, New York, 177 pp.

Gabrié, C. (1998) State of coral reefs in French overseas départements and territories. Ministry of spatial planning and environment report, 36 pp.

Guilcher, A. (1965) Coral reefs and lagoons of Mayotte Island, Comoro archipelago, Indian Ocean, and New Caledonia, Pacific Ocean. In Whittard, W.F., Bradshaw, R. (eds), Submarine geology and geophysics. *Proc. Symp. Colston Res. Soc. Bristol* **7**, 21–45.

Guilcher, A. (1971) Mayotte barrier reef and lagoon, Comoro islands, with other barrier reefs, atolls and lagoons in the world. In: *Regional variation in Indian ocean coral reefs, Symp. of the Zool. Soc. of Lond.* **28**, 65–86

Guilcher, A., Berthois, L., Le Calvez, Y., Battistini, R. and Crosnier, A. (1965) Les récifs coralliens et le lagon de l'île de Mayotte (Archipel des Comores, Océan Indien). *Mem. ORSTOM* **8**, 210 p.

Jocque, R. (1984) Quelques Invertébrés non-insectes terrestres et dulçaquicoles des Comores. *Africa-Tervuren* **30** (1–4), 68–74

Letourneur, Y. (1996) Réponses des peuplements et populations de poissons aux réserves marines: le cas de l'île de Mayotte, Océan Indien occidental. *Ecoscience*, 3(4).

Polunin, N.V.C. (1974) Devastation of a fringing reef by *Acanthaster*. *Nature* **249** (5457), 589–590

Polunin, N.V.C. and Frazier, J.G. (1974) Diving reconnaissance of twenty seven western Indian Ocean coral reefs. *Environmental Conservation* **1** (1), 71–72

Prosperi, F. (1957) *Vanished Continent. An Italian to the Comoro Islands.* Hutchinson, London.

Rao, T.S.S. and Griffiths R.C. (1998) *Understanding the Indian Ocean. Perspectives on Oceanography.* UNESCO Publishing, 187 pp.

Rouyer, P. (1988) Les récifs coralliens des Comores. *J. Nat. (Univ. Réunion)* **1** (1), 14–23.

Shah, N.J., Linden, O., Lundin, C.G. and Johnstone, R. (1997) Coastal management in Eastern Africa: status and future. *Ambio* **26** (4), 227–234.

Toussaint, A. (1966) *History of the Indian Ocean*. London, Routledge and Kegan Paul.

UNEP (1999) Western Indian Ocean environmental outlook. United Nations Environment Programme, Nairobi, 79 pp.

Vienne, E. (1900) Mayotte et les Comores, Paris, 200 p.

von Hentig, R. (1973) *Coral World*. British Broadcasting Corporation, London, 103 pp.

THE AUTHORS

Jean-Pascal Quod
ARVAM, 14, Rue du stade de l'Est,
97490 Réunion, France

Odile Naim
Laboratoire d'Ecologie Marine,
Université de la Réunion, BP 9151,Saint-Denis messag 9,
Réunion,France

Fouad Abdourazi
AIDE, Minizi, Mavouna, Moroni, Comoros

Chapter 70

THE MASCARENE REGION

John Turner, Colin Jago, Deolall Daby and Rebecca Klaus

The Mascarene Region in the southwest Indian Ocean consists of the deep Mascarene Basin, the banks and shoals of the Mascarene Plateau, the St Brandon archipelago, the volcanic islands of Mauritius, Reunion and Rodrigues, and the small oceanic islands of Agalega and Tromelin. Less than 10% of this 2×10^6 km^2 area lies within 200 m depth contours, and while the Mascarene Plateau has no land mass, it is effectively a shelf sea and a meridional barrier presenting a significant obstacle to deep flow in the western Indian Ocean.

Monsoon seasons bring northeasterly winds in December–February, and southeasterly winds in June–August, and the South Equatorial current flows west across the Mascarene Plateau. Tropical cyclones frequently cause severe damage.

The banks and shoals are best developed around St. Brandon. The volcanic islands are of different ages; the older Rodrigues and Mauritius have fringing reefs with wide lagoons, while on Reunion corals grow on steep volcanic rock. Mangroves are comparatively rare but 20 km^2 occur on Mauritius. The offshore islands are important seabird and turtle nesting sites, but many islands within lagoons have degraded habitats.

Major land clearance and decimation of fauna and flora followed soon after settlement in 1638. Sugar cane was planted, and fisheries in the region have been exploited since the 1830s. Mauritius has a population of 1.2 million and a booming economy, Reunion has 600,000 with a more constrained development, while Rodrigues (37,000) remains undeveloped and agricultural. Soil erosion and lagoonal sedimentation is significant in all of them.

Offshore waters support industrial-scale tuna fishing, and in lagoons artisanal fishing is influenced by pollution, sedimentation, mangrove removal and wetland reclamation. Fishing is not well controlled. Coastal erosion, sand extraction, sugar milling and other light industries pollute the lagoons, and waste-water treatment is recognised as a high priority. Tourism is important, but the islands face a water shortage and need to protect aquatic systems.

The Mascarene islands are committed to marine environmental protection, and several reserves have been declared. Management plans have been or are being developed, but despite protective measures, habitat degradation, pollution and over harvesting threaten the viability and productivity of the region. These priorities are recognised in many sectors, and are being addressed.

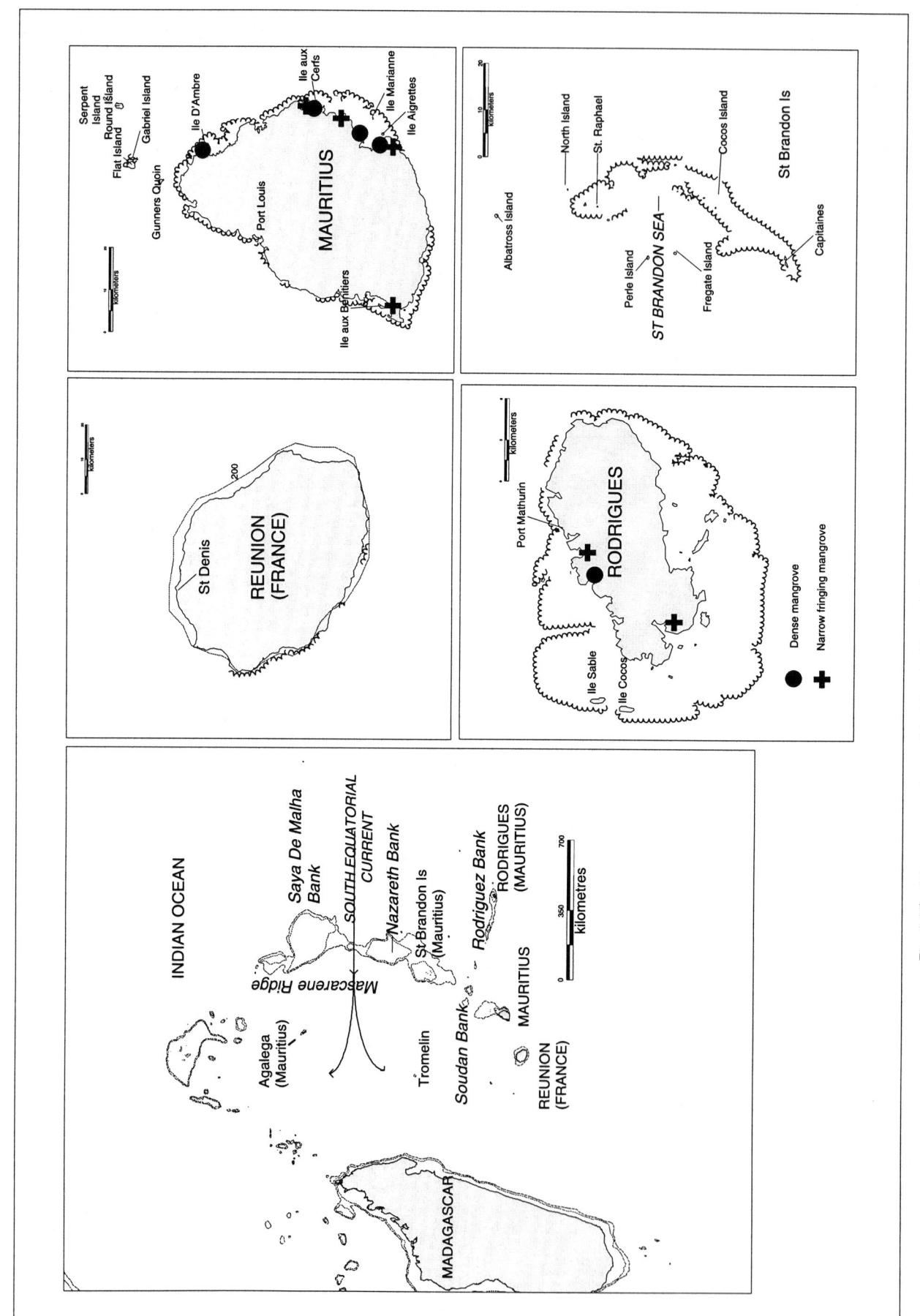

Fig. 1. The Mascarene region and its islands. The curly line around the islands is coral reef.

THE DEFINED REGION

The Mascarene region in the south western Indian Ocean extends approximately from 10 to 22°S, and 55 to 64°E (Fig. 1). This part of the Indian Ocean is affected by the Asian monsoon, annual cyclones and complex currents influenced by an oceanic ridge. The Mascarene Basin is over 4000 m deep, but is bordered by the Mascarene Plateau, a discontinuous ridge of about 200 m depth. The Plateau consists of the Saya de Malha, Nazereth, St Brandon, Soudon and Rodrigues submerged banks. The shoals and banks become more extensive in the north, and effectively form shelf seas, and with the exception of the St Brandon archipelago, have no associated land masses. The largest islands are volcanic, consisting of Mauritius, Reunion and Rodrigues, together with a number of small and scattered dependent coralline cays which harbour temporary human settlements (St Brandon, Agalega) or are uninhabited (Tromelin sea mount) (Table 1).

The extensive (28,350 km^2) Saya de Malha Bank is 1050 km north of Mauritius and is claimed by the Seychelles. The Mauritius banks include the Nazereth Bank (7625 km^2), St Brandon (1208 km^2), Soudan and Rodrigues Bank (1688 km^2).

Mauritius is the largest island (1869 km^2), lying 875 km east of Madagascar. Formed by three volcanic flows of 20,000 to 7–8 million years old, the basaltic island is 817 m high. The coastline is over 200 km in length, surrounded by 243 km^2 of lagoons bounded by fringing coral reefs. The French island district of Reunion is located at 21°7'S, 55°32'E, 175 km south west of Mauritius, and is the youngest of the Mascarenes, with a mountainous topography to 3069 m, with an active volcano and only 10–12 km of reef development on the southwest coast. Rodrigues is an island state of Mauritius, located 595 km east of Mauritius. It is the smallest (110 km^2) of the Mascarenes having formed 1.5 million years ago, and rises to 400 m. The island is situated on a shelf of 950 km^2 area, and is enclosed by a 200 km^2 fringing reef encompassing a very shallow lagoon extending to 13 km width in the south.

Table 1
Coastal zone data of the main island states

	Mauritius	Rodrigues	Reunion
Land area (×1000 km^2)	2.0	0.108	2.0
Coastal population (millions)	1.1	0.04	0.6
Percentage of total population	74	100	99
Shelf area (×1000 km^2) [0–200 m]	1.3	0.95	2.01
Coastline length (km)	320	90	207
Mean tidal amplitude (m)	0.5	0.4	1.2
Area of Exclusive Economic Zone (EEZ) (million km^2)	1.7	In Mauritius EEZ	2.7 (not all in Mascarene region)

St Brandon consists of 55 low elevation (<4.6 m) coralline islands and sand cays. The total area of reef is 190 km^2, and many of the islands are situated in the lee of a crescent-shaped reef, 38 km long by 5 km wide. Agalega consists of two coralline islands 24 km in total length and 1.6 km wide, with sand dunes reaching 7 m in height. About 100 km^2 of fringing reefs enclose a small lagoon. The island is thought to be a bank which emerged some 600 years ago. Tromelin is a sea mount claimed by France, Madagascar and Mauritius, with a flat sand island <2 km in length and 640 m wide, rising from depths of over 4000 m, surrounded by coral reefs.

SEASONALITY, CURRENTS, NATURAL ENVIRONMENTAL VARIABLES

The Mascarene Plateau is a major topographic, meridional barrier extending almost 2000 km from near the equator to 22°S, which presents a significant obstacle to deep water flow. Its shelf seas cover an area of some 300,000 km^2 which are exposed to strong winds, waves and tides.

Circulation

The surface circulation of this part of the Indian Ocean is dominated by a massive anticyclonic gyre, whose northern segment, flowing to the west, is the South Equatorial Current. It is at a minimum in the southern summer and at a maximum in the southern winter. The part between 12°S and 25°S encounters the Mascarene Plateau as it flows westwards. This is a major physical barrier to the flow below 200 m. In open water, the current reaches a maximum 0.5 m s^{-1} during the southern winter but it must accelerate as it passes through or over the Plateau. Kindle's (1991) model of the circulation of the Indian Ocean proposes that much of the flow is concentrated in the deep channels (such as the Nazareth Passage), but it is likely that part of the flow spills across the shallower parts of the Plateau, particularly near the equator. After passing through and over the Plateau, the current divides into two branches, one flowing around north Madagascar, the other flowing south. Tropical water of the South Equatorial Current forms a 100–200 m thick band of low-salinity water centred on 10°S which can be traced to the heavy equatorial rainfall near Sumatra and Java. Below this surface water is Subtropical Maximum Salinity water, a water mass which is formed to the south between 40°S and 20°S in the centre of the anticyclonic circulation. Cooling of this water induces convection and sinking along isopycnals. Although it has a maximum thickness of 600 m at 30°S, the thickness decreases northwards to 100 m at 10°S.

The base of the Subtropical Maximum Salinity water is marked by a permanent thermocline. Water immediately below the latter is comprised of Antarctic Intermediate Water, probably advected from the Atlantic Ocean, which occurs as layers of 300–800 m thickness (Wyrtki, 1971).

The Plateau sustains its own western boundary currents. Trade wind stress across the equatorial ocean causes an upward-sloping sea surface and a downward-sloping thermocline towards the west. The result is that waters on the western side of the Plateau are typically 2°C cooler than those on the east.

Tides

The Mascarene Plateau is close to a tidal amphidrome (Schwiderski, 1983) and there are relatively fast barotropic flows reaching 0.6 m s^{-1} through the channels. Mean tidal ranges at Reunion, Mauritius and Rodrigues are 0.4, 0.5 and 1.2 m, respectively. At Mauritius, open-sea shallow-water tidal currents locally reach 2.5 m s^{-1} (Fagoonee, 1990a). Tidal currents are much slower within the lagoons.

The impact of tides is probably not confined only to the Plateau. The shallow banks and shallow channels between banks (to 400 m) and the deep water around the Plateau (to 4000 m) create a geometry that is highly susceptible to internal wave generation (Morosov, 1995). Energy from the barotropic tide is transferred to an internal tide as stratified flow is forced over steep topography. Internal tides of 150 m height are generated over the Plateau region and lunar time-scale variations can be detected some 350 km from the Plateau (Konyaev et al., 1995).

Storms and Cyclones

Tropical cyclones form between 10 and 25°. They usually develop from small tropical depressions within the easterly Trades. When such air is warmed by the proximity of tropical waters, circular winds begin to blow and a storm develops. Cyclones last from 3 hours to 3 weeks, though most have lives of 5–10 days. They move westward and poleward at 5–40 km h^{-1} and occur predominantly between December and March. The southern part of the Mascarene Plateau is most susceptible to cyclones which frequently pass by the islands before running down over the cooler waters to the south. The high winds and heavy rainfall causes considerable destruction to shallow water, coastal, and land areas. Wind stress currents develop, sea levels are significantly elevated and there is the additional storm surge, a dome of water beneath the low pressure centre of the cyclone. There is little information on cyclones and their impacts on circulation and physical coastal processes, although the effects on reef communities have been studied, e.g. in Reunion (Letourneur et al., 1993).

At Mauritius, more than 70% of waves are driven by the Southeast Trades with typical periods of 7–10 s and heights up to 3 m; 11% of waves are generated in the extreme south of the Indian Ocean, reaching Mauritius after a few days, with typical periods of 15–18 s and heights of 3–5 m, although exceptionally they may be up to 20 s and 8 m (Sigma et al., 1991).

THE MAJOR SHALLOW WATER MARINE AND COASTAL HABITATS

Shoals and Lagoons

Lagoonal habitats are important in this region. Lagoons have good exchanges with the open ocean, display near to open ocean salinity, and are significantly influenced by tidal variability. Tides in the lagoons are mainly semi-diurnal, with ranges of 0.4–0.8 m, and weaker than those in adjacent shelf waters. Flood currents sweep approximately northwards, ebb currents southwards. Tidal flushing takes 4–7 days, though wind flushing may be faster, and lagoonal waters are vertically homogeneous. Lateral mixing times are 5–8 days so there are strong shore-normal and shore-parallel salinity and temperature gradients.

Sediments are predominantly carbonates (Bassias et al., 1993), and in Mauritian lagoons at least 50% of the grains are derived from broken coral debris. Lagoons fed by rivers have terrigenous sediments, especially around river mouths, but they can also constitute a significant proportion of the lagoonal sediments (Daby, 1999).

The distribution of bottom sediments in the lagoons influences water turbidity. In Mauritius, the most turbid waters occur above shoreline terrigenous sediments, or near areas of freshwater input. In Rodrigues, intense soil erosion has led to enhanced terrigenous sediment supply and turbidity in the lagoon.

The Banks

The habitats of the banks and shoals of the Mascarene Plateau are poorly known to the scientific community, but well known by the banks fishermen. The banks fishery is based on hand-lining and basket traps, suggesting extensive coral growth areas. Lebeau and Cueff (1976) report that the Saya de Malha Bank is covered by sandy areas separated by coral masses, and that green algae cover large areas. Twelve scleractinian coral genera were reported from the Bank by Stoddart (1984). The banks break the surface only at St Brandon, where Kaly and Brown (1998) describe in detail 11 types of coral reef, lagoon and shallow sea habitat, and two terrestrial habitat types: vegetated islands and sand cays. The marine habitats of St Brandon may give some indication of the habitats of the rest of the Mascarane Plateau.

St Brandon

The eastern sides of the St Brandon bank consist of reef slope extending down from a depth of between 10 and 40 m to the ocean floor, but has not been studied. Above the reef slope is a gently sloping, low-relief reef terrace at 3 to 40 m depth, dominated by coral rock, sand patches and variable cover of hard corals and encrusting coralline algae, with deep grooves parallel to the reef. Shallow spur and groove regions exist between 0 and 3 m, and are well populated with fish. The reef crest may be intertidal or subtidal and comprises reef rock and rubble, and may be several hundred metres wide, and composed mainly of calcareous algae on the lagoon side. A back reef is composed of large coral colonies, harbour clams, holothurians, *Halimeda* and sponges. Two large passes cut from the ocean into the lagoon are important for water exchange and fish migration, and are occupied by reef sharks and other larger reef fish. St Brandon has three main lagoons up to 7 m deep, and separated by reef crest walls. The lagoons have sand floors either heavily bioturbated by worms, crustaceans, bivalves and gastropods, or covered by cyanophyte mats. Patch reefs of 3 m diameter grow in the lagoons, often having dead or rubble tops. The St Brandon bank is less well developed at its western reef crest.

The St Brandon Sea is a shallow area 4–40 m deep bound by the western reef margin, lagoons and islands, with a sandy mud bottom with rocky patch reefs dominated by algae, foliose sponges and soft corals and some hard corals. To the west, the ocean floor slopes to the deep Mascarene Basin.

The 55 St Brandon islands are mostly vegetated and support large numbers of sea birds, including eight breeding species, and turtle nesting sites (Kaly and Brown, 1998). There are four types: the Guano islands (Albatross, Nord, Sirene, Pearle, Fregate); the sooty tern islands (Puits a Eau, Poulailler); islands of human habitation (Raphael, Avocaire, Cocos, Sud) and small islands and sand banks (Longue, Petite, Grande Dagorne, Capitaine and many others). Vegetation is low, usually an *Arguisia/Scaevola* association and a grassland in the centre of the islands used by nesting birds. Non-native plants such as coconut and casuarinas occur around current or former human habitations. Sooty terns *Sterna fuscata* are dominant in the north and the Lesser Noddy *Anous tenuirostris* and Common Noddy *Anous stolidus* occur in the south together with fairy terns *Gygis alba*. Birds vulnerable to extinction are Boobies *Sula dactylatra melanops*, Frigate Birds *Fregata ariel* and Greater Frigate *Fregata minor* and roseate terns *Sterna dougalli*. An estimated 1000 turtles (Green and Hawksbill) nest here. Kaly and Brown (1998) reported 87 species flowering plant, 10 species of algae, 3 species of sponge, 36 species of corals, and 4 species of soft corals/anemones, 4 species of holothuria, 12 species of birds, 2 species of turtles, 202 species of fish and 2 species of mammal (rat and mouse).

Tromelin

The sea mount of Tromelin is unique in the region, and is surrounded by coral reefs with a 150 m wide reef flat, and steep profile. Briefly described by Bouchon and Faure (1979), it remains unstudied. Stoddart (1984) reported 14 scleractinian coral genera. The beaches provide nesting for the Green turtle *Chelonia mydas*, and the sparsely vegetated island supports breeding colonies of Red-footed and Blue-faced boobies, Frigatebirds and fairy terns.

Another submerged sea mount, La Perouse at 19°40'S, 54°10'E, 332 km west of Mauritius arises from a depth of 4500 m to within 54 m of the surface, but remains unexplored. The mount is probably too deep to support coral reefs, but may harbour deep-water coral species and attract large pelagic fish.

Agalega

Agalega was described during the largely ornithological study by Cheke and Lawley (1983). The two small islands are separated by a 1.5 km sandy channel which dries at low tide. The island consists of consolidated coral debris, upon which are coconut plantations. The island was probably once used by turtles and sea birds, but these have been deterred by human settlement.

The Volcanic Islands

The coastlines of the volcanic islands were formed by seaward flows of lava which have since eroded. The islands are younger to the west, and are situated sufficiently far south to have rock shores, macro-algae, and coral reef at different stages of development. The islands have no continental shelf, and deep water exists close to their coasts. The major habitats are basaltic rock shores, coral reefs, lagoons, sand beaches, estuaries, saltwater wetlands and mangroves. All habitats do not occur on all islands, and the community structure of the habitats is often different on each island (e.g. Faure, 1975, 1977).

Mauritius

The greatest diversity of habitats is found on Mauritius. The south coast is the most exposed, and has a fairly uniform coastline of basaltic cliffs and boulders. Elsewhere there are beaches, dunes, saltwater wetlands, mangroves, estuaries, lagoons, sheltered bays, rocky shores and coral reefs (Hodgkin and Michel, 1961; Baissac et al., 1962; Pichon, 1971; Salm, 1976; Montaggioni and Faure, 1980; Fagoonee, 1990a,b; Daby, 1999) and most recently a series of nine unpublished MSc theses at the University of Wales, Bangor.

About 150 km (300 km^2) of fringing reefs surround the island, cut by surge channels and estuaries. Reefs are absent from two stretches (15.5 km in the south and 10.5 km on the west coast) where basalt rocky cliffs and boulders are found. Lagoon waters are generally shallow (2–3 m) and discontinuous, with a total area of 243 km^2, extending from 0 to 8 km from the beach. Lagoon morphology varies from single lagoons, to complex systems with a variety of habitats: beaches, mangroves and sand or mud swamps, channels, seagrass beds, corals, sand beds and a reef zone (Fagoonee, 1990a,b). Sediments are both terrigenous near river mouths and carbonates from the reefs.

The dominant organisms of exposed shores are the urchins *Echinometra mathaei*, *Stompneustes variolaris* and

Fig. 2. Mangrove infilling, Mauritius.

Colobocentrotus atratus, with the limpet *Patella chitonoides* and coralline algae at lower levels (Hodgkin and Michel, 1961). Sheltered shores are dominated by the barnacles *Cthamalus*, and *Tetralicta*, and the limpet *Siphonaria*, below which is a zone of algae *Laurencia* and *Ulva*, providing substrate for molluscs such as *Planaxis* and hermit crabs. Exposed limestone shores exhibit a midlittoral platform, and less exposed, undercut and overhanging rock faces have altered zonation (Baissac et al., 1962).

Sandy beaches and dunes occupy about 1.7% of the total surface area of the island (31 km^2) and nearly all are of white biogenic carbonate sand (Karisiddaiah et al., 1988). The beaches of Mauritius have moderate benthic biomass (159–1479 mg m^{-2} dry weight), with population densities ranging from 133 to 1187 meiofauna in 10 cm^{-2} to 5 cm deep (Ingole et al., 1998).

Estuaries are associated with the larger and longer rivers, and have yet to be studied. Terre Rouge Bird Sanctuary covers intertidal and estuarine habitats and is of value to migrating birds.

Mangroves are limited to narrow belts mainly on the southeast coast. Historically mangroves covered most of the shoreline of Mauritius, but have been largely destroyed by cutting for firewood and shoreline development (Fig. 2). Two species were identified by Sauer (1962) *Bruguiera gymnorrhiza* and *Rhizophora mucronata*, the latter being most prevalent. A mangrove planting programme has planted a total of 22750 m^2 of shore at nine sites around the island with over 12,000 seedlings having an estimated survival of 65% (MCFMRD, 1996) (Fig. 3).

There are three types of coral reefs around Mauritius: peripheral and sheltered fringing reefs, barrier reefs, and lagoon patch reefs (Salm, 1976). The peripheral fringing reef is a narrow rim, less than 25 m wide, almost surrounding the island. Sheltered fringing reefs border deep water channels or protected bays and most support branching and tabular *Acropora* corals, blocks of massive *Porites*, foliaceous *Montipora*, *Pavona* fields, and patches of loose sand or sand consolidated with seagrasses. A small barrier reef exists at Mahebourg on the south east coast, which is 0.5 to 2 km wide and approximately 10 km long, separated from the coast by 3–5 km of lagoon and channel, 15–30 m deep. Where reefs are absent, corals grow on the younger volcanic flows whose walls plunge vertically into the sea. Sheltered zones of the passes support extensive brown algae *Turbinaria*, *Sargassum* and thin populations of the seagrass *Halophila*.

The spur and groove zone is composed of many, mostly encrusting, corals and soft corals such as *Sarcophyton* and *Lobophyton*, and in the boulder zone the brown algae *Turbinaria ornata*, and *Sargassum* sp. dominate with the urchin *Echinometra mathaei*. At shallow depths (0–20 m), the spur and groove zone of the outer slope is sometimes replaced by dead coral 'flagstone', while at greater depths (20–50 m), volcanic flagstone can be found (Montaggioni and Faure, 1980). The narrow reef flat does not accumulate sediment deposits, which usually get swept into the lagoon. Most lagoons on the eastern coastline are characterized by reticulated and meandering channels. These are major depressions, usually close to the reefs and running parallel to the shore, reaching depths of 6–35 m.

Patch reefs in Mauritian lagoons develop directly from the reef flat and cover areas of up to 1 km^{-2}. They are exposed only during the lowest spring tides, and they support dense coral growth. Distribution of the patches is limited by terrigenous sediment, river freshwater or underground freshwater seepage.

Extensive and important seagrass beds occur in lagoons, with seven species (Montaggioni and Faure, 1980). A taxonomic survey of the lagoon algae (Mshigeni, 1985) indicated that the southwest, south and north east coasts have a high species diversity, with up to 48 species per site. Limestones favour the growth of *Laurencia*, *Centrocerus* and *Ceramium* species, though dense growth (up to 3000 g m^{-2} wet weight) of *Ulva* and *Enteromorpha* species on some shores are indicative of eutrophication and pollution. Some calcareous algae (e.g. *Halimeda* sp., *Jania* sp., *Cheilosporum* sp. and *Corallina* sp.) are also common around the coast and together with corals, they form a major source of the lagoon sediments.

Fig. 3. Mangrove plantation stabilising sediments, Mauritius.

There are about 600 species of fish, 244 crab species, 190 echinoderm species, 170 pelecypod species and some 1150 species of gastropods (Fagoonee, 1990b). A fish species list (Harmelin-Vivien, 1976) suggests that the majority of littoral fish species are common to East African Indo-Pacific faunas. However, 12 species of butterfly fish (Chaetodontidae) have been found to be endemic to Mauritius and Madagascar. There are only a few other endemic species; the echinoderm (*Acanthocidaris curvatispinis*) and gastropod (*Harpa costata*) are two.

Ecological and hydrographic surveys of Mauritian lagoons have been undertaken by Fagoonee (1990), Muller et al. (1991) and more recently by Torres (1997), Walley (1997), Muller (1998), Ingrams (1998) and Daby (1999). A series of studies have employed remote sensing to classify and quantify the coastal marine habitats (Daby, 1990; Klaus, 1995; Dykes, 1996; Orme, 1997; Taylor 1998), and the data has been embedded in a Geographical Information System (Eastwood, 1998; Hardman, 1999). Area cover estimates of some of the different benthic habitats within the lagoons (total area 240.35 km^2) were found to be as follows: sand (69.76 km^2); mixed sand and algae (15.09 km^2); seagrass (30.42 km^2); lagoonal corals (45.16 km^2) and dead coral (7.52 km^2).

A northern prolongation encompasses four small (<250 ha) uninhabited islands: Gunners Quoin, Gabriel Island, Flat Island, Round Island and Serpent Island, formed in the most recent volcanic episode. A small reef and lagoon join Gabriel and Flat Island (Salm, 1976), and contain abundant blue coral *Heliopora coerulea*. Round Island is of greatest biological importance, being the largest area in the Mascarenes to have remained free from introduced mammals and plants, probably because of its steep and inaccessible shores. The island still supports palm savannah, nine plants endemic to Mauritius, endemic reptiles (including three endemic to Round Island alone) and thousands of nesting sea birds (especially shearwaters), and the only known breeding ground in the Indian Ocean for a race of the Herald Petrel *Pterodroma arminjoniana*. The island supports the largest breeding population of wedge tailed shearwaters, white tailed tropic bird *Phaeton lepturus lepturus* and red tailed *P. rubicauda rubicauda* in the Mascarenes. Small islands occur within the larger lagoons around Mauritius, and most of these are heavily used by tourists and fishermen, and their habitats are degraded. Isle aux Cerfs and a number of smaller islets off the east coast harbour most of the remaining mangrove. Ile aux Aigrette is a 27 ha limestone island in Mahebourg Bay, surrounded by shallow lagoon. The island supports many native plants and birds which are now rare on the mainland, including the last remnant of coastal forest in the Mascarenes.

Reunion

Reunion is the youngest Mascarene island, and its steep volcanic surfaces support corals, but little reef growth. Coral reefs grow discontinuously on a 10–12 km stretch of fringing reef in the southwest (Bouchon, 1981; Montaggioni and Faure, 1980), being broken by rivers. The lagoon habitats are poor compared to those of other Mascarene islands, consisting of rubble, coral heads, and only one seagrass *Syringodium isoetifolium*. The reef flat consists of coral rock cut by sandy channels, with several species of corals. The reef front is about 20 m wide, lacks an algal ridge, and deep channels connect to the grooves of the outer slope, which stretch gently and evenly for about 400 m to a depth of 20 m. Branching corals are typical of this region. Beyond 20 m depth, volcanic rock is colonised by scattered coral colonies, and at 30 m, the rocks are embedded in a nodule field of calcareous algae and encrusting foraminifera to 65 m and beyond. Letourneur (1996) described the fish communities of the fringing reefs, reporting 217 species. The eastern rocky coast of Reunion has recently been explored, and the composition, structure and trophic networks of fish communities are described in Letourneur (1998).

Rodrigues

Rodrigues is the smallest of the Mascarenes, with a higher tidal amplitude (1.2 m) than the other islands. The reef platform has a maximum development in the west, and encompasses some small basaltic and sand cay islands (Montaggioni and Faure, 1980). The reef flat is covered by extensive areas of sand and rubble, and sparse corals, with greater living coral cover especially of branching *Acropora* close to the seaward edge. Small channels run parallel and perpendicular to the reef, and especially around the lagoon islands, and back reef channels run along the coast. These channels have soft corals on their walls and increasing hard coral cover to 30 m where rubble, sand and rock may form their bases. The lagoons are heavily silted, and soft-bottom communities of shrimps, crabs and polychaete worms are plentiful. Seagrass beds are uncommon and, together with algae, tend to grow on sand bars (Pearson, 1988).

Most of the coast is basaltic rock rubble on terrigenous sediments, and there are only five small sand beaches. Two lagoon islands support important nesting sea bird colonies and original vegetation. Ile aux Sables and Ile aux Cocos have large colonies of Brown and Lesser Noddies and Fairy Terns. The islands are no longer used by turtles. Mangroves (*Rhizophora mucronata*) occur in three bays (Baie aux Haitre, Baie Malgache and Baie Topaze) around the mainland where they have been planted to stabilise river-borne sediments.

OFFSHORE SYSTEMS

Offshore waters are generally not productive, except over the banks (Mordasova, 1980). Phytoplankton communities are nutrient-limited (Devassey and Goes, 1991), although enhancement does occur near Agalega due to an upwelling process caused when the South Equatorial Current reaches the Mascarene Ridge. Benthic macrofauna (Ingole et al., 1992) show a general decline with water depth and have a positive correlation with chlorophyll levels in surface waters.

The only industrial-scale fishing is for tuna, which are caught by foreign purse seiners and long-liners. Sanders (1988) indicates that 150,000 tonnes of tuna are caught in the south west Indian Ocean annually; in 1996, 14,715 tonnes of tuna, mainly *Thunnus alaunga* (81%) were transhipped in Port Louis, Mauritius. Both Mauritius and Reunion have recognised a potential for both a long-line swordfish fishery and a deep-water shrimp fishery (*Heterocarpus laevigatus*) using traps in 600–1000 m depth (MCFMRD, 1996; Renee et al., 1998). Sports fishing is an important attraction in the region, with about 80–100 boats engaged in this fishery in 1990. Total annual catch is around 400 tonnes, consisting of bill fish, tuna and shark (Dwivedi and Venkatasamy, 1991). Commercial fishing on the offshore banks of the Mascarene Plateau is undertaken by 15–17 mother ships from Mauritius, and one from Reunion, carrying small doreys from which hand-lining, basket traps and large nets are used. Six hundred and sixty-five fishermen and 95 frigo boys are registered in Mauritius to fish the banks. The whole production is consumed locally in Mauritius, accounting for 30% of total fish consumption, and 6% is exported in Reunion. Both Mauritius and Reunion fish on the Nazareth and Saya de Malha banks, and Mauritius also fishes on St Brandon and Chagos. The total annual catch is around 5000 tonnes, mostly of *Lethrinid* species (90%), and there is evidence that the banks fishery is overexploited, for catch per unit effort has fallen from 100 kg in early 1980s to 53.5 kg in 1996. Total allowable catch is being reduced by 5% per year. A semi-industrial chilled fishery operates around St Brandon and Soudan banks from 10 boats fishing by hand-lining to 300 m depth. Although a total allowable catch is set at 315 tonnes, 59 tonnes were taken in 1996, consisting of Lethrinids (66%) snappers (25%) and tuna (MCFMRD, 1996). Sanders (1988) also notes a demersal trawl fishery in the southwest Indian Ocean accounting for 16,000 tonnes of fish.

There are deep-sea mineral deposits, mainly of manganese polymetallic nodules at 3000 m in the Mascarene Basin over an area of 11,900 km^2 off Tromelin (Nath and Prasad, 1991). Nodules of 2–4 cm diameter consisting of hydrogenous or detrital minerals varied in density from 1.23 to 10 kg m^{-2}. Mining prospects are unlikely at present, and limited by expense and appropriate technology, but may be important in the future.

POPULATIONS AFFECTING THE AREA

All populations in the region are effectively coastal. Mauritius had the highest population growth in the world in the 1960s, and by 1972, fertility levels declined from 6 to 3 children per woman, which is the most rapid change in human history (Titmuss and Abel Smith, 1968). Mauritius was first colonised in 1638 by Dutch East India Company settlers, who cut and sold the ebony forests, and introduced alien species which severely damaged the island ecosystems, before abandoning the island in favour of the Cape of Africa. The Dutch also visited Rodrigues, but showed little interest in the island. The Mascarenes were re-colonised by the French in 1721. Sugar cultivation on Mauritius and Reunion began with slave workers, when an agricultural economy was established. Fishing probably started in Rodrigues a little later in 1792 and land clearance for agriculture and introduction of cattle began in the early 1800s. The British took over the Mascarenes in 1810, abolished slavery and indentured 100,000 labourers from India to work the sugar plantations. By independence in 1968, malaria, which had decimated coastal populations, had been eradicated from Mauritius, and improved health resulted in a population dominated by young people, and today the population is 1.2 million. A booming export-orientated economy and near full employment has since developed.

The Rodriguan population of 37,000 still depends on agriculture and fishing. Reunion has a population of 600,000, most of whom live in the narrow coastal strip, and work in agriculture, light industry and the service/tourism sector. The St Brandon archipelago was discovered by the Portuguese, and was important for exporting guano. The richness of the fishing grounds around the islands has been exploited since the 1830s, and in 1910, 100 fishermen were based on the islands. Temporary settlements still operate, together with a small coastguard and meteorological station staff. Agalega has been used for copra production since early Dutch and French occupation, and today the island is run by the Outer Islands Development Corporation who manage a population of 180 to 200 people, who work the plantations and fish. Tromelin was known in the 18th century for guano exploitation and was inhabited for 15 years by a group of shipwrecked slaves. The French from Reunion have cultivated turtles on the island and operate a meteorological station, and Mauritius claims fishing rights.

RURAL FACTORS

Agriculture

Mauritius

Mauritius was covered by dense forest until 1638, when native vegetation was reduced to just 5% of the land area, or 56,403 ha (Safford, 1998). Palm savannah along the coast was destroyed by fire and is now extinct, the *Diospyros/ Elaeodendron* low altitude forest was cut by the French for building and by the English for sugar cane production, and the plateau Sapotaceae forests have been invaded by exotics. About 63% of the remaining forest is in the southwest, and 44% of the total native vegetation is protected by the Black River Gorges National Park. Deforestation causes erosion of top soil which eventually reaches lagoons and reservoirs. Clearing of agricultural land along river banks has also caused siltation in river basins and freshwater systems, and residues of pesticides and fertilisers have damaged aquatic plants and caused the disappearance of most freshwater fish, shrimps and other aquatic life.

About 100,000 ha of land (48% of total land area) is devoted to agriculture, of which 90% is sugar cane, and 10% tea, coffee and other crops. Large quantities of fertiliser (600 kg year^{-1} ha^{-1}) and pesticides (7.4 kg ha^{-1} year^{-1}) are applied (Dwivedi and Venkatasamy, 1991). Nutrient enrichment and pesticide contamination of the sea occurs, causing hypernutrification and eutrophication of the lagoons. Nitrate concentrations vary throughout the year from 5 to 40 μg l^{-1} outside the reef, and 5 to 807 μg l^{-1} within the lagoon and 134–845 μg l^{-1} at a sea-bed seepage point (Daby, 1999). Phosphate concentrations were 5–9 μg l^{-1} outside the reef compared to 5–54 μg l^{-1} in the lagoon, and silicate 60–420 μg l^{-1} outside the reef compared to 6570 μg l^{-1} in the lagoon. Proliferation of filamentous algae and smothering of the corals occurs throughout the lagoons.

Reunion

Reunion's land area is 2500 km^{-2}, most of which is steep mountain, with 35% forest cover, containing important endemic plant species. Fertile lowlands are valuable agriculture areas, and 75% are used for sugar cane, creating conflict with urbanisation. Erosion is a serious threat (Perret et al., 1996), and sedimentation, eutrophication, bio-erosion, and run-off from urbanisation is causing reef degradation (Naim, 1993; Chabanet et al., 1995; Chazottes, 1996).

Rodrigues

Most of Rodrigues is steeply sloping and was stripped of much of its vegetation when planning efforts aimed to transform the island into a 'granary' for Mauritius to compensate for the sugar mono-crop in the latter. The nutrient-poor soil has eroded. Terracing was introduced in 1966, but the terraces have been poorly maintained, and overgrazing and damage to the terraces by cattle has caused further loss of soil, silting rivers, dams and lagoon channels and areas of lagoon, reducing the productivity of coastal waters (Pearson, 1988). Rodrigues has 38 endemic flora species, but the native vegetation is decimated and exists only on mountain tops where it is still threatened by exotics.

There is no agriculture in St Brandon, although a little 'gardening' has taken place around human habitations. On Agalega, coconut plantations are grown, while on Tromelin there is no agriculture.

Artisanal Fishing

Mauritius

Artisanal fishing in Mauritius involves lagoon and outer reef activities by 2863 fishermen using 1274 6–10 m long pirogues (Fig. 4). About 84% of the fleet is motorised, and the rest use punt poles, oars and sails. A variety of methods including hook and line, basket trap, seine nets, gill nets, canard nets, cast nets and harpoons are used. Siganids (13%), Scarids (12%), Lethrinids (11%), octopus (10%) and shark (0.2%) are caught. Annual catch is about 1600 tonnes, which is believed to be very close, if not above, maximum sustainable yield. A closed season operates for deployment of nets (MCFMRD, 1996), and 17 Fish Aggregating Devices (FADS) have been deployed over 400–3000 m water depth between 2 and 12 n.m. offshore, allowing about 300 fishermen to use vertical longlines and live bait.

Although there are no large-scale commercial sea cucumber fisheries in the Mascarene archipelago at present (Conand and Bryne, 1994), 2.7 metric tonnes of bêche-de-mer (dried/salted or in brine) were imported by Hong Kong from Mauritius between January and March 1996. Their total value was approximately US$ 56,000, making it a valuable artisanal fishery (Conand, 1990). Muller (1998) surveyed the abundance and distribution of holothuria in an east coast lagoon of Mauritius, recording 19 species and densities reaching 1.2 m^{-2} of lagoon bed.

Reunion

Artisanal fishing involves some 500–600 fishermen, catching 600–800 tonnes by hand lines, often operating around 29 FADs where the catch consists of swordfish (68%), albacore (15%) and yellowfin tuna (10%) (Renee and Poisson, 1996). Crayfish and lobster are also caught (3% of catch) off rocky areas. Ciguatera poisoning outbreaks, usually from consumption of Serrannidae, have been a problem in

Fig. 4. Lagoon fishing, Mauritius.

Reunion, with an incidence of 0.78 per 10,000 residents (Quod and Turquet, 1996).

Rodrigues

Artisanal fishing is important in Rodrigues, since the lagoon area is three times the land area. However, the channels rarely provide access to the outside sea due to siltation or sand bank formation, and hence small fishing boats (pirogues) are punted through channels and over deeper areas of lagoon. Fish are either caught on the outer reef slope or from the coral communities along the edges of the channels which support octopus, snappers, triggerfish, parrot fish and groupers, and shrimp and crabs are taken from the mud. Local consumption of fish is 500 tonnes per year, and 100–130 tonnes are exported (90% lobster and octopus) to Mauritius (Dwivedi and Venkatasamy, 1991). There is concern that dredging activities to deepen channels may further reduce catch, even though access and safety are improved (Pearson, 1988).

In St Brandon, 30–40 fishermen are employed by a private company to fish the lagoon, St Brandon Sea and other waters to a depth of 35 m using handlines and basket traps. About 350–600 tonnes of fish per year are caught, 80% of which are Lethrinids. The fish is frozen or salted and sent to Mauritius. There are no signs of overfishing at present (Kaly and Brown, 1998). In Agalega a small artisanal fishery produces some 30 tonnes of fin fish per year (Dwivedi and Venkatasamy, 1991), while in Tromelin there is no recorded artisanal fishing.

Effects of Fisheries Activities

The artisanal fisheries are directly influenced by pollution, sedimentation, mangrove removal and wetland reclamation. Artisanal fishing is also detrimental to coastal habitats because of various illegal fishing methods: dynamite and spearguns, nets with undersized meshes, anchoring and poling of boats which causes substantial coral breakage, poor enforcement of national legislation, capture of non-target species and fishing in prohibited areas, and by aquarium fish collectors who use chemicals to capture target species.

COASTAL EROSION AND LANDFILL

Coastal erosion is potentially a serious problem in many of the Mascarene islands. The marked seasonal cycle of rain, wind, and waves ensures that parts of the coast are in a delicate state of dynamic equilibrium so that changes in the frequency and intensity of physical forcings (e.g. cyclones) may produce significant changes in coastal morphology. As everywhere, the coasts are also affected by soil erosion accelerated by agriculture, and enhanced sediment removal from mining of lagoonal, beach and dune sands.

Although coastal sand accumulations are a vital asset to the tourist industry, the industry itself is often inadvertently responsible for degradation of sand bodies. Clearing of seagrass and deepening of channels in lagoons to improve their attraction and use may increase current velocities and erosion. Mining of lagoonal sand has the same effect. The result is locally enhanced sand transport and increased wave action on beaches. Structures such as groynes and jetties, often built for tourism, interrupt longshore drift of sand and give rise to local accretion and erosion. These effects have reached a critical state in parts of Mauritius. Attempts to halt the erosion have had mixed success; ill-advised sea walls have increased the problem. Limited beach nourishment has been applied with some success in Mauritius (e.g. Pointe Cannonier on the northwest coast) although the aesthetic impact of imported, exotic sand is questionable. Impacts on the coast are potentially greatest on those coasts most exposed to storms (Fagoonee, 1990a).

Solid wastes are dumped at sites adjoining lagoons causing heavy pollution from tyres, glass, plastic and metal. Incineration of material causes further air and sea pollution. In Mauritius, coastal dump sites are at Roche Bois and Poudre d'Or in the north, and there are plans to close all surface dumping sites by creating five landfill sites (Dwivedi and Venkatasamy, 1991).

EFFECTS FROM URBAN AND INDUSTRIAL ACTIVITIES

Artisanal and Non-industrial Uses of the Coast

About 3000 tonnes of salt are produced annually in Mauritius by evaporation of sea water pumped into gravity-fed tiered pans on the west coast at Tamarin and Black River. Coastal aquaculture is limited due to the destructive effects of cyclones. In Mauritius, barachois (fish ponds) are areas of lagoons enclosed by stone walls constructed for rearing fish (18 tonnes), crabs (4.1 tonnes) and oysters (108 tonnes). There are 24 barachois around the island located mainly on the east coast with a total area of 300 ha. Each barachois is between 0.6 to 51 ha in size. There is a continuous exchange of waters across the barachois walls. Fingerlings are caught by seine net and introduced into the barachois for later harvesting. Seed oysters are also collected from mangrove roots (Dwivedi and Venkatasamy, 1991) Production is limited by siltation, and poor management. Pilot projects have investigated sea bream *Rhabdosargus sarba* production in floating cages, and prawn farming (*Penaeus mondon* and *Macrobrachium rosenbergii*) in ponds on sugar estate lands, and rearing of mud crab *Scylla oceanica*, bivalves *Asaphis dicholoma*, and freshwater crayfish *Cherax quadricarinatus* (MCFMRD, 1996). Turtles and tilapia are farmed on Reunion for local consumption (Miard, 1997).

Two companies collect live aquarium fish from Mauritius for export. Over 4550 marine fish and 61,000 freshwater ornamental fish were exported in 1996, compared to 20,000 in 1990 (Dwivedi and Venkatasamy, 1991; MCFMRD, 1996). Mauritius is a contracting party to the CITES convention on trade in endangered species.

One of the major problems around the coast of Mauritius and the west coast of Reunion is residential ribbon development along the foreshore. In Mauritius, access to some of the best beaches is restricted by holiday houses and hotels. There is supposed to be public access to the sea every 500 m, but such access is not always available. Beach hotels are often constructed next to one another, and construction of commercial and shopping centres along coastal roads is prevalent. In Reunion, traffic congestion is a major problem in the tourist region of the west coast. Beach frontage has often been acquired at high cost, but coastal man-made structures in the path of longshore movements cause erosion, especially after cyclones, and there is no guarantee that such expensive land will still exist in the medium future. Wetlands, mangroves and marshes have been infilled and, as a result, such areas no longer help regulate tides and floods during heavy rains, and are becoming separated into small parcels of reduced ecological value (Dwivedi and Venkatasamy, 1991).

Industrial Uses of the Coast

Sand Mining and Lime Production

Medium and coarse-grained carbonate sands are required by the construction industry and for water purification, and demand, especially in Mauritius, has increased due to rapid industrialisation. Sand is effectively a non-renewable resource because its formation in the reef and lagoon is much slower than the rate of removal. In 1990, 800,000 tonnes of sand was extracted of which 300,000 tonnes came from the lagoon (Fig. 5). Sand is mined from lagoon beds manually by shovelling into pirogues. In an attempt to control activities and to protect beaches, four designated sites have been demarcated: Mahebourg 0.9 km^2, Grand Gaube 0.4 km^2, Grande Riviere South East 1.2 km^2 and Roche Noire 0.5 km^2. Sand is also removed from dunes, potentially weakening coastal defences against storm waves, flooding and future sea-level rise, and the flat areas are often used for residential development. Mauritius plans to phase out sand mining and to utilise rock sand from basalt crushing. About 8000 tonnes of lime for agriculture is also produced from 13,000 tonnes of fossil coral, eolionites and coral residues from sand quarries, through one kiln at Mahebourg. Coral sand and dune extraction is permitted in Reunion, where industrialisation has been constrained. In Rodrigues, fossil corals are cut into blocks for building.

The demand for basaltic products has increased dramatically, especially due to regulations to decrease use of coralline sand. There are 25 stone-crushing plants, including 10 basaltic plants on Mauritius, producing dust and suspended particles; sludge and related waste which may contaminate rivers and lagoons. Dust emission from the stone-crushing plants containing silica is damaging to human health and covers the environment in the vicinity of the plants (Dwivedi and Venkatasamy, 1991).

There are currently 17 sugar mills in operation on Mauritius and three on Reunion. These discharge large volumes of hot water, carbon column washing fly-ash, scrubbing liquid alkaline waste water and other organic effluent. High BOD effluents are directed into gutters between fields, streams and rivers. Most estates have no settling ponds or aeration systems to treat the effluent before it enters the lagoons. Similarly rum, alcohol and perfume distilleries produce highly organic and acidic waste water (Dwivedi and Venkatasamy, 1991).

There are over 568 textile firms currently operating in Mauritius. The total export accounted for 31% of the GDP in 1991. Effluent from dyeing activities and textile cuts, wastes and wrapping are dumped in rivers, or seep into the soil from absorption pits. Stone washing activities discharge detergents, and red reactive and chrome-based dyes are especially persistent and visible even at 0.05 ppm. Lagoons have become coloured with dye effluents, and large volumes of fresh water are used in the industry wastes (Dwivedi and Venkatasamy, 1991).

Light Industry

Industrial diversification, especially on Mauritius, has produced a range of industries, about 50% of which are located on the coast. These also have inadequate waste water treatment and solid waste disposal. The problems are enhanced by limited financial resources and lack of in-house expertise and environmental awareness (Dwivedi and Venkatasamy, 1991). Waste water treatment is recognised as a major priority in Mauritius, and this has been heightened by increasing water shortages.

Fig. 5. Lagoon sand mining, Mauritius.

Tourism

Tourism is increasingly important. Mauritius and Reunion have invested heavily in coastal infrastructure so that damage to coastal ecosystems is already evident. There is an urgent need to address the issue and policy review has started in Mauritius and Rodrigues with a 25-year tourism development plan. The present levels of awareness and corrective action are insufficient to prevent further degradation of the marine environment.

Tourism is well developed in Mauritius, heavily promoted in Reunion, and in its infancy in Rodrigues. Offshore islands such as St Brandon could be developed for tourism in the future. Tourism is the third largest source of foreign income for Mauritius, after agriculture and industry, and increased between 1983 and 1990 from 124,000 to 292,000 arrivals with an expected ceiling of 400,000 by year 2000. Gross earning increased from $34 million to $233 million. Mauritius has targeted high spending, beach resort tourism and has prevented charter flights. In contrast, Reunion receives charter flights, and offers both beach resorts and trekking in the mountain areas, especially along the south west coast beaches from St Gilles to La Saline. In Mauritius, unplanned urbanisation around the northern and eastern tourist zones has increased the impacts on the marine environment. Tourism also increases the cost of living in these zones, disturbing socio-cultural balances in local communities. Boat congestion, water sport activity, anchor damage, benthic habitat modification and coral breakage have also occurred. However, some abandoned sites have been transformed, areas reforested, flora and fauna protected, and artificial reefs established. Between 1980 and 1993, the Mauritius Marine Conservation Society sank nine ships in coastal waters originally to create artificial reef reserves, but without protection, these have become a focus for fishing and SCUBA diving. There is an opportunity for 'ecotourism' at St Brandon, based from live-aboard boats to observe bird colonies, turtle nesting and hatching, snorkelling, diving, subsistence fishing, and involvement in island restoration (Kaly and Brown, 1998).

Cities and Sewage Discharges

The human population density is high in the narrow coastal strips of the Mascarene islands and coastal urbanisation with its environmental implications remains a priority issue to be tackled. Although marine pollution can be considered to be relatively low, pollution hotspots cause much concern, especially near coastal towns. Waste treatment facilities are poor or virtually non-existent throughout the region.

The main urban centre of Mauritius extends from the capital city of Port Louis to the upland residential areas of Plaine Wilhelm where conditions are more temperate. About 33% of the rural population live in 12 large villages. Port Louis is only several centimetres above sea level and the port itself is built on reclaimed land. It is equipped for bulk and container handling and is one of the most modern harbours in the southwest Indian Ocean (Fagoonee and Daby, 1993). St Denis, capital of Reunion, and a narrow coastal region, houses 99% of the island's population. In Rodrigues, the population is scattered in some 137 hamlets, although some concentration occurs in the semi-urbanised zone of Port Mathurin and Oyster Bay with around 10,000 persons.

The residents of Mauritius produce 125,000 m^3 of waste water per day with a BOD load of 60 tonnes. Only 18% of households are connected to a sewage system, and in 1990 most industries were also unconnected to proper waste disposal systems. Three untreated and undiffused discharges, and one treated sewage discharge of domestic and industrial waste occur in shallow water close to the shore near Port Louis. Sewage discharges onto and over fringing reefs, along reclaimed land and is trapped in the lagoon, remaining near to the shore. These discharges have increased the turbidity of the water and led to an increased frequency of algal blooms.

Ciguatera fish poisoning remains a common problem and becomes more prominent after cyclonic episodes. Fish kills have been attributed to high BOD from sewage outfalls on two occasions (MCFMRD, 1996), and surface sediments have been found to be contaminated with coprostenol, a reliable marker of faecal pollution (Gendre et al., 1994) in all sites sampled, except an area of high wave energy at Tamarin Bay on the west coast. Mauritius has a Sewerage Master Plan developed in 1993, and currently being implemented. A new waste water treatment plant is planned, discharging at a depth of 30 m in a wave exposed rocky habitat. A novel alternative possibility currently being investigated involves injection through boreholes into lava tunnels which dissipate at 20 m depth out to sea through fissures in the basaltic rock bed.

A similar situation exists on Reunion, where nutrient-enriched submarine groundwater discharges cause algae to dominate corals in the hot season (Naim, 1993). In Rodrigues water consumption is very low at 47 l d^{-1}. Most households have pit latrines rather than flushed toilets, and the few hotels have cess-pits. There is no industry and impact is low.

Fresh Water

The sustainability of development is at stake due to a shortage of fresh water (Ramjeawon, 1994). Mauritius, for example, is coming under increasing 'water stress', defined as an annual water supply of between 1000 and 1700 m^3 per person, with a predicted fall from 1970 m^3 in 1995 to 1485 m^3 by 2025 (Hopkins, 1998). During the second half of 1998 and most of 1999, the Mascarene islands faced the most severe water stress situation for 20 years, impacting heavily on the economies. More than 75% of the water withdrawals are used for agriculture, mainly for irrigation. Domestic consumption accounts for 15–20% of withdrawals. Freshwater pollution is becoming increasingly evident in all countries, and saltwater intrusions into boreholes are evident.

Shipping and Offshore Accidents and Impacts

A major sea route passes along the west coast of Madagascar, along which passes an estimated 470 million tonnes of oil, plus a further 6.5 million tonnes to the islands themselves. Mauritius now has a National Oil Spill Contingency Plan, and is a contracting party to MARPOL, Ocean Dumping and a signatory to the Law of the Sea.

Over 1000 vessels visit Port Louis each year, of which one third are fishing vessels. The discharge of petroleum cargoes and bunkering represents a further threat to the marine environment. Extensive dredging has taken place to deepen the main channel, extend the fishing port, and construct a cargo handling area. In 1990, over 1287,000 m³ of dredging caused sediments rich in organic matter to be transported to the north and south of Port Louis (Dwivedi and Venkatasamy, 1991).

The channel at Port Mathurin in Rodrigues has also been dredged to allow the *Mauritius Pride* supply vessel easier access to the port. There are proposals to dredge more of the channels and to dispose of dredge spoil for coastal land reclamation. Increased turbidity and sedimentation is likely to affect coral communities, and loss of habitat will occur.

PROTECTIVE MEASURES

These small island states face disadvantages to their development from an interplay of factors such as remoteness, geographical dispersion, vulnerability to natural disasters, a high degree of economic openness, small internal markets, limited natural resources and fragile coastal ecosystems which are under increasing pressure. Their sustainable development is increasingly dependent on a thorough understanding of the marine resources and the ecological processes that control them, and the activities of their entire populations impact on these essential ecosystems.

The following environment issues are priorities:
- protection of the coral reefs and associated ecosystems;
- coastal erosion;
- unsound utilisation of the coastal and marine resources;
- protection of biodiversity;
- watershed management;
- coastal zone use conflicts;
- slow implementation of national environmental legislation;
- absence of a sustainable regional development policy for coastal and marine ecosystems.

The islands have established a number of Nature Reserves, Fishing Reserves and Marine National Parks (Table 2).

Table 2

Marine Protected areas of the Mascarene region

Country	Fishing Reserve	Nature Reserve	National Marine Park
Mauritius	Flacq Rivière du Rempart–Poudre d'Or Black River Grand Port–Mahebourg Port-Louis Trou d'Eau Douce	Flat Island Ile aux Aigrettes Round Island Serpent Island Ilot Gabriel Ilot Marianne Gunner's Coin	Blue Bay–Le Chaland Balaclava (Proposed)
Reunion	Cap la Houssaye–Ravine Trois Bassins Ravine Trois Bassins–Pte de Bretagne Pte de Bretagne–Pte Etang Sale	Ile Tromelin Non Mascarene Region sites: Ile Europa Iles Glorieuses Ilot de Bassas de India	
Rodrigues		Ile aux Cocos Ile aux Sables	

National environmental legislation exists in all the states to regulate and manage the coastal fisheries and the environment in general, although no specific legislation exists to date for Integrated Coastal Zone Management (ICZM).

In Mauritius there are six Fishing Reserves where no net fishing is allowed, and the Fisheries Act (1980) provides regulations related to number, types and sizes of nets as well as to the period of utilisation and allowable mesh size of basket traps in the lagoon fisheries. Nine islets around Mauritius are declared as Nature Reserves (no legislative protection) because they have important reef areas and/or are home to seabirds, endemic plants or animals. Most have management plans prepared by NGOs such as the Mauritius Wildlife Foundation working with the Conservation Unit. Blue Bay/Le Chaland has been designated as a Marine National Park, but Balaclava/Arsenal Bay is still awaiting statutory protection. Ile aux Cocos and Ile aux Sables of Rodrigues are Nature Reserves for their plants and seabird colonies. Round Island and Serpent Island, and St Brandon are nationally considered worthy of World Heritage Site status. A detailed management plan has been prepared for St Brandon (Kaly and Brown, 1998), which is recognised as a regional priority for marine protection by IUCN and the World Bank. Mauritius has a national Man and Biosphere Committee but no marine and coastal biosphere reserves. A National Environment Policy and an Environmental Investment Plan have been prepared by the Government of Mauritius with the World Bank, including preparation of a Marine Environmental Management Plan. The Environmental Protection Act, 1991 of Mauritius includes a section on coastal and maritime zone management. Integrated Coastal Zone Management has been proposed (Fagoonee and Daby, 1993).

There are three Fishing Reserves in Reunion which are closed to fishing (except on foot and by line) on a rotational basis for a period of two to three years (Sheppard and Wells, 1988). Three of the four island dependencies of Reunion (important for nesting seabirds, green and hawksbill turtles) are in the Madagascar Channel outside the Mascarene Region, and have been declared as Nature Reserves, but none has been formally designated by legislation because the islets are also claimed by Madagascar (Ile Europa, Iles Glorieuses and Ilot de Bassas de India) or by Mauritius (Tromelin).

The Indian Ocean Commission is conducting pilot operations in ICZM at the national level on Mauritius and Reunion, with a view to regional plans for sustainable development (IOC, 1997). The Mascarene islands participate in the Action Plan for the Protection, Management and Development of Marine and Coastal Environments in the East African Region. The Convention has one protocol on marine pollution issues, and another one on Protected Areas and Wild Fauna and Flora. Parties are required to establish protected areas for rare, threatened, endemic, migratory and economically important marine species, and a regional network for the protected areas. The Indian Ocean Sanctuary has been designated for the protection of cetaceans (15 species have been recorded in the region) and covers a large part of the Indian Ocean, and efforts are under way to propose the designation of the Mascarene Plateau as a Large Marine Ecosystem. The marine resources are of significant economic importance, and are protected in theory through the proclamation of EEZs covering several million square kilometres. Catch quotas for the banks fisheries have been imposed since 1994 in an attempt to control entry and safeguard the fish stocks.

Despite these measures, coastal degradation, marine pollution and over-harvesting remain among the most prominent pressures that threaten the long-term viability and productivity of the marine areas of this region. Such environmental deterioration is due to socio-economic forces as well as poor implementation of legislation and inadequate enforcement, which promote unsustainable resource utilisation. Considerable gaps in information regarding the status of many species, habitats and marine resources remain a major difficulty. There is an urgent need to promote marine research and monitoring, and current development policies should also be reviewed to respond to emerging environmental issues.

REFERENCES

Baissac, J. de B., Lubet, P.E. and Michel, C.M. (1962) Les Biocoenoses benthiques littorales de L'Ile Maurice. *Receuil Trav. Stn. Mar. Endourne* **25** (39), 253–291.

Bassias Y., Denisclocchiatti, M. and LeClair, L. (1993) The Mascarene Plateau - evolution of a neritic platform in the oceanic environment. *Comptes Rendus de L'Academie des Sciences II* **317** (4), 507–514.

Bouchon, C. and Faure, G. (1979) Apercu sur peuplement a base de scleractiniaires de recif de Tromelin (Ocean Indien). *Cashiers de l'Indo Pacifique* **1** (1), 25–37.

Bouchon, C. (1981) Quantitative study of the scleractinian coral communities of a fringing reef of Reunion island (Indian Ocean). *Marine Ecology Progress Series* **4**, 273–288.

Chabanet, P., Duour, V., and Galzin, R. (1995) Disturbance impact on reef fish communities in Reunion Island (Indian Ocean). *Journal of Experimental Marine Biology and Ecology* **168** (1), 29–48.

Chazottes, V. (1996) Experimental study of bioerosion and sedimentogenesis in a reefal environment: Eutrophication effects (Reunion Island, Indian Ocean). *Comptes Rendus de l'Academie des Sciences II* **323** (9), 787–794.

Cheke, A.S. and Lawley, J.C. (1983) Biological history of Agalega, with special reference to birds and other land vertebrates. *Atoll Research Bulletin* **273**, 65–108.

Conand, C. (1990) The fishery resources of Pacific island countries. Part 2. Holothurians. FAO Fisheries Technical Paper, 272.2, FAO, Rome, 143 pp.

Conand, C. and Bryne, M. (1994) A review of recent developments in the world sea cucumber fisheries. *Marine Fisheries Review* **55**, 1–13.

Daby, D. (1999) Structure and function of two lagoon ecosystems of Mauritius. PhD thesis, University of Mauritius, 904 pp.

Daby, D. (1990) A Coastal Zone Inventory of Mauritius. MSc thesis, School of Ocean Sciences University of Wales, Bangor, 105 pp.

Devassey, V.P. and Goes, J.I. (1991) Phytoplankton assemblages and pigments in the exclusive economic zone of Mauritius (Indian Ocean). *Indian Journal of Marine Science* **20**, 163–168.

Dwivedi, O.P and Venkatasamy, V. (1991) State of the Environment in Mauritius: A report prepared for presentation at the UN Conference of Environment and Development, Rio de Janeiro, Brazil. 403 pp.

Dykes, R. (1996) An evaluation of remote sensing using SPOT 3XS data for the classification and mapping of two Mauritian lagoonal ecosystems. MSc Thesis. University of Wales, Bangor. 151 pp.

Eastwood, P. (1998) Geographical Information System Development for Coastal Research, Education and Resource Management in Mauritius. MSc Thesis. University of Wales, Bangor. 91 pp.

Fagoonee, I. (1990a) *Coastal Zone of Mauritius—Sea level rise considerations*. Ministry of Agriculture and Natural Resources, Port Louis, 195 pp.

Fagoonee, I. (1990b) Coastal marine ecosystems of Mauritius. *Hydrobiologia* **208**, 55–62.

Fagoonee, I. and Daby, D. (1993) Coastal Zone Management in Mauritius. *Workshop and Policy on Integrated Coastal Zone Management in East Africa*, 21–23 April 1993, Arusha International Conference Centre, Tanzania. 58 pp.

Faure, G. (1975) Etude comparative des recifs coralliens de l'archipelago des Mascareignes (Ocean Indien). *Bulletin Mauritius Institute* **8** (1), 1–26.

Faure, G. (1977) Distribution of coral communities on reef slopes in the Mascareignes Archipelago, Indian Ocean. *Marine Research Indonesia* **17**, 73–97.

Gendre, F., Beck, C., Ruch, P. and Kubler, B. (1994) Human impacts on coral ecosystems at Mauritius island: Coprostonol in surface sediments. *Ecologae Geologicae Helvetiae* **87** (2), 357–367.

Hardman, E. (1999) A rapid assessment of the extent of coral bleaching in Mauritius after the 1998 seawater warming event. MSc Thesis. University of Wales, Bangor. 124 pp.

Harmelin-Vivien, M.L. (1976) Icthyofaune de quelques recifs corallines des iles Maurice et La Reunion. *Bulletin Mauritius Institute* **VIII**, 69–104.

Hodgkin, E.P. and Michel, C. (1961) Zonation of plants and animals on rocky shores. *Proceeding of the Royal Society Art and Sciences Mauritius* **2**, 121–145.

Hopkins, J. (1998) Solutions for a water short world. Population Report, XXVI (1) Johns Hopkins Population Information program, Baltimore, MD.

Ingole, B.S., Ansari, Z.A. and Parulekar, A.H. (1992) Benthic fauna around Mauritius Island, southwest Indian Ocean. *Indian Journal of Marine Sciences* **21** (4), 268–273.

Ingole, B.S., Ansari, Z.A. and Parulekar, A.H. (1998) Spatial variation in meiofaunal abundance of some coralline beaches of Mauritius. *Tropical Ecology* **39** (1) 103–108.

Ingrams, A. (1998) Survey of echinoderms in Trou D' Eau Douce lagoon, Mauritius and studies on the feeding behaviour of *Nardoa variolata* (Retzius) Asteroidea. MSc Thesis. University of Wales, Bangor. 115 pp.

IOC (1997) Integrated coastal area management: the role of the Regional Environment Programme of the Indian Ocean Commission. *Bull. EC Fish. Coop.* **10** (1) 13–17.

Kaly, U. and Brown, J. (1998) St Brandon Archipelago Management Plan. Volume 2. Technical Report. World Bank/Government of Mauritius. 60 pp.

Karisiddaiah, S.M., Veeryya, R., Guptha, S. (1988) Texture carbonate content and component composition of Mauritius beach sands, Indian Ocean. *Journal Coastal Research* **4** (3), 465–474.

Kindle, J.C. (1991) Topographic effects on the seasonal circulation of the Indian Ocean. *Journal of Geophysical Research—Oceans* **96** NC9, 16827–16837

Klaus, R. (1995) An evaluation of the use of a Landsat 4 TM satellite image for the qualitative and quantitative mapping and assessment of the coastal zone habitats of Mauritius (Indian Ocean). MSc Thesis. University of Wales, Bangor. 176 pp.

Konyaev, K.V., Sabinin, K.D. and Serebryany, A.N. (1995) Large amplitude internal waves at the Mascarene Ridge in the Indian Ocean. *Deep Sea Research I* **42** (11–12), 2075–2091

Lebeau, A. and Cueff, J.C. (1976) Biologie et peche des Lethrinids sur les haut-fonds de Saya de Malha. *Trav. Doc. ORSTOM* **47**, 333–348.

Letourneur, Y. (1996) Dynamics of fish communities on Reunion fringing reefs, Indian Ocean. 1. Patterns of Spatial distribution. *Journal of Experimental Marine Biology and Ecology* **195** (1), 1–30.

Letourneur, Y. (1998) Composition,, structures and trophic networks of the fish communities of the windward coast of Reunion Island. *Cybium*, **22** (3), 267–283.

Letourneur, Y., Harmelin-Vivien, M. and Galzin, R. (1993) Impact of hurricane Firinga on fish community structure on fringing reefs of Reunion Island, S.W Indian Ocean). *Environmental Biology of Fishes* **37** (2), 109–120.

MCFMRD (1996) Fisheries. Annual Report 1996. Ministry of Cooperatives, Fisheries and Marine Development. Republic of Mauritius. 68 pp.

Miard, R. (1997) Reunion: Turtle farm. *Oceanorama Inst. Oceanogr, Paul Ricard* **27**, 31–36.

Montaggionni, L. and Faure, G. (1980) Les recifs coralliens des Mascareignes (Ocean Indian) *Université Francaise de l'Ocean Indien Centre Universitaire de la Reunion.* 151 pp.

Mordasova, N.V. (1980) Chlorophyll in the southwestern Indian Ocean in relation to hydrological conditions. *Oceanology* **1**, 116–122.

Morosov, E.G. (1995) Semi-diurnal internal wave global field. *Deep Sea Research I* **42**, 135–198.

Mshigeni, K.E. (1985) Marine algal resources of Mauritius: a preliminary survey. Report to the Commonwealth Science Council, London, 72 pp.

Muller, J. (ed.) (1991) Etude des Ecosystems littoraux de Maurice, Rapport 5. CEC/ Université d'Aix-Marseille I et III/University of Mauritius. 255 p.

Muller, C.V. (1998) The role and distribution of Holothurians (Echinodermat: Holothuroidea) in a shallow coastal lagoon, Mauritius. MSc Thesis. University of Wales, Bangor. 92 pp.

Nath, B.N. and Prasad, M.S. (1991) Manganese nodules in the Exclusive Economic Zone of Mauritius. *Marine Mining* **10** (4), 303–335.

Naim, O. (1993) Seasonal responses of a fringing reef community to eutrophication (Reunion, western Indian Ocean). *Marine Ecology Progress Series* **99** (1–2), 137–151.

Orme, C.D. (1997) The remote mapping of Mauritian coral lagoon habitats using Landsat Thematic Mapper imagery. MSc Thesis. University of Wales, Bangor. 108 pp.

Pearson, M.P. (1988) Rodrigues. Rapid survey of the status of exploitation and environmental damage of the lagoon and coral reefs off Rodrigues. Report prepared for the project assistance to artisanal fishermen and development of outer-reef fishery. FAO, Rome. 49 pp.

Perret, S., Michellon, R., Boyer, J. and Tassin, J. (1996) Soil rehabilitation and erosion control through agro-ecological practices on Reunion Island. *Agriculture Ecosystems and Environment* **59** (3), 149–157.

Pichon, M. (1971) Comparative studies of the main features of some coral reefs of Madagascar, La Reunion and Mauritius. *Symposium of the Zoological Society London* **28**, 185–216.

Quod, J.P. and Turquet, J. (1996) Ciguatera in Reunion Island (SW Indian Ocean): Epidemiology and Clinical Patterns. *Toxicon* **34** (7), 779–785.

Ramjeawon, T. (1994) Water resources management on the small island of Mauritius. *International Journal of Water Resources Development* **10** (2), 143–156.

Renee, F.A. and Poisson, F. (1996) Recent evolution of the fisheries sector in La Reunion. *Proceedings of the Sixth Expert Consultation on Indian Ocean Tunas. Colombo, Sri Lanka.* pp. 110.

Renee, F., Poisson, F. and Teisser, E. (1998) Evolution of swordfish long line fishery *Xiphias gladus* operating in the Indian Ocean from Reunion. *Colloq. Semin. Inst. Fr. Rech. Sci. Dev. Coop. ORSTOM, Paris.* pp 287–312.

Safford, R.J. (1998) A survey of the occurrence of native vegetation remnants on Mauritius in 1993. *Biological Conservation* **80**, 181.

Salm, R. (1976) The structure and successional status of three coral reefs at Mauritius. *Proceedings of the Royal Society of Arts and Science, Mauritius* **3**, 227–240.

Sanders, M.J. (1988) Summary of the fisheries resources information for south west Indian Ocean. In: *Proceedings of the workshop on the assessment of the fishery resources in the south west Indian Ocean. Albion, Mauritius,* ed. S.C. Venema, pp. 187–229.

Sauer, J.D. (1962) Effects of tropical cyclones on the coastal vegetation of Mauritius. *Journal of Ecology* **50**, 275–290.

Schwiderski, E.W. (1983) *Atlas of Ocean Tides and Maps.* Naval Surface Weapons Centre, Dohlgran, Virginia.

Sheppard, C.R.C. and Wells, S.M. (1988) *Coral Reefs of the World. Volume 2: Indian Ocean, Red Sea and Gulf.* UNEP Regional Seas Directories and Bibliographics. IUCN. Gland, 389 pp.

Sigma, Soogreah and Codra (1991) Preparation of a master plan for the development of the northern tourist zone and the protection of the seaboard. Consultancy Report on the study of the coral reef from Cap Malheureux to Trou aux biches. 54 pp.

Stoddart, D.R. (1984) Coral reefs of the Seychelles and adjacent regions. In *Biogeography and Ecology of the Seychelles Islands,* ed. D.R. Stoddart. Junk, The Hague, pp. 63–81.

Taylor, E. (1998) Accuracy Assessment for remote sensing techniques of a Mauritian lagoon. MSc Thesis. University of Wales, Bangor. 125 pp.

Titmuss, R. and Abel Smith, B. (1968) *Social Policies and Population Growth in Mauritius.* Frank Cass & Co.

Torres, R.J. (1997) Spatial and temporal variability of turbidity and hydrography in a shallow tropical lagoon (Ile aux Cerfs, Mauritius). MSc Thesis. University of Wales, Bangor. 103 pp.

Walley, L. (1997) A critical evaluation using visual and video point sampling techniques to estimate mega-benthos from a lagoonal ecosystem at Trou d'eau Douce, Mauritius. MSc Thesis. University of Wales, Bangor. 176 pp.

Wyrtki K. (1971) *Oceanographic Atlas of the International Indian Ocean Expedition.* Natural Science Foundation, Washington, DC. 531 pp.

THE AUTHORS

John Turner
*School of Ocean Sciences,
University of Wales Bangor, Marine Science Laboratories,
Anglesey, Gwynedd LL59 5EY, U.K.*

Colin Jago
*School of Ocean Sciences,
University of Wales Bangor, LL59 5EY, U.K.*

Deolall Daby
*Faculty of Science, University of Mauritius,
Reduit, Mauritius*

Rebecca Klaus
*Department of Biological Sciences,
University of Warwick, Coventry CV4 7RU, U.K.*

Chapter 71

THE BAY OF BENGAL

D.V. Subba Rao

The Bay of Bengal has several remarkable features. It has a huge influx of fresh water from several major rivers, it is affected by intense northeast and southwest monsoon winds, of which the latter breed cyclones. In the bay there is a seasonal anticyclonic coastal current which is seasonally replaced by a cyclonic current, and unlike most coastal regions, there is a western boundary coastal upwelling.

The Bay supports various tropical biotopes such as brackish lakes, estuaries, mangroves, coral reefs, and offshore waters with a great diversity of marine fauna and flora. The steadily growing coastal population (~55 million) utilizes the coastal zone for several purposes. Increasing development pressure from urban settlements, industry, fishery, mariculture, ports and harbours has resulted in the alteration of coastline, loss of critical habitats, and pollution of the environment, giving rise to serious environmental and socio-economic problems. Massive education of the public about the environment, impending disasters, mitigation measures, and the need for conservation of resources and sustainable development are all crucial to our understanding or use of this unique sea. There is a need for the creation of a coastal zone management authority empowered to implement environmental regulations, and a need to establish regional scientific programmes.

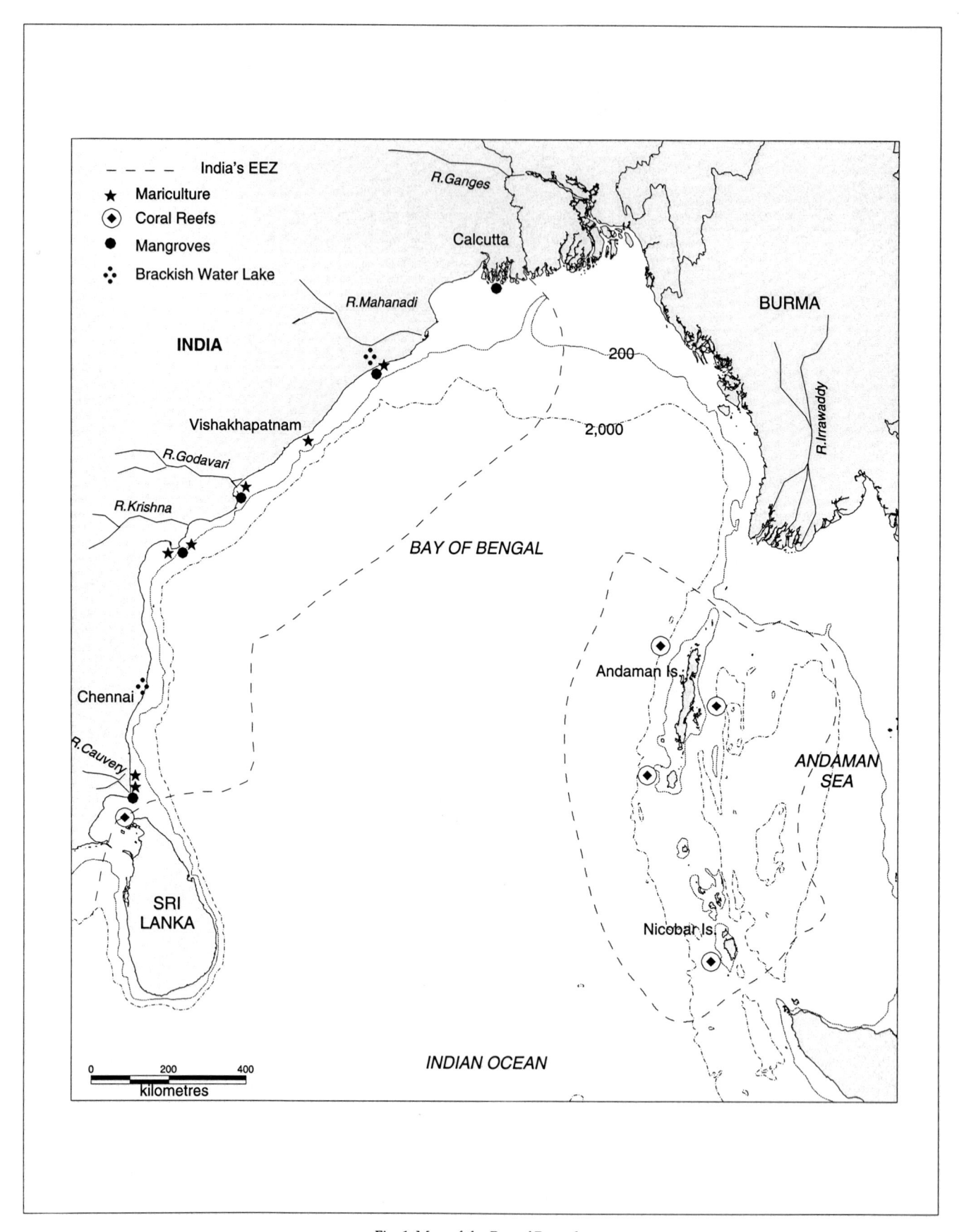

Fig. 1. Map of the Bay of Bengal.

THE BAY OF BENGAL

The Bay of Bengal stretches from the equator to 25°N and from 80°E to 100°E (Fig. 1) and has an area of about 4.09×10^6 km^2. Features such as alternating monsoon currents, upwelling of sub-surface waters and enormous river run-off provide an unusual combination of oceanographic features. Additionally, the Bay of Bengal offers diverse biotopes: estuarine waters and mangroves associated with several rivers, oceanic waters, brackish lakes and coral reefs around the 325 Andaman and Nicobar islands. The total population of the east coast of India i.e. Tamil Nadu, Andhra Pradesh, Orissa and West Bengal and the Andaman Nicobar Islands is 222 million, of which 25% live along the coast. The coastal population heavily utilizes the bay for fishing, mariculture, marine transport, recreation and waste disposal.

During the last 50 years India has acquired capabilities to study the Indian seas (Table 1) and also shows greater awareness of the value of marine sciences in this region. Most marine research publications address applications such as fisheries and aquaculture, and oil and gas exploration, but urbanization of the coast has also encouraged research into marine pollution, geology and geophysics.

The 0.44×10^6 km^2 shelf area (<200 m depth) of the Bay of Bengal (Table 2) approximates that off the British Isles or that off Peru and Chile. The total shelf and upper slope area of 0.62×10^6 km^2 approximates that of the Eastern Central Pacific (California to Colombia) or the Gulf of Mexico. Of the 2.02×10^6 km^2 Indian Exclusive Economic Zone, 59% is due to the Bay of Bengal (the eastern seaboard plus the Andaman, and Nicobar Sea) and covers 1.19×10^6 km^2.

NATURAL ENVIRONMENTAL VARIABLES

River Run-off

The Bay of Bengal receives an annual precipitation of 11,000 km^3 and a run-off of about 126×10^{10} m^3 from the major rivers Brahmaputra, Ganges, Mahanadi, Godavari, Krishna and Cauvery (Table 3). The Bay of Bengal and Nicobar fan is the world's largest, being 2800–3000 km long, 830–1430 km wide and >16 km thick in the north (Chauhan et al., 1993). Fresh-water run-off, monsoon winds, cyclones and coastal currents modify the bay in several ways. Transport of sediment widens the shelf; muds from the Gangetic fan extend thousands of kilometres, and the thickness of the sediment cones exceed 1.5 to 2.5 km (Ewing et al., 1969). The sediment brought by the rivers amounts to 4×10^9 m^3, enough to cover the entire bay to a depth of 2 mm annually. A north to south decrease in the width of the continental shelf along the east coast can be attributed to deposition of sediments by the major rivers in the north, and the seaward edge of the continental shelf seems to be steadily spreading annually about 1.8 km (LaFond, 1957). The annual particle flux patterns correspond with the fresh-water discharge

Table 1

Number of abstracts in *Aquatic Sciences and Fisheries* cited by Cambridge Scientific Abstracts

	1978–87	1988–96	1997–98
Indian Ocean	1853	2960	901
Andaman and Nicobar Sea	31	130	80
Bay of Bengal	397	856	188
Oceanography	38	161	37
Physical	14	38	6
Chemical	17	40	10
Biological	22	56	9
Geological and geophysical	8	78	10
Pollution	10	63	20
Fisheries	141	314	66
Aquaculture	24	102	15
Other	123	4	15

Table 2

Areas (10^6 km^2) in the Bay of Bengal based on Moiseev 1971

	Shelf <200 m	Continental Shelf >200–<300 m	Shelf and upper slope 0–1000 m	0–3000 m
Eastern India	0.19	0.84	0.33	1.03
Myanmar	0.25	0.20	0.29	0.45
Total	0.44	1.04	0.62	1.48

patterns of the Ganges-Brahmaputra rivers, being highest (>50g m^2 y^{-1}) in the central bay and least (37 g m^2 y^{-1}) in the south (Ramaswamy and Nair, 1994). The lighter illite-rich suspended sediments from Ganges–Brahmaputra are transported southwards along with the low salinity layer, an important sedimentary process in the Bay of Bengal. Some of these illite and chlorite rich clay sediments are transported as far as the southwestern peninsular continental margin on the Arabian Sea (Chauhan and Gujar, 1996).

Canyons

Opposite the mouths of the major rivers are deep canyons. At the head of the Bay, a deep Ganges Submarine canyon, known as 'Swatch of No Ground' cuts the plain in a northeast–southwest direction, and several others occur off Visakhapatnam (Andhra, Mahadevan and Krishna canyons). Their formation is attributed to turbidity currents (LaFond, 1957).

Shoreline Dynamics

Coastal currents and deposition of deltic muds affect the shoreline morphodynamics. Sediment transport can be very large (Fig. 2). Chilka inlet near Mahanadi is exposed to an annual littoral drift of about 1×10^6 m^3; as a result it

Table 3

Annual river inputs into the Bay of Bengal. Units for discharge are in 10×10 m^3 for P,N,U and ^{226}Ra in 10^6 g; for DOC and POC and particulate metals in 10^{12} g. (Based on Kumar et al., 1992)

	Ganges	Brahmaputra	Mahanadi	Godavari	Krishna	Cauvery	Pennar	Total
Discharge	39.3	60.9	6.7	10.5	6.8	2.1		126
Na	395.6	128.8	68.0	182.0	234.0	468.3	1477	
K	105.6	113.4	10.0	27.0	18.0	44.1		318
Ca	998.0	857.7	70.0	274.0	136.0	483.0	2819	
Mg	274.7	233.4	64.0	64.0	52.0	77.7		766
Cl	195.0	67.4	98.0	153.0	176.0	199.5		889
SO$_4$	336.0	614.4	210.0	109.0	211.0	63.0	1543	
HCO$_3$	4995.9	3556.3	410.0	1380.0	977.0	294.011613		
SiO$_4$	324.4	474.0	6.0	230.0	95.0	38.1		1167
TDS	7626.0	6045.0	1039.0	2436.0	1911.0	668.020725		
Phosphate	45981	–	3900	–	–	17220		
Nitrate	51090	–	1600	–	–	–		
Uranium	711	384	17	65	78	–		
226 Ra10^{10} dpm	2750	5480	–	420	340	–		
DOC	1.24							
POC	0.28							
Al	1.53		0.012	0.813	0.014	0.00019		
Fe	1.70		0.08	0.72	0.019	0.034		
Mn	0.035		0.007	0.022	0.003	0.0009		
Ni	0.005		0.00001	0.005	0.0003	–		
Cu	0.002		0.00008	0.001	0.0002	0.00002		
Zn	0.004		0.00009	0.0006	0.00003	0.00006		

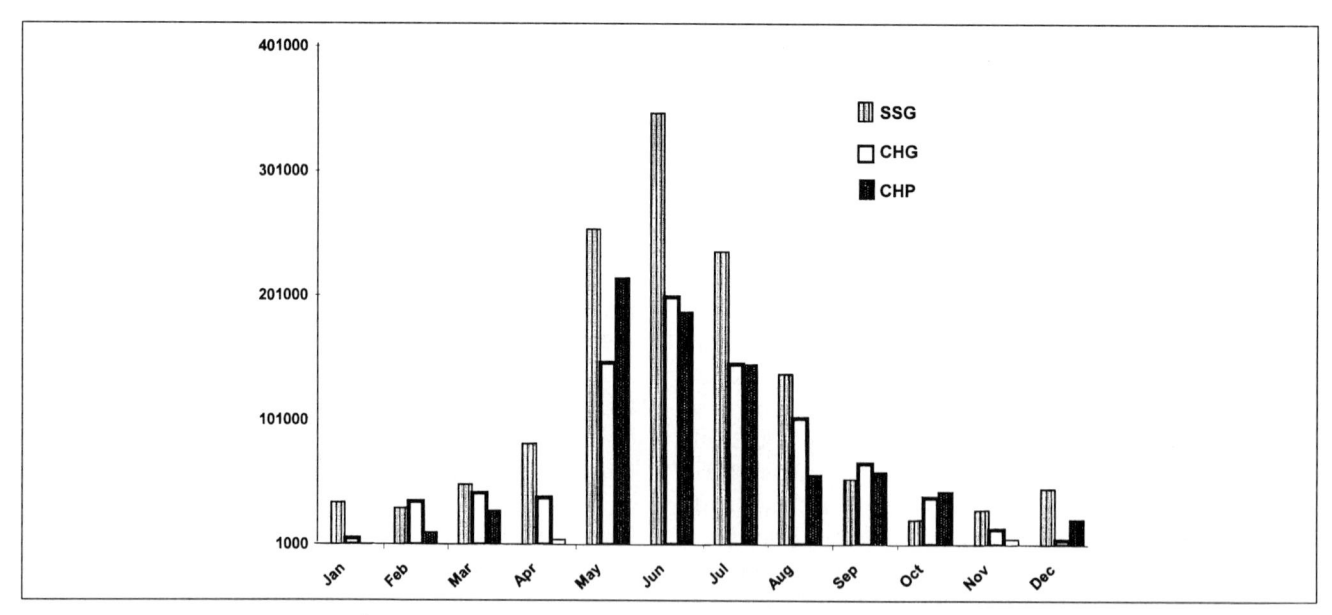

Fig. 2. Monthly sediment transport (m^3) off Orissa coast. SSG: off Gopalpur (Sundar and Sarma, 1992); CHG: off Gopalpur (Chandramohan et al., 1993); CHP: off Paradeep (Chandramohan et al., 1993).

migrates annually 500 m north (Chandramohan et al., 1993) and 0.16×10^6 m^3 south (Sundar and Sarma, 1992). The Puri coast experiences active erosion amounting to 5.37 m^3 m^{-1} y^{-1} due to littoral currents, in addition to another 70 m^3 m^{-1} y^{-1} sediments per cyclone (Chauhan, 1992). Further south near Kakinada, a 16 km long sand spit named after the river Godavari, runs north–south. The spit seems to continuously lengthen and ultimately the bay will be converted into a lagoon (Prasad, 1990). Due to littoral drift, further south the coastline of Tamil Nadu undergoes significant changes due to both erosion and accretion (Ramaiyan et al., 1997).

Sediments

The inner shelves are dominated by sandy terrigenous muds that grade to finer clayey and silty sediments down to the 50 m isobath. Medium to coarse sediments with local pebble zones occur on the inner and middle shelf. In the outer shelf, some calcareous oolitic sand might represent a regressive phase of the sea level during the Pleistocene (Sen Gupta et al., 1992). Regional differences in the characteristics of the sediments exist (Table 3) although kaolinite illite, chlorite, smectite and montmorillonite are common. Based on the clay minerals along the shelf, Raman et al. (1995) distinguished from north to south (a) a Himalayan zone dominated by illite followed by chlorite, (b) an Eastern Ghats zone with mixed sediments, and (c) a Deccan Provinces zone rich in smectite.

River Impacts

On a global scale the Ganges–Brahmaputra river system alone ranks fourth in sediment transport. They contribute 118×10^6 tonnes of dissolved solids annually (Sarin and Krishnaswamy, 1984) which control the water chemistry. Surface discharge of fresh-water, excluding that from precipitation during the Southwest Monsoon (June–September) into the Bay of Bengal, amounts to 50×10^3 m^3 s^{-1}. When this is evenly spread, it raises the water level by 70 cm, but circulation in the Bay is essentially density (geostrophic) controlled (Murty et al., 1992). At the head of the Bay, a substantial amount of flood water, dependent on the relative strength of the wind and river discharge, is held back inside Bangladesh by the Southwest Monsoon (Ali, 1995). Known as the 'back water effect', this causes flooding, loss of life and damage to property.

In coastal waters detrital matter dominates the particulate organic carbon. In oceanic waters detrital matter correlates with protein and lipid (Sreepada et al., 1995). At the head of the bay, Subarnamukha river contributes to the heavy metal flux (Senapati and Sahu, 1996) while off Orissa, the ratio between calcium, magnesium and fluoride with chlorinity decreases both off-shore and from north to south (Das and Sahoo, 1996). These ratios were high in areas with pronounced fresh-water. The highest concentrations of quartz (80–90%) are in sediments between Mahanadi and Kalingapatnam, and are lowest (50–60%) in the south i.e. Krishna–Godavari and in the outer shelf off Visakhapatnam (Purnachandra Rao and Vijay Kumar, 1990).

Monsoons, Currents and Gyres

Seasonal land–sea heat fluxes change the atmospheric pressure over the Asian continent and result in monsoons. The Northeast Monsoon, also known as the winter monsoon, blows from land to sea whereas the Southwest Monsoon blows from sea to land. The latter sets in as early as March and may extend to October. Based on 10-year sensible heat

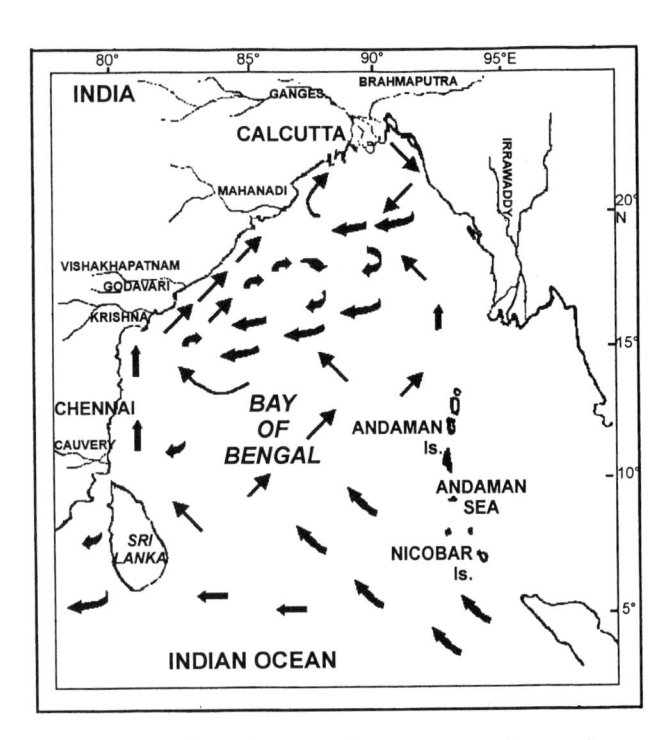

Fig. 3. Circulation during south west monsoon (summer).

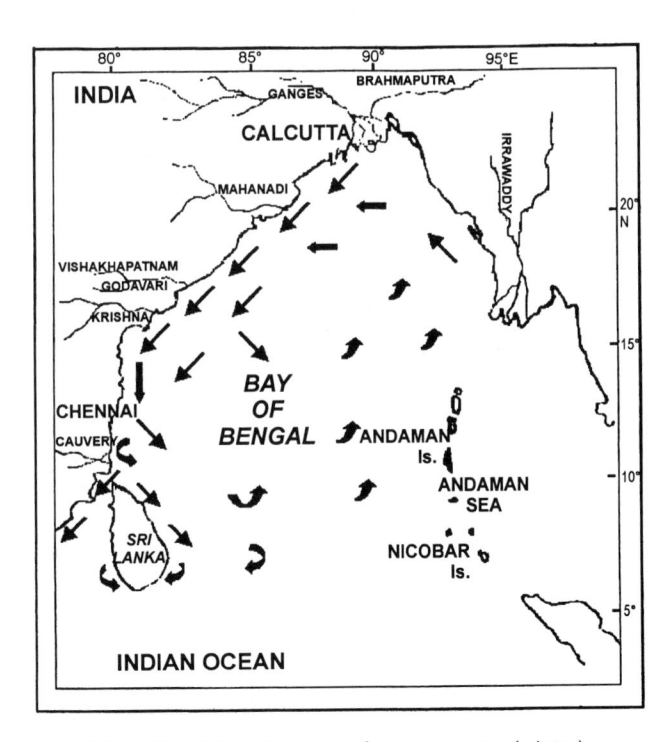

Fig. 4. Circulation during north east monsoon (winter).

(SHF) and latent heat (LHF) fluxes, Devi et al. (1994) showed that SHF of 40 W m^2 or more intensifies severe cyclonic storms. Good monsoon years have SHF of 20 W m^2 and poor years 10 W m^2 (Devi et al., 1994).

The east coast is influenced in a south to north sequence during the Southwest Monsoon and north to south during the Northeast Monsoon (Figs. 3 and 4). During the former, the Sri Lankan Coast will be influenced during February–

March, the Coramandel Coast during April, Orissa Coast during March–May and Calcutta Coast during April–July. A cyclonic gyre sets in at 5°N resulting in an eastward flow known as the Indian Monsoon Current (Molinari et al., 1990). At the head of the Bay the salinity remains as low as 26.0 psu due to freshwater influx, leading to a weak 2.0×10^6 m^3 s^{-1} northeast transport (Murty et al., 1992). The Northeast Monsoon passes along the east coast of India and Ekman pumping in the interior bay results in the East India Coastal Current (EICC) that flows southwards (Shetye et al., 1996). North of 15°, within 100 km from the coast, freshwater flow influences the circulation, and beyond this wind forcing is the dominant factor (Suryanarayana et al., 1992). During this monsoon the Calcutta, Orissa, Coramandel and Sri Lankan coasts will be influenced in September, October, November and December, respectively.

Numerical modelling studies (Potemra et al., 1991) agree with observations; temperature, salinity and flow data off the rivers Krishna, Godavari, Mahanadi and off New More Island during the Southwest Monsoon show south-southwest flows off Krishna and Godavari, whereas off Mahanadi it is north-northeast. Both flows are associated with gyres, clockwise off Mahanadi and anti-clockwise off Krishna and Godavari (Rao et al., 1987). Two anticyclonic gyres also develop in the central and southern parts of the Bay. The anticyclonic gyre during winter changes into a weak cyclonic gyre by late summer and is intensified into a western boundary current. Flow of river water results in a decrease in salinity and an increase in silicate, which is characteristic of these waters (Subba Rao, 1973).

Upwelling

During the Southwest Monsoon, a northerly current sets in and, due to Coriolis force, surface water is deflected to the right, away from the coast. As a result of Ekman transport, cooler subsurface waters upwell (LaFond, 1957; Gopalakrishna et al., 1996). In the southern part of the Bay the mean local temperature anomaly (LTA defined as the difference between coastal and mid-ocean sea surface temperatures) during the Southwest Monsoon is up to >2.0 due to upwelling (Naidu et al., 1999). However, in the northern Bay of Bengal it is negative (<–0.5) and so also is its correlation with Ekman transport ($r = -0.35, p = 0.05$). This is probably due to the presence of warm and freshwater discharges from the Brahmaputra and the Ganges (Naidu et al., 1999). The Hoogly and Mahanadi rivers (19°N) suppress coastal upwelling too (Sasamal, 1989; Shetye et al., 1991). Similarly, off the Godavari river, Johns et al. (1993) observed a marked cross-shelf spread of a plume of relatively low salinity water that weakens the local coastal upwelling.

Chemical Features

River discharge and the associated sediment and upwelling of bottom waters govern the chemistry of the Bay of Bengal. Mixed surface Bay waters (0–50 m) are saturated with oxygen, attributed to phytoplankton production (Satyanarayana et al., 1987; Sarma et al., 1988). Intermediate (100–500 m) waters have minimum oxygen suggesting active decomposition of organic matter (Satyanarayana et al., 1987). During the Southwest Monsoon, nutrients are brought up and tilt the isolines of oxygen-poor waters (Lafond, 1957; Sarma et al., 1988; Banse, 1990). Organic nitrogen and phosphorus were detected up to 3000 m deep indicating that supply of oxygen regulates oxidation processes in deeper waters (Sen Gupta et al., 1977). The rapidly sinking riverine matter contributes to biogenic silica that constitutes >80% of the total particulate silica (Gupta and Sarma, 1997). This reaches a maximum in mid depths of the euphotic layer, and its maxima deepens from north to south corresponding to the deepening of the thermocline and thickening of the surface layer. North of 15°N near the head of the Bay dissolved silicate levels were not high, suggesting that the contribution of the Ganges to the sea is insignificant (Rajendran et al., 1980).

Trace elements such as Ni, Zn, Cd, Cu, Mn, Pb and Fe are relatively higher in surface waters associated with low salinities than offshore (Satyanarayana and Murty, 1990). These grade from highest in the eutrophic harbour to lowest offshore (Satyanarayana et al., 1987).

COASTAL HABITATS AND BIODIVERSITY

The various biotopes in the Bay of Bengal offer a great biodiversity of fauna and flora (Table 4).

Table 4

Species diversity in the Bay of Bengal

Group	No. of species	Reference	Year
Diatoms	176	Subrahmanyan	1946
Dinoflagellates	108	Subrahmanyan	1971
Foraminifera	46	Reddy and Reddi	1994
Ciliates (Tintinnids)	47	Krishnamurty et al.	1979
Copepoda	58	Mishra and Panigraphy	1996
Hydromedusae	80	Santhakumari and Nair	1998
Mysidaceae	27	Tattersall	1922
Chaetognatha	19	Nair and Rao	1973
Euphausiaceae	22	Brinton and Gopalakrishnan	1973
Appendicularians	26	Fenaux	1973
Fouling organisms	54	Ganapati et al.	1958
Phytal fauna	80	Sarma and Ganapati	1975
Meiofauna	65	Sarma and Ganapati	1975
Sand dwelling fauna	30	Ganapati and Lakshmana Rao	1962
Coral reef fauna	120	Patterson and Ayyakkannu	1996
Bottom fauna	47	Ganapati and Lakshmana Rao	1962
Fish (commercially important)	45	James	1992

Phytoplankton Ecology

About 300 species of phytoplankton exist, mostly diatoms. Although tropical species dominate, a few temperate species such as *Chaetoceros affinae*, *Chaetoceros lorenzianum* also exist. There are at least eight floristic elements in the Bay (Table 5) (Thorrington Smith, 1971). The diatoms *Stephanopyxis palmeriana*, *Streptoptheca indica* and *Planktoniella sol* may be considered as indicators of upwelling off Visakhapatnam (Subba Rao, 1976).

Phytoplankton growth is bimodal, coinciding with the two monsoons. During summer, the crop is usually low but there are instances of a third, atypical non-seasonal bloom during June–August (Gopinathan et al., 1994). Observations in the Hoogly estuary (De et al., 1994) suggest exponential growth from February leading to a major peak during March. Off Visakhapatnam, initiation of phytoplankton growth coincides with upwelling (Subba Rao, 1974). The first signs of phytoplankton growth are in February in the 24–32 km zone, and it attains bloom proportions in the 0–16 km zone during early March. By April these blooms extend throughout the 0–32 km zone (Subba Rao, 1974). Heavy phytoplankton densities were reported in subsurface waters up to 122 m deep (Subba Rao, 1974). There is a time lag from north to south in the start of the maximum (Subba Rao, 1974; Gouda and Panigrahy, 1996; Gopinathan et al., 1994). Phytoplankton biomass is high and comparable to most productive coastal seas (Table 6) although in the northern Bay of Bengal it appears to be low (3.16 μg chl-a l^{-1}), probably because of the enormous river run-off.

Massive blooms of algae, mostly diatoms but sometimes with dinoflagellates, have been reported along the east coast (Murthy and Venkataramaiah, 1958; Mishra and Panigraphy, 1995; Subba Rao, 1969, 1973; Satpathy and Nair, 1996). Some are short lived and are a response to environmental perturbations such as dredging operations (Subba Rao, 1973), sudden lowering of temperature (Subba Rao, 1969) or dumping of fertilizer (Subba Rao, 1971), change in water quality (Satpathy and Nair, 1996) or due to organic pollution (Ganapati and Raman, 1979). Blooms cause red tides with 93.2×10^6 cells l^{-1} and 35.99 μg chl-a l^{-1} (Subba Rao, 1969). There were no fish kills or anoxic conditions associated with the algal blooms investigated, but the occurrence of euglenophyceae and cyanophyceae off river mouths and then in the brackish water lagoon at Chilka (Raman et al., 1990) flag a warning.

Primary production rates vary between 0.04–3852 mg C m^{-3} d^{-1} and 0.08–5.41 g C m^{-2} d^{-1} (Table 7). Annual primary production in coastal waters is about 250 g C m^{-2} while that of the coral reef waters off Mandapam and the Andamans has been measured at 2500 and 1200 g C m^{-2} (Nair, 1970). Surface production in the Bay is 4.9 t km^{-2} y^{-1} which is more than the 3.9 t km^{-2} y^{-1} for the Arabian Sea (Qasim, 1977). There is a significant increase from east to west (Pant, 1992). Qasim (1977) calculated that rivers contribute 33.6×10^6 tonnes of humus to the Bay of Bengal which would increase the phosphate in the upper 25 m by 0.03 μmol and nitrogen by 5.5 μmol. However, this enrichment and elevated production are limited to the top 25 m, unlike in the Arabian Sea. The calculated annual total column production for the Bay of Bengal is 394×10^6 t carbon compared to 1064×10^6 t for the Arabian Sea (Qasim, 1977).

Probably the enormous discharge of fresh water reduces total column production in the Bay of Bengal. Carbonate fluxes in the Bay of Bengal increase from north to south and seem to be controlled by input from rivers (Ittekkot et al., 1991). The spread of freshwater by the surface currents controls the total carbon dioxide and results in low pCO$_2$ (Dileep Kumar et al., 1996) and as a result the northwestern Bay of Bengal acts as a sink for atmospheric CO$_2$.

Zooplankton

Zooplankton in the northwestern bay is most productive in the Northeast Monsoon (Mathew et al., 1996). Eighty species of hydromedusae occur, of which several are limited to the east coast (Shanthakumari and Nair, 1998). Copepod

Table 5
Phytogeographical distribution of phytoplankton in the Bay of Bengal

Floral element	Taxa
South equatorial	*Ceratium pulchellum*, *Chaetoceros diversum*
Southern side of the equatorial under-current	*Ceratium vulture*
Centred on 5°S	*Ceratium tripos*, *Coscinodiscus lineatus*, *Grammatophora* sp., *Navicula* sp.,
Equatorial under-current	*Ceratium teres*
Equatorial subsurface	*Asteromphalus heptactis*, *Bacteriastrum hyalinum*, *Coscinodiscus marginatus*, *Rhizosolenia cylindrus*, *Rhizosolenia hebetata*, *Bacteriastrum delicatulum*, *Chaetoceros affinae*, *C. pendulum*, *C. peruvianum*, *C. lorenzianum*, *Coscinodiscus excentricus*, *Coscinodiscus lineatus*, *Nitzschia closterium*, *N. seriata*, *Planktoniella sol*, *Rhizosolenia alata*, *Thalassionema nitzschioides*
South-west monsoon	*Chaetoceros affine*, *Gossleriella tropica*, *Guinardia flaccida*, *Pleurosigma* sp. *Rhizosolenia alata gracillima*, *R. styliformis*, *Schroderella delicatula*,
Fluctuating conditions	*Ceratium axiale*, *Chaetoceros affine* var *circinalis*, *C. compressum*
Other West India Ocean	*Actinoptychus undulatus*, *Bacteriastrum comosum*, *B. varians*, *Ceratium buceros*, *C. carriense*, *C. declinatum*, *C. extensum*, *C. furca*, *C. fusus*, *C. macroceros*, *C. pentagonum*, *C. trichoceros*, *C. tripos*, *Chaetoceros affine*, *C. messanense*, *C. pelagicum*, *Coscinodiscus centralis*, *C. excentricus*, *C. oculis-iridis*, *Eucampia zoodiacus*, *Fragilaria oceanica*, *Gonyaulax* sp., *Oscillatoria* sp., *Rhizosolenia acuminata*, *R. alata*, *R. calcar-avis*, *R. hebetata*, *R. imbricata*, *R. osolenia setigera*, *R. stolterfothii*, *Thalassiothrix frauenfeldii*, *T. longissima*, *Triceratium alterans*, *Triceratium* sp.

Table 6

Phytoplankton biomass (Chl-*a*) in selected waters in the Bay of Bengal

Locality	Range mg m^{-3}	Range mg m^{-2}	Average mg m^{-2}	Reference	Year
Bay (Premonsoon)		1.0–34.0	8.1	Sarupria and Bhargava	1998
Bay (SW Monsoon)		1.1–93.2	14.8	Sarupria and Bhargava	1998
Bay			22.0	Sarupria and Bhargava	1998
N.E.Bay			23.2	Sarupria and Conkright	1998
Hoogly estuary	2.55–3.16			De et al.	1994
Chilka Lake	0.94–12.53			Raman et al.	1990
Gopalpur	0.19–17.22			Gouda and Panigrahi	1996
Tuticorin	1.2–11.8			Gopinathan et al.	1994
Chennai coastal waters	0.6–9.2			Nair	1990
Kalpakkam 12°33 N	<22.7			Satpathy and Nair	1996
Chennai backwaters	2.1–55			Nair	1990

Table 7

Primary production measurements in the Bay of Bengal

Locality	Range of production		Reference	Year
	mg C m^{-3} d^{-1}	g C m^{-2} d^{-1}		
Off Visakhapatnam	5–980	0.525–5.410	Subba Rao	1965
Visakhapatnam – Red water Dredging blooms	537–1609		Subba Rao	1969
Off Visakhapatnam	92–3380		Subba Rao	1973
Off Visakhapatnam	0.37–135.6	0.08–2.168	Ryther and Menzel	1964
Bay of Bengal	0.04–295.7	<3.495	Ryther and Menzel	1964
Bay of Bengal	2.1–44	0.19–0.63	Steeman Nielsen and Jensen	1957
Palk Bay	1.8–2342	1–8.68	Nair	1970
Gulf of Mannar	18–298		Raghu Prasad and Nair	1963
Vizinjam Bay	114–672		Jacob	1984
Vellar estuary	2940		Kawabata et al.	1993
Kalpakkam	36–3852		Nair	1990
Tuticorin	114–1600		Gopinathan et al.	1994
Shelf 16–20°N	74	2.9	Selvaraj and Srinivasan	1996
Slope 16–20°N	81	3.25	Selvaraj and Srinivasan	1996
Hoogly estuary	187–500		Bhunia and Choudhury	1982
Off Sagar Island	162–431		De et al.	1987
Bay	0.12–2988		Qasim	1977
Bay		0.3	Pant	1992

densities during the Southwest Monsoon are high (901 m^3) (Stephen, 1992). While some species avoid low salinity waters, members of the Centropagidae and Pontellidae seem to flourish in them.

Zooplankton biomass attains high densities, the maximum (92.25 cc10^6 l sea water) being around Andaman and Nicobars (Mathew et al., 1990), but in oceanic areas it is 36.55 ml in a million litres sea water. Maximum production is from July through October when 60–80% of the total occurs on the shelf. The balance between phytoplankton and zooplankton (Cushing, 1959) becomes overturned during periods of drastically lowered salinity, high temperature or lack of phytoplankton (Subba Rao, 1974).

In the southern region where hydrographical conditions are fairly stable, meroplankton and fishery production go hand-in-hand (Krishnamoorthy et al., 1999). In the northwestern Bay of Bengal peak zooplankton abundance coincides with maximum fishery landings (Mathew et al., 1996). Estimates of secondary production based on zooplankton biomass range between 0.5–5.98 g C m^{-2} y^{-1} on the East coast, and the maximum of 10.47 g C m^{-2} y^{-1} was around the Andaman and Nicobar Islands (Mathew et al.,

1990). Assuming a 10% conversion efficiency from secondary to tertiary production, the calculated annual fish production for the Bay of Bengal, and Andaman Nicobar Sea is 1.38 and 1.39 million tonnes (Mathew et al., 1990).

Macrobenthos and Meiobenthos

Of 29 major taxa, polychaetes are the most dominant, followed by bivalves and amphipods (Ganapati and Lakshmana Rao, 1962; Harkantra et al., 1982; Jegadeesan and Ayyakkannu, 1992). Maximum population density and biomass were 12,572 individuals m^{-2} (mean 839) and 150.6 g m^{-2} (mean of 10.61 g m^{-2}), comparable with that of the west coast of India. Diversity decreases with increasing depth. Based on biomass, Harkantra et al. (1982) calculated an annual benthic production of 21.22 g m^{-2}, an organic carbon equivalent of 1.805 g C m^{-2} y^{-1}.

Meiofaunal abundance is affected by the annual cycles (Suresh et al., 1992; Rao, 1994), as well as by discharges of sewage. Groups vary with the monsoons; nematodes dominate during the post Southwest Monsoon, foraminiferans in the Southwest Monsoon and ostracods during the summer (Sunita Rao and Rama Sarma, 1990). Sandy interstitial habitats at Gopalpur are dominated by harpacticoid copepods (70%) followed by nematodes (Patnaik and Lakshmana Rao, 1990) with faunal densities of 323–525 cm^{-2} during March–May. In mangrove systems, the same groups were positively correlated with high organic matter (Sarma and Wilsanand, 1994). Offshore, varied meiofaunal densities have been reported (Rudnick et al., 1985; Patnaik and Lakshmana Rao, 1990; Suresh et al., 1992).

Mangroves

In estuaries the mangroves are mostly fringing. The largest area (4262 km^2) is the Sunderbans, off Sagar Island, near Hoogly River. *Rhizophora mucronuta*, *R. apiculata*, *Avicennia officianalis*, *A. marina*, *Ceriops* sp., *Exoecaria agallocha*, *Acanthus ilicifolius*, and *Acristucyn* sp. are the most dominant of the 59 mangrove species. Near the water's edge and with the highest plant density, *Rhizophora*, *Bruguiera*, *Sonneratia* and *Avicennia* dominate, with *Ceriops decandra*, *Aegiceros* sp. and *Clarodendrum* sp. in the middle of the forest. *Suaeda* sp. and *Lumnitzera* sp. grow landward. Mangroves are a natural coastal green belt which offers protection from cyclones, storms, winds and tidal excursions. They are highly productive and the litter fall beneath *Avicennia marina* of the Sunderbans is 1603 g m^{-2} y^{-1} dry weight (Ghosh et al., 1990). Detritus from the mangrove litter is a food for a variety of organisms, and commercially important crustaceans, fishes and shellfish use mangroves as spawning grounds. Additionally, mangroves are a source of fire wood, construction material, tannin for the leather industry, and the leaves are fodder for cattle and are used in herbal medicines. Increased exploitation, however, exerts severe stresses on the mangrove ecosystem.

Coral Reefs

About 20 coral islands cover 683 h in the southern part of the coast. There are 120 coral species of which 110 are hermatypic and 10 ahermatypic (Patterson and Ayyakkannu, 1996). *Montipora* and *Acropora* constitute 39% of the species. Most are fringing reef corals and provide important biological resources like finfishes, shell fishes and seaweeds. Although corals are fragile and critical ecosystems, there has been active destruction due to indiscriminate use as raw material for construction of houses, carbide and the white cement industry, and they are affected by pollution, aquaculture, fishing and tourism.

MARINE FISHERIES

In the EEZ the fish fauna includes 242 species of 87 families (Balachandran and Nizar, 1990). On the east coast, mackerel, sardines, anchovies, catfish, other clupeoids, coastal tunas, carangids, seer fish, ribbon fish, and pelagic sharks constitute the major pelagic fisheries (Sudarsan, 1993). The identity of species in landings varies regionally (Raja and Philipose, 1977). Annual fish catches on the east coast are 1.762×10^6 t of which 0.656×10^6 t are demersal and 1.106×0^6 t are pelagic. Potential fish stocks off the east coast EEZ are estimated to be 2.180×10^6 t (Somvanshi, 1998).

Demersal finfish are an important resource off Orissa and west Bengal (John and Sudarsan, 1990) (Table 8) and scope exists for increasing the catch, particularly offshore (Devraj et al., 1996). Cyclones which often hit Orissa adversely affect gillnetting (Pati, 1982).

On the east coast of India, nearly 60% of the total catch is of demersal finfish, nearly 70% of which is from depths >50 m (Vijayakumaran and Naik, 1991) and dominated by more than 20 carangid species (Sivakami et al., 1996). The highest densities were in the 60–80 m depth range, yielding about 150 kg h^{-1}. Also viable is a long-line tuna fishery, particularly during December to April (John et al., 1988). The tuna catch is particularly good between 65 and 120 m depth which is the boundary between the surface uniform layer and the sharp gradient layer of temperature and dissolved oxygen (Kurita et al., 1991). This is the shallowest tuna fishing in the world and Morinaga et al. (1992) attribute this to the location of a subsurface dissolved oxygen minimum

Table 8

Changes in fishing operation between 1961 and 1991

Mechanised vessels	1228 (1961)	12,223 (1991)
Potential fishing area	53 (inshore)	4,261 (1991)
Fish production	0.187 (1960–64)	0.634 (1990–94)
Potential yield for the continental shelf		1.5×10^6 t
Trawler catch		48.8%

layer. Altogether, estimated potential stocks in the Bay of Bengal EEZ are 1.106×10^6 t fish stocks, of which 0.0656×10^6 t are demersal (Somvanshi, 1998).

Brackish Water Fisheries

On the east coast off India there are several brackish water areas with a potential for fish farming. In West Bengal examples are the Hoogly estuary, in Orissa, Chilka Lake (790 km^2), in Andhra Pradesh several areas totalling about 0.2×10^4 km^2, and in Tamilnadu, Edaiyur and Lake Pulicat a total area of 461 km^2. In Lake Pulicat, the second largest lagoon of India, grey mullets (*Mugil cephalus* and *Liza macrolepis*) were cultured without supplementary feeding, yielding 617 kg ha^{-1} in 7 months in pens and 556 kg ha^{-1} in 10.5 months in cages (Prasadam and Kadir, 1988). Lake Pulicat supports large edible crabs and three species of penaeids, *Penaeus indicus, P. monodon* and *P. semisulcatus* which yield a total of about 500 tonnes annually (Sanjeevaraj, 1981). Potential also exists for culturing molluscs, particularly the clam *Meretrix casta* (Thangavelu and Sanjeevaraj, 1985). The green mussel *Mytilus viridis* occurs along the east coast and is ideally suited for rope culture because it tolerates wide fluctuations in salinity (Qasim et al., 1977). Annually three harvests are possible, yielding 181% return on investment.

Approximately 26% of the brackish water is under cultivation, with an annual yield of 0.2 million tons of shrimp. To meet demand, a potential annual production of 5.6 million tons of shrimp is required, though there is a targeted estimated production of 8 million tons. Of ten species of shrimp, the Tiger shrimp, Indian white shrimp and the Green tiger shrimp are cultured intensively, and requirements for shrimp seed tripled from 2.2 billion in 1992–93 to 6.8 billion in 1996–97.

POPULATIONS AFFECTING THE AREA

With increased industrialization there is rapid urbanization in certain regions of the East coast of India. The coastal population is about 55×10^6 and growing. Very few monitoring stations specifically for the marine environment exist here. The Central Pollution Control Board (CPCB) has declared North Arcot (Tamil Nadu), Visakhapatnam and Patancheru (Andhra Pradesh), Talcher (Orissa) and Howrah (West Bengal) as critically polluted from numerous industries. As an example, Tamilnadu has a 100-year-old tanning industry with 573 tanneries. More than 700 tonnes or 40% of the country's leather is processed, generating 25 tonnes of solid wastes and over 30 million l of effluents daily. The warning that they are an environmental threat came as early as in 1939. Some tanneries are trying to reduce the total dissolved salt to 4000 mg l^{-1} in the effluent, almost twice the permissible amount (2100 mg l^{-1}) set by the World Health Organization.

Coming from numerous sources, there are no precise estimates of the quality and quantity of pollutants entering the Bay of Bengal and baseline studies do not exist for this region. Strict enforcement of environmental laws becomes secondary when most locals are dependent on the local industries. Yet several coastal cities, particularly harbours, are experiencing environmental problems due to unrestricted disposal of treated or partially treated sewage, leading, on occasion, to the formation of H_2S and anoxia.

RURAL FACTORS

Shrimp Farming and its Collapse

On the east coast of India several areas are suitable for shrimp farming. In Andhra Pradesh with a 974-km coastline shrimp farming was modest around 1980. By 1990–91, 6000 hectares were used for shrimp farming which increased to 34,500 ha by 1994–95. Peak yield was 82,850 tonnes in 1993, fetching about $100 million. About 70,000 aquaculturists were employed on 200 farms, ranging from 0.4 ha to several hectares. It was hoped to generate $300 million by the turn of the century; the farm-raised shrimp harvest, often referred to as "dollar crop", constituted 30% of shrimp exported. Projected exaggerated claims are to increase the shrimp farms to 1.5 million ha in Andhra.

Big farms are intensive or semi-intensive monoculture farms. They are cement-lined, and their wastewater often cross-contaminates farms downstream. Shrimp are grown along with Roho, and Catla, though recently introduced Chinese Carp are taking over and are becoming a pest. In 1995, 76,589 ha in Andhra were rushed into shrimp culture, on fertile land traditionally used for cultivation of rice. However, during 1995–96 several shrimp farms were closed due to diseases. Nearly twenty viral shrimp diseases caused catastrophic outbreaks. Bacterial and fungal diseases, specifically White Spot Disease (WSD) and Brown Spot Disease (BSD) resulted in a heavy loss. Lack of a holistic scientific approach and severe lapses in the management policy led to the collapse of shrimp farming in Andhra. Cost-benefit analysis did not include the cost of environmental damage.

Heavy Metals, Pesticides

Hoogly River, a branch of the Ganges, is the most polluted in India. Large quantities of heavy metals particularly Hg, Cd and Pb enter, of which less than 9% are precipitated in the estuarine region, 45–50% at the river mouth, leaving 40% which reach the Bay (Qasim, 1998). Metal "hot spots" exist (e.g. Sasamul et al., 1987). Ganges and Brahmaputra river sediments also contain barium and radium (Carroll et al., 1993). The uranium in the Hoogly estuary ranges between 3.5 and 3.9 μg l^{-1}, the highest for any estuary (Somayajulu, 1994). At the sediment–water interface very likely 25% of the uranium in the estuary is removed into sediments (Somayajulu, 1994). The weighted mean uranium input from Mahanadi estuary is 36 t y^{-1} out of which 18.3% is in dissolved form (Ray et al., 1995). Most of the uranium input takes place during the NE monsoon season.

Persistent organochlorine residues occur in sediments and biota. Despite a ban on its usage, total DDT levels in the sediments ranged between 0.02 and 720 ng g^{-1} (Shailaja and Sarkar, 1992). Total DDT in zooplankton from Vellar estuary ranged from 1.2 to 47 ng g^{-1} (Rajendran et al., 1990) and from 0.04 to 2.38 ng g^{-1} in 14 species of fishes (Babu Rajendran et al., 1992). Assuming 2% of wet weight as dry weight, in the bottom-feeding fish total DDT ranged from 0.026 to 2.318 ng g^{-1} (Shailaja and Singbal, 1994). In blubber samples from dolphins netted in coastal waters of Porto Novo, total DDT was between 0.09 and 3.3 ng g^{-1} (Tanabe et al., 1993).

MINING, EROSION AND LAND FILL

Along the East coast of India, loss of crucial coastal habitats is a serious problem. A case in point is Chilka Lake—Asia's largest brackish water lagoon. Chilka Lake is a national concern because of heavy siltation, macrophyte growth and shrinkage. Nearly 100,000 fisher-folk earn a livelihood from Chilka Lake (Lakshmana Rao et al., 1994). During 1978 and 1987 more than 60 papers were published on the potential for aquaculture (Satpathy, 1992; Lakshmana Rao et al., 1994), but sadly lacking were holistic studies on the functioning of the ecosystem and its management. Since its origin around 3500 years BP, Chilka Lake has suffered a 50% reduction up to the beginning of the present century. A further reduction of about 255 km^2 took place between 1929 and 1988 due to soil erosion caused by active deforestation (Hema Malini et al., 1993). The estimated annual silt input is 30×10^6 tonnes contributed by alkaline soils of Daya and Bhargavi rivers (Mangala, 1989). In the 18 km stretch of shore between Gopalpur and Rushikulya river mouth, 2900 ha of sand dunes contain an estimated 230×10^6 t of heavy minerals, one of the world's largest deposits. The Orissa Sand Complex (OSCOM) has designed mining operations of 500 t h^{-1}, to produce annually 220,000 t ilemnite, 10,000 t rutile, 2000 t zircon, 4000 t monazite and 30,000 t sillimanite (Lakshmana Rao et al., 1994). The *Casuarina* and cashew plantations along the sand dunes prevent the migration of sand to agricultural fields, residential areas and the national highway.

Remote sensing studies have shown that 68% of the mangroves in Orissa state are in this area but that the mangals are getting depleted at a rate of 2 km^{-2} y^{-1} (Sarma and Wilsanand, 1994). In 1987 alone, 2000 ha of pristine mangrove land was brought under prawn culture, with an export target of 1500 tonnes. Areas up to 200 ha in one site were banded for reclamation of mangroves for shrimp culture, but ironically this failed due to acidification of the soil (Hatcher et al., 1989).

EFFECTS FROM URBAN AND INDUSTRIAL ACTIVITIES

Desalination Plants

To supply fresh water to villagers, several desalination plants have been built which use reverse osmosis to create 50–100 thousand m^3 fresh water each day (Qasim, 1998). Ironically in India—the land of rivers—clean potable water has become a scarce commodity. The effluent from desalination plants is discharged via an earthen canal to the sea. Salinity has increased in ground water and several wells due to seepage of brine (Rao et al., 1990).

Ocean Thermal Energy Plants

Due to chronic power deficiency, the Department of Ocean Development, Government of India has commissioned an Indian wave energy programme at Chennai. A 10 m wide module operates on the Oscillating Water Column principle, generates 150 kW (Ravindran, 1994). Under the Indian Ocean Thermal Energy Conversion (OTEC) Programme in collaboration with Sea Solar Power, USA, several thermal power plants (100 MW) probably will operate. The impact of thermal pollution on the biota needs to be evaluated, however, and thermal outfall from Ennore thermal power station near Chennai has had profound impacts on the hydrobiological conditions (Subrahmanian et al., 1990). It has elevated mean levels of temperature, nutrients and BOD and decreased pH and gross production.

Nuclear Power Station

For cooling purposes Kalpakkam Atomic Power Station near Madras uses coastal water which is chlorinated to 0.2 g l^{-1} to control biofouling. *Perna virisdis, Balanus reticulatus, Modiolus striatulus* still seem to cause severe blockage (Thiyagarajan et al., 1997), and to control this, shock chlorination is done that leaves 1.10 to 1.50 mg l^{-1} residual chlorine which adversely reduces (80–83%) primary production (Ahmed et al., 1993).

Mechanised Craft

On the East coast, mechanized trawling and bottom gillnets reduced catfish catches from 43.7 tonnes in 1972 to 17 tonnes by 1980–81 (Lakshmi and Srinivasa Rao, 1992). *Arius tenispinis* which once contributed 61% of the catfish catch disappeared by 1984 (Roy, 1997). For the collective good, the fishing communities recently banned the use of a new kind of fishing gear, a snail net, in spite of its profitability (Bavinck, 1996).

Shipping Impacts: Exotic Biota Introductions

Faunal surveys in the Indian seas, initiated by the British naturalists in 1878 and subsequently continued by local researchers, have made the fauna and flora in the Bay of

Bengal quite well known. However, in recent years several species were reported to occur for the first time (Subba Rao et al., unpublished). These include the polychaete *Sigambara tentaculata* (Achari, 1975) and the pelecypod *Mytillopsis sallei* (Rao and Rao, 1975) which now seem to flourish. These have fully established themselves in Visakhapatnam harbour where they are pests. The commercially important wood-boring isopods *Limnoria insulae, L. unicornis, L. platycauda* (Ganapati and Lakshmana Rao, 1960) and the pelycypod *Xylophaga mexicana* (Ganapati and Lakshmana Rao, 1961) are also new records from the Bay of Bengal. Some of these taxa occur in the West Atlantic, Pacific, Hawaii and the Mediterranean, Madagascar, Malayan Archipelago and could be considered truly exotic.

These may have been introduced via ballast water. Since 1947 (post-independence) the nation has been advancing industrially, and with it the sea trade. It was even necessary to construct a new port, Paradeep, to meet shipping demands. Besides ballast water and sediments, other possible dispersal vectors of exotic organisms are the scrapings of fouling organisms resulting from maintenance and dry-docking operations. There are several ecological impacts of these invaders. They may out-compete the native fauna and flora for resources, colonize, and cause shifts in populations. This in turn will cause changes in the trophodynamics of the ecosystems, leading to elimination of desirable species and the newly introduced pests may prevail, as in the temperate seas (Carlton and Geller, 1993; Subba Rao et al., 1994).

Marine Pollution

Several rivers contribute to the pollutants in the Bay of Bengal; its coastal zone is used as a receptacle for a variety of pollutants i.e. domestic sewage, agricultural wastes and pesticides, petroleum hydrocarbons and PCBs, and heavy metals. Nearly 357 MT of Gulf oil are transported along the EEZ of India, around Sri Lanka, southern Bay of Bengal to the Far East and Japan (Qasim, 1998). In the Bay of Bengal, north of 10°N, during 1978 to 1980, 61 oil slicks were sighted and floating tar ranged between 0–69.75 mg m^{-2} with a mean of 1.52 mg m^{-2} (Qasim, 1998). Besides bilge washings, damaged oil tankers contribute to large oil slicks and floating tar in the Bay of Bengal (Sen Gupta et al., 1995). Due to an oil spill from the Danish VLCC Maersk Navigator, in the Andaman Sea, dissolved petroleum hydrocarbon levels ranged between 0.31 and 1.85 μg l^{-1} (Sen Gupta et al., 1995).

PROTECTIVE MEASURES

Central Pollution Control Boards (CPCB) and State Pollution Control Boards are responsible for enforcement of Acts for Prevention and Control of Pollution of Water (1974, 1977) and Environmental Protection Act (EPA, 1986). Laws for the prevention and abatement of pollution were adopted in 1992. Guidelines for technical support, training awareness programmes, waste minimization measures and promotion of clean technologies are provided. During 1997–98, 332 EPA projects were reviewed, 186 projects were granted site clearance and those concerned with water were mostly with fresh water resources. Under the Coastal Ocean Monitoring and Prediction Systems (COMAPS) along the Indian Coast 77 locations are now monitored. Of these, 32 are declared "hot spots" and were referred to State Pollution Control Boards. The Andaman and Nicobar Center for Ocean Development (ANCOD) is set up to monitor coastal ocean pollution, coral reef rejuvenation and enhancement of marine living resources through sea ranching and sea farming.

Public Awareness

Some of the developmental activities have resulted in environmental and socio-economic problems which have become issues of public concern and protest. In a few rare cases, regulatory measures were developed, for example the successful halting of the construction of a fishing jetty near Orissa which could have been a threat to the Olive Ridley sea turtle. About 50,000 Olive Ridleys (*Lepidochelys olivacea*) use 35 km of this sandy beach during December–January and March–April for nesting. These turtles enjoy protection under IUCN. Also, a major prawn farming venture in the periphery of Chilka Lake that would have necessitated clearance of prime mangrove forest was halted due to public pressure.

Diseases associated with shrimp farming on the East coast have cast long socio-economic shadows. Agitation by affected shrimp fishermen, championed by environmentalists in Tamil Nadu, Andhra Pradesh and Orissa, culminated in a few interim orders by the Indian Supreme Court on 11 December 1996. These include: (a) no shrimp farm may be placed within 500 m from high tide mark; (b) traditional and improved farmers are exempted; (c) any farm within 500–1000 m zone should be demolished; (d) a Government authority will enforce preparatory principles and the "polluter pays" principle; (e) fixed compensation for environmental damage, and (f) retrenched workers shall be compensated. Consequences of the above judgement are far reaching.

NEW MILLENNIUM: NEED FOR EAST COAST ZONE MANAGEMENT AUTHORITY

The key requirement for the management of the Bay of Bengal coastal waters is massive education of the public, mitigation measures and the need for conservation of resources and sustainable development. Various user agencies, advisory groups, joint task forces, and recognized NGOs should publicize that no coastal project will succeed without the willing participation and cooperation of the coastal communities. Cyclone and flood disaster mitigation is of prime importance. The large-scale alteration of the coastline due to human activity and the monsoons will raise the sea-level which will have serious impact. This expected sea-level rise (SLR) will submerge low-lying tracts of the Sunderbans and Chilka Lake. It may also enhance the frequency and magnitude of damaging storm surges, erosion of sandy shores away from river mouths, spread of saline waters in low-lying coastal lands and other effects of economic importance. Mangroves may flourish under rising sea level, but their wanton destruction in recent years may deprive the coast of defence against storms (Lakshmana Rao et al., 1994). The constitution of a coastal zone management authority and development of a regional scientific program to study the functioning of the ecosystems are essential.

DEDICATION

I dedicate this paper to the memory of the late M.V. Lakshmana Rao, my good friend, philosopher and a naturalist with a commitment to the coastal environment.

REFERENCES

Achari, G.P.K. (1975) Studies on new or little known polychaetes from Indian Seas. 4. On a new record of *Sigambra tentaculata* (Treadwell) (Pilargidae), from the southwest coast of India along with observations on its early larval stages. *Journal of the Marine Biologists Association of India* 17, 238–241.

Ahmed, M.S., Suresh, K., Durairaj, G. and Nair, K.V.K. (1993) Effect of cooling water chlorination on primary productivity of entrained phytoplankton at Kalpakkam, east coast of India. *Hydrobiologia* 271, 165–168.

Ali, A. (1995) A numerical investigation into the back water effect on flood water in the Meghana river in Bangladesh due to the south-west monsoon wind. *Estuaries, Coastal and Shelf Science* 41, 689–704.

Babu Rajendran, R., Karunagaran, V.M., Babu, S. and Subrahmanian, A.N. (1992) Levels of chlorinated insecticides in fishes from the Bay of Bengal. *Marine Pollution Bulletin* 24, 567–570.

Balachadran, K. and Nizar, M.A. (1990) A check list of fishes of the Exclusive Economic Zone of India collected during the research cruises of FORV Sagar Sampada. *Proceedings of the First Workshop on Scientific Results of FROV Sagar Sampada*, ed. K.J. Mathew, pp. 305–324. India Central Marine Fisheries Research Institute, Cochin.

Banse, K. (1990) Remarks on oceanographic observations off the east coast of India. *Mahasagar* 23, 75–84.

Bavinck, M. (1996) Fishery regulations along the Coramandel coast: a case of collective control of common pool resources. *Marine Policy* 20, 475–482.

Bhunia, A.B. and Choudhury A. (1982) Primary production of the estuarine water around Sagar Island, Sunderbans. *Indian Journal of Marine Sciences* 11, 87–89.

Brinton, E. and Gopalakrishnan, K. (1973) The distribution of Indian Ocean Euphausids. In *The Biology of the Indian Ocean*, eds. B. Zeitzschel and S.A. Gerlach, pp. 357–382. Springer-Verlag, New York.

Carlton, J.T. and Geller, J.B. (1993) Ecological roulette: The global transport of nonindigenous marine organisms. *Science* 261, 78–82.

Carrol, J., Falkner, K.K., Brown, E.T. and Moore, W.S. (1993) The role of Ganges–Brahmaputra mixing zone in supplying barium and radium-226 to the Bay of Bengal. *Geochimica et Cosmochimica Acta* 57, 2981–2990.

Chandramohan, P., Sanil Kumar, V. and Nayak, B.U. (1993) Coastal processes along shorefront of Chilka lake, east coast of India. *Indian Journal of Marine Sciences* 22, 268–272.

Chatterji, A., Ansari, Z.A., Mishra, J.K. and Parulekar, A.H. (1995) Seasonality in meiofaunal distribution on a tropical beach at Balramgari, northeast coast of India. *Indian Journal of Marine Sciences* 24, 49–55.

Chauhan, O.S. (1992) Sediment dynamics at Puri and Konark beaches, along northeast coast of India. *Indian Journal of Marine Sciences* 21, 201–206.

Chauhan, O.S. and Gujar, A.R. (1996) Surficial clay mineral distribution on the southwestern continental margin of India: evidence of input from the Bay of Bengal. *Continental Shelf Research* 16, 321–333.

Chauhan, O.S., Borole, D.V., Gukar, A.R., Mascarenhas, A., Mislanker, P.G. and Rao, Ch.M. (1993) Evidence of climatic variations during Late Pleistocene–Holocene in the eastern Bay of Bengal. *Current Science* 65, 558–562.

Cushing, D.H. (1959) On the nature of production in the sea. *Fish. Invest.Lond. Series II* 22, 1–40.

Das, J. and Sahoo, K. (1996) Influence of fresh water influx on the ratio of Ca, Mg and F with respect to Cl in the coastal water of Orissa, east coast of India. *Indian Journal of Marine Sciences* 25, 74–77.

De, T.K., Choudhury, A. and Jana, T.K. (1994) Phytoplankton community organization and species diversity in the Hugli estuary, north east coast of India. *Indian Journal of Marine Sciences* 23, 152–156.

Devi, K.V., Reddy, K.G. and Rao, G.R.L. (1994) A study on the energy flux in relation to meteorological systems over the Bay of Bengal during Southwest Monsoon. pp. 347–356 In *Ocean Technology: Perspectives*, eds. Sushilkumar, V.V. Agadi, V.K. Das, B.N. Desai. Publ. Inf. Dir., New Delhi, India.

Devraj, M., Raj, R.P., Vivekanandan, E., Balan, K., Sathiadhas, R. and Srinath, M. (1996) Coastal fisheries and aquaculture management in the east coast of India. *Mar. Fish. Inf. Tech. Ext. Ser.* 141, 1–9.

Dileep Kumar, M., Naqvi, S.W.A., George, M.D. and Jayakumar, D.A. (1996) A sink for atmospheric carbon dioxide in the northeast Indian Ocean. *Journal of Geophysical Research Oceans* 101, 18121–18125.

Ewing, M., Eittreim, S., Truchan, M. and Ewing, J.I. (1969) Sediment distribution in the Indian Ocean. *Deep-Sea Research* 16, 231–248.

Fenaux, R. (1973) Appendicularia from the Indian Ocean, the Red Sea and the Persian Gulf. In *The Biology of the Indian Ocean*, eds. B. Zeitzschel and S.A. Gerlach, pp. 409–414. Springer-Verlag, New York.

Ganapati, P.N. and Lakshmana Rao, M.V. (1960) On some crustacean wood-borers from Andamans. *Current Science* 29, 275–276.

Ganapati, P.N. and Lakshmana Rao, M.V. (1961) The marine woodborer *Xylophaga* from Bay of Bengal. *Current Science* 30, 464–465.

Ganapati, P.N. and Lakshmana Rao, M.V. (1962) Preliminary observations on the bottom fauna of the continental shelf of the north-east coast of India. *Proc. First All-India Congress of Zoology, Part 3*. The Zoological Society of India, pp. 8–13.

Ganapati, P.N. and Raman, A.V. (1979) Organic pollution and *Skeletonema* blooms in Visakhapatnam harbour. *Indian Journal of Marine Sciences* 8, 184–187.

Ghosh, P.B., Singh, B.N., Chakravarthy, C., Saha, A., Das, R.L. and Choudhury, A. (1990) Mangrove litter production in a tidal creek of Lothian island of Sundarbans, India. *Indian Journal of Marine Sciences* 19, 292–293.

Gopalakrishna, V.V., Pednekar, S.M. and Murty, V.S.N. (1996) T-S variability and volume transport in the central Bay of Bengal during Southwest Monsoon. *Indian Journal of Marine Sciences* 25, 50–55.

Gopinathan, C.P., Rodrigo, J.X., Kasim, H.M. and Rajagopalan, M.S. (1994) Phytoplankton pigments in relation to primary production and nutrients in the inshore waters of Tuticorin, Southeast coast of India. *Indian Journal of Marine Sciences* 23, 209–219.

Gouda, R. and Panigrahy, R.C. (1996) Ecology of phytoplankton in coastal waters off Gopalpur, east coast of India. *Indian Journal of Marine Sciences* 25, 81–84.

Government of India (1999) *India: A Reference Manual*. Publications Division, Ministry of Information and Broadcasting, Government of India, 773 pp.

Gupta, V.M. and Sarma, V. (1997) Biogenic silica in the Bay of Bengal during the southwest monsoon. *Oceanologica Acta* 20, 493–500.

Harkantra, S.N., Rodrigues, C.L. and Parulekar, A.H. (1982) Macrobenthos of the shelf off north eastern Bay of Bengal. *Indian Journal of Marine Sciences* 11, 115–121.

Hatcher, B.G., Johannes, R.E. and Robertson, A.I. (1989) Review of research relevant to the conservation of shallow tropical ecosystems. *Oceanography and Marine Biology Annual Reviews* 27, 337–414.

Hema Malini, B., Srinivasa Rao, K. and Nageswara Rao, K. (1993) Evolution and dynamics of the Chilka Lake. In *The Management of Coastal Lagoons and Enclosed Bays*, eds. J. Sorensen, F. Gable and F. Bandarin. American Society of Civil Engineers, New York, pp. 257–268.

Ittekkot, V., Nair, R.R., Honjo, S., Ramaswamy, V., Bartsch, M., Manganini, S. and Desai, B.N. (1991) Enhanced particle fluxes in Bay of Bengal induced by injection of fresh water. *Nature* 351, 385–387.

Jacob, R.M. (1984) Primary productivity in the nearshore waters of Vizhinjam. *Journal of the Marine Biological Association of India* 26, 66–70.

James, P.S.B.R. (1992) The Indian marine fisheries resources scenario—past, present and future. *Indian Journal of Fisheries* 39, 1–8.

Jegadeesan, P. and Ayyakkannu, K. (1992) Seasonal variation of benthic fauna in marine zone of Coleron estuary and inshore waters, south east coast of India. *Indian Journal of Marine Sciences* 21, 67–69.

John, M.E. and Sudarsan, D. (1990) Marine fishery resources off Orissa, west Bengal Coast. *Bulletin Fish. Surv. India* 19, 1–50.

John, M.E., Patil, S.M. and Somvanshi, V.S. (1988) Tuna resources off the east coast of India as revealed by charter operations. *Bulletin Fish. Surv. India* 17, 22–48.

Johns, B., Rao, A.D., Dube, S. and Sinha, P.C. (1993) The effect of freshwater discharge from the Godavari river on the occurrence of local upwelling off the east coast of India. *Estuaries, Coastal and Shelf Science* 37, 299–312.

Kawabata, Z., Magendran, A., Palanichamy, S., Venugopalan, V.K. and Tatsukawa, R. (1993) Phytoplankton biomass and productivity of different size fractions in the Vellar estuarine system, southeast coast of India. *Indian Journal of Marine Sciences* 22, 294–296.

Krishnamurty, K., Damodara Naidu, W. and Santhanam, R. (1979) Further studies on tintinnids (Protozoa: Ciliata). *Archiv fuer Protistenkunde* 122, 171–180.

Kumar, M.D., George, M.D. and Sen Gupta, R. (1992) Inputs from Indian rivers to the ocean: A synthesis. *Oceanography of the Indian Ocean*, ed. B.N. Desai. Oxford and IBH Publishing Pvt. Ltd., New Delhi.

Kurita, Y., Saotome, Y., Kasuga, I. and Hayashi, T. (1991) Relation between tuna catches and oceanic condition in Bengal Bay. *Bulletin of the Japanese Society of Fish. Oceanogr.* 55, 18–24.

LaFond, E.C. (1957) Oceanographic studies in the Bay of Bengal. *Proceedings of the Indian Academy of Science* B46, 1–46.

Lakshmana Rao, M.V., Mahapatra, K. and Subba Rao, D.V. (1994) The coastal zone of Orissa, Bay of Bengal: Threats and prospects for sustainable development. In: pp 304–319. *Coastal zone Canada '94. Cooperation in the Coastal Zone: Conference Proceedings, Vol. 1*, eds. P.G. Wells and P.J. Ricketts. Coastal Zone Canada Association, Bedford Institute of Oceanography, Dartmouth, Nova Scotia, Canada.

Lakshmi, K. and Rao, K.S. (1992) Trends in catfish catches at Visakhapatnam showing disappearance of *Arius teuispinis*. *Journal of the Marine Biologists Association of India* 34, 26–37.

Mangala, B. (1989) Chilka Lake: desilting Asia's largest brackish water lagoon. *Ambio* 18, 298–299.

Mathew, K.J., Naomi, T.S., Antony, G., Vincent, D., Anilkumar, R. and Solomon, K. (1990) Studies on zooplankton biomass and secondary and tertiary production of the EEZ of India. *Proc. First Workshop Scient. Resul. FORV Sagar Sampada*, 5–7 June, 1989. pp. 59–69.

Mathew, K.J., Kurup, K.N., Naomi, T.S., Antony, G. and Solomon, K. (1996) Zooplankton abundance in the continental shelf waters of the northeast coast of India. *Proc. Second Workshop Scient. Resul. FORV Sagar Sampada*, eds. V.K. Pilai, S.A.H. Abidi, V. Ravindran, K.K. Balachandran and V.V. Agadi. Dept. Ocean Development, New Delhi, pp. 257–269.

Mishra, S. and Panigrahy, R.C. (1995) Occurrence of diatom blooms in Bahuda estuary, east coast of India. *Indian Journal of Marine Sciences* 24, 99–101.

Mishra, S. and Panigrahy, R.C. (1996) Copepods of Bahuda estuary (Orissa), east coast of India. *Indian Journal of Marine Sciences* 25, 98–102.

Molinari, R.L., Olson, D. and Reverdin, G. (1990) Surface current distributions in the tropical Indian Ocean derived from compilations of surface buoy trajectories. *Journal of Geophysical Research, Oceans* 95, 7217–7238.

Morinaga, T., Imazeki, A., Tanada, S. and Arakawa, H. (1992) The environmental conditions of the tuna's maneuvering sphere in the Bay of Bengal. *Umi Mer.* 30, 5–16.

Murthy, V.S.R. and Venkataramaiah, A. (1958) The diatom *Asterionella* in the Krishna estuary region. *Nature* 181, 360–361.

Murty, V.S.N., Sarma, Y.V.B., Rao, D.P. and Murty, C.S. (1992) Water characteristics, mixing and circulation in the Bay of Bengal during the Southwest Monsoon. *Journal of Marine Research* 50, 207–228.

Naidu, P.D., Kumar, M.R.R. and Ramesh Babu, V. (1999) Time and space variations of monsoonal upwelling along the west and east coasts of India. *Continental Shelf Research* 19, 559–572.

Nair, P.V.R. (1970) Primary productivity in the Indian seas. *Bull. Cent. Mar. Fish. Res. Inst.* 22, 56 pp.

Nair, K.V.K. (1990) Primary productivity and photosynthetic pigments in the Edaiyur–Madras estuarine systems at Kalpakkam. *Proceedings of the National Symposium on Marine Resources, Techniques, Evaluation and Management*, eds. R. Vaidyanathan, D. Satyanarayana, P.C. Mohan, M.U. Rao, R. Varadarajulu and V.R.R.M. Babu, pp. 33–35. Visakhapatnam, India.

Nair, V.R. and Rao, T.S.S. (1973) Chaetognaths of the Arabian Sea. In *The Biology of the Indian Ocean*, eds. B. Zeitzschel and S.A. Gerlach, pp. 293–318. Springer-Verlag, New York.

Naqvi, S.W.A., Shailaja, M. S., Kumar, M.D. and Gupta, R.S. (1996) Respiration rates in subsurface waters of the northern Indian Ocean: Evidence for low decomposition rates of organic matter within the water column in the Bay of Bengal. *Deep Sea Research Topical Studies in Oceanography* 43, 73–81.

Pant, A. (1992) Primary productivity in coastal and off-shore waters of India during two south west monsoons, 1987 and 1989. In *Oceanography of the Indian Ocean*, ed. B.N. Desai. Oxford and IBH Publishing, New Delhi.

Pati, S. (1982) Observations on the effects of tropical cyclones on gillnetting in the Bay of Bengal. *J. Cons. Int. Explor. Mer* 40, 62–66.

Patnaik, A. and Lakshmana Rao, M.V. (1990) Composition and distribution of interstitial meiofauna of the sandy beach at Gopalpur, south Orissa coast. *Indian Journal of Marine Sciences* 19, 165–170.

Patterson, E.J.K. and Ayyakkannu, K. (1996) Changes in natural coastal systems in the Gulf of Mannar and the Palk Bay, southeastern coast of India with particular emphasis on coral reef ecosystem. Intergovernmental Oceanographic Commission of

UNESCO-IOC Paris 105, suppl. ed. E. Duursma, pp. 402–406.

Potemra, J.T., Luther, M.E. and O'Brien, J.J. (1991) The seasonal circulation of the upper ocean in the Bay of Bengal. *Journal of Geophysical Research, Oceans* **96**, 12667–12683.

Prasad, D.S. (1990) Morphodynamics of the sand spit, Kakinada Bay, East coast of India. *Proc. National Sym.marine Resources, Techniques, Evaluation and Management*, eds. R. Vaidyanathan et al. pp. 69–71.

Prasadam, R.D. and Kadir, P.M.A. (1988) Experimental pen and cage culture of Grey mullets in the Pulicat Lake. *The First Indian Fisheries Forum Proceedings*, ed. M.M. Joseph. Mangalore Karnataka, pp. 143–145.

Purnachandra Rao, V. and Vijay Kumar, B. (1990) Quartz and feldspar distribution in continental shelf sediments of east coast of India. *Indian Journal of Marine Sciences* **19**, 191–195.

Qasim, S.Z. (1977) Biological productivity of the Indian Ocean. *Indian Journal of Marine Sciences* **6**, 122–137.

Qasim, S.Z. (1998) *Glimpses of the Indian Ocean*. University Press (India) Ltd., Hyderabad, 206 pp.

Qasim, S.Z., Parulekar, A.H., Harkantra, S.N., Ansari, Z.N. and Nair, A. (1977) Aquaculture of green mussel *Mytilus viridis* L.: Cultivation on ropes from floating rafts. *Indian Journal of Marine Sciences* **6**, 15–25.

Raghu Prasad, R. and Nair, P.V.R. (1963) Studies on organic production I. Gulf of Mannar. *Journal of the Marine Biologists Association of India* **5**, 1–26.

Raja, S.K.D. and Philipose, V. (1977) Trends in the yields of major exploited fisheries of the east coast of India. *Indian Journal of Fisheries* **22**, 187–197.

Rajendran, A., Rajagopal, M.D. and Reddy, C.V.G. (1980) Distribution of dissolved silicate in the Arabian Sea and Bay of Bengal. *Indian Journal of Marine Sciences* **9**, 172–178.

Rajendran, K., Sampathkumar, P., Govindasamy, C., Ganesan, M., Kannan, R. and Kannan, L. (1993) Levels of trace metals (Mn, Fe, Cu and Zn) in some Indian Sea weeds. *Marine Pollution Bulletin* **26**, 283–285.

Raman, A.V., Satyanarayana, Ch., Adiseshasai, K. and Phani Prakash, K. (1990) Phytoplankton characteristics of Chilka Lake, a brackish water lagoon along east coast of India. *Indian Journal of Marine Sciences* **19**, 274–277.

Raman, C.V., Krishna Rao, G., Reddy, K.S.N. and Ramesh, M.V. (1995) Clay mineral distributions in the continental shelf sediments between the Ganges mouths and Madras, east coast of India. *Continental Shelf Research* **15**, 1773–1793.

Ramaswamy, V. and Nair, R.R. (1994) Fluxes of material in the Arabian Sea and Bay of Bengal—sediment trap studies. In *Biogeochemistry of the Arabian Sea: Present Information and Gaps*, ed. D. Lal, pp. 189–210. Proc. Indian Acad. Sci. Earth Planet. Sci.

Ramaswamy, V., Vijaya Kumar, B., Parthiban, G., Ittekkot, I., and Nair, R.R. (1997) Lithogenic fluxes in the Bay of Bengal measured by sediment traps. *Deep-Sea Research* **44**, 793–810.

Ramiayan, M., Prasad, E.K. and Suresh, P.K. (1997) Shoreline oscillation of Tamil Nadu Coast. *Proc. Second Indian National Conference on Harbour and Ocean Engineering*, **2**, pp. 1176–1181.

Rao, G.S. (1994) Environmental constraints and meiofaunal abundance in the Gosthani Estuary, India. *Journal of Ecobiology* **6**, 103–108.

Rao, K.V. and Rao, K.H. (1975) Macro- and micro faunal associates of the fouling dresseinid *Mytilopsis sallei* (Recluz) in the Visakhapatnam harbour. *Bulletin of the Dept. of Marine Sciences, University of Cochin* **7**, 623–629.

Rao, K.H., Antony, M.K., Murty, C.S. and Reddy, G.V. (1987) Gyres in the NW Bay of Bengal—some observed evidences. *Indian Journal of Marine Sciences* **16**, 9–14.

Rao, A.V.R.K., Bohra, A.K. and Rajeswara Rao, V. (1990) On the 30–40 day oscillation in southwest monsoon: a satellite study. *Mausam* **41**, 51–58.

Rao, C.K., Naqvi, S.W.A., Kumar, M.D., Varaprasad, S.J.D.,
Jayakumar, D.A., George, M.D. and Singbal, S.Y.S. (1994) Hydrochemistry of the Bay of Bengal: possible reasons for a different water-column cycling of carbon and nitrogen from the Arabian Sea. *Marine Chemistry* **47**, 279–290.

Ravindran, M. (1994) Ocean energy activities in India. *Marine Industrial Technology Monitor* **1**, 5–7.

Ray, S.B., Mahanti, M. and Somayajulu, B.L.K. (1995) Uranium isotopes in the Mahanadi River estuarine system, India. *Estuarine and Coastal Shelf Science* **40**, 635–645.

Reddy, A.N. and Reddi, K.R. (1994) Seasonal distribution of foraminifera in the Araniar river estuary of Pulicat, southeast coast of India. *Indian Journal of Marine Sciences* **23**, 39–42.

Roy, R. (1997) Nowhere to go. *Bay of Bengal News* **2**, 9–11.

Rudnick, D.J., Elmgren, R. and Frithson (1985) Meiofaunal prominence and benthic seasonality in a coastal marine ecosystem. *Oecologia* **67**, 157–168.

Ryther, J.H. and Menzel, D.W. (1964) Anton Brunn Cruise 1. Final Report. U.S. Program in Biology. International Indian Ocean Expedition, Woods Hole Oceanographic Institution, Woods Hole, MA.

Sanilkumar, K.V., Kuruvilla, T.V., Jogendranath, D. and Rao, R.R. (1997) Observations of the western boundary current of the Bay of Bengal from hydrographic survey during March 1993. *Deep Sea Research, Oceanographic Research Papers* **44**, 135–145.

Sanjeevaraj, P.J. (1981) Pulicat Lake as a giant pond for prawn culture under rural science and technology. Short term Training Programme in Brackish Water Prawn and Fish Culture, Kakinada, Central Institute for Fisheries Education, Bombay, India. pp. 105–107.

Santhakumari, V. and Nair, V.R. (1998) Distribution of hydromedusae from the exclusive economic zone of the west and east coasts of India. *Indian Journal of Marine Sciences* **28**, 150–157.

Sarin, M.M. and Krishnaswamy, S. (1984) Major ion chemistry of the Ganga–Brahmaputra river systems, India. *Nature* **312**, 538–541.

Sarma, A.L.N. and Ganapati, P.N. (1973) *Musculus strigatus* (Hanley) (Mollusca: Bivalvia) associated with littoral algae at Visakhapatnam. *Indian Journal of Marine Sciences* **2**, 146–147.

Sarma, A.L.N. and Wilsanand, V. (1994) Littoral meiofauna of Bhitarkanika mangroves of river Mahanadi system, east coast of India. *Indian Journal of Marine Sciences* **23**, 221–224.

Sarma, V.V., Narasimha Rao, T.V., Rama Raju, V.S., Vijaya Kumar, R. and Suguna, C. (1988) Interrelationships and distribution of hydrochemical parameters in coastal waters off Visakhapatnam, east coast of India. *Mahasagar* **21**, 197–207.

Sarupriya, J.S. and Bhargava, R.M.S. (1998) Seasonal distribution of chlorophyll *a* in the exclusive economic zone (EEZ) of India. *Indian Journal of Marine Sciences* **27**, 292–298.

Sasamal, S.K. (1989) Hydrography of the northern Bay of Bengal during southwest monsoon. *Mahasagar* **22**, 105–112.

Sasamal, S.K., Sahu, B.K. and Panigrahy, R.C. (1987) Mercury distribution in the estuarine and nearshore sediments of the western Bay of Bengal. *Marine Pollution Bulletin* **18**, 135–136.

Satpathy, D. (1992) Prospectus of the culture of mud crab *Scylla serrata* in Chilka Lake, Bay of Bengal. *Seafood Export Journal* **24**, 11–12.

Satpathy. K.K. and Nair, K.V.K. (1996) Occurrence of phytoplankton bloom and its effect on coastal water quality. *Indian Journal of Marine Sciences* **25**, 145–147.

Satyanarayana, D. and Murty, P.V.S.P. (1990) Distribution of dissolved trace metals in western Bay of Bengal. *Indian Journal of Marine Sciences* **19**, 206–211.

Satyanarayana, D. and Ramana, Y.V. (1996) Geochemical studies of major elements in central Bay of Bengal sediments. IAPSO publication Scientifique 36 and Andhra University Oceanographic Memoirs 3, ed. S.V. Durvasula, pp. 63–77. Andhra.

Satyanarayana, D., Rao, I.M. and Prasad Reddy, B.R. (1987) Chemical oceanography of harbour and coastal environment of Visakhapatnam (Bay of Bengal): Part 1. Trace metals in water and particulate matter. *Indian Journal of Marine Sciences* **14**, 139–146.

Selvaraj, G.S.D. and Srinivasan, V. (1996) Some observations on primary production and plankton biomass along the continental shelf and slope off the northeast coast of India during January 1989. *Proc. Second Workshop on Scientific Results of FOVR Sagar Sampada*, eds. V.K. Pillai, S.A.H. Abidi, V. Ravindran, K.K. Balachandran and V.V. Agadi, pp. 103–113. India Department of Ocean Development, New Delhi.

Sen Gupta, R., De-Sousa, S.N. and Joseph, T. (1977) On nitrogen and phosphorus in western Bay of Bengal. *Indian Journal of Marine Science* 6, 107–110.

Sen Gupta, R., Basu, P.C., Bandhopadhya, R.R., Bandhopadhyay, A., Rakshit, S. and Sharma, B. (1992) Geological Survey of India Special Publication No. 29, 29 pp.

Sen Gupta, R., Naik, S. and Varadachari, V.V.R. (1989) Environmental pollution in coastal areas of India. In: *Ecotoxicology and Climate*, eds. P. Bourdeau, J.S. Haines, W. Klein and C.R. Krishna Murti. SCOPE, pp. 235–246. John Wiley & Sons Ltd.

Sen Gupta, R., Fondekar, S.P., Shailaja, M.S. and Sankaranarayanan, V.N. (1995) *Maersk Navigator* oil spill in the Great Channel (Andaman Sea) in January 1993 and its environmental impact. 1995. *Spill Science & Technology Bulletin* 2, 113–119.

Shailaja, M.S. and Sarkar, A. (1992) Organochlorine pesticide residues in the northern Indian Ocean. In *Oceanography of the Indian Ocean*, ed. B.N. Desai, pp. 379–383. Oxford and IBH, New Delhi.

Shailaja, M.S. and Singbal, S.Y.S. (1994) Organochlorine pesticide compounds in organisms from the Bay of Bengal. *Estuarine and Coastal Shelf Science* 39, 219–236.

Shetye, S.R., Shenoi, S.S.C., Gouveia, A.D., Michael, M.S., Sundar, D. and Nampoothiri, G. (1991) Wind-driven coastal upwelling along the western boundary of the Bay of Bengal during the south west monsoon. *Continental Shelf Research* 11, 1379–1408.

Shetye, S.R., Gouveia, A.D., Shankar, D., Shenoi, S.S.C., Vinayachandran, O.P.N., Sundar, D., Michael, M.S. and Nampoothiri, G. (1996) Hydrography and circulation in the western Bay of Bengal during northeast monsoon. *Journal of Geophysical Research, Oceans* 101, 14011–14025.

Sivakami, S., Nair, P.V.R., Jayaprakash, A.A., Kasim, M., Yohannan, T.M., Sivadas, M., Koya, K.P.S. and Seetha, P.K. (1996) Distribution and abundance of carangids along EEZ India. *Proceedings of the Second Workshop on Scientific Results of FORV Sagar Sampada*, eds. V.K. Pillai, S.A.H. Abidi, V. Ravindran, K.K. Balachandran and V.V. Agadi, pp. 347–361. India Department of Ocean Development, New Delhi.

Somayajulu, B.L.K. (1994) Uranium isotopes in the Hoogly estuary, India. *Marine Chemistry* 47, 291–296.

Somvanshi, V.S. (1998) Fishery resources in the Indian EEZ: Recommendations for deep-sea fisheries development. *Indian Journal of Marine Sciences* 27, 457–462.

Sreepada, R.A., Bhat, K.L. and Parulekar, A.H. (1995) Particulate organic constituents of surface waters of east coast of India. *Indian Journal of Marine Sciences* 22, 132–134.

Steemann Nielsen, E. and Jensen, A.E. (1957) Primary oceanic production. *Galathea Reports* 1, 49–136.

Stephen, R. (1992) Copepod composition along southwest and southeast coasts of India. In *Oceanography of the Indian Ocean*, ed. B.N. Desai, pp. 121–127. New Delhi, India, and IBH, Oxford.

Subba Rao, D.V. (1965) Studies of hydrography, phytoplankton and primary production off Waltair Coast, Bay of Bengal. Ph.D. Thesis, Andhra University, Waltair, India 180 pp.

Subba Rao, D.V. (1969) *Asterionella japonica* bloom and discoloration off Waltair, Bay of Bengal. *Limnology and Oceanography* 14, 632–634.

Subba Rao, D.V. (1971) Effects of ammonium phosphate dumping on phytoplankton in Visakhapatnam harbour, Bay of Bengal. *Environmental Letters* 2, 65–73.

Subba Rao, D.V. (1973) Effects of environmental perturbations on short-term phytoplankton production off Lawson's Bay, a tropical coastal embayment. *Hydrobiologia* 43, 77–91.

Subba Rao, D.V. (1974) Monsoonal influences on production in the Bay of Bengal. IIe Colloque International sur l'Exploration des Océans. 4, 28 pp., Paris, France.

Subba Rao, D.V. (1976) Marine plankton diatoms as indicators of ocean circulation in the Bay of Bengal. *Botanica Marina* 19, 183–187.

Subba Rao, D.V., Sprules, W.G., Locke, A. and Carlton, J.T. (1994) Exotic phytoplankton from ships' ballast waters: risk of potential spread to mariculture sites on Canada's East Coast. *Canadian Data Report of Fisheries and Aquatic Sciences* 937, 51 pp.

Subrahmanyan, R. (1946) A systematic account of the marine plankton diatoms of the Madras coast. *Proceedings of the Indian Academy of Sciences* B24, 1–197.

Subrahmanyan, R. (1971) *The Dinophyceae of the Indian Seas*. Marine Biological Association of India, Cochin, India, 334 pp.

Subrahmanian, B., Prabhu, S.K. and Mahadevan, A. (1990) Influence of thermal power station effluents on hydrobiology of seawater. *Water Air and Soil Pollution* 53, 131–137.

Sudarsan, D. (1993) Marine fishery resources in the Exclusive Economic Zone of India. *Proc. National Workshop on Low Energy Fishing*. Fish. Technol. Soc. India, Special Issue. pp. 3–11.

Sundar, V. and Sarma, S.S. (1992) Sediment transport rate and its distribution across surf zone off Gopalpur port, east coast of India. *Indian Journal of Marine Sciences* 21,102–106.

Sunita Rao, G. and Rama Sarma, D.V. (1990) Meiobenthos of the Gosthani estuary. *Indian Journal of Marine Sciences* 19, 171–173.

Suresh, K., Ahamed, M.S., Durairaj, G. and Nair, K.V.K. (1992) Ecology of interstitial meiofauna at Kalpakkam coast, east coast of India. *Indian Journal of Marine Sciences* 21, 217–219.

Suryanarayana, A., Murty, C.S. and Rao, D.P. (1992) Characteristics of coastal waters of the western Bay of Bengal during different monsoon seasons. *Australian Journal of Marine and Freshwater Research* 43, 1517–1533.

Tanabe, S., Subramanian, A.N., Ramesh, A., Kumaran, P.L., Miyazaki, N. and Tatsukawa, R. (1993) Persistent organochlorine residues in dolphins from the Bay of Bengal, South India. *Marine Science Bulletin* 26, 311–316.

Tattersall, O. S. (1922) Indian Mysidaceae. *Rec. Indian Mus.* 24, 445.

Thangavelu, R. and Sanjeevaraj, P.J. (1985) Fishery and biology of the clam *Meretrix casta* (Chemnitz) in the Pulicat Lake. *Journal of the Marine Biologists Association of India* 27, 75–83.

Thiyagarajan, V., Venugopalan, V.P., Subramoniam, T. and Nair, K.V.K. (1997) Macrofouling in the cooling water conduits of a coastal power station. *Indian Journal of Marine Sciences* 26, 305–308.

Thorrington-Smith, M. (1971) West Indian Ocean phytoplankton: a numerical investigation of phytohydrographic regions and their characteristic phytoplankton associations. *Marine Biology* 9, 115–137.

Vijayakumaran, K. and Naik, S.K. (1991) Demersal finfish resources of the inner-continental shelf of the EEZ. *Proccedings of the National Workshop on Fishery Resources*, eds. D. Sudarsan and V.S. Somvanshi, pp. 71–79. India Fishery Survey of India, Bombay.

THE AUTHOR

D.V. Subba Rao

Mariculture and Fisheries Department,
Kuwait Institute For Scientific Research,
P.O. Box 1638,
Salmiya 22017, Kuwait

Chapter 72

BANGLADESH

Dihider Shahriar Kabir, Syed Mazharul Islam, Md. Giasuddin Khan, Md. Ekram Ullah and Dulal C. Halder

---

The Bangladesh coast is very low-lying. The continental shelf is 200 km wide and covers about 60,440 km², over half of which is no deeper than 50 m. Numerous rivers deposit 2.4 billion tons of silt per year, and reduce coastal salinity to 12 ppt during the monsoon period. However, partly because of the Farraka barrage in India and water withdrawal for irrigation, salinisation of agricultural land is extending further inland.

Cyclones can cause immense damage. The area experiences the most severe storm surges in the world, and over 30% of the country (almost all the coastal region) is prone to flooding, and has been repeatedly devastated.

The eastern region is the most settled and is fringed mainly by mud flats and submerged sands. The famous 145 km Cox's Bazar sand beach lies here. The central region runs east from the Tetulia River to the Big Feni River and includes the mouth of the combined Ganges, Meghna and Brahmaputra rivers. This active delta area is characterised by heavy sediment input, formation of islets and bank erosion, with an approximate yearly net accretion of 35.6 km² of land. The western region is relatively stable and mostly covered with dense mangrove forests, including the Sunderbans. Most offshore islands are deltaic in origin and low-lying, some of which are not yet fully consolidated.

The Sunderban is the largest continuous mangrove ecosystem in the world. Its area is expanding slightly due to deposition, but the biota is subjected to degradation and encroachment, and the vegetation on it is disappearing at an alarming rate. 500,000 to 600,000 people depend on the Sunderbans directly for their livelihood.

Coastal waters are biologically productive. Fisheries contribute 10% to agricultural GDP and 3% to total GDP, and involve about 190,000 fishermen. The catch of the most valuable penaeid shrimp is showing a gradual decline due to overfishing, and due to inadequate recruitment resulting from intensive collection of wild shrimp fry.

Some of the coastal area's groundwater has been severely contaminated by inorganic arsenic, notably in the Bengal delta. Sewage treatment plants are not yet available in cities or settlements; instead, drains, canals and rivers are used for waste removal. Numerous factories contribute to considerable pollution of both waterways and the Bay.

Estimating the effects of sea-level rise is very important but it is a difficult task in this low-lying coastal region; one fear is that the Sunderbans and some western parts might completely disappear. No protected areas have been declared in Bangladesh, and there seem to be no examples of successful land-use planning controls, or of offshore management or of integrated coastal protection, but it is essential that each sector of the user community understands the nature and benefits of planned development.

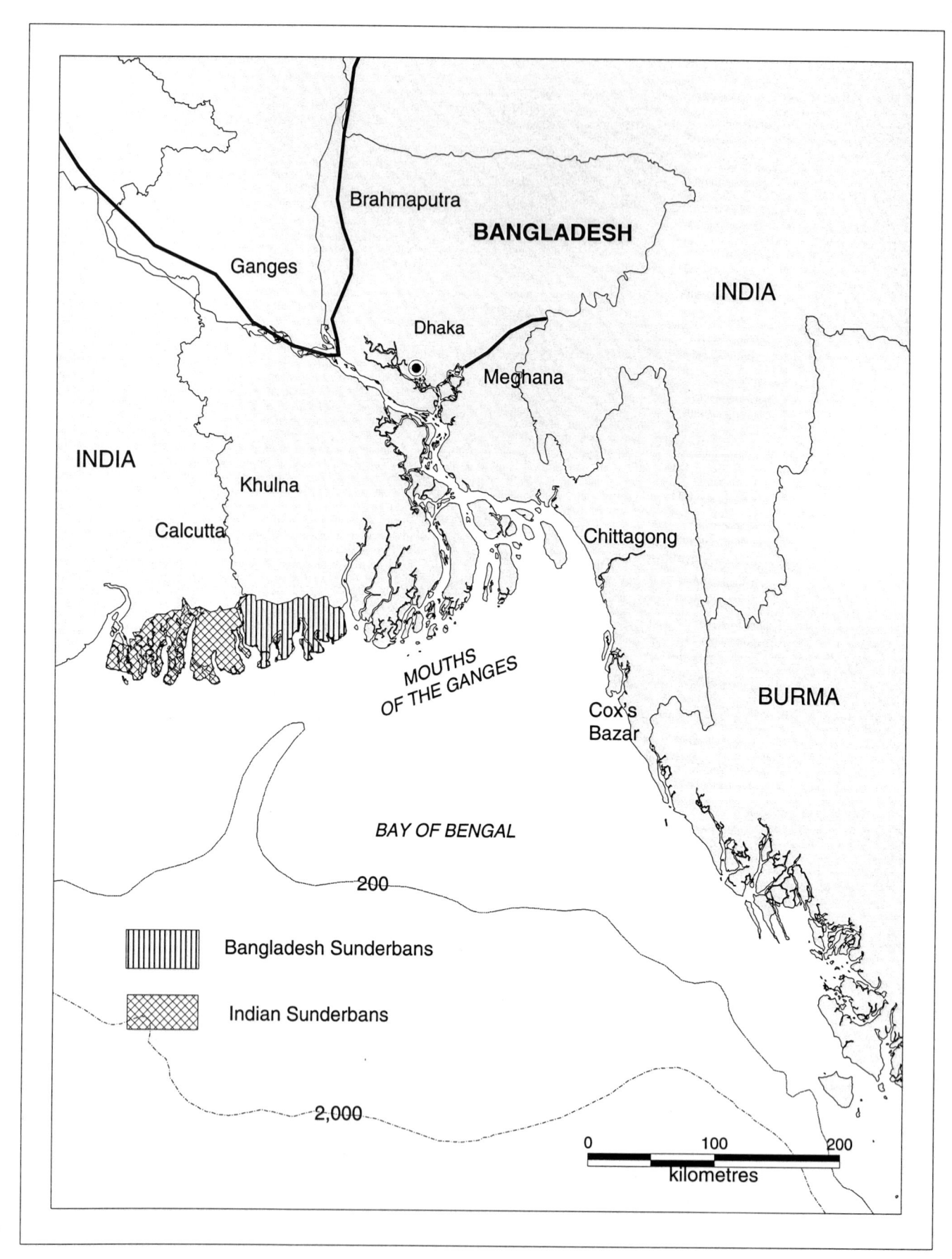

Fig. 1. Map of Bangladesh.

INTRODUCTION

Bangladesh has a low-lying coastline of 480 km which is dominated by three major river systems: the Ganges, the Meghna and the Brahmaputra (thereafter GMB). These flow into the Bay of Bengal, where sedimentation and the land accretion process contribute to deltaic land-formation on the shallow continental shelf. This ceaseless encroachment of land into the Bay of Bengal is still continuing, providing opportunity for settlement for the growing population. Sprawling estuaries, dense mangrove ecosystems, salt marshes, islands and coral reefs and sandy beach systems characterise the coastal zone of Bangladesh.

THE PHYSICAL SETTING OF THE BAY OF BENGAL

The coast falls within the tropical region and is affected by the monsoon system (Fig. 1). The Bay of Bengal is a Large Marine Ecosystem, whose continental shelf covers an area of 60,440 km^2, 37,000 km^2 of which is no deeper than 50 m (Khan et al., 1995; Rahman et al., 1995). The continental shelf of the Bay is about 200 km wide along its northern fringe. Fluvial sediments overlie the substrate in a gently sloping finger-like fashion (ESCAP, 1988; Pernetta, 1993), extending southwards to form the "Bengal Deep Sea Fan" for about 3000 km between 20°N and 7°S, and with a gradient of about 10 cm per km. According to Miah (1998) and Pernetta (1993), a series of turbid currents draw towards the south huge volumes of sediments brought down by rivers. In addition, the Bay of Bengal is situated at the junction of three major tectonic plates, so Bangladesh is also a highly tectonically active area.

The vast catchment area of the Himalayas plays the most significant role in the overall physical environment of the Bay of Bengal. Several large rivers, entering the northern narrow neck of the Bay form a large estuary at its apex, carrying large quantities of nutrients which mix with organic matters derived from the mangroves (Rahman et al., 1995). These tributaries deposit a high amount of silt with a total volume of 2.4 billion tons, and about 1300 million cubic feet of water every year (Khan, 1988; Eysink, 1983; ESCAP, 1988).

SEASONALITY, CURRENTS, AND NATURAL ENVIRONMENTAL VARIABLES

The seasonal oscillation of sea level in the Bay of Bengal is amongst the highest in the world. Moreover, monsoon winds of the region and the effect of the rivers determine the major oceanographic conditions of the coastal region of Bangladesh. The Bay of Bengal has a mean tidal height of 2.48 m at Chittagong in the east, increasing to 3.64 at spring tides. Along the Sunderban coast in the west the tidal range is much higher, reaching nearly 7 m during monsoon storms (FAO, 1968; Eysink, 1983). These factors have a significant bearing on the marine environment. Tides and currents originating in the colder Indian Ocean influence the temperature distribution in the Bay and the monsoon wind causes a seasonal reversal in the directions of currents along with a consequential change in its environment. This sometimes causes tidal surges along with cyclonic storms (Miah, 1998; Lamboeuf, 1987). Another important feature of the Bay is the substantial dilution of its surface water (FAO, 1968; Eysink, 1983) which causes salinity of the northern Bay to be amongst the lowest found in any sea (Fig. 2) Coastal salinity may be 12–33 ppt during the monsoon to 16–39 ppt in the dry season (Rahman et al., 1995), while the temperature of the surface water of the Bay varies from 29.5 to 38.8°C (Khan et al., 1997). In general, surface salinity is less than that of deeper, cooler water (Fig. 3).

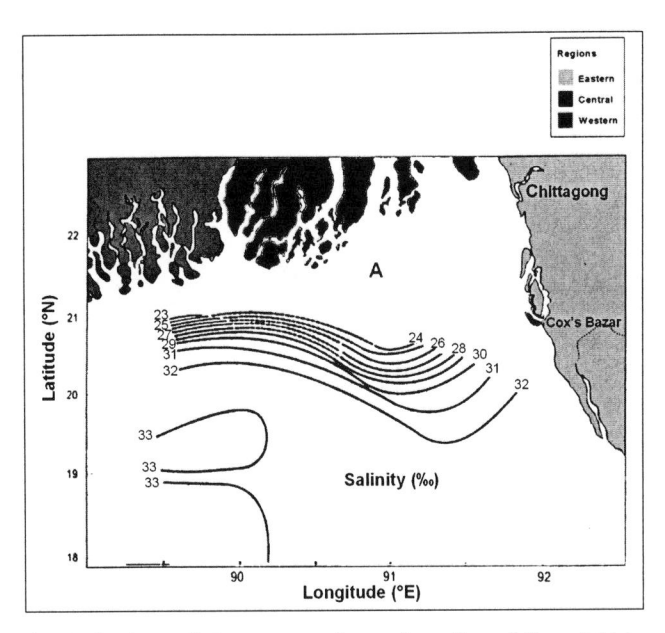

Fig. 2. Surface salinity contours in northern Bay of Bengal (data collected from Satere, 1981).

However, salinity intrusion in the southwestern region, especially close to the Sunderban, has been of great environmental concern for the past few years. It has been recognised since the early 1970s that salinity intrusion near the coast depends mostly upon the fresh water flow received from the Ganges through the Gorai–Madhumati river system. Because of the construction of the Farraka barrage on the upper Ganges in India, and also because of withdrawal of water upstream, mainly for irrigation and other reasons, a very low flow of fresh water now occurs further down during the dry season. Thus the tidal flow of the Bay extends further landward. During winter, water in the Ganges decreases sharply and shoals form. Efforts have been made to keep the Gorai as active as possible since 1982 through dredging operations, but still it dries almost every year. As a result, soil salinity has risen to between 1000 and 49,000 micro-mhos in the Sunderban (Hasan, 1990), creating a salinity problem which is affecting agriculture, aquaculture, forestry and industries.

Salinity in the Meghna estuary also has changed in recent years. The formation of numerous shoals and bars in

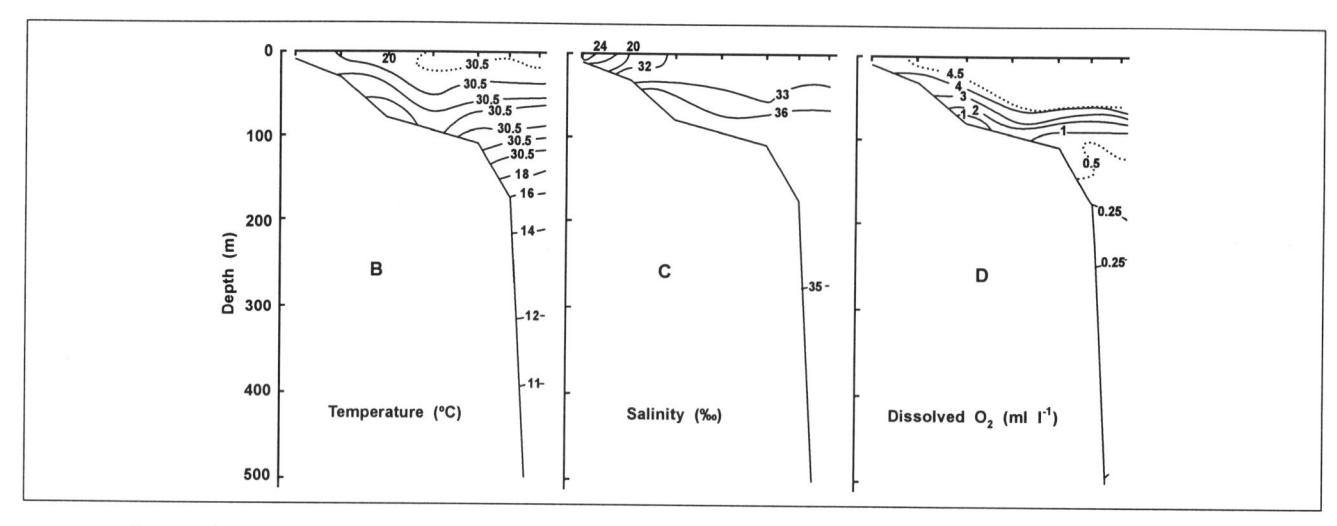

Fig. 3. Salinity, temperature and dissolved oxygen profiles with depth in northern Bay of Bengal (source: Satere, 1981).

the midstream of the lower Meghna river, resulting in fresh-water flow in the left bank, and a reduction in upland fresh-water flow in the dry months, have brought about instability in this area. The variation in salinity with depth has been found negligible. The tidal current in the area is strong enough to mix the salt with the fresh water and there is no clearly defined salt-water/fresh-water interface.

CYCLONES AND STORM SURGES

Cyclones and storm surges lasting for a few minutes to hours can cause immense changes in sea level and landmasses. The Bay of Bengal acts as a giant funnel of storms, creating the most severe storm surges in the world. As such, they are a substantial threat to the coastal people of Bangladesh. The most devastating cyclones occur in the late rainy season (October–November) originating from depressions which develop in the South China Sea and intensifying into tropical cyclones as they approach the Bay of Bengal. They proceed to the northwest and then commonly move northeast and hit the Bangladesh coast. Andhra Pradesh and Orissa of India are also heavily affected. Cyclones generate a low pressure centre and raise sea level up to 6–7 m, adding to the devastation in the low lying coastal areas. However, no satisfactory answer has ever been put forward (Islam, 1995) as to how the low pressure forms in the Bay of Bengal. Although the forces behind storms are still not well known it may be mainly due to the Coriolis force acting on the trade winds when the Inter-tropical Convergence Zone is displaced. In general, cyclonic circulations create wind speeds of 110 to 160 km/h or more, though in 1960 winds reached 193 km/h and in November 1970 they reached 241 km/h. Human mortality varies considerably and depends on various factors; it reached 100% in 1970 in Char Hare and Sonal Char, though usually it is not more than 10% (ESCAP, 1988).

These cyclones are regarded as the most intense storms affecting the surface of the earth. Heavy and often continuous rainfall occurs near the eye, and there are showers up to the limit of the cyclonic circulation (Islam, 1995). Air converges rapidly into the centre in the lower part of the storm. In addition a heavy swell known as the storm wave surges near the eye of the cyclone. On average, six typhoons occur every year in the Bay of Bengal, mainly during early summer or the late rainy season. Over 30% of the country (especially the coastal region) is prone to flooding and has been repeatedly devastated (Table 1). The damage caused by cyclones in 1970 and 1991 were the highest of all, and that in November 1970, was accompanied by massive flooding, causing half a million deaths.

To minimise cyclone casualties a few hundred cyclone shelters have been constructed. With the recent development of satellite communication and information technology and awareness among the people, the impact of cyclones has been decreasing sharply. However, success in reducing the effects of the disaster caused by cyclones has often been limited because of lack of adequate communication or early warning systems.

Table 1

Cyclones with tidal waves affecting Bangladesh

Year	Date	Area/district affected	Estimated loss (Tk million)
1958	October	Noakhali	N/A
1959	10 & 31 October	Chittagong	168.90
1961	10 May	Chittagong, Barisal	300.00
1963	29 May	Chittagong, Barisal	245.10
1965	12 May	Chittagong, Barisal	559.40
1966	1 October	Chittagong, Barisal, Bhola, Noakhali	121.00
1970	12 & 13 Nov.	Patuakhali	889.20
1985	May	Sandwip, Chittagong	N/A
1991	29 April	16 Districts	

Source: Islam (1995), BBS (1997).
Conversion rate 49.60 Taka = 1 US$).

COASTAL ENVIRONMENT AND HABITATS

The entire coast can be divided into three distinct geomorphologic regions: eastern, central and western regions (Fig. 2).

The eastern region is known as the Pacific type, and is the most settled. The shoreline, extending from the Big Feni River to Badar Mokam along Chittagong, is regular and unbroken and is protected along the sea by mud flats and submerged sands. The famous Cox's Bazar sand beach, about 145 km long, is part of the coastline. The region includes wetlands, such as the Chakaria Sunderbans, Teknaf Peninsula and the Naaf Estuary, the latter being of international importance and protected under the Ramsar convention. Only a few rivers such as Karnafuli, Matamuhuri, Shangu and Naf, traverse this region (ESCAP, 1988; Moudud et al., 1989).

The central region runs east from the Tetulia River to the Big Feni River and includes the mouth of the combined GBM rivers. This is the most active area of the delta, which is characterised by heavy sediment input, formation of Chars (small islands) and bank erosion. The general flow of water in this part of the Bay is westward, heading toward the 'Swatch of No Ground'. The result of these and other factors is an approximate yearly net accretion of 35.6 km² of land. New land has been formed in the south of Hatiya, Manpura and Bhola islands as well as near Char Clark, Char Balua and Uriis Char on the Noakhali coast (Moudud et al., 1989).

The western region is known as the Atlantic type and it covers the Bangladesh coastline westward from the Tetulia River to the Indian border located at the Hariabhanga River. This region is relatively stable and mostly covered with dense mangrove forests, which reduces bank erosion. It is characterised by many river tributaries (ESCAP, 1988).

Coastal soils generally consist of sediment deposits from the rivers. They are fertile, but salinity intrusion, both normal and cyclonic, offsets part of the fertility. One third of cultivated land can be classified as saline, Khulna and Patuakhali being the most affected regions. All the three coastal regions in Bangladesh are very rich in biodiversity.

Matamuhori Delta and Coastal Islands

The fast-flowing Matamuhori river deposits its silt on the plains at Kakhara, and later the river divides, forming its delta. Three main branches flow into the Mishkhal channel, two of which are known as Matamori khal and one as Bura Matamori. It supplies a large V-shaped lowland with its base between Mognama and Gomata and an apex at Kakhara. Tides are felt as far as Manikpur near Kakhara. It supports the small Chokoria Sunderban.

Most offshore islands are of deltaic origin and low-lying. There are six islands along the east coast, and in addition there are over a dozen Chars. Two of the islands, Ujanita and Koriardia, are not yet fully raised above high-water level and are still in the process of consolidation. Kutabdia is the northernmost island, covering an area of 55 km² and has a beach 10 km long. The Matarbari Channel, 0.5 km wide and 16 km long, separates Ujanita and Koriardia from Moishkhal islands. Southeast of Matabari, across the Kohalia channel is the large island of Moishkhali, the only hilly island in Bangladesh, covering over 161 km² in area. The greater part of the eastern half is a mass of low, tangled hills and ravines covered partly with forests and scrub.

To the southwest of Moishkhal is the small group of sandy, recurved compound and complex spits known as the Sonadia islands. At low tide, Moishkhal and Sonadia are joined together by extensive mud-flats which now contain a small mangrove forest. Another crescent island is forming to the southeast of Sonadia and threatens to push the Moishkhal channel towards Cox's Bazar town. There have been important accretions south of Kutubdia and Matarbari. On the western coast, Dublar Char is one of the largest islands (ESCAP, 1988).

In recent decades, vast shoals and off-shore islands have been formed and have surfaced in the shallow shelf region of the coastal areas of the country. "Nijhum Deep", "Urir Char" and "South Talpatti" are worth mentioning. It is hoped that by the turn of the century at least another 100 km² of newly-formed landmass will develop in the Bay of Bengal.

Beaches

Bangladesh has one of the largest beaches on earth: Cox's Bazar beach, about 145 km in length. It stretches along the eastern coast, and has great attractions for both domestic and foreign tourists. There are also beaches in Chittagong, Noakhali and Patuakhali districts, and in the offshore islands of the country. The Chittagong beach is being explored by local tourists, but several others, e.g. Kuakata, still remain unexplored.

St. Martin's Island: Virgin Jinjira Islets and Coral Reefs

St. Martin's island (20°36'N, 90°20'E) is sometimes said to be the only coral island of Bangladesh, but it is really a tectonic dumbbell-shaped island separated from the mainland Bangladesh by a channel of about 8 km. It is known as Jinjira in Bengali. Table 2 illustrates the land-use pattern on the island (Alam and Hasan, 1998). Its waters are thought to contain a huge biodiversity (Choudury et al., 1992), and the islets are a focus for eco-tourism. Its West Coast islet is an important nesting beach for the Olive Ridley and the Green turtle, and the islets support a wide variety of reef fishes. Corals are abundant, and the ten known genera include four species of *Acropora* (Mahmood and Haider, in press). The coral reef of St. Martin's island is also a habitat for 165 species of benthic algae (Mahmood and Haider, in press); mangrove formations on the southern intertidal mudflats

Table 2

Land use of St. Martin's Island

Land types	Area (acres)	Area (%)
Arable land	400	29
Settlements	100	8
Stony surface	150	12
Forested/bushes	30	2.5
Saline swamps	80	6
Sandy beaches	200	14
Coral reefs/shoals	400	29
Total	1400	100

Source: Land use survey (1992).

are also important. Nearly 150 species of shorebirds visit these during winter, including several globally threatened waders such as Spoon-billed Sandpiper, Nordman's Greenshank and Asian Dowitcher. To conserve these threatened species some measures have been taken by both Government and NGOs, but generally the measures are inadequate.

Seagrasses

Little is known about the seagrasses of Bangladesh. Generally, the shores of newly formed islands are seen to be carpeted with seagrasses. St. Martin's Island is reported to contain numerous seagrass beds, and it has recently been reported in daily newspapers that seaweeds are being smuggled into neighbouring Myanmar (The Daily Ittefaque, 1999). Seagrass is rich in protein, and the beds are being exploited in such a manner that the local extinction of some species is possible.

Mangroves

Bangladesh has the largest continuous block of mangrove ecosystem in the world. The main expanse is called Sunderban, and there is also another small Chokoria Sunderban. The Sunderban mangrove spreads across the ancient Gangetic delta bordering India and southwestern Bangladesh. Until recently these were viewed as being mosquito-infested wasteland of little or no value, but it has become clear more recently that the mangroves of Bangladesh are highly productive rainforests, extremely sensitive to changes within and outside their boundaries.

Chokoria Sunderban

Chokoria Sunderban, a relatively small tidal forest which comprises 20 species of trees, is a delta at the mouth of the river Matamuhari, and is a low saline swamp. It is in the Cox's Bazar Forest Division, between 21°36'N and 21°45'N and between 91°58' and 92°05'E (Choudhury et al., 1994). A management plan (Choudhury, 1967, 1968) was prepared for the area over 30 years ago, at a time when the forests were typical mangrove, resembling the larger Sunderban but with an abundance of Chulia Kanta (*Dalbergia spinosa*) and a profusion of Nunia (*Aegialities rotandifolia*). As observed in 1988, however, this mangrove forest had completely disappeared except for some patches on the banks measuring only 973 ha out of an original total of 7500 ha (Choudhury et al., 1994).

The Chokoria Sunderban is now little more than a distant memory. Another menace adding to the misery of this area is shrimp farming which conflicts with the farming of rice, greatly reducing its productivity. Saline water brought into the area as a result of shrimp cultivation increases soil salinity, degrading its quality and productivity.

The Sunderban

The name Sunderban has three possible derivations, but most probably derived its name from the proliferating sundari trees (*Heritiera fomes*), meaning beautiful. The mangrove forests are a harmonious combination of terrestrial and aquatic environments. It is the home of the Bengal tiger and is situated near Khulna Divisional headquarters which is about 320 km west of the capital city Dhaka. The Sunderban (including its Indian counterpart) is the largest single continuous mangrove ecosystem in the world. It is highly productive, and extremely sensitive to changing conditions. Sunderban mangroves give natural protection against cyclonic storms and tidal surges, stabilise the coastline, enhance land accretion and enrich the soil and the aquatic environment in the greater Khulna region in the southwestern corner of Bangladesh (Chaffey and Sandom, 1985a). This ecosystem is swampy, saline, and is criss-crossed by waterways. The flora of halophytic tree species contain about 65 of the world's 70 species of mangrove plants (Chaffey and Sandom, 1985b).

The north of the Sunderban is subjected to degradation, encroachment and vegetation changes, while the southern part is a zone of land accretion. The eastern part is connected with the fresh water from the Ganges river system while the western part forms the boundary with India.

In Bangladesh, this mangrove ecosystem covers an extensive area of the coast, but only a few studies have been made on it. The Sunderban of Bangladesh was apparently reduced to 1.4% of the total area of the country according to two inventories made in 1960 and 1985. However, this estimated reduction rate is not reliable. What is clear is that in a densely populated country like Bangladesh where land is a scarce commodity and where per capita land holding is only 0.38 acres, substantial land changes have taken place. What is unclear is how the increase of aquaculture and agricultural land, and dramatic increase of population growth is connected with this reduction (Kabir, 1997).

Land Area Changes and Biodiversity

Despite reduction or degradation of land in some places, the total land area of the Sunderban is, in fact, increasing.

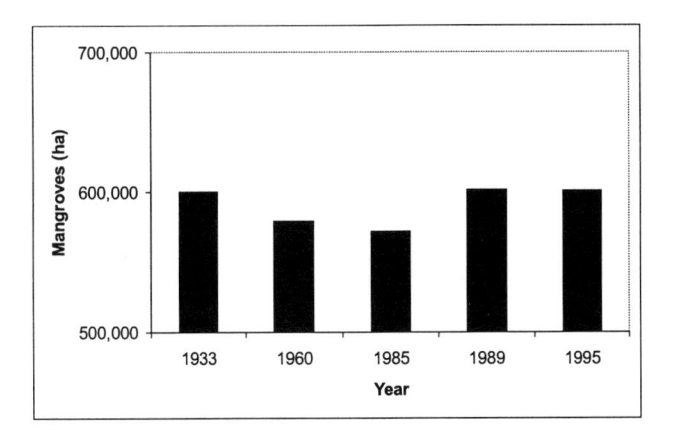

Fig. 4. Changes in area of Sunderbans over the past 25 years.

Using a GIS approach based on Landsat TM satellite data, Kabir (1997) recorded a total land area of the Sunderban as being 427,806 ha (Fig. 4). This is an increase in area of almost 5.07% in comparison to that in 1960, and of 6.52% in comparison to that in 1985. However, though the land area is increasing, the vegetation on it is still disappearing at an alarming rate.

The Sunderban teems with various forms of life. It is home to 300 species of birds, as many as 400 species of fish, at least 35 species of reptiles, and 42 species of other notable animals including the globally endangered saltwater crocodile *Crocodulus porosus* and the world's largest number of Royal Bengal Tigers *Panthera tigris*. There are large numbers of other threatened and endangered wildlife including pythons, King cobras, Adjutant storks, White bellied sea eagles, Clawless otters, Masked finfoots, ring-lizards, river terrapins etc. Some animals spend their entire lives in the mangroves, while others pass through them as temporary migrants. Mangrove animal communities include bottom-dwellers, such as crabs and worms, and tree dwellers, as well as a host of insects, birds and bats. During high tide, fish and crustaceans invade the mangrove to forage for food or to spawn. Animals such as birds, insects and bats go further to help mangrove reproduction by functioning as pollination agents and seed dispersers.

According to a recent ESCAP (1988) survey, 500,000 to 600,000 people depend on the Sunderbans directly for their livelihood. This includes both industrial and commercial enterprises. Besides generating a consumptive dollar value from fisheries, forestry, food, medicine, ecotourism and other products, the Sunderban also provides coastline protection, nutrient supply, sediment and toxicant traps and biomass production. A conservative estimate by the Forestry Department shows that Bangladesh has sustained a loss of over $US 100 million caused by the death of sundari trees alone, resulting in the loss of 1,440,000 m^3 of timber. Consequences of continuing degradation could be severe. As expressed by Burd (in Litherland, 1996) during the commencement of a project whose aim is salvaging what is left of the Sunderbans: "The disappearance of the Sunderbans, which is the only thing stopping Bangladesh from sliding into the Bay of Bengal, would be an ecological catastrophe.".

Chronic spillage of oil resulting from different port activities has now also endangered the Sunderban. The eventual impairment of the ecosystem could be irreversible. Oil and gas exploration in the Sunderban and in the Bay pose further threats.

The Sunderban's Biodiversity

Habitats: The mangroves are the basis for a high biodiversity, with 65 species of mangroves.

Mammals: About 40 species. The Rhesus macaque is the only primate existing in small groups along the waterways. The tiger is an endangered species. Deer are targets for hunters. Three species of otters exist, the flying fox (*Pteropus giganteus*), three species of wild cat (*Felis oengalensis, F. viverrina* and *F. chaus*) and the Gangetic dolphin (*Platanista gangeticac*).

Birds: About 300 species in total, including nine species of kingfishers.

Reptiles: Thirty-five species and eight of frogs and toads have been identified. The marsh crocodile and rock python are now extinct as a result of over-exploitation. Turtles, King Cobras, vipers.

Crustacea and fish: Sixty species of shrimp and crab, and 120 species of fish.

Source: Hendrich (1978); Chaffy et al. (1985); Bowler (1985); ESCAP (1988); IRMP, 1998.

POPULATION AND AGRICULTURE FACTORS IN THE COASTAL AREAS

According to the national census, Bangladesh had a total population in 1991 of 116.7 million, with a density of 938 km^2. The average annual growth rate in 1981–91 was 1.8%. It is feared that the population may rise to 154.4 million by 2010 (Moudud et al., 1989).

Population density however, varies from region to region (Table 3). It is high in major cities, but on the whole, is one-third less in coastal areas. Of the three coastal regions, the eastern region has a higher density than the Bangladesh average, while the western region has a much lower density because of the physical existence of the Sunderban. The coastal region as a whole had an estimated population of 10 million in the early 1980s (Moudud et al., 1989; Pernetta, 1993).

The coastal population is ethnically diverse. Both landless people and large land owners co-exist, and the literacy rate of 48% is higher than the national average. It may be noted that the rural people of the coastal areas are courageous and enterprising; they live in a very difficult

Table 3

Population density of the Coastal Area of Bangladesh

	Western region	Central region	Eastern region	Total
Population density (per km^2)	668	712	1029	724
Population per household	5.8	5.6	6.0	5.8
Population per village	1244	2330	1932	1556

Source: BBS (1994).

environment and have become used to struggling with the challenging coastal environment (Moudud et al., 1989).

About 30% of the net cultivable land of the country is located in the coastal area. Of the 2.85 million hectares of coastal land about 0.833 million hectares are arable. This constitutes about 52.8% of the net cultivable area in 64 Thanas (smallest administrative unit), though some is affected by varying degrees of soil salinity. The eastern and western regions are being used for a booming shrimp farming industry which affects agricultural crop production (ESCAP, 1988; Moudud et al., 1989).

Agriculture is the dominant sector in the central and western regions and roughly accounts for half the GDP, while in the eastern region industry is dominant; Chittagong port area has a considerable industrial sector, for example. In the western region, forestry accounts for about 17% of the GDP compared with 4% countrywide.

OFFSHORE SYSTEM AND FISHERIES RESOURCES

The coastal waters are biologically productive. The coast itself is rich in mangroves, which is a mixed evergreen forest. Tourism, maritime trade, marine and coastal fishery, aquaculture and exploitation of minerals and seas-bed resources are the most valuable of the resources here. Each biotic resource is multi species: the fisheries comprise 475 fish and 36 shrimp species (Khan et al., 1997).

In the coastal areas of Satkhira and Khulna districts, earthen embankments are constructed along the banks of rivers to allow sea-borne shrimps to enter and grow under natural conditions without supplementary feed. Output with this method has always been very poor, and the Government has recently begun to modernise the extensive shrimp culture in the coastal areas. However, the productivity of such shrimp farming is still relatively low compared to that in other Asian countries.

Fisheries include capture of hilsa, shrimp, lobster, Bombay Duck, pomfret, tuna, sardine, herring and mackerel. According to West (1973) the standing stock of fish is around 264,000 to 373,000 metric tons (mt) and is 9000 mt in the case of shrimps on the Bangladesh continental shelf. Depending on the survey, estimated demersal fishery biomass is between 55,000 mt and 373,000 mt. The highest figure was recorded by West (1973) and the lowest by Penn

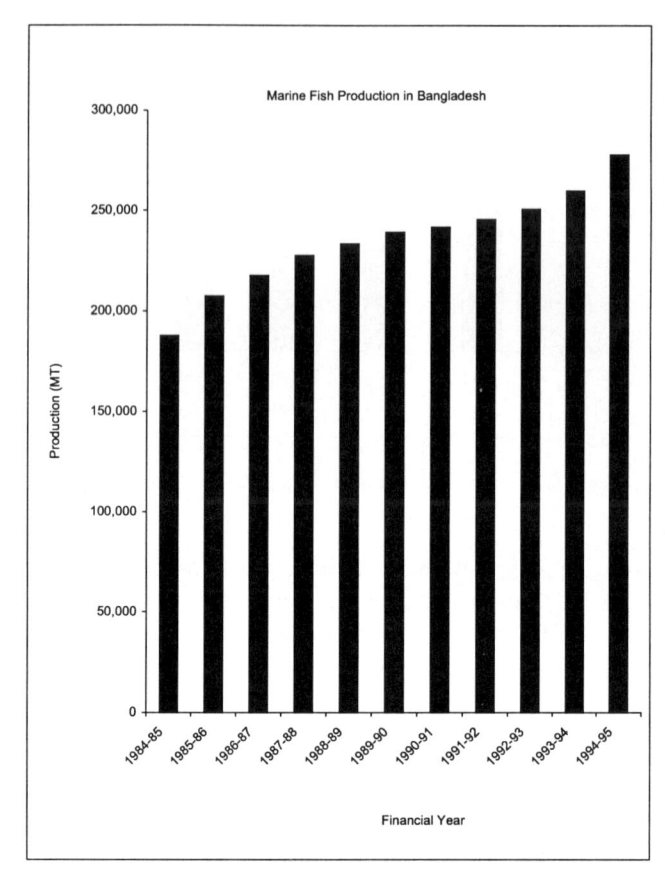

Fig. 5. Fisheries capture in Bangladesh.

(cited in Rahman et al., 1995). Tables 4 and 5 summarise fisheries resources of the Bay of Bengal (Khan et al., 1995). Marine fish accounts for 23% of the country's total fish production (Fig. 5).

On the basis of West's (1973) survey, the Government introduced commercial fin fish trawlers in 1974. Later, shrimp trawlers were introduced in 1978 (Rahman et al., 1995). As a result, the marine catch has increased by 265% between 1975–76 and 1992–93 (Rahman et al., 1995).

Fisheries in Bangladesh contribute 10% to agricultural GDP and 3% to total GDP. Its share in export earning in 1995–96 ($US 321.9 million) was 11%. Despite this large contribution, this sector has not received adequate attention from the government. Innovative policy, financial support and resource allocation in this sector are needed to increase employment opportunities.

Table 4

Stock of Marine fish

Marine fish	Standing stock (mt)	MSY
Demersal	152,000–160,000	47,500–88,500
Pelagic	60,000–120,000	–
Shrimp	4,000–9,000	7,000–8,000

Source: Khan et al. (1997); Khan (1988); Lambouef (1987); Satere (1981); West (1973).

Table 5

Commercially important fishes and shrimps exploited in Bangladesh

Scientific name	Common name
Demersal/small pelagics	
Pampus argenteus	Silver pomfret
P. chinensis	Chinese pomfret
Pomadasys argenteus	White grunter
Lutjanus johnii	Red snapper
Mene maculata	Moonfish
Tenualosa ilisha	Hilsa shad
Polydactylus indicus	Indian salmon
Lepturacanthus savala	Robbinfish/hairtail
Arius sp.	Catfish
Johnius belangerii	Croaker
Otolithes ruber	Croaker
Nemipterus japonicus	Japanese threadfin bream
Upeneus sulphurus	Goatfish
Saurida tumbil	Lizardfish
Ilisha filigera	Bigeye ilisha
Sphyraena barracuda	Great barracuda
Congresox talabonoides	Indian pike conger
Larger pelagics	
Euhynnus affinis	Eastern little tuna
Katsuwonus pelamis	Skipjack tuna
Thunnus maccoyii	Southern blue fin tuna
T. obesus	Bigeye tuna
T. tonggol	Longtail tuna
Auxis rochei	Bullet tuna
A. thazard	Frigate tuna
Shrimps	
Penaeus monodon	Giant tiger
P. semisulcatus	Tiger shrimp
P. jaaponicus	Tiger shrimp
P. indicus	White shrimp
P. merguiensis	Banana/White shrimp
Metapenaeopsis monoceros	Brown shrimp
M. brevicornis	Brown shrimp
M. spinulatus	Brown shrimp
Parapenaeopsis sculptilis	Pink shrimp
P. stylifera	Pink shrimp

Source: Khan et al. (1997).

Artisanal and Industrial Fishing

Information on the marine fishery resources of Bangladesh, including both industrial and artisanal fishery, is still insufficient. Various surveys over the past 20 years show that Bangladesh may be over-fishing.

Small-scale marine fisheries account for the bulk (85%) of the fish catch and about half of the shrimp catch. The small-scale fisheries, have both non-mechanized (75%) and mechanized boats (25%), involving nearly 185,000–190,000 fishermen. There were about 75 trawlers in 1983–84, this number dwindling to 45 by 1985–86. But in the last few years the number of shrimp trawlers dramatically increased, most operating deeper than 40 m in the EEZ. The major gear employed is gillnets, set bag nets, and tram nets, and as a result there is great concern about the future sustainability of this resource (White and Khan 1985).

Artisanal craft normally operate up to a depth of 40 m, but there is strong competition between the trawling and artisanal fishing operations up to 100 m deep. The catch of most valuable penaeid shrimp species *P. monodon* is showing a gradual decline due to overfishing, but also because of inadequate recruitment resulting from intensive collection of wild shrimp fry.

Deeper-water trawlers use small mesh nets to optimise catches and, moreover, pollution of river water, is making parts of the Bay of Bengal inhospitable for fish breeding. It is reported in the daily newspapers that the most important breeding places in the Bay no longer teem with fish and that fishing trawlers often return with very slender hauls. Sea turtles are also being killed in great numbers: the winter of 1999 alone has seen the death of several hundred in St. Martin's Island. In addition, there has been a remarkable fall in the *Hilsa* catch.

Domestic, Municipal and Industrial Wastes

Some of the coastal area's groundwater has been severely contaminated by inorganic arsenic compound, notably in the Bengal delta. Arsenic contamination in groundwater is a historical reality in Bangladesh. Unfortunately insufficient research has been done in this area, but the issue is of national concern and the media have widely reported that many people are suffering from arsenic poisoning. The coastal regions of Noakhali, Feni, Laxmipur and Bagerhat are severely affected by arsenic contamination. The sources and routes through food chains of arsenic are still unknown and under-researched.

Sewage treatment plants are not yet available in cities or settlements in the coastal region of Bangladesh. Instead, drains, canals and rivers are used for waste removal. A survey report of a waste-water expert mission in 1985 shows that the most populous coastal cities of Chittagong and Khulna have poor sanitary conditions due to lack of facilities or improper functioning of the existing ones. Septic tank effluents are also dumped into the rivers directly or indirectly, causing localized water pollution surrounding the drainage outfalls (ESCAP, 1988).

Rivers including the Karnaphuly and Passur directly receive raw excreta from the huge population living beside them. The total BOD of the domestic waters was estimated at some 3.5 tons/day. Also 760 tons of solid wastes are produced by the 3.3 million population daily. The Chittagong City Corporation (CCC) has its only dumping ground at Halishahar, and recently has begun dumping wastes at Patenga Beach and along the Karnaphuly Shah Amanat

> ## Bangladesh Fishery Resources
>
> Penaeid shrimps account for the greatest part of demersal catches. Artisanal catches mainly include pre-adults, post-juveniles, juveniles, as well as larvae, while commercial catches are mainly of the adult Tiger shrimp, *Penaeus monodon*. About 63% of total catch is of brown shrimp, *Metapenaeus monoceros* (Rahman et al., 1995).
>
> Caridean prawns (Family Palaemodiae) are important; these are mainly freshwater crustaceans which inhabit estuaries during the breeding season. Five species of lobster abound on the continental shelf, and are harvested commercially near St. Martin's Island as by-catches of existing bottom set gillnet and shrimp trawl. Among 25 species of crabs caught, the most exploited is the Mud Crab *Scylla serrata*, caught by traps and hooks. This fetches a high price in the international market. Other crabs are smaller in size and valued less. In Bangladesh, most people do not eat crabs for religious reasons (Rahman et al., 1995).
>
> Squid, cuttlefish and octopus are abundant. Though they have high protein value and low fat content, they are not popular among the Bangladeshi people.
>
> Fin-fish presently exploited are mainly demersal marine fishes, though almost 100 species of shallow, brackish-water estuarine and mid-water species are also caught (*The Daily Janokanto*, 1999). The pelagic Hilsa (River shad) are caught by drift gillnet. *Hilsa ilisa* accounts for about 13% of inland, 46% of marine and 22% of the total fish production, and *Hilsa* accounts for 78% of the fish caught by mechanized gillnets (Rahman et al., 1995).
>
> The Bangladesh continental shelf is rich in 15 species of anchovies, four species of sardines and two species of herring. They are harvested commercially, and are also caught as by-catch of small gillnet, beach seines and set bag nets (Rahman et al., 1995; Khan and Latif, 1995). Similarly 30 species of carangid are reported from the Bay of Bengal, caught both commercially and as by-catch of gillnets, and bottom trawls.
>
> Seven species of tuna and four species of mackerel exist (Khan, 1992) but these are not yet popular among the Bangladeshi people. Fifty-three species of shark, skate and ray exist (Hussein, 1970). Several are commercially important and appear to be abundant. They are in high demand in the international market, which exports shark-fins and shark-liver oil. Caught by longlines, gillnets and bottom trawl, they are considered by-catches and are not exploited commercially.
>
> ### REFERENCES
>
> Hussain, M.M. (1970) The marine and estuarine fishes of the north-eastern part of Bay of Bengal. Scientific Researches, East Regional Laboratories, Dhaka, Pakistan). Vol. Vii, No. 1:54.
>
> Khan, M.G. and Latif, M.A. (1995) Potentials, constraints and strategies for conservation and management of open water and marine fisheries resources. Paper presented at the National Seminar on Fisheries Resources Development and Management organised by MOFL, Bangladesh, with FAO and ODA, 29 October to 31 November, 1995. Dhaka, Bangladesh.19 pp.
>
> Khan, M.G. (1992) Bangladesh sustainable development of Marine fishery resources and assessment requirement of trained manpower. Paper presented at the first Asia and Pacific seminar on sustainable ocean development and management, July 27–30, 1992AIT Bangkok, Thailand.
>
> Rahman, A.K.A. et al. (1995) Economically important marine fishes and shell fishes of Bangladesh. Department of Fisheries, Dhaka.
>
> *The Daily Janokanto* (Bengali newspaper), 30th July, 1999. Dhaka.

Bridge Approach Road. The domestic waste load in Chalna and Mongla Ports is estimated to be approximately 2.2 tons BOD/day. A substantial amount of animal waste from the slaughterhouses in Chittagong finds its way into the Karnaphuly and hence the Bay of Bengal. Domestic and kitchen wastes, as well as wastes from ships anchored in the area, all contribute to the creation of unhealthy living conditions in this area (ESCAP, 1988). Moreover, the waterways carrying these wastes reverse their flows at high tide and return to the coastal areas, causing pathogenic and microbial pollution and serious health hazards, which are increased during monsoon downpours and floods.

Pollution from Shrimp Farming and Agrochemicals

After harvesting, shrimps are brought to the shrimp processing centres where they are processed, iced and packaged for transportation to the freezing plants in Khulna, Satkhira and Chittagong. Discards are mostly thrown into the rivers or canals. The coastal environment has also suffered from embankments built from 1962 to 1973, as well as through unplanned shrimp cultivation. Shrimp cultivation has also reduced the stock of other indigenous fish varieties, and destroyed mangrove flora and fauna. In addition, it has had impacts on rice cultivation and cattle farming. Moreover, agro-chemical use has risen steeply, and about 1800 tons/year of these residues enter the coastal water annually (Mahmood et al., 1994).

Oil and Shipping

Oil pollution is said to be heavy in the vicinity of Chittagong and Chalna harbours. However, more than 50% of oil pollutants in the marine environment are usually from urban and river run-off. Nearly 1000 ships and 40–50 oil tankers use Chittagong Port, and nearly 500 ships use Khulna Port at Mongla annually. Other contributors are the diverse and abundant river craft, launches and steamers which ply the waterways and which discharge waste oil, spillage and bilge washings. These vessels are the main sources of water pollution in the marine environment in the coastal regions of Bangladesh. Crude oil and its derivatives are among the most catastrophic pollutants that enter the Chittagong coastal area. Dripping of oil from hoses, overfilled sumps and deteriorating containers contribute to chronic oil pollution within the harbour (ESCAP, 1988).

Another source is ballast water regularly dumped into the sea and into the Karnaphuly and Passur rivers.

Discharges of sewage from ships are spreading infectious diseases. In addition, rotting food grains, cement dust, fertiliser etc. are also dumped into the marine water near the port areas. Ship breaking has begun on an industrial scale in Bangladesh in recent years, especially on the seashore from Kumira to Fauzderhat in Chittagong and near Mongla Port in Khulna. About 50 ship-breaking units have started operations. Residual heavy oil sludges and engine oils cause a considerable amount of spillage during washing and dismantling operations. They are not only detrimental to marine life, but also pose a threat to the beauty of the beach (ESCAP, 1988).

Chemical Wastes

The arsenic-containing compound 'geomarkovet rock', in Ghorasal (near Dhaka), is dumped into the Bay a few hundred kilometres away from the source.

Toxins discharged from the tanneries, the Karnaphuly pulp and paper industries, Khulna newsprint mills and the Barabkunda pesticide manufacturing plant all ultimately end up in the Bay of Bengal. Nearly 40,000 kg BOD/day is dumped into the Karnaphuly and 45,000 kg BOD/day enters the Bay of Bengal. Portions of the river near outfall pipes at Kalurghat, Chandraghona and Patenga are seriously polluted. The dissolved oxygen level of the river has been found to drop down to 0.1 ppm, a lethal level for aquatic flora and fauna (ESCAP, 1988).

There are a considerable number of factories in Khulna which actively contribute to pollution. Match factories, shipyard and fish-processing units, as well as other industries, directly dump their effluents into the Rupsa River. The Khulna newsprint mills, Goalpara Power Station and some jute and steel factories from the Khalispur industrial belt all discharge their effluents directly into the Bhairab. The DO varies from 3.1 to 7.9 ppm (ESCAP, 1988). The eastern region is relatively more industrially developed, but contributes much greater pollution compared to the Khulna region.

Sea-level Rise and Climate Change

To estimate the possible impact of a rise of sea level is an important but very difficult task in this low-lying coastal region. One fear is that the Sunderban on the western coast of Bangladesh might completely disappear. The eastern coast will be relatively less affected, but cyclonic storms and floods are assumed to have increased both in frequency and in intensity. Projections made by different sources are that half of coastal agricultural and a quarter of industrial activities would be affected. According to Rahman et al. (1990) a 1 m sea-level rise would cause a total loss of 22% of industrial and 18.18% of other sectors. In a country like Bangladesh, retreat is not possible, and there is little choice but to continue shoreline protection. It is also necessary to conduct research in the areas of ecological forecasting, monitoring environmental changes, development of appropriate socio-economic plans and, of course, to implement actions (Moudud et al., 1989).

THE LEGAL REGIME

Bangladesh has extended its territorial limit to 12 nautical miles in the Bay. This was possible by enacting the 1974 Territorial Water and Maritime Zones Act, while the contiguous and exclusive economic zone extends 200 nautical miles into the Bay, although overlapping areas are claimed by India. The 1974 act and 1977 rules have provisions for maritime pollution control, but enforcement of such rules cannot be established because of institutional weaknesses. Moreover, illegal fishing and 'robbing' takes place by neighbouring countries. During the past few decades the Navy has been the sole authority to maintain law and order in the Bay, but recently the coast guard has been formed to maintain this role. A few new regulations are being enacted also; for example, a trawler with cold-storage facilities is allowed to stay 30 days continuously at sea, whereas a simple trawler is allowed only 15 days.

THE NEED FOR INTEGRATED COASTAL MANAGEMENT

In the Bangladeshi part of the Bay no protected area has yet been declared. Strategies for the protection, enhancement and sustainable use of the marine and coastal environment and resources should be based on comprehensive observations, and on a sound and scientifically based understanding of marine ecology, ocean processes and their interactions with the terrestrial and atmospheric systems. Unfortunately, little data is available in this area. Further, the problems in these zones are often administered by different agencies reporting to a variety of government departments, with a potential for conflict. For example, the responsibility for managing the coast involves two ministries and two political jurisdictions. The coast itself is the responsibility of the Ministry of Agriculture, Fisheries and Food and its agency, the Water Development Board, whereas the coastal zones are also administrated by the district councils and the Department of the Environment (DOE). In addition, the Sunderban mangroves are managed by the Forest Department.

There would seem to be no examples of a successful triple alliance of land-use planning controls, offshore management controls and integrated coastal protection in Bangladesh. It is essential that each sector of the user community understands the nature of planned development of every other sector.

In view of the above, what is needed at the moment is an umbrella administration, integrating and co-ordinating all these sectoral activities of conserving natural habitats and species, controlling pollution and the alteration of shorelines and beachfronts, controlling watershed activities that adversely affect coastal zones, controlling excavation, mining and other alteration of coral reefs, water basins and sea floors. If this could be done, it would provide a mechanism for rational resources management in the next century.

REFERENCES

Alam, M. and Hassan, M.Q. (1998) The origin of beach rock of St Martin's Island of Bay of Bengal. *Bangladesh Oriental Geographer (Journal)* **42** (2), 21–32.

Anon. (1998) Integrated development of the Sunderbans Reserved Forest (IRMP), Bangladesh Volume 1, Project BGD/84/056. UNDP/FAO.

BBS (Bureau of Statistics) (1994) *Statistical Pocketbook of Bangladesh.* Ministry of Planning, Government of Bangladesh, Bangladesh.

BBS (Bureau of Statistics) (1997) *Statistical Pocketbook of Bangladesh.* Ministry of Planning, Government of Bangladesh, Bangladesh.

Chaffey, D.R. and Sandom, J.H. (1985a) ODA main Report on the Sunderbans Forest inventory project, Dhaka, Bangladesh.

Chaffey, D.R. and Sandom, J.H. (1985b) A glossary of vernacular names and a field key to the trees. The Sunderbans Forest inventory project, Dhaka, Bangladesh.

Choudhury, A.M. et al. (1994) "Study of Chokoria Sunderbans using remote sensing techniques". *ISME: Mangrove Ecosystems Technical Reports, Vol. 4.*

Choudhury, M.U. (1967–68) Working Plan of the Sunderbans. For the period from 1967–68 to 1979–80, Vol. 1. Forest Department, Government of East Pakistan.

Choudhury, S.Q. et al. (1992) Coastal geomorphology of St. Martin's Island. *Bangladesh Oriental Geographer (Journal)* **36** (2), 30–43.

ESCAP (1988) Coastal environment management plan for Bangladesh. UNDP/ESCAP Vols. 1 and 2. Bangkok.

Eysink, W.D. (1983) Basic consideration on the morphology and land accretion potentials in the estuary of lower Meghna river. Technical report 15. Land Reclamation Project, Bangladesh Water Development Board.

FAO (1968) Report to the government of East Pakistan on Oceanography of the North Eastern Bay of Bengal, useful to fisheries development and research in East Pakistan. Based on the work of Dr. Z. Popovioi. United Nations Development Programme. PEEK/FI. 102 p. FAO/BGD.FE:DP/BGD/80/075. 26 p. Marine Fisheries and Research Management and Development project.

Hasan, M.M. et al. (1990) Soil hydrology and salinity in the Sunderbans. In *Proceedings of the Seminar on Top Dying of Sundri Heritiera Fomes Trees*, eds. M.A. Rahman et al. Bangladesh Agricultural Research Council, Dhaka.

Islam, A. (1995) *Environment Land Use and Natural Hazards in Bangladesh.* Dhaka University Press.

Kabir, D.S. (1997) An areal and ecological extent of the mangrove Sunderbans using GIS analysis. MSc Dissertation, Department of Biological Sciences, University of Warwick. U.K.

Khan, M.G., Alamgir, M. and Sada, M.N. (1997) The Coastal Fisheries of Bangladesh. In *Status Management of Tropical Coastal Fisheries in Asia*, eds. G. Silvestre and D. Pauly, pp. 26–37. ICLARM Conf. Proc. 53, 208 pp.

Khan, M.G. (1988) Offshore marine fisheries resources of Bangladesh, its potentiality and status of exploitation. In Compilation of Reports on the Marine Fisheries Survey and Research. MFSMDP. Ctg. 304–310 pp.

Khan, M.S. et al. (1995) *Wet Lands of Bangladesh.* Bangladesh Centre for Advanced Studies, Dhaka.

Lambouef, M. (1987) Bangladesh demersal fish resources of the continental shelf. FAO/BGD Marine Fisheries Research, Management and Development Project, FI:DP/BGD/80/075, 26 pp.

Litherland, S. (1996) Is mangrove dying? *The Daily Observer* Nov. 1996. Dhaka.

Mahmood, N. and Haider, S.M.B. A note on corals of the St. Martin Island. Bangladesh. Chittagong University Studies (in press).

Mahmood, N., Chowdhury, N.J.U., Hossain, M.M., Haider, S.M.B. and Chowdhury, S.R. (1994) Bangladesh. In: An assessment in the Bay of Bengal region. Swedish Centre for coastal Development and Management of Aquatic Resources. Bay of Bengal programme. BOBP/Rep/67. pp. 75–129.

Miah, M.M. (1998) *Flood in Bangladesh.* Academic Publishers, Dhaka.

Moudud, H.J. et al. (1989) The Greenhouse Effect and Coastal Area of Bangladesh. Proceedings of an International Conference held in Dhaka, Bangladesh, 5 March 1989.

Pernetta, J.C. (ed.) (1993) *Marine Conservation and Development Report on Marine Protected Area Needs in the South Asian Region, Vol. 1. Bangladesh.* IUCN, Gland, Switzerland.

Rahman, A.K.A. et al. (1990) Environmental aspects of agricultural development in Bangladesh, UPL, Dhaka, 25 pp.

Rahman, A.K.A. et al. (1995) Economically important marine fishes and shell fishes of Bangladesh. Department of Fisheries, Dhaka.

Satere, R. (1981) Survey of the marine fisheries resources of Bangladesh, Nov.–Dec. 1979 and May 1980. Reports on surveys with the RN Dr. Fridtjof Nansen, Institute of Marine Research, Bergen.

The Daily Ittefaque (Bengali newspaper), 31st July, 1999. Dhaka.

West, W.Q.B. (1973) Fishery resources of the upper Bay of Bengal. Indian Ocean programme, Indian Ocean Fisheries Commission, Rome, FAO.IOFC/DEV/73/28.

White, T.F. and Khan, M.G. (1985) Marine fisheries resources survey. Demersal trawling survey Cruise Rep.1. FAO/BGD/80/025/CRI. 67 pp. Chittagong.

THE AUTHORS

Dihider Shahriar Kabir

*School of Environmental Science and Management,
Independent University, Bangladesh (IUB),
Plot # 3 & 8, Road 10,
Baridhara, Dhaka-1212, Bangladesh*

Syed Mazharul Islam

*School of Liberal Arts and Science,
Independent University, Bangladesh (IUB),
Plot # 3 & 8, Road 10,
Baridhara, Dhaka-1212, Bangladesh*

Md. Giasuddin Khan

*Centre for Environment and Geographical Information
Systems Support,
House 49, Road 27,
Banani, Dhaka-1212, Bangladesh*

Md. Ekram Ullah

*Environment Section,
Water Resources and Planning Organisation (WARPO),
House 4 A, Road 22,
Gulshan-1, Dhaka-1212, Bangladesh*

Dulal C. Halder

*Independent University, Bangladesh (IUB),
Plot 3 & 8, Road 10,
Baridhara, Dhaka-1212, Bangladesh*

Chapter 73

THE GULF OF THAILAND

Manuwadi Hungspreugs, Wilaiwan Utoomprurkporn and
Charoen Nitithamyong

The Gulf of Thailand is a semi-enclosed shallow sea bordered by four countries in Southeast Asia: namely Thailand, Malaysia, Cambodia and Vietnam. Twenty-one Thai rivers discharge directly into the Gulf and the Mekong River, which discharges into the adjacent South China Sea and has a major influence on the circulation and salinity of the Gulf water. Several important habitats such as mangrove, coral reefs and seagrasses exist in the coastal area of the Gulf. These habitats have degraded to a certain extent as a result of anthropogenic causes. Political conflict prevented co-operation among the littoral States for almost 40 years before joint surveys could resume in 1998. The Gulf used to be very productive for fisheries but has declined greatly in recent years. Non-living resources, e.g. oil and gas production, have replaced fisheries since 1990. Co-operative scientific research to gain more understanding of the Gulf is progressing with the help of international agencies like the IOC/WESTPAC and SEAPOL.

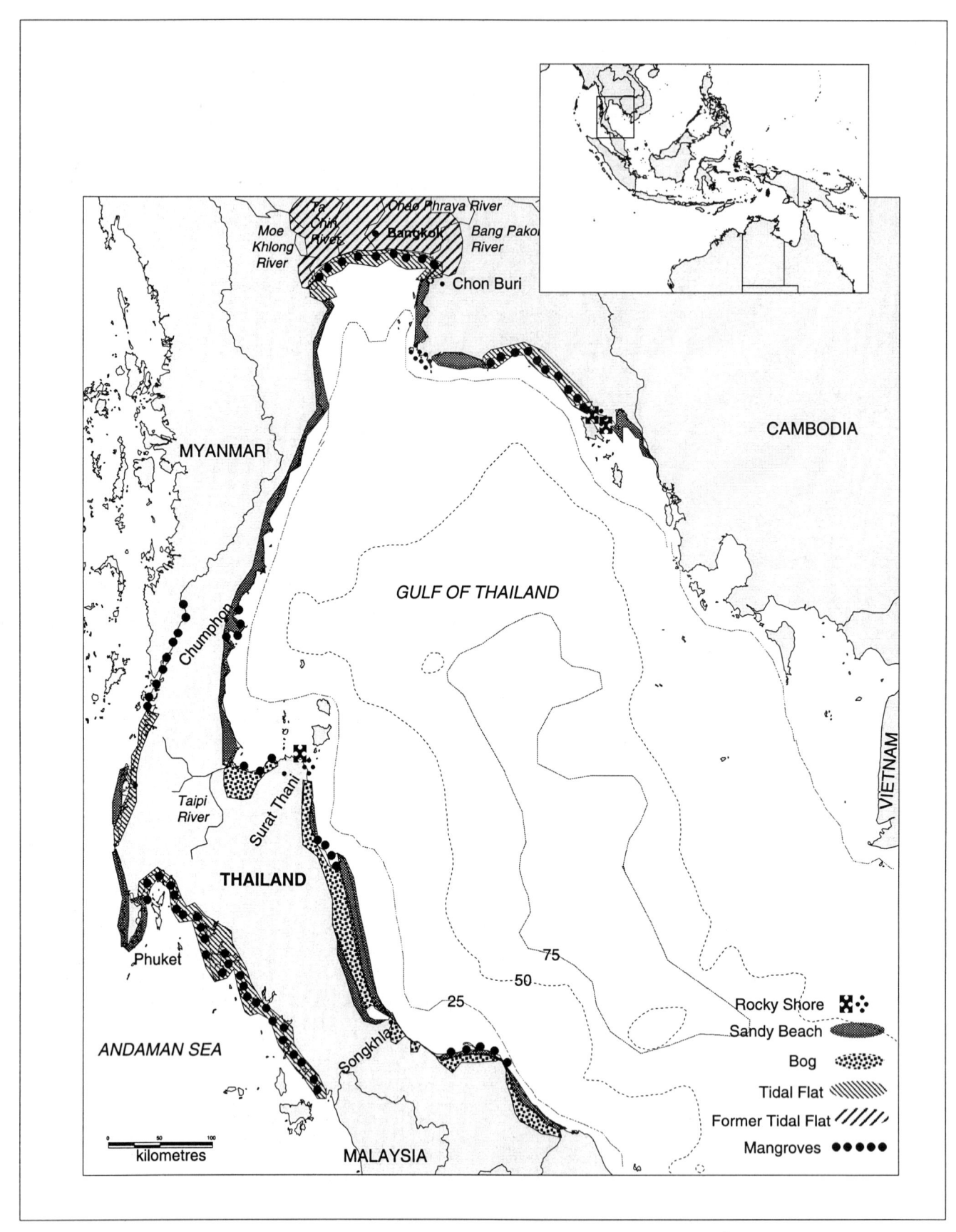

Fig. 1. The Gulf of Thailand.

THE DEFINED REGION

The Gulf of Thailand is a semi-enclosed tropical basin on the Sunda Shelf between latitudes 6° and 14° N and longitudes 99° and 105° E with the maximum depth of about 80 m and a sill depth of 67 m at the entrance (Fig. 1). The geographical mouth of the Gulf is a line drawn from just below the Thai–Malaysia border to Cape Camau of Vietnam, opening to the South China Sea. The lengths of the coastlines bordering the Gulf belonging to each country, namely Thailand, Malaysia, Vietnam and Cambodia are 1450 km, 20 km, 340 km and 280 km, respectively. The total sea area is 320,000 km² with the exclusive economic zones claimed by the four countries overlapping to a certain extent. The political difficulties between 1960 and the early 1980s prevented any meaningful co-operation in the management of resources, living and non-living. Only in the mid 1990s did international co-operation on the Oceanography of the Gulf of Thailand start with the Intergovernmental Oceanographic Commission, Sub-Commission for the Western Pacific (IOC/WESTPAC) Project and the European Union (EU INCO-DC) Cuulong Project on the Mekong Delta. Although Cambodia is not a member of the IOC/WESTPAC, its representatives have always been invited to participate in most of the activities of these organisations through the partial sponsorship of the Southeast Asian Programme in Ocean Law Policy and Management, SEAPOL which includes the Gulf of Thailand among its priority areas. Lack of infrastructure and an uncertain political climate prevents any real participation with Cambodia.

Due to the above reasons, large sections of the eastern part of the Gulf of Thailand have been largely unexplored since the NAGA Expedition in 1959–1960 and before the concept of the Exclusive Economic Zone was implemented. There have been some Russian oceanographic surveys of the nearshore Vietnamese coastal area.

Geological Description

The Gulf of Thailand is overlaid with a thick layer of sediment deposited since the Tertiary Period. Basement rocks beneath the sediments are believed to have formed during the Oligocene Period or earlier (Bunopas and Vella, 1983).

Studies of the recent sediment, from Holocene to the present, were done using sediment grabs for surface sediments and on grain size composition (sand–silt–clay) and total organic content (Emery and Niino, 1963; Umnuay, 1984) but more recent studies worked on trace metal contents and clay mineral speciation (Sompongchaiyakul, 1989). The Hydrographic Department, in a more recent study (1993), surveyed 39 stations from the upper Gulf of Thailand and found that although the near river-mouth stations consisted of clay, the grain size increased with distance away from the river mouth (Fig. 2). The contour lines of the sediment grain size distribution are aligned in the southwest–northeast direction which probably relates to the physical oceanographic condition of this area (Snidvongs and Vongprommek, 1997). For the Lower Gulf, Umnuay (1984) found mostly clay and sandy clay, and sediments along the western coast are usually finer than along the eastern coast.

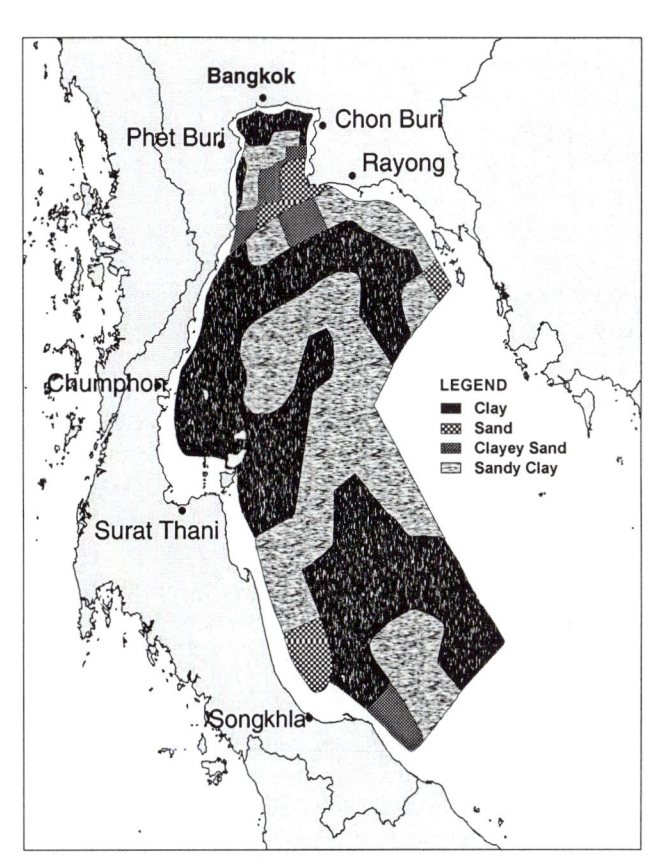

Fig. 2. The sediment grain size distribution of the Gulf of Thailand (after Umnuay, 1984).

Calcareous particles of shell fragments are widely distributed throughout the Gulf with the average carbonate content in silt and clay fraction being about 13% in weight, with no relationship with the sediment grain size. A higher percentage was found offshore (Sompongchaiyakul, 1989). Biogenic opal in the Upper Gulf was about 2–3%

Most of the sediments brought in by the major rivers into the Upper Gulf are trapped there (Windom et al., 1984) with a small fraction going into the Lower Gulf. Sedimentation rate, based on the excess Pb-210 activity, ranged from around 1 g/cm²/year near the mouth of the Chao Phraya River to less than 0.05 g/cm²/year near the centre of the Gulf (Windom et al., 1984; Srisuksawad et al., 1997).

Physical Oceanography

The understanding of the hydrography and circulation of the Gulf of Thailand has been quite limited due to the political barrier and suspicion among the bordering countries. Data from the 1995–1997 Thai–Malaysia Co-operative Study of Fisheries Oceanography of the Gulf of Thailand and the South China Sea under Southeast Asian Fisheries Development Centre (SEAFDEC) have not been fully analysed and interpreted, and the IOC/WESTPAC Co-operative Project on Oceanography of the Gulf of Thailand recently completed a joint survey across the mouth of the Gulf in 1999. It is hoped that, after the interested parties have had time to work on these hydrographic data, much more understanding will result. In the past 40 years, only the NAGA Expedition, conducted by the Scripps Institution under joint Thai–US–Vietnam co-operation during 1959–1961, covered the whole Gulf but that survey was made before the current CTD instrumentation was available.

Forcing mechanisms in the coastal seas include freshwater input (precipitation, river discharge from 21 rivers, sub-surface freshwater input and land overflow), winds, tides and open ocean influence. The Gulf of Thailand is subject to large seasonal change in the wind system (the two Monsoons, which reverse direction) and large freshwater fluxes.

Stansfield and Garrett (1997) used the historical NAGA data set to examine the physical oceanography of the Gulf of Thailand:

(a) *Tidal stirring versus buoyancy input.* Pukasab and Pochanasomburana (1957) used data from 12 tide gauges in the Gulf to show that tides vary from purely diurnal to mixed diurnal/semi-diurnal with inequalities. Siripong (1985), presumably using various information from this paper and from Wyrtki (1961), presented a simplified version of the tides in Fig. 3. Near Bangkok, tidal range is about 1.5 m, but amplitude declines to 0.5 m or less in the south.

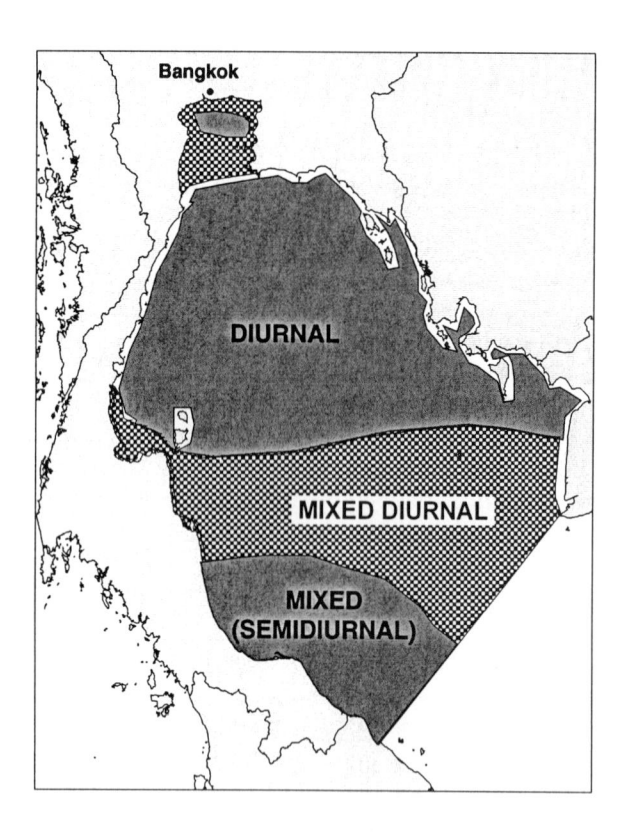

Fig. 3. Tidal type in the Gulf of Thailand (data from Siripong, 1985).

(b) *Wind forcing.* The Southwest Monsoon lasts from May/June to September bringing rain to the whole Southeast Asia land mass and sea, while the Northeast Monsoon lasts from November to February bringing the dry cool wind from the northern land mass. The speed and direction of the monsoon winds in the Gulf are variable due to the surrounding land and coastal mountains. Nasir, Rojana-anawat and Snidvongs (1999) analysed the temperature, salinity and density profiles to determine the effect of the Northeast Monsoon and found it to be quite small with regard to the stratification of the offshore water. The pycnocline moved a little deeper due to the higher wave energy caused by the stronger Northeast wind. In the nearshore area, the pycnocline was broken down. The average wind speed of the Northeast Monsoon was 9 m/s while that of the Southwest Monsoon was 6 m/s. Heavy rainfall in the Lower Gulf during the Northeast Monsoon decreased the surface salinity.

(c) *Currents.* Wyrtki (1961) showed the circulation in the Gulf of Thailand to be weak and anticyclonic from May to September and cyclonic from October to January, but there are several other versions of circulations suggested, and several numerical models have been proposed. No single concept is widely accepted yet.

Snidvongs and Sojisuporn (1999) made numerical simulations of the net current using a hydrological model using tri-monthly averages of observed temperature, salinity, depth and wind. They worked on the data collected from the Thai-Malaysian Collaborative Survey (SEAFDEC) in 1995–1996 for the western part of the Gulf, supplemented by the long-term data from WDC-A (Oceanography) and JODC for the same months. They reported that, along the western half of the Gulf, surface current was westward toward the coastline but the mid-depth water flows in the opposite direction. The water balance indicated a net gain of water in the Gulf and this was in agreement with the observation that the sea level was higher during the Northeast monsoon. Deep water temperature and salinity below the pycnocline suggested an intrusion of intermediate water from the South China Sea along the shore of Vietnam and Cambodia. Deep water circulated anticlockwise and exited from the Gulf along the West Coast.

As for the Southwest Monsoon period, the sea surface and mid-depth temperatures of the Gulf are higher near the centre of the Gulf and spread along the Southwest coast, towards the Thai–Malay border. Surface and mid-depth salinity suggests that the major sources of freshwater input comes from the head of the upper Gulf and along the East coast through Cambodia and Vietnam. A mid-depth clockwise eddy is evident in the Northern part of the Gulf. The general flow pattern in the mid-depth layer near the mouth is southwest, from the Cambodia–Vietnam EEZ i.e. to the Thai–Malay Peninsula. Inflow of intermediate water from the South China Sea into the sub-pycnocline layer was extended to 10°N, with considerable velocity.

SEASONALITY, CURRENTS, NATURAL ENVIRONMENTAL VARIABLES

Seasonality

Southeast Asia is under the strong influence of the monsoon system when the Southwest monsoon blows towards the continent during May to October, and the Northeast monsoon comes from the northern land in November to January. This system is unique to Asia because there is an unequal distribution of land mass in the two hemispheres. The northern half is mainly land which is heated up more easily, while the southern half is mainly ocean (except the Australian Continent). Although the general direction of the monsoon in the Gulf of Thailand is mainly northeast and southwest, the actual directions can vary from place to place due to coastal morphology.

Other seasonal factors affecting the climate of the Southeast Asian region are the tropical cyclones or typhoons whose season starts in May and ends in December, though most occur between August and October. However, the Gulf of Thailand is usually not in the direct path of these storms, being deflected by the Philippines and the mountains of Vietnam.

Freshwater Discharge

The major rivers draining into the Upper Gulf of Thailand are (from left to right in Fig. 1) the Mae Klong, Ta Chin, Chao Phraya, and Bang Pakong, accounting for more than 80% of the total discharge of river water into the Gulf, with approximately 10^8 m^3/y. However, summing up, all the Thai river discharge into the Gulf is not enough to justify the surface salinity distribution actually observed. Hence, there must be a considerable amount of freshwater discharge, either by sub-surface discharge or overland flooding from the coastal area, or from another river. Stansfield and Garrett (1997) showed that, if half of the Mekong discharge found its way into the Gulf of Thailand, it would make the predicted monthly depth-average salinity distribution more in agreement with the observed values (Fig. 4). In support of this hypothesis, Landmann et al. (1998) confirmed that in both seasons, most of the Mekong water found its way into the Gulf.

COASTAL HABITATS

The first marine-related study in the Gulf of Thailand dated back to 1899 when Johannes Schmidt, on a Danish Expedition in Siam during 1899–1900, identified phytoplankton and algae off Chang Island (Suvapepun, 1994). Later in 1951, the Galathea Deep Sea Expedition also made an oceanographic survey in the Gulf. During 1959–1961, a co-operative oceanographic survey between the United States, Thailand and Vietnam, namely the NAGA Expedition, made a complete survey in the Gulf and the South China Sea and initiated the interest in oceanographic studies in Thailand and Vietnam.

Several types of coastal habitats are found in the Gulf of Thailand, namely rocky shore, sandy beach, mud flat, bog flat, mangrove, seagrasses, and coral communities (Fig. 1). Mangrove and coral communities have been intensively studied recently as they are important nursery grounds for economically important species.

Mangrove

Mangrove was the first coastal habitat intensively studied in the Gulf of Thailand. The study was started with the UNESCO Workshop on Mangrove Ecology at Phuket Marine Biological Station in January 1976. Not long after that meeting, several international co-operative projects were launched and resulted in better understanding of the role of mangrove in the ecosystem. While the knowledge grew and people learnt of its usefulness, the resource itself soon diminished at a rapid rate. In the past, a high percentage of the coastal area was covered by mangrove forest: in 1961, for example, mangrove forest covered 36% of the shoreline. However, within 30 years the coverage decreased by 47% due to human encroachment and shrimp-farming. At present, the remaining mangrove is only as shown in Fig. 1. There have been efforts to transplant mangrove trees in the Inner Gulf of Thailand.

Seagrasses

Seagrasses are flowering plants found in the shallow waters. Data on the seagrass in the Gulf are few (Aryuthaka and Poovachiranon, 1994) and only done on specific areas. Five genera of seagrasses are found in the Gulf: *Halophila* spp., *Halodule* spp., *Enhalus* spp., *Thalassia* spp., and *Ruppia* spp. (Liewmanomont et al., 1991; Aryuthaka and Poovachiranon, 1994). The most widely distributed and abundant species is *Halophila ovalis*. Seagrass beds near Samui Island are in a better state than those on the East coast.

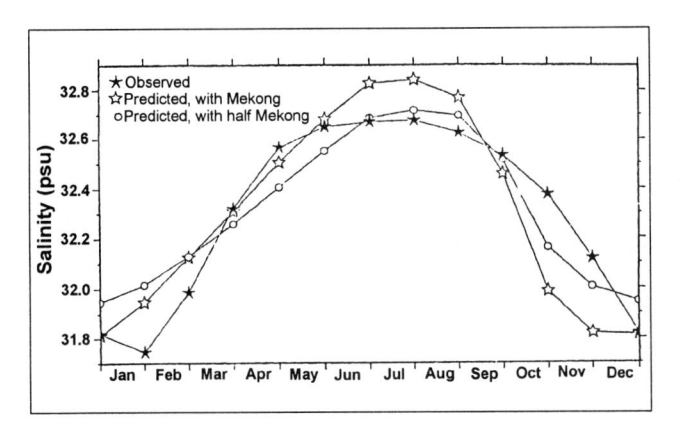

Fig. 4. Observed and predicted depth-averaged salinity (psu) in the Gulf of Thailand (Stansfield and Garrett, 1997).

Coral Reefs in Thailand

Niphon Phongsuwan

Phuket Marine Biological Center, P.O. Box 60, Phuket, 83000, Thailand

Coral reefs are one of the most productive marine systems in Thai waters. Reef status has been monitored under a coral reef resource management programme between 1995 and 1999 by the Phuket Marine Biological Center, Department of Fisheries. Maps, with detailed description of almost all coral reefs in Thailand, have been recently published (Chansang et al., 1999). Reefs in Thailand are separated into two main regions: the Andaman Sea and Gulf of Thailand (Fig. 1).

ANDAMAN SEA

The Thailand coastline on the Andaman Sea is approximately 700 km in length. In this region the width of the continental shelf varies from 27 km to 130 km, the sea is strongly influenced by the southwest monsoon during May–October, and the coastline is subjected to a semi-diurnal tide, with a maximum tidal range of 2.8–3 m. The northern half of the coastline is more open to oceanic water, with a salinity of about 32–33 ppt, while the southern half, which is more influenced by run-off from the mainland, has a salinity of about 29–32 ppt. (Limpasaichol et al., 1991). On the shorelines of both the mainland and its small islands, varied habitats are found including coral reefs, mangroves, seagrass beds and rocky shores. The coral reefs are generally found in clear water where the shoreline is sheltered from the southwest monsoon; the major reefs usually fringe the east coasts of the islands, while most of the west-facing coasts harbour only sparse communities of corals. Soft corals are found on rocky areas in reef and non-reefal areas.

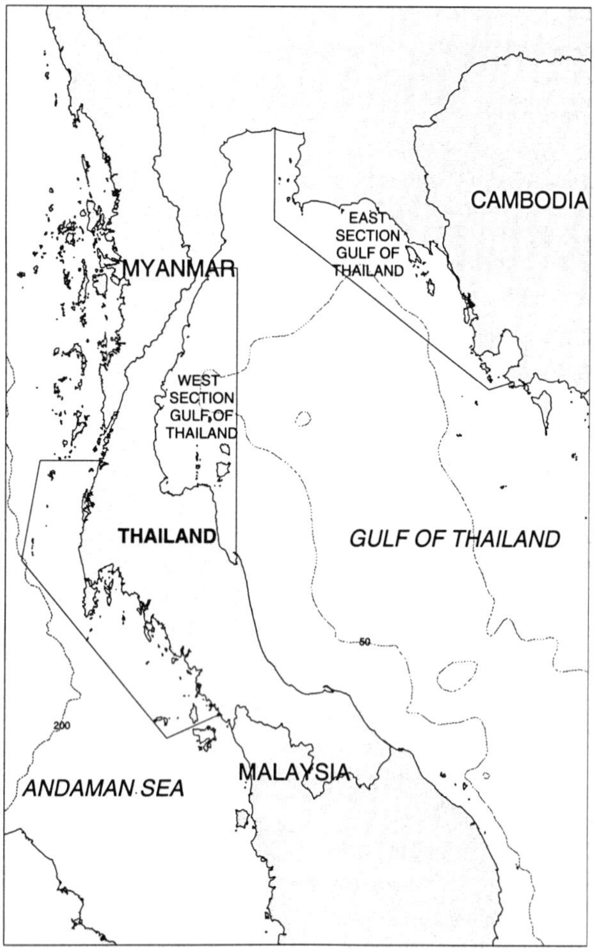

Along the Andaman Sea there are hundreds of islands. The Similan and the Surin Archipelagos are located near the continental edge, and thus have a water clarity which is much higher than in inshore areas. Coral reefs around each of these archipelagos extend to 20–30 m deep. In contrast, coral reefs nearer shore live in turbid waters and extend only to depths of 3–10 m. Many of both kinds of reefs show distinct zonation, and have a wide reef flat ranging from 20–300 m, and most of the reefs located in near-shore areas are usually aerially exposed during the lowest tides. In some cases reef zonation is not clearly marked; the reef may begin at the shore and from this point slope gently down towards the base. This type of reef is usually found on the small islands where shelter is limited. Reefs located near mangroves usually consist of reef flats only, without a slope. About 240 species of hard corals have been collected from the Andaman Sea.

GULF OF THAILAND

The coastline along the Gulf of Thailand is approximately 1600 km in length. This sea is generally shallower than the Andaman Sea, with the deepest part being only 86 m. The Gulf is influenced by both diurnal and mixed tides, as well as by the Southwest (May–September) and Northeast monsoon October–February). The northern part of the Gulf is affected by river run-off, causing high primary productivity. Coral reefs are mainly found fringing islands, of which there are two major groups: an eastern group of about 100 islands and a western group of about 150 islands. Due to the influence of the Southwest monsoon, the islands in the eastern part of the Gulf support coral reefs mainly on the sheltered east and north coasts. Conversely, for islands in the west part of the Gulf, coral reefs are mainly found on the west and south coasts due to the influence of the Northeast monsoon. The windward sides of the islands are rocky with sparse communities of corals. Reefs around the eastern islands are usually shallow, extending to depths of only 2–5 m, while those around western islands develop into deeper water of about 8 m depth. Further offshore, reefs may extend down to depths of 15 m. Because of their limited extent, zonation of these reefs is usually not distinct, and diversity of scleractinian corals is not as high as in the Andaman Sea, with only 90 species reported in a preliminary survey (Jirawat, 1985).

Fig. 1. Map showing the main areas in the Gulf of Thailand and in the Andaman Sea where coral reefs of Thailand are located.

The status of entire coral reefs on the Andaman Sea coasts was surveyed on over 130 islands and on a few additional places on the mainland shore by the "Manta-tow method" (English et al., 1994), in which the ratio of cover of live to dead coral was measured, following which sites were ranked into status categories (Table 1). The areal size of the reef was analyzed after the boundary of the reef was mapped. It is estimated that the total area of these reefs is about 78 km^2. In the same manner, in the Gulf of Thailand, reefs around 250 islands were surveyed with the total area of the reefs in the Gulf being estimated to be about 75 km^2. Communities of corals growing on rocky coasts were not included in this estimate.

From this survey it appeared that reefs in the Gulf of Thailand were in better condition than those in the Andaman Sea. Poor reef status could be attributed to many causes—to natural disturbances and catastrophic storms. In 1986 a Southwest monsoon storm damaged reefs in the Andaman Sea (Phongsuwan, 1991) The southern site of Rawi Island has shown a slow recovery rate after the storm which reduced live coral cover from 59% on average (Phongsuwan, 1986) to 10% in 1988 (Chansang et al., 1989); this then increased to 16% in 1994 (Chansang et al., 1999). Similarly, on a small reef on the southeast cape of Phuket Island, live coral which had almost disappeared after the storm, increased to 3% cover in 1988 and to 52% in 1991. The southern part of the Gulf of Thailand was severely damaged by Typhoon Gay in 1989. Here too, there is some recovery, with fast-growing species such as *Pocillopora damicornis* and *Acropora formosa* extensively covering the shallow reef zone.

Table 1

Status of the edge–slope zone of coral reefs measured as % of total area of this zone. A ratio of live:dead corals of ≥3:1 = very healthy reef, 2:1 = healthy reef, 1:1 = fair condition, 1:2 poor condition, and 1:≥3 = a very poor condition.

Region	Very healthy	Healthy	Fair	Poor	Very poor
Andaman	4.6	12.0	33.6	26.5	23.3
Gulf	16.4	29.0	30.8	10.9	12.9

OTHER IMPACTS

An outbreak of Crown-of-thorns starfish *Acanthaster planci* damaged the reefs over a wide area, mostly in the Andaman Sea, during 1984–1986 (Chansang et al., 1986). The abundance of *Acanthaster planci* and degree of damage inflicted differed from place to place.

Abnormal sea surface temperature increases during the dry season in 1991, 1995 and 1998 caused corals to bleach in the Andaman Sea. In May 1995, sea water temperature exceeded the seasonal maximum by 0.66°C (Brown et. al., 1996). In the Gulf of Thailand, this phenomenon occurred in 1998. Coral bleaching events in the Andaman Sea in 1991 and 1995 generally caused only 10% coral mortality at each site, since most corals recovered once temperature declined. A total of 94 Scleractinian coral species showed bleaching. However, at some sites where *Acropora* was abundant, mortality of this most sensitive species reached levels of 15–20%. About 80% of *Acropora* at sites at Surin and Similan Islands were dead in 1995 (Phongsuwan, 1998). In 1998 coral bleaching was not severe in the Andaman Sea due to a cool upwelling early in the year, in contrast to the Gulf of Thailand where there was severe damage to corals from bleaching.

In the Andaman Sea, sedimentation effects have been marked at Phuket Island due to off-shore mining activity in the 1980s. Other impacts have come from tourism and land reclamation at some sites both in the Andaman Sea and Gulf of Thailand, and from anchoring from tour boats, though mooring buoys have now been installed to limit this. In Thailand, trampling on corals in shallow water by snorkellers and boat groundings are still major factors causing damage to reefs.

In the past, dynamite blasting for fish was common. However, at present this type of illegal fishing is rare. Impacts from trash fishing, nets covering coral reefs and coral trampling by fishermen searching for shellfish are still common and give cause for concern.

REFERENCES

Brown, B.E., Dune, R.P. and Chansang, H. (1996) Coral bleaching relative to elevated sea surface temperature in the Andaman Sea (Indian Ocean) over the last 50 years. *Coral Reefs* **15**, 151–152.

Chansang, H., Boonyanate, P., Phongsuwan, N., Charuchinda, M. and Wangboonkong, C. (1986) Infestation of *Acanthaster planci* in the Andaman Sea. Paper presented at the 2nd International Symposium on Marine Biology in Indo-Pacific, Guam. (Abstract published in *Bulletin of Marine Science* **41**(2), 634.)

Chansang, H., Phongsuwan, N. and Boonyanate, P. (1989) Inventory and monitoring on coral reefs. In ASEAN-Australia Cooperation Program on Marine Science. Technical report submitted to National Environmental Board. 132 pp.

Chansang, H., Satapoomin, U. and Poovachiranon, S. (eds.) (1999) *Maps of Coral Reefs in Thai's Waters, Vol. 1. Gulf of Thailand*. Coral Reef Management Project, Department of Fisheries. 284 pp.

Chansang, H., Satapoomin, U. and Poovachiranon, S. (eds.) (1999) *Maps of Coral Reefs in Thai's waters, Vol. 2. Andaman Sea*. Coral Reef Management Project, Department of Fisheries. 198 pp.

English, S., Wilkinson, C.R. and Baker, V. (eds.) (1994) *Survey Manual for Tropical Marine Resources*. Australian Institute of Marine Science, Townsville. 368 pp.

Limpasaichol, P., Khokiattiwong, S. and Bussarawich, N. (1991) Water quality along the Andaman coastline, Thailand, in 1982–1986. Technical Report no. 2/2534, Phulet Marine Biological Center. 28 pp.

Phongsuwan, N. (1986) Coral reef resources of the Tarutao National Park, Thailand. *Proceedings on Coral Reef Management in SE-Asia*. Borgor, Indonesia, Biotrop Special Publication no. 29, pp. 141–156.

Phongsuwan, N. (1991) Recolonization of coral reef damaged by a storm on Phuket Island, Thailand. *Phuket Marine Biology Centre Research Bulletin* **56**, 75–83.

Phongsuwan, N. (1998) Extensive coral mortality as a result of bleaching in the Andaman Sea in 1995. *Coral Reefs* **17**, 70.

Coral Communities

The most studied coastal habitat in the Gulf is probably the coral communities. The ASEAN–Australian Co-operative Programme on Marine Science during 1986–1994 has supported research on taxonomy, distribution, and coral mapping in the Gulf. Due to human disturbance such as dynamite fishing, boat anchoring and commercial coral collection, coral degradation in several tourist areas has occurred. The Gulf of Thailand, although not as famous as the Andaman sea side of the southern coast, is well known for its beautiful beaches and coral communities. The water in the Gulf is not as clear as the Andaman Sea, but Ko Tao, Ko Nang Yuan, Ko Phangan, Ang Thong Archipelago National Marine Park and Ko Samui are popular tourist sites. All these cover an area of approximately 75 km^2. Some islands off Pattaya and Rayong on the East coast have some good coral too, but not as unspoiled as the southern islands. Ko Tao, being 80 km offshore, has clear water, and about 79 species of corals have been recorded there. Some smaller islands with good beaches are swallows' nest concession areas so they are well protected from the tourists. Branching corals, especially of the *Acropora* group are easily damaged by dynamite fishing, although they grow fairly quickly. In water made turbid from freshwater discharge, *Porites lutea* is dominant due to its tolerance. This is also the case of Ko Samui and Ang Thong Archipelagos. Coral transplantation in several areas has been attempted in order to revive the damaged coral communities, with some success. Coral bleaching in large areas has recently been observed in the Gulf as well as in the Andaman Sea, possibly due to an increase in water temperature from the El Niño phenomenon. As part of the Gulf of Thailand rehabilitation programme, the Department of Fisheries has completed the mapping of coral communities in Thai waters which will be useful as a management tool.

Biological Studies

Most studies on plankton in the Gulf emphasised species composition, abundance, distribution and seasonal changes, but very few looked at the ecology or relationship with the environmental factors. Boonyapiwat (1999) regularly studied phytoplankton in the Gulf for over 20 years, first with 80 μ-mesh plankton nets, and more recently with 20 μ nets, so more species were discovered. In 1995–1996, she found 260 taxa comprised of 2 species of blue-green algae, 133 species of diatom and 107 species of dinoflagellates. Cell counts for the flood season of September–October, Southwest Monsoon, were 214–33,520 cells/l while the April–May gave 178–14,223 cells/l, and these are lower than are found in the Chao Phraya Estuary. The dominant species found were *Oscillatoria erythraea*, *Thalassionema frauenfeldii*, *Chaetoceros lorenzianus* and *C. compressus*. Toxic dinoflagellates were also found at extremely low cell densities, and evenness indices were high in the coastal areas in the April–May survey. Although much has been reported on the occurrence of toxic blooms in Southeast Asian waters, in the Gulf of Thailand only one toxic bloom was recorded (and on a very small scale) in 1983 at the mouth of Pranburi River, although there was some eutrophication. On-going research is trying to find the reason for this exceptional case.

As a nation that depends heavily on fisheries, Thai scientists are interested in studies on economically important larvae in the coastal areas, in order to determine fisheries potential. Studies on pathogenic bacteria and fish diseases also receive much attention. Most benthic studies concentrate on the near-shore areas or river mouths, and the offshore area was monitored only irregularly, and only received attention when possible pollution effects from the oil/gas platforms made it compulsory to examine benthos as an indicator of impact. So far, no noticeable effect has been found.

OFFSHORE SYSTEMS

The central area of the Gulf of Thailand consists of two permanent layers of water separated by a thermocline and pycnocline at about 45 m deep during the Southwest Monsoon. The pycnocline moves deeper to 60 m during the Northeast Monsoon (Snidvongs, 1998). The Gulf of Thailand used to be very productive in fisheries but in the past 30 years there has been a drastic decline in fisheries catches due mainly to overfishing, since the number of fishing boats has increased about 20 times at the same time that the fishing methods also became more destructive. Although some people laid the blame on deteriorating water quality, the latter really played a small part. All the offshore water quality is normal although certain river mouth areas are low in dissolved oxygen and high in dissolved organic content and nutrients from domestic wastes (Utoomprukporn, 1997). There had been an erroneous report on high concentrations of trace metals in sea water and river water, but several others reported normal levels of trace metals in nearshore and offshore waters of the Gulf of Thailand (Hungspreugs et al., 1998, 1989; Hungspreugs and Utoomprukporn, 1998; Utoomprukporn, 1999).

Since the mid 1980s, there has been considerable petroleum exploration and production in the Central Gulf of Thailand and in the adjacent South China Sea. The main operators in the Gulf now are Unocal, the Petroleum Authority of Thailand, Thai Shell, Chevron and Triton. Several more companies are exploring, and recently Thailand and Malaysia signed a landmark agreement to buy natural gas from the Joint Development Area of the Gulf of Thailand.

Oil and Gas in the Gulf

During the Second World War, when a large part of Thailand was briefly occupied by the Japanese Forces, foreign oil companies ceased operation in Thailand, leading to shortages. They returned after the war ended but the Thai Government, in 1959, began to involve itself in the petroleum industry once more. In 1960, the Government extended an invitation to the private sector to explore for and refine petroleum in Thailand. This marked the beginning of the modern petroleum industry.

The exploration was made on land and offshore. For the offshore exploration, the first discovery in the Gulf was made by Union Oil (later Unocal) in the Pattani Basin in January 1973 when the deposits of natural gas were named Erawan, after the three-headed elephant of Thai folklore. Later, another significant discovery was found 160 km away, and named Bongkot, which came on stream in 1990, and which was operated by Total until 1998. There had been a theory that the Gulf only contained gas fields but in 1987, Shell struck oil at Nang Nuan-I exploration well. Production soon followed but was terminated after six months due to early influx of water. However, another oil discovery was found in the nearby Nang Nuan-B Block in 1993. From the report "Inside Petroleum" made by the Petroleum Institute of Thailand in 1998, the oil reserves in the Gulf were estimated at about 40 m bbl of proven reserves and 205 m bbl of probable and possible reserves. The gas reserves in the Gulf (Thai territory only) were 6026 bcf of proven and 13,915 bcf of probable and possible reserves.

One problem with the petroleum production in the Gulf of Thailand, was that the sediment contained mercury at 10 times the world average. The heavy metal concentrations in the sediments are shown in Table 1 (TOTAL MEMEP report, 1997). From the analyses of 'reactive' and 'total' mercury in seawater, of total mercury in sediment and of methyl mercury and total mercury in fishes in both the production platform area and reference area of the Bongkot Field (112°E and 8°N), no unusually high concentration of mercury was discovered. In the water, the content was about 1 ng per litre although the 'produced water' emulsion from the platform contained very high amounts of mercury in 'reactive' and 'total' forms. Since the mercury released did not stay in the water, biota or sediment, the mass balance estimate suggested that the excess mercury produced must have escaped into the air above the sea. Soon afterwards, the company re-injected the wastewater back into the empty holes on the seafloor after the initial treatment to remove excess mercury. The 1998 environmental campaign around the Bongkot Field confirmed the result obtained in 1997.

FISHERIES

The Gulf of Thailand has been a major source of marine fishery products for the four countries bordering it. The fisheries around the Gulf are multi-gear, multi-species fisheries mostly conducted by small-scale fishermen estimated at more than four million (Menasveta, 1998; National Statistical Office, 1997). In these countries, even though commercial fisheries constitute only a small part of the fishery sector, they provide a considerable economic return (Menasveta, 1998). Modern-day marine fisheries development in the Gulf started in 1962 when the otterboard trawl net fishing was introduced and quickly gained acceptance from local fishing communities in Thailand and Malaysia, while in Cambodia and Vietnam rapid development took place in the eighties (Menasveta, 1998). At present, trawl is still the major fishing gear type in Thailand, accounting for 56% of total fishing gears used while purse seine (32%) is also popular for harvesting pelagic resources (Fisheries Economics Division, 1998). Important pelagic fishery developments were the introduction of luring purse seines in the 1970s and anchovy fisheries with luring light in 1983 (Ibrahim, 1999). During the past three decades, development in fisher related technology and increasing demand for fishery products, among other factors, have caused a decline in the fishery resources as fish diversity (Vidthayanon, 1999) and abundance (Longhurst and Pauly, 1987) decrease. Consequently, conflicts among fishermen using different fishing gears have increased due to the depleted resources. Changes in species and size compositions are also observed, as cephalopods and anchovies constitute major catches at present, instead of demersal fishes as in the past (Longhurst and Pauly, 1987).

The major fishery resources in the Gulf can be categorised into 4 major groups: demersal fishes, coastal pelagic fishes, shrimps, and cephalopods, as described in Chullasorn and Martosubroto (1986).

Six dominant demersal fishes commercially harvested in the Gulf are threadfin breams (*Nemipterus hexodon, N. japonicus, N. mesoprion, N. nematophorus,* and *N. peronii*), big-eyes (*Priacanthus tayenus*), lizardfishes (*Saurida undosquamis,* and *S. elongata*), snappers (*Lutjanus* spp.), groupers (*Epinephelus* spp.), and rays (Fisheries Economics Division, 1996). Vidthayanon (1999), working on the joint Thai-Malay otterboard trawling survey during 1995–1996, found 300 species of fish from 18 orders and 89 families, showing a

Table 1

Mercury and some selected metals in the surficial sediments from the Lower Gulf of Thailand (after TOTAL MEMEP, 1997)

Location	%Al	Trace metals ppm dry wt.									
		V	Cu	Ni	Zn	AS	Cd	Pb	Ba	Sn	Hg
Near Bongkot platform	3	31.2	19	19	66	5	0.2	49	3117	1.4	119
Other stations	3–7	36–68	15–24	22–38	51–57	5–7	0.1–0.2	17–20	132–374	1–2	0.01–1.50

diversity decline from 380 (Wongratana, 1968, 1985). The maximum catch per hour was 175.3 kg in the Malaysian waters and the minimum was 4.33 kg from Thai waters, but the average percentage of economic fish in Thai waters was higher at 55–82%.

Dominant pelagic groups of fish are anchovies (*Encrasicholina heteroloba* as a dominant species), sardines (*Sardinella gibbosa*, *S. frimbriata* and *S. albella*), Indo-Pacific or short-body mackerel (*Rastrelliger brachysoma*), Indian mackerel (*Rastrelliger kanagurta*), Caragids (*Decapterus maruadsi*, *Megalaspis cordyla*, *Selar crumenophthalmus*, *Atule mate*, and *Selaroides leptolepis*), king mackerel (*Scomberomorus commerson*, *S. guttatus*), and coastal tunas (*Euthynnus affinis*, *Auxis thazard* and *Thunnus tonggol*) (Fisheries Economics Division, 1996).

The important shrimp species in the Gulf are *Penaeus* spp. and *Metapenaeus* spp. The commercially important species are banana prawn (*Penaeus merguiensis*), Indian white prawn (*P. indicus*), green tiger prawn (*P. semisulcatus*) and black tiger prawn (*P. monodon*), Jinga shrimp (*Metapenaeus affinis*), yellow shrimp (*M. brevicornis*), and greasyback shrimp (*M. ensis*). There are also small-size pelagic shrimps of the genus *Acetes* which are used for shrimp paste production.

Important cephalopods in the Gulf are squid and cuttlefishes while octopus is given less interest by the fisherman and consumer. Three commercially important species of squid are Indian squid (*Loligo duvauceli*), mitre squid (*L. chinensis*), and big fin reef squid (*Sepioteuthis lessoniana*) while two species of cuttlefish commonly found in the commercial catch are pharaoh cuttlefish (*Sepia pharaonis*) and curvespine cuttlefish (*S. recurvirostra*).

Other species that are commercially important but constitute smaller catches are crabs (*Portunus* spp. and *Scylla* spp.), molluscs (*Perna viridis*, *Anadara granosa*, *Crassostrea* spp., *Paphia undulata*), Jellyfishes (*Rhopilema* spp. and *Stomolphus* spp.), and flathead lobster (*Thenue* spp.) (Fisheries Economics Division, 1996).

POPULATIONS AFFECTING THE AREA

History shows that the Thai people have always been linked to the sea. There has been trading with ships since at least the Sukhothai Period over 700 years ago, according to the records written in stone and from evidence from china-ware in sunken ships in the Gulf. In the Southern Province of Nakorn Si Thammarat (formerly Srivichai Dynasty) there was regular sea trade with the Indonesian Archipelago. Later on, during the Ayutthaya Period, the Portuguese arrived and set up a Portuguese Village. In the Bangkok Period, beginning over 200 years ago, the English and the French arrived.

In recent years, mariculture has expanded in all aspects; fishes, shrimp, cockles, oysters, mussels. Shrimp farming has been expanding so fast, ahead of the controlling authorities, that a lot of damage is being done to the coastal environment. While the shrimp export was at its peak, long stretches of shrimp ponds lined the coastline of the Gulf, discharging wastewater directly to the sea or nearby freshwater canals or rivers. In the early stages of shrimp culture expansion, most of these culture ponds used the mangrove forest area whose soil became acidic after exposure to the air, which in turn acidified the water, so that finally the ponds had to be abandoned. There are now long stretches of abandoned shrimp farms on the southern coasts. Now these shrimp farmers have moved inland where there is a freshwater supply, to try their hand at raising tiger prawns, a saltwater species, by transporting seawater inland. This procedure is causing much damage to the rice farmers and to the fruit orchards in the vicinity. At present, the government prohibits this harmful practice.

Currently, there is also a major conflict between light luring anchovy fishermen and subsistence fishermen. Subsistence fishermen claim that their catches are affected by the light luring operation which uses a small net mesh size, as the light attracts not only anchovies but other small economic species as well, and results in low or no recruitment of their target species. This is a tragic sign of competition for depleted fishery resources, and more conflicts amongst fishermen using different gears can be expected in the near future.

RURAL FACTORS

In past times, Thailand and Myanmar were famous for their teak and other valuable forest products but these days are long gone for Thailand. Wood cutting has gone on, legally and illegally, so rapidly that Thailand has had to stop all forms of the cutting of natural forest.

It is well known among river scientists that Southeast Asia has the most turbid rivers, and this region produces over 25% of the world input of suspended matter to the ocean. On top of the increased erosion of land, water is becoming scarce for the growing population, so that a water storage system had to be implemented. This resulted in several dam reservoirs, built across most river systems, for irrigation, domestic use and hydroelectric generation. But, more recently, NGOs have become more powerful in protesting against dam building. One side-effect of these dams is that they reduce the flux of nutrients, especially silicates, into the coastal seas. Other forms of nutrients like phosphate and nitrate can be replaced by discharge of agricultural and domestic wastes downstream, but silicate can only come from natural sources. River systems show this decrease in silicate by a reduction of diatoms in the estuaries, but for the Gulf of Thailand, it was seen as a decrease of biogenic silica in bands of the coral *Porites lutea*, after construction of the first large dam across the Chao Phraya River system in the early 1960s.

In the coastal towns where fisheries are the main livelihood of the people, there are conflicts of interest due to the declining resources, both on the coast and in the sea. Large commercial vessels are taking away the fish, clams, etc. from the traditional fisherfolk who operate with small vessels. Companies cut down the mangrove forest to build culture ponds, and discharge aquaculture waste into the public waterways, possibly causing eutrophication.

COASTAL EROSION, LAND SUBSIDENCE AND SEA-LEVEL RISE

Generally, the west coast of the Gulf of Thailand is experiencing coastal erosion. Land subsidence in the Bangkok area has been quite considerable, in parts, due to the heavy pumping of underground water for new housing developments which, in their early stages, were not served by public tap water. But when the artesian wells were later prohibited, the sinking of the land almost stopped. This problem is serious for the metropolis because, on average, the city is only 1 m above sea level, and flooding is frequent after heavy rain. Because of land subsidence, it is difficult to say if the sea level is rising.

EFFECTS FROM URBAN AND INDUSTRIAL ACTIVITIES

During the past 40 years when housing demand has increased, the public utility services have not been able to keep up with the population. Traffic jams are the norm, leading to increased pollution in the inner city area of the Bangkok metropolis. Here, as with many economically constrained countries, wastewater treatment never catches up with increasing domestic demand, or with small agro-based industries which discharge high BOD water into the public waterways. During the low flow season, these waterways are very low in dissolved oxygen and high in organic content, nitrogen and phosphorous nutrients. Eutrophication occurs in coastal waters more frequently than in former times. The Pb and Cd concentrations in the upper layers of sediment cores in the Upper Gulf of Thailand are higher than the levels 40 years ago, (Hungspreugs and Yuangthong, 1983) although in the water itself, metals are not yet a problem (Hungspreugs et al., 1989).

PROTECTION MEASURES

It is widely known that coalition government is never effective in the implementation of any long-term policies for the good of future generations, if these policies run against the benefit of the present generation, especially if the present generation are politically influential. So each government tends to deal with immediate problems that confront them and leave the long-term plans for the unforeseeable future. Problems often grow larger with time until they become extremely hard to solve, for example, in Thailand, the conflicts in fisheries resources, coastal land use, forest conservation, industrial wastes etc. It is hoped that, once there is a majority government, these long-term plans will become reality.

REFERENCES

Aryuthaka, C. and Poovachiranon, S. (1994) Status of seagrasses in Thai waters. A paper presented at the Fifth National Seminar on Marine Science: Status of Thai Seas and the Future Trend 22–24 August 1994, National Research Council of Thailand. 23 pp.

Boonyapiwat, S. (1999) Distribution, abundance and species composition of phytoplankton in the South China Sea, Area I: Gulf of Thailand and East Coast of Peninsular Malaysia. In *Proceedings of the First Technical Seminar on Marine Fishery Resources survey in the South China Sea Area I: Gulf of Thailand and East Coast of Peninsular Malaysia, 24–26 November 1997, Bangkok, Thailand.* pp. 111–134.

Bunopas, S. and Vella (1983) Opening of the Gulf of Thailand—Rifting of Continental Southeast Asia, and late Cenozoic tectonics. *Journal of the Geological Society of Thailand* 6, 1–12.

Chullasorn, S. and Martosubroto, P. (1986) Distribution and important biological features of coastal fish resources in Southeast Asia. FAO Fish. Tech. Paper # 278. 84 pp.

Emery, K.O. and Niino, H. (1963) Sediments of the Gulf of Thailand and adjacent continental shelf. *Geological Society of America Bulletin* 74, 541.

Fisheries Economics Division (1996) The Marine Fisheries Statistics 1994 based on the sample survey. Department of Fisheries, Ministry of Agriculture and Cooperatives. 58 pp.

Fisheries Economics Division (1998) The Marine Fisheries Statistics 1995. Department of Fisheries, Ministry of Agriculture and Cooperatives.

Hungspreugs, M. and Yuangthong, C. (1983) A history of metal pollution in the Upper Gulf of Thailand. *Marine Pollution Bulletin* 14 (12), 465–469.

Hungspreugs, M., Utoomprurkporn, W., Dharmvanij, S. and Sompongchaiyakul, P. (1989) The present status of the aquatic environment of Thailand. *Marine Pollution Bulletin* 20 (7), 327–332.

Hungspreugs, M., Utoomprurkporn, W., Snidvongs, A. and Ratanachongkiat, S. (1998) A comparative study of some trace metals in the Mekong estuary (Vietnam) with the Chao Phraya estuary (Thailand). *Proceedings of the International Symposium on the Mekong Delta, 23–27 February 1998, Chiang Rai, Thailand.* pp. 150–168.

Hungspreugs, M. and Utoomprurkporn, W. (1998) Trace metal contamination in Thai rivers. *Proceedings of the International Workshop on Environmental Biogeochemistry, Jawaharial Nehru University, New Delhi, India. 13–18 December 1998.* pp. 207–226.

Hydrographic Department (1983) Report on the Analysis of Oceanographic Data from the Gulf of Thailand, Royal Thai Navy, 85 pp.

Ibrahim, H.M. (1999) Overfishing in the Gulf of Thailand: Issues and resolution. In *SEAPOL Integrated Studies of the Gulf of Thailand. Vol. II,* ed. D.M. Johnston, pp. 55–93. Southeast Asian Programme in Ocean Law, Policy and Management, Bangkok.

Landmann, G., Hutfils, V., Hagermann, F. and Ittekkot, V. (1998) Distribution and behaviour of suspended matter and sediments in the Mekong River and the adjacent sea. In *Proceedings of the International Workshop on the Mekong Delta, Chiang Rai, Thailand.* pp. 100–115.

Liewmanomont, K., Deetae, S. and Srimanobhas, V. (1991) Taxonomic and Ecological Studies of Seagrasses of Thailand. A research report submitted to the National Research Council. Faculty of Fisheries, Kasetsart University. 77 pp.

Longhurst, A.R. and Pauly, D. (1987) *Ecology of Tropical Oceans.* Academic Press, New York, 407 pp.

Menasveta, D. (1998) Fisheries management needs and prospects for the countries bordering the Gulf of Thailand. In *SEAPOL Integrated Studies of the Gulf of Thailand. Vol. I*, ed. D.M. Johnston, pp. 205–224. Southeast Asian Programme in Ocean Law, Policy and Management, Bangkok.

Nasir, M.S., Rojana-anawat, P. and Snidvongs, A. (1999) Physical characteristics of water-mass in the South China Sea. In *Proceedings of the First Technical Seminar on Marine Fishery Resources Survey in the South China Sea Area I: Gulf of Thailand and East Coast of Peninsular Malaysia, 24–26 November 1997, Bangkok, Thailand*. pp. 1–5.

National Statistical Office (1997) 1995 Marine Fishery Census: Whole Country.

Office of the Prime Minister and Department of Fisheries, Ministry of Agriculture and Cooperatives. Thailand. 209 pp.

Pukasab, P. and Pochanasomburana, P. (1957) The Types of Tide and Mean Sea Level in the Gulf of Thailand. Hydrographic Department, Royal Thai Navy. 7 pp.

Robinson, M.K. (1974) The Physical Oceanography of the Gulf of Thailand. *Naga Report* **3**(1), 5–110.

Sinsakul, S. (1997) Coastal Zone in Thailand. In *Status of Marine Scientific Study in Thai Waters*, eds. M. Hungspreugs, pp. 97–116. National Research Council of Thailand.

Siripong, A. (1985) The Hydrography of the South China Sea and the Gulf of Thailand. Waves, Tides and Currents, United Nations Environment Programme. 253 pp. (unpublished manuscript).

Snidvongs, A. (1998) The Oceanography of the Gulf of Thailand: Research and Management Priority. In *SEAPOL Integrated Studies of the Gulf of Thailand Vol. 1*, ed. D.M. Johnston, pp. 1–8. Southeast Asian Programme in Ocean Law, Policy and Management.

Snidvongs, A. and Sojisuporn, P. (1999) Numerical Simulation of the net current in the Gulf of Thailand under Different Monsoon Regimes. In *Proceedings of the First Technical Seminar on Marine Fishery Resources Survey in the South China Sea, Area I: Gulf of Thailand and East Coast of Peninsular Malaysia, 24–26 November 1997, Bangkok, Thailand*. pp. 52–72.

Snidvongs, A. and Vongprommek, R. (1997) Tertiary and Quaternary Sediment in the Gulf of Thailand. Status of Marine Science Studies in Thailand, Chemical Oceanography and Geological Oceanography, Sub-Committee on Marine Science, National Research Council of Thailand. pp. 129–146.

Sompongchaiyakul, P. (1989). Analysis of Chemical species of Trace metals in Near-shore Sediments by Sequential Leaching Method. M.Sc. Thesis. Department of Marine Science, Chulalongkorn University, Bangkok.

Srisuksawad, K., Porntepkasemsan, B., Nouchpramool, S., Yamkate, P., Carpenter, R., Peterson, M.L. and Hamilton, T. (1997) Radionuclide activities, geochemistry and accumulation rates of sediments in the Gulf of Thailand. *Continental Shelf Research* **17**, 925–965.

Stansfield, K. and Garrett, C. (1997) Implications of the salt and heat budgets of the Gulf of Thailand. *Journal of Marine Research* **55**, 935–963.

Supongpan, M. (1996) Marine capture fisheries in Thailand. *Thai Fisheries Gazette* **49**(2), 154–162.

Suvapepun, S. (1994) Present status of marine biology in Thai sea waters. A paper presented at the Fifth National Seminar on Marine Science: Status of Thai Seas and the Future Trend 22–24 August 1994, National Research Council of Thailand. 5 pp.

TOTAL Exploration and Production Thailand (1997) Bongkot Field Environmental Field Survey, August 1997. Marine Environmental Mercury Expert Panel. 38 pp.

Umnuay, G. (1984) The behaviour of some trace elements in the Chao Phraya Estuary. In *Proceedings of the Third Seminar on the Water Quality and the Quality of the Living Resources in Thai Waters*. National Research Council of Thailand. pp. 304–334.

Utoomprurkporn, W. (1997) Chemical oceanography review, Part 1: General water quality and nutrients. In *Status of Marine Scientific Study in Thai Waters*, eds. M. Hungspreugs, pp. 1–46. National Research Council of Thailand.

Vidthayanon, C. (1999) Species composition and diversity of fishes in the South China Sea, Area I: Gulf of Thailand and East Coast of Peninsular Malaysia. *Proceedings of the First Technical Seminar on Marine Fishery Resources Survey in the South China Sea, Area I, Gulf of Thailand and East Coast of Peninsular Malaysia*. pp. 172–240.

Windom, H.L., Silpipat, S., Chanpongsang, A., Smith, R.G. and Hungspreugs, M. (1984) Trace metal composition of and accumulation rates of sediments in the Upper Gulf of Thailand. *Estuarine, Coastal and Shelf Science* **19**, 133.

Wongratana, T. (1968) A checklist of fishes caught during trawl surveys in the Gulf of Thailand and off the east coast of the Malay Peninsula. Department of Fisheries. Bangkok. Thailand. 96 pp.

Wongratana, T. (1985) Ichthyological result of the otter-board bottom trawl operated by M.V. Nagazaki-Maru at off shore Songkhla and Nakorn Si Thamarat, Gulf of Thailand, 28–29 Oct. 1982 with account of ecology of the region, pp. 365–390. *Proceedings of the Aquatic Resources Research Institute*, Chulalongkorn University, Bangkok, Thailand.

Wyrtki, K. (1961) Scientific Results of Marine Investigations of the South China Sea and the Gulf of Thailand 1959–1961. Physical Oceanography of the Southeast Asian Waters. NAGA Report Vol. 2, The University of California Scripps Institution of Oceanography, La Jolla, California.

THE AUTHORS

Manuwadi Hungspreugs
*Department of Marine Science,
Chulalongkorn University,
Bangkok 10330,
Thailand*

Wilaiwan Utoomprurkporn
*Department of Marine Science,
Chulalongkorn University,
Bangkok 10330,
Thailand*

Charoen Nitithamyong
*Department of Marine Science,
Chulalongkorn University,
Bangkok 10330,
Thailand*

Chapter 74

THE MALACCA STRAITS

Chua Thia-Eng, Ingrid R.L. Gorre, S. Adrian Ross,
Stella Regina Bernad, Bresilda Gervacio and Ma. Corazon Ebarvia

---

The Malacca Straits have long been an important trade route linking the Indian Ocean to the South China Sea and Pacific Ocean. From the 7th to the 11th century, the Srivijaya empire controlled them, followed in the 15th century by the port kingdom of Malacca. Western maritime powers also recognised the strategic importance of the Straits, and in 1511, the Portuguese captured Malacca. In 1641, the Dutch occupied what is now known as Jakarta, and from the 17th to the 18th century, the Dutch East India company controlled the trade in the Straits (Ross et al., 1995). The British also recognised the need to control the Straits to ensure the safe passage of British merchant ships on their way to China, and in 1819 established a colony in Singapore. In 1824, the British and the Dutch ended their rivalry with a treaty whereby Britain agreed to "safeguard the Straits and keep them open for other friendly nations" (Chia, 1998).

In recent years, the Straits have become a very important trade route. In 1993 and 1995, over 100,000 oil and cargo vessels traversed through each year, carrying 3.23 million barrels of crude oil through the Straits each day (Sakura Institute of Research, 1998). Shipping accidents have been occurring more frequently recently, which is attributed to the heavy traffic in the Straits combined with shallow, narrow channels and shoals. Despite these hazards, economic efficiency dictates that vessels continue to use the Straits.

The Straits are also rich in renewable and non-renewable resources, including productive coastal ecosystems, extensive capture fisheries, aquaculture, coastal tourism, mining and valuable natural gas reserves.

This chapter looks at the natural environmental conditions and the status of the coastal resources, the sustainability of existing activities, critical environmental problems and management. It is based on the Malacca Straits Environmental Profile (Chua et al., 1997) and other studies undertaken by the GEF/UNDP/IMO Regional Programme for Marine Pollution Prevention and Management in the East Asian Seas, which is referred to as the Regional Programme in this document.

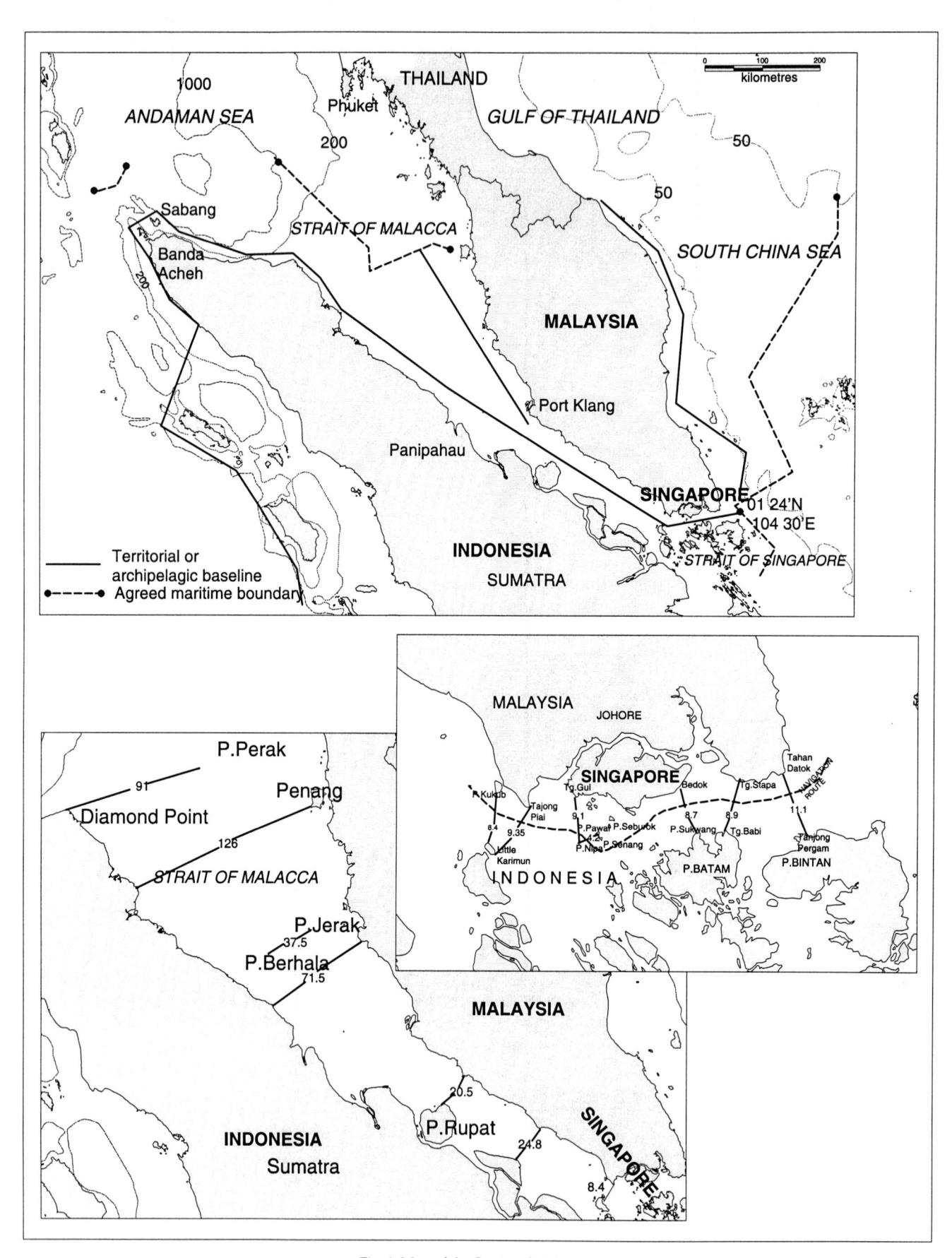

Fig. 1. Map of the Straits of Malacca.

NATURAL ENVIRONMENTAL CONDITIONS

Geography

The Strait of Malacca is located between the east coast of Sumatra Island in Indonesia and the west coast of Peninsular Malaysia, and is linked with the Strait of Singapore at its southeast end. The Strait of Malacca is bordered on the northwest by a line from Ujung Baka (5°40'N, 95°26'E), the northwest extremity of Sumatra, to Laem Phra Chao (7°45'N, 98°18'E), the south extremity of Ko Phukit Island, Thailand, and on the southeast by a line from Tahan (Mount) Datok (1°20'E, 104°20'N) and Tanjung Pergam (1°10'E, 104°20'N) (Hamzah, 1997; Shaw, 1973).

Three smaller straits are found within this waterway: the Bengkali Strait, which is located between the islands of Bengkali and Sumatra; the Rupat Strait, which is between Rupat and Sumatra; and the Johore Strait, which is situated between the southern tip of Peninsular Malaysia and the north coast of Singapore. Together, these five straits form part of an important international shipping route linking the Indian Ocean (via the Andaman Sea) to the South China Sea and the Pacific Ocean and are collectively referred to as the Malacca Straits or the Straits of Malacca (Fig. 1). The Malacca Straits are bordered by four littoral States, namely Thailand, Indonesia, Malaysia and Singapore. However, the navigational channel passes through the territorial seas of Indonesia, Malaysia and Singapore. For the purpose of this chapter, the littoral States shall be limited to the latter three only. Their length is around 600 nautical miles (nm), with the widest section (220 nm) near the northwest entrance, narrowing gradually to around 8 nm at the southeast entrance near the Riau Archipelago. The depth of the water is irregular, varying from 17 to 55 m.

Topography

The West Coast of Peninsular Malaysia is dominated by coastal plains and basins formed by alluvial deposits. Sandy beaches are found in a few areas. The eastern islands of Sumatra, mainly the Riau Archipelago, are generally made of granitic and old sedimentary materials that are rich sources of tin and bauxite. The broad alluvial coastal plain of central Sumatra consists largely of thick deposits of sediments where Indonesia's main deposits of oil and gas are located. Offshore, there is a mix of 241 islands, coral reefs, estuaries, deltas and lagoons.

Around 14 rivers in Sumatra drain into the Straits of Malacca with an estimated annual discharge of 0.94×10^{11} m^3 (Indra Jaya et al., 1998), and another 12 major rivers enter from Malaysia, including the Perak River, which is the second largest (Chua et al., 1997). Singapore has around 40 smaller drainage basins.

Populations Affecting the Area

In 1994, the population of the coastal states in the east coast of Sumatra was 10.9 million, at a density of 135 per km^2; that of the states in the west coast of Peninsular Malaysia was 13.2 million (MPP-EAS, 1999a). Within the peninsula, the population tends to concentrate in the major urban centres of Penang, Ipoh and Johore Bahru, and this distribution can be attributed to the presence of rich natural resources which encouraged greater infrastructure development there. Singapore has an estimated population of 2.85 million with a population density of 4608 per km^2 (DESIPA, UN, 1995).

The coastal economy of these areas used to be mainly agricultural. In recent years, however, there has been a shift in the employment pattern towards non-agricultural production and services. In 1990, for example, the agricultural labour force was 49.9% of the total while the industrial and service sector was 50.1% of the labour force. In Singapore, only 0.2% are now employed in the agricultural and fisheries sector.

Climatology and Oceanography

The Malacca Straits have a tropical climate. They are strongly influenced by the Northeast Monsoon which brings rain from December to February, and by the dry Southwest Monsoon from June to August. In the two inter-monsoon periods the weather becomes unpredictable.

The direction of the wind varies with the monsoon seasons. Strong squalls, locally known as "Sumatras" can occur in any month. There is abundant rainfall and torrential rains of short duration may occur any time of the year. Visibility in the Straits is clear (10 km or more about 95% frequency and less than one km below 0.5% frequency) except during heavy rain and when haze occurs. More recently, visibility in the area has been affected by a regular occurrence of the haze (Hamzah, 1997).

Tides in the Straits are mainly semi-diurnal with a tidal range of 1.6 to 3.7 m depending on location. In certain localities along the East Coast of Sumatra, tidal range may reach 4 to 5 m due to sea-bottom conditions.

Currents follow the topographic configuration of the sea bottom. The flow of surface current is stronger during the northeast monsoon when the dominant direction of the surface current is from the South China Sea to the Andaman Sea. Surface current also enters from the Andaman Sea but turns northwest off Pulau Penang and the Perak coast. There is an undercurrent flowing from the Andaman Sea towards the Straits during monsoon periods which causes a light upwelling near the One Fathom Bank (Uktolseya, 1988). Tidal currents in the open waters are generally weak, though stronger tidal currents occur in narrow parts and passages around Pulau Iyu Kecil. Wave strength is strongest during the Northeast Monsoon.

Surface water temperature is warmer during the Southwest Monsoon when it ranges from 28 to 30°C, which drops by 1 to 2°C during the Northeast Monsoon.

Salinity is influenced by high rainfall and by the numerous rivers. Fresh water from the rivers is largely responsible for lowering the surface salinity, while the monsoons affect the annual salinity variation. Coastal salinity may range as much as 6.8 to 31.66 ppt off Malaysia, much more than off Singapore where the fluctuation is only 29–32 ppt (Din et al., 1996). In open water, surface salinity ranges from 30.80 to 31.83 ppt, which is similar to bottom values.

Turbidity around river mouths and the coasts may be high, though offshore transparency may range from 10 to 30 m. Primary productivity is constant the whole year but varies geographically. There appear to be higher phytoplankton counts in the south where water is shallower, and where there is both vertical mixing and high nutrient input from rivers from Sumatra. Surface chlorophyll ranges from 0.51 to 0.95 mg/m^3 without any distinct seasonal variation, but this varies depending on location; the northern, deeper and more open areas have lower chlorophyll than do the shallower and narrower southern areas. Zooplankton, on the other hand, varies with the monsoon seasons, with plankton counts of $0.38 \times 10^3/m^3$ to $0.50 \times 10^3/m^3$ in the Northeast Monsoon and $0.67 \times 10^3/m^3$ to $0.84 \times 10^3/m^3$ in the Southwest Monsoon (Chua et al., 1997).

Coastal and Marine Ecosystems

Southeast Asia is the centre of marine biodiversity, and the Malacca Straits are located centrally within it. The Straits are one of the largest estuarine environments in the region, characterised by soft-bottom habitats, fringing coral reefs, seagrass beds and mangroves lining the coastlines (Fig. 2).

In East Sumatra, some protected species of reptiles, sea birds, and mammals are found in coastal areas and in Riau Island. Numerous endangered species occur in this area and, in 1991, a number of protected species of reptiles, sea birds, and mammals were listed, along with the designation or proposal of around 17 marine conservation areas on the East Coast of Sumatra (Indra Jaya et al., 1998). There are also important nesting sites in this area for species such as *Chelonia mydas* (green turtle) and *Eretmochelys imbricata* (Hawksbill sea turtle), for sea birds such as the brown booby, brown noddy, black-naped, bridled and roseate terns and milky terns, and some waterfowl (Chua et al., 1997; MPP-EAS, 1999a).

Coastal Forests

The coastal forests along the Malacca Straits consist of an estimated 447,680 ha of mangroves, of which 385,000 ha occur in Riau province of Sumatra. In East Sumatra as a whole there are at least 28 recorded genera of mangrove species, and their dominant fauna is generally crustaceans and other invertebrates, though a limited number of mammals are also found, such as the long-tailed macaque (*Macaca memestrina*). In Malaysia, there are an estimated 74,022 ha of mangroves, but in the last few decades there has been a substantial loss because of coastal reclamation projects (Chong et al., 1998).

Distinct zonation of mangrove species in Sumatra is common depending on tidal regimes. At mean and low tide is *Avicennia* and *Sonneratia* which grow on waterlogged substrate. Further up the shore occurs the *Rhizophora–Bruguiera* forest, and towards the upper shore are a greater variety, with *Ceriops tagal* and other less common species. Along the estuarine banks grow formations of *Sonneratia alba* and *Avicennia alba*, while along creek banks within the mangroves, *Rhizophora mucronata* is the dominant tree (Chong et al., 1998).

Peat swamps are estimated to cover 3.3 million ha in the East of Sumatra. They may have been substantially reduced by the rapid conversion to other land uses. Species of *Callophylum, Ganoa, Durio, Goystylus,* and *Trustania* are common. Certain species with horticultural or pharmacological potential are also found, e.g. *Alstonia*, for its latex and *Melaleuca,* for its essence (Indra Jaya et al., 1998).

Mudflats and Beaches

The Malacca Straits have extensive mudflats which contain rich organic substances and fine silt particles that support rich benthic communities. In East Sumatra, heavy sediment loads from the rivers result in deep and broad mudflats which eventually grade into coastal land. Mangroves and prevailing intertidal current systems also influence their formation. The entire alluvial plain of the East Coast of Sumatra and many adjacent islands were developed through this process (Chua et al., 1997).

Beaches in the Malacca Straits amount to a total of 272,289 ha, most being found in Sumatra. In general, these grade into broad mudflats, except where they have been developed into ports, industrial estates, housing, and tourist resorts.

Almost the entire East Coast of Sumatra can be considered to be estuarine. In addition to the above habitats there are also important nursery grounds for commercial species of fishes and shrimps. These areas have become highly vulnerable to environmental threats, such as siltation, pollution, oil spills and other negative impacts of human activities.

The Malacca Straits, together with the Java Sea, the southern South China Sea and the Gulf of Thailand, are part of the extensive and shallow Sunda Shelf, joining western Southeast Asia to the continental shelf of mainland Asia. Offshore therefore, shallow, soft-bottom habitats are extensive, and support a high biodiversity of commercial species, which include fishes, seaweeds, horseshoe crabs, shrimps, bivalves, gastropods, sea cucumbers and sea urchins.

Seagrass Beds

Seagrass beds are mostly found on the Malaysian side, particularly in Cape Rochado, Port Dickson and around Langkawi Island. The largest known Malaysian seagrass

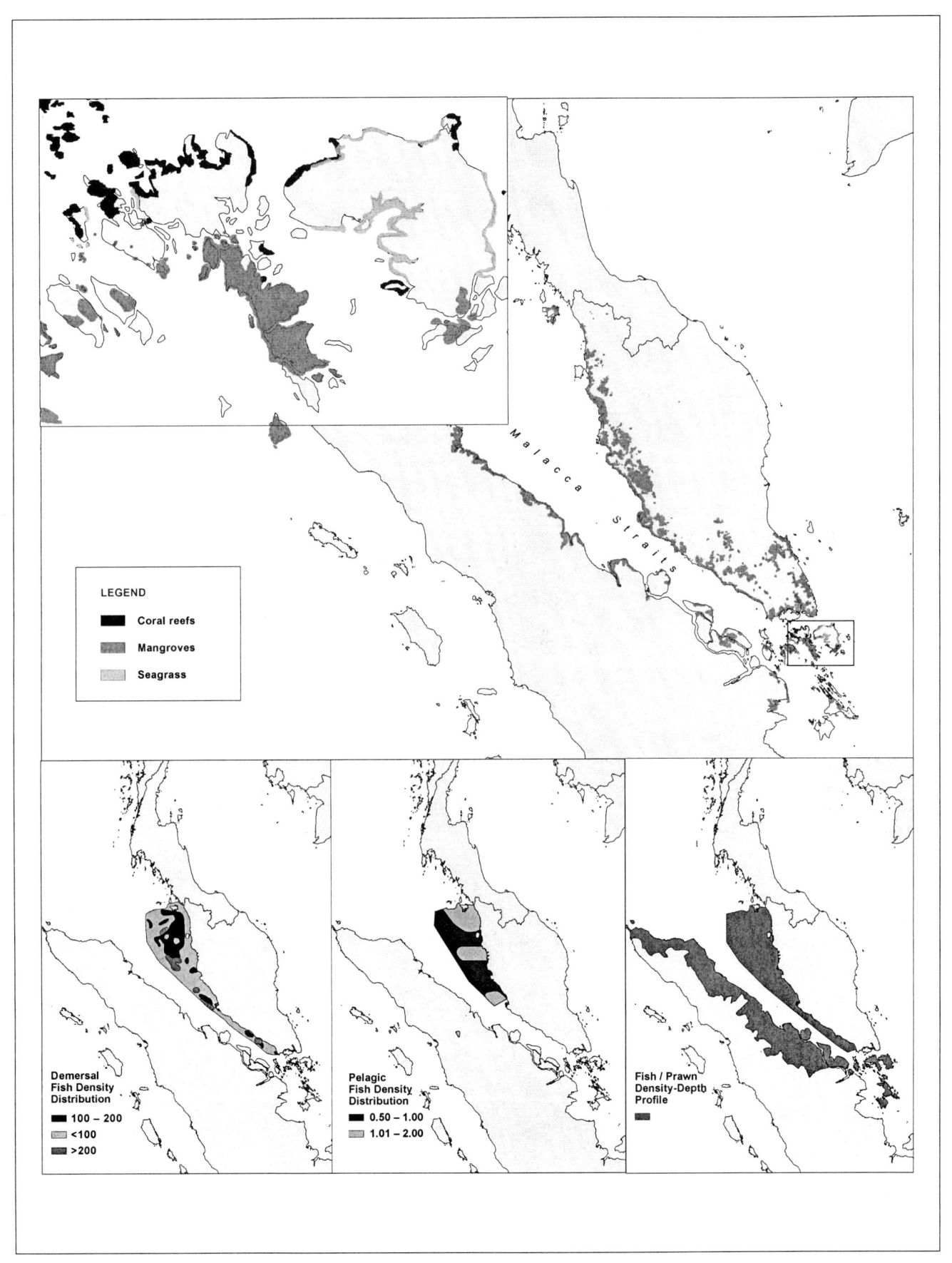

Fig. 2. Coastal ecosystems and natural resources in the Straits of Malacca. Source: MPP-EAS (1999a).

bed can be found on Meranmbong Shoal with a length of 1.2 km and a width of 80–100 m (Japar, 1994). Dense beds occur also in the Riau Archipelago of Indonesia and in some isolated patches off the southern islands of Singapore where there is considerable loss caused by land reclamation (Chua et al., 1998). Fourteen species of seagrass were recorded in Indonesia, nine in Malaysia and 11 in Singapore. Of the latter, only seven remain. The common genera include *Enhalus, Halophila, Thalassia* and *Cymodocea,* whose beds serve as the habitats of some endangered species such as the sea cow (*Dugong dugon*) and green turtles (*Chelonia mydas*).

Coral Reefs

The total coral reef area in the Straits is around 56,000 ha. (Low et al., 1998). In East Sumatra, they mostly occur in the Riau province and in the northeastern tip of Sumatra. A large area of coral reefs was reported off the Asahan district but had apparently already been degraded following the conversion of mangroves to shrimp ponds (Indra Jaya et al., 1998). Only about 1300 ha of coral reefs are found on the eastern side of the Straits, mainly around offshore islands or rocky outcrops, notably around the Langkawi islands. Off Singapore, coral reefs have also been found along the southern islands in particular.

On the Indonesian side, the best reef development is on the southeastern entrance to the Straits and at the remote island groups of Natuna, Anabas, and Tambelan. The best coral reefs in terms of cover are found in the Riau Islands; coral cover in Batam ranges between 48 and 66%, while that in Senayang-Lingga is 64–90% (MPP-EAS, 1999a). However, these reefs are deteriorating. On the Malaysian side, cover ranges from 25–46% (MPP-EAS, 1999a), and reefs in Singapore are severely stressed because of sedimentation and deteriorating water quality resulting from port activities and land reclamation (Goh and Chou, 1991; Chou et al., 1994).

This sparse distribution is due to the relatively turbid conditions, fluctuating salinity and high sediment loads discharged from land sources. Other contributory factors of coral reef degradation in the Straits include deforestation, land reclamation and dredging, intensive fishing, and industrial, ship-based and domestic sources of pollution (Chua et al., 1998).

RESOURCE EXPLOITATION, UTILIZATION AND CONFLICTS

During the last five decades, the three littoral States have experienced rapid economic development and burgeoning population. These have affected the pattern of resource exploitation and utilisation, resulting in resource-use conflicts.

Capture Fisheries and Coastal Mariculture

Capture fisheries are extensive, giving employment to 245,161 fisherfolk. Around 45% of Malaysian fishers are dependent on the Straits. In 1991, their fish catch was 410,900 tonnes, or 57% of the total Straits fish catch (Hamzah, 1997). The area is estimated to hold a potential annual stock of 253,000 tonnes of pelagic fish, 220,000 tonnes of demersal fish, and 88,400 tonnes of shrimp (Moosa, 1988). In 1993, total fish landing in the Straits' coasts of Malaysia and Sumatra amounted to 884,159 tonnes.

The Straits have a rich, multi-species fish stock, consisting mainly of demersal and pelagic finfish, crustaceans and molluscs. The abundance of each species varies with the season, but pelagic species dominate the total catch (62% in Sumatra).

The fisheries are mostly small-scale. Traditional gear such as the bottom gill net, push net and tidal trap used to catch shrimp are used on the Indonesian side while synthetic-fibre fishing nets and mechanized fishing boats are used in Malaysia. In the early 1960s, trawl fishing was introduced. Coastal aquaculture and mariculture are growing industries here. Brackish pond culture of shrimps and fish dominate aquaculture activities. The total volume of brackish water aquaculture production in Sumatra reached 38,725 tonnes in 1993. Other coastal aquaculture practices include floating net-cage culture of groupers, seabass (*Lates calcarifer*), and snappers, and the culture of cockle (*Anadara granosa*).

Exploitation of Natural Resources

Construction of brackish ponds has been made at the expense of mangroves. This has created conflicts in the competing use of the mangrove resources, and has affected their nursery function, thus severely affecting the sustainable recruitment of fish stocks. Aside from their ecological functions, mangroves support several economic activities in the area. Many mangrove species are used for firewood, charcoal, tanning, dyes, timber and boat construction. *Nypa* leaves can be used for roofs, baskets and cigarette "paper." The stalks of the flower are cut off and the sap is tapped for making brown sugar or the fermented palm wine or "arak". In the past, conflicts have arisen in North Sumatra between large concessionaires of mangroves and charcoal producers who no longer had sufficient areas to practise a sustainable pattern of harvesting.

Mangrove drainage channels form a unique ecosystem supporting a variety of aquatic species, caught with a variety of traditional fishing gears. Landings of mangrove-dependent species from the Malaysia coast are around 87,806 tonnes per year, which is 25% of the total fish landings. Extensive mudflats in the vicinity of mangroves along the Malaysia coast, particularly off Selangor and Perak, are utilised for semi-intensive culture of cockles. In the East Coast of Sumatra, large areas of mangroves have also been cleared and converted into coastal ponds for shrimp or fish

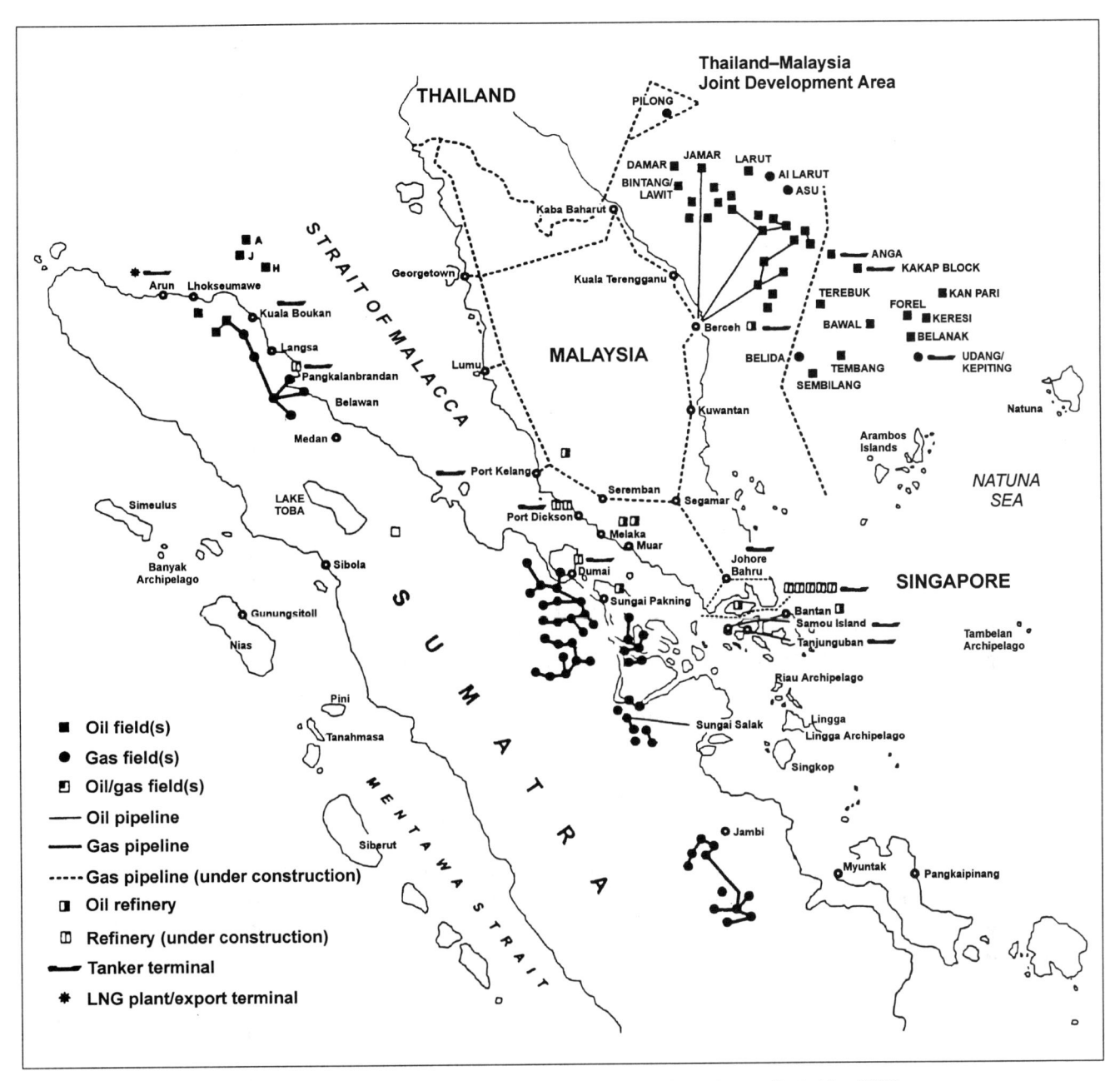

Fig. 3. Oil and gas fields and refineries in the Straits of Malacca. Source: Burbridge (1988).

culture. As of 1993, shrimp farms accounted for approximately 7% of current land use on what was formerly primary forest cover in North Sumatra province (McPadden, 1993).

This exploitation of mangroves clashes with its ecological functions of coastline protection, support for the yield of fisheries and biodiversity conservation.

Coral reefs also provide physical protection to coastal and island ecosystems. However, coastal communities have mined corals, coral gravel and coral sands for construction materials and for lime, and many species are also bleached and dried for sale to tourists. Coral reefs are also tourist attractions, and many parts of Indonesia and Malaysia have developed marine tourism resorts based on coral reefs.

Oil and Mineral Resources

Almost all oil fields in the Malacca Straits are located on the East Coast of Sumatra. Around 8 million barrels of potentially recoverable oil have been identified in Sumatra (Fig. 3), and offshore oil production from the Malacca Straits amounts to 55,000 barrels per day. Indonesia is a major liquefied natural gas (LNG) producer. In 1993, Indonesia had a 38.4% share of the LNG traded internationally and exported more than 23.1 million tonnes. A major field is the Arun field in Aceh, near the northwest entrance to the Malacca Straits. This gas field produces around 12 trillion cubic feet of LNG annually or 47% of Indonesia's gas production. So far, no oil is produced along the Malaysia coast

Table 1

Inventory of important coastal tourist spots in the Malacca Straits, and numbers of visits and revenue earned. Source: DOE/Malaysia (1989); TDC/Malaysia (1995); Wong, 1991b, 1995 cited in Chua et al. (1997).

Country/State	Coastal Tourist Spots	Number of tourists	Revenues from beach use
Peninsular Malaysia		2,928,800	$ 115,382,063
Kedah	Langkawi Island: Tanjung Rhu, Pantai Cenang, Pasir Hitam, and Kuah Pantai Merdeka Payar, Segantang, Kala, and Lembu Islands		
Penang	Penang Island: Batu Feringgi, Teluk Bahang, Gurney Drive and Padang Kota		
	Seberang Perai: Pantai Bersih		
Perak	Pangkor Island: Pasir Bogak, Pulau Pangkor Laut		
	Lumut: Pantai Teluk Bato		
Selangor	Pantai Morib		
Negeri Sembilan	Port Dickson		
Malacca	Tanjung Bidara, Pantai Kundur, Pantai Klebang, Pulau Besar		
Indonesia		1,167,480	$ 248,072,782
Riau Province	Nongsa Beach in Batam Island		
Sumatra	Trikora Beach in Bintan Island		
Singapore		3,965,000	$ 413,810,397
	Sentosa Island		
	East Coast Resort Area		
	Changi Resort Area		
	Pasir Ris-Loyang Resort Area		

but the area has already been subdivided into "contract areas" for exploration.

The oil and gas fields support a number of refineries. The total refining capacity along the Malacca Straits in 1995, was estimated to be around 1.54 million barrels per day, of which over two thirds is refined in Singapore.

Minerals such as tin ore and bauxite are also found here (Burbridge, 1988) and are mined along the Indonesian coast. Other resources include urea, kaolin, and granite, and sand mining is also undertaken in places such as Johore and Riau and exported to Singapore for reclamation.

Coastal Tourism

Coastal tourism is a thriving industry in the Malacca Straits, and white sandy beaches and coral reefs are some of the main attractions (Table 1). Large-scale integrated resorts are being developed in areas along the Straits, such as the joint-venture project between the Indonesian government, the Singaporean government and the private sector for the development of the 23,000-ha Bintan Beach International Resort. Coastal tourism has contributed to the development of several localities (Table 1). Of the revenues from beach use of $777 million about $221 million comes from diving and snorkelling (MPP-EAS, 1998).

Expansion of coastal tourism is not without cost. Impacts of inadequately planned tourism development include erosion, displacement of local residents and pollution from boats, untreated sewage and solid waste. This continued decline in quality of the beaches will eventually affect the coastal tourism industry itself.

Ports, Trade and Navigation

The Straits are a vital economic component and trade link for the littoral States. For example, 75% of Malaysia's trade passes through them. The rapid industrial development of many East Asian countries has led to an increase in shipping traffic (Fig. 4), and today vessels from around 20 flag States pass through. Traffic increases are both in oil tankers and in cargo vessels carrying manufactured goods from the East Asian countries to West Asia and beyond. It has been estimated that on average about 36.75% of the vessels passing through the Straits are general cargo vessels and container vessels (MPP-EAS, 1999g).

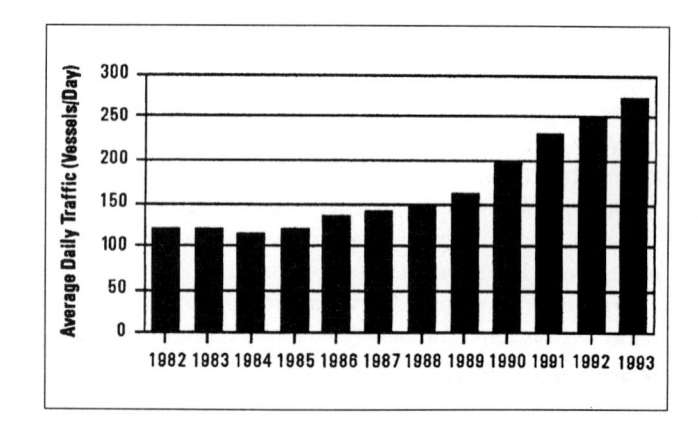

Fig. 4. Average daily shipping traffic through the Straits of Malacca. Source: Lee (1994); Naidu, 1997, cited in Chua et al. (1997).

Carriers of crude oil and petroleum products form about 35% of the vessels passing through the Malacca Straits carrying, in 1995, about 7 million barrels of oil daily. Traffic in vessels carrying LNG is also growing and is expected to increase in the future as an increasing number of countries favour the use of LNG as a source of energy.

The growth of trade and industry along the Straits has led to the development of ports and urban centres to service it. Indonesia has about 300 registered ports scattered throughout the archipelago. Among those located in the Straits are Belawan in North Sumatra, Dumai in Riau and Llokseumawe in Aceh, located in the vicinity of large oil fields. These also handle large volumes of exports, in addition to their main function as centres of crude and refined oil distribution. Another port, Kabil in Batam Island, Riau Archipelago, is also being expanded to serve as a major cargo port.

In Malaysia, the Straits are the main transport channel for trade. Thus, the major ports in Malaysia, together with associated infrastructure are found here; the major ports include Port Klang, Port Dickson, Ipoh Cargo Terminal, Malacca Port and Penang Port.

The Port of Singapore is the biggest in the Malacca Straits. Since 1986, it has been rated as the world's busiest port in terms of shipping tonnage and, since 1989, has also been regarded as the world's largest container port. It provides all major port services, including environmental control services such as the cleaning of oil and debris from the sea. Singapore has also grown into a major oil distribution centre and bunkering port with a very large storage capacity for crude oil and crude oil products. Singapore is also one of the major shipbuilding, rig-building and ship repair centres in the world with a total of 70 large and small shipyards. It is also the salvage and towage centre in the region.

CRITICAL ENVIRONMENTAL PROBLEMS

The various activities in the Straits have exposed the coastal and marine resources to environmental risks such as pollution and degradation of critical ecosystems.

Coastal Erosion, Sedimentation and Coastline Modification

There is increasing erosion of the coastline. As of 1985, 1972 km of the coastline of West Malaysia had been affected by erosion. Some of these areas, including the coastline from Kuala Perlis to Sungai Udang, the West Johore and Northern Kuala Selangor coast, were in a critical or significant stage of degradation. Coastal erosion has also been reported in the Indonesian side, particularly in the Bengkalis, Anak, and Setahan islands. High rates of shoreline erosion are attributed to mangrove extraction and its conversion to agricultural purposes.

Sedimentation is a growing problem in the Straits. In southeastern Sumatra, there is an increasing rate of sedimentation because of the heavy influx from the rivers, and dredging has become necessary to ensure the safe passage of larger vessels. Serious siltation problems have also been reported in some river mouths and harbours along the West Coast of Malaysia, especially around the ports and harbours at Klang, Kuala Perlis and Kuala Kedah. Sedimentation in the Straits arises from both natural processes and human activities such as deforestation, land reclamation, onshore mining and dredging.

In Singapore, coastline modification is primarily the result of land reclamation and port development. Since the 1970s, dredged material has enlarged several offshore islands, as well as creating new islands such as Pulau Ular and Terumbu Retan Laut. Between 1965 and 1985, Singapore's total land area increased by some 6.71% as a result of land reclamation.

Pollution

The Straits of Malacca have received waste discharged from both land- and sea-based sources. Many of the wastes contain hazardous substances harmful to human health, as well as to the marine ecosystem. Pollutant sources include discharges from agricultural activities, human settlements, industry and shipping.

Land-Based Pollution

Wastes are either directly discharged into the Straits or into the coastal rivers emptying into the Straits. The rapid population growth in the littoral States has resulted in the steady increase of sewage discharges into the rivers and the coastal waters resulting in high biochemical oxygen demand (BOD). In 1989 alone, the estimated BOD loading from domestic sewage discharges in the coastal areas of Indonesia, Malaysia and Singapore was 5014 tonnes per day (Table 2). BOD loading is expected to increase to over 6000 tonnes per day by the year 2000. High levels of BOD are usually found close to industries or sewage outfalls, such as in the Juru estuary, Chuah (Negeri Sembilan) and the Penang and Johore Straits.

There is significant industrial pollution in the Straits, generated from agro-based and manufacturing industries, such as factories for food, beverage, palm oil, rubber, tapioca and starch and for the manufacture of fertilizers, textiles, pulp and paper, tanneries and sugar (Table 3). Industrial sources of hazardous waste are mining, energy production, oil refining and pharmaceutical sectors, as well as hospital and healthcare facilities and research laboratories.

In Sumatra, these industries are mostly located in coastal areas or along a major river system. Most of them discharge untreated effluents directly to receiving waters. As a result,

Table 2

Estimated BOD loading from sewage discharges in coastal areas of the littoral states bordering The Malacca Straits (tonnes per day). Source: Low et al. (1998).

Year	Parameter	Indonesia	Malaysia	Singapore
1989	National population (million)	184.6	20.14	2.7
	Total BOD load (tonne)	9230	870	324
	Coastal population (million)	110.76	12.8	2.7
	% of national population	65	60	100
	Daily BOD load in coastal area			
	Generated (tonne)	5538	565	324
	Primary treatment (%)	60	70	10
	Secondary treatment (%)	0	5	90
	Daily BOD removal (tonne)	997	144	272
	Residual daily BOD disposal (tonne)	4541	421	52
2000	National population (million)	222	20.9	2.9
	Total BOD load (tonne)	11100	1045	348
	Coastal population (million)	133.2	13.58	2.9
	% of National Population	60	65	100
	Daily BOD load in coastal area			
	Generated (tonne)	6600	679	348
	Daily BOD removal (tonne)	1199	173	313
	Residual daily BOD disposal (tonne)	5461	506	35

Table 3

Major industries of Indonesia and their potential impact. Source: Noor et al. (1998)

Type of Industry	Potential Pollutant
Aceh	
1. Palm Oil	Organic
2. LNG, Fertilizer, LPG, Paper Kraft and Olefin	Mercury, ammonia and sulphur
3. Oil and Gas (drilling and processing)	Oil, organic, heavy metals
4. Forestry (sawmill and plywood)	Organic, chlorinated hydrocarbons
North Sumatra	
1. Power generation	Thermal, oil
2. Oil and gas	Oil, sulphur
3. Aluminum	Heavy metals
4. Mining of sand quartz	Coastal erosion and sedimentation
5. Palm oil	Organic
6. Forestry (pulp mills)	Organic
Riau	
1. Plastic, electronics, glass, ashes and porcelain, chemicals, textile, tannery, iron and metal works, food and organic material, shipping and shipyard construction	Organic, heavy metals, chlorinated hydrocarbons, thermal
2 Oil refinery	Oil
3. Palm oil	Organic
4. Pulp, rubber and sago	Organic

many river systems in Sumatra, including the Aceh, Deli, Belawan and Asahan rivers, are moderately to heavily polluted, and others, such as Lhokseumawe, are already heavily polluted.

Heavy metals, primarily from the manufacturing sector, include cadmium, copper, lead, mercury and nickel, all of which were reported in the coastal waters of West Malaysia, particularly the coastal waters of Perak and Penang (Choo et al., 1994). In addition, most rivers of West Malaysia contain concentrations of heavy metals which exceed standards.

There are reports of heavy metal contamination in some locations in Sumatra, such as at Llokseumawe in North Aceh and Asahan and Deli Serdang (Dahuri and Pahlevi, 1994). Mercury, lead, cadmium and copper were found to exceed national water quality standards for bathing, swimming and mariculture, and metal content was also high in the sediments of Pakning River in the Riau Province where oil refining, offshore oil exploitation and transportation activities occur.

High concentrations of copper, cadmium, cobalt, nickel, lead and zinc were found in the waters off the southern coast of Singapore, especially in Keppel harbour and the main port (Grace et al., 1987). Also, high concentrations of lead, nickel and cobalt were reported in waters near petroleum refineries, and metals were also detected in bottom sediments, especially in areas experiencing heavy shipping traffic.

Agricultural Waste

The lowland areas of East Sumatra have been used extensively for agriculture, agro-industries and plantations for rubber, palm, coconut and pineapple. North Sumatra and Riau Provinces have the largest areas of palm plantations in Indonesia.

The use of agricultural chemicals has increased substantially in recent years. About 3780 tonnes of insecticides, 110 tonnes of fungicides, 291 tonnes of rodenticides and 22 tonnes of herbicides are used annually. There is no research or information on the presence or impact of those chemicals on the coastal waters of the East Coast of Sumatra, but it is highly possible that some are leached to rivers and to coastal and estuarine environments.

In Malaysia, swine farming, which mostly occurs along river banks, is a major source of agricultural waste. In 1989, pig farms near the Linggi and Langat rivers generated waste with an estimated 218 tonnes/day, or 46% of the total national pollution load. Fertilisers and pesticides are also sources of water pollution in the river systems of Malaysia. Residues of organochlorine pesticides, particularly DDE, liquid and powder DDT and heptachlor were commonly detected in nearly all the rivers of Peninsular Malaysia (Abdullah et al., 1998). Pesticide residues were also found in the marine biota in West Malaysia but these were in low levels except for lindane in some samples of mussels (Abdullah et al., 1998). Recent studies show that there is a decline in organochlorine residues detected in Malaysian waters because of the increasing use of less persistent organophosphorous and carbamate pesticides (Abdullah, 1995, cited in Abdullah et al., 1998).

In Singapore, farmlands occupy only 3% of the total land area. Most of the agricultural processes occur in the northern portion of the main island and on Pulau Ubin. These are small-scale operations and the use of toxic and persistent chemicals is not allowed. The only chlorinated pesticide allowed is chlordane, of which an estimated 47 tonnes were used in 1987. Farmers are educated in methods of pesticide and fertiliser application, and many farms have drainage systems that collect the run-off into ponds, from which the water is pumped back to the field for irrigation.

Coliform Contamination

The discharge of untreated or partially treated domestic sewage has resulted in an increase in bacterial contamination, specifically faecal coliform bacteria. In many areas, coliform levels have exceeded interim national standards, and most water samples exceed the interim water quality standard for recreation (100 MPN/100 ml) in the West Coast of Peninsular Malaysia (DOE/Malaysia, 1990). In 1988, *E. coli* levels in Penang and Selangor reached 100,000 MPN/100 ml. Documented health-related impacts of high coliform level in seawater include skin disease and eye irritation. Furthermore, the consumption of raw or inadequately cooked seafood coming from coastal waters with a high coliform level is the main cause of cholera, typhoid and other gastrointestinal disorders.

Sea-Based Sources of Marine Pollution

Sea-based sources of marine pollution occur in the normal course of operation of a vessel and of gas platforms or as a consequence of accidents. A large percentage of oil pollution incidents are operational discharges from normal tank operations. These occur during deballasting, oil tank and cargo cleaning, and operation of fishing boats and small vessels. The residue of oil in a tank is estimated to be 0.1% of its total load. These residues are washed off with seawater and eventually discharged directly into the sea. Deballasting can cause other environmental risks aside from the discharge of oily waste. The ballast water contains harmful aquatic organisms which may cause the introduction of harmful microscopic toxic aquatic plants such as dinoflagellates that can cause algal bloom or pathogens such as the bacterium *Vibro chlorea* that causes cholera (Anon, 1998). An estimated two tonnes of oily waste is discharged daily into the Straits, mainly as a consequence of tanker ballasting (Tahir, 1996).

Collectively, small fishing vessels also contribute a large volume of oily discharge. The 10,000 small fishing boats operating in the Malaysian side of the Straits discharge around two tonnes of oil daily. A proportionally greater volume of discharge is expected from the 50,000 fishing vessels operating along the Indonesian side (Tahir, 1996).

Relatively high concentrations of tri-butyl tin (TBT) from antifouling paints have been observed along the shipping lane. Recent surveys indicate an increasing frequency of imposex in areas with shipping traffic (Hashimoto et al., 1998). Imposex was found in muricid snails and neogastropods in Singapore and Port Dickson in 1989, and in 1992, studies showed high levels (20.5–61.3 ng/l) in the seawater at Port Klang and in the sediments (27.6 ng/g wet weight) of Port Dickson (Tong et al., 1996, cited in Abdullah et al., 1998; Chua et al., 1997). The levels of TBT in the tissues of mussels and cockles were between 0.5 and 23.5 ng/g wet weight. TBT stays active even outside the target area and influences non-target animals. The effects of TBT are found within shellfish culture plots, threatening not only biodiversity but also food safety and human health.

Recent studies also show an increasing sulphur deposition in areas bordering the Malacca Straits (Streets, 1997), mainly from shipping emissions. Tar balls have also been reported along the beaches lining both sides of the Straits (Bilal, 1980). Beaches are considered to be adversely polluted when tar levels reach 10 grams per metre of shoreline (UNEP, 1990). In Malaysia, tar pollution has exceeded the UNEP standards in the Pantai Pasir Panjang, Perak, the Tanjung Rhu, Kedah, and in the Kepulan Riau islands, which lie 16 km south of Singapore.

Accidental oil and chemical spills are major sources of marine pollution. Vessel accidents are caused by collision or grounding of ships; others are due to accidental operational discharges and accidents. There were around 476 vessel accidents in the Straits from 1978 to 1994. The most recent oil spill in the Singapore Strait took place in 1997 with the collision of Evoikos and Orapin Global when an estimated 29,000 tonnes of oil were spilled.

These oil spill incidents can have a serious impact on the fragile ecosystem of the Straits, especially the intertidal zone. Recovery of mangroves affected by major oil spills can take 50 to 80 years. Fisheries, mariculture, coastal tourism, and crops are also affected, and other impacts of oil spills on marine fisheries include loss of revenues for fishermen because of fouled nets or closure of fishing areas, reduction in fish stocks, tainting of fish tissue and consumers refusing to purchase fish products from affected areas.

The aquaculture industry of cockles, oysters, and mussels is also particularly vulnerable to damage from oil spills.

For example, the aquaculture industry in Malaysia was expected to lose an estimated RM 66.5 million (US$ 26 million) as a consequence of an oil spill in Johore (Tahir, 1996).

The fouling of beaches, coastlines and visitor facilities can have a profound effect on tourism. For example, a 1993 collision between an oil tanker and an LPG carrier one mile from Sentosa Island, a popular tourist resort, resulted in a financial loss estimated at US$ 1.5 million to the hotel business. Popular tourist destinations such as Pangkor, Penang and Langkawi, the sandy beaches of Port Dickson and Lumut in Malaysia and Batam and Bintan in Indonesia are also at risk.

The shipment of hazardous wastes around the world is also increasing. There is no published record of dangerous goods being shipped through the Straits. However, it is apparent that as the region become more industrialised, the demand for such goods will increase.

Declining Fish Resources

Various indicators show that there is a decline of fish stock in the Malacca Straits, especially on the Indonesian side. More effort is needed to catch pelagic fish in the Strait, so fishers have moved their activities to other waters. On the Malaysian side, there has been a decline in total fish catch, a decline in catch per unit effort and a fall in the ratio of commercial fish to fish. Indigenous species in the Straits are also slowly disappearing.

There is evidence that some waters around Malaysia, including the Malacca Straits, are being heavily fished or overfished. Overfishing here is mainly attributed to several factors: (a) introduction and rapid spread of trawling; (b) fishery development to meet rising domestic and export demands; (c) increase in the number of fishers; and (d) participation of the government in some commercial fishing activities. As early as 1971, fish catch (428,000 tonnes) in the Straits exceeded the maximum sustainable yield of 400,000 tonnes per year (Burbridge, 1988). Trawling was thought to have led to the depletion of demersal stock, and controls over trawling were eventually instituted by the Malaysian government. In 1983, a complete ban on trawling was introduced by the Indonesian government.

The annual fish catch in Singapore has substantially dropped from 40,000 tonnes in the 1940s to the early 1950s to only 11,290 tonnes in 1994 (Chua et al., 1997). Some causes of the decline include: (a) increased sedimentation resulting from the extensive coastal reclamation; (b) removal of mangrove; (c) canalising of river courses which reduces the amount of fine sediments and nutrients; (d) increased shipping activities; (e) increased land-based activities and the resulting deliberate or inadvertent discharge of oil and other pollutants into coastal waters; (f) restrictions in giving licenses to operate traps; and (g) availability of more lucrative alternative employment opportunities resulting in fewer fishers.

Harmful Algal Blooms and Fish Kills

High nutrient levels in the Straits' waters underpin a high primary productivity and sustain the relatively high levels of aquatic standing crops. However, hyper-nutrification could cause a gradual change of species composition such as the increased frequency of harmful algal blooms and red tides. The occasional blooms of these organisms are reported to cause fish and shrimp mortality as a result of oxygen depletion. Red tides have been reported in the coastal waters of both sides of the Straits and in Singapore waters. Paralytic shellfish poisoning (PSP) in Malaysia was first reported in 1993, followed by a second incident in April 1994 in Sg. Sebatu and Sg. Rambai in a shellfish culture area. Monitoring results showed that dinoflagellates constitute a small fraction of the total phytoplankton populations (Anton et al., 1995). In Singapore, the first PSP occurrence was in 1987, and in 1989, fatalities of farmed fish associated with red tide were reported. Studies have not detected toxic dinoflagellates fatal to humans but have identified the species *Cochlodinium catenatum*, which causes fatalities to farmed fish. This species has been found to bloom in the Strait of Johore probably because of the warm temperature (above 30°C) and low salinity (28 ppt) with peak densities at neap tide (Khoo, 1992). Some of the possible causes of algal bloom include the introduction of harmful toxic species spread from ships during deballasting, and the increasing organic waste in the area.

Risks to Ecosystems and Human Health

It is evident that the various human activities in the Straits impact both ecosystems and human health. To find out the extent of the harmful consequences of human activities in the Straits, the Regional Programme has conducted an environmental risk assessment for the Malacca Straits, to study the effect of land- and sea-based pollution to living and non-living resources in them. This included a retrospective and prospective risk assessment.

Risk pathways (Fig. 5), identified during the retrospective assessment, illustrated the complex relationships between the potential causes of the problems for human and ecosystem health (habitat and species) and their consequences. Findings showed that sources of hazards were ultimately attributable to economic and social factors non-uniformly distributed among the littoral States (Calow and Forbes, 1997; MPP-EAS, 1999b).

The retrospective risk assessment showed the declining health of the marine ecosystems (Table 4). In terms of human health, the history of gastrointestinal problems in the region, shows that the problem could not be totally attributed to the consumption of contaminated seafood but is also caused by lack of potable water, improper waste disposal and poor food hygiene.

In a prospective risk assessment, the analysis of likelihood of adverse effects of environmental conditions to

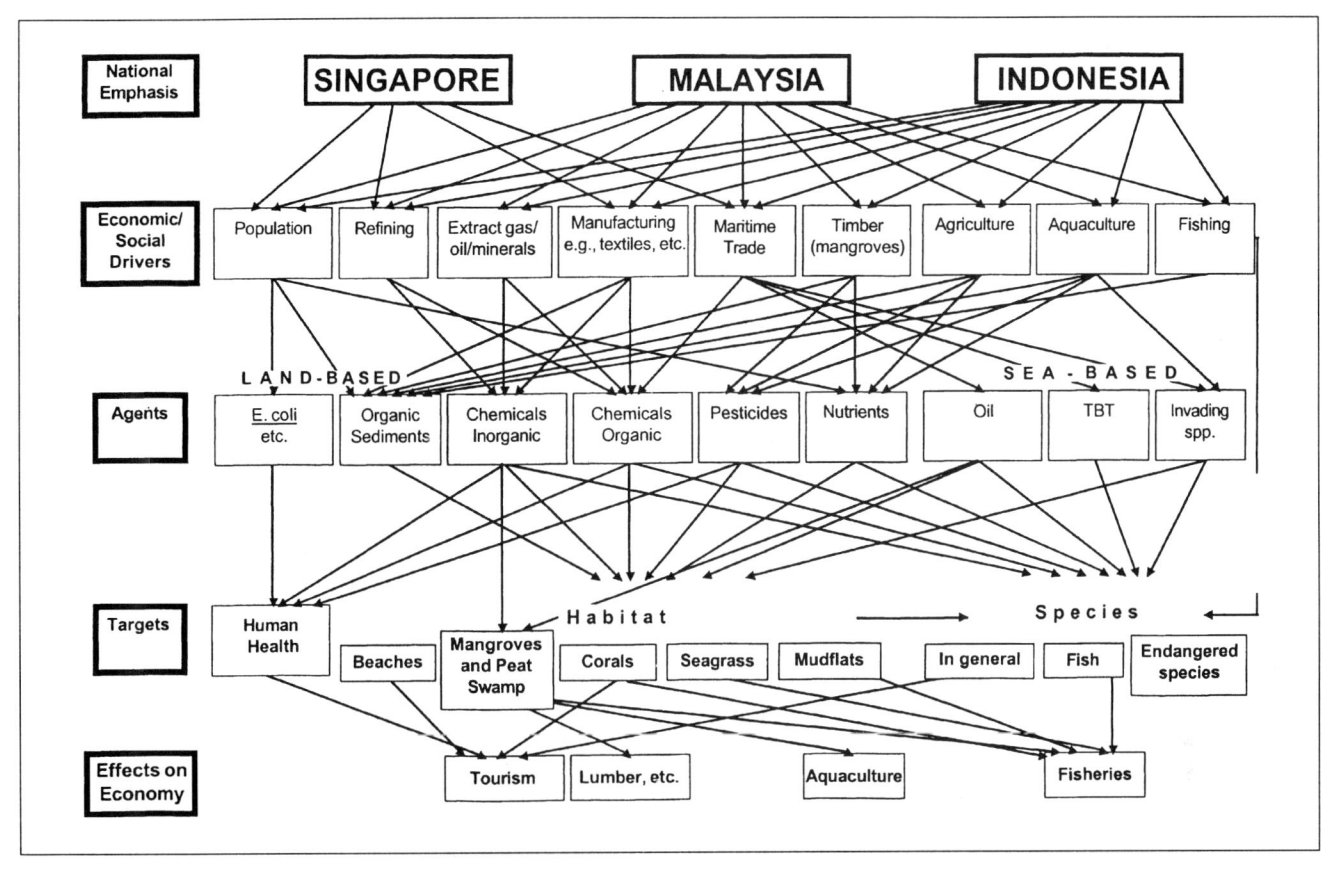

Fig. 5. Risk pathways illustrating relationships between potential causes of human health and environmental problems and their consequences in the Straits of Malacca. Source: MPP-EAS (1999b).

Table 4

Retrospective analysis of decline in key habitats for the Straits. Source: MPP-EAS, 1999a. Areal extent is an estimation of the relative abundance of each habitat type as large, moderate or small; evidence indicates decrease in habitat quantity (i.e., areal extent) and quality indicates a large decrease, moderate decrease and minimal decrease or no decrease. Judgments on the relative seriousness of consequences for the ecology of the Straits are indicated by number of asterisks, (i.e., more asterisks, more serious), NI indicates that no information was provided. (S) indicates information for Singapore only.

Habitat type	Areal extent	Decrease in quantity	Decrease in quality	Ecological consequences	Economic consequences
Mangroves	Large	Large	Moderate (S)	***	**
Peat swamps	Large	Large	NI	***	**
Coral reefs	Small	NI	Moderate to large	**	*
Seagrass beds	Moderate	NI	Moderate (S)	**	*
Soft bottoms	Large	No decrease	Moderate	**	**

human health, habitat and species in the Straits was estimated (MPP-EAS 1999a,b,c). The effects of the human activities were estimated through risk quotients (RQs). The study showed that heavy metals in the water column and sediments, pesticides in the sediments and suspended solids, TBT and oil and grease were likely to cause harm to ecological systems. For human health, likely problems from consumption of heavy metal- and pesticide- contaminated seafood were identified. Coliform bacteria also presented risks, primarily for bathers. The extent of the risk varied from place to place.

PROTECTIVE MEASURES

There are two regulatory frameworks which have to be considered in the management of the Malacca Straits: (1) the national legal and regulatory framework of the three littoral States and (2) the international legal regime governing straits used for international navigation.

National Management Systems and Legislations

Indonesia

The framework legislation of Indonesia on the environment is the Law Regarding Environmental Management (EMA-1997) or Act No. 23 which repealed Act No. 4 of 1982 (MPP-EAS, 1999d). The legislation on the marine environment is Act No. 5 of 1983 on the EEZ and its Implementing

Table 5

Protected Marine Areas in the east coast of Sumatra. Source: Salm and Halim, 1984, cited in Chua et al. (1997).

Province	Name	Status	Major interest
Aceh	Kuala Jambu Air	p.CA	Mangrove
	Kuala Langsa	p.CA	Mangrove
North Sumatra	SM. Karang Gading	Ministerial Decree of the Dept. of Forestry 11/Kpts/ Urn/i 1/80	Mangrove/ nesting turtles
	Sei Prapat	p.CA	Mangrove
Riau	Tg. Sinebu - P. Alang Besar	p.SM	Mangrove/mangrove island
	CA P. Berekeh	Ministerial Decree of the Dept. of Forestry 13/Kpts/ Um/3168	Mangrove/nesting water birds
	Bakau Selat Dumai	p.SM	Mangrove/crocodiles
	P. Pasir Panjang	p.TW	Mangrove islet
	Bakau Muara Kaupas	p.CA	Fringing mangrove/swamp forest
	Muara Gunting	p.CA	Fringing mangrove/swamp forest
	CA P. Burung	Ministerial Decree of the Dept. of Forestry 13/Kpts/ Um/3168	Mangrove/nesting water birds
	P. Bulan	p.SM	—
	P. Penyengat	p.TW	Mangrove island
	Tg. Datuk	p.CA	Mangrove
	CA P. Laut	Ministerial Decree of the Dept. of Forestry 13/Kpts/UM/3/68	Mangrove island/nesting water birds and turtles

p = Proposed.
CA = *Cagar Alam Laut* – **Strict Marine Reserve** for the protection of nature and maintenance of natural processes in an undisturbed state in order to have an ecologically representative example of the natural environment available for scientific study, environmental monitoring, education and for the maintenance of genetic resources.
SM = *Suaka Margasarwa Laut* – **Managed Marine Reserve/Marine Sanctuary** for the protection of significant species, biotic communities or physical features of the environment, where these may require specific human manipulation for their perpetuation.
TW = *Taman Wisata Laut* – **Protected Seascape** for the maintenance of nationally or provincially significant natural areas which are characteristic of the harmonious interaction of the mainland with islands, coasts and sea, while providing opportunities for public enjoyment through recreation and tourism within the normal lifestyle and economic activities of these areas.

Government Regulation 15 of 1983 on the management of the living resources within the Indonesian EEZ. The delineation and preservation of nature conservation and protected areas is another strategy of the Indonesian government in managing the marine environment (Table 5).

Land-based sources of marine pollution are regulated through the establishment of effluent quality standards. Industrial waste is controlled from the planning phase (EIA process) to actual production (effluent standards for each type of activity e.g., fertiliser, tapioca and palm oil). Measures to prevent and manage pollution are required to be integrated in the workplan of mining operations. The regulation on toxic waste in Indonesia utilises the cradle-to-grave approach. Procedures and a permit system for production, storage, collection, transportation and processing of such wastes are imposed by the regulation. Pesticides and fertilisers are also regulated under Act No. 12/1992.

Sea-based marine pollution is governed by the EMA-1997, Act No. 5 of 1983, and the Continental Shelf Act, respectively. Pollution from vessels is governed by Chapter 5 of Act No. 21 of 1992, which is the comprehensive legislation on all aspects of shipping including shipping management, prevention of and response to pollution. Ships are required to have pollution prevention equipment and are prohibited from disposing of wastes except in accordance with requirements. In particular, ships are required to have oily-water separators and an oil discharge monitoring system in accordance with MARPOL 73/78. The four gateway ports of Indonesia have already been instructed to establish ship waste reception facilities; however, only Tanjung Priok and Tanjung Perak have done so and neither facility is operating effectively.

As a policy, Indonesia does not allow ocean dumping. However, it is not entirely prohibited but only regulated through a licensing system covering the place, method and frequency of dumping, and the type, content and volume of wastes to be dumped.

The major oil spill caused by the Showa Maru accident in 1975 prompted the Indonesian government to establish the Standard Operation Procedures (SOP) for oil spill response in the Straits of Malacca and Singapore. The SOP is a Joint Decree between the Director-General of Sea Communication and the Director-General of Oil and Gas, signed in 1981 (Joint Decree No. DKP/1/2/27/ Kpts/DM/MIGAS/1981). The

Directorate-General of the state-owned oil company is tasked to operate the equipment for oil pollution control.

A draft Presidential Decree on Emergency Control of Oil Pollution at Sea is pending. This would define the composition and structure of response teams at national, regional and local levels. The DGSC would be the lead agency while Pertamina would be the supporting agency. After the decree is issued, it will become the basis for the preparation of the National Oil Spill Contingency Plan. However, the operations prescribed by it have already been put into use during actual accidents and exercises. Further regulation on the matter was provided in the Decision of the Minister of Communication Number: KM 86 of 1990 regarding the Prevention of Oil Pollution from Ships.

At national level, the Office of the State Minister for Environment (MENLH) has primary jurisdiction over environment and development issues. This co-ordinates national efforts on environmental matters. The National Committee on Marine and Coastal Environment advises the MENLH, and the Environmental Impact Management Agency (BAPEDAL) acts as the environmental enforcement agency, being involved in all aspects of environmental regulation. There is no single agency responsible for marine environmental management, but for marine pollution, the responsible agency is the Director-General of Sea Communication (DGSC) under the Ministry of Communications.

Malaysia

The Environment Quality Act of 1974, as amended (EQA), is the framework legislation of Malaysia on environmental protection. It prohibits the pollution of the atmosphere, soil, inland waters, and discharge of environmentally hazardous substances, pollutants or wastes into Malaysian waters, as well as noise pollution. The EQA is complemented by the framework of policies for sustainable development provided under the Second Outline Perspective Plan, the National Development Policy Plan and the Sixth Malaysia Plan.

Vessel-related marine pollution is further regulated under the Malaysian Exclusive Economic Zone Act of 1984 (EEZ Act), the Merchants' Shipping Ordinance of 1952 and the Merchant Shipping Act of 1994. Dumping of wastes at sea is also prohibited under the EEZ Act unless it is authorised by the Director General of the Department of Environment.

Land-based marine pollution is regulated under the Environmental Quality (EQ) Regulations on several subjects, including 'Sewage and Industrial Effluents' under which any plan to carry out any construction activity which may result in a new source of effluent discharge or cause a material change in the quantity or quality of the discharge from an existing source must first be approved by the Director General for Environment.

Marine pollution from sea-bed activities is regulated by several Acts, and the Petroleum Act of 1984 governs the transportation of petroleum by pipes. The pipeline owner is required to submit a Written Emergency Plan, which will be implemented in case there is a system failure. The plan includes prompt and remedial measures to protect the environment, minimise property damage and limit the accidental discharge from the pipeline.

In the regulation of toxic and hazardous wastes, there are the Guidelines for the Storage of Scheduled Wastes (Guidelines) and the Customs Orders of 1993 on Prohibition of Imports and Exports. The Guidelines prohibit the disposal of any waste listed in the schedule on land or into Malaysian waters except in prescribed premises. The Customs Orders of 1993, on the other hand require the prior written approval from the Director General of Environment for the export or import of toxic or hazardous wastes into or out of Malaysia.

Malaysia combines the use of command and control and market-based approaches to prevent and control marine pollution. Market-based approaches have been adopted for the major polluting industries such as the agro-based industries and toxic and hazardous wastes generators. This approach includes the imposition of license fees, the phasing of standards, the assessment of an effluent-related fee based on the 'polluter pays principle', the full or partial waiver of the effluent-related fee for industries conducting research on effluent treatment, the granting of tax incentives, especially to industries that reduce emissions and the conferment of pioneer status for five years to companies engaged in an integrated operation for the storage, treatment and disposal of toxic and hazardous wastes (MPP-EAS, 1999d). Civil or criminal penalties are also used as a deterrent for violation of marine pollution laws. With specific reference to oil pollution, a legal regime of liability for damages was enacted under the Merchant Shipping Act of 1994. This regime is applied together with the CLC and the FUND Convention.

Malaysia has a National Oil Spill Contingency Plan which is supported by the private sector. The initiative is subscribed to by companies involved in downstream processing and upstream production of petroleum in Malaysia. The responsibility for the enforcement of marine environmental legislation in Malaysia does not rest with one single department. The Director-General of the Department of Environment under the Ministry of Science and Technology has extensive powers under the 1985 Exclusive Economic Zone Act to deal with marine pollution which threatens to damage Malaysia's coastline, or any part of the environment or related interests, such as fishing interests, in the EEZ. The Director-General may issue directives or take necessary action to respond to or mitigate the damage or the threat to damage.

Singapore

Although Singapore is one of the world's busiest ports, it has no framework regulation on the environment, but has adopted specific measures to control and prevent both land- and sea-based sources of marine pollution.

The Prevention of Pollution of the Sea Act of 1991 (PPSA) regulates pollution from ships and contains the main provisions of MARPOL 73/78 on sea-based pollution sources. The discharge of any ashes, solid ballast, sludge or any other matter into the waters of the Port of Singapore without permission is also prohibited. In addition to the permit requirement, Singapore requires the treatment of ballast water prior to discharge. Other regulatory measures include the setting-up and use of reception facilities and the requirement of keeping oil and cargo record books. Another measure regulating the transport of petroleum and other potentially hazardous substances is the Petroleum Act and the Hazardous Waste Act of 1997.

Land-based pollution is regulated under the PPSA, the Water Pollution Control and Drainage Act of 1975, as amended (WPDA) and the Environmental Public Health Act of 1987, as amended (EPHA). The PPSA prohibits the discharge of oil or oily mixture into Singapore waters from any place on land, or from any apparatus used for transferring oil from or to any ship. It also prohibits the deposit of any oily mixture, refuse, garbage, plastic waste matter, carcass, noxious liquid substances and marine pollutant. Trade effluents and sewage collection and treatment is regulated under the Water Pollution Control and Drainage Act. The discharge of toxic substances into any inland water is also prohibited under the Water Pollution Control and Drainage Act, and the sewage collection and treatment facilities, as well as the effective solid waste system in Singapore are major factors in the control of land-based marine pollution.

The two government offices that play major roles in the prevention and control of marine pollution are the Maritime and Port Authority of Singapore (MPA) and the Ministry of Environment (ENV) which have general jurisdiction over marine and ship-based sources of pollution and land-based pollution, respectively.

International Legal Regime Governing Malacca Straits

It is a general principle of international law that each state shall have the duty to ensure that the activities within its territory do not cause transboundary harm. Specifically applied to shared resources such as the Malacca Straits, the littoral States have the "general duty to cooperate on the basis of a system of information and prior consultations in order to achieve the optimum use of such resources without causing damage to the legitimate interests of the others." (U.N. General Assembly Resolutions 3129 (XXVIII) of December 1973 and 3281 or the Charter of Economic Rights and Duties of States, 1974). For example, the littoral States must ensure that the level of pollution that they discharge into the Straits will not substantially affect the interests of other littoral States in the Straits. International law and UNCLOS consider the Malacca Straits as an international strait. Table 6 shows other relevant international conventions that the three littoral States have ratified.

Vessels have the right of transit passage, but in order to ensure safety of navigation and the protection of the marine environment, UNCLOS requires that ships comply "with generally accepted international regulations, procedures and practices for safety at sea, including the International Regulations for Preventing Collisions at Sea". Furthermore, ships are required to "comply with generally accepted international regulations, procedures and practices for the prevention, reduction and control of pollution from ships". UNCLOS also allows the littoral States to undertake measures for navigation safety and marine environment protection. The littoral States are authorised to take appropriate enforcement measures against any foreign ship, other than a warship, auxiliary vessel, or vessel owned by a State and used for non-commercial purposes, which threatens major damage to the marine environment of the Straits and violates regulations adopted by the littoral

Table 6

Ratification of littoral states in the Malacca Straits of international conventions relating to marine pollution

Convention/ Protocol	Year	Year of Ratification/Accession		
		Indonesia	Malaysia	Singapore
UNCLOS	1982	1986	1996	1994
MARPOL	1973/78			
Annex I/II		1986	1997	1990
Annex III				1994
Annex IV				
Annex V			1997	1999
Annex VI				
London Convention				
Convention	1972			
Protocol	1996			
Intervention				
Convention	1969			
Protocol	1973			
CLC				
Convention	1969	1978	1995	D*
Protocol	1976			1981
Protocol	1992			1997
Fund				
Convention	1971	1978	1995	
Protocol	1976			
Protocol	1992			1997
Salvage	1989			
OPRC	1990		1997	1999
Basel	1989	1993	1993	1996
HNS	1996			

*D = Denunciation in accordance with the 1992 Protocols.

State for the safety of navigation and for the prevention, reduction and control of pollution.

As coastal States, Malaysia, Indonesia and Singapore are authorised to undertake enforcement measures when there are clear grounds for believing that a vessel has violated provisions and obligations. Flag states of vessels are also required to ensure compliance, but flags of convenience are a problem in the Straits because these have very lax vessel safety regulations and do not enforce IMO Conventions.

Coordination in the Management of The Malacca Straits

The complexity of the issues and problems in the Malacca Straits makes it impossible for the littoral States to individually manage the Straits. They have thus engaged in various levels of co-operative activities. One important initiative has been the preparation of the present Traffic Separation Scheme and development of an Electronic Chart Display and Information System (ECDIS). In 1998, a hydrographic survey was completed by the three littoral States and Japan, which was used in the formulation of a new routing system (Hamzah, 1999).

Oil spills are examples of transboundary hazards that have to be coordinated by the littoral States of the Malacca Straits, as well as by the private sector. In 1975, oil companies/agencies in each ASEAN country formed the ASEAN Council on Petroleum (ASCOPE) to promote and extend cooperation. The Council consists of heads of each national company/agency, and in 1980 ASCOPE initiated the ASCOPE Plan for the Control and Mitigation of Marine Pollution (ASPCMMP) in order to enhance individual national plans. In 1986, ASCOPE and ASEAN unified their plans into a regional oil spill contingency plan. In 1993, the then six ASEAN countries agreed to establish the ASEAN Oil Spill Response Action Plan (ASEAN-OSRAP) to provide a cooperative plan for mutual assistance from member states and organisations for oil spill response.

Oil spill equipment is maintained by the littoral States in co-ordination with industry. At present, oil spill contingency planning has been done at the sub-regional level under the auspices of the Association for South East Asian Nations. The Petroleum Association of Japan (PAJ) also extends assistance by stockpiling oil spill response equipment in the Straits which can be used in an oil spill.

In 1994, the Regional Programme initiated the Malacca Straits Demonstration Project (MSDP) to assist the littoral States in their efforts to identify existing and potential pollution risks, and to strengthen regulatory mechanisms and instruments for managing pollution. The Regional Programme, together with the littoral States, prepared an environmental profile of the Malacca Straits for use in environmental management and assessment of the natural resources (Straits of Malacca Environmental Information System, SMEIS). A comprehensive economic assessment of the market and non-market values of the marine natural resources was made.

The Regional Programme has identified a verified process for assessing and managing transboundary marine pollution in sub-regional sea areas. In particular, the MSDP has demonstrated the usefulness and effectiveness of risk assessment/management as a tool for addressing the transboundary pollution problem, particularly marine pollution risk assessment arising from land- and sea-based sources. An assessment of marine pollution risks in the Straits has already been completed. The risk assessment provides additional input in identifying options for managing risks in the Straits. In order to echo the lessons learned from the risk assessment study in the Straits, the MSDP organised a regional training and developed a training manual on environmental risk assessment and natural resource damage appraisal.

The Regional Programme has also undertaken a study on the need and feasibility of designating the Malacca Straits as a special area under MARPOL 73/78, wherein more stringent restrictions exist on the disposal of harmful substances. Oceanographic and ecological conditions, vessel traffic characteristics, availability and adequacy of port reception facilities, are among the criteria for determining a special area. The ecological conditions and the vessel traffic in the Malacca Straits make it a potential candidate, which will ensure stricter protective measures for the marine environment (MPP-EAS, 1999f).

To sustain the activities in the Straits after the project, the Regional Programme has conducted capacity building activities on marine pollution prevention measures and the implementation of marine pollution-related international conventions. In particular, the Regional Programme has conducted training workshops on the implementation of international conventions, risk assessment, damage claims, OPRC, and effective shore reception facilities.

CONCLUSION AND RECOMMENDATIONS

The integrity of the Malacca Straits as safe and clean international waterways can only be maintained if both user States and littoral States co-operate. The users of the coastal resources, international sea-lanes and the littoral States have the responsibility to ensure safe passage and a clean marine environment in the Straits.

The littoral and user States should develop and implement sustainable sub-regional marine pollution prevention and management policies, strategies and action plans to address marine pollution arising from sea-based activities. The latter action plans need to include ratification and implementation of marine pollution conventions.

The existing environmental issues and potential environmental threats demand serious consideration. Pollution management in the Straits must address four main problem areas: (1) minimising pollution; (2) enforcement of

Table 7

Total economic value, by ecosystem, by country (US$, in 1995 prices)

Ecosystem	Indonesia		Malaysia		Singapore		Total	
	Gross benefits	Net benefits	Gross benefits	Net benefits	Gross benefits	Net benefits	Gross benefits	Net benefits
Mangroves	2,260,475,206	1,841,575,576	1,695,275,129	1,379,014,506	31,229,604	27,798,929	3,986,979,940	3,248,389,011
Mudflats	26,250	26,250	49,680,083	28,131,904	10,289	10,289	49,716,623	28,168,444
Beach	348,033,632	248,804,519	115,408,583	69,255,758	413,810,397	248,286,238	877,252,612	566,346,515
Coral reefs	567,387,522	466,360,403	47,503,305	26,145,178	118,068,560	70,863,522	732,959,386	563,369,103
Seagrass		n.d.	10,765,355	2,333,656	71,702	71,702	10,837,057	2,405,358
Seaweeds	16,382,625	1,933,575	3,345,767	1,037,413		n.d.	19,728,392	2,970,988
Fisheries	875,050,250	458,445,077	686,340,025	154,429,860	8,400,000	1,657,320	1,569,790,275	614,532,258
from ecosystems	875,050,250	458,445,077	425,551,661	102,976,316	4,200,000	828,660	1,304,801,911	562,250,053
Aquaculture	170,940,387	17,094,039	97,852,203	38,973,538	18,600,000	11,160,000	287,392,590	67,227,576
Sea lanes[3]							600,210,000	600,210,000
Total	3,363,245,622	2,575,794,362[1]	2,280,618,789	1,596,345,497[2]	585,990,552	359,019,340[2]	6,830,064,963	5,131,369,199

[1] The total figure excludes the value of fisheries since this has been already accounted for under the ecosystems.

[2] The total figure excludes the value of fisheries from the ecosystem since this has been already accounted for under the ecosystems.

[3] The value of sea lanes refers to the shipping costs saved by Japanese petroleum vessels from using the Straits of Malacca instead of alternative routes.

regulations for controlling vessel-based pollution; (3) oil spill prevention and control; and (4) marine environment rehabilitation and compensation for damage (MPP-EAS, 1999g).

For coastal tourism areas, the following management measures should be considered: (a) coral reef protection by limiting the number of divers; (b) provision of setback zoning for beaches to allow natural coastal processes and vegetation growth to occur; and (c) pollution prevention. An important concept in recreational planning is the determination and upholding of carrying capacity in beach and coral reef areas. For beach protection, adequate buffer zones between development areas and shorelines must be established and setback zones established to protect life and property against erosion, preserve coastal habitats such as sea turtle nesting grounds, and enhance the scenic values. According to one study, tourism and recreational benefits gained from maintaining biodiversity and the state and quality of coastal and marine water and resources far outweigh the costs of implementing the management measures (MPP-EAS, 1998).

Pollution control by vessels traversing the Straits is essential. Vessel accidents are another major source of pollution, so improvement of navigational services is also essential.

Table 8

Expenditures for navigational safety

	RM		US$
Navigational aids	40,955,000		16,382,000
Surveillance cost	9,590,000		3,836,000
Emergency and rescue services:			
search and rescue (operational cost)	4,884,372	or 14,076 per incident	1,953,749
oil pollution control	77,000	per oil spill incident	30,800
traffic separation	6,500,000		2,600,000
vessel traffic services	110,500,000	per year	44,200,000
hydrographic survey	2,350,000	(1993)	940,000
Institutional cost:			
Marine Department (operational expense)	10,690,000	(ave. 84-93)	4,276,000
Light Dues Board (expense)	6,210,000	(ave. 84-93)	2,484,000
Hydrographic Directorate	7,550,000	(ave. 84-93)	3,020,000
Total	199,306,373		79,722,549

Source: Hamzah (1997).

Total net economic value of marine and coastal resources in the Straits of Malacca is estimated to be about $5.1 billion (Table 7). Revenues from tourism and fishery activities alone account for 11.7% and 10.9%, respectively, of the total (net) economic value of the Straits (MPP-EAS, 1999h). At present, however, these resources are at risk because of widespread deforestation leading to siltation and sedimentation, pollution, and destruction of habitats. These problems, if not mitigated, would lead to the loss of economic values. For example, the deterioration of the ecosystems, water quality and decline of biodiversity will lead tourists to other sites and will push fishing boats into other fishing areas, thereby reducing the economic benefit that the tourism and fisheries industries derived from the Straits.

Measures to prevent accidents, to reduce discharge of wastes, and to respond and control the consequences must be in place to ensure that there can be continued beneficial use of the Straits. These services do not come free (Table 8). While many user States enjoy the benefits of using the Straits, only Japan has made significant contributions for the maintenance of navigational safety and prevention of pollution (Hamzah and Nizam Basiron, 1997).

With the increase in traffic and activities in the Straits, the littoral States will be unable to subsidise the cost of maintaining them without assistance from the users. Under the "polluter pays" principle, the shipping community should also shoulder the cost of marine environmental protection (Tan Sri Razali Ismail, 1999). This is a necessary expense if the beneficiaries of the Straits would like to continue deriving both market and non-market benefits from its resources. It becomes imperative for the littoral States and the user States to cooperate and formulate an innovative financing scheme, which will distribute the financial burden on all parties that benefit from the Straits. There may be several methods of collecting revenues for this, but in addition, non-monetary means of cooperation, such as capacity building and cooperative research on transboundary environmental issues should be explored.

During the International Conference on the Straits of Malacca held on 19–22 April 1999, the establishment of an inter-ministerial mechanism was proposed. The proposed mechanism shall be tasked with reaching consensus on priority programmes and policies within the Straits, including the design of a financial mechanism within the Straits, generated from revenues from users and beneficiaries of the Straits.

Over the long term, a Strategic Environment Management Plan (SEMP) in the Malacca Straits must be developed by the littoral States in order to guide future development in the area. At present, it is more likely that cooperative activities in the Straits will proceed on an *ad hoc* or a per issue basis. Eventually, the numerous issues "may form a web-like framework upon which can be constructed a broader, multi-sectoral management institution" (MPP-EAS, 1999f).

REFERENCES

Abdullah, A.R., Mohd. Tahir, N., Tong, S.L. and T. Mohd Hoque (1998) Marine pollution sources database and GIS mapping of the Straits of Malacca, GEF/UNDP/IMO MPP-EAS, Quezon City, Philippines (unpublished).

Anon. (1998) MEPC adopts in principle HNS protocol. *IMO News* **4**, 8–11.

Anton, A., Normawaty, Mohammed Noor and Tambi, H. (1995) Studies on harmful algal bloom species in the Malacca Straits, Malaysia. In *Proceedings of the ASEAN-Canada Midterm Technical Review Conference on ASEAN Criteria and Monitoring: Advances in Marine Environmental Management and Human Health Protection, 24–28 October 1994*, eds. D. Watson, K.S. Ong and G. Vigers, pp. 357–362. Singapore, EVS Environmental Consultants and National Science and Technology Board.

Bilal, J. (1980) Status report on research, monitoring and assessment of oil pollution and its impact on living aquatic resources in Southeast Asia. Technical paper presented at the FAO/UNEP/UNESCO Expert Consultation Meeting on Aquatic Resources in Southeast Asia, February 1980, Manila, Philippines. SCS/Oil/Poll/80/8. 38 p.

Burbridge, P. (1988) Coastal and marine resource management in the Strait of Malacca. *Ambio* **17** (3), 170–177.

Calow, P. and Forbes, V. (1997) Malacca Straits: Initial risk assessment. MPP-EAS/Info/91/117. GEF/UNDP/IMO MPP-EAS, Quezon City, Philippines. 82 p.

Chia, L.S. (1998) The importance of the Straits of Malacca and Singapore. *Singapore Journal of International and Comparative Law* **2**, 301–322.

Chia, L.S. (1997) Alternative routes for oil tankers: a financial, technical and economic analysis. In *The Straits of Malacca: International Co-operation in Trade, Funding and Navigational Safety*, ed. A. Hamzah, pp. 103–122. In Kuala Lumpur, Pelanduk Publications (M) Sdn. Bhd. and Maritime Institute of Malaysia.

Chong, V.C., Sasekumar, A., Phang, S.M. and Jaafar, M.N. (1998) Marine and coastal resources mapping for the Straits of Malacca. GEF/UNDP/IMO MPP-EAS, Quezon City, Philippines (Unpublished).

Choo, P.S., Ismail, I. and Rosly, H. (1994) The west coast of Peninsular Malaysia, In An Environmental Assessment of the Bay of Bengal region, ed. S. Homgren, pp. 33–54. Bay of Bengal Report No. 67.

Chua T.E., Ross, S.A. and Yu, H. (eds.) (1997) Malacca Straits environmental profile. MPP-EAS Technical Report 10. GEF/UNDP/IMO MPP-EAS, Quezon City, Philippines, 259 pp.

Chua T.E., Ross, S.A., Yu, H., Jacinto G. and Bernad, S.R. (1999) Sharing lessons and experiences in marine pollution management. MPP-EAS Technical Report No. 20. GEF/UNDP/IMO Regional Programme for the Prevention and Management of Marine Pollution in the East Asian Seas, Quezon City, Philippines. 94 pp.

Chua, T.E. and Ross, S.A. (eds.) (1996) Marine pollution prevention and management in the East Asian Seas: from planning to action. Annual Report. GEF/UNDP/IMO Regional Programme for the Prevention and Management of Marine Pollution in the East Asian Seas, Quezon City, Philippines. 40 pp.

Chua, T.E. and Ross, S.A. (eds.) (1997) Pollution prevention and management in the East Asian Seas: a paradigm shift in concept, approach and methodology, Annual Report. GEF/UNDP/IMO Regional Programme for the Prevention and Management in the East Asian Seas, Quezon City, Philippines. 45 pp.

Chua, T.E., Natarajan, R. and Ross, S.A. (1998) Analysis of the state of the marine environment of the Straits of Malacca and Singapore. *Singapore Journal of International and Comparative Law* **2**, 323–349.

DESIPA/UN (Department for Economics and Social Information/United Nations) (1995) World population prospects: the 1994 revision, United Nations Secretariat, United Nations, New York.

DOE-Malaysia (Department of Environment, Malaysia) (1990) Environment quality report, 1989. Department of Environment, Ministry of Science, Technology and Environment, Malaysia, Kuala Lumpur. 240 pp.

Din, Z., Ahamad, B.A., Jamaliah, M.R.S.N. and Wan Mustapha, W.D. (1996) Water and sediment quality of West Coast Peninsula Malaysia. Paper presented at the Conference on ASEAN marine environmental management: quality criteria and monitoring for aquatic life and human health protection, 24–28 June 1996, Penang, Malaysia. 21 pp.

General Assembly Resolutions 3129 (XXVIII) of December 1973 and 3281 or the Charter of Economic Rights and Duties of States, G.A. Res. 3129, 28 UN GAOR Supp. (No. 30), UN Doc. A/Res./ 3050-3199; G.A. Res. 3281, 29 U.N. GAOR Supp. (No. 31), U.N. Doc. A/9631 (1974).

Goh, B.P.L. and Chou, L.M. (1991) Coral reef-associated flora and fauna of Singapore. In Proceedings of the Regional Symposium on Living Resources in Coastal Areas, ed. A. Alcala, pp. 47–53. Marine Science Institute, University of the Philippines, 30 January to 1 February 1989.

Chou, L.M., Wilkinson, C.R., Licuanan, W.R.Y., Aliño, P., Chesire, A.C., Loo, M.K.G., Tangjatrong, S., Ridzwan, A.R. and Soekarno (1994) Status of coral reefs in the ASEAN region. In *Proceedings Third ASEAN-Australia Symposium on Living Coastal Resources. Volume 1: Status Review*, eds. S. Sudara, C.R. Wilkinson and L.M. Chou, pp. 1–10. Department of Marine Sciences, Chulangkorn University, Bangkok, Thailand.

Grace, L.M., Woo, K.H. and Chou, L.M. (1987) Singapore Country/ Status Report, pp. 188–274. In Development and management of living marine resources workshop on pollution and other ecological factors in relation to living marine resources. Report of ASEAN-Canada Cooperative Programme on Marine Science.

Hamzah, A. (1997) The Straits of Malacca: A Profile. In *The Straits of Malacca: International Co-operation in Trade, Funding and Navigational Safety*, ed. A. Hamzah, pp. 3–14. Kuala Lumpur, Pelanduk Publications (M) Sdn. Bhd. and Maritime Institute of Malaysia. 359 pp.

Hamzah, A. (1999) International law and funding of specific services in Straits used for international navigation: A case study of the Straits of Malacca and Singapore. In *Challenges and Opportunities in Managing Pollution in the East Asian Seas*, eds. T.E. Chua and N. Bermas, pp. 257–275. MPP-EAS Conference Proceedings 12/PEMSEA Conference Proceedings 1, GEF/UNDP/IMO MPP-EAS, Quezon City, Philippines, 567 pp.

Hamzah, B.A. and Nizam Basiron, M. (1997) Funding partnership for safer nagivation and cleaner environment in the Straits of Malacca: some preliminary thoughts. In *Sustainable Financing Mechanisms: Public Sector–Private Sector Partnership*, eds. S.A. Ross, C. Tejam and R. Rosales, pp. 87–103. Proceedings of the Regional Conference on Sustainable Financing Mechanisms for the Prevention and Management of Marine Pollution: Public Sector–Private Sector Partnership. MPP-EAS Conference Proceedings No. 6. GEF/UNDP/IMO MPP-EAS, Quezon City, Philippines, 352 pp.

Hashimoto, S., Watanabe, M., Noda, Y., Hayashi, T., Kurita, Y., Takasu, Y. and Otsuki, A. (1998) Concentration and distribution of butyltin compounds in a heavy tanker route in the Strait of Malacca and in Tokyo Bay. *Marine Environmental Research* 45 (2), 169–177.

Indra Jaya, Siregar, V., Sondita, M.F. and Rustandi, Y. (1998) Marine and coastal resources mapping for the Malacca Straits. GEF/ UNDP/IMO MPP-EAS, Quezon City, Philippines (Unpublished).

International Convention for the Prevention of Pollution from Ships, 1973/1978 (MARPOL 73/78).

Japar, S.B. (1994) Status of seagrass resources in Malaysia. In *Proceedings, Third ASEAN–Australian Symposium on Living Coastal Resources. Volume 1: Status Review*, eds. S. Sundar, C.R. Wilkinson and L.M. Chou, pp. 283–289. Department of Marine Sciences, Chulangkorn University, Bangkok, Thailand.

Khoo, H.W. (1992) Red tide management issues: a Singapore perspective. In *Proceedings of the ASEAN-Canada Midterm Technical Review Conference on Marine Environmental Quality: Perspectives on ASEAN criteria and monitoring, Vol. 1. Jakarta, Indonesia*, eds. D. Watson, D. Setiapermana and G. Vigers, pp. 131–133.

Low, K.S., Phua, Y.N., Lee, C.K., Jaafar, M.N. and Wan Zahari, W.A. (1998) Final Report for the Malacca Straits Development Project on the development of a regional database system and GIS for the Straits of Malacca. GEF/UNDP/IMO MPP-EAS, Quezon City, Philippines. (Unpublished).

MPA (Maritime and Port Authority of Singapore). Singapore Electronic Navigational Chart (visited 29 September 1999). http:// www.mpa.gov.sg/homepage/services/sg_enc.html).

McPadden, C.A. (1993) The Malacca Straits coastal environment and shrimp aquaculture in North Sumatra province. 190 pp.

Moosa, M.K. (1988) Strait of Malacca: environment, living resources and transnational problems. In *Coastal Zone Management in the Strait of Malacca. Proceedings of a Symposium on Environmental Research and Coastal Zone Management in the Strait of Malacca, 11-13 November 1985. Medan, Indonesia*, eds. P. Burbridge, Koesoebiono, H. Dirschl and B. Patton, pp. 132–141. School for Resource and Environmental Studies, Dalhousie University, Halifax, Nova Scotia.

MPP-EAS (1995) Bi-Annual Report. GEF/UNDP/IMO Regional Programme for the Prevention and Management of Marine Pollution in the East Asian Seas, Quezon City, Philippines. 27 pp.

MPP-EAS (1997) Proceedings on the consultative meeting on the Malacca Straits Demonstration Project. Workshop Proceedings 4, GEF/UNDP/IMO MPP-EAS, Quezon City, Philippines. 50 pp.

MPP-EAS (1998) Benefit–cost analysis of tourism development and sustainability in the Malacca Straits. MPP-EAS Technical Report 17. GEF/UNDP/IMO MPP-EAS, Quezon City, Philippines. 53 pp.

MPP-EAS (1999a) Straits of Malacca environmental information system (CD-ROM). GEF/UNDP/IMO MPP-EAS, Quezon City, Philippines.

MPP-EAS (1999b) Malacca Straits: refined risk assessment. MPP-EAS Technical Report No. 23. GEF/UNDP/IMO MPP-EAS, Quezon City, Philippines. 89 p.

MPP-EAS (1999c) Environmental risk assessment manual: a guide for tropical ecosystems. MPP-EAS Technical Report No. 21, GEF/ UNDP/IMO MPP-EAS, Quezon City, Philippines. 88 pp.

MPP-EAS (1999d) Assessment of national legislation on marine pollution in the East Asian Seas. MPP-EAS Technical Report 26. GEF/ UNDP/IMO MPP-EAS, Quezon City, Philippines. 363 pp.

MPP-EAS (1999e) Sustainable financing for ship-based pollution prevention and management in the Malacca Straits. GEF/UNDP/ IMO MPP-EAS, Quezon City, Philippines. 42 pp.

MPP-EAS (1999f) Malacca Straits: Special area? MPP-EAS Information Series 194. MPP-EAS, Quezon City, Philippines.

MPP-EAS (1999g) Marine pollution management in the Malacca/Singapore Straits: lessons learned. MPP-EAS/Info/99/195. GEF/ UNDP/IMO MPP-EAS, Quezon City, Philippines. 168 pp.

MPP-EAS (1999h) Total economic valuation: coastal and marine ecosystems in the Straits of Malacca. Technical report No. GEF/ UNDP/IMO MPP-EAS, Quezon City, Philippines.

Noor, E., Hariyadi, S. and Dahuri, R. (1998) Marine pollution sources database & GIS mapping for the Malacca Straits. GEF/UNDP/ IMO MPP-EAS, Quezon City, Philippines (Unpublished).

Ross, A., Lintu, S. and Hachmoller, B. (1995) Navigational safety in the Malacca Straits. *Tropical Coasts* 2 (1), 7–10.

Shaw, K.E. (1973) The Straits of Malacca: in relation to the problems of the Indian and Pacific Ocean. University Education Press, Singapore. 174 pp.

Streets, D.G., Carmichael, G.R. and Arndt, R.L. (1997) Sulphur dioxide emissions and sulphur deposition from international shipping in Asian waters. *Atmospheric Environment* 31 (10), 1573–1582.

Tahir, N.M. (1996) Fate of spilled oil and ecological and socio-economic impacts of oil pollution in the Straits of Malacca. Paper presented at the Regional Workshop on Operational Oil Spill Modelling 31 May–3 June 1996, Pusan, Republic of Korea.

Tan, A.K.J. (1996) Country Report on Marine Pollution Legislation and Implementation. In GEF/UNDP/IMO MPP-EAS Inception Workshop of the Regional Network of Legal Experts on Marine Pollution, 18–19 March 1996. Manila.

Tan Sri Razali Ismail (1999). Keynote Address at the International Conference on the Straits of Malacca, April 19–22, 1999.

Trail Smelter Arbitration (U.S. vs. Canada), 3 R. Int'l. Arbitral Awards, 1938, 1941.

Uktolseya, H. (1988) Physical and biological characteristics of the Straits of Malacca in the Framework of coastal resource management. In *Coastal Zone Management in the Strait of Malacca. Proceedings of a Symposium on Environmental Research and Coastal Zone Management in the Strait of Malacca, 11–13 November 1985, Medan, Indonesia*, eds. P. Burbridge, H. Koesoebiono, H. Dirschl and B. Patton, pp. 118–131. School for Resource and Environmental Studies, Dalhousie University, Halifax, Nova Scotia.

UNEP (1978) Draft Principles of Conduct on Shared Natural Resources.

UNEP (1990) GESAMP: the state of the marine environment. UNEP Regional Seas Reports and Studies No. 115, 111 pp.

Chua Thia-Eng
GEF/UNDP/IMO Regional Programme on Partnerships in Environmental Management for the Seas of East Asia (PEMSEA), P.O. Box 2502, Quezon City, 1165 Metro Manila, Philippines

Ingrid R. L. Gorre
GEF/UNDP/IMO Regional Programme on Partnerships in Environmental Management for the Seas of East Asia (PEMSEA), P.O. Box 2502, Quezon City, 1165 Metro Manila, Philippines

S. Adrian Ross
GEF/UNDP/IMO Regional Programme on Partnerships in Environmental Management for the Seas of East Asia (PEMSEA), P.O. Box 2502, Quezon City, 1165 Metro Manila, Philippines

Stella Regina Bernad
GEF/UNDP/IMO Regional Programme on Partnerships in Environmental Management for the Seas of East Asia (PEMSEA), P.O. Box 2502, Quezon City, 1165 Metro Manila, Philippines

Bresilda Gervacio
GEF/UNDP/IMO Regional Programme on Partnerships in Environmental Management for the Seas of East Asia (PEMSEA), P.O. Box 2502, Quezon City, 1165 Metro Manila, Philippines

Ma. Corazon Ebarvia
GEF/UNDP/IMO Regional Programme on Partnerships in Environmental Management for the Seas of East Asia (PEMSEA), P.O. Box 2502, Quezon City, 1165 Metro Manila, Philippines

Chapter 75

MALACCA STRAIT INCLUDING SINGAPORE AND JOHORE STRAITS

Poh Poh Wong

The Malacca Strait extends about 430 nautical miles between Sumatra and Peninsular Malaysia and is contiguous with the Singapore Strait and the Johore Strait. All of these straits have extensive coastal areas with rich diverse ecosystems. As a drowned estuary of the Sunda Shelf, the region is one of the largest estuarine environments in Southeast Asia, characterized by soft-bottom habitats, mangrove swamps, peat swamp forests, seagrasses, and fringing corals reefs. Economic development in three littoral states of Indonesia, Malaysia and Singapore has a marked impact on the habitats, coastal resources and water quality of the area. Along its coastal areas are a population of 11 million in Sumatra, 10 million in Malaysia, and 3 million in Singapore.

The conversion of mangroves and peat swamp forest for other uses, especially on the Sumatran coast, has led to siltation, inundation and erosion along the coast, the destruction of habitats for fish nurseries, and an increase in the level of total suspended solids. Few mangrove areas are being exploited sustainably, and an inherent conflict exists between the use of mangroves to protect capture fisheries and for aquaculture.

The vast soft-bottom habitats of the area offer a variety of demersal, pelagic and shrimp species. Overfishing, from as early as the mid-1970s, and conflicts between traditional and commercial fishermen are the major problems. Proper fisheries management cannot be carried out as stock assessment is not completed for all waters and species. In the meantime, several measures have been taken. These include better licensing, the control of trawling, management zones in Malaysia, and the use of marine parks to exclude fishing. However, there are still conflicts between commercial fishing, subsistence fishing and aquaculture.

Within the straits, a major problem is the cross-boundary water quality issue. Pollution comes mainly from land-based sources (70%). Suspended sediments come from forest clearance and coastal landfill activities, while agriculture and domestic sewage are responsible for the high levels of coliforms. From 1986 to 1991, 60% of samples from west Malaysia exceeded the proposed interim standard of 100 MPN/100 ml. From Indonesia, the discharged sewage from the population generates loadings of BOD, COD, total nitrogen and total phosphorus estimated at 167,000, 381,000, 74,000 and 7000 tonnes per year, respectively. In the Johore Strait, both Johore and Singapore have implemented a joint plan for cleaning up the strait by 2006.

Oil pollution is related to the high shipping traffic. In 1974–94 there were 17 major spills (>5000 barrels per spill) in the strait and its approaches. This is likely to increase as shipping traffic increases. The implementation of traffic separation schemes (TSS) and under keel clearance (UKC) in the Singapore Strait and its northern approach have been effective in reducing shipping accidents.

Seas at The Millennium: An Environmental Evaluation (Edited by C. Sheppard)
© 2000 Elsevier Science Ltd. All rights reserved

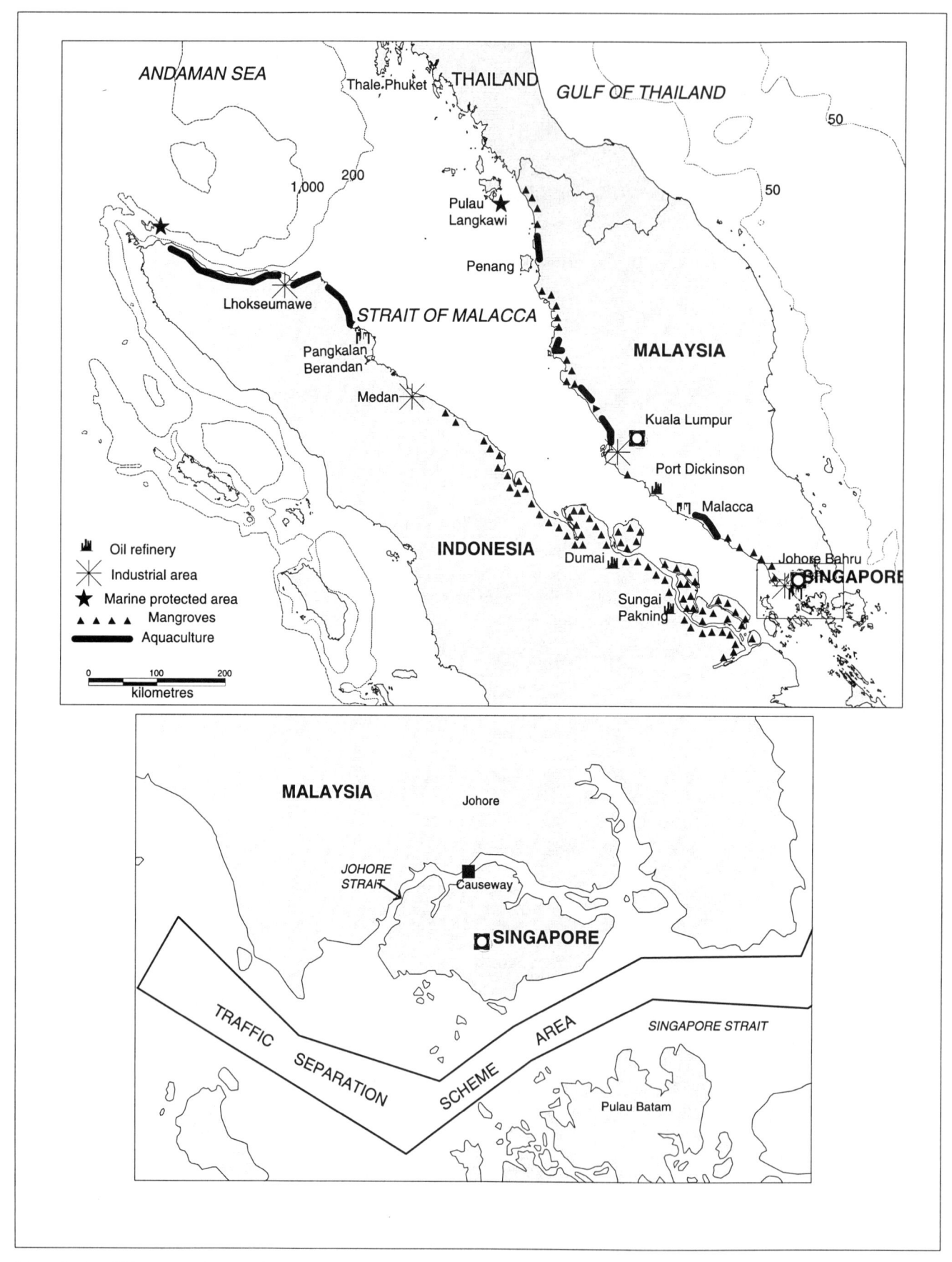

Fig. 1. Straits of Malacca, with inset of Straits of Singapore and Johore, showing major urban centre, industrial area, mangrove distribution, aquaculture areas, marine protected areas.

Marine pollution in these straits is often seen in fish kills, algal blooms and paralytic shellfish poisoning. Hydrocarbons are one of the contributors. A less evident impact comes from tributyltin (TBT) antifoulant.

In future, there will be more economic development on the west coast of Peninsular Malaysia and at both ends of the strait. The SIJORI growth triangle linking Singapore, Johore and Riau Province is at the southern end of the strait. It focuses on industries on Batam and a large integrated resort on the northern coast of Bintan, both islands lying just south of Singapore. Economic growth at the northern end of the strait is strongly tied to the IMT growth triangle linking North Sumatra and Aceh of Indonesia, five provinces of southern Thailand and four northern states of Malaysia. The area under natural habitats will decrease as economic activities expand. The protected areas remain limited and are at Pulau Weh in North Sumatra, four islands off Kedah and an area of wetlands at Kuala Selangor. This is not encouraged by the lack of sandy beaches which can put a premium on the need to keep the coastal waters clean for the growing tourism industry.

INTRODUCTION

The Malacca Strait situated between the east coast of Sumatra and the west coast of Peninsular Malaysia is contiguous with the Strait of Singapore (Fig. 1). It is funnel-shaped with its northwest entrance at 126 nautical miles wide from Ujung Taming, Sumatra, to Penang, Malaysia (or 220 nautical miles from Sabang, Sumatra, to Phuket) and tapers to 8 nautical miles near the Riau archipelago (Chua et al., 1997). According to Morgan and Valencia (1983) the strait has a length of 431 nautical miles, a perimeter of 1252 nautical miles and an area of 45,882 nautical miles^2. It is one of the world's major shipping lanes and contains at least three other straits, Bengkali Strait, Rupat Strait and Johore Strait, the latter separating Singapore from Malaysia. Physically, the series links the Andaman Sea and South China Sea.

Geologically, the Malacca Strait is a drowned estuary of the Sunda Shelf, an extensive low-relief surface joining the western part of Southeast Asia to the continental shelf of Asia at a depth of less than 200 m. Much of the Sunda Shelf was dry until inundated by the Pleistocene postglacial sea-level rise. Within the 60 m submarine contour are extensive coastal areas, home to several major habitats and ecosystems, mangroves, peat swamp forests, coral reefs, seagrasses, soft-bottom habitats, rocky and sandy coasts.

GENERAL ENVIRONMENTAL SETTING

The area is strongly influenced by monsoons. The north monsoon (December to February) brings more rain while the south monsoon (June to August) is generally dry. The transition periods are rather unstable, punctuated by the occasional "sumatras" or squalls from the direction of Sumatra with winds of up to 50 knots (Chua et al., 1997). Waterspouts are common in the area (The Hydrographer, 1997).

Waves are more influenced by the North monsoon and by the following transition period, when 27% and 60% of waves are below 1 m and 2 m respectively. During the South monsoon and the following transition, 52% and 82% of the waves are below 1 m and 2 m, respectively (Uktolseya, 1988). The change in monsoons does not influence the direction of current flow which generally follows the topographic configuration of the sea bottom. The current flow is predominantly northwest at 1–1.25 knots in all seasons but increases to 5 knots in some localities by the effects of tidal streams (Kamaruzaman, 1996). The total volume of the strait is about 10^{12} m^3 (Calow and Forbes, 1997) and the monsoons have an influence on the velocity of surface currents. A stronger flow of surface current occurs during the North monsoon (Chua et al., 1997) as water is pushed from the South China Sea. Large sandwaves with amplitudes of 4–7 m and wavelength of 250–450 m at right angles to the water flow are found at several areas, such as northwest of One Fathom Bank and southwest of Tanjung Tuan (Cape Rachado) (The Hydrographer, 1997). Coastlines show cycles of accretion and erosion as a result of the monsoon cycle (Fig. 2).

Water temperature regimes in the strait are relatively uniform with a slightly lower surface water temperature prevailing during the north monsoon. The strait is greatly diluted by river discharges from both sides. Recent surveys show that salinity varies from 6.8 ppt to 31.6 ppt and pH varies from 6.2 to 8.4 (Chua et al., 1996), according to the state of the monsoons (Mossa, 1988). Dissolved oxygen in the nearshore water varies according to the influence of land drainage and freshwater run-off from large rivers, particularly on the Sumatran coast. There is no strong seasonal variation in dissolved oxygen in the open waters (Chua et al., 1997).

Tides in Malacca Strait are mainly semi-diurnal, and strongly influenced by the tides from the Indian Ocean. The tidal range varies from 1.6 to 3.7 m depending on locality. The local sea-bottom topography causes tides to rise 4–5 m at Asahan and Bagansiapi-api (Uktolseya, 1988). Tidal currents do not exceed 3 knots (Kent and Valencia, 1985).

The waters of the straits are relatively fertile, with an abundance of phytoplankton (Mohsin and Ambak, 1996).

Fig. 2. Impact of monsoons on the coast. The impact of the monsoons is very evident on the beaches. On the north coast of Pulau Bintan (Riau) the beach is cut back during the North monsoon with accretion occurring during the south monsoon. Over the years, flotsam has also accumulated on the unutilized beach.

Surface run-off and river discharge of the rivers from both sides contribute to its fertility, but the wastes also cause organic and chemical pollution. There is no significant seasonal variation in surface chlorophyll content and phytoplankton counts. The primary productivity of the water is higher in the southern part than the northern part of the strait due to shallower water with vertical mixing and high nutrient input from the rivers of Sumatra. A seasonal impact is evident on the zooplankton with lower counts during the North monsoon than the South monsoon (Chua et al., 1997).

SINGAPORE STRAIT

South of Singapore lies the Singapore Strait which connects the Malacca Strait with the South China Sea (see Fig. 1). The Singapore Strait merges with the coastal and port waters of Singapore. Daily tidal currents keep the port waters clear of silt, making the whole of the southern coastal area an ideal natural harbour. Also, the area is well protected by small islands.

The physical characteristics of the Singapore Strait differ slightly from those of the Malacca Strait. For example, it has a lower salinity. Towards the east, the tidal characteristics are increasingly influenced by the water from the South China Sea, and the diurnal component increases rapidly to be a diurnal tide at about Tanjung Ayam and P. Kapalajernih (The Hydrographer, 1987).

The loss of habitats has been most severe in Singapore waters. Land reclamation and port development were the main driving forces for this. Massive alterations of the coastal area have resulted in an increase in land size at the expense of the sea. Many islets south of Jurong Industrial Estate were formerly ringed by reefs; they were completely destroyed by landfill through enlargement of the islands (Chia, 1992).

Land reclamation continues, especially on the southern coast and at the eastern end of Singapore.

To some extent, the loss of habitats is compensated by the development of the port. Today Singapore has five gateways. Since 1986 it has been rated as the world's busiest port in terms of shipping tonnage, and in the 1990s it became the world's largest container port as well. The port waters are divided into navigational channels (fairways) and anchorages for various purposes.

Navigation, traffic separation and oil pollution are the main concerns in the Singapore Strait. The grounding of the 244,000-DWT *Showa Maru* in this strait in 1975 eventually led to the implementation of the traffic separation scheme (TSS) and under keel clearance (UKC) (Purwaka, 1996). Since 1993, the implementation of TSS at both ends of the strait has been effective in curbing accidents within its limits (Kamaruzaman, 1996).

JOHORE STRAIT

The Johore Strait lies between Singapore and the state of Johore at the southern end of the Malay Peninsula. A causeway, 1.2 km long, was built in 1924 between Johore Bahru and Woodlands. Originally, ten culverts were built below the causeway but this has been reduced to only two during its widening over the years (Anon, 1997). The causeway has, in effect, resulted in two separate bodies of water, reduced the water circulation and encouraged the build-up of pollutants. A distinct area has developed several kilometres on either side which consists of anoxic, watery and black muds smelling of hydrogen sulphide with no macrobenthic organisms. This is primarily due to the discharges of sewage wastes into waters with little capacity for dispersion (Wood et al., 1997).

This strait has very low wave energy, other than those generated by passing ships going to the ports at Sembawang in Singapore and Pasir Gudang in Johore. As a result of the causeway, the tide levels are elevated 0.2 m above that on the south coast of Singapore (Chua and Lim, 1986).

In studies carried out on the Singapore side (Lim, 1983; 1984a; 1984b) the East Johore Strait shows temperature and salinity mainly influenced by wind and rainfall, currents influenced by the monsoons and temperature and salinity changes caused by tides. In general, the monthly variations of temperature and salinity are quite similar to those of Singapore Strait (Lim, 1984a). In contrast, the West Johore Strait shows two areas where, north of West Reach, there is an inner area of high temperature, low salinity, strong stratification of the surface layer and a nearly anoxic bottom layer (Lim, 1984b).

Except for the area near the causeway, the western half of the strait fares better than its eastern half, in terms of natural habitats. Extensive seagrass beds at the Merambong Shoal were discovered recently (1992) and these extend 1.2 km long and over a width of 80–100 m (Japar, 1994). The west half of the strait has low traffic and little socio-economic development along its shores. The Singapore side has streams dammed and converted into reservoirs and is largely a military restricted area with a small area of mangroves protected in the Sungei Buloh Nature Park. The Johore side has mainly tidal mudflats and mangroves. The only major development is the new bridge at Tuas (Second Link), Singapore, connected to the opposite coast.

The Johore Strait was an important source of local fisheries but this has been drastically reduced due to decreasing fish population, increasingly polluted water and increased shipping traffic. Small areas in Singapore waters are used for floating fish farms, which sometimes suffer from sudden fish kills due to water pollution. The water quality has also not favoured the strait as a recreational area either. Some holiday resorts for the local population have been established at Pasir Ris and Changi where the beaches are being upgraded.

With a significant population on the north coast (Johore Bahru, capital of Johore state), an increase in industrial activities, pig farms and other activities, there has been an increase in sewage and industrial waste discharges into the waterways and Johore Strait. The mean values of BOD, COD, AN, NN, SS, DO and pH for 1987 are available for Skudai, Tebrau and Johore rivers (ASEAN/US CRMP, 1991). In the 1985–87 mean faecal coliform (FC) results from nine stations in the strait, only two sites at the estuaries of Melayu and Masai satisfied the proposed Malaysian criterion for bathing water. Waters closer to the causeway are consistently more polluted than the eastern portion due to the discharge of urban wastes from Johore Bahru which has reduced the recreational potential of the western end as well (Lim and Leong, 1991).

Some idea of heavy metal pollution is given in a study in 1990–92 on the marine environmental levels in sediments of copper, zinc, lead and cadmium (Goh and Chou, 1997). All five sample stations show very high levels of metals in fine sediments which are associated with the heavy boat traffic and shipping industry at Pulau Ubin and Sembawang respectively.

The Department of Environment, Johore and the Ministry of Environment, Singapore, carry out joint monthly monitoring of Johore Strait seawater. The existing water quality is generally within normal variations for seawater. However, some sites show high levels of Total Organic Carbon (TOC) concentration, DO, phytoplankton counts, chlorophyll-*a* contents, faecal coliforms and refuse and flotsam. A ten-year programme was undertaken in 1997 for a complete clean-up of the strait (Anon, 1997).

MARINE AND COASTAL HABITATS OF THE MALACCA STRAIT

The Malacca Strait is one of the biggest estuarine environments in this region, characterised by soft-bottom habitats, mangrove swamps, seagrasses and fringing coral reefs. It has high productivity and diversity, and a rich mix of fauna and flora. For example, the mangroves on the Malaysian side of the straits are believed to have reached their optimal development (Chew, 1996); Sungai Merbok in Kedah is perhaps the most biodiverse mangrove (Ong and Gong, 1996).

Information on the marine and coastal habitats of the strait is available in several accounts (e.g. Morgan and Valencia, 1983; and a special issue of *Ambio*, vol. 17, no. 3, 1988). For the Sumatran side, the coastal systems are described by Whitten et al. (1987), Silvius (1987), BAPPENAS (1993) and Burbridge et al. (1988). On-going biological research on Southeast Asia and ASEAN continues to describe the strait (e.g. Sudara et al., 1994). Perhaps the most recent and most comprehensive account on the Malacca Strait is by Chua et al. (1997). This summarises the literature and provides new information from Indonesia. Statistics are available for the provinces of Aceh, North Sumatra and Riau. In the light of the difficulty of obtaining Indonesian data, this document provides a useful supplement to the earlier review by Burbridge et al. (1988). The Malaysian Institute of Maritime Affairs (MIMA) has released a number of useful overviews (e.g. Ibrahim, 1995; Dow, 1995) and recent conferences on the Malacca Strait have also generated useful overviews but not all are easily available. The papers presented in April 1999 at the International Conference on Malacca Straits are to be published by MASDEC (Malacca Straits Research and Development Centre) which was established in 1998. In future, the MASDEC is a potential source to update information on the Malacca Straits.

Mangroves

Coastal forests along the Malacca Strait are mainly mangroves and peat swamps. Mangroves thrive best where the tidal regime is normal with a significant amplitude of 1–2 m or more and where seawater mixes with freshwater from land run-off. According to Chua et al. (1997) mangroves cover 386,100 ha on the Sumatran coast (Aceh, North Sumatra and Riau provinces), 111,409 ha on the Malaysian coast (Matang, K. Selangor and Klang islands) and 600 ha in Singapore. One problem has been the unreliability of figures on mangroves as the rate of conversion of mangroves to other uses has been rapid in recent years, especially on the Sumatran side, but the area is clearly decreasing.

Mangroves trap and stabilise sediments, prevent siltation and damage to coral reefs and seagrass beds and offer protection against coastal erosion due to waves and currents. These roles are heightened since the strait is an important navigation channel where waves from ship wakes can be damaging to the coast.

Mangroves are productive and important spawning and nursery grounds for many economically important species of finfish, crustaceans and molluscs. Available evidence points to mangroves as being important to coastal fisheries but obtaining a quantitative link is difficult (Ong and Gong, 1996). For example, 42% of capture fisheries along the west coast of Peninsular Malaysia are mangrove-related species (Ibrahim, 1995). A variety of endangered species depends on them for survival, such as the proboscis monkey, estuarine crocodile, milky stork, and lesser adjutant stork (Chua et al., 1997).

Mud flats are usually associated with accreting mangroves; good examples are Matang in Perak and the Klang islands in Selangor. These can extend several km offshore (Chew, 1996). Similarly, the Sumatran alluvial coast has been extended by mudflats which have been developed into coastal land. On the inner boundary of mangroves and extending inland into brackish water, distinct nipa swamps with monospecific stands of palm *Nypa fructicans* may develop.

Peat Swamp Forest

Peat swamp forests occur landward of mangroves and are more influenced by fresh water. They also have a high biodiversity. In Malaysia, they develop over clay which derived from sediments eroded and weathered from the Main Range. Peat releases tannin and organic acids into water, resulting in a high acidity with pH 3–4 and a typical tea-colour (Chew, 1996). As with mangroves, figures on areas of peat swamp are not accurate. An estimated 299,000 ha are found along the west coast of Malaysia, though about 80% is disturbed and logged (Ibrahim, 1995). Peat swamps along the three eastern provinces of Sumatra have been reduced by rapid conversion to agricultural land and other land uses. The major mangrove and peat swamp forests in the Malacca Strait occupy 69,721 ha for the Malaysian side and 210,965 ha for the Sumatran side (Chua et al., 1997).

Coral Reefs

Coral reefs are limited to much smaller patches in the Malacca Strait due to unsuitable conditions for coral growth. On the west coast of Peninsular Malaysia, corals are found around islands and off the coast of Port Dickson (Chua et al., 1997). They are in fair condition with live corals ranging from 25–46% cover (Ridzwan, 1994). On the east coast of Sumatra, corals are mainly around the islands of the Riau archipelago with small patches around the northeastern tip of Sumatra (outside the Malacca Strait proper). Their condition is not known.

Singapore has only small patches of corals remaining, confined to the southern islands (Chou, 1991). Much of the reefs have been lost as a result of reclamation, shipping activities, oil-related activities, dumping of solid wastes and other activities (Hilton and Manning, 1995).

Seagrasses

Seagrasses are less extensive here than they are in other waters in the region, such as the Philippines. Nevertheless,

Table 1

Species of seagrasses bordering the Malacca Strait (Chua et al., 1997)

Species	Indonesia	Malaysia	Singapore
Round-tipped seagrass (*Cymodocea rotundata*)	X	X	X
Toothed seagrass (*C. serrulata*)	X	X	X
Tropical eelgrass (*Enhalus acoroides*)	X	X	X
Fibre-strand grass (*Halodule pinifolia*)	X	–	X
Fibre-strand grass (*H. uninervis*)	X	–	X
Estuarine spoon-grass (*Halophila beccarii*)	x	X	X
Veinless spoon-grass (*H. decipiens*)	X	–	–
Small spoon-grass (*H. minor*)	X	X	X
Spoon-grass (*H. ovalis*)	X	X	X
Curled-base spoon-grass (*H. spinulosa*)	X	X	X
Syringe grass (*Syringodium isoetifolium*)	X	X	X
Dugong grass (*Thalassia hempricii*)	X	X	X
Woody seagrass (*Thalassodendron ciliatum*)	X	–	–
Ruppia maritima	x	–	–
Total	12	9	11

x = Limited record.

many species have been recorded (Table 1). Seagrasses are grazed directly by green turtles and dugong (*Dugong dugon*), and they protect benthic fauna from direct contact with oil slicks; even if their blades are in contact with oil, the rhizomes and roots remained undamaged (Fortes, 1988).

Off the east coast of Sumatra, seagrasses are found in the estuaries of the Riau archipelago. These are mixed species meadows, dominated in particular by *Enhalus acoroides* and *Thalassia hemprecii* or are pure stands depending on substrate, although Whitten et al. (1987) observed some distinct zonation. In the north of the west coast of Malaysia, seagrasses cover the substrate from sandy-muddy to sand-covered corals from Langkawi Island in the north to Teluk Kemang in Negeri Sembilan (Chua et al., 1997). Seagrasses off the west coast of Malaysia cover five main areas (Ishak and Man, 1996; Japar, 1994). Singapore has isolated patches of seagrasses which are limited to reef flats and reef crest zones (Loo et al., 1994).

Rocky Shores and Sandy Shores

Rocky shores are present where headlands extend to the coast and are associated with fringing coral reefs. They have distinctive profiles, influenced by geological structure.

Sandy shores are limited (Fig. 3), found along the west coast of Malaysia, but virtually absent on the Sumatran side.

Fig. 3. Typical sandy beach between rocky headlands. Rocky and sandy shores are limited in the Malacca Strait. On the north coast of Pulau Bintan, the beaches are found between granite headlands which are often fringed by coral reefs.

They are more abundant on islands where the sediments are derived from weathered rocks and coral reefs, such as the north coast of Bintan island. Some are found as narrow intertidal sandy strips above broad tidal flats.

Soft-bottom Habitats

Soft-bottom habitats are extensive over large areas of the Sunda Shelf with muddy bottoms commonly near the major river mouths (Chou, 1994). They are characterized by a high biodiversity of marine animals and are economically important for commercial species of fishes, seaweeds, horseshoe crabs, shrimps, bivalves, gastropods, sea cucumbers and sea urchins (Chua et al., 1997). They form the basis of a thriving cockle industry in Malaysia.

The fisheries stock of the Malacca Strait has not been assessed properly. According to one estimate, the strait holds a potential stock of 253,000 tons of pelagic fish, 220,000 tons of demersal fish and 88,400 tons of shrimp and is a critical fishing ground for small pelagic fish and shrimp (Mossa, 1988). The Indonesian side has more than 45 commercial species, with pelagic fish accounting for more than 60% of the catch. The estimated sustainable potential is more than 270,000 tonnes/year which represents 5.9% of the fishery potential for the whole of Indonesia (Chua et al., 1997).

The Malaysian part, up to 12 miles off the coast, has an estimated potential yield of 321,800 tons, or 55% of the catch from Peninsular Malaysia (Chua et al., 1997). The Department of Fisheries identifies six fishing areas, and the highest catch is off Johore and Perak, with the lowest off Malacca.

COASTAL POPULATION

The Malacca Strait has a long history in which coastal waters and habitats have been used for artisanal fishing and extraction of resources. Before the arrival of the European colonial powers, the strait was already an important navigation and trading route for traders from the Mediterranean, Middle East, India and China. The Riau islands at the southern end of the strait provided a safe haven for boats, and kingdoms were established within the coastal area. The best known was Malacca which was established at the beginning of the 15th century and which became the centre of a spice trade. From the 16th century, European colonial powers, first the Portuguese, then the Dutch and the English, established their presence in the strait. The division of the strait between Indonesia, Malaysia and Singapore dated from the division of Dutch and English interests in 1824. The colonial powers introduced large-scale plantations and mining which have had a great impact. The plantations were supplemented by logging, industrial estates, aquaculture, and large settlements. Important growth triangles have established at both ends of the strait. At the southern entrance, the SIJORI growth triangle links Singapore, State of Johore (Malaysia) and Riau Province (Indonesia) and focuses mainly on Bintan (tourism and industries) and Batam (industries). At the northern entrance of the strait, the IMT (Indonesia–Malaysia–Thailand) growth triangle, links North Sumatra and Aceh of Indonesia, five provinces of southern Thailand and four northern states of Malaysia.

In the last five decades, the rapid economic development of the three littoral states has had a marked impact on the marine environment. The population is about the same on either side of the strait but unevenly distributed. On the Sumatran side, Aceh, North Sumatra and Riau are the provinces fronting the strait, and they have a total population of 10.9 million (1993–94) (Chua et al., 1997). On average, the east coast of North Sumatra has the highest, and Riau Province, the lowest population density. Batam as part of the Growth Triangle is likely to attract more migrants from other parts of Indonesia as it develops. On the Malaysian side, 9.62 million or 49% of the country's population live along the strait (Chua et al., 1997). Agriculture is the most dominant land-use (49%), followed by forest land (47%) and urban areas (3.5%) (Ibrahim, 1995).

Singapore has a population of 2.8 million (1995) with a density of 4608 persons/km^2 (Chua et al., 1997) and has a highly urbanised population and economy. The national increases for the last two decades vary from 1.77–2.41%, 2.32–2.60% and 1.03–1.74% in Indonesia, Malaysia and Singapore respectively. A significant feature in Singapore's population is the gradual increase in the 'greying population'. After 2020, more than 40% of the population will be more than 60 years old (Chua et al., 1997).

IMPACTS OF HUMAN ACTIVITIES

Several activities affect the coastal and marine ecosystems and coastal water quality of the strait.

Forest Clearance

The catchment areas on both sides of the strait have suffered from continuous and intensive land clearing, uncontrolled development and logging activities. Mounting pressure to convert mangrove and peat forests has an impact on the coastal waters. Substantial levels of suspended solids are found in the estuaries of S. Kedah, S. Juru in Penang, S. Kurau and S. Sepetang in Perak, S. Melaka in Malacca and S. Lumut in Johore (Ishak and Man, 1996). From 1985–1991, 57% of the water samples from the west coast exceeded the proposed interim standard for total suspended solids of 50 mg/l (Ibrahim, 1995).

Forest clearance has destroyed important fish breeding and nursery grounds, wildlife sanctuaries and threatens the depletion and extinction of important species (Ibrahim, 1995). The problem is made difficult by the fact that land is a state matter and the Forestry Department has no jurisdiction over state land forests.

In Sumatra, uncontrolled conversion of naturally vegetated upland areas has led to an acceleration of natural erosion, and is a primary cause of down-coast flooding and increased siltation, both of which impact navigation, agricultural production and fisheries. Asahan, Kuala and Barumun are examples of the most seriously affected (Burbridge et al., 1988).

Mangrove forests have been cut for firewood, charcoal, tanning dyes and timber, and Riau province has been an important centre for charcoal. The peat swamp forests provide ramin wood species, rattans and resins (Chua et al., 1997). Massive deforestation started in the late 1960s for the production of wood chips for export to Japan. In the transition of mangroves from traditional uses to large-scale woodchip production, there are conflicts between traditional users and the concessionaires and there has been a failure to regenerate desired species (Burbridge et al., 1988).

Agriculture

Rivers flowing into the Malacca Strait have been contaminated to varying degrees by agricultural activities such as wastes from pig farms, oil-palm and rubber processing and pesticides. In the early 1990s, more than 3,200 pig farms with 2.3 million pigs on the west coast of Malaysia generated an estimated 218 tonnes/day of pollution load (Chua et al., 1997). Large quantities of waste and some of the effluents discharge directly into the estuaries and tidal rivers without any pre-treatment, contaminating the water with bacteria and parasites, exerting a high BOD and causing eutrophication (Ishak and Man, 1996). This problem has been somewhat reduced with the recent outbreak of nipah virus which led to the culling of 890,000 pigs.

Agro-based effluents also contaminate the rivers and coastal waters. These effluents contain very high concentrations of organic matter, suspended solids, nitrogen and phosphorus, leading to a rapid depletion of oxygen in the water and resulting in the destruction of aquatic life and natural ecology of the rivers (Ishak and Man, 1996). A similar situation exists with the processing of oil palm and coconut palm products in North Sumatra and Riau (Burbridge et al., 1988).

Pesticides such as organochlorines have contaminated some of the rivers on the west coast of Peninsular Malaysia. DDE, DDT and heptachlor are present. Organochlorines and PCBs are present in cockles, oyster and green mussels but still are well within the acceptable limits for human consumption (Ishak and Man, 1996).

Fishing

A variety of demersal, pelagic and shrimp species are harvested, mainly down to about 100 m contour. The main pelagic fish are listed in Burbridge et al. (1988). Overfishing is a problem and is widespread (Burbridge et al., 1988; Ooi, 1990).

There are 139 fishing villages on the Malaysian side of the strait (Fig. 4). A wide variety of gear is used by Malaysian fishermen. Successful trawling started in the mid-1960s and is centred at Pangkor but this has given rise to conflict between traditional fishermen and trawlers (Ibrahim, 1995).

Almost 80% of Indonesian fishing is artisanal, using a variety of inshore fishing gear and depending very much on human skill rather than advanced technology. Tidal traps are used where the strait is narrow and tides are strong (Burbridge et al., 1988). Shrimp trawling was introduced in the 1960s, but with the ban on trawling from 1981, this has actually stimulated the conversion of mangroves to fishponds as a means of replacing the loss of shrimp exports. Strong conflicts exist between artisanal and commercial fisheries.

Fig. 4. Typical fishing hut with a landing platform. Fishing villages are found on the Malaysian and Indonesian sides of the Malacca Strait. They are often sited to take advantage of the sheltered waters of estuaries, mangrove coasts or the lee side of islands. The houses are typically built on stilts with landing platforms, as shown in this example of a fishing hut located in a mangrove environment on Pulau Karimum Besar.

Aquaculture

Brackish water pond (*tambak*) culture, mainly for prawns and milkfish and characterized by low productivity, dominates the coastal areas of Sumatra. Its extension has been at the expense of mangrove forests and covered an estimated 39,800 ha in 1993 (Chua et al., 1997). Mariculture, including the culture of seaweeds, is developed around Batam and Bintan islands.

Aquaculture is a fast-growing industry on the west coast of Malaysia. This includes semi-intensive brackish-water pond culture of prawns, floating net-cage culture of groupers, seabass and snappers, cockle culture in coastal mudflats, and mussels and oysters in coastal waters (Chua et al., 1997). In 1993, fish pond and cockle rearing occupied 5700 ha, while fish cages, mussel and oyster beds took up nearly 7 ha (Kamaruzaman, 1996).

Aquaculture has several impacts on the coastal waters. The main problem is associated with the clearing of mangroves for brackish-water ponds utilised for penaeid prawn culture, leading to the eutrophication of the coastal waters and disease (Ishak and Man, 1996). A lesser problem comes from cage culture of grouper (*Epinephelus* sp.), seabass (*Lates calcarifer*) and snapper (*Lutianus* sp.) introduced in the mid-1970s, whose wastes reduce the dissolved oxygen in the water and cause high BOD (Ishak and Man, 1996).

Coastal Erosion and Landfill

Serious coastal erosion occurs at the northern end of Sumatra, the west end of Bengkalis Island and Anak Setahan, due to mangrove felling and land clearance (Chua et al., 1997). In contrast, soil erosion has brought a large amount of silt to the coast, causing coastal accretion and rapid siltation at some ports, e.g. Bagan-siapi-api, one of the largest fishing ports (Burbridge et al., 1988).

Stanley Consultants (1985) found that 36 out of 43 critical areas of coastal erosion in Peninsular Malaysia were along the coast of Perlis, Kedah, north Selangor, south Selangor and south Johore. In south Johore, the erosion of mangrove coast is significant where the upland area has been bunded for agriculture or aquaculture, thus depriving the coastal area of sediments and freshwater supply and contributing to mangrove loss (Zamali and Lee, 1991).

Siltation and sedimentation seriously affect Port Klang, Kuala Perlis and Kuala Kedah, restricting navigation and requiring dredging work (Stanley Consultants, 1985). Accretion at river mouths causes flooding, hampers navigation, destroys natural habitats and results in the degradation of ҉ (Ibrahim, 1995). Land reclamation has been carried out ҉sively along the west coast of Peninsular Malaysia and ҉ ҉end is increasing. This has led to the destruction of spawning and nursery grounds, contamination related to construction, reduced primary production and changes to the circulation pattern (Ishak and Man, 1996).

In Singapore, coastline modification is primarily the result of land reclamation and port development. Landfill is still carried out at the eastern and southwest coasts and the dumping of dredged materials has enlarged several islands. The impacts include the loss of the natural coastline, and smothering of coral reefs due to siltation (Hilton and Manning, 1995). Increased sedimentation is associated with areas of reclamation or dumping (Goh and Chou, 1997).

Shipping and Related Activities

The Malacca Strait is one of the world's busiest sea lanes and a major shipping route for petroleum oil tankers. The shipping traffic is increasing and changing in composition. The daily average of vessels has increased from 119 in 1982 to 274 in 1993 and is projected to reach 598 in 2000 (Chua et al., 1997).

Although oil spills are low-frequency events, they pose a considerable potential threat to fisheries, tourism and aquaculture. Oil contamination in the strait occurs mostly through oil spills arising from collision between vessels. The period 1974–94 saw 17 major oil spills (>5000 barrels per spill) in the strait and its approaches (Meyer, 1996). In contrast, most spills in the Port of Singapore occur during bunkering operations (Chua et al., 1989). Studies following major spills indicate that oil spills have an adverse impact on the fragile intertidal ecosystems of the strait, and mangroves recovered only slowly from major spills (Chua et al., 1997). The strait is affected by sulphur dioxide emissions from international shipping (Streets et al., 1997).

Hydrocarbons are only one contributor to marine pollution, where the results are often seen in fish kills, algal blooms and paralytic shellfish poisoning. A less evident impact is tributyltin (TBT) from antifouling paints used by tankers. Despite indications that the concentrations of TBT in water and sediments are relatively low, there is evidence of their deleterious effects on aquatic organisms, e.g. imposex in female gastropods (Hashimoto et al., 1998; Swennen at al., 1997).

Industrial Activities

Many industrial activities along the strait are highly site-specific in terms of their influence. For example, industrial pollution on the Sumatran coast (Burbridge et al., 1988) is related to the discharge of heated waste water and waste products, sawmill refuse and wastes from the plywood industry and motor-powered river craft in rivers. Mining is associated with problems of land degradation, erosion and the destruction and pollution of the environment (Ibrahim, 1995).

Singapore is also a major industrial centre in the strait with a large industrial estate at Jurong. Many islands south of the Jurong Industrial Estate have been converted for the oil refining industry.

Urban Activities

An estimated 60% of the sewage from the Sumatran coast reaches the strait. This generates loadings of BOD, COD, total nitrogen and total phosphorus estimated at 167,000, 381,000, 74,000 and 7000 tonnes per year, respectively (Chua et al., 1997). The west coast of Malaysia has eight major urban settlements with a population of 10,000–300,000 each, about 20 industrial estates and 139 fishing villages (Ibrahim, 1995). An estimated 4500 tons of BOD per day is derived from domestic sewage and discharged into the waters near Penang island. Coliform counts of 4.9×10^7/100 ml and *E. coli* at 1.6×10^7/100 ml are reported at some distance from the Penang sewage outfall (Sivalingam, 1988) and high faecal coliform counts exist off Port Dickson.

The major ports and harbours along the strait are the Port of Singapore; Pasir Gudang, Malacca Port, Port Dickson, Port Klang, Lumut and Penang in Malaysia; Tanjung Uban, Sambu, Bengkalis, Sungai Pakning, Dumai, Belawan, Pangkalan Susu and Lhokseumawe in Indonesia (see Fig. 1). Tourism development is a relatively new activity along the strait. In Malaysia the main tourist centres are at Langkawi, Penang, Pangkor, Pangkor Laut, Malacca and Port Dickson (Wong, 1990). The main Indonesian tourist centres are on the islands of Bintan and Batam (Fig. 5) and there are considerable negative impacts on the coastal environment on the latter. In contrast, the integrated resort on Bintan has proper sewage treatment plants. In Singapore the holiday resorts at Pasir Ris and Changi are for the local population.

WATER QUALITY

The water quality of the Malacca Strait is greatly affected by sediment load from deforestation, land clearing, coastal reclamation, unsound agricultural and forestry practices, mining activities, and dumping of dredged materials. Pollutants are from land- and sea-based sources and these include pesticides, organic wastes and fertilizers from agricultural activities; sewage and garbage from human settlements; heavy metals, hydrocarbons, phenol, etc. from industries; and sewage, garbage, oil, and chemicals from shipping. Land-based activities pose the most serious impact, contributing as much as 70% of the pollution originating from urban, agricultural and industrial activities. Sea-based pollution is from operational discharges of vessels and potential oil spills from tanker accidents. Atmospheric pollution is difficult to measure (Basiron, 1996).

Agricultural, industrial and domestic sources release a wide variety of pollutants to the rivers which act as conduits to the coastal waters. The persistence of high values of total suspended solids (TSS), *E. coli* and oil and grease are indicative of widespread water pollution in the strait. This reflects the high population density, rapid transformation of land use, industrial growth and shipping traffic. In recent years, harmful algal blooms, commonly known as red tides, have occurred and increased their frequency.

Fig. 5. Typical modern beach resort, Pulau Bintan (Riau). In contrast to unplanned tourism development, planned tourism development is one of the more recent activities along the Malacca Strait. Beach resorts with a variety of facilities and planned environments are being established on the western half of the north coast of Pulau Bintan (Riau) to make it the largest integrated resort in this region.

Faecal coliform count is normally used as an indicator for sewage contamination arising from domestic and animal wastes. *E. coli* contamination is a serious problem. For the west coast of Malaysia, the coastal waters show substantially high levels of faecal coliform (Ishak and Man, 1996). From 1986–91, 60% of the west coast samples exceeded the proposed interim standard of 100 MPN/100 ml (Ibrahim, 1995). The beaches in Penang were seriously affected until the construction of a sewage treatment plant in 1992 (Sivalingam, 1988).

Environmental baseline studies are required to establish the benchmarks of environmental quality and to quantify the quality in economic terms. However, few such surveys are available for this region. Until recently, there was very little reliable data on the levels of heavy metals in marine sediments; they are of doubtful quality as there is no quality assurance (Hungspreugs, 1988).

Malaysia's Department of Environment (DOE) has a water quality monitoring programme with 254 sites on 33 rivers and 127 coastal sites on the west coast states. These provided the only data base without a 'shifting baseline syndrome'. Table 2 shows the range of mean values of some selected parameters for the monitoring stations from 1985–1991. More details and an overview including data for 1992 are given in a review by Dow (1995).

Various studies and routine surveys by the DOE showed the TSS values between 100 and 200 mg/l which are above the interim standard of 50 mg/l (Chua et al., 1996). The DO levels are also low according routine surveys by DOE. The BOD varies from 1.33 mg/l to 9.95 mg/l, most of them around 3–4 mg/l. In recent years, there have been improvements in the bacterial levels (Chua et al., 1996).

Table 2

Water quality data of the west coast of Peninsular Malaysia, 1985–91

Parameter	Interim standard for marine quality	Mean value	% Samples exceeding standard
TSS	50 mg/l	16–265	7–63.2
E. coli	100 MPN/100 ml	599–24,452	36.4–76.1
Lead	0.1 mg/l	0.108	36.9
Mercury	0.001 mg/l	0.068	15.4
Copper	0.100 mg/l	0.034	5.0
Cadmium	0.010 mg/l	0.049	39.8
Arsenic	0.100 mg/l	0.008	0.8
Chromium	0.500 mg/l	0.062	0.3

Source: Ibrahim (1995).

Almost all samples collected from the coastal waters off the west coast of Peninsular Malaysia had values of lead, copper and cadmium exceeding the proposed standards for these metals. However, the levels of heavy metal in Malaysian fish and shellfish have not yet posed a major threat to public health (Ishak and Man, 1996). In a survey of pesticides in Malaysian waterways, S. Selangor and S. Bernam are the most polluted rivers along the west coast (Tan et al., 1991). Organochlorine pesticides found in cockle, oyster and green mussel are generally low and within acceptable limits (Ishak and Man, 1996).

A picture of total hydrocarbons in water and sediments along the coastal areas of the Malaysian coast is provided by Abdullah et al. (1994). Tahir (1997) summarized this survey and another survey to provide some idea of baseline data. The DOE conducted a tar ball survey in 1991 (Chua et al., 1997).

There is no systematic monitoring of pollution on the Indonesian side of the Malacca Strait. Data which exist are from various sources and some preliminary results for BOD and COD are given in Chua et al. (1997). There are no comprehensive estimates of pesticide residues in Sumatran coastal waters. Heavy metal contamination occurs in some locations, especially near the ports, oil refineries and offshore oil exploration. There is no systematic monitoring of hydrocarbon content in coastal waters but studies show the hydrocarbon content level higher around offshore oil fields and refineries, especially in the Riau islands (Chua et al., 1996).

Singapore has an extensive network of sewers and measures to clean the polluted rivers. Rapid improvements to the water quality were reported from 1980–1985 (Hungspreugs, 1988). Treated effluent meets the standards of 20 mg/l of BOD and 30 mg/l of SS. From 1993, a scheme was implemented to prevent indiscriminate dumping of sludge and slop oil in Singapore waters (Loh, 1996).

There is no regular monitoring for heavy metals in the Singapore marine environment. A two-year study in 1990–92 on the levels of metals copper, zinc, lead and cadmium in sediments shows relatively high levels of metals in finer materials. This was the result of the shipping industry which releases anti-fouling paints into the water (Goh and Chou, 1997). Tang et al. (1997) confirmed an earlier survey that the port of Singapore has the highest concentration of metallic pollutants, followed by Johore Strait.

PROTECTIVE MEASURES AND SUSTAINABLE USE

The main problems and issues of the Malacca Strait relating to the coastal and marine resources include the following: over-exploitation and degradation of fishing grounds; loss of habitats; pollution from sediments, sewage, agricultural and industrial effluents, hydrocarbons and heavy metals; and erosion and accretion of shoreline (Burbridge, 1988; Ibrahim, 1995). The marine environment of the strait will worsen with expanding human activities if no management actions are taken to reduce pollution. Unfortunately, relevant information is not adequate and readily available for planning and management (Chua et al., 1997).

In fisheries management, the basic problem is the lack of comprehensive studies on factors that have a direct bearing upon sustained development of the fishery resources, such as the accelerated rate of siltation and conversion of mangrove forests (Burbridge, 1988). Stock assessment has not yet been completed for all waters and species. More than 40 major commercial species have to be assessed and managed individually. Statistics on catches are misleading as they include catches outside the strait (Chua et al., 1997). The fisheries in the shallow waters of the strait had approached or surpassed the level of maximum sustainable yields probably by the mid 1970s (Martosubroto and Naamin, 1988). Overfished resources include prawns, fish and crabs (Chua et al., 1997). Conflicts exist between commercial fishing, subsistence fishing and aquaculture, and the promotion of one inadvertently affects the others.

In Malaysia, management zones have been established to reduce conflicts between traditional and commercial fishermen. Marine parks, marine reserves and fisheries protected areas are now possible under the 1985 Fisheries Act (Ishak and Man, 1996). In North Sumatra, traditional management schemes have evolved over the course of generations and are based on the widely accepted concept of resource user rights. No provision for customary rights is made in the Basic Fisheries Law No. 9/85. Also, no fishing fees are levied on domestic fishermen, depriving the authorities of one method of control. The Indonesian government imposed a ban on trawling in the strait in 1981, which led to the growth of small-scale demersal fisheries and created resource management problems. The development of appropriate coastal zone management strategies are at the incipient stage and have yet to resolve multiple use conflicts (World Bank, 1994).

There is a strong need for mangrove ecosystem management on a sustainable basis. The Fisheries Department, Malaysia, faces the thorny issue of how to protect coastal

Table 3

List of protected and threatened species along the east coast of Sumatra

English Name	Scientific Name
Corals	
Black coral	*Anthipates* spp.
Crustaceae	
King crab	*Tachlypleus tridentatus*
Molluscs	
Horned helmet	*Cassi cornuta*
Triton trumpet	*Charonia tritonis*
Chambered nautilus	*Nautilus compilius*
Green mussel	*Turbo manoratus*
Top shell	*Trochus niloticus*
Bear's paw clam	*Hippopus hippopus*
Small giant clam	*Tridacna maxima*
Boring clam	*Tridacna corcea*
Southern giant clam	*Tridacna derasa*
Giant clam	*Tridacna gigas*
Scaly clam	*Tridacna squamosa*
Reptiles	
Estuarine crocodile	*Crocodylus porosus*
Green turtles	*Chelonia mydas*
Hawksbill turtles	*Eretmochelys imbricata*
Aves/Birds	
Greater egret	*Egretta alba*
Pacific reef egret	*Egretta sacra*
White-collared kingfisher	*Halcyon chloris*
White-bellied sea-eagle	*Haliaetus leucogaster*
Lesser adjutant	*Leptoptilos javanicus*
Milky stork	*Mycteria cinerea*
Tern	*Sterna* sp.
Mammalia	
Macaque monkey	*Macaca fascicularis*
Malayan sun bear	*Helarctos malayanus*
Dugong	*Dugong dugong*
Barking deer	*Munticus muntjak*

Source: Chua et al. (1997).

capture fisheries and encourage mariculture since the two are incompatible. The conversion of mangroves for aquaculture is no longer considered valid (Chua et al., 1997; Ong and Gong, 1996). An example of good conservation and sustainable exploitation of mangroves is the Matang Mangrove Forest Reserve in Perak where timber extraction has been sustainably managed since the 1920s (Chua et al., 1997). The value of timber products per ha on a 30-year rotation is reported to be substantially higher than upland forest production on a 60-year rotation (Ibrahim, 1995).

In Indonesia, the conversion of peat swamp forest is not based on thorough assessment of the value of production nor on whether alternative uses can be sustained (Burbridge, 1988). A change in the leasing system for mangroves was made in 1970 to encourage foreign investment in large concessions. This prevented traditional charcoal producers from access to sufficient areas for sustainable production. Some confusion was present in the maintenance of a mangrove buffer belt along the coast for the protection of fish stocks. The fisheries authority's recommendation of 400 m differs from the 50 m recommended by the forestry authority. Eventually, based on ecological studies, the width of the buffer belt was set at 130 times the tidal range in metres (Soemodihardjo and Soerianegara, 1989). Some recent modification has been introduced in Aceh and North Sumatra provinces. The *tambak empang parit* now allows local residents to use the broad channels around rehabilitated mangrove areas for fishing or fish culture (Chua et al., 1997).

There are not enough protected areas in the strait. On the Malaysian coast, there is a marine park covering four islands (P. Paya, P. Segantang, P. Kala and P. Lembu) off the Kedah coast (Chua et al., 1997). The mangroves have been identified as a potential area for ecotourism and some mangroves at P. Langkawi are used for this. The Malayan Nature Society, a local NGO, maintained a small area of wetlands at the Kuala Selangor Nature Park (Chew, 1996).

Indonesia has a national park programme in which marine reserves are used for preserving coral reefs, mangroves and other specific coastal zones. Within the strait, a marine park of 2600 ha covers P. Weh at the northern end of Sumatra and eleven others have been proposed (Haeruman, 1988). Table 3 summarizes the protected and threatened marine species and mangrove-associated species along the east coast of Sumatra from various sources, compiled by Chua et al. (1997).

Various strategies have been taken to solve coastal erosion and siltation. In Malaysia, the National Erosion Control Council and Coastal Engineering Technical Centre were established in 1987. The construction setback line is set at 60 m from MHSW for a sandy coast and 400 m for a mangrove coast (Ibrahim, 1995).

For Singapore, with the virtually complete loss of mangrove and reef habitats, it seems more imperative to establish the commitment to and the concept of sustainable development for these resources (Hilton and Manning, 1995).

REFERENCES

Abdullah, A.R., Tahir, N.M. and Lee, K.W. (1994) Hydrocarbons in seawater and sediments from the west coast of Peninsular Malaysia. *Bulletin of Environmental Contamination and Toxicology* **53**, 618–626.

Anon (1997) Framework to clean up Straits of Johor. *Asian Water and Sewage* **13** (3), 33–34.

ASEAN/US CRMP (1991) *The Coastal Environmental Profile of South Johore, Malaysia*. ICLARM Technical Reports 24. ICLARM, Manila.

BAPPENAS (1993) Biodiversity: Action Plan for Indonesia. Jakarta.

Basiron, M.N. (1996) Managing marine pollution in the Straits of Malacca. *Coastal Management in Tropical Asia* **6**, 22–25.

Burbridge, P. (1988) Coastal and marine resource management in the Strait of Malacca. *Ambio* **17**, 170–177.

Burbridge, P., Koesoebiono and Dahuri, R. (1988) Problems and issues in coastal resources management and planning in eastern Sumatra and the Straits of Malacca. In *Coastal Zone Management in the Strait of Malacca*, eds. P. Burbridge et al., pp. 3–117. School for Resource and Environmental Studies, Dalhousie University, Halifax.

Calow, P. and Forbes, V.E. (1997) Malacca Straits: Initial Risk Assessment. MPP-EAS/Info/97/117. GEF/UNDP/IMO Regional Programme for the Prevention and Management of Marine Pollution in the East Asian Seas. Quezon City, Philippines.

Chew, Y.F. (1996), Wetland resources in Malaysia. In *State of the Environment in Malaysia*, pp. 111–116. Consumers' Association of Penang, Penang.

Chia, L.S. (1992) *Singapore's Urban Coastal Area: Strategies for Management*. ICLARM Technical Report 31. International Center for Living Aquatic Resources Management, Manila.

Chou, L.M. (1991) Artificial reefs in Singapore : development potential and constraints. In *Urban Coastal Area Management: the Experience of Singapore*, eds. L.S. Chia and L.M. Chou, pp. 47–52. ICLARM Conference Proceedings 25. ICLARM, Manila.

Chou, L.M. (1994) Marine soft bottom benthic habitats—an invaluable but unappreciated resource. In *Living Coastal Resources of Southeast Asia: Status and Management*, ed. C.R. Wilkinson, pp. 123–128. Australian Institute of Marine Sciences, Townsville.

Chua, N.F.W. and Lim, W.K. (1986) Tides and currents in Singapore. In *Biophysical Environment of Singapore and It Neighbouring Countries*, eds. L.S. Chia et al., pp. 77–92. Singapore.

Chua, T.E., Natarajan, R. and Ross, S.A. (1996) Analysis of the state of the marine environment of the Straits of Malacca and Singapore. Paper presented at Conference on Navigation Safety and Control of Pollution in the Straits of Malacca and Singapore: Modalities of International Co-operation, 2–3 September 1996, Singapore. Unpublished.

Chua, T.E., Paw, J. and Guarin, Y.F. (1989) The environmental impact of aquaculture and the effects of pollution on coastal aquaculture development in Southeast Asia. *Marine Pollution Bulletin* **20**, 335–343.

Chua, T.E., Ross, S.A. and Yu, H. (1997, editors) Malacca Straits Environmental Profile. MPP-EAS Technical Report 10. GEF/UNDP/IMO Regional Programme for the Prevention and Management of Marine Pollution in the East Asian Seas. Quezon City, Philippines.

Dow, K.M. (1995) *An Overview of Pollution Issues in the Strait of Malacca*. MIMA Issue No. 5/95. Maritime Institute of Malaysia, Kuala Lumpur.

Fortes, M.D. (1988) Mangrove and seagrass beds of East Asia : habitats under stress. *Ambio* **17**, 207–213.

Goh, B.P.L. and Chou, L.M. (1997) Heavy metal levels in marine sediments of Singapore. *Environmental Monitoring and Assessment* **44**, 67–80.

Haeruman, H. (1988) Conservation in Indonesia. *Ambio* **17**, 218–222.

Hashimoto, S. et al. (1998) Concentration and distribution of butylin compounds in a heavy tanker route in the Strait of Malacca and in Tokyo Bay. *Marine Environmental Research* **45**, 169–177.

Hilton, M.J. and Manning, S.S. (1995) Conversion of coastal habitats in Singapore: indications of unsustainable development. *Environmental Conservation* **22**, 307–322.

Hungspreugs, M. (1988) Heavy metals and other non-oil pollutants in Southeast Asia. *Ambio* **17**, 178–182.

Ibrahim, I. (1995) Coastal Resources of the Malacca Strait and Their Development. *MIMA Issue Paper 1/95*. Maritime Institute of Malaysia, Kuala Lumpur.

Ishak, Z. and Man, A. (1996) Malaysian marine ecosystems and fishery resources. In *State of the Environment in Malaysia*, pp. 90–99. Consumers' Association of Penang, Penang.

Japar, S.B. (1994) Status of seagrass resources in Malaysia. In *Proceedings, Third ASEAN-Australian Symposium on Living Coastal Resources, Vol. 2. Status Review*, eds. S. Sudara, C.R. Wilkinson and L.M. Chou, pp. 283–289. Department of Marine Sciences, Chulalongkorn University, Bangkok.

Kamaruzaman, R.J. (1996) Navigational safety in the Straits of Malacca and Singapore. Paper presented at Conference on Navigation Safety and Control of Pollution in the Straits of Malacca and Singapore: Modalities of International Co-operation, 2–3 September 1996, Singapore. Unpublished.

Kent, G. and Valencia, M.J. (eds.) (1985) *Marine Policy in Southeast Asia*. University of California Press, Berkeley, CA.

Lim, L.C. (1983) Coastal fisheries oceanographic studies in Johore Strait, Singapore. 1. Current movement in the East Johore Strait and its adjacent waters. *Singapore Journal of Primary Industry* **11** (2), 83–97.

Lim, L.C. (1984a) Coastal fisheries oceanographic studies in Johore Strait, Singapore. 2. Hydrological condition in the East Johore Strait. *Singapore Journal of Primary Industry* **12** (1), 17–39.

Lim, L.C. (1984b) Coastal fisheries oceanographic studies in Johore Strait, Singapore. 3. Hydrological condition in the West Johore Strait. *Singapore Journal of Primary Industry* **12** (2), 128–146.

Lim, P.E. and Leong, Y.K. (1991) Water quality status and management in South Johore, Malaysia. In *Towards an Integrated Management of Tropical Coastal Resources*, eds. L.M. Chou et al., pp. 95–101. ICLARM Conference Proceedings 22, National University of Singapore and ICLARM.

Loh, A.T. (1996) Panel discussion on marine pollution. Paper presented at Conference on Navigation Safety and Control of Pollution in the Straits of Malacca and Singapore: Modalities of International Co-operation, 2–3 September 1996, Singapore. Unpublished.

Loo, M.G.K. et al. (1994) A review of seagrass communities in Singapore. In *Proceedings, Third ASEAN-Australia Symposium on Living Coastal Resources, Vol. 2, Research Papers*, eds. S. Sudara, C.R. Wilkinson and L.M. Chou, pp. 311–316. Department of Marine Sciences, Chulalongkorn University, Bangkok.

Martosubroto, P. and Naamin, N. (1988) Fisheries of the Strait of Malacca with special reference to the environment. In *Coastal Zone Management in the Strait of Malacca*, eds. P. Burbridge et al., pp. 225–236. School for Resource and Environmental Studies, Dalhousie University, Halifax.

Meyer, T.A. (1996) Panel discussion on marine pollution. Paper presented at Conference on Navigation Safety and Control of Pollution in the Straits of Malacca and Singapore: Modalities of International Co-operation, 2–3 September 1996, Singapore. Unpublished.

Mohsin, A.K.M. and Ambak, M.A. (1996) *Marine Fishes and Fisheries of Malaysia and Neighbouring Countries*. University Pertanian Malaysia Press.

Morgan, J.R. and Valencia, M.J. (eds.) (1983) *Atlas for Marine Policy in Southeast Asian Seas*. University of California Press, Berkeley, CA.

Mossa, K.M. (1988) Strait of Malacca : environment, living resources and transnational problems. In *Coastal Zone Management in the Strait of Malacca*, eds. P. Burbridge et al., pp. 142–141. School for Resource and Environmental Studies, Dalhousie University, Halifax.

Ong, J.E. and Gong, W.K. (1996) Mangroves, fish and chips. In *State of the Environment in Malaysia*, pp. 121–124. Consumers' Association of Penang, Penang.

Ooi, J.B. (1990) *Development Problems of An Open-Access Resource: the Fisheries of Peninsular Malaysia*. Occasional Paper No 86. ASEAN Economic Research Unit, Institute of Southeast Asian Studies, Singapore.

Purwaka, T.H. (1996) Panel discussion on marine pollution. Paper presented at Conference on Navigation Safety and Control of Pollution in the Straits of Malacca and Singapore: Modalities of International Co-operation, 2–3 September 1996, Singapore. Unpublished.

Ridzwan, A.R. (1994) Status of coral reefs in Malaysia. In *Proceedings, Third ASEAN–Australia Symposium on Living Coastal Resources, Volume 2, Research Papers,* eds. S. Sudara, C.R. Wilkinson and L.M. Chou, pp. 49–56. Department of Marine Sciences, Chulalongkorn University, Bangkok.

Silvius, M.J. (1987) Wetlands—Indonesia. In *A Directory of Asian Wetlands,* pp. 981–1109. IUCN.

Sivalingam, P.M. (1988) Impact of domestic, industrial and agricultural waste disposal on the coastal zone environment of the Strait of Malacca. In *Coastal Zone Management in the Strait of Malacca,* eds. P. Burbridge et al., pp. 179–207. School for Resource and Environmental Studies, Dalhousie University, Halifax.

Soemodihardjo, S. and Soerianegara, I. (1989) Country report: Indonesia. The status of mangrove reports in Indonesia. In *Symposium on Mangrove Management: Its Ecological and Economic Considerations,* I. Soerianegara et al., editors, pp. 73–114. BIOTROP Special Publication No. 37, Bogor.

Stanley Consultants (1985) National Coastal Erosion Study. Final Report, 2 volumes.

Streets, D.G., Camichael, G.R. and Arndt, R.L. (1997) Sulfur dioxide emissions and sulfur deposition from international shipping in Asian waters. *Atmospheric Environment* 31, 1573–1582.

Sudara, S., Wilkinson, C.R. and Chou, L.M. (eds.) (1994) *Proceedings, Third ASEAN–Australia Symposium on Living Coastal Resources. Volume 2, Research Papers.* Department of Marine Sciences, Chulalongkorn University, Bangkok.

Swennen et al. (1997) Imposex in sublittoral and littoral gastropods from the Gulf of Thailand and strait of Malacca in relation to shipping. *Environmental Technology* 18, 1245–1254.

Tahir, N.M. (1997) Fate of spilled oil and ecological and socioeconomic impacts of oil pollution in the Straits of Malacca. In *Oil Spill Modelling in the East Asian Region,* eds. H. Yu et al., pp. 242–263. MPP-EAS Workshop Proceedings No. 5, Quezon City.

Tan, G.H., Goh, S.H. and Vijayaletchumy, K. (1991) Analaysis of pesticide-residues in Peninsular Malaysian waterways. *Environmental Monitoring and Assessment* 19, 469–479.

Tang, S.M. et al. (1997) A survey of levels of metallic and organic pollutants in Singapore coastal waters and marine sediments. In ASEAN Marine Environmental Management: Quality Criteria and Monitoring for Aquatic Life and Human Health Protection, eds. G. Vigers et al., pp. II-51–60. Proceedings of ASEAN–Canada CPMS-II End-of-Project Conference, 24–28 June 1996, Penang, Malaysia. EVS Consultants Ltd., Canada, and Department of Fisheries, Malaysia.

The Hydrographer (1997) *Malacca Strait and West Coast of Sumatera Pilot.* 6th edition. Hydrographic Department, Taunton.

Uktolseya, H. (1988) Physical and biological characteristics of the Strait of Malacca. In *Coastal Zone Management in the Strait of Malacca,* eds. P. Burbridge et al., pp. 118–131. School for Resource and Environmental Studies, Dalhousie University, Halifax, Nova Scotia.

Whitten, A.J. et al. (1987) *The Ecology of Sumatra.* 2nd edn. Gajah Mada University Press, Yogyakarta.

Wong, P.P. (1990) Coastal resources management: tourism in Peninsular Malaysia. *ASEAN Economic Bulletin* 7, 213–221.

Wood, A.K.H. et al. (1997) Geochemistry of sediments in Johor Strait between Malaysia and Singapore. *Continental Shelf Research* 17, 1207–1228.

World Bank (1994) *Indonesia: Environment and Development.* World Bank, Washington, DC.

Zamali, B.M. and Lee, S.C. (1991) Analysis of coastal protection works along the southwestern coast of Johore, Malaysia. In *Towards An Integrated Management of Tropical Coastal Resources,* eds. L.M. Chou et al., pp. 109–117. ICLARM Conference Proceedings 22. ICLARM, Manila.

THE AUTHOR

Poh-Poh Wong
Department of Geography,
National University of Singapore,
Singapore 119260

Chapter 76

EAST COAST OF PENINSULAR MALAYSIA

Zelina Z. Ibrahim, Aziz Arshad, Lee Say Chong, Japar Sidik Bujang,
Law Ah Theem, Nik Mustapha Raja Abdullah and
Maged Mahmoud Marghany

East Coast Peninsular Malaysia borders the South China Sea and faces the continental shelf of the Sunda Platform. Its climate is dominated by the ITCZ which moves through the area in a seasonal cycle, bringing the Northeast Monsoon between November and March and the Southwest Monsoon between May and September. Annual rainfall is greater than 300 cm on the coast and 250 cm on the islands.

The nearshore area is dominated by material transported by rivers and by wave interaction in an environment with low tidal range. The whole length of the coast is lined with a series of beach ridges extending up to 12 km inland. Rocky headlands, and rock exposures provide greater productivity. There are numerous estuaries, most of which have good water quality. All estuaries and many lagoons have mangrove vegetation along their banks, and extensive examples have been gazetted as Mangrove Forest Reserves. Soft substrates in relatively sheltered lagoons, bays, and mangroves of Kelantan and Terengganu also support seagrass communities, which are also found in association with fringing coral reefs which occur around the islands. The beaches also attract several migratory shore birds.

Offshore investigations have mainly been carried out for fisheries purposes. Nutrient levels are low. Human populations are fairly low and dispersed, especially in the south where the terrain is rugged. Almost all of the fishing on the East Coast is artisanal though there is a more organized deep-sea fishing. Generally 80 to 90% of the fishermen are coastal and use traditional gear, though this is now mostly motorised.

The greatest impact of agriculture is the change in the catchment land use to oil palm, rubber plantation or silviculture. The rate of mangrove destruction is quite alarming. Sections of the coastline suffer from erosion or coastal change, and among measures taken to overcome erosion and river mouth sedimentation problems are beach nourishment, shore armouring, and the construction of breakwaters and jetties.

The discovery of oil on the continental shelf has led to an increase in oil pollution and tar-balls, and there also appears to be some seepage of oil from the sea floor. Problems of the coastal area have been recognised since the early 1970s when the rapid expansion of the fishing industry on the west coast and other factors led to a sudden drop in catch. The main problems identified were overfishing and the reduction in habitats supporting the resources. The creation of Marine Parks for many of the coral reefs has greatly reduced destructive practices which previously affected them, and two main initiatives instituted in the 1980s resulted in a Fisheries Act and a coastal erosion survey. Since then the Government has also strengthened regional ties for the investigation and management of the seas in this region.

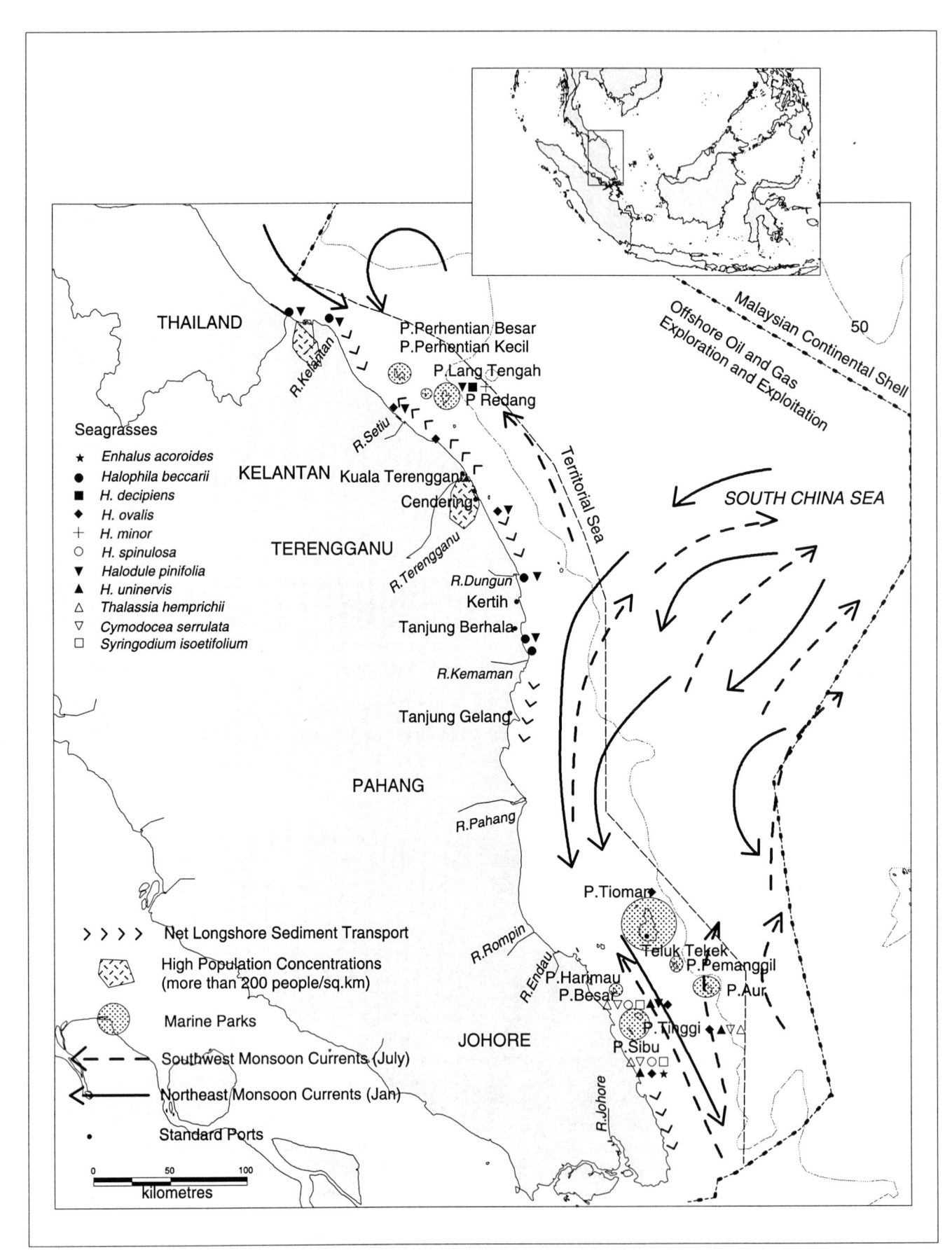

Fig. 1. Map of the East Coast of Malaysia.

THE DEFINED REGION

East Coast Peninsular Malaysia borders the South China Sea, the largest water body in Southeast Asia (Fig. 1) and faces the continental shelf of the Sunda Platform which has water depths not exceeding 100 m. Four states border the East Coast: Kelantan, Terengganu, Pahang and Johor. Although the area described is based on political boundaries, it contains clear environmental characteristics, and forms part of a larger system of interspersed islands and seas in the Malay Archipelago. It differs, for example, from several other Sunda shelf environments in that it includes an edge of the shelf with direct access to a large and long exposure of open sea. This physical setting has a major influence on the habitats present.

SEASONALITY, CURRENTS, NATURAL ENVIRONMENTAL VARIABLES

Climate

Peninsular Malaysia lies between 7° and 1°N in the equatorial belt. Air temperature is fairly uniform throughout the year, varying from 24 to 28°C with an average relative humidity of approximately 80%. The average amount of cloud cover over the sea (50–75%) is also fairly constant. The Inter-tropical Convergence Zone (ITCZ) moves through the area in a seasonal cycle. This diffuse frontal zone, of varying width, separates the trade wind regions of the northern and southern hemispheres and has dense clouds and thunderstorms along its borders. In June the ITCZ lies to the north at 20°N. It moves south in September and lies over southern Peninsular Malaysia in November. From December to April it lies south of $0°$ and starts to move north in May. This movement is related to the location of the pressure systems which result in the monsoons, which in turn are the main forcing factors on the coast.

The Northeast Monsoon occurs between November and March and the Southwest Monsoon occurs between May and September. Annual rainfall is high, greater than 300 cm along coastal areas and 250 cm on the offshore islands. Most falls in heavy showers, especially during the Northeast Monsoon. Tropical cyclones affect the northeastern portion of the South China Sea, but mainland Malaysia rarely feels the effects of these phenomena.

The Northeast Monsoon causes the greatest sea and swell conditions experienced in the South China Sea. The winds blow over a long reach of sea, bringing winds, waves and rain. The Southwest Monsoon is a period of recovery, with calmer winds and smaller waves. The presence of offshore islands reduces the effects of sea and swell on the coast.

Wind-induced waves, and currents resulting from them, are the most important factors affecting coastal transport of water and materials. Where river estuaries are present, they also influence the nature of the coastline.

Tides

Tides are mainly diurnal with a strong diurnal influence. The semidiurnal component becomes more important during the equinoxes, resulting in mixed tides. Ranges are between 1 to 2 m, and the tidal wave approaches the coast from the north and moves southward. Tidal streams are weak, usually less than 0.8 m/s, setting south on the rising tide and north on the falling tide. During the Northeast Monsoon, when streams and currents flow in the same direction at flood, speeds of 1.5–1.8 m/s may be reached on a rising tide. In river mouths the tidal streams are stronger, and higher freshwater discharges or strong onshore winds, such as during the Northeast Monsoon, can alter the speed and direction of the flood and ebb currents.

Winds, Waves and Currents

There are four seasons: the two monsoons and the two transitional intermonsoon periods (Table 1) summarised by Morgan and Valencia (1983), Chua and Charles (1984), and the China Sea Pilot Vol. 1 (NP 30) (Hydrographer of the Navy, 1987). More recent work has been carried out by Kawamura (1986), Liew et al. (1987), Saadon and Rosnan (1989), Sakri (1991), Maged (1994), Saadon and Maged (1996), Nasir et al. (1997), among others.

During the Northeast Monsoon, the northeasterly winds induce surface flow into the South China Sea which generally approaches the East Coast from the northeast and east at 0.3–0.5 m/s, moving southward, at much reduced speeds, along the coast. Over 65% of the waves arriving during this period are from 30° to 60°, but on the northern coast, waves mainly approach from the east. Winds produce considerable swell and during the peak of the Monsoon, the seas are mainly moderate with occasional rough seas. In February, moderate to higher sea-state conditions occur half of the time. The China Sea Pilot notes that waves from the north and northeast may exceed 3.7 m in height.

In the Intermonsoon changeover period, between April and May, the sea state improves and is mainly smooth to light with some calm periods. Current velocities reduce to 0.1–0.5 m/s, more than half of the time, and come mainly from the north. Saadon and Maged (1996) have also found surface currents of 1.35 m/s moving east-northeast in April.

During the Southwest Monsoon, southerly and southwesterly winds prevail, and a reversed current stream of 0.2–0.3 m/s flows northward up to the Terengganu coastline where it is deflected towards Vietnam. Some waters also enter from the Malacca Strait and the Java Sea. Along the Kelantan coast and north Terengganu, a reduced southward coastal current is maintained which appears to originate from the Gulf of Thailand (Kawamura, 1986). Sea

Table 1

Summary of seasonal conditions on East Coast, Peninsular Malaysia

Season	Nominal period	Coastal current	Sea state	Wave height	Wave direction
Northeast Monsoon	November to March	southeasterly; 0.1–0.9 m/s	≥ moderate	1.8 m; >3.7 m possible	from northeast to east
Intermonsoon	April	southward reversing to north; 0.1–0.5 m/s	smooth to light		mixed
Southwest Monsoon	May to September	in the north, southward; in the south, northeasterly; 0.5–0.8 m/s	smooth to light	<2.8 m	from south
Intermonsoon	October	variable	moderate to slight	0.4–2 m	mixed

state is often smooth to slight, and during May to October, waves are less than 2.8 m in height and travel from the south and southwest. From radar data, the significant wave height ranged from 0.4 m to 1.24 m in August (Maged et al., 1998). During this period, squalls are frequent, occurring on average 1–2 days a month, resulting in rougher seas.

In the subsequent Intermonsoon, the currents are similar in strength but run south less than half of the time. The sea state is moderate to slight. In September and October, the significant wave height, from radar, ranged 0.4 to 2 m (Maged et al., 1998).

Around the larger river mouths, coastal currents are also modified by river outflow and tides (Saadon and Rosnan, 1991; Maged, 1994). The high river discharges may result in outflow current velocities of more than 1 m/s.

Temperature, Salinity and Density

Average sea surface temperature ranges from 27 to 29°C. Minima occur in February with maxima in August. Bottom water temperatures of 25°C have been reported. Lateral gradients of surface temperature occur across the South China Sea, in winter between 23°C in the north and 28°C in the south, and in summer this weakens (Liong, 1974; Sivalingam, 1977; Levitus, 1982; Soegiarto, 1981; Husain et al., 1986; Saadon and Liew, 1988; Maged, 1994; Maged et al., 1997).

Salinity in the South China Sea is extremely variable due to the large volumes of river freshwater inputs. Between June and October all coastal areas have lower salinity and density. Levitus (1982) reported surface salinity decreased from 34 ppt, in the north, to 32.4 ppt, in the south. Values of 34.6 ppt have been measured by Sivalingam (1977).

Vertical profiles of the waters at around 100 km offshore, in water depths of 60 to 80 m, indicate the presence of both thermocline and halocline at depths of from 10 to 60 m (Husain et al., 1986; Liew et al., 1987; Saadon and Liew, 1988). Vertical variations of temperature were from 2°C off Terengganu (May 1985) to 5°C off Pahang (May 1986). Salinity variations were about 1 ppt. Off Terengganu, the thermohaline that exists in deeper waters offshore during the Southwest Monsoon weakens during the Northeast Monsoon (Maged et al., 1996), which is consistent with Soegiarto (1981) who reported the depth of mixing in the South China Sea to be about 30–40 m during the Southwest Monsoon and about 70-90 m during the Northeast Monsoon.

Increase in the mixing depth may be brought about by upwelling. Wrytki (1961) had postulated upwelling during the Southwest Monsoon and downwelling during the Northeast Monsoon. Measurements at 16 km offshore in water depths of 35 m to 50 m off the coast of Kuala Terengganu (Maged, 1994) showed isothermal and isohaline contours, suggestive of an upwelling, especially during the Southwest Monsoon, and of downwelling during the Northeast Monsoon.

THE MAJOR SHALLOW-WATER MARINE AND COASTAL HABITATS

The whole coastline is predominantly an exposed sandy shore, interspersed with rocky coasts, mangrove-lined estuaries and lagoons, river deltas and sheltered tidal flats. The shallow seas support coral reef communities around the offshore islands. The area is dominated by material transported by rivers and wave interaction in a low tide-range environment. The larger rivers which are important are the Kelantan River and distributaries, the Terengganu, Kemaman, Kuantan, Pahang, Endau, Rompin, and the Johore Rivers. The difference in the coastline, from north to south, is mainly in terms of the intensity of the coastal processes affecting the shoreline. In general, the southern coastline tends to experience less severe weather effects than the north. In sheltered areas seagrass and mangrove communities are able to develop. The most comprehensive information on the coastline is from studies which mapped the geomorphology, flora, fauna, wildlife, and fisheries of the whole coastline on a scale of 1:25,000 between 1982 to 1993 by teams from Universiti Putra Malaysia (Ridzwan and De Silva, 1983; Ibrahim, 1988; Zauyah and Arshad, 1992; Ibrahim et al., 1993) sponsored by oil companies which extract oil and gas from the continental shelf. A general review of the macrobenthos is found in Chou et al. (1994). Most of these studies focus on the more productive major ecosystems of coral reefs, mangroves and seagrasses, but it should be remembered that most of the coast comprises low-productivity sand beaches.

Sand Beaches, Spits and Shoals

The beaches are mostly of Quaternary alluvium deposits of both marine and continental origin. Raised beaches are common and shallow deposits also occur inland along the river valleys. The landward portion of the whole length of the coast is lined with a series of beach ridges, locally termed "permatang", extending, in some places, up to 12 km inland. These are remnant barrier-lagoon complexes probably formed 5000 to 6000 years ago after the Holocene marine transgression (Teh, 1993).

Sands are fine to very coarse, with finer sand beaches in the south. The beach slope is generally moderate (5–10°) with some very steep (>20°) and sands are generally well sorted. Spits form from entrainment of river flow along the coast, and shoals and bars occur across river mouths and along the coastline.

During the Northeast Monsoon the beaches are generally much shorter and steeper than during the calmer Southwest Monsoon. Net littoral drift is to the south, but net drift direction in specific localities may differ: generally northward in the lee of headlands and southward on exposed stretches of beaches. On beaches that are sheltered by offshore islands, beach progradation may be seen.

Beach vegetation forms a narrow strip of woodland, rarely more than 70 m in width along the coastal berm. This is often disturbed due to natural processes. Woodland is frequently found on low sandy ridges in association with cultivated coconut palms and sea oaks (*Casuarina equisetifolia*) and there are often localised occurrences of trees (*Terminalia catappa, Barringtonia asiatica, Calophyllum inophyllum, Eugenia grandis*), shrubs (*Scaevola taccada, Pandanus odoratissimus, Desmodium umbellatum*) and scattered creepers (*Ipomoea pes-caprae, Canavalia maritima, Vigna marina*), and herbs (*Vitex ovata, Ischaemum muticum*).

The coastal processes create unstable conditions which inhibit many intertidal dwellers. Active forms of life are restricted to such as ghost crabs (e.g. *Ocypode ceratophthalma*), gastropods (e.g. *Nassarius suturalis, Natica tigrina* and *Mactra mera*) and donax (*Donax faba*), razor (*Solen delesserti*) and venus clams (e.g. *Meretrix meretrix*) and polychaete worms (e.g. *Glycera* spp., *Perenereis singaporensis*).

Headlands and Rocky Shores

Rocky headlands, and rock exposures on the beach are dominated by crabs, barnacles (*Tetraclita porosa*), limpets (*Patella* sp., *Acmaea* sp.) and gastropods (e.g. *Morula musiva, Littoraria ventricosa, Monodonta lasio* and *Nerita chamaeleon*). These sessile or sedentary animals tend to have a higher species diversity. Limpets and barnacles (Carripedia) are commonly observed on jetty pillars, seawalls and rocks.

River Estuaries, Tidal Flats, Lagoons, Seagrasses and Mangroves

A number of rivers discharge from the lee of headlands, while some discharge onto straight, beach coastlines. Estuaries are more productive than the coastal waters (Lokman et al. 1984); most are stratified, have good to very good water quality, with moderate to high benthic diversity (Ibrahim et al., 1996).

Perhaps only the Kelantan, Terengganu, Kuantan and Pahang Rivers discharge substantial amounts of sediment. In the north, tidal flats are relatively narrow except when they occur as part of a lagoon system, though in sheltered areas, 200 to 1000 m of tidal flats may be exposed at low water. Sand spits are often found in association with lagoons formed by blocked river outlets, and many small rivers may be blocked from entering the sea by high storm beaches for one season or even for a few years. This often results in eutrophication of the enclosed waters. In some cases artificial exits have been made across the storm beach.

All of the estuaries and many lagoons have mangrove vegetation along their banks. Where these are extensive they have been gazetted as Mangrove Forest Reserves. Their extent and development are strongly influenced by seasonal processes (Gong et al., 1984), though mangrove usually terminates 0.5 to 1 km from the river mouths (Wong et al., 1984). Mangroves also occur on some of the bigger offshore islands. The mainland mangrove vegetation is both patchy and isolated and often not diverse. Isolated stands comprise mainly *Nipa fruticans, Lumnitzera racemosa, Ceriops decandra, Sonneratia caseolaris, Excoecaria agallocha, Rhizophora apiculata* and *Hibiscus tiliaceus*. Extensive growths of *Nipa fruticans* are common in all areas and particularly along river banks in association with other mangrove species. These are mainly medium-sized *Sonneratia caseolris, Rhizophora apiculata* and large *Avicennia alba*. Other mangrove species, such as *Lumnitzera racemosa, Hibiscus tiliaceus* and *Excoecaria agallocha, Avicennia intermedia, Bruguiera gymnorrhiza*, and *Thespesia populnea*, are usually found in low densities and tend to occupy the drier back mangrove. The mangrove vegetation, whether existing in patches or in a long narrow belt, does not show clear zonation of individual species distribution.

The sandy to silty-sand substrates in relatively sheltered lagoons, bays, and mangroves of Kelantan and Terengganu also support seagrass communities. Seagrass also inhabit fine sandy substrate in association with fringing coral reefs of offshore islands such as Tioman, Tengah, Besar, Tinggi and Sibu Islands. Eleven species occur: *Halophila beccarii, H. ovalis* and *Halodule pinifolia* are common in lagoons, bays, and mangroves. Others such as *Enhalus acoroides, Halophila decipiens, H. minor, H. ovalis, H. spinulosa, Halodule pinifolia, H. uninervis, Cymodocea serrulata* and *Thalassia hemprichii* are found on offshore islands.

This highly rich and productive habitat attracts a wide range of marine organisms including several gastropods notably Trochids, Pyrenids (e.g. *Pyrene versicolor*, *P. philippinarum*), strombids and nassarids. The bivalves include cockles (*Anadara* spp.) Penshells (*Atrina* sp.), mussels (e.g. *Perna viridis*) and venus clams (e.g. *Meretrix* sp., *Gafarium* sp.). Holothuroids (*Penctacta quandrangularis*), small gastropods (*Nassarius*) and crabs (mainly *Portunus pelagicus* and *Matuta lunaris*) are regular visitors to the site, especially during their spawning periods. Several commercially important penaeid shrimps, such as *Penaeus merguiensis*, are also widely reported to be present, confirming the nursery role played by the seagrass ecosystem (Arshad et al., 1994).

Offshore Islands and Coral Reefs

There are several islands off the coast, both inhabited and uninhabited. The larger islands are important tourist islands. The islands are mountainous outcrops from the continental shelf with fringing coral reefs, with a steep topography with narrow coastal strips. Sandy pocket beaches, crescent-shaped bays and sheer cliff walls are the normal features here. In the larger islands, rivers have small areas of mangrove vegetation.

Fringing reefs occur around the islands, in isolated patches in water depth of up to 40 m. De Silva et al. (1984) recorded over 170 different species of hard corals belonging to 47 genera, notably *Acropora*, *Montipora*, *Pocillopora*, *Porites*, *Fungia* and *Galaxea*. The status of coral reefs in Malaysia, including the east coast of Peninsular Malaysia, has been compiled by Ridzwan (1994). They support numerous species typical of coral reefs in this particularly rich part of the world's coral seas.

The Migratory Communities

The beaches attract several migratory shore birds. A favoured site is the northern delta of the Kelantan River (Malaysian Wetland Working Group, 1987), but small groups of shore birds can be observed elsewhere along the coastline. Common species are the Malayan plover, Mongolian plover, lesser golden plover, Kentish plover, terek sandpiper, common sandpiper and little green heron. Otters can be seen near the river mouths, along lagoons, and among mangrove vegetation.

Turtles use the beaches as nesting sites throughout the year. A freshwater turtle, the painted terrapin (*Callagur borneoensis*) occurs, with three species of marine turtle; the green turtle (*Chelonia mydas*), the hawksbill turtle (*Eromochelys imbricata*), and leatherback (*Dermochelys coriacea*). The seasons for turtle landings on the beach are between April and September in the north, and February to June in the south. The turtles land on a wide range of beaches both on the mainland and on islands (Chan, 1993; Chan and Liew, 1999). Many places on the beach have been identified by the Department of Fisheries as protected turtle nursery and hatching grounds as part of their turtle conservation programme.

THE SHALLOW SEAS

Offshore investigations have mainly been carried out for fisheries purposes (Jothy et al., 1975; Lim, 1978; Mohsin et al., 1986, 1987) (Table 2). Nutrient levels are low and are reflected in the low chlorophyll *a* levels. Hydrocarbon values off Terengganu are relatively high (>100 ppb) indicating some pollution. Considering the presence of several oil wells in the area this should not be unexpected, but values in sediments also indicate pollution. Secchi disk values indicate that the whole water column, down to 40–90 m, lies within the euphotic zone. The heavy metal levels in sediment were low, indicative of a non-polluted environment.

Fish trawls show that the demersal resources of the area closely reflect the species landed by local fishermen. The average catch rate was 72.08 kg/h in 1985 and 51.3 kg/h in 1986. Earlier studies gave higher catch rates of 166 kg/h (Jothy et al., 1975) and 428 kg/h (Anon, 1967). The lower catch rate was attributed to the gear size and the short trawl time of only 1 hour. The dominant fish families caught were Lutjanidae, Nemipteridae, Synodontidae, Priacanthidae, Ballistidae, Carangidae, Mullidae, Loliginidae, Sepiidae and Ariidae.

POPULATIONS AFFECTING THE AREA

The major towns here are Kota Bahru, Kuala Terengganu, and Kuantan. Other townships are small, consisting of a few thousand persons. There are many important tourist areas with seasonally higher populations, such as Mersing. Elsewhere the population is rural and dispersed, and much of the coastline in the south is uninhabited due to its rugged terrain (Table 3). The intercensal population increase is low,

Table 2

Summary of water quality in the South China Sea, East Coast Peninsular Malaysia

Cruise year	Temp.	Salinity	NH_4-N (μg-at/l)	NO_3-N (μg-at/l)	NO_2-N (μg-at/l)	PO_4-P (μg-at/l)	Chlorophyll *a* (μg/l)	Petroleum hydrocarbon (ppb)
1985	27.5–30.2	32.7–33.5	0.97–1.12	0.90–1.36	0.05–0.09	0.07–0.17	0.095–0.39*	960–980
1986	24.5–29.0	33.0–33.8	0.62–0.74	0.57–0.81	<0.01–0.10	0.18–0.47	0.026–0.059*	37.77–38.42

Values for nutrients are range of means at different depths. *Increasing with depth.

Table 3

Population on the East Coast

State	Population within 30 km of coast (1991)	Coast length (km)	Persons/km of coast
Kelantan	742,100	73	10,166
Terengganu	674,044	244	2,762
Pahang	241,464	204	1,184
Johor (East Coast)	61,896	250	248
Total	1,719,504	771	2,230

Population based on subdistrict (Maxim) level (Department of Statistics, 1995).

Table 4

Average annual growth rates, urban and total population by state

State	Urban population		Total population	
	1970–80	1980–91	1970–80	1980–91
Kelantan	8.50	4.50	2.30	2.90
Terengganu	7.20	3.80	2.60	3.40
Pahang	7.40	4.00	4.20	2.80
Johor	5.10	4.70	2.10	2.50

Source: Teh et al. (1998).

with the urban population growth rate about double that of the total population (Table 4), indicating rural–urban migration.

The population densities of the East Coast states are low compared to the rest of Peninsular Malaysia. The total coastal population, along an area approximately 30 km inland of the coast, is only 1.7 million, with the greater number located along the north, where the larger coastal towns are also found. The towns of Kota Bahru, Kelantan, and Kuala Terengganu have populations of about a quarter of a million people, though the largest town in Johor, Mersing, has only about 30,400 people.

Thus, notwithstanding the presence of the two very populous urban areas at Kota Bahru and Kuala Terengganu, the whole of the East Coast may be regarded as rural in nature.

RURAL FACTORS

Land Development

Along the coast itself there has not been much change in land development over the past twenty years. The major changes have been in the growth of existing towns, increase in residential housing and the introduction of some mainly small industrial estates, particularly in the northern section of the coastline. In addition there have also been many small and scattered developments to cater to the newly growing tourism industry. This is mainly of a rural nature and the main attraction of the East Coast for local and foreign tourists is the natural state of the environment and the relative isolation from urban development. There are new highways constructed to improve accessibility to the region.

The greatest impact of agricultural practices on the coast is the change in the catchment land use. Much of the forested hinterland has been converted to oil palm, rubber plantation or silviculture. There are also cocoa, tobacco, and mulberry plantations. This has changed the hydrological and water quality regime, but, as these developments are normally far upstream, their effects are greatly reduced at the coast. This is evident from the results of Ibrahim et al. (1993) where all the East Coast estuaries surveyed showed much higher water quality than the estuaries on the more developed West Coast. Nevertheless the concern here is the degree of sanitation facilities available to the less developed urban and rural population. De Silva (1985) commented that in many coastal towns in the East Coast, untreated sewage is discharged directly into the sea.

An assessment of the liquid and airborne pollutant loading to the coastal area of the East Coast of Peninsular Malaysia (Nakamura, 1985) gave a value of 25,000 tons BOD_5/annum (less than 12% of the Malaysian total input) and an airborne pollutant load of 65,500 tons/annum (less than 5% of the Malaysian total input). The BOD_5 input is primarily from domestic sources (85%), with industry discharging only 15%. The major polluting industries are palm oil, rubber, food and beverage, and pig farming. The airborne pollutant load is mainly from power plants (7%), industry (34%), transport (53.2%), wood burning (5.3%) and domestic sources (0.5%). It is expected that at present the petroleum and petro-chemical related activities account for perhaps less than 3%.

Measurements by the Department of Environment (1998) on suspended sediment and *E. coli* (Table 5) in the coastal waters indicate that much of the time the values exceed standards (Table 6). There has been much improvement in reduction of bacterial counts in the last few years, and the higher values of suspended sediment indicate the effects of hinterland soil erosion on the coast due to urban development and construction. De Silva et al. (1984) have reported that sediment and freshwater run-off have resulted in damage in coral reefs of several islands on the

Table 5

Percentage exceedance of interim standards for marine water quality for suspended sediment (SS) and *E. coli*

	Kelantan		Terengganu		Pahang		Johor	
Year	SS	E. coli	SS	E. coli	SS	E. coli	SS	E. coli
1992	91.67	0	90.91	0	97.6	38.78	15.97	80.53
1993	88.9	0	79.3	0	88.4	45.5	16.3	65.9
1994	80	36.7	52.8	53.8	97.7	45	23.8	60
1995	87.5	42.5	62.5	35.4	75	42.9	42.9	28.6
1996	90	30	77.4	16.1	53.8	17.3	11.4	33.3
1997	65	15	94	18	100	41.4	15.3	35.3

Table 6

Malaysian Department of Environment Interim Standards for Marine Water Quality

Parameter	Unit	Value
Suspended solids	mg/l	50
Oil and grease	mg/l	0
E. coli	MPN/100 ml	100
Cadmium	mg/l	0.1
Chromium	mg/l	0.5
Mercury	mg/l	0
Lead	mg/l	0.1
Arsenic	mg/l	0.1
Copper	mg/l	0.1

East Coast. At sites where development has occurred such as on Redang Island, much of the live coral and mangrove have been destroyed. The effects of development include destruction of coral by sea transport: for example, the corals at Perhentian Island were in better condition than those at Redang Island because smaller boats and engines are used and houses were also located further back on the beach (Ibrahim et al., 1993).

Artisanal Fishing

Almost all of the fishing on the East Coast is artisanal. The more organized business of deep-sea fishing is new to Malaysia (Fig. 2). Generally, 80–90% of the fishermen population are inshore coastal fishermen and use traditional gear, though this is now mostly motorised (Fig. 3). The main fishing gear used are purse-seines, trawl net, pull/lift net, drift net, gill net, encircling net, traps and hook and line. In estuaries, the main gear used are fish traps, hook and line, cast net, drift net, gill net, and spears. Rotenone, a pesticide, has been used illegally for fishing in the rivers, and in the Kuantan River the effect has been a decline in the amount of giant freshwater prawns. Explosives have also been used, though this is illegal and its use is rare now.

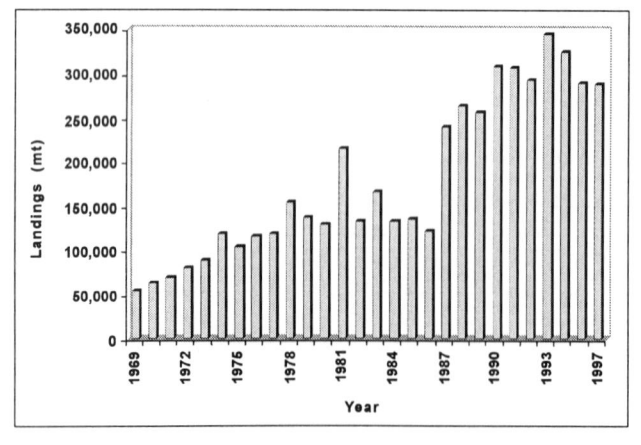

Fig. 2. Total marine fish landings on the East Coast, Peninsular Malaysia.

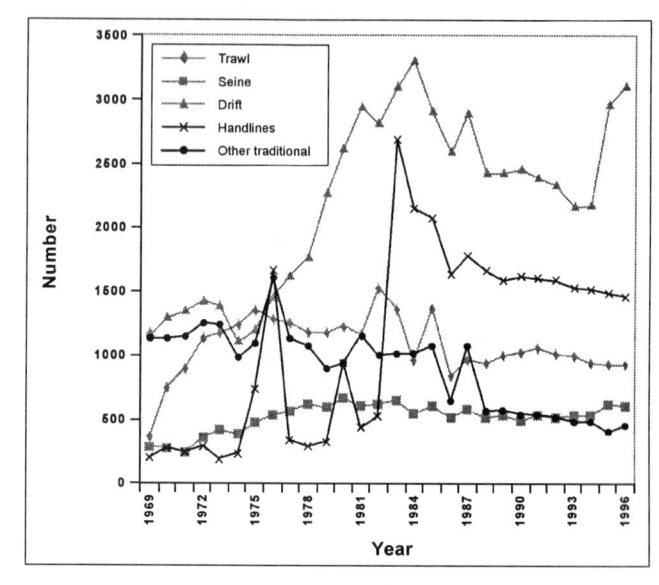

Fig. 3. Trends for the type of various fishing gear licensed in 1969–1996.

The East Coast fishery is characterized by the presence of a large number of commercial fish species. The total fish catch is dominated mainly by the pelagic fishes which contribute around 65% to 70% of the total marine landings. The major species landed in this region include Selar (*Selar* spp.), Selayang (*Decapterus* spp.), Bilis (*Stolephorus* spp.), Tamban (*Sardinella* spp.), Aya/Tongkol (*Thunnus*) and Kembong (*Rastrelliger* spp.). Marine fish landings have been increasing from 1989 to 1996 (see Fig. 2). The marked increase in 1987 is attributed mainly to deep-sea fishing operations, and this masks a decline in the artisanal catch. The highest catch was recorded in 1993 where some 345,103 mt of fish were landed but the catch has stabilized in recent years at about 290,000 mt.

Drift nets seem to be the most popular net used in the fishing operations which mostly target pelagic species. There were some 3,134 drift nets operating in this area in 1996. This is followed by handlines and trawlers. Since 1969, the number of traditional gears has been declining, as has the artisanal fish catch; government policy has been to replace this with deep-sea fishing and fish culture.

With the exception of the year 1977, the total number of licensed fishing vessels was relatively constant during the 1974 to 1996 period (Fig 4). The majority of these boats were small in size (<70 GRT) and not suitable for deep-sea fishing. This trend in types of boats is consistent with fish landings, where total fish catch increased with the inclusion of deep-sea fishing catches.

Artisanal uses of the coast now usually revolve around processing, and there has been more effort directed towards aquaculture. Fish culture can result in income more than twice as much as that of traditional fishing. The problems include higher initial capital cost, higher education, and conversion of land from mangrove. Other legal fishing methods do not impose any great burden on the

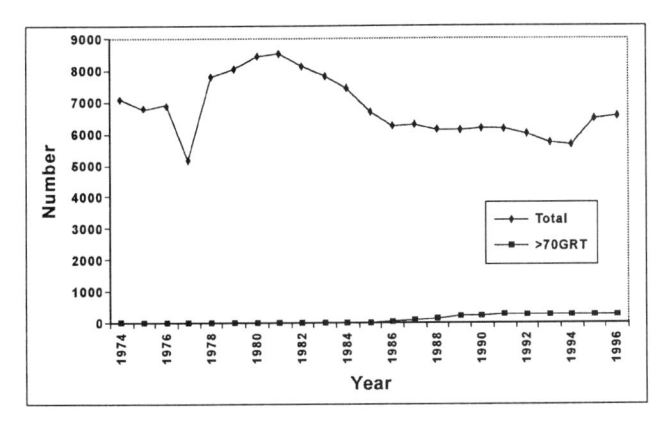

Fig. 4. Total Number of Fishing Vessels and Vessels of Greater than 70 GRT

ecosystem resources, but trawlers have been known to operate illegally in inshore waters.

COASTAL EROSION AND LAND RECLAMATION

Sections of the coastline suffer from erosion or coastal change. It is difficult to assess if many of the changes are man-induced or part of the natural cycle of processes. The historical record is scarce and, since village houses are normally of wood, there is little evidence of lost settlements. The low coastal population density also allows coastal changes to go unnoticed. Several studies have been carried out on beach changes (Wong, 1981; Raj, 1982; Phillips, 1985; Mastura, 1987; Husain et al., 1995b) and in 1985 the National Coastal Erosion Report (Stanley Consultants et al., 1985) classified the coastline into three classes of coastal erosion (Table 7). The net longshore drift is based on this (see Fig. 1).

Among measures taken to overcome erosion and rivermouth sedimentation problems are beach nourishment, shore armouring, and the construction of breakwaters and jetties. Rubble-mound jetties have been placed along some medium-sized river mouths in attempts to prevent spit formation from closing the river. River sand mining has been allowed where shoaling of the river has occurred. Dams have resulted in some hydrological changes, possibly

Table 7

Coastal erosion categories

Category	Erosion condition	Definition
1	Critical	Areas suffering from coastal erosion where shore-based facilities are in imminent danger of loss or damage.
2	Significant	Areas where shore-based facilities are expected to be endangered within 5 to 10 years if no remedial action is taken.
3	Acceptable	Erosion areas that are generally undeveloped with consequent minor economic loss if erosion continued unabated.

including beach erosion or sedimentation (Teh and Shamsul Baharin, 1994; Universiti Kebangsaan Malaysia, 1999). Traditional use of the beach for festivals also results in disruption (Husain et al., 1995a).

Land reclamation is not a major activity on the East Coast. Most is for the construction of ports and marinas in deeper water and with some sea exposure. The more sensitive habitats are not disturbed. There have been proposals to reclaim some tidal flat areas, such as at the Kuantan rivermouth, but these have not yet materialized. On offshore islands some construction activities occur right to the beach berm, and jetties are constructed over coral, resulting in their destruction and erosion. Many hectares of mangroves on Pulau Redang were destroyed with ill-managed development.

At present the major cause of habitat loss is through conversion of reserve land, such as mangrove forest that is close to urban centres, into housing or aquaculture projects. Some of the mangrove conversion activities are illegal but their relative isolation and lack of accessibility make enforcement difficult.

EFFECTS FROM URBAN AND INDUSTRIAL ACTIVITIES

The main commercial activities on the East Coast are related to the fish industry, land cultivation, tourism, and the petroleum industry. Land cultivation on the coast is locally managed except for oil palm and rubber plantations. Tourism is mainly local although there are some international companies investing in the region. The petroleum industry is the main industrial activity causing change to the coast. Except for the potential effect on water quality, that too is very localised to a small area in Terengganu, Kertih, which serves the offshore oil platforms. The effects of human activities have been more apparent on the more sensitive habitats of mangrove, seagrasses, and coral reefs, and on the turtle community. In some areas, levels of both *E. coli* and suspended sediments remain high (Figs. 5 and 6).

Aquaculture

Most aquaculture activities are artisanal in nature, though the pond culture method is expected to cause slow changes to the water quality due to waste discharges, and to habitat area, due to conversion of mangrove. In Johore, however, large-scale conversion of coastal land for aquaculture pond development has occurred and water quality around discharge points has been reported by local fishermen to be of poor quality in the initial stages of operation. There are no documented studies.

Tourism

The tourism industry on the East Coast concentrates on the offshore islands. This industry has resulted in degradation

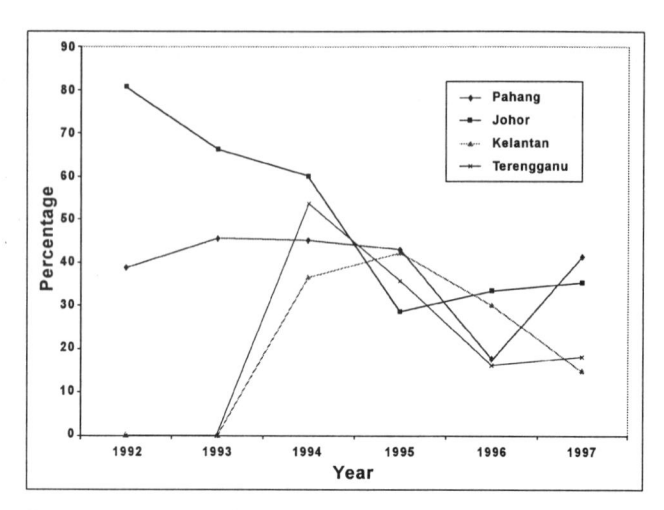

Fig. 5. Percentage of water samples exceeding *E. coli* interim standards.

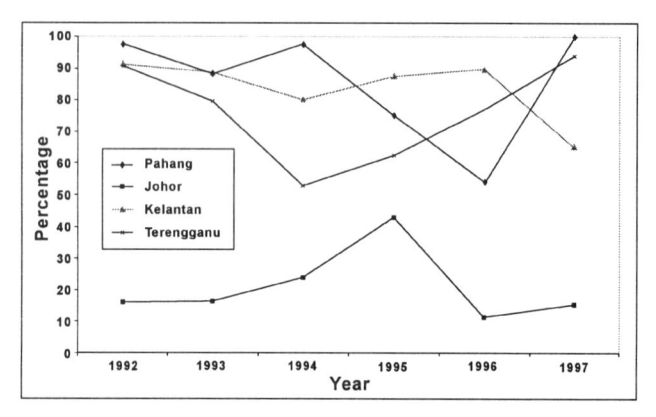

Fig. 6. Percentage of water samples exceeding suspended sediment interim standards.

of coral reefs, mangroves and island coastal water quality. The development of hotels, piers, marine transport facilities, the growth of water-based sports such as scuba-diving, snorkelling, spearfishing, since the early 1970s has resulted in appreciable degradation of coral reefs (Lulofs, 1977). Since then such activities have been much controlled.

Petroleum Industry

The discovery of oil on the continental shelf has led to a localised growth in service industries. There are 56 oil fields and 89 gas fields in the continental shelf, of which 32 oil fields and 6 gas fields are in production. The present gas and oil production are 3500 MMcfd gas/day and 650,000 barrels oil/day, respectively (Law, 1994; Tong et al., 1997). Martin (1984) calculated, from wind data, that oil spilling during the Northeast Monsoon could possibly arrive on the coast south of Kuantan after a period of about five weeks. From wind and current data, the oil slick would likely to affect Tioman Island or Mersing in about eleven days or to track further and miss the coast.

Marine pollution in the South China Sea has been reviewed by Law (1994). The East Coast seas are also vulnerable because of the huge number of ships that pass by and which sometimes discharge oily water, and from land-based activities of bordering countries. There are not less than 40,000 oil tankers sailing past annually, and the amount of crude oil spilled is estimated at 1–5 metric tonnes per day.

Oil pollution seems to be increasing as judged from the increasing amounts of tar-balls along East Coast beaches. Apparently, there is also some seepage of oil from the sea floor of the South China Sea. High hydrocarbon content was found in the sediments of the continental shelf off Terengganu where the highest value found was up to 1332 mg/kg dry sediment. In the vicinity of these highly oiled sediments, high levels of oil in water, up to 1750 μg/l, were also detected (Law et al., 1986). The levels in an unpolluted location would be expected to be below 100 mg/kg in dry sediment and 100 μg/l in water.

Some work on the impact of oil on marine organisms has been performed on crustacean embryology and larvae (Fisher and Foss, 1993; Lee and Nicol, 1980; Law, 1997; Lai and Kessler, 1992). Water-soluble fraction of crude oil ranging from 1 to 20 mg/l can have lethal effects on marine organisms. Toxicities in Malaysian waters have been summarised by Law (1997). The 96hLC$_{50}$ values of the water-soluble fraction of a Malaysian crude oil for penaeid prawn larvae, *Mugil* juveniles, *Lates* juveniles, *Scylla* and cockle *Andara granosa* were 20.3 mg/l, 22.8 mg/l, 23.1 mg/l, >23.6 mg/l, and 26.2 mg/l, respectively.

In the sediment surrounding Redang Island, the mean levels of oil in sediment and water were 3.71 mg/kg dry sediment and 15.69 μg/l respectively (Law et al., 1997). In the coastal waters off Pahang, low hydrocarbon levels were found in water; 17.9–243.95 μg/l, and in sediment; 7.66–14.08 mg/kg dry sediment respectively (Law et al., 1987). Thus, except for some areas, the level of oil pollution in the South China Sea is still low, with mean values below 150 μg/l in water and 50 mg/kg dry sediment in sediment.

IMPACT ON HABITATS AND COMMUNITIES

Mangroves and Seagrasses

The rate of mangrove destruction is quite alarming (Singh et al., 1994). Most of the potential reasons for the decline of both diversity and abundance are organic and metal pollution that are discharged through rivers, as well as from direct clearing. The future of mangroves on the East Coast of Peninsular Malaysia and the associated fauna will depend on the prudence of policy makers. Government policies for mangrove management in the South East Asia region are outlined in Atmadja et al. (1994).

Several threats to the seagrasses have been identified. Natural stress such as waves and sand drift have buried *H. ovalis* and *H. pinifolia* under 5 to 8 cm of sand at Telaga

Table 8
Estimated numbers of nesting leatherback turtles at Rantau Abang, Terengganu

Decade	Estimated number of females
1950s	1779
1960s	1424
1970s	613
1980s	100–300
1990s	30

Source: Salleh (1986); Liew et al (1995); Chan and Liew (1996).

Simpul and Paka, Terengganu, and prolonged, excessive freshwater exposure from the Northeast Monsoon rains temporarily wiped out almost all *H. ovalis* and *H. pinifolia* at Kuala Setiu and Merchang, Terenggau. Human-induced activities also pose threats to seagrasses. Sand extraction by suction at Pengkalan Nangka, Paka and Telaga Simpul resulted in heavy sedimentation. The use of seagrass areas as transport routes by local fishermen, as in the lagoons of Pengkalan Nangka, Kuala Setiu, Merchang and Paka, also disrupt the system. Aquaculture activities such as fish culture, at Pengkalan Nangka and Kuala Setiu, and oyster farming, at Merchang, apparently had no adverse impacts on the seagrass communities there, and seagrass beds around off-shore islands are also seemingly undisturbed.

Coral Reefs

The impacts of human activities on coral reefs have been reviewed by De Silva (1985). Shells, corals, and aquarium fish species have been collected on a small scale, mainly on the offshore islands. The use of poison to catch fish is uncommon here and effects can generally be ignored. Explosives have been used, illegally, to catch fish in coral reefs and this has resulted in extensive damage to the reefs. Trawl fishing is not normally carried out in coral reef areas but several massive trawl nets have been observed in reef areas indicating that some fishermen come too close to the reef. Spearfishing has also been carried out. These activities resulted in stress on the fish population in several island coral-reef areas.

The creation of Marine Parks over many of the island coral reefs has greatly reduced these destructive practices. However, Marine Park areas only encompass the coastal or marine waters and not the island land mass itself. Since most of the coral-fringed islands, such as Redang, Tioman and Perhentian, are inhabited, they have been partially developed mainly for tourism. In general such developments have been kept well managed by the Marine Park authorities to regulate their possible negative impacts.

Degradation of major coral reef sites along the East Coast has also resulted from infestation of 'crown of thorns' starfish (*Acanthaster plancii*), sedimentation, fishing by explosives, and boat anchoring. Impact of sediment on the growth of coral polyps in the region has been well established (Sudara et al., 1991).

Turtles

Turtle egg collection used to be a fishery activity. However, the increase in egg collection, and in the disruption of turtle egg laying by tourists, have caused a decline in the number of turtle landings (Chan, 1988). This has resulted in the Fisheries Department imposing a ban on the collection of eggs of endangered turtles and in guidelines of conduct for turtle watchers. The previously large leatherback population, in particular, has almost collapsed (Table 8) (Salleh, 1987).

PROTECTIVE MEASURES

Problems of the coastal area have been recognised since the early 1970s when the rapid expansion of the fishing industry on the west coast and other factors, led to a sudden drop in catch. The main problems identified were overfishing and the reduction in the habitats supporting the resources. As a result of this, two main initiatives were instituted in the 1980s. These initiatives resulted in the Fisheries Act 1985 and in the National Coastal Erosion Study Report (Stanley et al., 1985) mentioned previously. Since then the Government has moved forward in strengthening regional ties for the investigation and management cooperation in the seas of the region. In the 1980s, the development of ASEAN–USAID cooperative projects resulted in a coastal resources management plan for south Johore, and growing education and awareness of coastal management issues. In the mid 1980s, the ASEAN–Australia cooperative programme on marine resources resulted in surveys of coral reef, benthic and mangrove communities. In early 1990 the Economic Planning Unit of the Prime Minister's Department initiated the development of a Coastal Zone Resources Management Policy. The development of this Policy is currently at the draft stage with the assistance of the Danish DANCED fund.

The control of activities in the coastal and marine environment is governed by several legal requirements and guidelines (Table 9). Some of the major problems faced are the implementation of the mitigation measures prescribed and the continuous monitoring of the project development.

The Fisheries Act 1985

The most effective legislation in the conservation of coastal resources is the Fisheries Act 1985 which deals in a commercial commodity for fisheries as well as for tourism. The Act requires the preparation and review of fisheries plans "based on the best scientific information available and designed to ensure optimum utilization of fishery resources, consistent with sound conservation and management principles and with avoidance of overfishing, and

Table 9

Legislation and Guidelines Relevant to the Coastal Zone

No.	Legislation
1	Environmental Quality Act 1974
	– Environmental Quality (Prescribed Activities) (Environmental Impact Assessment) Order 1987
	– Environmental Quality (Sewage and Industrial Effluents) Regulations 1979
2	Town and Country Planning Act 1976
3	Land Conservation Act 1960
4	Merchant Shipping (Oil Pollution) Act 1994
5	Fisheries Act 1985
6	Continental Shelf Act 1966
7	Exclusive Economic Zone Act (EEZ) 1984
8	National Forestry Act 1984
9	Protection of Wildlife Act 1972
10	Port Authorities Act 1963
11	Fisheries Act 1985

Guidelines

Department of Environment

1. Environmental Impact Assessment Guidelines for Coastal and Land Reclamation
2. Environmental Impact Assessment Guidelines for Petroleum Industries
3. EIA Guidelines For The Management and Disposal of Waste in DownStream Petroleum Industries
4. EIA Guidelines For The Management and Disposal of Waste in UpStream Petroleum Industries
5. Environmental Impact Assessment Guidelines for Fishing Harbours and/or Land Based Aquaculture Projects
6. Environmental Impact Assessment Guidelines for Dams and/or Reservoir Projects
7. Environmental Impact Assessment Guidelines For Coastal Resort Development Projects
8. EIA Guidelines For Development of Tourist and Recreational Facilities on Islands in Marine Parks

Drainage and Irrigation Department

1. Controlling Coastal Erosion Arising from Coastal Development
2. General Guidelines for Hydraulic Study Using Computer Model
3. Sand Mining Guidelines
4. Guidelines for Erosion Control for Development Projects in the Coastal Zone

in accordance with the overall national policies, development plans and programmes".

It allows the Department to create Marine Parks. Several areas have already been designated (see Fig. 1). The purpose of the Parks is to:

(a) afford special protection to the aquatic flora and fauna of such area or part thereof and to protect, preserve and manage the natural breeding grounds and habitat of aquatic life, with particular regard to species of rare or endangered flora and fauna;

(b) allow for the natural regeneration of aquatic life in such area or part thereof where such life has been depleted;

(c) promote scientific study and research in respect of such area or part thereof;

(d) preserve and enhance the pristine state and productivity of such area or part thereof; and

(e) regulate recreational and other activities in such area or part thereof to avoid irreversible damage to its environment.

Within Marine Parks it is an offence to fish, take or remove any aquatic animal or plant whether dead or alive, collect any coral, dredge materials, pollute or alter the natural aquatic habitat, construct any building or structure within a marine park or marine reserve, and to anchor any vessel other than to a designated mooring structure.

Turtle Hatcheries

The Department of Fisheries maintains turtle hatcheries along the coast and gazettes beaches for turtle nesting. In 1990 the Marine Fishery Resources Development and Management Department (MFRDMD) was established under the Southeast Asian Fisheries Development Center (SEAFDEC). SEAFDEC is a regional institution incorporated in 1967 for the development of the fisheries industry in Southeast Asia.

The MFRDMD has a Turtle Conservation Unit, established in 1992, responsible for turtle conservation and marine animal research. The Unit administers the Rantau Abang, Redang, Perhentian, Setiu, Paka, Kijal and Kemaman Hatcheries in Terengganu; the Chendor and Cherating Hatcheries in Pahang; and the Mersing Marine Parks Hatcheries in Johore. Sanctuary Centres for public education and activities are established at Rantau Abang, Terengganu and at Pahang.

Coastal Erosion

Following the National Coastal Erosion Study Report, the Department of Irrigation and Drainage (DID) established the Coastal Engineering Technical Centre (presently Coastal Engineering Division) in 1987. The Division is responsible for implementing the National Coastal Erosion Control Plan involving construction of coastal protection and regulation of development through land-use controls. Various methods of mitigating coastal erosion have been implemented, ranging from hard engineering solutions such as revetments and groin fields, to soft engineering approaches such as beach nourishment, as well as controls such as setback lines and buffer zones. There has been a conscious shift toward the use of beach nourishment as opposed to hard structures. The DID has made significant progress in the implementation of coastal protection works in Category 1 (critical) areas and it is presently envisaged that the short-term, construction-focused, strategy for all Category 1 areas will be fully achieved by the end of 2000.

The Division has also issued Guidelines on Erosion Control for Development Projects in the Coastal Zone as part of the management approach to erosion control. The DID has also been working closely with the Department of Environment in standardizing the requirements for hydraulic studies for impact assessment resulting from coastal works through the issuance of the General Guidelines for Hydraulic Study Using Computer Model, the contents of which have largely been absorbed into the Environmental Impact Assessment Guidelines for Coastal and Land Reclamation.

The Division has further expanded the management focus on coastal protection by initiating the preparation of an Integrated Shoreline Management Plan (ISMP) for a section of the state of Pahang in late 1999. While the underlying purpose of the ISMP is coastal defence, the terms of reference for the project also consider the hinterland development and its effect on the waters of the coast. The production of ISMP is to include clear strategic guidelines for the future management of the coast.

Prognosis

Great changes to the coast from the 1970s at a time of economic growth drew attention to the problems faced by the coastal zone and its resources. The measures described above formed the basis for subsequent actions taken to safeguard the coastal resources. The formation of Marine Parks, the conservation of turtles and the control of fishing by gear and area represent the more positive steps taken. The East Coast has generally been isolated from the pressure of West Coast development activities due to its remoteness and reduced accessibility. However, new roads will ensure that development will reach the remoter areas. This will bring, in addition to economic wealth, the attendant misfortunes of pollution. Due to the slower pace of development here the East Coast may be better prepared to meet the challenges of the new Millennium. It has the experience of the West Coast to look to. In addition, coastal tourism represents an important economic aspect of the region, therefore the naturalness of the environment will be directly linked to its economic value.

At present, the major focus of coastal and marine research has been in relation to fisheries but there is growing realization that habitat destruction or degradation and excessive catch is resulting in depleted resources. The policy of moving the concentration of fisheries into the deep sea will alleviate the stress on the coastal ecosystem (but with so far unknown effects offshore) but this will need to be balanced by control of water quality degradation from the expected development of residential and industrial areas.

Investigations into habitats other than coral reefs also need to be stressed. Seagrass beds, for example, are facing destruction from various anthropogenic factors such as coastal erosion, sedimentation and fishing activities. The relatively scarce information has meant that less attention has been given to their conservation needs.

The South China Sea, too, is exposed to pollution from sources beyond the control of the littoral states. Tankers and offshore petroleum development pose a potential threat, and the management of the area needs to occur in a cooperative forum, with all countries achieving a greater capability and capacity of control in the management of resources and pollution abatement. The seasonal circulation pattern of the shelf seas shows clearly the influence of extra-regional pollution. The long-term effect of pollution from remote sources on the coasts, river mouths, and townships are yet unknown. In view of the continuous physical development taking place in the region, it is anticipated that organic and chemical pollution, through land-clearing activities, domestic and industrial wastes disposal, could represent a threat to the aquatic life in these seas.

REFERENCES

Anon (1967) The Results of the Joint Thai–Malaysian–German Trawling Survey off the East Coast of the Malay Peninsula, 1967. The Marine Fisheries Laboratory, Department of Fisheries, Ministry of Agriculture, Bangkok, Thailand and the Fisheries Research Institute, Fisheries Division, Ministry of Agriculture and Cooperatives, Kuala Lumpur, Malaysia. 64 pp.

Arshad, A., Siti Sarah, M.Y. and Japar Sidek, B. (1994) A comparative qualitative survey on the invertebrate fauna in sea grass and non sea grass beds in Malaysia. In *Proceedings of the Third Australia Symposium on Living Coastal Resources, Vol. II: Research Papers*, eds. S. Sudara, C. Wilkinson and L.M. Chou. Chulalongkorn University, Bangkok. pp. 337–347.

Atmadja, W.S., Alias, M., De Leon, R.O. and Tantichodok, P. (1994) Government policy for mangrove management in ASEAN. In *Proceedings of the Third ASEAN-Australia Symposium on Living Coastal Resources. Vol. I: Status Reviews*, eds. C.R. Wilkinson, S. Sudara and L.M. Chou. Chulalongkorn University, Bangkok, Thailand. pp. 171–177.

Chou L.M., Paphavasit, N., Kastoro, W.W., Nacorda, H.M.E., Othman, B.H.R., Loo, M.G.K. and Soedibjo, B.S. (1994) Soft bottom macrobenthic communities of the ASEAN Region and the influence of associated marine ecosystems. In *Proceedings Third ASEAN-Australia Symposium on Living Coastal Resources. Vol. I: Status Reviews*, eds. C.R. Wilkinson, S. Sudara and L.M. Chou. Chulalongkorn University, Bangkok, Thailand. pp. 325–331.

Chua T.-E and Charles, J.K. (eds.) (1984) *Coastal Resources of East Coast Peninsular Malaysia, An Assessment in Relation to Potential Oil Spills*. Universiti Sains Malaysia, Penang. 306 pp.

Chan, E.H. (1988) An update on the leatherback turtles of Rantau Abang, Terengganu, Malaysia. *Marine Turtle Newsletter* 42, 4.

Chan, E.H. and Liew, H.C. (1996) Decline of the leatherback population in Terengganu, Malaysia, 1956–1995. *Chelonian Conservation and Biology* 2 (2), 196–203.

De Silva, M.W.R.N. (1985) Enviromental controls for offshore oil exploration and production in Malaysia waters. In *Enviromental Protection and Coastal Zone Management in Asia and the Pacific*, eds. I. Kato, N. Kumamoto, W.H. Matthews and A. Suhaimi. University of Tokyo Press. pp. 135–144.

De Silva, M.W.R.N., Betterton, C. and Smith, R.A. (1984) Coral reef resources of the east coast of Peninsular Malaysia. In *Coastal Resources of East Coast Peninsular Malaysia. An Assessment in Relation to Potential Oil Spills*, T.-E. Chua and J.K. Charles. Universiti Sains Malaysia, Penang. pp. 57–90.

Fisher, W.S and Foss, S.S. (1993) A simple test for toxicity of Number 2 fuel oil and oil dispersants to embryos of grass shrimp, *Palaemonetes pugio*. *Marine Pollution Bulletin* **26**, 385–391.

Gong, W.K., Ong, J.E. and Leong Y.K. (1984) The mangrove resources of the east coast of Peninsular Malaysia. In *Coastal Resources of East Coast Peninsular Malaysia, An Assessment in Relation to Potential Oil Spills*, eds. T.E. Chua and J.K. Charles. Universiti Sains Malaysia, Penang. pp. 91–109.

Husain, M.L, Rosnan, Y. and Shazili, N.A. (1986) Some measurements of temperature and salinity on a portion part of the South China Sea. In *Ekspedisi Matahari '85*, eds. A.K.M. Mohsin, M.I. Mohamed and M.A. Ambak. Faculty of Fisheries and Marine Science, Universiti Pertanian Malaysia, Serdang. pp. 49–79.

Husain, M.L., Ismail, K.A. and Yaakob, R. (1995a) Short-term impact of beach fest on the topography, vegetation coverage and sediment distribution of the Mengabang Telipot Beach, Terengganu. *Pertanika Journal of Science and Technology* **3**(2), 325–336.

Husain, M.L., Yaakob, R. and Saad, S. (1995b) Beach erosion variability during a Northeast Monsoon: The Kuala Setiu coastline, Terengganu, Malaysia. *Pertanika Journal of Science and Technology* **3**(2), 337–348.

Hydrographer of the Navy (1987) *China Sea Pilot Vol. 1* (NP 30). The Hydrographic Department Ministry of Defence, U.K.

Ibrahim, Z.Z. (ed.) (1988) Environmental Sensitivity Index Mapping of the Chukai to Penor Coastline. Universiti Pertanian Malaysia, Serdang.105 pp.

Ibrahim, Z.Z., Ramdzani, A., Aziz, A., Japar Sidek, B., Ismail, A., Yaacob, R. and Nik Mustapha, R.A. (1993) Enviromental Sensitivity Index Mapping of the Coastline between Tumpat to Tanjung Jara, Pulau Perhentian, Pulau Redang and Pulau Kapas. Universiti Pertanian Malaysia, Serdang. 129 pp.

Ibrahim, Z.Z., Low Soo Moi, Abdullah, R. and Arshad, A. (1996) Classification of Malaysian estuaries for development planning. *Aquatic Conservation: Marine and Freshwater Ecosystems* **6**, 195–203.

Jothy, A.A, Ranck, G., Mohd Shaari bin S.A Latiff, Sing, O.K., Chong, L.P. and Carvalho, J.L. (1975) Dimersal fish resources in Malaysian waters. *Fisheries Bulletin* **4**, 36 pp.

Kawamura, G. (1986) A Preliminary survey of the sea surface in Malaysian waters and the Gulf of Thailand with the aid of satellite data utilization system. In *Ekspedisi Matahari '85*, eds. A.K.M. Mohsin, M.I. Mohamed and M.A. Ambak. Faculty of Fisheries and Marine Science, Universiti Pertanian Malaysia, Serdang. pp. 45–48.

Lai, H.C and Kessler, A.O. (1992) Acute Effect of Crude and Chemical-dispersed Oil on Marine Fauna of Commercial Importance. Universiti Sains Malaysia, Penang. 38 pp.

Law, A.T. (1993) Marine pollution in the South China Sea. Paper presented at the ISIS/SEAPOL Seminar on Overview Workshop on Ocean Policy and National Sub-regional and Extra-Regional Perspectives. Seminar on Overview Workshop on Ocean Policy and National Sub-regional and Extra-Regional Perspectives.

Law, A.T. (1995) Oil pollution in the Malaysian Seas. *Fishmail Malaysia Fisheries Society* **6**, 5–23.

Law, A.T. (1996) Petroleum hydrocarbons in the South China Sea and Straits of Malacca and their toxicities to organisms. *Proceeding of UNEP Workshop on Land-based Oil Discharges to Coastal Waters: Ecological Consequences and Management Aspects*. Penang, Malaysia. pp. 22–26.

Law, A.T. (1997) Acute and chronic toxicity of the water soluble fraction (Wsf) of a Malaysian crude oil to *Penaeus monodon* postlarvae. In *ASEAN Marine Enviromental Management: Quality Criteria and Monitoring Conference on Marine Science, Penang, Malaysia*, eds. G.A. Vigers, K.S. Ong, C. McPherson, N. Milson, J. Watson and A. Tang. EVS Enviroment Consultants, North Vancouver and Department of Fisheries Malaysia.

Law, A.T. and Rahimi bin Yusuf (1986) Hydrocarbon Distribution in the South China Sea. In *Matahari Expedition '85*, eds. A.K.M. Mohsin, M. Ibrahim and M.A. Ambak. Faculty of Fisheries and Marine Science, Universiti Pertanian Malaysia, Serdang. 233 pp.

Law, A.T. and Zulkifli b. Mahmood (1987) Distribution of petroleum hydrocarbon in the South China Sea. In *Matahari Expedition '86*, eds. A.K.M. Mohsin, M.A. Ambak and Abd. Ridzwan. Faculty of Fisheries and Marine Science, Universiti Pertanian Malaysia, Serdang. 197 pp.

Law, A.T., Japar, S.B., Mohd. Ibrahim Hj. Mohamed, Noor Azhar, M.S., Lim, Li Ching and Mohd. Nasir Abdul Salam (1997) Some physical, chemical and microbiological properties of Pulau Redang coastal waters. Paper presented at Science and Technology Developments in East Coast States with Special Reference to Chemistry. November 1997. Kuala Terengganu.

Lee, W.Y. and Nicol, J.A.C. (1980) Toxicity of a field oil to the eggs of *Parhyale hawaienci* and *Amphithoe valida* (Arrphipcda). *Marine Environmental Research* **3**, 297–305.

Levitus (1982) Climatological Atlas of the World Ocean. NOAA. Professional Paper No. 13. U.S.A. Govermnent Printing Office, Washington D.C.

Liew, H.C. and Mohd Rafiee Mohd Saufi (1987) Zooplankton Biomass, Caloric Equivalent and Composition in the South Western Portion of the South China Sea. In *Ekspedisi Matahari '86*, eds. A.K.M. Mohsin, R.A. Rahman and M.A. Ambak. Faculty of Fisheries & Marine Science, Universiti Pertanian Malaysia, Serdang. pp. 85–92.

Lim, L.C. (1978) MFRD oceanographic data report. Marine Fisheries Research Department, South East Asia Fisheries Development Centre.

Liong Pit Chong (1974) Hydrographic conditions off the East Coast of the Peninsular Malaysia during August 1972. *Malaysian Agriculture Journal* **59**, 492–513.

Lokman Shamsudin, M., Mansor, I. and Husin, J. (1984) Primary productivity and mineral nutrient status of some estuarine and coastal waters along the East Coast of Peninsular Malaysia during the off Monsoon Period. *Pertanika* **7**, 79–87.

Lulofs, R.B. (1977) Conservation of the Marine Enviroment—Coral Reefs. *Proceedings of the First Annual Seminar of the Malaysian Society of Marine Sciences, Penang*. The Malaysian Society of Marine Sciences. pp 35–49.

Maged, M.M. (1994) Coastal water circulation off Kuala Terengganu. M.Sc. Thesis, Universiti Pertanian Malaysia.

Maged, M.M., Saadon, M.N., Hussain, M.L. and Mohamed, M.I. (1997) Seasonal Thermohaline Variation in Coastal Waters off Kuala Terengganu, Malaysia. In *Climate Change in Malaysia: Proceedings of the National Conference on Climate Change*. Faculty of Science and Environmental Studies. Universiti Putra Malaysia Press, Serdang.

Maged, M.M., Ibrahim, Z.Z. and Ibrahim, M.M. (1998) ERS-1 and Modeling of Shoreline Change. *International Science Conference '98*. Serdang. Malaysia.

Malaysian Wetland Working Group (1987) Malaysian Wetland Directory. Department of Wildlife and National Parks, Peninsular Malaysia. Kuala Lumpur.

Martin, P. (1984) Prediction analysis of oil slick drift from tapis field—a model study. In *Coastal Resources of East Coast Peninsular Malaysia, an Assessment in Relation to Potential Oil Spills*, eds. T.-E. Chua and J.K. Charles. Universiti Sains Malaysia, Penang. pp 11–51.

Mastura, S. (1987) *Coastal Geomorphology of Desaru and Its Implication for Coastal Zone Management*. Monograph No. 13. Faculty of Social Sciences and Humanities. Universiti Kebangsaan Malaysia, Bangi.

Mohsin, A.K.M., Ibrahim, M., Mohamed, H. and Ambak, M.A. (eds.) (1986) *Ekspedisi Matahari '85*. Faculty of Fisheries and Marine Science, Universiti Pertanian Malaysia, Serdang.

Mohsin, A.K.M., Rahman, R.A. and Ambak, M.A. (eds.) (1987) *Ekspedisi Matahari '86*. Faculty of Fisheries and Marine Science, Universiti Pertanian Malaysia, Serdang.

Morgan, J.R. and Valencia, M.J. (1983). The natural environmental setting. In *Atlas for Marine Policy in Southeast Asian Seas*, eds. J.R. Morgan and M.J. Valencia. University of California Press. pp. 18–22.

Nakamura, M. (1985) Development of coastal zones and its impact on the environment: the Singapore experience. In *Enviromental Protection and Coastal Zone Management in Asia and the Pacific*, eds. I. Kato, N. Kumamoto, W.H. Matthews and A. Suhaimi. University of Tokyo Press. pp. 183–200.

Nasir, M.S. and Camerlengo, A.L. and Kadir, W.H.W. (1997) Coastal current in the northern region of the East Coast of Peninsular Malaysia. *Sains Malaysiana* 26 (2), 5–14.

Phillips, R.P. (1985) Longshore transport of sediment during August and September on the Terengganu Coast. *Pertanika Journal* 8 (2), 273–275.

Raj, J.K. (1982) Net directions and rates of present-day beach sediment transport by littoral drift along the East Coast of Peninsular Malaysia. *Geological Society of Malaysia Bulletin* 15, 57–82.

Ridzwan, A.R. (1994) Status of Coral Reefs in Malaysia. In *Proceedings of the Third ASEAN-Australia Symposium on Living Coastal Resources. Vol. I: Status Reviews*, eds. C.R. Wilkinson, S. Sudara and L.M. Chou. Chulalongkorn University, Bangkok. pp. 49–56.

Ridzwan, A.R and De Silva, M.W.R.N. (eds.) (1983) Coastal Resources of the Chukai-Dungun Coastline Sensitive to Impacts of Oil Spills. Universiti Pertanian Malaysia, Serdang.

Saadon, M.N. and Liew, H.C. (1988) Salinity and temperature vertical profiles. In *Ekspedisi Matahari '87*, eds. A.K.M. Mohsin and M.I. Mohamed. Faculty of Fisheries and Marine Science, Universiti Pertanian Malaysia, Serdang. pp. 29–36.

Saadon, M.N. and Maged, M.M. (1996) Surface Circulation off Kuala Terengganu in the Transitional Period between the Northeast and Southwest Monsoons. *Pertanika Journal of Science and Technology* 4 (1), 141–148

Saadon, M.N. and Rosnan, Y. (1989) A study on the water movement due the tides off the Kuala Terengganu Coast. Paper presented in Workshop on Tides and Tidal Phenomena. Kuala Terengganu, Malaysia.

Saadon, M.N. and Rosnan, Y. (1991) The Relationship of Tide and Wind to the Coastal Water Movement of Kuala Terengganu. Proceedings IEM/ICM Joint Conference on Coastal Engineering in National Development, Kuala Lumpur, Malaysia. pp. C1–C15.

Sakri, I. (1991) Proper construction and set up of Malaysian fish aggregating devices (UNJAM). Ph.D. Thesis, Universiti Pertanian Malaysia.

Salleh, U. (1986) Rancangan pemuliharaan penyu di Terengganu. Universiti Pertanian Makaysia, Kuala Terengganu.

Singh, H.R., Chong, V.C., Sasekumar, A. and Lim, K.H. (1994) Value of mangroves as nursery and feeding grounds. In *Proceedings of the Third ASEAN-Australia Symposium on Living Coastal Resources. Vol. I: Status Reviews*, eds. C.R. Wilkinson, S. Sudara and L.M. Chou. Chulalongkorn University, Bangkok. pp. 105–122.

Sivalingam, P. (1977) Chemical oceanography observations of the South China Sea. *Sains Malaysiana* 6 (2), 139–154.

Soegiarto, A. (1981) The oceanographic features of South East Asian waters. In *South East Asian Seas, Frontiers and Development*, eds. L.S. Chia and C. MacAndrews. McGraw-Hill. pp. 20–47.

Stanley Consultants, Moffat and Nichol Engineers and Jurutera Konsultants (1985) National Coastal Erosion Study Final Report, Vols. 1 & 2. Unit Perancang Ekonomi, Kuala Lumpur.

Sudara, S., Sanitwongs, A., Yeemin, T., Moordee, R., Panutrakune, S., Suthanaluk, P. and Natekanjanalarp, S. (1991) Study of the impact on growth of the coral *Porites lutea* in the Gulf of Thailand. In *Proceedings of the Regional Symposium on Living Resources in Coastal Areas*, ed. A.C. Alcala. Marine Science Institute, University of Philippines, Quezon City. pp: 107–112.

Teh Tiong Sa (1993) Potential impacts of sea level rise on the Permatang coasts of Peninsular Malaysia. *Malaysian Journal of Tropical Geography* 24, 41–55.

Teh Tiong Sa, Voon Phin Keong, Chan Kok Eng, Tan Wan Hin and Tan Lee Seng (eds.) (1988) Background paper, GEF-ISIS Project on Assessment of Impacts of Climate Change on Key Economic Sectors in Malaysia. Department of Geography, Universiti Malaya and Malaysian Meteorological Services, Kuala Lumpur.

Teh Tiong Sa and Shamsul Bahrin, T. (1994) Socio-economic Impacts of Erosion at Pantai Sabak, Kelantan. *Malaysian Journal of Tropical Geography*, 129–142.

Tong, S.L., Goh, S.H., Rani Abdullah, A., Tahir, N.M. and Wang, C.W. (1997) Development of ASEAN marine water quality criteria for oil and grease. In *ASEAN Marine Environmental Management Quality Criteria and Monitoring for Aquatic Life and Human Health Protection. Proceedings of the ASEAN-Canada Technical Conference on Marine Science, June 1996, Penang, Malaysia*, eds. G.A. Vigers, K.-S. Ong, C. McPherson, N. Millson, I. Watson and A. Tang. EVS Environmental Consultants, North Vancouver, and Department of Fisheries, Malaysia.

Universiti Kebangsaan Malaysia (1999) Draft Final Report, Benefit Monitoring And Evaluation of The National Coastal Erosion Control Sector Project (Volume II : Main Project). Bureau of Consultancy and Innovation, Universiti Kebangsaan Malaysia, Bangi.

Wong, T.M., Charles, J.K. and Khoo, T.T. (1984) The mangrove invertebrate resources of the East Coast of Peninsular Malaysia. In *Coastal Resources of East Coast Peninsular Malaysia. An Assessment in Relation to Potential Oil Spills*, eds. T.-E. Chua and J.K. Charles. Universiti Sains Malaysia, Penang. pp. 110–130.

Wong P.P. (1981) Beach Changes on a monsoon coast of Peninsular Malaysia. *Geological Society of Malaysia Bulletin* 14, 59–74.

Wrytki, K. (1961) Physical Oceanography of the South-East Asian Waters. NAGA Report Vol. 2. University of California Scripps Institute of Oceanography, La Jolla, CA.

Zauyah, S.and Arshad, A. (eds.) (1992) Environmental Sensitivity Index Mapping of the Kuala Penor, Pahang, to Johore Bharu, Johore, Coastline. Universiti Pertanian Malaysia, Serdang. 320 pp.

THE AUTHORS

Zelina Z. Ibrahim
National Hydraulic Research Institute of Malaysia, Km 7 Jalan Ampang, 68000 Ampang, Malaysia

Aziz Arshad
Universiti Putra Malaysia, 43400 UPM Serdang Malaysia

Lee Say Chong
National Hydraulic Research Institute of Malaysia, Km 7 Jalan Ampang, 68000 Ampang, Malaysia

Japar Sidik Bujang
Universiti Putra Malaysia, 43400 UPM Serdang Malaysia

Law Ah Theem
Universiti Kolej Terengganu, Universiti Putra Malaysia, 21030 Kuala Terengganu, Malaysia

Nik Mustapha Raja Abdullah
Universiti Putra Malaysia, 43400 UPM Serdang Malaysia

Maged Mahmoud Marghany
Universiti Putra Malaysia, 43400 UPM Serdang, Malaysia

Chapter 77

BORNEO

Steve Oakley, Nicolas Pilcher and Elizabeth Wood

The seas of Borneo lie in one of the most biodiverse areas of the world, and have a range of environments, species and habitats of global significance. The island sits astride the Equator, and is divided into Brunei, East Malaysia (Sabah and Sarawak) and Indonesia (Kalimantan). Only in Brunei are large areas of original forest still intact; elsewhere they are being heavily logged for timber which, aside from terrestrial consequences, has dramatically increased sediment loads in rivers and has smothered and killed nearshore marine communities. Mangrove forests are also being heavily logged and many have been destroyed by conversion and landfill (up to 80% loss in Kalimantan).

Petrochemicals are important throughout Borneo but there is limited industrial development, so that human and agricultural wastes are the most significant sources of pollution. Human populations are growing while waste treatment infrastructure is lagging behind. Population pressures are also the cause of widespread overfishing. All nearshore resources are over exploited, while huge areas of coral reefs are being systematically destroyed by both blast fishing and cyanide fishing. Only reefs protected by ecotourism are in good condition and support breeding populations of the largest fish. There is a desperate need for well managed marine protected areas where fishing is prohibited, and where adult fish and other resources can mature and breed.

Borneo is still a place of unspoilt wild natural beauty, containing huge areas of undisturbed mangrove swamps and reduced areas of pristine coral reefs, large populations of river dolphins and important turtle nesting areas. However, the threats to these are increasing, large areas being destroyed, and management measures mostly do not protect the diversity. This chapter describes the physical environment, the issues and threats, as well as the steps that concerned people in the region are taking towards a more sustainable future.

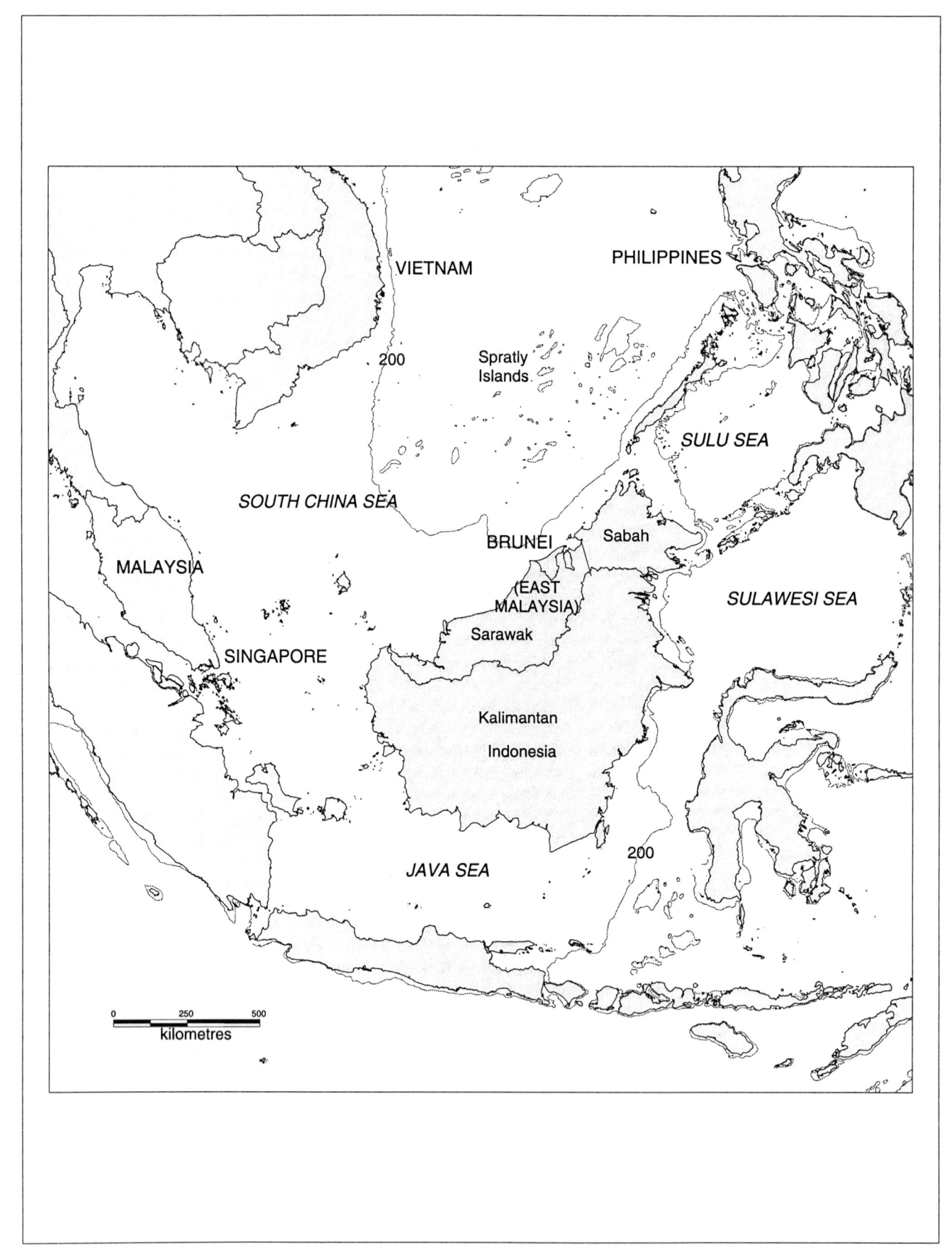

Fig. 1. Borneo and surrounding seas.

REGIONAL EXTENT

The seas of Borneo lie at the heart of the Malay and Indonesian archipelago, and include the Java, Sulawesi, Sulu and the South China Sea (Fig. 1). The island is shared by three countries: Indonesia (Kalimantan), the Sultanate of Brunei, and Malaysia (the States of Sarawak and Sabah and the Federal Territory of Labuan). These three countries have varied ethnic groups, religions, population densities and incomes, which are reflected to a large extent in the environmental issues which dominate in each state. Brunei is small (approx. 5800 km^2) and wealthy with a petrochemical-based economy. Sarawak covers 125,000 km^2, but has a small population. Its natural resources include timber and petrochemicals. Sabah covers 74,000 km^2 but has a larger population than Sarawak. Its natural resources include oil and a mixed agricultural economy in which palm oil is important. Kalimantan is the largest state (approx. 532,000 km^2), largely undeveloped and with a small population, and with petrochemical reserves off the Natuna islands between Sarawak and Peninsular Malaysia. It has significant timber resources which, along with palm oil, are a major export. Large forested areas of Kalimantan, Sarawak and Sabah have been degraded in the last three decades, much being converted to plantations.

Many of the international borders around Borneo remain to be formalised. After UNCLOS, and under some bilateral agreements, Indonesia's claim to the waters among its islands was recognised. Its territorial waters extend into the South China Sea to include the Natuna islands under an agreement with Malaysia. However, the Spratly Island bank is claimed by most countries bordering the South China Sea, and the border across the Sulu Sea between Malaysia and the Philippines is disregarded by many fishermen, resulting in a continuous stream of fishery violations.

The Java Sea from Western Kalimantan towards the Natuna islands and Peninsular Malaysia is shallow, and in the recent geological past Borneo was connected to mainland Asia. The area lies at the epicentre of tropical biodiversity, including that for coral reefs (Wood, 1979), such that some of the best coral reefs in the world have developed in the clear water parts of this region.

South China Sea

The South China Sea extends 2800 km north from the Equator to the Tropic of Cancer, and averages 1000 km in width, with an area exceeding 2.8 million km^2. The sea is bordered by several countries and is a major 'connection' between the Pacific Ocean and the Straits of Malacca and Indian Ocean. Much of the area overlies the Sunda shelf (100–200 m depth) which is partially divided by a deep central trench (>5000 m). It is from this deep abyssal plain that the Spratly and Paracel islands and reefs arise. There are approximately 200 isolated oceanic islands, along with numerous partially submerged cays, rocks and reefs, many of which are not suitable for habitation.

There is considerable speculation about oil and gas reserves which may exist in the South China Sea, especially in the Spratly islands, which has led to tensions and some conflict. The sea is also the world's second busiest international sea lane, carrying more than half of the world's supertankers.

Sulu Sea

The Sulu Sea between Borneo, Palawan and the Philippines, has a southern border at the Sibutu–Basilan Ridge. The southern half is a deep basin, but between this and the Palawan Shelf there is a broad slope. An oceanic ridge runs parallel to Palawan and the Sulu archipelago, from which rise oceanic atolls and platform reefs. The reefs have a gradually sloping underwater terrain leading to vertical walls that drop to 1000 m depth. The Sulu archipelago divides the Sulu and the Celebes seas, and consists of over 400 scattered islands extending from Zamboanga southwest to Borneo. Sulu abounds in timber resources, although fishing is the most important industry, in which sea cucumbers, endangered sea turtles, fish and pearls are extensively gathered.

Java Sea

Situated within the epicentre of marine biological diversity, the Java Sea is home to numerous fishes, corals and other invertebrates, many as yet unidentified. The Java Sea is particularly fragile since it is a semi-enclosed sea bordering relatively dense human settlements. Specific threats and damage include extensive large-scale fisheries, disappearance of some fish species, coastal erosion as a result of the removal of mangroves, fish farming and uncontrolled development of prawn hatcheries along the coast, pollution from industrial wastes and heavy use of pesticides on coastal lands.

Fishery production in the Java Sea has risen steadily although not all of it has been sustainable. Indonesia reported a catch of 3.1 million tons of marine fish in 1991, ranking eighth in the world. Most came from coastal waters and remains artisanal, despite government-sponsored attempts to modernise the fishing industry. After the ban on trawling in the Java Sea, purse seines became one of the main fishing gears used in the area. The purse seine fishery for coastal pelagic fish such as sardinella, scads, and mackerels expanded significantly after 1980, when landings increased from 30,000 tons to over 120,000 tons in 1991.

There is little official fishery monitoring or regulation. Sharks appear to migrate through the Java Sea area and may be a significant part of the Southeast Asian stocks. Dolphins are thought to include the finless porpoise, Irrawaddy dolphin, endemic dwarf spinner dolphin, and the hump-backed dolphin, which occur in relatively small, localised and vulnerable populations.

> ## Spratly Islands
>
> The Spratly islands lie in the centre of the South China Sea near several primary shipping lanes. Their 100 or so coral cays cover less than 5 km^2, but are scattered over a very large area. There are no indigenous inhabitants, but China, Malaysia, the Philippines, Taiwan, and Vietnam claim and have armed garrisons on about 50 of the islands and reefs. Brunei has established a fishing zone at Louisa Reef in the south, but has not publicly claimed the island. Importantly, there are multiple overlapping claims for most of the Spratly and Paracel reefs and islands and all claimants to the islands allege that their claims are fully supported under international law and UNCLOS.
>
> Oil claims lie at the heart of the dispute. Reserves are unproven, but it has been speculated that there is the equivalent of 6 billion barrels of oil in the Spratly Islands area, of which 70 percent would be natural gas. In fact, there is little evidence outside of Chinese claims to support this speculation.
>
> The most accessible and most studied island is Layang Layang, meaning Swallows in Malay, occupied by Malaysia. These and other birds are resident in large numbers. From July to October, populations swell to more than 10,000 birds from northern climates, including many terns, providing a rich avifauna (Pilcher et al., 1999). The reefs of the Spratlys have traditionally supported a limited fishery but in recent years, fishing pressure has increased, especially in areas that are not strictly controlled by the military. Destructive fishing practices are used, and most of the large fish sought by the live fish trade had been removed by the mid-1990s (Johannes and Riepen, 1995).
>
> The underwater habitats of the Spratlys are probably some of the region's most pristine environments, and most are still largely unexplored. The reefs that have been studied harbour a rich diversity of corals. In sheltered waters, table corals grow to several meters in diameter, and soft corals and gorgonians abound. During a recent survey at Pulau Layang Layang, over 140 species of scleractinian corals, belonging to 58 genera and 17 families, were recorded on the outer reef slope. The genus *Acropora* was represented by at least 23 species (Pilcher and Oakley, 1997).
>
> The reefs may be a vital source of larvae for the overfished areas of the region (McManus, 1994; Oakley and Pilcher, 2000; Pilcher and Oakley, 1997). The mechanisms are far from understood, but because most invertebrate larvae remain in the water column for periods that vary from a few days to a few weeks, and because larval fish are able to find reefs from open water and swim long distances even against normal oceanic current speeds, there is evidence that the Spratly islands are acting as a source of fishery recruits. For this reason, and due to the lack of conclusive evidence on the presence of petroleum reserves, the Spratlys were proposed as a Marine Sanctuary from which the remainder of the South China Sea could derive its recruitment, so that all countries would benefit economically. At the moment, the Spratlys do not produce noticeable quantities of resources for any country, but if an agreement could be reached and management implemented, the region could also support a valuable tourist and tuna fishing industry as well as continuing to supply fishery recruits.
>
> **REFERENCES**
>
> Johannes, R.E. and Riepen, M. (1995) Environmental, economic and social implications of the live reef fish trade in Asia and the West Pacific. The Nature Conservancy. 11 16 Smith Street, Honolulu, HI, 87 pp.
>
> McManus, J.W. (1994) The Spratly Islands: A marine park? *Ambio* 23, 181–186.
>
> Oakley, S. and Pilcher, N. (2000) The role of marine protected areas as a strategy for sustainable fisheries management. Proceedings of the Workshop on Aquaculture and Sustainable Reef Fisheries, Kota Kinabalu, Sabah, Malaysia, Dec. 1996, In press.
>
> Pilcher, N.J. and Oakley, S.G. (1997) Unsustainable fishing practices: Crisis in coral reef ecosystems of Southeast Asia. Procs. Oceanology International 97 Pacific Rim, Singapore, 12–14 May 1997. Spearhead Exhibitions, Surrey, UK, pp. 77–87.
>
> Pilcher, N.J., Oakley, S.G. and Ismail, G. (1999) *Layang Layang: A Drop in the Ocean*. Kota Kinabalu, Natural History Publications. Borneo. 142 pp.

Pressures from a rapidly growing human population and industrial development have brought about changes to coastal ecosystems. For example, turtle populations in the Java Sea have decreased in size, in particular the hawksbill turtles which have suffered up to 88% reduction in numbers during the past 10–15 years (Yusuf et al., 1999). These declines were driven by the increase in coastal development, the increase in egg collection and the export of shells and stuffed turtles to Japan prior to 1991 when the trade became illegal. At the same time, there has been massive loss of marine mammals, turtles and fish in trap nets in pelagic migratory channels (Rossiter, 1997). For example, two nets operating during 1997 (in the pelagic migratory channel at Tangkoko, Manado area), caught some 1424 manta rays, 18 whale sharks, 312 other sharks, 4 minke whales, 326 dolphins, 577 pilot whales, 789 marlin, 84 turtles, and 9 dugong (Rossiter, 1997). The minke whales were probably the small form of Bryde's whales, which may be a separate species (possibly *B. edeni*) endemic to shallow waters of Southeast Asia. Since then, the operation is reported to have increased its fleet from one fibreglass skiff to include three much larger ships with crews of 15–20 each, making trips two or three times a day to each net, some days working on a 24-hour basis due to large catches.

Habitat Types

The habitat types around Borneo range from sediment-dominated environments to coral reefs. These vary from inshore to offshore, and from Southwest to Northeast. The land is predominantly covered by tropical forest, and much of the area is influenced by large rivers. These carry heavy sediment loads which have created extensive estuaries dominated by mangrove growth. The continental shelf is wide (over 100 km) along Southern Kalimantan and Sarawak, and the large offshore seabed is dominated by muddy environments.

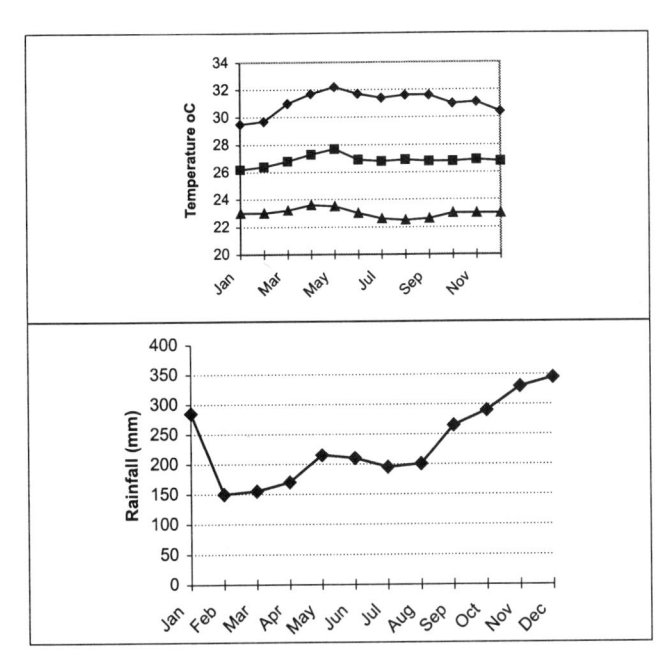

Fig. 2. Temperatures and rainfall on west coast of Borneo: Maximum, mean and minimum temperatures (top), rainfall (bottom). Monthly averages for 12 years of data from Kimanis, Brunei.

The coastal habitats of the Sulawesi and Sulu Seas, and northeastern Borneo, are predominantly coral reef environments with limited mangrove development. In Sabah and Northeastern Kalimantan, rivers are narrower and shorter, generally with less sediment, so that coral reefs have flourished both around nearshore islands and as isolated patch reefs. Overall, the presence of coastal and nearshore reefs and islands reflects the sediment output of nearby rivers: small coral patches occur in southern Sarawak where rivers carry high sediment loads; small, poorly developed coral reefs similarly occur in Brunei; while coral islands and reefs are abundant in Sabah where sedimentation is significantly lower. The terrestrial influence on habitat type diminishes with distance from the shoreline.

NATURAL ENVIRONMENTAL VARIABLES

Borneo has an equatorial climate characterised by uniformly high temperature and rainfall throughout the year. Rainfall can exceed 4,000 mm per annum in many parts of the region, with relative humidity invariably above 70%, with air temperatures that fluctuate between 20 and 36°C (Fig. 2). Surface waters in the South China Sea range from 21 to 29°C.

Weather

There is no distinct seasonality on land but the seas, especially the South China Sea with its long fetch, are noticeably affected by the monsoon winds. The Sulu Sea and the Sulawesi Seas rarely experience large waves and the major climatic influence is the daily land–sea breeze cycle with its associated rainfall. Over the South China Sea, the monsoons are generated by the low-pressure troughs in the intertropical convergence zone which moves north or south following the sun through the seasons (Cooper, 1990). During the Northeast Monsoon from November to March, winds in the South China Sea blow from the northeast. During the early part of the monsoon, up to the end of January, northeasterly winds from between Japan and mainland China bring stormy weather. During this time, the Northeast Monsoon is subject to surges in strength and these bring rapid deterioration of weather and sea conditions. A northerly swell generated in the northern part of the South China Sea will usually take 24 h to reach Borneo, but it will produce rough seas very quickly upon reaching shallow water.

The second type of weather experienced during the Northeast Monsoon occurs during February and March when the winds come predominantly from the Pacific. These dry winds bring dry and sunny weather (see Fig. 2), which makes February the driest month of the year in Northern Borneo. During the transition between monsoons in March–April, winds are light and weather is controlled mainly by the land and sea breezes, with localised showers and thunderstorms. During the Southwest Monsoon from April–October, the equatorial trough moves north as low pressure forms over China. The trade winds then blow from Australia, curving as they cross the Java Sea and equator to blow from the southeast along the coast of Borneo. These winds are dry and subsiding, producing hot weather with calm seas. However, this is the time for cyclones to form in the northern parts of the South China Sea, and these create westerly winds from the Indian Ocean bringing unsettled wet weather with frequent squalls.

Localised weather effects may be important. Sea breezes combined with the Northwest Monsoon create heavy rain or thunderstorms wherever the breeze is forced over hills. The land-breeze at night blows at right angles to the Northwest Monsoon wind, creating a line of convergence parallel to the coast which creates a line of showers or thunderstorms about 8 km off the coast. This rain rapidly dies away at dawn as the land-breeze weakens. During the Southwest Monsoon, the land and sea breezes are both at right angles to the monsoon wind so, while the rain dies away at dawn, further showers form near the coast as the sea breeze develops late in the morning. These rain showers move inland with the sea breeze and often develop into thunderstorms over high land or later in the afternoon.

Typhoons rarely have any great effect on Borneo weather. During June to November, the typical typhoon tracks far to the North (Fig. 3), leaving the Southern South China Sea with calm, fine conditions. Occasionally, typhoons cross the South China Sea and create severe storm conditions in the Spratly Islands, but in general, cyclones bypass Borneo; indeed Sabah is known as the 'Land Below The Wind'. However, cyclones occasionally wander far enough South to affect its northern tip.

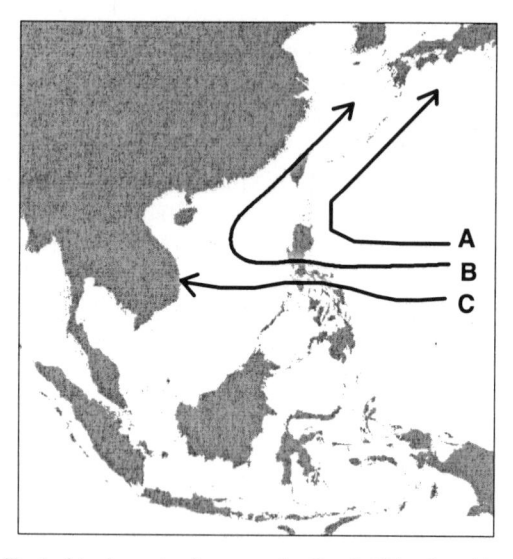

Fig. 3. Typical typhoon tracks across the South China Sea: (a) early in the season (June–August), (b) mid season (August–September) and (c) late in the season (November–December). Map from Cooper (1991).

Currents

Surface currents resulting from the Southwest Monsoon rotate coastal waters counter-clockwise up the northern coast of Borneo toward the Philippines and the Spratly Islands. The Northeast Monsoon rotates the surface currents clockwise down the Northern coast of Borneo and upward along the Eastern coast of peninsular Malaysia and Thailand, toward Vietnam (Fig. 4). Diverging and converging currents occur at the edge of the continental shelf and north of Labuan at the edge of the 2500 m-deep Sunda trough.

MAJOR COASTAL HABITATS

Alluvial plains are the most characteristic coastal feature of Borneo. Southern and southeastern coasts are dominated by major rivers, each of which drains a large catchment area and transports large amounts of sediment seaward. High sedimentation rates shape the coastal communities. Long stretches of shore are lined with mangrove forest, seaward of which are extensive mud flats. Most muddy shores are an interim feature in the succession process as mangroves colonise and spread. Coastal accretion has been rapid over the last decade due to increased sediment loads as a result of logging and land conversion in the catchment areas; parts of Kuching bay, Sarawak, have been accreting by up to 70 m per year. The mudflats at the mouth of the Rajang river estuary are composed of alluvial deposits (fine clays) which are being rapidly colonised by mangroves. In East Sabah, most of the long, slow-flowing rivers carry high sediment loads downstream (Murtedza and Ti, 1993).

Terrigenous sediments dominate in coastal areas with higher rainfall. Fast-flowing rivers carry sediment in suspension as river load, while heavier particles are transported as bed load. The short, fast-flowing rivers in Sabah generally carry larger particles than do the longer, slower flowing rivers in Sarawak and Kalimantan. In the latter areas, large particles are absent on the seabed, and terrigenous muds and sands (rich in clay minerals) form extensive coastal intertidal mudflats which support diverse and productive soft-bottom benthic communities. In the drier north and northeast of Borneo, shallow coastal sediments are predominantly biogenic carbonates and chemically precipitated carbonates. These form vast intertidal and shallow subtidal flats of a variable mixture of silt, sand and calcareous gravel that frequently support seagrass or reef communities.

Mangroves

Mangroves are abundant at the mouths of all major rivers in southern Borneo, and form a coastal belt in nearly all sheltered coastal conditions (Fig. 5). Mangrove development is limited along the more exposed western coast, where mud or silt swamps are only found in the lower reaches of the rivers. The distribution of mangrove forests in Borneo is controlled by physical forces, with biotic factors

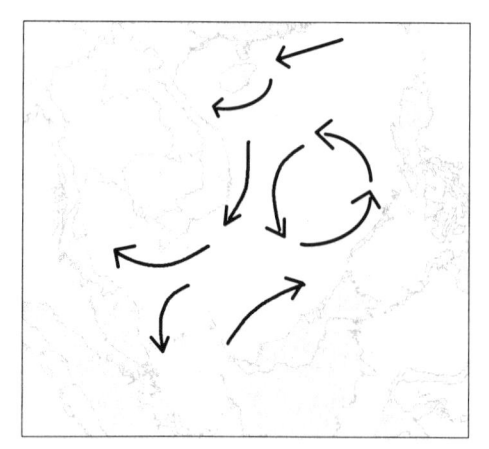

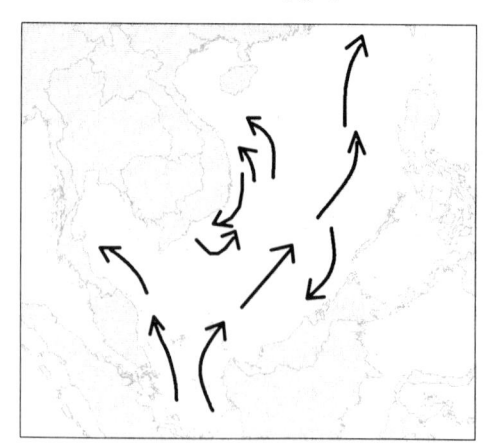

Fig. 4. Surface water current patterns in the South China Sea during the Northeast Monsoon (left) and Southwest Monsoon (right).

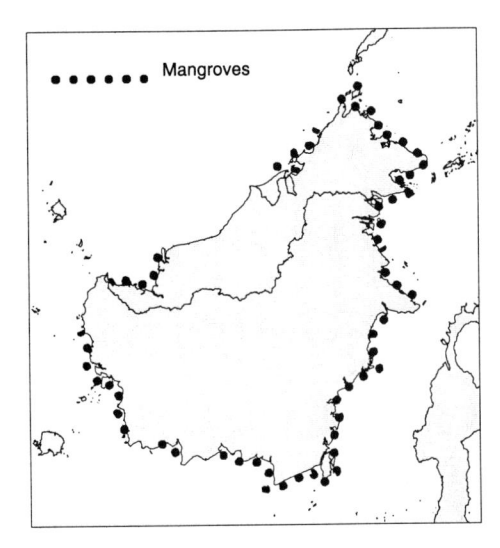

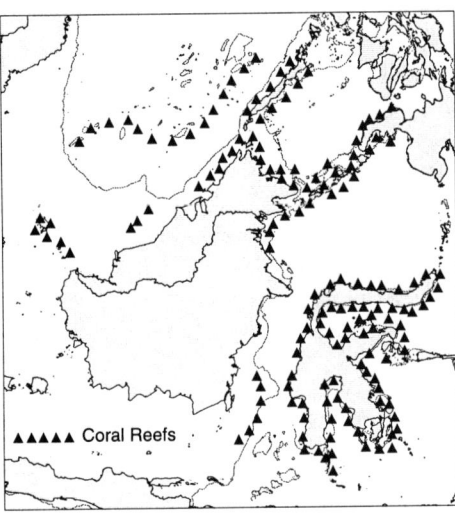

Fig. 5. Distribution of mangrove and coral reef around Borneo. Mangrove distribution partly from Tomascik et al. (1997).

being less important. In the Mahakam delta in East Kalimantan, Dutrieux (1991) recognised four environmental zones: the freshwater apex zone, middle distributary zone, river-mouth zone and the central non-distributary zone. Various species of mangrove flora occupy most of these environmentally distinct zones but the abundance and diversity of aquatic flora and fauna varies greatly. Highest diversity occurs in the most stable environments at the river mouth (mainly marine organisms), there is moderate diversity at the upper extent of the estuary (mainly freshwater organisms), with the lowest diversity in the environmentally stressful mid-estuary zone.

Sedimentation is probably the greatest controlling factor in mangrove development. In Southern Borneo, the Mahakam river discharges up to 1500 m^3 s^{-1}, with a suspended organic matter load reaching 80 mg l^{-1}, with sediment discharges of 10×10^6 tonnes year^{-1}. The estuary is continuously accreting across an extensive river delta covering 1500 km^2. The seaward margin of the intertidal areas, with a salinity 20–32 of psu, is dominated by mangrove forests (*Avicennia* sp., *Bruguiera gymnorrhiza*, *Aegiceras corniculatum* and *Acrostichum aureum*). In more freshwater conditions behind the *Avicennia* zone, thick monospecific stands of *Nypa fruticans* are dominant (Dutrieux, 1991). Further upriver, where the salinity drops to 0–10 psu, the paired association of *Heritiera littorals* and *Oncosperma tigillarium* replace the *Nypa* swamp (Dutrieux et al., 1990a). Tidal range is high, up to 2.5 m, and turbidity is also high, with SPM concentrations as high as 160 mg l^{-1}. Sediment load drops from 25 mg l^{-1} in the lower estuary to less than 1 mg l^{-1} in the adjacent waters. The fluvial sediments extend to the 30 m isobath 40–50 km offshore, while a low-salinity plume is identifiable 400 km to the southeast of the delta. This river plume extends 50–100 m deep, with an SPM of 0.3–0.5 mg l^{-1}, compared with the much clearer coastal waters (0.1 mg l^{-1}).

Mangrove forests on accreting shores in Borneo are successional communities. *Sonneratia* and *Avicennia* are pioneer genera which establish initially on intertidal mudflats. As the sediment becomes stabilised, genera such as *Rhizophora* and *Bruguiera* succeed the pioneers. *Bruguiera* and *Xylocarpus* are found in areas that are subject to only a few tidal washes each month.

Nypah swamps are dominated by the mangrove palm *Nypa fruticans* and occur in habitats with salinities that range between that of mangrove and peat swamps, but are predominantly freshwater environments, which are flooded at spring tides. Inland of the Nypah swamps, other salt-tolerant freshwater species become established on the landward edge of the mangroves which are flooded only by a few tides a year. This distinct zonation of dominant species, arranged more or less parallel to the coastline, is reflected by the morphological and physiological adaptations of each species.

On open coasts with wave-protected shores such as Kuching bay and north Sabah, the outer margins of the mangrove stands are dominated by *A. marina* and *S. caseolaris* with *Rhizophora* and other genera forming a diverse, mixed community behind the seaward zone. Mixed mangrove communities may extend long distances upriver, and these delta mangroves are highly productive. In East Kalimantan, fixation rates vary from 32–40 kg C ha^{-1} d^{-1} during the SE monsoon (rainfall <100 mm month^{-1}), to 48–51 kg C ha^{-1} d^{-1} during the NW monsoon (rainfall >300 mm month^{-1}) (Sukardjo, 1995). These figures are comparable with the highest primary productivity levels recorded, and also influence offshore fisheries.

In north Sabah and Kalimantan, where mangroves are associated with coral cays and coarser sediments, *Rhizophora stylosa* replaces *Avicennia marina* as the pioneering mangrove. It often precedes sand build-up on protected reefs. *Sonneratia alba* is also characteristic of coral cays, although it rarely becomes established until after the formation of a stable beach.

Mangrove stands are included in several protected areas. In Sarawak, these include Bako National Park, Samunsan and Sibuti Wildlife Sanctuaries, while in Sabah they are found in the Pulau Tiga National Park and Klias Forest Reserve. In Eastern Kalimantan, the Kutrai NP includes a very large area of mangroves.

Mangrove Fauna

In Sarawak, the proposed Pulau Bruit National Park is being gazetted to protect one of the most important sites in Eastern Asia for migratory birds. The large, fish-eating Brahminy kite and the White Bellied Sea Eagle are common over large rivers and mangrove areas, while arboreal mammals, especially flying foxes (*Pteropus*) and fruitbats (*Macroglossus*), dominate the terrestrial component. Cave nectar bats (*Eonycteris spelaea*) fly long distances from inland caves to feed on *Sonneratia* nectar when the Durian (*Durio* spp.) is not in flower. Proboscis monkeys *Nasalis larvatus*, are found in undisturbed mangrove areas along several major rivers. They are endemic to Borneo and feed extensively on the young leaves of mangroves. They are easy targets for hunters which, combined with major losses of coastal habitat, has led to severe population declines so that the species is now severely endangered (Bennett, 1991; Bennett and Gombek, 1993). Long-tailed macaques *Macaca fascicularis* can be seen in many areas foraging at low tide, while the silvered leaf monkey *Presbytis cristata* feeds entirely on leaves with some flowers and seeds but is a much shyer animal, seen rarely on river banks and mangrove edges. Other protected mangrove fauna include estuarine crocodiles (*Crocodylus porosus*), monitor lizards (*Varanus*) and otters (*Lutra* and *Aonyx* spp.). Mud flats associated with mangroves are very important for resident and migrating shore birds and several areas in Borneo are protected specifically for birdlife.

Invertebrates are dominated by fiddler crabs (*Uca* sp.), Sesarmid and Grapsid crabs. Mudskippers (*Pteropthalmus* sp.) are also abundant, especially at the mangrove edge and on soft mudflats. The cockle *Anadara granosa* is common on the harder sediments of the mid-littoral zone, and is harvested commercially in large numbers. The clams *Geloina expansa* and *Mactra veneriformis* are found in the soft organic mud in mangrove channels at the top of the littoral zone, and are also harvested at a subsistence level (S. Zulkifli, pers. comm.). Many Gastropods, especially *Cerithedia obtusa*, and littorinids are harvested and can be seen in large numbers in local markets. Mud lobsters *Thalassina anomala* are very important mangrove sediment mixers.

Estuaries and coastal waters along the west coast of Sarawak from Kuching to Miri are home to large numbers of three small marine mammals: the finless porpoise *Neophocaena phocaenoides*, Irrawaddy dolphin *Orcaella brevirostris*, and the Humpbacked dolphin *Sousa chinensis*. Irrawaddy dolphin are abundant in the rivers at the mouth of the Rajang delta. In most of Sarawak, fishers catch dolphins only accidentally but in the Mukah district, villagers actively hunt for them even though they are protected by law. This seems to be a localised practice. Coastal dolphins are not regularly seen in Sabah. Dugong (*Dugong dugon*) are suspected to be present in South Kalimantan (de Longh, 1997), have been seen off Brunei (Elkin, 1990) and there have been several unconfirmed reports of individuals in other parts of Sarawak and Sabah.

Mangrove Clearing for Shrimp Farms

Mangrove productivity is closely linked with the life cycle of many economically important Penaeid prawn species, and the dependence of prawn fisheries on adjacent mangroves is indicated by the strong positive correlation between total area of mangrove forests and commercial catch of prawns (Martosubroto and Naamin, 1977; Sasekumar and Chong, 1987).

The total mangrove area of ASEAN countries was 3,650,395 ha in 1984 (FAO, 1984), at which time Borneo contained 40% of the total. Unfortunately, overfishing and mangrove destruction have occurred concurrently in most Southeast Asian countries (Pauly, 1989; Primavera, 1999), and it is generally not possible to separate the effect of each of these on reduced shrimp catches. However, in Sarawak, which has converted the smallest area of mangrove (Table 1), there has been very little fall in shrimp catches. These have fluctuated between 9000 and 16,000 tonnes since 1982 (Sarawak Fisheries Department annual reports, 1982–1997). The Sarawak Ministry of Agriculture has shown a very

Table 1

Estimates of mangrove area (ha) and amount lost to land conversion in recent years for all the states of Borneo

Region	Remaining mangroves	Original area	Percentage lost
Sarawak	130,000[3] (1999)	174,000[2] (1975)	25%
Sabah	Unknown	350,342[6] (1979)	Unknown
Brunei	No recent estimate[5]	18,418[4] (1982)	Unknown
W Kalimantan[1]	40,000 (1986–90)	213,400 (1982)	81%
Central Kalimantan[1]	20,000 (1986–90)	84,000 (1982)	76%
South Kalimantan[1]	66,650 (1986–90)	115,000 (1982)	42%
East Kalimantan[1]	266,800 (1986–90)	680,000 (1982)	61%
Total		1,461,160	

Based on:
1 Giesen (1993). 2 Chai (1975). 3 Sarawak Forest Dept. annual report 1996 reports permanent forest of 36,900 ha for mangroves while 131,000 ha are state land forest of which an unknown percentage has been converted. An unpublished report from the Sarawak State planing unit in 1999 estimates 130,000 ha. of mangroves remaining.
4 DOF-MIPR (1992) based on aerial photos taken in 1982.
5 Davies (1998) uses the 1982 estimate (Chua et al., 1987).
6 Anon (1979) quoted by Chua et al. (1987).

conservative attitude in respect to prawn farming, following FAO guidelines (Garcia et al., 1999). This was developed after the environmental lessons learnt in other countries in the region (Rajendran and Kathiresan, 1996).

The conversion of mangrove into prawn farms still has environmental consequences. One major problem has been the acid sulphate soil conditions associated with most mangroves which limit the life of the shrimp farming operations. A second problem of the tiger prawn industry is viral disease and its likely introduction into wild prawn stocks. Prawn farms undoubtedly contribute large amounts of nutrient to the environment, and between 2 and 22 ha of undamaged mangrove are required to neutralise the effects of 1 ha of prawn pond. The Sarawak regulations certainly head in the right direction with a 20% maximum area of ponds within a catchment, and it is hoped that the tough environmental protection attitude will prevail. To date, very few licences for prawn farms have been issued, and these are subject to strict environmental protection regulations, including buffer zones of undisturbed mangrove, settlement reservoirs and mangrove areas for effluent clean-up.

Mangrove Use

Within Borneo, primary direct non-conversion uses include mangrove tree bark collected and exported for tannin, and Nypah for sugar and alcohol production, but the most important product is timber. In Malaysia, 60% of the mangrove areas are designated as forest reserves to be used for sustainable harvest of mangrove products, protected in theory from destructive conversion. The states of Sabah and Sarawak use a harvesting system based on minimum girth of 21–23 cm; there is no clear felling and smaller trees are left behind (Chai and Lai, 1984). However, there continues to be an alarming rate of destruction and degradation. In 1982, 1% of Malaysian mangrove (Borneo and Peninsular Malaysia) was estimated to be destroyed each year by conversion or destructive harvesting (Ong, 1982).

One of the most destructive uses of mangroves in Borneo was the Japanese-based mangrove wood chip industry. In Sabah, 40% of the mangroves (120,000 ha) were allotted for this and there was a mean annual clear felling of 4000 ha. In Sarawak, 1600 ha were clearcut annually (Ong, 1995). This industry still continues to decimate the Kalimantan mangrove forests, at an unknown rate.

It has been suggested that 5% of mangroves in Malaysia (30,000 ha) should be conserved (Ong and Gong, 1991). Progress towards this ambitious aim has been made, especially in Sarawak where in recent years there has been a better understanding and awareness, such that approximately 2500 ha are formally protected within various National Parks and another 10,000 ha are in the process of being gazetted (Forestry Dept., 1997). These conservation gains are significant but unfortunately are dwarfed by the huge areas that have been converted and land-filled for industrial and residential development. Estimates of mangrove areas in Borneo are hard to obtain, but evidence shows that 50–80% of the mangroves have been lost in 20 years (Table 1), most in Kalimantan, where the Japanese wood chip industry obtained licences in 1990 for 200,000 ha.

There is still a critical shortage of reliable information on Borneo's mangrove forests, despite the loss of up to 80% in 20 years. There are still large areas of logged but otherwise undamaged mangrove, but whether these continue to be eroded is unknown. In Brunei there are forest reserves, but mangrove outside these areas are in effect unprotected since they are simply classified as State Land (Davies, 1998). The Integrated Management Plan for the Coastal Zone of Brunei (Anon., 1989) recommended 99% of the mangroves be protected and proposed conversion of only 200 ha, while 4141 ha were to be protected in National Parks and 6545 ha maintained for coastal protection. This was ambitious, apparently never implemented, and is clearly being ignored. Piecemeal clearance of the forests throughout Borneo continues although most information is out of date. For example, in Brunei, the most recent area estimates available are based on aerial photos taken in 1982 (Chua et al., 1987). In Sarawak, the most recent estimate was made in 1975 (Chai, 1975), while in Sabah, the most recent figure is from Anon (1979) although the GIS used by the Dept. of Fisheries Sabah is based on maps made in 1985. Proper management of mangroves, and of the national shrimp fisheries, thus depends on data which are over 18 years out of date. Sarawak's shrimp fisheries fluctuate around 14,000 tonnes per year (Table 2), yielding US$ 42 million, yet there are no fisheries biologists with exclusive responsibility to manage the resource or the environment that sustains it, and there is still very poor understanding on the part of planners and politicians.

Rocky Shores

Rocky shores are relatively uncommon in Borneo, except on windward sides of headlands or rocky islands, particularly in Western Sabah. Rocky littoral shores are dominated by barnacles and gastropods. Seaweed or oyster dominated rocky shores are rare. The green mussel *Perna viridis* is locally abundant, especially in West Sabah. In exposed areas, rocky shores merge sublittorally with veneering or poorly developed fringing reefs. Rocky shore zonation follows classical patterns but there are few studies on the habitat.

Sandy Shores

Terrigenous or riverine sandy shores predominate along the west-facing coast of Borneo. From Pulau Bruit at the eastern end of Kuching Bay to the northern end of Brunei, a distance of 600 km, the coast is a long continuous beach. There is a predominantly southwest sand drift, and most river mouths have a sand bar across the mouth, with an

Table 2
The 1998 fisheries resources of Sarawak and Sabah, East Malaysia

	Biomass (MT)	Potential Yield (MT)	Current yield (MT)	Status
Sarawak				
Coastal shrimp	22,000	15,000	16,000	Overfished
Coastal demersal fish	80,000	54,000	57,000	Overfished
Offshore demersal fish	145,000	73,000	20,000	Slightly exploited
Rough ground fish	68,000	34,000	23,840	Slightly exploited
Small pelagic fish	1,088,000	435,200	18,000	Exploited
Tuna	90,000	45,000	20,000	Slightly exploited
Sabah				
Coastal crustaceans	42,000	21,000		Heavily overfished
Coastal demersal fish	130,000	65,000	N/A	Heavily overfished
Offshore demersal fish	22,000	11,000	N/A	Heavily overfished
Coral reef fish	N/A	N/A	N/A	Heavily overfished
Offshore small pelagic fish	612,000	244,800	N/A	Exploited
Tuna	80,000	40,000	N/A	Exploited

Data from Dept. of Fisheries unpublished reports.

elongated sand spit on their northeastern shores. Beach sediments grade vertically from coarse sand in the upper littoral, to muddy sand in the subtidal zone and mud offshore. Beaches are relatively depauperate. *Ocypode* crabs on the upper shore and *Dutila* crabs in the mid-shore are the most common species. The clam *Meretrix meretrix* and various razor clams of the genus *Solen* may be locally abundant.

Most sandy shores are immediately backed by a line of *Casuarina equisetifolia* trees which follow the line of a very low (0.5 m) supratidal sand dune. In Sarawak, behind the high tide line, the sandy belt is often very narrow and merges imperceptibly with peatswamp or other forest, while in western Sabah, the forest has been cut to the shore in most places and the sand dune belt is more distinct.

Coralline white sand beaches do not occur on mainland Borneo but are associated with offshore coral cays where they are narrow (5–20 m) with sand deposits 1 to 2 m thick. The sands are fine to medium grained, well sorted, with scattered coral and shell rubble. On high islands with rocky substrate, such as Pulau Banggi, P. Matawali and P. Mantanani, the beach sediments originate in part from erosion of the island.

Seagrass and Algae

In northeastern Borneo where there are no mangroves, seagrasses commonly occur behind many fringing reefs. They frequently dominate deeper parts of the moat that separates the reef flat from the shore. On islands such as Mabul, leeward reef slopes may have dense mixed beds of *Thalassia hemprichii*, *Cymodocea rotundata* and *Halodule uninervis*. *Halophila ovalis* is also common and appears to extend deeper than other seagrasses.

Mangrove, seagrass and coral communities occur together in several locations, mainly in areas with limited or low rainfall and small rivers. The best example is in the Berau islands in East Kalimantan. The mainland near the Berau islands supports extensive mangrove swamps which lead first into broad mud flats, then to subtidal seagrass beds and finally to fringing reefs.

Fringing reefs with intertidal flats are covered with thick turfs of fleshy corallines (*Galaxaura*, *Amphiora* as well as the brown algae *Cytoseria*, *Turbinaria*, *Sargassum*). Coral reefs and *Halimeda* banks are mutually exclusive in the eastern Java Sea. In areas typically occupied by coral reefs, reefs are confined to depths shallower than 15 m, whereas extensive *Halimeda* bioherms occur in depths between 20 and 100 m, including shelf and carbonate platform environments (Roberts and Phipps, 1988). Extensive areas composed primarily of disarticulated *Halimeda* plates occur on a large carbonate bank (Kalukalukuang Bank) located 50–70 km east of the central Sunda Shelf margin (eastern Java Sea, Indonesia) (Roberts et al., 1987).

Coral Reefs

Most shelf reefs around Borneo are influenced primarily by the adjacent landmass in terms of geology and lithology as well as by river run-off and coastal circulation. Southern Kalimantan and Sarawak are dominated by major river systems, and coral reefs are found only in areas with relatively clear water. They include small patch reefs in very shallow water between the major rivers e.g. Talang Talang Islands, Sarawak, and offshore patch reefs such as around the Samarang and Vernon banks which loosely follow the edge of the continental shelf (Fig. 5).

Turbid waters from rivers and coral reefs are found together on the eastern coast of Kalimantan where the Berau Barrier reef system occurs immediately adjacent to the major Berau River and Bulungan River 60 km to the north. These release large quantities of silt year round. The Berau Barrier reef complex includes a large, flourishing delta front barrier reef, with numerous patch and platform reefs such as Pulau Panjang (145 km^2), as well as an extensive 40–60 km long fringing reef complex. There are three atolls (Kakaban, Maratua and Muaras) offshore, though these are sufficiently far offshore (60 km) that they experience mostly oceanic conditions. The mainland has a wide fringing reef with a mixed community of corals growing at depths of up to 5–8 m consisting of massive *Porites* and *Goniastrea* with various soft corals including *Alcyonium*, *Lobophytum*, *Sinularia*, and *Sarcophyton*. The reef flats have strongly lagoonal characteristics supporting large numbers of burrowing macroinvertebrates, especially *Callianassa* and

Fig. 6. A sediment plume from a river affecting an island and smothering its coral reefs.

holothuroids. The seaward edge of the reef flat and reef crest has numerous soft corals with the gently shelving reef slope being dominated by foliaceous corals *Pavona, Montipora, Pectinia* and *Leptoseris* spp. The net flow through the channels is southwards, and it seems likely that both the strong coastal currents of up to 50 cm s^{-1} and the tidally induced, frequent upwelling are likely to be responsible for reef development (Tomascik et al., 1997).

Pulau Sangalaki is a remote coral cay directly in front of the Berau barrier reef, but 60 km offshore. Conditions for reef development are ideal and there are many well-developed patch reefs, as well as very diverse fringing reefs. There is a large productive seagrass bed in the moat. Strong tidal currents (2.5 m s^{-1}) reduce sedimentation but transport nutrient-rich coastal water into the area. Tidally induced upwelling has been observed, and it is an important Manta Ray (*Manta birostris*) feeding area. It also contains one of the largest turtle nesting populations in the Malay–Indonesian archipelago.

The northeastern seas of Borneo have the best conditions for reef development. In northern Kalimantan and eastern Sabah there are extensive patch reefs and even a few stretches of fringing reef. All offshore islands are surrounded by fringing reefs with high species diversity, although destructive fishing and increased sedimentation (Fig. 6) has dramatically reduced coral cover and fish abundance. The most diverse reef development is found in the Mabul, Kapali and Ligitan group on the edge of the continental shelf. The most dramatic reef in the area is Sipadan, located 8 nm off the continental shelf; a needle-like pinnacle reef with almost vertical sides down to the sea floor at over 1,000 m deep (Wood and Wood, 1987).

The whole of the northern coastline of Sabah and northern Kalimantan has vigorous coral growth in a series of offshore reefs and islands, albeit in turbid conditions. As an example, reefs on the Banggi Islands in north Sabah have a relatively high species diversity and very high coral cover (80%) in turbid conditions (secci disc 1–3 m) near large mangrove swamps. Live coral cover extends only to a depth of 8 m where fine silt deposits take over. Thus the reefs of Borneo exhibit a wide diversity of form and species, mainly because of very complex factors which control their development. Unfortunately, not much information is available for most reefs of the region and anthropogenic activities have severely damaged them.

An extensive survey of the reefs around most of north Sabah was made in the late 1970s (Wood, 1979) together with general surveys of the corals of other reefs (Lulofs, 1973; Johnston, 1984). Pulau Sipadan was surveyed in detail by Wood et al. (1987) and Wood (1994) and there have been several studies in Tunku Abdul Rahman Park. For Kalimantan, there are brief descriptions of visits to Kakaban, Sangalaki and Derawan by Tomascik et al. (1997).

The most recent information on Sabah was made in 1997 using 'reefcheck' methods (Hodgson, 1999). The results showed devastation; reefs continue to be destroyed by destructive fishing practices, especially blast fishing (Fig. 7) and cyanide fishing (Fig. 8). In addition to this, invertebrates were being heavily collected: sea cucumbers, clams, triton shells and lobsters were all absent or severely reduced on all but a few reefs. Other invertebrate populations had expanded in the absence of competition or predators: *Diadema* urchins had reached plague

Fig. 7. *Top*: A blasted reef crest, note the absence of any living corals or fish, and the rubble/sand substrate. *Bottom*: A typical unblasted reef.

silty mud and is inhabited by shrimp and other soft-bottom communities. Carbonate sands dominate in coral reef areas and become more frequent to the north on both coasts. The major fishing effort is on muddy bottoms in nearshore waters, and the offshore resources are only lightly fished. In Brunei shrimps are taken with 22% of total fishing effort, but pelagic resources are also important at 23% of effort. The presence of offshore oil installations of Brunei are of vital importance, since these structures act as aggregating devices for both migratory and residential fish stocks. Fish aggregating devices (FADs) are commonly used in both Sarawak and Brunei to concentrate fish for handline fisheries. Unfortunately, the fish resources of FADs are relatively easily caught; a series of FADs in Brunei were opened for fishing two years after establishment, and were completely fished out within a few months.

In contrast to the muddy waters of Sarawak, reef fisheries are important in Sabah, but these are heavily overfished (see Table 2) in 1998; the 130,000 tonnes caught required almost twice the fisheries effort of the maximum economic yield, and needed much greater effort than the optimum required for maximum sustainable yield (201,000 Mt). Sabah used to have the highest catch rate for all states in Malaysia, but the 1998 survey shows that catch rate has declined by nearly 74% since 1974 (Shaari et al., 1976). Around Sabah, all of the known resources are exploited; there are a number of small artisanal fisheries for pelagic resources, light and pole fisheries for squid, liftnet fisheries for anchovies and small clupeoids as well as a small deep-water trap fishery for *Nautilus* shells for the tourist trade. The major mollusc fishery is for the blood cockle (*Anadara granosa*) from Labuk Bay and Tawau, although large numbers of giant clams (*Tridacna* and *Hippopus*) and others are caught and sold locally.

In the more exposed South China Sea and Java Sea, the size of fish landings is related to the monsoon. Generally, fish catches are low due to the rough weather during the Northeast Monsoon whereas catches in the Java Sea are reduced during the Southwest Monsoon. The monsoon climate is the main factor governing the environmental characteristics of the Java Sea. Seasonal water exchange with the Flores Sea affects the abundance and occurrence of pelagic fishes, whereas coastal fish populations tend to stay in the Java Sea throughout the year.

In common with most tropical seas fisheries, there is a very high percentage of low-value species in the demersal catches, high-value fish being less than 12% of catch. Cephalopods are an important local fishery resource, though few data exist. Octopus, squid and cuttlefish are increasingly dominant in the demersal catches and it appears that the fishers are proceeding to fish down the food chain as described by Pauly et al. (1998).

Around Sarawak and Sabah, carangids and scombrids form the backbone of the pelagic fishery although mackerels (*Rastrelliger* spp.), round scads (*Decapterus* spp.) and tunas are seasonally important. In general, the small neritic

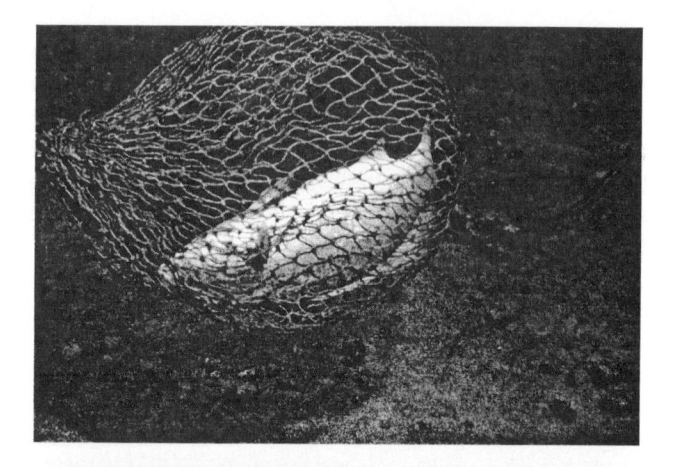

Fig. 8. *Top*: A grouper recovering after being anaesthetized and caught with cyanide. *Bottom*: Humphead wrasse in a live fish restaurant, for sale at RM18/100 g or US$55/kg. Note the fish is mislabelled as Parrotfish.

proportions on a number of reefs while crown-of-thorns starfish were locally abundant on some.

Offshore Systems

The continental shelf is very wide in the Java and South China Seas and becomes very narrow in the Sulawesi Sea. The Sarawak continental shelf has an area of 125,000 km^2 of which 97,000 km^2 is trawlable, while on the West coast of Sabah the continental shelf has an area of 28,000 km^2 with 14,000 km^2 trawlable. The seabed of the south and west is

> ## Cyanide and the Live Reef Fish Food Trade
>
> The live reef fish trade supplies live, large fish to gourmet markets in major Asian cities. It specifically targets a few high-priced species, using cyanide as an anaesthetic. Boats travel far afield since shallow, nearby reefs are badly overfished and blasted by bomb fishing.
>
> There are usually 4–6 divers per boat with rotating shifts of two divers underwater searching the reef throughout the day. Divers breathe from air flowing continuously from a car tyre compressor, and dives usually last 40 min to one hour, at depths of up to 50 m. Suitable fish are chased into a hole, then all entrances are sprayed with cyanide. When captured, the fish is enclosed in a net to recover and its swim bladder is punctured to allow the gas to expand as it is brought to the surface. In both Indonesia and Sabah, boat owners generally employ the poorest groups of local fishermen for this trade. Once catches decline, the fishers move on. In Sabah, fishers are currently concentrating on the Saracen bank and the islands of Banggi in the far north.
>
> Cyanide is lethal in small doses to a wide range of marine organisms, so the method destroys reef inhabitants with unknown long-term ecological effects. The social cost of the trade is of concern too; divers regularly suffer crippling decompression injuries or deaths, and mixing cyanide by hand in open containers has also been observed, but high economic returns encourage the practice.
>
> The fish fetch premium prices and for social reasons the trade is unaffected by normal market forces; in Chinese culture, rare, large, expensive fish offered to guests confer status. Taste seems to be less of an issue; gourmets could not readily differentiate between farmed and wild caught fish (Fox, 1997). In 1995, prices in Hong Kong ranged from $30 to 90 per kilo (Oakley and Pilcher, 1996), while today prices may exceed US$ 55/kg (see Fig. 8b). With this level of financial incentive, fishers can spend considerable time searching reefs; fishers from north Sabah often search for more than a day for a single fish.
>
> The trade swept through most reefs in the South China Sea and Sulu Sea several years ago, and has overfished almost all populations from Sri Lanka to Micronesia and south to Australia (Johannes and Riepen, 1995). During surveys in Sabah from 1997 to 1999, searches on over 45 reefs along 2,500 km of coast found reproductive sized fishes of targeted species on only four reefs.
>
> The trade has prompted the classification of some species as vulnerable. They have very wide ranges but may be locally extinct. Regulations exist in several countries (Vincent and Hall, 1996) although these are ineffective. Cyanide detection laboratories now set up at export centres in the Philippines (Barber and Pratt, 1997) are an important controlling measure, but the large coastlines make customs avoidance easy.
>
> Vulnerability of groupers and Humphead wrasse is compounded by their sex change during growth. With severe fishing pressure focused on the largest (male) fish, there are probably very few surviving males, and spawning aggregations are selectively targeted. Within the huge areas recently surveyed, spawning aggregations, even spawning individuals, no longer exist. Fish below 40 cm are also caught, and are transported to cage farms where they are grown to a market size of 1–1.5 kg. These fish could be the first marine fish to become extinct in the new millennium.
>
> **REFERENCES**
>
> Barber, C.V. and Pratt, V.R. (1997) Sullied Seas: Strategies for Combating Cyanide Fishing in Southeast Asia and Beyond. World Resources Institute, Washington, D.C.
>
> Fox, C. (1997) Asian gourmets taste fish to help save coral reefs. *Live Reef Fish Information Bulletin* no. 2,
>
> Johannes, R.E. and Riepen, M. (1995) Environmental, economic and social implications of the live reef fish trade in Asia and the West Pacific. The Nature Conservancy, 11 16 Smith Street, Honolulu, HI, 87 pp.
>
> Oakley, S. and Pilcher, N. (1996) The role of marine protected areas as a strategy for sustainable fisheries management In Procs. Workshop on Aquaculture and Sustainable Reef Fisheries, Kota Kinabalu, Sabah, Malaysia.
>
> Vincent, A. and Hall, A. (1996) The threatened status of marine fishes. *Tree* 11.

tuna species dominate the catches along the western coastline of Sarawak, Sabah and Brunei. Coastal tuna landings are mainly represented by five species: *Euthynnus affinis*, *Sarda orientalis*, *Thunnus tonggol*, *Thunnus allalunga* and *Katsuwonus pelamis*. Oceanic species such as skipjack are found more frequently along the eastern coast where the continental shelf is narrow and consequently deep-sea conditions and larger pelagic fish are found closer to shore. The larger species, bigeye (*Thunnus obesus*) and yellow fin (*T. albacares*) are concentrated over the edge of the continental shelf, especially in the South China Sea, where suspected localised upwelling raises productivity.

Human Populations

Borneo is multi-cultural with several large ethnic groups. In Sarawak, there are 25 ethnic communities in a population of 1.8 million. The Iban tribe is the biggest, making up 29.6%. The second biggest is the Chinese with about 29.1%. Malays represent 20.7%, the Bidayuh people (8.3%), the Melanau (5.8%), the Orang Ulu (5.4%) and the Indians (1.1%).

Sabah has the highest population density. Officially, there are 1.8 million people but illegal immigrants boost this to well over 2.5 million. There have always been close cultural ties between Sabah and the Sulu Archipelago of the Philippines, and Sabah offers lower population densities, much better prospects for employment and better minimum wages (US$3 per day for illegal immigrants working in plantations) than do the Philippines or Indonesia. Several problems associated with illegal immigrants include the marine version of slash-and-burn, with entire villages raiding reefs until little fauna is left, before moving on. Enforcement of marine and fisheries laws invariably takes a

> ## Blastfishing
>
> For many years, the reefs over all the seas of South East Asia have been subjected to blast fishing, using fertiliser for home-made bombs of approximately 1 kg. Blasts kill fish in a 15–25 m radius (~500 m²) forming a 3–4 m diameter crater. Blast fishers prefer schooling fish, and choose sites for maximum catch. After detonation, they use dip nets to collect the stunned and dying fish. Many larger boats use divers also, so banning their air compressors has been suggested as a simple regulation (Pet and Djohani, 1998).
>
> In northern Sabah, blasts were heard at rates of up to 15 per hour and most reef crests are now continuous bands of rubble and sand, with over-turned and fractured corals (Oakley et al., 1999). Coral boulders dislodged by blasts create underwater avalanches. Blast fishing kills all the fish in an area; after one blast, over 2500 dead fish were counted in an area <100 m², mostly damsel fish, small fusiliers and numerous juveniles. Blasts also reduce the reefs' three-dimensional structure, so it can no longer provide food or shelter, and becomes much more vulnerable to erosion from waves.
>
> Economic models (Pet-Soede et al., 1999) show that blast fishing was initially four times as rewarding as non-destructive fishing. After a few years, non-destructive fishing is no longer an option, and by year 20 income declines to one fifth of what would have been available by sustainable methods. In southwest Sulawesi, an area directly comparable to Sabah or Kalimantan, net annual income per fisher dropped from US$ 6,450 to less than US$ 550. The opportunity cost of this widespread destruction is high: fisheries production has dropped and net income has fallen by 80%. All reefs that may have been suitable for new tourism developments have been severely damaged, such that destructive fishing has cost the state over US$ 55,000 per year in lost revenue for each potential facility (Pet-Soede et al., 1999).
>
> Governments have tried to combat blast fishing with command and control regulations, or with community-based marine protected areas. In Sabah, there is still a lack of commitment by authorities, even though fisheries landings of the state have fallen by 44% in 10 years (Rajali et al., 1999). Blast fishing is banned, yet magistrates do not take cases seriously. It costs much more to catch culprits than is yielded by fines, and most officials are unaware of the economic losses. There is also the difficulty of enforcing laws. Some blast fishers understand that their activities destroy the habitat that the fish depend on, but are not aware that their activities threaten their own livelihood. All knew that their reefs had deteriorated, but most were convinced that there were better reefs further out, and that they had to fish these before others took the fish.
>
> Detection and regulation are clearly not the whole answer, because if fishers cannot use sustainable methods or find alternatives, regulations will be ignored. Alternative income sources are essential, though finding these is not simple. The Fisheries Department of Sabah has an aquaculture master plan which will guide some developments, but there is no structural plan to improve enforcement of the Fisheries Laws. The general lack of funds, staff and facilities for enforcement, coupled with the lack of knowledge and awareness and a shortage of political will, means that the destruction will continue for the foreseeable future.
>
> ### REFERENCES
>
> Oakley, S., Pilcher, N., Atack, K., Digges, C., Enderby, S., Mackey, G., Clubb, R., Stapelton, K., Mei, T.S., Huet, C. and Morton, T. (1999) Reefs under attack: the status of coral reefs of Sabah, East Malaysia. Presented at 4th International Conference on the Marine Biology of the South China Sea, Quezon City, Philippines, 20–22 Oct. 1999. In press.
>
> Pet, J.S. and Djohani, R.H. (1998) Combating destructive fishing practices in Komodo National Park: Ban the hookah compressor! *SPC Live Reef Fish Information Bulletin* **4**, 17–28.
>
> Pet-Soede, C., Cesar, H.S.J. and Pet, J.S. (1999) An economic analysis of blast fishing on Indonesian Coral reefs. *Environmental Conservation* **26**, 83–93.
>
> Rajali, H.B., Gambang, A., Awang, D.B. and Basran, R.B.H. (1999) Fisheries resources survey in the EEZ of Malaysia 1997/98. Survey report for demersal fish resource survey on the west coast of Sabah 13 Oct–6 Nov 1998 Fisheries Research Institute, Kuching, Sarawak.

back seat to issues of citizenship, with the defendants frequently being deported before facing trial.

Sarawak and Brunei have much smaller population densities than Sabah and have tight immigration controls. Sarawak's 1.8 million people for the most part reside in a few large urban centres. Brunei's 270,000 people have the second highest per capita income on earth. Very few people actively fish full-time but the major coastal fishery is located in Brunei Bay. Because of the low population densities only a few commercial species are overfished. The coastal and marine environments are largely unexploited and undisturbed, except for the activities related to oil extraction. However, the Sarawak population is growing at 4–5% per year which will undoubtedly put additional pressure on the coastal resources.

The majority of people in Kalimantan are Iban, with significant numbers of Chinese in urban areas. The low population density prompted the previous Indonesian government to promote a transmigration programme to relocate Muslim people from overcrowded Java and Sumatra. This programme was very unpopular in Kalimantan and was the root cause for the civil unrest. Possibly, recent events will bring more democracy for the Kalimantan ethnic majorities.

Rural Factors

Until very recently, Borneo was covered in dense tropical rainforest. This forest has been systematically cut and cleared across huge tracts of land. Logging operations, reduced vegetation cover and the logging roads have all contributed to the increased run-off and the high sediment run-off. Slash-and-burn agriculture has followed the roads. Small-scale agriculture is still dominant, with rice, fruit and pepper being important crops. Hill rice is very important, although wet (paddy field) rice is also grown in suitable areas. Many logged areas have been converted into oil palm plantations, which cover very large areas and are the most important cash crop. Large-scale burning of secondary forests in preparation for planting of oil palms was responsible for widespread haze during the 1997/1998 ENSO, which stretched from Indonesia to Thailand and across Borneo, with the air pollution index at a dangerously high level for many weeks. While satellite photographs clearly pinpointed which areas were burnt, those responsible have not yet made reparations for lost tourist revenue, detrimental health effects or their contribution to global warming. Air pollution notwithstanding, sediment and agricultural fertiliser run-off are the most important rural land issues that affect the coastal seas of Borneo.

Mangrove crab culture in the natural forest seems to be a good example of sustainable community aquaculture. Throughout southeast Asia the mangrove crabs *Scylla olivaceous, S. tranquebarica* and others are caught in traps placed in the mangroves. Catches in Sarawak vary with tide and are greatest at 14–17 kg per day during spring tides and 5–7 kg during neap tides. Crabs larger than 10 cm carapace width are considered market sized, and the small crabs are used as bait or included in the family meal. In Sematan and other areas of Sarawak, small crabs are now purchased by the community to be stocked in aquaculture pens built within the mangroves (Ikwannuddin and Oakley, 1999). The pens are small, 10×20 m and are built with minimum disturbance, so that dense *Rhizophora* mangrove trees may continue to grow within the pens. Small crabs are typically stocked at 1000 per pen, fed trash fish daily and grow to market size in 4–6 months. Profits are then shared; the operation is profitable and has the advantages of maintaining the mangrove, minimising the loss of small crabs, and even of allowing female crabs to breed within the pens and release larvae back into the natural environment.

Erosion and Landfill

Most coasts are soft sediment-depositing shores so that generally there is coastal accretion rather than erosion. However, north of Miri, the shore is being eroded at a rate of 4–8 m per year and at the mouth of the Baram River, nearly one km of shore has been lost since 1985.

The west coast of Brunei and Sarawak features a long sand beach, with sediments carried southwest. All major rivers contribute to this deposition while minor rivers are diverted south to flow for long distances behind sandbars (as much as 10 km at Miri) before reaching the sea. In an effort to stabilise the Miri river mouth, many millions of dollars are currently being spent to divert the river outflow.

Further north, in Brunei, a large number of wave barriers have been built to reduce wave action, and the sand has developed large beaches behind the barriers. Natural longshore sand drift has been stopped and consequently beaches further south in the Miri area are being starved of sand. Coastal erosion is exacerbated by sand mining operations at the mouth of the Baram River (between Brunei and Miri), which remove over 1000 m^3 of sand per day from the river bed, depleting the sand bar. Thus the coastline between Kuala Baram and Miri is being rapidly eroded. In an effort to curb this, beach protection works are being carried out in Miri while, just a few km up the coast in Kuala Baram, and in conflict with these works, sand is being pumped ashore for landfill and for the construction of the Kuala Baram industrial area. These projects alone are estimated to require 25 million m^3 of sand by the year 2002. Hotels are already losing their beaches to erosion caused by sand mining, and the continuing sand extraction will cost beachfront properties large sums for additional protection works. Possibly the cost of erosion will be passed back to the sand mining companies through the implementation of an ICZM plan for Sarawak.

The opposite process occurs in most of the major towns in Sabah, where shallow coastal seas are land-filled. Significant areas off Kota Kinabalu in Sabah have been land-filled and the city has grown out into the bay by at least 1 km. There is currently heated debate about the merits of additional large landfill for municipal buildings from unspoiled Likas Bay.

Urban and Industrial Pollution

Sedimentation is a major concern throughout most of Borneo. Rivers which originate in forest reserves or national parks are clear, whereas rivers from land which is being converted are murky brown with very high suspended sediment. The concentration of sediments in streams or rivers has a direct relationship with the land use and topography. Streams draining areas with steep slopes and high human activity have high average TSS concentrations (3000 mg/l) which is 20 times higher than the relevant Malaysian standard. When construction removes the vegetation cover, the nearby rivers receive average TSS concentrations of 6000–7000 mg/l. Not surprisingly, this causes major changes to the river ecology as well as the estuary and nearshore coastal waters.

Pollution is a growing problem affecting most towns in Borneo. Solid wastes, especially in the poorer districts, are neither collected nor properly disposed, ending up in urban drains or rivers. Litter and floating wastes from all types of industries are common in the rivers and coastal seas. Solid waste collection in urban areas is woefully inadequate and a major concern near all towns (Willoughby et al., 1997); the seas within the Tunku Abdu Rahman Park in Sabah are frequently filled with floating garbage from Kota Kinabalu. In Miri, one of the largest towns in Sarawak, only 75% of the houses or commercial centres have waste collection services. There is no municipal collection or disposal of industrial wastes, while the 4000 families in squatter areas are only provided with limited waste-disposal facilities. Not surprisingly, the banks of the Miri river are lined with garbage.

Liquid wastes and domestic sewage receive little or no treatment in most areas of Borneo. In Sarawak towns, less than 28% of homes have modern flush toilet systems. Some domestic sewage is treated in septic tanks, but most sewage enters directly into rivers untreated. Another issue concerns pollution of groundwater from liquid wastes seeping out of landfill disposal sites. Throughout Borneo, industries are required to provide special treatment, but even the most casual observations show that this is rare. None of the industries of the Kuching Industrial area had any treatment facilities in 1997 (unpublished data) and industries like the Bau gold mines in Sarawak casually discharge metals and other chemicals into water courses. An EIA is required only for industries larger than 50 ha, so known polluting industries either occupy smaller areas or, in most cases, simply ignore regulations.

Agricultural wastes are also a problem. Pesticides leach out of soils. Larger pig and poultry farms have oxidation ponds although these are frequently too small, leading to washout, with residues concentrating downstream. Palm oil processing mills produce high temperature water with high BOD, which is normally treated in retention ponds, but impacts on the rivers are obvious especially during low flow periods. In Sarawak even rivers in areas with low population densities are moderately polluted, and generally all rivers have been deteriorating since 1995.

Pollution from the petroleum industry is a major threat to coastal ecosystems. Also of concern are organochloride compounds from herbicide and pesticide use, but the concentrations and fates of these pollutants are unknown. Areas with high levels of hydrocarbons include waters of the Riau Archipelago and Java Sea (Wisaksono, 1972).

Red tides created by blooms of the toxic algae *Pyrodinium* were unknown in East Malaysia prior to 1976 (Ming and Wong, 1989). Since then, periodic blooms have occurred and it has been postulated (Seliger, 1989) that coastal lagoons, such as Kuala Penyu, act as the culture vessel where agricultural fertiliser run-off is the key trigger.

Marine Conservation Areas

Integrated Coastal Zone Management Planning has yet to be successfully applied in Borneo, and despite several multi-million dollar projects, recommendations have not progressed further. The CRMP for Brunei Darussalam was completed in 1989 (Anon, 1989) and this along with other projects (White, 1991; ONEB-MSTE, 1992) were demonstration projects for the region. However, many years later few, if any, of the key recommendations have been put into practice.

Throughout Borneo, there are a number of protected areas that range from well protected and enforced areas through to 'paper' parks. In Sarawak, all the coastal National Parks and Protected Areas are well managed and

---

Tunku Abdul Rahman Park

Many reefs around Borneo have developed under conditions of high natural levels of sediment and turbidity. Since much human coastal development causes more turbidity, relatively minor activities are increasingly becoming key factors determining the survival of many coral reefs. In Sabah, the Tunku Abdul Rahman Park (TARP) was created to protect reefs and islands off Kota Kinabalu. This area perfectly demonstrates the need for consideration of wider issues than simply park management. The reefs are under stress from blast fishing, anchors, line fishing, shell collecting, reef gleaning and tourists, as well as pollution and coastal landfill.

Recent surveys in TARP and other areas of Sabah show that the *Diadema* urchins are extremely abundant in lagoons, caused both by removal of competitors and of the predators of the urchins. The urchins can erode reef substrate at rates of up to 4 kg/m^2/yr, or more than the rate of reef accretion, even on healthy reefs. Increased sediment also affects coral and reef building algae, and indirect impacts also come from petrochemicals, heavy boat traffic, sewage, garbage, silt from construction sites, as well as industrial wastes discharged into every storm drain, river or stream. The latter appears to be toxic enough to kill larvae needed for a giant clam hatchery, and it is not likely that coral planulae or other pollution-sensitive larvae can survive either. Data on coral recruitment, growth or survival within TARP are not available but there is a broad consensus of opinion that the reefs are deteriorating.

The net result in Kota Kinabalu bay and TARP is that the reefs are being eroded faster than they are growing, and the long or even medium term outlook for reef survival is bleak. Reefs adjacent to other capital cities in the region are already functionally dead (Tomascik et al., 1997), so the question that must be asked is not: Can we reverse the effects of Kota Kinabalu city development on the TARP, because the technology is available if the political will can be found? The relevant question is: "Can Kota Kinabalu afford the price paid by Jakarta or Manila or Bangkok?" The Tunku Abdul Rahman Park is one of the major tourist attractions of the State of Sabah, the direct economic benefits are easily measured, yet the reefs which maintain the beaches and attract the tourists are being destroyed and will never recover unless action is taken.

REFERENCES

Tomascik, T., Mah, A.J. et al. (1997) *The Ecology of the Indonesian Seas*. Periplus Editions (HK) Ltd.

---

attract a number of tourists. Bako National Park, northeast of Kuching, has representative rocky, mangrove and mudflat coastal habitats, and is a common destination for tourists and educational field trips. The Talang Talang islands off the southwest tip of Sarawak support a small turtle-nesting population, but are not open to visitors. A recent project has positioned small artificial reefs around the islands to promote additional coral growth and to discourage trawl fishing. The Samunsan, Sibuti and Simunajan protected areas are all managed to protect mangrove and beach habitats. These protected areas in Sarawak have muddy seas and, while there are no fishing restrictions, there are also minimum conflicts. This is not the case for any of the protected areas within the Seas of Borneo that include coral reefs. Excluding the Talang Talang islands, all coral reefs within national parks are fished and have been subjected to illegal destructive fishing techniques (Figs. 7a and b).

In East Kalimantan: Derawan, Sangalaki, Kakaban and Meratua are primary SCUBA diving destinations. Both Derawan and Sangalaki support significant marine turtle nesting. Sangalaki is home to a resident population of manta rays, which form the backbone of the diving attraction. Kakaban contains an inland saltwater lake renowned for its non-stinging jelly fish, while Meratua has steep drop-offs at which hammerhead sharks are commonly observed. Villagers on the islands collect and sell turtle eggs, and engage in destructive fishing techniques that have not yet been curbed either by the presence of two dive resorts or by the fact that Sangalaki is a National Park.

In Sabah, the islands of Sipadan, and Layang Layang are internationally renowned SCUBA diving destinations. Dive operators at Sipadan have effectively banned blast and cyanide fishermen from the reefs, and the vigilance and presence of the operators on the islands has dissuaded nearby villagers from encroaching on the recovering reefs. This was not always the case: as recently as 1982, blast fishing on Sipadan was commonplace. Today, very little, if any, blastfishing occurs on Sipadan, although blasts can still be heard, probably originating from reefs around Mabul and Kapalai.

Sipadan is also a special security area, being the cause of an international dispute between Indonesia and Malaysia, both of which claim ownership of the island, and for this reason is frequented more regularly by the Malaysian Navy and Marine Police. This, along with the strict protection by the dive operators, has brought about the recovery of the once-bombed reefs. In contrast, the reefs off Pulau Tiga National Park on the Western coast of Sabah which lacks such a deterrent, have been virtually completely destroyed, as have several sites on the protected reefs of Tunku Abdul Rahman Park, off Kota Kinabalu, Sabah.

Layang Layang is one of the Spratly islands claimed by Malaysia. Its resort also enforces strict rules against touching the reef or harassing the wildlife. Remoteness and the presence of a military base deters blast and cyanide fishermen, although fishing boats of this type can sometimes be found seeking shelter in the lagoon during the monsoon season. Lankayan, off the east coast of Sabah, boasts a wide array of marine life. Here, only the resort operator accords protection, and on Mantanai, the resort owner has negotiated a 3 km no-blasting zone with the police, community leaders and the blast fishers.

The proposed Semporna Islands Park off the east coast of Sabah is recognised, both nationally and internationally, as one of the prime marine conservation sites in South East Asia. The area covers over 325 km² and encompasses high rocky islands, low-lying sand cays, lagoons and open water. This park illustrates many of the problems and possible management solutions common to most of the coral reef areas within the seas of Borneo. It contains a wide range of marine and terrestrial ecosystems, with high biodiversity. Several of the islands are inhabited. Tranquil surroundings and outstanding natural beauty provide enormous potential for tourism development in the area.

A management plan for the proposed Semporna Marine Park is being produced through the Semporna Islands Project (1998–2000). Involvement and input of local communities, other users and interested parties is a vital component of this process, and an education programme is ensuring that conservation issues and potential benefits of the proposed park are appreciated. There are many issues that have to be addressed in order to ensure a healthy and productive future for the environment and the resources on which people depend (Table 3).

In general, the management objectives for Semporna and other marine protected areas can be summarised as follows:
– Conserve and maintain representative habitats and species, ensuring that populations of rare, endangered or endemic species are protected.
– Introduce appropriate fishery management strategies, ensuring that local people fish in a way that is sustainable.
– Develop strategies for dealing with 'natural' ecological disturbances, and consider ways to rehabilitate sites which have been degraded.
– Maximise opportunities for sustainable use (e.g. recreation, aquaculture, artisanal fishing), and economic development in line with estimated carrying capacity. This may involve creating or developing alternative activities for local communities.
– Maintain and promote cultural and social values, and involve local communities in development and management of the area.
– Develop and maintain an awareness and education programme, and develop research and monitoring programmes that support the management plan.

These are applicable to other areas, whether already gazetted or unregulated. Above all, the inclusion of local people in the development of protected areas will prove the most successful measure. The employment of local villagers is essential.

Dive operators on some islands have played a beneficial role in conserving reef habitats, but several factors remain unchecked. On Sipadan, for instance, the presence of six dive resorts is believed to have eroded the fresh water table to the extent that operators import water. The resorts have now agreed to import water and/or use desalinated water. Disposal of waste products on many islands is invariably by open burning or by ocean dumping, although a few operators are conscientious enough to transport solid waste back to the mainland. On Layang Layang, a tertiary treatment

Table 3
Current and potential threats to the proposed Semporna Marine Park

Threats to reefs and marine resources
- Overharvesting of fish and shellfish
- Degradation of coral reefs due to use of explosives for fishing
- Degradation of reefs from other harvesting methods
- Mortality of corals due to increased water temperature (coral beaching)
- Damage caused by large populations of coral-feeding starfish, *Acanthaster planci*
- Damage caused by 'black sponge' infestation
- Loss of biodiversity due to these factors

Issues dealing with shoreline and land use
- Land claims, customary rights and ownership issues
- Illegal settlements and immigration
- Unregulated use of mangrove and forest resources
- Unregulated land clearance and agricultural development
- Loss of biodiversity

Water quality and wastes
- Threats from river-borne pollution and run-off from Sabah mainland
- Vulnerability of reef lagoons to sediment run-off from island slopes
- Limited fresh water supply
- Lack of proper disposal of sewage and rubbish

Recreation and tourism
- No environmental guidelines for visitors
- Lack of a plan for tourism development

Aquaculture
- Established seaweed farming operations are not part of integrated plan
- Lack of a plan for potential aquaculture projects

Poverty and lack of opportunities for local communities
- Low income and standard of living
- Reliance on marine resources which are in a state of decline
- Few other job opportunities
- Lack of education and training for alternative jobs

Gaps in information and education
- Lack of awareness of the need for—and benefits of—conservation and management.
- Incomplete understanding of natural systems and processes.

plant exists for treating sewage, while less satisfactory cesspit treatment systems are used on all other islands.

In summary, protected areas have served to protect habitats in some places, but failed in others, due in a major part to the lack of participation of local villagers. The development of protected areas must be based on a need to conserve these areas not only for the sake of tourism or economic development, but also for the betterment of local populations. The provision of alternative livelihoods for those who are affected, along with the consultation of local inhabitants in decision-making, will ultimately result in areas that are well managed and in which regulations are enforced through people-participation programmes, rather than through traditional and commonly ineffective enforcement methods.

REFERENCES

Anon (1989) The Integrated Management Plan for the Coastal Zone of Brunei Darussalam. Dept of Fisheries, Min. of Industry & Primary Resources, Bandar Sri Begawan, Brunei Darussalam.

Bennett, E.L. (1991) Diurnal primates. In *The State of Nature Conservation in Malaysia*, ed. R. Kiew. Malayan Nature Society, Kuala Lumpur, Malaysia, pp. 150–172.

Bennett, E.L. and Gombek, F. (1993) Proboscis Monkeys of Borneo. Natural History Publications (Borneo) and KOKTAS Sabah Berhad, Kota Kinabalu.

Chai, P.P.K. (1975) Mangrove forest of Sarawak. *The Malaysian Forester* **38**, 108–134.

Chai, P.P.K. and Lai, K.K. (1984) Management and utilization of mangrove forests in Sarawak. In *Proceedings of the UNESCO Asian Symposium on Mangrove Environment: Research and Management*, eds. E. Soepadmo, A.N. Rao and D.J. Macintosh. University Malaya, Malaysia, pp. 785–795.

Chua, T.E., Ming, C.L. and Sadorra, M.S.S. (1987) The coastal environmental profile of Brunei Darussalam: Resource assessment and Management issues. Dept of Fisheries, Min of Industry & Primary resources Bandar Sri Begawan, Brunei Darussalam, and ICLARM, Manila Philippines.

Cooper, R. (1990) Forecasting and understanding local weather in Brunei Darussalam. *Brunei Museum Journal* **7**, 88–97.

Davies, J. (1998) The mangrove forests of Brunei Darussalam. *Borneo* **4**, 40–49.

de Longh, H.H. (1997) Current status of dugongs in Indonesia. In *The Ecology of the Indonesian Seas Series, Vol. 8*, eds. T. Tomascik, A.J. Mah, A. Nontji and M.K. Moosa, pp. 1158–1166. Periplus Editions (HK) Ltd.

DOF-MIPR (1992) The integrated management plan for the coastal zone of Brunei Darussalam Department of Fisheries, Ministry of Industry and Primary Production, Bandar Seri Begawan, Brunei Darussalam, .

Dutrieux, E. (1991) Study of the ecological functioning of the Mahakam Delta (East Kalimantan, Indonesia). *Estuarine, Coastal and Shelf Science* **32**, 415–420.

Dutrieux, E., Denis, J. and Populus, J. (1990a) Application of SPOT data for the study of the Mahakam delta mangrove (East Kalimantan, Indonesia). Definitions of sensitive areas to oil pollution. *Oceanologica Acta* **13**, 317–326.

Elkin, J. (1990) Observations of Marine animals in the coastal waters of western Brunei Darussalam. *Brunei Museum Journal* **7**, 74–87.

FAO (1984) Mangrove forests in Asia-Pacific Region: A Summary of Available Information. Special study on forest management, afforestation and utilization of forest resources in developing regions. GCP/RAS/106/JPN, pp. 38.

Forestry-Dept. (1997) Annual Report, Sarawak Forestry Dept., Kuching, Sarawak, Malaysia.

Garcia, S.M., Cochrane, K., Van Santen, G. and Christy, F. (1999) Towards sustainable fisheries : a strategy for FAO and the World Bank. *Ocean & Coastal Management* **42**, 369–398.

Giesen, W. (1993) Indonesia's mangroves: an update on remaining area and main management issues. Asian Wetland Bureau, Bogor, Indonesia In Unpublished manuscript presented at the International Seminar on Coastal Zone Management of Small Island Ecosystems, Ambon.

Hodgson, G. (1999) A global assessment of human effects on coral reefs. *Marine Pollution Bulletin* **38**, 345–355.

Ikwannuddin, A.M. and Oakley, S.G. (1999) Culture of mud crabs in Mangrove areas: The Sarawak experience. In *Regional Workshop on Integrated Management of Mangrove/coastal Ecosystems for Sustainable Aquaculture development, 23–25 March 1999*, ed. D. Mackintosh. Kuching, Sarawak, Malaysia. In press.

Johnston, N.A. (1984) *Coral Reefs of Pulau Tiga*. West Sabah, University of Kebangsaan Malaysia, Bangi.

Lulofs, R.B. (1973) A reef survey of Pulau Gaya and associated islands Sabah National Park Dept., Kota Kinabalu, pp. 16.

Martosubroto, P. and Naamin, N. (1977) Relationship between tidal forests (mangroves) and commercial shrimp production in Indonesia. *Marine Research Indonesia* **18**, 81–86.

Ming, T.T. and Wong, J.T.S. (1989) Summary of Red tide and paralytic shellfish poisonings in Sabah, Malaysia. In *Biology, Epidemiology and Management of Pyrodinium Red Tides, Vol. 21*, eds. G.M. Hallegraef and J.L. and Maclean. ICLARM Conference Proceedings, Brunei, 286 pp.

Murtedza, M. and Ti, T.C. (1993) Managing ASEAN's forest: Deforestation in Sabah. In *Environmental Management in ASEAN: Perspectives on Critical Regional Issues*, ed. M. Seda. Institute of Southeast Asian Studies, Singapore, pp. 111–140.

ONEB-MSTE (1992) The integrated management plan for Ban Don Bay and Phangnga Bay, Thailand. In ICLARM Tech. Rep., Vol. 30. Office of the National Environment Board, Ministry of Science, Technology and Environment, Thailand, 161 pp.

Ong, J.E. (1982) Mangroves and aquaculture in Malaysia. *Ambio* **11**, 252–257.

Ong, J.E. (1995) The ecology of mangrove conservation and management. *Hydrobiologia* **295**, 343–351.

Ong, J.E. and Gong, W.K. (1991) Mangroves. In *The State of Nature Conservation in Malaysia*, ed. R. Kiew. Malayan Nature Society, Kuala Lumpur, pp. 22–28.

Pauly, D. (1989) Fisheries management in Southeast Asia: Why bother? In *Coastal Area Management in Southeast Asia: Policies, Management Strategies and Case Studies*, eds. T.E. Chua and D. Pauly. ICLARM, Philippines, pp. 1–9.

Pauly, D., Christensen, V., Dalsgaard, J., Froese, R. and Jr., F.T. (1998) Fishing Down Marine Food Webs. *Science* **279**, 860–863.

Pauly, D. and Ingles, J. (1986) The relationship between shrimp yields and intertidal vegetation (mangrove) areas: a reassessment. In *IOC/FAO Workshop on Recruitment in Tropical Coastal Demersal Communities*, eds. A. Yañez-Arancibia and D. Pauly, pp. 277–282. IOC Workshop Report No. 44, Supplement.

Primavera, J.H. (1999) Impacts of Coastal Aquaculture development on Mangrove ecosystems: Lessons for Southeast Asia. In Regional Workshop on Integrated Management of Mangrove/Coastal Ecosystems for Sustainable Aquaculture Development, 23rd–25th March 1999, ed. D. Mackintosh. Kuching, Sarawak, Malaysia. In press.

Rajendran, N. and Kathiresan, K. (1996) Effect of effluent from a shrimp pond on shoot biomass of mangrove seedlings. *Aquaculture Research* **27**, 745–747.

Roberts, H.H. and Phipps, C.V. (1988) Proposed oceanographic controls on modern Indonesian reefs: A turn-off/turn-on mechanism in a monsoonal setting. In *Proceedings of the 6th International Coral Reef Symp., Vol. 3*, eds. J.H.B. Choat et al. Townsville, Australia, 8–12 Aug. 1988., pp. 529–534.

Roberts, H.H., Phipps, C.V. and Effendi, L. (1987) Halimeda bioherms of the eastern Java Sea, Indonesia. *Geology* **15**, 371–374.

Rossiter, W. (1997) Massive destruction of marine mammals, sea turtles and fish reported from trap nets in pelagic migratory channels. Unpub. Report, Cetacean Society International.

Sasekumar, A. and Chong, V.C. (1987) Mangroves and prawns: further perspectives. Proc. 10th Annual Seminar of the Malaysian Society of Marine Science, pp. 10–21.

Seliger, H.H. (1989) Mechanisms for red tides of *Pyrodinium bahamensis* var *compressum* in Papua New Guinea, Sabah and Brunei Darussalam. In *Biology Epidemiology and Management of Pyrodinium Red Tides, Vol. 21*, eds. G.M. Hallegraef and J.L. Maclean. ICLARM Conference Proceedings, Brunei, pp. 286.

Shaari, S.A.L.M., Chong, B..J. and Pathansali, D. (1976) Assessment of Marine Fisheries Resources of Malaysia Ministry of Agriculture, Malaysia.

Sukardjo, S. (1995) Structure, litter fall and net primary production in the mangrove forests in East Kalimantan. In *Vegetation Science in Forestry*, eds. T. Fujiwara and E.O. Box. Kluwer Academic publishers, The Netherlands, pp. 585–611.

Tomascik, T., Mah, A.J., Nontji, A. and Moosa, M.K. (1997) *The Ecology of the Indonesian Seas*. Periplus Editions (HK) Ltd.

White, A.T. (1991) Coral reef management in the ASEAN/US Coastal Resources Management Project In *Towards An Integrated Management of Tropical Coastal Resources. ICLARM Conference Proceedings 22*, eds. L.M. Chou et al. National University of Singapore, Singapore; National Science and Technology Board, Singapore; and International Center for Living Aquatic Resources Management, Philippines, pp. 343–353.

Willoughby, N.G., Sangkoyo, H. and Lakaseru, B.O. (1997) Beach litter: an increasing and changing problem for Indonesia. *Marine Pollution Bulletin* **34**, 469–478.

Wisaksono, W. (1972) Sea pollution in Indonesia: the role of man and technology in facing the challenges of pollution. In *Procs. Indonesian Petroleum Association First Annual Convention, Jakarta, June 1972*.

Wood, C.R. and Wood, E.M. (1987) The coral reefs of the Bodgaya Islands (Sabah: Malaysia) and Pulau Sipadan. 5. Fishes. *Malay. Nat. Journal* **40**, 285–310.

Wood, E.M. (1977) Coral reefs in Sabah: present damage and potential dangers. *The Malay. Nat. Journal* **31**, 49–57.

Wood, E.M. (1979) Ecological Study of Coral Reefs in Sabah. WWF Project MAL/15. Kuala Lumpur: World Wildlife Fund Malaysia. 163 pp.

Wood, E.M. (ed.) (1994) Pulau Sipadan: Reef Life and Ecology. WWF Project Number MYS 233/92. 160 pp.

Wood, E.M., George, J.D., George, J.J. and Wood, C.R. (1987) The coral reefs of the Bodgaya Islands (Sabah: Malaysia) and Pulau Sipadan. 6. Scientific summary, conservation issues and recommendations. *Malayan Nature Journal* **40**, 311–324.

Yusuf, A., Tanaka, S., Hamid, S., Kamezaki, N. and Suganuma, H. (1999) Recent conservation of sea turtles in the Java Sea, Indonesia. In *Second ASEAN Symposium and Workshop on Sea Turtle Biology and Conservation*, ed. N. Pilcher. Kota Kinabalu, Sabah, 15–17 July, 1999.

THE AUTHORS

Steve Oakley
Institute of Biodiversity and Environmental Conservation, University of Malaysia, Kota Samarahan 93400, Sarawak, Malaysia

Nicolas Pilcher
Institute of Biodiversity and Environmental Conservation, University of Malaysia, Kota Samarahan 93400, Sarawak, Malaysia

Elizabeth Wood
*Marine Conservation Society
9 Gloucester Road, Ross on Wye HR9 5BU, U.K.*

Chapter 78

CONTINENTAL SEAS OF WESTERN INDONESIA

Evan Edinger and David R. Browne

The continental seas of western Indonesia cover the Indonesian areas of the Sunda Shelf, and include the Java Sea and the southern portion of the South China Sea. Adjacent areas covered in this chapter include the west coast of Sumatra, the Flores Sea, and the western edges of the Makassar Strait. This region is surrounded by Java and Bali to the south, Sumatra to the west, and Indonesian Borneo to the north, and covers approximately 1 million km^2. The waters of the Sunda shelf are less than 100 m deep throughout, and most of the region is microtidal. The mountainous islands surrounding the Sunda shelf are home to approximately 150 million people. The monsoonal climate with high rainfall and runoff leads to high rates of terrigenous sedimentation. The Sunda shelf is dominated by fine siliciclastic sediments, except around coral reefs or straits with strong currents. Most coastlines are prograding rapidly, with localized cases of coastal erosion related to human activities.

The marine biodiversity of western Indonesia is high, because of its proximity to the Southeast Asian biodiversity centre in eastern Indonesia and New Guinea. The dominant nearshore habitats are mangroves, seagrass beds, and coral reefs, all of which are threatened by land-based pollution and marine resource extraction. Of the 2.1 million ha of mangrove area present in western Indonesia in 1980, nearly 60% have since been destroyed for timber, plantations, or shrimp ponds. Similarly, about 50,000 km^2 of coral reefs occur in western and central Indonesia, but half of these are in poor condition.

Major environmental threats to the seas of western Indonesia stem from rural, industrial and domestic land-based pollution, and from a variety of marine resource extraction activities. Agricultural runoff, deforestation, and mining shed nutrients and sediments into nearshore waters. Likewise, untreated sewage and poorly controlled industrial effluents are discharged into rivers, and thence into nearshore marine environments. Landfill and over-extraction of groundwater lead to ground subsidence and flooding in coastal districts of major cities, causing a variety of health problems. Marine debris from both urban and rural sources causes a serious litter problem on many beaches in western Indonesia. The effects of land-based pollution on coral reefs are most severely demonstrated on the reefs of Jakarta Bay and the Thousand Islands.

Marine resource extraction activities with major effects in western Indonesia include oil production, shipping and refining, capture fisheries, aquaculture, and coral mining. Western Indonesia produces vast amounts of oil and natural gas, and major petroleum shipping lanes run through the Strait of Malacca, South China Sea, Makassar and Lombok Straits. The major shallow water habitats of western Indonesia—mangroves, seagrass beds, and coral reefs—are all highly

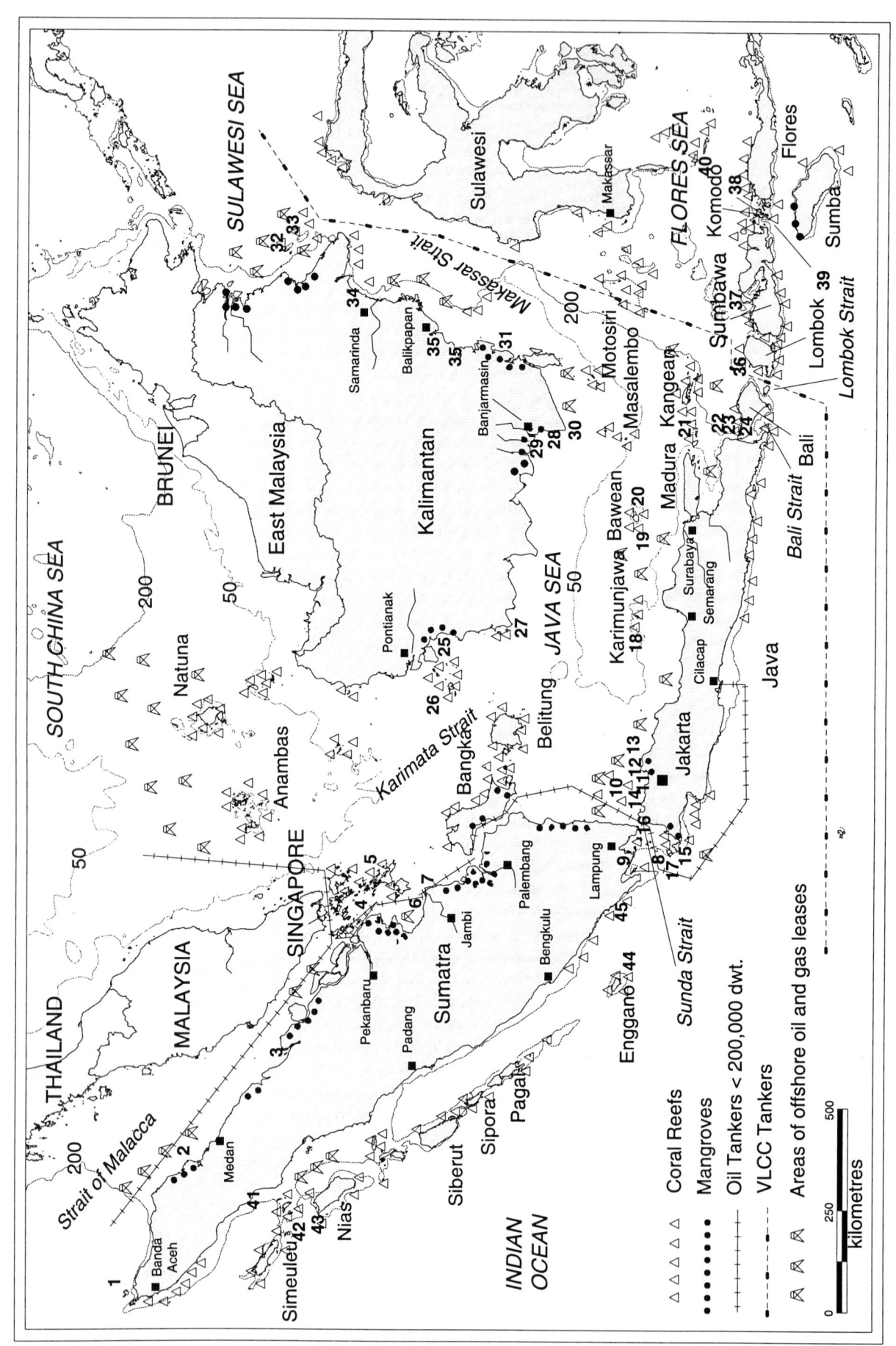

382   SEAS AT THE MILLENNIUM: AN ENVIRONMENTAL EVALUATION

Fig. 1. Map of the region. International boundaries, Indonesian provinces, and major cities are shown. Major rivers, large tracts of mangrove forests, coral reefs, offshore oil and gas leases in Indonesian waters, and oil shipping lanes are indicated. Marine protected areas are numbered according to Table 6. (Note: mangrove areas do not include large tracts of coastal swamp in Sumatra and Kalimantan).

sensitive to oil pollution. Capture fisheries for both pelagic and demersal fish are severely overexploited throughout western Indonesia, despite two decades of warnings and the 1980 ban on trawlers in most of Indonesia. Fisheries' decline is further exacerbated by mangrove clearing, and by destructive fishing practices including blasting and cyanide fishing. Aquaculture has grown explosively in western Indonesia, and releases large fluxes of nutrients, organic detritus, antibiotics and other chemicals, and sediment into nearshore waters. Shrimp farms are the primary cause of mangrove destruction, contributing to coastal erosion. Coral mining for building stone remains a threat to coral reefs.

Protective efforts include several large government projects, endangered species legislation, and a slowly growing system of marine protected areas. The Clean Rivers program (PROKASIH) has reduced discharge of industrial pollutants into nearshore waters, but is limited to major industries, ignoring pollution from small industries and domestic sources. Endangered species legislation nominally protects giant clams, marine turtles, dugongs, and other species, but implementation and enforcement are insufficient. Indonesia has established a system of Marine Protected Areas (MPAs), with 1.1 million ha of MPAs in western Indonesia. Indonesia remains far behind its target of 30 million ha of MPAs by the year 2000. Furthermore, most MPAs do not have management plans; existing management efforts suffer from poor implementation. The recent economic and political turmoil in Indonesia bodes ill for marine environmental conditions in western Indonesia, as rapid economic return takes priority over environmental protection and conservation.

THE DEFINED REGION

The broad shallow seas of western Indonesia, overlying the Sunda Shelf, include the Java Sea, the southern portion of the South China Sea, the Malacca Strait, the Makassar Strait, and the Sunda Strait (Fig. 1). The area covered in this chapter includes the entire Java Sea and extends into adjacent waters immediately to the west, and east. The Java Sea is bounded by Sumatra on the west, by Java on the south, and by Kalimantan on the north. To the east, the Java Sea opens onto the relatively deep waters of the Makassar Strait and the Flores Sea, which reach 1500 m in depth. The northwestern margin of the Java Sea extends through the Karimata Strait to the beginning of the Strait of Malacca in the west and opens onto the deep basin of the South China Sea to the north.

The continental seas of western Indonesia are unique as broad shallow seas in the tropics, in areas of high runoff, sediment load, and nutrient availability. They are of great interest geologically for their similarity to the epeiric sea environments that were common habitats through much of earth history. The continental seas of western Indonesia are also influenced by intense human activities, both on land and at sea. Approximately 150 million people live in land areas whose watersheds drain into the waters of the Sunda Shelf. The major cities, their population and major industries, as well as provincial population densities, are presented in Table 1. Six major cities, namely, Medan (North Sumatra), Kuala Lumpur (Malaysia), Singapore, Jakarta (West Java), Semarang (Central Java), and Surabaya (East Java) are the dominant urban-industrial zones affecting the region. Smaller cities bordering the Sunda Shelf include Melaka, Batam, Palembang, Pekanbaru, Cirebon, Pekalongan, Jepara, Rembang, Pontianak, Banjarmasin, Balikpapan, and Samarinda. The population densities on habitable or cultivable lands (<40° slope) of the provinces whose watersheds drain into this region are high, particularly in Java and Bali (Table 1). The Sunda Shelf region has experienced intensive human land use for at least seven thousand years, and was the route of migration of several waves of people colonizing Australia, Melanesia, and Micronesia.

OCEANOGRAPHY

The Sunda Shelf covers an area of 1.85 million km^2, of which roughly 1 million km^2 lie within the continental seas of western Indonesia. The western Indonesian seas region lies in the core tropics and spans the equator between approximately 3°N and 6°S. The Shelf is less than 100 m deep throughout, and was exposed during Pleistocene lowstands and incised by rivers draining northward to the South China Sea or eastward to the Makassar Strait (Tjia, 1980). The oceanographic conditions in the region are determined by the shallow depth of the Sunda Shelf, the high rainfall and topography of the landmasses draining into it, and the monsoonal climate.

The climate of the Java Sea region is monsoonal, with winds from the east during the east monsoon (April–November), and from the west during the west monsoon (December–March). During both the east and west monsoons, Pacific ocean water flows through the region into the Indian Ocean through the Makassar Strait and Lombok Strait. Surface currents in the Java Sea are determined mostly by these winds, and flow west and north through the Sunda Shelf during the east monsoon, and flow east

Table 1

Major Indonesian cities adjacent to the Sunda Shelf, city population, provincial population density, and major industries (source: BAKOSURTANAL, 1998)

City	Population (10^3)	Province	Provincial pop. density (people/km^2)	Major industries
Medan	1500	North Sumatra	228	Manufacturing, electronics, textiles
Pekanbaru	250	Riau	33	Petroleum
Jambi	230	Jambi	57	Timber, oil
Palembang	800	South Sumatra	68	Petroleum, timber, shipping, pulp and paper
Pontianak	350	West Kalimantan	32	Timber, shipping
Banjarmasin	400	South Kalimantan	93	Timber, oil
Balikpapan	300	East Kalimantan	19	Timber, oil, mining
Jakarta	10000	West Java	1185	Manufacturing, pharmaceuticals, pesticides, metal processing, paper processing, shipping, financial centre
Semarang	1500	Central Java	1080	Textiles, tanneries, metal processing, food processing
Surabaya	3000	East Java	879	Heavy industries, oil refinery

and south during the west monsoon (Tomascik et al., 1997). Current speeds generally range between 0.25 to 0.5 m/s. Surface currents through the Strait of Malacca and the Sunda Strait always transport surface water out of the Sunda shelf and into deeper waters of the Indian Ocean (Wyrtki, 1961; Burbridge et al., 1988). Because the Sunda Shelf is protected from large oceanic swell, most waves are short period wind-driven waves, and the storm weather wave base is shallow, about 25 m (Wyrtki, 1961). The Sunda shelf region lies outside the subtropical cyclone belt, and is not affected by major tropical storms, although local storms can cause considerable damage to shallow coral reefs and other marine habitats.

The region is microtidal; the normal tide range is generally less than 1 m (Wyrtki, 1961), with local exceptions due to resonance phenomena in the Strait of Malacca (Burbridge et al., 1988) and near the Madura Strait (Hoekstra, 1993). Tides are diurnal, or diurnal-dominated mixed tides, in the Java Sea, and semidiurnal or semidiurnal-dominated mixed tides in the Strait of Malacca and the South China Sea portion of the Sunda Shelf (Wyrtki, 1961). Sea surface temperatures are relatively constant, ranging between 26 and 30°. SST follows day length, such that highest water temperatures in the Strait of Malacca (north of the equator) occur in June–July, while highest water temperatures in the southern and eastern Java Sea (south of the equator) occur in December (Wyrtki, 1961). Episodic 2–3°C increases in SST are related to droughts and low cloud cover during ENSO events (Brown and Suharsono, 1990). Runoff and precipitation greatly exceed evaporation in all areas of the Sunda Shelf. Salinity varies between 30 and 35 S, with generally lower salinities in nearshore waters and during the wet season.

Total river discharge on the Indonesian portions of the Sunda shelf excluding the Strait of Malacca is approximately 9.09×10^{11} m^3/year, of which roughly 28% drains from Sumatra, 61% from Kalimantan, and 11% from Java. Java has shorter rivers with smaller catchment areas than Sumatra or Kalimantan. The major rivers of each province, their catchment area, discharge, and main anthropogenic impacts are presented in Table 2. Active tectonism and uplift, steep slopes, heavy rainfall, rapid weathering, and deforestation result in very high sediment loads. Sediment yield per drainage area in the Indonesian archipelago averages 3000 t/km^2/yr, among the highest rates in the world (Milliman and Meade, 1983). As a result, the Sunda Shelf is mainly covered by siliciclastic sediments, primarily silts and clays (Tjia, 1980), except in areas immediately adjacent to coral reefs. Rivers deliver strong pulses of sediment during the wet season (December to March in Java), with much less sedimentation occurring during the dry season (April to November in Java; Hoekstra, 1993).

As a result of naturally high rates of sedimentation often exaggerated by human activities, most coastlines of the Sunda shelf are prograding. Coastline extension rates of up to 100 m/yr have been reported in some parts of Jambi and South Sumatra (Burbridge et al., 1988). Progradation rates of up to 30 m/yr have been recorded near Demak, Central Java (Tomascik et al., 1997), with similar rates found in the Cimanuk delta of West Java, which advanced ca. 1 km over 30 years. High sedimentation rates throughout the region are often the result of anthropogenic land conversion and deforestation, either for agriculture or forestry (Burbridge, et al., 1988). Because the sediments delivered by many rivers are mud-dominated, compaction and subsidence rates are high, contributing to flooding in coastal portions of some cities.

The waters of the Sunda Shelf have the highest primary productivity in the Indonesian seas (Polunin, 1983). The euphotic layer is <50 m and waters are generally mesotrophic, with locally eutrophic environments nearshore, and hypertrophic conditions in areas subject to intense sewage pollution. Coastal phytoplankton biomass concentrations in the wet season are typically double those of the dry season. Primary productivity is fed primarily by nutrients in coastal runoff, although upwelling contributes to

Table 2

Major rivers draining into the Sunda Shelf. All rivers in Kalimantan carry heavy sediment loads due to poorly managed logging operations. Rivers in Sumatra are heavily loaded with sediment due to erosion from poorly managed forest operations, conversion of forest to oil palm plantations, and shifting cultivation agriculture. (Data courtesy of Dr. Ir. Imam Soeseno, Bogor Agricultural Institute).

River	Province	Catchment area (km^2)	Discharge (m^3/s)	Anthropogenic impacts
Siak	Riau	17474	499	Wood processing industry
Kampar	Riau	31232	1054	Wood processing, deforestation
Indragiri	Riau	31594	971	Wood processing, deforestation
Batanghari	Jambi	44554	1485	Wood processing and rubber pollution
Musi	South Sumatra	77234	2482	Pollution from wood processing, rubber, and oil refineries.
Sugihan	South Sumatra	10619	351	Wood processing and rubber
Baturusa Cecurut	South Sumatra (Island of Bangka)	16727	541	Tin mining runoff.
Tulangbawang	Lampung	16610	449	Pollution from Cassava processing industry
Seputih Sekampung	Lampung	7060	219	Cassava processing
Ciliwung	Jakarta, DKI	6510	214	Heavily polluted from many sources (see box)
Cisadane	West Java	1440	110	Textile industry pollution
Cisadek-Cikuning	West Java	10121	436	Industrial pollution
Ciwulan	West Java	6510	214	Industrial pollution
Citarum	West Java	7250	275	Industrial pollution
Serang	Central Java	3860	130	Industrial pollution and agricultural run-off
Jratun Seluna	Central Java	10465	2366	Industrial pollution and agricultural run-off
Bengawan Solo	Central and East Java	15400	340	High sediment, pollution from textile industries
Brantas	East Java	11050	290	High sediment
Pekalen Sampean	East Java	14969	366	High sediment load, agricultural run-off
Kapuas	West Kalimantan	95557	3857	High sediment load
Pawan	West Kalimantan	33641	1215	High sediment load
Sambas	West Kalimantan	14145	445.4	High sediment load
Kahayan	Central Kalimantan	17303	706	High sediment load
Mendawi	Central Kalimantan	23626	734	High sediment load
Sampit	Central Kalimantan	16169	514	High sediment load
Pembuang	Central Kalimantan	41000	1216	High sediment load
Barito	Central Kalimantan	80536	2695	High sediment load
Kayan	East Kalimantan	33005	1254	High sediment load
Sesayap	East Kalimantan	31568	1167	High sediment load
Mahakam	East Kalimantan	92641	2613	Chemicals from wood processing industry, mining
Berau-Kelai	East Kalimantan	19115	394	High sediment load
Karangan	East Kalimantan	20315	405	High sediment load
Cengal Batulicin	South Kalimantan	16219	437	High sediment load

productivity at the eastern edge of the shelf, where deep water from the Makassar Strait penetrates the Java Sea during the dry season. In general, the standing phytoplankton crop is smaller in the middle of the Java Sea than along the coast. Java Sea zooplankton is dominated by five species of large copepod with the number of common species increasing towards the Sunda Strait (Polunin, 1983).

MAJOR MARINE AND COASTAL HABITATS

Mangroves, seagrass beds, and coral reefs form the main marine and coastal habitats of western Indonesia along with the open sea habitat described above. The biota represented in the seas of western Indonesia is less diverse than that of eastern Indonesia due to the relatively recent colonization of the area following the last glacial period. The high human population and associated resource extractive and industrial development activities threaten each of the main habitats throughout the Java Sea and adjacent waters.

Mangroves

Mangrove forests were common along most of the low-gradient coastlines of the Sunda shelf. The area covered by mangrove forests in western Indonesia before and after intensive human interference is listed by province in Table 3. Before 1980, Western Indonesia accounted for 51% of Indonesian mangrove forests, but widespread clearing

Marine Ecosystems of Western Sumatra

Andreas Kunzmann

ZMT Bremen, Germany

OVERVIEW

The island of Sumatra comprises 470,000 km² (6°N to 6°S; 96°E to 106°E). In contrast to the eastern coast, where coastal plains with large rivers, peat swamps and mangroves dominate, the western coast is characterised by rocky and sandy shores and coral islands.

The climate of western Sumatra is characterised by more than nine consecutive wet months (up to 6000 mm/y) and a maximum of two consecutive dry months. The average surface water temperatures range from 28°C in December to 30°C in May. The salinity and density values are very stable throughout the year ranging from 33–34 ppt and 1020–1021 kg/m³ (DHI, 1990) for most of the area. The regional oceanography is largely unknown. It is characterised by exchange processes between the Andaman Sea and the Malacca Strait in the north and between the Pacific and Indian Oceans, via Sunda Strait, in the south. Upwelling develops off the southwest coast during the northeast monsoon from December through March.

In contrast to most western Indonesian seas, western Sumatra is characterised by a narrow shelf with steep continental slopes and steep islands. A 1700 m deep trench is less than 80 km west off the mainland. Only 100 km west of the Mentawai Islands, depths of 5000 m are common. From a geosciences perspective, the very active Sumatra subduction zone shows characteristic features, such as trench, ridge, fore- and back-arc basin and volcanic arc (island). The large sedimentary basins of North Sumatra, Sibolga and Bengkulu are little studied areas, but are thought to host considerable demersal fish resources (Lohmeyer, 1983).

The territorial waters of western Sumatra (70,000 km²) constitute 2% (8% including EEZ) of the total for Indonesia. In 1995, 10% of Indonesia's total catch or 250,000 t of marine fish were produced here, including 15,000 t reef fishes. In contrast to the Fisheries Department, the Ministry for Environment claims that shrimps are fully overexploited and reef fishes are heavily overexploited.

MARINE ECOSYSTEMS

The marine ecosystems of western Sumatra are poorly documented. Exceptions are the reefs of West Sumatra (Kunzmann, 1997; Steffen, 1998) and Lampung (Wiryawan, 1999).

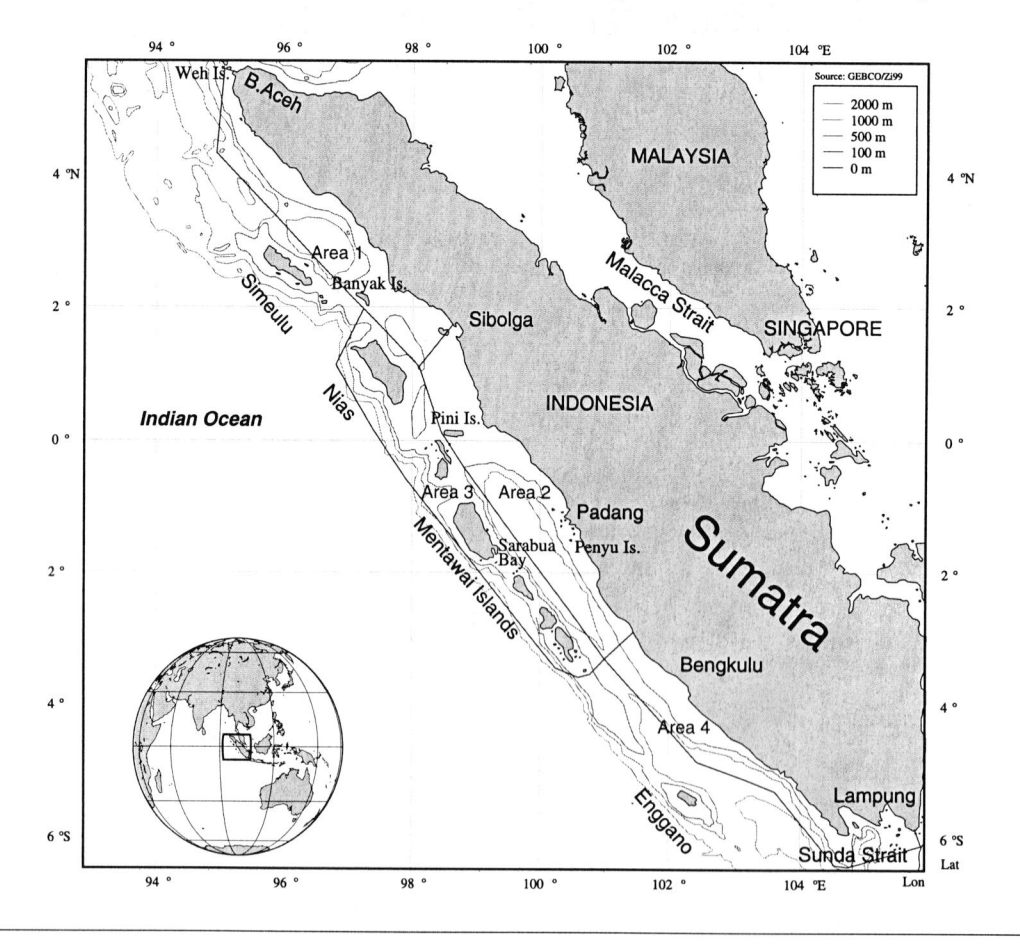

Apart from coral reefs, mangroves, seagrass/seaweed, mudflats, rocky shore, beaches, swamps, estuaries, we also find deep-sea environments. Especially in West Sumatra, all these very different systems occur in a comparatively small area, with a mixed fauna of Pacific and Indian Ocean species, which more closely resembles the Pacific Ocean fauna rather than that of the Indian Ocean (Kunzmann et al., 1999; Wallace, 1997). The Joint Indonesian–German Sumatra Expedition JIGSE planned for 2001 is expected to produce a comprehensive understanding of the area.

The sandy beaches, north of Padang reaching lengths of hundreds of km, are colonised by *Pescaprae*—and sometimes *Barringtonia*—formations. Sand and ghost crabs (*Ocypode, Dotilla*), as well as molluscs (*Tellina, Pinna, Arca*) inhabit these beaches. Nesting sites of sea turtles are common in the south, but only two species (*Chelonia* and *Eretochelys*) are seen regularly.

The rocky shores, in contrast to the sandstone of the East, are formed of old limestone (Aceh) or volcanic rock (Padang). Barnacles, oysters, limpets and Nerita snails are common and crabs (*Grapsus*) and rockpool blennies occur. The volcanic rock south of Padang hosts few corals.

Approximately 60% of the 1.5 million ha of mangroves in Sumatra lie on the east coast. The only large mangrove area in the west is the Tomak estuary. Few publications exist on this area or on the smaller areas of Painan and Air-Hadji further south. The Mentawai Islands also host mangroves, which are increasingly and illegally deforested. For more details, see Whitten et al. (1984) and Mastaller (1992).

Little work has been done on macroalgae and seagrass. *Ulva, Halimeda* and *Padina* are found on reefs. Mixed meadows of *Thalassia* and *Enhalus* are found on a few fringing islands. Larger shallow areas with seagrass are only documented from the Tanahmasa and Pini Islands, where dugongs are occasionally reported. As seagrass is grazed on by green turtles, their abundance close to Padang suggests there must remain as yet unreported seagrass resources in the area.

Coral reefs are most developed in the province of West Sumatra. Few fringing reefs occur along the main shoreline and most reefs are breaking or submerged patch reefs, or fringing reefs surrounding high islands lying between 1 and 26 km offshore. The Mentawai islands (Siberut, Sipura and Pagai), 160 km to the west, host many coral reefs and mangroves with the roots of *Rhizophoras* sometimes reaching large *Porites* blocks in locations such as Sarabua Bay. Further north along the coast, preliminary studies have been completed on the reefs of Banyak (Stehmann, unpubl.) and Weh Islands, both in Aceh.

DISTURBANCE AND STRESS

Unfortunately, most systems are under heavy human pressure. Destructive fishing with dynamite, poison and trawls (Kunzmann, 1998), coral and shell mining, pollution and sediments from settlements, logging and mining activities, as well as natural disasters such as the recent occurrence of red tides (Praseno et al., 1998) pose a major threat to the coastal ecosystems.

Due to repeated disturbance, succession is halted and the reefs are in a process of constant resettlement. This seems to favour either fast-growing species (small island areas) or species with a high tolerance for siltation (inshore areas).

Mangrove areas are reduced to small patches of degraded vegetation dominated by Nipah. This frequently leads to coastal erosion with narrow beaches. The replacement of mangroves with tambak ponds for shrimps and fish stopped only after dramatic drops in Penaeid shrimp yield. Many jobs were lost, since other species of economic importance, such as *Scylla* crabs, *Acetes* shrimps and fishes (*Chanos, Mugil* and *Lates*) also disappeared due to loss of habitat.

The over-exploitation of turtle eggs and adults has resulted in considerable reduction of turtle populations. Of the four species formerly abundant, only two are still regularly observed. Even on Pulau Penyu, a Marine Protected Area specifically dedicated to turtles, where sometimes the rare *Dermochelys* is seen, eggs are exploited regularly in large quantities.

While the prognosis is poor for the future, the area is of special interest to national and international projects, like MAREMAP, COREMAP and JIGSE and it is hoped that, with more attention focused on the west coast of Sumatra, the mounting human impacts on coastal and marine systems can be reduced and the resources and unique habitats of the area preserved.

REFERENCES

DHI (1990) *Handbuch der Suedkueste Asiens No. 2036*, Deutsches Hydrographisches Institut, Hamburg, 396 pp.

Kunzmann, A. (1997) The coral reefs of West Sumatra. In *The Ecology of the Indonesian Seas, Part Two*, eds. T. Tomascik, A.M. Mah, A. Nontji, and M. Kasim Moosa. Periplus Editions (HK) Ltd, Singapore, pp. 1249–1262

Kunzmann, A. (1998) "Management" of marine ecosystems in West Sumatra: theory and everyday practice. *Image of Indonesia* IV (5), 19–22

Kunzmann, A., Randall, J.E. and Suprihanto, I. (1999) Checklist of the shore fishes of the Mentawai Islands, Nias Island and the Padang region of West-Sumatra. *Naga* 22 (1), 4–10.

Lohmeyer, U.P. (1983) Jetindofish, Final Progress Report. GTZ, Eschborn, 31 pp.

Mastaller, M. (1992) Marine Survey of Lorentz, Togian Island and Siberut Island Management and Conservation of Tropical Forest Ecosystems and Biodiversity. Specialist Reports, ADB and PHPA Project T.A. 1430.

Praseno, D.P., Fukuyo, Y., Wiadiarti, R., Badrudin, Efendi, Y. and Pain, S.S. (1998) The 1997/1998 HAB/red tide blooms in Indonesian waters (unpublished; presented at BPPT/UBH Workshop in Padang, August).

Steffen, J.H. (1998) Comparative assessment of coral reefs in West-Sumatra. Thesis, Bremen University, 106 pp.

Tomascik, T., Mah, A.M., Nontji, A. and Kasim Moosa, M. (eds.) (1997a) *The Ecology of the Indonesian Seas, Part I*. The Ecology of Indonesia Series, Vol. VII. Periplus Editions, Singapore, 642 pp.

Tomascik, T., Mah, A.M., Nontji, A. and Kasim Moosa, M. (eds.) (1997b) *The Ecology of the Indonesian Seas, Part II*. The Ecology of Indonesia Series, Vol. VIII. Periplus Editions, Singapore, 746 pp.

Whitten, A. (1984) *The Ecology of Sumatra*. Gadjah Mada Univ. Press, pp. 89–149

Wiryawan, B., Marsden, B., Susanto, H.A., Mahi, A.K., Ahmad, M. and Poespitasari, H. (eds.) (1999) Lampung Coastal Resources Atlas. Government of Lampung Province and Coastal Resources Management Project (Coastal Resources Center, University of Rhode Island and Center for Coastal and Marine Resources Studies, Bogor Agricultural University). Bandar Lampung, Indonesia (in Indonesian), 109 pp.

Table 3

Area of mangrove forest in western Indonesia, by province. Source: Tomascik et al. (1997)

Province	Pre-1980 area of mangroves (1000 ha)	1990 area of mangroves (1000 ha)	% Loss since 1980
Aceh	60	<20	>67
North Sumatra	95	30.75	68
Riau	260	184.4	29
Jambi	18.5	4.05	78
South Sumatra	355	231	35
Lampung	56.5	11	81
West Java	66.5	<5	92
Central Java	32.2*	2.0*	94
East Java	57.5	0.5	99
Bali	1	0.5	50
West Kalimantan	213	40	81
Central Kalimantan	84	20	76
South Kalimantan	115	66.7	42
East Kalimantan	680	267	61
Total western Indonesia	2104	885	58

*Central Java figures exclude the Segara Anakan estuary on the south coast of Java.

for wood-pulp, aquaculture, tidal agriculture, and plantations has reduced mangrove cover in western Indonesia by approximately 60%. For example, 95% of the mangrove forests in Kalimantan have been allocated as timber concessions (MacKinnon et al., 1996). Most severe losses are along the north coast of Java, where more than 90% of mangroves present in 1980 have now been cleared, nearly all converted into aquaculture ponds (Tomascik et al., 1997).

Indonesian mangroves are immensely productive, with primary production rates averaging 26.2 kg C/ha/day (Tomascik et al., 1997). The high productivity of mangroves supports an abundant and diverse marine invertebrate and fish fauna, as well as many bird species. Mangrove forests in Indonesia trap large amounts of siliciclastic muds and play an important role in coastal protection along non-reefal coastlines. Extensive clearing of mangroves for brackish water shrimp pond development in western Indonesia has contributed to growing coastal erosion problems. Mangroves also function to export particulate organic matter to the adjacent coastal waters, mainly in the form of floating leaves and seeds. Mangrove detritus is exported to nearshore waters, and helps sustain nearshore fisheries. Mangroves are important feeding and nursery grounds for many offshore fisheries, including the prawn fishery, and Indonesian prawn catches are strongly positively correlated with mangrove area ($r = 0.89$; Tomascik et al., 1997). The widespread destruction of mangroves in the Sunda Shelf region poses a grave threat to the region's capture fisheries (Fortes, 1988).

Seagrass Beds

Seagrass beds are well-developed in peri-reefal areas in western Indonesia, and adjacent to mangroves in Sumatra and Kalimantan. Southeast Asia represents the centre of diversity for tropical seagrasses, with 12 of 16 known Southeast Asian species present in Indonesia. There are no published estimates of the area of seagrass beds in Indonesia, but Sloan (1993a) suggests that seagrass beds are at least as extensive as coral reefs. Like mangroves, seagrasses are important habitat for juveniles of many demersal fish and crustaceans and sustain high populations of zooplankton. Seagrasses are also among the primary food sources for green turtles and dugongs.

Seagrass beds thrive in mesotrophic conditions without heavy siliciclastic sedimentation. Excess nutrients damage seagrass beds by promoting the growth of epiphytic algae on the seagrass blades, preventing access to light by the seagrasses themselves. Seagrass beds export significant fluxes of carbon to adjacent environments as detritus (dead seagrass blades), or through various animals that feed on seagrasses or the epiphytic algae on seagrasses. Export rates reach as high as 90% of net primary production (Tomascik et al., 1997).

Seagrass beds exist in depth ranges from the low water line down to approximately 10 m, depending on water clarity. Increased turbidity from nutrient loading or sedimentation decreases the depth range of seagrass beds by decreasing light penetration (Sloan, 1993a). Intertidal seagrass beds are highly susceptible to damage from oil spills. Subtidal seagrass beds are better able to resist oil pollution than intertidal seagrasses, because the underground rhizomes of the seagrasses often survive, even if the aboveground shoot portions of the seagrass are killed. Nonetheless, seagrass beds, along with mangroves and coral reefs, are among the coastal habitats most endangered by oil spills within western Indonesia.

Coral Reefs

There are approximately 50,000 km^2 of coral reefs in western and central Indonesia, slightly more than half of the estimated 85,000 km^2 total reef area in Indonesia (Tomascik et al., 1997; see Fig. 1). Barrier reefs are well developed off the northeast coast of Kalimantan, at the edge of the Sunda shelf, covering an area of 20,600 km^2. A further 9450 km^2 of barrier reefs occur around the islands off the west coast of Sumatra. Platform reefs are well developed around sets of islands in the Sunda shelf, such as the Natuna and Anabas Islands, the Riau Archipelago, the islands of Bangka and Belitung, Pulau Seribu Islands, Karimunjawa Islands, Bawean, and Kangean (Fig. 2). Such offshore reefs cover an area of about 8000 km^2. These islands are far enough from coastal sources of runoff and clastic sedimentation that nutrient and turbidity conditions are favourable for reef development, and coral growth reaches depths of 25 m (Suharsono, 1998). Most of these reefs are coral cays rising

Fig. 2. Aerial view of fringing reefs and coral cays in the Karimunjawa islands, Central Java.

from water depths of 30 to 60 m. While water temperatures are generally well suited for coral growth, increased water temperatures during ENSO-related droughts have caused extensive coral bleaching throughout the region (Brown and Suharsono, 1990).

Fringing reefs and nearshore coral cays are developed along portions of the clastic sediment-dominated coastlines of the Sunda Shelf, particularly in the provinces of Riau, Jambi, South Sumatra, West Java, Central Java, and along the coast of Madura (East Java). These reefs cover approximately 2000 km². Many of these reefs are developed adjacent to seagrass beds or mangroves. Coral growth on fringing reefs is limited to less than 10 m by turbidity associated with sedimentation and nutrient loading from land-based pollution (Edinger et al., 1998). In general, the coral reefs of western Indonesia have suffered greatly from anthropogenic impacts. Nearly half (49%) of the reefs surveyed in western Indonesia during the 1990s were in poor condition (i.e. had less than 25% live coral cover; Suharsono, 1998). Comparable figures for central and eastern Indonesia are 37% and 29%, respectively.

Several atolls occur in the Flores Sea (central Indonesia), covering approximately 8440 km². The largest and best known of these is Taka Bone Rate, off South Sulawesi. Taka Bone Rate, one of 6 National Marine Parks in western Indonesia (see Table 6), rises from clear waters of the Flores Sea near the island of Selayar. The reefs of the atoll have been considerably damaged by destructive fishing practices (see Box 2), as well as sustained high fishing pressure by traditional means. Recent surveys indicated that 68% of the sites surveyed in Taka Bone Rate had less than 25% live coral cover (Suharsono, 1998). Giant clams have been largely hunted to extinction within the atoll, although cultivation efforts are sustaining small populations.

Biogeography and Biodiversity

The southeast Asian archipelago, including Indonesia, the Philippines, and Papua New Guinea, is a zone of megabiodiversity, with centres of diversity for many marine organisms, including mangrove trees (Polunin, 1983), seagrasses (Sloan, 1993a), marine algae, corals (Best et al., 1989), molluscs (Paulay, 1997), reef fish (Randall, 1998), and others. The region has approximately 80 genera and >400 species of scleractinian corals, 50 species of mangrove trees, 16 seagrass species, 400 marine algal species, many hundreds of molluscs, and over 2000 reef fish species (Randall, 1998). The greatest diversity of these organisms is found in the waters of eastern Indonesia.

The Sunda Shelf lies in the Asian floral and faunal realm west of Wallace's Line, the major biogeographic division between Asian biota in continental southeast Asia and Australasian biota in New Guinea and Australia. Wallace's line runs through the Makassar Strait and the Lombok Strait, the deep-water channels which prevented migration of land plants and animals during Pleistocene lowstands. From the perspective of marine organisms, Wallace's Line represents the division between shallow areas exposed during Pleistocene lowstands, in western Indonesia, and areas in deep water where local habitat migration allowed marine communities to persist relatively undisturbed during Pleistocene lowstands. Thus most marine organisms on the Sunda Shelf have recolonized the region within the last 10–15,000 years since deglaciation, either from the high diversity refuge areas of eastern Indonesia, or from the Indian Ocean (Wallace, 1997).

The diversity of scleractinian corals in western Indonesia is slightly less than that in eastern Indonesia (Wallace, 1997). About 75% of the coral genera and species recorded on reef surveys in eastern Indonesia have also been documented in the Java Sea (Hoeksema, 1997). Within the highly speciose genus *Acropora*, the greatest species diversity occurs in northeastern Indonesia, and the least in western Indonesia. Levels of endemism in the west are much lower than in the east (Wallace, 1997). No endemic *Acropora* species are known from the Sunda Shelf region.

The high diversity of molluscs in eastern Indonesia is well-represented on the Sunda shelf (Paulay, 1997). Several reef-dwelling molluscs with direct development or brooded larvae are known from eastern Indonesia, but have not been recorded on the Sunda shelf. By contrast, more mangrove-associated gastropod and bivalve species are found in the Sunda Shelf and Philippines region than in the Moluccas and Irian Jaya (Tomascik et al., 1997). The global centre of biodiversity for reef fishes is in the Philippines and eastern Indonesia (Randall, 1998). This reef fish diversity was well represented in western Indonesia, but intense fishing pressure and habitat destruction have reduced populations and may have caused local extirpations.

Threats to marine biodiversity within the region stem mainly from habitat destruction, with human hunting posing a threat to some of the large marine vertebrates, such as sea turtles and dugongs. In western Indonesia, land-based pollution presents the greatest threat to coral reef biodiversity (Edinger et al., 1998). Clearing of mangroves poses a great threat to the flora and fauna of mangrove habitats, as well as contributing to higher nutrient and sediment fluxes to nearshore waters (Chua et al., 1989).

MARINE RESOURCE EXTRACTION

Oil Production and Shipping.

Indonesia is one of the world's major petroleum-producing nations. Total average daily oil production in 1997 was 1,364,000 barrels/day, produced from 8535 wells (Oil and Gas Journal, 1998). More than 90% of this production comes from 10 major sedimentary basins in western Indonesia, specifically along the north and east coasts of Sumatra, in the Java Sea, and in coastal regions of Kalimantan (Tomascik et al., 1997). Most of this oil and gas production is exported, and petroleum continues to provide Indonesia's largest source of export earnings. Both domestic and foreign tanker traffic is concentrated in three major shipping lanes: the Strait of Malacca, for traffic between the Indian Ocean and the South China Sea, and the Makassar Strait and Lombok Strait for traffic between the Pacific and the southern Indian Ocean. Large volumes of domestic oil are shipped through the Sunda strait from production fields in Sumatra and the Java Sea to the refinery at Cilacap, on the south coast of Java (see Fig. 1).

The threat of oil spills in these shipping lanes is high, and the major shallow marine habitats in western Indonesia—mangroves, seagrass beds, and coral reefs—are all highly sensitive to oil spill damage (Sloan, 1993b). The number and volume of spills in Indonesian and adjacent waters, as well as the percent of total spills accounted for by each area and the average volume of oil released per spill, is presented in Table 4. Between 1974 and 1994, 36 major tanker spills were reported in Indonesian waters, spilling a total of 1.17 million barrels, about half of which was crude oil. Two thirds of the spills occurred in the Strait of Malacca. Other high risk areas for oil spills in Indonesian waters are the Makassar Strait, the Lombok Strait, the Sunda Strait, and the Indonesian portions of the South China Sea (Table 4). The Indonesian, Malaysian, and Singapore governments have established an oil spill response network for the Strait of Malacca, but the ability of the Indonesian network to respond to a major spill has not yet been tested. Indonesia and Malaysia prefer mechanical responses to spills, because dispersants often increase the impact of oil spills on coral reefs and mangroves (Sloan, 1993b). Singapore, however, relies heavily on dispersants in its response to oil spills.

Despite the high tanker traffic through Indonesian waters, chronic oil pollution from oil refineries and production facilities probably presents a greater threat of oil contamination to many western Indonesian mangroves, seagrass beds, and coral reefs (Sloan, 1993b). Eight major oil or gas refineries are located on Indonesian coastlines bordering the Sunda Shelf, including four in Sumatra, two in Java, and two in Kalimantan. Oil leakage from the refineries near Palembang, South Sumatra, has been blamed for extensive pollution of the coastal rivers in eastern Sumatra (Whidden et al., 1987; Burbridge et al., 1988).

Table 4

Oil tanker spills reported in/adjacent to Indonesian waters, 1974–1994. Number of spills, volume spilled in barrels, % of total spills by volume and average volume per spill. Oil discharge from blowouts, storage tank failures, and on-land operations not included.

Location	No. of spills	Volume spilled (barrels)	% Total spills	Average volume
Malacca Strait (all)	24	961,511	82.4	40,021
Malacca Strait (Indonesia)	11	771,614	66.2	64,301
Malacca Strait (Malaysia)	3	97,330	8.3	32,443
Malacca Strait (Singapore)	9	91,567	7.9	10,174
Makassar Strait	2	36,650	3.1	18,325
Lombok Strait	1	7,330	0.6	7,330
Sunda Strait	1	3,358	0.3	3,548
South China Sea	8	157,947	13.5	19,743
Total	36	1,165,986		32,389

Indonesia is also the world's leading producer of liquefied natural gas (LNG), much of which is processed at the LNG plants at Lhoksumawe, Aceh, and at Bontang Bay in South Kalimantan (Burbridge et al., 1988; MacKinnon et al., 1996). Hot water discharges from the Bontang plant have killed corals on the adjacent fringing reefs, along with other localized impacts of thermal pollution. Mercury extracted from natural gas at liquid natural gas refineries in Aceh poses a continuing contaminant hazard, and may contribute to high dissolved mercury levels in the Malacca Strait (Burbridge et al., 1988). Further mercury flushed from production waters in offshore wells has been blamed for mercury contamination in the northern Java Sea and Singapore region.

Increases in offshore oil production in the Sunda Shelf coastal regions has apparently led to an increased number of spills from production platforms, including blowouts near Balikpapan (E. Kalimantan) and Madura (E. Java). In a general study of marine debris in the Pulau Seribu region, north of Jakarta, oil patches on beaches were the only form of marine debris whose frequency increased away from Jakarta (Uneputty and Evans, 1997). The most likely source of this oil is the series of production platforms in the Sunda and Northwest Java basins, north of the Pulau Seribu (Willoughby et al., 1997; see Fig. 4). Various heavy metals, including Cr, Cu, Ni, Pb, Sn, and Zn incorporated into coral skeletons in the Pulau Seribu, are thought to be derived partially from offshore oil platforms in the Sunda and Northwest Java Sea basins (Scott and Davies, 1997).

Coral Mining

Illegal mining of live corals for building materials continues in areas of the Sunda shelf with ready access to coral reefs. Most mining occurs on reef flats, where the corals can be easily loaded into a canoe or other shallow draft vessel. There are no quantitative estimates of the volume of coral currently mined in western Indonesia. Coral and sand mining on the inner islands of Jakarta Bay contributed to the ultimate demise and submergence of those islands (Tomascik et al., 1993). Coral mining persists because corals remain the building material of choice in many coastal communities (Yates, 1994), even when alternative materials are available. For example, 75% of construction (1994) on Karimunjawa island, Central Java, used corals as the primary building material, even though this island is a high island composed of sandstone suitable for building. To discourage coral mining, the local government has provided machines for making cement bricks from carbonate sand; these efforts have had moderate success, despite some distribution problems.

Fisheries

Indonesian fisheries account for approximately one third of the total catch of ASEAN countries. Between 1986 and 1996, the total Indonesian catch increased from 2 to 3 million tonnes with the bulk of this catch coming from the highly productive seas of the Sunda Shelf (Tomascik et al., 1997). Over 90% of the total catch is landed by artisanal fishermen. The fisheries sector provided employment for 1.4 million people in 1986 directly supporting over 5.6 million people if dependants are included (Rice, 1991). The fishing fleet is based primarily in Central and East Java and along the northeast coast of Sumatra with the coral reef fishing fleet centred in South Sulawesi. Marine fish is a primary source of dietary protein in Indonesia, accounting for between 30 and 50% of protein intake across the country and higher percentages in coastal communities of the outer islands (Polunin, 1983). In 1988, the fisheries sector accounted for 2% of Indonesia's GDP and US$626 million in exports, mainly from shrimp and tuna sales (Rice, 1991). Extensive areas of the Sunda Shelf are relatively barren with only specific fishing grounds yielding large catches including the Strait of Malacca, the Karimata Strait, the seas surrounding the islands of Karimunjawa, Bawean, Masalembu and Matasiri, and the Bali Strait (Fig. 1). In general in the Java Sea, fish standing stocks are highest inshore and decrease with depth. Upwelling areas at the edge of the shelf such as the Bali Strait are extremely productive and support large pelagic fisheries (Pet et al., 1997).

Exploitation of the fishery resources of the seas of western Indonesia is characterized by extremely high fishing pressure on major stocks in an environment of limited regulatory enforcement, minimal scientific understanding of the fisheries, and incomplete and inaccurate catch and effort data (Marr, 1981; McElroy, 1991; Pet et al., 1997). Important species in the fishery are anchovy, mackerel, sardine, scad, trevally, skipjack, ponyfish, blood cockle, banana prawn, drumfish, marine catfishes, coral fish, and sharks (Polunin, 1983). However, over 50% of the pelagic catch from the Java Sea consists of just three genera and six main species, namely scads (layang, *Decapterus macrosoma* and *D. russelli*), Indo-Pacific mackerel (kembung, *Rastrelliger brachysoma, R. kanagutta*), and sardine including both Fringescale sardinella (tembang, *Sardinella fimbriata*), and Indian oil sardine (lemuru, *Sardinella longiceps*) (McElroy, 1991). Most stocks in western Indonesia have already reached or exceeded their production limit including reef fish, penaeid shrimp, sea cucumber, small pelagics, and sharks (Tomascik et al., 1997; Pet et al., 1997).

Significant commercial exploitation of the offshore fisheries in the western seas of Indonesia began in the 1960s with the introduction of the otter trawl (1966) and the purse seine (1968). Inshore fisheries along the north coast of Java and northeast Sumatra were reported as overfished by the late 1970s and by 1979 the demersal and penaeid shrimp resources were considered fully exploited throughout western Indonesian seas with significant fisheries potential remaining in pelagic species (Bailey et al., 1987). Due to the rapid over-exploitation of demersal and prawn fishery resources by trawlers in the 1980s and clashes between artisanal fishers and commercial trawlers, trawlers were banned in the western seas of Indonesia in 1981 (McElroy, 1991). Following the ban on trawls, effort shifted to pelagic fishing, primarily using purse seines (Fig. 3). By 1988, exploitation of the Java Sea pelagic resources were considered above the maximum sustainable yield as evidenced by a declining total catch despite a significant increase in fishing effort. Furthermore, along the north coast of Java where artisanal fishing effort in the nearshore environment is very high, catches have been at or above the maximum sustainable yield since 1986 (McElroy, 1991).

Fig. 3. Purse-seines are the dominant gear in the Java Sea pelagic fisheries. Hand-drawn purse-seines such as this are used for anchovy and sardine, while mechanically drawn seines are used for jack, mackerel, and *tongkol*.

Destructive Fishing Practices in Indonesian Seas

Mark V. Erdmann

Dept. of Integrative Biology, University of California, Berkeley, Berkeley, CA 94720, U.S.A.

Destructive fishing practices (DFP), defined here as any fishing technique that results in direct damage to either the fished habitat or to the primary habitat-structuring organisms in the habitat (e.g., scleractinian corals in a reef fishery), are widespread throughout the Indonesian archipelago and constitute one of the greatest threats to Indonesian reefs. As reviewed in Erdmann and Pet-Soede (1998), the most common DFP encountered in Indonesia include blast and cyanide fishing, *muro-ami* (drive-in net fishing), *bubu* trap fishing, and inshore trawling. This summary will focus on blast and cyanide fishing, widely considered the most destructive in Indonesia.

BLAST FISHING

First introduced by Japanese soldiers during WWII, blast fishing is now so pervasive in Indonesian coral reef fisheries that it might in some respects be considered "traditional". While the explosive used has evolved from actual dynamite (first from WWII munitions, then from international development aid civil engineering projects) to home-made kerosene and fertilizer bombs in beer bottles, the technique remains the same. Schooling reef fishes (caesionids, acanthurids, siganids, and lutjanids) are visually located by a snorkeling fisherman, who directs the capture boat to within 5 m of the school. A "bomb" is then thrown into the middle of the school, and after it has exploded, divers enter the water to retrieve the fish that have been killed or injured by the shock wave.

Though explosive fishing is illegal in Indonesia, it is still common throughout the islands due to the highly lucrative nature of the fishery. Fishermen involved in medium and large-scale blasting operations can earn up to THREE times the salary of government officials or University professors. With these incentives, it is not surprising that in some areas of Indonesia, it is estimated that up to 15% of the fishers rely heavily on blasting, while 10–40% of the total landings in these areas are supplied by blast fishing operations (Pet-Soede and Erdmann, 1998).

Unfortunately, this is an extremely habitat-destructive fishery. Damage to the reef framework is well-documented; branching, tabulate and foliose hard corals are shattered while massive and columnar corals are often fractured. Explosions can leave craters of rubble of 1–2 m in diameter, and reefs subject to repeated blasting are often reduced to little more than shifting rubble fields. Edinger et al. (1998) document decreases in coral species diversity on bombed Indonesian reefs of ca. 50%, and the prospects for recovery on these reefs seem dim indeed. Not only is the unconsolidated rubble a poor substrate for hard coral recruitment, but the greatly-reduced three dimensional structure of blasted reefs make them less attractive to emigrating adult and settling larval fishes—reducing their fishery potential for years to come. Previously blasted reefs in the Bunaken National Park, protected from further blasting since the late 1980s and otherwise relatively unimpacted by land-based pollution or sedimentation, have shown virtually no hard coral recovery, dominated instead by zoanthids and the soft corals *Xenia* and *Sarcophyton* spp. The unsustainable nature of this DFP is obvious; in many areas of western Indonesia, blast-fishing has ceased altogether because the heavily damaged reefs no longer support the fish biomass necessary for the practice to be economically viable (Erdmann, 1999).

CYANIDE FISHING

The use of potassium or sodium cyanide solution to stun fish and invertebrates desired for live collection is also extremely widespread throughout Indonesia (Erdmann and Pet-Soede, 1996). In this technique, divers (often supported by hookah compressors) use bursts of cyanide solution from squirt bottles to stun their targets. While overdose results in death of the target organism, a "controlled" squirt allows the diver to easily remove the stunned animal from its refuge in the reef framework, often after some breakage of the surrounding coral. The captured organism is then placed in a live well in the fisher's boat, transported to a central collection pen, and eventually exported.

While this technique was at one time used almost exclusively to capture ornamental fishes and invertebrates for the aquarium trade, since the late 1980s it has been the method of choice for capture of spiny lobsters (*Panulirus* spp.) and live reef food fish (primarily grouper and the Napoleon wrasse *Cheilinus undulatus*). These animals are exported live and sold in upmarket Chinese and Japanese restaurants, where wealthy diners pay exorbitant prices for the privilege of picking their seafood live from display aquaria. The incredible prices paid for these species (reportedly up to US$ 5000 for a single adult wrasse!) translate into a very lucrative fishery; for a live grouper or lobster, fishers receive up to 25 times what they would for the same animal dead. The huge sums of money involved in these fisheries are not only irresistible for fishers, but also ensure that any officials

Fig. 1. Overview of a thoroughly blasted reef, Morotai Island, northeast Halmahera. The reef structure has been reduced to rubble and is almost completely devoid of lifeforms other than algae and bioeroding invertebrates. With the current intensity of blast fishing this scene is becoming more and more common across the Indonesian Seas.

inclined to enforce the laws against cyanide fishing are easily persuaded otherwise.

The widespread use of cyanide in these fisheries has resulted in significant reef degradation in Indonesia. Besides the direct mechanical breakage of corals when collecting the live target species, cyanide solution in concentrations used to capture large reef fishes has proven lethal to most reef organisms, including smaller fishes, mobile reef invertebrates, and most germane to this discussion, hard corals. Perhaps even more importantly, these fisheries are extremely efficient in decimating target species' populations.

Predictably, cyanide fishing for live reef food fish and lobster is currently in sharp decline throughout Indonesia due to economic overfishing (Erdmann and Pet-Soede, 1998). In its place, fishers are now using the less capital-intensive techniques of fish traps and hook and line to take the remaining individuals of these target species. In many parts of western Indonesia, lobster and adult grouper are simply no longer encountered by fishermen, though grouper fingerlings are still captured for grow-out aquaculture.

Unfortunately, this decline in the live reef food fish and lobster fisheries may be leading unemployed cyanide divers to focus their attentions once again on the marine ornamentals trade, which has the potential for even greater reef destruction. With a much wider range and abundance of target species, significantly larger volumes of cyanide are used in collecting marine ornamentals, often with daily-repeated exposures. Moreover, because this fishery is relatively high value, with little capital investment required, it is virtually impervious to economic overfishing.

Without a much stronger commitment to enforcement on behalf of the Indonesian government, these reefs may be doomed to the sad fate of being squirted (and blasted) until they are reduced to barren carbonate skeletons supporting little more than a community of bioeroders and algae. Current reef management efforts in Indonesia, particularly the Coral Reef Rehabilitation and Management Project (COREMAP), have identified DFP prevention and enforcement as a key component of the national coral reef management strategy. The effectiveness of these efforts may well determine the long-term survival of Indonesia's extensive and once magnificent reefs.

REFERENCES.

Edinger, E.N., Jompa, J., Limmon, G.V., Widjatmoko, W. and Risk, M.J. (1998) Reef degradation and coral biodiversity in Indonesia: effects of land-based pollution, destructive fishing practices and changes over time. *Marine Pollution Bulletin* **36** (8), 617–630.

Erdmann, M.V. (1999) Destructive fishing practices in the Pulau Seribu archipelago. In *UNESCO Study No. 10: Proceedings of the Coral Reef Evaluation Workshop*, Pulau Seribu, Jakarta, Indonesia, 11–20 September 1995, pp. 84–89.

Erdmann, M.V. and Pet-Soede, L. (1998) $B^6 + M^3$ = DFP; An overview of destructive fishing practices in Indonesia. In *Proceedings of the Workshop on the Impacts of Destructive Fishing Practices on the Marine Environment*, Hong Kong, pp. 25–34.

Erdmann, M.V. and Pet-Soede, L. (1996) How fresh is too fresh? The live reef food fish trade in Eastern Indonesia. *NAGA, ICLARM Quarterly* **19** (1), 4–8.

Reef fisheries, while a minor portion of the reported total catch, are highly valuable and production has increased rapidly since approximately 1980. Landings of Lutjanidae species increased from 18,000 to over 47,000 tonnes between 1980 and 1990 and ornamental fish export increased from 23 tonnes in 1968 to 2300 tonnes in 1991 (Tomascik et al., 1997; MLH, 1996). Groupers (Serranidae; kerapu), snappers (Lutjanidae, kakap), and fusiliers (Caesionidae, ekor kuning) are the three main reef fisheries. The most valuable portion of the catch is the export trade in live reef fish to Hong Kong and other major Asian cities (Fig. 4). The majority of this catch is unreported as fishing methods are often illegal and the live fish is exported directly without landing at an Indonesian port. A significant catch of reef fish is also taken for local markets most often using blast fishing techniques. Unfortunately, the majority of reef fish is taken using destructive fishing practices and large tracts of reef have been either completely denuded of large fish and valuable molluscs or destroyed entirely. The reefs of Pulau Seribu off Jakarta and the Spermonde archipelago off South Sulawesi have been heavily fished by destructive methods and fishing effort has now shifted to more remote areas of reef development in both western and eastern Indonesia.

Overall, the Java Sea pelagic resource base is considered heavily exploited throughout its range and for many stocks throughout much of their life-cycle by a variety of fishing gears. This situation exists partly because a large subsistence and artisanal economy depends upon these resources for its own nutrition and as a source of income (McElroy, 1991). Reef resources throughout western Indonesian seas have been fished to the point of severe stock depletion and habitat destruction (Erdmann and Pet-Soede, 1998). The

Fig. 4. Napoleon Wrasse in a holding pen awaiting transport to Hong Kong. The majority of these fish are caught using sodium cyanide to stun the fish and fetch prices of up to thousands of dollars per fish in Hong Kong restaurants.

FAO (1997) lists shad, ponyfish, bream, scads, sardine, anchovy, yellowfin tuna, skipjack tuna, and banana prawn as likely fully exploited in Southeast Asian seas, including western Indonesia.

Despite these assessments, the currently reported annual catch of approximately 3 million tonnes is less than half the estimated sustainable yield for the entire EEZ of Indonesia and the government expects to increase the annual catch to 6.6 million tonnes by 2018 with most of the increase expected in the waters of eastern Indonesia (MLH, 1996). Marr (1981) has characterized the problems facing fisheries in Indonesian seas as stemming from over-capitalization of certain fisheries, overfishing due to unlimited entry, problems of allocation of fishing rights, a lack of management strategy and theory on which to base it, a lack of enforcement capability, and a lack of management mechanisms at the intra and international levels. Artisanal fishermen and their families comprise one of the most impoverished sections of the Indonesian population and in the many areas where traditional fisheries are already seriously overfished, the only way to increase the yield per fisherman is to reduce the number of fishermen by offering alternative employment. The reduction of waste through improvements in handling, such as more effective use of ice at sea, improved surfaces in fish landing markets, and a greater use of freezing and refrigerated land transport could significantly increase the economic yield of the fisheries. Greater use of quality standards and inspection as well as upgrading the use to which landings are put, i.e. from dried to fresh, could also result in significant improvements. The Indonesian Directorate General of Fisheries has been upgrading port and landing facilities as well as increasing the availability of ice and refrigeration across the north coast of Java to address the problem of fisheries waste due to poor handling practices. These improvements, however, address neither the problem of commercial overfishing and destructive fishing practices which requires, among other measures, an increased capacity for enforcement of fisheries regulations, nor the problems facing subsistence fisheries which require the inclusion and participation of coastal communities in comprehensive coastal zone planning and development.

Aquaculture

Aquacultural development in western Indonesia has expanded explosively since 1980, when Indonesia began promoting intensive shrimp aquaculture, primarily aimed at increasing export earnings (Hannig, 1988). Total aquaculture production in the Southeast Asian region has more than doubled since 1984, and shrimp farms in south and southeast Asia (692,000 tonnes in 1994) account for 75% of total world production of farmed shrimp (Kongkeo, 1997). The environmental consequences of shrimp aquaculture have caused great concern in Thailand, Malaysia, and Indonesia (Chua et al., 1989; Philips et al., 1993). Indonesia produced 80,000 tonnes of farmed shrimp in 1995 (Kongkeo, 1997). Shrimp pond development has primarily proceeded in areas previously covered by mangrove forests (Fig. 5). In Java alone, 128,700 ha (82%) of north coast mangrove forests were cleared for shrimp ponds by 1991 (Tomascik et al., 1997). Shrimp ponds are among the principal causes of mangrove forest destruction along the coasts of Sumatra, Kalimantan, and South Sulawesi.

Intensive chemical use and high stocking rates in shrimp aquaculture shed high nutrient, organic carbon, and bacterial loads into nearshore waters (Philips et al., 1993). Figure 6 presents the nutrient budget for shrimp production using intensive tambak methods. Nitrogen and phosphorus loading from Indonesian shrimp ponds produces the equivalent of sewage from approximately 8 million people (cf. Briggs and Funge-Smith, 1994). Most semi-intensive shrimp ponds in Indonesia are stocked at densities ranging from 10–30 individuals/m^2, such that pond operators are racing to harvest the crop before bacterial and fungal infections destroy it (Tomascik et al., 1997). High biological oxygen demand requires that operators use electrical aerators to mix oxygenated surface waters down to the oxygen limited zones at the bottom of the 1–2 m deep ponds. Use of antibiotics and fungicides is widespread and intensive. Organo-metal compounds, including organo-tin complexes, are used as molluscicides in shrimp ponds (Philips et al., 1993). In some areas with intensive shrimp aquaculture, such as Jepara, 100 km northeast of Semarang, Central Java, bacterial and ciliate concentrations in nearshore waters are so high that operators of shrimp hatcheries can no longer use local seawater, and must purchase seawater trucked in from other sites further up the coast.

Particularly high fluxes of nutrients and sediment from shrimp ponds to coastal waters occur when ponds are drained for harvest. Release of pesticides and antibiotics

Fig. 5. Aerial view of semi-intensive shrimp ponds (*tambak*) west of Semarang, Central Java. Extensive areas of mangrove have been cleared for shrimp farming along the coasts of the Java Sea, especially on Java. Note the line of remnant mangrove marking the previous coastline. White spots are floating aerators.

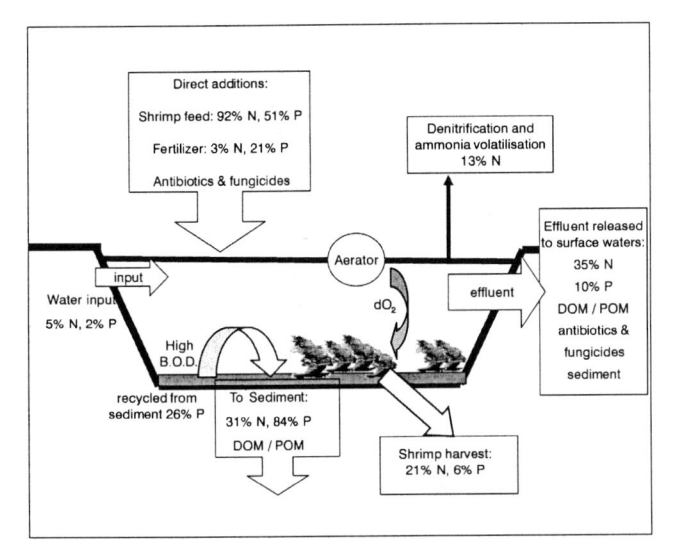

Fig. 6. Nitrogen and phosphorus loading from intensive Indonesian shrimp ponds, following Briggs and Funge-Smith (1994). Intensive shrimp farms in Thailand release to the environment inorganic N and P equivalent to 80 people, and 130 people, respectively, per tonne of shrimp produced (Briggs and Funge-Smith, 1994). Indonesian farmed shrimp production in 1995 was 80,000 tonnes, of which 63% was produced in intensive shrimp ponds, and most of the remainder in semi-intensive ponds (Kongkeo, 1997). Extrapolating these ratios, and assuming that semi-intensive ponds release 50% the N and P loadings of intensive ponds, Indonesian intensive and semi-intensive shrimp farming produces N and P runoff approximately equivalent to 5.2 million people (N) and 8.4 million people (P) annually, or about 5% of the human population in the region.

into coastal waters have raised considerable concern, especially reports of antibiotic resistance in some pathogens of shrimp, including some potential human pathogens. Introductions of exotic shrimp breeds and species have also raised concerns regarding genetic biodiversity conservation (Philips et al., 1993).

Development of shrimp farms in western indonesia is further complicated by acid-generating soils where farms are developed in mangroves or peat swamps (Burbridge et al., 1988). High organic content in mangrove soils and peats produces high sulphur content in the soils, mainly in the form of authigenic pyrite (FeS_2). Aeration of the soils for development of shrimp ponds liberates sulphuric acid, which then leaches Fe and Al from clay minerals in the mangrove soils. The concentration of metals in the ponds may reach toxic levels, reducing productivity of the shrimp ponds (Philips et al., 1993), and poisoning the soil, making rehabilitation difficult or impossible. Problems with acid-generating soils have been particularly severe in South Sumatra, where transmigration projects have promoted immigration of settlers from Java into lowland peat forests, with shrimp farms and tidal rice cultivation as primary economic activities (Burbridge et al., 1988). Similar problems have been experienced in the extensive tidal peat forests of Kalimantan (MacKinnon et al., 1996).

The effects of shrimp ponds on coastal erosion throughout western Indonesia have been dramatic. Serious coastal erosion problems have been reported in all the Indonesian provinces bordering the Sunda shelf, where mangrove forests have been cleared for shrimp ponds (Tomascik et al., 1997). For example, the village of Bulak, south of Jepara, Central Java, has experienced such high rates of coastal erosion and loss of farmland that it has been relocated inland, but remains endangered by erosion, as loss of shoreline has already reached the new village location. Coastline retreat in these areas has been estimated at 500 m since 1946, with higher rates of coastline retreat between 1972 and 1990 than between 1946 and 1972. Mangrove clearing for shrimp ponds in the late 1970s and 1980s made the muddy coastline more easily eroded, and further increased rates of shoreline retreat (Fig. 5). Efforts to stem coastal erosion, focused on mangrove restoration, have had limited success.

LAND-BASED PROCESSES AFFECTING WESTERN INDONESIAN SEAS

Rural Processes

Impacts of rural land use on western Indonesian seas are primarily due to run-off from deforestation, mining, and agriculture. These terrestrial activities have led to large increases in fluxes of sediment, nutrients, and organic carbon from the land to coastal waters, such that eutrophication is a problem in many of the coastal waters of Indonesia.

Deforestation

Deforestation rates in Sumatra and Kalimantan accelerated dramatically from the 1970s onward, and large areas of forest on both islands have been converted to cultivation or palm oil plantations. Loss of forest cover typically leads to lower rates of water infiltration, higher proportions of run-off, and dramatic increases in rates of soil erosion. The main impact of high rates of deforestation in Kalimantan and Sumatra is increased sediment loads to the rivers draining those watersheds and a resulting increase in turbidity of coastal waters near the river mouth (Whidden et al., 1987; MacKinnon et al., 1996; see Table 2).

Mining

Surface and submarine strip mining for placer tin deposits takes place on several islands in Jambi and Riau provinces of Sumatra. Mining is most concentrated near the islands of Bangka and Belitung, Jambi, where total reserves have been estimated at 3.48 million tonnes (Burbridge et al., 1988). Mining for similar tin deposits near Phuket, Thailand, has been implicated in the demise of fringing reefs in that region. The primary concern emanating from undersea mining of placer tin deposits is high turbidity from dredging; potential tin pollution itself poses a lesser threat. Bauxite is mined in Riau province, principally on the island of Bintan (Burbridge et al., 1988). Aluminum contamination from mining activities may influence mangroves and fringing reefs around Bintan island. Mining activities on land are most concentrated in Kalimantan, where extensive placer gold deposits are mined within the rivers. Mercury and cyanide contamination in Kalimantan rivers from these mining activities has had significant impacts on river and estuarine biota (MacKinnon et al., 1996).

Agriculture

Since the mid-1970s, Indonesia's agricultural sector has undergone exponential growth. In Java and southern Sumatra, much of this expansion has been fuelled by more intensive cultivation of traditional agricultural lands using chemical-intensive methods. The tonnage of inorganic N and P fertilizers applied has increased by more than 260% since 1970. In Kalimantan and Sumatra, large areas of rainforest have been cleared for timber operations, and converted into cultivation areas for rice, sugar cane, palm oil, or other commercial crops. In Jambi and South Sumatra, much of this land conversion has been associated with Indonesia's controversial transmigration program, which seeks to alleviate high population densities in Java and Bali by sending colonists to relatively sparsely occupied regions elsewhere in the archipelago (Whidden et al., 1987).

Agricultural runoff is one of the principal origins of non-point source pollution affecting coastal waters of western Indonesia. Chemical fertilizers applied to rice fields and other crops wash into rivers, and ultimately drain into coastal marine habitats. This increased agricultural runoff has contributed to eutrophication of estuaries and coastal waters throughout southeast Asia, including Java, Sumatra, the Malacca Strait, and Kalimantan (Nixon, 1995). Incidence of dinoflagellate blooms and paralytic shellfish poisoning in Indonesia may be related to the increased flux of terrigenous nutrients into coastal waters, as a result of agricultural and aquacultural runoff (MacLean, 1993).

Urban and Industrial Activities

Marine Pollution

Marine pollution in western Indonesia is concentrated around the industrial cities/complexes of Batam, Padang, Medan, Palembang, Balikpapan, Jakarta, Semarang, and Surabaya (Table 1). Knowledge of the extent of the pollution problem in coastal areas is very limited, as studies to date have been scarce and sporadic and have not been undertaken for periods long enough to indicate definite trends. Jakarta Bay is the most studied coastal area in the region suffering from urban and industrial pollution and will be used as an example (see Fig. 7). The pollution trends of the Jakarta area are repeated in urban areas across the coastline of western Indonesia although the problems in Jakarta are significantly larger in scale.

The dumping of untreated liquid waste into rivers is one of the area's most pressing pollution problems. Pollution entering the rivers and coastal seas includes organic matter, heavy metals, and organochlorines. Domestic sewage is the major source of organic pollution. Seawater functions largely as a sewer for the over 100 million inhabitants of the region since treatment facilities either operate far below the necessary capacity or are simply not present. The volume of sewage released to Jakarta rivers and coastal waters is estimated at 202,400 m^3/day with an estimated BOD pollution load of 93 tonnes per day. Overall, approximately 79% of wastewater and 73% of BOD pollution derives from domestic sources (World Bank, 1994). Of the 15% of organic pollution contributed by industry in Jakarta, food processing, textile industries and tanneries are the major contributors. Organic pollution from domestic and industrial sources leads to eutrophication of coastal areas off the mouth of affected rivers and has been attributed to the increased occurrence of red tides and toxic bacteria. Industry is the primary source of heavy metal contamination of the riverine and coastal environment. Tanneries contribute high chromium loads, metal processing releases large amounts of untreated waste high in zinc, iron, and nickel and battery and paper manufacturers contribute significant loads of mercury to the coastal waters.

The total heavy metal loading into Jakarta Bay from the nine major rivers in the city is given in Table 5. Similar problems ecist in Semarang and Surabaya. Heavy metals accumulate in the marine sediments of Jakarta Bay posing a toxic threat to benthic biota. Through processes of biomagnification mercury can reach hazardous levels in biota at higher trophic levels. Fish, shrimp and molluscs of Jakarta Bay have been found to contain mercury levels above the level safe for human consumption (Hutagalung, 1987) and incidences of mercury poisoning (Minimata disease) have

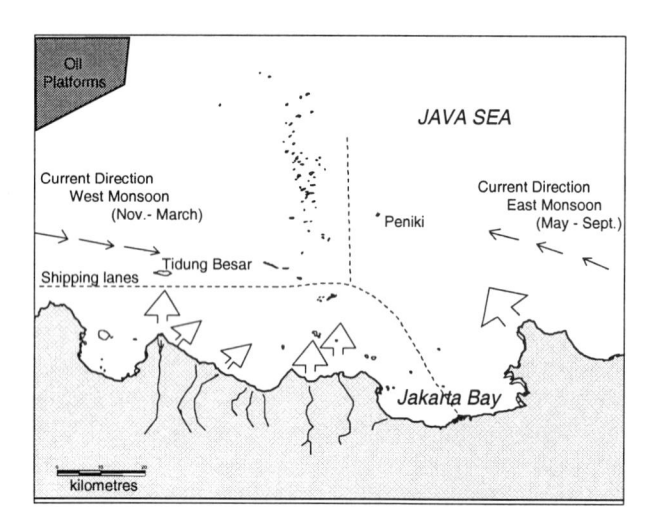

Fig. 7. Map of the Jakarta Bay – Pulau Seribu region, north of Jakarta. Shaded area indicates most highly polluted zone; shaded arrows indicate river runoff. Runoff from coastal rivers carries heavy industrial effluent, sewage, sediment, and garbage from Jakarta to the islands of Jakarta Bay and the Pulau Seribu. All of these contaminants decline exponentially with distance from Jakarta (from Tamascik et al., 1997).

Where Have All the Reefs Gone? The Demise of Jakarta Bay and the Final Call for Pulau Seribu

Anmarie J. Mah and Tomas Tomascik

The coral reefs of Jakarta Bay and Pulau Seribu, located just north of Indonesia's capital city of Jakarta, were once an underwater paradise. The magical beauty of Jakarta Bay reefs inspired a Dutch scientist in 1928 to write the following prose:

"The unrivaled splendor and wealth of forms and the delicate tints of the coral structures, the brilliant colours of fishes, clams, sea anemones, worms, crabs, star fishes and the whole rest of the reef animals are so attractive and interesting that it seems impossible to give an adequate description of such a profusion of serene and fascinating beauty."

Umbgrove (1939) was describing the reefs of Nyamuk Besar, a small "resort island" located a mere 6.6 km from Jakarta. Today, Nyamuk Besar is no longer a resort island. Its reefs and those of Jakarta Bay are now, in large part, functionally dead systems, no longer able to supply the various resources and amenities they provided in the past.

Verwey (1931a) was among the first to conduct environmental studies in Jakarta Bay, demonstrating an inverse relationship between water turbidity and the maximum depth of the living reef. He went on to suggest that reef development in the bay was made possible by the unique combination of sedimentation and hydrologic conditions. Verstappen (1953) further demonstrated that water transparency in the bay fluctuated from 6 to 12 m during the dry season and from 4 to 6 m during the wet season. These studies suggest that the reefs are well adapted to the natural seasonal fluctuations in water turbidity and that land runoff has played a significant, not necessarily detrimental, role in the structuring of these coral reef communities.

Despite this inherent adaptability, however, many reefs in Jakarta Bay have succumbed to the fatal impacts of recent anthropogenic activities. The influence of Java on water quality of Jakarta Bay and Pulau Seribu is reflected in a strong ($r^2 = 0.76$) inverse relationship between water transparency and distance from the mainland. The comparison of Verwey's 1929 water transparency data to present-day conditions (Brown, 1986; Tomascik et al., 1993) demonstrates a dramatic reduction in water transparency from 1929 to 1993. Land runoff from mainland Java, and effluents from Jakarta, have increased to such an extent that river plumes, laden with an assortment of agricultural, urban and industrial pollutants, now reach the outermost reefs of the bay. The negative impact of increased land runoff is reflected in an inverse relationship between water turbidity and percent coral cover and coral diversity. High water turbidity has also decreased coral growth rates and significantly reduced the maximum depth of functional coral communities (Tomascik et al., 1997). Nutrient enrichment of coastal waters is now causing massive phytoplankton blooms that extend up to 12 km from Jakarta, compared to 2 km in 1986 (Harger 1992). Harmful algal blooms (mainly *Noctiluca* sp.) are now a major concern to the coastal aquaculture industry.

In 1929, the reefs of Jakarta Bay supported thriving and diverse coral communities. Ninety-six coral species were recorded from Nyamuk Besar alone (Umbgrove, 1939). Today, only 16 species remain and these are considered very rare. In 1929, Nyamuk Besar reef had a prominent *Montipora digitata* (*ramosa*) and *M. foliosa* facies. Today, none of these remain. The large moat of *Acropora aspera* present in 1928 also disappeared. It is a sad reflection of "modern development" that the once thriving reefs of Jakarta Bay are now functionally dead systems. Clearly, something other than, or in addition to, the natural turbidity must have played a significant role in their demise.

Pollution and increased land runoff are not the only threats facing the reefs. Verwey (1931b) observed that approximately 8500 m^3 of coral was extracted from the shallow reef flats of Jakarta Bay. He further estimated that approximately 20,000 m^3 was being removed annually from the reefs of Pulau Seribu. In 1982 alone, approximately 840,000 m^3 of coral rock was extracted from the reefs (Ongkosongo and Sukarno, 1986). The construction of Jakarta's international airport was facilitated by large quantities of coral rock from the now defunct coral cays of Air Kecil and Ubi Kecil.

The physical destruction of the reefs is mirrored by the reduced abundance and diversity of scleractinians and reef-associated fauna and flora. Reef-associated fishes in Jakarta Bay are very low in abundance. The reefs support an unusually high abundance of benthic filter- and suspension-feeding invertebrates, indicating a major ecosystem-wide trophic shift. The reef matrix itself is now perforated by a great profusion of boring organisms dominated by boring bivalves (*Lithophaga* spp.). Bioerosion is currently the main biophysical process that is contributing to a rapid physical destruction of the reef matrix. Coralline algae, important structural components of reefs, are in low abundance and their functional role of cementing the reef matrix has been greatly reduced. The destruction of reefs in Jakarta Bay, and the continuing degradation of reefs in Pulau Seribu, is also having a drastic impact on reef-associated fisheries (Fig. 1).

Healthy coral reefs provide a wide range of services and amenities to coastal communities world-wide. If human activities on mainland Java, Jakarta and Pulau Seribu are properly managed, coral reefs may recover and will continue to

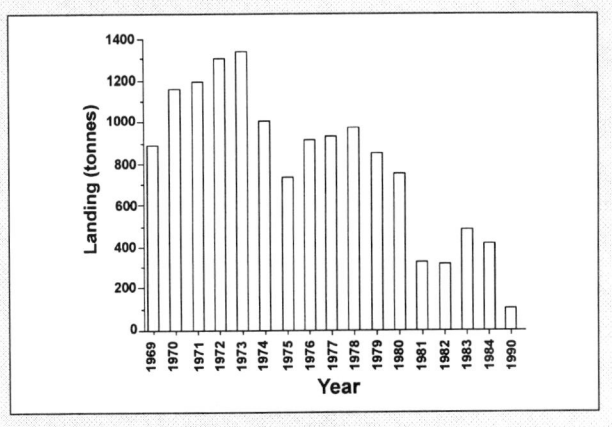

Fig. 1. The recent history of declining coral reef fish landings from the destructive Muro-ami reef fishery in Kepulauan Seribu (total catch per year in tonnes).

support an economically important tourism industry, as well as provide a livelihood for the resident population. There is little that can be done to prevent impacts from natural perturbations such as storms, ENSO events or *Acanthaster* outbreaks. However, if the coral reefs of Pulau Seribu are to continue to provide the wealth of resources and amenities to future generations of islanders, as they do today, an ecosystem-based management program needs to be implemented to address:
- Resource over-exploitation
- Destructive fishing practices
- Coral mining
- Sewage pollution
- Solid waste disposal
- Urban and agricultural runoff
- Dredging
- Industrial pollution
- Aquaculture
- Tourism
- Marine transportation

REFERENCES

Brown, B.E. (ed.) (1986) *Human Induced Damage to Coral Reefs*. UNESCO Reports in Marine Science 40, 180 pp.

Harger, J.R.E. (1992) Environment trends and reef monitoring strategies. Prepared for the 7th Int. Coral Reef Symp., Guam. UNESCO/ROSTSEA, Jakarta, pp. 1–69.

Ongkosongo, O.S.R. and Sukarno (1986) Background to the study sites in the Bay of Jakarta and Kepulauan Seribu. In *Human Induced Damage to Coral Reefs*, ed. B.E. Brown. UNESCO Reports in Marine Science 40, pp. 56–79.

Tomascik, T., Mah, A.J., Nontji, A. and Moosa, M.K. (1997) *The Ecology of the Indonesian Seas. Parts I and II*. The Ecology of Indonesia Series, Vols. VII and VIII. Periplus Editions, Singapore.

Tomascik, T., Suharsono and Mah, A.J. (1993) A historical perspective of the natural and anthropogenic impacts in the Indonesian Archipelago with a focus on the Kepulauan Seribu, Java Sea. In *Proceedings of the Colloquium on Global Aspects of Coral Reefs: Health, Hazards and History, 1993*, compiled by R.N. Ginsburg, pp. 304–310. Rosenstiel School of Marine and Atmospheric Science, University of Miami.

Umbgrove, J.H.F. (1939) Madreporaria from the Bay of Batavia. *Zoölogische Mededeelingen* **XXII**, 1–64.

Verstappen, H.T. (1953) Djakarta Bay, a geomorphological study on shoreline development. Ph.D. Thesis. Utrecht, 101 pp.

Verwey, J. (1931a). Coral reef studies. II. The depth of coral reefs in relation to their oxygen consumption and the penetration of light in the water. *Treubia* **13** (2), 169–198.

Verwey, J. (1931b) Coral reef studies. III. Geomorphological notes on the coral reefs of Batavia Bay. *Treubia* **13** (2), 199–216.

been reported from Jakarta. Organochlorines have received much less study than heavy metals due to the high cost of analyses and the lack of laboratory facilities. In Jakarta Bay, PCB concentrations reach 9 μg/l while DDT reaches 13 μg/l exceeding the maximum allowable limit of 0.5 μg/l (Chua et al., 1989). Jakarta Bay pollution, coupled with intensive fishing, has essentially eliminated the inshore demersal fishery of the Bay and has degraded or destroyed much of the coastal associated mangrove, seagrass and coral reef ecosystems in the vicinity (MLH, 1996).

Marine pollution of the Java Sea resulting from urban and industrial expansion is expected to increase over the next 20 years. The off-Java share of industrial production is expected to increase from 25% in 1990 to about 35% by 2020; thus the threat of industrial pollution is expected to fan out from Java's present urban centres to those of the relatively undisturbed outer islands (World Bank, 1994).

Responses to the threat posed by urban and industrial marine pollution must begin with the development of an inventory of pollution sources, threatened habitats, and particular contaminants leading to immediate environmental degradation or threats to public health. An action plan with effective monitoring, enforcement and financial support can then be developed (MLH, 1996). The general lack of monitoring, baseline data, and capacity to analyze samples for a wide variety of pollutants makes control of urban pollution difficult. Effective enforcement of existing regulations would result in major improvements in water quality. The greater challenge is to find effective ways of reducing massive domestic waste inputs to coastal waters.

Land Reclamation, Coastal Subsidence and Flooding

Many of the major coastal cities in western Indonesia have built harbour facilities on reclaimed land. In Semarang, for example, the city is currently creating approximately 20 ha of new harbour land by filling in nearshore areas of the old port facility (Fig. 8). In many cases, the areas surrounding the new reclamation projects experience flooding particularly at high tides due to resulting changes in drainage and tidal amplification in narrow channels. For example, the low-income coastal districts of Jakarta and Semarang are flooded approximately twice monthly on the high spring tides (Fig. 9). The seawater flooding these neighbourhoods is contaminated with sewage and industrial waste, contributing to skin diseases and other ailments among the coastal populations. Uncontrolled extraction of groundwater from deep wells also contributes to ground subsidence and flooding.

Marine Debris

Marine debris in Indonesia comes from both land-based urban sources and from shipping and fishing vessels (Fig.

Table 5

Total heavy metal and organic (in BOD equivalents) loading into Jakarta Bay (kg/h). Adapted from MLH (1996) and World Bank (1994).

Pollutant	As	Cr	Cu	Hg	Ni	Pb	Zn	BOD
Loading	46	325	191	372	154	757	42,983	13,984

Fig. 8. Aerial view of reclaimed land in Semarang Harbour. Major sections of the coastline have been infilled to an elevation of 1–2 m above adjacent land to create a large container port facility. Such reclamation is occurring in most cities along the north coast of Java. This has exacerbated urban flooding and displaced thousands of coastal residents.

Fig. 9. Residential streets in the low-income coastal neighbourhoods of Semarang are flooded with sewage-contaminated seawater twice monthly at the high spring tides. Ground subsidence is exacerbated by groundwater over-extraction, resulting in coastal flooding and salt-water intrusion of shallow wells.

Fig. 10. Urban solid waste accumulated at the mouth of a drainage canal along the north coast of Central Java. Plastic forms a significant portion of the catch of inshore fishermen in urban areas.

10). Approximately 30–40% of solid waste goes uncollected or inappropriately disposed in urban centres and it is estimated that 860,000 tonnes of garbage is left uncollected each year in Jakarta (MLH, 1996). Accumulation of marine debris has been best studied in the Pulau Seribu area north of Jakarta (see Fig. 7). Here, debris abundance on beaches declines exponentially with distance from Jakarta, and is dominated by land-derived plastic garbage (Uneputty and Evans, 1997). The amount of litter per beach area doubled between 1985 and 1995 (Willoughby et al., 1997). Plastic bags are very abundant in the litter flux, but frequently sink, such that Jakarta Bay "is carpeted in plastic bags". In the northern parts of the island chain, lost or abandoned fishing gear dominates the composition of marine debris. Marine debris has received little attention elsewhere in western Indonesia, but qualitative observations suggest that floating garbage is a widespread problem throughout the region (Willoughby et al., 1997).

PROTECTIVE MEASURES

Integrated Coastal Zone Management

Instituting a program of integrated coastal zone management across the Indonesian archipelago has been identified as an important step in reducing human impacts on the marine environment. Several projects have recently been initiated to improve the resource management of Indonesia's coastal habitats in the new Millennium. Of particular note are the USAID-sponsored Coastal Resources Management Project (CRMP), the CIDA-sponsored Collaborative Environmental Project in Indonesia (CEPI) and the World Bank/Asian Development Bank initiated Coral Reef Rehabilitation and Management Project (COREMAP). The CRMP project focuses on small-scale community-based management of coastal resources. Resource management projects are focused on the province of Lampung, Sumatra and North Sulawesi. It is hoped that successful examples of locally controlled resource management practices will be repeated throughout the archipelago. The CEPI project, on the other hand, focuses on improving the capacity of the environmental management sector of the Indonesian government to carry out environmental impact assessment, monitoring and coastal zone planning. Together, the demonstration projects resulting from these two initiatives will help define the future of coastal resource management in Indonesia and hopefully lead to more sustainable use of coastal habitats.

Recognizing the importance of conserving its coral reef resources, Indonesia has sought international funding for COREMAP (Coral Reef Rehabilitation and Management Project). This project directs assessment and monitoring of reef conditions throughout Indonesia (Suharsono, 1998), and plans active interventions in 10 priority regions of the country including West Sumatra and Riau. Reef rehabilitation measures planned under COREMAP include construction of artificial reefs from concrete on sand flats, and transplantation of branching corals into areas damaged by blast fishing and other mechanical agents. The project will also attempt to instate a more sustainable system of coral reef resource use through both community-based resource management and national/regional improvements in planning and development practices. COREMAP recognizes the major threat to reefs from land-based sources of pollution, but addressing this threat is not part of the COREMAP mandate.

The survey methods used in Indonesia's national coral reef monitoring program have been questioned, in particular the analysis of non-taxonomic life-form benthic surveys. Data on reef status are reported as a linear scale based exclusively on live coral cover, ignoring intermediate disturbance effects on coral diversity and habitat complexity (Tomascik et al., 1997). Furthermore, the benthic life-form surveys do not distinguish between different sources of reef degradation. Reefs may be more effectively classified from life-form survey data using triangular diagrams based on coral morphology (Edinger and Risk, 1999), combined with mortality indices. Sources of degradation can be identified using bioindicators or simple geochemical analyses. COREMAP will address these and related shortcomings of past coral reef management practice. The project has the potential to be the single largest reef management initiative in the world. The health of Indonesia's vast coral reef resources will partially depend on the success or failure of COREMAP in bringing about effective conservation of coral reefs across the archipelago.

PROKASIH Program

The threat to marine resources from land-based pollution is particularly acute in western Indonesia, where population, industry, and shipping are much more concentrated than in eastern Indonesia. The national marine park in Pulau Seribu suffers not only from poorly regulated and illegal activities within the park, but from large scale land-based pollution from Jakarta (Tomascik et al., 1993). The degree of pollution damage to the reefs of Pulau Seribu has prompted the Worldwide Fund for Nature's Indonesia Program to focus its conservation efforts elsewhere, considering the Pulau Seribu too severely damaged for conservation and rehabilitation.

To address these problems, the government has pursued an incentives-based program to reduce industrial water pollution, known as PROKASIH (Clean Rivers Program). PROKASIH targeted key industries responsible for the greatest loads of organics and heavy metals to the nation's rivers, and focused on the 18 most polluted rivers in the country, 12 of which drain into the Java Sea (Jardine, 1993). Because the program is voluntary and incentives-based, the effluent standards are targets for industries to meet, rather than obligatory compliance measures. Initial results at improving water quality by reducing industrial pollution have reduced biological oxygen demand in many rivers, but heavy metal fluxes from many industries remain high (Jardine, 1993). The Clean Rivers program has made some progress in reducing pollutant loads from industrial sources, but has largely ignored domestic wastes which contribute up to 80% of the BOD load to some rivers. The lack of adequate sewage treatment anywhere in the country remains a major pollution problem.

Endangered Species Protection

Indonesia is a signatory to several UN conventions relating to marine resources and endangered species protection, including the Convention on International Trade in Endangered Species (CITES), the Convention on the Protection of Biodiversity (COPD), and the UN Law of the Sea. Indonesia has enacted legislation consistent with these international conventions, including marine protected areas, and penalties for use of explosives or poisons in fishing. How a new (1998) law legalizing and regulating harvest of wild corals for the aquarium trade conforms to CITES is unclear. Enforcement of these laws, however, is generally lacking, and many abuses continue. Endangered species have been largely fished out of some areas, and harvesting of food fish, ornamental fish, and corals by illegal methods is common.

Of six species of sea turtles known to occur in Indonesian waters, only the green turtle (*Chelonia mydas*) and the hawksbill turtle (*Eretmochelys imbricata*) occur on the Sunda Shelf. These species are best represented on sparsely populated offshore islands in the Natuna, Riau, Karimunjawa, Bawean, Kangean, and small islands off the coast of Borneo. Human predation and egg collection remain the greatest threats to marine turtles (Sloan et al., 1994). Green turtles have been extirpated around Bali, where they are hunted for ceremonial feasts (Tomascik et al., 1997). Although all sea turtles are now protected by Indonesian law, it is not

Fig. 11. Sea turtles, corals, and other CITES-protected organisms for sale at a curio booth, Jepara, Central Java.

uncommon to see stuffed hawksbill turtles for sale in coastal regions (Fig. 11), and continuing illegal trade in turtles remains a major obstacle to sea turtle conservation efforts.

Dugongs (*Dugong dugon*) were once common on the Sunda shelf, but severe hunting pressure has reduced their populations to small relicts in the Sunda Straits, Bangka island (Sumatra), and off the southern coast of Kalimantan. Cetaceans are present in western Indonesia, but in lower diversity than in the deeper water environments of the east. Porpoises, dolphins and beaked whales occur in the Java Sea and Malacca Straits (Tomascik et al., 1997). Black and southern right whales and the sei whale have been reported from the Java Sea. Although artisanal hunting of some whales continues in eastern Indonesia, commercial whaling does not pose a major threat to cetaceans in Indonesia (Tomascik et al., 1997).

Conservation efforts for these species include public education about the protected status of these species, establishment of protected sea turtle nesting beaches, and establishment of giant clam (*Tridacna*) hatcheries to promote cultivation and restocking of giant clam populations. Of these efforts the giant clam hatchery on Barang Lompo, South Sulawesi, run by Hassanuddin University, has been particularly successful. This hatchery donates juvenile *Tridacna* for conservation efforts in marine parks and protected areas throughout Indonesia, and sells juveniles abroad and for domestic giant clam farming enterprises and the aquarium trade. Despite these conservation measures, illegal harvest of wild giant clams continues, and *Tridacna* remain severely threatened.

Marine Protected Areas

Indonesian marine protected areas (MPAs) are selected according to eight criteria: diversity, naturalness, representativeness, uniqueness, rareness, size, accessibility, and effectiveness. MPAs within Indonesia are divided into four types: strict nature reserves, wildlife reserves, national parks, and recreation parks (Alder, 1995). Strict nature reserves and wildlife reserves set aside areas for total protection, with access limited to research. National marine parks are areas with outstanding natural value, and are zoned to allow some development. Recreation parks, areas with high recreation value, may be developed in ways that do not permanently threaten the natural resource. The marine conservation areas and terrestrial parks with marine components are presented in Table 6 along with the type of reserve, the total area, major habitats, and the year the park was created.

Indonesia set a marine conservation target of 10 million hectares of marine protected areas by 1990, and 30 million hectares for the year 2000. Existing protected areas total around 2.7 m ha, and the majority of this area is in eastern Indonesia. Established MPAs in the Sunda Shelf region total approximately 1.1 m ha. (Table 6). Relatively few MPAs have been established in the last decade, and many more MPAs are proposed than are established (BAKOSURTANAL, 1998).

Many coastal populations in Indonesia are dependent on exploitation of marine resources for their livelihoods, and are often among the poorest portions of Indonesian society. Therefore, most marine protected areas are managed through a zoning system intended to allow local residents to maintain traditional resource exploitation in some areas of the park. Zoning schemes for national marine parks include four zones: core sanctuary zone, wilderness zone, development zone, and buffer zone. In core sanctuary zones, access is restricted to research only, while in wilderness zones, non-destructive tourism is permitted, but construction is not permitted. In development zones, construction of facilities for tourism, and traditional resource uses such as fishing are permitted. The buffer zone allows most traditional forms of resource use, such as fishing and aquaculture (Alder, 1995). This general zoning scheme recognizes the high economic value of Indonesia's coral reefs (Cesar et al., 1997), and strives to combine compatible multiple uses within marine protected areas. Legislation requires that local residents be consulted in design of management plans for marine protected areas, although the effectiveness of this consultation is sometimes questionable (Yates, 1994).

Implementation remains a major difficulty for marine conservation throughout southeast Asia, and in Indonesia in particular (Yates, 1994; Wilkinson and Chou, 1997). Of the various national marine parks established in Indonesia, only three have management plans, and none of these has been approved or implemented (Alder, 1995; Yates, 1994; Tomascik et al., 1997). Thus, many of the declared MPAs in Indonesia are considered "paper parks" (Wilkinson and Chou, 1997). For example, coral reefs in Karimunjawa national marine park are threatened by unregulated construction for foreign-owned resorts, construction of shrimp ponds in coastal mangroves, and frequent fishing within sanctuary and wilderness zones. Unclear and overlapping jurisdictions among various government agencies responsible for various activities in the coastal zone, and between federal, provincial, and regional governments, have hindered efforts at effective management of marine protected areas (Alder, 1995).

Furthermore, even in those marine protected areas which do have management plans, local populations have often been largely excluded from economic benefits from tourism, making them more resistant to park restrictions on fishing and other extractive activities. Inadequate staffing and funding for equipment hinders efforts at enforcement and public education, such that various illegal and destructive activities often continue within the boundaries of marine protected areas (Sloan and Sugandhy, 1994). For example, illegal poison fishing for aquarium fish and the live food fish trade both continue in several national marine parks, and very few people, islanders or external fishers, have been apprehended and charged for these activities.

Table 6

Established marine protected areas and terrestrial national parks containing marine zones in the Sunda shelf region, western Indonesia. Protected areas in eastern Indonesia are not listed (see Dhahuri, this volume). Marine protected areas on the west coast of Sumatra are listed at the end of the table. There are an additional 13 small parks established along the south coast of Java. *Taka Bone Rate, in the Flores Sea, is administered by South Sulawesi, and is the easternmost protected area discussed in this chapter.

No.	MPA Name	Type	Province	Area(ha)	Primary marine and coastal habitats	Year created
1	Pulau Weh	Recreation Park	Aceh	2600	Coral reef	1982
2	Karang Gading	Nature reserve	North Sumatra	15,765	mangrove	1980
3	P. Berkeh	Nature reserve	Riau	500	mangrove	1968
4	P. Burung	Nature reserve	Riau	200	mangrove	1968
5	P. Laut	Nature reserve	Riau	400	Coral reef, mangrove	1968
6	Kelompok Hutan	Nature reserve	Jambi	6,500	Mangrove	1981
7	Berbak	Nature reserve	Jambi	170,000	Mangrove	1935
8	P. Krakatau	Nature Reserve	Lampung (Sunda Strait)	11,200	Coral reef	1990
9	Way Kambas	Nature reserve	Lampung (Sunda Strait)	130,000	mangrove	1937
10	Kep. Seribu	Nat'l. Marine Park	DKI Jakarta	108,000	Coral reef	1982
11	P. Bokor	Nature reserve	DKI Jakarta	15	Beach	1921
12	P. Rambut	Nature reserve	DKI Jakarta	18	mangrove	1939
13	Muara Angke	Nature reserve	DKI Jakarta	25	Estuary, mangroves	1978
14	P. Dua	Nature Reserve	West Java	30	Mangrove, bird habitat	1974
15	Ujung Kulon	National Park	West Java (Sunda Strait)	78,359	Mangroves, coral reefs	1980
16	P. Sangiang	Recreation Park	West Java (Sunda Strait)	70,035	Mangroves, coral reefs	1985
17	P. Panaitan	Nature reserve	West Java (Sunda Strait)	17,500	Coral reefs	1937
18	Karimunjawa	National Marine Park	Central Java	111,625	Coral reef, mangroves, seagrasses	1986
19	Bawean	Wildlife reserve	Central Java	3,832	Coral reef	1979
20	P. Noko, P. Nusa	Nature reserve	East Java	15	Sea cliffs, seabird nesting	1926
21	P. Saubi	Nature reserve	East Java	430	Beach, turtle nesting	1926
22	Baluran	Wildlife reserve	East Java	25,000	Beach, mangrove	1962
23	P. Menjangan	Wildlife reserve	Bali	n/a	Coral, mangrove	1978
24	Bali Barat	National Park	Bali	77,727	Mangrove, reef	1947
25	Gunung Palung	Nature reserve	West Kalimantan	30,000	Mangrove	1936
26	Kep. Karimata	Nature reserve	West Kalimantan	77,000	Coral reef, seagrass beds	1985
27	Tj. Keluang	Nature reserve	Central Kalimantan	2000	Beach forest	1984
28	P. Kembang	Recreation Park	South Kalimantan	n/a	Beach, mangrove	1976
29	P. Kaget	Nature reserve	South Kalimantan	n/a	mangrove	1976
30	Pleihari	Wildlife reserve	South Kalimantan	35,000	Beach, turtle nesting	1975
31	Tl. Kelumpang	Nature reserve	South Kalimantan	13,500	Mangrove, estuary	1981
32	P. Semama	Wildlife reserve	East Kalimantan	220	Beach	1982
33	P. Sangalaki	Wildlife reserve	East Kalimantan	280	beach	1982
34	Kutai	National Park	East Kalimantan	200,000	mangrove	1982
35	Tl. Apar & Adang	Nature reserve	East Kalimantan	46,900	mangrove	1971
36	Gili Meno	Marine recreation park	Lombok (NTB)	2,954	Beach, coral reef	
37	P. Moyo	Wildlife reserve	Sumbawa (NTB)	22,500	Beach, coral reef	1986
38	Tujuhbelas Pulau	Nature reserve	Flores (NTT)	11,900	Coral reef	1987
39	Komodo	National Park	Komodo (NTT)	75,000	Coral reef, mangrove, Komodo dragon	1965
40	Taka Bone Rate	Nat'l. Marine Park	South Sulawesi*	530,765	Coral reef, seagrasses	
41	P. Bangkaru	Wildlife reserve	Aceh	400	Sea turtle nesting	
42	Kluet	Wildlife reserve	Aceh	23,425	Sea turtle nesting	1936
43	Taitabatti	Wildlife reserve	North Sumatra	56,500	Sea cliff	1976
44	Nanuua	Hunting park	Bengkulu	10,000		1978
45	Bukit Barisan Selatan	Nature reserve	Lampung	21,600	Coral reef	1982
	Total	All types	All western Indonesia	1,989,720		

The live food fish trade, in particular, is very lucrative, such that there are strong incentives for local fishers to engage in illegal destructive fishing activities.

Summary and Prospects for the Future

Western Indonesia is a tremendously important region for coastal marine habitats, but these environments are seriously threatened by human activities, both on land and at sea. The recent economic and political turmoil in Indonesia promises further threats to marine conservation, as the government emphasizes export earnings over conservation. The recent economic crisis has severely impacted the commercial and industrial economic sectors and has prompted a surge in natural resource exploitation, including the live food fish, lobster, sea-cucumber, and tuna fisheries. There are disturbing accounts of villages carrying out a 'clean sweep' of grouper or lobster stocks. The road to economic recovery for Indonesia is expected to be a long slow climb, and poses a serious threat to marine resource conservation.

Indonesia has been in a state of reform since former President Suharto resigned in May 1998. This climate of reform has encouraged more awareness, consciousness and openness and has created more opportunities to critique various policies and laws. Formal and informal discussions are concerned with decentralization and greater autonomy for local governments at the provincial and sub-district levels, especially in managing their own natural resources. For example, new legislation under consideration would give regional governments authority to manage their own marine and coastal territories. Viewed positively, this bill has the potential to clarify the mandate for various sectors of coastal resource management, empower local communities, restore traditional management systems and give coastal communities a greater sense of stewardship of their natural resources. On the other hand, this legislation may effectively decentralize the corruption and lack of enforcement to the lower echelons of regional government, and encourage even more rapid, unsustainable marine resource extraction in the rush to maximize profits before natural resources are completely exhausted. Despite many sincere efforts by government agencies, conservation NGO's, and researchers, the status and prospects of western Indonesian seas at the Millennium appear bleak indeed.

ACKNOWLEDGEMENTS

The authors thank M. Erdmann, G. Llewellyn, A. Mah, J. Rendell, I. Soeseno, and T. Tomascik for supplying data and/or critiquing early drafts of this chapter. Field work in Indonesia during 1994–1999 was supported by the Universitas Diponegoro–McMaster University Coastal Ecodevelopment Project, CIDA University Partnerships in Cooperation and Development agreement # 098/S47074-(099).

REFERENCES

Alder, J. (1995) Marine park management in Indonesia. Environmental Management and Development in Indonesia (EMDI) Report #47, Dalhousie University, Halifax, Canada, 239 pp.

Bailey, C., Dwiponggo, A. and Marahudin, F. (1987) *Indonesian Marine Capture Fisheries*. ICLARM, Manila, 196 pp.

BAKOSURTANAL (National Coordination Agency for Surveys and Mapping) (1998) *Indonesian Marine Resource Atlas*. Bogor, West Java.

Best, M.B., Hoeksema, B.W., Moka, W., Moll, H. and Sutarna, N. (1989) Recent Scleractinian coral species collected during the Snellius-II Expedition in eastern Indonesia. *Netherlands Journal of Sea Research* 23 (2), 107–115.

Briggs, M.R.P. and Funge-Smith, S.J. (1994) A nutrient budget of some intensive marine shrimp ponds in Thailand. *Aquaculture and Fisheries Management* 25, 789–811.

Brown, B.E. and Suharsono (1990) Damage and recovery of coral reefs affected by El Nino related seawater warming in the Thousand Islands, Indonesia. *Coral Reefs* 8, 163–170.

Burbridge, P., Koesoebiono and Dahuri, R. (1988) Problems and issues in coastal resources management and planning in eastern Sumatra and the Strait of Malacca. In *Coastal Zone Management in the Strait of Malacca*, eds. P. Burbridge, Koesoebiono, H. Dirschl and B. Patton. School for Resource and Environmental Studies, Dalhousie University, Halifax, Canada, pp. 8–117.

Cesar, H., Lundin, C.G., Bettencourt, S. and Dixon, J. (1997) Indonesian coral reefs—an economic analysis of a precious but threatened resource. *Ambio* 26, 345–350.

Chua, T.E., Paw, J.N. and Guarin, F.Y. (1989) The environmental impact of aquaculture and the effects of pollution on coastal aquaculture development in Southeast Asia. *Marine Pollution Bulletin* 20(7), 335–343.

Edinger, E.N., Jompa, J., Limmon, G.V., Widjatmoko, W. and Risk, M.J. (1998) Reef degradation and coral biodiversity in Indonesia: effects of land-based pollution, destructive fishing practices, and changes over time. *Marine Pollution Bulletin* 37, 617–630.

Edinger, E.N. and Risk, M.J. (1999) Reef classification by coral morphology predicts coral reef conservation value. *Biological Conservation* 92, 1–13.

Erdmann, M.V. and Pet-Soude, L. (1998) B^6 and M^3 = DFP; an overview of destructive fishing pracices in Indonesia. In *Proceedings of the Workshop on the Impacts of Destructive Fishing Practices on the Marine Environment*. Hong Kong, pp. 25–34.

FAO (1997). *Review of the State of the World Fishery Resources: Marine Fisheries*. FAO Fisheries Circular No. 920. Rome.

Fortes, M.D. (1998) Mangrove and seagrass beds of east Asia: habitats under stress. *Ambio* 17, 207–213.

Hannig, W. (1988) *Towards a Blue Revolution: Socioeconomic Aspects of Brackishwater Pond Cultivation in Java*. Gadjah Mada University Press, Yogyakarta, Indonesia. 404 pp.

Hoekstra, P. (1993) Late Holocene development of a tide-induced elongate delta, the Solo delta, East Java. *Sedimentary Geology* 83, 211–233.

Hoeksema, B.W. (1997) Generic diversity of Scleractinia in Indonesia. Box 7.2 in *The Ecology of Indonesian Seas*, eds. T. Tomascik, A.J. Mah, A. Nontji and M.K. Moosa. Periplus Editions, Singapore, pp. 308–310.

Hutagalung, H.P. (1987) Mercury content in the water and marine organisms in Angke Estuary, Jakarta Bay. *Bulletin of Environmental Contamination and Toxicology* 39, 406–411.

Jardine, C.G. (1993) PROKASIH: Strategy for Water Quality Monitoring and Assessment. Environmental Management and Development in Indonesia (EMDI) Report #43, Dalhousie University, Halifax, Canada, 156 pp.

Kongkeo, H. (1997) Comparison of intensive shrimp farming systems in Indonesia, Philippines, Taiwan and Thailand. *Aquaculture Research* **28**, 789–796.

MacKinnon, K., Hatta, G., Halim, H. and Mangalik, A. (1996) *The Ecology of Kalimantan*. Periplus Editions, Singapore, 802 pp.

MacLean, J.L. (1993) Developing country aquaculture and harmful algal blooms. In: *Environment and Aquaculture in Developing Countries*, eds. R.S.V. Pullin, H. Rosenthal and J.L. MacLean. ICLARM, Manila, pp. 252–283.

Marr, J.C. (1981). Southeast Asian marine fishery resources and fisheries. In: *Southeast Asian Seas: Frontiers for Development*, eds. C.L. Sien and C. MacAndrews. McGraw-Hill International Book Company, Singapore.

McElroy, J.K. (1991) The Java Sea purse seine fishery: A modern-day 'tragedy of the commons'? *Marine Policy* July, 255–217.

Milliman, J.D. and Meade, R.H. (1983). World-wide delivery of river sediment to the oceans. *Journal of Geology* **91**, 1–21.

MLH (Menteri Negara Lingkungan Hidup, Ministry of State for Environment) (1996). Indonesia's Marine Environment: A Summary of Policies, Strategies, Actions and Issues. Division of Policy Development for Environmental Management, Ministry of State for Environment, Jakarta, 86 pp.

Nixon, S.W. (1995) Coastal marine eutrophication: a definition, social causes, and future concerns. *Ophelia* **41**, 199–219.

Paulay, G. (1997) Diversity and distribution of reef organisms. In: *Life and Death of Coral Reefs*, ed. C.E. Birkeland. Chapman and Hall, New York, pp. 298–353.

Pet, J.S., van Densen, W.L.T., Machiels, M.A.M., Sukkel, M., Setyohadi, D. and Tumuljadi, A. (1997). Catch, effort and sampling strategies in the highly variable sardine fisheries around East Java, Indonesia. *Fisheries Research* **31**, 121–137.

Philips, M.J., Lin, Kwei Lin, C. and Beveridge, M.C.M. (1993) Shrimp culture and the environment: lessons from the world's most rapidly expanding warmwater aquaculture sector. In: *Environment and Aquaculture in Developing Countries*, eds. R.S.V. Pullin, H. Rosenthal and J.L. MacLean. ICLARM, Manila, pp. 171–197.

Polunin, N.C.V. (1983) The marine resources of Indonesia. *Oceanography and Marine Biology Annual Review* **21**, 455–531.

Randall, J.E. (1998) Zoogeography of shore fishes of the Indo-Pacific region. *Zoological Studies*, Bishop Museum, Honolulu, Hawaii, USA, **37**, 227–268.

Rice, R.C. (1991) Environmental degradation, pollution, and the exploitation of Indonesia's fishery resources. In *Indonesia: Resources, Ecology, and Environment*, ed. J. Hardjono. Oxford University Press, Singapore.

Scott, P.J.B. and Davies, M. (1997) Retroactive determination of industrial contaminants in tropical marine communities. *Marine Pollution Bulletin* **34**, 975–980.

Sloan, N.A. (1993a) Science and Management Review of Tropical Seagrass Ecosystems in Support of Integrated Coastal Zone Management in Indonesia. Environmental Management and Development in Indonesia (EMDI) Report #39, Dalhousie University, Halifax, Canada, 25 pp.

Sloan, N.A. (1993b) Effects of Oil on Marine Resources: A Worldwide Literature Review Relevant to Indonesia. Environmental Management and Development in Indonesia (EMDI) Report #32, Dalhousie University, Halifax, Canada, 65 pp.

Sloan, N.A. and Sughandy, A. (1994) An overview of Indonesian coastal environmental management. *Coastal Management* **22**, 215–233.

Sloan, N.A., Wicaksono, A., Tomascik, T. and Uktolseya, H. (1994) Pangumbahan sea turtle rookery, Java, Indonesia: Toward protection in a complex regulatory regime. *Coastal Management* **22**, 251–264.

Suharsono (1998) Condition of coral reef resources in Indonesia. *Pesisir dan Lautan (Indonesian Journal of Coastal and Marine Resources, Bogor Agricultural Institute)* **1**, 44–52.

Tjia, H.D. (1980) The Sunda Shelf, Southeast Asia. *Zeitschrift Geomorphologie* **24**, 405–427.

Tomascik, T., Suharsono and Mah, A.J. (1993) Case histories: a historical perspective of the natural and anthropogenic impacts in the Indonesian archipelago with a focus on the Kepuluauan Seribu, Java Sea. In: *Global Aspects of Coral Reefs: Health, Hazards and History*, ed. R.N. Ginsburg, pp. J26–31, University of Miami.

Tomascik, T., Mah, A.J., Nontji, A. and Moosa, M.K. (1997) *The Ecology of the Indonesian Seas Parts 1 and 2*. Periplus Editions, Singapore, 1388 pp.

Uneputty, P.A. and Evans, S.M. (1997) Accumulation of beach litter on Islands of the Pulau Seribu Archipelago, Indonesia. *Marine Pollution Bulletin* **34**, 652–655.

Wallace, C.C. (1997) Separate ocean basin origins as the explanation for high coral species diversity in the central Indo-Pacific. *Proceedings of the 8th International Coral Reef Symposium, Panama* Vol. 1. Smithsonian Tropical Research Institute, Balboa, Panama, pp. 365–370.

Watson, R.A. (1995) Marine biodiversity management. Environmental Management and Development in Indonesia (EMDI) Report #44, Dalhousie University, Halifax, Canada, 127 pp.

Whidden, A.J., Damadik, S.J., Anwar, J., Hisyam, N., (1987) *The Ecology of Sumatra*. Gadjah Mada University Press, Yogyakarta, Indonesia, 583 pp.

Wilkinson, C.R. and Chou, L.M. (1997) The role of science in the establishment and management of marine protected areas in Southeast Asia. *Proceedings of the 8th International Coral Reef Symposium, Panama*, Vol. 2: 1949–1954. Smithsonian Tropical Research Institute, Balboa, Panama.

Willoughby, N.G., Sangkoyo, H. and Lakaseru, B.O. (1997) Beach litter: an increasing and changing problem for Indonesia. *Marine Pollution Bulletin* **34**, 469–478.

World Bank. (1994) Indonesia: Environment and Development. Washington, D.C.

Wyrtki, K. (1961) Physical Oceanography of Southeast Asian Waters. NAGA Report, No. 2, La Jolla, California 195 pp.

Yates, B.F. (1994) Implementing coastal zone management policy: Kepuluan Seribu Marine Park, Indonesia. *Coastal Management* **22**, 235–249.

THE AUTHORS

Evan Edinger
Department of Geology, St. Francis Xavier University, P.O. Box 5000, Antigonish, Nova Scotia B2G 2W5, Canada
Current address: Department of Earth Sciences, Laurentian University, Ramsey Lake Road, Sudbury, Ontario, P3E 2C6, Canada

David R. Browne
Department of Biology,
McGill University, 1205 Docteur Penfield Avenue, Montreal, PQ, H3A 1B1 Canada

Chapter 79

THE PHILIPPINES

Gil S. Jacinto, Porfirio M. Aliño, Cesar L. Villanoy,
Liana Talaue-McManus and Edgardo D. Gomez

The 7150 islands of the Philippines form one of the largest island groups in the world, which lie near the centre of marine biodiversity. The region experiences distinct seasons which are predominantly associated with the Asian Monsoon System, and an average of 19 typhoons pass through every year. The archipelago is also in a geologically active and volcanic zone.

Several of its marine habitats are the most extensively surveyed in the South East Asian region; fringing reefs are among the most prominent features of the coastal areas, especially in the Palawan and Visayas groups and Sulu archipelago. Mangrove forests are also important, but of an estimated 500,000 ha of mangrove cover in 1918, there are now only about 139,725 ha, and around a third of the country's seagrass areas is also estimated to have been damaged. Seaweed resources have a good economic potential.

Offshore surface waters are extremely poor in nutrients, though important upwelling occurs, especially about 100 km off northwest Luzon. The Philippines is an important fish producer, producing 2.766 million tonnes in 1997. It is a populous country of 68 million people, and its growth rate of 2.3% will, if unabated, cause the population to double by 2025. Utilization of upland areas has been very extensive; about 14 million ha are classified as severely or moderately eroded. Pesticide sales grew tremendously until integrated pest management was introduced. Forest cover has been reduced from more than 50% in 1948 to less than 20% in 1996, and siltation is the most important factor influencing conditions in coral reefs, mostly in reef areas near rivers. Many reefs are also overfished, resulting in changes to fish populations.

However, the Philippines is perhaps one of the most advanced countries in the East Asian region with respect to its appreciation and practice of coastal management. Numerous initiatives on the management of the country's coastal areas and resources have been undertaken since the 1980s. There have been at least 45 coastal management programs and projects involving about 150 sites, carried out by various institutions and groups, which it is hoped will greatly improve the situation.

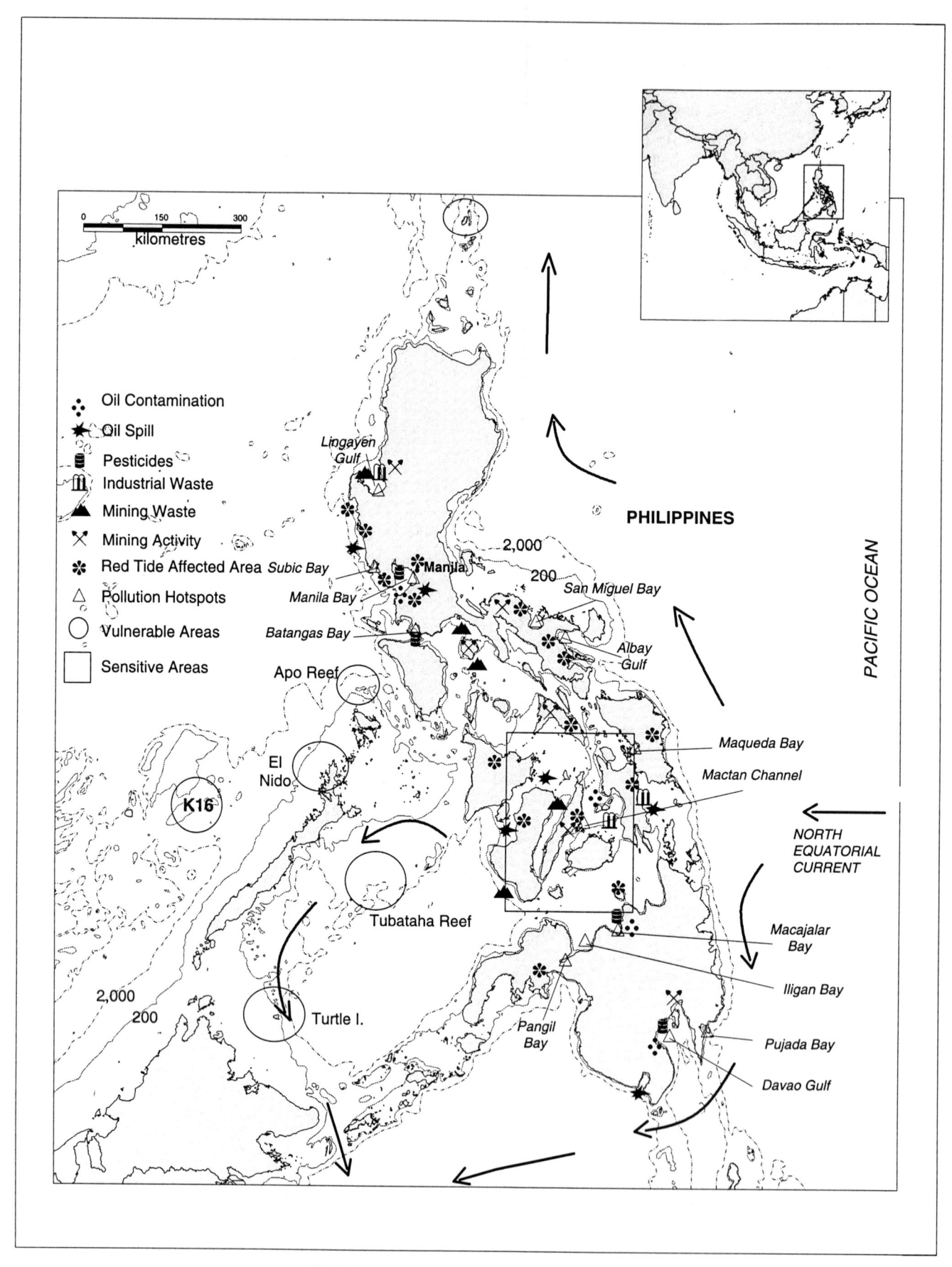

Fig. 1. Seas within and adjacent to the Philippines.

THE AREA AND ITS NATURAL ENVIRONMENTAL VARIABLES

The Philippines is composed of 7150 islands and is one of the largest island groups in the world (Fig. 1). It lies at the centre of marine biodiversity for many marine taxa, and this combination makes it an especially important location at the junction of the Pacific and Indian Oceans.

The country experiences distinct seasons which are predominantly associated with the Asian Monsoon System. The Northeast Monsoon (locally called 'amihan') usually starts developing during October in the northern part of the Philippines above 15°N and reaches its peak around January and February. April and May is a transition period with weak easterly trades. The Southwest Monsoon ('habagat') develops during June and prevails until September. The typhoon (tropical cyclone) season usually occurs from June to September, and an average of 19 typhoons pass through the country every year. Tropical cyclones from the western Pacific intensify the Southwest Monsoon, and precipitation rates in the western and southern parts of the country. Climate types (Fig. 2) exhibit northeast–southwest gradients, and peak rainy seasons often vary depending on whether a particular area is more exposed to the Northeast or Southwest Monsoon.

The eastern side of the Philippines experiences more severe weather conditions, being exposed to incoming tropical cyclones from the Pacific Ocean and large waves generated from distant North Pacific storms. Tropical storms are most frequent and strongest in Northern Mindanao, specifically in the Siargao group of islands. Unlike Mindanao, most coastal areas in northwest Luzon are strongly influenced by medium and high storm intensities.

The major current found east of the Philippines is the westward flowing North Equatorial Current which splits into two upon reaching the Philippines at around 5–15°N (see Fig. 1). The northward branch forms the root of the Kuroshio Current while the southern branch forms the Mindanao Current. Upon moving past southeast Mindanao, the Mindanao current further splits into two; the western branch enters the Sulawesi Sea while the eastern branch turns eastward to feed into the North Equatorial Counter Current. Since these currents are part of the North Pacific Subtropical gyre, its seasonality is not readily influenced by the Monsoons that prevail mostly in the west Pacific. Ship's drift data used to infer surface currents show these flow patterns east of the Philippines persisting throughout the year.

Surface currents in the South China Sea show more seasonal variability owing to its semi-enclosed configuration and seasonally reversing wind forcing. Flow patterns in the western part of the South China Sea basically follow prevailing wind conditions and exhibit seasonal reversals in direction. Inflow from the Pacific through the Luzon Strait appears to be strongest during the Northeast Monsoon months. The narrow and shallow opening of the South China Sea in the Java Sea constricts the flow, resulting in recirculating currents which characterize the persistent northward current off western Luzon. Surface circulation in the Sulu Sea is characterized by cyclonic flow during the Northeast Monsoon and anticyclonic during the Southwest Monsoon.

The distribution of surface temperatures and salinities in the South China Sea is dominated by northwest–southeast gradients due to advection by the prevailing currents. During the Northeast Monsoon, the southwestward flow off the Chinese and Vietnamese coasts advects cold and saltier water from the north through the Taiwan and Luzon Straits, while the Southwest Monsoon brings warm, fresh

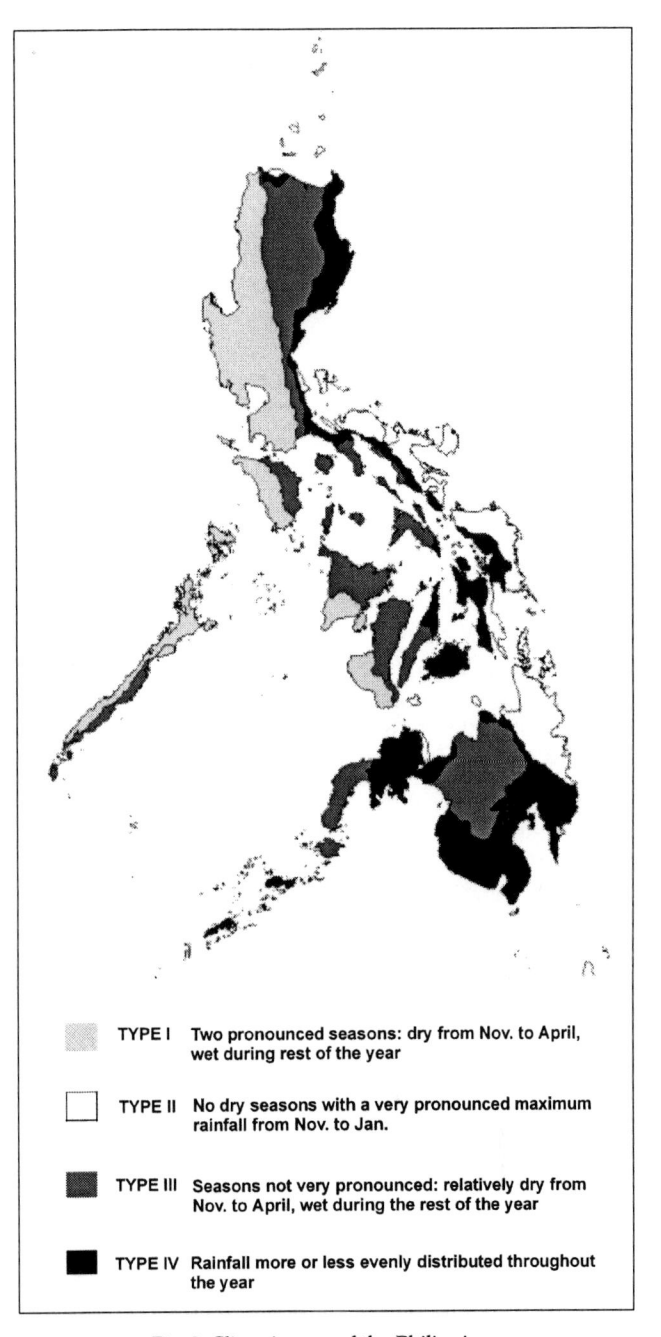

Fig. 2. Climatic map of the Philippines.

TYPE I Two pronounced seasons: dry from Nov. to April, wet during rest of the year

TYPE II No dry seasons with a very pronounced maximum rainfall from Nov. to Jan.

TYPE III Seasons not very pronounced: relatively dry from Nov. to April, wet during the rest of the year

TYPE IV Rainfall more or less evenly distributed throughout the year

water from the south. At any time during the year, the eastern part of the South China Sea is always warmer and less saline than the western part.

The manifestation of El Niño events in the Philippines is drought in many parts, and generally higher-than-normal sea surface temperatures, particularly in the South China Sea. The effect of this natural phenomenon on marine ecosystems in the country is little known. However, during the 1998 El Niño, massive bleaching was observed in various reefs throughout the Philippines, beginning early June until late November 1998 (Arceo et al., 1999). Their studies detected a significant reduction of live coral cover of over 40% and a concomitant increase of dead coral cover of over 40%. Increasing evidence suggests that despite the possible link with natural global climate change, the detrimental effects of bleaching might be exacerbated by anthropogenic factors. Aside from coral reefs, there were also suggestions that harmful algal blooms (HAB) in the region, including those in the Philippines, may be associated with El Niño-Southern Oscillation (ENSO) events (Maclean, 1989).

The archipelago is also in a geologically active zone and is part of the Pacific "ring of fire". Thus, earthquakes are common and there are numerous dormant and some active volcanoes in the country. The eruption of Mt. Pinatubo in June 1990, for instance, was apparently associated with the subduction of the South China Sea beneath the Manila trench (Punongbayan, 1998). Mt. Pinatubo's eruption blanketed most of Southeast Asia with ash and created a dust cloud that slightly cooled the earth. This resulted in geomorphic changes within a few weeks. Among the determined effects of this volcanic eruption was the major reduction of coral cover in some reef areas of the Zambales coast (e.g., from 60–70% before the eruption to 10–20% just a week after eruption) (Atrigenio et al., 1992). Initial estimates of annual fisheries losses for that period were around $0.5 million (Ochavillo et al., 1992). Another study was done by Siringan et al. (1999) on the impact of the Mt. Pinatubo eruption on sedimentation in Manila Bay. They found that the sediment load of coastal rivers along Pampanga Bay, adjacent to Manila Bay, increased and that surface sediments in Manila Bay had become finer-grained relative to pre-eruption periods. They also noted that the sediment input rate into the bay had increased by at least three orders of magnitude, from 10^7 t y^{-1} to 10^{10} t y^{-1}.

MAJOR SHALLOW-WATER MARINE AND COASTAL HABITATS

Main and Important Habitats

With the adoption of UNCLOS and the extension of the maritime zone of coastal states due to the additional 200-mile exclusive economic zone (EEZ), the Philippine EEZ now covers approximately 2.2 million km^2, of which 226,000 km^2 is coastal. The highly productive shelf ecosystems where the bulk of fishing activities are conducted cover 184,620 km^2.

Fringing reefs are among the most prominent features of the coastal areas. The Philippines holds the distinction of being the first country in the world to conduct a nationwide assessment of coral reefs (Gomez et al., 1994a) and they are the most extensively surveyed in the South East Asian region (Aliño et al., 1995). Estimates on the areal extent of these areas range from 25,000 km^2 (Gomez et al., 1994) to 33,500 km^2 (Carpenter, 1977) depending on the assumptions of where the maximum depth limits of corals are. The geographical concentration of coral reefs are as follows: Palawan group of islands (37.86%), Sulu archipelago (27.31%), Visayas group (21.7%), Northern Luzon (7.63%), central and southern Mindanao (3.21%), and the Turtle Islands Group (1.74%).

In conjunction with coral reefs are the mangrove forests, which serve as nursery grounds for fish and invertebrates. Of the estimated 500,000 ha of mangrove cover in 1918, there are now only about 139,725 ha. Seagrass beds add to the stability of many of the country's coastal areas and also serve as critical habitat for many invertebrates and fish as well as endangered/threatened marine life. Around a third of the country's seagrass areas is estimated to have been damaged.

Biodiversity

There are at least 4951 identified species of marine plants and animals in Philippine coastal and marine habitats (DENR, 1997). Fishes, non-coral invertebrates and seaweeds constitute the greatest numbers catalogued. Of these, 1396 species (28%) are economically important, 403 (10%) are flagship species, while 145 species (2.4%) are under threat. Fifteen species are listed as endangered or threatened, and 16 species (0.3%) are endemic.

Of the major habitats found in the country, coral reefs are, by far, the most diverse with 3967 species recorded, followed by seagrass beds with 481 species, and mangroves with 370 species. Soft-bottom communities have the lowest recorded species richness with 70 species. With 410 coral species and at least 1030 species of coral reef fish recorded, Philippines reefs are recognized as one of the most diverse marine habitats in the world. The 16 taxa of seagrasses recorded in the Philippines makes the country the second highest in seagrass species richness in the world too (Fortes, 1994).

Coral Reefs

The most recent countrywide status reviews for coral reefs in the Philippines were undertaken by Gomez and co-workers in collaboration with the group of Alcala (Alcala et al., 1987; Gomez and Alcala, 1979; Yap and Gomez, 1985; Gomez et al., 1981, 1994a, 1994b). Reef health condition was based on the arbitrary % quartile classification of live coral

Table 1

Overall condition of coral reefs in the Philippines. Reef health condition was based on the arbitrary % quartile classification of live coral cover (i.e., 0.1–24.9% = poor; 25–49.9 = fair: 50–74.9% = good; and 75–100% = excellent).

Source	No. of transects (Stations)	Excellent (75–100%)		Good (50–74.9%)		Fair (25–49.9%)		Poor (0–24.9%)	
		No.	%	No.	%	No.	%	No.	%
Yap and Gomez (1985)	632	35	5.5	153	24.2	242	38.3	202	32.0
ASEAN-Australia MSP; LCR	103	4	3.9	32	31.1	46	44.7	21	20.4
ASEAN-US CRMP	40	0	0.0	18	45.0	17	42.5	5	12.5

cover (i.e., 0.1–24.9% = poor; 25–49.9 = fair; 50–74.9% = good; and 75–100% = excellent). The overall condition of the reefs in the country has been evaluated based on samplings gleaned from over 700 transects sampled in over 14 provinces (Table 1).

The nationwide surveys made in the 1970s showed 5% of the reefs in excellent condition, 25% good, 39% fair and 30% poor (Gomez and Alcala, 1979). Later studies in the 1980s showed 70% of the reefs in poor to fair condition, with less than 50% living coral. Despite the ongoing debate on how to come up with a better indicator of the condition of reefs (e.g., quadrat method and % live cover; belt transect and coral mortality index), there remains an urgent need to address the unabated destruction of coral reefs in the country (Gomez et al., 1994a; Aliño et al., 1994; Aliño et al., 1999 in press).

In general, coral reefs have primary productivity in the order of 1500–3500 gC m^{-2} y^{-1}, producing valuable resources of fishes, invertebrates and seaweeds. In the Philippines, coral reef fisheries yield (reef fish and invertebrates) ranges between 1–2 t km^{-2} y^{-1} in damaged reefs and 37 t km^{-2} y^{-1} in pristine or healthy reefs (Alcala and Gomez, 1985). However, fisheries yields up to 78–105 t km^{-2} y^{-1} have been estimated for pristine reef areas such as the "Kalayaan Islands" or the Spratly Islands Group (Aliño et al., 1998). Groupers, snappers, carangids, and caesionids are among the commercially sought fish harvested from reef areas along with various species for the tropical aquarium fish industry. Of the macroinvertebrates, lobsters, crabs, prawns and cephalopods are highly priced. Other than these, some 160 species of macroinvertebrates are commercially exploited, as has been done on the reef flat in Bolinao, Pangasinan (McManus et al., 1992).

Associated with coral reefs are seaweeds of which a total of 1,062 species have been recorded in the country. Of these, 538 species have been identified to have economic potential (e.g., as food, medicine, and raw material for industrial colloids). Four genera are cultured (*Kappaphycus, Eucheuma, Caulerpa, Gracilaria*) while several species are exported (*Kappaphycus alvarezii, Eucheuma denticulatum, Caulerpa lentillifera, Gelidiella acerosa, Gracilaria* spp. and *Sargassum* spp.). Natural stocks of *Halymenia, Gelidiella, Acanthophora, Gracilaria, Codium,* and *Caulerpa* are heavily exploited.

Mangroves

In the early 1900s, mangrove coverage in the country was estimated at 400,000–500,000 ha (Brown and Fischer, 1918). In 1984, only half (51%) of the mangrove forest remained at ~230,000 ha (BFD, 1984). The rate of exploitation in the mid 1980s in the different regions in the country was highest in the Visayas (at 72%) and Luzon (at 64%) (Table 2). Mangroves were intensively utilized in these regions compared to Mindanao (10%) and Palawan (21%) where old growth stands are still found (PNMC, 1986). In the mid 1990s, mangrove forests were estimated to be only 200,000 ha, a loss of 60% from the 1920 estimate.

Mean production rates measured at Philippine mangrove sites range from 1–2 gC m^{-2} d^{-1}. Production rates are directly proportional to interstitial salinity, temperature and canopy cover, and inversely proportional to soil erosion and light penetration through the canopy (Fortes, 1991). About 54 species of crustaceans, 63 species of molluscs and 110 species of fish have been reported in Philippine mangroves (PNMC, 1987 in PCAMRD, 1991; De la Paz and Aragones, 1985), a number of which are commercially important (Camacho and Bagarinao, 1987).

Mangroves are important sources of fishery products (e.g., fish, shrimps, molluscs, crabs, fry) and forest products. Mangrove areas are also used for aquaculture, salt production and human settlement.

Table 2

Mangrove forests in the Philippines (× 1000 ha) from 1920–1994 in Luzon, Visayas, Mindanao, and Palawan

Year	Total	Luzon	Visayas	Mindanao	Palawan
1920	450				
1968	448	47	228	125	47
1970	288	42	78	124	44
1975	254	28	67	117	42
1980	242	23	66	115	39
1988	228				
1994	200				

Data Sources: BFD Statistics; Ronquillo (1988); Calumpong (1994).

Seagrasses

Seagrass beds are highly productive ecosystems. Tropical seagrass beds have an average gross primary productivity of 465 g C m^{-2} y^{-1} or an average of 1.27 g C m^{-2} d^{-1} (Phillips and Meñez, 1988). A study in the Philippines recorded production of 0.9 g C m^{-2} d^{-1} (Fortes, 1996). Fish yield from seagrass beds indicates more than 10 t km^{-2} y^{-1} of fish alone, with a production potential of 20 t km^{-2} y^{-1} in biomass of fish, invertebrates and seaweeds (McManus et al., 1992). Siganids are the most abundant fish and occur in seagrass beds as adults and juveniles; others are mostly represented by juveniles. Seagrasses support the siganid fry industry. Estimated annual catch of siganid fry from one sampling area in northwestern Philippines is 2.6 million pieces in 1986 and 12.2 million in 1987 (Ungson, 1990). In Bolinao, northern Philippines, a dominant siganid, *Siganus fuscescens* in seagrass beds had an estimated mean biomass of 2 t km^{-2} and a fish yield of 4 t km^{-2} y^{-1} (del Norte and Pauly, 1990).

No comprehensive evaluation of the state of seagrass communities has been undertaken but the depth range of seagrass communities in the intertidal and subtidal is extensive (to 30 m). Studies which have been conducted were mainly in conjunction with environmental impact assessments (EIA) of industries along the coast, ecological assessments of degraded fishing grounds, and resource assessments related to coastal management issues. One study (Table 3) shows the only estimates done. Carried out in connection with the ASEAN–Living Coastal Resources Project on seven sites using remote sensing techniques (Fortes, 1995), most of the sites were seen to have fair to good seagrass beds except in the sites in Southeast Luzon, i.e., Pagbilao, Calauag and Calancan Bays, where most of the seagrass beds are in poor condition.

Endangered Species

Giant Clams

The family Tridacnidae is included in Appendix II of CITES, and of the seven species of giant clams found in the Philippines, six are included in the IUCN Red List of Threatened Invertebrates (*Tridacna gigas*, *T. derasa*, *T. squamosa* Lamarck, *T. maxima* (Roding), *Hippopus hippopus* (Linnaeus) and *H. porcellanus*). The seventh, *Tridacna crocea* Lamarck, was removed from the list in 1996 when sufficient data had been gathered about its abundance. However, at present, Philippine laws restrict the export of all species of giant clams.

Marine Turtles

The Philippines prohibits the collection, gathering, utilization, possession, transport and disposal of marine turtles. A major breakthrough in conservation was the establishment of a Turtle Island Heritage Protected Area in Region 9 (Southern Philippines) through a Memorandum of Agreement between Malaysia and the Philippines in May 1996.

Table 3
Seagrass assessment in seven sites in the Philippines using remote sensing techniques (from Fortes, 1995). Status numbers are % of sites in each category

Site	Area (km^2)	Status		
		Good	Fair	Poor
Cape Bolinao	25.0	38	49	13
Calauag Bay	7.7	33	–	67
Pagbilao Bay	1.89	–	–	100
Puerto Galera	1.14	70	–	30
Ulugan Bay	2.97	17	51	32
Banacon Island	7.81	73	27	–
Calancan Bay	0.07	29	–	71

The five species of marine turtles recorded in the Philippines are: *Caretta caretta* (loggerhead), *Chelonia mydas* (green sea turtle), *Eretmochelys imbricata* (hawksbill sea turtle), *Lepidochelys olivacea* (olive ridley turtle), and *Dermochelys coriacea* (leatherback turtle).

Marine Mammals

Marine mammals are protected in the Philippines by a Fisheries Administrative Order which prohibits the taking, catching, sale, purchase, possession, transport and export of dolphins. An amendment was later issued to cover whales and porpoises as well, and the sea cow, *Dugong dugon*, is also protected (DEN/DILG/DA-BFAR/CRMP, 1997). Twenty-two species of marine mammals have been recorded in Philippine waters.

Transboundary Straddling Stocks/Species

The country's migratory fish species were determined from catches of fishing gear for tuna. These are: (1) tunas and tuna-like species—*Thunnus albacares*, *T. obesus*, *T. tonggol*, *T. alalunga* and *Katsuwonus pelamis*; (2) seerfishes—*Scomberomorus commerson*, *Gymnosarda unicolor* and *Grammatorcynus bicarinatus*; (3) billfishes—*Makaira mazara* and *M. indica*; and (4) dolphinfish—*Coryphaena hippurus*.

The Philippine Tuna Research Project (1991–1993), studied the movement of tuna within Philippine waters and between the Philippines and adjacent areas. Local movement is significant from the Celebes Sea to the Sulu Sea. Movement to adjacent areas suggests mixing of stocks between the Philippines and eastern Indonesia and the large purse seine fishing grounds to the east.

OFFSHORE SYSTEMS

In contrast to the high productivity of coastal ecosystems, nutrients and productivity of Philippine offshore waters are low. Data taken from the previous oceanographic expeditions, such as the Dana, Albatross and the Galathea, described the surface layers of Southeast Asian (SEA) waters as extremely poor in nutrients (Wyrtki, 1961). Similarly, Megia (1952) described the surface waters in and around the Philippines as nutrient-poor. Low organic production of less than 0.5 g C m^{-2} d^{-1} were reported for the China Sea, Philippine waters and the Celebes Sea (Nielsen and Jensen, 1957) without considering the vertical distribution of productivity and depth of the phytoplankton layer. A more recent study by San Diego-McGlone et al. (1999) found depth-integrated primary production rates in the oceanic region of the Sulu Sea at 195 ± 2 g C m^{-2} y^{-1} and that in the oceanic region of the SCS at 147 ± 3 g C m^{-2} y^{-1}.

Upwelling and Internal Waves

Two oceanographic features in Philippine offshore waters have generated interest in recent years: upwelling off the west coast of Luzon and internal waves in the Sulu Sea. Offshore fisheries distribution and abundance associated with these features have not been determined.

A major upwelling region has been shown to be present about 100 km off northwest Luzon in the South China Sea (Fig. 3), between 16 and 19°N (Hung et al., 1986; Shaw et al., 1996; J. Udarbe, pers. comm.). This feature seems to occur during the Northeast Monsoon (between October and January) and is, apparently, not driven by local winds. Rather, Shaw et al. (1996) suggest that the upwelling is sustained by an offshore Ekman drift from above and convergence in the northward undercurrent from below. This upwelled water may be an important nutrient source for the interior South China Sea during the monsoon transition period (April–May) in the Philippines.

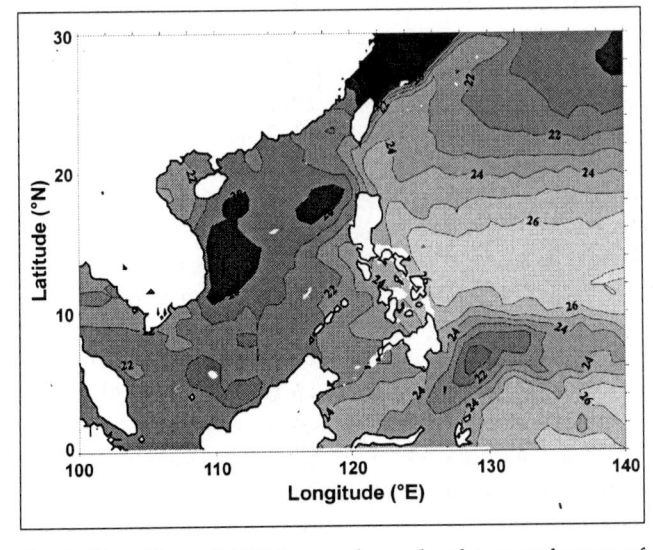

Fig. 3. Upwelling off NW Luzon shown by the annual mean of objectively analyzed monthly temperature field at 100 m depth using a grid resolution of 0.5 × 0.5 degrees. The data used in the objective analysis was obtained from the National Oceanographic Data Centre (NODC) World Ocean Atlas CD-ROM 1994. A description of the objective analysis theory is in Carter and Robinson (1987) and the objective mapping routine was supplied by E.F. Carter. Figure courtesy of J. Udarbe.

Apel et al. (1985) and Liu et al. (1985) described the generation and propagation of large-amplitude internal solitons in the southeastern Sulu Sea by intense tidal flow over the Pearl Bank sill every fortnight. These solitons have wavelengths of 5–16 km, a period of 35–55 min, a phase speed of 1.8–2.6 m s^{-1} and reach the coast of Palawan after about 2.5 days. Chapman et al. (1991) observed internal swash at the shelf edge off Puerto Princesa, Palawan, and attributed this to the breaking of the Sulu Sea solitons. Energy from these internal wave packets was also found to drive coastal seiches in Puerto Princesa (Giese et al., 1998) and may induce upwelling episodes on the northwestern section of the Sulu Sea, off Palawan.

Oil and Gas

In the 1990s, more than 80 wells were drilled in offshore areas within Philippine territory: 56 wells in North Palawan, 20 in South Palawan and seven in the Reed Bank (Austria and Rillera, 1993). Commercial deposits of oil and gas have been found in northwest Palawan. West Linapacan is estimated to contain more than 100 million barrels of recoverable crude oil. The Camago–Malampaya offshore area is a major natural gas deposit, estimated to contain 2.5–3.2 trillion cubic feet of recoverable gas reserves (Balce and Pablico, 1997)

Offshore Fisheries

The Philippines is an important fish producer, ranking 12th in 1995 (BFAR, 1997). It is the second biggest tuna and tuna-like producer in the Indian Ocean and in the ASEAN region. Its total fish production in 1997 reached about 2.766 million tonnes, valued at PHP 80.7 billion or US$2.73 billion (BFAR, 1997). The contribution of the fishery sector to GNP in 1997 was 2.8 and 3.8% in current and constant prices, respectively. Table 4 shows a comparison of the 1975 and 1995 commercial fisheries production by fishing grounds. The waters west of Palawan (i.e., the South China Sea) and the Sulu Sea (South and East) contributed about 43% (~387,000 Mt) of the country's total commercial fisheries production in 1995. About 5% (44,140 t) of the commercial fisheries production was obtained from international waters.

On the Pacific seaboard, there are indications that the area east of Mindanao is also a major fishing ground as indicated by upwelling features. The area off eastern Luzon, between 14 and 16°N, is, likewise, believed to contribute to the major commercial fish catch of Lagonoy Gulf and Lamon Bay.

POPULATIONS AFFECTING THE AREA

The Philippines ranks as the 9th most populous country in Asia and the 14th in the world. A recent estimate of the

Table 4
Commercial fisheries production by major fishing grounds

Fishing Ground	1995 Total (t)	1995 %	1975 Total (t)	1975 %
West Palawan Waters	187,728	21.02		
South Sulu Sea	180,532	20.21	6,738	1.60
Visayan Sea	120,267	13.46	181,030	42.96
Moro Gulf	83,352	9.33	12,690	3.01
Lamon Bay	55,325	6.19	16,782	3.98
International Waters	44,140	4.94		
Cebu/Bohol Sea	36,161	4.05	12,351	2.93
Samar Sea	25,602	2.87	14,518	3.45
Manila Bay	25,046	2.80	51,743	12.28
Guimaras Strait	24,163	2.71		
Sibuyan Sea	20,065	2.25		
East Sulu Sea	18,994	2.13		
Tayabas Bay	17,140	1.92	1,379	0.33
Others	54,717	6.13	12,803	3.04
North Sulu Sea			111,386	26.43
Total	893,232	100	421,420	100

country's population is at 68 million with an average density of 207 per km^2 (NSCB, 1997). While the country's population growth rate has gone down from 3.3% to 2.3% over the past 30 years, the current growth rate, if unabated, will cause the population to double to 128 million by 2025. About 60% of the population resides in some 10,000 coastal barangays (villages) and urban centres.

At present, there are 16 regions, each made up of local government units, from provinces down to the barangays. As of June 30, 1998, there were 78 provinces, 82 cities, 1,525 municipalities and 41,939 barangays (NSCB, 1999). Twenty-five cities are considered highly urbanized. Of these, 14 are located in Luzon, five in the Visayas and six in Mindanao. The National Capital Region (NCR), which includes Metro Manila, is the most populous region, accounting for 13.8% of the country's inhabitants.

The last 25 years have seen continued rural-to-urban migration, affecting the quality of life both in cities and rural communities. The open-access nature of the country's coastal waters has contributed heavily to the burgeoning in the number of coastal dwellers. In many urbanized coastal municipalities/cities, the influx of migrants is very high, being from 1 to over 4% annually during the 1990s.

Due to population pressure, utilization of upland areas for settlement, livelihood and other uses has been very extensive since the 1980s. Previous studies cited that the shortage of livelihood opportunities in the lowlands results in upland migration. For instance, in 1995 and 1996 alone, a total of 29,288 ha of forest were destroyed by natural and man-made activities, with forest fires, slash-and-burn farming and illegal logging as the major causes of destruction (NSCB, 1997).

In terms of total municipal fisher population, a study by Padilla and Cortez (1996) projected an increase in the number of fishers in the country. Fishers were also found to engage in several activities to augment their incomes and therefore put more pressure on other natural resources. Studies in Lingayen Gulf indicate an annual increase of 2.3% of total number of fishers per km^2 of fishing ground (Padilla and Morales, 1997) and is apparently similar to the status of other major fishing grounds in the country.

RURAL FACTORS

Agriculture

The agriculture sector is the largest single contributor to the national economy, generating 36% of the country's export earnings and 28% of its GNP. About 32% (9.73 million ha) of the country's total land area of 30 million ha is devoted to agriculture and includes ~3.42 million farms, most of which are less than 3 ha in size (NEDA, 1998). The total agricultural lands converted to non-agricultural purposes for the period 1987–1997 was 50,719 ha or an average conversion rate of 5635 ha y^{-1}.

Soil Erosion

As of 1993, some 5.2 million ha were classified as severely eroded, 8.5 million ha as moderately eroded and 8.8 million ha as slightly eroded. Mindanao had the largest eroded area at 2.4 million ha (Francisco, 1994). The human causes of accelerated soil erosion were attributed to improper land uses, shifting cultivation practices, extensive deforestation of sloping areas and improper use of chemicals.

Eroded soil from agricultural areas, including slash-and-burn agriculture, amounted to about 399 million t in 1988 and 342 million t in 1993 for an area of ~10 million ha (Franciso, 1994). Thus, about 600,000 t of soil are being eroded annually. The estimate of the volume of sediment load from farming was 100 million t y^{-1}. The corresponding annual average nutrient loss from sediments, expressed in fertilizer equivalent, was estimated to be at least 290,000 t of N (45–0–0), 6,000 t of P (0–20–0) and 45,000 t of K (0–0–60).

Fertilizers are also used in aquaculture areas. However, little is known about the amounts and rates of fertilizer discharge in the country's coastal waters, principally arising from run-off and the aquaculture practices.

Pesticides

While virtually no published data are available on pesticide residues in the country's marine environment, an indication of the problems may be gleaned from pesticide importation, production and use. There are about 200 agricultural pesticide companies, including about 30 formulation plants (Fertilizer and Pesticide Authority, 1998). They include manufacturers, formulators, repackers, importers, distributors, traders and suppliers.

Pesticide sales in the Philippines grew tremendously until the introduction of the integrated pest management (IPM) concept in the 1990s. From 1987 to 1989, the major pesticide companies in the Philippines put on the market 20,000 t of pesticides, and were growing at an annual average of about 17.5%.

The majority of pesticide applications by rice farmers in Leyte were insecticides (Heong et al., 1994), about half of which were organophosphates. About 22% of the chemicals in the Philippines were classified as 'extremely hazardous' (Category 1a) by the World Health Organization and another 17% of the insecticides were classified as 'highly hazardous' (Category 1b). High proportions of the sprays were targeted at leaf-feeding insects and accounted for 28% of insecticide sprays.

Another study (GEF/UNDP/IMO MPP-EAS and FAO, 1998) carried out a preliminary environmental risk assessment of pesticides in the Batangas Bay Region. In 1996, the major crops grown in the area (30,000 ha) were rice, corn, mango and vegetables. The study found that pesticide consumption in the area was relatively low, with the total load of pesticides at 0.7 kg ha^{-1} on the agricultural soil and an average of 0.24 kg ha^{-1} on the whole basin. However, no actual values were available in this study for pesticide residue levels in the tributaries, the bay area, or in marine organisms.

Forest Clearance

Forest cover has been reduced from more than 50% in 1948 to less than 20% in 1996. Estimates of total forest cover in 1990 were 6.1 million ha or 20% of total land area (DENR, 1997). Old growth dipterocarp forests have dwindled to ~800,000 ha from about 4.6 million ha during the last 40 years. Agroecosystems, both lowland and upland, expanded during the past 40 years to about 10 million ha. Urban ecosystems have grown at a much faster rate. During the period 1980–1990, urban areas grew by 142,000 ha.

Siltation is the most important factor influencing conditions in coral reefs, caused either by natural processes or by anthropogenic activities. Upland deforestation causes siltation mostly in reef areas near rivers, e.g., Bacuit Bay, Palawan (Hodgson and Dixon, 1988). The sediments physically smother the corals, and further reduce recruitment success, due to poor larval settlement and increased post-larval mortality (Gomez et al., 1994a).

Aquaculture Practices

Effluents discharged from aquaculture farms also contribute significantly to high nutrient loads in the receiving marine waters. Levels of nitrogen, phosphorus and other water quality parameters are generally higher in effluent than inflowing water (Macintosh and Phillips, 1992). While the pollution potential of aquaculture pond effluents is minimal compared to domestic or industrial waste water (Macintosh and Phillips, 1992), problems arise because of the large volumes of water discharged from intensive farms, compounded by the high concentration of farm units in areas with limited water supplies and inadequate flushing.

A related issue with respect to intensive aquaculture practices is the use of chemicals and biological products to solve the self-polluting characteristic of intensive ponds. In a study of intensive prawn farms in the Philippines, Primavera et al. (1993) found that chemical and biological products used included therapeutants and disinfectants, soil conditioners, bacteria–enzyme preparations, algicides and piscicides, plankton growth promoters and feed additives. Chemical and biological pollution by shrimp farms results from disposal in coastal waters of pond effluents and sludge, misuse of antibiotics and other chemicals; and introduction of exotic shrimp species and diseases (Primavera, 1998; Naylor et al., 1998).

COASTAL EROSION AND LANDFILL

Fortes (1995) made an inventory on the conversion and uses of seagrass areas (Table 5). Industrial developments, ports and recreation are the major and widespread cause of problems. The impacts of waste disposal and boat traffic are widespread but moderate. Aquaculture farms of fish, prawns and crabs from seagrass areas are still confined to small areas. Localized but major habitat modifications to seagrass beds result from urban development and mining, followed by conversion to solar salt beds, use for sugar-cane fields and airports.

Table 5

Conversion and utilization of seagrass beds in the Philippines (source: Fortes, 1994)

Activity	Use	Ranking
Industrial Development	widespread	major use
Ports	widespread	major use
Recreation	widespread	major use
Waste disposal	widespread	moderate use
Boat traffic	widespread	moderate use
Aquaculture-fish	localized	major use
Urban development	localized	major use
Mining	localized	major use
Aquaculture	localized	moderate use
Solar salt	localized	moderate use
Flood run-off	localized	moderate use
Aquaculture-crabs	localized	minor use
Sugar-cane	localized	minor use
Airports	localized	minor use

Seagrass beds are also subject to unsustainable practices of collection of the other economically important invertebrates that abound in this habitat. The beds are commonly raked to collect shells for the ornamental trade, trawled to harvest shrimps and prawn, and covered with fish corrals to catch rabbitfishes. Sea cucumbers are easily harvested, as are sea urchins for the gonads, and sea hares for their eggs.

In the Philippines, the effect of land clearing and deforestation in the watershed areas and from slash-and-burn agriculture or 'kaingin' has resulted in sedimentation of seagrass beds. Apart from that, coastal eutrophication from industrial, residential and commercial facilities adds further stresses to the seagrass community.

The over-exploitation and degradation of seagrass beds also threaten populations of some endangered organisms like dugongs and marine turtles (Fortes, 1995) which have been hunted for subsistence and for commercial trade.

EFFECTS FROM URBAN AND INDUSTRIAL ACTIVITIES

Artisanal and Non-industrial Uses of the Coast

Between 1920 and 1994, around 60% of the mangroves in the country were lost (PNMC, 1986; Evangelista, 1992). Mangroves are economically important as sources of timber, firewood, charcoal, tannin, tanbark, nipa sap and shingles. However, another major cause of the reduction of mangrove forests is their conversion into fishponds (Fig. 4). This more than doubled from 88,000 ha in 1952 to 176,000 ha in 1975 (Datingaling, 1977; PNMC, 1986), and was mainly for the culture of milkfish (*Chanos chanos*) and prawns (*Penaeus monodon*). In 1982, brackishwater fishponds still increased to >195,000 ha (Gomez et al., 1990), while in 1988, mangrove areas converted to fishponds totalled 224,000 ha which was almost equal to the remaining mangrove area of 228,000 ha (Ronquillo, 1988; Primavera, 1993).

This loss of mangrove corresponds to a loss of coastal productivity affecting coastal fisheries (Primavera, 1993; Camacho and Bagarinao, 1987; Silvestre, 1989). Fishponds bring about pollution due to the use and disposal of organic/inorganic fertilizers, chemical toxins, and antibiotics, among others (PNMC, 1986; Primavera, 1991). At the same time, it affects coastal dwellers dependent on the mangrove swamps who are more numerous than those working in the aquaculture farms (Primavera, 1991; Velasco, 1979). The loss of mangrove habitats brings about the loss of its commercial value as an important breeding, nursery and feeding ground of economically important fish (mullet, milkfish, tilapia, eel, kitang) and invertebrates (shrimp, prawns, molluscs and sea cucumber) (see Zamora, 1982; de la Paz and Aragones, 1985; Alcala, 1982; Calumpong, 1994). The fry gathered in the mangrove areas, especially of milkfish (*Chanos chanos*), shrimp (*Penaeus monodon*), and prawn (*Penaeus merguiensis*), contributed greatly to the country's fry-gathering industry and have become important species in aquaculture (Gonzales, 1977).

The salinization of the water aquifer from over-abstraction for fishpond use is a problem in some areas e.g., Central Visayas. However, the impact of salinization from shrimp pond practices is still localized in the Philippines, compared to other countries like Bangladesh, Ecuador, and Indonesia where rice production decreased due to the loss of soil fertility (Primavera, 1993).

The shrinking mangrove resources may become less easily renewable (Jara, 1983; Hamilton, 1984; Calumpong, 1994). This brings about social inequity between the marginalized fishers and the few fishpond entrepreneurs (Primavera, 1993). A large percentage of the coastal population depends on coastal fisheries for subsistence and income, compared to a seasonal and undependable employment in aquaculture farms (Primavera, 1991). However, the increasing contribution of aquaculture to the total fishery income in the country indicates its economic viability (BFAR Statistics; Camacho and Bagariano, 1987). Hence, fishpond development has been the recipient of government support since its inception, with loans, tax privileges, and other incentives being provided (Primavera, 1993).

Yet, the most important but least appreciated value of mangroves is their role in coastal erosion and sediment stabilisation. The importance of this has now been recognized because of the loss of property and lives in some areas. The major catastrophes in the country exacerbated by the loss of this habitat include approximately 3000 deaths in Zamboanga in 1996; 1000 deaths in Northern Panay in 1984; and 7000 deaths in Ormoc, Leyte in 1991 (Primavera, 1993).

Small-scale Mining

Small-scale mining activities, mainly of gold, take place in various parts of the country. A classic example of toxic effects associated with this type of activity are mercury poisoning in Mt. Diwalwal, Davao del Norte, on the Agusan River basin on eastern Mindanao. An estimated 600 million m³ of untreated mine tailings have been discharged over

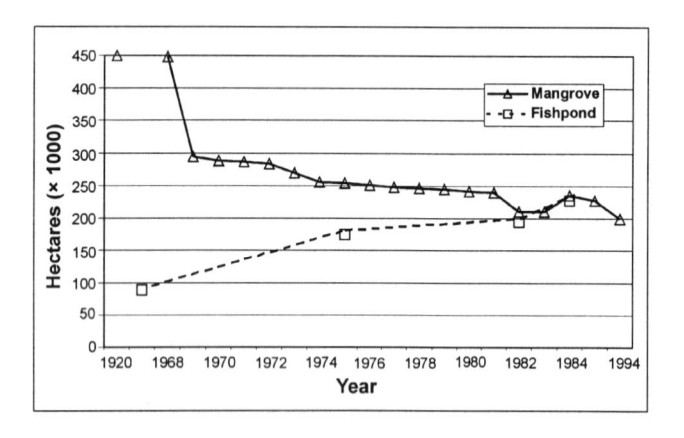

Fig. 4. Fishpond development and decline in mangrove forest cover.

the last 14 years of illegal gold mining. A recent study (Appleton et al., 1999) showed that drainage downstream has extremely high levels of Hg both in solution (maximum 2906 μg l^{-1}) and in bottom sediments (>20 mg kg^{-1}). Filtered surface water Hg levels exceeded the WHO Drinking Water guideline and the US-EPA Water Quality Criteria for the Protection of Aquatic Life for a downstream distance of more than 14 km, including channel sections utilized for fishing and potable water supply. This same study suggests that operators processing Hg-contaminated tailings may be at risk of high occupational Hg exposure

Mining of beach sand has increased with the proliferation of beach resorts in parts of the country, especially in the Visayas (Cebu, Bohol). Sand is obtained from beaches that are undeveloped and moved to supply and "enhance" the beaches of newly-developed high-end resorts, catering principally to foreign tourists. The extent of this practice remains unknown but anecdotal accounts indicate major changes in coastlines and coastal features (e.g., Mactan Island in Cebu).

Effects on Coral Reefs

Three factors commonly affect coral reefs: sedimentation, overfishing and destructive fishing (Aliño et al., 1995). These were investigated by Aliño et al. (1995) based on the frequency distribution of reefs experiencing each (Table 6).

Many of the country's reefs are overfished, resulting in shifts in fish size, abundance, and fish composition within reef communities. For example, in the 1960s and 1970s, several giant clam species became locally extinct due to overharvest for their large shells. The sea urchin, *Tripneustes gratilla*, was found in dense populations across a 24 km^2 seagrass bed in Bolinao, Pangasinan in 1987. With the arrival of commercial traders in the area, the sea urchin was believed to have become locally extinct in 1995 (Talaue-McManus and Kesner, 1995).

Destructive fishing practices include blast fishing, fishing with cyanide and other poisonous chemicals, muro-ami (pounding reefs with weighted bags to scare fish out of crevices), and in deeper waters, trawling. Reefs destroyed by these fishing practices can take between 5 and 50 years to partly recover, although other factors may prolong this period further (Gomez and Yap, 1985).

Cyanide is used to stun and capture live coral-reef fish. The practice began in the 1960s to supply the growing market for aquarium fish in Europe and North America (Bryant et al., 1998). In the 1970s, cyanide was also used to capture larger live reef fish (especially groupers) for sale to specialty restaurants within and outside the country. Studies on the impact of cyanide on reefs are few, but anecdotal accounts indicate that cyanide kills corals and is toxic to fish. Moreover, cyanide fishing poses threats to fishers through accidental exposure.

Industrial Uses of the Coast

Coastal Mining

The country is one of the world's leading producers of gold, copper, metal and chromite. As of 1996 it was reported that the country had a total of 6.671 B t of metallic and 78.472 B t of non-metallic reserves. Seventy-two percent of the metallic reserves are copper, followed by nickel with 16%. Limestone and marble account for 47% and 37% of the non-metallic reserves, respectively.

Mine wastes and tailings are the more serious threats to the marine environment. A total of 47.44 million t of mine wastes was generated in 1991 alone. However, for the period 1990–1996 the volume of mine wastes decreased by 17%, except for 1990–1991. Coastal waters affected by mine tailings include Calancan Bay (Marinduque, 200 km south of Manila) and Tañon Strait, between Cebu and Negros.

The discharge of mine tailings in Calancan Bay was exacerbated in March 1996, when an accident released 1.6 million m^3 of mine tailings into the Boac River and Bay at a rate of 5–10 m^3 s^{-1}. Heavy sedimentation clogged river channels and flooded banks. Major fish kills were reported and costs to date exceed $US 71 million.

Problems have also arisen from cinnabar (HgS) mines in Palawan, 14 km away from the City of Puerto Princesa. From 1955 to 1975, mine tailings of approximately 100,000 t y^{-1} were produced and deposited at the nearby sea, enough to build a 600 meters long artificial peninsula. At present, the mine site is home to a thriving community of at least 200 households and the tailings have long been used by the community, including the City of Puerto Princesa, as filling materials, although samples collected from the peninsula

Table 6

Critical coral reef areas in terms of siltation, pollution, fishing and storm impacts

Impact	Type	Areas
Siltation	Agricultural	Palawan Southern Mindanao Southeast Luzon
	Industrial	Southwest Luzon Northern Mindanao Visayas
Pollution	Agricultural	Northwest Luzon Southeast Luzon Southwest Luzon
	Industrial	Southern Mindanao Northern Mindanao Central Visayas
Fishing	Overfishing	Palawan = Northern Mindanao Central Visayas = NE & SE Luzon
	Destructive	Southern Mindanao Southwest and Northeast Luzon

and sediments from the surrounding waters had elevated mercury levels as high as 570 ppm (Benoit et al., 1994). However, results of a subsequent study of the Mines and Geosciences Bureau have not shown any direct link between apparent manifestations of mercury poisoning in people living within or in the vicinity of the artificial peninsula and the high levels of HgS in the substrate.

Trace metal pollution in the country's marine environment is generally localized, with high concentrations of trace elements found in water and sediments in areas where mine tailings reach the sea (e.g., Honda Bay, Calancan Bay and Macajalar Bay) or in the immediate vicinity of effluent outlets. In coastal areas next to major cities such as Manila Bay, studies by Jacinto et al. (1999a) and Prudente et al. (1994) suggest that generally high levels of trace metals in sediments are limited to specific metals (e.g., Pb, Cd) and remain in the vicinity of major tributaries (e.g., Pasig River). Data on trace metals in the country's coastal and offshore waters remain sparse, in part due to the difficulty in accurately determining very low concentrations. Reports in the 1980s or earlier of extremely high metal concentrations in certain coastal areas may be suspect and may have been biased due to sampling and analytical problems. Recently, greater attention to the issues of contamination and even better analytical techniques have afforded more reliable estimates of ambient trace metal concentrations even in non-contaminated areas (see e.g., Watson et al., 1995).

Large-scale fishing

There are issues with the intrusion of commercial fishers into municipal waters. Commercial fishers use ships of 3.1 GWT or more, while municipal fishers use much smaller vessels. Under present laws, commercial fishers are not supposed to fish within municipal waters (15 km from shore). However, commercial vessels do enter municipal waters and this results in unfair competition between commercial and municipal fishers, with the municipal fishers losing (Lacanilao, 1998). Commercial fishing in municipal waters has reduced the municipal catch yearly since 1991, with corresponding increases in commercial harvest until 1995 (Fig. 5). These two groups not only compete for fishing grounds but also for their main target fishes—roundscads, anchovies, tunas, sardines, and mackerel. These fishes made up 60% of commercial catch and 31% of municipal catch in 1995 (BFAR, 1997).

Another factor that contributes to economic losses in the fisheries sector is illegal fishing (e.g., poaching in offshore areas, smuggling, use of prohibited gears, etc.). While the landed value of Philippine fisheries is about US$ 2 billion per year, harvests by illegal fishing activities have been valued at more than US$ 1.5 billion per year.

Cities

Based on the 1995 population census, Metro Manila alone, with only 0.2% of the country's land area, accounts for almost 14% of the total Philippine population and almost 30% of the urban population. Rapid urbanization has brought with it the proliferation of informal and overcrowded settlements and associated problems.

The absence of systematic waste collection and disposal facilities in the country's urban areas and coastal municipalities has largely contributed to the decline in river water quality and nearshore areas. In 1995, per capita solid waste generation in Metro Manila was estimated at 0.66 kg d^{-1}. This amounts to 6,102 t of solid wastes generated by the metropolis per day. The current rate of collection of domestic solid waste in Metro Manila is 85% or 5187 t per day with the remaining 15% either burned or disposed into the waterways. Fifty percent of the total domestic waste is disposed of at the different disposal facilities, while 40% is dumped illegally and the remaining 10% is apparently recycled.

Even in primary urban centres, there is no adequate sewerage system. In 1992, 82% of households in urban areas had access to sanitary toilet facilities. Only 13% of the population of Metro Manila is served by a centralized sewage collection. About 70% of the households used septic tanks, many of which are inadequate.

Water pollution of river systems has also become more conspicuous, and about 50 major river systems in the country are polluted. Pasig River, a major river traversing metropolitan Manila, that empties into Manila Bay, is a classic example. According to a DENR report (DENR-DANIDA, 1991), 327 t of pollutants are dumped daily into Pasig River. The pollutants in the river consist of 45% liquid domestic wastes, 45% industrial wastes, and 10% solid wastes. From the study, it was determined that sources of pollution could be attributed to the residents of about 367 riverside barangays in the vicinity, the 63,000 squatters living along the sides of the river, and the 820 companies surrounding the river, 320 of which are known major polluters. Without

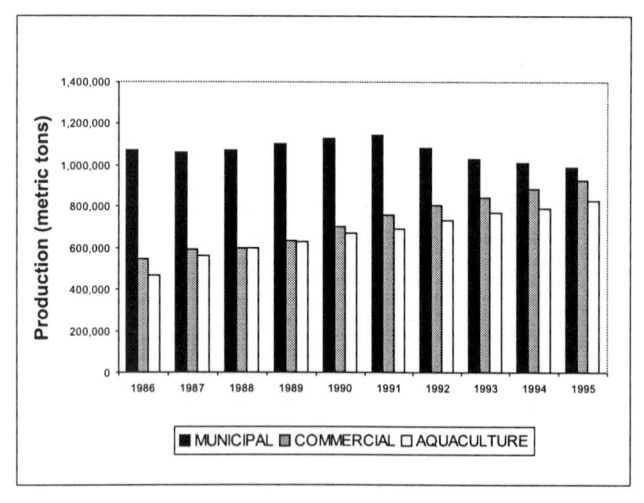

Fig. 5. Fisheries production by sector (1986–1995).

Table 7

Estimated BOD (t d^{-1}) of Pasig River and major tributaries assuming no intervention is initiated (source: DENR, 1991)

	1990	1995	2005
Liquid domestic waste	34	183	250
Solid domestic waste	145	41	55
Commercial and industrial waste	327	135	110
Total BOD	327	359	415

any rehabilitation effort by government, particularly with respect to domestic sewage, while maintaining the existing moderate abatement of industrial and commercial pollution, the BOD loads by pollution source of the river system were projected to reach 415 t d^{-1} (Table 7).

The major industrial regions in the Philippines are Metro Manila and Regions III and IV. Most industrial activities are concentrated near urban areas. Information on toxic and hazardous wastes was obtained by an IMO (1995) study and are summarized as follows:

"Liquid hazardous wastes are currently discharged in an uncontrolled manner to sewers and watercourses. Chlorinated organics, trace metals, pathogens and oily wastes enter the inland water system and seas. Toxic wastes are dumped at municipal landfills (where they occur), posing a health risk to local scavengers. Ineffective monitoring and control of industrial waste occurs because of a lack of government resources. Conflicting land uses and illegal development of open spaces occurs and contributes to further environmental problems. There is also a general lack of awareness among waste generators. Although larger companies may be aware of the adverse health and environmental impacts of hazardous wastes, small and medium-sized generators are ignorant of the potential harm that is associated with their wastes."

"Industrial wastes (both hazardous and non-hazardous) are poorly handled for the most part and are either dumped on land within the plant area or discharged with the wastewater. In Metro Manila, an estimated 25 million m^3 of acid and alkaline liquid waste is disposed annually, primarily from the electronics industry. In addition, almost 2000 m^3 of solvents and 22,000 t of heavy metals, infectious wastes, biological sludges, lubricants and intractable wastes are disposed on land or into watercourses. About 4000 t of solid wastes are generated daily in Metro Manila; of these, only about 3400 t are collected and transported to existing dumpsites."

The problem of eutrophication is probably most pronounced in bay areas, not only where there is a very large human population discharging untreated sewage but also where the tributaries and watersheds that drain into such bays have significant agriculture and aquaculture (e.g., Manila Bay). However, aside from the studies by San Diego-McGlone et al. (1995) that considered the relative importance of nitrogen and phosphorus as indicators of the influence of sewage and run-off into a coastal lagoon, and the work of Jacinto et al. (1999b) that sought to determine the nutrient regimes in Manila Bay, few other assessments have been done on the effect of nutrient inputs on coastal areas in the country.

Shipping and Offshore Accidents and Impacts

Because oil spills are particularly striking and visual, the perception is that they are the major cause of marine pollution in the country. However, long-term studies are indicating that, while a spill in restricted water could be a local disaster, and in some circumstances oil residues could persist in patches for as long as 10 years, operational discharges from ships contribute a greater overall volume of oil to the world's oceans and could be a more widespread threat to birds and beaches than the annual quota of shipping accidents (McIntyre, 1995). Most of the oil spills in the country's waters were in the order of less than 1000 l and were mostly ship-generated or associated with terminal/depot operations. Anecdotal reports of tar balls in Batangas Bay, and in the middle of the South China Sea among the islands, reefs and shoals are consistent with the heavy traffic of oil tankers that go through the area daily.

The assessment of other organic compounds associated with petroleum, such as polycyclic aromatic hydrocarbons (PAH), remain sparse. However, PAHs are ubiquitous marine sediments in Manila Bay (Santiago, 1997) and tend to show highest concentrations near point sources of pollution such as near outfalls of petroleum refineries. Urban run-off appears also to contribute alkylated naphthalenes and phenanthrenes which are associated with petrogenic sources.

Sea-bed Exploration

In the Philippines, offshore seabed development at the moment is focused in the northwest Palawan area. No major spill incidents have been reported involving offshore oil operations at this site. However, initial environmental impact studies on the development of an underwater gas pipeline from the Palawan offshore site crossing towards the Mindoro Strait into Batangas Bay and Bataan suggests some possible areas of concern. An oil blowout in these areas, though a remote possibility, may threaten the environmentally critical areas of northwest Palawan within five days of a major oil spill.

Ports, Harbours and Maritime Transport

Bryant et al. (1998) developed a risk criteria classification to evaluate potential risks from port and harbour activities on reefs. Manila Bay, Subic Bay, Batangas Bay and Puerto Galera Bay are considered high-risk areas. Those considered of medium risk are Bacuit Bay and the Malampaya-

Camago Shell–Oxy exploration site in Palawan. The Mindoro Strait area is considered to be medium risk as well as similar narrow passageways in central and eastern Visayas.

Pollution Hot Spots

Pollution hot spots, derived from overlaying areas considered to be Regional Growth Centres (RGCs) and where there have been incidences of red tide, include Manila Bay and the Calabarzon (Cavite, Laguna, Batangas, Rizal, Quezon) area, the Zambales area where the Subic Bay freeport is located, and the Mindoro strait where Batangas Bay and Puerto Galera Bay interact.

Areas with red tide occurrences are considered high-risk areas (see Fig. 1). Since the first documented report of red tide occurrence in the Philippines in 1983 (Gonzales, 1989), a total of 1992 Paralytic Shellfish Poisoning (PSP) cases and 116 deaths have been reported. The principal causative organism is *Pyrodinium bahamense* var. *compressum*. More recently, however, two other toxic species, *Alexandrium tamayavanichi* and *Dinophysis caudata*, have been reported (Azanza, 1999). While the highest occurrence of red tide has been in Manila Bay, numerous other areas have also experienced these events.

PROTECTIVE MEASURES

In recent years, coastal and marine areas throughout the country have been experiencing serious resource depletion and environmental degradation. These changes have been brought about by resource over-exploitation, pollution from land-based and maritime activities, rapid population growth, and lack of clear government policies and management actions.

However, the Philippines is perhaps one of the most advanced countries in the East Asian region with respect to its appreciation and practice of coastal management as a strategy to address the problems in this sector. Numerous initiatives on the management of the country's coastal areas and resources have been undertaken since the 1980s. These have been promoted under different labels, among them: coastal area management (CAM), coastal zone management (CZM), community-based coastal resources management (CB-CRM), integrated coastal zone management (ICZM), and more recently, integrated coastal management (ICM).

One of the most dominant of these is CB-CRM, defined as "a process by which residents of a coastal community are provided with the opportunity and responsibility of managing their own resources; defining their needs, goals and aspirations; and making decisions and taking actions affecting their well-being". CB-CRM has traditionally focused on areas where people are dependent primarily on natural resources for their livelihood (e.g., fisheries).

More recently, ICZM or ICM systems, defined as "a resource management system which employs an integrative, holistic approach and an interactive planning process in addressing the complex management issues in the coastal area", have been pursued (Chua, 1992). Management measures centre on the promotion of sustainable development by cost-effectively addressing three concerns, namely: (1) adverse environmental and natural resource impacts; (2) resource-use conflicts; and (3) human welfare (Scura et al., 1992). ICM has started to focus on areas not just dependent on natural resource-based sectors but on economies that are more diversified and industrializing, where industrial, manufacturing and service activities are present.

An important feature of ICM as applied to specific localities is the development, adoption, implementation, monitoring and evaluation of environmental management plans. Such plans are meant to employ an integrated, multisectoral, strategic approach for the efficient allocation of scarce resources among competing uses, and to minimize unintended natural resource and environmental effects (Scura et al., 1992).

Since the mid-1980s, there have been at least 45 coastal management programs and projects involving about 150 sites all over the Philippines. These have been carried out by various institutions and groups, and vary according to sites, interventions, and target groups.

In the 1980s, coastal management programs of small island systems (e.g., Sumilon Island in Central Visayas; Apo Island in Negros Oriental) were developed, and a few have been sustained (Kent and Valencia, 1985; Alcala, 1988). However, these models were found inadequate for the management of major coastal areas which subsume more than one town or province. The coastal management of Lingayen Gulf was a pioneering effort to manage a coastal area larger than small islands. Among the lessons learned were: (a) law enforcement programs through punitive measures and vigorous public environmental education through the mass media are easy to implement and have significant potential to create immediate positive impact among local communities; and (b) a program for training resource managers from local government units is crucial for long-term impact.

The devolution of the management of municipal waters to local government units (with the passage of the Local Government Code in 1991) up to 15 km from the shoreline increased interest in coastal resource management. Two major government programs facilitated CRM, the Fisheries Sector Program of the Department of Agriculture (DA) and the Coastal Environment Program (CEP) of the Department of Environment and Natural Resources (DENR).

The Fisheries Sector Program (FSP) under the Department of Agriculture was the largest and most comprehensive attempt to address declining productivity in the artisanal fisheries sector and environmental deterioration in the coastal zone (A. Isidro, pers. comm.). With coastal resources management as its core component, the Program was implemented from 1990 to 1995 by an institutional network of government agencies, non-government organizations and academic institutions in 12 priority bays in the country. Policy reforms were instituted by the Program and have initiated a shift from the open access nature of nearshore fisheries to managed fisheries regimes based on the principle of sustainable yield. Apparently, the CRM component has led to significant improvements in catch rates and biodiversity. However, the imposition of controls on fishing effort was found extremely difficult to establish (A. Isidro, pers. comm.)

The Coastal Environment Program (CEP) was started by the DENR and integrates programs, projects and initiatives related to or concerning coastal environments. The program aims to promote community-based sustainable use of resources in coastal areas by encouraging the use of environment-friendly technologies, providing livelihood opportunities to coastal communities, promoting equitable access to resources, and building DENR capabilities in the management of coastal areas. The strategies of CEP are community organizing, involving communities in the protection and management of coastal ecosystems, mobilizing financial and administrative resources from the public and private sectors, and using contingent approaches in identifying issues, problems and opportunities for human and environmental details.

In 1993, GEF and UNDP approved a project to prevent and manage marine pollution in the East Asian Seas. Implemented by IMO with the cooperation of participating countries, the project sought to demonstrate the effectiveness of various approaches to integrated coastal management (ICM) in tackling marine pollution. Among the project's major activities was a demonstration project at Batangas Bay in the Philippines. The project's management strategy focused on the preventive aspects of ICM and emphasized collaboration between the public and private sectors to anticipate waste management problems. The project has included the private sector as a stakeholder in framing major policies within the management process. It has also actively sought opportunities for developing private/public partnerships, particularly in the area of waste management and marine pollution monitoring. Because of the significant progress made in applying the principles of ICM to prevent and mitigate marine pollution in the East Asian region using Batangas Bay as one of its demonstration sites, UNDP has expressed interest in developing other parallel sites in the Philippines.

Other coastal management initiatives include a USAID-funded Coastal Resource Management Program (CRMP) implemented by the DENR covering six sites: Palawan, Cebu, Negros Oriental, northwest Bohol, Davao del Sur, and Sarangani. This five-year (1996–2001) project provides technical assistance and training to local governments and communities in coastal resource management. Community participation is an inherent and integral part of the program.

In many of these coastal resource management initiatives, the use of marine protected areas (MPAs) as a tool and strategy has gained a lot of support. An MPA is defined as "any area of intertidal or subtidal terrain, together with its overlying water and associated flora, fauna, historical and cultural features, which has been reserved by law or other effective means to protect part or all of the enclosed environment" (Kelleher and Kenchington, 1992). The MPA categories recommended by UNEP include: marine sanctuary, marine reserve, marine park, protected seascape and world heritage site, depending on the purpose, degree of protection and management design (Salm and Clark, 1984).

The number of MPAs in the Philippines was only 16 in 1980 (White, 1988) and rose to 160 in 1994 (GBRMPA/WB/IUCN, 1995). Most MPAs appear to have been designed primarily to improve fishery yields and stocks. Regulations include no-fishing zones, regulated fishing zones or a combination of these, as well as open- and closed-season fishing zones. Enforcement was best accomplished in small areas, while larger MPAs had only moderate enforcement. The Marine Science Institute of the University of the Philippines, with assistance from UNDP, has started a community-based MPA monitoring project aimed to provide training and technical assistance, and systematize and consolidate technical information on MPAs.

In all of the initiatives described, there has been no single government agency coordinating all these efforts, nor a national ICM framework plan being referred to in the implementation of these initiatives. To help address this need, the DENR is soon to undertake a project to prepare a national master plan for the sustainable management of coastal and marine resources and environment. The intent is to take stock of all available information related to the coastal and marine environment, delineate the country's coastal zone into coastal planning units and to develop a typology of coastal planning units. Also expected from this activity is the assessment and classification of coastal areas across various uses to match this "supply" of goods and services from the coast with the projected future demand for coastal resources.

The Philippines was the first country to set up a national council for sustainable development. In 1996, after 20 months of extensive multisectoral consultations, the country completed its own Philippine Agenda 21 as its blueprint for sustainable development.

To address issues on fishery depletion, the National Marine Policy of the Philippines was formulated. This policy underscores the archipelagic nature of the country and highlights the conservation and sustainable use of marine resources as major agents of economic development.

Complementing these coastal management initiatives are laws, executive orders, administrative orders and municipal ordinances that relate to the protection and sustainable use of the marine environment (Table 8).

In October 1998, a DENR Administrative Order entitled, "Guidelines for the Designation of DENR Recognized

Table 8

Laws, decrees, executive orders and administrative orders pertinent to the utilization, conservation, and management of Philippine coastal areas

Law, Decree, Executive Order or Administrative Order	Year enacted
National Water and Air Pollution Control Commission Act	1964
Revised Coast Guard Law	1974
Marine Pollution Decree	1976
National Pollution Control Commission	1976
The Water Code of the Philippines	1976
Environmental Impact Statement System	1978
Regulations for the Conservation of Marine Turtles	1979
The Coral Resources Development and Conservation Decree	1980
Environmental Impact Statement System—Areas/Types of Projects	1981
Small Scale Mining Law	1984
Philippine Environment Code	1988
Toxic Substances and Hazardous and Nuclear Wastes Control Act	1990
National Integrated Protected Areas System Act	1992
Strategic Environmental Plan for Palawan Act	1992
Guidelines on Biological and Genetic Resources	1995
Philippine Mining Act	1995
The Water Crisis Act	1995
Preferential Treatment of Small Fisherfolks	1996
Agriculture and Fisheries Modernization Act	1997
Philippine Environmental Policy	1997
Philippine Fisheries Code	1998
Philippine Clean Air Act	1999

Source: http://www.chanrobles.com/legal9.htm

Environmental Laboratories" (DAO 98-63) was signed. This provides a mechanism to improve the standards of laboratories engaged in generating environmental data. At the same time, the guidelines are meant to encourage private laboratories and groups to participate in environmental monitoring efforts. A complementary initiative of the DENR is to consider the privatization of environmental compliance monitoring.

REFERENCES

Alcala, A.C. (1982) Why conserve Philippine mangroves: Economics and ecology. In *Proceedings of the National Mangrove Research Coordination Symposium Workshop*, pp. 1–6. Cavite, Philippines.
Alcala, A.C. (1988) Effects of marine reserves on coral fish abundance and yields of Philippine coral reefs. *Ambio* 17, 194–199.
Alcala, A.C. and Gomez, E.D. (1985) Fish yields of coral reefs in Central Philippines. In *Proceedings of the Fifth International Coral Reef Congress*, eds. V.M. Harmelin and B. Salvat, pp. 521–524. Antenne Museum, French Polynesia.
Alcala, A.C., Gomez, E.D. and Yap, H.T. (1987) Philippine coral reefs: Status and human responses to changes. *Resource Management Optim.* 4, 297–340.
Aliño, P.M., Juinio-Menez, M.A. and Gomez, E.D. (1995) Developments in planning and management of coastal resources in the Philippines. In *Reg. Workshop on Planning and Management of Coastal Resources*, pp. 21–31. Sabah, Malaysia.
Aliño, P.M., Uychiaco, A.J. and Campos, R. The relevance of coral reef community ecological research to coastal zone and resource management issues and challenges. In *Proceedings of the National Symposium-Workshop on Coral Reef Resources Management Conservation*, April 19–21, 1995, Olongapo City. In press.
Aliño, P.M., Nañola, C.L. Jr., Ochavillo, D.G. and Rañola, M.C. (1998) The fisheries potential of the Kalayaan Island Group, South China Sea. In *Proceedings of the Third International Conference on the Marine Biology of the South China Sea, Hong Kong*, ed. B. Morton, pp. 219–226. Hong Kong University Press.
Aliño, P.M., Pamintuan, I.S., Nañola, C.L. Jr. and Miclat, R.I. (1994) Considerations of spatial scaling in artificial reef monitoring. In *Proceedings of the 3rd ASEAN-Australia Symposium on Living Coastal Resources, Vol. 2: Research Papers*, eds. S. Sudara, C. R. Wilkinson, and L.M. Chou, pp. 187–192. Chulalongkorn University, Bangkok, Thailand.
Apel, J.R., Holbrook, J.R., Liu, A.K. and Tsai, J.J. (1985) The Sulu Sea internal soliton experiment. *Journal of Physical Oceanography* 15, 1625–1651.
Appleton, J.D., Williams, T.M., Breward, N., Apostol, A., Miguel, J. and Miranda, C. (1999) Mercury contamination associated with artisanal gold mining on the island of Mindanao, the Philippines. *Science of the Total Environment* 228, 95–109.
Arceo, H.O., Quibilan, M.C., Alino, P.M., Lim, G. and Licuanan, W.Y. (1999) Coral bleaching in Philippine reefs: Coincident evidences with mesoscale thermal anomalies (Abstract) In *Proceedings of the International Conference on Scientific Aspects of Coral Reef Assessment, Monitoring, and Restoration*. Fort Lauderdale, Florida, p. 48.
Atrigenio, M.P., Bina, R.T., Opena, F. and Valderrama, R. (1992) Evaluation of the environmental impacts of the Mt. Pinatubo eruption on the terrestrial habitats of Zambales, Western Philippines. In *Third ASEAN Science and Technology Week Conference Proceedings, Vol. 6, Marine Science Living Coastal Resources*, eds. L. M. Chou and C. R. Wilkinson, pp. 387–389. Department of Zoology, National University of Singapore and National Science and Technology Board, Singapore.
Austria, B.S. and Rillera, F.G. (1993) The hydrocarbon potential of the South China Sea within Philippine territory and neighboring areas. In *Problems, Prospects, and Policies: Non-living Marine Resources of the Philippines. Scientific Concerns. A Roundtable Discussion*, eds. T.M. Santos and I.M.V. Fernandez, pp. 119–142. Institute of International Legal Studies, University of the Philippines Law Centre, Diliman, Quezon City.
Azanza, R.V. (1999) The ASEAN-Canada Red Tide Network: 1993–1998. In *ASEAN Marine Environmental Management: Towards Sustainable Development and Integrated Management of the Marine Environment in ASEAN. Proceedings of the Fourth ASEAN-Canada Technical Conference on Marine Science*, eds. I. Watson, G. Vigers, C. Ong, C. McPherson, N. Millson, A. Tang, and D. Gass, pp. 341–345. Vancouver and Malaysia: EVS Environment Consultants and Department of Fisheries.
Balce, G.R. and Pablico, E.F. (1997) Philippine natural gas resources: Maximizing their potential. (Unpublished manuscript).
Benoit, G., Schwantes, J.M., Jacinto, G.S. and Goud-Collins, M.R. (1994) Preliminary study of the redistribution and transformation of HgS from cinnabar mine tailings deposited in Honda Bay, Palawan, Philippines. *Marine Pollution Bulletin* 28, 754–759.
BFAR (1997) 1997 Philippine Fisheries Profile. Bureau of Fisheries and Aquatic Resources, Department of Agriculture, Philippines. 51 pp.
BFD (1984) *Forestry Statistics Yearbook*. Bureau of Forest Development, Diliman, Quezon City, Philippines.

Brown, W.W. and Fischer, A.F. (1918) Philippine mangrove swamps. *Bull. Bur. For. Philipp. Is.* **22**, 9–125.

Bryant, D., Burke. L., McManus, J. and Spalding, M. (1998) *Reefs At Risk—A Map-Based Indicator of Threats To The World's Coral Reefs.* World Resources Institute.

Calumpong, H.D. (1994) Status of mangrove resources in the Philippines. In *Proceedings of the 3rd ASEAN-Australia Symposium on Living Coastal Resources, Volume 1: Status Reviews*, eds. C. Wilkinson, S. Sudara and L. M. Chou, pp. 291–310. Australian Institute of Marine Science, Townsville.

Camacho, A. S. and Bagarinao, T. (1987) Impact of fishpond development on the mangrove ecosystem in the Philippines. In *Mangroves of Asia and the Pacific: Status and management*, eds. R.M. Umali et al., pp. 383–405. Natural Resource Centre and National Mangrove Commission, Ministry of Natural Resources, Philippines, Quezon City, Metro Manila.

Carpenter, K.D. (1977) Philippine coral reef fisheries resources. *Philipp. J. Fish.* **17**, 95–125.

Carter, E.F. and Robinson, A.R. (1987) Analysis methods for the estimation of oceanic fields. *Journal of Atmos. Oceanic Technology* **4**, 49–74.

Chapman, D.C., Giese, G.S., Collins, M.G., Encarnacion, R. and Jacinto, G.S. (1991) Evidence of internal swash associated with Sulu Sea solitary waves. *Continental Shelf Research* **11**, 591–599.

Chua, T.-E. (1992) The ASEAN/US coastal resources management project: initiation, implementation and management. In *Integrative Framework and Methods for Coastal Area Management. ICLARM Conference Proceedings*, eds. T.E. Chua and L.F. Scura, pp. 71–92. ICLARM, Philippines

Datingaling, B. (1977) Aquaculture potentials of mangrove areas. In *Proceedings of the National Symposium/Workshop on Mangrove Research and Development*, pp. 42–52. Paranaque, Rizal.

De la Paz, R.M. and Aragones, N. (1985) Mangrove fishes of Pagbilao (Quezon Province, Luzon Island), with notes on their abundance and seasonality. *Natur. Appl. Sci. Bull.* **37**, 171–190.

Del Norte, A.G.C. and Pauly, D. (1990) Virtual population estimates of monthly recruitment and biomass of rabbitfish, *Siganus fuscescens* off Bolinao, Northern Philippines. In *Proceedings of the Second Asian Fisheries Forum*, eds. R. Hirano and I. Hanyu, pp. 851–854. Asian Fisheries Society, Manila.

DENR (Department of Environment and Natural Resources) (1997) *Philippine Biodiversity: An Assessment and Plan of Action.* Bookmark, Makati City, Philippines. 298p.

DENR (Department of Environment and Natural Resources) (1998) Philippine Asset Accounts: Forest, Land/Soil, Marine Fishery, Mineral and Water Resources 1998. Environment and Natural Resources Accounting Sub-Programme. Integrated Environmental Management for Sustainable Development, National Statistical Coordination Board.

DENR-DANIDA (Department of Environment and Natural Resources–Danish International Development Assistance) (1991). Feasibility Study on the Pasig River Rehabilitation Program. DENR, Quezon City, Philippines

DENR/DILG/DA-BFAR/CRMP (Department of Environment and Natural Resources/Department of Interior and Local Government/Department of Agriculture–Bureau of Fisheries and Aquatic Resources/Coastal Resources Management Project) (1997) *Legal And Jurisdictional Guidebook For Coastal Resource Management in the Philippines.* Coastal Resource Management Project, Manila, Philippines, 196 pp.

Evangelista, D. (1992) Management of mangrove areas in Calauag Bay, Quezon Province. *NAGA: The ICLARM Quarterly* **15**, 48–49.

Fertilizer and Pesticide Authority (1998) Fertilizer Handlers per Activity, 1998 Report. Fertilizer and Pesticide Authority, Quezon City, Philippines.

Fortes, M.D. (1991) Structure and productivity of mangroves at five study sites in the Philippines (Abstract). In *Proceedings of the Regional Symposium on Living Resources in Coastal Areas*, ed. A.C. Alcala, p. 474. Marine Science Institute, University of the Philippines.

Fortes, M.D. (1994) Philippine seagrasses: Status and perspectives. In *Proceedings of the 3rd ASEAN-Australia Symposium on Living Coastal Resources, Vol. 1: Status Reviews*, eds. C. Wilkinson, S. Sudara and L. M. Chou, pp. 291–310. Australian Institute of Marine Science, Townsville.

Fortes, M.D. (1995) Seagrasses of East Asia: Environmental and management perspectives. PCU/EAS Technical Report Series No. 6, UNEP.

Fortes, M.D. (1996) Seagrass research in Southeast Asia: where does it lead to? In *Seagrass Biology: Proceedings of an International Workshop*, eds. J. Kuo, R.C. Philips, D.I. Walker and H. Kirkman, pp. 359–366. Faculty of Sciences, The University of Western Australia, Nedlands, Western Australia.

Francisco, H.A. (1994) Upland soil resources of the Philippines: Resource assessment and accounting for soil depreciation. Philippine Environmental and Natural Resources Accounting Project. (ENRAP Phase II). Pasig City. Philippines.

GBRMPA/WB/IUCN (Great Barrier reef Marine Park Authority/The World Bank/The World Conservation Union) (1995) *A Global Representative System Of Marine Protected Areas: Vol. III Central Indian Ocean, Arabian Seas, East Africa And East Asian Seas*, eds. G. Kelleher, C. Bleakley and S. Wells. The International Bank for Reconstruction and Development and The World Bank, Washington D.C., U.S.A.

GEF/UNDP/IMO MPP-EAS and FAO (Global Environment Facility/United Nations Development Programme/International Maritime Organization Regional Programme for the Prevention and Management of Marine Pollution in the East Asian Seas and the Food and Agriculture Organization) (1998) Initial environmental risk assessment of pesticides in the Batangas Bay Region, Philippines and the Xiamen Seas, China. MPP-EAS Technical Report No. 16, 49 p. GEF/UNDP/IMO MPP-EAS, Quezon City, Philippines and FAO, Rome, Italy.

Giese, G.S., Chapman, D.C., Collins, M.G., Encarnacion, R. and Jacinto, G. (1998) The coupling between harbour seiches at Palawan Island and Sulu Sea internal solitons. *Journal of Physical Oceanography* **28**, 2418–2426.

Gomez, E.D. and Alcala, A.C. (1979) Status of Philippine coral reefs. In *Proceedings of the International Symposium on Marine Biogeography and Evolution in the Southern Hemisphere*, pp. 645–661. Auckland, New Zealand.

Gomez E.D. and Yap, H.T. (1985) Coral reefs in the Pacific - their potentials and their limitations. In *Environment and Resources in the South Pacific: A Regional Approach.* UNEP Regional Seas Reports and Studies, pp. 89–106.

Gomez, E.D., Alcala, A.C. and San Diego, A.C. (1981) Status of Philippine coral reefs 1978. In *Proceedings of the 4th International Coral Reef Symposium, Manila, Volume 1*, eds. E.D. Gomez, C.E. Birkeland, R.W. Buddemeier, R.E. Johannes, J.A. Marsh, Jr. and R.T. Tsuda, pp. 275–282. Marine Sciences Centre, University of the Philippines, Diliman, Quezon City.

Gomez, E.D., Alino, P.M., Licuanan, W.R. and Yap, H.T. (1994) Status report on coral reefs in the Philippines 1994. In *Proceedings of the 3rd ASEAN-Australia Symposium on Living Coastal Resources, Vol. 1: Status Reviews*, eds. C. Wilkinson, S. Sudara and L.M. Chou, pp. 57–76. Australian Institute of Marine Science, Townsville, Australia.

Gomez, E.D., Alino, P.M., Yap, H.T. and Licuanan, W.Y. (1994a) A review of the status of Philippine reefs. *Marine Pollution Bulletin* **29**, 62–68.

Gomez, E.D., Deocadiz, E., Hungspreugs, M., Jothy, A.A., Kuan, K.J., Soegiarto, A. and Wu, R.S. (1990) State of the marine environment in the East Asian Seas Region. UNEP Reg. Seas Rep. Stud.

Gonzales, F.R. (1977) Mangrove and estuarine area development in the Philippines. Presented in the International Workshop on Mangrove and Estuarine Area Development for the Indo-Pacific Region, 14–19 November 1977. Unpublished.

Gonzales, C.E. (1989) Pyrodinium blooms and paralytic shellfish poisoning in the Philippines. In *Biology, Epidemiology and Management of Pyrodinium Red Tides. ICLARM Conference Proceedings 21*, eds. G.M. Hallegraeff and J.L. Maclean, pp. 39–47. Fisheries Department, Ministry of Development, Brunei Darussalam, and International Centre for Living Aquatic Resources Management, Manila, Philippines.

Hamilton, L.S. (1984) A perspective on forestry in Asia and the Pacific. In *Wallaceana (A Global Newsletter for Tropical Ecology)*, ed. A. Sasekumar, pp. 3–8. University of Malaya, K.L., Malaysia.

Heong, K.L., Escalada, M.M. and Mai, V. (1994) An analysis of insecticide use in rice—case studies in the Philippines and Vietnam. *International Journal of Pest Management* 40, 173–178.

Hodgson, G. and Dixon, J. (1988) Logging versus fisheries and tourism in Palawan: An environmental and economic analysis. East-West Environment and Policy Institute Occasional Paper No. 7.

Hung, T.C., Tsai, C.C. and Chen, N.C. (1986) Chemical and biomass studies. 1. Evidence of upwelling off the southwestern coast of Taiwan. *Acta Oceanographica Taiwanica* 17, 29–44.

IMO (1995) Global Water Survey (1995). Final Report, International Maritime Organization. Manila, Philippines.

Jacinto, G.S., Duyanen, J.P., Narcise, C.I.S. and Santiago, E.C. (1999a) Approaches in assessing contamination of sediments by trace elements and PAHs in Manila Bay. In *ASEAN Marine Environmental Management: towards sustainable development and integrated management of the marine environment in ASEAN. Proceedings of the fourth ASEAN CANADA Technical Conference on Marine Science*, eds. I. Watson, G. Vigers, K.S. Ong, C. McPherson, N. Millson, A. Tang and D. Gass, pp. 352–363. EVS Environment Consultants, North Vancouver and Department of Fisheries, Malaysia.

Jacinto, G.S., San Diego-McGlone, M.L., Velasquez, I.B. and Smith, S.V. (1999b) N and P budget of Manila Bay, Philippines. In *ASEAN Marine Environmental Management: towards sustainable development and integrated management of the marine environment in ASEAN. Proceedings of the fourth ASEAN CANADA Technical Conference on Marine Science*, eds. I. Watson, G. Vigers, K.S. Ong, C. McPherson, N. Millson, A. Tang and D. Gass, pp. 341–345. EVS Environment Consultants, North Vancouver and Department of Fisheries, Malaysia.

Jara, R.S. (1984) Aquaculture and mangroves in the Philippines. In *Proceedings of the Workshop on Productivity of the Mangrove Ecosystem: Management Implications*, eds. J.-E. Ong and W.-K. Gong, pp. 97–107. Universiti Sains Malaysia.

Kelleher, G. and Kenchington, R. (1992) *Guidelines for Establishing Marine Protected Areas*. A Marine Conservation and Development Report. IUCN, Gland, Switzerland, 79 pp.

Kent, G. and Valencia, M.J. (1985) *Marine Policy In South East Asia*. University of California Press.

Lacanilao, F. and Fernandez, P. (1998) Ignoring sustainability and social equity in fisheries development. *UP-CIDS Chronicle* 3, 129–130.

Liu, A.K., Holbrook, J.R. and Apel, J.R. (1985) Nonlinear internal wave evolution in the Sulu Sea. *Journal of Physical Oceanography* 15, 1613–1624.

Macintosh, D.J. and Phillips, M. J. (1992) Environmental issues in shrimp farming. In *Proceedings of the 3rd Global Conference on the Shrimp Industry*, eds. H. de Saram and T. Singh, pp. 118–145. Infofish, Kuala Lumpur, Malaysia.

Maclean, J.L. (1989) Indo-Pacific red tides, 1985–1988. *Marine Pollution Bulletin* 20, 304–310.

McIntyre, A.D. (1995) Human impact on the oceans: the 1990s and beyond. *Marine Pollution Bulletin* 31, 147–151.

McManus, J., Nañola, C.L., Reyes, R.B. Jr. and Kesner, K.N. (1992) Resource ecology of the Bolinao coral reef system. *ICLARM Studies and Review* 22, 117 p.

Megia, T. (1952) Oceanographic background of Philippine fisheries. In *Bureau of Fisheries Handbook*, pp. 10–19. Bureau of Fisheries, Department of Natural Resources, Quezon City, Philippines.

Naylor, R.L., Goldburg, R.J., Mooney, H., Beveridge, M., Clay, J., Folke, C., Kautsky, N., Lubchenco, J., Primavera, J. and Williams, M. (1998) Ecology—nature's subsidies to shrimp and salmon farming. *Science* 282, 883–884.

NEDA (National Economic Development Authority) (1998) The Philippine National Development Plan—Directions for the 21st Century. NEDA. Manila, Philippines

Nielsen, E.S. and Jensen, E.A. (1957) Primary oceanic production. The autotrophic production of organic matter in the oceans. "*Galathea*" *Rep.* 1, 49–120.

NSCB (1997) Philippine Statistical Yearbook. National Statistical Coordination Board, Sen. Gil J. Puyat Ave., Makati City, Philippines.

NSCB (1999) Philippine Statistical Yearbook. National Statistical Coordination Board, Sen. Gil J. Puyat Ave., Makati City, Philippines.

Ochavillo, D.G., Hernandez, H.B. and Aliño, P.M. (1992) The effects of Mt. Pinatubo eruption on the coral reef fish community of Zambales. In *Third ASEAN Science and Technology Week Conference Proceedings, Vol. 6, Marine Science Living Coastal Resources*, eds. L. M. Chou and C. R. Wilkinson, pp. 163–165. Department of Zoology, National University of Singapore and National Science and Technology Board, Singapore.

Padilla, J.E. and Cortez, A.C. (1996) Implications of a fishery collapse in the Philippines: Applications to small-pelagic fishery. In *Philippine Environmental and Natural Resources Accounting Project (ENRAP Phase III): Policy Studies*. DENR and USAID, Philippines.

Padilla, J.E. and Morales, A.C. (1997) Evaluation of fisheries management alternatives for Lingayen Gulf: An options paper. In *Philippine Environmental and Natural Resources Accounting Project*. DENR and USAID, Philippines.

PCAMRD (1991) Management of nearshore fishery resources. In *Proceedings of the Seminar-Workshop on Management of Nearshore Fishery Resources*. (Book Series No. 10/1991).

PNMC (Philippine National Mangrove Committee) (1986) Philippine Report. In *Mangroves of Asia and the Pacific: Status and management*. Technical Report of the UNDP/UNESCO Research and Training Pilot Programme on Mangrove Ecosystems in Asia and the Pacific (RAS/79/002), pp. 175–210.

Philipps, R.C. and Meñez, E.G. (1988) Seagrasses. In *Smithsonian Contributions to the Marine Sciences* 34, 104 pp.

Primavera, J.H. (1991) Intensive prawn farming in the Philippines: Ecological, social and economic implications. *Ambio* 20, 28–33.

Primavera, J.H. (1993) A critical review of shrimp pond culture in the Philippines. *Reviews in Fisheries Science* 1, 151–201.

Primavera, J.H., Lavilla-Pitogo, C.R., Ladja, J.M. and Dela Peña, M.R. (1993) A survey of chemical and biological products used in intensive prawn farms in the Philippines. *Marine Pollution Bulletin* 26, 35–40.

Primavera, J.H. (1998) Tropical shrimp farming and its sustainability. In *Tropical Mariculture*, ed. S. de Silva, pp. 257–289. Academic Press, London.

Prudente, M.S., Ichihashi, H. and Tatsukawa, R. (1994) Heavy metal concentrations in sediments from Manila Bay, Philippines and inflowing rivers. *Environmental Pollution* 86, 1–6.

Punongbayan, R. (1998) Geologic History. In *Environment and Natural Resources Atlas of the Philippines*, ed. G. Magdaraog. Environment Centre of the Philippines Foundation, Manila.

Ronquillo, I.A. (1988) The fishery resources in the Philippines and its relation to mariculture development. In *Proceedings of the Workshop on Appropriate Applied Mariculture and Fisheries*, pp. 93–117. Silliman University Press, Dumaguete City, Philippines.

Salm, R.V. and Clark, J.R. (1984) *Marine and Coastal Protected Areas: A Guide for Planners and Managers*. International Union for the Conservation of Nature and Natural Resources, Gland, Switzerland.

San Diego-McGlone, M. L., Villanoy, C. L. and Aliño, P. M. (1995) Nutrient-mediated stress on the marine communities of a coastal lagoon (Puerto Galera, Philippines) *Marine Pollution Bulletin.* **31**, 355–366.

San Diego-McGlone, M.L., Jacinto, G.S., Dupra, V.C., Narcise, I.S., Padayao, D.O. and Velasquez, I.B. (1999) A comparison of nutrient characteristics and primary productivity in the Sulu Sea and South China sea. *Acta Oceanographica Taiwanica* **37**, 219–229.

Santiago, E.C. (1997) The levels and distribution of polycyclic aromatic hydrocarbon contaminants in bottom sediments in Manila Bay. *Science Diliman.* Vol. 10.

Scura, L., Chua, T.-E., Pido, M.D. and Pao, J.N. (1992) Lessons for integrated coastal zone management: the ASEAN experience. In *Integrative Framework and Methods for Coastal Area Management. Proceedings of the Regional Workshop on Coastal Zone Planning and Management in ASEAN: Lessons Learned,* eds. T.-E. Chua, and L. F. Scura, pp. 1–70. ICLARM, Manila.

Shaw, P.T., Chao, S.Y., Liu, K.K., Pai, S.C, and Liu, C.T. (1996) Winter upwelling off Luzon in the northeastern South China Sea. *Journal of Geophysical Research* **101**, 16,435–16,448.

Silvestre, G.T. (1989) Philippine marine capture fisheries—exploitation, potential and option for sustainable development. Working Paper No. 48, Fisheries Stock Assessment, Collaborative Research Support Program.

Siringan, F., Ringor, C., Berdin, R. and Santos, M.G. (1999) Impact of the 1991 eruption of Mount Pinatubo on Manila Bay sedimentation. Presented at the 4th International Conference on Asian Marine Geology: Geology of the Asian Oceans in the 21st Century, 14–18 October 1999, Quingdao, China.

Talaue-McManus, L. and Kesner, K.P.N. (1995) Valuation of a municipal sea urchin fishery and implications of its collapse. In *Philippine Coastal Resources Under Stress. Selected Papers from the Fourth Common Property Conference,* eds. Juinio-Meñez and Newkirk. Manila, Philippines.

Ungson, J.R. (1990) The fry fishery of Ilocos Norte, Philippines. In *Proceedings of the Second Asian Fisheries Forum,* eds. R. Hirano and I. Hanyu, pp. 777–781. Asian Fisheries Society, Manila.

Velasco, A.B. (1979) Some socio-cultural factors influencing mangrove utilization: Their implications on conservation and economic policies in the Philippines. In *Proceedings of the Symposium on Mangrove and Estuarine Vegetation in Southeast Asia,* eds. P.B.L. Srivastava, A.M. Ahmad, G. Dhanarajan and I. Hamzak, pp. 181–189. Universiti Pertanian Malaysia, Serdang, Selangor, Malaysia.

Watson, D., Ong, K.S. and Vigers, G. (Eds.) (1995) ASEAN Criteria and Monitoring: Advances in Marine Environmental Management and Human Health Protection. In *Proceedings of the ASEAN-Canada Midterm Review Conference on Marine Science, Singapore.* EVS Environment Consultants, Vancouver and National Science and Technology Board, Singapore. 422 pp.

White, A.T. (1988) Marine parks and reserves: management for coastal environments in Southeast Asia. ICLARM Education Series 2, 36 pp. International Centre for Living Aquatic Resources Management, Manila, Philippines

Wyrtki, K. (1961) Physical Oceanography of the Southeast Asian Waters. Scientific results of marine investigations of the South China Sea and the Gulf of Thailand, *NAGA Report,* Vol. 2, Scripps Institution of Oceanography, La Jolla, California. 196 pp.

Yap, H.T. and Gomez, E.D. (1985) Coral reef degradation and pollution in the East Asian Seas Region. In *Environment and Resources in the Pacific,* eds. A.L. Dahl and J. Carew-Reid, pp. 185–207. UNEP Regional Seas Rep. Studies No. 69.

Zamora, P.M. (1982) Conservation and management of Philippines mangroves. In *Proc. First Nat. Conserv. Conf. Nat. Res. Manila* 1981, pp. 85–104.

THE AUTHORS

Gil S. Jacinto
*Marine Science Institute,
University of the Philippines,
1101 Diliman, Quezon City, Philippines*

Porfirio M. Aliño
*Marine Science Institute,
University of the Philippines,
1101 Diliman, Quezon City, Philippines*

Cesar L. Villanoy
*Marine Science Institute,
University of the Philippines,
1101 Diliman, Quezon City, Philippines*

Liana Talaue-McManus
*Marine Science Institute,
University of the Philippines,
1101 Diliman, Quezon City, Philippines*

Edgardo D. Gomez
*Marine Science Institute,
University of the Philippines,
1101 Diliman, Quezon City, Philippines*

Chapter 80

THE CORAL, SOLOMON AND BISMARCK SEAS REGION

Michael E. Huber and Graham B.K. Baines

The Coral, Solomon, and Bismarck Seas lie in the western equatorial Pacific, bounded by Australia, New Caledonia, Vanuatu, the Solomon Islands and Papua New Guinea (PNG). PNG and the Solomon Islands are the coastal countries discussed.

North–south migration of the Inter-Tropical Convergence Zone brings about alternating Trade Wind and monsoon conditions. The South Equatorial Current (SEC) drives the major ocean circulation. The region's major land mass is mainland Papua New Guinea (PNG). All remaining land is grouped into archipelagos. Major coastal ecosystems are coral reefs, mangroves, seagrass beds, soft estuary and lagoon bottoms, inland saltwater lagoons, and freshwater coastal wetlands. The biodiversity of these coastal ecosystems is of global significance in extent and diversity and, overall, their condition is generally good. Little of the baseline knowledge needed to monitor trends and to assess environmental change is available, and the level of research and survey is very low.

Though the open seas of the region are characterised by low primary production, it is likely that upwelling and turbulent mixing around reefs and islands results in localised areas of high primary productivity. The region's large pelagic tuna resource is being exploited by both PNG and the Solomon Islands. Both receive support in this from the South Pacific Forum Fisheries Agency (FFA).

The majority of the population of the countries of the region (about 1.4 million) lives along the 22,400 km coastline. Port Moresby is the largest urban population (250,000). Urban centres are characterised by inadequate provision for treatment and disposal of human and industrial wastes, localised overfishing, coastline disturbances arising from land sediment discharge, and from poorly designed and positioned engineering structures, and chronic low-level pollution.

Outside urban areas the level of material development is very low and most people live a mixed subsistence and artisanal lifestyle. While traditional agriculture is the main source of food and livelihood there is a high degree of dependence upon coastal fisheries for both subsistence and income. Few coastal villagers are solely farmer or fisher.

Local fisheries are diverse and, apart from the wide range of finfish and invertebrates taken for food, there is an export trade in sea cucumbers (to produce bêche-de-mer), molluscs (for mother-of-pearl), and a few high-value food species. Not only are these stocks inherently difficult to manage but coastal fisheries management capacity in the countries of the region is very low. Over-exploitation is common. Recently, a trade in live reef food fish for export has emerged. Characterised by a quick depletion of stocks, and the use of toxic substances for stunning fish, it is considered to be a significant threat to the sustainability of coastal marine resources and to coral reef ecosystems.

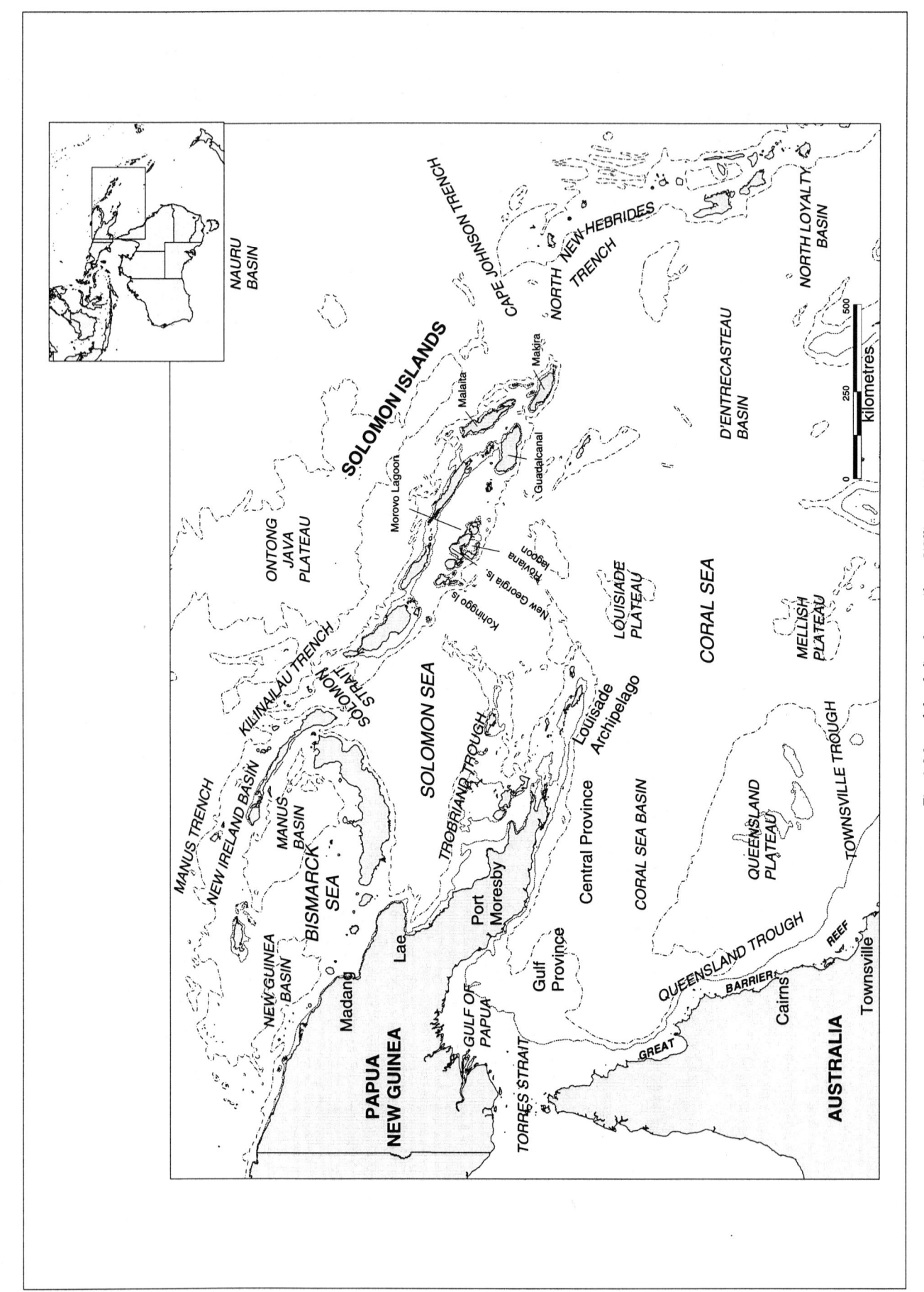

Fig. 1. Map of the defined area (from Williams, 1994).

Land-slips and their sediment plumes in nearby coastal waters are a natural feature of the dynamic landscape. Where land is disturbed by logging and agriculture, additional sediment stress on aquatic ecosystems is superimposed on this natural background stress.

Poor logging practices are effecting major changes in the nature of the land and the ecosystems from which freshwater enters the sea. Sediment loads are increased, stressing coral reefs and seagrass beds. This is accompanied by increases in the amount of freshwater runoff, adding further stress.

Domestic sewage is the most important source of point-source pollution, not only in urban areas but sometimes in rural lagoons. The low level of industrial development has precluded widespread industrial pollution but localised pollution from mines, breweries, food and agricultural processing plants, and probably other industries has occurred. There is no systematic water quality, biological, or public health monitoring of coastal waters.

Though both countries in the region have a legislative and administrative base for environmental and resource management, effective implementation is compromised by a lack of political will and weak—and declining—capacity in the management agencies. Increased attention is being paid to the promotion and support of community-based approaches that incorporate elements of traditional land and sea tenure systems and management practices.

THE CORAL, SOLOMON, AND BISMARCK SEAS REGION

At a broad scale the region's seas can be considered as three partially enclosed, deep basins at the western extreme of the southern equatorial Pacific (Fig. 1). They are bounded by Papua New Guinea (PNG), which consists of the main island of New Guinea and several hundred smaller islands, and the Solomon Islands, another archipelagic nation (Table 1). The Coral Sea is also bounded by Vanuatu (discussed by Vuki, this volume) and New Caledonia (discussed by Labrosse et al., this volume) to the east and south, the Great Barrier Reef (GBR) to the west (discussed by Zann, this volume), and Torres Strait and the Gulf of Papua (including the Western and Gulf Provinces of PNG) to the northwest (discussed by Huber, this volume). Although only the southern coasts of the Solomon Islands and Bismarck Archipelago border directly on the Bismarck and Solomon Seas, the entire coastal area of these island groups is included in the discussion.

Table 1

Physical characteristics of Papua New Guinea and Solomon Islands (sources: Davis and Huber, 1995; Wright, 1993)

	Papua New Guinea	Solomon Islands
Land area (km^2)	462,243	27,556
EEZ area (km^2)	3,120,000	1,340,000
Length of coastline	17,100	5,300
No. of high islands	83	37
No. of low islands	>350	65
No. of atoll islands and cays*	47	2

*These are probably underestimates.

The region's seas are of a general depth of the order of 4000–4500 m in the Coral Sea Basin, reducing to 2000 m above plateaus, and these depths extend eastward to New Caledonia and Vanuatu. In the vicinity of the Solomon Islands, and north of Papua New Guinea in the Bismarck Sea, depths are much less; typically of the order of 1000–1500 m. However, a marked seabed feature of the latter area is the New Britain Trench, whose bottom approaches 8000 m. The island of Guadalcanal (Solomon Islands) is the first of a long line of islands, extending east and south, and eventually along the full length of Vanuatu, which emerge from a ridge adjacent to a series of narrow ocean trenches which are mostly 5000 m in depth. The area lies at the convergent boundary of the Pacific and Australian lithospheric plates and is geologically complex (Pandolfi, 1992; Vedder and Coulson, 1986). The New Britain Trench delineates the small Bismarck lithospheric plate. The San Cristobal Trench marks the eastern edge of the small Solomon Plate.

The southern, or Papuan, coast of mainland PNG is largely of sedimentary origin overlying continental material. The northern coast was formed by the accretion of material as New Guinea collided with a Pacific Island arc commencing in the late Oligocene (Pigram and Davies, 1987). The islands north and east of the eastern tip of the island of New Guinea are of continental and raised-reef origin, while most of the northern islands of PNG, and of the Solomon Islands, are primarily volcanic, often with thick overlying raised reef deposits.

Most coastal areas of mainland PNG and the high islands are backed by mountains, and the coastline is often steep with a narrow, even non-existent, continental shelf. Important exceptions are the broad coastal plain of the Sepik River, the world's 18th largest in terms of water discharge (Alongi, 1990) and an extensive shelf area in Milne Bay Province. Most coastal watersheds are forested, although there are some areas of apparently human-induced grasslands in areas with a pronounced dry season (Leary, 1993a; Paijmans, 1982).

It is doubtful that the Large Marine Ecosystem (LME) concept fits this seas region. Environmental variability is driven largely by global-scale oceanographic and climatic systems. The area can be considered as having several distinct ecosystems that probably have stronger interactions with adjoining systems outside the region than with each other. The Papuan Barrier Reef on the south coast of PNG, for example, has greater geological and biogeographic affinities with the GBR to the south than with the north coast of PNG (e.g., Hoeksma, 1992).

A DYNAMIC TROPICAL ENVIRONMENT: SEASONS, CURRENTS, SEISMICITY, VOLCANICITY AND CYCLONIC STORMS

Climate and Currents

The overall regional climate is "wet tropical", with average minimum and maximum temperatures of around 22° and 32°C, respectively, and relatively little annual variation (McAlpine et al., 1983; Spenceley, 1982). Annual rainfall generally ranges between 2500 mm and 5000 mm and, though there is seasonal variation, (see below), there is considerable rainfall throughout the year in most areas. Significant exceptions are "rain shadow" areas such as the central Papuan coast (which receives less than 1500 mm per year with almost no rain during the pronounced dry season of May to December) and north Guadalcanal in the Solomon Islands.

From May or June to October or November, depending on the location, the weather is dominated by southeasterly Trade Winds that blow steadily at average speeds of 10–15 knots, somewhat higher in the western Coral Sea (Fig. 2). Over most of the region this represents a dry season of varying degree; although the southeastern, windward, coasts of the islands of Guadalcanal, Malaita, and Makira experience rainfall maxima during the Trades. From December or January to March or April monsoonal conditions prevail as the Inter-Tropical Convergence Zone migrates southward over the area. It is the monsoons which bring most areas their rainfall maxima and the winds, lighter and more variable than the Trade Winds, are predominantly from the northwest. The effects of the northwesterly monsoon decrease progressively from north to south. The doldrums between the Trade and monsoon wind periods are characterised by light, variable winds and relatively frequent calms.

Ocean circulation is driven by the South Equatorial Current (SEC), the westward path of which is blocked by the islands of New Guinea and Australia (Williams, 1994). The southernmost component of the SEC bifurcates in the southwestern Coral Sea, off the coast of Australia. The northern branch of this bifurcation moves northwesterly in the Coral Sea as the Coral Sea Coastal Current before turning into the Solomon Sea as the New Guinea Coastal Undercurrent. This pattern constitutes a true western boundary current (Burrage et al., 1995). In the Solomon Sea a small countercurrent flows southeasterly along the southern coast of the Solomon Islands. This general circulation pattern is substantially modified by seasonal shifts in wind patterns and by north–south movement of the Equatorial Countercurrent system in relation to bathymetry and land masses (Fig. 2). During the monsoon the SEC moves southward, bringing the Bismarck and Solomon Seas under the influence of the North Equatorial Counter Current flowing from the northwest and causing seasonal reversal of flows around the periphery of the Bismarck Sea. Surface currents in the Bismarck and Solomon Seas are also strongly influenced by seasonal wind patterns, those in the Coral Sea less so.

The oceanography of the region is of some global significance. Its waters are one of the sources of the Indonesian Throughflow from the Pacific to the Indian Ocean, aided by the SEC. Also, the Bismarck and northern Solomon Seas lie within a large, semipermanent pool of warm surface water believed to have a critical role in global climate regulation. A further point of interest is the suggestion that tectonic processes in this region may have played a significant role in enhancing primary productivity in the iron-limited eastern Pacific by injecting iron into the source waters of the Equatorial Undercurrent (Wells et al., 1999).

El Niño

El Niño events typically bring drought to the region. At the ocean basin scale El Niño is associated with lower sea-surface temperatures as the western Pacific pool of warm surface water moves eastward in response to reduced trade winds. At local scales, however, there may be surface warming. Elevated surface temperatures, for example, were observed on reefs in New Britain during the 1997–98

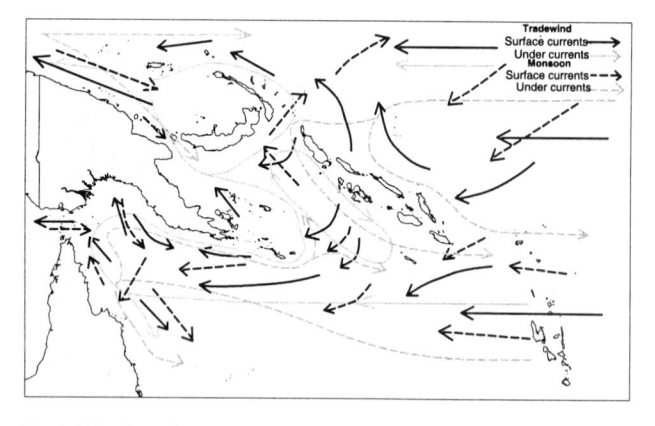

Fig. 2. Winds and current patterns during Trade Wind and monsoon conditions (from Williams, 1994).

El Niño and coincided with mass coral bleaching and mortality (Wilkinson, 1998). Bleaching events, however, are not necessarily associated with unusually high water temperatures. For example, bleaching was reported from Milne Bay during 1997–98 in the absence of abnormally high water temperatures. Neither does bleaching occur only during El Niño events. The first published report of mass bleaching in the region, again from Milne Bay, was from early 1996 when El Niño was not underway (Davies et al., 1997).

Earthquakes, Volcanoes, and Cyclones

The area is tectonically active and experiences occasional volcanic eruptions, and numerous earthquakes. Recent dramatic events include a major volcanic eruption which obliterated much of the town of Rabaul in 1994 and, in 1998, a seismic sea wave generated by a submarine earthquake in the floor of the Bismarck Sea which killed at least 2200 coastal villagers. Wells and Jenkins (1988) have noted tectonic activity as a significant natural disturbance to the region's coastal ecosystems. This point is well taken. Over the period 1960–80, for instance, the island of Guadalcanal (Solomon Islands) experienced 34 strong earthquakes (Richter magnitude 5.5 and greater) and some of these produced mild tsunamis and localised coastline change. Perhaps of greater significance for coastal shallow water ecosystems were the land-slips triggered by these events. For instance, those from a series of four 1977 earthquakes on Guadalcanal dramatically increased stream bedloads and steepened stream gradients. Offshore sediment plumes from these streams persisted for many years (Baines, pers. obs.). There have presumably been significant ecological changes resulting from such long-term sediment impacts. Coral community structure, for example, has been shown elsewhere to be affected by elevated sediment loads (Hodgson, 1994). It is important to note that, while some of these land-slips were associated with subsistence food gardens on steep slopes, many occurred where there had been no human disturbance. A general point to be drawn is that these land-slips and their sediment plumes in nearby coastal waters are a natural feature of a dynamic landscape (Baines and Morrison, 1990).

Elsewhere in the Solomon Islands one of us (Baines) has examined cracking, subsidence, and the breaking off of reef edges on narrow fringing reefs alongside the islands of Vella Lavella and Santa Ysabel arising from earthquake events. Though some short-term effects arise from differential coral species damage, the physical changes have given rise to a diversification of physical habitat which can be expected to contribute positively to species diversity. On a larger scale, at the time of a severe Guadalcanal earthquake in 1977 a reef said to have been at a depth of 10 m was forced upwards to the sea surface. As sand deposits formed on it and built up into cays, it was claimed as new land under local customary law.

Tropical cyclones form in the Coral Sea or in the southwestern Pacific east or northeast of the Solomons (McAlpine et al., 1983). Their path after formation is unpredictable, but generally includes a southwesterly component, so that most cyclones originating in the Coral Sea move away from PNG, although they regularly impact remote reefs in the Coral Sea and occasionally move northward to affect the Papuan coast. The only part of PNG subject to major cyclonic events is its southeastern tip, including the Milne Bay area and the archipelagoes to the east. In the Solomon Islands, too, most cyclones do not cross over the country's major islands although the country still experiences an average of 1.2 cyclones a year (Leary, 1993a). Cyclone impacts on coastal shallow water ecosystems are primarily episodic sediment and freshwater fluxes resulting from high-intensity rainfall and subsequent soil erosion. No cyclone-induced wave impacts on coral communities have been studied, though it is reasonable to expect these to be a natural source of disturbance.

SHALLOW WATER MARINE AND COASTAL HABITATS

The region's major shallow-water coastal biomes are coral reefs, mangrove forests, and seagrass beds. Other important habitats are soft estuary and lagoon bottoms. These biomes and the habitats which they provide are ecologically linked at a number of locations with coastal swamps and floodplains.

Reef–Mangrove–Seagrass Complex

Most of the coastline of PNG has associated reefs, though there are extensive stretches of shoreline which are devoid of significant fringing reef development where natural freshwater and sediment levels are high, as along the Gulf of Papua (see Huber, this volume). A standard, but probably high, estimate is that there are 40,000 km^2 of coral reef and associated shallow habitats to a depth of 30 m (Dalzell and Wright, 1986). A more recent estimate of 12,000 km^2 (Bryant et al., 1998) is probably an underestimate: a map based on the same database fails to indicate the presence of reefs in many areas (e.g., most of the north coast) where in fact they occur. Even this low estimate represents nearly 5% of the world's coral reef area, and the fourth highest of any country (Bryant et al., 1998). About a third of PNG's reef area occurs around its southeast tip, in Milne Bay Province. Though coral reefs occur in all coastal provinces, their development is restricted along the northwestern mainland coast due to the influence of the Sepik and Ramu rivers.

Reef development in the Solomon Islands is less extensive. Bryant et al. (1998) estimate that there are 6000 km^2 of coral reef in the Solomon Islands. Stoddart (1969a) believed the paucity of living coral on reef flats could be the result of tectonic upheaval of reefs, a small eustatic fall in sea level, or seasonal fluctuations. Since this condition persists, the latter possibility can probably be ruled out. Wells and Jenkins (1988) suggest that scarcity of suitable substrate on the steep coastline, or the effects of storms, tectonic activity, and/or the depression of sea level during extreme ENSO events may be responsible.

Fringing and patch reefs are most common, but the full spectrum of reef types occurs, including barrier reefs, atolls, and coral pinnacles in both countries. The Papuan Barrier Reef extends from Yule Island for some 900 km along the south coast of PNG to merge with the extensive reef systems of Milne Bay. Geologically and biogeographically it is essentially an extension of the GBR. Another large barrier reef is found off Madang, on the north coast. Some of the many islands of the Solomon Islands have associated barrier reefs. Those at Marovo Lagoon on New Georgia Island and at Kohinggo Island along Blackett Strait are of particular interest because they are considered to be among the best-defined double barrier reefs in the world. The barrier reef at Marovo Lagoon, in fact, is in part a triple barrier (Stoddart, 1969 a,b). There are many smaller barrier reefs. More detail on the coral reefs of PNG and the Solomon Islands is provided by Wells and Jenkins (1988).

Mangroves are found on most coasts where there is some protection from wave exposure and the shoreline is of sufficiently low gradient to allow accumulation of suitable sedimentary substrate. They are best developed in bays, lagoons, and estuaries, where there is more fresh water input, soft substrate, and protection from waves. Mangroves cover more than 4100 km^2 in PNG and 640 km^2 in the Solomon Islands (Saenger et al., 1983). Most of PNG's mangroves are found in Western and Gulf provinces (see Huber, this volume), but there are also significant areas of mangrove in other parts of the country. Central Province, for example, has nearly 600 km^2 of mangroves (Osborne, 1993). Other large areas of mangrove in PNG include the mouth of the Sepik–Ramu river system, Madang lagoon, the mouth of the Markham River at Lae, and the Orangerie Bay/Mullins Harbour area of Milne Bay Province (Percival and Womersley, 1975; Huber, pers. obs.) The largest mangrove areas in Solomon Islands are at Santa Isabel, New Georgia, Choiseul and South Malaita (Maramasike Passage; Hansell and Wall, 1976).

In PNG, the genera *Rhizophora, Bruguiera, Sonneratia,* and *Avicennia* typically dominate the seaward parts of mangrove forests, while common genera to landward include *Ceriops, Osbornia, Lumnitzera,* and *Sonneratia* (Cragg, 1987; Johnstone and Frodin, 1982; Percival and Womersley, 1975). Well-developed estuaries with enough freshwater input to produce very low (<1 ppt) salinities usually have stands of the nypa palm (*Nypa fruticans*). There are extensive stands on the Sepik floodplain. *Rhizophora, Bruguiera,* and to a lesser extent *Avicennia* are the dominant genera of the Solomon Islands (G. Pillai, pers comm; Leary, 1993a).

Seagrasses are widespread on reef flats and shallow areas of lagoon floors in the Coral, Solomon, and Bismarck Seas. Coastlines that slope steeply into deep water have relatively little seagrass. Seagrass beds support a diverse and productive flora and fauna, are breeding and nursery grounds, and are feeding grounds for dugong (*Dugong dugon*) and the green turtle (*Chelonia mydas*). The distribution of seagrass beds in the region has not been mapped, and no quantitative estimates of the area of seagrass coverage are available. There have been some geographic surveys of seagrass PNG (Balat, 1993; Heijs and Brouns, 1986; Johnstone, 1982), and detailed studies of the associated flora and fauna, trophic relationships, and community structure of seagrass beds near Port Moresby (Brouns and Heijs, 1991; Hattori, 1987; and references therein). Seagrass beds occur in the Solomon Islands, but have not been the subject of scientific investigation.

The coral reefs, mangroves, and seagrass beds rarely exist as isolated ecological units. Usually they function as interacting components of an integrated coastal ecosystem. In a typical barrier reef–lagoon–estuary system, for example, the barrier reef and associated sand cays enclose a lagoon with extensive development of patch reefs. A fringing reef along the shoreline is backed by mangroves, and seagrass beds are on parts of the reef flats and shallow (typically 30–50 m) lagoon floor. Prominent examples of this type of coastal complex include Marovo and Roviana Lagoons in the Solomons, which are among the world's largest, and the Papuan and Madang Lagoons in PNG. Wells and Jenkins (1988) provide some detail of Marovo Lagoon's coral reefs, while Osborne (1993) details the coastal swamp forests through which freshwater feeds into the lagoon. Extensive soft-bottom areas are also associated with these systems, but the soft-bottom communities here, and elsewhere in the region have been poorly studied.

Coastal Swamps

A number of enclosed coastal saltwater lagoons are of interest. The largest, the Murik Lakes at the mouth of the Sepik, cover about 90 km^2 with 185 km^2 of associated mangroves (Dalzell et al., 1996). Similar but smaller lagoons include Sissano Lagoon to the west of the Murik Lakes and, in the Solomon Islands, Lauvi Lagoon on the south coast of Guadalcanal.

In many parts of the region there are lowland freshwater swamps that are important sources of nutrients and organic matter to coastal marine waters. Most notable, because of the large area involved, is the Sepik–Ramu river system. These rivers meander through a swampy coastal floodplain with more than 1500 oxbow and other lakes (Osborne, 1993), and the above-mentioned extensive nypa thickets. Also important are smaller areas of "big tree" swamp forests dominated by *Terminalia* and *Campnosperma*, and swamps dominated by sago palm (*Metroxylon sagu*) or *Pandanus* spp.

Detail on these swamps is given by Osborne (1993). Similarly constituted swamp forests occur in the Solomon Islands.

Because of the high freshwater outflow from the Sepik (75 km^3/yr; Alongi, 1990) there is little estuarine development upstream of the Murik Lakes. Even so, a number of marine fish species occur several hundred kilometres upstream. The depauperate, highly endemic native fish fauna is largely unsuitable to support fisheries. Some two-thirds of Sepik–Ramu fisheries production is based on tilapia (*Oreochromis mossambica*) and carp (*Cyprinus carpio*), which were initially introduced to PNG for pond culture and subsequently escaped or were released into rivers (Kailola, 1995). Several other species have been deliberately introduced to further enhance fisheries despite environmental concerns about their possible suppression of indigenous species (Coates, 1993; Allen, 1991).

Biodiversity

The Coral, Solomon, and Bismarck Seas region lies within or immediately adjacent to the global centres of diversity of corals, mangroves, and seagrasses. Shallow-water biological diversity here is among the highest in the world. In general, the diversity is highest in PNG, Indonesia and the Philippines, and decreases moving eastward across the Pacific as well as to the north and south (McCoy and Heck, 1976; Mukai, 1993; Veron, 1993; Woodroffe, 1987). Though the current estimates of species diversity, below, are high, these should be regarded as conservative because of very low sampling effort.

More than 70 coral genera have been recorded from PNG's reefs (Veron, 1993) which are as, or more, diverse than any in the world. Coral diversity on the Papuan Barrier Reef is higher than on the well-studied Great Barrier Reef to the south (Veron and Kelley, 1988). With more than 60 coral genera, Solomon Islands' reefs are also very diverse by global standards. The species diversities of reef-associated fishes and invertebrates are also among the world's highest (e.g., Allen and Swainston, 1993; Ghiselin, 1992; Gosliner, 1992; Wells, 1990).

While in nearby coral species-rich Indonesia and the Philippines a high proportion of reefs is threatened by human activities (Bryant et al., 1998; Chou et al., 1994) the reefs of PNG and the Solomon Islands are still in relatively good condition (Huber, 1994; Maragos, 1998) and so are of great significance for the conservation of global biodiversity.

There are more than 30 recorded species of mangroves in PNG (Percival and Womersley, 1975; Woodroffe, 1987), while in the Solomons 27 have been identified (G. Pillai, pers comm). As with coral reefs, the mangrove forests support a diverse associated biota. A total of more than 130 species of fishes, for example, has been recorded from the mangrove habitat of PNG and the Solomon Islands; at any one site, from 20 to more than 50 species have been found (Blaber and Milton, 1990; Collette, 1983).

Seagrass species diversity reaches a maximum in western PNG and the Torres Strait, where 13 species have been identified (Huber, this volume). Elsewhere on the PNG coast there are 9–11 seagrass species (Johnstone, 1982; Mukai, 1993). The diversity of seagrasses, like that of other shallow tropical communities, declines moving eastward across the Pacific (McCoy and Heck, 1976; Mukai, 1993). Accordingly, the Solomon Islands probably has fewer species, but there have been no studies of seagrasses there.

Six of the world total of eight species of marine turtle are found in PNG seas, and five of these have been recorded in the Solomon Islands. The estuarine crocodile (*Crocodylus porosus*) population of the Solomons has been severely depleted by hunting (Messell and King, 1989). A ten-year ban on export of crocodile skins, implemented in 1989, appears to have been followed by some recovery (M. Lam, pers. comm.) The largest population now is at Lauvi Lagoon. Though substantially reduced in PNG, management measures introduced for this species (including farming) have had some success in maintaining population levels while also catering for a village-based industry in skins. Dugong populations in both countries have been much reduced by hunting.

Several coastal endemic animals are known from the region. An anemone fish (*Amphiprion leucokranos*) is known only from northern PNG and the Solomon Islands (Fautin and Allen, 1992). An endemic subspecies of the mangrove monitor lizard, *Varanus indicus spinulosus*, has been found only in the small area of San Jorge and Thousand Ships Bay, at the southern end of Santa Isabel. For another rare species, an endemic giant rat, *Uromys ponceleti*, the preferred habitat is believed to be *Terminalia brassii* coastal swamp forest (Leary, 1993b). In Lake Tenggano, on the small Solomons island of Rennell, there is an endemic sea krait *Laticauda crockeri* (McCoy, 1980). Undoubtedly there are more discoveries to be made of interesting endemics. A comprehensive account of coastal biodiversity in PNG is to be found in Sekhran and Miller (1994). The biodiversity of coastal ecosystems in the Solomon Islands has been less studied.

OFFSHORE SYSTEMS

The Environment

The Coral, Solomon, and Bismarck Seas are oligotrophic. At oceanic stations in the Coral and southern Solomon Seas, Furnas and Mitchell (1996) found deep (generally >60 m) surface mixed layers with low surface nutrient concentrations. They estimated primary production in the oceanic Coral Sea, which was dominated by picoplankton, to be 9–180 gC/m^2/yr.

Upwelling and turbulent mixing around reef and archipelagic systems, however, may result in much higher primary production in coastal areas. For instance, Furnas and Mitchell (1996) measured daily production rates of 1–3 g/C/m²/day at sites adjacent to the Papuan Barrier Reef and in the Louisiade Archipelago. This, coupled with the area's complex bathymetry and circulation patterns, probably gives rise to high spatial and temporal variability in productivity. Very little is known, however, about such variability, or its relationship to climatic and oceanographic forcing functions. Furnas and Mitchell (1996) found high variability in productivity estimates derived from two cruises undertaken in 1985 and 1988. They were unable to determine whether this reflected natural variability or whether it was a methodological artefact. High cloud cover and inadequate ground truthing limit the utility of satellite observations in oceanographic studies of the region (e.g., Furnas and Mitchell, 1996).

Tuna Fisheries

Pelagic tunas are by far the largest fisheries resource for both PNG and the Solomon Islands. In 1995 the combined catch of over 340,000 mt (Table 2) in these countries' EEZs accounted for about 11% of the global catch of tuna (Gillett, 1997). About 40% of this catch, which is dominated by skipjack and yellowfin tuna, comes from the Coral, Solomon, and Bismarck Seas (Fig. 3). (The catch from these seas within the EEZs of PNG and the Solomons cannot be calculated precisely; statistics provided by the South Pacific Commission are tabulated on the basis of grids of five degrees of latitude and longitude. The data for the Coral Sea in Fig. 3 include a small catch from the EEZs of Australia, Vanuatu, and New Caledonia.) Catches are well below the estimated maximum sustainable yield for all species, a situation rare among the world's fisheries.

The Solomon Islands tuna fishery was established in 1971 with a catch of 4666 tonnes by a small pole-and-line fleet. The tuna fishing fleet has since expanded and now includes purse seiners and long liners. The annual domestic catch, 1990–96, averaged over 48,000 tonnes. Continued development of the industry led to a 1998 domestic fleet catch of 116,438 tonnes (S.I. Fisheries Department files). The economic returns from the tuna fishery, including royalties earned from distant water fishing vessels that do not land their catch in the Solomon Islands, make it a key industry for the country. The domestic fleet takes 65–95% of the total tuna catch of the Solomon Islands' EEZ, and around half the catch is consumed locally or processed prior to export (Gillett, 1997; FAO, 1998; Solomon Islands, 1992). Tuna fishing and processing account for over 8% of the country's formal employment (and a higher proportion of private sector employment) and a third of export revenue; foreign access fees provide 5% of government revenue (Gillett, 1997; Leary, 1993a). In the face of opposition from some distant water fishing nations, the Solomon Islands has taken the initiative to introduce a far sighted "Tuna 2000" policy as a basis for sustainable management of the nation's tuna resources. Under this regime, the nation is to receive fairer economic returns.

In PNG the situation is different. Since a pole-and-line fleet based in New Ireland ceased operations in 1981 (Doulman and Wright, 1983, describe the domestic tuna fishery at that time), tuna fisheries in PNG have been exploited almost entirely by distant water fishing fleets from other countries. Other than access fees of about US$20 million annually, or about 2% of government revenue (Gillett, 1997; Gumoi and Sekhran, 1994), tuna fisheries provided almost no domestic benefit during most of the 1980s and 1990s. Tuna vessels did not call or land product at PNG ports and, in 1995, tuna fisheries, with a catch approaching US$500 million in value, provided only an estimated 125 jobs in PNG (Gillett, 1997).

Recent government policy has given high priority to the development of domestic participation in tuna fishing and processing, as well as to support services for domestic and foreign vessels. In 1995 the long-line industry was reserved exclusively for domestic vessels, and small longline and purse seine fleets have since developed. A cannery opened in Madang in 1997. Despite such initiatives, domestic benefits from the large tuna resource remain largely unrealised. Tuna exports during the first half of 1998 amounted to less than 9400 mt with a value of about US$12.5 million; 70% of the exports were unprocessed frozen fish (NFA, 1998). There are plans to continue to develop the embryonic domestic industry, but previous efforts have often failed to materialise and it remains to be seen whether the potential of PNG's tuna stocks can be managed so as to make a significant contribution to its economy.

The resources of the region's seas are important for the nations who control them, and not least because of their very long coastal fishing tradition. Inshore tuna schools have always been an integral part of this tradition, and for many coastal communities the skipjack tuna (*Katsuwonus pelamis*, locally called "bonito") had special status, symbolised in cultural practice through art, song and dance.

Table 2

Nominal tuna catches in the EEZs of Papua New Guinea and Solomon Islands in 1995 by gear type (from Gillett, 1997)

Gear Type	Papua New Guinea	Solomon Islands
Longline[1]	406	8,248
Pole and line	66	31,943
Purse seine	264,531	37,053
Total catch	265,003	77,244
Estimated value (US$ million)[2]	452.3	131.8

1. Gillett (1997) suggests that longline catches should be adjusted upward by a factor of 1.7 to allow for under-reporting.
2. Calculated from an average value of US$1707/mt for tuna catches in the South Pacific Commission statistical area (Gillet, 1997).

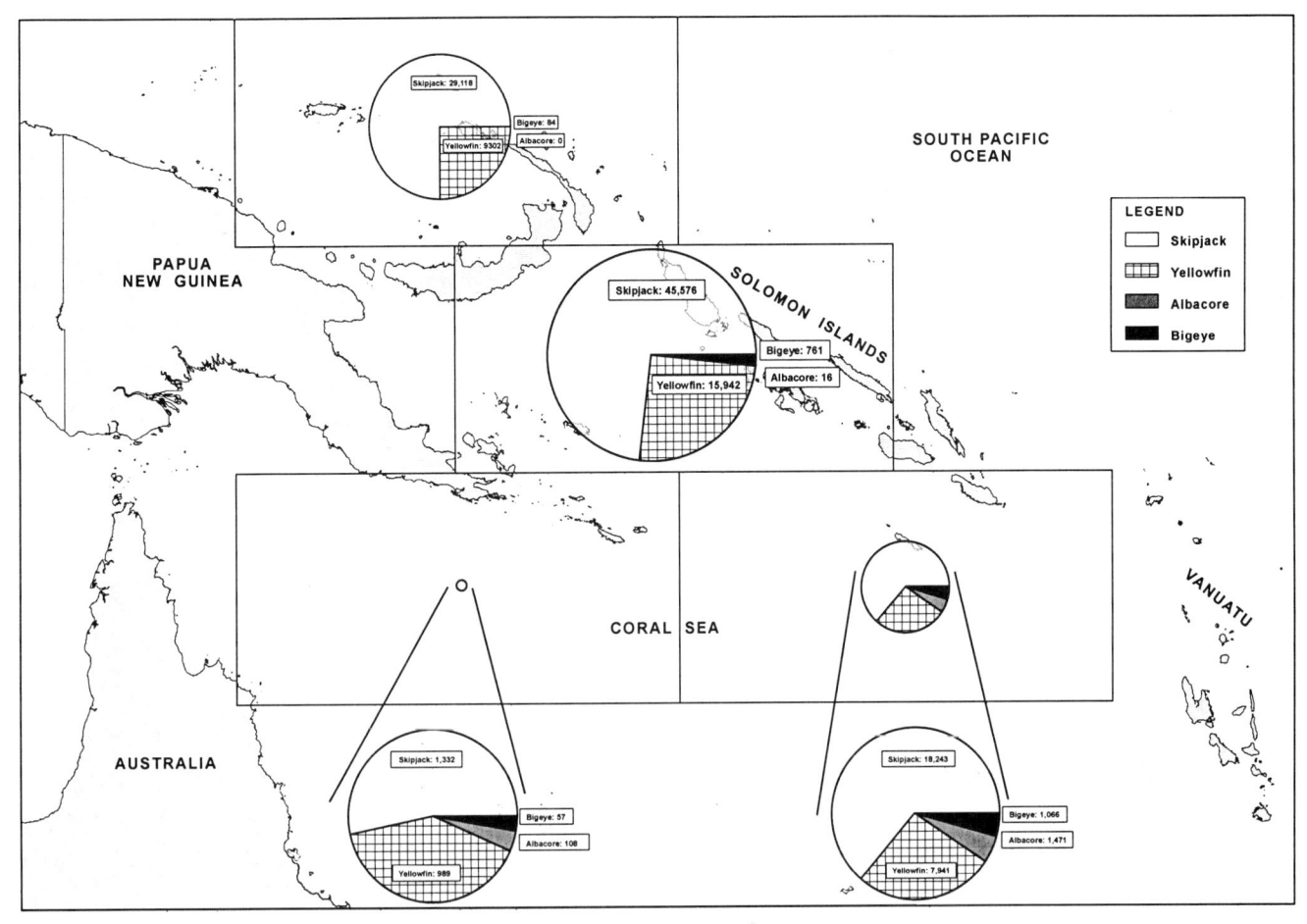

Fig. 3. Tuna catch and catch composition in the Coral, Solomon, and Bismark Sea region. Size of circles inside the 5° latitude and longitude grids is proportional to total catch (source: SPC database).

PEOPLE, DEVELOPMENT AND CHANGE

Population and Demography

Population data for PNG and the Solomon Islands are shown in Table 3. These are presented as national aggregates, since data specific to the coastal population are not available. A large component of the PNG population lives inland. The five highlands provinces isolated from the coast account for 38% of the national population. Furthermore, most of the "coastal" provinces, i.e. those that have some coastline, also contain significant inland populations. On the basis of 1980 census data, Frielink (1983) estimated that the rural coastal population of PNG was 23.1% of the national total, which if unchanged would currently represent some 1 million people. For PNG as a whole only 16% of the population lives in urban areas (UNDP, 1994). The 1980 census revealed that about 44% of the coastal population lived in urban areas (Frielink, 1983). The country's largest two cities, Port Moresby (195,000; 1990 census) and Lae (88,000), account for a great part of the coastal urban population and are undergoing rapid growth as a result of in-migration from rural areas. Recent population estimates are about 250,000 for Port Moresby and 140,000 for Lae. Other important coastal towns, most of which are growing much less rapidly, include Madang (27,000; 1990 census), Wewak (23,000) Rabaul (17,000), Kimbe/Hoskins (9000), Vanimo (7800), Kavieng (6800), Alotau (6400), and Lorengau/Lombrum (5800). The Kieta/Arawa/Panguna urban area on Bouganville island had an estimated 18,000 people in 1980 but the population has declined dramatically as a result of civil unrest.

In the Solomon Islands, where there are no large land masses and no place as much as 20 km from the coast, virtually everyone is a coastal dweller. Except for a couple of urban areas, virtually every family harvests coastal resources. Thus, the data in Table 3 do reflect the coastal population of the Solomon Islands. Most of the urban population lives in Honiara, with an estimated 1997 population of 65,000 growing at about 10% annually (Commonwealth, 1998). Gizo (2331; 1986 census) and Auki (948; 1986 census) are the other towns of significance.

In both countries, although the crude population density (number of people per unit land area) is low on average, the population per unit of arable land is considerably higher, especially on smaller islands. There are also

Table 3

Selected Social Indicators for Papua New Guinea and Solomon Islands (sources: AusAID, 1997; Leary, 1993; SPC, 1998; UNDP, 1989, 1994, 1998)

	Papua New Guinea[1]	Solomon Islands
Estimated population in 1998	4,412,400	417,800
Annual population growth rate (%)	2.3[2]	3.4
Population density (persons/km^2)	10	15
Fertility (no. children per woman)	4.7	5.8
Population under 15 (%)	42	47
Dependency ratio[3]	85	109
Life expectancy	54	65
School attendance by school-age children (%)	28	35
Adult literacy (%)	28	30
GDP per capita (US$)	1196	926
Participation in cash economy (males/females; % of population aged 15–64[4])	54/37	36/13
Formal employment (% of workforce)	10	15
Public sector employment (% of formal employment)	27	36
% Urban population	15	13
% Population in main centre	5	11
UNDP Human Development Index	0.341	0.371

1. Excludes North Solomons Province.
2. Unisearch (1991) suggest that this is an underestimate
3. Dependency ratio = (population under 15 + population over 60)/(population 15–60).
4. For Solomon Islands females, aged 14 and over.

wide differences in crude population density among islands. High fertility in both countries has produced a youthful population structure and high population growth rates.

The Nature of Development

The level of social and economic development is very low. Indeed, PNG and the Solomon Islands have the lowest "human development indices" (an index calculated by UNDP for all countries, from a range of social and economic parameters) of Pacific Islands nations, by a considerable margin (UNDP, 1998). There has been little progress during the 1990s. The national aggregate statistics presented in Table 3 mask wide disparities in development among provinces, and between urban and rural areas. For the most part the rural majority lives an essentially traditional lifestyle in small villages constructed in large part of locally available "bush materials". They often have little access to education, health care, safe water, and other basic services. Electricity supplies are restricted to urban centres, except in a few rural communities where a nearby industrial-scale development provides limited access. The level of formal employment is low and is dominated by the public sector. Over 80% of the workforce is engaged in agriculture, primarily in traditional subsistence and market gardening. The commercial sector is dominated by extractive primary resource industries—mining, logging, agriculture, and fisheries.

Use of the Sea and its Resources

In PNG and the Solomon Islands the main use of coastal marine resources is subsistence and artisanal fishing. There is a high degree of rural reliance upon seafood and other marine resources (see below). Marine resources were first exploited at a commercial level early in the 19th century when American whalers began to take humpback whales. An export trade in "tortoiseshell" (the carapace of the hawksbill turtle, *Eretmochelys imbricata*) then developed. In the Solomon Islands tortoiseshell was exchanged with visiting Europeans for firearms and ammunition. This trade enabled communities with access to good stocks of hawksbill turtle to gain superiority in the inter-island warfare endemic at the time (McKinnon, 1975). In the western Solomons these gains are still reflected today in the distribution of people and the rights of communities to access land and sea resources. Foreign interest in marine resources subsequently extended into mother-of-pearl and bêche-de-mer. A similar sequence of commercial trade developed over this period in PNG. As described below, present-day commercial trade in coastal marine resources is dominated by bêche-de-mer and mother-of-pearl.

Though in PNG and the Solomon Islands the State has ultimate jurisdiction over coastal sea areas, most of these are subject to a long tradition of customary ownership and resource allocation. A typical traditional coastal resource management system is a community-based, participatory system for the integrated management of both land and sea. It is not a management system specifically for fisheries. It is based on social relationships among people, and is expressed as rights to exploit resources (Baines, 1985; Hviding, 1996). Rights to access marine resources are mediated through a system of customary marine tenure (CMT). These rights may apply not only to specific places such as reefs, spawning grounds, or fish migration routes, but also to particular species or fishing methods. Though CMT is widespread, and profoundly affects marine resource use, only limited information about it has been documented (see Baines, 1985; Carrier and Carrier, 1989; Hviding, 1996; Polunin, 1984, Wright, 1985 and references therein).

The Information Baseline

Fisheries agencies in both PNG and the Solomons have generated considerable information on coastal fisheries stocks (see Dalzell et al., 1996; Kailola, 1995; Skewes, 1990), though there are important gaps, questions about reliability, and often a lack of continuity through time. The information is largely limited to stock assessments and production figures for export products. There are few catch data for subsistence and non-export artisanal fisheries, and catch-effort data in all coastal fisheries is very limited.

Aside from fisheries, there is inadequate information on the marine environment to establish any form of baseline against which trends might be identified as a basis for management. The situation in PNG is somewhat better than in the Solomon Islands. Some of the earliest work on shallow water ecosystems in PNG was undertaken from the Kanudi Research Station near Port Moresby in support of fisheries development. Coastal research by staff at the University of Papua New Guinea commenced in the early 1970s. Around the same time the University established a research station on Motupore Island, near Port Moresby, and the Royal Belgian Institute of Natural Sciences established another at Laing Island on the north coast; the latter has recently closed. The Christensen Research Institute operated a research station at Madang from the mid-1980s until 1998. Another small research facility opened at Kimbe on the north coast of the island of New Britain in the late 1990s. These and other efforts have provided considerably more information about the coastal biota of PNG than is available for the Solomon Islands, but it is still very limited, geographically and taxonomically. Large areas of the country, and major ecotypes such as soft-bottom communities, remain essentially unknown. Nearly all work has consisted of "one off" studies at one or a very few sites, and there is no information about spatial and temporal variability.

Reports arising from a 1965 Royal Society Expedition (Stoddart, 1969a,b), remain the primary source of published information about Solomon Islands reefs. A coordinated range of ecological, resource and anthropological studies has been made, in support of a community resource management initiative, in the Marovo Lagoon (Baines, 1985; Hviding, 1996). Increased knowledge is emerging from local research on giant clam (*Tridacna* spp.), pearl oysters and bêche-de-mer (various holothurians) at an International Centre for Living Aquatic Resource Management coastal aquaculture centre (ICLARM-CAC) at Aruligo, Guadalcanal. Elsewhere, research is underway on trochus and green snail. A research centre of the University of the South Pacific's Institute of Marine Resources is under establishment.

In PNG, water quality, habitat, and related environmental surveys and monitoring have been undertaken in connection with large mining projects that impact on coastal waters. A few monitoring sites have been established as part of the Global Coral Reef Monitoring Network established in the late 1990s but the viability of this effort is questionable.

A considerable source of knowledge is yet to be accessed and presented in a form that makes it useful for advancing scientific understanding. This is local traditional knowledge of coastal marine systems and resources, which underpins traditional coastal resource use systems. Such knowledge may not be overtly linked with conservation objectives, but its application often serves to conserve resources. Nor is this knowledge "complete" in an ecological sense. Knowledge of the behaviour of target species, for instance, is extensive—of other species, very limited. Further, it appears that an appreciation of ecological processes is only now developing as information from scientific studies is made available to traditional fishers. Nonetheless, the body of traditional knowledge represents an important store of scientifically useful information. Johannes (1988), for example, made use of traditional knowledge in a study of spawning behaviour of coral trout, *Plectropomus leopardus*.

LAND AND SEA USE FACTORS IMPACTING ON COASTAL WATERS

Land Use

Food gardens for subsistence, and for limited marketing, are widespread in coastal PNG and the Solomon Islands. Some of this cropping is undertaken as shifting cultivation, which is a source of forest disturbance (Leary, 1993a; Levett and Bala, 1994; Louman and Nicholls, 1994). It can also give rise to increased soil erosion. Large-scale agriculture is conducted mostly as plantations of export tree crops, mainly coconuts, oil palm, and cocoa. Rubber, and coffee, and sugar, are also grown in parts of PNG, primarily inland but in some cases also in coastal areas. Much of the coastal tree cropping is on flat or gently sloping land and, since ground cover is maintained in these plantations, soil losses to coastal waters may not be significant.

A desk study of fertiliser and pesticide loading of the coastal environment suggests that this is relatively low by world standards, though the Solomon Islands is estimated to have among the highest in the Pacific (Convard, 1993; Table 4). A more significant problem for coastal waters arises from organic processing wastes (see below).

A crude form of rainforest logging, using heavy machinery and with little effort to minimise collateral damage, has been an economic mainstay which in recent years has outstripped fisheries as the leading source of export earnings in the Solomon Islands. In PNG, logging revenues are second only to those of mining. Nearly all production is exported as unprocessed round logs. The extraction of logs has greatly exceeded sustainable rates, and environmental practice has been poor (Hughes and Sullivan, 1989; Louman and Nichols, 1994; Leary, 1993a; Marshall, 1990; Nadarajah,

Table 4

Estimated total loadings of fertilisers and pesticides from agricultural runoff per kilometre of coastline (calculated from data in Convard 1993 and Table 1)

	Papua New Guinea	Solomon Islands
Fertilisers (kg/km/yr)	0.1	22.6
Pesticides (kg/km/yr)	1.8	4.7

1993; Solomon Islands, 1992; UNDP, 1994). Increased sedimentation arising from forest canopy removal and soil disturbance by logging is widely regarded as a major threat to coral reefs and other coastal ecosystems in the area (Asquith et al., 1994; Huber, 1994; Leary, 1993a; Maragos, 1998). However, there are inadequate data with which to assess the problem. Baines and Morrison (1990) reported that, for the logged island of Kolombangara in the Solomons, "Within a few years of the cessation of logging, except following heavy rain, these streams were flowing clear, into coastal seas where no polluting suspended sediment was apparent.". This does not rule out the possibility that damage was inflicted while logging was in progress, or that accumulated sediment on the reef may have had subsequent effects. It has been suggested (e.g., Asquith, 1994) that the generally narrow continental shelf, which may reduce sediment retention in the coastal zone, and high rainfall and steep slopes, which are likely to produce high natural sediment fluxes, might enhance ecosystem resilience, but to what extent, if any, is not known. Neither is there information on possible interactions between sedimentation and other chronic stresses and ecosystem capacity to recover from natural disturbances such as storm damage and mass bleaching events.

Subsistence and Artisanal Fisheries

There are few full-time fishers and, where land is available, agriculture is the more important subsistence activity. Nevertheless, subsistence and artisanal fisheries are by far the most important use of coastal marine resources in the region. Seafood consumption is very high in the Solomon Islands (Richards et al., 1994), and Pernetta and Hill (1981) have pointed out the importance of seafood as a source of protein in PNG. In both production and value the subsistence catch in both countries greatly exceeds the commercial (mostly artisanal) catch (Table 5).

Subsistence fishers typically use the full range of available resources, directing effort to estuarine, mangrove, lagoon, reef, and/or nearshore pelagic species according to season or weather conditions, resource availability, and social factors such as customary access rights. Their methods vary widely with location, target species, season, and

> Subsistence fishing is often defined as fishing for the direct consumption of fishers and their families and artisanal fishing as fishing for products that will be sold or traded. In practice it may be difficult to distinguish between subsistence and artisanal catches. The exchange of fisheries products in the area is part of a complex system of social exchange and obligation, and a distinction between consumption by a fisher's family and consumption by others in the community is artificial. It is also common for fishers who lack access to agricultural land to trade part of their catch for plant foods; in the case of Western and Gulf provinces of PNG, sago. Subsistence catches have been estimated from seafood consumption rates (e.g. Dalzell et al., 1996), in which case locally-traded catches, even if bought and sold in markets, are included in subsistence catch estimates. The basis of reporting of "subsistence" and "artisanal" catch categories in statistics is often poorly defined. In this chapter "artisanal" catches are primarily those of products that are exported beyond the local area in exchange for cash. "Subsistence" catches refer in general to those that are consumed locally. This distinction is, however, neither precise nor necessarily consistent.

the age and sex of the fisher. They include drive nets, gill nets, beach seines, droplines, handlines, troll lines, spears (used both from the surface and while underwater), bow and arrow, hand traps, stone traps and weirs, hand gathering (both by gleaning shallow flats and by diving), derris root and other natural poisons, and explosives. Hviding (1996) provides a comprehensive account for Marovo Lagoon. Turtles, dugong, and porpoises are sometimes herded with boats to facilitate their capture with spears or nets. There are also some unusual traditional methods such as "calling" sharks with coconut shell rattles and then lassoing them, and catching "garfish" (Hemirhamphidae) by entangling their teeth in lures of spiders' web suspended from kites.

Catches are highly diverse. Several hundred species of fishes may be involved, of more than 20 families (Dalzell et al., 1996; Lock, 1986a; Richards et al., 1994; Wright and Richards, 1985). A variety of molluscs, crustaceans, echinoids, and other invertebrates is also taken, as well as sea turtles, dugong, crocodiles, and porpoises. A number of other marine products are harvested for traditional purposes. Though a trade in sea salt with inland communities is no longer of importance in PNG, certain shells are still traded between communities as a form of currency and as a demonstration of cultural links. Demand for one of the species involved (*Nassarius camelus*) is so great that some supplies are imported from the Solomon Islands, where this species does not have the same cultural significance. In 1990, over 12 tonnes of this tiny shell were subject to this international trade (SI Fisheries Department data). There are other mollusc species harvested for cultural purposes by both PNG and Solomon Islander communities, often physically processed into small discs from which "shell money" strings are made. This product is of vital social importance for some

Table 5
Estimated mean annual coastal fisheries production, 1989–1992 (from Dalzell et al., 1996)

	Papua New Guinea		Solomon Islands	
	Production (t)	Value (US $)	Production (t)	Value (US $)
Subsistence fisheries	21,014	42,028,000	10,000	8,405,660
Commercial fisheries	4,966	22,096,078	1,150	4,343,811

communities, for example on Malaita, Solomon Islands and the Trobriand Islands, PNG. Other marine products collected include *Acropora* coral, which is burned to produce lime to be chewed with betel nut.

Commercial activity in inshore fisheries is dominated by artisanal fisheries, which are an important source of village incomes. Export fisheries for bêche-de-mer, the dried product of various holothurians, and shells used to produce mother-of-pearl, most importantly trochus (*Trochus niloticus*), black- and gold-lipped pearl oyster (*Pinctada margaritifera* and *P. maxima* respectively), and green snail (*Turbo marmoratus*), dominate inshore fisheries in both volume and value. These products are a major source of cash income for many coastal communities but, being poorly managed, stocks are rapidly depleted.

Other targets of artisanal fisheries include sharks, spiny lobsters (*Panulirus* spp.), mangrove crab (*Scylla serrata*), and snappers taken by deeplining along reef fronts (including high priced market favourites such as *Pristipomoides*). Corals, shells, and reef fish are collected, the former for processing into decorative products for sale to tourists, and the latter for an export aquarium trade. Fisheries for giant clams (*Tridacna*), hawksbill turtle shell (*Eretmochelys imbricata*), and the estuarine crocodile (*Crocodylus porosus*) have been curtailed by local regulation both in response to the listing of these species in the CITES Convention (Convention on International Trade in Endangered Species) because of their endangered status, and through local recognition of population declines (Hair, 1994; Kailola, 1995; Richards et al., 1994). The effectiveness of controls is uncertain, however.

Threats to Sustainability

Overfishing is a primary concern in coastal fisheries of the Coral, Solomon, and Bismarck Seas. While, overall, coastal living marine resources are regarded as under-utilised, this results from a lack of markets rather than from effective management. Wherever there has been access to receptive markets, over-exploitation has inevitably followed. Export coastal fisheries typically expand until the resource is depleted, or a drop in the market price provides a disincentive to harvest. Fisheries for bêche-de-mer, mother-of-pearl, giant clams, and other sedentary species typically undergo "boom-and-bust" cycles; these resources appear to be inherently difficult to manage and susceptible to over-exploitation (see chapters in Wright, 1993).

Data presented in Table 6 demonstrate this. Through the 1980s, the primary sources of cash income for coastal villagers in the Solomon Islands were trochus shell and copra (dried coconut flesh). Copra prices, however, were depressed and, where the alternative of fishing for trochus was available, a typical villager response was to concentrate on that. The resultant over-harvesting is evident in the subsequent rapid decline in exports after 1990. A renewed interest in bêche-de-mer emerged in 1991. Not only was this a convenient alternative source of income, but there had been a surge in prices offered for this product. Again, a steady production decline followed, culminating in a collapse of interest as stocks became depleted. A similar sequence is evident in the export tonnage of pearl shell. For pearlshell, the "nil" entries beginning in 1995 mark the introduction of an export ban. This pattern of "boom, bust and ban" reflects an inability to manage stocks for sustainability. PNG fisheries for sedentary marine resources have followed a similar pattern.

In the Solomon Islands hunting for skins and shells for export quickly drove crocodiles to near extinction and severely reduced breeding populations of hawksbill turtle (*Eretmochelys imbricata*; Richards et al., 1994).

The complex social and political difficulties faced by Solomon Islands and PNG fisheries authorities in effecting sustainable management of coastal resources can to some extent be avoided through export bans on commodities for which there is no local market. However, bans cannot be applied to resources such as shellfish and most fishes other than tunas, for which the primary market is domestic. This is troubling, particularly as artisanal fisheries near urban centres are expanding to meet local demand. Overfishing of finfish and other food species has also occurred near urban areas (Solomon Islands, 1992; Lock, 1986b). In at least some subsistence and rural artisanal fisheries of PNG, human population growth has been identified as a major factor in overfishing (Apinia, 1986; Swadling, 1982).

Habitat degradation from subsistence and artisanal fishing is not widespread, but there are some localised

Table 6

Exports (tonnes), of selected non-fish marine products arising from artisanal fisheries in the Solomon Islands, 1990–1997 (source: Solomon Islands Fisheries Department files)

Product	1990	1991	1992	1993	1994	1995	1996	1997
Trochus shell (*Trochus niloticus*)	307	87	51	24	67	nil	nil	113
Bêche-de-mer (various holothuria)	119	622	715	316	285	219	nil	nil
Green snail shell (*Turbo marmoratus*)	3.6	3.3	2.0	0.6	0.9	nil	nil	nil
Black lip pearl shell (*Pinctada margaritifera*)	31.4	43.7	27.9	26.0	0.3	nil	nil	nil
Gold lip pearl shell (*P. maxima*)	25.7	13.9	4.8	1.2	nil	nil	nil	nil

concerns. One of these is blast fishing, involving explosives diverted from construction and mining projects, or from World War II munitions caches still found in many remote areas. Though illegal, enforcement capacity is minimal. Blast fishing breaks up corals, so damaging habitat, and brings about collateral damage to non-target species. The use of poisons to stun fish is an occasional but not widespread problem (Veitayaki et al., 1995). Substances used include cyanide, some pesticides, and timber preservative chemicals. The use of cyanide in the live reef-fish trade is a particular concern (see below). Though not yet widespread there is a possibility that the practice, once introduced, could spread to other fisheries and involve other poisons. This dangerous practice may be more readily accepted at a community level in the light of the long-established traditional use of *Derris* and other plant poisons. Traditional poisons are typically used on a small scale, and on special occasions, and it has not been established that they cause significant damage. There are anecdotal reports of localised pollution from bêche-de-mer processing, which produces organic matter and in some cases natural toxins.

Some forms of netting and gathering cause physical disturbance to reefs and seagrass beds. The collection of corals to produce lime also causes physical disruption. Leary (1993a) believes that the collection of corals for the aquarium and shell trade has contributed to reef degradation in some areas in the Solomon Islands. While no evidence of serious widespread impact has been sighted, these practices are expanding to an extent that makes them a concern.

COASTLINE CHANGE

In a region of high levels of tectonic activity and of geological complexity, it is not surprising that different sections of the coastline of a single island may be subject to both long-term subsidence, and uplift at the same time. Superimposed on these long-term adjustments are the short-term shifts brought about by earthquakes. An example of reef uplift is described earlier for Guadalcanal. Both earthquakes (by inducing land-slips), and periods of unusually intense rainfall attendant on tropical cyclones, result in peaks of sediment transfer to coastal waters. As these sediments are reworked by waves and currents, they can bring about large short-term changes in coastline.

Sea transport is of such importance for the coastal population centres of PNG and the Solomon Islands that there are many examples of small-scale, sometimes improvised, coastal infrastructure development. Local coastal dynamics often are not closely studied, nor is local knowledge used in siting decisions or infrastructure design. In some cases the result is that a port structure is damaged by wave action or scouring, or its function impaired by sediment accumulation.

Most urban ports have undergone significant development involving foreshore reclamation, and channel dredging. Many PNG coastal villages have landing facilities for small, drop-front freighters (landing craft). A small area of reclamation may be involved, though more commonly the freighter offloads directly on to the foreshore. Larger reclamation projects for log and agricultural produce loading facilities are also widespread. These rural "port facilities" are usually very basic. At a few locations modern loading facilities have been developed for mines. With the exception of the mine project ports, there is minimal environmental management of this port development, but it affects only a small fraction of the coastline.

HUMAN IMPACTS ON COASTAL SEAS

Village-level Impacts

Corals and beach sand are widely used as building materials in the Polynesian Pacific Island nations, but this practice is not widespread in PNG and the Solomons. More common is the use of mangroves for building materials and firewood, and the clearing of mangroves for housing sites. Considerable areas of mangrove have been deforested in this way near urban areas, but the greater impact on mangrove habitat has been from clearance alongside coconut plantations in the mistaken belief that this is necessary for the health of coconuts (Baines, pers. obs). Mangrove deforestation has also occurred where mangrove fuel is used for drying bêche-de-mer or copra (Huber, Baines, pers. obs.; Leary, 1993a). Removal of mangroves has led to coastal erosion (e.g., Hughes et al., 1986) and to loss of important fisheries habitat which mangroves provide.

Though the ecological conditions required for small-scale aquaculture are present, the social requirement for sustained effort is not. A generally satisfactory supply of wild stock removes much of the incentive for aquaculture for subsistence or small-scale commerce. Small-scale aquaculture has been conducted on a trial basis in north Guadalcanal. Here, villagers grow out giant clams for the aquarium trade from spat produced at the nearby ICLARM coastal aquaculture facility, which is also investigating the suitability of pearl oysters, sea cucumbers, and other species for village-level aquaculture enterprises. Various industrial-scale aquaculture enterprises have been established but some have been short-lived. Those currently operating include two small penaeid prawn farms. Prospects for aquaculture remain promising, and a recent boost for this in the Solomon Islands has been a 1999 crop of high quality pearls harvested from a culture trial at Gizo.

In the Solomon Islands there are concerns about pesticide pollution in rural areas from spraying for malaria control (Convard, 1993; Leary, 1993a; Solomon Islands, 1992). The ecologically notorious chemical used (DDT) was introduced regularly into the coastal village environment over a forty year period in line with World Health Organization advice to the Solomon Islands Department of Health.

Industrial-scale Impacts

Mining in PNG is easily the largest industry in the area. The massive excavation of surface and/or subsurface material that accompanies mineral development poses a pollution threat to coastal waters. Baines and Morrison (1990) provide a dramatic example of this—the Bougainville copper mine. Until its closure in 1989 by the actions of militant customary landowners, for 18 years the mine discharged 150,000 tonnes of tailings a day into the Jaba River, and from there into Empress Augusta Bay. In the Jaba River valley, the sediments built up to a depth of 27 m, spread out across the valley floor. These sediments are reported by Brodie et al. (1990) to have created a 7 km^2 delta and affected more than 100 km^2 of sea floor. An open cut gold mine on the island of Misima for 20 years has discharged tailings directly into the sea at a depth of 75–100 m on the outer edge of a fringing coral reef. In addition, some 20,000 tonnes of soft waste rock have been tipped daily into the sea. This ocean outfall was positioned below the thermocline, the objective being to minimise chances of sediment-laden water rising to the surface, where its ecological effects would have been more damaging. An ocean outfall of similar design is in use at another island mine developed later at Lihir. Both mines were subject to environmental assessment prior to mining. Following on the disastrous Bougainville example, mining industry interest in establishing acceptable environmental practice has resulted in assessments which are among the best undertaken in the Pacific island region, with most impacts being predicted, and addressed. Environmental monitoring has been conducted at both mines. An independent assessment of the impacts of these discharges would be valuable as more coastal mines are expected to open.

Environmental aspects of the one and only mine in the Solomon Islands—at Gold Ridge, Guadalcanal, have been independently assessed. Its waste is contained in storage areas which in 1999 attracted an Australian mining industry award for excellence in tailings dam design. This mine has been described as "a model for future development of a responsible mining industry in the South Pacific region." (Islands Business 25 (7), 18. July 1999).

In both PNG and the Solomon Islands the manufacturing base is small. Convard (1993) estimated marine contaminant loadings from major agricultural and food processing industries including palm and coconut oil milling, sugar refining, brewing, soft drink manufacture, and meat and fish processing (Table 7). Other small industries, such as paint manufacturing, metal plating and fabrication, and vehicle and ship repair, also generate marine contamination of unknown quality and quantity (Convard, 1993). Various proposals for new canneries, oil and mineral refineries, and other coastal industrial infrastructure have been prepared. Only a few of these are expected to materialise. The contaminant loads in Table 7 are low in terms of average coastal loadings, especially in PNG, but of course are concentrated near discharges. There have been cases of serious localised pollution from fish canneries, oil palm mills, and other industries, especially when discharges are located in waters with restricted circulation (Convard, 1993; Huber, 1994; Solomon Islands, 1992). There is no routine water quality or other pollution monitoring, however, and the status of the problem is not known.

There is no mass tourism. Both countries have a small tourism industry and there is some focus on marine tourism—small island and beach resorts and guest houses, and live-aboard dive vessels. A few cruise ships also visit. At present levels tourism does not appear to have major environmental impacts. There is considerable potential for expansion of tourism, and this prospect has given rise to concerns that adverse impacts on coastal ecosystems may result. At the same time, there is hope that tourism-focused attention to natural attractions may help deflect coastal developments that would otherwise damage the coastal environment and its resources.

The Impacts of Large Urban Areas

Rapid population growth and poor urban planning, especially in Port Moresby, Lae, and Honiara, have clearly had significant adverse impacts on the surrounding coastal areas. These include overfishing, soil erosion and other

Table 7

Estimated waste loads (mt/yr) on the marine environment in Papua New Guinea (PNG) and Solomon Islands (SI) (from Convard, 1993)

Source	Country	BOD	SS	Oil and grease*	N	P
Domestic wastewater	PNG	5666	2425	–	3107	374
	SI	2137	1763	–	979	139
Industrial wastewater (selected industries	PNG	509	1083	765	0	0
	SI	513.6	494.8	560.8	18.7	0.1

*Loads of oil and grease from domestic wastewater not available.

consequences of intensified agriculture, and deforestation. Part of the problem arises from the fact that many urban dwellers live in unserviced squatter settlements. They retain aspects of the rural lifestyle that do not suit urban areas, such as partial dependence on subsistence and market gardening, the use of wood as a cooking fuel, and direct disposal of human wastes to the sea.

Domestic sewage is the largest urban source of marine contamination. The population is not well served with sanitary services (Table 7). Convard's (1993) data for PNG contain obvious errors (e.g., the population of Lae is reported as only 33,000, entirely served by sewerage, while Madang's population is reported at over 90,000) and overestimate the provision of sanitary services. Simple bush or over-the-water latrines are not uncommon in urban areas of PNG, and in rural areas are the norm. Even where septic or reticulated sewerage systems are available, they are often poorly designed, maintained, and operated. Except in Port Moresby, where much of the sewage is treated by settlement in a natural swamp, primary treatment, where it exists, usually consists only of screening. Secondary treatment is confined to oxidation ponds serving some of the populations of Lae, Popondetta, and a few hospital and other facilities (Waterboard of PNG files). A few other hospitals have packaged secondary treatment plants. There is no systematic monitoring of water quality, much less environmental or human health impacts, but microbiological contamination and other evidence of sewage pollution has been reported in both urban and rural areas (Moore, 1982; Naidu et al., 1991; Waterboard of PNG, 1987).

Solid waste disposal is a persistent problem. In both urban and rural areas household waste may be dumped directly into watercourses or the sea, and rubbish dumps are often located in mangrove or other foreshore areas, or near watercourses. There are no waste management provisions to prevent leaching. While this problem has not been assessed in PNG or the Solomon Islands, coastal heavy metal and other pollution has been observed near waste dumps in other Pacific island countries (Naidu et al., 1991).

Fisheries

There are few industrial fisheries in coastal areas. A small trawl fishery in Orangerie Bay, Milne Bay Province, produces around 20 mt/yr of prawns, mostly banana prawn (*Penaeus merguiensis*). The fishery is considered fully exploited and access is limited to two small (15 m) vessels. Other PNG industrial fisheries, in the Gulf of Papua and in Torres Strait, are discussed by Huber (this volume).

The main inshore industrial fishery in the Solomon Islands is a bait fishery conducted at several locations by the pole-and-line tuna fleet, harvesting some 2000 mt annually, mostly of anchovies (*Stolephorus*) and sprats (*Spratelloides*). Based on limited data this appears to be around the maximum sustainable yield. Fishing grounds are rotated as catches decline (Tiroba, 1993). Concerns have been expressed about possible adverse trophic and by-catch effects of the bait fishery on subsistence fisheries. Blaber et al. (1990) concluded that the species most likely to be affected are nearshore pelagics caught primarily by trolling. Trolling is a relatively minor method in the subsistence fishery (Blaber et al., 1990; Leqata et al., 1990). So long as this remains the case baitfish harvests are not expected to adversely affect local catches. However, should increasing wealth in local communities bring about an increase in the use of fishing techniques targeting nearshore pelagic species, there is a possibility that a classic industrial–subsistence fisheries clash could develop.

There is potential for small-scale industrial development of a few other coastal wild fisheries, but little development activity. Studies and pilot projects are underway regarding the culture of pearl oysters, bêche-de-mer, and tridacnid clams.

In both PNG and Solomons an unsustainable form of live reef food fish trade has been introduced. This focuses on the capture of a few species of grouper, wrasse, stonefish, and spiny lobster for transport, live, to Hong Kong, China, Taiwan and other Asian markets, where they command very high prices in the restaurant trade. Overfishing, the targeting of spawning aggregations, the use of cyanide and other poisons, physical disruption of the reef matrix to extract fish, and high mortality of both target species and by-catch are typical in this trade (Aini and Hair, 1995; Johannes and Riepen, 1995; Johannes and Lam, 1999; Richards, 1993). To date, most operations have been short-lived, often because of rapid resource depletion, withdrawal of harvest approvals by customary reef owners concerned about the adverse environmental consequences, or dissatisfied with benefits they derive from the operations, or breaches of licensing conditions. With its poor environmental track record, high profitability, and the prospect of market expansion in southern Asia, this trade continues to pose a threat to the reef ecosystems of the Coral, Bismarck, and Solomon Seas. National capacity to contain and manage the threat is low.

Shipping and Offshore Accidents and Impacts

There are no port reception facilities for solid or oily wastes in either PNG (Envision, 1993) or the Solomon Islands. Ship-based sewage is, at best, discharged into municipal systems. Pollution controls are not enforced and pollution from ships in harbours and in passage is common. The extent of the problem has not been quantified.

Though reef groundings of ships are not unusual there is no systematic reporting of such accidents. Neither are there procedures for assessing or addressing threats of physical damage to habitats or pollution from this cause.

Concern has been expressed in the Solomon Islands about prodigious quantities of bunker oil which lie in the deteriorating fuel tanks of battleships, cruisers and numerous smaller warships sunk in deep water close to the coast of Guadalcanal during World War II. While gradual seepage over the years appears not to have had significant environmental effects, this legacy of war could become a marine environmental issue for the new millennium if these corroding tanks were to rupture.

Both PNG and the Solomon Islands have addressed the need for oil spill contingency planning, and PNG has a national plan, and practical procedures, developed in close consultation with the petroleum industry. Yet very little oil spill control equipment is in place, and only limited training has been carried out. Plan implementation relies heavily on support from nearby countries, notably Australia.

PROVISIONS FOR THE MANAGEMENT AND PROTECTION OF COASTAL SEAS

National Administrative and Legal Arrangements

The relatively healthy coastal environments of PNG and the Solomon Islands reflect low development and population pressures rather than effective environmental management. When industrial and urban development has taken place, environmental problems have followed.

Environmental protection and sustainable resource use are enshrined as one of the national goals of PNG's constitution. This is supported by generally adequate environmental legislation (Hedemark and Sekhran, 1994). However, implementation of PNG's innovative environmental legislation has been very weak. The Solomon Islands has less comprehensive legislation than PNG but here, too, it is of limited effectiveness because implementation is hindered by the inadequate number and experience of enforcement officials and by a limited general awareness of the legislation and of the rationale for it.

PNG has at times been prominent on the international stage in advocating and making commitments to environmental management initiatives. However, in the absence of the required political support, and in the face of a heavy economic development emphasis on resource extraction with minimal control, it has not been possible to sustain these initiatives at home. Those advocating a more responsible approach to management of the coastal environment in the Solomon Islands have faced similar pressures. A useful participative multisectoral National Environmental Management Strategy (Solomon Islands, 1992, 1993) has been prepared. However, it has not been possible to translate this into effective action. As parties to relevant international conventions (Table 8) both countries have entered into important commitments to protect and manage the shared regional marine environment. Yet neither country has been able to demonstrate that its good international intentions can be translated into rigorous attention to growing marine environmental problems within national boundaries.

Protected Species, Habitats and Areas

Some marine protected areas have been formally established in this region, mostly in PNG. However, in line with the general deficiency in support for environmental management activities, few have any form of management regime. A major problem has been the approach to protected area identification and management, which has been "top-down", driven by conservation criteria which are often not appreciated by those who have long-established customary rights to access and harvest those areas. Much of the consultation by protected area proponents with local communities has been on the basis of persuasion rather than participation, making local communities feel the protective regime is being imposed on them. Lessons have been learned. For instance, a late 1970's effort to establish a marine sanctuary focused on marine turtle habitat at the Arnarvon Islands, Santa Isabel, Solomon Islands, failed dramatically with the destruction of infrastructure and equipment by a group claiming to have customary rights to the area and resentful that they had no role in the exercise. A new project established in the 1990s after extensive consultation with all groups with a history of access and use rights, and continuing that consultation throughout, appears to have been successful. PNG legislation provides for local communities to establish and manage Wildlife Management Areas (WMAs). Some of the WMAs declared have the objective of protecting marine species such as turtles and dugong. The concept is sound. Yet most WMAs have not succeeded, because the prevailing political lack of enthusiasm for conservation means that the Department of Environment and Conservation has not been funded to support the WMA trustees with the technical and financial support they need. In late 1999 the Department of

Table 8

Participation in Selected Conventions. ✓ indicates party to the convention. (Source: CIA, 1998 and World Wide Web pages of administering agencies of some agreements.)

	Papua New Guinea	Solomon Islands
MARPOL	✓	
London Dumping	✓	✓
Civil Liability for Oil Pollution	✓	
Basel (Hazardous Wastes)	✓	
Law of the Sea	✓	✓
CITES (Trade in Endangered Species)	✓	
Ramsar (Wetlands)	✓	
Climate Change	✓	✓
Montreal Protocol (Ozone Layer)	✓	✓
Biodiversity	✓	✓
Straddling and Migratory Fish Stocks		✓
Apia (Conservation in the South Pacific)	✓	✓
SPREP (South Pacific Environment Programme)	✓	✓

Environment and Conservation was downgraded to the status of an "Office", with an attendant loss of Cabinet status and a dramatic reduction in already low levels of funding and staffing.

Responding to signs of depletion of coastal fisheries stocks, the two countries have introduced a range of fisheries management measures including closed areas and seasons, minimum sizes, licensing of fishers and traders, restriction of certain areas and/or stocks to domestic participants, harvest quotas and moratoria, and export bans on certain species. Though the National Fisheries Authority of PNG has developed management plans for several commercial stocks, a great deal more surveying and planning is needed.

Because of their multispecies nature, what seems to be the inherent vulnerability of some stocks (for instance, sedentary species), and a limited scientific understanding of them, tropical fisheries are notoriously difficult to manage even in developed countries. Given the limited capacities of management authorities, and the eagerness of some outside interests to exploit these limitations, prospects for sustainable coastal fisheries management in PNG and the Solomon Islands are uncertain.

Regional Cooperation

In light of the limited capacity for sustainable development and environmental management, not only in PNG and the Solomon Islands but throughout the South Pacific, several regional organisations have been established to assist with technical support, information exchange, and capacity building. These include the South Pacific Environmental Programme (SPREP), the South Pacific Community (SPC, formerly known as the South Pacific Commission), the South Pacific Forum Fisheries Agency (FFA), the South Pacific Applied Geoscience Commission (SOPAC), and the University of the South Pacific (USP), in particular its Institute of Marine Resources (IMR). A variety of international aid and non-governmental organisations have undertaken sustainable development and conservation projects in the area and some of these initiatives, too, are regional.

Community-based Management

It is no surprise that a western-style, "top-down" approach to resource management and conservation has been ineffective. Self-management of coastal marine areas has long been carried out by small kinship-based groups living in villages. The first attempt at centralised administration in PNG and the Solomon Islands was barely 100 years ago and only gradually, over decades, took hold. The concept of nation is still being developed among disparate communities who speak many languages—it is estimated that about 700 distinct languages are spoken in PNG, and 65 in the Solomon Islands. Most languages also have dialect variations. As declining funding increasingly compromises the national governments' capacity to undertake, or even to oversee, resource management in rural areas—despite the existence of provincial governments established to address local needs—so, management responsibilities revert to those who held them before. However, this is by default, not by plan, and so is not happening smoothly or efficiently.

Though Government fisheries agencies have attempted to enforce "modern" fisheries management systems, these have been largely unsuccessful. The coastal resources and ecosystems in PNG and the Solomons are dispersed over a vast, remote area with poor transportation and communication infrastructure. The financial, technical, and human resources required to centrally manage them far exceed those that could ever be made available to the relevant agencies. More importantly, government-imposed management measures have not been widely accepted among rural communities. Evidence of this emerges from the discussion, above, on artisanal fisheries. Fisheries policy makers are coming to realise that these problems are best addressed by community-based initiatives that incorporate traditional management systems. In PNG and the Solomon Islands the best approach to coastal fisheries management is for governments to provide the legal and administrative framework to support traditional management systems (Baines, 1995). Though a new Solomon Islands Fisheries Act, 1998 makes provision for this, PNG missed an opportunity to incorporate appropriate provisions in its most recent Fisheries legislation.

An examination of traditional coastal area management systems is revealing. As Johannes (1978) has pointed out, there are many parallels with modern fisheries management. Taboos that prohibit harvests in certain places, at certain times, or by certain people, are analogous to area or seasonal closures, or limited entry. Customary marine tenure has been described as a form of "limited entry". In most areas traditional systems continue to operate, though they have been weakened by the centralisation of power, which has had the effect of undermining local-level leadership. Increasing economic pressures on fishers to exploit their resources at a commercial scale, together with the pressures of rapid population growth, also threaten the integrity of traditional coastal-area management systems. Further, rural communities are profoundly disadvantaged in dealing with external pressures, as in negotiations with proponents of commercial fishing enterprises in their areas. The practicalities of one of these systems, the stresses on it, and the adaptations it has successfully made, are described for Marovo, Solomon Islands by Hviding and Baines (1994).

Traditional management systems are common and widespread in the region, though it has not been established that they are ubiquitous. Some have questioned their effectiveness (Aswani, 1998; Polunin, 1984; Swadling, 1982). They do not, in their present form, guarantee sustainable resource management in the modern context, a point made by Adams (1998), Baines (1989), Johannes (1978), and others. Nevertheless, there is growing consensus that, since local communities organised on socially meaningful lines are "on site", have extensive knowledge of local ecology and conditions, and have a vested interest in sustainability, the best management option is government support and empowerment of communities to manage their own resources (Adams, 1998; Dalzell et al., 1996; Hviding and Baines (1994)).

REFERENCES

Adams, T.J.H. (1998) The interface between traditional and modern methods of fishery management in the Pacific Islands. *Ocean and Coastal Management* **40**.

Aini, J. and Hair, C. (1995) Live fishing and export in Northern Papua New Guinea. Unpublished Report. Papua New Guinea Department of Fisheries and Marine Resources, Fisheries Research Station, Kavieng, New Ireland.

Allen, G.R. (1991) *Freshwater fishes of New Guinea*. Christensen Research Institute, Madang, Papua New Guinea.

Allen, G.R. and Swainston, R. (1993) *Reef fishes of New Guinea*. Publication No. 8. Christensen Research Institute, Madang, Papua New Guinea.

Alongi, D. (1990) The ecology of tropical soft-bottom benthic ecosystems. *Oceanography and Marine Biology Annual Review* **28**, 381–496.

Apinia, J. (1986) The declining mangrove resources of Wom. *Ples* **2**, 29–32.

Asquith, M., Kooge, F. and Morrison, R.J. (1994) *Transporting sediments via rivers to the ocean, and the role of sediments as pollutants in the South Pacific*. SPREP Reports and Studies No. 72. SPREP, Apia, Western Samoa.

Aswani, S. (1998) Patterns of marine harvest effort in south-western New Georgia, Solomon Islands: resource management or optimal foraging? *Ocean and Coastal Management* **40**, 207–235.

AusAID (1997) *Economic survey of Papua New Guinea*. Australian Agency for International Development, Canberra.

Baines, G.B.K. (1985) A traditional base for inshore fisheries development in the Solomon Islands. In *The traditional knowledge and management of coastal systems in Asia and the Pacific*, eds. K. Ruddle and R.E. Johannes, pp. 39–52. United Nations Educational Scientific and Cultural Organisation, Regional Office for Science and Technology for Southeast Asia, Jakarta Pusat, Indonesia.

Baines, G. (1989) Traditional resource management in the Melanesian South Pacific: a development dilemma. In *Common Property Resources: Ecology and Community-Based Sustainable Development*, ed. F. Berkes, pp. 274–295. Belhaven Press, London.

Baines, G.B.K. (1995) Sustaining livelihoods in Coastal Fisheries: Lessons for modern management from the South Pacific. *Appropriate Technology* **22** (2), 6–8.

Baines, G.B.K. and Morrison, R.J. (1990) Marine pollution from land use in the Pacific island region. *Marine Pollution Bulletin* **21** (11), 506–515.

Balat, N. (1993) A preliminary observation on the ecological distribution of northern New Guinea seagrasses. In *Seagrass Resources in Southeast Asia*, eds. M.D. Fortes and N. Wirjoatmodjo, pp. 221–226. UNESCO, Jakarta, Indonesia.

Blaber, S.J. and Milton, D.A. (1990) Species composition, community structure, and zoogeography of fishes of mangrove estuaries in the Solomon Islands. *Marine Biology* **105**, 259–267.

Blaber, S.J., Milton, D.A. and Rawlinson, N.J. (1990) Diets of lagoon fishes of the Solomon Islands: Predators of tuna baitfish and trophic effects of baitfishing on the subsistence fishery. *Fisheries Research* **8**, 263–286.

Brodie, J.E., Arnould, C., Eldridge, L., Hammond, L., Holthus, P., Mowbray, D. and Tortell, P. (1990) *State of the marine environment in the South Pacific region*. UNEP Regional Seas Reports and Studies No. 127 and SPREP Topic Review No. 40. UNEP and South Pacific Regional Environment Programme, Nairobi.

Brouns, J.W.M. and Heijs, F.M.L. (1991) Seagrass systems in the tropical west Pacific. In *Intertidal and Littoral Ecosystems*, eds. A.C. Mathieson and P.H. Nienhuis, pp. 371–390. Elsevier, Amsterdam.

Bryant, D., Burke, L., McManus, J. and Spalding, M. (1998) *Reefs at risk. A map-based indicator of threats to the world's coral reefs*. World Resources Institute, Washington, D.C.

Burrage, D., Hughes, R., Bode, L. and Williams, D. (1995) Dynamic features and transports of the Coral Sea circulation. In *Recent Advances in Marine Science and Technology '94*, eds. O. Bellwood, H. Choat and N. Saxena, pp. 95–105. James Cook University of North Queensland, Townsville, Queensland, Australia.

Carrier, A.H. and Carrier, J.G. (1989) Marine tenure and economic reward on Ponam Island, Manus Province. In *A Sea of Small Boats*, ed. J. Cordell, pp. 94–121. Cultural Survival Inc, Cambridge, Mass.

Chou, L.M., Wilkinson, C.R., Licuanan, W.R.Y., Alino, P., Cheshire, A.C., Loo, M.G.K., Tangjaitron, S., Sudara, A., Ridzwan, A.R. and Soekarno (1994) Status of coral reefs in the ASEAN region. In *Proceedings, Third ASEAN-Australia Symposium on Living Coastal Resources*, eds. C. Wilkinson, S. Sudara and L.M. Chou, pp. 1–10. Australian Institute of Marine Science, Townsville, Queensland, Australia.

CIA, 1998. *The World Factbook 1998*. United States Central Intelligence Agency, Washington D.C.

Coates, D. (1993) Fish ecology and management of the Sepik-Ramu, New Guinea, a large contemporary tropical river basin. *Environmental Biology of Fishes* **38**, 345–368.

Collette, B. (1983) Mangrove fishes of New Guinea. In *Biology and Ecology of Mangroves*, ed. H.J. Teas, pp. 91–102. W. Junk Publ., The Hague.

Commonwealth, 1998. *The Commonwealth Yearbook 1998*. Great Britain Foreign and Commonwealth Office, London.

Convard, N. (1993) *Land-based pollutants inventory for the South Pacific Region*. SPREP Reports and Studies no. 68. SPREP, Apia, Western Samoa.

Cragg, S. (1987) Papua New Guinea. In *Mangroves of Asia and the Pacific: Status and Management*, pp. 299–309. Natural Resources Management Center and National Mangrove Committee, Ministry of Natural Resources, Quezon City, The Philippines.

Dalzell, P., Adams, T.J.H. and Polunin, N.V.C. (1996) Coastal fisheries in the Pacific Islands. *Oceanography and Marine Biology Annual Review* **34**, 395–531.

Dalzell, P. and Wright, A. (1986) An assessment of the exploitation of coral reef fishery resources in Papua New Guinea. In *The First Asian Fisheries Forum*, eds. J.L. Maclean, L.B. Dizon and L.V. Hosillos, pp. 477–481. Asian Fisheries Society, Manila.

Davies, J.M., Dunne, R.P. and Brown, B.E. (1997) Coral bleaching and elevated sea-water temperature in Milne Bay Province, Papua New Guinea, 1996. *Marine and Freshwater Research* **48**, 513–516.

Davis, D. and Huber, M. (1995) Melanesia. In *Coastal management in the Asia-Pacific*, eds. K. Hotta and I.M. Dutton, pp. 319–335. Japanese International Marine Science and Technology Federation, Tokyo.

Doulman, D.J. and Wright, A. (1983) Recent developments in PNG's tuna fishery. *Mar. Fish. Rev.* **45**, 47–59.

Envision (1993) *Establishment of port waste facilities and oil pollution control in Papua New Guinea*.

FAO (1998) *Solomon Islands*. Food and Agriculture Organization of the United Nations, Rome.

Fautin, D.G. and Allen, G.R. (1992) *Field Guide to Anemonefishes and their Host Sea Anemones*. Western Australian Museum, Perth. 160 pp.

Frielink, A.B.J. (1983) *The coastal population of Papua New Guinea*. Report No. 83–11. Papua New Guinea Department of Primary Industry, Fisheries Research and Surveys Branch, Port Moresby.

Furnas, M.J. and Mitchell, A.W. (1996) Pelagic primary production in the Coral and southern Solomon Seas. *Marine and Freshwater Research* 47, 695–706.

Ghiselin, M.T. (1992) How well known is the opisthobranch gastropod fauna of Madang, Papua New Guinea. *Proceedings of the Seventh International Coral Reef Symposium, Guam* 2, 697–701.

Gillett, R. (1997) *The importance of tuna to Pacific Island countries*. FFA Report 97/15. Forum Fisheries Agency, Honiara, Solomon Islands.

Gosliner, T.M. (1992) Biodiversity of tropical opisthobranch gastropod faunas. *Proceedings of the Seventh International Coral Reef Symposium, Guam* 2, 702–709.

Gumoi, M. and Sekhran, N. (1994) An overview of the Papua New Guinean economy: The implications of conservation. In *Papua New Guinea Country Study on Biological Diversity*, eds. N. Sekhran and S. Miller, pp. 41–57. Department of Environment and Conservation, Conservation Resource Centre, Waigani, Papua New Guinea.

Hair, C. (1994) Fisheries in Papua New Guinea. In *Papua New Guinea Country Study on Biological Diversity. A report to the United Nations Environment Program*, eds. N. Sekhran and S. Miller, pp. 169–185. Department of Environment and Conservation, Conservation Resource Centre, Waigani, Papua New Guinea.

Hansell, J.R. and Wall, J.R. (1976) *Land Resources of the Solomon Islands*. Land Resource Study 18 Vol. 1. Ministry of Overseas Development, Land Resources Division, Surrey, England.

Hattori, A. (1987) *Studies on the dynamics of the biological community in tropical seagrass ecosystems in Papua New Guinea: the second report*. Ocean Research Institute, University of Tokyo, Tokyo.

Hedemark, M. and Sekhran, N. (1994) Overview of current conservation-oriented policies, legislation, and interventions. In *Papua New Guinea Country Study on Biological Diversity. A report to the United Nations Environment Program*, eds. N. Sekhran and S. Miller, pp. 339–359. Department of Environment and Conservation, Conservation Resource Centre, Waigani, Papua New Guinea.

Heijs, F.M.L. and Brouns, J.W.M. (1986) A survey of seagrass communities around the Bismarck Sea, Papua New Guinea. *Proceedings of the Koninklijke Nederlandse Akademie van Wetenschappen* C89, 11–44.

Hodgson, G. (1994) Sedimentation damage to reef corals. In *Proceedings of the colloquium on global aspects of coral reefs: Health, Hazards, and History, 1993*, ed. R.N. Ginsburg Compiler, pp. 298–303. Rosenstiel School of Marine and Atmospheric Science, University of Miami, Miami.

Hoeksma, B.W. (1992) The position of northern New Guinea in the center of marine benthic diversity: a reef coral perspective. *Proceedings of the Seventh International Coral Reef Symposium, Guam* 2, 710–717.

Holland, A. (1994) The beche-de-mer industry in the Solomon Islands: recent trends and suggestions for management. *SPC Beche-de-mer Information Bulletin* 6, 2–9.

Huber, M.E. (1994) An assessment of the status of the coral reefs of Papua New Guinea. *Marine Pollution Bulletin* 29, 69–73.

Hughes, P. and Sullivan, M. (1989) Environmental impact assessment in Papua New Guinea: lessons for the wider Pacific region. *Pacific Viewpoint* 30, 34–55.

Hughes, P., Sullivan, M. and Asigau, W. (1986) Coastal erosion at Gabagaba village, Central Province, Papua New Guinea—a case study. *Ples* 2, 33–40.

Hviding, E. (1996) *Guardians of Marovo Lagoon*. Pacific Island Monograph Series 14. University of Hawaii Press, Honolulu.

Hviding, E. and Baines, G.B.K. (1994) Community-based Fisheries Management, Tradition and the Challenges of Development in Marovo, Solomon Islands. *Development and Change* 25, 13–39.

Johannes, R.E. (1978) Traditional marine conservation methods in Oceania and their demise. *Annual Review of Ecology and Systematics* 9, 349–364.

Johannes, R.E. (1988) Spawning aggregation of the grouper, *Plectromus areolatus* (Rüppel) in the Solomon Islands. Proc. 6th Int. Coral Reef Symposium, Vol. 2, pp. 751–755.

Johannes, R.E. and Lam, M. (1999) The live reef food fish trade in the Solomon Islands. *SPC Live Reef Fish Information Bulletin* 5, 8–15.

Johannes, R.E. and Riepen, M. (1995) *Environmental, economic, and social implications of the live reef fish trade in Asia and the western Pacific*. Report to The Nature Conservancy and the South Pacific Forum Fisheries Agency.

Johnstone, I.M. (1982) Ecology and distribution of seagrasses. In *Biogeography and Ecology of Papua New Guinea*, ed. J.L. Gressit, pp. 497–512. Dr. W. Junk, The Hague, Netherlands.

Johnstone, I.M. and Frodin, D. (1982) Mangroves of the Papuan sub-region. In *Biogeography and Ecology of New Guinea*, ed. J.L. Gressit, pp. 513–528. Dr. W. Junk, The Hague, Netherlands.

Kailola, P.J. (1995) *Fisheries Resources Profiles: Papua New Guinea*. FFA Report No. 95/45. Forum Fisheries Agency.

Leary, T. (1993a) *Solomon Islands State of the Environment Report*. South Pacific Regional Environment Programme, Apia, Western Samoa.

Leary, T. (1993b) Solomon Islands. In *A Directory of Wetlands in Oceania*, ed. D.A. Scott, pp. 331–361. IUCN, Cambridge.

Leqata, J.L., Rawlinson, N.J., Nichols, P.V. and Tiroba, G. (1990) Subsistence fishing in Solomon Islands and the possible conflict with commercial baitfishing. In *Tuna Baitfish in the Indo-Pacific Region: Proceedings of a Workshop, Honiara, Solomon Islands, 11–13 December 1989*, eds. S.J. Blaber and J.W. Copland, pp. 169–179. Australian Council for International Agricultural Research, Canberra, Australia.

Levett, M. and Bala, A. (1994) Agriculture in Papua New Guinea. In *Papua New Guinea Country Study on Biological Diversity. A report to the United Nations Environment Program*, eds. N. Sekhran and S. Miller, pp. 125–153. Department of Environment and Conservation, Conservation Resource Centre, Waigani, Papua New Guinea.

Lock, J.M. (1986a) *Study of the Port Moresby artisanal reef fishery*. Technical Report 86/1. Department of Primary Industry, Papua New Guinea, Port Moresby.

Lock, J.M. (1986b) *Effects of fishing pressure on the fish resources of the Port Moresby barrier and fringing reefs*. Technical Report 86/3. Department of Primary Industry, Papua New Guinea, Port Moresby.

Louman, B. and Nicholls, S. (1994) Forestry in Papua New Guinea. In *Papua New Guinea Country Study on Biological Diversity. A report to the United Nations Environment Program*, eds. N. Sekhran and S. Miller, pp. 155–167. Department of Environment and Conservation, Conservation Resource Centre, Waigani, Papua New Guinea.

Maragos, J. (1998) Status of coral reefs of the southwest and east Pacific: Melanesia and Polynesia. In *Status of Coral Reefs of the World: 1998*, ed. C. Wilkinson, pp. 89–107. Australian Institute of Marine Science, Townsville, QLD, Australia.

Marshall, G. (1990) The political economy of logging: The Barnett Inquiry into corruption in the Papua New Guinea timber industry.

The Ecologist **20**, 174–181.

McAlpine, J.R., Keig, G. and Falls, R. (1983) *Climate of Papua New Guinea*. Commonwealth Scientific and Industrial Research Organisation, Canberra, Australia.

McCoy, E.D. and Heck, K.L. (1976) Biogeography of corals, seagrasses, and mangroves: an alternative to the center of origin concept. *Systematic Zoology* **25**, 201–210.

McCoy, M. (1980) Reptiles of the Solomon Islands. Wau Ecology Institute Handbook No. 7, Wau, Papua New Guinea.

McKinnon, J.M. (1975) Tomahawks, turtles and traders. *Oceania* **45**, 290–307.

Messell, H. and King, W. (1989) Report on CITES and Solomon Islands Government national survey of the crocodile populations of the Solomon Islands. Unpublished report to the Ministry of Natural Resources, Honiara, Solomon Islands.

Moore, N.H. (1982) Oceanographic Observations in the Western Sector of the Papuan Coastal lagoon, PNG. Motupore Island Research Department, University of Papua New Guinea, Port Moresby.

Mukai, H. (1993) Biogeography of the tropical seagrasses in the Western Pacific. *Australian Journal of Marine and Freshwater Research* **44**, 1–17.

Nadarajah, T. (1993) *Effectiveness of the Papua New Guinean Environmental Planning Act 1978 in forestry timber projects*. NRI Discussion Paper 69. The National Research Institute, Boroko, Papua New Guinea.

Naidu, S., Aalsberg, W.G., Brodie, J.G., Fuavo, V.A., Maata, M., Naqasima, M., Whippy, P. and Morrison, R.J. (1991) *Water quality studies on selected South Pacific lagoons*. UNEP Regional Seas Reports and Studies No. 136. UNEP, Nairobi.

NFA (1998) PNG Exports of fish and marine products for the period January to June 1998. *Papua New Guinea National Fisheries Authority Fisheries Newsletter* **4**, 13–14.

Osborne, P.L. (1993) Papua New Guinea. In *A Directory of Wetlands in Oceania*, ed. D.A. Scott, pp. 301–320. IUCN, Cambridge.

Paijmans, K. (1982) Vegetation. In *Papua New Guinea Atlas*, eds. D. King and S. Ranck, pp. 92–93. Robert Brown and Associates, Bathurst, NSW, Australia.

Pandolfi, J. (1992) A review of the tectonic history of New Guinea and its significance for marine biogeography. *Proceedings of the Seventh International Coral Reef Symposium, Guam* **2**, 718–728.

Percival, M. and Womersley, J. (1975) *Floristics and ecology of the mangrove vegetation of Papua New Guinea*. Botany Bulletin No. 8. Papua New Guinea National Herbarium, Department of Forests, Brisbane.

Pernetta, J. and Hill, L. (1981) A review of marine resource use in coastal Papua. *J. Soc. des Oceanistes.* **71–73**, 175–191.

Pigram, C.J. and Davies, H.L. (1987) Terranes and the accretion history of the New Guinea Orogen. *BMR J. Aust. Geol. Geophys.* **10**, 193–211.

Polunin, N.V.C. (1984) Do traditional marine "reserves" conserve? A view of Indonesian and Papua New Guinean evidence. *Senri Ethnological Studies* **17**, 15–31.

Richards, A.H. (1993) Live reef-fish exports to S.E. Asia from the South Pacific. *South Pacific Commission Fisheries Newsletter* **67**, 34–36.

Richards, A.H., Bell, L.J. and Bell, J.D. (1994) Inshore fisheries resources of the Solomon Islands. *Marine Pollution Bulletin* **29**, 90–98.

Saenger, P., Hegerl, E.J. and Davie, J.D.S. (1983) Global status of mangrove ecosystems. *The Environmentalist* **3(1983)**, 3–88.

Sekhran, N. and Miller, S. (1994) Papua New Guinea Country Study on Biological Diversity. A Report to the United Nations Environment Programme, Waigani, Papua New Guinea, Department of Environment and Conservation, Conservation Resource Centre; and Nairobi, Kenya, Africa Centre for Resources and Environment (ACRE). 438 pp.

Skewes, T. (1990) *Marine Resource Profiles: Solomon Islands*. FFA Report No. 90/61. South Pacific Forum Fisheries Agency, Honiara, Solomon Islands.

Solomon Islands (1992) *Country Report for UNCED*. South Pacific Regional Environment Programme, Apia, Western Samoa.

Solomon Islands (1993) *National Environmental Management Strategy*. South Pacific Regional Environment Programme, Apia, Western Samoa.

SPC (1998) *Selected Pacific economies—a statistical summary, number 14*. South Pacific Commission, Noumea, New Caledonia.

Spenceley, P. (1982) Rainfall and temperature. In *Papua New Guinea Atlas*, eds. D. King and S. Ranck, pp. 94–95. Robert Brown and Associates, Bathurst, NSW, Australia.

Stoddart, D.A. (1969a) Geomorphology of Solomon Islands coral reefs. *Philosophical Transactions of the Royal Society of London B* **255**, 355–382.

Stoddart, D.A. (1969b) Geomorphology of the Marovo elevated barrier reef, New Georgia. *Philosophical Transactions of the Royal Society of London B* **255**, 383–402.

Swadling, P. (1982) Shellfishing in Papua New Guinea, with special reference to the Papuan coast. In *Traditional Conservation in Papua New Guinea: Implications for Today*, eds. L. Morauta, J. Pernetta and W. Heaney, pp. 307–310. Institute of Applied Social and Economic Research, Boroko, Papua New Guinea.

Tiroba, G. (1993) Current status of commercial baitfishing in Solomon Islands. In *Tuna baitfish in Fiji and Solomon Islands: Proceedings of a workshop, Nadi, Fiji, 17–18 August 1993*, eds. S.J. Blaber, D.A. Milton and N.J. Rawlinson, pp. 113–116. Australian Centre for International Agricultural Research, Canberra.

UNDP (1989) Fishery Sector Review, Papua New Guinea. Final Report.

UNDP (1994) *Pacific Human Development Report*. United Nations Development Programme, Suva, Fiji.

UNDP (1998) *Human Development Report 1998*. United Nations Development Programme, New York.

Vedder, J.G. and Coulson, F.I. (1986) Introduction to geology and offshore resources of Solomon Islands. In *Geology and Offshore Resources of Pacific Island Arcs—Central and Western Solomon Islands*, eds. J.G. Vedder, K.S. Pound and S.Q. Boundy, pp. 1–15. Circum-Pacific Council for Energy and Mineral Resources, Houston, Texas.

Veitayaki, J., Ram-Bidesi, V., Matthews, E., Gibson, L. and Vuki, V. (1995) *Overview of destructive fishing practices in the Pacific Islands region*. SPREP Reports and Studies No. 93. SPREP, Apia, Western Samoa.

Veron, J.E.N. (1993) *A biogeographic database of hermatypic corals*. Australian Institute of Marine Science Monograph Series Volume 10. Australian Institute of Marine Science, Townsville, Australia.

Veron, J.E.N. and Kelley, R. (1988) Species stability in reef corals of Papua New Guinea and the Indo Pacific. *Mem. Assoc. Australas. Palaeontologists* **6**, 1–69.

Waterboard of PNG (1987) *Environmental plan Joyce Bay sewage outfall*. 1. Waterboard of Papua New Guinea, Port Moresby.

Wells, F.E. (1990) Comparative zoogeography of marine mollusks from northern Australia, New Guinea, and Indonesia. *The Veliger* **33**, 140–144.

Wells, M.L., Vallis, G.K. and Silver, E.A. (1999) Tectonic processes in Papua New Guinea and past productivity in the eastern equatorial Pacific ocean. *Nature, Lond.* **398**, 601–604.

Wells, S.M. and Jenkins, M.D. (1988) *Coral Reefs of the World. Volume 3: Central and Western Pacific*. UNEP Regional Seas Directories and

Bibliographies. UNEP/IUCN, Nairobi.

Wilkinson, C. (1998) The 1997–1998 mass bleaching event around the world. In *Status of Coral Reefs of the World: 1998*, ed. C. Wilkinson, pp. 15–38. Australian Institute of Marine Science, Townsville, QLD, Australia.

Williams, D. McB. (1994) *Coral Sea Region Billfish Atlas*. Australian Institute of Marine Science, Townsville, QLD, Australia.

Woodroffe, C.D. (1987) Pacific island mangroves: distribution and environmental setting. *Pac. Sci.* **41**, 166–185.

Wright, A. (1985) Marine resource use in Papua New Guinea: can traditional concepts and contemporary development be integrated. In *The Traditional Knowledge and Management of Coastal Systems in Asia and the Pacific*, eds. K. Ruddle and R.E. Johannes, pp. 79–99. United Nations Educational Scientific and Cultural Organization, Regional Office for Science and Technology for Southeast Asia, Jakarta Pusat.

Wright, A. (1993) Introduction. In *The Nearshore Marine Resources of the South Pacific*, eds. A. Wright and L. Hill, pp. 1–13. Institute of Pacific Studies/Forum Fisheries Agency, Suva, Fiji/Honiara, Solomon Islands.

Wright, A. and Richards, A.H. (1985) A multispecies fishery associated with coral reefs in the Tigak islands, PNG. *Asian Marine Biology* **2**, 69–84.

THE AUTHORS

Michael E. Huber
Global Coastal Strategies,
P.O. Box 606, Wynnum, Qld. 4178,
Australia

Graham B.K. Baines
Environment Pacific,
3 Pindari St., The Gap,
Brisbane, QLD 4061,
Australia

Chapter 81

ASIAN DEVELOPING REGIONS: PERSISTENT ORGANIC POLLUTANTS IN THE SEAS

Shinsuke Tanabe

This chapter provides an overview of the status of contamination and toxic impacts of persistent organic pollutants (POPs) such as butyltins (BTs) and organochlorines (OCs) in the coastal waters and open seas of the Asian developing regions. BTs were detected in a wide range of environmental media and biota from Asian coastal waters, indicating a widespread contamination by these compounds even in developing countries. Several studies have also made clear the existence of significant OC contamination in the coastal waters and open seas of tropical Asian countries in recent years. It has been also pointed out that OCs used in the tropical regions disperse on a global scale by long-range atmospheric transport and deposit in colder regions. This feature of contamination has led to high accumulation and toxic effects in marine mammals. Continuous monitoring studies of POPs in the aquatic environment of Asia-Pacific region are needed to effectively address pollution problems affecting the health of oceans.

Seas at The Millennium: An Environmental Evaluation (Edited by C. Sheppard)
© 2000 Elsevier Science Ltd. All rights reserved

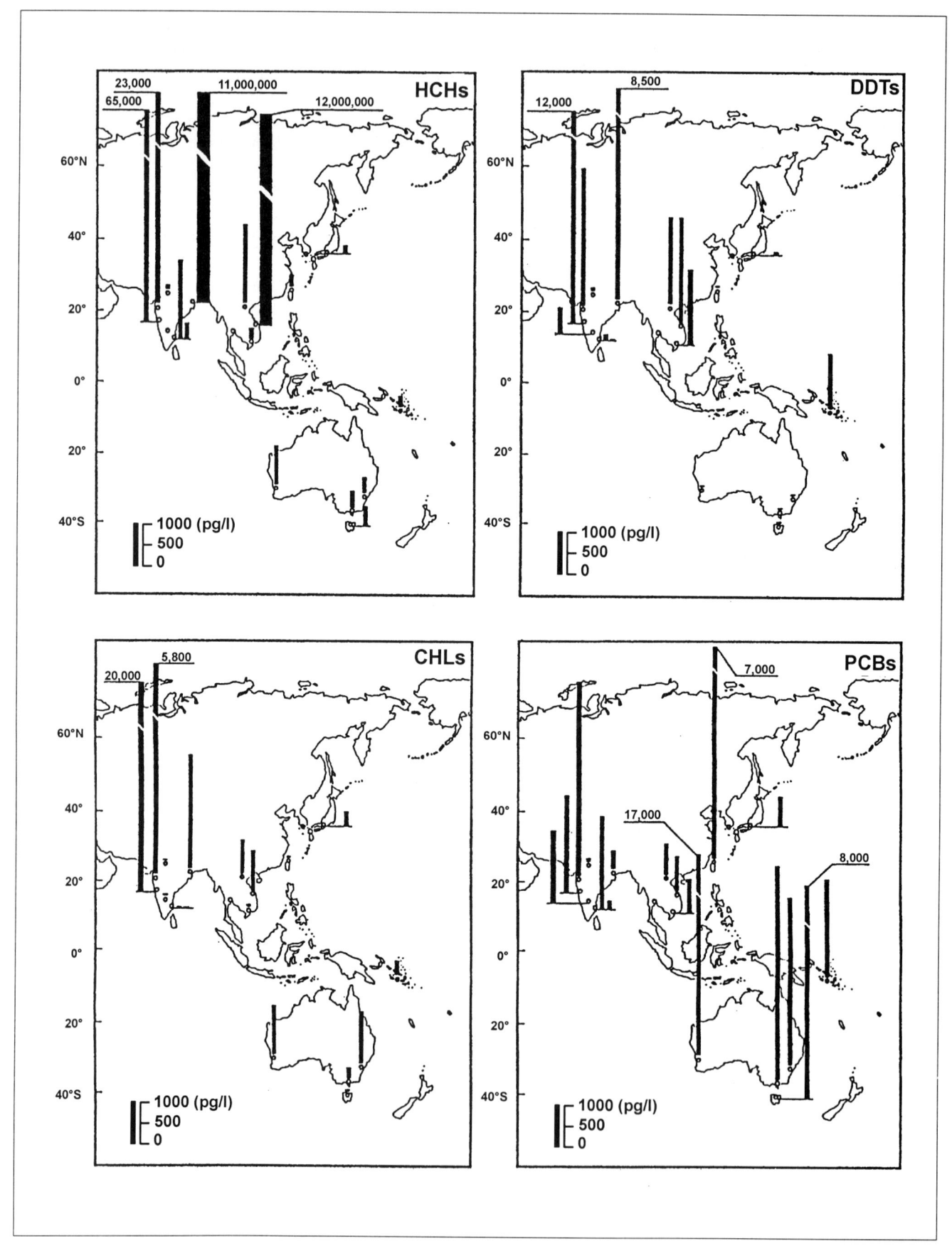

Fig. 1. Distribution of persistent organochlorines in river and estuarine air from the eastern and southern Asia and Oceania.

INTRODUCTION

The rapid increase in the chemical trade in Asian developing countries in recent years implies greater production and usage of toxic chemicals including organochlorines (OCs) and butyltins (BTs). Consequently, environmental problems associated with toxic contaminants such as BTs and OCs are of great concern.

BTs, represented by tributyltin (TBT), have been used widely for pleasure boats, large ships and vessels, docks and fishing nets, lumber preservatives and slimicides in cooling systems, and as an effective anti-fouling agent in paints. Its derivatives, dibutytin (DBT) and monobutyltin (MBT) have been used mainly as stabilizers in polyvinyl chloride and as catalysts in the production of polyurethane foams, silicones, and in other industrial processes (Fent, 1996). Aquatic pollution by TBT arising from antifouling paints has been of great concern in many countries because of their reported effects on non-target marine organisms, i.e., shell malformation in oysters (Alzieu and Heral, 1984; Alzieu and Portmann, 1984), mortality on the larvae of mussels (Beaumont and Budd, 1984), and imposex in gastropods (Bryan and Gibbs, 1991). These findings led to the restriction of TBT-based antifouling paints in most developed nations; however, in most developing countries in Asia, such control is yet to be implemented.

Persistent OCs such as PCBs (polychlorinated biphenyls), DDTs, HCHs (hexachlorocyclohexane isomers), CHLs (chlordane compounds), and HCB (hexachlorobenzene) have been reported to show estrogenic properties (Soto et al., 1994), in addition to the well-documented environmental problems and human health hazards associated with these toxic compounds. Although most developed nations have banned or restricted the production and use of these compounds during the last decade, some developing countries still use these chemicals for agricultural and public health purposes (Dave, 1996; Li et al., 1996).

The widespread use of these compounds in Asia has implications for human and wildlife exposure to the so-called Persistent Organic Pollutants (POPs). This chapter gives an overview of the status of contamination by POPs with special reference to OCs and BTs in coastal waters and open seas of Asian developing regions.

COASTAL WATERS

Air, Water and Sediments

In Asia, the comprehensive monitoring surveys have been conducted on the contamination by POPs such as PCBs (polychlorinated biphenyls) and organochlorine insecticides (e.g. DDTs, HCHs [BHC] and HCB) in air, water and sediments from the latter half of the 1980s to the early 1990s (Iwata et al., 1994a). These surveys reported the presence of higher concentrations of DDTs and HCHs in air and water samples collected from coastal and estuarine areas in the developing countries of tropical and subtropical regions (e.g., India, Thailand, Vietnam) rather than the developed nations such as Japan and Australia (Figs. 1 and 2). Unexpectedly, PCBs and CHLs which have been largely produced and used in industrialized countries were also found to be higher in lower latitude areas, reflecting that the developing countries in the tropics are also using these chemicals, which were prohibited for domestic usage in the developed nations. As evidence, India produced 4200 tons of DDT and 32,000 tons of HCH during 1994–1995 (Dave, 1996). In the suburb of Bangkok, Thailand—a dumpsite for secondhand transformers and capacitors imported from industrialized countries—caused serious PCB pollution near this area (Watanabe et al., 1996). Similar situations are more or less expected in some other Asian countries.

Unlike the pollution pattern found in air and water, the OC residue levels in coastal and estuarine sediments revealed smaller variation between northern and southern countries in Asia (Fig. 3) (Iwata et al., 1994a). This is attributable to the shorter residence time of these contaminants in the water phase because of their active evaporation due to prevailing high temperatures in the tropics. In a field study of HCH application in a paddy of tropical India, more than 90% of this insecticide applied evaporated into the atmosphere within two weeks (Tanabe et al., 1991). Further studies which investigated the fate of this insecticide in the watershed of Vella River, South India, demonstrated its large-scale atmospheric evaporation (more than 99% used in this area in a year) and less transfer into the water phase (less than 1%) of riverine and estuarine environment (Takeoka et al., 1991).

Regarding the organotin compounds, data on the atmospheric pollution is not available for Asian coastal regions. Reports on water pollution are also few, but high concentrations of organotins were recorded in surface seawater from Malacca Strait in Malaysia (Tong et al., 1996; Hashimoto et al., 1998). Butyltin compounds have been measured in Ganges River waters in India (Ansari et al., 1998). Concentrations of MBT, DBT and TBT were in the ranges of 2.1–70 ng Sn/l, 1.7–101 ng Sn/l and 2.9–20 ng Sn/l, respectively. Monitoring studies on sediment pollution by organotins have been conducted in Malaysia (Tong et al., 1996), Hong Kong (Ko et al., 1995), Thailand (Kan-atireklap et al., 1997) and Korea (Kim et al., 1998), in which the maximum concentration (130 μg/g dry wt.) of TBT (tributyltin) was detected in the sediment sample from Hong Kong. In Thailand, 20 locations in coastal and estuarine areas were monitored for BT pollution and higher concentrations were reported in high boating ports (Table 1) (Kan-atireklap et al., 1997a).

Fish

An extensive study of OC pollution in fish from Asian waters was conducted by Kannan et al. (1995a). The OC

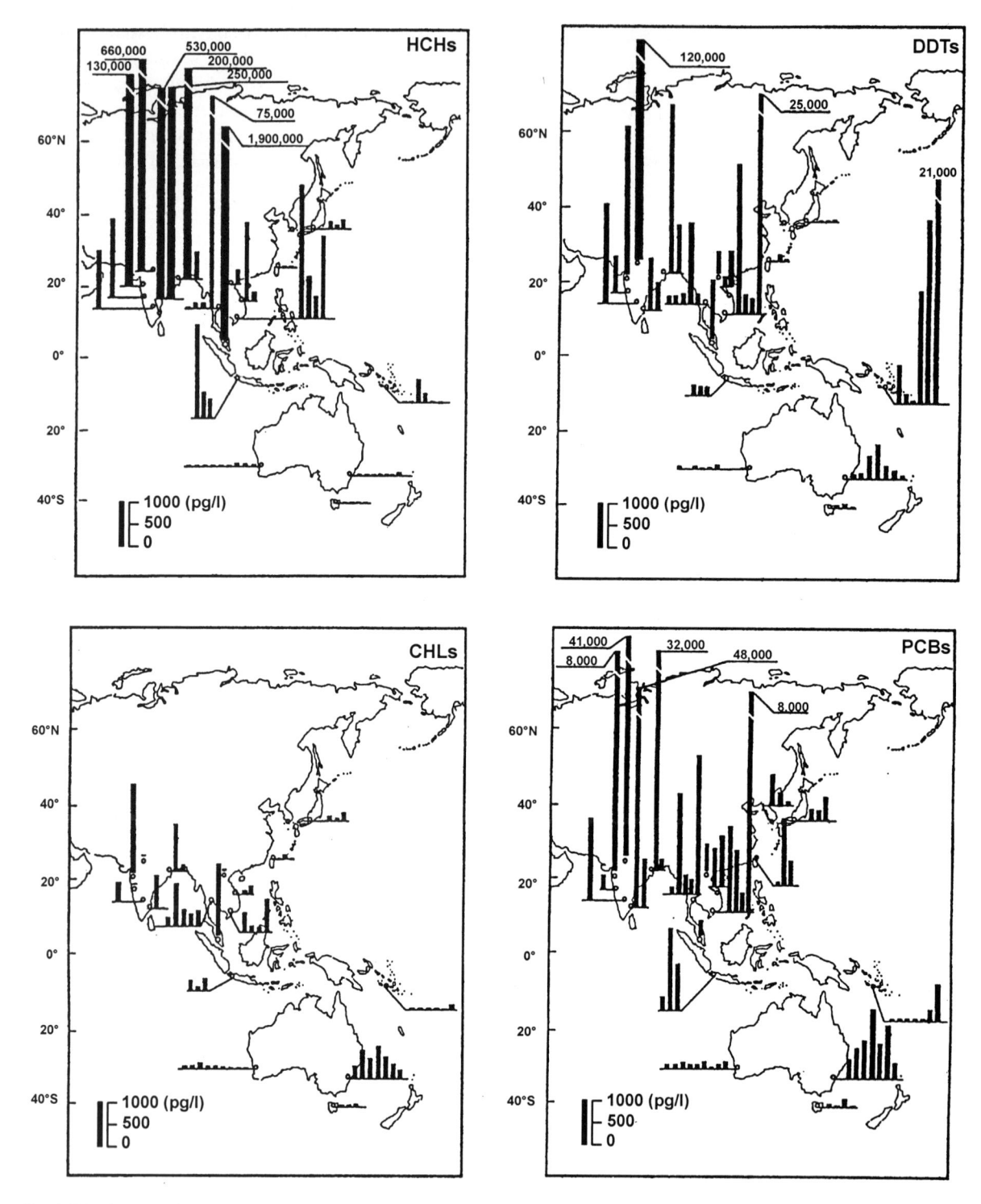

Fig. 2. Distribution of persistent organochlorines in river and estuarine water from the eastern and southern Asia and Oceania.

contamination in fish was prominent over the whole region of Asian coastal waters with smaller variations in residue levels between northern and southern countries (Fig. 4) similar to sediments, again suggesting the nature of these contaminants having shorter residence times in the water phase in the tropics. The occurrence of OC residues in fish are pointed out to pose a great concern in food pollution in Asian developing countries (Kannan et al., 1997a). In addition, marine pollution monitoring using green mussels as bioindicators has also made clear the status of contamination by OCs in some Asian countries, which suggested serious contamination by HCHs and DDTs in Indian coastal waters and by PCBs and CHLs in Philippines coastal waters (Fig. 5) (Tanabe et al., 1998).

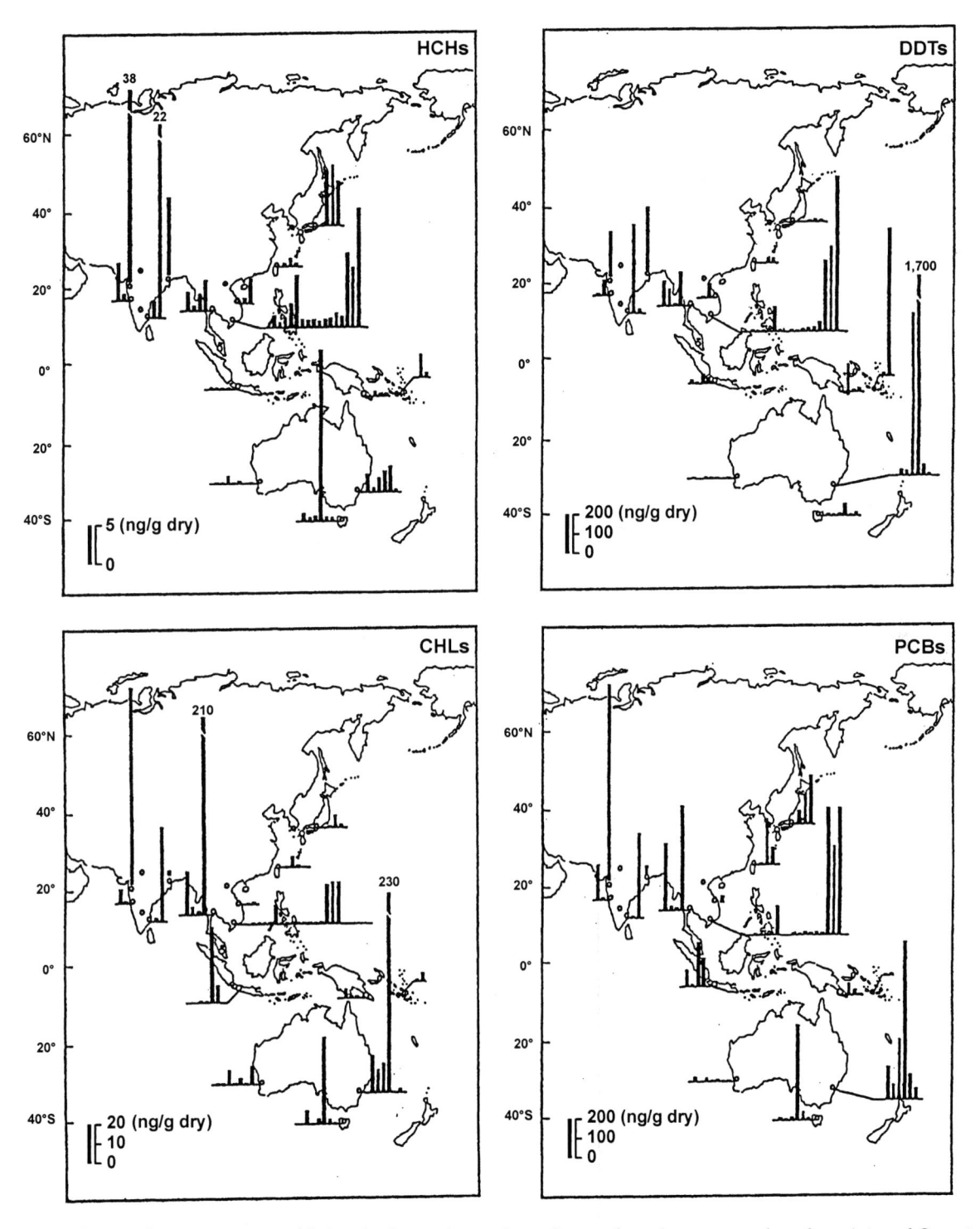

Fig. 3. Distribution of persistent organochlorines in river and estuarine sediments from the eastern and southern Asia and Oceania.

BT pollution in fish has also been studied and relatively lower residue levels were found in Asian developing countries as compared with developed nations (Fig. 6) (Kannan et al., 1995b). However, continuous monitoring of organotins is strongly needed in Asian coastal waters, because of the increasing demand for anti-fouling paints in Asian developing countries (Reish, 1996). The Mussel Watch Project using bivalves as bioindicators has elucidated relatively higher contamination by BTs in the coastal waters of Thailand and Philippines than that of India (Table 2), and suggested by the usage of organotins in aquaculture in Thailand (Kan-atireklap et al., 1997b, 1998; Prudente et al., 1999).

Table 1

Concentrations (ng g^{-1} dry wt) of butyltin compounds in sediment samples collected from coastal areas of Thailand in 1995

Sampling Location	Code*	Date#	MBT	DBT	TBT	S BTs
Coastal mariculture sites						
Kung Kra Baen, Chanthaburi	THTBKB	950306	7	2	4	13
Ang Sila, Chonburi	THCBAL	950307	14	13	81	108
Ban Laern, Phetchaburi	THPBBL	950312	77	21	32	130
Mean ($n = 3$)			33	12	39	84
Fishing boat piers/coastal mariculture sites						
Muang, Prachuap Khiri Khan	THPKMU	950312	8	2	7	17
Yong Star, Trang	THTRYS	950318	24	66	93	183
Mean ($n = 2$)			16	34	50	100
Fishing boat piers/ports						
Khlong Yai, Trat	THTTKY	950306	9	9	12	30
Trat River, Trat	THTTTR	950305	190	190	480	860
Pak Nam Prasae, Rayong	THRYPS	950305	120	53	100	273
Rayong River, Rayong	THRYRR	950304	130	82	120	332
Thachin River, Samut Sakhon	THSKTC	950322	26	14	430	470
Pak Phanang, Nakhon Si Thammarat	THNSPP	950314	25	13	10	48
Yaring, Pattani	THPNYR	950316	300	450	880	1630
Tham Ma Lang, Satun	THSTTL	950317	12	11	9	32
Muang, Krabi	THKBMU	950318	110	140	370	620
Bang Toey, Phang Nga	THPGBT	950319	43	35	84	162
Paknam, Ranong	THRNPN	950321	53	40	70	163
Mean ($n = 11$)			93	94	233	420
Far seas vessel harbours						
Chao Phraya River, Samut Prakan	THSPCP	950323	400	1600	4500	6500
Paknam, Chumphon	THCPPN	950321	200	210	3600	4010
Songkhla Lake, Songkhla	THSHSK	950317	7	9	36	52
Ao Makham, Phuket	THPKAM	950320	410	1900	3900	6210
Mean ($n = 4$)			254	930	3009	4193

*First, second and last two alphabets indicate country, state or city and site name, respectively.
#First, second and last two digits indicate year, month and date, respectively.

Birds and Mammals

To our knowledge, only a few studies have reported on OC pollution in Asian avifauna, comprising resident and migrant birds from India (Ramesh et al., 1992; Tanabe et al., 1998b) and Vietnam (Yen et al., 1998). The HCH residue levels in Indian birds were quite high, while PCB and DDT concentrations were lower than in the birds from developed nations (Fig. 7). One of these studies also made clear that the migrating birds that stop over in India and the Middle Eastern countries are heavily exposed to HCHs and PCBs, respectively (Tanabe et al., 1998b). Butyltin residues were also detected in Indian (Senthilkumar et al., 1999) and Philippines birds (Senthilkumar et al., 1998), but those levels were very low.

Contamination by OCs and BTs in dolphins and whales from Asian developing countries was relatively low as compared with those from developed nations (Prudente et al., 1997; Tanabe et al., 1998c), which seems to be due to less accumulation in tropical fish, the prey of dolphins. River dolphins collected from the Ganges River in India have been analyzed for organochlorines and butyltins (Kannan et al., 1993a, 1994, 1997b). DDTs were the most predominant contaminants, followed by PCBs and HCHs. A recent study has suggested that there are signs of increasing concentrations of PCBs in river dolphins (Senthilkumar et al., 1999b).

In India, human breast milk is highly contaminated by HCHs (Tanabe et al., 1990). DDT residue levels are also high, while PCBs are low in Indian human breast milk. The analytical data on Chinese human breast milk have also been reported and extremely high concentrations of HCHs and DDTs were noted (Slorach and Vaz, 1983).

Table 2

Concentrations (ng g^{-1} wet wt) of butyltin compounds in green mussel (*Perna viridis*) collected from coastal waters of Thailand, Philippines and India

Location	MBT	DBT	TBT	BTs	Location	MBT	DBT	TBT	BTs	Location	MBT	DBT	TBT	BTs
India					Philippines					Thailand				
INWBDH	<3	1	2	3	PHFIPQ	15	13	76	104	THTTTR-1	<3	1	8	9
INOSSK	16	1	1	18	PHEMRP	47	100	640	787	THTTTR-2	<3	2	11	13
INOSKR	11	12	85	108	PHMLCC	51	43	200	294	THTBKB	42	80	680	802
INAPVP	20	1	1	22	PHBLBA-1	5	4	13	22	THRYBP	38	10	25	73
INAPKD	<3	1	1	2	PHCVBC-1	9	8	34	51	THCBSI	45	66	200	311
INAPMN	<3	1	2	3	PHSMJB	<3	3	1	4	THCBAL	3	5	24	32
INTNMD	9	21	150	180	PHSMVR	<3	2	28	30	THSPCP-1	7	10	56	53
INTNKK	3	1	2	6	PHLTDT	<3	<1	<1	nd	THSPCP-2	8	9	48	65
INPCPC-1	3	2	4	9	PHCZSP	<3	1	<1	1	THSKTC-1	<3	3	25	28
INPCPC-2	6	1	2	9	PHBTSL	<3	<1	<1	nd	THSKTC-2	<3	2	9	11
INTNCD-1	<3	<1	3	3	PHCVBC-2	<3	16	43	59	THPBBL-1	5	7	7	19
INTNCD-2	24	1	1	26	PHMMMB	3	15	44	62	THPBNBL-2	<3	4	23	27
INTNPP	250	110	18	378	PHBLBA-2	<3	12	34	46	THPKMU-1	9	20	130	159
INTNMP	11	4	24	39						THPKMU-2	7	16	210	233
INKLCH	11	3	16	30						THCPPN	3	3	11	17
INKLCC	8	<1	<1	8						THSRPD	4	8	49	61
INKTML	<3	7	10	17						THSHSK	3	5	27	35
INGOGA	7	1	<1	8						THPNMU	5	6	41	52
INMRMH	10	1	2	13						THTRYS	<3	8	89	97
										THPGBT	<3	1	3	4
										THPKAM	<3	4	28	32

Nd: less than detection limit (3 ng g^{-1} for MBT and 1 ng g^{-1} for DBT and TBT); BTs: MBT+DBT+TBT

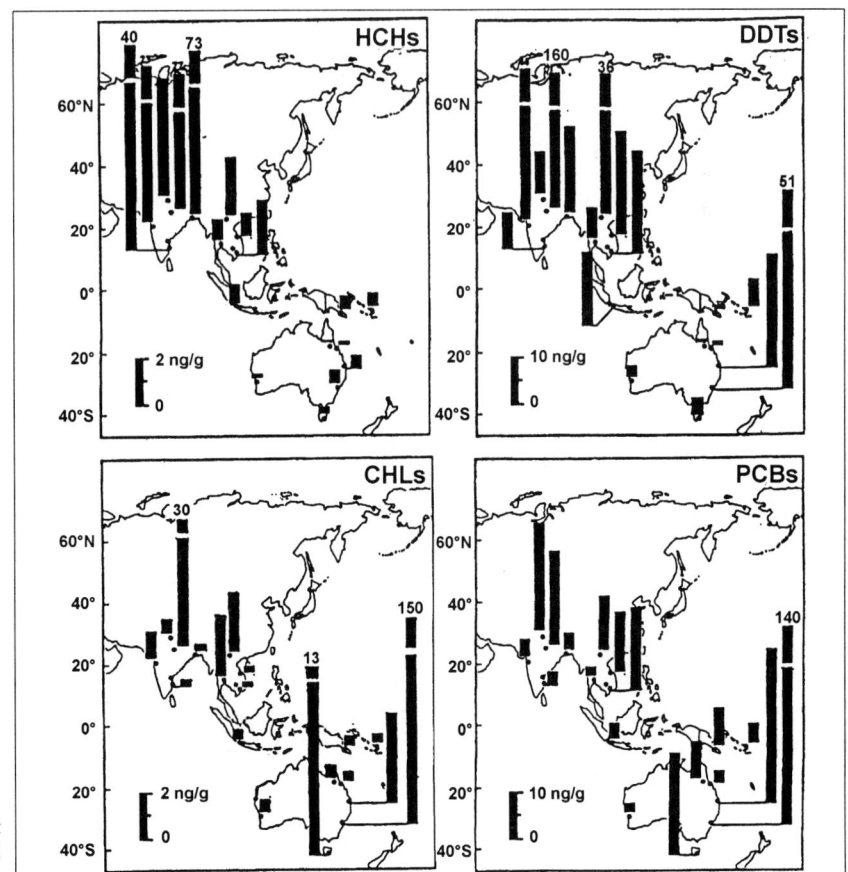

Fig. 4. Contamination by persistent organochlorines in fish from eastern and southern Asia and Oceania.

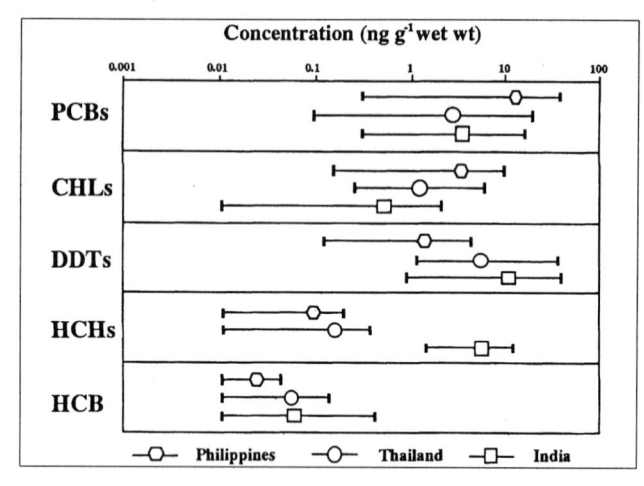

Fig. 5. Comparison of organochlorine residue levels in green mussel (*Perna viridis*) from India, Philippines and Thailand.

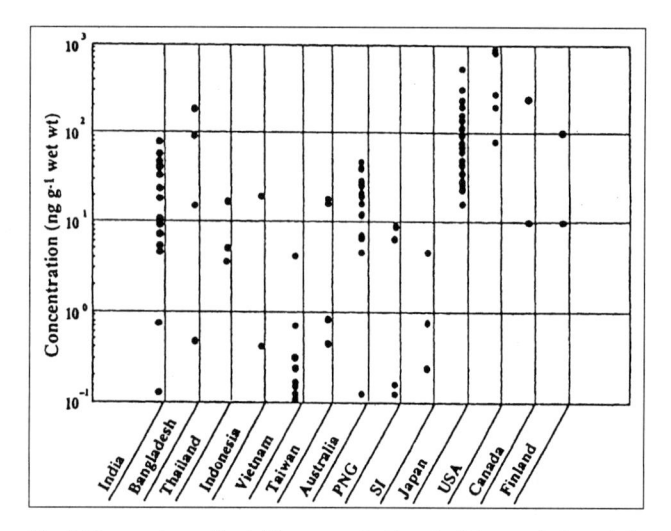

Fig. 6. Comparison of butyltin concentrations in fish muscle from Asia and Oceania with those in some developed countries.

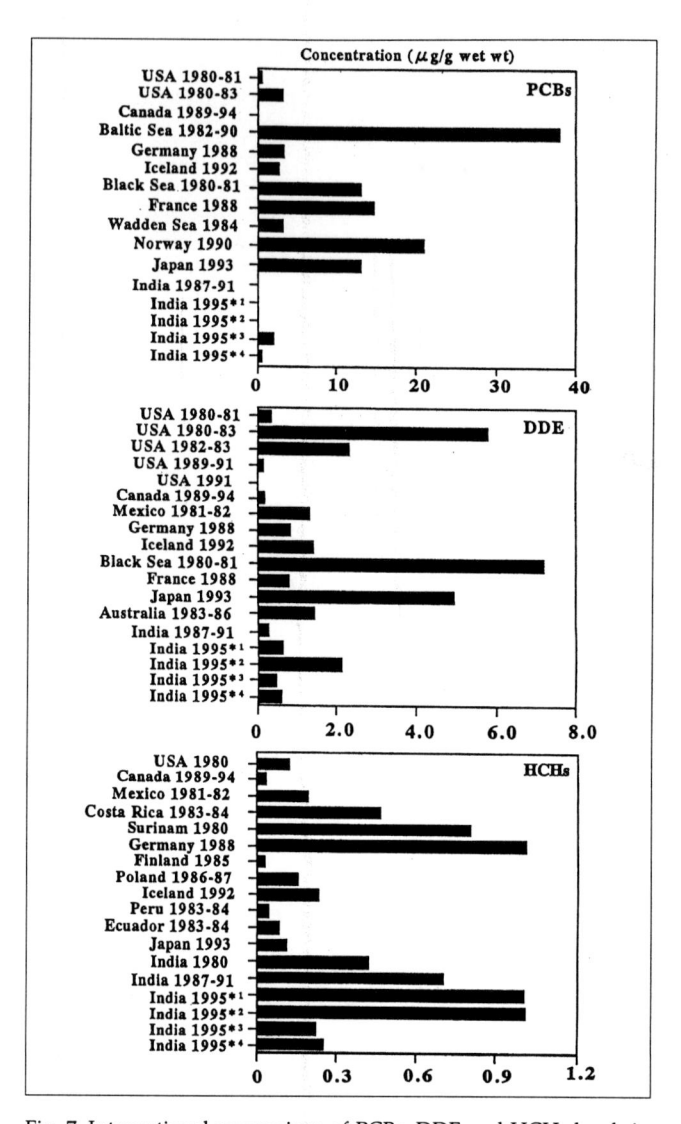

Fig. 7. International comparison of PCBs, DDE and HCHs levels in birds. (1) Strict residents, (2) local migrants, (3) short-distance migrants, (4) long-distance migrants.

OPEN SEAS

Air and Seawater

Studies on contamination by persistent OCs in the North and South Pacific began in the latter half of the 1970s. Predominant contamination in the northern hemisphere relative to the southern hemisphere and the presence of the highly contaminated zone in mid-latitudes of the northern hemisphere, which was affected by the extensive environmental release of OCs from developed nations, were discerned from these studies (Tanabe et al., 1982, 1983). It was further demonstrated by analysing deep sea water and biota that pollution by OCs had penetrated to the bottom of the oceans (Tanabe and Tatsukawa, 1983; Melzian et al., 1987; Lee et al., 1997).

A comprehensive investigation into organochlorine contaminants in open-sea air and surface seawater was initiated in 1989/1990. It was found that OC pollution expanded over not only the Pacific, but also the Atlantic, Indian, Arctic and Antarctic Oceans (Iwata et al., 1993). The highly contaminated zone found to be in the mid-latitudes had disappeared in a recent survey, but the low latitude oceans were significantly contaminated by OCs, particularly HCHs and DDTs (Figs. 8 and 9). Such a temporal change in the pattern of OC pollution is attributable to a decline in their usage by industrialized nations. In the 1970s, Goldberg (1975) had predicted the transfer of the principal regions of DDT usage from developed to developing countries. This prediction turned out to be true not only for DDT usage, but also for other OC consumption such as HCH and PCBs. The fact that PCB pollution was found over the world's oceans in uniform concentrations (Fig. 10) seems to be derived both from the continuous environmental discharge of PCBs from various equipment used in industrialized countries and from the release from dumping of the older transformers and capacitors exported to developing countries.

Table 3

Estimated PCB loads in the global environment

Environment	PCB load (t)	Percentage of PCB load	Percentage of world production
Terrestrial and coastal			
Air	500	0.13	
River and lake water	3500	0.94	
Seawater	2400	0.64	
Soil	2400	0.64	
Sediment	130,000	35	
Biota	4300	1.1	
Total (A)	143,000	39	
Open ocean			
Air	790	0.21	
Seawater	230,000	61	
Sediment	110	0.03	
Biota	270	0.07	
Total (B)	231,000	61	
Total load in the environment (A+B)	374,000	100	31
Degraded and incinerated	43,000		4
Land-stocked*	783,000		65
World production	1,200,000		100

*Still in use in electrical equipment and other products, and deposited in landfills and dumps.

Global Fate

In order to evaluate the global fate of OCs, an analysis was conducted to estimate the environmental loads of PCBs (Table 3) (Tanabe, 1988). Global PCB production was estimated to be 1200 thousand tons and around 30% (37 thousand tons) of the total production was estimated to be released into the environment. Out of the 37 thousand tons which escaped into the environment, 35% appear to be present in coastal sediments and 60% were retained in the open sea water. Although no such mass-balance analysis has been conducted for other OCs, the marine environmental loads of OC insecticides such as HCH and DDT which were used in the open lands were likely to be greater than those of PCBs used largely in the closed systems.

In an estimation of fluxes of OCs in air and surface seawater interface in open seas, these chemicals were shown to be scavenged from air to water, suggesting that the open sea water body plays a role as a sink for OCs (Iwata et al., 1993). The colder waters such as the Arctic Ocean had higher rates of OC fluxes from air to water.

At present, the major pollution sources of OCs are located in the tropical and sub-tropical regions and thus severe contamination has been observed in coastal and estuarine environments in the low latitude countries (Iwata et al., 1994a). However, the pattern of contamination of open oceans varied depending on the OCs. DDT distribution in open sea water was predominant in the tropics (Fig. 9), whereas HCHs were abundant in colder oceans (Fig. 8) (Iwata et al., 1993). This difference is ascribed by the atmospheric transportability of contaminants. DDTs, which are less volatile, remain in the water near to the sources of emission. On the other hand, HCHs which are relatively more volatile, evaporate and disperse following the long-range atmospheric transport, and finally deposit in colder oceans. The open sea water eventually functions as the sink for OCs, but the destination and temporal trend vary according to the nature of the contaminants (Wania and Mackay, 1996).

Contamination and Bioaccumulation in Marine Mammals

Marine pollution by OCs has been recorded not only in air and water, but also in higher trophic animals such as cetaceans and pinnipeds. In particular, the coastal species of marine mammals from developed nations are known to be heavily contaminated by OCs. The bottlenose dolphins in the Mediterranean Sea (Corsolini et al., 1995), finless porpoise in the Seto Inland Sea (Kannan et al., 1989) and Beluga whale in the St. Lawrence River (Martineau et al., 1987) contained with several hundred parts-per-million of PCBs in blubber. The OC contamination in marine mammals was extended over the animal populations in polar regions where there is less industrial and human activity (Tanabe et al., 1986; O'Shea and Brownell, 1994; Norstrom and Muir, 1994). Exposure from sources arising out of long-range atmospheric transport and high rates of bioaccumulation through food-webs predispose these animals to accumulate noticeable concentrations of OCs. A recent study measured OCs in dolphins and whales from the western Pacific (Fig. 11) and elucidated that the pattern of contamination in cetaceans was similar to water pollution (Prudente et al., 1997).

It has been well documented that marine mammals accumulate high concentrations of OCs in their bodies. A food-chain accumulation study for the western North Pacific ecosystem (Kurosho region) showed that striped dolphins, a top predator, retained OCs with more than 10 million-fold of concentration in water (Tanabe et al., 1984). Moreover, cetaceans inhabiting even cleaner oceans accumulated PCBs at much higher concentrations than terrestrial mammals which live near to pollution sources (Fig. 12) (Tanabe et al., 1994a). The reasons for this extraordinarily high accumulation were made clear in the follow-up studies. Marine mammals have thick subcutaneous fat that retains large quantities of organochlorines (reservoir for contaminants) (Tanabe et al., 1981), transfers organochlorines from mother to newborn through lactation (Tanabe et al., 1994a), and has a low capacity to degrade OCs due to the lack of some cytochrome P-450

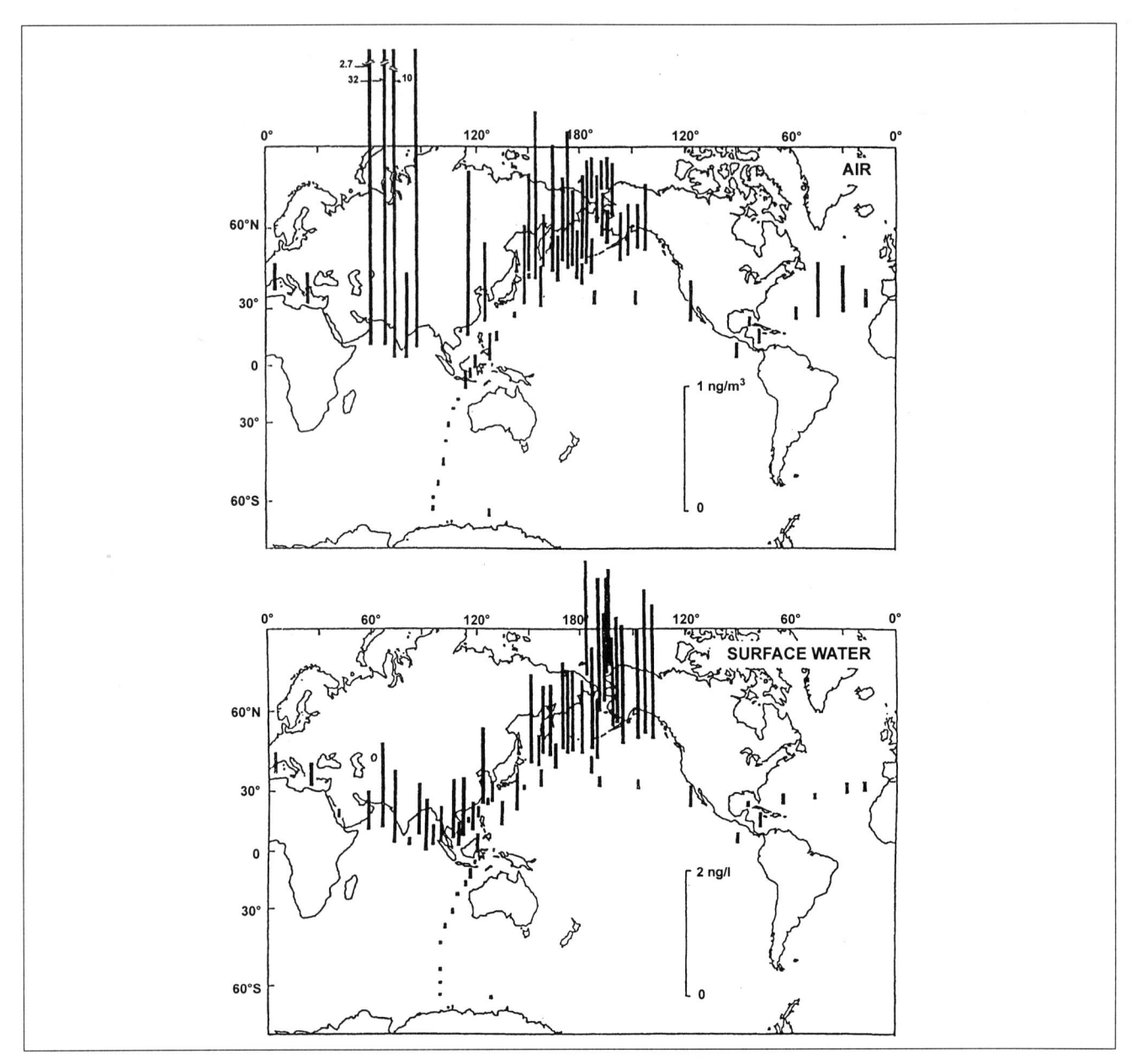

Fig. 8. Distribution of HCH concentrations in air and surface seawater.

drug-metabolizing enzyme system (Fig. 13) (Tanabe et al., 1988; Watanabe et al., 1989). Such specific physiological aspects led to high accumulation of OCs in cetaceans, and therefore experience toxic effects.

Toxic Effects

In marine mammals, many abnormalities have been suspected to be due to the effects of toxic chemicals. Colborn and Smolen (1996) listed 65 examples of perturbations (population decline or dieoffs, reproduction and endocrine impairment, immune dysfunction, organ damage, infection health impairment, tumours etc.) observed in marine mammals since 1968. Simmonds (1991) also reported 11 incidents of mass mortality of marine mammals in the 20th century. Interestingly, out of 11 incidents, 9 mass mortalities took place after the 1970s, indicating adverse effects of toxic chemicals originating from rapid material civilization. Therefore comprehensive studies describing mutual interactions of contaminant accumulation, the activity/concentration of specific enzymes and hormones and the disease/abnormality found in the body are needed in marine mammals. Although there are few studies examining the toxic effects of contaminants in marine mammals, Dall's porpoise and northern fur seal have been investigated along these lines.

In Dall's porpoise, inhabiting cold waters nearby Japan, a significant negative correlation was observed between PCB/DDE concentrations in the blubber and testosterone levels in the blood, suggesting possible endocrine

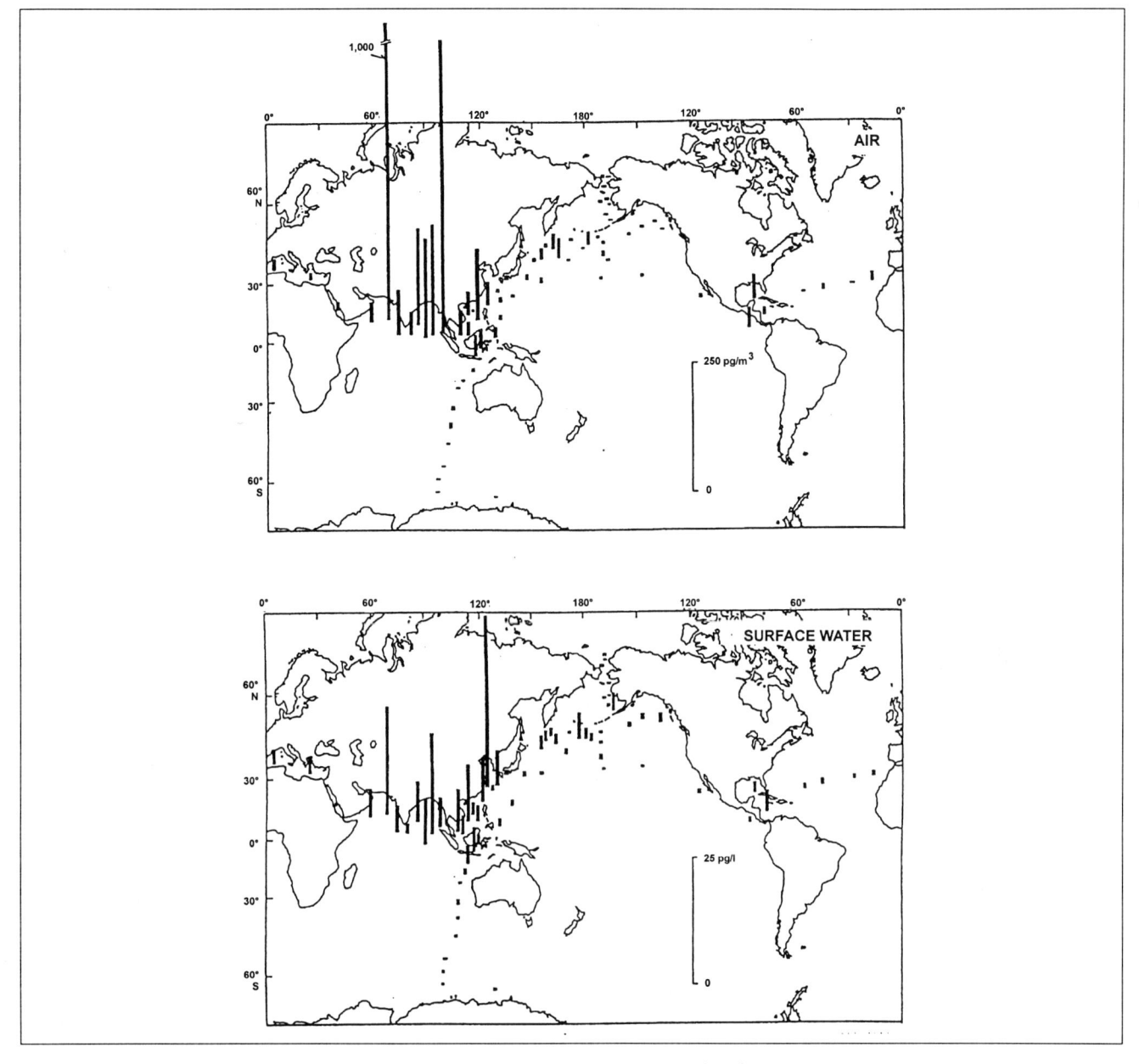

Fig. 9. Distribution of DDT concentrations in air and surface seawater.

disruption by these contaminants (Fig. 14) (Subramanian et al., 1987). In another study examining the effect of PCBs in northern fur seal from coastal waters in Japan, a significant positive correlation between PCB residue levels and activities of some cytochrome P-450 enzymes was noted, indicating enzyme induction by PCBs (Fig. 15) (Tanabe and Tatsukawa, 1992). Similar results have also been observed in short-finned pilot whales from the Pacific coast of Japan (Tanabe and Tatsukawa, 1992) and striped dolphins from the Mediterranean Sea (Kannan et al., 1993b). The enzyme induction by toxic contaminants can disrupt reproductive cycles since they share a common/overlapping pathway. These observations imply that some OCs can act through enzyme induction causing hormonal imbalance, which can lead to abnormalities in marine mammals.

In the course of these studies on adverse health effects, coplanar PCBs, a group of highly toxic PCB congeners, are the most suspected contaminants in addition to polychlorinated dibenzo-p-dioxins (PCDDs) and dibenzofurans (PCDFs) (Kannan et al., 1989). PCDDs and PCDFs have much lower vapour pressure and high particle affinity compared to PCBs. They are less dispersible through air and water, and thus contamination by these pollutants on a global scale is a great deal less and their toxic impact is also less for the open sea ecosystem. On the other hand, coplanar PCBs are relatively volatile and dispersible. They are, therefore, ubiquitous over the globe and bioaccumulate significantly in marine mammals through the food-webs. Coplanar PCBs are believed to pose much greater toxic impacts to marine mammals than PCDDs and PCDFs.

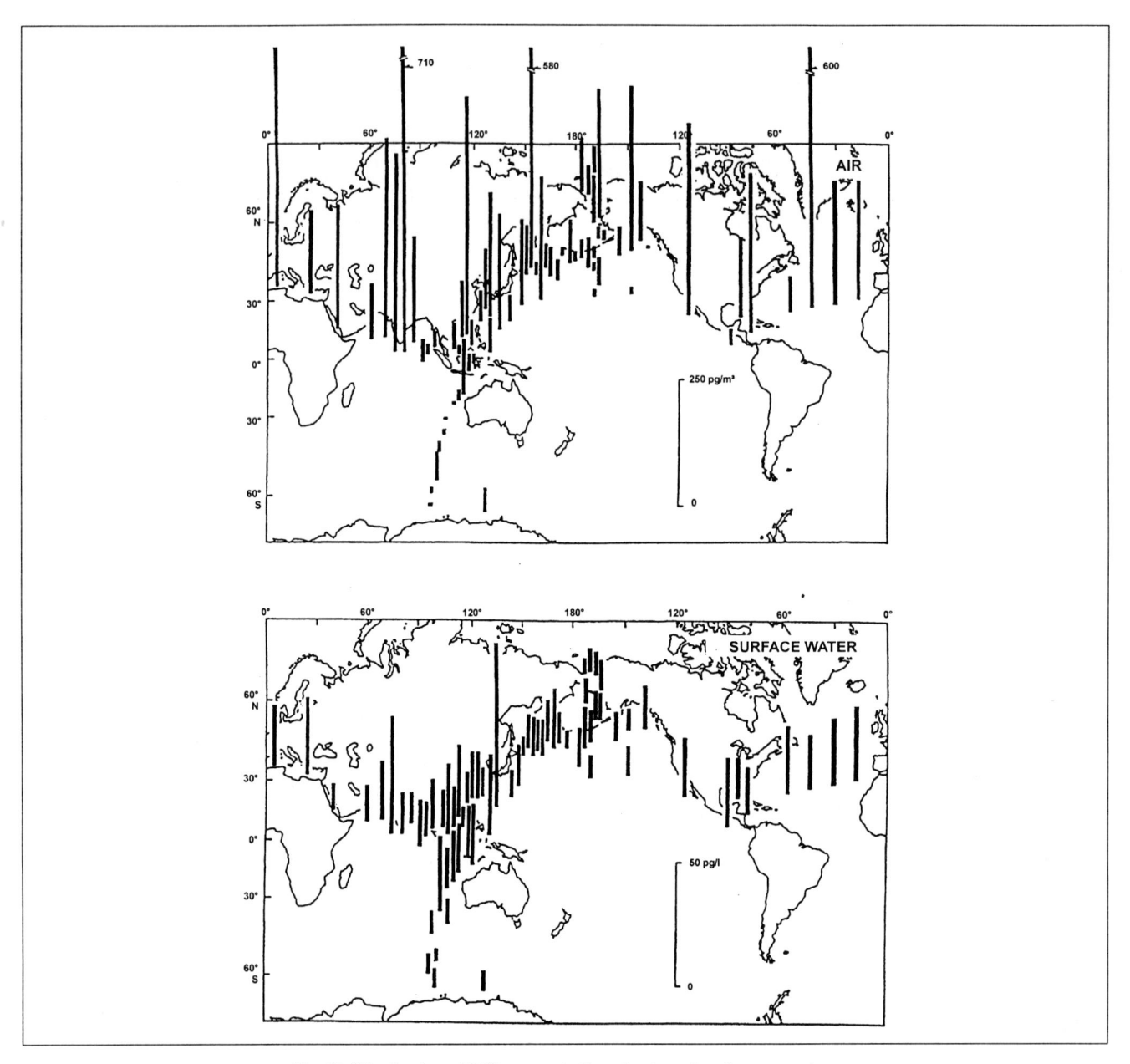

Fig. 10. Distribution of PCB concentrations in air and surface seawater.

Temporal Trend of Contamination

Four studies have reported temporal pollution of OCs in the open seas. The determination of OCs in sediment core samples from the Bering Sea clearly indicated continuous environmental input of these contaminants by long-range atmospheric transport (Iwata et al., 1994b). A study using blubber samples of the striped dolphin from the Pacific coast of Japan suggested less declining pattern in the concentrations of PCBs and DDTs during the period 1978–1986 (Loganathan et al., 1990). Archived fat tissues of northern fur seals collected from the northern Pacific of Japan showed maximum contamination of PCBs and DDTs in the mid 1970s and then showed a declining trend (Tanabe et al., 1994b). However, in this study PCBs levels were steady in the 1980s and HCH concentrations were constant in the whole period examined. Interestingly, the blubber samples of male minke whale from the Antarctic Ocean represented the invariable temporal pattern in the residue levels of organochlorine insecticides and the increasing trend of PCBs during a recent period of 10 years (Fig. 16) (Aono et al., 1997). All the observations mentioned above suggest a continuing significant long-term contamination of organochlorines in the future. This also means that the oceanic ecosystem is the ultimate repository for persistent pollutants and marine organisms, particularly marine mammals, are at risk from exposure to such pollutants.

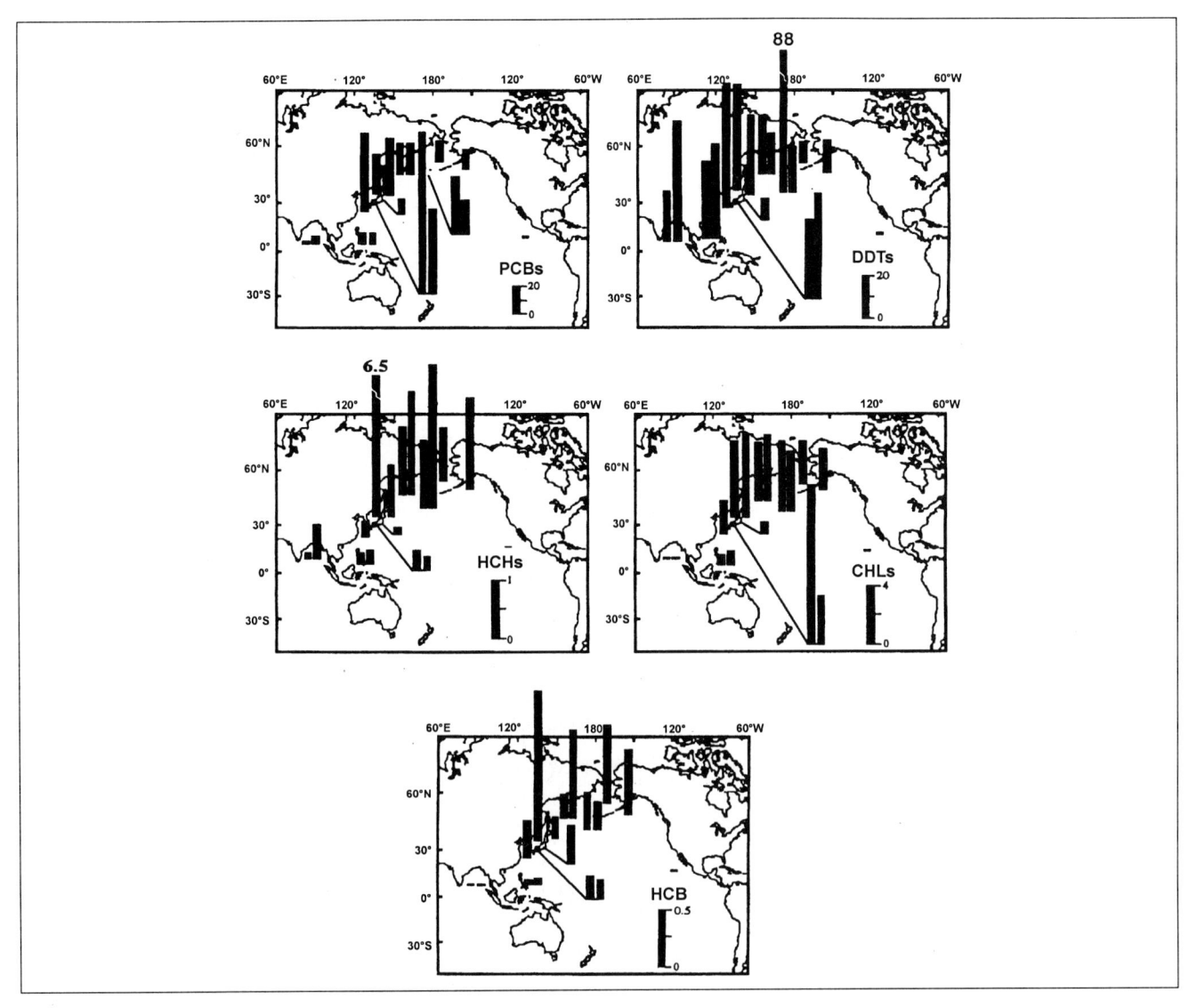

Fig. 11. Distribution of organochlorine residue levels (μg/g wet wt) in cetaceans from western North Pacific and Bay of Bengal.

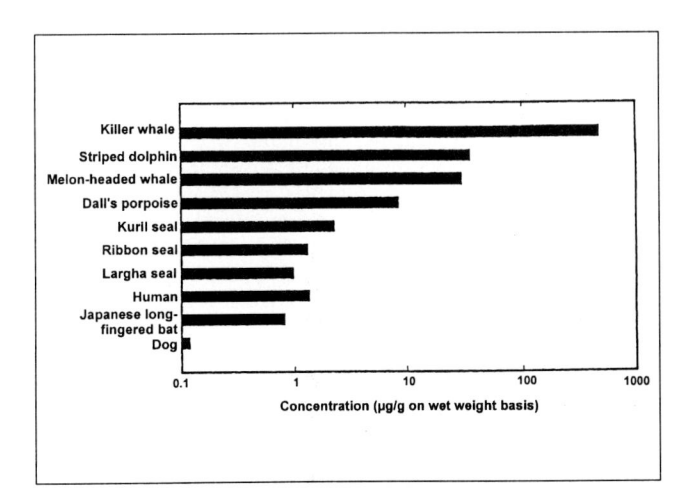

Fig. 12. Comparison of PCB concentrations in fat tissue of cetaceans from western North Pacific, and coastal and terrestrial mammals from Japan.

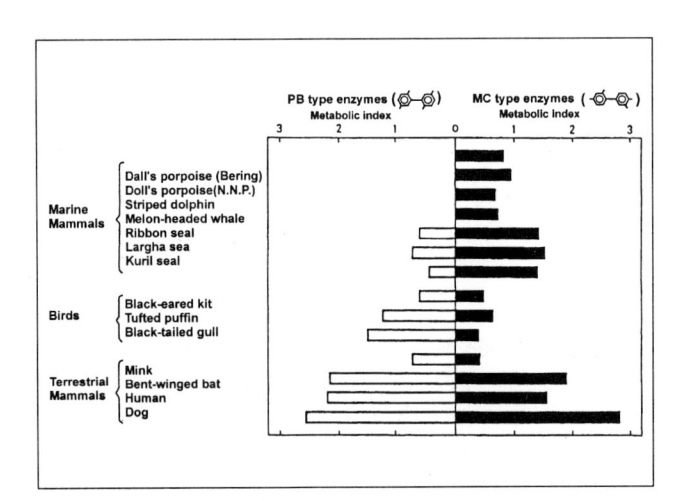

Fig. 13. PB(phenobarbital)- and MC(methylcholanthrene)-type enzyme activities in higher animals as estimated by Metabolic Index (MI) of 2,2',5,5'- and 2,3',4,4'-tetrachlorobiphenyl isomers which indicate the capacity of PCB metabolism.

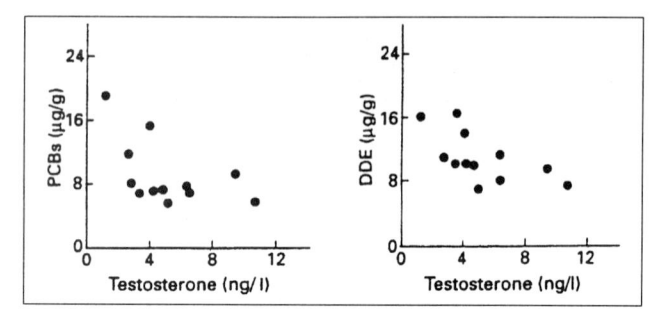

Fig. 14. Relationship between the concentrations of PCBs and DDE in the blubber and levels of testosterone in blood of adult male Dall's porpoises from the northwestern North Pacific.

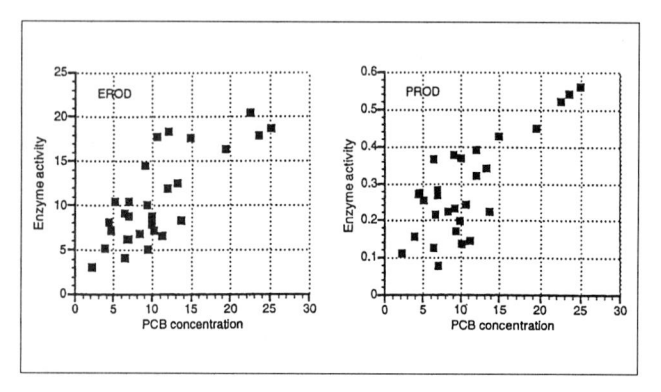

Fig. 15. Relationship between PCB residue levels (μg/g wet wt) and EROD(7-ethoxyresorufin O-deethylase)/PROD(pentoxyresorufin O-deethylase) activities (nmol/min/g liver) in northern fur seals from the Pacific coast of northern Japan.

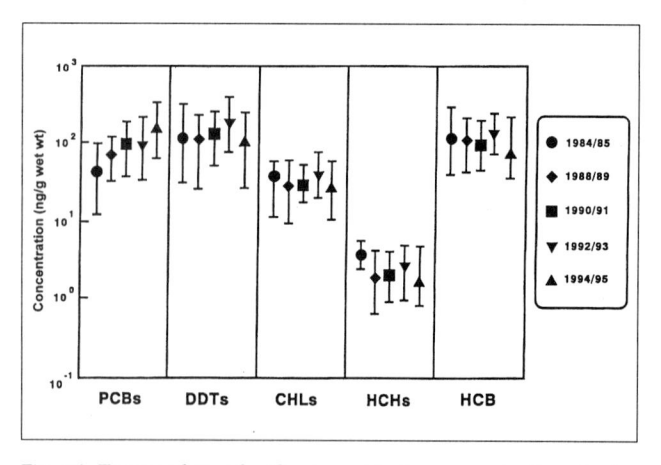

Fig. 16. Temporal trends of organochlorine concentrations in the blubber of minke whale from the Antarctic Ocean.

ACKNOWLEDGEMENTS

The author thanks Dr. K. Kannan (Michigan State University) and Dr. M.S. Prudente (De La Salle University System, Philippines) for critical reading of the manuscript. This study was partly supported by a Grant-in-Aid for the Scientific Research Program from the Ministry of Education, Science and Culture of Japan (Project No. 09041163 and 10559015).

REFERENCES

Alzieu, C. and Heral, M. (1984) Ecotoxicological effects of organotin compounds on oyster culture. In *Ecotoxicological Testing for the Marine Environment*, eds. G. Persoone, E. Jaspers and C. Claus, Vol. 2, State University, Belgium, pp. 187–195.

Alzieu, C. and Portmann, J.E. (1984) The effect of tributyltin on the culture of *C. gigas* and other species. In *Fifteenth Annual Shellfish Conference Proceedings*. The Shellfish Association of Great Britain, London, pp. 87–104.

Ansari, A.A., Singh, I.B. and Tobschall, H.J. (1998) Organotin compounds in surface and pore waters of Ganga plain in the Kanpur-Unnao industrial region, India. *The Science of the Total Environment*, 223, 157–166.

Aono, S., Tanabe, S., Fujise, Y., Kato, H. and Tatsukawa, R. (1997) Persistent organochlorines in minke whale (*Balaenoptera acutrostrata*) and their prey species from the Antarctic and the North Pacific. *Environmental Pollution* 98, 81–89.

Beaumont, A.R. and Budd, M.D. (1984) High mortality of the larvae of the common mussel at low concentrations of tributyltin. *Marine Pollution Bulletin* 15, 402–405.

Bryan, G.W. and Gibbs, P.E. (1991) Impact of low concentrations of tributyltin (TBT) on marine organisms: a review. In *Metal Ecotoxicology Concepts and Applications*, eds. M.C. Newman and A.W. McIntosh. Lewis Publishers, 323–361.

Colborn, T. and Smolen, M.J. (1996) Epidemiological analysis of persistent organochlorine contaminants in cetaceans. *Reviews of Environmental Contamination and Toxicology* 146, 91–172.

Corsolini, S., Focardi, S., Kannan, K., Tanabe, S., Borrell, A. and Tatsukawa, R. (1995) Congener profile and toxicity assessment of polychlorinated biphenyls in dolphins, sharks and tuna collected from Italian coastal waters. *Marine Environmental Research* 40, 33–53.

Dave, P.P. (1996) India: a generic giant. *Farm Chemical International*, November 10, 36–37.

Fent, K. (1996) Ecotoxicology of organotin compounds. *Critical Reviews of Toxicology* 26, 1–117.

Goldberg, E.D. (1975) Synthetic organochlorines in the sea. *Proceedings of the Royal Society of London* Series B 189, 277–289.

Hashimoto, S., Watanabe, M., Noda, Y., Hayashi, T., Kurita, Y., Takasu, Y. and Otsuki, A. (1998) Concentration and distribution of butyltin compounds in a heavy tanker route in the Strait of Malacca and in Tokyo Bay. *Marine Environmental Research* 45, 169–177.

Iwata, H., Tanabe, S., Sakai, N. and Tatsukawa, R. (1993) Distribution of persistent organochlorines in the oceanic air and surface seawater and the role of ocean on the global transport and fate. *Environmental Science and Technology* 27, 1080–1098.

Iwata, H., Tanabe, S., Sakai, N., Nishimura, A. and Tatsukawa, R. (1994a) Geographical distribution of persistent organochlorines in air, water and sediments from Asia and Oceania, and their implications for global redistribution from lower latitudes. *Environmental Pollution* 85, 15–33.

Iwata, H., Tanabe, S., Aramoto, M., Sakai, N. and Tatsukawa, R. (1994b) Persistent organochlorine residues in sediments from the Chukchi Sea, Bering Sea and Gulf of Alaska. *Marine Pollution Bulletin* 28, 746–753.

Kan-atireklap, S., Tanabe, S. and Sanguansin, J. (1997a) Contamination by butyltin compounds in sediments from Thailand. *Marine Pollution Bulletin* 34, 894–899.

Kan-atireklap, S., Tanabe, S., Sanguansin, J., Tabucanon, M.S. and Hungspreugs, M. (1997b) Contamination by butyltin compounds and organochlorine residues in green mussel (*Perna viridis*, L.) from Thailand coastal waters. *Environmental Pollution* 97, 79–89.

Kan-atireklap, S., Yen, N.T.H., Tanabe, S. and Subramanian, An. (1998) Butyltin compounds and organochlorine residues in green mussel (*Perna viridis*, L.) from India. *Toxicological and Environmental Chemistry* **67**, 409–424.

Kannan, K., Sinha, R.K., Tanabe, S., Ichihashi, H. and Tatsukawa, R. (1993a) Heavy metals and organochlorine residues in Ganges River dolphins from India. *Marine Pollution Bulletin* **26**, 159–162.

Kannan, K., Tanabe, S., Borrell, A., Aguilar, A. Focardi, S. and Tatsukawa, R. (1993b) Isomer-specific analysis and toxic evaluation of polychlorinated biphenyls in striped dolphins affected by an epizootic in the western Mediterranean Sea. *Archives of Environmental Contamination and Toxicology* **25**, 227–233.

Kannan, K., Tanabe, S., Tatsukawa, R., Sinha, R.K. (1994) Biodegradation capacity and residue pattern of organochlorines in Ganges River dolphins from India. *Toxicological and Environmental Chemistry*, **42**, 249–261.

Kannan, K., Tanabe, S. and Tatsukawa, R. (1995a) Geographical distribution and accumulation features of organochlorine residues in fish in tropical Asia and Oceania. *Environmental Science and Technology* **29**, 2673–2683.

Kannan, K., Tanabe, S., Iwata, H. and Tatsukawa, R. (1995b) Butyltins in muscle and liver of fish collected from certain Asian and Oceanian countries. *Environmental Pollution* **90**, 279–290.

Kannan, K., Tanabe, S., Giesy, J.P. and Tatsukawa, R. (1997a) Organochlorine pesticides and polychlorinated biphenyls in foodstuffs from Asian and Oceanic countries. *Reviews of Environmental Contamination and Toxicology* **152**, 1–55.

Kannan, K., Senthilkumar, K. and Sinha, R.K. (1997b) Sources and accumulation of butyltin compounds in Ganges River dolphin, *Platanista gangetica*. *Applied Organometallic Chemistry* **11**, 223–230.

Kannan, N., Tanabe, S., Ono, M. and Tatsukawa, R. (1989) Critical evaluation of polychlorinated biphenyl toxicity in terrestrial and marine mammals: increasing impact of non-*ortho* and mono-*ortho* coplanar polychlorinated biphenyls from land to ocean. *Archives of Environmental Contamination and Toxicology* **18**, 850–857.

Kim, G.B., Tanabe, S. and Koh, C. (1998) Butyltins in surface sediments of Kyeonggi Bay, Korea. *Journal of the Korean Society of Oceanography* **33**, 64–70.

Ko, M.M., Bradley, G.C., Neller, A.H. and Broom, M.J. (1995) Tributyltin contamination of marine sediments of Hong Kong. *Marine Pollution Bulletin* **31**, 249–253.

Lee, J.S., Tanabe, S., Takemoto, N. and Kubodera, T. (1997) Organochlorine residues in deep-sea organisms from Suruga Bay, Japan. *Marine Pollution Bulletin* **34**, 250–258.

Li, Y., Mcmillan, A. and Scholtz, M.T. (1996) Global HCH usage with 1° × 1° longitude/latitude resolution. *Environmental Science and Technology* **30**, 3525–3533.

Loganathan, B.G., Tanabe, S., Tanaka, H., Watanabe, S., Miyazaki, N., Amano, M. and Tatsukawa, R. (1990) Comparison of organochlorine residue levels in the striped dolphin from western North Pacific, 1978–79 and 1986. *Marine Pollution Bulletin* **21**, 435–439.

Martineau, D., Beland, P., Desjardins, C. and Lagace, A. (1987) Levels of organochlorine chemicals in tissues of beluga whales (*Delphinapterus leucas*) from the St. Lawrence Estuary, Quebec, Canada. *Archives of Environmental Contamination and Toxicology* **16**, 137–147.

Melzian, B., Zoffmann, C. and Spies, R.B. (1987) Chlorinated hydrocarbons in low continental slope fish collected near the Farallon Islands, California. *Marine Pollution Bulletin* **18**, 388–393.

Norstrom, R.J. and Muir, D.C.G. (1994) Chlorinated hydrocarbon contaminants in Arctic marine mammals. *The Science of the Total Environment* **154**, 107–128.

O'Shea, T. and Brownell, R.T.Jr. (1994) Organochlorine and metal contaminants in baleen whales: a review and evaluation of conservation implication. *The Science of the Total Environment* **154**, 179–200.

Prudente, M., Tanabe, S., Watanabe, M., Subramanian, A., Miyazaki, N., Suarez, P. and Tatsukawa, R. (1997) Organochlorine contamination in some odontoceti species from the North Pacific and Indian Ocean. *Marine Environmental Research* **44**, 415–427.

Prudente, M., Kan-atireklap, S., Watanabe, I. and Tanabe, S. (1999) Butyltins, organochlorines and metal levels in green mussel (*Perna viridis*, L.) from the coastal waters of the Philippines. *Fisheries Science* **65**, 441–447.

Ramesh, A., Tanabe, S., Kannan, K., Subramanian, A.N., Kumaran, P.L. and Tstsukawa, R. (1992) Characteristic trend of persistent organochlorine contamination in wildlife from a tropical agricultural watershed, South India. *Archives of Environmental Contamination and Toxicology* **23**, 26–36.

Reisch, M.S. (1996): Paints and coatings. *Chemical and Engineering News*, October 14, 44–63.

Senthilkumar, K., Kannan, K., Tanabe, S., Prudente, M. (1998) Butyltin compounds in resident and migrant birds collected from Philippines. *Fresenius Journal of Environmental Bulletin* **7**, 561–571.

Senthilkumar, K., Tanabe, S., Kannan, K. and Subramanian, An. (1999a) Butyltin residues in migratory and resident birds collected from South India. *Toxicological and Environmental Chemistry* **68**, 91–104.

Senthilkumar, K., Kannan, K., Sinha, R.K., Tanabe, S. and Giesy, J.P. (1999b): Bioaccumulation profiles of polychlorinated biphenyl congeners and organochlorine pesticides in Ganges river dolphins. *Environmental Toxicology and Chemistry* **18**, 1511–1520.

Simmons, M. (1991) Marine mammal epizootics worldwide. In *Proceedings of the Mediterranean Striped Dolphins Mortality International Workshop*, eds. X. Pastor and M. Simmonds, Greenpeace International Mediterranean Sea Project, Madrid, Spain, pp. 9–19.

Slorach, A.S. and Vaz, R. (1983) Assessment of human exposure to selected organochlorine compounds through biological monitoring. Global Environment Monitoring System (GEMS) UNEP/WHO. Presented by the Swedish National Food Administration, Uppsala, Sweden, 134.

Soto, A.M., Chung, K.L. and Sonnenschein, C. (1994) The pesticides endosulfan, toxaphene and dieldrin have estrogenic effects on human estrogen-sensitive cells. *Environmental Health Perspectives* **102**, 380–383.

Subramanian, An., Tanabe, S., Tatsukawa, R., Saito, S. and Miyazaki, N. (1987) Reduction in the testosterone levels by PCBs and DDE in Dall's porpoise of northwestern North Pacific. *Marine Pollution Bulletin* **18**, 643–649.

Takeoka, H., Ramesh, A., Iwata, H., Tanabe, S., Subramanian, An., Mohan, D., Magendran, A. and Tatsukawa, R. (1991) Fate of the insecticide HCH in the tropical coastal area of south India. *Marine Pollution Bulletin* **22**, 290–297.

Tanabe, S. (1988) PCB problems in the future: foresight from current knowledge. *Environmental Pollution* **50**, 5–28.

Tanabe, S. and Tatsukawa, R. (1983) Vertical transport and residence time of chlorinated hydrocarbons in the open ocean water column. *Journal of the Oceanographical Society of Japan* **39**, 53–62.

Tanabe, S. and Tatsukawa, R. (1992) Chemical modernization and vulnerability of cetaceans: increasing toxic threat of organochlorine contaminants. In *Persistent Pollutants in Marine Ecosystems*, eds. C.H. Walker and D.R. Livingstone. Pergamon Press, Oxford, 161–177.

Tanabe, S., Tatsukawa, R., Tanaka, H., Maruyama, K., Miyazaki, N. and Fujiyama, T. (1981) Distribution and total burdens of chlorinated hydrocarbons in bodies of striped dolphins (*Stenella coeruleoalba*). *Agricultural and Biological Chemistry* **45**, 2569–2578.

Tanabe, S., Tatsukawa, R., Kawano, M. and Hidaka, H. (1982) Global distribution and atmospheric transport of chlorinated hydrocarbons: HCH (BHC) isomers and DDT compounds in the western Pacific, eastern Indian and Antarctic Oceans. *Journal of the Oceanographical Society of Japan* **38**, 137–148.

Tanabe, S., Hidaka, H. and Tatsukawa, R. (1983) PCBs and chlorinated hydrocarbon pesticides in Antarctic atmosphere and hydrosphere. *Chemosphere* **12**, 277–288.

Tanabe, S., Tanaka, H. and Tatsukawa, R. (1984) Polychlorobiphenyls, DDT, and hexachlorocyclohexane isomers in the western North Pacific ecosystem. *Archives of Environmental Contamination and Toxicology* **13**, 731–738.

Tanabe, S., Miura, S. and Tatsukawa, R. (1986) Variations of organochlorine residues with age and sex in Antarctic minke whale. *Memoirs of National Institute of Polar Research* **44**, 174–181.

Tanabe, S., Watanabe, S., Kan, H. and Tatsukawa, R. (1988) Capacity and mode of PCB metabolism in small cetaceans. *Marine Mammal Science* **4**, 103–124.

Tanabe, S., Gondaira, F., Subramanian, A.N., Ramesh, A., Mohan, D., Kumaran, P., Venugopalan, V.K. and Tatsukawa, R. (1990) Specific pattern of persistent organochlorine residues in human breast milk from South India. *Journal of Agricultural and Food Chemistry* **38**, 899–903.

Tanabe, S., Ramesh, A., Sakashita, D., Iwata, H., Mohan, D., Subramanian, An. and Tatsukawa, R. (1991) Fate of HCH (BHC) in tropical paddy field: application test in South India. *International Journal of Environmental Analytical Chemistry* **45**, 45–53.

Tanabe, S., Iwata, H. and Tatsukawa, R. (1994a) Global contamination by persistent organochlorines and their ecotoxicological impact on marine mammals. *The Science of the Total Environment* **154**, 163–177.

Tanabe, S., Sung, J., Choi, D., Baba, N., Kiyota, M., Yoshida, K. and Tatsukawa, R. (1994b) Persistent organochlorine residues in northern fur seal from the Pacific coast of Japan since 1971. *Environmental Pollution* **85**, 305–314.

Tanabe, S., Kan-atireklap, S., Prudente, M.S. and Subramanian, An. (1998a) Mussel watch: marine pollution monitoring of butyltins and organochlorines in coastal waters of Thailand, Philippines and India. In *Proceedings of Fourth International Scientific Symposium "Role of Ocean Sciences for Sustainable Development"*, UNESCO/IOC/WESTPAC, 331–345.

Tanabe, S., Senthilkumar, K., Kannan, K. and Subramanian, An. (1998b) Accumulation features of polychlorinated biphenyls and organochlorine pesticides in resident and migratory birds from South India. *Archives of Environmental Contamination and Toxicology* **34**, 387–397.

Tanabe, S., Prudente, M., Mizuno, T., Hasegawa, J., Iwata, H. and Miyazaki, N. (1998c) Butyltin contamination in marine mammals from North Pacific and Asian coastal waters. *Environmental Science and Technology* **32**, 193–198.

Tong, S.L., Pang, F.Y., Phang, S.M. and Lai, H.C. (1996) Tributyltin distribution in the coastal environment of peninsular Malaysia. *Environmental Pollution* **91**, 209–216.

Wania, F. and Mackay, D. (1996) Tracking the distribution of persistent organic pollutants. *Environmental Science and Technology* **30**, 390–396.

Watanabe, S., Shimada, T., Nakamura, S., Nishiyama, N., Yamashita, N., Tanabe, S. and Tatsukawa, R. (1989) Specific profile of liver microsomal cytochrome P-450 in dolphin and whales. *Marine Environmental Research* **27**, 51–65.

Watanabe, S., Laovakul, W., Boonyathumanondh, R., Tabucanon, M.S. and Ogaki, S. (1996) Concentrations and composition of PCB congeners in the air around stored used capacitors containing PCB insulator oil in a suburb of Bangkok, Thailand. *Environmental Pollution* **92**, 289–297.

Yen, N.T.H., Tanabe, S., Hue, N.D. and Qui, V. (1998) Pollution by polychlorinated biphenyls and organochlorine pesticides in resident and migratory birds from North Vietnam. In *Proceedings of The Fourth International Scientific Symposium "Role of Ocean Sciences for Sustainable Development"*, UNESCO/IOC/WESTPAC, 444–451.

THE AUTHOR

Shinsuke Tanabe
Center for Marine Environmental Studies,
Ehime University,
Tarumi 3-5-7,
Matsuyama 790-8566, Japan

Chapter 82

SEA OF OKHOTSK

Victor V. Lapko and Vladimir I. Radchenko

The Sea of Okhotsk is a marginal sea in northeast Russia. It is almost entirely bordered by the Russian Federation, with its southern part bordered by Japan. Climatic conditions, especially in the north, are the most severe of the far-eastern seas and are similar to those of arctic seas, as a result of the extensive continental coast with cold polar conditions, and strong prevailing winds.

It has a general cyclonic current system with peculiar conditions of water exchange with the Pacific. Northward water transportation occurs in the east along the Kamchatka peninsula, with a southward transfer in the west. In winter the Sea of Okhotsk is mostly covered by ice, except for waters near Kamchatka and the Kuril Islands. In summer, surface water temperature reaches 11–13°C, and 18–19°C near Hokkaido.

The region is very productive, and is exceptionally important for fisheries. Recent assessments suggest there are about 430 million tonnes (Mt) of plankton, 230 Mt of benthic biota, 35 Mt of fishes, 3.5 Mt of squids and 0.5 Mt of marine mammals. This region provides about 65–70% of the total Russian catch in the Far East. The main target species and groups include pollock, herring, pacific salmon, cod, flatfishes and crabs. To preserve these resources at high levels, it is necessary to keep commercial fisheries within sustainable catch limits.

Currently, the region is relatively safe from effects of pollutants because of the relatively low population and low industrial activity. However the Okhotsk shelf contains oil. Mining operations have been started near northeastern Sakhalin Island, and the west Kamchatka shelf is the next area to be considered for prospecting. Oil extraction on the shelf causes inevitable dredging and oil pollution, and it is recognised that the prevention of serious accidents requires a high priority. There is no doubt that the Sea of Okhotsk will remain the primary fishery region in the Far East. The multispecies nature of the fisheries is providing stable catches of fishes and invertebrates, and it is anticipated that this will remain the case.

Seas at The Millennium: An Environmental Evaluation (Edited by C. Sheppard)
© 2000 Elsevier Science Ltd. All rights reserved

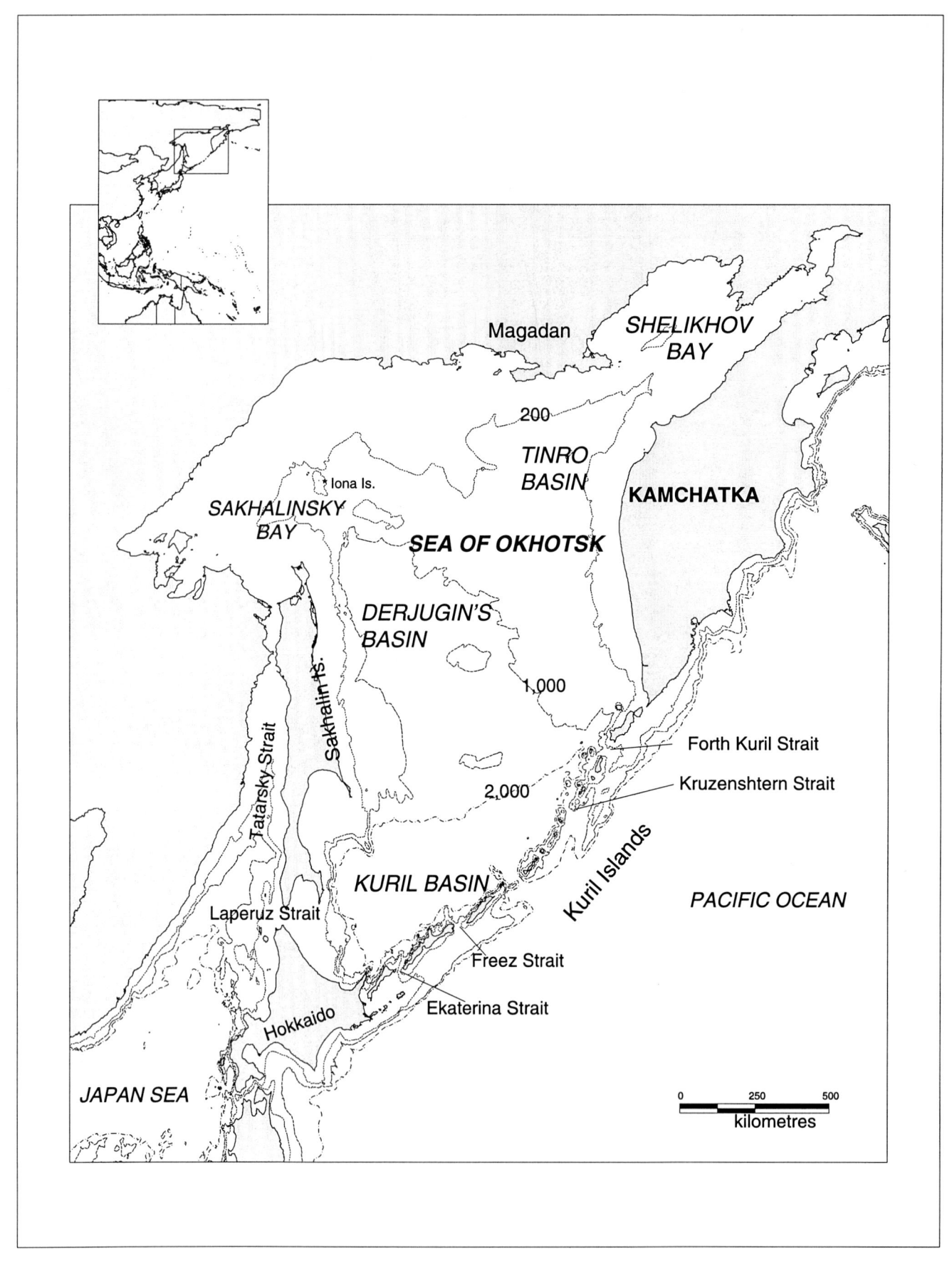

Fig. 1. Map of the Sea of Okhotsk (Udintsev, 1957, simplified).

THE DEFINED REGION

The Sea of Okhotsk is on the northeastern margin of the Asian continent and runs deeply inland (Fig. 1). It is connected to the Pacific Ocean by 30 large Kuril straits and more than 20 lesser ones, over about 500 km. Its average depth is 821 m with a maximum of 3916 m (Dobrovolsky and Zalogin, 1982). The most important water exchange between the sea and ocean takes place in the Forth Kuril Strait (depth 1700 m), Kruzenshtern's Strait (1400–1900 m), Bussol (2500–3000 m), Freez's Strait (800 m) and Ekaterina's Strait (437 m), the rest being substantially shallower. Straits connecting the Sea of Okhotsk and the Sea of Japan are much shallower; the minimum depth of the Tatarsky Strait is 10 m, and that of Laperuz's Strait is 40 m (Leonov, 1960; Balev, 1992).

The benthic relief is rugged. There is the deep Kuril basin in the south, which is connected by troughs with Dejrugin's hollow and TINRO basin. From these hollows, deep-water troughs extend to both Iona island and Shelikhov's Bay. There are two underwater rises, named Academy of Science of USSR Rise and Oceanology Institute Rise in the southern part (Udintscv, 1957). The proportion of seabed area at different depths is shown in Fig. 2 (Larina, 1968). The shelf is widest in the northern part (up to 400 km), where the two large Shelikhov and Sakhalinsky Bays are situated. The shelf is much narrower elsewhere.

Climatic conditions in the Sea of Okhotsk, especially its northern part, are similar to those of arctic seas. This results from the extensive continental coast, its high latitude, and prevailing pressure gradients and winds. The total energy balance of the sea is negative: -25 kcal/cm^2 year^{-1} (Yakunin, 1974), and this deficit is balanced by warm water advection from the Pacific Ocean and Japan Sea. In this sense, the Okhotsk Sea can be divided into two approximately equal parts: northwestern and southeastern. The Sea absorbs warmth in the northwest and gives it back in the southeast, especially near the West Kamchatka coast. The net result is that these conditions build up a characteristic feature of a residual cold water layer with a core of $-2°C$, which remains in several areas even during the warm season.

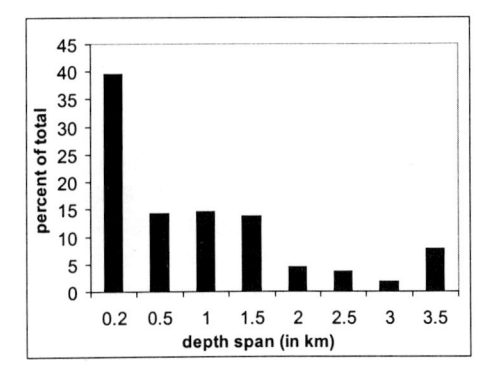

Fig. 2. Percent of the total area of the Sea of Okhotsk at different depths.

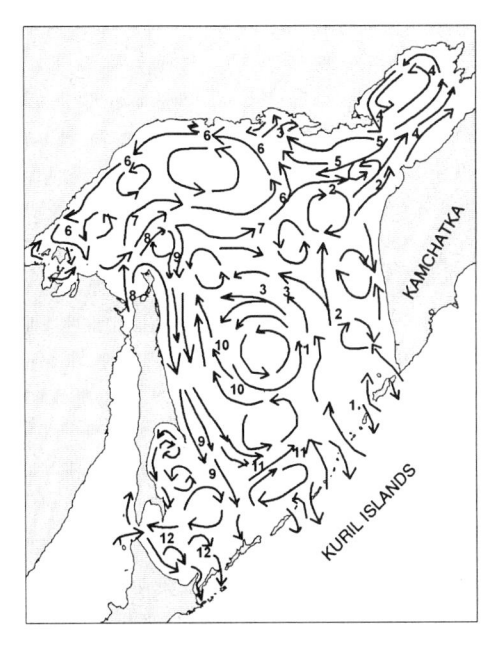

Fig. 3. Scheme of general water circulation in active layer in the Sea of Okhotsk in summer. 1: West Kamchatkan current; 2: its Northern Branch; 3: Middle current; 4: Penzhinskoe; 5: Yamskoe; 6: Northern Okhotsk's; 7: Northern Okhotsk's concurrent; 8: Amurskoe; 9: East Sakhalin's; 10: East Sakhalin's concurrent; 11: North-Eastern; 12: Soya.

Currents

The Sea of Okhotsk has two major flows: northward in the eastern part and southward in the west (Fig. 3) (Leonov, 1960; Moroshkin, 1966; Kitani, 1973; Favorite et al., 1976). The interaction of these with the coastline, with bottom relief and with each other, generates several smaller currents (Chernyavsky, 1981; Markina and Chernyavsky, 1984; Chernyavsky et al., 1981; Luchin, 1982; Anon, 1995). The main currents are the West Kamchatkan current in the east and the East Sakhalin current in the west. With others, these generate a general cyclonic water movement (Stepanov, 1974).

The main driving force of Okhotsk Sea water movement is created by large-scale atmospheric circulation over the sea in winter (Moroshkin, 1966; Leonov, 1960), when powerful and directionally stable winds predominate. These winds are the most pronounced in the northeastern part where, throughout the cold period, a high-pressure gradient prevails, built up from the interaction between the Siberian High Pressure Core (Siberian Maximum, SM) and the Aleutian Low Pressure region (Aleutian Minimum, AM). The locations of the SM and AM cause the highest pressure gradient and most powerful winds to develop over the northern Okhotsk coast and Shelikhov Bay. The winds are significantly strengthened by the arrival of cyclonic eddies (Chernyavsky, 1981), the resultant winds frequently exceeding 15 m/s over Shelikhov Bay (Koshinsky, 1978). The winds over the rest of the northern coast are not so strong but are very stable in direction, thus driving water drift. Originally flowing westward, drift currents deflect southward following the shore, leaving the Sea of Okhotsk through the Kuril straits. One branch of the East Sakhalin current turns east off eastern Sakhalin and carries its water and ice out into the Pacific through the Bussol Strait.

In winter, the Sea of Okhotsk is mostly covered by ice, except near Kamchatka and the Kuril Islands. There are open water sections in the northern part of the sea during the whole winter, created by the strong northern and northwestern winds. There is always a stable ice-hole, a so-called "polynia", near Kashevarov Bank (Alfultis and Martin, 1987; Mitnik and Kalmykov, 1992), generated by the powerful upwelling coming from comparatively warm Middle Current water which flows up and over the Bank's slope.

Water exiting from the sea is compensated by inflowing Pacific water through the northern Kuril straits. However, strong northern winds in winter prevent the development of a warm West Kamchatkan current, but they do promote and strengthen the southerly flow off Sakhalin. In the warm season, atmospheric circulation changes and southerly winds dominate, but these are weaker (Bokov, 1995). Accordingly, the West Kamchatkan current strengthens while the East Sakhalin current weakens.

SEASONALITY, CURRENTS, NATURAL ENVIRONMENTAL VARIABLES

Seasonal changes are very sharp and distinctive. First of all, as mentioned above, in winter most of the sea is covered by ice, and advection of relatively warm Pacific waters establishes the hydrologic regime (Chernyavsky, 1981). In winter, Pacific waters can be clearly tracked as an elongated section of open water, not covered by ice, along the West Kamchatka coast, a so-called "eastern channel". Waters around the Kuril Islands stay uncovered due to intensive horizontal and vertical mixing in the Kuril straits. Nevertheless in winter, surface waters cool to –1.5 or –1.8°C, and only in the southeastern part do surface temperatures not fall below zero (near the Kuril Islands 1–2°C). In summer, insolation, vertical mixing and increased currents from the Pacific raise water temperature to about 11–13°C over most of the area and to 18–19°C near Hokkaido. Summer warming extends to depths of 30–75 m (Shuntov, 1985).

Seasonal changes in biota of the Okhotsk Sea mirror the hydrologic processes. In the warm season, the Sea of Okhotsk "blooms". Species permanently inhabiting the sea migrate from deep layers, where they spend the winter, to shallow shelf waters and near-surface layers for reproduction and feeding. Also in summer many diverse species migrate to the Okhotsk Sea either actively (fishes, squids, marine birds and mammals) or passively (planktonic organisms, jellyfishes, small mesopelagic fishes etc.) with the West Kamchatkan and Soya currents. The most striking examples of active migrants are common subtropical fishes such as saury (*Cololabis saira*), sardine (*Sardinops sagax*

Table 1

Biomass and proportion (× 1000 t/%) of common fishes in epipelagic layer of the Sea of Okhotsk in the second half of 1980s

Species and groups	Summer	Winter
Walleye pollock	11300/75.0	3609/77.0
Herring	500/3.3	450/10.0
Deep-sea smelt	2457/16.3	500/11.0
Salmons	150/1.0	62/1.0
Sardine	500/3.3	–
Capelin	15/0.1	39/0.8
Other fishes	135/0.9	10/0.2
Total	15057	4670
Concentration (t/km^2)	10.0	3.5

melanosticta), anchovy (*Engraulis japonicus*) as well as marine birds (Shuntov, 1998). Seasonal fish biomass in the epipelagic layer (0–200 m) may be high (Table 1) (Lapko, 1996).

In winter, the general concentration of fish in the upper 200 m layer decreases markedly, especially in the case of pollock, deep-sea smelt (also northern smoothtongue, *Leuroglossus shmidti*), sardine and salmon. All these species have seasonal migrations. In winter the most abundant fish—pollock—migrates from the shelf areas to the continental slope, and from the upper layers to deeper ones (Shuntov et al., 1993). This is stimulated by strong cooling of surface waters, and by the seasonal descent to greater depths of planktonic crustaceans which are their main food. In winter, most fishes move to waters deeper than 200 m.

The small deep-sea smelt are the most numerous mesopelagic fish living in the deep layer, mainly from 200 to 1000 m (Balanov and Ilyinsky, 1992; Ilyinsky and Gorbatenko, 1994). Daily they undertake vertical migrations to feed on the ascending plankton, creating substantial concentrations in the epipelagic layer over the deep water basins (Lapko, 1995, 1996; Shuntov et al., 1993). In summer, biomass at night of deep-sea smelt was estimated as approximately 2.5 million tonnes (Mt), while in winter it was considerably less. The main reasons for a decrease of migratory activity of deep-sea smelt in winter are the same as with pollock, namely a cooling of surface waters and concentration of food organisms (macroplankton) in deep layers.

The most pronounced example of horizontal migrations in summer is that of sardines. In the 1980s, sardine was very abundant in the northwestern Pacific. Assessments of their annual biomass ranged from several million to several tens of million tons (Shuntov and Vasilkov, 1982; Shuntov et al., 1993). In summer this fish usually leaves its spawning grounds around Japan and moves north for feeding, becoming widespread in the Sea of Okhotsk, Japan Sea and in the Pacific waters of the Kuril Islands, reaching Sakhalin and Kamchatka. Sardine enter the Sea of Okhotsk in July, and by October migrate back to warmer waters outside the Okhotsk Sea. When present in the southern part of the

Okhotsk Sea, sardine biomass has been estimated as several hundreds of thousand tons, making it one of the most numerous fish.

Pacific salmon (pink, chum, coho, sockeye, chinook and cherry) spend approximately half of each year inside the Okhotsk Sea, and they spawn in freshwater rivers of the mainland. In general, in spring and summer, salmon juveniles migrate to the sea, where they feed and grow intensively. Forced out by winter cooling, young fishes then leave the Okhotsk Sea for the ocean where they continue to feed and mature until the next spring. When the water warms, salmon return, and over 1–2 months gradually migrate to the river mouths for spawning. Thus, in the warm season, both juveniles and mature fishes co-exist inside the Okhotsk Sea (Takagi et al., 1981; Heard, 1991; Salo, 1991; Neave et al., 1976; Sandercock, 1991).

Thus, in the warm season, spawning of local species as well as migration of animals from outside greatly increases the biomass, in what appears to be a characteristic feature of arctic and temperate seas.

THE MAJOR SHALLOW-WATER MARINE AND COASTAL HABITATS

As noted in Fig. 2, areas having less than 200 m depth, which could be defined as shelf, comprise about 40% of the total area. The shelf is the widest in the north (up to 400 km) and much narrower in other parts; about 75 km off eastern Sakhalin and up to 90 km off western Kamchatka. The hydrological regime of the shelf waters usually develops under the influence of the intermediate cold layer, the layer of residual winter cooling (depth approx. 50–150 m). In general, areas covered by near-bottom waters with temperatures below 0°C are smallest on the West Kamchatka shelf and greatest off the northern shore. Therefore, the northern and northwestern parts are the most inclement, not only in this Sea but throughout all far-east Asian seas.

Despite severe natural conditions, shelf areas are extraordinarily productive. For instance, in the depth range 16–200 m the total benthos biomass varies between 15–3500 g/m^2, more frequently 100–500 g/m^2. Areas with highest biomass, exceeding 1000 g/m^2, are observed on the northern shelf (up to 3500 g/m^2), off eastern and southeastern coasts of Sakhalin (up to 2700 g/m^2) and in Shelikhov Bay. The minimum biomass recorded (up to 50 g/m^2) lies in central parts of the northern shelf and in the south of Sakhalinsky Bay. Average benthic biomass over the whole shelf is 394 g/m^2 (Koblikov et al., 1990), more than half (63.1%) of this value being composed of species from the most common groups: bivalves, sea urchins, polychaetes and sponges. Detailed analysis of the composition and distribution of these groups in several areas is given by several authors (Neyman, 1969; Pavluchkov, 1982, 1987; Nadtochy, 1984; Koblikov, 1985, 1988).

The Okhotsk shelf in general and West Kamchatka in particular are known to be the most productive fish areas in the far-eastern seas. About 309 species of demersal fishes are recorded (Borets, 1997), mostly species belonging to the families Zoarcidae, Cottidae, Liparididae, Stichaeidae and Pleuronectidae. However, flatfishes, sculpins and cod are the most abundant. For example, in the 1980s these contributed more than 90% of the total biomass of all demersal fishes on the West Kamchatkan shelf (Table 2).

The most numerous bottom fishes of the Okhotsk shelf are cod, saffron cod, yellowfin sole, flathead sole, Alaska plaice, and 2–3 species of sculpins. Pelagic fishes are much more abundant than demersal ones in shelf areas. Common pelagic species such as pollock and herring form particularly dense concentrations. For example, in the 1980s, pollock biomass was several million tonnes in the northeast (Shuntov et al., 1993; Shuntov and Dulepova, 1996), and in the warm season pelagic fishes such as capelin and salmon also form considerable aggregations in coastal waters. The total biomass of pelagic fishes exceeds that of demersal species several-fold (Table 3, from Shuntov and Dulepova,

Table 2

Composition (%) and biomass of demersal fishes on the West Kamchatka shelf in 1980s according to trawl surveys (Borets, 1997)

Family	1982	1986
Rajidae	0.6	1.5
Gadidae	38.7	28.5
Zoarcidae	0.2	0.3
Cottidae	12.3	15.7
Agonidae	1.2	0.8
Liparididae	0.6	0.4
Pleuronectidae	45.9	52.7
Others	0.5	0.1
Total (× 1000 t)	916	1365

Table 3

Biomass and production in Okhotsk Sea ecosystem

Groups	Biomass (× 10^6 t)	Production (× 10^6 t)
Phytoplankton	–	15100
Bacteria	–	5200
Protozoa	–	2100
Herbivorous zooplankton	314	2520
Predatory zooplankton	115	480
Nonpredatory zoobenthos	208.6	318
Predatory zoobenthos	21.4	22.1
Pelagic fishes	31.5	15.7
Demersal fishes	3.5	1.7
Demersal invertebrates	1.5	0.5
Squids*	3.5	12.5
Seabirds	0.012	0.004
Marine mammals	0.5	0.1

*Corrected according to Lapko (1995a).

1996). In the 1970–80s on the West Kamchatka shelf, the annual average catch ranged from 8 to 22 t/km^2.

OFFSHORE SYSTEMS

There are three major and permanent zones of intensive vertical intermixing in the Sea of Okhotsk (Kitani, 1973). One of them, the Yamskoy upwelling, is at the mouth of Shelikhov Bay, created by the West Kamchatkan current flowing along the side of the TINRO basin where the sea bed rises from 600–750 to 200–300 m. The second upwelling is observed over the Kashevarov Bank, generated when middle current water flows up onto the Kashevarov Bank slope. The third permanent upwelling is on both sides of the middle part of the Kuril Islands, generated by interaction of opposing currents and tidal events. All these areas are highly productive and greatly affect the surrounding pelagic layers, as well as the shelf and continental slope waters. For example, Shelikhov Bay and Kashevarov Bank are important fishing grounds for fish and shellfish, and the former is also the breeding area for the most abundant population of red king crab.

Waters over deep areas are much more sparsely populated. Salmon and squid predominate in the upper 0–200 m. Mesopelagic fishes are also abundant at night. In the 1980s, pollock were discovered to migrate annually into the waters of deep-sea basins where they form dense concentrations. The total biomass of fishes in the epipelagic layer over deep-sea areas was assessed as being several million tonnes, most of which consists of mesopelagic fishes (Shuntov et al., 1993; Lapko, 1995).

The mesopelagic layer in the Sea of Okhotsk from 180 to 1500 m corresponds to the warm intermediate waters in the subarctic Pacific. This stratum is densely populated by both resident mesopelagic species and transient migrants. About 61 species of fishes belonging to 53 genera and 33 families have been recorded there (Balanov and Ilyinsky, 1992; Ilyinsky, 1995), most from one species (20 families) or two (6 families). In general, species composition of the mesopelagic fishes community differs little from that of the subarctic Pacific. According to data collected in trawl surveys in 1987–91, the total biomass of mesopelagic fishes ranged from 19 to 30 Mt. Deep-sea smelt always predominated and constituted about half of these values (42–66%). This is a characteristic feature here; in other Pacific regions, fishes such as *Stenobrachius leucopsarus* and *Diaphus theta* are the most numerous. Additionally, *Lipolagus ochotensis*, another common species belonging to the same family as deep-sea smelt, and two myctophids (*Stenobrachius leucopsarus* and *S. nannochir*) are very abundant in the Sea of Ohhotsk. These species undertake daily vertical migrations to the surface and form significant concentrations and sound scattering layers in the upper horizons at night (Shuntov et al., 1993; Lapko, 1995).

Sixteen species of squid, belonging to nine genera and six families, are known. The family *Gonatidae* contains ten species, two of which (*Gonatopsis borealis* and *Berryteuthis magister*) are the most abundant cephalopods (Nesis, 1989; Lapko, 1995a; Iljinsky, 1991). Squid are a very important item in the food web of the deep-sea communities. Their total biomass reaches 3.5–4.0 Mt, and the total consumption of squid is about 12 Mt per year, a figure made possible by the highly productive character of this group. Common squids also migrate to the surface at night, participating in a sound scattering layer.

Thus, areas over the deep-sea basins are relatively poorly inhabited and are less attractive to commercial fisheries, compared with shelf waters.

POPULATIONS AFFECTING THE AREA

The Okhotsk Sea coast includes part of five administrative territories of the Russian Federation: Khabarovsky, Sakhalinsky, Magadansky, Kamchatsky and Koryaksky regions. The southern coast belongs to Hokkaido Island (Japan).

The Regions

Sakhalinsky region is a unique Russian administrative territory, covering 59 islands. It consists of Sakhalin Island, adjacent small islands and the Kuril Islands. The total area of this region is 87,100 km^2, with a population of about 608,000 (January 1999). Yzhno-Sakhalinsk is the main city. Sakhalin Island is the most densely populated part, especially in the south where relatively big towns such as Yuzhno-Sakhalinsk, Korsakov, Poronaysk, Shakhtersk and Nevelsk are located. About 80% of the total population live in southern Sakhalin, the north being poorly inhabited. There are several villages on the larger, southern Kuril Islands, Kunashir and Iturup, and on the northern Paramushir. Most of the middle islands are uninhabited. Because of the island character of the Sakhalinsky region, all residents can be regarded as coastal, although about one third lives on the west coast of Sakhalin facing the Sea of Japan.

Table 4 details the important features of the regions. These administrative territories were established comparatively recently, most in 1932–38 but Magadansky region in 1953. This illustrates the relatively short history of the industrial use of the regions. The main development in all of them is based around mineral and primary resources. Parts of the territories are extremely remote from large industrial and cultural centres and depend on good communications and supplies from the mainland of food, energy and other materials, which are delivered largely by ships during summer. In recent years, these distant territories have begun to experience difficulties in obtaining supplies in the necessary quantities.

Thus, about 800,000 people permanently inhabit the Russian coast of the Sea of Okhotsk, mainly in only two centres of population: Sakhalin and Magadan. In general

Table 4
The five administrative territories of the Russian Federation

Sakhalinsky region
Minerals: Coal, oil, natural gas, native sulphur. Well known coal fields are: Gornozavodskoe, Vakhrushevskoe, Solntsevskoe, Uglegorskoe. Oil: Mongi, Dagi, Ust-Tomi, Katangli Yubileynoe, Okruzhnoe. Gas: Vostochno-Lugovskoe. Oil and gas condensate: Odoptu, Chaivo, Lunskoe. Sulphur: Novoe (Iturup Island).
Main industry: Fishery and fish processing, pulp and paper, fuel and ship repair.

Khabarovsky region
Main city: Khabarovsk. Total area is about 790,000 km^2. Population: 1,634,000. There are 17 districts, 7 towns and 31 villages in this territory. Only about 10% of total population are coastal.
Minerals: Tin, mercury, iron ore, coal and lignite, graphite, manganese etc.
Industries: Machinery construction, metal working, ship repair, ferrous and non-ferrous metallurgies, forestry, woodworking, pulp and paper, chemical, refinery, fishery etc. None of the substantial industrial activities relates to the Okhotsk Sea coast.

Kamchatsky and Koryaksky regions
Main city: Petropavlovsk-Kamchatsky. Total area: 170,000 km^2. In 1996, total population was estimated at 411,000. Koryaksky region administrative centre is Palana town. Total area: 3,015,000 km^2, population: 38,000. Only about 10% of the population inhabits the coast.
Minerals: Coals, ores of gold, silver, mercury, polymetals, native sulphur, building materials. Regions are being prospected for oil. There are numerous thermal and mineral springs, geysers, boiling lakes, mud volcano.
Industries: Fishery and fish processing, forestry, ship repair, electroenergetics, metallurgy, food.

Magadan region
Main city: Magadan. Total area: 4,614,000 km^2, population: 363000.
Minerals: Gold, non-ferrous metals, building materials. A well-known gold field is Karamken, and tungsten is mined at Omsukchan.
Industries: Mining, fishery, electroenergetics, machine repair, building materials, food and light industries.

this is a poorly populated area. This can be partly explained by the general worsening of the economic situation in the country, causing greater difficulties in these most distant regions.

RURAL FACTORS

There is relatively poorly developed agriculture in these regions, due to severe climatic conditions. For example, the number of warm days with air temperature >20°C ranges from one to six on the Kamchatka coast and does not exceed 30 on the peninsula. There are two main types of agriculture. First, the traditional type by the aboriginal population which raise deer. Second is active farming, which provides more than 90% of the total agricultural production.

Almost all production is consumed inside the regions, and substantial quantities of food are also imported. Thus, agriculture has not reached a level of development where it may make much significant impact on the environment. In several areas, the farming industry has actually declined (Muzurov and Korolev, 1999).

However, forestry may cause serious problems on Kamchatka and especially on Sakhalin. There is only one place on Kamchatka where coniferous forest grows: the Kamchatka river basin (the river itself flows into the Pacific Ocean). The ecological importance of the forests for preserving rivers and fish populations is incomparably higher than their industrial importance, but from the beginning only the latter has been considered. As a result there has been a significant decrease in the area of coniferous forests, from 1,200,000 ha in the first quarter of the current century to less than 350,000 ha by 1996 (Lazarev, 1999).

The forests of Sakhalin are more abundant than those in Kamchatka. Sakhalin is situated in the forested "taiga" zone, and covers about 6.9 million ha, or 87% of the area of Sakhalin. Coniferous forest (fir and pine) predominates, providing a commercially valuable wood. The total stock of wood is estimated at 600 million m^3, 340 million of which comprise mature and old trees (>100 years). Assessed annual yield is about 3.5 million m^3. The timber industry is one of the most important in Sakhalin, forming social and commercial infrastructures of towns and villages, consisting of timber enterprises, pulp, paper and wood processing. For example, in 1998, 3,669,000 m^3 of commercial wood and 56,200 m^3 of sawn-timber were produced, with 2393 tons of cardboard and 1700 tons of paper. More than half of production is exported abroad.

Thus, this industry is profitable. However, the forests of Sakhalin are one of the key natural ecosystems, supplying unique rivers where pacific salmon spawn. Over-harvesting is providing cause for anxiety today.

The aboriginal population is very numerous on this coast. Approximately 4000 people live relatively compactly around Shelikhov Bay, and more are scattered further west. These are the Koryaks, Itelmens and Evenks. Nivchs live on the northern Sakhalin (several thousand). Their traditional activities are deer raising, hunting and fishing, the latter having a seasonal character related to the spawning run of pacific salmon in the rivers. Sweep net and drift nets are usually used for fishing. We have no definitive data regarding any impact on the environment that could be affected by aboriginal populations; most likely it is insignificant.

EFFECTS FROM URBAN AND INDUSTRIAL ACTIVITIES

Undoubtedly, the large-scale commercial fishery causes a pronounced effect on the ecosystem of the Okhotsk Sea.

Russian explorers discovered the Sea in the 17th century, and after that, commercial use of marine resources started, insignificantly at first. Exploitation slightly increased in the 18th century and the main attention was focused on marine mammals: sea otters, seals and whales. Later in the 19th century, fish such as salmon and herring were also targeted. However, by the 20th century total fisheries can be described as being low, excluding that for marine mammals. Over long periods, the total catch and its composition were not constant. Pacific salmon, pacific herring and red king crab were the main items caught. This situation remained stable until the mid-1950s, until industrialization of the fishery began. By the early 1960s total catch reached 1 Mt and started to rise much faster later. In the mid 1960s it was about 1.5 Mt, and in 1975 it reached 2.7 Mt. At the beginning of the 1960s the target species were mainly flatfishes, herring, crabs and salmon, and an expansion of the walleye pollock fishery also began. In the second half of the 1960s, catches of salmon, flatfishes and some other species decreased while herring and especially pollock substantially increased (Fig. 4). By the late 1970s, total catch significantly decreased, until in the mid 1980s it increased again and became stable at 2.4–2.6 Mt (1.8–2.0 Mt in Russian waters). This growth arose from an increase of fish abundance that, in turn, resulted from fishery regulations and favourable reproductive conditions for some, notably pollock (Shuntov, 1985).

During recent years, annual catches have been about 1.4–1.5 t/km² (1.2–1.3 t/km² inside the Russian EEZ) (Shuntov, 1987; Shuntov, Dulepova, 1996) (Table 5). These values indicate a relatively high fishing intensity; the harvest in shelf waters of the northern Atlantic and Pacific ranges from 0.56 to nearly 2 t/km², although in limited shelf

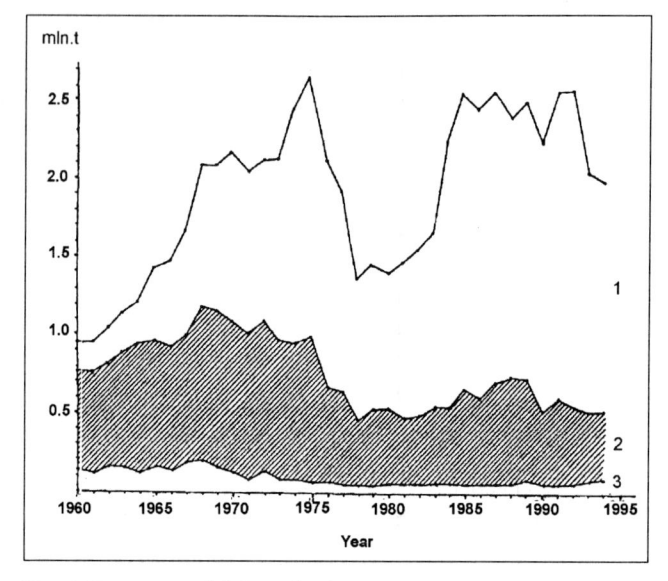

Fig. 4. Dynamics of fish and other targets removal in the Sea of Okhotsk since 1960 (million tonnes). 1: Pollock; 2: other fishes; 3: other targets.

areas the catch is higher: over 5 t/km² off the western coast of Greenland and in the Gulf of Maine, for example (Moiseev, 1989).

However, in some areas of the Okhotsk Sea fisheries are much more intensive. For example, average catch ranged from 8 up to 22 t/km² in the West Kamchatka shelf in 1970–80s, reflecting the high productivity there.

Early in the 1990s there was a decrease of pollock and sardine stocks. Total catch dropped from 2.5–2.6 to 2 Mt in 1993. The main reason for this was the directed fishery aimed at the most common and fluctuating species (pollock and sardine) while stocks of flatfishes, cod and herring were being under-exploited. Nevertheless, pollock remains the main target species and provides the majority of the total catch. The multispecies base for the fisheries in this sea secures against abrupt drops of the catches of any one species, and it is predicted that in the future, the Okhotsk Sea will remain the most important fishery area in Russia.

An important resource is oil and gas on the Sakhalin shelf. Currently, oil operations have been started near the northeastern coast of Sakhalin Island, and the west Kamchatka shelf is considered as a prospective area. Recently three gas fields and one oil field were opened. The "Odoptu-Kupol" field, started in 1998, is located 4 km off-shore, but is exploited by a horizontal oil well drilled from the coast.

Recently, three new projects were created with the participation of foreign investors to exploit oil and gas fields on the shelf of northeastern Sakhalin. Sakhalin 1 is estimated to contain 310 Mt of oil, 425 billion m³ of gas and 33 billion m³ of gas condensate. The estimated investment here is about US$ 15 billion. Sakhalin 2 is estimated to contain 140 Mt of oil and 408 billion m³ of gas. Total investment cost is about US$ 10 billion. Commercial oil extraction has already started about 18.5 km from the northeastern coast, with a

Table 5

Fish catch in the Sea of Okhotsk after 1960, t/km²

Year	Whole sea	Without Japanese EEZ
1960–1969	0.87	0.55
1970–1979	1.25	0.99
1980	0.88	0.63
1981	0.92	0.63
1982	0.98	0.73
1983	1.0	0.8
1984	1.40	1.20
1985	1.60	1.26
1986	1.52	1.20
1987	1.61	1.29
1988	1.50	1.18
1989	1.53	1.44
1990	1.39	1.29
1991	1.59	1.50
1992	1.59	1.50
1993	1.27	1.17
1994	1.27	1.13

platform, a terminal for oil shipment and a floating oil tank with a capacity of one million barrels. The first buyers of this oil, which is light, low in sulphur and similar to that from the Middle East, will be Asian countries. Participants plan to create a combined infrastructure for oil and gas processing and transporting. The third project is Sakhalin 3, again a multi-national project, created for searching and exploration work.

Unfortunately, oil mining on the shelf causes dredging, leaking oils and oil pollution. Already at the end of September 1999 an accident on one production complex resulted in a spill of about 3.5 tons of oil. No doubt other cases will occur in future. This kind of industrial activity, as well as the commercial fishery, can seriously degrade the marine ecosystem. Domestic and industrial waste from towns and villages is estimated to be relatively insignificant.

PROTECTIVE MEASURES

In general, the Sea of Okhotsk and its coast can be regarded as a relatively safe region regarding pollution. Nevertheless, pollutant concentrations are subject to special sanitary control like everywhere in Russia. At present the environment seems to be little affected.

Fisheries are regulated by means of "fishery regulations" which define time limits and areas for fishing as well as the quantities (wet weight) or so-called quotas and minimum catch size. To determine quotas, research fisheries institutes, including TINRO-center, conduct annual surveys in the far-eastern seas to monitor the state of the diverse species stocks and their ecosystems as a whole. Further, fishery biologists estimate current population and community characteristics to calculate values for feasible catches in the various areas. Feasible catches for the forthcoming year need to be approved by the state Government, and have the authority of federal law. The special federal organization RYBVOD supervises the observance of fisheries regulations, and the Federal Frontier Service also recently obtained authority for this.

However, poaching is significant in this region. Both Russian and foreign poachers catch the most commercially valuable species, overfishing them and exhausting populations. Their main targets are crabs, shrimps, shellfishes, sea urchins etc., which are in steady and high demand in fishmarkets of neighbouring Asian countries: Japan, Korea, China. Because of this illegal fishery, the feasible catch of some species is becoming substantially exceeded. For example, according to Japanese statistics during the last few years, the real import of red king crab from Russia is double its official catch. A similar situation occurs with other invertebrates. Unfortunately, defects of legislation and a general worsening in the state economy impede the solution of this problem.

Natural reserves and prohibited territories should be mentioned. "Yamskoy" preserve is situated on the islands with that name near the western coast of the mouth of Shelikhov Bay. It is intended for the protection of a unique habitat for marine birds, the so-called "birds' bazaar", and a rookery of Steller's Sea lions. A similar reserve has now been established in the Shantarsky Islands in the northwest. To protect species of marine mammals, prohibited areas have been established around Tjuleny Island near eastern Sakhalin where a famous seal-rookery is situated, and a similar reserve has been established around the Kuril Islands to preserve unique habitats of sea otters.

Additionally, there is a cryobank in Petropavlovsk-Kamchatsky. This was founded in 1994 and is a part of the world system of low temperature banks, preserving the genetic fund of nature. Kamchatsky cryobank intends to create a representative cryocollection of the genetic materials from rare and valuable far-eastern marine species, including fishes and invertebrates. Work has been started to form cryocollections of salmon genomes of Kamchatka (Zheltonozhko and Zheltonozhko, 1999).

REFERENCES

Anon. (1995) The Okhotsk Sea and Oyashio Region. PICES Scientific Report No. 2. Institute of Ocean Science, P.O.Box 6000, Sidney, B.C., Canada. 227 pp.

Alfultis, M.A. and Martin, S. (1987) Satellite passive microwave studies of the Sea of Okhotsk ice cover and its relation to oceanic processes, 1978–1982. *Journal of Geophysical Research* 92, 13013–13028.

Balanov, A.A. and Iljinsky, E.N. (1992) Species composition and biomass of mesopelagic fishes in Okhotsk and Bering Seas. *Voprosy Ichthyologii* 32 (1), 56–63.

Balev, B.F. (1992) The straits of Asiatic–Pacific region. In *Main Straits and Channels of the World Ocean and Their Role in International Relations*. Institute of World Economy RAS, Moscow, Vol. 4, Part 2, 95 pp.

Bokov, V.N. (1995) Spatial–temporal regime of wind over the Barents and Okhotsk Seas. *Meteorologia and Hydrologia* 2, 46–54.

Borets, L.A. (1997) *Demersal Shelf's Ichthyocens of Russian Far-Eastern Seas: Composition, Structure, Elements of Functioning and Importance for Fisheries*. TINRO Center, Vladivostok. 217 pp.

Chernyavsky, V.I. (1981) Circulation Systems of Okhotsk Sea. *Izvestia TINRO* 105, 13–19.

Chernyavsky, V.I., Bobrov, V.A. and Afanasiev, N.N. (1981) Main productive zones in Okhotsk Sea. *Izvestia TINRO* 105, 20–24.

Dobrovolsky, A.D. and Zalogin B.S. *The Seas of USSR*. Moscow State University, Moscow, 192 pp.

Favorite, F., Dodimead, A.J. and Nasu, K. (1976) Oceanography of the Subarctic Pacific region, 1960–71. *Bull. INPFC* 33, 187 pp.

Heard, W.R. (1991) *Life History of Pink Salmon (Oncorhynchus gorbuscha). Pacific Salmon Life Histories*. UBC Press, Vancouver, pp. 119–230.

Iljinsky, E.N. (1991) Distribution of squids in mesopelagic layer of Okhotsk Sea in summer. *Okeanologia* 31 (1), 151–154.

Iljinsky, E.N, (1995) Composition and structure of mesopelagic nekton in Okhotsk Sea. Master's thesis. TINRO Center, Vladivostok, 25 pp.

Iljinsky, E.N. and Gorbatenko, K.M. (1994) Main trophic relations of nekton in mesopelagic layer. *Izvestia TINRO* 116, 91–104.

Kitani, K. (1973) An oceanographic study of the Okhotsk Sea particularly in regard to cold waters. *Bull. Far Seas Fish. Res. Lab.* 9, 45–77.

Koblikov, V.N. (1985) Qualitative and quantitative description of macrobenthos on the shelf and upper slope off the eastern Sakhalin. Master's thesis. Moscow, 20 pp.

Koblikov, V.N. (1988) Quantitative description of the demersal population in near Sakhalin waters of the Sea of Okhotsk. In *Quantitative and Qualitative Distribution of Benthos: Food Base of Benthivorous Fishes*. VNIRO, Moscow, pp. 4–22.

Koblikov, V.N., Pavluchkov, V.A. and Natdtochy, V.A. (1990) Benthos of the continental shelf in Okhotsk Sea: composition, distribution and biomass. *Izvestia TINRO* **111**, 27–38.

Koshinsky, S.D. (1978) *Regime Characteristics of the Powerful Winds in the Seas of Soviet Union. Part 2: Northern Japan, Okhotsk and Bering Seas)*. Hydrometeoizdat, Leningrad, 392 pp.

Lapko, V.V. (1996) Composition, structure and dynamics of epipelagic ichthyocen in the Sea of Okhotsk. Master's thesis. TINRO, Vladivostok, 25 pp.

Lapko, V.V. (1995) Role of mesopelagic fishes in epipelagic ichthyocen in the Sea of Okhotsk. *Voprosy Ichthyologii* **35** (6), 782–787.

Lapko, V.V. (1995a) Role of squids in communities of Okhotsk Sea. *Oceanologia* **35** (5), 737–742.

Larina, N.I. (1968) Calculation of the areas in Pacific ocean, its seas and some basins. *Oceanologia* **8** (4), 646–658.

Lazarev, G.A. (1999) Problems of rational exploitation of Kamchatkan coniferous forests. Abstract. *Conference of Problems of Protection and Rational Use of Kamchatka Bioresources*, Petropavlovsk-Kamchatsky, 10–12 June 1999, p. 30.

Leonov, A.K. (1960) *Regional Oceanography*. Hydrometeoizdat, Leningrad, 765 pp.

Luchin, V.A. (1982) The diagnostic account of water circulation in Okhotsk Sea in summer. *Proc. of Far-Eastern Res. Hydr.-Met. Ins.* **96**, 69–77.

Markina, N.P. and Chernyavsky, V.I. (1984) Quantitative distribution of plankton and benthos in Okhotsk Sea. *Izvestia TINRO* **109**, 109–119.

Mitnik, L.M. and Kalmykov, A.I. (1992) Structure and dynamics of the Sea of Okhotsk marginal ice zone from "Okean" satellite radar sensing data. *Journal of Geophysical Research* **97**, 7429–7445.

Moiseev, P.A. (1989) *Biological Resources of the World Ocean*. Agropromizdat, Moscow, 368 pp.

Moroshkin, K.V. (1966) *Water Masses of Okhotsk Sea*. Nauka, Moscow, 67 pp.

Muzurov, e.L. and Korolev, M.P. (1999) Problems of man's impact on the water bioresources and their environment in Kamchatka in modern socio-economic conditions. Abstract. *Conference of Problems of Protection and Rational Use of Kamchatka Bioresources*, Petropavlovsk-Kamchatsky, 10–12 June 1999, pp. 119–120.

Nadtochy, V.A. (1984) About long-term variability and quantitative distribution of benthos on West Kamchatka shelf. *Proceedings of TINRO Center* **109**, 126–129.

Neave, F., Yonemori, T. and Bakkala, R. (1976) Distribution and origin of chum salmon in offshore waters of the North Pacific ocean. *Bull. INPFC* **35**, 79 pp.

Nesis K.N. (1989) Teuthofauna of Okhotsk Sea. Biology of squids Berryteuthis magister, Gonatopsis borealis (Gonatidae). *Zoologichesky Zhurnal* **68** (9), 45–56.

Neyman, A.A. (1969) Benthos of the West Kamchatka shelf. *Proceedings of VNIRO* **65**, 223–232.

Pavluchkov, V.A. (1987) Macrobenthos of northern Okhotsk shelf and upper part of slope. Master's thesis. TINRO Center, Vladivostok, 20 pp.

Pavluchkov, V.A. (1982) Quantitative description of zoobenthos in the northern part of Okhotsk Sea. In *Ecology and Spawning Conditions for Fishes and Invertebrates in Far-Eastern Seas of North Western Pacific*. TINRO Center, Vladivostok, pp.114–119.

Pavluchkov, V.A. (1990) Benthos of continental shelf in Okhotsk Sea: composition, distribution, stocks. *Proceedings of TINRO Center* **111**, 27–38.

Salo, E.O. (1991) *Life History of Chum Salmon (Oncorhynchus keta)*. Pacific Salmon Life Histories. UBC Press, Vancouver, pp. 231–310.

Sandercock, F.K. (1991) *Life History of Coho Salmon (Oncorhynchus kisutch)*. Pacific Salmon Life Histories. UBC Press, Vancouver, pp. 395–446.

Shuntov, V.P. (1987) About fish productivity of the far-eastern seas. *Voprosy Ichthyologii* **27** (5), 747–754.

Shuntov, V.P. (1985) *Biological Resources of the Sea of Okhotsk*. Agropromizdat, Moscow, 224 pp.

Shuntov, V.P. (1998) *Birds of the Far-eastern Seas of Russia, Vol. 1*. TINRO Center, Vladivostok, 423 pp.

Shuntov, V.P. and Dulepova, E.P. (1996) Modern status and interannual dynamics of demersal and pelagic communities in Okhotsk Sea ecosystem. *Izvestia of the Pacific Research and Fisheries Centre (TINRO-Centre)* **119**, 3–32.

Shuntov, V.P. and Vasilkov, V.P. (1982) Periods of atmospheric circulation and cyclicity of Japanese and Californian sardines. *Voprosy Ichthyologii* **22** (2), 187–199.

Shuntov, V.P., Volkov, A.F., Temnykh, O.S. and Dulepova E.P. (1993) *Walleye Pollock in the Ecosystems of Far-eastern Seas*. TINRO, Vladivostok, 310 pp.

Stepanov, V.N. (1974) *World Ocean*. Znznie, Moscow. 256 pp.

Takagi, K., Aro, K.V., Hart, A.C. and Dell, M.B. (1981) Distribution and origin of pink salmon (Oncorhynchus gorbuscha) in offshore waters of the North Pacific ocean. *Bulletin INPFC* **40**, 195 pp.

Udintsev G.B. (1957) Bottom relief of the Sea of Okhotsk. *Proc. of Inst. Ocean. of Academy of Science of USSR* **22**, 3–76.

Yakunin L.P. (1974) Estimation of Okhotsk Sea energy balance considering ice cover. *Proc. of Far-Eastern Res. Hydr.-Met. Ins.* **45**, 36–45.

Zheltonozhko, O.V. and Zheltonozhko, V.V. (1999) Reserving of biodiversity and genofund of Kamchatkan hydrobionts with help of cryoconservation of gametes and larvas in cryobank. Abstr. of Conference on Problems of Protection and Rational Use of Kamchatka Bioresources. 1999. Petropavlovsk-Kamchatsky, June 10–12, p. 101.

THE AUTHORS

Victor V. Lapko
Pacific Fisheries Research Centre,
TINRO Centre,
Vladivostok, Russia

Vladimir I. Radchenko
Pacific Fisheries Research Centre,
TINRO Centre,
Vladivostok, Russia

Chapter 83

SEA OF JAPAN

Anatoly N. Kachur and Alexander V. Tkalin

The Sea of Japan is a semi-enclosed sea whose shores and islands are inhabited by more than 50 million people. The Russian section supports 1.6 million people, including 1.2 million in Primorsky Krai and 217,000 in Sakhalin Island. However, as a result of currents and airflows, another 60 million people living beyond the basin boundaries also affect its condition.

Vegetation cover has been significantly reduced over the last 150 years. Water and wind erosion of soils is widespread and there is coastal erosion as well, such that more than half the arable land in this territory urgently needs erosion control measures. There is little control of pesticide levels, and pollution of surface and ground waters is widespread; all the rivers flowing into the Sea of Japan are polluted to varying degrees.

Annual discharges of polluted waste waters per citizen are nearly 30% more than the Far East average due to inefficient wastewater treatment, though metal pollution of surface waters is not severe. Where agricultural production is developing, organic substances pollution is growing. In the northern region there are several local sources of significant pollution from ore-mining and chemical production, including large quantities of metals. A build-up of pollutants has made most marine species inedible.

In the central region, pollution is continuous, especially where there are highly developed industries and agriculture. Practically no coastal waters can be used for recreation purposes. Water pollution in the southern region is sporadic and with careful selection marine species can be consumed. Radioactive pollution appears not to be an issue, despite accidents and the location of the Russian Pacific fleet base.

In the last decade, bottom communities in the Peter the Great Gulf have begun to show visible changes due largely to increased volumes of waste waters. High calorie communities of polychaetes are being replaced with lower calorie communities of basket stars and bivalves, which affects the feeding of demersal fish. There is also a loss of seagrass, leading to a shrinkage of the spawning grounds of Pacific herring and to lower stability.

From north to south there is a change in marine biota, from boreal to sub-tropical species. This area and its adjoining waters have numerous unique characteristics. A mild climate, warm sea, coastal landscapes, presence of many relict species, exotic features and ancient historical monuments, all make it attractive for tourists, especially towards the south where the summer season is longer. A system of protected natural areas makes up a framework for nature conservation activity, and together these now encompass an area covering many tens of thousands of hectares.

Seas at The Millennium: An Environmental Evaluation (Edited by C. Sheppard)
© 2000 Elsevier Science Ltd. All rights reserved

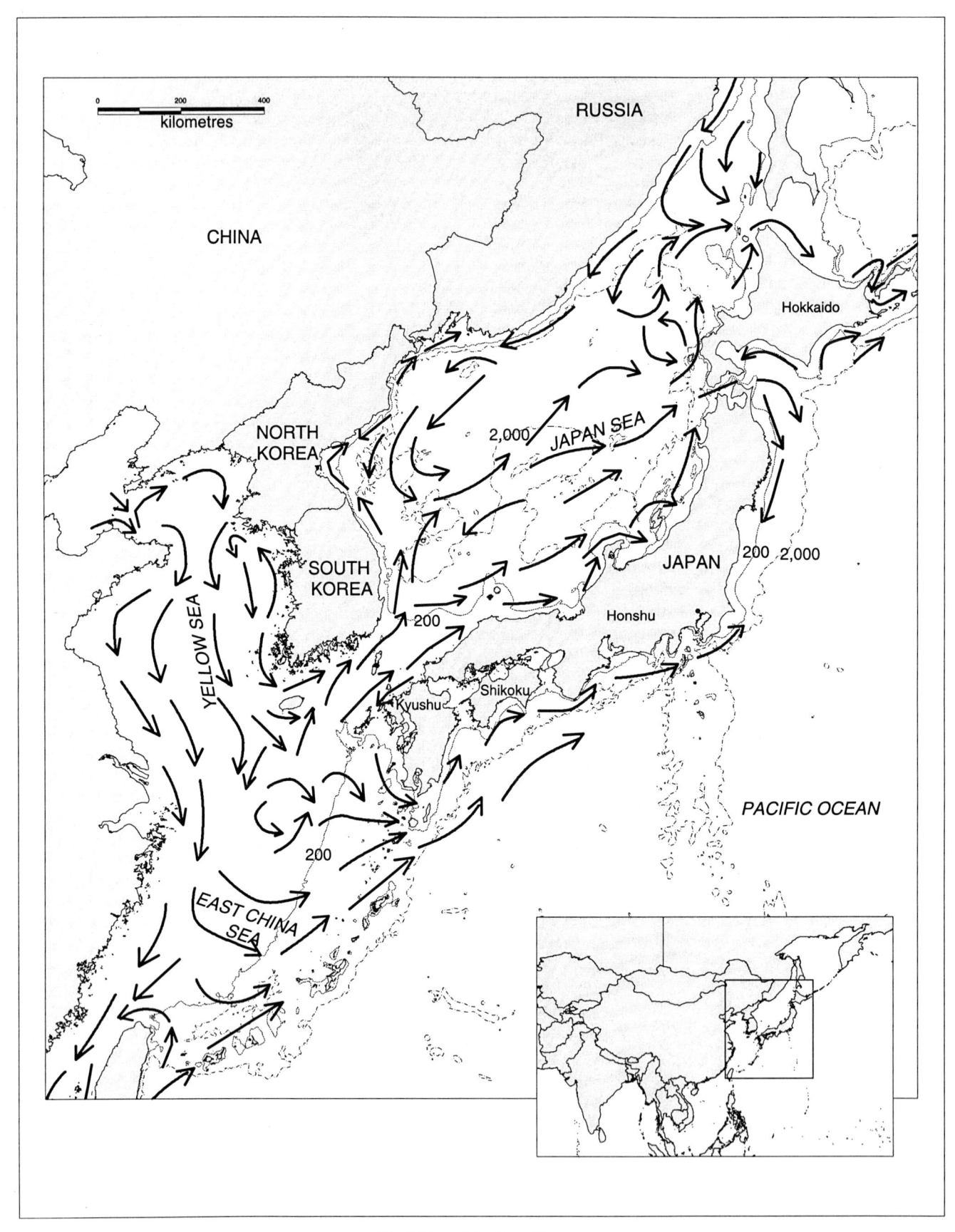

Fig. 1. Map of the Sea of Japan.

INTRODUCTION

The Sea of Japan is a semi-enclosed sea surrounded by Russia, Japan and Korea. It is connected with the East China and Yellow Seas through the Korea Strait in the south, and with the Sea of Okhotsk through the Tatarskyi Strait and Laperuz Strait in the north (Fig. 1). The former is the major connection, accounting for 97% of the total annual water exchange or supply.

The Russian section of the Sea of Japan basin contains several rivers draining the eastern slope of the Sikhote-Alin and Eastern Manchurian Mountains. Rivers are generally steep, so carry substantial particulate matter, some of them from basins as large as 1900 km². Also, a small portion of the Amur River flows into the Sea of Japan via the Tatarskyi Strait.

More than 50 million people live on the mainland and islands. The Russian section of the basin supports 1.6 million people, including 1.2 million in Primorsky Krai, 139,000 in the Khabarovsk oblast and 217,000 in Sakhalin (Table 1). However, as a result of waters brought with the Tsushima current, another 60 million people living beyond the basin

Table 1
Industrial centres in the Russian section of the Sea of Japan basin (by number of industrial personnel, 1992)

Industrial centre	Large		Medium		Small	
	>10,000	10,000–5000	5000–1000	1000–500	500–100	<100
City Artiom	+					
City Vanino	+					
City Vladivostok	+					
City Dalnegorsk	+					
City Nakhodka	+					
City Partizansk	+					
City Ussuriisk	+					
City Bolshoy Kamen		+				
City Sovetskaya Gavan		+				
Set. Zarubino			+			
Set. Kavalerovo			+			
Set. Preobrazhenie			+			
Set. Slavynaka			+			
Set. Tavrichanka			+			
Set. Mikhailovka				+		
Set. Plastun				+		
Set. Popova				+		
Vil. Sergeyevka				+		
Set. Valentin					+	
Set. Kraskino					+	
Set. Pazdolnoye					+	
Set. Shkotovo					+	
Set. Olga						+

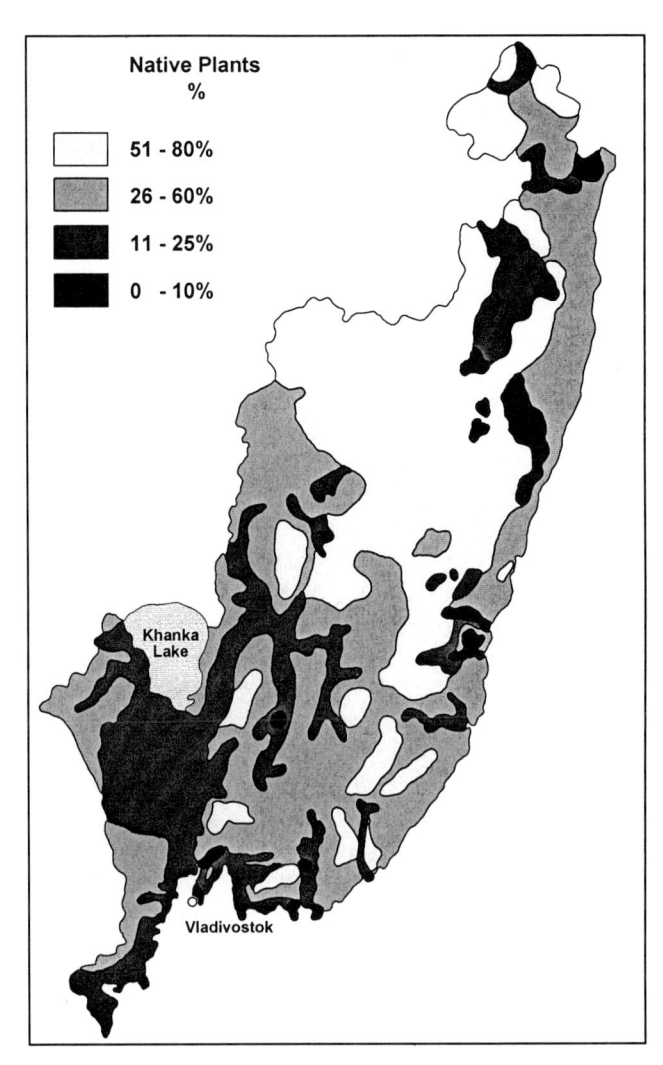

Fig. 2. Condition of natural vegetation cover around and north of Vladivostok.

boundaries also affect its condition; this includes those in the Amur River watershed. Although the region has been developing for many centuries, most environmental change has occurred during the last century, and the mainland has suffered most.

Vegetation cover (Fig. 2) and especially of primary forest in the south of the Russian Far East, has been significantly reduced over the last 150 years. Water and wind erosion of soils is widespread. On paddy fields there are signs of sheet erosion and scouring, and there is coastal erosion as well. Areas growing dry crops on slopes suffer intensive sheet erosion and scouring; on fields there is scouring, river erosion and destruction of canals; on grasslands and pastures there is grass-cutting and pasture burning, followed by wind erosion. More than half the arable lands in this territory urgently need close attention and erosion control measures. Also, there is little control of pesticide levels in the soil, so that paddy fields, vegetable-growing areas, potato fields, gardens and fruit growing areas have significant loads of pesticides.

The economy of the Far-Eastern regions has been developing without consideration of ecological consequences, and without a conservation policy. The economic development has aggravated the ecological situation, as a result of which, primary ecosystems over almost the whole of the territory of the Russian section have changed dramatically.

STATE OF MARINE, COASTAL AND FRESHWATER ENVIRONMENT

Air Pollution

Air pollution in the Sea of Japan basin comes from a combination of global, regional and local sources. Researchers from the Primorskiy Kray (initially FERHRI and the Pacific Institute of Geography of FEB RAS) have appraised the terrestrial air composition in the central part of the Ussuriisk taiga, in the Sikhote-Alin Biosphere Nature Reserve (Fig. 3). More than 73% of pollutants arrive from China, and the most heavily polluted air masses come from the east. Principal pollutants are SO_2, NO_2, CO and dust (Fig. 4). On these figures, maximum admissible concentrations are marked by a ring, so it is evident that average annual concentrations of NO_2 and solid substances exceed this in practically all cities in the Russian section. The situation is most unsatisfactory in Vladivostok, Dalnegorsk and Rudnaya Pristan, and the distribution of pollutants in the air emitted by the largest cities in the Primorskyi Krai stretch for hundreds of km². The regional air pollution comes mainly from China and the Korean Peninsula.

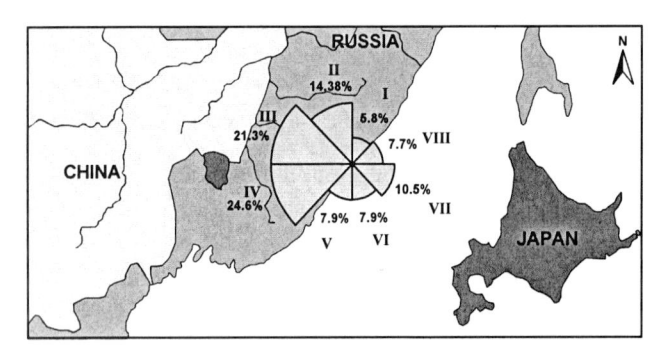

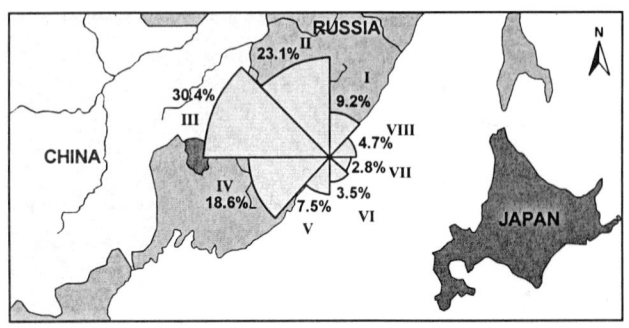

Fig. 3. *Top*: Directions of contributions of different air masses to the formation of pollution over the Sea of Japan area. *Bottom*: Ratio of different contributions of different air masses to the pollution over the Sea of Japan area.

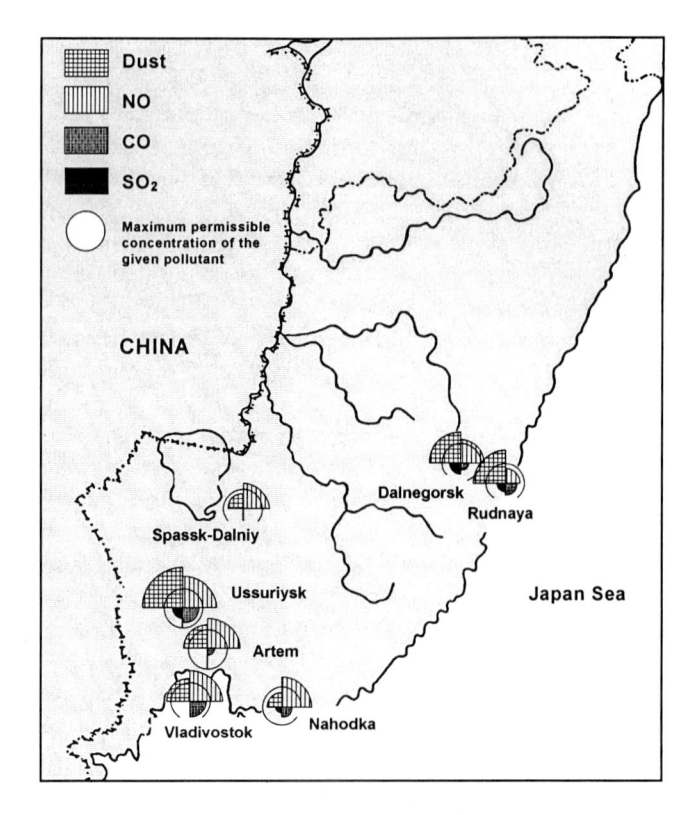

Fig. 4. Air pollution in the main settlements of Primorye Territory. 1: dust; 2: NO; 3: CO; 4: SO_2. Circles show maximum admissible concentration of the given pollutant.

The greatest local contributor of air pollution (up to 51%) is the fuel-energy and mineral-ore complex. In power plants, 97% of solid pollutants are scrubbed, but gaseous emissions vary from 6 to 8.6 tons per million kW/h of generated electricity. Emissions of sulphur anhydride, the most toxic substance, vary from 4.6 to 7.7 tons/million kW/h.

Metal pollution is also very prevalent. In cities and settlements in the south of the Russian Far East maximum permitted concentrations of lead were exceeded in several places (an average daily admissible concentration is 0.3 kg/cu m), and in Rudnaya Pristan the annual average level of lead is twice as high. This reflects all the large cities in the Japan Sea basin. For example, in the late 1980s and early 1990s, Vladivostok—the largest city and industrial centre on the Russian coast of this basin—had pollutants from stationary sources equal to 79,900 t/year, including solid matter (42.8), sulphur anhydride (27.4), nitrogen oxides (3.6), carbon oxides (4.6), hydrocarbons (1.3) and others (0.09). Emissions including vehicle transport were 124,300 t/year.

The main contributors to emissions are industries related to fuel-energy 68.3%, construction organizations 8.4%, marine fleet 12.1%, and ship-repair industry 4.8%. In the last five or six years air pollution in the south of the Russian Far East has somewhat diminished, but this is not connected with an improved treatment of emissions, but with a decline in production. When the economy revives, air pollution will increase.

Table 2
Comparative content of elements in dissolved and suspended phases of river waters

Region	Fe		Mn		Zn		Cu		Ni	Cr	Pb	Cd
	µg/l	%	µg/l	%	µg/l	10^{-3}	µg/l	10^{-3}	10^{-3}	10^{-3}	10^{-3}	10^{-3}
Eastern Sikhote-Alin	22.9	3.1	3.5	103.2	52.7	142.0	1.2	14.7	16.4	4.1	31.9	0.94
Watershed of Peter the Great Gulf	57.8	4.0	11.6	130.7	41.9	107.2	1.5	12.2	8.2	7.2	20.5	0.35
Average for world rivers (according to Gordeyev, 1983)	410	5.1	10	100	20	31	7.0	8.3	8.4	13	14.7	0.07

It should be noted that some elements (Pb, Cu, Zn, Mn), which are monitored in the atmosphere, are insufficient for an objective appraisal of the environment. Seasonal dynamics also affect the concentrations. For example: in Vladivostok, Nakhodka and partially in Sovetskaya Gavan and Vanino, concentrations of pollutants are the highest in spring and autumn; in Artem, pollutant concentrations are lower in summer; in Dalnegorsk and Rudnaya Pristan concentrations of solid matter and some other pollutants are constantly high and exceed maximum admissible levels. These situations are dependent to some extent on emission sources in a particular city, but in all cities, concentrations of many pollutants are higher than is admissible, which is indicative of a dangerous level of pollution. In general, the sulphur load is 0.25 to 1.0 t/ km^2 and nitrogen from 0.5 to 2.0 t/ km^2.

Surface and Ground Waters

Pollution of surface and ground waters is of great consequence, not only for the localities where these waters flow, but also for the whole basin in general. Water pollution in the Sea of Japan basin is made up of atmospheric fall-outs, discharges and effluents from industries and settlements, and run-off from land. All the rivers in the Sea of Japan basin are polluted to varying degrees (Table 2). This is also true of the Amur River basin that partially contributes to the pollution of the northern part of the Sea of Japan.

Dissolved and suspended matter from the Amur basin is about 20 t/km^2 a year for each. Surface waters are subject to pollution, but mineral and ion composition does comply with drinking water requirements, except for several rivers in the north and east of the Gulf of Peter the Great, and rivers in the main ore-mining regions in the Central Primorskyi Krai. Here the rates of run-off are much higher and equal to 30 to 50 t/km^2 a year. Annual releases of polluted waste waters per citizen in the Primorskyi Krai in 1989–1990 averaged 183 m^3/person, which is nearly 30% more than the Far East average. In agricultural regions, water pollution is caused by seepage from fields treated with chemicals and fertilizers, and waste waters from animal farms. Water pollution is very high because of inefficient wastewater treatment (an average figure for the Primorskyi Krai is not more than 15%).

Metal pollution of surface waters is not severe. Northern regions of the Primorskyi Krai, regions of the Khabarovsk Territory, some highlands and watersheds of rivers and sparsely populated areas in the south of the Primorskyi Krai are not subject to perceptible metal pollution. A great part of the southern half of the Primorskyi Krai can be classified as an area with a moderate metal pollution of waters, as is revealed in a higher level of one or several parameters, or else in likely pollution by nearby ore-mining settlements.

Pollution of rivers with metals and other chemical components is connected with sanitary and industrial waste waters of several towns, and with areas of intensive ore-mining. Pollution of surface waters is heavier in industrial zones, such as the basins of Zerkalnaya, Rudnaya Rivers and others (Fig. 5). Where agricultural production is developing, pollution by organic substances is growing.

The problem becomes more serious in the face of the planned development of a 'free zone' of Tumen river. Already, this region suffers the most from pollution, while only 15–20 years ago pollution in this basin was insignificant. According to available information, almost none of the adjacent industries in China or the Korea Democratic Republic are provided with treatment facilities. The Marine Nature Reserve in the Peter the Great Gulf is endangered, as are other coastal areas in this region.

Marine Waters

The western coast of the Sea of Japan can be divided into five regions according to pollution loads: Amur Lagoon, northern, central, and southern regions of Primorskiy Kray, coastal marine waters of the Korean Peninsula.

Water composition in the Amur lagoon is controlled to a great extent by the Amur River. In the lagoon waters there is a raised content of detergents, petroleum hydrocarbons and metals, and a fairly high turbidity. The Amur flow can be traced (by metal content) to coastal waters of the Sikhote-Alin Nature Reserve. In recent years, development of oil production in the territory of the Northern Sakhalin and its shelf has caused serious concern regarding it possible effects on marine ecosystems.

In the northern region from the Zolotoy Cape to Povorotny Cape there are several local sources of significant pollution to coastal waters, largely from ore mining

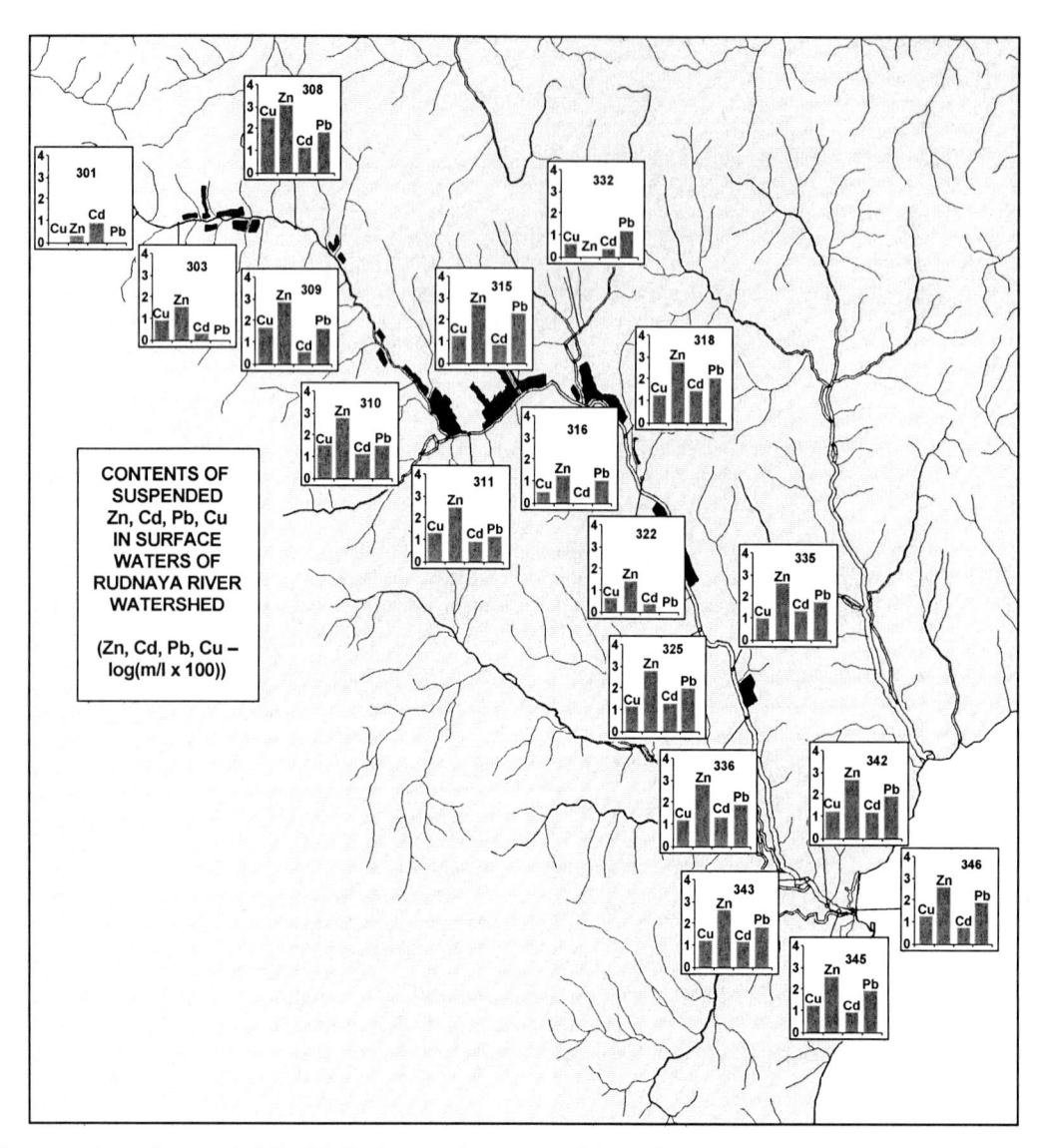

Fig. 5. Concentrations of suspended Zn, Cd, Pb, Cu in surface waters of the Rudnaya River watershed (values in log(mg/l × 100)).

and ore-chemical production. The largest is located near Rudnaya and Zerkalnaya Bays. Pollution includes large quantities of Pb, Cu, Zn, Cd, As, B and others in dissolved and suspended forms. A build-up of pollutants is revealed in the marine species, making them inedible. No elevations above background levels are found around Olga Bay with its harbour for the Pacific Fleet of Russia.

In the central region from Povorotny Cape to Gamov Cape, pollution is continuous, especially in Amursky, Nakhodka and Ussuriysky Bays, which have highly developed industries and agriculture. There is a wide range of pollutants here, as reflected in marine biota and water. Practically all coastal waters cannot be used for recreation purposes due to the great volume of waste water discharged here. Therapeutic mud-baths in Amursky Bay have been virtually ruined.

The southern region includes coastal waters of Posjet Bay from Gamov Cape to the Tumen River. Water pollution is sporadic here and is connected with disposal of sewage and, to a lesser extent, industrial waste waters. Build-up of pollutants in marine organisms is not high. With careful selection they can be consumed, although constantly increasing pollution from the Tumen River basin has become an issue here (Fig. 6).

Coastal marine waters of eastern Korea are subject to pollution, with waste waters from the peninsula on the one hand, and with the heavy impact of pollutants brought with the Korean current from Liaotung Bay, on the other.

Water pollution of the Japan Sea varies markedly between areas (Fig. 7). Most contaminated are coastal waters in the south of the Primorskyi Krai and the Korean Peninsula.

Radioactivity

Of special significance is the problem of radioactive pollution. Several sites here have been used for dumping of radioactive wastes (Fig. 8). The average level in the Japan

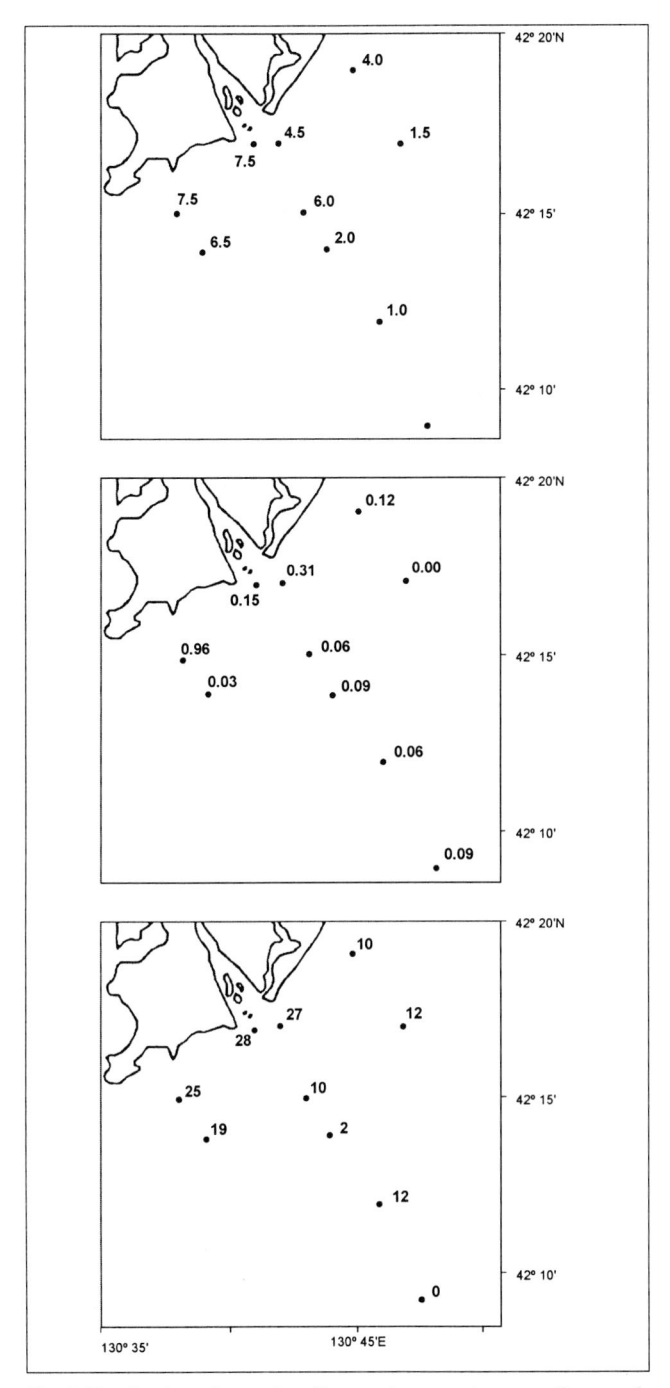

Fig. 6. Distribution of organic pollutants in sea waters near the mouth of the Tumen River. Top nitrite (mg l^{-1}), middle = aromatic hydrocarbons (μg l^{-1}, bottom = anionic detergents (μg l^{-1}). (Tkalin and Shapovalov, 1991).

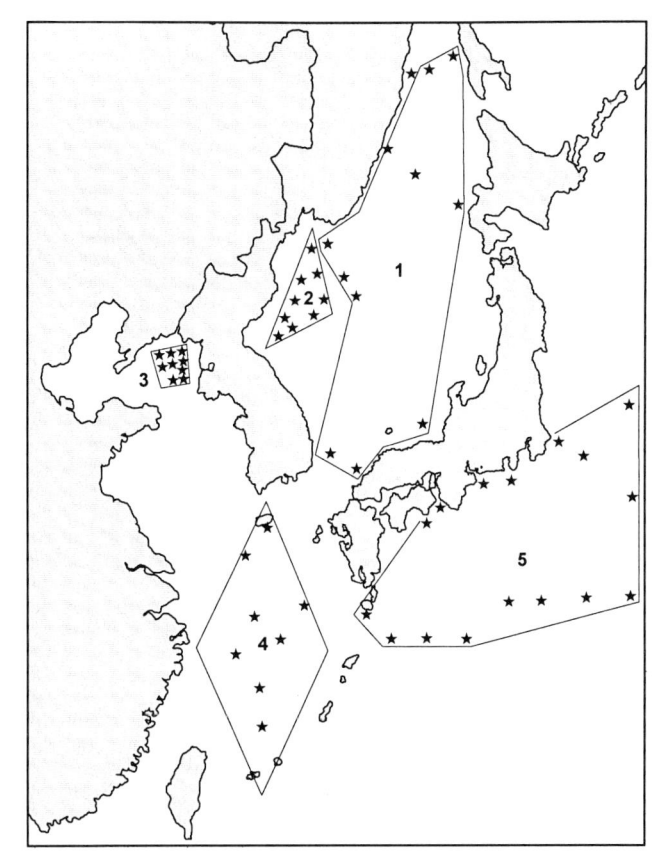

Fig. 7. Organic pollution of the Japan Sea. Values are average/maximum (Tkalin, 1991a).

Pollutant	Group 1 (μg/l)	Group 2 (μg/l)	Group 3 (μg/l)	Group 4 (μg/l)	Group 5 (μg/l)
Phenol	1/6	1/2	0/1	0/1	0/1
Anionic detergents	10/24	10/29	1/2	3/9	5/10
Aromatic hydrocarbons	–/–	0.07/0.13	0.09/0.13	0.18/0.11	0.07/0.14
Non-polar petroleum hydrocarbons	–/–	6/11	2/10	1/8	3/6

Sea of caesium-137 is 0.004 Bq/l. Recent information shows that one of the main sources of radioactive pollution in the northwest of the Japan Sea is solid radioactive wastes. This region has bases for nuclear submarines and surface ships, ship repair facilities, points for harbouring written-off nuclear submarines, and on-land maintenance bases. In 1985, in Chazhma Bay, the environment was polluted with radionuclides as a result of an accident on nuclear submarine K-431. What resulted was an uncontrolled runaway of a reactor, with the destruction of a core, release of fission products (together with effluents formed in the course of fire fighting in the reactor compartment of the submarine). From the atmosphere and sea water of the bay, wind dispersed a fine aerosol over the industrial zone (shipyards of the Navy Fleet) and further on over the coast of the Dunai Peninsula, bypassing settlements "Shkotovo-22" (now Fokino) which were located here. Recent research has indicated that, generally, the situation in the water areas does not cause any special concern, as the levels of ^{137}Cs and ^{90}Sr are within the background values 1.7–5.7 and 1.7–3.5 Bq/m^3, respectively. Differences between the levels of these elements in surface and bottom layers of water were not found. In general, a ratio of ^{137}Cs to ^{90}Sr levels varies within 0.8 to 2.4, an average value being 1.54, which agrees well with the ratio and levels of these radionuclides globally.

The contribution of fleet nuclear ships and their support facilities to the general radioactivity in the Gulf of Peter the

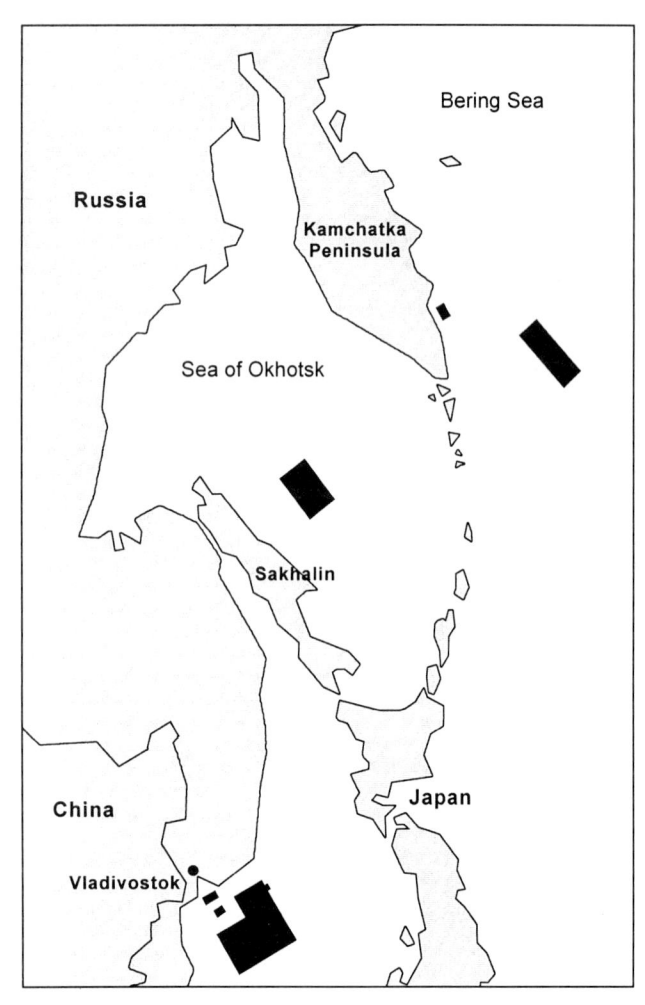

Fig. 8. Location of nuclear waste dumping sites in Russian Far Eastern waters.

Great does not exceed 0.1%. As a result of mixing processes in the upper layer to 20–50 m depth, radionuclides become diluted. In surface run-off, their concentrations are much higher, leading to their higher content in areas with river inflow.

NATURAL RESOURCES, SPECIES AND PROTECTED AREAS

By virtue of geological and historical factors, this area and the adjoining waters of the Sea of Japan, especially Primorskyi Krai, have unique characteristics. Its mild climate, relatively high degree of land development, warm sea, aesthetic coastal landscapes, presence of many relict species of animals and plants, exotic features, and ancient historical monuments, all make this area very attractive for Russian and foreign tourists.

The mainland may be sub-divided into three natural resource regions: Severoprimorsky (including some regions of the Khabarovsk Territory and the southwestern coast of Sakhalin Island), Sredneprimorsky and Yuzhnoprimorsky.

The Severoprimorsky region experiences severe climatic conditions which make construction and development difficult. The absence of convenient bays, except for Vanino-Sovetskaya Gavan Bay, hinders marine transport. Combined with difficulties in road construction on the mainland, this creates serious obstacles to economic development. Biological resources (commercial fish, including migratory fish) are its most important natural resource. Recreation potential is low.

The mainland part of the Sredneprimorsky region (Mosolov Cape–Povorotny Cape) is distinguished by very steep slopes with moderately high mountain landscapes. The rough topography here also makes road building and economic development difficult, and the prevailing erosion and steep slopes leave only narrow river valleys for building on. The shore is open with few bays, which prevents much development of marine transport. There is little recreation potential in its coastal zone, and only in the south where climate is milder and where there are sandy beaches in relatively large bays can recreation activities develop on the seashore.

Amongst the marine organisms in these two regions are widespread boreal forms. For some of them, the southern border of distribution is Povorotny Cape. More than 500 species of macrobenthos are found here, of which over 200 species are macrophytic algae. The commonest groups of animals are bristle worms, bivalve and obelis shells, and amphipods. In general, the potential production and resources of the region have been studied only occasionally. In recent years, there has been commercial harvesting of *Laminaria* at a few sites on the coast, though the main biological resources collected and fished here are fish, algae and echinoderms.

The southernmost Yuzhnoprimorsky region practically coincides with the borders of the Peter the Great Gulf. There are three sub-regions:

(a) *The Western region* (from the Tumen River mouth to the Amursky Bay) has a high natural resource potential. Landscapes here are mainly low mountains with river valleys going out to the sea. The coastline is heavily rugged. Sources of sediment run-off in combination with an active wave regime has resulted in the formation of sea deposits of titanium–magnesium and building sand. Accessibility of shores and large bays allows easier port construction here and the vast shallow-water bays and favourable hydrological conditions give rise to valuable fishery resources, providing a good base for mariculture development. Rivers have maintained their fishery potential only partially, however. The aesthetic beauty of the shores, favourable climate and sandy bays and beaches make this region suitable for recreation. Deposits of curative mud in the south of this sub-region have led to development of curative sanatorium establishments.

(b) *The Central sub-region* (coastal zone of Amursky and Ussuriysky Bays) is the most developed part of the Primorskyi Krai and it has a relatively high natural potential.

Dominant here are landscapes with vast river valleys, fertile soils, a smooth relief and favourable climate. The coastal zone is fairly rugged. Sheltered parts of bays are suitable as rural recreation zones; the shallowness of the bays provides good warm water and they have a high biological potential.

Marine organisms in these two sub-regions have low-boreal species; Povorotny Cape forms the northern boundary. At least 300 species of macrobenthos are found. At a depth of 0–5 m in open and semi-closed bays there are widespread communities of Japanese and chicory blade kelp, Irish moss and Miabe seagrass giving biomass up to 12 kg/m^2. The heavily silted bottoms of bays are populated with bivalves. Greatest diversity is at depths of 5–20 m, where biomass of the communities reaches 1–10 kg/m^2.

In the last decade, bottom communities in the Peter the Great Gulf have begun to show visible changes as a result of the impact of anthropogenic factors. High calorie communities of polychaetes are being replaced with lower calorie communities of basket stars and bivalves, which affects the feeding of plaice and other fish. There is a progressive reduction of stenobiontic species of benthos and plankton; populations of some species of polychaetes, sea-lettuce and other organisms being classic indicators of an increase in pollution. These increased anthropogenic pressures are due largely to increased volumes of industrial waste waters and sewage from the coastal zone and agricultural lands in the river watersheds. Degradation of the marine grass *Zostera* inhabiting the northeastern part of Amursky Bay deserves special mention. This leads to a shrinkage of the spawning grounds of the Pacific herring and to lower stability of the ecosystems.

(c) *The Eastern sub-region* (Strelok Bay to Povorotny Cape) is also characterized by a rugged coastline. Unlike the western sub-region it is more open to southern storms. Many bays are suitable for port construction and many placer deposits of, for example, building sand, zirconium and gold are found. Deeply cut bays (Nakhodka, Vostok) are promising for mariculture. Picturesque shores and favourable climatic conditions, in particular in semi-closed bays, offer great recreation possibilities.

Fauna and flora in this sub-region feature a higher percentage of subtropical species. Rough estimations of plankton and benthos include about 519 species of bottom animals, including 85 species of macrophytic algae, three seagrasses, 190 phytoplankton, 100 fish and three marine mammals. Coastal waters have high productivity supported by organic detritus brought in rivers flowing into semi-closed bays. More than once warm-water tropical sea snakes, turtles and sharks have been seen here.

In the last 25 years there has been a reduction in the density of the macrobenthos, notably autotrophic species and a growing quantity of heterotrophic species. Increasing areas are occupied by bivalves resistant to pollution and silting. Evidently, this is a result of industrial development of this bay. Especially severe are effects of bottom "ploughing" by trawlers.

PROTECTED AREAS

A system of protected natural areas makes up a framework for nature conservation activity in the Russian section of the Sea of Japan basin. There are four principal groups:

1. A system of specially protected natural territories (water areas) (SPNT), of high scientific and conservation significance and ensuring preservation of representative areas, unique natural complexes, genetically rich collections of plants and animals, and scenic formations. This includes state nature reserves (including biosphere), state national parks, monuments of nature, natural preserves, and some scientific stations.
2. Cultural and historical zones comprising localities with the highest concentration of monuments of antiquity and from the Middle Ages.
3. A system of "ethnic territories" ensuring preservation of national cultures and ways of life of indigenous peoples having no autonomy of their own, in their traditional locations, within territories run by the Soviets of People's Deputies who represent the interests of national minorities.
4. A system of 'normative' territories, having special scientific or cultural values. Normative territories ensure quality of the natural environment and rational management of natural resources. A system of these territories includes forests, water-conservation zones in river valleys and, in the coastal marine belt, territories ensuring the safety of groundwater reserves.

Specially Protected Natural Territories

Large forests growing on the eastern slopes of the Sikhote-Alin cover the Russian section of the Sea of Japan basin. Many rivers flow into the Sea of Japan down steep slopes across different landscapes and varied combinations of dark and light taiga, cedar and oak forests. Endemic plants (Korean larch, Sikhote-Alin rhododendron) are widespread here, in some localities small populations of goral are preserved. Ginseng grows in the southern part of the Sikhote-Alin eastern slopes. Populations of spotted deer inhabit this region too.

Ridges and basalt plateaux near the coast of Sikhote-Alin are characterized by spruce-fir taiga, usually with cedar and yellow birch. In a transitional zone from dark coniferous forests to subgolets forests in the south of Sikhote-Alin is a species of tea plant, and there is also a single representative of an endemic genus of cypress shrub.

In general, of the species in the IUCN Red Data Books, there are 26 varieties of vascular plants, five species of mammals, eight species of birds and 25 invertebrates which deserve special protection on the eastern slope of Sikhote-Alin. There are a great number of additional rare and endemic species. In comparison to other regions this area has the greatest number of protected areas. There are two nature reserves: Sikhote-Alin Biosphere Reserve covering an area of 347,050 ha, and the Lazovsky Reserve with an area of 120,024 ha; five natural preserves: Losiny (26,000 ha), Goralovy (8170 ha), Cerhnye skaly (2,920 ha), Vasilkovsky (28,700 ha); a forest-park zone of Vladivostok (16,500 ha); and about 60 monuments of nature. But even these protected territories do not cover all of the main natural sites and conditions, and their size is inadequate for the preservation of some species such as tiger, leopard, Mandarin duck, merganser, black stork and others.

The ecosystem of black fir–hardwood forest is confined to low-mountain areas with less severe winters and a longer growing season. In such forests, mesothermal Manchurian flora and fauna are well represented. Indicative of virgin forests are needle (black) fir and various broadleaved species, typically with abundant flora and fauna (heartleaf hornbeam, maple, Manchurian maple, Japanese yam, sika deer and others), as well as some rare species (Manchurian aristolichia, bog birch, Ussuri currant, leopard, and others). Special scientific significance is attributed to this ecosystem. Included in IUCN and Russian Red Data Books are 16 varieties of vascular plants, four species of mammals, six species of birds, and 35 species of invertebrate which need special protection.

Within this natural complex is the "Kedrovaya pad" nature reserve covering 17,896 ha, a natural preserve "Barsovy" of 71,100 ha, and two monuments of nature. However, existing protection measures are not enough for even a partial restoration of the black fir-hardwood forests. Remnants of primary ecosystems are permanently threatened by destruction from fires and felling. The whole southern-maritime forest of needle fir needs special protection.

In 1996 the natural preserve "Borisovskoye plato" was established, covering 103,400 ha. The main protected species are black fir–broadleaved forests with Khingam fir and Yeddo spruce, larch, Schmidt's birch, pine forests with needle juniper and several mammals (leopard, sika deer, tiger, goral and others); birds, and rare and unique species of insects. In the southern part of the Khasan Region are several other rare species, and the only nesting places in the southern Far East of the golden eagle.

Of all species included in the Red Books of IUCN and Russia, 21 varieties of vascular plants, two species of mammals, seven species of birds, and 16 species of insects need special protection in this complex. A significant number are rare thermophilic or kserophyte plants.

Marshlands

In saline lakes, brackish lagoons and estuaries of rivers are other specific flora and fauna which flourish in conditions of high productivity, some of which (e.g. bivalves or shrimps) are promising for mariculture. These brackish waters are also crucial to the most valuable migratory commercial fish—sturgeon and salmon. In comparison to other brackish water complexes in Russian Siberia and the Far East, those in the Sea of Japan basin are distinguished by a great variety of species, especially in Posjet Bay. Numerous molluscs, nemertines, bristle worms and sand hoppers form the basis of the brackish communities. Some are unique to Primorskyi Krai and need special protection, but in the Russian Far East brackish-water complexes are not, in fact, included in protected areas. In the Primorskyi Krai a desalted bay of Posjet Gulf and a lagoon of the Razdolnaya River are declared monuments of nature, but this does not guarantee their effective protection. In the Khasan Region a large territory of marshlands is given the status of a natural park "Ptichyi" as a refuge for waterfowl.

Historical Sites

According to the Institute of History, Archaeology and Ethnography of the Peoples of the Far East, Far-Eastern Branch of RAS, the territory of the Russian Far East around the Sea of Japan contains more than two thousand monuments of history and ancient culture, of which about 1000 date back to the primitive communal system and to the Middle Ages. It should be kept in mind that investigations in many regions are far from complete, but we can note the localities of the greatest concentration of those which have been detected. Territories with the highest concentrations will be singled out as protected cultural–historical zones. These territories include the Posjet Bay basin (about 120 monuments), the Razdolnaya River valley, several sites on the coast of the Sea of Japan, and others.

LANDSCAPES AND TOURISM

In the Primorskyi Krai there are 295 monuments of nature, many of which represent a basis for tourism development. Most valuable in recreational terms are mineral springs, caves, landscapes of mountains, waterfalls, bays and cliffs on the coast, and islands in the Peter the Great Gulf with rare varieties of plants.

At least four types of coastal areas can be ranked for recreational suitability:
– Most suitable is in the south of the Primorskyi Krai in Posjet-Slavyanka, Livadia-Vostok, Nakhodka and Preobrazhenia Regions. The coast is heavily rugged, with wide sand beaches and forests. The bathing season here lasts from 70 to 106 days. This coastal zone is most promising for development of thalassotherapy, water sports and underwater tourism.
– Suitable coast is located largely in the Peter the Great Gulf. It is rugged with small beaches, otherwise is similar to the most suitable category (above).
– Partially suitable coast is located between Titov Cape and Ryndaya Bay. The coastline is less rugged, and steep escarpments occur. There are few bays and gulfs and the bathing season lasts less than a month. There is a possibility for development of underwater tourism.

– Unsuitable coast is that located north of Rynda Bay. The coastline is nearly straight, no bays and gulfs of considerable size are found here, and steep escarpments tower over the coast. Flat areas occur in river valleys. Some areas, for instance Arka Cape, are very picturesque cliffs, but severe natural conditions exclude much development.

Natural curative mineral waters along the coast are important, unique, and include mineral springs of cold carbon dioxide, thermal nitrogen and cold sulphate waters. Some deposits of curative mud are found on the coast of Primorskyi Krai as well; according to some authors, their reserves are estimated at over 100 million tons. In the Russian section of the Sea of Japan basin there are known to be more than 2,000 archaeological and historical monuments of cultural significance; in the Primorskyi Krai alone more than 800 monuments date back to settlements of Palaeolithic, Mesolithic and Neolithic periods and to the Middle Ages.

There are, however, some specific constraints to recreational development. These include: a monsoon climate involving frequent typhoons; high relative air humidity; a great number of days with strong winds and mists; a short period of steady snow cover in the southern and piedmont regions; widespread bloodsucking insects in the Primorskyi Krai and especially on the coast of the Khabarovsk Territory; and finally, great quantities of poisonous jellyfish which appear during the bathing season.

CLIMATE CHANGE

Effects of processes in previous warm and cold climate epochs in the Pacific and in inland regions have been dissimilar. In past periods of warming, an intensification of oceanic and weakening of continental influences have been recorded. The reverse was observed in epochs of Late Pleistocene cooling. Climate became more severe in cooling epochs due to lowered ingress of warm oceanic waters, considerable shrinking of the area of water and their nearly complete freezing. In warming epochs, the influence of warm currents became enhanced resulting in a milder climate. A rise and lowering of the sea level brought about restructuring of coastal landforms, with the coastline becoming more complicated, erosion becoming weaker, but with abrasion rates increasing. During the transgression the whole coastal zone, including some inland depressions, was subject to the most acute erosion.

Detailed layer-by-layer studies of Holocene profiles and their correlation using radiocarbon dating, has revealed a quick response of the natural environment to even small and short-term climate fluctuations. This response manifested itself in a changed structure of vegetation, rising or lowering of the upper border of a forest, changes in the degree of waterlogging and changed patterns of forest growth. Some changes in vegetation structure have been detected corresponding to temperature deviations of 1.0–1.5°C from the present level, for climatic phases of 250–300 years duration. Terrestrial scientists, therefore, have documented numerous and marked changes to vegetation in response to relatively modest temperature changes of relatively short duration.

Against this background of major changes to the natural environment in the Far East one can see some shorter-period climate variations lasting for dozens to one hundred of years. To determine climate variations in the recent 100 to 200 years, specialists from the Pacific Institute of Geography, FEB RAS, carried out investigations of multi-year, including century-long, fluctuations of a hydro-thermal regime in the south of the region. The length of the time series at Seoul-Vladivostok is 200 years, and if we start counting from the years with the actual minimum precipitation (approximately the 1890s), then data show that periods with inadequate precipitation are 20 to 30 years shorter than previously, and a return to the initial condition takes at least 70 years (Fig. 9). This allowed a forecast of diminished precipitation beginning from the 1970s, and the actual situation confirmed fully this forecast. Annual precipitation in low periods is lower than the average yearly figures by only 15–20%, but is lower by 30–40% compared with the rainy part of the cycle. It should be noted that a decrease of precipitation by 20–30% below the normal level leads to profound ecological consequences.

ECOLOGICAL PROBLEMS AND THEIR CAUSES

Amursky Bay

Amursky Bay is subject to an anthropogenic load from five administration regions in the Primorskyi Krai. Also, waters of the Razdolnaya River bring pollutants from China. According to official information, more than 120 million m^3 of waste waters are discharged annually. Of this quantity nearly 78 million m^3 are untreated and more than 26 million m^3 are inadequately treated. Only 8.1% of wastewater is treated properly. The main sources of pollutants are concentrated discharges of wastewater, flow from polluted surfaces of watersheds during rainstorms and snow melting, and scattered discharges into the sea. A total of 27 wastewater outfalls into coastal waters of the bay have a daily total discharge of at least 130,000 m^3 (excluding the Razdolnaya River flow).

Waste waters bring into Amursky Bay about 104,690 tons of organic substances, 110,050 tons of suspended matter, 1540 tons of fats, 880 tons of petroleum products, 980 tons of detergents, 4.6 tons of phenols and 1.2 tons of pesticides. Pollution is officially assessed at a network of Goskomgidromet stations (Table 3). It is seen that from time to time the maximum admissible concentrations are substantially exceeded (2–38 times) for all controlled parameters. For phenols the level is exceeded 3–4-fold practically all the time.

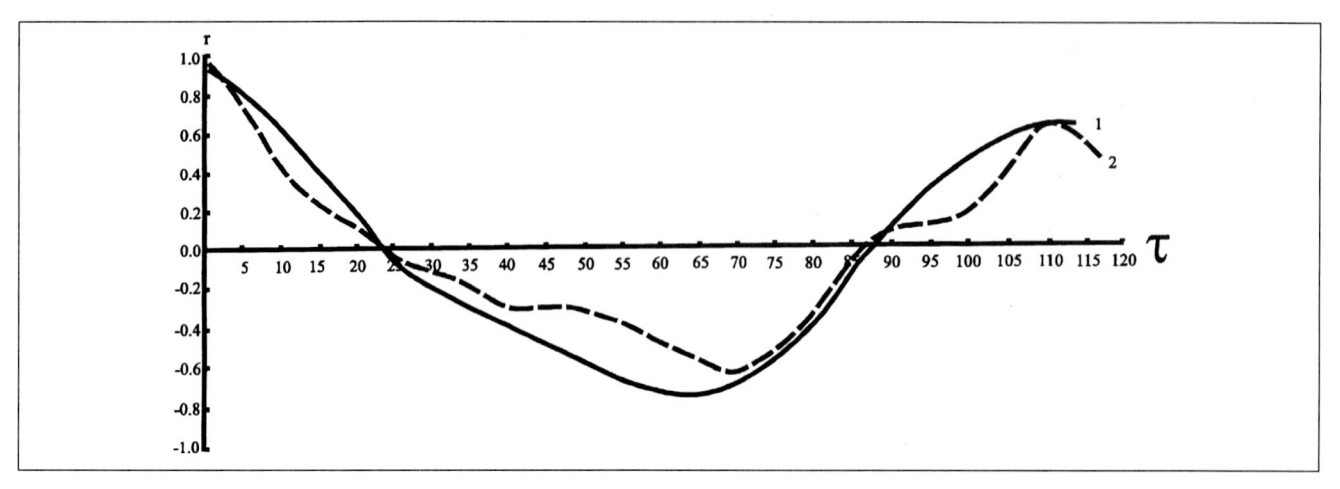

Fig. 9. Autocorrelation of sliding 26 years annual means (1) and sliding 12 years annual means (2) of precipitation in Seul-Vladivostok.

Table 3

Chemical pollution of Amursky Bay (Roskomgidromet data)

	Year	1986	1987	1988	1989
Petroleum hydro-carbons (mg/l)	average	0.03	0.04	0.03	0.02
	maximum	0.57	0.40	4.61	0.20
Phenols (mg/l)	average	0.03	0.001	0.003	0.004
	maximum	0.036	0.008	0.025	0.017
Detergents (mg/l)	average	0.01	0.08	0.04	0.07
	maximum	0.37	0.28	0.116	0.15
DDT (ng/l)	average	8.9	2.6	6.5	1.8
	maximum	97.2	35.0	251	39.5
Oxygen (mg/l)	average	9.16	9.48	9.05	9.53
	maximum	1.90	1.54	2.04	3.47

The oxygen regime in the Bay waters in the warm season is satisfactory. However, in July–September dissolved oxygen is halved. Ratios of background concentrations of suspended matter in coastal water of Amursky Bay and oceanic waters show the anthropogenic loading factor:
– dissolved forms of heavy metals Pb(93), Co(19), Ni(6.1), Cd(5.9), Cu(1.9), Ag(1.6), Zn(1.0)
– suspended forms of heavy metals Cd(115), Ag(63), Ni(54), Pb(42), Co(36), Zn(14), Cu(3.0).
(Figures in brackets denote the ratio of concentrations in coastal waters to that in oceanic waters. Elements are arranged by decreasing concentrations.)

Metal concentrations are especially high in suspended matter. Analysis of the material assembled for several years has revealed some zones with higher concentrations of pollutants. Kirpichny Zavod Bay is most severely affected, receiving metals with waters of the Pervaya and Vtoraya Rivers and from sewage outlets. In this region there are also high recorded concentrations of cadmium, zinc, petroleum hydrocarbons, and a high coverage in the water surface by films of organic pollutants, and pathogenic microflora. Bottom sediments here contain high concentrations of petroleum hydrocarbons, cadmium, zinc, mercury and others in comparison to a nearby marine nature reserve.

A geochemically abnormal zone in respect to some metals (Pb, Zn, Cu) is found in bottom sediments in the vicinity of Egersheld Cape, where bottom sediments taken in the Zolotoy Rog Bay in the course of bottom dredging are dumped.

From material obtained by the institutes of the Far-Eastern Branch of RAS and TINRO the following conclusions can be reached:

1. Between 1925–1933 and 1970–1972 the Bay to 80 m depth was subject to a profound restructuring of bottom communities as revealed by the replacement of highly productive species by silt-adapted species. In the past 40 years, a general reduction of species populations was witnessed. Average biomass declined three-fold, while the area inhabited by communities with low biomass expanded 28-fold.
2. Observations of the distribution of urchins (grey and black), which are used as good indicators of water quality, have shown that in the Bay areas with a higher level of metals and petroleum hydrocarbons in water and bottom sediments (Vtoraya Rechka River, waste outfall on the De-Frize Peninsula) urchins completely disappeared. At stations with lower pollution levels, the reproductive functions of animals are impaired.
3. Surveys of algae distribution have shown that in the recent decade the species that require clear water have practically disappeared. They have been replaced with sea-lettuce.

The results suggest that the Amursky Bay ecosystem is decaying. Probably, in order to lessen the anthropogenic load on the Amursky Bay, it is necessary to carry out a wide range of infrastructural and technical measures using considerable capital investment, and this will require changes in the structure of the industrial complexes in Vladivostok and Ussuriisk, and agriculture in the Razdolnaya River basin.

In the past three to four years some improvement in the Bay has been observed, but this is a temporary event connected with economic decline.

Rudnaya River Valley

At present, the situation in the Rudnaya River valley can be characterized as pre-crisis. The reasons for this are:

1. An intensive growth of industry, especially of "ecologically" hazardous kinds—poly-metal and chemical—that has entailed an increased input of pollutants, gas and dust emissions, industrial and sanitary waste waters. In some cases unique technologies are being developed which have no analogies in the rest of the world and, consequently, no existing models for their treatment.
2. A growth of atmospheric pollution from motor transport—the main transport means in the region.
3. Development of ore deposits, without giving proper consideration to local conditions. At present the main processing enterprises are found in parts of the valley which experience a great number of inversions, both in winter and summer. Settlements are located so that in the summer they are subject to prevailing pollution. Industry chimney heights are such that air discharges especially affect these settlements, and wind speeds prevailing at times of inversions correspond to the most dangerous speeds for pollutant transfer.
4. Settlements are usually located downstream from ore mines, dressing factories and others, so the population cannot use water from nearby surface water streams.

As a result of a combined effect of these factors, degradation and decay of terrestrial and aquatic ecosystems is taking place in some parts of the Rudnaya River basin, including Rudnaya Bay.

In Rudnaya Pristan and Dalnegorsk, analysis of average annual concentrations of metals in air over settlements has indicated that of all analyzed metals only lead exceeds acceptable levels. In Rudnaya Pristan, the average annual concentration of lead is twice as high as the maximum acceptable level, and is the highest atmospheric lead level in all controlled cities in the Russian section of the Sea of Japan. A nonferrous metallurgy industry is the major contributor to atmospheric pollution in this settlement.

Surface and ground waters over the whole territory of the Rudnaya River valley from Krasnorechesnk to Rudnaya Pristan are heavily polluted. Moreover, in many places their composition endangers people's health with normal contact with water. Pollution of the surface waters makes them unsuitable for use, and the Rudnaya River has completely lost its fishery significance. Sickness levels in the Dalnegorsk region are 1.2 times higher than the territory average, and include frequent occurrence of such diseases as cancer of the lymphatic tissue and acute infections of respiratory organs. It follows that the health condition of the population living in the Rudnaya River valley requires urgent measures for its improvement.

CONCLUSION

The Sea of Japan basin is subject to ever-increasing industrial pollutant loads arising not only from its own surrounding territories, but from neighbouring territories that are closely connected with it via a system of marine currents and air transfer. Improvement of the state of its ecosystems is possible only if the region follows a concept of sustainable development.

For this purpose it is necessary to establish a system of protected natural areas that will help not only to preserve a genetic fund of natural ecosystems, but will become a basis for their restoration in areas that are now partially or completely destroyed. Finally, it is necessary to change outdated production technologies, starting with those which have modern treatment facilities.

REFERENCES

(*This review was based on numerous publications, government reports, internal working papers and conference documents, many in Russian. A selection of these articles is listed here.*)

Anon. (1992) Long-term program of protection of a nature and rational land-use of natural resources of Primorye till 2005 (Ecological program). Part 2. *FEB RAS, Vladivostok*, 297 pp. (In Russian).

Fairbridge, W. (ed.) (1966) *Encyclopedia of Oceanography.* Reinhold Publishing Corporation, New York, pp. 626–631.

Ikeuchi, Y., Amano, H., Aoyama, M., Berezhnov, V.I., Chaykovskaya, E., Chumichev, V.B., Chung, C.S., Gastaud, J., Hirose, K., Hong, G.H., Kim, C.K., Kim, S.H., Miyao, T., Morimoto, T., Nikitin, A., Oda, K., Povinec, P.P., Pettersson, H.B.L., Tkalin, A., Togawa, O. and Veletova, N.K. (1999). Anthropogenic radionuclides in seawater of the Far Eastern seas. *Science of the Total Environment* 237–238, 203–212.

Kachur, A.N. (1996) Regional pollution background for the southern Far East of Russia. Resources, environment and sustainable development. International Symposium, Shenyang, P.R. China. pp. 202–203.

Kachur, A.N. (1998) State of marine and coastal environment and connected with it freshwater environment of Japan basin. Resources, environment and sustainable development. 1998 International Symposium. Changchun, P.R. China. pp. 31–35.

Kachur, A.N. 1999. The system of natural and international nature parks in the Russian Far East: issues and perspectives. *"Beringian Notes". National Park Surveys. Alaska Region* 7 (1), 4–5.

Kachur, A.N. (1998) Ecological problems in the Russian part of Japan sea basin. *Metals & Technology KINZOKY* 68 (9), 65–69. (In Japanese).

Kachur, A.N. (1997) Pollution of a basin of the Japanese sea and its adjacent regions. Academic Forum for Northeast Asia. Kyoto, Japan, pp. 38–44.

Kitamura Moritsugu (1996) How to select Acid Precipitation Monitoring Points. Strategy for air pollution control in East Asia. Tokyo, Japan, pp. 71–81.

Min-Hee-Lee (1996) Situation of Acid Rain in Korea. Strategy for air pollution control in East Asia. Tokyo, Japan, pp. 25–30.

Petterson, H.B.L., Amano, H., Berezhnov, V.I., Chaykovskaya, E., Chumichev, V.B., Chung, C.S., Gastaud, J., Hirose, K., Hong, G.H., Kim, C.K., Kim, S.H., Lee, S.H., Motimoto, T., Nikitin, A., Oda, K., Povinec, P.P., Suzuki, E., Tkalin, A., Togawa, O., Veletova, N.K., Volkov, Y. and Yoshida, K. (1999). Anthropogenic radionuclides in sediments in the NW Pacific Ocean and its marginal seas. Results of the 1994–1995 Japanese–Korean–Russian Expeditions. *Science of the Total Environment* **237–238**, 213–224.

Sevis, G.D., Baklanov, P.Ya., Kachur, A.N. and Voronov, B.A. (1996) A sustainable land use and allocation program for the Ussury/Wusuli river watershed and adjacent territories (northeastern China and the Russian Far East) and others. ESD Inc. USA; FEB-RAS PGI, FEB-RAS IAEP, RUSSIA; HPTS, CHINA; National Committee on U.S.-China Relation. 97 pp.

Sivintcev, Yu.V., Visotckiy, V.L. and Danilyan, V.A. (1993) Radiation circumstances in Peter Great Gulf for want of draining of liquid radioactive scraps in October. *Atomic Energy* **82** (4), 309–314. (In Russian).

Soyfer, V.N., Danilyan, V.A., Malkin, S.D. and Chaykovskaya, E.L. (1997) Radiation condition of a water environment of northern part of the Japanese sea. *Bulletin of FEB RAS, Vladivostok* **4**, 86–104 (in Russian).

Svinuhov, V.G. (1997) *Ecology of Primorye Cities*. Far East University, Vladivostok. 140 pp. (In Russian).

Tkalin, A.V. (1991). Chemical pollution of the NW Pacific. *Marine Pollution Bulletin* **22**, 455–457.

Tkalin, A.V. (1991a) Present state of the organic chemical pollution of the western Pacific Ocean surface waters. *Ocean Research* **13**, 103–108.

Tkalin, A.V. (1992) Present status of the Japan Sea chemical pollution: An overview. *La Mer* **30**, 1–4.

Tkalin, A.V. and Shapovalov, E.N. (1991) Influence of typhoon Judy on chemistry and pollution of the Japan Sea coastal waters near the Tumangan River mouth. *Ocean Research* **13**, 95–101.

Tkalin, A.V., Lishavskaya, T.S. and Veretshak, A.A. (1996) Persistent organochlorines in the East Sea (Japan Sea). *Ocean Research* **18**, 159–163.

Yablokov Commission (1993) Facts and problems related to radioactive waste disposal in seas adjacent to the territory of the Russian Federation (Data from Government Commission on Matters Related to Radioactive Waste Disposal at Sea). Moscow, Russia.

THE AUTHORS

Anatoly N. Kachur

Pacific Geographical Institute,
Far East Branch,
Russian Academy of Sciences
Vladivostok 690022,
Russia

Alexander V. Tkalin

Far Eastern Regional Hydrometeorological Research Institute (FERHRI),
Russian Academy of Sciences
24 Fontannaya Street, Vladivostok 690600,
Russia

Chapter 84

THE YELLOW SEA

Suam Kim and Sung-Hyun Kahng

The Yellow Sea is a semi-enclosed shelf-type shallow basin, which has relatively low water exchange with the open ocean. The residence time of water in the Yellow Sea is long, i.e. 5–6 years. This region is where the Siberian High and the subtropical Pacific Low weather systems meet producing cold, dry winters and warm, wet summers. Typhoons accompanied by heavy rains during the summer and autumn cause large quantities of fresh water runoff. The Yellow river carries 40.4 km^3 of fresh water and approximately 1.0 billion tonnes of silt into the Yellow Sea every year, and the Yangtze river annually transports a further 900 km^3 of fresh water and 0.47 billion tonnes of sediment.

The Yellow Sea is regarded as one of the most heavily exploited seas in the world, with Korean, Chinese and Japanese fishermen targeting numerous species with multigear fisheries. Along the coastal provinces of China and Korea, there are countless small and large-scale mariculture farms rearing fish, invertebrates and seaweeds. Over the last several decades, the living resources in the Yellow Sea ecosystem have changed greatly, stressed by overfishing, environmental changes and pollution. The Yellow Sea's water quality has been impaired in many locations by contaminants, waste material, and disturbances from human activities. Many of these threats to marine environmental quality originate from land.

The coastal areas on either side of the Yellow Sea are highly populous, with approximately 600 million people residing in the area. Increasing coastal population coupled with rapid industrialization and urbanization exerts severe pressure on the marine environment. Large quantities of industrial wastewater and domestic sewage are directly discharged into the coastal waters with either little or no treatment. Nutrient loads in coastal waters have contributed to eutrophication and frequent outbreaks of red tides. Land-use change, deforestation, soil erosion and intensive farming are closely interrelated with the heavy use of chemical fertilizer and pesticides. Massive reclamation projects have created land for agriculture, industrial complexes, and coastal cities, and significant decreases of coastal wetlands have become a matter of serious concern in relation to the reduction of natural purification capabilities, loss of coastal habitats and biodiversity, and adverse effects on fisheries. The first and most serious obstacle to be overcome is a national development strategy that considers income growth as the primary national priority. In addition, the management of the Yellow Sea is especially complicated by being surrounded by nations with differing political systems and levels of economic development. International co-operation between them has begun to facilitate the prevention of damage to the environment and its resources.

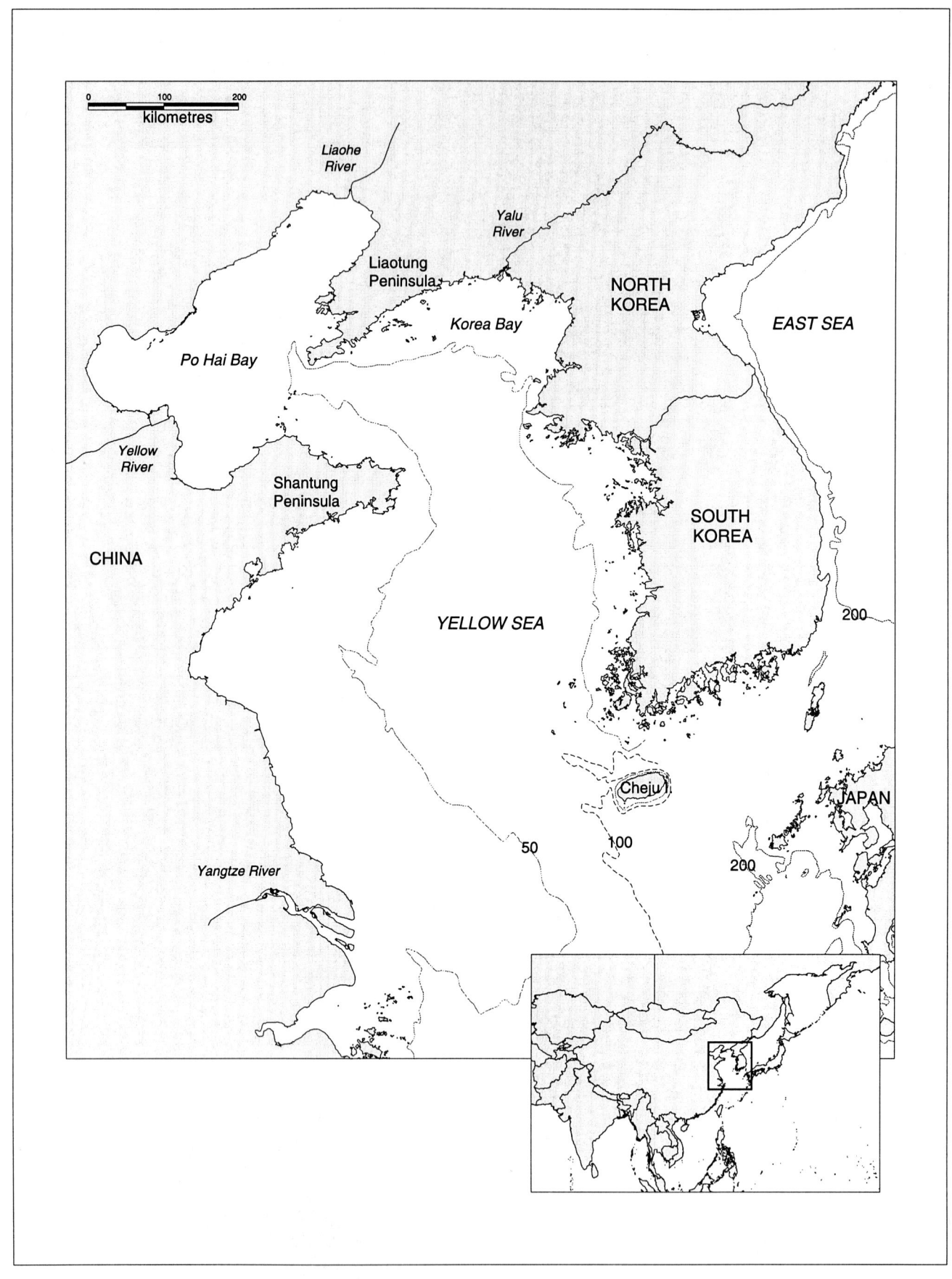

Fig. 1. Geography of the Yellow Sea.

THE DEFINED REGION

The Yellow Sea is a semi-enclosed sea between mainland China and the Korean Peninsula in the mid latitudes of the northwestern Pacific. It borders the Republic of Korea (hereafter South Korea) and the Democratic People's Republic of Korea (hereafter North Korea) on the east, and the People's Republic of China (hereafter China) on the north and west. It extends to around 40°N to the North, and the Korea Bay occupies most of the shallow (<50 m) northeastern section. Another inlet, the Po Hai Bay (Bohai Sea), is separated along the line between the Liaotung Peninsula and the Shantung Peninsula in the northwest (Fig. 1). To the south, it joins the East China Sea near 31°N, and the boundary between the Yellow Sea and the East China Sea runs from the mouth of the Yangtze River (Chang Jiang) to Cheju Island. In general, the sea bottom is shallow and relatively flat. The mean and maximum depths are 44 m and 103 m, respectively, and the area of the Yellow Sea (excluding the Po Hai Bay) is 404,000 km^2, which embodies about 120,000 km^3 of seawater. An elongated trough running in a north-south direction is found along its central part. Tidal flats and ria coasts are comparatively well developed in the eastern Yellow Sea along the Korean Peninsula.

Several rivers and small tributaries supply fresh water as well as sediment to the Yellow Sea with the maximum supply occurring during the summer. The Yellow River (Huang Ho) and the Yangtze River in China are major sources compared to rivers in the Korean Peninsula (e.g. the Yalu and Taedong River in North Korea, and the Han and Kum River in South Korea). The Yangtze River, which is the longest river joining the Yellow Sea and the world's third longest river, has an annual transport capacity of 900 km^3 of fresh water and 0.47 billion tonnes of sediment (Sheng et al., 1997). The Yellow River also carries 40.4 km^3 of fresh water annually; however the body of freshwater contains a large amount (25 kg/m^3) of silt, transporting a total of approximately 1.0 billion tonnes to the Yellow Sea every year (Yang et al., 1997). It is regarded as the world's muddiest river. It has been reported, though, that the water and sediment discharges from the Yellow River have been decreasing in recent years as a result of increased water consumption and the dam constructed for water conservation (Yang et al., 1997). The Yalu River, located at the boundary between the Korean Peninsula and the Manchurian region, is the longest river in the Korean Peninsula transporting 34.7 km^3 of fresh water annually.

Even though there are high concentrations of organic matter and nutrients supplied from the land, primary productivity in the Yellow Sea is neither very fertile nor productive. A wide range of primary productivity has been reported. Zhu et al. (1993) measured 425 m gC/m^2/day. Previous researchers observed a wide range of productivity from 80–213 g C/m^2/yr (Choi et al., 1988; Kang, 1991). At present, 166 zooplankton and 276 fish species are reported as resident species of this large marine ecosystem, forming a relatively rich biodiversity of fish and marine organisms. Zooplankton biomass seems to be lower than in other adjacent marginal seas. The Yellow Sea ecosystem, however, supports substantial populations of fishery resources including marine mammals and seabirds, and its primary and secondary productions can sustain about one million tonnes of pelagic and ground fish catch every year. Approximately 100 commercial species have been identified, comprising demersal fish (66%), pelagic fish (18%), cephalopods (7%), and crustacea (7%). Nevertheless, the recorded fish catch indicates overfished and overburdened stocks. In addition, many fish stocks are threatened by pollution and habitat losses brought on by improper management of coastal areas. Some fish species are transient, seasonally migrating to and from neighbouring seas, while others show onshore–offshore seasonal migration inside the Yellow Sea. Mudflats with a total area of approximately 1700 km^2 are exposed and drained during low tide. These mudflats are regarded as important habitats for countless benthic organisms, as well as providing a naturally filtering and purging treatment system for the polluted waters dumped from the land.

PHYSICAL PARAMETERS AND ENVIRONMENTAL VARIABLES

Circulation in the Yellow Sea and the East China Sea is a continuous system. A minor divergent branch of the Kuroshio Current from the East China Sea penetrates into the Yellow Sea, bringing a warm and saline water mass, which slowly (<0.25 m/sec) flows northward over the deep trough area in the central Yellow Sea. The process of mixing with river runoff dilutes and cools the water as it moves towards the north. The flow bifurcates toward the east and west at the northern end of the Yellow Sea. Both of the bifurcated flows turn southward along the coast on both sides of the Yellow Sea forming two gyres. There are several coastal currents including the Yellow Sea Warm Current, Jiangsu Coastal Current, Zhejiang–Fujian Coastal Current, and occasionally small gyres are formed at nearshore areas. For example, in the south of Haeju of the Korean Peninsula, there is a southward countercurrent near the coast that is strongly developed during winter. In addition, freshwater input from the Yangtze River varies seasonally, which in turn influences the seasonal patterns of the Yellow Sea. The strong Kuroshio Current pushes the major portion of the water mass containing freshwater inputs from the Yangtze River towards the northeast during summer, and mixes with the Yellow Sea. However, a large portion of the northerly outflows is formed in winter. Given the relatively limited water exchange with the open ocean, the residence time of water in the Yellow Sea is long, i.e. 5–6 years (Hong et al., 1997).

Tidal currents are very strong, reaching 0.4–1.0 m/sec, and large tides of 4–8 m range are common on the west coast of the Korean Peninsula. Typhoons frequently pass through the Yellow Sea during summer and autumn, and strong storms are occasionally observed in winter also, with high waves of 4–5 m. As a result of the strong tidal currents and shallow bottom topography, water columns are well mixed and turbulent except during the rare calm and sunny days in summer. The silts suspended by the strong turbulence are reflected in the high turbidity measured in the waters.

The Yellow Sea region is where the Siberian High and the subtropical Pacific Low collide producing cold, dry winters and warm, wet summers. Snow, rain, and typhoons, which influence river runoff, act as major determining factors for the changes in seawater temperature and salinity. High precipitation in the west coast of the Korean Peninsula is observed in the summer monsoon seasons ranging between 100–700 mm during June–August, while low precipitation is the norm in winter with 50–60 mm during January–March. Typhoons accompanied by heavy rains, which form in the subtropic area of the west Pacific, frequently occur during summer and autumn. On average, there are twenty-eight typhoons every year, with as many as nine hitting China's coastal belt. However, a mean frequency of three typhoons per year is known to occur in the Yellow Sea and west coast of the Korean Peninsula.

Seawater temperatures vary with the seasons. Influenced by the high pressure formed from the continent, the cold strong northwesterly wind prevails in winter, lowering the water temperature to 2–8°C. As mentioned above, the winter water columns of the Yellow Sea are homogeneous as a result of seasonal cooling and turbulent processes. Sea ice is seldom formed in the open waters during the winter, but it is frequently found in the innermost part of the coastal embayment. As spring approaches, the surface water temperature rises with the increasing insolation. During spring, surface temperatures near the coastal area measure up to 15°C, while the water temperatures in the central Yellow Sea remain cooler at below 9°C, thus forming a front between two regions. Surface water temperatures in summer fluctuate between 24 and 28°C. A strong thermocline with relatively constant temperature measurements of 11–15°C, at a depth of 10–30 m, separates the surface water from the bottom water. In the trough area, mean temperatures measured at 75 m do not show any significant difference through the seasons (Kim et al., 1997).

Salinities usually show relatively wide ranges in the Yellow Sea, depending on spatial and temporal variabilities—except cool bottom water over the trough which is a constant 32–33‰ all through the year. Higher surface salinities in the south (33–34‰) and lower salinities in the north (31–32‰) are observed during winter. Sometimes a salinity of 22‰ is observed when large quantities of river runoff are dumped in the Yellow Sea during rainy seasons.

SHALLOW WATER HABITATS

Sand beaches and tidal flats are ubiquitous features along the west coast of the Korean Peninsula. Although only small and sparse coastal sand dunes are present, they are more common in some regions. The dunes are aligned north–south or northwest–southeast along the directions of prevailing winds. The eastern coast of the Yellow Sea (i.e. the west coast of the Korean Peninsula) is characterized by long stretches of tidal flats reminiscent of sea level rises in the Holocene. Surface sediments on the tidal flats generally consist of mud, mud mixed with fine sand, and sand. Tidal flats with muddy sediments are found mainly either in shallow (and therefore calm) embayments or within estuaries. Sandy flats occur in the eroded coasts or in locations with strong tidal currents or waves. However, many of the tidal marsh zones in Korea disappeared long ago, because supratidal flats have been reclaimed for agricultural uses since 1900.

Quantitative surveys of subtidal soft-sediment benthos, including detailed examinations of spatial and temporal variations, are carried out in several places in the west coast of the Korean Peninsula. The biotic assemblages of the tidal flat habitats are well described by several studies (e.g. Choi et al., 1998; Kim et al., 1998; Lim and Je, 1998; Lee et al., 1998).

Large-scale tidal flats occurring along the Chinese coastal areas are associated with two different types of coasts. The total length of Chinese coastline of the Yellow Sea supporting tidal flats is about 4000 km. Major tidal flats in China have developed along the fringing zone of the North China plain, where the tidal range is about 3 m and where the velocity of tidal currents is less than that found in the estuaries. The tidal flats observed in this area are between 3 and 18 km in width and are much more extensive than in estuarine areas. Given that the slopes in the coastal zone are very gentle, tidal currents are the dominant force in dynamic processes shaping the flats. Sediments of tidal flats in China are fluvial in origin, and silt (ranging from fine to coarse) is the major constituent of the deposits. Very fine sands and clays have been observed in many regions of tidal flats. They were largely supplied to the coastline during the lowered sea level period via large rivers such as the Yellow River and the Yangtze River (Wang and Zhu, 1994).

OFFSHORE SYSTEMS

The Yellow Sea occupies a relatively broad area; 960 km south to north, and 690 km east to west. As the entire area sits on a continental shelf, the Yellow Sea is fairly homogeneous from an ecological point of view. Neritic diatoms including *Skeletonema* and *Chaetoceros* dominate the phytoplankton community, while *Calanus*, *Euphausia*, *Sagitta*, and *Thermisto* are the major components of the zooplankton. The area also provides grounds for spawning, nesting and nursery areas for various economically important animals such as mackerel, anchovy, and jack mackerel during spring through fall (KORDI, 1994). However, small degrees of heterogeneity in water properties are formed resulting from local weather and/or bottom topography, and in turn these result in different seasonal patterns in fish distribution. For example, in the Yellow Sea, two groups of fish communities exist, resident species near coastal areas and migratory species. The deep trough area where the water temperature is comparatively warmer in winter and cooler in summer than the coastal areas, is occupied by both of these two fish groups. In winter, the temperature in coastal areas is very low, so that coastal species such as skate and pomfret escape to deeper water for overwintering.

Warm water species leave the cold waters of the Yellow Sea during winter. Some of the major fish species such as the small yellow croaker and hairtail spawn in the coastal areas in the Yellow Sea, but migrate to the south for overwintering. Recent statistical analyses on the small yellow croaker between fish recruitment and environmental conditions demonstrated that warm and stable oceanographic conditions favour the survival of younger fish more than cooler and more turbulent waters (Kim et al., 1997). On the other hand, cold water masses over the trough during summer provide a good refuge area for the cold water fish species such as Pacific cod, which migrated to this area in the past from the east during severe cold periods over the Korean Peninsula. However, the overwintering species in the confined trough area have become overfished and face complete depletion by the motorized artisanal fishing fleets. The annual catch of demersal fishes which comprises about 40% of the total in the Yellow Sea and the East China Sea has increased continuously, though the CPUE trend is the reverse of this. The overall catch doubled between 1970 and 1990 (Zhang and Kim, 1999); China's catch increased by the largest proportion, Korea's catch was stable, but Japan's decreased (Fig. 2).

POPULATION

Increasing coastal human population surrounding the Yellow Sea is especially significant and exerts severe pressures on the marine environment. The coastal areas on either side of the Yellow Sea are very populous—approximately 600 million people reside in the area—and all the by-products of human habitation are drained into the Yellow Sea. Several large cities such as Shanghai, Qingtao, Dalian, Pyungyang and Seoul have grown up close to the Yellow Sea: at least 15 million people reside in the most populous of these cities, i.e. Shanghai and Seoul.

China has the largest national population in the world which continues to increase at a rate of 15 million annually. Its population was 400 million in 1900, 600 million in 1950, and reached 1.2 billion in 1995; it is expected to reach 1.4–1.5 billion by 2030 (Brown, 1995; Fischer and Heilig, 1997). Currently, it is estimated that 41% of its population reside in the coastal provinces.

This ever-growing population places a severe strain on every type of environment in China. Meeting the needs of the vast population with China's limited natural and economic resources is an enormous challenge. Over the last 40 years, China's agriculture, cities, energy supplies and forests have all come under increasing stress as its population has more than doubled. To keep up with demands for housing and buildings, coastal land that was previously cultivated for food is now being developed for human occupancy at a rate of 3400 km^2/yr (Zhijie and Côte, 1990). Coastal waters are also being forced to absorb increasing volumes of runoff from increasing industrial complexes as well as the population explosion.

Estimated coastal population density in China is about 405 people per square kilometre. With an estimated 40% of the industrial plants located along the coast, more workers and their families are tempted to move to coastal areas in search of work. The coastal provinces and municipalities account for just 13% of the country's land area, but 41% of the population, 51% of the GNP, 71% of import and export values, and 50% of the state's tourism revenues (SOA, 1986; Zhijie and Côte, 1990; KORDI, 1994).

The Korean Peninsula, by contrast, is characterized by high and rugged mountainous areas in the east and relatively flat plains in the west. By virtue of necessity, the majority of the population resides near the coast of the Yellow Sea. The population of South Korea was 45 million in 1995 with a density of 454 people per square kilometre (NSO, 1996), while the population of North Korea was estimated at 25 million (UN, 1997). The population of North Korea, however, is currently at risk of a steady decline. Recent studies have shown that 15% of the towns in North Korea are suffering from famine. Many people are dying, not only from starvation, but from diseases associated with malnutrition (NAPSNet, 1996). As a whole, the Korean Peninsula currently supports a total population of approximately 70 million.

Geographical distribution of the population is a matter of concern when managing marine pollution. Rapid industrialization in South Korea since the 1960s has been accompanied by drastic changes in the spatial distribution of human settlements. The urban population doubled to 57.3% in 1980 and further increased to 74.4% in 1990,

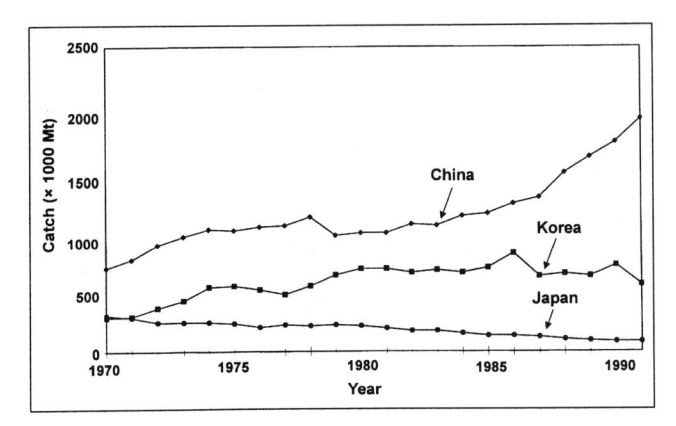

Fig. 2. Annual catches of demersal fishes by Korea, China and Japan in the Yellow Sea and East China Sea, 1970–1991 (Zhang and Kim, 1999).

primarily as a result of continuous rural-to-urban migrations (Kim and Oever, 1992). Metropolitan growth and urbanization have been most pronounced in the capital city of Seoul and consequently, the Seoul metropolitan area supported 42.8% of the total national population in 1996 (NSO, 1997).

Rapid industrialization and urbanization in South Korea and China are similar in many aspects, although there is a time gap of a decade or more in the history of economic development. South Korea achieved the rapid economic development as a result of the successful implementation of its five-year economic development plans. The Korean Government promoted export-oriented light manufacturing in the initial stage of industrialization and heavy and chemical industries in the later stage, given the nation's weak agriculture base. To accommodate the influx of people from rural areas, major cities expanded their boundaries; furthermore, industrial complexes grew and evolved into new coastal cities, accelerating urbanization and resulting in environmental pollution (Hong, 1991).

RURAL FACTORS

China, with its growing population and shrinking arable lands, faces an extraordinary challenge in feeding its people. The amount of Chinese arable land per capita is only one-third of the global per capita average. While China has only 7% of the world's arable lands, these lands have been responsible for feeding 22% of the world's population. Agricultural production must increase by over 20–40% in order to meet the steadily increasing demand for grain (Brown, 1995; Huang and Rozelle, 1995; Chen et al., 1996). However, China has experienced a dramatic loss of arable lands during the post-reform economic boom period of 1978–1995. Net decreases during 1988–1995 amount to 2.2 million hectares (ha), about 1.6% of China's cultivated lands (Fischer et al., 1998). If this trend continues, then by the year 2030, Chinese arable lands will have decreased by another 10.3 million ha and per capita arable lands will have dropped to 0.053 ha (Li and Sun, 1997). In addition, deforestation and desertification have caused further degradation of croplands. In the coastal areas of eastern China, there has been an unprecedented conversion of arable lands for non-agricultural uses following rapid industrialization. Preferential human settlement in coastal regions has also accelerated the loss of croplands (Heilig, 1997; Rozelle et al., 1997).

In spite of tripled grain production from 113 million metric tonnes in 1949 to 454 million metric tonnes in 1993, China's capacity to feed its population is questionable. It is estimated that achieving the goal of delivering 400 kg of grain per capita will be very difficult as a result of the limited productivity gain (US Embassy Beijing, 1996a). Soil erosion is still a serious problem in China. Between 1985 and 1994, 360,000 ha of farmland annually suffered from topsoil losses. The total area of soil erosion in China is estimated at 1.79 million km^2, accounting for almost one fifth of China's total land area (NEPA, 1995). The erosion has reduced agricultural productivity by washing away vital minerals such as nitrogen, phosphorus and potassium, which then serve as natural fertilizers in the sea. The total amount of natural fertilizers washed away as a result of soil erosion is almost equivalent to China's total annual consumption of chemical fertilizers (US Embassy Beijing, 1996c).

Intensive use of the arable soils made the use of fertilizers imperative in replacing the depleted nutrients and in helping to increase the grain yields. The use of chemical fertilizers has been rising since the 1960s because of the limited natural reserves for the production of organic fertilizers. Use of chemical fertilizers is measured at more than 150 kg/ha. In addition, the majority of small scale farmers still do not understand the concept of appropriate timing of fertilizer application—or that overuse leads to serious environmental and marine pollution. Only 30–40% of fertilizers are used at appropriate times during the farming seasons, so that the fertilizers are only used half as efficiently as in advanced industrialized countries. Consequently, enormous waste occurs with increasing eutrophication in coastal waters.

It was estimated that 17,000 tonnes of organochlorine pesticides were also spread every year on land along Chinese coastlines during 1980–1987 (Zheng, 1997). About 85% of these pesticides applied in the field are transported into the sea. Pesticide residues and wastes also pose serious problems because of lack of information and proper education among farmers in proper application methods. Hand sprayers commonly used in China rely on an old technology employing poor quality raw materials. Low atomizing properties, severe leakage, and uneven spraying result in an application effectiveness of 10–15% lower than in developed countries. The effective utilization rate is only 30% compared with the world average of 50% (US Embassy Beijing, 1996b). In addition, China faces the dual challenges in overcoming insecticide-resistant pests such as bollworms in cotton fields as well as preventing food and water contamination by pesticide residues.

In North Korea, a widespread famine following a series of floods, droughts, and typhoons since 1995 is attributed to deforestation, soil erosion and natural disasters, which reflect other aspects of environmental deterioration in Northeast Asia. According to the United Nations, North Korea was able to produce only about half the food it needed to feed its population in 1997–1998, resulting in a food shortfall of 2–2.5 million tonnes (UN WFP, 1998). Despite the use of improved seed varieties, expansion of irrigation, and the heavy use of fertilizers, North Korea has yet to become self-sufficient in food production. Indeed, a shortage of arable lands, several years of poor harvests, and deforestation have worsened the chronic food shortages.

Two thirds of land territory in Korea is made up of mountainous areas. Forest regions cover 77.2% of South Korea, 51.2% of North Korea but only 14.3% of China (FAO, 1997). Here too, intensive farming with high inputs for maximum yields is indispensable to maintain the national food supply level. Therefore, use of chemical fertilizers and pesticides in South Korea has doubled since the 1970s. The average application rate of 615 kg/ha in South Korea is much greater than the average of 155 kg/ha in China. In order to raise the grain productivity as well as reduce pest and insect damages, use of pesticide in paddy fields increased from 3.8 kg/ha in 1978 to 6.8 kg/ha in 1990 (NSO, 1997). Rapid decreases in rural populations since the 1980s also caused the increase of herbicide use in South Korea.

COASTAL EROSION AND LANDFILL

Since the end of the 1950s, extensive and escalating coastal erosion has been taking place in China. Nearly all of the muddy coasts and 70% of the sandy coasts are subject to erosion (Xia et al., 1993). Coastal erosion has threatened villages, roads, factories, coast-protection forests and tourism resources. The speed of erosion of muddy coasts varies considerably from place to place, while that of the sandy coasts ranges between 1 and 2 m/yr. The main causes of coastal erosion in China are the quarrying of sands and the decrease of sediment discharges from the rivers into the sea.

The largest sources of modern sediments to the Yellow Sea have been from the Yellow and Yangtze rivers. Annually, these rivers discharge about a billion tonnes and half a billion tonnes of sediments respectively, and rank second and fourth in sediment discharge among the world's rivers (Milliman and Mead, 1983). However, most of the sediments from the Yangtze River are transported southward and accumulate in the East China Sea (Beardsley et al., 1985). Thus, the dominant fine-grained deposits in the Yellow Sea are from the Yellow River.

The mouth of the Yellow River has frequently alternated between the north and south side of the Shantung Peninsula (Wang and Aubrey, 1987). The shifting of the river mouth has been recorded eight times since 2278 BC. The Yellow River has made debouches into Po Hai Bay, and its sediment was transported to the Yellow Sea through Po Hai Bay. However, in 1126 the river channel shifted southward into the Yellow Sea and its sediment discharged directly into the Yellow Sea, greatly increasing the amount of sediment passing into the Yellow Sea. In 1855, the Yellow River channel shifted to the north and entered into the Po Hai Bay again, reducing the Yellow River sediment supply to the Yellow Sea.

Strong coastal erosion in the abandoned deltas of the Yellow River is significant. For example, a total of 4.7×10^9 tonnes of sediment has been eroded away in the three erosion centres of sub-deltas along the river channel (Yang et al., 1997). Sediment discharges of the Yellow River delta to the sea have been constantly decreasing in recent years as a result of the increasing water consumption and water conservation constructions. This contributes to the acceleration of coastal erosion in the river mouth.

Significant environmental issues in the west coast of the Korean Peninsula, on the other hand, result from land fill. In both South and North Korea, planned land fill of estuaries, shallow bays, and intertidal mudflats threaten enormous areas of highly productive coastal habitats. South Korea has initiated unprecedented large-scale land fill projects in its western coasts since 1962 to replace the crop lands lost to human encroachment as well as creating land for constructing industrial complexes and supporting coastal cities. The total area of mudflats decreased from 82,621 to 53,856 ha between 1945 and 1996. A loss of as much as 65% of total coastal wetlands is anticipated if the current development plans are implemented (Ahn et al., 1998). Massive coastal land fill and river modification are also underway in North Korea, with little consideration for the impact on the coastal ecosystem.

There are positive contributions to social and economic development in the land fill projects of South Korea. These massive projects have created land that can be used for constructing residential houses, industrial estates, recreational facilities, parks, expressways and the expansion of airport and seaport facilities. However, some negative impacts on the coastal environment were inevitable including the loss of natural coastlines and coastal habitats, adverse effects on fisheries and a decrease in the self-purification capacity of mudflats.

Large-scale land fill projects have removed mudflats, reducing the natural pollution-absorption capabilities while increasing the pollution loads. Dam construction in estuaries has produced several artificial lakes in the west coast of South Korea; however, most of the fresh and brackish lakes have suffered from severe eutrophication because of heavy nutrient inputs (Jung et al., 1997). For example, the Shiwha artificial lake—a major reclamation development project site in South Korea—is so heavily polluted that even treated waters from the lake cannot be used for irrigation of agricultural lands as a result of high concentrations of COD, e.g. 30.9 mg/l in 1997 (Kim, 1998a). Costly clean-up and restoration projects, such as large-scale dredging projects or construction of artificial wetlands, will be essential to maintain at least a moderate water quality in the Shiwha lake.

In Korea, various harbour and reclamation engineering projects have altered natural sediment transport patterns along the nearshore and caused coastal erosion and sedimentation. The nearshore land fill may alter the coastal water and current flow regimes and coastal residents living within the vicinity may frequently experience unusual intrusion of seawater, particularly at spring tides.

Decreases of coastal wetlands have become a matter of serious concern in relation to the loss of habitat for migratory birds. The wetlands of the Yellow Sea support over 150 species of water birds, including ducks, geese, and cranes. According to the IUCN's Red Data Book, 27 of these are listed as threatened or endangered, and some are nearly extinct (IUCN, 1987; Collar et al., 1993). The Han River estuary, for example, which is located 50 km west of Seoul, is one of Northeast Asia's most important stopping points for migratory birds (Yoo and Lee, 1998). As well as hunting and pollution, pressures on these birds come from the loss of wetland habitats for nesting, feeding, and migratory stopovers.

Estuary wetland ecosystems have been strongly affected by oil and agricultural exploitation in North China. The Liaohe Delta originally had 3660 km^2 of wetland, but about 50% has been converted to paddy fields and salt fields, leaving less than 2000 km^2 of natural and seminatural landscape. Within the Shuangtaizi Nature Reserve, the area for oil fields, roads, agricultural fields and residential areas increased from 4.1% to 8.5% during the period from 1977 to 1986. The natural wetland is still decreasing at a rate of 0.43% per year (Xiao et al., 1996).

URBAN AND INDUSTRIAL ACTIVITIES

Aquaculture and Coastal Industries

Aquaculture has been a major use of the coastal waters adjoining the Yellow Sea. Along the coastal provinces of China and Korea, there are countless small- and large-scale mariculture farms rearing fish, invertebrates and seaweeds. Major species of invertebrates cultured are oysters, mussels, razor clams, cockles, short-necked clams, pearl oysters, scallops, abalone, ark shells, pen shells and hard clams (Valencia, 1987).

Aquaculture production in the Yellow Sea has significantly increased in the last 20 years, with all the major natural fish stocks being heavily exploited, and the catch-per-unit area has decreased to 2.3 tonnes/km^2. Aquaculture has become a mainstay of many local coastal economies, with steadily increasing production and strong demand. However, aquaculture farms in the coastal area have been threatened by red tides, oil spills, chemical contaminants, and other disturbances from human activities.

Furthermore, marine fish farms using floating net-cages, culturing rock fish and flounder, suffer from direct, heavy losses of unconsumed animal feeds through the net-cages, which accelerate organic pollution and eutrophication. Since 1990, natural oyster spat fall was very low or failed completely in Korean coasts, which may be related to TBT contamination and the parasitic organism *Perkinsus* sp. (Kahng et al., 1996; Choi and Park, 1997; Shim et al., 1998). Because of these widespread spat fall failures, in recent years Korean oyster farms have solely depended on cultured spats, mainly imported from Japan and USA.

Seaweeds, such as *Sargassum*, *Porphyra* and *Undaria* sp., are also important crops in the Yellow Sea. The most important cultivated seaweed in China is the brown algae, *Laminaria japonica*, which is grown in more than 3000 ha of Chinese coastal water with a production rate of 10,000 dry tonnes/year (Valencia, 1987). The seaweed production in the west coast of South Korea was valued at US$ 165 million in 1996 (MOMAF, 1997). However, dam construction in the river mouth resulted in nitrogen deficient waters, and the latter, as well as sporadic fresh water discharges, have caused decreased production of *Porphyra* cultures in enclosed areas such as Cheonsoo Bay (Kim, 1998b).

Littoral wetlands of northern China are mainly distributed along the coastal belt on the northern region of the Yangtze River. These lands have a total area of 21,000 km^2, including the extensive beaches and estuary wetlands on the deltas of several large rivers, such as the Yellow River and Liaohe River. The reed field in the Liaohe Delta is the second largest in the world, with an area of 70,000 ha, which constitutes the main part of the Liaohe Estuary wetland. Average reed productivity is 5.7 tonnes/ha, and the primary production of reed communities is measured at 14,150 (avg. 8300~20,000 range) kg/ha/yr (Xiao et al., 1996).

Numerous prawn ponds along the coasts bring a large quantity of nutrients to the marine environment. It is estimated that 160×10^8 m^3 of wastewater from mariculture ponds is discharged into the sea each year. In the Liaohe Delta, there are 5000 ha of artificial shrimp ponds, with a production of 8000 tonnes/yr, which give enormous economic benefits to local residents (Xiao, 1994). Shrimp ponds, however, have become a serious pollution source, which contribute to occurrences of red tides in coastal waters. It is reported that 7470 tonnes/ha of COD are discharged from shrimp ponds (Xiao et al., 1996).

Sea-salt production in China is probably the largest in the world. In 1992, the area for sea-salt production was 352,000 ha and the output was 19.8 million tonnes. There are more than 30 salt chemical plants now in operation throughout the country which can also produce potassium chloride, bromide, mirabilite, magnesium chloride etc., with an annual output of over 0.3 million tonnes and value of 1.51 billion Yuan. In Korea, whatever remaining salt-production is still in operation has greatly declined since 1980, because land for even the small-scale salt ponds has been converted into other uses.

Wastewater and Solid Waste Discharges

Since 1978, there have been rapid societal and economic developments in the coastal regions of China as a result of the implementation of the reform policy which opened its doors to the outside world. The rapid economic growth was accomplished at great cost to the marine environment, especially with regard to the quality of coastal waters, with industrial wastewater and domestic sewage being directly discharged into the coastal waters. At present, of the 22.7 billion metric tonnes of industrial wastewater discharged annually in China, less than 30% is treated and only about 50% of the treated waters meet the national effluent standards (World Bank, 1997b). Chinese coastal waters are estimated to have approx. 66.5×10^8 tonnes of wastewater annually discharged into them. There are 91 major outfalls discharging sewage into the coastal seas of China. The amount of discharged sewage increased from 2.5 billion tonnes in 1990 to

3.2 billion tonnes in 1995, and it is estimated that 4.0 billion tonnes will be discharged into the sea in 2000 (UNEP, 1998). In addition, coastal agriculture will use 24×10^4 tonnes of pesticides and 268×10^4 tonnes of chemical fertilizer. The annual total quantity of domestic wastes dumped into the coastal environment without pretreatment is estimated to be 2×10^7 tonnes. A water quality assessment showed that 18.7% of coastal waters met Grade I standards, 21.4% met Grade II, 6.5% met Grade III, and 53.4% exceeded Grade III. Consequently, China's coastal areas are subject to great pressures from the improperly treated discharges of land-based pollutants carried by rivers, and the pressures from environmental impacts introduced by coastal economic development activities and coastal construction projects.

China is rapidly expanding its requirements for energy, with coal being a major source of fuel. Increased combustion of coal brings increased quantities of fly ash that require a disposal system. China is considering disposal of its fly ash in the Yellow Sea, alongside calcium carbonate residues from the production of fertilizer (Valencia, 1995). At the end of 1986, the State Council in China had approved three areas as possible waste disposal sites, two in the East China Sea and one in the Yellow Sea.

In 1995, total national wastewater production in Korea was about 8.0 billion tonnes including 3.2 million tonnes of industrial wastewater. In coastal areas, about 77% of wastewater was untreated in 1995 and was released directly or indirectly to the sea. Since 1988, South Korea has dumped human wastes as well as industrial wastes from food and beverage facilities at a designated site in the Yellow Sea. The total quantity of dumped materials involved so far is 6 million tonnes, which is relatively small compared with other regional seas (MOE, 1996), but is still an enormous quantity.

Nutrient loads in coastal waters have contributed to frequent outbreaks of red tides with increasing cases of paralytic shellfish poisoning. China has recently recognised the importance of ever-increasing red tides. There were 27 incidents of large-scale harmful algal blooms in the Yellow Sea and Po Hai Bay from 1970 to 1991. In 1989, harmful algal blooms of *Gymnodinium* sp. occurred in the vicinity of Huanghua District in the Western part of the Po Hai Bay, and caused enormous economic losses to shrimp cultures, approximately US$ 40 million (Tseng et al., 1993; Zhang and Zou, 1997). Total economic losses due to 256 red tides from 1972 to 1994 were estimated at US$ 0.62 billion.

Korean fishermen have also suffered increased frequencies of harmful algal blooms since the 1980s. There were 65 outbreaks of red tides in 1995, and economic losses were approximately US$ 9 million. The blooms of *Cochlodinium polykrikoides* in 1995 occurred along 150 km of the southern coasts of Korea, resulting in mass mortality of fish in aquaculture farms.

Oil and Oil Spills

Rapid industrial and economic development in South Korea and China mean that these countries are highly dependent on oil as their major source of energy. Oil exploitation, trans-shipment and off-loading at terminals often result in oil spills. China experienced 21 major oil spills (>50,000 l) from 1990–1997 (Xu and Cheng, 1998). The largest tanker spill occurred along the Guangdong coast in 1976, discharging 8000 tonnes. This incident polluted more than 1000 km of coastline and seriously damaged numerous mariculture farms (Cai, 1988). South Korea is one of the major oil refiners in the region, processing crude oil imported mainly from the Middle East. Demand for oil in South Korea during the next decade is expected to increase from 2.2 million barrels per day (bpd) to 2.7 million bpd in the 2005. Also, the oil demands of China, which recently became a net oil importer, were 2.7 million bpd in 1993 and are now increasing.

Offshore outputs, which are from relatively new fields, still account for only 3% of overall Chinese oil production. Annual offshore oil production will increase to 50 million tonnes from the present 5 million tonnes by the year 2000. Nearshore oil fields (Liaohe, Dagang and Shengli oil fields) mainly located in the Po Hai Bay discharge 12,730 tonnes of oil into Po Hai Bay.

The number of reported marine spills in South Korea increased from 268 in 1984 to 374 in 1996, although most of the spills were classified as minor. In South Korea, there have been 42 major spill accidents (>50,000 l) during the period 1990–1997 and the total quantity of oil spilled was 39,000 tonnes (KMPA, 1997). In July 1995, the super tanker *Seaprince* was grounded off Kwangyang Bay and spilled 4150 tonnes of crude oil, which severely damaged marine aquaculture farms. The economic losses caused by the incident were estimated at $900 million.

Large volumes of oil transported, combined with heavy shipping traffic and poor navigation conditions, make the Yellow Sea a high risk area. The estimated annual increase of shipping traffic through the Yellow Sea is 7.6%. With increasing demands for energy accompanied by high shipping traffic, the risk of oil spills in the Yellow Sea is likely to increase in the future.

PROTECTION AND CONSERVATION MEASURES

Factors Inhibiting Conservation

The Yellow Sea is semi-enclosed and therefore is particularly sensitive to the effects of land-based pollution including agricultural runoff, sewage, industrial emissions, and atmospheric deposition. High population densities, linked with rapid urban growth and industrialization in coastal areas, pose major threats to natural resources and sustainable uses of the Yellow Sea. Resulting ecological and economic damage includes commercial losses of fisheries and aquaculture, habitat modification, destruction of flora and fauna, and frequent occurrences of red tides.

There are many obstacles to successful implementation of land-based pollution prevention programmes in the Yellow Sea. The first and most serious obstacle to overcome is national development strategies that consider income growth as the main national priority. The final costs to each citizen will be higher in the development of environmentally blind economic growth than it would be to environmentally friendly development with higher initial costs. The "grow now, pay later" philosophy adopted by South Korea, which imposes high financial, social and ecological costs on the next generation, is still preventable in China and North Korea. Another impediment is the socio-economic conditions of Northeast Asia. In order to improve capacities for handling increasing wastes, there needs to be enhanced investment for expanding coastal infrastructures. In the case of China and North Korea, however, it may be difficult to increase environmental investment to prevent and minimize pollution and ecosystem degradation within the anticipated period.

Economic problems have been the major impediments to pollution-prevention projects and to the controlling of point sources. Since the early 1980s, China and Korea have enacted various environmental laws and regulations. Implementation of new rules, however, has proven to be a difficult task when immediate livelihoods are at stake. Governments face strong resistance in enforcing regulations. Company owners are cutting corners in environmentally friendly systems and are unwilling to run wastewater treatment facilities in order to save money. The ineffective and loosely enforced system of imposing severe penalties and fines cannot stop frequent illegal discharges of industrial wastewater. In addition to incomplete and inadequate surveillance, penalties are too low to stop violations in most cases. It is more profitable for companies to pollute the environment and pay the fines and penalties than carry out environmentally friendly operations.

Non-point-source pollutions are yet to be a major environmental issue in the states bordering the Yellow Sea. In order to address non-point-source pollution problems as well as to achieve national goals for better water quality, watershed management planning should be introduced; this involves an integrated and holistic approach which considers all possible impacts to water quality in watersheds and which can help successfully implement the regulatory tools (US EPA, 1991). Development of a GIS-based watershed information system including a database, models, and decision-supporting tools would also be very helpful in supporting and promoting better watershed management (Yoon, 1996; Leon et al., 1997).

Land-based pollution originates from diverse activities carried out within each nation's jurisdiction, making it especially difficult to control pollution sources from the other coastal states. Crucial environmental issues such as the uses of rivers and watersheds, the emission of industrial and household pollutants, and the management of farmlands, forests, and wetlands, remain largely within the regulatory regimes of an individual state. In order to address these problems properly and effectively, institutional frameworks should be developed by creating regional co-ordination structures to select the priorities which would provide the purposes and policies needed as catalysts for project implementation.

Current Conservation and Marine Protection Measures

In 1992, a co-operative process to establish a regional governance structure was initiated under the auspices of UNEP's Regional Seas Programme—the Northwest Pacific Action Plan (NOWPAP). The five participating nations of NOWPAP are China, Japan, The Russian Federation, South Korea and North Korea. Focused initially on marine pollution, the NOWPAP may eventually embrace many facets, including fisheries, tourism and coastal management (Verlaan, 1995). The NOWPAP action plan was adopted in 1994, and a regional overview report on pollution from land-based activities is under preparation. The NOWPAP members will work towards the development of a regional convention or agreement for the protection and management of the coastal and marine environment. It will also explore the formulation of appropriate protocols that could be adopted to formalize the commitment to mutual cooperation, assistance and collaboration in protecting regional marine environments.

International co-operation among the NOWPAP countries is currently under progress, and is working to develop a regional legislative framework and other agreements for a contingency plan of mutual support in case of oil spill emergencies. Development of a NOWPAP contingency plan will improve the oil spill response capability of the region as well as facilitate the rapid mobilization of equipment and personnel among involved countries (UNEP/IMO/NOWPAP, 1998). The NOWPAP countries are also preparing a comprehensive regional database and information management system for the affected region. This will serve as a repository of all relevant, available data, act as the authority for decision-making, and serve as a source of information and education.

The lack of complementary, formal, and long-term monitoring programmes in the Yellow Sea has resulted in severe shortages of reliable and consistent data. The lack of a formal arrangement limits or prevents analyses of the status and trends in marine resources and environmental qualities for most coastal and offshore waters of the Yellow Sea. There is an urgent need to co-ordinate and integrate monitoring and data-gathering systems on a regional basis, which would make the best use of the expertise and facilities available within the region on a consistent and collective basis.

The first step towards controlling marine pollution and sustaining resources in the Yellow Sea is a comprehensive and complete regional monitoring programme to determine its ecological status. A typical example is the Yellow Sea Large Marine Ecosystem (YSLME) programme being conducted by China and South Korea. The overall objective of the programme is to monitor and decrease sources of stresses on the YSLME (She, 1993; KORDI, 1994; Cyr, 1995).

The Northeast Asian Regional Global Ocean Observing System (NEAR GOOS) will also contribute in measuring relevant information on the evaluation of anthropogenic impacts on the marine coastal environment, including contaminant releases, sediment mobilization and physical alterations. In addition, natural environmental changes of long- and short-term variations will be monitored on delayed, near real-time and real-time bases. Another key issue requiring regional environmental governance is to develop and adopt a harmonious approach towards the integrated management of coastal and marine resources, in a manner which combines protection, restoration, conservation and sustainable uses of the Yellow Sea. The most urgent concerns are the joint management of the fisheries, conservation of biodiversity and prevention of habitat destruction.

The loss of biodiversity stems in large part from habitat destruction, as well as introduction of exotic species, hunting, and over-harvesting. Protection of migratory animals, especially birds and fish, requires regional commitments. Tidal wetlands are also important in supporting marine ecosystems and fisheries. The intertidal organisms living in mudflats may support up to 90% of the fish living in coastal waters.

In order to conserve and restore critical habitats, the establishment of an extensive network of nature reserves of many different types and status in the Yellow Sea is essential. Additionally, there is an urgent need to prevent the mudflat reclamation in the west coasts of Korea. Preserving mud-flats requires governments to change national policies which determine the fate of coastal wetlands. Without more integrated and sustainable approaches to coastal zone management, reclamation will further disturb the habitats of marine biodiversity and high productivity which, in the long run, will benefit human society.

REFERENCES

Ahn, T.M., Lee, Y.H., Lee, W.H., Park, S.Y. and Je, J.G. (1998) Some recommendations for establishment of legal systems for conservation and wise use of tidal flats. *Ocean Research* 20(2), 209–219.

Beardsley, R.C., Limeburner, R., Yu, H. and Cannon, G.A. (1985) Discharge of the Yangtze River into the East China Sea. *Continental Shelf Research* 4, 57–76.

Brown, L.R. (1995) *Who Will Feed China? Wake-up Call for a Small Planet*. W.W. Norton, New York.

Cai, G.Y. (1988) Some problems of setting up the oil-spill emergency system. *Haiyang Huanjing Kexiu* 6(2), 9–18.

Chen, X., Wang, Z. and Guo, J. (1996) China's grain supply and demand in the 21st century. *Bulletin of the Chinese Academy of Science* 10, 136–142 (in Chinese).

Choi, J.K., Park, Y.C., Kim, Y.C., Lee, S.K., Wang, H.J., Han, B.S. and Jung, C.S. (1988) The study on the biological productivity of the fishing ground in the western coastal area of Korea, Yellow Sea. *Bulletin of the National Fisheries Research & Development Agency* 42, 143–168.

Choi, K.S. and Park, K.I. (1997) Report on the occurrence of *Perkinsus* sp. in the manila clams, *Ruditapes philippinarum* in Korea. *Journal of Aquaculture* 10, 227–237.

Choi, J.W., Kim, D.S., Shin, S.H. and Je, J.G. (1998) Spatial distribution of macrobenthos in the sandflat of Taebudo, Kyonggi Bay, the west coast of Korea. *Ocean Research* 20(2), 97–104.

Collar, N.J., Gonzaga, L.P. and Wege, D.C. (1993) *IUCN Red Data Book*. 3rd ed., Smithsonian Institute Press.

Cyr, N. (1995) An LME monitoring and assessment program for the Yellow Sea The Large Marine Ecosystems of the Pacific Rim. A Report of a Symposium. Tang, Q. and Sherman, K. (eds.). Gland-Switzerland IUCN 1995. 117 p.

FAO (1997) State of the World's Forest 1997. 200 pp.

Fischer, G., Chen, Y. and Sun, L. (1998) The balance of cultivated land in China during 1988–1995, IIASA Report IR-98-047.

Fischer, G. and Heilig, G.K. (1997) Population momentum and the demand on land and water resources. *Philosophical Transactions of the Royal Society London* 352, 869–889.

Heilig, G.K. (1997) Anthropogenic factors in land use change in China. *Population and Development Review* 23(1), 139–168.

Hong, G.H., Kim, S.H. and Chung, C.S. (1997) Contamination in the Yellow Sea proper: A review. *Ocean Research* 19(1), 55–62.

Hong, S.Y. (1991) An assessment of coastal zone issues in the Republic of Korea. *Coastal Management* 19(4), 391–415.

Huang, J. and Rozelle, S. (1995) Environmental stress and grain yields in China. *American Journal of Agricultural Economics* 77(4), 853–864.

IUCN (1987) Directory of wetlands of international importance: sites designated under the Convention on Wetlands of International Importance especially as Waterfowl Habitat.

Jung, H.S., Choi, K.W., Kim, D. and Kim, C.S. (1997) Status of environmental conditions of Shihwa lake in the west coast of Korea and strategies for its remediation. *Ocean Policy* 12, 123–163.

Kahng, S.H., Oh, J.R., Shim, W.J. and Shim, J.H. (1996) Imposex as an evidence of TBT pollution in the Chinhae Bay. *J. Malaco. Soc. Korea* 12(2), 123–131.

Kang, Y.S. (199). Primary productivity and assimilation number in the Kyonggi Bay and in the eastern coast of Yellow Sea. M.S. Thesis. Inha Univ.

Kim, E.S. (1998a) A study on environmental changes in Shihwa lake. BSPE 97610-00-1035-4. 230 pp.

Kim, J.K. (1998b) Environmental monitoring in Cheonsoo Bay area. BSPG 96256-1063-3. 900 pp.

Kim, O.K. and Oever, P. (1992) Demographic transition and patterns of natural resources use in the Republic of Korea. *Ambio* 21(1), 56–62.

Kim, S., Jung, S. and Zhang, C.I. (1997) The effect of seasonal anomalies of seawater temperature and salinity on the fluctuation in yields of small yellow croaker, *Pseudosciaena polyactis*, in the Yellow Sea. *Fisheries Oceanography* 6, 1–9.

Kim, D.S., Choi, J.W., Je, J.G. and Lee, J.H. (1998) Community structure of meiobenthos in tidal flats at Taebudo, west coast of Korea. *Ocean Research* 20(2), 81–87.

KMPA (1997) Marine pollution statistics. Korea Maritime Police Agency.

KORDI (1994) Yellow Sea Large Marine Ecosystem: a project development. BSPE 00406-725-3. 340 pp.

Lee, S.W., Lee, H.G., Shin, S.H., Kim, D.S, Lee, J.W. and Je, J.G. (1998) Marine biota of the macro-tidal flat of Namsari, Taebudo in the west coast of Korea. *Ocean Research* 20, 105–119.

Leon, L.F., Lam, D.C.L., Swayne, D., Farquhar, G. and Soulis, R. (1997) Integration of a nonpoint source pollution model with a decision support system. In: Proc. Second International Symposium on Environmental Software Systems, Whistler, BC Canada.

Li, X. and Sun, L. (1997) Driving forces of arable land conversion in China. IR-97-076/Sept, International Institute for Applied Systems Analysis, Laxenburg, Austria.

Lim, H.S. and Je, J.G. (1998) Macrobenthic communities on the macro-tidal flats of Taebudo and Tando in the west coast of Korea. *Ocean Research* 20(2), 121–130.

Milliman, J.D. and Mead, R.H. (1983) World-wide delivery of river sediment to the oceans. *Journal of Geology* **91**, 1–21.

MOE (1996) Environmental statistics yearbook. Ministry of Environment. Republic of Korea. 648 pp.

MOMAF (1997) Korean fisheries yearbook. Ministry of Maritime and Fisheries Affairs. Republic of Korea.

NAPSNet (1996) Why the world should help North Korea, Northeast Asia Peace and Security Network Daily Report. February 6, 1996, Berkeley, California, USA.

NEPA (1995) Report on China's Environmental Condition.

NSO (1996) Population and housing census report, National Statistical Office, Republic of Korea. CD-ROM.

NSO (1997) Korea statistical yearbook, National Statistical Office, Republic of Korea. CD-ROM.

Rozelle, S., Veeck, G. and Huang, J. (1997) The impact of environmental degradation on grain production in China: 1975–1990. *Economic Geography* **73**(1), 44–66.

She, J. (1993) Monitoring and researches on pollution/environment and their ecological effects in YSLME(Yellow Sea Large Marine Ecosystem). *J. Oceano. of Huanghai & Bohai Sea* **11**(3), 61–69.

Sheng, H., Zhang, C., Xiao, C. and Zhu, J. (1997) Change of the discharge and sediment flux to estuary in Changjiang River. In *Proceedings of the International Symposium on the Health of the Yellow Sea*, 30 Sept. 1997, Seoul, Korea. pp. 79–89.

Shim, W.J., Oh, J.R., Kahng, S.H. and Shim, J.H. (1998) Accumulation of tributyl- and triphenyltin compounds in Pacific oyster, *Crassostrea gigas*, from the Chinhae Bay system, Korea. *Archives of Environmental Contamination and Toxicology* **35**, 41–47.

SOA (1986) Marine pollution prediction in the year 2000. State Oceanic Administration.

Tseng, C.K., Zhou, M. and Zou, J. (1993) Toxic phytoplankton studies in China. In: *Toxic Phytoplankton Blooms in the Sea*, eds. T.J. Smayda and Y. Shimizu. Elsevier, Amsterdam, pp. 347–352.

UN (1995) World Population Prospects 1994. United Nations, New York.

UN (1997) World Population Prospects, 1950–2050. UN Population Division, New York.

UNEP (1998) Land-based sources and activities affecting the marine environment in the East Asian Seas. UNEP Regional Coordinating Unit. Draft 17 June 98. p. 106.

UNEP/IMO/NOWPAP (1998) Northwest Pacific action plan forum on marine pollution preparedness and response. In *The Second Meeting of NOWPAP Forum on Marine Pollution Preparedness and Response*, April 15–17, 1998, Taejon, Korea. pp. 49–50.

UN WFP (1998) 1998 Estimated Food Needs and Shortfalls for WFP-assisted: Emergency Operations, Protracted Relief and Recovery Operations, Development Projects, Special Operations (updated September 1998).

US Embassy Beijing (1996a) Can China feed itself in the 21st century: land use patterns may provide some answers.

US Embassy Beijing (1996b) Agricultural pesticide use in China.

US Embassy Beijing (1996c) Chemical fertilizers—blessing for Chinese agriculture or curse for China's environment.

US EPA (1991) The Watershed Protection Approach: an Overview. EPA 5039-92-001. Office of Water. Washington, DC.

Valencia, M.J. (1987) International Conference on the Yellow Sea: Transnational Ocean Resource Management Issues and Options for Cooperation. Occasional Papers of the East-West Environment and Policy Institute. Paper No. 3. 166 p.

Valencia, M.J. (1995) Northeast Asian Marine Environmental Quality and Living Resources: Transnational Issues for Sustainable Development (draft).

Verlaan, P. (1995) Current legal developments. UNEP: now we are twelve—the Northwest Pacific joins UN Environment Programme's Regional Seas Programme. *International Journal of Marine Coast. Law* **10**(3), 426–430

Wang, Y. and Aubrey, D.G. (1987) The characteristics of the China coastline. *Continental Shelf Research* **7**, 329–349.

Wang, Y. and Zhu, D. (1994) Tidal flats in China. In *Oceanology of China Seas*, eds. D. Zhou et al. Vol. 2, pp. 445–458.

World Bank (1997a) China 2020: development challenges in the new century. 162 pp.

World Bank (1997b) China 2020: Clear water, blue skies, 114 pp.

Xia, D., Wang, W., Wu, G., Cui, J. and Li, F. (1993) Coastal erosion in China. *Acta. Geogr. Sin. Dili-Xuebao* **48**(5), 468–476.

Xiao, D. (1994) Natural resource and regional exploitation in Liaohe Delta. *J. Nat. Res.* **9**, 43–50.

Xiao, D., Li, X., Hu, Y. and Wang, X. (1996) Protection of the littoral wetland in northern China: ecological and environmental characteristics. *Ambio* **25**(1), 2–5.

Xu, C. and Cheng, Z. (1998) National report on the identification of sources of existing data on fate and effects of oil spills within the NOWPAP member states. In *Second Meeting of NOWPAP Forum on Marine Pollution Preparedness and Response*, April 15–17, 1998, Taejon, Korea.

Yang, Z., Sun, X. and Chen, Z. (1997) Sediment discharge of the Yellow River to the seas: its past, present, future and human impact on it. In *Proceedings of the International Symposium on the Health of the Yellow Sea*. Sept. 30. 1997. Seoul, Korea. pp. 67–78.

Yoo, J.C. and Lee, K.S. (1998) Current status of birds on the west coast of Korea and a recommendation for conservation. *Ocean Research* **20** (2), 131–143.

Yoon, J. (1996) Watershed-scale nonpoint source pollution modelling and decision support system based on a model-GIS-RDBMS linkage. In *AWRA Symposium on GIS and Water Resources*, Sept. 22–26. 1996. Ft. Lauderdale, FL.

Zhang, C. and Zou, J. (1997) Recent progress of study on harmful algal blooms in China: an overview. In *Sources, Transport and Environmental Impact of Contaminants in the Coastal and Estuarine Areas of China*, ed. J. Zhou. SCOPE China Publication Series 4. China Ocean Press, Beijing. pp. 105–110.

Zheng, J. (1997) Organic pesticides. In *Sources, Transport and Environmental Impact of Contaminants in the Coastal and Estuarine Areas of China*, ed. J. Zhou. SCOPE China Publication Series 4. China Ocean Press, Beijing. pp. 91–96.

Zhijie, F. and Côte, R.P. (1990) Coastal zone of PR China: management approaches and institutions. *Marine Policy* **14**, 305–314

Zhu, M., Mao, X., Lu, R. and Sun, M. (1993) Chlorophyll a and primary productivity in the Yellow Sea. *J. Oceanogr. Huanghai, Bohai Seas Huangbohai Haiyang* **11** (3), 38–51 (in Chinese).

Zhang, C.I. and Kim, S. (1999) Living marine resources of the Yellow Sea ecosystem in Korean waters: Status and perspectives. In *Large Marine Ecosystems of the Pacific Rim: Assessment, Sustainability, and Management*, eds. K. Sherman and Q. Tang. Blackwell Science, Inc. pp. 163–178.

THE AUTHORS

Suam Kim
Korea Ocean Research and Development Institute, Ansan P.O. Box 29, Seoul, 425-600, Korea
Current address: Dept. of Marine Biology, Pukyong National University, 599-1 Daeyeon 3-Dong, Nam-Gu, Pusan, 608-737, Korea

Sung-Hyun Kahng
Korea Ocean Research and Development Institute, Ansan P.O. Box 29, Seoul, 425-600, Korea

Chapter 85

TAIWAN STRAIT

Woei-Lih Jeng, Chang-Feng Dai and Kuang-Lung Fan

Taiwan Strait is a passage between the East China Sea to the north and the South China sea to the south. The strait is a shallow shelf with an average depth of about 40 m, but the southern quarter is steep. The Penghu Islands are located in the middle of the strait, the Taiwan Shoal in the southwest, and Kaoping (submarine) Canyon in the southeast. Seventeen rivers in Taiwan flow into the strait. Steady currents in the offshore area and tidal currents along the western coast of Taiwan are two major physical features. In winter, the NE monsoon drives the water in the north strait southward to encounter the northward current from a branch of the Kuroshio, and the merging current flows to the South China Sea. In summer, when the SW monsoon prevails, South China Sea water flows northward and passes all the way through the strait. Along the western coast of Taiwan, the tidal currents flow northward in the southern section and southward in the northern section during flood tides. The current direction reverses during ebb tides. The semi-diurnal tides dominate the currents along the coast.

Five types of habitat, including estuaries, sandy beaches, mud flats, lagoons, and coral reefs, exist in the coastal areas of western Taiwan. Estuaries and lagoons are important nursery grounds for many commercial species of fishes, shrimps, and prawns. The wetlands have been exploited for aquaculture; oysters, hard clams and purple clams are the main cultured organisms. Coral reefs in the Penghu Islands and Hsiaoliuchiu are dominated by reef-building corals. The reef areas in the strait are the major fishing grounds of anchovy larval fishery, demersal fish and shrimps. The offshore areas are important fishing grounds for coastal fisheries; major target fishes are mackerel, jacks, grey mullet and clupeoid larvae. The grey mullet fishery is an important seasonal fishery in the strait, generally in December every year. On the other hand, clupeoid larvae are the most abundant fishes in the coastal waters; the peak fishing season of larval anchovies is in March and April. Two popular methods for artisanal fishing are beach seine and spear fishing.

Major wave directions in the strait induce longshore currents to flow southward in the north strait and northward in the south strait. Generally, Taiwan's coasts have been eroded in the north and south sections of the western coast, while in the middle section, sand is deposited reducing water depth, or even forming new land. In recent years, the impact of groundwater extraction for mariculture has caused serious land subsidence along the littoral zone. There are seven districts on the western coast which suffer from serious land subsidence problems.

Most of the population on Taiwan resides on the western side bordering the strait; the western coast with dense population is therefore seriously affected by human activities. Over the last two to three decades, rapid industrialization and population concentration in cities have led to runoff of domestic, industrial and agricultural wastewater, thermal effluents from power plants, pesticides and garbage via rivers to the sea. This has seriously polluted the coastal environment of

Seas at The Millennium: An Environmental Evaluation (Edited by C. Sheppard)
© 2000 Elsevier Science Ltd. All rights reserved

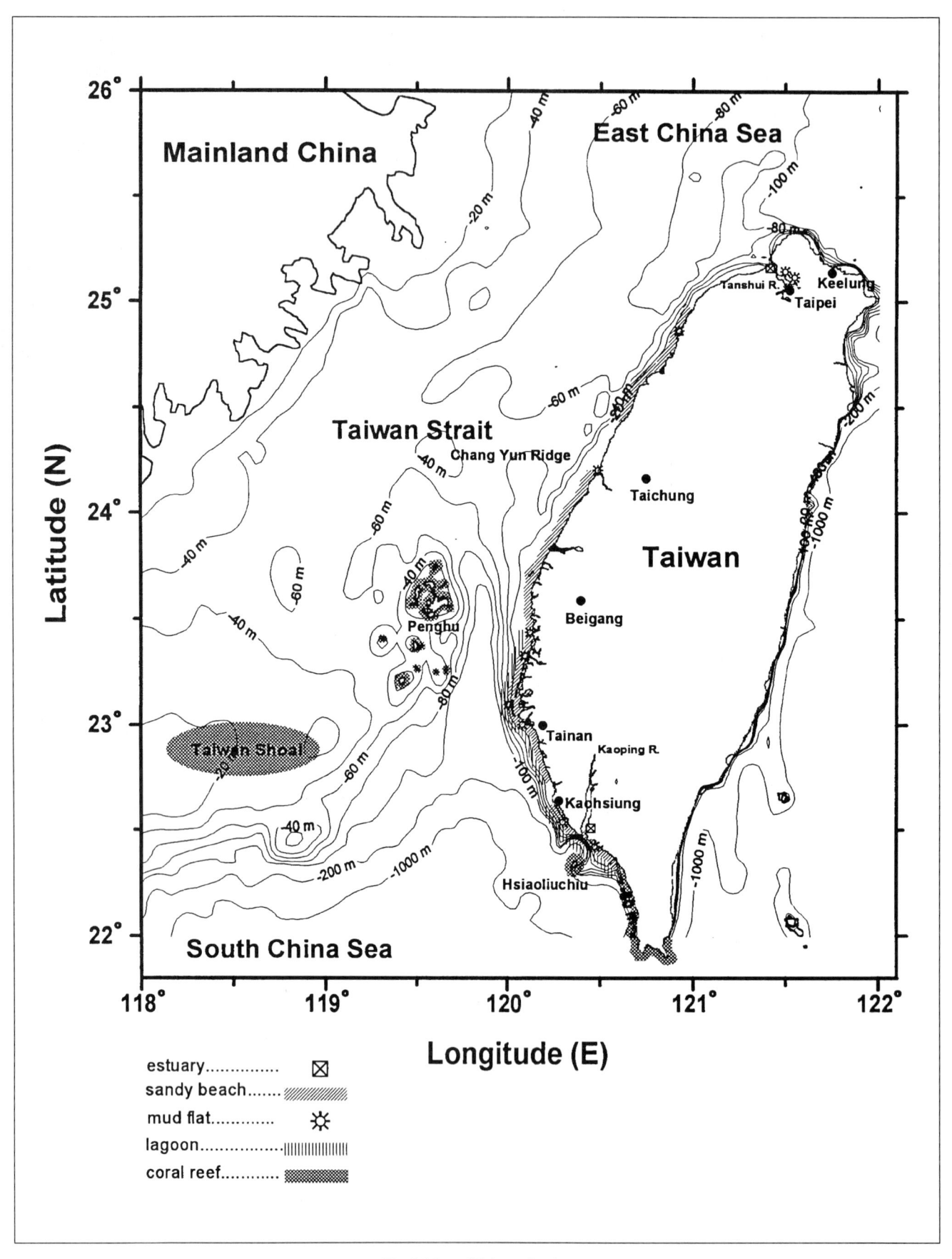

Fig. 1. Map of Taiwan Strait.

Taiwan, especially the western coast. In addition, increasing fishing intensity and improved fishing methods have over-exploited the coastal fishery resources, leading to resource degradation.

The coastal zone is severely abused by both the private and public sectors, causing serious damage to existing land and water resources, environmental pollution, adverse geographical changes in coastal areas, etc. Currently a wide spectrum of measures, including preventive environmental impact assessment and environmental monitoring programs, have been employed. Coastal resources are protected under a special plan, and seven coastal conservation zones (CCZ) have been established. In addition, three societies are dedicated to promoting public awareness, environmental education, research and conservation activities in Taiwan.

THE DEFINED REGION

Taiwan Strait is an area between Taiwan Island and mainland China (Fig. 1). It is a passage between the East China Sea to the north and the South China Sea to the south. Tectonically, the strait is situated on the margin of the Eurasia plate which has obliquely collided with the Philippine Sea plate since Plio-Pleistocene. The northern three quarters of the strait is a shallow shelf, while the southern quarter is a steep slope (Chen, 1991). The shelf can be divided into western, eastern, Penghu-Beigang Uplift, Taiwan Shoal and submarine canyons—located off southwestern Taiwan (Yu and Wen, 1992). The average depth of the continental shelf of the strait is about 40 m. The topography of the western part of the strait is usually gentler than that of the eastern part where it is characterized by small, shallow basins and low broad ridges with about 20 m relief. These topographic features are indicated by the contour lines of the 60 and 80 m depths. The Penghu-Beigang Uplift which extends from the Penghu Islands to Beigang of Taiwan is unconformably overlain by Miocene and Plio-Pleistocene deposits (Chen, 1991). The Taiwan Shoal is a shallow bank located southwest of the Penghu Islands. The shoal is about 10 to 20 m in depth with an area of about 6,500 km². The Penghu Islands are a cluster of 64 islands located about 50 km west of Taiwan in the strait. Kaoping (submarine) Canyon is well aligned with the Kaoping River on land, runs across the continental shelf, continues its course southwestward onto the continental slope, and terminates at a depth of about 3000 m (Ma, 1963; Liu et al., 1993). It is a V-shaped valley with a steeper slope on the eastern bank. There are several alternate basins and uplifts or platforms in the strait, including Taihsi-Taichung Basin, Penghu Basin, Tainan Basin, and Penghu Platform (Sun, 1982).

The coastal area along the eastern part of the strait has been developed in the past for human activities including agriculture, aquaculture, and harbour construction. The reclamation of the coastal area for saltpans as well as for rice and fish farming was an extensive government and private practice in the seventeenth century after the mass immigration from mainland China (Hwang, 1985). During the past three hundred years, the low land along the western coastal area has been reclaimed and developed into arable land for agriculture and aquaculture. Rapid development of aquaculture and intensive use of underground water along the west coast of Taiwan has resulted in land subsidence and environmental pollution in many areas.

SEASONALITY, CURRENTS, NATURAL ENVIRONMENTAL VARIABLES

Steady currents in the offshore areas and tidal currents along the coast are the two major physical features dominating the marine environment in Taiwan Strait.

Offshore Currents

The surface current patterns in the strait have been studied by Chu (1963) and Fan and Yu (1981). During winter time, the NE monsoon drives the water in the north strait southward to encounter the northward current near the Penghu Islands; then the merging current flows to the South China Sea (Fig. 2). The northward current actually comes from a branch of the Kuroshio (Wang and Chern, 1988; Shaw, 1989). This current encounter may result in more suspended matter deposition which is supported by a geochemical observation that more land-derived organic matter has been deposited in the surface sediments just off the northern Penghu Islands (Jeng, 1979). In summer, when the SW monsoon prevails, South China Sea water flows northward and passes all the way through the strait.

Generally speaking, the currents in the southern strait flow northward at least from the surface to 100 or 200 m depth for the whole year (Fan, 1984). Chuang (1986) measured the currents in the strait and found a steady northward current with a speed of about 27 cm/sec. Even in the southern strait, the NE monsoon also affects the surface current during winter time (Wang et al., 1993).

The topography of the strait is characterized by the presence of the Chang Yun Ridge (Fig. 1), almost normal to the west coast of Taiwan, separating the strait into its north and south parts. Chern and Chen (1994) used mathematical models to study the diffusion of the pollutants discharged from the southwestern coast and carried by the current. Model results show that the advective transport of ocean currents is the most effective way to disperse the pollutants in the downstream direction. As a result, the northward flow turns toward northwest in front of this ridge, then turns toward east immediately after passing the western end of the ridge and transports to the area off the northern coast. This is a typical flow pattern in the strait most of the year. So, the discharged pollutants from the southwestern coast can remain near the discharged site for a long period, and then some of the pollutants might also be transported to the northern coast. Unfortunately, many pollutant sources are located along the southwestern coast of Taiwan.

Tidal Currents

Along the western coast of Taiwan, the tidal currents flow northward in the southern section and southward in the northern section during flood tides. The current direction reverses during ebb tides (Fan, 1991) as shown in Fig. 3. The semi-diurnal tides dominate the currents along the western coast of Taiwan (Fan, 1995). The current velocity ranges from 10 to 30 cm/sec with the current directions mainly parallel to the coastline. Figure 4 shows the tidal ranges along the western coast of Taiwan: the highest range, more than 3 m on average, is in the middle strait, and the lowest range, about 1 m on average, is at both ends. The quantity of sea water flowing in and out of the rivers is small, and the discharge from the rivers is small except during the summer flood period. Most of the pollutants in the river flow to the ocean only during ebb tides.

Tidal currents are the main force for dispersing the pollutants discharged from river mouths or other pollutant sources. The polluted river water mainly flows into the sea during ebb tides. Since semi-diurnal tides dominate the currents in this area, the tidal currents change direction every six hours or so. With a tidal current velocity of 20 cm/sec, during the six-hour ebb tide period, the polluted water may travel a distance of about 4 km along the coast in the current direction with ebb tides. During the next six-hour flood tide period, the polluted water in the ocean, together with some newly discharged river or harbour water if any, may travel a distance of about 4 km along the coast in the reverse direction. So, the polluted water, in general, extends to an area of about 4 km upcoast and 4 km downcoast from the discharge source, and within about 3–4 km from the coast. In some estuaries, the water discharged from the river is seriously polluted, and it is easy to distinguish from the clean sea water. Sometimes, the boundary between harbour water and sea water can be recognized easily even five or six hours after the harbour water has been discharged to the ocean. In Taiwan, most rivers on the western coast are seriously polluted, especially on the southwestern coast, and there are many pollutant sources along the western coast. Sometimes, in one area, sea water contains many different pollutants discharged from a variety of sources.

The weather type in Taiwan also makes the situation worse, On average, 76% of the rainfall in Taiwan is in the period from May to October. The other six-month period, from November to April, is considered the dry season. The

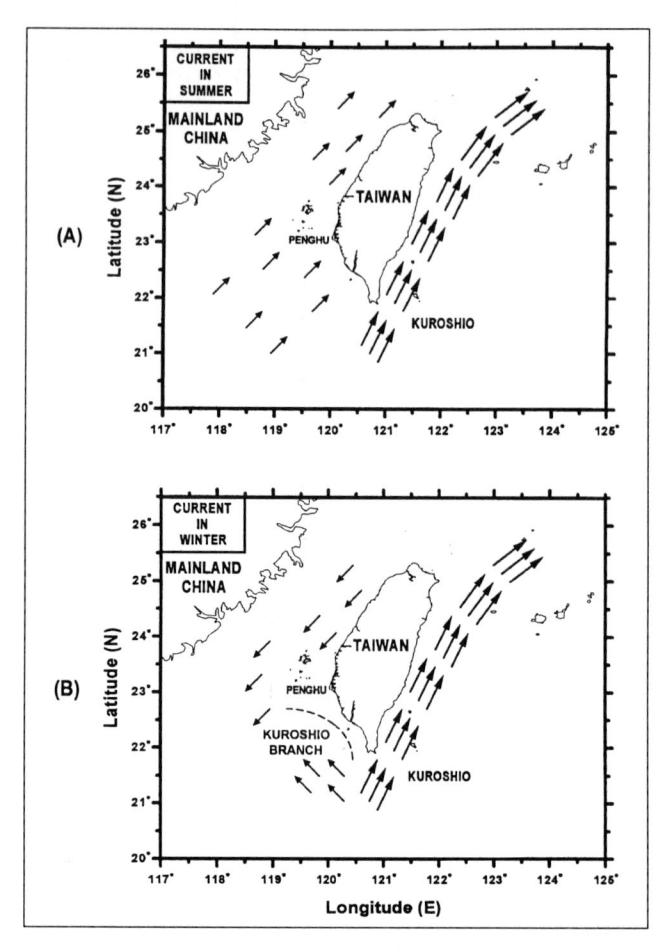

Fig. 2. Surface current patterns in the vicinity of Taiwan in summer (A) and winter (B).

contrast of winter and summer is even more evident in southern Taiwan, with about 90% of the rainfall in the wet season (Fan, 1987). Industrial effluents discharged from riverside factories into the almost dry river bed are accumulated during winter. When the spring rainy season begins, the pollutants are released to the coastal areas endangering the cultured oysters, clams and fishes.

THE MAJOR SHALLOW WATER MARINE AND COASTAL HABITAT

Five types of habitat, including estuaries, sandy beaches, mud flats, lagoons, and coral reefs exist in the coastal areas in western Taiwan. All of these habitats, except coral reefs, are different types of wetland. The total area of wetlands in Taiwan is about 11,896 hectares (Hsueh, 1996). Estuaries occur erratically on the western coast of Taiwan where the majority of large rivers drain into the strait. They are continuously supplied with freshwater and excess terrestrial nutrients which support high primary productivity. Sandy beaches, mud flats and lagoons are three major types of habitat occur on the coastal plain of western Taiwan. Estuaries and lagoons are important nursery grounds for many commercial species of fishes, shrimps and prawns. These soft bottom habitats are often occupied by mangroves, salt

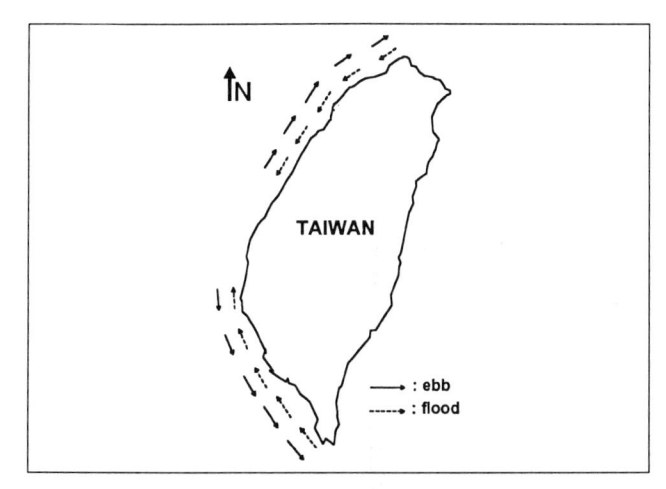

Fig. 3. General tidal current patterns along the western coast of Taiwan.

marshes or sea grasses. Coral reefs are a habitat with high biodiversity and productivity; they occur mainly in the shallow water areas of offshore islands including the Penghu Islands, Hsiaoliuchiu, and Taiwan Shoal (Fig. 1).

Mangroves were once widely distributed along the west coast of Taiwan from the Tanshui River estuary in the north to Dapong Bay (just to the south of the Kaoping River mouth) in the south during the early history of human colonization. They occupied the estuaries, mud flats and lagoons. Due to land reclamation and deforestation for agriculture and aquaculture development, most of the mangroves have been destroyed and only seven localities with a total area of 287 hectares remain less disturbed (Hsueh, 1996). In addition, two of the six major mangrove species in Taiwan, namely *Ceriops tagal* and *Bruguiera conjugata*, have become almost extinct. The dominant species in the mangroves in northern Taiwan is *Kandelia candel* and that in southern Taiwan is *Avicennia marina* (Chou et al., 1987). Mangroves in the estuaries are continuously supplied with freshwater and terrestrial nutrients which support high primary productivity and biodiversity. They produce large amounts of biomass which contribute a detritus food web to the ecosystem and support many organisms. Forty-six species of crabs, 42 species of molluscs, 79 species of fish (Lin et al., 1995) and 243 species of plants have been recorded from the mangroves in Taiwan (Hsueh, 1996). Fishes in the early stages of life often utilize this habitat for feeding and avoidance of predators. Salt marshes and sea grasses occur sporadically in central and southern coastal areas. These habitats are also characterized by high productivity and abundant biomass. The high productivity and abundant biomass of wetlands provide rich food and good habitats for migratory birds.

Traditionally the coastal wetlands have been intensively utilized and have resulted in various exploitative activities in coastal areas. Among them, aquaculture and saltpans are the most common activities. The exploitation of wetland for aquaculture has expanded rapidly since 1971. As a consequence, most of the wetlands on the west coast of Taiwan have been exploited. Oysters, hard clams and purple clams are the main cultured organisms. The oyster culture area is distributed along the western coast between Hsiangshan in Hsinchu County and Tungkang in Pingtung County. The main culture areas are in Changhua, Yunlin, and Chiayih Counties (Yang and Ting, 1988). The remaining unexploited wetlands have been allocated for industrial and recreation uses or harbour construction.

Coral reefs in the Penghu Islands and Hsiaoliuchiu are dominated by reef-building corals. About 1000 species of reef fish and 200 species of scleractinian corals have been recorded from these areas (Dai, 1997). Coral reefs are important for both fisheries and tourism. Fish and invertebrates collected from waters adjacent to the reefs comprise a considerable portion of the total catch of nearshore and coastal fisheries. The reef areas in the strait are one of the major fishing grounds of anchovy larval fishery, demersal fish and shrimps. Due to the practices of illegal fishing methods using dynamite and cyanide in the past two decades, reef habitats in the strait have been severely damaged. In addition, reef areas are major attractions for tourists and are popular for recreational fishing. Various marine recreational activities, such as Scuba diving, snorkeling, glass bottom boat trips and recreational fishing, depend on coral reefs. Many local people rely on coral reefs for their subsistence.

OFFSHORE SYSTEMS

The offshore systems of Taiwan Strait are important fishing grounds for coastal fisheries. Major target fishes caught in this area are mackerel, jacks, grey mullet and clupeoid larvae (Hsieh, 1993). Mackerel and jacks are migratory fish; they migrate to the strait for over-wintering and spawning from late winter to early spring. They are mainly caught by purse seine.

The grey mullet (*Mugil cephalus*) fishery is an important seasonal fishery in the strait. The main fishing period of

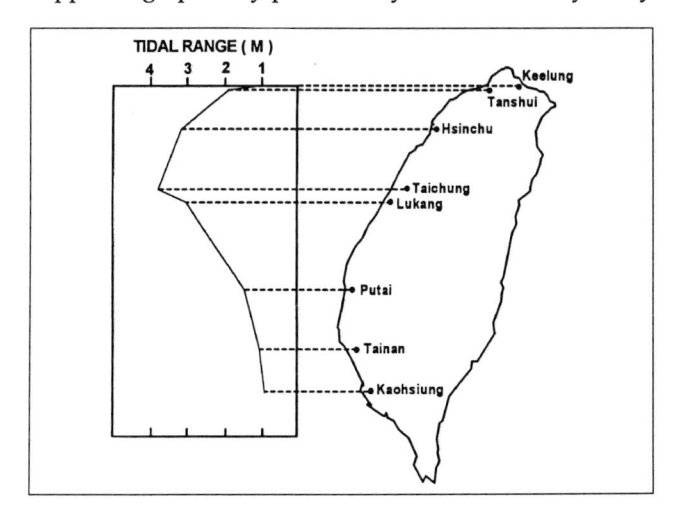

Fig. 4. Average tidal ranges along the western coast of Taiwan.

> ## Coral Reefs in Taiwan
>
> Corals are found in all the waters around Taiwan except in the sandy area on the west coast. The main reef area is around the southern tip of the island, where well developed fringing reefs are found (Dai, 1988, 1991). These reefs are characterized by diverse and abundant scleractinians and alcyonaceans. Scleractinians dominate most of the areas except on two headlands (Maobitou and Oluanpi) where the current is relatively strong and alcyonaceans dominate (Dai, 1991, 1993). The deepest extent of coral reefs in southern Taiwan is usually between 20 and 30 m, and none are found below 40 m where the majority of substrates are covered by sand. The depth limit of reef-building corals in southern Taiwan is between 70 and 80 m.
>
> The north, northeast and east rocky coasts have flourishing or patchy scleractinian coral communities, but reef development is generally absent (Randall and Cheng, 1977, 1979; Yang and Dai, 1982).
>
> Coral reefs and coral communities are also well developed around offshore islands including Lutao (Green Island), Lanyu (Orchid Island), Hsiaoliuchiu and Penghu (Pescadores). Lutao and Lanyu, located off southeast Taiwan, are in the pathway of the Kuroshio current. These reefs are densely covered by abundant scleractinians and alcyonaceans; alcyonaceans (mainly Xeniidae) dominate the west and southwest coasts while scleractinians dominate the north and east coasts (Dai, 1997). Hsiaoliuchiu, to the southwest, is an emergent reef island. The coral fauna of this island is dominated by stony corals (Yang et al., 1975). The Penghu Islands are a cluster of 64 islands located about 50 km west of Taiwan in Taiwan Strait, dominated by stony corals especially in the shallow subtidal areas (Dai, 1997).
>
> Species diversity of reef organisms on the coral reefs around Taiwan is relatively high. Approximately 300 species of scleractinians, 50 species of alcyonaceans, 30 species of gorgonians, 130 species of decapod crustaceans, 90 species of echinoderms, 1,200 species of reef fishes and 150 species of algae have been recorded from coral reefs in southern Taiwan, Lutao and Lanyu (reviewed in Dai, 1997). About 200 species of scleractinian corals and 1000 species of fishes have been reported from Hsiaoliuchiu and the Penghu Islands (Shao et al., 1994). Approximately 100 species of scleractinian corals and 800 species of fishes have been reported from the coastal areas on the north and northeast coasts of Taiwan.
>
> Coral reefs in Taiwan are important for both fisheries and tourism. Fish and invertebrates collected from reef waters provide a considerable portion of the total catch of near shore fisheries. Coral reefs are major attractions for tourists and are popular for recreational fishing, and most coral reefs are within National Parks or National Scenic Areas. These include the Kenting National Park in the south, the Northeast Coast National Scenic Area, the East Coast National Scenic Area, and Penghu National Scenic Area. Various marine recreation activities in these areas include scuba diving, snorkeling, glass bottom boat trips and recreational fishing.
>
> Coral reefs in Taiwan have suffered increasing threats during the past two decades due to rapid economic and human population growth. Heavy sedimentation and wastewater pollution associated with rapid coastal development are increasing in all regions. Other anthropogenic impacts including tourist abuse, dynamite fishing, aquarium fish collection and coral collection are also threatening many reef systems. In addition, typhoon disturbances in summer and autumn often cause high coral mortality over large areas. In general, living coral cover in many areas has been reduced from approximately 50–30% (Dai, 1997). Some species of reef fishes, gastropods and crustaceans have become locally extinct.
>
> ### REFERENCES
>
> Dai, C.F. (1988) Coral communities of southern Taiwan. Proceedings of the 6th International Coral Reef Symposium, Australia, 2, 647–652.
>
> Dai, C.F. (1991) Reef environment and coral fauna of southern Taiwan. *Atoll Res. Bull.*, **354**, 1–28.
>
> Dai, C.F. (1993) Patterns of coral distribution and benthic space partitioning on the fringing reefs of southern Taiwan. *P.S.Z.N.I: Mar. Ecol.*, **42**, 185–204.
>
> Dai, C.F. (1997) Assessment of the present health of coral reefs in Taiwan. In: R.W. Grigg and Birkeland (eds), *Status of Coral Reefs in the Pacific*. Sea Grant Program, University of Hawaii, pp. 123–131.
>
> Randall, R.H. and Y.M. Cheng (1977) Recent corals of Taiwan. Part I. Description of reefs and coral environments. *Acta Geol. Taiwanica*, 19, 79–102.
>
> Randall, R.H. and Y.M. Cheng (1979) Recent corals of Taiwan. Part II. Description of reefs and coral environments. *Acta Geol. Taiwanica*, **20**, 1–32.
>
> Shao, K.T. (1994) Biodiversity and conservation of marine fishes in Taiwan with emphasis on the coral reef fishes. In: C.I. Peng and C.H. Chou (eds.) *Biodiversity and Terrestrial Ecosystem*. Institute of Botany, Academia Sinica, Monograph Series, 14, pp. 307–318.
>
> Yang, R.T., K.S. Chi, S.C. Hu and H.T. Chen (1975) Corals, fishes, and benthic biota of Hsiaoliuchiu. Institute of Oceanography, National Taiwan University, Special Publication No. 7, 53 p.
>
> Yang, R.T. and C.F. Dai. (1982) Coral communities in Yen-Liao Bay, Taiwan. *Acta Oceanogr. Taiwanica*, **13**, 167–180.

grey mullet is generally in December every year. The schools of grey mullet seem to move with the protruding cold front resulting from a branch of the warm Kuroshio current from the south and the cold coastal current along the western coast of Taiwan (Huang and Su, 1989; Su and Kawasaki, 1995). The fishing ground of grey mullet is mainly in the coastal waters off southwestern Taiwan, from Tainan to Kaohsiung. Purse seines are the most effective fishing gear by which at least 87% of the total catches are obtained. The grey mullet fishery started in 1951 with sporadic catch. The total catch increased rapidly in 1959 when purse seine was applied for this fishery, peaked in 1979–1986, and has declined since then.

Clupeoid larvae (anchovies, sardines and round herring) are the most abundant fishes in the coastal waters off Taiwan (Cheng, 1988; Yu and Chiu, 1994). The fish larvae become the target of this "bull-ard" fishery—the name given to a collection of small fish, generally <3 cm in length—about one month after being hatched with a body length of less than 3 cm. They are harvested for local

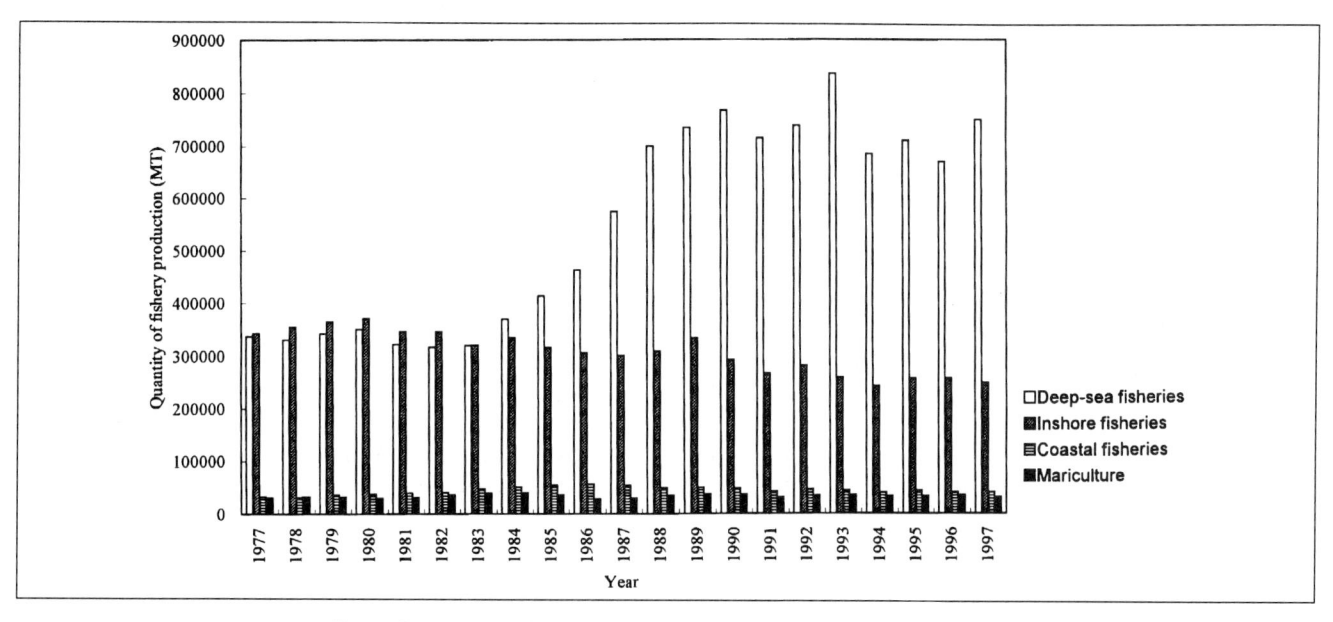

Fig. 5. Quantity of fishery production in Taiwan over time (Anon., 1998).

consumption and constitute the bulk of the catch in the local fisheries of Taiwan. The peak fishing season of larval anchovies is in March and April. The lag of three months of catch per unit effort behind sea surface temperature suggests that its peak spawning season is in December and January, corresponding to the spawning and fishing season of grey mullet (Tsai et al., 1995). Apparently, the abundance of both larval anchovies and grey mullet is related to the intensity of coastal upwelling, which is regulated by sea surface temperature and the intensity of the northeast monsoon in December and January.

Seventeen out of 21 major rivers in Taiwan flow into the strait. Inland sources of nutrients fertilized the strait water and made the strait an important fishing ground. Fish and invertebrates collected from the strait comprise a considerable portion of the total catch of coastal fisheries. Traditionally, many local people rely on this fishery for their subsistence. However, due to overfishing, illegal fishing practices and marine pollution, fishery resources in the strait have declined rapidly in the past two decades (Hsieh, 1993) (Fig. 5).

POPULATIONS AFFECTING THE AREA

A brief description of the historical use of the Taiwan coast is given by Hwang (1996). The following is exctracted from his paper. In Taiwan, there was no record of when people started to utilize reclaimed tidal land. However, it actually started on the southwestern coast about 300 years ago when Dutch people were driven out of Taiwan followed by an immigration from mainland China. Before 1895, there were 520 hectares of saltpans producing table salts and other uses. During the Japanese era (1895–1945) the sea-salt-producing fields from Lukang (near Taichung) to Kaohsiung increased to 4,967 hectares and a number of culture ponds. Beginning in 1937, planned reclamation of tidal land for paddy fields has been greatly increased. By 1945, there were 2,212 hectares of paddy fields from reclaimed tidal land. On the Tainan coast, 484 hectares of reclaimed tidal land were developed for fish culture. Then, integrated and planned development of reclaimed tidal land began in 1959. To date, 12,269 hectares of reclaimed tidal land have been developed, and 5,035 hectares are planned for future development.

Reclaimed tidal land ranges from Nankang Creek (south of the Tanshui River) to the Tsengwen River (near Tainan) covering an area of 54,028 hectares, of which about 8,000 hectares are developed or are being developed. Large amounts of crabs, shells and fishes occupy the natural habitats on the undeveloped reclaimed tidal land.

Most of the population of Taiwan is found on the western side bordering the strait; the western coast with dense population is therefore seriously affected by human activity. According to statistics (Anon., 1997), fishery population was 471,819 (about 83% in deep-sea fisheries including inshore fisheries, coastal fisheries and inland water fishing and about 17% in inland water culture including mariculture) in 1953, and the peak population was 675,715 in 1964. Detailed information about fishery population in inshore and coastal fisheries and mariculture is available starting in 1989. In 1996, the population engaged in fisheries was 504,317 (about 4.4% in deep-sea fisheries, 28.1% in inshore fisheries, 32.1% in coastal fisheries, 6.3% in mariculture, 1.5% in inland water fishing, and 27.6% in inland water culture). It is important to note that the population in both inshore fisheries and mariculture was decreasing from 1989 to 1996, suggesting decreasing resources and decreasing space for mariculture. On the other hand, the population in both inland water fishing and inland water culture was increasing during the same period. Another statistic shows that the total area of shallow sea culture decreased from 11,831 hectares in 1970 to 5,578 hectares in 1990 (Anon., 1995).

RURAL FACTORS

There have been many cases of mass mortality of shellfish in the spring in the coastal areas from Taichung to Tainan. The first investigation into the possible cause of these mortalities was conducted by Hung et al. (1975), and they speculated that the mortality of oysters and clams in April 1975, immediately after the heavy rain, may have been due to the low oxygen content in association with the high concentrations of heavy metals and nutrients in the water.

One of the major industries in southern Taiwan is pig-farming along the Kaoping River; some two million animals are being raised along the river banks. Excretions from pigs, treated and untreated, are discharged directly into the river and eventually carried to the sea off southwestern Taiwan. A study by Jeng et al. (1996) shows that geographically the percent coprostanol is highest around the river mouths, and decreases to about 1% at the shelf break and to nearly zero at a distance of approximately 40 nautical miles (74 km) from the river mouth. They also show that a sediment core exhibits comparatively higher percent coprostanol in the top 15 cm, indicating an increased input of coprostanol over the past 20 years.

Jan and Chang (1993) indicated that two popular methods for artisanal fishing are beach seine and spear fishing. On the rocky coast, beach seine is the major method of fishing. The net is generally set along the path where fish migrate, resulting in a good harvest with a number of fish species. On the other hand, spear-fishing is generally conducted in spring when plankton bloom. In the rocky reef areas, herbivorous molluscs (e.g., abalones) and echinoderms are abundant. In the Penghu area, the harvest is 123 Mt for 1991 (Taiwan Fisheries Bureau, 1992).

COASTAL EROSION AND LANDFILL

Two major factors affect the coastal environment.

Longshore Currents

When waves come close to the coastline and where the water depth becomes shallow, the waves break. If the propagating direction is not perpendicular to the coastline, then there will be alongshore movement generated as the wave breaks. The longshore currents are strong enough to displace sand near the shoreline. The sand then drifts along the coastal area. On some coasts, the sand accumulates to form reclaimed land, while elsewhere the beaches are eroded. The major waves, mainly induced by wind, approach northern Taiwan from the N-NNE direction, and in southern Taiwan from the WSW-SW direction. These major wave directions in the strait induce the longshore currents to flow southward in the north, and northward in the south (Lee and Hou, 1979). The major wind and sand drift directions are shown in Fig. 6. The currents can not only transport the pollutants in seawater near the shore but also move the coastal sand to a degree which changes the coastline morphology. Generally, the coasts have been eroded in the northern and southern sections of the western coast, while in the middle section, the drift sand, either from the littoral zone or suspended particles in the river discharges, is deposited reducing water depth, or even forming new land (Fan, 1995). In the past two decades, coastal erosion has become serious, especially along the western coast. Coastal dykes have been positioned in most areas to protect the land. Figure 7 shows six coastal districts with serious erosion problems, three are located on the western coast. They are protected by the coastal dykes, but are destroyed frequently (Chang, 1993). The main reason for recent serious coastal erosion can be attributed to:

1. construction of many dams upstream in the rivers in the past 20 years, reducing the transfer of some sand to the littoral zones;
2. over-extraction of groundwater by fishermen for mariculture making the coastal zones unstable and inducing erosion (described below).

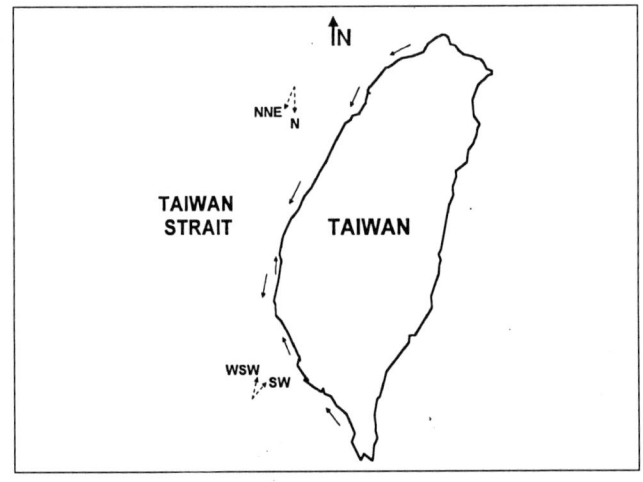

Fig. 6. Major wave directions (dotted arrows) in Taiwan Strait and the longshore current directions (solid arrows) along the western Taiwan coast (after Fan, 1995).

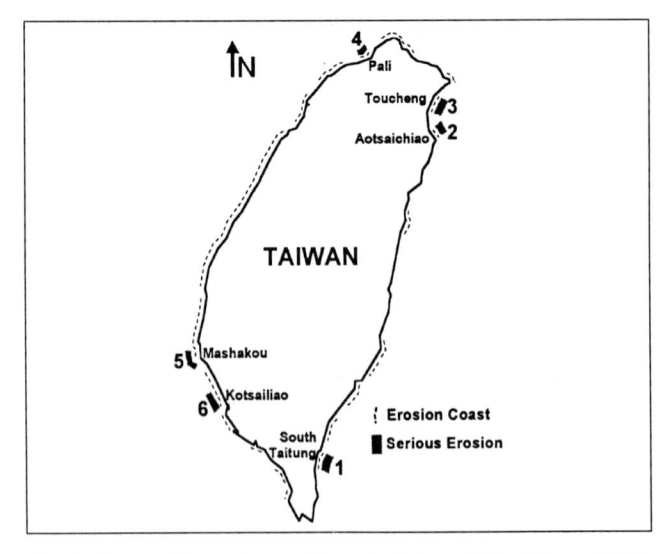

Fig. 7. Coasts with erosion problems in Taiwan (after Chang, 1993).

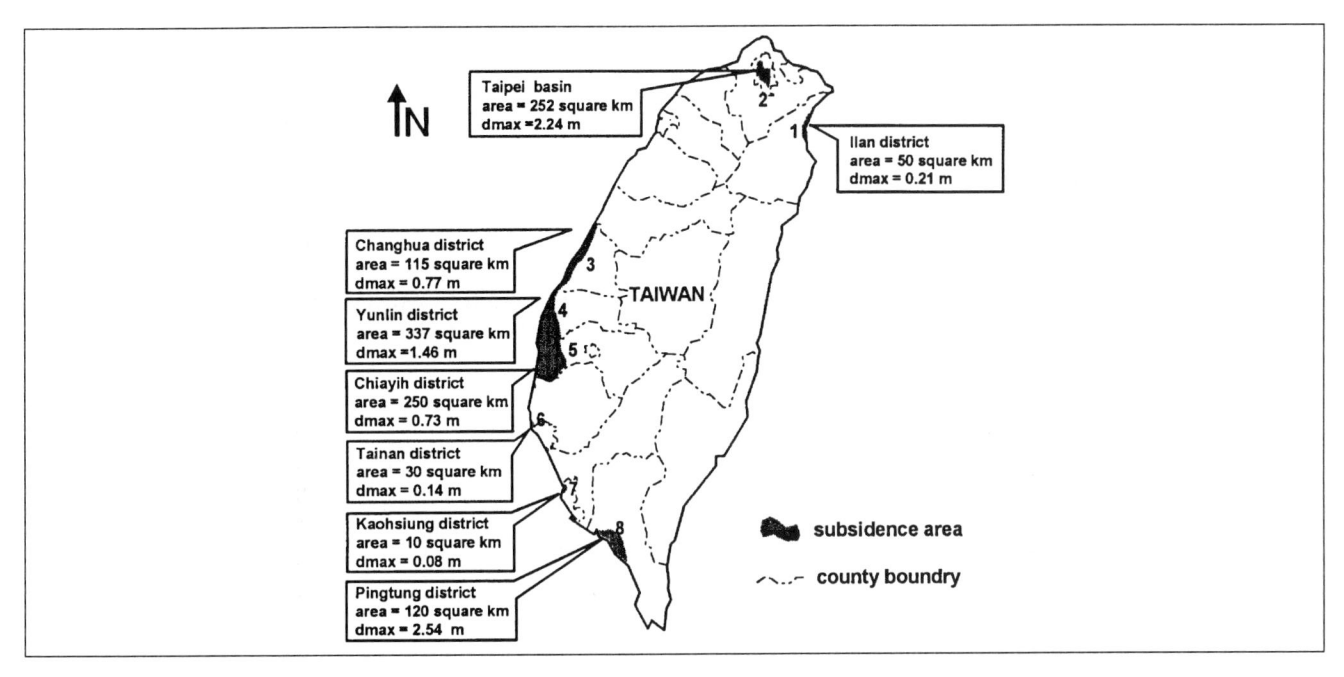

Fig. 8. Map of land subsidence districts in Taiwan (after Tseng and Yen, 1993).

Coastal Land Subsidence

Since 1983, the impact of groundwater extraction for mariculture has caused a serious land subsidence problem along the littoral zone in Taiwan (Tseng and Yen, 1993). There are eight districts which suffer from serious land subsidence problems. Among these, only the Taipei basin is located inland. The other seven are located along the littoral zone in seven counties (Fig. 8). Figure 8 shows that:

1. The biggest subsidence area covering three counties, Changhua, Yunlin and Chiayih, is located in the middle part of the western coast, an area reclaimed in the last one hundred years.
2. Two districts with a subsidence depth of over 2 m are the no. 8, Pingtung district in southern Taiwan, where the maximum subsidence depth is 2.54 m and no. 2, Taipei basin, located in northern Taiwan where the maximum subsidence depth is 2.24 m. The government has restricted pumping groundwater in the Taipei basin in 1978; since then the land has subsided no more than 1 cm a year (Chang, 1993).
3. Groundwater extraction has not only caused land subsidence but also resulted in a progressive deterioration of groundwater quality. Seawater intrusion was first detected in the Yunlin district, which is now the largest land subsidence area in Taiwan. The county has more than 6,000 deep wells for pumping groundwater, most of which are for fishery business (Tsao, 1983). Once the land subsidence occurs, even if the groundwater level recovers in the future, the ground-layer will not be able to recover. This will lead to the loss of national land resources, the exhaustion of groundwater resources, problems of seawater intrusion and flooding of coastal villages. Almost every year in the summer typhoon season, storm surges often destroy the coastal dykes, and seawater floods the coastal villages, especially along the western and southwestern coasts. (Fan, 1993). In some villages, the flood water rises to 2 m deep or more during flood tides. Some villages even have to evacuate their inhabitants to other safe places. These phenomena have happened almost every year during the summer typhoon season, especially in the no. 4–7 districts.
4. The total land subsidence area is about 1,164 km^2. The area of Taiwan is about 36,000 km^2, so this means that 3.23% of Taiwan suffers from land subsidence. In fact, most land area of Taiwan is mountainous. Plains have a total area of only 11,000 km^2, so that land subsidence covers 11% of the productive plains in Taiwan.
5. In no. 8 district, the land subsidence phenomenon is even more peculiar; some old houses and tombs subside into the ground, while some newly built houses are 1–3 m above ground.

EFFECTS FROM URBAN AND INDUSTRIAL ACTIVITIES

Over the last two to three decades, rapid industrialization and population concentration in cities have led to runoff of domestic wastewater, industrial wastewater, agricultural wastewater, thermal effluents from power plants, pesticides and garbage via rivers to the sea. This has seriously polluted the coastal environment of Taiwan, especially the western coast (Tables 1 and 2). In addition, increasing fishing intensity and improved fishing methods have over-exploited the coastal fishery resources, leading to a state of resource degradation (Chen et al., 1991).

Eastern Taiwan

Off the north Taiwan coast is the southern East China Sea shelf which borders the southern Okinawa Trough (Fig 1). The slope area of the trough near Taiwan has relatively high sedimentation and is considered a major site for exchange of material between the East China Sea and the Kuroshio. Further south, the Ilan Ridge extends eastward to connect with Yonakuni Island. This ridge separates the Okinawa Trough from the deep west Philippine Sea. Off southern Taiwan, the west Philippine Sea and the South China Sea are connected by the Bashi Channel located between Taiwan and Luzon.

The eastern coast of Taiwan is bounded on the west by the Central and Coastal Ranges, and habitable space is very limited. The coast is rocky with high cliffs except for a couple of places with sandy beaches. Rock and gravel (mainly boulders and cobbles) are the major feature of the coast. The sediment dispersal pattern off the eastern coast extends offshore eastward to the west Philippine Sea; the inshore transport may represent the transport of the countercurrent from the main Kuroshio current (Chen, 1980).

The Kuroshio current, a strong surface current of warm and saline water, originates in Luzon Strait, flows by Taiwan, Ryukyu, and Japan, and then slows down as a drift current of the north Pacific Ocean after meandering off southeast Japan. When reaching southeastern Taiwan, the Kuroshio divides into two: the main one flows by eastern Taiwan and the other flows through Taiwan Strait. While flowing by Taiwan, the main Kuroshio reaches its maximal surface velocity and total volume transport (Nitani, 1972). Chu (1971) reports that the width of the main Kuroshio east of Taiwan ranges from 110 to 150 km, its axis shifts from 50 to 150 km offshore, and maximum velocity ranges from 59–100 cm/sec in different areas and at different times of the year. The Kuroshio shoals up across the Ilan Ridge and turns northeastward along the East China Sea continental slope. A persistent cold water dome, resulting from the intrusion and upwelling of the cold subsurface water that is topographically induced at the continental margin, located at around 25°30′N, 122°12′E is clearly indicated by satellite photographs and hydrography (Fan, 1980; Chern et al., 1990). Another coastal upwelling, first observed in a Satellite Gemini X photograph taken on July 20, 1966, occurs along the southeast coast of Taiwan in the area around latitude 22°45′N to 23°N, and at depths down to several hundred meters (Fan, 1979 and references therein).

Unlike the western coast of Taiwan, tidal ranges on the eastern coast are about 1.2 m during spring tides and about 0.6 m during neap tides, averaging 0.9 m.

A survey off southeastern Taiwan along the 21°45′N parallel shows that plankton assemblages were dominated by species of blue-green algae, *Trichodesmium* and *Pelagothrix*, which formed patches, particularly in the surface and near-surface water (Huang et al., 1988). Another survey conducted across the Kuroshio near 21°50′N shows that firstly, the most dominant species of living planktonic forams is *G. sacculifer* followed by *G. ruber*, secondly, living planktonic forams are concentrated highly at the 100 m water depth (near 120°20′E), up to 6,937 specimens per 1000 m^3, which may be associated with the position of the main Kuroshio axis, and thirdly, these living assemblages can be divided into neritic, deep water and Kuroshio groups (Shieh and Chen, 1990). Ichthyoplankton in offshore waters off eastern Taiwan (123°E westward to the coast and in between 22 and 25°N) are dominated by engraulids (37.7%), myctophids (19.3%), scombrids (15.5%), carangidis (3.6%), and acanthurids (2.3%) (Chiu and Liu, 1989).

Kuroshio and its neighboring waters consist of 243 fish larvae and juveniles representing 35 families and 58 species; they can be grouped into oceanic epipelagic, mesopelagic and coastal species categories, and among them the mesopelagic fishes especially Myctophids and Gonostomatids were found to be the most dominant, constituting up to ca. 60% of the total catch (Tseng, 1989). An investigation of the pelagic fishing grounds in the waters off eastern Taiwan during April indicates that of all the catch dorado was the most abundant species and among billfishes most of the catch was sailfish (Huang and Yang, 1980).

Most common marine organisms along the eastern coast of Taiwan (Chen and Yuan, 1987) are as follows.

(1) *Fish and shrimps*. Most economic fishes are epipelagic, including tunas, bonitos, billfishes, dorado, mackerel, jacks, flying fish and juvenile fishes. In deep waters are sharks, shrimps, and various demersal fishes. Spiny lobsters are abundant on rocky coastal areas, but they have declined in recent years due to overfishing. Squids and cuttlefish are abundant in the area near Ilan. The production of milkfish fry in Taitung County comprises about 42% of total catch in Taiwan.

Reef fishes are widely distributed along the rocky coast of eastern Taiwan. However, due to the special environment of deep water and strong current, the species composition is different from that around the Nanwan area. They are abundant, and many of them are collected for aquaria.

(2) *Corals*. Stony corals are widely distributed along the rocky coast, with abundant areas found at Shicheng (north of Ilan), Suao, Sansientai (north of Taitung) and Tawu (south of Taitung). The species diversity of stony corals is less than that in the Nanwan area. Soft corals are less common. Red corals are collected below 300 m in the areas adjacent to Lutao and Lanyu, but these fishing grounds have been destroyed by over-harvesting.

(3) *Molluscs* The east coast from Tachi to Tawu (between Taitung and Nanwan) is the most important fishing ground for abalones in Taiwan. The shallow water area with boulders is the most favorable habitat for this, and the largest abalones are found in the Taitung area. Production of the green snail shell, *Turbo marmoratus*, in Taitung has declined rapidly in recent years. Oyster farming is increasing in the coastal area of Taitung.

(4) *Seaweeds*. Seaweeds are abundant on the east coast, and several species of green algae and red algae are commonly utilized.

(5) *Other animals*. The east coast is one of the major migratory routes for marine mammals in the west Pacific. Sea urchins are abundant along the east coast.

REFERENCES

Chen, D.M. and H.L. Yuan (1987) Report on the exploitation of the Taiwan coasts. Taiwan Fisheries Bureau Series No. 2, Taiwan Fisheries Bureau, p. 16. (In Chinese).

Chen, M.P. (1980) Modern and redeposited coccolithophores for the indication of sediment dispersal pattern around Taiwan offshore region. *Acta Oceanogr. Taiwanica*, 11, 129–156.

Chern, C.S., J. Wang and D.P. Wang (1990) The exchange of Kuroshio and East China Sea shelf water. *J. Geophys. Res.*, **95**, 16017–16023.

> Chiu, T.S. and H.C. Liu (1989) Investigation on the kinds and occurrence of ichthyoplankton in the waters off eastern Taiwan. *Acta Oceanogr. Taiwanica*, **23**, 53–62.
> Chu, T.Y. (1971) Environmental study of surrounding waters of Taiwan. *Acta Oceanogr. Taiwanica*, **1**, 15–31.
> Fan, K.L. (1979) On upwelling along the southeastern coast of Taiwan. *Acta Oceanogr. Taiwanica*, **10**, 155–163.
> Fan, K.L. (1980) On upwelling off northeastern shore of Taiwan. *Acta Oceanogr. Taiwanica*, **11**, 105–117.
> Huang, R., L.L. Jan and C.H. Chang (1988) A preliminary analysis of phytoplankton variability in the western Philippine Sea. *Acta Oceanogr. Taiwanica*, **21**, 82–91.
> Huang, C.C. and R.T. Yang (1980). A survey of the pelagic fishing grounds in the waters east off Taiwan. *Acta Oceanogr. Taiwanica*, **11**, 228–237. (In Chinese with English abstract).
> Nitani, H. (1972) Beginning of the Kuroshio. In: H. Stommel and K. Yoshida (eds.) *Kuroshio—Its Physical Aspects*, University of Tokyo Press, pp. 129–163.
> Shieh, Y.T and M.P. Chen (1990) Planktonic foraminiferal assemblages within the surface 200 M water, across the Kuroshio current, June, 1989. *Acta Oceanogr. Taiwanica*, **25**, 87–98. (In Chinese with English abstract).
> Tseng, W.N. (1989) Preliminary studies on species composition of fish larvae and juveniles in the Kuroshio waters adjacent to Taiwan with reference to water mass and diurnal variation. *Acta Oceanogr. Taiwanica*, **22**, 102–115.

Artisanal and Non-industrial Uses of the Coast

Filling the low land along the coast for reclaiming land, construction of culture ponds, felling mangroves, lagoon disappearance or conversion for other uses etc. have wiped out the nursery ground for larval fish, which results in decreased fishery resources. The development of social economics over the last few decades has led to ever-increasing forms of management for increasing production values. Land utilization along the coastal areas has been converting from traditional agriculture to fish culture, and the number of culture ponds is on the increase. The operation of those culture ponds needs a supply of fresh water for both mariculture and aquaculture. Due to lack of surface water, groundwater is the first choice, given that it is convenient and inexpensive. Long-term pumping of groundwater without any control over it has led to the over-extraction of groundwater and lowering of the groundwater levels, which in turn leads to land subsidence. Seawater flooding in the land subsidence areas during high tide periods has already frequently occurred, causing serious damage to these areas (Hwang, 1995).

Industrial Uses of the Coast

Besides the several areas of reclaimed land which have been in use for industry, seven major development plans for industrial uses along the western coast of Taiwan will reclaim a further 35,000 hectares of land. The second largest one (planned area over 12,000 hectares) is for the development of industrial parks along the coastal lagoons of southwestern Taiwan which may cause problems. Originally, the tidal area was muddy with an abundance of biological resources. Pumping sand from offshore and dumping it on the muddy tidal coastal area is the present method of artificial land reclamation. To protect industrial parks from wave attack, jetties will be built. This is expected to wipe out the original ecosystem completely.

Cities

There are three ocean outfalls on the coast of Kaohsiung. The water quality in the area has recently been surveyed by Lin et al. (1994). Their results are as follows: (1) The organic matter contamination becomes apparent from the analysis of pH, dissolved oxygen, and BOD_5; (2) The concentration of NO_3-N shows an increasing trend with time, but that of PO_4-P shows a decreasing trend with time; (3) The concentration of phenolic compounds is far lower than the standard of second-class seawater set by the EPA(ROC) and lower than that of the previous year; (4) Among Cr(VI), Cd, Ni, Zn, Pb and Hg examined, only one case shows an increase in Zn, and the concentration of Hg decreases substantially compared to that of the previous year, all other metals have the same levels as those of the previous year.

Of the six phthalates currently being used in the domestic plastic industry, three phthalates—dibutyl phthalate (DBP), di-(2-ethylhexyl) phthalate (DEHP), and di-n-octyl phthalate (DOP)—have been detected in the coastal sediments by Jeng (1986b). His results show that, in general, DEHP is virtually ubiquitous the ppb level and very seldom have extraordinarily high levels (ppm) of DEHP been encountered. Furthermore, DBP and DOP have only been found in sediments off the Kaohsiung coast; this can probably be attributed to heavy industrial activities around the Kaohsiung area.

Sedimentary coprostanol in Kaohsiung Harbour and the Tanshui Estuary has been investigated by Jeng and Han (1994). In the Kaohsiung Harbour sediments, coprostanol has a concentration range between 0.58 and 128 $\mu g/g$ dry wt with a mean of 20.8 $\mu g/g$ dry wt; higher concentrations are found near the river mouths where some untreated sewage from Kaohsiung city is discharged into the harbour. In Tanshui Estuary sediments, coprostanol ranges in concentration from 1.00 to 230 $\mu g/g$ dry wt with an average of 63.5 $\mu g/g$ dry wt; relatively high levels of coprostanol (>10 $\mu g/g$) with $5\beta/5\beta+5\alpha$ cholestanol ratios >0.7 indicate positive sewage pollution. This probably results from an input from the sewage outfall, anoxicity, shallow water depths, inadequate tidal flushing, etc.

The area around the Erhjin Chi estuary is especially affected by large discharges of heavy metals from acid cleaning of metal scrap on the riverbanks. For example, the first incident of "green oysters" in the Charting coastal area was observed in January 1986. Han and Hung (1990)

Table 1

Metal concentrations (mg/kg dry wt) in surficial sediments (<63 μm) from the southwestern coast of Taiwan (Lee et al., 1998)

Location*	Cu	Zn	Pb	Ni	Cr	Cd
Dadu Chi	25.0±2.9	112.8±6.7	32.1±2.5	65.0±8.0	33.0±3.8	0.2±0.1
Choshui Chi	29.9±3.3	108.2±2.3	31.8±2.5	70.3±3.0	36.7±0.7	0.4±0.3
Jeeshui Chi	26.2±2.5	109.6±6.8	33.3±9.3	68.3±4.4	34.6±2.0	0.2±0.1
Tainan coast	24.6±7.4	105.2±14.6	27.3±3.9	65.1±7.0	36.7±4.8	0.1±0.1
Kaohsiung coast	34.3±15.2	142.0±50.9	34.3±8.3	66.7±11.7	35.6±7.9	0.2±0.1
Fengkung Chi	26.2±1.4	88.8±1.6	25.5±3.8	71.0±1.5	37.7±2.6	0.1±0.0
S. Bay**	2.6±1.4	11.3±3.7	4.1±5.6	8.5±2.6	20.0±13.4	0.06±0.06

*Refer to original paper. **Sediment grain size < 1000 μm.

Table 2

Average concentrations (ng/g dry wt) of organochlorine pesticides in bivalves from the southwestern coast of Taiwan (Hung et al., 1997)

Location	Organism	Date	p,p'-DDE	p,p'-DDD	p,p'-DDT
Charting (south of	oyster*	1994/01	13.8	12.6	8.5
Tainan coast)		1995/03	6.4	Tr	Tr
		1995/09	9.4	11.5	Tr
Charting	mussel**	1994/12	21.3	57.3	124
		1995/09	Tr	Tr	Tr
Yunlin (north of Beigang coast)	oyster*	1995/09	5.4	Tr	Tr
Makung (Penghu)	oyster*	1995/09	Tr	Tr	6.2

*Crassostrea gigas. **Perna viridis.

pointed out that the green color was found to be due to a high concentration of copper in the oyster tissue and that the green oysters collected from the Erhjin Chi estuary on January 26, 1989 gave the highest copper concentration of 4,401 ± 79 μg/g.

Shipping and Offshore Accidents and Impacts

An investigation of hydrocarbons in surface marine sediments off southwestern Taiwan has been conducted by Jeng (1986a). His results are as follows. Total contents of n-alkanes range from 337 to 3767 ng/g with a mean of 2019 ng/g, which is the highest among the coastal areas of Taiwan. Carbon preference indices (C_{25}~C_{31}) vary from 1.3 to 4.4 with an average of 1.8; these low CPIs indicate that land-derived organic matter doesn't play a major role. In addition, there are rather high levels of both pristane and phytane—88 ng/g (mean) and 27 ng/g (mean), respectively—suggesting inputs of fossil hydrocarbons. Total aromatic hydrocarbons range in concentration from 0.49 to 12.57 μg/g having an average of 1.95 μg/g, the highest value found at the station nearest to Kaohsiung Harbour. The results can be attributed to heavy tanker traffic and operations, other ship traffic, and industrial activities.

PROTECTIVE MEASURES

Because of an intensive and rapid development in Taiwan's coastal zone during the past several years, the quality of Taiwan's marine and coastal environment has been declining dramatically. Concerning the degradation of the marine environment, Taiwan is actively undertaking several programs for better management of its coastal zone.

The coastal policy in Taiwan in the 1960s and 1970s focused on tidal land reclamation to meet land requirements. Regarding the damage or loss of natural resources, conservation awareness had grown significantly in the 1980s. Conserving coastal resources seemed to be the theme of that decade. Since the early 1990s, the coastal policy has wandered between facilitating development and promoting conservation in the coastal zone. The three-stage history of coastal development implies the instability of Taiwan's coastal policy (Chiau, 1994). It also reflects the reality that the coastal policy often mirrors the level of development and economic situation in a country.

The authorities dealing with coastal zone management in Taiwan are spread among many agencies at each level of government. According to a recent review, more than eight agencies and nearly 30 laws and regulations related to coastal affairs (Chiau, 1994). For example, the Ministry of Interior (Construction and Planning Administration) manages land use, the Council of Agriculture is in charge of fisheries and natural resource conservation, the Ministry of Economic Affairs is actively involved in tidal land reclamation for industrial uses, the EPA takes care of water quality, and the Ministry of Finance (National Property Bureau) is responsible for managing the national land in the coastal area. However, several other affairs such as marine pollution cleanup and wetland protection are not the exclusive responsibility of any of the existing agencies. Within the existing administrative system, the lack of coordination among all related affairs becomes an inherent weakness in coastal zone management. The situation reveals the need for an integrated coastal management in the future. In addition, an effective and exclusive law for strengthening coastal zone management is urgent in Taiwan.

The problems regarding the overuse and abuse of coastal zone resources in Taiwan mainly come from the distorted land policy, the misleading land tax system, the weakness and loopholes in the allocation and operation of property rights governing the land and sea areas within coastal zones, the poor efficiency of the administrative authorities, and essentially, the unsound and unbalanced national land use program (Ma, 1993). Because of these shortcomings, the coastal zone is severely abused both by the private and public sectors, causing serious damage to existing land and water resources, environmental pollution, adverse geographical changes in coastal areas, etc. In order to improve the current situation, it is suggested that the most urgent mission is to correct the distortions in land policy, land tax system, and national land use programs so that the utilization of coastal areas can be significantly improved (Ma, 1993). Currently a wide spectrum of measures, including preventive environmental impact assessment and environmental monitoring programs, have been employed. Coastal resources are protected under the Coastal Area Environment Protection Plan which is administered by the National Park Department, Bureau of Construction and Planning, Ministry of Interior. Under the plan, seven coastal conservation zones (CCZ) have been established: Northeast Coast CCZ, Langyang River Mouth CCZ, Suhua CCZ, Huatung CCZ, Kenting National Park, Chanyunchia CCZ and Tanshui River Mouth CCZ. The last two CCZs lie on the sandy west coast, and the others include coral reefs. They are managed by the Ministry of the Interior and Government of Taiwan Province. Activities permitted depend on the type of zones. In addition, based on the Fishery Law, twenty-five Fishery Conservation Areas have been established which include three on the north coast, eleven on the east coast, one at Lutao, one at Hsiaoliuchiu, one bordering Kenting National Park, two in the Penghu Islands, and six on the west coast. Nineteen of them include or are adjacent to coral reefs. These areas are managed in the same way as CCZ, but commercial fishing is prohibited as their aim is to protect nursery areas for fisheries. Within the conservation areas, environmental impact assessments are required before approval of any development project.

The ignorance of effects of human impacts on wetlands and coral reefs and the lack of public awareness of the significance of these ecosystems are two of the main issues for the management of coastal ecosystems in Taiwan. The Wetland Conservation Association, the Wild Bird Society of Taiwan and the Taiwanese Coral Reef Society are three societies dedicated to promoting public awareness, environmental education, research and conservation activities in Taiwan.

It is suggested that the legislation of the Coast Law be accomplished and the administrations improve their ability in implementing laws and orders (Tsai, 1993). An agency should be designated to achieve coordination and cooperation among different levels of government. The investment for exploitation and preservation of the coastal area should be a joint venture of national and local government fulfilled by achieving a good balance between development and conservation.

REFERENCES

Anon. (1995) Statistical Abstract of Agricultural Censuses, Taiwan, Republic of China. Compiled by Statistics Office, Council of Agriculture, Executive Yuan, Taipei, Taiwan, R.O.C. (In Chinese).

Anon. (1997) Agricultural Production Statistics Abstract, Republic of China 1996. Edited by Statistics Office, Council of Agriculture, Executive Yuan, Taipei, Taiwan, R.O.C. (In Chinese).

Anon. (1998) Agricultural Production Statistics Abstract, Republic of China 1997. Edited by Statistics Office, Council of Agriculture, Executive Yuan, Taipei, Taiwan, R.O.C. (In Chinese).

Chang, S.C. (1993) Land-subsidence-water resources problem in Taiwan. Proceedings of Symposium on Sustainable Development of Water Resources, Taipei, Taiwan, R.O.C., Vol. 3, pp. 39–62. (In Chinese).

Chen, C. (1991) Physiographic and tectonic provinces of Taiwan Strait. Proceedings of Symposium on Geology and Seismology of Taiwan Strait and its Coast, Fuzhou, Fujian, China, pp. 180–187.

Chen, C.C., T.W. Shieh and C.C. Li (1991) Protection issues and measures on the present fishery resources of coastal seas. In: *Environment Protection of Coastal Ecosystem*. Council of Agriculture, Executive Yuan, Taipei, Taiwan, R.O.C., pp. 3–11. (In Chinese).

Cheng, T.S. (1988) Study and investigation of bull-ard and anchovy fisheries in coastal waters of Taiwan. *Bull. Taiwan Fish. Res. Inst.*, **32**, 219–233. (In Chinese).

Chern, C.S. and J.C. Chen (1994) Model study of oceanic diffusion ability related to leakage of low level radwaste. *Nuclear Sci. J.*, **31**(4), 264–273. (In Chinese).

Chiau, W.Y. (1994) Coastal policy and administration in Taiwan: a review. Proceedings of Coastal Zone Canada '94, Halifax, Nova Scotia, Vol. 1, pp. 72–84.

Chou, C.H., F.C. Chang and Y.H. Hwang (1987) Ecological research of mangrove forests: a review. In: C.H. Chou, C.I. Peng and S.M. Chaw (eds.), *Plant Resources and Conservation in Taiwan*. Society of Wildlife and Nature, Taipei, Taiwan, R.O.C., pp. 23–48. (In Chinese)

Chu, T.Y. (1963) The oceanography of the surrounding waters of Taiwan. Report of the Institute of Fishery Biology of Ministry of Economic Affairs and National Taiwan University 1(4): 29–44.

Chuang, W.S. (1986) A note on the driving mechanisms of current in the Taiwan Strait. *J. Oceanogr. Soc. Jpn.*, **42**, 355–361.

Dai, C.F. (1997) Assessment of the present health of coral reefs in Taiwan. In: R.W. Grigg and C. Birkeland (eds.), *Status of Coral Reefs in the Pacific*. Sea Grant Program, University of Hawaii, pp. 123–131.

Fan, K.L. (1984) The branch of Kuroshio in the Taiwan Strait. In: *Ocean Hydrodynamics of the Japan and the East China Seas*. Oceanography Series, 39: 77–82.

Fan, K.L. (1987) The pollutant movements along the southwestern coast of Taiwan. Proceedings of a Workshop on Ocean Outfalls, Taipei, Taiwan, R.O.C., pp. 12–23.

Fan, K.L. (1991) Current affecting the marine environment along the Taiwan southwestern coast. Proceedings of International Marine Pollution Symposium, Tainan, Taiwan, R.O.C., pp. 19–26.

Fan, K.L. (1993) The coastal environmental status and problems in Taiwan. *Engng. Environ.* (R.O.C.), **5**, 1–27. (In Chinese).

Fan, K.L. (1995) The coastal environmental characteristics of Taiwan. *Chem. Ecol.*, **10**, 157–166.

Fan, K.L. and C.Y. Yu (1981) A study of water masses in the seas of southernmost Taiwan. *Acta Oceanogr. Taiwanica*, **2**, 94–111.

Han, B.C. and T.C. Hung (1990) Green oysters caused by copper pollution on the Taiwan coast. *Environ. Pollut.*, **65**, 347–362.

Hsieh, M.H. (1993) Coastal and nearshore fisheries. In: C.Y. Sha, P.N. Sun and H.C. Kuo (eds.), *Forty Years of Fisheries in Taiwan, Council of Agriculture.* Executive Yuan, R.O.C., pp. 87–110. (In Chinese).

Hsueh, M.L. (1996) The status of wetland conservation on western Taiwan. In: A.C.T. Chen (ed.), *Proceedings of Conservation and Management of Coastal Areas Conference, Taipei, Taiwan, R.O.C.*, pp. 158–168 (In Chinese).

Huang, C.S. and W.C. Su (1989) Studies on the fluctuations of fishing conditions for grey mullet (*Mugil cephalus* Linnaeus) from the western coast of Taiwan. *J. Fish. Soc. Taiwan*, **16**, 47–83.

Huang, J.B. and T.S. Chiu (1991) Eco-geographic difference of larval fish assemblage in the coastal waters of the western central Taiwan. *J. Fish. Soc. Taiwan*, 18, 241–256.

Hung, T.C., J.C. Chen, L.P. Lin and N.K. Liang (1975) Pollution studies on shellfish cultivating area of Taiwan western coast. Special Publication No. 6, Institute of Oceanography, National Taiwan University, Taipei, Taiwan, R.O.C. (In Chinese with English abstract).

Hung, T.C., Y.C. Ling, W.L. Jeng, C.C. Huang and B.C. Han (1997) Marine environmental monitoring and QA/QC system in Taiwan. *J. Chin. Environ. Prot. Soc.*, **20** (1), 69–90.

Hwang, C.S. (1985) A review study of tidal land reclamation along the western coast of Taiwan. Proceedings of the 8th Ocean Engineering Symposium, Hsinchu, Taiwan, R.O.C., pp. 527–539.

Hwang, J.S. (1995) Land subsidence and its induced sea water flooding in Taiwan. Special Issue of *Engineering Environment* (R.O.C.), no. 6, pp. 48–61. (In Chinese with English abstract).

Hwang, J.S. (1996) Overall assessment of the development and utilization of Taiwan's reclaimed tidal land. In: A.C.T. Chen (ed.), *Symposium on the Assessment Techniques of Environmental Impact—Protection and Development of Coastal Areas.* Environment Protection Administration (R.O.C.), pp. 294–310. (In Chinese)

Jan, R.Q. and K.H. Chang (1993) Coastal ecosystems around Taiwan with an assessment of the productivity. *J. Engng. Environ.* (R.O.C.), **13**, 59–84. (In Chinese with English abstract).

Jeng, W.L. (1979) Hydrocarbons in southern Taiwan Strait surface sediments. *Acta Oceanogr. Taiwanica*, **12**, 94–111.

Jeng, W.L. (1986a) Hydrocarbons in marine sediments off southwestern Taiwan. *Proc. Natl. Sci. Counc. ROC(A)*, **10**, 123–129.

Jeng, W.L. (1986b) Phthalate esters in marine sediments around Taiwan. *Acta Oceanogr. Taiwanica*, **17**, 61–68.

Jeng, W.L. and B.C. Han (1994) Sedimentary coprostanol in Kaohsiung Harbour and the Tan-shui Estuary, Taiwan. *Mar. Pollut. Bull.*, **28**, 494–499.

Jeng, W.L., J. Wang and B.C. Han (1996) Coprostanol distribution in marine sediments off southwestern Taiwan. *Environ. Pollut.*, **94**, 47–52.

Lee, J.P. and H.S. Hou (1979) The study of littoral drift vs. wave energy along the Taichung coast from Tachia River to Tatu River. Proceedings of the 3rd Ocean Engineering Workshop, Suao, Taiwan, R.O.C., pp. 193–208. (In Chinese).

Lee, C.L., M.D. Fang and M.T. Hsieh (1998) Characterization and distribution of metals in surficial sediments in southwestern Taiwan. *Mar. Pollut. Bull.*, **36**, 464–471.

Lin, C.S., L.C. Shen, W.C. Chang and T.H. Wen (1994) Survey of the influence of ocean outfalls on fisheries. In: Reports on Fish Environmental Protection (VIII), Council of Agriculture Fisheries Series No. 45, pp. 30–87, Executive Yuan, Taipei, Taiwan, R.O.C. (In Chinese).

Lin, H.C., K.T. Shao and S.R. Kuo (1995) Fish fauna of the mangrove ecosystems in Taiwan. Proceedings of Mangroves Ecosystem Symposium, Taichung, Taiwan, R.O.C., pp. 181–197. (In Chinese).

Liu, C.S., N. Lundberg, D.L. Reed and Y.L. Huang (1993) Morphological and seismic characteristics of the Kaoping Submarine Canyon. *Mar. Geol.*, **111**, 93–108.

Ma, K. (1993) The environmental resources and development and management of the coastal zone in Taiwan: the economic aspect. *J. Engng. Environ.* (R.O.C.) **12**, 23–30. (In Chinese).

Ma, T.Y.H. (1963) Twin origin of the submarine canyons around Taiwan and the Quaternary glaciation in Taiwan as basis for refutation of the turbidity current theory and the so called 'Great Ice Age' as due to universal lowering of temperature. *Petrol. Geol. Taiwan*, **2**, 209–219.

Shaw, P.T. (1989) The intrusion of water masses into the sea southwest of Taiwan. *J. Geophys. Res.*, **94**, 18213–18226.

Su, W.C. and T. Kawasaki (1995) Characteristics of the life history of grey mullet from Taiwan water. *Fish. Sci.*, **61**, 377–381.

Sun, S.C. (1982) The Tertiary basins of offshore Taiwan. Proceedings of ASCOPE Conference, Manila, pp. 125–135.

Taiwan Fisheries Bureau (1992) Annual Report on Fisheries 1991.

Tsai, C.Y. (1993) Coastal land reclamation and management in Taiwan. Proceedings of Coast Environmental Management Conference, Taipei, Taiwan, R.O.C., pp. 1–19. (In Chinese).

Tsai, C.F., P.Y. Chen, C.P. Chen, M.A. Lee, G.Y. Shiah, K.T. Lee, C.S. Huang and W.C. Su (1995) Periodicity of the abundance of fish stocks in coastal waters of southwestern Taiwan and its causal factor. Proceedings of Climate Change and Agricultural Production Symposium, Taipei, Taiwan, R.O.C., pp. 223–225.

Tsao, Y.S. (1983) Change of groundwater quality in the Yunlin basin due to over-pumping. In: *Relation of Groundwater Quantity and Quality* (Proceedings of the Hamburg Symposium, August 1983) IAHS Publication no. 146, pp. 33–43.

Tseng, R.H. and C.P. Yen (1993) The study of regional land subsidence problems in Taiwan. Proceedings of the 6th Conference on Environmental Planning and Management in R.O.C., Taiwan. pp. 409–422. (In Chinese).

Wang, J. and C.S. Chern (1988) On the Kuroshio branch in the Taiwan Strait during wintertime. *Prog. Oceanogr.*, **21**, 469–491.

Wang, J., C.S. Chern, N.K. Liang and M.Y. Shyu (1993) A preliminary study on the response of subtidal flows to the winter monsoon in the vicinity of Formosa banks. *Acta Oceanogr. Taiwanica*, **31**, 1–43.

Yang, H.S. and Y.Y. Ting (1988) Studies on life cycle and the habitat environment of bivalves along the west coast of Taiwan. *Bull. Taiwan Fish. Res. Inst.*, **45**, 67–81.

Yu, H.S. and Y.H. Wen (1992) Physiographic characteristics of the continental margin off southwestern Taiwan. *J. Geol. Soc. China*, **35**, 337–351.

Yu, P.T. and T.S. Chiu (1994) Fishery target species of larval anchovy fishery in the western central Taiwan. *J. Fish. Soc. Taiwan*, **21**, 227–239.

THE AUTHORS

Woei-Lih Jeng
Institute of Oceanography,
National Taiwan University,
Taipei, Taiwan, Republic of China

Chang-Feng Dai
Institute of Oceanography,
National Taiwan University,
Taipei, Taiwan, Republic of China

Kuang-Lung Fan
Institute of Oceanography,
National Taiwan University,
Taipei, Taiwan, Republic of China

Chapter 86

XIAMEN REGION, CHINA

Chua Thia-Eng and Ingrid Rosalie L. Gorre

Xiamen is located in the southern part of the Fujian Province of the People's Republic of China (PRC), west of the Taiwan Strait. This chapter discusses the physical environment of the Xiamen coastal and marine waters, the activities that have an impact on these waters, and the measures adopted by the Xiamen Municipal Government to ensure the sustainable management and development of the coastal and marine resources.

The coastal resources of this region are used for fisheries, mariculture, tourism, port, navigation and other activities. Fisheries and mariculture are the two major sources of income in the coastal areas of Xiamen. In 1993, marine fish production reached 42,229 tonnes, while annual mariculture production reached 25,862 tonnes in 1992. Conflicts among the users are not uncommon.

Despite the rapid economic growth of Xiamen in recent years, marine pollution has been kept within the National Seawater Quality Standards, except for some areas, such as the Maluan Bay.

In order to address these coastal and marine problems, the Municipal Government have adopted an integrated approach to coastal management. New management concepts have been introduced to transform the short-term and single-sectoral perspectives into more rational and sustainable policies. Inter-agency co-ordinative mechanisms have been established to facilitate coastal and marine management. The Municipal Government of Xiamen adopted a marine functional zonation scheme to allocate the coastal resources among the users.

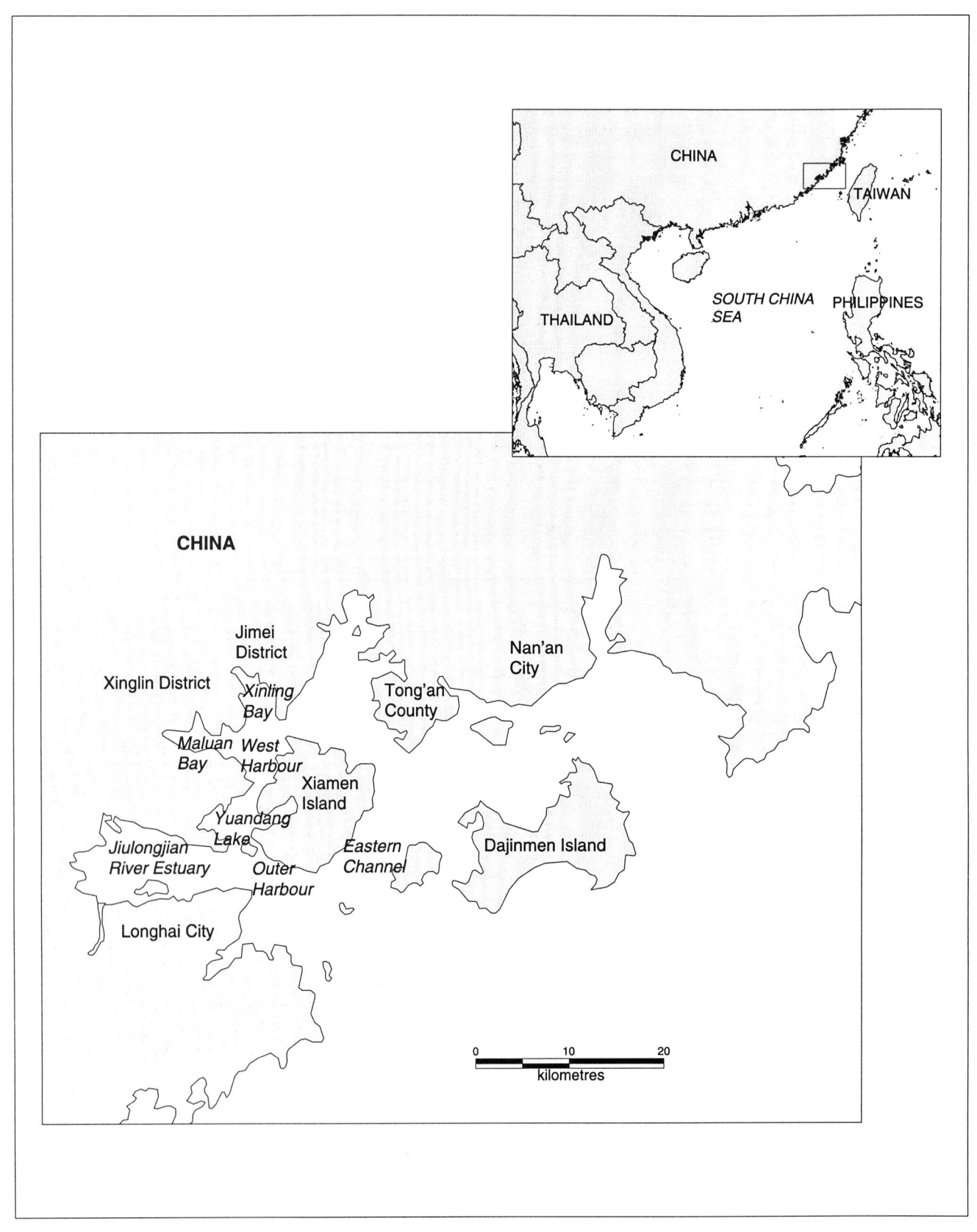

Fig. 1. Xiamen and its coastal waters.

GEOGRAPHY OF THE XIAMEN REGION

Xiamen Municipality is located in the southern portion of the Fujian Province of the People's Republic of China (PRC), west of the Taiwan Strait (Fig. 1). It is composed of Xiamen Island proper, Gulangyu Islet, and the part of the southern Jiulongjiang River estuary and its adjacent continental parts at longitude 117°53'W 118°25'E and latitude 24°24'N 24°5'S. The Municipality has six districts and one county, namely: Kaiyuan, Siming, Gulangyu, Huli, Jimei and Xinglin Districts and Tongan County (ITTXDP, 1996). It has a coastline of 234 km and total sea area of 334 km^2 (Chua et al., 1999).

The coastal waters of Xiamen Municipality cover a portion of the Jiulongjiang River Estuary; the West Harbour, Maluan Bay, and Xinglin Bay; the Outer Harbour, Tong'an Bay, the Eastern Channel of Xiamen Island, and the Northern Channel to Jinmen Island (ITTXDP, 1996). The boundaries of the Eastern Channel are Xiaojinmen Island, Dada Island, and Xiamen Island. The boundaries of the Outer Harbour are Gulangyu Island, Dada Island, Erda Island and Qingyu Island. The Eastern Channel and the Outer Harbour link the Tong'an Bay, the West Harbour, and Jiulongjiang River Estuary to the Taiwan Strait.

Around nine streams from the northwest hilly areas of Xiamen Municipality drain into Maluan Bay. Maluan Bay and Xinglin Bay eventually meet the Jiulongjiang River Estuary in the south and form the peninsulas of Jimei, Xinglin and Haicang.

The main rivers in the Tong'an County are the Xiqi, Dongqi, Guanxunqi and Litouqi. The Dongqi and Xiqi are the largest rivers, though the main source of freshwater in the Xiamen area is still the Jiulongjiang River, which supplies about 3.75×10^8 m^3 of water per annum (ITTXDP, 1996).

SEASONALITY, CURRENTS, NATURAL ENVIRONMENTAL VARIABLES

Climate

Xiamen has a subtropical climate characterized by a long period of warm weather with very minor changes over the rest of the year. It has no severe winter nor intense summer. The average temperature in Xiamen is 20.9°C (ITTXDP, 1996). The warmest average monthly temperature is 28.4°C which occurs in July. The coldest weather occurs in January and February with an annual variation of 15.8°C and a diurnal variation of 7°C. The warmest temperature which has been recorded is 38.5°C and the coldest is 2.0°C (Fig. 2) (ITTXDP, 1996).

Xiamen's weather is humid and wet. The average precipitation is 1143 mm while the average relative humidity is 74%. The level of humidity and precipitation is high in spring and summer and low in autumn and winter. The

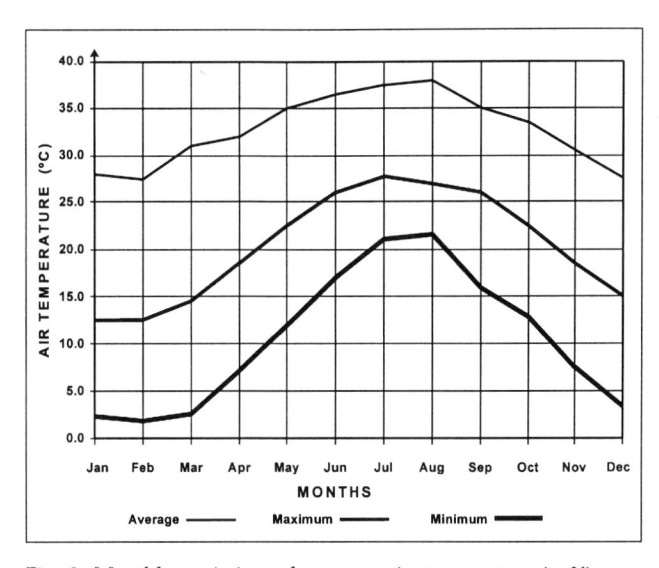

Fig. 2. Monthly variation of average air temperature in Xiamen Municipality. Source: Office of the Steering Group for Comprehensive Investigation of Xiamen Island Resources et al. (1994)

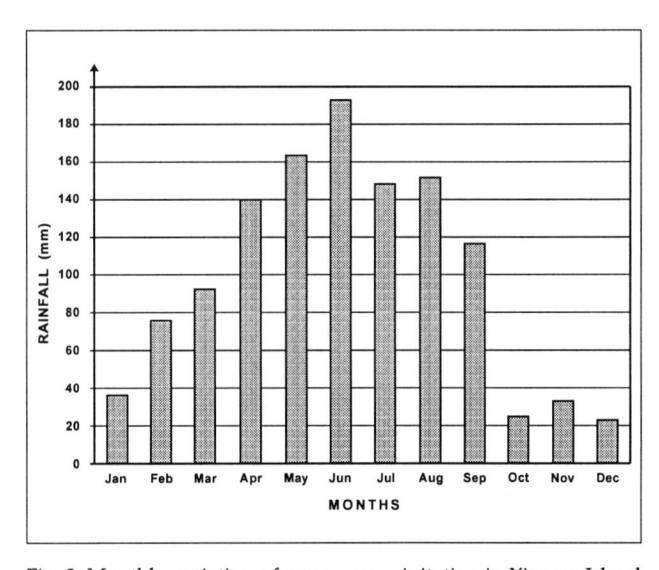

Fig. 3. Monthly variation of average precipitation in Xiamen Island (1952–1990). Source: Office of the Steering Group for Comprehensive Investigation of Xiamen Island Resources et al. (1994)

average evaporation is 1910 mm. The highest evaporation rate is in summer and autumn, and the lowest is in winter and spring (Fig. 3) (ITTXDP, 1996).

The average wind speed is 3.4 m/s, occurring from the northeast 15% of the time. Monsoonal circulation varies with the season. Wind velocity is high during the northeast monsoon from September to March and decreases during the southeast monsoon from April to August (ITTXDP, 1996).

Hydrology

The average annual temperature range of Xiamen coastal waters is from 14.8°C to 27.8°C. The coldest water temperature is in February, with temperatures ranging from 14.2°C to 16.2°C. The warmest water temperature is in September, with temperatures ranging from 27.2°C to 28.9°C. Water salinity ranges from 28.7 ppt to 30.1 ppt. Salinity of the coastal waters can fluctuate from 22.9 parts per thousand (ppt) to 31.1 ppt in February and from 13.9 ppt to 32.4 ppt in September (ITTXDP, 1996).

The average sea level of Xiamen coastal waters is 0.30 meters above the Huanghai Sea. Its coastal waters have a normal semi-diurnal tide of high range (Table 1) (ITTXDP, 1996). The tidal current of Xiamen's coastal waters is strong, bifurcated and of the semi-diurnal type, and both its direction and velocity are influenced by the coastal geomorphology. For example, the velocity decreases from the opening to the upper portion of the bay, and also from the deep channel to the shore. Xiamen has a strong tidal current with a maximum of 60–90 cm/s and an occasional velocity above 100 cm/s. Strong currents occur along the narrow sections of the Dongdu Channel in the West-Harbour, between Songyu and Gulangyu islands, between Gulangyu and Xiamen islands, between the channels in the east coast of Houyu Island, along the north coast of Jiyu Island, and in the waters of Autou-Wutongdao at the entrance of Tong'an Bay. The highest tidal current recorded in the Xiamen waters is 152 cm/s at the Dongdu Channel while the tidal current in the rest of the area is 40 cm/s. Slowest current speeds occur in the waters around Baozhuyu Island in Xinglin Bay (ITTXDP, 1996).

The current velocity at flooding and ebbing phases in Xiamen waters varies. Generally, waters with a strong tidal current have an ebbing velocity faster than the flooding tidal current resulting in the seaward transportation of the suspended materials or pollutants. In the weak tidal areas in Xiamen, however, there is very little variation between the flooding and the ebbing tidal currents and surface layer ebbing tidal current in the Xiamen waters generally lasts 20–50 minutes longer than the flooding current, with a maximum time difference of 1 hour and 50 minutes between flooding and ebbing (ITTXDP, 1996).

Residual currents in the Outer Harbour, in the estuary and in the West Harbour are strong and often higher than 10 cm/s and are attributed mainly to the strong effect of the water flowing from the Jiulongjiang River. In contrast, the residual currents near Xinglin, in the upper part of the West Harbour, and in Tong'an Bay are weak.

In the Dongdu channel, the direction of the residual current is similar to its ebbing tidal current, unlike the condition in northern Jiyu Island, western Songyu-Gulangyu Channel, the western Huoshaoyu Island and off the northern coast of Tong'an Bay where the residual current direction is similar to the flooding tidal currents. In other parts of Xiamen, the residual current of the upper layer follows the flooding direction of the lower layer. The direction of the residual current at the culvert of Gao-Ji Causeway flows from the West Harbour towards Tong'an Bay at 10–40 cm/s (ITTXDP, 1996).

As waves reach the coastal waters of the Outer Harbour, the current gradually weakens, especially towards the

Table 1

Tidal patterns (in metres)

Average high tidal level	5.68	Maximum tidal range	7.78
Average sea level	3.58	Average tidal range	3.99
Average low tidal level	1.89	Minimum tidal range	0.99
Maximum tidal level on record	7.78	Average flood duration	6 h 8 min
Minimum tidal level on record	–0.06	Average ebbing duration	6 h 10 min

Source: Office of the Steering Group for Comprehensive Investigation of Xiamen Island Resources, et al. (1994) cited in ITTXDP (1996).

inner waters of Xiamen. This is attributed to the sheltering effects of the island chain of Dajinmen, Xiaojinmen, Dada, Erda, Qingyu, and Wuyu.

In the West Harbour, the dominant wind wave has a wind wave to swell ratio of 89:11. However, waves are usually weak because of the short fetch (ITTXDP, 1996). The prevailing wave direction is towards the northeast, while the secondary prevailing wave direction is east-northeast. The strong wave direction is north-northeast, having a maximum height of 1.3 meters. The secondary strong wave direction is toward south-southeast with a maximum wave height of 1.2 m. The yearly average wave height reaches 1.2 m. There are stronger waves in the summer and NNE/ENE waves in the other three seasons (ITTXDP, 1996).

The opening of Xiamen Bay in the southeast allows waves from that direction to dominate Xiamen waters. Hence, during typhoons, southerly winds and southeasterly wind waves usually affect a large part of the region.

The offshore waters of Xiamen are affected by the Zhejiang–Fujian coastal current, the Kuroshio branch and the South China Sea warm current (ITTXDP, 1996). The prevailing direction of the current is toward the northeast. However, there are variations during the different seasons. From June to September, for example, the strait is influenced by the northeast warm current and by the high salinity and temperature from the southern warm current and the Kuroshio branch entering the strait in the south. From October to May, the southward Zhejiang–Fujian coastal currents affect the upper layer of western strait up to the Xiamen Harbour and Lishi Islands in the south (ITTXDP, 1996).

Natural Hazards

Earthquake

Xiamen Municipality sits on the Quanzhou–Shantou fault zone, which is the most active fault in the southeast coast of China. It must be noted that seismic movement is weak in the western and middle parts of the fault zone where Xiamen is located. However, historical records show that earthquakes with an epicentre outside of Xiamen occurring in the eastern part of the Taiwan Strait and above intensity 7 on the Richter scale have caused damage to Xiamen (ITTXDP, 1996). Medium to strong earthquakes with a magnitude of about 6 in the Richter scale are predicted to occur in this particular fault zone in the next 100 years, and may cause extensive damage in the urban areas.

Tropical Storm

Typhoons are the most severe and dangerous weather disturbances in the Xiamen region. Typhoons usually occur in July, August and September at an average rate of 5.4 times a year (Table 2). The strongest wind force can reach 8 on the wind scale with a maximum speed of 60 m/s and with a maximum rainfall of 510 mm (ITTXDP, 1996).

Storm Surges

Storm surges occur in summer and autumn, increasing water level by an average of 50 cm with an amplitude from 1.50 to 2.00 m. The maximum level relative to high tide can reach 1.31 m. The storm surges in Xiamen are often accompanied by rising tides and waves causing seawater flooding. This results in loss of life and damage to properties (ITTXDP, 1996).

Coastal Erosion

Nearly half of the shorelines of Xiamen have been eroded or have receded at varying rates (Table 3). Wave action, sand quarrying, and poor planning of construction projects have all caused coastal erosion. The erosion rates of these shorelines range from 1–2 m per year and can reach 2–3 m per year in some places, e.g. the Tapu–Huangchuo span in Xiamen Island. Coastal erosion has resulted in damage to infrastructure, large-scale soil and field depletion, accelerated port siltation, farmland flooding and the reduction of the environmental integrity of the Xiamen area.

MAJOR SHALLOW WATER MARINE AND COASTAL HABITATS

The coastal waters of Xiamen provide an important habitat for various marine species. Studies show that there are 181 species of phytoplankton, 192 species of zooplankton, 248 species of nekton, 731 species of benthos, and 817 intertidal zone species in the Xiamen area (see Table 4) (ITTXDP, 1996). Over time, certain marine species in Xiamen have become rare and endangered, particularly the Chinese white dolphin (*Sousa chinensis*), lancelet (*Branchiostoma belcheri*), and egret (*Egretta eulophotes*).

Tidal Flats

Xiamen has extensive mud or tidal flats extending over a total area of 136.7 km². These tidal flats are distributed in the following areas: 15.1 km² in Xiamen Island; 98.3 km² in

Table 2

Frequency of tropical storms in Xiamen

Month	5	6	7	8	9	10	11	Total
No. of typhoons	7	18	53	61	101	6	209	455

Source: Office of the Steering Group for Comprehensive Investigation of Xiamen Island Resources, et al. (1994), cited in ITTXDP (1996).

Table 3

Some eroded coastal areas in Xiamen

Location	Length of retreated coastline (km)
Gaoqi-Hechuo	19
Yinchuo-Aoguan	3
Western Tong'an Bay	2
Hechuo-Xiamen University	16

Source: ITTXDP (1996).

Table 4

Numbers of marine species in Xiamen coastal waters

	Group	No.
Phytoplankton	Bacillariophyta	164
	Pyrrophyta	16
	Cyanophyta	1
	Total	181
Zooplankton	Medusa	75
	Copepoda	73
	Ostracoda	3
	Decapoda	4
	Amphipoda	1
	Mysidacea	8
	Euphausiacea	1
	Chaetognatha	11
	Tunicata	5
	Others*	11
	Total	192
Nekton	Fish	157
	Shrimp	42
	Crab	32
	Squid	4
	Cephalopoda	13
	Total	248
Benthos	Polychaeta	248
	Crustacea	175
	Mollusca	146
	Echinodermata	43
	Demersal fish	84
	Others	35
	Total	731

*Including larval plankton and ichthyoplankton.
Source: Third Institute of Oceanography, SOA (1988) and Office of the Steering Group for Comprehensive Investigation of Xiamen Island Resources, et al. (1994), cited in ITTXDP (1996).

Tong'an, Xinglin and Jimei; 24.4 km² in Dadeng Island; and about 1 km² in Gulangyu Island (ITTXDP, 1996). Some can be found in the northern coast of the West Harbour; the coast of Haicang, the southwest, north and northeast coasts of Tong'an Bay, and in the north coast of Xiamen Island. Most of the mud flats have been used for mariculture, especially those in the northern coast of West Harbour, Tong'an Bay including the north of the Dongdu port, the northern coast outside Maluan Causeway, around Paitou and west of Dayu Island and the north of Songyu pier. Several tidal flats have also been used for port construction and coastal engineering projects. In the past, disputes have arisen between the mariculture industry and the port authorities as a result of conflicting interests in this resource.

Freshwater Resources

Freshwater resources are scarce in Xiamen. Groundwater has a high saline content and is not suitable for drinking.

Mangroves

Mangroves are found in small patches in Haicang, Dongyu and west of Jiyu Island (Wu et al., 1988). Six mangrove species grow in Xiamen: *Kandelia candel*, *Avicennia marina*, *Aegiceras comiculatum*, *Acanthus ilicifolius*, *A. ebracteatus* and *Excoecaria agallocha* (ITTXDP, 1996).

Estuarine Habitats

Jiulongjiang River Estuary tapers from the middle towards its mouth. In the middle lies an underwater shoal around Jiyu Island. Deep-water channels, with an average depth of 10 m, are found along the south and north banks of the estuary. The harbour of the estuary is known as the Outer Harbour, and has a water depth of 7–20 m. Its northern side faces the beaches of Gulangyu Island, the Hulishan and the Baishi fort.

The estuary is characterized by hyposaline surface water with saline bottom water, with the mid water column being a mixture of offshore water (55%) and freshwater (45%) (ITTXDP, 1996). The water quality of the estuary, generally, falls within Classes I and II of the National Seawater Standards, particularly with regard to the parameters of dissolved oxygen, COD, inorganic phosphorus, and oil contents. The concentration of inorganic nitrogen exceeds Class III of the National Seawater Standard, while the faecal coliform count at the mouth of the estuary exceeds Class IV of the National Seawater Standard (Wu et al., 1992).

The phytoplankton cell count in the estuary and the Outer Harbour has an average of 1.06×10^6 ind/m³ (Wu et al., 1987). The dominant species include *Chaetoceros curvisetus*, *Skeletonema costatum* and *Thalassionema nitzschioides*. The zooplankton cell count has a maximum total biomass and density of 626.7 mg/m³ and 462.0 ind/m³, respectively in the month of May. The dominant species include *Labidocera euchaeta*, *Schmacheria poplesia* and *Pleurobrachia globosa* (ITTXDP, 1996). The most abundant of the zooplankton species are the nearshore warm water species.

Jiulongjiang River is the main source of sediments for the estuary. The sedimentation patterns of the estuary vary from west to east. Sediment quality in the estuary conforms to the Standards of Sediment Environmental Quality, except those found in the northern coast of Jiyu Island and in the southern coast of Haicang. The movement of these sediments towards the West Harbour is one of the main causes of siltation in the West Sea.

RESOURCE EXPLOITATION, UTILISATION AND CONFLICTS

Table 5 summarizes the environment and resource issues in Xiamen (Chua, 1998). At the end of 1993, the population of Xiamen reached 1.1749 million. This population is expected to reach 1.5 million and 2.3 million by 2000 and 2010, with annual growth rates of 3.71% and 3.9%, respectively. The population density is about 11,700 persons/km² (ITTXDP, 1996).

The Xiamen area began as a rural settlement during the Yuan Dynasty and has subsequently been used both for commercial and military purposes. In 1394, a military base was constructed in Xiamen for the purpose of resisting Japanese invasion. The town eventually evolved into a small trading town and served as the outer harbour of Yue Harbour in Zhangzhou. Commercial and trade activities began in 1684 and by 1842 Xiamen had become one of the five major trade ports in China.

In time, Xiamen developed into an industrial city. Economic development was furthered by the declaration of

Table 5

Resource utilization conflicts in Xiamen

Kind of conflict	Specific problem
Use conflicts	Multiple use of coastal zone and marine space
	Oyster culture and navigation
	Eel fry gathering and navigation
Conservation and economic activities	Amphioxus fishing
	Coral reef fish harvesting
	Protection of white dolphin and egret
	Coastline preservation and sand/tin mining
	Coastline preservation and reclamation
Degradation of water quality	High nutrients loading
	High level of heavy metals
	High level of coliform bacteria or *E. coli*
	High concentration of hydrocarbon
	Oil spill occurrence
	High level of pesticides
Habitat degradation	Destruction of spawning/nursery grounds of marine organisms
Environment disaster	Red tide outbreaks
	Major fish kills
Overexploitation	Overfishing

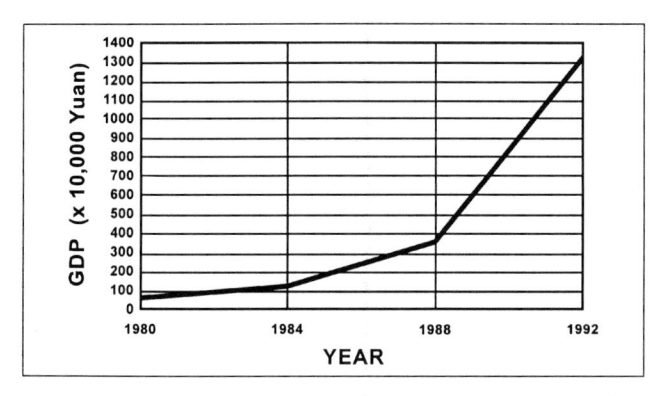

Fig. 4. Gross Domestic Product in 1980–1993. Source: China Statistics Press (1994).

Xiamen and Gulangyu Islands as a Special Economic Zone (SEZ) in 1984. During this time, there was a rapid growth in business and industry, tourism, and the real estate sector in the Xiamen area (Chua et al., 1997). Since its declaration as a SEZ, Xiamen has experienced increased socio-economic growth. From 1980–1993, Xiamen's economic development grew by an annual rate of more than 20% (Fig. 4) and in 1993, the Gross Domestic Product (GDP) reached 13.2 billion Yuan (ITTXDP, 1996).

Light industries are central to the Xiamen economy. In 1993, light industry accounted for 66% of Xiamen's production; heavy industry accounted for 26%; and agriculture accounted for 8%. Xiamen's industrial base includes manufacture of machinery, electronics, textiles, chemicals, construction materials and processed food products (ITTXDP, 1996).

The rapid economic development of Xiamen is attributed to its natural port advantages, hospitable climate, scenic attractions, strategic location within the Xiamen–Zhangzhou–Quanzhou trade triangle and its proximity to Hongkong and Taiwan, and to capabilities in scientific research and technological development.

Fishing and Mariculture

Capture fisheries and mariculture are the two major sources of income in the coastal areas of Xiamen. Capture fisheries are mainly undertaken in the offshore waters with the use of the set net, drifting gill net and push net methods. Small trash fish, shrimps and cephalopods are the main catch in summer and autumn while juveniles and fingerlings comprise the fish catch during winter and spring. In 1989, marine fish production reached 41,193 tonnes. In the early 1990s, however, this declined to around 40,000 tonnes. Production bounced back to 42,229 tonnes in 1993 (ITTXDP, 1996).

Mariculture production, on the other hand, gained momentum in the late 1980s, reaching a total output of 7,162 tonnes in 1988 (ITTXDP, 1996). This was mainly a result of the sharp increase in seafood prices and open access to coastal resources. Annual production more than tripled in the early 1990s and reached 25,862 tonnes in 1992 (ITTXDP, 1996). The growth of the mariculture industry stimulated not only areal expansion but also the development of new farming species and methods, such as oyster farming using stone stack and suspended feeding in shallow waters as well as large-scale caged-fish farming (Table 6).

Some other marine products and derivatives are being used for food, pharmaceutical and health purposes. For example, pharmaceutical products derived from the Chinese horseshoe crab have been produced and marketed.

Beach Reclamation and Coastal Construction Activities

The construction of the Gaoqi-Jimei and the Maluan causeways, and the large-scale reclamation of Maluan Bay, Xinglin Bay, Dongyu Bay and Yuan Dang Lake not only changed the natural contours of Xiamen but also caused environmental problems such as siltation of some navigable channels. The serious pollution of the Yuan Dang Lagoon is partly attributable to the construction of the Maluan causeway.

Coastal Tourism

Xiamen is a major tourist area. Attractions include the Gulangyu–Wanshisan national park, the beach resorts, natural scenic spots and some artificial attractions (Tables 7 and 8). The environmental quality of some coastal attractions, such as the beach areas in Gulangyu Island, has suffered only slightly from tourism. In other coastal tourism areas, the environmental quality of the beaches has deteriorated as a result of pollution from land-based wastewater discharges, dumping of rubbish, and oily wastewater discharges from ships. Consequently, approximately one third of the sandy beaches around Xiamen Island are affected by microbial contamination and oil pollution (ITTXDP, 1996) (Table 9)

Port Development

Xiamen port has a geographical advantage, given that it is prominently and strategically located within three bustling economic and trade centres. Firstly, in the Fujian Province, it is located in the Xiamen–Zhangzhou–Quanzhou trade triangle, which is the most important port for foreign trade. Secondly, Xiamen lies between Shanghai and Guangzhou, China's two leading trade economic centres. Shanghai is the centre of economic development in the Changjiang River Delta. Thirdly, Xiamen is a part of the East Asian Region and is China's important gateway to the bigger Asia-Pacific economic belt.

Deep harbours—an important consideration for port development—are located mainly in the coasts west of Xiamen Island proper, near Paitou, Dayu Island, and from Songyu to Qianyu Island. Deep channels are located between Xiamen and Gulangyu, Sonyu and Gulangyu,

Table 6
Mariculture production in 1988–1992

Species	1988			1989			1990			1991			1992		
	Area	P	kg/u	Area	P	kg/u	Area	P	kg/u	Area	P	kg/u	Area	P	kg/u
Shellfish															
Razor clam	155	108	98.2	2190	2105	984	2585	2600	1046	2620	2627	1003	4055	3289	811
Ruditapes philippinarum	6335	3290	755	555	4576	973	5605	5183	1025	6295	6024	1029	7508	6640	884
Blood clam	20	10	500	300	290	1000	320	290	906	320	290	906	250	200	800
Oyster	17670	2649	150	55709	8449	158	56371	9187	165	60198	95006	158	60633	9211	152
Mussel				120	120	1000	150	165	1100	380	470	1238	540	1510	2790
Fan shell mussel							17	567		2000	367	184			
Algae															
Sea Tangle	80	13	5	5	1000	5	10	2000	10	10	1000	110	165	1550	134
Laver			163	5320	636	120	5470	682	125	6120	814	133	9000	1208	
Fish	1477	62	42	1228	104	87	1275	98	77	1445	100	69	1596	161	101
Soya	70	60	1045	155	63	417	168	63	377	204	98	492	699	157	224
Prawn	9547	971	102	25970	3317	128	25385	3262	129	26167	4112	157			
Total	35314	7162		96552	19770		97364	21557		105759	24412		113571	25862	

P = production (kg); kg/u = kilogram per unit. Fish includes rock, cod, perch and red porgy which are raised mainly in cage nets. Source: Center of Environmental Science Studies, Xiamen University (1992) cited in ITTXDP (1996).

Table 7
Natural scenic sites

Type	Subtype	Location
Island	1. national level mountain	Wanshiyan, Sunrock
Hill	2. city level mountain	Hongjishan, Wulaofeng, Yupingshan, Hongshan, Huweishan
Island scenic spot	3. sea stacks	Southeast of Xiamen Island, Gulangyu, Jiyu
	4. sea caves	Southeast of Xiamen Island, Gulangyu, Huoshaoyu, Baozhuyu
	5. sea-erosion-arch-bridge	Gulangyu, Jiyu, Eyuyu
	6. sea erosion group	Southeast of Xiamen Island, Eyuyu
	7. wind shaking rock	Xianyueshan at Xiamen Island, Huajing Garden
Beach dune	8. sandy beach (swimming)	East coast of Xiamen Island, Gulangyu, Dadeng Island
	9. dune	
Island reef	10. islands	
	11. reefs	Gulangyu, Huoshaoyu, Jiyu, Baozhuyu, Eyuyu
Chann el bay	12. channel	East coast of Xiamen Island, south coast of Gulangyu
	13. bay	Xiamen-Gulangyu channel
Spring	14. mineral spring	Xinglin Bay, Yuan Dang Lake, Xiamen Bay, Tong'an Bay
	15. hot spring	East Longsheshan, southeast Wulaofeng
Geological site	16. tectonic landscape	Zhongzhai
	17. sea eroded landscape	Xiamen Island, Gulangyu, Huoshaoyu, Jiyu
Ecosystem	18. coastal protected forest	Dadeng Islands, east coast of Xiamen Island,
		East coast of Xiamen Island, Wanshi rock at Xiamen Island, Subtropic Plant Institute
	19. coastal ornamental plants	Gulangyu Plant Introduction, northwest of Dayu, Eyuyu, Jiyu, Dayu, Xiamen Island and Yuan Dang Lake, Xinglin Bay
	20. mangroves	Huoshaoyu
	21. egret	Xiamen Harbour
	22. dolphin	Amphioxus (Eyuyu, Dadeng, Xiaodeng, Huangcuo)
	23. valuable seafood	Mitella (Tuyu)
Lake	24. lake	Yanwu pool and confederate rose lake (ancient Lagoon), Yuan Dang Lake
		Shangli, dongshan, Wanshiyan, Hubian Reservoirs
Reser voir	25. reservoir	

Source: Office of the Steering Group for Comprehensive Investigation of Xiamen Island Resources et al., 1994 cited in ITTXDP (1996).

Table 8
Historical sites in Xiamen

Type	Subtype	Location
Archaeological ruins	1. neolithic ruins	Xiangyu coast, Shijantou
	2. museum of mankind	Museum of Mankind of Xiamen University
Historical Sites	3. stone tablet, stonecutting	Nanputuo Temple, Wutong ferry, Wanshiyan, Sunlight rock
	4. ancient town/barbett e	Hulishan barbette, Baishi barbette
	5. temple, church	Nanputuo Temple, masjid, Xinjie church, Qingjiaochiji Temple
Garden	6. cemetery and park	Cemetery of Xiamen revolutionary martyr, Shuzhuang Garden and Ming Garden at Gulangyu
Cultural Sites	7. memorial hall	Museum of Overseas Chinese, Zhengchenggong memorial hall, Luxum memorial hall, Jimei cultural village
Island Attractions	8. salt ponds	Zhongzhai at Xiamen Island and Dadeng Island
	9. mariculture sites	North and northeast coast at Xiamen Island and Huoshaoyu, Dadeng, Xiaodeng, Dalimuyu
Seaport	10. modern port	Dongdu port, Heping berth
Seafood	11. fresh seafood	Xiamen Island, Gulangyu
	12. local flavour snacks	Xiamen Island
Tourist Shopping	13. carved stone handicrafts	Stone-carving factory
	14. native products	Xiamen carved lacquerware, pearl embroidery sites
Causeway	15. causeway	Gao-Ji causeway, Dadeng causeway
Bridge	16. bridge	Xiamen Bridge

Source: Source: Office of the Steering Group for Comprehensive Investigation of Xiamen Island Resources et al., 1994 cited in ITTXDP (1996).

Table 9
Status and impacts of seaside resorts and marine parks in Xiamen

Area	Capacity per 10,000 persons	Status per 10,000 persons	Environmental impacts
Gulangyu	0.4/d, 146/a	>1/d at peak	minimal
Aoyuan park	0.35/d	0.16/d	minimal
Wanshi park	800/a	0.3–0.5/d	deteriorating
Nanputuo	6/d/580/a	0.3/d	5.5/d at peak
Hulishan barbette	2d, 73/a	0.1–0.2/d	minimal
Yundingyan	n.d.	dozens of hundred/d	minimal
Qingjiao	n.d.	4/month in 1993	minimal
Seaside:			
Gulangyu	0.2–0.6/d	n.d.	deteriorating
Xiamen University	0.2/d	>1/d at swimming season	sand beach erosion, seawater pollution
Hulishan-Baishi barbette	1/d, 190/a	underdeveloped	high potential impacts

n.d. = No data. Source: Center of Environmental Science Study, 1994 cited in ITTXDP (1996).

Huoshaoyu and Paitopu, Jiyu and Qianyu (Table 10) (ITTXDP, 1996).

At present, Qingyu Island Channel serves as the navigational entry to Xiamen Port. It has a minimum water depth of 14–18 m and a maximum depth of 26 m (ITTXDP, 1996). Port construction and development began in the 1980s. The first phase of Dongdu port project was finished in 1984 and the second phase in 1994. Since then, port-handling capacity has doubled and the number of berths has been increased to 66 to handle the yearly increase in the volume of passengers and goods (Fig. 5) (ITTXDP, 1996).

Port development activities include the third phase of the Dongdu port project, construction of special berths at Songyu and Haicang and medium-small berths at Tong'an. Xiamen's long-term plan is to have seven ports, with a total handling capacity of 50 million tonnes.

CRITICAL ENVIRONMENTAL PROBLEMS

Human activities such as mariculture, construction of causeways and land-based waste discharge have resulted in pollution of the marine waters and changes in the ecosystem of the waters of Xiamen (Table 11). Despite the accumulation of pollution in some areas, the Xiamen coastal waters are still in good condition, not exceeding

Table 10

Distribution of deep-water harbours in Xiamen shoreline

Location of deep-water harbours			Shoreline length (km)
Xiamen	Xiamen Island	Dungshang	1.4
		Wutong	5.0
		Shihushan–south fishing port	6.0
		Xiamen Island facing Lujiang	3.0
	Jimei Xinglin	West to Xiangbishan–Neikeng	4.0
		North to Xiangbishan	0.3
		Songyu–Dayu	1.0
		North of Shuitou, north to Paitou	0.6
	Islands	North of Dayu	0.6
	Tong'an	Liuwudian	0.5
Longhai	Gangwei Village	Dashikeng	1.6
		Yuziwei	2.7
		Total	26.7

Source: Office of the National Land Planning Management Division (1993) cited in ITTXDP (1996).

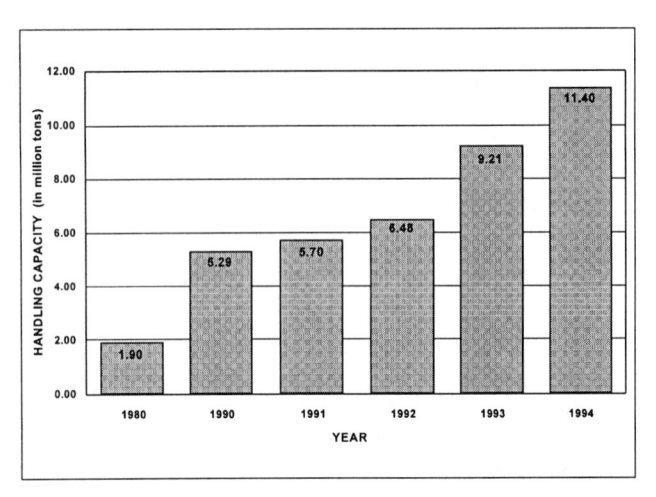

Fig. 5. Development of handling capacity of the Xiamen Port. Source: Xiamen Statistics Yearbooks, 1991–1994, cited in ITTXDP (1996).

Classes I and II of the Seawater Quality Standards of the country—with the exception of the waters in Maluan Bay and those near Baozhuyu Island.

It must be pointed out, however, that the rapid economic development of Xiamen poses a threat to the quality of the coastal waters. Economic development has resulted in an increase in the use of petroleum and fuel products (e.g. LPG) and chemicals. This has heightened the potential for environmental damage. Areas where the risk of environmental damage has increased include the Zhaishang Village where jetties and depots of fuel oil and petroleum are located, the airport fuel depot area, the chemical jetties in Xinglin District, and the petroleum trans-shipment jetties in Haicang.

Organic Pollution, Eutrophication and Red Tides

Organic pollution in the coastal waters, mainly from domestic, industrial and mariculture wastes, is one of the most critical problems in Xiamen. This is evidenced by an increase in Chemical Oxygen Demand (COD), an indicator for organic pollution. This is particularly evident in the Maluan Bay area. This bay has the most serious level of organic pollution in these coastal waters due to high concentrations of COD, sulphide and organic matters. The high concentration of sulphide and organic materials in the surface sediment is a result of accumulated organic materials from the water and reflects the long-term organic pollution in the water column.

The COD in Maluan Bay is within the range of Classes III and IV of the Seawater Quality Standard while the rest of the coastal waters of Xiamen have a smaller concentration of sulphide and organic materials, and lower COD levels. The COD contents of Tong'an Bay and Outer Harbour are the lowest and meet Class I of the standard while the COD level in the West Harbour falls between Classes I and II.

In general, Xiamen coastal waters exhibit a moderate level of eutrophication caused by the presence of inorganic nitrogenous nutrients (Hong et al., 1995). Most of Xiamen coastal waters fall within Class I of the Seawater Quality Standards for dissolved inorganic phosphate except for the bottom water of Maluan Bay which has a high concentration of dissolved inorganic phosphate.

In 1986 and 1987, major red tide incidents due to *Gymnodinium* and *Eucampia zoodiacus* occurred at the West Harbour, particularly on the north coasts of Dongdu-Huoshaoyu Island. In 1986 a *Gymnodium* red tide outbreak in the Xia-Gu Channel even extended to the offshore waters of Hulishan Fortress, forming a red tide belt in Xiamen coastal waters (Fig. 6). Studies of Xiamen waters indicate that red tide occurrences follow a complex process triggered by one or several of the following factors: eutrophication, extreme weather variation (particularly in the summer and spring), excessive nutrient loading during storms, rapid transformation of nutrients, subsequent transportation of sedimentary organic matter during the warm season, and photosynthesis of blooming red tide species aided by clear or sunny weather after the rains (ITTXDP, 1996).

Bacterial Pollution

Bacterial pollution has also been found in varying degrees in Xiamen coastal waters. The concentration of faecal coliform is highest in the West Harbour where the faecal coliform count reaches 3914 individuals/l which exceeds the standard for recreational seawater (2000 ind/l) (ITTXDP,

Table 11
Concentration of some pollutants in Xiamen coastal waters

Date	Areas	DO (mg/l)	COD (mg/l)	BOD (mg/l)	RP (mg/l)	TIN (mg/l)	S^2 (mg/l)	OM (%)	Oil (mg/l)	Faecal coliform (ind/l) Planning	Faecal coliform (ind/l) Sea Islands
Jun–Dec 1982	West Harbour		1.24								
Jan–Dec 1987	West Harbour	7.16	1.38		0.015	0.21					
Mar 1987–Feb 1988	West Harbour	7.22	1.33		0.015	0.22	118	1.82		TC < 54,000 30% exceeding standard	
Mar–Oct 1989	West Harbour	6.68	1.50		0.013	0.26		2.40	0.012		3914
Apr 1989–Mar 1990	West Harbour		1.81		0.011	0.34	284			69% exceeding standards**	
1992	West Harbour								0.026		
	Maluan Bay: shallow water	9.37	4.32	3.29	0.003	0.29			0.058	646	
	Maluan Bay: bottom water	3.45*	7.72	3.50	0.161	0.08	1636	3.51		(20–5420)	
	Estuary and Outer Harbour: shallow water	7.24	1.53	0.88	0.016	0.41			0.015	5136+	1800+
	Estuary and Outer Harbour: bottom water	6.87	1.03		0.015	0.24	147	1.74	0.015	1610+	+1187++
	Tong'an Bay: shallow water	8.15	1.08	0.89	0.012	0.21	31	0.99	0.012	108	
	Tong'an Bay: bottom water	8.19	0.93		0.012	0.17				(20–1700)	
	Eastern Waters										2806

DO = dissolved oxygen; COD = chemical oxygen demand; BOD = biological oxygen demand; RP = reactive phosphate; TIN = total dissolved inorganic nitrogen; S^2 = sulphide at surfacial sediment; OM = organic matter at surfacial sediment; ** = based on the FC standard of 2000 ind/l; * = dissolved oxygen was not detected for some stations (data in brackets are range of determination).
Sources: Institute of Environmental Science Study, Xiamen City (1990); Office of the Steering Group for Comprehensive Investigation of Xiamen Island Resources et al. (1994); Third Institute of Oceanography, SOA 1983, 1993a, 1993b, 1993c, 1994; and Xiamen Management Region, SOA, 1990 cited in ITTXDP (1996).

1996). This is the result of large quantities of domestic wastes being dumped into coastal waters.

The faecal coliform content in the Eastern Channel and the Jiulongjiang River Estuary is lower than in the West Harbour. Tong'an Bay has the lowest faecal coliform content which makes it suitable for shellfish culture. The level of faecal coliform content in Maluan Bay is not as high as the organic pollution in the area. The faecal coliform count varies from 20 to 5240 ind/l, with an average of 646 ind/l. Bacterial pollution in Maluan Bay mainly comes from mariculture rather than from domestic wastes.

Oil Pollution

Generally, oil pollution is not a major problem in Xiamen coastal waters except in Maluan Bay where there is a concentration of 0.058 mg/l of oil in the water column, exceeding the standard of Class I of the Seawater Quality Standard (ITTXDP, 1996). The oil contamination in this area is attributed to the oily wastewater discharge from the numerous small boats used for mariculture in combination with limited water exchange between the bay and the West Harbour.

The West Harbour is second to Maluan Bay in terms of oil contamination. Oil pollution has been directly caused by oil spill from the oily water discharges from sea vessels and motorized fishing boats. In recent years, however, prevention and management of oil pollution have improved considerably.

From 1984 to 1992, several oil spills occurred in Xiamen. It was discovered that some of these oil spills originated from factories along the coastlines and illegal ship discharges. In 1989, 21 oil slick incidents covering an area over 500 m² were recorded from Dongdu Port to southeast Gulangyu Island. Similar oil slick incidents occurred again in 1990 and 1991 in the same area.

Sea Dumping and Disposal of Solid Waste

The major wastes dumped in the Xiamen coastal waters are the dredged materials from jetties, port basins and navigation channels, and the waste from the production of monosodium glutamate (MSG). The volume of wastes dumped at sea vary according to levels of production but the volume has never exceeded one million tons per year (Programme Group for Waste Management, 1996).

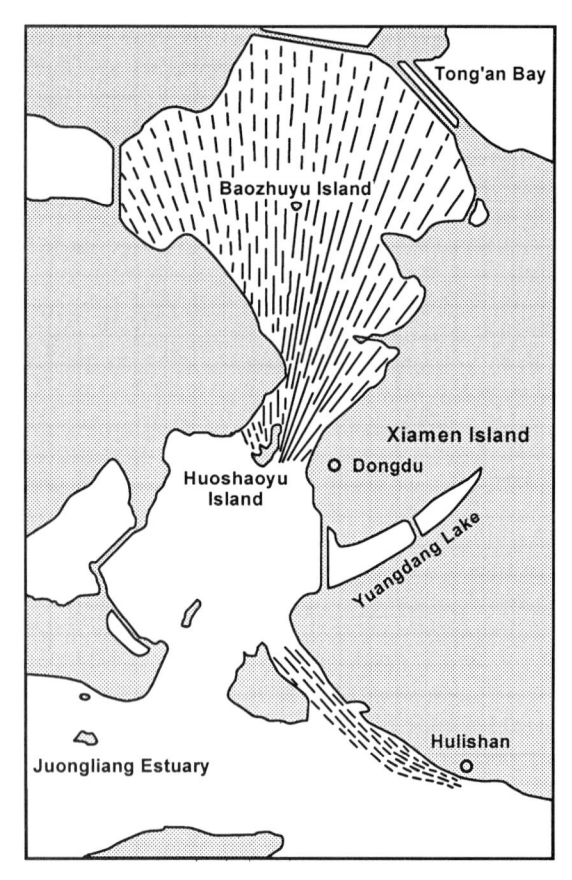

Fig. 6. Red Tide areas in 1986 and 1987. Source: Third Institute of Oceanography, SOA (193) cited in ITTXDP (1996).

In 1990, the Xiamen Government prohibited inshore dumping of wastewater and relocated the dumpsite to Zhenghaijiao in accordance with the 1972 London Dumping Convention. Since 1990, around 0.325 million tonnes of wastewater from MSG production has been dumped. Alternative solutions were eventually developed by the industry and ocean disposal of MSG was terminated in 1995 even prior to the global ban on ocean dumping of industrial wastes.

All other industrial solid wastes are disposed of at designated landfills except for rubbish such as plastic bags, bottles, etc., mainly from households, recreational and tourist facilities and sea vessels which litter the shore and offshore waters of Xiamen.

Sediment Deterioration and Secondary Pollution

The polluted sediments of Maluan Bay are becoming a secondary pollution source. Its sulphide content exceeds 1636 mg/kg giving off a foul smell during both summer and winter, and its organic matter content reaches 3.51%, both parameters exceeding the Seawater Quality Standards of the country (ITTXDP, 1996). The high content of organic matter in the sediment poses a potential source for eutrophication. The mercury concentration in the surface sediment has reached 0.107 mg/kg, also becoming a secondary pollution source (ITTXDP, 1996).

Pollution from Pesticides

The situation with respect to pesticide use in the Xiamen agricultural area is of major concern. The total load of pesticides on agricultural soil is 27 kg/ha on average and it is dominated by organophosphates which are both very toxic to aquatic fauna and very mobile (MPP-EAS and FAO, 1998).

ECOSYSTEM CHANGES

Over the past 40 years, the coastal ecosystems of Xiamen have changed significantly. An indicator of ecosystem change is the decline of the quantity of some phytoplankton-sensitive species. The populations of once common species such as the lancelet and the Chinese white dolphin have likewise declined. In fact, only few marine species can now be found in the seriously polluted Maluan Bay and the sea areas around Baozhuyu Island. In addition, the habitat and community structure in Yuan Dang Lake have significantly changed because of pollution.

Maluan Bay

Maluan Bay became an enclosed water body as a result of the construction of the Maluan Causeway and the intensive mariculture activities in the bay. The causeway itself has reduced the surface area of the bay by half. The level of pollution is aggravated by the lack of water exchange between the bay and the West Harbour because the sluice gates of the causeway are usually closed throughout the year. In addition, freshwater runoff diluted the salinity of the waters in the bay. The low salinity in the bay may suggest poor water exchange and pollutant accumulation.

The water between the causeway and Baozhuyu Island is stagnant because of the weak tidal currents of only 10 cm/s in the upper portion of the West Harbour and 20 cm/s in the channel southwest of Baozhuyu Island. In the place of weak tidal current, a local eddy current allows suspended particulate materials to deposit in the sea area (ITTXDP, 1996).

There are only four groups of species recorded thus far in Maluan Bay. Phytoplankton biomass is recorded at 1.62×10^9 ind/m^3. The dominant species are *Nitzchia delicatissima*, *Cyclotella comta*, *Skeletonema costatum* and *Chaetoceros affinis* (ITTXDP, 1996). The dominant benthic species include *Neanthes donghaiensis*, *Nectoneanthes oxypoda*, *Nectoneanthes multignatha*, *Dendronereis pinnaticirris*, *Mytilopsis sallei*, *Musculista senhausia*, *Ruditapes philippinarum*, *Potamocorbula ustulata* and *Sermyla riqueti* (ITTXDP, 1996).

Tong'an Bay

Before the Gaoqi–Jimei Causeway was constructed in 1956, West Harbour and Tong'an Bay used to be one water body.

After the causeway was built, the West Harbour became a semi-enclosed bay, with entrance in the south where it combines with the Jiulongjiang River Estuary as they flow into the Outer Harbour.

The construction of the Gao–Ji Causeway, the reclamation of mud flats and mariculture activities have caused significant environmental problems in the Tong'an Bay ecosystem. The construction and its accompanying reclamation have reduced the navigable area by 30 km². Mariculture has likewise reduced the navigable area by another 35 km². These activities resulted in pollution of the coastal waters causing some adverse effects on the marine ecosystem.

Annual fish catches in the area have drastically declined. It has been observed that the population of the lancelet (*Branchiostoma belcheri*) has declined and its habitats have moved outward. Prior to 1956, the average annual production of fish ranged from 70–150 tonnes. In the 1970s, fish catch declined to 1 tonne. The decline in fish catch was attributed to overfishing and the reclamation near Liuwidian.

PROTECTIVE MEASURES

In the past, coastal management measures in Xiamen were sectoral in approach. Recently, the Municipal Government adopted an integrated approach to coastal management. New management concepts have been introduced to transform the short-term and single-sectoral perspectives into more rational and sustainable policies.

Laws and Regulations

General policies, guidelines and standards on marine environmental protection are governed by the environmental provisions of the 1982 Constitution of the People's Republic of China, laws promulgated by the National People's Congress and its Standing Committees, national standards issued by the State Council and governmental agencies, as well as local standards issued by the local governments, local codes promulgated by the Local People's Congress and its Standing Committees, and administrative stipulations issued by the local government (ITTXDP, 1996).

The framework legislation of the People's Republic of China on environment is the Environmental Protection Law. The Marine Environmental Protection Law and the Seawater Quality Standard provide the regulatory framework for marine environmental protection. Specific measures for the management of the marine environment include the following:
– Environmental quality standards;
– Marine protected areas;
– Environmental Impact Assessment System;
– Requirement of three-concurrency (i.e. development projects and their facilities for pollution prevention and

Table 12

Laws and regulations on marine environment

	Title	Date of Promulgation
1.	Regulations Concerning Environmental Protection in Offshore Oil Exploration	29 December 1983
2.	Regulations Concerning the Prevention of Pollution in Sea Areas by Vessels	29 December 1983
3.	Fisheries Law	1 January 1986
4.	Regulations for the Implementation of the Fisheries Law	14 October 1987
5.	Regulations Concerning the Dumping of Wastes at Sea	6 March 1985
6.	Regulations Concerning the Prevention of Environmental Pollution by Ship-Breaking	18 May 1988
7.	Rules for the Implementation of Laws on Water Pollution Prevention and Treatment	12 July 1989
8.	Regulations Concerning Pollution Damage to the Marine Environment by Coastal Construction Projects	25 May 1990
9.	Regulations Concerning the Prevention of Damage to the Marine Environment by Land-Based Pollutant	25 May 1990
10.	Interim Regulations Concerning Management and Use of State Seawaters	31 May 1993

Source: ITTXDP (1996).

mitigation must be simultaneously designed, constructed, and put into production);
– Requirement of application, registration, and licensing of pollutant discharge;
– Fees for excessive discharge of pollutants;
– Requirement of eliminating pollution within a given period of time;
– Identification of compulsory emergency measures, such as production scale limitation or production stoppage to minimize pollutant damage;
– Monitoring, surveillance, and inspection; and
– A system of economic incentives.

Administrative, civil, and criminal liabilities are incurred for violating these environmental laws. In addition to the general environmental laws, detailed rules have also been enacted to govern specific activities affecting the marine environment, such as offshore oil exploration and exploitation, traffic of marine vessels, dumping of waste at sea, ship breaking, fisheries, and coastal construction projects (Table 12).

The Fujian Province and Xiamen Municipality People's Representative Congress and its Standing Committee have also enacted several local marine environmental regulations for the Fujian Province and the Xiamen Municipality. These include the Interim Provisions Concerning Resources Protection and Management of Sand, Stone and Soil, Provisions Concerning Management of Storm Shelter at Shapowei Fishing Port, and the Measures Concerning the Management of Yuan Dang Lake in Xiamen City.

Institutional Arrangements for Marine Environment Management in Xiamen

Prior to 1994, the management approach in the Xiamen Municipality was largely sectoral. Around 14 government agencies, both national and local, are responsible for the various aspects of coastal management. Often, concerned agencies function only in accordance with their mandates and are limited by their administrative jurisdictions despite the cross-sectoral concerns of coastal and marine management. Unclear mandates or indistinct delineation of duties among maritime agencies hampered the enforcement of laws and regulations.

At the national level, the State Environmental Protection Administration (SEPA), the State Oceanic Administration (SOA), the Environment Protection Departments, the Harbour Superintendency Administration, the state agency in charge of fishery harbour and superintendency, and the environmental protection department of the armed forces are all involved in marine environmental protection.

The national administrative framework adopted the traditional dichotomy between marine management and development and environmental protection. The SEPA has the mandate for environmental protection in general, while SOA is responsible for "comprehensive marine management", which includes organizing scientific investigation, monitoring, surveillance and research on the marine environment. SOA is also in charge of protecting the marine waters against damage caused by offshore oil exploration and exploitation and waste dumping at sea.

The Harbour Superintendency Administration is responsible for overseeing, investigating and dealing with the discharge of pollutants from vessels and for keeping under surveillance the waters of the port areas. It is also in charge of environmental protection against pollution damage caused by vessels.

At the local level, the protection and management of Xiamen's coastal waters is the primary responsibility of the Municipal Government. Several local agencies also have specific mandates in relation to the marine waters of Xiamen, which include the following:
- The Commission of the Sea/Frontier Defense of Xiamen Municipality which is the co-ordinating body for the comprehensive management of marine frontier affairs;
- A branch of the Office of Port Affairs which is responsible for the management and operation of the seaport and airport in Xiamen;
- Xiamen Customs, the field agency of the State General Customs which is tasked to collect customs duties;
- Xiamen Frontier Inspection, which supervises the entry and exit of passengers and inspects passports or certificates, baggage and transportation vehicles;
- Xiamen Maritime Safety Superintendency of Fuijian Province, an agency of the Transportation Department of the Fujian Province, which is responsible for the maritime safety and management of vessels;
- Xiamen Sanitary Quarantine, an agency of the Ministry of Health, which implements rules on quarantine of animals and plants including the containers and vehicles used within the area;
- Xiamen Commercial Commodity Inspection, which inspects imported and exported commodities;
- Xiamen Marine Management Region, which carries out marine surveillance, investigation, scientific research and supervision of ocean dumping;
- Marine Management Division of Xiamen Municipality, which is tasked to co-ordinate the various marine management sectors for the comprehensive management of the coastal waters;
- Port Bureau of Xiamen Municipality, which is responsible for the development and management of ports and jetties, as well as the management of cargo transportation by ship;
- Environmental Protection Agency of Xiamen Municipality, which is mandated to monitor and manage Xiamen coastal waters and is responsible for preventing damage to the coastal marine environment arising from coastal engineering projects and land-based pollution;
- Fishery Agency of Xiamen Municipality, which is responsible for the management and surveillance of mariculture in coastal waters and beaches, and the surveillance and management of waste discharges from fishing jetties and fishing boats; and
- Xiamen Navy, which has the function of providing marine environmental protection to military ships and ports (ITTXDP, 1996).

In the past, no coordinating structure for marine management existed at the national level or the local level.

Major Management Problems

There are about 12 sectoral agencies from the central, provincial and local governments engaged in coastal management. They carry out their respective functions without integrated planning and coordination. This results in fragmented policy-making and sometimes coastal-use conflicts.

Multiple-use Conflicts and the Lack of Integrated Planning

Conflicting uses in the marine waters of Xiamen have arisen in the past. Some of the major types of multiple use conflicts include the following:
- shipping vs. fisheries;
- shipping vs. biodiversity protection;
- shipping vs. coastal engineering;
- reclamation vs. habitat protection;
- waste disposal vs. human health; and

– current use vs. future use.

In the past, the Municipal Government of Xiamen had no mechanism for cross-sectoral conflict resolution. There was no integrated plan that could prioritize the use of the waters and address the issue of interaction among the sectors. The legal system was likewise inadequate to effectively resolve multiple-use conflict because the laws and regulations only reflected the sector-oriented management strategy and did not consider the overall benefit from multiple coastal uses.

Low Capability for Integrated Management

Integrated coastal management requires a combination of consultative mechanisms, various types of expertise, competent personnel, effective administration and communication, etc., which are all deficient in Xiamen.

Weak Capability for Pollution Prevention and Mitigation

Facilities for pollution prevention and mitigation are inadequate. For example, there is no comprehensive sewage pipeline system covering the entire Xiamen Municipality. Factories also suffer from a shortage of waste treatment facilities and are left without any choice but to use waste treatment facilities that are poorly designed and technologically outdated. The total volume of waste water in Xiamen Municipality is estimated at 100×10^6 t/a. In contrast, the capacity of the waste treatment facilities is only 35×10^6 t/a.

Weak Marine Environmental Consciousness Among Policy-makers and Lack of Awareness by the General Public

At present, there is a tendency to place marine environmental protection behind marine resource development and exploitation. This is partly due to low environmental public awareness. In order to ensure the sustainable development of Xiamen Municipality, there is a need to enhance environmental education and awareness.

Insufficient Funding for Integrated Environmental Management

Shortage of funds make it difficult to achieve the objectives of environmental protection. The fund for environmental protection in Xiamen Municipality has increased in recent years, accounting for 0.8% of the GDP. However, it is still insufficient as 2.19% would be necessary for effective pollution prevention and environmental protection. Environmental funding is primarily dependent on government appropriation. Without sufficient funding, many important environmental projects remain unimplemented.

Poor Law Enforcement

Marine pollution laws and marine management policies remain largely unenforced in Xiamen. One reason is the absence of enforcement mechanisms. Procedures for coastal environmental management are yet to be established.

Lack of Sound Scientific Knowledge

In the past, scientific knowledge on coastal and marine ecosystems was hardly used as a factor in decision-making in the development and management of the coastal zone in Xiamen Municipality. For example, the Yuan Dang and the Maluan causeways were constructed despite the considerable adverse effects on the coastal ecosystem. These effects include changes in hydrodynamics, siltation of nearby shore areas and navigation channels, reduction of environmental carrying capacity and marine pollution.

In addition, there were few studies concerning marine pollution and management, for example: impacts of vehicular exhausts, mariculture, port and ship pollution to the quality of the marine waters, design of a system for emergency management of oil spill and chemical accidents, effect of sand and pollutants from the Jiulongjiang River flowing into the marine waters of the West Harbour, forecasting impacts of socioeconomic development on the environmental quality of the West Harbour, and studies on the long-term variability of the marine ecosystem.

Lack of a Sound Information System

A considerable amount of marine environmental data has been accumulated from the marine environmental studies carried out by research institutions, marine monitoring activities, and environmental impact assessments conducted by marine management and protection agencies. However, these studies could not provide sufficient scientific information useful for integrated coastal management for several reasons.

First, there has been no thorough study of the coastal environment conditions of the different areas of Xiamen. Most of the studies were on West Harbour and the Estuary. Second, there has been no thorough study of sources of pollution in Xiamen marine waters except for one conducted on land-based sources of pollution in the West Harbour in 1989. Information on sea-based pollution sources, pollutants entering the coastal waters through the atmosphere and urban storm runoff is deficient. Finally, there is an absence of a cohesive information management system for the data. Information is scattered among various research institutions and government agencies making the systematic use of the data difficult. In addition, an information service system for decision-making in environmental emergency response is lacking.

Transboundary Problems

Pollution from sources outside the jurisdiction of Xiamen Municipality also has a significant effect on the environmental quality of the marine waters of Xiamen. For example, the waters under the jurisdiction of Xiamen Municipality receive pollutants from the Jiulongjiang River and adjacent regions.

Towards Integrated Coastal Management: The Impacts of the Partnership of Xiamen Municipality and the GEF/UNDP/IMO Regional Programme

Xiamen's continued economic development compounded with population growth, resource-use conflicts, pollution, and other coastal problems has caused stress in the marine ecosystem. In response to these multifaceted problems, the Municipal Government explored innovative strategies and management options for the sustainable development of coastal and marine areas of Xiamen.

In 1993, the Municipality of Xiamen, together with the Global Environment Facility/United Nations Development Programme/International Maritime Organization (GEF/UNDP/IMO) Regional Programme for the Prevention and Management of Marine Pollution in the East Asian Seas (the Regional Programme) launched the Xiamen Demonstration Project (the Project) with the objective of building the capability of Xiamen Municipality to apply the integrated coastal management system (ICM) in the prevention, mitigation, and management of marine pollution problems on a self-reliant basis (MPP-EAS, 1995).

The municipal government undertook capacity-building activities to strengthen its planning and management capability, eventually formulating and adopting the Strategic Management Plan for Marine Pollution Prevention and Management in Xiamen. An interagency multi-sectoral coordinating mechanism was established to integrate coastal management in the municipality. Environmental quality monitoring programmes were also developed as part of the new management mechanism (Chua and Ross, 1996).

The introduction of ICM in Xiamen has significantly affected the management regime of its sea space and marine resources. There is a greater public awareness of environmental concerns, a stronger political will to address environmental problems, and improved perception of the integrated management system among the resource managers and concerned scientific community. Interagency conflicts were reduced as a result of the new co-ordinating mechanisms and use conflicts were resolved through the adoption of a functional marine zonation scheme.

The management interventions of the municipal government arrested the increasing environmental degradation in the waters of Xiamen. In the Western Sea, for example, variations in the average concentration of nutrients have occurred in the past ten years. In 1996, however, average concentrations of total nitrogen, nitrate and phosphate were lower than those in 1986. In addition, no red tide event was observed in the Western Sea of Xiamen after 1990. The levelling off of nutrient levels is believed to be a result of pollution management interventions.

Strategic activities of the project included a coastal environmental survey, the preparation of a strategic management plan, the establishment of a marine management co-ordinating mechanism, and the preparation of a marine functional zonation scheme for Xiamen waters.

Strategic Management Plan (SMP) and its Implementation

The SMP builds on the overall policy of the Xiamen Government to "develop Xiamen into a socialist, modern, international, and scenic port city", taking into account, Xiamen's coastal environmental status and management problems (ITTXDP, 1997).

The SMP document assessed the causes and effects of the identified environmental problems and the associated risks, evaluated existing management measures, if any, and prioritized types of management actions that could be undertaken to resolve them. The SMP contains a strategic action plan and guidelines for both short- and long-term management of the coastal resources of Xiamen, as well as arrangements for the implementation, monitoring and evaluation of the plan.

The SMP was prepared after a series of consultation meetings with various agencies and experts. The general objective of the SMP was to ensure that economic development is achieved within the context of sustainable development. The SMP proposed the following policy measures to implement the strategic management objective:
- integrated planning and management;
- resource evaluation and environmental accounting;
- principles of marine area use and pollution damage compensation;
- adoption of precautionary principles;
- public participation; and
- sustainable financing mechanisms for environmental management (e.g. market-based instruments).

New Management Measures

The Municipal Government of Xiamen recognized the need to move away from the sectoral approach in the management of the coastal and marine resources and established the Marine Management Co-ordination Committee (MMCC or the Leading Group) as the integrated mechanism for coastal management (Chua and Ross, 1997). The MMCC is a high-level inter-agency committee set up for coordination and consultation on coastal development policies. It is chaired by the Executive Vice-Mayor and includes four other vice-mayors in charge of fisheries, marine, ports and harbours, tourism, industries, science and technology, and city planning and construction. The MMCC was created to address cross-agency management issues related to utilization of sea-space and marine resources. It also provides an organization structure that can implement the adopted recommendations of the SMP.

Table 13
Development of legal instruments for the marine environment in Xiamen

Year	Major Project Activities	Legal Instruments
1994	Strengthening local government commitments Public awareness campaign	Regulation for Environmental Protection
1995	Integrated management committee/office established Profile/environmental management plan prepared Marine laws reviewed and new legal instruments proposed	Regulation for Managing the Resources of Sands Rocks and Soils Regulations for the Management of Navigation Municipal Ordinance for Egret Nature Reserve in Dayu Island Administrative Rules on the Relocation of Aquaculture in the Marine Area for the Siting of Xiamen Shipyard Administrative Rules for Strengthening the Management of Catching Marine Eel Larvae Regulations for the Management of Water Resources
1996	Yuan Dang Lagoon case study Waste management problems assessed Aquaculture impact study Integrated monitoring system established	Municipal Ordinance for Managing Yuan Dang Lagoon Area Municipal Ordinance for Urban Landscaping and Environmental Health Administrative Rules for Aquaculture in Shallow Seas and Tidal Flats (1996) Regulations for Marine Environment Protection
1997	Integrated environmental impact assessment Functional zoning scheme developed studies on sustainable financing mechanisms	Regulations for the Uses of Sea Areas (1997) Regulations for the Protection of Chinese White Dolphin (1997) Regulations for the Management of Tourism (1997) Government Notice on the Implementation of Xiamen Marine Functional Zoning Scheme

The functions and responsibilities of the MMCC include the following:
- preparation of a functional zonation scheme;
- protection and management of the marine environment;
- enactment of local coastal and marine regulations;
- enforcement of laws jointly with relevant agencies to address coastal use and conflicts; and
- management of activities covering the use of coastal resources.

The implementing arm of the MMCC is the Marine Management Co-ordination Office (MMCO) under the Secretary-General of the Municipal Government. The MMCO is the reorganized Marine Management Division formerly under the Municipal Science and Technology Commission. The MMCO is tasked with planning, development, construction, management and co-ordination of the areas below the high tide line.

The co-ordinating mechanisms established by the Municipal Government have proven to be effective and efficient in resolving use-conflicts (Kho and Agsaoay, 1999). Furthermore, the MMCC and the MMCO are instrumental in the enactment of marine environment regulations covering such subjects as a functional zonation scheme, utilization of maritime space, environmental protection and aquaculture in marine waters.

Laws and Regulations

In the past, legislation was ineffective and unable to respond to the various issues and concerns pertaining to the management of Xiamen marine waters. Most laws were enacted to respond to specific sectoral concerns. Hence, it became imperative to establish an integrated legal regulatory system that would respond to these various issues in a coordinated manner, consistent with the framework of integrated coastal management.

Xiamen Municipality enacted the Regulation on the Use and Management of the Xiamen Seas, which is a comprehensive regulation on marine environmental management and protection addressing cross-sectoral concerns and multiple use of the coastal waters. The new regulation provided measures for inter-agency co-ordination in the coastal project review and permit process, scientific decision-making and the use of market-based instruments. The regulation institutionalized the interagency co-ordinating mechanisms for integrated coastal management and in particular identified the former Marine Management Division as the office tasked with the comprehensive management of the Xiamen waters. The new regulation also requires individuals to procure permits and to pay the corresponding users' fee for the use of the marine waters. Table 13 shows other relevant local measures on marine management and protection in Xiamen (Chua et al., 1999).

A Functional Marine Zonation Scheme

In 1997, the Xiamen Municipal Government adopted a functional zonation scheme (Fig. 7). This integrated the ecosystem and socioeconomic functions of coastal land and waters. The main purpose of the zonation is to reduce multi-use conflicts, maximize net social benefits in the coastal area, conserve biodiversity, and ensure long-term sustainable growth of Xiamen marine waters. The marine waters were classified according to the priority of the uses, taking into account the dominant use of the area, possible compatible uses, and activities which should be restricted in the area.

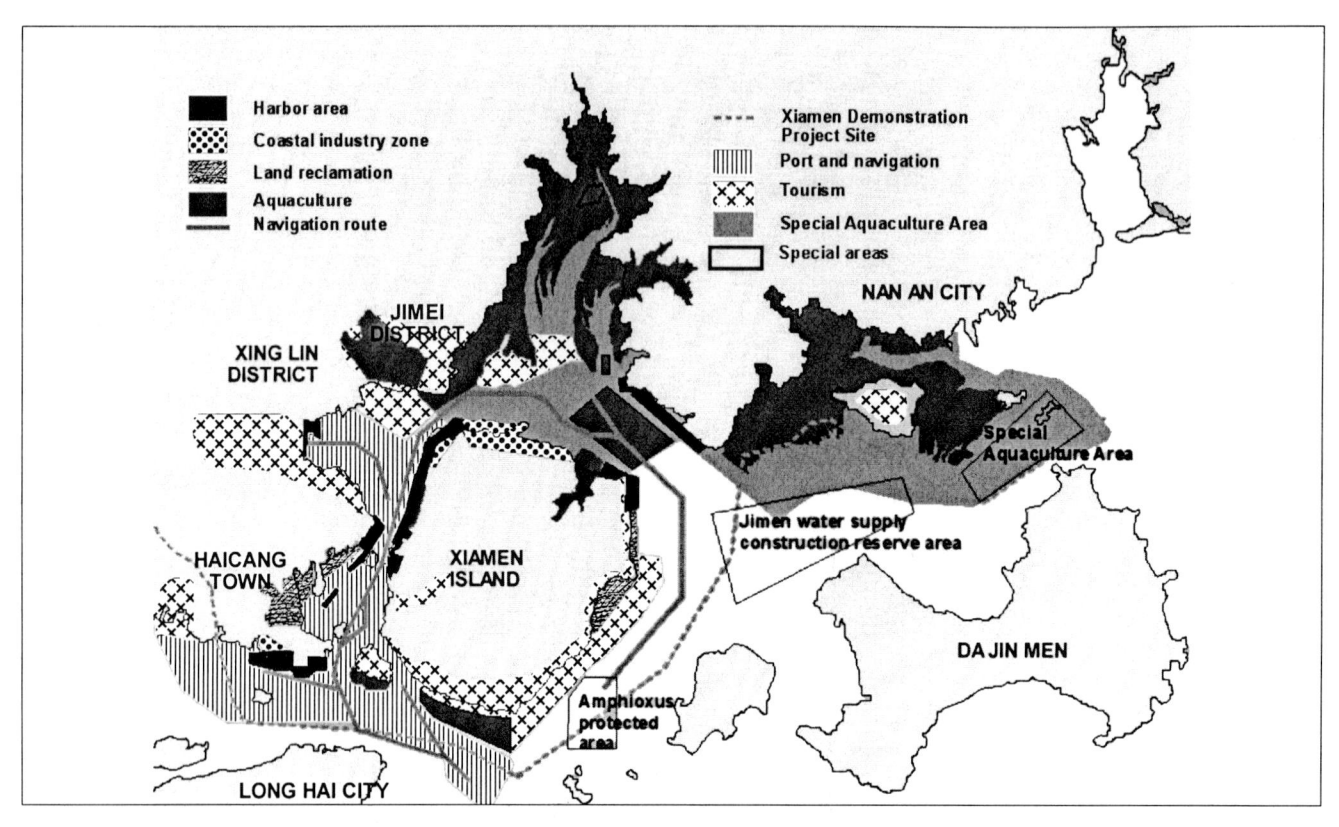

Fig. 7. Sea-use zoning scheme in Xiamen.

Xiamen marine waters were classified mainly into various zones: a shipping/port zone, tourism zone, aquaculture zone, coastal industrial zone, ocean engineering zone, mining zone, nature reserve zone, special function zone, and rehabilitation zone (MPP-EAS, 1997) (Table 14). To ensure compliance with the zonation scheme, the implementing legislation required that all development of coastal and marine resources of Xiamen must be consistent with the functional zonation scheme.

One of the positive effects of the functional zonation scheme was the reduction of multiple use conflicts by identifying, *a priori*, allowable compatible uses of a particular area. For example, in the West Sea, shipping and port development are the dominant uses of the marine waters. At the same time, the West Sea is an important habitat of the Chinese white dolphin (*Sousa chinensis*). In the past, this created a conflict. The functional zonation scheme resolved this problem by designating a core protected area of 5500 hectares and establishing other special regulations to protect the Chinese white dolphin. Navigation is allowed outside the protected area because it became unnecessary to completely prohibit navigation in the West Sea (Wuqi and Yu, 1999).

Xiamen also adopted the following regulatory measures in the West Sea for conservation of the Chinese white dolphin:

- regulation of ship cruising speeds below 8 knots, except in emergencies;
- prohibition of bottom trawling or use of gill nets;
- prohibition of surfing and the use of high speed recreational boats;
- prohibition of discharge of effluents without a permit and beyond the allowable limits;
- requirement of special permits for reclamation; and
- prohibition of underwater explosions and other activities that would disturb the dolphin habitat or increase suspended sediment loads in the water column.

Table 14

Functional zonation scheme in Xiamen

Area/Function	West Harbour	Tong'an Bay	East Sea	Dadeng Sea
Dominant	Shipping/Port	Aquaculture	Tourism	Aquatic Resource Enhancement
Compatible	Tourism/Reserves	Tourism/Port/Reserves	Shipping/ Engineering/ Reserves	Shipping/ Tourism
Restricted	Aquaculture	Waste disposal	Aquaculture	Waste disposal

Table 15

Network of marine monitoring in Xiamen

Institution	Activity
Third Institution of Oceanography	Routine water quality monitoring. Poisonous residuals in organisms
Environmental Monitoring Station	Surface seawater quality monitoring
Environmental Research Center, Xiamen University	Sea bathing water monitoring
Monitoring Station of Fujian Fishery Institute	Aquaculture area monitoring
Fujian Institute of Oceanography	Sediment monitoring
Monitoring Station of Xiamen Port	Port environment monitoring

Source: Xu (1997).

Monitoring, Surveillance and Emergency Preparedness

Marine pollution monitoring is one of the important links in the chain of environmental management. In the past, there were many isolated but inconsistent efforts at marine environmental monitoring. Cross-sector marine monitoring is necessary to meet the needs of integrated coastal management (ITTXDP, 1997). The Xiamen Municipal Government developed a collaborative network and programme under which monitoring efforts are optimized, resources are shared and methods, standards and results are exchanged (Xu, 1997). A panel composed of representatives from the participating institutions manage and guide the activities of the network.

The new monitoring network did not replace the existing sectoral monitoring plans of the participating institutions but sought co-ordination and standardization among them. The network sought to integrate, rationalize and improve upon the activities of the various agencies tasked with monitoring the Xiamen Bay. Tasks were distributed among the agencies and each agency focused on parameters that they were most comfortable with and competent in determining (Table 15). Each agency is required to periodically submit monitoring results. These results are validated and eventually aggregated into a consolidated report.

Present Limitations and Recommendations

The sustainable development of the Xiamen waters necessitates the consolidation of the new measures adopted for integrated coastal management during the pilot phase of the Xiamen Demonstration Project. Co-ordinating mechanisms established must be strengthened and institutionalized to ensure long-term viability. Capacity-building activities must be undertaken to ensure adequate and effective support for coastal and marine management. Awareness building activities must also be undertaken to consolidate the support of local communities and society in general.

Other measures that need to be undertaken include the consolidation of land- and sea-use plans, the adoption of precautionary measures to respond to hazards relating to port activities and navigation, and the consolidation of financing mechanisms for the sustainable management of the Xiamen waters.

Land–Sea Integration

The adoption of an integrated land- and sea-use framework is the optimal approach to coastal management. While the marine functional zonation scheme is a great step forward for coastal management in Xiamen, it is unable to regulate land-based activities, which also have significant

---

Cleanup of Yuan Dang Lagoon

Yuan Dang Lagoon was originally a fishing harbour connected to the West Sea, covering an area of more than 10 km^2. In 1970, a causeway was built at the mouth of the lagoon. The construction of the causeway cut off the water flow to the lagoon and converted it into an enclosed body of water. The area was eventually used as a catchment of industrial and municipal waste discharges. The surface area of the lagoon was further reduced as a result of reclamation projects.

Yuan Dang Lagoon became seriously polluted and eventually biologically dead. The waters of the lagoon became black and smelly. Rubbish and weeds were everywhere. The pollution problem in the lagoon began to pose a danger to the health of the city dwellers and started to drive investors away from the area.

The local government sought to remedy the environmental problem by launching a cleanup project in the late 1980s. The cleanup involved the construction of sewage treatment facilities, restructuring of surrounding drainage systems, dredging, increasing water exchange and developing landscaping of the embankment (Xiamen Yuan Dang Lake Administration Office and the Xiamen Wastewater Preparatory Office, 1998).

Social, recreational and aesthetic benefits resulted from the clean-up. Water quality improved and fell within the national standards for fisheries. There were also economic benefits from the cleanup. By 1994, the total economic gain from the cleanup, through sewage fees, land upgrading and revenue increase exceeded the total cost of the cleanup. An analysis of long-term benefits versus costs of the cleanup showed that investment in the environment had paid off (Chua et al., 1999).

More investments poured in to the Yuan Dang Lagoon area. Based on a survey conducted by the Xiamen Demonstration Project, the majority of the investors chose the Yuan Dang Lagoon area as the location of their business for aesthetic reasons (Chua et al., 1999). In 1997, the area emerged as a new city centre for international and domestic investments, tourism, and residential development.

impacts on the coastal environment and should be taken into account in coastal management.

For example, the use of land for agricultural production affects the coastal environment because pesticide run-offs pollute the coastal waters. This poses a high risk to the aquatic environment and human health and requires relevant remedial measures to be undertaken.

Marine Pollution Management

Xiamen needs to strengthen its capability for prevention and management of marine pollution from ports and vessels. At present, only the Dongdu Port is constructing pollution control facilities. Other port areas have no such facility. Eventually, the facilities in Dongdu Port will also be inadequate to meet the requirements for pollution prevention and mitigation because of rapid port expansion.

Preparedness and Response Systems

The projected increase in port handling capacity by 20 million tonnes per year for the next five years combined with the intensified sea traffic in the West Harbour without any adequate sea traffic control significantly increases the risk of oil and dangerous chemical spill-related accidents (Roos, 1995). The risk is further increased due to the non-existence of monitoring of dangerous cargoes in the port and adjacent areas and the lack of adequate trained personnel in the handling of dangerous goods (Roos, 1995).

At present, emergency response procedures and an accident prevention infrastructure for oil and chemical spills are lacking and legislation on emergency response planning and response, as well as specific port regulations for the handling and storage of dangerous goods, are likewise inadequate. The emergency response organizations are presently neither equipped nor trained to cope with chemical accidents. It may seem that this is not needed because of the small amount of chemicals handled but it should be stressed that the situation may change in the future because of the increasing global traffic of dangerous chemicals. Development of the country's economy as well as the planned expansion of the port facilities will most likely attract additional cargo in the future. Dangerous cargoes account for approximately 10% of the total cargo shipped worldwide and it is likely that some of these will either be handled in or transit through Xiamen ports (Roos, 1995). Therefore, it becomes important to monitor closely the volume and composition of cargoes handled in or transiting through the port.

Finally, appropriate economic instruments need to be established to create new funding mechanisms and diversify funding sources for coastal and marine environmental protection and management. The user fee system established under the Regulation for the Management of Uses of Xiamen Seas is an important economic instrument which may be utilized for sustainable financing and at the same time ensure the rational utilization of the marine resources (ITTXDP, 1997). Financing mechanisms must be institutionalized to provide funds for the effective implementation of the SMP, marine environmental protection activities and integrated coastal and marine management.

REFERENCES

Chua, T.E. (1998) Lessons learned from practising integrated coastal management in Southeast Asia. *Ambio* **27** (8), 599–610.

Chua, T.E. (1999) Integrated coastal management: An effective mechanism for local implementation of coastal and marine environment related international conventions. In *Challenges and Opportunities in Managing Pollution in the East Asian Seas*, eds. T.E. Chua and N. Bermas, pp. 3–23. MPP-EAS Conference Proceedings 12/PEMSEA Conference Proceedings 1, 567 pp.

Chua, T.E. and Ross, S.A. (eds.) (1996) Marine pollution prevention and management in the East Asian Seas: from planning to action. Annual Report. GEF/UNDP/IMO Regional Programme for the Prevention and Management of Marine Pollution in the East Asian Seas, Quezon City, Philippines. 40 pp.

Chua, T.E. and Ross, S.A. (eds.) (1997) Pollution Prevention and Management in the East Asian Seas: A paradigm shift in concept, approach and methodology, Annual Report. GEF/UNDP/IMO Regional Programme for the Prevention and Management in the East Asian Seas, Quezon City, Philippines. 45 pp.

Chua, T.E., Ross, S.A., Yu, H., Jacinto, G. and Bernad, S.R. (1999) Sharing Lessons and experiences in marine pollution management. Final Report. GEF/UNDP/IMO Regional Programme for the Prevention and Management in the East Asian Seas, Quezon City, Philippines, 94 pp.

Chua, T.E., Yu, H. and Chen, G. (1997) From sectoral to integrated coastal management: a case in Xiamen, China. *Ocean and Coastal Management* **37** (2), 233–251.

Hong, H. Xu, L., Zhang, L., Chen, J.C., Wong, T.S. and Wan, T.S.M. (1995) Environmental fate and chemistry of organic pollutants in the sediment of Xiamen and Victoria Harbours. *Marine Pollution Bulletin* **31** (4–12), 229–236.

ITTXDP (Integrated Task Team of the Xiamen Demonstration Project) 1996. Coastal environmental profile of Xiamen. MPP-EAS Technical Report No. 6. GEF/UNDP/IMO Regional Programme for the Prevention and Management of Marine Pollution in the East Asian Seas, Quezon City, Philippines. 117 pp.

ITTXDP (Integrated Task Team of the Xiamen Demonstration Project) (1997) Strategic management plan for marine pollution prevention and management in Xiamen. MPP-EAS Technical Report 7. GEF/UNDP/IMO Regional Programme for the Prevention and Management of Marine Pollution in the East Asian Seas, Quezon City, Philippines. 60 pp.

Kho, J. and Agsaoay, E. (1999) ICM Implementation at the local level: the Batangas Bay and Xiamen experience. national coastal policy for the East Asian Seas: status review and model policy development. MPP-EAS/Info/99/190. GEF/UNDP/IMO Regional Programme for the Prevention and Management in the East Asian Seas, Quezon City, Philippines, pp. 29–57.

MPP-EAS (1995) Bi-Annual Report. GEF/UNDP/IMO Regional Programme for the Prevention and Management of Marine Pollution in the East Asian Seas, Quezon City, Philippines. 27 pp.

MPP-EAS (1997) Proceedings of the regional workshop on partnerships in the application of integrated coastal management, 12–14 November 1997, Burapha University, Bangsaen, Chonburi, Thailand. Workshop Proceedings No. 10. GEF/UNDP/IMO Regional Programme for the Prevention and Management in the East Asian Seas, Quezon City, Philippines. 167 pp.

MPP-EAS and FAO (1998) Initial environmental risk assessment of pesticides in the Batangas Bay Region, Philippines, and the Xiamen Seas. MPP-EAS Technical Report No.16. GEF/UNDP/IMO Regional Programme for the Prevention and Management in the East Asian Seas, Quezon City, Philippines and Food and Agriculture Organization of the United Nations, Rome Italy. 49 pp.

Programme Group for Waste Management (Programme Group of Subactivity 1.2.3.3) (1996) Waste management and prevention in Xiamen coastal zone. GEF/UNDP/IMO Regional Programme for the Prevention and Management of Marine Pollution in the East Asian Seas, Quezon City, Philippines. 111 pp. (Unpublished).

Roos, H.G. (1995) Feasibility study on a comprehensive preparedness and response plan to oil and chemical spills, integrating national, sub regional and regional plans, Mission Report for Xiamen, China. GEF/UNDP/IMO Regional Programme for the Prevention and Management of Marine Pollution in the East Asian Seas, Quezon City, Philippines. 29 pp. (Unpublished).

Wu, Y., Lu, C., Huang, J. and Lin, Y. (1988) Removal of heavy metals from sediments by mangroves in Jiulong Estuary, Xiamen Harbour, China. *Water Science and Technology* 20 (6/7), 49–54.

Wu, Y., Lin, Y., Guo, T., Wang, L. and Zheng, Z. (1992) Mechanisms of phosphorus released from the sediment–water interface in Xiamen Bay, Fujian, China. *Science of the Total Environment*, Supplement, 1087–1097.

Wu, Y., Chen, C. and Wang, L. (1987) Relationship between heavy metal pollution and water productivity in Xiamen Estuarine Harbour. *Chinese Journal of Oceanology and Limnology* 5 (3), 205–216.

Wuqi, R. and Yu, H. (1999) Design and implementation of marine functional zoning scheme in Xiamen, China. In *Challenges and Opportunities in Managing Pollution in the East Asian Seas*, eds. T.E. Chua and N. Bermas, pp. 341–354. MPP-EAS Conference Proceedings 12/PEMSEA Conference Proceedings 1, 567 pp.

Xu, Kuncan (1997) Integrated marine pollution monitoring system in Xiamen, People's Republic of China. Highlights of the Second Technical Workshop of the Regional Network for Marine Pollution Monitoring and Information Management. MPP-EAS/Info/98/157. GEF/UNDP/IMO Regional Programme for the Prevention and Management in the East Asian Seas, Quezon City, Philippines. Annex C.

Xiamen Yuan Dang Lake Administration Office and Xiamen Wastewater Treatment Preparatory Office (1998) Terminal Report of the Yuan Dang Lake integrated treatment project. GEF/UNDP/IMO Regional Programme for the Prevention and Management of Marine Pollution in the East Asian Seas, Quezon City, Philippines. 8 pp. (Unpublished).

THE AUTHORS

Chua Thia-Eng
GEF/UNDP/IMO Regional Programme on Partnerships in Environmental Management for Seas of East Asia (PEMSEA), P.O. Box 2502, Quezon City, 1165 Metro Manila, Philippines

Ingrid Rosalie L. Gorre
GEF/UNDP/IMO Regional Programme on Partnerships in Environmental Management for Seas of East Asia (PEMSEA), P.O. Box 2502, Quezon City, 1165 Metro Manila, Philippines

Chapter 87

HONG KONG

Bruce J. Richardson, Paul K.S. Lam and Rudolf S.S. Wu

The Hong Kong Special Administrative Region of southeast China has a total land area of 1042 km² and encompasses 235 islands of varying sizes. It is affected by an unusual combination of water masses which not only keeps the coastal waters of Hong Kong warm and maintains the local sub-tropical/tropical biotic communities, but also allows temperate species to survive. It is greatly affected by freshwater dilution from both heavy rainfall during monsoons, and from the Pearl River which has an annual flow of 308 billion m³, from a drainage area of 228,000 km². These may lower the salinity of the western and northwestern approaches to 1 to 7‰, while the eastern and northeastern waters remain virtually unaffected.

These conditions create sheltered, estuarine environments to the west, exposed, oceanic environments to the east, and a transitional zone in the middle, typified by the Victoria Harbour region. Species surviving are both euryhaline and eurythermal, and able to tolerate highly sedimented waters. These natural stresses also make the marine fauna and flora much more susceptible to pollution stresses. Nevertheless, there is considerable plant life, including mangroves, and the bird life associated with sheltered shore and mangroves is of considerable international significance.

There is heavy use of the marine environment. A pronounced decline in demersal fish stocks in the South China Sea due to over-fishing has occurred. The sea receives over 2 million m³ of waste water per day, and at present about 80% of Hong Kong's sewage is largely untreated. About 75% enters Victoria Harbour directly, representing a daily BOD loading of over 300 tonnes. As a result, many of Hong Kong's beaches are contaminated, especially along the western shores of the New Territories. There are some signs of a trend of improving beach water quality in general, but there are also increasing signs of harmful algal blooms.

The huge volume of sea traffic in Hong Kong, with considerable paint stripping and painting, results in the release of chemical contaminants and biocides, e.g. tributyltin (TBT); the latter leading to concentrations of 53,000 and 18,300 ng Sn g^{-1} in sediments, which exceed typical levels recorded elsewhere. Farming practices have become more specialized, resulting in large increases in application of chemical fertilizers and pesticides. Crop and livestock farms are large sources of nutrient-rich and pesticide-laden run-off, the former being considered the culprit in an increasing number of red tides. Some pesticides include compounds that have been banned in Hong Kong (e.g. DDT), which are attributable to contamination sources in the Chinese mainland. Indeed, the annual discharge of the Pearl River accounts for over 85% of the total nutrient input to the local waters of Hong Kong.

Rapid urban development has exerted an unprecedented pressure on the local marine environment. Large-scale reclamation has altered current patterns, reduced tidal flushing and caused hypoxia/anoxia in many areas. During 1991 to 1999, a total of 366 million m³ of dredged

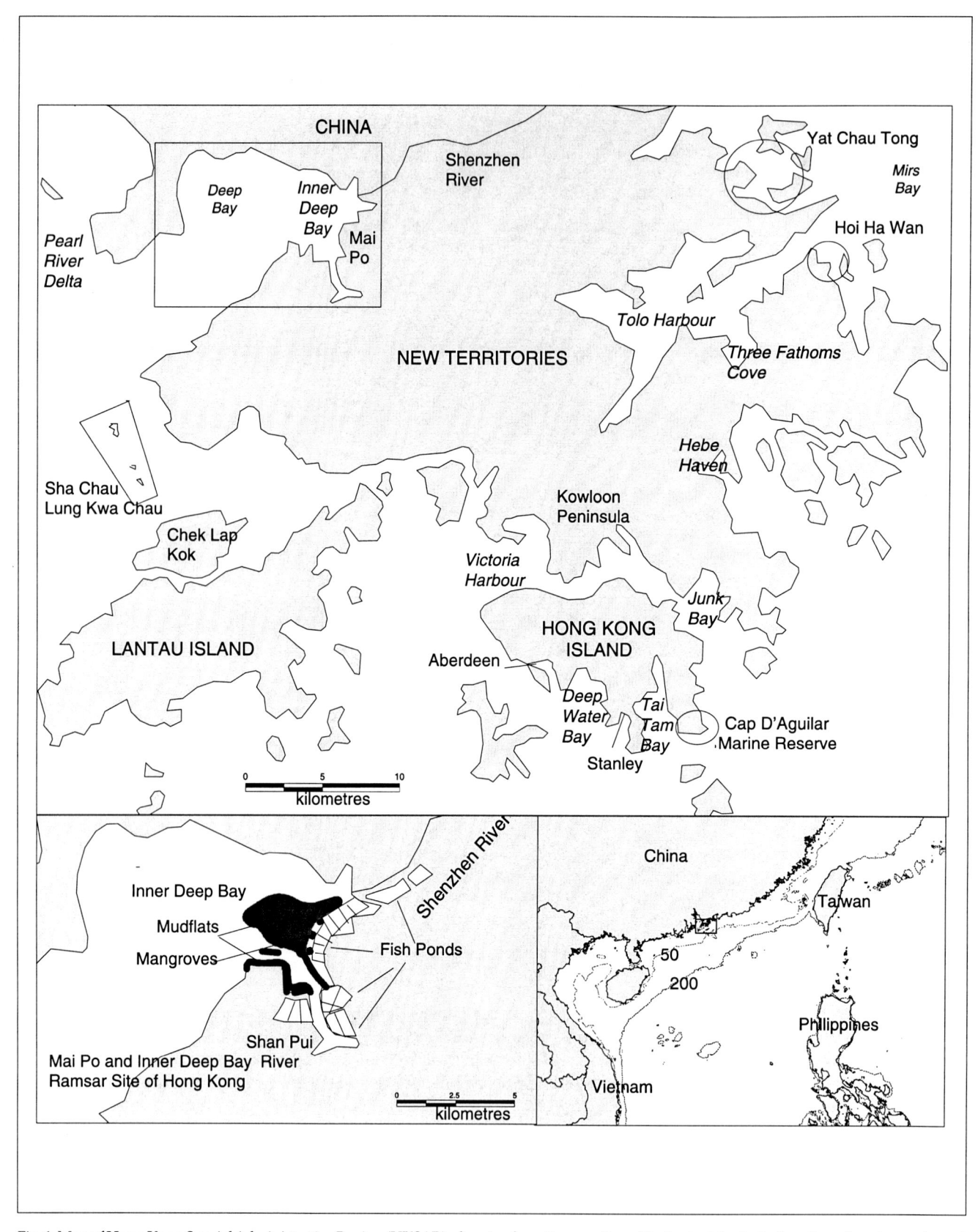

Fig. 1. Map of Hong Kong Special Administrative Region (HKSAR), showing key sites mentioned in the text. Includes location of marine parks and reserves.

spoil was disposed into coastal waters, and the new port and airport development involved the dredging of 880 million m³ of fill materials, and the dumping of 380 million m³ of dredged materials within a very small coastal area, affecting fisheries resources, coral, seagrass, benthos and water quality. Health issues, including PSP and ciguatera intoxication are a problem, and some 3 to 7% of shellfish samples contain unacceptably high levels of PCB and Σ DDT.

Some important wildlife areas are now protected. Habitats such as mudflats are increasingly scarce in south China; there are only 23 remaining sites with intertidal mudflat habitats within 800 km of Hong Kong. The large Mai Po/Deep Bay Ramsar site is one such, and is described.

THE DEFINED REGION

The Hong Kong Special Administrative Region (HKSAR), with a total land area of 1042 km², lies on the coast of mainland China to the south of Guandong Province (Fig. 1). It is situated between 22°9' and 22°37'N and 113°52' and 114°30'E, and includes Hong Kong Island, which is separated from the Kowloon Peninsula by Victoria Harbour (Fig. 2). The Kowloon Peninsula adjoins the New Territories, beyond which lies the hinterland of mainland China. In addition, the HKSAR encompasses 235 islands of varying sizes.

Hong Kong has an extensive coastline and its coastal waters extend to the South China Sea. Its hydrography has been described in detail by Morton and Wu (1975) and Morton (1982). In short, the region is predominantly influenced by the Hainan current, which is characterized by high salinity (34.4–36‰) and variable temperature (12°C at 300 m depth to 29°C at the surface). During winter months, the Hainan current is replaced by the Kuro Shio current which has high salinity (34.4–35‰) and high temperature (26–29°C) and by the Taiwan current of reduced salinities (31–33‰) and temperatures (19–23°C). This combination of water masses not only keeps the coastal waters of Hong Kong warm and maintains the local sub-tropical/ tropical biotic communities, but also allows certain temperate species to survive.

Fig. 2. Victoria Harbour, showing the densely populated areas of Hong Kong Island and the Kowloon Peninsula.

Local hydrography is also considerably affected by freshwater dilution from two sources. There is heavy rainfall (average 217 cm year⁻¹) associated with monsoon conditions which prevail from May to September, and there is also the large spate of turbid freshwater exiting from the Pearl River (the third largest river in China). The annual flow of this river is 308 billion m³, and it drains an area of 228,000 km².

The salinity of the western and northwestern approaches of Hong Kong may be lowered to 1–7‰ during summer as a result of these influences. However, the eastern and northeastern waters remain virtually unaffected by freshwater dilution, and the salinity in these localities is typically oceanic all year round.

Because of these hydrographic influences, Hong Kong's territorial waters can be divided into three zones: a sheltered, estuarine environment to the west; an exposed, oceanic environment to the east; and a transitional zone in the middle (typified by Victoria Harbour and its adjacent waters). Such an environmental gradient, combined with the large seasonal variations in hydrographic conditions affect the local marine communities to a considerable extent, sometimes making it difficult to differentiate pollution effects from natural perturbations.

Seasonal variations in water temperature are also typically large, ranging from a low of 13°C in winter to a high of 30°C in the summer, and a thermocline often occurs at a water depth of about 3 m. The high water temperatures in summer, especially in shallow waters, often approach the thermal tolerance limit of the marine biota. Marine communities in the western approaches are further subjected to large seasonal changes in salinity. Thus, species surviving in this locality are both euryhaline and eurythermal, and able to tolerate waters highly laden with sediment. These natural physical stresses also make the marine fauna and flora much more susceptible to the pollution stresses which prevail in Hong Kong's coastal waters.

MAJOR SHALLOW WATER MARINE AND COASTAL HABITATS

Hong Kong's shallow water habitats and their biological communities have been extensively documented and reviewed by Morton and Morton (1983). As a result of climate and hydrography, a mixed subtropical fauna and flora from tropical corals to temperate species are found in Hong Kong's coastal waters. Tropical species tend to be maintained by warmer Pacific waters; the Taiwan Current and the Hainan Current bring species from the shores of Japan and east China, and from the South China Sea respectively. The north-western areas of Hong Kong are influenced by the influx of freshwater during the summer (see above), which may impair breeding of certain intolerant species. As a result, first spawning in this locality occurs in spring, followed again in autumn (due largely to higher salinity and moderately high temperatures at these times).

Hong Kong's nearshore environments exhibit a considerable range of habitats, ranging from hard through soft shores, including mangroves. Within Hong Kong Island itself, in open, south-facing coastal areas, exposed shores are subject to high wave action as typified by the example of Cape D'Aguilar, on the extreme southern tip of the island. Sheltered and semi-sheltered hard shores occur in Deep Water Bay in the south, where water movement is high enough to remove loose debris, but protection from high wave action is sufficient. Boulder and cobble beaches also occur, as typified at Stanley and Tai Tam, respectively. The infrastructure of Hong Kong also offers ample examples of specialised and protected hard surfaces; for example wharf piles and other man-made structures that project into coastal waterways.

Soft shores are also amply represented in Hong Kong. The south eastern facing beaches are exposed to considerable wave action, but such areas are not common in the territory. In contrast, in parts of Lantau Island, more moderate wave action occurs and the resultant beaches are wider, and a richer fauna exists. Protected sand flats consisting of silty sand occur in the east, for example at Tai Tam, Three Fathoms Cove and Hebe Haven. In such localities, species diversity on sheltered shores is at its maximum.

Enclosed shores, which are to all intents and purposes landlocked, can also be found in Hong Kong. Many such areas have been subject to human alterations, either for aquaculture purposes (Fig. 3), or for building. Excellent examples can be found in the northwest, in Deep Bay and near Mai Po marshes, where extensive cultivation of various fish has occurred. In addition, these sheltered shorelines show considerable plant life, including mangroves. Hong Kong is relatively rich in this regard, given its latitude, which is at the northern extremity of the mangrove zone. The bird life associated with the sheltered shore zones in Hong Kong, and with the mangrove areas, is also of considerable significance internationally.

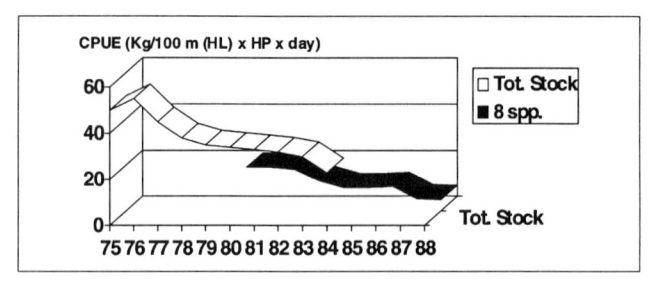

Fig. 4. Catch per unit effort (CPUE) of fish stock in the South China Sea. (Data from the Hong Kong Agriculture and Fisheries Department, personal communication).

OFFSHORE SYSTEMS

The landing of the Hong Kong fishing fleet is approximately 200,000 t per annum. About 90% of landings are harvested from the South China Sea, and the remaining 10% from coastal waters. A pronounced decline in demersal fish stocks in the South China Sea due to over-fishing is clearly evident. Fisheries statistics collected by the Agriculture and Fisheries Department indicate that Catch per Unit Effort (kg 100 m head line^{-1} horsepower^{-1} d^{-1}) of the demersal stock in the South China Sea has declined from 52 to 18 (i.e. by 65%) from 1975 to 1984. A recent study carried out in 1996–97 showed that 12 out of 17 commercial species in coastal waters are heavily over-exploited. Likewise, fish fry production has decreased by 90% within 7 years (from a total of 68 million tails in 1990 to 6.4 million in 1997) due to a combination of pollution and over-exploitation (AFD, 1998) (Fig. 4).

POPULATIONS AFFECTING THE AREA

Hong Kong is one of the most densely populated cities in the world, with over 6.7 million people, mostly living along the waterfront and in its immediately connecting hinterland. The marine environment has been acting as the ultimate sink of the majority of wastewater (estimated to be over 2 million m^3 per day) generated by the people as well as the industrial and service sectors. At present, about 80% of Hong Kong's domestic wastewater is discharged into coastal waters with only primary screening. The remaining

Fig. 3. Coastal fish ponds in the Deep Bay region of Hong Kong and the Pearl River Delta.

Mai Po Marshes

Louise Richards, Ed Gmitrowicz and Tom Dahmer

Hyder Consulting Ltd, Hong Kong; Ecosystems Ltd, Hong Kong

Mai Po Marshes was designated a Site of Special Scientific Interest (SSSI) in 1976. It encompasses 172 ha of mangroves and 209 ha of traditional and modified shrimp ponds and fish ponds within the 535 km^2 catchment of Deep Bay (Young and Melville, 1995). With a maximum tidal range of 2.8 m (the average depth in Deep Bay is 2.9 m), a mudflat of 1.8 km^2 is exposed during low tide. The diverse wetland habitats within the Reserve support an abundant bird life, such that Mai Po has become a site of regional and global importance, and was incorporated within the boundaries of Hong Kong's only and China's sixth Ramsar Site in 1995.

The Ramsar Site supports more than 330 bird species, representing over 70% of those recorded in Hong Kong since the Second World War. The high diversity is not only due to the mosaic of habitats, but also to intensive management and the site's location on the East Asian/Australasian Flyway.

In recent years more than 50,000 waterfowl over-wintered in the Deep Bay area annually, including significant populations of globally endangered species such as the Black-faced Spoonbill (*Platalea minor*), and Saunders' Gull (*Larus saundersi*). The Deep Bay area also supports the largest over-wintering population of Cormorants (*Phalacrocorax carbo*) in East Asia (Peking University, 1996) and large flocks of the near threatened Silky Starling *Sturnus sericeus* (Leven et al., 1994).

Numerical hydrodynamic and water quality modelling of Deep Bay (Hyder Consulting Ltd CES (Asia) Ltd. JV, 1997) has shown that in both the wet (summer) and dry (winter) seasons the residual water circulation and flushing of Deep Bay is controlled by exchanges with the Pearl River Estuary. In both wet and dry seasons, Pearl Estuary water is dominant, except in the near vicinity of individual local river discharges. The high pollution loads carried by the local rivers mean that water quality is strongly influenced by freshwater discharges. This is the case particularly within the Ramsar Site, which is situated between the two major freshwater inflows to the Bay, the Shenzhen River and the Shan Pui River. The pollution loads carried by these rivers lead to failure of local dissolved oxygen, BOD and *E. coli* water quality standards within the Ramsar Site in both wet and dry seasons. Nutrient concentrations also exceed the water quality standards, due to high concentrations in both the Pearl Estuary and the local rivers.

Pollution loads arise from urban, industrial and agricultural sources within the catchment of Deep Bay and the wider Pearl River catchment. The marine ecology of the Ramsar Site is influenced by the local water quality, particularly during episodes of oxygen depletion in the waters of the Shenzhen River.

MUDFLATS

Mudflats are increasingly scarce in south China due to coastal development. There are only 23 remaining sites with intertidal mudflat habitats within 800 km of Hong Kong (Aspinwall, 1997). The mudflat at the Ramsar Site is an important foraging habitat for ducks and waders (Peking University, 1996). The number of ducks feeding on the mudflat often exceeds 12,000 (Leven et al., 1994). The mudflat is also a preferred winter foraging habitat of Saunders's Gull, which is listed in Appendix I of the Bonn Convention, and is classified as endangered by IUCN (1996). Deep Bay supports 5% of the world population (approx. 100 birds).

The mudflat is the most important habitat in the Ramsar Site for mudskippers, which contribute to nutrient cycling, and as prey for birds. The intertidal benthos is a key component of the ecosystem supporting the large numbers of over-wintering birds in the Deep Bay area. McChesney (1993; 1997) described the benthic infauna on the Inner Deep Bay mudflats in relation to migratory bird-carrying capacity at the Ramsar Site, concluding that the Mai Po mudflat may be at carrying capacity.

The mudflat ecosystem is also prone to the effects of extreme water quality events, and algal blooms have resulted in severely depleted oxygen levels, and concomitant declines in benthos abundance and biomass. A decline in the 1997 winter waterfowl count was possibly a result of such an occurrence.

FISHPONDS

Historically, Inner Deep Bay was dominated by mangrove wetlands. With the advent and spread of human populations, mangroves were converted to man-made wetlands (paddies, shrimp ponds and fish ponds) or to dry land (Irving and Morton, 1988). Fish ponds in the Deep Bay catchment, which were very limited in extent before World War II, began to replace paddies, shrimp ponds and other less profitable types of agriculture around the middle of the 20th century. Fish ponds now account for the largest wetland area in the North-west New Territories (NWNT), excluding Deep Bay itself. Fish ponds in the Deep Bay area probably peaked in extent during the 1970s or 1980s. In recent years, these ponds have been increasingly subject to the more lucrative conversion to open storage of shipping containers and residential development.

However, fish ponds in Deep Bay perform some of the ecological functions of the natural wetlands that preceded them. Fish ponds are almost certainly less useful to most native species than the natural habitats that originally occupied this area, but they are now among the only significant areas of wetland remaining in the Deep Bay catchment, and are hence extremely important for wetland-dependent species (Melville et al., 1994; Chu, 1995).

Fish ponds are of low botanical interest but the pond bund vegetation (Fig. 1) constitutes important habitats for many bird species. At least 136 species of birds (more than 25% of the species recorded in Hong Kong) have been recorded on the Deep Bay fish ponds. These records cover a broad taxonomic range (13 orders) and include many globally vulnerable species such as Imperial Eagle (*Aquila heliaca*) and Spoon-billed Sandpiper (*Eurynorhynchus pygmeus*). The annual draining of fishponds for commercial fish harvest is also ecologically important as the non-harvested fish biomass available from pond drainage in winter could support as many as 9000 ardeids within the Deep Bay fish pond system (Hyder Consulting Ltd, 1999).

Fig. 1. Mai Po Gei Wais showing the replacement of mangroves by vegetation dominated by reedbeds.

SHRIMP PONDS

The tidal shrimp pond (*gei wai*) is a traditional method of shrimp culture in Deep Bay. *Gei wais* in Hong Kong were excavated in the early 1940s. The ponds are built in the inter-tidal zone and the sluice gate is opened during high tide to wash in juvenile shrimp from Deep Bay. Shrimp are harvested at low tide by opening the sluice gates to drain the pond and by placing a narrow-mesh net in the gate to capture the mature shrimp (Irving and Morton 1988). Mangrove communities inside the *gei wais* were retained when the ponds were excavated, and these provide shelter and food for the maturing shrimp. *Gei wai* shrimp culture was formerly widespread in Hong Kong, but is now only practised at the Mai Po Marshes Nature Reserve in some 16 gei wais covering 175 ha. In recent years the production of shrimp using this method has declined, largely due to water quality impacts and increased sedimentation (Irving and Morton, 1988; Melville et al., 1989). Metal concentrations in crabs and shrimps were also found to be elevated (Cheung and Wong, 1992) suggesting that reduced production of shrimp larvae in Deep Bay may be attributable to industrial pollution inputs.

The gei wais are actively managed by the World Wide Fund for Nature to attract and support the large and diverse numbers of migratory and resident birds (Young and Melville, 1995).

THREATS TO THE RAMSAR SITE

Water pollution is one of the key threats to the ecology of Deep Bay. Deep Bay is sheltered and relatively slow-flushing, and an increase in pollution from the Shenzhen River and other local feeders has been documented in recent decades (Leung et al., 1975; Hyder Consulting and CES (Asia) Ltd JV, 1997; Young and Melville, 1995). The most recent water quality report (EPD, 1998a) indicated that although major rivers entering the Deep Bay catchment were still heavily polluted in 1997, signs of improvement were found. Livestock waste discharges from the HKSAR are believed to be declining as a result of government livestock waste collection and treatment initiatives. Other sources, notably industrial discharges from mainland China, have been increasing. Human population levels are also increasing on both sides of the boundary, though both the HKSAR and Shenzhen Special Economic Zone governments are making efforts to keep pace by installing or upgrading sewage treatment systems (HKGDEPLG, 1992; 1995).

Habitat loss is a further threat to the ecology of Deep Bay. Between 1985 and 1994, 500 ha or 25% of the fish ponds in Hong Kong and 50% of the fish ponds on the Shenzhen side were lost (Aspinwall, 1997). Key land-use changes over the past few decades which have resulted in the filling of fish ponds and conversion of agricultural land to other uses include the construction of new towns and industrial estates since the 1970s, widespread increases in open storage and car repair yards, and Government infrastructure and drainage works.

The Deep Bay area is still under intense development pressure due to the decrease in economic returns from agriculture and fish farming, as well as human population growth in the catchment. Furthermore, the invasive climber *Derris trifoliata* is recognised as a potential threat to the mangrove habitat at Mai Po. Hyder Consulting Ltd (1998) have mentioned *Mikania micrantha* as a threat to all vegetation at Mai Po and in the Northwest New Territories of Hong Kong. *M. micrantha* is particularly threatening to egretries because it degrades the quality of nesting habitats.

PROTECTIVE MEASURES

Despite the potential threats to the ecology of Deep Bay, the HKSAR Government is taking several significant measures to protect the area. Planning controls now prevent development within this area and establish a buffer area within which no net loss of fishponds and wetlands is permitted (TPB, 1999).

The HKSAR government is currently implementing its Conservation Strategy and Management Plan for the Ramsar Site (Aspinwall 1997), and is developing a long term ecological monitoring programme for the site. The monitoring programme will quantify the baseline conditions of the area and detect any changes in the ecological character of the site (Hyder Consulting, 1999).

The HKSAR government is also implementing a number of improvement measures to protect and improve water quality in the NWNT and Deep Bay catchment. These include tightening of discharge standards for livestock farms; implementation of the Chemical Waste Control Scheme; implementation of the North District Sewerage Master Plan (SMP) works; and implementation of the revised SMP works in the Yuen Long/Kam Tin District.

To ensure the protection of the ecology of Mai Po and Deep Bay into the next millennium, it was concluded (Hyder Consulting and CES (Asia) Ltd JV, 1997) that pollution control measures which are already in place or planned for the future will not be adequate, and that even more stringent measures will be required. Various options have been considered, preferably the exportation of all wastes from the Deep Bay catchment, or the selective elimination of waste streams from the catchment.

REFERENCES

Aspinwall (1997) Agreement No. CE 47/95, Development of a comprehensive Conservation Strategy and a Management Plan in Relation to the Listing of Mai Po and Inner Deep Bay as a Wetland of International Importance under the Ramsar Convention. Reports prepared for the Hong Kong SAR Department of Agriculture and Fisheries.

Cheung, Y.H. and Wong, M.H. (1992) Comparison of trace metal contents of sediments and mussels collected within and outside Tolo Harbour, Hong Kong. *Environmental Management* **16** (6), 743–751.

Chu, H.F. (1995) Planning considerations for conservation and development within Deep Bay Buffer Zones in North West New Territories Hong Kong. Thesis (MSc, Urban Planning), University of Hong Kong.

EPD (Environmental Protection Department) (1998) River Water Quality for Hong Kong, 1997. Hong Kong Government, Hong Kong.

HKGDEPLG (1992). Hong Kong–Guangdong Environmental Protection Liaison Group Technical report on the environmental protection of Deep Bay and its catchments. Technical Sub-Group.

HKGDEPLG (1995). Hong Kong–Guangdong Environmental Protection Liaison Group. Designation of Mai Po Marshes and Inner Deep Bay as a Wetland of International Importance Under the Ramsar Convention. TDD, Hong Kong-Guangdong Environmental Protection Liaison Group.

Hyder Consulting Ltd and CES (Asia) Ltd JV. (1997) Deep Bay Water Quality Regional Control Strategy Study. Reports prepared for the Hong Kong SAR Environmental Protection Dept. Water Policy and Planning Group, Agreement No. CE17 95.

Hyder Consulting Ltd (1998, 1999) Development of an Ecological Monitoring Programme for the Mai Po and Inner Deep Bay Ramsar Site. Study reports prepared for Agriculture and Fisheries Department, HKSAR.

Irving, R. and Morton, B. (1988) *A Geography of the Mai Po Marshes*. World Wide Fund for Nature Hong Kong, Hong Kong.

IUCN (1996) IUCN Red List of Threatened Animals. Compiled by the World Conservation Monitoring Centre for IUCN The World Conservation Union. Accessed on world-wide website www.wcmc.org.uk

Leung, C., Shortridge, K.F., Morton, B. and Wong, P.S. (1975) The incidence of faecal bacteria in the tissues of the commercial oyster *Crassostrea gigas* Thunberg 1793 correlated with the hydrology of Deep Bay, Hong Kong, eds. B. Morton and Li Mingfeng, pp. 114–127.

Leven, M.R., Carey, G.J. and Picken, V.B. (1994) Systematic list. The Hong Kong Bird Report 1993, pp. 16–90.

McChesney, S (1993) Mai Po mudflat invertebrate assemblage-changes through one year (abstract only). Asia-Pacific Symposium on Mangrove Ecosystems; Hong Kong. Hong Kong University of Science and Technology, pp. 104–105.

McChesney, S. (1997) The benthic invertebrate community of the intertidal mudflat at the Mai Po Marshes Nature Reserve, with special reference to resources for migrant shorebirds. Thesis (M. Phil.), University of Hong Kong.

Melville, D.S., Young, L. and Leader, P.J. (1994) The importance of fish ponds around Deep Bay to wildlife especially waterbirds, together with a review of potential impacts of wetland loss and mitigation measures. World Wide Fund for Nature, Hong Kong.

Peking University (1996) Environmental impact assessment study on Shenzhen River Regulation Project: final EIA study. Shenzhen River Regulation Office of Municipal Government.

TPB (1999) Town Planning Board guidelines for application for developments within Deep Bay area under Section 16 of the Town Planning Ordinance. TPB PG-No. 12B (Revised April 1999), Planning Department, Hong Kong Special Administrative Region Government.

Young, L. and Melville, D. (1995) Environmental research at the WWF Hong Kong Mai Po Marshes Nature Reserve. In *Proc. Environmental Research in the Pearl River and Coastal Areas*, eds. C.K. Wong, K.H. Chu, Q.C. Chen and X.L. Ma, pp. 58–68. Guangdong Higher Education Press, Guangzhou, PRC.

---

wastewater receives mainly secondary treatment. Of all the wastewater discharges into the coastal waters of Hong Kong, about 75% enters Victoria Harbour directly, representing a daily BOD loading of over 300 tonnes (EPD, 1997). The fast currents and high flushing rate in the harbour have long been taken as an effective, natural means of dilution and dispersion of the essentially raw sewage.

Along the coastline of Hong Kong, there are some 41 gazetted bathing beaches and numerous ungazetted ones (EPD, 1993). On average, 1.6 million people visited the beaches each month of the bathing season in 1997 (Broom et al., 1999). The waters of many of Hong Kong's beaches are contaminated by sewage containing human and animal faecal materials, especially along the western shores of the New Territories. Previous studies have indicated a clear relationship between swimming-associated illness (in terms of the combined gastrointestinal illness and the symptom rate) and *E. coli* concentration in beach water (EPD, 1987), indicating a substantial health risk to people bathing on contaminated beaches. Since 1993, the Environmental Protection Department has undertaken regular monitoring of the bacteriological water quality (primarily based on geometric mean *E. coli* counts) of all gazetted and a number of selected ungazetted beaches. All gazetted beaches are graded weekly in terms of their water quality, and their suitability for bathing purposes promulgated to the public (EPD, 1998). The monitoring data indicate that many Hong Kong beaches are still polluted (*E. coli* counts >180 per 100 ml) to an extent that they were unsuitable for swimming. Notwithstanding, the percentage of gazetted beaches ranked "fair" and "good" (i.e. complying with the Water Quality Objectives) increased from 63% in 1997 to 78% in 1998. In contrast to the trend of improving beach water quality in general, there are signs that incidents of harmful algal blooms are on the rise in local coastal waters (Anderson et al., 1999). During the bathing seasons in 1998, 12 algal blooms were reported at beaches, resulting in closure of these facilities.

Hong Kong is one of the world's busiest container ports, handling more than 13 million twenty-foot-equivalent units (TEUs) in 1996 and rising by an average of 8% per annum (Broom et al., 1999). Vessel arrivals and departures number about half a million each year (Marine Department, 1997). A number of important impacts on the local coastal environment have resulted from the huge amount of marine traffic and their associated activities. These include: construction of port facilities and related infrastructure, domestic refuse from local and ocean-going vessels, discharge of ballast water and wastes generated by hull stripping and cleaning, oil pollution due to accidental spills, leakages during shipping operations, and cleaning of vessels.

Developments along the shores have caused a rapid loss of ecological habitats. The threats to local seagrass (*Zostera japonica, Halophila beccarii, H. ovalis* and *Ruppia maritima*) communities have raised some concern, and resulted in calls for the development of conservation plans for these species. Loss of shallow sheltered waters, elevated pollution levels (particularly of persistent trace organics and organometals), and increased sea traffic have been considered the main threats to marine animals, most notably the Chinese White Dolphin, *Sousa chinensis* (Parsons, 1998a,b).

Given the huge volume of sea traffic in Hong Kong, vessel stripping and painting at dockyards may result in the release of chemical contaminants, e.g. tributyltin (TBT), into the sea. These compounds are widely used as biocides for controlling fouling organisms on the hulls of vessels and net cages of fish farms. A survey commissioned by the EPD revealed maximum concentrations of 53,000 and 18,300 ng Sn g^{-1} in sediments collected in Causeway Bay and Aberdeen Marinas respectively (Ko et al., 1995). These levels well exceeded typical TBT levels (500 to 1000 ng Sn g^{-1}) recorded elsewhere. Also of concern is the introduction of exotic marine species via hull fouling and ship's ballast water, although this problem has, so far, received little attention by local authorities. The ecological impacts of these introductions are virtually unknown. A recent study has recommended that guidelines on the discharge of ballast water be established (Chu et al., 1997).

There have been numerous oil spills in Hong Kong waters over the past ten years, but the majority of them were minor, causing contamination of limited sections of the shoreline (Wu, 1988; Morton, 1989). Other more major spills have resulted in substantial losses of cultured fish, and closures of contaminated swimming beaches on Hong Kong Island. Subsequent monitoring of these events, however, revealed that the effects were largely transient, with effects on benthic and planktonic communities becoming undetectable after about five months (Wu, 1988).

RURAL FACTORS

Almost all marine pollutants in Hong Kong have a land-based origin. They are either directly dumped or released into the sea in the form of screened sewage via outfall pipes, or indirectly via storm-water drains or streams, especially following heavy rains. Traditional agricultural activities have rapidly declined in Hong Kong over the past decades, and farming practices in the territory have become more specialized, requiring more intensive use of agrochemicals. Indeed, application of chemical fertilizers and pesticides, on a per unit area basis, have increased tenfold and fivefold respectively over the past two decades (Broom et al., 1999). The remaining croplands and poultry/pig farms, particularly in the north-western New Territories, are still sources of nutrient-rich run-off that finds its way into coastal waters via inland watercourses. These discharges are typically loaded with high amounts of nitrates and phosphates, as well as pesticides. The nutrients cause localized eutrophication in areas with relatively poor flushing, and have been considered the culprit in an increasing number of "red tides" in local waters. Records in Hong Kong from 1983 to 1997 revealed a total of 457 "red tide" incidents, averaging approximately 30 incidents per year (Anderson et al., 1999). Although the introduction of water pollution control regulations in the late 1980s appeared to coincide with the levelling off of the number of red tide incidents after 1988, the number of red tides in the first six months of 1998 was higher than that in any full year since 1988. It is clear that red tides will continue to be an important environmental problem in Hong Kong in the foreseeable future.

The more persistent pesticides and other substances detected locally, e.g. the organochlorine compounds, tend to adsorb onto sediment particles, and accumulate to very high concentrations in marine sediments (Richardson et al., in press). There is also evidence that the high levels of pesticides in the marine sediments around the Deep Bay area, including compounds that have been banned in Hong Kong (e.g. DDT), are attributable to contamination sources in the Chinese mainland. The lower reaches of the Pearl River, receiving two million tonnes of various types of wastes and wastewater annually, are heavily polluted by domestic, industrial and livestock waste and agrochemicals (Neller and Lam, 1994). Indeed, the annual discharge of the Pearl River (340 billion m^3), with an average sediment load of 84 million tonnes, accounts for over 85% of the total nutrient input to the local waters of Hong Kong (Hills et al., 1998). This, together with inputs from the heavily polluted Shenzhen River, have been of major concern to the long-term sustainability of important wetland systems (Mai Po and Deep Bay areas fringing the northwestern coast of the New Territories) which are protected under the Ramsar Convention.

In the northeast, Mirs Bay, which receives discharges from Hong Kong and the Special Economic Zone of Shenzhen, is also showing signs of eutrophication and localized hypoxic stress. Despite this, the environmental quality in general, and water quality in particular, is much better than that in western waters.

COASTAL EROSION AND LANDFILL

Since the early 1960s, large-scale reclamation has occurred in Hong Kong. Some 3600 hectares of land have been reclaimed in the years up to 1985 (compared with a total land area of only 1042 km^2), and the natural coastline fringing Victoria Harbour has been completed reclaimed. This large-scale reclamation has altered the current flow patterns and reduced tidal flushing, further augmenting pollution problems in local waters and causing hypoxia/anoxia in many areas (Wu, 1988). A study by Yip (1979) showed that it takes some 8 to 10 years for a new shoreline community to reach a climax stage. During 1981–1986, about 100 million m³ of marine spoil and sludge, generated by various development projects, have been dumped into the coastal waters of Hong Kong. During 1991 to 1999, a total of 366 million m³ of dredged spoil were disposed into coastal waters. A huge, but unknown amount of sand was extracted from local and nearby Chinese waters to support the booming building industry. According to the statistics of the Agriculture and Fisheries Department, fish fry production was seriously affected by these dumping and dredging activities.

In recent years, the rapid urban development in Hong Kong has exerted an unprecedented pressure on the local marine environment. Enormous amounts of waste, wastewater and contaminated mud have been disposed into the coastal environment. For example, the port and airport development (one of the largest civil projects in the world), involved the reclamation of 600 hectares of land, the dredging of 880 million m³ of fill materials, and the dumping of 380 million m³ of dredged materials within a very small coastal area. Fisheries resources, coral, seagrass, benthos and water quality were all seriously affected. As compensation, the government has paid some US$26 million to fishermen, in the form of an ex-gratia payment, for fisheries/habitat damage. Parallel with these urban developments, a new sewage disposal scheme is now being implemented. This scheme involves the discharge of wastewater from some 3.5 million people into the South China Sea, although the impact of the discharge on the coastal environment is virtually unknown. Further to this, Hong Kong has recently launched another large urban construction plan, intended to take place during the coming four years. The capital cost of this plan will exceed US$30 billion, and another 665 hectares of coastal area will be reclaimed before the year 2002.

EFFECTS FROM URBAN AND INDUSTRIAL ACTIVITIES

Contaminants originating from anthropogenic inputs have been found to be widespread in Hong Kong's coastal waters, sediments and biota. Several factors in combination have resulted in these inputs (Blackmore, 1998a,b). These include the dense population in Hong Kong, especially on the northern seaboard of Hong Kong Island and in the southern Kowloon Peninsula, together with considerable industrialization by both major and minor industries and a lack of controls, especially prior to the 1980s, on the release of agricultural and industrial effluents. The fact that many inputs are skewed in areas of densest population and industrial activity, especially in Victoria Harbour and Tolo Harbour has not helped the situation; this is especially true for organic (i.e. sewage-based) materials. Finally, there are severe cross-border influences, especially from chemicals (e.g. DDT) which have been previously banned in the HKSAR.

The introduction of organic wastes via sewage and run-off has been rated by Wu (1988) as the most problematic pollution concern in Hong Kong coastal waters. For example, sewage from approximately 3.5 million people is discharged directly into Victoria Harbour via 13 screening plants, which do little more than remove solids of >6 mm in size (Connell et al., 1998a). It is only waters with poor flushing, such as Tolo Harbour, that receive secondary treated sewage, but even in this locality other inputs from rivers, streams, rain storm events, urban run-off and agriculture contribute a considerable organic load.

Concomitant with these inputs, a number of effects have been observed. These include potentially detrimental changes in a number of physical and biotic parameters, such as % oxygen saturation, chlorophyll a, BOD_5, total nitrogen and inorganic phosphate (Wu, 1988). However, despite these factors, changes to biotic communities tend to be limited to localized areas, and are in many cases associated with areas of especially poor flushing, such as typhoon shelters (Fig. 5).

Nonetheless, the occurrence in recent years of an increasing number of red tides in Hong Kong's coastal waters has been reason for concern, especially where these have impacted upon local fish production through mariculture activities (Anderson et al., 1999). In 1998, a major red tide outbreak in Hong Kong (caused by *Gyrodinium digitatum*) killed over 80% (3400 tonnes) of mariculture fish, and the total loss was estimated at over US$40 million. There have been 35 cases of PSP intoxication reported thus far, and over 100 people have been intoxicated. According to the Department of Health, PSP outbreaks, in which toxicities exceeded the WHO limit (i.e. 4000 MU/kg), have been reported in shellfish in 1989, 1990, 1993 and 1997. Likewise, 69 cases of ciguatera intoxication were reported during 1993–1996, and 305 people were affected (Anderson et al., 1999). It is particularly worrying to note that there has been an increase in the presence and variety of toxic algal species in local waters (Hodgkiss and Ho, 1997), and the problem of

Fig. 5. A typical typhoon shelter (at Aberdeen), characterised by high concentrations of contaminants, largely originating from uncontrolled discharges of boats and surrounding housing and industrial developments.

HAB can be expected to escalate in the coming years (Anderson et al., 1999).

In addition, the loss and/or mortality of phytoplankton, fish, benthic communities and coral populations in such localities as Tolo Harbour has been reported (Horikoshi and Thompson, 1980; Wu and Richards, 1979; Wu, 1982, 1988), although summer-based mortalities of benthos tend to be short-term, the status quo being restored during a relatively rapid winter recolonization.

Trace metals in Hong Kong's coastal waters, sediments (Table 1) and biota have been the subject of a number of studies spanning the 1970s to the 1990s, and the resultant data have been extensively reviewed by Wu (1988), Phillips (1989) and Blackmore (1998a,b). In general, two particular areas have been studied—Victoria and Tolo Harbours. Levels of metals in Victoria Harbour during the 1970s and 1980s were typically high. For example, Wu (1988) reported that Cu and Zn in this locality were some 5 and 2.5 times higher respectively in sediments than in other Hong Kong localities. Monitoring of living organisms (e.g. *Saccostrea glomerata* and *Perna viridis*) also supports sediment data for this period (Phillips, 1979; Wu, 1988).

The development of the New Territories as a site for light industrial activities during the late 1970s has resulted in a shift in trace metal distribution within Hong Kong coastal waters. Tolo Harbour has been the recipient of trace metal inputs from this development, and Blackmore (1998a,b) has noted that the long retention times of waters within this harbour have resulted in a steady elevation of trace metal concentrations in sediments and biota throughout the 1980s. Most recently, economic factors have resulted in another shift in Hong Kong's industrial infrastructure—this time to the Special Economic Zone of Shenzen, immediately across Hong Kong's northern border, in mainland China. This has resulted in a fall in trace metals in biota from Hong Kong, reflecting a lowering of trace metal bioavailability locally (Blackmore 1998a,b).

Nonetheless, trace metal concentrations remain high in many parts of Hong Kong, and are reason for concern, especially regarding the consumption of certain seafoods which have been shown to have elevated levels of certain metals (especially Cd; Wu, 1988). The situation in waters in semi-enclosed inlets and bays—and especially in typhoon shelters, where considerable human populations may live aboard boats—is of particular concern.

Trace organic contaminants, including the pesticides, polychlorinated biphenyls (PCBs) and petroleum hydrocarbons, have received less attention than the trace metals in Hong Kong research studies (Tables 2 and 3). Although Hong Kong has little arable land, pesticides (including HCHs, DDT and metabolites, dieldrin, aldrin and chlordane) may be found in coastal waters, sediments and biota (Richardson and Zheng, 1999), and it is likely that cross-border influences play a significant role in their distribution. In the case of PCBs, however, local inputs almost certainly play an important role. For instance, Tanabe et al. (1987) showed that a localised industrial input in the Junk Bay area was responsible for elevated concentrations of PCBs which they observed in transplanted mussels (*Perna viridis*).

Elevated concentrations of petroleum hydrocarbons and polycyclic aromatic hydrocarbons (PAHs) have been observed since the 1970s (Wu, 1988), and in a few recent studies of Hong Kong waters, their potential as toxicologically significant contaminants has been assessed. Hong et al. (1995) and Zheng and Richardson (1999) have reported high concentrations of petroleum hydrocarbons and PAHs locally, the latter authors noting that nearshore concentrations were indicative of serious pollution in certain areas, including Victoria and Tolo Harbours. Zheng and Richardson (1999) also suggest that oil and its products are the main contributors to these inputs, and that sources appear to be stable and continuing (the result of well established human and industrial activities in Hong Kong's most densely populated areas).

Connell et al. (1998a) recently reviewed local Environment Protection Department data on the occurrence and distribution of persistent organic contaminants in Hong Kong waters, and concluded that Hong Kong marine sediments contained elevated levels, especially within Victoria and Tolo harbours, and within typhoon shelters. They concluded that the contaminants concerned, including

Table 1

Maximum concentrations of Cd, Cr, Cu, Mn, Ni, Pb and Zn in Hong Kong marine sediments. After Blackmore (1998b).

Site	Metal						
	Cd	Cr	Cu	Mn	Ni	Pb	Zn
Southern Waters	NM	10.2	NM	NM	NM	69.2	13
North Western	NM	27.2	28.8	739	NM	50.0	98.5
Tolo Harbour	8	2.35	150	352	NM	161	182
Victoria Harbour	NM	601	3789	NM	NM	138	610
Deep Bay	2.96	34.2	41.1	438	NM	35.4	146
Port Shelter	0.36	51.1	27.2	NM	9.34	35.9	100
Typhoon Shelters	NM	2220	6250	NM	550	550	1000

NM = Not measured.

Table 2

Concentrations[a] of total polychlorinated biphenyls (PCBs) in Hong Kong waters during 1995–1996. After Connell et al. (1998a)

Zone	Stations							Average (SD)
	S1	S2	S3	S4	S5	S6	S7	
Deep Bay	10.0 (7.0)	8.75 (3.5)	6.75 (2.3)	15.5 (16.6)	NA	NA	NA	10.25 (3.7)
Eastern Buffer	6.25 (2.5)	5.25 (0.5)	33.0 (27.4)*	7.25 (2.6)	NA	NA	NA	12.9 (13.3)
Junk Bay	NA	9.0 (2.7)	NA	NA	NA	NA	NA	9.0 (2.7)
Mirs Bay**	6.75 (3.5)	6.5 (3.0)	7.75 (3.2)	5.0 (0)	5.0 (0)	9.75 (9.5)	6.25 (2.5)	6.7 (1.5)
North Western	NA	6.7 (2.0)	7.0 (2.8)	7.25 (3.8)	14.7 (11.8)*	9.75 (4.2)	NA	9.1 (3.4)
Port Shelter	12.6 (11.5)*	6.75 (3.5)*	5.75 (2.25)	11.75 (12.2)	5.3 (0.5)	5.75 (1.5)	5.0 (1.6)	7.5 (3.2)
Southern	5.25 (0.5)	5.0 (0)	5.5 (1.0)	8.3 (5.7)	8.25 (3.9)	5.0 (0)	16.0 (14.9)*	7.6 (3.9)
Tolo Harbour	NA	7.0 (2.6)	7.7 (7.2)	5.0 (0)	10.0 (9.2)	NA	15.25 (16.6)*	8.99 (3.9)
Western	10.5 (4.5)	7.75 (5.5)	NA	NA	NA	NA	NA	9.1 (1.9)

**Mirs Bay Stations continued: S8 5.3(0.5); S9 6.6 (2.8); S10 5.0 (0); S11 7.6 (4.6); S12 6.3 (2.3); S13 4.6 (0.5); S14 6.6 (2.0); S15 8.5 (7.0); S16 6.0 (1.7); S17 9.0 (6.9).
[a] $\mu g\ kg^{-1}$ wet weight with standard deviations in parenthesis.
*Typhoon shelters and sheltered anchorages.
NA = not available.

Table 3

Concentrations[a] of total polyaromatic hydrocarbons in Hong Kong waters during 1995–1996. After Connell et al. (1998a)

Zone	Stations							Average (SD)
	S1	S2	S3	S4	S5	S6	S7	
Deep Bay	231 (213.7)	64 (29.1)	51 (15.3)	47 (13.1)	NA	NA	NA	98.8 (88.9)
Eastern Buffer	42 (5.2)	40 (1.4)	230 (158.3)*	60 (13.2)	NA	NA	NA	93.6 (92.5)
Junk Bay	45 (9.9)	NA	NA	NA	NA	NA	NA	45.5 (9.9)
Mirs Bay**	41 (1.5)	41 (3.6)	40 (1.4)	40 (1.2)	40 (1.2)	44 (3.2)	40 (1.4)	40.4 (1.5)
North Western	NA	62 (19.7)	44 (4.9)	49 (9.9)	125 (51.9)*	61 (28.1)	NA	68.6 (32.6)
Port Shelter	120 (117.7)*	42 (2.0)*	43 (5.3)	48 (11.3)	40 (1.15)	43 (6.9)	51 (11.8)	55.7 (28.8)
Southern	43 (5.4)	49 (8.5)	170 (174.4)	104 (53.6)	60 (5.7)	41 (2.8)	156 (44.8)*	89.5 (55.0)
Tolo Harbour	NA	40 (2.5)	40 (2.1)	39 (0.9)	41 (2.6)	NA	50 (13.9)*	42.41 (4.4)
Western	60 (13.5)	55 (22.4)	NA	NA	NA	NA	NA	58.1 (3.4)

**Mirs Bay Stations continued: S8 44 (8.6); S9 39 (0); S10 40 (2.3); S11 40 (2.8) S12 39 (0); S13 40 (1.7); S14 40 (1.7); S15 39 (1.0); S16 39 (0); S17 39 (0).
[a] $\mu g\ kg^{-1}$ wet weight with standard deviations in parenthesis.
*Typhoon shelters and sheltered anchorages.

petroleum hydrocarbons, PAHs, PCBs and certain organochlorine pesticides were likely derived from a combination of industrial discharges, stormwater run-off, sewage and combustion. They also noted the declining values of PCBs and PAHs in most areas during the period 1978–1996.

However, in a fate and risk evaluation of these substances, Connell et al. (1998b) indicate that the concentrations present in sediments are of a long-term significance to the health of Hong Kong waters, especially those within Victoria Harbour. The levels of PAHs were found to be such that they posed a risk not only to the marine ecosystem, but also to consumers of seafoods. They also noted that the risks posed by other contaminants, including total alkanes, non-aromatic hydrocarbons, linear alkyl benzenes and chlorohydrocarbons, were likely to be additive to those of other contaminants present, and required further investigation to elucidate their distribution and ubiquity in local waters.

Results of the shellfish surveillance program carried out by the Department of Health in 1995–1997 showed that some 3 to 7% of shellfish samples contained an unacceptable high level of PCBs (>2 ppm) and $\sum$DDT (>5 ppm). The estimated daily ingestion of total BHC and total DDT from consumption of local food stuffs, including fish, was 10 and 9 times higher, respectively, than that found in USA (Ip, 1990). Epidemiological surveys also showed that the concentrations of DDT, DDE and βHCH in Hong Kong breast milk were amongst the highest reported in the literature, and the high level of contamination was attributable to consumption of seafood in Hong Kong (Ip, 1983; Ip and Phillips, 1989). Likewise, high body burdens of organochlorines have been reported in local dolphins and porpoises, which have been related to the high neonatal mortality in local marine mammals (Parsons and Chan, 1998). The above evidence clearly indicates that the health

Fig. 6. Reclamation work along Hong Kong coastal waters.

of both human and marine populations in Hong Kong is seriously under threat.

In addition to the conservative contaminants, other pollution-related factors of importance in the Hong Kong coastal environment include the effects of reclamation and dredging (Fig. 6), which have had a considerable influence on water flows, flushing rates and the burial of benthic communities. Large-scale reclamation has been undertaken in Hong Kong since 1960, reclaiming in excess of 3600 ha for major infrastructure projects in Kowloon, Hong Kong Island and as part of the new airport construction at Chek Lap Kok. Dumping and dredging activities have important effects, which have similar endpoints to the effects of reclamation. Wu (1988) reports that some 100 million m^3 of spoil have been generated from local developments, urban, sewage and industrial activities. In addition, toxic, hazardous and difficult wastes (THD), which include acids and alkalis, oil and oil–water emulsions, solvents and various other wastes of local metal, plastic and electronic industries (Wu, 1988) are important, as is floating refuse, which has long been a problem in Hong Kong, and which constitutes materials such as plastics, polystyrene foams, wood, paper, etc.

PROTECTIVE MEASURES

Oil pollution is controlled by three ordinances in Hong Kong:
1. the Shipping and Port Control Ordinance;
2. the Merchant Shipping (Prevention and Control of Pollution) Ordinance; and
3. the Merchant Shipping (Liability and Compensation for Oil Pollution) Ordinance.

Waste from vessels is controlled by the Waste Disposal Ordinance and the five annexes of the MARPOL convention—the 1978 Protocol of the 1973 Convention on Prevention of Pollution from Ships.

The deteriorating water quality along Hong Kong shores, particularly bathing beaches, has partly been kept in check as a result of the Water Pollution Control Ordinance and the Waste Disposal Ordinance. Under the former (enacted in 1980, and amended in 1990 and 1993), Hong Kong as a whole is divided into ten Water Control Zones (WCZs), each having its own Water Quality Objectives (WQO), based on the specific characteristics and sensitivity of individual receiving environments. The discharge of polluting wastes into a WCZ is subject to licensing control. Local environmental managers are of the view that significant improvements in water quality along Hong Kong shores will only be possible with the full implementation of the Sewerage Master Plans (SMPs), which set out to collect sewage on a catchment-by-catchment basis and direct it to respective treatment facilities. SMPs have been produced for all the 16 catchment areas. However, there is still some debate as to the appropriate degree of treatment that should be applied to the wastewater before it is discharged offshore into the South China Sea. In addition to the above two Ordinances, the Public Health and Municipal Services Ordinance and the Summary Offenses Ordinance also provide additional regulatory tools for controlling littering of beaches, and the former Ordinance also governs the illegal discharge of hazardous materials into sewers.

With regard to nature conservation, the Marine Parks Ordinance provides a legal framework for the designation and protection of marine parks and marine reserves; e.g., Hoi Ha Wan Marine Park, Yan Chau Tong Marine Park, Sha Chau and Lung Kwu Chau Marine Park, as well as a marine reserve at Cape D'Aguilar. The Wild Animals Protection Ordinance, on the other hand, allows the designation and protection of Restricted Areas, e.g. the Mai Po and Inner Deep Bay area.

REFERENCES

AFD (1998) Fisheries resources and fishing operations in Hong Kong waters. Agriculture and Fisheries Department, Hong Kong Government.

Anderson, D.M., Anderson, P., Bricelj, V.M., Cullen, J.J., Hodgkiss, I.J., Ho, K.C., Rensel, J.E., Wong, J.T.Y. and Wu, R.S.S. (1999) Red tide—HAB monitoring and management in Hong Kong. Final report submitted to the Agriculture & Fisheries Department, Hong Kong SAR Government.

Blackmore, G.R. (1998a) The Importance of Feeding Ecology in Investigating the Heavy Metal Burden Accumulated by *Thais clavigera* (Kuster) (Mollusca: Neogastropoda: Muricidae) in Hong Kong. Thesis (PhD). University of Hong Kong.

Blackmore, G.R. (1998b) An overview of the trace metal pollution in coastal waters of Hong Kong. *Science of the Total Environment* **214**, 21–48.

Broom, Y.S., Leung, S.H., Sze, W.C. and Wong, Y.H. (eds.) (1999) *Hong Kong's Present Challenge: Profile of the Environment in the 1990's*. WWF Hong Kong, Hong Kong.

Chu, K.H., Tam, P.F., Fung, C.H. and Chen, Q.C. (1997) A biological survey of ballast water in container ships entering Hong Kong. *Hydrobiologia* **352**, 201–206.

Connell, D.W., Wu, R.S.S., Richardson, B.J., Leung, K., Lam, P.K.S. and Connell, P.A. (1998a) Occurrence of persistent organic contaminants and related substances in Hong Kong marine areas: an overview. *Marine Pollution Bulletin* **36** (5), 376–384.

Connell, D.W., Wu, R.S.S., Richardson, B.J., Leung, K., Lam, P.K.S. and Connell, P.A. (1998b) Fate and risk evaluation of persistent organic contaminants and related compounds in Victoria Harbour, Hong Kong. *Chemosphere* **36** (9), 2019–2030.

EPD (Environmental Protection Department) (1987) Epidemiological studies of beach water pollution in Hong Kong, phase II. Technical paper EPD/TP14/88, Hong Kong Government, Hong Kong.

EPD (Environmental Protection Department) (1993) Bacteriological water quality of bathing beaches in Hong Kong. Hong Kong Government, Hong Kong.

EPD (Environmental Protection Department) (1997) Marine Water Quality in Hong Kong for 1996. Hong Kong Government, Hong Kong.

EPD (Environmental Protection Department) (1998) Environment Hong Kong. Hong Kong Government, Hong Kong.

Hills, P., Zhang, L. and Liu, J. (1998) Transboundary pollution between Guangdong Province and Hong Kong: threats to water quality in the Pearl River estuary and their implications for environmental policy and planning. *Journal of Environmental Planning and Management* **41**, 375–396.

Horikoshi, M. and Thompson, G.B. (1980) Distribution of subtidal molluscs collected by trawling in Tolo Harbour and Tolo Channel, Hong Kong, with special reference to two venerid bivalves. In *Proceedings of the First International Workshop on the Malacofauna of Hong Kong and Southern China, Hong Kong, 1977*, ed. B. Morton. Hong Kong University Press, Hong Kong, pp. 149–162.

Hodgkiss, I.J. and Ho, K.C. (1997) Are changes in N:P ratios in coastal waters the key to increased red tide blooms? *Hydrobiologia* **852**, 141–147.

Hong, H., Xu, L., Zhang L., Chen, J.C., Wong, Y.S. and Wan, T.S.M. (1995). Environmental fate and chemistry of organic pollutants in the sediment of Xiamen and Victoria Harbours. *Marine Pollution Bulletin* **31**, 229–236.

Ip, H.M.H. (1983). Breast milk contaminants in Hong Kong. *Bulletin of the Hong Kong Medical Associations* **35**, 1–16.

Ip, H.M.H. (1990) Chlorinated pesticides in foodstuffs in Hong Kong. *Archives of Environmental Contamination and Toxicology* **19**, 291–296.

Ip, H.M.H. and Phillips D.J.H. (1989) Organochlorine chemicals in human breast milk in Hong Kong. *Archives of Environmental Contamination and Toxicology* **18**, 490–494.

Ko, M.M.C., Bradley, G.C., Neller, A.H. and Broom, M.J. (1995) Tributyltin contamination of marine sediments of Hong Kong. *Marine Pollution Bulletin* **31**, 249–253.

Marine Department (1997) Statistical Tables 1996. Government Printer, Hong Kong.

Morton, B. (1982) An introduction to Hong Kong's marine environment with special reference to the north-eastern new territories. In *Proceedings of the First Marine Biology Workshop. The marine flora and fauna of Hong Kong and southern China, Hong Kong, 1980*, eds. B.S. Morton and C.K. Tseng. Hong Kong University Press, pp. 25–53.

Morton, B. (1989) Pollution of the coastal waters in Hong Kong. *Marine Pollution Bulletin* **20** (7), 310–318.

Morton, B. and Morton, J. (1983) *The Sea Shore Ecology of Hong Kong*. Hong Kong University Press. 350 pp.

Morton, B. and Wu, R.S.S. (1975) The hydrology of the coastal waters of Hong Kong. *Environmental Research* **10**, 319–347.

Neller, R.J. and Lam, K.C. (1994) The environment. In *Guangdong: Survey of a Province Undergoing Rapid Change*, eds. Y.M. Yeung and D.K.Y. Chu, pp. 401–428. The Chinese University Press, Hong Kong.

Parsons, E.C.M. (1998a) Strandings of small cetaceans in Hong Kong territorial waters. *Journal of the Marine Biological Association of the UK* **78** (3), 1039–1042.

Parsons, E.C.M. (1998b) Trace metal pollution in Hong Kong: implications for the health of Hong Kong's Indo-Pacific hump-backed dolphins (*Sousa chinensis*). *Science of the Total Environment* **214** (1–3), 175–184.

Parsons, E.C.M and Chan L.H.M. (1998) Organochlorines in the Chinese white dolphins (*Sousa chinensis*) and finless porpoises (*Neophocaena phocaenoides*) in Hong Kong. In *The Marine Biology of the South China Sea III*, ed. B.S. Morton. pp. 423–438. Hong Kong University Press, Hong Kong.

Phillips, D.J.H. (1979) The rock oyster *Saccostrea glomerata* as an indicator of trace metals in Hong Kong. *Marine Biology* **53**, 353–360.

Phillips, D.J.H. (1989) Trace metals and organochlorines in the coastal waters of Hong Kong. *Marine Pollution Bulletin* **20**, 319–327.

Richardson, B.J. and Zheng, J.S. (1999) Chlorinated hydrocarbon contaminants in Hong Kong surficial sediments. *Chemosphere* **39**(9), 61–71.

Richardson, B.J., Zheng, G.J. and Tse, E.S.C. (In press) A comparison of mussels and SPMDs for monitoring organochlorine pesticides and total petroleum hydrocarbons in Hong Kong waters. In *Fish Physiology, Toxicology and Water Quality*. Proceedings of the Fifth International Symposium, Hong Kong, November 9–13, 1998.

Tanabe, S., Tatsukawa, R. and Phillips, D.J.H. (1987) Mussels as bioindicators of PCB pollution: A case study on uptake and release of PCB isomers and congeners in green-lipped mussels (*Perna viridis*) in Hong Kong waters. *Environmental Pollution* **47**, 41–62.

Wu, R.S.S. (1982) Periodic defaunation and recovery in a sub-tropical epibenthic community, in relation to organic pollution. *Journal of Experimental Marine Biology and Ecology* **11**, 117–130.

Wu, R.S.S. (1988) Marine pollution in Hong Kong: a review. *Asian Marine Biology* **5**, 1–23.

Wu, R.S.S. and Richards, J. (1979) *Mass Mortality of Benthos in Tolo Harbour*. Hong Kong Fisheries Occasional Paper No. 21. Agriculture and Fisheries Department, Hong Kong.

Yip, S.S.Y. (1979) The ecology of coastal reclamation in Hong Kong. Thesis (M.Phil.), University of Hong Kong.

Zheng, J.S. and Richardson, B.J. (1999) Petroleum hydrocarbons and PAHs in marine sediments of Hong Kong. *Chemosphere.* **38**(11), 2625–2632.

THE AUTHORS

Bruce J. Richardson
*Department of Biology and Chemistry,
City University of Hong Kong,
83 Tat Chee Avenue, Kowloon,
Hong Kong*

Paul K.S. Lam
*Department of Biology and Chemistry,
City University of Hong Kong,
83 Tat Chee Avenue, Kowloon,
Hong Kong*

Rudolf S.S. Wu
*Department of Biology and Chemistry,
City University of Hong Kong,
83 Tat Chee Avenue, Kowloon,
Hong Kong*

Chapter 88

SOUTHERN CHINA, VIETNAM TO HONG KONG

Zhang Gan, Zou Shicun and Yan Wen

The rapid economic development in South China, especially in the Pearl River delta, during the last two decades, has had a significant human impact on the coastal ecosystem. Increasing discharges of industrial wastewater, domestic sewage, agricultural wastewater, and particularly pig farm wastewater, have severely polluted the coastal water bodies by N, P and petroleum hydrocarbons. As a result, red tide events are occurring more frequently in the Pearl River mouth. The biodiversity of phytoplankton has also decreased due to increasing pollution.

Contamination of petroleum hydrocarbons in commercially significant fishes and trace toxic organic compounds such as BHC, DDT, PCBs and PAHs in sediments are also widely detected in this area. Sediments in Macao estuary have been heavily contaminated by organochlorine compounds.

Crude fishing methods, as well as uncontrolled fishing activities before 1997, have seriously reduced the fishery resources in this region. The increasing disregard of environmental impact of mariculture has also contributed to the eutrophication of the nearshore waters, and uncontrolled and unsustainable land reclamation has regularly destroyed excellent harbours, bays, spawning habitats of fishes and the natural coastal landscape on this coast of the South China Sea.

The commencement of new "Management Regulations for Utilization of Marine Area in Guangdong" in 1996 and closed seasons for fishing from 1999 may prove to be effective in protecting the coastal environment. However, further control of water pollution in river catchments as well as the restoration of currently impacted river water environments are needed urgently before overall improvement of the coastal environment in the region can be achieved.

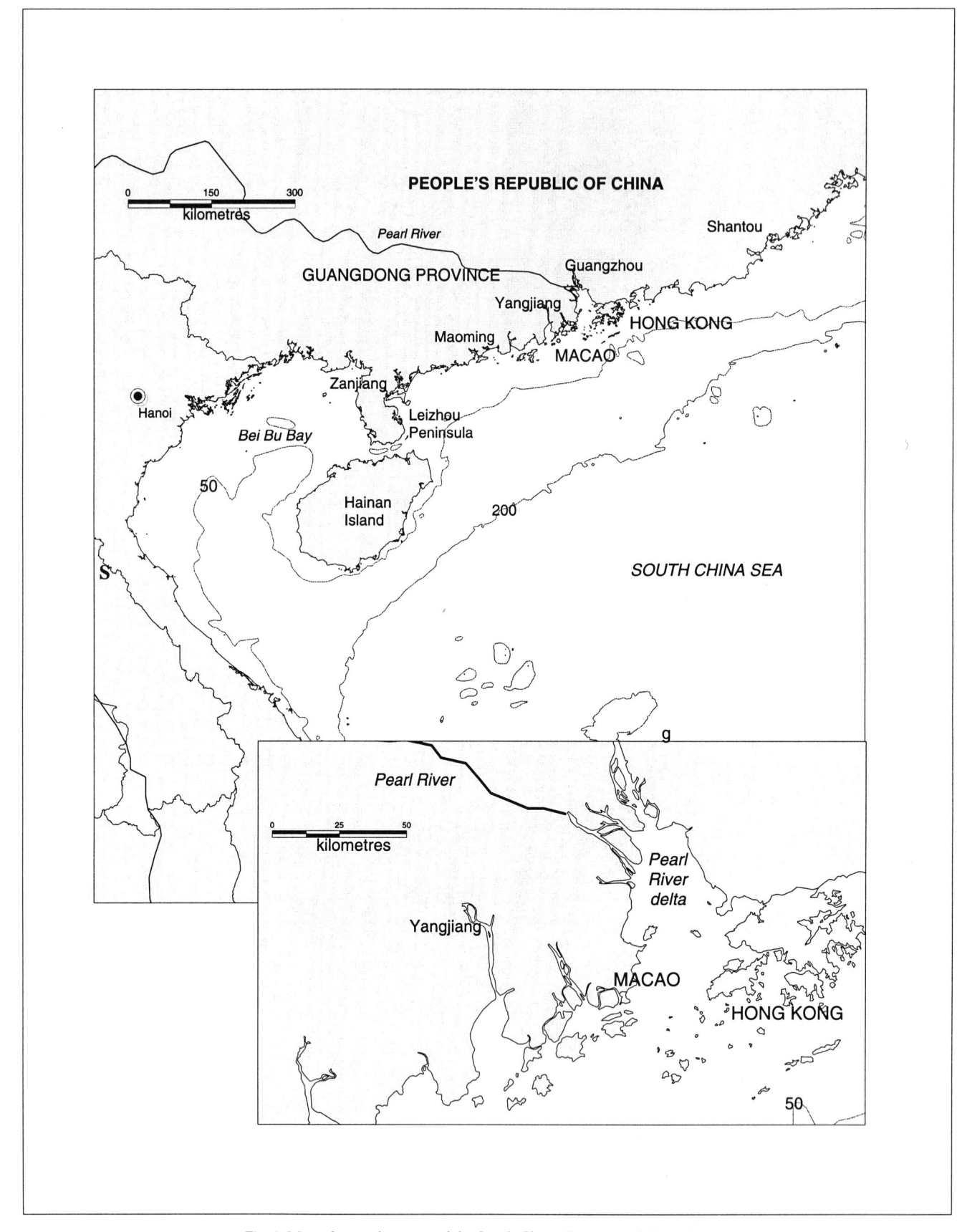

Fig. 1. Map of coastal regions of the South China Sea, west of Hong Kong.

THE DEFINED REGION

The coastal areas of the South China Sea (West of Hong Kong) include four regions distinguished by economic development, population density and natural geographic condition. These are: the Pearl River Mouth and adjacent areas, the coast of West Guangdong province (Yuexi Coast), Bei Bu Bay and the coast of Hainan Island (Fig. 1).

Hainan Island was formerly a special administrative region of Guangdong province and became a new province in 1987. The island is located in the tropical zone with a sub-tropical monsoon climate, possesses a land area of 3.39 million ha, a population of 7.71 million and a coastline of 1,528 km in length. Most parts of the island remain undeveloped with abundant tropical coastal and fishery resources.

Bei Bu Bay is surrounded by Leizhou Peninsula of Guangdong province, the southern part of Guangxi Autonomous Region of Zhuang Minority (province) and the coast of northeast Vietnam. Abundant fishery resources, coastal mangroves as well as marine, oil and gas resources are found in the bay.

The only systematic investigation of the coastal ecology and environment of the above two regions was organized and carried out by Guangdong and Guangxi provinces respectively in 1984–1986 under the State Investigation Scheme for Coastal and Beach Resources, launched by the State Council of China in 1979 (GDCRIP, 1988; GXCRIP, 1986). There are no more recent detailed ecological or environmental reports on these two regions. In addition, most of the statistical data was based on administrative regions, especially by province, so in some parts, the coast of Guangdong province is examined as a whole, including the coast of the South China Sea, east of Hong Kong.

The coast of West Guangdong including Leizhou Peninsula is 1976 km in length (GDCRIP, 1988). It is the main fishery region in Guangdong province, and its economy is characteristically based on traditional fishery industries and large-scale artisanal mariculture. The exception is Maomin which is an oil-shale based petrochemical city.

The Pearl River plays an important role in the geography and eco-environmental conditions of the Pearl River Mouth. It is the largest river system in South China and the second largest river in China. It runs across six provinces of China (Yunnan, Guizhou, Guangxi, Guangdong, Hunan and Jiangxi) and northeast Vietnam as well, covering a watershed area of 453,700 km². The average annual rainfall in the watershed is 1470 mm, and the maximum annual rainfall is 2.0–2.5 times higher than the minimum. The summer (April–September) brings 80% of the annual rainfall. The average annual amount of surface run-off from this watershed is 331.9 billion m³ (391.7 billion m³ in 1994), and the high–low difference over the last 40 years has been 261.2 billion m³. Theoretically, this difference is capable of causing an annual 7 cm change in seawater level in the South China Sea (Chen and Yang, 1998). The water level change over the last decade in the Pearl River Delta as calculated by Yang et al. (1998) is 3.1 mm/yr. In 1950–1989, there were 8731 reservoirs constructed in the watershed with a total capacity of 42.9 billion m³, which effectively adjusted the surface run-off (Shi Xieguo, 1990).

The Pearl River Delta (including Hong Kong and Macao), comprises three main tributaries, i.e. East River (Dong Jiang), North River (Bei Jiang) and West River (Xi Jiang). The waters merge and flow into the South China Sea via eight river outlets, which defines the outline of the Pearl River mouth. The coastline of the Pearl River delta is 522 km in length. The main silt and sand load of the watershed is carried by West River (80–87%) and North River (10–13%), while in the East River the silt and sand are precipitated and adjusted by large reservoirs in the upper reaches. About 20–35.5% of silt and sand from the upper reaches deposits in the river network, while the rest enters the river mouth (Mou et al., 1988). The north bay of Pearl River mouth, including Hong Kong, Macao and Humen, is geographically called the Lingding Yang, with an area of 2110 km².

The Pearl River delta has a history of civilization of over 3500 years (Situ, 1996). In August 1980, Shengzhen and Zhuhai became the first two special economic zones of China and opened their doors to the world. From then on, the economy has developed rapidly with an annual regional GDP increasing at an average rate of >14%. The region has been urbanized from being an agricultural area to an integrated industrial region, and is now one of the most prosperous economic regions in China and it is the heart of the coast of the South China Sea. However, with the expanding economy, urbanization and the consequent increase in population, the environmental quality, including the coastal environment in the Pearl River delta, has deteriorated significantly.

SEASONALITY, CURRENTS, NATURAL ENVIRONMENTAL VARIABLES

The coastal region of the South China Sea is located south of the Tropic of Cancer. The most remarkable characteristics of the climate result from the monsoon. Most parts of the region display a southern subtropical monsoon pattern, while the Leizhou Peninsula and Hainan Island show a tropical monsoon pattern. In winter (October–March), the winds come from inland areas of Asia, from the northeast, with an average wind strength of 6–10 m/s. In summer (May–August) winds come from the southwest with an average strength of <7 m/s.

The annual average air temperature is 21.2–23.3°C in the coastal areas of Guangdong and 23.2–25.5°C on the Hainan coast. The isogram of annual average air temperature parallels the coastal line, displaying an increasing gradient from north to south and from east to west. Highest air temperature is generally 37.0–40.0°C observed in July–August (April–June in Leizhou Peninsula and Hainan Island), whereas lowest air temperature usually occurs in January, falling to −1.3°C in the Pearl River mouth, −1.4°C in West Guangdong and 0.4°C in Hainan.

The average annual water temperature is 21.2–26.3°C. The isograms of water temperature on the coast of the South China Sea parallel the depth contours. The average surface water temperatures in summer and winter are 28–30°C and 17–18°C respectively in the Pearl River mouth, and 29–31°C and 13–17°C on the coast of West Guangdong. When seasons change, coastal surface water temperature changes at a rate of 3–4°C/month. Extreme surface water temperatures are 33°C (high), 11.6°C (lowest) in the Pearl River mouth. An even greater range of 35°C and 11.8°C has been seen in Hainan Island, and 33.3°C and 10.0°C on the coast of West Guangdong (GDCRIP, 1988).

The salinity of the South China Sea is a result of mixing by high-salinity offshore water and low-salinity land run-off. Thus horizontal variations of water salinity are more evident than vertical variations, the latter often being less than 0.5 ppt. In the delta of the Pearl River, salinity displays remarkable changes together with changes in river level. Average surface water salinity in the delta is 14.89 ppt in flood season (April–September), and 32.69 in the low water level season. On the coast of West Guangdong, surface salinity is affected by coastal water all the year, and remains low at 19–33.6 ppt, except in winter when it remains above 30 ppt.

The frequency of typhoons in the South China Sea is one of the highest in the world. The number of annual typhoons ranges from 7 to 22, with an average of 14.1 in the last 30 years. Typhoons often occur in July–September, but are rare in January–April. Up to 70% of the typhoons in the South China Sea come from the West Pacific, while others are generated in the northern or eastern parts between 10–20°N, 112–120°E. Statistics showed that about 41% of the typhoons landed on Hainan Island, 20% on the Pearl River mouth, while 25% hit the coast of West Guangdong (Fig. 2).

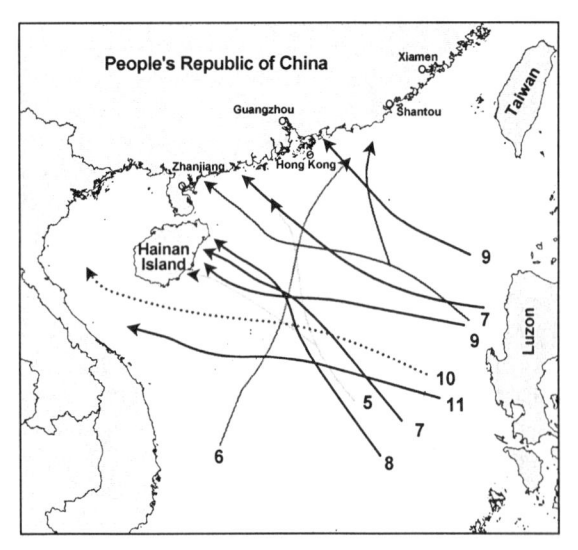

Fig. 2. Paths of typical typhoons in the South China Sea (after Situ, 1996).

The direction and speed of ocean currents in the South China Sea is affected by monsoons and displays seasonal variations. In winter, the northeast monsoon drift-current occurs with an average speed of 0.5–1 mile/h (maximum 3 mile/h). Seawater comes from the Java Sea through the Karimata Strait and Gaspar Strait, and turns northeastward, merging into the Pacific through Bashi Channel and Balingtang Channel, or entering the East Sea of China through the Taiwan Strait. In summer, the Southwest Monsoon drift-current flows at 0.2–0.5 (maximum less than 2) mile/h. Seawater comes from the East Sea of China, and flows southwestward and enters the Java Sea. Each year in April and September, the current of the South China Sea becomes mixed before changing with the monsoons. In addition, local currents can be observed in Bei Bu Bay (clockwise in winter). This is the basis of the South China Sea current (Situ, 1996).

In the waters of the Pearl River mouth, the irregular semi-diurnal tides dominate. The tides show unequal solar tides and anti-clockwise alternating currents. Generally, the tidal current reaches the maximum six hours after high tide. The current speed on the surface layer is 153 cm/s, and is 104 cm/s on the bottom. During the spring tides, current speed on the surface layer reaches 190 cm/s, which is close to the theoretical value. Residual flow is much weaker, and the maximum current speed of the residual flow is less than one third of the maximum tidal speed (Lin Zuheng and Liang Shunhua, 1996).

The major environmental variables controlling the distribution of biota in the region are water salinity and temperature. In the Pearl River mouth, the phytoplankton is a mixture of near-shore low-salinity species, salinity-tolerant estuarine species and some freshwater species. This contrasts with the coast of West Guangdong where the phytoplankton is comprised mainly of species acclimatised to high water temperatures and salinity; here the phytoplankton biomass is lower than that in the Pearl River mouth due to the relative stability of water temperature and salinity in the region (Guo et al., 1994).

THE MAJOR SHALLOW-WATER BIOTA

Intertidal Zone

Five types of intertidal habitat are found on the coast of Guangdong: sand, mud, gravel, reef and mangrove (GDCRIP, 1988).

Mangrove is mainly distributed in Hainan, along the coast of West Guangdong, and also can be found in the Pearl River Mouth. Mangroves in the area are typically rich and diverse, including 12 families, 14 genera and 22 species, comprising 13 vegetation assemblages in the region. There are a further four vegetation assemblies of mangrove associates, with 10 families of plants, 12 genera and 14 species (GDCRIP, 1988).

Table 1
Average biomass in intertidal zones in mainland as well as in islands of Guangdong. Biomass in g/m², densities in individuals/m²

Item	Total	Polychaetes	Mollusc		Shellfish	Echinoderms	Algae	Others	Details (region)
			Univalve	Bivalve					
Biomass	580.93	1.29	57.27	258.45	83.52	8.23	169.15	3.02	1980–1985
Density	469.92	12.04	170.62	175.70	105.17	1.36	–	5.03	(total intertidal)
Biomass	53.13	0.23	7.75	39.52	3.28	0.10	–	2.25	1980–1985
Density	213.35	5.97	36.67	159.26	8.07	0.04	–	3.25	(Pearl River mouth)
Biomass	2053.33	1.67	44.59	1512.77	380.15	0.02	109.11	5.05	1980–1985 (island,
Density	1381.71	58.25	158.08	728.16	413.37	0.06	–	23.79	Pearl River mouth)
Biomass	231.66	0.71	80.40	53.82	25.02	15.63	53.98	2.10	1980–1985
Density	382.60	6.10	282.2	71.30	17.20	1.40	–	4.40	(West GD)
Biomass	709.34	0.34	121.90	154.15	187.06	4.68	236.08	4.95	1980–1985
Density	345.20	3.70	196.00	37.00	97.00	2.60	–	8.90	(island, West GD)
Biomass	506.77	2.71	44.46	105.91	27.71	2.16	318.69	5.13	1980–1985
Density	391.25	13.87	122.67	128.33	119.43	1.07	–	5.87	(Hainan Island)
Biomass	958.24	0.37	59.48	159.73	223.72	2.97	511.79	0.18	1990–1991
Density	282.74	4.67	58.01	156.48	62.45	0.54	–	0.59	(island)

The intertidal biota on the coast of the South China Sea has been systematically investigated in 1980–1985 (GDCRIP, 1988) for coastal resources, and in 1989–1991 for island resources (Guo et al., 1994). More recently, only random investigations have been carried out. Results of these investigations are listed in Table 1. The biomass of intertidal zones on the mainland was much higher than on islands within the same region; the intertidal zones on islands are mainly reef which contain many more molluscs, especially oysters.

Phytoplankton and Red Tide Organisms

Phytoplankton identified on the coast of Guangdong include 406 species in 93 genera, in six phyla (1989–1991). Diatoms and pyrrophyta were the most popular groups, accounting for 96.9% of the total, whereas chrysophyta, cyanophyta, chlorophyta and xanthophyta together accounted for the rest.

In the Pearl River delta, phytoplankton were mainly low-salinity and wide-tolerance species affected by the high horizontal salinity gradients caused by the freshwater run-off from Pearl River. The average densities of phytoplankton were 194×10^4 ind/m³ in spring and 3121×10^4 ind/m³ in autumn. On the coast of West Guangdong, the phytoplankton consisted of species acclimatised to high water temperature and salinity. The average biomass of phytoplankton there is 199×10^4 ind/m³ in spring and 154×10^4 ind/m³ in autumn (Guo et al., 1994).

Dominant diatoms displayed clear seasonal changes. In spring, the dominant species in different coastal regions were similar, and mainly comprised of species of Chaetoceros, Rhizosolenia and Leplocylindrus. In autumn, these changed to Skeletonema, Thalassiothrix, Thalassionema and Nitzschia. Although pyrrophytes accounted for only a small portion of the total phytoplanktons, their dominant species such as Ceratium tripos, Ceratium trichoceras, Ceratium furca, Ceratium fusus, Noctiluca scintillans and Ceratium breve showed a relatively high biomass. Under optimum conditions, these may greatly increase to cause red tide events. In addition, Ceratium massiliense, Perdinium depressumm, and Perdinium ocrannicum are distributed widely in the region, despite their low biomass.

In 1988, 98 species of red tide phytoplankton were observed in the Pearl River mouth and adjacent waters where red tide events frequently occur (Qian Honglin and Liang Song, 1999). Among them were several species known to be toxic: Alexandrium catenella, A. tamarense, Gonyaulax polyedra, G. polygramma, Gymnodinium, Cochlodinium and Pyrodinium bahauehse. P. pungens was the main red tidal diatom widely distributed in the Pearl River mouth, and detailed work is needed to identify whether it is a species of Pseudonitzschia pungens f. multiseries which may also be toxic. No Garmbierdiscus toxicus was found in the Pearl River mouth, although Hook (after Qian Honglin and Liang Song, 1994) considered that the Ciguatera in Hong Kong might originate from fishes from mainland China. However, Notiluca scintillans was the predominant red tide organism in the Pearl River mouth. It was reported that 28 out of 53 red tide records in 1981–1992 were caused by Notiluca scintillans, mainly in the spring (March–May). A recent red tide event which occurred in March–April, 1998 lasted for 15 days; this was invoked by Gymnodinium mikimotoi, and the biomass of red tidal algae in a "net case" mariculture area was 0.63×10^6–7.60×10^6 ind/dm³ with an average of 3.6×10^6 ind/dm³ (Pearl River Environment Post, Zhujiang Huanjing Bao, 25/5/1998).

Huang et al. (1997) analysed the basic biomass and biodiversity of phytoplankton in the Pearl River mouth. They pointed out that the biodiversity of phytoplanktons decreased from low water level season to high water level season, due to the poorer water quality in the high water season.

Zooplankton

In a survey in 1990–1991, 208 species of 15 classes of zooplankton were identified in the region. This included 99 species of copepod, 39 coelenterates (especially jellyfish), 15 species of chaetognatha, 12 species of pteropoda and heteropoda, 11 species of tunicate, seven ostracods, six amphipods, three species of cladocera and others (Guo et al., 1994) (Table 2).

Benthic Organisms

Altogether 607 benthic species in 339 genera have been identified on the Guangdong coast, including 29 species of alga, 19 species of soft corals, 242 molluscs, 212 crustaceans and 105 echinoderms (GDCRIP, 1988). The regional distribution of biomass is shown in Table 3.

Pelagic Organisms

The coast of Guangdong is located in the transition between tropical and subtropical zones. Pelagic organisms in the region generally belong to the Indian-West Pacific type, and a total of 282 species are known from a detailed survey carried out in 1989–1991. Among the fishes and Cephalopods, nine species were global in their distribution, 169 species were distributed in the Indian and Pacific Ocean, with 104 species being Pacific Ocean only. Diversity was high, especially in autumn. Most coastal species are relatively small and widely dispersed (Guo et al., 1994). Biomass and basic composition of pelagic species is listed in Table 4, the taxonomic distribution of fish species is listed in Table 5, and the degree of diversity (calculated after the formula Shannon–Wiener) in Table 6.

MARINE POLLUTION PARAMETERS

Pollution Sources and Waste Products

Water pollution in the Pearl River delta area mainly originates from industrial discharges, municipal sewage, harbours, shipping and agriculture. Investigations have shown that the total amount of industrial wastewater and municipal sewage in the early 1980s was approximately 1 billion and 1 million tons respectively, which contained 717 and 99 thousands of tons of pollutants. It was estimated that 1.406 million tons of pollutants flowed into the whole coastal belt via rivers annually in the 1980s.

The accelerated urbanization process in the Pearl River delta in the 1990s resulted in the rapid growth of pollutants

Table 2

Dominant species of plankton in the coast of Guangdong, west of Hong Kong

Region	Spring	Autumn
Pearl River mouth	Acartia spinicauda Acartilla sinensis Labidocera euchaeta Calanus sininus Sagitta enflata Planktonic larvae	Eucalanus subcrassus Torttanus gracilis Canthocalanus pauper Lucifer intermidius Planktonic larvae Sagitta enflata Saggita bedoti
Coast of west Guangdong	Planktonic larvae Penilia avrirostis Calanus sinicua Centropages tenuiremis Canthocalanus pauper Saggita enflata Temora discaudata	Eucalanus subcrassus Canthocalanus pauper Candacia bradyi Labidocera minuta Saggita enflata Temora discaudata Temora turbinata

Table 3

Biomass (g/m^2) composition of benthic biota in coast of South China Sea west of Hong Kong (1982 in PRM, 1984 in West GD and Hainan)

Class	Pearl River mouth	Coast of West GD	Hainan Island
Molluscs	11.1	35.1	8.4
Echinoderms	2.6	5.1	2.2
Crustaceans	2.1	3.1	1.4
Polychaetes	1.2	2.6	0.6
Coelenterates	–	–	8.2
Others	0.8	13.4	4.7
Algae	–	–	2.4

Table 4

Biomass composition of pelagic organisms in coastal waters of the South China Sea west of Hong Kong

Item	Fish		Cephalopods		Crustaceans	
	Spring	Autumn	Spring	Autumn	Spring	Autumn
Average biomass (kg/h)	54.17	60.19	1.58	4.29	1.94	44.26
% of total	93.90	87.56	2.74	6.24	3.36	6.20
Average density (ind/ha)	2647.90	5001.58	30.61	175.42	101.46	304.30
% of total pelagic	95.25	91.25	1.10	3.20	3.65	5.55

Table 5

Distribution of fish species in coast of South China Sea, west of Hong Kong

Order	Families	Genera	Species
Orectolobiformes	1	1	1
Carcharhiniformes	1	2	2
Rajiformes	3	3	4
Myliobatiformes	3	3	6
Torpediniformes	2	2	2
Clupeiformes	3	11	26
Salmoniformes	1	1	1
Myctophiformes	2	4	6
Anguilliformes	6	9	10
Siluriformes	2	2	4
Gadiformes	1	1	1
Gasterosteiformes	2	3	4
Mugiliformes	3	5	9
Perciformes	46	87	149
Scorpaenigormes	6	17	21
Pleuronectiformes	4	7	18
Tetraodontiformes	4	11	17
Lophiformes	1	1	1

Table 6

Diversity of pelagic organisms (1990–1991) (Shannon–Wiener diversity index)

Season	Pearl River Mouth	West Guangdong
Spring	2.18	2.97
Autumn	3.12	3.51

Table 7

Waste products (tons) in Guangdong Province from 1995–1998 (data from annual report on environment status of Guangdong, GDEPB)

	1995	1996	1997	1998
Industrial waste water tons $\times 10^9$	12.82	15.09	12.5	11.77
Sanitary waste water tons $\times 10^9$	22.06	22.05	29.4	31.64
Solid industrial wastes tons $\times 10^4$	1260	315.44	1506	1025
Solid sanitary wastes tons $\times 10^4$	207	565.1	744	1771
Toxic compounds*	218	402.1	283.7	222
COD tons $\times 10^4$	34.98	40.1	41.4	36
Oil	1756	2446.1	1800	1400

*Total amount of Cd, Hg, Cr, Pb, As, As, volatile phenols and cyanides.

entering the environment (Table 7). Compared to the 1980s, the waste production has increased sharply, and correspondingly the pollutants entering the river catchment and coastal waters will increase as a result of delays in pollution treatment. It should be emphasized that the proportion of municipal sewage has further increased, and yet most of the wastewater is discharged directly into the environment without any treatment.

Conventional Water Quality Parameters

The official, conventional water quality parameters comprise COD, DO, dissolved inorganic N and P, total petroleum hydrocarbons (TPHs) and heavy metals, according to the country's water quality monitoring standards. Statistically computed results for the COD, TPHs, phosphate and inorganic nitrogen along the coast of Guangdong between 1985 and 1995 were determined by Li Zhongqing and Chen Kouping (1998). In general, the coastal water pollution had characteristic levels of DIN, PO_4-P and TPHs. These three parameters, as well as COD, have all displayed an increasing trend from 1985 to 1995. The highest contents of phosphate–P, inorganic nitrogen and TPHs in seawater were up to 12.9, 295.9 and 70 $\mu g/l$ respectively. The lower values were within the country's permitted standards. The concentrations of some heavy metals, such as Hg, Cu, Pb, and Cd, were lower than permitted values and they have changed little over time.

Using monitoring data obtained from 1990 to 1995 as an example, the mean concentration sequences in different regions were as follows:

- Organic pollutant (COD): Pearl River Mouth > West Guangdong > Bei Bu Bay > Hainan Island
- Inorganic phosphate: Pearl River Mouth > West Guangdong > Hainan Island > Bei Bu Bay
- Inorganic nitrogen: Pearl River Mouth > West Guangdong > Beibu Gulf > Hainan Island.
- TPHs: West Guangdong > Beibu Gulf > Pearl River Mouth > Hainan Island.

Trace Toxic Organic Contaminants in the Pearl River

Trace toxic contaminants such as DDTs, BHCs, PAHs, PCBs in waters in the Pearl River mouth have been tentatively studied in recent years (e.g. Cai et al., 1997; Mai et al., in press; Zhang et al., 1999). Concentrations of BHC and DDT in surface water were 0.087 $\mu g/l$ and 0.080 $\mu g/l$, whereas those in bottom waters were 0.117 $\mu g/l$ and 0.506 $\mu g/l$, respectively. BHC and DDT concentrations in sediments were 11.15 ppb (dry weight) and 33.46 ppb respectively, and a reciprocal relationship between microorganisms and BHC and DDT concentrations was observed. Mai et al. (in press) found that surface sediment in Macao estuary was heavily contaminated by organochlorinated compounds, in which the concentrations of DDTs, PCBs were 1629, 339 ppb respectively.

It is interesting that the spatial distribution of toxic organic compounds in sediment (Cai et al., 1997; Mai et al., in press) showed an increasing trend from the east (Hong Kong side) to the west (Macao side). While tentative data of heavy metals showed a similar trend (Li Xiangdong, personal communication), it is still unknown whether the concentrated pollutants in the west of the Pearl River Mouth resulted from mainland China as carried by land run-off, or from Hong Kong as transported by ocean currents in a northwesterly direction.

Zhang et al. (1999) studied the time trend of BHCs and DDTs in a sediment core in the Macao estuary, and found that the highest BHCs and DDTs concentrations were at a depth corresponding to a ^{210}Pb time in 1992–1993. It was suggested that the large-scale land transformation (see below), driven by the so-called "economic over-heating" in the Pearl River region had caused significant transportation of pesticides from soil residues to the sedimentary system.

Chemical Contaminants in Biota

Toxic heavy metals in oysters and commercially significant fishes were also studied on the coast of the South China Sea (e.g. Lu Chaohua et al., 1991, 1995; Gan Juli and Jia Xiaoping, 1998; Lin Yantang and Yang Meilan, 1995). Lu Chaohua et al. (1991) pointed out that the concentrations of heavy metals in fishes in West Guangdong and the Pearl River delta (Table 8) were very low compared to fishes in seriously contaminated sea areas in the world, and have no significant effects on human health.

Gan Juli and Jia Xiaoping (1998) analysed the petroleum hydrocarbons in 471 fishes of 48 species in the north South China Sea, and found all were contaminated with concentrations between 1–15 μg/g (dry weight), with the highest concentrations found in the Pearl River Mouth. This was identical to the high concentrations of TPHs.

POPULATIONS AFFECTING THE AREA

The population affecting the coastal areas in the South China Sea, West of Hong Kong, is mainly distributed in the Pearl River Delta (Guangzhou, Shengzhen and Zhuhai are directly on the coast) and West Guangdong (Table 9). The population in Hainan Island (3.39 million hectare land area) is only 6 million. Detailed data on the natural population growth rate in Guangdong (coastal populations) displayed a decreasing trend from 1.65% in 1990 to 1.15% in 1997. The population is strictly controlled according to the country's family planning policy.

Activities and use of marine resources by people in Guangdong includes oil and gas exploitation, land reclamation, mariculture, fishery and marine transportation. Until 1998, the total value of this was 105.3 billion RMB Yuan, which included fishery, transportation, tourism, oil and gas, salt industry, shipbuilding and repair, marine electricity etc. The portion of domestic GDP in the marine industry increased from 8% in 1990 to 13% in 1996.

Table 8

Concentrations of heavy metals in fishes in West Guangdong and the Pearl River Delta (Lu Chaohua et al., 1991)

Region	Species	Sample nos.	Concentration (ppm dry weight)					
			Cu	Pb	Zn	Cd	Cr	Ni
West Guang-dong	5	40	3.15±1.41	1.27±0.58	49.28±24.74	0.27±0.10	0.99±0.20	0.64±0.24
Pearl River Mouth	15	169	3.80±1.64	0.72±0.24	21.42±8.49	0.08±0.05	1.07±0.60	0.71±0.68

Table 9

Population and land area of the Pearl River Delta and Coast of West Guangdong[a]

Region	Land area				Population				
	km^2	1980	1985	1990	1994	1995	1996	1997	1998
Guangzhou	7434.4	501.55	544.98	594.25	637.02	646.71	656.05	666.49	674.14
Shengzhen	327.50	32.09	47.86	68.65	93.97	99.16	103.38	109.46	114.6
Zhuhai	121.00	36.53	41.17	50.25	60.66	63.24	65.37	67.34	69.48
Pearl River Delta[b]	25274.8[c]				2095.26	2137.73	2170.4	2208.7	2237.68
Yangjiang	7813.4	193.11	205.62	224.55	237.94	239.97	242.51	244.35	246
Zhanjiang	12471	439.77	477.26	546.48	593.77	603.86	616.02	627.32	636.51
Maomin	11459	426.02	457.87	521.42	562.23	571.57	581.36	590.19	606.87
Coast West Guangdong[d]	31743.4	1058.9	1140.75	1292.45	1393.94	1415.4	1439.89	1461.86	1489.38

(a) Data from the *Yearbook of Guangdong* (1996, 1998, 1999). Only administrative regions directly on the coastline were included, other regions in the Pearl River Delta are not included.
(b) Summation of the whole Pearl River Delta, including Guangzhou, Shengzhen, Zhuhai, as well as other cities not directly on the coastline in the integrated region affecting the Pearl River Mouth.
(c) After the definition of the Pearl River Delta by Mou et al. (1988).
(d) Summation of Yangjiang, Zhanjiang and Maomin, the coast of northern Bei Bu Bay not included.

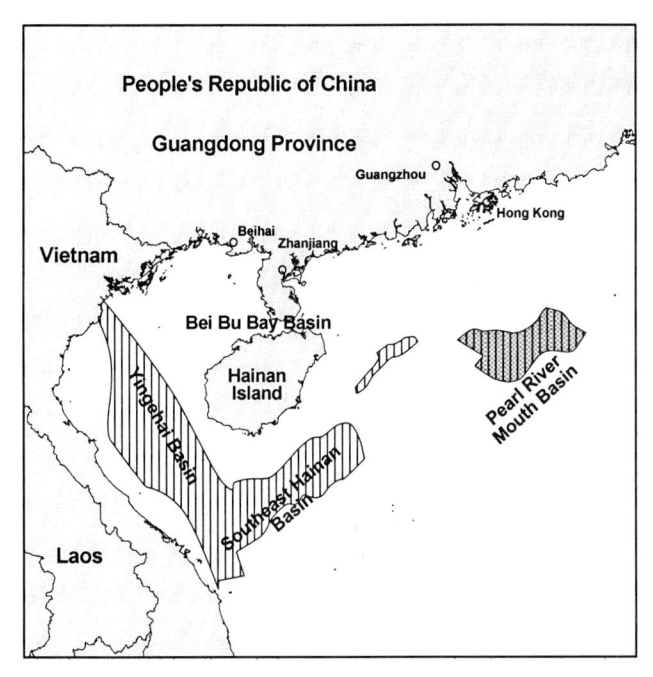

Fig. 3. Map showing the oil and gas bases in northern South China Sea (redrawn after China Environment Post, Zhongguo Huanjing Bao, 27/7/1999).

Oil and Gas Exploitation

The continental shelf possesses abundant oil and gas resources. According to recent estimations, the quantity of natural gas in the north of the South China Sea is about 3321 billion m^3 (Fig. 3). Up to 1995, there were four oil fields in the Pearl River basin, two in Bei Bu Bay Basin and one marine natural gas field in Yingehai Basin (Situ, 1996). The production of oil in the Guangdong in 1995 (Li Yihua, 1998) was 5.6 million tons, accounting for 67% of the country's total offshore oil. The Ya-13 natural gas field located in Yinggehai Basin to the southwest of Hainan Island has reserves of 92.7 billion m^3. It provides Hong Kong with 2.9 billion m^3/yr for power generation, and Hainan Island with 0.5 billion m^3/yr for both power generation and the fertilizer industry. Recently, Zhanjiang on the coast of West Guangdong, and Shengzhen on the Pearl River Delta have become bases for the marine oil and gas industries in the South China Sea.

Land Reclamation

Land reclamation on the coast of Guangdong began in the Song Dynasty, 960 A.D., and developed rapidly throughout the Ming and Tsing Dynasties. In 1934, the area of reclaimed land was about 16.7 million ha, accounting for 10% of the total cultivated land in the Guangdong province (Situ, 1994).

Until 1983, the total reclaimed coastal land of the coast of the South China Sea (West of Hong Kong, including Hainan) was 7.04 million ha. In 1989–1993, over 200 km^2 of bays and beaches were reclaimed at a rate of 4000 ha/yr (Li Yihua, 1998). Since the 1980s, reclamation activities have been concentrated in the Pearl River mouth as a result of the rapid economic development in the region. More than 80% of the area is cultivated within a year of reclamation in the Pearl River delta, while the use of reclaimed land on the coast of West Guangdong is still low (Situ, 1994).

However, excellent harbour bays, habitats of spawning fishes and the natural coastline have been extensively destroyed as a result (Li Yihua, 1998). According to a survey in 1997, reclamation without any suitable management has caused a significant decrease in the area of unaffected shore, affecting mariculture, and speeding up the eutrophication of economically important water bodies on the coast of West Guangdong, especially in the Leizhou Peninsula (Huang, 1997).

Fishery Activity

Mariculture and fishing are the dominant marine biological resources in Guangdong. In 1990–1996, the average annual increased rate of fishing was around 10%, and the total marine fishery products kept on increasing (Fig 4). In 1996, the fish products in Guangdong totalled 171.9×10^4 tons, accounting for 15.3% of the total marine catch in China. Today, marine fishing in Guangdong has progressed from coastal and offshore systems to ocean and deep-sea areas. In 1995, the ocean fishing product in Guangdong was 12×10^4 tons.

In the meantime, however, excessive fishing has brought many traditional fishery resources to extinction. In 1997, the fishing fleets in West Guangdong had outstripped the sustainable fishing capacity in the region. For example, the fishing capability of fleets in Yangjiang (a county on the coast of West Guangdong) in 1995 had increased by 25% compared to that in 1994, but the fishing product had only increased by 10%. Until 1997, there was no spatial and temporal limitation or restriction of coastal and offshore fishing.

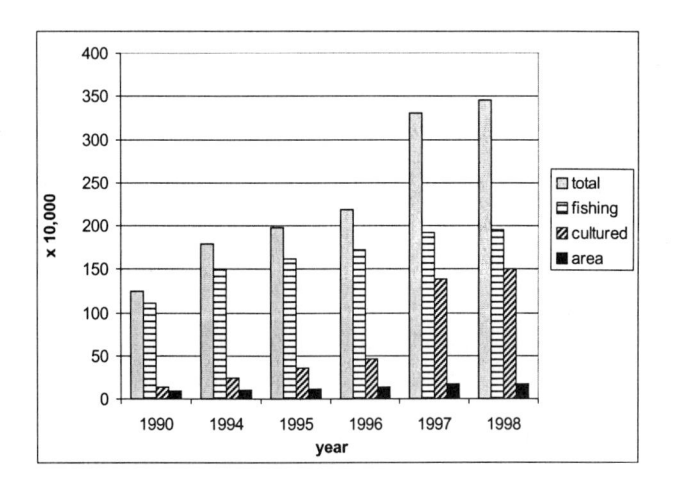

Fig. 4. Fishery products of Guangdong in 1990–1998 (compiled after Guangdong Yearbooks 1994, 1996, 1997, 1998). Fishery yields in tons (×10^4). Culture area in ha (×10^4).

Table 10

Sizes and numbers of harbours in coast of South China Sea west of Hong Kong (after Situ, 1994)

Region	Sizes and numbers of harbours				
	>10 million ton	10–5 million ton	5–1 million ton	<1 million ton	Total
Pearl River Mouth	3	0	2	14	19
West Guangdong	2	0	1	20	23
Hainan Island	1	–	5	16	22
Bei Bu Bay	4	–	–	12	12

Some crude fishing methods such as electric shock, poisons, explosives and small mesh nets were still widely used in the region (Huang, 1997). In shallow fishing fields within 200 m of the shoreline in the South China Sea, available fishing products were estimated to be less than 1.2 million tons, together with a decrease in fish quality (60% being small fish) (Li Yihua, 1998).

The scale of mariculture also kept on expanding and reached 17.64×10^4 ha in 1998, nearly double that of 1990. The average annual increased rate of mariculture in Guangdong was 7.0% in 1990–1998.

The main mariculture methods on the coast of the South China Sea (west of Hong Kong) are the "beach pool" and "net box". Due to the lack of scientific management and planning, these methods have caused serious pollution problems to the aquaculture water bodies themselves, and have speeded up the processes of eutrophication and aging of water bodies around the aquaculture areas (Huang, 1997).

Marine Transportation

Marine transportation in Guangdong developed rapidly in the 1990s. In 1996, there were 77 harbours and 936 berths (84 of which were larger than 10 kton) along the coast of Guangdong. The annual throughput of these harbours reached 1.45×10^8 tons (Table 10).

RURAL FACTORS

Guangdong is one of the provinces which has decreased its cultivated land area. In 1978–1996, 47.45×10^4 ha of agricultural land in Guangdong was lost, which accounted for 17.07% of the cultivated land area. The sharpest decline in cultivated land area occurred in 1991–1996 (22.45×10^4 ha was lost), accounting for 47.31% of the total area lost between 1978–1996 (Huang Ningsheng, 1999). The main driving force was the speed and level of regional economic development, while the impact from the intensified population was not obvious (Huang Ningsheng, 1999). In the Pearl River delta where rapid urbanization has taken place during the last 20 years, large areas of cultivated land were transformed for non-rural usage, and, in the meantime, many land resources were unnecessarily wasted. In a typical city in the Pearl River delta, most of the lost agricultural land was around the city and towns or along highways, and this loss occurred mainly after 1990, especially in 1992–1993 with large-scale real estate development in the region (Li Xia et al., 1998). According to the government's "Overall Planning of Land Use in Guangdong", annual reduction in agricultural land will be 1×10^4 ha until 2010. It was estimated that the lost agricultural land due to industries, transportation, real estate and tourism development might cause severe soil run-off, the rate of which is often 10–20 times higher than that caused through agriculture.

From the 1970s, 92,000 ha of agricultural land has been endangered by soil run-off in the province. The total land run-off area of Guangdong was 17,070 km² in 1980s, and 8,650 km² in the last decade. In the Guangxi province at the upper reach of Pearl River, the total land run-off area has doubled from 11,999.9 km² in the 1950s to 38,200 km² in the 1980s.

Between the 1950s and 1980s, the forests in Guangdong were drastically destroyed by human activity. For example, from 1975 to 1984, the forests were reduced from 38% to 27%. From 1986, the government took steps to restore the forests, which have thus increased in area to 42% in 1990 and 56.6% in 1997. However, each year throughout the 1990s, statistical data also showed that about 500–600 km² of new land is subjected to run-off through urbanization (Xia Hanping, 1999).

The land run-off in the watershed of Pearl River has raised the riverbed of Humen Section of the river by 0.5 m, and that of Nanxiong to Shaoguan Section by 2 m during the past 50 years (Zhang Shengcai, 1997). The coastal water turbidity was found to be high in the 1990s due to the soil run-off caused by the large-scale land transformation in this period, especially in the river mouth of West River and Macao estuary. This greatly affected the coastal environment of the Pearl River mouth and has had a significant negative impact on related habitats.

In addition, the fertiliser applied to cultivated land in Guangdong also doubled from 333.54×10^4 tons in 1980 to 601.09×10^4 tons in 1995. Application of fertiliser has also greatly increased (Liao Jinfeng, 1998). Severe soil run-off linked with the increased use of fertiliser to river catchments has affected and accelerated the eutrophication process in estuaries and coastal waters in the region.

EFFECTS OF URBAN AND INDUSTRIAL ACTIVITIES

Artisanal and Non-industrial Uses of the Coast

The major artisanal uses of the coast involve conversion of mangrove and shores to small aquaculture ponds for oysters. This has occurred generally on the coast of West

Table 11

Total products of marine industries in Guangdong in 1990–1998

	1991	1992	1993	1994	1995	1996	1997	1998
Product (billion Yuan)	23.5	33.0	40.0	50.0	64.7	84.3	92.6	105.3
Increasing rate (%)	29	40	21	25	29	30	9.8	13.7

Table 12

Composition of marine industries in Guangdong in 1998

	Fishery	Transportation	Coastal tourism	Oil & gas	Salt	Sand	Ship building & repair	Marine pharmacy	Marine electricity	Total
Product (billion Yuan)	26.42	13.1	33.5	13.4	0.07	0.03	3.15	0.09	15.44	105.3
Portion (%)	25.1	12.4	31.8	12.7	0.07	0.03	2.99	0.08	14.66	100

Guangdong, especially in Leizhou Peninsula, which is the shrimp base of South China. A survey of the marine resources of West Guangdong in 1997 reported that drastic unplanned uses of beach land had critically destroyed mangroves (Huang, 1997).

Industrial Uses of the Coast

The average rate of increase of marine industries was faster than that of GDP in Guangdong (Tables 11 and 12). Compared to the marine industry in the 1980s, high-tech marine industries, such as marine pharmaceuticals and marine electricity generation, have developed rapidly during the 1990s, especially since 1996. According to statistical data in 1998, 23% of the coast in Guangdong province is used.

Cities

Most of the large cities in Guangdong are located along the coastline, such as Guangzhou, Shengzhen, Zhuhai, Maoming and Zhanjiang. As a result of the rapid urbanization of the Pearl River delta, it is now difficult to differentiate rural areas from urban cities. It was reported recently that parts of the river are seriously polluted and that only 36.2% of the total river can be considered as source water complying with Grade III of the country's water quality standard. The rivers in West Guangdong are more seriously affected, with only 25% of the sections attaining Grade III water quality. The water pollution is characteristic organic pollution. In 1998, municipal sewage discharges were three times greater than industrial wastewater, and only 12.5% was treated before discharge. Pig farming is another important source of organic pollution. In 1998 there were 41.90 million pigs bred in 403 farms in Guangdong. However, only six farms had water treatment facilities authorized by local EPBs.

Shipping and Offshore Accidents and Impacts

Shipping activity on the coast of the South China Sea west of Hong Kong is very intensive, especially in the Pearl River mouth. Offshore accidents in the region were mainly due to oil tankers. In 1993–1995, more than 10 serious oil spill accidents were reported. Among them were six cases in 1995, which, according to estimates, resulted in an economic loss of 0.2 billion Yuan. According to annual reports on the province's environment during the 1990s, petroleum hydrocarbons (TPHs) are an important contaminant which exceeded the country's coastal water quality standards. Undoubtedly, intensive shipping activities in the region have greatly contributed to TPHs pollution of the coastal water bodies.

PROTECTIVE MEASURES

Local Legislation

The first local legislation on the management of marine resources and eco-environmental protection commenced in Guangdong in 1996, i.e. "Management Regulations for Utilization of Marine Area in Guangdong (MRUMAG)". This was followed by "Regulations for Prohibition of Electronic Shock, Explosive and Poison as Fishing Methods" in 1996. These two regulations have proved to be very effective in promoting the protection of marine fishery resources and eco-environment in the province.

In 1997, more detailed regulations were established to execute MRUMAG, including "Execution Guidance for MRUMAG" and "Temporary Standards for Changing of Utilisation of Marine Areas in Guangdong". Generally, MRUMAG constitutes the basis of the local marine legislation system.

In 1998, allied to the "International Year of the Sea", the province launched an integrated programme for execution of MRUMAG, and established the "Guangdong Monitoring Centre for Marine and Fishery Environment" and the "Guangdong Research Centre for Red Tides".

Closed Fishing Seasons

In April–July 1999, the first closed fishing season in the South China Sea was commenced, which affected 17,000 fishing ships, 300,000 fishery workers and a total population of 1 million in Guangdong, Guangxi and Hainan provinces. It was found that only 0.041% of the related fishing ships violated the regulations. It is expected that this closed fishing season will prove to be most effective in protecting the fishery resources of this part of the South China Sea.

REFERENCES

Cai Fulong, Linzhifeng, Chen Ying, Yang Jiadong, Chen Shumei, Cai Feng and Qian Lumin (1997) The study on the behaviour character of BHC and DDT in tropical marine environment. *Marine Environmental Science* **16** (2), 9–14.

Chen Tegu and Yang Qingshu (1998) Surface runoff of the Pearl River watershed in the last 40 years and its impact on sea level in the Pearl River Mouth. *Research and Development of South China Sea (Nanhai Yanjiu Yu Kaifa)* **3–4**, 12–18.

Gan Juli and Jiao Xiaoping (1998) Characteristics of frequency and regional distribution of petroleum hydrocarbons in marine fishes from northern South China Sea. *Tropical Oceanology* **17** (2), 53–58.

GDCRIP (1988) *Investigation Report on Coastal Resources of Guangdong Province.* Ocean Press, Beijing.

Guangdong Yearbook House (1995) Yearbook of Guangdong 1994.

Guangdong Yearbook House (1997) Yearbook of Guangdong 1996.

Guangdong Yearbook House (1998) Yearbook of Guangdong 1997.

Guangdong Yearbook House (1999) Yearbook of Guangdong 1998.

Guo Jingfu, Li Maozhao and Yu Mianyu (eds.) (1994) *Marine Biological and Fishery Resources in Island Sea Areas of Guangdong.* Guangdong Sci. & Tech. Press, Guangzhou.

GXCRIP (1986) *Investigation Report on Coastal Resources of Guangxi Autonomous Region (Province).* Ocean Press, Beijing.

Huang Changjiang (1997) Marine environmental and resources status of West of Guangdong and protective suggestions. *Research and Development of South China Sea (Nanhai Yanjiu Yu Kaifa)* **4**, 56–59.

Huang Liangmin, Chen Qingchao, Yin Jianqiang et al. (1997) Basic organisms composition and environment status of the Zhujiang estuary and adjacent waters. *Marine Environmental Science (China)* **16** (3), 1–7.

Huang Ningsheng (1999) Spatial distribution of cultivated land area of Guangdong and its relationship with economic development and population growth. *Tropical Geography (China)* **19** (1), 29–33.

Li Xia, Peng Peiquan and Liao Qifang (1998) Analysis of the spatial distribution of agricultural land loss and its determinants in Dongguang with remote sensing and GIS. *Tropical Geography (China)* **18** (2), 97–102.

Li Yihua (1998) Sustainable development of marine resources in Guangdong province. *Tropical Geography (China)* **18** (2), 151–155.

Liao Jinfeng (1998) Problems of the cultivated land resources in Guangdong province. *Tropical Geography (China)* **18** (2), 113–116.

Lin Yantang and Yang Meilan (1995) Heavy metals in animals from the Pearl River Estuary. In: *Environmental Research in Pearl River and Coastal Areas*, eds. Wong Chong Kim, Chu Ka Hou, Chen Qing Chao and Ma Xiao Ling. Guangdong Higher Education Press, Guangzhou, pp. 109–115.

Lin Zhongqing and Chen Kouping (1998) Water quality and its trend in coast of South China Sea. *Research and Development of South China Sea (Nanhai Yanjiu Yu Kaifa)* **1**, 15–23.

Lin Zhuheng and Liang Shunhua (1996) A study on the tidal currents in the waters of the Pearl River Mouth. *Marine Science Bulletin (China)* **15** (2), 11–22.

Lu Chaohua, Jia Xiaoping and Zhou Guojun (1995) The distribution and temporal and spatial fluctuations of Cd concentration in oysters from the coastal waters along Guangdong province. *Marine Environmental Science (China)* **14** (4), 27–32.

Lu Chaohua, Lin Yantang and Yang Meilan (1991) Concentration and assessment of heavy metal contaminants in economic marine fishes in Guangdong. *Marine Environmental Science (China)* **10** (2), 46–50.

Mai Bixian, Lin Zheng and Zhang Gan (1999) The spatial distribution of PAHs and organochlorinated pesticides in sediments from the Pearl River Delta, south China. *Environmental Science in China*, in press.

Mou Hongji, Shen Canyan and Huang Guangyao (1988) *Water and Soil Resources in the Pearl River Delta.* Zhongshan University Press, Guangzhou. pp. 30–50

Qian Honglin and Liang Song (1999) Study on the red tide in the Pearl River Estuary and its near waters. *Marine Environmental Science (China)* **18** (3), 69–74.

Shi Xieguo (1990) Achievements in flood control and prevention in the watershed of Pearl River. *People's Pearl River* **50**, 2–6.

Situ Shangji (1996) *Ocean Territory of Linnan.* Guangdong People's Press, Guangzhou.

Xia Hanping (1999) On the flood disaster, land runoff and plant restoration engineering in the watersheds of Changjiang and Pearl River. *Tropical Geography (China)* **19** (2), 124–129.

Yan Qingshu, Luo Zhangren and Zhang Xiujie (1998) Secular trend of water level change in Zhujiang River Delta in recent decades. *Tropical Geography (China)* **17** (2), 9–14.

Zhang Gan, Min Yusun, Mai Bixian, Sheng Guoying, Fu Jiamo and Wang Zhishi. 1999. Time trend of DDTs and BHCs in a sedimentary core from Macao estuary, Southern China. *Marine Pollution Bulletin* **39** (1–12), 325–329

Zhang Shengcai (1997) '94 flood disasters in Guangdong and strategies for flood prevention and control. *Tropical Geography (China)* **17** (1), 30–35.

THE AUTHORS

Zhang Gan

Guangzhou Institute of Geochemistry, Chinese Academy of Sciences, Guangzhou 510640, People's Republic of China

Zou Shicun

School of Chemistry and Chemical Engineering, Zhangshan University, Guangzhou 510301, People's Republic of China

Yan Wen

South China Sea Institute of Oceanology, Chinese Academy of Sciences, Guangzhou 510301, People's Republic of China

Chapter 89

VIETNAM AND ADJACENT BIEN DONG (SOUTH CHINA SEA)

Dang Duc Nhan, Nguyen Xuan Duc, Do Hoai Duong, Nguyen The Tiep and Bui Cong Que

This chapter presents an overview of the main features of the Bien Dong (South China) Sea off the coast of Vietnam and the current status of its environment. This is still not a well studied marine area, despite its large size.

The formation and circulation of water masses are mainly determined by the seasonally reversing monsoon winds in the region. As a consequence its salinity varies strongly with the seasons. The shallow seas support a high biodiversity, containing varied habitats including littoral forest, mangrove forest and coral reefs. So far large numbers of different species have been recorded, including 537 species of phytoplankton, 657 species of zooplankton, 853 species of algae, 92 mangroves and about 6,000 species of zoobenthos. The region is known to contain at least 2,000 species of marine fishes.

The coastal and shallow marine environment is under pressure from industrialization and from general development, and current low levels of pollution reflect mainly this low level of industrialisation since pollution control is very limited. The coastal region has intensive agricultural activities to support the population growth, and fertilizer run-off and agrochemical residues from agricultural activities seem to be of increasing concern. Other problems are conversion and loss of large areas of estuarine habitat, including the rich mangroves, which are converted for shrimp and fish farming, which has also led to salt water intrusion inland.

A system of legislation is being developed. Foremost of these are Presidential Ordinances, Governmental decrees and Inter-ministerial Circulars which both direct and implement laws. However, a wider cooperation among scientists and managers in Vietnam, as elsewhere, is obviously needed to solve many of the existing problems.

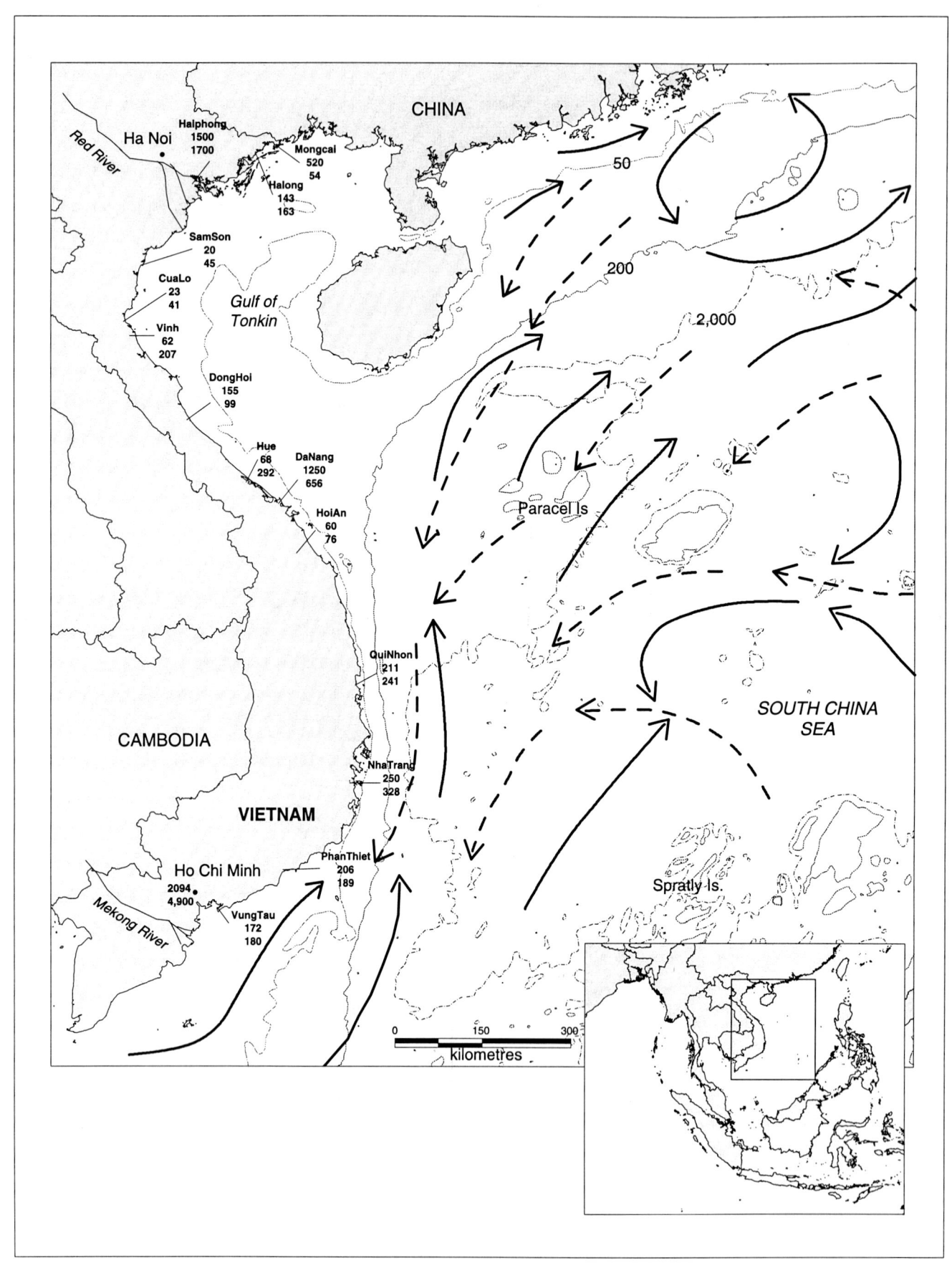

Fig. 1. Map of Vietnam and the Bien Dong (South China) sea. Surface currents in February and May are shown. The main cities on the coast are shown (numbers attached to cities are its area (km^2) and population (thousands)

REGIONAL SETTING

The Bien Dong Sea (BDS) is one of the world's largest continental semi-enclosed seas, with an area of about 3,600,000 km². Vietnam is one of the 10 countries belonging to the Association of South East Asian Nations (ASEAN) Block and its coast extends for about 3000 km along the western BDS (Fig. 1). In its coastal region live 17 million people, or more than 20% of the population of the country. The coastal people are highly dependent on the sea, and about half of the export currency gained in 1998 was from seafood. However, the major activity of inland Vietnamese farmers is rice growing, and in 1998, the rice yield reached about 30 million tonnes, making the country the third largest rice exporter in the world. To maintain this high productivity, farmers use large quantities of fertilizers and pesticides. These agricultural activities, together with the intensive fishery, will undoubtedly directly affect the marine ecosystem.

In Vietnam at present, scientists from several institutions are conducting research into various aspects of the BDS system. At the Department of Oceanography (Hanoi University) studies are being made on the variation of currents, temperature and salinity; the Institute of Ecology and Biological Resources continues to study biodiversity in the marine coastal zone, island and open sea ecosystems; while the Institute of Hydrometeorology in Hanoi, and of Oceanology in Haiphong and Nhatrang are studying factors which can affect the BDS ecosystem. This chapter summarises results of these studies.

PHYSICAL PARAMETERS

Based on bimonthly observations over a long time series from 1907 to 1995, oceanologists from the Vietnam National University (VNU) have drawn a series of maps of the temperature and salinity fields as well as water masses (Uu and Brankart 1997); Fig. 2 shows examples. During winter, isotherms follow the coastline. An upwelling in summer is observed simultaneously in both the temperature and salinity fields, and it is clear that this sea contains two types of water masses, the first permanent and the second seasonal.

In the upper mixed layer there are two permanent masses: water of the open sea and water of the continental shelf. The open sea, where the depth is greater than 50 m, is characterized by a salinity between 33.0 and 34.0 p.s.u. and a temperature between 25 and 27°C during winter, rising to 28 and 29°C during summer. This water mass occupies 60% of the surface during winter and 80% during summer in the central part of the sea.

Continental shelf water has a salinity less than 33.0 p.s.u. and temperature is between 18 and 20°C in the Bac Bo Gulf in winter, and between 27 and 30°C in summer. The mean upper mixed layer thickness is from 10 to 50 m, depending upon the season.

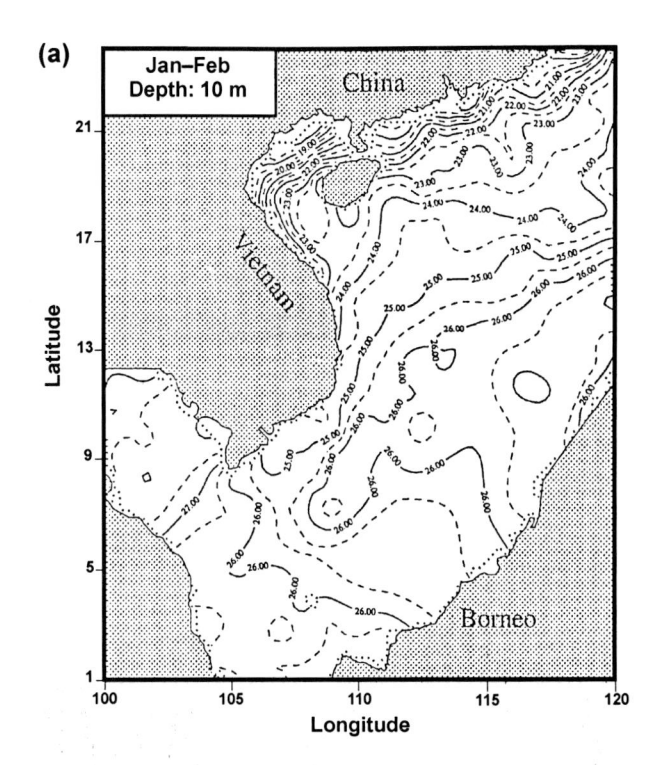

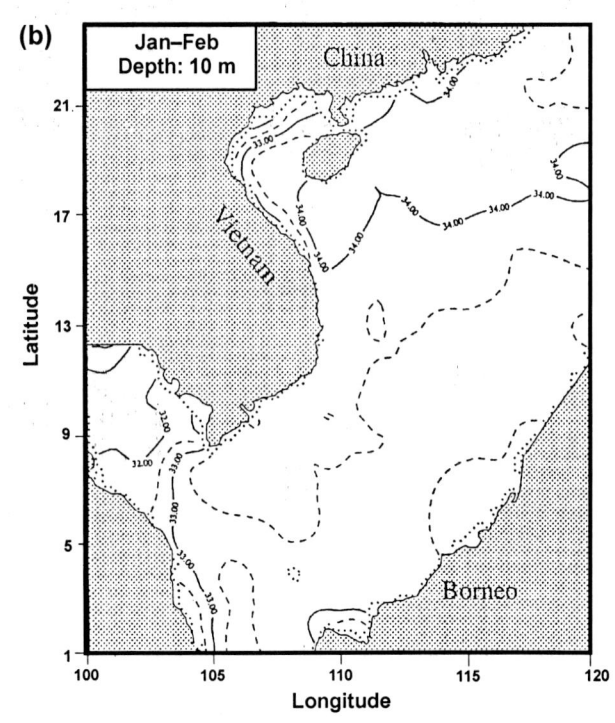

Fig. 2 (a) Temperature of water in the BDS at 10 m depth in January–February. (b) Salinity of water in the BDS at 10 m depth in January–February.

A third permanent water mass is made of maximum salinity water (>34.5 p.s.u.) with a temperature of 15 to 17°C. This water is located under the seasonal thermocline, the depth of its centre varying from 150 to 200 m.

A fourth permanent water mass is located at depths greater than 1000–1500 m; this deep water has a salinity greater than 34.5 p.s.u., and a temperature of 2 to 5°C.

Table 1
Characteristics of the water masses in the BDS

No.	Water masses	Salinity (p.s.u.)	Temperature (°C)	Depth (m)
1	Open sea water	33–34	25–27 in winter 28–29 in summer	0–30
2	Continental shelf water	<33	17–20 in winter (north) 27–30 in summer (north) and whole year (south)	0–50 0–30
3	Maximum salinity water	34.5–35.0	13–15	150–200
4	Deep sea water	34.0–35.0	2–5	>1000
5	Northern open sea during winter	34.0–34.5	23–25	0–100
6	Pacific Ocean water	34.0–35.0	25–27	0–100

There are two seasonal water masses. One from the northern part of the open sea during winter is formed by the strong vertical wind mixing process and thermohaline convection, produced by heat and water flux from sea to atmosphere in the winter monsoon. The second seasonal water mass is from the Pacific Ocean; its salinity is similar, but its temperature is greater than 25°C. Table 1 describes the principal characteristics of these water masses.

Transformation and Spreading of the Water Masses

The seasonally reversing monsoon wind and the variation of net air–sea fluxes play an important role in determining the transformation and spreading of the water masses. Generally, the open sea water mass occupies most of the region deeper than 50 m. During the summer monsoon, water from the Thailand Gulf mixes with low salinity waters from the Mekong River and moves along the coast of Vietnam to 10°30–11°N. The region north and west of this is occupied by higher salinity and lower temperature water rising from the seasonal thermocline. This upwelling is partly caused by the ITCZ (To and Uu, 1995). The hydrological front formed between those two water bodies in the southern shelf of Vietnam creates a zone of high biological production and rich fish stocks.

The summer monsoon season is also the rainy season; the large excess of rainfall over evaporation causes a decrease of salinity of the upper mixed layer of about 0.25 p.s.u. The area occupied by such waters increases from May to November, after which the upwelling decreases.

BIODIVERSITY

Marine biologists in Vietnam have defined three main regions: the continental shelf, the continental slope, and the open sea. Each has coastal, island and coral reef ecosystems. In the continental shelf there are also typical coastal systems such as estuaries, mangroves, a large man-made pond ecosystem, and lagoons.

Vietnam appears to be a region with one of the highest levels of biodiversity in the world. There are 537 known species of phytoplankton (Nguyen Van Tien, 1994), of which 318 species are found in the northern Tonkin Gulf, the remaining species being found south of 17°N (Ho Thanh Hai, 1998). Different assemblages occur in estuarine brackish water, low salinity water along the coast, high salinity water in open sea, and there is a mixed community in both low and high salinity waters. Similarly there are 853 known species of algae, about a third being in the north, with more in the south (Nguyen Van Tien, 1994), and it is certain that this number is considerably fewer than those which remain to be recorded.

Mangroves grow abundantly on tidal mud flats and form large forests. About 25 species have been recorded, with a greater number of mangrove associates found in secondary, planted forests at higher elevations where the tide rarely reaches (Phan Nguyen Hong, 1994).

Zooplankton is similarly rich, with 657 species so far recorded, a third of which are in the Tonkin Gulf (Ho Thanh Hai, 1998). However, it is in the sublittoral benthos that the highest diversities of animals are found, with over 6000 recorded species to date (Nguyen Xuan Duc, 1997), including about 2500 species of molluscs and 1500 species of crustaceans. About 20% of these are found in Tonkin Gulf, in the central part there are 30% and in the south there are about 50%. There is thus a gradient of increasing richness southwards, though 30% of the number of species are common for all the three zones.

About 36 of these species have been placed in the list of endangered, threatened or vulnerable species (*The Red Book of Vietnam*, Vol. I, 1992).

Marine Fishes

About 2038 species of 717 genera have been recorded, and the pattern again shows a gradient of increasing diversity southwards (Nguyen Nhat Thi, 1994). Two groups have been defined, one inhabiting a narrow range of temperature and distributed in a narrow area in the west and Tonkin Gulf only, and a larger group distributed widely in the Indo-Pacific. Of all these, the most diverse groups are demersal and benthic; coral reef species here are much less diverse, and a total of 37 species are listed as rare, endangered or threatened.

MARINE AND COASTAL HABITATS

Littoral habitat

Along the coast are more than 100 estuaries, covering at least 649,000 ha, most of which are located either in a region close to the Chinese border or in another large region near

Cambodia in the southeast. These tidal wetlands have high densities of numerous invertebrates, including important shrimp and molluscs.

The rich mangroves include stands of *Avicennia marina, Kandelia candel, Rhizophora stylosa, Bruguiera gymnorrhira, Aegiceras comiculatum*, etc. with high diversity of associated species, including 21 reptiles and 41 bird species, along with high numbers of species associated with mangroves throughout the southeast Asian region (Doan Canh and Pham Van Mien, 1994).

Coral Reefs

Coral reefs are distributed widely in the Vietnam sea, from the coast to open sea and from the north to the south. According to Nguyen Huy Yet (1995), 147 species of corals have been recorded, a number which further sampling is likely to increase considerably. Three important reef areas are Cat Ba-Ha Long (20°8 N, 107°E), whose reefs are the most developed in the northern coastal area, The Tran Island (Quang Ninh province) area (21°13′30″-21°15′30″N, 107°56′-107°58 E) covering about 3 km^2 but a little less rich because of the shallow water conditions with a wide seasonal variation of temperature, and a coral reef in Son Tra (Thua Thien-Hue, 16°11 N, 108°13 E). All of these have communities of corals, other invertebrates and fishes typical of the region.

WATER QUALITY

The pH value of seawater has been continuously measured for several years and it remains between 8.0 and 8.1 which is within the TCVN-5943 (Vietnam Standard, 1995). Suspended or particulate material is between 14 and 58 mg/l, with highest values in the northern and southern regions due to high river water discharge from the Red and Mekong Rivers. Dissolved oxygen is around 6–7 mg/l. Chemical Oxygen Demand (COD) ranges between 2 and 26 mg/l (Do Hoai Duong and Nguyen Dinh Chan, 1998) with highest values found in the south; the reason for this might be higher concentrations of oil or detergents. Biological Oxygen Demand (BOD_5^{20}) is relatively low, ranging from 0.5 to 3 mg/l.

A survey of organochlorine pesticide residues was carried out during 1995–1997 in the Ha Long bay (Tonkin Gulf) and for the Balat estuary (Red River mouth) (Dang Duc Nhan et al., 1998). Concentrations of Lindane and pp′-DDE ranged from 0.1 to 1.2 μg/l. Other members of the DDT pesticide family were below the ECD detection limit (a few ng/l). Another survey for organochlorine residues in marine sediment and clam (*Meretrix meretrix*) for the same region in 1997 showed that the content of the pp′-DDT pesticide members: pp′-DDE, pp′-DDD, pp′-DDT varies between 4–6, 0.5–1.0, and 1–3 ng/g dry sediment and between 7–15, 3.0–5.5, and 2–3 ng/g dry soft tissue, respectively (Dang Duc Nhan et al., 1999a). The ratio of pp′-DDT/Σ(DDE + DDD + DDT) ranges from 0.2 to 0.3 in the sediments studied. DDT is still being used in the country, but for sanitary purposes only, and is now found only with very low concentrations in surface soils from rice paddy (up to 8.3 ng of Σ(DDE + DDD + DDT)/g dry soil), although it is found in rather high concentrations in sediment from canals located close to densely populated areas in Hanoi (almost 70 ng of Σ(DDE + DDD + DDT)/g dry sediment) (Dang Duc Nhan et al., 1999b). Phuong et al. (1998) have surveyed the level of the pp′-DDT pesticides members in sediment from canals of Ho Chi Minh city, and found that the concentration of Σpp′ isomers of DDT has a wide range of between 1.7 and 253.6 ng/g dry sediment. The results obtained in this study suggested the recent use of the pesticides in the Ho Chi Minh areas, too. However, the purpose of this DDT use is not known.

Inorganic Nutrition Content in Seawater

Phosphate content in seawater ranges from 4 to 6 mg P/l, while nitrate content is 2–59 mg N/l. In the south of Vietnam, rice is farmed all year round, and farmers apply a high rate of inorganic nitrogen fertilisers for this, which explains the higher nutrient run-off concentrations there.

Pollution from Industrial Activities

Industrialisation is in its early stages in the development of Vietnam, and at present there are very few measures in place to reduce environmentally harmful impacts of industrial activities. Almost all the factories functioning in Vietnam, including paints, agrochemicals, paper, food processing etc. use very primitive technology to process or treat liquid, solid and gas wastes. This can lead to the release of a large amount of waste into the environment which, by run-off or dispersion, reaches the marine environment.

Recently, a survey was undertaken on PCB concentrations in marine sediment and in the clam (*Meretrix meretrix*) collected from five stations along the northeast coast, from Mong Cai close to the Vietnam China border, to Balat on the Red River mouth, by Vietnamese scientists in cooperation with the IAEA-MEL (Monaco) (Dang Duc Nhan et al., 1999a). The level of Aroclor 1254 and Aroclor 1260 in sediment is between 0.5–28.2 and 0.2–6 ng/g dry weight of sediment respectively, but the concentration of the same Aroclors in the biota range from 5.1–25.3, and 1.6–13.7 ng/g dry weight of soft tissue. In sediments collected from canals of Ho Chi Minh City, the concentration of a total of six CB congeners (28, 52, 101, 153, 138, and 180) was found to vary between 0 to 122.8 ng/g dry sediment (Phuong et al., 1998). The ratio of Σ6CBs to total PCB was 4.8 which is in the range of those in other southeast Asian countries. For instance, in the Thailand and Malaysian seas, PCBs content in sediment is 0.01–2.00 μg/g dry wt (Everaarts et al., 1991). On the other hand, the PCBs in marine sediment are much lower than those in the aquatic environment of industrialised countries in Europe. For instance in the Marseilles harbour in the Mediterranean the concentration of PCBs was up to 2500 ng/g sediment (Villenueve et al., 1998). The marine environment of the Vietnam coast is still relatively clean from PCBs, and this probably is representative of most of the additives which are widely used in industry.

Metal analysis of marine sediment in Tonkin Gulf shows that Zn and Cu are between 12 and 14 mg/kg, but that the concentrations of Pb and Hg are below AAS detectable limits (Nguyen Chu Hoi, 1996).

Oil Spills and Shipping

Crude oil is found with a concentration ranging from 0.1 to 2 μg/l, averaging 1 μg/l. This varies with the seasons. In summer it might reach up to 75 μg/l, but in winter it usually falls (Pham Van Ninh, 1996). This could be explained by the currents. In summer, the current flows from the north bringing water with high oil content from the Japan Sea, whereas in winter the current is from the south, bringing cleaner water.

Annually about 200 million tonnes of crude and refined oil are transported through the region. Harbours in Vietnam annually receive 24–25 millions tonnes of imported goods, and export nearly the same quantity, though plans exist to increase this to 70–75 million tonnes. These are potential sources of marine pollution.

River Systems and Estuaries

In Vietnam there are eight river systems with a total of 2360 rivers longer than 10 km. On average, there is an estuary every 20 km of the coast. It is estimated that each year these rivers discharge into the sea about 900 billion m³ of water, carrying 200–300 million tonnes of suspended matter including man-made organic compounds. The total amount of metals and inorganic ions annually carried into the Vietnam Sea by rivers has been estimated as follows: Cu 18,084; Pb 2,063; Cd 1,082; Zn 21,739; Co 504; Ni 523; As 2,407; Hg 134; PO_4^{3-} 54,221; NO_3^- 230,710 (values in tonnes). These pollutants mainly come from land transportation and agricultural activities (Ta Dang Minh, 1996).

IMPACTS FROM DEVELOPMENT

The population in Vietnam now is about 80 million, with a rate of increase of about 2%. It is estimated that more than 20% depend on the sea (17 million). The population density along the coast and islands is 276/km² and so it is expected that if the population increase rate is maintained, then by 2010 the coastal population could be 22 million. In the Vietnam sea there are more than 3000 islands whose total area is estimated to be 1600 km². Living on 66 of these are farmers, with a total population of 150,000 inhabitants (Nguyen Chu Hoi, 1997).

This population increase coupled with poor environmental management is causing a serious deterioration of the marine environment. Destruction of mangrove forests for fuel causes both landslides along the coast and deep salt intrusion into fields, affecting rice growing. To survive, the farmers presently can see only short-term benefits, so there has also been widespread conversion of mangrove for intensive shrimp farming. This has also led to salt intrusion and coastal flooding.

In order to fish, farmers use explosives or even poisons (cyanide). The total resource of sea fish in the BDS has been estimated as 4.8 million tonnes, and the quantity within 100 miles from the coast is 2.8 million tonnes. The threshold of the fishery is estimated to be of 1.1 million tonnes per year, and since 1994 the fishery in the Vietnam sea has been kept around this figure.

Farmers in some coastal provinces also produce mortar for construction from coral.

Urbanization and its Impact on the Marine Environment

Halong City provides an example of the impact of urbanization on the marine environment. Halong City is a tourist city in the northeast, to which each year about half a million tourists come to enjoy the bay. At present, all liquid and part of the solid wastes are discharged directly into this famous bay, so nutrient enrichment has been marked. Now, rapid increases in algae and other undesirable effects have become intolerable to tourists. This is not one of the larger cities. It has been estimated that every day about 1 million m³ of effluent is discharged into the Vietnam Sea, of which Ho Chi Minh City supplies 550,000 m³ and Hanoi another 350,000 m³ (Nguyen Thuong Hung, 1996).

LEGISLATION

The problem of sustainable exploitation of marine resources in Vietnam has been discussed at many meetings of state managers and scientists. In 1994 the Law on Environmental Protection including the Marine Environment was approved by the National Assembly.

Recognizing the importance of the necessity to preserve marine biodiversity for its sustainable development, Vietnam has joined the International Convention of Biodiversity. In June 1991, the Chairman of the Ministers Council, now the Prime Minister of the SR Vietnam, signed the National Action Plans on the Environment and Sustainable Development. In the framework of the Plans, several programmes have been designed on biodiversity protection, protection of mangrove forests, and a programme for improvement of the management for National Parks, Protected Areas and maintenance of rare fauna and flora.

The Environmental Protection Law of Vietnam was enacted in 1994. One of the main elements of the Law is to

Table 2
Number of species in need of protection in Vietnam

	Endangered	Vulnerable	Threatened	Rare	Insufficient knowledge
Fauna	68	97	71	124	6
Flora	24	61	83	187	27

preserve biodiversity. In 1995, the Prime Minister approved the National Action Plans on the preservation of biodiversity, and for the marine environment, and the Government and the Ministry of Aquaculture of Vietnam have issued necessary documents aimed at the protection of marine resources. These documents include the Presidential Ordinance on protection and development of aquacultural resources approved by the President in 1989, the Decree of the Government to implement the Ordinance, and Circular No. 04/TS/TT of the Ministry of Aquaculture which directed the implementation of all the above.

To better implement all the Governmental Documents on protection and development of aquacultural resources, management systems have been designed. The Department for Protection of Aquacultural Resources which belongs to the Ministry of Aquaculture leads other bodies and local forces, and can intervene where actions could cause harm to the marine environment. One of the effective measures to further sustainable development is to organise the fishery in order to reduce over-exploitation in the coastal zone. To this end, in 1997 the Government invested US$ 400 million for farmers to build boats to enable them to go further out into the open sea.

After many years of investigation and studies, scientists in Vietnam have published two volumes of the *Red Book for Vietnam*. The first volume (1992) includes fauna, and the second volume (1994) includes flora (Table 2).

CONCLUDING REMARKS

The Bien Dong, South China Sea has great natural resource potential. Its ecosystems are diverse, but the area is far from well studied to date. At present, it appears that population increase is one of the threats that could affect the marine environment. Cutting of mangroves for fuel and intensive fish and shrimp farming have led to serious problems of flood and salt intrusion. Artisanal fishing by farmers using nets of small mesh size and explosives is of great concern, and urbanization in some cities on the coast is leading to an increasing amount of waste discharged into the sea.

Industrial activities are not marked in this region. Metals and PCBs in sediment and biota are at lower levels compared to developed countries, which reflects the low level of industrial development in the region. Nitrogen and phosphorus fertilizer residues from agricultural activities are higher. In conclusion, a proper policy to manage the environment in general, and the marine environment in particular, is urgently needed for the sustainable development of the Bien Dong Sea.

REFERENCES

Dang Duc Nhan, Nguyen Manh Am, Nguyen Chu Hoi, Luu Van Dieu, Carvalho, F.P., Villeneuve, J-P. and Cattini, C. (1998) Organochlorine pesticides and PCBs in the Red River Delta, North Vietnam. *Marine Pollution Bulletin* **36**, 742–749.

Dang Duc Nhan, Nguyen Manh Am, Carvalho, F.P., Villeneuve, J-P. and Cattini, C. (1999a) Organochlorine pesticides and PCBs along the coast of North Vietnam. *The Science of the Total Environment* **237/238**, 363–371.

Dang Duc Nhan, Carvalho, F. P., Nguyen Manh Am, Nguyen Quoc Tuan, Nguyen Thi Hai Yen, Villeneuve, J-P. and Cattini, C., (1999b) Chlorinated pesticides and PCBs in sediments and molluscs from freshwater canals in the Hanoi region. *Environmental Pollution* (in press).

Do Hoai Duong (1995) Recommendation on a monitoring Network for the Environment in Vietnam. Final Report of the Natl. Project KT-02.02 to the MOSTE. MOSTE, Hanoi, 1995.

Do Hoai Duong and Nguyen Dinh Chan (1998) Marine environment and it's management for the marine coastal zones in Vietnam. An overview. The National Centre for Scientific and Technological Informations (NCNTI), Hanoi.

Doan Canh and Pham Van Mien (1994) Impacts of the socio-economical development on the biodiversity in mangroves forests adjacent to estuaries in the south Vietnam. Final Report of the Subproject 2E.2: Biodiversity preservation in Vietnam. NCNST, Hanoi.

Everaarts, J.M., Nasreen, B., Swennen, C. and Hillebrand, M.T.J. (1991) Cyclic chlorinated hydrocarbons in benthic invertebrates from three coastal areas in Thailand and Malaysia. *Journal of the Science Society Thailand* **17**, 31–49.

Ho Thanh Hai (1998) Technico-Economical Report to the Ministry of Science, Technology and Environment. Present status of the Biodiversity preservation in the Vietnam Sea. MOSTE, pp. 301–325.

Nguyen Chu Hoi (1996) The marine coastal zone ecosystem in Vietnam and it integrated management. MOSTE-SIDA Project. MOSTE, Hanoi.

Nguyen Huy Yet (1995) Scleractilia in the west of the Tonkin Bay. Ph.D Thesis, Vietnam National University, VNU, 153 pp.

Nguyen Nhat Thi (1994) In *The Handbook on the Sea of Vietnam, Vol. IV: Biological Resources and the Marine Ecosystems*. NCNST, Hanoi, 587 pp.

Nguyen Thuong Hung (1996) The marine resources and environment of Vietnam—the present status. In *Proceedings of Sci. Reports of a Natl. Programme on Environmental Protection*. NCNST, Hanoi.

Nguyen Van Tien (1994) In *The Handbook on the Sea of Vietnam, Vol. IV: Biological Resources and the Marine Ecosystems*. NCNST, Hanoi, 587 pp.

Nguyen Xuan Duc (1997) Exploitation of living marine resources of Vietnam. Report of UNEP-EAS/RCU, NCNST, Hanoi, 50 pp.

Red Book of Vietnam (1992) Volume I: Fauna. Publishing House Science & Techniques, Hanoi.

Red Book of Vietnam (1992) Volume II: Flora. Publishing House Science & Techniques. Hanoi.

Pham Dinh Trong (1996) Zoobenthos in the mangroves forest ecosystem of the west of Tokin Bay. Ph.D Thesis, Vietnam National University, VNU, 147 pp.

Pham Van Ninh (1996) Final Report of the national Project KT-03.07: Marine pollution from river discharge. MOSTE, Hanoi.

Phan Nguyen Hong (1994) Environmental assessment of the impact of aquaculture and fishery to biodiversity and the marine environment in the sea of the north Vietnam. Final report of the subproject 2E.2: Biodiversity preservation in Vietnam. NCNST, Hanoi.

Phuong, P.K., Son, C.P.N., Sauvanin, J-J., Tarradellas, J. (1998) Contamination by PCBs, DDTs and heavy metals in sediments of Ho Chi Minh City's canals, Vietnam. *Bulletin of Environmental Contaminant Toxicology* **60**, 347–354.

Ta Dang Minh (1996) Final report of the national Project KT-03.07: Oil spillage pollution in the Vietnam sea. MOSTE, Hanoi.

To, L.D. and Uu, D.V. (1995) The meterological and oceanographical regime in The Bien Dong sea and the possibility of predict marine fishery conditions. Final Report of Vietnam National Marine Programme Project KT-03.10, NCNST, Hanoi.

Villeneuve, J-P., Carvalho, F.P., Fowler, S.W. and Cattini, C. (1998) Levels and trends of PCBs, chlorinated pesticides and petroleum hydrocarbons in mussels from the N.W. Mediterranean Coast. *The Science of the Total Environment* (in press).

Vietnam Standards for Seawater (1995) TCVN-5943/1995, Vietnam Standardization Authority.

Uu, D.V. and Brankart, J-M. (1997) Seasonal variation of temperature and salinity fields and water masses in the Bien Dong (South China) Sea. *Mathematical Computer Modelling* **26**, 97–113.

THE AUTHORS

Dang Duc Nhan
Institute of Nuclear Sciences and Techniques,
P.O. Box 5T-160, Hoang Quoc Viet,
Hanoi, Vietnam

Nguyen Xuan Duc
Institute of Ecology and Biological Resources,
Nghia Do, Hanoi, Vietnam

Do Hoai Duong
Institute of Hydrometeology,
Lang Thuong, Dong Da, Hanoi, Vietnam

Nguyen The Tiep
Institute of Oceanology, Hoang Quoc Viet, Nghia Do,
Hanoi, Vietnam

Bui Cong Que
Institute of Oceanology, Hoang Quoc Viet, Nghia Do,
Hanoi, Vietnam

Chapter 90

CAMBODIAN SEA

Touch Seang Tana

Located in the northeastern part of the semi-enclosed Gulf of Thailand, the Cambodian Sea contains several diversified coastal ecosystems. The area is dominated by a wet monsoon (summer) and by a dry monsoon (winter), which alternately drive the oceanographic character and influence the area's marine habitats. The Cambodian Sea described in this chapter is that of Cambodia's Exclusive Economic Zone, which was officially claimed in January 1978. The Sea has an average depth of 50 m, and receives a number of freshwater rivers and streams originating from both the Elephant Chain and Cardamomes Mountains. These have greatly increased volumes during the rainy season and flow into large estuaries along the coastline. The estuarine systems and predominantly wet monsoon winds create rich and varied biological habitats in all the near shore waters.

The inshore waters contain three major estuarine bays: Kompong Som is the largest, covering about 40% of the overall coastline, where upwelling occurs during the wet and early dry monsoons. Koh Kong bay, is dominated by the large river Dong Tong and forms a large estuary covered by mangrove forest and seagrass beds. The southeastern edge of the coastline includes Kampot bay which contains important seagrass habitats. Offshore the sea floor structure from 40 m deep is a rugged structure with numerous ridges 3–4 m high, some being elevated up to 10 m.

Cambodian waters include 474 known fish species in 105 families, including 30 commercially important Mackerel, Scad, Anchovy and Snapper. These are abundant from September to January, while peak periods of fishing for Penaeus and Metapenaeus shrimps is May–August. Blue swimming crab, Squid and Cuttle fish are fairly abundant throughout the year. Commercially important Green mussel and oysters are found mainly in Koh Kong estuary, while the Blood cockle is important in Kompong Som Bay and Kampot Bay. Dugong, Sea turtle and Dolphin also seasonally inhabit the Cambodian Sea.

In recent years there has been greatly increased development and settlement along the Cambodian coastline. Concerns focus on the environmental impact of logging in both watershed forests and mangroves, coastal watershed damming, widespread poaching and destructive fishing. Shrimp farming is expanding in the mangrove forest. Poor management generally has led to drastic degradation of many marine habitats. There is an absence both of natural resource policy and a lack of legislation for environmental protection and resource conservation, which has allowed poor enforcement of the natural habitats and resources. There is collusion among state bureaucracy and authority, which has allowed rampant and anarchic natural resource exploitation, which has led to serious deterioration of the natural environment and of its resources.

Seas at The Millennium: An Environmental Evaluation (Edited by C. Sheppard)
© 2000 Elsevier Science Ltd. All rights reserved

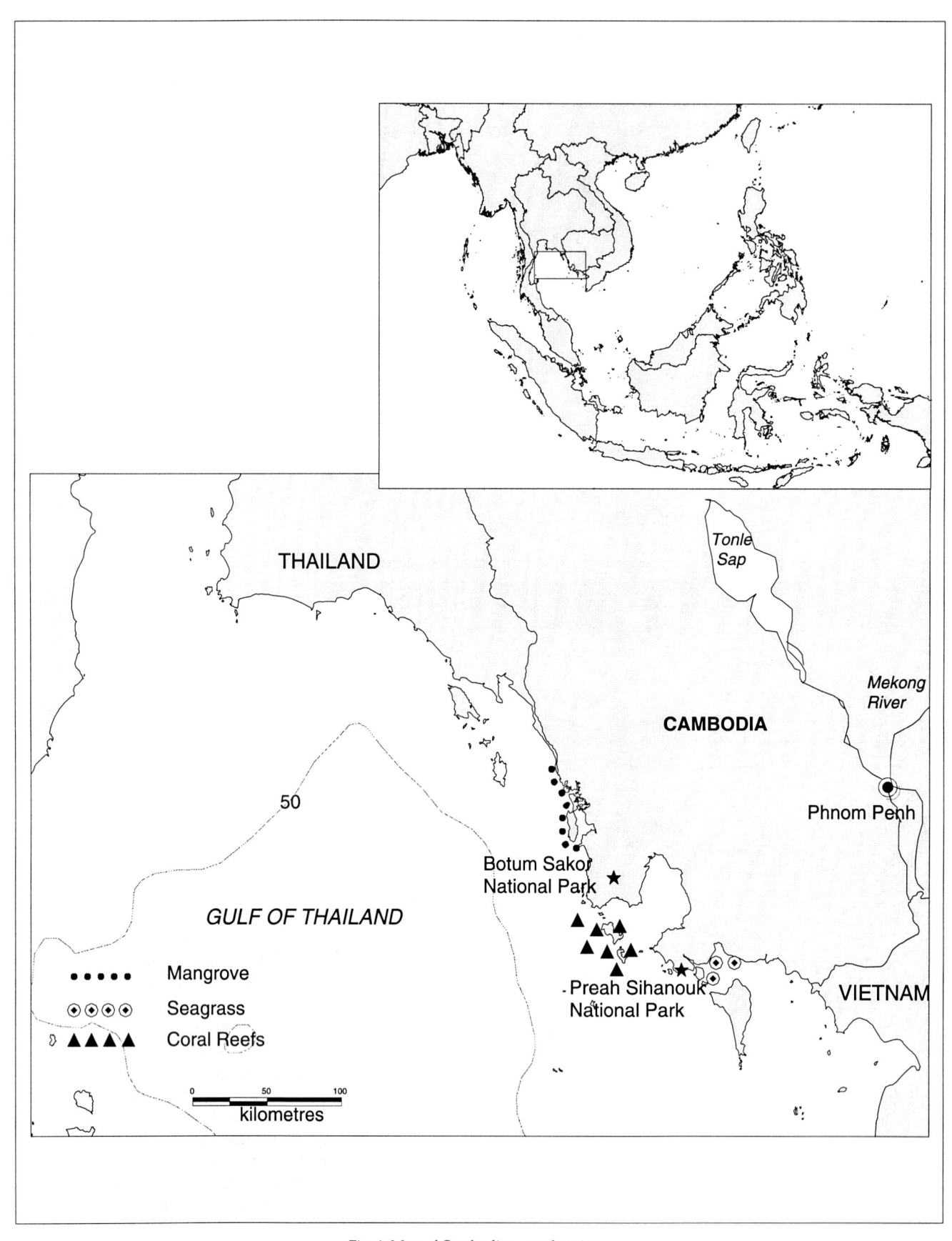

Fig. 1. Map of Cambodia coastal region.

THE DEFINED LOCAL MARINE ENVIRONMENT

The Cambodian seas included in the Exclusive Economic Zone (EEZ) are located in the Gulf of Thailand between 8 and 12°N and between 101 and 104°E. The region is very diversified in terms of physical, chemical and biological characteristics, both because of freshwater inputs and interactions between inshore and offshore waters (Fig. 1).

The northern estuary in Koh Kong bay is influenced by several freshwater rivers originating from the Cardamomes Chain of hills. These seasonally form very large mixing areas of freshwater and seawater, creating huge and productive habitats for numerous species. During the rainy season when the freshwater flow is enormously increased, the salinity of the estuary falls to zero while during the dry season, the salinity varies with an average of about 29.5 ppt. Part of this area, including that covering the most important mangrove forest of 23,750 ha, has already been defined as the Peam Krasob Wildlife Sanctuary. Further south in this region, a pristine ecosystem, defined by the same Royal Kret as Botum Sakor National Park, covers a 171,250 ha area of evergreen forested watershed, which meets the sea along white sandy and undisturbed rocky beaches (see Fig. 1).

Kompong Som, the largest bay of the Cambodian coastline, is partly enclosed by two big islands at its mouth, and behaves like a small semi-enclosed sea within the Gulf of Thailand. The massive currents of the South China Sea, pushed by southerly and southwesterly winds toward the Cambodian shore, mix with the huge volumes of freshwater from several rivers during the wet monsoon (June–October) and early dry monsoon (December–January). This phenomenon creates large mixing areas in the northern part of the bay and several turbulent areas at the mouth of the bay. This creates an area which is rich with microorganisms and both vertebrate and invertebrate species. Several species of small Cetaceans and sea turtles often appear during this period also, with occasional Baleen whales. The entire shoreline of the bay is mainly covered by mud or white sandy beaches.

The southeastern part of the coastline from the naval base at Ream consists of the small Kampot Bay, leading to the border with Vietnam. The muddy and rocky beaches of this area support rich seagrass beds. A small part of this bay is bordered by terrestrial lowland evergreen forest, and together with some mangrove and small islands this area is gazetted as the 21,000 ha Preah Sihanouk National Park. As in other areas of the Cambodia coast, massive water currents from the South China Sea, influenced by the southerly winds of the wet monsoon, make the water of this area a rich but turbulent environment.

The offshore area extending from 20 m depth out to the international border of the EEZ is a very dynamic zone in terms of physical, chemical and biological properties. Major influences come from the interaction of oceanic currents, winds, an unusual sea floor structure and seasonal variations of freshwater outflow from the rivers. In the central part within 30 and 60 m deep, cyclonic turbulence currents causing upwelling occur in February and July; substantial benthic and pelagic biomass occurs at this time. The muddy sea floor has a structure of rocky ridges, and with their well developed sponge species these create suitable spawning grounds for many fish, especially mackerel and scad during the dry monsoon period. Penaeus shrimp use the same spawning grounds in the wet monsoon when oceanic currents flow toward the shore. Coral reefs are well developed on most of the offshore islands except Polowaii Island, where disturbed water exists all year round.

SEASONAL VARIABILITY OF NATURAL ENVIRONMENT

The natural environment of the Cambodian Sea is unusual compared with other regions in the Gulf of Thailand due to the specific interaction of inshore and offshore water, both of which are influenced by seasonal oceanic currents and monsoon winds.

Meteorology

Cambodia is located in the climatic sub-equatorial zone. It is therefore greatly influenced by the Monsoons, during which the difference in temperature between the air masses over the sea and the continent creates substantial energy exchanges. In summer, the Asian continent is hotter than the sea creating a low-pressure region and strong monsoon winds, while in the winter this is reversed.

During the period of February–April, southeast and south winds are pre-dominant. In May, when the monsoon is changing its direction, the wind regime is extremely unstable but southwest winds are very significant, if not dominant. During June and the first half of October, the west and the southwest winds are dominant, while from the second half of October to December, the monsoon winds change direction. Typhoons sometimes occur during this time and create a large upwelling in the central part of the Cambodian coast, leading to increased densities of marine organisms. Calm periods then occur during May and in September–October.

Water Currents and Upwelling

The water currents of Cambodia, as with the whole Gulf of Thailand, are basically formed by the interaction of winds, freshwater inflow, sea floor relief and by salinity and density gradients.

In February, there is a large upwelling that dominates from 20 miles off shore to 80 miles at the central part. This causes complex surface currents in the outer part of the zone where high salinity water from the South China Sea dominates (see Fig. 2.1).

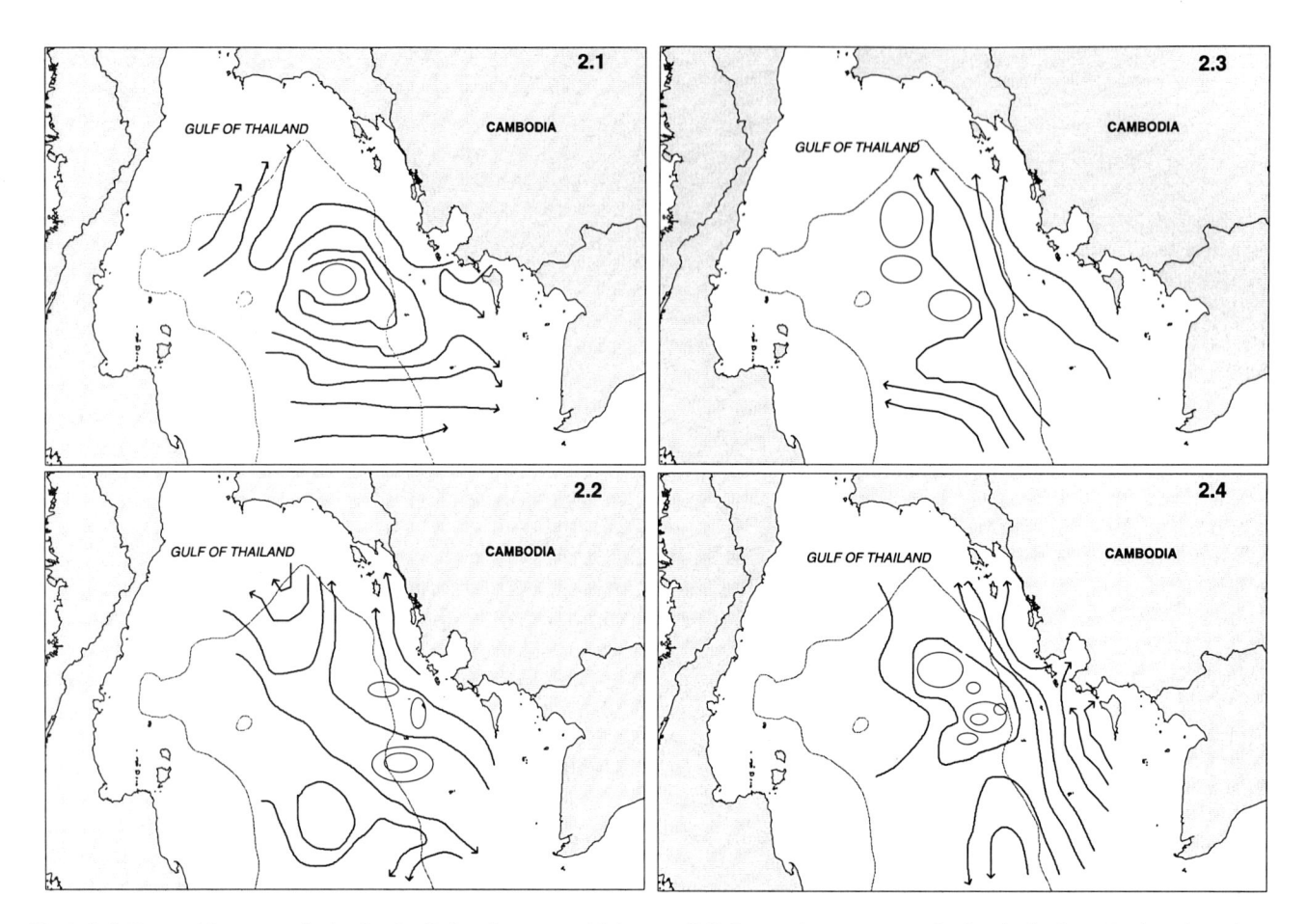

Fig. 2. (2.1) Geotrophic currents in the Cambodia Sea: January and February. (2.2) Geotrophic currents in the Cambodia Sea: March and April. (2.3) Geotrophic currents in the Cambodia Sea: June. (2.4) Geotrophic currents in the Cambodia Sea: July and August.

The water transport in April as well as in February is cyclonic. Northwestern currents are high, with surface velocities of 0.15–0.35 knots, found near Polowaii and Tang islands and in Koh Kong bay. Deeper velocities are much less (Fig. 2.2). In May, the central part of the region is under the influence of the southern and southeastern currents, which are deflected to the southwest with a velocity of about 0.3–0.5 knots (Fig. 2.3). In June, the current changes drastically due to the regular winds of the southwest summer monsoon. The winds, with speeds of from 2 to 15 m/s, force northward currents from the South China Sea, and these dominate the southern and eastern parts of the Cambodia Sea. In the southwestern Cambodian waters, meandering westward currents occur (Fig. 2.4).

In June and July as well as during the entire wet season, the northward and shoreward currents from the south dominate the inshore water. Upwelling occurs around the larger off-shore islands. Further west off Koh Tang Island, the southwestward current from the northern region has a maximum speed of not greater than 0.3 knots (Fig. 2.4). In deep water at 50 and 60 m depth, bottom currents are stronger and flow from the north to the south, generating turbulence at the western border. These bottom currents are seasonal and do not exist during December–February.

Physical and Chemical Conditions

In general, temperatures of surface waters of the Cambodia Sea do not fluctuate significantly, especially offshore. Table 1 below shows that temperature at the water surface ranges only between 27 and 30.5°C throughout the year even though air temperature may fall considerably below this to 23.5°C.

Salinity varies widely during the rainy season. Nearshore surface water salinity can be as low as 27.4 ppt (entirely fresh in the estuaries) but it can also be as high as 32.6–33 ppt at the southern border of the Cambodian Sea which is dominated by water from the South China Sea.

Surface oxygen concentrations remain almost stable throughout the year. In the early dry season, saturation at the surface varies between 102 and 104% with dissolved oxygen levels (DO) between 4.2 and 4.8 ppt. Maximum saturation of about 105% is found in Koh Kong bay where DO increases up to 5 ppt. Saturation steadily decreases from nearshore to northern, northwestern and southeastern parts of the region. Oxygen concentration remains similar to 20 m depth, then saturation decreases to 50–60% at 50 m. In the wet season, maximum concentration of oxygen (4.5–4.6 ppt) is found near shore between 0 to 30 m

Table 1
Physical and chemical conditions in surface waters of the Cambodian Sea

Period	Surface water temperature (°C)		Salinity (ppt)		Silicon (ppb)		Phosphorus (ppb)		Oxygen (ppt)		Air temperature (°C)	
	Min	Max	Min	Max	Min	Max	Min	Max	Min	Max	Min	Max
December	27.0	30.5	29.5	30.5	–	–	–	–	–	–	24.0	32.0
February	27.1	28.4	30.7	32.4	10	30	0	0.3	4.2	4.8	25.0	34.6
April	29.9	30.1	31.8	33.8	10	20	0	0.3	4.5	4.9	27.1	37.3
May	30.0	30.9	31.3	33.0	6	9	0	2.0	4.2	4.9	25.1	32.8
June	28.5	29.0	31.5	33.5	0	10	0.1	2.2	4.3	4.7	25.4	32.8
July	28.5	29.7	27.4	32.6	0	24	0.1	3.8	4.8	4.8	23.5	33.0
September	28.0	29.5	29.0	32.0							24.5	32.5
October											26.0	33.5

deep, especially at Koh Kong bay, though there is a decreasing gradient to about 2.3–3 ppt (50–70% saturation) in the north of the region. There is sometimes a reduced minimum concentration of oxygen (3.8 ml/l) in June in near shore waters, especially in Koh Kong bay, probably due to intense oxygen uptake at this time.

During the dry season, phosphorus distribution at the water surface is patchy in character, varying from 0 to 2 ppb (see Table 1), and this remains the same to the bottom. Maximum concentrations are found in northern and southwestern parts of the region. Silicon in surface water varies from 30 ppb in the western part to 6–20 ppb in the rest of the region (see Table 1). Near the sea bed, silicon varies between 20 and 35 ppb. In the wet season, phosphorus concentration at the surface is a little higher, varying between 0.1 and 3.8 ppb, increasing slightly at the bottom.

Primary Production

In general, highest biomass concentrations are found in estuarine waters due to the flow of organic matter from the rivers and streams in Koh Kong, Kompong Som and Kampot bays. Planktonic and detritus biomass varies seasonally; in February–April it averages 4500 mg/m^3, while in May–June this rises to about 4600 mg/m^3, while in July–August, it increases tremendously to about 21,800 mg/m^3 before decreasing gradually again. In the coastal zone, diatoms dominate in depths between 20 and 40 m. It has been recorded that juvenile shrimp occur abundantly in northern and eastern parts of the coastal zone as a result. Further offshore, concentrations of phytoplankton decline to 3–5% of the total biomass, as the proportion of detritus increases. The zooplankton represent almost 30% of the total biomass in most of the coastal zone, but their average contribution to biomass in the whole area does not generally exceed 5–10%. The main zooplankton groups are from the Decapoda, Amphipoda, and Chaetognata.

COASTAL HABITATS

The Cambodian coast contains four main interacting ecosystems: the Koh Kong bay ecosystem, the Botum Sakor National Park, the Kompong Som semi-enclosed bay ecosystem, and the Kampot bay ecosystem.

Koh Kong Bay

This is the largest estuarine ecosystem of the Cambodian coastline and is much influenced by huge inflows of freshwater during the wet monsoon. The two major freshwater rivers are Dong Tong and Trapeang Roung. The estuarine water forms a large delta in waterways where mangrove vegetation spreads throughout a vast inter-tidal swamp of about 60,000 ha. This pristine mangrove forest is very diversified with 74 species including plants that occur in the limit of salinity influence. *Rhizophora mucronata* and *Rhizophora conjugata* are particularly important as their roots are a major substrate for green mussel and oyster during the dry season when the freshwater from the streams diminishes. Seagrass, especially *Enhalus* species, is found at the delta of Trapeang Roung river and in the muddy beaches at the eastern part of the bay. Halodule species grow well in the area between the shoreline and Koh Kong Island, developing particularly during the dry season. The mangrove forest and the seagrass beds form a very important habitat for mud crab, cuttle fish, Penaeus and Metapenaeus shrimp, and many juvenile and young estuarine and euryhaline fish species. Shallow water mammals such as the dolphin *Ocaella brevirostris* are also found in this habitat almost all year round.

Botum Sakor National Park

Botum Sakor National Park is the only extensive lowland evergreen forest in good condition, and the only coastal Dacrydium/Podocarpus swamp forest in Cambodia. The lowland evergreen forest is a good habitat for resident and migratory birds, and is also a safe habitat for the endangered brackish water crocodile (*Crocodilus porosus*). The coastal side of the National Park which comprises rocky and white sandy beaches is a very favourable habitat for coral reef development due to a greatly reduced input of freshwater from the continent. The coral reefs extend from both the rocky shoreline of the mainland as well as around a number of islands. Fish species associated with the coral reefs number at least 50 species, mostly being found between November and January. Penaeus shrimp, especially 'white shrimps' accumulate in large numbers in this area, avoiding storms and strong northern winds.

Kompong Som Semi-enclosed Bay Habitat

Konpong Som Bay is located in the central part of the Cambodian coastline. Its deepest part is about 20 m. The northeastern coastal habitat is defined as Dong Peng Multiple-Use Area where two major estuaries and some mangrove/Melaleuca wetland forests are located. The two estuaries are fed by the Andong Tuk and Sre Ambel rivers during the wet season when low salinity exists. There are other freshwater streams in the eastern and southern parts of the bay that also contribute freshwater. A large mixing area occurs at the narrow part (along some 20 km) of the bay that makes this area more productive for fisheries, especially fish and shrimp. This unique mixing characteristic of the bay occurs during the dry season, when northern and northeastern winds push the freshwater from the continent into the bay environment. Dolphin and octopus species occur seasonally in large numbers in the mouth of the bay during this mixing period. The Hawksbill and Loggerhead Sea turtles commonly enter the bay to lay eggs on the eastern beaches. Abundant jellyfish occur during the wet monsoon, and mollusc species such as the blood cockle, and a number of cone shells are also found in the shallow water.

Kampot Bay habitat

Kampot bay is characterized by a swampy and rocky marine habitat. There is only little effect of freshwater here, since most continental streams entering it are small. The biggest stream into the bay is the Kampot that forms a small delta covered with land shrubs and mangrove. Salinity of the nearshore water varies between 30.5 and 32 ppt during the wet season and increases up to 32.5–33.4 ppt during the dry season. This has led to the development of some salt ponds used for national consumption. The deepest area is found near Koh Tral island (presently called Phu Quoc island) and does not exceeded 20 m. Upwelling occurs in the wet season and early dry season (November–December) due to turbulence caused by monsoon storms coming from the eastern and western sides of Phu Quoc Island, and from the oceanic current from the South China Sea. Turbidity is very high during the wet monsoon in shallow water less than 10 m, except in the area between the Trapeang Ropov and Stung Kampot estuaries where transparency remains high almost all year round. The unique characteristic of this bay is its extended seagrass communities; beds of *Enhalus* species which project from the Trapeang Ropov and Stung Kampot estuaries and which grade into pristine Halodule seagrass meadows growing on the sandy sea floor. These beds host endangered Dugong species during November to January. This unique bay habitat also supports a number of mollusc species such as blood cockle, clam and cone shell, and is a feeding ground for a number of resident and migratory fish, squid, octopus and crustaceans.

THE OFFSHORE HABITAT

The offshore boundary extends from 20 m depth to the international boundary of the Cambodian EEZ. At present the Cambodian EEZ claims overlap with those of neighbouring countries, so the full extent of the Cambodia offshore habitat cannot be fully defined at present.

The offshore ecosystem is strongly affected by interactions of currents, wind, substrate and sea floor structures, by the locations of coral reefs and by the effects of inshore water. Sea bed characteristics at different depths are described in Table 2.

A substantial coral barrier is found along the 'reef beaches' of Koh Rong, Koh Rong Salem, Koh Tang and Koh Pring. These are well populated by grouper, coral cod, wrasse, and other large schooling fish species, especially during the dry season. Mollusc species such as numerous bivalves, cone shells, Strombus and other gastropods are abundant in these pristine habitats. Shark and cuttle fish are fairly abundant in this region, especially in deeper areas. Mackerel and scad are the most productive species, occurring abundantly in shallow water in the region during September; their peak spawning period is in late January–February when fish move to deep water between 40 and 60

Table 2

Cambodian sea floor structure

Depths (m)	Sea floor characteristics
20–30	The sea floor is entirely flat and consists of sand with some inclusion of broken shell (small pieces).
40–60	The sea floor is wavy with irregular mounds and ridges of 2–4 m which in some locations are up to about 10 m high. The floor of this area consists of viscous mud, often with broken shell deposits.
60–70	The floor is fairly flat and consists of very soft mud with inclusions of plastic mud or clay. In the north-western part of this area, the floor is covered by pointed and high rock (which damages the bottom trawls).

m to spawn. A different pattern is seen with shark and crustaceans, which congregate during the wet season, especially in the area within Koh Rong, Koh Rong Sanlem, Koh Tang, Koh Pring and Phu Quoc island.

To date, 472 fish species have been identified from 105 families in the offshore zone (FAO, 1974, 1984, 1988, 1991). Recent observations found that the composition of species in the total catches (estimated about 30,000 and 50,000 tons in 1998) mainly consisted of *Atule mate* (scad), *Sela crumennophthalmus* (big eyes scads), *Decapterus maruadsi* (yellow-tail round scads) and others such as *Leiognathidae* (pony fishes), *Scombridae* (tunas, mackerels) and *Lutjanidae* (snappers). Other sizeable catches are of rays, sharks and invertebrates (cephalopods and crustacea), and large fish of the *Lutjanidae* (snappers), *Pleoctorhynchidae* (grunters), tunas, *Formionidae* (black pomfrets), *Platax pinnatus*, and *Rachycentron canadum* (trevally). These latter species are abundant in shallow water area (20–30 m depth). Mackerel and *Clupeidae* are concentrated mainly in the northeastern part of the region, while in the southeastern part the main composition is *Leignathidae*. Stingray and octopus are found and caught almost everywhere in the Gulf.

The biologically important offshore zone lies in the southeastern and the central parts of the region where upwelling occurs during the wet monsoon and early dry season. Typhoons occur in September–November, whose mixing effects make this region very productive for fisheries. Other visitors such as sea turtle, dolphin and Baleen whale are often seen throughout the year.

THE COASTAL POPULATION

For most Cambodians, the marine environment is of much less interest than the inland fisheries of the Mekong River and its wetland ecosystems. In the past, 90% of its people rely on agriculture and freshwater fisheries. Reportedly, marine fisheries started commercially around 1957 (Blanc, 1958) using fishing weir and ramie fibre bag nets at the time of establishment of commercial ports in Sihanouk Ville. Trawler and gill net operations were introduced to Cambodian coastal waters during 1958 (Eng, 1972) through joint operation with Thai fishermen, while the freshwater fisheries started commercially in 1864. Indigenous coastal people in the past lived in small communities, usually settling around freshwater tributaries. It is known that small fishing villages of about 10 to 30 households have long settled the Kompot delta vicinity and along the banks of the rivers Sre Ambel, Andong Tuk, Peam Krasob, Pakloang and Dong Tong, and around Koh Kapik Island. Coastal settlements and fishing activities in Ream Bay, Koh Takiev Island, Chak Sre Cham and Koh Rong of Sihanouk Ville port city were developed during 1958–64. In other provinces such as Kampot and Koh Kong, new fishing village settlements were located along almost every stream and around the inshore islands. Employment in the area, mainly of officials, port workers, salt pond manufacturers, fishermen and so on, was not higher than 2% of the total population of about 5 million inhabitants in 1960. The coastal and islander people, especially fishermen, increased steadily until the civil war period in 1975, but did not exceed 3% of the population.

In recent years, public awareness of the diverse marine habitats has led to a large migration of urban and rural inland people into coastal zones, including the islands, where exploitation of natural resources such as fishing, mangrove logging, upland forest logging and employment in different public and private services and the industrial sector has increased substantially. Within a five-year period from 1993 to 1998 the coastal people in two provinces and two secondary cities rose from 675,000 to 845,540 inhabitants; the Cambodian population was 9,870,000 in 1993 and 11,426,223 in 1998 (Anon., 1998). A census result of 1998 showed that the number of females was higher than that of males (51.2%) in the coastal zone, and that the population density was 49 inhabitants per km^2, while the first census in 1962 found only 23 inhabitants km^2.

Although the population in coastal provinces and cities increased by 25% within a five-year period, the number of coastal people is still less than 10% of the total population. Most are farmers (about 80%) followed by fishermen including those involved in aquaculture and fish trading (about 10%). Other groups are officials, soldiers, policemen, loggers (including non-timber forest gatherers), employees and workers for tourist services, transport services and the industrial sector. These latter groups have increased over the last five years, following the establishment of factories in Sihanouk Ville city.

THE EFFECTS OF THE RURAL SECTOR

According to the second census of 1998, the rural population represented 70.6% of the total coastal population. Moreover, the rural poor of the area also often engaged in small-scale fishing, agriculture and non-timber forest gathering and hired labour in both the private and public spheres. If this latter is taken into consideration, the percentage of rural people should be about some 80% of the total coastal population.

The coastal rural sector has increased over the last 10 years. Upland logging, mangrove logging, shrimp farming, small-scale fishing and foreign industrial investment have become more and more attractive in coastal socio-economic terms. These activities have brought with them the development of the service sector and an intensified state authority interest in supporting and controlling these activities. However, degradation of the coastal forested watershed, the mangrove forest and especially the fisheries resource and its habitats is increasing markedly, partly due to anarchic exploitation and collusion of logging interests with irresponsible authorities.

Agricultural practices in the coastal zone as well as inland are mostly still traditional and undeveloped, and do not impose serious environmental impacts at the present time. But unregulated claims of private land developments and the recently developed plantations of industrial crops such as palm oil, fruit trees and so on, may possibly cause future crucial environmental problems. Not surprisingly, except in some controllable protected areas, most of the coastal forested watershed has already been cleared for agriculture and plantations, and almost all the rest has been changed in quality into secondary forest.

The pristine mangrove forests in Koh Kong bay have been seriously degraded by the active operation of some hundred charcoal kilns during 1989–99 and by the claims for several hundred hectares of shrimp farm constructions. The shrimp farm industry no longer survives, due to a spread of disease that caused the abandonment of most farms.

The increase of small-scale inshore fisheries over the last 10 years is depleting inshore resources, which in the future may cause a decline in the offshore catch. Foreign poaching occurred heavily during the above period because of collusion of and poor control by local authorities. This has changed local fishers' behaviour; large-scale offshore fisheries holders gave up their businesses and shifted to small-scale inshore operations in order to survive. With the addition of new arrivals from the rural inland sector, the originally small number of fishing boats operating in Cambodian coastal waters has risen to more than 5000 of less than 1 ton (or less than 50 Hp). Most of the fleet use different multi-fishing practices as they follow seasonal appearances of the different marine resources; purse seiner, gillnetters and pushnetters are all used, though most of the time trawling is undertaken. The two latter practices are the most ecologically damaging since they destroy the shallow sea floor environment where seagrass and other important habitats are located. Mono-filament gill net is the most popular gear for small fishing boats due to its low cost investment. Reportedly, about 65% of small boats use this gear for catching shrimp, crab and fish, and every one of them appears to loose or damage at sea about 10–20 kg of net every year. Thus some dozen tons of mono-filament gill net are lost every year which may cause a serious problem of ghost net mortality in the future.

Foreign poachers—according to complaints of officials and local fishermen—usually operate with large gears, and most of the time they also use prohibited gear such as large bottom trawls, long drift nets, pair trawlers, light fishing, as well as fishing with explosives. These have caused a huge destruction of sea floor habitat and coral reef, and have caused a serious decline in some marine biodiversity resources.

In recent years the demand for coral fish by the aquarium trade and by exclusive restaurants has risen in both international and local markets. This has persuaded fishermen to use cyanide fishing in the coral reefs. During the last two years, the export to Hong Kong of live coral fish such as grouper, snapper, coral trout, mantis shrimp, Strombus shell etc. was at least 5 tons per month. In 1999, when the Hong Kong market price dropped drastically, the demand by local exclusive Chinese restaurants increased to stabilize this fishing activity. No one authority in Cambodia is concerned with the impact of this fishing activity on the natural environment.

COASTAL DETERIORATION DUE TO EROSION AND LANDFILL

Erosion is still limited but is increasing. Future landslides may damage the on-going hotel constructions on the very degraded hill opposite the maritime port of Sihanouk Ville city. This may be due to a lack of unawareness by both the local authority and the developer of the long-term environmental impact of such developments. However, flooding in the coastal watershed systems due to this deterioration occurred as early as October 1991 in this site. Recently, in early August 1999, a monsoon storm in the coastal Cambodian area caused floods over Kampot town and the watershed of Koh Kong province. This flood cut off the communication between the capital city and the coastal region, especially Sihanouk Ville city, for at least a week. Other remote coastal watersheds are also under threat but are less socially affected due to their smaller populations.

Natural and man-made landfill will become one of the major concerns of coastal habitat loss in the near future. Long-term natural erosion and the rapid conversion of coastal wetland for development is presently being undertaken by irresponsible developers in the commercial city Sihanouk Ville, which may reduce both the area of estuarine habitat and underground water, the latter being used to supply both domestic water and irrigation for agriculture.

EFFECTS OF URBAN AND INDUSTRIAL DEVELOPMENT

There is no clear perception of environmental problems among local communities, or even in the public sectors or state institutions, except for the Ministry of Environment. Communication and awareness is very weak in this country. Dirt, garbage and waste dumping is widespread. Solid wastes such as cans, plastic bags and so on are found everywhere, especially in the urban areas, towns, market places, road sides, in the sea, and also around homesteads and even in remote coastal villages and communities. There is a long-term concern regarding solid waste loads at sea, which seriously affect both the landscape and marine environment. This may increase as the coastal population itself is likely to increase drastically in the future.

At the present time, industrial development in the coastal zone does not cause a serious environmental pollution problem, except for the shrimp farming industry, since there are only a few factories and domestic tourists (Monirith et al., 1999). But the attitude of local people towards waste disposal is a major concern. Shrimp farming started sometime in 1991 with a few farms in Koh Kong province, and then expanded to almost one hundred farms operating along the entire 500 ha inter-tidal mangrove area. The pristine mangrove forest of Koh Kong bay, which was already exploited by logging for charcoal production, was cleared for shrimp farm construction, and is now heavily polluted. The shrimp farm industry itself did not survive; the spread of disease caused most farms to be abandoned after hundreds of hectares of mangrove habitat had been destroyed.

PROTECTED MEASURES

Cambodian perception of environmental impact is very limited, especially at the grass-roots level. The establishment of the Ministry of Environment is very recent and has limited authority on living and non-living natural resources since it is subordinate to the Ministry of Agriculture, Forestry and Fisheries, and to the Ministry of Industry, Mines and Energy. A number of Acts, decrees and sub-decrees on living and non-living natural resources were in place during the former regime before 1993. Interestingly, with strong support of a number of international agencies such as the World Bank, Asian Development Bank, IDRC-Canada, Danida-Denmark, Wetlands International, IUCN and so on, the Royal Decree for the creation of Protected Areas was promulgated in 1993, and an Environment Management Law was prepared for implementation in 1997. To enforce this law required a number of sub-laws and sub-decrees enacted by the Royal Government. However, the legislative system in Cambodia is very incomplete and is burdened by old legislation from former regimes that has to be changed, so that an environmental impact assessment system is not yet in place. Moreover, the performance of resource and environment monitoring and surveillance by the legal authorities is still complicated in practice. In recent years, international pressure to stop serious environmental destruction pushed the government to issue strict measures to stop anarchic natural resource exploitation, beginning in 1999. Even so, the protection and conservation

Table 3

Status of marine environment and habitats protection measures

Marine environment and defined habitat and natural resource	Legal frame work	Institutional arrangement	Enforcement status
Forested watershed excluding protected areas	Forestry Decree (Dept Forestry and Wildlife 1988)Environme nt Decree (Ministry of Environment 1997)	Department of Forestry and Wildlife, Ministry of Environment	Many monitoring control surveillance groups from different military and police units involved with forestry official that create more collusion
Mangrove forest excluding protected areas	Sub-decree on management of inundated forest	Department of Fisheries	Poor monitoring control surveillance due to collusion
Peam Krasop Wildlife Sanctuary in Koh Kong province	Royal decree of Creation of Protected Areas (1993)	Ministry of environment, Wildlife Sanctuary authority	Poor monitoring control surveillance due to less experience and collusion
Botum Sakor National Park in Koh Kong province	Royal Decree on Creation of Protected Areas (1993)	Ministry of Environment. (Park authority has yet established)	Different military and police units involved with central and provincial forestry officials that create more collusion
Dong Peng Multiple-Use Area in Koh Kong province	Royal Decree on Creation of Protected Area (1993)	Ministry of Environment (Authority has yet established)	Poor monitoring control surveillance due to lack of management authority
Preah Sihanouk National Park in Ream	Royal Decree on Creation of Protected Area (1993)	Ministry of Environment Park authority	Poor monitoring control surveillance due to lack of fund and inexperience
Sea grass habitats	absent	undefined	absent
Coral reef habitats	absent	undefined	absent
Inshore habitat	Fisheries Decree (1987) Sub-decree of hired of inland and marine fishing ground for exploitation	Department of Fisheries	Poor monitoring control surveillance due to incompatibility of the fisheries legislation
Offshore habitat	Fisheries Decree (1987) Sub-decree of hired of inland and marine fishing ground for exploitation	Department of Fisheries	Poor monitoring control surveillance due to lack of fund and collusion
Marine pollution	Environment Decree Lack of Environmental Impact Assessment Sub-decree	Ministry of Environment	Negligible monitoring due to lack of expertise and power

of marine endangered species such as dolphin, dugong, sea turtle, whale shark and so on is still undermined by local fishermen due to a lack of both legislation and interest on the part of authorities and public institutions. As is evident, these endangered animals have been accidentally caught by local fishermen without any restriction. However, during the 1990s, several laws and decrees may allow the situation to improve (Table 3).

REFERENCES

Anon. (1998) General Population Census of Cambodia. National Institute of Statistics, Pnom Penn.

Blanc, M. (1958) Mission hydrobiologique et oceanographique au cambodge. *Cahiers de Pacifique* **1**.

Eng, C.M. (1972) Contribution a l'Etude de la Faune Benthique des Codes du Cambodge. Memoire a la fin d'etude.

FAO (1974) Eastern Indian Ocean (Fishing Area 57) and Western Central Pacific (Fishing Area 71). FAO Species Identification Sheets for Fishery Purposes. Four Volumes.

FAO (1984) Western Indian Ocean (Fishing Area 51). FAO Species Identification Sheets for Fishery Purposes. Five Volumes.

FAO (1988) Evaluation de la Situation Alimentaire et Agricole. Kampuchea Rapport de la Mission FAO/PAM/ONU.

FAO (1991) Situation of the Fisheries Subsector in Cambodia and Urgent needs for its rehabilitation and development.

Monirith, I., Nakata, H., Tanabe, S. and Touch Seang Tana (1999) Persistent Organochrorine in Marine and Freshwater Fish in Cambodia. *Marine Pollution Bulletin* **38**, 604–612.

THE AUTHOR

Touch Seang Tana
*Department of Fisheries,
186 Norodom Blvd., P.O.Box 582,
Phnom Penh, Cambodia*

Chapter 91

THE AUSTRALIAN REGION: AN OVERVIEW

Leon P. Zann

Australia has a 200-mile EEZ of over 11 million km^2, one of the largest in the world. Its waters span almost 60 degrees in latitude, and include all ocean temperature zones, and communities from corals reefs to temperate kelps. The coastline of the mainland and associated islands is around 70,000 km long, and there are over 12,000 islands. The shelf area is generally narrow, and waters are limited in nutrients. It has a high marine biodiversity, with two major bioregions: the tropical north (Indo-Pacific), and the temperate south (with a high proportion of endemics). Australia has the largest areas of coral reefs, and seagrasses, and the third largest area of mangroves in the world. While human populations are low by world standards, over 85% live in the coastal zone, mostly in metropolitan areas. Much of the land has been cleared for agriculture, resulting in massive land degradation. This has increased run-off of sediments and nutrients, degrading estuaries, threatening inshore coral reefs and causing massive die-back of temperate seagrass. Pollution from heavy metals and chlorinated compounds is locally severe around urban estuaries and bays, and adjacent industrial areas. The region has limited fisheries, and virtually all areas are subject to commercial fishing. Marine environmental management is undertaken by State agencies within the three-mile limit, and by the Commonwealth (national) government to the 200 mile EEZ. Australia has the largest area of marine protected areas in the world, and in 1998 produced a comprehensive oceans policy. The state of the marine environment is regarded as generally good (largely because of the sparse rural populations) but declines in estuaries, seagrass and fisheries are of major concern.

Seas at The Millennium: An Environmental Evaluation (Edited by C. Sheppard)
© 2000 Elsevier Science Ltd. All rights reserved

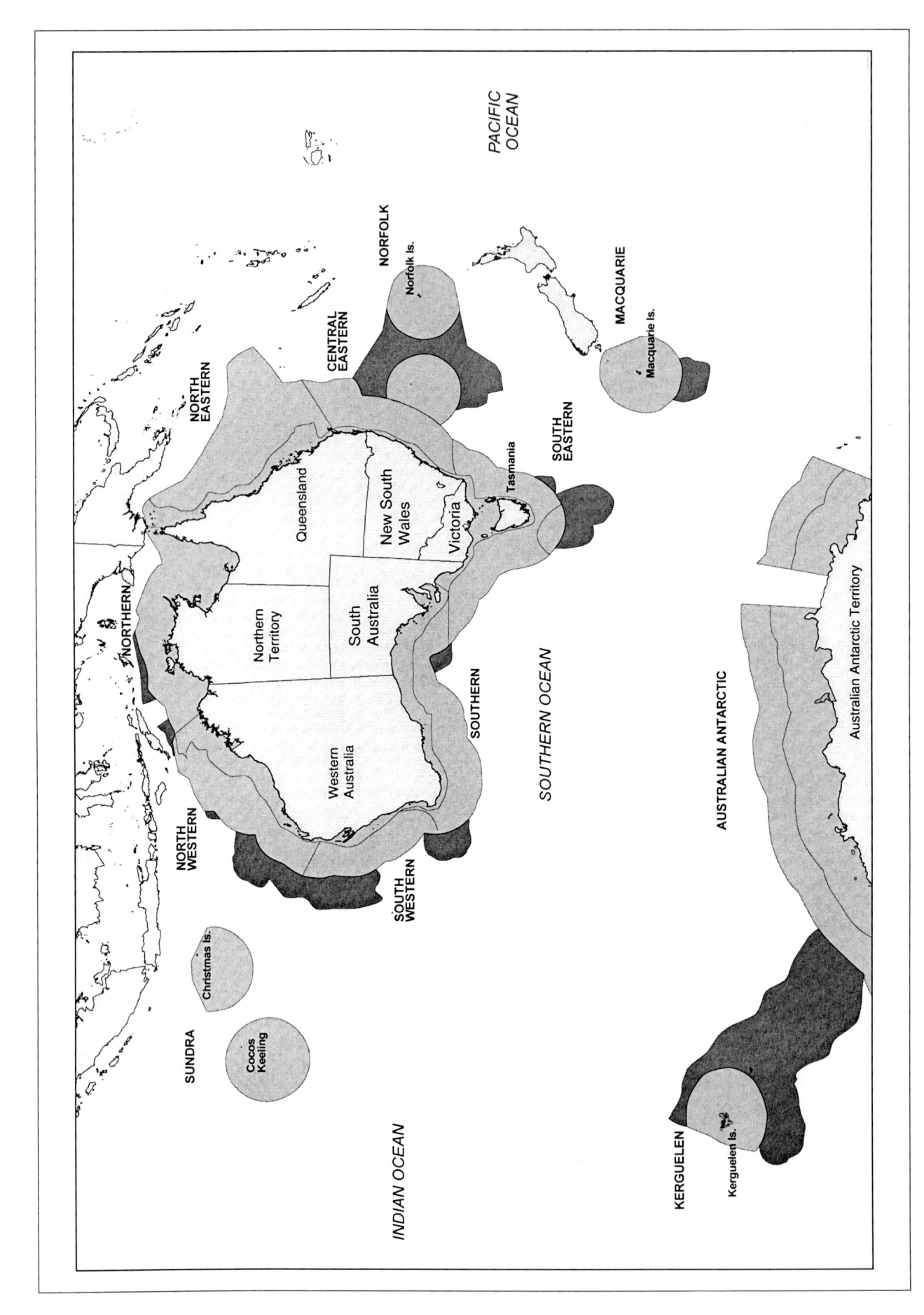

Fig. 1. Australia's large marine ecosystems or marine domains, EEZ (light stipple) and claimable shelf and EEZ (dark stipple) (from Lyne, CSIRO).

THE DEFINED REGION

Australia is an island continent surrounded by three oceans: the Pacific, Indian and Southern Oceans (Fig. 1). Australia's long coastline and its geographically scattered External Territories (Antarctica and subantarctic islands in the Southern Ocean, Cocos (Keeling) Atolls and Christmas Island in the Indian Ocean and the Coral Sea Territories and Norfolk Islands in the Pacific Ocean) make Australia's Exclusive Economic Zone one of the largest and most diverse in the world. Australian waters span almost 60° in latitude, from Torres Strait in the north, to Antarctica in the south, and 72° in longitude from Norfolk Island in the east to Cocos (Keeling) Island in the west. They include a great range of geographic, geologic and oceanographic features, and over 12,000 islands. The length of the coastline of the mainland and that of the larger islands is 69,630 km (Zann, 1995).

This chapter provides a general overview of Australia's marine environment and national environmental issues and management initiatives. The main marine domains or large marine ecosystems are specifically described in the following chapters. A national overview is needed because ocean management is the responsibility of the Commonwealth (national) Government and an integrated approach is required for the scientific understanding and management of the interconnected oceans. This review has been compiled largely from the State of the Marine Environment Report for Australia (Zann, 1995; Zann, 1996a) and State of the Environment, Australia (State of the Environment Advisory Council, 1996).

Australia's Exclusive Economic Zone

Australia's 200 nautical mile Exclusive Economic Zone (EEZ), proclaimed in August 1994, is over 11 million km^2 in area (Fig. 1). Australia's 200 nautical mile Fishing Zone, which excludes the Australian Antarctic Territory, is 8.74 million km^2 in area, and is the third largest in the world. The total area of the claimable continental shelf around Australia, as defined by the 1982 United Nations Convention on the Law of the Sea, is 14.8 million km^2 (Zann, 1995).

The six Australian States have jurisdiction over marine areas within three nautical miles of their shores, while the Commonwealth has jurisdiction from there to the edge of the 200 nautical mile EEZ, and for the Great Barrier Reef and Australian Territories. However there are agreements between Commonwealth and States for the latter to manage specified activities within Commonwealth waters (Zann, 1995).

Biogeography

Australia has been long isolated geologically and climatically, and many of its marine species are as distinctive as its unique terrestrial species. Northern Australia has a very high species diversity but with a low proportion of endemism (around 10% in some groups), while southern Australia has a lower species diversity but very high proportion of endemism (over 90% in some groups). There is an overlap of biota in the east and west (Poore, 1996a).

While there is no agreed bioregionalisation for Australian waters, an interim ecosystem-based classification has been recently developed as a basis for maritime planning and management. This hierarchical classification comprises: two pelagic provinces with two biotones; nine demersal provinces with eight biotones; and 60 meso-scale bioregions (IMCRA, 1998). Seven 'marine domains' or large marine ecosystems (LMEs) have also been proposed (Lyne, pers. comm.) and form the bases of the following chapters (Fig. 1).

SEASONALITY, CURRENTS, NATURAL ENVIRONMENTAL VARIABLES

Australia's seas encompass all five of the world's ocean temperature zones: tropical (25 to 31°C); subtropical (15 to 27°C); temperate (10 to 25°C); subpolar (5 to 10°C); and polar (–2 to 5°C). The main ocean currents affecting the marine environment around the continent are the East Australian Current which brings warm equatorial and Coral Sea water down the east coast, and the Leeuwin Current bringing warm, low-salinity water down the west coast. These are not major currents by world standards. Other major ocean features are the sub-tropical and antarctic convergences (Middleton, 1996) (Fig. 2).

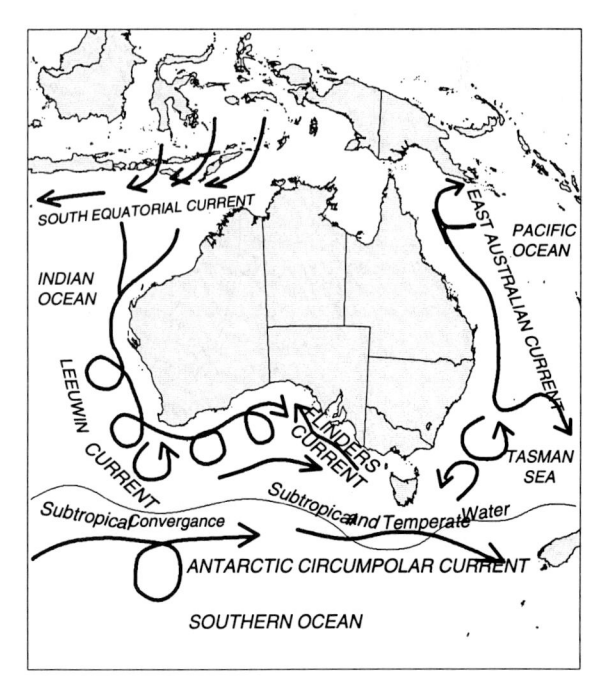

Fig. 2. Major currents and water bodies around Australia (from Middleton, 1996).

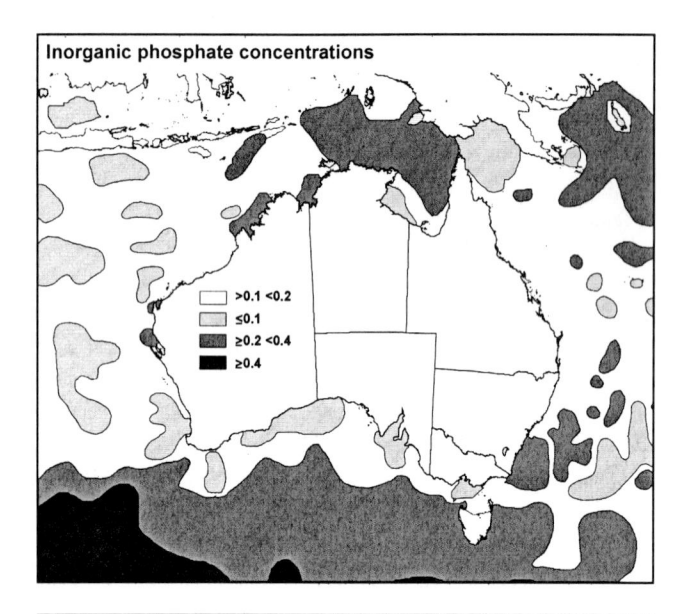

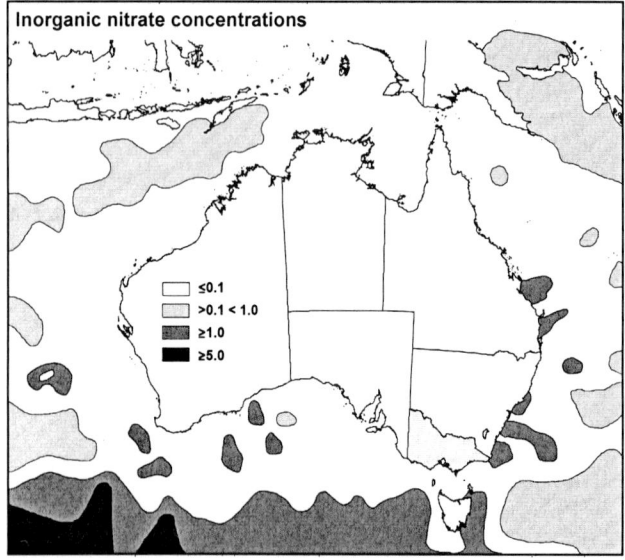

Fig. 3. Nutrient concentrations in Australian waters (from Middleton, 1996).

Australian waters are generally low in nutrients and therefore have a relatively low biological productivity (Fig. 3). This is because they are largely dominated around the continent by low-nutrient tropical water masses; there are no major upwellings of nutrient-rich deep water in the region; and the run-off from the ancient, leached land is naturally low in nutrients (Middleton, 1996). Paradoxically, some inshore areas tend to be dominated by highly productive mangroves, seagrass and coral communities which are adapted to low-nutrient waters. The generally low-nutrient status also contributes to the relatively low fisheries production of Australian waters (Zann, 1995).

The periodic influence of the El Niño/Southern Oscillation has a great effect on ocean temperature and biological productivity, and on Australia's terrestrial climate and agricultural production. Similarly, there are the periodic influences of strong winds, swirls and storm surges associated with tropical cyclones and mid-latitude low pressure systems (Middleton, 1996).

MAJOR SHALLOW WATER MARINE AND COASTAL HABITATS

Australia's continental shelf is 15–400 km wide, and is 2.5 million km^2 in area. The shelf is connected to Papua New Guinea in the north, and to Tasmania in the south. With the exception of the complex Great Barrier Reef in the northeast, and the undersea cliffs and pinnacles off Tasmania, the sea floor is generally rather featureless except for occasional reefs, terraces and plateaus (Harris, 1996).

Australia's inshore waters support a diverse and rich biodiversity. Northern Australia has the largest areas of coral reefs in the world, one of the largest areas of tropical seagrass, and the third largest area of mangroves. It has some of the world's last secure populations of vulnerable and endangered species such as giant clams, turtles and dugongs. Australia's tropical marine ecosystems are relatively little affected by human activities and will have an increasing global value as many of the other tropical ecosystems in the tropical Indo-Pacific region become degraded (Zann, 1995).

Southern Australia also has a high marine biodiversity. It has the largest areas of temperate seagrasses in the world, and the highest diversity of seagrass and marine algae. Southern Australia also has significant populations of fur seals and sea lions, and important breeding grounds for endangered Southern Right and Humpback whales. However, much of the south is also far more densely populated and developed than the north, and is under greater environmental pressure (Zann, 1995).

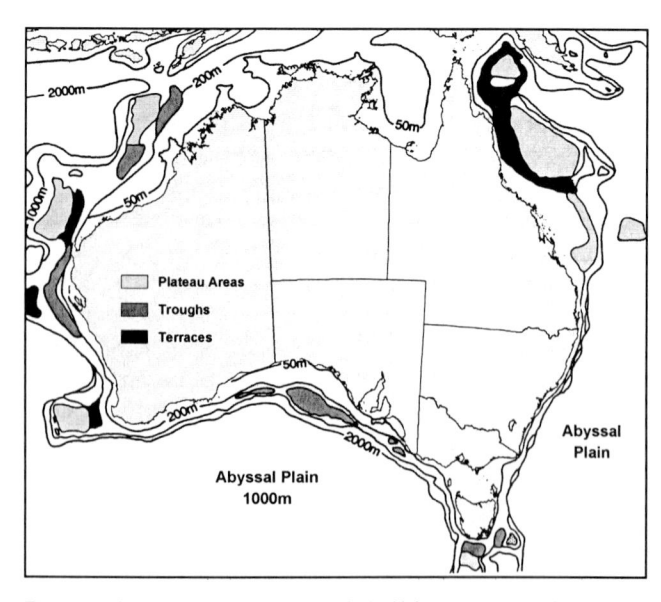

Fig. 4. Bathymetry and continental shelf features around Australia (from Harris, 1996).

OFFSHORE SYSTEMS

The offshore systems in Australia's EEZ are less well known because of the region's vast size, remoteness and a relatively low level of scientific capability in the area.

Geomorphology

The Australian continental slope is continuous with that of Indonesia and New Guinea in the north. In parts it is steeper than slopes elsewhere in the world; gradients up to 40° may extend down to the abyssal plain at 4000 m depths. The slope is intersected by canyon systems (particularly around Bass Strait) and terraces, major plateaus and troughs (particularly off the north west of the mainland, and southern Tasmania) (Poore, 1996b). Major offshore features include: the Queensland trough in the Coral Sea; the Lord Howe Rise and Norfolk Ridge in the Southern Coral Sea and Tasman Sea; deepwater volcanic pinnacles off eastern Tasmania; the Macquarie Rise and Kerguelan Plateau in the Southern Ocean; the Exmouth Plateau off north Western Australia; and Vening Meinesz Seamounts in the Indian Ocean which give rise to Cocos (Keeling) Atoll (IMCRA, 1998).

Oceanography

The prevailing westerly winds in mid-latitudes, and easterly winds in the tropics drive the ocean currents in the major ocean basins. Below the surface waters to about 800 m, and extending from the equator to the Subantarctic Convergence, lies the South Pacific Central Water in the east, and the Indian Central Water in the west. The low salinity, cold Antarctic surface waters meet subantarctic waters at the Antarctic Convergence and the resulting mixture sinks to around 1000 m. Below this, and flowing southwards at depths of 1000–3000 m lies the Pacific Deep Water which is of Atlantic and Antarctic origin (Middleton, 1996).

The ocean around Australia is generally low in nutrients, especially nitrates and phosphates. The east and west are dominated by low-nutrient subtropical waters, while the south is largely isolated from the rich subantarctic waters. Significant upwellings are lacking. Nutrient inputs from land run-off are also limited because of the poor nutrient status of the continent's ancient, leached soils, and the low rainfall. As a result, large areas of the oceans around Australia are virtual deserts (<0.1 μmole/l nitrate). Off most of the coastline, levels are low (>0.1–1.0 μmole/litre nitrate), and typical of subtropical belts of the world ocean. South of Tasmania at the Subtropical Convergence, levels are higher (1–5 μmole/l nitrate) but these are significantly lower than levels in the world's more productive waters, for example 20–25 μmole/l nitrate off southwest Africa (Middleton, 1996).

Biodiversity

There are two main offshore pelagic provinces in the Australian region: the North (from ca. 10°S to 20°S), and the South (Gulf of Carpentaria to southern Tasmania (45°S), with two extensive biotones between: the Eastern, and Western Biotones (IMCRA 1998). Some 5000 species of diatoms and 2000 species of dinoflagellates have been described from Australian waters. There are three distinct phytoplankton assemblages around Australia: temperate neritic off the southeast; tropical neritic in the Gulf of Carpentaria to the Northwest Shelf; and tropical in the waters of the Coral Sea and Indian Ocean (Hallegraff, 1996). Little is known of deepwater seafloor communities around Australia. Studies off the Great Barrier Reef, Bass Strait and the Southeast Slope, and the Northwest Shelf indicate that continental slope communities are highly diverse (Poore, 1996b).

Offshore Territories

The offshore territories lie in separate biogeographic provinces. The Indian Ocean Territories of Cocos (Keeling) Atolls and Christmas Island have been placed in the Sunda Province (Indonesia region). Norfolk Island, Macquarie Island and Kerguelen Islands are placed in separate provinces (IMCRA, 1998) (Fig. 1).

HUMAN POPULATIONS AFFECTING THE AREA

The Australian mainland has been inhabited by Aboriginal peoples for the past ca. 50,000 years. The Torres Strait Islands between Australia and New Guinea are inhabited by people of Melanesian origin. Australia was colonised by Britain in 1788 and today the majority of the population are of European descent.

Indigenous Communities

Traditionally, coastal Aboriginal clans relied seasonally on marine foods, and their sea domains were indistinguishable parts of their clan estates. Despite dispossession by Europeans, the cultural associations and concerns of coastal Aboriginal people for the sea and its resources remains strong. Major issues and concerns of coastal Aboriginal peoples centre around their dispossession from their traditional land/sea estates; the threats, desecration and injury to sites of cultural significance; the loss of ancient fishing and hunting rights; their lack of commercial fishing opportunities; and their general lack of participation in coastal environmental planning and management (Smyth, 1996).

However, Aboriginal land and sea rights have been increasingly recognised in recent years. In the Northern Territory, 84% of the coastline to the low tide mark is Aboriginal land and beyond the low tide mark legislation provides for the closure of waters up to 2 km offshore for Aboriginal usage. In the Torres Strait, indigenous fisheries are specifically protected under the Torres Strait Treaty. In 1992 the historic 'Mabo decision' of the Australian High Court recognised native title on land, and in 1997 the Federal Court ruled in the Croker Island Decision that native title can also exist in the sea (Smyth, 1996; Federal Court of Australia, 1998).

General Community

Australia is today a complex, dynamic multicultural society of over 18 million people. Despite its great size, Australia is essentially a coastal nation. Over a quarter of the population live within 3 km of the sea, three-quarters live within 50 km of the coast, and Australians are increasingly moving to live along the coast from inland rural areas as farming declines. All State capitals and all major cities (excepting the national capital, Canberra) lie on the coast. Although the average population density is very low, over 80% of the population living in coastal metropolitan areas, largely in the south east (Zann, 1996b).

Economic Importance of the Sea

The value of Australia's marine resources is around Aust $20 billion per year, and is expected to double in the next 20 years. Offshore oil and gas is valued at around $5.5 billion per year. The tourist industry, largely based on the coast and sea, contributes over 5% of the gross domestic product and employs 5.8% of the workforce. Commercial fisheries production is worth around $1.3 billion each year and recreational fishers spend over $2 billion each year on their activities. Marine transport is essential for the island continent and Australia ranks fifth in the world in terms of frequency and volume of shipping (Zann, 1995).

RURAL FACTORS

The major land uses in Australia are agriculture (67%), forestry (2%) and nature conservation (3.5%). Urban and transport uses occupy only 1.3% of the land area. Around 26% is desert or inaccessible mountainous terrain (DAHE, 1991). SOMER (Zann, 1996b) identified that the major issues affecting Australia's marine environment, elevated nutrients and sediments, were largely the result of inappropriate catchment land use practices.

Agriculture

In the 200 years since European colonisation, Australia has become one of the world's leading agricultural nations. It is the largest exporter of wool, the largest supplier to the free market of sugar, the second largest exporter of meat, the fourth largest exporter of wheat and a major supplier of

---

Status of Scientific Knowledge of Australia's Marine Environment

A major finding of the State of the Marine Environment Report (SOMER) (Zann, 1995; 1996) was that there are serious gaps in scientific knowledge and understanding of the marine environment in Australia, both geographically and by issue (Zann, 1996c). In particular, there is a serious lack of long-term data on water quality, environmental health and human uses which could be used for quantitative assessment of the state of the environment (Zann and Oliver, 1996).

Reasons for the poor state of knowledge include: (a) the vast size of Australia's EEZ, and the large proportion of coastline which is sparsely inhabited; (b) a relatively small scientific population; (c) limited funding in marine science (particularly from the private sector); (d) a lack of a strategic, long-term approach to marine sciences, particularly inventory, monitoring and data management (Zann, 1996c).

However, Australia's marine sciences have grown very rapidly over the past 20 years, and have achieved pre-eminence in areas of tropical marine ecology and management. Despite its small population, Australia is now amongst the world's top ten nations in marine research effort and output (McKinnon, 1989).

---

dairy produce, fruits, grains, cotton and other commodities. Of agricultural lands, 91% are used for grazing, 6% pastures and 3% cropping (DAHE, 1991; State of the Environment Advisory Council, 1996).

Poor agricultural practices have resulted in soil erosion; loss of soil fertility, structure, and increasing acidity; utilisation of marginal lands; increasing salinity; effects of agricultural chemicals; and feral and native animals (State of the Environment Advisory Council, 1996).

Fertilisers are a significant source of nutrients in agricultural areas. Because Australia's ancient leached soils are particularly low in phosphorus, superphosphate fertilisers are applied to crops and pastures. Superphosphate use has steadily increased since 1950, and is currently about 350,000 tonnes a year. Application of nitrogenous fertilisers has also steadily increased and is around 370,000 tonnes a year (Brodie 1996). Sewage and storm-water run-off are also major sources of nutrients in urban areas.

Forestry

About 30% of Australia's forest is managed primarily for wood production. Timber production in 1989-90 was around 16.5 million cubic metres (63% native hardwoods and 37% exotic softwoods) and was worth $548 million. Major environmental consequences include conservation of forest biodiversity; loss of rainforests; and clear-felling for wood chip (DAHE, 1991).

Mining

Australia is among the world's leading exporters of coal, gold, iron ore, bauxite, alumina, zinc, manganese, lead and mineral sand, with an annual value of around $30 billion. Major environmental consequences are local site disturbance, and air and water pollution. This includes air pollution from coal burning stations, motor vehicles, smelters, and greenhouse emissions, particularly carbon dioxide (DAHE, 1991).

Land Degradation

Of about 226 million ha of forest and woodland at the time of European settlement, only 135 million ha remains in a close-to-original state. Almost half of Australia's rainforests have been cleared. Around 52% of agricultural and pastoral lands are considered degraded and in need of reclamation. Major types of degradation include soil erosion (sheet and rill, gully, wind); soil surface scalding; soil structure decline; soil compaction; soil acidity; nutrient decline; salinisation of dry land and irrigated land; waterlogging; and coastal degradation (State of the Environment Advisory Council, 1996).

Declining Water Quality

Australia's rainfall and run-off are less than those of other continents, and twice as variable. Floods and droughts are an integral feature of the climate and have moulded both the terrestrial and marine environments. Rainfall is concentrated on the coasts. Australia has around 250 rivers, with a total annual discharge of about 440,000 million m^3. Vast areas of the arid and semi-arid interior have no external discharge. Most run-off is into the Timor Sea and Gulf of Carpentaria (48%), and to the northeast (21%). The quality of Australia's lakes and inland waterways is declining because of over clearing; intensive irrigation and fertilisation; raising water tables and increasing salinities; and inappropriate cultivation practices contributing to soil erosion, which increases turbidity and salinity. Blooms of toxic algae and fish kills are increasingly common in inland waters (State of the Environment Advisory Council, 1996).

Declining Estuaries and Inshore Environments

As Australian coral reefs and seagrass communities evolved under low-nutrient and sediments regimes, they are particularly vulnerable to eutrophication and sedimentation. Increased sedimentation is the major contributor of nutrients entering the sea (Zann, 1995). For example, it is estimated in Queensland that four times more sediments, nitrogen and phosphorus now enter the sea each year than in pre-European times (Brodie, 1996).

Many areas around Australia have experienced major changes to marine ecosystems through elevated nutrients. Very large areas of seagrass in southern Australia have suffered die-back from eutrophication and sedimentation. Nuisance and toxic blooms of seaweeds and phytoplankton are also now common in many southern bays and estuaries. Amongst the worst affected areas are the Peel–Harvey system and Cockburn Sound (WA); Holdfast Bay and Barker Inlet (SA); Gippsland Lakes and Port Phillip Bay (Vic); and Lake Illawarra, Lake Macquarie, Tuggerah Lakes, and Georges River (NSW) (Brodie, 1996). There is a growing concern that the Great Barrier Reef lagoon, the body of water between the Reef and mainland, is becoming eutrophic, and that corals on inner reefs may be declining (Brodie, 1996).

COASTAL EROSION AND LANDFILL

Because the great majority of Australia's population and development is concentrated along some 10% of the coastal zone in the southeast and southwest, that area has been particularly affected by urban, industrial and port development, and by tourism and recreation. However, most of the remaining coastline has been little affected (Zann, 1996b).

Major coastal modifications in developed areas include engineering structures such as breakwaters and seawalls associated with ports, harbours, canal estates and marinas, and reclamations. Estuaries and the coastal lakes and lagoons in the southeast have been particularly affected by seawall construction and there have been significant local losses in saltmarsh, mangrove and seagrass habitats (Bird, 1996).

Coastal Erosion and Sea-level Change

Beach erosion, which has been occurring naturally in many parts of Australia in geologically recent times, is an increasing problem in many areas. Erosion is expected to accelerate in the following decades if sea-level rises resulting from the enhanced greenhouse effect do eventuate (Bird, 1996).

Possible greenhouse scenarios, or outcomes for Australia, are that by 2030 AD global atmospheric warming will be between 0 and 1.5°C in the north, and between 0.5 and 2.0°C in the south. Based on 1990 sea levels, the resultant global rise in sea level by 2030 AD could be between 5 and 35 centimetres, and by 2100 AD could be between 15 and 120 centimetres. The lower estimates would cause minor problems around Australia; the higher estimates would result in inundation of low-lying areas and serious erosion of coastlines (CSIRO, 1992). Changes in patterns of rainfall and cyclones due to global warming may have a greater impact on coastal environments than sea-level rise. Coastal saltmarshes and wetlands, beach processes, coral growth and fisheries' productivity may also be affected (CSIRO, 1992).

EFFECTS OF URBAN, INDUSTRIAL AND OTHER ACTIVITIES

Urban Impacts

Although Australia's population density is the lowest of any continent other than Antarctica, and the actual urban area is very small (0.1% total area), the population is one of the most urbanised in the world, with 85% living in large cities (Zann, 1996b). Australia's population has grown rapidly over the past 20 years, mainly in the coastal zone. While the Australian population grew by 32% between 1971 and 1991 (from 12.6 million to 16.6 million), growth in the non-metropolitan coastal zone grew by 95% (from 2.1 million to 4.1 million). The growth has been most rapid in the coastal zone of Queensland (increasing from 33% to 46% of population), NSW (from 14% to 20%) and Western Australia (from 19% to 32%) (RAC, 1993).

Key environmental issues include urban sprawl (poorly planned expansion into adjacent lands of agricultural and environmental value); access to open space (for recreational, environmental and aesthetic purposes); transportation management (high dependence on private road transport, with high gaseous emissions of carbon monoxide and nitrogen oxides); sewage and waste management (water quantity and quality problems); and air, noise and water pollution resulting from the above (RAC, 1993; State of the Environment Advisory Council, 1996).

Although over 95% of urban households are sewered, most of Australia's sewage is only secondarily treated and remains high in nutrients. Each year around 10,000 tonnes of phosphorus and 100,000 tonnes of nitrogen are discharged through sewage, much of which finds its way into the marine environment. Tertiary treatment which removes nutrients is now being introduced in inland towns because of eutrophication of rivers. However, stormwater run-off from cities is also high in nutrients from animal and other wastes and may equal that generated from the city sewage (Brodie, 1996).

Marine Industries and Their Environmental Impacts

The impacts of major industries in the coastal zone are outlined here and summarised in Table 1.

Marine and Coastal Tourism

Australia's marine environment is the focus for domestic and international tourism. There are over 2.5 million overseas visitors each year and around half the visitors participate in some coastal or marine outdoor activity. Marine ecotourism is a fast-growing industry. For example, the Great Barrier Reef receives over 10 million visitor days per year, and is valued at over $1 billion p.a. As the quality of the marine environment deteriorates elsewhere in the world, Australia's significant areas of undisturbed coasts,

Table 1

Major uses of marine environment and their effects. Developed areas (urban, high density rural) and undeveloped areas (low populations, remote) are treated separately (from Zann, 1995)

Uses	Effects*	Issues
Tourism and recreation	A–C (mainly developed)	Coastal strip development; over-development; channel dredging and marinas; loss of habitat; shore and beach erosion; social impacts in communities; loss of amenity
Marine transport	C–D (developed)	Chronic spills of oil and hazardous chemicals; introductions of foreign organisms in ballast waters; occasional large spills; dredging of channels etc.
Offshore oil	A–B (undeveloped)	Operational discharges; occasional oil spills and shipping accidents
	A minus (undeveloped)	Localised effects of exploration and operational discharges. Insignificant oil spills to date but some risk of significant spills exists. Leases may affect establishment of marine protected areas
Fisheries	A–D (most areas)	Declines of significant stocks; declines in coastal catches; increasing conflicts between commercial and recreational sector in coastal fisheries; widespread serious concerns on effects of trawling on sea floor; concerns on waste of by-catch and effects on ecosystem.

*Scores. A: No apparent effects; B: Some effects (general slight effects, or few sites with moderate effects); C: Moderate effects (general moderate effects, or some sites with serious effects); D: Serious effects (general serious effects, or many very serious effects).

seas and reefs will assume even greater importance for international ecotourism (Driml, 1996).

Fisheries

Fishing is the major extractive use of the marine environment. Although fish are a renewable resource, fisheries' production of a growing number of species has been declining since the late 1980s (Kailoa et al., 1993). The Australian Fishing Zone is the third largest in the world. However, Australia's fish catches are relatively low, at about 200,000 tonnes each year, and only rank around 50th in the world. This is the result of the generally low productivity of Australia's waters because of naturally limited run-off of nutrients from land, a relatively small area of continental shelf, and the absence of major upwellings of nutrient-rich deep waters. Australia has a number of high value export fisheries such as abalone, rock lobsters and prawns, and a large cultured pearl industry. Annual exports of marine products are valued at over $1 billion per annum. Australia's commercial fishing fleet consists of almost 10,000 vessels and some 200 different species of fish, 60 species of crustaceans and 30 species of molluscs are fished (Kailoa et al., 1993; DPIE, 1996).

Australia has experienced declines in some important commercial fisheries, particularly southern bluefin tuna, southern sharks and gemfish. There are also now serious concerns that the high catches of the long-lived, deep-sea orange roughly cannot be sustained. Of 100 main fisheries, nine are considered to be overfished, 23 are fully or heavily fished, nine are under-fished, and 59 are of unknown status (Kailoa et al., 1993). Reasons for declines in some fisheries include overfishing, use of non-selective fishing gear, loss of habitat, pollution and Australia's marine jurisdictional complexity which hinders management of a fish stock or population. While it is considered that many of Australia's fisheries have not been managed in a conservative manner in the past, fisheries management is now focusing more on fisheries ecosystem management (State of the Environment Advisory Council, 1996).

Aquaculture

Australia has generally lagged in aquaculture development with the exception of oyster and pearl farming. However, aquaculture began to grow rapidly in the mid 1980s, and has increased in value from around $50 million in 1985, to $300 million in 1995. Sixty aquatic species, from seaweeds to crocodiles, are now farmed in Australia (Kailoa et al., 1993; State of the Environment Advisory Council, 1996).

Recreational Fisheries

Almost 30% of Australians go fishing for recreation each year. The recreational fishing sector's significance was not recognised until this decade but the sector remains largely unmanaged. Details on recreational fisheries remain sketchy. A 1984 survey estimated that over 4.5 million Australians went fishing at least once a year, and over 800,000 of these went fishing at least 20 times a year. Spending on fishing and related activities was estimated at $2.2 billion each year and later surveys suggest this figure is much higher. Some 100,000 people are directly employed in servicing Australia's recreational fishery. Australia's recreational fisheries are little managed, although size limits (often imposed without adequate knowledge of the life history of the species) have existed for many years. Because of increasing pressure on stocks, bag limits have been imposed in recent years. Australian recreational and commercial fishing sectors are increasingly competing for the same, often dwindling stocks (Coleman and Shorthouse, 1996).

Environmental Effects

Fishing has had direct and indirect effects on marine ecosystems. In some fisheries, particularly those using trawl and gill nets, large numbers of other species, termed the 'by-catch', are also taken but discarded dead. In 1997, gillnetting was attributed to an alarming decline in the globally endangered dugong in the southern Great Barrier Reef. Steady declines in green and loggerhead turtles in northern Australia have been attributed in part to trawling. Australia has been slow in the introduction of by-catch excluders in trawl nets but they are now to be phased in over the next few years in Queensland (State of the Environment Advisory Council, 1996).

Fish Kills

Fish diseases are also a significant problem. Introduced toxic marine dinoflagellates, thought to have been introduced in ship's ballast waters, are responsible for periodic fish and shellfish poisoning in southern Australia (Hallegraeff, 1996). Runoff from disturbed acid soils in coastal areas of northern and eastern Australia have resulted in periodic acidification of estuaries, and massive fish kills and diseases (Anon., 1998). In 1995 a major epidemic caused by a herpes virus killed millions of pilchards along thousands of kilometres of coast in southern Australia and a second mass mortality began in October 1998.

Shipping

Australia is an isolated island continent with a long coastline and shipping is a major economic use of its seas, estuaries and coastlines. In terms of tonnage carried and distance travelled, Australia ranks as the fifth largest user of shipping in the world. Each year there are around 12,000 overseas shipping arrivals and almost 380 million tonnes of freight carried in Australian waters (Raaymakers and Zann, 1996).

Shipping and port operations produce a variety of environmental impacts such as pollution from oil and from hazardous cargoes; tributyl tin from antifouling paints; litter and sewage; and introductions of exotic or foreign organisms in ships' ballast waters or attached to ships' hulls. Australia has 68 main ports. The impacts of ports include loss of habitat from reclamation and dredging; physical alteration of coastal environments; increased sediment from dredging; and pollution by oil, toxic chemicals, litter and antifouling paints. While ports are amongst the most disturbed marine environments in Australia, technical, engineering and management solutions do exist to prevent or minimise many of their environmental impacts (Raaymakers and Zann, 1996).

Oil Spills

Australia has been fortunate in that there have been only three large spills (over 1000 tonnes), the *Oceanic Grandeur* in Torres Strait and the *Kirki* off Western Australia, and *Iron Monarch* off Tasmania in 1995 but none of these appeared to result in lasting environmental damage (Raaymakers and Zann, 1996). The risk of a major spill from shipping around Australia is considered to be high. It has been estimated that the risk of a major spill from shipping was 49% in any five year period, and 84% in any 20-year period (BTCE, 1991).

Offshore Petroleum Exploration and Production

Offshore petroleum production is of considerable economic and strategic importance to Australia. Some 72% of Australia's liquid fuels and natural gas comes from offshore wells in Bass Strait, the Timor Sea and the North West Shelf. Current production of oil is over 3.5 million barrels per year (Beck, 1996). Over the past 30 years more than 1100 wells have been drilled offshore and around 2800 million barrels of oil have been extracted (Swan et al., 1993).

The Australian offshore petroleum industry has a very good environmental record and only about 800 barrels of oil have been spilt since drilling began (Swan et al., 1993). However, a risk study estimated that the theoretical probability of a major spill (that is, greater than 1000 tonnes) from offshore petroleum sources was around 26% in any five year period, and the risk of a pipeline spill was around 17% for the same period (BTCE, 1991). Oil drilling is prohibited in marine protected areas such as the Great Barrier Reef World Heritage Area and in Ningaloo Marine Park. In the Australian Antarctic Territory, any activity relating to mineral resources, other than scientific research, is prohibited (Beck, 1996).

Marine Pollution and Other Issues

Because of the very low human populations around most of Australia's coastline, the levels of marine pollution are, on average, very low. However, pollution around the major cities and industrial areas may be locally severe. The status of pollutants is summarised in Table 2.

Hydrocarbons

In Australia the greatest sources of marine oil pollution are terrestrial outflows from sewage systems and drains, which are estimated at 16,000 tonnes a year, and operational discharges from shipping (Swan et al., 1993). The largest number of spills at sea result from accidents during fuelling of vessels in ports (Raaymakers and Zann, 1996). Waters from areas in Port Phillip Bay and Western Port (Vic), and the Brisbane (Qld), Parramatta (NSW) and Yarra Rivers (Vic) have been contaminated from urban run-off, sewage and petrochemical industries. Sediment concentrations similarly range from background levels in remote areas (e.g. on the Great Barrier Reef) to localised contamination in urban areas (e.g. in Yarra, parts of Western Port) (Connell, 1996).

Heavy metals

There are isolated 'hot spots' of heavy metal contamination in sediments off some State capitals and industrial areas (e.g. Derwent River (Tasmania); Port Pirie (South Australia), Lake Macquarie (New South Wales (Batley, 1996)). However, significant advances have been made in reducing the problem through implementation of water quality guidelines and effluent controls and heavy metals now pose only localised problems, or potential problems. There are naturally high levels of some heavy metals (e.g. cadmium) off northern Australia (State of the Environment Advisory Council, 1996).

Tributyl tin (TBT) has been widely used in Australia as a poison in antifouling paints since the 1970s. During the 1980s TBT was found to affect the growth of oysters and other molluscs and levels of concentration in Australian dockyards and marinas were frequently 50 times the guidelines. It has been banned in most vessels under 25 m in length and levels have dropped in most areas to below the guidelines (Batley, 1996).

Organochlorines

A range of organochlorine compounds have been very widely used in Australia as herbicides (e.g. 2,4-D, 2,4,5-T), insecticides (e.g. DDT, lindane, chlordane), fungicides (e.g. hexachlorobenzene, chlorinated phenyls) and as insulating fluids (e.g. poly-chlorinated biphenyls, or PCBs). Little is known of levels in Australia's marine environment. PCBs have been detected in offshore waters of Australia, with increasing levels towards the coast, but at lower levels than in equivalent Atlantic waters. PCBs have been detected at very low levels on the Great Barrier Reef and Brisbane River, and at higher levels off Sydney ocean sewage outfalls and off Melbourne (Richardson, 1996).

Ocean and Beach Litter

Australia's beaches are increasingly littered with plastic bottles, plastic bags, tangled fishing lines, nets and other rubbish. Litter comes from 'tourist trash' left by beach-goers or is washed there. The latter comes from land litter washed from catchments and stormwater drains, from ships' garbage, from discarded fishing gear from anglers and fishing boats, and from remote sources far across the ocean. Urban beaches are worst affected, but even the most remote coastal and island beaches are not free from litter (Wace, 1996).

Litter not only reduces aesthetic values, but may also endanger marine life. In Australia the incidence of entanglement of fur seals in net fragments and other litter is alarmingly high. Surveys indicate that at any one time, around 500 seals off Tasmania, and 45 seals off Victoria have 'collars' of plastic litter (Marsh et al., 1996).

Introduced Species

Introduced or exotic species of plants and animals have had a catastrophic effect on Australia's unique terrestrial environment. While impacts of exotic organisms on the

Table 2

Issues in Australia's marine environment. Developed areas (urban, high density rural), and undeveloped areas (low populations, remote) are treated separately. (A: low, not detectable, to D: severe, very high)

Issue	Level	Sources and effects
Coastal strip development	C–D (developed)	Decline in water quality; loss of habitat from reclamations; dredging, marinas; erosion of beaches; over-development; conflicts amongst users; loss of amenity; poor integrated planning
Nutrients (nitrogen, phosphorus)	C–D (developed)	Source: inland waters; developed sewage and run-off. Effects: widespread locally serious eutrophication; algal blooms; toxic algae; die-back of temperate seagrass; loss of bathing water quality
	B–C (undeveloped)	Source: inland waters (erosion, fertilisers etc.) Effects: locally low to moderate eutrophication; occasional algal blooms; die-back of temperate seagrass; threats to nearshore corals
Hydrocarbons	B–C (developed)	Source: developed run-off; chronic small spills while fuelling in harbours; occasional large shipping spills Effects: chronic effects not known; large spills cause concentrated, moderate to long term environmental damage
	A–B (undeveloped)	Source: operational discharges from ships, occasional spills Effects: large spills as above
Heavy metals (mercury, cadmium, copper etc)	B–D (developed)	Source: localised industrial discharges; refineries, antifouling paints; sewage; mining ports. A few areas seriously contaminated (e.g. Derwent Estuary, Tasmania; Port Pirie, SA) Impacts little known; may be very toxic; bioaccumulate in food chains; contamination of seafood
	A–B (undeveloped)	Source: natural concentrations and mining (e.g. Torres Strait) Effects: little known; possible bioconcentration in food chain
Chlorinated compounds (pesticides etc)	B–C? (developed)	Source: industrial discharges; dumps; developed run-off, sewage
	A–B (undeveloped)	Effects: little known; bioaccumulates in food chains; may be very toxic Source: agricultural chemicals; dumps; distant sources Effects: as above
Litter	B–C (developed)	Source: local catchments; dumps; visitors; fishers; shipping Effects: loss of visual amenity; ingestion by wildlife
	A–B (undeveloped)	Source: vessels; fishers; visitors Effects: as above. Locally serious in seal populations
Sewage	C–D (developed)	Source: primary treatment; ocean sewage outfalls Effects: pathogenic microorganisms; bathing water standards frequently exceeded in some developed beaches; important source of nutrients (above)
	A–B (undeveloped)	Localised problem
Introduced species	B (potentially A)	Currently localised problems (e.g. North Pacific sea star in eastern Tasmania) but potentially very widespread and serious
Population outbreaks	B (?)	Currently localised serious problems (e.g. crown-of-thorns starfish on the Great Barrier Reef) but not known whether they are natural or human-induced.

marine environment appear to have been relatively minor in the past, there is now great concern about the threats posed by the introductions of exotic marine species via ships' ballast waters (Lehane and Arundell, 1996).

At least 55 species of fish and invertebrates and a number of seaweeds have been introduced into Australia either intentionally, for aquaculture, or accidentally, in ships' fouling and ballast waters. Six species are regarded as pests. Principal organisms of concern are the toxic alga *Gymnodinium catenatum*, which causes red tides, the seaweed *Undaria pinnatifida*, which smothers native kelps, the Northern Pacific seastar *Asterias amurensis*, and fish pathogens such as *Myxosoma cerebralis* (Lehane and Arundell, 1996).

PROTECTIVE MEASURES

Marine environmental conservation in Australia involves a large number of international, regional, Commonwealth, State and Territory, and local government agreements, arrangements and agencies and involves a large number of different management strategies (RAC, 1993).

International Arrangements and Responsibilities

Australia is a party to numerous international treaties, conventions and agreements that are relevant to the coastal zone. Many of these deal with general global issues such as climate change, biological diversity, and world heritage (Herriman et al., 1997). In 1993 Australia signed the United Nations Law of the Sea Convention which allows nations to claim six zones including territorial seas, (which extend 12 nautical miles from the coastal baseline), and a 200 nautical mile exclusive economic zone (EEZ). Australia's EEZ came into force on the 16 November 1994 (Herriman et al., 1997).

Government Responsibilities in Ocean Management

Australia has a three-tiered system of government, consisting of Commonwealth, State/Territory (7), and Local (almost 1000) Governments. Management of the coastal and marine environments involves each sector. The legislative basis for planning and management of the land area of the coastal zone is primarily provided by the States. State and Local Governments are generally responsible for day-to-day decision-making in the coastal zone. States generally have primary jurisdiction over marine areas to three nautical miles from baselines (except in the Great Barrier Reef, and the Commonwealth from there to the 200 nautical mile line (Bergin et al., 1996).

The Offshore Constitutional Settlement established jurisdictions between the Commonwealth and States over marine areas. There are 'agreed arrangements' on the management of oil and gas, and other seabed minerals, the Great Barrier Reef Marine Park, other marine parks, historic shipwrecks, shipping, marine pollution and fishing (Bergin et al., 1996). However jurisdictional problems remain serious and many government inquiries have identified the fragmented and often duplicatory responsibilities in the coastal zone as severe impediments to effective planning and management (Zann and Earley, 1997).

General Marine Environmental Management Strategies

The main strategies for marine environmental management in Australia include:
- maintaining water quality through controlling disposal of wastes and emissions entering catchments, the atmosphere and the sea;
- prohibiting or regulating destructive and unsustainable activities;
- protecting important habitats and areas;
- zoning for particular uses to separate and control incompatible uses;
- requiring environmental impact studies to minimise effects of developments;
- protecting certain organisms, particularly vulnerable and threatened species; and
- regulating fisheries through licences, size limits, quotas (total allowable catches), seasons and other mechanisms (Zann, 1995).

Marine Protected Areas

Marine protected areas (MPAs) are a very important 'catch all' tool for marine conservation and management, particularly in protecting biodiversity, and achieving sustainable use of marine resources. Australia is a world leader in using MPAs for marine conservation and management and has around 310 MPAs, totalling around 0.75 million km^2. This is 24% of the world total number of MPAs. Over 7% of Australia's marine environment is protected in MPAs, a very large proportion of this is within a single MPA, the Great Barrier Reef Marine Park. However while the number of MPAs has increased by over 60% in the past decade, large sections of Australia's marine environment still have few or no MPAs (Bleakley et al., 1996).

Australia has widely adopted the concept of large, multiple-use marine protected areas intended to operate at the large marine ecosystem level, conserve biodiversity in core areas, and allow sustainable resource use. These include: Torres Strait Protected Zone (between Queensland and New Guinea); Great Barrier Reef Marine Park (off Queensland; area: 360,000 km^2; Coringa-Herald and Lihou National Nature Reserves (Coral Sea; areas: 8856 km^2 and 8436 km^2 (respectively)); Elizabeth and Middleton National Nature Reserve (Northern Tasman Area: 1880 km^2); Solitary Islands Marine Reserve (northern New South Wales; area 1000 km^2); Jervis Bay Marine Park (southern New South Wales); Great Australian Bight Marine Park (off South Australia); Ningaloo Marine Park (off Western Australia; area 4500 km^2); Shark Bay Marine Park (off Western Australia; area 7487 km^2); Hamelin Pool Marine Nature Reserve (off Western Australia; area 1320 km^2); Ashmore Reef National Nature Reserve (off Western Australia; area 583 km^2); Mermaid Reef National Nature Reserve and Rowley Shoals Marine Park (off Western Australia; area 232.5 km^2); and Australian Antarctic Territory (6 million km^2); and Territory of Heard Island and McDonald Island (declared 1998).

Australia's Oceans Policy

The lack of a national policy for Australia's oceans has been a major impediment to their sustainable use (Zann, 1995). In 1998 the Commonwealth Government announced a comprehensive and integrated Oceans Policy for Australia as a means of asserting sovereign rights over Australia's EEZ and as a critical step in ensuring the ecologically sustainable development of ocean resources both for wealth creation and environmental protection (CoA, 1998).

The Oceans Policy provides the strategic framework for the planning management and ecologically sustainable development of fisheries, shipping, petroleum, gas and seabed resources, while ensuring the conservation and protection of the marine environment. The Policy complements existing and planned programs and policies, for example, the Commonwealth Coastal Policy; the National Strategy on Ecologically Sustainable Development; the National Biodiversity Strategy; and the Marine Science and Technology Plan (CoA, 1998).

The Oceans Policy particularly addresses the cross-jurisdictional and cross-sectoral issues which have previously plagued marine environmental management in Australia. It includes a national representative system of marine protected areas and focuses on reducing diffuse sources of marine pollution, particularly from agricultural run-off. It focuses also on sustainable fisheries and increased involvement of the fishing industry in reducing fishing by-catch (CoA 1998).

CONCLUSIONS: STATUS OF MARINE ENVIRONMENT AND MAJOR ISSUES

The State of the Environment Report for Australia (Zann, 1996) and State of the Environment Advisory Council (1996) concluded that Australia's marine and estuarine environments and habitats are generally in good condition, to the extent that this can be measured. However, only a few areas can be regarded as pristine, because of wide-ranging pressures such as nutrient loading, pollution with persistent chemicals and fishing, which have some effect on nearly all parts of Australia's marine and estuarine systems. Near many of the cities and in other localised inshore areas, the condition of some habitats is poor. Of particular concern are the coral reefs of the north-east coast and the temperate seagrass beds of southern Australia. However, for many systems, particularly those offshore or distant from population centres, we do not have enough information to make even a first estimate of their state of health. Remote areas are assumed to be in good health because they face few pressures, but the data to confirm this are scarce.

REFERENCES

Anon (1998) Draft national strategy for the management of acid sulfate soils. National Working Party on Acid Sulfate Soils, Department of Primary Industries and Energy, Canberra.

Batley, G.E. (1996) Heavy metals and organometallic. In *The State of the Marine Environment Report for Australia. Technical Summary*, ed. L.P. Zann. Great Barrier Reef Marine Park Authority and Department of the Environment, Sport and Territories, Canberra, pp. 250–253.

Beck, J. (1996) Offshore petroleum exploration and production. In *The State of the Marine Environment Report for Australia. Technical Summary*, ed. L.P. Zann. Great Barrier Reef Marine Park Authority and Department of the Environment, Sport and Territories, Canberra, pp. 209–214.

Bergin, A., Prescott, V. and Haward, M. (1996) Australia's maritime zones, international borders, and intergovernmental arrangements. In *The State of the Marine Environment Report for Australia. Technical Summary*, ed. L.P. Zann. Great Barrier Reef Marine Park Authority and Department of the Environment, Sport and Territories, Canberra, pp. 341–347.

Bird, E. (1996) Coastal modifications and impacts. In *The State of the Marine Environment Report for Australia. Technical Summary*, ed. L.P. Zann. Great Barrier Reef Marine Park Authority and Department of the Environment, Sport and Territories, Canberra, pp. 226–232.

Bleakley, C., Ivanovici, A. and Ottesen, P. (1996) An assessment of marine protected areas in Australia. In *The State of the Marine Environment Report for Australia. Technical Summary*, ed. L.P. Zann. Great Barrier Reef Marine Park Authority and Department of the Environment, Sport and Territories, Canberra, pp. 403–414.

Brodie, J. (1996) Nutrients and eutrophication in coastal waters. In *The State of the Marine Environment Report for Australia. Technical Summary*, ed. L.P. Zann. Great Barrier Reef Marine Park Authority and Department of the Environment, Sport and Territories, Canberra, pp. 238–254.

BTCE (1991) Major marine oil spills risk and response, Report 70, Bureau of Transport and Communications Economics, Australian Government Publishing Service.

CoA (1998) Australia's oceans policy. Department of the Environment, Canberra.

Coleman, G. and Shorthouse B. (1996) Recreational fisheries and the catch-sharing issue. In *The State of the Marine Environment Report for Australia. Technical Summary*, ed. L.P. Zann. Great Barrier Reef Marine Park Authority and Department of the Environment, Sport and Territories, Canberra, pp. 188–192.

Connell, D.W. (1996) Hydrocarbons in the marine environment. In *The State of the Marine Environment Report for Australia. Technical Summary*, ed. L.P. Zann. Great Barrier Reef Marine Park Authority and Department of the Environment, Sport and Territories, Canberra, pp. 247–249.

CSIRO (1992) Climate change scenarios for the Australian region, CSIRO Division of Atmospheric Research, Aspendale.

DAHE (1991) Australian national report to the United Nations Conference on Environment and Development, Department of Heritage and Environment, AGPS, Canberra.

DPIE (1996) Fishery status reports, AGPS, Canberra.

Driml, S. (1996) Coastal and marine tourism and recreation. In *The State of the Marine Environment Report for Australia. Technical Summary*, ed. L.P. Zann. Great Barrier Reef Marine Park Authority and Department of the Environment, Sport and Territories, Canberra, pp. 159–165.

Federal Court of Australia (1998) Mary Yarrmirr & Ors v The Northern Territory of Australia & Ors. 771 FCA. http://www.austlii.edu.au

Galloway, R.W. (1981) An inventory of Australia's coastal lands. *Australian Geographical Studies* **19**, 107–116.1.

Hallegraff, G.M. (1996) Phytoplankton: the pastures of the sea. In *The State of the Marine Environment Report for Australia. Technical Summary*, ed. L.P. Zann. Great Barrier Reef Marine Park Authority and Department of the Environment, Sport and Territories, Canberra, pp. 66–69.

Harris, P.T. (1996) The sea floor. In *The State of the Marine Environment Report for Australia. Technical Summary*, ed. L.P. Zann. Great Barrier Reef Marine Park Authority and Department of the Environment, Sport and Territories, Canberra, pp. 15–19.

Herriman, M., Tsamenyi, M., Ramli, J. and Bateman, S. (1997). Australia's Oceans policy. International Agreements. Background Paper 2. Environment Australia, Canberra.

IMCRA (1998) Interim marine and coastal regionalisation for Australia: an ecosystem-based classification for marine and coastal environments. Version 3.3. Environment Australia, Department of Environment, Canberra.

Kailoa, P.J., Williams, M., Stewart, P.C., Reichelt, R., McNee, A. and Grieve, C. (1993) Australian fisheries resources, Bureau of Resource Sciences and Fisheries Research and Development Corporation, Canberra.

Lehane, L. and Arundell, L. (1996) Exotic marine species and the ballast water problem. In *The State of the Marine Environment Report for Australia. Technical Summary*, ed. L.P. Zann. Great Barrier Reef Marine Park Authority and Department of the Environment, Sport and Territories, Canberra, pp. 268–274.

McKinnon, K. (1989) Oceans of wealth? Department of Industry, Technology and Commerce, AGPS, Canberra.

Marsh, H., Cockeron, P.J., Limpus, C.J., Shaughnessy, P.D., and Ward, T.M. (1996) The status of marine reptiles and mammals. In *The State of the Marine Environment Report for Australia. Technical Summary*, ed. L.P. Zann. Great Barrier Reef Marine Park Authority and Department of the Environment, Sport and Territories, Canberra, pp. 88–97.

Middleton, J.H. (1996) The ocean. In *The State of the Marine Environment Report for Australia. Technical Summary*, ed. L.P. Zann. Great Barrier Reef Marine Park Authority and Department of the Environment, Sport and Territories, Canberra, pp. 9–14.

Nicholson, A. and Cane, S. (1996) Pre-European use of the sea by Aboriginal people: an archaeological perspective. In *The State of the Marine Environment Report for Australia. Technical Summary*, ed. L.P. Zann. Great Barrier Reef Marine Park Authority and Department of the Environment, Sport and Territories, Canberra, pp. 99–103.

Poore, G. (1996a) Biogeography of Australia's marine biota. In *The State of the Marine Environment Report for Australia. Technical Summary*, ed. L.P. Zann. Great Barrier Reef Marine Park Authority and Department of the Environment, Sport and Territories, Canberra, pp. 25–28.

Poore, G. (1996b) Deepwater communities. In *The State of the Marine Environment Report for Australia. Technical Summary*, ed. L.P. Zann. Great Barrier Reef Marine Park Authority and Department of the Environment, Sport and Territories, Canberra, pp. 64–65.

Raaymakers, S. and Zann L. (1996) Shipping and ports. In *The State of the Marine Environment Report for Australia. Technical Summary*, ed. L.P. Zann. Great Barrier Reef Marine Park Authority and Department of the Environment, Sport and Territories, Canberra, pp. 202–208.

RAC (1993) Coastal Zone Inquiry—Final Report, Resource Assessment Commission, Australian Government Publishing Service, Canberra.

Richardson, B. (1996) Organochlorines. In *The State of the Marine Environment Report for Australia. Technical Summary*, ed. L.P. Zann. Great Barrier Reef Marine Park Authority and Department of the Environment, Sport and Territories, Canberra, pp. 255–257.

Smyth, D. (1996) Aboriginal maritime culture. In *The State of the Marine Environment Report for Australia. Technical Summary*, ed. L.P. Zann. Great Barrier Reef Marine Park Authority and Department of the Environment, Sport and Territories, Canberra, pp. 105–115.

State of the Environment Advisory Council (1996). The State of the Environment Australia. 1996. Department of the Environment, Sport and Territories, Canberra.

Swan, J.M., Neff, J. and Young, P.C. (eds.) (1993) Environmental implications of oil and gas development in Australia—the findings of an independent scientific review, Australian Petroleum Exploration Association, Sydney,

Wace, N. (1996) Ocean and beach litter. In *The State of the Marine Environment Report for Australia. Technical Summary*, ed. L.P. Zann. Great Barrier Reef Marine Park Authority and Department of the Environment, Sport and Territories, Canberra, pp. 258–263.

Zann, L.P. (1995) Our sea, our future. The State of the Marine Environment Report for Australia. Department of Environment, Sport and Territories, Commonwealth of Australia. 112 pp.

Zann, L.P. (1996a). *The State of the Marine Environment Report for Australia. Technical Summary*. Department of Environment, Sport and Territories, Commonwealth of Australia. 515 pp.

Zann, L.P. (1996b). The coastal zone: an introduction. In *The State of the Marine Environment Report for Australia. Technical Summary*, ed. L.P. Zann. Great Barrier Reef Marine Park Authority and Department of the Environment, Sport and Territories, Canberra, pp. 2–8.

Zann, L.P. (1996c) The status of marine environmental science in Australia and a critical review of its contribution to management. In *The State of the Marine Environment Report for Australia. Technical Summary*, ed. L.P. Zann. Great Barrier Reef Marine Park Authority and Department of the Environment, Sport and Territories, Canberra, pp. 371–379.

Zann, L.P. and Oliver, J. (1996) Marine environmental monitoring, information management and reporting. In *The State of the Marine Environment Report for Australia. Technical Summary*, ed. L.P. Zann. Great Barrier Reef Marine Park Authority and Department of the Environment, Sport and Territories, Canberra, pp. 380–387.

Zann, L.P. and Earley, O. (1997). Australia's Oceans Policy. Background Paper 4. Reviews and recommendations. Department of Environment, Commonwealth of Australia.

THE AUTHOR

Leon P. Zann
*Southern Cross University,
Lismore, New South Wales 2480,
Australia*

Chapter 92

TORRES STRAIT AND THE GULF OF PAPUA

Michael E. Huber

The very shallow Torres Strait separates Australia and Papua New Guinea (PNG). The Gulf of Papua, to the northeast, is the northern extremity of the Coral Sea. The major physical driving forces affecting the area are strong tidal currents through Torres Strait and very large inputs of freshwater and sediment from rivers flowing into the Gulf of Papua. These rivers form large deltas along the Gulf of Papua.

Torres Strait has about 600 fringing and patch coral reefs, which are characterised by an abundance of soft sediment on the reef tops. Reefs at the easternmost edge of the Strait represent the northern extension of the Great Barrier Reef and are less affected by sediment inputs. There are approximately 300,000 ha of relatively undisturbed mangrove forest in the deltas. Mangroves also occur on the Torres Strait islands, especially the northern ones, and in a fringe along the PNG coast west of Daru. There are large areas of seagrass in Torres Strait, both on reef tops and in inter-reefal areas. Soft-bottom benthic communities are found in the Gulf of Papua. Torres Strait has a variety of hard- and soft-bottom inter-reefal benthic communities, determined in large part by tidal current stress.

The area has a total human population of about 60,000, mostly rural. Western Province has a high population growth rate, while out-migration restrains population growth in Gulf Province. Population growth in Torres Strait is concentrated on the outer islands. Torres Strait Islanders, as Australian citizens, enjoy a dramatically higher living standard than people living on the adjacent coastline of PNG, but a low one by Australian standards. Government allowances, government-related jobs, remittances from relatives, and artisanal fishing are the main sources of income in Torres Strait, where subsistence fishing is still an important source of food. The rural population in PNG lives a largely subsistence lifestyle; government jobs, remittances from relatives, and artisanal fishing are the main sources of cash incomes.

Development pressure in the area is low, and the coastal and marine habitats remain largely intact, except for localised degradation around towns. The disposal of mine tailings into the Fly River system has caused concern, but studies reveal no adverse effects on coastal and marine environments. Upstream logging and agricultural development are potentially significant threats to the mangals of the Gulf of Papua deltas. There is potential for oil spills from a pipeline, offshore loading facility, and tanker operations.

The bilaterally managed Torres Strait Protected Zone, established in part to provide for sustainable fisheries, encompasses most of Torres Strait. The main commercial fisheries, for prawns and rock lobster, appear to be sustainable under present conditions. Torres Strait stocks

Seas at The Millennium: An Environmental Evaluation (Edited by C. Sheppard)
© 2000 Elsevier Science Ltd. All rights reserved

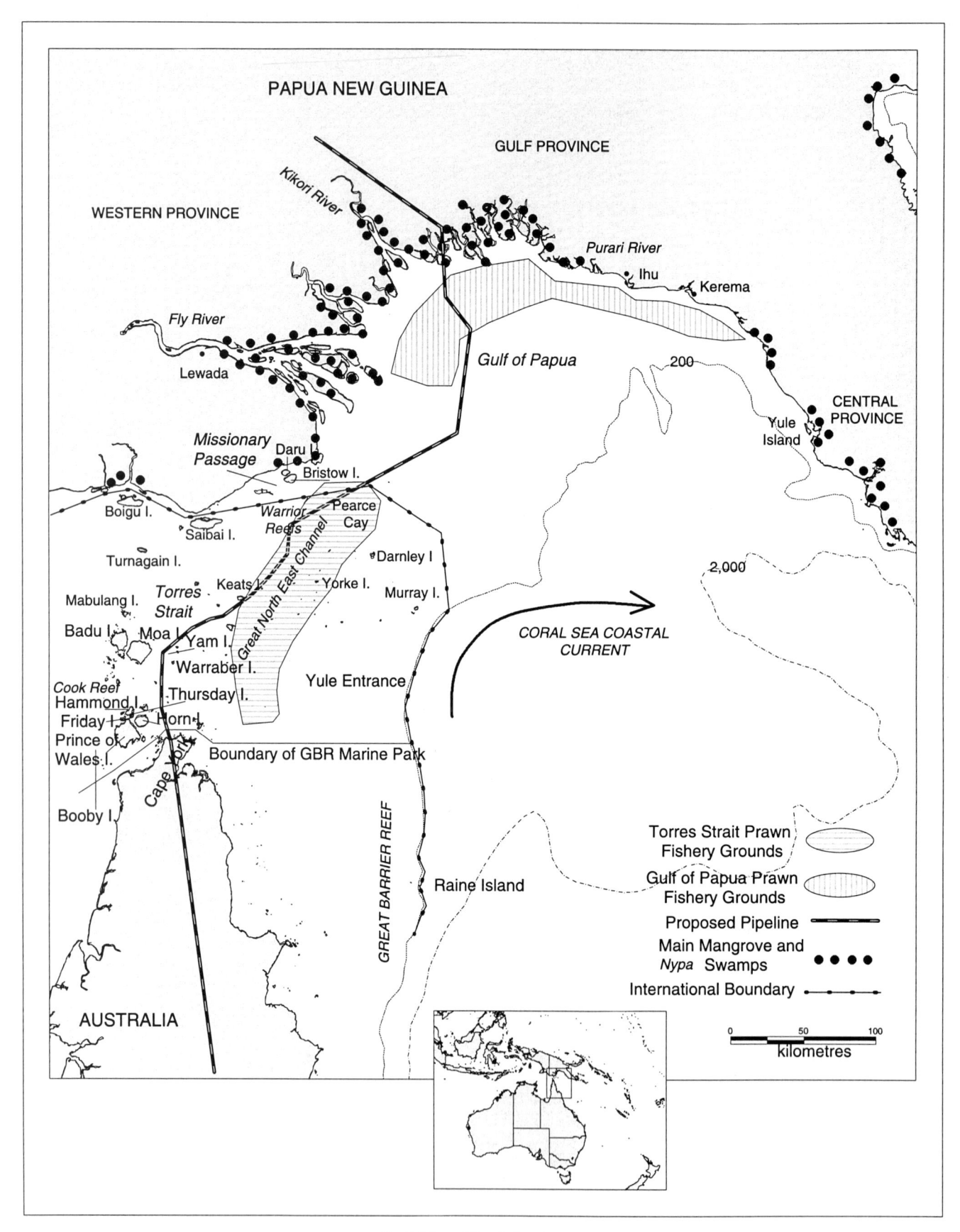

Fig. 1. Map of the defined area.

of pearl oysters, bêche-de-mer, and trochus were severely overfished in the past, and bêche-de-mer and barramundi have been severely overfished in PNG in recent years. These fisheries have for the most part not recovered. Other fisheries stocks are probably healthy, but this results from a lack of development more than effective management. The sustainability of traditional harvests of sea turtles and dugong is uncertain.

THE DEFINED REGION

Torres Strait is a shallow area of submerged continental shelf separating Cape York, Australia, from New Guinea (Fig. 1). The Gulf of Papua lies to the northeast. The boundary of the area described in this chapter has been set as the 20 m depth contour on the west side of Torres Strait. The southern boundary includes the islands immediately offshore of Cape York, but not Cape York itself; continuing eastward, the southern boundary corresponds to the northern boundary of the Great Barrier Reef Marine Park at 10°41'S. The eastern boundary is the 200 m depth contour, marking the upper continental slope, which extends in an arc to the northeast. The boundary extends south from the border of Gulf and Central Provinces, but there are references in the text to Yule Island, which lies in Central Province. No precise northern boundary has been defined. The chapter concentrates on coastal and estuarine areas but the watersheds draining into the Gulf of Papua are considered in parts of the discussion.

Torres Strait is about 150 km wide and contains a highly complex system of more than 100 islands, about 600 reefs, and many shoals and channels. The western and central Strait is less than 20 m deep. The eastern Strait is somewhat deeper , but still generally less than 60 m. The Strait has formed a land bridge between Australia and New Guinea during times of low sea level, resulting in high biogeographic affinities among the biota in and adjacent to the Strait (Barham and Harris, 1983). In broad terms the islands can be subdivided into a southwestern group of granitic and volcanic high islands, a northwestern group of mud and peat islands overlying coralline platforms, a central group of low-lying sand islands, and an eastern group of basic volcanic origin (Barham and Harris, 1983). Torres Strait is administratively part of the State of Queensland.

The mainland area of Papua New Guinea (PNG) considered in this chapter comprises the coastal areas of Western and Gulf Provinces. Unlike most of PNG, Western Province is predominantly lowland (<300 m) with little relief; there are large areas of grassland. From the mouth of the Fly River east to Ihu is a huge delta system, formed by the inputs of a series of large rivers. Delta development gradually diminishes east of Ihu due to lower rainfall and reduced river input. Upstream of the deltas are large tracts of lowland rainforest.

The area falls under the jurisdictions of Australia and PNG. The international border in Torres Strait extends to within 3 km of mainland PNG and turns to the south in the Gulf of Papua (Fig. 1). A bilateral agreement, the Torres Strait Treaty, has established jurisdictional boundaries that do not necessarily coincide with the border and there are provisions for co-management of some resources.

The Torres Strait and Gulf of Papua do not have typical characteristics of an LME in terms of being largely self-contained within well-defined geographic boundaries. Torres Strait is itself a boundary, between New Guinea and Australia to the north and south and between the Coral and Arafura Seas to the east and west. Furthermore, the reefs at the eastern edge of the continental shelf in Torres Strait are geologically and biogeographically the northern extension of the Great Barrier Reef (GBR); the boundary of the GBR Marine Park is an administrative rather than natural one. Similarly, there is no clear distinction between the Gulf of Papua and the Coral Sea, and the 200 m isobath used herein is purely arbitrary.

Nonetheless, for management purposes many aspects of the LME concept are applicable. As described elsewhere in the chapter, the physiographic characteristics of the area, while driven by outside forces, are largely determined by the area's internal characteristics. Tidal circulation, for example, is driven by the tidal regimes of adjacent seas, but the resultant currents are determined primarily by local topography. More importantly, the reproductive cycles, migrations, and other life history of much of the biota, including the species most important to fisheries, take place largely, if not entirely, within the defined area.

SEASONALITY, CURRENTS, NATURAL ENVIRONMENTAL VARIABLES

Average annual rainfall is around 2000 mm in Torres Strait and coastal areas of PNG as far east as Daru (McAlpine et al., 1983). This rainfall is highly seasonal, with 80% or more occurring during the wet season of December to April. The dry season, from May to November, is dominated by strong, steady, southeasterly trade winds. Lighter and more variable northwesterlies predominate during the wet season. Coastal areas of Gulf Province experience higher rainfall (generally 3–4000 mm/yr, in the Kikori area over 6000 mm/yr). There is much less seasonality, though unlike the adjacent coastline to either side the area from the mouth of the Turama River east to about Bereina (Central Province) receives its rainfall maximum during May to August, and rainfall is heavy throughout the year (McAlpine et al., 1983). Cyclones occasionally pass through Torres Strait, but are relatively rare; two weak cyclones transited the Strait in the period 1969–1997.

Currents in Torres Strait, which constitutes the junction of the tide regimes of the Coral Sea and Arafura Sea/Gulf of Carpentaria systems, are predominantly tide-driven. These tidal regimes, both dominated by the semidiurnal lunar (M_2) component, are effectively 180° out of phase across the Strait, with large phase differences occurring over very small distances. The M_2 component, for example, is about 120° out of phase between Twin and Booby Islands, which are separated by only 55 km (Bode and Mason, 1995). This and the large (4 m) tidal range results in extremely strong tidal currents through the shallow, narrow channels. Tidal streams peak at up to 8 knots, for example, in the Prince of Wales channel. Tidal circulation is strongly affected by the many islands and reefs in the Strait, and by the secondary influences of wind and oceanic circulation. The resulting current pattern, which has been modeled by Bode and Mason (1995) is extremely complex. Net transport through the Strait, however, is very low; water in the Strait essentially moves back and forth with the tides (Wolanski et al., 1988, 1995a).

The strong tidal flows have a strong influence on the biota of Torres Strait. Seismic data indicate that tidal currents control modern growth of reefs in the southern Strait, which have developed as long, narrow platforms parallel to the east–west tidal flow through a process of patch reef formation and sediment deposition in the current lees at the reef ends (Jones, 1995). Long et al. (1997a) found that seabed current stress calculated from tidal current models is positively correlated with the density of sessile epibenthos, and successfully tested the predictive power of the relationship. The presence of hard (rock and rubble) substrata is also positively correlated with epibenthos density, independent of the positive correlation between hard substrata and current stress.

The Coral Sea Coastal Current is the dominant forcing agent of circulation in the Gulf of Papua, and appears to generate a net counter-clockwise eddy in the Gulf of Papua (Wolanski et al., 1995a) (Fig. 2a). Seasonal wind patterns, however, have significant influences on nearshore surface currents (Fig. 2b).

Estimates of freshwater and sediment inputs to the Gulf of Papua vary widely, but by any estimate are very high. Wolanski et al. (1995a) estimate the total freshwater input to be about 15,000 m^3 s^{-1}, with little seasonal variation. In terms of water flow, the Fly is the world's tenth-largest river, and three other rivers flowing into the Gulf, the Purari, Kikori, and Bamu, are in the world's top 25 (Table 1). The Turama, Vailala, and Era are also large rivers.

This high sediment discharge is a major determinant of the area's geomorphology, particularly the vast series of deltas in the Gulf of Papua. The line of islands along the PNG coast east of the Fly river mouth was formed by the

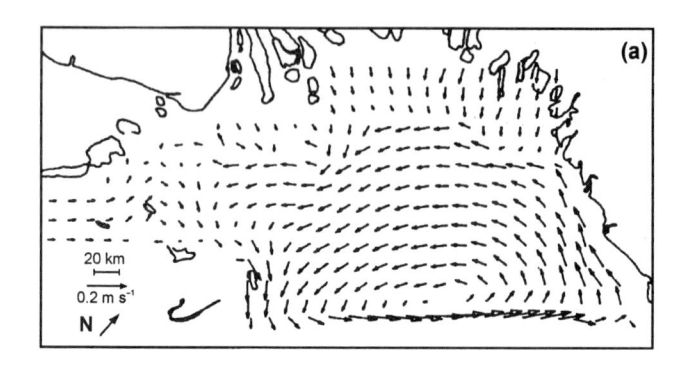

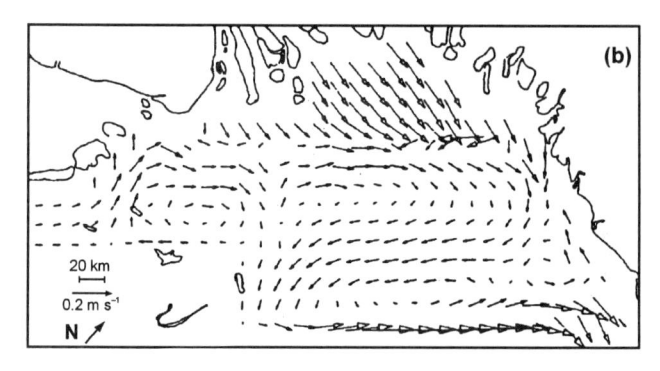

Fig. 2. Model predictions of surface currents in the Gulf of Papua under conditions of (a) no wind and (b) steady southeasterly trade winds (from Wolanski et al., 1995a).

deposition of sediment on pre-existing reef platforms (Barham and Harris, 1983). The muddy reefs in central and western Torres Strait are also strongly affected by sediment.

Surface currents trap low-salinity (<30 ppt) surface water against the PNG coast for much of the year (Harris et al., 1993; Wolanski et al., 1995a), and surface salinities in even the offshore regions of the gulf are generally less that 34 ppt (Robertson et al., 1990; Wolanski et al., 1984). The entire Gulf is salinity-stratified in the upper 20 m. Residence time for freshwater runoff in the Gulf is about 2 months (Wolanski et al., 1995a). A semi-permanent tongue of reduced-salinity water intrudes from the mouth of the Fly southwest through the Bligh Entrance into Torres Strait,

Table 1

Estimates of water and sediment discharge from the four largest rivers flowing into the Gulf of Papua. Data for the Amazon and Nile are given for comparison (from Alongi, 1990)

River	Water discharge (km^3/yr)	Sediment discharge (10^6 ton/yr)	World ranking (water discharge)
Fly	238	74	10
Purari	100	105	15
Kikori	43	30	21
Bamu	32	2	22
Amazon	6300	900	1
Nile	10	3	23

where tidal mixing maintains vertical homogeneity (Wolanski et al., 1984, 1995a).

Runoff and other terrestrial influences have important effects on the biota of the area. The distributions of mangroves and coral reefs are strongly influenced by terrestrial influences, notably salinity and sedimentation rates. Long and Skewes (1997b) found that the abundance of seagrasses on Torres Strait reefs was correlated with proximity to the Fly River and other major freshwater and sediment sources; they proposed that this is a response to nutrient input. Similarly, Milton et al. (1997) found that species composition in reef fishes is affected by terrestrial runoff.

THE MAJOR SHALLOW WATER MARINE AND COASTAL HABITATS

The area contains typical tropical shallow marine habitats: coral reefs, mangroves, and seagrass beds. Inter-reefal areas in Torres Strait have a number of distinct epibenthic seabed communities. The Gulf of Papua has a muddy bottom. The area's geologic history, geography, and connections to global centres of faunal and floral diversity in Indonesia, Malaysia, and the Philippines have resulted in extremely high physical and biological diversity.

Nearly all of the coral reef in the area is found in Torres Strait, where there are approximately 600 reefs covering more than 240,000 ha (Long and Skewes, 1997a). The foundations of these reefs date from the mid-Miocene and are the oldest in the entire Great Barrier Reef region, but their present structure has been determined primarily by growth during the last 8000 years, i.e. since Torres Strait was drowned by the most recent rise in sea level and reef growth re-commenced on pre-existing limestone platforms (Barham and Harris, 1983). The biological and structural diversity of these reefs exceeds that found on the Great Barrier Reef to the south.

The reefs in central and western Torres Strait are little more than vast mud banks with a surrounding fringe of coral, although a few reefs, such as Bet Reef north of Warraber (Sue) Island, do have significant areas of consolidated carbonate reef platform (Long et al., 1997c). The Warrior Reefs in central Torres Strait, together with Dungeness and Warraber reefs to the south, form the largest reef complex in the central and western Strait, and the one most important to fisheries. The Orman Reefs, another large reef complex north of Mabuiag I., are poorly known. Interspersed among the islands in the southwestern Strait are a number of long, narrow platform reefs, which are oriented east–west because of the influence of the strong tidal currents (Jones, 1995). Most of the islands also have fringing reef platforms of varying degrees of development.

The line of reefs along the easternmost edge of the continental shelf is quite distinct from those in the rest of Torres Strait. They represent the northern extension of the Great Barrier Reef. Bathed in clear Coral Sea waters, they have less soft sediment and much more carbonate "pavement" than other Torres Strait reefs (Long et al., 1997c). These reefs have been described by Veron (1978a,b). At the northernmost extreme they grow in narrow strips bisected by deep channels. Further south they develop a pattern similar in appearance to a series of river deltas, again bisected by deep, narrow channels. Between this outer barrier and the reefs of central Torres Strait is a large, shallow lagoon with numerous platform and patch reefs, which like those in the central and western Strait are predominantly soft sediment on the reef top (Long et al., 1997c). Many of these "Eastern Patch Reefs" are quite large, and several have associated sand cay islands. The largest islands, Murray and Darnley, have fringing reefs.

The northern extension of the Great Barrier Reef terminates just south of 9°S, probably because of present or past inputs of freshwater and sediment from the rivers draining into the Gulf of Papua. Reef development is very restricted along the coast of Western and Gulf Provinces although, contrary to some accounts, small patch and fringing reefs do occur (pers. obs.). The extensive Papuan Barrier Reef commences at Yule Island and runs east along the southern coast of PNG.

The series of vast river deltas that make up the New Guinea coastline between Daru and Kerema support some of the largest remaining more or less pristine mangrove forests. Gulf Province has around 200,000 ha of mangrove forest (Osborne, 1993). The Fly River delta, where mangroves extend some 300 km upstream, has another 87,000 ha. of mangal and there is also extensive mangrove development on the Bamu River delta between the Fly and the Gulf Province border (pers. obs.) These forests are highly diverse. Robertson et al. (1990) recorded 29 mangrove tree species in the Fly River delta, and Cragg (1983) recorded 30 species in that of the Purari. These authors concluded that the community structure of these forests is determined largely by salinity, sediment deposition rates, and the age of the mud bank, island, or other coastal feature upon which the mangroves grow. Established forests in areas with relatively high (>10 ppt) salinities are typically dominated by *Rhizophora apiculata* or *Bruguiera parviflora*, while the nypa palm (*Nypa fruticans*) dominates where salinities are reduced to 1–10 ppt. *Sonneratia lanceolata* is the most abundant mangrove species along river banks, and on newly formed islands and banks, where salinities are very low (<1 ppt). Where salinities are higher (>5 ppt), rapidly accreting banks typically support stands of *Avicennia marina* and/or *S. alba*. Forests that are transitional between these major types generally have the highest diversity (Cragg, 1983). Mangrove species common in adjacent high-salinity environments, including *Aegialitis annulata*, *Ceriops australis*, *C. tagal*, *Lumnitzera* spp., and *Osbornia octodonta*, are rare or absent in the deltas, which do not have high-salinity habitats.

There are some 5000 ha of mangroves on the PNG coast west of Daru, as determined from LandSat images (D. Milton, pers. comm.), growing mainly in a narrow fringe along the shoreline. Mangroves grow on less than a third of the islands in Torres Strait, covering a total area of about 18,000 ha (Long and McLeod, 1997). The largest areas of mangrove occur on Boigu (4744 ha), Saibai (3209 ha), Moa (1484 ha), Turnagain (1142 ha), Sassie (1006 ha), and Zagai (908 ha) islands. Like Boigu and Saibai, the smaller islands near the New Guinea coast are low and muddy, and have high mangrove coverage. Mangroves occur in patches and/or as a narrow fringe on most other islands in central Torres Strait. The islands in the eastern Strait have little or no mangrove.

The mangrove communities along the New Guinea coast west of Daru and in Torres Strait generally lack the low-salinity species of the delta communities such as *N. fruticans* and *S. lanceolata*, but high-salinity species such as *A. annulata, C. australis, C. tagal, Lumnitzera* spp., and *O. octodonta* are common (Long and McLeod, 1997; pers. obs.) The mangrove diversity of the Torres Strait–Gulf of Papua region as a whole is among the highest in the world.

Mangrove areas, both on the PNG mainland and on the large muddy islands immediately offshore, typically grade into wooded freshwater swamps (Osborne, 1993; TSPZJA, 1999). These swamps are the source of sago, the dietary staple, and are critical habitat for barramundi (*Lates calcarifer*; Kare, 1995), an anadromous fish important to subsistence and artisanal fisheries.

Seagrass beds are the most extensive marine habitat in Torres Strait and are critical feeding, breeding, and nursery habitat for dugong (*Dugong dugon*), green sea turtles (*Chelonia mydas*), and several commercially important species including shrimps (*Penaeus esculentus, Metapenaeus ensis*), rock lobster (*Panulirus ornatus*), and pearl oysters (*Pinctada maxima* and *P. margaratifera*). Seagrass grows both on the coral reefs and on the sea floor in inter-reefal areas to depths of 40 m. Torres Strait seagrass communities have recently been mapped and quantitatively described by Long and Poiner (1997) and Long and Skewes (1997b). Altogether there are some 1.4 million ha of seagrass in Torres Strait. The most extensive development is in the western and central Strait, where about half of the inter-reefal sea bed is covered by seagrass. In the eastern Strait, by contrast, seagrass covers <3% of the sea bed, and is much less diverse (Long and Poiner, 1997). This may be due to the greater depths and higher turbidity in eastern Torres Strait, and possibly also the effects of shrimp trawling. These factors do not explain the pattern of seagrass on reefs, however, which parallels that in inter-reefal areas, i.e. seagrass is much less common on reefs in eastern Torres Strait than in the central and western Strait. Seagrass abundance on reefs is correlated with proximity to the main sediment sources, the Fly and Mai Rivers and the large continental islands in the southwestern Strait. Thus, terrestrial influences seem to generally favour seagrass development (Long and Skewes, 1997b). Long et al. (1997b), however, report an extensive dieback in seagrass in northwestern Torres Strait following an unusually large runoff event from the Mai River. They were unable to determine the causal relationship, if any, between the high runoff and seagrass mortality, although they concluded that reduced salinity was not a direct cause. They suggested that runoff pulses may be important determinants of seagrass and epibenthic community structure in the area.

Little is known about seagrass on the adjacent PNG coast. Given the low abundance of seagrass in the Torres Strait east of the Warrior Reefs and the high turbidity and sediment input, seagrass development is probably minimal in the Gulf of Papua. Robertson et al. (1990) do not report seagrasses in their intertidal and subtidal benthic samples in and offshore of the Fly River delta. Seagrass is probably common on the coast to either side of the deltas. There are extensive beds at Daru and Bristow Islands, and the coast west of Daru presumably has seagrass beds similar to those of Torres Strait islands immediately offshore. Seagrass development increases at some point east of the Purari Delta; seagrass beds in the Port Moresby area are well-developed and generally similar to those in the Torres Strait.

As for coral reefs and mangroves, the seagrasses, with 11 reported species, are among the world's most diverse. The most abundant species is turtlegrass, *Thallassia hemprichii*, but there are a number of distinct assemblages and most species can occur as dominants (Long and Poiner, 1997). For example, *Halophila ovalis* is numerically dominant in deep-water assemblages (mean depth = 12.1 m; Long and Poiner, 1997). Seagrasses typically occur in mixed beds of up to eight species, and monospecific beds are uncommon. Community structure appears to be determined by complex interactions of depth, tides, currents, turbidity, substrate, micro-habitat and other physical factors (Long and Poiner, 1997). In the central and western Strait total seagrass abundance decreases exponentially with depth, primarily because of a similar relationship for the dominant *T. hemprichii*. Several other species have abundance maxima at intermediate depths (2–6 m), where species diversity is also highest. In inter-reefal areas the eastern Strait seagrass is found at depths of 25–40 m (Long and Poiner, 1997).

The bottom of the Gulf of Papua is soft sediment with a diverse and productive fauna and high rates of bioturbation (Robertson et al., 1990). It supports a prawn trawl fishery that is the most important commercial fishery in the area. A variety of subtidal benthic habitats are found in Torres Strait, ranging from rocky bottoms with dense epibenthos to soft mud (Long and Taranto, 1997).

OFFSHORE SYSTEMS

Because the 200 m contour was set as the boundary for the Gulf of Papua, there are no deep habitats in the area. Water-column processes in the neritic Gulf of Papua are poorly studied, but appear to be dominated by the large river inputs of sediment, organic material, and nutrients, especially nitrate and silicate (Furnas, 1991; Robertson et al., 1990; Wolanski, 1984). The oceanographic characteristics, and presumably biophysical processes, vary along a gradient from the river estuaries to the oligotrophic waters of the Coral Sea. The remainder of this section refers only to the relatively inshore waters inside the 200 m isobath; nearly all the available information is from the northwestern Gulf of Papua off the mouth of the Fly River.

Based on very few measurements, typical mean water column nitrate concentrations are 0.5–1 μM, relatively high for tropical waters. Robertson et al. (1990) measured water-column averaged primary production rates of 0.13–0.34 g C m^{-2} d^{-1} in the northwestern Gulf of Papua. This is again relatively high for tropical waters, but lower than other sites on the far northern GBR (Furnas, 1991). Robertson et al. (1990) suggest that turbidity may reduce primary production in the Gulf of Papua. The primary production is dwarfed by the high (12–36 g C m^{-2} d^{-1}) respiration, reflecting the high riverine inputs of organic carbon.

Small amounts of nutrients are input to surface waters along the far northern GBR by tidally-forced upwelling along the shelf break (Furnas, 1991; Wolanski, 1991). Though the fate of these nutrients is not known with certainty they are probably rapidly assimilated locally and contribute to highest measured rates of primary productivity on the far northern GBR (0.61–0.68 g C m^{-2} h^{-1} at mid-day) (Furnas, 1991).

There is very little information about plankton communities in the Gulf of Papua. Phytoplankton standing stock averages around 0.5 μg l^{-1} over the water column (Furnas, 1991; Robertson et al., 1990). Robertson et al. (1990) measured zooplankton standing stocks of 44–83 mg m^{-3} at three stations in the northwestern Gulf of Papua; copepods numerically dominate the zooplankton.

POPULATIONS AFFECTING THE AREA

The human population is small (Table 2). On a province-wide basis, Western and Gulf Provinces have population densities of 1–2 persons/km^2, the lowest of all PNG provinces (Bakker, 1996). Coastal population densities are higher, but still low. Though the Torres Strait population is small in absolute numbers, population densities are generally higher than on the PNG mainland due to the small size of the islands, only 16 of which are inhabited. The Torres Strait population is concentrated on the continental "Inner Islands", which are relatively large and nearest the Australian mainland; some 35% of the Islands' population lives on Thursday Island, the administrative centre, and the adjoining Horn Island, the airport site (Tallegalla/Sanbis, 1998). The largest urban centres are Daru (8500), Kerema (4000), Ihu (3300) and Thursday Island (2500).

High fertility results in a youthful population structure (Table 2). Western Province has a high population growth rate but in Gulf Province out-migration results in low population growth despite the high fertility (Bakker, 1996). Population growth in the Torres Strait Islands is concentrated in the outer islands (Tallegalla/Sanbis, 1998).

The area has a very low level of development. Gulf and Western Provinces have among the lowest levels of human development in PNG, which has the lowest Human Development Index among Pacific island nations (UNDP, 1994a, 1994b). Few people living outside the urban centres have ready access to communications, electricity, healthcare, education, or running water. Malaria and other tropical diseases are a significant problem. Most villagers live in houses built largely from traditional bush materials. Rural people are highly dependent upon gardening, hunting, and fishing, while formal employment and participation in the cash economy are low (Bakker, 1996). Sago is the dietary staple, but seafood, including green turtles and dugong, is important both nutritionally and as a trade commodity (Pernetta and Hill, 1981). Government-related enterprises are the largest employer. Remittances from relatives employed elsewhere are a significant source of income, particularly in Gulf Province. There are few roads. Some coastal villages are serviced by coastal freighters. Traditional canoes and other small boats remain an important means of transportation. English, Motu, Melanesian Pidgin, and a variety of indigenous languages are spoken.

Table 2

Population statistics for Western and Gulf Provinces and Torres Strait

	Total pop.	Children /youth[5] (%)	Annual growth rate (%)	Rural coastal pop.	Urban coastal pop.
Western Province[1]	110,420	51.5[3]	2.8[3]	16,000[10]	8,501[11]
Gulf Province[1]	68,737	45.3[3]	0.7[3]	23,000[10]	6,415[12]
Torres Strait[2] (Inner Is.[6])	3,396[4,7]	42.4[4,8]	0.3[8,9]	298[4]	2,948[4,13]
Torres Strait[2] (Outer Is.)	4,966[4,7]		1.2[8,9]	3,278[4]	0[4]

1: 1990 census.
2: 1996 census.
3: Source: Bakker (1996).
4: Source: Tallegalla/Sanbis (1998).
5: Figures for Gulf and Western Provinces include persons under age 15; figures for Torres Strait include age 15 and under.
6: The inner islands are those nearest Cape York: Thursday, Horn, Friday (Prince of Wales), and Hammond.
7: Island populations only; does not include populations of Cape York, Australia, which are normally included in the Torres Strait Local Area for census purposes.
8: includes aboriginal populations of northern Cape York (1,508 persons).
9: mean for 1991–1996 for indigenous residents only (from figures in Tallegalla/Sanbis, 1998).
10: % rural coastal population in 1980 (Frielink, 1983a) multiplied by total provincial population in 1990.
11: Daru.
12: Kerema, Ihu, Baimuru, Kikori.
13: Thursday Island, Horn Island.

By the developed-country standards of Australia, Torres Strait also has a very low level of economic and human development. The standard of living is, however, much higher than that of adjacent PNG villagers. Indeed, lines of sovereignty in Torres Strait were drawn to accommodate the desire of Torres Strait Islanders to remain Australian and thus maintain their relatively higher living standard (Johannes and MacFarlane, 1991). Most Islanders have access to electricity, education, health care, communications, and transport, and are eligible for government benefits and subsidies.

Despite the availability of imported foods and money to buy them, seafood consumption in Torres Strait is among the world's highest; sea turtles (mainly the green turtle), dugong, and reef fish are the most important foods (Johannes and MacFarlane, 1991; NSR, 1998). Unemployment rates are lower than for mainland Queensland, but most employment is in low-wage, unskilled jobs, especially away from the inner Islands. More than two-thirds of all jobs are government-related (Tallegalla/Tanbis, 1998). Remittances are an important source of income. Light aircraft, coastal freighters, and small power boats are the most important means of transportation.

The population is primarily indigenous. Non-citizens make up only about 4% of the population of Western and Gulf Provinces (Bakker, 1996); there are also an unknown but small number of PNG citizens from other parts of the country. The non-citizen population, mostly Indonesian citizens, is concentrated in the Kiunga district in the Western Province highlands. About 80% of the Torres Strait population is indigenous, with the non-indigenous population concentrated on Thursday Island. Some 85% of the indigenous population identify themselves as Torres Strait Islander. The remainder are Australian Aboriginal or mixed Aboriginal–Torres Strait Islander (Tallegalla/Tanbis, 1998).

Prior to European contact there were extensive cultural and trade linkages among Torres Strait Islanders and the Papuans of the south PNG coast. Cultural links between Torres Strait Islanders and the Kiwai-speaking peoples around the Fly River estuary, though diminished by Western influences, remain strong (Johannes and MacFarlane, 1991). Europeans first sighted the area in the late 16th century but contact was mostly restricted to passing vessels. European exploration and colonisation did not commence in earnest until the late 18th century in Torres Strait and the mid-19th century in the Gulf of Papua. Outside contact remains limited on much of the coast of PNG to the present day. In Torres Strait, the establishment of fishing industries for sedentary resources (see below) and the arrival of missionaries brought in an increasing number of migrants from the Pacific islands and elsewhere during the latter half of the 19th century. At the same time, the indigenous populations were reduced by the availability of alcohol and firearms, measles and smallpox, and reprisals for the killings of shipwreck victims (Tallegalla/Sanbis, 1998).

Commercial use of marine resources commenced in Torres Strait in the mid-19th century with the exploitation of bêche-de-mer (the cleaned and dried product of various holothurians), pearl shell, and trochus (*Trochus noctiluca*) (Ganter, 1994)). By 1870 these and related industries employed 25% of Torres Strait Islander males and, except for the interruption of World War II, they remained the main source of employment until the 1960s. Ganter (1994) has described these industries as "resource raiding"; they essentially involved exploiting the marine resources of an area to exhaustion and then moving on to a new area. Prices for pearl shell and other commodities began to decline in the 1950s and by the late 1960s these industries had essentially collapsed. The PNG coastal area around Daru participated in the Torres Strait sedentary resource boom, and Daru is still the centre of seafood marketing and processing in Western and Gulf provinces. Commercial use of marine resources elsewhere in these provinces is limited and essentially a 20th-century phenomenon.

Historical baseline information on the area is nonexistent. Except for a few scattered reports with very limited geographical scope, information about the biota and resource use has only become available in the last few decades at best, generally via environmental assessments of development projects. Assessments of the potential effects of a hydroelectric scheme on the Purari River, conducted in the late 1970s, provide detailed information about the Purari Delta's biota, patterns of resource use, and ecological linkages with adjacent habitats, in particular the Gulf of Papua prawn fishery (Petr, 1983; Gwyther, 1980; and references therein). The Torres Strait Baseline Study was initiated by Australia in response to environmental concerns about the development of mines at Ok Tedi, and subsequently Panguna, which discharge large amounts of tailings into highland tributaries of the Fly River. This study primarily addresses the potential impacts on Torres Strait of elevated inputs of sediment and heavy metals. It did not commence until 1991, and few marine baseline data were collected prior to the commencement of operations at Ok Tedi in 1984. Most recently, environmental impact assessment of a proposed natural gas pipeline from fields in the PNG highlands through the Kikori River delta and Torres Strait to mainland Australia, and an ancillary Australian government conservation planning initiative, has provided detailed, quantitative descriptions, mapping, and analysis of the marine biota of much of Torres Strait (NSR, 1998; Poiner et al., 1997 and appendices).

RURAL FACTORS

Agriculture

Agricultural development is largely restricted to subsistence and market gardening. In Gulf Province the most important cash crop is betel nut, grown in village gardens and sold locally or in Port Moresby (R.M. Bourke, pers. comm.) Very small amounts of copra, rubber, and cocoa are grown, mostly by smallholders. Bananas, other fruits, vegetables, and, in swampy areas, sago are grown for local consumption. Shifting cultivation is a source of forest disturbance in most of the lowland forests of PNG (Levett and Bala, 1994; Louman and Nicholls, 1994), but the nature and severity of disturbance is not documented. No information is available on the downstream effects of agriculture, but given the low levels of population, development, and technology (mechanisation and pesticide and fertilizer use are very low), and the very high natural runoff and sediment loads, the coastal impacts of village-based agriculture are probably small and sustainable in their present form. In Torres Strait traditional gardens have fallen into disuse and most produce is imported (Johannes and MacFarlane, 1991).

Subsistence and Artisanal Fisheries

Subsistence and artisanal fishing is the primary use of marine resources in Torres Strait and the Gulf of Papua. Subsistence fisheries in the area have been documented in detail by Frielink (1983b), Haines (1978/79), Johannes and MacFarlane (1991), and Pernetta and Hill (1981). Seafood consumption in Torres Strait is among the world's highest (Dews and Harris, 1995; Johannes and MacFarlane, 1991). In PNG, where suitable land is available more time is spent upon, and calories derived from, agriculture than fishing, but fishing is still a very important activity and nutritional source (Harris, 1977; Frielink, 1983b; Pernetta and Hill, 1981). Dalzell (1993) estimates that for Papua New Guinea the subsistence catch is four times the total commercial catch. The proportion is probably higher for Western and Gulf Provinces because of the predominantly subsistence lifestyle and because these provinces produced little if any of the large commercial catch of tuna included in Dalzell's figures. Reliable current catch data for subsistence and artisanal fisheries in these provinces, however, are not available.

Traditional methods vary with target species, location, and the age and sex of the fisher. They include gathering by hand (either by gleaning mud and reef flats or by free diving), handlines, spearing, harpooning, bow and arrow, hand traps, traps and weirs, and the use of derris root and other natural poisons (Frielink, 1983b; Haines, 1978/79; Johannes and MacFarlane, 1991; Pernetta and Hill, 1981). Nylon gill nets and other introduced methods are increasingly supplanting traditional ones. Blast fishing and the use of cyanide, bleach, or formalin do not appear to be widespread.

The subsistence catch is highly varied. A notable feature of subsistence fishing in Torres Strait, including the adjacent coastal areas of mainland Western Province, is that sea turtles, mostly the green turtle (*Chelonia mydas*), and dugong are typically consumed in larger amounts than reef fish (Fig. 3). Dugong and turtles are also important to the Kiwai people who live around the mouth of the Fly River in PNG (Johannes and MacFarlane, 1991). They were traditionally hunted by harpooning them from sail- or paddle-powered canoes or from platforms built over seagrass feeding grounds, but now they are herded and harpooned from outboard-powered dinghies. Dugong and turtles also have great social importance as ceremonial foods, in bride price and other social exchanges, and in conveying status upon successful hunters. This has been formally recognised in legislation and management regimes. Reef fisheries in Torres Strait are not as diverse as in some other parts of Oceania but may still include over 50 fish species (Johannes and MacFarlane, 1991). Turtle, dugong, and reef fish make up about 95% of the subsistence catch in Torres Strait (Johannes and MacFarlane, 1991; Dews and Harris, 1995).

In the delta region of Gulf Province the subsistence catch is dominated by finfish, mud crab (*Scylla serrata*), and, depending upon the local salinity, freshwater (*Macrobrachium*) or marine (Penaiedae) prawns (Frielink, 1983b; Haines, 1978/79). The finfish catch is dominated by barramundi, catfishes (Arridae), threadfins (Polynemidae), and sharks and rays, but a large number of other fishes are caught in small amounts. Traditional diets tend to be protein-deficient, but the consumption of seafood, though it is readily available, is apparently much lower than in Torres Strait (Haines, 1978/79; Pernetta and Hill, 1981). Haines (1978/79) estimated that the average catch of crabs and fish in villages in the Purari delta amounted to about 40 and 80 g/person/day respectively, significant amounts of which were traded for sago.

Potential over-exploitation of dugong and sea turtles is the primary environmental concern regarding subsistence fisheries in the area. Because these species are long-lived, with a long time to first reproduction and low fecundity, they are vulnerable to over-exploitation, there is likely to be a time lag between over-exploitation and its detection via catch statistics, and recovery from over-exploitation is likely

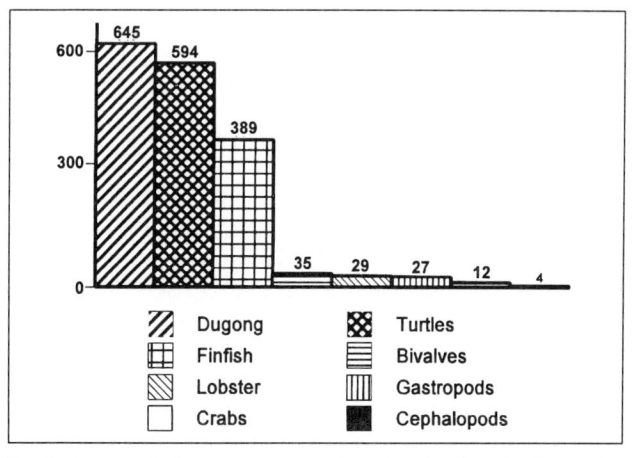

Fig. 3. Average daily non-commercial catches (kg) by Islanders in the Torres Strait Protected Zone, 1991–93 (from data in Dews and Harris, 1995).

to be difficult. Population growth and technological improvements in hunting have probably increased the subsistence harvest, and it is uncertain whether present harvests are sustainable (Marsh et al., 1997; Williams, 1994). In Torres Strait there is no evidence of overfishing of reef fish and it is unlikely that it will occur as long as the harvest of reef fish remains a primarily subsistence fishery (Johannes and MacFarlane, 1991; Williams, 1994). While recent subsistence catch data from PNG are scarce, it is also unlikely that over-exploitation of other species as a result of subsistence harvests is a widespread problem. The potential for overfishing of species in the subsistence harvest arises primarily from artisanal fishing of the same species for export product.

Artisanal fisheries are an important realised or potential source of village incomes. Because coastal people in the area have limited terrestrial resources, an increased use of coastal marine resource exploitation represents the primary opportunity for economic development. Unfortunately, most artisanal fisheries in the area have histories of over-exploitation and decline.

The largest artisanal fishery in the area at present is for rock lobster, and does appear to be sustainable under present conditions. The fishery is based on the ornate rock lobster (*Panulirus ornatus*), which lives in reef and rocky areas in Torres Strait. Between August and November each year, mature adults migrate to spawning grounds on the northern Great Barrier Reef and on reefs between Yule Island and Port Moresby. The larvae are apparently carried back into Torres Strait by the Coral Sea Coastal Current (Pitcher et al., 1995).

Lobsters are collected by hand or with hand-held scoop nets from the reef crests and slopes either by free diving or with hookah equipment. The most productive grounds are the Orman and Warrior Reefs (Williams, 1994). The divers work from small (usually 4–6 m) dinghies and canoes, usually in conjunction with larger (9–15 m) freezer vessels. The fishery involves about 600 boats from Cape York and the Torres Strait Islands (TSPZJA, 1999), while estimates of the number of boats involved in the Daru-based fishery range from 150 to more than 800 (ANZDEC, 1995; Evans and Polon, 1995). Most fishers are indigenous, and regulations prevent expansion of non-indigenous participation. Catch is landed at Thursday Island and Daru. The Australian-based catches average around 200 mt per year, with landing of 244 mt, worth about US$ 6 million, in 1997 (TSPZJA, 1999). Catch landed at Daru is 30–90 mt per year (Evans and Polon, 1995; Kailola, 1995). There is a small fishery based at Yule Island that targets spawning aggregations; no production statistics are available. The fishery appears stable, with room for expansion (Kailola, 1995; TSPZJA, 1999).

Artisanal fisheries for sedentary resources operate in the area on a periodic boom-and-bust cycle due to lack of management and resultant over-exploitation. The primary targets are pearl shell, bêche-de-mer, and trochus shell. Early European involvement in Torres Strait, and subsequently Western Province, centred around commercial fisheries for these species, which by the early 20th century dominated the local economy and, in the case of pearl shell the world market. By the 1960s these fisheries had collapsed, in large part because of over-exploitation (Ganter, 1994; Williams, 1994). A bêche-de-mer fishery was re-established around Daru in the late 1980s, expanded rapidly, and crashed due to overfishing by 1993, when the fishery was closed by government regulation until 1995.

Today, these resources support only small, sporadic artisanal fisheries. Torres Strait pearl shell stocks remain depleted and the only pearl shell fishery involves a few boats collecting live shells to supply pearl farms in the Torres Strait (TSPZJA, 1999). The catch fluctuates markedly and is difficult to measure. In 1997 the fishery involved only two boats, fishing in PNG waters. They collected about 15,500 shells. Lobster divers recorded an incidental catch of 620 shells. Western Province is considered to have large unexploited pearl grounds (ANZDEC, 1995). About 45 reefs in eastern Torres Strait are fished for trochus. The catch varied from around 50–200 mt per year during the 1990s, with the recommended total allowable catch declining from 150 mt in 1985 to 60 mt in 1993 (NSR, 1998; Williams, 1994). In 1990 Islanders were reported to be harvesting trochus on the northern GBR, having depleted stocks in Torres Strait (Johannes and MacFarlane, 1991). The apprehension of Indonesian fishermen (Ganter, 1994) indicates illegal fishing pressure on trochus stocks in western Torres Strait. The Daru-based bêche-de-mer fishery does not appear to have recovered despite the closure of the fishery from September 1993 to March 1995 and the subsequent imposition of minimum size limits and an annual catch limit of 40 mt (Long and Skewes, 1997a). The Australian catch, primarily from the Warrior Reefs, is about 20 mt (NSR, 1998).

Barramundi are an important resource for villagers, especially in the delta region of the Gulf of Papua, where the population is concentrated (Moore, 1982). Barramundi are catadromous, living in rivers and freshwater swamps but migrating annually along the coast to inshore marine spawning grounds (Moore, 1982). They are taken in lakes, swamps, rivers, and along the coast, primarily with nylon gill nets. Barramundi are important in subsistence catches in Gulf and Western Provinces and on the Torres Strait islands near PNG (Frielink, 1983b, Haines, 1978/79, Pernetta and Hill, 1981; TSPZJA, 1999). On the Torres Strait islands barramundi fishing is primarily for subsistence use (TSPZJA, 1999) but Western and Gulf provinces have artisanal fisheries that until the early 1990s were an important source of village income. Unfortunately, these fisheries targeted juveniles and migrating adults, and were not sustainable. Management measures such as seasonal closures and minimum size limits have not been effective, largely because of a lack of enforcement, and the fishery has declined (Fig. 4) as a result of over-exploitation and possibly other factors (ANZDEC, 1995; Kare, 1995; Kailola, 1995). Reliable catch statistics are not available but the total

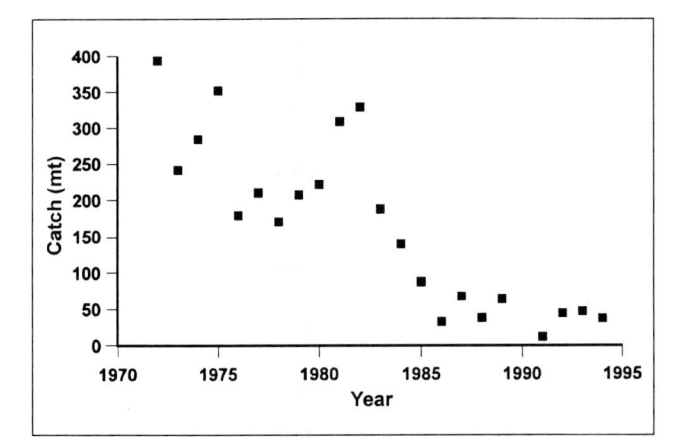

Fig. 4. Landings of barramundi (*Lates calcifer*) at Daru, 1971–1994. Data from Kare (1995).

artisanal catch is now no more than a few tens of tons (ANZDEC, 1995; Kailola, 1995) and commercial processing at Daru has essentially ceased. The artisanal fishery now revolves around the sale of juvenile fish at local markets.

Throughout coastal Western and Gulf Provinces there are small artisanal fisheries for a variety of other species, including mud crab, prawns, sharks, and various bony fishes. The catch is sold either at local markets or to small processing plants or boats. There are few reliable recent data on the catch or status of these fisheries; what data are available indicate that catches are highly variable. Small amounts of some products, including mud crabs, prawns, and shark fin, are processed and exported from the area, but much of the catch is sold locally. These fisheries are probably limited primarily by lack of processing and marketing infrastructure and are unlikely to be presently over-exploited except perhaps occasionally at the local level. Experience indicates, however, that over-exploitation is likely if ready cash markets develop for these resources. Crocodile farming is a significant source of income in some villages; the skins are processed at Daru.

Published information does not suggest that artisanal fishing in Torres Strait and the Gulf of Papua has widespread detrimental effects on coastal and marine habitats. A notable exception to this is the Daru area, where the use of mangroves as fuel for drying bêche-de-mer has contributed to mangrove deforestation (Kailola, 1995).

COASTAL EROSION AND LANDFILL

Because of the low level of economic development, problems of coastal erosion and landfill are restricted to a few locations. At Daru and on some of the Torres Strait Islands there have been small reclamation projects to provide housing sites and airstrips, small-scale port development involving dredging and/or jetty construction, and some use of beach sand for construction (pers. obs.), but these are relatively minor activities. There have been some coastal erosion problems at Daru as a result of mangrove deforestation.

EFFECTS FROM URBAN AND INDUSTRIAL ACTIVITIES

Artisanal and Non-industrial Uses of the Coast

On the Torres Strait Islands there are no significant artisanal activities other than fishing. The primary source of environmental alteration is the construction of houses, airstrips, and a few small businesses and roads. The primary effects are the loss of lowland and mangrove forest from the developed areas.

The effects of small-scale use of coastal resources in rural areas of Western and Gulf Provinces are probably negligible because these areas are so sparsely populated, and the level of economic development is so low. Small amounts of forest are cut for house sites and materials, and there is disturbance of forest by shifting cultivation for subsistence and market gardens. There are also a few roads and airstrips. The impact of these activities is relatively small.

Industrial Uses of the Coast

Mining and petroleum extraction are by far the largest industrial activity in the area in terms of revenue. These activities take place in the highlands and are not uses of the coast *per se*, but they have significant impacts on coastal areas. The Ok Tedi gold and copper mine has been in operation at Mt. Fubilan, about 1000 km inland from the Gulf of Papua, since 1984. The mine discharges tailings into the Ok Tedi River, a tributary of the Fly, increasing the sediment load of the Fly from about 85 to 120 million tons per year (Salomons and Eagle, 1990) and is expected to remain in operation until 2008. The Porgera Gold mine, at the headwaters of the Porgera river, commenced operation in 1990 and discharges about 4 million tons of tailings per year into the Fly system, again until 2008.

Concerns about the possible environmental impacts of mining upon Torres Strait and the Gulf of Papua, particularly with regard to fisheries, have stimulated a series of detailed studies on distribution and fate of sediment and metals from the Fly River. At least 50% and perhaps up to 90% of the Fly's sediment load is retained in the Delta (Harris et al., 1993; Wolanski et al., 1995b). Less than 2% is transported to the southwest, into Torres Strait, and the remainder is inferred to be transported into the Gulf of Papua (Harris et al., 1993). Monitoring data indicate little or no effect of mine discharges on the fish fauna of the middle and lower Fly (Williams, 1994), and no evidence of increased levels of metals in seafood (Evans-Illidge, 1997).

Production of oil from fields at Lake Kutubu, in Southern Highlands Province approximately 265 km northwest of Kikori, commenced in 1992, and produced 27.6 million barrels in 1997, and nearby fields are being developed. The oil is transported via a pipeline running along the Kikori River to a terminal some 50 km offshore, where it is loaded onto tankers. To date no significant spills have been reported. There are plans for a pipeline to deliver natural gas from the oil fields to markets in Australia (Fig. 1). Extensive impact assessment and other studies have been carried out in Torres Strait (NSR, 1998; Poiner et al., 1997 and Annexes thereto) to select a pipeline route that reduces environmental impacts and disruption of trawling and other uses.

Logging is another major industrial activity in the watersheds draining into the Gulf of Papua. Well over half of the forested areas in Western and Gulf Provinces have been allocated to industrial logging, including 750,000 ha of forest upstream of the Purari–Kikori delta area (Louman and Nicholls, 1994). Environmental management of the logging industry has been poor (Hughes and Sullivan, 1989; Louman and Nicholls, 1994). Clear-fell logging is usually practised only in conjunction with plantation development, but selective logging practice in PNG is not good and causes high tree mortality (Louman and Nicholls, 1994). Declining export prices caused a sharp downturn in logging in PNG in 1997, and many operations have ceased at least temporarily. There is no information about the downstream effects of industrial logging on marine and coastal environments in PNG, but it is likely to represent a significant threat, particularly to the mangrove swamps of the delta (Louman and Nicholls, 1994).

Industrial agriculture in the area is insignificant. Gulf and Western provinces once had significant coastal plantations for copra and rubber, but these have largely fallen into disuse and production is low (e.g., Levett and Bala, 1994). The Torres Strait islands have little if any scope for commercial agricultural development because there is so little land. Much of the land in both Western and Gulf provinces is unsuitable for commercial agricultural development, although some areas in Western Province are moderately suitable (Eaton, 1982; Hanson et al., 1998). There are proposals for the development of oil palm and rubber plantations, and some 600,000 ha upstream of the delta, representing most of the forest not allocated to logging, has been allocated for plantation development (Louman and Nicholls, 1994). Some observers have suggested that these proposals are in fact clear-fell logging projects in disguise. These plans are not proceeding at the moment due to economic and political circumstances.

Fisheries

Prawn trawling is the primary industrial fishing activity in the area. There are two main fisheries, a Torres Strait fishery east of the Warrior Reefs and a Gulf of Papua fishery operating from the coast to approximately the 70 m depth contour (Fig. 1). These fisheries are by far the largest source of fisheries revenue in the area. Both fisheries are multi-species fisheries, and in both there are catch-sharing arrangements between Australia and PNG.

Torres Strait fishery operates in the TSPZ east of the Warrior Reefs. It is based primarily on blue endeavour and brown tiger prawns (*Metapenaeus endeavouri* and *Penaeus esculentus*, respectively), with small catches of several other species. Seagrass nursery grounds, especially on the large Warrior Reef complex, are essential to the health of the fishery (Williams, 1994). The annual catch varies in the 1000 to 2000 mt range; in 1997 the Australian catch was 1626 mt valued at around US$16 million (TSPZJA, 1999). Under catch-sharing arrangements PNG is entitled to an annual quota of 200 mt but the actual catch is only a few mt (Hair, 1994).

The Gulf of Papua fishery operates to about the 70 m depth contour between the Fly river delta and Iokea, an area of about 9000 km^2. Banana prawns (*P. merguiensis*, *P. indicus*) make up about half the catch, the remainder being tiger (*P. monodon*, *P. semisulcatus*), endeavour (*M. ensis*, *M. demani*, *M. eboracensis*) and coral (*Parapenaeus sculptilis*, *P. cornuta*) prawns. The fishery is dependent upon the recruitment of a single cohort into mangrove areas in the Gulf of Papua (Gwyther, 1983; Kailola, 1995). Annual production varied between 650 and 900 mt during 1990–93. The Gulf of Papua prawn fishery represents the second largest source of fisheries revenue to the PNG government. Unlike the larger tuna industry, which is operated mostly by foreign vessels, the prawn fishery generates local economic benefits other than government taxes and licensing fees.

As in most trawl fisheries, the large majority of the catch is by-catch, which is dominated by finfish, but also includes cephalopods, crustaceans, turtles, and a range of other taxa (Kailola, 1995; Harris and Poiner, 1990). Scallops, scyllarid lobsters, large fishes, and other valuable by-catch species are retained, and a small proportion of "trash fish" from Gulf of Papua by-catch is sold for domestic consumption in PNG, but most is discarded. There is some turtle mortality and evidence of detrimental impacts on demersal fish populations from intensive trawling (e.g., Poiner and Harris, 1996; Poiner et al., 1997), and Long and Poiner (1997) suggest that the scarcity of seagrass in inter-reefal areas west of the Warrior Reefs may result in part from trawling. Assessment of the impact of trawling, however, requires further research.

There is a small hook-and-line fishery for Spanish mackerel (*Scomeromorus commerson*) in eastern Torres Strait. There are more than 300 vessels participating in the fishery, the majority of them based locally, but most of the annual catch of 100 mt is taken by about 15 operators based in mainland Queensland (TSPZJA, 1999). The 106 mt taken in 1997 had a value of about US$ 500,000.

A trawl fishery for rock lobster formerly operated in the Gulf of Papua, targeting lobsters during their spawning migration. The trawl fishery was discontinued in 1984 because of its negative impact on the Yule Island artisanal fishery and potential impacts on the Torres Strait fishery. The PNG trawl industry, however, seeks to re-establish a trawl fishery (ANZDEC, 1995), and Kailola (1995) notes that though prawn trawlers are required to release lobster by-catch alive, there is no monitoring to ensure that this occurs.

Small industrial fisheries for sharks, mackerel, tuna, other nearshore pelagic fishes, and other species come and go in the Gulf of Papua with fluctuating markets and fisheries policy, aid initiatives, and entrepreneurial circumstances. The most promising emerging fishery at the moment is tuna long-lining; no data on the fishery are available.

Cities

Urban development in the area is minimal. Daru, the largest town by far, has a population of only 8500 (1990 census). The main industries are government (Daru is the provincial seat) and fisheries, including harvesting, processing, and related services such as fuel and gear sales. There is net immigration from surrounding villages because Daru offers modern infrastructure such as education, health care, transport links, electricity, and radio and telephone communications. Though the standard of these in Daru is generally low, they are to a large extent not available at all in the villages.

Wage earners and their dependants generally live in relatively modern housing provided with roads, electricity, and other agencies at some distance from the foreshore. Sewage is collected via a night-bucket system and dumped without treatment on the north side of the island, away from the town. Solid waste is disposed of in a crude landfill. Small quantities of seafood waste are discharged along the foreshore from processing plants. Settlers from surrounding villages typically clear housing sites in the mangroves along the foreshore and construct stilt houses from a combination of bush materials (including mangrove posts) and modern materials such as sawn timber and corrugated iron roofing. The provision of electricity, water, sewage collection, and other amenities is variable. Housing construction, along with the use of mangrove fuel in bêche-de-mer processing, has caused substantial mangrove deforestation. In addition to permanent residents, there is usually a transient population of several hundred people, who have come to sell seafood and produce and purchase supplies, living on boats and canoes long the waterfront. No waste reception facilities or other amenities are provided for them.

The reefs and reef flat at Daru are heavily impacted by food collection, trampling, and solid waste dumping (pers. obs.). Several hectares of mangroves and reef have been killed at the site where sewage is dumped from night-buckets (pers. obs.) No water quality data are available, but the strong tidal circulation probably provides adequate flushing away from the immediate discharge site.

The other coastal towns in Gulf and Western Provinces, all centres of provincial or district government, have populations of a few thousand. Septic tanks are the main source of sewage disposal, and solid waste is generally disposed of in crude dumps and landfills, or thrown into rivers or coastal waters. Impacts are probably confined to the immediate local area due to the small size of the towns.

Thursday Island, with a population of 2500, is the main focus of settlement in Torres Strait. Like Daru, government and seafood processing are the main industries; there is also a small tourist industry. Sewage has been disposed of in septic tanks or via night-bucket, but in 1999 a reticulated sewerage system and activated sludge treatment plant was commissioned. Tidal circulation provides good flushing. The other communities in Torres Strait consist of townships of a few hundred people. Solid waste is typically disposed of in dumps, sewage is disposed of in septic tanks, via night-bucket, and occasionally directly into coastal waters. Disposal of night-buckets typically degrades several hectares of mangrove and/or reef at the disposal site (pers. obs.). Water quality data are not available but there is usually strong tidal flushing.

Shipping and Offshore Accidents and Impacts

Some 2000–3000 ships traverse the Torres Strait each year, including several hundred oil tankers. Though pilotage through Torres Strait is not compulsory, most large ships and tankers are piloted. Nonetheless, the strong, unpredictable tidal currents and numerous reefs and shoals make navigation difficult, and there is a significant risk of maritime accident. Indeed, some 130 shipwrecks occurred in Torres Strait between 1803 and 1991 (Loney, 1993, after NSR, 1998), and groundings and other incidents are not uncommon. One of the larger spills in Australian waters took place in the Strait in 1970, when the oil tanker *Oceanic Grandeur* struck an uncharted reef and spilled about 1000 tons of oil. The oil was dispersed with detergent. Most accounts indicate that there was little environmental impact (e.g., Zann, 1995) but there were unsuccessful court proceedings claiming significant damage to the pearl culture industry (Ganter, 1994).

The development of the oil pipeline from the Kutubu oil fields, and operation of the offshore Kumul marine terminal to load tankers, present the risk of a spill offshore or in the sensitive mangrove areas of the delta. The pipeline operators have developed a technically sound spill contingency plan, but it is not set in the context of a good understanding of the ecosystem or its uses, and the ability to avert ecological damage in the event of a spill is uncertain (G. Baines, pers. comm.).

PROTECTIVE MEASURES

Fisheries Management

The Torres Strait Treaty, a bilateral agreement between Australia and PNG ratified in 1985, provides the primary framework for environmental management in Torres Strait, particularly with regard to fisheries. The treaty was negotiated to:

- clearly define lines of sovereignty and jurisdiction in the Strait;
- protect the traditional way of life of the indigenous inhabitants, including movement across the present national border to harvest marine resources;
- protect and preserve the marine environment in the strait.

The Treaty established a complex set of boundaries and associated regulations and management provisions. Two main boundaries were defined with regard to marine resource use and management (Fig. 5). The seabed jurisdiction line defines the boundary of jurisdiction over seabed resources, including sedentary living marine resources. It also defines the border of international sovereignty, with the exception that Australia maintains sovereignty over fifteen islands and cays north of the line and the surrounding waters within 3 nautical miles. The fisheries jurisdiction line sets the boundaries of jurisdiction over free-swimming fisheries resources.

The Torres Strait Treaty also established a protected area, the Torres Strait Protected Zone (TSPZ), that includes sovereign waters of both countries. The Treaty contains provisions for the co-management of marine resources within the TSPZ to ensure sustainable use, and catch sharing between Australia and PNG (Table 3). It also recognises the right of traditional inhabitants to move freely within the TSPZ for fishing, hunting, and other traditional activities. Both countries have enacted complementary national fisheries acts to provide the legislative basis for fisheries management within the TSPZ.

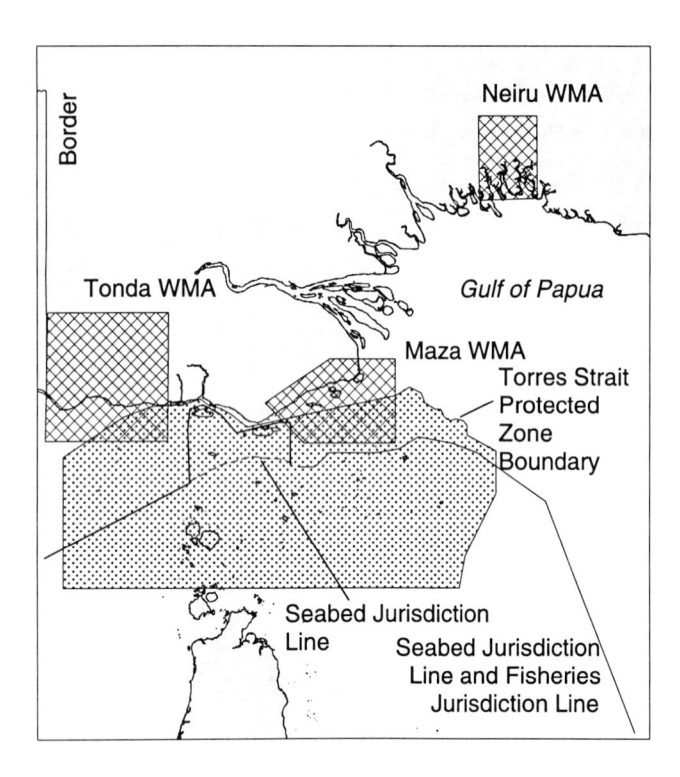

Fig. 5. Protected areas in Torres Strait and the Gulf of Papua.

Table 3

Agreed provisions for catch sharing in the TSPZ, 1997–2000

Fishery	Allowances
Torres Strait Prawn	7 PNG vessels in the Australian sector
Pearl Shell	5 Australian vessels (3 at any one time) in the PNG sector
Rock Lobster	27 PNG dinghies and associated freezer boats in the Australian sector
Spanish Mackerel	10 PNG dinghies and associated freezer boats in the Australian sector

Torres Strait fisheries on the Australian side are managed by either the Torres Strait Protected Zone Joint Authority (TSPZJA) or the State of Queensland, which was responsible for fisheries prior to the Torres Strait Treaty. The TSPZJA consists of one minister each from the national and Queensland governments, with guidance from a multi-tiered set of committees comprising various government agencies, the Island Coordinating Council, and fishing and scientific interests. In 1998 the TSPZJA managed the prawn, Spanish mackerel, pearl shell, rock lobster, dugong, turtle fisheries, barramundi and other traditional fisheries. Queensland managed all other commercial fisheries in the TSPZ until April 1999, when the TSPZJA assumed management responsibility for finfish, crab, trochus, and bêche-de-mer. On the PNG side, fisheries both in Torres Strait and the Gulf of Papua are managed by the National Fisheries Authority (NFA).

The management measures applied to these fisheries are summarised in Table 4. In the Torres Strait prawn fishery both the size of the fleet and the total fishing effort have been gradually reduced since 1985, when the fishery was first managed separately from the broader Queensland prawn fishery. The fleet has been reduced from about 110 to 86 vessels and effort capped at 13,570 access days since 1993, when analysis (Reid et al., 1993) indicated that the fishery was overcapitalised. Effort has also been reduced in the Gulf of Papua fishery, from 73–95,000 trawl-hours in the 1980s to 41–46,000 hours in 1991–93 (Evans and Opnai, 1995), and entry is limited to 13 vessels. These and other management measures applied to the prawn, rock lobster, and mackerel fisheries appear to be effective. The prawn and rock lobster industries appear stable, while the Spanish mackerel fishery is under-exploited. Reef fish, sedentary resources, and other fisheries in the Australian fishing zone in Torres Strait are for the most part harvested sporadically or for subsistence purposes, and ongoing management measures appear to be adequate under these circumstances.

Management of other coastal fisheries resources in Gulf and Western Provinces has not been as successful. The human and financial resources of the NFA are severely limited, and are typically concentrated on the high-value prawn and tuna fisheries, and on ongoing organisational

Table 4
Summary of management measures applied to major commercial fisheries

Fishery	Sector	Measures	Status
Gulf of Papua prawn	PNG	Limited entry, seasonal closure, gear restrictions, area closure within 3 nm of shore	Fully exploited
Torres Strait prawn	Australian	Seasonal and area closures, limited entry, effort quota, gear restrictions	Fully exploited
Torres Strait rock lobster	Australian	harvest only by hand or with hand-held implements, seasonal ban on use of hookah, minimum size, bag limits for traditional use, ban on retention of by-catch by trawlers, no expansion of non-indigenous capacity or effort	Approaching full exploitation
Spanish mackerel	Australian	gear restrictions, minimum size, expansion limited to traditional inhabitants	under-exploited
Pearl shell	Australian	taking of dead shell prohibited, minimum and maximum sizes, methods restricted to diving, expansion limited to traditional inhabitants	poor recovery from severe over-exploitation
Dugong	Australia, PNG	sanctuary in western Torres Strait, subsistence and traditional use only, commercial sale banned, harvest with traditional spear only	uncertain, population appears stable
Turtle	Australia, PNG	subsistence and traditional use only; traditional methods only; commercial sale banned	uncertain
Reef fish	Australia	none (primarily a subsistence fishery)	under-exploited
Bêche-de-mer	PNG	catch quota, minimum sizes, seasonal closure, no harvest by Scuba, hookah, nets, or at night with lights, participation restricted to PNG citizens, limits on no. of buyers and exporters	over-exploited, recovery apparently poor
Barramundi	PNG	minimum sizes, minimum net mesh sizes	over-exploited

restructuring and policy and programme development. Management measures that are in place, for example minimum size limits for barramundi and bans on the sale of dugong and turtle meat, are not enforced. Even if management measures on sedentary resources are observed, these fisheries are notoriously difficult to manage on a sustainable basis (see chapters in Wright and Hill, 1993). Many smaller fisheries, e.g. for shark, tuna, and coastal pelagics, are essentially unmanaged. In many instances resource stocks are prevented from over-exploitation by lack of investment capital and marketing infrastructure. The restriction of trade in sedentary marine product to PNG nationals has, in some cases, led to catch reductions (Kailola, 1995). Where there has been ready access to markets, however, severe overfishing has occurred. Notable examples are the collapses of the Daru barramundi and bêche-de-mer fisheries.

Environmental Management

Environmental legislation in PNG is adequate but the capacity and political will to implement it is generally lacking (Hedemark and Sekhran, 1994; Osborne, 1993). Large-scale mining projects do comply with requirements for environmental impact assessment and monitoring. Large-scale environmental degradation, notably severe impacts of mine tailing disposal in the Ok Tedi river, has been considered acceptable, relative to the economic benefits generated by the projects. Environmental impact legislation has not been effective in other industries. The logging industry in particular has been plagued by corruption and controversy (e.g., Marshall, 1990). As of 1990 only about a quarter of active logging projects had submitted environmental plans (Nadarajah, 1993). New forestry guidelines promulgated in 1993 sought to address these deficiencies but their implementation has been limited, although compliance with environmental plan submission does appear to have increased (Hedemark and Sekhran, 1994). Forestry environmental plans are sometimes written from a generic template with little site-specific information, and provisions such as the retention of buffer zones of intact forest along water courses are routinely violated (pers. obs.). Monitoring and enforcement are essentially non-existent. Large-scale agricultural projects are often exempted from environmental planning requirements (Hughes and Sullivan, 1989).

Primary responsibility for environmental management in PNG rests with the Department of Environment and Conservation (DEC), whose capacity is even more limited than that of the NFA. DEC suffers from a crippling lack of financial and human resources. Given the low level of, and resultant political pressure for, economic and human development, the government's high reliance on revenues from mining and logging, and the general perception within government that environmental management is inimical to development, DEC is given a very low priority in funding and other decisions. There has been a recent move to downgrade it from a line department to an "office", thereby depriving DEC of cabinet representation, and the administrative status of the department is unclear. In late 1999 the

Department was in fact downgraded, and staffing reduced by about half.

The PNG constitution recognised traditional land ownership rights, and some 97% of the land is privately held by the customary owners. Traditional sea rights are also implicitly recognised. A central component of DEC's conservation strategy developed in this context is the designation of Wildlife Management Areas (WMAs). These are areas reserved at the request of the customary landowners for the purpose of conserving specified wildlife species. WMAs have formal status as conservation areas, and DEC assists local committees in establishing and administering a set of management rules. There are three coastal WMAs in Gulf and Western Provinces (Fig. 5). Only the Maza WMA, established to manage turtles and dugong, is specifically marine; the Tonda and Neira WMAs were established to manage terrestrial wildlife. Though WMAs are designed to protect given species, rather than providing broader habitat conservation and environmental management, they have the potential to serve as useful frameworks for community-based management. Unfortunately, their effectiveness has declined, in part because of limited support and monitoring from government. Initiatives are underway to revitalize them.

The Australian sector of Torres Strait is subject to the environmental management legislation and policies of Australia and Queensland. Environmental assessment is required for new developments. The degree of assessment and level of approval (Island councils, Queensland government, Australian government) depend upon the scale of the project. Mineral and oil extraction is prohibited in the TSPZ, a provision which is periodically reviewed. Oil spill contingency plans have been developed for Torres Strait and for the port at Thursday Island. In 1996–1997 oil spill response equipment consisted of 300 m of general purpose boom and one small weir-type skimmer based at Thursday Island. Ships over 70 m and oil tankers are required to report to the Australian Maritime Safety Authority when entering and leaving Torres Strait, and there have been initiatives to improve navigational aids in the Strait including the installation of differential GPS beacons. A reticulated sewage system and treatment plant have been commissioned in Torres Strait. Given the low development pressure and absence of major shipping accidents, the environmental management regime appears adequate, although local degradation of reef, mangrove, and seagrass areas has occurred.

International Agreements

Australia and PNG are parties to a number of relevant international agreements, including:
- Convention on Biological Diversity
- RAMSAR Convention on Wetlands of International Importance
- Convention on the Prevention of Marine Pollution by Dumping of Wastes and Other Matter (London Dumping Convention)
- Convention on International Trade in Endangered Species of Wild Flora and Fauna (CITES)
- South Pacific Regional Environment Programme (SPREP) Convention

REFERENCES

Alongi, D. (1990) The ecology of tropical soft-bottom benthic ecosystems. *Oceanography and Marine Biology Annual Review* **28**, 381–496.

ANZDEC (1995) Asian Development Bank TA No. 2258-PNG. Fisheries management project marine fisheries sector plan and provincial fisheries profiles. Asian Development Bank, Port Moresby, Papuan New Guinea.

Bakker, M.L. (1996) The provincial populations of Papua New Guinea: profiles based on 1990 census data. UNFPA/ILO Project PNG/94/P01. Project Research Report No. 1. United Nations Population Fund/International Labour Organisation, Port Moresby, Papua New Guinea.

Barham, A.J. and Harris, D.R. (1983) Prehistory and palaeoecology of Torres Strait. In *Quaternary Coastlines and Marine Archaeology*, eds. P.M. Masters and N.C. Flemming, pp. 529–557. Academic Press, London.

Bode, L. and Mason, L. (1995) Tidal modelling in Torres Strait and the Gulf of Papua. In *Recent Advances in Marine Science and Technology '94*, eds. O. Bellwood, H. Choat and N. Saxena, pp. 55–65. James Cook University of North Queensland, Townsville, Australia.

Cragg, S. (1983) The mangrove ecosystem of the Purari delta. In *The Purari: Tropical Environment of a High Rainfall River Basin*, ed. T. Petr, pp. 295–324. W. Junk, The Hague.

Dalzell, P. (1993) Coastal fisheries production in the South Pacific. *SPC Fisheries Newsletter* **66**, 27–29.

Dews, G. and Harris, A. (1995) The islander-based seafood catches in the Torres Strait Protected Zone. In *Recent Advances in Marine Science and Technology '94*, eds. O. Bellwood, H. Choat and N. Saxena, pp. 107–117. James Cook University of North Queensland, Townsville, Queensland, Australia.

Eaton, P. (1982) Land Use Potential. In *Papua New Guinea Atlas*, eds. D. King and S. Ranck, pp. 46–47. Robert Brown and Associates, Bathurst, NSW, Australia.

Evans, C.R. and Opnai, L.J. (1995) Research and management of the commercial prawn fisheries of the Gulf of Papua and Orangerie Bay, Papua New Guinea. *Sci. New Guinea* **21**, 89–99.

Evans, C.R. and Polon, P. (1995) A preliminary stock assessment of the ornate rock lobster, *Panulirus ornatus*, in reefs of the Torres Strait protected zone, Papua New Guinea. *Sci. New Guinea* **21**, 59–68.

Evans-Illidge, E. (1997) Heavy metals in commercial prawn and crayfish species in Torres Strait. Report Series 5b. Great Barrier Reef Marine Park Authority, Townsville.

Frielink, A.B.J. (1983a) Coastal fisheries in Papua New Guinea, the current situation. Research Report 83-10. Department of Primary Industry, Papua New Guinea, Port Moresby.

Frielink, A.B.J. (1983b) A socio-economic study of the artisanal fisheries in the delta of Gulf Province Papua New Guinea. Report No 83-12. Fisheries Research and Surveys Branch, Department of Primary Industry, Papua New Guinea, Port Moresby.

Furnas, M.J. (1991) Biological oceanographic measurements in the Torres Strait and far northern Great Barrier Reef. In *Sustainable Development for Traditional Inhabitants of the Torres Strait Region. Proceedings of the Torres Strait Baseline Study Conference, Kewarra Beach, Cairns, Australia, 19–23 November 1990*, eds. D. Lawrence and T. Cansfield-Smith, pp. 197–212. Great Barrier Reef Marine Park Authority, Townsville, Australia.

Ganter, R. (1994) *The Pearl-Shellers of Torres Strait. Resource Use, Development, and Decline, 1860s–1960s*. Melbourne University Press, Carlton, Victoria, Australia.

Gwyther, D. (1983) The importance of the Purari River delta to the prawn fishery of the Gulf of Papua. In *The Purari: Tropical Environment of a High Rainfall River Basin*, ed. T. Petr, pp. 355–365. Junk Publ., The Hague.

Haines, A.K. (1978/79) The subsistence fishery of the Purari delta. *Sci. New Guinea* 6, 80–95.

Hair, C. (1994) Fisheries in Papua New Guinea. In *Papua New Guinea Country Study on Biological Diversity. A report to the United Nations Environment Program*, eds. N. Sekhran and S. Miller, pp. 169–185. Department of Environment and Conservation, Conservation Resource Centre, Waigani, Papua New Guinea.

Hanson, L.M., Bourke, R.M. and Yinil, D.S. (1998) *Cocoa and Coconut Growing Environments in Papua New Guinea. A Guide for Research and Extension Activities*. Australian Agency for International Development, Canberra.

Harris, A.N. and Poiner, I.R. (1990) By-catch of the prawn fishery of Torres Strait: Composition and partitioning of the discards into components that float or sink. *Australian Journal of Marine and Freshwater Research* 41, 37–52.

Harris, D.R. (1977) Subsistence strategies across Torres Strait. In *Sunda and Sahul: Prehistoric Studies in Southeast Asia, Melanesia, and Australia*, eds. J. Allen, J. Golson and R. Jones, pp. 421–463. Academic Press, London.

Harris, P.T., Baker, E.K., Cole, A.R. and Short, S.A. (1993) A preliminary study of sedimentation in the tidally dominated Fly River Delta, Gulf of Papua. *Continental Shelf Research* 13, 441–472.

Hedemark, M. and Sekhran, N. (1994) Overview of current conservation-oriented policies, legislation, and interventions. In *Papua New Guinea Country Study on Biological Diversity. A report to the United Nations Environment Program*, eds. N. Sekhran and S. Miller, pp. 339–359. Department of Environment and Conservation, Conservation Resource Centre, Waigani, Papua New Guinea.

Hughes, P. and Sullivan, M. (1989) Environmental impact assessment in Papua New Guinea: lessons for the wider Pacific region. *Pacific Viewpoint* 30, 34–55.

Johannes, R.E. and MacFarlane, J.W. (1991) *Traditional Fishing in the Torres Strait Islands*. Commonwealth Scientific and Industrial Research Organisation, Canberra.

Jones, M.R. (1995) The Torres Reefs, North Queensland, Australia: Strong tidal flows a modern control on their growth. *Coral Reefs* 14, 63–69.

Kailola, P.J. (1995) *Fisheries Resources Profiles: Papua New Guinea*. FFA Report No. 95/45. Forum Fisheries Agency.

Kare, B. (1995) A review of research on barramundi, reef fish, dugong, turtles and Spanish mackerel and their fisheries in the Torres Strait adjacent to Papua New Guinea. *Sci. New Guinea* 21, 43–55.

Levett, M. and Bala, A. (1994) Agriculture in Papua New Guinea. In *Papua New Guinea Country Study on Biological Diversity. A report to the United Nations Environment Program*, eds. N. Sekhran and S. Miller, pp. 125–153. Department of Environment and Conservation, Conservation Resource Centre, Waigani, Papua New Guinea.

Long, B.G. and McLeod, I. (1997) Distribution of mangroves in Torres Strait. Torres Strait Conservation Planning Final Report Annex 15. Commonwealth Scientific and Industrial Research Organisation. Marine Research Division, Cleveland, Queensland, Australia.

Long, B.G. and Poiner, I.R. (1997) Seagrass communities of Torres Strait, Northern Australia. Torres Strait Conservation Planning Final Report Annex 11. Commonwealth Scientific and Industrial Research Organisation. Marine Research Division, Cleveland, Queensland, Australia.

Long, B.G. and Skewes, T. (1997a) Distribution and abundance of commercial beche-de-mer on Torres Strait. Torres Strait Conservation Planning Final Report Annex 9. Commonwealth Scientific and Industrial Research Organisation. Marine Research Division, Cleveland, Queensland, Australia.

Long, B.C. and Skewes, T. (1997b) The potential influence of land and runoff on distribution and abundance of seagrass on the reefs in Torres Strait. Torres Strait Conservation Planning Final Report Annex 14. Commonwealth Scientific and Industrial Research Organisation. Marine Research Division, Cleveland, Queensland, Australia.

Long, B.G. and Taranto, T.J. (1997) Habitat classification of seabed areas of Torres Strait, Northern Australia. Torres Strait Conservation Planning Final Report Annex 17. Commonwealth Scientific and Industrial Research Organisation, Australia. Marine Research Division, Cleveland, Queensland, Australia.

Long, B.G., Bode, L. and Pitcher, R. (1997a) Seabed current stress as a predictor of the distribution and abundance of epibenthos in Torres Strait. Torres Strait Conservation Planning Final Report Annex 13. Commonwealth Scientific and Industrial Research Organisation. Marine Research Division, Cleveland, Queensland, Australia.

Long, B., Skewes, T., Taranto, T., Thomas, M., Isdale, P., Pitcher, R. and Poiner, I. (1997b) Seagrass dieback in north-western Torres Strait. Torres Strait Conservation Planning Final Report Annex 12. Commonwealth Scientific and Industrial Research Organisation. Marine Research Division, Cleveland, Queensland, Australia.

Long, B., Skewes, T., Taranto, T., Jacobs, D. and Dennis, D. (1997c) Torres Strait reef resource inventory and reef habitat mapping. Torres Strait Conservation Planning Final Report Annex 5. Commonwealth Scientific and Industrial Research Organisation. Marine Research Division, Cleveland, Queensland, Australia.

Louman, B. and Nicholls, S. (1994) Forestry in Papua New Guinea. In *Papua New Guinea Country Study on Biological Diversity. A report to the United Nations Environment Program*, eds. N. Sekhran and S. Miller, pp. 155–167. Department of Environment and Conservation, Conservation Resource Centre, Waigani, Papua New Guinea.

Marsh, H.M., Harris, A.N., and Lawler, I.R. (1997) The sustainability of the indigenous dugong fishery in Torres Strait. *Conservation Biology* 11, 1375–1386.

Marshall, G. (1990) The political economy of logging: The Barnett Inquiry into corruption in the Papua New Guinea timber industry. *The Ecologist* 20, 174–181.

McAlpine, J.R., Keig, G. and Falls, R. (1983) *Climate of Papua New Guinea*. Commonwealth Scientific and Industrial Research Organisation, Canberra, Australia.

Milton, D.A., Long, B.G. and Thompson, A.A. (1997) Influence of coastal processes on large scale patterns in reef fish communities of Torres Strait, Australia. Torres Strait Conservation Planning Final Report Annex 10. Commonwealth Scientific and Industrial Research Organisation, Marine Research Division, Cleveland, Queensland, Australia.

Moore, R. (1982) Spawning and early life history of the barramundi, *Lates calcarifer* (Bloch), in Papua New Guinea. *Australian Journal of Marine and Freshwater Research* 33, 803–813.

Nadarajah, T. (1993) Effectiveness of the Papua New Guinean Environmental Planning Act 1978 in forestry timber projects. NRI Discussion Paper 69. The National Research Institute, Boroko, Papua New Guinea.

NSR (1998) PNG Gas Project. Draft environmental impact statement and impact assessment study. Main Report. NSR Environmental Consultants Pty. Ltd., Hawthorn East, Victoria, Australia.

Osborne, P.L. (1993) Papua New Guinea. Introduction. In *A Directory of Wetlands in Oceania*, ed. D.A. Scott, pp. 301–320. IUCN, Cambridge.

Pernetta, J. and Hill, L. (1981) A review of marine resource use in coastal Papua. *J. Soc. des Oceanistes* 71–73, 175–191.

Petr, T. (1983) *The Purari: Tropical Environment of a High Rainfall River Basin.* Junk, The Hague.

Pitcher, R., Skewes, T. and Dennis, D. (1995) Biology and fisheries ecology of the ornate rock lobster in Torres Strait. In *Recent Advances in Marine Science and Technology '94*, eds. O. Bellwood, H. Choat and N. Saxena, pp. 311–321. James Cook University of North Queensland, Townsville, Queensland, Australia.

Poiner, I.R., Blaber, J.M., Brewer, D.T., Burridge, C., Caesar, D., Connell, M., Dennis, D., Dews, G.D., Ellis, N., Farmer, M., Glaister, J., Gribble, N., Hill, B.J., O'Connor, R., Milton, D.A., Pitcher, R., Salini, J.P., Taranto, T., Thomas, M., Toscas, P., Wang, Y., Veronise, S. and Wassenburg, T.J. (1997) Final report on the effects of prawn trawling in the far northern section of the Great Barrier Reef: Final report to the Great Barrier Reef Marine Park Authority and Fisheries Research and Development Corporation on 1991–96 (Years 1–5) Research. Commonwealth Scientific Industrial Research Organisation, Marine Research Division.

Poiner, I.R. and Harris, A.N.M. (1996) Incidental capture, direct mortality and delayed mortality of sea turtles in Australia's northern prawn fishery. *Marine Biology* **125**, 813–825.

Poiner, I., Skewes, T., Pitcher, R., Taranto, T.J. and Long, B.G. (1997) Torres Strait Conservation Planning Project Overview Report. Torres Strait Conservation Planning Final Report. Commonwealth Scientific and Industrial Research Organisation. Marine Research Division, Cleveland, Queensland, Australia.

Reid, C., Collins, P. and Battaglene, A. (1993) Torres Strait prawn fishery: An economic analysis. ABARE Research Report 93.15. Australian Bureau of Agriculture and Resource Economics, Canberra.

Robertson, A.I., Alongi, D.M., Christoffersen, P., Daniel, P.A., Dixon, P. and Tirendi, F. (1990) *The influence of freshwater and detrital export from the Fly River system on adjacent pelagic and benthic systems.* Australian Institute of Marine Science Report No. 4 No. 4. Australian Institute of Marine Science, .

Salomons, W. and Eagle, A.M. (1990) Hydrology, sedimentology, and the fate and distribution of copper in mine-related discharges in the Fly River system, Papua New Guinea. *The Science of the Total Environment* **97/98**, 315–334.

Tallegalla/Sanbis (1998) PNG Gas Project. Review of socio-economic impacts. Interim report. Report prepared by Tallegalla Consultants Pty. Ltd, and Sanbis Pty. Ltd. as Appendix 6 in: PNG Gas Project Draft Environmental Impact Statement (Commonwealth)Impact Assessment Study (Queensland). Annexures and Appendices. Report prepared by NSR Environmental Consultants Pty Ltd for South Pacific Pipeline Company Pty Ltd.

TSPZJA (1999) Torres Strait Protected Zone Joint Authority. Fisheries Management Annual Report 1997–1998. Australian Fisheries Management Authority, Brisbane.

UNDP (1994a) Yumi Wankain. Report of the United Nations joint inter-agency mission to Papua New Guinea on sustainable development. United Nations Development Programme, Port Moresby.

UNDP (1994b) Pacific Human Development Report. United Nations Development Programme, Suva, Fiji.

Veron, J.E.N. (1978a) Deltaic and dissected reefs of the far Northern Region. *Philosophical Transactions of the Royal Society of London B* **284**, 23–37.

Veron, J.E.N. (1978b) Evolution of the far northern barrier reefs. *Philosophical Transactions of the Royal Society of London B* **284**, 123–127.

Williams, G. (1994) *Fisheries and Marine Research in Torres Strait.* Australian Government Publishing Service, Canberra.

Wolanski, E., Pickard, G.L. and Jupp, D.L.B. (1984) River plumes, coral reefs and mixing in the Gulf of Papua and northern Great Barrier Reef. *Estuarine, Coastal and Shelf Science* **18**, 291–314.

Wolanski, E., Ridd, P. and Inoue, M. (1988) Currents through the Torres Strait. *Journal of Physical Oceanography* **18**, 1535–1545.

Wolanski, E., Norro, A. and King, B. (1995a) Water circulation in the Gulf of Papua. *Continental Shelf Research* **15**, 185–212.

Wolanski, E., King, B. and Galloway, D. (1995b) Dynamics of the turbidity maximum in the Fly River estuary, Papua New Guinea. *Estuarine, Coastal and Shelf Science* **40**, 321–337.

Wright, A. and Hill, L. (1993) *The Nearshore Marine Resources of the South Pacific.* Institute for Pacific Studies/Forum Fisheries Agency, Suva, Fiji/Honiara, Solomon Islands.

Zann, L.P. (1995) State of the Marine Environment Report for Australia. Technical Summary. Ocean Rescue 2000, Dept. of the Environment, Sport, and Territories, Canberra.

THE AUTHOR

Michael E. Huber
Global Coastal Strategies,
P.O. Box 606,
Wynnum, QLD 4178,
Australia

Chapter 93

NORTHEASTERN AUSTRALIA: THE GREAT BARRIER REEF REGION

Leon P. Zann

The Great Barrier Reef on Queensland's continental shelf is around 2000 km long, consists of over 2800 reefs and is the largest complex of coral reefs in the world. It is also one of the most diverse large marine ecosystems, with tropical estuaries, seagrass beds, continental and coral islands, shelf benthos, and fringing, platform and ribbon reefs. Offshore in Australia's Coral Sea Islands Territory are oceanic coral reefs, plateaus and deep water basins. The human population of the GBR Catchment is relatively low (one million), and the far north is largely uninhabited. Most of the reefs lie considerable distances offshore, and human impacts are largely focused around coastal population centres. However large-scale clearing of catchments for cattle grazing and sugarcane has resulted in major land erosion and increases of sediments and nutrients entering into coastal waters. This may be causing degradation of the distinctive inner-shelf reefs in some areas. Direct human uses include recreational and commercial fishing, tourism and shipping, but these are not large by world standards. The GBR Marine Park (area 344,000 km^2) is a multiple-use marine protected area, the largest in the world. The GBR Marine Park is a model for multiple-use, large ecosystem management. There is a good scientific information base and the GBR Marine Park is generally well resourced and managed. Most of the Coral Sea reefs are Australian Nature Reserves but are not actively managed. Major issues in the GBR region are: effects of terrestrial run-off on inner-shelf reefs; effects of fishing, and particularly of prawn trawling on shelf benthos; effects of concentrated tourism in some areas; a recent rapid decline in dugongs in the southern part of the GBR; and threats to GBR breeding turtles in neighbouring countries. Of great concern is the massive scale of coral mortality from repeated outbreak episodes of crown-of-thorns starfish (e.g. 17% of reefs in the 1979–92 episode) and from the unprecedented 1998 coral bleaching episode (affecting 87% of inner-reefs and 28% of offshore reefs). The dilemma facing management authorities is that many of the major issues are outside their control.

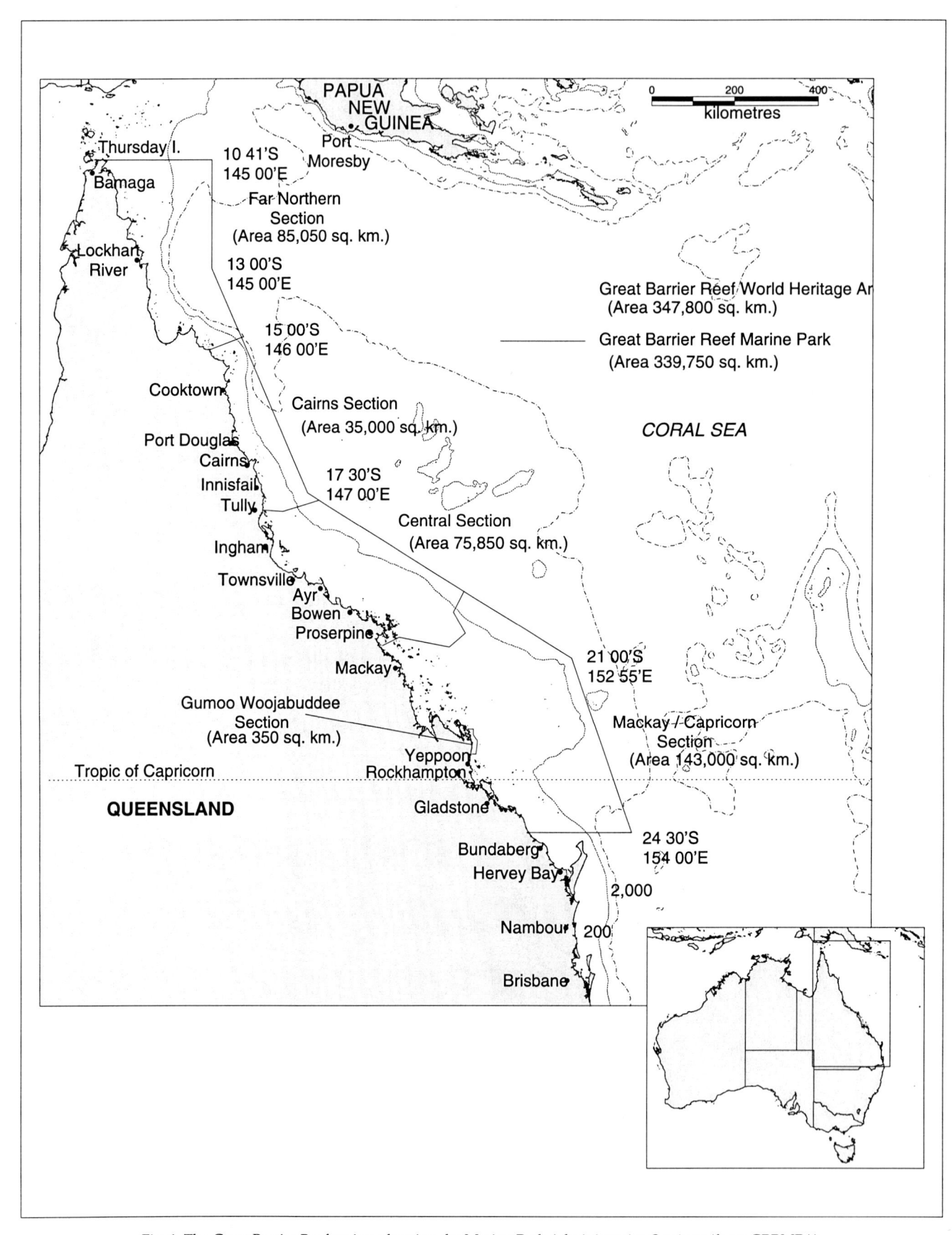

Fig. 1. The Great Barrier Reef region, showing the Marine Park Administrative Sections (from GBRMPA).

THE DEFINED REGION

The Great Barrier Reef (GBR) is the largest complex of coral reefs in the world. It is around 2000 km in length, 347,000 km² in area, and consists of 2,800 separate coral reefs (Figs. 1 and 2). It includes diverse marine habitats and communities, including inshore estuaries, soft and hard shores, saltmarsh, mangroves and seagrass; mid-shelf coral reefs, shoals, islands and shelf benthos; and outer-shelf ribbon reefs and continental shelf slopes and oceanic waters. The GBR is also one of the most biodiverse bioregions in the world. For example, it supports 14 species of seagrass (around one third of the world's total), 37 species of mangroves (over half of the world's total), 350 species of coral (over two thirds of the world's total) and up to 2,000 species of fish. It also has secure populations of threatened species such as giant clams, turtles and dugongs. The GBR was established as a Marine Park in 1975, and was inscribed in the World Heritage List in 1981 (Wachenfeld et al., 1998).

The region examined in this chapter comprises the Queensland mainland catchment and coast adjacent to the GBR, the Great Barrier Reef World Heritage Area (which includes the GBR Marine Park, plus islands and adjacent coastal waters), and the offshore Coral Sea reefs. It is referred to here as the GBR region (Fig. 1). It is continuous with the Torres Strait region to the north. The GBR region is relatively well known scientifically, and this review has largely been compiled from *State of the GBR World Heritage Area, 1998* and its sources (Wachenfeld et al., 1997; 1998).

Biogeography

The GBR region lies in the Northern Australian Marine Bioregion. It is part of the Indo-Pacific Marine Province, the largest and most diverse marine province in the world (Poore, 1995). It has been placed in the 'North Eastern Australia' marine domain. The biodiversity of the GBR varies along its north–south axis, and across the continental shelf. It has been divided into three latitudinal sectors: north (16–10°41'S), central (16–20°S) and southern (20–24°30'S); and three cross-shelf categories: inner-, mid- and outer-shelf (Wolanski, 1994). A tentative meso-scale bioregionalisation has divided it into eight meso-scale regions (IMCRA, 1998). The offshore Coral Sea reefs are not well studied but are typically Indo-Pacific in biodiversity.

Jurisdiction

The GBR lies on the eastern continental shelf of the State of Queensland. The coastline, islands and seas within a three nautical mile limit are part-Queensland. The GBR lies within Australian Commonwealth (national) waters and the Commonwealth GBR Marine Park. Most of the offshore reefs lie within the Commonwealth Coral Sea Islands Territory.

Fig. 2. The Great Barrier Reef consists of over 2800 separate reefs. A platform reef in the Capricorn Group (L. Zann).

SEASONALITY, CURRENTS, NATURAL ENVIRONMENTAL VARIABLES

Climate

Because of the great latitudinal range of the GBR region, the climate ranges from subequatorial in the north, to subtropical in the south. Major influences are the monsoonal wind and rainfall patterns which extend south to around 16°S. Annual rainfall ranges from 1700 mm in the north, to 900 mm in the south. Most rainfall falls in summer (>70% between October and March). Strong southeasterly trade winds dominate during the drier winter months (April–October), while weaker northerlies dominate the wetter summer months (Lough, 1997).

Cyclones

Tropical cyclones are major influences on rainfall and wave height, and have a major influence on coral reefs in the region. The belt from 16-20°S is most affected, and an average of four cyclones (with five cyclone 'days') affect the region each year (Lough, 1997). Over the past 28 years, at least 135 cyclones have affected the GBR region, and most reefs have probably been affected in some way. Much of the great community variability or 'patchiness' within and between reefs is now attributed to the different stages in recovery from cyclone disturbances (Wachenfeld et al., 1998).

ENSO

El Niño/Southern Oscillation events have a marked longer term influence on climate, cyclones, oceanography and biota of the region. The summer monsoon effect is notably enhanced during anti-ENSO (La Niña) years. Cyclone frequency is also affected; during anti-ENSO years an average 7.5 cyclones occur; during ENSO years an average of 2.5 cyclones occur. River flow and freshwater influences are much higher in anti-ENSO years (e.g. 688% higher for the Burdekin River, the major river in the region). Over the past three decades, major ENSO events occurred in 1997, 1992, 1990, 1987, 1983, 1978, 1975, and 1966 (Lough, 1997).

Currents

The circulation in the Western Coral Sea is strongly dominated by the inflowing South Equatorial Current, which comprises the northern limb of the South Pacific sub-tropical gyre. When this reaches the GBR shelf edge, at around 14°S, it bifurcates into the northwards flow, the Hiri current, and the stronger southwards flowing boundary current, the East Australia Current. These may induce localised upwellings of nutrient-rich deep waters onto the shelf. Below 14°S, the East Australia Current crosses the shelf to induce a net southerly flow but this can be modified or even reversed during periods of high south-east trade winds in which northerly flows are common. Current flows are complex at the meso-scale within the reef matrix and closely spaced reefs may locally influence or even block prevailing currents and isolate GBR lagoonal waters. Tidal flows may also superimpose a 12-hour oscillatory flow on the prevailing currents. Currents are also strongly affected by ENSO influences (Burrage et al., 1997).

Ocean Temperature and Salinity

In the Western Coral Sea a uniform salinity is found in the top 100 m, while a temperature stratification may exist in summer. These surface waters prevail over the shallow GBR shelf, but salinity declines towards the coast with terrestrial runoff (Wolanski, 1994). Annual ocean temperatures range from 26–29°C in the north, to 20–26°C in the south. Temperatures are higher in shallow inshore areas than in oceanic outer reefs (Lough, 1997).

Nutrients and Sediments

Coral Sea waters are generally oligotrophic but waters below the thermocline are nutrient-rich and may be carried to the surface in localised upwellings (Wolanski, 1994). Major inputs of nutrients are from terrestrial sources. Thirty-five drainage basins, with a total catchment of 450,000 km^2, discharge an average of 60 cubic kilometres of water into the GBR region each year. These input around 77,000 tonnes of N, 11,000 tonnes of P, and 23,000,000 tonnes of sediments each year. A large proportion of P (ca. 80%) and N (ca. 40%) is associated with fine sediments. Nutrient and sediment loads have increased by around four-fold in the 130 years since European colonisation of the adjacent coast. Strong southeasterly trade winds and occasional cyclones resuspend large volumes of shallow water sediments, releasing nutrients. These stimulate phytoplankton growth with chlorophyll concentrations reaching 18 μg/L (Wachenfeld et al., 1998).

In contrast with most other tropical regions, river flows are highly erratic from year to year, according to the strength of the summer monsoon, which in turn is coupled to global ENSO variability. River flow, occurring mainly in the summer, depresses inshore salinities in inshore areas, and following major floods, plumes may reach mid-shelf coral reefs. Inner-shelf waters (<20 m depth) are most affected (Furnas et al., 1997).

Tides and Waves

Tides are mixed diurnal and semidiurnal and show marked latitudinal and cross-shelf changes. Tidal ranges are 3–4 m in most areas, but near Broad Sound (23°S) the semidiurnal tides are amplified by northern and southern topographic influences to produce macrotidal conditions up to 9 m in range. Tidal currents vary and may be very strong in reef channels, to 2 m/s (Pickard et al., 1977; Wolanski, 1994).

The main wave-driving mechanisms are the forcing from the Coral Sea (characterised by prevailing south-east trades and long fetch), local winds (mainly affecting inshore waters) and tropical cyclones (extreme wave heights, storm surges). The reef matrix greatly filters waves entering the GBR lagoon. Waves breaking on steep-sided coral reef edges generate circulation over the reef platforms which may produce and maintain coral cays (Wolanski, 1994).

MAJOR SHALLOW WATER MARINE AND COASTAL HABITATS AND BIOTA

The following briefly describes the characteristics of the main inshore habitats, communities and selected biota; environmental pressures; and their general status.

Estuaries

There are around 190 estuaries along the adjacent Queensland coastline and larger islands. These range in size and importance, but around 20% of these are large and are considered to have high conservation and fisheries values (Bucher and Saenger, 1991). Estuaries support a highly productive saltmarsh, mangrove and seagrass communities, as well as many invertebrates, fish, reptiles and mammals.

Pressures and status: Major threats are increased erosion and sedimentation following clearing of catchments. Over half the catchments have been at least partially cleared (Bucher and Saenger, 1991).

Mangroves

Estuaries and mainland- and island-protected shores are dominated by highly productive mangrove wetlands. There are 37 mangrove species of 19 families in the region. There are around 4000 km^2 of saltmarsh and mangroves

(Duke, 1997). Floral communities vary greatly according to salinity gradients and vertical height in the intertidal gradient. In high rainfall humid regions in the north, there are tall (to 40 m) closed canopies dominated by a high diversity, mainly of *Rhizophora* and *Bruguiera* species. In arid areas, water and salinity stress increases intertidally and more open canopy woodlands or short shrublands develop (Robertson and Alongi, 1996).

Pressures and status: Major threats are reclamation and draining, mainly for agricultural purposes. There are no estimates of area lost, but this is not expected to be great. In many areas there has been a net increase in area because of increased sedimentation of estuaries because of land erosion. For example, in the Johnson River mangroves increased from 166 ha to 202 ha between 1951 and 1992, but there was a 65% loss of freshwater wetlands (Duke, 1997).

Island Communities

There are 940 islands and innumerable smaller islets in the GBR. Of these, 618 are continental islands and 87 are vegetated sand cays. These support a wide range of plant communities, from rainforests to rare native grasslands, and have high floral biodiversity. Around 2200 species of plants have been identified. This is around 25% of the total for Queensland, even though the island area represents only 0.1% of the land area. Island plant communities tend to be dominated by rainforest species (48%) and open-forest species such as eucalypts (46%), with littoral species (6%). Coral cays have a lower diversity (350–400 species in the north, and around 140 species in the south). Jurisdictionally, islands are part of the state of Queensland. Most are protected as national parks and are managed by Queensland Department of Environment and Heritage (Wachenfeld et al., 1998).

Pressures and status: Two islands are urbanised and 27 have tourist resorts. Pressures include introduced weed species, and feral animals such as rats and cats which prey on native wildlife and breeding seabirds. Some islands have been partially cleared for farming and been disturbed by grazing cattle. Increased tourist visitation rates have increased weed introductions, trampling of soils, disturbances to nesting seabirds and turtles, and bush fires (Wachenfeld et al., 1998).

Seagrasses

Seagrasses are important in primary production and support complex marine food webs in the region. They are critical nursery habitats for many species, including the juveniles of some commercial crustaceans and fish, and threatened species such as turtles and dugongs. Around 3000 km^2 of coastal seagrass habitat (<15 m) have been mapped in the region. It is estimated that there is also at least 2000 km^2 of deep-water habitat (15–58 m depths) which is important pasture for dugongs. Some 14 species of seagrass have been identified to date; two *Halophila* species may be endemic (Lee Long and Cole, 1997). *Pressures and status*: Seagrass beds may be smothered by sediments from rivers and dredging. Some beds may have expanded because of elevated nutrients. Little is known of the large natural, year-to-year variability in pastures, or of the newly discovered deep-water shelf communities (Lee Long and Coles, 1997).

Algae

The GBR region has a highly diverse algal flora, reflecting the great latitudinal extent, diversity of reef and other substrata and water conditions, and consequent great habitat diversity. Despite their great ecological importance, they are not well studied. Around 230 species have been identified (largely Indo-Pacific in origin, with low endemism), and it is likely that there are around 400–500 species in total. There are overt cross-shelf differences in algal communities, with an abundance of brown algae (Phaeophyta, e.g. *Sargassum*) in inshore areas, and red algae (Rhodophyta) offshore. There is a marked seasonal blooming of many algae, particularly the planktonic blue-green *Trichodesmium*, which may form vast slicks thousands of hectares in area. During summer, *Sargassum* blooms in inshore areas, while blue-green algae, red algae and green algae species bloom on mid- and outer-shelf reef flats. Little is known of natural variability and it is not possible to determine whether many blooms are natural or the result of human activity (McCook and Price, 1997).

Pressures and status: It has been suggested that the high abundance of *Sargassum* on inner-shelf reefs is the result of eutrophication and reef degradation. There is no strong scientific evidence to support this and studies indicate it may be more the result of reduced grazing pressure by fish (McCook and Price, 1997).

Shelf Benthos

Around 95% of the GBR region is the seafloor of the continental shelf, which ranges in depth from subtidal to around 60 m. The seafloor, almost entirely soft sediments, is ecologically diverse and complex. There is a marked, cross-shelf zonation in sediments and biota. Close to shore, sediments are fine, and of terrestrial origin; offshore they are coarser, calcareous marine sediments. While biota in inshore muddy areas are generally less diverse than offshore areas, both areas support rich benthic communities, including prawns and other species of commercial importance. In shallow inshore waters, and clearer deeper waters, benthic communities are dominated by seagrass and macroalgae (above). Macroalgae are particularly abundant in the deep-water, inter-reefal areas in the Northern Section. Here meadows of the green calcareous alga *Halimeda* cover up to 2000 km^2 in area, and form mounds up to 20 m in depth (Wachenfeld et al., 1998).

Pressures and status: While shelf benthos is an intrinsic part of the GBR ecosystem, its connectivities with coral reef communities and particular biodiversity values have not been adequately recognised in the past, and it has not been adequately studied or protected. The main pressures are probably increased terrestrial runoff and the effects of trawling. Clearing of coastal vegetation for agriculture and other developments have greatly increased runoff of nutrients and sediments. Trawling for prawns may remove or dislodge sessile and mobile benthos.

Coral Reefs

Coral reefs make up 6% of the GBR region's area. Of the 2900 separate reefs in the GBR, 566 are submerged shoals, 466 are patch reefs, 254 are cracentic reefs, 270 are lagoonal reefs, 544 are planar reefs, 545 are fringing reefs and 213 are incipient fringing reefs. Reefs show a marked latitudinal and cross-shelf variation. In the north where the continental shelf is narrow and there are fewer major rivers, there are well developed coastal fringing reefs and a semi-continuous line of ribbon reefs along the shelf edge. In the centre where the shelf is wider and there are major river influences, there are few inshore reefs and sand cays, and well developed mid- and outer-shelf patch reefs. In the south there are many cays, and well developed offshore reef complexes (Pompeys, Capricorns, Swains) (Zann, 1996).

The GBR has 350 different species of corals, compared with some 450 species in the Indonesian–Philippine region, at the centre of global coral species diversity. Most species have a wider Indo-Pacific distribution, but 10 species are considered endemic. Diversity varies latitudinally along the reef, with around 350 species in the north and 244 species in the south. Diversity and species assemblages also vary across the shelf. There is a lower species diversity and higher proportion of non-Acroporid forms such as *Galaxea*, *Montipora* and *Goniopora* in inner-shelf reefs, with more plate-forming *Acropora* in mid-shelf reefs, and more digitate and sub-massive species on outer-shelf reefs. Coral cover also varies greatly from place to place, according to local environmental factors and past disturbances (Wachenfeld et al., 1998).

Pressures and status: Natural factors affecting coral reefs include elevated freshwater and sediments during flooding, physical destruction during storms and cyclones, predation by outbreaks of crown-of-thorns starfish and *Drupella* snails, and coral bleaching events. Some of these disturbances may be exacerbated by human activities, or may lower the resilience of coral reefs to more direct human stresses. More direct human pressures include dredging and reclamation, trampling, collecting and harvesting, damage by anchors, shading by pontoons and other structures, pollution and shipping accidents.

Although there are anecdotal reports of declining corals reefs on the GBR, particularly in inshore areas, there is little long-term scientific monitoring to confirm this. Considerable natural variability on reefs may mask human impacts. For example, in a 30-year study on Heron Island in the Southern Section, coral cover varied from 0–80%, and most declines in cover were due to cyclones. A study of six reefs in the Central and Northern Sections between 1980 and 1995 showed major community changes, deteriorating in some cases and improving in others. An analysis of coral banding spanning 230 years indicated a recent decline in growth rates, but also that there had been previous decreases. Analysis of early photographs taken on 14 reefs showed that there had been declines in coral cover at four sites, no obvious changes at six sites, and partial declines at four sites (Wachenfeld et al., 1998). While there is little scientific evidence of long-term declines on monitored reefs, massive mortality of corals due to crown-of-thorns starfish outbreaks and coral-bleaching events are well documented.

Crown-of-thorns Starfish Outbreaks

The first documented outbreak of the corallivore *Acanthaster planci* (Fig. 3) on the GBR began at Green Island off Cairns (19°S) in 1962. This outbreak spread slowly southwards in the next 14 years due to larval dispersal, and affected many as far south as Mackay (21°S). A second outbreak episode was also first recorded at Green Island, in 1979, but modelling suggests that both outbreaks originated further north, on reefs around 15°S. A third outbreak episode was detected around the predicted source reefs (around 15°S) in 1993, and has progressively moved southwards, to around 17°S by 1998. Effects on reefs vary from partial to almost complete coral mortality. In the second episode, around 17% of the entire GBR Reefs (around 500 reefs) were affected. Almost 60% of these reefs suffered 30–50% coral mortality. Mid- and outer-shelf reefs are most affected (Zann and Eager, 1987; Engelhardt et al., 1997).

Fig. 3. Outbreaks of the crown-of-thorns starfish (*Acanthaster planci*) have caused massive damage to parts of the GBR (L. Zann).

Outbreak Cause(s): As the extent of coral mortality in these outbreaks is on a massive scale, the possibility of human influences has been a highly controversial issue. Despite decades of research, the causes of outbreaks, and whether they are natural phenomena or human-induced, have not been determined. Studies of GBR sediments indicate that high densities were present in the past (Engelhardt and Lassig, 1993) and recent studies suggest that outbreak episodes may be correlated with ENSO events (Zann unpubl.). However, while it is claimed that recent scientific opinion favours natural causes (Engelhardt et al., 1997; Wachenfield et al., 1998), there is still no scientific evidence to preclude human influences.

Management Dilemma: In the absence of definitive evidence on causation, the GBR Marine Park Authority has adopted a policy of minimal intervention, and only strategic controls are undertaken in areas of importance to tourism or scientific research (Engelhardt et al., 1997). In practice, manual controls are not feasible at a scale beyond a few hectares. The great economic importance of some the contributing factors proposed (e.g. fishing and run-off from mainland agriculture) makes unequivocal scientific evidence necessary, and an effective management response very difficult.

Coral Bleaching Events

Coral polyps may eject their symbiotic zooxanthellae when under temperature, osmotic or other stress, and lose their colours, or 'bleach'. They may later regain zooxanthellae, or may die. Mass coral bleaching events occur on whole reefs and even reef systems. Events of varying intensity occurred on the GBR in 1980, 1982, 1987, 1992, 1994 and most recently, in early 1998 (Birkelman and Oliver, in press).

The 1998 episode was on a vast scale. Aerial surveys found that 87% of inner-shelf reefs had at least moderate bleaching (>1% of corals bleached, 66% had high levels (>10% bleached) and 25% had extreme bleaching (>60% bleached). Some 28% of offshore reefs had moderate bleaching, 14% had high levels and none had extreme levels. Bleaching was more intense in the Central and Southern Sections. Ground surveys by divers of 23 reefs found the aerial surveys actually underestimated the extent of the phenomenon, and that it was most intensive on the reef flat and upper 4 m. The overall mortality rate has not yet been determined, but preliminary surveys indicate that it has been high in inner- and mid-shelf reefs. For example, on Orpheus Island fringing reef 60–80% of coral on the reef flat had died, and on Otter Reef, 70% of coral on the reef flat to 6 m depths had died (Birkelman and Oliver, in press).

Cause(s): The massive 1998 event is attributed to high ocean temperatures, intense solar radiation and, in the Central Section, lowered salinities after flooding. Massive coral bleaching was also experienced in widespread reefs of the Pacific, Indian and Caribbean in late 1997 and early 1998, indicating a global, probably ENSO-related influence (Birkelman and Oliver, in press). The massive, global scale of the 1998 bleaching event has led to the suggestion that it may result from Greenhouse Effect global warming. In many parts of the world, 1998 was the warmest year recorded, and the last decade has been the warmest recorded. It also indicates that in many part of the world, corals live very close to their upper temperature tolerances, and that any rise in ocean temperature will have major, widespread consequences on coral reefs (Harrison, pers. comm.).

Invertebrates

The GBR also supports many thousands of other invertebrates (e.g. >4000 species of molluscs). Most species (90–95% in most taxa) are Indo-Pacific in origin. Little is known of the pressures on, and status of, invertebrates. Trawling may affect populations of benthic invertebrates, particularly sessile sponges. There have been concerns that the Heron Island volute (*Cymbiolacca pulchra*) is threatened in the Capricorn group (Zann, 1996).

Reef Fish

There are between 1500 and 2000 species of fish in the GBR (Sweatman et al., 1998). In the better studied Capricorn Group, there are around 920 species, belonging to 123 families. The most speciose families there are Gobiidae (104 species); Labridae (69 species), Pomacentridae (69 species), and Blenniidae (40 species), Serranidae (32 species), Chaetodontidae (32 species) (Russell, 1983; Lowe and Russell, 1990). There are few endemic species (e.g. only four of 91 Pomacentrid species) (Sweatman et al., 1997). While many reef species are locally uncommon, they have a very wide geographic distribution and large overall populations.

Reef fish have a strong cross-shelf distribution pattern; there is a weaker latitudinal pattern, with more species found in the north. Research has found that many reef fish species are long-lived and have highly variable annual recruitment, resulting in a high natural variability of populations on reefs. This has made assessment of the effects of fishing more difficult. Distributions of pelagic and inshore species are not well known (Wachenfeld et al., 1998).

Pressures and status: Recreational and commercial line fishing, and to a lesser extent aquarium fish collecting and spearfishing, are the direct pressures on fish (Wachenfeld et al., 1998). Degradation of habitats, particularly inshore wetlands, are indirect pressures. Fisheries target the larger predatory reef fish species (e.g. coral trout and snappers), and the abundant smaller species and herbivores (e.g. parrotfish) are unfished. While there are many anecdotal reports of declines in the fished species, results of catch and effort studies and underwater visual surveys are unclear. Little is known of the effects of loss of habitat, but surveys on reefs affected by crown-of-thorns indicate a decline only in the obligatory coral feeders (Sweatman et al., 1998).

Reptiles

One species of crocodile, 17 species of sea snakes, and six species of turtles are found in the GBR region (Wachenfeld et al., 1998). Estuarine crocodiles (*Crocodylus porosus*) occur between Cape York and the Tropic of Capricorn, but are most common in mainland estuaries and rivers of Cape York Peninsula. Populations south of Cairns are fragmented and under pressure. Populations were greatly reduced by commercial hunting until the 1960s, but they are now protected (Miller and Bell, 1997). Sea snakes are common in muddy inner-shelf waters and around reefs but little is known of their ecology and status. Four species of turtles (flatback, hawksbill, loggerhead and green) have internationally significant populations in the region; the flatback nests only in northern Australia (Wachenfeld et al., 1998).

GBR turtles: pressures and status

The sea turtle species are characterised by being slow-growing, slow to mature (decades), long-living, long-distance migratory, non-annual breeding, with great fidelity to feeding and nesting sites. Their life history makes them vulnerable and they are globally threatened. They are protected in Australia except for traditional hunting by indigenous people (Limpus, 1997). The following summarises the status of GBR breeding populations.

Green turtles (Chelonia mydas)

There are two breeding units, the northern (>30,000 females annually), and the southern (ca. 8000 females annually). Annual nesting numbers vary greatly, according to ENSO events (with a two-year lag). Adults migrate up to 2,600 km between nesting and feeding sites. Tagging indicates that over half the nesting population have feeding grounds in neighbouring countries (South Pacific and Indonesia). The largest source of mortality is hunting, mainly in neighbouring countries. Additional sources of mortality are boat strikes, fisheries' by-catch and entanglements in marine litter. GBR populations are unsustainable, largely because of the very high mortality due to hunting outside Australia. It is listed as 'vulnerable' in Australia (Limpus, 1997).

Loggerhead Turtle (Caretta caretta)

This species is also highly migratory (to 2600 km), and most breeding within the South Pacific stock occurs within the GBR region and Southern Queensland (ca. 1000 females annually). There has been a 50-80% decline in nesting females since the late 1970s. This has largely been attributed to fisheries by-catch, boat strikes, entanglements and hunting in Papua New Guinea. Most anthropogenic mortality occurs within Australia. The regional populations are unsustainable. It is listed as 'endangered' (Limpus, 1997).

Hawksbill Turtle (Eretmochelys imbricata)

The Northern Australian stock is distinct from others in the Indo-Pacific, and is the largest remaining population in the Pacific. It is extensively hunted for its shell and meat in neighbouring countries, and its populations have been greatly reduced. There is limited census and demographic data. Populations are unsustainable. It is listed as 'vulnerable' (Limpus, 1997).

Flatback (Natator depressus)

The eastern Australian stock (ca. 1000 females annually) breeds mainly on continental islands in the GBR and do not migrate outside the GBR region and are more easily managed. Anthropogenic mortality is probably mainly fisheries by-catch and boat strikes. Census data suggests the population may have commenced a decline. It is listed as 'vulnerable' (Limpus, 1997).

Seabirds

The region supports many land, shore and seabirds. There are 55 species of seabirds, and 1.4–1.7 million breeding seabirds. There are over 55 nesting islands, mainly in the north and south; the Capricorn-Bunker islands in the south support 73–75% of the total seabird biomass of the region. Populations are highly variable, largely due to El Niño fluctuations in food availability (Hulsman et al., 1997).

Pressures and status: Natural pressures on seabirds include El Niño events which limit food, and cyclones and storms which kill and disturb breeding birds and destroy nesting sites. Human pressures include loss of habitat (through tourist developments), and disturbance of nesting birds by tourists, boats and aircraft. As most nesting sites are on low coral cays, sea level rise due to Greenhouse climate change would have a major effect. Although populations are naturally variable from year to year, long-term monitoring indicates that populations are stable at most sites. However, between 1984 and 1994 there was a 46% decline in common noddy terns at Michaelmas Cay, probably due to a rapid rise of visitors to the island (to a peak of 70,000 p.a.). Human use of the cay has now been restricted to a small area, and visitation restricted to 50 people at a time. Seabirds are totally protected in the region, and access to many nesting islands is seasonally restricted (Hulsman et al., 1997).

Mammals

One species of sirenian (the dugong), and at least 26 species of cetaceans are resident or visitors to the GBR. These include humpback whales, which migrate annually from Antarctic feeding grounds to breed in the Southern and Central GBR, minke whales (including the 'dwarf' minke), sperm whales, pilot whales, killer whales, false killer whales and various dolphins (Corkeron, 1997).

Pressures and status: Populations of the great whales are slowly recovering since the cessation of whaling in the 1960s. The eastern Australian humpback population has increased from a low of 200–500 in 1961, to over 2500 in 1997. While GBR populations of Indo-Pacific humpback dolphins and Irrawaddy dolphins appear secure, there are concerns regarding their wider status, particularly in the Asian area. There are also considerable concerns on the status of dugongs, particularly in the southern GBR (below) (Wachenfeld et al., 1998).

Dugongs: Pressures and status

Dugongs (*Dugong dugon*) are large, herbivorous marine mammals found in the Indo-Western Pacific. They are long-lived and slow to reach maturity, with a low reproductive rate. It is estimated that for populations to be sustained, total mortality cannot exceed 5% per year, and the maximum rate of hunting cannot exceed 1–2% of females per year. Populations outside Australia have been severely depleted by over-hunting and they are globally 'vulnerable'. They are protected in Australia, but traditional hunting by Aboriginal and Torres Strait Islander communities may be permitted (Marsh and Corkeron, 1997).

Monitoring by aerial surveys since 1984 indicates that populations in the northern GBR (north of 15°30'S) are around 8190 (±11720) and are considered secure. However, those in the south have been rapidly declining, from 3470 (±459) in 1986/87, to 1682 (±236) in 1994, and are now regarded as 'critically endangered' in this area (Marsh and Corkeron, 1997).

Pressures: Dugongs are an important traditional festive food for Aboriginal communities. There are no accurate estimates of kills, but in the Hopevale Aboriginal community, one of the main Aboriginal communities, 74 dugongs were taken between 1984 and 1987. Since the southern decline became evident, most communities have voluntarily agreed to cease hunting (Marsh and Corkeron, 1997).

Dugongs may become entangled and drown in nets deployed to keep beaches safe from sharks, and in commercial fish gill nets. Since shark meshing has been employed in Queensland in the mid-1960s, 540 dugongs have been caught, most of which died. Because of the high take of dugongs and turtles, many nets have been replaced by hook and lines. Gill-netting for inshore fish is a significant fishery in the region, and there are around 1000 net licences in Queensland. While there are no estimates of by-catch of dugongs, it is considered high in some areas (e.g. in 6 months of 1996, 14 dugong carcasses were reported, at least eight of which has evidence of net marks) (Marsh and Corkeron, 1997).

While most of the known seagrass beds in the GBR are protected from trawling, seagrass are susceptible to dieback as a result of increased sediments and algal overgrowth due to elevated nutrients. In Harvey Bay, immediately south of the GBR region, around 1000 km^2 of seagrass beds died in 1992-93 after floods, and there was a major mortality and emigration of dugongs. Noise from boats may deter dugongs from some areas (Marsh and Corkeron, 1997).

Management: A management plan has been developed as a response to the recent rapid decline in dugongs. This includes bans on hunting in the southern GBR, and controls or total bans on commercial gill netting in 16 new 'Dugong Protection Areas' along the coast. Shark meshing has also been reduced (Wachenfeld et al., 1998).

OFFSHORE SYSTEMS

Major sea-floor features in the Coral Sea include the Queensland Plateau (to 2000 m depths) which has extensive coral reef development in shallower waters, and the Coral Sea Basin (>4500 m depths). The climate is oceanic, with a rainfall 1250–1500 mm p.a., and annual air temperatures are 26–28°C in the north, and 18–24°C in the south. Surface sea temperatures are 25-30°C in the north, and 20–26°C in the south. Surface salinity ranges from 34.6–35.4 ppt. The main currents are the South Equatorial Current in the north and consequent Hiri and East Australian Currents along the GBR, and the Western Central South Pacific Current in the south. The nutrient status of surface waters is very low (Pickard et al., 1977). Jurisdictionally, most of this area lies within Australia's Coral Sea Island Territory.

The main coral reefs on the Queensland shelf are Holmes and Flinders (west), and Willis, Herald, Coringa and Tregosse (east). These reefs are protected in the Coringa-Herald and Lihou Reef National Nature Reserves, managed by the Commonwealth Australian Nature Conservation Agency. Few detailed scientific studies have been conducted in the area. Reefs have less habitat complexity than the GBR, and tend to have a lower coral cover, and a high coralline algae and sponge cover. The cays are important seabird roosting and nesting sites for 24 species of seabird (Griffin, 1996).

POPULATIONS AFFECTING THE AREA

Indigenous People

The GBR region has been inhabited by Aboriginal people for around 50,000 years. This long pre-dates the Holocene sea-level rise and even the formation of the present GBR. The coastal and island Aboriginal people traditionally have a maritime culture and fished, hunted and harvested seafood from shores and inner reefs. Tribal and clan estates extended from the shoreline to include islands, coastal and reef waters. The sea was (and remains) very important for subsistence, and culturally and spiritually. In the far north, Torres Strait Islanders, of Melanesian origin, traditionally hunted and fished off Cape York reefs (Benzaken et al., 1997).

With European colonisation, Aboriginal people were dispossessed from their land/sea estates, and populations were reduced by warfare and diseases. Early this century, the survivors (many from the inland regions) were forcibly relocated into coastal and island settlements away from their traditional lands. Today there are 11 Aboriginal and Torres Strait Islander communities in the North, with a total resident population of 11,000. In addition, 8000 Aboriginals and 3850 Torres Strait Islanders reside in urban areas. Around 11% of indigenous people still engage in hunting and fishing activities. In some Cape York communities, up to 80% of dietary protein comes from the sea (Benzaken et al., 1997).

Despite dispossession, the indigenous people continue to identify themselves as traditional owners and custodians of their land/sea estates. While Aboriginal interests were not considered in the *GBR Marine Park Act 1975*, and minimally in the initial establishment of the GBR Marine Park, their interests are increasingly recognised. Indigenous communities are now represented in the GBR Consultative Committee and the Authority (the governing body). The *GBR Marine Park Act* was specifically amended in 1995 to give representation of indigenous interests in management. The Commonwealth *Native Title Act 1993* recognised customary law and customary land tenure, and while it did not rule on sea rights, this was recently recognised in the Federal Court's 'Croker Island Decision' in 1998. There are now nine Native Title claims by Aboriginal communities over parts of the GBR Marine Park. Indigenous peoples aspire to greater self-determination and a meaningful role in the economic development and management of the GBR region (Benzaken et al., 1997; Wachenfeld et al., 1998).

Trends in Populations

The GBR catchment was colonised by Europeans between 1860 and 1880. The GBR coastal population has grown rapidly and is today around 700,000. That of the entire GBR catchment is 1.1 million (Fig. 4a). The region also receives an additional one million visitors each year. While the overall population density in the catchment is low (1 person/38 ha), the population is concentrated in coastal towns and cities. The fastest growing area is Cairns (4.3% pa). The Northern Section (eastern Cape York Peninsula) has a very low coastal population of around 8000 (Doyle, 1998). The Coral Sea Islands are uninhabited.

RURAL FACTORS

The GBR Catchment consists of 28 river catchments, and is 42 million ha in area (Fig. 4). The major rural land uses are cattle grazing (77%) and cropping (3%). Catchments have been extensively modified since European colonisation some 130 years ago (Pulsford, 1993).

Agriculture and General Impacts in the Region

The catchments vary greatly in topography, rainfall, soils and agriculture. Most of the land has been partially cleared for cattle grazing (ca. 4.5 million head of cattle). Some 3.4 million ha is improved pasture containing introduced sown species, and 1.3 million ha is used for cropping. The main crop is sugarcane (390,000 ha) which has been very rapidly growing in recent decades (Fig. 4b). Bananas and cotton are also important (Pulsford, 1993).

Land Clearing

Extensive areas of wetlands, rainforest and woodlands in coastal plains, ranges and tablelands have been cleared for agriculture, particularly in central and southern areas. For example, in the lower Herbert catchment *Melaleuca* wetlands have been reduced from 30,000 ha to less than 5000 ha, and rainforest from 12,500 ha to less than 4000 ha. In the upper Fitzroy catchment 4 million ha of brigalow

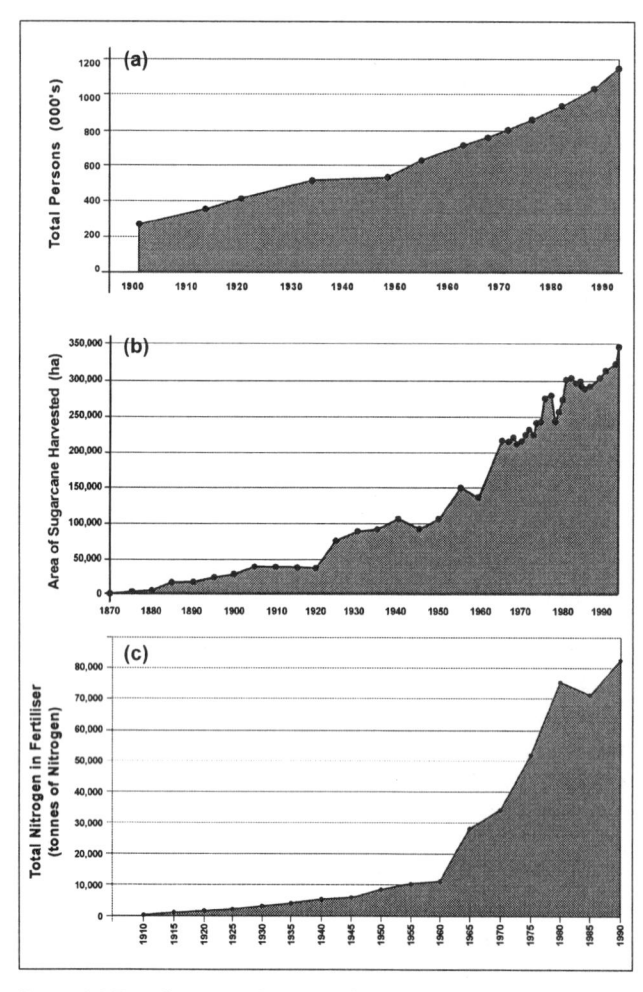

Fig. 4. (a) Trends in population in GBR Catchment since 1900. (b) Trends in sugar cane cultivation in GBR Catchment since 1870. (c) Trends in nitrogen fertilizer application since 1910. (from Wachenfeld et al., 1998)

bushland was cleared for cattle grazing between 1950 and 1975 (Wachenfeld et al., 1998).

Sedimentation and Nutrients

There have been major changes in hydrology, soil erosion and sedimentation of rivers and coastal areas as a consequence of land clearing (Fig. 5). Higher rates of water runoff have resulted in more serious flooding, and increased freshwater, sediments and nutrients entering inner GBR waters. It is estimated that sediment and nutrient levels have increased four-fold, and each year around 15 million tonnes of sediments, 77,000 tonnes of nitrogen and 11,000 tonnes of phosphorus are discharged. The greatest source of this has been land clearing for cattle grazing, and most of this increase has occurred in the past 40 years (Brodie, 1997).

Fertiliser Application

Regional soils are naturally limited in nutrients, and 3.3% of the sown pastures and 37% of the crops are fertilised. Application is increasing rapidly. For example, in 1990, 83,000 tonnes of nitrogen and 13,500 tonnes of phosphorus were applied. This was equal to the total applied between 1900 and 1945 (Fig. 4c). Most of this (71%) was applied to sugarcane (Pulsford, 1993)

Acid Sulfate Soil Run-off

Drainage of acid sulfate soils, which are formed in anaerobic conditions along tropical shores, may result in acid run-off and acidification of estuaries. This kills fish and other organisms and is the probable cause of 'red-spot' disease which affects estuarine fishes in the region (Wachenfeld et al., 1998). Although there has been extensive drainage of potential acid soils in the region for sugar cane, and fish kills are common, the extent of the problem has not been adequately recognised in the region.

Recreational Fisheries and Impacts

The region supports small indigenous subsistence fisheries, commercial fisheries, and major recreational fisheries. Around 30% of the coastal population go fishing at least once per year. Of these, 29,000 fish in offshore reef waters (37% making more than five trips per year). There are around 24,300 private boats in the region, most of which are outboard skiffs used for recreational fishing. The recreational fleet and fishing effort is fast growing, increasing five-fold between 1968 and 1994 (Higgs, 1997; Blamey and Hundloe, 1993).

The annual number of recreational fishing trips is 210,000–270,000, and the catch is 3500–4300 tonnes. Catch per unit effort is around 0.78 kg/person/h, and the average catch is 1.4 kg. The major fish are coral trout (*Plectropomus* spp.), lethrinids and lutjanids. The main techniques are line

Fig. 5. The possible decline of inner-shelf reefs due to increased sediments and nutrients is a major issue in the GBR. *Top*: Flood plumes from the Tulley and Murray Rivers. *Bottom*: Smothering of coral by sediments, Magnetic Island. (L. Zann)

fishing from private boats on reefs and open water, fishing from charter boats and fishing from shores. Spearfishing is of some significance. Because of the limited range of the recreational fleet, fishing effort is concentrated on reefs closer to the coast (Higgs, 1997; Blamey and Hundloe, 1993).

The status of the fishery is not adequately known. Surveys of the recreational catch and effort suggest that the inshore catch and effort is declining; 59% of fishers interviewed also considered that catch rates had declined in the previous five years, and 41% thought mean sizes had also declined (Blamey and Hundloe, 1993). Studies of records of fishing clubs found a decline in mean size in some species in some areas, but that there had not been a significant decline in overall catches over 30 years of records. While it was concluded that there was no evidence to support public opinion on declining catches (Higgs, 1997), interpretation of catch and effort does not consider the great technological advances in reef fishing over 30 years (e.g. GPS, depth sounders, fish finders), and increasing visitation by fishing clubs to more distant, less fished reefs. Because the recreational and commercial line fisheries often compete for the same fish, it is necessary to consider both fisheries jointly.

COASTAL EROSION AND LANDFILL

Coastal erosion is a localised problem in the region. The GBR mainland coastline is generally protected from prevailing seas by the matrix of mid- and outer-shelf reefs, while many islands' shores are protected by fringing reefs. Some areas are periodically prograding and receding. For example, the shores of Flying Fish Point near Innisfail prograded up to 100 m between 1883 and 1922, and receded by 90 m between 1922 and 1968 (Bird, 1985). Coastal engineering structures (breakwaters, sea walls, boat ramps etc.) have created localised erosion problems in some places. Shore erosion is a particular problem in Green and Heron Island sand cays where natural sand dynamics has been altered through the construction of buildings and seawalls (e.g. Hacker and Gouley, 1997).

Land fill is only a localised problem. In some ports and urban areas, shores have been dredged or reclaimed. More serious has been drainage of freshwater and estuarine swamps for sugarcane cultivation in some catchments. Coastal progradation, due to increased fluvial sediment yields from cleared catchments, has been extensive. For example, the Barron Delta near Cairns prograded by 300 m between 1942 and 1960, and the Burdekin River delta prograded up to 3 km between 1942 and 1980 (Bird, 1985).

EFFECTS OF URBAN AND INDUSTRIAL ACTIVITIES

Urbanisation

Around 3% of the GBR Catchment is occupied by urban and rural development. The major towns and cities in the region are: Gladstone (27,000), Rockhampton (60,000), Mackay (71,500), Townsville (132,000), and Cairns (113,000). These are growing at 1.1–4.3% pa (Doyle, 1998).

Mining

The region's hinterland is very rich in coal and minerals and there are a number of large refineries, power stations and ports along the coast. There are over 30 major metallic mines in the GBR catchment (gold, copper, zinc, silver etc.), including one of the largest open cut gold mines in the world (Mount Leyshon, producing 6000 kg p.a.). Central Queensland has amongst the world's largest coal deposits (reserves over 10,000 million tonnes), and supplies most of the State's coal exports (100 million tonnes p.a., worth $4.4 billion pa) (Doyle, 1998). The major industrial centre is Gladstone, in the south. Gladstone port handles over 20 million tonnes of coal, bauxite, alumina, cement and grain each year. Other industrial centres are Mackay (sugar processing, coal mining and port) and Townsville (copper and nickel refineries, sugar and mineral port) (Doyle, 1998). The major issues from industrial development are localised loss of coastal habitat and loss of water quality (see pollution).

Tourism

Each year, over 10 million visitor nights are spent in the region, and tourism are brings in over $1 billion. In the GBR Marine Park in 1997 there were 742 tourist operations, 23 reef structures or pontoons, 1674 vessels and 1.6 million visitor days were spent on the GBR. Tourism is growing very rapidly (averaging 10% p.a.), particularly since the introduction of high speed vessels which have greatly extended the range of day trip operations. Most tourism (95%) is centred in the offshore Cairns and Whitsunday areas (Dinesen and Oliver, 1997; Wachenfeld et al., 1998).

The impacts of tourism are localised, and include damage to corals by divers, reef walkers, anchors and tourist pontoons; effects of fishing and collecting by visitors; waste discharges from vessels and tourist resorts; damage to island vegetation and disturbances to nesting turtles and seabirds, and social and cultural impacts. Tourist operators are subject to an Environmental Management Charge equivalent to $4 per visitor per day (Dinesen and Oliver, 1997).

Shipping and Ports

Major shipping routes running along the GBR lagoon (Inner Route) and off the eastern boundary (Outer Route) are used by over 2000 ships annually. For example, over a six month period in 1997 and 1998, there were 723 transits along the Inner Route, and 850 transits through the four main passages through the outer GBR. Vessels included bulk carriers, ore carriers, container vessels, gas tankers, oil tankers, RORO cargo and other ships, of up to 200,000 dwt in size. Bulk cargoes included metal ore (bauxite, alumina, manganese, nickel etc.), sugar, coal and oil. Oil tankers (to 60,000 dwt) make up 10% of shipping, and other vessels may carry up to 10,000 tonnes of fuel oil (Wachenfeld et al., 1998).

Fig. 6. Large-scale clearing and reclamation of salt marsh and mangroves for industrial development, Gladstone (L. Zann).

Between 1985 and 1997, there were 28 ship groundings and 19 collisions in the region, but none resulted in major oil or chemical spills. However, small chronic spills, illegal cleaning of tanks and ballast water and operational discharges have increased hydrocarbon levels in many channels and ports (Wachenfeld et al., 1998). Introduction of exotic species through ballast water discharges is a serious problem in southern Australia and a potential problem in the GBR region. As the biodiversity of inshore waters is not well known, it is not possible to identify exotic species.

There are many small ports, and 10 major ports along the GBR coastline. The main ports and approximate annual throughputs are: Hay Point (36 million tonnes), Gladstone (32 million tonnes), Abbot Point (6 million tonnes), Townsville 94.5 million tonnes, Mackay (1.6 million tonnes) and Cairns (1.2 million tonnes) (Quirk, 1994). Environmental impacts include localised loss of coastal habitats, increased turbidity from ship movements and dredging, and hydrocarbon and heavy metal pollution.

Pollution in GBR region

Hydrocarbons

The main source of hydrocarbons is from urban and industrial discharges, operational discharges from ships and minor spills while loading. Hydrocarbons are present in most harbours examined, and toxic polyaromatic hydrocarbons are high in sediments in Townsville and Gladstone harbours (Wachenfeld et al., 1998).

Heavy metals

The main source of heavy metals is from terrestrial runoff (naturally high in some catchments), agri-chemicals, metal refineries, ships' hulls, and spills of ore in some ports. Sediments in Townsville are contaminated by nickel, chromium, iron and zinc. High concentrations have been found in tissues of some organisms. Corals off Townsville had relatively high levels, and a range of bivalves had extremely high levels of arsenic (probably natural). Seagrass sampled along the coast had high levels near population centres. Dugongs had unusually high concentrations of zinc in liver tissues, and cadmium in kidneys; levels detected in Cape York and Torres Strait are sufficiently high to have health implications for human consumers (Wachenfeld et al., 1998).

Chlorinated Compounds

The main sources of chlorinated compounds are herbicides and pesticides, mainly from agricultural use, and particularly from sugar cane farms. Pesticide residues have been detected in rivers (e.g. significant levels of atrazine and endosulphan in Fitzroy River, and lower levels in Johnstone River) and in sediments around river mouths. Trace levels of DDT and lindane have been detected in offshore waters in the Coral Sea. Dioxans and PCB congeners were detected in seagrass along the coast. DDE and dieldrin have been detected in fish livers off Townsville. Lindane, dieldrin and other chlorinated compounds have also been detected in dugong. In most cases levels have been very low and it is not known whether these pose any threat to the biota or ecosystem (Wachenfeld et al., 1998).

Commercial Fishing

Commercial fishing is the major extractive industry within the GBR region, and has an economic value of around $143 million p.a. The main fisheries are prawn trawling, reef line fishing, and inshore netting and crabbing. There are minor fisheries for aquarium fish, beche-de-mer, trochus shell and coral. There are 840 commercial fishing vessels (most 12–25 m in length) licensed for offshore otter trawling. The main landings are prawn (5000 tonnes p.a.), other crustaceans (500 tonnes p.a.), scallops (1000 tonnes p.a.) and fish (1500 tonnes p.a.) (Healy, 1997). There is little long-term spatial and temporal information on catch and effort. Catches monitored since 1988 indicate considerable year-to-year variation, but catch per unit effort declined for some species (scallops and eastern king prawns), despite increased gear efficiency (e.g. sonar and GPS). The major issues are impacts of trawling on benthic species (see box) (Wachenfeld et al., 1998).

There are 250 licences for line fishing, mainly for reef species such as coral trout (1200 tonnes p.a.), emperors (1200 tonnes p.a.) and pelagic Spanish mackerel (300 tonnes p.a.). Fishers generally operate a mother vessel (20 m) with freezer or live-holding tanks, and smaller tenders or dories for fishing. There is a growing live food-fish export to Asia. While there are anecdotal reports of declining catches in some areas, long-term information on catch and effort is limited. Monitoring since 1988 indicates a generally stable trend, but with possible declines in some species (e.g. sweetlip, *Lethrinus miniatus* and mackerel, *Scomberomorus commerson*) (Elmer, 1997).

There are around 300 net licences for GBR inshore waters. The main techniques are beach seining and mesh netting (for mullet, flathead, whiting etc.), and set nets (for barramundi, threadfin salmon and grunter). There is little long-term and spatial data on trends. The major issue is by-catch, particularly of dugongs.

Effects of trawling

The effects of prawn trawling on the benthic communities in the GBR have been a controversial issue. Around 95% of the GBR region is shelf communities which support commercially valuable prawns and other species. Most trawling is non-selective, and much of the haul (50–90% in many

Fig. 7. Effects of trawling on sea floor communities is a major issue in the GBR region (L. Zann).

cases) consists of detached sessile and mobile benthic organisms, fish, and occasionally sea snakes and turtles. This by-catch is discarded, and generally dies (Wachenfeld et al., 1998).

Long-term studies on the effects of trawling by CSIRO indicate that a single trawl sweep removes around 10% of biomass attached to the sea bed, and after 13 passes up to 90% of the biomass is removed (Fig. 7). Northern GBR by-catch includes 245 species of fish. Trawling may also affect food pathways, favouring scavengers (sea birds, dolphins, sharks, some fish, crabs) and benthic detrital feeders. In 1998 the study concluded that trawling has significant impacts, and that there may be significant environmental impacts from repetitive trawling (Wachenfeld et al., 1998).

There is little long-term spatial information on catch-and -effort to assess trawling impacts on the GBR. Recent spatial databasing of log-book catch-and-effort data indicates that effort is patchy (reflecting the distribution of target species), most of the GBR shelf is trawled and in half the trawled area the effort was high (>2000 hours per 6′ lat/long grid monitored). There is also considerable illegal trawling of protected areas (Wachenfeld et al., 1998).

Evaluation of Management

The importance of shelf benthos and the need for a 'systems' approach in management was not well recognised in earlier zoning plans, and minimal areas were protected from trawling. Spatial protection is also poor, and only one large area is protected, in the remote Far Northern Section. Policing is minimal but compulsory satellite vessel monitoring devices will greatly aid in monitoring and enforcement. By-catch reduction devices are to phased in over the next few years following threats of US embargoes on Australian prawns caught without by-catch excluders. The issue is politically sensitive as commercial fisheries are under the jurisdiction of the State of Queensland and are beyond the immediate control of the GBR Marine Park Authority. However, as a result of the recent findings by CSIRO, the Commonwealth has threatened to intervene if Queensland does not improve fisheries management.

PROTECTIVE MEASURES

GBR Marine Park

Around 98.5% of the GBR region is included in the Commonwealth GBR Marine Park. This is a multiple-use marine protected area, and covers an area of 344,000 km^2, with significant additional inshore areas to be included in 1999. In addition, most of the Queensland coastal waters and islands are Queensland marine and terrestrial protected areas. The entire area is managed collaboratively, and together makes up the GBR World Heritage Area. The offshore Coral Sea Island Territory is protected in two Commonwealth National Nature Reserves (Griffin, 1996).

Management Framework

The GBR Marine Park was established under the Commonwealth *GBR Marine Park Act 1975*, and is managed by the GBR Marine Park Authority, based in Townsville. The management philosophy of the Authority is based on: management at the ecosystem level; conservation while allowing sustainable use; community involvement in management; and monitoring and performance evaluation of management. A 25-year strategic plan has been developed by GBR 'stakeholders' to provide long-term goals and objectives in planning, management and research (Wachenfeld et al., 1998).

The GBR Marine Park Authority is responsible for the overall planning and management of the Marine Park, while day-to-day management is undertaken by Queensland Department of Environment and Heritage (QDEH). Fisheries are managed by the Queensland government (above).

Human and Financial Resources

The GBR Marine Park Authority has a staff of 110, and QDEH marine section has a staff of 100. Total funding in 1998/99 is Au $27.2 million. Of this, $17.8 million is provided by the Commonwealth; $3.98 by the Queensland Government; and $8.0 million through an Environment Management Charge on visitors to the Marine Park (of $4 per person per day) (Wachenfeld et al., 1998).

The great size and ecological complexity of the GBR necessitate a strong scientific base for management. While the Authority and QDEH coordinate research, most GBR research is undertaken at the Australian Institute of Marine Science (AIMS) and the Cooperative Research Centre (CRC) Reef Research Centre, in Townsville, and in universities. AIMS with a staff of 180, and an annual budget of $23 million, undertakes core scientific research in reef ecosystems, oceanography and other areas. The CRC, which is funded by government and industry (through the Environment Management Charge and government contributions) supports applied research in ecologically sustainable use of the GBR (e.g. effects of tourism, fishing, shipping).

Management Mechanisms

The GBR Marine Park is divided into four administrative sections, each with a comprehensive zoning plan. Within each zone, reefs or groups of reefs are designated for different purposes (Table 1). Some reefs may also be subject to seasonal and short-term closures, while more sensitive or heavily used areas (e.g. Low Isles, Green Island, Whitsunday Islands) have detailed, strategic management plans (Zann, 1996).

Policies have also been formulated, in consultation with industry and other groups, to provide guidelines for specific types of activities (e.g. offshore tourism development, pontoons, floating hotels). Activities such as commercial tourism developments, commercial fishing and specimen collecting require a permit, to which specific conditions are attached (e.g. on place, size, duration). If the nature of a development's impacts are potentially significant, alterative options are considered. If there are none, an Environmental Impact Assessment is required under the Commonwealth *Environment Protection (Impact of Proposals) Act 1974* (Wachenfeld et al., 1998).

Management Issues

The major issues identified by the GBR Marine Park Authority are: (a) protection of the natural values of the GBR (by protecting threatened species and habitats through a system of protected representative areas, and determining effects of disturbances (such as crown-of-thorns starfish); (b) reducing the adverse impacts of tourism while providing diverse tourist opportunities; (c) maintenance of ecologically sustainable fisheries (through research on impacts, monitoring of catch and effort, reducing by-catch); (d) maintaining water quality (through reducing impacts of coastal development and pollution, and risks of shipping spills and accidents) (Wachenfeld et al., 1998). The following summarises management mechanisms in these areas.

Biodiversity Conservation

Biodiversity management has been mainly through habitat protection, and management of threatened species. Large areas of coral reefs and inshore seagrass are protected, but some habitats (e.g. inner-shelf reefs and shelf benthos) are not adequately protected. There is no agreed bioregionalisation and no formal process for identifying representative areas (see 'Evaluation of management', below). The conservation status of most species is unknown; protection of threatened species has focused on

Table 1

Main zones in GBR Marine Park, by Section, number of reefs, area, % section area and total area within GBR (from Whitehouse, 1992)

Zone type/section	Mackay/Capricorn	Central	Cairns*	Far Northern	Total GBR
GUA	279 reefs	16 reefs		219 reefs	
	126,737 km^2	61,023 km^2	25,600 km^2	61,247 km^2	274,607 km^2
	85.1%	79.3%	73%	74.0%	80.0%
GUB	642 reefs	531 reefs		304 reefs	
	20,111 km^2	13,873 km^2	7,800 km^2	9,960 km^2	51,744 km^2
	13.5%	18.0%	22%	12.0%	15.7%
MNPA	10 reefs	5 reefs		13 reefs	
	103 km^2	597 km^2	600 km^2	135 km^2	1,435 km^2
	0.1%	0.8%	2%	0.2%	0.4%
MNPB	115 reefs	32 reefs		183 reefs	
	1,940 km^2	1,460 km^2	700 km^2	11,1113 km^2	15,213 km^2
	1.3%	1.9%	2%	13.4%	4.4%
Sci/Pres	8 reefs	5 reefs		9 reefs	
	109 km^2	44 km^2	100 km^2	52 km^2	206 km^2
	–	–	<1%	0.3%	0.06%
Total section	1,052 reefs	589 reefs		735 reefs	
	149,000 km^2	77,000 km^2	35,000 km^2	83,000 km^2	343,265 km^2**

GUA: General Use A (trawling and other legal activities permitted).
GUB: General Use B (trawling not permitted but other legal activities permitted).
MNPA: Marine National Park A (or Conservation Park Zone) (commercial fishing not permitted, recreational fishing permitted).
MNPB: Marine National Park B (or National Park Zone) (no collecting).
Sci/Pres: Scientific or Preservation Zone (no entry).
*Equivalent to new categories in 1992 zoning plan. **Does not include buffer zones (total ca. 700 km^2).

the charismatic macrofauna (dugongs and turtles). The effects of disturbances such as crown-of-thorns starfish and coral-bleaching is well known, but the underlying causes are not.

Impacts of Tourism

Tourism management is largely through pre-existing zoning and permits, with specific conditions. Permits for tourist developments are assessed on a 'first-come, first served' basis with a tenure of six years. As this is *ad hoc* in nature, and does not consider long-term and cumulative impacts, a more strategic planning approach has been taken in the fast-growing Cairns area. Carrying capacities may necessitate controls on certain places or activities (Dinesen and Oliver, 1997).

Fisheries Management

Recreational fisheries management is through zoning restrictions, catch and size limits on the most popular species, and prohibitions on commercial sales. Commercial fishing is managed by the Queensland Fisheries Management Authority and Department of Primary Industries. It is regulated through limited entry, licences and gear restrictions, and some spatial and seasonal closures (Robertson, 1997).

Water Quality

Water quality management is difficult as most sources of contaminants are from outside the GBR Marine Park. Within the Marine Park, sewage discharges must be tertiary treated, or secondary treated with land re-use. While no coastal urban areas discharge directly into the Marine Park, most treat sewage to secondary level and discharge into coastal water ways. Discharges of industrial wastes are regulated by the new Queensland *Environment Protection Act 1994*. Dredging and sea dumping is regulated by GBR Marine Park permit. Management of agricultural run-off (nutrients and sediments)—the major problem—is very difficult and beyond the control of the GBR Marine Park Authority (Brodie, 1997; Wachenfeld et al., 1998). While the scale of the problem is now widely recognised, and better land-use practices are being promoted, it will require a massive, long-term commitment to reverse a century of land degradation.

Shipping and Oil Spills

The major management mechanism is prevention. The International Maritime Organisation has identified the GBR as a 'Particularly Sensitive Area', the first in the world. This enabled the introduction of compulsory pilotage for all vessels over 70 m, or carrying bulk oil or chemicals in some passages. An oil spill contingency plan, REEFPLAN, exists and stockpiles of oil spill response equipment are based in Townsville and Brisbane. However, the reality of large distances and difficulties in dealing with oil spills are such that a major spill outside of port limits would pose great problems. Oil drilling is prohibited under the *GBR Marine Park Act* (Zann, 1996).

Evaluation of Management of the GBR

The GBR Marine Park is considered a model in large marine ecosystem conservation, and multiple-use management (e.g. RAC, 1993). The management model entails a large-scale, ecosystem approach, and long-term strategic objectives. There is a very strong legislative, education and scientific resource base, and strong community support. Jurisdictional difficulties are overcome through a collaborative approach and agreed responsibilities. The GBR Marine Park is relatively well resourced financially, although the effort must be spread over a very large area, and a multitude of issues.

While the GBR model is an appropriate one in its context, it is not necessarily appropriate to other coral reef areas, particularly those in undeveloped nations with a limited scientific and management base, subsistence economies, limited economic resources and alternatives and large populations.

There are, of course, problems with the model, and its practical implementation. It relies very heavily on zoning plans which were developed in the early 1980s as a matter of political urgency. There was then very limited knowledge on biodiversity, functional ecology and bioregionalisation. Zoning was based on the coral reefs (only 5% of the total area), with a strong bias towards offshore reefs of high aesthetic value, and low existing patterns of use. While it was intended that zoning plans would be revised every five years to accommodate changes in scientific understanding and increasing pressures, this has not occurred for bureaucratic reasons. As a result, zoning plans are now rather arbitrary and *ad hoc* in nature; do not consider the imperatives of water quality maintenance and ecosystem connectivities; under-represent the more distinctive and more threatened inner-shelf communities; and minimally protect shelf benthos. There is still no detailed inventory or bioregionalisation for biodiversity conservation (Zann and Brodie, 1997).

The multiple-use management model predisposes itself to criticism from user groups. There have been a number of parliamentary inquiries on controversial issues, and the effectiveness of the GBR Marine Park Authority. Inquiries include crown-of-thorns starfish outbreaks and controversial tourist developments such as floating hotels, Magnetic Keys and Port Hinchenbrook. A recent audit by the Australian National Audit Office was critical of the Authority's structure and effectiveness, resulting in a restructuring in 1998 to better address the major issues (biodiversity conservation, water quality, effects of fishing, etc.) (GBRMPA, 1998).

Jurisdictional arrangements are of concern. The relatively benign tourist industry is managed by the well-resourced Commonwealth. Fisheries, by far the greatest extractive use of the GBR, is the responsibility of Queensland, is poorly resourced, and has not been well managed. This resulted in the threat in early 1999 of Commonwealth intervention.

Evaluation of Management in Coral Sea Island Territory

The Commonwealth Coringa-Herald and Lihou Reef National Nature Reserves are infrequently visited and are not effectively monitored or managed. It has been suggested that the GBR Marine Park be expanded to include the Coral Sea Island Territories (e.g. Whitehouse, 1993).

CONCLUSIONS: STATUS OF GBR REGION AND MAJOR ISSUES

The 1998 'State of the Great Barrier Reef World Heritage Area', while noting many potential threats, indicated that these were being addressed by the management agencies, and expressed a cautious optimism on the state of the GBR region (McPhail, 1998). However, the massive scale of coral mortality in repeated outbreak episodes of crown-of-thorns starfish and the 1998 coral bleaching episode; possible degradation of inshore reefs from terrestrial run-off; the rapid decline of dugongs in the southern GBR; and the threats to GBR breeding turtles in neighbouring countries are of great concern. The dilemma facing management authorities is that most of these issues are outside their immediate control.

ACKNOWLEDGEMENTS

I would like to thank Associate Professor Vicki Harriott for reviewing this chapter.

REFERENCES

Benzaken, D., Smith, G. and Williams, R. (1997) A long way together: the recognition of indigenous interests in the management of the Great Barrier Reef World Heritage Area. In *State of the Great Barrier Reef World Heritage Area Workshop*, eds. D. Wachenfeld, J. Oliver and K. Davis. Workshop series No. 23, Great Barrier Reef Marine Park Authority, Townsville pp. 471–497.

Birkelmans, R. and Oliver, J. (in press) Large scale bleaching of corals on the Great Barrier Reef. *Coral Reefs* 18, 55.

Bird, E.C.F. (1985) *Coastline Changes*. John Wiley and Sons, London

Blamey, R.K. and Hundloe, T.J. (1993) *Characteristics of Recreational Boat Fishing in the Great Barrier Reef Region*. Institute of Applied Environmental Research, Griffith University, Queensland.

Brodie, J. (1997) The water quality status of the Great Barrier Reef World Heritage Area. In State of the Great Barrier Reef World Heritage Area Workshop, eds. D. Wachenfeld, J. Oliver and K. Davis. Workshop Series No. 23, Great Barrier Reef Marine Park Authority, Townsville, pp. 69–96.

Bucher, D. and Saenger, P. (1991) An inventory of Australian estuaries and enclosed marine waters: an overview of results. *Australian Geographic Studies* 9, 370–381.

Burrage, D., Steinberg, C., Bode, L. and Black, K. (1997) Long-term current observations on the Great Barrier Reef. In *State of the Great Barrier Reef World Heritage Area Workshop*, eds. D. Wachenfeld, J. Oliver and K. Davis. Workshop series No. 23, Great Barrier Reef Marine Park Authority, Townsville, pp. 21–45.

Corkeron, P. (1997) The status of cetaceans in the Great Barrier Reef Marine Park. In *State of the Great Barrier Reef World Heritage Area workshop*, eds. D. Wachenfeld, J. Oliver and K. Davis. Workshop series No. 23, Great Barrier Reef Marine Park Authority, Townsville, pp. 283–287.

Dinesen, Z. and Oliver, J (1997) Tourism impacts. In *State of the Great Barrier Reef World Heritage Area Workshop*, eds. D. Wachenfeld, J. Oliver and K. Davis. Workshop series No. 23, Great Barrier Reef Marine Park Authority, Townsville, pp. 414–427.

Doyle, B. (1998) *Queensland Yearbook 1998*. No. 56. Australian Bureau of Statistics, Brisbane.

Duke, N.C. (1997) Mangroves in the Great Barrier Reef World Heritage Areas: current status, long-term trends, management implications and research. In *State of the Great Barrier Reef World Heritage Area Workshop*, eds. D. Wachenfeld, J. Oliver and K. Davis. Workshop series No. 23, Great Barrier Reef Marine Park Authority, Townsville, pp. 288–299.

Elmer, M. (1997) Reef fish fisheries in the Great Barrier Reef World Heritage Area. In *State of the Great Barrier Reef World Heritage Area Workshop*, eds. D. Wachenfeld, J. Oliver and K. and Davis. Workshop series No. 23, Great Barrier Reef Marine Park Authority, Townsville, pp. 346–351.

Engelhart, U. and Lassig, B. (1993) *The Possible Causes and Consequences of Outbreaks of the Crown-of-thorns Starfish*. Workshop Series 18, Great Barrier reef Marine Park Authority, Townsville.

Engelhardt, U., Miller, I., Lassig, B.R. and Bass, D. (1997) Crown-of-thorns starfish (Acanthaster planci) populations in the Great Barrier Reef World Heritage Area: status report 1995-96. In *State of the Great Barrier Reef World Heritage Area Workshop*, eds. D. Wachenfeld, J. Oliver and K. Davis. Workshop series No. 23, Great Barrier Reef Marine Park Authority, Townsville, pp. 158–184.

Furnas, M., Mitchell, A. and Skuza, M. (1997) River inputs of nutrients and sediments into the Great Barrier Reef. In *State of the Great Barrier Reef World Heritage Area Workshop*, eds. D. Wachenfeld, J. Oliver and K. Davis. Workshop series No. 23, Great Barrier Reef Marine Park Authority, Townsville, pp. 46–68.

GBRMPA (1998) Great Barrier Reef Marine Park Authority Annual Report 1997/98, GBRMPA, Townsville.

Griffin, M. (1996) National nature reserves. In *The State of the Marine Environment Report for Australia. Technical Summary*, ed. L.P. Zann. Great Barrier Reef Marine Park Authority and Department of the Environment, Sport and Territories, Canberra, pp. 441–447.

Hacker, J.L.F. and Gouley, M.R. (1997) Harbour bund wall construction at Heron Island: coral response. In *The Great Barrier Reef. Science, Use and Management. Vol. 2*. Great Barrier Reef Marine Park Authority, Townsville, pp. 55–60.

Healy, T. (1997) Queensland east coast trawl fisheries. In *State of the Great Barrier Reef World Heritage Area Workshop*, eds. D. Wachenfeld, J. Oliver and K. Davis. Workshop series No. 23, Great Barrier Reef Marine Park Authority, Townsville, pp. 352–362.

Higgs, J. (1997) Fishing club activities on the Great Barrier Reef. In *State of the Great Barrier Reef World Heritage Area workshop*, eds. D. Wachenfeld, J. Oliver and K. Davis. Workshop series No. 23, Great Barrier Reef Marine Park Authority, Townsville, pp. 335–345.

Hulsman, K., O'Neill, P. and Stokes, T. (1997) Current status and trends in seabirds on the Great Barrier Reef. pp. 259–282.

IMCRA (1998) Interim marine and coastal regionalisation for Australia: an ecosystem-based classification for marine and coastal environments. Version 3.3. Environment Australia, Department of Environment, Canberra.

Lee Long, W.J. and Coles, R.G. (1997) Status of seagrasses in the Great Barrier Reef region. In *State of the Great Barrier Reef World Heritage Area Workshop*, eds. D. Wachenfeld, J. Oliver and K. Davis. Workshop series No. 23, Great Barrier Reef Marine Park Authority, Townsville, pp. 185–193.

Limpus, C. (1997) Marine turtles of the Great Barrier Reef World heritage Area. In *State of the Great Barrier Reef World Heritage Area Workshop*, eds. D. Wachenfeld, J. Oliver and K. Davis. Workshop series No. 23, Great Barrier Reef Marine Park Authority, Townsville, pp. 256–258.

Lough, J.M. (1997) recent climate variation on the Great Barrier Reef. In *State of the Great Barrier Reef World Heritage Area Workshop*, eds. D. Wachenfeld, J. Oliver and K. Davis. Workshop series No. 23, Great Barrier Reef Marine Park Authority, Townsville, pp. 7–20.

Lowe, G.R. and Russell, B.C. (1990) Additions and revisions to the checklist of fishes in the Capricorn-Bunker Group Great Barrier Reef Australia. Great Barrier Reef Marine Park Authority Technical memorandum 19, Townsville.

Marsh, H. and Corkeron, P. (1997) The status of the dugong in the Great Barrier Reef Marine Park. In *State of the Great Barrier Reef World Heritage Area Workshop*, eds. D. Wachenfeld, J. Oliver and K. Davis. Workshop series No. 23, Great Barrier Reef Marine Park Authority, Townsville, pp. 231–247.

McCook, L. and Price, I.R. (1997) The state of the algae in the Great Barrier Reef: what do we know? In *State of the Great Barrier Reef World Heritage Area Workshop*, eds. D. Wachenfeld, J. Oliver and K. Davis. Workshop series No. 23, Great Barrier Reef Marine Park Authority, Townsville, pp. 194–204.

McPhail, I. (1998) Foreword. In *State of the Great Barrier Reef World Heritage Area, 1998*, eds. D. Wachenfeld, J. Oliver and J. Morrissey. Great Barrier Reef Marine Park Authority, Townsville.

Miller, J.D. and Bell, I.P. (1997) Crocodiles in the Great Barrier Reef World Heritage Area. In *State of the Great Barrier Reef World Heritage Area Workshop*, eds. D. Wachenfeld, J. Oliver and K. Davis. Workshop series No. 23, Great Barrier Reef Marine Park Authority, Townsville, pp. 248–254.

Pickard, G.L., Donguy, J.R., Henin, C., and Rougeries, F. (1977) *A Review of the Physical Oceanography of the Great Barrier Reef and Western Coral Sea*. Australian Institute of Marine Science Monograph Series Vol. 2. AGPS, Canberra.

Poore, G. (1995) Biogeography of Australia's marine biota. In State of the Marine Environment Report for Australia. Technical Annex, eds. L.P. Zann and P. Kailola. Great Barrier Reef Marine Park Authority and Department of the Environment, Sports and Territories, Canberra, pp. 75–84.

Pulsford, J.S. (1993) *Historical Nutrient Usage in Coastal Queensland River Catchments Adjacent to the Great Barrier Reef Marine Park*. Great Barrier Reef Marine Park Authority Research publication No. 40, Townsville.

Quirk, P. (1994) Development in ship safety standards: implications for protection of the Great Barrier Reef. In *Hulls, Hazards and Hard questions*. Great Barrier Reef Marine Park Authority Workshop Series No. 19, Townsville, pp. 25–43.

RAC (1993) Coastal Zone Inquiry—Final Report, Resource Assessment Commission, Australian Government Publishing Service, Canberra.

Robertson, A.I. and Alongi, D.M. (1996) Mangrove systems. In *The State of the Marine Environment Report for Australia. Technical Summary*, ed. L.P. Zann. Great Barrier Reef Marine Park Authority and Department of the Environment, Sport and Territories, Canberra, p. 39.

Robertson, J. (1997) The management of fisheries in the Great Barrier Reef Marine Park. In *State of the Great Barrier Reef World Heritage Area Workshop*, eds. D. Wachenfeld, J. Oliver and K. Davis. Workshop series No. 23, Great Barrier Reef Marine Park Authority, Townsville, pp. 395–401.

Russell, B.C. (1983) Checklist of fishes Great Barrier Reef Marine Park Capricorn section. Great Barrier Reef Marine Park Authority, Special Publication Series (1), Townsville.

Smyth, D. (1997) Recognition of Aboriginal maritime culture in the Great Barrier Reef Marine Park: an evaluation. In *State of the Great Barrier Reef World Heritage Area Workshop*, eds. D. Wachenfeld, J. Oliver and K. Davis. Workshop series No. 23, Great Barrier Reef Marine Park Authority, Townsville, pp. 496–501.

Sweatman, H., Bass, D., Cheal, A., Coleman, G., Miller, I., Ninio, R., Osbourne, K., Oxley, W., Ryan, D., Thompson, A. and Tomkins, P. (1998) Long-term monitoring of the Great Barrier Reef. Status Report 3. AIMS, Townsville.

Wachenfeld, D., Oliver, J. and Davis, K. (eds.) (1997) *State of the Great Barrier Reef World Heritage Area Workshop*. Workshop series No. 23, Great Barrier Reef Marine Park Authority, Townsville.

Wachenfeld, D., Oliver, J. and Morrissey, J. (eds.) (1998) *State of the Great Barrier Reef World Heritage Area. 1988*. Great Barrier Reef Marine Park Authority, Townsville.

Whitehouse, J.F. (1992) *A Review of the Great Barrier Reef Marine Park Authority*. Department of Environment, Sport and Territories, Canberra.

Wolanski, E. (1994) *Physical Oceanographic Processes of the Great Barrier Reef*. CRC Press, Boca Raton, Florida.

Zann, L.P. (1996) The Great Barrier Reef Marine Park: the world's largest multiple use managed area. In: The State of the Marine Environment Report for Australia. Technical Summary (Zann, L.P.). Great Barrier Reef Marine Park Authority and Department of the Environment, Sport and Territories, Canberra, pp. 416–424.

Zann, L.P. and Eager, E. (1987) The crown of thorns starfish. *Australian Science Magazine* 3, 14–55.

Zann, L.P. and Brodie, J. (1997) Towards a scientifically based implementation plan for ecologically sustainable use and biodiversity conservation in the Great Barrier Reef World Heritage Area. In: *The Great Barrier Reef: Science, Use and Management. Vol. 2*. GBRMPA, Townsville, pp. 128–138.

THE AUTHOR

Leon P. Zann
*Southern Cross University,
Lismore, New South Wales 2480,
Australia*

Chapter 94

THE EASTERN AUSTRALIAN REGION: A DYNAMIC TROPICAL/TEMPERATE BIOTONE

Leon P. Zann

Eastern Australia lies in an overlap of tropical Indo-Pacific and temperate bioregions because of the influence of the southward-flowing East Australian Current. It has a very high biodiversity, from tropical coral reef and seagrass in the north, to coastal lakes and kelp communities in the south. Offshore, Norfolk and Lord Howe Islands, and Elizabeth and Middleton Reefs, are the highest latitude coral reefs in the world. The coastal region includes South East Queensland and New South Wales (NSW), and supports almost half of Australia's population. Catchments have been extensively cleared for agriculture and water quality in many rivers and estuaries is poor. There has been major loss of wetlands and acid soil runoff causes fish kills and diseases in estuaries. The coastline is exposed, and much has been mined for heavy minerals and is subject to erosion. Urbanisation and industrialisation, particularly in the Sydney area, has resulted in sewage and heavy metal pollution. Many of the coastal fisheries are declining, largely because of loss of water quality and wetlands habitats, and in some cases overfishing. Marine environmental management (the responsibility of the two States to the three-mile limit, and the Commonwealth to the boundary of the 200 mile EEZ) is variable, and largely focused on the urban centres. Queensland has a system of marine protected areas while NSW has minimal protected areas. The major issues are declining water quality, loss of estuarine habitats, degradation of coastal lakes, and localised pollution from heavy metals and sewage in the Sydney area.

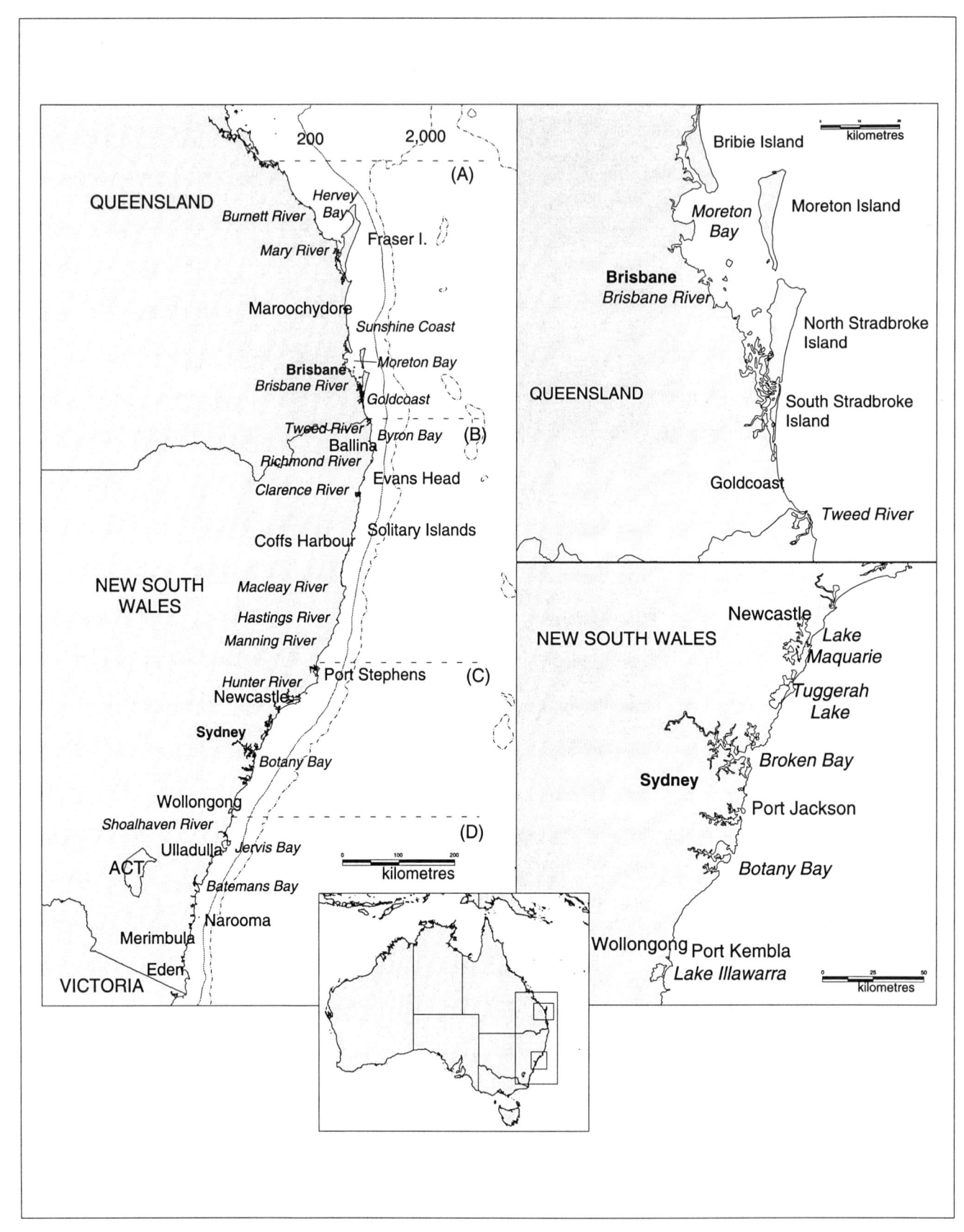

Fig. 1. Eastern Australia (South Eastern Queensland to NSW Victorian border) showing major coastal river catchments (italics) and urban areas. Insets (right): detail of Brisbane and Sydney metropolitan areas. The area is a tropical/tropical ecocline and has been divided into five sectors for this review: (A) Southeast Queensland; (B) Northern New South Wales; (C) Central New South Wales; (D) Southern New South Wales; and (E) Norfolk region (offshore).

THE DEFINED REGION

The eastern coast of Australia (from latitude 24°S to 37°S) lies in a overlap of the tropical Indo-Western Pacific and temperate bioregions, and therefore has a particularly rich biodiversity (Poore, 1996). The region, which adjoins the southern Great Barrier Reef in the north, includes the largest sand barrier islands in the world; large sheltered bays dominated by tropical seagrasses; coral reefs; open coastlines of long sandy beaches; river estuaries with well developed wetlands; drowned river valleys; rocky shores; and coastal lakes dominated by temperate seagrasses. Offshore lie the highest latitude fringing and platform coral reefs in the world.

The coastal zone in this region includes the south-eastern part of the State of Queensland, and all of the State of New South Wales (NSW). It supports around eight million people, almost half Australia's population, and includes two State capitals (Brisbane and Sydney), as well as Australia's major tourist and industrial centres. As a consequence, considerable pressures are placed on the region's unique marine environment.

Biogeography

A proposed bioregionalisation of Australia places this region in the Central Eastern Biotone, the warm temperate Central Eastern Province (shelf area 22,800 km^2), and the northern part of the South Eastern Biotone. The offshore reefs and islands are in the Norfolk province. The coastal bioregions are further separated into four meso-scale bioregions: Tweed-Moreton (the tropical/temperate overlap), Manning Shelf, Hawkesbury and Bateman's Shelf (IMCRA, 1998). However, as this region includes separate State jurisdictions and many local districts, for the purposes of this chapter the region has been divided into five geographic sectors: (a) South-East Queensland, (b) Northern NSW, (c) Central NSW; (d) Southern NSW) and (e) Norfolk Island, Lord Howe Island and Elizabeth and Middleton Reefs.

Fig. 2. Eastern Australia is characterised by long, high energy sandy beaches (L. Zann).

SEASONALITY, CURRENTS, NATURAL ENVIRONMENTAL VARIABLES

As the region covers 13 degrees of latitude and is a biogeographic ecotone, it has a range of climates and complex oceanography.

Climate

The climate in the northern sector is subtropical with hot summers, mainly summer–autumn rainfall (1400–2000 mm) and occasional cyclones. In the south it is moist cool–temperate, with warm summers, but no seasonal pattern of rainfall (IMCRA, 1998).

Oceanography

The water masses in the region are largely South Pacific Central Water in origin. These are driven southwest by prevailing winds through the Coral Sea where they are deflected southwards along the outer Great Barrier Reef and mainland coastline to become a moderately strong western boundary current, the East Australian Current (EAC) (Middleton, 1996).

Currents

The main stream of the EAC meanders and eddies southwards before finally separating from the coast, usually in the vicinity of Sugarloaf Point (300 km north of Sydney), although it may sometimes continue as far south as Ulladulla (400 km south of Sydney). Off Northern NSW, the EAC influences the continental shelf oceanographic circulation 90% of the time while the waters off Sydney are influenced 75% of the time (Ortiz and Burchmore, 1992). The EAC then turns easterly into the Tasman Sea towards New Zealand, forming a southern closure of the South Pacific sub-tropical gyre. Meanders of the EAC may form loops which can detach to form large eddies up to 300 km in diameter. Under the influence of strong southerly winds, the EAC may be accompanied by a northward counter-current, or even vanish completely (Burrage, 1993). As the EAC passes along the coast it leads to the formation of even smaller-scale eddies and recirculation zones which are determined by the size and bathometry of headlands and coastal embayment. The direction of the current flow may be reversed due to northward flowing coastally trapped waves which originate in Bass Strait and southern NSW (Middleton, 1996). The current's velocity is variable, depending on prevailing winds, but may flow at up to 5 knots (CSIRO, 1987).

Surface Sea Temperatures

Ocean temperatures are highly variable because of seasonal and short-term meandering of the EAC (Fig. 3). The annual range is around 20-28°C in the north (latitude 24°S) to 14–21°C in the south (latitude 37°S). The temperature

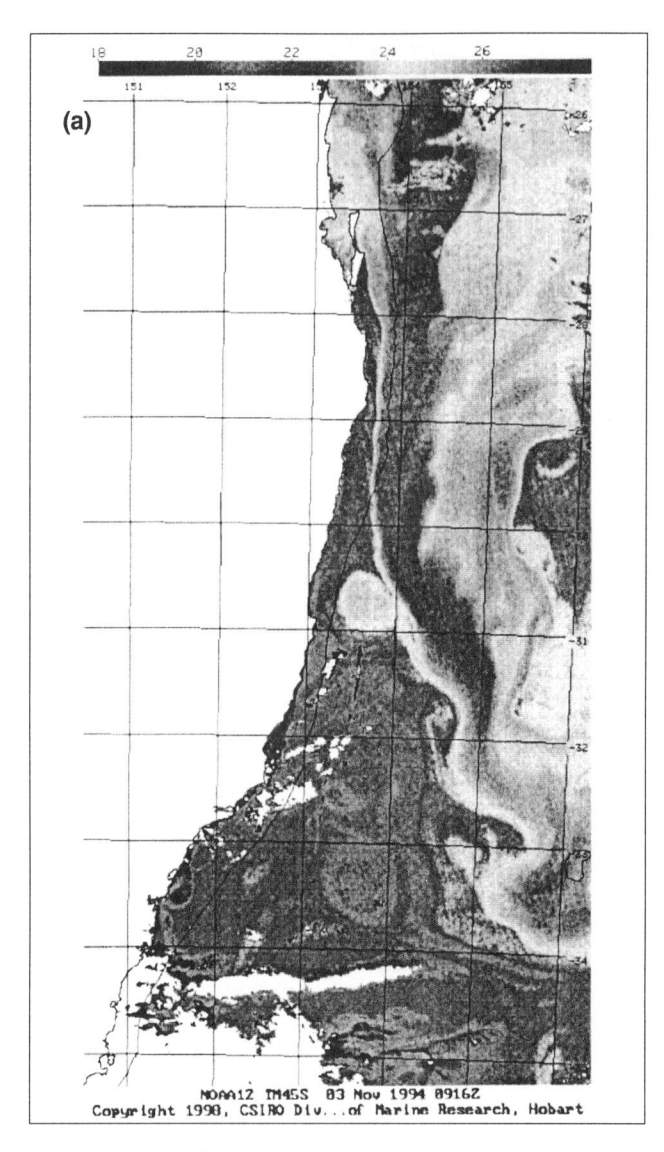

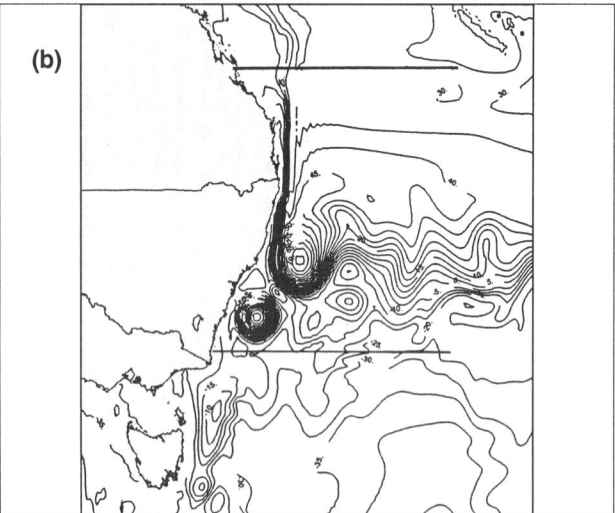

Fig. 3. The Eastern Australian Current brings warm tropical waters down the Eastern coast. (a) Typical NOAA 12 sst satellite image (3 Nov. 1994) showing coastal influence in north and coastal eddy formation. (b) Surface pressure streamlines of Coral and Tasman Seas showing EAC and detached eddies (from Burrage et al., 1997).

difference between the EAC and surrounding waters may be over 5°C (Davis and Zann, unpubl. data).

Nutrients, Productivity

The ocean off Central-eastern Australia is generally low in nutrients because of the tropical origins of the water masses, the lack of major upwellings, and the low nutrient status of many Australian catchments. Upwellings of cooler shelf waters may occur where the EAC separates from the coast (around Evans Head (29°17'S), bringing nutrient-rich waters onto the shelf (Davies and Zann unpubl. data).

Tides, Waves and Storms

The tidal range is 1–3 m and the wave clime is characterised by breakers 1.5–3.0 m in height. The northern sector experiences around 1.5 cyclones per year (IMCRA, 1998).

ENSO Events

Temperature, rainfall, flooding and ocean circulation vary seasonally and year-to-year and are strongly influenced by El Niño/Southern Oscillation (ENSO) events. For example, during the very strong 1996–97 ENSO, ocean temperatures exceeded 30°C in the north, and 26°C off Northern NSW for several months. ENSO events also strongly influence rainfall and cyclones, and terrestrial runoff.

MAJOR SHALLOW WATER AND COASTAL HABITATS

As most of the region lies in a dynamic ecotone under the influence of the tropical East Australian Current, inshore habitats vary greatly according to latitude and local geomorphology, and by season and year. The following firstly briefly describes the major community types, and secondly the major habitats and communities in the region's five sectors.

Community Types

Saltmarsh

Saltmarsh is well developed above the mean high water marks of protected shores. There is a higher species diversity in the temperate south (30 spp.) than the tropical north (14 spp.), contrary to the general trend in species diversity. The saltmarsh flora are not as unique as the Australian terrestrial flora and there are strong floral affiliations with saltmarsh elsewhere in the southern hemisphere, and linkages with the northern hemisphere (Adam, 1996).

Mangroves

Mangroves, mainly of tropical Indonesian-Malay origin, dominate protected intertidal shores in bays and estuaries in the region. They are very diverse and well developed in

South East Queensland (20 spp.; 485 km² cover) (Robertson and Alongi, 1996) but decline rapidly in diversity and cover southwards along the NSW coastline. There are five species in Northern NSW and one species in Southern NSW. The total coverage in NSW is 99 km² (West et al., 1985).

Rocky Shores

Rocky shores of the region have a high biodiversity. While there are some tropical influences in the northern sector, shores are largely dominated by more temperate species. Typically, there is an overlapping zonation of foliose and encrusting algae, barnacles, molluscs, polychaetes and other organisms. Communities are very patchy in time and space because of variables in the physical environment, irregular recruitment and complex ecological interactions such as competition and predation (Underwood and Chapman, 1995)

Sandy Beaches

Much of the exposed coastline in the region consists of sandy beaches. Around 70% of beaches in NSW (sectors b–d) are intermediate beaches which develop in areas of moderate waves (0.5–2 m) and fine to medium sediments. There is a distinct zonation of beach fauna, and species diversity and abundances increase as wave exposure decreases. Sandy shores have a rich meiofauna (nematodes, copepods, mites, gastrotrichs, oligochaetes, polychaetes, nemerteans, tardigrades, rotifers, protozoans and turbellarians) and macrofauna (crabs, hermit crabs, mysid shrimps, isopods, amphipods, insects, polychaete worms, gastropods, bivalves) (Jones and Short, 1995).

Soft Shores

In more protected bays and estuaries, the lower shores consist of mud flats composed of fine sediments. These have a marked vertical zonation, from a surface aerobic layer, to deeper anaerobic layers dominated successively by nitrate-reducing, sulphur-reducing and methane-producing bacteria. The soft sediments are typically dominated by a microflora of bacteria, diatoms and flagellates; a meiofauna of gastrotrichs, nematodes, platyhelminths and other small worms, and a macrofauna of burrowing species such as shrimps (e.g. the thalassinid *Trypaea australiensis*, crabs, polychaetes, gastropods and bivalves). These support a high diversity of fish, shore birds and other organisms (Inglis, 1995).

Seagrasses

Seagrasses are well developed in shallow, protected bays, lower estuaries and coastal lakes. In South Eastern Queensland there are 14 tropical species, with a total area of over 1300 km² (Poiner, 1996). In NSW there are seven species (dominated by *Zostera* spp. in the north and *Posidonia australis* in the south), showing a decreasing tropical influence from north to south. The total area is about 155 km² (West et al., 1985). Studies in NSW indicate that around 70% of all fish caught by recreational and commercial fishers are associated with seagrass and mangroves at some stage of their life histories (EPA, 1993)

Subtidal Rocky Reefs

In more temperate areas, shallow rocky reefs are typically dominated by canopy-forming large brown algae (e.g. *Ecklonia, Sargassum, Phyllospora*), and a high diversity of sessile colonial animals such as sponges, hydroids, soft corals, bryozoans and ascidians. Urchins (e.g. *Centrostephanus, Heliocidaris*) are important algal grazers (Underwood and Kennelly, 1991). In deeper water, reefs are dominated by sponges. Sponge assemblages studied off Sydney generally increase in species diversity with depth (>15 m), with an increase in erect and massive species, but a decrease in encrusting species, presumably because of reduced water turbulence and scouring (Roberts and Davis, 1996).

Coral Communities

Because the EAC brings warm tropical waters and tropical larvae down the coast from the Great Barrier Reef, rich coral communities are found on shallow, sheltered subtidal rocky reefs away from freshwater influences, between 24°S (Fraser Island) to 31°S (South West Rocks). However, while hermatypic or reef-building corals are generally the dominant benthic species, they do not form a limestone based reef but a veneer over the existing rocky substratum (Harriot et al., 1994). Offshore, Norfolk and Lord Howe Islands have extensive true fringing coral reefs, and Elizabeth and Middleton Reefs are platform reefs.

Coral species diversity is often high, but turn-over of species is rapid. Ninety species of coral, in 28 genera from 11 families, have been recorded in the Solitary Islands (ca. 30°S). This compares with 356 species from the Great Barrier Reef (Harriott et al., 1994). However, 21 (38%) of 55 species initially recorded from the Solitary Islands in 1974 (Veron et al. (1974) were not found in re-surveys 20 years later (Harriott et al., 1994). Many coral species are locally rare, and are replaced in the population by different species over time. These coral communities contain unique associations of tropical species near their southern latitudinal range (77 species or 85% of total), and sub-tropical species that are absent or rare in the Great Barrier Reef area (11 species or 12%). Notably, subtropical and temperate coral species dominate in terms of percentage cover (Harriot et al., 1994).

Shelf Benthos

The seafloor on the continental shelf is primarily detrital with relic siliceous sediments such as beach ridges and river deltas from times of lowered sea level. Since the Holocene, around 100 cm of sediments have accumulated on the shelf. However, rates of accumulation since European colonisation have increased between 10 and 100-fold because of increased land erosion (Harris, 1996).

Little is known of the deeper shelf communities in the region. Cross shelf transects off Northern NSW (lengths 5–9 km, depths 5–70 m) found that rocky reefs constituted 20% of the seafloor in the Solitary Islands Marine Park, 23% off South Evans Head and 2% off Byron Bay. In shallower reefal areas (to 20 m) macroalgae dominated, but hard corals were common off the Solitary Islands. In deeper waters (20–70 m) there was a high diversity of sponges, with gorgonians and antipatharians. The mid-shelf was almost entirely soft sediments, with a sparse community of sponges, with locally very high densities of crinoids (Mau, 1997; Byrnes, 1998). Almost nothing is known of the fauna in abyssal areas (Poore, 1996).

Plankton

The planktonic communities in the region are tropical oceanic in the north, and temperate oceanic in the south. The former are carried southwards down the coast and into the Tasman Sea by the East Australian Current and its eddies. Depth distribution is limited by light penetration, ranging from metres in turbid estuaries, to 200 m in clearest oceanic conditions (Hallegraeff, 1996).

Coastal Fishes

The fish fauna in the region is highly diverse. The northern sector has a high proportion of tropical species from the Great Barrier Reef region, the south is mainly temperate Australian in origin. The former has a very high species diversity (over 1500 species) of tropical Indo-Pacific origin and distribution. The latter has a lower species diversity (680 species) but is highly distinctive in composition, and unique to Australia. Reef fishes in the region are a distinctive mixture of species from tropical families such as wrasses and damselfish, with southern-temperate families such as weed whitings, rock cales and morwongs. Soft bottoms are typically dominated by more temperate species of sardines, sprats and other small baitfish; sand mullet and whitings; and predatory flatheads and stingrays (Smith and Jones, 1995).

Major Habitats in Each Sector

South East Queensland (Hervey Bay–Gold Coast)

This sector comprises the northern part of the Tweed-Moreton bioregion proposed by IMCRA (1998) and is characterised by large sand barrier islands and shallow sub-tropical bays. As the coastline to the north is very protected by the Great Barrier Reef, it has been a sink for quartz sands carried northwards by longshore currents, creating a series of large barrier islands.

Fraser Island, the northern-most island, forms the eastern edge of Hervey Bay. It is the largest sand island in the world. It is 123 km in length, 184,000 ha in area, 250 m in height and extends to 60 m below sea level. Hervey Bay is around 300,000 ha in area, is fringed by mangroves, has over 100,000 ha of seagrass and supports major populations of endangered turtles and dugongs. Major estuarine inputs into the bay are from the Mary and Burram Rivers. Fraser Island is a World Heritage Area and Hervey Bay is a Queensland Marine Park (Sinclair, 1995).

Moreton Bay, just off the Queensland capital of Brisbane is also semi-enclosed by large barrier sand islands (Moreton, and North and South Stradbroke Islands). The bay has a very high biodiversity, including saltmarsh, mangroves (6 spp., area 14,000 ha), seagrass (7 spp,. area 26,700 ha) and coral reefs (Robertson and Alongi, 1996; Poiner, 1996). It supports significant populations of turtles and around 600 dugongs, the only population near a major city. It also supports important commercial and recreational fisheries, and water sports. Moreton Bay is a Queensland multiple-use Marine Park (Eager and Campbell, 1996).

The seaward coastline, from Fraser Island to the Queensland/New South Wales border, consists of high energy sandy beaches, interrupted by rocky headlands and sand barrier estuaries. The main river estuaries are Burram, Burnett, Noosa, Pimpana, Brisbane, Logan and Nerang. There are well developed coastal lakes in the Noosa estuary. There are around 485 km² of mangroves in the region (Robertson and Alongi, 1996).

Northern NSW (Tweed-Hunter)

This sector includes the southern part of the Tweed-Moreton bioregion and the Manning Shelf bioregion (IMCRA, 1998). It is characterised by long, high energy sandy beaches interrupted by headlands and large sand barrier estuaries. It is strongly influenced by the tropical East Australian Current which brings the larvae of coral and other tropical species. Rich coral reef communities have developed around rocky islands and reefs away from estuarine influences at Cooks Reef (near Tweed Heads), Julian Rocks (near Byron Bay), the Solitary Islands (near Coffs Harbour) and Smoky Cape (near South West Rocks). Julian Rocks is a NSW Fisheries Reserve and Solitary Islands is a NSW Marine Park (Pollard, 1996).

There are around 40 estuaries in the sector, the larger ones being the Tweed, Richmond, Clarence, Macleay and Hastings River estuaries. The Clarence River is one of the largest estuaries in Australia (115 km long) and has well developed coastal lakes and wetlands. The total area of estuaries in this sector is 252 km². These support 18.6 km² of saltmarsh, 29.4 km² of mangroves and 29.8 km² of seagrass (West et al., 1985).

Central NSW (Hunter to Wollongong)

This sector lies in the warm temperate Central Eastern Province (Hawkesbury Shelf Region). Coral and Tasman Sea water masses meet in this area, forming the Tasman Front, and it is occasionally influenced by meso-scale eddies of the EAC. It is characterised by an open high energy sandy coastline, barrier sand estuaries with very well developed coastal lakes, rocky headlands and cliffs, and long drowned sandstone river valleys (IMCRA, 1998).

There are 50 estuaries in the sector with a total area of estuaries of 943 km^2. These support 32.5 km^2 of saltmarsh, 70 km^2 of mangroves and 111.5 km^2 of seagrass. The major coastal lakes are Wallis, Myall, Macquarie, Tuggerah, and Illawarra. Wallis and Macquarie have some of the largest seagrass beds in NSW (30.8 km^2 and 13.4 km^2, respectively) (West et al., 1985). The major bays and drowned river valley estuaries are Hawkesbury, Port Stephens, Port Jackson (Sydney), Botany Bay, Port Hacking and Jervis Bay.

Southern NSW (Wollongong–Victoria border)

This sector coincides with the South Eastern Biotone and the Batemans Bay Shelf, and is cool-temperate in nature. The shelf is steep and narrow (e.g. 17 km off Montague Island) and oceanographic circulation is influenced mainly by coastal trapped waves setting northwards. The coastline is characterised by short sandy and rocky beaches, rocky bays and saline coastal lagoons (IMCRA, 1998). There are 44 estuaries, most of which are small in size, with a total area of 181 km^2. These support 6.6 km^2 of saltmarsh, 6.7 km^2 of mangroves 6.7 km^2 and 13.5 km^2 of seagrass (West et al., 1985).

OFFSHORE SYSTEMS

The coastal continental shelf is relatively narrow in this region. Offshore, the major geological features are the Norfolk Ridge and the Lord Howe Rise, upon which lie Norfolk and Lord Howe Islands, and oceanic Elizabeth and Middleton Reefs (IMCRA, 1998).

Norfolk Island

Norfolk Island lies 1600 km off Northern NSW, at 29°S latitude. The Norfolk Ridge, which runs north west from New Zealand to New Caledonia, was formed with the break up of the Australian-Antarctic Plate in late Cretaceous to Early Tertiary. Norfolk Island is around 8 km long, is of volcanic origin and was formed around three million years ago. Because of its distinctive origins and biodiversity, it has been placed in a distinctive biogeographic zone, the Norfolk Province (IMCRA, 1998).

Norfolk Island has a subtropical oceanic climate (16–26°C), and annual rainfall of 1320 mm (Douglas and Douglas, 1994). It experiences moderate seas in all seasons, and periodic tropical cyclones between November and April. It is affected by warm currents from the New Caledonia area which flow from October to May, bringing migratory species such as humpback whales and seabirds (IMCRA, 1998). Norfolk Island has a rich marine biodiversity of tropical and subtropical species, and a significant number of endemic species because of its isolation. These includes 231 species of algae (some at their northernmost distribution), 400 species of molluscs (including 160 species of opisthobranchs, 25 of which are endemic) and 250 species of inshore fish (several endemic) (IMCRA, 1998).

Lord Howe Island

Lord Howe Island lies 630 km off Northern NSW at around 31'S, on a platform of the Lord Howe Rise. The island, and associated Admiralty Islands and Ball's Pyramid, are volcanic in origin and formed around 10 million years ago. Because of its biogeographic similarities with Norfolk Island, it is also placed in the Norfolk Province (IMCRA, 1998).

Lord Howe Island has well developed fringing coral reefs and lagoon, and a high marine biodiversity. This includes 235 species of marine algae (28 spp. endemic), seagrass and small stands of mangroves (IMCRA, 1998). Some 83 species of scleractinian corals have been recorded. Coral communities contain a rare association of tropical species at their southern limits of distribution, and subtropical species. Many of the species are rare, and probably result from chance recruitment of only a few larvae (Harriott et al., 1995). Other fauna includes 65 species of echinoderms (70% tropical, 24% temperate and 6% endemic) and 447 species of fish (60% tropical, 10% temperate and 4% endemic) (IMCRA, 1998).

Elizabeth and Middleton Reefs

Some 600 km off Northern NSW, these platform reefs (each ca. 8 km × 6 km) lie on volcanic seamounts rising steeply off the Lord Howe Rise. The reefs receive waters from the warm East Australian Current and its eddies, and the cool temperate Southern Ocean. Air temperatures range from 13.5–25°C. Because of similarities in biodiversity with the Norfolk and Lord Howe Islands, they are placed in the Norfolk Province. Biodiversity includes 122 species of coral; 266 species of molluscs (96.4% Indo-Western Pacific and 3.6% endemic to the Norfolk Province); around 500 species of crustaceans; and around 500 species of fish (IMCRA, 1998).

HUMAN POPULATIONS AFFECTING THE AREA

The central-eastern Australia coast has been inhabited by Aboriginal people for the past 50,000 years. Populations were generally low, and clans lived as hunter-gatherers in their estates which generally extended from the sea to the inland ranges. Shellfish and fish were very important seasonally and semi-permanent fishing camps were established in rich areas such as Moreton Bay and the Clarence estuary. In the century following the arrival of European colonists in 1788, Aboriginal populations were decimated by warfare and diseases, and they were dispossessed from their lands. However, following recent legal recognition of customary native law and title, their descendants are re-asserting their claims over traditional coastal lands and adjacent waters (Smyth, 1996; Schnierer et al., 1996).

European colonisation of Australia commenced in 1788 in Sydney. Over the following century all of the fertile central-eastern coast was colonised, and much of the coastal forests were cleared for agriculture. Australia's hinterland is dry and infertile, and the fertile Central Eastern coastal region today supports over 7.8 million people (46% of the total Australian population). The population is rapidly growing and in the sub-tropical north, ribbon or 'sunbelt' development has spread along the coastal strip. High populations in the coastal strip around bays and estuaries have created locally severe environmental impacts (Newman et al., 1996). The offshore islands of Norfolk and Lord Howe, formerly uninhabited, now have small populations.

South-East Queensland

This area was first colonised around the 1840s and supports most of Queensland's population. It is the fastest growing part of Australia and between 1971 and 1991 grew by around 225% (RAC, 1993). The major coastal population centres are Hervey Bay (38,000), the Sunshine Coast (170,000), Brisbane (1,930,000) and the Gold Coast (250,000). Growth rates along the coastal 'sunbelt' are very high; between 1991 and 1993, the Sunshine Coast and Gold Coast grew by 5% p.a. (Newman et al., 1996).

The major industries in the coastal strip are tourism, sugarcane and light manufacturing industries. The sector includes the most important coastal tourist destinations in Australia. Visitor nights each year are: Brisbane 10.7 million, Gold Coast 9.3 million and Sunshine Coast 7.3 million (Driml, 1996).

The inshore waters are important for recreation and tourism and fisheries. Hervey Bay is Australia's major whale-watching centre as humpback whales rest here during their annual migration between Antarctica and the tropics. Moreton Bay is the epicentre for Brisbane's recreational and commercial fishing. The Sunshine and Gold Coasts are major national and international tourist destinations.

Northern NSW

Northern NSW was first colonised by Europeans around the 1850s, and by 1900 virtually all of the unique coastal sub-tropical rainforest had been cleared for agriculture. The major industries in the coastal zone are tourism, cropping (sugarcane, tea-tree etc.) and cattle grazing. The population is around 400,000. The main coastal towns (populations >15,000) are Tweed Heads, Ballina, Coffs Harbour and Port Macquarie. The region is an important retirement and tourist destination for Sydney residents. It has experienced moderate 'sunbelt' development, and between 1991 and 1993 grew at 5.2% p.a. (Newman et al., 1996).

Central NSW

This sector was first settled in 1788. It includes the Hunter, Gosford-Wyong, Sydney and Illawarra Regions of NSW, and has the highest population in Australia. It is largely metropolitan in nature, with extensive industrial development (coal and steel works, powerhouses, refineries, heavy manufacturing industries etc.). The major cities are Newcastle (464,000), Sydney (4,486,00) and Wollongong (245,000), which make up 88% of the NSW population. Sydney, the largest city in Australia, grew very rapidly in the 1950s–70s following post-World War II international migration, and is rapidly approaching the boundaries of the industrial cities of Newcastle and Wollongong (Newman et al., 1996). Sydney is also the largest single tourist destination in Australia (13 million visitor nights per year) (Driml, 1996). The bays and inshore waters are very popular for recreational boating, water sports and fishing, while the coastal lakes and inlets support an important oyster aquaculture industry.

Southern NSW

The southern coast of NSW is largely rural in nature and has a population of only 124,000. There are no cities on the coast; the main towns are Ulladulla, Batemans Bay and Tathra. The major industries are tourism, fisheries, dairying, grazing and forestry. Shelf waters support important tuna fisheries.

Offshore Islands and Reefs

Norfolk Island has a population of 1922, largely of re-settled Pitcairn Islanders of HMS *Bounty* origin. The main industry is tourism (around 27,000 visitors p.a.). Lord Howe Island has a population of 300 and the main industry is also tourism (around 7000 visitors p.a.) (Douglas and Douglas, 1994). Elizabeth and Middleton Reefs are uninhabited.

RURAL FACTORS

The coastal zone in this region includes some of the most fertile areas in Australia. All major coastal catchments have been at least partially cleared for agriculture, including sugarcane (in the north), mixed cropping, dairying, and cattle grazing. Much of the inland soils are thin and nutrient-limited, and have been greatly degraded by compaction and erosion through clearing of catchments (particularly slopes and river banks) and overgrazing and overcropping. This has resulted in massive soil erosion and sedimentation of rivers and estuaries. Other issues in inland waters include river impoundments and interference with hydrological regimes; eutrophication resulting from agricultural fertilisers and urban sewage and stormwater; loss of drinking and bathing water quality; and acid soil runoff which is responsible for mass mortalities of fish (Tarte et al., 1996; EPA, 1997).

Commercial and recreational fishing is very important in the region, and coastal fisheries have been affected by loss of habitat, declining water quality, overfishing and use of destructive fishing techniques (Zann, 1995; 1996).

The following summarises impacts of agriculture and small-scale commercial and recreational fisheries in each sector. Large-scale commercial fisheries are discussed later.

South-East Queensland

Major issues in the sector's catchments include erosion from cropping lands, acid-soil runoff, construction of river impoundments and alterations of hydrological regimes, barriers to fish migration, stream channel instability, and in-stream and off-stream water use conflicts (Tarte et al., 1996). Sediment loads have increased five-fold in coastal catchments since colonisation and have had significant impacts on the marine environment (Brodie, 1996). For example, because of extensive flooding of the Mary and Burrum Rivers in 1992, and turbulence from Cyclone Fran three weeks later, Hervey Bay was very turbid for a prolonged period. This resulted in a massive die-back of around 90,000 ha of seagrass in 1992-93, and the subsequent mortality or migration of most dugongs (Poiner, 1996).

This sector supports major recreational and commercial fisheries. For example, Moreton Bay is used by around 300,000 recreational fishers who land approximately 2000 tonnes of fish a year (Tarte et al., 1996). Catches are declining in the Bay, increasing competition between the sectors. Recreational fishers are particularly concerned by the commercial use of gill nets and by-catch of commercial trawlers, while commercial fishers are concerned about the cumulative impacts of the recreational fishers on stocks, the lack of regulation of recreational fishers, and their illegal selling of fish (Shorthouse, 1996).

Northern NSW

Most of the catchments in this sector have been cleared for agriculture (Fig. 4), and all of the major rivers have been seriously affected by agricultural run-off. Major issues are soil erosion and sedimentation of inland waters and estuaries, loss of water quality, drainage of wetlands, acid

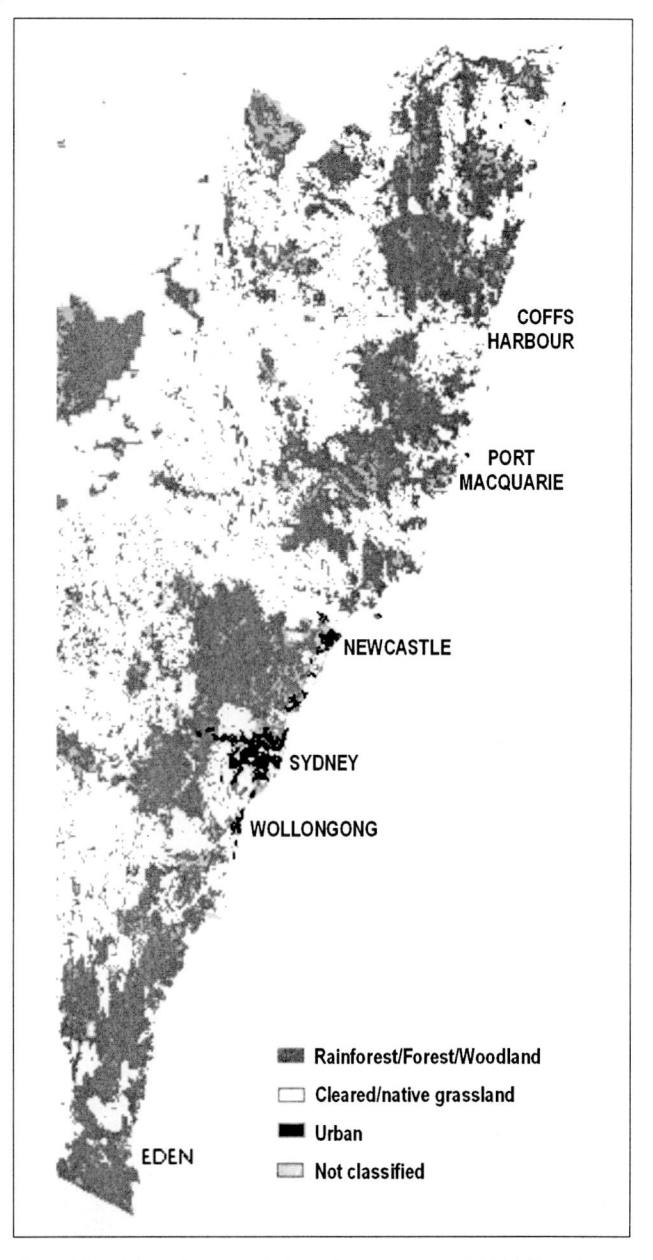

Fig. 4. Vegetation types and cleared areas in coastal NSW (redrawn from EPA, 1997)

soil run-off, fish diseases and mass mortalities, and loss of seagrass and other estuarine habitats. Chemical contamination (DDT, arsenic etc.) from former cattle dip sites is widespread in the catchments and 1624 contaminated sites have been identified (EPA, 1997).

Water quality in the river and estuaries of the Tweed, Brunswick and Richmond Rivers has been rated poor, to very poor, while only the Clarence was generally good (EPA, 1997). Major losses have occurred in seagrass in this area (e.g. 75% loss in the Tweed between 1947 and 1986; 84% loss in the Clarence between 1942 and 1986). Only the small Sandon and Esk River catchments are uncleared (mainly lying in National Parks and State Forests); these are considered to be the most pristine estuaries in the region (Bucher and Saenger, 1991). Overall, NSW has lost up to

Case Study
Acid Soil Run-off: A Critical Issue in Eastern Australia

Acid sulfate soils (ASS), initially formed in estuaries by anaerobic bacteria in seawater, contain iron sulfides, principally as the mineral iron pyrite. They are very widely distributed in coastal eastern Australia, below 5 m above sea level. Left undisturbed, ASS are benign. However, when exposed to air (through channelisation, reclamation etc.) oxidation takes place and sulfuric acid is ultimately produced. This may enter rivers and estuaries, killing aquatic flora and fauna. The effects of acid run-off were not recognised in Australia until 1987 when a massive fish kill decimated 23 km of the Tweed River. It is now known that there are 30,000 km^2 of ASS in Australia, containing over one billion tonnes of sulfuric compounds. Over $10 billion of coastal infrastructure may be at risk (Anon., 1998a).

Mass mortalities of fish, aquacultured oysters and other organisms are frequent following heavy rains after a dry spell. There is evidence that acidification is responsible for a prevalent fungal fish disease, epizootic ulcerative syndrome, or red spot disease. This is caused by an *Aphanomyces* fungus which invades fish through acid-damaged skin and scales. Up to 80% of fish catches in some estuaries are affected (EPA, 1997). Acid run-off may also dissolve concrete and steel piles, bridges, buildings and other structures.

The Tweed, Richmond, Macleay, Hastings and Manning Rivers in this sector are so seriously affected by acid that nearly all the major and medium tributaries show evidence of acid pollution (EPA, 1997). Queensland, particularly the south east, is also seriously affected (Anon., 1998a).

On the Richmond River, where there are 71,000 ha of ASS, sulfuric acid production from the Tuckean Swamp, a drained tributary, is about 300 kg/ha/year. It is estimated that a major flood event in 1994 released the equivalent of 1,000 tonnes of concentrated sulfuric acid, 450 tonnes of aluminium and 300 tonnes of iron from the 4000 ha sub-catchment. This 2.6 pH slug swept back-and-forth by the tide for seven weeks, and acidified over 90 km of river (Anon, 1998a).

MANAGEMENT

Management of ASS is difficult and costly. A national strategy is being developed to improve management, and restore water quality in flood plains and embayments. Objectives are to: (1) identify and define ASS; (2) minimise disturbances to ASS by draining and excavation through introduction of planning and development controls; (3) to mitigate disturbances to ASS through water treatment, water table management etc.; and (4) Rehabilitate disturbed ASS and acid damage through cost-effective new technologies (Anon., 1998).

70% of its coastal wetlands, largely because of draining, and 50% of its seagrass, largely due to elevated nutrients and sediments (State of Environment Advisory Council, 1996).

Fisheries are important in this sector. Small-scale commercial fishing occurs in river estuaries for prawns (using trawls and pocket nets) and fish (using gill nets) and on the inner continental shelf for prawns (using larger trawls) and bottomfish (using lines and traps). The Clarence River estuary is the largest fishery in NSW with around 200 commercial licences for river prawn trawling, and fish netting (NSW Fisheries, 1997).

Central NSW

While much of this area is urban, cattle grazing occurs on the northern coastal strip and the fertile Hunter valley is intensely cultivated. The region also supports small-scale commercial fisheries for prawns and bottomfish, and offshore tunas, and is the largest oyster producer in NSW. Fishing is a major recreational activity in the region; for example, in Botany Bay the recreational catch far exceeds the commercial catch (EPA, 1993).

Southern NSW

The south coast is rural, with beef and dairy cattle and cropping in the fertile valleys and coastal strip, and forestry (native hardwoods, wood chipping) on the ranges.

Offshore Islands and Reefs

Much of Norfolk Island was cleared for cultivation last century. Catastrophic soil erosion and loss of unique flora resulted from the introduction of rabbits on nearby Philip Island.

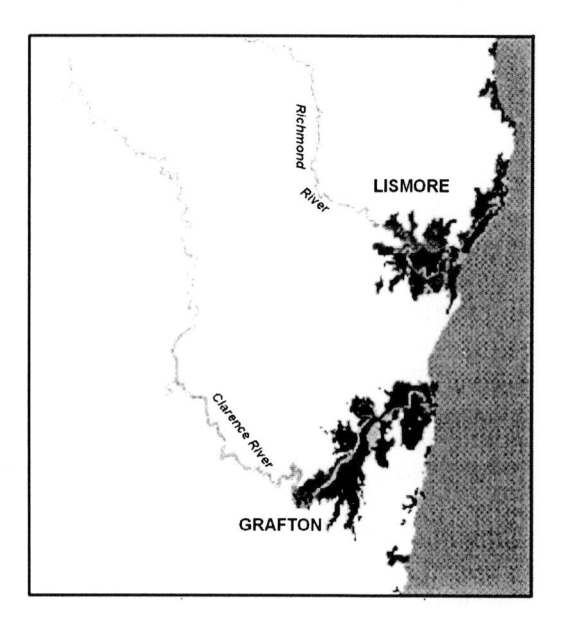

Fig. 5. Acid sulphate soils in Northern New South Wales.

COASTAL EROSION AND LANDFILL

Most of the region's open coastline is dynamic, with high wave energy sandy shores backed by low sand dunes. It is particularly vulnerable to natural disturbances such as cyclones and gales, and human disturbances from ports, seawalls and other coastal engineering, and foreshore development.

South-East Queensland

Beach erosion is a major issue in this sector, particularly in the tourist mecca of the Gold Coast (Fig. 6). Here the construction of the Tweed River breakwaters (in NSW) has interfered with the northern long-shore transport of sand, robbing the famous Gold Coast beaches (across the border in Queensland). Beaches disappear after high seas, and rock walls have been constructed to save foreshore developments such as high-rise hotels and apartments (Bird, 1996). The recent introduction of sand diversion dredges on the Tweed and Southport estuary bars has alleviated the problem.

Coastal dunes in the sector have been extensively degraded by mining for heavy metals such as zircon and rutile since the 1960s. This is continuing on a very large scale on the Stradbroke Islands. Impacts include loss of sensitive dunal and back-barrier marsh habitats, and invasion of unaffected dunes and undisturbed heathlands by exotic plants such as the African Bitou bush which was used in revegetation. Dune blowouts and erosion are caused by off-road vehicles and walking tracks on Fraser, Moreton and Stradbroke Islands (Bird, 1996).

The widespread construction of residential marinas in the Gold Coast's Southport estuary, Moreton Bay and Maroochydore has resulted in significant loss of mangrove and saltmarsh habitat (e.g. 10–20% in Moreton Bay), loss of fisheries, localised pollution from small boats, interference with sediment transport, acid-soil run-off and invasions of biting midges (RAC, 1993).

Northern NSW

Beach erosion is also a significant problem in this sector. Construction of breakwaters in each of the main estuaries has significantly altered beach processes, robbing many downstream (northern) beaches of sand. This has been exacerbated by natural beach erosion due to storms and high sea level, altered winds and rainfall associated with ENSO events (Bird, 1996). Beaches which are protected by engineering structures may undergo severe downdrift erosion. Byron Bay has experienced the largest recorded landward movement in NSW (EPA, 1993). Many of the beaches and dunes were also mined for sand and suffer similar problems to those in Queensland. While there have been fewer marina developments, these have created localised problems.

Central NSW

Beach erosion and sand drift are also issues in the open coastlines in this sector. The proximity of urban developments to dune areas has required coastal local councils to regularly remove sand drifts from roads and car parks. In Botany Bay, important wetlands have been inundated at Kurnell. In Port Stephens, a 10–30 m drift is threatening dwellings at Ana Beach (EPA, 1993).

This sector is the most urbanised area in Australia. In Port Jackson (Sydney) and Botany Bay there have been extensive reclamations of bay foreshores for ports, airports, roads, industrial and urban development. Many shores have been replaced by seawalls, swamps have been infilled, and streams have been replaced by storm water drains, with significant loss of saltmarsh and mangroves.

Southern NSW

There are fewer beaches and low coastal development in the sector, but sand drift is a localised problem. For example, at Wairo Beach near Ulladulla large migrating sand dunes have threatened to close the coastal highway (EPA, 1993).

EFFECTS OF URBAN AND INDUSTRIAL ACTIVITIES

Urban developments, from small coastal holiday villages to the large metropolitan areas of Brisbane and Newcastle/Sydney/Wollongong, industrial developments, commercial fisheries and shipping have had locally severe to widespread impacts on the region's coastal waters. The following describes the general impacts of these developments, and briefly examines these in each sector.

Fig. 6. The Gold Coast in southern Queensland is Australia's 'sunbelt'. High-rise constructions built on the shores are periodically threatened by beach erosion (L. Zann).

> # Case Study
> ## Beach Erosion and Sand Drift in NSW
>
> The 1700 km long NSW coastline has 700 beaches. In the north they are very long and sweeping, and in the south they are small pocket beaches between rocky headlands. The coastline is very dynamic and sensitive to natural and human factors. Hazards include beach erosion, shoreline recession, coastal entrance stability, vegetation degradation and sand drift, coastal inundation, slope and cliff instability and stormwater erosion (PWD, 1990). Beach erosion varies greatly because of differences in sediments, aspect of the coast, wave climate and evolutionary history (EPA, 1993).
>
> Although data are limited, it is thought that most NSW beaches have been eroding for the past 40 years at around 0.2 m per year (and up to 1.2 m in some areas). This may be related to rising sea levels (Gordon, 1988). Greenhouse sea level rises in NSW may result in higher peak run-off in estuaries, changes in the frequency and magnitude of oceanic storm surges and changes in wave patterns which would increase the incidence and severity of flooding in low-lying coastal areas and increase rates of coastal recession (EPA, 1997).

Coastal Settlements

While most of the scattered coastal settlements in rural areas are relatively small (<10,000 population), most have been constructed around estuaries and coastal lakes and have had disproportionately severe impacts on the local environment. These include clearing of saltmarsh and mangroves, and dredging of local ports and boat channels. Storm-water, septic runoff and sewage discharges have elevated nutrients and created eutrophic conditions in some estuaries, particularly the poorly flushed, intermittent estuaries and semi-enclosed coastal lakes. Human pathogens may also be a problem; in 1996 a major hepatitis outbreak was traced to aquacultured oysters from largely rural Wallis Lake in NSW, the nation's largest oyster producer. In the past, local wetlands were considered 'wastelands' and used for refuse tips but developments in mangroves and saltmarsh are now relatively well controlled by local councils. Beach litter, much from the coastal refuse tips, is a common problem in all rural beaches. There is very limited pond aquaculture in the region, and mining and dredging is regulated (Zann, 1995, 1996).

Cities

Around 85% of the region's population lives in metropolitan areas. These have experienced very rapid, *ad hoc* growth in recent decades which has placed great pressure on their sewerage, drainage and other services. As the major cities in the region are all built around bays, sewage, stormwater and industrial discharges have had relatively serious impacts on local water quality. In some cases sewage remains only primary treated, and is discharged through increasingly long ocean outfalls. Storm water systems also flush petroleum products, heavy metals, animal wastes and street litter into the sea. Because of the concentration of populations, recreational pressures (water sports, boating, fishing and collecting) are also intense. The State of the Marine Environment Report for Australia found that, although the condition of Australian waters was on average, 'generally good', around the cities it was often 'very poor' (Zann, 1995; 1996).

Industrial Uses

Industrial uses of coastal resources include commercial fisheries, shipping, ports and related infrastructure, mining of minerals and sediments, and dumping of wastes and industrial discharges. In the region most industries occur in the metropolitan centres (above), compounding environmental problems.

Commercial Fisheries

By world standards, the region is not rich in fisheries. Ocean waters are tropical in origin and low in nutrients, there are no major upwellings, and the shelf area is relatively small. There are no large-scale industrial fisheries, and most of the commercial fishing vessels are owner-operated and small (e.g. in NSW 80% of commercial vessels are <8 m in length) (EPA, 1993). Larger deep-water fish

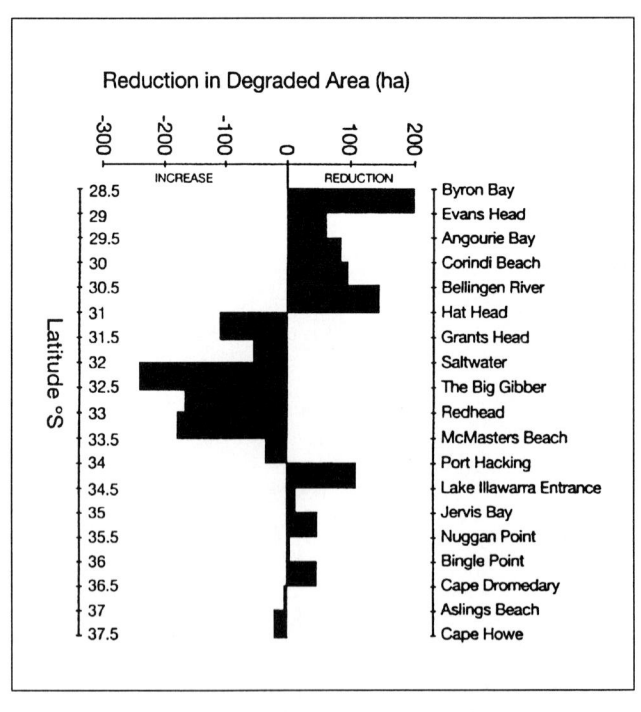

Fig. 7. Trends in sand drift along the NSW coastline (PWD, 1990).

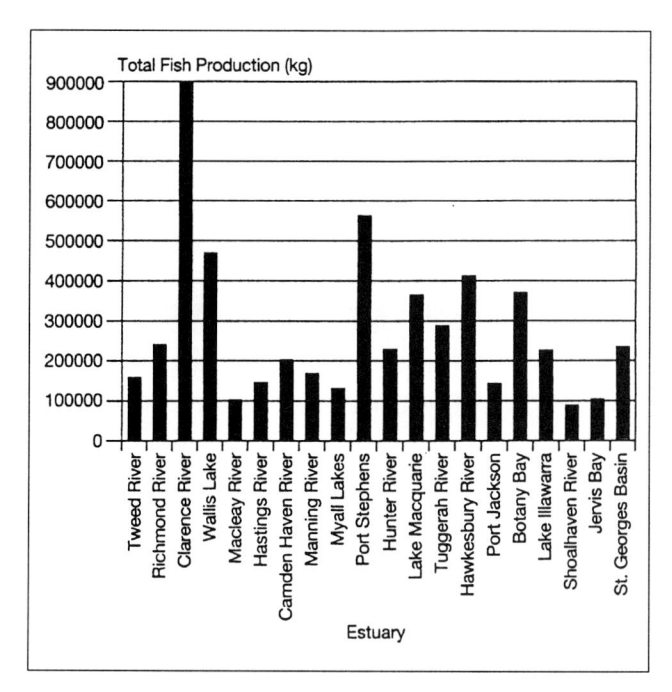

Fig. 8. Fisheries production from main NSW estuaries (from EPA, 1993).

trawlers and long-line and purse-seine tuna vessels operate from Southern NSW.

In South East Queensland, the major inshore commercial fisheries are for mullet, mackerel, snappers and prawns and crabs. Mullet comprises the largest landing by weight (around 1800 tonnes p.a.) while prawns are the most valuable. In Moreton Bay the commercial fish landings are around 700 tonnes p.a., compared with the recreational catch of around 2000 tonnes. There are serious concerns about declining landings in the bay; between 1988 and 1995, there have been 50% declines in mullet, squid, crabs and prawns (Anon., 1998b).

In NSW there are around 2000 commercial fishing operators, landing around 30,000 tonnes of estuary and ocean fish per year (Fig. 8). Fisheries' production has declined in some stocks in recent years. Gemfish, yellowtailed kingfish, blue mackerel, southern bluefin tuna, redfish, snapper and eastern rock and abalone have been seriously over-fished and there are serious concerns about the sustainability of other coastal and estuarine fisheries. Reasons for declines include overfishing, use of non-selective fishing gears, pollution and loss of estuarine habitats. There is inadequate information on catch effort, life histories and other biological information in many fisheries to effectively manage the fisheries. In general, it is considered that the region's fisheries have not been conservatively managed (Kearney, 1996).

Oyster-farming is the only large-scale aquaculture in the region. This is declining (currently around 100 million oysters p.a. from a peak of 170 million p.a. in 1977) because of declining water quality and diseases (EPA, 1993; NSW Fisheries, 1997).

Ports and Shipping

Australia is an island continent and one of the major users of shipping in the world. The region's principal ports are Brisbane (1200 arrivals p.a.), Newcastle (713 p.a.), Sydney (1115 p.a.), Botany Bay (572 p.a.) and Port Kembla (280 p.a.). Smaller fishing ports are situated on all of the major river estuaries. The main export cargoes are agricultural products and minerals; the main imports are manufactured goods, petroleum and chemicals (Raaymakers and Zann, 1996).

Chronic oil pollution occurs in ports from operational discharges and while bunkering. In NSW there are around 150 oil spills each year, mainly minor. Since 1980 there have been two large spills, in Sydney Harbour and Botany Bay (ca. 30,000 and 150,000 litres, respectively) (EPA, 1993). Fortunately there have been no major (Tier 1) spills in the region.

The introduction of exotic marine species in ships' ballast waters, largely from Japan, is a serious problem. Species introduced into NSW include toxic dinoflagellates, algae, gobies (three species), Japanese sea bass, molluscs (three species), and polychaetes. The introduced green alga *Caulerpa scalpelformis* has been observed over-growing seagrass beds in Botany Bay (Bohm, 1998).

Mining and Dredging

Much of the northern coastline has been mined for heavy metals. Limestone for cement manufacture, and sand and gravel for aggregate are mined in some areas. The Brisbane River is extremely turbid as a result of urban run-off and gravel dredging (Abal et al., 1998). Major channel dredging has occurred at Botany Bay and Port Kembla (EPA, 1997). All ports require continuous maintenance dredging.

Regional Issues

South-East Queensland

The State capital and largest city, Brisbane, lies on the Brisbane River estuary, and the western side of Moreton Bay. Most catchments have been cleared, increasing soil erosion and the Brisbane River's sediment load by up to 30-fold, and necessitating constant dredging of channels. Around 10% of the Moreton Bay's saltmarsh and 20% of mangroves have been cleared for developments. Brisbane's sewage is secondary and tertiary treated and is discharged into the bay; western parts of the bay show signs of nutrient pollution and high *E. coli* levels, and increasing toxic algal blooms (cyanobacteria *Lyngbya majuscula*) since 1990. There has been some associated loss of seagrass. Surveys of heavy metals and pesticides in 1979 found that although these were high compared with background levels, they were not sufficiently high for concern at that time. Surveys in 1997 found that dieldrin levels exceeded water quality guidelines in sediments and are close to the limit for water quality (Brisbane City Council, 1996; Tarte et al., 1996; Abal et al., 1998).

Because of the biodiversity importance of Moreton Bay and its proximity to a major population centre, there are growing concerns on declining water quality and inshore habitats. In the west and southwest near Brisbane there have been significant seagrass losses, and localised algal blooms. A multi-disciplinary study funded by local councils is currently investigating hydrology, nutrient status, plankton, mangrove, seagrass, coral reefs, microbenthos, and other communities (Abal et al., 1998).

Northern NSW

There are no major cities in the sector.

Central NSW

This sector is the most urbanised area in Australia, and is discussed in some detail. Newcastle, Sydney and Wollongong have a combined population of 5.2 million. The major issue is poor water quality, largely from sewage and stormwater discharges from metropolitan areas. In the Sydney region, 98% of properties (1.3 million) are sewered. There are five major treatment works with ocean outfalls of largely primary treated sewage at Malabar, North Head, Cronulla and Warriewood. During the 1980s sewage affected beaches along 40 km of the coastline, and Sydney beaches exceeded coliform bacteria for much of the time. Levels of chlorinated compounds such as chlordane and dieldrin were very high in fish and oysters. The commissioning of new deep-water outfalls in 1990 resulted in a marked improvement in inshore waters. Beaches now pass bathing water guidelines around 90% of the time, and chlorinated compounds in coastal fish have been greatly reduced. However, offshore water quality has been reduced, and more distant beaches are now occasionally affected (EPA, 1993; Macdonald, 1996).

Urban stormwater runoff contributes an equivalent amount of nutrients and significant levels of heavy metals and other contaminants to coastal waters. Monitoring indicates that water quality from each catchment is poor; levels of nutrients were very high, and 98% of samples exceeded faecal coliform guidelines. Stormwater is also largely responsible for very high levels of lead in upper harbour sediments (EPA, 1993, 1997). Levels of lead, copper, zinc, cadmium and mercury, and dioxin are high in sediments in some areas (e.g. Parramatta River where commercial fishing is prohibited; Rozelle Bay, Morrison's Bay, Port Kembla) (EPA, 1993, 1997). However, generally lower levels in surface sediments indicate the efficacy of recent controls (Macdonald, 1996; Batley, 1996).

Coastal lagoons with limited flushing are particularly affected. Sydney's smaller lagoons (Narrabeen, Dee Why, Curl Curl and Manly) are contaminated by heavy metals, organochlorines, nutrients and bacteria. Lake Macquarie, which receives discharges from industry, sewage and urban and agricultural run-off is contaminated with copper, lead, zinc and cadmium. Tuggerah Lakes and Lake Illawarra are highly eutrophic and regularly experience massive algal blooms, including toxic algae (EPA, 1993; Macdonald, 1996). There have been major declines of seagrass in these lakes (Poiner, 1996).

Channel dredging and sea dumping are extensive, and each year over one million tonnes of sediments are dumped at sea. In Botany Bay over 1,000 ha of seagrass have been covered by dredge spoil. In Port Kembla over 13 million tonnes of material have been dredged, exposing heavy metals in contaminated sediments (EPA, 1993).

Southern NSW

There are no cities in this area.

PROTECTIVE MEASURES

The waters in this region are managed by the States of Queensland and NSW (from the shores to the three nautical mile limit), and the Commonwealth government (from there to the edge of the 200 nautical mile EEZ). A complex and overlapping range of State and Commonwealth legislation and policies and international agreements and responsibilities related to coastal and marine environmental protection operate in the region (RAC, 1993).

Legislation and Responsibilities

A range of Australian Commonwealth legislation and international agreements relate to protection of marine species and areas, and controls on threatening processes. In Queensland the *Nature Conservation Act 1992* and *Environmental Protection Act 1995* are intended to provide a coherent framework for coastal protection. Other relevant legislation includes the *Marine Parks Act 1982–88*, *Fauna Conservation Act*, and the *Fisheries Act 1994* (Eager and Campbell, 1996). Management is undertaken by the Department of Environment and Heritage (Environmental Protection, Marine Parks), Department of Primary Industries (Fisheries, Boating and Fisheries Patrol) and Queensland Fisheries Management Authority.

In NSW over 80 Acts relate to coastal and marine environmental management, the major ones being the *Pollution Control Act 1970*, *Clean Waters Act 1970*, and *Waste Disposal Act 1970*. Important recent legislation includes the *Fisheries Management Act 1994*, and the *Marine Park Act 1997* (EPA, 1993, 1997). Management is undertaken by the Environment Protection Authority, NSW Fisheries and NSW National Parks and Wildlife Service. The recent NSW Coastal Policy (1997) provides important guidelines for coastal development.

Protected Species

Threatened species and a variety of other important marine species are protected under Commonwealth and State legislation. This includes all cetaceans, dugongs, turtles, seabirds and some fish and invertebrates such as corals and some molluscs (Ivanovici et al., 1996). In NSW 11 fish species are protected. These include endemic species such as sea dragons, wrasse and butterflyfish and nurse sharks and white pointer sharks (Pollard, 1996).

Marine Protected Areas

Maintenance of habitats for threatened and other marine species is critical, and marine protected areas (MPAs) are an important strategy in marine biodiversity conservation. Queensland has the largest area of MPAs in Australia and includes parts of the Great Barrier Reef Marine Park. Major MPAs in this sector are Hervey Bay, Woongarra, Pumicestone Passage and Moreton Bay Marine Parks which are multiple-use MPAs. In addition there are 17 Fish Habitat Reserves, 15 Wetland Reserves and two Fish Sanctuaries in this sector (Eager and Campbell, 1996).

NSW has 17 MPAs, the major ones being Solitary Island and Jervis Bay Marine Parks which are multiple-use MPAs (Pollard, 1996). However, the actual areas of effective protection are very small, the MPAs are very poorly resourced and are probably largely ineffectual. The NSW Marine Park Authority was therefore established in 1997 for the conservation of marine biodiversity and the management of the Marine Parks.

The offshore reefs have a high level of protection. Lord Howe Island reefs are Marine Reserves and part of the Lord Howe World Heritage Area. Elizabeth and Middleton Reefs are National Nature Reserves and are managed by the Commonwealth National Parks and Wildlife Service (Griffin, 1996).

Evaluation of Protection Measures

The effectiveness of marine environmental management varies greatly by jurisdiction and locality. Queensland has not had an effective environment protection legislation and capability until recently, and long-term monitoring data is very limited. Planning approvals have been liberal, and large-scale coastal developments, including marinas and beach front high rise developments are still permitted. However, Queensland has had a comprehensive system of marine protected areas in place for several decades, and has a national capability in MPA management. But given their large number and size, and the emphasis on the more glamorous Great Barrier Reef Marine Park in the north, MPAs in this sector are not adequately resourced.

NSW has had an effective and relatively well resourced Environment Protection Agency for two decades and there is relatively good, long-term monitoring in certain key areas. However, most emphasis and resources have been placed on issues in the metropolitan areas, particularly in Sydney, and rural areas have generally been neglected. NSW has had a stricter developmental approvals process and progressive local and state environmental protection plans, and now has a comprehensive coastal policy. Large-scale beachfront development is controlled, and an important system of coastal National Parks has been established over the past 20 years. However, marine biodiversity conservation has not been well undertaken in the past, and resources for marine conservation have been minimal. This has improved with the recent establishment of the NSW Marine Park Authority in 1997.

Lord Howe Island and Norfolk reefs appear to be adequately managed, but Middleton and Elizabeth Reefs are very remote and are not actively policed.

CONCLUSIONS: STATUS OF MARINE ENVIRONMENT

The region is large in size and has a very high marine biodiversity because it lies in a tropical/temperate overlap zone. It is also subject to a great diversity of pressures. The major issues include: (1) declining water quality and loss of habitat in estuaries, and in particular, the unique coastal lakes; (2) declining coastal fisheries; and (3) pollution in metropolitan areas, particularly in the Newcastle/Sydney/Wollongong area. While the great majority of the population lies in the metropolitan areas and human impacts are concentrated there, clearing of catchments in rural areas and disturbance of acid soils have had a disproportionately great impact on estuaries.

REFERENCES

Abal, E.G., Halloway, K.M. and Dennison, W.C. (1998) Moreton Bay catchment interim stage 2 scientific report, Brisbane and Moreton Bay wastewater management study, Brisbane.

Adam, P. (1996) Coastal saltmarsh. In *State of the Marine Environment Report for Australia. Technical Summary*, ed. L.P. Zann. Great Barrier Reef Marine Park Authority, Department of Environment Sport and Territories, Townsville, pp. 33–35.

Anon. (1998a) Draft national strategy for the management of acid sulfate soils. National Working Party on Acid Sulfate Soils, Department of Primary Industries and Energy, Canberra.

Anon. (1998b) Catch trends in Moreton Bay since 1988. Queensland Fisherman June 1988, pp. 16–17.

Batley, G.E. (1996) Heavy metals and organometals In *State of the Marine Environment Report for Australia. Technical Summary*, ed. L.P. Zann. Great Barrier Reef Marine Park Authority, Department of Environment Sport and Territories, Townsville, pp. 250–254.

Bird (1996) Coastal modifications and impacts, In *State of the Marine Environment Report for Australia. Technical Summary*, ed. L.P. Zann. Great Barrier Reef Marine Park Authority, Department of Environment Sport and Territories, Townsville, pp. 226–232.

Bohm, C. (1998) Introduced species in Botany Bay. *Waves* 5, 1–8.

Brisbane City Council (1996) State of the environment report, Brisbane. 1996. Brisbane City Council, Brisbane.

Brodie, J. (1996) Nutrients and eutrophication in coastal waters. In *State of the Marine Environment Report for Australia. Technical Summary*, ed. L.P. Zann. Great Barrier Reef Marine Park Authority, Department of Environment Sport and Territories, Townsville, pp. 238–246.

Bucher, D. and Saenger, P. (1991) An inventory of Australian estuaries and enclosed marine waters: an overview of results. *Australian Geographic Studies* 9, 370–381.

Burrage, D.M. (1993) Coral Sea currents. *Corella* 17, 135–145.

Burrage, D., Steinberg, C., Bode, L. and Black, K. (1997) Long-term current observations in the Great Barrier Reef. In *State of the Great Barrier Reef World Heritage Area Workshop*, eds. D. Wachenfeld, J. Oliver and K. Davies, Series No. 23, Great Barrier Reef Marine Park Authority, Townsville, pp. 21–45.

Byrnes, T. (1998) An assessment of coastal continental shelf communities of Northern New South Wales. Unpubl. Honours thesis, Southern Cross University, Lismore, Australia.

CSIRO (1987) Oceanography. CSIRO Research for Australia. 14, CSIRO, Canberra.

Douglas, N. and Douglas, N. (1994) *The Pacific Islands Yearbook*. Fiji Times, Fiji.

Driml, S. (1996) Coastal and marine tourism and recreation. In *State of the Marine Environment Report for Australia. Technical Summary*, ed. L.P. Zann. Great Barrier Reef Marine Park Authority, Department of Environment Sport and Territories, Townsville, pp. 33–35.

Eager, E. and Campbell, J. (1996) Marine conservation and marine protected areas in Queensland. In *State of the Marine Environment Report for Australia. Technical Summary*, ed. L.P. Zann. Great Barrier Reef Marine Park Authority, Department of Environment Sport and Territories, Townsville, pp. 458–465.

EPA (1993) New South Wales State of the environment, Environment Protection Authority, New South Wales, Chatswood.

EPA (1997) New South Wales State of the environment, Environment Protection Authority, New South Wales, Chatswood.

Gordon, A.D. (1988) A tentative but tantalising link between sea-level rise and coastal recession in NSW, Australia. In *Greenhouse: Planning for Climate Change*, eds. L. Pearman and J. Brill. Leiden.

Griffin, M. (1996) National nature reserves. In *State of the Marine Environment Report for Australia. Technical Summary*, ed. L.P. Zann. Great Barrier Reef Marine Park Authority, Department of Environment Sport and Territories, Townsville, pp. 441–447.

Hallegraeff, G.M. (1996) Phytoplankton: the pastures of the sea. In *State of the Marine Environment Report for Australia. Technical Summary*, ed. L.P. Zann. Great Barrier Reef Marine Park Authority, Department of Environment Sport and Territories, Townsville, pp. 66–69.

Harriott, V.J., Smith, S.D.A. and Harrison, P. (1994) Patterns of coral community structure of subtropical reefs in the Solitary Islands Marine Reserve, Eastern Australia. *Marine Ecology Progress Series* 109, 67–76.

Harriott, V.J., Harrison, P.L. and Banks, S.A. (1995) The coral communities of Lord Howe Island. *Marine and Freshwater Research* 46, 457–465.

Harris, P.T. (1996) The sea floor. In *State of the Marine Environment Report for Australia. Technical Summary*, ed. L.P. Zann. Great Barrier Reef Marine Park Authority, Department of Environment Sport and Territories, Townsville, pp. 15–19.

Inglis, G.J. (1995) Intertidal muddy shores. In: *Coastal Marine Ecology of Temperate Australia*, eds. A.J. Underwood and M.G. Chapman. UNSW Press, Sydney, pp 171–186.

IMCRA (1998) Interim marine and coastal regionalisation for Australia. Version 3.3. IMCRA Technical Group, Environment Australia, Canberra.

Ivanovici, A., Anderson, G., Antram, F., Crennan, J., Male, B., Mooree,R., and Weaver, K. (1996) Protection of marine species: national and international responsibilities. In *State of the Marine Environment Report for Australia. Technical Summary*, ed. L.P. Zann. Great Barrier Reef Marine Park Authority, Department of Environment Sport and Territories, Townsville, pp. 353–358.

Jones, A.R. and Short, A.D. (1995) Sandy beaches. In *Coastal Marine Ecology of Temperate Australia*, eds. A.J. Underwood and M.G. Chapman. UNSW Press, Sydney, pp. 136–151.

Kearney, R.E. (1996) Coastal fisheries: a critical review. In *State of the Marine Environment Report for Australia. Technical Summary*, ed. L.P. Zann. Great Barrier Reef Marine Park Authority, Department of Environment Sport and Territories, Townsville, pp. 181–184.

Macdonald, R. (1996) Issues in New South Wales' marine environment. In *State of the Marine Environment Report for Australia. Technical Summary*, ed. L.P. Zann. Great Barrier Reef Marine Park Authority, Department of Environment Sport and Territories, Townsville, pp. 295–303.

Mau, R. (1997) A preliminary survey of continental shelf habitats of the Solitary Islands Marine Park, New South Wales. Unpubl. Honours thesis, Southern Cross University, Lismore, Australia.

Middleton, J.H. (1996) The ocean. In *State of the Marine Environment Report for Australia. Technical Summary*, ed. L.P. Zann. Great Barrier Reef Marine Park Authority, Department of Environment Sport and Territories, Townsville, pp. 9–14.

Newman, P., Birrell, B., Holmes, C., Newton, P., Oakley, G., O'Connor, A., Walker, B., Spessa, A. and Tait, D. (1996) Human settlements. In *State of the Environment Australia. 1996*. State of the Environment Advisory Council, CSIRO Publishing, Collingwood, Australia pp. 3/1–57.

NSW Fisheries (1993) The oyster industry of New South Wales. NSW Fisheries, Sydney.

NSW Fisheries (1997) Annual report 1996/97, NSW Fisheries, Sydney.

Ortiz, E. and Burchmore, J. (1992) The development of a representative system of marine and estuarine protected areas for New South Wales. Ocean Rescue 2000. Unpubl. report, NSW Fisheries.

Poiner, I. (1996) Seagrasses. In *State of the Marine Environment Report for Australia. Technical Summary*, ed. L.P. Zann. Great Barrier Reef Marine Park Authority, Department of Environment Sport and Territories, Townsville, pp. 40–45.

Pollard, D. (1996) Marine conservation and protected areas in New South Wales. In *State of the Marine Environment Report for Australia. Technical Summary*, ed. L.P. Zann. Great Barrier Reef Marine Park Authority, Department of Environment Sport and Territories, Townsville pp. 366–471.

Poore, G.C.B. (1996) Biogeography of Australia's marine biota. In *State of the Marine Environment Report for Australia. Technical Summary*, ed. L.P. Zann. Great Barrier Reef Marine Park Authority, Department of Environment Sport and Territories, Townsville, pp. 33–35.

PWD (1990) Coastline management manual, NSW Government, Sydney.

Raaymakers, S. and Zann, L.P. (1996) Shipping and ports. In *State of the Marine Environment Report for Australia. Technical Summary*, ed. L.P. Zann. Great Barrier Reef Marine Park Authority, Department of Environment Sport and Territories, Townsville, pp. 202–208.

RAC (1993) Coastal zone inquiry, Resource Assessment Commission, Canberra.

Roberts, D.E. and Davis, A.R. (1996) patterns in sponge (Porifera) assemblages on temperate coastal reefs off Sydney, Australia. *Journal of Marine and Freshwater Research* 47, pp. 897–906.

Robertson, A.I. and Alongi, D.M. (1996) Mangrove systems. In *State of the Marine Environment Report for Australia. Technical Summary*, ed. L.P. Zann. Great Barrier Reef Marine Park Authority, Department of Environment Sport and Territories, Townsville, pp. 36–39.

Sinclair, J. (1995) *Fraser Island and Cooloola*. Lansdowne, Sydney, Australia.

Schnerier, S., Robinson, S., Heron, R., and Nayutah, J. (1996) Case study: Aboriginal use of the coastal environment in northern New South Wales. In *State of the Marine Environment Report for Australia. Technical Summary*, ed. L.P. Zann. Great Barrier Reef

Marine Park Authority, Department of Environment Sport and Territories, Townsville, pp. 116–126.

Shorthouse, B. (1996) Recreational fisheries and the catch-sharing issue. In *State of the Marine Environment Report for Australia. Technical Summary*, ed. L.P. Zann. Great Barrier Reef Marine Park Authority, Department of Environment Sport and Territories, Townsville, pp. 188–194.

Smith, M.P.L. and Jones, G.P. (1995) Fishes of shallow coastal habitats. In *Coastal Marine Ecology of Temperate Australia*, eds. A.J. Underwood and M.G. Chapman. UNSW Press, Sydney, pp. 241–253.

Smyth, D. (1996) Aboriginal maritime culture. In *State of the Marine Environment Report for Australia. Technical Summary*, ed. L.P. Zann. Great Barrier Reef Marine Park Authority, Department of Environment Sport and Territories, Townsville, pp. 104–115.

State of the Environment Advisory Council (1996) State of the Environment, Australia. 1996. CSIRO, Collingwood.

Tarte, D. Hall, M. and Stocks, K. (1996) Issues in Queensland's marine environment. In *State of the Marine Environment Report for Australia. Technical Summary*, ed. L.P. Zann. Great Barrier Reef Marine Park Authority, Department of Environment Sport and Territories, Townsville, pp. 285–294.

Underwood, A.J. and Kennelly, S.J. (1991) Ecology of marine algae on rocky shores and intertidal reefs in temperate Australia. *Hydrobiologia* **192**, 3–20.

Underwood, A.J. and Chapman, M.G. (1995) Rocky shores. In *Coastal Marine Ecology of Temperate Australia*, eds. A.J. Underwood, and M.G. Chapman. UNSW Press, Sydney, pp. 55–82.

Veron, J.E.N., How, R.A., Done, T.J., Zell, L.D., Dodkin, M.J. and O'Farrell, A.F.(1974) Corals of the Solitary Islands, Central New South Wales. *Australian Journal of Marine and Freshwater Research* **25**, 193–208.

West, R.J., Thorogood, C. Walford, T. and Williams, R.J. (1985) An estuarine inventory of New South Wales, Fisheries Bulletin, NSW Department of Agriculture, Sydney.

Zann, L.P. (1995) Our sea, our future. Major findings of the state of the marine environment report for Australia. Great Barrier Reef Marine Park Authority, Department of Environment Sport and Territories, Townsville.

Zann, L.P. (ed.) (1996) *State of the Marine Environment Report for Australia. Technical Summary*. Great Barrier Reef Marine Park Authority, Department of Environment Sport and Territories, Townsville.

Zann, L.P. and Alcock, D. (1996) The community and marine conservation. In *State of the Marine Environment Report for Australia. Technical Summary*, ed. L.P. Zann. Great Barrier Reef Marine Park Authority, Department of Environment Sport and Territories, Townsville, pp. 145–152.

THE AUTHOR

Leon P. Zann
*Southern Cross University,
Lismore, New South Wales 2480,
Australia*

Chapter 95

THE TASMANIAN REGION

Christine M. Crawford, Graham J. Edgar and George Cresswell

The Tasmanian region includes the eastern, southern and western coastal areas of Tasmania (the Tasmanian Province) and the inshore coastal waters of the northern Tasmanian coast and Bass Strait Islands (part of the Bassian Province). The biota of the Tasmanian Province is characterised by a number of endemic species but low species diversity. By contrast, marine communities in the Bass Strait area have high species richness but negligible endemism, largely because the north coast was connected to mainland Australia <10,000 years ago. The major currents east and west of Tasmania, the East Australian Current and the Zeehan Current, respectively, both flow consistently southward. The strength of the East Australian current, in particular, has a major influence on the distribution of flora and fauna. During La Nina years of strong southerly current flow, species associated with warm high-salinity waters are transported well down the Tasmanian east coast, whereas highly productive pelagic fish stocks tend to associate with the subantarctic water mass and move south or into deeper water during these years.

Industrial and urban development in Tasmania has largely occurred in the coastal zone, resulting in substantial changes to the natural environment along the northern and eastern coasts. By contrast, little human activity has occurred on the south and west coasts, with much of this region protected within a large National Park. Estuarine and coastal waters have been regularly used as a dumping ground for wastes, and the three largest estuaries in Tasmania are severely degraded with high levels of heavy metals and other industrial and urban wastes. However, very few baseline studies of the Tasmanian marine environment have been conducted, greatly limiting the identification of change.

Although issues of coastal and marine ecosystem health have historically attracted little public concern, this attitude is changing at both community and government levels. Recent changes to legislation have resulted in a more integrated approach to management of the marine environment, with land-based activities becoming increasingly accountable for downstream effects. Local communities are also showing greater interest and involvement in the management of their estuarine and marine neighbourhoods.

Nevertheless, major gaps remain in our knowledge of Tasmanian coastal and marine environments. Baseline conditions are poorly documented but such information is essential to ensure effective management and sustainable development. Detailed information is needed on how marine and estuarine ecosystems function and the processes affecting them. The effects on marine communities and habitats of human activities, including land-based and extractive industries, are also largely unknown and require investigation and monitoring.

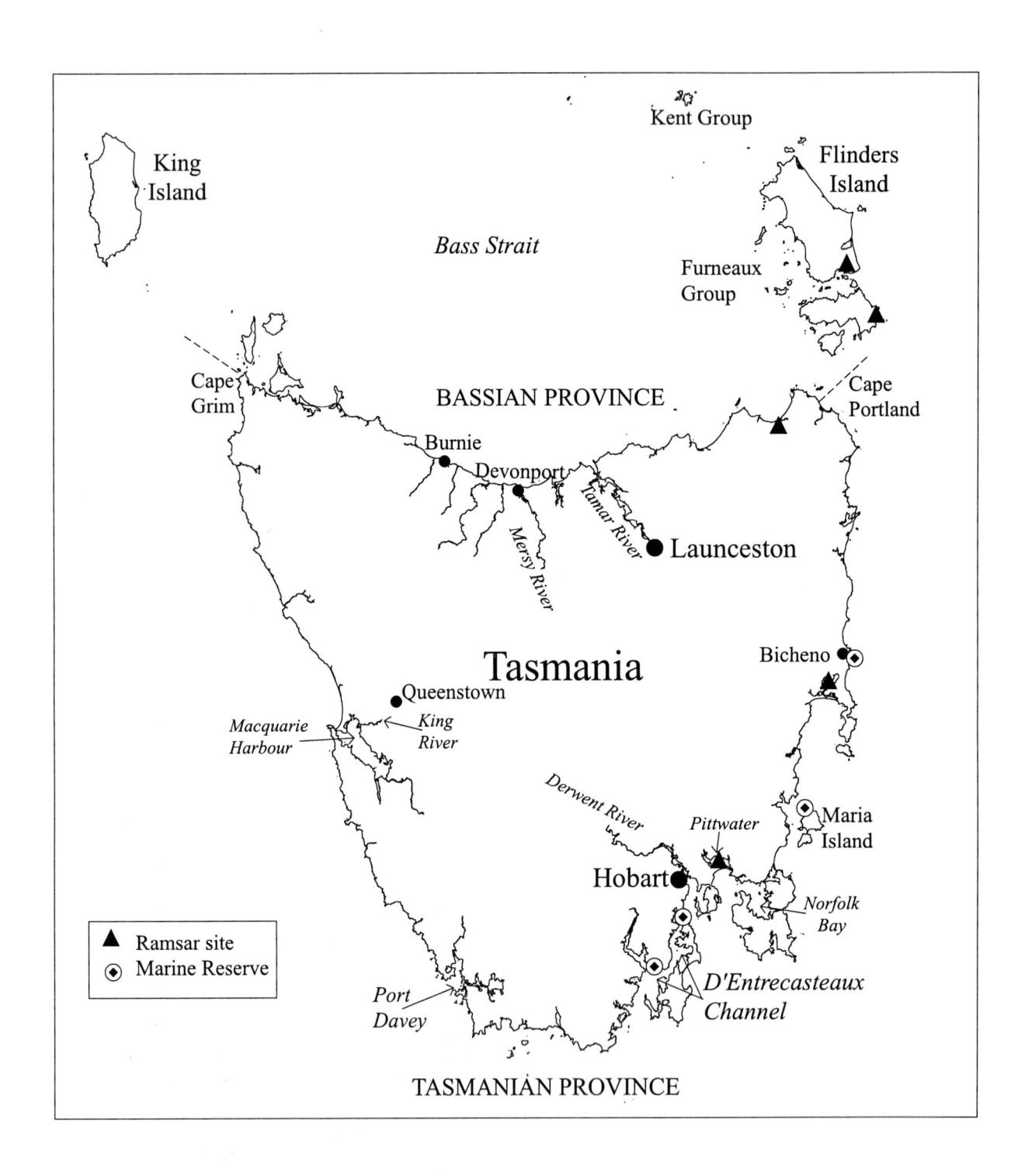

Fig. 1. Map of Tasmania.

THE DEFINED REGION

The Tasmanian region, as discussed in this chapter, includes waters in two subregional groupings: (i) the eastern, southern and western coasts of Tasmania to the 200 nautical mile offshore limit of the Exclusive Economic Zone, and (ii) inshore waters associated with the northern coast and Bass Strait islands (Fig. 1). These two groupings possess substantial ecosystem differences, with rapid changes in predominant plants and animals occurring over distances of less than 20 km at the northeastern (Cape Portland) and northwestern (Cape Grim) corners of Tasmania. Perhaps the most noticeable of the ecosystem changes is the abrupt cessation of extensive seagrass meadows south of Bass Strait, due to the absence of *Posidonia australis* and *P. angustifolia*.

The eastern, southern and western coasts of Tasmania comprise a distinctive biogeographic region, known as the Tasmanian Province (Interim Marine and Coastal Regionalisation for Australia Technical Group, 1998). By contrast, the northern Tasmanian subregion from Cape Grim to Cape Portland forms part of a larger Bassian Province that also includes the central sector of Victoria. The flora and fauna in northern Tasmanian waters, particularly around King Island in the western sector, are strongly influenced by elements of a widespread southern Australian biota. Additionally, the flora and fauna associated with the Furneaux and Kent Groups in eastern Bass Strait strongly interact with eastern Australian elements (Edgar et al., 1997).

When considered on geological time scales, the Bassian Province is a recent entity. During the last glacial epoch, which peaked between 15,000 and 20,000 years ago, sea levels were at least 130 m below present levels and Tasmania formed a southern promontory of Australia. The Bassian region remained fully terrestrial as a consequence of eustatic sea level decline as recently as 10,000 years ago. Consequently, few species are endemic to the Bassian region compared to the Tasmanian region further south. Remnants of the former landbridge can still be seen in the distribution of sibling species pairs, such as for the Australian salmon *Arripis trutta* and *A. esper* (Edgar, 1986). The emergence during the last ice age of a southern barrier separating a formerly widespread population presumably allowed disjunct evolution of eastern and western populations. These have remained distinctive as different species despite the overlap of ranges following water flow through Bass Strait.

SEASONALITY, CURRENTS, NATURAL ENVIRONMENTAL VARIABLES

Tasmania is set in the Southern Ocean near the subtropical convergence, with major influences coming from the Tasman Sea to the east and the Indian Ocean to the west. Three major current systems, the East Australian Current, the Zeehan Current and the Antarctic Circumpolar Current, converge around Tasmania, resulting in a complex series of interactions between water bodies. This complexity remains poorly understood because Tasmanian waters have been little studied compared to those off other Australian coasts.

In the past there has been some debate over the seasonal direction of currents off western Tasmania. Drift bottle studies in that region initially suggested southward flow in winter and northward flow in summer (Vaux and Olsen, 1961). More recently, a survey of temperature sections between Melbourne and Antarctica in summer showed warm southward flow above the continental slope at the western entrance to Bass Strait (Edwards and Emery, 1982), and this was named the Zeehan Current by Baines (Baines et al., 1983). The current was thought to be about 40 km wide and permanently southward, rather than seasonally reversing as previously claimed. Long-term moored

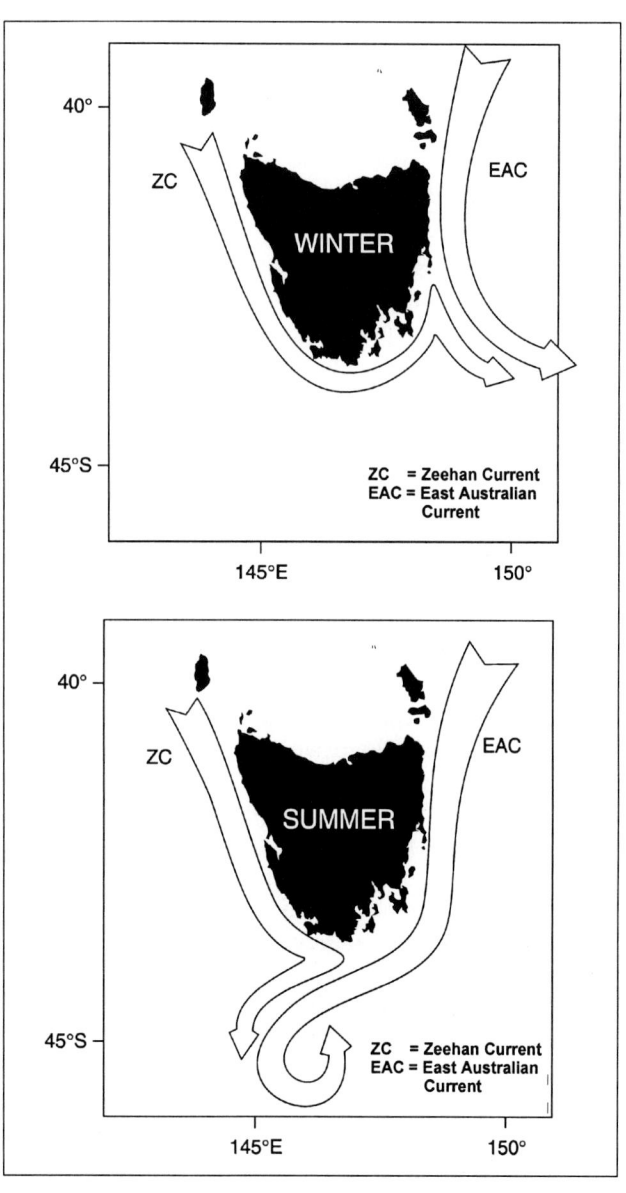

Fig. 2. Schematic pictures for winter and summer showing the Zeehan Current off western Tasmania and its interaction with the East Australian Current.

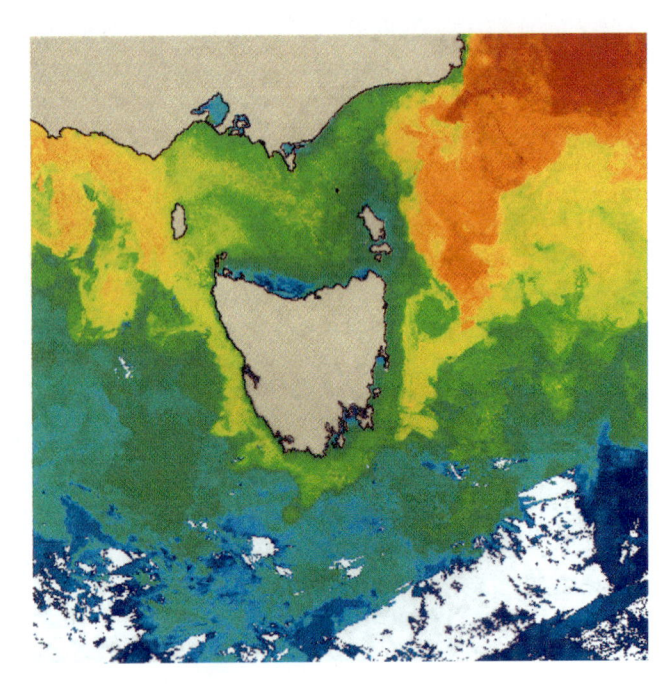

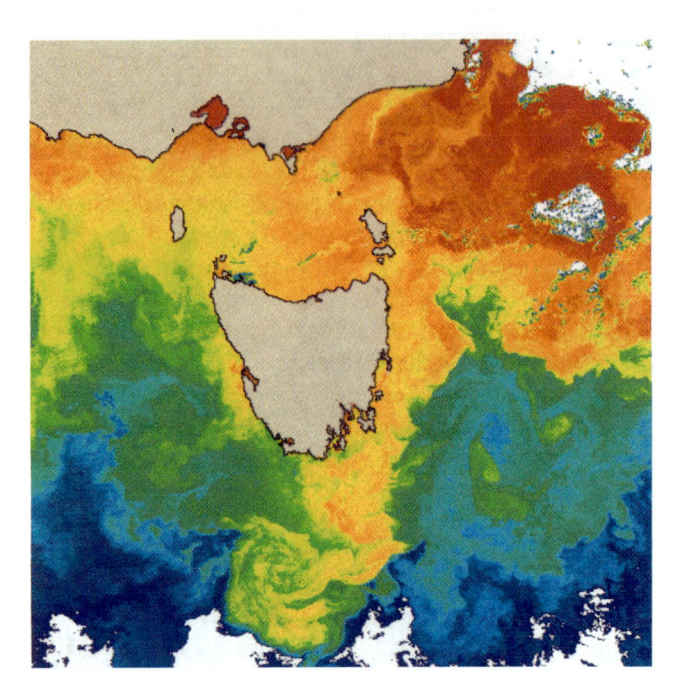

Fig. 3. Winter and summer satellite sea surface temperature images from the NOAA11 satellite. Black is warm ranging through grey to white, which is cold. Clouds are white. The images include a 5° latitude–longitude grid and the shelf edge (200 m isobath).

current meter data at four sites across the western Tasmanian shelf and slope between June and November 1988 also indicated currents were consistently southward and strong at the shelf edge, averaging 0.3 ms^{-1} (Lyne and Thresher, 1994).

Off eastern Tasmania, Harris et al. (1987) concluded that seasonal and interannual variability in water properties at coastal stations could be explained by the strength of the East Australian Current. This current generally varies in accordance with El Niño/Southern Oscillation cycles, with its strength controlling the latitude of the subtropical convergence off the central eastern Tasmanian coast.

Recent cruises by CSIRO confirm that the major currents east and west of the island, the East Australian and Zeehan Currents, respectively, (Fig. 2), both flow southward (Cresswell, in press). These two currents are out of phase in their strengths; the East Australian Current is strongest and reaches farthest southward in summer, while the Zeehan Current is strongest and extends farthest around southern Tasmania and up to the mid-east coast in winter. Figure 2 was drawn from the movements of satellite tracked drifters; however, this pattern can also be seen in satellite images (Fig. 3).

The two major currents differ in their positions relative to the coast. The East Australian Current, which brings water from as far north as the Coral Sea, is an open ocean current that can influence the continental shelf. By contrast, the Zeehan Current, possibly a tenuous extension of the Leeuwin Current that brings subtropical waters down the western Australian coast and then eastward south of Australia, is a near-persistent feature of the continental shelf off western Tasmania. It is strongest at the outer shelf and not present beyond the upper continental slope. Somewhat remarkably, the greater strength and depth of the Zeehan Current in winter causes the bottom temperatures on the western Tasmanian continental shelf to be 2°C warmer than at other times. Wind-driven upwelling events can occur on the west coast, but are not common.

Strong oscillatory tidal currents prevail in Bass Strait. A slow net eastward transport of water intensifies during winter under the influence of water pushed into western Bass Strait by the Zeehan Current (Tomczak, 1981). Eastern-flowing saline bottom water flows off the edge of the continental shelf in eastern Bass Strait during winter, when it is known as the Bass Strait Cascade (Godfrey et al., 1980).

The Antarctic Circumpolar Current possesses relatively little strength off southern Tasmania, compared to both its strength further south in latitudes around 50°S and the strength of the East Australian and Zeehan Currents. Nearly all satellite drifters released in Tasmanian waters have eventually meandered eastward towards New Zealand after leaving the influence of the East Australian Current, but required travel times in excess of 16 months to reach New Zealand waters. Off western Tasmania, well seaward of the Zeehan Current, the residence time for satellite-tracked drifters was as much as one year (Creswell, in press).

Because of the pronounced influence of the East Australian Current, summer sea surface temperatures around Tasmania are substantially warmer (19°C) in the northeastern Bass Strait region, particularly around the Kent and Furneaux Groups, than elsewhere, while cold water (16°C) intrudes onto the southwestern coast. During winter, waters are warmest at King Island and the northwest coast,

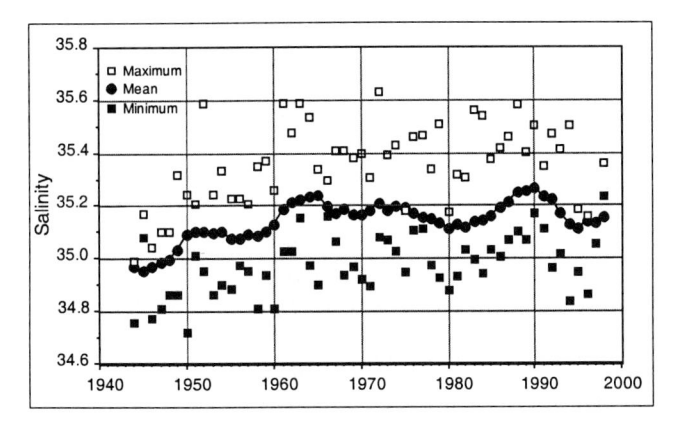

Fig. 4. Maximum and minimum salinities recorded at 20 m depth each year from 1944 to 1998 off northeastern Maria Island (CSIRO, unpublished data). Seasonally-corrected 5-year mean salinity is also shown.

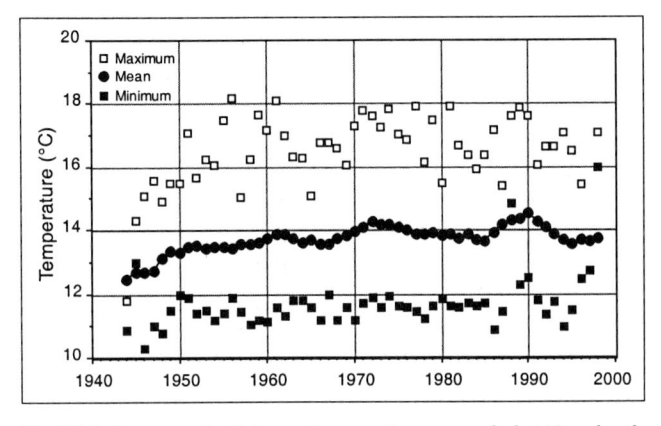

Fig. 5. Maximum and minimum temperatures recorded at 20 m depth each year from 1944 to 1998 off northeastern Maria Island (CSIRO, unpublished data). Seasonally-corrected 5-year mean temperature is also shown.

although they generally vary little throughout inshore Tasmanian waters from the south coast to the Furneaux Group (11°C). Localised cooling down to 7°C occurs in large sheltered embayments, particularly Great Oyster Bay, Storm Bay and D'Entrecasteaux Channel on the east coast, with temperatures declining further in estuaries that are subject to ice-melt runoff.

Nutrient concentrations around Tasmania are generally higher that those recorded around mainland Australia, albeit at levels amongst the lowest for temperate latitudes worldwide (inorganic nitrate 1–2 μM). Highest nutrient levels are associated with the south coast and intrusions of subantarctic water. Central Bass Strait waters are very low in nutrients, with nitrate concentrations generally <1 μM (Rochford, 1993).

Long-term hydrological records from the central east coast at Maria Island indicate that substantial changes in oceanographic climate have occurred off Tasmania over the past 50 years, presumably as a consequence of the East Australian Current increasing in strength and extending further south. Salinity increased from an average <35.0 in the late 1940s to >35.2 in the early 1990s, while temperature increased >1°C over the same period (Figs. 4 and 5). Nitrate concentrations are strongly negatively correlated with water temperature and salinity, and declined substantially over this period (Harris et al., 1987). High nitrate concentrations (>3 μM) have only been recorded during periods when salinity was <34.9 (Harris et al., 1987). Salinities >35.2 are considered indicative of subtropical water associated with the East Australian Current. The trends for increase observed during the second half of the twentieth century showed signs of reversing during the late 1990s, a period when mean annual salinity declined to 35 (Fig. 4).

Reflecting the interannual variability in ocean climate, annual catches of the most abundant inshore pelagic fish species, the jack mackerel *Trachurus declivis*, have oscillated between 9,000 and 42,000 tonnes during the period from 1985 to 1994 (Stewart and Pullen, 1997). In conjunction with overfishing, the trend for gradually increasing water temperatures over the past 50 years, with associated reductions in nutrient concentrations and biological productivity, has presumably contributed to documented declines in Tasmanian catches of important fishery species, such as the barracouta (*Thyrsites atun*).

THE MAJOR SHALLOW WATER MARINE AND COASTAL HABITATS

Tasmanian marine habitats are principally separated into pelagic and benthic systems, with well-defined boundaries also separating hard-bottom and soft-bottom habitats. Multivariate analyses indicate that Tasmanian inshore benthic ecosystems can be subdivided into eight regional groupings on the basis of biogeographic consistencies (Edgar et al., 1997). These 'bioregions' are located along the northern, northeastern, southeastern, southern and western coasts of Tasmania, and around the Kent Group, the Furneaux Group and King Island in Bass Strait. The primary determinants of these bioregions are large scale oceanographic processes, particularly strength of oceanic currents, hydrology of water masses, tidal range and sea state. For example, the distinctive character of ecosystems around the Kent Group in northeastern Tasmania is largely attributable to low nutrient waters, relatively high summer water temperatures, strong tidal currents, moderate sea state, and the advection of numerous larvae spawned in eastern Australia and transported south in association with the East Australian Current. By contrast, the biota in southern Tasmania is adapted to relatively high-nutrient water, large oceanic swells and cool water temperatures, hence few of the common species in that region also occur abundantly near the Kent Group (Edgar, 1984).

Because elements from several biogeographic regions overlap within Bass Strait, species richness in this area is extremely high for macroalgae, invertebrates and temperate fishes. However, despite this great diversity, very few species are endemic to the region. The southeastern

Tasmanian bioregion, by way of contrast, contains relatively few fish and invertebrate species, but includes a large component of plants and animals that are not distributed elsewhere. Notable amongst these species are four of the eight known species in the fish family Brachionichthyidae (*Brachionichthys hirsutus, B. politus* and two undescribed *Brachionichthys* species; P.R. Last, pers. comm.), seastars (*Patiriella vivipara, Marginaster littoralis* and *Smilasterias tasmaniae*), sea urchin (*Pachycentrotus bajulus*) and several macroalgae. Most of the endemic faunal species lack a pelagic larval stage.

Numerous soft-sediment and reef habitat types occur within each of the Tasmanian bioregions. Seagrass meadows typically dominate soft-sediments in shallow sheltered environments, particularly along the northeastern and northwestern Tasmanian coast, and in the Furneaux Group. The most prevalent seagrass species in these areas are *Posidonia australis, Posidonia angustifolia* and *Amphibolis antarctica* (Jordan et al., 1998). Numerous large embayments are also present in southeastern Tasmania; however, within this region the local seagrass species *Heterozostera tasmanica* and *Halophila australis* tend to occupy estuaries and sheltered fringes of bays rather than carpeting the seabed. Notable exceptions to this pattern are extensive seagrass meadows within Norfolk and Blackman Bays.

Very little work has been undertaken in Tasmania on the characterisation of soft-sediment habitat types. Fish and invertebrate assemblages tend to show gradients with depth and wave exposure rather than corresponding closely with the particle size–distribution or origin of sediments. Sediments on the Tasmanian continental shelf consist predominantly of calcareous bryozoan and shell fragments, although siliceous sediments are notably abundant adjacent to granitic coastlines in northeastern Tasmania and around the Furneaux Group (Harris, 1995). Shallow inshore sediments and beaches around Tasmania are all predominantly composed of siliceous particles (Davies, 1978). The few macrofaunal samples collected from shelf sediments possess exceptionally high species richness, with numbers of species collected in northeastern shelf samples amongst the highest recorded worldwide (Coleman et al., 1997).

Shallow reef habitats around eastern, southern and western Tasmania are more highly vegetated by macroalgae than reefs elsewhere in Australia, presumably as a consequence of the relatively high nutrient levels and cool water temperatures. Macroalgal-dominated habitats are most conspicuous along exposed coasts, where the massive alga *Durvillaea potatorum* usually monopolises space in the lower littoral zone and shallow subtidal. The *Durvillaea* habitat extends to depths exceeding 10 m at southern sites subjected to long rolling swells, but declines to insignificance on moderately-exposed and sheltered shores (Edgar, 1984).

Below the *Durvillaea* habitat on exposed coasts, two other large brown algae, the laminarian kelp *Lessonia corrugata* and the fucoid *Phyllospora comosa*, form extensive beds. These plants possess huge sweeping fronds that dislodge other benthic organisms on settlement, creating an almost monospecific habitat type. Numerous macroinvertebrate species are, however, associated with the holdfasts of *L. corrugata*.

Extensive beds of the giant kelp *Macrocystis pyrifera* often form adjacent to *Lessonia* and *Phyllospora* habitats off moderately exposed shores. *Macrocystis* plants possess long stipes that extend from holdfasts on the seabed in water depths of 5–25 m to fronds at the sea surface. The distinctive surface canopy buffers wave action and intercepts most solar radiation, and thus has far-reaching effects on the physical and biological environment in the near vicinity. The total extent of *Macrocystis* habitat within Tasmanian waters has decreased massively over the past four decades, with the almost complete loss of forests from sites north of Tasman Peninsula on the east coast (Sanderson, 1987). This decline has corresponded with a substantial decline in seagrass beds (Rees, 1993), and with a regional increase in temperature and decline in dissolved nitrate concentrations (Harris et al., 1987).

Below the *Lessonia/Phyllospora* habitat, in depths ranging from 1 m on sheltered shores to 20 m at wave-exposed sites, an extremely diverse assemblage of macroalgae is generally present. Included within this habitat type are numerous fucoid algae in the genera *Acrocarpia, Sargassum, Cystophora* and *Caulocystis*, green algae in the genus *Caulerpa*, and red algae in the genera *Plocamium, Phacelocarpus* and *Jeannerettia*.

As depth increases and light levels decrease, the diverse algal habitat merges into habitat dominated by the kelp *Ecklonia radiata*. *Ecklonia* plants predominate to depths of 35 m on clear exposed coasts, with the occasional plant reaching 40 m. Sites below 35 m along the open coast are typically dominated by sponges, octocorals, ascidians and other sessile animals. Although the diversity of invertebrates is exceptionally high, no studies have been conducted on these deep reefs and the majority of species remain poorly known.

Reef locations near the mouths of estuaries in the south and west of Tasmania often show pronounced compression of the habitat types described above. Waters above the halocline in these areas are highly tannin-stained as a result of fluvial drainage through peat deposits, hence light is rapidly absorbed near the water surface (Edgar and Cresswell, 1991). A consequence of the low ambient light in shallow water is that few plants survive below 5 m depth, and sessile animals that more typically live below the euphotic zone in depths greater than 50 m can occupy shallow reefs.

The tannin-stained estuaries of southern and western Tasmania are also unusual in possessing lower nutrient concentrations than the sea, very low faunal productivity and low faunal species richness in their upper riverine reaches (Edgar et al., 1999). By comparison, the species richness of estuarine fishes and macrofauna is high in the Furneaux Group and northeastern and southeastern Tasmania, and moderate elsewhere. These patterns primarily reflect differences in estuary type between regions rather than concentrations of locally endemic species. Coastal inlets and drowned river valleys typically accommodate large numbers of species, whereas river estuaries and coastal lagoons are depauperate. Very few estuarine species (1% of fishes and 3% of invertebrates) are endemic to Tasmania.

With the exception of fjords, all major types of estuaries are present around Tasmania. Annual rainfall in much of the south and west of the island exceeds 4 m, hence river estuaries with little marine dilution in outflow to the sea predominate in these areas. By contrast, barrier estuaries and lagoons, many of which are hypersaline during summer, occur commonly in the northeastern and eastern regions, which lie in a rainshadow and have annual rainfall <1 m. Estuaries in northern Tasmania are nearly all open as a consequence of mesotidal ranges (> 2 m) and high river runoff (2–3 m annual rainfall), and generally show pronounced salinity and temperature fluctuations during the tidal cycle (Edgar et al., 1999). Southeastern Tasmania possesses a highly dissected coast with numerous inlets and open estuaries that possess little freshwater dilution. The five large drowned river valleys present in Tasmania (Macquarie Harbour, Tamar estuary, Derwent Estuary, Huon Estuary and Port Davey) are evenly distributed around the Tasmanian coast.

Mangrove habitats do not extend south from Victoria into the Tasmanian region; however, saltmarshes occur extensively in association with estuaries along all Tasmanian coasts and on the larger Bass Strait islands. Numerous migratory wading birds are dependent on saltmarsh and associated mudflats, with five local estuaries listed under the Ramsar Convention (Pittwater, Moulting Lagoon, Ringarooma estuary, Thirsty Lagoon, Logan Lagoon).

OFFSHORE SYSTEMS

As is the case elsewhere in the world, species associated with pelagic habitats off Tasmania generally possess wide-ranging distributions that are primarily bounded by oceanographic features such as convergences, upwellings and warm core eddies. With respect to offshore species distributions, the most important oceanographic feature in the region is the location of the front between subtropical and subantarctic water masses, a seasonally fluctuating boundary known as the Subtropical Convergence that oscillates from south of Tasmania to the central coast. Nutrient levels, with associated phytoplankton, zooplankton and fish productivity, are substantially higher south of this front (Harris et al., 1987; 1991; 1992). Catches of schooling pelagic fishes drop dramatically during periods when subtropical water intrudes south and animals follow the subantarctic water mass or move into deeper waters (Harris et al., 1988; 1992; Young et al., 1993).

Valuable offshore fishery resources occur in deep water on the continental slope off eastern and western Tasmania. Recently the deepwater demersal trawl fishery has extended to the Cascade Plateau and South Tasmanian Rise and includes seamounts—undersea features unusually prevalent on the Tasmanian continental slope. Approximately 70 seamounts rising 200–500 m above the seabed from depths between 1000 and 2000 m occur in a field 100 km southeast of Tasmania (Hill et al., 1997; Koslow and Gowlett-Holmes, 1998). Several other more-isolated seamounts are also present around the Tasmanian coast.

Bottom currents intensify in the vicinity of seamounts and discontinuities on the seabed, concentrating food and enabling dense aggregations of filter feeding invertebrates such as gorgonians to survive (Genin et al., 1986). On the mid continental slope in water deeper than 700 m several commercial fish species, including orange roughy (*Hoplostethus atlanticus*) and oreo dories, aggregate and are targeted by trawlers. The deepwater fishery also exploits shallower areas of trawlable bottom on the upper slope, approximately 400–600 m depth, where blue grenadier (*Macruronus novaezelandiae*), gemfish (*Rexea solandri*) and ling (*Genypterus blacodes*) are captured in large quantities. Near the edge of the continental shelf species such as jackass morwong (*Nemadactylus macropterus*), blue warehou (*Seriolella brama*) and tiger flathead (*Neoplatycephalus richardsoni*) are trawled, and several species of shark caught using nets and hooks. Other commercially valuable deepwater species, such as blue eye (*Hyperoglyphe antarctica*), are targeted using droplines.

POPULATIONS AFFECTING THE AREA

Between 5000 and 10,000 aboriginal inhabitants are estimated to have occupied Tasmania prior to European settlement. Britain occupied the island in 1803, primarily to secure the region from the French, and partly to develop a penal colony. By 1996 the Tasmanian population had risen to 474,600. This population comprised only 2.6% of the total Australian population and is currently declining (ABS, 1998).

The Tasmanian coastline extends for 4900 km—more coastline per unit land area than for any other Australian state. Nowhere in Tasmania is located more than 115 km from the sea, with approximately two thirds of Tasmanians living in the coastal zone (Edgar et al., 1998). Tasmania is also the most decentralised state of Australia, with nearly 60% of the total population living outside the Greater Hobart statistical division. The population is widely distributed around the northern and eastern Tasmanian coastlines, while much of the south and west coasts are uninhabited, lying within the Tasmanian Wilderness World Heritage Area.

Intensive exploitation of coastal resources commenced at the time of European settlement, with seals and whales providing much of the financial basis for the developing colony. Rock lobsters (*Jasus edwardsii*) and native flat oysters (*Ostrea angasi*) were also extensively exploited during the nineteenth century. Fishing for finfish was of lesser importance, but expanded with little regulation until the 1880s when it was recognised that stocks of several inshore fishes, rock lobsters and native oysters were severely depleted (Harries and Croome, 1989). Fisheries legislation was introduced at that time. Fishing in marine and coastal waters has continued to be important to the economy and social fabric of Tasmania. Aquaculture also has developed into an important activity in estuaries and coastal waters since the 1970s.

Industrial and urban development has also expanded disproportionately in the coastal zone, due to shipping access, favourable climate and productivity of estuarine and inshore waters. Most of this development, and the use of estuaries and coastal waters as a dumping ground for industrial and urban waste, has occurred on an *ad hoc* basis, resulting in substantial changes to the natural environment. Little regard was given to the state of the marine environment until recently, even when grossly obvious pollution was evident. For example, industrial dumping of paint pigment wastes on the north-west Tasmanian coast stained waters a dark red-brown colour for a distance of 20 km along the coast and up to 8 km offshore (SDAC, 1996). Tasmanian seas were widely considered sufficiently large to absorb any anthropogenic activity and capable of rapidly returning to original conditions after an impacting activity ceased. During the last two decades, however, the general community and management agencies have become increasingly concerned about the state of the marine environment and the assimilative capacity of coastal waters for pollutants.

Very few baseline surveys have been conducted in Tasmanian marine waters due to logistic difficulties and costs, and the general perception that waters were sufficiently large to be able to absorb human impacts. Consequently, very few data are available for assessment of environmental conditions and trends. This lack of baseline data is now recognised as a major impediment to management decisions, and was listed as a key issue in the first State of the Environment Report (SDAC, 1996).

RURAL FACTORS

Tasmania's economy is principally dependent on primary production, especially highly intensive agriculture and forestry. Just under one third of the total area of Tasmania (almost 20,000 km^2) is utilised for farming, and 20,000 km^2 of forests are available for wood production, either from native forests or plantations. Pastures and crops are routinely improved using phosphorus-based fertilisers, and herbicides and pesticides are used in large quantities to protect crops and areas of timber regrowth. Estimates of the rate of clearance of natural vegetation, based on satellite images, for forestry, hydroelectricity and agricultural activities, averaged 100 km^2/year between 1972 and 1994 (Edgar et al., 1999). Runoff associated with catchment activities has caused large sediment and nutrient loadings in estuaries and coastal waters. Edgar et al. (1999) found that estuaries with catchments of moderate to high population densities (>10 persons per km^2) consistently had muddy sediments, whereas estuaries with low population densities in catchments possessed sandy sediments and different invertebrate assemblages to the high population-density estuaries. Increased siltation and nutrification of waters within catchments, and resultant declines in light penetration, are also considered to have contributed substantially to a major decline in seagrass beds around the Tasmanian coastline since 1960 (Rees, 1993).

Containment of freshwater systems for hydroelectricity generation, irrigation of agricultural lands and domestic or industrial water supply has also affected many estuarine and, to a lesser extent, coastal waters. Edgar et al. (1999) identified 39 major river basins in Tasmania that are affected by hydroelectricity developments. Large impoundments and diversions between catchments, such as the Mersey/Forth diversion, generally produce a regulated flow pattern which differs from natural cycles. Large irrigation dams also reduce freshwater flows into estuaries, which can affect overall productivity of the system. For example, an almost complete reduction in water flow by the Craigbourne Dam and South-East Tasmanian Irrigation Scheme appears to have affected the growth of oysters (*Crassostrea gigas*) in the downstream Pittwater estuary (Crawford 1997, unpublished report). The downstream ecological effects from altered freshwater systems, however, are largely unknown.

Mining activities have also had a major impact on Tasmanian coastal waters. Acid mine drainage and sediment runoff from large metalliferous mines in the western and northeastern regions have greatly depressed pH and increased heavy metal concentrations in particular rivers. High levels of suspended solids are also generally present in waste waters. The most severe problem is associated with tailings of the Mount Lyell copper mine, which has affected the Queen River and downstream King River to the extent that they are considered biologically dead at downstream sites. Macquarie Harbour, the largest estuary in Tasmania and recipient of King River waters, has also been severely affected. Over 100 million cubic metres of tailings, slag and topsoil have been deposited in a 2.5 km^2 delta at the mouth of the river in Macquarie Harbour (O'Connor et al., 1996) and heavy metals originating from Mt Lyell continue to affect the estuary.

The distribution of catchment impacts around the Tasmanian coastline is far from random, with virtually all estuaries along the northern and eastern coastline badly degraded by land clearing and siltation but with most estuaries along the southern and western coasts remaining relatively pristine. Nearly all estuaries located in the south and west are river estuaries or small open estuaries that occur in high-rainfall areas remote from human activity, with the majority of these estuaries and catchment areas included in the South West National Park (Graddon, 1997). The three largest estuaries in the state, the Derwent estuary (Coughanowr, 1997), Macquarie Harbour (O'Connor et al., 1996) and Tamar estuary (Pirzl and Coughanowr, 1997) are very badly degraded by urbanisation and heavy metal pollution, with the Derwent estuary having been considered one of the most polluted estuaries by heavy metals worldwide (Bloom and Ayling, 1977).

Nevertheless, the impacts of rural and industrial activities are expected to decrease over the next decade with the development of an Integrated Catchment Management Policy for Tasmania, and new water management and quality requirements, including allocating water for environmental flows and establishing Protected Environmental Values for Tasmanian waterways.

COASTAL EROSION AND LANDFILL

As elsewhere in the world, human activity around the Tasmanian coastline has altered natural cycles, with detrimental environmental effects now evident in many cases. For example, the introduction of marram grass (*Ammophila arenaria*) to stabilise mobile sand dunes has caused reductions in sand supply and reduced the width of adjacent beaches (SDAC, 1996). Similarly, the introduction of ricegrass (*Spartina anglica*) into the Tamar estuary early this century stabilised mud flats and navigational channels in that area, but the plant also trapped pollutants and spread to other estuaries where waterways became choked and land drainage and navigation were affected (DPIF, 1997). Eradication trials are currently underway in an attempt to clear affected estuaries.

Urban sprawl, ribbon development and recreational use of the coastal strip, especially on northern and eastern coasts, has contributed significantly to coastal erosion. Small holiday dwellings known as shacks, for example, are generally built close to the waters edge, altering the protective cover of native vegetation. This problem is exacerbated by an associated proliferation of vehicle tracks. Marine structures such as jetties or breakwaters have also been built in some areas, resulting in disruption of longshore sand drift and alterations to sand erosion and depositional processes.

Land reclamation has occurred on a relatively small scale to provide additional area for industrial development and port facilities. For example, at Emu Bay, Burnie, land infilling to expand industrial and port facilities has caused extensive alterations to the shoreline. Dredging also occurs in a few sites to maintain navigation channels, most notably in the Tamar River. Urban and industrial wastes are discarded in landfills in many municipalities, some of which are located close to waterways and seepage into coastal waters may occur. However, the environmental effects of these activities are poorly understood and remain largely unstudied, except for obvious major impacts. For example, an infilled causeway across the mouth of an enclosed bay at Orielton Lagoon near Hobart reduced tidal flushing to a negligible level, thereby causing waste containment from a sewage treatment plant and land runoff. The lagoon eventually developed a major bloom of blue-green algae, producing obnoxious odours and killing fish and other aquatic life. Remediation works have recently been completed to promote water circulation and reduce nutrient inputs. Several large wetlands around the coast, particularly in the northeastern and northwestern regions, have also been drained to expand agricultural land.

EFFECTS FROM URBAN AND INDUSTRIAL ACTIVITIES

Artisanal and Non-industrial Uses of the Coast

The presence of numerous large shellfish middens around the Tasmanian coast indicates that abalone (*Haliotis rubra* and *H. laevigatus*), turbo (*Turbo undulatus*), oysters (*Ostrea angasi*) and other molluscs provided a major component of the diet of coastal aboriginal tribes, but that scalefish were rarely utilised. Since European settlement, aboriginal exploitation of the coastal zone has declined substantially.

Recreational use of the coastal area is, however, high and increasing. Tasmanians are frequent users of the coast due primarily to the close proximity of most population centres and to technological advances in fishing gear and boats. A survey by the Australian Bureau of Statistics in 1983 estimated that a quarter of Tasmania's population participates at least occasionally in recreational fishing, and that well in excess of US$ 35 million is spent annually in Tasmania by recreational fishers. The main groups targeted recreationally are rock lobster, abalone, scallops and scalefish. The capture of large quantities of targeted species possibly causes substantial impacts to coastal marine ecosystems through functional loss of predatory and large grazing species; however, no studies have been conducted on this topic.

Commercial Usage of Coastal Resources

Commercial exploitation of wild fish, crustacean and mollusc stocks continues to be a major activity in Tasmanian coastal waters, with the total value of the catch in Tasmania in 1996/97 estimated to be US$70 million (ABARE, 1997). The total inshore fishery in terms of catch weight is divided approximately evenly between abalone, rock lobster and finfish, with the first two of these fisheries providing 90% of catch value. The rock lobster and abalone fisheries are both managed using individual transferable quota (ITQ) systems. In contrast to most abalone fisheries worldwide, stocks of Tasmanian blacklip abalone (*Haliotis rubra*) are considered healthy, with the total allowable catch increased in 1997 by 20%. Rock lobsters, on the other hand, are probably overfished in northern and eastern areas, with declining catch per unit effort and a reduction in the mean weight of animals taken in those regions.

Catches of scale fish are relatively small compared to catches in other geographic areas of similar size worldwide. The major species caught in terms of biomass is the jack mackerel (*Trachurus declivis*), most of which is used as fish meal in the salmonid aquaculture industry. Overall this fishery is thought to be operating at a sustainable level although intensity of fishing varies greatly between regions (Stewart and Pullen, 1997). In the offshore trawl fishery, managed by the Commonwealth Government, orange roughy (*Hoplostethus atlanticus*), blue grenadier (*Macruronus novaezelandiae*), blue eye trevalla (*Hyperoglyphe antarctica*) and blue warehou (*Seriolella punctata*) are important species to Tasmania. Catches of shark, particularly school shark (*Galeorhinus galeus*) and gummy shark (*Mustelus antarcticus*) have also been significant in both volume and value. Although these fisheries are tightly controlled by individual transferable quotas (ITQs), introduced in 1992 (or soon to be introduced for shark) to reduce overfishing of some species, fishing effort has increased overall. Based on available stock assessment information, orange roughy, blue eye trevalla, blue warehou and gummy shark were considered to be fully fished and school shark overfished (Tilzey and Chesson, 1997). Few southern bluefin tuna (*Thunnus maccoyi*) are now caught in Tasmanian waters, and the status of this species has been rated as overfished with severely depleted spawning stock (Caton et al., 1997).

Recent catches for species under Tasmanian jurisdiction have been around 1500–1700 tonnes per annum, with Australian salmon (*Arripis* spp.) accounting for approximately 400 tonnes p.a. Other important species include blue warehou, wrasse (*Notolabrus tetricus* and *N. furicola*) and banded morwong (*Cheilodactylus spectabilis*). The latter three species are inshore reef fishes collected for the premium live fish trade, and have been heavily fished in recent years. Limited restrictions have traditionally been applied to these inshore state-managed fisheries, although more restrictive 'input' controls have recently been introduced.

The effects of fishing activities on the marine environment have been little investigated in Tasmania. Even scallop dredges, which have an obvious impact by overturning upper layers of the seabed, have not been assessed for environmental degradation in Tasmanian waters. A major fishery for commercial scallops (*Pecten fumatus*) and, to a lesser extent, queen scallops (*Equichlamys bifrons*) and doughboy scallops (*Chlamys asperrimus*) developed in Tasmania in the 1930s. This fishery has dramatically declined in recent years due to over-exploitation, climate change and habitat alterations.

Fishing gear, in particular longlines and gill nets, are thought to cause high mortality amongst seabirds and mammals, especially albatrosses, gannets and seals (Caton et al., 1997).

Aquaculture developed rapidly during the final two decades of the twentieth century, and was valued at US$ 45 million in 1996–97. The two main species cultured, Atlantic salmon (*Salmo salar*) and Pacific oysters (*Crassostrea gigas*), were originally imported from overseas. Native species such as mussels, abalone and scallops are farmed on a small scale.

A total of 145 marine farms were registered in Tasmania in 1999, occupying almost 20 km^2 of seabed. Disputes over resource usage have increased with the development of the aquaculture industry, and been exacerbated because this industry is concentrated within the area of highest population density in the southeast. The level of conflict should decline with the recent development of a marine farming planning scheme that takes into account different users of inshore waters. Marine farming is recognised to potentially cause environmental degradation through the discharge of food-related organic wastes and nutrients in finfish culture, and by the increased removal of suspended particulate matter and deposition of faeces and pseudofaeces by shellfish. Although low ambient levels of nutrients in Tasmanian waters are likely to make ecosystems relatively sensitive to organic loadings, studies to date indicate ecosystem effects of aquaculture are largely confined to the immediate vicinity of farms (Woodward et al., 1992; Ye et al., 1991).

No significant mining activities occur in Tasmanian coastal waters.

Industrial Uses of the Coast

Tasmanian coastal waters are widely used industrially as a means of waste disposal. Although most industries now treat wastewater to some degree, impacts on the environment remain. The areas most affected by waste discharge are the Derwent and Tamar estuaries, Macquarie Harbour and the northwest coast around Burnie. The main contributors of industrial waste to the sea are mineral processors, paper and wood-chip mills, and food processors.

Several large mineral processing operations exist in Tasmania. Nearly all of these discharge contaminants into estuaries or the sea. The largest processors, located in the Tamar (Comalco and Temco) and Derwent (Pasminco Metals-EZ) estuaries, discharge metals, suspended solids, chemicals and other contaminants. Heavy metals accumulated in sediments in the Derwent estuary remain at amongst the highest levels in Australia (particularly zinc, lead, cadmium and mercury), and this estuary remains classified as highly degraded (Coughanowr, 1997). Ocean dumping of industrial wastes also occurred until recently. Jarosite, a chemical residue from electrolytic zinc refining, was dumped on the continental shelf 60 km east of Hobart until 1997 when exemptions were withdrawn as a consequence of international commitments given by the Federal Government to the London Convention on sea dumping.

Pollutants in waste waters discharged from pulp and paper mills in Tasmania include wood-derived organic wastes, suspended solids and resin acids. Pulp and paper mills now use primary waste treatment to remove much of the load of suspended solids. However, because of a lack of secondary treatment to reduce the level of biological oxygen demand, environmental problems continue to arise in areas of low water exchange such as the Derwent estuary. A number of food processing facilities, including several fish processing factories, also discharge primary-treated waste waters. Again, the lack of secondary treatment can result in high local biological oxygen demand and elevated levels of suspended solids and nutrients, although waste disposal from most food related industries is generally considered adequate. Several large food processing plants on the north coast discharge wastes to municipal treatment plants, with the discharges thus included in municipal sewage.

As is the case with other industries, the impact of manufacturing industries on Tasmania's marine environment cannot be reliably determined because virtually no baseline studies were conducted before the industries developed. Additionally, the sharing of management of waste treatment approximately equally between the Tasmanian Government and local councils has resulted in uncoordinated management and very little information available on wastes from numerous small operations. Large discharges (>1000 tons) are managed by the State Government and small discharges (<1,000 tons) by local councils.

Urban Use (Cities)

An increasing proliferation of human settlements around the Tasmanian coastline places increased pressure on the marine environment, partly because of diffuse pollution from storm water drains and also because of increased sewage discharges. Many municipal sewage treatment plants in Tasmania are currently operating at or above designed capacity and numerous illegal stormwater connections enter sewerage systems. Storm events periodically cause treatment plants to overflow, with diluted sewage then discharging directly into estuaries or the sea. This problem is expected to diminish during the first decade of the twenty-first century as sewage systems around the state are upgraded and waste waters increasingly recycled. However, similar to industrial wastes, sewage systems management is uncoordinated because it is split between State and Local Government depending on the size of the system.

Discharge of pollutants in storm water from urban areas also causes substantial environmental problems. A general expansion of roofs, roads, unvegetated soils and other impermeable ground coverings associated with urban sprawl results in elevated and channelised water runoff to estuaries and the sea. These runoffs also contain pollutants such as dog faeces, domestic wastes, herbicides, pesticides, nutrients, vehicle wastes and organic matter, which cause significant declines in estuarine water quality.

Studies of the Derwent estuary indicate high levels of pollution from urban and industrial wastes, and significant associated changes in the geomorphology and biota of the river. Nevertheless, no baseline data are available to quantify the magnitude of change. Although upgraded sewage treatment plants and improved treatment of industrial waste water are now causing a decline in end-of-pipe emissions, large nutrient loads continue to be released from sewage treatment plants, and substantial loads of faecal bacteria are deposited by urban run-off into the estuary.

Shipping and Offshore Accidents and Impacts

As an island State, Tasmania is heavily reliant on sea transport of goods. The potential for shipping accidents that affect the environment is therefore relatively large.

A moderate oil spill occurred in July 1995 when the BHP bulk carrier 'Iron Baron' ran aground on Hebe Reef, a shallow offshore reef 5 km northwest of the mouth of the Tamar River in northern Tasmania. Approximately 325 tons of Bunker C oil were discharged into Bass Strait, resulting in the largest oil spill and clean up operation in Tasmania's history. Subsequent surveys and research have shown that the major impact from the oil spill has been on populations of marine mammals and sea birds, particularly little penguins. Other marine life appears to have been little affected, other than at the hull grounding site on Hebe Reef (Edgar and Barrett, 2000).

Degradation of estuarine and coastal environments by marine oil spills is not generally considered a major problem in Tasmania. Most wastes from commercial shipping and recreational boating are so small that they are broken down and dispersed by natural processes with minimal impact on the environment (SDAC, 1996). Hydrocarbon discharges in marine waters primarily arise from urban wastes, such as oils and paints discharged in storm water, rather than from shipping spills.

The major impact of shipping on the Tasmanian environment has been the introduction of exotic taxa in ballast water and fouled hulls. Over thirty species have successfully colonised the marine environment and dispersed around the coast, disturbing ecosystems by displacing native species or altering predatory balances. Ecosystem effects caused by introduced species are, however, difficult to ascertain because of the lack of baseline data. Species presumed to have been introduced in ballast water and likely to have significant environmental impacts include the northern Pacific seastar *Asterias amurensis*, the European shore crab *Carcinus maenas*, the Japanese kelp *Undaria pinnatifida* and the toxic dinoflagellate *Gymnodinium catenatum*. These species are all relatively recent introductions with economic effects and their distributions and impact are currently under investigation. In the case of *G. catenatum*, periodic blooms of this alga result in the closure of shellfish farms to harvesting in the D'Entrecasteaux Channel and significant economic hardship to marine farmers.

A number of other exotic species were probably introduced near the beginning of the twentieth century on oysters shipped to Tasmania from New Zealand. The most important of these pest species are the screw shell *Maoricolpus roseus*, the venerid clam *Venerupis largillierti*, the porcellanid crab *Petrolistes elongatus*, the cancrid crab *Cancer novaezelandiae*, the chiton *Amaurochiton glaucus*, and the asterinid seastar *Patiriella regularis*. Despite the long period since these invertebrates were introduced, very little biological information is available on these species in Tasmania.

PROTECTIVE MEASURES

Management of Tasmania's marine resources is shared between the Australian and State Governments. The State generally has jurisdiction over land and sea within 3 nautical miles of the coast, and as a result of the Offshore Constitutional Settlement also controls some fishery species such as rock lobster and abalone out to the 200 nautical mile limit of the Australian Exclusive Economic Zone or to the Tasmanian/Victorian border at 39°12'. The Australian Government manages the marine area from 3 to 200 nautical miles, including most mobile pelagic species. Cooperative management arrangements between the Australian and Tasmanian Governments have also been agreed to in relation to petroleum, minerals, mammal and seabird conservation and offshore dumping.

Because the Australian Government is a signatory to a number of international treaties and conventions, it has broad management functions but little direct management responsibility in Tasmania. International conventions include the MARPOL Convention (International Convention on Pollution from Ships), Ramsar Convention (Convention on Wetlands of International Importance Especially as Waterfowl Habitat 1971) and treaties with Japan and China on the protection of migratory birds.

State legislation introduced during the last five years of the twentieth century signalled a major policy shift in coastal and marine resource management. The State Coastal Policy 1995, which is legally binding on all State agencies and local government, aims for sustainable development of the coastal zone, protection of natural and cultural values of the coast, and shared management and protection of the coastal zone between different users. Similarly, the Water Quality Policy 1997 and the proposed Integrated Catchment Management Policy provide new strategies for improved environmental management. Management of land and water areas are becoming increasingly integrated, with land-based activities increasingly accountable for downstream effects on estuarine and marine environments.

Recently introduced legislation, notably the Living Marine Resources Management Act 1995 and the Marine Farming Planning Act 1995, has changed the emphasis of fisheries management from largely commercial fisheries development to sustainable management of all living marine resources. Management Plans have been developed for the major commercial fisheries and stock assessments are conducted annually. However, although the Act provides for habitat protection plans, management of marine habitats has been ignored to date. Marine Farming Development Plans have been prepared for each major farming area and incorporate requirements for baseline environmental surveys and environmental monitoring so that any detrimental impacts, should they occur, will be noticed and managed appropriately.

Other relevant legislation includes the Environmental Management and Pollution Control Act 1995 which controls pollution of Tasmania's waters plus some aspects of marine environmental management. The Threatened Species Protection Act 1995 has been developed to conserve Tasmania's threatened flora and fauna. Of the 58 vertebrate species classified as rare, vulnerable or endangered under this Act, approximately half occur in marine or coastal environments. The majority of endangered animal species in Tasmania are birds, mammals and reptiles, but increasing numbers of echinoderms and fish are also becoming listed (SDAC, 1996). Of particular concern is the spotted handfish (*Brachionichthys hirsutus*), which faces possible extinction.

A number of Marine and Estuarine Protected Areas (MEPAs) have been declared in Tasmania, with the great majority historically declared as Nature Reserves to protect wading birds but with no protection afforded to aquatic animals or plants. More recently, a system of 'no-take' MEPAs is under development to protect and preserve representative and unique marine habitats, for educational purposes and scientific research, and for recreational activities. Four 'no-take' MEPAs, covering a total area of 22 km^2, were declared in 1991. Three of these MEPAs, located off the eastern and southeastern coast (Governor Island off Bicheno, and Tinderbox and Ninepin Point in the D'Entrecasteaux Channel), are relatively small, and were declared for scientific or recreational purposes. The largest reserve is also located off the east coast at Maria Island, and was declared to conserve representative Tasmanian east coast marine habitats.

Biological changes within this and the other MEPAs and at associated reference sites have been monitored since 1992 (Edgar and Barrett, 1997; 1999). Six years after the declaration of the MEPAs, substantial changes amongst reef communities protected from fishing were evident. The number of fish, invertebrate and macroalgal species recorded on transects, and the average size and abundance of exploited fishery species increased rapidly at the Maria

Island reserve, and to a lesser extent in the three smaller reserves. Increases of an order of magnitude in the biomass of rock lobsters, and increases of two orders of magnitude in the abundance of the bastard trumpeter *Latridopsis forsteri*—the major target of recreational fishers—were particularly noticeable. Ecosystem effects were also evident within the Maria Island MEPA, with the dominant macroalga on reefs changing over the period of study. Detailed proposals for the declaration of MEPAs in other coastal regions are well underway. Current research is evaluating several areas proposed by the fishing industry as MEPAs, and also to progress the development of MPA's in all Tasmanian marine bioregions.

One large MEPA managed by the Australian Government has also been declared in offshore waters south of Tasmania. This MEPA, which encompasses an area of 370 km^2, was declared to protect a representative subset of an unusual seamount field that arises from water depths between 1000 m and 2000 m. Exploratory surveys by CSIRO indicated that trawling for orange roughy rapidly eliminated unique benthic communities of sessile invertebrates associated with these seamounts (Koslow and Gowlett-Holmes, 1998). Trawl fishers voluntarily agreed to a closed area, which included eight seamounts rising to depths between 1100 m and 1600 m that had not been previously fished. This area is now protected.

REFERENCES

ABARE (1997) *Australian Fisheries Statistics*. Australian Bureau of Agricultural and Resource Economics, Canberra.

ABS (1998) *Year Book Australia 1998*. Australian Bureau of Statistics, Canberra.

Baines, P.G., Edwards, R.J. and Fandry, C.B. (1983) Observations of a new baroclinic current along the western continental slope of Bass Strait. *Australian Journal of Marine and Freshwater Research* **34**, 155–157.

Bloom, H. and Ayling, G.M. (1977) Heavy metals in the Derwent estuary. *Environmental Geology* **2**, 3–22.

Caton, A., McLoughlin, K. and Staples, D. (1997) *Fishery status reports 1997: Resource assessments of Australian Commonwealth fisheries*. Bureau of Resource Sciences, Canberra.

Coleman, N., Gason, A.S.H. and Poore, G.C.B. (1997) High species richness in the shallow marine waters of south-east Australia. *Marine Ecology Progress Series* **154**, 17–26.

Coughanowr, C. (1997) State of the Derwent Estuary: a review of environmental quality data to 1997. *Supervising Scientist Report* **129**, Supervising Scientist, Canberra.

Cresswell, G. (in press) Currents of the Continental Shelf and Slope of Tasmania. Proceedings of a symposium '98 UNESCO Year of the Ocean in the Tasmanian and Southern Ocean Context. 30 September– October 1998, Hobart Tasmania.

Davies, J.L. (1978) Beach sand and wave energy in Tasmania. In *Landform Evolution in Australia*, eds. J.L. Davies and M.A.J. Williams, pp. 158–167. ANU Press, Canberra.

DPIF (1997) Strategy for the Management of Rice Grass (*Spartina anglica*) in Tasmania, Australia. Draft for Public Comment. Department of Primary Industry and Fisheries, Marine Resources Division, Tasmania.

Edgar, G.J. (1984) General features of the ecology and biogeography of Tasmanian rocky reef communities. *Pap. Proceedings of the Royal Society of Tasmania* **118**, 173–186.

Edgar, G.J. (1986) Biogeographical processes in the Southern Hemisphere marine environment. In *Actas II Congreso Algas Marinas Chilenas*, pp. 29–46. Universidad Austral de Chile, Valdivia, Chile.

Edgar, G.J. and Barrett, N.S. (1997) Short term monitoring of biotic change in Tasmanian marine reserves. *Journal of Experimental Marine Biology and Ecology* **213**, 261–279.

Edgar, G.J. and Barrett, N.S. (1999) Effects of the declaration of marine reserves on Tasmanian reef fishes, invertebrates and plants. *Journal of Experimental Marine Biology and Ecology* **242**, 107–144.

Edgar, G.J. and Barrett, N.S. (2000) Impact of the Iron Baron oil spill on subtidal reef assemblages in Tasmania. *Marine Pollution Bulletin* **40**, 36–49.

Edgar, G.J., Barrett, N.S. and Graddon, D.J. (1999) A classification of Tasmanian estuaries and assessment of their conservation significance using ecological and physical attributes, population and land use. *Tasmanian Aquaculture and Fisheries Institute, Hobart, Technical Report Series* **2**, 1–205.

Edgar, G.J. and Cresswell, G.R. (1991) Seasonal changes in hydrology and the distribution of plankton in the Bathurst Harbour estuary, south-western Tasmania. 1988–1989. *Pap. Proceedings of the Royal Society of Tasmania* **125**, 61–72.

Edgar, G.J., Moverley, J.S., Barrett, D., Peters, N.S. and Reed, C. (1997) The conservation-related benefits of a systematic marine biological sampling program: the Tasmanian bioregionalisation as a case study. *Biological Conservation* **79**, 227–240.

Edwards, R.J. and Emery, W.J. (1982) Australian Southern Ocean frontal structure during Summer 1976–1977. *Australian Journal of Marine and Freshwater Research* **33**, 3–22.

Genin, A., Dayton, P.K., Lonsdale, F.N. and Spiess, F.N. (1986) Corals on seamounts provide evidence of current acceleration over deep sea topography. *Nature* **322**, 59–62.

Godfrey, J.S., Jones, I.S.F., Maxwell, J.G. and Scott, B.D. (1980) On the winter cascade from Bass Strait into the Tasman Sea. *Australian Journal of Marine and Freshwater Research* **31**, 275–286.

Graddon, D.J. (1997) Characteristics of Tasmanian estuaries and catchments: physical attributes, population, and land use. Unpublished Master of Environmental Studies thesis, University of Tasmania, Hobart.

Harries, D.N. and Croome, R.L. (1989) A review of past and present inshore gill netting in Tasmania with particular reference to the Bastard Trumpeter *Latridopsis forsteri* Castelnau. *Pap. Proceedings of the Royal Society of Tasmania* **123**, 97–110.

Harris, G., Nilsson, C., Clementson, L. and Thomas, D. (1987) The water masses of the East Coast of Tasmania: seasonal and interannual variability and the influence on phytoplankton biomass and productivity. *Australian Journal of Marine and Freshwater Research* **38**, 569–590.

Harris, G.P., Davies, P., Nunez, M. and Meyers, G. (1988) Interannual variability in climate and fisheries in Tasmania. *Nature* **333**, 754–757.

Harris, G.P., Griffiths, F.B. and Clementson, L.A. (1992) Climate and the fisheries off Tasmania—interactions of physics, food chains and fish. *South African Journal of Marine Science* **12**, 585–597.

Harris, G.P., Griffiths, F.B., Clementson, V., Lyne, L.A. and Van der Doe, H. (1991) Seasonal and interannual variability in physical processes, nutrient cycling and the structure of the food chain in Tasmanian shelf waters. *Journal of Plankton Research* **13**, 109–131.

Harris, P.T. (1995) Marine geology and sedimentology of the Australian continental shelf. In *The State of the Marine Environment Report for Australia. Technical Annex: 2. Pollution*, eds. L.P. Zann and D.C. Sutton, pp. 11–24. Great Barrier Reef Marine Park Authority, Townsville, Qld.

Hill, P.J., Exon, N.F. and Koslow, J.A. (1997) Multibeam sonar mapping of the seabed off Tasmania: results for geology and fisheries.

Third Australasian Hydrographic Symposium, Maritime Resource Development, Symposium Papers, Special Publication **38**, 9–19.

Interim Marine and Coastal Regionalisation for Australia Technical Group (1998) Interim Marine and Coastal Regionalisation for Australia: an ecosystem-based classification for the marine and coastal environments. Version 3.3. Environment Australia, Canberra.

Jordan, A.R., Mills, D.M., Ewing, G. and Lyle, J.M. (1998) Assessment of inshore habitats around Tasmania for life-history stages of commercial finfish species. Tasmanian Aquaculture and Fisheries Institute, Hobart.

Koslow, J.A. and Gowlett-Holmes, K. (1998) The seamount fauna off southern Tasmania: benthic communities, their conservation and impacts of trawling. CSIRO, Division of Marine Research, Hobart, Tasmania.

Lyne, V.D. and Thresher, R.E. (1994) Dispersal and advection of *Macruronus novaezealandiae* (Gadiformes: Merlucciidae) larvae off Tasmania: Simulation of the effects of physical forcing on larval distribution. In *The bio-physics of marine larval dispersal. Coastal and Estuarine Studies*, Vol. 45, eds. P.W. Sammarco and M.L. Heron, Chapter 6, pp. 109–136. American Geophysical Union, Washington, DC.

O'Connor, N.A., Cannon, F., Zampatti, B., Reid, M. and Cottingham, P. (1996) A Pilot Survey of Macquarie Harbour, Western Tasmania. Department of Environment and Land Management, Environmental Assessment Division Water Ecoscience and Monash University.

Pirzl, H. and Coughanowr, C. (1997) State of the Tamar Estuary: a review of environmental quality data to 1997. *Supervising Scientist Report* **128**, Supervising Scientist, Canberra.

Rees, C. (1993) Tasmanian seagrass communities. Unpublished MSc thesis, University of Tasmania.

Rochford, D. (1993) Nutrients. In *Australina Fisheries Resources*, eds. P.J. Kailola, M.J. Williams, P.C. Stewart, R.E. Reichelt, A. McNee and C. Grieve, pp. 29–31. Bureau of Resource Sciences, Department of Primary Industries and Energy, and the Fisheries Research and Development Corporation, Canberra.

Sanderson, J.C. (1987) A survey of the *Macrocystis pyrifera* (L.) C. Agardh stocks on the east coast of Tasmania. *Dept Sea Fisheries Technical Reports* **21**, 1–11.

SDAC (1996) *State of the Environment Tasmania Volume I: Conditions and Trends*. Sustainable Development Advisory Council, Department of Land Management, Tasmania.

Stewart, P. and Pullen, G. (1997) Jack Mackerel. In *Fishery Status Reports 1997: Resource Assessments of Australian Commonwealth Fisheries*, eds. A. Caton, K. McLoughlin and D. Staples, pp. 75–79. Bureau of Resource Sciences, Canberra.

Tilzey, R. and Chesson, J. (1997) South East Fishery quota species. In *Fishery Status Reports 1997: Resource Assessments of Australian Commonwealth Fisheries*, eds. A. Caton, K. McLoughlin and D.E. Staples, pp. 41–58. Bureau of Resource Sciences, Canberra.

Tomczak, M. (1981) Bass Strait water intrusions in the Tasman Sea and mean temperature-salinity curves. *Australian Journal of Marine and Freshwater Research* **32**, 699–708.

Vaux, D. and Olsen, A.M. (1961) Use of drift bottles in fisheries research. *Australian Fisheries Newsletter* **20**, 17–20.

Woodward, I.O., Gallagher, J.B., Rushton, M.J., Machin, P.J. and Mihalenko, S. (1992) Salmon farming and the environment of the Huon Estuary, Tasmania. *Division of Sea Fisheries Tasmania, Technical Report* **45**, 1–58.

Ye, L., Ritz, D.A., Fenton, G.E. and Lewis, M.E. (1991) Tracing the influence of sediments of organic waste from a salmonid farm using stable isotope analysis. *Journal of Experimental Marine Biology and Ecology* **145**, 161–174.

Young, J.W., Jordan, A.R., Bobbi, C., Johannes, R.E., Haskard, K. and Pullen, G. (1993) Seasonal and interannual variability in krill (*Nyctiphanes australis*) stocks and their relationship to the fishery for jack mackerel (*Trachurus declivis*) off eastern Tasmania, Australia. *Marine Biology* **116**, 9–18.

THE AUTHORS

Christine M. Crawford

*Tasmanian Aquaculture and Fisheries Institute,
University of Tasmania, Nubeena Crescent,
Taroona, Tasmania 7053, Australia*

Graham J. Edgar

*Zoology Department,
Tasmanian Aquaculture and Fisheries Institute,
University of Tasmania, GPO Box 252-05,
Hobart, Tasmania 7011, Australia*

George Cresswell

*CSIRO Marine Research,
Castray Esplanade,
Hobart, Tasmania 7000, Australia*

Chapter 96

VICTORIA PROVINCE, AUSTRALIA

Tim D. O'Hara

This chapter describes the state of the marine environment in Victoria and adjacent waters. The open coast is dominated by rocky reef, sandy beaches and a range of offshore soft sediment habitats. There is a diverse benthic biota on the continental shelf and slope, with extensive sponge beds occurring in Bass Strait. The marine embayments and estuaries support extensive seagrass beds, mudflats, mangroves and saltmarsh.

The bays and many estuaries have been profoundly changed by human activities, including the opening of new entrances to the sea, draining of wetlands, clearing of vegetation in the catchment, by urbanisation, and the increased use of agricultural fertilisers. Increased nutrient levels and turbidity have led to seagrass dieback and algal blooms in some inlets. Seagrass loss was extensive during 1975–1984, when Western Port Bay lost 70% of its cover and Corner Inlet 25%. The Gippsland Lakes have been adversely affected by reduced freshwater input, increased levels of nutrients, algal blooms and damage to aquatic vegetation from the introduced European Carp.

In Port Phillip Bay nutrient levels have declined since the 1960s following additional sewerage treatment and diversions. Toxicant levels are also declining but remain a localised problem around some outlets. Some fisheries are over-exploited, most notably for scallops, Southern Rock Lobster and Snapper. Trends in other commercial fisheries are masked by the natural variability of fish abundance and changes over time in fishing effort and technologies. Several fish species, favoured by recreation fishers, have declined in numbers in Port Phillip Bay since 1972, probably due to over fishing. Offshore there are concerns about damage to seafloor habitats from trawling and dredging. Over 175 introduced and cryptogenic species have been reported from Port Phillip Bay, more than any other port in the Southern Hemisphere. Several of these species now occur in large numbers and are considered significant pests or weeds.

Seas at The Millennium: An Environmental Evaluation (Edited by C. Sheppard)
© 2000 Elsevier Science Ltd. All rights reserved

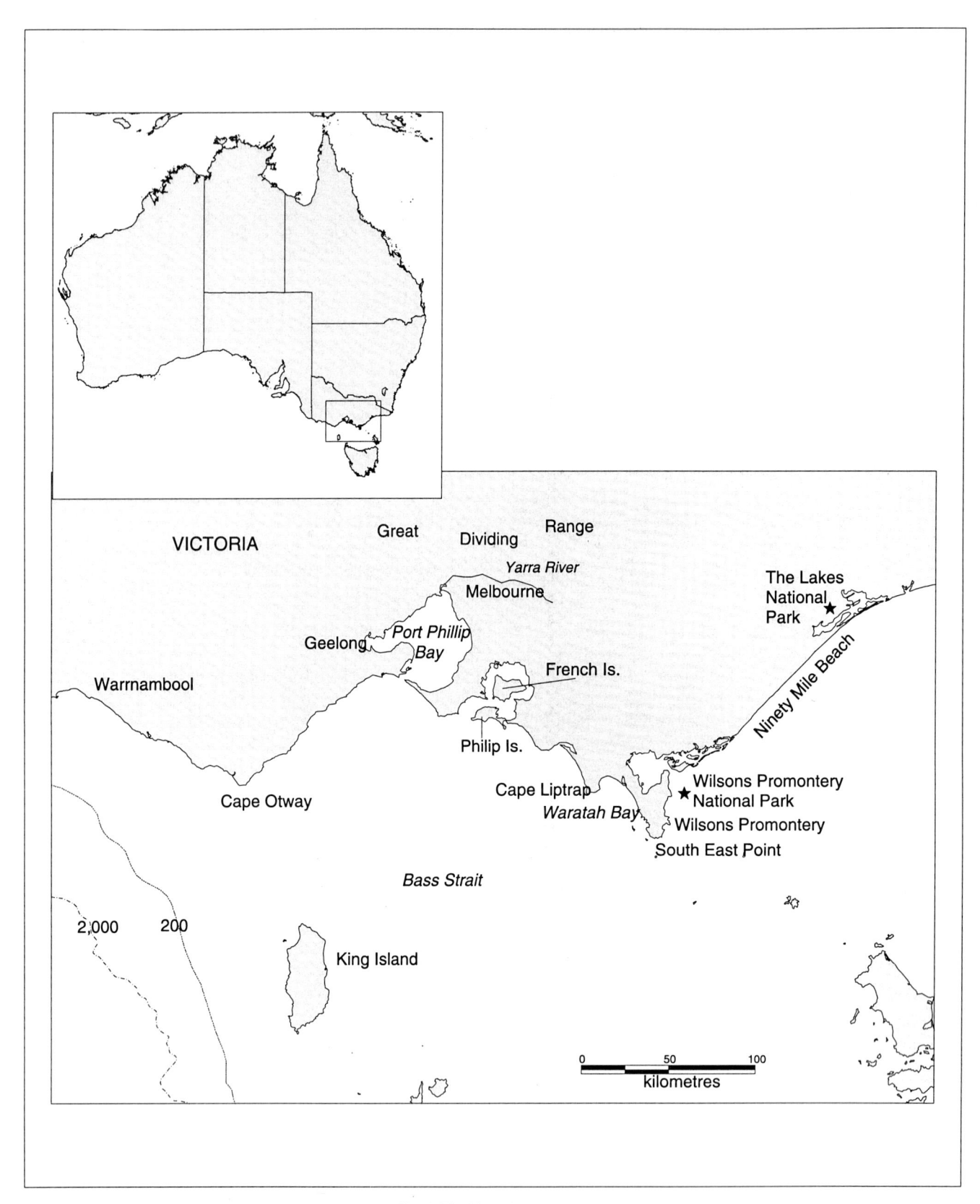

Fig. 1. The Victorian region.

THE DEFINED REGION

The region includes the coastline of Victoria, northern Bass Strait and adjacent offshore areas, ranging in longitude from 141°E to 150°E and latitude from 37.5°S to 39°S (Fig. 1). The area is dominated by the shallow Bass Strait shelf (maximum depth 85 m). On the east and west sides of the Strait the seafloor rapidly descends along the continental slope to the abyssal plain of the Southern Ocean which lies approximately 4 km deep. The coastline of Victoria, which is over 2000 km long, consists mainly of a series of rocky promontories or headlands interspersed by open sandy bays. A long stretch of sandy beach (136.8 km) is present along the mid Gippsland coast. There are three major embayments (Port Phillip Bay, Western Port Bay and Corner Inlet/ Nooramunga), an extensive estuarine lagoon system (Gippsland Lakes) and numerous smaller estuaries scattered along the coast. No large rivers enter the region (Land Conservation Council, 1993; Bird, 1993).

A recent biophysical classification, generated predominantly from oceanographic data, recognised a series of meso-scale bioregions throughout the Victorian marine environment (Thackway and Cresswell, 1998). The Otway region in the west is a high energy coastline fully exposed to southwesterly swell waves generated in the Southern Ocean. The shorelines of the Central Victorian region are dominated by sedimentary cliffs, orientated in various ways to the prevailing swell. It shares similar cool-temperate conditions as northern Tasmania. The Flinders region includes Wilsons Promontory: a large granite peninsula that forms the southernmost tip of the Australian mainland. This is a transition zone between the cooler waters of central Bass Strait and the warmer waters of the east. The Twofold region in the far east of Victoria extends into southern New South Wales, recognising the shared warm-temperate conditions. The muddy seafloor of central Bass Strait is also recognised as a distinct region.

From a biological perspective, this area is a part of the temperate southern region of Australia, and a high proportion of species are endemic (Wilson and Allen, 1987). Marine species occurring in Victoria have four principal distribution patterns: (1) species ranging across Southern Australia, often from southwest Western Australia to southern Queensland; (2) South Australian species having an eastern distribution limit within Victoria; (3) New South Wales species having a western distribution limit within Victoria; and (4) species endemic to the region or having a shared distribution only with Tasmania. For example approximately 50% of decapods and echinoderms that occur in Victoria have a distribution limit somewhere within the region. These species distribution limits are spread out along the central and eastern coastlines; there are no precise biogeographical end-points (Handreck and O'Hara, 1994; O'Hara, unpublished data).

There have been few studies of the deep-sea fauna of offshore regions. Poore et al. (1994) and Kornicker and Poore (1996) found that community composition of isopods and ostracods taken in four transects along the southeast Australian continental slope (200–2250 m) varied with both depth and latitude. There is little data on the biota of deep-sea reefs in the region, but rocky seamounts off Tasmania are known to support a fauna that is distinct from the neighbouring slope (Koslow and Gowlett-Holmes, 1998). There have been no comparative studies on the deep-water benthic fauna off east and western Bass Strait, but there do appear to be species restricted to either side (O'Hara, unpublished data).

Historically, Bass Strait and the Victorian coastline have only experienced marine conditions during interglacial periods. Bass Strait was last fully exposed 16,000 years ago, and re-flooded approximately 8000 years ago (Blom, 1988). There have been approximately 60 glacial cycles over the last three million years (Tiedemann et al., 1994) and the shallow-water marine communities have had to evolve in an environment of constant sea-level change. An exposed Bass Strait could have acted as a barrier to species dispersal, creating the opportunity for allopatric speciation (Dartnall, 1974).

SEASONALITY, CURRENTS, NATURAL ENVIRONMENTAL VARIABLES

Topography, Geology and Sediments

The topography of Bass Strait consists of a flat central plain, approximately 500×350 km in size, flanked on either side by low ridges that occasionally surface to form islands. The narrow continental shelf off western Victoria is rocky inshore, and covered in sand and shell offshore. A large shelf covered in sand lies off Gippsland in Eastern Victoria, punctuated on its eastern boundary by a series of submarine canyons. Victorian bays and estuaries are shallow, rarely exceeding 20 m in depth.

The coast of Victoria is predominantly composed of Mesozoic and Tertiary sedimentary rocks and Quaternary dune sandstone. There are outcrops of granite, principally at Wilsons Promontory and off eastern Gippsland, and basalt, in far western Victoria and around Western Port Bay. Seafloor sediments vary in size from the fine mud in central Bass Strait to the coarse sand near the shore and at the entrances to the Strait.

Sediments in western Victoria are predominantly calcareous, derived from bryozoans and molluscs, those to the east are siliceous, while those in central Bass Strait and in coastal embayments are largely derived from terrestrial sources (Land Conservation Council, 1993).

Tides, Currents, and Waves

Wave conditions vary considerably across the Victorian coast. The western coast to Cape Otway is maximally exposed to the southwest swell. Central, and to a greater extent eastern, Victoria are protected from the southwest by Tasmania and King Island. These areas are however still influenced by storm waves generated in Bass Strait and off southeastern Australia. Near shore the mean average wave power can vary from 46 kW/m at Portland to 4 kW/m on the eastern side of Wilsons Promontory (Water Power Consultants et al., 1992).

Tidal ranges along the Victorian coast are small, generally between 0.9 and 2.4 m in height. Some of the larger bays, with restricted entrances, have slightly higher tidal ranges. Peak tidal ranges in Western Port Bay can reach 3.1 m. The tidal regime along the open coast is complex due to it being on the north side of the shallow Bass Strait and the varying speed of tidal components moving around Tasmania. Tidal currents are an important cause of water movement in Bass Strait, and the relative strength of tidal currents is reflected in the average size of sediment grains on the seafloor. Tidal currents are strongest at the entrance to major bays, around the promontories and Bass Strait islands that form the entrances to the Strait.

Victoria and northern Bass Strait are subject to several oceanographic currents. During summer the Eastern Australian Current brings warm, saline waters from tropical areas to eastern Victoria. A cool saline body of water also moves in from the west into northern Bass Strait. Net flow through Bass Strait in summer is probably towards the west. During winter, climatic conditions force cool subantarctic water into Bass Strait and net flow is eastward. This water "cascades" down the continental slope off eastern Gippsland forming a weak undercurrent that flows eastward into the Tasman Sea. Central Bass Strait waters are stratified during summer whereas in winter they are wind-mixed.

Sea Surface Temperature, Salinity and Nutrients

Sea surface temperature (SST) varies from east to west Victoria by 2–3°C. There is a gradient running north-south throughout Bass Strait of approximately 1°C. In summer, sea surface temperatures reach 17–18°C in the west and 19–20°C in the east. In winter average temperatures reach 13–15°C. Water temperatures in coastal bays can be more extreme as shallow water gains or loses heat to the atmosphere at a greater rate. Temperatures in Port Phillip Bay can range from 10–21°C.

Salinity varies across Bass Strait from 35.1 to 35.7‰. Nutrient levels in Bass Strait are generally low throughout the year (<1 μM of nitrates). Low salinity nutrient-rich waters (7 μM) upwell in winter off the eastern shelf break associated with the Bass Strait Cascade, and to a lesser extent off the western shelf break southwest of King Island (Gibbs et al., 1986). These upwelling are of limited duration and increased nutrient levels do not persist into summer. Phytoplankton levels reach a maximum during winter just inshore of the upwellings. There is little transfer of the nutrients from these upwellings into the centre of Bass Strait (Gibbs et al., 1991). Zooplankton abundance is not correlated with areas of enriched nutrients or phytoplankton (Kimmerer and McKinnon, 1984; Gibbs et al., 1991).

THE MAJOR SHALLOW WATER MARINE AND COASTAL HABITATS

Rocky Shores

Rocky shores along the Victorian coast range from the basalt headlands of the west, through the sedimentary cliffs and wave cut platforms of the central region, to the granite boulders and slopes of Wilsons Promontory and the far east. They support a diverse assemblage of animals and plants which varies slightly according to the type of rock, the exposure to waves and the water temperature. Rocks such as basalt weather to small rocks and crevices with a range of microhabitats. Smooth granite rocks offer fewer protected surfaces. Rock platforms in the cool high energy zone off western Victoria are fringed by Bull Kelp, *Durvillaea*. The high energy of these regions scours many other organisms from the rocks. An unusual, highly diverse assemblage occurs at San Remo in Western Port Bay which supports over 600 species of larger invertebrates, including over 130 opisthobranch molluscs which is one quarter of the known southern Australian opisthobranch fauna (O'Hara, 1995).

Subtidal Rocky Reefs

There are a variety of subtidal rocky reef communities in Victoria. Kelp-dominated bommies and ridges occur in the nearshore environment (Fig. 2). *Phyllospora comosa* is the most common of the large brown algae but is usually replaced by *Ecklonia radiata* in deeper water. *Macrocystis angustifolia* occurs in isolated patches. Offshore there are sand-scoured cobble reefs and low limestone cliffs that generally support a cover of red algae or sessile invertebrates. In deeper water, especially where there is a strong current, exposed rock surfaces support a dense invertebrate assemblage dominated by colourful sponges (Fig. 3), bryozoans, ascidians, colonial anemones and gorgonians. Good examples occur at the "walls" near the Entrance to Port Phillip Bay and the "arches" off Port Campbell.

Seagrasses

Six of the sixteen seagrasses that occur in temperate Australia (Fig. 4) are found in Victoria. *Amphibolis antarctica* is the typical seagrass of the open coast. It occurs in dense beds around sheltered subtidal reefs from the west to as far as Wilsons Promontory. The eelgrasses *Heterozostera tasmanica* and *Zostera muelleri* dominate seagrass beds in the bays and inlets, with *Zostera* usually occurring in more intertidal and

Fig. 2. Forest of Kelp on high energy rocky surfaces.

Fig. 4. Seagrass beds of Port Phillip Bay.

Fig. 3. 'Sponge Gardens' of southern Australia.

estuarine habitats. There are also extensive beds of *Posidonia australis* in Corner Inlet/Nooramunga. *Halophila australis* is a small species that does not form large monotypic beds in Victoria and the subtropical species *Zostera capricorni* occurs in Mallacoota Inlet. The four main seagrasses support distinct assemblages of associated species. The perennial stems of *Amphibolis* support a diverse range of epiphytic algae, bryozoans and invertebrates. Several infaunal echinoderm species appear to occur only in *Posidonia* beds (O'Hara, unpublished data).

Mangroves and Saltmarsh

Saltmarshes occur on sheltered coastlines throughout Victoria. Extensive areas occur in the Barwon estuary, Western Port Bay and Corner Inlet. The White Mangrove (*Avicennia marina*) reaches its southerly distribution in Victoria. It occurs as isolated stands in intertidal areas on the western side of Port Phillip Bay, in Western Port Bay, Corner Inlet and various smaller estuaries. The co-existence of saltmarsh and mangroves in Victorian bays is significant, as on a global scale these two habitats are normally separated into temperate and tropical regions, respectively (Land Conservation Council, 1993; May and Stephens, 1996).

Beaches and Soft Substrates

The biological assemblages in soft-sediment habitats vary according to sediment size and water movement. Coarse sands dominate the nearshore oceanic beaches. While these beaches occur along the length of the Victorian coast they are particularly extensive in the east—including the Ninety Mile Beach from Woodside to Lakes Entrance. Offshore, coarse sands are replaced by fine sand and, in areas of low water movement (e.g. central Bass Strait), by mud. The diverse communities occurring on the sand off east Gippsland have been recently studied and over 800 invertebrate species identified from a single survey from Lake Tyres to Cape Conran (Coleman et al., 1997). Bass Strait is known to support extensive sponge beds, significant on an international scale (Ray and McCormick-Ray, 1992).

Sediment size has been identified as the most significant ecological factor stratifying benthic communities. In Port Phillip Bay, the eastern and western flanks are covered with sand, with thick mud occurring in the centre. Extensive intertidal mud and sand flats occur in Western Port Bay and Corner Inlet/Nooramunga and support large populations of migratory wading birds such as the Red-necked Stint or the Eastern Curlew. These wetlands are listed under the Ramsar Convention. Tidal water movement cuts deep channels into the mud-banks. These deeper channels support extensive populations of invertebrates normally restricted to the continental shelf. In Western Port Bay these include the bivalve *Neotrigonia margaritacea* and the brachiopod *Magellania australis*.

OFFSHORE SYSTEMS

Pelagic System

Pelagic biomass and diversity is low compared with other regions, reflecting the generally nutrient-poor waters. The zooplankton is dominated by a few abundant species. Pelagic fish species include Silver Trevally, Barracoota, Jack Mackeral, pilchards and anchovies. Other pelagic animals include Arrow Squid, the Paper Nautilus, various jellyfish, comb-jellies and salps. Large mammals include the Southern Right Whale which calves off the coast of Warrnambool, the Common and Bottle-nosed Dolphins, and the Australian Fur Seal which breeds on a few offshore islands along the coast. The Little Penguin also breeds and feeds off pilchards and anchovies in these waters.

A number of fish regularly migrate through the area. King George Whiting breeds in South Australia; the larval fish then make their way along the western Victorian coast before settling in the various bays and inlets. School Sharks also migrate from South Australia to Bass Strait as part of their life cycle. Eels migrate from Victorian streams to breed in the Coral Sea off Queensland.

Slope Communities

The small amount of research conducted on the southeastern Australian continental slope suggests that it supports a diverse biological community (Koslow et al., 1994; Poore et al., 1994). The sediment is biogenic and largely derived from the nearby shelf. The bottom fauna is dominated by mud-dwelling species, except in isolated rocky areas where a sessile invertebrate reef fauna is present. Above the sea floor there is a range of deep-water fish, prawns and cephalopods, several of which are of commercial importance.

HUMAN POPULATIONS AFFECTING THE AREA

Indigenous peoples have occupied the Victorian coast for millennia. At the time of white settlement. nine language groups occupied the coastal zone. These coastal peoples appear to have utilised food resources in the wetlands and open woodlands behind the coast more than the rather unproductive shoreline. Resources that were gained from the coast included shellfish from rock platforms and sandy beaches, seals and mutton birds if colonies were locally available, and the occasional beached whale. Shell middens abound along the coast.

The first white settlements in Victoria were short-lived. Sealing and whaling gangs set up temporary settlements along the coast and the first official settlement at Sorrento lasted only seven months in 1802. These first settlers reported catching up to 500 crayfish in a single evening from the shoreline of Point Nepean. Pastoralists from Tasmania increasingly showed interest in Victoria, and a settlement at Portland was established in 1828. Melbourne was unofficially founded in 1835 and, within a year, the arrival of 12,000 colonists forced official recognition. The colony rapidly expanded. By 1851 there were 77,345 Europeans, 391,000 cattle and 6,590,000 sheep in Victoria. A Gold Rush prompted large numbers of new arrivals and by 1861 the population had grown to 540,000. The influx of settlers had a devastating effect on the indigenous peoples, who numbered only 1067 by 1877 (Barwick, 1984; Land Conservation Council, 1993).

In 1997 there were over 4,605,100 people in the State of Victoria, including 22,598 indigenous peoples. The vast majority of Victorians live within 25 km of the coastline. Victoria has the highest population density per kilometre of coast of any state in Australia (Winstanley, 1994), and over 3,321,700 people live in Melbourne and surrounding suburbs. Other coastal cities include Portland (population 19,000), Warrnambool (28,000) and Geelong (153,100). The concentration of population at sites in the coastal zone imposes severe pressures on the neighbouring marine environment. There are also many smaller coastal townships that survive on coastal or marine activities including tourism, fishing and the offshore oil and gas industry.

RURAL FACTORS

The environment of many of Victoria's bays, inlets and estuaries has been profoundly changed over the last 120 years. The native vegetation has been removed from catchments, increasing erosion and the amount of sediment moving down the rivers into coastal systems. *Melaleuca* woodlands and saltmarshes have been drained and cleared, changing freshwater input into bays from low levels spread across the breadth of the swamp to a few major point sources concentrated around the main drains. This has removed the natural filtering function of coastal swamps and further facilitated the transport of sediment into bays, particularly after floods. Some of Victoria's estuaries, once blocked by a sand bar across the entrance, are now permanently kept opened to allow the passage of boats and to protect property from rising waters. This increases salinity and shifts the environment from freshwater to marine, killing fringing riparian vegetation. Other estuaries have suffered from the reduction of fresh water as

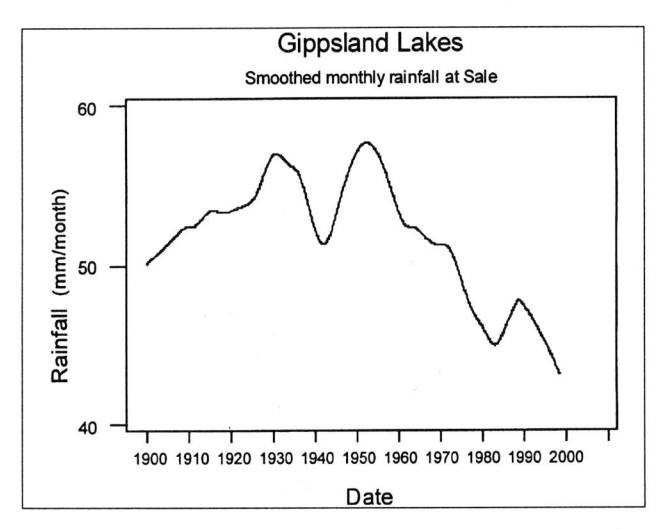

Fig. 5. Smoothed plot of rainfall at Sales in the Gippsland Lakes showing the overall decline since 1950 (reproduced from Harris et al., 1998).

rivers have been diverted for urban, rural or industrial use. Nutrient levels in the rivers and bays have become a problem, with the increased use of modern chemical fertilisers in agriculture and from additional sewage from a rapidly urbanising population.

The Gippsland Lakes are still adjusting to the opening of the permanent entrance in 1889. The resulting shift in the aquatic environment from freshwater to marine has been further exacerbated by a water diversion to the Thompson Dam for urban use, the opening of an industrial drain at Delray beach for industry, a long-term decrease in rainfall since the 1950s (Fig. 5), and saline ground-water drawn up by irrigation. Fringing riparian woodland, marshes and reed beds have declined. The water is now more turbid and nutrient levels have increased. Although there is marked inter-annual variability, there has been a long-term shift in aquatic vegetation from seagrass to planktonic algae, and algal blooms are now common. The bloom in 1999 is the most severe to date. The introduced fish Carp (*Cyprinus carpio*), which uproots aquatic vegetation and adversely affects water quality by increasing turbidity, is slowly expanding its range across the Lakes (Harris et al., 1998).

The hydrological regime of Western Port Bay has been fundamentally changed by the draining of the large Koo-wee-rup swamp at the north of the bay from 1885 until the 1930s. Before this, fresh-water flows to the bay were limited to a few small streams around the top of the bay and the Bass River which enters from the southeast. The waters of the Bay were largely clear and supported extensive beds of seagrass. Fresh water now flows freely into the northeast section of the Bay from several large drains, stressing nearby marine communities and raising nutrient levels and sedimentation throughout the Bay. Sediments have been generated from stream erosion, land clearing and urban development in the catchment. Over 100,000 m^3 of sediment per year currently enter the bay from the northern drains. There was a large die-off of seagrass from 1974–1984, possibly precipitated by massive erosion in the catchment associated with the building of roads and the Cardinia Dam. Over 70% of seagrass beds disappeared over this period, particularly on the intertidal mudflats at the top end of the Bay. The exact cause of the dieback is uncertain but is probably linked to increased turbidity, smothering of seagrass by blooms of epiphytic algae, or the raising of mud-banks through increased sedimentation. Turbidity levels have improved since 1984 and seagrass has recovered in small patches although the density is still low (May and Stephens, 1996).

Extensive clearing of native vegetation has occurred in the Corner Inlet–Nooramunga catchments and drains cut into the sea. There was a major *Posidonia* seagrass dieback during 1976–84 (Morgan, 1986). Port Phillip Bay is discussed below under urban use. Many of Victoria's smaller estuaries are degraded by reduced flow, increased levels of nutrient and sediment (Winstanley, 1994).

COASTAL EROSION AND LANDFILL

The development of artificial structures that limit the seasonal longshore movement of sand has caused beach erosion in several areas in Victoria, including near Portland and at several popular beaches adjacent to marinas and jetties in Port Phillip and Western Port Bays. Several exotic plants have been introduced into coastal areas to prevent erosion or to stabilise river banks. Marram Grass (*Ammophila arenaria*) was deliberately introduced in 1875 by the Government Botanist to stabilise denuded sand dunes in the west of Victoria. Ricegrass (*Spartina anglica*), planted along rivers to stabilise banks, is now regarded as a major weed which clogs rivers and estuaries. It is now a major problem around the Bass River in Western Port Bay, Andersons Inlet, and Corner Inlet/Nooramunga. There have been various attempts to locally eradicate it using herbicides or smothering it with black plastic; recent herbicide trials appear encouraging.

EFFECTS FROM URBAN AND INDUSTRIAL ACTIVITIES

Commercial Fisheries

The hunting of whales and seals began shortly after Bass and Flinders rowed into Bass Strait in 1797/8. These animals were quickly exploited to commercial extinction. A small fishing industry began in the bays shortly after European settlement and expanded offshore as transport and storage technologies improved. Fisheries now exist in almost all Victorian marine and estuarine waters.

A specialised bay and inlet fishery exists for various finfish, including Snapper, King George Whiting and various flathead. Fishing gear includes beach seines, gill nets and long lines. Catches have increased steadily since 1914,

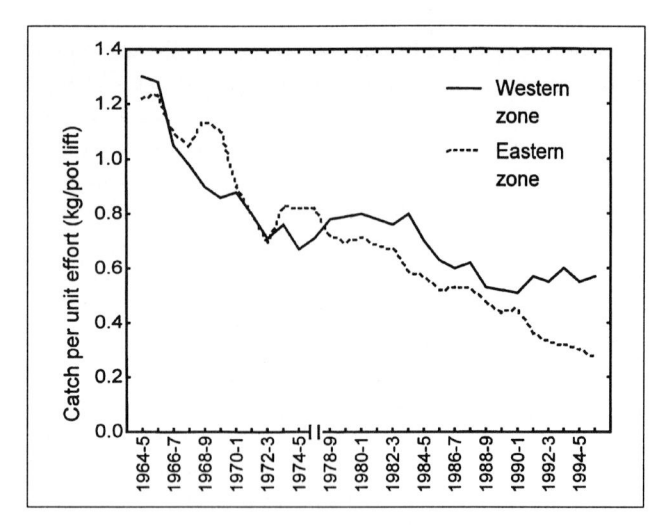

Fig, 6. Decline in catch per unit effort (kg/pot) for Southern Rock Lobster from the western (west of Cape Otway) and eastern Victorian management zones, data is unavailable for 1976–77 (reproduced from Victorian Fisheries, 1996).

peaking in the 1970s. About that time there was a reduction in the catch of King George whiting in Western Port Bay associated with seagrass die-off (May and Stephens, 1996). Since that time there has been concern over black bream stocks in the Gippsland Lakes resulting from decreasing water quality (Harris et al., 1998). A recent review has found some evidence of long-term decline of snapper stocks due to over-harvesting but little evidence of damage to habitat or by-catch from current netting practices (Hall, 1997). Fishing has gradually shifted towards species further down the food-chain. New fisheries for pilchards and calamari have rapidly developed over the past 20 years and now account for more of the total catch in Port Phillip Bay than the more traditional species. Jellyfish are now being exploited.

An offshore trawl fishery exists in eastern Bass Strait. It forms part of the South-East Trawl fishery managed by the Commonwealth Government. Gear includes Danish seines and Otter Board bottom trawls. A variety of different fish and crustacean species are caught from near-shore, mid-continental slope and offshore locations, including Tiger Flathead, School Whiting, John Dory, Silver Trevally and Royal Red Prawns. Problems include the catching of vulnerable slow-growing species using non-selective fishing techniques (e.g. Oreos, deep-sea sharks) and the impact of trawling on seafloor habitats. There has been minimal research into these issues.

The modern dive fishery for Abalone commenced in Victoria in the early 1960s. By 1968 entry into the fishery was limited and by 1988 a quota system was introduced. Today poaching is a major problem, possibly accounting for as large a catch as the legal fishery. Nevertheless the legal fishery is Victoria's most valuable, earning over $50 M in 1998. The Black-lip Abalone has escaped the decline suffered by abalone species elsewhere in the world through a combination of effective management and the naturally cryptic habit of the juveniles. There is some concern about the status of Green-lip Abalone stocks which may have to be managed separately in the near future.

The Southern Rock Lobster is caught along Victoria's open coast in beehive-type pots. The fishery is currently over-exploited, catch per unit effort has been in steady decline since the early 1960s (Fig. 6), and the fishery will require urgent restructuring in the near future (Victorian Fisheries, 1996). King Crab, Red-velvet Crab, octopus and wrasse were historically caught as by-catch in crayfish pots. King Crab and wrasse have become established fisheries in their own right, wrasse being caught for the live restaurant trade. Both species are now fully exploited and require management action.

A dredge fishery for native Mud Oysters developed in the major bays last century but stocks were extinguished by the 1920s. Scallop dredging began in Port Phillip Bay in 1964 and spread to eastern Bass Strait in the early 1970s. The Port Phillip Bay fishery was already over-exploited by 1967 and the Government moved to restrict entry to the fishery. Extreme variability in recruitment led to the closure of the fishery on several occasions. Concern about the impact of dredging on bottom habitats and water turbidity led to the permanent closure of the dredge fishery in Port Phillip Bay in 1997. The offshore fishery is currently closed (May 1999) due to low numbers of scallops. There has been no study on the impact of dredging on offshore communities. Of particular concern is the impact of dredging on the extensive sponge beds reported from Bass Strait.

Recreational Fisheries

Telephone surveys indicate that over 500,000 Victorians go fishing at least once a year. Overall numbers are declining gradually, however better fishing equipment and the increased availability of locality and species data is possibly still increasing catch efficiency. The recreational catch in Port Phillip Bay from 1989–1994 was likely to have been greater than the commercial catch of similar species, with boat anglers taking far more of the catch (88%) than shore anglers (12%) (Coutin et al., 1995). At some point additional catch limits may have to be imposed to protect some fish stocks. Gear normally used includes rods, hand lines and spear guns; nets are sometimes used illegally.

Current issues concerning recreational fishing include possible over-harvesting of some species, litter and damage to habitat through boat activity. Several fish species in Port Phillip Bay significantly declined in abundance between 1972 and 1991. The likely cause of decline for Toothy Flathead (99.9% decline), Sand Flathead (47.2%), Yank Flathead (33.6%) and Common Gurnard Perch (80.7%) is increased fishing pressure in the Bay (Harris et al., 1996).

Aquaculture

Aquaculture of marine species is currently limited in Victoria to Blue Mussel cultivation and some on-land farms growing Flounder, Abalone and Pacific Oyster. There have been various attempts to cultivate native Mud Oysters but these have been unsuccessful due to disease. Large amounts of debris still exist in Swan Bay from one such venture. The industry and Government is keen to develop industries based on fin-fish and algae, and expand shellfish production. Blue Mussels are grown in Port Phillip Bay and near the western entrance into it. Issues with Blue Mussel production include local modification of the seafloor communities and the accidental transfer of exotic species from Port Phillip Bay to Western Port Bay. The exotic Pacific Oyster was deliberately introduced into Victorian estuaries during the 1950s by CSIRO. At least one feral population still exist in Andersons Inlet. A new proposal to cultivate the Pacific Oyster in Victoria was rejected by the State Government in 1996 on the grounds that it readily forms feral populations on nearby shores.

Urban Use (Cities)

Port Phillip Bay has escaped many of the problems usually associated with proximity to major urban centres (Harris et al., 1996). Fortunately, the foresight of early water board planners resulted in the creation of a large sewage treatment facility at Werribee. Nevertheless nutrient levels continued to increase until the late 1960s when sewerage from the southeastern suburbs was diverted to an ocean outfall. The cycling of nutrients in Port Phillip Bay is now well understood and the aim is to keep nutrient levels below that which causes eutrophication. Turbidity in the Bay has increased since European settlement, particularly in the north of the Bay and there has been a local die-off of seagrass in Swan Bay and the Geelong Arm to the southwest. This die-off of seagrass is the likely cause of the documented decline in several fish species in the bay including King George Whiting, Cobbler, Greenback Flounder, Six-spined Leatherjacket, Mosaic Leatherjacket and Long-finned Pike (Harris et al., 1996). Additional environmental problems now associated with Port Phillip Bay are over-harvesting, particularly of intertidal shellfish, and introduced species (see below).

There are seventeen outfalls on the Victorian open coast. Treatment of sewage is improving but still only half of Victoria's sewage is treated to break down organic wastes (secondary treatment). No outfall attempts to remove nutrient and all toxicants (tertiary treatment). Some outfalls still pump untreated sewage out to sea during peak holiday periods. Other waste is carried out to sea via stormwater drains including litter, dog faeces, and residues from car emissions. These outfalls have a local effect; there is no overall nutrient enrichment of the open coastline.

Industrial Uses of the Coast

The impact of industrial development on the coast includes habitat modification through coastal engineering and degraded water quality around industrial outfalls. Industrial wastes (including heavy metals and organic toxicants) were originally discharged directly into the marine environment, predominantly into Port Phillip and Western Port Bays. Most industrial discharges are now routed through treatment plants, however, toxicants persist in seafloor sediments, particularly in Hobsons and Corio Bays (Land Conservation Council, 1993). The Gippsland Lakes are still affected by high levels of mercury generated by gold mining operations in the catchments in the nineteenth century.

Offshore oil and gas fields exist in northern Bass Strait. The eastern oil fields have been developed since the mid 1960s and the western gas fields are currently being developed. To date the operation of offshore rigs has been apparently without major incident. There has been local pollution of the sea floor from toxic drilling mud, low-level hydrocarbon contamination (Lavering, 1994) and the dumping of equipment. Much of the debris was cleaned up after supreme court action by commercial fishers in 1983 (Winstanley, 1994). There is currently no offshore mining in Victoria.

Shipping and Offshore Accidents and Impacts

Four major ports exist in Victoria, including Melbourne, Geelong (Port Phillip Bay), Hastings (Western Port Bay) and Portland. Environmental issues associated with ports and shipping include clearing of coastal habitat for industrial development, dredging operations required to open channels and keep them clear of sediment, oil spills and other ship-based pollution, toxic hull fouling paints, and the introduction of exotic species through ballast water discharges and hull fouling. Tributyl tin (TBT) anti-fouling paints were banned on vessels under 25 m in 1989. Low levels of TBT residues are still detectable in the sediments of Port Phillip Bay (Foale, 1993).

Victoria, and in particular Port Phillip Bay, has been a major recipient of exotic marine species. Over 175 introduced and cryptogenic species have now been recorded from Port Phillip Bay, the most for any port in the Southern Hemisphere (Hewitt et al., 1999). Several of these are now pest species and dominate the seafloor, including the seastar *Asterias amurensis*, the fan worms *Sabella spallanzanii* and *Euchone* sp., the bivalves *Corbula gibba* and *Theora fragilis*, the crabs *Carcinus maenas* and *Pyromaia tuberculata*, the macroalgae *Undaria pinnatifida* and *Codium fragile tormentosoides*, and various toxic dinoflagellates. The combined effect of these introductions may change the ability of the Bay's benthic community to recycle excess nutrients from sewerage outfalls. The abundance of several fish species (Globefish, Little Rock Whiting) has increased, possibly due to the increase in exotic species (Harris et al., 1996). Toxic dinoflagellate blooms have caused red-tides, fish kills, and temporary suspension of mussel harvesting in Port Phillip Bay (Winstanley, 1994).

PROTECTIVE MEASURES

Ecologically sustainable development principles are now enshrined in many pieces of Victorian legislation. However there is no integrated management regime for both Victoria's marine and coastal environments. Nevertheless, the importance of protecting the environment is recognised in various Government coastal strategies and action plans which address development on both private and public land. In theory, strip development along the coastline is discouraged and development proposals concentrated in existing "nodes". In practice, the current management boards have inherited inappropriate planning decisions made in the past, and emphasis on coastal protection varies with the priorities and experience of each local board. The population of the south-east growth corridor of Melbourne increased by 200% between 1970 and 1991 (RAC, 1993).

At present, Victoria has 12 marine protected areas (MPAs) that cover some 55,000 ha of marine waters. These MPAs are concentrated in the bays and inlets in the central region of the State and only 350 ha are equivalent to IUCN categories 1 and 2, fully protecting all plants and animals, including fish. A government process developing a representative marine protected area system has been in progress since 1991. Despite numerous interim proposals, a final report has still to be prepared for Government approval. There has been no new marine park declaration since 1991.

Catchment Boards are developing catchment management plans to reduce nutrient and sediment input into the bays and estuaries. A state-wide nutrient-reduction strategy was launched in 1995 and some $A 3 Million has been spent up to 1999 on various nutrient-abatement projects. A special catchment levy raises money to mitigate environmental problems. The Victorian Environment Protection Authority sets the water quality objectives for marine and freshwater waters. These are currently being progressively revised in line with Government preference for community cooperation rather than prescriptive regulation.

Statutory Fisheries Management Plans are now planned for major Victorian fisheries. These plans will outline sustainable fishing practices and catches and how to protect the important environmental values that support the fishery. However, to date only one Fishery Management Plan, for the inland eel fishery, has been finalised. Recreational bag limits are generally employed to share the resource between users rather than being based on ecological principles.

Endangered and vulnerable marine species and communities can be protected under the Victorian Fauna and Flora Guarantee legislation or as Protected Aquatic Biota under the Fisheries Act. Both pieces of legislation also set out a framework for the sustainable use of fauna and flora and the abatement of threatening processes. Two rare marine invertebrates, one invertebrate community, various sharks, seahorses and seadragons are listed under these Acts. Noxious species can be listed under the Fisheries Act. The Victorian Biodiversity Strategy, released in 1997, emphasises community awareness, cooperative programs and self-regulation to achieve biodiversity protection. However, there has been no allocation of funds to implement the strategy.

CONCLUSION

The main problems associated with Victoria's marine environment are in the bays and estuaries which in many instances are suffering from poor water quality, incremental coastal development, seagrass dieback, algal blooms and the proliferation of introduced species. The effect of proposed aquaculture developments on coastal ecosystems will increase and needs to be carefully monitored. Along the open coast there are concerns with over-harvesting of some species and the damage to seafloor habitats from trawling and dredging. The current system of marine protected areas is small, *ad hoc*, and not representative of the range of habitats present in the region. Many of these problems are understood by Governments and have been identified in various coastal strategies and action plans. However it is yet to be seen whether the good intentions contained in glossy Government brochures will translate into effective outcomes.

REFERENCES

Barwick, D.E. (1984) Mapping the past: an atlas of Victorian clans 1835–1904. *Aboriginal History* **8**, 100–131.

Bird, C.F. (1993) *The Coast of Victoria: The Shaping of Scenery*. Melbourne University Press, Victoria.

Blom, W.M. (1988) Late Quaternary sediments and sea-levels in Bass Basin, southeastern Australia—a preliminary report. *Search* **19**, 94–96.

Coleman, N., Gason, A.S.H. and Poore, G.C.B. (1997) High species richness in the shallow marine waters of south-east Australia. *Marine Ecology Progress Series* **154**, 17–26.

Coutin, P., Conron, S. and MacDonald, M. (1995) *The Daytime Recreational Fishery in Port Phillip Bay 1989–94*. Victorian Fisheries Research Institute, Queenscliff, Victoria.

Dartnall, A.J. (1974) Littoral biogeography. In *Biogeography and Ecology in Tasmania*, ed. W.D. Williams, pp. 171–194. Junk, The Hague.

Foale, S. (1993) An evaluation of the potential of gastropod imposex as a bioindicator of tributyltin pollution in Port Phillip Bay, Victoria. *Marine Pollution Bulletin* **26**, 546–552.

Gibbs, C.F., Cowdell, R.A. and Longmore, A.R. (1991) Seasonal variations of density patterns in relation to the Bass Strait cascade. *Australian Journal of Marine and Freshwater Research* **37**, 21–25

Gibbs, C.F., Tomczak, M. Jr. and Longmore, A.R. (1986) The nutrient regime of Bass Strait. *Australian Journal of Marine and Freshwater Research* **37**, 451–466

Hall, D. (1997) Review of the net and line fisheries in Victorian Bays and Inlets: Background report. Unpublished report for the Victorian Fisheries Co-management Council.

Handreck, C.P. and O'Hara, T.D. (1994) Occurrence of selected species of intertidal and shallow subtidal invertebrates at Victorian locations. A report to the Victorian Land Conservation Council by the Marine Research Group Inc., Melbourne.

Harris, G., Batley, G., Fox, D., Hall, D., Jernakoff, P., Molloy, R., Murray, A., Newell, B., Parslow, J., Skyring, G. and Walker, S.

(1996) Port Phillip Bay Environmental Study Final Report. CSIRO, Canberra, Australia.

Harris, G., Batley, G., Webster, I., Molloy, R., and Fox, D. (1998) Gippsland Lakes Environmental Audit: Review of Water Quality and Status of the Aquatic Ecosystems of the Gippsland Lakes. A report for the Gippsland Coastal Board prepared by CSIRO Environmental Projects Office, Melbourne.

Hewitt, C.L., Campbell, M.L., Thresher, R.E. and Martin, R.B. (1999) *Marine Biological Invasions of Port Phillip Bay, Victoria.* Centre for Research on Introduced Marine Pests. Technical Report No. 20, CSIRO Marine Research, Hobart, 344 pp.

Kimmerer, W.J. and McKinnon, A.D. (1984) Zooplankton abundance in Bass Strait and western Victorian shelf waters, March 1983. *Proceedings of the Royal Society of Victoria* 96, 1161–167.

Kornicker, L.S. and Poore, G.C.B. (1996) Ostracoda (Myodocopina) of the SE Australian continental slope, Part 3. *Smithsonian Contributions to Zoology* 573, 1–186.

Koslow, J.A., Bulman, C.M. and Lyle, J.M. (1994) The mid-slope demersal fish community off south-eastern Australia. *Deep-Sea Research* 41, 113–141.

Koslow, J.A. and Gowlett-Holmes, K. (1998) The seamount fauna off southern Tasmania: Benthic communities, their conservation and impacts of trawling. Final report to Environment Australia and the Fisheries Research Development Corporation. CSIRO Marine Fisheries, Hobart, 104 pp.

Land Conservation Council (1993) Marine and Coastal Special Investigation Descriptive Report. Government Printer, Melbourne.

Laverling, I.H. (1994) Marine environments of southeast Australia (Gippsland shelf and Bass Strait) and the impact of offshore petroleum exploration and production activity. *Marine Georesources and Geotechnology* 12, 201–226.

May, D. and Stephens, A. (eds.) (1996) The Western Port Marine Environment. Environment Protection Authority, State Government of Victoria, Melbourne.

Morgan, G. (1986) A survey of macrobenthos in the waters of Corner Inlet and Nooramunga, southern Victoria, with an assessment of the extent of Posidonia seagrass. *Fisheries and Wildlife Paper* No. 31.

O'Hara, T.D. (1995) Marine invertebrate conservation at San Remo. *Victorian Naturalist* 112, 50–53.

Poore, G.C.B., Just, J. and Cohen, B.F. (1994) Composition and diversity of Crustacea Isopoda of the southeastern Australian continental slope. *Deep-Sea Research* 41, 677–693.

RAC (1993) *Coastal Zone Inquiry.* Resource Assessment Commission, Australian Government Publishing Service, Canberra.

Ray, G.C. and McCormick-Ray, M.G. (1992) Marine and estuarine protected areas: A strategy for a national representative system within Australian coastal and marine environments. Australian National Parks and Wildlife Service, Canberra.

Thackway, R. and Cresswell, I.D. (eds.) (1998) Interim and Coastal Regionalisation for Australia (IMCRA): An Ecosystem-based classification of coastal and marine environments. Version 3.3. Environment Australia, Commonwealth Department of Environment, Canberra.

Tiedemann, R., Sarnthein, M. and Shackleton, N.J. (1994) Astronomic timescale for the Pliocene Atlantic $\delta 18O$ and dust flux records of Ocean Drilling Program site 659. *Paleoceanography* 9, 619–638.

Victorian Fisheries (1996) *Victorian Rock Lobster Fishery Future Management.* Victorian Department of Natural Resources and Environment, Melbourne.

Water Power Consultants and Lawson and Treloar (1992) Wave Energy Study. Report prepared for State Electricity Commission of Victoria and Renewable Energy Authority Victoria.

Wilson, B.R. and Allen, G.R. (1987) Major components and distribution of marine fauna. In *Fauna of Australia. Vol. 1A: General Articles,* eds. G.R. Dyne and D.W. Walton, pp. 43–68. Australian Government Publishing Service, Canberra.

Winstanley, R. (1994) Issues in the Victorian marine environment. The State of Marine Environment Report Technical Annex 3.

THE AUTHOR

Tim D. O'Hara
*Zoology Department,
University of Melbourne,
Parkville, Vic., Australia*

Chapter 97

THE GREAT AUSTRALIAN BIGHT

Karen Edyvane

The Great Australian Bight is an area of international conservation significance, containing globally significant breeding populations of rare and endangered marine mammals, and also, some of the highest levels of endemism and marine biodiversity in Australia (and the world). Much of this unique biota has resulted from the relatively long period of geological isolation and extensive, arid, east–west extent of the coast and the associated wide open, swell-dominated continental shelf, which is dominated by Holocene bioclastic carbonate sediments, particularly bryozoans. However, the fauna and flora of the inshore and offshore regions of the Bight, particularly the seabirds, fish and invertebrates, remain poorly known. Until recently, existing marine biodiversity research and conservation management efforts in the region have been low, with only 260 hectares of the 18.6 million hectares of the Bight, being formally protected and managed as Marine Protected Areas. Despite the risk of increasing conflicts with marine biodiversity in the region from existing uses, such as commercial fisheries, and also, increasing activity in the region, from marine mammal-based ecotourism and sea-based aquaculture (in WA and SA), regional, multiple-use management arrangements or management plans are generally lacking. The recent establishment of the Great Australian Bight Marine Park (State and Commonwealth waters) represents the first multiple-use management regime and the first formal reservation of the ecosystems and habitats of the Great Australian Bight.

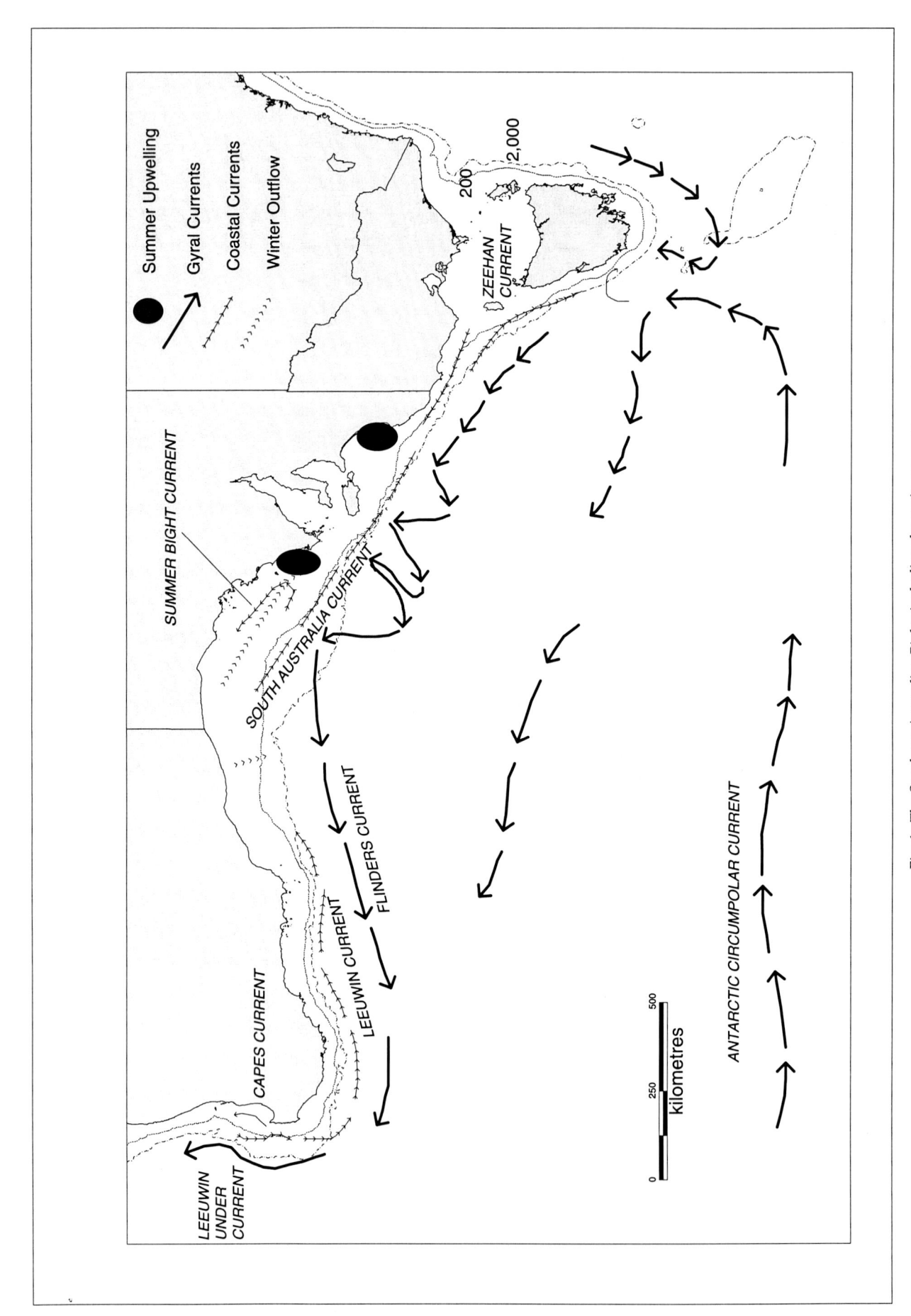

Fig. 1. The Southern Australian Bight, including the main currents.

THE DEFINED REGION

The Great Australian Bight forms part of the southern shelf of Australia, which is the northern boundary of the South Australian Basin of the Southeast Indian Ocean (Fig. 1). The inshore regions of the Bight extend over 1200 km, from Cape Pasley (near Esperance), in Western Australia to Cape Catastrophe, at the entrance of Spencer Gulf in South Australia (ACIUCN 1986). Along this highly variable coastline there are spectacular cliffs and rocky headlands, numerous offshore islands, surf-pounded beaches and large sheltered embayments (Short et al., 1986).

The Great Australian Bight region falls within the Southern Pelagic Province of Australia (IMCRA, 1998), which extends from near Albany (WA) in the west, along the southern coast, to Lakes Entrance (Victoria) in the east and enclosing Bass Strait and the Tasmanian waters. The southern coast of Australia, and Southern Pelagic Province encompass a major biogeographic region, the Flindersian Province, which extends from southwest Western Australia to southern New South Wales and includes the waters of Victoria and Tasmania. Within this broad region, the coastal waters of the Great Australian Bight are recognised as warm to cool temperate (Womersley, 1990) (Fig. 2).

Under a recent macroscale classification of marine ecosystems (based on demersal fish species diversity and richness), the Great Australian Bight is defined as a single demersal biotone known as the Great Australian Bight Biotone (GABB) (IMCRA, 1998). As such, the Great Australian Bight region represents a zone of transition between two core provinces, the South West Province (SWP) in southwestern Western Australia (extending from Perth to Israelite Bay), and the Gulfs Province in South Australia (comprising Spencer Gulf and Gulf St Vincent, and Kangaroo Island, east to Cape Jervis, near the mouth of the Murray River). The South Western Province represents the western limit of the Flindersian Province, and comprises warm temperate elements in the biota, while the Gulfs Province contains unique subtropical elements (particularly in the gulfs), and acts as a major biotone for warm temperate species and also, cool temperate species from Victoria and Tasmania (i.e. Tasmanian Province and Bassian Province) (IMCRA, 1998). At a finer, mesoscale scale, the nearshore regions of the Great Australian Bight encompass three major mesoscale, biophysical regions or bioregions (i.e. Eucla, Murat and Eyre Bioregions) (IMCRA, 1998; Edyvane, 1998).

SEASONALITY, CURRENTS, NATURAL ENVIRONMENTAL VARIABLES

The coast from Cape Leeuwin to Tasmania is the longest stretch of east–west, ice-free coastline in the Southern Hemisphere and is characterised by high deep-water wave energies, with no true rivers or streams arriving at the coast. As such, there are no true estuarine environments along the coast and the extensive shallow swell-dominated shelf, has no significant fluvial or riverine input. Adjacent to the only circumpolar ocean, the swell-dominated coast of the Great Australian Bight experiences some of the world's highest and most persistent waves (Chelton et al., 1981).

Climate

The climate of the Great Australian Bight is typically semi--arid or 'Mediterranean' and as such, is characterised by hot, dry summers and cool, moist winters. Occasional heavy rainfalls occur in mid to late summer from the remnants of tropical cyclones, but these events are exceptions. The climate of the Great Australian Bight is largely influenced by mid-latitude anticyclones or high pressure systems which pass from west to east across the continent. Winter generally brings southerly to southeasterly winds and low pressure systems which travel across the Southern Ocean between 40 and 50°S, bringing frontal activity and rain. Summer brings northerly to northwesterly winds. Along the Great Australian Bight and the western coast of Eyre Peninsula, strong westerly, onshore winds have reworked the coast, resulting in extensive dune development.

Waves

The Great Australian Bight is located within the 'west coast swell environment' where coastal processes are dominated by a persistently high southwest swell, generated by the westerly moving low pressure cyclones south of the mainland. The sea-state rapidly worsens during the passage of fronts and low pressure systems, and then gradually moderates as anticyclonic conditions return. The wave climate is

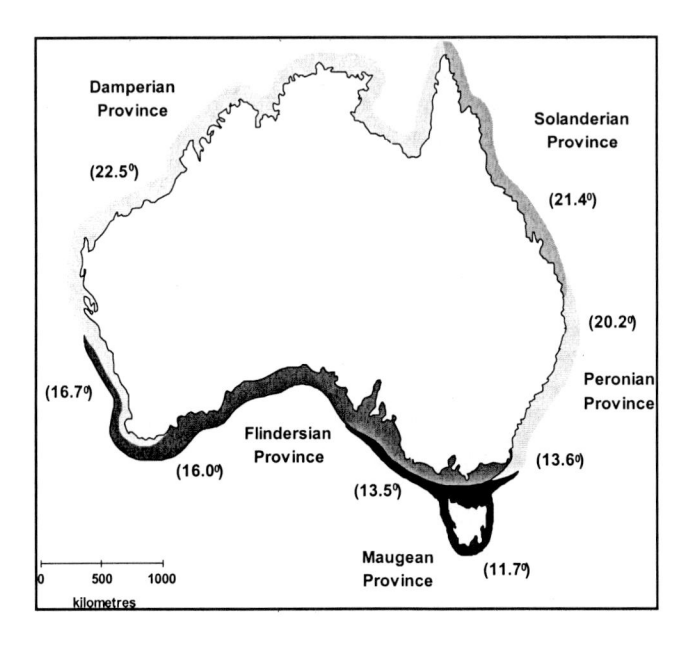

Fig. 2. Marine biogeographical provinces of Australia, and average surface sea temperatures (from Womersley, 1990).

typified by southwest to westerly swells which range from less than 2 m for 50% of the year, to 2–4 m for 30–45% of the year and exceed 4 m approximately 10% of the year. Wind-generated sea conditions also provide an additional source of wave energy, with seas averaging 0.5 to 1.25 m and may exceed 2 m for 10–15% of the year. Swell and wind waves are usually unimodal, since the storm belt lies sufficiently close to the coastline that a bimodal spectrum does not have the opportunity to develop (Bye, in press).

Tides

Tides along the Great Australian Bight coast are microtidal in range and are predominantly semi-diurnal with a marked diurnal inequality between the two daily tides. Tides in the eastern Great Australian Bight are semi-diurnal, with a mean tidal range of between 0.8 and 1.2 m.

Currents

Oceanographically, the Great Australian Bight represents the northern part of the southeast Indian Ocean. Circulation within the Great Australian Bight is dominated by a wind-driven anti-cyclonic gyre with strong upwelling in the east (Herzfeld et al., 1998) (see Fig. 1). The southern limit of this circulation (and the northern boundary of the Southern Ocean) is the Subtropical Front, or the northern boundary of the Southern Ocean (39–40°S, and closer to 47°S in the vicinity of Tasmania). Four major water masses or currents influence the oceanography of the Great Australian Bight region:
- the Leeuwin Current, a poleward eastern boundary current which flows along the continental shelf edge of Western Australia and brings warm, nutrient poor equatorial waters of low salinity (35.0‰) from the tropical waters of the Indian Ocean to the southern parts of Western Australia and the Great Australian Bight (Rochford, 1986; Cresswell, 1991). It is strongest in April–May, when it rounds the south western tip of the Western Australia and enters the Bight.
- the central Bight water mass from the southeast Indian Ocean. This warm (17–21°C) and very high salinity (35.4–36.0‰) water mass, occurs in the central and eastern half of the Great Australian Bight for most of the year, and drifts to the southeast and occupies much of the shelf and slope region east of 135°E, particularly in winter (Rochford, 1986).
- the West Wind Drift cold water mass, which has low salinities (35.0–35.6‰) and lower temperatures (9–14°C). This cold water mass of lowest salinity, is found throughout the year off the slope region of southern Australia and periodically intrudes into the shelf break, especially when the Leeuwin Current is weakly developed (Rochford, 1986).
- the surface-flowing Flinders Current, which has a mean salinity of 35.35‰ and a mean temperature of 14°C. This water mass originates from the gyre south of South Australia (Bye, 1972). These mixed waters have an average surface flow velocity of approximately 5 cm s^{-1}.

The Leeuwin Current is critical to several ecological processes and helps determine patterns of marine biodiversity in the Great Australian Bight. The warm low-salinity waters of the Leeuwin Current are thought to be responsible for the distribution, migration and patterns of dispersal of many pelagic marine organisms from the warm waters of the northwest of Australia to the southern seaboard. This has resulted in a unique 'tropical' or Indo-Pacific element both in the demersal and pelagic fauna of the Great Australian Bight (Maxwell and Cresswell, 1981). Among the benthic invertebrate fauna, many of the echinoderm species recorded from South Australia originate from the Indo-Pacific region. For instance, some 20 of the 84 species of hydroid recorded from South Australia have their principal distribution in the warm waters of the Indo-Malay region. Other echinoderms of Indo-Pacific origin include the Basket Star (*Euryale aspera*), and the holothurians, *Pentacta anceps* and *Pentacta quadrangularis* (Maxwell and Cresswell, 1981).

The warm water of the Leeuwin Current is also responsible for the dispersal of pelagic marine organisms from waters northwest of Australia to the southern seaboard (Maxwell and Cresswell, 1981). Tropical pelagic species, such as the Oriental Bonito and the Southern Bluefin Tuna move with the Leeuwin Current in their migration from the spawning grounds in the Java Sea. These same warm equatorial waters are also thought to be responsible for the significant tropical element in the phyto- and zooplankton of the Great Australian Bight (Markina, 1976) and the suspected relic tropical assemblage in the marine flora (i.e. *Sargassum* spp.). The influence of the Leeuwin Current is also reflected in the Holocene sediments of the Eucla Shelf, which are characterised by a lack of bryozoans, and an abundance of coralline algae and the large foraminifer, *Marginopora* (James et al., 1994).

The Leeuwin Current is intimately linked to the population dynamics of many of Western Australia's and to a lesser extent, South Australia's, commercially important pelagic species (Lenanton et al., 1991), Southern Bluefin Tuna (*Thunnus maccoyii*), Mackerel (*Scomber australasicus*), Horse Mackerel (*Trachurus declivis*), Australian Salmon (*Arripis truttaceus*), and Australian Herring (*Arripis georgianus*), with their distribution and abundance influenced by the seasonality, strength and timing of the Leeuwin Current. The Leeuwin Current has also been suggested as a mechanism for bringing the tropical Bryde's Whale (*Balaenoptera edeni*) into southern waters, while the warm waters in the nearshore regions of the Great Australian Bight, particularly at the Head of Bight (Petrusevics, 1991), may partly explain the significance of this region as a key calving site for the endangered Southern Right Whale (Kemper and Ling, 1991).

Temperature and Salinities

Open coast salinities and water temperatures along the coast of the Great Australian Bight generally vary according to the water masses that prevail, for all or part of the year, within the continental shelf and slope region off southern Australia.

Open coast sea temperatures in the Great Australian Bight vary from a mean summer sea surface temperature of 18°C to a mean winter sea surface temperature of 14°C (decreasing to 11–12°C under the influence of upwellings). Water temperatures (and salinities) vary markedly within shallow coastal embayments and other sheltered areas in the eastern Bight region. Generally, high salinity is a feature of the Bight, with levels of 35.7% being recorded at 100 m depths (Rochford, 1979).

During the summer months (February–March), the waters of the Great Australian Bight are subject to warm surface water (2–3°C higher than surrounding waters) which originates in the west and spreads slowly eastward and, in the eastern Great Australian Bight, localised periodic cold-water, nutrient-rich upwellings (Wenju et al., 1990). The circulation of water within the Great Australian Bight is dominated by a wind-driven anticyclonic gyre with strong upwelling in the east. The dominant surface exchange process is evaporation, which in the absence of any major freshwater inflows (including the River Murray), causes permanently elevated salinity levels such that the Southern Shelf constitutes one extended inverse estuary (Bye, in press). In Spencer Gulf and Gulf St Vincent high salinities result in annual, seasonal (i.e. April–December) outflows or pulses of saline water in a gravity current on their eastern sides. A gravity outflow also occurs in the eastern Bight over the wide shelf and on to the continental slope, south of Eyre Peninsula (Hammat, 1995).

The high temperature surface water (21–22°C) within the Bight was traditionally believed to be advected into the Bight from the warm water Leeuwin Current (Rochford, 1986). However, recent studies (Petrusevics, 1991; Herzfeld, 1997; Herzfeld and Tomczak, 1997; Herzfeld et al., 1998) have shown that the origin of the warm surface water mass is an extensive shallow region (<30 m deep) in the northwest of the Bight (between 124 and 129°E). During winter this region is of comparable temperature to the open ocean (i.e. 18–19°C). First warming over shallow topography occurs in early summer (i.e. November) and intensifies over summer and autumn. During late summer and autumn the locally formed water mass is advected eastward across the shelf (to 136°E) towards the shelf break. The hot summer climate of the Great Australian Bight contributes to the heating of the shallow surface waters. Winds carry hot continental air across the Bight, and air–sea temperature difference during "hot events" can reach up to 15°C, and last 2–3 days. The Leeuwin Current enters the region in autumn (i.e. late April), well after the warm water mass is established (see Fig. 1). This current moves along the shelf break (to 130°E), with only minor eddies into the Great Australian Bight. However, by winter (i.e. July), the Leeuwin Current dominates the sea surface temperature field.

The process of local water mass formation and eastward advection of the warm water corresponds to the west to east passage of anticyclones. With regard to the vertical structure of the water column, in summer the water is stratified with a well defined near mid depth temperature and salinity discontinuity. By winter, this stratification is eroded and the water column is near homogeneous in terms of temperature and salinity. Some additional mixing is brought about by the reversal in direction of an easterly–southeasterly flow in winter, to a westerly–northwesterly flow during November to March.

Nutrients and Productivity

Localised periodic cold-water, nutrient-rich upwellings occur off the coastal regions of the eastern Great Australian Bight (east of 134°E), from Baird Bay to western Kangaroo Island) (Wenju et al., 1990). While summer weather anticyclonic conditions over the Bight are favourable for coastal upwelling along the entire coastline, upwellings only occur along the east coast of the Bight. The upwelling is not driven by the coastal wind but is a result of conservation of potential vorticity over a sloping bottom, which produces a shift of the gyre centre relative to the centre of the wind system, resulting in upwelling through the bottom boundary layer (Herzfeld et al., 1998). As such, the process responsible for the observed coastal upwelling along the eastern coast of the Bight and its absence along the central and western coast is the mismatch between the atmospheric anticyclone and the resulting anticyclonic oceanic circulation in the Bight.

These cold coastal upwellings occur over a narrow strip of water (less than 100 km in offshore extent) and by providing nutrients to surface waters, are sites of significant productivity. Recent studies have indicated a close relationship between these upwellings and pilchard egg abundance in the region (Ward and McLeay, 1998; Ward et al., 1998). This area is also a region of very high benthic biodiversity, seabird abundance and marine mammal abundance (Edyvane, 1998)

Overall however, productivity in the shallow shelf waters of the Great Australian Bight is very low and is considered one of the most nutrient-poor marine environments in the world (Rochford, 1979). The low nutrient levels are most likely the result of the lack of river runoff, the prevailing stable oceanographic conditions of the region, and the Leeuwin Current acting as a barrier to the rich sub-Antarctic waters (Lenanton et al., 1991). A maximum biomass of 500–600 mg/mm of phytoplankton has been recorded for the region (Makarov and Pashkin, 1968). The seasonal variability of the circulation of the shelf waters determines the seasonal aspect of the fishing capabilities of the Bight, with concentrations of pelagic fish occurring in areas of increased plankton abundance (Makarov and Pashkin, 1968).

MAJOR SHALLOW WATER MARINE AND COASTAL HABITATS

Coastal Habitats

The coast of the Great Australian Bight consists of an ancient bedrock geology overlaid by younger Cainozoic, Tertiary and Quaternary deposits. The major open coastal landforms of the Great Australian Bight can be divided into rocky and sandy sections. The rocky sections consist of Tertiary limestone cliffs (with the largest section comprising the Nullarbor Cliffs, which extend over 179 km and average 90 m high) (Figs. 3 and 4), Precambrian bedrock (usually capped by dune calcarenite); and Pleistocene dune calcarenite which is exposed in cliffs up to 150 m high, usually fronted by well developed shore platforms and reefs (Parker et al., 1985; Short et al., 1986; Curry, 1987). The sandy sections include numerous beaches with backing foredunes and transgressive dunes. At the Head of Bight, the Yalata dunes form an extensive active dune transgression. Historic records indicate that the dunes are transgressing inland 11 metres per year. At the Merdayerrah Sandpatch, Pleistocene calcarenite and Holocene marine sediments front the Nullarbor cliffs to form a narrow coastal plain which stretches 30 km to the State border.

Many offshore islands and reefs are scattered within the Great Australian Bight region. Of particular significance are the Recherche Archipelago (WA) and Nuyts Archipelago (SA) and of geological significance is the Investigator Group of Islands, particularly the Pearson Islands, which are true granitic inselbergs (island mountains).

Inshore Habitats

The inshore waters of the Great Australian Bight encompass three major mesoscale, biogeographical regions (i.e. Eucla, Murat and Eyre Bioregions) (IMCRA, 1998). Within these three bioregions, a further 11 finer-scale biounits have

Fig. 3. The vast, remote and featureless Nullarbor Plain lies adjacent to the Great Australian Bight and contains one of the world's largest semi-arid karst landscapes

Fig. 4. The Tertiary limestone Nullarbor Cliffs are a major coastal geological feature of the Great Australian Bight, extending unbroken for 179 km of coast.

been recognised within the Bight (Edyvane, 1998). Inshore habitats range from the high energy cliffs and warm temperate waters of the Nullarbor Cliffs region (i.e. Eucla Bioregion), to the shallow, seagrass-mangrove dominated embayments of Murat Bay and offshore islands of Nuyts Archipelago (i.e. Murat Bioregion), to the high energy, cool temperate waters and offshore islands of southwestern Eyre Peninsula (i.e. Eyre Bioregion). The swell-dominated cliff habitats of the Eucla Bioregion are dominated by extensive sand habitats, interspersed with offshore parallel reefal systems and the absence of seagrass meadows. In contrast, the rocky crenulate coast of the Murat Bioregion provides numerous sheltered embayments for seagrass, and also, abundant reefal habitat. The exposed rocky coast of the Eyre Bioregion comprises mostly sand habitats and reefal habitats, with seagrass habitats principally confined to the sheltered embayments of Port Douglas, Venus Bay and Baird Bay.

Marine Biodiversity

The southern temperate coast of Australia is a region of high marine biodiversity and high levels of endemicity (Edyvane, 1996a,b). This is in part due to the long east–west extent of the southern coastline and the long period of geological isolation (Edyvane, 1998). Within the Flindersian Province, approximately 1200 species of macroalgae, 17 species of seagrasses, 110 species of echinoderms and 189 species of ascidians (Shepherd, 1991) and over 200 ascidians (Greenwood and Gum, 1986) and over 300 species of fish (Glover, 1982) have been reported.

Of these, approximately 85% of fish species, 95% of molluscs and 90% of echinoderms are endemic (Poore, 1995). Similarly, the marine macroalgal diversity in the region is among the highest in the world, with over 75% endemism in the red algae (Womersley, 1990). Of the 1200 species of macroalgae recorded, around 32% are endemic and only 10% are widespread throughout the oceans of the world. In contrast, approximately 13%, 10% and 13% of fish, mollusc and echinoderms, respectively, are endemic in the tropical regions of Australia (Poore, 1995).

Marine biodiversity and biogeography of the Great Australian Bight is strongly influenced by physical factors such as the shallow, broad continental shelf, which has provided abundant shallow-water habitat; the semi-arid climate of the Bight and lack of fluvial inputs (i.e. rivers, estuaries, creeks), which has provided clear oceanic waters for the growth of organisms; and importantly, the oceanographic features of the Bight (i.e. Leeuwin Current, coastal upwellings). In this latter regard, the Eucla Bioregion (and to a lesser extent, the Murat Bioregion) is an area of relatively low productivity and benthic biodiversity, but is influenced by the warm waters of the Leeuwin Current and coastal heating (exhibiting clear warm temperate affinities in the marine biota). In contrast, the inshore habitats of the Eyre Bioregion are under the influence of localised nutrient-rich, cold-water, seasonal upwellings, which has resulted in very high levels of benthic biodiversity (particularly in the macroalgae) and productivity. Similarly, this region is also a key area of pilchard abundance and one of the most important sites for seabirds and marine mammals in temperate Australia.

Rocky Reef Habitats

In general, the marine benthic habitats of the Great Australian Bight have been the subject of few studies. Recent surveys of the inshore regions of the central Great Australian Bight region indicate benthic communities' assemblages are typical of warm to cool temperate waters, and high swell wave conditions (Edyvane, 1998). The coastal nearshore marine habitats of the Great Australian Bight Marine Park comprise mostly sand out to at least 2–3 km, interspersed with small narrow patches of low profile limestone reef (Edyvane and Baker, 1996). Subtidal macroalgal communities on rocky reefs are dominated by the kelp, *Ecklonia radiata* and the fucoid *Scytothalia dorycarpa* and species of *Cystophora*. Of particular interest is the presence of a suspected relic tropical assemblage in the macroflora, indicated by the presence of an undescribed fucoid, *Sargassum* spp. (Edyvane, 1998). Based on the area of reefal habitat available for reef-associated species, existing and potential demersal fishing grounds for species, such as blacklip abalone and southern rock lobster are likely to be limited within the central Great Australian Bight region.

Seagrass Habitats

Seagrass communities in the region are restricted to the lee of headlands (e.g. Point Fowler, Point Sinclair and Point Bell) in the central Bight, and the large sheltered embayments of the eastern Great Australian Bight (e.g. Murat Bay, Streaky Bay, Smoky Bay, Port Douglas, Venus Bay and Baird Bay) and generally comprise large meadows of *Posidonia* (*P. sinuosa*, *P. angustifolia*) and *Amphibolis* (*A. antarctica*, *A. griffithii*). Some of the largest meadows occur in the South Australian gulfs, Spencer Gulf (5520 km^2) and Gulf St Vincent (2440 km^2), and the shallow sheltered embayments on the west coast of the Eyre Peninsula (880 km^2) (Edyvane, 1998).

Fish

The fish fauna of the Great Australian Bight (particularly non-commercial species) is poorly known and there is an urgent need for detailed fish studies, for both inshore and offshore areas. However, the relatively few fish studies which have been conducted in the region indicate that the marine fish fauna of the Great Australian Bight is typical of the Flindersian Province of southern Australian coastal waters (Glover and Olsen, 1985), with approximately 85% of the inshore species endemic to the region (Wilson and Allen, 1987). Of the 344 species of fish listed for southwestern Western Australia, 61% extend eastwards to, at least, off South Australia. There are some 300 species of known marine fish recorded off the Great Australian Bight down to the base of the continental slope, which represents approximately 67% of the species recorded from all South Australian marine waters (Glover, 1982).

Midwater and pelagic fish, some of which are commercially important, are generally widespread in the Bight. Many of them are migratory, especially tropical species that visit cooler southern waters during summer. Regular migratory visitors include commercial species such as Southern Bluefin Tuna (*Thunnus maccoyii*), Mackerel (*Scomber australasicus*), Horse Mackerel (*Trachurus declivis*), Australian Salmon (*Arripis truttaceus*), and Australian Herring (*Arripis georgianus*). The distribution and abundance of these species is strongly influenced by the timing and strength of the easterly flowing Leeuwin Current (Glover and Olsen, 1985). Occasional oceanic vagrants include oceanic Sunfish (*Mola* sp.), Basking Shark (*Cetorhinus maximus*), Black Marlin (*Makaira indica*), and the Lizardfish (*Saurida undosquamis*). Of particular interest is the relative abundance of the White Shark (*Carcharodon carcharias*) in the Great Australian Bight and western Eyre Peninsula region, which may be due to the abundance of prey species, such as pinnipeds (Bruce, 1992). The status of other shark species in the Bight (i.e. gummy sharks, school sharks, bronze whalers, dogfish) is poorly known.

Seabirds

Very little information on seabirds has been collected in the Great Australian Bight, particularly feeding and nesting aggregations (Edyvane, 1998). Breeding colonies of Little Penguins (*Eudytula minor*) occur at the base of the Nullarbor Cliffs in the region (Reilly, 1974), Many seabird species, such as Short-tailed Shearwater (*Puffinus tunuirostris*) and White-faced Storm Petrel (*Pelangodroma marina*), probably feed in the area and/or breed wholly or largely on the offshore islands of the Great Australian Bight (Copley, 1995). Other species include the Black-faced Shag, Eastern Reef Egret, Cape Barren Goose, White-bellied Sea-eagle, Sooty Oystercatcher, Pacific Gull, Fairy Tern, Crested Tern, Rock Parrot, and the Fleshy-footed Shearwater (Smith Island) (Eckert et al., 1985). Non-breeding migratory seabirds such as Albatrosses, Petrels and Prions are known to frequent the coastal and shelf regions of the Great Australian Bight, but to an unknown degree (Copley, 1995).

Marine Mammals

Southern Right Whales (*Eubalaena australis*) seasonally migrate to Australian coastal waters in the winter months between May and October to calve and mate, mainly on a three-year cycle (Fig. 5). The summer feeding grounds and routes of migration to winter calving and breeding areas are not precisely known (Kemper et al., 1994), but probably comprises a broad front rather than a narrow migratory path. Very recently, direct evidence of the presumed link between animals wintering on the southern Australian coast and those feeding in sub-Antarctic waters has been observed (Bannister et al., 1996).

The breeding and calving aggregation or nursery area at the Head of Bight is the largest, densest and most consistent aggregation in Australia and represents one of the two major breeding and calving areas for this species in the world—with approximately one-third of the calves born in Australian waters being born in this single discrete area (Bannister and Burnell, unpubl. data, cited in Burnell and Bryden, in press; Edyvane, 1998). Every year, between the months of May and November, approximately 50–60 whales visit the Head of Bight region to calve, nurse their young and breed. Low numbers of individual Southern Right Whales are also reported irregularly in areas off the New South Wales and Victorian coasts, with sporadic appearances of several individuals in favoured areas such as Warrnambool (Victoria), northwest and east Tasmania and eastern South Australia. The world population of Southern Right Whales is estimated at approximately 1500 to 3000 individuals, with an Australian population of about 600 to 800 (Bannister et al., in press). The populations off South Australia have shown no apparent increase (Ling and Needham, 1991).

At least 17 species of cetaceans have been recorded, including migratory Blue Whales (*Balaenoptera musculus*), Sperm Whales (*Physeter macrocephalus*), Minke Whales (*Balaenoptera acutorostrata*), Rorquals and Humpbacks (*Megaptera novaeangliae*) (Kemper and Ling, 1991). Many cetaceans prey upon cephalopods (squids, octopus, cuttlefish), which are thought to be diverse and abundant in the Ceduna canyons and on the edge of the continental shelf. Possibly resident cetaceans include the Beaked Whale, Killer Whale (*Orcinus orca*) and Risso's Dolphin (*Grampus griseus*). Bottlenose Dolphins (*Tursiops truncatus*) have been seen on many occasions playing amongst Southern Right Whale groups (Ling and Needham, 1988; Bannister, 1993a,b). Little information is available on the distribution, abundance and ecological requirements of resident cetaceans, although in the case of Killer Whales their presence is probably related to the abundance of pinnipeds in the region upon which they feed (Kemper and Ling, 1991).

Two species of seals or pinnipeds are found in the Great Australian Bight region (Fig. 6). These are the rare Australian Sea Lion (*Neophoca cinerea*) and the New Zealand Fur Seal (*Arctocephalos forsteri*) (Edyvane and Andrews, 1995). The region contains many breeding colonies of these species and as such, is becoming increasingly recognised as an

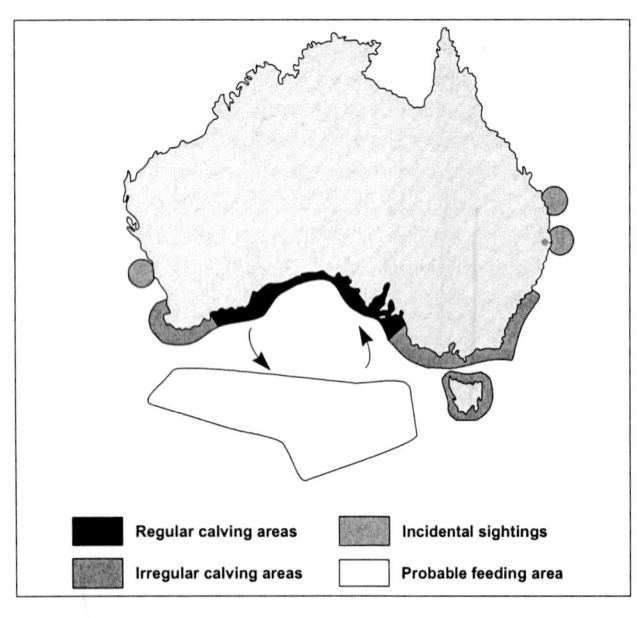

Fig. 5. Distribution of the Southern Right Whale *Eubalaena australis* (after Marsh et al., 1995)

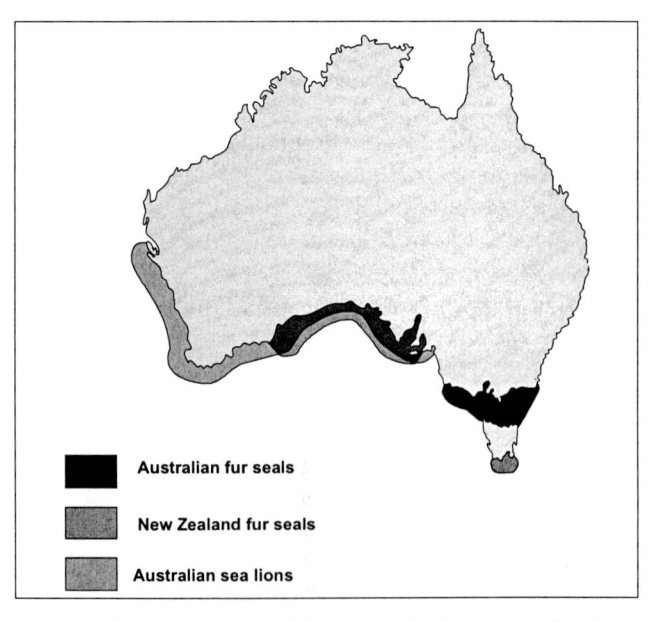

Fig. 6. Distribution of seals and sea lions in the Great Australian Bight (after Marsh et al., 1995).

area of global conservation significance for these species, particularly the rare Australian Sea Lion which is one of Australia's most endangered marine mammals, one of the rarest and most endangered pinnipeds in the world and is endemic to Australia (Gales, 1990).

The Australian Sea Lion is currently recognised as 'rare' under South Australian legislation; a 'Special Protected Species' in Western Australia; and classified as 'rare' under the IUCN Red List of Threatened Animals (WCMC, 1993). However, Shaughnessy (in press) has recently proposed that the species be reclassified as 'near threatened' in the 'lower risk' IUCN category. Prior to seal-hunting, this species occurred along the whole of the southern coastline, but is now confined to the waters of South Australia and Western Australia. The estimated world population for this species is 9300–11,700 individuals, with estimated population sizes of 7500 sea lions in South Australia and 3100 in Western Australia (Gales, 1990; Gales et al., 1994).

The populations of Australian Sea Lions at the head of the Great Australian Bight, which comprise approximately 7% of the total world population, are highly significant because it is very likely that the populations were never commercially harvested—unlike many other populations along the southern coasts and islands of Australia last century which were generally harvested to the point of extinction (Dennis and Shaughnessy, 1996). This is principally because of the isolation and general inaccessibility of the Great Australian Bight coast, both from land and sea. Thus the populations have remained intact, providing probably one of the greatest sources of genetic diversity for this species in the world, and also, a very important genetic and geographic bridging population between the South Australian and Western Australian sea lion populations (Dennis and Shaughnessy, 1996).

The New Zealand Fur Seal does not presently fall into any of the IUCN 'threatened' categories (i.e. critically endangered, endangered or vulnerable), but is currently recognised as 'conservation dependent' in the 'lower risk' IUCN category (Shaughnessy, in press). The total population size for this species in Australia has been estimated at 34,700 individuals in 1989/90 (Shaughnessy et al., 1994).

OFFSHORE SYSTEMS

The waters of the Great Australian Bight region submerge continental crust of the ancient Gondwana supercontinent. In the central Great Australian Bight, that crust extends oceanwards for up to 500 km. The continental margin itself is considerably younger, having formed by continental extension and rifting within Gondwanaland in Jurassic and Cretaceous times (~150 to 95 Ma). The offshore geology is dominated by three sedimentary basins, the Bremer, Great Australian Bight, and Otway Basins (Willcox et al., 1988).

The Great Australian Bight forms part of the Southern Shelf of Australia which is the northern boundary of the South Australian Basin of the South East Indian Ocean. The Southern Shelf is a key bathymetric feature of the Great Australian Bight. This extensive shallow continental shelf has a maximum width of about 200 km in the central Great Australian Bight and narrows to about 20 km south of Western Australia and on the Bonney Coast of South Australia. The continental shelf is almost featureless, forming a gently sloping plain out to the shelf break at about 125–165 m depth. Minor changes in slope also occur at about 25 m and 90 m depth. Beyond the shelf are two large marginal terraces (Eyre and Ceduna Terraces) with a water depth range of about 200–4000 m (Willcox et al., 1988). Along the Southern Shelf of the Great Australian Bight and the Bonney Coast, in particular south of the Eyre Peninsula, about 25 very large (length 50 km, width 5 km) and steep (slope 1:10) canyons connect the continental slope and the abyssal plain over the depth range 1–5 km. The South Australian Basin has an abyssal plain of depth about 5.5 km.

The wide, swell-dominated, open shelf waters of southern Australia have also allowed some of the largest modern, cool-water, open shelf accumulations of carbonate sediments in the world (Connolly and von der Borch, 1967; James et al., 1994). Export of continental terrigenous sediment to the wide continental shelf is low (because of the low continental relief and a predominantly arid climate), and together with the cold water upwelling ocean waters, has resulted in luxuriant growths of carbonate-producing bryozoans and coralline algae, sponges, molluscs, asteroids, benthic and some planktonic foraminifera (Gostin et al., 1988).

Commercial pelagic fish species such as Pilchards (*Sardinops neopilchardus*), Jack Mackerel (*Trachurus declivis*) and Blue Mackerel (*Scomber australasicus*), have been recorded in inshore regions of the Bight, while mature fish were located near the shelf (Stevens et al., 1984). Recent studies have also shown a strong link between the upwellings in the eastern Great Australian Bight and pilchard egg abundance in the region (Ward and McLeay, 1998). Pilchards, as well as being commercially significant as baitfish, are one of the major prey species of a range of fish, bird and mammalian predators, including the Little Penguin (*Eudyptula minor*), the Southern Bluefin Tuna (*Thunnus maccoyii*) and Australian Salmon (*Arripis truttaceus*) (Fletcher, 1990).

Shuntov (1969) observed that pelagic fish in the Bight were generally concentrated along the shelf break, possibly due to the increased productivity of the region. As such, the irregular catches made by fishing vessels in the region are probably the result of the high variability in the abundance of pelagic fish, and are possibly associated with the timing and duration of enrichment events in the Bight (Makarov and Pashkin, 1968; Shuntov, 1969; Stevens et al., 1984). A survey of offshore fish at depths of 400–1200 m recorded 166 species of deep-sea fishes, and some new species (Glover and Newton, 1991). A survey of the crustacean isopod fauna from the Australian southeastern continental slope between 200 and 3150 m depth found a very rich fauna which is largely undescribed (Poore et al., 1994).

POPULATIONS AFFECTING THE AREA

The Great Australian Bight region is one of the most remote and uninhabited regions of southern temperate Australia. This is due largely to the arid landscape and the marginal or limited potential for grazing and agricultural activities. The closest and largest coastal townships in the region include Ceduna (population 2877) and Streaky Bay (population 992) to the east, and Eucla (population <500) to the west. A large proportion of the coastal land in the region is reserved in coastal national parks and conservation reserves. This includes Cape Arid National Park (279,832 ha) and Cape Le Grand National Park in the western Bight, to the Nullarbor National Park (588,300 ha) and Wahgunyah Conservation Reserve (15,555 ha) in the central Bight region, amongst others. The Nullarbor National Park protects the world's largest semi-arid karst (limestone) landscapes. Similarly, in the eastern Bight, several large national parks, such as the Coffin Bay National Park (28,106 ha) and Lincoln National Park (29,060 ha) reserve significant areas of pristine coastal wilderness.

Aboriginal People

In the past, the Aboriginal people lived on the southern and northern fringes of the Nullarbor Plain and avoided the waterless, treeless interior. Population density was probably low and small bands of Aboriginal people may have used unusually good weather to explore the heartland. The ability of the environment to support more permanent populations has improved over the last 10 000 years. Hunting and gathering focused on the coast and the hinterland where kangaroo, wombat and small marsupials were exploited. The coastal Mirning people occupied the southern portion of the Nullarbor plain and the Wirangu people lived from the Head of Bight to the Gawler Ranges and Streaky Bay.

Aboriginal subsistence was dominated by seasonal weather patterns. The inhabitants were described as living on the coast throughout the spring and summer, travelling inland when the sea was rough and cold and the inland rock holes were full. Much of the coastline was inaccessible because of the cliffs, so occupation focused around fresh water soaks at the Head of Bight, Eucla and Merdayerrah Sandpatch. Coastal resources included seals, shell fish, fish, birds, wombats and other large game. With the exception of the saline water table of some of the larger limestone caves, the Nullarbor contains no known permanent supplies of water. Temporary supplies in rockholes were used on the plains and these were connected by Aboriginal pathways or dreaming trails which formed major highways across the plains from the Head of Bight to Eucla, from Ooldea to Fowlers Bay, and from Muckera Rockhole to Koonalda.

Present coastal Aboriginal settlements in the region are confined to the Yalata Aboriginal Land Lease (456,000), which abuts the eastern section of the Great Australian Bight Marine Park, adjacent to the critical breeding and calving areas of the Southern Right Whale.

European Colonisation

Europeans first saw the Nullarbor from the sea in the seventeenth century and called it Nuytsland but shied away from the formidable cliffs. Early terrestrial visitors to the Nullarbor were interested in its grazing potential. Explorers Eyre and Warburton reported the Nullarbor Plain unfit for settlement while others reported enthusiastically about the grazing opportunities. European pioneers who attempted settlement of the Great Australian Bight in the 1800s found limited opportunities for grazing. However, in the southern Nullarbor region, pastoralists settled in areas where palatable borewater was located. A relatively small area is now occupied by pastoral leases and none of the Great Australian Bight Marine Park borders pastoral leases.

The first commercial industry to be established in the Great Australian Bight was the whaling and sealing industry in the 1800s. The proximity of the seals and whales to the shore was seen as a great advantage to prospective sealers and whalers. Large numbers of United States and French whalers visited the waters around Streaky Bay and Encounter Bay in the 1830s and 1840s, and also helped chart the coast of the eastern Bight. Albany was a major centre for commercial whaling activity in Australia, and in the 1970s was also one of the last stations to cease commercial whaling activities.

RURAL FACTORS

The current commercial use of the Great Australian Bight is largely restricted to commercial fishing activity. Until recently, there were no mineral or petroleum exploration or development proposals. However, in 1999, gas and petroleum acreage was released in the southern waters of the Bight (i.e. approximately 75.6 nautical miles offshore, south to the edge of the Exclusive Economic Zone) (Commonwealth of Australia, 1999). Of this, areas S99-2 and S99-4 pass through a portion of the Benthic Protection Area of the Great Australian Bight Marine Park.

Commercial Fisheries

The offshore waters of the Bight are the focus of three major commercial fisheries: the Southern Bluefin Tuna fishery, the Southern Shark Fishery and the Great Australian Bight Trawl Fishery. The Southern Bluefin Tuna Fishery and the Great Australian Bight Trawl Fishery are Commonwealth-managed fisheries and are managed under the Commonwealth's Fisheries Management Act 1991, while the shark fishery is managed under the provisions of the South Australian Fisheries Act 1982. The inshore commercial fishing activities in the Bight are restricted to a scalefish industry concentrating on shark gillnetting; limited Mulloway and

Australian Salmon netting; and several licensed Southern Rock Lobster fishers. In South Australia, commercial fisheries are primarily managed under provisions and regulations under the Fisheries Act 1982 and through a system of Integrated Management Committees, which provide for co-management of the fisheries.

The Southern Shark Fishery has operated for more than 60 years, targeting several temperate species of shark, and other species of marine scalefish. However, the majority of this catch (90%+) is made up of Shark. Large mesh gill netting (>15 cm mesh) is by far the most important method of capture of marine scalefish species in the Great Australian Bight, followed by handlines and rod and line (Jones, 1991). The main marine scalefish species taken in inshore (<50 m) waters were Gummy (*Mustellus antarcticus*) and Bronze Whaler (*Charcharinus brachyurus*) sharks, Sweep (*Scorpus aquepinnis*), Mulloway (*Argyrosomus hololepidotus*) and Australian Salmon (*Arripis truttaceus*), and in offshore waters, School Shark (*Galeorhinus galeus*), Ocean Leatherjackets (*Nelusetta ayraudi*) and Deep Sea Trevalla (*Hyperoglyphe antartica*) (Jones, 1991). This fishery also includes such species as Pilchards, Whiting, Tommy Ruff, and Redfish. Marine scalefish catches in the central Bight have varied from a maximum of 143 tonnes in 1991/92 to 34 tonnes in 1996/97 (Edyvane, 1998). Increases in shark and marine scalefish catches in recent years (i.e. since 1991/92) with the Great Australian Bight region, are not apparent. Records indicate that pregnant School Sharks move from southeastern Australia to the Great Australian Bight for the period of gestation, and then return eastwards to give birth.

Extensive Pilchard (*Sardinops neopilchardus*) fisheries occur in bays to the west of the Great Australian Bight (Esperance, Albany, WA), whilst in South Australia, operations target areas east of the Bight. In 1992/93, the total allowable catch (TAC) for the pilchard fishery (3450 tonnes) constituted half of the South Australia's commercial marine scale fishery. Most of the present fishing activity for Southern Rock Lobster (*Jasus edwardsii*) occurs in the eastern and western Bight, but is limited in central Great Australian Bight due to the lack of reefal habitat and exposed nature of the coast.

The Great Australian Bight Trawl Fishery extends from Kangaroo island off South Australia, to Cape Leeuwin in Western Australia, a distance of over 2000 km. The southern boundary of the fishery is the edge of the Australian Fishing Zone, which extends 200 nautical miles offshore. The northern boundary follows the 200 m depth contour about 20–90 nautical miles offshore. Fishing activity in the Great Australian Bight Trawl Fishery is currently confined to a fairly narrow margin off the continental shelf and slope, in depths of less than 1200 m, mostly within the 100–200 m depth zone (BRR, 1993). Species caught at depths less than 200 m (i.e. the continental shelf) comprise mostly Deepwater Flathead (*Neoplatycephalus conatus*) and Bight Redfish (*Centroberyx gerrardi*), while at depths greater than 200 m (i.e. the continental slope), Orange Roughy and various Oreo Dories (family Oreosomatidae) are the main species. The slope fishery has contracted in recent years and, in 1994, yielded the lowest catch on record (BRR, 1994).

Southern Bluefin Tuna (*Thunnus maccoyii*) is a highly migratory species that spawns between September to March, in the warm waters of the Indian Ocean, south of Indonesia and migrates through the southern oceans between 30°S and 50°S. They mature at approximately 8 years of age and may live as long as 40 years (BRR, 1995). Young tuna (1–4 years) tend to stay in relatively shallow waters associated with coasts and continental shelves where they are fished by domestic purse seine, pole and troll vessels. However, once they reach maturity, they move to deeper oceanic waters, live a pelagic existence and have an almost circumpolar distribution, where they are taken by long-liners. Because of their long exposure to fishing activity prior to spawning, the species is highly vulnerable to overfishing, and then slow to recover. In recent years an active purse seine and pole fishery for this species has been confined to the South Australian sector of the Great Australian Bight (Jones, 1991). With the shift away from long-line fishing to purse seining, most fish in this Australian fishery are now taken south of 32°00'S and east of the Great Australian Bight.

Recreational Fisheries

The major recreational fishery is a shore line fishery targeting Mulloway (*Argyrosomus hololepidotus*) and Australian Salmon (*Arripis truttaceus*). Fishers include members of fishing clubs, organised fishing safaris and four-wheel-drive fishing enthusiasts (Jones, 1991). Fishing is beach-based with major fishing areas occurring east of the Head of Bight and at Coymbra. Although total recreational fishing effort has not been quantified in the region, its importance for local and regional tourism for western Eyre Peninsula is well accepted.

Whale Watching

The Head of Bight region is increasingly being recognised as a major site for whale-watching. It is one of the few areas in the world where the breeding and general behaviour of the Southern Right Whales can be observed closely by the public. Whale-watching within the region is restricted to the area immediately adjacent to the Head of Bight, particularly at Collosity Point, and is a land-based activity. This area affords spectacular close views of Southern Right Whales calving, nursing and mating, often within 100 m of shore. These views are further enhanced by the unspoiled coastline.

Other Activities

Key tourist and recreational activities in the region (other than whale-watching) principally include low-key camping and recreational fishing, with major sites of activity, many occurring at beach sites, such as the Merdayerrah Sandpatch, and the beach from Coymbra west to Twin Rocks. Other water-based activities, such as swimming and surfing, also occur within the region. Diving and yachting opportunities are limited in the region due to the exposed nature of the coast and the limited reef habitats. Camping infrastructure is generally very limited or absent within the coastal areas adjacent to the Marine Park (Josef and Mingatjuta, 1993a,b). There are presently no facilities for launching recreational boats along the entire coast of the Great Australian Bight Marine Park. It is possible, however, to launch vessels from Coymbra beach and Merdayerrah Sandpatch, weather permitting.

COASTAL EROSION AND LANDFILL

Due to the very sparse settlement along the Great Australian Bight coast there are very few coastal management issues compared with the rest of the coast. As such, erosion, seawater flooding and sand drift are not perceived as problems, as they pose no risk to any coastal development. The most serious concern along the Great Australian Bight coast however, is uncontrolled vehicle access. Four-wheel-drive vehicles are currently accessing most parts of the far west coast of South Australia along tracks largely created over the last 10–15 years. Extensive track development has occurred in the dunes, on headlands and along cliff tops and inland (for access to water holes). Vehicle access through the Nullarbor National Park is essentially via established tracks running off the Eyre Highway. The track development has caused major vegetation loss in some regions and contributed to soil erosion. As such, access management is urgently required along the coast to prevent further degradation. Damage to dune vegetation and litter is a significant problem at many of the popular campsites. With the likely rapid increase in visitors to the region, particularly whale watchers, increased planning and resources for visitor management at the Head of Bight is urgently required. Results from systematic observations now clearly depict nodes of concentrations for both vehicles and pedestrians (Reid, in press).

The visitor experience at the Head of Bight has to do with more than just whale-watching. In a recent visitor survey of the region (Reid, in press), the tranquillity of the area and avoidance of crowds were considered by visitors to be important parts of the whale-watching experience at the Head of Bight. An appreciation of the surrounding environment was equally important.

Feral animals, particularly rabbits, are also having a significant impact on the coastal vegetation in the region. Feral camels migrate seasonally to these coastal regions and are also present at the Head of Bight, primarily in the dunes. Feral cats and dogs are a danger to wildlife. Due to the scale of rabbit infestation in the region generally, control measures would be very costly and difficult to implement. However, without effective controls, coastal vegetation in the region will continue to be seriously damaged, with long-term effects on the structure and composition of the coastal plant communities and scenic amenity.

EFFECTS FROM URBAN AND INDUSTRIAL ACTIVITIES

Commercial fisheries have the potential to significantly affect the conservation values within the Great Australian Bight through over-exploitation (reducing fish populations), fisheries by-catch, habitat modification (through the action of equipment such as demersal trawls and nets), and fisheries entanglements from litter. Marine litter in particular has the potential to cause mortality in marine mammals in the region. In recent years, results from annual beach litter surveys conducted at Anxious Bay (on the remote far west coast of South Australia) indicate that commercial fisheries have been a major contributor to ocean litter in the Great Australian Bight (Wace, 1995, Edyvane, 1998). Much of the marine litter can be sourced directly to trawling, lobster and commercial netting operations (Table 1). The amount of marine litter has decreased significantly between 1991 and 1998, with the highest levels recorded in 1992 (i.e. 344 kg) and the lowest recorded in 1995 (i.e. 150 kg) (Edyvane, 1998) (Fig. 7). Despite the long history of the Great Australian Bight Trawl Fishery and Southern Shark Fishery in the region, almost nothing is known of the extent of by-catch and the extent and nature of the disturbance to the benthos in the Bight.

Potential threats to seabirds in the Great Australian Bight region include fisheries entanglements (longline, squid fisheries), competition for prey species (particularly pilchards), marine litter entanglements and human disturbance to nesting sites. Potential impacts of longline and squid fisheries could be minimised by the introduction of night fishing practices, bird lines, weighted hooks and thawed bait, and bait-throwing devices for long line fisheries (Flaherty, 1996). There may also be an economic advantage to fishers in avoiding interactions with seabirds

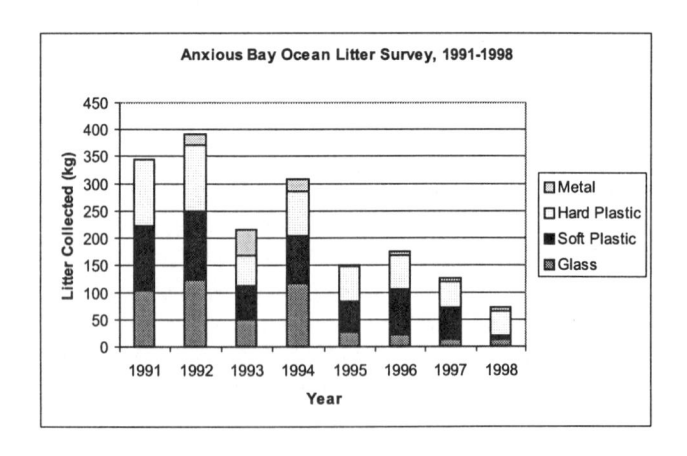

Fig. 7. Marine litter in the Great Australian Bight, as recorded from the Anxious Bay ocean litter survey, 1991–1998.

Table 1

Major categories of marine litter in the Great Australian Bight, as recorded from the annual Anxious Bay ocean litter survey, 1991–1998. Weights in kg

Type of Litter	1991	1992	1993	1994	1995	1996	1997	1998
Hard Plastic (moulded)	122	121	56	81.6	64.7	62.3	46.3	45.7
Soft Plastic (flexible)/Rope	119	127	64	89.2	57.5	84	59.5	6.9
Glass	103	123	49	116	26	21.7	12.5	12.6
Metal (steel/aluminium)	n/a	20.5	47	21.5	2.5	6.8	7.3	5.9
Totals	344	391	216	308	150	174	125	71
kg/km	13.2	15.0	8.3	11.8	5.8	6.7	4.8	2.7

(and other non-target species). In the case of interactions between the long line fishery for Southern Bluefin Tuna and Albatrosses (which take bait from hooks), Brothers (1991) estimated costs to the industry at $7.2 million annually from decreased catch and $4.9 million in lost bait.

While competition between seabirds and commercial fisheries (for prey species) in the Great Australian Bight is presently undocumented, competition could potentially arise due to increased fishing effort (Flaherty, 1996). For instance, pilchards are an important food item for a number of seabird species. Extensive pilchard fisheries occur in bays to the west of the Great Australian Bight (Esperance, Albany, WA), whilst in SA, operations target areas east of the Bight. Other fisheries which may affect seabird populations include calamari and cuttlefish, which represent important prey for some albatrosses and other seabirds (Flaherty, 1996). Nesting seabird populations in the Great Australian Bight region may also be affected by increased human visitation and use of coastal areas. For instance, foot, vehicle and vessel traffic disturbance could result in decreases in populations and local extinction of populations of White-bellied Sea Eagles, which nest along the cliffs in the region (Flaherty, 1996). Little Penguins favour rocky shorelines which provide suitable breeding sites and are known to occur within the area bounded by the Great Australian Bight Marine Park (Reilly, 1974). Specific threats to Little Penguins include fisheries entanglement in nets and predation from feral animals such as foxes and cats.

Southern Right Whales, like all marine mammals, are protected under Commonwealth legislation under the Commonwealth Whale Protection Act 1980 from direct disturbance or harm. However, this legislation does not protect marine mammals from indirect disturbance (e.g. aircraft, fisheries entanglements, seismic drilling) or direct habitat disturbance (e.g. dredging, pollution). With the cessation of commercial harvesting, the greatest potential threats to the recovery of Southern Right Whales occur in the nearshore calving and breeding areas and during migrations to these sites. In the Great Australian Bight these threats include noise, oil spills, entanglements and collision with vessels (Edyvane, 1998; Commonwealth of Australia, 1999). Cetaceans are particularly vulnerable to disturbance to underwater noise because of their reliance upon sound for communication, prey detection and orientation (Reeves, 1992). The major threats to cetaceans in Australian waters have been recently prioritised and summarised in the draft National Cetacean Action Plan (Bannister et al., in press). Types of direct disturbance and threats possible to Southern Right Whales include:

– whale watching and research vessels/aircraft, pleasure craft (e.g. acoustic disturbance, vessel crowding), swimming and divers;
– low flying aircraft (e.g. acoustic disturbance);
– coastal industrial activity (e.g. seismic, drilling, sand-mining and shipping operations);
– defence operations;
– collision with large vessels, particularly on shipping routes on eastern seaboard (e.g. in Bass Strait, across the Great Australian Bight);
– entanglements in fishing gear—at least three recent examples;
– pollution including increasing amounts of plastic debris at sea, oil spills and dumping of industrial waste into waterways and the sea (leading to bio-accumulation of toxic substances in body tissue), though less serious for species rarely feeding in low latitudes;
– impacts on critical migratory pathways, calving and mating areas from inshore habitat degradation, such as from fisheries exploitation and petroleum and mineral exploration and exploitation.

Threats to the recovery of Australian Sea Lion populations in Australia include human disturbance to breeding colonies and haul out sites, fisheries entanglements, competition with fisheries for common prey species and the effects of oil spills (Marsh et al., 1995). Breeding populations of both *Neophoca cinerea* and *Arctocephalus forsteri* are highly susceptible to disturbance by humans (Gales, 1990; Shaughnessy, 1990). As such, affording breeding colonies prohibited area status has been recommended as the most straightforward method of protecting these colonies (Gales, 1990; Shaughnessy, 1990).

It is not currently possible to assess the level of direct competition between Australian Sea Lions and fishers. Sea lions are known to rob lobster pots, as well as nets set for

School Shark (Shaughnessy, in press). Commercial and recreational fishermen often, rightly or wrongly, regard seals and sea lions as competitors and as pests. In South Australia, records of the Department of Environment and Natural Resources indicate that sea lions have been shot ashore or taken for bait. As the Australian Sea Lion population increases, interaction between sea lions and fishers can be expected to increase (Shaughnessy, 1990). Gales (1990) reported instances of conflict with fisheries operations in the form of net and bait-band entanglements and drownings in lobster pots. In South Australia, sea lion entanglements in monofilament netting of 150 mm mesh (used in the shark fishery) is the most common form of entanglement found on Australian Sea Lions (Robinson and Dennis, 1988). Entanglement rates vary from 0.2% in a survey of Western Australia and South Australia populations (Gales, 1990) to 0.3% at Seal Bay and The Pages. While the present rate of entanglements for Australian Sea Lions appears low, the effect on a rare species can be significant. For instance, studies on the closely related Hooker's Sea Lion in New Zealand, indicate that an increase in mortality of only 1% is sufficient to cause the population to decrease (Woodley and Lavigne, 1993). A further potential threat is a reduction in food supply through, for example, enhanced development of a pelagic squid fishery, as occurred in the New Zealand Subantarctic (Shaughnessy, in press).

Mineral and petroleum exploration and extraction activities also have the potential to significantly affect the conservation values of the region through reduced visual amenities; acoustic disturbance to marine mammals from seismic testing, underwater seismic pulses and offshore drilling and collisions with ships; and also, through risk of oil spills and ballast water introductions. Oil spills pose a threat to all seal populations, especially those at breeding colonies near major fishing lanes (Shaughnessy, in press). Two month old pups from fur seal colonies in the Recherche Archipelago (WA) were affected by a major oil spill from the Sanko Harvest in February 1991 (Commonwealth of Australia, 1999).

The need for greater research and conservation management efforts in the Great Australian Bight is highlighted by the increasing interest in the economic potential of Australia's EEZ and the need to resolve potential user-group conflicts, particularly between fisheries and mineral and petroleum interests and significant marine biodiversity values in the region. While economic activity in the Bight is presently low, being essentially confined to offshore commercial fisheries and inshore oyster aquaculture along western Eyre Peninsula, there are signs of increasing activity in marine-mammal-based ecotourism in the region, particularly at Esperance and Albany (WA), and at Yalata (SA). Further, there is increasing activity in sea-based aquaculture in the region, with aquaculture proposals at Esperance and also, Fowlers Bay and Point Sinclair (west of Ceduna), in addition to expanding areas for oyster aquaculture along western Eyre Peninsula (Ashman, 1995). Such activities, including proposals for intensive fish-farming, have the potential to cause serious and significant conflicts with fisheries and conservation values of the Great Australian Bight through: reduction in coastal scenic amenity; site-based pollution (i.e. sedimentation, eutrophication, hypoxia); entanglements of marine mammals and sharks in predator nets; and potential spread of feral populations (such as Pacific Oysters).

PROTECTIVE MEASURES

Despite the level of marine biodiversity and endemism along the southern coast of Australia, the management of ecosystems, habitats, and species of this region, particularly the Great Australian Bight, has until recently, been significantly under-represented in terms of Marine Protected Areas (or MPAs). It has been estimated that 21 out of the 32 biogeographic regions around Australia lack any significant protection as protected areas (Ivanovici, 1993). Notable regions included the Great Australian Bight, the Gulf of Carpentaria, and deep offshore regions.

Great Australian Bight Marine Park

In June 1995, the Government of South Australia proclaimed the Great Australian Bight Marine Park Whale Sanctuary at the Head of Bight (175 sq. km), specifically to protect the critical breeding and calving areas of the Southern Right Whale, as part of an interim step in considering the establishment of a larger, multiple-use Great Australian Bight Marine Park (Edyvane and Andrews, 1995). This was followed in May 1996, by the proclamation of the larger, multiple-use Great Australian Bight Marine National Park (Government of South Australia, 1998; Edyvane, 1998) (Fig. 8). The proposal comprised the existing Great Australian Bight Whale Sanctuary, and also, the proclamation of the Great Australian Bight Marine National Park (totalling 26,750 ha), including six Sanctuaries for the protection of breeding colonies of Australian Sea Lions. The Great Australian Bight Marine National Park and the prohibited areas were proclaimed on 26 September 1996 under the National Parks and Wildlife Act 1972. The establishment of the Great Australian Bight Marine Park protects the largest and densest breeding aggregation of Southern Right Whales in Australia, and one of the two major breeding and calving areas for this species in the world. As such, it has been identified as a major step in assisting the global recovery of this endangered and vulnerable species. Although all whales, dolphins, porpoises and seals are completely protected in Australian waters under the Commonwealth Whale Protection Act 1980, there is a world-wide recognition of the need to protect the key calving and breeding areas of these marine mammals. As such, the establishment of the Great Australian Bight Marine Park will protect the key critical breeding and calving areas of the endangered Southern Right Whale and rare Australian Sea Lion, while allowing

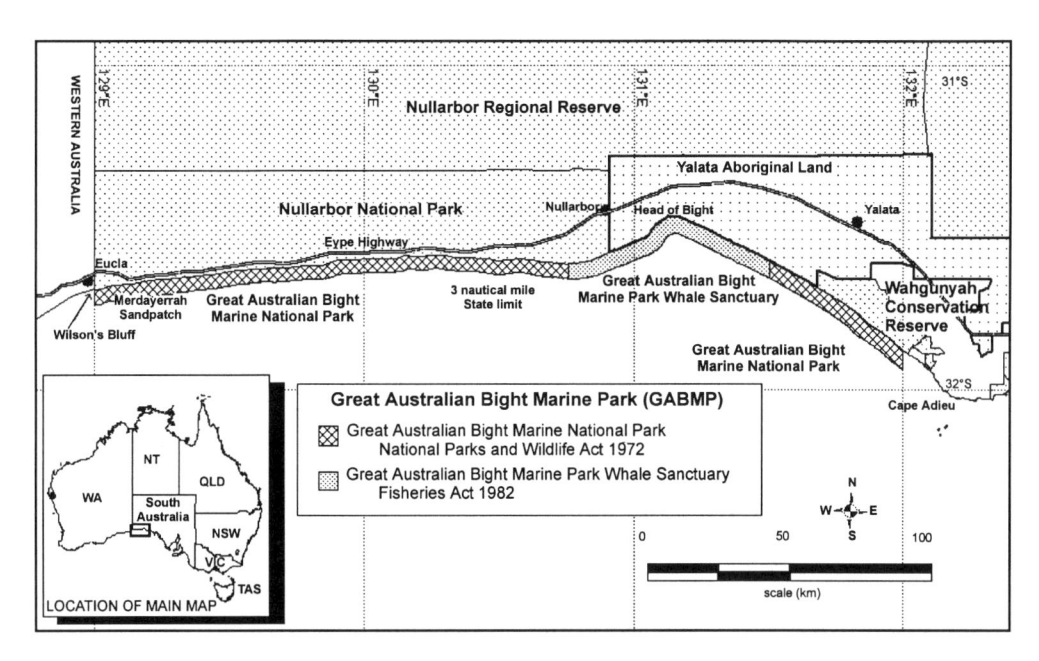

Fig. 8. Boundaries of the State waters of the Great Australian Bight Marine Park (from Edyvane, 1998).

and managing a range of human activities (such as commercial fishing, recreation, tourism, mining, and research).

On 22 April 1998, the Commonwealth Government proclaimed a 1.9 million hectare Marine Park under the National Parks and Wildlife Conservation Act 1975 in Commonwealth Waters of the Great Australian Bight. Environment Australia is the Commonwealth agency responsible for managing the Commonwealth Waters component of the Marine Park. The Commonwealth Marine Park complements the Marine Park established by the South Australian Government in 1995/96, and the intention of both Governments is to manage both Parks cooperatively as much as is practicable. It is a statutory requirement under the National Parks and Wildlife Conservation Act 1975 (Cth) for Environment Australia of the Commonwealth Government, to prepare a plan of management for the Park. Management planning for the Commonwealth waters of the Marine Park is currently in progress (Commonwealth of Australia, 1999).

In developing a cooperative State–Commonwealth approach to the management of the Great Australian Bight Marine Park, there is a critical need for complementary levels of protection and standardised guidelines and processes for defining conservation and multiple-use management regimes, particularly prescribing activities within management zones (across jurisdictions) and also, in defining activities which conflict with primary conservation management objectives for a protected area. For instance, while the State waters of the Great Australian Bight Marine Park afford significant levels of protection (IUCN Category II), the Commonwealth waters of the park currently prescribe for multiple-use and relatively low levels of protection for offshore ecosystems (IUCN Category VI). At the time of proclamation, current management intentions prohibit one activity in the waters of the Commonwealth Park—bottom trawling (for Deepwater Flathead, Bight Redfish) (Commonwealth of Australia, 1999). This is despite the known impacts of mid-water trawling, demersal shark netting and exploration/extraction on benthic communities. Significantly, the lack of high conservation or IUCN Category I and II areas in the Commonwealth waters of the park, will result in a difficulty in assessing, in any research and monitoring program, the effects of trawling (mid-water trawling, demersal shark), fishing (shark, lobster), mineral and petroleum extraction and also, demonstrating the effects/benefits of the management zones, as there are no reference or control sites (Slater, 1999).

There are also several activities permitted in the proposed management zones of the Commonwealth park that potentially conflict with the management objectives of the zones. For instance, the primary goal of the Benthic Protection Area is to "preserve a representative sample of benthic flora and fauna and sediments" (Commonwealth of Australia, 1999). Yet activities such as demersal shark, mid-water trawling and exploration/extraction, activities which are likely to cause "disturbance" to the seabed, and the taking of lobster, abalone, demersal sharks (i.e. biota which are part of the unique and diverse bottom-dwelling fauna representative of the Bight), are currently permitted in the management prescriptions. Similarly, the Marine Mammal Protection Area, which has the designated primary goal "to protect the Australian Sea Lion", currently permits shark netting (on a six-month seasonal basis), although this activity is known to negatively interact with Australian Sea Lions (and is a recognised major cause of juvenile mortality). Mineral and petroleum exploration and extraction are currently permitted in both the Marine Mammal Protection Area (on a 6-month seasonal basis) and the Benthic Protection Area, subject to permit approval and an environmental impact assessment (Commonwealth of Australia, 1999). The

Table 2

Summary of potential pressures on marine conservation values in the Great Australian Bight Marine Park (Commonwealth waters) (from the Commonwealth of Australia 1999).

Pressures	Southern Right Whale	Australian Sea Lion	Benthos	Cultural Heritage
Commercial Fishing	noise, particularly underwater, oil spills, entanglements in nets and ropes, collisions with vessels	Noise, human visitation, oil spills, entanglement in nets and ropes	disturbance from trawl nets and weighted shark nets, from lobster pots (dependent upon density)	no pressures currently perceived
Mining/ Petroleum	oil spills, collisions with vessels, seismic noise and other noise	oil spills, seismic noise and other noise	direct disturbance to benthos through drilling, anchorage and infrastructure, pollution from lubricants and cuttings, seepage from drill holes	no pressures currently perceived
Tourism	noise, particularly from aircraft, collisions from boats.	noise, particularly from aircraft, human visitation	no pressures currently perceived	no pressures currently perceived

management intentions at proclamation for the Commonwealth waters of the Marine Park are currently subject to review under the Commonwealth's formal management planning process. A summary of the potential pressures on marine conservation values in the Great Australian Bight Marine Park (Commonwealth Waters) is outlined in Table 2.

From a management perspective, the Great Australian Bight Marine Park has very high wilderness and coastal scenic values. This is due largely to the remoteness and inaccessibility of the coastline. While this may in some respects protect the natural and cultural values of the Marine Park (and adjacent terrestrial lands) from the types of human impacts found close to highly urbanised areas or impacts associated with high levels of recreational boating, it also presents its own unique management difficulties, particularly as regards compliance and monitoring.

Research

Ongoing research projects within the Great Australian Bight region include: research into the biology, behaviour and status of the Southern Right Whale at the Head of Bight by the University of Sydney; and regular monitoring of Southern Right Whales by SA and WA Museums using aerial surveys. More recently, specific research and monitoring needs and priorities for the State and Commonwealth waters of the Great Australian Bight Marine Park have been identified (Slater, 1999).

Despite the high level of biodiversity and endemism recorded in the Great Australian Bight based on existing studies, very little is known of the ecosystems, habitats, species and ecological processes within this remote region. In particular, there is a need for systematic surveys of the marine fauna and flora of both the offshore and inshore regions of the Bight. Floral, seabird, fish and invertebrate distributions are particularly poorly known. While the limited biodiversity studies have, in the past, concentrated on cetaceans and marine mammals in the region, information is still required on migratory routes, behavioural ecology and general population biology. Spatial information on the diverse faunal and floral assemblages will be particularly important with increasing activity and development in the region, and also to implement an ecosystem-level approach to the management of the region.

REFERENCES

Ashman, G. (1995) Far West Coast Aquaculture Management Plan. Public Consultation Draft. Primary Industries South Australia, Adelaide, South Australia.

ACIUCN (1986) Australia's Marine and Estuarine Areas—a Policy for Protection. Occasional Paper No. 1. Australian Committee for International Union for Conservation of Nature and Natural Resources, Canberra, 32 pp.

Bannister, J.L. (1993a) Report on aerial survey for southern right whales off southern Western Australia, 1992. Unpublished report for the Australian Nature Conservation Agency.

Bannister, J.L. (1993b) Report on aerial survey for southern right whales off southern Australia, 1993. Unpublished report for the Australian Nature Conservation Agency.

Bannister, J.L., Burnell, S., Burton, C. and Kato, H. (1996) Right whales off southern Australia: direct evidence of a link between onshore breeding and offshore probable feeding grounds. Paper presented to Scientific Committee of the International Whaling Commission, no. 48.

Bannister, J.L., Kemper, C.M. and Warneke, R.M. (in press). The Action Plan for Australian Cetaceans. Australian Nature Conservation Agency, Canberra.

Brothers, N. (1991) Albatross mortality and associated bait loss in the Japanese longline fishery in the Southern Ocean. *Biological Conservation* 55, 255–268.

Bruce, B. (1992) Preliminary observations on the biology of the White Shark, *Carchardon carcharias*, in South Australian waters. In Sharks: Biology and Fisheries, ed. by J.G. Pepperell. *Australian Journal of Marine and Freshwater Research* 43, 1–11.

BRR (1993) Great Australian Bight Trawl. Bureau of Rural Resources Fishery Status Reports 1993, 81–85.

BRR (1994) Great Australian Bight Trawl. Bureau of Rural Resources Fishery Status Reports 1994, 85–90.

BRR (1995) Southern Bluefin Tuna. Bureau of Rural Resources Fishery Status Reports 1995, 49–90.

Burnell, S.M. and Bryden M.M. (in press). Coastal residence periods and reproductive timing in southern right whales, *Eubalaena australis*. Journal of Zoology (London).

Bye, J.A.T. (1972) Ocean Circulation South of Australia. Antarctic Research Series. 19. *Antarctic Oceanology 2. The Australian–New Zealand Sector*, ed. D.E. Hayes. American Geophysical Union, Washington, DC. pp. 95–100

Bye, J.A.T. (in press). The South East Indian Ocean and Great Australian Bight: a brief oceanographic survey. In *Procs. South East Indian Ocean and Great Australian Bight USA–Australia Bilateral Workshop, held in Port Lincoln, South Australia, 28 August–3 October, 1998.*

Chelton, D.B., Hussey, K.J. and Parke, M.E. (1981) Global satellite measurements of water vapour, wind speed and wave height. *Nature* **294**, 529–532.

Connolly, J.R. and von der Borch, C.C. (1967) Sedimentation and physiography of the sea floor south of Australia. *Sedimentary Geology* **1**, 181–220.

Commonwealth of Australia (1999) Great Australian Bight Marine Park (Commonwealth Waters) Plan of Management. Environment Australia.

Copley, P. (1995) Status of South Australian Seabirds. Unpublished report, Department of Environment & Natural Resources.

Cresswell, G.R. (1991) The Leeuwin Current—observations and recent models. *Journal of the Royal Society of Western Australia* **74**, 1–14.

Curry, G. (1987) Climate. In *A Biological Survey of the Nullarbor Region South and Western Australia in 1984*, eds. N.L. McKenzie and A.C. Robinson. South Australian Department of Environment and Planning, Western Australian Department of Conservation and Land Management. Australian National Parks and Wildlife Service, pp. 6–16.

Dennis, T.E. and Shaughnessy, P.D. (1996) Status of the Australian Sea Lion, *Neophoca cinerea*, in the Great Australian Bight. *Wildlife Research* **23**, 741–54.

Edyvane, K.S. (1996a) The role of Marine Protected Areas in temperate ecosystem management. In *Developing Australia's Representative System of Marine Protected Areas: Criteria and Guidelines for Identification and Selection*, ed. R. Thackway. Proceedings of a technical meeting held at the South Australian Aquatic Sciences Centre, West Beach, Adelaide, 22–23 April 1996. Department of Environment, Sport and Territories, Canberra.

Edyvane, K.S. (1996b) The 'Unique South': marine biodiversity in the Great Australian Bight. In *Workshop on Multiple-Use in Marine Environments—Proceedings*, eds. Australian Petroleum Production and Exploration Association (APPEA), held at the Australian Academy of Science, Canberra, 18 October 1995. APPEA, Canberra, pp. 176–196.

Edyvane, K.S. (1998) *Great Australian Bight Marine Park Management Plan. Part B. Resource Information*. Department of Environment, Heritage and Aboriginal Affairs, South Australia.

Edyvane, K.S. and Andrews, G.A. (1995) Draft Management Plan for the Great Australian Bight Marine Park. Prepared for Primary Industries South Australia. South Australian Research & Development Institute, West Beach, South Australia.

Edyvane, K.S. and Baker, J. (1996) Major Marine Habitats of the Proposed Great Australian Bight Marine Park. Unpublished report, South Australian Research and Development Institute, Adelaide, 7 pp.

Flaherty, A.F. (1996) Conservation and community concerns for present and future multiple-use of the Great Australian Bight region. In *Workshop on Multiple-Use in Marine Environments—Proceedings*, eds. Australian Petroleum Production and Exploration Association (APPEA), held at the Australian Academy of Science, Canberra, 18 October 1995. APPEA, Canberra, pp. 200–211.

Fletcher, W.J. (1990) A Synopsis of the Biology and the Exploitation of the Australian Pilchard, *Sardinops neopilchardus* (Steindachner). Part 1. Biology. Fisheries Department of Western Australia, Fisheries Research Report No. 88, Perth, WA.

Gales, N.J. (1990) Abundance of Australian Sea Lions, *Neophoca cinerea*, along the Southern Australian Coast, and Related Research. Unpublished report to the Western Australian Department of Conservation and Land Management, South Australian National Parks and Wildlife Service and the South Australian Wildlife Conservation Fund.

Gales, N.J., Shaughnessy P.D. and Dennis, T.E. (1994) Distribution, abundance and breeding cycle of the Australian sea lion *Neophoca cinerea* (Mammalia: Pinnipedia). *Journal of Zoology, London* **234**, 353–370.

Glover, C.J.M. (1982) A provisional checklist of marine fishes (Amphioxi, Petromyzones, Myxini, Elasmobranchii, Holocephali, Teleostomi) recorded in South Australian coastal waters. *South Australian Museum Information Leaflet* **70**, 1–25.

Glover, C.J.M. and Newton, G. (1991) Denizens of the deep: deep-sea fishes in the Great Australian Bight. *Australian Fisheries* May, 30–35.

Glover, C.J.M. and Olsen, A.M. (1985) Fish and major fisheries. In *Natural History of the Eyre Peninsula*, eds. C.R. Twidale, M.J. Tyler and M. Davies, pp. 169–181. Royal Society of South Australia.

Gostin, V.A., Belperio, A.P. and Cann, J.H. (1988) The Holocene non-tropical coastal and shelf carbonate province of southern Australia. *Sedimentary Geology* **60**, 51–70.

Government of South Australia (1998) Great Australian Bight Marine Park Management Plan. Part A. Management Prescriptions. Department of Environment, Heritage and Aboriginal Affairs, South Australia.

Greenwood, G. and Gum, E. (1986) *The State of Biological Resources in South Australia*. Dept. of Environment and Planning, Adelaide, South Australia.

Hammat, J. (1995) Surface Current Structure on the Southern Shelf and in the Adjacent Southern Ocean: A Comparison of Geostrophic Velocities and ADCP Measurements. B.Sc.(Hons.) Thesis, FIAMS, Flinders University, South Australia.

Herzfeld, M. (1997) The annual cycle of sea surface temperature in the Great Australian Bight. *Progress in Oceanography* **39**, 1–27.

Herzfeld, M. and Tomczak, M. (1997) Numerical modelling of sea surface temperature and circulation in the Great Australian Bight. *Progress in Oceanography* **39**, 29–78.

Herzfeld, M., Schodlok, M. and Tomczak, M. (1998) Water mass formation, upwelling and fronts in the Great Australian Bight. A poster presented at the TOS and IOC Meeting on Coastal and Marginal Seas at the UNESCO Headquarters in Paris, France, 1–4 June 1998.

IMCRA (1998). Interim Marine and Coastal Regionalisation of Australia: an Ecosystem Classification for Marine and Coastal Environments. Version 3.3. Interim Marine & Coastal Regionalisation of Australia Technical Group, Environment Australia, Commonwealth Department of the Environment. Canberra.

Ivanovici, A. (1993) Planning for a national, representative system of Marine and Estuarine Protected Areas: identification of priority areas. In *Protection of Marine and Estuarine Areas—A Challenge for Australians*, eds. A. Ivanovici, D. Tarte and M. Olson. Proceedings of the Fourth Fenner Environment Conference Canberra, 9–11 October 1991, Occasional Paper No. 4. Australian Committee for International Union for Conservation of Nature and Natural Resources, Canberra, pp. 56–60.

James, N.P., Boreen, T.D., Bone, Y. and Feary, D.A. (1994) Holocene carbonate sedimentation on the west Eucla Shelf, Great Australian Bight: a shaved shelf. *Sedimentary Geology* **90**, 161–177.

Jones, G.K. (1991) Fin fishery considerations in the management of the proposed Great Australian Bight Marine Park. *Safish* **15** (4), 11.

Josif, P. and Mingatjuta (1993a) Yalata Land Management Plan. *Ngura Nganampa Atunymankutjaku.* (Part 1). Prepared for the Yalata Community Council.

Josif, P. and Mingatjuta (1993b) Yalata Tourism Feasibility Study. (Part 2). Prepared for the Yalata Community Council.

Kemper, C.M. and Ling, J.K. (1991) Possible influences of oceanic features of GAB on cetaceans (abstract). In *Collection of Abstracts: The Great Australian Bight; A Regional Perspective*, Adelaide, 2 May 1991. South Australian Department of Fisheries, Australian National Parks and Wildlife Service and the Australian Marine Sciences Association.

Kemper, C.M., Mole, J., Warneke, R., Ling, J.K., Needham, D.J. and

Wapstra, H. (1994) Southern Right Whales in South-Eastern Australia During 1991–1993. Report to BHP Petroleum Pty. Ltd., Melbourne.

Lenanton, R.C., Joll, L., Penn, J. and Jones, G.K. (1991) The influence of the Leeuwin Current on coastal fisheries of Western Australia. *Journal of the Royal Society of Western Australia* 74, 101–114.

Ling, J.K. and Needham, D.J. (1988) Final Report on Southern Right Whale Survey, South Australia, 1988. Unpublished report to the Australian National Parks and Wildlife Service.

Ling, J.K. and D.J. Needham (1991) Southern Right Whale Survey: South-Eastern Australia—1991. Final Report. Report to BHP Petroleum Pty. Ltd., Melbourne.

Makarov, V.N. and Pashkin, V.N. (1968) General features of fish distribution in the Great Australian Bight (in Russian). *Rybnoe Khozyaistvo (Moscow)* 3, 14–16.

Markina, N.P. (1976) Biogeographic regionalisation of Australian waters of the Indian Ocean. *Oceanology* 15, 602–4.

Marsh, H., Corkeron, P.J., Limpus, C.J., Shaughnessy, P.D. and Ward, T.M. (1995) The reptiles and mammals in Australian seas: their status and management. In *The State of the Marine Environment Report for Australia. Technical Annex 1: The Marine Environment*, ed. by L Zann. Department of Environment, Sport and Territories, Canberra, pp. 151–166.

Maxwell, J.G.H. and Cresswell, G.R. (1981) Dispersal of tropical marine fauna to the Great Australian Bight by the Leeuwin Current. *Australian Journal of Marine and Freshwater Research* 32, 493–500.

Parker, A.J., Fleming, C.M. and Flint, R.B. (1985) Geology. In *Natural History of the Eyre Peninsula*, eds. C.R. Twidale, M.J. Tyler and M. Davies. Royal Society of South Australia, Adelaide, pp. 21–45.

Petrusevics, P. (1991) Oceanography of the Great Australian Bight (abstract). In *Collection of Abstracts: The Great Australian Bight: A Regional Perspective*, Adelaide, 2 May 1991. South Australian Department of Fisheries, Australian National Parks and Wildlife Service and the Australian Marine Sciences Association.

Poore, G.C.B. (1995) Biogeography and diversity of Australia's marine biota. In *The State of the Marine Environment Report for Australia. Technical Annex 1: The Marine Environment*, ed. by L Zann. Department of Environment, Sport and Territories, Canberra, pp. 75–84.

Poore, G.C.B., Just, J. and Cohen, B.F. (1994) Composition and diversity of crustacea isopoda of the southeastern Australian continental slope. *Deep-sea Research* 41, 677–693.

Reeves, R.R. (1992) *Whale Responses to Anthropogenic Sounds: a Literature Review*. Science and Research Series, no. 47. Department of Conservation, Wellington, New Zealand.

Reid, E. (in press). *Whale Watchers of the Head of the Bight: A 1995 Visitor Profile and Implications for Management*. Mawson Graduate Centre Occasional Paper No. 11, 70 pp.

Reilly, P.N. (1974) Breeding of Little Penguins along the Great Australian Bight. *Emu* 74, 198–200.

Robinson, A.C. and Dennis, T.E. (1988) The status and management of seal populations in South Australia. In *Marine Mammals of Australasia. Field Biology and Captive Management*, ed. M.L. Augee. Royal Zoological Society of New South Wales, pp. 87–110.

Rochford, D.J. (1979) *Nutrient Status of the Oceans Around Australia*. Report 1977–1979. CSIRO Division of Fisheries and Oceanography, Hobart.

Rochford, D.J. (1986) Seasonal changes in the distribution of Leeuwin Current waters off southern Australia. *Australian Journal for Marine and Freshwater Research* 37, 1–10.

Shaughnessy, P.D. (1990) Distribution and Abundance of New Zealand Fur Seals, *Arctocephalus forsteri*, in South Australia. Unpublished report to the South Australian Wildlife Conservation Fund. CSIRO, Sydney.

Shaughnessy, P.D. (in press). Seal Action Plan. Incomplete Revised Draft Report. October 1996. Unpublished report to the Australian Nature Conservation Agency.

Shaughnessy, P.D., Gales, N.J., Dennis, T.E. and Goldsworthy, S.D. (1994) Distribution and abundance of New Zealand fur seals, *Arctocephalus forsteri*, in South Australian and Western Australia. *Wildlife Research* 21, 667–95.

Shepherd, S.A. (1991) Biogeography of the GAB Region (abstract). In *Collection of Abstracts: The Great Australian Bight: A Regional Perspective*, Adelaide, 2 May 1991. South Australian Department of Fisheries, Australian National Parks and Wildlife Service and the Australian Marine Sciences Association.

Short, A.D., Fotheringham, D.G. and Buckley, R.C. (1986) Coastal Morphodynamics and Holocene Evolution of the Eyre Peninsula Coast, South Australia. Coastal Studies Unit Technical Report No. 86/2. Department of Geography, University of Sydney.

Shuntov, V.P. (1969) Some features of the ecology of pelagic fishes in the Great Australian Bight. *Problems of Ichthyology* 9, 801–809.

Slater, J. (ed.) (1999) *Proceedings of the Workshop on Research and Monitoring of the Great Australian Bight Marine Park: Past, Present and Future*, 19–21 October 1998, Kangaroo Island, South Australia. Biodiversity Group, Environment Australia.

Stevens, J.D., Hausfeld, H.F. and Davenport, S.R. (1984) Observations on the Biology, Distribution and Abundance of *Trachurus declivis*, *Sardinops neopilchardus* and *Scomber australisicus* in the Great Australian Bight. CSIRO Marine Laboratories Report 164. CSIRO, Cronulla, NSW.

Wace, N. (1995) Ocean litter stranded on Australian coasts. In *The State of the Marine Environment Report for Australia. Technical Annex 2: Pollution*, eds. L. Zann and D. Sutton. Department of Environment, Sport and Territories, Canberra, pp. 73–87.

Ward, T.M., Jones, G.K. and Kinloch, M. (1998). *Baitfisheries of Southeastern Australia*. Final report to FRDC.

Ward, T. and McLeay, L. (1998) Use of the Daily Egg Production Method to Estimate the Spawning Biomass of Pilchards (*Sardinops sagax*) in Shelf Waters of Central and Western South Australia in 1998. Report to the South Australian Pilchard Working Group. South Australian Research and Development Institute (Aquatic Sciences).

WCMC :World Conservation Monitoring Centre in association with IUCN Species Survival Commission and Birdlife International (1993) *1994 IUCN Red List of Threatened Animals*. IUCN, Gland, Switzerland.

Wenju, C. Schahinger, R.B. and G.W. Lennon (1990) Layered models of coastal upwelling: a case study of the South Australian region. In *Modelling Marine Systems*, ed. A.M. Davies, Vol. 1, pp. 73–91. CRC Press, Boca Raton, FL, USA.

Willcox, J.B., Stagg, H.M.J. and Davies, H.L. (1988) Rig seismic research cruises 10 and 11: structure, stratigraphy and tectonic development of the Great Australian Bight region—preliminary report. Bureau of Mineral Resources, Australia, Report No. 88/13.

Wilson, B.R. and Allen, G.R. (1987) Major components and distribution of marine fauna. In *Fauna of Australia. Volume 1A. General Articles*, ed. G.W. Dyne. Australian Government Publishing Service, Canberra, pp. 43–68.

Womersley, H.B.S. (1990) Biogeography of Australasian marine macroalgae. In *Biology of Marine Plants*, eds. M.N. Clayton and R.J. King. Longman Cheshire, Melbourne, pp. 367–381.

Woodley, T.H. and Lavigne, D.M. (1993) Potential effects of incidental mortalities on the Hooker's sea lion (*Phocarctos hookeri*) population. *Aquatic Conservation* 3, 139–48.

THE AUTHOR

Karen Edyvane

SA Research and Development Institute,
P.O. Box 120, Henley Beach,
South Australia 5022, Australia

Chapter 98

THE WESTERN AUSTRALIAN REGION

Diana I. Walker

Western Australia has some 12,500 km of coastline, and about four million square kilometres of Australia's EEZ lies off its coastline. This long coastline has a diversity of environments, ranging from corals and mangroves in the tropics to seagrasses and cool, temperate macroalgal reefs on the south coast. Much of this marine environment is poorly described or understood. This chapter provides a brief description of the coastal geomorphology, habitats and their biogeography, and assesses human impacts on this diverse marine environment. The low population base in Western Australia has resulted in relatively low scales of impacts on the marine environment, but there are examples of degradation, particularly associated with centres of population in the southwest. The dominant factors affecting the marine environment in Western Australia are degradation of habitat, contamination, the introduction of exotic species and harvesting of marine species. More detailed studies of the West Australian marine environment are required if a sound basis for management is to be developed, both within and outside the marine park and reserve system. There have been few coherent, broad-based studies (both in time and space) that have researched the cumulative impact of pollution, siltation, habitat fragmentation and introductions of invasive species on community structure of marine communities. There are some 15 different government agencies which have some responsibility for management of the West Australian marine environment and a more coherent approach to management is required by government agencies.

Seas at The Millennium: An Environmental Evaluation (Edited by C. Sheppard)
© 2000 Elsevier Science Ltd. All rights reserved

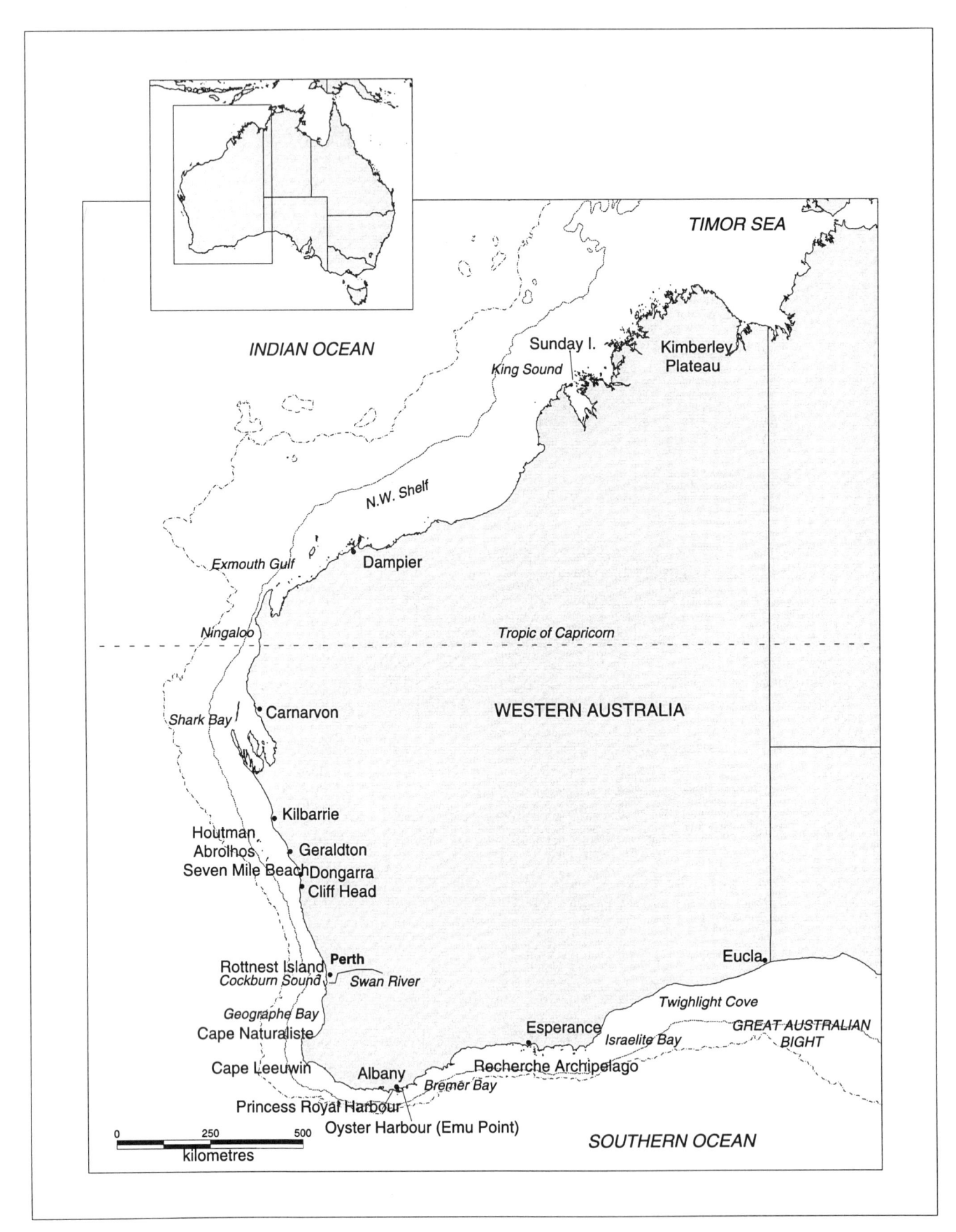

Fig. 1. Map of the coast of Western Australia.

THE DEFINED REGION

The coastline of Western Australia extends 12500 km from the temperate waters of the Southern Ocean at 35°S, to the tropical waters of the Timor Sea at 12°S (Fig. 1). This long coastline has a diversity of environments, ranging from corals and mangroves in the tropics to seagrasses and cool temperate macroalgal reefs on the south coast. These are exposed to different tidal conditions (amplitudes 9 m in the north to <1 m on the west and south coast) (Anon., 1994), substratum types, and exposure to wave energy. Although some areas of the West Australian coast, such as Cockburn Sound, have been the subject of much research, much of the rest of the marine environment is poorly described or understood. The adjacent area under Australia's Exclusive Economic Zone (EEZ) is approximately one quarter of Australia's total of approximately 12 million km^2 (Anon., 1998), and has been very little studied.

This chapter provides a brief description of the coastal geomorphology, habitats and their biogeography. Current uses will be described and current and potential threats to these habitats/uses considered. Recommendations to address these issues will be made. Extensive use of the 1998 Western Australian State of the Environment Report (Anon., 1998) and of the State of the Marine Environment (Zann and Kailola, 1995; Zann, 1996), has been made in compiling the latter section of this review.

Geomorphology of the Coast

The underlying geology of the coast consists of granitic rocks in the south and southwest, with extensive mantling of tertiary limestone, and sandstones in the north west and north. In the southeast of the state, the vertical limestone cliffs of the southern edge of the Nullarbor Plain delimit a narrow coastal plain. For almost 300 km, offshore reefs protect sandy beaches and high foreshore sand dunes from oceanic swell, producing a calmer habitat between the reefs and the shore, suitable for seagrass growth. These reefs are colonised by extensive macroalgal and invertebrate communities. At Twilight Cove the cliffs again approach the sea and follow the coastline to just east of Israelite Bay. From there to Esperance, beaches and seagrass beds are sheltered by the granitic islands of the Recherche Archipelago, 5–50 km offshore.

From Esperance to Albany, sheltered beaches are broken by granite outcrops although occasionally limestone reefs and eroded cliffs occur. Small rivers flow into a number of bays along this 500 km coastline, but they have relatively low discharge rates, particularly during the summer dry season. Their estuaries are also generally small, unlike those further west, and may be seasonally closed by sand bars which gives rise to extreme fluctuations in temperature and salinity, resulting in low biodiversity within them. Offshore of these estuaries, seagrasses which can withstand swell and sediment movement commonly occur, and reef communities are extensive.

From Albany to Cape Naturaliste, limestone overlies granitic rocks for much of the coast. Again the reef communities are extensive. Seagrasses occur in this region in sheltered inshore lagoons protected by offshore reefs.

Geographe Bay, east of Cape Naturaliste, is north-facing and the prevailing southwesterly swell is refracted into the relatively sheltered embayment. The coastal reefs again support extensive reef communities. The embayment has a thin sediment veneer (mean thickness, 1 m) overlying Pleistocene limestone (Searle and Logan, 1978). This provides an ideal habitat for seagrasses (McMahon et al., 1997), and extensive meadows are found to depths of 25 m. A number of estuaries, larger than those further east, also afford habitat for seagrasses and other submerged aquatic plants such as *Ruppia* (Congdon and McComb, 1979), and their associated invertebrates.

The western coastline, from Geographe Bay to Kalbarri, is relatively straight and continuous, as it has been eroded by the action of winds and currents which have built up sand-dunes and bars parallel to the coast. There is also a fringe of limestone reefs running parallel to the coast (Fig. 2) which are relict Pleistocene dune systems composed of aeolianite. These break the Indian Ocean swells, forming relatively calm, shallow (4–10 m deep) lagoons up to 10 km wide, in which the tidal range is small (<1 m), and the waters generally clear. These lagoons are dominated by seagrasses.

From Kalbarri to Steep Point (the most westerly point of the mainland), along Dirk Hartog Island, Bernier and Dorre Islands and up to Point Quobba, there are high cliffs composed of sandstone to the south and limestone to the north. These continue below sea level, and are exposed to full oceanic swell with largely undescribed reef communities, with extensive macroalgae and filter feeders. These cliffs shelter Shark Bay, a large (13,000 km^2) shallow, semi-enclosed embayment. This is an area of intense carbonate sedimentation, which is affected by wind and tidal driven water movement, leading to high turbidity. It also has relatively low water temperatures in winter (down to 13°C, Walker, 1989).

North of Point Quobba, the Pilbara coastline has a low relief with gently sloping beaches, numerous headlands and many small offshore islands. Headlands are composed of isolated patches of very hard haematite-bearing quartzite, which is more resistant to erosion than the surrounding rocks. Normal erosion processes, combined with submergence, have led to a broken, rough coastline. Mangroves become conspicuous. Coral reefs and atolls occur north of the Tropic of Capricorn, where tropical seagrasses are found in lagoons, as well as in mangrove swamps and around islands (Walker and Prince, 1987). There is a progressive increase in tidal amplitude with decreasing latitude. Large tides affect seagrass distributions by resuspending sediments; the high turbidity limits seagrass growth to shallow water. On broad intertidal flats, seagrasses are restricted to those species which can tolerate

Fig. 2. Kimberley coast–Mermaid Island with corals at low tide, showing tidal extent (9 m).

Fig. 3. The seagrass, *Enhalus acaroides*, at high tide at Sunday Island.

high temperatures and desiccation, as well as periodic freshwater inundation from rainfall.

The Kimberley coast is a typical ria (drowned river valley) system, characterised by resistant basement rock, with faults oriented at angles to the shore, creating a rugged coastline. The area is subject to large tidal amplitudes and is remote and sparsely populated, with little information available about the marine habitats. Embayments and sounds grade shorewards into mangrove-covered tidal flats, and there are many offshore islands. Extensive terracing of these expanses of intertidal often results in seagrass, particularly *Enhalus acaroides* (Walker and Prince, 1987) high in the intertidal just below the mangroves (Fig. 3).

Much of the Kimberley landscape is of extraordinary natural beauty. With a vast land area and a small population, the Kimberley has been, until recently, largely unexplored by biologists. Its isolated coastline is devoid of settlement along the 2000 km stretch between Derby and Wyndham. The area is receiving increasing attention from tourists, with increasing activity by small private boats and charter operators. As part of the development of a marine park and reserve system in Western Australia, several areas are being considered as potential marine parks. In addition, some of the areas have been designated as potential Aboriginal reserves. These designations have been based on severely limited data available from the few scientists and other people who have travelled in the area. The only substantial data on marine organisms in the Kimberley relate to salt water crocodile populations and turtles. Marine plants, fish and invertebrates are largely unknown. Recent surveys by the West Australian Museum, the University of Western Australia and the Northern Territory Museum, and by CSIRO, have yet to be published, but will help provide a basis for future research

SEASONALITY, CURRENTS, NATURAL ENVIRONMENTAL VARIABLES

Algal and zoological biogeographers have divided the western coastline into two main provinces, the tropical Dampierian and the temperate Flindersian (Womersley, 1990). The boundary line between them is diffuse for both marine algae and animals; the seagrasses are no exception. The Flindersian Province, which commences at the Abrolhos Islands, extending southwards and eastwards into the Great Australian Bight, is rich in biodiversity. Despite the wide latitudinal range of the coast, the annual sea temperature range is remarkably small. This is due partly to the Leeuwin current, which transports warm tropical water southwards in winter (Rochford, 1984; Pearce and Cresswell, 1985) (Fig. 4), and partly to the absence of any upwelling on the west coast of the continent. There is some influence on the distribution of invertebrates (Wells, 1985) and on algae (Huisman and Walker, 1990) and perhaps on seagrass distributions (Walker, 1991).

The Leeuwin Current, which brings warm tropical water down the west coast of Australia in winter may affect these processes through its transport of reproductive material (both vegetative and sexual) and its effect on water temperatures (Pearce and Walker, 1991; Edyvane, this volume).

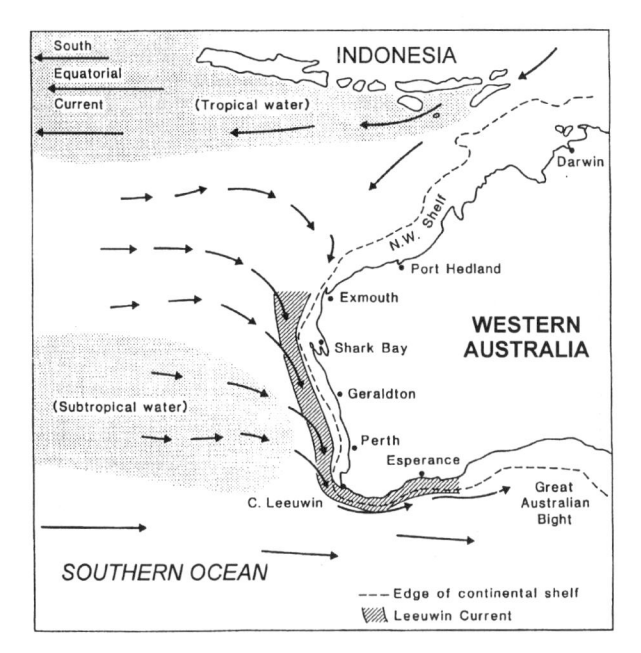

Fig. 4. Typical position of the Leeuwin Current (after Kirkman and Walker, 1989).

Biogeography

On the West Australian coast, marine invertebrates (Morgan and Wells, 1991) and fish (Hutchins and Pearce, 1994) show strong tropical influences, particularly clearly demonstrated on the west and south sides of Rottnest Island. In contrast, the algae and seagrasses show much less strong trends. This may be an effect of habitat availability, but it probably also results from a difference in the dispersal mechanisms of the different types of organisms. There is evidence for long-range dispersal by drift algae, particularly for positively buoyant algae such as *Sargassum* and any attached epiphytes. *Sargassum* species are often widely-distributed (Womersley, 1987) and, for example, *Sargassum decurrens* occurs at Rottnest Island (32°S) and around the northern Australian coast to Keppel Bay (23°S) in Queensland. This may be a consequence of the southerly flow of the Leeuwin Current. The potential for transport of algal propagules by the Current is low in comparison to that for invertebrate larvae, but there is the potential for algal drift to result in range extensions as a consequence of the Leeuwin Current.

MAJOR SHALLOW WATER MARINE AND COASTAL HABITATS

Mangroves

Mangrove forests extend 2500 km around the coast from the Northern Territory border and cover an area of approximately 2517 km^2. Most of the mangrove forests are in good condition. In some places—for example, the Pilbara—the mangrove forests have both local and global significance. Some areas have been lost as a result of salt farms, ports and road works and industrial/urban landfill (Anon., 1998).

Algal Assemblages

The southern Australian algal flora is well documented (Womersley, 1984, 1987) and has one of the richest algal floras in the world (400 genera and 1100 species, Womersley, 1984) (see also Chapter 97). However, the western coast has a reduced diversity in comparison to the south coast, despite the transition from the cold temperate flora to the subtropical and typically Indo-Pacific tropical flora of the northern waters. The relative absence of intensive collections and taxonomic studies of algae in the north of Western Australia does make biogeographical analyses of algal distributions difficult, but recent works by Kendrick et al. (1988, 1990) for Shark Bay, and collections from the Abrolhos Islands (Huisman, 1997) have also improved distribution records.

Macroalgae occur on rocky substrata persisting under a range of water movement conditions. Their depth range is from the intertidal to about 50 m, although this may be greater in clear offshore areas, and less in more turbid inshore waters. In general, the rocky subtotal environment of southwest Western Australia has extensive populations of the kelp *Ecklonia radiata* and *Sargassum* spp. Mixed macroalgal assemblages of foliose red, green and brown algae also occur, particularly on convoluted limestone reef substrata. As the substratum changes from limestone to granite at Cape Naturaliste, cold temperate brown alga become much more conspicuous (Walker, 1991).

Soft Bottom Benthos

There are limited data on Western Australia's soft sediment sea floor communities other than in Cockburn and Warnbro Sounds and on the North West Shelf. Data indicate that the North West Shelf is an area of high biodiversity (Ward and Rainer, 1988).

Fig. 5. The seagrass *Posidona sinuosa* in the lee of a limestone reef at Rottnest Island.

Seagrass

The main habitats for seagrasses are very extensive shallow sedimentary environments that are sheltered from oceanic swell, such as embayments (e.g. Shark Bay, Cockburn Sound), protected bays (e.g. Geographe Bay, Frenchman's Bay) and lagoons enclosed by fringing reefs (e.g. Bunbury to Kalbarri). Seagrasses occupy approximately 20,000 km^2 on the Western Australian coast (Kirkman and Walker, 1989), ranging in depth from the intertidal to 45 m (Cambridge, 1980), making up a major component of nearshore ecosystems. The diversity of seagrass genera (10) and species (25) along this coastline is unequalled elsewhere in the world (Walker and Prince, 1987).

Large, mainly monospecific meadows of southern Australian endemic seagrass species form about one third of the habitat in the coastal regions of Western Australia. These meadows have high biomasses (500–1000 g m^{-2}) and high productivities (>1000 g m^{-2} yr^{-1}: Hillman et al., 1989). Southern Australian seagrasses are unusual in that they occur in water bodies exposed to relatively high rates of water movement—in most of the world, Zostera and other seagrasses are confined to estuaries and lagoons. Nevertheless, Australian species also occur where there is some protection from extreme water movement and most are found in habitats with extensive shallow sedimentary environment, sheltered from the swell of the open ocean, such as embayments (e.g. Shark Bay and Cockburn Sound, Western Australia), protected bays (e.g. Geographe Bay and Frenchman's Bay, Western Australia) and lagoons sheltered by fringing reefs (e.g. the western coast from 33° to 25°S).

Coral Reefs

Extensive coral reefs occur from the Kimberley down as far as the Abrolhos Island, although coral may be found as far south (and east) as Esperance. The remoteness of many of Western Australia's coral reefs has protected them from degradation. Most reefs are considered to be in pristine condition (Anon., 1998).

Compared with those on Australia's east coast, the western reefs have been much neglected scientifically. The coastal reefs are of particular interest as they are distributed down the coast in a series of stepping stones, connected by the southward flowing Leeuwin Current, and resulting as a chain of geographically and environmentally discrete settings for long-distance dispersal of reefal fauna from Indonesia (Veron, 1995).

The reefs off Kimberley are only superficially known as the sea is very turbid and currents strong. Those of Dampier off the Pilbara coast are better described. With an environment ranging from muddy inshore waters to clear offshore waters and strong tidal currents, their diversity of habitats is unmatched in Australia. The 230-km-long Ningaloo Reefs are by far Australia's longest fringing reefs. They have around 300 species of coral, nearly 500 species of fish and over 600 species of mollusc. Lying along the mainland and readily accessible to Perth, until recently they were heavily fished. They were seriously damaged by an outbreak of coral-eating *Drupella* snails in the 1980s. Ningaloo was declared a Marine Park in August 1987, under the Western Australian Department of Conservation and Land Management, in collaboration with the Australian Nature conservation Agency (Veron, 1995). The Houtman Abrolhos reefs, situated 400 km north of Perth, are the most southerly reefs in the Indian Ocean and comprise the south limit of distribution of most Western Australian coral species. However, they show few signs of environmental stress and are amongst the most luxuriant in Australia (Veron, 1995).

Reefs in Western Australia, whether granite in the south, limestone on the west coast or coral reefs in the north, support diverse invertebrate communities. Important commercial species are associated with these reefs, such as the Western Rock Lobster, *Panurilus cygnus*, which is taken between 28 and 32°S, an annual value of more than A$ 200 million (Phillips et al., 1991).

OFFSHORE SYSTEMS

Ashmore Reef, situated 350 km off the Western Australian Kimberley coast on the outer edge of the Sahul Shelf, is a large sedimentary accumulation of reef patches. It has the highest diversity of corals and probably other reef taxa in the west. Scott and Seringapatam Reefs and Rowley Shoals are 'shelf-edge atolls', a reef type not represented elsewhere in Australia. These are visually spectacular due to clear oceanic water and a high tidal range. Apart from a few faunistic studies, they are not well-described (Veron, 1995).

Ashmore Reef is a National Nature Reserve administered by the Australian Nature Conservation Agency. Situated only 120 km from Indonesia, the Reserve is closed to fishing and collecting under a Memorandum of Understanding between the Australian and Indonesian Governments. Poaching of turtles and seabirds and their eggs occurs from time to time (Veron, 1995).

Cocos (Keeling) in the eastern Indian Ocean is Australia's only true atoll, and of interest faunistically because of its isolation. It has significant populations of seabirds and its marine ecology is relatively well known. Christmas Island to the east is a high mountainous island with a plunging shoreline, and a similar reefal fauna (Veron, 1995).

POPULATIONS AFFECTING THE AREA

Despite its great size, Western Australia is amongst the most urban of the Australian states. Around 72% of the population of 1.7 million live within the Perth metropolitan area, and 80% between Geraldton and Esperance in the southwest. Outside the metropolitan area, the coast supports an average of less than 30 people/km, and most of its ecosystems are regarded as virtually pristine. By contrast, around the metropolitan areas human influence has severely degraded some marine ecosystems (Stoddart and Simpson, 1996).

According to the 1998 Western Australian State of the Environment Report (Anon., 1998), human activities most affecting coastal habitats in Western Australia are:

(i) direct physical damage caused by port and industrial development, pipelines, communication cables, mining and dredging, mostly in the Perth Metropolitan and Pilbara marine regions;

(ii) excessive loads of nutrients from industrial, domestic and agricultural sources, mostly in the Lower West Coast, Perth Metropolitan and South West Coast marine regions;

(iii) land-based activity associated with ports, industry, aquaculture and farming, mostly in the Pilbara, Central West Coast, Lower West Coast, Perth Metropolitan and South West Coast marine regions;

(iv) direct physical damage caused by recreational and commercial boating activities including anchor and trawling damage, mostly in the Kimberley, Pilbara, Shark Bay, Perth Metropolitan and Geographe Bay areas. Trawling nets remove sponges and other attached organisms from the sea floor.

The marine environment provides the receiving environment for most surface water from land. The quality of this water is affected by activities and the environment of the catchments through which it flows. Soil and nutrients can be carried by river discharges to coastal waters, causing water quality deterioration. Ground water can also carry terrestrial pollution into the marine environment. Direct discharges such as sewage and/or treated wastewater and industrial outfalls, and accidental discharges such as spills and shipping accidents also influence coastal water quality (Anon., 1998).

These land-based activities, their impacts on ground and surface water and the ultimate movement of these waters into nearshore marine environments are the major human influence on the Western Australian coast. They result in most pollution of the marine environment and the resulting chronic degradation of marine habitat. Degradation of the marine environment leads to reductions in the area of seagrass, corals, mangroves or other habitats.

Growing land and marine-based tourism development in Western Australia and the centralisation of population growth will cause these impacts to increase unless adequate protection and management of the coast occurs.

RURAL FACTORS

Commercial Fisheries

Most fishing methods in Western Australia are suggested to have limited effect on the environment (Anon., 1998). Methods that may significantly affect the environment, for example, dredging and pelagic drift gill-netting, are banned. Other methods, such as trawling, that alter the benthic environment are restricted to prescribed areas. Currently, many of these impacts cannot be quantified (Anon., 1998), thus their effects on biodiversity in the long term are unknown.

At present there are less than 100 trawlers operating in a series of discrete managed fisheries within the total Western Australian fishing fleet of around 2000. The numbers of these trawl licences will be reduced over time. Areas available to trawling within each trawl fishery management area are also restricted. There are significant demersal gill netting closures in areas of high abundance of vulnerable species such as dugong (for example, Shark Bay and Ningaloo).

Fishing does change the overall abundance of the target species and in some cases species which are closely related in the food chain. Quantifying these changes is extremely difficult as the effects from fishing are difficult to isolate from other human-induced effects and from natural variations. For example, the strength of the Leeuwin Current is a natural environmental factor which has been shown to have a significant effect on many of Western Australia's fish stocks (Lenanton et al., 1991).

Pollution, loss of habitat, sedimentation from dredge spoil and agricultural run-off can impact heavily on fish stocks, primarily in nearshore waters and estuaries. Nutrient enrichment of some Western Australian estuaries continues to be a problem. Construction of dams and other physical barriers threatens the abundance and distribution of many freshwater native fish species, particularly in the southwest of the State.

The introduction of exotic marine organisms from ballast water and via the aquarium industry remains an area of concern.

Annual assessments of fish stocks, levels of fishing activity and trends in fish catches indicate that, with only a few exceptions, Western Australia's major commercial fisheries stocks are being managed at sustainable levels. In 1995/96, 16 of the 21 commercial fisheries had breeding stock levels that were adequate to maintain present stocks (Anon., 1998)

Fishing pressure from recreational fishing in Western Australia is increasing and attempts to quantify its impact are being made, so that it can be taken into account in future management decisions. Gaps in information are being addressed in order to assess whether the State's fisheries are being managed in an ecologically sustainable manner. Indicators of ecologically sustainable fisheries are being developed at a national level in consultation with the States (Anon., 1998). Table 1 presents a summary of the status of each major fishery in Western Australia.

The West Australian government agency responsible for fisheries management, Fisheries WA, is investigating environmental management measures such as by-catch reduction techniques and reduction of commercial fishing

Table 1

Stock and exploitation status for major commercial fisheries (Source: Fisheries Western Australia in Anon., 1998).

Fishery	Assessment complete	Exploitation status	Breeding stock levels	Catch (t)	Year
Invertebrates					
A. *Rock lobster*					
Western rock lobster	Yes	Fully exploited	Increasing	9784	95/96
Esperance	Yes	Limited data	Adequate	100	95/96
B. *Prawns*					
Shark Bay	Yes	Fully exploited	Adequate	1880	96
Exmouth Gulf	Yes	Fully exploited	Adequate	771	96
Onslow	No	Limited data	Limited data	94	96
Nickol Bay	No	Fully exploited	Limited data	164	96
Broome	No	Not available		83	96
Kimberley	No	Not available	Limited data	477	96
Fin Fish					
A. *Tropical fin fish*					
Kimberley gillnet and barramundi	Yes*	Fully exploited	Adequate	46	95/96
Pilbara trap and line	No	Fully exploited	Limited data	727	96
Kimberley trap and demersal	No	Fully exploited	Limited data	955	96
Lake Argyle catfish	Yes	Fully exploited	Decreasing	129	95/96
Pilbara trawl	Yes*	Over exploited**	Limited data	3201	96
B. *Estuarine and coastal embayments*					
Shark Bay beach seine and mesh net	Yes*	Fully exploited*	Adequate*	99 (whiting)	96
Exmouth Gulf beach seine	No	Not available	Not appropriate	70	96
Estuarine fisheries	Yes*	F/O exploited	Not appropriate	882	96
Cockburn Sound	No	Not available	Not appropriate	872	96
Princess Royal Harbour/King George Sound	No	Not available	Not appropriate	71.3	96
C. *Temperate marine*					
Western Australian salmon	Yes	Fully exploited	Adequate	2523	96
South west trawl	No	Not Available	Not available	85	96
Shark Bay snapper	Yes	Fully exploited	Adequate	454	96
Southern demersal gillnet and longline	Yes*	F/O exploited	Decreasing	883	95/96
Australian herring trap	No	Fully exploited	Adequate	804	96
West coast demersal gillnet and longline	Yes*	F/O exploited	Decreasing	421	95/96
D. *Pelagic*					
West coast purse seine	Yes	Fully exploited	Adequate	3920	96
Albany/King George Sound purse seine	Yes	F/O exploited	Decreasing (low)	4370	96
Bremer Bay purse seine	Yes	Fully exploited	Decreasing	2340	96
Esperance purse seine	Yes	Under exploited	Adequate	1330	96
Mid-west purse seine	No	Not available	Not available	Cannot report	96
Molluscan					
A. *Pearl oyster fishery*	Yes	Fully exploited	Increasing	592,000 oysters	96
B. *Abalone*					
Abalone Zone 1 and Zone 2	No	Fully exploited	Adequate	216	96
Roe's abalone	No	Fully exploited	Adequate	121.2	96
C. *Scallops*					
Shark Bay	Yes	Fully exploited	Adequate	364	96
Abrolhos Islands and Mid-west trawl	No	Fully exploited	Adequate	229	96

* For key species only.
**Some species only.
F/O Fully or over exploited.

effort through mechanisms such as licence buy-back schemes. Fisheries WA has identified growing recreational fishing effort as an important issue that will be taken into account in the overall management of fish stocks.

Sustaining fish habitat is considered an integral part of sustaining fisheries. One of the management tools made available by the *Fish Resources Management Act* 1994 (WA) is the establishment of Fish Habitat Protection Areas which provide special protection and management for fish and their habitats. This will complement provisions in the *Conservation and Land Management Act* 1984 (WA) for the establishment and management of a marine reserves system.

The extent of illegal fishing by commercial fishers or illegal and inappropriate fishing by recreational fishers is difficult to quantify. It is thought that significant amounts of the higher value species such as abalone, rock lobster and dhufish are sold illegally. Fishing for marron out of season, illegal gill netting and the taking of high value reef fish are currently issues for compliance officers and recreational fishing managers (Anon., 1998).

Loss of Habitat

Loss of habitat for reef communities, especially macroalgae, often results from changes to the environment in which they occur. Macroalgae are limited to the photic zone (Kirk, 1994), and usually require the presence of hard substratum (Round, 1981). Reductions in water quality can lead to a reduction in the depth of the photic zone (Walker and McComb, 1992), and hence to a direct loss of macroalgal habitat. Increasing population pressure in Western Australia leads to increasing pressure on the coast. Development of the coastal zone in the form of construction of marinas, port facilities and canal estates, for example, results in alienation of coastline causing direct destruction of existing communities and indirect changes in hydrodynamics and sedimentation.

All large macroalgae provide habitats for other organisms both *in situ* and as beach cast wrack. Changes to macroalgal distributions resulting from human interference may change the amount and composition of the wrack. Beach wrack is a food source and habitat for many organisms (Koop and Griffiths, 1982; Lenanton et al., 1982; Hansen, 1984; Robertson and Lenanton, 1984; McLachlan, 1985), including at least one vulnerable bird species in Australia, the Hooded Plover (Schulz, 1992; Schulz and Bamford, 1987). Consequences of changes to macroalgal communities may have ramifications for other aspects of biodiversity in associated communities.

Habitat Alienation

The development of the coastline, particularly related to increased population pressure in coastal areas, leads to alienation and fragmentation of habitats available for macroalgae on the coast. Coastal development in Western Australia is localised to centres of population, and takes the form of construction of ports, marinas and groynes. Housing developments alienate coastal terrestrial communities impacting on coastal water quality, whereas canal estates have greater direct impact on the marine environment. All these developments have potential consequences for macroalgal habitats and hence for biodiversity.

Some developments have resulted in direct destruction of reef invertebrate and macroalgal communities by smothering or deterioration in water quality. For example, construction of the causeway at the southern end of Cockburn Sound (Cambridge et al., 1986) destroyed existing reef environments, as well as loss of seagrass habitat due to reduced flushing. The construction of ports and marinas in the Perth Metropolitan area has alienated existing reef habitats, as well as fragmenting the remaining distributions. Subsequent dredging and sediment infill reduces the water quality and results in further losses of reef communities.

Canal estates, where convoluted rock walls surround a coastal waterbody, are increasingly common in Western Australia, as the demand for real estate with water frontage has increased. Such developments could be considered as increasing potential reef habitat, and as providing fish 'nursery grounds' (Morton, 1992). Water quality within such developments, however, is often poor due to inadequate design, resulting in limited flushing capacity (Morton, 1992).

There are many instances in Western Australia where indirect changes in hydrodynamics and sedimentation resulting from coastal development have altered light availability as well as having direct effects on reef environments. Groynes alter sediment transport in the nearshore zone. Consequences for reef communities especially macroalgae involve smothering or scour (Kendrick, 1991; Airoldi et al., 1995).

Some additional solid substratum may be provided by these constructions, some of which will be within the photic zone. These new surfaces are then available for colonisation by macroalgae and invertebrates, providing some additional habitat. These habitats may not be of the same quality (texture) as those they replace, and the processes of colonisation and recruitment may take some time, with fully mature communities often taking more than six years to develop on new surfaces such as artificial reefs (Carter et al., 1985; Palmer-Zwahlen and Aseltine, 1994; Jara and Cespedes, 1994). These processes have not been studied in Western Australia, but overseas studies suggest they are slow. In conjunction with other consequences of development, particularly related to reduced water quality and light availability, these 'new' habitats are subject to extra environmental stress, reducing their suitability as habitat (Walker and Kendrick, 1998). There is unlikely to be an overall net gain in habitat, even with comprehensive mitigation efforts (Cheney et al., 1994).

Table 2

Summary of major human-induced declines of seagrass in Australia. The principal apparent cause is given, but in almost every case other factors are implicated.

Place	Seagrass community	Extent of loss	Cause	Reference
Cockburn Sound, Western Australia	Posidonia sinuosa	3300 ha lost	Increased epiphytism blocking light	Cambridge et al. (1986)
	P. australia			Silberstein et al. (1986)
Princess Royal and Oyster Harbours, Western Australia	Posidonia australis	8.1 km^2 lost (46%)	Decreased light, increased epiphyte and drift algal loads	Bastyan (1986)
	Amphibolis antarctica	7.2 km^2 lost (66%)		Wells et al. (1991)

Areas of the coast of Western Australia are being targeted for aquaculture developments. Aquaculture for algal biomass or fish farms has being shown to result in major environmental impacts both locally (Sommer et al., 1990), but also extensively overseas (e.g. Friedlander and Ben-Amotz, 1991). The effects of fish farms and other aquaculture developments are of concern in tropical areas, such as the Philippines (Fortes, pers. comm., 1996) and the Gulf of Thailand (Ruangchoy, pers. comm., 1996). As yet these impacts are unstudied in Western Australia, but aquaculture ventures are being proposed for sections of the Western Australian coast from the Kimberley to the south coast in areas for which no systematic biodiversity surveys have been undertaken.

COASTAL EROSION, LANDFILL AND EFFECTS FROM URBAN AND INDUSTRIAL ACTIVITIES

Pollution of coastal environments can result in major changes to water quality, either from point or diffuse sources which can influence algal community structure. Marine disposal of sewage from coastal developments, in the form of more (secondary or, rarely, tertiary) or less (primary or none) 'treated wastewater' can contribute nutrients and other chemicals to coastal environments (Lord, 1994). Industrial developments also rely on marine disposal of their effluents.

Declining Water Quality

The Western Australian coastal environment is particularly sensitive to cultural eutrophication. The effects of cultural eutrophication include an increase in frequency, duration and extent of phytoplankton and macroalgal blooms (Lukatelich and McComb, 1989; Malone, 1991), low oxygen concentrations in the water column (Dubravko et al., 1993), shifts in species composition (Lukatelich and McComb, 1986; Lavery et al., 1991), loss of seagrass and benthic vegetation (McMahon et al., 1997; McMahon and Walker, 1998), decrease in diversity of organisms present (Walker and McComb, 1992) and an increase in diseases in fish and waterfowl. Western Australian marine waters are generally low in nutrients and biological productivity. Serious seagrasses losses resulting from increased nutrient loading have occurred in the Albany Harbours, Cockburn Sound and parts of Geographe Bay. Cockburn Sound is the most degraded marine environment in Western Australia, having experienced the second largest loss of seagrass in Australia (80%) (Zann and Kailola, 1995).

The major human-induced declines of seagrass in Australia are summarised in Table 2, with suggested principal causes—in most cases, other factors interact to make the process of loss more complex. The general hypothesis for all these instances of seagrass decline is that a decrease in the light reaching seagrass chloroplasts reduces effective seagrass photosynthesis. The decrease may result from increased turbidity from particulates in the water, or from the deposition of silt or the growth of epiphytes on leaf surfaces or stems (Walker and McComb, 1992). Seagrass meadows occur between an upper limit imposed by exposure to desiccation or wave energy, and a lower limit imposed by penetration of light at an intensity sufficient for net photosynthesis. A small reduction in light penetration through the water will therefore reduce the depth range of seagrass meadows, while particulates on leaves could eliminate meadows over extensive areas of shallower water (e.g. Princess Royal Harbour, Western Australia) (Bastyan, 1986; Walker et al., 1991; Wells et al., 1991).

Increasing turbidity of water above seagrasses may occur directly, by discharge or resuspension of fine material in the water column from, for example, sludge dumping. Indirect effects on attenuation coefficients occur through increased nutrient concentrations resulting from the discharge of sewage and industrial wastes, or from agricultural activity in catchments, which in turn increase phytoplankton biomass reducing light penetration significantly (Chiffings and McComb, 1981; Lukatelich and McComb, 1986). The extent of phytoplankton blooms associated with nutrient enrichment will be determined by water movement, and mixing will dilute nutrient concentrations. Deeper seagrass beds further from the sources of contamination may show no influence of turbidity.

In Cockburn Sound, nutrient enrichment has lead not only to enhanced phytoplankton growth, but also to enhanced growth of macroscopic and microscopic algae on leaf surfaces (Silberstein et al., 1986). Macroalgae dominate over seagrasses under conditions of marked eutrophication, both as epiphytes and as loose-lying species (e.g. the genera *Ulva, Enteromorpha, Ectocarpus*) which may originate as attached epiphytes (Bastyan, 1986). Increased epiphytic growth results in shading of seagrass leaves by up to 65% (Silberstein et al., 1986), reduced photosynthesis and hence leaf densities (Walker and McComb, 1992). In addition, the epiphytes reduce diffusion of gases and nutrients to seagrass leaves.

As photosynthetically active radiation passes through water, it is attenuated by both absorption and scattering. Attenuation is increased by the presence of suspended organic matter (e.g. phytoplankton) and inorganic matter, particularly in eutrophic systems when phytoplankton concentrations are high (Kirk, 1994), thus reducing light penetrating to benthic primary producers. In Cockburn Sound, where this continued for extended periods of time, reduction in density and loss of benthic macrophytes resulted (Cambridge et al., 1986; Silberstein et al., 1986).

The requirement of light by benthic macrophytes makes the presence of submerged aquatic vegetation an indicator of water quality (adequate light penetration) and hence, nutrient status (i.e. low nutrient concentrations) (Walker and McComb, 1992). Light reduction for extended periods, which is common in eutrophic systems, causes loss of benthic macrophyte biomass (Silberstein et al., 1986).

Siltation

Changes in land use practices often results in increased sediments in run-off from land. Larger sediment loads reduce light penetration. Increased sedimentation can result in changes in the abundance and percent cover of macroalgae due to increased sediment deposition or scour (Kendrick, 1991). On shores south of Lirorno, western Mediterranean Sea, which are characterised by high rates of sediment deposition and movement, macroalgae were dominated by turf forming species, whereas erect macroalgae were rare (Airoldi et al., 1995).

These sediments can carry pollutants, which affect benthic communities. Contaminants may be derived from industrial wastes. Marinas are prone to accumulation of toxic chemicals (McGee et al., 1995). The combined effects of sedimentation and pollution affect survival of reef organisms, but can also have chronic effects on growth and reproduction.

Toxic Chemicals

In general, the West Australian coastal environment is not subjected to large-scale inflows of toxic chemicals. The 1998 West Australian State of the Environment report does not consider them a threat (Anon., 1998). Awareness of toxic, human-produced chemicals and their impacts on marine organisms has increased, and such industrial inflows are controlled by Licence Conditions from the WA Department of Environmental Protection. Urban run-off may include such chemicals, but in Western Australia, the runoff is separated from the sewage system. Some direct run-off may still influence groundwater or the coastal environment, and increasing population pressure will result in increased risk of contamination.

Fortunately, the aquaculture industry in Western Australia has avoided the use of antibiotics in fish foodstuffs. The potential effects of antibiotics (Coyne et al., 1994; Kerry et al., 1994). may result in widespread changes in microbial activities, with consequences up the food webs, as well as for nutrient recycling in coastal sediments.

The effects of antifouling compounds are of concern. Tributyl tin (TBT) has been recorded from Western Australian locations (Kohn and Almasi, 1993), highest near marinas and ports. Tributyl tin contamination is present at various levels in all major ports in Western Australia. Tributyl tin contamination is widespread throughout Perth metropolitan marine environment (Anon., 1998). The use of TBT has been banned in Western Australia on vessels longer than 25 m.

Chronic low level effects of pollutants are difficult to detect and may result in long-term changes which may be hard to separate from natural spatial and temporal variability.

Exotic Species

Exotic marine organisms have been introduced to Western Australia via ballast water and hull fouling from shipping. It is estimated that 100 million tonnes of ballast water are discharged into this region's marine waters each year. Currently, controls are only voluntary. Introduced marine species may threaten native marine flora and fauna and human uses of marine resources such as fishing and aquaculture. Knowledge of species introduced and their distribution has recently been updated. The risk of damage to marine biodiversity is largely unknown but international experience suggests that the potential for significant environmental impact is high (Anon., 1998). Displacement of existing flora and fauna by introduced species, intentional or accidental, has been widely reported (Sindermann, 1991).

The 1998 Western Australian State of the Environment Report estimated that over 27 exotic species have been introduced to Western Australia (Anon., 1998). Twenty-one of these are known to have been introduced into Perth metropolitan waters, the most highly visible being a large polychaete worm (Sabellidae Family). This worm occupied up to 20 ha of the sea floor and most of the structures in Cockburn Sound, but its incidence may be declining. It is capable of excluding and out-competing native species (Chaplin and Evans, 1995).

Petroleum Exploration and Production

Petroleum exploration and development in the State is concentrated on the North West Shelf. The area of most concern is the shallow waters surrounding the reefs and islands in the west Pilbara. Petroleum development is routinely assessed by WAEPA. Approval is usually conditional on the companies having adequate Environmental Management Plans, which include monitoring and remedial management components for routine operations, and oil spill contingency plans for accidental spills.

PROTECTIVE MEASURES

The low population base in Western Australia has resulted in relatively low scales of impacts on the marine environment. No marine issues are included in the top two categories of highest priority for the environment of Western Australia (Table 3). The understanding of its marine environment is still patchy and care needs to be taken, especially as the surrounding Indo-West-Pacific region has much higher population pressures, and suffers from much greater human impacts (see elsewhere in this volume).

The establishment of the West Australian Marine Parks and Reserves Authority, in which Marine Conservation Reserves are vested, should help facilitate the development of a comprehensive series of reserves. This process is, however, slow and serious current issues such as extensive plans for Aquaculture developments being implemented by one section of government (Fisheries, WA) whilst another part of government is trying to develop these reserves may compromise its effectiveness. The development of Marine Conservation Reserves within Western Australia must form part of the framework being developed federally for Australia, and it must be assessed to see if it provides the necessary comprehensiveness, adequacy and representativeness for marine conservation to be effective.

On an urgent basis, more detailed studies of the West Australian marine environment are required if a sound basis for management is to be developed, both within the marine park and reserve system and outside it. There have been few coherent, broad-based studies (both in time and space) that have researched the cumulative impact of pollution, siltation, habitat fragmentation and introductions of invasive species on community structure of marine communities (Anon., 1998). Further effort is needed on the influence of these human activities on the whole community, although this will take a long-term commitment to fund these multi-disciplinary studies.

A more coherent approach to management of the Marine Environment is required by government agencies. There are some 15 different government agencies which have some responsibility for management of the West Australian marine environment. As a start, the 1998 State of the Environment Report (Anon., 1998) recommends that State Government should establish a formal framework to co-ordinate environmental management within Perth's metropolitan marine region and between these waters and their land catchments. This should be used as a pilot program for expansion to other areas under pressure from domestic and rural discharges.

ACKNOWLEDGEMENTS

This chapter was reviewed by L. Zann who included some additional material.

REFERENCES

Airoldi, L., Rindi, F. and Cinelli, F. (1995) Structure, seasonal dynamics and reproductive phenology of a filamentous turf assemblage on a sediment influenced, rocky subtidal shore. *Botanica Marina* 38, 227–237.

Airoldi, L., Fabiano, M. and Cinelli, F. (1996) Sediment deposition and movement over a turf assemblage in a shallow rocky coastal area of the Ligurian Sea. *Marine Ecology Progress Series* 133, 241–251.

Anon. (1994) *Australian National Tide Tables 1995*. Aust. Govt. Publ. Serv. Canberra 256 pp.

Anon. (1998) Environment Western Australia 1998: State of the Environment Report. Department of Environmental Protection, Western Australia.

Bastyan, G. (1986) Distribution of seagrasses in Princess Royal Harbour and Oyster Harbour on the southern coast of Western Australia. Dept of Conservation and Environment, Perth, Western Australia. Technical Series 1, 50 pp.

Cambridge, M.L. (1980) Ecological studies on seagrass of south Western Australia with particular reference to Cockburn Sound. PhD Thesis, University of Western Australia. 326 pp.

Cambridge, M.L., Chiffings, A.W., Brittan, C., Moore, L. and McComb, A.J. (1986) The loss of seagrass in Cockburn Sound, Western Australia. II. Possible causes of seagrass decline. *Aquatic Botany* 24, 269–285.

Carter, J.W., Carpenter, A.L., Foster, M.S. and Jessee, W.N. (1985) Benthic succession on an artificial reef designed to support a kelp-reef community. *Bulletin of Marine Science* 37, 86–113.

Chaplin, G. and Evans, D.R. (1995) The status of the introduced marine fanworm *Sabella spallanzanii* in Western Australia: a preliminary investigation, Technical Report 2, Centre for Research on Introduced Marine Pests, Division of Fisheries, Hobart, TAS.

Cheney, D., Oestman, R., Volkhardt, G. and Getz, J. (1994) Creation of rocky intertidal and shallow subtidal habitats to mitigate for the construction of a large marina in Puget Sound, Washington. *Bulletin of Marine Science* 55, 772–782.

Table 3

Priority environmental issues for the Western Australian Marine Environment (Anon., 1998)

Five stars	Four stars	Three stars
Contamination of the marine environment	Degradation of marine habitats	Introduction of exotic marine species
Maintaining biodiversity		

Chiffings, A.W. and McComb, A.J. (1981) Boundaries in phytoplankton populations. *Proceeding of the Ecological Society of Australia* **11**, 27–38.

Commonwealth of Australia (1995) Our Sea our Future: Major Findings of the State of the Marine Environment Report for Australia, Great Barrier Reef Marine Park Authority: Brisbane, QLD.

Congdon, R.A. and McComb, A.J. (1979) Productivity of *Ruppia* : seasonal changes and dependence on light in an Australian estuary. *Aquatic Botany* **6**, 121–132.

Coyne, R., Hiney, M., O'Connor, B., Kerry, J., Cazabon, D. and Smith, P. (1994) Concentration and Persistence of Oxytetracycline in Sediments Under a Marine Salmon Farm. *Aquaculture* **123**, 31–42.

Dubravko, J., Rabalais, N.N., Tyrner, R.E. and Wiseman, W.J. (1993) Seasonal coupling between riverborne nutrients, net productivity and hypoxia. *Marine Pollution Bulletin* **26**, 184–189.

Friedlander, M. and Ben-Amotz, A. (1991) The effect of outdoor culture conditions on growth and epiphytes of *Gracilaria conferta*. *Aquatic Botany* **39**, 315–333.

Hansen, J. (1984) Accumulations of macrophyte wrack along sandy beaches in Western Australia: biomass, decomposition rate and significance in supporting nearshore production. Unpublished Ph.D. thesis, University of Western Australia.

Hillman, K., Walker, D.I., McComb, A.J. and Larkum, A.W.D. (1989) Productivity and nutrient limitation. In *Seagrasses: A treatise on the Biology of Seagrasses with Special Reference to the Australian Region*, eds. A.W.D.Larkum, A.J. McComb and S.A. Shepherd. Elsevier/North Holland, Amsterdam and New York, pp. 635–685.

Huisman, J.M. (1997) Marine benthic algae of the Houtman Abrolhos Islands, Western Australia. In *The Marine Flora and Fauna of the Houtman Abrolhos Islands, Western Australia*, ed. F.E. Wells. Western Australian Museum, Perth. Volume 1, pp. 177–237.

Huisman, J.M. and Walker, D.I. (1990) A catalogue of the marine plants of Rottnest Island, Western Australia, with notes on their distribution and biogeography. *Kingia* **1**, 349–459.

Hutchins, J.B. and Pearce, A.F. (1994) Influence of the Leeuwin Current on recruitment of tropical reef fishes at Rottnest Island, Western-Australia. *Bulletin of Marine Science* **54**, 245–255.

Jara, F. and Cespedes, R. (1994) An experimental evaluation of habitat enhancement of homogeneous marine bottoms in southern Chile. *Bulletin Marine Science* **55**, 295–307.

Kendrick, G.A. (1991) Recruitment of coralline crusts and filamentous turf algae in the Galapagos archipelago: effect of stimulated scour, erosion and accretion. *Journal of Experimental Marine Biology and Ecology* **147**, 47–63.

Kendrick, G.A., Walker, D.I. and McComb, A.J. (1988) Changes in distribution of macro-algal epiphytes on stems of the seagrass *Amphibolis antarctica* along a salinity gradient in Shark Bay, Western Australia. *Phycologia* **27**, 201–208.

Kendrick, G.A., Huisman, J. and Walker, D.I. (1990) Benthic macroalgae in Shark Bay, Western Australia. *Botanica Marina* **33**, 47–54.

Kerry, J., Hiney, M., Coyne, R., Cazabon, D., Nicgabhainn, S. and Smith, P. (1994) Frequency and distribution of resistance to oxytetracycline in micro-organisms isolated from marine fish farm sediments following therapeutic use of oxytetracycline. *Aquaculture* **123**, 43–54.

Kirk, J.T.O. (1994) *Light and Photosynthesis in Aquatic Ecosystems*. Cambridge University Press, Great Britain.

Kirkman, H. and Walker, D.I. (1989) Western Australian seagrasses. In *Biology of Seagrasses: A Treatise on the Biology of Seagrasses with Special Reference to the Australian Region*, eds. A.W.D. Larkum, A.J. McComb and S.A. Shepherd. Elsevier/North Holland, Amsterdam and New York, pp. 157–181.

Kohn, A.J. and K.N. Almasi (1993) Imposex in Australian *Conus*. *Journal of the Marine Biology Association, U.K.* **73**, 241–244.

Koop, K. and Griffiths, C.L. (1982) The relative significance of bacteria, meio- and macrofauna on an exposed sandy beach. *Marine Biology* **66**, 295–300.

Lavery, P.S., Lukatelich, R.J. and McComb, A.J. (1991) Changes in the biomass and species composition of macroalgae in a eutrophic estuary. *Estuarine and Coastal Shelf Science* **33**, 1–22.

Lenanton R.C.J., Robertson, A.I. and Hansen, J.A. (1982) Nearshore accumulations of detached macrophytes as nursery areas for fish. *Marine Ecology Progress Series* **9**, 51–57.

Lenanton, R., Joll, L., Penn, J. and Jones, H. (1991) The influence of the Leeuwin Current on coastal fisheries in Western Australia. *Journal of the Royal Society of Western Australia* **74**, 101–114.

Lord, D.A. (1994) Coastal eutrophication: Prevention is better than cure. The Perth Coastal Water Study. *Water* **45**, 22–27.

Lukatelich, R.J. and McComb, A.J. (1986) Distribution and abundance of benthic microalgae in a shallow southwestern Australian estuarine system. *Marine Ecology Progress Series* **27**, 287–297.

Lukatelich, R.J. and McComb, A.J. (1989) Seasonal changes in macrophyte abundance and composition in a shallow southwestern Australian estuarine system. Waterways Commission, Perth, Western Australia.

Malone, T.C. (1991) River flow, phytoplankton production and oxygen depletion in Chesapeake Bay. Modern and ancient continental shelf anoxia. *Geological Society of America Special Publication* 83–93.

McGee, B.L., Schlekat, C.E., Boward, D.M. and Wade, T.L. (1995) Sediment contamination and biological effects in a Chesapeake Bay marina. *Ecotoxicology* **4**, 39–59.

McLachlan, A. (1985) The biomass of macro- and interstitial fauna on clean and wrack-covered beaches in Western Australia. *Estuarine and Coastal Shelf Science* **21**, 587–599.

McMahon, K. and Walker, D.I. (1998) Fate of seasonal, terrestrial nutrient inputs to a shallow seagrass dominated embayment. *Estuarine Coastal and Shelf Science* **46**, 15–25.

McMahon, K., Young, E., Montgomery, S., Cosgrove, J., Wilshaw, J. and Walker, D.I. (1997) Status of a shallow seagrass system, Geographe Bay, south-western Australia. *Journal of the Royal Society of Western Australia* **80**, 255–262.

Morgan, G.J. and Wells, F.E. (1991) Zoogeographic provinces of the Humboldt, Benguela and Leeuwin Current systems. *Journal of the Royal Society of Western Australia* **74**, 59–70

Morton, R.M. (1992) Fish assemblages in residential canal developments near the mouth of a subtropical Queensland estuary. *Australian Journal of Marine Research* **43**, 1359–1371.

Palmer-Zwahlen, M.L. and Aseltine, D.A. (1994) Successional development of the turf community on a quarry rock artificial reef. *Bulletin Marine Science* **55**, 902–923.

Pearce, A.F. and Walker, D.I. (1991) The Leeuwin Current: An influence on the coastal climate and marine life of Western Australia. *Journal of the Royal Society of Western Australia* **74**, 1–140.

Pearce, A.F. and Cresswell, G. (1985) Ocean circulation of Western Australia and the Leeuwin Current. CSIRO Division of Oceanography Information Service Sheet 16, 4 pp.

Phillips, B.F., Pearce, A.F. and Litchfield, R.T. (1991) The Leeuwin Current and larval recruitment to the rock (spiny) lobster fishery off Western Australia. *Journal of the Royal Society of Western Australia* **74**, 93–100.

Robertson, A.I. and Lenanton, R.C.J. (1984) Fish community structure and food chain dynamics in the surf-zone of sandy beaches: The role of detached macrophyte detritus. *Journal of Experimental Marine Biology and Ecology* **84**, 265–283.

Rochford, D.J. (1984) Effect of the Leeuwin Current upon sea surface temperatures off south-western Australia. *Australian Journal of Marine Freshwater Research* **35**, 487–489.

Round, F.E. (1981) *The Ecology of Algae*. Cambridge University Press, Cambridge.

Schulz, M. (1992) Hooded plover *Charadrius rubricollis*. Dept. of Conservation and Environment - Victoria. Action Statement 9, 1–5.

Schulz, M. and Bamford, M. (1987) The hooded plover, A RAOU

Conservation Statement. RAOU Report 35, pp. 1–11.

Searle, J.D. and Logan, B.W. (1978). A report on sedimentation in Geographe Bay. Research Project R T 2595. Department of Geology, University of Western Australia. Report to Public Works Department, W.A.

Silberstein, K., Chiffings, A.W. and McComb, A.J. (1986) The loss of seagrass in Cockburn Sound, Western Australia. III. The effect of epiphytes on productivity of *Posidonia australis* Hook. F. *Aquatic Botany* 24, 355–371.

Sindermann, C.J. (1991) Case histories of effects of transfers and introductions on marine resources—Introduction. *Journal du Conseil* 47, 377–378.

Sommer, T.R., Potts, W.T. and Morrissy, N.M. (1990) Recent progress in the use of processed microalgae in aquaculture. *Hydrobiologia* 204, 435–443.

Stoddart, J.A. and Simpson, C.J. (1996) Issues in the Western Australian marine environment. The State of the Marine Environment Report for Australia, Technical Annex 3. The Marine Environment, eds. L.P. Zann and D. Sutton. Department of Environment, Sport and Territories, Commonwealth of Australia.

Trowbridge, C.D. (1996) Introduced versus native subspecies of *Codium fragile*: how distinctive is the invasive subspecies *tomentosoides*. *Marine Biology* 126, 193–204.

Veron, J.E.N. (1995) Coral reefs—an overview. The State of the Marine Environment Report for Australia, Technical Annex I. The Marine Environment, eds. L.P. Zann and P. Kailola. Department of Environment, Sport and Territories, Commonwealth of Australia.

Walker, D.I. (1989) Seagrass in Shark Bay—the foundations of an ecosystem. In *Seagrasses: A Treatise on the Biology of Seagrasses with Special Reference to the Australian Region*, eds. A.W.D. Larkum, A.J. McComb and S.A. Shepherd. Elsevier/North Holland, Amsterdam and New York, pp. 182–210.

Walker, D.I. (1991) The effect of sea temperature on seagrasses and algae on the Western Australian coastline. *Journal of the Royal Society of Western Australia* 74, 71–77.

Walker, D.I. and McComb, A.J. (1992) Seagrass degradation in Australian coastal waters. *Marine Pollution Bulletin* 25, 191–195.

Walker, D.I. and Kendrick, G.A. (1998) Threats to macroalgal diversity: Marine habitat destruction and fragmentation, pollution and introduced species. *Botanica Marina* 41, 105–112.

Walker, D.I. and Prince, R.I.T. (1987) Distribution and biogeography of seagrass species on the north-west coast of Australia. *Aquatic Botany* 29, 19–32.

Walker, D.I., Hutchings, P.A. and Wells, F.E. (1991) Seagrass, sediment and infauna—a comparison of *Posidonia australis*, *Posidonia sinuosa* and *Amphibolis antarctica*, Princess Royal Harbour, South-Western Australia I. Seagrass biomass, productivity and contribution to sediments. In *Proc. 3rd International Marine Biological Workshop: the flora and fauna of Albany, Western Australia*, eds. F.E. Wells, D.I. Walker, H. Kirkman and R. Lethbridge. WA Museum Vol. 2, pp. 597–610.

Ward, T.J. and Rainer, S.F. (1988) Decapod crustaceans of the North West Shelf, a tropical continental shelf of north-western Australia. *Australian Journal of Marine and Freshwater Research* 39, 751–65.

Wells, F.E., Walker, D.I. and Hutchings, P.A. (1991) Seagrass, sediment and infauna—a comparison of *Posidonia australis*, *Posidonia sinuosa* and *Amphibolis antarctica* in Princess Royal Harbour, South-Western Australia III. Consequences of seagrass loss. In *Proc. 3rd International Marine Biological Workshop: The Flora and Fauna of Albany, Western Australia*, eds. F.E. Wells, D.I. Walker, H. Kirkman and R. Lethbridge. Western Australian Museum Vol. 2, pp. 635–639.

Wells, F.W. (1985) Zoogeographical importance of tropical marine mollusc species at Rottnest Island, W.A. *W.A. Naturalist* 16, 40–45.

Womersley, H.B.S. (1984) *The Marine Benthic Flora of Southern Australia Part 1*. Government Printer, Adelaide, 329 pp.

Womersley, H.B.S. (1987) *The Marine Benthic Flora of Southern Australia Part 2*. Government Printer, Adelaide, 384 pp.

Womersley, H.B.S. (1990) Biogeography of Australian marine macroalgae. In *Biology of Marine Plants*, eds. M.N. Clayton and R.J. King. Longman, Cheshire, pp. 367–382.

Zann, L.P. and Kailola, P. (eds.) (1995) The State of the Marine Environment Report for Australia, Technical Annex I. The Marine Environment. Department of Environment, Sport and Territories, Commonwealth of Australia.

Zann, L.P. (1996) The State of the Marine Environment Report for Australia. Technical Summary. Department of Environment, Sport and Territories, Commonwealth of Australia.

THE AUTHOR

Diana I. Walker
Department of Botany,
The University of Western Australia,
Perth WA 6907, Australia

Chapter 99

THE SOUTH WESTERN PACIFIC ISLANDS REGION

Leon P. Zann and Veikila Vuki

The south Western Pacific Islands region consists of 1170 islands and thousands of islets and coral reefs. The major island groups are Vanuatu, New Caledonia, Fiji, Tonga and Samoa (including American Samoa and Samoa). The total land area is 55,700 km^2, EEZ area is 4.89 million km^2, and population is 1.3 million. Marine biodiversity is rich and relatively similar throughout, but island groups vary geomorphologically, politically and socio-economically, necessitating separate descriptions in this chapter. Populations are small but growth and urbanisation has been rapid, placing acute pressures on the limited natural resources of the small islands. The major issues include: the effects of increased sedimentation resulting from changing land-use on estuarine and lagoonal environments; loss of wetlands, lagoons and reefs through land reclamations, draining, seawalls, ports and other constructions; declining inshore fisheries production through over-fishing, use of destructive fishing practices and loss of fish habitats; environmental degradation, overfishing in urban areas; declines in endangered turtles, giant clams and other wildlife through overfishing and poor water quality in urban areas; and effects of cyclones and crown-of-thorns starfish on coral reefs. Industrialisation is limited, and pollution is confined largely to ports and industrial sites. Impacts vary greatly within and between groups, with the urbanised, smallest and most densely inhabited islands most affected. Environmental management capabilities are very limited in most countries but regional fisheries and environmental management agencies are relatively effective, but under-resourced. Western conservation mechanisms such as marine national parks are culturally inappropriate and there are few marine protected areas. Because of traditional land/sea tenure, co-management models (government and village) are more appropriate in this region.

Seas at The Millennium: An Environmental Evaluation (Edited by C. Sheppard)
© 2000 Elsevier Science Ltd. All rights reserved

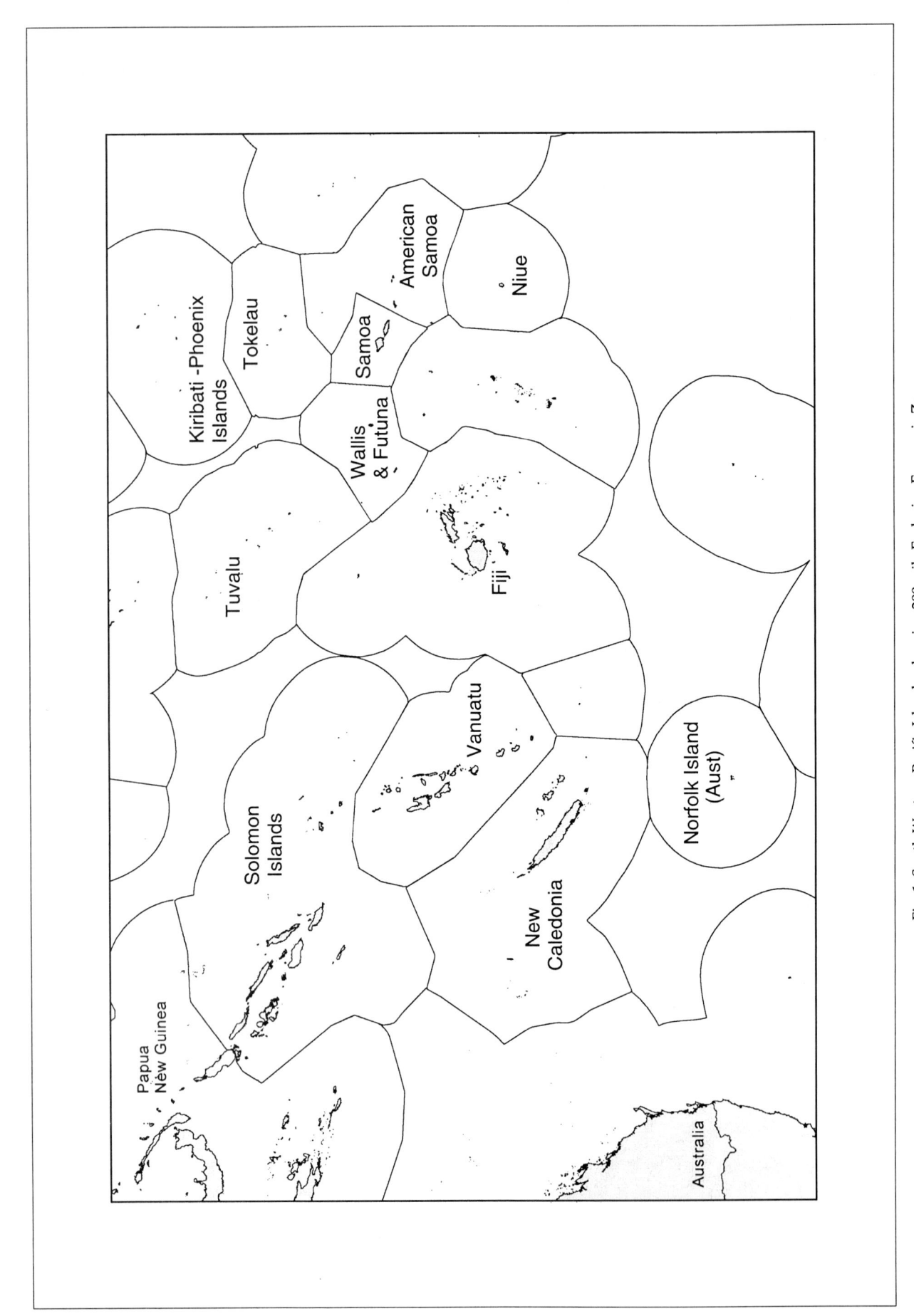

Fig. 1. South Western Pacific Islands, showing 200 mile Exclusive Economic Zones.

THE DEFINED REGION

The region between 12°S–22°S, and 160°W–170°E, termed here the 'South Western Pacific Islands region' (Fig. 1) is strewn with some 1170 major islands, and many thousands of islets and coral reefs. It includes the major island groups of Vanuatu, New Caledonia, Fiji, Tonga and Samoa (Anon, 1984). As it lies within the tropics, coral reefs are very well developed and most of the high islands are surrounded by fringing coral reefs. Platform, patch and barrier reefs have developed in the shallow and generally narrow insular shelf waters. In oceanic waters there are near-atolls and atolls of different ages (IUCN, 1988).

The islands have been inhabited for at least 3500 years. In the west the indigenous people are of Melanesian origin, in the east they are of Polynesian origin and they overlap in the centre, in the Fiji Group. Island sizes, populations and natural resources vary greatly, from the 400 km long, Le Grande Terre (population ca. 170,000) in New Caledonia in the west, to uninhabited Rose Atoll in American Samoa in the east. Most islands are small, with limited natural resources, and most of the rural dwellers continue to live at a semi-subsistence level. Politically, four groups are independent nations and two are colonies or territories (Douglas and Douglas, 1994).

Although the region is relatively homogeneous biogeographically, the island groups are diverse socio-economically and politically, necessitating separate descriptions of each nation or territory. More detailed case studies are provided on specific issues.

Island Groups

Vanuatu

Vanuatu (formerly the condominium of the New Hebrides) (Fig. 2a), is a Y-shaped archipelago of 80 islands, the largest of which is Espiritu Santo. Its land area is around 11,880 km^2 and its EEZ is 650,000 km^2 (Douglas and Douglas, 1994).

New Caledonia

New Caledonia (Fig. 2b) comprises the main island, Le Grande Terre (area: 16,750 km^2), and smaller islands including the Loyalty Group (Lifou, Mare and Ouvea; area: 1970 km^2), and the remote, uninhabited Chesterfields in the west and Walpole in the south. It has a total land area of 19,100 km^2 and an Exclusive Economic Zone of 1,740,000 km^2 (Douglas and Douglas, 1994).

Fiji

The Fiji Group comprises about 844 high islands, cays and islets, 106 of which are inhabited. Fiji (Fig. 2c) has a land area of 18,500 km^2 and an EEZ of 1,290,000 km^2. The major islands are Viti Levu and Vanua Levu (87% of the land area), Taveuni, and Kadavu (Douglas and Douglas 1994).

Tonga

Tonga consists of three main island groups (Vava'u, Ha'apai and Tongatapu Groups) (Fig. 2d). Tonga comprises 171 islands, 37 of which are inhabited. It has a total land area of 700 km^2, and an EEZ of 700,000 km^2 (Douglas and Douglas, 1994).

Samoa Group

The Samoa Group in the sub-equatorial Central Pacific consists of the independent nation of Samoa (known until 1997 as Western Samoa) and the United States Territory of American Samoa.

Samoa (Fig. 2e) consists of the oceanic volcanic islands of Savaii and Upolu, and six smaller islands. It has a land area of 2935 km^2, and an Exclusive Economic Zone of 120,000 km^2, one of the smallest in the South Pacific (Douglas and Douglas, 1994).

American Samoa (Fig. 2f) comprises six islands in the Samoan Archipelago (the Tutuila and Manua Groups), and remote Swains and Rose Atolls. The largest island is Tutuila, which is almost bisected by Pago Pago harbour. The land 197 km^2, and the EEZ is 390,000 km^2 (Douglas and Douglas, 1994).

Biogeography

Biogeographically, the tropical Western Pacific lies within the Indo-West Pacific Marine Province. This is the world's largest and most diverse bioregion, and is of Pan-Pacific Tethyan origin. Species diversity declines eastward from the epicentre of coral reef diversity, the Phillipine–Indonesia region, and declines markedly from west to east through this region (e.g. corals: Veron, 1986). The archipelagos were formed along past and present boundaries of the Pacific and Indo-Australian lithospheric plates and the separate Fiji microplate. Some islands are still tectonically active and are uplifting (Nunn, 1990).

There is no accepted bioregionalisation for the region. Holthuis and Maragos (1992) defined the 'South Pacific Marine Region' as extending east from eastern Australia to Chile, and from about 30°S to 30°N. Dahl (1979) divided this area into 20 biogeographic zones, mainly on national borders. Holthuis and Maragos (1992) proposed a more ecosystem-based classification of 'Island Domains': atoll (low island); with mangrove and seagrass; without mangrove and seagrass; high limestone; high non-limestone.

The defined region might be broadly classified as a large marine ecosystem (LME) or marine domain. The different groups are characterised by similarities in oceanography, plate tectonics, and biogeography. Although there are very strong biotic affinities with the surrounding LMEs in the west and north west (Coral Sea, Great Barrier Reef, New-Guinea–Solomon Islands archipelago), the region tends to be isolated from these by the prevailing southeasterly flowing Sub Equatorial Current. Conversely, although it is downstream from

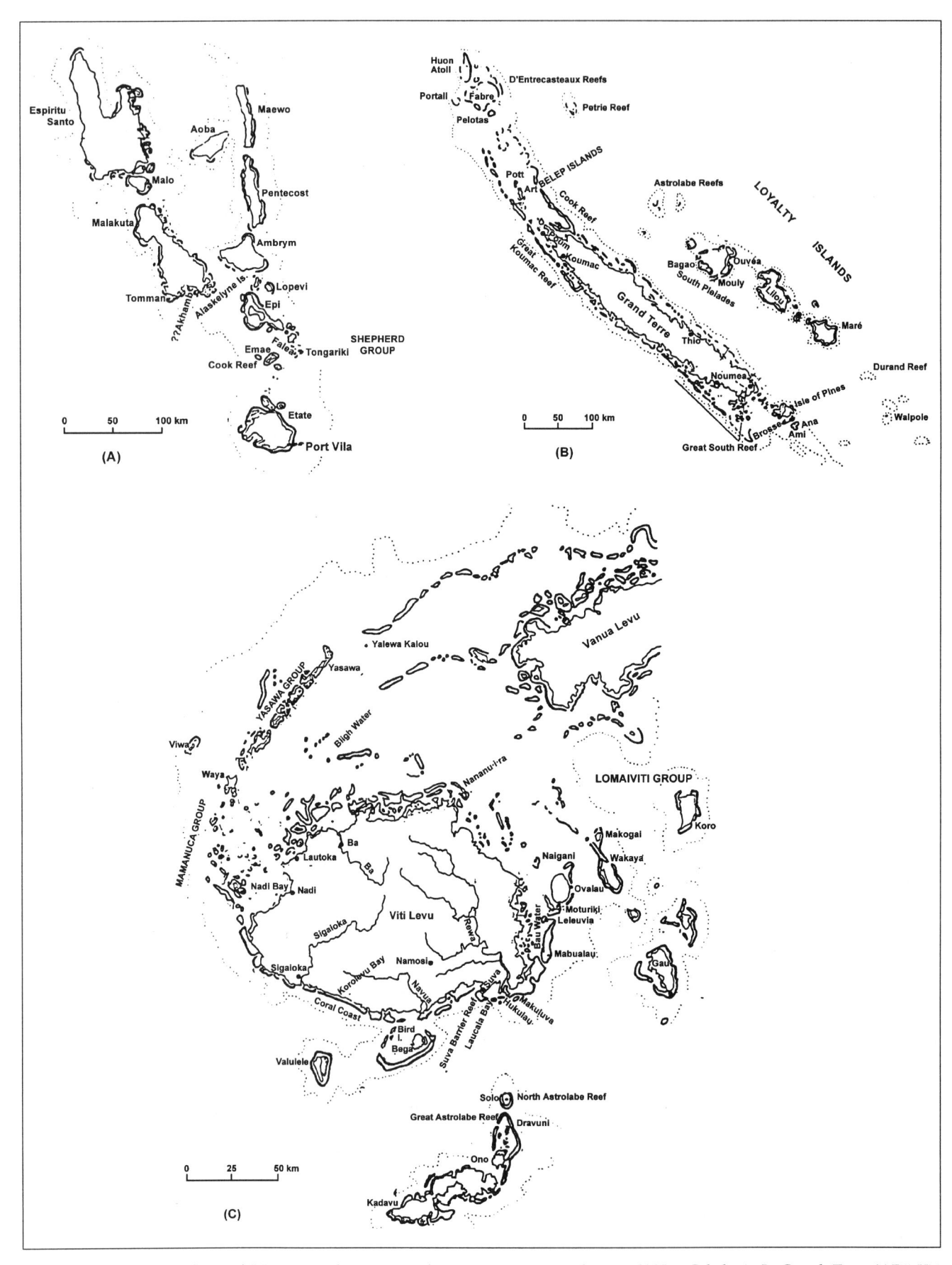

Fig. 2. The main islands in the South Western Pacific countries. *Above*: (a) Vanuatu, central group. (b) New Caledonia, Le Grande Terre. (c) Fiji, Viti Levu. *Opposite page:* (d) Tonga, Tongatapu. (e) Samoa, Upolu and Savaii. (f) American Samoa, Tutuila (after IUCN, 1988)

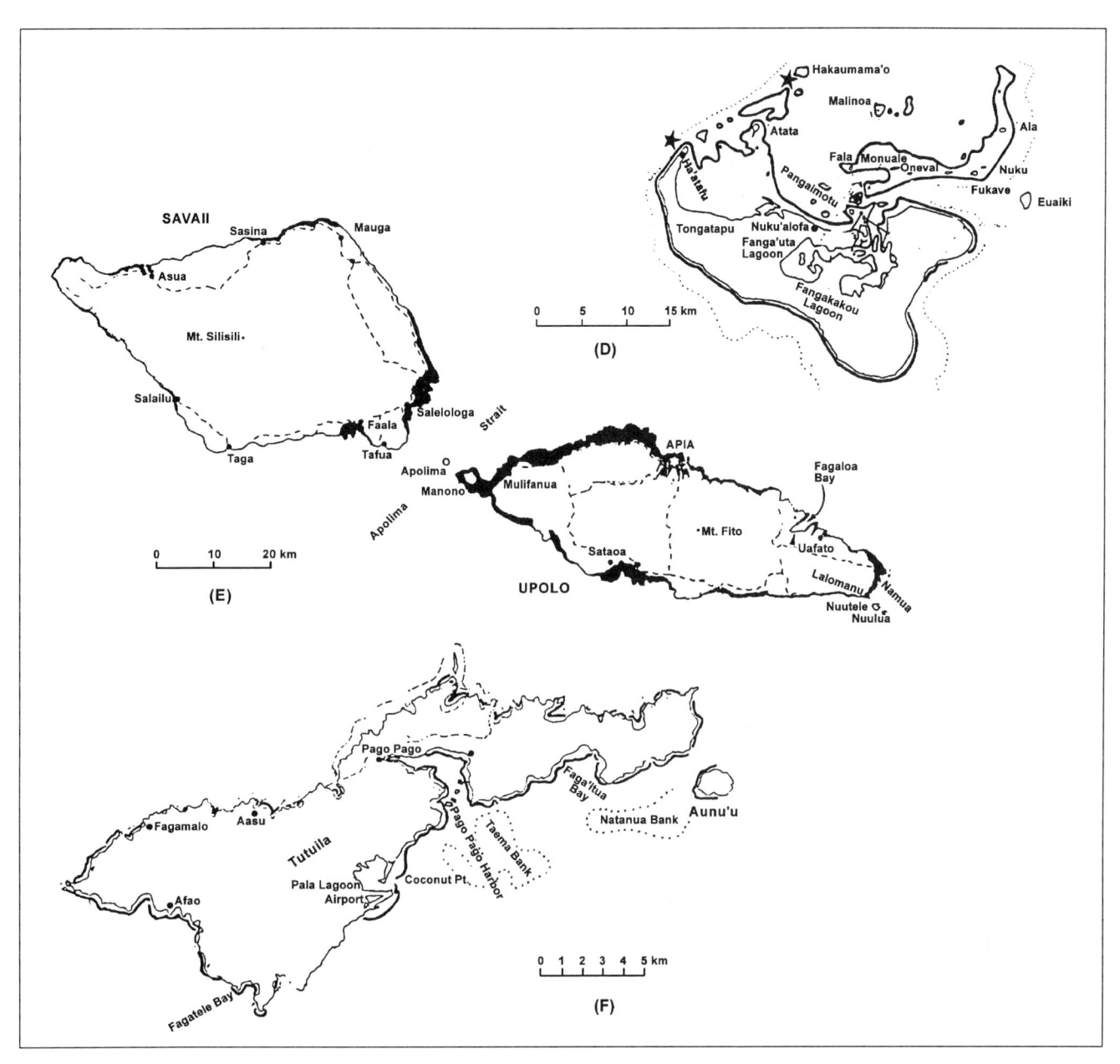

Fig. 2 (continued).

LMEs in the east (Polynesia), it has a much higher coral species diversity as eastern species dispersal has been against the prevailing ocean flow. East of Samoa, species diversity very rapidly declines (e.g. corals: Veron, 1986). There are very low biotic affinities and oceanographic connectivities with most of the shallow water biota to the south (temperate northern New Zealand). As there is no agreed nomenclature for the defined region, it is here termed the 'South Western Pacific Islands Bioregion' or LME.

SEASONALITY, CURRENTS, NATURAL ENVIRONMENTAL VARIABLES

The region's climate is tropical, with island air temperatures rising to 35°C in summer. The surrounding sea (temp. range 21–28°C) and prevailing winter south east trade winds have a moderating effect on the temperature. The east to southeast trade winds are generally light to moderate and predominate in all seasons. They are most persistent in the period June to November (Anon, 1984).

The islands have variable rainfalls according to their topographies; these range from moderate to extremely high rainfall (to 5 m pa). Most have a distinct dry season (May–October) and wet season (November–April). All are subject to very high 24 h rainfall events, often associated with cyclones (Anon, 1984; IUCN, 1988). These cause sudden and severe flooding and severe soil erosion in cultivated areas and sometimes on pristine forested slopes. The high sediment loading and low salinity may have a destructive effect on coral reefs (IUCN, 1988).

Cyclones

Tropical cyclones, characterised by strong winds at the edge of a relatively calm eye, abnormally high sea levels and heavy rain periodically influence the region. Most start as low pressure areas in the monsoonal trough (intertropical convergence zone) in the summer months (November–April). They generally move south and reach a peak between 17 and 22°S, after which they lose intensity. Fiji, for example, experiences an average of one severe cyclone (Saffir–Simpson category 3) every 3–4 years, and one mild to moderate (categories 1–2) every year (Dickie et al., 1991). Vanuatu is particularly prone to cyclones; an average of two cyclones affects the group each year (Anon, 1984). Cyclones have a major effect in sculpting coral reefs.

ENSO

El Niño/Southern Oscillation (ENSO) events have a great effect on the region's oceanography and climate, and on its biota. A recent model developed in the region suggests that during El Niño trade winds weaken and westerly winds commence and intensify with atmospheric convection. The winds move the western edge of the Pacific warm-water pool, located at the equator, some 3000 km eastward as a series of equatorial waves. The rebounding of the equatorial waves off the South American coast pulse back towards the centre of the Pacific basin and water temperatures become cold, giving rise to La Niña (Anon, 1997a). Major changes in water temperatures, currents and nutrients affect migratory tuna and billfish movements in the region. ENSOs also affect cyclone frequency and intensity and rainfall and coastal runoff (Anon, 1997b). Coral bleaching events in the region appear to be related to ENSO.

Currents

The water in the region comes from the South Equatorial and Central South Pacific area. The trade wind drift through the area produces warm westward and southwestward flowing currents (Fig. 1). These average 0.2–0.3 m/s, reaching a maximum velocity in March/April. Locally, currents are tidally induced. At high tide, waves overtopping the reef crests send ocean water across the reefs. At low tide most of this exits through reef passes (Penn, 1982).

Ocean Temperature and Salinity

Ocean temperature ranges from about 21–28°C. Lagoonal temperatures vary, reaching highs of 31°C in late summer. Ocean surface salinity ranges from about 34.9–35.3‰. Lagoonal salinity is lowered by freshwater input but generally remains well stratified. During floods the surface water may be almost fresh, while the bottom layers are oceanic (e.g. Penn, 1982; Dickie et al., 1991).

Tides and Waves

Tides are semi-diurnal with a pronounced diurnal inequality with the lower tides falling during the night in summer, and during the day in winter. For example, in Suva in Fiji, the mean range is 1.1 m, mean high water spring is 1.90 m and mean low water spring is 0.60 m (Dickie et al., 1991).

The prevailing South East trade winds generate prevailing swells from that direction at all times of the year. The high energy deep-water waves have a major effect on coral reef growth in exposed reef fronts (i.e. southwest to east). The windward reefs are subject to high wave energy and good flushing by locally enriched and well oxygenated ocean water. By contrast, inshore reefs protected by a barrier and those on the sheltered (leeward) coastlines have reduced water exchange and are more prone to disturbance from land-based pollutants and siltation (Zann, 1994).

Waves associated with cyclones cause considerable damage to normally protected reefs and coastlines. For example, in Fiji, cyclone storm surges of 2–4 m frequently occur in association with category 2 and 3 cyclones and cause major damage to reefs and coastlines (Dickie et al., 1991).

MAJOR SHALLOW WATER MARINE AND COASTAL HABITATS

The high islands consist mainly of volcanic and plutonic rocks of various ages, which have been extensively eroded under wet tropical conditions. Low islands are coral cays or highly eroded, raised reef limestones. Vegetation varies greatly with biogeography, geomorphology, size and climate, and ranges from beach strand communities to high altitude tropical rainforests. As with the marine biota, terrestrial biota is far more diverse in the larger, continental islands in the west, which are closer to the Australian region, and in Fiji which has geological and floral connectivities to the unique New Zealand region (Watling et al., 1992).

Mangroves

Mangroves are well developed along protected shores and estuaries of the larger islands. Dominant forms are *Rhizophora* and *Bruguira* in seaward areas, with associated *Xylocarpus* and *Exoceocaria* in landward areas. Species diversity is moderately high in New Caledonia, Vanuatu and Fiji, e.g. five species in Fiji (Morton and Raj, 1980), and low in Tonga (three species) (Zann et al., 1984) and Samoa (two species) (Zann, 1991). This contrasts with 35 species in northeastern Australia (Robertson and Alongi, 1995). Fiji has the largest area (45,000 ha) and they are best developed along northern and eastern Viti Levu, the Rewa delta in southern Viti Levu, and along the western coastline and Buca Bay in Vanua Levu (Watling and Chape, 1992). Samoa has the eastern-most mangrove forests in the Pacific (Morton, 1989).

Seagrass

Seagrass beds are well developed in protected subtidal soft shores and lagoons and back reefs. Species diversity is again higher in the west. For example, there are 14 species in adjacent north East Australia (Lanyon, 1986), nine in Vanuatu (Chambers, 1990), four in Fiji (Morton and Raj, 1980), three in Tonga (Zann et al., 1984) and two in Samoa (Zann, 1991). Seagrass beds are very important habitats for fish, turtles and dugongs. Small populations of dugongs are found only in Vanuatu, the most southeastern limit of their distribution (Chambers et al., 1990).

Coral Reefs

Coral reefs are very well developed in each group. Because of their biodiversity and fisheries importance, coral reefs are described separately. The general status of reef communities is discussed here, and specific anthropogenic impacts are discussed later.

Vanuatu

Vanuatu's reefs are little studied. A survey of 16 islands found that reef development depends on aspect (protected to exposed), the tectonic history of the area (some areas are rapidly uplifting, killing reef-top corals), and the impacts of disturbances such as cyclones, coral bleaching events and crown-of-thorns starfish. Outer reefs are dominated by coralline algae and robust plating and branching corals (*Acropora* and Pocilloporidae) with sheltered areas (reef flats, lagoons) dominated by Acropora (especially *A. palifera*) and *Montipora* (especially *M. digitata*). Embayments are dominated by massive *Porites*, branching *P. cylindrica*), various *Acropora* and *Turbinaria* (Done and Navin, 1990). Some 295 species of scleractinian corals (Veron, 1990) and 469 species of fish (Williams, 1990) were identified in the survey, but more are likely to be found.

Pressures and status: Despite low human population pressures, reefs were generally considered to be in poor condition. Severe cyclone damage was evident at 50% of sites and coral bleaching was widespread (Done and Navin, 1990). Active crown-of-thorns starfish outbreaks were seen at 10% of sites, and major mortality of corals was seen at another 20% of sites (Zann et al., 1990).

New Caledonia

New Caledonia's reefs are relatively well studied. There are around 24,000 km^2 of reefs. Le Grande Terre is surrounded by an almost continuous 1600 km barrier reef with a 16,000 km^2 lagoon with depths to 40 m. This is second only to the Great Barrier Reef in size. The west is bordered by two main reef systems with well defined barrier reefs and numerous patch reefs, some with vegetated sand cays. The southern reef, from Noumea to the Isles of Pines, is well developed. The Loyalty Islands in the east are surrounded by extensive fringing and barrier reefs. The isolated, volcanic islands of Hunter and Matthew have some reef development. Chesterfield Islands and Bellona reef in the Coral Sea have very well developed reefs (IUCN, 1988; Coudray et al., 1985). Some 108 species of corals have been described (IUCN, 1988) but the number is expected to be higher. There is a high diversity of invertebrates, of Indo-Western Pacific origins.

Pressures and status: With the exception of reefs off Noumea, reefs appear to be little affected by human activities. There have been chronic outbreaks of crown-of-thorns starfish off Noumea (Conand, 1984) and there are concerns on over-fishing in this area. The effects of increased sedimentation from nickel mining are not well known.

Fiji

There are probably over 1000 reefs of all types in Fiji: fringing reefs, barrier reefs, platform reefs (with and without sand cays), oceanic ribbon reefs, drowned reefs, atolls and near atolls. The largest fringing reef is the Coral Coast reef of Viti Levu, and the largest barrier reef is Mamanuca/Yasawa/Great Sea Reef along the shelf edge of northern Viti Levu and Vanua Levu. Reef assemblages vary according to their aspect (windward or leeward) and position on the shelf. Fiji's marine biota includes almost 200 species of scleractinian corals, 15 zoanthids, 600 shelled gastropods, 250 nudibranch gastropods, 145 bivalves, 60 ascidians, 1198 species of fish from 162 families, five sea turtles, and three species of sea snakes (Zann et al., in press).

Pressures and status: Coral reefs off Suva have been significantly degraded by pollution, elevated nutrients and crown-of-thorns starfish outbreaks. A major outbreak of crown-of-thorns starfish affected the group in 1965–70, and smaller outbreaks in 1978–82, 1983–85 and 1993–present (Zann et al., 1990b, and unpubl. data). Leeward reefs in northern Viti Levu and some other areas have probably been degraded by increased sedimentation from agricultural areas. Most reefs are moderately to heavily fished and stocks of reef fish and invertebrates such as giant clams, trochus and beche-de-mer have been reduced (Zann et al. in press). Mass coral bleaching is occasional. A massive mortality of corals, algae, invertebrates and fish recently occurred between Savu Savu and Buca Bay on Vanua Levu in December/January 1999 and may be more related to extreme sea temperatures and flooding.

Tonga

Tonga's reefs have been little studied. Coral reefs are well developed and most reef types are represented: fringing reefs (from incipient growth on recent volcanic shores, to wide, well developed platforms), platform reefs (on the shelves off each major group), wave-cut raised reefs, and barrier reefs on outer shelves. Around Tongatapu, a raised reef island, there are high energy limestone platforms in the windward south, and well developed fringing and platform reefs with 23 coral cays on the leeward north. A large, almost land-locked, estuarine embayment, Fanga'uta lagoon, partially bisects the island. A broken line of reefs and coral islands extends around 300 km northwards from Tongatapu, to the Ha'apai Group and to Vava'u along the upthrust margin of the Indian–Australian plate. The Ha'apai Group contains the largest area of coral reefs in Tonga. Vava'u is an uplifted and highly eroded reef structure consisting of a maze of islands and lagoons. The recent volcanic islands east of Ha'apai (Late, Fonualei and Toku) and Nuiafo'ou, Niuatopotapu and Niuafo'ou) north of Vava'u are developing fringing reefs. Isolated Minerva Reef, south of Tonga, is an oceanic coral reef (ESCAP, 1990).

No comprehensive collections have been made in Tonga, but species diversity is overtly lower than Fiji and the western groups.

Pressures and status: Reefs off Tongatapu have been significantly degraded by reclamations, runoff and overfishing. Fanga'uta lagoon is particularly affected (Zann et al., 1984). Most reefs in Tonga are subject to moderate to intensive fishing pressure and there are major declines in some species (e.g. giant clams, mullet, deep-water snappers) (ESCAP, 1990).

Samoa Group

Samoa and American Samoa lie only 40 km apart in the same group. Biodiversity is similar, but Samoa, being geologically older, has more developed reef systems.

Samoa

Because of its volcanic origins and recent activity, Samoa has a relatively small area of shelf, and a small area of coral reef (ca. 30,000 ha to the 40 m isobath). Much of the coastline is recent volcanic lava (85–600 years BP) and without reefs, but narrow barrier and fringing reefs have developed around the geologically older parts of most islands. Reef development also varies according to aspect (windward, leeward), and disturbances (cyclones, crown-of-thorns starfish predation, terrestrial runoff, etc.) (Zann, in press).

Pressures and status: Many of Samoa's reefs have been degraded by cyclones, crown-of-thorns starfish outbreaks and human activities (e.g. dynamite and poison fishing, and elevated nutrients and sediments). Cyclone Ofa in 1990 caused massive physical destruction of Samoa's coral reefs and formation of long high rubble banks; these are to 3 m high, 41 km in total length, and comprise 15% of the total outer reef edge. The north western Upolu lagoon becomes periodically eutrophic, and extensive algal blooms have occurred since 1990. Outbreaks of crown-of-thorns starfish have seriously affected Samoan reefs in recent decades. Major outbreaks began off Upolu (1969, 1979 and 1992) and Savaii (1978, 1983 and 1993), causing localised high coral mortality which was great in some areas (over 90%). Recovery has been very slow in some leeward bays, possibly due to disturbances from cyclone and elevated sediments and nutrients (Zann, 1991b; in press).

American Samoa

American Samoan reefs are well studied and a detailed inventory has been produced for all islands (AF and AECOS, 1980). The high islands have extensive areas of fringing reefs, broken by lava cliffs. On Tutuila, two thirds of the coast, largely in the south, have fringing reefs. These are generally narrow (<300 m), but there is a broad reef off Nu'u'uli (1000 m), and an associated, well developed lagoon at Pala. Around 20% of Tutuila's reefs occur around Pago Pago harbour, but most of these have been degraded. There are two submerged coral banks (Taema and Nafanua) off Pago Pago entrance which are the remains of a drowned barrier reef. There are well developed fringing reefs off Ofu and Olosega, in the Manua Group (AF and AECOS, 1980). Rose Atoll in the east is very small (5 ha cay and 640 ha reef) (IUCN, 1988). There have been few detailed collections from the Samoa Group but species diversity in most groups is overtly less than in western groups. Around 1000 species of fish have been identified (Wass, 1980).

Pressures and status: Much of American Samoa's coral reef has also been seriously degraded by cyclones, crown-of-thorns starfish outbreaks and human activities. A massive crown-of-thorns starfish outbreak devastated reefs around Tutuila from 1978–81, and although some 487,000 starfish were eradicated, coral damage was great (Birkeland and Randall, 1979). A massive coral bleaching event occurred in 1994, affecting 60% of corals. Over the past 20 years coral cover off Tutuila has declined from 60% to 10%, and numbers of reef fish have declined by 70%. Pago Pago harbour is heavily polluted by heavy metals and fishing is prohibited (Craig et al., 1995).

OFFSHORE SYSTEMS

Offshore systems are not well known in the region. Island shelves are generally small or absent, and outer reef slopes generally descend rapidly to depths of 1300–3500 m to the continental margins and abyssal plain and deep seabed at 3500–5000 m. Major geological features include the Norfolk Rise which supports New Caledonia and Norfolk Island, the Rotuma Ridge north of Fiji, the Tonga–Kermadec Ridge which supports the Fiji, Tonga and Kermadec Groups, isolated seamounts of different heights, and the New Caledonia, North and South Fiji and Lau Basins, and the Tonga Trench (SOPAC, 1998).

Information on deep-sea communities is very limited. Deep sea thermal vents and ore fields have recently been found in the North Fiji Basin by the German RV *Sonne*. These support rich abundant bacteria, and mussels, shrimps, snails and other species, in densities of up to 250 macro-organisms/m^2 (Anon, 1998a).

Pelagic communities are better understood, particularly in the New Caledonia area. Satellite imagery has been employed to monitor ENSO effects in the region, and a network of telemetry buoys (TRITON, JAMSTEC, GOOS) have been deployed (Anon, 1998b). Studies on the distribution and abundances of tuna and billfish indicate that they migrate many thousands of kilometres, and through many countries' EEZs, and that they are particularly affected by ENSO events (Anon, 1998c).

POPULATIONS AFFECTING THE AREA

The region's total land area is 55,700 km^2; its sea area (combined EEZs) is 4,890,000 km^2; and its population is around 1.3 million (Douglas and Douglas, 1994). Population densities average 23 persons/km^2 of land, and 0.26 persons/km^2 of sea. However, terrestrial and marine resources, and human population densities, vary greatly throughout the region.

Population Trends

Human populations have been rapidly increasing in most countries during the past 50 years. Before European colonisation, island populations were small because of their small areas and limited resources, and populations were limited by famines, birth control, infanticide and warfare. Following European contact, populations initially declined because of introduced diseases and warfare. However, these subsequently recovered with pacification, improved medical care, and more effective fisheries and agriculture (Douglas and Douglas, 1994). Populations are now up to four to five times those of pre-European levels, placing great pressures on limited land and sea resources. Many islands are now no longer self-sufficient and rely on imported foodstuffs and overseas aid. While development and urbanisation has generally been restricted to the national capitals, and while urban populations are small compared to most places around the world, environmental impacts have been locally severe (Zann, 1994).

Group Populations

Each group differs greatly in natural resources, human populations, economic development and marine scientific capabilities.

Vanuatu

Vanuatu (pop. 151,000) is among the least developed nations in the Pacific, and in the world. The indigenous Melanesian people, belonging to 40 different language groups, today make up 94% of the population. Following European contact, Vanuatu's population was greatly reduced by introduced diseases, but has been rapidly increasing in recent decades (ca. 3% per year). Formerly the British and French condominium of the New Hebrides, Vanuatu achieved independence from Great Britain and France in 1980 and is today a parliamentary democracy (Douglas and Douglas, 1994). However, customary lands and laws are protected under the Constitution, and in rural areas customary laws are generally pre-eminent.

In rural areas the people live as subsistence farmers and fishers in small villages or hamlets in customary language areas, and there are few roads and other services. Only the capital, Port Vila (pop. 21,000) on the island of Efate, Luganville (pop. 6,000), on Espiritu Santo, are urbanised. The national income is derived from tourism, fisheries and limited timber and agricultural products (Douglas and Douglas, 1994). The major port and light industrial area is at Port Vila.

Vanuatu does not have a capability in marine environmental management, and very few scientific studies have been undertaken on its marine resources.

New Caledonia

New Caledonia (pop. 173,300) is a French Overseas Territory. The indigenous Melanesian (kanak) people make up around 45% of the population and mainly live in their customary villages in rural areas. The French (38%), other Pacific Islanders (11%) and Vietnamese (5%) mainly live around the capital, Noumea (pop. 60,000) (Douglas and Douglas, 1994).

Case Study: Nickel Mining in New Caledonia

Le Grande Terre contains 40% of the world's known nickel deposits. This lies in ultramafic rocks under around 30 m of weathered material which is scraped off in open-cut mining. In the past century there have been 330 mines dug, 110 million tonnes of ore have been extracted, and 280 million tonnes of overburden removed. There were 130 open-cut mines at the height of the industry in 1970. The open-cut mines are largely based at Thio, Kouaoua and Poro on the east coast of Le Grande Terre, and Nepoui on the west coast (SPREP, 1986).

Nickel mining has destroyed large areas of unique and environmentally sensitive vegetation and massive dumping of wastes down steep slopes has occurred. Sediments have infilled 40 streams and rivers and their estuaries, and smothered agricultural lands. Turbidity has been greatly increased in bays and lagoons. Environmental degradation has been particularly serious around Thio and Dothio Rivers where fine reddish clays clog channels in the deltas, and discolour coastal waters. The Barrier Reef around the island has increased environmental impacts by entraining the silt in the lagoonal waters (Bird et al., 1984).

Environmental impacts assessments have been required for new mines since the 1970s. These require spoil dumps to reduce erosion, and prevent slippage of sediments; terracing; dams and settlement ponds; canalising polluted rivers; and erosion controls on access roads. However, the environmental impacts of abandoned mines will continue for decades to centuries; a huge investment would be needed to stabilise and restore all sites (SPREP, 1986).

New Caledonia is one of the largest nickel producers in the world. As a consequence, Noumea is the most developed and industrialised city in the South Pacific Islands. New Caledonia also supports a major international tourism industry, and small fisheries industries (Douglas and Douglas, 1994). Because of the French expertise, the waters around New Caledonia are well studied.

Fiji

Fiji (pop. 746,000) is the largest and wealthiest island group in the region. Indigenous Fijians constitute 49% of the population. The remainder are Indian (42%), and Chinese, European and other Pacific Islander in origin (9%). Fiji was formerly a British Protectorate and achieved independence in 1970 (Douglas and Douglas, 1994). The capital and only city is Suva (pop. ca. 120,000), on Viti Levu, and there are eight smaller towns. The majority of the indigenous Fijians live in rural villages in customary tribal groups. Fiji Indians live in towns and on small farms leased from Fijians. The major industries are tourism (ca. 270,000 visitors pa), sugar, fisheries, timber and gold (Douglas and Douglas, 1994).

Fiji's marine environment has been better studied than most other groups, although in a generally *ad hoc* manner. Fiji has a relatively good marine research and training capability. The international University of the South Pacific based in Fiji undertakes marine studies programmes, and visiting scientists undertake much of their research within Fiji waters. Reef monitoring programmes are underway off Suva, Astralabe Lagoon (Kadavu), Ovalau Island and other sites.

Tonga

Tonga (pop. 95,000) is an independent Polynesian kingdom and was never colonised. The population is almost entirely Polynesian. The capital town is Nuku'alofa (pop. 30,000) on the main island of Tongatapu Island (pop. 61,000). The Ha'apai (pop. 9,000) and Vava'u (pop. 15,000) Groups have small administrative centres (Douglas and Douglas, 1994).

Tonga has limited natural resources and few exports, and the majority of the rural population are semi-subsistence farmers and fishermen. The major overseas earnings come from remittances from Tongans working abroad (mainly in New Zealand); exports of agricultural products to New Zealand, Hawaii, and Japan; and from a small tourist industry (ca. 20,000 visitors pa). Marine resources are important for the subsistence of many Tongans (ESCAP, 1990).

Tonga's waters are not well studied. Tonga has a small Fisheries Department, which has largely focused on commercial fisheries development, and a small Environment Section.

Samoa (160,000)

Samoa (formerly Western Samoa) is an independent Polynesian nation. The resident population is around five times that of pre-European levels, while at least 100,000 more Samoans reside in New Zealand and the US. Around 110,000 people live on Upolu Island and 50,000 live on Savaii. The capital and only town is Apia (population 30,000), on Upolu. Although overall population density is not high, the recent volcanic activity renders much of the land uninhabitable and the arable coastal fringes are densely populated (Douglas and Douglas, 1994).

With few natural resources and no significant exports, Western Samoa has the lowest per capita GDP in the region, and most of its population live at a subsistence level as farmers and fishermen. The main foreign exchange comes from remittances from Samoans working overseas, tourism (47,000 visitors pa) and agricultural products. In 1990 and 1991 catastrophic cyclones caused massive damage to the environment, buildings and infrastructure. In 1992 a fungal blight appeared which has destroyed the main staple and cash crop, taro (Ward and Ashcroft, 1998). Samoa has also suffered major degradation of coral reefs, and major declines in coastal fisheries over the past two decades (Zann, 1991; in press).

Samoa has a small Fisheries Division and Department of Environment which are largely dependent on international aid.

American Samoa

Neighbouring American Samoa (pop. (51,000) is geographically and ethnically part of Samoa but is an American unincorporated Territory. The majority of the population (47,000) live on the largest island, Tutuila, in rural villages or in the capital, Pago Pago. Around 4,000 live in the Manua Group to the east. Although American Samoa has fewer natural resources than Samoa it is moderately developed. The major income is from tuna canneries (Pago Pago is the base for many of the distant-water tuna vessels in the South Pacific) and remittances from American Samoans in the US and US Social Security (Douglas and Douglas, 1994).

Fisheries and marine environmental management are undertaken by a relatively well resourced Department of Fisheries and Wildlife, largely with US scientific capability. There are few American Samoan marine scientists. Because of its US affiliation, American Samoan waters are well investigated (IUCN, 1988).

RURAL FACTORS

Agriculture and fisheries development varies greatly throughout the region, necessitating a brief summary of each group.

Vanuatu

Vanuatu has the least developed agriculture and fisheries in the region. Land is held communally by language groups. The main commercial crops are copra, cocoa and

Fig. 3. Clearing of catchments and slopes has resulted in serious sedimentation of rivers and lagoons in all groups (L. Zann).

coffee, and cattle grazing (over 110,000 head). Subsistence crops are bananas, cassava and breadfruit.

Fish are only of moderate importance in the subsistence diet: average fish consumption is 18 kg/person/year (Dalzell et al., 1995). There is a small artisanal fishery based on deep-water species (Brouard and Grandperrin, 1985). Total landings of inshore fish are around 2400 tonnes/year (David, 1985). Despite relatively low levels of land use and fish landings, Done and Navin (1990) considered that reefs were moderately impacted by human activities, and that increased siltation from logging of catchments, and nutrient enrichment from agricultural lands and urban sewage, and over-exploitation of reef fish threatened some reefs.

New Caledonia

Around half the land is held as unalienable Melanesian reserves, and half is private rural properties. Main rural land-use is for agricultural crops (copra and coffee) but cattle grazing is more important (120,00 head). Nickel mining is the largest source of income (see case study) (Douglas and Douglas, 1994).

Coral reefs are important for subsistence fisheries in rural areas (ca. 500 tonnes/year), and for recreational or sports fisheries (ca. 4000 tonnes/year), and small-scale commercial fisheries (ca 170 registered boats, 1600 tonnes/year) (Anon, 1985). Fish consumption in rural areas is around 28.6 kg/person/year (ITSEE, 1993). Fisheries yields indicate that the southwest of Le Grande Terre is overfished but that the north is under-fished (Labrosse, 1998).

Fiji

Around 85% of land is owned communally by Fijians, and 10% is freehold. Fijian's lease some agricultural lands to Indians for commercial cultivation. In rural areas Fijians cultivate some of their lands for subsistence for taro, cassava and other crops. The main commercial crops are sugarcane (ca. 400,000 tonnes/year), copra (15,000 tonnes/year), rice (30,000 tonnes/year), ginger (6500 tonnes/year) and other crops. Timber production from native forests has been unsustainable, but in the 1970s Fiji commenced large-scale plantation pine production on the degraded native lands which has now become a major export industry (Douglas and Douglas, 1994).

Clearing of coastal and slope vegetation has caused extensive soil erosion, and sedimentation of rivers, estuaries and lagoons. Mayer (1924) noted that reefs in the Suva area had deteriorated by the 1920s, and attributed this to silting following the clearing of the Rewa watershed. Fish kills are relatively frequent in the major river estuaries. Anecdotal information from fishermen of a characteristic clear blue/green water in the areas of kills suggests acid soil runoff (Zann, unpubl.).

Mangroves, lagoons and coral reefs provide fish for subsistence and sale. Fijians consume around 40 kg of seafood per capita per year. In 1996 the subsistence fisheries landed around 6500 tonnes. The small-scale commercial or artisanal fishery of 1900 small craft landed around 4700 tonnes of fined fish and 2000 tonnes/year of invertebrates.

Fig. 4. Artisanal or small scale commercial fisheries in Fiji supply fresh seafood to the urban market.

Turtles and some giant clam species have been fished to a level which endangers their survival, and stocks of trochus and beche-de-mer species have been fished far past their sustainable yields in the past decade. Stocks of some inshore fish (e.g. mullet, Indian mackerel) have probably also been overfished. The use of illegal fishing techniques (explosives and poisons) remains widespread (Zann, 1992).

Tonga

Tonga traditionally had a feudal system under which the nobility owned the estates and the commoners were farmers and fishers. Upon emancipation in 1875, bush lands and a town allotment were allotted to each taxpayer, but population increases over recent decades have created land shortages. Virtually all coastal forest and arable lands are used for subsistence (taro, coconuts etc.) and cash crops (bananas, copra, taro, pumpkins and vanilla) (Douglas and Douglas, 1994). Most islands are flat and without water courses. Localised erosion and runoff around Fanga'uta lagoon on Tongatapu have resulted in sedimentation and infilling. Elevated nutrients have resulted in local eutrophication, and expansion of seagrass and mangroves, and high mortality of corals in the lagoon (Zann et al., 1984).

Fishing is an important source of protein for Tongans, particularly in the smaller outer islands. Fish consumption ranges from 20 kg/person/year in urban areas, to 50 kg/person/year in some rural islands. There are around 6000 fishermen and 500 motorised fishing craft. The annual catch is estimated to be around 3100 tonnes, of which the majority (1920 mt) comes from shallow reefs. Tonga's subsistence fisheries for invertebrates and fish are based on inshore coral reefs and slopes. The small scale commercial or artisanal fisheries are based on deep-water snappers on the offshore slopes and sea mounts (ESCAP, 1990).

Although fisheries monitoring is very limited, it is apparent that inshore landings have declined over the past two decades. Mullet, previously a major component of landings, collapsed in the late 1970s (Zann et al., 1984). A fishery for deep-water snapper which was developed in the late 1970s declined in the 1980s (ESCAP, 1990).

Samoa Group

Samoa

Around 60% of Samoans are subsistence farmer/fishers. Customary lands are collectively owned by clans. The traditional crops are taro (decimated by a fungal blight in 1992), bananas, coconuts and sweet potato. The main commercial crop is copra (slowly recovering from the 1980s slump), with small amounts of cocoa (Ward and Ashcroft, 1998). The coastal catchments were at least partially cleared for commercial copra plantations (1870s–1960s), for commercial exotic forest plantations (1960s–1970s), and for commercial taro plantations (1960s–1992). The current rate of clearance of forest is 3000 ha/year. Small remnants of coastal rainforest remain on Upolu, and significant stands remain on Savaii. Because of the high seasonal rainfall, erosion is severe, and sedimentation of stream beds and inshore lagoons has been extensive (Taule'alo, 1993). For example, between 1975 and 1981, Apia harbour infilled by an average of 1.5 m due to sediments from the Vaisagano River (Gauss, 1982).

Elevated sediments and nutrients are thought to be a major cause of progressive coral mortality around Apia. Aerial photographs between 1954 and 1991 show an increase in seagrass (particularly *Syringodium*) and a major decline in *Porites* assemblages in outer lagoons around Apia, suggesting the impacts of elevated sediments and nutrients (Zann, in press).

Inshore fisheries are essential in the subsistence economy. There are 9600 fishers, and 55% of households fish for subsistence at least once per week. The average annual per capita consumption is 36 kg in rural areas on Upolu, 50 kg on Savaii and 19 kg in urban areas. Around 2800 tonnes of inshore fish are landed per year; 200 tonnes of inshore fish and 400-500 tonnes/year of offshore fish are sold (Zann, 1991). Inshore fisheries have declined since the 1970s. Market landings of all inshore fish and invertebrate groups have declined since monitoring began in 1986. Between 1984 and 1991, fish consumption, fishing effort and the catch-per-unit effort in surveyed villages declined by 30–35%. In the urban villages the frequency of fish meals remained similar, but meal sizes were smaller (Zann, 1991).

American Samoa

As in Samoa, coastal forests have been cleared for cultivation. Subsistence agriculture production is declining because of the cash economy, and devastation by cyclones and taro blight. Increased sedimentation from poor agricultural and urban land use practices, and eutrophication from sewage and pollution from the fish canneries and slipways is considered responsible for a massive decline of corals which has occurred in the past 20 years. Both fishing effort and catch-per-unit effort have greatly declined in recent decades. As in Samoa, giant clams and turtles have declined because of overfishing (Craig et al., 1995).

COASTAL EROSION AND LANDFILL

Sea-level Rise

Coastal zones of high islands and the low-lying coral islands in the region would be seriously affected by a Greenhouse sea level rise, and climate changes (e.g. in ENSO intensity, cyclone frequency and intensity, rainfall). Low-lying Tonga would be most affected. Predicted rises of 5–35 cm in the next 50 years fall within the existing natural variation of sea level (10–20 cm during strong trade winds,

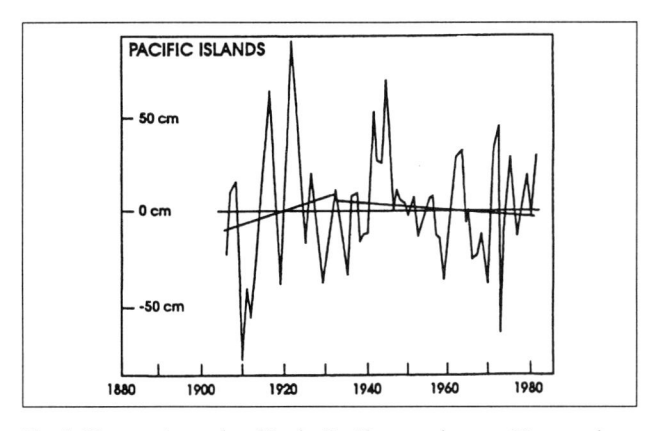

Fig. 5. Changes in sea level in the Pacific over the past 80 years show cyclic changes of up to 50 cm with ENSO events, but no net changes (from SOPAC, 1996).

and up to 50 cm during ENSO) (SOPAC, 1996). Cumulative effects, especially with cyclone storm surges, would be extremely serious (Fig. 5).

Shore Modifications

In all urbanised areas in the region there have been significant reclamations of shores and wetlands for building and other infrastructure, and dredging and blasting of lagoons and reefs for deep-water ports. Low-lying islands have been particularly affected. Coastal modifications have caused significant loss of wetlands, lagoons and coral reefs, with consequent loss of biodiversity and fisheries production. The following summarises coastal modifications and their consequences in the region. A more detailed case study is provided on Samoa.

General Coastal Modifications

In Vanuatu, coastal modifications have been minimal, and are largely confined to the capital, Port Vila. Several kilometres of seawalls have been constructed in the port area, and along Erakor Lagoon. In the short term, natural impacts from cyclones and seismic movements of the coastline have had far greater impacts on coral reefs (Done and Navin, 1990).

In New Caledonia, coastal modifications (e.g. harbour construction, channel dredging, and seawalls) are largely restricted to the Noumea area in southwest Le Grande Terre (IUCN, 1988).

In Fiji coastal modifications are largely restricted to the Suva area, and the other coastal towns. On Suva Reefs coral sands have been mined for several decades for cement manufacture. Around 3000 ha of an original 45,000 ha of mangroves have been cleared for sugar cane, housing and industrial sites (Watling et al., 1992).

In the low-lying islands of Tonga, coastal modifications have been extensive in some areas. On Tongatapu, almost the entire fringing reef off Nuku'alofa was reclaimed in the 1980s for a major port, and much of the coastline of Fanga'uta lagoon has been reclaimed for building. Beach sand mining is a major problem around the island as sand for construction purposes is lacking. In Ha'apai and Vava'u, solid-fill causeways have been constructed to link some adjoining islands, creating barriers to migrating fish, and causing beach erosion (Zann et al., 1984).

In American Samoa there have been extensive port works, dredging and filling in Pago Pago Harbour with consequent loss of fringing reef. Many of the original coral transects surveyed by Mayor in 1918 have completely disappeared as the result of dredging and filling (Dahl and

Case Study: Coastal Modifications in Samoa

Coastal modifications have had serious environmental impacts in Samoa. There are few natural harbours and major engineering and dredging has been necessary in Apia on Upolu, and Asau on Savaii, and minor works at Aleipata and Saleaologa. Because of inundation of coastal areas during cyclones Ofa (1990) and Val (1991), high sea walls have been constructed around the capital, Apia, and lower walls around many rural villages. These have altered long-shore drift and created adjacent beach erosion, as well as occluded tidal movement into wetlands. Of 11.5 km of shores around Apia, 5.5 km have been reclaimed (Zann, 1991).

Although ample elevated lands are available inland for building in Apia, there have been major reclamations of lagoons and fringing reefs for building (particularly the New Market, government administrative buildings, and hotels). Mulinu'u Peninsula has been actively eroding, necessitating continuing seawall construction. Vaiusu Bay is mined for coral sand while long causeways have restricted water flows. Extensive areas of mangroves have been lost at Moata'a and Vaiusu Bay, the largest wetlands in the Samoa Group (Zann, 1991).

Coastal road construction has degraded most of the nation's estuaries. Because of problems acquiring lands from the customary owners, the modern coastal roads constructed since the 1970s often follow Crown Lands along the upper shore. As the estuaries are small and insufficient funds were available for bridges, most were spanned by solid-fill causeways with either several small drain pipes, or with a high concrete culvert (0.9 m high, where tidal range is 1.2 m). Some occluded estuaries became very stagnant and were completely filled by the villagers. Of eight estuaries examined, six were almost completely occluded by causeways. Cross-sectional areas and tidal flows were reduced to below 10%. As a consequence of tidal occlusion, above the causeways salinity is lowered, algal blooms are frequent and mangrove wetlands have been replaced by freshwater marshes, inhabited by introduced tilapia fish (Gaskell, 1998). Loss of wetlands habitat and access for juvenile fish has probably substantially contributed to the dramatic decline of mullet and several other species in Samoa over the past 20 years (Zann, 1991).

Lamberts, 1977). Likewise, large areas of Pala Lagoon, the only estuary in American Samoa, were reclaimed during construction of the present airport (IUCN, 1988).

EFFECTS OF URBAN AND INDUSTRIAL ACTIVITIES

In most Pacific Islands urbanisation is restricted to the capital, and rural areas are generally little developed. While urban areas and populations are small by world standards, because of the small size of many islands, environmental impacts may be locally very severe.

There are only two cities in the region, Noumea (pop. 60,000) in New Caledonia and Suva (pop. 120,000) in Fiji. There are also relatively few large-scale industrial activities along the region's coastlines. All national capitals have ports and some light industrial development (e.g. boat-building, engineering and steel fabrication, petroleum storage and distribution, coconut oil production, sand dredging, quarries) (Fig. 7). Heavy industries, in Noumea and Suva include sugar mills and refineries, mineral processing, timber wood-chipping, cement manufacturing, soap and paint manufacturing, ports and shipyards. Environmental regulations are limited throughout the region, and pollution and environmental impacts range from localised to extensive. (See case study examining the major pollution issues in Fiji.)

Shipping

Shipping plays a major role in communications and commerce in the island nations. Shipping includes traffic transiting through the region on international shipping routes, international traffic into the major ports, inter-island vessels, industrial fishing vessels, and pleasure craft such as liners and yachts.

As the region is strewn with submerged coral reefs, groundings are not infrequent. Reported groundings of

Fig. 6. Construction of coastal roads has affected most of Samoa's estuaries. This road on Savaii has occluded the estuary (lower left) and caused beach erosion (lower right).

Fig. 7. In urban areas wetlands and reefs have been degraded through construction of port works and sea walls. In Samoa sediment runoff infilled the harbour by 1.5 m over 6 years. Sediments from port reclamations and runoff have degraded the only marine park, at Palolo Deep (upper right) (L. Zann).

ships in the region between 1976 and 96 were: Vanuatu 3; New Caledonia 2; Fiji 34; Tonga 0; Samoa 18. Groundings largely occurred on reefs at harbour entrances, with major causes being unfamiliarity with passages, excess speed, failure of gear and heavy weather (Preston et al., 1997).

Fortunately, there have been no major oil spills in the region. The nearest oil spill response groups are located in Australia and Singapore but several regional countries have local oil spill contingency plans (Preston et al., 1997). However, given the lack of equipment and great distances involved, an effective response to a large spill is unlikely.

Industrial Fisheries

Almost one million tonnes of tuna and billfish are caught in the South Pacific each year by Korean, Taiwanese, Japanese, US and domestic vessels. Major canneries are based in Fiji and American Samoa. In Fiji in 1996, 4700 tonnes of yellowfin, big-eye and other tuna were caught by long line for export, largely to Japan and the US. A further 4400 tonnes of tuna, mainly skipjack, were canned by PAFCO for export, although not all of these were caught in Fijian

> ### Case Study: Pollution in Suva Harbour
>
> The enclosed, protected nature of harbours makes them prone to pollution as emissions of all kinds from urban and industrial areas eventually find their way into the adjacent waters. Because of the number of light and heavy industries with uncontrolled discharge of wastes into Suva harbour, levels of pollution are moderate to very severe.
>
> Studies have found elevated biochemical oxygen demand (BOD), elevated nutrients (nitrates and phosphates), high suspended solids, pH, and high coliform bacterial levels in discharges from a large number of light and medium industries in the city (Cripps, 1992). Levels of heavy metals are also high, and are equal to the most polluted harbours in Australia (Naidu et al., 1989). Levels of tributyl tin (TBT), an extremely toxic compound used in antifouling paints, were higher in Suva Harbour than in any port reported in the literature. Indiscriminate use of TBT in anti-fouling and anti-mildew paints may also have contributed to the high levels reported (Stewart and deMora, 1990).
>
> Suva's dump for domestic and industrial wastes is situated in mangroves in the harbour. Leaching of a broad spectrum of hazardous chemicals into the adjacent water course is inevitable with the high tropical rainfall, posing a potentially very serious health problem for consumers of shellfish and fish in the area. Studies of sediments and shellfish from near the dump showed high levels of faecal coliform, mercury, zinc and lead (Cripps, 1992).
>
> About 40% of urban residences lack proper sanitation (Watling et al., 1992). Because of the geology of the Suva area (impervious clay marl), overflow of septic tanks and contamination of local streams is common. Studies in Suva Harbour, Laucala Bay, Vatuwaqa River and Samabula River have shown extremely high coliform levels along beaches and in local shellfish (which are frequently consumed raw) (Cripps, 1992; Zann, 1992).

waters (MAFF, 1996). Pago Pago in American Samoa is the major base for US purse seiners and canneries produce around 15 million cases of canned tuna, valued at over US $250 million per year (Douglas and Douglas, 1994).

Most tuna vessels are large distant-water fishing vessels between 30–80 m in length, and use long-lines, pole-and-lines and purse seines. Ocean drift netting, which had a large by-catch of turtles, cetaceans and fish, has been banned because of environmental impacts. Collection of live-bait used in pole-and-lining may localised environmental impacts as juvenile reef species may be caught. The by-catch of the present tuna fisheries is generally unknown.

PROTECTIVE MEASURES

Island and coral reefs are environmentally very sensitive and over thousands of years Pacific Islanders evolved practical techniques to conserve the resources of their coastal and sea estates. The most important of these was sea tenure under which the land-owning group owned, used and defended their marine estates. Overfished areas could be reserved ('tabooed') or spelled from fishing. Some species prone to overfishing (e.g. turtles) were reserved for priests and chiefs. Typically, clans were responsible for looking after their totems, often marine species. Although many of these customs broke down with colonisation, some have persisted. With declines in inshore species, villages in some areas (Samoa, parts of Fiji) are re-asserting sea tenure to protect their dwindling resources from outsiders (Zann and Vuki, 1994).

Modern Conservation Practices

At the national level, all countries have developed fisheries and environmental legislation based on Western models. Typically, fisheries legislation controls commercial fisheries through licences, restricts the use of certain gears and mesh sizes, imposes size limits on certain species, prohibits the use of poisons, prohibits the taking of certain species at certain times (e.g. nesting turtles), or in certain areas (marine protected areas). In some places clearing of mangroves is prohibited under fisheries or forestry legislation.

American Samoa and New Caledonia are covered by the comprehensive US and French environmental legislation and other environmental legislation. Environmental legislation has progressively been developed in the other countries to control certain developments, protect certain species and areas, and regulate the use, disposal and discharges of certain substances. Some form of environmental impact assessment is required for designated developments, but this is generally less comprehensive than in Western nations.

Marine Protected Areas

Marine protected areas are an important 'catch-all' strategy to conserve marine biodiversity, particularly in the absence of scientific knowledge and monitoring. The following marine protected areas have been declared in the region (Kelleher et al., 1995).

Vanuatu

President Coolidge and Million Dollar point Reserves (WW II ship wreck and wreckage, Espiritu Santo); Narong Marine Reserve (Malakula area, 160 ha).

New Caledonia

Reserve Speciale de Faune et Flore de I'Ilot Maitre; Reserve Speciale Tournante de Faune Marine; Reserve Speciale Marine Yves Merlet; Parc Territorial du Lagon Sud (Amedee, Bailly, Canard, Laregnare, Signal); Reserve Speciale de Faune se I'ile de pam; Reserve Speciale de Faune de I'Ilot Lepredour; Southern Botanical Reserve (sections contain marine habitats).

Fiji

None

Tonga

Fanga'uta and Fangakakau Lagoons Marine Reserve; Ha'atafu Beach Reserve; Hakaumam'o Reef Reserve; Malinoa Island Park and Reef Reserve; Pangaimotu Reef Reserve.

Samoa Group

Samoa: Palolo Deep Marine Reserve; 20 village fishing reserves. *American Samoa*: Fagatele Bay National Marine sanctuary; Rose Atoll National Marine Sanctuary; American Samoa National Park.

These MPAs are too few in number and too small to be effective. In most cases there is no education and enforcement, and MPAs are protected in name only. Marine protected areas have probably not been accepted by Pacific Islanders because of customary sea tenure or fishing rights of local communities. For example, in Fiji, Dahl (1979) recommended 50 MPAs but none has ever been established, largely because of customary fishing rights issues.

Evaluation of Marine Environmental Protection Measures

While legislation and policies for marine environmental protection exist, there are inadequate resources for their effective implementation and enforcement. Most central governments have limited human and financial resources for effective governance, and have little real influence in more isolated rural areas. Bureaucracies are inefficient, and a level of corruption in some nations prevents good government. Environmental conservation is not a national priority in most countries. Government environment protection agencies are small, generally with only a few local and foreign aid staff, and have very limited resources. There are also few Pacific Islanders with specialist training and skills in marine environmental management.

Given the limited national capabilities, international fisheries and environmental support agencies have been established to service the South Pacific Islands. The University of the South Pacific, South Pacific Regional Environment Program, Secretariat for the South Pacific and Forum Fisheries Agency provide an important regional and international framework for island governments, and a technical and information base. These agencies are funded by former colonial powers and Pacific rim countries but there are insufficient resources to effectively service the increasing demands of the region.

Because of a generally low level of local awareness on environmental issues, there are few grass-root environment conservation organisations or lobby groups in the island nations. However, in recent years international conservation organisations such as Greenpeace and Worldwide Fund for Nature (WWF) have established Pacific offices and are active in raising local awareness on issues such as pollution, overfishing and loss of biodiversity.

Most environmental projects in the region are funded by international aid. While these have generally been beneficial in raising local awareness of environmental issues, foreign advisers often try to impose Western conservation ethics, and environmental protection models such as national parks. Many advisers also have little cultural awareness, and little appreciation of the realities of life for subsistence dwellers. Most aid projects are of short duration, and often languish when the overseas experts leave.

Given the lack of government resources and the pre-eminence of customary and local government, recent environmental initiatives have focused at the community level. For example, in Samoa a co-management arrangement was established in 1995 between the national government and village fishing rights holders for fisheries and marine environmental management. Villages were given responsibility for fisheries management and assisted by government and international aid (AusAID) to develop plans of management for village waters and were assisted financially to develop alternative resources. By mid-1998, some 40 villages had entered into the arrangements, and 20 village fishing reserves had been established (King and Faasili, 1998).

CONCLUSIONS: STATUS OF MARINE ENVIRONMENT AND MAJOR ISSUES

It is not possible to accurately assess the status of the region's marine environment as an insignificant area has ever been surveyed, and there are very few monitoring programmes in place. However, it is evident that coastal environments have been significantly degraded in many areas through clearing of forests and poor agricultural practices, and that over-fishing is widespread. Moreover, the current level of protection is poor, and the deterioration can only continue in the foreseeable future.

REFERENCES

Anon (1984) *Pacific Islands Pilot*. Vol. 2. Hydrographer of the Navy.
Anon (1985) *Pacific Islands Year Book*. Fiji Times, Suva.
Anon (1986) *Pacific Islands Year Book*. Pacific Publications, Sydney.
Anon (1997a) New theory clarifies the complex mechanisms of El Nino. *SPC Fisheries Newsletter* 82, 19–20.
Anon (1997b) Tuna follow el Nino. *SPC Fisheries Newsletter* 82, 17.
Anon (1998a) Large ore field revisited by German RV Sonne in the North Fiji basin. *SOPAC News* 15, 11.
Anon (1998b) International initiatives. *SOPAC News* 15, 7.
Anon (1998c) Ocean fisheries program. *SPC Fisheries Newsletter* 85, 11–16.
AF and AECOS (1980) American Samoa coral reef inventory. A & B. US Army Corps of Engineers and American Samoa Development

Planning Office, Honolulu.

Bird, E.C., Dubois, J.P. and Iltis, J.A. (1984) The impacts of opencast mining on the rivers and coasts of New Caledonia. United Nations University, Tokyo.

Birkeland, C. and Randall, R.H. (1979) Report on the *Acanthaster planci* (Alamea) studies on Tutuila, American Samoa. Office of Marine Resources, Government of American Samoa, Pago Pago, American Samoa.

Brouard, F. and Grandperrin, R. (1985) Deep-bottom fishes of the outer reef slope in Vanuatu. SPC/Fisheries 17/WP. ORSTOM Centre, Noumea.

Chambers, M. (1990) Seagrass communities. In: *Vanuatu Marine Resources*, eds. T.J. Done and K. Navin, pp. 92–103. Australian Institute of Marine Science, Townsville.

Chambers, M., Nguyen, F. and Navin, K. (1990) Seagrass communities, In: *Vanuatu Marine Resources*, eds. T.J. Done and K. Navin, pp. 92–103. Australian Institute of Marine Science, Townsville, pp. 37-65.

Conand, C. (1984) Distribution, reproductive cycle and morphometric relationships of *Acanthaster planci* (Echinodermata, Asteroidea) in New Caledonia, western tropical Pacific. In: *Proceedings of the 5th International Echinoderm Conference, Galway, 24–29 September 1984*, eds. B.F. Keegan and B.D.S. O'Connor, pp. 499–506. A.A. Balkema, Rotterdam.

Coudray, J., Thomassin, B.A. and Vasseur, P. (1985) Comparative geomorphology of New Caledonia and Mayotte barrier reefs (Indo-Pacific Province). *Proc. 5th Int. Coral Reef Congr. Tahiti* 6, 427–432.

Craig, P., Green, A. and Saucerman, S. (1995) Coral reef troubles in American Samoa. *SPC Fisheries Newsletter* 72, 21.

Cripps, K. (1992). Survey of point sources of industrial pollution entering the port waters of Suva. Engineering Dept., Ports Authority of Fiji, Suva, Fiji.

Dahl, A.L. and Lamberts, A.E. (1977) Environmental impact on a Samoan coral reefs: a resurvey of Mayor's 1917 transect. *Pacific Science* 31, 309–319.

Dahl, A. (1979) Marine ecosystems and biotic provinces in the South Pacific Area. In *Proceedings of the International Symposium on Marine Biogeography in the Southern Hemisphere*. New Zealand DSIR Information Series No 137.

Dalzell, P.T., Adams, T. and Polunin, N. (1995) Coastal fisheries in the South Pacific Islands. In Background Paper, South Pacific Inshore Fisheries Workshop Manuscript Collection, eds. P. Dalzell and T. Adams. South Pacific Commission, Noumea, New Caledonia. Vol. 1, No. 30.

David, G. (1985) Le perche villageoise a Vanuatu: recensement 1. Moyens de production et production globale. Report Mission ORSTOM, Vanuatu.

Dickie, R.M., Fenney, R.M. and Schon, P.W. (1991) *The Planning Manual*. Ports Authority of Fiji, Suva, Fiji.

Done, T.J. and Navin, K. (1990) Shallow-water benthic communities on coral reefs. In: *Vanuatu Marine Resources*, eds. T.J. Done and K. Navin, pp. 10–36. Australian Institute of Marine Science, Townsville.

Douglas, N. and Douglas N. (1994) *Pacific Islands Yearbook*. Fiji Times, Suva.

ESCAP (1990) *Environmental Management Plan for the Kingdom of Tonga*. ESCAP, Bangkok, Thailand.

FAO. (1991). Environmental guidelines for dredging and river improvement in Fiji. Draft. Project TCP/FIJ/0154.

Gaskell, I. (1998) The impacts of causeway construction on estuarine hydrology, flora and fauna in Samoa. Unpubl. Integrated Project Report, Southern Cross University, Lismore, Australia.

Gauss, G. (1982) Apia Harbour survey, Samoa., 19 January–6 February 1981, 20–31 March 1992. CCOP/SOPAC Cruise Report 55. CCOP/SOPAC. Suva, Fiji.

Holthuis and Maragos (1992) Marine biological diversity conservation in the Central/South Pacific region. Report to CNPPA.

ITSEE (1993) Budget consomation de menages 1991, principaux resultats. Vol. 1. Insitute of Statistics and Economic Research, New Caledonia.

IUCN (1988) *Coral Reefs of the World. Vol. 3. Central and Western Pacific*. UNEP, IUCN, Gland, Switzerland and Cambridge, UK.

Kelleher, G., Bleakley, C. and Wells, S. (1995) A global representative system of marine protected areas. Great Barrier Reef Marine Park Authority, World Bank, IUCN, Washington, USA.

King, M. and Fa'asili, U. (1998) Community-based management of fisheries and the marine environment. In: *Proceedings of Pacific Science Congress 1997, University of South Pacific, Suva*, pp. 115–125.

Labrosse, Y. (1998) Assessment of commercial fish resources in the lagoon of the northern province, New Caledonia. *SPC Fisheries Newsletter* 85, 22–31.

Lanyon, J. (1986) *Seagrasses of the Great Barrier Reef*. Great Barrier Reef Marine Park Authority, Special Publication Series 3, Townsville, Australia.

MAFF (1996) Fiji Fisheries Division Annual Report 1996. Ministry of Agriculture, Fisheries and Forests, Fiji.

Mayer, A.G. (1924). Structure and ecology of Samoan reefs. *Pub. Carnegie Inst.* 340, 1–25.

Morton, J. (1989) Western Samoa's shores. Western Samoa Fisheries Division, Apia, Western Samoa.

Morton, J. and Ray, U. (1980) The shore ecology of Suva and south Viti Levu. Institute of Marine Resources, University of the South Pacific, Fiji.

Nunn, P. (1990) Coastal processes and landforms in Fiji: their bearing on Holocene sea level changes in the South and western Pacific. *Journal of Coastal Research* 6, 279–310.

Naidu, S., Aalbersberg, W.G.L., Brodie, J.E., Fuavao, V.A., Maata, M., Naqasima, M., Whippy, P. and Morrison, R.J. (1989). Water quality studies on selected South Pacific lagoons. UNEP regional Seas Reports and Studies No. 136, UNEP; and SPREP reports and Studies No. 49.

Penn, N. (1982). The environmental consequences and management of coral sand dredging in the Suva region, Fiji. PhD Thesis., University of Wales.

Preston, G.L., Gillett, R.D., McCoy, M.A. and Lovell, E.R. (1997) Ship groundings in the Pacific Islands region. Issues and guidelines. SPREP, Apia, Western Samoa.

Robertson, A.I. and Alongi, D.M. (1995) Mangrove ecosystems in Australia: structure, function and status. In *State of the Environment Report for Australia. Technical Annex 1*, eds. L.P. Zann and P. Kailola, pp. 119–134. Dept. Environment, Sport and Territories, Canberra.

SOPAC (1996) Coasts of Pacific Islands. SOPAC Miscellaneous Report 222, Suva, Fiji.

SOPAC (1998) Coastal states and the continental shelf. South Pacific Applied geoscience Commission, Number 11, Suva, Fiji.

Stewart, C. and deMora, S.J. (1990). A review of the degradation of tri(n-butyl) tin in the marine environment. *Environmental Technology* 11, 565–570.

SPREP (1986) The effects of mining on the environments of high islands: a case study of nickel mining in New Guinea. Environmental Case Study 1. South Pacific Regional Environment Program.

SPREP (nd). Coral reef survey: Vava'u, Tonga. SPREP/Royal Geographic Society of Great Britain. (unpubl. ms)

Taule'alo, T. (1993) Western Samoa. State of the environment. SPREP, Apia, Western Samoa.

Veron, J.E.N. (1986) *Corals of Australia and the Indo-Pacific*. Angus and Robertson, Sydney.

Veron, J.E.N. (1990) Hermatypic corals. In: *Vanuatu Marine Resources*, eds. T.J. Done and K. Navin, pp. 37–65. Australian Institute of Marine Science, Townsville.

Ward, G.G. and Ashcroft, P. (1998) Samoa: mapping the diversity. Institute of Pacific Studies, University of the South Pacific, Suva and

National University of Samoa, Apia.

Wass, R.C. (1980) The shoreline fishery of American Samoa—past and present. UNESCO Seminar, Motopure Research Centre, University of Papua New Guinea. Regional Office for Science and Technology for South East Asia, Jakarta, Indonesia.

Watling, R. and Chape, S. (1992) Environment Fiji—the national State of the Environment Report. IUCN, Gland, Switzerland.

Watling, R. et al. (1992) A state of the environment report for Fiji. National Environmental Management Project, Department of Town and Country Planning, Fiji.

Williams, D.McB. (1990) Shallow-water reef fishes. In: *Vanuatu Marine Resources*, eds. T.J. Done and K. Navin, pp. 66–76. Australian Institute of Marine Science, Townsville.

Zann, L.P. (1991). The inshore resources of Upolu, Western Samoa. Coastal inventory and fisheries database. Field report 5. SAM/89/002 FAO, Rome.

Zann, L.P. (1992) The state of the marine environment of Fiji. Unpubl. report to National Environmental Management Project, Environmental Management Unit, Suva, Fiji.

Zann, L.P. (1994) The status of coral reefs in the South Western Pacific Islands. *Marine Pollution Bulletin* 29, 52–61.

Zann, L.P. (1998) The inshore resources of Savaii, Western Samoa. Samoa Fisheries/AusAID, Apia, Samoa.

Zann, L.P., Kimmerer, W.J. and Brock, R.E. (1984) The ecology of Fanga'uta lagoon, Tongatapu, Tonga. Uni. South Pacific and Uni, Hawaii Seagrant Cooperative Program, Hawaii.

Zann, L.P., Ayling, A. and Done, T.J. (1990) Crown-of-thorns starfish. In: *Vanuatu Marine Resources*, eds. T.J. Done and K. Navin, pp. 104–118. Australian Institute of Marine Science, Townsville.

Zann, L.P., Brodie, J. and Vuki, V. (1990) History and dynamics of the crown-of-thorns starfish *Acanthaster planci* (L.) in the Suva area, Fiji. *Coral Reefs* 9, 135–144.

Zann, L.P., Vuki, V.C., Lovell, E., and Nyeurt, A. and Seeto, J. (in press). Biodiversity of Fiji's marine environment. In: Proceedings of Pacific Science Congress, University of South Pacific, Suva.

Zann, L.P. and Vuki, V. (1994) Marine environmental management and the status of customary marine tenure in the Pacific Islands. In *Traditional Marine Tenure and Sustainable Management of Marine Resources in Asia and the Pacific*, Proceedings of the International Workshop 4–8 July, 1994, eds. G.R. South, D. Gouley, S. Tuqiri and A. Church, pp 62–70. International Ocean Institute, University of South Pacific, Suva, Fiji.

Zann, L.P. and Vuki, V. (1999) The status and management of subsistence fisheries in the South Pacific. Ocean Yearbook 1999 (eds. Borgeses, E.M. Chircop, A., McConnell, M.L. and Morgan, J.R.). Halifax, Nova Scotia.

THE AUTHORS

Leon P. Zann
*Southern Cross University,
Lismore, NSW 2480,
Australia.*

Veikila Vuki
*Marine Studies Program,
University of the South Pacific,
Suva, Fiji*

Chapter 100

NEW CALEDONIA

Pierre Labrosse, Renaud Fichez, Richard Farman and Tim Adams

New Caledonia lies on the edge of the tropics 1500 km east of Australia. Its total land covers 19,100 km^2, though its EEZ and territorial waters extend to over 1,450,000 km^2. The main island, Grande Terre, is surrounded by a barrier reef 1100 km in length enclosing a lagoon approximately 23,400 km^2 in area and up to 50 m deep. The coral reefs around New Caledonia can be considered to be largely pristine because of low anthropogenic pressure resulting from low population density. However, urbanisation is the main factor affecting demand for land on the coastline and mangroves have suffered both deliberate destruction and the indirect effects of development. Generally, there is a growing acknowledgement of the economic and environmental value of most habitats, and protective regulations are now in place.

New Caledonia has one of the lowest overall population densities in the South Pacific region, with 10.4 inhabitants per km^2. Traditionally, the people (first Melanesian, then European) depended mostly on agriculture for subsistence. Most people live on the coast, and more than 60% live in the Greater Noumea area. Subsistence fishing is traditionally important, but its practice has changed considerably with the advent of efficient equipment and gears, and professional fisheries also target high value species for export. Impacts on fishery stocks and habitats have not been extensively researched. More recent is the development of prawn aquaculture and industrial tuna long-lining. Seafood is the second export commodity, after nickel ore, but the economy generally is highly dependent on financial support from France.

Coastal erosion has been apparent all around New Caledonia, more markedly on eastern shores and smaller islands. The phenomenon is too widespread however to be attributed to human activity alone, so that several natural or climatic effects are also seen.

In the past few years, numerous regulations have been enacted to protect economically or ecologically important species such as rock lobsters, mullets, mangrove crabs, rock oysters, coral, aquarium fish, sea turtles, dugongs and trochus. These are usually based on size limits, closures or exclusion zones. Also regulated are fishing techniques, together with the conditions for operating commercial vessels. These, with development regulations, help to maintain the generally good environmental conditions and sustainable practices that exist in New Caledonia.

Seas at The Millennium: An Environmental Evaluation (Edited by C. Sheppard)
© 2000 Elsevier Science Ltd. All rights reserved

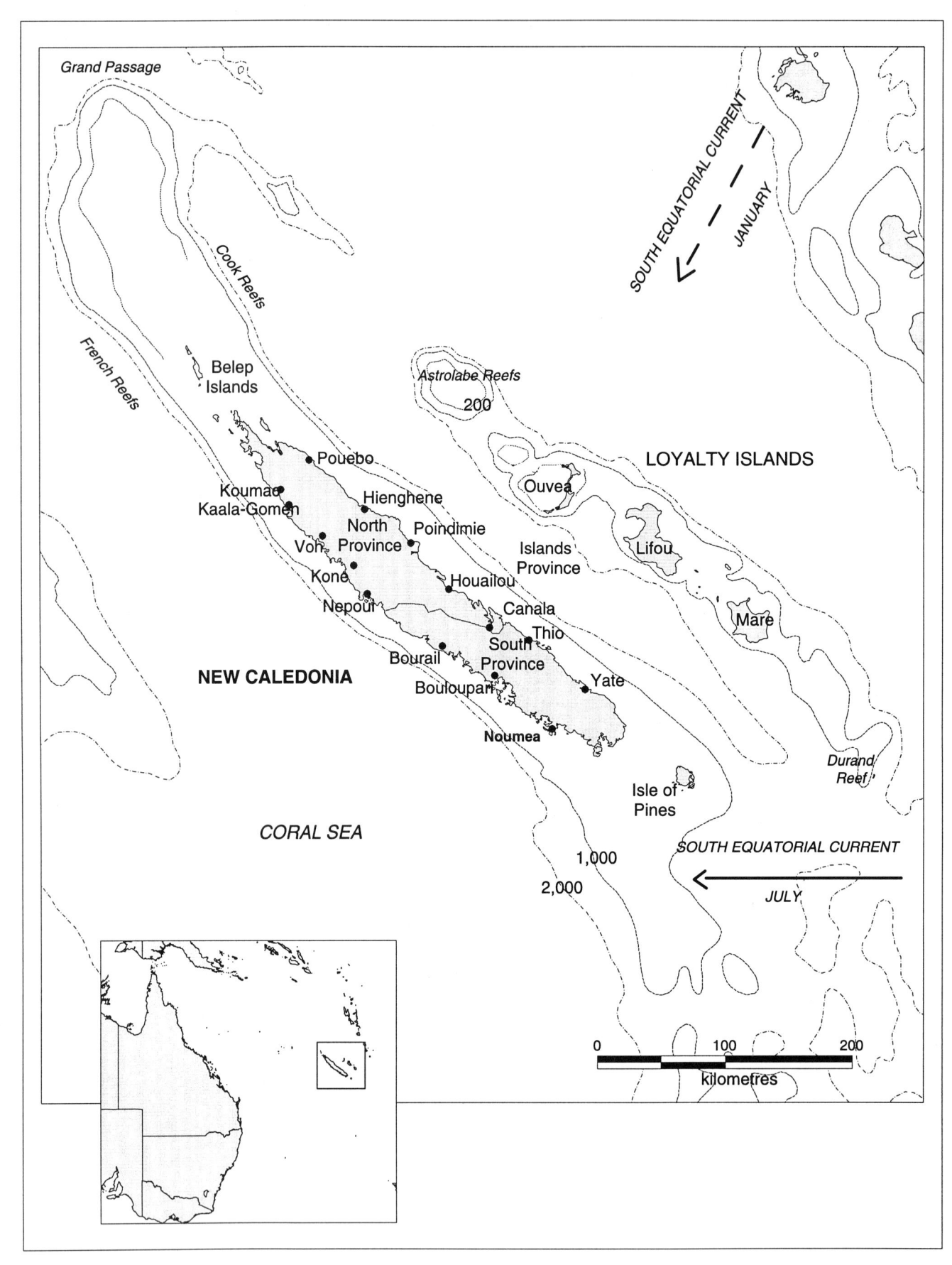

Fig. 1. Map of New Caledonia.

THE DEFINED REGION

New Caledonia lies just to the north of the Tropic of Capricorn in the South Pacific Ocean (Fig. 1) between latitudes 19–23°S and longitudes 163–168°E. It is situated 1700 km northwest of New Zealand, 1500 km east of Australia and 18,000 km from France with which it is politically affiliated (Jost, 1998). It comprises Grande Terre (the main island), three island groups (Loyalties, Isle of Pines and Belep) and some uninhabited islets.

Its total land area is 19,100 km^2, though its maritime area (EEZ and Territorial Waters) extends over 1,450,000 km^2. The main island, Grande Terre, is large (400×50 km) and is surrounded by a barrier reef 1100 km in length around a lagoon approximately 23,400 km^2 in area (Testaud and Conand, 1983) and up to 50 m deep. Its width can vary from a few hundred metres to 65 km. On the west coast the continuous reef reaches the sea surface and clearly defines the boundary between the lagoon and the ocean; on the east coast the barrier reef is not so continuous as on the west coast and is non-emergent in places, resulting in a less distinct separation between ocean and lagoon. The various reef and coastal systems bordering the lagoon around *Grande Terre* present a wide variety of biological and geomorphological characteristics (Richer de Forges, 1998). Lagoon water quality is chiefly determined by the rate of renewal resulting from ocean inflow and outflow, and the lagoon also undergoes changes due to material discharged by rivers.

SEASONALITY, CURRENTS, NATURAL ENVIRONMENTAL VARIABLES

Climate and Ocean

New Caledonia stands on the edge of the tropics and the climate is considered to be semi-tropical. Two distinct seasons result from the displacement of the inter-tropical convergence zone (ITCZ), a major oceanographic feature in the Pacific. In summer (November–March) the ITCZ is centred on the 15th southern parallel, bringing hot and wet air over New Caledonia and raising ocean temperatures to 26–27°C. High ocean water temperatures are responsible for the build-up of the tropical depressions and cyclones that affect New Caledonia with an average frequency of 1.5 cyclones per year. In winter (June–August) the ITCZ shifts to the northern side of the equator allowing polar depressions to extend north and these occasionally reach New Caledonia. During this period, ocean temperatures decrease to 22–23°C. During transition periods the climate is usually drier especially from September to November. Trade winds from the southeast are a common feature and are more frequent (250 days per year in Noumea) and sustained (5–10 m s^{-1}) in the southern part than in the northern part and the Loyalty Islands. During winter, occasional westerly gales are accompanied by rainfall.

The ocean surface circulation around New Caledonia is under the influence of the easterly equatorial counter-current. However superficial currents generated by the dominant trade winds partly oppose this general circulation scheme.

In addition to these general influences, climate in this area of the South Pacific is strongly impacted by the climatic process known as the El Niño southern oscillation (ENSO). During El Niño events the ITCZ shifts along the equator and the associated water mass known as the West-Central Pacific warm pool is displaced eastward (Delcroix and Lenormand, 1997). As a consequence, lower average temperatures (1–1.5°C) are observed during summer but the main ENSO effect in the area is the occurrence of long periods of dryness and a significant decrease in cyclone frequency.

As a consequence of environmental conditions, ocean waters in the New Caledonia basin area are poor in nutrients, and chlorophyll concentrations are below 0.2 mg m^{-3} (Dandonneau and Gohin, 1984). The New Caledonia EEZ is oligotrophic.

Lagoon

The coral barrier reef of New Caledonia encloses a large lagoon. Water temperature in the lagoon tends to be 1–2°C lower in winter and 1–2°C higher in summer than that of the ocean (Rougerie, 1986). Salinity is generally close to ocean values but rainfall can occasionally lower the local salinity of the surface water in protected embayments. On the west coast, where rainfall is moderate and the reef is more than 10 km from the coast, it is extremely rare to observe estuarine sediment plumes extending toward the reef, in contrast to the east coast where high rainfall occurs and the reef is close to the coast; here it is much more common to observe estuarine plumes extending offshore. Terrigenous sediments have been retrieved from the reef slopes down to 100 m depth. The semi-diurnal tide, with amplitudes of 1.5 m at spring tide and 0.3 m at neap tide, propagates from the south to the north but its influence on lagoon circulation is largely dependent on geomorphology (Douillet, 1998). In general, the lagoon is subject to strong oceanic influence, and oligotrophic conditions prevail. Trade winds further favour frequent water renewal in the lagoon and recent estimates using modelling suggest an average residence time of no more than 11 days for an area of 2000 km^2 in the southwest part of the lagoon (Bujan et al., 1999).

THE MAJOR SHALLOW WATER MARINE AND COASTAL HABITATS

Coral Reefs

Around the main island of New Caledonia coral reefs have been developing since the Miocene and there is a major barrier reef delimiting the lagoon. Other reefs are present in the Loyalty islands and more remote islands, atolls or submerged reefs. The number of coral (*Madrepora*) species has been estimated to total 300 (Gabrié, 1998) but a comprehensive study of coral diversity in New Caledonia is still needed.

Corals from New Caledonia have been largely free from coral bleaching. The only significant coral bleaching event was reported in 1995, a year marked by temperature anomalies and bleaching events worldwide. Crown-of-thorns starfish (*Acanthaster planci*) outbreaks are reported occasionally but generally appear to be confined to limited areas. Fringing reefs are locally exposed to anthropogenic threats such as an increase in terrigenous inputs or discharge of waste water (detailed below). Patch reefs located in the vicinity of urban areas or frequented by tourists can be endangered by effects such as anchor damage. However, relative to the rest of the world, the coral reefs of New Caledonia can be considered as largely pristine, mainly because of the low anthropogenic pressure resulting from low population density.

Mangroves

There are 16 species of mangrove in New Caledonia and mangrove ecosystems cover an estimated area of 200 km^2 (Thollot, 1996). Mangroves have been traditionally used for wood, medicinal plants, and as fishing grounds at a level that did not significantly alter their balance. Stretching along the land–sea interface, mangroves have been much more affected by modern economic development. Urbanisation is the main factor affecting demand for land on the coastline and mangroves have suffered both deliberate destruction or the indirect effects of other nearby development. However, there is a growing acknowledgement of the economic and environmental value of mangrove swamps and protective regulations are now in place.

Seagrass and Green Algal Beds

Seagrasses are well represented, although their distribution is poorly known except for the southwest lagoon (Garrigue, 1995). A total of 12 species have been sampled from various depths, but dense seagrass beds are mostly located in shallow waters (1–6 m) on the leeward side of the patch reefs and on the landward face of the barrier reef.

Significant beds are also formed by green algae of the genera *Caulerpa* and *Halimeda*. The species from the *Halimeda* group produce calcareous skeletons and the regular carbonate inputs from these productive algae contribute significantly to lagoon sediment deposition.

Lagoon Sediments

It is important to emphasise that, compared with the previously described communities, sediment beds are by far the most common habitat in the lagoon of Grande Terre. Three major subsystems stretching from the land to the back reef have been defined on the basis of various criteria such as sediment composition, benthic community composition, or metabolic activity (Chardy et al., 1988). Muddy substrates are located close to the shore and largely influenced by terrigenous inputs; grey sands cover the middle of the lagoon, while white sands are confined to the vicinity of the back reef. It has been demonstrated that lagoon sediments are essentially of local origin and that, even for the white sands, coral carbonates contribute to only a small proportion of deposited sediments (Chevillon, 1996).

Biodiversity

New Caledonia stands close to the Pacific marine biodiversity hot spot in the Philippines and its lagoon thus has relatively high species biodiversity (Table 1). Furthermore, the remoteness of the islands has promoted the appearance and encouraged the maintenance of species which are not found elsewhere in the world. Endemic species account for 5% of all marine aquatic species. In groups with low dispersal capacities like sponges, up to 71% of the identified species might be endemic (Levi et al., 1998).

OFFSHORE SYSTEMS

The pelagic zone of New Caledonia is relatively important by Pacific island standards. The economic zone covers over 1 million km^2. Its physical environment has been studied over recent decades but the knowledge base was quite heterogeneous until the various administrative and research bodies of New Caledonia decided to undertake a comprehensive research and exploratory programme (see ZoNéCo Programme box). This programme, covering both living and non-living resources, compiles existing data and coordinates all new data acquisition.

Today, bathymetrical and geophysical information has been collected over about 20% of New Caledonia's area

Table 1
Diversity of several groups in New Caledonia

Group	Identified no. of species	Estimated no. of species	Reference
Algae	336	1000	Garrigue and Tsuda (1988)
Sea grasses	12	12	Garrigue (1995)
Sponges	300	600	Levi et al. (1998)
Corals		300	Gabrié (1998)
Echinoderms	246	400	Guille et al. (1986)
Molluscs	513	6500	Bouchet (1979)
Crustaceans	544	3000	Richer de Forges (1998)
Ascidians	199	300	Monniot et al. (1991)
Fishes	1610	1800	Kulbicki and Rivaton (1997)

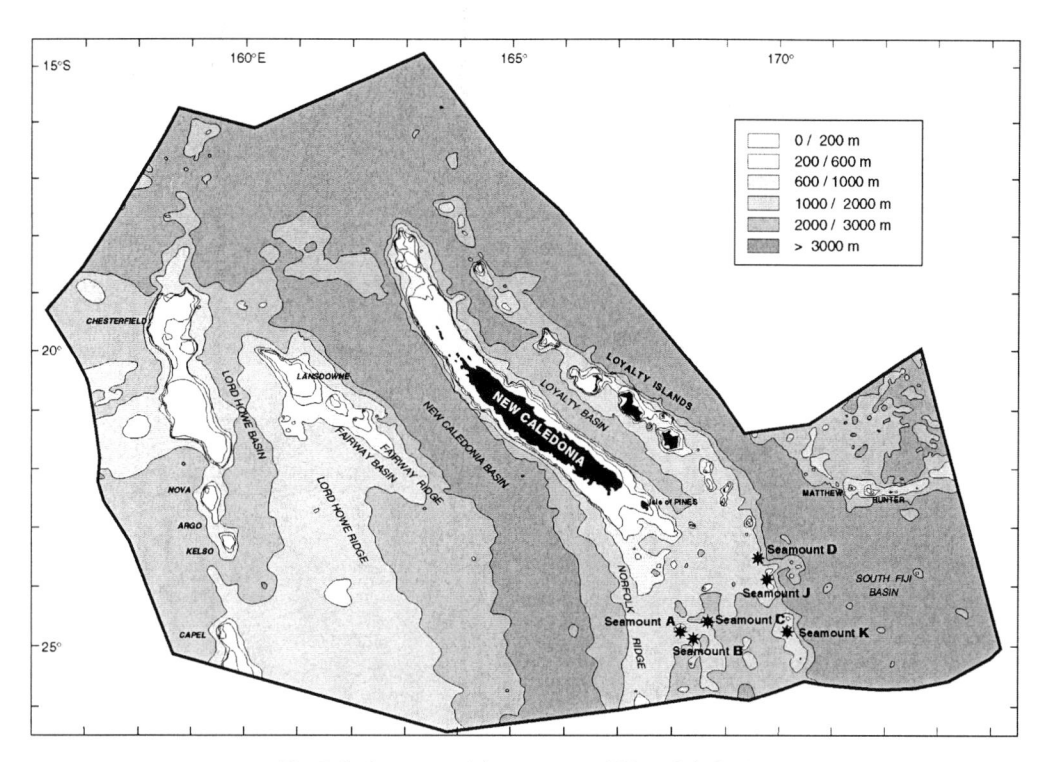

Fig. 2. Bathymetry of the sea around New Caledonia.

The ZoNéCo Programme

Initiated in 1991, the objective of the ZoNéCo Programme is to survey and assess living and mineral marine resources in New Caledonia's Exclusive Economic Zone (Anon., 1997). It is a grouping of many partners representing the French State, New Caledonia, the three Provinces and the research organisations (IFREMER, IRD and University of New Caledonia). This long-term and multidisciplinary programme (geology, geophysics, biology, physical oceanography and fisheries) aims to identify marine resources of economic interest and contribute to their sustainable development.

This programme is based on a survey-type strategy and comprises three phases:
- a "strategic phase" of reconnaissance and mapping designed to identify favourable areas;
- a "tactical phase" of qualitative and quantitative assessments of resources available at the various relevant sites, as identified during the strategic phase;
- a "targeting phase" designed to yield an economic appraisal of the target areas identified during the previous phases.

After being entirely focused on the oceanic portion of the EEZ, the Steering Committee meeting held on Lifou in 1998 addressed the extension of ZoNéCo's work to the coastal zones. The purpose is to supplement and update current knowledge on reef and lagoon resources and their exploitation and then to determine how they can be sustainably managed.

The ZoNéCo web site may be consulted at: http://www.territoire.Nouvelle-Calédonie/zoneco.

(focusing on the 0–1500 m depth range) and over 70 reports and articles have been produced, including historical fisheries data compilations (Virly, 1996; 1997). A full bathymetric chart has also been produced (Fig. 2). Sea surface temperature maps from satellite imagery are produced routinely.

As far as offshore resources are concerned, only pelagic stocks (tuna and associated species) are currently exploited and support a developing industry. There are no comprehensive sustainable yield estimates yet but the current catch levels (under 2000 tons) can probably be increased several-fold with no risk to the stocks. Demersal stocks (mainly alfonsino *Beryx splendens*) have also been identified, principally in the most southern part of the EEZ (Norfolk and Loyalty ridges) but do not appear to be able to support large-scale fishing (MSY estimated around 400 tons).

XBT casts and ADCP data demonstrate a deep counter-clockwise gyre reaching 700 m depth on these southern fishing grounds, which is believed to play a major role in larval fish and juvenile dispersion (Anon, 1997). These data also indicate an upwelling along the south west coast of the main island.

POPULATIONS AFFECTING THE AREA

Population Structure

The structure of New Caledonia's population stems from its history and colonisation. The 1996 census recorded a population of 196,836 in New Caledonia (Ahmed-Michaux and

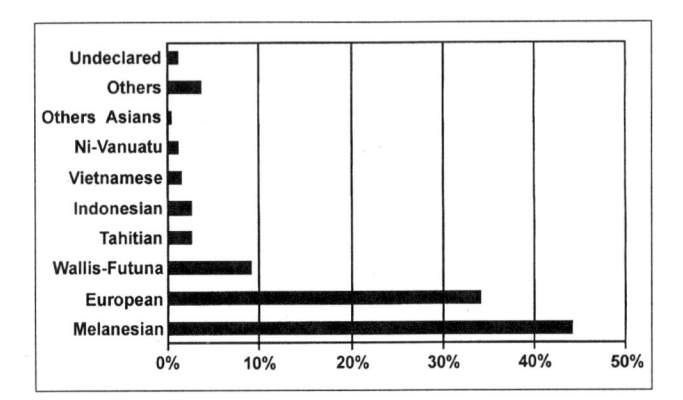

Fig. 3. Ethnic groups as percentages of the population in 1996.

Roos, 1997). Despite a high natural growth rate (19 per thousand), New Caledonia still has one of the lowest overall population densities in the South Pacific region: 10.4 inhabitants per km². Before the arrival of Europeans in 1840, its population was estimated at approximately 50,000. Although the Melanesian ethnic group is constantly increasing in numbers in absolute terms, its relative proportion of the total population was decreasing until the 1970s, mainly due to the growth of the other ethnic groups. The nickel "boom" and government policy of the day favoured European immigration. Immigrants also arrived from other French Pacific territories (Wallis and Futuna; French Polynesia) and former French possessions (Vietnam), joining the descendants of immigrant Asian workers brought to New Caledonia from 1865 onwards to develop the sugarcane industry and work in the mining and timber industries. These factors explain the multi-racial nature of the New Caledonian population today (Fig. 3), in which the Melanesian and European ethnic groups are the largest.

The population is unevenly distributed. The geographical distribution of the population by ethnic groups shows that there is a large Melanesian majority in the Northern and Loyalty Islands Provinces and that other ethnic groups, Europeans in particular, are in a majority in the Southern Province and especially in the 'Greater Noumea' area (Table 2).

Virtually the entire population lives on the coast and more than 60% of the country's inhabitants live in the Greater Noumea area (Noumea and adjacent Communes of Dumbea, Mont-Dore and Païta), which makes this the only real urban area of Grande Terre and the island group as a whole (Table 3).

Activities Affecting the Sea

As in all island settings, cultural and economic ties with the sea are strong in New Caledonia, whose marine environments are receiving increasing attention from scientists, especially within IRD (formerly ORSTOM), who have been doing research in physical and biological oceanography in New Caledonia for over 50 years (Grandperrin and Hénin, 1996; Fromaget and Richer de Forges, 1992). This research involved exploring marine areas and, in some key areas, performing situation assessments of immense value not only for modern oceanography, but also economic development.

For centuries, because of the subsistence nature of their activities, local communities lived in close harmony with their marine environment and its resources. Shipping services long remained the one and only means of communication with the outside world. Today, they continue to represent an important link with the neighbouring islands and continents, mainly for the transport of goods, commodities and raw materials (ore in particular). The development of modern transport facilities, increasing demographic pressure and economic development have led to a greater use of coastal zones. Close to the Greater Noumea area, urban development has also led to environmental pollution and habitat degradation. Occasional other activities linked to marine non-living resources exploitation have been developed, such as sand quarrying in the lagoon (now prohibited) which has had only local impact.

Fishing is traditionally important in Melanesian culture (Leblic, 1993). It has changed considerably in recent decades with the advent of efficient equipment and gears. With the much more recent development of prawn aquaculture and industrial tuna long-lining, it remains the only significant coastal marine resource exploitation activity. From the economic point of view, seafood is the second export commodity after nickel ore, far ahead of other agricultural goods. It is nevertheless secondary and accounted for less than 0.6% of GDP in 1995 (ITSEE, 1997). New Caledonia's economy depends heavily on government

Table 2

Percentage distribution of the population in terms of the main ethnic groups in 1996

Region	European	Others	Melanesian
Islands Province	2	1	97
Northern Province	17	4	79
Southern Province	30	53	17
Greater Noumea	46	32	22

Table 3

Populations and population densities in New Caledonia in 1996

Region	Land area (km²)	Population	%	Density (pop./km²)
Loyalty Islands	1981	20877	10.6	10.5
Northern Province	9853	41413	21	4.2
Southern Province (SP)	7012	134546	68.4	19.2
Greater Nouméa (GN)	1643	118823	60.4	72.3
S.P.–G.N.	5369	15723	8	2.9
New Caledonia	18846	196836	100	10.4

transfers (more than a quarter of GDP), an export-oriented nickel industry, an industrial sector made up of small businesses, an under-developed agricultural sector and a services sector not tuned towards the productive sector. It is highly dependent on direct and indirect financial support from France.

The value of coastal fisheries lies more in its importance for rural and tribal coastal communities, to whom it represents a major subsistence food source. Sea fishing and aquaculture are priority development options for the Provinces in New Caledonia and incentives are, therefore, being offered in order to stimulate their growth. Leisure activities for tourists are expanding and are an integral part of the uses made of the maritime and particularly lagoon areas. They are also a priority development focus, as recently reaffirmed by the Government of New Caledonia.

RURAL FACTORS

Agriculture

Traditionally, the populations of New Caledonia (first the Melanesians, then the Europeans) have depended mostly on agriculture for subsistence. Still, only about 2% of the land is used for crop culture (40%) and cattle ranching (extensive: 29%). The major crops are listed in Table 4, together with the current volume of production of livestock.

Habitat alterations have been more noticeable on the terrestrial environment than on the marine. However, the use of pesticides and fertilizers is not widespread (about 3000 tons of all types combined were used in 1998) and is likely to have only a localized impact. Biotic indices have been developed to monitor water quality, and stream watches will soon be done routinely.

Table 4

Major crops and livestock of New Caledonia in 1998. Source: Memento Agricole 1998, Gouvernement de la Nouvelle-Calédonie, Direction de l'Economie Rurale)

Crop	Tonnage
Beef production	4080
Poultry	581
Pig	1490
Apiculture	54.6
Fruits	4222
Vegetables	5055
Root crops	250
Coffee	50
Copra	171
Vanilla	0.4
Cereals	2030
Forest products	2310

Because water is, or will be, the limiting factor for agriculture development in the coming years, the main impact might very well be the decrease of freshwater input into the coastal ecosystems. Indeed, streams are often impounded to prevent saltwater incursions upstream in order to provide a greater capacity for irrigation. EIAs have shown that these types of infrastructures can indeed have an impact on estuarine ecosystems, particularly mangroves.

Artisanal Fishing

Coastal fishing in New Caledonia can be considered an entirely artisanal activity apart from the special case of the semi-industrial exploitation of a bivalve mollusc (*Amusium balloti*) in the part of the lagoon located to the North of *Grande Terre*, which recently ceased.

Reef and Lagoon Fisheries

Reef and lagoon fisheries are focused inside the lagoon, from the shore to the barrier reef. Activities are carried out on foot on reefs, estuaries and mangroves, or from boats generally smaller than 10 m in length and fitted with outboard motors in open water or near the reef. Various fishing methods are used. The main ones are hand-lining, trolling, gathering by hand, gillnetting, seining, cast-netting, free diving, spearing and, to a lesser extent, prawn dragnetting and trawling (under derogatory measures for only one fishery).

Two types of fishing activity can be distinguished. Firstly, subsistence fishing, which mainly targets species such as mullets, emperors, sea breams, groupers, snappers and rabbit-fish. Catch quantities are difficult to assess. The most recent estimate was based on an analysis of household consumer budgets (ITSEE, 1993), which reported over 3000 tonnes of catches annually (Paddon, 1997), accounting for almost half the total production from sea fisheries and aquaculture in New Caledonia in 1996 and 1997 (Territoire de Nouvelle-Calédonie, 1998). Recreational fisheries were included in these estimates; more specific research is needed on Greater Noumea where most of the recreational fleet is concentrated. Developments in the subsistence fishery are influenced by growing population pressure, accompanied by increasing demand for seafood, but also by the monetarisation of the Melanesian rural economy (integration into the cash economy). It is dependent on the ethnic structure (EGC, 1991).

Secondly, professional fisheries target either high-added-value species for export (mainly beche de mer, trochus), or species for marketing locally (fish, crabs, lobsters and shellfish to a lesser extent). In 1997, there were 273 boats fitted out for professional lagoon fishing. Catch quantities were reported as 1,254 tons for an estimated value at first sale of approximately US$ 10 million. Although the "professional fisher" status is clearly defined in the regulation, many fishermen are still engaged in other activities as well and would not qualify as professionals *per se*. For example, only a very low proportion of fishers registered in the Northern Province make their living from that activity alone (Labrosse et al., 1997b; Paddon, 1997). Others fish part-time or as a supplementary activity. Also, people belonging to fishing clans in the Melanesian environment are often not declared but, through their social function within the community, could be considered as professionals (Leblic, 1993). The registrar in the Southern Province on the other hand only includes professionals under the legal definition, all other fishers being considered as subsistence fishermen.

Coastal Fishery

The professional coastal fishery outside the lagoon comprises a fleet of approximately 20 multipurpose vessels 9 to 12 metres in length, of 15 gross registered tonnes on average. These boats generally operate less than 12 miles offshore and basically target demersal species of the outer slope and seamounts, and pelagic species, including tunas. This activity also includes FAD (fish aggregation device) fishing. The main types of gear used are trolling lines, traps, long-lines and hand-lines. In 1997, 56 tonnes were caught, representing a value of half a million US dollars at first sale, including 18 tonnes of tunas and associated species and 38 tonnes of deep-swimming fish (chiefly snappers).

Impact of Artisanal Fishing

The impact of the artisanal fishery on stocks and habitats is difficult to assess and has not been extensively researched. Most fishing gears currently in use are not destructive and are relatively selective, except the trawls and dragnets in limited use. Three boats previously used this type of gear in the reef passages or on soft bottoms, particularly to harvest *Amusium* in the northern lagoon, but only one is still operating in reef passages outside the lagoon. Fishing on foot probably has a destructive effect on the most accessible reefs (especially the fringing reef), but there is no scientific data available on this. Fisheries research has yielded a better biological and ecological knowledge of some species of economic interest and has made it possible to determine stock exploitation parameters for those species. This is the case for Ouvea (Kulbicki et al., 1994) and Northern Province (Labrosse et al., 1996; Labrosse et al., 1997a; Letourneur et al., 1997a), lagoon fish, lagoon fish that can be used as tuna bait fish (Conand, 1988), lagoon bottom fish (Kulbicki, 1988; Wantiez, 1992, 1993), beche de mer (Conand, 1989), trochus (Bour, 1992), corals (Joannot, 1990), mud crab *Scylla serrata* (Delathière, 1990), coconut crab *Birgus latro* on Ouvéa (El Kadiri-Jan, 1995), lobsters (Chauvet and Weil, 1991; Joop, 1992; Grandin and Chauvet, 1994; Chauvet and Farman, 1994) and the scallop *Amusium balotii* (Clavier and Laboute, 1987; Clavier, 1991). Generally speaking, fishing pressure remains low at the scale of the various coastal biotopes and

> ### Beche-de-Mer and Trochus Exploitation
>
> Artisanal beche-de-mer and trochus production is one of the main sources of monetary income for coastal communities. Virtually all the country's output is exported to southeast Asian countries. Trochus shells are used to manufacture pearl buttons, while beche-de-mer is exported in dried form and is considered a choice dish. Their export values were estimated at US$ 640 and 360 thousand respectively in 1997, taking third and fourth places in the list of seafood product exports after aquaculture prawns and tuna.
>
> The trochus fishery in New Caledonia became significant in 1907, and for a long time was the main source of cash income for local communities. The fishery was one of the biggest in the Pacific, but with increases in the number of fishers, the unexploited habitat of *Trochus niloticus* shrank, and access difficulties hampered its exploitation. Further, the fragility of this species, which had already become apparent, led to regulations from 1911, with minimum size being restricted first to 8 cm, then to between 9 and 12 cm and fishing being closed during the breeding period (from January to April), or even for a whole year (1956–1957). From 1994 to 1997, shell production dropped from 273 to 124 tons when the MSY was estimated to be 400 tons.
>
> After a period of irregular catches up to 1940, commercial beche-de-mer production in New Caledonia virtually ceased until about 1983. This situation meant that stocks could be considered as unexploited during that time. From over 100 tonnes per year dry weight at the beginning of the fishery, exports have fallen by half over the last six years to below 50 tonnes at present. This equates to a significant reduction in production depending mainly on the areas exploited, the number of fishermen and its price, which has noticeably increased recently. Also, from the end of the 1980s, changes in exploitation areas and species were observed, and 'grade two' beche-de-mer began to be collected. Two-thirds of the current production occurs in the Northern Province.

fisheries resources would not appear to be threatened for the moment. For most of the species studied and exploited, quantities currently caught would appear to be less than the limits of safe exploitation. However, there are a few exceptions. Beche-de-mer and trochus have registered significant production drops in recent years. They are probably showing a stock reduction due to the intensification of fishing effort. This is prompting fishers and neighbouring communities to temporarily cease their activities in some areas; the authorities are considering new management and regulatory measures.

Experimental long-lining surveys in the southwestern lagoon showed differences in fish population compositions between the area lying in a radius of 20 miles from Noumea and the rest of the lagoon, with special reference to a drop in the number of species of economic interest (Kulbicki,

1988). Research on demersal reef and lagoon fish of the Northern Province lagoons which are of interest as food fish and/or commercial fish, revealed the impact of fishing on reef stocks in the most highly populated areas (Labrosse et al., 1997a) and also showed a terrigenous influence on the structure of their populations (Letourneur et al., 1997b; 1999). These few examples show that fisheries resources are not safe from certain threats principally associated with population growth, urbanisation and economic development. Apart from species protected from international commerce (turtles, corals, dugongs, etc.), specific territorial or provincial regulations have been drawn up for species deemed to be ecologically or economically sensitive to exploitation. Also, on-going or proposed research programmes should be able to provide the information necessary for sustainable fisheries management and therefore make it possible to avoid overfishing or habitat degradation situations comparable to those found in other Pacific islands.

COASTAL EROSION AND LANDFILL

Landfill has been used since the early days of European settlement to claim land for urban development, mainly in the urban centre of Noumea where 230 ha of mangroves have been reclaimed between 1960 and 1989 (Thollot, 1996). Today, landfill claims are subjected to administrative approval which is often subject to an Environmental Impact Assessment (EIA), thereby reducing this practice mainly to public infrastructure constructions. Furthermore, sand and coral extraction is prohibited in the lagoon.

Coastal erosion has been apparent all around New Caledonia, more markedly so on the eastern shores and smaller islands. The phenomenon is too widespread however to be attributed to human activity. The main impact so far has been principally the loss of beaches and ocean front which, in some cases, has necessitated the use of sea walls.

EFFECTS FROM URBAN AND INDUSTRIAL ACTIVITIES

Because of its large size, oceanic setting, active water circulation and low population density, urban and industrial impacts on the New Caledonian lagoon are essentially of local significance. Nevertheless, economic development and population are rising and it is necessary to ensure accurate scientific assessments and sustained management efforts to ensure the conservation of present environmental conditions.

Artisanal and Industrial Uses of the Coast

Due to the low population density, individual or small community-based activities are of very limited impact.

Fishing is by far the most scattered activity and its influence is detailed above. Coral mining for lime is almost non-existent and subject to strict regulation; as an example, gravel and sand extraction has been totally forbidden in the southern half of the lagoon. Forest felling and conversion has been practised at the beginning of European settlement to provide cattle land but is now insignificant. Bush fires of natural or human origin can occasionally run out of control, especially during ENSO dry periods, and result in soil erosion.

Mining

Mining is one of the main local economic resources of New Caledonia. Minerals exploited since 1875 include some chromium, a little iron, cobalt, manganese, gold, copper, lead, silver and coal (Fig. 4), but nickel is by far the main mineral resource. New Caledonia is currently the third largest nickel producer in the world. It is estimated that a total of 300 million m^3 of soil has been displaced by direct extraction since the beginning of mining activities. Open-cast mining on hilltops has been the main method and has resulted in major alteration to landscapes and habitats (Bird et al., 1984). Not only mining activities but also prospectors have caused deforestation and soil erosion. A most important consequence is the destruction of terrestrial habitats for endemic species. Eroded soils are also transported through river catchments, occasionally filling in river mouths and contributing to delta formation. Bird et al. (1984) reported that 40 rivers in New Caledonia were significantly affected.

The main environmental consequence for the lagoon is sedimentation induced by the additional input of particles. Fringing reefs are especially sensitive to the increase in sediment deposition and decrease in light penetration but all communities, both pelagic or benthic, are affected to some extent. A second consequence is the impact of metals discharged in the coastal zone. Surprisingly, the potential of this discharge has not been rigorously investigated and studies are currently being conducted to bridge this enormous gap. As with their terrestrial counterparts, it is very likely that coastal habitats have been adapting to naturally occurring high levels of metal inputs but a very significant effort is needed to assess the fate of metal in the lagoon, including the key issue of metal accumulation in organisms.

In the past, ores were either directly exported or smelted locally, whereas new mining projects designed to exploit hitherto unexploited ore deposits with nickel content lower than 2% (laterite) should now rather use chemical extraction procedures using hydrochloric or sulphuric acid. It is likely that such processes will radically modify the geochemical characteristics and subsequent biological availability of metals in the discharged residues, requiring scientific information for the adaptation of environmental management plans. As part of industrial prospective investigation, a pilot plant has been constructed to study the

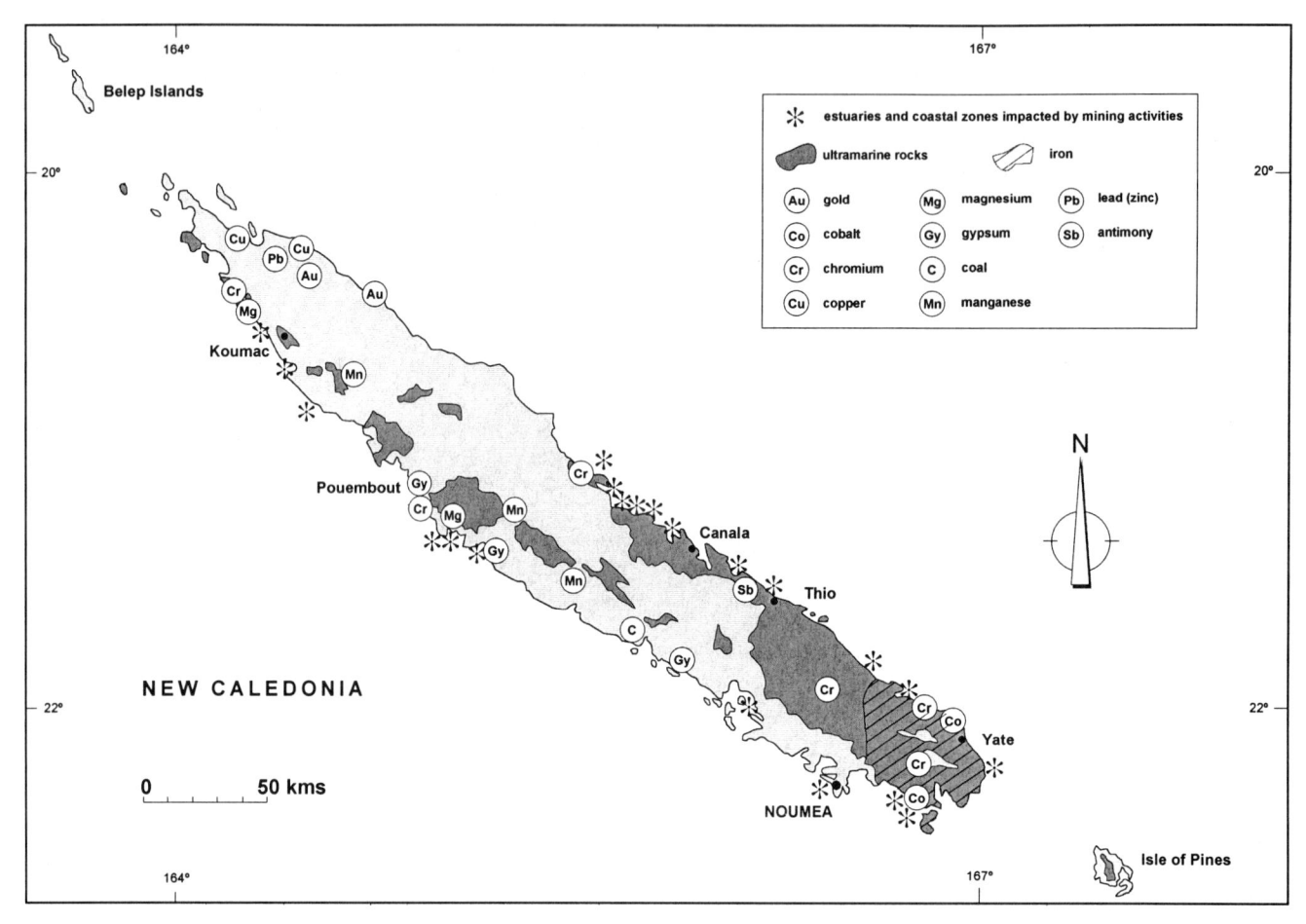

Fig. 4. Distribution of mineral deposits and of mining effects.

technico-economic feasibility of this process and to provide information on the possible environmental impacts.

Aquaculture

Recent progress in aquaculture of tropical shrimp (*Litopenaeus* spp.) has led to the development of shrimp farms on the west coast of Grand Terre. This activity is presently carried out only on a modest scale by international standards, with a total production of 1500 tons for a farming surface of 422 ha in 1999, but is on a greater scale than shrimp farming in any other Pacific Island. Furthermore, with artisanal fishing and agriculture it offers one of the major sources of rural employment and it is the first export product after the ores and before tunas. Shrimp farms are located on land in bays just behind mangrove swamps. The development of aquaculture was linked early on with environmental awareness, and protective measures have been taken and are regularly improved. An independent scientific study of a new pilot farm was conducted over three years (Guelorget et al., 1998; Chevillon, 1999; Garrigue et al., 1999) to assess its impact on the coastal environment and water quality. No significant impact on the benthic system was noted during this time. Additional information from larger-scale environmental surveys is needed as much additional aquaculture development is planned for the future, mainly to assess the risk of pollution and disease propagation.

Tourism

Tourism is a developing sector in New Caledonia but with around 100,000 visitors per year, its impact is currently minimal. As tourism has been defined as the number one development priority, special attention has been given to environmental aspects and EIA, and water treatment plans are now routinely required for new projects. Considering the rapid increase of the recreational fleet, the growing impact of boats anchoring near the shores of the smaller islands of the southern lagoon on fringing reefs and seagrass beds prompted local authorities to initiate a mooring buoy program. Today, over 100 buoys are available for safe anchoring on these spots but environmental information and awareness is still considered as a key issue to achieve some significant improvements in public behaviour.

Cities

Noumea, with its surrounding conurbation, is the only significant urban centre, hosting 60% of the 200,000

> ### Shrimp Aquaculture
>
> The New Caledonian shrimp farming industry took advantage of favourable conditions but also had to face several constraints (Coatanea et al., 1997). These include a geographical isolation with high operating, processing and shipping costs. There is also a relatively high labour cost of US$ 4.50 per hour for a non-qualified worker, which is much higher than that for most of SE Asia. There is a high cost of connection to primary infrastructure, and sometimes difficult access rights to the coast. Further, there is a limited tradition in primary production in this region (besides nickel ore).
>
> Advantages of shrimp farming here include favourable climatic conditions, and a very low population pressure (9 inhabitants km^{-2}). There are numerous suitable areas, however. About 6000 ha of potential space have been identified using satellite imagery, of which only around 400 ha are currently exploited. Areas which are used operate with ecologically sound practices, and there is a 20-year research and development effort by IFREMER, backed by French and New Caledonian Authorities, which is wholly dedicated to shrimp farming research, training and operation.
>
> Government incentives, through subsidy and tax alleviation, encourage the development of new farms and reduce the financial risk. In recent years, public sector involvement has increased to about 40% of the productive investment. This has led to an integrated, self-sufficient industry, with hatcheries, feed production and processing plant, all co-managed by the farms in a high quality approach (HACCP and 1509002 standards). This results in the production of 350 t for the local market and 1150 t for the export market, mainly Europe, Japan and Australia.

inhabitants of New Caledonia. Other villages around New Caledonia generally contain from 1000 to 4000 inhabitants, never exceeding 7000. Ignoring very local problems related to point source discharges, Noumea is the only place where significant urban effects are evident. The Noumea peninsula is flanked by the Bay of Boulari and the Bay of Dumbea where small river estuaries favour the presence of mangrove swamps. One of the most important urban effects results from coastline modification due to earthworks (building, roadways, harbours, etc.) which, from 1955 to 1993, accounted for the loss of 380 ha of natural habitat, notably mangroves and fringing reefs.

Industrial pollution has been poorly assessed in the past and the presence of a growing industrial area at the entrance of the city with a large diversity of activities represents a significant environmental issue. Today, all polluting industries are subject to a very stringent regulation (Installations Classées pour la Protection de l'Environnement) and European standards are usually applied. However, a proper assessment of environmental conditions in the coastal embayments located on both sides of the industrialized peninsula still has to be completed. The proximity of mangrove swamps that provide a complementary food source for the locals is of major concern, even though fishing has been officially forbidden in the area. Atmospheric discharge and industrial practice at the Doniambo nickel ore processing plant has been regularly improved in the past but the impact of this major plant on the coastal system of Grande Rade, a subsystem of Dumbéa Bay, has been poorly assessed and surveyed. The smelting process is based on ore melting and therefore involves limited addition of chemicals (mostly sulphur). Hot water, discharged at 45°C, is cooled down along a channel and the temperature signature at sea is of limited importance and extent. The effect of the processing plant on metal distribution, nature and fate in the lagoon is largely unknown. Preliminary results from the Ecotrope research program launched in 1997 by IRD provided evidence for metal enrichment in the Grande Rade but also demonstrated that the extension of contamination in the south west lagoon was limited and that metal concentrations in benthic organisms were comparable to those found at the vicinity of some estuaries discharging significant terrigenous inputs to the lagoon.

Solid domestic wastes are deposited in a controlled dumpsite located on the coastline in the Bay of Koutio. Except for controlled storage and burial no specific treatment is presently applied and it is very likely that leakage through the substratum occurs. This 'old generation' dumpsite is coming to saturation and a new site will be opened in 2001 according to European standards for solid waste disposal.

Finally, the discharge of waste water is a major issue for the Noumea urban area. Because of its complex geomorphology, the city is separated into numerous waste water collection systems, and whilst some treatment plants have been built during the past decade, a great deal more remains to be accomplished before satisfactory environmental standards are reached. This issue has been classified as the highest priority for the 2000–2004 development plan. One of the most critical sites is the Bay of Sainte-Marie where raw sewage is discharged in the inner part of the bay on the fringing reef and after destruction of mangroves. This bay is part of a pilot study under national research programmes on coral reefs (PNOC) and on the coastal environment (PNEC). Eutrophication resulting from this discharge has resulted in the development of green algae (*Ulva* spp.) around the sewage point sources and a general alteration in benthic populations. These alterations are mostly confined however to the inner part of the bay, due to prevailing currents.

Shipping and Offshore Accidents and Impacts

Despite some significant ship traffic, essentially directed to and from Noumea harbour, few wrecks have been recorded in the recent past. The traffic is mainly due to the delivery of nickel ore to the processing plant and in October 1992 a ship ran aground on a patch reef in the Havannah channel and an estimated amount of 4500 tons of ore was

Table 5

Marine Protected Areas of New Caledonia

Reserve name	Date of creation	Area (ha)	Special environmental interest	Tourist facilities	Scientific facilities
Signal Island	Jan 89	243	yes	yes	no
Larégnère Island	Jan 89	649	no	yes	no
Maître Island	Jul 81	765	no	yes	no
Amédée Island (and Aboré barrier reef)	Jul 81	15,070	yes	yes	yes
Canards Island	Jan 89	176	no	yes	no
Bailly Island	Jan 89	215	no	yes	no
Fausse passe de Uitoé	Dec 92	110	no	no	no
Ténia Island	Apr 98	101	yes	no	no
Kuendu point	Apr 98	55	yes	yes	no
Roche Perçée and Baie des Tortues	Jun 93	120	yes	yes	no
Verte Island	Jun 93	84	no	no	no
Poé	Jun 93	2,800	yes	yes	no
Casy Island	Jun 93	145	yes	yes	no
"L'Aiguille"	Jun 93	12.5	yes	yes	no
Complete MPA "Yves Merlet"	Jul 70	17,200	no	no	no
Séche-Croissant	Aug 94	0	yes	no	no
Goéland Island	Nov 95	0.5	yes	no	no
Total		37,746			

lost. A study conducted to assess the impact found limited alteration, mainly because a large part of the discharge was immediately dispersed by very active currents (Chevillon, 1992).

New Caledonia is also dependent on imports and in January 1997 a small vessel went aground spilling around 100 tons of petrol. Fortunately, this fuel is much more volatile than crude oil and the impact was very limited.

PROTECTIVE MEASURES

The earliest regulations pertaining to the conservation of marine species date back to the late 1800s and were usually inspired by French regulations.

A whole array of species-specific regulations have since been enacted to protect economically or ecologically important species such as rock lobsters, mullets, mangrove crabs, rock oysters, coral, aquarium fish, sea turtles, dugongs and trochus. These are usually based on size limits, seasonal or permanent closures (dugongs), or exclusion zones (coral, oysters). In addition, a whole array of fishing techniques, professional or recreational, are also regulated separately together with the conditions for operating commercial vessels. For instance, there is a 1000 m exclusion zone for professional gill netting all around the coast line of New Caledonia. Furthermore, spearfishing at night is prohibited as well as spearfishing with scuba gear. Recreational or subsistence catches are limited to 50 kg/boat/day. The use of poisons or any chemical substances for catching marine and freshwater species is also prohibited. The aquarium fish trade is carefully monitored and restricted to commercial operators fishing with specially authorized nets. Trawling and dredging is prohibited but a scallop fishery was authorized in a limited area of the northern lagoon. This fishery has since stopped its operations. As far as protective measures for the environment are concerned, the successive governments of New Caledonia have started, as early as 1975, to create marine protected areas (MPA). Today, 37,746 ha are protected, 17,200 ha of which are accessible only for traditional fishing. Table 5 lists the different MPA and their respective areas.

Since 1991, the Southern Province has operated two patrol boats which are in charge of surveillance, public information and general management of these MPA. The number of poachers has been declining ever since. Finally, several of these MPA include small islands or cays which often shelter marine bird colonies. Boardwalks have been provided to keep the public clear of nesting sites and in two instances, the islets have been closed to the public during breeding seasons.

A reef-monitoring project has been initiated to evaluate the status of reef communities near the more active rural centres. This project has been carried out only in the southern Province so far but should be generalized to the whole of New Caledonia in the near future. Data are routinely transmitted to the Global Reef Monitoring Network. Several international treaties or conventions are enforced in New Caledonia, namely: the Apia and Noumea conventions of regional scope, and the RAMSAR and CITES conventions. In that respect, only the CITES convention has direct application to New Caledonia. The local representative of the French Government is in charge of issuing CITES permits and the scientific authority is the research institute IRD. Annual reports are produced, summarizing the permits issued by species.

ACKNOWLEDGEMENTS

We are grateful to Marika Tortelier (IRD computer graphic artist), Jipé Lebars (SPC graphic artist), Roy Benyon (SPC translator), Sheryl Mellor (SPC translator), and Sabine Brocero (IRD) for their valuable help.

REFERENCES

Ahmed-Michaux, P. and Roos, W. (1997) Images de la population de la Nouvelle-Calédonie, principaux résultats du recensement 1996. Institut Territorial des Statistiques et des Études Économiques, 64 pp.

Anon. (1997) Marine resources of New Caledonia: the ZoNeCo program. Grandperrin R., Richer de Forges B. and Auzende J.M. (eds.), 90 p.

Bird, E.C.F., Dubois, J.P. and Iltis, J.A. (1984) The impact of opencast mining on the rivers and coasts of New Caledonia. The United Nations University, NRTS-25/UNUP-505, 53 pp.

Bouchet, P. (1979) How many molluscan species in New Caledonia? *Hawaiian Shell News* 27, 10.

Bour, W. (1992) Un mollusque nacrier du Pacifique, biologie, écologie et gestion rationnelle du troca (*Trochus niloticus* L.) de Nouvelle-Calédonie. Collection Études et Thèses. Editions de l'ORSTOM. 174 pp.

Bujan, S., Grenz, C., Fichez, R. and Douillet, P. (1999) Modélisation préliminaire de l'évolution saisonnière du cycle biogéochimique du lagon sud-ouest de Nouvelle-Calédonie. C.R. Acad. Sciences, Paris, submitted.

Chardy, P., Chevillon, C. and Clavier, J. (1988) Major benthic communities of the south-west lagoon of New Caledonia. *Coral Reefs* 7, 69–75.

Chauvet, C. and Farman, R. (1994) Les langoustes *Panulirus penicillatus* et *Panulirus longipes* de l'île des Pins (Nouvelle-Calédonie). Croissance, mortalité et rendement par recrue. 25ème conférence technique régionale sur les pêches, Nouméa, 14–18 mars 1994, CPS. 5 pp.

Chauvet, C. and Weil, S. (1991) Étude des pêcheries de langoustes de l'île des Pins. Université Française du Pacifique, 38 pp.

Chevillon, C. (1992) Echouage du minéralier "Manyland U" et déversement de minerai en mer. Rapport d'expertise. 9 pp.

Chevillon, C. (1996) Skeletal composition of lagoonal modern sediments in New Caledonia: Coral, a minor constituent. *Coral Reefs* 15 (3), 199–207.

Chevillon C. (1999) Influence des rejets d'une ferme aquacole sur l'environnement littoral "La Pénéide de Ouano", sédimentologie. Rapports Conventions Sciences de la Mer, Biologie Marine, Vol. 24, 35 p.

Clavier J. and Laboute P. (1987) Connaissance et mise en valeur du lagon nord de la Nouvelle-Calédonie: premiers résultats concernant le bivalve pectinidé *Amusium japonicum balloti* (étude bibliographique, estimation de stock et données annexes). Nouméa: ORSTOM. *Rap. Sci. tech.: Sci. Mer: Biol. Mar.*, 48, 73 p.

Clavier J. (1991) État des connaissances sur *Amusium balloti* (Bivalve, Pectinidé) dans les lagons de Nouvelle-Calédonie. Nouméa: ORSTOM. *Conv. Sci. Mer: Biol. Mar.*, 4, 54 p.

Coatanea D., Farman, R. and Galinié, C. (1997) Shrimp aquaculture in New Caledonia: socio-economical aspects. In Proceedings of the Conference on Island and Tropical Aquaculture, Martinique, FWI, 4–9 May 1997.

Conand, C. (1989) Les holothuries aspidochirotes du lagon de Nouvelle-Calédonie, biologie, écologie et exploitation. Collections Études et Thèses, Editions de l'Orstom, 393 pp.

Conand, F. (1988) Biologie et écologie des poissons pélagiques du lagon de Nouvelle-Caledonie utilisables comme appât thonier. Collection Études et Thèses. Editions de l'ORSTOM., 239 pp.

Dandonneau, Y. and Gohin, F. (1984) Meridional and seasonal variations of the sea surface chlorophyll concentration in the south-western tropical Pacific (14–32°S, 169–175°E). *Deep Sea Research* 31, 1377–1393.

Delathière, S. (1990) Biologie et exploitation du crabe de palétuviers *Scylla serrata* en Nouvelle-Calédonie. Th. Doct.: Océanogr. biol. Univ. Bretagne Occid., 291 pp.

Delcroix, T. and Lenormand, O. (1997) ENSO signals in the vicinity of New Caledonia, south western Pacific. *Oceanologica Acta* 20, 481–491.

Douillet, P. (1998) Tidal dynamics of the south-west lagoon of New Caledonia: observations and 2D numerical modelling. *Oceanologica Acta* 21, 69–79.

E.G.C. (École de Gestion et de Commerce de Nouméa) (1991) La consommation des produits de la mer à Nouméa, 173 pp.

El Kadiri-Jan, T. (1995) Biologie des populations de crabes de cocotiers, *Birgus latro* (L.), dans trois îles du Pacifique: Lifou et Vauvilliers: Archipel des Loyauté, Nouvelle-Calédonie; Taiaro: Archipel des Tuamotu, Polynésie Française. Thèse de Doctorat de l'Université Française du Pacifique, Nouméa. 149 pp.

Fromaget M. and Richer de Forges B. (1992) *Catalogue indexé du milieu marin de Nouvelle-Calédonie*. 2ème édition. ORSTOM, Nouméa. Cat.: Sci. Mer., 274 pp.

Gabrié, C. (1998) L'état des récifs coralliens en France Outre-Mer. Ministère de l'Aménagement du Territoire et de l'Environnement, Paris, 136 pp.

Garrigue, C. (1995) Macrophyte association on the soft bottoms of the south-west lagoon of New Caledonia: description, structure and biomass. *Botanica Marina* 38, 481–492.

Garrigue, C., Bach, C., Richer de Forges, B., Bargibant, G., Hamel, P., Laboute, P. and Lapetite, A. (1999) Influence des rejets d'une ferme aquacole sur l'environnement littoral "La Pénéide de Ouano", organismes benthiques. Rapports Conventions Sciences de la Mer, Biologie Marine, Vol. 25, 44 pp.

Garrigue, C. and Tsuda, R.T. (1988) Catalog of marine benthic algae from New Caledonia. *Micronesica* 21, 53–70.

Grandin, P. and Chauvet, C. (1994) La pêche à la langouste à Lifou. 25ème conférence technique régionale sur les pêches, Nouméa, 14–18 mars 1994, CPS. 5 pp.

Grandperrin, R. and Hénin, C. (1996) Un demi-siècle de recherches en Nouvelle-Calédonie—La mer. ORSTOM Actualités no. 51.

Guelorget, O., Lefebvre, A., Martin, J.L.M., Lemonier, H., Fuchs, J. and Favry, A. (1998) Characterisation of ecosystems and follow-up of their evolution under the impact of shrimp aquaculture development in 3 geographic areas Mekong delta (Vietnam) Lampung Province (Indonesia) and North and Central parts of New Caledonia. In Assessment of tropical shrimp aquaculture impact on the environment in tropical countries, using hydrobiology, ecology and remote sensing as helping tools for diagnosis. Rapport final du contrat RS3-CT 94-00284. Rapport interne de la direction des ressources vivantes de l'IFREMER, DRV/RA/RST/98-05, 262 pp.

Guille, A., Laboute, P. and Menou, J.-L. (1986) Guide des étoiles de mer, oursins et autres échinodermes du lagon de Nouvelle-Calédonie. Faune Tropicale, ORSTOM, Paris, Vol. 25, 238 p.

ITSEE (Institut Territorial des Statistiques et des Études Économiques), (1993) Budget consommation des ménages 1991, principaux résultats, tome 1, 200 pp.

ITSEE (1997) Tableaux de l'économie calédonienne. Supplément au

"Bulletin de conjoncture de Nouvelle-Calédonie. 224 pp.

Joannot, P. (1990) Étude d'un récif exploité pour son corail: le récif Tétembia, Nouvelle-Calédonie. Recrutement et croissance des madréporaires ; dynamique du stock et exploitation rationnelle. Thèse de doctorat de 3ème cycle de l'Université d'Aix-Marseille II. 125 pp.

Joop, S. (1992) Étude préliminaire à l'évaluation de l'état des stocks de langoustes en Nouvelle-Calédonie. DEA Université de Nouvelle-Calédonie. 32 pp.

Jost, C. (1998) *The French-speaking Pacific: Population, Environment and Development Issues*. Géopacifique Association, Nouméa, 271 pp.

Kulbicki M. (1988) Bottom longlining in the south-west lagoon of New Caledonia. *Australian Fisheries* Sept. 88, 41–44.

Kulbicki, M., Bargibant, G., Menou, J.L., Mou Tham, G., Thollot, P., Wantiez, L. and Williams, J. (1994) Évaluation des ressources en poissons du lagon d'Ouvéa. 3ème partie: les poissons. Rapport Conv. Sci. Mer Biol. Mar. ORSTOM, Nouméa, 448 p.

Kulbicki, M. and Rivaton, J. (1997) Inventaire des poissons lagonaires et récifaux de Nouvelle-Calédonie. *Cybium*, 21 suppl., 81–98.

Labrosse P. and Letourneur Y. (1998) Définition et mise en oeuvre de méthodes de suivi des stocks et de la pression de pêche des poissons d'intérêt commercial des lagons de la Province Nord de la Nouvelle-Calédonie. Rapport Conv. Sc. Mer, Biol. Mar. ORSTOM Nouméa, 21, 52 pp.

Labrosse, P., Letourneur, Y., Audran, N., Boblin, P. and Kulbicki, M. (1996) Évaluation des ressources en poissons démersaux commerciaux des lagons de la Province Nord de la Nouvelle-Calédonie: résultats des campagnes d'échantillonnage de la zone nord. Rapport Conv. Sc. Mer, Biol. Mar. ORSTOM Nouméa, 16, 118 pp.

Labrosse, P., Letourneur, Y., Audran, N., Boblin, P., Malestroit, P. and Kulbicki, M. (1997a) Evaluation des ressources en poissons démersaux commerciaux des lagons de la Province Nord de la Nouvelle-Calédonie: résultats des campagnes d'échantillonnage de la zone ouest. Rapport Conv. Sc. Mer, Biol. Mar. ORSTOM Nouméa, 17, 110 pp.

Labrosse, P., Letourneur, Y., Paddon, J. and Kulbicki, M. (1997b) Incidences de la pression de pêche sur les stocks de poissons démersaux commerciaux du lagon ouest de la Province Nord. Rapport Conv. Sci. Mer Biol. Mar. ORSTOM Nouméa, 19, 15 pp.

Leblic, I. (1993) *Les Kanaks face au développement—La voie étroite*. ADCK Presses Universitaires de Grenoble, 441 pp.

Lehodey, P. (1994) Les monts sous-marins de Nouvelle-Calédonie et leurs ressources halieutiques. Thèse de l'Université Française du Pacifique. 402 pp.

Letourneur, Y., Labrosse, P., Audran, N., Boblin, P., Paddon, J. and Kulbicki, M. (1997a). Évaluation des ressources en poissons démersaux commerciaux des lagons de la Province Nord de la Nouvelle-Calédonie: résultats des campagnes d'échantillonnage de la zone est. Rapport Conv. Sc. Mer, Biol. Mar. ORSTOM Nouméa, 20, 130 pp.

Letourneur, Y., Kulbicki, M. and Labrosse, P. (1997b) Spatial structure of commercial reef fish communities along a terrestrial run off gradient in the Northern lagoon of New Caledonia. *Environmental Biology of Fishes* 51, 141–159.

Letourneur, Y., Labrosse, P. and Kulbicki, M. (1999) Comparison of commercial fish assemblages of New Caledonian fringing reefs subjected to different levels of ground erosion, *Oceanologica Acta* (in press).

Levi, C., Laboute, P., Bargibant, G. and Menou, J.-L. (1998) *Sponges of the New Caledonian lagoon*. Orstom, Paris, 214 pp.

Monniot, C., Monniot, F. and Laboute, P. (1991) Coral reef Ascidians of New Caledonia. Faune Tropicale, ORSTOM, Paris, 247 pp.

Paddon, J.R. (1997) Fishing pressure on demersal lagoon fish in the North Province of New Caledonia, South Pacific: evaluation, impact on stock assessment and implications for management. Master of Arts Thesis, RSMAS, University of Miami, 158 pp.

Richer de Forges, B. (1998) Le diversité du benthos de Nouvelle-Calédonie: de l'espèce à la notion de patrimoine. Thèse Museum Nat. Hist. Nat., Paris, 326 p.

Rougerie, F. (1986) Le lagon sud-ouest de Nouvelle-Calédonie: spécificité hydrologique, dynamique et productivité. Etudes et Thèses, ORSTOM Editions, Paris, 234 p.

Territoire de Nouvelle-Calédonie (1996) Pêches maritimes et aquaculture: les chiffres de 1994 et 1995. STMMPM, 19 p.

Territoire de Nouvelle-Calédonie (1998) Pêches maritimes et aquaculture: les chiffres de 1996 et 1997. STMMPM, 15 p.

Testaud, J.L. and Conand F. (1983) Estimations des surfaces des différentes zones des lagons de Nouvelle-Calédonie. ORSTOM, Nouméa, 5 p.

Thollot, P. (1996) Les poissons de mangroves du lagon sud-ouest de Nouvelle-Calédonie. Etudes et Thèses ORSTOM, ORSTOM Editions, Nouméa, 321 p.

Thollot, P. and Wantiez, L. (1995) Les mangroves du littoral du Grand Nouméa: inventaire, caractérisation écologique et pressions anthropiques. Unpublished report.

Virly, S. (1996) Synthèse halieutique des données thonières de la zone économique de Nouvelle-Calédonie (années 1956–1994). Nouméa: programme ZoNéCo d'évaluation des ressources marines de la zone économique de Nouvelle-Calédonie, février 1996, 215 pp.

Virly, S. (1997) Les pêches profondes réalisées dans la zone économique de Nouvelle-Calédonie: synthèse des données de 1970 à 1995. Nouméa: programme ZoNéCo d'évaluation des ressources marines de la zone économique de Nouvelle-Calédonie, mai 1997, 200 pp.

Wantiez, L. (1993) Les poissons des fonds meubles du Lagon Nord et de la baie de St-Vincent de Nouvelle-Calédonie. Description des peuplements, structure et fonctionnement des communautés. Thèse Univ. Aix-Marseille 2, 444 pp.

Wantiez, L. (1992) Importance of reef fishes among the soft-bottom fish assemblages of the north lagoon of New-Caledonia. *Proc. 7th Intern. Coral Reef Symp., Guam*, 2, 942–950.

THE AUTHORS

Pierre Labrosse
Secretariat of the Pacific Community (SPC),
B.P. D5,
98848 Nouméa Cedex, New Caledonia

Renaud Fichez
IRD (Institute of Research for Development),
B.P. A5,
98848 Nouméa Cedex, New Caledonia

Richard Farman
Southern Province, Department of Natural Resources,
B.P. 3718,
98846 Nouméa Cedex, New Caledonia

Tim Adams
Secretariat of the Pacific Community (SPC),
B.P. D5,
98848 Nouméa Cedex, New Caledonia

Chapter 101

VANUATU

Veikila C. Vuki, Subashni Appana, Milika R. Naqasima and Maika Vuki

This chapter reviews and assesses the condition of the marine environment in Vanuatu mainly from unpublished sources. It is in generally very good condition. However, severe damage from cyclones and crown-of-thorns starfish outbreaks have caused considerable damage to reefs in recent years, and reefs near major population centres show signs of degradation. Some reefs are threatened by increased siltation from logging in catchment areas, nutrient enrichment from sewage pollution and from over-exploitation of reef fish. Sewage disposal, domestic and solid wastes are serious environmental concerns in urban areas and, as a result, water quality is generally poor. There are generally no refuse dumps in many rural areas, and beaches, streams and lagoons are heavily littered. Industrial marine pollutant loadings appear to be increasing.

Trochus, green snails and land crabs (*Birgus latro*) have declined because of high rates of exploitation and the giant clam, *Tridacna gigas*, has been driven into extinction through excessive harvesting in Vanuatu. Turtles and dugongs are endangered and are sensitive to over-hunting. The huge leatherback turtle which also occurs here is threatened with extinction because of egg collection. Seabirds have also been significantly depleted through egg collection at rookeries and through disturbance of nesting sites through clearing.

There is limited management of the marine environment in Vanuatu and very little environmental awareness, technical or financial resources for this. Recent marine awareness-raising through workshops, drama, radio broadcasts has been successful in establishing community-based marine sanctuaries for replenishing overfished trochus and giant clams. Some village-based fishing restrictions on fishing methods, fishing ground closures and size restrictions have been initiated and implemented. But there is generally a lack of finances and technical expertise to support monitoring programmes or to implement and enforce legislation and policies pertaining to the marine environment. Vanuatu has declared a moderate number of marine protected areas (MPA) but the involvement of customary fishing right owners is integral to the success of MPAs in Vanuatu. These MPAs are an important strategy to conserve marine biodiversity in the absence of scientific knowledge and monitoring.

Seas at The Millennium: An Environmental Evaluation (Edited by C. Sheppard)
© 2000 Elsevier Science Ltd. All rights reserved

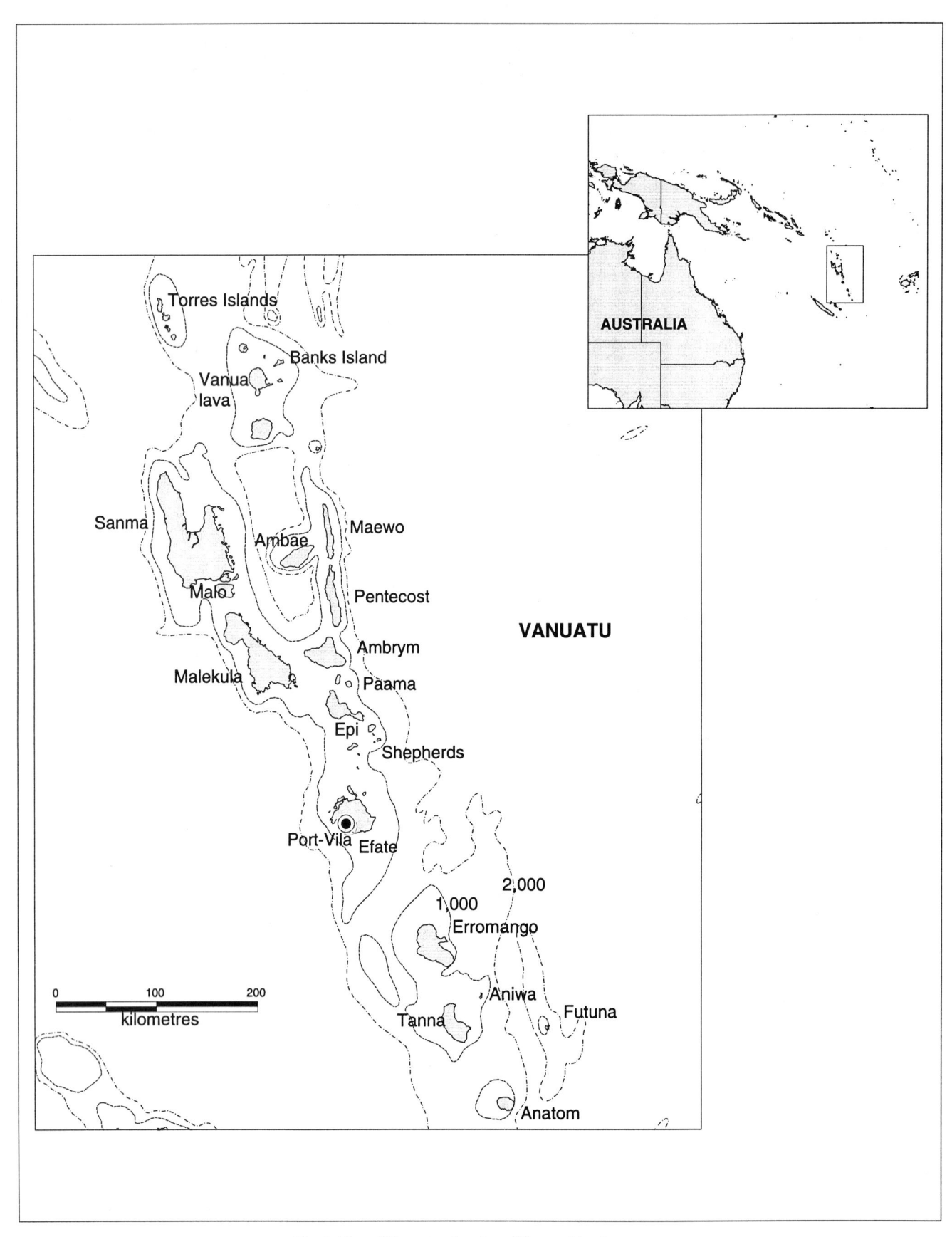

Fig. 1. Map of Vanuatu showing different island groups.

THE DEFINED REGION

Vanuatu (formerly known as the New Hebrides) is an archipelago of dispersed islands in the shape of a 'Y' stretching over about 850 km of the South Western Pacific Ocean (Fig. 1). It lies on the outer Melanesian ridge which extends from New Guinea through the Solomons, Vanuatu, Fiji, Tonga and the Kermadec Islands to the North of New Zealand (Mallick and Ash, 1975). Its closest island neighbours are New Caledonia to the South, Fiji to the East, and Solomon Islands to the Northwest. It has an Exclusive Economic Zone of 650,000 km^2 (Douglas and Douglas, 1994) which is about fifty times greater than its total land area.

The islands include both volcanic rocks and marine limestones derived from fringing coral reefs. There are nine active volcanoes, four of which are submarine. Approximately half of the islands and islets of Vanuatu are inhabited. The indigenous people are Melanesian in origin and are known as ni-Vanuatu. About 80% of ni-Vanuatu live in rural communities and are involved in subsistence agriculture and fisheries.

According to the Constitution (Government of Vanuatu, 1980) all land in Vanuatu belongs to customary landowners, and this ownership carries with it the right to own adjacent reef and other near-shore areas (Fairbairn, 1992).

Island Groups

The Republic of Vanuatu comprises over 80 islands, 67 of which are inhabited. These islands lie between 12°S and 21°S latitude and 166°E and 172°E longitude. The total land area is 12,270 km^2 of which 5500 km^2 (45%) is considered to be potential arable land (Environments of Vanuatu, 1997). The four main islands are Espiritu Santo, which at 3947 km^2 is the largest island, Malekula, Erromango and Efate, which together make up about 66% of the total land area. The capital of Vanuatu, Port Vila, is located on the principal island of Efate. The most northerly islands are the Banks Group and the Torres Group which lie 60 km northwest of the Banks.

Vanuatu consists almost entirely of 'high islands' of volcanic origin. Most of the larger islands rise above 700 m, most of them are mountainous and are relatively young geologically. There is an abundance of rainfall averaging about 2200 mm per year, but there are few perennial streams. This is probably the result of their small size and rugged topography (Convard, 1993).

The interaction of glacially controlled sea level and tectonic uplift throughout Quaternary times has produced locally extensive flat areas of reef limestone on most islands. The older islands typically consist of volcanic hill ranges surrounded by these limestone plateaus (Weightman, 1989).

Biogeography

The biogeography of the tropical Pacific has been studied extensively (e.g. Kay, 1980; Dahl, 1979; Stoddart, 1984; Holthus and Maragos, 1992). Dahl (1979) included Vanuatu in the "South Pacific Marine Region" which is the world's largest and most diverse bio-region. This was further subdivided into 20 biogeographical regions and according to Dahl's (1979) classification, Vanuatu is in Zone V.

The area has very strong biotic affinities with the New Guinea, Solomon Islands and the New Caledonia archipelagos. Most of the Vanuatu coral reef species are believed to have originated in south-east Asia, the centre of coral reef biodiversity (Veron, 1986). The number of coral species in the group decreases from west to east (Veron, 1986); many coral species may have been transported throughout the Pacific either as larvae or as colonies attached to floating material (Jokiel, 1990). High biogeographic affinities exist between New Caledonia mangrove fish and those from Fiji and Vanuatu (Thollot, 1996). Zann and Vuki (see Chapter 99) have included Vanuatu in the "South-Western Pacific Islands Bioregion" or Large Marine Ecosystems (LME).

SEASONALITY, CURRENTS, NATURAL ENVIRONMENTAL VARIABLES

Climate

Vanuatu lies in the south-east trade-winds which predominate for most of the year. The climate varies from tropical in the north to sub-tropical in the south. However, the majority of the Vanuatu archipelago has a wet and tropical climate. There are two distinct seasons: a hot and rainy season from November to April (also known as the cyclone season), and a cool and dry season from May to October (Environments of Vanuatu, 1997). In some years, the "dry season" can be very pronounced and extended, with little rain falling between June and December.

The average sea temperature ranges between 25 and 29°C, and average humidity ranges from 75 to 80%. Prevailing south-east trade winds have a moderating effect on this temperature. The mean annual rainfall is normally greater than 2000 mm for all islands, with the northern islands reporting a higher annual rainfall than the south. Vanuatu is subject to 24 hour rainfall events, often associated with cyclones (IUCN, 1988). These cause flooding and soil erosion in cultivated areas, and sometimes the damage extends to the forest slopes as well.

Cyclones

Vanuatu lies in one of the most active cyclone belts within the South Pacific and the Coral Sea (Republic of Vanuatu, 1985). Thus it is particularly prone to cyclones, and an average of two cyclones affect the group each year (Anon., 1984). November to April is the period generally termed the rainy, hot or cyclone season.

Tropical cyclones are characterised by strong winds, heavy swells and high rainfall. The effects are generally localised, affecting only one or two islands, but effects can be very severe causing extensive damage to gardens, tree crops and forest, and can cause the destruction of buildings, wharves and other infrastructure. Vanuatu experienced severe cyclones in 1951, 1959, 1972 and 1985, and economic losses to development activities because of cyclones has been estimated at more than 300 million Vatu in crops, material and equipment (Weightman, 1989).

Vanuatu's coastal marine ecosystems have developed under the regular influence of cyclones. The high energy and severe impacts of the storms are detrimental to coral reefs in the short term. Large swells resulting from the winds can cause considerable breakage, coastal erosion and re-suspension of bottom sediments, resulting in widespread mortality of reef organisms.

ENSO

Heavy rainfall associated with El Niño (ENSO) years can cause major coral mortality on reef flats. These events have a great effect on Vanuatu's oceanography and climate, and on its biota. Major changes in water temperatures, currents and nutrients affect migratory tuna and billfish movements in the region, and of course ENSOs also affect cyclone frequency and intensity, rainfall and coastal runoff (Anon, 1997). Coral bleaching events and outbreaks of crown-of-thorns starfish (*Acanthaster planci*) in Vanuatu appear to be related to ENSO. Sea-level depression associated with El Niño years can cause major coral mortality on shallow reefs and seagrass beds, as organisms are exposed to desiccation and/or rainfall. Any increase in storm frequency and intensity because of ENSO will probably have an impact on reefs.

Oceanography

The trade-wind drift through the area produces warm westward and southwestward flowing currents. These average 0.2–0.3 m/s, reaching a maximum velocity in March/April. Eddy currents are found nearshore which often flow in the opposite direction to that of the currents a little further from shore (Carter, 1983). In general, stronger currents are often found further offshore.

However, most currents are generally localised and tidally induced. At high tides, waves overtopping the reef crests pump ocean water across the reefs. At low tide most of this exits through reef passages (Carter, 1983). Currents are generally slow. Measurements at the entrance of Vila Harbour indicated that surface currents range in speed from 0.007–0.125 m/s. Mid-depth currents were generally <0.02–0.25 m/s. The flood currents were in the range of 0.11–0.26 m/s (Carter, 1985).

Sea-surface temperatures in lagoons vary, reaching a maximum of 31°C in January and February (Carter, 1985). Sea surface temperatures in open waters at Vanuatu are warm and range between 25–29°C (Anon, 1984). The coolest sea surface temperature recorded was 23°C and this occurred between July and August.

Surface salinity ranges from about 34–35‰. In general, salinity in lagoons is lowered by freshwater input during heavy rainfall. The annual rainfall is high ranging from 1700 mm at Tanna, in the south, to almost 3000 mm at Santo, in the north (Anon., 1984). Tides are semi-diurnal with a pronounced diurnal inequality with the lower tides falling during the night in summer, and during the day in winter.

MAJOR SHALLOW WATER MARINE AND COASTAL HABITATS

Vanuatu has a very large coastline in relation to its area (Environments of Vanuatu, 1997). However, much of the mountainous coastline is an abrupt transition from land to sea and possesses narrow or discontinuous beaches and mangrove fringes.

Coastal and shallow water habitats include wetlands (salt marsh, mangroves), estuaries, seagrass beds, hard and soft shores, fringing coral reefs, lagoons, barrier reefs and outer reef slopes. The nearshore shelves are very small because of Vanuatu's volcanic origin and small size. Most islands have emerged reef flat which is dissected by shallow pools, spurs and grooves, and some low islands or coral cays exist.

Mangroves

Mangroves are generally scarce in Vanuatu, but where they do occur even as small areas, they can be of great local importance for fisheries (Baines, 1981). An estimated 2500–3500 ha of mangroves is found in total (David and Cillaurren, 1992). On a regional basis, mangroves are unevenly distributed with more than half in Malakula alone (about 2000 ha). A further 22% is found in Santo/Malo, with the remainder occurring on Efate and in the Banks/Torres Groups.

Mangroves are well developed along protected shores and estuaries of the larger islands, for example, Big Bay in Espiritu Santo. The dominant mangrove species are *Rhizophora stylosa* and *Bruguiera gymnorhiza* in seaward areas, with associated *Xylocarpus granatum* and *Excoecaria agallocha* in landward areas. Total species diversity is much higher than this (Zann, 1991) with 17 species of mangroves recorded from Malakula by the Royal Society Expedition in 1971 (Table 1). These mangrove areas are found on low-lying areas of mud, silt and sand deposits, subjected to daily inundation by salt or brackish water. Brackish swamp communities occur on tidal flats.

Seagrass

Seagrass beds associated with the coral reefs are occasionally very dense and extensive, usually limited to shallow waters. A good description of seagrass in Vanuatu is provided by Chambers (1990). Most seagrass species are widely

Table 1
Vanuatu's known marine biodiversity

Marine plants	Description	Sources	Species
Algae	Coralline	Done and Navin (1990)	numerous
Seagrass	9 species	Chambers et al. (1990)	Cymodocea rotundata; C. serrulata; Enhalus acoroides; Halodule pinifolia; H. uninervis; Halophila ovalis; Syringodium isoetifolium; Thalassia hemprichii; Thalassodendron ciliatum
Mangroves	17 species	Marshall & Medway (1976)	Acrostichum aureum L.; Heritieera littoralis Ait.; Xylocarpus granatum Koenig.; Derris trifoliata Lour.; Rhizophora mucronata Lam.; R. stylosa griff.; R. apiculata Bl.; Bruguiera gymnorhiza (L.) Lam.; B. parviflora (Roxb.) W. & A. (from Hou, 1958); Ceriops tagal (Perr.) C. B. Rob.; Lumnitzera littorea (Jack) Voigt From Van Balgooy, 1971); Sonneratia caseoloris (L.); S. alba J. Smith (from Backer & van Steenis, 1951); Avicennia marina (Forsk.) Vierh. var. australisica (Forst. F.) Moldenke; Clerodendron inerme (L.) Gaertn.; Excoecaria agallocha L
Scleractinian corals	295 species	Veron (1990)	
Beche-de-mer	19 species	Done and Navin (1990)	
Turtles	5 species	Tacconi and Bennett, (1997)	
Fish	469 species	Williams (1990)	

distributed throughout the islands. Nine species were recorded from Vanuatu (Table 1), but only four species (*Cymodocea serrulata*, *Halodule pinifolia*, *Syringodium isoetifolium* and *Thalassodendron ciliatum*) have been recorded in Malakula.

Thalassia hemprichii occupies a wide range of habitats and, in contrast, *H. pinifolia* is restricted to a narrow range of distribution. The main habitats of seagrass are reef crest, reef flat, reef passage, lagoon, backreef, bays and intertidal areas. Diversity and abundance are lowest in the reef habitats and greatest in the intertidal areas, and intermediate in lagoons and bays.

Sandy and sheltered intertidal areas provide the most suitable conditions for seagrass growth in Vanuatu. Seagrass beds are, however, well developed in protected subtidal soft shores and lagoons and back reefs of Vanuatu. They are very important habitats for fish, turtles and dugongs. Small populations of dugongs (*Dugong dugon*) are found in Vanuatu, which forms the most southeastern limit of their distribution, and these seagrass beds provide their staple diet. The beds are also important spawning, nursery and feeding grounds for a variety of fish and turtles. However, the greatest financial values that seagrass beds provide may be that of protection and stabilization of Vanuatu's coastlines.

Coral Reefs

Most of the islands of Vanuatu have fringing coral reefs, and as a result of tectonic activity, have been uplifted above present sea level. The importance of coral reefs is clearly seen in that, although the reef-fringed shores only occupy 5% of the area of Vanuatu, they support 26% of its population.

Coral reefs also provide protection of Vanuatu's coastlines. Coral terraces are predominantly found in Santo/ Malo (40%), Malakula (29%) and Efate (18%) (Environments of Vanuatu, 1997). Traditional ownership of nearshore areas and particularly of the coral reefs is hereditary. Customary law in Vanuatu states that coral reef flats are owned by clans or larger communal groups and are not subjected to open access fisheries (Fairbairn, 1992).

Vanuatu's reefs are exposed to heavy oceanic swells and wind-generated waves. Most coastal regions have narrow fringing reefs, generally 100–200 m wide with shallow lagoons and intertidal reef flats. At the reef crest, the seabed drops rapidly to depths of 1000 m. A comprehensive survey of the reefs (35 sites on 16 islands) was undertaken in 1988 (Done and Navin, 1990). These surveys were carried out from Aneityum in the south to Ureparapara in the north. Survey results indicated that Vanuatu has a very high biodiversity of marine organisms and there are great similarities between corals of Vanuatu and those on the Great Barrier Reef (Done and Navin, 1990). Some 295 species of scleractinian corals (Veron, 1990) and 469 species of fish (Williams, 1990) have been identified from Vanuatu. The extent of reef development depends on different aspects (protected to exposed), the tectonic history of the area (some areas are rapidly uplifting, killing reef top corals), and the impacts of disturbances such as cyclones, coral bleaching events and crown-of-thorns starfish predation.

A summary of the survey results reported by Done and Navin (1990) and the island-by-island account of some reefs in Vanuatu is shown in Table 2. In general, outer reefs are dominated by coralline algae and robust plating and

Table 2

Condition or special features of coral communities in some of the islands in Vanuatu (source: Done and Navin, 1990)

Island	Description
Aneityum	– one of the most beautiful reef sites (sheltered lagoon)
	– shallow coral communities
	– crown-of-thorns-starfish (*A. planci*) damage
Tanna	– diverse coral community on the slopes
	– low coral cover with patches of attractive coral
	– localised sedimentation
Erromango	– robust reef adjacent to sandy bay
	– evidence of high rate of recolonisation
Efate and offshore islands	– massive coral destruction due to *A. planci* and cyclone
	– high cover of massive *Porites* (Port Vila)
	– good subtidal coral growth (Lelepa)
	– rich coral reef community inside Port Havannah (Moso)
	– resilient image of reefs with no signs of destruction from *A. planci*/cyclones (Paul Rock)
Epi	– high mortality due to *A. planci*
	– little reef development
	– dense populations of heavily pigmented massive corals
Malakula and offshore islands	– attractive coral knolls (Uripiv)
	– gradient of spectacular diversity observed near deepwaters (Port Stanley)
	– corals grow on muddy sand and showed extensive bleaching (Crab Bay)
	– seagrass bank surrounded by damaged fringing reef (Vulai)
	– spectacular reef community with dense beds of *Halimeda opuntia* (Nanamoru)
	– corals in bay areas appear very pristine and beautiful while those in reef flats are bleached (Gaspard Bay)
	– rich coral diversity (Akhamb island)
Pentecost	– cyclonic damage, colonies of massive *Porites* (Wanuru)
	– widespread bleaching of shallow water corals (Loltong)
Santo and offshore islands	– minor bleaching (Big Bay)
	– communities established on fossil reefs with no signs of damage (Hog Harbour)
	– exceptionally rich and beautiful coral communities (Lathu)
	– floods and cyclones have caused major physical damage to reefs (Turtle Bay)
	– western shore virtually has no coral colonies (Malao)
	– well stratified patch reefs and slope areas showed rich fauna despite the little damage and wave-beaten topography
	– *A. planci* and *Drupella* spp. are the major cause of collapse of beautiful coral colonies (Palikulo)
Reef islands	– spectacular reef slopes and lagoonal holes. The tops of reefs appear smothered and lack corals
Ureparapara	– recent beds of *Goniastrea* corals with dead, dome-shaped tops.

branching corals (*Acropora* and Pocilloporidae) with sheltered areas (reef flats, lagoons) dominated by *Acropora* (*A. palifera*) and *Montipora* (*M. digitata*). These shallow communities change abruptly or gradually to dominance by a mixture of massive and branching corals 3–5 m below the level of the reef flat. Massive corals, notably *Diploastrea heliopora* and *Goniopora* spp. are locally abundant, both on steep slopes and on ledges. Embayments are dominated by massive *Porites* (branching *P. cylindrica*), and by various *Acropora* and *Turbinaria*.

Various species of *Acropora* and *Montipora* are characteristic of the various partially sheltered habitats, for example, reef flats, (subtidal and intertidal), and lagoon slopes. Forests of thick branching colonies of *Acropora palifera* are characteristic of subtidal flats (about 1 m below low water level). Staghorn and plate *Acropora* are characteristic of lagoonal slopes, and thickets of *Montipora digitata* are characteristic of intertidal flats near the shore.

Massive *Porites*, though present in most reef habitats, are particularly characteristic of open embayments, and by branching *P. cylindrica*, various *Acropora* and *Turbinaria*. In sheltered parts of open bays, soft corals, (especially *Sinularia flexibilis*) are co-dominants with massive *Porites*.

Most sheltered reefs in Vanuatu are strongly dominated by soft corals and *Porites*. On shallow rubble and sand flats, *Acropora aspera* and *Montipora stellata* frequently form extensive beds. On muddy slopes, *Turbinaria frondens*, *Porites cylindrica* and various branching *Acropora* frequently reach large size.

Status of Reefs in Vanuatu

Vanuatu is endowed with many beautiful reefs and lagoons. But despite low population pressures, reefs were generally considered to be in poor condition. Severe cyclone damage was evident at 50% of sites surveyed in 1988 and coral bleaching was widespread (Done and Navin, 1990). Active crown-of-thorns starfish outbreaks were seen at another 20% of sites (Zann et al., 1990).

Coral reefs have provided a harvest of valuable protein for ni-Vanuatu villagers (David and Cillaurren, 1992). However to date, exploitation of reefs has been of little importance to the national economy (Chambers, 1990). A small but economically significant dive-tourism industry has been based on Vanuatu's coral reefs and on wreck diving. The reefs have also been heavily exploited.

Vanuatu's location in the cyclone belt and on the margin of the New Hebrides plate (Done and Navin, 1990), is of particular relevance with respect to disturbances to coral reefs. So too is the increasing urbanisation of the capital, Port Vila. Done and Navin (1990) considered that reefs were moderately impacted by human activities. Some reefs were threatened by increased siltation from logging of catchments, nutrient enrichment from agriculture lands and urban sewage, and over-exploitation of reef fish.

OFFSHORE SYSTEMS

Offshore systems and information on deep-sea communities are very limited in Vanuatu. Island shelves are generally small or absent, and outer reef slopes generally descend rapidly to depths of 1000–3000 m to the continental margins and abyssal plain and deep seabed at 3000–5000 m.

Upper reef slopes (to 500 m depths) support a diversity of shrimps and molluscs (including *Nautilus*). Deep sea snappers caught on the outer reef slope of the islands of Vanuatu are of commercial importance. A total of 107 species are caught from depths ranging from 100 to 400 m, and deep-sea fish belonging to the family Etelidae form the bulk of the catch (Brouad and Grandperrin, 1985). Some deep-sea surveys have been conducted in some areas for precious corals and for deep-sea minerals, manganese nodules and hydrothermal deposits, at depths of 2–3 km (Howarth, 1998).

The Secretariat for the Pacific Community has undertaken research investigating the distribution and abundances of migratory tuna and billfish in the Pacific region. Results indicate that these species may migrate many thousands of kilometres and that they are particularly affected by ENSO events (Anon., 1998).

POPULATIONS AFFECTING THE AREA

The people of Vanuatu are Melanesian in origin. The ni-Vanuatu comprise 94% of the total population, and the remainder are European, Chinese and other Pacific Islanders. Over half of Vanuatu's population live on the islands of Efate, Espiritu Santo and Malakula (Baines, 1981).

The village of Mele, some 10 km outside Port Vila, with a population of about 2000 people, is the largest village in Vanuatu (Republic of Vanuatu, 1985). In 1989, a national census recorded the total population at 142, 944 (Government of Vanuatu, 1991). The rural population is 116,600. Vanuatu's population is expanding at an average rate of 2.9% annually. The urban population density is 11.6 persons/km^2 and the rural density is 9.5 persons/km^2 (Environments of Vanuatu, 1997).

Vanuatu has a dual economic structure with a substantial subsistence sector co-existing with the modern cash economy. The proportion of the population that depends on the subsistence economy is estimated to be 80%, these people being subsistence farmers and fishers who live in small villages. The cash economy is derived from primary activities, notably fisheries, timber, copra and beef, and by services such as government and tourism (Douglas and Douglas, 1994).

The economy is primarily agriculture-based, and beef, copra and fish remain the primary exports. Industrial scale logging also occurs as well as a small industrial sector mainly in Port Vila. A brewery, a soft drink manufacturer, biscuit factory and other manufacturers and commercial operations also contribute to the economy (Fairbairn, 1992).

RURAL FACTORS

Agriculture and Impacts

Agriculture in Vanuatu is characterised by an increasing diversity of crops contributing to both subsistence and commercial production. Some of the subsistence crops include yam, taro, sweet potato, manioc, banana, breadfruit and other minor subsistence crops. Other cash crops include arrowroot, cotton, maize, coffee, and cocoa. Non-traditional export crops are kava, pepper, and vanilla. Rainforests provide timber in Vanuatu but these have been overexploited and production is declining (Douglas and Douglas, 1994).

The goal of the Vanuatu Department of Agricultural Research is to encourage the more efficient use of land and to increase production. This is done through the association of crops and livestock, plant and animal breeding, and the control of pests and diseases. For many years copra has been Vanuatu's most important export commodity. Thus, coconut plantations have replaced lowland rainforest ecosystems. The increase in production of copra as a single commodity followed the collapse of cocoa and coffee production, and a decline in the cattle sector. There is a renewed interest in the development of cocoa, coffee and cattle.

Agrochemical use in Vanuatu is small considering the large importance and size of the agricultural sector. Only some fertilizers and pesticides are imported each year. The Agriculture Department does not encourage the use of agrochemicals on its Pilot Plantation Project which uses integrated pest management procedures using the right crops and grasses to minimize weeds (Stone, 1992). The malaria prevention program may contribute greater quantities of pesticides to the environment than agriculture because of the heavy use of DDT and other mosquito-control pesticides that are sprayed as part of the programme.

Fisheries and Impacts

Inshore mangroves, lagoons and coral reefs traditionally provide most of the protein, and they remain important for subsistence in rural areas. Some hundreds of species of fish and invertebrates are taken using a large number of fishing techniques. Normally women and children collect shellfish from mangroves and shores while men fish from canoes or skin-dive in the lagoons or reefs (Fig. 2).

The major impacts from fisheries are a loss of inshore fish stocks resulting from a high demand for fresh fish from urban areas and more effective fishing techniques. For example, modern nylon fishing lines, steel hooks, imported lures, nylon gillnets, bigger boats, depth sounders and freezers have increased catches (Zann, 1994). Second order impacts of fishing are also unknown. For example, it has been suggested that outbreaks of crown-of-thorns starfish result from overfishing or collection of natural predators (Endean, 1976).

Vanuatu has an unsatisfied domestic demand for fresh fish. It has limited baitfish resources and little capital to exploit the offshore tuna fishery. Japan, and to a lesser extent other nations, have entered into a number of joint ventures which earn the island state additional revenue and other gains e.g. a South Pacific Fishing Company set up in Malakula. The chances of Vanuatu significantly increasing its tuna catches or establishing a cannery with further processing facilities is low, however. Foreign-based fishing fleets have shown little interest in Vanuatu's waters and the Vanuatu government has itself concentrated its efforts on its inshore fisheries (Waugh, 1986).

The present level of fisheries activity in Vanuatu is therefore relatively low and the fisheries sector remains undeveloped in comparison with other island states in the South Pacific Region. Traditionally, there has been little fishing, and knowledge of stocks is poor. Although fishery activity is low, small-scale fisheries for beche-de-mer, green snail and *Trochus* provide sufficient output for export markets. A small factory in Port Vila which manufactures buttons from the trochus shell has had difficulties in remaining operational due to inconsistent supplies. Small fisheries for rock lobster and deep-bottom snapper exist for local supply. *Trochus* resources in Vanuatu have been harvested for subsistence and commercial purposes since the beginning of the 19th century, but not until recently have they been under high fishing pressure as a result of the increase in demand for the shells by overseas markets.

Vanuatu is perhaps an exception among island states. It may need greater government assistance over a longer period of time to sustain the momentum of fisheries development (Waugh, 1986). Fisheries, both coastal and oceanic, are viewed as a major area for development. The Fisheries Department plays a key role in the development of commercial projects in fisheries and fish licensing.

Fig 2. A traditional outrigger canoe used in the artisanal fishery.

Vanuatu's population is relatively small, and inshore marine resources are generally not seriously over-exploited when compared to other islands in the South Pacific Region. Commercial fisheries largely target deep-water species beyond the reef, so conflict between commercial and subsistence fishermen appears to be infrequent. Potential exists for conflict over access to beche-de-mer, *Trochus* and green snail.

Fish are only of moderate importance in the subsistence diet and the average fish consumption is 18 kg/person/yr (Dalzell et al., 1996). There is a small artisanal fishery based on deep-water species (Brouard and Grandparrin, 1985). The total landings of inshore fish are around 2400 tonnes/year (David and Cillaurren, 1992).

COASTAL EROSION AND LANDFILL

In Vanuatu, coastal modifications have been minimal, and are largely confined to the capital, Port Vila. Several kilometres of seawalls have been constructed in the port area, and along Erakor Lagoon. In the short term, natural impacts from cyclones and seismic movements of the coastline have had far greater impact on coral reefs (Done and Navin, 1990). In 1965, parts of the northwest coast of Santo and Malakula were forced upwards almost 2 m, as a result of earthquake activity. At the same time a tsunami (5 m in height) hit the southwest coast of Santo.

Most of Vanuatu's population lives along the coast and infrastructural development on coastal sand deposits is at risk from storm surges and from tsunami waves. The only practical approach to coastal development is to avoid unstable areas, establish buffer strips and encourage natural processes of coastal protection by mangroves (Baines, 1981).

Coastal erosion has been a recurring problem at Mele Bay, Efate for a number of years (Temakon and Harrison, 1988) and has become a major concern. Estimates of land loss vary, but signs of coastal erosion are evident and include erosional scarps at the beach edge, exposed tree roots, fallen trees that litter portions of the beach, short and steep beach faces and undercut building structures. Coastal erosion at Mele Bay has been aggravated in recent years by the unregulated mining of beach sand at two sites, one to the east of Swango point and one to the west. Both areas show signs of coastal erosion. Surveys show that the beaches of the two areas are relatively steep and narrow (Douglas, 1990).

Due to the narrow continental shelf and steep continental slope of the nearshore zone of Mele bay, sand that is transported offshore during storms is probably permanently lost to the sediment budget system. When sediment lost is added to the sand that is extracted by beach mining, an average of 450–500 m^3 is lost per month and the coast becomes even more susceptible to erosion (Douglas, 1990). It appears that sediment input into the bay is unable to supply enough sediment to compensate for the amount extracted by miners. A possible solution to the problem of mining of beach sand is to find several other areas in which to mine sand.

EFFECTS OF URBAN AND INDUSTRIAL ACTIVITIES

Urban development in Vanuatu, as elsewhere in Melanesia, has been clearly associated with European colonial expansion during the 19th century. The two major towns are located in Port Vila and Luganville (Cairo, 1994). In Vanuatu, environmental issues that primarily affect the rural population include marine degradation, erosion and sedimentation and in the urban area of Port Vila, sewage contaminated lagoons. Industrial contribution to marine environmental problems is still quite small.

Sewage disposal in Vanuatu is one of the more serious environmental concerns, particularly in urban areas. Water-related diseases are common and this suggests that public health and the environment are both affected by the type of sanitation facilities available. In all villages in Vanuatu, sewage and wastes are untreated. Previous studies identified poorly functioning sanitary facilities together with rapid groundwater flow as the causes of the microbial and nutrient population in the lagoons around Port Vila (Sinclair-Knight, 1991). The uncontrolled flow of nutrient and bacterial contamination combined with the poor natural flushing in these lagoons results in decreased oxygen concentrations, high turbidity and contaminated marine food resources. In Vanuatu, Erakor lagoon near the capital, Port Vila, is severely eutrophic (Naidu et al., 1991).

Vanuatu's industrial sector is relatively small. With the exception of the beef, fish-freezing, and timber industries, the manufacturing industry is almost exclusively for local consumption. Industries for the local market include, soft drink manufacturing, printing, small cement works, brewery, soap manufacturing, garment manufacturing, printing, baked goods, milk production and boat building.

Waste management in the industrial sector is minimal. The brewery and soft drink manufacturing facilities utilize septic tanks for their waste effluents. Simple treatment is possible through septic tanks but overflow and seepage can be a problem for these types of facilities. Overflow from septic tanks enters a marshy area and eventually enters the nearby bay. Small wastewater-producing activities such as laundries, printers, restaurants, and photo shops discharge their wastes to overflowing septic tanks or directly to storm drains.

Solid waste disposal, as elsewhere throughout the region, is poor in the urban centres of Vanuatu. The Port Vila dump at Fres Wata is of particular concern, as it is located over a water supply. In Port Vila, storm debris and litter are sometimes dumped directly into the harbour area as landfill (Sinclair-Knight, 1991). There is no refuse collection or designated refuse dumps; beaches, streams and lagoons are often heavily littered.

PROTECTIVE MEASURES

Marine conservation in Vanuatu includes both traditional and introduced modern practices. Customary law prevails throughout, except in the urban areas of Port Vila and Santo. Vanuatu has many cultural and language groups, and although details of customary sea tenure vary greatly from place to place, the legal status of customary ownership of land and sea is enshrined in the national constitution. Some traditional practices include placing of curses on reefs so villagers are threatened with supernatural retribution and fines. Some curses are still being practised in some villages, and fines are in the form of pigs, cattle or cash, or both livestock and cash (Amos, 1993).

In Vanuatu, the traditional approach to resource management is similar to other countries in the South Pacific Region and is based on chiefly authority, custom and "tabu" restrictions. Although these approaches are widely practised, they have declining authority and influence because of social changes and especially of shifting emphasis from the communal to the individual (Whyte et al., 1998). Johannes (1998) documented several villages (Marae, Erakor, Mangaliilu, Emau, Ebau, Uripi and Analgawat) which have now abandoned customs in which reef taboos were traditionally related to certain occasions such as a chief's death or initiation, a family member's death and circumcision ceremonies.

At the national level, the Vanuatu Fisheries Department now realizes that developing its fisheries legislation and managing most of its coastal fisheries from Port Vila are impossible. The fisheries legislation controls commercial fisheries through licences, restricts the use of certain gears and mesh sizes, imposes size limits on certain species and prohibits the use of poisons.

The research, monitoring and enforcement in a vast area of ocean with many scattered islands such as exist here would not only be costly but would outweigh the benefits. The Fisheries Department is now playing a vital role in providing awareness programmes to help villagers to combine traditional knowledge of their fishing grounds with modern management practices, to improve village-based marine resource management. A major drawback is that the Fisheries Department has very little manpower to promote awareness. However, the Fisheries Department began assisting its fishing villages in community-based marine resource management in the early 1990s (Amos, 1993). The assistance by government was initially for the conservation of one species of *Trochus*. As a result of the benefits of these community-based initiatives, fishing controls of many species of fish and invertebrates were put in place by the villagers themselves (see Tables 3 and 4).

The fishing controls were in several forms and ranged from closure of fishing grounds to species-specific closures (Tables 3 and 4). There were specific closure periods for *Trochus*, beche-de-mer, limpets, parrotfish, shoe crabs, and rudderfish. In addition, restrictions were also imposed on

Table 3

Village-based fishing restrictions (source: Johannes, 1998)

Practice Species tabooed	Area
(1) All species tabooed	Marea, Lamen, Emua, Ebau, Pelong, Uri, Litslits, Uripi, Tongamea, Tabakoro, Analgawat, Lamap
Trochus and green snail	Lamen, Erakor, Eton, Mangaliilu, Seviri, Pelong, lutes, Marae, Tabakoro, Utche, Analgawat
Trochus, greensnail and beche-de-mer	Tongamea
Trochus, green snail and lobster	Aneitcho (Port Patrick)
Octopus	Sangava, Marae, Pescarus
Beche-de-mer	Mangaliilu, Emua
All shellfish	Marae
Mangrove crabs	Uri
Shore crabs, limpets, parrotfish and rudderfish	Anawanjei
Rock lobsters	Analgawat
	Areas
(2) Methods tabooed	
All methods	Marea, Lamen, Emua, Ebau, Pelong, Uri, Litslits,Uripi, Tongamea, Tabakoro, Pescarus
Night fishing	Pelong, Uri
Gillnetting	Pelong, Tabakoro, Utche, Uri
Dropline fishing	Tabakoro
All methods except bow and arrow or throwing spear	Lamap
All methods except line fishing and daytime spearfishing	Pelong
Breaking coral or not replacing rocks properly while reef gleaning	Erakor, Marae
Gillnetting	Utche
Commercial gillnetting	Uripi
Spearfishing	Utche
Night Spearfishing	Mangaliilu
(3) Size restrictions	
Beche-de-mer	Uri
Mangrove crabs	Uri, Sangava

Table 4

Village-based fishing ground closures (source: Johannes, 1998)

Practice Fishing grounds closures	Areas
(1) Extent of areas affected by closure	
Throughout village	Mangaliilu, Erakor, Lamen, Emau, Paunangisu, Uri, Norsup, Tautu, Tabakoro, Aneitcho (Port Patrick)
About half the village	Marae, Ebau, Litslits, Sangava.
Smaller portion	Uri, Pelong, Tabakoro, Tongamea
Individual subdivisions owned by different descent groups i.e clans or families	Lamen, Uripi, Pescarus, Pelong, Lamap, Lutes, Tautu, Marae, Analgawat
	Areas
(2) Duration of closures	
One to seven months	Sangava, Uri
One year	Erakor, Mangaliilu, Paunangisu, Ebau, Pelong, Lamap, Norsup, Tautu, Tabakoro, Marae, Pescarus, Analgawat, Anawanjei
Two years	Lamap, Uripi, Pelong, Sangava, Analgawat
Three years	Marae, Lamap, Tabakoro
Five years	Uri, Pelong, Seviri, Pescarus, Aneitcho (Port Patrick)
"Until it is decided that the area is ready"	Tongamea
Indefinitely	Mangaliilu, Emau
(3) Villages in which reef taboos were traditionally related to	
Chief's death or initiation	Marae, Lamen, Erakor, Mangaliilu, Emau, Ebau, Uripi, Analgawat
Family members death	Lamen
Circumcision ceremony	Uripi

specific methods of fishing, for example, gillnetting and night spearfishing. By late 1993, 26 coastal villages of Vanuatu were reported to have introduced fishing controls for the purpose of marine resource management in their fishing grounds (Johannes, 1998).

There are several reasons why there have been so few marine protected areas in Vanuatu in the past. Many ni-Vanuatu are not aware of the importance of marine protected areas and western ideas of marine national parks are not culturally appropriate. Many ni-Vanuatu rely on subsistence fishing as their major source of protein and are not willing to give up ownership of fishing grounds for a national park. They are also spiritually linked to their lands and seas.

However, marine protected areas are an important strategy to conserve marine biodiversity, particularly in the absence of scientific knowledge and monitoring. Vanuatu has a moderate number of marine protected areas namely: President Coolidge and Million Dollar Point Reserve (WW II shipwreck and wreckage, Espiritu Santo); Narong Marine Reserve (Malakula, 160 ha) (see Table 5).

In particular, the President Coolidge and Million Dollar Point Reserve covers the area off the south coast of Espiritu Santo Island and is located 6 km from Luganville. It includes the wreck of the American wartime troopship *President Coolidge* and the area known as Million Dollar Point (Crossland, 1984). The *President Coolidge* sank in 1942 and the wreck is over 198 m long and 24.4 m wide. The wreck was declared a marine reserve on 18 November 1983 by the Minister of Land and Natural Resources. The wreck supports a large population of fish, notably, groupers and cod. It is also covered with corals. Sand and gravel cannot

Table 5

Marine conservation activities in Vanuatu (source: Whyte et al., 1998)

Island	Location	Village Community	Marine Conservation Activity
Efate	Emua	Emua, Saama	Tabu on reef
Efate	Efate	Emua, Saama	Tabu on turtle harvesting
Efate	Erakor Reef	Erakor Village	Trochus restocking/tabu area
Efate	Hideaway Is.	Mele Village	Trochus restocking/marine reserve
Efate	Ifira	Ifira Village	Trochus restocking/tabu area
Efate	Mangaliliu	Mangaliliu Village	Protect marine resources
Efate	Onesua	Ulei Village	Tabu on marine resources
Efate	Paunangisu	Paunangisu Village	Protect beche-de-mer
Efate	Siviri	Siviri Village	Trochus restocking/tabu area
Efate	Takara	Takara Village	Tabu on turtle harvesting
Efate	Utanlang	Utanlang Village	Tabu to maintain reef resources
Efate	Utanlang	Utanlang Village	Tabu on turtle harvesting
Emae	Marae	Marae Village	Trochus restocking/tabu area
Emae	Sasake	Sasake Village	Trochus restocking/tabu area
Epi	Ponkovio	Ponkovoi Village	Custom tabu protecting marine resources
Maewo	Kerepei	Kerepei Village	Tabu protecting marine resources
Maewo	Lolorugu	Naumumu Village	Manage resource use
Malaku la	Nagha & Pineia Protected Area	Wiawi Village	Marine and forest conservation
Malaku la	Narong Marine Reserve	Selanamboro Village	Protect marine nursery area
Malaku la	Nawo Marine Conservation Area	Uripiv Island	Conserve reef and marine area
Malaku la	Nevnal	Leviamp Village	Conserve marine and forest area
Malaku la	Patpang Res	Patpang Village	Protect coconut crab
Malaku la	Vendik Pik	Bamboo Bay Village	Conserve marine area
N. Efate	Unakap	Unakap Village	Trochus restocking/tabu area
N. Efate	Woraviu	Pele Village	Trochus restocking/tabu area
Paama		Lironesa Village	Protect custom/marine resources
Pentec ost	Lekavik	Lebwibwi Village	Tak tabu on marine area
Pentec ost	Lekavik	Lebwibwi Village	Tak tabu on marine area
Santo			Tabu to maintain coconut crab stocks
Santo	Hulu Wildlife Protection Area	Khole Village	Protect bird and flying fox habitat
Santo	Lope Lope Reef	Landholders	Maintain coastal resources
Santo	Loru Conservation Area	Khole Village	Protect forest/marine resources
Santo	President Coolidge Marine Reserve	None	War grave and dive site
Santo	Sarete	Sarete Village	Protect forest/marine resources
Santo	Vatthe Conservation Area	Sara and Matantas Villages	Forest conservation/community development

be removed from the reserve and removal of wreck souvenirs is prohibited. Fishing in the marine reserve for all types of marine organisms is banned.

There is a recent upsurge in awareness-raising and conservation activities in Vanuatu (see Table 5). In particular, marine conservation activities have been initiated in response to marine-awareness workshops, radio broadcasts and drama activities of government departments, non-government and regional organisations. For example, the Fisheries and Environment Departments, Wan Smol Bag Theatre Company, Foundations for the People of the South Pacific—Vanuatu and the South Pacific Regional Environment Programme. A result of community response to awareness raising is a 10-year "tabu" or closure placed on marine turtles at Utanlang (Whyte et al., 1998).

CONCLUSIONS

The status of Vanuatu's marine environment is relatively good, although around population centres it shows signs of degradation. Many of Vanuatu's villages, towns and tourist resorts are constructed close to the seashore and these have degraded the coastal environment through clearing of forests and poor agriculture practices. Overfishing is also widespread on most of Vanuatu's inshore reefs and those close to major population centres.

Domestic wastes, sewage and solid wastes are the major contributors to marine pollutants. As in other countries of the region, industrial contributions appear to be growing. Waste management practices should continue to be improved. There is no regular monitoring programme for the marine environment, and it is critical that data on industrial activity be collected and analysed routinely to monitor environmental quality.

There is generally a lack of medium to long-term planning for the marine environment in Vanuatu, as is the case elsewhere in the region. There needs to be forward planning and incorporation of environmental considerations into the earliest phases of development. This will help avoid one of the most common problems in the region, that is, acting to solve problems only after they have arisen. A major drawback in formulating long-term planning is the problem of insufficient environment legislation to cope with problems. In addition, there are also very few trained marine scientists in Vanuatu and generally a lack of finances to support monitoring, effective implementation and enforcement of legislation and policies.

REFERENCES

Amos, M.J. (1993) Traditionally based marine management system in Vanuatu. *Tradition Marine Resource Management and Knowledge Information Bulletin* 2, 14–17.

Anon (1984) *Pacific Islands pilot*. Vol. 2. Hydrographer of the Navy.

Anon (1997) Tuna follow El Niño. *SPC Fisheries Newsletter* 82, 17.

Anon (1998) Ocean fisheries program. *SPC Fisheries Newsletter* 85, 11–16.

Baines, G.B.K. (1981) *Mangrove Resources and Their management in the South Pacific*. South Pacific Commission, New Caledonia.

Brouard, F. and Grandperrin, R. (1985) Deep-bottom fishes of the outer reef slope *in Vanuatu*. SPC/Fisheries 17/WP. ORSTOM Centre, Noumea, New Caledonia.

Cairo (1994) Pacific island populations. Report prepared by the South Pacific Commission for the International Conference on population and development 5–13 September, 1994.

Carter, R. (1983) Baseline studies of Port Vila and Erakor Lagoons, Vanuatu. Cruise report No. 82. Economic and Social Commission for Asia and the Pacific. UNDP.

Carter, R. (1985) Baseline current and water density studies in Mele Bay and Teuma Bay, Vanuatu. Cruise Report no.99. South Pacific Applied Geoscience Commission (SOPAC), Suva, Fiji. Pp. 14.

Chambers, M. (1990) Seagrass communities, In *Vanuatu Marine Resources*, eds. T.J. Done and K. Navin, pp. 92–103. Australian Institute of Marine Science, Townsville.

Convard, N. (1993) Land-Based Pollutants Inventory for the South Pacific Region. Envirosearch International. SPREP, Apia, Western Samoa.

Crossland, J. (1984) Vanuatu's first marine reserve. *Naika* 14, 2–3.

David, G. (1985) Le perche villageoise a Vanuatu: recensement 1. Moyens de production et production globale. Report Mission ORSTOM, Vanuatu.

Dahl, A. (1979) Marine ecosystems and biotic provinces in the South Pacific Area. In *Proceedings of the International Symposium on Marine Biogeography in the Southern Hemisphere*. New Zealand DSIR Information series No. 137.

Dalzell, P.T., Adams, T. and Polunin, N. (1996) Coastal fisheries in the Pacific Islands. *Oceanography and Marine Biology: An Annual Review* 34, 395–531.

David, G. and Cillaurren, E. (1992) National fisheries development policy for coastal waters, small scale village fishing, and food self-reliance in Vanuatu. *Man and Culture in Oceania* 8, 35–38.

Done, T.J. and Navin, K.F. (1990) Vanuatu Marine Resources: Report of a Biological Survey. A Project of the Australian International Development Assistance Bureau. Australian Institute of Marine Science, Townsville.

Douglas, M.R.(1990). Baseline study of coastal erosion at Mele Bay, Efate, Vanuatu. South Pacific Applied Geoscience Commission (SOPAC), Suva Fiji. SOPAC Technical Report 116.

Douglas, N. and Douglas, N. (1994) *Pacific Islands Yearbook*, 17th Edition. Suva, Fiji Times.

Endean, R. (1976) Destruction of coral communities. In *Biology and Geology of Coral Reefs. Vol 3: Biology* 2, eds. O.A. Jones and R. Endean, pp. 215–254. Academic Press, New York.

Environments of Vanuatu (1997) A classification and atlas of the natural resources of Vanuatu and their current use as determined from VANRIS. Prepared by the CSIRO Brisbane and the Department of Primary Industries Forest Service for the Australian International Development Assistance Bureau.

Fairbairn, T.J. (1992) Reef and lagoon tenure in the Republic of Vanuatu and prospects of mariculture development. Research Reports and Papers in Economics of Giant Clam Mariculture. Department of Economics, University of Queensland, Australia.

Government of Vanuatu (1980) Land reform regulation No. 31 of 1980. Port Vila.

Government of Vanuatu (1991) Vanuatu National Population Censuses, May, 1989. Statistics Office, Port Vila, Vanuatu.

Holthus and Maragos (1992) Marine biological diversity conservation in the Central/South Pacific region. Report to CNPPA.

Howarth, R. (1998) A review of non-living resources and threats to the Pacific region. South Pacific Applied Geoscience Commission, Suva Fiji, Technical Report 247.

IUCN (1988) *Coral Reefs of the World, Volume 3: Central and Western Pacific*. IUCN, Gland.

Johannes, R.E. (1998) Government-supported, village–based management of marine resources in Vanuatu. *Ocean and Coastal Management* 40 (2–3), 165–186.

Jokiel, P.L. (1990) Transport of reef corals into the Great Barrier Reef. *Nature* 347, 665–667.

Kay, E.A. (1980) *Little Worlds of the Pacific: An Essay on Pacific Basin Biogeography*. University of Hawaii, Hawaii.

Mallick, D.I.J. and Ash, R.P. (1975) Geology of the Southern Banks Islands, New Hebrides Condominium. Geological Survey. British Service, New Hebrides. 33 pp.

Marshall, A.G. and Medway, L. (1976) A mangrove community in the New Hebrides, south-west Pacific. *Biological Journal of the Linnean Society* 8, 319–336.

Naidu, S., Aalbersberg, W.G.L., Brodie, J.E., Fuavao, V.A., Maata, M., Naqasima. M., Whippy, P. and Morrison, R.J. (1991) Water quality studies on selected South Pacific lagoons. UNEP regional Seas Reports and Studies No. 136, UNEP; and SPREP reports and Studies No. 49.

Republic of Vanuatu (1985) Report of the Disaster Preparedness Workshop for Non-governmental Organisations. Held in the Republic of Vanuatu 19–23 August, 1985.

Sinclair-Knight (1991) Report on water quality improvements for Port Vila, Vanuatu. World Bank, Government of Vanuatu. June, 1991. Sinclair-Knight and Partners, Port Vila, Vanuatu.

Stoddart, D.R. (1984) *Biogeography and Ecology of the Seychelles Islands*. W. Junk Publishers, Netherlands.

Stone, A. (1992) Contingent valuation of the Barmah wetlands, Victoria. In *Valuing Natural Areas—Applications and Problems of the Contingent Valuation Method*, eds. M. Lockwood and T. DeLacy. Albury, New South Wales, The Johnstone Centre of Parks, recreation and heritage, Charles Sturt University.

Tacconi, L. and Bennett, J. (1997) Protected area assessment and establishment in Vanuatu: A socio-economic approach. ACIAR, Canberra, Australia.

Temakon, J.S. and Harrison, W. (1988) Nearshore and coastal mineral resources of Vanuatu: Manuscript presented at CCOP/SOPAC Nearshore Minerals Workshop, Savusavu, Fiji, September-October, 1998.

Thollot, P. (1996) *Les Poissons de Mangrove du Lagon Sud-ouest de Nouvelle-Caledonie*. ORSTOM, Paris.
Veron, J.E.N. (1986) *Corals of Australia and the Indo-Pacific*. Angus and Robertson, Sydney.
Veron, J.E.N. (1990) Hermatypic corals. In *Vanuatu Marine Resources*, eds. T.J. Done and K. Navin, pp. 37–65. Australian Institute of Marine Science, Townsville.
Waugh, G. (1986) *The Development of Fisheries in the South Pacific Region with Reference to Fiji, Solomon Islands, Vanuatu, Western Samoa and Tonga*. National Centre for Development Studies, Canberra, Australia.
Weightman, B. (1989) *Agriculture in Vanuatu. A Historical Review*. The British Friends of Vanuatu, Surrey.
Whyte, J., Siwatibau, S., Tapisuwe, A., Kalotap, J. and Fraser, T. (1998) Participatory resource management in Vanuatu : research report. Report on research conducted by FSPI for ACIAR project ANRE96224 "Participatory planning and management of resource development in Vanuatu".
Williams, D.McB. (1990) Shallow-water reef fishes. In *Vanuatu Marine Resources*, eds. T.J. Done and K. Navin, pp. 66–76. Australian Institute of Marine Science, Townsville.
Zann, L.P., Ayling, A. and Done, T.J. (1990) Crown-of-thorns starfish. In *Vanuatu Marine Resources*, eds. T.J. Done and K. Navin, pp. 104–118. Australian Institute of Marine Science, Townsville.
Zann, L.P. (1991) The inshore resources of Upolu, Western Samoa. Coastal inventory and fisheries database. Field report 5. SAM/89/002 FAO, Rome.
Zann, L.P. and Muldoon, J.W. (1992) Management of marine resources in the Kingdom of Tonga. Tonga National Tourism Plan. Nicholas Clark and Associates, Australia.
Zann, L.P. (1994) The status of coral reefs in the South Western Pacific Islands. *Marine Pollution Bulletin* **29**, 52–61.
Zann, L.P. and Vuki, V. (1994) Marine environmental management and the status of customary marine tenure in the Pacific Islands. In *Traditional Marine Tenure and Sustainable Management of Marine Resources in Asia and the Pacific*. Proceedings of the International Workshop 4–8 July, 1994, eds. G.R. South, D. Gouley, S. Tuqiri and A. Church, pp. 62–70. International Ocean Institute, University of South Pacific, Suva, Fiji.
Zann, L.P. (1998) The inshore resources of Savaii, Western Samoa. Samoa Fisheries/AusAID, Apia, Samoa.

THE AUTHORS

Veikila C. Vuki
Marine Studies Programme,
University of the South Pacific,
P.O. Box 1168, Suva, Fiji

Subashni Appana
Marine Studies Programme,
University of the South Pacific,
P.O. Box 1168, Suva, Fiji

Milika Naqasima
Marine Studies Programme,
University of the South Pacific,
P.O. Box 1168, Suva, Fiji

Maika Vuki
Chemistry Department,
University of the South Pacific,
P.O. Box 1168, Suva, Fiji

Chapter 102

THE FIJI ISLANDS

Veikila C. Vuki, Leon P. Zann, Milika Naqasima and Maika Vuki

---

Fiji is one of the largest and wealthiest island groups in the South Western Pacific Island Region and has a very large Exclusive Economic Zone (EEZ) of 1.26 million km². Its many reefs support a substantial subsistence population and a broad-based economy. Rapid population growth has resulted in locally severe overfishing of reefs and lagoons in high population centres in the urban areas. Growth and stock overfishing have occurred with many inshore species, particularly mullet, sea cucumbers and giant clams. There is also significant loss in major coastal habitats because of unplanned development in urban areas.

Turtles, giant clams, seabirds, sea cucumbers, giant triton shell and some species of fish are endangered and the world's largest parrotfish, *Bolbometapon muricatum* (humphead parrotfish) is in danger of being fished to extinction in Fijian waters. Water quality is pristine in rural areas but is poor in urban areas because of sewage pollution and poor land-use practices which cause sedimentation and eutrophication. Outbreaks of crown-of-thorns starfish have caused moderate damage to some reefs, and it is evident that Fiji's coral reefs are under increasing pressure from human activities and will require protection and effective management. There is, however, a recent interest in marine protected areas because customary fishing right owners are becoming aware of the value of marine eco-tourism, habitat protection and fisheries conservation and management.

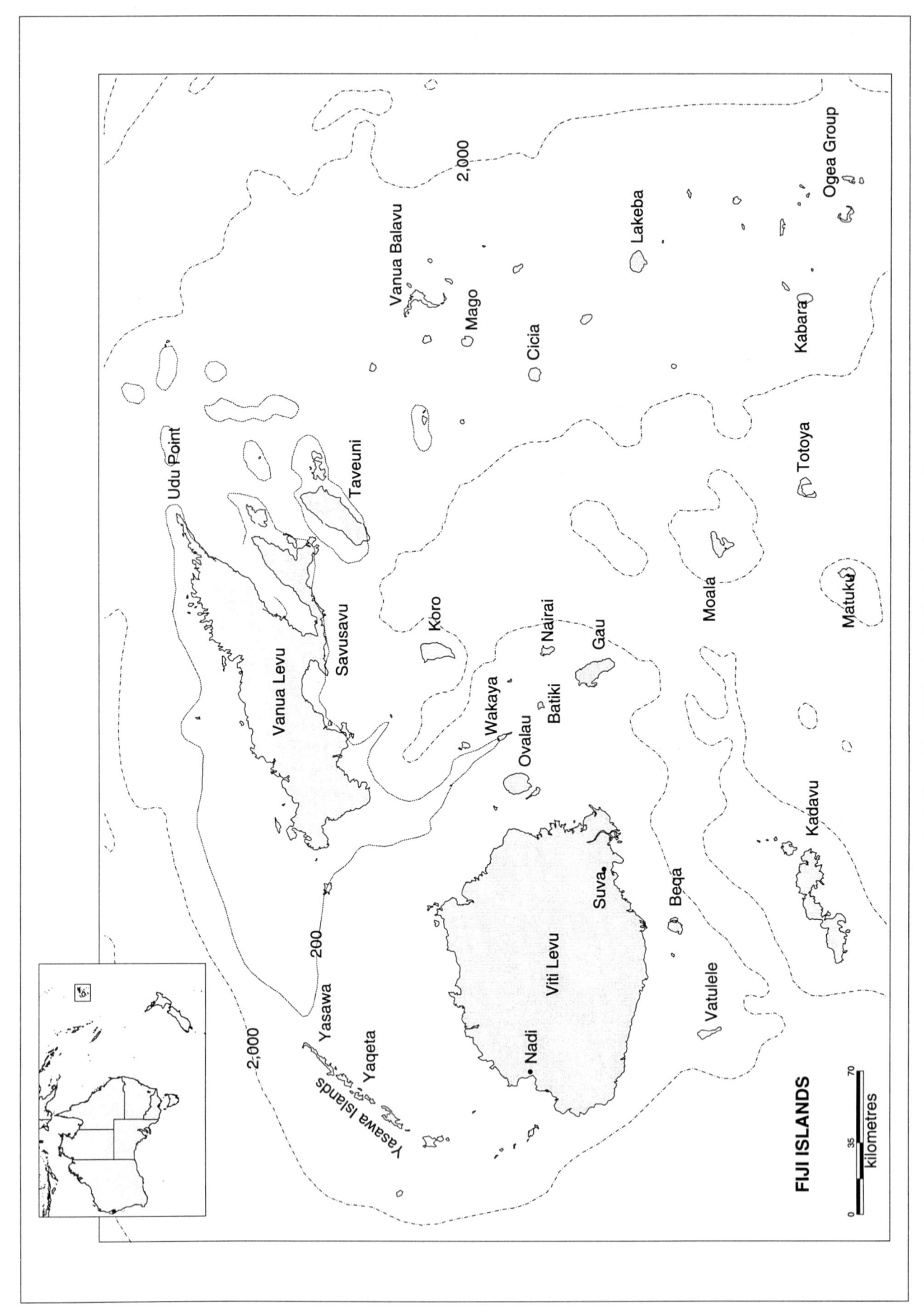

Fig. 1. Map of the Fiji Islands with Exclusive Economic Zones.

THE DEFINED REGION

Fiji occupies a central position in the South Pacific and is one of the largest and most scattered archipelagic groups in the region. It lies close to the Kingdom of Tonga, Western Samoa, Vanuatu and New Caledonia. The nearest metropolitan countries are New Zealand, which lies approximately 2100 km to the south, and Australia, which is 3100 km to the southwest.

Located between 15–23°S and 177–178°W (Fig. 1), the Fiji Islands comprise about 844 high islands, cays and islets, 106 of which are inhabited. It has a land area of 18,500 km^2, and an Exclusive Economic Zone of 1.29 million km^2. The Fiji Group lies within the tropics and so coral reefs are well developed. Most of the islands are surrounded by fringing reefs and many of the reef systems include barrier, platform and patch reefs (IUCN, 1988).

Island Groups

Fiji is one of the largest and wealthiest island groups in the South Western Pacific Island Region, and is economically self-sufficient. Its major islands are Viti Levu (10,386 km^2) and Vanua Levu (5534 km^2), which together account for 87% of the total land area. The other large islands are Taveuni, Kadavu, Gau and Ovalau (Douglas and Douglas, 1994).

The islands are volcanic with some sedimentary rocks, and some atolls exist in the Lau Group. The two large submerged platforms are the Viti Levu and Vanua Levu platforms, which form the Fiji Plateau. The islands may be subdivided into several distinct groups, which are: Viti Levu and adjacent islands; Vanua Levu, Taveuni and adjacent islands; the Lau Group; the Lomaiviti Group; the Yasawas; and the Kadavu and associated islands.

Biogeography

Fiji lies within the Indo-West Pacific Marine Province. Its species diversity is generally high with strong affinities with the South East Asian–Great Barrier Reef Region, which is the centre of Indo-Pacific marine species diversity. Fiji is geographically isolated and, as a result, the number of species is significantly lower than in, for example, the Philippines (Veron, 1995). The reduced number of species compared to Southeast Asian sites may also be related to the position of the Fiji Islands in relation to the Pacific Plate, the largest of the Earth's lithospheric plates.

The Fiji Group has a small number of central Pacific marine endemics. Although a number of marine species are currently known only from Fiji, this probably reflects more the lack of collections than a real feature of endemism. Low endemism is common in most marine groups because of the prevalence of larval transport mechanisms.

SEASONALITY, CURRENTS, NATURAL ENVIRONMENTAL VARIABLES

Climate

Fiji has a moderate tropical climate and is influenced by the Southeast Trade Winds. These are more persistent from July to December. Average air temperature is about 25°C and the hottest months (when air temperatures reach 35°C) are from December to April (Penn, 1982).

The pattern of rainfall within the Fiji Group is highly variable. The windward sides of the large islands have higher rainfall than the leeward sides; the average rainfall ranges from 3000 to 4000 mm on the windward side and is usually about 2000 mm on the leeward sides of islands. The mountainous southeastern part of Viti Levu island has a high annual rainfall associated with tropical cyclones, and the annual rainfall is unevenly distributed because of the rain shadow caused by mountains (Dickie et al., 1991).

High rainfall associated with cyclones often leads to severe floods, land erosion and high sediment loads in local rivers and coastal waters. Rainfall is seasonal with distinct dry (May–October) and wet (November–April) seasons. About 67% of the annual rainfall occurs in the wet season (Dickie et al., 1991).

Cyclones, Tsunamis and ENSO

Tropical cyclones are common in Fiji and can be hundreds of kilometres in diameter. They are characterised by strong winds circling a relatively calm eye and very low atmospheric pressure in the eye. This combination creates elevated sea levels and extremely high rainfall. Most cyclones start in the Inter-Tropical Convergence Zone (ITCZ) during the summer months of November to April when the ITCZ lies between 10 and 15°S (Fenney, 1989), and generally move SSE or SSW reaching peak intensity between 17 and 22°S. The areas in Fiji most affected by cyclones are the Yasawas, West Viti Levu, Kadavu and North West Vanua Levu, Cikobia and the Lau Group (Dickie et al., 1991).

Over the past century, Fiji appears to have experienced an average of one cyclone per year, with a severe cyclone every 3–4 years. There is significant annual and decadal variability; Carter (1990) found there were 15 cyclones which affected Fiji in the period from 1970 to 1989. ENSO events and sea temperature change could also affect the frequency, intensity and location of formation of cyclones in this area.

Eleven tsunamis have been recorded in the Fiji Group since 1877. Three produced waves of nearly 2 m in Viti Levu, and during the 1953 tsunami, considerable reef damage occurred (IUCN, 1988).

The El Niño/Southern Oscillation Index (ENSO) also has far-reaching effects on the climate and oceanography of Fiji. During an El Niño event there is a reversal or a weakening of trade winds, the ITCZ moves south and east, and warm surface waters spread eastward into the Eastern Pacific. Fiji experiences major changes in its oceanography during an El Niño year and this affects the migration of tuna and billfish. During an El Niño event, there are weaker trade winds, lower western Pacific oceanic temperatures and less precipitation (Nunn, 1994). El Niño years are characterised by increased frequencies of cyclones and higher rainfall. Cyclones that coincided with high tide caused havoc in the Yasawas in 1997 because of saltwater intrusion.

Oceanography

Southeasterly swells predominate through the year with significant easterlies occurring during July to December. The wave and swell records show positive correlation with wind data (Dickie et al., 1991). Tidal range is very small, with an annual mean range of 1.1 m. Neap tides have a mean range of 0.9 m and springs reach an amplitude of 1.3 m (Ryland, 1981). Fiji experiences predominantly semidiurnal tides, with the lower low water springs falling during the night in summer and during the day in winter.

Surface current flows southwesterly through the Fiji Group. The annual sea surface temperatures vary from 24°C to 31°C during the year with an average annual variation of 6°C. The normal surface salinity is 35‰ but may drop to 10–15‰ after heavy rainfall (Zann et al., 1987).

In lagoons, strong tidal currents occur three hours before and after low and high tides. There is high variability of currents within the lagoon and these are usually influenced by strong and continuous trade winds and high tides. The quantity of oceanic water entering the lagoon over the reef top and through different passages also depends on tidal height, which influences the currents within the lagoon (Vuki, 1994).

MAJOR SHALLOW WATER MARINE AND COASTAL HABITATS

Marine Algae and Seagrass

The most complete list of the algal flora to date has listed 422 taxa for the Fiji Islands (N'Yeurt et al., 1996). The best known areas are Suva Lagoon in Viti Levu and the Great Astrolabe Reef in Kadavu. The Fiji algal flora contains very few endemic species and the majority of species are pantropical in distribution. Large areas of Fiji remain unexplored, and these include the Lau Islands and much of Vanua Levu and Taveuni.

The development of seaweed aquaculture has resulted in the import of the exotic alga *Euchema* into Fiji. Recently, the Fiji Fisheries Department has encouraged the cultivation of *Euchema* as part of its development projects in rural areas, and seaweed farms have been established in remote rural areas such as the Southern Lau and Rotuma.

Several other species of algae are edible and are an important part of the Fijian diet; South (1993) has made an ethnobotanical study of edible seaweeds in Fiji. Algae such as *Ulva* and *Enteromorpha* are key indicators of pollution in coastal and estuarine environments (Tabudravu, 1995).

There are four species of seagrasses recorded in Fiji (Table 1) whose beds create common and important features of inshore areas. They are well developed in protected subtidal soft shores and lagoons and back reefs, and are also found intertidally and in the shallow subtidal and soft shores throughout the area. Seagrass have high biological productivity and support a large number of fish and invertebrates that are of fisheries importance.

Table 1

Summary of Fiji's marine biodiversity

	Information known	Sources (references)
Marine plants		
Algae	422 taxa: 39 Cyanophyceae 113 Chlorophyceae 42 Phaeophyceae 228 Rhodophyceae	N'Yeurt et al. (1996)
Seagrass	4 species *Halodule uninervis* *Halophila ovalis* *Syringodium isoetifolium* *Halophila ovata*	Morton and Raj (1980)
Mangrove	9 species *Bruguiera gymnorrhiza* *Rhizophora stylosa* *R. mangles* *R. samoensis* *R. x selala* *Lumnitzera littorea* *Xylocarpus granatum* *X. moluccensis* *Excoecaria agallocha* *Heritiera littoralis*	Whippy-Morris and Pratt (1998)
Invertebrates		
Stony Corals	198 species	Zann (1992)
Gorgonians	5 species	Muzik and Wainwright (1977)
Zoanthids	15 species	Muirhead and Ryland (1981)
Molluscs		
Gastropods	123 species	Parkinson (1982)
Opisthobranch	12 families	
Bivalves	253 species 102 species 25 families	Brodie and Brodie (1990) Parkinson (1982)
Ascidians	60 species	Kott (1981); Ryland et al. (1984)
Vertebrates		
Bony fish	162 families, 1198 species	Baldwin & Seeto (1986)
Reptiles	3 species	Guinea (1980)
Seabirds	10 species	Clunie (1985)
Whales	4 species	Zann (1992)

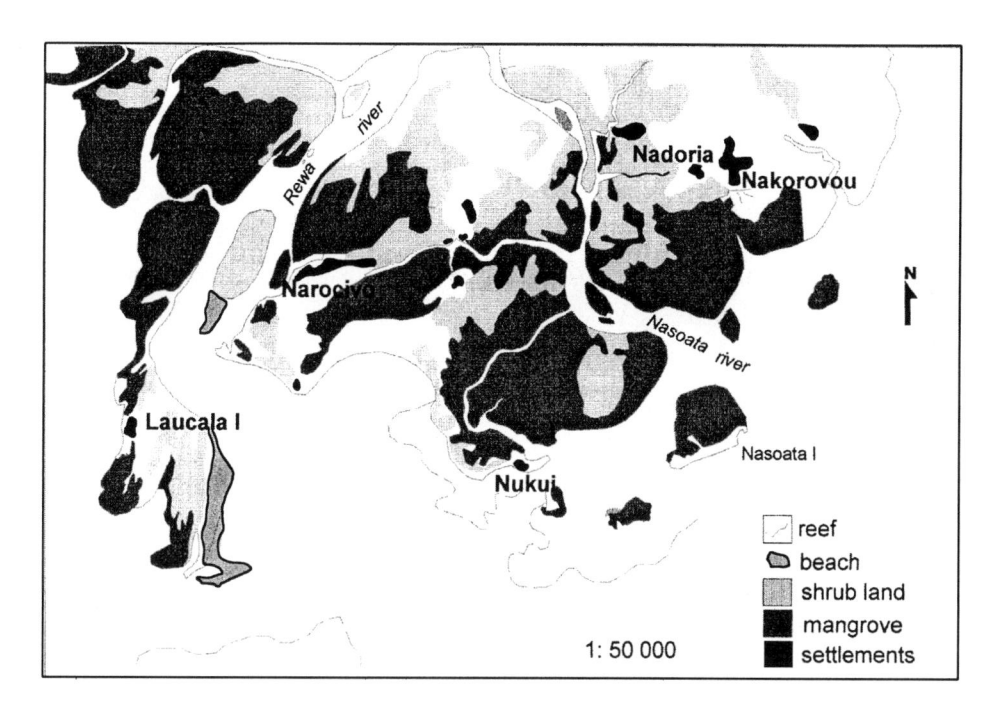

Fig. 2. Distribution of mangrove habitat in the Rewa Delta, near Suva.

Long-term ecological studies of seagrass beds on Suva Reef revealed that losses occurred in some years because of major disturbances such as tsunami, cyclones and flood (Vuki, 1994). Analysis of spatial patterns of seagrass beds from airborne images showed clearly that there were oscillations in abundance on Suva Reef; seagrass beds extended towards the lagoon in some years and regressed in others. Regressions in seagrass beds on Suva backreef areas were attributed to high turbidity and siltation caused by foreshore reclamations (Vuki, 1994).

Mangroves

Fiji has one of the largest formations of mangrove forest (45,000 ha) in the South Pacific (Watling and Chape, 1992). They are well developed in deltas along the mouths of the larger rivers in Rewa (Fig. 2), Ba and Nadi in Viti Levu. They are also developed along the Labasa river, Dreketi river and Bua Bay in Vanua Levu (Watling and Chape, 1992). The growths of mangrove on the southeastern Viti Levu are luxuriant, but are short and stunted on the western and drier parts. On Viti Levu, the Suva-Navua mangroves and the Nadi Bay mangroves are considered to be the most threatened because of development pressure (Watling, 1985). The mangrove of the Rewa Delta in Viti Levu are the most diverse and have been identified as one of the sites important for biodiversity conservation (Whippy-Morris and Pratt, 1998).

Fiji has a considerable area of mangrove but its floristic community is relatively simple by comparison with those in southeast Asia. Species diversity is moderate with five species and a unique hybrid. There are no endemic mangrove species (Morton and Raj, 1980). Fiji's mangrove flora is dominated by *Bruguiera gymnorrhiza*, *Rhizophora mangle* and *Rhizophora stylosa*. The sterile hybrid *Rhizophora × selala* is a cross between *Rhizophora stylosa* and *Rhizophora samoensis*. The mangrove assemblage shows a clear zonation. Seaward zones are of *Rhizophora* and *Bruguiera*, the former especially occupying the most seaward zone where they are adapted to tidal immersion and soft, muddy conditions. *R. stylosa* is common along sandy tidal flats while *R. samoensis* occurs along river channels. The *Bruguiera* grows behind the *Rhizophora* thickets, and the associated *Xylocarpus* and *Exoceocaria* occur to landward of these. These mangrove ecosystems are important traditional sources of food, dyes, medicine and building materials (Watling, 1985). Mangrove wood has excellent burning properties and is a good source of fuel-wood for domestic and industrial uses. Mangrove wood is usually sold in service stations and along roadside stalls.

More than 60% of the commercially important fish species and 83% of subsistence fish species are mangrove-associated at some point in their life cycle (Lal, 1983). Several species of crustaceans, *Cardisoma cardifex*, *Scylla serrata*, *Sesarma erythrodactyla* and *Thalassina anomala* are vital to the subsistence fishery. In areas where extensive reclamation has taken place, there has often been an associated decrease in fish and crustacean production.

Coral Reefs

There are around one thousand coral reefs in Fiji (Zann, 1992) . These reefs are geologically recent structures, being generally younger than 10,000 years and forming a cap of

Fig. 3. Luxuriant coral growth on the well known Great Astrolabe Reef, Kadavu.

biogenic limestone over previous reef formations. The major reef types are fringing reefs, which surround almost all high islands, and barrier reefs that lie at the edges of island shelves. Sizes range from patch reefs which are mostly less than 50 m long, to the 100 km long Coral Coast fringing reef system on Viti Levu and the 370 km long broken barrier reef chain of the Mamanucas/Yasawas/Great Sea Reef System.

Fiji's present coral reefs (Fig. 3) have a range of geomorphological features which reflect their Pleistocene and Holocene histories, during which there has been tectonic uplift and submergence, as reflected in the steep sided nature of its reefs. There are old 'drowned' reefs in Ovatoa, Bua on the Great Sea Reefs, and deep reef shoals on the edge of Viti Levu's shelf. Reef passes along the coast of Viti Levu and other parts of Fiji are former river canyons, while the inner reefs off Lautoka–Ba also have narrow channels which were once meandering river beds. Reticulate features in lagoons off Ra are due to erosion on what were once limestone hills. The reefs are diverse in both their geomorphological structure and origin. All of the different, classical reef types are represented (Table 2). They can be further categorised by their position on the insular shelf (outer-, mid- and inner shelf) and by their exposure to the prevailing wind and seas. Zann and Vuki (see Chapter 99) have described their reef assemblages in more detail. A tentative classification based on position, reef type and geomorphology (Zann, 1992) divided the group into 17 distinct areas of reef provinces based on similarities in geomorphology (Table 3).

There have been very few long-term ecological studies on the reefs in Fiji because of lack of local technical expertise. However, several studies have been carried out on Suva reef because of its proximity to the University of the South Pacific (Zann et al., 1987; Zann et al., 1990; Vuki, 1994). These studies have reported that major disturbances have influenced the structure of reef flat communities on this reef. In particular, excavations of the sea urchin *Echinometra mathaei* have caused massive changes to the reef

Table 2

Reef types represented in Fiji

Fringing reefs	partially surrounding or fringing a high island
Patch reefs	small patches of coral in lagoon
Barrier reefs	elongate reefs forming a wall or "ribbon" offshore along the edge of the continental shelf
Platform reefs	rising to sea level on the insular shelf
Oceanic ribbon reefs	partially enclosed wall or ribbon reefs growing on a submerged feature or seamount
Drowned reefs	deep-water reefs, not in an active growth phase
Atolls	circular reefs, with small sandy islets or motus
Near atolls	circular reefs, with small sandy islets or motus, but with a part of the volcanic basement protruding from the reef as a rocky islet

Table 3

Reef provinces in Fiji based on similarities in geomorphology

Distinct areas	Reef types
South Viti Levu	windward, outer-shelf barrier reefs
Coral Coast	windward, mid- and outer shelf fringing reefs
Beqa and Vatulele	windward, isolated shelf (uplifting) barrier reefs
North/Western/Eastern Viti Levu	leeward, mid- and inner-shelf platform reefs
Mamanucas and Yasawas	leeward, midshelf platform reefs
Outer Mamanucas/Yasawas/ Great Sea Reef/Northern Vanua Levu reef line	leeward, outer-shelf reefs, shoals and northern barrier reef system
Northern Vanua Levu reefs	leeward, midshelf platform reefs
Northern and Eastern Vanua Levu reefs	leeward, innershelf
South Vanua Levu Barrier Reef	moderate windward, outershelf
South eastern Taveuni Fringing reefs	windward, isolated, uplifted
Cikobia	fringing and barrier reefs, isolated, uplifted
Lomaiviti reefs	moderately windward, mid- and outer-shelf
Eastern Vanua Levu reefs, atolls and near atolls	moderately windward, outer-shelf
Kadavu	isolated shelf barrier and fringing reefs
Yasayasa Moala	isolated, barrier and fringing reefs
Lau Ridge	uplifting fringing, barrier, platform and oceanic ribbon reefs
Rotuma	isolated, shelf, fringing and platform reef

substratum which were associated with an increase in turf algae. Other disturbances that made significant contributions in shaping the reef community structure on Suva Reef were tsunami, cyclone and flood damage, *Acanthaster planci* predation and the effects of human activities (for example fishing activities, dredging, pollution and reclamation).

About 100 species of stony corals were identified from the Great Astrolabe Reef, Kadavu (Paulay, 1990). However, the most detailed description of corals from Fiji to date was 198 species from the Mamanucas and the Southern Viti Levu (Zann, 1992). Five species of gorgonian corals or sea fans have been described (Muzik and Wainwright, 1977) and 15 species of zoanthids have been described from Viti Levu (Muirhead and Ryland, 1981).

Biodiversity of Other Groups

A large range of recorded animals on Fijian reefs include all major invertebrate groups that would be expected in a rich reef environment. Vertebrates include sharks and bony fish, reptiles (turtles and seasnakes), seabirds and whales.

The lower invertebrates (common reef sponges, polychaete worms etc.) are not well known. Molluscs are well represented and are scientifically well described; Cernohorsky's books and monographs (1968, 1972, 1977) describe many of the Fijian species. Parkinson (1982) collected 123 species of gastropods belonging to 12 families from deeper water in southern Viti Levu while Brodie and Brodie (1990) listed 253 species of opisthobranchs, mainly from southern Viti Levu. The bivalves are of moderate importance in the subsistence and artisanal fisheries in Fiji and Parkinson (1982) collected 102 species of bivalves from 25 families from Viti Levu and adjacent islands. The nautilus *Nautilus pompilius*, a 'living fossil', is the best known of the Fijian cephalopods and has been the subject of intensive research (Ward et al., 1977; Tanabe et al., 1983; Zann et al., 1984).

Crustaceans are moderately well studied and collections of shallow water amphipods have been described by Myers (1985). Copepods have been collected by Wells (1977) and Yeatman (1983) and parasitic fish isopods by Bruce (1984). Echinoderms are not well known, though *Acanthaster planci* is well studied because of the damage done by this animal to Fijian reefs (Owens, 1971; Zann et al., 1987, 1990). The sea squirts are relatively well known; Kott (1981) and Ryland et al. (1984) have described 60 species including 14 diademnids from reefs in Viti Levu and Kadavu.

A preliminary listing of reefal, pelagic and deep-water bottom fish by Baldwin and Seeto (1986) gives a total of 1198 species from 162 families. The green turtle, *Chelonia mydas*, and the hawksbill turtle, *Eretmochelys imbricata*, nest in Fiji (Bustard, 1970). The loggerheads (*Cate caret*) are present but uncommon. Flatbacks (*Chelonia depressa*) and Ridleys (*Lepidochelys olivacea*) are occasional visitors. The three known species of seasnake in these waters are *Laticauda colubrina*, *Laticauda laticauda* and *Hydroplus melanocephalus* (Guinea, 1980).

Sensitive and Endangered Marine Species

The giant clams, *Hippopus hippopus* and *Tridacna gigas* have become recently extinct in Fiji. *Tridacna gigas* was seen 35 years ago while *Hippopus hippopus* can be seen as dead shells and subfossils along many shores (Lewis et al., 1988). *H. hippopus* have been reintroduced to the Makogai clam hatchery from the Great Barrier Reef for aquaculture purposes. Giant clams are of special significance in Fiji and four species are present today: *Tridacna derasa*, *T. tevoro*, *T. squamosa*, and *T. maxima*.

Over-harvesting on most islands in Fiji has seriously depleted the giant coconut crab *Birgus latro*. It is today common on the limestone islands of Northern Lau and eastern Vanua Levu (Zann, 1992).

Fijian breeding populations of green turtles (*Chelonia mydas*) and hawksbill turtles (*Eretmochelys imbricata*) are very seriously endangered and will become extinct unless urgent action is taken. The major turtle rookeries in Fiji are in the Ringgold Group and the Mamanuca Group. Fiji's turtle populations have been depleted by commercial hunting for meat and shells. Some turtle rookeries have been impacted by tourist development, for example in the Mamanuca Group. Despite the large, and apparently sustained, harvests of turtles in Fiji, the breeding population is very small. Tag returns from turtles killed in Fiji indicate that they are turtles migrating from French Polynesia, American Samoa and the Great Barrier Reef in Australia.

Two species of banded seasnake are recorded from Fiji: *Lauticauda colubrina* and *Lauticauda lauticauda* (Zann, 1992). The Indian mongoose *Herpestes auropunctatus*, introduced in 1933 to control rats in sugar cane, is a major threat. This is clearly evident by the lower populations of seasnake on islands inhabited by the mongoose when compared with those not colonised by mongoose (Guinea, 1980).

Major threats to seabird rookeries in Fiji come from mongoose predation, egging at rookeries by fishermen and the disturbance of colonies by the construction of tourist resorts on low islands (Clunie, 1985). Seabird nesting sites are also significantly disturbed through clearing for gardens in villages and the introduction of modern diesel-powered fishing vessels, which has enabled fishermen to visit isolated seabird rookeries more frequently.

REFERENCES

Clunie, F. (1985) Seabird breeding colonies on Namena Island and the Ringgold Isles of north-eastern Fiji. *Domodomo* III, 3, 90–109.

Guinea, M. (1980) The Seasnakes of Fiji. *Proc. 4th International Coral Reef Symposium, Manila 1981*, Vol. 2, pp. 581–585.

Lewis, A.D., Adams, T.J.H. and Ledua, E. (1988) Fiji's giant clam stocks—A review of their distribution, abundance, exploitation and management. In *Giant Clams in Asia and the Pacific*, eds. J.W. Copland and J.S. Lucas. ACIAR Monograph, No. 9, 274 pp.

Zann, L.P. (1992) The state of the marine environment of Fiji. Unpubl. report to National Environmental Management Project, Environmental Management Unit, Suva, Fiji.

Seabirds are well described by Watling (1982) and Clunie (1985), the latter also describing the nesting seabirds of the Ringgold Islands. The marine mammals of Fiji are poorly known. Pods of humpback whales (*Megaptera novaeangliae*) and bottlenose dolphins (*Tursiops truncatus*) are occasionally seen in the open sea in the Lomaiviti Group, Mamanucas and Natewa Bay (Zann, 1992).

OFFSHORE SYSTEMS

Oceanographic studies are limited. Currents are driven westward by the trade winds at speeds of 0.19 ms^{-1} to 0.32 ms^{-1} just off Viti Levu (Wyrtki, 1990). Immediately outside the reef, currents are tidally driven, eastward on the flood and westward on the ebb at a maximum of 0.1 ms^{-1} (Webster, 1979). Barstow and Haug (1994) described the deep-water wave climate off Fiji. Waves were measured using a waverider buoy located off Kadavu where wave heights were 2–4 m with peak periods of 10–15 seconds.

Surveys of the deep slope by the Fiji Fisheries Division and the Secretariat for the Pacific Community have established the presence of stocks of deep-water snappers, groupers, emperors and amberjacks (Dalzell and Preston, 1992). The most important commercial tuna species in Fiji waters are yellowfin (*Thunnus albacares*), bigeye (*T. obesus*) and albacore (*T. alalunga*). There is some evidence that interchange of yellowfin between the eastern and western Pacific is limited (Hampton, 1993). Other tuna species caught by local fishermen for subsistence are skipjack (*Katsuwonus pelamis*) and dog-toothed tuna (*Gymnosarda unicolor*). These species are highly migratory and are capable of unrestricted movement throughout the Pacific Ocean. The cephalopod, *Nautilus pompilius* is very common in the upper continental slope off Southeast Viti Levu, at depths of between 200–600 m, and is an abundant by-catch of the deep water shrimp fishery.

POPULATIONS AFFECTING THE AREA

Fiji is an independent nation which became a republic in 1987. Its population is dominated (51%) by indigenous people of Melanesian and Polynesian origins. The Indo-Fijians (who are mostly descendants of indentured labour brought from India in the last century to work on sugar plantations) account for 44% of the total population; the remaining 5% are other Pacific Islanders, Europeans, Chinese and people of mixed ethnic origin. At the 1996 census, the population was 775,077. Ninety percent of the population lives on the two main islands of Viti Levu and Vanua Levu (Bureau of Statistics, 1996). The capital city is Suva (pop. 170,000), located on Viti Levu, and there are eight smaller towns. The majority of Fiji's population lives in the coastal areas because of the inaccessibility of mountainous areas which are unsuitable for settlement.

RURAL FACTORS

Agriculture and Impacts

The economy is based on the development of Fiji's varied natural resources as well as on service industries, including tourism (Douglas and Douglas, 1994). The sugar industry dominates the agriculture sector and accounts for over 50% of the country's export earnings. Notable agricultural commodities are ginger, fruit juices, vanilla, pawpaw, coconut products and cocoa. Tourism and related activities comprise a leading sector, which has potential for expansion. Exports are dominated by sugar but ginger, wood products, gold and processed fish are also significant. Much of the agricultural activities in Fiji are associated with the subsistence economy.

Morrison (1990) has described the environmental impacts of agriculture in Fiji. Soil erosion and land degradation were significant problems, and the intensive farming of sugar cane, ginger, pineapples and pine seedlings on steep slopes have contributed to severe soil erosion in Viti Levu and Vanua Levu. Erosion from sugar cane plots ranged from 68 to 77 tonnes/hectare/year, erosion from pineapple plots is approximately 71 t/ha/year, while erosion from pine seedlings is up to 4877 t/ha/year (Convard, 1993). About 85.7 t/ha/year of soil loss was estimated for the typical ginger cultivation in the eastern Viti Levu using the universal soil loss equation (Morrison, 1981).

Intensive agriculture has contributed to greatly increased sediment loads deposited into Fiji's rivers and marine environment. The sediment load in the Waimanu River in southeast Viti Levu indicates that the average soil loss was about 53 tonnes/hectare/year. This corresponds to a loss of 2–2.5 mm of soil per year (Watling and Chape, 1992). As a result of high sediment loads in Fiji's rivers, dredging of sediments from rivers in Viti Levu and Vanua Levu alone has cost the Fiji government more than $50 million to date. A significant proportion of this sediment is attributed to agricultural run-off.

Fisheries and Impacts

Subsistence fisheries in Fiji provide the rural population the bulk of their protein and support traditional rituals and culture. Fijians consume approximately 40 kg per capita per year of seafood and the subsistence catch from reefs is estimated to be 17,000 tonnes per year (Zann and Vuki, 1998). Prior to the Deed of Cession to the British in 1874, the Fijian system of ownership was based on communal lifestyle. This enabled Fijians to control the exploitation of fisheries resources. At present, Fijians have the right to fish for their own consumption but the ownership of the seabed belongs to the State.

Fishing for subsistence purposes is restricted to within the inshore areas using a variety of fishing methods. The most common method is hand line fishing. Other methods

Overfishing of Marine Invertebrate Species

Reefs close to urban areas have experienced heavy fishing pressure for shellfish, and stocks of invertebrates subject to export fisheries have been overfished. These include *Trochus niloticus* and holothurians such as *Actinopyga miliaris*, *Holothuria nobilis*, *Holothuria fuscogilva*, *Holothuria scabra*, and *Thelenota ananas*.

The meat of *Trochus*, pearl oysters and giant clams are important food items for subsistence and are sold in urban markets. The shells of *Trochus* and pearl oysters are used for the production of buttons locally. Although shells of giant clams are not commercially important, the meat is a delicacy and highly priced in the Southeast Asian markets. Collection of shells for sale is common and has resulted in depletion of giant triton shell, *Charonia tritonis*.

The introduction of Scuba apparatus, coupled with lucrative prices offered by overseas business people, are often seen as the major causes of shellfish over-exploitation. A survey on Fiji's giant clam population was conducted by the Fiji Fisheries Division between 1984 and 1988 under the auspices of the Australian Centre for International Agricultural Research (ACIAR) International Giant Clam Project. The survey results show that giant clam stocks on most reefs around Fiji have been exploited beyond sustainable level. In 1988, the Fiji government passed a new regulation banning the export of giant clam meat. However, despite this, giant clam meat continued to be exported through exemption powers of the Minister of Agriculture, Fisheries and Forest. The exemption power of the Minister has not only contributed greatly to the devastation of the already overfished giant clam stock, but also disregarded the collaborative attempts of the Fisheries Division and ACIAR to prevent further decline of the giant clam wild stock.

Pinctada margaritifera, commonly known as black-lip pearl oysters, have been over-fished in most reefs around Fiji. Murray (1992, in Ledua and Vuki, 1998), concluded that Fiji's pearl oyster stocks have been depleted, and attributed the poor condition of the stocks to heavy fishing pressure, environmental stress caused by agricultural runoff and to general pollution of inshore areas from effluents. Passfield (1995, in Ledua and Vuki, 1998), concluded after a five-week survey at four locations in Fiji (Totoya, Beqa, Makogai and Bua waters) that the number of pearl oysters was low, and insufficient to consider establishing commercial pearl farming based on wild stock. He considered that the number of pearl oysters was also too low for spat collection to be considered economically viable.

Trochus niloticus are easily collected and thus vulnerable to over-exploitation. However, with the 1988 peak in raw shell exports, coupled with the additional demand from button factories, recruitment overfishing may have occurred on heavily fished reefs, followed by a steady decline in major stocks. As a result of overfishing, fishermen no longer fish specifically for *Trochus* but they are collected as part of an incidental catch.

Stocks of commercial beche-de-mer holothurians in Fiji have declined (Ledua and Vuki, 1998). Because of the rapid decline in the stocks of sea cucumbers, the Government intervened in December 1988 when Cabinet legislated to prevent the export of any beche-de-mer less than 7.6 cm in length (in any form). The export of *Holothuria scabra* was banned unless a permit is obtained from the Minister for Agriculture, Fisheries and Forest. The protection of *H. scabra* reflects the importance of this species as a local and emergency food item. Unfortunately, the ban on the export of *H. scabra* did not prevent the decline of the already heavily fished wild stock. The permits issued by the Minister for Agriculture, Fisheries and Forest continued to allow large amounts of *H. scabra* to be exported. As a result, most of the coastal areas in the Northern Vanua Levu (Bua and the Great Sea Reef) have been overfished. The main target species (*H. scabra*, *H. fuscogilva* and *H. nobilis*) have been fished beyond sustainable level and will take a long time to recover.

REFERENCE

Ledua, E. and Vuki, V.C. (1998) The inshore fisheries resources of Fiji. In: Fisheries and marine resources. Papers presented at Symposium 8, VIIIth Pacific Science Inter-Congress. The University of the South Pacific, Fiji, 13–19 July 1997. Marine studies technical report No. 98/3. The University of the South Pacific.

---

of fishing include the use of fish traps (both traditional and modern traps), fish fences, gill nets, seine nets, hand nets, fish drives, spears, use of poisonous plants (such as *derris* roots), line trawling, reef gleaning and skin diving (especially for collecting shellfish and sea cucumber). Women carry out reef gleaning during low tide and target shellfish, sea cucumbers, octopus, worms, sea urchins, eels and small fish. Men dominate skin diving and spear fishing. Approximately 23,253 tonnes of inshore fisheries products are harvested annually with an estimated value of US$ 64.1 million. The volume of resources collected for subsistence purposes constituted 71.4% (16,600 tonnes, valued at US$ 45.7 million) of the total inshore resources harvested annually (Ledua and Vuki, 1998).

Inshore fin-fish have been over-exploited because of the high dependence of Fiji's increasing population on these resources. Fin-fish resources closer to heavily populated centres are declining rapidly due to high fishing pressure. In spite of the high fishing pressure on fisheries resources, there is an over-emphasis on the export fisheries and in 1998 alone, a budget of $3.17 million was allocated to support export fisheries (World Bank, 1999).

There is no systematic monitoring to detect early signs of overfishing. Although the Fiji Fisheries Division have data on the volume of fish landed on various market outlets, there is very little information on fish sizes caught, fishing effort, fishing locations, catch per unit effort and volume of fish caught for subsistence needs.

The commercial catch landing data collected by the Fiji Fisheries Division have provided some information on long-term production trends of various fin fish species. Stocks of highly targeted species such as *Lethrinus harak*, *Lethrinus xanthochilus*, *Lethrinus nebulosus*, *Bolbometopon muricatus* and all the Mugilid species have been overfished. Overfishing by commercial fishermen residing in urban centres may be responsible for the decline in mullet landing (Lal, 1983; Fong, 1994). Further decline in abundance of mullet was noted by Sasa villagers in Macuata on the island of Vanua Levu (Fong, 1994). Fishers have reported a decline in mullet, stout chub mackerel and trevallies, and they also noted that they were spending more hours catching fish, and had to travel further away from the village to catch them. They attributed the cause of stock decline to gillnet fishing.

Jennings and Polunin (1996) carried out studies on the effects of fishing effort and catch rate on targeted reef fish communities subjected to different levels of exploitation. They found that decreases in carnivorous fishes such as groupers and emperors are among the most detectable effects of fishing pressure. For the six fishing grounds examined, the main targeted species were from the Seranidae (grouper family) and Lethrinidae (emperor family). Low yields indicate that these areas are overfished. This study confirmed that stocks of highly targeted fish species such as *Lethrinus harak*, *Lethrinus xanthochilus*, *Lethrinus nebulosus*, *Bolbometopon muricatus* and all the Mugilid species have been overfished.

Fong (1994) reported that commercial fishermen, especially Indian fishermen, agree that catch rates as well as sizes of fin-fish have declined rapidly within the Macuata Province. In the Lau Group, the Lau Provincial Council imposed a total ban on tuna bait fishing in 1991 as it was implicated as the cause of fish stocks' decline on some islands. It was reported to the Council that stocks of fish were abundant prior to the arrival of tuna pole-and-line vessels. The people of Lau believed that tuna baits, and other important subsistence food fish were being caught in large numbers in bouki-ami nets of tuna pole-and-line vessels.

COASTAL EROSION AND LANDFILL

Coastal modifications are largely restricted to major urban areas and other coastal towns. Reclaimed land occupies most of the Suva waterfront. Coastal modifications in Suva include three harbours and several km of seawalls. Around 3000 ha of an original 45,000 ha of mangroves have been cleared for sugar cane, housing and industrial sites (Watling and Chape, 1992).

Most of Suva's downtown area is built on reclaimed land. Many of the modern buildings have been built along the low coastal fringe which is also reclaimed land. This area includes infrastructure of buildings, wharves, port facilities and tourist facilities. Concrete and pavement are the dominant ground cover. Seawalls and other artificial materials have replaced Suva's foreshore. The main constituents of the current shoreline are 51% seawall, 11% revetment, 8% wharf and structures, 8% beach and 22% unprotected reclamation. At least 110 hectares have been reclaimed in Suva since 1881. Most of the existing shore protection is degraded and the eastern coast of Suva's shoreline shows evidence of erosion.

In most villages in Vanua Levu and Viti Levu, concrete seawall has been constructed because of beach erosion. Most beaches have become narrower and thinner because of the low sediment supply. This is of great concern to hotel operators because one of the main tourist attractions to Fiji is its beaches. Furthermore, the response of these eroded beaches to cyclones is not known. It is likely that significant wave and storm events have led to beach erosion, and the abilities of Fiji's beaches to respond to and recover from wave and storm events are highly localised and will depend on wind and wave directions and on local sediment input.

Human-related causes of beach erosion in rural areas might be related to building houses too close to the shore and reef blasting. In urban areas, the removal of mangroves and land reclamations are major causes of beach erosion. The Ministry of Fijian Affairs is supporting engineering projects such as village seawalls to prevent erosion but most are being built without any investigation on the nature of the problem.

EFFECTS OF URBAN AND INDUSTRIAL ACTIVITIES

As populations grow in Fiji, urbanisation and development expand and pressure on the coastal and the marine environment increases. There are additional demands for land and, as a result, mangrove areas are reclaimed. There is also a high demand for coral sands for cement construction material. Potential sources of point source pollution in Fiji include: mining, shipyards and slipways, moorings, tourist developments, sugar mills, timber mills, cement factory, litter refuse disposal sites, sewage, agricultural pesticides and herbicides, changing land use, and small industries (Cripps, 1992).

Urban areas are served by both sewered and individual wastewater systems. These individual facilities are inadequate and often discharge overflows directly into the sea, streams or storm drains. In Suva, areas that are not served by sewerage include the older areas of the city (Toorak and Nasese) and squatter settlements. Toorak and Nasese areas have inadequate septic tanks that can often discharge overflows into storm drains during heavy rainfall. The septic systems in the Suva region are further hampered by the local clay marl which does not allow for absorption of the septic tank overflows. In Suva, faecal coliform levels have been found to be high and of concern to public health. In the Suva area, the fact that 95% of mangrove oysters collected at eight sites exceed World Health Organization limits for human consumption is cause for serious alarm (Naidu et al., 1989).

Litter is a conspicuous source of pollution in the marine environment in Fiji. Solid wastes such as plastic bags, metal cans, glass and bottles are often discarded indiscriminately in urban areas, on roadsides, on beaches, in mangroves and in the sea. This litter is not only visually offensive but also dangerous (for example, broken bottles) and the environment (for example, plastic bags ingested by turtles and sea birds). All shores around Suva are seriously littered; the average cover of litter in some areas is >50% (Naidu et al., 1991).

In 1991, an extensive survey of solid waste disposal sites in Fiji was carried out (Bronders, 1991). It was concluded that the major rubbish dumps were not adequately maintained and were located in environmentally sensitive areas, for example in mangrove areas, riverbanks and coastal areas. Bronders noted environmental problems such as smoke, smell, insects, water pollution and loss of natural beauty and also highlighted the need for improved solid waste management. The large Suva dump is located in mangroves within the city boundaries and poses a serious public health risk.

A survey of point source pollution in Suva indicated alarming levels of many pollutants due to industrial pollution (Cripps, 1992). Studies found that the major industries contributing to the problem included: a brewery and food manufacturing industry (high BOD, pH, oils/grease, suspended solids), paint factories (extremely high lead), electroplating (zinc, cyanide), service stations (motor oils), rubbish dump (wide spectrum of hazardous chemicals) and slipways (TBT). There are also high levels of TBT in Suva Harbour.

Industries outside Suva which pose pollution problems include mines, sugar and timber mills. Liquids from the cyanide tailing ponds at the Emperor Gold Mine at Vatukoula are periodically discharged into the Navisi River. Although these comply with WHO standards, bund failure at the site could result in a serious problem (Cripps, 1992).

Fiji sugar mills discharge large quantities of organic wastes resulting from cane crushing and caustic soda from washing water. Waste waters have a high BOD and fish kills from anoxic conditions have been frequently reported from Qawa river, Vanua Levu and Ba River, Viti Levu.

PROTECTIVE MEASURES

Fiji has a poor success rate in establishing marine protected areas because of customary fishing rights ownership. The National Trust of Fiji is currently endowed with the responsibility for the management of protected areas in Fiji. The various recommendations for marine environmental management made in the past have concentrated on marine

Table 4

Proposed Marine Protected Areas in Fiji (source: Dunlap and Singh, 1980)

General area	Specific areas
Viti Levu, Kadavu and Yasawas	Draunibota, Labiko and Vuo Islets in Suva's Bay of Islands Makuluva and reefs Suva barrier reef and cays Bird Islets Beqa, Beqa lagoon Coral Coast Reefs Makogai Reefs Wakaya and Reefs Cakau Momo and Reef Taqa rocks Nadi Bay reefs Mamanuca Group North Astrolabe Reef Great Astrolabe Reef
Lau Group	Qilaqila Fulaga Bay Sovu Islands Cakau Lekaleka Barrier Reef Nukutolu Islet and reefs
Rotuma	Rotuma Islands and Reefs
Vanua Levu and Taveuni	Namenalala Nanuka Great Sea Reef Qelelevu Atoll Rainbow Reef

protected areas. The marine protected areas recommended by Dunlap and Singh (1980) have never been established and there is no established system of protected areas. Marine protected areas recommended by Dunlap and Singh (1980) are shown in Table 4.

Several additional areas have been recommended by Dahl (1986). These were Leleuvia; Ra coast; Mana; Yasawa-i-Rara, Vuaqava; Yasawa (Sawailau); Vatulele (red prawn pool); Koro (turtle pool and cave); Balolo Point, Ovalau (balolo rise area); Moturiki (land crabs); Rewa Delta (seagrass and mangroves).

However, various resorts and privately owned islands have established private sanctuaries on the adjacent reefs through agreements with the customary fishing right holders. "Beachcomber Cruises" established the first marine sanctuary in the early 1970s on the reefs around Beachcomber and Treasure Islands. Most resorts on the Coral Coast and Mamanucas now have private arrangements with the customary fishing right holders for the "lease" of the waters and reefs adjacent to their resort but these reef areas are all very limited in size. The largest private sanctuary is that of the US religious community on Naitauba Island in Northern Lau which encompasses the entire reef on the island.

Yadua Tabu in Bua, off western Vanua Levu is now protected as an iguana reserve by the National Trust for Fiji in cooperation with the traditional landowners. The Sigatoka sand dunes were gazetted as a National Park in 1991. A joint terrestrial and marine park is suggested for Monuriki in the Mamanucas (Zann, 1992).

Fiji's first National Marine Park was to be Makogai Island in the Lomaiviti Group, to give protection to the Fisheries Division's clam hatchery. Just prior to the final ratification in late 1991, however, negotiations with the traditional owners broke down because of inadequate compensation (US $250 per annum).

The fact that none of these areas suggested have yet been protected and that only one National Park has been established in the entire country after 30 years of discussion, indicates that the concepts of resource management through a system of protected areas is a very slow process in Fiji. This is because most land is communally owned and the adjacent sea is the group's customary fishing ground.

There is, however, recent resurgence of interest in marine protected areas because customary fishing right owners are becoming aware of the value of marine eco-tourism, habitat protection and fisheries management and conservation (Whippy-Morris and Pratt, 1998). Several government departments have a strong interest in promoting marine protected areas and conservation. The setting up of marine protected areas is integral to the responsibilities of Fiji as a partner in the UN Convention on Biological Diversity, UNCLOS III, Convention on the Conservation of Nature, the South Pacific Regional Environment Programme Convention and the Convention on International Trade on Endangered Species of Wild Fauna and Flora (CITES–Convention).

The importance of marine protected areas in attracting dive eco-tourists from around the world is currently recognised. The new guidelines for village-level eco-tourism have further enhanced the establishment of protocols focusing on marine protected areas and marine conservation as important elements in the tourism industry. The Fiji Dive Operators, Fisheries Department, Environment Department, Fijian Affairs and the University of the South Pacific are co-operating on the promotion of protected areas. The development of national policies to support village-level eco-tourism and dive eco-tourism and the tourism industry has been proposed (Whippy-Morris and Pratt, 1998). International conservation organisations such as Greenpeace and the World Wildlife Fund for Nature (WWF) and Foundation of the People of the South Pacific are active in raising local awareness on environmental issues such as pollution and overfishing.

The US based Biodiversity Support Programme (BSP) and its partners (University of the South Pacific, Rainforest Alliance, South Pacific Action Committee for Human Ecology and Environment, WWF-South Pacific and the Fiji Department of Environment) have been working with community members in Verata, Tailevu to monitor their marine resources. BSP and its partners have developed biological prospecting agreements between the Verata community and pharmaceutical companies to enable the community to benefit directly from any monies paid from chemical screening of marine organisms and possible drug development. They have succeeded in persuading the Verata community to establish marine protected areas or no-take zones and these areas have shown recovery of marine organisms.

CONCLUSION

Fiji's marine environment is critical to its subsistence and economic development. There are widespread concerns about the degradation of this marine environment because of increasing populations and development and declines in fisheries. The two major islands, Viti Levu and Vanua Levu which are more urbanised and developed have been seriously affected, while the more remote rural islands are less affected.

The major issues affecting Fiji's marine biodiversity include: loss of water quality in estuaries and lagoons as a result of poor land-use practices; consequent declines of coral reefs because of elevated sediments and nutrients; loss of coastal wetlands due to urban and related infrastructure developments; widespread coral mortality from crown-of-thorns starfish outbreaks; decline of inshore fisheries due to loss of critical habitats, overfishing and destructive fishing techniques. It is difficult accurately to assess the extent of environmental degradation because there are so few comprehensive scientific studies. Most studies have been descriptive and there is very little long-term quantitative monitoring, and the status of scientific knowledge is poor. However, it is clear that Fiji's marine environment is under increasing pressure from human activities and will require protection and effective management.

REFERENCES

Baldwin, W. and Seeto, J. (1986) A checklist of the fishes of Fiji. Unpublished manuscript, University of the South Pacific.

Barstow, S.F. and Haug, O. (1994) The wave climate of the southwest Pacific, SOPAC Technical report 206, SOPAC, Suva, 25 pp.

Brodie, G.D. and Brodie, J.E. (1990) A checklist of the opisthobranch molluscs of Fiji. *J. Malac. Soc. Aust.* **11**, 53–63.

Bronders, J. (1991) *Waste Disposal in Fiji. Part 2: Field Visits*. Mineral Resources Department, Suva, Fiji.

Bruce, N.L. (1984). A new family for the isopod crustacean genus *Tridentella* Richardson, 1905, with description of a new species from Fiji. *Zoological Journal of the Linnean Society* **80**, 447–455.

Bureau of Statistics (1996) 1996 Fiji Census of Population and Housing, Bureau of Statistics, Suva, Fiji.

Bustard, H.R. (1970). Turtles and an iguana in Fiji. *Oryx* **10**, 317–322.

Carter, R. (1990) Predicted storm and wave set-up for Suva and Laucala Bay harbours on Viti Levu in Fiji. CCOP/SOPAC. SOPAC Technical Report, 115 pp.

Cernohorsky, W.O. (1968). *Marine Shells of the Pacific. Vol. 1*. Pacific Publications, Sydney.

Cernohorsky, W.O. (1972). *Marine Shells of the Pacific. Vol. 2.* Pacific Publications, Sydney. 411 pp.

Cernohorsky, W.O. (1977). Report on the molluscan fauna of the Lay Group, Fiji islands. Lau-Tonga 1977. *The Royal Soc New Zealand, Wellington, Bulletin* **17**, 41–52.

Clunie, F. (1985) Seabird breeding colonies on Namera Island and the Ringgold Isles of north-eastern Fiji. *Domodomo III*, **3**, 90–109.

Convard, N. (1993) *Land-based Pollutants Inventory for the South Pacific Region.* SPREP, Apia, Western Samoa, 165 pp.

Cripps, K. (1992) *Survey of Point Sources of Industrial Pollution Entering the Port Waters of Suva.* Engineering Dept., Ports Authority of Fiji, Suva, Fiji.

Dahl, A.L. (1986) *Review of the Protected Areas in Oceania.* IUCN, Gland, Switzerland.

Dalzell, P. and Preston, G.L. (1992) Deep reef slope fishery resources of the South Pacific: a summary and analysis of the dropline fishing survey data generated by the activities of the South Pacific Commission Fisheries Programme between 1974–1988. South Pacific Commission, Noumea, New Caledonia, 299 pp.

Dickie, R., Fennery, R.M. and Schon, P.W. (1991) Ports Authority of Fiji, Planning, Engineering Dept. Ports Authority of Fiji, Suva.

Douglas, N. and Douglas, N. (1994) *Pacific Islands Yearbook.* 17th Ed. Fiji Times, Suva.

Dunlap, R.C. and Singh, B.B. (1980) A national parks and reserves systems for Fiji. A report to the National Trust of Fiji.

Fenny, R.M. (1989) Suva foreshore study, unpublished report for the Ports Authority of Fiji, Suva.

Fong, G. (1994) case study of traditional marine management system: Sasa village, Macuatu Province, Fiji. In: FAO Field Report, No. 94/1, FAO, Rome, 85 pp.

Guinea, M. (1980) The Seasnakes of Fiji. *Proc. 4th International Coral Reef Symposium, Manila, 1981*, Vol. 2, 581–585.

Hampton, J. (1993) Sixth standing committee on tuna and billfish. Working paper five: work programme review 1992–93 and work plan 1993–94. South Pacific Commission, Noumea, New Caledonia.

IUCN (1988) *Coral Reefs of the World, Vol. 3: Central and the Western Pacific,* IUCN.

Jennings, S. and Polunin, N.V.C. (1996) Impacts of fishing on tropical reef ecosystems. *Ambio* **25**, 44–49.

Kott, P. (1981) The ascidians of the reef flats of Fiji. *Proceedings of the Linnean Society of New South Wales* **105**, 147–212.

Lal, P.N. (ed.) (1983) Mangrove resource management. Proceedings of an interdepartmental workshop. Ministry of Agriculture and Fisheries (Technical report No. 5, Fisheries Division).

Ledua, E. and Vuki, V.C. (1998) The inshore fisheries resources of Fiji. In Fisheries and marine resources. Papers presented at Symposium 8, VIIIth Pacific Science Inter-Congress. The University of the South Pacific, Fiji, 13–19 July 1997. Marine studies technical report No. 98/3. The University of the South Pacific. pp. 45–59.

Meyers, A.A. (1985) Shallow water, coral reef, and mangrove Amphipoda (Gammaridea) of Fiji. *Records of the Australian Museum Suppl.* 0812-7387.

Morrison, R.J. (1990) The efforts of land use changes on soil properties and the impact of soil erosion in the South Pacific. Unpublished manuscript, University of the South Pacific, Suva, Fiji.

Morrison, T.M. (1981) Report to the Vice-Chancellor, University of the South Pacific. Assessors report, School of Agriculture, January 1981. University of the South Pacific, Suva, Fiji.

Morton, J. and Ray, U. (1980) The shore ecology of Suva and south Viti Levu. Institute of Marine Resources, University of the South Pacific, Fiji.

Muirhead, A. and Ryland, J.S. (1981) A review of the henus Isaurus Gray 1828 (Zoanthidea) including new records from Fiji. *Journal of Natural History* **19**, 323–335.

Muzik, K. and Wainwright, S. (1977) Morphology and habitat of five Fijian sea fans. *Bulletin of Marine Science* **27** (2).

Naidu, S., Aalbersberg, W.G.L., Brodie, J.E., Fuavao, V.A., Maata, M., Naqasima. M., Whippy, P., and Morrison, R.J. (1989) Water quality studies on selected South Pacific lagoons. UNEP Regional Seas Reports and Studies No. 136, UNEP; and SPREP Reports and Studies No. 49.

Nunn, P. (1994) Coastal processes and landforms in Fiji: their bearing on Holocene sea level changes in the South and western Pacific. *Journal of Coastal Research* **6**, 279–310.

N'Yeurt, A.D.R., South, G.R. and Keats, D.W. (1996) A revised checklist of the benthic marine algae of Fiji (including the island of Rotuma). *Micronesica* **29** (1), 49–96.

N'Yeurt, A.D.R. and South, G.R. (1997) Biodiversity and biogeography of marine benthic algae in the Southwest Pacific, with specific reference to Rotuma and Fiji. *Pacific Science* **51**, 18–28.

Owens, D. (1971). Acanthaster planci starfish in Fiji: survey of incidence and biological studies. *Fiji Agricultural Journal* **33**, 15–23.

Paulay, G. (1990) Astrolabe corals. Unpublished manuscript, University of the South Pacific, Suva, Fiji.

Parkinson, B. (1982). *The Specimen Shell Resources of Fiji.* South Pacific Commission, Noumea, New Caledonia.

Penn, N. (1982) The environmental consequences and management of coral sand dredging in the Suva region, Fiji. Ph.D. Thesis., University of Wales.

Ryland, J.S. (1981). Reefs of south-west Viti Levu and their tourism potential. *Proc. 4th Int. Coral Reef Symp., Manila* **1**, 293–298.

Ryland, J.S., Wrigley, R.A. and Muirhead, A. (1984) Ecology and colonial dynamics of some Pacific reef flat Didemnidae (Ascidiacea). *Zoological Journal of the Linnean Society* **80**, 261–282.

South, G.R. (1993) edible seaweeds-an important source of food and income to indigenous Fijians. In: International Center for Living Aquatic Resources Management, Manila, Philippines. *NAGA, ICLARM Quarterly* **16** (2–3), 4–6.

Tabudravu, J.N. (1995) Experimental and field evaluation of *Enteromorpha flexuosa* as an indicator of heavy metal pollution by zinc, lead and copper in coastal waters of Lami, Fiji. Masters Thesis. University of the South Pacific. 146 pp.

Tanabe, K.S., Hayasaka, T., Saisho, T., Shinomiya, A., and Aoki, K. (1983) Morphologic variation of Nautilus pompilius from the Phillipines and Fiji Islands. Occas. Pap. No. 1, Research Centrre for the South pacific, Kagoshima University, Japan.

Veron, J.E.N. (1995) *Corals in Space and Time: The Biogeography and Evolution of the Scleractinia.* University of the New South Wales Press, New South Wales.

Vuki, V.C. (1994) Long term changes of Suva reef flat communities from conventional *in situ* survey and remote sensing methods. Ph.D thesis, University of Southampton.

Ward, P., Stone, R., Westermann, G and Martin, A. (1977). Notes on animal weight, cameral fluids, swimming speed, and color polymorphism of the cephalopod *Nautilus pompilius* in the Fiji islands. *Paleobiology* **3**, 377–388.

Watling, R. (1982). *Birds of Fiji, Tonga and Samoa.* Millwood Press, Wellington.

Walting, D. (1985) A mangrove management plan for Fiji. Zonation requirements and a plan for the mangroves of Ba, Labasa and Rewa deltas, South Pacific Commission, Fiji, 67 pp.

Watling, R. (1992) A state of the environment report for Fiji. National Environmental Management Project, Department of Town and Country Planning, Fiji.

Watling, R. and Chape, S. (1992) Environment Fiji—The National State of the Environment Report. IUCN, Gland, Switzerland.

Webster, I.T. (1979) Preliminary oceanographic study, Namosi copper project, Fiji—report. Viti Copper Ltd.

Wells, J. (1977). A preliminary report on the littoral copepods of Moce island, Lau group, Fiji. *Royal Society of New Zealand Bulletin* **17**, 91–93.

Whippy-Morris, C. and Pratt, C. (eds.) (1998) Marine Biodiversity Technical Group Report. Fiji biodiversity strategy and action plan project, Environment Department, Fiji Government.

World Bank (1999) Voices from the village: A comparative study of coastal resource management in the Pacific Islands: World Bank Report. Washington, DC, 67 pp.

Wyrtki, K. (1990) Sea level rise: the facts and future. *Pacific Science* **44** (1).

Yeatman, H.C. (1983). Copepods from microhabitats in Fiji, Western samoa and Tonga. *Micronesica* **19**, 57–90.

Zann, L.P. (1992) The state of the marine environment of Fiji. Unpublished report to the National Environmental Management Project, Environmental Management Unit, Suva, Fiji.

Zann, L.P., Kimmerer, W.J. and Brock, R.E. (1984) The ecology of Fanga'uta lagoon, Tongatapu, Tonga. Uni. South Pacific and Uni, Hawaii Seagrant Cooperative Program, Hawaii.

Zann, L., Broadie, J., Berryman, C. and Naqasima, M. (1987) Recruitment, ecology, growth and behaviour of juvenile *Acanthaster planci* (L.). (Echinodermata:Asteroidea). *Bulletin of Marine Science* **41**(2), 561–575.

Zann, L.P., Brodie, J. and Vuki, V. (1990) History and dynamics of the crown-of-thorns starfish *Acanthaster planci* (L.) in the Suva area, Fiji. *Coral Reefs* **9**, 135–144.

Zann, L.P., Ayling, A. and Done, T.J. (1990) Crown-of-thorns starfish. In *Vanuatu Marine Resources*, eds. T.J. Done and K. Navin, pp. 104–118. Australian Institute of Marine Science, Townsville.

Zann, L.P. and Vuki, V.C. (1998) Subsistence fisheries in the South Pacific. In: Fisheries and Marine Resources. Papers presented at Symposium 8, VIIIth Pacific Science Inter-Congress. The University of the South Pacific, Fiji. 13–19 July 1997. Marine Studies Technical Report No. 98/3. The University of the South Pacific. pp 103–114.

THE AUTHORS

Veikila C. Vuki
Marine Studies Programme,
University of the South Pacific, P.O. Box 1168,
Suva, Fiji

Leon P. Zann
School of Resource Science and Management,
Southern Cross University, P.O. Box 57,
Lismore, NSW 2480,
Australia

Milika Naqasima
Marine Studies Programme,
University of the South Pacific, P.O. Box 1168,
Suva, Fiji

Maika Vuki
Chemistry Department,
University of the South Pacific, P.O. Box 1168,
Suva, Fiji

Chapter 103

THE CENTRAL SOUTH PACIFIC OCEAN (AMERICAN SAMOA)

Peter Craig, Suesan Saucerman and Sheila Wiegman

This chapter describes several environmental issues in a small portion of Oceania, the vast expanse of open ocean and small tropical islands that are scattered across the South Pacific Ocean. American Samoa consists of seven islands, ranging from a small uninhabited atoll to the densely populated high island of Tutuila (145 km^2). Mean air and sea surface temperatures (27.0 and 28.3°C, respectively) vary little year-round, although average air temperatures have risen sharply (2°C) in the 1990s. The main islands are volcanic mountains that descend rapidly to depths of 1000 m within 1–3 km from shore. Offshore ocean waters are 4000–6000 m deep and stratified, with cold water of 5–6°C below 600 m. The fringing coral reefs around the islands support over 200 coral and 890 fish species. The corals are recovering from a series of natural disturbances over the past two decades, but at least some reef resources are overfished for local consumption. The most serious environmental problem facing American Samoa is its uncontrolled population growth rate (2.5%). The current population of 63,000 in 2000 is already straining the environment, with extensive harbour pollution, loss of coastal habitats by urban expansion, and coastal sedimentation from poor upland management practices. Enforcement of environmental regulations is not widespread, and environmental educational programs have difficulty keeping pace with population growth. Several marine protected areas have been established, but illegal fishing is a general problem.

Seas at The Millennium: An Environmental Evaluation (Edited by C. Sheppard)
© 2000 Elsevier Science Ltd. All rights reserved

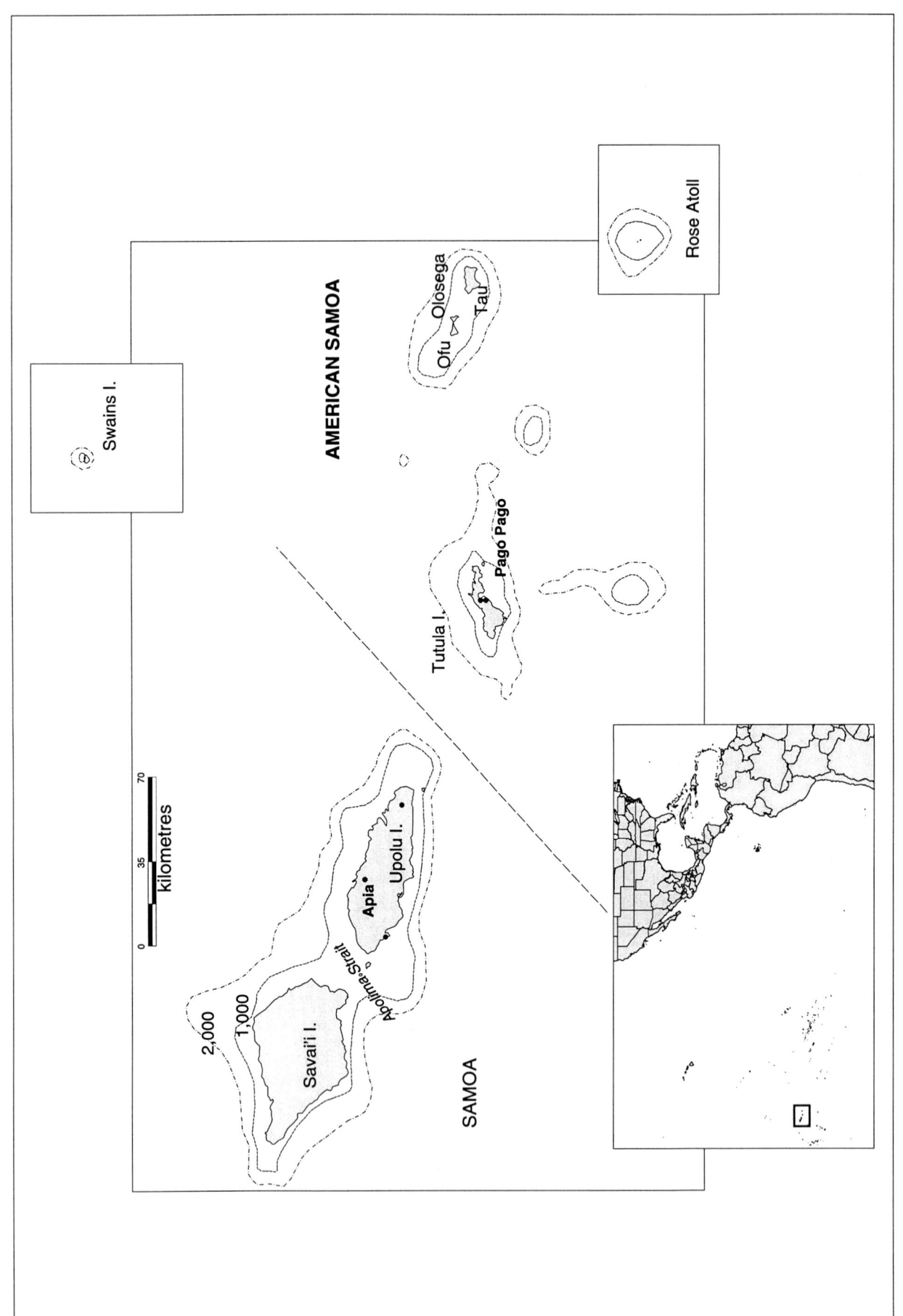

Fig. 1. Regional location of American Samoa in the South Pacific Ocean. The Samoan archipelago is comprised of the independent country of (Western) Samoa and the United States Territory of American Samoa.

THE REGION

The South Pacific Ocean spans 15,000 km (9000 miles) between Australia and South America, and contains some 10,000 islands in its central and western portions which have collectively been called Oceania. It is difficult to subdivide Oceania into ecosystem units based on distinctive physical or biotic characteristics, but one designation of convenience is the group of isolated oceanic islands in the central South Pacific Ocean. This chapter focuses on one group of islands in that region: American Samoa (Fig. 1).

American Samoa consists of seven islands in the eastern portion of the Samoan Archipelago (14°S, 168–173°W): Tutuila, Aunu'u, Ofu, Olosega, Ta'u, Swains and Rose (Fig. 1). The islands are small, ranging in size from the populated high island of Tutuila (145 km^2) to the remote and uninhabited Rose Atoll (4 km^2). In that the seven islands include examples of high/low islands and densely populated/uninhabited islands, many of the environmental issues occurring in the study area are shared by other island nations in the central South Pacific region.

THE ENVIRONMENT

American Samoa lies in the broad, westward-flowing South Equatorial Current. It has a maritime climate of year-round tropical heat and rain, with ocean surface temperatures generally higher than air temperatures (Fig. 2). Tutuila Island receives 300–500 cm of rain annually. There is a wet season (October–May) and a slightly drier and cooler season (June–September) which is characterised by 2°C cooler air temperatures and increased easterly trade winds.

Despite American Samoa's open ocean location near the equator, there are seasonal signals in the marine environment. Sea surface temperatures, both nearshore and offshore, drop 2–2.5°C during the cooler season (Fig. 3). Day length varies 1 hour through the year. The tidal range is 1 m.

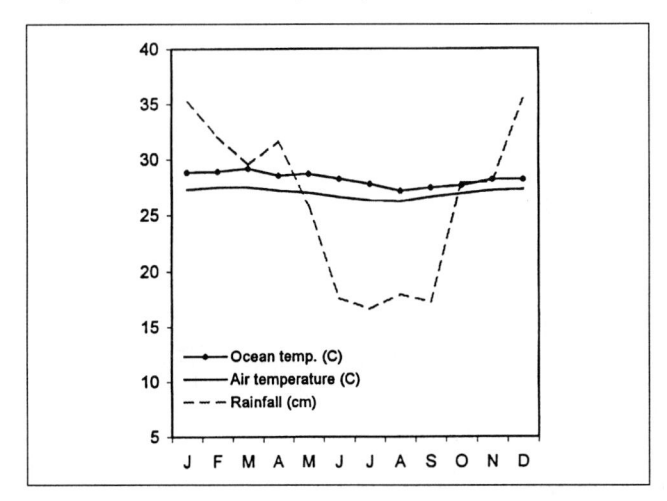

Fig. 2. Monthly rainfall and air temperature (1965–1997), and offshore ocean surface temperature (1943–1990) in American Samoa. Source: National Oceanic and Atmospheric Administration (1997) and NOAA ship data.

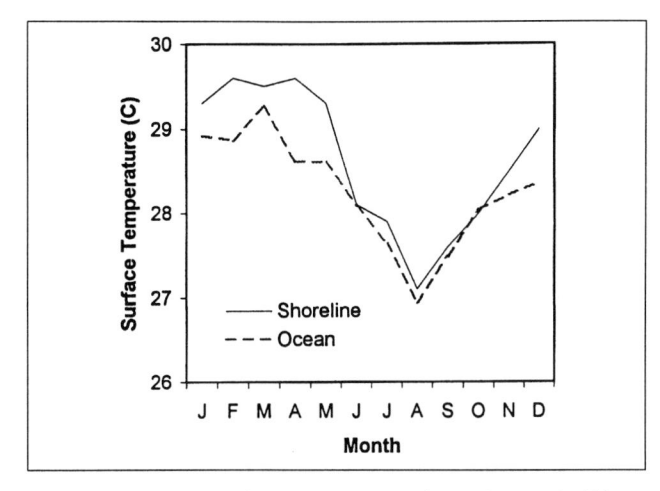

Fig. 3. Monthly sea surface temperatures of nearshore and offshore waters in American Samoa. Nearshore temperature data were made near the reef crest at Afao Village, 1990–1995 (n = 295 daily measurements, Craig 1998). Offshore SST data were collected at deep ocean locations (>2000 m), 1985–1991 (n = 102 measurements in the region of 14–16°S and 169–172°W, NOAA ship data).

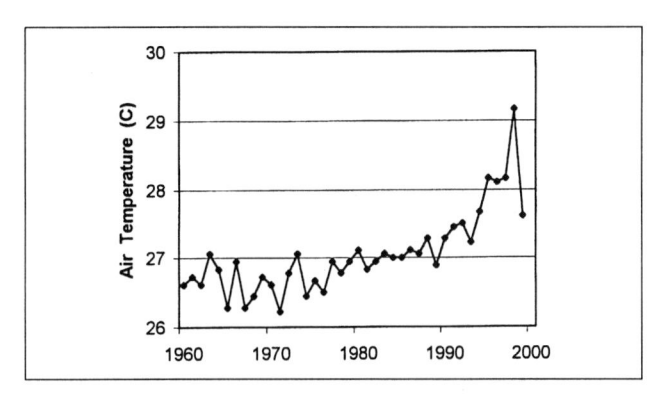

Fig. 4. Annual air temperature in American Samoa (NOAA, 1999).

Severe storms occur almost annually and hurricanes hit periodically, the last four occurring in 1966, 1986, 1990, and 1991. A sharp rise in air temperature over the past decade (Fig. 4) suggests future climatic uncertainty and a probable increase in the frequency of hurricanes in the region. The record high temperature in 1998 was due in part to El Niño which also caused drought conditions on land and unusually low tides that caused mortalities among exposed corals.

The islands have latent volcanic activity. Secondary volcanism occurred as recently as 90 and 120 years ago in the archipelago, presumably in association with the subduction of a portion of the Pacific Plate (which Samoa rests upon) down into the nearby Tongan Trench.

COASTAL HABITATS

Coral Reefs

Due to the steepness of the main islands and their relative geological youth, shallow water habitats around the islands are limited in size and consist primarily of fringing coral reefs (Fig. 5). The reefs have narrow reef flats (50–500 m) that drop to a depth of 3–6 m and descend gradually thereafter to 40 m. Depths of 1000 m are reached within 1–3 km from shore. As might be expected with this limited shallow area, nearshore water temperatures measured at the edge of the reef flat are nearly the same as offshore sea surface temperatures (Fig. 3).

Over 200 coral species occur in American Samoa, representing over half of all coral species found throughout the Indo-Pacific region (Hunter et al., 1993; Maragos et al., 1994; Veron, 1995; Mundy, 1996). Dominant genera at the 30-m depth are *Montipora* and *Porites*, followed by *Pavona*, *Pocillopora*, *Psammocora* and *Acropora* (Mundy, 1996).

The reefs are currently recovering from a series of natural disturbances over the past two decades: a crown-of-thorns invasion (1978), three hurricanes (1986, 1990, 1991), and mass coral bleaching (1994), as well as chronic human-induced impacts in areas like Pago Pago Harbour (described later) (Fig. 6). By 1995, the reefs were beginning to recover, as evidenced by an abundance of young corals (Mundy, 1996; Birkeland et al., 1997), and the recovery has continued through 1999.

The reefs support a diverse assemblage of 890 fish species (Wass, 1984), which is about twice the number of fishes occurring in Hawaii but half the number in the Indo-Australia region. Dominant families are damselfish (Pomacentridae), surgeonfish (Acanthuridae), wrasse (Labridae) and parrotfish (Scaridae) (Green, 1996a,b; Green and Hunter, 1998). Spawning for some, and perhaps most species occurs year-round, although peak spawning may be seasonal (Craig et al., 1997; Craig, 1998).

Coral reef fishes and invertebrates are harvested in subsistence and small-scale artisanal fisheries where catches are sold locally. In 1994, the only year when both components of this fishery were measured, catches were 86 and 76 mt, respectively, and consisted primarily of surgeonfish, parrotfish, groupers, octopus and sea urchins (Saucerman, 1996; Craig et al., 1997). A decreasing trend in catches was apparent in the early 1990s (Ponwith, 1992a; Saucerman, 1995, 1996), and a number of management reports indicate that overfishing is a continuing problem (ASCRTF, 1999).

Other Coastal Environments on the Main Islands

The occurrence of other types of coastal habitats is limited in the territory. Although numerous steep, short creeks (0.5–3 km) drain the landscape on Tutuila Island, many are intermittent due to porous volcanic substrates. Streams are virtually absent on the other six islands in the territory. Freshwater fishes and macro-invertebrates inhabiting local streams are generally species that are widely distributed throughout the South Pacific (ACOE, 1981; Cook, 1999).

Mangroves and other coastal wetlands were not historically abundant in these oceanic islands, and the few present are steadily being encroached upon by urbanization. Whistler (1976) estimated that there were 51 ha (127 acres)

Fig. 5. Volcanic mountains and fringing reefs on Ofu and Olosega Islands, American Samoa. Population levels here are low (25 people/ km^2) and declining slightly as people move to the main island (Tutuila) for jobs or schooling.

Fig. 6. Urban and industrial expansion around Pago Pago Harbour on the main island of Tutuila, where 96% of the territory's population dwells, mostly along the shoreline of this steeply sloped island.

of mangroves in the territory, found only on two of the seven islands (Tutuila, Aunu'u). The largest remaining mangrove area is at Nu'uuli Pala Lagoon (34 ha, 85 acres) on Tutuila Island. This lagoon supports a variety of lagoon-resident fishes and visitors of species more commonly found on the nearby coral reefs, but catch rates in the lagoon were generally low (Iose and McConnaughy, 1993), and subsistence catches there were an order of magnitude lower than that on the nearby coral reef flats (Ponwith, 1992b). Kluge (1992) sampled for larval fish in the lagoon over a one-year period and concluded that the Pala Lagoon was not a nursery area for the non-resident coral reef fish that swim into the lagoon to feed.

Given the limited occurrence of mangrove habitat in American Samoa, it is not surprising that coastal reef fishes and invertebrates here have developed little direct dependence on using mangroves as nursery areas, as occurs elsewhere in the world. Similarly, there are virtually no seagrass beds in the territory, thus further limiting the diversity of coastal environments among these islands.

Atolls

Two of the islands in the archipelago are remote atolls: Swains and Rose atolls. They differ from the islands described above in that they are geologically much older, and all that remains of these sunken mountains are small atolls. Rose is a coralline algae-dominated atoll that has received intermittent scientific attention over the past 60 years (Rodgers et al., 1993), particularly since a vessel ran aground there in 1994 causing some damage to the atoll (Green et al., 1998). Rose Atoll is important for several reasons—it is a seabird nesting area (USFWS, unpub. data), the site of a successful rat eradication program (Morrell et al., 1991), a refuge for giant clams (*Tridacna maxima*) that have been heavily exploited elsewhere in the archipelago (Green and Craig, 1999), and the principal nesting site for the few green sea turtles remaining in the territory (Tuato'o-Bartley et al., 1993). Tagging data show that most turtles nesting at Rose migrate 1500 km to Fiji to feed (Balazs et al., 1994).

OFFSHORE SYSTEMS

The central South Pacific region consists of blue ocean waters 4000–6000 m deep, punctuated by occasional volcanic eruptions that emerge from the seafloor as seamounts or small islands. Ocean surface waters are warm year-round (27–30°C) but vary 2°C seasonally (Fig. 3). The water column is stratified, with cold water of 5–6°C below 600 m (Fig. 7).

Pelagic fishes (primarily tuna, marlin, mahimahi) migrate through these waters, but not in significant commercial quantities. Local fishermen, using small 10-m boats, caught 400 mt in 1997 (WPRFMC, 1998a). The large volumes of commercially caught tuna that are delivered to the two canneries in American Samoa (about 160,000 mt annually) are caught elsewhere in the South Pacific region (Craig et al., 1993; WPRFMC, 1998a). A much smaller fishery occurs for bottomfish (primarily emperors, snappers) around the islands and seamounts—the artisanal catch in 1997 was 12 mt (WPRFMC, 1998b).

A few humpback whales from Antarctica are regular visitors to local waters where they calve and mate, mostly during September–October. Little is known about the number of whales present, their specific migration routes, or their residency time in local waters.

POPULATIONS AFFECTING THE AREA

Perhaps the most serious environmental and social problem facing American Samoa is its uncontrolled population growth (Fig. 8) (Craig, 1995; Craig et al., 2000). The

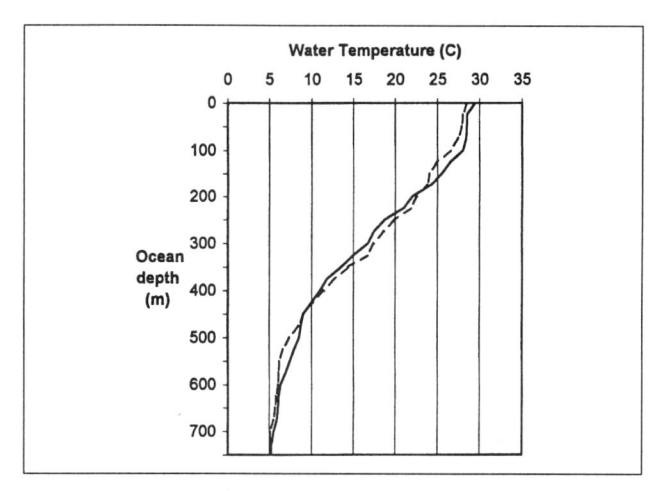

Fig. 7. Ocean temperature profiles at two locations north of Tutuila Island, Nov. 16, 1993: 2 km offshore (dashed line) and 80 km offshore (solid line). Source: NOAA ship data.

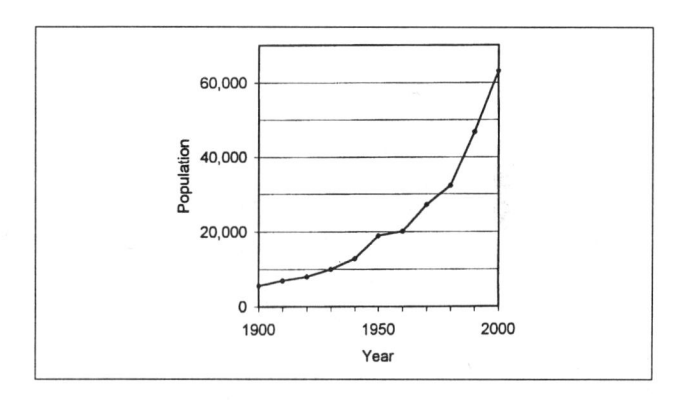

Fig. 8. Population growth in American Samoa (Craig et al., 2000).

population estimate of 63,000 in 2000 is increasing at a rate of approximately 2.5%, which equates to a doubling time of 28 years. This is the third highest growth rate in the South Pacific region (SPC, 1995). A continued increase is expected given the high birth rate (4.5 children per female) and high proportion of pre-reproductives in the population (nearly 50% of the population is younger than age 20).

The population is unevenly distributed, with 96% living on Tutuila Island (ASG, 1996). Even there, however, most of the island landscape is uninhabited due to its steepness, except for the populated band of relatively flat land along the southern coast. Population densities there will jump from about 620 people/km^2 (1600/sq mile) in 1990 to 980/km^2 (2,570/sq mile) at the millennium.

The population has increased far beyond what the local environment can currently support. In 1996, American Samoa imported $200 million of food, fuel, oil, textiles, clothing, machinery, and miscellaneous goods (ASG, 1996). The economy is based primarily on federal grants and exports of canned tuna (a product that is not caught within the EEZ of American Samoa but at distant South Pacific locations). Most of the workforce of 14,000 is employed by the local government (31%) and canneries (33%). Over half of these workers (62%) were not born in American Samoa.

RURAL FACTORS

Land use practices on Tutuila Island are determined in large part by the steep topography of the island. With 50% of the land area having a slope of greater than 70%, there is relatively little level land to use. However, as the human population expands, hillsides are being developed, particularly for agriculture and housing.

Samoans have traditionally farmed and fished for subsistence, although a cash-based economy is now firmly in place. Presently, 16% (3160 ha) of the total land area is in agriculture, but farming has declined with 5% less land and 41% fewer farmers since 1970 (ASG, 1996). As some of this agriculture occurs on mountain slopes, impacts to water quality from soil erosion are a major concern. Additionally, many families raise pigs and the waste washes directly into streams, contributing directly to eutrophication and other water quality impacts (ASEPA and ASCMP, 1995). Feral pigs that root-up vegetation also contribute to hillside erosion.

Local fishing patterns are changing as well. There is less reliance on subsistence fishing as many Samoans obtain full-time employment. At present, there are inter-related nearshore fisheries for subsistence, recreation and artisanal purposes where the catch is sold locally. There is evidence that marine fish and invertebrate populations have diminished due to fishing pressure (ASCRTF, 1999), and at least one favoured invertebrate (giant clams) has been overfished (Green and Craig, 1999).

Mariculture, as currently practised, consists of giant clam grow-out 'farms' on reef flats that probably have little impact on coastal waters because there are few of them. They are generally situated in shallow sandy areas unsuitable for coral growth, and there is no contamination by excess food, because the clams do not have to be fed. A number of alien fish and aquatic invertebrate species have been introduced to local waters (Eldredge, 1994), but they do not seem to have fared well in the wild.

EFFECTS FROM URBAN AND INDUSTRIAL ACTIVITIES

Habitat Loss

Over the years, there have been extensive alterations of coastal habitats (sandy beaches, wetlands, coral reefs) due to highway construction and urban expansion, particularly along the south shore of Tutuila Island where the majority of the territory's population is concentrated. Much of the coastal highway, built across beaches and landfills, requires shoreline protection with concrete revetments and boulders (SEI & BCH, 1994). Coastal erosion is amplified by the removal of large quantities of beach sand and coral rubble from the shoreline by villagers for use around their homes (Volk et al., 1992). Together, these shoreline alterations have largely eliminated the use of the south coast by nesting sea turtles.

Cumulative wetland losses over the last 30 years have been substantial (28%), due mainly to urban expansion. The largest mangrove areas on the island were either reclaimed (inner Pago Pago Harbour) or significantly reduced by urban development (Nu'uuli Pala Lagoon).

For coral reef habitats, direct losses occurred where harbours have been built in Pago Pago and Faleasao (Ta'u Island), and where the reefs were dredged to construct the airport runway. Other unquantified losses have occurred due to chronic pollution and natural damages (e.g., hurricanes).

Other Urban Activities and Impacts

In addition to agricultural erosion, soil runoff originates from many sources, including quarries, road projects and other ground-disturbing activities. Because of the island's steep terrain and high rainfall, storm events easily wash loose soil into streams and coastal waters, but the impacts of sedimentation to adjacent reefs have not been quantified. Coastal waters outside Pago Pago Harbour generally meet local water quality standards, although the effects of non-point source pollution are found in all areas occupied by humans (ASEPA and ASCMP, 1995).

Sewage treatment is provided only in the densely populated areas of Pago Pago and Tafuna near the airport. Three million gallons of primary treated sewage per day are pumped into the ocean in outer Pago Pago Harbour and two million gallons into Vai Cove, adjacent to the airport. For the remainder of the territory, septic tanks and illegal cesspools are used for sewage disposal.

Other urban activities that impact the coastal environment are landfills and small business activities. Although landfills adjacent to the shoreline are illegal, that regulation is not well known or enforced, and families often fill areas adjacent to their property to reclaim the land. Small businesses situated near the shoreline may have a considerable environmental impact by releasing contaminated runoff and/or dumping solid waste directly into streams and coastal waters.

Westernization and an increased consumption of imported material goods has also increased the amount of solid waste generated, with the associated problems of increased litter and indiscriminate waste disposal. Recent improvements in the solid waste disposal and collection system will assist with this problem, but the volume of waste that has accumulated over the years presents a formidable problem.

Industrial Activities and Impacts

Several industrial activities, all located within Pago Pago Harbour, have had a major impact on water quality in the harbour. There are two tuna canneries, a sewage treatment plant, ship repair yard, fuel tank farm and power plant whose collective discharge and runoff have caused several kinds of pollution.

Heavy metal contamination of fish and substrates in the inner harbour caused the government to issue a health advisory warning in 1991 for people not to eat harbour fish (AECOS, 1991). Fortunately, a similar level of contamination was not found in residents living in the vicinity, presumably because most did not eat fish from the harbour (ATSDR, 1993). Sources of this contamination are unknown, though they may be related to historical operations and poor environmental practices. Heavy metal contamination in fish and shellfish was also found at several other locations around Tutuila Island (ESI, 1994), but again, the source is not known.

Stormwater runoff from the industries located around the harbour are potentially contaminated with petroleum products, paint chips and other toxic materials. There are also frequent diesel and bilge spills in port, most of which occur during hours when surveillance is decreased. Additionally, nine derelict fishing vessels that remained beached on harbour reefs for nine years after being grounded there during a hurricane in 1991 were finally removed in 2000.

Nutrient loading from cannery and sewage disposals has also been an important environmental issue. For the past 45 years, cannery discharges of fish wastes into the inner harbour have caused extensive eutrophication, resulting in perpetual algal blooms and occasional fish kills due to oxygen depletion. However, in the early 1990s the canneries were required to construct a pipe to dispose of their treated cleaning waste at a 47-m depth in the outer harbour where there is better water circulation, and to haul high strength fish waste 8 km offshore. These actions significantly decreased the nutrient levels in the inner harbour, from 0.63 mg/l to 0.04 mg/l total nitrogen, and from 0.07 mg/l to 0.04 mg/l total phosphorous (ASEPA, 1998).

PROTECTIVE MEASURES

On-island expertise in the fields of environmental and habitat protection has increased in recent years, and legislation is in place for water quality standards, land use regulations, waste disposal, fishery management, habitat protection, endangered species, protected areas, ship pollution and other environmental issues. Environmental violations are more frequently detected and prosecuted, but enforcement of these regulations is not widespread and many problems persist. Local environmental agencies have also undertaken aggressive education programs to increase community understanding of environmental issues. This effort is commendable, but it is difficult to keep pace with the territory's rapidly growing population and concurrent development pressures.

Several examples of Indo-Pacific coral reefs have been designated marine protected areas in the territory (Table 1). The National Park of American Samoa, located on three islands (Tutuila, Ofu, Ta'u), and Fagatele Bay National Marine Sanctuary (Tutuila) allow subsistence fishing by villagers but not commercial fishing. Fishing is prohibited altogether at Rose Atoll National Wildlife Sanctuary. However, poaching is a general problem and there is little enforcement capability at present.

Table 1

Marine protected areas in American Samoa. Fishing is prohibited at Rose Atoll; subsistence fishing by villagers is permitted at the other locations.

Protected area	Year established	Island	km^2	Acres
Rose Atoll National Wildlife Sanctuary	1973	Rose Atoll	158.8	39,251
Fagatele Bay National Marine Sanctuary	1986	Tutuila	0.7	161
National Park of American Samoa	1993	Tutuila, Ofu, Ta'u	42.6*	10,520*
Vaoto Territorial Marine Park	1994	Ofu	0.5	120

*79% of this amount consists of terrestrial land adjacent to coral reefs.

ACKNOWLEDGEMENTS

We thank Pat Caldwell (NOAA/National Ocean Data Center/ Pacific Liaison, Honolulu) for his assistance with oceanic data.

REFERENCES

ACOE (Army Corps of Engineers) (1981) American Samoa stream inventory, Island of Tutuila, American Samoa Water Resources Study. US Army Corps of Engineers (Hawaii). 122 pp.

AECOS (1991) A preliminary toxic scan of water, sediment and fish tissues from inner Pago Pago Harbour in American Samoa. Prepared for American Samoa Government. 75 pp.

ASCRTF (American Samoa Coral Reef Task Force) (1999) A 5-year plan for management of coral reefs in American Samoa (FY00-FY04). Prepared for American Samoa Government. 20 pp.

ASEPA (American Samoa Environmental Protection Agency) (1998) Territory of American Samoa water quality assessment and planning report. American Samoa Government. 23 pp.

ASEPA & ASCMP (American Samoa Environmental Protection Agency & American Samoa Coastal Management Program) (1995) American Samoa coastal nonpoint pollution control program. American Samoa Government. 141 pp.

ASG (American Samoa Government) (1996) *Statistical Yearbook, 1996.* Prepared by Dept. Commerce, Statistical Division, American Samoa. 185 pp.

ATSDR (Agency for Toxic Substances Disease Registry). 1993. Biological indicators of exposure to heavy metals in fish consumers in American Samoa, June 1993. Prepared by US Dept. of Health and Human Services for American Samoa Government. 11 pp.

Balazs, G., Craig, P., Winton, B. and Miya, R. (1994) Satellite telemetry of green turtles nesting at French Frigate Shoals, Hawaii, and Rose Atoll, American Samoa. Proceedings of 14th annual symposium on sea turtle biology and conservation (Georgia). 4 pp.

Birkeland, C., Randall, R., Green, A., Smith, B. and Wilkins, S. (1997) Changes in the coral reef communities of Fagatele Bay National Marine Sanctuary and Tutuila Island (American Samoa) over the last two decades. Rept. to National Oceanic and Atmospheric Admin. 225 pp.

Cook, R. (1999) Survey of Laufuti Stream, Ta'u Unit, National Park of American Samoa. Rept. by National Park of American Samoa. 30 pp.

Craig, P. (1998) Temporal spawning patterns of several surgeonfishes and wrasses in American Samoa. *Pacific Science* 52, 35–39.

Craig, P. (1995) Are tropical nearshore fisheries manageable in view of projected population increases? Workshop on Management of South Pacific Inshore Fisheries, New Caledonia, 26 June–July 7, 1995. Joint Forum Fisheries Agency–South Pacific Comm. Biol. Paper 1. 6 pp.

Craig, P. and 11 others (2000) Impacts of rapid population growth in American Samoa. Report by Advisory Group to Governor's Task Force on Population Growth. American Samoa Government. 28 pp.

Craig, P., Choat, J., Axe, L. and Saucerman, S. (1997) Population biology and harvest of the coral reef surgeonfish *Acanthurus lineatus* in American Samoa. *Fisheries Bulletin* 95, 680–693.

Craig, P., Ponwith, B., Aitaoto, F. and Hamm, D. (1993) The commercial, subsistence and recreational fisheries of American Samoa. *Marine Fisheries Review* 55, 109–116.

Eldredge, L. (1994) Introduction of commercially significant aquatic organisms to the Pacific Islands. Vol. 1, Perspectives in aquatic exotic species management in the Pacific islands. South Pacific Commission, New Caledonia. 127 pp.

ESI (EnviroSearch International) (1994) Human health risk assessment for the consumption of fish and shellfish contaminated with heavy metals and organochlorine compounds in American Samoa. Prepared for American Samoa Government. 104 pp.

Green, A. (1996a) Status of the coral reefs of the Samoan Archipelago. Dept. Marine & Wildlife Resources (American Samoa). Biological Report Series. 125 pp.

Green, A. (1996b) Fish communities. 66 p. In Birkeland (ed.). Changes in the coral reef communities of Fagatele Bay National Marine Sanctuary and Tutuila Island (American Samoa) over the last two decades. Report to National Oceanic and Atmospheric Admin. 225 pp.

Green, A., Burgett, J., Molina, M. and Palawski, D. (1998) The impact of a ship grounding and associated fuel spill on Rose Atoll National Wildlife Refuge, American Samoa. US Fish & Wildlife Service (Honolulu). 64 p.

Green, A. and Craig, P. (1999) Population size and structure of giant clams at Rose Atoll, an important refuge in the Samoan Archipelago. *Coral Reefs* 18, 205–211.

Green, A. and Hunter, C. (1998) A preliminary survey of the coral reef resources in the Tutuila Unit of the National Park of American Samoa. Rept. for National Park of American Samoa, Pago Pago, American Samoa. 42 pp.

Hunter, C., Friedlander, A., Magruder, W. and Meier, K. (1993) Ofu reef survey: baseline assessment and recommendations for long-term monitoring of the proposed National Park, Ofu, American Samoa. Rept. for National Park of American Samoa, Pago Pago, American Samoa. 90 pp.

Iose, P. and McConnaughy, J. (1993) Fish resources in Pala Lagoon. Dept. Marine & Wildlife Resources (American Samoa). Biological Report Series 37. 42 pp.

Kluge, K. (1992) Seasonal abundance of zooplankton in Pala Lagoon. Dept. Marine & Wildlife Resources (American Samoa). Biological Report Series 36. 33 pp.

Maragos, J., Hunter, C. and Meier, K. (1994) Reefs and corals observed during the 1991-92 American Samoa Coastal Resources Inventory. Rept. for Dept. Marine & Wildlife Resources (American Samoa). 50 pp.

Morrell, T., Ponwith, B., Craig, P., Ohasi, T., Murphy, J. and Flint, E. (1991) Eradication of Polynesian rats (*Rattus exulans*) from Rose Atoll National Wildlife Refuge, American Samoa. Dept. Marine & Wildlife Resources (American Samoa). Biological Report Series 20. 10 pp.

Mundy, C. (1996) A quantitative survey of the corals of American Samoa. Dept. Marine & Wildlife Resources (American Samoa). Biological Report Series. 75 pp.

NOAA (National Oceanic and Atmospheric Administration) (1999) Local climatological data, annual summary with comparative data, Pago Pago, American Samoa. National Climatic Data Center, Asheville, North Carolina. 8 pp.

Ponwith, B. (1992a) The shoreline fishery of American Samoa: a 12-year comparison. Dept. Marine and Wildlife Resources (American Samoa). Biol. Rept. Series 22. 51 pp.

Ponwith, B. (1992b) The Pala Lagoon subsistence fishery. Dept. Marine and Wildlife Resources (American Samoa). Biol. Rept. Series 22. 28 pp.

Rodgers, K., McAllan, I., Cantrell, C. and Ponwith, B. (1993) Rose Atoll: an annotated bibliography. Tech. Rept. Australian Museum. ISSN 1031-8062. 37 pp.

Saucerman, S. (1996) Inshore fishery documentation. Annual Rept. by Dept. Marine & Wildlife Resources (American Samoa). Biol. Rept. Series. 29 pp.

Saucerman, S. (1995) Assessing the needs of a fishery in decline. Workshop on Management of South Pacific Inshore Fisheries, New Caledonia, 26 June–July 7, 1995. Joint Forum Fisheries Agency -South Pacific Comm. Biol. Paper 18. 26 pp.

SEI & BCH (Sea Engineering Inc. & Belt Collins Hawaii) (1994) American Samoa shoreline inventory update II. Prepared by US Army Corps of Engineers for American Samoa Government.

SPC (South Pacific Commission) (1995) South Pacific economies, pocket statistical summary. SPC, Noumea, New Caledonia.

Tuato'o-Bartley, N., Morrell, T. and Craig, P. (1993) Status of sea turtles in American Samoa in 1991. *Pacific Science* 47, 215–221.

Veron, J. (1995) *Corals in Space and Time: The Biogeography and Evolution of the Scleractinia*. Cornell Univ. Press, Ithaca, NY. 321 pp.

Volk, R., Knudsen, P., Kluge, K. and Herdrich, D. (1992) Towards a territorial conservation strategy and establishment of a conservation areas system for American Samoa. Rept. prepared by Le Vaomatua, Inc. for American Samoa Natural Resources Comm. 114 pp.

Wass, R. (1984) An annotated checklist of the fishes of Samoa. National Oceanic and Atmospheric Administration, Tech. Rept. SSRF-781. 43 pp.

Whistler, A. (1976) Inventory and mapping of wetland vegetation in the Territory of American Samoa. Rept. prepared for Army Corps of Engineers, Pacific Ocean Division, Fort Shafter, Hawaii. 94 pp.

WPRFMC (Western Pacific Regional Fisheries Management Council) (1998a) Pelagic fisheries of the western Pacific region, 1997 annual report. WPRFMC (Hawaii). 243 pp.

WPRFMC (Western Pacific Regional Fisheries Management Council) (1998b) Bottomfish fisheries of the western Pacific region, 1997 annual report. WPRFMC (Hawaii). 170 pp.

THE AUTHORS

Peter Craig
National Park of American Samoa,
Pago Pago, American Samoa 96799, U.S.A.

Suesan Saucerman
Environmental Protection Agency,
EPA Region IX – WTR-5, 75 Hawthorne St.,
San Francisco, CA 94105-3901, U.S.A.

Sheila Wiegman
American Samoa Environmental Protection Agency,
Pago Pago, American Samoa 96799, U.S.A.

Chapter 104

THE MARSHALL ISLANDS

Andrew R.G. Price and James E. Maragos

The Republic of the Marshall Islands (RMI) is a low-lying atoll nation within Micronesia, comprising a western (Ralik) chain of 16 atolls and three small islands, and an eastern (Ratak) chain of 13 atolls and two small islands. Wake is included in this chapter as it is geologically, climatically, and oceanographically a part of the Marshall Islands, although separate politically. Like other coral islands, the Marshall islands are influenced by episodic events, such as damage from typhoons (e.g. 'Paka' in 1997), ENSOs, crown-of-thorns starfish infestations, and coral bleaching. The 1998 El Niño was severe, bringing extensive droughts and causing much suffering. Threats posed by climatic changes and impending sea level rise are taken very seriously by the RMI, particularly as mean island height is only 2 m above sea level.

Land area of the islands is small (180 km^2), in contrast to the marine environment and vast EEZ (1.2 million km^2). Coral reefs are the most important and biologically diverse ecosystem, providing a dynamic and unstable environment, yet underpinning the country's physical and economic existence. On Bikini atoll alone, some 250 species of corals have been recorded, mostly due to the collections there since the 1950s. This represents more than half the number known for the entire Pacific or Indian Oceans. Of the marine flora, algae are moderately speciose (238 species), whereas seagrasses and mangroves (four species) are uncommon and of low diversity. Among species of importance to conservation are endemics (e.g. the Wake rail, which went extinct in 1945) and 'threatened' species including five sea turtle species.

The human population of the Marshall Islands may reach around 70,000 by the year 2000. Tourism is increasing and may expand further, but is constrained by freshwater availability, remoteness, heavy wave action, and other factors, especially in the northern Marshalls.

Coastal uses and pressing environmental concerns include: earth-moving and dredging, particularly in association with military activities (e.g. on Kwajalein atoll) as well as civilian land expansion and seawall and causeway construction (e.g. Ebeye Island and Majuro Atoll); severe droughts and water problems; hazardous wastes, in particular radionuclides on Bikini and Eniwetak from nuclear tests conducted between 1946 and 1958, resulting in increased leukaemia rates which have hindered permanent resettlement of Bikini; and limited consideration of environmental or cross-sectoral issues in development planning. Resource use problems include over-harvesting in some areas, in particular of giant clams, coconut crabs and turtles. Some fish (e.g. big-eye tuna) are possibly being fished close to their maximum sustainable yield throughout their range, because of high demand in Asian markets.

Appreciable national and international environmental legislation is now in place. However, implementation of regulations, enforcement of compliance and monitoring remain major constraints. A number of environmental measures for the Marshall Islands were identified in the 1992 National Environmental Management Strategy. Progress has been made, although many activities have yet to be implemented.

Seas at The Millennium: An Environmental Evaluation (Edited by C. Sheppard)
© 2000 Elsevier Science Ltd. All rights reserved

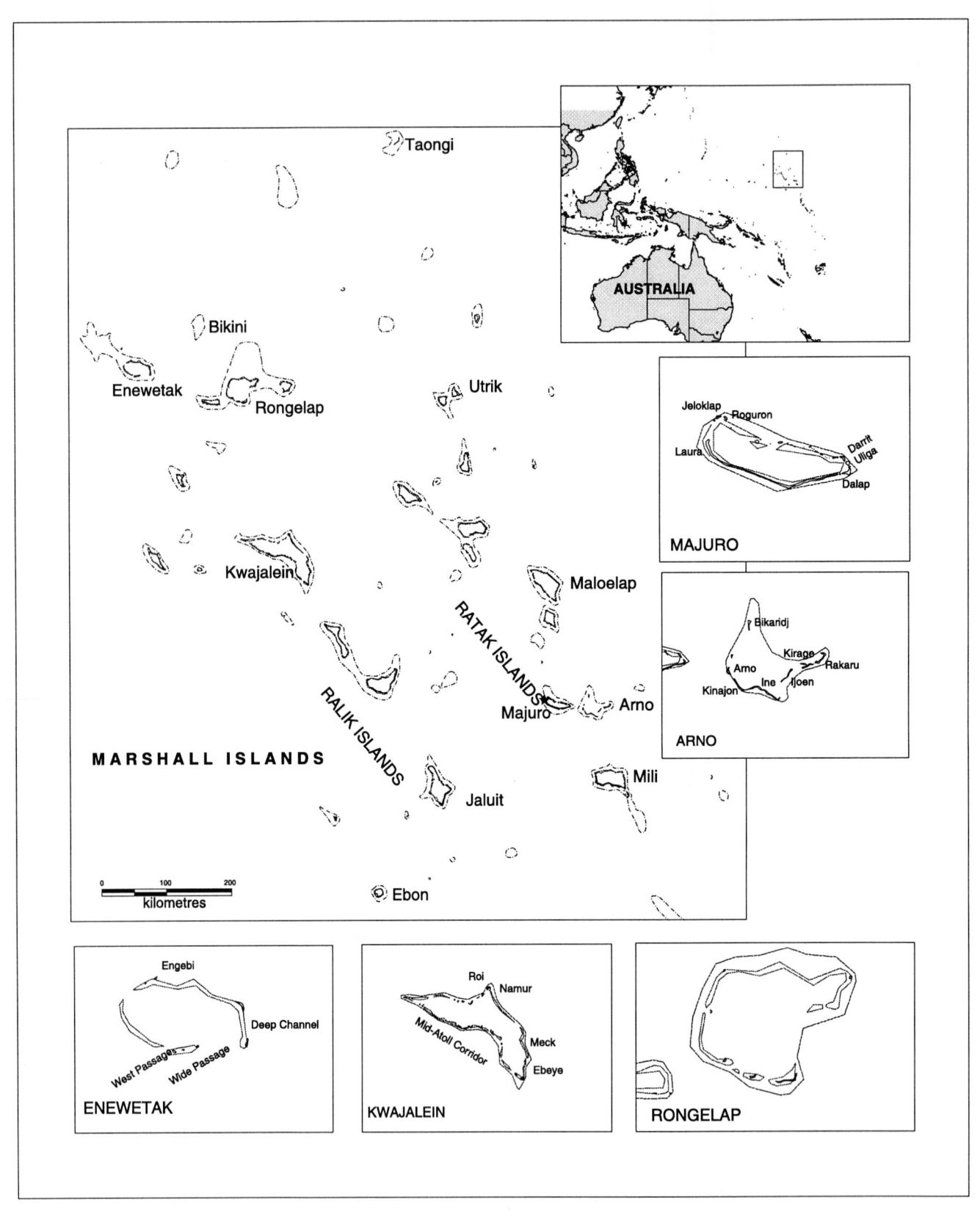

Fig. 1. Map of the Marshall Islands.

Environmental measures required include the following, which build on earlier and ongoing initiatives: development of a system of protected areas and fishery reserves; development of a national coastal zone management plan; use of EIA and cost benefit analysis in development projects; and development of contingency plans for environmental hazards and accidents. Traditional maritime skills and customary practices are valuable elements of Marshallese life, despite dilution by western influence during recent decades. Customary marine practices may have some parallels with contemporary approaches to protected area management (i.e. marine parks, 'no-take' fishery reserves). A blend of the two management systems may help optimise marine resource use and minimise conflicts, engender cooperation, and help align economic development more closely with environmental sustainability.

THE DEFINED REGION

Isolated by ocean, the Republic of the Marshall Islands (RMI) is a Pacific Island nation that is large in extent (total lagoon area 6511 km^2, EEZ 1.2 million km^2) but small in terms of land area 180 km^2) (Fig. 1). The archipelago is composed of a western ('Ralik') chain of 16 atolls and three small islands, including Kwajalein, enclosing the largest atoll lagoon in the world (Wells and Jenkins, 1988); and an eastern ('Ratak') chain of 13 atolls and two small islands (Fig. 1). Of the total, 22 atolls and four islands are inhabited. Wake Island is included in this chapter, where information is available. It is geologically, climatically, and oceanographically a part of the Marshall Islands, although separate politically.

The Marshall Islands are culturally a part of Micronesia and a single language, Marshallese, was spoken throughout the islands prior to western contact in the 18th century. However, the country has undergone considerable political administrative change since the end of World War II. In 1947 control of RMI was officially placed under the responsibility of the US by the UN, whereby the country was administered as part of the Trust Territory of the Pacific Islands (TTPI). In 1986 a treaty, the Compact of Free Association, was signed between the RMI and US. This compact recognises RMI as a sovereign nation, with its own constitution; it defines political, military and economic relations with the US; and the compact also provides funds from the US until 2001, with special compensation for residents of the four islands most affected by nuclear testing (Bikini, Eniwetak, Rongelap, Utrik). Hence, while the Marshall Islands are clearly defined geographically, the country's sociopolitical status has altered considerably over the past 50 years and this is likely to continue.

SEASONALITY, CURRENTS, NATURAL ENVIRONMENTAL VARIABLES

Physical environmental features of the Marshall Islands, including geographical variations, are summarised in Table 1. These, mostly latitudinal, gradients are of scientific interest, several of which also have implications to future economic development plans.

Atolls and Islands

Like other atolls and coral islands, those in the Marshall Islands are (naturally) highly dynamic platforms, often unstable and can alter on a seasonal basis. Inadequate recognition of this has sometimes led to serious erosion and even loss of houses. Further, the mean island height above sea level is only 2 m. Not surprisingly, the government of the RMI and other low-lying island states in the Pacific (e.g. Kiribati, Tuvalu, Tokelaus, FSM, Cooks, French Polynesia) and elsewhere (e.g. Maldives) view climate change and impending sea level rise very seriously.

Climate and Currents

Trade wind patterns differ between the northern and southern atolls, with the 5°N latitude meridian serving as the approximate boundary between the two trade wind belts. This boundary shifts seasonally, resulting in the occurrence of two different regimes (northeast and southeast trade winds) and a seasonal shift in exposure of different parts of an island to onshore winds. In some years, westerly winds can prevail during certain months (e.g. during El Niño periods). Episodic events, including typhoons, storm surges, and El Niños are considered below.

Lying within the equatorial belt of high rainfall, the southern atolls receive considerably more precipitation than the northern atolls (Table 1), while further north, Wake receives less than 10 cm y^{-1} (Wells and Jenkins, 1988). As a result the southern atolls are the most heavily populated, while the northern atolls are sparsely inhabited or uninhabited. Within the Marshall Islands there is geographic variation in the amount of water available from catchment and the rate at which aquifers are recharged. This is likely to be an important factor in future urban development plans, unless water produced by desalination becomes widespread. Larger, wider islands tend to have larger reservoirs of potable water.

The Marshall Islands are influenced by three different ocean currents (Table 1): the wide belts of the westward-moving North and South Equatorial Currents and the Equatorial Countercurrent along the 5°N latitudinal meridian and moving eastwards between them. Marine

Table 1

Some geographical gradients in physical features of the Marshall Islands (from Wells and Jenkins, 1988 and other sources; ? indicates that information is preliminary)

Physical feature	Northern atolls	Southern atolls
Island height	Increases towards north (?)	Decreases towards south (?)
Island area	Decreases towards north (?)	Increases towards south (?)
Winds	NE trades dominant	NE trades (Dec–Apr) SE trades (May–Nov)
Typhoons/ storms	Frequency relatively high	Frequency relatively low
Rainfall	10–175 cm y^{-1}	300–430 cm y^{-1}
Currents	North of 9°N: N. Equatorial Current (west-moving; av. 0.4 knots up to 2 knots)	5–8°N: Counter- current (east-moving; av. 0.4–2 knots) South of 4°N: S. Equatorial Current (west-moving; av. 3–4 knots)

primary productivity rates of these water bodies vary both spatially and temporarily, and influence the abundance and availability of fishery resources.

Upper Water Column Characteristics

Mean annual surface salinities are 35.5 ppt (Lalli and Parsons, 1993) and typical of values for the world's oceans. Despite spatial and temporal variability, planktonic primary productivity is low (<100 mg C m^{-2} d^{-1}) and is reflected by generally low abundance of zooplankton (<50 mg m^{-3}; Couper, 1989). The lagoons within the southern atolls may have slightly depressed salinities during periods of heavy rainfall, and salinities within the lagoons of the northern atolls may have slightly raised salinities due to higher rates of evapotranspiration from heavier winds and less rainfall.

Water residence time within the lagoons can vary from a few days to several weeks depending upon the extent to which open reef flats and passes through the reefs are available to promote tidal exchange between the lagoon and ocean, and the degree to which island formations on the reefs block circulation. Aside from tides, wave set-up on windward reefs contributes substantially to water exchange between the ocean and lagoon. Unlike tidal exchange which moves seawater in both directions between the ocean and lagoon, wave driven currents tend to move uni-directionally into the lagoon no matter what the tidal or wind state. The cooler ocean waters flooding into the lagoon are denser, sinking to the bottom of the lagoon and displacing shallow lagoon waters warmed by the sun.

Natural Events, Hazards and Uncertainty

Unpredictable environmental events have a critical impact on human well-being in the Marshall Islands. For example, construction of homes or other infrastructures near hazard-prone areas (e.g. exposed beaches, causeways/island interconnections) can result in extensive damage or even loss of structures. Droughts and water shortage are also of critical concern. However, the distinction between, and relative influence of, natural vs. human-induced events is not always clear-cut.

Typhoons, Wave Surges and Other Episodic Events

Although of infrequent occurrence, the impacts of these phenomena can be catastrophic. Only six typhoons have been reported between 1900 and 1998 (1905, 1918, 1951, 1958, 1967 and 1997—'Paka'). However, the severe ones caused untold physical damage to the islands. This extended to loss of resources (e.g. food crops) and probably also impact to coral reefs. The frequency of typhoons and storms is greater towards the north and west of the Marshall Islands. Extensive damage can also be caused by wave surges, which result from an ocean swell arising from distant storms. One such event occurred in 1992, which inundated Majuro for several hours. Much causeway sand linking islands of Majuro was washed away leading to erosion and loss of many houses. The 1979 flooding of Majuro also caused major damage.

ENSOs

The phenomenon of El Niño Southern Oscillations (ENSO) is addressed separately, although many climate-related events are thought to be linked to it. Under normal conditions the trade winds moving west cause a westerly flow of tropical water as part of an overall clockwise gyre (anticlockwise in the Southern Hemisphere; Table 1). During an El Niño event, the usual pattern of water flow is slowed or stops and sometimes is reversed temporarily causing an easterly flow of tropical water. This is characterised by (i) higher temperature and (ii) reduced primary productivity, coral bleaching, lower fish abundance. However, the effects of El Niños extend to major droughts, storms and other weather abnormalities which can be felt across the globe.

El Niños occur at an average frequency of about four years, although very strong events occur less frequently. Many have occurred over the past 50 years and they have been known since earliest times. The extent to which human activities influence El Niños is not clear. A major El Niño event occurred in 1982/83. However, the recent El Niño (1997/98/99) may have been of even greater magnitude and was associated with widespread drought in the Marshall Islands.

Coral Bleaching

Coral bleaching is a stress response by corals resulting in the expulsion of their symbiotic single-celled plants (zooxanthellae). Since zooxanthellae give corals most of their pigmentation, their expulsion results in the corals having a bleached appearance. This phenomenon has been linked to warm sea temperatures resulting from El Niños (above). However, other stress factors can also be involved (e.g. reduced light, pollution, and sedimentation). Coral bleaching has been reported in the Marshall Islands, but its overall extent is not clear. Infestations of the crown-of-thorns starfish can lead to heavy predation on corals resulting in the prevalence of white coral skeletons on the reef, as occurred on the reefs of Majuro and Arno in 1969–1971. In turn, these patches of white corals could be misconstrued as coral bleaching.

Sea-level Rise

Small island ecosystems are among the most vulnerable to sea level rise and other natural events and excessive human perturbations. Damage to agricultural crops on entire islands is of real concern. The best guess for predicted global average sea level rise is now around 0.5 cm per year (see Houghton, 1997). A tidal gauge has been established recently in Majuro through a project involving the Republic of the Marshall Islands Environmental Protection Authority (RMIEPA) and the Office of Planning and Statistics with assistance from NOAA and SPREP. The time series obtained to date (1998) is insufficient for determining any clear trends in sea level change. In the absence of local estimates, several points are significant:

- atolls and reefs are not static but living and have the capacity for vertical growth; sustained maximum rates of vertical reef accretion of 10 mm per year have been recorded (Wilkinson and Buddemeir, 1994);
- heavily degraded reefs have less capacity to respond to sea level rise, associated events and other impacts than healthy ones, and linked to this;
- high nutrient levels (e.g. phosphates and nitrates—in certain liquid effluents) result in eutrophication: this often leads to algal increase and coral degradation, and coral calcification rates are also impaired by high phosphate levels;
- the elevation of the reef flats and islands of some northern atolls of the Marshall Islands are reported to be higher than those in the south, although this does not necessarily imply that reefs are more robust in the north (or better able to withstand sea level changes); however, reefs in the north may be more robust, but mainly due to the preponderance of crustose coralline red algae.

Crown-of-Thorns (CoT) starfish

Outbreaks of crown-of-thorns (CoT) starfish occur in the Marshall Islands, but the extent of damage has not been recently assessed in detail. CoT outbreaks may be largely a natural phenomenon, although there is evidence that over-harvesting of its predators (e.g. pufferfish, triggerfish, triton shells) can also be an important factor. In 1969–1971 CoT starfish infestations were studied at Arno Atoll and Iroij Islet and other nearby islets at Majuro Atoll. The reef coral communities in both places subsequently recovered quickly (J. Maragos, unpubl.).

THE MAJOR SHALLOW WATER MARINE AND COASTAL HABITATS

Islands and Beaches

Islands in the Marshalls are derived from the weathered carbonate remains of reef building organisms such as corals, sand-producing algae, echinoderms, molluscs, and foraminiferans, and are often called coral islets or islands. These remains are either cast on the shallow reefs by large waves from local or distant storms, or are residues of reef materials formed when sea level was higher. For example, many geologists believe that sea level was up to 2 m higher about 4000 years ago, during the so-called Holocene sea-level stand. When sea level subsequently receded, much of the reef materials remained on the reef flats to form the foundations for many modern islets. Tropical cyclones and other tropical storms can also generate powerful storm surges and waves that can mobilize loose submarine deposits and cast them up on the reefs, contributing to island formation (see Maragos et al., 1973).

Normally the edges of coral islets are raised beach berms maintained by episodes of heavy wave action which carry fresh reef debris including sand up on the beaches. Fine beach sands, including the spectacular pink foraminiferal sand beaches of many atolls in the Marshalls, tend to accumulate where wave regimes are less turbulent, such as along lagoon shorelines. Along more exposed reaches facing towards the open sea, beaches tend to consist of larger gravels, cobbles and boulders, often eroded and rounded by the constant breaking of waves and flood tides.

The centres of larger atoll islets are often depressed, in some places with marshy soils and in others with open brackish water ponds and pools. These environments often support beach forest vegetation, and in the southern Marshalls, mangrove stands. Leaf fall and the decay of the vegetation promote acidic conditions which dissolve reef carbonates and help to maintain the depressions and their often saturated soil conditions.

Terrestrial Vegetation

Soils are of coral/sand origin, nutrient-poor and of high salinity, restricting the nature and abundance of terrestrial resources. Currently some 60% (20,000 acres) of the Marshall Islands are covered with coconut plantations

(*Cocos nucifera*) as a result of extensive plantations developed this century (Crawford, 1993a). Other cultivated plants include taro, grown in shallow pits and patches, and other marsh plants for food, medicine and weaving. Natural vegetation has been described in several works (e.g. Dahl, 1980; Crawford, 1993a). The following assemblages are important:
- mixed broad leaf forest (typically of low to medium stature, with closed canopy and few species (e.g. *Tournefortia argentea*, *Pisonia grandis*);
- monospecific forest (rare but includes: *Neisosperma oppositifolia*, *Pisonia grandis* and *Tournefortia argentea* forests);
- shrubs and herbaceous plants (several species of shrubs: e.g. *Scaevola ericea* and *Suriana maritama* on sandy shores, and *Pemphis acidula* often associated with arid, high energy areas with salt spray, and rocky areas; herbaceous plants occur mostly under forests and in open areas)'
- mangroves (e.g. *Bruguiera* probably introduced from high islands in the Pacific, and *Rhizophora*; see also above), especially in the central depressions of islets in the wetter southern atolls and along protected lagoon shores.

Fresh/Brackish Water Lakes and Pools

As noted above, brackish water pools, ponds, and lakes are most often found in the central depressions of islets where the surface of the ground dips below the level of the water lens. These water bodies are usually covered with fine sands or muds and support fish and shellfish adapted to variable salinity conditions. Salinity levels are lower in water bodies of the wetter atoll islets in the south, and where wind energy, salt spray, and evapotranspiration rates are less.

Land Invertebrates

The principal land invertebrates, all of which depend on the ocean for part of their life-cycles, are crabs, including the coconut crab (*Birgus latro*) the world's largest land invertebrate, burrowing land crabs (*Cardisoma* and others) and hermit crabs (*Coenobita* and others). The burrowing land crabs are most abundant on the wetter southern atolls and are absent on the dry northern atolls, but the other crabs are found throughout the Marshalls. The land crabs are small but are seasonally collected and eaten where they are abundant. The coconut crab is a favoured delicacy, which has led to its depleted status except on a few uninhabited atolls in the north.

Land Birds

There are very few land birds in the Marshalls. Birds in general (including sea birds) are an important component of the Marshallese diet and so are depleted on most inhabited atolls. Micronesian pigeons and introduced fowl (especially chickens) are the most common. Of interest was the recent extinction of the Wake Rail (*Rallus wakensis*), a ground bird endemic to Wake atoll, which probably was driven to extinction during the war years when starving Japanese soldiers were trapped on the atoll and survived on any available source of food.

Mammalian Pests and Insects

Many mammals have been introduced to the Marshall Islands during the 19th century (Crawford, 1993a), some of which have become pests in some areas. Among the non-native mammals are: domestic dogs, cats, goats, house mice and rats (Norway rat and Roof rat). In some areas, particularly abandoned human habitations, pigs, dogs, cats and goats have become feral. Dogs and cats have also become feral in villages and towns.

Coral Reefs

Corals

Coral reefs of the Marshall Islands are the major repository of national biodiversity. From just one well-studied atoll (Bikini) nearly 250 species of reef corals have been recorded (Wells, 1954), which is comparable to the coral diversity of another well-studied atoll, Eniwetak (Wells and Jenkins, 1988). For comparison, approximately 200 species of reef corals are known for the entire Red Sea, and 187 species for the Maldives. Coral species richness on Bikini is also more than half that reported for the entire Pacific Ocean, or Indian Ocean (491 species; Sheppard, 1998). The diversity of coral reef fish and invertebrates in the RMI is also high. In contrast, coral diversity and development in Wake, which has not been well-studied, are reported to be low. This is a result of several factors: geographic isolation, storms, shallowness of lagoon, small size and limited habitat variety, and other environmental conditions (Wells and Jenkins, 1988).

Reef Invertebrates

Important reef invertebrates in the Marshalls include several species of giant clams (*Tridacna* spp.), lobsters, many crabs, sea cucumbers, sea urchins, octopi, starfishes and marine shrimps. Many are foraged and an important component of the diet of the Marshall islanders.

Reef Fish

Reef fish are numerous and diverse off rural and uninhabited atolls, but are showing signs of depletion in the lagoon environments of the most populous atolls: Majuro and southern Kwajalein. The most common varieties are surgeonfish, parrotfish, wrasses, damselfish, sharks, jacks, butterflyfish, angelfish, goatfish, mullets, rudderfishes, baracudas, eels, and fusiliers. Rabbitfish are highly prized as a culturally important delicacy. Skipjack tuna, yellowfin tuna, and wahoo are migratory pelagics found in abundance close to the outside reefs, and are prized food fish and sport fish.

Marine Flora

Seagrasses

Extensive seagrass beds appear to be rare in the Marshall Islands, although seagrass litter is often seen washed ashore. Only three taxa are reported by Crawford (1993a): *Thalassia hemprichii*, *Cymodocea rotundata* and *Halophila*. Large beds of seagrasses have been reported at Majuro near the northern islet of Laura, and off Kwajelein and Roi-Namur islands at Kwajelein atoll.

Mangroves

Mangroves are another functionally important ecosystem, but are neither common nor diverse in the Marshall Islands, and only four species have been reported. In general mangrove species diversity declines from the Western to the Eastern Pacific (Kelleher et al., 1995), and the Marshalls occur within the low diversity belt. For Micronesia as a whole 14 species of mangroves are known (Spalding et at., 1997), with most found in the Carolines, especially Palau.

Marine Algae

Based largely on extensive taxonomic studies during the 1950s, a total of 238 marine algal species are listed for the country (McDermid, 1989). This is virtually the same number reported for the Maldives (237 taxa), an extensive atoll group in the central Indian Ocean.

Species, Habitats and Sites of Conservational Importance

The RMI harbours a number of conservationally important species, several of which are listed as 'threatened' in the IUCN Red Data Books. Among these species are the following:
- marine reptiles; five species believed to occur, important nesting sites for Green and Hawksbills being located in the northern atolls of Bikar, Bikini and Bokaak (Taongi); Thomas et al. 1989; Crawford, 1993a; Maragos, 1994). Only the green turtles and hawksbill turtle are common. Sea snakes occasionally arrive via the Equatorial Countercurrent from the west, and one was reported on the southern atoll of Namdrik (Namorik) in 1971 (Maragos, unpubl.);
- seabirds (31 species, at least 15 species breeding especially on uninhabited northern atolls), particularly Bikar and Bokaak (Taongi)
- migratory pelagic seabirds and shorebirds (many species of the latter nesting in the Arctic);
- terrestrial and freshwater birds (30 species; Crawford, 1993a);
- marine mammals (at least 20 species of whales and several dolphin species; see also below);
- endemic species (e.g. Ratak Micronesian pigeon, *Ducula oceanica*, on Wotje and Arno; grass species, *Lepturus gassaparicensis*, on Bokaak (Taongi) (Dahl, 1980); and the Wake Rail, *Rallus wakensis*, a endemic bird on Wake Island, which went extinct in 1945 (Wells and Jenkins, 1988).

Cultural and Historical Resources

Among the valuable resources of the Marshall Islands are archaeological sites (e.g. Majuro, Bikini, Kwajalein, Arno), giving clues to the origin of the people, most likely from southeast Asia several thousand years ago. Cultural sites include coral grave scatters and shell middens, while historic sites include military ships sunk during and after World War II and pillboxes, artillery, and armour on several atolls fortified by the Japanese, especially on Mili, Jaluit, Likiep, Enewetak, and Kwajalein atolls. Traditional skills in canoe building, navigation, fishing and customary marine practices are also seen as valuable elements of Marshallese life, despite increasing western influence during recent decades. Customary marine practices may have some parallels with contemporary approaches to protected area management (i.e. marine parks, 'no-take' fishery reserves). Bikar, Bokaak (Taongi), and perhaps Wake (which has a Marshallese name, Enen Kio) were earlier considered as "pantry reserves". A blend of the two management systems may provide one useful mechanism to help optimise marine resource use and minimise conflicts on at least certain atolls.

OFFSHORE SYSTEMS

Oceanography

The atolls and islets of the Marshall Islands are the emergent tops of the Ratak Submarine Ridge, the northern end of a near continuous series of seamounts that extend for thousands of kilometres to the southeast past the Gilbert Ridge and the Ellice Ridge to the Samoan and Fijian archipelagoes. The ridges jut up from the deep sea floor more than five kilometres below the sea surface. The Ratak Ridge consists of ancient drowned volcanoes, most of which at one time were emergent volcanoes. As the volcanoes began to subside roughly more than fifty or sixty million years ago, the coral reefs fringing the volcanoes maintained upward growth, eventually most surviving as atolls, and a few as the reefs supporting the isolated coral reef islets (Jemo, Mejit, Lib, Kili, and Jabat).

As noted previously, two major boundary currents run past the Marshall Islands, the broad westward-flowing North Pacific Equatorial Current (NPEC), through the central and northern atolls) generally north of the 5°N parallel of latitude, and the narrower Equatorial Countercurrent (ECC) which meanders to the east between latitudes of 4° and 6°N, depending on season. The ECC defines the location of the dynamic oceanographic equator in the Pacific, and is situated below the weather phenomenon known as the Intertropical Convergence Zone (ICZ). The wet atolls of Majuro, Arno, Mili, Jaluit, Ebon, Namdrik and the islet Kili are all within the influence of the ICZ and the ECC, while the rest are influenced more by the NPEC and northeast trade winds.

The ICZ is also known as the doldrums, or a zone of slack winds and high humidity and rainfall, and the boundary between the northeast trade winds and the southeast trade winds at any one time. As these wind systems collide, they rise, dropping immense quantities of rainfall. Within this zone the ECC flows eastward, bringing migratory fish, marine reptiles, marine mammals, and the spores and seeds of algae and plants from the west, and larvae of countless species of reef organisms from the higher species diversity areas in the western Pacific. In a similar manner the NPEC transports the adults, larvae, seeds, and spores of marine plants and animals from the east, ensuring the connectivity of the island and reef ecosystems within the region.

Marine Mammals

Although not well studied, the Marshall Islands support a rich marine mammal fauna, which includes the following (Crawford, 1993a):
- 20 species of whales (including blue, fin, sei, minke, killer and short-finned pilot whale); and
- several dolphin species (e.g. bottlenose dolphin, common dolphin, striped dolphin and spotted dolphin).

Dugongs appear to be absent from the Marshall Islands (Kelleher et al., 1995), as are other marine mammals (Crawford, 1993a). Marine mammals in the Marshalls are all pelagics and consist of dolphins, porpoises and whales. Occasionally the Hawaiian Monk Seal, an endangered species is sighted, most recently off Wake Atoll.

Benthos

Deep oceanic sediments in the Marshall Islands are of mainly calcareous origin and red clays. Benthic biomass in the region is low (0.05–0.1 g m^{-2}), which is the second-lowest or lowest category in a ranked scale of eight compiled for the world's oceans (Couper, 1989).

POPULATIONS AFFECTING THE AREA

This section summarises the main human population trends in the Marshall Islands, as well as the main concerns arising from coastal use and other human activities.

Demographic Patterns

Between 1980 and 1988 the national population increased by more than 3% annually. The current population of the Marshall Islands (1998) is probably in the mid 50,000s, and projections estimate a total population of around 70,000 by the year 2000. Many environmental problems in the Marshall Islands, where land is scarce, relate directly or indirectly to population pressures. Hence Majuro and Ebeye, where two-thirds of the population live, face the most overt problems. The high population density on Majuro is not actually due to shortage of land, but to crowding around the limited infrastructures and facilities, which are concentrated in relatively few areas.

Tourism

In addition to the resident population, there is increasing ecotourism and other forms of tourism in the Marshall Islands. Although still in its infancy, the industry might expand further (Douglas and Douglas, 1994), although shortage of freshwater and other factors may constrain its large-scale development in the foreseeable future. During 1994, 564 tourists were reported to have visited the RMI, whereas in 1997 the number was an estimated 2000 visits (Maxwell Stamp, 1998). Although the figures may partly reflect improvement in the recording system, it is likely that there has been some real increase in tourism over recent years.

Summary of Coastal Uses and Environmental Issues

Environmental issues linked to coastal uses and economic development in the Marshall Islands are summarised in Table 2. Issues are grouped into natural (and unnatural) events, hazards and uncertainty, practical problems and cross-sectoral issues. However, there is clearly overlap between the categories. An attempt has been made to score the severity of each issue by sub-region and for the country as a whole. Despite the qualitative and somewhat tentative data, greatest problems for the country as a whole include the following:
- severe droughts and water problems,
- hazardous wastes (e.g. radioactive materials),
- earth-moving and dredging (impacting e.g. reefs),
- limited understanding of cross-sectoral issues and incorporation of environmental issues into development planning,
- depletion of renewable natural resources (including reef fish, sea turtles, sea birds, giant clams, pearl oysters, coconut crabs, native *Pisonia* beach forest, etc.).

Individual issues and concerns are considered further in the context of (1) natural environmental variables (above); and (2) human-induced problems in terms of rural factors, coastal erosion and landfill, together with effects from urban and industrial activities, and other environmental concerns (below).

Table 2

Summary of major environmental concerns for Marshall Islands by sub-region. Perceived relative magnitude of the problem is indicated by: 0 none or minor, + minor or moderate, ++ moderate to major, +++ very major or serious; 'nuclear islands' include Bikini, Eniwetak, Utrik, Rongelap; ? knowledge incomplete (from Price, 1998).

Environmental Issue	Majuro	Ebeye	Outer Islands	'Nuclear' Islands	Marshall Islands
Natural Events, Hazards and Uncertainty					
Severe droughts and water shortage (e.g. 1998)	+++	+++	?	?	+++
Typhoons, wave surges and other episodic events creating physical damage and loss to resources (e.g. crops)	++	+	+	+	++
El Niños (causing reduced marine productivity and linked to climate related problems above)	++	++	++	++	++
Other *environmental hazards* and events (e.g. coral bleaching)	+/++	+/++	?	?	+
Practical Problems					
Rapid human *population growth*	+++	+++	++	0	++
Non-sustainable consumption and/or degradation of freshwater	+	++	+	+	++
Solid waste disposal problems	+++	+++	+	+	++
Hazardous waste disposal problems	+++	+++	+	+++	+++
Sewage/liquid waste disposal problems	+	+++	++	+	++
Sand/coral mining (due to shortage of building materials), which can cause erosion and impair the capacity of reefs to act as natural sea defences, and undermine their role as fishery and biodiversity areas	++	+	0	0(?)	++
Earth-moving and dredging causing degradation of reefs/habitats, exacerbating impacts from wave damage and other events (e.g. erosion and deposition)	+++	++	+	+	+++
Over-harvesting of some marine resources (e.g. giant clams, coconut crabs, turtles, fish)	+++	+++	+	0	++
Sea level rise, in particular the potential threat to urban centres (Majuro, Ebeye), populated atolls and actual/potential tourism areas	++	++	++	++	++
Cross-Sectoral Issues					
Limited understanding and data *on cross-sectoral environment development issues*	++	++	++	++	+++
Need for greater *incorporation of environmental concerns* (e.g. EIA, protected areas) into development planning, which is limited by shortage of human and physical resources	+	++	++	++	+++
Conflicts from different user interests in the same area (e.g. land owner and hotel developer)	+++	+++	+	+	++
Community vs. individual conflicts linked to land tenure and governance system	++	++	0	0	++

RURAL FACTORS

Agriculture and Freshwater

This section reviews concerns facing the rural environment, some of which also apply to other settings in the Marshall Islands, such as the urban environment. The widespread transformation of land to coconut plantations in the RMI appears not to have halted, as there is at least one proposed new major coconut replanting scheme. Such schemes clearly involve both environmental costs as well as socio-economic benefits.

Shortage of freshwater is a serious problem in both rural and urban areas. This results from droughts (above) and non-sustainable use of freshwater. The situation is compounded by contamination from faecal coliform bacteria in sewage and burial of the dead near to freshwater sources. Cremation is very uncommon, with only two reported for Majuro up to 1998. Heavy utilisation of aquifers can also lead to ingress of seawater (saline intrusion) into the freshwater lens of aquifers. The same problem can result from sea level rise. Water-borne diseases, including cholera, typhoid, as well as bacterial and protozoan induced intestinal disorders, are extensive in the Marshall Islands (Crawford, 1993a).

Aquaculture

The Marshall Islands Marine Resources Authority (MIMRA) is engaged in several aquaculture schemes, involving in particular the giant clam and pearl oyster. The commercial potential for algae is also currently being explored. Private-sector aquaculture includes enterprises on Mili and Majuro (Fig. 2). The hatchery farm on Majuro has been operational for 10–12 years and is reported to be breaking even financially (1998). The main species under culture include the following:

- clams (four species of *Tridacna*, bred in captivity and sold for aquarium trade, e.g. in USA),
- pearl oysters (*Pinctada margaritifera*),
- *Trochus* shells (introduced to the Marshalls by the Japanese from the western Pacific) mainly for ornaments/buttons).

Souvenir or Curio Species

A number of ornamental marine species are on sale, for example in hotels on Majuro. Examples include cowries, other mollusc shells, corals and turtle shells. Reef flats are sometimes systematically combed by 20 individuals or more, possibly for corals, shells, urchins or other species. If so, this can not only reduce local populations, but also adversely affect reef structure and possibly even function. Should the RMI become signatory to CITES (Convention on International Trade in Endangered Species), the curio species trade will require further scrutiny. Corals and jewellery made from hawksbill turtle shell ('tortoiseshell'), are both subject to CITES.

COASTAL EROSION AND LANDFILL

Sand/Coral Mining

Sand/coral is mined from lagoons and other inshore areas and used for construction. The practice is most widespread in Majuro where building materials are needed most. Problems include unforeseen sedimentation and erosion in adjacent areas. Mined sand also serves as infill for causeways which, without provision for through-flow of water, often results in similar problems and reduced lagoon circulation. Both coral mining and sedimentation impair the capacity of reefs to act as natural sea defences, and undermine their biological role as fishery areas and repositories of biodiversity. Mining activity too close to beaches can cause beach erosion, as occurred near the new Outrigger Hotel. Virtually no reef mining or dredging operations are based on considerations of sustainability, yet most reefs have substantial amounts of carbonate sands in some areas that are not critical to reef functioning.

Earth-moving and Dredging

Earth-moving and dredging activities (Fig. 3) cause problems similar to those described for mining (above). Significant levels of earth-moving require a permit from RMIEPA, but this is sometimes by-passed. Kwajalein atoll was subject to saturation bombing before the 1944 US invasion by armed forces and Mili, Jaluit and Enewetak suffered similar bombing and shoreline development. Many of the larger islets have been altered by military activities, dredging and infilling (i.e. earth-moving), and Wake was similarly impacted (Wells and Jenkins, 1988).

Fig. 2. Private-enterprise aquaculture hatchery on Majuro, where species under culture include *Tridacna* clams sold in the USA and elsewhere for aquarium trade, pearl oysters (*Pinctada margaritifera*) and *Trochus* shells mainly for ornaments/buttons (A.R.G. Price).

Fig. 3. Earth-moving on Majuro. The practice, together with dredging, often causes degradation of reefs/habitats, exacerbating impacts from wave damage and other events, such as erosion and deposition. Many larger islets of the Marshalls have been altered by earth-moving associated with military activities, and Wake was similarly impacted. Natural resources of Ebeye have been particularly degraded (A.R.G. Price).

Fig. 4. Disposal of solid wastes is a major problem in the Marshall Islands, as shown here in Majuro, and for small island states in general (A.R.G. Price).

Natural resources of Ebeye have been particularly degraded. The cumulative impact of all activities and development has never been determined, although there are now strict environmental standards in place for USAKA military activities (below). The extensive impact of coastal construction on shallow water ecosystems in US-affiliated Pacific islands is described by Maragos (1993).

EFFECTS FROM URBAN AND INDUSTRIAL ACTIVITIES

Solid Wastes

Disposal of solid wastes is a major problem in the Marshall Islands (Fig. 4) and for small island states in general. There are some solid waste dump sites (e.g. Majuro), but extensive junk and refuse are a common sight. Layers of sand are often put over solid waste at the dump site, a practice constrained by the shortage of sand and the lack of readily available land in the Marshall Islands. Old equipment and shortage of funds for maintenance are further problems. An additional problem is the need for burning of waste from the Majuro hospital. This has become a pressing concern, particularly since the hospital incinerator burned down recently.

In a recent US EPA report, the use of the following in the RMI was recommended: crushers, incineration and the recycling of aluminium cans. These recommendations have yet to be implemented. As with other projects, there will be costs associated not only with equipment purchase, but also repair, maintenance and other recurrent costs.

Hazardous Wastes

Common hazardous materials include batteries, pesticides, toxic wastes and PCBs (e.g. from transformers). Fuel and batteries from cars are supposed to be removed before dumping. However, at present there are currently no special facilities for dealing with hazardous wastes.

Bikini and Eniwetak atolls were subject to numerous nuclear tests by the US between 1946 and 1958. This caused inestimable damage and social disruption to many reef and island communities (Wells and Jenkins, 1998). Some inhabited downwind islands (e.g. Rongelap, Utrik) were also affected by radiation fallout (IAEA, 1998). Some evacuees from Bikini returned to the island in 1969, but became contaminated by radioactivity and had to leave the same year. Rondik Atoll, although now uninhabited, was used for resettlement for two years by the Enewetakese people following their removal from Enewetak atoll to make way for other nuclear weapons testing. Subsequently Rondik became contaminated from the fallout following test Bravo at Bikini in 1954. Billions of dollars of nuclear research has been conducted at Bikini, but the atoll has not yet been cleaned up sufficiently to allow permanent resettlement of the Bikini islanders. The cleanup costs of Bikini atoll are estimated to be over 100 million dollars, while the cleanup of Rongelap and Utrik and possibly other atolls has not yet been planned. As a result of exposure to radiation, leukaemia rates and other radiation-associated medical disorders increased, for which compensation is still being paid by the US Government.

Liquid Wastes

Untreated liquid wastes and other effluents discharged into the sea often cause environmental problems. This is particularly acute in Ebeye, whereas in Majuro there are better waste treatment facilities. However, infrastructures for dealing with sanitation, water and liquid wastes are planned for Ebeye, pending a loan from an outside agency. Many health issues are also environmental concerns. Sewage and waste water discharged into the sea contain plant nutrients which, if in sufficient quantity, cause enhanced growth of algae to the detriment of corals or in extreme cases can lead to deoxygenation of bottom waters and the exclusion of most marine life. Untreated sewage may also present a health hazard to users of the sea by the transmission of pathogens. In the face of a rising population, environmental issues are generally of lower priority compared with health and education.

Accidental spills of oil or other hazardous waste are increasing. Major users of oil and/or diesel fuel are Majuro, Ebeye and Jaluit, for example for engines, generators and power for desalination plants (Majuro and Ebeye). Small oil spills (less than 1000 gallons) occur frequently.

Tourism Infrastructures and Activities

Environmental degradation resulting from tourism is not currently a major concern in the RMI. However, the situation could change if the industry expands. The following groups of concerns are identified for the Maldives, a group of islands sharing certain features in common with the Marshall Islands and so these same values could become important here: (1) impacts from tourism infrastructure; (2) impacts from diving and snorkelling; and (3) boating, fishing and other effects of tourism.

Fisheries

The fisheries are the principal renewable resource of the Marshall Islands, contributing 7.3% to GDP. The RMI has jurisdiction of fishery stocks within its EEZ (1.2 million km^2). Of particular importance are tuna, a resource shared with other Central West Pacific Island States. The value of the catch of vessels fishing within the RMI's EEZ is conservatively estimated to be $50 million annually. However, the RMI receives only a small fraction of this amount via treaties or fishing agreements with the foreign fishing fleets and nations. The maximum sustainable yield (MSY) is considered to be appreciably more than current harvests.

The main fishery sectors in the Marshall Islands are the pelagic fisheries, artisanal fisheries, aquarium fish trade and cultured invertebrates.

Pelagic Fisheries

Of particular importance are tuna. Among the countries licensed to fish in the Marshall Islands' EEZ are the Japanese, and earlier a large fleet of Chinese vessels were fishing in the Marshalls. In 1997 the number of vessels operating dropped to less than thirty. The number is expected to increase during 1998 due to the addition of purse-seiners to the fleet of long-liners and pole-and-line vessels normally used by the Japanese. Total catches by Japanese and other foreign vessels from 1989 to 1995 have fluctuated enormously, ranging from a reported 2562 tonnes (1995) to 111,198 (1993). No clear patterns are available. Stock assessments are being undertaken by the Forum Fisheries Agency (FFA) but are not yet completed. However, preliminary analysis indicates that big-eye tuna (caught by longline) is being fished close to the MSY. In contrast, yellowfin tuna and skipjack do not yet appear to be heavily exploited.

The average revenue to the Marshall Islands from the Japanese during the 1990s to date has been $1.2 million per year, based on 5% of the landed value of their catch. (Hence 95% of the value of the resource is not retained in the Marshall Islands). Revenues during 1998 are expected to be greater than the $1.75 million high set in 1994, largely due to an increase in number of vessels.

Artisanal Fisheries

These are mostly subsistence-based reef fisheries, but are expected to be developed as part of the national plan for 'spatially integrated development'. It is anticipated that this will help meet the domestic demand for fish, and help to stem urbanisation by encouraging economic development on the outer atolls. With Japanese assistance, artisanal fishery projects are being implemented for Likiep, Namu and Ailinglaplap atolls.

Aquarium Fish

These have been commercially harvested from Majuro, using slurp guns, for around 20 years. In 1992 at least 3000 fish per month were exported to Hawaii. Growth of the industry is expected. Efforts are made to follow a well defined reef rotation scheme harvesting fish from different reefs each day, and allowing exploited reefs time to recover. The scheme should provide an input to the future development of regulations to manage the resource.

Other Exploited Species Groups

Other species targeted include giant clams, coconut crabs and marine turtles, of which all five species are reported in the Marshall Islands. However, Green and Hawksbill turtles are the commonest species. In a recent project on the outer atolls, harvested *Trochus* shell was sold for $2.11 per pound (totalling $20,000), with producers receiving $1.78 per pound and agents $0.07 per pound. In the same project sharks caught as by-catch in ocean and lagoon tuna fishing were obtained for their fins. Prices paid to producers ranged from $10–27 per pound, while wholesale selling price to buyers ranged from $15–30 per pound. Shark finning is becoming increasingly controversial because the rest of the shark is normally discarded following removal of the fins, frequently when the sharks are still alive. Desirable species of sea cucumbers (holothurians) were difficult to harvest and quantities were not considered sufficient for a commercial operation.

Resource Over-harvesting

Several renewable resources have been over-harvested in some areas of the Marshall Islands, in particular giant clams, coconut crabs, sea birds, and sea turtles. It is also possible that some fish (e.g. big-eye tuna) are being fished close to their MSY. This indicates moderately high levels of exploitation. However, knowledge on the status of most fishery stocks in the Marshall Islands is incomplete. Without such information, optimal (i.e. sustainable) fishing levels cannot be easily determined.

OTHER ENVIRONMENTAL CONCERNS

A range of environmental problems related mainly to cross-sectoral issues are examined in the following sections.

Limited Understanding on Cross-sectoral Issues

In general, understanding of the interconnections between environment and development is extremely limited in the Marshall Islands. The value of the environment is not recognised widely, or is ignored by developers and local people, or is not considered in some aid projects. There is a widely held view that environmental interests are always against development. Coastal ecosystems are therefore increasingly threatened by growing development pressures, and also by management policies which stress economic development and focus on human activities and not on the systems which sustain them. During recent ship groundings, overriding concerns were for damage to cargo and financial compensation, while environmental damage was not considered.

Limited environmental awareness is linked to the widespread consideration of only short time scales in the Marshall Islands. Both may partly originate from the perception that environmental and other problems will somehow be alleviated by outside intervention. It is partly for these reasons that there is only limited data on cross-sectoral environment development issues in the Marshall Islands. However, shortage of physical and human resources are also critical constraints.

Need for Greater Incorporation of Environmental Concerns into Development Planning

This issue relates to limited environmental understanding (above). However, shortage of technical staff, resources, field equipment, analytical tools and skills in environmental management are recognised difficulties. The Asian Development Bank earlier invested in capacity building for the RMIEPA during 1993–1994, but the benefits cannot be sustained unless there is more political will to do so and technical assistance is sustained over a longer time period.

While an EIA is now needed for major projects, implementation and monitoring compliance of the development with EIA requirements can be problematic. It is particularly important that environmental and economic factors are combined in cost–benefit analyses (multidisciplinary assessments), to determine the true costs and benefits. These include not only direct ones, but also indirect ones or 'externalities'. An example is the adverse effects of sedimentation on a coral reef and associated fishery caused by earth-moving needed for construction.

There have been recent proposals for an extensive, multi-million dollar coconut tree replanting project in Majuro. The immediate socioeconomic benefits are likely to be considerable. However, impacts caused by the massive equipment and soil excavation are also likely to be very significant. It is therefore critical that a long range planning process be instituted and comprehensive cost–benefit analysis is undertaken, to demonstrate whether or not the gains are likely to be greater than the losses. Similarly, a large fish plant is under development (or shortly to be initiated) at Majuro for processing and cooking fish caught by purse seiners. This will employ 200–300 people and probably generate considerable income. However, environmental (and economic) costs associated with its construction and operation can also be expected. All possible socioeconomic and environmental issues therefore require close examination and need to be incorporated into a cost–benefit analysis. Such assessments should be undertaken prior to project implementation. It requires consideration of not only short-term (1–3 years) expected positive and negative consequences, but also of longer-term ones. The pros and cons of desalination plants is another example of the need for thorough cost–benefit analysis.

The absence of a protected area system and a national coastal management plan is a very real constraint, and would help to integrate the needs of economic development and the environment. A coastal management plan initiated in 1993 for Majuro is still not yet completed (1999).

Conflicts from Different User Interests

Majuro and Ebeye in particular are associated with a broad and growing spectrum of human activities. Conflicts have arisen largely due to difficulties with integrating complementary activities and segregating conflicting ones. For example, the extraction of sand in lagoons/inshore area may limit options for fishing and other marine-dependent activities by degrading nursery grounds or inshore fishing areas and may exacerbate shoreline erosion. (However, the present lack of alternative to sand for building in itself is a very real problem, unless equipment and training and development begin to assess removal of sand from offshore sites.) The significance of this potential conflict may only become apparent if the RMI wishes to expand the inshore fisheries to provide food and jobs for Marshallese people, and/or as a source of fish for tourists, should the industry ever take off.

Conflicts between a land owner and hotel developer also came to light recently in Majuro. Often, however, a prospective developer is also the land owner. More often conflicts arise between government regulatory agencies and the developer. A further point is that traditionally in the RMI island chiefs (iroij) have had authority for conservation related issues. At the same time they are generally major land owners, and some chiefs are now becoming the proponent of economic development projects such as hotels. Pristine coral reefs are clearly a major tourist attraction for hotels. There is clearly at least the potential for a chief's recommendations on conservation to be influenced by personal as well as communal interests. This might become a significant issue in the development of a protected area system for the RMI.

In 1988 a team of American and New Zealand scientists conducted a survey of seven northern atolls and islands, all uninhabited (Bokaak, Bikar, Taka, Jemo, Rondik and Erikub) except for sparsely inhabited Wotho. The team concluded that all island areas warranted greater levels of protection, particularly Bokaak, Bikar, Taka and Jemo. The South Pacific Regional Environment Programme's South Pacific Biodiversity Support Programme has introduced community-based designation and management over protected areas, and have initiated efforts via working with the landowner to provide Bokaak and Bikar with more protection.

Community and Individual Conflicts Linked to Land Tenure and Governance

Incompatibilities between longstanding land tenure rights and contemporary governance systems in the Marshall Islands has been mentioned earlier. Of particular significance is landowner resistance to environmental regulation by government mentioned above. Resolution of such problems requires the integration of the interests of different users of the coast. In some instances, this will require adjustment of individual activities to optimise the use of coastal resources to meet society's needs in the Marshall Islands. Without national protected areas and coastal management plans and their implementation, resource-use conflicts are likely to escalate further. Consideration of land tenure and customary marine practices will undoubtedly remain of paramount importance.

PROTECTIVE MEASURES

This section reviews the management of environmental affairs in the Marshall Islands. It examines the main environmental agencies, existing environmental legislation and other governance arrangements which have a bearing on the environment.

Institutions

The main institutions involved with environmental management in the Marshall Islands include: (1) ministries (Health and Environment, previously Health Services; Resources, Development and Public Works; and the Ministry of Justice); (2) statutory authorities, agencies and public/private joint ventures (Environmental Protection Authority: RMIEPA; Marshall Islands Marine Resources Authority: MIMRA; Marshall Islands Development Authority: MIDA; Kwajalein Atoll Development Authority: KADA; Office of Planning and Statistics; Marshall Islands Visitor Authority: MIVA); (3) government offices, e.g. Majuro Atoll Local Government (MALG) is heading a Coastal Zone Management Working Group for Majuro; and (4) coalitions (e.g. Women United Together in the Marshall Islands (WUTMI), which has developed a policy for increasing the involvement of women in development). The activities carried out by these institutions are assessed by Price (1998) and the overall status of governance is considered below.

Land Tenure and Customary Marine Practices

Under the traditional Marshallese system of land tenure, all land is privately owned. Hence there has been no provision for non-private (e.g. government) recreation areas, national parks or space for other public land-use. Landowners are also accustomed to exercising ultimate control over land use, and often oppose regulations that may restrict usage of their property. Land tenure rights have major environmental and economic implications across virtually all development sectors. Conflicts can also apparently arise if a landowner also happens to be influential in key development decisions. This may lead to interests of an individual overriding those of the community. An analysis of land tenure and other aspects of environmental law is given by Harding (1992).

Of much significance is recent legislation developed, 'eminent domain', whereby government can take over private land in the public interest. Social and wider repercussions from this are likely to be considerable.

Traditional property rights extend out to sea. Some unpopulated islands and reefs used to be set aside as "pantry" reserves for the protection of fish, crayfish, coconut crabs, seabirds, eggs, and turtles (Wells and Jenkins, 1988). Fishing was only permitted on certain occasions. During the Japanese occupation these and other customary marine practices were not honoured. However, there is renewed interest in, and possibly some re-emergence of, marine customary law in the Marshall Islands.

Table 3

Key environmental and marine regulations for the Marshall Islands and their legal status (1998; regulations are based on modified Trust Territory Regulations; ++ = enacted, + = in draft, and still to be enacted: 1998)

Regulation	Status
Clean Air Regulations (1995 draft)	+
Earthmoving Regulations (1989, Incorporating 1994 Amendments)	++
Environmental Impact Assessment Regulations (1994)	++
Marshall Islands Pollutant Discharge Elimination System (MIPDES) Regulations (1995 draft)	+
Pesticide Regulations (1995 draft)	+
Public Water Supply Regulations (1994)	++
Solid Waste Regulations (1989)	++
Toilet Facilities and Sewage Disposal Regulations (1990)+	++
Regulations for Sustainable Development of the Coastal Zone (1994 draft)	+
Marine Water Regulations (1992)	++

National Legislation

A considerable body of national environmental legislation is now in place in the Marshall Islands. Details of acts having a direct or indirect bearing on the environment are summarised below and examined in more detail elsewhere (Price, 1998):
- National Environment Protection Act (NEPA), 1984 (and supporting environmental regulations) (Table 3);
- Planning and Zoning Act, 1987;
- Endangered Species Act, 1975;
- Coast Conservation Act (CCA), 1988;
- Marshall Islands Marine Resources Authority (MIMRA) Act (1988) (coordination of exploration, exploitation and management of living and non-living resources);
- *Trochus* act of 1983 (licensing system for harvesting);
- Marine Resource Act (prohibits killing of turtles on land and the collection of eggs; limits for harvesting of cultivated sponges and the pearl oyster).

US Army, Kwajalein Atoll (USAKA) Procedures and Standards

Stringent environmental standards (1995) are in place to regulate the environmental impacts from US army training and ballistic missile testing activities on the 11 islands on Kwajalein Atoll leased from Marshallese landowners and used by the US Army for military activities, as well as other areas leased or used by it. The same standards apply to non-USAKA users of USAKA territory. The standards, which are being updated (1998), are divided into (i)

Table 4

Major international environmental conventions and regional agreements for Marshall Islands. (++ Signed, ratified, accepted, approved, or acceded; + signed, but not ratified, accepted or approved)

Environmental Convention or Agreement	Status
Atmosphere	
Aircraft Engine Emissions (ICAO)	+
Climate Change (FCCC)	++
Ozone Layer Convention	++
Montreal Protocol	++
London Amendment	++
Copenhagen Amendment	++
Marine environment	++
MARPOL 73/78	++
Civil Liability for Oil Pollution (CLC)	++
UNCLOS	++
SPREP	++
Nature conservation and terrestrial living resources	
Convention on Biological Diversity	++

permissible procedures and (ii) standards for seven sets of environmental concerns (air quality, water quality and reef protection, drinking water quality, endangered species and wildlife resources, ocean dumping, material and waste management and cultural resources). A coastal resource atlas has also been compiled for Kwajalein as well as for Majuro and Arno atolls by the US Army Corps of Engineers (US Army Corps of Engineers, 1989–1990).

International Legislation and Regional Programmes

The RMI is party to about 11 international and regional treaties, conventions and other agreements relating to the environment (Table 4). Upholding these is particularly important in a maritime state whose trans-boundary resources (e.g. migratory tuna) constitute a 'regional commons' shared by several Pacific Island nations and utilised by some other countries (e.g. China, Japan). The agreements can be divided into the following categories:
- atmosphere (six conventions, one not yet concluded);
- marine global environment (three conventions);
- marine regional environment (one convention);
- nature/terrestrial conservation (one convention).

There are also at least six international conventions for dealing with hazardous substances and at least three concerning nuclear safety. However, the RMI is not party to these, nor to several major nature conservation agreements such as the World Heritage Convention, the Bonn Convention on Migratory Species and the Ramsar Convention (wetlands). The UN Convention on the Law of the Sea (UNCLOS) serves as useful umbrella legislation for a wide range of marine environmental, resources and other concerns, and the RMI is party to this agreement.

The RMI is also a member of major regional programmes including the South Pacific Forum (SPF), the South Pacific Commission (SPC), the Alliance of Small Island States (AOSIS), the South Pacific Regional Environment Programme (SPREP) and the Forum Fisheries Agency (IFA). Membership of these programmes can benefit Pacific Island states such as the Marshall Islands, for example during negotiations or debates over economic and environmental affairs such as fishing licenses and catch revenues.

Responses to Environmental Issues: Assessment and Status

Progress has been made in some aspects of environmental management, despite constraints. This section reviews the status of programmes for which funding was sought in the National Environment Strategy (1992). It also considers future opportunities and prospects relating to the environment. This provides the context and setting for the policies and environmental measures outlined in concluding parts of the chapter.

Strategies and Programmes identified in 1992 Environmental Management Strategy

A number of strategies (policies) to help improve the environmental situation in the Marshall Islands were identified in the 1992–96 National Environmental Management Strategy (Crawford, 1993b). Several were ranked as high priority and considered appropriate for seeking funding from outside agencies. The extent to which activities have been funded and implemented is highly variable (Table 5). Some progress has been made, many activities have yet to be implemented.

Environmental Opportunities and Prospects

Environmental measures to date in the RMI have been largely reactive in an attempt to deal with the more pressing concerns. Considerable opportunities may become available for more proactive, integrated environmental management. While this need has been recognised, resources for implementation have been meagre. In view of possible new economic opportunities (e.g. fisheries and tourism) which might help finance environmental work, and possibly also increasing assistance from outside agencies, prospects for environmental management may improve. In 1988 the US Department of Defense, Office of Economic Adjustment, developed a Water Resources Plan of Action, which integrated economic development and environmental protection and planning needs (U.S. D.O.D., 1989). As with many other planning documents produced for the RMI over the past two decades, very few, if any, were implemented. The proposed actions focused on development and protection measures to promote local fishing, aquaculture, and eco-tourism, the same sorts of

Table 5

Action strategies and programmes of high priority recommended in 1992 National Environmental Management Strategy: Status of Implementation (from Price, 1998)

Action strategy	Extent of implementation*
1. Anticipating sea level rise	1 out of 2 projects
2. Enhancing freshwater supply	2 of 5 projects
3. Improving disposal of solid and hazardous waste	3 of 4 projects
4. Managing marine and coastal resources for sustainability	2 or 3 of 5 projects
5. Improving sewage disposal and management	2 of 3 projects
6. Strengthening environmental education programmes	1 of 3 projects
7. Strengthening environmental legal instruments	3 of 3 projects
8. Protecting special areas and species	1 of 2 projects
9. Protecting cultural values and practices	2 of 3 projects
10. Managing terrestrial resources for sustainability	3 of 3 projects
11. Anticipating environmental emergencies	0 of 3 projects

*Number of programmes/projects with funding and/or at least some implementation.

compatible development options still being debated now, but still without the political will to implement them.

Overall Assessment of Environmental Governance

A number of environmental governance structures are now in place in the Marshall Islands, together with a reasonable body of national and international environmental legislation. In many instances, implementation of regulations, enforcement of compliance and monitoring remain major constraints and lack local political and community support and understanding. Better coordination between the various development and environmental agencies is also needed. This will help to ensure that (i) environmental concerns are given more prominence in development planning, and related to this (ii) short-term economic projects are not undertaken at the expense of the environment and sometimes also community interests.

There is a complete lack of national parks and protected areas in the Marshall Islands, which is seen as a serious constraint, an issue that has been brought to the attention of the RMI government almost continuously over the past decade. However, the conservational and socioeconomic value of protected area systems seems quite widely recognised. Neither the Department of Public Works (within the Ministry of Resources, Development and Public Works), nor the Office of Planning and Statistics, are represented in the CZM Working Group for Majuro. Given their influence in many aspects of development, their representation in the group perhaps should be reconsidered.

The principal resistance to designation of protected areas stems from strong traditional ownership and control over land, and the lack of sufficient financial incentives to take land out of future potential development options. There is strong cultural sentiment to maintain controls over land and not empower government authorities via managing protected areas. There are at least 8 uninhabited atolls and islands in the RMI, any of which would be prime candidates for protected area designation (Bokaak, Bikar, Taka, Jemo, Ailinginae, Rondik, Erikub, and Knox). It would be advantageous to pursue protection of these areas because they pose higher levels of natural diversity and fewer existing population and land-use constraints (see Thomas et al., 1989; Maragos, 1994).

Proposed Environmental Policies and Strategies

Policies and mechanisms for responding to the perceived environmental problems and opportunities in the Marshall Islands are considered below (see also Price, 1998). These build on earlier and ongoing initiatives, including those in the 1992 National Environmental Management Strategy and include: (1) development of system of protected areas and fishery reserves: Bikar atoll and Bokaak atoll and others have already been proposed as protected areas (above references and Kelleher et al., 1995); (2) development of national coastal zone management plan; (3) greater incorporation of EIA and cost–benefit analysis into development projects; (4) development of contingency plans for environmental hazards and accidents. Through these and other measures the multiple benefits from natural systems of the Marshall Islands should sustain both present and future generations long into the 21st century and beyond.

For several of the above areas, planning and advanced measures (protected area recommendations, EIA procedures, and CZM regulations) have been completed, but they are frequently ignored or not implemented. Lack of political will and support for anything that can be construed as a hindrance to economic development, is the main barrier to further progress. There is still strong traditional tenure over land and water ownership and use, and government actions to promote sustainable development and conservation of natural resources that are under traditional control will not be supported unless there is financial compensation and other incentives accrue to the affected traditional landowners and leaders.

ACKNOWLEDGEMENTS

One of us (ARGP) would like to express thanks to Jorelik Tibon of Republic of the Marshall Islands Environmental Protection Authority, Majuro, for his advice and valuable information provided during a visit to the Marshalls in 1998.

REFERENCES

Couper, A. (ed.) (1989) *The Times Encyclopaedia and Atlas of the Sea*. Times Books Ltd., London, 272 pp.

Crawford, (1993a) Republic of the Marshall Islands. Part A. State of the Environment Report (1992). South Pacific Regional Environment Programme (SPREP), 90 pp.

Crawford, (1993b) Republic of the Marshall Islands. Part B. Action Strategy for Strengthening Environmental Management (1992–1996). South Pacific Regional Environment Programme (SPREP), 70 pp.

Dahl, A.L. (1980) Regional ecosystem survey of the South Pacific area. South Pacific Commission Tech. Paper 179, 1–99.

Douglas, N. and Douglas, N. (1994) *Pacific Islands, Yearbook*. Fiji Times Ltd., Fiji.

Harding, E. (1992) Strengthening environmental management capabilities in Pacific Island developing countries. Regional Environment Technical Assistance (RETA) 5403. South Pacific Regional Environment Programme (SPREP).

Houghton, J.T. (1997) *Global Warming: The Complete Briefing*. Cambridge University Press, Cambridge, 251 pp.

Kelleher, G., Bleakley, C. and Wells, S. (1995) A global representative system of marine protected areas: Vol. IV. South Pacific, Northeast Pacific, Northwest Pacific, Southeast Pacific and Australia/New Zealand. The Great Barrier Reef Marine Park Authority/The World Bank/The World Conservation Union (IUCN).

IAEA (1998) Radiological conditions at Bikini atoll: prospects for resettlement. Radiological Assessment Reports Series, IAEA, Vienna.

Lalli, C.M. and Parsons, T.R. (1993) *Biological Oceanography: An Introduction*. Pergamon Press, Oxford.

Maragos, J.E., Baines, G.B.K. and Beveridge, P.J. (1973) Tropical cyclone Bebe creates a new land formation on Funafuti Atoll. *Science* **181**, 1161–1164.

Maragos, J.E. (1993) Impact of coastal construction on coral-reefs in the United States-affiliated Pacific Islands. *Coastal Management* **21** (4), 235–269.

Maragos, J.E. (1994) Description of reefs and corals for the 1988 protected area survey of the northern Marshall Islands. *Atoll Research Bulletin* **419**, 88 pp. + 5 app.

Maxwell Stamp (1998) Report on tourism as input from Policy Advisory Team on Economic Management for the Marshall Islands.

McDermid, K.J. (1989) An annotated list of marine algae from Majuro and Arno atolls, Republic of the Marshall Islands. Appendix E in Coastal Resource Inventory of Majuro Atoll, Republic of the Marshall Islands. Sea Grant Extension Service, University of Hawaii, Manoa, Hawaii.

Price, A.R.G. (1998). The environment and sustainable development in the Marshall Islands. Report by Maxwell Stamp plc to the Asian Development Bank.

Sheppard, C.R.C. (1998) Biodiversity patterns in Indian Ocean corals, and effects of taxonomic error in data. *Biodiversity and Conservation* **7**, 847–808.

Spalding, M., Blasco, F. and Field, C. (eds.) (1997) *World Mangrove Atlas*. The International Society for Mangrove Ecosystems. Okinawa, Japan, 178 pp.

Thomas, P.E.J. et al. (1989) Report of the Northern Marshall Islands Natural Diversity and Protected Areas Survey, 7–24 September 1988. South Pacific Regional Environment Programme, Noumea, New Caledonia, and Test-West Center, Honolulu, Hawaii, 133 pp.

US Army Corps of Engineers, Pacific Ocean Division (1989–1990) Coastal Resources Atlases for Majuro, Arno and Kwajalein Atoll, Republic of the Marshall Islands. Ft. Shafter, Honolulu, Hawaii.

US Dept. of Defense, Office of Economic Adjustment (1989) Republic of the Marshall Islands, Water Resources Plan of Action. Prepared by the US Army Corps of Engineers, Pacific Ocean Division. 64 pp. + 21 App.

Wells, J.W. (1954) Recent corals of the Marshall Islands. *US Geol. Survey Prof. Paper* **260-I**, 285–486.

Wells, S.M. and Jenkins, M.D. (1988) *Coral Reefs of the World. Vol. 3: Central and Western Pacific*. IUCN/WCMC/UNEP.

Wilkinson, C.R. and Buddemeir, R.W. (1994) Global climate change and coral reefs: implications for people and reefs. Report of the UNEP-OOC-ASPEI-IUCN Global Task Team on the implications of climate change on coral reefs. IUCN, Gland, Switzerland. 124 pp.

THE AUTHORS

Andrew R.G. Price
*Ecology and Epidemiology Group,
Department of Biological Sciences,
University of Warwick,
Coventry, U.K.*

James E. Maragos
*US Fish and Wildlife Service,
Pacific Islands Ecoregion,
300 Ala Moana Blvd.,
Box 50167,
Honolulu, HI 96850, U.S.A.*

Chapter 105

HAWAIIAN ISLANDS (U.S.A.)

James E. Maragos

The Hawaiian Islands are the fifth largest in terms of land area (16,643 km^2), and with 1.3 million people, the second most populous of the archipelagos in the tropical Pacific. The principal islands are in two groups: eight large and several smaller volcanic islands in the warmer southeast end of the chain where the human population is concentrated, and very small volcanic and coral islands and atolls to the northwest which are uninhabited or sparsely inhabited. Marine biodiversity is high in terms of habitat quantity, variety, and degree of endemism but lower in terms of species richness. Hawai'i is geographically the most isolated island chain in the world.

The people rely heavily on air and sea transportation for maintaining an economy dependent on tourism, military expenditures, and agriculture. The population has expanded rapidly since World War II, with three-fourths residing on O'ahu, and most living in or near the main city of Honolulu.

During the earlier part of this century, plantation agriculture (sugar cane, pineapple), ranching, grazing of feral animals, military construction, and transportation projects changed the marine environment via the draining of wetlands, dredging and filling coastal areas, and the flushing of eroded soils. Urbanization now fuels coastal construction, resort and golf course development, increased coastal pollution, coastal flooding, sedimentation, and overfishing in nearshore waters. Alien and invasive species, derelict fishing gear, shark finning, and depletion of fish stocks are now emerging as major marine environmental concerns, along with poorly regulated aquarium fish collection, ship groundings, anchor damage, and overuse of popular coastal parks, beaches, and nearshore reefs. The endemic monk seal and many seabirds are being threatened by commercial fishing off remote reefs. Green sea turtles are afflicted with a poorly understood disease. The incidence of coral bleaching and diseases appears to be increasing, although existing levels are low.

Hawai'i contains 84% of all reefs under U.S. jurisdiction, accounting for 17,520 km^2 of habitat above a depth of 200 m. The northwest islands account for 82% of the Hawaiian total. Together with adjacent beaches and waters, the reefs support millions of dollars of economic activity, including marine and coastal tourism, sport-fishing, subsistence fishing, commercial fishing, commercial and recreational boating, scientific research, and mariculture development. The coral reefs also afford natural protection and anchorages for boats, and protect coastal property and beaches from the damaging effects of tropical storms and large waves.

Extensive local, state, and federal environmental legislation and regulations exist, and there are adequate controls over solid waste, coastal water pollution, further draining or loss of wetlands, and coastal construction. The state and the four county governments, with federal support, have active programs that regulate development in the coastal zone. The network of

Seas at The Millennium: An Environmental Evaluation (Edited by C. Sheppard)
Published by Elsevier Science Ltd.

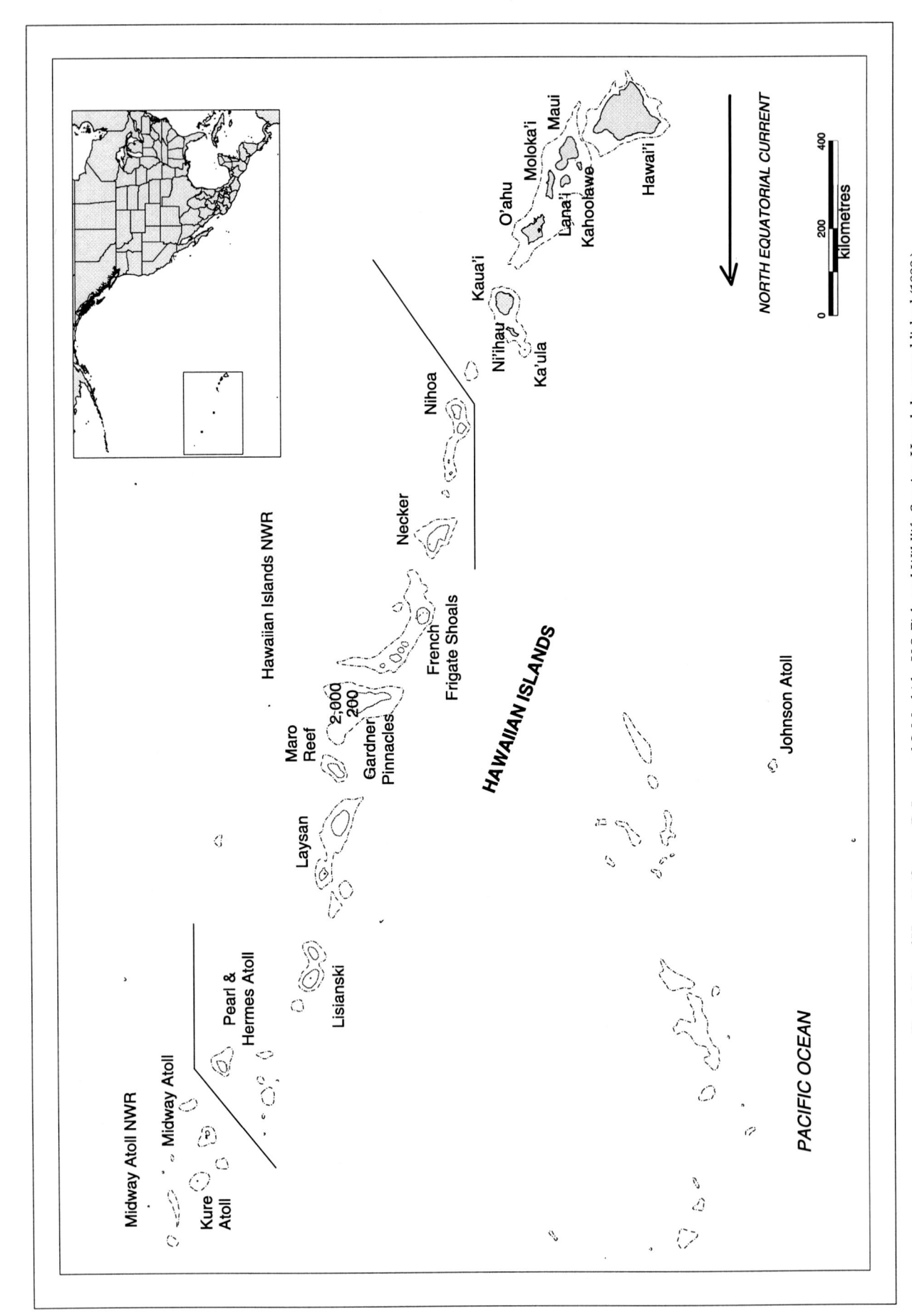

Fig. 1. Map of Hawaii. (Source: R. Low and S. Machida, U.S. Fish and Wildlife Service, Honolulu, unpublished (1999.)

federal and state marine and coastal protected areas is substantial, especially in the northwest islands. Sustainable fisheries management, increased community-based management of coastal areas in the main islands, public education, additional marine protected areas, and increased monitoring and enforcement pose as the major future challenges for conserving marine ecosystems in Hawai'i.

THE DEFINED REGION

The principal Hawaiian Islands consist of 13 larger volcanic islands, three coral islands and five atolls, together with numerous smaller islets that straddle the Tropic of Cancer in the central northeast Pacific Ocean (Fig. 1, Tables 1 and 2). At 3000 km in length, Hawai'i is the second longest island chain in the tropical Pacific after the Caroline Islands. The northwest end of the chain lies at 29°N latitude and 179°W longitude, and the southeast end of the chain lies at 19°N and 155°W (UNEP/IUCN, 1988; Juvik and Juvik, 1998).

Hawai'i is the most isolated archipelago in the world. All reefs and islands are U.S. sovereign territory, and all except Midway Atoll, are also under the jurisdiction of the State of Hawai'i out to a distance of 3 miles from shore (see UNEP/IUCN, 1988). Of the eight main islands at the southeast end of the chain, six are much larger, wetter, and heavily populated (O'ahu, Kaua'i, Maui (Fig. 2), Moloka'i (Fig. 3), Lana'i and Hawai'i), with the other two uninhabited (Kaho'olawe) or sparsely inhabited (Ni'ihau, Fig. 4). Of the northwest islands, three are sparsely populated (Midway, French Frigate Shoals, Laysan) and the rest uninhabited (Kure, Pearl-and-Hermes, Maro, Lisianski, Gardner Pinnacles, Nihoa, and Necker). The latter are generally smaller, drier, and subtropical (Juvik and Juvik, 1998; Amerson, 1971; Amerson et al., 1974; Clapp, 1972; Clapp et al., 1977; Clapp and Wirtz, 1975; Dana, 1971; Ely and Clapp, 1973; Woodward, 1972).

The geological origin for all of the islands is basaltic volcanoes that formed over a "hot spot" beneath the ocean floor (19°N, 154°W) near the centre of the Pacific tectonic plate (Coulbourn et al., 1974; Grigg, 1988; Stearns, 1966; Clague in Juvik and Juvik, 1998). Over the past 70 million or more years, the hot spot has remained stationary while the crust of the Pacific plate has moved north and more recently northwest. As older volcanic islands formed on the crust and moved away from the hotspot, new volcanoes and islands formed in their wake. The older islands supported reef growth along their upper flanks, and later some

Fig. 3. Waterfalls off the 600 m high sea-cliffs of northeast Moloka'i, 1983 (photo James E. Maragos).

Fig. 2. Rocky shoreline and sea-cliffs off Ke'anae, northern Maui, 1981 (photo James E. Maragos).

Fig. 4. Aerial view of Ni'ihau Island, at the northern end of the main Hawaiian Islands, 1999 (photo James E. Maragos).

Table 1

Hawaiian Islands listed from northwest to southeast and their corresponding land and contiguous ocean areas to a depth of 100 fathoms (approximately 200 m). Asterisk (*) indicates that the Moloka'i Island total includes Penguin Banks. Sources: Rod Low and Susan Machida, U.S. Fish and Wildlife Service, Honolulu, unpublished (1999), Juvik and Juvik (1998), and Hunter (1995).

Island name	Emergent land area (km^2)	Submerged reef area (km^2)
Kure Atoll	0.9	323.77
Midway Atoll	6.3	1,201.70
Pearl and Hermes Atoll	0.3	784.75
Lisianski Island	1.5	1,250.38
Laysan Island	4.1	588.40
Maro Reef	0.004	1,934.88
Gardner Pinnacles	0.02	2,447.72
French Frigate Shoals	0.3	942.89
Necker Island	0.2	1,556.70
Nihoa Island	0.7	574.57
Ka'ula Rock	0.6	28.0
Ni'ihau Island	179.9	60.0
Lehua Island	1.0	4.0
Kaua'i Island	1,430.5	266.0
O'ahu Island	1,546.5	504.0
Moloka'i Island*	*673.5	*998.0
Lana'i Island	364.0	95.0
Kaho'olawe Island	115.5	58.0
Maui Island	1,883.7	270.0
Molokini Island	0.1	1.0
Hawai'i Island	10,433.1	252.0
Totals: 21	*16,642.7	*14,141.8

Table 2

Additional submerged reef areas to a maximum depth of 200 m that are not contiguous with the principal islands. Source: Rod Low and Susan Machida, U.S. Fish and Wildlife Service, Honolulu, unpublished (1999) and Cynthia Hunter (1995). Asterisk (*) indicates that the Penguin Banks estimate is included in the Moloka'i Island total in Table 1.

Name	Area (km2)
Unnamed: SW of Kure Atoll, area	0.28
Unnamed: SW of Kure Atoll, area 2	0.51
Unnamed: S of Kure Atoll	1.02
Unnamed: SE of Kure Atoll	1.40
Nero Seamount	76.57
Ladd Seamount	153.97
Gambia Shoal	15.84
Unnamed- E of Pearl and Hermes Atoll	4.98
Salmon Bank	163.69
Unnamed: S of Pearl & Hermes Atoll	5.67
Unnamed: W of Lisianski Island, area 1	13.14
Unnamed: W of Lisianski Island, area 2	93.50
Pioneer Bank	436.28
Northampton Seamounts, area 1	193.67
Northampton Seamounts, area 2	6.82
Northampton Seamounts, area 3	13.55
Northampton Seamounts, area 4	187.26
Raita Bank	571.79
Unnamed: between Gardner Pinnacles and St. Rogatien Bank	68.30
St. Rogatien Bank	384.58
Brooks Bank, area 1	3.66
Brooks Bank, area 2	158.35
Brooks Bank, area 3	144.63
Brooks Bank, area 4	29.83
Unnamed: E of French Frigate Shoals	12.02
Unnamed: N of Necker	6.70
Unnamed: between Necker & Nihoa Is., area 1	3.46
Unnamed: between Necker & Nihoa Is., area 2	69.17
Unnamed: between Necker & Nihoa Is., area 3	5.56
Unnamed: between Necker & Nihoa Is., area 4	65.06
Unnamed: W of Nihoa Island	339.50
Unnamed: E of Nihoa Island	147.12
Penguin Banks	*
Totals: 33	3,377.88

formed barrier reefs and lagoons as they subsided during their multi-million year journey to the northwest. Eventually the volcanic portion of the oldest islands submerged beneath the sea surface with the reefs forming atolls in accordance with Darwin's theory. Eventually the atolls and islands moved completely out of the warmer Pacific, resulting in reductions in upward reef growth to the point that it could no longer keep pace with the rate of subsidence.

As a result, all the former Hawaiian Islands are now submerged, consisting mostly of "drowned" atolls (guyots) of the Emperor Seamounts, a vast chain stretching northwest past Midway and Kure for thousands of miles towards the Kamchatka Trench off Siberia. The oldest have already been subducted. None of the present Hawaiian islands are more than 25 million years old (e.g. Midway and Kure Atolls at the northeast end of the chain). However, biological evolution has been going on for a much longer period of time on both the shallow marine habitats encircling the islands and the terrestrial habitats on the islands. Marine and terrestrial species were also able to "hop" from older to younger islands and reefs, allowing continuous evolution in isolation from other island groups for as long as 70 million years or more (see Nullet et al. in Juvik and Juvik, 1998).

Hawai'i's geographic isolation has shaped its unusual marine biodiversity. Although only 52 species of shallow-water stony corals have been reported, 25% are endemic or unique to Hawai'i and a few nearby islands. Similar proportions of the nearly 500 species of nearshore fishes, about 1000 marine mollusc species, and 450 marine algae species in Hawai'i are also endemic (Devaney and Eldredge, 1977; Kay, 1972, 1979, 1980, 1994; Maragos, 1995; J. Juvik in Juvik and Juvik, 1998; Randall, 1996; Hoover, 1998; Magruder and Hunt, 1979; Maragos et al., 1995).

SEASONALITY, CURRENTS, NATURAL ENVIRONMENTAL VARIABLES

Climate and Currents

The Hawaiian islands lie in the broad oceanic zone of the North Pacific Equatorial Current (NPEC), which moves west at a rate of one to two knots. The northeast trades are the dominant wind system, prevailing for 80% of the year. The trades cool the air temperatures, generate wind-driven surf along "windward" coasts (north and east facing shorelines), and bring moisture. As the trades pass over the higher volcanic islands, they release rainfall in large amounts on the windward sides of the islands, resulting in dryer "rain shadows" on the leeward sides. As a result, rivers, streams, surf, and estuaries are more common on the windward sides of the islands and greatly influence the character and distribution of marine habitats, including beaches, reefs, and mangroves (Giambelluca and Schroeder in Juvik and Juvik, 1998; Sanderson, 1993).

Because Hawai'i lies near the periphery of the northern tropics, its climate is subtropical in winter months and tropical in summer months. Seasons may be defined by humidity, temperature, and winds. In the main southeast islands, the annual air temperatures vary between 20° and 30°C, while surface seawater temperatures vary between about 22° and 27°C. The cooler northeast trade winds prevail between March and October, and tend to slacken in the latter months. Between November and February, the trades are generally weak, non-existent, or are temporarily replaced with warm, humid "kona" winds from the south and southwest. Kona winds are often strong and accompanied by heavy rainfall. Tropical cyclones achieving hurricane strength pass over the islands about once every ten years, usually from the southeast between June through December. Two recent cyclones, "Iwa" in November 1982, and "Iniki" in September 1993, caused extensive damage (Schroeder in Juvik and Juvik, 1998; Fletcher in Juvik and Juvik, 1998).

Upper Water Column Characteristics

The oligotrophic pelagic waters surrounding the Hawaiian Islands have high transparency and low primary productivity. Hawai'i is located near the centre of the north Pacific subtropical anti-cyclonic gyre where marine productivity is very low (Flament et al. in Juvik and Juvik, 1998). The thermocline occurs at an average depth of about 100 m, but in summer months it descends and in winter months it rises. Marine productivity is enhanced in the neritic or coastal waters of the islands, especially off the downstream reaches of higher islands. Groundwater and surface water run-off are rich in nutrients and can elevate nearshore primary production, especially in confined bays and off stream mouths and wetlands.

Wake eddies also form off the downstream (southwest or west) sides of the islands, and some tend to spin off in unpredictable directions, at times moving warmer and nutrient-rich surface waters trapped in the eddies away from the islands. These eddies serve the important function of transporting marine larvae between reefs and islands, and perhaps between Hawai'i and neighbouring islands (e.g. Johnston, Kingman, and Palmyra atolls to the south and Wake Atoll to the west, etc.).

Gradients in Biophysical Features

The winter minimum temperatures for surface seawaters are sub-optimal for coral growth, and these minimums are even lower and marginal for coral growth in the northwestern islands (Grigg and Maragos, 1974; Grigg and Pfund, 1980; Grigg, 1988). The islands are also vulnerable to damage from large waves approaching from distant polar winter storms whose large waves traverse the North Pacific unimpeded and strike the north and west facing coasts. From June to September, Antarctic storms also generate waves that strike the southern coasts, but these are generally smaller because of the longer distances and the need to pass through the islands and reefs of French Polynesia and the Line Islands before reaching Hawai'i (see Maragos in Juvik and Juvik, 1998).

The combination of exposure to wave action and lower water temperatures, especially in winter, exerts major control over the formation of reefs and maintenance of islands. Few contemporary shallow reefs are found along northern facing coasts of the main islands due to exposure to heavy surf during winter (Dollar, 1982; Dollar and Tribble, 1993; Grigg, 1983). Only coral reefs along sheltered coasts, within embayments, and in deeper water below the influence of large surface waves have high coral cover and are actively accreting at this time. Reef growth rates in the northwest islands are even lower due to unfavourably low temperatures, strong winds, and heavy surf, especially during winter months (Grigg, 1988). Consequently, the eastern end of the Hawaiian chain has larger, warmer, higher, and wetter, basaltic islands with fewer reefs and beaches. The islands tend to become smaller, lower, cooler, and dryer with more carbonate rock, reefs, beaches, lagoons, and atolls towards the northwest.

THE MAJOR SHALLOW-WATER MARINE AND COASTAL HABITATS

About 20 shallow-water and coastal marine habitats have been described for Hawai'i at depths of 50 m or less (Maragos in Juvik and Juvik, 1998; Kay and Chave, 1984). Tidepools, anchialine ponds, sandy beaches, rocky basalt beaches (Fig. 5), limestone benches, ancient Hawaiian fishponds (Fig. 6), boat harbours, constructed seawalls and revetments, estuaries, coastal marshes, mangrove forests

Fig. 5. The black sand beach at Kalapana, southern Hawai'i Island, in 1973 before the beach was covered over by lava flows during the subsequent decade (photo James E. Maragos).

Fig. 6. Kaloko Fishpond in 1977, off the west coast of Hawai'i Island, one of many fishponds constructed by the early Hawaiians (photo James E. Maragos).

(Fig. 7), seagrass beds, algal beds, coral communities, fringing reefs (Fig. 8), sand and mud flats, coral rubble flats, and natural and dredged channels are found off volcanic islands. Further offshore are found embayments, lagoons, patch reefs, pinnacles, offshore islets, and barrier reefs. More prevalent in the northwestern islands are basaltic sea stacks, white sand beaches, coral rubble flats, cobble beaches, low coral islets, lagoons, patch reefs, pinnacles, sand flats, natural and dredged passes through reefs, and atoll perimeter reefs. The water column habitats include the nearshore neritic zone, the offshore pelagic zone, and the deep water or "twilight" zone at depths between 50 and 200 m, and the "midnight" zone at greater depths.

Coral Reefs

Coral reefs, beaches, and associated ecosystems are the most common of the native coastal habitats. Introduced invasive mangroves have spread quickly to many shores in the main islands during the past century. A single endemic seagrass species occurs in Hawai'i, and seagrass beds are rare. Anchialine pools, with only subsurface connections to the sea, are marine or brackish, and found along the basaltic coasts of younger volcanic islands (Maui, Hawai'i) and where coral reefs have been uplifted (O'ahu). Fleshy algal beds dominate the shallow inner reaches of fringing reef flats, and stony coral communities dominate the outer flats and upper reef slopes. Crustose coralline algae are dominant reef builders too, and are particularly prevalent in wave-exposed littoral and sublittoral habitats (Fig. 9). Stony corals are normally not abundant in wave-exposed habitats, but are well developed on reef slopes at depths greater than 10 m, within marine lagoons, and within protected embayments. Coral communities are especially well developed off the Kona (or western) coast of the "Big Island" (Hawai'i).

Many coral reefs in the main islands have been inventoried during the past two decades and several coral reef

Fig. 7. Aerial view of southwest Moloka'i coastline showing spread of alien mangroves (*Rhizophora mangle*) into subtidal coral reef habitats, 1990 (photo James E. Maragos).

areas have been well studied during the past several decades. Kane'ohe Bay off the northeast coast of O'ahu supports one of only two barrier reefs in the islands, and extensive coral and benthic algae communities in the lagoon. The Bay has been the site of monitoring before, during, and after the removal of sewage discharges from the bay over the past 30 years, and represents the site of the first successful large-scale coral reef restoration in the Pacific (Hunter and Evans, 1995). Other important and well-studied reefs in Hawai'i include those off Mamala Bay (Grigg, 1994) and Hana'uma Bay (O'ahu), the Kona Coast (Hawai'i), Lahaina and Ma'alaea (Maui), south Moloka'i, and Hanalei Bay and adjacent reefs off north Kaua'i. Structural reefs are well developed off the south and northeast coasts of O'ahu, the south coast of Moloka'i, the west coast of Maui, the north coast of Kaua'i,, and the northeast coast of Lana'i. The atolls at the northeast end of the chain (Midway, Kure, Pearl and Hermes, Maro, and French Frigate Shoals) and the submerged reefs off Necker, Gardner

Fig. 8. Submerged reef flat community off Ha'ena Reef, north Kaua'i, 1997 (photo James E. Maragos).

Fig. 9. Necker Island from the air (photo James E. Maragos).

Pinnacles and Lisianski are major reef systems supporting commercial fishing grounds in the northwest islands (Grigg and Pfund, 1980).

Aside from the position of the archipelago at the margins of the northern tropics, the North Pacific Equatorial Current flowing through the archipelago tends to keep shallow-water marine species upstream of the western tropical Pacific where marine species richness is much higher. This factor serves to further isolate marine ecosystems in Hawai'i from their more abundant and diverse neighbours to the south and west (Jokiel, 1987).

Offshore Islands and Beaches

About 30 small basaltic and coral islets fringe the larger eight main islands at the southeast end of the chain. These islets provide sheltered habitat for coral communities and support important nesting beaches for seabirds. Rare endemic plant life is found on several of these islets, and virtually all are designated as State Offshore Island Seabird Sanctuaries. All but one (Moku O Loe in Kane'ohe Bay) of the islets are uninhabited, and sea turtles and seals may also haul out on the beaches of some of the islets.

Sandy beaches are, of course, prime resort and recreational sites, and attract many of the 7 million tourists visiting Hawai'i each year (Sakai in Juvik and Juvik, 1998). However, beaches are also important habitat for sand dwelling crustaceans, molluscs and fish, and also important sites for resting and nesting seabirds, sea turtles, and the Hawaiian monk seal. Beaches are most abundant along the lagoon reaches of atoll islets, and along the coasts of several of the main southeast islands, especially the west and south sides of Kaua'i, O'ahu, Moloka'i, Maui, Lana'i, and Hawai'i. Four kinds of beaches have been recognized, and white sand or carbonate beaches are the most common. Black sand beaches derived from recent lava flows are common off Hawai'i and Maui, and a pink volcanic cinder beach is located off east Maui. Green sand beaches consisting of the hardy remains (olivine crystals) of lava rock, are found off south Hawai'i, and off northeast and southeast O'ahu (at Mokapu Peninsula and Hanauma Bay). Beach sand tends to be coarser off wave-exposed and windy reaches of islands, and finer within lagoons, embayments, mangroves, harbours and fishponds (Maragos in Juvik and Juvik, 1998).

Mangroves and Seagrasses

Geographic isolation also prevented the natural colonization of mangroves in Hawai'i and only a single endemic seagrass species (*Halophila hawaiiana*) has become established. Many other families and groups of marine organisms have failed to reach the islands. Over the past century mangroves (*Rhizophora mangle* and *Bruguiera gymnorhiza*) were intentionally introduced to O'ahu and Moloka'i, and the former have since spread throughout most of the main volcanic islands. Seagrass beds are rare and limited to shallow, sandy bottoms on coasts protected by reefs and embayments.

Habitats, Sites and Species of Conservation Importance

Seabirds

About 22 species of seabirds nest here, although populations of some ground-nesting species have been decimated or extirpated in the populous main islands (Table 3). In contrast, the largest nesting populations of many resident seabirds occur in the uninhabited northwest Hawaiian islands where they have not been exposed to rats and other predators (Flint in Eldredge et al., 1999). Near the beginning of the 20th century, President Theodore Roosevelt initiated the establishment of one of the most important seabird refuges in the world, the Hawaiian Islands National Wildlife Refuge, now administered by the U.S. Fish and Wildlife Service. This refuge protects millions of nesting seabirds and, together with the adjacent Midway Atoll National Wildlife Refuge and the State's Wildlife Reserve at Kure Atoll, protects all the seabird habitats of the northwest Hawaiian islands except Ka'ula, a distance of 1020 nautical miles (2000 km) (U.S. Dept. of Interior, 1999).

Important nesting seabird species are two albatrosses, three petrels, three shearwaters, two storm petrels, two tropic birds, three boobies, the great frigate-bird, and six terns and noddies.

Seasonal migratory shorebirds winter and rest in Hawai'i and reside in the Arctic. Important species include Pacific golden plovers, ruddy turnstones, wandering tattlers, sandpipers, and bristle-thighed curlews (Flint in Eldredge et al., 1999).

Seabirds are protected by migratory bird treaties and a few by the U.S. Endangered Species Act (ESA). However, several endemic seabird and shore species have become

Table 3

List of the native or indigenous seabirds and waterbirds that nest in Hawai'i. One asterisk (*) indicates the species is listed as vulnerable, threatened, or endangered, and two asterisks (**) indicates that it is also endemic to Hawai'i. Sources: E. Flint, U.S. Fish & Wildlife Service (unpublished); E. Flint in Eldredge et al. (1999); Hawaii Audubon Society (1993)

Common name	Scientific name	Weight (g)	Wingspan (cm)	Longevity (year)
Seabirds				
Black-footed Albatross	*Diomedea nigripes*	3000	210	35
Laysan Albatross	*D. immutabilis*	2800	203	43
Bonin Petrel	*Pterodroma leucoptera*	176	68	>21
Hawaiian Petrel**	*P. phaeopygia sandwichensis*	434	91	7
Bulwer's Petrel	*Bulweria bulwerii*	100	58	22
Wedge-tailed Shearwater	*Puffinus pacificus*	340	97	33
Christmas Shearwater	*P. nativitatis*	356	81	17
Newell's Shearwater**	*P. auricularis newelli*	388	82	?
Tristram's Storm-petrel*	*Oceanodroma tristrami*	90	56	14
Band-rumped Storm-petrel* (possibly endemic)	*O. castro cryptoleucura*	44	35	?
Red-tailed Tropicbird	*Phaethon rubricauda*	660	112	28.5
White-tailed Tropicbird	*P. lepturus*	?	93	?
Masked Booby	*Sula dactylatra*	2000	152	27
Brown Booby	*S. leucogaster*	1340	140	16
Red-footed Booby	*S. sula*	1110	100	22
Great Frigate-bird	*Fregata minor*	1500	229	34
Gray-backed Tern	*Sterna lunata*	146	74	31
Sooty Tern	*S. fuscata*	198	86	34
Blue-gray Noddy	*Procelsterna cerulea*	53	46	11
Brown Noddy	*Anous stolidus*	205	84	25
Black Noddy	*A. minutus*	108	71	24
White Tern	*Gygis alba*	111	71	36
Waterbirds				
Black-crowned Night Heron	*Nictocorax nictocorax hoactli*	?45?		
Hawaiian Duck**	*Anas wyvilliana*	670	?	?
Laysan Duck**	*A. laysanensis*	?	?	?
Hawaiian Gallinule**	*Gallinula chloropus sandvicensis*	? ??		
Hawaiian Coot**	*Fulica americana alai*	?	?	?
Hawaiian Stilt**	*Himantopus mexicanus knudseni*	?	??	

Table 4

Provisional list of Hawaiian marine species worthy of listing as threatened or endangered. The importance score in the right column is an indicator of justification for listing, and is based on the following criteria: range or depth restriction, limited population size and abundance, importance to commercial and recreational fishing, importance in the aquarium trade, dependent on declining critical habitat, importance for scientific research, biologically dependent on other species, displaying territorial behaviour, excluded by competition from other species, and showing restrictive spawning behaviour. Compiled by D. Gulko, with input of other marine specialists at a State Division of Aquatic Resources workshop, September 1999. Final list will be subject to outside review and approval by relevant management agencies.

Common Name	Scientific Name	Importance score
Hawaiian Sea Grass	*Halophila hawaiiana*	35
Hawaiian Green Seaweed	*Boodleopsis hawaiiensis*	26
Hawaiian Sand Anemone	*Heteractis malu*	39
Mushroom Coral	*Fungia scutaria*	20
Hawaiian Plate Coral	*Montipora dilatata*	41
Duerden's Finger Coral	*Porites duerdeni*	31
Puko'o Lobe Coral	*P. pukoensis*	30
Orange Tube Coral	*Tubastraea coccinea*	10
Large Hawaiian Table Coral	*Acropora cytherea*	34
Small Hawaiian Table Corals	*A. gemmifera, humilis, nasuta* and *valida*	36
Moloka'i Leather Coral	*Sinularia molokensis*	29
Hawaiian Sea Fan	*Acabaria bicolor*	26
Black-lipped Pearl Oyster	*Pinctada margaritifera*	30
Hawaiian Sea Grass Snail	*Smaragdia bryanae*	36
Mushroom Coral Snail	*Epitonium ulu*	27
Hawaiian Bobtail Squid	*Euprymna scolopes*	41
Reef Squid	*Sepioteuthis lessoniana*	4
Hawaiian Octopus	3 unidentified endemic species	25
Triton's Trumpet	*Charonia tritonis*	18
Tiger Cowry	*Cypraea tigris*	29
Giant Opihi	*Cellana talcosa*	37
Polished Nerite	*Nerita polita*	28
Anchialine Pond Nerites	*Nerita* (2 unidentified species)	33
Northwest Hawaiian Islands' Nerite	*N. plicata*	20
Inarticulate Brachiopod	*Lingula reevii*	33
Anchialine Pond Shrimp	*Halocaridina rubra*	38
Harlequin Shrimp	*Hymenocera picta*	28
Giant Grouper	*Epinephelis lanceolatus*	28
Hawaiian Grouper	*E. quernus*	35
Spectacled Parrotfish	*Scarus perspicillatus*	31
Hawaiian Turkeyfish	*Pterois sphex*	25
Masked Angelfish	*Genicanthus personatus*	34
Hawaiian Flame Angelfish	*Centropyge loriculus?*	28
Bandit Angelfish	*Holacanthus arcuatus*	22
Hawaiian Seahorse	*Hippocampus* sp.	25
Hawaiian Pipefish	*Doryhampus baldwini*	28
Giant Trevally	*Caranx ignoblis*	28
White-saddle Goatfish	*Parupeneus porphyreus*	19

extinct during historic times (Pratt, 1994), and countless species of ground-nesting seabirds and flightless birds must have gone extinct in prehistoric times during the era of colonization by the Hawaiians and other Polynesians (Steadman, 1995; Olson and James, 1982).

Benthic Reef Life

Other marine groups of conservation importance include molluscs, reef corals, reef fish, and crustaceans associated with coral reefs and which are prized by collectors and aquarium enthusiasts or heavily sought after by fishers (Kay, 1979; Randall, 1996; Gulko, 1998; Hoover, 1998). Those warranting protection as endangered and threatened species are listed in Table 4. Virtually all of these are endemic to Hawai'i.

Waterbirds

Five endemic and endangered species of waterbirds (the Hawaiian Stilt, Hawaiian and Laysan Ducks, Hawaiian Gallinule, and Hawaiian Coot) also rely heavily on coastal habitats, especially wetlands, fishponds, lakes, and mudflats for foraging, nesting and resting (Table 3) (Hawaii Audubon Society, 1993).

Sea Turtles

At least five species of sea turtles have been reported from Hawai'i (Table 5). Green turtle (*Chelonia mydas*) populations increased dramatically after 1978 when the U.S. Endangered Species Act (ESA) was extended to protect them. One of the largest green turtle nesting sites in the Pacific and the principal breeding habitat for the monk seal are the sandy islets at French Frigate Shoals and adjacent atolls and islets. A few remaining nesting sites for the endangered hawksbill turtle (*Eretmochelys imbricata*) in the insular Pacific occur on a few beaches in the main islands (Maui, Hawai'i), and feeding hawksbills are also commonly seen on Hawaiian reefs. Large groups of green turtles are regularly seen off many coral reefs in the main islands and are popular visitor attractions. Other sea turtles less frequently observed in Hawai'i are the leatherback, loggerhead, and olive ridley (NMFS and USFWS, 1998a,b; Balazs in Juvik and Juvik, 1998).

During the past several decades, the frequency of a tumorous disease, (fibropapillomatosis) on the necks, heads, and fins of green turtles has increased. Although now under intensive research, the cause of and the cure for the tumours are not known. Outside of Hawai'i these types of tumours are known only from Florida. Since these tumours show higher incidence off the main Hawaiian Islands, agricultural and urban pollution is suspected to be a possible cause for the disease.

Table 5

Marine mammals and reptiles commonly reported in Hawaiian waters. An asterisk (*) indicates the species is listed as endangered or threatened, and a second asterisk (**) indicates the species is endemic to Hawai'i. Sources: L. Eldredge, G. Balazs and J. Maragos, in Juvik and Juvik (1998); National Marine Fisheries Service and U.S. Fish and Wildlife Service (1998a,b); Reeves et al. (1999)

Common Name	Scientific Name
Marine Reptiles	
Green Turtle*	*Chelonia mydas*
Hawksbill Sea Turtle*	*Eretmochelys imbricata*
Leatherback Turtle*	*Dermochelys coriacea*
Loggerhead Turtle*	*Caretta caretta*
Olive Ridley Turtle*	*Lepidochelys olivacea*
Marine Mammals,	
Hawaiian Monk Seal**	*Monachus shauinslandi*
Humpback Whale*	*Megaptera novaeangliae*
Minke Whale	*Balaenoptera acutorostrata*
Bryde's Whale	*B. edeni*
Fin Whale	*B. physalus*
Right Whale*	*Eubalaena glacialis*
Rough-toothed Dolphin	*Steno bredanensis*
Spinner Dolphin	*Stenella longirostris*
Bridled Dolphin	*S. attennata*
Striped Dolphin	*S. coeruleoalba*
Risso's Dolphin	*Grampus griseus*
Melon-headed whale	*Peponocephala electra*
Pygmy Killer Whale	*Feresa attenuata*
False Killer Whale	*Pseudora crassidens*
Shortfin Pilot Whale	*Globicephala electra*
Killer Whale	*Orcinus orca*
Sperm Whale*	*Physeter macrocephalus*
Pygmy Sperm Whale	*Kogia breviceps*
Cuvier's Beaked Whale	*Ziphius cavirostris*

Marine Mammals

Of interest is the endemic Hawaiian monk seal (*Monachus shauinslandi*), now confined to the northwest islands. The seal is the only surviving marine mammal completely dependent on coral reefs for its existence. Despite its designation as an endangered species, its populations have declined steadily over the past several decades to about 1,500 individuals. Adult males greatly outnumber females at some overcrowded mating sites, resulting in mobbing behaviour by the males, death to some pups and females, and reduced reproductive capacity of the species. The seal and pups are often malnourished from the lack of prey food and from competition with commercial fishers for the same prey species, including lobsters and eels. Seals and seabirds are also accidentally hooked by long-liners fishing just offshore from the main breeding and pupping beaches at French Frigate Shoals. The seal is easily disturbed by humans at resting and pupping beaches and are especially sensitive to noise and movement. At one time the seal had a wider distribution range and is still occasionally seen hauling out on beaches in the main Hawaiian islands, as well as at nearby atolls (Johnston, Palmyra, Wake).

The other prominent marine mammal in Hawai'i is the endangered Humpback Whale (*Megaptera novaeangliae*). As part of their annual migration cycle, the whales pass through the islands between November and May to give birth and rear young whales and possibly to escape the colder seasons in their far north Pacific feeding grounds. Increased federal and international protection has allowed populations to increase dramatically in recent decades, and pods travel close to the coastlines of the populated islands where whale watching is now a major visitor industry. Other common marine mammals in Hawai'i include the spinner dolphin (*Stenella longirostris*) and the endangered sperm whale (*Physeter macrocephalus*). Fifteen other marine mammals are reported from Hawai'i (Table 5) (Reeves et al., 1999; Maragos in Juvik and Juvik, 1998).

Endemic Species

About one-fifth to one-fourth of the nearshore marine biota of Hawai'i (including corals, fish, algae, molluscs) consists of endemic species, the highest level of endemism reported for any comparable marine area of the world (Kay and Palumbi, 1987; Kay, 1994). In addition, Hawai'i is home to several endemic terrestrial species including many birds, insects, plants, spiders and an insectivorous bat. Streams and anchialine pools are also habitat for endemic species of fish and shellfish. Tables 3, 4, and 5 collectively list important endemic marine species from Hawai'i.

OFFSHORE SYSTEMS

Benthic deep-sea environments have been described as part of explorations to study palaeo-climates, evaluate the scope and impacts of deep-sea mining, and assess the environmental effects of ocean dumping of harbour-dredged spoils. The University of Hawai'i, other out-of-state research institutions, the National Marine Fisheries Service and the U.S. Geological Survey all operate ships and conduct deep-sea research in Hawai'i. Fossil coral reefs have been extensively studied along the Emperor Seamounts and off some of the main islands to understand the palaeo-history of the archipelago. Some precious coral beds have been located and studied along the deep flanks of the islands and seamounts using deep-diving submersibles and remotely operated vehicles equipped with video and other sensors. Deep-sea cobalt manganese crusts and manganese nodules have been reported on seamounts and the ocean floor adjacent to Hawai'i, but there are no plans to move forward with any commercial exploitation of the deposits.

Deep-sea pelagic and bottom environments extend to depths of more than 6000 m, and include seamounts, guyots, drowned reef terraces, abyssal plains, and sea floor fracture zones. An actively growing submerged volcano, Loihi, is situated 30 nm southeast of the Hawai'i island and is now within 1000 m of the sea surface. Eventually Loihi may become the newest volcanic island in the chain.

POPULATIONS AFFECTING THE HAWAIIAN ISLANDS

Before the arrival of Captain Cook in 1778, the resident Hawaiian population was estimated at 250,000 to 300,000. Exposure to diseases brought to Hawai'i by foreigners and other factors (emigration, war, famine, high infant mortality) reduced the native population to 54,000 by 1876. The need for adequate labour to work on agricultural plantations compelled plantation owners to import additional foreign labour, and boatloads of indentured labourers and sometimes their families were brought to Hawai'i. In succession, Chinese, Japanese, Portuguese, Puerto Ricans, Koreans, Filipinos, and others arrived in Hawai'i between 1852 and 1907. The Hawaiian population rebounded after 1876, and population as a whole increased steadily until the end of World War II. Including the armed forces and defense workers, the population in Hawai'i was 859,000 in 1945. Troop withdrawals after the war reduced the population to 622,000 by 1954. Afterwards tourism spending, increased military expenditures, and statehood status in 1959 have resulted in a steady population increase to the present level of about 1.3 million people (Schmitt in Juvik and Juvik, 1999; Daws, 1974; Howe et al., 1994).

Demographic Patterns

In 1990 about 75% of the Hawaii population resided on O'ahu, and about 45% of the population of O'ahu resided in Honolulu, the only large city in the state. Smaller cities and towns are found on the neighbouring main islands of Maui, Hawai'i, Kaua'i, Moloka'i, and Lana'i. These other islands are more rural and agricultural in character, with Hawai'i Island supporting a 1990 population of 120,000; Maui 92,000; Kaua'i 51,000; Moloka'i 6000, and Lana'i about 2500, in contrast to O'ahu's 1990 population of 836,000. About 300 people (native Hawaiians) live on Ni'ihau, and less than 100 workers and visitors reside on Midway at any one time. French Frigate Shoals supports about 10 staff and visitors, and Laysan seasonally supports a staff of about two workers. At any one time about 19,000 tourists and visitors are in the islands. The remaining northwest Hawaiian islands are uninhabited (Schmitt in Juvik and Juvik, 1998).

Tourism Trends

Romantic tales about the South Pacific lured the first visitors to Hawai'i at the end of the 19th century. Tourism expanded rapidly after World War II and the advent of air travel, and by 1959 there were about 250,000 annual visitors to the islands. Tourism continued to grow rapidly; by 1970 the visitor count was nearly two million and by 1990 nearly seven million a year. Tourism has remained steady at about 6.5 million visitors per year over the past decade (Sakai in Juvik and Juvik, 1998).

Most current visitors are from the continental United States (3.6 million), Japan (2 million) and Canada (300,000). Most tourists visit the City and County of Honolulu (O'ahu) followed by Maui, Hawai'i, and Kaua'i counties in descending order of magnitude. Recently the U.S. Fish and Wildlife Service has opened Midway Atoll for ecotourism, and other remote islands are less frequently visited by tourists and adventurers. Six of the 20 top visitor destinations, and nine of the 20 most popular visitor activities in Hawai'i are ocean related. The coral reefs, beaches, clean water, and ocean recreation opportunities are among the principal reasons for Hawai'i being the top visitor destination in the world over the past several decades.

RURAL FACTORS

Deforestation and Soil Erosion

The rural landscapes of the main islands of Hawai'i have been extensively modified over the past two centuries. At the time of James Cook's discovery of the islands in 1778, terrestrial vegetation dominated by endemic species covered all upland and most coastal areas. Villages, fishponds, *kalo* (taro) ponds, and other agricultural plots and gardens were mostly concentrated in lowland and coastal areas, and woodlands and forests dominated the upper slopes of the high islands. Within a century, many of the lowland areas were damaged from sandalwood exploitation, and the clearing of vegetation for roads, sugar and pineapple plantations, logging, and livestock ranches. Game animals and livestock were also released in the wild to propagate on their own, especially cattle, deer, and goats, adding to the populations of pigs brought earlier by the Hawaiians. These populations quickly grew, leading to overgrazing of vegetation, soil loss, and downstream sedimentation. Many of the Hawaiian plants discovered by Cook's botanist in 1778 have never been seen again, and hundreds of other endemic plants are now being listed as endangered or threatened species.

The loss of groundcover, displacement of native species, fires, overgrazing by livestock and feral ungulates, and extensive plantation agriculture all worked to mobilize and erode massive amounts of soil and to increase sedimentation discharge to coastal areas. Although poorly documented at the time, sedimentation into coastal waters must have damaged coral reefs, wetlands, streams, estuaries, and beaches downstream from concentrated sources of soil erosion and in confined bays. The activities on land had major impacts on downstream coastal and marine areas, particularly from soil erosion, sediment discharge, and the blanketing of coral reefs, stream mouths, and beaches with terrigenous sediments.

Overgrazing by livestock on ranches (cattle, sheep, horses) and feral game animals (in conservation areas) is still a major problem on several islands. Years of overgrazing by goats and deer along the southeast coast of Moloka'i and north Kaua'i has led to major soil loss and sedimentation to coastal waters that persist to this day. For over a century, the island of Kaho'olawe was severely overgrazed by goats to the point that very little soil and larger woody vegetation persisted. Offshore reefs and beaches were covered with terrigenous sediments. In the early 1990s the U.S. Navy decided to terminate use of the island as a bombing range and took measures to eradicate all goats from the island. Complete eradication was successful, and vegetation is now rapidly recolonizing many landscapes, including native species never before recorded from the island. Sediment discharges on adjacent reefs also appear lower.

Alien and Invasive Species

Many alien species were intentionally introduced to Hawai'i, and many other aliens were unintentionally established, such as rodents and weed plants. These have led to the extinction of many endemic terrestrial plants and animals (particularly land snails, insects, and birds) and have endangered many others (Warshauer in Juvik and Juvik, 1998). Mangroves, intentionally introduced a century ago from Florida to O'ahu and Moloka'i, have now spread to most of the other main islands, displacing stream, fishpond, native beach and coral reef habitats. Sediment discharges and the spread of invasive species of aquatic weeds has also clogged many streams, ponds, canals, and lakes, and filled in many coastal wetlands and fishponds. In recent years many invasive alien marine algae and several reef fish species have invaded offshore marine habitats in Hawai'i, and are spreading and displacing resident reef species including corals, such as off west Maui, Kaneohe Bay on O'ahu, and south Moloka'i (Walsh, 1967; Maragos in Juvik and Juvik, 1998).

COASTAL USES

Subsistence, Artisanal and Recreational Fishing

The past two centuries have seen a transition from subsistence fishing in nearshore coastal waters practised by the Hawaiians to small-scale commercial, artisanal, and recreational fishing by present residents. However, subsistence fishing is still practised by small isolated Hawaiian settlements. Most local residents now use mostly wooden and fibreglass skiffs with outboard motors launched from boat ramps. Modern gear includes metal fishhooks, monofilament line, wire traps, large fishnets, spearguns, snorkelling gear, scuba gear, underwater flashlights, and other technologies which have extended harvesting to greater depths with improved efficiencies. Today many thousands of residents engage in surf casting, throw netting, surround netting, trolling, spearing, bottom fishing, night fishing, and other techniques. At present, most easily accessible nearshore areas are heavily fished with many species depleted (goatfish, parrotfish, surgeonfish, soldier fish etc.). Small-scale artisanal and commercial fishers prefer to use gillnets, large surround nets, and traps made from wire mesh. The use of ice chests and coolers also allowed fishers to extend their time at sea without the loss of spoiled fish. Virtually all nearshore fish around all the main islands of Hawai'i are now depleted.

Licensing and regulation of recreational and aquarium fish collection is presently very weak, although the state government is now investigating options for licensing all fishers and improving the monitoring of fish stocks. Sport fishing is also popular for visitors.

Mariculture

Research and development has been active in Hawai'i over the past three decades. Mariculture ventures and research have concentrated on marine shrimp (*Penaeus* spp.), dolphin fish or mahimahi, threadfin (*Polydactyus*), mullet, milkfish, and other tropical and pelagic species. Nutrient-rich cold water discharges from an ocean thermal energy conversion plant (OTEC) off the west coast of Hawai'i island are being used for experimental rearing and some commercial production of abalones, Maine lobsters, and other cold-water species. Several fishponds off O'ahu and Moloka'i are also being used for the production of mullets, milk fish, flagtails, threadfin, and edible marine algae, called *limu* in Hawaiian. Research is actively being conducted at government laboratories in Honolulu, the Oceanic Institute in windward O'ahu, and by the University of Hawaii Institute of Marine Biology at Coconut Island (*Moku o Loe*) in Kane'ohe Bay (Pooley in Juvik and Juvik, 1998).

Environmental concerns associated with the mariculture industry are many and include the escape of non-native mariculture species and the diseases they carry into the wild where they can, and often do, proliferate and displace native species. Other concerns are the quality of mariculture effluent discharges, and genetic mixing of mariculture and wild stocks of species. In general, alien species introductions to the marine environment are a growing concern in Hawai'i and other Pacific islands (see Eldredge, 1994).

Commercial Fisheries

Commercial fisheries using long lines, purse seines, hook-and-line, drop-line, trap, and trolling gear have been active in Hawai'i for most of the past century. In general, pelagic fisheries focus on yellowfin and bigeye tunas, skipjack tuna, swordfish and other billfish, wahoo, sharks, mahimahi, and others. Two seamount fisheries target armorheads and alfonsins. The offshore bottom fishery (at depths of 150–750 m) targets kona crab, deep-water pink and red snappers, and deep water groupers. The principal inshore commercial fisheries using scuba, spear, trap, and net gear target aquarium fish, coastal scombrids, particularly *akule*, jacks, and a variety of reef fish and shellfish. Spiny lobsters and other crustaceans are fished commercially, particularly in the northwest Hawaiian Islands. Recent (1998) data reveal that the spiny lobster fishery in the NW Hawaiian Islands is severely over-fished and may soon be closed. Recently sharks are being fished in large numbers to supply Asian demand for shark fin soup. Shark finning is presently very controversial in Hawai'i because mostly only the fins are taken, with the rest of the shark carcass dumped, often while still alive and bleeding. Inshore stocks of the native sardine *nehu*, are captured using surround nets and then kept in live bait wells and chummed during capture of offshore tunas using pole-and-line gear (Pooley in Juvik and Juvik, 1998).

Some deep-water precious corals are commercially important, but only black corals (*Antipathes* spp.) are being collected actively at the present time. Resident divers descend to depths of 60 m or more to manually collect the corals. The other precious corals (pink, bamboo, and gold corals) occur at depths of 400 m or more and are collected by manned submersibles and remotely operated vehicles. At present no active collection of the deeper precious corals is occurring, except experimental harvesting and life history data collection by researchers.

Inshore commercial fish stocks seem to be declining, and catch per unit effort has also declined. Commercial fisheries in nearshore waters and reefs are depleted for all groups, except for *akule* stocks, which have been traditionally harvested and managed for many decades. To date government management of inshore commercial fisheries has been ineffective, although fishers are subject to gear, seasonal, and area restrictions.

Fisheries Management Areas (FMAs) and Other Limited-take Marine Areas

The State of Hawai'i Division of Aquatic Resources has established a network of Fisheries Management Areas (FMAs) in the main islands including one off east Kaua'i, two off O'ahu, and 10 off the Big Island. A series of additional FMAs are being designated off the west coast of the Big Island to regulate aquarium fishing and promote sustainable use of nearshore fisheries. The State also prohibits fishing around the waters of Coconut Island (site of the Hawai'i Institute of Marine Biology), and in partnership with the Kaho'olawe Island Reserve Commission, regulates fishing around the uninhabited island of Kaho'olawe. The State has also established a series of Marine Life Conservation Districts, where fishing is prohibited at two of them (Hana'uma Bay on O'ahu and Kealakekua Bay on Hawai'i) and generally restricted to pole-and-line fishing within the rest (Table 6). The State has also established a Natural Area

Table 6

Marine protected areas. Sources: Clark and Gulko (1999); U.S. Dept. of Interior (1999); Maragos and Holthus, in Eldredge et al. (1999).

Marine Protected Area	Coverage
Federal	
Hawaiian Islands National Wildlife Refuge (NWR)	islands and reefs to 10 fathoms around Nihoa, Necker, French Frigate Shoals, Gardner, Maro, Laysan, Lisianski, & Pearl and Hermes
Midway Atoll NWR	50 sq. mile area around Midway Atoll
Kaloko-Honokohau National Historic Park	coastal waters off Honokohau, W. Hawai'i
Kalaupapa NHP	coastal waters off Kalaupapa, N. Moloka'i
Hawaiian Humpback Whale National Marine Sanctuary	coastal waters to 100 fa. off Kaua'i, Maui, Moloka'I, Maui, O'ahu, and Hawai'i.
State	
Kealakekua Bay Marine Life Conservation District (MLCD)	coastal waters off Kealakua Bay, SW Hawai'i
Lapakahi State Historical Park MLCD	coastal waters off Lapakahi, N.W. Hawai'i
Wailea Bay MLCD	coastal waters off Wailea, N.W. Hawai'i
Old Kona Airport MLCD	coastal waters N. of Kailua, W. Hawai'i
Hulopoe Bay & Manele Bay MLCD	coastal waters off S. Lana'i
Honolua &Mokuleia Bay MLCD	coastal waters off N.W. Maui
Molokini Shoals MLCD	Molokini Islet off S. Maui
Hana'uma Bay MLCD	off S.E. coast of O'ahu
Waikiki MLCD	off S. coast of O'ahu
Pupukea Beach Park MLCD	off N. coast of O'ahu
Kure Atoll State Wildlife Sanctuary	islets and reefs around Kure Atoll
Cape Kinau, Ahihi and La Perouse Bays State Natural Area Reserve	coastal waters between Ahihi & La Perouse Bay, S. Maui
Kaho'olawe Island Restricted Area	waters surrounding Kaho'olawe
Hawai'i Marine Laboratory Refuge	waters around Coconut Is, N.E. O'ahu
West Hawaii Fishery Management Area (FMA)	multiple coastal areas along W. Hawai'i
Kailua Bay FMA	coastal waters off Kailua, W. Hawai'i
Hilo Bay FMA	coastal waters off Hilo Bay, E. Hawai'i
Waimea Bay & Recreational Pier FMA	coastal waters off Waimea, S. Kaua'i
Hanamaula Bay & Ahukini Recreational Pier FMA	coastal waters off E. Kaua'i
Manele Boat Harbour FMA	harbour waters off S. Lana'i
Kahului Harbour FMA	harbour waters off N. Maui
Waikiki-Diamond Head FMA	coastal waters off S. O'ahu
State Offshore Island Seabird Sanctuaries	3 off Hawai'i; 3 off Kaua'i; 4 off Lana'i; 3 off Kaua'i; 6 off Maui; 2 off Moloka'i, 13 off O'ahu; 2 off northwest Hawaiian islands

Reserve at Ahihi-Kinau off south Maui and a Wildlife Sanctuary around Kure Atoll, where fishing in adjacent marine waters is prohibited (Loope and S. Juvik in Juvik and Juvik, 1998; Clark and Gulko, 1999).

Disturbances to the Northwest Islands

The smaller northwest islands were spared the urbanization effects in the main islands, although substantial military and navigational facilities were established at Midway, and less so at Kure Atoll and French Frigate Shoals. At the turn of the last century, Laysan island was heavily mined for guano, and its vegetation denuded by feral rabbits. The disturbance led to the documented extinction of two endemic species, the Laysan rail and Laysan millerbird, and reduced populations of other endemic species (Laysan finch, and Laysan duck).

Derelict Fishing Gear

Purse seiners, gill/drift netters, and trawlers normally do not operate near Hawai'i, although vast quantities of their derelict fishing nets are found snagged on reefs and beaches, especially in the northwest islands. Large commercial fishing fleets from several Pacific-rim nations operate in the open Pacific north of Hawai'i. Fishing gear is regularly lost, discarded, or abandoned by the fleets including lines, nets, floats, traps, and hooks which may drift around in the north Pacific for many years. Aside from being a nuisance to marine mammals, sea turtles, and seabirds, floating derelict fishing gear can accumulate into drifts and eventually become snagged on reefs and beaches.

The northwest Hawaiian Islands jut out towards the centre of the circular ocean circulation system known as the north Pacific sub-tropical anti-cyclonic gyre, and the North Pacific Equatorial Current is the southern component and westward-moving current in the gyre. Because of their location near the middle of the gyre, the Hawaiian islands have snagged many thousands and perhaps millions of tons of fishing gear on beaches and reefs. Although organized large-scale removal of derelict fishing gear was initiated last year in the northwest islands, the several tons of collected debris amounts to less than one percent of the gear that remains. Adequate resources and manpower to remove remaining gear and to restore reefs damaged by the gear remain a huge challenge.

URBANIZATION FACTORS

Over the past half century, urbanization has eliminated or modified many rural landscapes in coastal areas, particularly around the larger cities and plantation towns in the main islands (McGranaghan and Goss in Juvik and Juvik, 1998). Construction of ports, airfields, roads, and military bases further reduced rural areas, particularly on O'ahu. Many of the ancient coastal fishponds were modified, dredged, or filled for land expansion, housing, harbours, commercial fish propagation etc. Parts of O'ahu, Hawai'i, and most of Kaho'olawe were used as bombing and live fire ranges by the military, while other rural lands, primarily on O'ahu and Hawai'i, were used as military training areas. Later, flood control projects and urban settlements encroached on stream courses or led to the modification and diversion of many streams.

Eutrophication

Sewage impacts are more pronounced in bays and lagoons where water circulation is more restricted. Sewage discharges off other open coastal areas seem to have less or poorly understood effects, such as off south O'ahu (servicing Honolulu), and west Maui (servicing Ka'anapali to Lahaina). Aside from ecological effects on coral reefs, other concerns over sewage discharges include public health effects, undesirable growths of algae, and aesthetically displeasing odours and marine water quality. The local governments of Hawai'i have opted for strategies that include some treatment of sewage and location of discharge outfalls far offshore in deeper waters where mixing will be greater and effects on coral reefs minimized. In the future the effects of sewage discharges on coral reefs will increase in proportion to population increases and will require constant monitoring and possibly treatment upgrades to keep adverse effects at tolerable levels. Clean coastal waters are especially important in a major tourist destination like Hawai'i, where visitors expect to find clean ocean waters and healthy coral reefs.

A major difficulty has been scientific documentation of the effects of sewage discharges on coral reefs. Although there is considerable circumstantial evidence showing that reefs are stressed near the sites of sewage discharges, to date there is little water quality evidence to substantiate a cause-and-effect relationship between the two. In general, substantial water quality data collection at three sewage discharge locations in Hawai'i (Kane'ohe Bay and Mamala Bay on O'ahu and Lahaina-Ka'anapali off west Maui) has led to largely inconclusive results. These results demonstrate that traditional water quality monitoring techniques are not sensitive enough to detect the changes, and that new techniques, perhaps using a combination of appropriate ecological, pathogenic, and sewage indicators may need to be devised, through research, to substantiate the relationships.

Coastal Erosion, and Seawalls

The geological subsidence of most of the main volcanic Hawaiian islands results in steady rates of natural coastal erosion over long time periods. Sea level in Hawai'i underwent substantial fluctuations during the transitions between the glacial and inter-glacial periods. Natural cycles of contemporary beach erosion and accretion are also attributed to periodic fluctuations in coastal wave energy striking the coasts, and rare natural events such as tsunamis, tropical cyclones, and sea-level fluctuations attributed to tides and thermal heating of surface ocean waters.

Coastal erosion only became a socio-economic problem when people began to occupy the coasts and build permanent buildings and other structures within the zones of natural shoreline fluctuations. The early Hawaiians probably found it much easier to shift huts and village areas out of harm's way when threatened by natural cycles of coastal erosion, but the "modern" age of concrete, lumber, steel, glass, plumbing, and electricity has made this option costly, unpopular, and rarely pursued. Seawalls, groins, offshore breakwaters, and revetments have been erected along many beach areas to protect residences and public facilities (roads, airfields, parks, etc.) Although the government may erect shore structures to protect public use areas, private individuals have also constructed seawalls and rubble-mound revetments to protect their property. Virtually all coastal towns and cities in Hawai'i are now threatened with shoreline erosion (Curtis in Juvik and Juvik, 1998; Fletcher in Juvik and Juvik, 1998).

Many modern residential and urban areas were established too close to many Hawaiian shorelines and flood-prone areas before residents were aware of the dynamic state of the coastline and before government could regulate development in coastal erosion-prone areas. The present 40-foot shoreline setback ordinance was established by the state in the 1970s to restrict shoreline development, but exemptions are still granted. The state and federal governments and lending institutions have sponsored the development of flood and coastal hazard maps and have imposed insurance requirements as a means to discourage occupation of hazardous areas, including the shoreline and transfer the cost of protection to those most responsible for its need. Nevertheless, a much wider setback zone and more restrictions against shoreline development and fortification are needed.

The development of Waikiki as the principal visitor destination in Hawai'i warrants mention. During the first half of the 20th century, many hotels were built on the beach, perhaps the most beautiful in the islands at the time, and the main thoroughfare was placed too close to the beach. During the ensuing decades, the threat of shoreline erosion has forced the government and hotels to spend millions of dollars to build seawalls, crib walls, groins, and replenish lost beach sand. The sand lost to the beach has moved seaward, smothering coral reefs and degrading surfing sites. Very little of the original Waikiki beach now remains. A shoreline setback zone of about 100 m would have preserved the beach and saved vast resources.

Coastal Construction

Substantial coastal construction including the dredging basins and the filling of beach, wetland, and reef areas for shoreside facilities and protective structures was required to establish most commercial ports and some airfields in Hawai'i. These include the deep draft harbours at Honolulu on O'ahu; Kaunakakai on Moloka'i; Kahului on Maui; Nawiliwili and Port Allen on Kaua'i; and Hilo and Kawaihae on Hawai'i. Smaller marinas and docks have been constructed on all the main islands except Ni'ihau and Kaho'olawe, but including Tern Islet at French Frigate Shoals in the northwest islands. Airfields extending seaward of the shoreline have also been constructed off Honolulu and Ka'anapali. The era of military development in Hawai'i began with the dredging and filling of Pearl Harbour during the first decade of the 20th century and accelerated during the decades before, during, and after World War II. Extensive coastal construction was required for the military bases at Midway, Kane'ohe, Honolulu, and other sites. Road construction along many coasts has encroached upon beach areas, including many coastlines on O'ahu, southwest Kaua'i, south Maui, west Hawai'i, and southeast Moloka'i.

Coastal wetlands and fishponds were drained and filled, primarily on O'ahu in Honolulu, Waikiki, Kailua, Kane'ohe, Salt Lake, and Hawaii Kai to control flooding and provide land for housing and hotels. A century ago alien species of mangroves were intentionally planted off southwest Molokai by agricultural plantations to stabilize sediment and expand coastal land. During the past century mud flats and mangroves have replaced inner tidal and reef flats west of Kaunakakai.

Industrial Discharges

Earlier during the 20th century the Hawaiian islands experienced industrial discharges from military installations (Pearl Harbour and Hickam), the pineapple and fish canneries in Honolulu, meat packing plants, and wood preservative plants in Honolulu and Hilo. At present most of these industries have gone and all coastal discharges terminated, and military and industrial installations are subject to rigid water pollution controls. The principal remaining discharges of concern are fuel spills and leaks from power plants, offshore pipelines, refuelling facilities in harbours, ship ballast water discharges and ship groundings and collisions. Many minor fuel spills have occurred, mostly in the Honolulu area. The U.S. Coast Guard, with its local headquarters in Honolulu Harbour is primarily responsible for enforcing prevention and coordinating cleanup of industrial maritime spills, especially fuel, ballast and sewage discharges and spills from vessels.

Most municipal power plants use ambient seawater for cooling. Before passage of the Federal Water Pollution Control Act in 1972 (and later the Clean Water Act in 1977), heated effluent waters from the large Kahe power plant in southwest O'ahu were discharged on nearshore shallow reefs, killing and injuring corals. As a result of the legislation, effluent temperatures were lowered and the discharge pipe moved further offshore. The result is that further injury to reef communities was prevented, and now the moderately heated effluent waters encourage coral recolonization with colonies actually growing faster than normal because of the more favourable warmer waters.

Tourism and Related Ocean Recreation

Hawai'i is the most popular visitor destination in the world, averaging nearly seven million visitors a year. Over the past century Waikiki Beach on O'ahu and east of Honolulu has been the primary visitor destination in the islands, but tourism has gradually expanded on the neighbouring islands of Maui, Kaua'i, Hawai'i, and to a lesser extent Lana'i, and Moloka'i. The transfer of Midway Atoll from the U.S. Navy to the U.S. Fish and Wildlife Service has opened up small-scale eco-tourism in the northwestern islands over the past decade. During recent years many tourists have expanded their interests beyond traditional beach hotel stays and seek out more outdoor and adventuresome activities, especially ocean and coastal recreation.

Infrastructure

Hawai'i supports an extensive network of visitor accommodations totalling more than 70,000 hotel rooms, condominiums, and other lodgings generally situated on O'ahu (Sakai in Juvik and Juvik, 1998). Hotels are concentrated in Waikiki (south O'ahu), but resorts have also been established at Punalu'u, Kahuku, Makaha, and the Ewa-Kahe areas over the past three decades in northeast, north, west, and southwest O'ahu respectively. Maui has more recently emerged as an important tourism destination with resorts concentrated in Lahaina-Ka'anapali (west Maui), Kahului-Wailuku (north Maui), and Wailea-Kihei-Makena (south Maui) areas. On Kauai, resort infrastructure is concentrated along the northeast quadrant between Lihu'e and Hanalei. On Lana'i the two resorts are located on the south coast and centre of the island. On Moloka'i small resorts are situated off the west and south coasts. On the Big Island, resort infrastructure is concentrated along the northern half of the west coast between Keauhou and Kawaihae, and to a lesser extent in Hilo town along the northeast coast. Local docks, marinas, and a small airfield are located near most of the resort centres and provide visitor access to remote destinations and ocean recreational opportunities.

Golf Courses

Over 90 golf courses are located in the main Hawaiian islands with at least two on all of the above six tourism islands. Over 15 are found on Maui, over 20 found on Hawai'i, and over 40 on O'ahu (S. Juvik et al. in Juvik and Juvik, 1998). Golfing in Hawai'i is extremely popular to both visitors and residents. Resorts operate most of these courses, and many are situated along scenic coasts. In the past some residents have expressed concerns over sediment run-off during construction and discharge of nutrients and pesticides during the operation of golf courses. However, evidence for documented impacts of the discharges on coral reefs and other coastal waters is scant or inconclusive. Still, some golf course effluents discharge into waters near stressed reefs, raising circumstantial evidence of a cause-and-effect relationship, such as the discharges from golf courses into southern Hanalei Bay off north Kaua'i.

Diving and Snorkelling

All six of the main tourism islands and Midway Atoll provide snorkelling and SCUBA diving opportunities for visitors. However, many coasts along the main islands are exposed to seasonally heavy waves and wind conditions, limiting access to many suitable dive sites. For example, O'ahu, with the most visitors, has only a few accessible diving and snorkelling sites, and one, Hana'uma Bay, is so popular that there are controls in place to limit access. Other popular dive and snorkelling sites include the north and south coasts of O'ahu, Lana'i, Kaua'i, and Maui; the south coast of Moloka'i; most of the west coast of the Big Island; the east and northwest coast of Ni'ihau; Molokini Island off south Maui, and Midway Atoll. There are some conflicts between sport diving advocates and fishers, especially aquarium collectors, at popular dive sites. Anchor damage from boaters is also a concern, such as at Molokini. The State however, is developing management plans to resolve user conflicts (such as off west Hawai'i), and advocating the use of mooring buoys to limit anchor damage to corals.

Sport Fishing

Blue water trolling for game fish has been a popular pastime for many visitors along calmer leeward coasts, including the west and south coasts of Hawai'i, O'ahu, and Maui. The sport fish include *mahimahi* (dolphin fish), *ono* (wahoo), *aku* (skipjack tuna), *ahi* (yellowfin tuna), *au* (swordfish), marlin, *ulua* (trevally), and *papio* (small jacks). Certain reef fish are also prized by mostly resident sport fishers, including *weke*, *moano*, and *kumu* (goatfish); *kala* and *palani* (surgeon fish); *u'u* (soldierfish), and *uhu* (parrotfish). Popular shellfish include the spiny lobster, samoan or mangrove crab, seven-eleven crab, and native grouper (*hapu'upu'u*). The abundance of the above reef fish and shellfish has been steadily declining over the past several decades, while the remaining species (generally open-ocean or pelagics) appear to be holding their own. Anchor damage and the accumulation of lost fishing gear are concerns to divers, snorkellers, resource managers, and researchers.

Boating, Surfing, and Submarines

Commercial boating other than for diving and fishing is also popular for whale watching, water skiing, and para-sailing. Speed boat racing is also pursued by residents and some visitors in calm water locations such as Ke'ehi Lagoon and Kane'ohe Bay off O'ahu. Jet skis are popular rentals at some resorts, although generally disliked by locals, especially kayakers, divers, and snorkellers. Commercial jet skis are now used for transporting surfers to offshore surfing sites, especially where access is difficult or strenuous, such as to the famous "Jaws" surfing site off north Maui. Ocean kayaking, wind-surfing, and sailing are popular to many residents and visitors. Hawaiian outrigger canoe rides have been a long-time favourite attraction for visitors at Waikiki, and many local residents are active in traditional Hawaiian canoe racing throughout the main islands. Commercial submarine dive operations are active off south Oahu and west Hawai'i and are very popular visitor attractions. The submarines dive to depths of over 30 m, and depend upon access to coral reefs, artificial reefs or shipwrecks to attract customers. Several marine aquariums are now established on O'ahu and Maui and stimulate education, interest, and use of marine areas.

Board surfing and body surfing are also traditional pastimes in Hawai'i and now attract many visiting surfers. The best surf sites tend to be located in rougher waters away from divers and boaters, especially along north, northwest and south facing coasts. Surfing is seasonal, with larger waves attracting surfers to northerly coasts during winter months and to southerly locales during summer (austral winter) months.

Other Environmental Concerns

Crown-of-Thorns Starfish

Infestations of the crown-of-thorns sea star (*Acanthaster planci*) have been major concerns in other coral reef locales in the Indo-Pacific region, but have not been an issue in Hawai'i over the past three decades. In 1968 a large aggregation of the sea star was reported off the south coast of Moloka'i and monitored by scientists for two years. The coral predator was feeding actively on corals of the common genus *Montipora* and to a lesser extent *Pocillopora*, but avoided predation on the dominant corals of the genus *Porites*. Nevertheless, the State government elected to eradicate it in 1970, using divers to inject sodium hydroxide into them. Since that time there have been no other reported infestations or aggregations from Hawai'i.

Coral Bleaching and Freshwater Flooding

Until 1995, coral bleaching was never a concern in the islands, having been reported only sporadically over the previous several decades. However, in late 1996, coral bleaching was widespread in some reef localities, affecting most of the common coral species. The bleaching episode diminished during the following year with mainly *Pocillopora* suffering some mortality. Corals in the Hawaiian Islands were not affected during the 1998 coral reef bleaching event that damaged many other coral reef areas around the globe.

Heavy rainfall events in 1965 and 1988 in Kane'ohe Bay watershed caused unusual freshwater kills of corals in confined reef localities in the lagoon. Apparently flood waters discharged into the bay through several stream mouths, with the buoyant freshwater plumes submerging to depths that injured and killed the corals. After the initial 1968 freshwater kill, other corals succumbed during periods of low dissolved oxygen concentrations, caused by the decomposition of dead corals and associated reef biota. After both events, corals gradually recolonized the shallow reef substrates affected by the earlier kills. Some scientists (including the author) believe that the flooding was exacerbated by land use changes in the surrounding watersheds attributed to urbanization, which resulted in higher run-off rates (due to the expansion of impervious surfaces such as sealed roads, roofs, and concrete-lined flood control channels).

Coral Diseases

Hawai'i is relatively free of diseases that affect corals. However, local researchers are concerned because there are now several types of diseases or tumours observed on common corals, including dominant species, and they appear to be reported with increasing frequency. Evidence is not yet conclusive on the origin or cause of tumours and other diseases on Hawaiian corals, nor can it yet be substantiated that any of the diseases have an anthropogenic origin. Local scientists and conservation agencies and organizations, however, are concerned over the possible introduction of alien coral species to Hawai'i for mariculture propagation. These non-native corals may bring diseases that could attack native corals (Hunter in Maragos and Grober-Dunsmore, 1999).

PROTECTIVE MEASURES

Environmental Legislation

All three levels of government have passed extensive environmental legislation. The counties, in concert with the state, have passed and implemented coastal zone management legislation, primarily to guide and control development within Special Management Areas, a continuous strip of coastal land averaging about 100 m around all islands. The counties are also responsible for grading and sediment control ordinances and compliance with effluent standards for municipal sewage treatment plants.

The state government has passed legislation and regulations to manage fisheries in state waters, require permits for use of conservation-designated lands and territorial waters, protect wildlife, conserve state forests, manage state parks, monitor and enforce water quality standards (both effluents and receiving waters), designate and protect endangered species, manage the state's historic preservation program, and administer the state's coastal zone management program, including review of federal actions for consistency with the program (Parks in Juvik and Juvik, 1998). The state has also passed laws prohibiting the removal of nearshore sand and coral, the harvest of live corals, the introduction of alien species without a permit, and the protection of important streams and their adjacent estuaries (Clark and Gulko, 1999).

The federal government has passed legislation and regulations to protect endangered and threatened species, air and water quality, and support state coastal zone management programs. Other statutes control solid waste pollution, toxic and hazardous waste, ocean dumping, and construction, dredging and filling in of wetlands and other tidal waters. The federal government also has procedures to ensure that development projects do not injure marine mammals, migratory birds, important fish and wildlife resources, and sites eligible for listing on the National Register of Historic Places. Many state and federal environmental statutes allow citizen suits against polluters and other suspected violators of environmental standards and procedures.

As can be surmised from the above, the county, state and federal governments have established a plethora of environmental ordinances, laws, and regulations that apply to marine and coastal areas. However, it is a commonly held perception that Hawai'i's natural resources are over regulated but under managed. The flood of environmental legislation and corresponding statutes that emerged primarily in the 1970s and 1980s has been effective in controlling point source water pollution, including discharges of sewage and other industrial effluents. It has also been effective in controlling coastal construction, reducing soil erosion, and limiting further loss of wetland habitat, and the new Presidential Executive Order shows promise in increasing protection of the nation's coral reefs. Establishment of marine protected areas, as discussed below and earlier, has also been impressive. On the down side, government has demonstrated limited capacity to enforce existing statutes, including established fisheries, surveillance, and monitoring programs. Presently some local governments, ocean-dependent businesses and communities are dissatisfied and want to play greater roles in managing coral reefs, beaches and other marine areas and address the lack of an adequate fisheries management and enforcement regime.

Development Plans

The county and state governments are primarily responsible for development and land use planning, although the federal government is responsible for similar planning on its own land holdings and in ocean waters beyond the territorial seas. In Hawai'i, federal jurisdiction begins beyond three nautical miles (nm) from shore with the State exerting jurisdiction in waters within three nm. The county governments manage land-use zoning for residential and urban areas, and administer local ordinances on shoreline setbacks, and development in Special Management Areas. The counties are also responsible for planning and managing sewage treatment and other public works and ordinances. The state government has developed a State Plan to guide long-range development and has established a land-use districting system and review process. The state is directly responsible for managing agricultural and conservation districts, the latter including offshore territorial waters. The State has also developed a Hawaii Ocean Resources Management Plan in 1991 (Hawaii Ocean and Marine Resources Council, 1991), but many of its provisions have yet to be funded and implemented. The state is also responsible for managing the Hawai'i Coastal Zone Management Plan, including special plans for Areas of Particular Concern. The state is responsible for planning and operating state highways and civilian harbours and airports.

Environmental Impact Assessment

Two levels of environmental impact assessment procedures have been established. The state's environmental impact statement (EIS) law covers proposed state, county, and private actions that may result in significant negative environmental impacts on lands and waters within state jurisdiction. A similar but independent federal process authorized by the National Environmental Policy Act of 1970 requires an EIS for any proposed federal project or action requiring a federal permit that may significantly affect the environment. Both of these programs require proponents to develop and evaluate feasible alternatives for each proposal, and to avoid or reduce negative impacts through mitigation or implementation of less damaging alternatives. Environmental assessment regulations have been very effective in reducing development impacts in sensitive environmental areas and protecting important and vulnerable species.

Marine Protected Areas

Both the state and federal governments have jurisdiction in marine waters and manage marine protected areas (MPAs) in Hawai'i. State government-managed MPAs are extensive and include approximately 13 FMAs, one Hawai'i Marine Laboratory Refuge, one Natural Area Reserve, one State Wildlife Sanctuary, 11 Marine Life Conservation Districts, and the Kaho'olawe Island Restricted Area, which is jointly managed with the Kaho'olawe Island Reserve Commission. Of these, the last four categories are more restrictive with respect to fishing and harvesting. The State also manages 31 Offshore Island Seabird Sanctuaries for the small islets offshore from the main volcanic islands and several State Parks which afford protection to shoreline areas and wildlife (Table 6) (Clark and Gulko, 1999; Loope and S. Juvik in Juvik and Juvik, 1998; Maragos and Payri, 1997).

The federal government has also established several categories of MPAs, including the Hawaiian Humpback Whale National Marine Sanctuary managed by NOAA, and three National Parks and Monuments administered by the National Park Service. The U.S. Fish and Wildlife Service administers the Midway Atoll National Wildlife Refuge and the (Northwest) Hawaiian Islands National Wildlife Refuge (Complex) which collectively is the largest MPA in the United States and the second largest MPA in the Pacific, after Australia's Great Barrier Marine Park. All existing federal MPAs prohibit fishing except for small-scale catch-and-release sport fishing at Midway (Table 6).

The Nature Conservancy of Hawai'i (TNCH) is a nongovernment organization that manages or co-manages many private ecological reserves in the U.S. and other countries. One of their reserves protects beach and coastal areas at Mo'omomi, along the northwest coast of Moloka'i, although reserve jurisdiction does not extend into offshore waters. However, the State of Hawaii, Department of Land and Natural Resources has an ongoing initiative to establish community-based marine reserves, including an existing one at Mo'omomi. This initiative should be expanded to other candidate Hawaiian communities and valued marine areas because of the potential for more effective local management and surveillance and the role these kinds of reserves can play in preserving the subsistence lifestyles of the traditional Hawaiian culture.

Marine Environmental Restoration

The effects of sewage discharges into Kane'ohe Bay, O'ahu over the past 35 years has been documented by marine scientists affiliated with the University of Hawaii Institute of Marine Biology which is located on a small island in the southern bay. Sluggish water circulation in the south bay exacerbated the adverse effects of municipal sewage discharges from two outfalls servicing urban populations in the surrounding watershed in the early 1960s. The longer residence time of water in the south lagoon magnified the growths of marine phytoplankton stimulated by the nutrients in the sewage and nearly denuded all live coral. Nutrients from the sewage also stimulated bottom algal growths in the middle bay lagoon that smothered live corals. The studies compelled the county government to remove the sewage outfalls from the bay in 1977–78. Studies conducted 6 years after the diversion of sewage documented improved lagoon water quality, rapid recolonization of corals in the south bay, and higher coral cover and much lower benthic algal (bottom seaweed) cover in the mid bay lagoon (Maragos et al., 1985; Hunter and Evans, 1995).

The removal of sewage from the lagoon was the first successful large-scale restoration of a coral reef ever attempted. In recent years non-point sources of nutrients into the bay may be increasing, and may explain the subtle rise of bottom algae in the mid-lagoon of the bay. The state government passed legislation to establish a special coastal management project for Kane'ohe Bay, and a local citizen council is given the responsibility to execute the plan.

CHALLENGES FOR THE NEW MILLENNIUM

Establishment of new MPAs and better management of all MPAs and fishery management areas (FMAs) is the principal challenge and goal for strengthening protection of ocean areas around the Hawaiian Islands. The strategies below appear to be the most effective in achieving this goal.

Community-based Management of Coastal Areas

Government has demonstrated limited capacity to monitor compliance with, and enforcement of, important environmental and conservation controls protecting marine areas in Hawai'i. Furthermore, there is no clear indication on the status of coral reefs and other nearshore marine areas, due to the lack of long-term monitoring and other quantitative information (Maragos and Grober-Dunsmore, 1999; Wilkinson, 1998). Citizen groups and local communities have the interest and resources to assist government manage coastal and marine resources. Volunteers, such as members of sport diving clubs, can be trained and certified to monitor marine resources and habitats and provide early warnings to government regulators to take corrective action. The government itself will need to support volunteer programs and delegate some management authorities and responsibilities to them. This strategy would be particularly useful for stewardship of established MPAs and surveillance of FMAs for detecting illegal or unauthorized fishing activity. The communities and volunteers themselves should be involved in all stages, beginning with planning of, as well as the monitoring and surveillance of, the management regimes.

Public Education

Many citizens and public servants are not aware of the depleted status of many resources and the destructive effects of many actions such as anchor damage, destructive fishing methods, and dumping of pollutants and derelict fishing gear and refuse into the ocean. Many fishers are unaware of the depleted status of many fisheries and the implications of continued over-fishing. There is tremendous political and public resistance, for example, to the designation of additional MPAs and the imposition of licensing and monitoring requirements for recreational and artisanal fishing. It is very clear that fishery resources are heavily depleted, and that an effective management regime is needed. Public education is an important strategy to convince legislators and the fishers themselves of the efficacy of and need for fisheries management reform, especially for fishing grounds around the main islands.

Controls Over Fishing

Additional restrictions are needed to render fisheries management more effective. All commercial fishing vessels should be equipped with vessel monitoring systems in order to plot continuously the location of the vessels and determine if they are fishing in authorized waters. Fishing gear should be marked with identifying "signatures" or tags so the origin of the gear can be determined, if it is lost or becomes derelict. Large vessels and ships should be required to deposit a monetary bond for the privilege of fishing in federal or state waters. Should the ship or vessel be grounded or wrecked on a reef or other coastal habitat, the bond could be immediately tapped by the U.S. Coast Guard or the fishery management agency to cover the cost of timely vessel removal.

Better Management in the Northwest Hawaiian Islands

Most of Hawaiian coral reefs are in the remote northwest islands, in uninhabited localities beyond the watchful eyes of residents and resource managers. Evidence is compelling that unauthorized and destructive fishing is occurring in the remote refuges, particularly by fishers fetching species with high prices in U.S. and Asian markets: sharks, aquarium fish, lobsters, pearl oysters, and large food fish for the live fish trade. Authorized commercial fishery vessels operating or transiting near the refuge should be equipped with vessel monitoring systems (VMS) and use labelled gear as discussed earlier. In addition, remotely operated surveillance cameras equipped with radar, video warning sirens and warning lights should be developed, prototypes tested, and eventually installed on towers or platforms on all the remote islands in the refuge. These systems would help to detect and discourage unauthorized vessels from approaching no-take reefs. Vessels entering no-take areas would be warned and management authorities notified. If necessary, the U.S. Coast Guard could then be launched to track and apprehend illegal fishers. Additional punitive actions should also be taken for illegal or destructive fishing including confiscation of gear and vessels and forfeiture of vessel bonds and commercial fishing licences. The U.S. Fish and Wildlife Service (USFWS) is presently working with the U.S. Department of Defense to develop remotely operated surveillance systems for its remote National Wildlife Refuges in the Pacific.

Additional Marine Protected Area Coverage

Existing boundaries of some MPAs are too close to shallow reef areas (especially French Frigate Shoals and adjacent atolls) and sensitive species (sea turtles, seabirds, monk seals). The boundaries need to be moved further offshore and made more defensible from the standpoint of surveillance and enforcement. The USFWS is presently working with the National Marine Fisheries Service and the State of Hawaii to extend protection of coral reef and marine areas out to 12 miles beyond an island in the Northwest Hawaiian Islands. These islands and reefs are the most important coral reefs in the United States.

Additional MPAs in the main islands are also needed, especially a system of no-take preserves where fish stocks would be allowed to reproduce without threat of harvest and disturbance. In turn these replenishment areas would provide larval fish recruits to adjacent fishing grounds which are presently overfished or depleted. The location and size of these replenishment reserves should be scientifically based and emphasize selection of fishing grounds where currents are favourable for outside larval dispersal, and where habitat is conducive to support viable breeding populations of depleted fishery species.

Partnerships With Other Nations and Educational Institutions

The State of Hawai'i and the several federal agencies responsible for ocean management, must work cooperatively to ensure adequate monitoring, protection and conservation of marine resources, and set a good example for neighbouring nations, many of which lack the technology and funding to adequately protect their own resources. Through exchange training and internship programs, the U.S. and the State could empower the other nations with similar marine resource management goals. In particular the Republic of Kiribati and the Republic of the Marshall Islands are both close to Hawai'i and should welcome and benefit from such technical assistance.

Preservation of Natural Tourism Amenities

Many marine resources and sites are valued by tourists, attracting many millions of dollars to Hawai'i. The visitor industry should work more closely with marine resource management agencies and local communities in sponsoring or supporting efforts to protect or restore marine areas and resources attractive to visitors and tourism.

REFERENCES

Amerson, A.B., Jr. (1971) The natural history of French Frigate Shoals, Northwestern Hawaiian Islands. *Atoll Research Bulletin* **150**, 1–383.

Amerson, A.B. Jr., Clapp, R.B. and Wirtz II, W.O. (1974) The natural history of Pearl and Hermes Reef, Northwestern Hawaiian Islands. *Atoll Research Bulletin* **174**, 1–306.

Clapp, R.B. (1972) The natural history of Gardner Pinnacles, Northwestern Hawaiian Islands. *Atoll Research Bulletin* **163**, 1–29.

Clapp, R.B. and Kridler, E. (1977) The natural history of Necker Island, Northwestern Hawaiian Islands. *Atoll Research Bulletin* **206**, 1–102.

Clapp, R.B., Kridler, E. and Fleet, R.R. (1977) The natural history of Nihoa Island, Northwestern Hawaiian Islands. *Atoll Research Bulletin* **207**, 1–147.

Clapp, R.B. and Wirtz, II, W.O. (1975) The natural history of Lisianski Island, Northwestern Hawaiian Islands. *Atoll Research Bulletin* **186**, 1–196.

Clark, A.M. and Gulko, D. (1999) *Hawaii's State of the Reefs Report, 1998*. Department of Land and Natural Resources, Honolulu, Hawaii. 41 pp.

Coulbourn, W.T., Campbell, J.F. and Moberly, R. (1974) Hawaiian submarine terraces, canyons, and quaternary history evaluated by seismic reflection profiles. *Marine Geology* **17**, 215–234.

Dana, T.F. (1971) On the corals of the world's most northern atoll (Kure: Hawaiian Archipelago). *Pacific Science* **25** (1), 80–87.

Daws, G. (1974) *Shoal of Time: A History of the Hawaiian Islands*. Univ. Hawaii Press, Honolulu.

Devaney, D.M. and L.G. Eldredge (eds.). *Reef and Shore Fauna of Hawaii, Section 1: Protozoa through Ctenophora*. Bernice P. Bishop Museum Special Publication 64 (1), pp. 1–278.

Dollar, S.J. (1982) Wave stress and coral community structure in Hawaii. *Coral Reefs* **1**, 71–81.

Dollar, S.J. and Tribble, G.W. (1993) Recurrent storm damage and recovery: a long-term study of coral communities in Hawaii. *Coral Reefs* **12**, 223–233.

Eldredge, L.G. (1994) *Perspectives in Aquatic Exotic Species Management in the Pacific Islands*. Vol. 1: Introductions of Commercially Significant Aquatic Organisms to the Pacific Islands. South Pacific Commission, Noumea, New Caledonia. 127 pp.

Eldredge, L.G., Maragos, J.E., Holthus, P.F. and Takeuchi, H.F. (1999) *Marine and Coastal Biodiversity in the Tropical Island Pacific Region: Vol. 2. Population, Development, and Conservation Priorities*. East-West Centre and Pacific Science Association c/o Bishop Museum, Honolulu. 456 pp.

Ely, C.A. and Clapp, R.B. (1973) The Natural History of Laysan Island. Northwestern Hawaiian Islands. *Atoll Research Bulletin* **171**, 1–361.

Grigg, R.W. (1983) Community structure, succession and development of coral reefs in Hawaii. *Marine Ecology Progress Series* **11**, 1–14.

Grigg, R.W. (1988) Paleoceanography of coral reefs in the Hawaiian-Emperor chain. *Science* **240**, 1737–1743.

Grigg, R.W. (1994) Effects of sewage discharge, fishing pressure and habitat complexity on coral ecosystems and reef fishes in Hawaii. *Marine Ecology Progress Series* **103**, 25–34.

Grigg, R.W. and Maragos, J.E. (1974) Recolonization of hermatypic corals on submerged lava flows in Hawaii. *Ecology* **55**, 387–395.

Grigg, R.W. and R. Pfund (eds.). (1980) Proceedings of the Symposium on the Status of Resource Investigations in the Northwestern Hawaiian Islands. Univ. Hawaii Sea Grant College Pub. UNIHI-SEAGRANT-MR-80-04. Honolulu, 333 pp.

Gulko, D. (1998) *Hawaiian Coral Reef Ecology*. Mutual Publishing, Honolulu. 245 pp.

Hawaii Audubon Society (1993) *Hawaii's Birds. 4th edn*. Hawaii Audubon Society, Honolulu.

Hawaii Ocean and Marine Resources Council (1991) Hawaii Ocean Resources Management Plan. Department of Business, Economic Development and Tourism, State of Hawaii. Honolulu. 49 pp. + Tech. Sup., 159 pp.

Hoover, J.P. (1998) *Hawai'i's Sea Creatures, A Guide to Hawai'i's Marine Invertebrates*. Mutual Publishing, Honolulu, 366 pp.

Howe, K.R., Kiste, R.C. and Lal, B.V. (eds.) (1994) *Tides of History: The Pacific Islands in the Twentieth Century*. Univ. Hawaii Press, Honolulu, 475 pp.

Hunter, C.L. (1995) Review of Coral Reefs around American Flag Pacific Islands and Assessment of Need, Value, and Feasibility of Establishing a Coral Reef Fisheries Management Plan for the Western Pacific Region. Final Report. Western Pacific Regional Fisheries Management Council, Honolulu, 30 pp.

Hunter, C.L. and Evans, C.W. (1995) Coral reefs in Kaneohe Bay, Hawaii: Two centuries of western influence and two decades of data. *Bulletin of Marine Science* **57**, 501–515.

Jokiel, P.L. (1987) Ecology, biogeography and evolution of corals in Hawaii. *Trends in Ecology and Evolution* **2** (7), 179–182.

Jokiel, P., Hunter, C.L., Taguchi, S. and Watarai, L. (1993) Ecological impact of a fresh-water "reef kill" in Kaneohe Bay, Oahu, Hawaii. *Coral Reefs* **12**, 177–184.

Juvik, S.O. and Juvik, J.O. (eds.) (1998) *Atlas of Hawaii, 3rd edn*. University of Hawai'i Press, Honolulu, 333 pp. [51 multiple authored reports, maps and statistical tables]

Kay, E.A. (ed.) (1972) *A Natural History of the Hawaiian Islands. Selected Readings*. Univ. Press of Hawaii, Honolulu. 653 pp.

Kay, E.A. (1979) *Hawaiian Marine Shells*. Reef and Shore Fauna of Hawaii Section 4: Mollusca. Bishop Museum Press. Bernice P. Bishop Museum Special Publication 64(4), pp. 1–653, Honolulu.

Kay, E.A. (1980) *Little Worlds of the Pacific*. H.L. Lyon Arboretum, University of Hawaii, Honolulu.

Kay, E.A. (ed.) (1994) *A Natural History of the Hawaiian Islands. Selected Readings II*. Univ. Hawaii Press, Honolulu.

Kay, E.A. and Chave, E.H. (1983) Reef and Shore Communities. In *Atlas of Hawaii, 2nd edn*. Dept. Geography, Univ. Hawaii. Univ. Hawaii Press, Honolulu.

Kay, E.A. and Palumbi, S.R. (1987) Endemism and Evolution in Hawaiian Marine Invertebrates. *Trends in Ecology and Evolution* **2** (7), 183–186.

Magruder, W.H. and Hunt, J.W. (1979) *Seaweeds of Hawaii: A Photographic Identification Guide*. Oriental Publishing Company, Honolulu,

Maragos, J.E. (1995) Revised Checklist of Extant Shallow-water Stony Coral Species from Hawaii (Cnidaria: Anthozoa: Scleractinia). *Bishop Museum Occasional Paper* **423**, 54–55.

Maragos, J.E. and Grober-Dunsmore, R. (eds.) (1999) *Proceedings of the Hawai'i Coral Reef Monitoring Workshop*, June 9–11, 1998, Honolulu, Hawai'i. East-West Centre and the Hawaii Department of Land and Natural Resources, Honolulu, 334 pp.

Maragos, J.E. and Payri, C. (1997) The Status of Coral Reef Habitats in the Insular South and East Pacific. *Proc. 8th Int. Coral Reef Symp.* **1**, 307–316.

Maragos, J.E., Evans, C. and Holthus, P. (1985) Reef corals in Kaneohe Bay six years before and after termination of sewage discharges (Oahu, Hawaiian Archipelago). *Proc. 5th Int. Coral Reef Congr.* **4**, 189–194.

Maragos, J.E., Peterson, M.N.A., Eldredge, L.G., Bardach, J.E. and Takeuchi, H.F. (eds.) (1995) *Marine and Coastal Biodiversity in the Tropical Island Pacific Region: Vol. 1, Species Systematics and Information Management Priorities*. East–West Centre, Ocean Policy Institute, and the Pacific Science Association, Honolulu. 424 pp.

NMFS and USFWS (National Marine Fisheries Service and U.S. Fish and Wildlife Service) (1998a) Recovery Plan for U.S. Pacific Populations of the Green Turtle (*Chelonia mydas*). National Marine Fisheries Service, Silver Spring, Maryland. 84 pp.

NMFS and USFWS (National Marine Fisheries Service and U.S. Fish and Wildlife Service) (1998b) Recovery Plan for U.S. Pacific Populations of the Hawksbill Turtle (*Eretmochelys imbricata*). National Marine Fisheries Service, Silver Spring, Maryland. 82 pp. [also similar plans developed for the 3 other less common sea turtle species reported from Hawai'i]

Olson S.L. and James, H.F. (1982) Prodromus of the Fossil Avifauna of the Hawaiian Islands. *Smithsonian Contributions to Zoology* **365** 1–59.

Pratt, H.D. (1994) Avifaunal change in the Hawaiian Islands, 1893–1993. *Studies in Avian Biology* **15**, 103–118.

Reeves, R.R., Leatherwood, S., Stone, G.S. and Eldredge, L.G. (1999) Marine Mammals in the Area Served by the South Pacific Regional Environment Programme (SPREP). South Pacific Regional Environment Programme, Apia, Samoa. 48 pp.

Randall, J.E. (1996) *Shore Fishes of Hawai'i*. Natural World Press, Vida, Oregon. 216 pp.

Sanderson, M. (ed.) (1993) *Prevailing Trade Winds: Climate and Weather in Hawai'i*. Univ. Hawaii Press, Honolulu.

Stearns, H.T. (1966) *Geology of the State of Hawaii*. Pacific Books, Palo Alto, California. 266 pp.

Steadman, D.W. (1995) Prehistoric Extinctions of Pacific Island Birds: Biodiversity Meets Zooarchaeology. *Science* **267**, 1123–1131.

UNEP/IUCN (1988) *Coral Reefs of the World. Vol. 3: Central and Western Pacific*. UNEP Regional Seas Directories and Bibliographies. IUCN Gland, Switzerland and Cambridge, U.K./UNEP Nairobi, Kenya. xlix + 329 pp. + 30 maps. [Hawaii pp. 143–165]

U.S. Department of the Interior (1999) Protecting the Nation's Coral Reefs. U.S. Department of the Interior, Washington D.C. 10 pp.

Walsh, G.E. (1967) An Ecological Study of a Hawaiian Mangrove Swamp. In *Estuaries*, ed. G.H. Lauff. American Association for the Advancement of Science, Washington D.C., pp. 420–431.

Wilkinson C.W. (ed.) (1998) *Status of Coral Reefs of the World: 1998*. Australian Institute of Marine Science, Cape Ferguson, Queensland. 184 pp. [Hawai'i: pp. 97–98].

Woodward, P.W. (1972) The Natural History of Kure Atoll, Northwestern Hawaiian Islands. *Atoll Research Bulletin* **164**, 1–318.

THE AUTHOR

James E. Maragos
U.S. Fish and Wildlife Service,
Pacific Islands Ecoregion,
300 Ala Moana Blvd.,
Box 50167,
Honolulu, HI 96850, U.S.A.

Chapter 106

FRENCH POLYNESIA

Pat Hutchings and Bernard Salvat

This chapter summarises the available information on the physical and biological characteristics of the coral reefs and associated areas in French Polynesia. Reefs vary from almost pristine to heavily degraded, although the biological diversity of these relatively isolated reefs is low compared to reefs in the Western Pacific. The chapter then discusses the various anthropogenic factors impacting these reefs and how these are distributed within the region. Commercial uses of the coral reef resources are reviewed, as are the various management options being developed to manage and conserve them. However it is apparent that with a rapidly growing population, far more political will is needed in order to develop and manage these reefs, which are probably the best studied in the area.

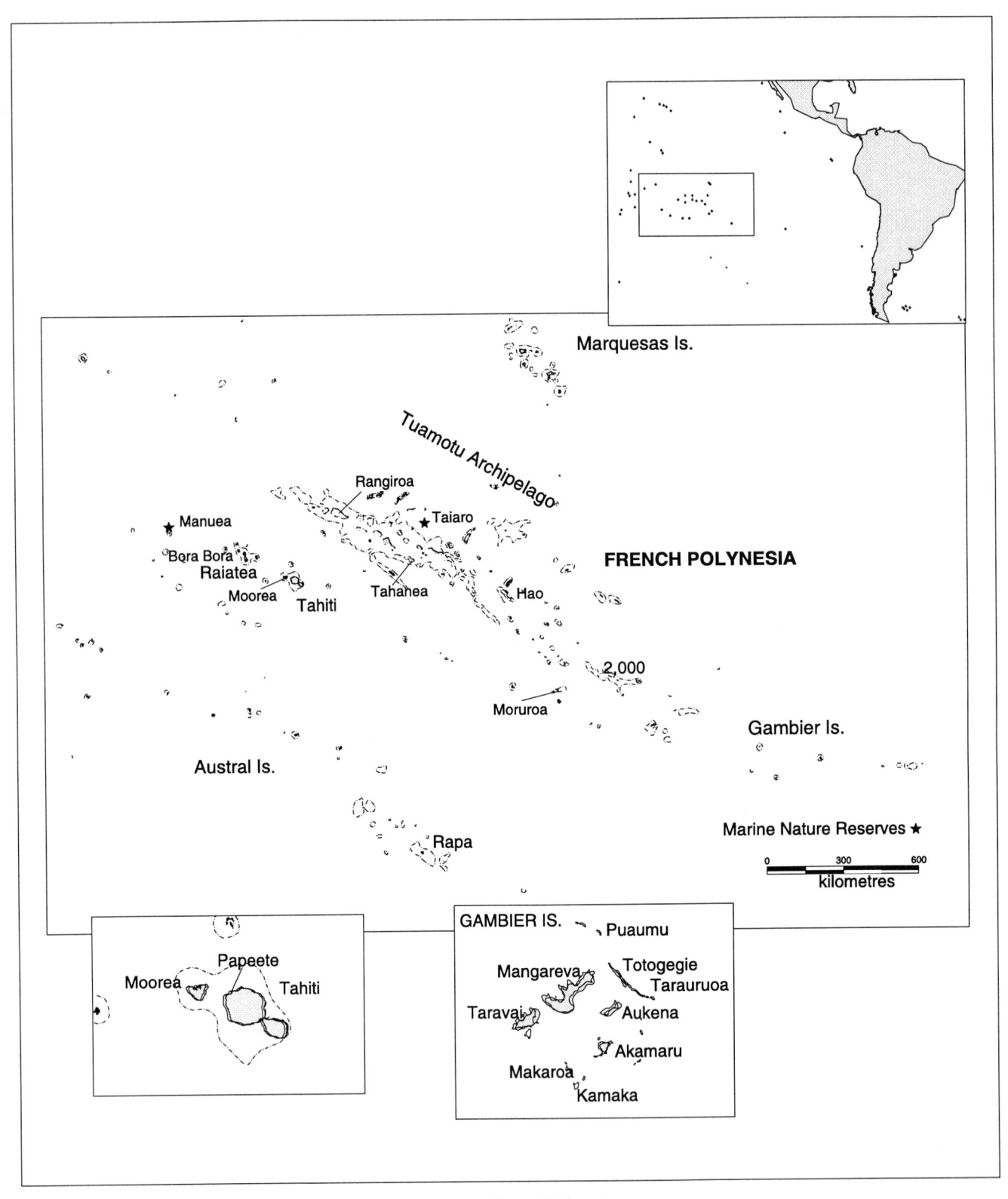

Fig. 1. Map of French Polynesia.

THE DEFINED REGION

French Polynesia lies in the South Pacific and extends over about 2,500,000 km² of ocean from 134°28 W (Temoe) to 154°40'W (Manuae or Scilly) and from 7°50'S (Moto One) to 27°36'S (Rapa) (Fig. 1). Emergent lands total 3430 km², with about 12,800 km² of reef formation totalling more than 2000 km in length (Gabrié, 1998) and there is about 7000 km² of lagoon (Wells, 1988), giving a land-to-sea ratio of 0.16%. In addition, the region is surrounded by an Exclusive Economic Zone (EEZ) of 5,500,000 km².

French Polynesia is part of the larger Polynesian Province which covers 8733 km² of land and 13,200,000 km² of sea (Wauthy, 1986), giving a land-to-sea ratio of 0.066%. Polynesia, together with Micronesia and Melanesia, constitute Oceania. All these areas, while characterised by numerous small islands, have very small land masses compared to their surrounding waters (Bleakely, 1995). Thus the waters are of great significance to the people of Oceania both in terms of culture and economies. The marine environments of French Polynesia are probably the best studied of this Province.

French Polynesia consists of 118 islands, of which 84 are atolls, most of the remainder being high volcanic islands, many of which are mountainous with inaccessible interiors. The islands form five archipelagoes (Society, Tuamotu, Gambier, Marquesas and Austral), each situated along a southeast to northwest axis. This is due to their origin from a hot spot on the sea floor towards the northwest. Thus the age of the islands increases from the southeast to the northwest. Most of the Tuamotu islands are atolls, due to subsidence of high volcanic islands on a plateau formed on the East Pacific ridge, which is still separating. Tahiti the largest, highest island in French Polynesia (1042 km², 2241 m) is in the middle of the South Pacific, 6100 km from Sydney, 8000 km from Santiago, Chile, 6400 km from Los Angeles and 9500 km from Tokyo.

A classification of the islands based upon geomorphological characteristics is given by Salvat (1985). The islands vary greatly in size with Rangiroa the largest, 1800 km², and some of the smallest being only 2 km². Additional information is given by Wells (1988) together with many relevant references to the geomorphology, fauna and flora (see also Gabrié and Salvat, 1985). Reports on the status of coral reefs in French Polynesia have been published in which many references are given (Hutchings et al., 1994; Payri and Bourdelin, 1997). This chapter attempts to update these, as well as commenting on future environmental developments for the area. An overview of French Polynesian reefs was given by Gabrié and Salvat (1985) and a comprehensive bibliography on the biology–ecology of these reefs up until 1990 was collated by Salvat (1991).

POPULATION AND POLITICS

French Polynesia became a French colony in 1840, and was declared a French Overseas Territory in July 1972, with New Caledonia, and Wallis and Futuna, with Papeete as its capital. The Statute for Self Government was enacted in September 1984, and granted the Territory of French Polynesia complete responsibility for its environmental protection policy (Fontaine, 1993; Hutchings et al., 1994). The Territory is working towards more autonomy and is becoming more responsible for international matters.

The population of French Polynesia was estimated in 1996 to be about 220,000, which is, however, not evenly distributed. Of the 118 islands only 76 are inhabited and of these Tahiti and Moorea in the Society Islands have 75% of the entire population. In fact, 91% of the entire population of the Territory lives in the large urban zone of Papeete, in an area about 100 km² representing about 3% of the total land mass of the entire Territory. The current rate of population increase is 1.9% annually (Gabrié, 1998) and 43% of the population is under 20 years of age. Basically, the entire population of French Polynesia lives along the narrow coastal fringes of the atolls and the high volcanic islands. So while the population of French Polynesia has always traditionally used the coastal resources, with increasing urbanisation and population increases over the past 20–30 years, these uses have increased and with it consequential impacts on the marine environment. A similar pattern is occurring throughout the South Pacific as people move to towns in search of employment.

Since the end of nuclear testing in 1991, which provided a lot of economic support to the Territory from France, a development contract has been agreed which maintains the flow of funds from France and which should allow the Territory to become more independent.

SEASONALITY, CURRENTS, AND NATURAL ENVIRONMENTAL VARIABLES

Weather

The region experiences two seasons—a warm rainy season from November to April and a relatively cool dry season from May to October. Eastern trade winds predominate from October to March. April to June tend to be characterised by long calm periods, occasionally broken by cyclones, which generally arrive from the northeast and northwest. However within this basic pattern, there are significant differences between archipelagoes (Teissier, 1969; SPREP, 1980). Cyclones have typically been rare in the past, averaging from one per century to the north of the Marquesas, between one and three per century from the Marquesas to the region north of the Tuamotu to the Gambiers and one every 2–3 years in the Austral group (Gabrié and Salvat, 1985). However this pattern changed during 1982–1983, when six cyclones occurred, related to the abnormal El Niño of that period. These cyclones caused considerable damage and, at Tikehau, destruction of coral communities occurred to depths of 75 m. Other cyclones occurred during 1991 and 1997, but the most devastating occurred early this century. Accompanying at least one of the cyclones in 1983, was an abnormally low tide, 20–25 cm below low water during the middle of the day which exacerbated the damage done to coral reef communities by the cyclone (Rougerie and Wauthy, 1983; Harrison and Cane 1984). This was particularly evident on Moorea in the Society Islands.

Ocean Waters

Sea surface temperatures decrease southward and eastward to Rapa where the minimum temperature for coral growth is found. Summer temperatures are 26–30°C and winter temperatures are 20–22°C (Wells, 1988). Sea surface water salinity throughout the area varies from normal salinities to salinities above 36‰, associated with subtropical gyres east of French Polynesia where evaporation is intense and exceeds precipitation (Donguy and Henin, 1978).

A synopsis of the circulation patterns in the Polynesian South Pacific is given by Rougerie and Rancher (1994, 1995) (Fig. 2). Data collected around atolls and on barrier reefs indicate that oceanic conditions (i.e. oligotrophic mixed layer) dominate all around these structures, in spite of periodic thermal oscillations related to the tidal regime. The vortex and eddies created around the islands are insufficient to modify the vertical stratification or to pump up nutrients from the nutricline, which explains the absence of local upwelling or "island mass effect".

Tides are semi-diurnal with an amplitude rarely exceeding 40 cm.

Reef and Lagoonal Waters

Reef and lagoonal waters have been studied in relatively few French Polynesian lagoons. Detailed studies have been undertaken all around Moorea and in the urban area of Tahiti (both high islands) and in some atolls, primarily Takapoto, the site of a pearl fishery, and Tikehau and Mataiva, all in the Tuamoto Archipelago.

The waters surrounding Tikehau atoll and lagoonal areas have been studied by Charpy-Roubaud and Charpy (1994). This atoll is open and there is a large pass through which water exchange occurs. The average nutrient concentrations of lagoonal waters are in the same order of magnitude as reported for other coral reef areas (Crossland, 1983). Nutrient concentrations vary over time, although they are homogenous throughout the lagoon at a specific time (Charpy et al., 1997; Dufour and Harmelin-Vivien, 1997). Oceanic water as it crosses the reef flat or enters the lagoon through the pass, loses NO_2 and NO_3 and becomes enriched in NH_4 and organic nitrogen. The total nitrogen concentration of lagoonal waters is about twice as high as surrounding oceanic waters. In summary, high primary production over the reef and in the lagoon results from an input of nitrate and phosphate from oceanic waters, and by

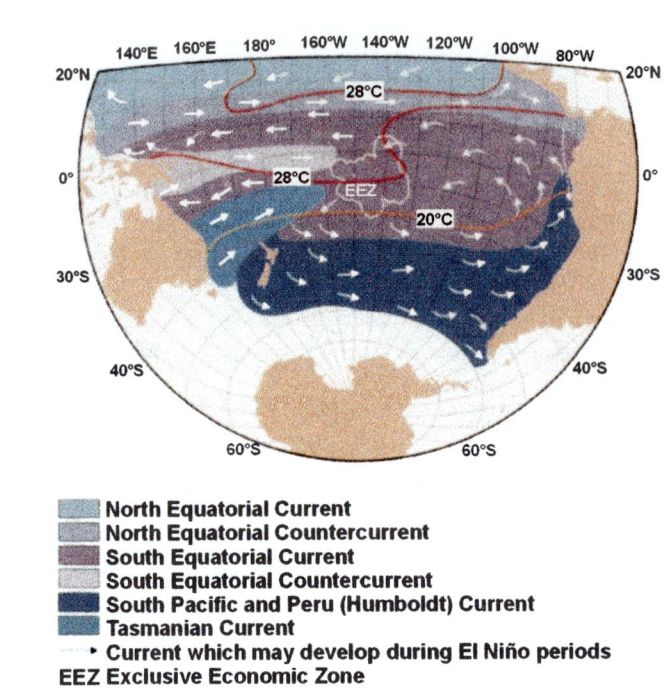

Fig. 2. Synopsis of the circulation patterns in the South Pacific after Rougerie and Rancher (1995) but taken from Gout et al. (1997).

nitrogen fixation by cyanobacteria on reef substrates and from the mineralisation of organic compounds in the lagoon. They discuss the trophic structure and the productivity of the lagoonal communities which are dominated by macrophyte production.

A recent study in Moorea has measured directly the export of carbon carbonate from the coral reef ecosystem in Moorea (Delasalle et al., 1998), and estimated that 47% and 21% of the total organic and inorganic carbon produced by the reef is exported through the pass during normal atmospheric oceanic conditions. Obviously, in order to complete the carbon model for the reef, a precise estimation of the inputs from both oceanic water and from land-based sources needs to be made.

Water Residence Time in Lagoons and Their Planktonic Communities

Delesalle (1990) synthesised the available information on lagoonal waters in French Polynesia, and the residence time of lagoonal waters varies from 0.25 to 10 h for lagoons of high islands, and from one week to 4.4 years for lagoonal atolls. Closed atolls (Fig. 3) show longer water exchange times. The longest rate of exchange was for the closed atoll of Takapoto (Sournia and Ricard, 1976). Studies of 11 lagoonal waters from high islands and atolls in the Pacific, showed that a linear relationship exists between the residence time (when less than 50 days) and phytoplankton biomass (Delesalle and Sournia, 1992). For residence time greater than 50 days, the data show a much lower phytoplankton biomass which is the case in Takapoto and Tikehau atolls in the Tuamotu. Obviously the rate of exchange is dependent upon the degree of openness of the lagoon, and this will influence the type of lagoonal phytoplankton communities present.

Recently, Dufour and Berland (1999) have investigated the nutrient control of phytoplankton biomass in the atolls within the Tuamoto archipelago. Although the atolls within this archipelago are located in an oligotrophic oceanic area, some of their lagoons have experienced exceptionally harmful phytoplankton blooms during the past 30 years and these were sometimes followed by massive fish, coral and cultivated pearl oyster mortality (Harris and Fichez, 1995). By adding specific nutrients, Dufour and Berland (1999) were able to show that atoll lagoons and ocean waters in the Tuamoto archipelago are highly susceptible to eutrophication with the addition of both nitrogen and phosphorus. They suggest that inputs of nitrogen, phosphorus and silicate into these atoll lagoons should be considered when developing management plans.

A comprehensive model of the planktonic food web has been developed for Takapoto atoll where extensive farming for the black pearl oyster (*Pinctada margaritifera*) occurs. The oysters are grown on suspended ropes in the water column, and feed on the lagoonal plankton. A knowledge of the pelagic food web is essential in determining the standing stock of oysters which can be supported (Niquil et al., 1998). Takapoto is a closed lagoon which is isolated from the ocean by an almost continuous reef rim. Although some exchange occurs through three channels ("hoa"), it has been estimated that the renewal time of the lagoonal water is 2.5 years for 50% renewal and 18 years for 99% (Magnier and Wauthy, 1976). Because this lagoon is almost completely closed, the planktonic system is sensitive to local variations in the lagoon, which would in an open atoll be buffered by access to the relatively unchanging surrounding oceanic waters (Niquil et al., 1999). This sensitivity can lead to algal blooms (Harris and Fichez, 1995).

Water Quality

A study monitoring water quality has been conducted in Papeete Harbour and around Tahiti since 1980 (Cariès, 1991). High levels of various pollutants occur in the urban areas of Papeete, related to the discharge of two rivers, and industries such as the power station and the brewery, with the highest levels occurring in the harbour. Results from water quality studies carried out in Papeete Harbour from between 1990 and 1995 are given by Adjeroud et al. (submitted). Water quality is monitored once a year all around Tahiti at 73 stations (39 in urban and 34 in rural zones). In 1998, 74% of urban sites exhibited good and mid-quality water and 26% exhibited poor or occasionally polluted water quality; the values for rural zones were 71 and 29% respectively. The reasons for the poor water quality in the rural zone are the increasing number of piggeries in the Territory and the disposal of their wastes into rivers as well as domestic sewage.

According to Salvat (1982) and Poli et al. (1983), the central part of the lagoon around Papeete which consists of soft sediment has no live coral colonies remaining as heavy siltation has occurred. This decline they attribute to the concreting of the barrier reef between the Papeete and Taaone passes, declining water quality, pollution from ship traffic and hydrocarbon spillage, discharge of urban sewage, sedimentation caused by increasing land run-off and dredging of lagoonal sediments. Although a recent survey by Adjeroud et al. (submitted) has shown that 24 coral species are present and 104 species of fish, and they suggest that strong currents replace the entire volume of water in the inner harbour in 4–12 hours, which allows these communities to survive. It therefore appears that the survival of coral communities around Papeete is highly dependent on local conditions and perhaps a comprehensive survey needs to be undertaken to see if the conditions reported by Salvat in 1982 have changed.

MAJOR SHALLOW WATER MARINE AND COASTAL HABITATS

Almost all of the 118 islands of French Polynesia have fringing coral reefs, variously developed lagoons and an outer barrier reef (Figs. 4 and 5). Exceptions are found in the Marquesas islands, Mehetia (a recent Society island) and Rapa in the Austral archipelago. While a large number of marine studies have been conducted, not all the islands have been visited by scientists, and Salvat (1982) estimates that around 70 islands have been visited by scientists from the Antenne de Musée National d'Histoire Naturelle et de L'Ecole Pratique des Hautes Etudes (Antennae Museum-EPHE) based in Moorea, since it was established in 1971 and publications have appeared on about 30 of these. However it should be stressed that most information is available on the reefs of Moorea and Takapoto on which was launched in 1971, a research project within the Man and Biosphere programme of UNESCO. In addition the reefs of Tahiti, and the atolls of Tikehau and Mataiva, are well studied for a variety of reasons. Mururoa and Fangataufa atolls have been extensively studied as a result of the nuclear testing which was carried out on these atolls from 1967 to 1991 (Gout et al., 1997). Immediate and long-term effects on coral reef gastropod assemblages on the reef flat have been studied on Fangataufa atoll after their exposure to atmospheric nuclear tests (Salvat et al., 1995; Lanctot et al., 1997). Results suggest that, even if the densities of several species dropped immediately after the nuclear tests, most species were able to quickly recolonise the perturbed reefs. The least known areas in the Territory are the Gambier, Marquesas and the Austral Archipelagos (Gabrié, 1998) and it is these groups which contain most of the 50 odd islands, the reefs of which have never been investigated. A bibliography of recent theses and dissertations carried out in the South Pacific has been collated by Angleviel et al. (1991),

Fig. 3. Anu Anurunga atoll, Tuamotu archipelago. An enclosed atoll (without pass) of about 7 km². A total of 84 atolls are in French Polynesia, more than half are enclosed (B. Salvat).

Fig. 6. Algal ridge, Fangataufa atoll, Tuamotu archipelago. These algal ridges around all atolls in French Polynesia have greatest development where the swell is heavy. (B. Salvat)

Fig. 4. Reef complex on Moorea island, Society archipelago. All Society, Gambier and most of the Austral Islands are surrounded by reefs which generally are no more than about 100 m wide. (C. Rives).

Fig. 7. In lagoons, either of atolls or around high volcanic islands, the coral reef community is dominated by the genus *Porites*. (B. Salvat)

Fig. 5. Atoll rim of Marutea du Sud, Tuamotu archipelago. Islets are covered with vegetation (mainly Coconut trees). Channels ("hoa") allow ocean waters to come into the lagoon. (B. Salvat).

Fig. 8. On the outer slope, coral community cover is normally about 50 to 60% and is well developed in some atolls down to 90 m. These communities suffered from bleaching events in 1991, 1994 and 1998. (R. Hayes)

and those dealing specifically with French Polynesia are given by Salvat (1991).

The importance scientifically of French Polynesia is the high geomorphological diversity present within the region, and some of the atolls are often used to illustrate atoll reef formation. In addition the narrow width of the reefal system between the outer slope and the land, often less than 1 km typically around the high volcanic islands, provides unique opportunities for research and they have been the focus of several global reef research programmes.

The distribution of coral reef communities around the high island of Moorea has recently been studied by Adjeroud (1997), and he found two major gradients, one along the bays and one from the fringing reef to the outer reef slope. Although he identified factors such as depth, sand and algal coverage which have been commonly identified elsewhere, he also noted that the high densities of sea urchins also have had a major effect on the reefs, a factor absent on the Great Barrier Reef in Australia (Done et al., 1991). The low diversity of the major reefal organisms such as corals, molluscs, sponges, echinoderms and macroalgae present in Moorea is characteristic of French Polynesian reefs and is strongly correlated with regional diversity (Caley and Schulter, 1997). With a total of 1159 species of molluscs, 346 of algae, 168 of corals and 30 echinoderms recorded from French Polynesia, this represents a low diversity area of the Pacific Province (Richard, 1985) (recently updated by Gabrié (1998) who lists about 1500 species of molluscs, 170 corals (representing 30 genera) and 800 of fish). In comparison, the Great Barrier Reef has 350 species of corals (representing 110 genera), 4000 species of molluscs and over 1500 species of fish (Veron, 1993). Two main reasons suggested for this are the distance from the 'colonisation source' which is located in the maximum diversity area defined by the Ryukyu Islands, Indonesia and New Guinea (Rosen, 1984). The distance is too great for larval dispersal (Scheltema, 1986). Salvat (1967) noted that most of the molluscs present in French Polynesia also occur in the Western Pacific and have larvae that are adapted for long-distance dispersal. The second reason suggested is the small size of the coral reef ecosystems in French Polynesia, and that there is no equilibrium between rates of immigration and extinction since relatively few species are present (MacArthur and Wilson, 1967). In general immigration is generally higher on a large island than on a small island, and generally small areas provide fewer different habitats compared to larger land masses.

Although Adjeroud and Salvat (1996) have shown that local conditions such as salinity can influence the species composition in the innermost parts of the bays in Moorea, presumably similar factors operate on reefs elsewhere in French Polynesia.

The soft-bottom benthic infaunal communities have recently been investigated around the island of Tahiti by Frouin and Hutchings (in review) as part of a study to investigate the effects of terrigenous inputs on these communities (Frouin, in press). Other studies of these communities have included those on Moorea (Thomassin et al., 1982) and in the atoll lagoon of Mururoa (Salvat and Renaud Mornant, 1969). A characteristic of the reefal communities is the low level of endemism present within the Territory, although there are differences between the Archipelagos, with the highest rates of endemism being recorded in the Marquesas and the Gambier Archipelagos (Gabrié, 1998).

Interesting biological questions can be asked using some of the diverse geomorphological features present in French Polynesia. An example of this, investigated during the Franco-Australian expedition to Taiaro Atoll in the Tuamotu group in 1994, was whether the fish populations in this slightly uplifted atoll (which is hypersaline) were self-sustaining despite being in limited contact with the ocean. While the lagoon contained few oceanic species of zooplankton, confirming its general isolation, some species of fish may depend upon infrequent colonisation from the ocean during storm conditions (Galzin et al., 1998).

OFFSHORE SYSTEMS

Deep-water habitats below 80 m on the outer slope have been investigated only once with a submersible to depths of 1100 m off Tahiti. A subvertical cliff of dead corals from 100 to 200 m depth revealed a community which is completely distinct from the shallow reefal community and differs completely in species composition. Gorgonians, as well as actinarians, antipatharians, and crinoids, which are almost completely absent on shallow coral reefs in French Polynesia, were abundant (Salvat et al., 1985).

EXPLOITATION OF RESOURCES

Fisheries and Pearl Culture

Fishing in French Polynesia consists of both subsistence and commercial fisheries, and includes lagoonal species and the pelagic tuna. Landings of lagoonal species are about 4000 tons annually, consisting of about 3500 tons from subsistence fishing with the rest being sold. Of the fish sold, 100 tons is caught in Rangiroa and Tikehau in the Tuamotu where this commercial fisheries is the sole source of income for many communities. About 2300 tons of tuna are caught for the domestic market and about 3300 tons were exported in 1998. Tuna exported are caught mainly by the Koreans in the Exclusive Economic Zone. However there is a great variation in landings between years: in 1991, 4730 t were landed but this dropped to 1700 t in 1992 (Hutchings et al., 1994).

A comprehensive study on the fisheries in Tikehau atoll was carried out by Caillart et al. (1994), where a major fishing industry exists with large quantities being exported to

Papeete for sale. While over 276 species of fish were collected in the lagoon and outside the atoll on the reef slope, the fishing industry is based on about 14 species, none of which are dominant. About 200 metric tons of fish were caught annually, and an additional 40 metric tons were taken for subsistence with another 40 tons being caught by occasional fishermen for commercial purposes, so that 280 tons was landed in 1993. This equates to an annual harvest of 0.7 t km^{-2}. The fish were caught by bottom-fixed fish traps, all located in the pass, and were primarily carnivorous species. While additional species could be caught by diversifying the types of fishing gear employed and fishing elsewhere in the lagoon, Caillart et al. (1994) were reluctant to recommend this until a far more detailed study was undertaken, focusing on target species. Little is known as to the source of new recruits to this fishery and how much exchange occurs between atolls in this region.

Aquaculture

As of 1995, there were two hatcheries in Tahiti for producing larval prawns and undertaking research on raising the gastropod *Turbo marmoratus*, one run by EVAAM (Local Government) and the other run by IFREMER (French Government). Four farms in Moorea and Tahiti raise prawns and produce about 21–30 tonnes/ha/year and this is sufficient for the local market both as fresh and frozen product. It seems unlikely that the acreage available for raising prawns can be increased as more suitable land is not available. A similar amount of land is used for rearing freshwater prawns and again there are few possibilities for expansion, in 1995 about 1 ton/ha/year was being produced in Tahiti and Moorea.

Some research is being undertaken on raising barramundi, and other tropical fin species. Possibilities of using cages to rear barramundi are being explored around Tahiti, and currently larvae are being reared in earthen ponds. About 30 people are employed in such activities.

The trochus shell *Trochus niloticus* was introduced from Vanuata to Tahiti in 1957, and the species was introduced to many other islands, both volcanic and atolls, all through French Polynesia, and it now occurs on more than 70 of the island coral reef ecosystems. The green snail *Turbo marmoratus* was introduced from Vanuata in 1967, and reefs in Tahiti and the Outer Islands atolls were stocked. Two factories were established in 1992 for cutting the shells of *Turbo marmoratus* and *Trochus* for local craft jewellery, most of which is exported. The flesh of these two species is also eaten. It appears that it provides supplementary income for fishermen and Outer Islands people. However there are still problems with ensuring an adequate supply of these two mollusc species and maintaining their populations.

Some interest has been raised in the feasibility of establishing a beche-de-mer industry and raising clams, but to date no commercial operations exist.

In French Polynesia, an extensive pearl fishing industry exists, using the black-lipped oyster *Pinctada margaritifera* which is used to culture large black pearls. Currently more than 98% of all black pearls are cultured in French Polynesia. The production of these pearls has increased over 10-fold during the past 10 years and in 1997 was responsible for 95% of export revenue from the Territory (130 million US$). In addition, it is the main source of income for over 5000 people in about 100 cooperatives on about 50 island atolls, as of 1995. Since the time French Polynesia became a French Protectorate in 1842, the nacreous layer inside oyster shells has been used to make jewellery and buttons, and any natural pearls found were also used. However the first attempts to culture pearls were made in 1961 in the lagoon of Bora Bora, and by 1972 an export industry had developed, although major expansion of the industry did not occur until around 1983. While the number of permits given for pearl culture is known (5137 at the end of 1997), the impact of these large numbers of oysters suspended in lagoons, and how they modify the phytoplankton regimes of these lagoons is poorly understood. The development of a model of the phytoplankton communities in the lagoon at Takapoto, one of the main pearl-growing lagoons, is the first step in this process (Niquil et al., 1998). Also the development of extensive racks of oysters effectively excludes many other activities in the lagoon, as well as removing larvae of other organisms through predation. Currently, while there are regulations restricting the transport of oysters from one lagoon to another, and with it the possibility of transfer of pathogens, these are not enforced. Mass mortality of oysters and with it the eutrophication of the lagoon, albeit temporary, has been reported on several occasions (Gabrié, 1998).

Since the mass mortality of the oysters in Takapoto in 1985, considerable research into the oyster has been undertaken to ensure the long-term viability of these commercial operations. The development of an extensive pearl industry within an atoll lagoon impacts the lagoonal fisheries, agriculture and perhaps tourism, but it does provide rewarding employment in the atolls and helps to prevent migration of people to the already overcrowded island of Tahiti. The industry has developed a unique grading system for "Tahitian Culture Pearls" and is limiting the production of low-quality pearls.

EFFECTS OF HUMAN ACTIVITIES

Savoie (1989a,b) has listed 80 environmental publications related to development which have appeared in the previous 20 years, of which 61 are related to reef-lagoonal systems. Aubanel and Salvat (1990) analysed these studies and found that from 1981–1989, studies have increased 10-fold on the previous decade, and many of these were related to a regular monitoring program of Papeete Bay associated with the port development.

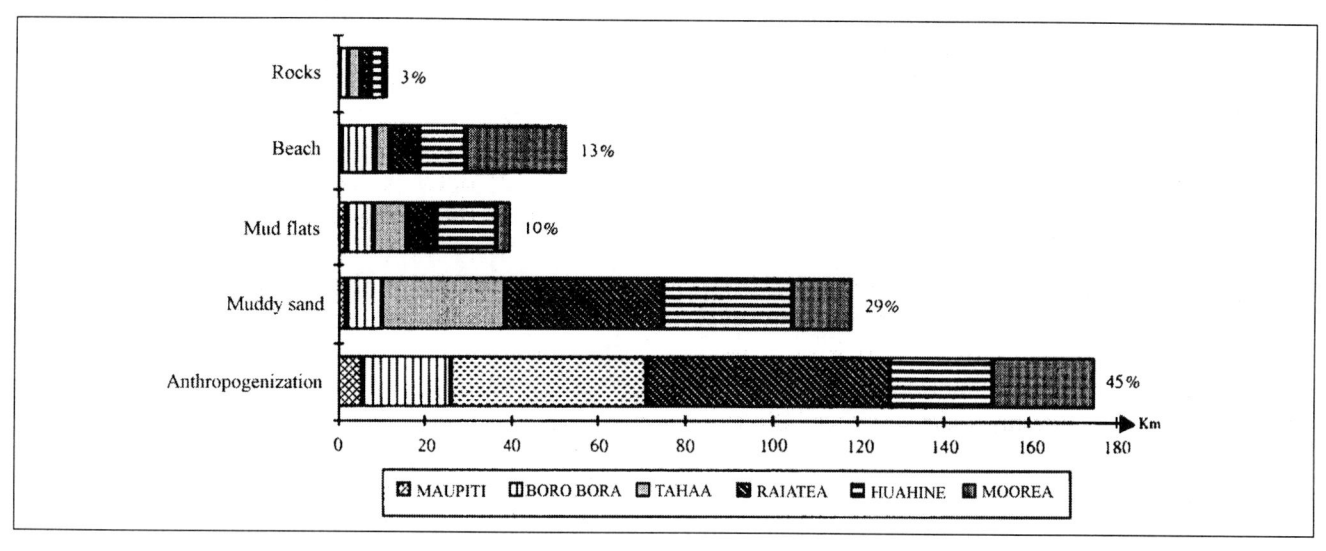

Fig. 9. Total lengths of shore line categories in high islands of the Society Archipelago (from Aubanel et al., 1999)

Modification of Coastlines

Extensive modification of the coastline has occurred in the urban areas of Tahiti and Moorea, often in areas where resorts are located. From the 1950s onwards, a lot of dredging occurred around Tahiti and Moorea, for projects such as the construction of the harbour and the airport, however in 1984 such dredging activities were banned. Nevertheless dredging and filling still occurs and with the high cost of taking material from the mountains (with other environmental impacts) it may continue to occur. Up to now, no rehabilitation of these dredged areas on fringing reefs has occurred and often such areas are then reclaimed. A recent study (Aubanel et al., 1999) on the extent of shoreline modification in the Society islands has shown that almost 50% has been modified by human activities (Fig. 9).

Far less modification has occurred in the atolls, except perhaps for construction of airport runways which are present on most atolls, but in most cases this has not impacted on the coral reef ecosystem.

Land Run-off

Extensive land run-off occurs from the high islands of Tahiti and Moorea during the rainy season and Demougeot (1989) has estimated that about 1000 t of sediment are washed into the lagoon of Tahiti annually. Frouin (in press) has shown that this sediment is transported across Tahiti lagoon and some even reaches the barrier reef. Although developers are increasingly clearing steep hill sides on the high islands of Tahiti and Moorea, for residential and agricultural purposes, there appears to be little action taken to minimise soil erosion, so that as the vegetation is cleared, plumes of water filled with sediment can be seen flowing across the reef after heavy rains. Such sediment impacts both the coral and soft-bottom communities (Frouin, in press).

The sediment may also contain pesticides and fertilisers. Levels of pesticides in lagoonal waters have been measured several times around Tahiti and levels vary both in the sediments, water column and in sessile organisms such as mussels around the island (see Hutchings et al., 1994, for details).

Obviously land run-off is almost non-existent in the atolls, but in lagoons which are poorly flushed, run-off from terrestrial activities (including rubbish tips), should be carefully controlled to prevent water chemistry changes in the lagoon, and with it the potential for dangerous algal blooms.

Blooms of the toxic benthic dinoflagellate *Gambierdiscus toxicus* have been recorded from time to time in different places in French Polynesia but mainly on the inshore reefs in the Gambier Islands (Bagnis et al., 1990). They suggest that the occurrence of blooms in various parts of the Archipelago can be correlated with several natural disturbances (e.g. cyclones, high coral mortality) and man-made disturbances such as the construction of the airport that damaged the inshore reefs from 1965 to 1974. The toxins of this dinoflagellate are absorbed into the flesh of herbivorous fish as they feed on these dinoflagellates and, if eaten by humans, cause ciguatera. The decline in levels of the dinoflagellate since 1980 appears to be related to an increasing cover of live coral on these reefs.

Disposal of Sewage

Disposal of sewage in French Polynesia varies from individual septic tanks to private sewage treatment works which may be used by large apartment blocks or hotels, or disposing of untreated sewage into the lagoon or creek. Currently there are no large-scale sewage treatment works in any urban area, although there are plans for building one in Papeete.

In addition to human health concerns, water quality of creeks and of groundwater entering the lagoons impacts the coral reef communities. For example, increased rates of nutrients in the water column around the airport at Faaa, Tahiti, have allowed high levels of algal cover on the reefs to develop, which encourages high populations of sea urchins, which feed on the algae by scraping dead coral reef substrates. Previously, populations of the sea urchins were controlled by fish, but widespread overfishing in the area has allowed large populations of urchins to develop. Extensive grazing by the sea urchins prevents any new coral recruits from settling and so the reef is being eroded at much higher rates than those at which it is growing. Net rates of loss by grazing are estimated at 6.87 ± 2.16 kg m^{-2} yr^{-1} for this site (Pari et al., 1998). On other reefs studied in the region not subjected to eutrophication, net rates varied from 0.89 ± 1.24 kg m^{-2} yr^{-1} to 0.33 ± 0.20 kg m^{-2} yr^{-1}.

Mining

In the Tuamotus, the island Makatea was intensively mined for phosphate from 1917–1966, and was the most populated island in the group (3000), but now the population is only about 40 (Montaggioni, 1985). The plateau-like surface of the island at an average elevation of 70 m is deeply dissected by a karstic system after the extraction of phosphate, but the fringing reefs all around the islands appear to still be in good health. There is a potential for phosphate extraction in Mataiva atoll but so far this has not occurred because of opposition to mining by the local residents.

Bleaching

Significant mass bleaching events, where corals lose their symbiotic zooxanthellae and may then die, have been recorded in French Polynesia in 1991 (Salvat, 1991, 1992), in 1994 on Moorea (Hoegh-Guldberg and Salvat, 1995) and on the barrier reef in Tahiti (Fagerstrom and Rougerie, 1994), resulting in extensive coral death. It is estimated that 20% of the bleached corals subsequently died on the outer slopes of Moorea during the 1991 event. In April 1994, extensive bleaching occurred again at a wide variety of sites along the outer reef slope of Moorea, although it was less severe than in 1991. Bleaching varied as a function of depth and included a wide variety of species. Species of *Acropora* were most subjected to bleaching with over 89–100% of colonies completely bleached and *Porites* spp. the least with 12.9–42.5% of all colonies partly bleached. The extent of the 1994 bleaching was similar to that observed in 1991. Temperatures recorded during April 1994 were 29.5–30°C, about 1.5°C above that normally expected for this time of the year, and preceded the onset of bleaching by about 2–3 weeks,

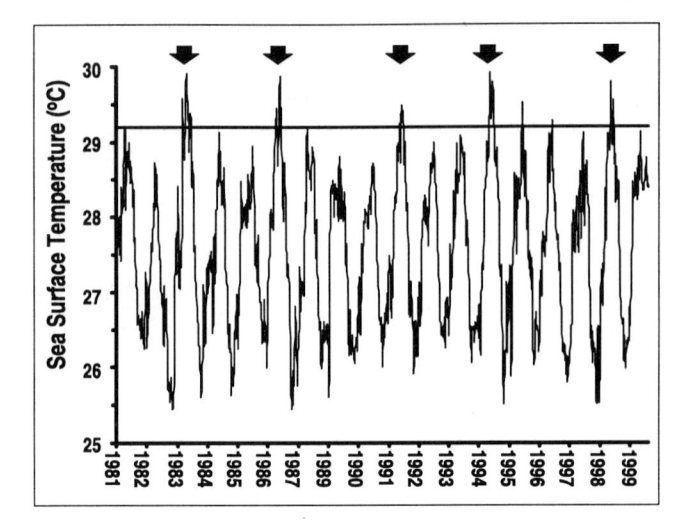

Fig. 10. Weekly sea surface temperature data for Tahiti (149.5°W 17.5°S). Arrows indicate bleaching events reported in the literature. Horizontal line indicates the minimum temperature above which bleaching events occur (threshold temperature). IGOSS-nmc blended data courtesy of the Lamont-Doherty Climate Centre at Columbia University (from Hoegh-Guldberg, 1999).

which is consistent with the hypothesis that thermal anomalies are responsible for these beaching events (Fig. 10). Bleaching also occurred in 1998, but was variable among atolls and in locations on the atolls. Strong bleaching was seen in some areas like Takapoto, where 20% coral cover in 1994 was reduced to 12% after the 1998 bleaching. Severe bleaching also occurred on Rangiroa and Manihi with significant mortality (Wilkinson, 1998).

Thus the reefs of French Polynesia are being subjected to rising sea temperatures, as are most coral reefs worldwide. However the isolation of the reefs from other reefal systems, combined with great distances between individual reefs, may mean that coral recruitment could take many years. If the frequency of bleaching events increases, this may jeopardise the recovery of these reefs.

Nuclear Testing

Mururoa and Fangataufa atolls have been extensively studied as a result of the nuclear testing which was carried out on these atolls from 1967 to 1991 (Gout et al., 1997). These atolls were selected in part as they are isolated and it is therefore difficult to compare them with other atolls in the region. The tests have now been concluded, as have the biological studies, which are summarised by Gout et al. (1997). As detailed studies were not carried out prior to testing (except surveys of molluscs on Fangataufa reef flats, Salvat et al., 1995; Lanctot et al., 1997), it is difficult to really assess the impact these tests have had on the marine communities. Also, our knowledge of coral reefs has increased substantially over the past 30 years, as well as methods by which we investigate reefal processes especially with regard to sampling design, so that the initial studies carried out in the late 1960s seem very basic to us today.

Collection of Coral

No collecting of live coral for the aquarium industry occurs in the Provence.

Status of Reefs

A recent survey of coral reefs in French Polynesia (Gabrié, 1998) reports that 20% of fringing reefs in urban areas of Tahiti, primarily around Papeete, have been destroyed. The quality of reefs around the island varies considerably from healthy to considerably damaged reefs. In Moorea, reef quality has declined over the past 25 years on both the fringing and barrier reefs. On the Leeward Islands, at least 6% of the fringing reef has been totally destroyed, and 7–11% of coral reefs have been disturbed by coral extraction and reclamation operations. In Bora Bora, 44% of the reef is intact, with 56% under threat of damage and 75% of the fringing reefs moderately to severely disturbed. However, for most reefs such figures are not available, although one would presume that on many islands the reefs are intact except where damage from cyclones, bleaching or Crown-of-Thorns has occurred, from which they are expected to recover although slowly.

Tourism

Tourism represented 7% of the GDP for the Territory in 1998 and employs about 10% of the local working population. While 190,000 people visited the Territory in 1998, it is predicted that 350,000 will visit by 2005. The majority of tourists participate in lagoonal-based activities, and most hotels are in Tahiti, Moorea and Bora Bora, with 3500 beds available at the end of 1999. Construction of hotels often impacts on the coastal zone, with development of bungalows on wharfs extending out into the lagoon, or on reclaimed projections into the lagoon. While hotels are required to have their own sewage treatment works, excess water may be used to water the surrounding gardens, and as studies by Wolanski et al. (1993) in Moorea revealed, this water may still reach the lagoon via groundwater, leading to pollutants accumulating in the nearby lagoon. Such problems may be exacerbated on the atolls where there is a narrow fresh water lens from which drinking water is extracted; polluting this would have serious consequences for all the inhabitants of the atoll, where rainfall is often intermittent.

As tourism is predicted to increase over the next 10 years, careful planning is needed to ensure that the lagoonal coral reef is maintained in a healthy condition. This means minimising impacts of construction, the management of drinking water and the disposal of wastes. Certainly on the atolls, where the local population lives at a subsistence level, the construction of a five-star resort will have major impacts on both the terrestrial and marine environments. Ecotourism is developing in atolls such as Rangiroa as well as on many Society Islands.

Sea-level Rises

The effects of a sea-level change of 1 cm per year over a 30-year period on the high island of Moorea have been discussed by Salvat and Aubanel (1993). Assuming population increases of 1.6 and 3.8%, this would increase the island's population from 8000 to 14,000 or 35,000 by 2020, most of whom would be engaged in the tourist industry. While not actually calculating the loss of land area by rising sea levels, the growing population would increasingly be forced to occupy less area, increasingly forming an urbanised community. They suggest that the outer barrier which encircles the island and which protects the coastal fringe from wave action during storms would be maintained, but the quality of the coral reef communities would deteriorate with increasing human activities in the lagoonal area. While the area occupied by barrier reef would remain similar, the area occupied by the fringing reefs would increase. About 30% of the shore line of Moorea has already been modified by development, and with sea level rises would be expected to increase to up to 70%, in order to prevent flooding to surrounding areas.

EFFECTS OF CROWN-OF-THORNS

Several outbreaks of the Crown-of-Thorns starfish *Acanthaster planci* occurred in French Polynesia in the late 1970s and early 1980s (Faure, 1989) in Moorea which reduced the percentage of live coral cover in shallow water coral communities. Recovery of the coral communities from these outbreaks was further hindered by cyclone damage in the early 1980s and by the occurrence of some abnormally low tides during the day during the same period. This was probably related to the abnormal El Niño which occurred at this time (Pirazzoli, 1985). Since then, an occasional Crown-of-Thorns is seen on the reefs of Moorea, but no additional outbreaks have been reported.

PROTECTIVE MEASURES

Since 1984, with the enactment of the statute law of the French Polynesian autonomous Territory, all environmental issues have become the responsibility of the local Government. In 1985 the 'Délégation à l'Environnement' was formed and is currently responsible to the Minister of Environment. Other Ministers are responsible for seas, land management, research, tourism etc. The influence of the Department of Environment has varied between governments. Currently about 15 associations have formed a federation which is concerned with the protection of the environment; this federation plays an important role in decision-making processes and in the implementation of environmental measures. Increasingly Environmental Impact Assessments are being undertaken, although they are not always taken into account when making final decisions regarding that development.

Recently an agreement has been signed between France and French Polynesia that will ensure that about 12% of the annual budget of the Territory will be spent on a range of environmental projects. Two projects of relevance to coral reefs are programmes related to developing a sewage system for domestic and industrial wastes. The European Union will also contribute funds to these projects. Currently, French Polynesian inhabitants do not pay personal income tax, although this may have to change if the Territory wishes to have sufficient funds to protect and manage its environment.

Other positive developments include the development of Marine Area Management Plans which should shortly be adopted for Moorea and Bora Bora. These will help the resolution of conflicting activities based on lagoonal reefal resources such as fishing, tourism, diving and boating. Similarly some progress is being made towards improving land-use management practices and the disposal of wastes from piggeries leading to improved river quality. Reef restoration schemes have been initiated in Bora Bora (Porcher and Salvat, 1999). A French Polynesian Coastal Conservation Agency has been established which is gradually purchasing littoral areas for protection, however they have been allocated few resources with which to do this. Updating of environmental laws and regulations is occurring, and it is helping to ensure that they are compatible with French and European legislation and an 'Environmental Code of Practice' is being discussed.

In addition, international conventions which have been ratified by France are also applicable to French overseas territories. These include RAMSAR on wetlands (1971), World Heritage (1972), although no sites have been nominated in French Polynesia, CITES on endangered species (1973), Bonn on migratory species (which includes birds, turtles and marine mammals) (1979) and Rio on biodiversity (1992). There are also regional conventions, including those signed in 1976 in Apia and in 1996 in Noumea, regarding the protection of natural resources in the South Pacific. The South Pacific Regional Environmental Program based in Apia Samoa (SPREP) is assisting all the countries in the region to protect and manage their marine environmental resources. While many marine species are already protected in French Polynesia, including species of molluscs, crustaceans, fish, corals, black coral, turtles and birds, little enforcement of the relevant legislation appears to occur.

Marine nature reserves have been declared in the Scilly and Bellinghausen atolls which are both uninhabited. Taiaro Atoll, in the Tuamoto archipelago, a closed and privately owned atoll, has been declared a Biosphere Reserve, and there are plans to restructure and enlarge the reserve to encompass surrounding atolls. Another reserve is the islet of Motu One, a sand cay with a surrounding coral platform in the north of the Marquesas. However these reserves represent only about 1% of French Polynesian reefs, and obviously do not adequately cover the diversity of coral reef habitats present in the region.

Many monitoring networks on coral reefs in French Polynesia have been established. The most comprehensive is the one on Moorea, at north west Tiahura led by CRIOBE-EPHE which was established in 1970, and which collects quantitative data annually from the outer barrier and the outer slope (Augustin et al., 1997). In addition CRIOBE-EPHE undertakes a quantitative survey of coral cover on outer slopes of about 15 islands which are distributed throughout all the Archipelagoes. The data are transferred to Reef Base at ICLARM in Manilla as a contribution to the Global Coral Reef Monitoring Network (GCRMN) of the International Coral Reef Initiative (ICRI). At regular intervals, the "Delegation à l'Environnement" undertakes monitoring programs around Tahiti and the Office of Marine Resources undertakes surveys in all the atoll lagoons where pearl culture occurs, although the reports are not easily available.

CONCLUSIONS

In summary, while there is a considerable amount of information available on the diverse reefal systems present in French Polynesia, and they are probably the best known in the region, relatively little active management of these resources occurs. Yet the reefs and their resources are critical to the economy of French Polynesia by providing both food and income from fisheries, aquaculture and tourism. With a rapidly increasing population the threats to these resources are also increasing and there needs to be a more active program of reef and coastal zone management and a greater commitment by the government to implement these programs. Sufficient data exists on which to develop and implement these programs, but time is running out, especially for reefs associated with centres of population.

ACKNOWLEDGEMENTS

We would like to thank the following people who provided data, papers in press or directed us to relevant unpublished reports etc, Annie Aubanel, Rene Galzin, Mireille Harmelin-Vivien, Bruno Delesalle, Ove Hoegh-Guldberg, Chris Bleakeley, Clive Wilkinson, Bernard Thomassin, Mehdi Adjeroud, Serge Planes, Nathalie Niquil, Christian Monier, Mireille Peyrot–Clausade and Georges Remoissenet. Pat Hutchings would like to thank the staff of the Centre de Biologie Écologie Tropical Méditerranéenne for their hospitality during her visit in May 1999.

REFERENCES

Adjeroud, M. (1997) Factors influencing spatial patterns on coral reefs around Moorea, French Polynesia. *Marine Ecology Progress Series* **159**, 105–119.

Adjeroud, M. and Salvat, B. (1996). Spatial patterns on biodiversity of a fringing reef community along Opunohu Bay, Moorea, French Polynesia. *Bulletin Marine Science* **59**, 175–187.

Adjeroud, M., Planes, S. and Delesalle, B. (submitted) High diversity of coral and fish communities in a disturbed environment, Papeete harbour, Tahiti. Submitted to *Pacific Science*.

Angleviel, F., Charleaux, M. and Coppell, W.G. (Doumenge, J.-P. ed.) (1991) Le Pacifique Sud bibliographie des thèses et mémoires récents. *Collection Iles et Archipels no 13*.

Aubanel, A., Marquet, N., Colombani, J.M. and Salvat, B. (1999) Modifications of the shore line in the Society islands (French Polynesia). *Ocean Coastal Management* **42**, 419–438.

Aubanel, A. and Salvat, B. (1990) French Polynesian government and research institution involvement in coral reef environmental studies during the last 20 years. *Proceedings of the International Society for Reef Studies*, Noumea, pp. 9–16.

Augustin, D.R., Galzin, R., Legendre, P. and Salvat, B. (1997) Variation interannuelle des peuplements récifaux du récif barrière de Tiahura (île de Moorea, Polynésie française). *Oceanologica Acta* **5**, 743–756.

Bagnis, R., Inoue, A. and Pascal, H. (1990) Dynamic of the dinoflagellate *Gambierdiscus toxicus* in Gambier islands (French Polynesia). *Proceedings of the International Society for Reef Studies*, Noumea, pp. 35–40.

Bleakely C. (1995) Marine Region 14 South Pacific. In *Global Representative System of Marine Protected Areas. Volume 4. South Pacific, Northeast Pacific, Northwest Pacific, Southeast Pacific and Australia/ New Zealand*, eds. G. Kelleher, C. Bleakley and S. Wells.The Great Barrier Reef Marine Park Authority, The World Bank, The World Conservation Union (IUCN), Washington DC.

Caley, M.J. and Schulter, D. (1997) The relationship between local and regional diversity. *Ecology* **78**, 70–80.

Caillart, B., Harmelin-Vivien, M., Galzin, R. and Morize, E. (1994) An atoll of the Tuamotu Archipelago (French Polynesia) part 11. Reef fish communities and fishery yields of Tikehau atoll (Tuamotu Archipelago, French Polynesia). *Atoll Research Bulletin* **415**, 1–36.

Cariès, J.C. (1991) Marine Environment monitoring of Papeete harbour and of coastal ecosytem of Tahiti as part of Territorial monitoring network. *Proceedings of the International Society for Reef Studies*, Nouméa 1991, 17–24.

Charpy-Roubaud, C.J. and Charpy, L. (1994) An atoll of the Tuamoto Archipelago (French Polynesia). Part 11. Nutrients, particulate organic matter, and planktonic and benthic production of the Tikehau Atoll (Tuamoto Archipelago, French Polynesia). *Atoll Research Bulletin* **415**, 1–30.

Charpy, L., Dufour, P. and Garcia, N. (1997) Particulate organic matter in sixteen Tuamotu atoll lagoons (French Polynesia). *Marine Ecology Progress Series* **151**, 55–65.

Crossland, C.J. (1983) Dissolved nutrients in coral reef waters. In *Perspectives on Coral Reefs*, ed. D.J. Barnes. Australian Institute for Marine Science, Townsville, pp. 49–56.

Delesalle, B. (1990) Ecologie du phytoplancton des lagons de Polynesie Francaise. These de Doctorat présentée a l'Ecole Pratique des Hautes Etudes.

Delesalle, B., Buscail, R., Carbonne, J., Courp, T., Dufour, V., Heussner, S., Monaco, A. and Schrimm, M. (1998) Direct measurements of carbon carbonate export from a coral reef ecosytem (Moorea Island, French Polynesia). *Coral Reefs* **17**, 121–132.

Delesalle, B. and Sournia, A. (1992) Residence time of water and phytoplankton biomass in coral reef lagoons. *Continental Shelf Research* **12** (178), 939–949.

Demouget, P. (1989) Etude préliminaire des conséquences des processus érosifs anthropiques sur les riviéres et les lagons et de la mise en place des techniques de lutte dans les îles hautes de Tahiti et Moorea. DEA, Université Française du Pacifique. 70 pp.

Done, T.J., Dayton, P.K., Dayton, A.E. and Steger, R. (1991) Regional and local variability in recovery of shallow coral communities: Moorea, French Polynesia and central Great Barrier Reef. *Coral Reefs* **9**, 183–192.

Donguy, J.R. and Henin, C. (1978) Anomolous navifacial salinities in the tropical Pacific Ocean. *Journal of Marine Research* **34**(3), 355–364.

Dufour, P. and Berland, B. (1999) Nutrient control of phytoplanktonic biomass in atoll lagoons, Pacific ocean waters: Studies with factorial enrichment bioassays. *Journal of Experimental Marine Biology and Ecology* **234**, 147–166.

Dufour, P. and Harmelin-Vivien, M. (1997) A research program for a typology of atoll lagoons: strategy and first results. In *Proceedings of the 8th International Coral Reefs Symposium, Panama*, ed. H.A. Lessios. Allen Press, New York, Vol. 1, pp. 843–848.

Fagerstrom, J.A. and Rougerie, F. (1994) Coral Bleaching Event, Society Islands, French Polynesia. *Marine Pollution Bulletin* **29** (1–3), 34–35.

Faure, G. (1989) Degradation of coral reefs at Moorea island (French Polynesia) by *Acanthaster planci*. *Journal of Coastal Research* **5** (1), 295–305.

Fontaine, Y. (1993) French Polynesia. In *A Directory of Wetlands in Oceania*, ed. D.A. Scott, pp. 105–128. International Waterfowl and Wetlands Research Bureau (IWRB), Slimbridge, UK.

Frouin, P. (in press) Anthropogenic disturbances of tropical soft-bottom benthic communities. *Marine Ecology Progress Series*.

Frouin, P. and Hutchings, P.A. (in review) Macrobenthic communities in a tropical lagoon (Tahiti, French Polynesia, Central Pacific) *Coral Reefs*.

Gabrié, C. (1998) L'État des récifs coralliens en France outre-mer. Nouvelle Calédonia, Wallis et Futuna, Polynésie française, Clipperton, Guadeloupe, Martinique, Mayotte, La Réunion, Iles Éparses de l'Océan Indien. Ministère de L'Aménagement du Territoire et de L'Environnement Secrétariat d'état à l'Outre-Mer. 136 pp (English summary also available).

Gabrié, C. and Salvat, B. (1985) General features of French Polynesian islands their coral reefs. *Proc. Fifth International Coral Reef Congress, Tahiti* **1**, 1–16

Galzin, R., Planes, S., Adjeroud, M., Chauvet C., Doherty, P.J. and Poupin, J. (1998) Objectives and background to the 1994 Franco-Australian expedition to Taiaro Atoll (Tuamotu Archipelago, French Polynesia) *Coral Reefs* **17**, 15–22.

Gout, B., Bablet, J-P. and Goutière, G. (1997) *The Atolls of Muroroa and Fangataufu (French Polynesia) 111. The Living Environment and its Evolution*. CEA/DAM Direction des Essais, 305 pp.

Harris, P. and Fichez, R. (1995) Observations et mécanismes de la crise dystrophique de 1994 dans le lagon de l'atoll d'Hikuera (Archipel des Tuamotu, Polynésie Française). Océanographie, Notes et documents, Volume 45, ORSTOM, Tahiti ed., 25 pp.

Harrison, D.E. and Cane, M.A. (1984) Changes in the Pacific during the 1982–1983 event. *Oceanus* **27**, 21–28.

Hoegh-Guldberg, O. (1999) Climate change, coral bleaching and the future of the world's coral reefs. *Marine Freshwater Research* **50**, 839–866.

Hoegh-Guldberg, O. and Salvat, B. (1995) Periodic mass-bleaching of outer reef-slope communities, Moorea, French Polynesia. *Marine Ecology Progress Series* **57**, 173–186.

Hutchings, P., Payri C. and Gabrié, C. (1994) The current status of coral reef management in French Polynesia. *Marine Pollution Bulletin* **29**, 26–33.

MacArthur, R.H. and Wilson, E.O. (1967) *The Theory of Island Biogeography*. Princeton University Press Princeton.

Lanctot, J.L. Legendre, P. and Salvat, B. (1997) How do coral reef gastropods feel about nuclear testing? A long term study of the effects of man-made perturbations. *Oceanologica Acta* **20**(1), 243–257.

Magnier, Y. and Wauthy, B. (1976) Esquisse hydrologique du lagon de Takapoto, Tuamotu. *Cahiers ORSTOM, Serie Océanographie* **14**, 279–287.

Montaggioni, L.F. (1985) Makatea island, Tuamotu archipelago. In

Proceedings of the Fifth International Coral Reef Congress, Tahiti, Vol. 1, eds. B. Delesalle, R. Galzin and B. Salvat, pp. 103–158.

Niquil, N., Jackson, G.A., Legendre, .and Delesalle, B. (1998) Inverse model analysis of the planktonic food web of Takapoto Atoll (French Polynesia). Marine Ecology Progress Series 165, 17–29.

Niquil, N., Aris-González, J.E., Delesalle, B. and Ulanowicz, R.E. (1999) Characterisation of the planktonic food web of Takapoto Atoll lagoon, using network analysis. Oecologia 118, 232–241.

Pari, N., Peyrot-Clausade, M., Le Campion-Alsumard, T., Hutchings, P., Chazottes, V., Golubic, S., Le Campion, J. and Fontaine, M.F. (1998). Bioerosion of experimental substrates on high islands and on atoll lagoons (French Polynesia) after two years of exposure. Marine Ecology Progress Series 166, 119–130.

Payri, C. and Bourdelin, F. (1997) Status of Coral Reefs in French Polynesia. In Status of Coral Reefs in the Pacific, eds. R.W. Grigg and C. Birkeland, Hawaii, pp. 43–57.

Pirazzoli, P.A. (1985) Leeward islands Maupiti, Tupai, Bora-Bora, Huahine, Society Archipelago. In Proceedings of the Fifth International Coral Reef Congress, Vol. 1, eds. B. Delesalle, R. Galzin et B. Salvat. pp. 17–72.

Poli, G., Delesalle, B., Gabrié, G., Montaggioni, L., Monteforte, M., Naim, O., Payri, C., Richard, G. and Trondle, J. (1983) Etude de l'environnement lagonaire du secteur urbain de Papeete. Evolution des pollutions et des dégradations. Rapport Muséum/EPHE RA 10, 1–110.

Porcher, M. and Salvat, B. (1999) A restoration of damaged coastal zone and reef flat in Bora Bora island (Society, French Polynesia). International Conference Scientific Aspects of Coral Reef Assessment, Monitoring and Restoration, 14–16th April 1999, Fort Lauderdale, Florida, Abstract, pp. 154.

Richard, G. (1985) Fauna and flora: a first compendium of French Polynesian sea-dwellers. In: 5th International Coral Reef Congress, Tahiti, Vol. 1, eds. B. Delesalle, R. Galzin and B. Salvat, pp. 379–520.

Rosen, B. (1984) Reef coral biogeography and climate through the late Cainozoic; just islands in the sun or a critical pattern of islands. In Fossils and Climate, ed. Brenchley. John Wiley & Sons. pp. 201–262.

Rougerie F. and Rancher, J. (1994) The Polynesian South Ocean: Features and Circulation. Marine Pollution Bulletin 29 (1–3), 14–25.

Rougerie F. and Rancher, J. (1995) L'environnement océanique de l'archipel des Tuamotu (Polynésie Française. Oceanologica Acta 18, 43–60.

Rougerie, F. and Wauthy, B. (1983) Anomalies de l'hydroclimat et cyclogenèse en Polynésie en 1982 et 1983. Met-Mar. 121, 27–40.

Salvat, B. (1967) Importance de la faune malacologique dans les atolls polynésiens. Cahiers du Pacifique 11, 7–49.

Salvat, B. (ed.) (1982) OFAI, Bulletin de Liaison 1. Muséum National d'Histoire Naturelle, Ecole Pratique des Hautes Etudes Antenna de Tahiti.

Salvat, B. (1985) An integrated (geomorphological and economical) classification of French Polynesian atolls. In 5th International Coral Reef Congress, Tahiti, Vol. 2, eds. B. Delesalle, R. Galzin et B. Salvat, pp. 337.

Salvat, B. (1991) Biologie et écologie des récifs et lagons coralliens du Pacifique (1960–1990). Bulletin de l'Institut océanographique de Monaco 74 (1439), 231–266.

Salvat, B. (1992) Blanchissement et mortalité des scléractinians sur les récifs de Moorea (Archipel de la Société) en 1991. Compte rendu de l'Académie des Sciences, Paris 314, Série 111, 105–111.

Salvat, B. and Aubanel, A. (1993) Conséquences d'une élévation du niveau de la mer pour un littoral à récif corallien: le cas d'une Ile haute volcanique, Moorea, Polynésia française. Sémaire "Elévation du niveau de la mer le long des côtes de France "DRAIE-Ministere de l'environnement 6–7 Décembre 1993.

Salvat, B., Legendre, B., and Lanctot, J.L. (1995) Atmospheric nuclear tests: immediate and long-term effects on coral reef gastropod assemblages. International Conference. Long term changes in marine ecosystems, Arcachon.

Salvat, B. and Renaud-Mornant, J. (1969) Etude écologique du macrobenthos et du meiobenthos d'un fond sableux du lagon de Mururoa (Tuamotu Polynésie). Cahiers du Pacifique 13, 159–179.

Salvat, B., Sibuet, M. and Laubier, L. (1985) Benthic megafauna observed from the submersible Cyanaon the fore-reef slope of Tahiti (French Polynesia) between 70 1100 metres. In 5th International Coral Reef Congress, Tahiti, Vol. 1, eds. B. Delesalle, R. Galzin and B. Salvat, p. 338.

Savoie, A. (1989a) Les études environnementales en Polynésia francaise. Mémoires de D.E.A, Université française du Pacifique, Papeete, Tahiti 70 pp.

Savoie, A. (1989b) Les études environnementales. Journées de la Recherche, Tahiti 1989. Rapport final, Ministère de la Recherche scientifique, Tahiti, Polynésie française: 145–154.

Scheltema, R.S. (1986) Long-distance dispersal of planktonic larvae of shoal-water benthic invertebrates among central Pacific islands. Bulletin of Marine Science 39, 241–256.

Sournia, A. and Ricard, M. (1976) Phytoplancton and its contribution to primary productivity in two coral reef areas of French Polynesia. Journal of Experimental Marine Biology and Ecology 21, 129–140.

SPREP (1980) French Polynesia. Country Reports 5. South Pacific Commission, Noumea, New Caledonia.

Tessier, R. (1969) Les cyclones en Polynésia française. Bulletin de la Société d' Océanographiede France 14, 1–48.

Thomassin, B., Jouin, C., Renaud Mornant, J., Richard, G. and Salvat, B. (1982) Macrofauna and microfauna in the coral sediments on the Tiahura reef complex, Moorea island (French Polynesia). Tethys 10 (4), 392–397.

Veron, J.E. (1993) A Biogeographic Database of Hermatypic Corals, Species of the Central Indo-Pacific, Genera of the World. Australian Institute of Marine Science Monograph Series 9.

Wauthy, B. (1986) Physical ocean environment in the South Pacific Commission Area. UNEP Regional Seas Report Studies No 83. UNEP

Wells, S.M. (1988) Coral Reefs of the World. Vol. 3. Central and Western Pacific. UNEP, Nairobi; International Union for Nature and Natural Resources, Switzerland.

Wolanski, E., Delesalle, B., Dufour, V., and Aubanel, A. (1993) Modelling the fate of pollutants in the Tiahura lagoon, Moorea, French Polynesia. Proceedings of the Eleventh Australian Conference Coastal Oceanography Engineering, Townsville, Vol. 2, pp. 583–587.

Wilkinson, C.R. (ed.) 1998. Status of Coral Reefs of the World. Global Coral Reef Monitoring Network. Australian Institute of Marine Science, Townsville.

THE AUTHORS

Pat Hutchings
*The Australian Museum,
Sydney, NSW 2010,
Australia*

Bernard Salvat
*Ecole Pratique des Hautes Etudes, URA CNRS 1453,
Université de Perpignan, France, and
Centre de Recherches Insulaires et Observatoire de
l'Environnement, BP 1013, Moorea, Polynésia française*

INDEX

Page numbers in bold refer to tables, in italics to illustrations. Roman numerals indicate volume number.

abalone, recreational harvesting, South Africa II-139
abalone fishery II-27, II-137
— Australia II-586
— dive fishery, Victoria Province II-668
— Tasmania II-655–6
Aboriginal people, Australia II-583–4, II-619–20
— dugongs a festive food for II-619
— southern and northern Nullabor Plain II-682
aboriginal subsistence whaling III-74
Abrolhos Archipelago and the Abrolhos Bank, Brazil *I-732*
— coral reefs *I-720*, *I-724*, *I-726*
Abrolhos Bank–Cabo Frio, Southern Brazil I-734
— Brazil Current I-734
— environmental degradation I-740
— eucalyptus plantations for pulp production I-738
— — industrial pollution from pulp production I-740
— major activities I-737–8
— shallow water marine and coastal habitats I-735–6
— Vitória Bay
— — environmental degradation I-740
— — oil spill risk for surrounding fragile areas I-739
Abrolhos National Marine Park, Brazil *I-720*, I-726–7, *I-728*
abundance, change in may be due to new competitor I-257
abundances, damped oscillations in III-275
accidental introductions *see* alien/accidental/exotic/introduced organisms/species
accretion, Gulf of Guinea coast I-778
acid mine drainage, Tasmania II-654
acid sulphate soil conditions, associated with mangrove clearance II-369, II-395
acid sulphate soil run-off
— a critical issue in eastern Australia II-638
— Great Barrier Reef region II-620–1
adaptive management III-400
Aden, Gulf of *II-36*, II-38, II-47–61, *II-48*
— coastal erosion and landfill II-58
— defined II-49
— effects from urban and industrial activities II-58–9
— major shallow water marine and coastal habitats II-51–4
— offshore systems II-54–5
— populations affecting the area II-55–6
— ports **II-55**
— production higher II-42
— protective measures II-59–60
— — conventions signed relating to the Marine Environment **II-59**
— rural factors II-56–8
— seasonality, currents, natural environmental variables II-49–51
Aden port II-55
— sand and mud flats II-53
Adriatic Sea *I-268*
— agricultural pollution load I-275
— circulation and water masses I-270
— described I-269–70
— differences between east and west coasts I-269, I-272
— effects of urban and industrial activities I-277–8
— — lagoon ports and canal harbours I-277
— — nuisance diatom bloom III-298

— salinity and tidal amplitude I-269
— seasonality, currents and environment variables I-270–1
adult biomass, minimum viable, required for stock replacement III-382–3, *III-383*
adverse health effects, of toxic chemicals II-456–8
— coplanar PCBs most suspect contaminants II-457
Aegean Sea I-233–52
— Aegean basins I-235
— — partitioned by islands I-237
— behaviour of water masses I-236–7
— coastal erosion and landfill I-245
— defined I-235–6
— — Miocene—present evolution I-235
— effects from urban and industrial activities I-245–9
— human populations I-241–2
— northern, Turkish, sediment decreased through dam construction III-352
— offshore systems I-240–1
— protective measures I-245–9
— rural factors I-242–5
— seasonality, currents and natural environmental variables I-236–7
— shallow water ecosystems and biotic communities I-237–40
aerial surveys, for bird distributions III-109
Africa, Eastern, coral bleaching and mortality events III-48
Africa, southwestern I-821–40
— Benguela Current Large Marine Ecosystem I-823
— commercial fisheries I-833
— effects of urban and industrial activities I-835
— major shallow water marine and coastal habitats I-828–30
— the next millennium I-835–7
— — fisheries management, past, present and future I-836–7
— — oil spill contingency plans I-837
— — other development and commercial activities I-837
— offshore systems I-830–5
— physical environment I-823–8
— status of selected Convention, Agreement and Codes of Conduct **I-836**
African dustfall
— an ecological stressor in the Florida Keys I-410
— potential cause of other Caribbean coral diseases I-410
Agalega (Mascarenes) II-255
— coconut plantations II-257
— population II-260
— upwelling increases nutrients II-259–60
Agenda 21 III-351
— and promotion of ICZM III-352
— — evolution of III-352
— on Protection of the Oceans I-58
— on rights and duties of fishing III-158
— using two UNCED principles I-459
aggregate extraction
— offshore sites, Irish Sea I-95–6
— sand for beach nourishment I-53–4, I-96
— sand and gravel, off UK coasts I-53, I-74
aggregate mining, wadi beds II-27
agrarian reform, Nicaragua, created environmental problems I-538–9
agricultural pollution II-376
— Malacca Strait II-338
— Mauritius II-261
— problem of, Adriatic and Tyrrhenian basins I-275–6

— southern Spain I-174
— Sri Lanka II-184
agriculture
— American Samoa, Tutuila Island II-770
— Argentina I-758
    — Buenos Aires Province, fertilisers and pesticides I-762
— Australia II-584
    — poor agricultural practices II-584
— in the Bahamas I-423–4
— Bangladesh II-292
— Belize I-508
    — commercial, effects of I-508
    — potential impacts on the marine system I-508
— Borneo II-374–5
— Cambodia, traditional II-576
— Chesapeake Bay
    — best management practices I-347
    — contributes N and P to pollution I-345
— China's Yellow Sea coast, increased output II-492
— coastal
    — Mozambique II-107
    — subsistence, Gulf of Aden II-57–8
— coconut plantation, Marshall Islands II-781
— Colombian Pacific Coast I-682
— the Comoros II-247
— Coral, Solomon and Bismark Seas region II-435
— Côte d'Ivoire, basis of the economy I-816
— creating problems, Andaman, Nicobar and Lakshadweep Islands II-193
— eastern Australian region II-636–8
    — central and southern New South Wales II-638
    — northern New South Wales II-637–8
    — south-east Queensland II-637
— effects of in Tanzania II-88–9
— El Salvador I-551
    — poor land-use and agricultural practices I-551–2
— extensive cattle industry I-479
— Fiji Islands, and its impact II-758
— Great Barrier Reef region II-620–1
    — land clearing II-620–1
    — sedimentation and nutrients II-621
— Gulf of Guinea, use of agrochemicals I-784–6
— Gulf of Papua II-600–1, II-604
— Hawaiian Islands II-801
— Hong Kong, declining II-542
— influenced by water availability, Arabian Peninsula southern coast II-25–6
— intensive
    — Greece and Turkey I-242
    — Guadalquivir valley I-174
— Jamaica I-568
— Lesser Antilles I-635
    — export crops I-635
— Madagascar II-123–4
    — hill rice planting leads to soil loss II-123
— the Maldives **II-208**
    — environmental effect of II-207
— Marshall Islands II-778
— Mascarene Region
    — Mauritius II-260–1
    — Reunion II-261
    — Rodrigues II-261
— Mexican Pacific coast I-489
    — and the green revolution I-492
— Mozambique coast, inappropriate techniques II-109
— N and P to Baltic Sea I-104, *I-104*
— New Caledonia II-729, **II-729**
— Nicaragua
    — Caribbean coast I-523
    — Pacific coast I-538–9, I-540
— Peru, use of organochlorine pesticides I-696

— the Philippines II-412–13
— plant nutrients from I-93
— poor farming practice leads to estuarine sedimentation and turbidity II-136
— Samoa II-716
— Sea of Okhotsk, poorly developed II-469
— the Seychelles
    — deforestation for coconut plantations II-238
    — poor agricultural practice and unwise development activity II-238–9
— Somalia
    — few crops 179, II-77
    — nomad/semi-nomad II-77
— southern Brazil
    — subsistence I-738, I-739
    — sugar cane cultivation I-737–8
— Southern Gulf of Mexico, plantations I-476
— in the Sundarbans II-153
— Taiwan west coast II-503
— Tonga II-716
— Turks and Caicos Islands, small amount only I-591–2
— Vanuatu
    — and its impacts II-743
    — Pilot Plantation Project II-743
— Vietnam, rice growing II-563
— western Indonesia II-396
agrochemical pollution
— Malacca Strait II-318–19
— south coast, southern Brazil I-739
agrochemicals
— use of
    — China II-492
    — Vanuatu II-743
— western Indonesia II-396
Agulhas Bank II-135, II-137
Agulhas current *I-822*, I-824, *II-101*, *II-134*, II-135
Agulhas Gyre *II-101*
Agulhas Province *II-134*
air masses
— main winter transport routes over the Arctic *I-13*
— transport of from industrialised areas to the Arctic I-5
air pollution
— by ships, prevention of III-337
— Chesapeake Bay I-341
— locally high, Jamaica I-570
— Rudnaya River Valley, eastern Russia II-485
— Sea of Japan II-476–7
    — industrial contributions II-476–7
    — sources of II-476
— southern Brazil
    — Rio de Janeiro metropolitan area I-741
    — Santos Bay I-743
— through forest burning, Borneo II-375
air temperatures
— Côte d'Ivoire I-809
— Gulf of Guinea I-776
Air-Ocean Chemistry Experiment (AEROCE) I-224, I-229
airborne remote sensing, use of multiple digital video cameras III-284
airmasses, contribution to pollution over the Sea of Japan *II-476*
airport construction, Bermuda, effects of I-228
airports, environmental problems of II-214
Airy's theory, to predict energy in a wave III-312
Al Batinah *II-18*, II-19, II-26
— coastal erosion acute II-28
— dams across wadis keep sediment from coast II-27
— a wadi plain II-27
Alaska, protection for high relief pinnacles III-386
Alaska Coastal Current *I-374*, I-375
— and Ekman drift I-378
Alaska Current *I-374*, I-375, *III-180*
Alaska, Gulf of I-373–84, III-128

— anomalous along-shore flow III-184
— baseline studies, pollutants I-381-I-383
— climatic changes III-179-86
    — changes in physical properties III-181-3
    — impact of El Niño events III-183-4
    — implications of observed changes III-184-5
    — physical properties III-181
    — possible biological effects: salmon as an example III-185-6
— effects from urban and industrial activities I-383
— the *Exxon Valdez* oil spill I-382
— Forest Practices Act I-383
— geographic setting I-375
— human populations I-381
— major shallow water marine and coastal habitats I-375
— overfishing of virgin groundfish stocks, change to fishing other species III-120-1
— oxygen depletion impacts fisheries III-219
— physical oceanography I-375, I-378-9
    — shelf hydrography and circulation vary seasonally I-378
— primary productivity and nutrient cycles I-379-80
    — biomass doubling round perimeter, hypotheses I-380
— rural factors I-383
— 'strip-mining' of Pacific Ocean Perch III-120, III-121
— tidal propagation III-188-90
— trophic shift in marine communities I-376-8
Alaska Gyre *III-180*
Alaska Peninsula *I-374*
Alaskan Stream *I-374*, I-375, *III-180*
albacore I-141
albatrosses, wandering I-757
Albermarle Sound *I-352*, I-360
— environmental problems for bottom communities I-360
Albermarle—Pamlico Estuarine System *I-352*, I-353, I-362
— low benthic diversity and high seasonality in abundances I-360
— supports extensive shellfisheries I-360
Albufeira Lagoon *I-152*, I-154
— phytoplankton community I-156
— purification of mussels I-158
— RAMSAR site I-162, *I-163*
Aleutian Islands *I-374*
Aleutian Low Pressure region/system I-376, II-465
— linked to Alaskan Basin and shelf I-375, I-378
Alexander Archipelago *I-374*, I-375
algae I-590, I-649, I-754
— Andaman and Nicobar Islands II-192
— benthic, degraded, Black Sea I-291
— blue-green I-106
— brown I-47, I-70, I-109, I-125, I-159, I-290, I-669, II-23, II-42, II-664
— calcareous, maërl I-69-70, I-140
— calcareous, exploitation of, Abrolhos Bank–Cabo Frio, Southern Brazil I-737
— coralline II-370, II-711, II-741
    — crustose II-72, II-796
— drift, long-range dispersal, Western Australia II-695
— encrusting I-21
— epigrowth I-21
— epiphytic II-665, III-2
    — reduce diffusion of gases and nutrients to seagrass leaves II-701
— fucoid I-70
    — sensitive to oil III-275
— Galician I-137
— Great Barrier Reef region II-615
    — cross-shelf difference in communities II-615
— green I-70, I-71, I-194
    — *Halimeda* II-615
— Gulf of Maine, comparable to northern Europe I-311
— Gulf of Mannar II-164
— marine
    — in the Azores I-205-6
    — Fiji Islands II-754
    — Marshall Islands II-779
— Mexican Pacific coast, diversity of I-490
— red I-47, I-70, I-109, I-125, I-159, I-290, I-754
    — endemism, Australian Bight II-678
— reefal I-599-600
— the Seychelles II-236
— The Maldives II-204
— toxic
    — in English Channel I-69
    — Irish Sea I-90
— Vietnam II-564
— Wadden Sea, changes in I-49
algal assemblages, Western Australia II-695
algal beds
— green, New Caledonia II-726
— Hawaiian Islands II-796
algal blooms I-743, I-821, III-198
— Baltic Proper I-126, I-128
— Baltic Sea I-125
— Bay of Bengal II-275
— Black Sea, intensifying I-290
— detection of III-298-300
    — coccolothophores III-298
    — cyanobacteria III-298-300
    — red and white tides (HABs) III-298
— English Channel I-69
— entrainment blooms II-20
— harmful (HAB) I-827
    — and fish kills II-320
    — Hong Kong coastal waters II-541
    — increase in I-20
    — may be associated with ENSO events II-408
    — Yellow Sea II-495
    — *see also* harmful algal blooms (HABS)
— lakes, central New South Wales II-642
— novel III-259
— nuisance, creating anocic/hyppoxic disturbances III-219
— potentially harmful II-10-11, II-13
— red, Zanzibar II-92
— toxic
    — Australia's inland waters II-585
    — Carolinas coast I-356
    — increase may be due to ballast water I-26
    — Portuguese coastal waters I-160
— toxic and nuisance
    — Irish Sea I-90
    — North Sea I-53
algal mats I-131
— cyanophyte mat formers I-357
— may cause anaerobic condition I-53
algal reefs, Red Sea II-41
algal ridges, Fangataufa atoll, French Polynesia *II-818*
algal turf I-205, I-583
— Red Sea II-42
algal-vermetid reefs (boilers) I-227
alien/accidental/exotic/introduced organisms/species I-26, II-587, II-588-9, II-623, II-641
— accidental introduction of III-86
— Adriatic and Tyrrhenian seas I-282
— Argentine coastal waters I-765
— the Azores I-209
— Baltic Proper I-110
— Baltic Sea I-129
— Bay of Bengal II-279-80
— Black Sea I-290, I-291
— carried in ballast water II-319, III-227
— a coastal transboundary issue III-355
— a concern in Hong Kong II-542
— effects on seabirds III-114
— English Channel I-70
— Gulf of Maine I-312
    — and Georges Bank I-310

— Hawaiian Islands II-802
— Irish Sea I-89
— a major impact on the Tasmanian marine environment II-657–8
— Mediterranean mussel, southwestern Africa I-830
— the Mediterranean via the Suez Canal I-257–8
— North Sea I-47–8
— Victoria Province, Australia II-661, II-669
— Western Australia II-701
— *see also* new, novel occurrences and invasive disturbances
Alisios Winds *see* Trade Winds
alkyl lead I-95
alluvial fans II-37
alluvial plains, Borneo II-366
Altata-Ensenda del Pabellón, Mexican Pacific coast I-493
— pesticides causing environmental stress I-493
aluminium contamination, from bauxite mining II-395–6
Amazon River water, entering the Caribbean I-579
American flamingo, at Laguna de Tacarigua, Venezuela I-652
American Samoa II-765–72
— coastal habitats II-767–9
— coral reefs II-712, II-767–8
  — degradation of II-712
— effects from urban and industrial activities II-770–1
— environment II-767
— offshore systems II-769
— population II-714, II-769
  — must import to support population II-769
— protective measures II-771
  — marine protected areas II-771, **III-771**
— rural factors II-770
Americas, Pacific Coast, coral bleaching and mortality events III-53–4
americium I-94
Amirantes Bank *II-234*, II-235
ammonia, exports from Trinidad and Tobago I-638
amnesic shellfish poisoning (ASP) III-221
Amnesic Shellfish Toxin contamination, Scottish scallop fishery I-90
Amoco Cadiz oil spill
— change to *Fucus* III-272
— migration of oil layers downward within beach sediments III-270–1
amphidromic points *I-44*, I-45
amphipods, highly sensitive to oil III-274–5
Amursky Bay, eastern Russia
— chemical pollution **II-484**
— decline in annual biomass II-484
— ecological problems II-483–5
— the ecosystem is decaying II-484–5
— heavy metals in bottom sediments II-484
— metal concentrations, high in suspended matter II-484
— organic substances in wastewater II-483
anchialine pools, Hawaii II-796
anchor damage
— fishing and diving II-806
— ship groundings, Florida Keys I-409
— to corals I-605–6, I-637
anchovy
— beach seining, southern Brazil I-739
— Peru, replaced by sardines during an El Niño event I-693
anchovy fishery I-141
— Black Sea
  — collapse of I-290
  — Turkey I-292
— larval fishery, Taiwan Strait II-504–5
— Mexican Pacific waters I-490
— Peru I-691, I-693
— southwestern Africa I-832, I-833
Andaman Islands *II-270*
Andaman and Nicobar Center for Ocean Development (ANCOD) II-280
Andaman, Nicobar and Lakshadweep Islands II-189–97
— biodiversity II-192

— climate and coastal hydrography II-191
— coastal ecosystems II-191–2
— conservation measures II-195–6
— fish and fisheries of the Andamans II-192
— impacts of human activities on the ecosystem II-193, II-195
— islands of volcanic origin II-191
— Lakshadweep Islands II-194–5
— national parks and wildlife sanctuaries II-195
— population II-193
Andaman Sea *II-270*, *II-298*, *II-310*
— coral reefs II-302
— poor reef status II-303
— surfaces water influenced by freshwater continental runoff II-191
Andros Island *I-416*, I-431
Angola, development limited by civil war I-837
Angola Current *I-822*, I-824
Angola–Benguela front *I-822*, I-825
— southward displacement I-827
anguilla I-226, *I-616*, I-617
— beaches I-620
— climate I-617
— coastal protection I-624
— coral reefs I-618
— Exclusive Fisheries Zone I-621
— hurricane damage I-622
— impact of urban development I-622–3
— increased tourism I-620
— mangroves I-619
— marine parks I-625
— rocket launching site, environmental impact assessment criticised I-623
— seagrass beds I-618, *I-619*
— sustainable yields for fisheries I-621
— tourism I-620, I-622
Anguilla Bank I-618
Anjouan (Ndzouani/Johanna)
— coconut and ylang-ylang II-246
  — fringing reefs II-246
— reef front changed by quarrying II-248
— young island II-246
Annapolis Royal, Nova Scotia, tidal energy project III-316
Anole, Lac and Lac Badana *II-67*, II-68
— ecological profiles *II-76*
anoxia I-337, III-366
— due to decay of large plant biomass III-259
— from HABs III-298
— summer, Thessaloniki I-241
anoxic basins, confined, for dumping of contaminated sediments I-27
anoxic water
— Oslofjord *I-27*
— shallow-silled fjords I-17
anoxic/hypoxic disturbances III-219–21, *III-220*
— fish kills due to hypoxia III-220
Antarctic Circumpolar Current *II-581*, II-649, II-650, *II-674*
Antarctic Convergence II-583
Antarctic ecosystem, diversity and abundance in I-706–7
Antarctic Intermediate Water I-579, I-632
— Mascarene Region II-255
Antarctic polar front I-721
Antarctic Treaty system **III-335**, III-342
— Madrid Protocol III-342
Antarctica
— birds of III-108
— the grave of whaling III-75
anthropogenic disturbance, imposed on seagrasses III-10–11
anthropogenic influences, Norwegian coast I-24
anti-cancer/anti-infective agents, natural or modelled on natural products III-38–9
antibiotics, used in marine fish cages III-369
anticyclones
— Azores I-188, I-203

— Azores—Bermuda High I-439
— Bermuda High I-223
— Great Australian Bight II-675
— Mascarene II-116
— North Pacific High I-378
— Siberian High I-287, I-376, II-465
— south Atlantic I-823
antifouling paints I-788
— *see also* marine antifoulants; organotins; tributyltin (TBT)
Antilles Current I-419, *I-628*
Apo Reef *II-406*
Aqaba, Gulf of *II-36*, II-39
— minerals as pollutants *II-36*, II-39
aquaculture I-52, I-162, I-366, I-835, II-558–9, III-368–9
— Adriatic coasts I-275
—— lagoons and small lakes I-277
— Asian, significance of III-166
— Australia II-587
— the Bahamas, unsuccessful I-427
— Cambodia II-576–7
— can be a perfect environment for epidemics III-221
— Chilean fjords I-712
— the Comoros II-249
— Côte d'Ivoire I-818
— diseases, from and among facilities III-226
— east coast, Peninsular Malaysia II-352–3
—— changing water quality II-353
— effective planning and management to control pollution and disease III-170
— French Polynesia II-820
—— raising barramundi II-820
— Galicia I-144
— global, overview III-166
— Godavari-Krishna delta II-171
— limited in the Mascarenes II-262
— Malacca Straits, vulnerable to oil spill damage II-319–20
— marine, Tasmania II-656
— marine, Argentina I-764–5
—— biological species introduction I-764
—— environments: characteristics and adaptability I-764
—— harmful algae I-764–5
— marine, overview III-166–71
—— culture facilities III-170
—— fish aquaculture, development of III-168–9
— Marshall Islands II-781–2
— New Caledonia II-729
— oyster farming, eastern Australia II-641
— the Philippines, issues associated with II-413
— ponds in marine and brackish water, high demands on water III-368–9
— scenarios for opportunities to secure and increase production III-171
— seaweed, Fiji Islands II-754
— small-scale II-438
— sustainable II-375
— sustainable, Chinese and Thai experience III-171–7
—— national perspectives on issues and challenges III-171–2
—— national support for III-172
— The Maldives II-207
— Tyrrhenian coast I-279
— Venezuela I-653-4
— Victoria Province, Australia II-668–9
— west coast of Malaysia II-339
— Western Australia II-700
— western Indonesia II-394–5
—— high fluxes of nutrients and sediments into nearshore waters II-394–5
— Yellow Sea coastal waters II-494
— *see also* mariculture; shrimp aquaculture; shrimp farming
aquarium fish
— from Mauritius II-263
— trade in I-510
aquarium fish collection
— Cambodia II-576
— Great Barrier Reef II-617
— Hawaiian Islands II-803
— Marshall Islands II-784
— western Indonesia II-400
aquarium fish trade, New Caledonia II-734
aquasports, non-consumptive resource use II-141
aquatic ecosystems, atmospheric pathway a significant pathway of pollutant and nutrient fluxes III-201, III-207
aquatic organisms, farming of *see* aquaculture
aquatic plant farming III-167–8
aquatic vegetation
— decline of, Tangier Sound I-345, I-347
— improving, Chesapeake Bay I-347
Arabian Gulf II-1–16, *II-18*
— autumn seabird breeding season III-110
— coral bleaching and mortality events III-47
— corals little affected by Gulf War oil spills III-274
— defined II-3–4
— development issues II-9–11
— evaporation in excess of precipitation II-3
— future marine studies II-13
— major coastal habitats and biodiversity II-6–9
— marine studies II-3–4
— natural environmental variables II-4–6
— oil II-3
—— oil spills and the Gulf War II-11–13
— protective measures II-13
Arabian Gulf Co-operative Council (AGCC), Marine Emergency Mutual Aid Centre (MEMAC) II-29
Arabian Oryx Sanctuary II-31
Arabian Plate II-37
Arabian Sea
— abundance of meso-pelagic fish II-24–5
— coral bleaching and mortality events III-47–8
— crossed by major trading routes II-25
— a cyclone-generating region II-20
— erosion–deposition cycle II-27
— extreme marine climate II-20–1
— floor of II-19
— low pressure, ambient-temperature water flushing removes oil from mangroves III-277
— mangrove stands II-23
— northern, no freshwater inflow from Arabian Peninsula II-23
— suboxic conditions II-21
Arcachon Bay, French coast, serious environmental impacts of TBT III-250
arctic haze I-13, I-14
Arctic Mediterranean I-33
Arctic Monitoring and Assessment Programme (AMAP) I-9, I-26
— POPs ubiquitous III-361
areal surveillance, in fisheries management III-160
Argentina
— co-operation with Uruguay I-463
— laws and decrees undertaken by **I-766**, I-767
— main coastal management concerns I-751
— provincial and municipal ordinances **I-767**
Argentine coast
— estuaries and salt marshes I-755–6
—— Mar Chiquita coastal lagoon I-755
Argentine Sea *see* Southeast South American Shelf Marine Ecosystem
Argentinian continental shelf
— circulation I-754
— coastal system I-753
— Malvinas/Falklands system I-753
— subantarctic shelf waters system I-753
arid soil extraction, for beach regeneration, southern Spain I-177
Arkona Basin I-101, I-106
— deterioration of oxygen conditions I-105

armouring
— of beaches
    — and oil penetration III-269–70
    — and oil persistence III-276
Arrow oil spill, effect on *Fucus* spp. III-272
arsenic II-12
arsenic contamination, Bangladeshi groundwater II-293
artificial islands, for airport construction I-54
artificial reef structures I-347–8
artisanal fishing, impact of, New Caledonia II-730–1
Aruba *I-596*, I-597
— coastal and marine habitats *I-599*
— coastal urbanization I-606
— fishery I-603
— landfill I-602
— limited reef development I-599
— mass tourism I-598
— sewage discharge I-606
ASEAN Council on Petroleum (ASCOPE) II-325
Ashmore Reef National Nature Reserve, Western Australia II-696
Asia Minor Current I-236
Asian developing regions: persistent organic pollutants in the seas II-447–62
— coastal waters II-449–54
— open seas II-454–60
asphalts, formation of III-270
astronomical tidal forcing III-190, III-193
Aswan Dam, effects of I-256
Atacama Desert I-704
Atchafalaya River *I-436*, I-443
Athens *I-234*, I-241
— primary sewage treatment I-245–6
Atlantic basin
— and ENSO events I-223–4
— and tropical cyclones I-224
Atlantic Conveyor, postulated reversal of III-86–7
Atlantic Niños, related to Pacific El Niños, effects of I-781–2
Atlantic Rainforest
— Abrolhos Bank–Cabo Frio, Southern Brazil I-735
— Brazilian tropical coast I-725–7
Atlantic Water I-255
— entering the Tyrrhenian circulation I-272, I-274
— flowing into the North Sea I-45, *I-45*
atmosphere—surface fluxes, determined using surface analysis methods III-204
atmospheric nuclear experiments, fallout from I-11
'atmospheric particles'
— described III-199
— origins of III-199
— primary and secondary particles III-199
— size distribution in space and time III-199–200
    — can change as a result of physical and chemical processes III-200
Australia
— an island continent II-581
— biogeography II-581
— coastal erosion and landfill II-585
— continental shelf II-582
— coral bleaching and mortality events III-52–3
    — Western Australia (1998), variable III-3
— degradation of the Great Barrier Reef III-37
— EEZ II-581
— effects of urban, industrial and other activities II-586–9
— human populations affecting the area II-583–4
— laws protect flatback turtle III-64
— major shallow water marine and coastal habitats II-582
— mangroves III-20
— marine biogeographical provinces *II-675*
— offshore systems II-583
— protective measures II-589–91
    — general marine environmental management strategies II-590
    — International arrangements and responsibilities II-590
    — marine protected areas II-590
    — oceans policy II-590–1
— a regional overview II-579–92
    — status of scientific knowledge of the marine environment II-584
— rural factors II-584–5
— seasonality, currents, natural environmental variables II-581–2
— status of the marine environment and major issues II-591
Australia, eastern: a dynamic tropical/temperate biome II-629–45
— biogeography II-631
— coastal erosion and landfill II-639
— effects of urban and industrial activities II-639–42
— human populations affecting the area II-635–6
— major shallow water and coastal habitats II-632–5
— protective measures II-642–3
    — evaluation of protected areas II-643
    — legislation and responsibilities II-642
    — marine protected areas II-643
    — protected species II-643
— rural factors II-636–8
— seasonality, currents, natural environmental variables II-631–2
— status of the marine environment II-643
Australia, northeastern, Great Barrier Reef region II-611–28
— biogeography II-613
— coastal erosion and landfill II-622
— effects of urban and industrial activities II-622–4
    — protective measures II-624–7
— major shallow water marine and coastal habitats and biota II-614–19
— offshore systems II-619
— populations affecting the area II-619–20
— rural factors II-620–1
— seasonality, currents, natural environmental variable II-613–14
Australian Sea Lion II-687
— on IUCN Red List II-681
— threats to recovery of II-685–6
AVHRR imagery
— detection of cyanobacteria blooms in the Baltic III-299–300
    — algorithm for bloom detection III-299
— sea surface temperature (SST), Venezuela *I-647*
— for trends in water clarity III-296
The Azores I-201–19
— climate I-203–4
— coastal erosion and landfill I-211
— effects from urban and industrial activities I-212–13
— major shallow water marine and coastal habitats I-205–9
    — importance of harbours I-209
— offshore systems I-209–10
— populations I-210–11
— protective measures I-213–17
    — in word but not deed I-214–15
— the region defined I-203
— rural factors I-211
— seasonality, currents, natural environmental variables I-203–5
Azores anticyclone I-188, I-203
Azores Current I-204–5, *I-204*
Azores Microplate I-203
Azores—Bermuda High I-439
Azov, Sea of *I-286*, I-287
— suffering from hypoxia I-291

Bab el Mandeb *II-36*, II-38
— flow through not clear II-39–40
— traversed by oil tankers II-59
back water effect, Bangladesh II-273
bacteria
— on the Southern Californian shoreline I-399
— sulphur-oxidizing III-263, *III-263*
— zooplankton and phytoplankton as reservoirs for III-221
bacterial decomposition, Arabian Sea II-24

Baffin Bay *I-6*, I-7, I-8
— increased mercury in upper sediments I-9
Bahama Bank Platforms, calcium carbonate I-419
Bahama Banks
— biota of Caribbean origin I-418
— fossil reefs, used to date sea-level change I-417
— islands formed during sea-level lowstands I-417
— origins of debated I-589
The Bahamas I-415–33
— agriculture I-423–4
— aquaculture I-437
— artisanal and commercial fisheries I-424–5
— coral bleaching and mortality events III-54
— effects from urban and industrial activities I-427–30
— Landsat TM–seagrass biomass relationship III-286
— major shallow water marine and coastal habitats I-419–22
— Marine and Coastal National Parks of the Bahamas **I-431**
— offshore systems I-422–3
— origin of I-417–18
— other fisheries resources I-426–7
— population I-423
— protective measures I-430–2
— seasonality, currents, natural environmental variables I-418–19
— tourism and its effect on the population I-423
Bahamas National Trust I-432
Bahamas National Trust Park *I-416*
Bahamas Reef Environment Education Foundation I-432
Bahamian Banks *see* Bahama Banks
Bahamian Exclusive Economic Zone (EEZ) I-422
Bahrain *II-2*
— coral bleaching and mortality events III-47
Baie de Seine *I-66*, I-70
— dumping of metals I-75
— PCBs and organochlorines in meio- and macrofauna I-75
Baie de Somme *I-66*
— marsh with ponds I-69
baiji (Yangtze river) III-90
— will be affected by Three Gorges Dam III-97
Baird's beaked whale III-76, III-92, III-93
Baja California I-486
— marine habitat I-487
Baja California Sur I-486, I-487
Bajuni Islands (Archipelago) *II-67*, *II-68*, *II-75*
— coral carpet development II-72
— development of shelf since isotope stage 5e *II-73*
— fringing reefs II-72, II-74
— shows features of barrier island complex II-69, *II-73*, *II-75*
— build-up of sand bodies II-69, *II-71*
Bajuni Sound
— coral knobs, patch reefs and seagrass beds II-74
— intertidal abraded flats facing channels II-77
— mixture of habitats *II-71*, *II-75*, II-76, *II-78*
"balance of nature" III-86
baleen (rorqual) whales III-74, III-76
— estimates of numbers III-82–3
— "safe catch" limit calculations III-79
ballast water
— dumping of II-295
— environmental risks from II-319, III-226–7
Baltic ecosystem, pollution sensitivity of I-129–30
Baltic Proper *I-122*
— biota
— alien species I-110
— birds and mammals I-109
— main coastal and marine biotopes I-108–9
— pelagic and benthic organisms I-109
— central parts permanently stratified I-123
— chemical munitions dumped in I-117
— eutrophication I-103–8
— biological effects of I-109–10
— inputs from land and atmosphere I-103–4
— oxygen I-104–6
— temporal and spatial variability in nutrients I-106–8
— major oceanic inflows I-101–2, *I-102*
— environmental conditions dependent on I-124
— natural immigrants I-108
— persistence of anoxic zones I-105–6
— reasons for problems I-101
— shallow banks providing spawning and nursery areas I-124
— standing stock and carbon flow I-126, *I-127*
Baltic Sea *I-100*, *I-122*, III-205
— anoxic areas III-258
— basins *I-121*–I-123
— cyanobacteria blooms III-299
— investigation using AVHRR imagery III-299
— described I-123
— freshwater immigrants I-125
— including Bothnian Sea and Bothnian Bay I-121–33
— coastal and marine habitats I-124–8
— effects of pollution I-130–1
— environmental factors affecting the biota I-123–4
— fish and fisheries I-128–9
— pollution sensitivity of the Baltic ecosystem I-129–30
— protective measures I-132
— southern, eutrophication of coastal inlets III-260–3
— southern and eastern regions I-99–120
— biota in the Baltic Proper I-108–10
— condition of the Baltic Proper I-103–8
— environmental pollutants I-112–17
— fish stocks and fisheries I-110–12
— regional setting I-101–3
— a young sea I-101, I-108
Baltic Water *I-18*, I-23
band ratio, for retrieval of chlorophyll content from spectral radiance III-297
Bangladesh II-285–96
— coastal environment and habitats II-289–91
— cyclones and storm surges II-288
— fisheries resources II-294
— legal regime II-295
— Naaf estuary Ransar site II-289
— need for integrated coastal management II-295–6
— offshore system and fisheries resources II-292–5
— physical setting of the Bay of Bengal II-287
— population and agriculture factors in the coastal areas II-291–2
— seasonality, currents, and natural environmental variables II-287–8
bank reefs, Brazil I-723, I-723–4
banks and shoals
— Madagascar II-118
— Mascarene Plateau, habitats poorly known II-256
Bar al Hickman *II-18*
Barbados Marine Reserve I-639
Barcelona Convention I-249, I-262, III-363
Barents Sea, effects of collapse of capelin stock III-129
barnacles I-71, I-194
— barnacle belt, Spanish north coast I-140
— bioindicators of Cu, Zn, and Cd concentrations I-213
— Deltaic Sundarbans II-156
— Mauritius II-258
Barnegat Inlet *I-322*
barramundi fishery, Gulf of Papua II-602–3, II-607
barrier island lagoons I-437
barrier island-sound systems, diverse communities I-361
barrier islands I-353
— provide shelter for seagrass communities I-358
barrier reefs II-370
— Andros Island I-421
— Belize *I-502*, I-503, I-504
— affected by hurricanes I-504
— Barrier Reef Committee I-512–13
— coral mortality and macroalgae increase I-505

— system may be damaged by sediment, nutrients and contaminants I-508
— Fiji Islands II-756, **II-756**
— French Polynesia II-817
— Madagascar II-118
— Mauritius II-258
— New Caledonia II-725
— off northeast Kalimantan II-388
— San Andrés and Providencia Archipelago I-668
— *see also* Great Barrier Reef
basin water exchange processes I-21
basking sharks, Irish Sea I-91
Basque Country
— heavy industry decline, leaving environmental damage I-142
— recovery of coastal water quality I-142
Bass Strait *II-648, II-662*, II-663
— local sea floor pollution from offshore rigs II-669
— oscillatory tidal currents II-650
— species richness II-651
— tidal currents II-664
— topography II-663
Bassian Province II-649
bathing water quality I-51, I-76
— Aegean Sea I-248
— Israeli coast I-261
— Turkish Black Sea coast I-302-3
Bay of Biscay Central Water, off Ortegal Cape I-139
Bay of Fundy, Canada, tidal barrage III-316
bays, mudflats and sand spits
— southwestern Africa I-828-9
— — saltpans I-828
— — Walvis Bay and Sandwich Bay, Ramsar wetlands I-829
Bazaruto Islands, Mozambique II-104-5
— fishing a key activity II-109
— shallow and shelf waters II-104
beach angling I-835
beach armour, or beach nourishment III-66
beach erosion
— Australia II-585
— the Bahamas I-428
— Belize I-510
— Fiji Islands II-760
— human-induced I-211
— and sand drift, NSW, case study II-640
— South Florida I-428
— south-east Queensland and New South Wales II-639
— Tanzania, possible reasons for II-93
— — buffer zone principle II-93
— through sand mining II-782
beach forest vegetation II-777
beach formation, and longshore drift, east coast of Madagascar II-117
beach loss, in the Comoros II-248
beach mining *see* sand mining
beach nourishment/replenishment I-604, II-356, III-352
— Gulf of Guinea I-787
— sand for I-53-4, I-96, I-176, I-366, I-428
beach sands, quarrying of II-248
beach seining, Great Barrier Reef region II-623
beach vegetation, east coast, Peninsular Malaysia II-349
beach-rocks habitat, possible origin of I-273
beaches
— Adriatic, more severe storm patterns I-276
— Anguilla I-620
— artificial, for tourists I-604
— Curaçao, littered I-607
— impermeable layers affect oil penetration and persistence III-270
— loss of sand from, Jamaica I-569
— Malacca Straits II-312
— Maldivian II-203
— not meeting EU standard, Irish Sea coast I-95
— oil contamination of III-364

— and soft substrates, Victoria Province, Australia II-665-6
— Southern California, recreational shoreline monitoring I-399
— Sri Lanka, squatters on II-183
— Turkish Black Sea coast I-302, **I-302, I-303**
— *see also* sandy beaches
beaked whales III-92
Beaufort Sea, summer ice thickness and residual circulation III-190, *III-190*
Beaufort's Dyke *I-84*, I-85
bêche-de-mer fishery, Coral, Solomon and Bismark Seas region II-437
bêche-de-mer production, New Caledonia II-730
Belfast Lough *I-84*
Belgium, offshore wind farm planned III-309
Belize I-501-16
— after the 1998 bleaching and hurricane Mitch III-55
— coastal erosion and landfill I-509-10
— Coastal Zone Management Authority I-512, I-514
— Coastal Zone Management Unit I-512
— effects from urban and industrial activities I-510-12
— geology of I-504
— major shallow water marine and coastal habitats I-504-6
— Marine Protected Areas Committee I-513
— National Coral Reef Monitoring Working Group I-513
— offshore systems I-506
— populations affecting the area I-506-7
— — population and demography I-506-7
— — use of the coastal zone I-507
— protective measures I-512-14
— — challenges I-514
— — policy development and integration I-512-13
— — regulation of development I-513
— rural factors I-508-9
— seasonality, currents and natural environmental variables I-503-4
— source area for fish, coral and other larvae I-503
Belize Barrier Reef *I-502*, I-503, I-504
Belize City, habitat destruction during growth of I-510
Belle Tout Lighthouse, Beachy Head, relocation of III-353
beluga whales III-76, III-90
— contaminant-induced immunosuppression III-95
— hunted in Arctic and sub-Arctic III-92
Bengal, Bay of (northwest coast) and the deltaic Sundarbans II-145-60
— effects from urban and industrial activities II-154-6
— major rural activities and their impact II-153-4
— major shallow water marine and coastal habitats II-149-51
— protective measures II-156-9
— — captive breeding programmes II-158-9
— — conservation of biological resources II-156-7
— — conservation policies II-157-8
— seasonality, currents and natural environmental variables II-148-9
— social history and population profile II-151-3
Bengal, Bay of II-269-84
— chemical features of the water II-274
— coastal habitats and biodiversity II-274-7
— defined II-271
— effects from urban and industrial activities II-279-80
— marine fisheries II-277-8
— mining, erosion and landfill II-279
— natural environmental variables II-271-4
— new millennium: need for east coast zone management authority II-280-1
— physical setting II-287
— populations affecting the area II-278
— protective measures II-280
— — Coastal Ocean Monitoring and Prediction System, Indian Coast II-280
— — public awareness II-280
— rural factors II-278-9
Bengal Deep Sea Fan II-287
Bengal tiger II-150, II-290
— Project Tiger II-156, II-157

Bengkali Strait II-311
Benguela Current *I-822*, I-825
— eastern boundary current I-823
— fish kills, from upwelling, blooms and oxygen depletion III-129
— red tides III-218
Benguela ecosystem *II-134*
— upwelling II-135
Benguela Environment Fisheries Interaction and Training Programme I-837
Benguela Niño years I-825, I-826, I-827
benthic assemblages, offshore, Côte d'Ivoire I-814
benthic biomass, South China Sea coast **II-554**
benthic communities
— affected by turbidity and oxygen depletion III-259
— Gulf of Guinea I-782, **I-782**
— Irish Sea
    — linked to sediment type *I-86*, I-89
    — threats to I-89, *I-94*
— macrofaunal, English Channel bed I-71
— offshore, Carolinas coast I-362
— Portuguese coastal waters, temporal changes due to natural causes I-159–60
benthic fauna
— Aegean Sea I-240
— Baltic Proper, adverse changes below the halocline I-110
benthic microbial communities, effects of eutrophication on, the Bodden III-261–2
benthic monitoring III-243
benthic organisms
— as biomonitors of radionuclides I-116
— West Guangdong coast II-554
benthic species, high mortality rate from trawling III-123
Benthic Surveillance Project (USA) I-452
benthic vegetation, Portuguese coastal waters I-159–60
benthos
— Campeche Sound I-474
— Côte d'Ivoire I-813
— infaunal, filtering ability, Chesapeake Bay I-342
— Marshall Islands II-780
— Southern California Bight, effects of anthropogenic inputs I-395–6
Bergen *I-18*, I-27
Bering Sea *I-374*
— by-catch as an issue III-146
— decline in some marine mammal populations through the pollock fishery III-129
— Eastern, change in groundfish species composition III-128
— residual currents III-190
Bering Sea ecosystem, example of cumulative and cascading impacts III-98–9
Bermuda
— acid rain I-229
— Bermuda Atlantic Time-series (BATS) program I-224–6
— conservation laws I-230
— Hydrostation S program I-224–5
— threats to reef environment I-228
Bermuda High I-223
Bermuda Platform I-227
Biddulphia sinensis, an Asian introduction I-26
Bien Dong Sea II-563
— great natural resource potential II-567
— *see also* South China Sea
Bijagos Archipelago, Guinea-Bissau I-781
— breeding and nursery ground, fish and crustaceans I-781
— diverse tidal habitats I-781
Bikini Atoll, nuclear tests at, costs of clean up II-783
Bimini *I-416*
— deep water sport fishing I-422
— mangroves stunted I-422
— sand mining/dredging I-428
bioaccumulation
— of heavy metals II-156

— in seabirds and fish I-212
— of pesticides, northern Gulf of Mexico I-444
— in the Southern California Bight I-394
    — of DDTs and PCBs in seabirds I-398
    — effects of in fish I-397
    — in marine mammals I-399
— of TBT III-250
bioavailability, of contaminants in sediments I-25, I-26
biocides
— antifouling, the ideal III-249
— in marine antifoulant paints III-248
biodiversity II-336
— Andaman, Nicobar and Lakshadweep Islands II-192, **II-193**
— Anguilla I-620
— Arabian Gulf II-6–7
— Australia II-583
— of the Baltic Sea I-124–5
    — marine species decrease to the north I-124–5
— Bay of Bengal **II-274**
— benefits to tourism and recreation II-326
— British Virgin Islands I-620
— coral reefs III-34
    — endangered III-36–7
— Coral, Solomon and Bismark Seas region II-431
— Deltaic Sundarbans II-150, **II-150**
— English Channel
    — fauna I-70–1
    — flora I-69–70
— Fiji Islands II-757
    — marine **II-754**, II-762
— Godavari-Krishna delta II-170
— Great Barrier Reef II-613
— Gulf of Aden II-49
— Gulf of Guinea LME I-794
— Gulf of Mannar II-164, **II-164**
— high
    — macroalgae and invertebrates, the Quirimbas II-102, II-104
    — New Caledonia II-726, **II-726**
— hunted in Arctic and sub-Arctic III-92
— increase from fjord head to coast I-22
— Lakshadweep Islands II-194
— Malacca Straits II-312
— marine
    — Australian Bight II-678–9
    — Hawaiian Islands II-794
    — southern Australia II-582
    — Vanuatu **II-741**
    — Vietnam II-564
    — western Indonesia, threats to II-389–90
— Palk Bay **II-166**
— Palk Bay–Madras coast II-169
— Patagonian shores I-756
— Peru I-691–3
    — biological effects of El Niño I-692–3
    — wetlands and protected areas I-691–2
— the Philippines II-408–10
— potential loss of in the Comoros II-251
— range of effects of fishing III-378–9
— rocky shores, eastern Australian region II-633
— South African shores II-135
— in the Sundarbans II-291
— super-K species III-39
— and system integrity, mangroves III-28
— in terms of higher taxa, mine tailings III-241–2
— threatened by by-catch III-136
— western Indonesia II-389–90
— Wider Caribbean I-589
— within the Belize coastal zone I-506
— Yellow Sea, loss of II-497
biodiversity management, Great Barrier Reef Marine Park II-625–6
biodiversity recovery, mine tailings III-241

bioerosion II-69
— contributes to destruction of the reef matrix II-397
— the Maldives, after coral reef bleaching event II-214
biofilms
— as indicators for eutrophication III-263
— and microorganism habitats III-261
— oxygen supply to III-263
— photoheterotrophic III-263
— toxic III-219
biogenic species, effects of oil spills III-273–4
biogeochemistry, around Bermuda I-225
biogeography
— Australia II-581
— eastern Australian region II-631
— equilibrium theory of III-380
— Fiji Islands II-753
— Vanuatu II-739
— Western Australian region II-695
biological communities
— and coastal stability III-352
— effects of oil spills III-272–5
    — alterations in pattern of succession and dominance III-275
    — on consumers (predators and herbivores) III-274
    — on prey species III-274
    — on sensitive species with localized recruitment III-274–5
    — on structuring communities III-2734
biological factors, the key to the Sundarban coast II-148
biological invasions *see* alien/accidental/exotic/introduced
    organisms/species
biological oxygen demand (BOD)
— direct increases in through nutrient loading I-367
— high, leads to hypoxia/anoxia I-364
biological production
— increased, Baltic Sea I-130
— Portuguese coastal waters I-153–4
biology, evolutionary, and "punctuated equilibrium" III-86
biomagnification I-5
— of Cd and Hg, Greenland I-9, *I-10*
— of mercury in the biota, Jakarta Bay II-396
"biophile" elements I-116
biophysical features, gradients in
— Hawaii II-795
— the Maldives II-202, **II-202**
biopollution *see* alien/accidental/exotic/introduced organisms/species
bioregions, Tasmania II-651
biosphere, value of ecosystem services III-395
Biosphere Reserves
— Gulf of Mannar II-163
— Mananara Nord, Madagascar II-127
— Nancowrie Biosphere Reserve II-196
— Odiel saltmarshes I-170
— Rocas Atoll, Brazil I-724
— Sikhote-Alin Biosphere Reserve, eastern Russia II-481–2
— Sundarban Biosphere Reserve II-147, II-156–7
biota
— influence on oil persistence III-271
— Madagascan, effects of climate on II-116–17
    — migratory patterns of some species II-116
    — shallow-water assemblages differ from North to South II-116
biotoxin and exposure disturbances III-218–19, *III-220*, **III-220**
biotoxins
— causing mortalities III-218
— cyanobacterial, implicated in chronic diseases III-218
— effects of direct exposure to III-219
birds
— Asian, organochlorine pollution in II-452, *II-454*
— commercially reared III-225
— Doñana National Park as breeding site and migratory stop I-171–2
— El Salvador I-549
— Gulf of Mexico Coast *I-473*, **I-473**, I-474
— land and sea, Marshall Islands II-778

— Mai Po marshes, Hong Kong, high species diversity II-539
— migratory
    — disease among III-225
    — exposed to HCHs and PCBs in India II-452
    — Sarawak II-368
— *see also* marine birds; seabirds; shorebirds
Biscay, Bay of, weak circulation I-138
Biscayne Bay *I-406*, I-407, III-2, III-9
Biscayne National Park I-412
Bismark Sea *II-426*
Bitter Lakes, kept Suez Canal salinity high II-38
bivalve molluscs, particle-feeding III-240
bivalve mortalities III-226
bivalves
— Aegean I-239
— Bothnian Bay I-125
— filter feeding activity reduced by brown tides III-219–20
— Portuguese coastal waters, some bacterial problems I-158
— southern Baltic I-109
BKD (kidney disease) I-35
Black River Morass, Jamaica I-565
— mangroves I-565
Black Sea I-285–305
— anoxic interface I-287
— climate I-287
— coastal development expected to continue I-303
— effects of recurring hypoxic conditions III-219
— environmental policy difficult to implement I-304
— land-based pollution I-296–303
— major shallow water marine and coastal habitats and offshore
    systems I-290–4
— residence time of waters I-303
— seasonality, currents, natural environmental variables I-287–9
— southern Black Sea, Turkey I-294–6
— surface circulation I-287–8, *I-288*
    — upper layer general circulation *I-289*
— world's largest anoxic water mass I-287
Black Sea Environmental Programme I-290, I-294–5
— Black Sea Action Plan I-295, I-303
— financing of I-295
Black Sea Water I-236, I-243
blacklisting, of fishing vessels III-160–1
blast fishing *see* destructive fishing
blast fishing, Indonesian seas II-382
blue crab pot fishery I-362
— Pamlico and Neuse estuaries I-360
Blue Mountains, Jamaica, effects of I-561
blue mussels I-52, I-131
— Baltic Proper I-126
— Greenland, high lead concentrations I-12
blue whales III-82, III-82–3
— secretly killed III-77, III-86
BOD *see* biological oxygen demand (BOD)
Bodden, southern Baltic Sea
— changes of ecosystem structure and function following
    eutrophication III-264, **III-265**
— characteristics of III-260
    — high filter and buffer capacity III-260
— eutrophication of III-260–1
    — buffer capacity exhausted III-261
    — causes of III-260
    — effects on benthic microbial communities III-261–2
    — investigations of the impact on the nitrogen cycle III-262–3
— Nordrügensche Bodden, effects of increasing eutrophication
    III-261–2
— remediation possibilities III-263–4
bolide impact, effects of Chesapeake Bay I-337
"Bolivian winter" I-705
Bonaire *I-596*, I-597
— coastal development I-606
— coastal and marine habitats *I-599*

— dive tourism I-597
— excavation for construction, ruined groundwater quality I-602
— fishery I-603
— Flamingo Sanctuary and Washington Park I-609
— sewage discharge I-606
— turtle grass I-601
Bonifacio Strait I-273
Bonn Convention III-340
Borneo II-361–79
— erosion and landfill II-375
— habitat types II-364–5
— human populations II-373–4
— major coastal habitats II-366–72
— management objectives for marine protected areas II-377
— marine conservation areas II-376–8
— natural environmental variables II-365–6
— offshore systems II-372–3
— regional extent II-363–5
— rural factors II-374–5
— urban and industrial pollution II-375–6
Bornholm Basin I-101
— deterioration of oxygen conditions I-105
Bornholm Deep I-102, I-106
Bosphorus I-198, *I-286*
— connects Black Sea to the Mediterranean I-287
Boston *I-308*, I-312, I-313
Boston Harbor I-314, I-317
— a history of lead I-315
Bothnian Bay *I-100*, *I-122*, *I-123*, I-126
— heavy metals I-131
— standing stock and carbon flow I-126, *I-127*
Bothnian Sea *I-100*, *I-122*
— fresh water species I-125
— heavy metals I-131
— standing stock and carbon flow I-126, *I-127*
bottlenose whales III-76, III-83, III-92
bottomland forest wetlands, Gulf Coast, USA I-440
boulder/cobble shores, the Azores I-207
Boundary Current, North and East of Madagascar *II-114*
boundary-layer flow III-206
— *see also* internal boundary layers
bowhead whales III-76
— catch limits III-83
— distinct stocks III-81
— occasional Canadian aboriginal kills III-81
brackish-water conditions, adaptation to in Baltic Sea I-125
Brahmaputra River *II-146*
braided channels, Somalia
— biota in II-77
— channel levees II-77, *II-78*
— encrusted hard bottoms *II-76*, II-77
Brazil
— degradation of coastal areas I-463
— education to solve some problems I-463
— mangroves III-20
— Special Management Zone, Bahia de Caraquez I-463
Brazil Current I-722, I-751, I-753
— Abrolhos Bank–Cabo Frio region I-734
Brazil Current–Falklands/Malvinas Current confluence zone I-751
Brazil, southern I-731–47, I-744–5
— activities I-737–40
— degradation I-740–3
— east coast I-733
— Abrolhos Bank–Cabo Frio I-734
— habitats I-735–6
— environmental laws I-744–5
— territorial sea defined I-744
— historical setting I-733–4
— major shallow water marine and coastal habitats I-735–7
— physical description I-734–5
— south coast I-733
— habitats I-737
— large coastal plain I-737
— Southern Brazil Shelf I-735
— southeast coast I-733
— habitats I-736–7
— South Brazil Bight I-734–5
Brazil, tropical coast of I-719–29
— major coastal habitats I-725–7
— major environmental concerns and preservation I-727–8
— major marine habitats I-722–5
— oceanographic parameters I-721–2
— the region I-721
— Rocas Atoll I-723
— Biosphere Reserve I-724
— Southern Bahia, coral reefs I-724
breaking wave forces, increased with rising sea level III-190
bridges, may obstruct water flow in estuaries II-135
brine pools
— contain hydrogen sulphide I-442
— hot, Red Sea II-39
brine rejection, due to freezing I-33
British Indian Ocean Territories (BIOT)
— included in UK's ratification of conservation and pollution Conventions II-231
— *see also* Chagos Archipelago, Central Indian Ocean
British Virgin Islands *I-616*, I-617
— Coast Conservation Regulations I-624
— coral communities I-618
— effects from urban and industrial activities I-623
— Exclusive Fishing Zone I-621
— expansion of tourism I-620
— hurricane damage I-622
— mangroves I-619
— marine protected areas I-625
— mooring system I-624
— National Integrated Development Plan I-624
— tourism I-620
Brittany coast *I-66*, I-69, I-74
Broad River Estuary *I-352*, I-354
brominated chemical, widespread contamination by III-362
brown tides I-439, III-1298
— anoxic impact III-219–20
— cause persistent economic fisheries losses III-220
— high bivalve mortality III-220
Browns Bank *I-311*
Brunei *II-362*, II-365
— population II-374
Bryde's Whale II-676, III-78
bubble burst activity 202
— change in deposition velocity due to III-202–3, *III-203*
Buckingham Canal, southeast India, a health hazard II-170
Buenaventura Bay *I-678*
— high hydrocarbon levels in bivalves I-683
Buenos Aires Province I-751
— coast *I-752*
— coastal erosion I-762
— effects of harbour construction I-762
Burmeister's porpoise III-90
burning
— of grassland, Madagascar, leads to topsoil loss II-123–4
— of secondary forest, produced air polluting haze, Borneo II-375
— *see also* fire
burrowing animals, effects on oil spills III-271
Busc Busc Game Reserve II-80
Busc Busc, Lac *II-67*, II-68
by-catch I-445, II-57, II-59
— black-browed albatross as I-758
— by-catch reduction programs III-140–1
— reductions in BPUE III-141
— reductions in effort III-140–1
— as a component of fishing mortality III-140

— creating conservation problems III-136
— defined III-137
— and discard mortality III-123–5
    — reasons for discard III-123
— Great Barrier Reef region II-623
— history of the issue: some early examples III-141–7
    — coastal gillnets and seabirds III-146
    — discards in shrimp and prawn trawls III-143
    — gillnets and cetaceans III-144–5
    — high seas drift nets III-145
    — longlines and sea turtles III-146
    — longlines and seabirds III-145–6
    — Northeast Pacific groundfisheries III-146–7
    — shrimp–turtle problem III-143
    — trawls and cetaceans III-145
    — tuna–dolphin problem III-141–3
— and incidental take III-379
— includes many dolphins III-91
— indirect results of III-379
— a main fisheries issue III-161
— of non-target organisms III-368
— not incorporated in most fisheries management models III-367
— originally ignored, now important III-136–7
— prawn trawling Torres Strait and Gulf of Papua II-604
— problems and solutions III-135–51
    — by-catch classification: why is it useful? III-139–40
    — definitions III-137
    — into the 21st century III-147–8
    — reasons for discarding III-137–8
    — regulations and guidelines III-138–9
— of seabirds III-113
— shrimp fishing/trawlers I-490, I-536, I-539, I-554, II-121
— solutions to III-147
— squids, southern Brazil I-739
— and technology III-148
— wasteful I-3
by-catch-per-unit effort (BPUE), reductions in III-141
— deployment and retrieval changes III-141
— management action III-141
— training III-141
by-catch-reduction devices (BRDs) III-144, III-145, III-146
bycatch quotas III-156
Bylot Sound, Thule, nuclear weapon accident, plutonium contamination I-11

Cabo de Santa Marta Grande-Chui, southern Brazil
— environmental problems I-743
    — primary problems I-743
— major activities I-739–40
    — tourism I-739
— marine and coastal habitats I-737
— Port of Rio Grande I-740
Cabo Delgado, dividing point for South Equatorial Current II-101
Cabo Frio–Cabo de Santa Marta Grande, southern Brazil I-736–7
— bays
    — Rio de Janeiro coast I-736
    — Sao Paulo State I-736
— environmental degradation I-738
    — Cubatao Pollution Control Project I-743
    — from unregulated urbanisation I-740
    — in Guanabara Bay I-741
    — Santos estuary I-742–3
    — in Sepetiba Bay I-741
— estuarine–lagoon complex, Iguape–Cananéia–Paranaguá
    — important littoral ecological system I-736–7
    — subsistence agriculture and fisheries I-739
— major activities I-738–9
— mangroves degraded, south of Caraguatatuba Bay I-736
cachelot see sperm whales/whaling
Cadiz Bay *I-168*
— commercial port I-176
— saltmarshes in decline I-177
— urban–industrial development environmentally hazardous I-174
Cadiz, Gulf of *I-168*
— connects Atlantic and Mediterranean trading ports I-181
— contamination related to heavy industrial activity I-179, **I-180**, *I-181*
— meeting of water masses I-169
— monitoring quality of coastal waters I-179
— rich fishing grounds I-173
— use of traditional fishing techniques I-175–6
Cadiz–Tarifa arc
— rocky I-172
— rugged bottoms, Tarifa I-172
cadmium (Cd) I-5, I-75, I-260–1, I-301
— Baja California I-495, **I-496**
— concentrations in the Azores I-213
— in Greenland seabirds and mammals I-9, *I-10*, I-14
— levels unsafe in Norwegian mussels and fish liver I-25
— off Cumbrian coast I-94
— in whales I-230
caesium (Cs), conservative behaviour I-13
caesium-137(^{137}Cs)
— from Chernobyl I-11
— from Sellafield
    — by long-distance marine transport I-11
    — in Irish Sea water I-94, **I-95**
— outflow from the Black Sea I-249
— Sea of Japan II-479
Caicos Bank *I-588*
— reef areas I-590
Caicos Passage *I-416*, I-417
calamari fishery, Victoria Province, Australia II-668
calcification, corals
— enhanced by algal turf communities III-38
— and nutrient uptake III-35–6, *III-36*, III-38
— physiology of III-35, *III-36*
California Current *I-374*, *I-386*, I-486, I-487, I-488, I-489, I-543, III-113, *III-180*
— affected by El Niño and La Niña events I-387
California sea lion I-399
Californian Coastal Province, Mexico I-486–7
— environmental importance of I-487
— features of I-486
— habitats I-486–7
— monitoring work I-487
Calvados Coast *I-66*, I-69
Calvert Cliffs *I-336*, I-346
Cambodia II-299
Cambodian Sea II-569–78
— coastal deterioration due to erosion and landfill II-576
— coastal habitats II-573–4
— coastal population II-575
— defined local marine environment II-571
— effects of the rural sector II-575–6
— effects of urban and industrial development II-576–7
— offshore habitat II-574–5
— physical and chemical conditions in surface waters II-572–3, **II-573**
— protective measures II-577–8
    — limited perception of environmental impact II-577
    — status of marine environment and habitats protection measures **II-577**
— seasonal variability of the natural environment II-571–3
Cameroon
— conservation concerns I-790, *I-791*
    — coastal environment assessment I-790
— increasing waste contamination I-789
— rivers I-777
Campeche Sound *I-468*
— oil industry I-476
— primary production I-474
Canadian Archipelago I-5, *I-6*

Canary Current I-204, *I-204*, I-223
Canary Islands I-185–99
— climate I-188
— development pressures and protective measures I-196–8
    — ecosystem characteristics and impacts on coastal communities **I-197**
— fishing resources I-195–6
— marine ecosystems I-193–5
— physical and geological background I-187–9
    — geological evolution of I-187
— seasonality, currents and natural environmental variables I-189–92
Canary Islands Counter Current I-189
Canary Islands Stream I-189
canneries, Fiji and American Samoa II-718, II-770
canning industry, Venezuela I-653
Cantabrian Sea *I-136*
— summer subsurface chlorophyll maximum I-141
Cantabrian Shelf, sand and silt *I-136*, I-137
Cap-Breton Canyon *I-136*
Cape Cod *I-308, I-311*, I-314
Cape Fear Estuary I-353, I-362
— source of chronic BOD load I-364
— well flushed I-363
Cape Fear River *I-352*, I-354, I-361, I-364
— turbidity, faecal coliforms and BOD, correlation I-364, **I-365**
Cape Hatteras *I-352*
— associated with geographic division of plankton I-356
Cape Horn Current I-701
Cape Johnson Trench *II-426*
Cape Sable *I-308*
capelin I-376
— industrial fishery based on I-18
capelin fishery, collapse of affected small cetaceans III-97
Capelinhos Mountain, Faial I-203, I-207
captive breeding programmes, Sundarbans
— estuarine crocodile II-158
— horseshoe crabs II-158–9
— Olive Ridley turtle II-158
carbon dioxide (CO_2)
— anthropogenic, elevated III-38
— atmospheric
    — increase in III-188
    — reduces oceanic $CaCO_3$ supersaturation III-36
— disposal at sea to mitigate climate change III-366
— from conversion of bicarbonate III-35, III-35–6
carbon dioxide pollution, transferred to deep sea I-7
carbonate platforms, the Bahamas I-417
carbonate sediments
— dominate tropical Brazilian middle and outer shelves I-722–3
— open shelf, Australian Bight II-681
carbonates, biogenic and chemically precipitated, Borneo II-366
Cardigan Bay *I-84*
— shore communities I-88
— Special Area of Conservation I-90–1
Cariaco Gulf *I-644*
— *Thalassia* beds I-650
Caribbean, collapse of coral reefs III-37, *III-37*
Caribbean and Atlantic Ocean, coral bleaching and mortality events III-54–6
Caribbean Basin, eddies within I-580
Caribbean Coastal Marine Productivity (CARICOMP), research and monitoring network I-634
Caribbean Current I-580, I-617–18, *I-628*, I-665
Caribbean Lowlands, Nicaragua I-524
— ethnic hierarchy I-524
— home to Miskito, Creoles and mestizos I-524
Caribbean Oceanographic Resources Exploration (CORE) I-634
Caribbean Sea I-579
Caribbean Small Island Developing States I-617
Caribbean Surface Water, density, temperature and salinity I-579
Caribbean–North Atlantic convergence I-497, **I-497**

Carlsberg Ridge II-49
Carolina Coasts, north and south I-351–71
— better management on non-point source runoff essential I-367
— coastal rivers, problems of I-361
— environmental concerns in estuarine and coastal systems I-363–6
— flora and fauna of the coastal waters I-354–63
— limits needed on nutrient inputs to rivers and estuaries I-367
— physical setting I-353–4
    — sources of pollutants I-354
— prognoses for the future I-366–7
carrying capacity III-86
Cartagena Convention
— protocol on Specially Protected Areas and Wildlife (SPAW) I-608, I-640
— SPAW protocol III-68
— and whale protection III-85
Carysfort Reef
— continuing decline in deep and shallow waters I-410, I-412
— coral vitality: long-term study I-410–12
— reaching stage of ecological collapse I-412
— sediments of I-409
"cascade hypothesis" III-99
catchment impacts, Tasmanian coastline II-654
categorical correlation matrix, from HEED survey *III-217*
catfish, mysterious mortalities III-223
Cay Sal Bank I-417, I-421
Cayman Islands, coral bleaching and mortality events III-54
cays
— Belize I-505
    — shifting populations I-507
CC:TRAIN III-327–8
— operates under TRAIN-X principles III-327
— role of III-327
— Vulnerability and Adaptation Assessment (V&A) COURSE III-327–8
CDOM (coloured dissolved organic matter) III-297
cellulose industry, Chile, disposal of liquid waste I-712–13
Central Adriatic I-269
— surface circulation I-271
— water column divisions I-270–1
Central America
— analysis of coastal zones I-463
— described I-460
— institutional issues in coastal resource management I-460
— major coastal resource management issues **I-460**
Central Bass Strait Waters II-664
central Bight water mass, salinity of II-676
Central Equatorial Water I-679
central south Pacific gyre ecosystem I-706
Central South Pacific Ocean *see* American Samoa
cephalopods, important in Gulf of Thailand II-306
'Certain Persistent Organic Pollutants', negotiations on III-340
cetaceans
— caught by gillnets III-144–5
— caught in midwater trawls III-145
— in the English Channel I-70
— Great Australian Bight II-680
— Gulf of Aden II-54
— increased strandings in the North Sea I-50
— organochlorine residue levels, western Pacific II-455, *II-459*
— Patagonian coast I-756–7
— PCB concentrations II-455, *II-459*
— Southern California I-399
— vulnerable to undersea noise II-685
— *see also* dolphins; whales
Chagos Archipelago, Central Indian Ocean II-221–31
— available for defence purposes, with conservation provisions II-230
— biogeographic position in the Indian Ocean II-224–5
— a British Indian Ocean Territory II-223
— geographical and historical setting II-223–4

— importance of Chagos II-231
— major shallow water marine and coastal habitats II-226–9
— offshore systems II-229–30
— population, urban and industrial activities II-230
— a pristine environment II-230
— protective measures II-230–1
— reef studies II-223
— seasonality, currents, natural environmental variables II-225–6
Chagos Bank *see* Great Chagos Bank
Chain Ridge *II-64*, II-67
chalk cliffs, erosion of I-68
Challenger Bank I-227
Chang Yun Ridge, Taiwan Strait, effects on pollutants II-501–2
chank fishery, Sri Lanka II-183
Channel Islands I-71
Char Bahar *II-18*, II-24
Charleston Harbor *I-352*, I-353
chemical cargoes, loss of from shipping III-364
chemical contamination/pollution
— Amursky Bay, eastern Russia **II-484**
— Bay of Bengal II-295
— English Channel, and its impacts I-74–6
— Faroes I-31, I-36
— northern New South Wales catchments II-637
— Norway I-24–5
— southern Brazil, Paranaguá Port I-743
chemical energy, from natural oil seeps I-442
chemical industry, southern Brazil I-738
chemical spills, toxic I-72–3
chemical warfare agents, dumped in the Baltic Proper I-117
chemical wastes, Bangladesh, disposal of II-295
Chernobyl
— contamination from I-11, I-55
— — in Baltic seawater I-116–17
Chesapeake Bay I-335–49, III-207
— C and D canal *I-336*, I-339
— coastal erosion and landfill I-346
— the defined region I-337–9
— eutrophication and remediation III-205, III-206
— land use *I-338*
— largest estuarine system in the USA I-337
— major shallow water marine and coastal habitats I-341–3
— offshore systems I-343–4
— pesticides affected eelgrass III-9
— populations affecting the area I-344–5
— principal rivers entering I-337, I-339, **I-339**
— protective measures I-346–8
— — nutrient management plans for manures I-347
— — pollution clean up I-346
— removal of eelgrass by rays III-8
— rural factors I-345
— seasonality, currents, natural environmental variables I-339–41
Chesapeake Bay ebb tidal plume I-344
— USEPA Chesapeake Bay Program I-346
Chesapeake Bay ecosystem, loss of keystone species leads to ecosystem shift III-227
Chesil Beach *I-66*, I-68
Chichester Harbour *I-66*, I-74
Chile
— industrial fishery development I-463
— possibilities for coastal management plan I-463
— small cetaceans as fish bait III-92–3
Chile–Peru Current I-705
Chilean Coast I-699–717
— advances in control and pollution abatement I-715
— coastal marine ecosystems I-705–9
— — major determinants of distribution/abundance of marine species I-708–9
— human coastal activity I-709–15
— large and mesoscale natural variability I-702–5
— — long-term variations: El Niño Southern Oscillation I-702–4
— — seasonal variations I-704–5
— physical setting I-701
— — oceanic islands I-701
— water masses I-701–2
Chilean Coast oceanic islets I-701
Chilean Trench I-701
Chilka Lake, India, and environmental problem II-279
China
— aquaculture
— — allocation and utilisation of natural resources III-174
— — changes in living standards and consumer preferences III-172–3
— — changing scenarios in development III-172–5
— — freshwater aquaculture expansion III-173
— — production diversification III-174–5
— — provincial expansion III-173–4
— — short comings in the sector III-175
— coastal population, Yellow Sea II-491
— increased agricultural output, Yellow Sea coast II-492
— legislation and regulations concerning water quality III-175, **III-175**
— mortalities in mariculture attributed to red tides III-219
— oil spills II-495
— species used in aquaculture III-171, III-174–5
China Sea, effects of sea level rise, cases considered III-190, *III-192*, III-193
chlorinated compounds, Great Barrier Reef region II-623
chlorinated polycyclic aromatic hydrocarbons (Cl-PAHs), in the Baltic I-114
chlorobiphenyl congeners, decreased concentrations, Baltic Sea I-114
chlorofluorocarbons (CFCs) III-188
chlorophyll
— Argentine Sea I-757, *I-757*
— concentrations, Côte d'Ivoire I-814
— deep chlorophyll maximum, Levantine Basin I-256
— southern Aegean Sea I-240
chlorophyll *a* III-297
chlorophyll content
— Case 2 waters, determination of III-297
— distinguished from CDOM III-297
chlorophytes I-821
Chocó Current I-679
Chokoria Sundarbans II-289, II-290
Christiaensen Basin, New York Bight, highly contaminated I-325
Christmas Island II-583, II-696
Chrysochromulina leadbeteri bloom I-26
Chrysochromulina polylepis bloom (spring 1988) I-20
Chumbe Island, Tanzania
— Marine Park II-94
— — environmental education programme II-94
Chwaka Bay, Tanzania, effects of herbicide use II-89
ciguatera poisoning, Mascarene Region II-261–2, II-264
Ciguatoxic Fish Poisoning (CFP) III-221
circulation
— Aegean I-236
— Argentinian continental shelf I-754
— Bay of Bengal, monsoonal II-273, *II-273*
— English Channel I-67
— fresh water-induced, off estuaries I-46
— Gulf of Mexico, Loop Current system I-469
— largest Baltic Sea basins I-123
— northern Gulf of Mexico I-439
— Yellow Sea and East China Sea II-489
circulation patterns
— Coral, Solomon and Bismark Seas Region II-428, *II-428*
— the Maldives II-201
— Polynesian South Pacific II-816, *II-816*
CITES III-340
— Convention on Migratory Species III-68
cities
— Guangdong II-559

— Indonesia **II-384**
— the Philippines II-416–17
clams, Maputo Bay, *Vibrio* contamination II-109
clay–oil flocculation, reduces oil retention in fine sediments III-271
clear water, assists remote-sensing III-284
cliffs, high, central Peruvian coast I-689
climate
— Arabian Gulf II-4
   — winter and summer monsoons II-4
— Arabian Sea coastal areas II-20
— Bahamas archipelago I-418–19
— Chesapeake Bay I-339–41
   — tornadoes and hurricanes I-340–1
— and coastal hydrography, Andaman and Nicobar Islands II-191
— coastal, tropical Brazil I-721
— Coral, Solomon and Bismark Seas Region II-428
— Côte d'Ivoire I-808–9
— east coast, Peninsular Malaysia II-347
— eastern Australian region II-631
— effect on Madagascar's biota II-116–17
— El Salvador I-547
— Fiji II-753
— Florida Keys, hurricanes I-408
— Great Australian Bight II-675
— Great Barrier Reef region II-613
— Gulf of Guinea I-775–6
— Hawaiian Islands II-795
— leeward and windward Dutch Antilles *I-596*, I-598
— Mexican Pacific Coast I-486, I-487, I-488, I-489
— Nicaragua I-519, I-534
— northern Gulf of Mexico I-439
— the Philippines II-407, *II-407*
— Red Sea, driven by migration of the Inter-Tropical Convergence Zone II-39
— Sea of Okhotsk, similar to arctic seas II-465
— and seasonal rainfall, Chagos Islands II-225
— seasonal variation, Chilean coast I-704–5
— the Seychelles
   — controlling factors II-235
   — humid tropical II-235
— Somalian Indian Sea coast
   — bi-modal rainfall II-65–6
   — temperature II-66
— Vanuatu II-739
— (weather), Borneo II-365
— (weather), French Polynesia II-815
— western Sumatra II-386
— Xiamen region, China II-515
climate change III-369–70, III-394
— and changes in marine mammals and seabirds I-9
— and changes in thermohaline circulation I-33
— combined with stressors, possible effects of III-230
— and coral reef degradation II-44
   — *see also* coral reef bleaching
— Global Seagrass Declines and Effects of Climate Change III-10–11
— and greenhouse gas emissions III-47
— Gulf of Alaska III-179–86
— monitoring for Caribbean Planning for Adaptation to Climate Change project, Jamaica I-571–2
— and the North Sea I-47
— past, west coast, Sea of Japan II-483
— and sea level change, effects on coastal ecosystems III-187–96
— and warming, Chagos Islands II-225
Climate Change Convention III-327
— implementing the challenges III-327
Climate Change Project (GEF), the Maldives II-215, II-216
climate disturbances, significant III-224
climate influences, act alongside global-scale environmental change III-225
climate shift, Gulf of Alaska I-377
cloud cover, limiting factor for remote sensing III-286

co-management
— in fisheries III-159
— Fisheries Master Plan, Mozambique II-111
— from management to co-management: the *Pomatomus saltatrix* fishery II-138
co-occurring biological anomalies III-214, III-217
coagulation, of particles III-200
coal mining
— Great Barrier Reef hinterland II-622
— southern Brazil I-739, I-743
coastal accretion
— Borneo II-366, II-375
— western Indonesia II-384
coastal area management, integrated, need for III-170–1
coastal area management systems, traditional, Coral, Solomon and Bismark Seas region II-442–3
coastal areas
— changes, tides and long waves through sea level rise III-188
— propagation of tides in III-188
— sea breezes III-207
— showing signs of eutrophication III-199
— Venezuela
   — areas under Special Regulation **I-658**
   — relevant planning regulations **I-657**
Coastal Biodiversity Action Plans I-77
coastal cold water *I-688*
coastal construction
— Hawaiian Islands II-805
— may provide additional solid substratum II-699
— Xiamen region II-519
coastal currents, Puerto Rico and US Virgin Islands I-580
coastal data coordination III-355
coastal defences
— English Channel coast I-78–9
— soft and hard, loss of, Mozambique II-108
coastal development
— in the Comoros II-247–8
— impacts of, the Bahamas I-427–9
— Western Australia, leading to habitat loss and alienation II-699
coastal dunes, heavy metal mining II-639
coastal ecosystems
— Andaman and Nicobar Islands II-191–2
— changes in, Xiamen region II-524–5
   — Maluan Bay II-524
   — Tong'an Bay II-524–5
— complex, Grande Island Bay with Septiba Bay, southern Brazil I-738
— Coral, Solomon and Bismark Seas region II-430
— effects of climate change and sea level on III-187–96
   — coastal effects III-188
   — effects on coastal storm surges and estuarine flood risk III-190, III-193–4
   — effects on tides and tidal currents III-188–90
   — other effects III-194–5
— Marshall Islands, threatened II-784
— Palk Bay II-166–7
— Palk Bay–Madras coast II-168–9
   — lagoon ecosystem II-169
— southeast India, degradation of II-172
— Taiwan, problems of ignorance and lack of public awareness II-511
coastal environment, and habitats, Bangladesh II-289–91
— beaches II-289
— mangroves II-290–1
— Matamuhuri delta and coastal islands II-289
— St Martin's Island II-289–90, **II-290**
— seagrasses II-290
Coastal Environment Program, the Philippines II-419
coastal erosion
— American Samoa II-770
— Ancash, Peru I-689

— Australia
    — and landfill II-585
    — and sea-level change II-585
— the Azores I-211
— the Bahamas I-428–9
— Bay of Bengal II-272
— Brunei II-375
— Cambodia II-576
— Chesapeake Bay I-337
— China's Yellow Sea coast, causes of II-493
— Colombian Caribbean Coast I-672–3, **I-673**
— Colombian Pacific Coast I-682
— Côte d'Ivoire I-816
— and deposition, El Salvador I-552, *I-552*
— east coast, Peninsular Malaysia II-353, II-356–7
— English Channel
    — English coast I-68–9
    — French coast I-69
— Great Australian Bight II-684
    — problem of uncontrolled vehicle access II-684
— Gulf of Guinea
    — and sediment supply I-777
    — through anthropogenic activities I-786
— Gulf of Mexico I-476
— Gulf of Thailand, west coast II-307
— Hawaiian Islands
    — now a socio-economic problem II-805
    — through subsidence II-804–5
— and landfill
    — Adriatic Sea I-276
    — Argentine coastlines I-762–3
    — around the North Sea I-53–4
    — Belize I-509–10
    — Bermuda I-228–9
    — Chesapeake Bay I-346
    — eastern Australian region II-639
    — Great Barrier Reef region II-622
    — Gulf of Aden II-58
    — Hong Kong II-542–3
    — Irish Sea Coast I-93
    — Jamaica I-569
    — Lesser Antilles I-636
    — Madagascar II-124–5
    — Marshall Islands II-782–3
    — Mozambique II-108
    — northern Gulf of Mexico I-445–8
    — northern Spanish coast I-144–5
    — Oman II-27–8
    — The Seychelles II-239
    — South Western Pacific Islands II-716–18
    — southern Spanish coast I-176–7
    — Sri Lanka II-184
    — Torres Strait II-603
    — Turks and Caicos Islands I-592
    — Tyrrhenian Sea I-276–7
— landfill, and effects from urban and industrial activities, West Australia II-700–2
— and landfill, Fijian Islands II-760
— Malacca Straits II-317
— The Maldives II-208–10
    — from pleasure boats II-209–10
— Mediterranean coast of Israel I-258–9
— New Caledonia II-731
— northern end of Sumatra II-339
— Peru I-696
— Portuguese coast I-161
— potentially a serious problem in the Mascarenes II-262
— Puerto Rico I-584
— Tanzania II-93
— Tasmania II-655
— Turks and Caicos Islands I-592

— Vanuatu II-744
— Victoria Province, Australia II-667
— western Indonesia, effects of shrimp ponds on II-395
— western Taiwan II-506
— Xiamen region II-517, **II-517**
coastal flooding, and sea level rise III-190
coastal forests
— Madagascar east coast, alleviate mangrove problem II-125
— Malacca Straits II-312
— Tanzania II-88, II-89–90
coastal habitats
— affected by livestock II-26
— American Samoa II-767–9
    — changes to II-770
— Australian Bight II-678
— Bay of Bengal II-274–7
    — loss of on east coast of India II-279
— and biodiversity, Arabian Gulf II-6–9
— Borneo II-366–72
    — coral reefs II-370–2
    — mangroves II-366–9
    — rocky shores II-369
    — sandy shores II-369–70
    — seagrass and algae II-370
— Brazilian tropical coast I-725–7
    — Abrolhos National Marine Park I-726–7
    — Atlantic Rainforest (maritime forest) I-725–7
    — restinga I-727
    — wetlands I-727
— Cambodia II-573–4
    — Botum Sakor National Park II-573–4
    — Kampot Bay habitat II-574
    — Koh Kong Bay II-573
    — Kompong Som semi-enclosed bay habitat II-574
— Côte d'Ivoire, depletion and degradation of I-817–18
— Dutch Antilles **I-597**
    — under environmental pressure I-612
— Gulf of Thailand II-301–4
— Oman and Yemen, effects of cold nutrient rich upwellings II-50
coastal lagoons I-645
— Adriatic Sea I-272, I-276
    — intrusion of allochthonous species I-282
— Baltic I-124
— and basin estuaries, Sri Lanka II-178–9
— and coastal lakes, Baltic Proper I-109
— Côte d'Ivoire I-811, *I-811*
    — Ebrié lagoon I-811–12
    — and estuaries, depletion and degradation of habitats I-817–18
— Curaçao, habitat destruction I-602
— and estuaries, Tasmania, depauperate II-653
— forming New River Estuary I-353
— Gulf of Guinea coast, suffering from eutrophication I-788–9
— Mar Chiquita coastal lagoon, Argentina I-755
    — changes made by alien reef-builder I-762
    — experimental aquaculture hatcheries I-754
    — heavy metals I-762
— Mexican Pacific Coast I-482, I-486, I-490
    — Huizache y Caimanero lagoon system I-493–4
    — Teacapán—Agua Brava lagoon system I-494–5, I-497
— Muthupet Lagoon, Palk Bay II-167
— Nicaragua, Caribbean coast I-522
    — depth preferences, fish and shrimp I-522
— Pulicat Lake, Palk Bay–Madras coast II-169
— saltwater, Coral, Solomon and Bismark Seas region II-430
— Venezuela I-651–3
    — affected by natural and anthropogenic factors I-651
    — Laguna de Tacarigua I-652
    — legal protection for some I-651
— west Taiwan II-502
    — industrial parks II-509
coastal lowlands, Nicaraguan Pacific coast I-533

coastal management
— Argentina, main concerns I-751
— Latin America, regional examples I-463–4
— rational, lack of, Adriatic and Tyrrhenian seas I-281
— southern Brazil I-744
    — active environmental groups I-745
    — Coastal Management Law not entirely successful I-745
    — National Programme of Coastal Management, proposed de-centralization I-744
    — preservation and conservation areas I-745
    — Sao Paulo I-744–5
— southern Brazil Rio de Janeiro State Environmental Agency I-745
— techniques, classification of I-464–5
— working with nature III-352–3
— Xiamen region, an integrated approach to II-525–32
coastal management, in the future III-349–58
— coastal management and policy evolution III-350–2
— data and inclusivity III-355
— financing III-356
— integration in III-350
— legal issues III-353–5
— physical systems and management III-352–3
— relocation of historic buildings III-353
— solitary waves III-355–6
— success needs political will III-351
— transboundary issues III-355
coastal management and planning
— effectiveness depends on access to relevant information III-356
— legal issues III-353–5
coastal management professionals, training programmes for III-329–30
coastal management programmes, integration with national climate change action plans III-330
coastal management schemes, at the planning stage III-351–2
coastal marine areas, self-management of II-442
coastal and marine ecosystems
— Malacca Straits II-312–14
— nearshore, El Salvador, ecological important I-547
coastal marine ecosystems, Chile I-705–9
— biogeography of the pelagic system I-707–8
— a biological perspective I-707–9
— south Pacific eastern margin I-705–7
coastal and marine problems, frequently transboundary III-355
coastal marine waters
— eastern Korea, wastewater pollution II-478
— western coast, Sea of Japan II-477–8
    — Amur lagoons II-477
    — Northern Sakhalin II-477
    — pollution
    — in Amursky, Nakhodka and Ussuriysky Bays II-478
    — from ore mining and chemical production, Zolotoy Cape–Povorotny Cape II-477–8
    — southern region, sporadic water pollution II-478
coastal models, for areas associated with sea-level rise III-195
coastal morphology, Colombian Caribbean coast I-666–7
coastal plains, fertile, Al Batinah and Salalah II-26
coastal platform, raised I-170
coastal pollution, Chile, related to geography I-709
Coastal Protected Areas (Proposed), Mozambique II-110, **II-110**
coastal protection, Sri Lanka II-184
coastal reclamation I-93
coastal reef terrace, Somalia, diversified ahallow marine environments II-69, *II-70*
coastal reefs, Western Australia, macroalgal and invertebrate communities II-693
coastal region, Southern Gulf of Mexico, recognition of importance of swamps I-472
Coastal Regulation Zones (CRZ), West Bengal II-157–8, **II-158**
Coastal Resource Management Program, the Philippines II-419
coastal resources
— Australian Bight II-682

— east coast, Peninsular Malaysia, conservation legislation II-355–6
— southern Brazil
    — destructive use of I-740–3
    — economic, social and environmental differences causing problems I-744
— Tasmania, commercial usage of II-655–6
— use in Coral, Solomon and Bismark Seas region II-434
coastal seas
— Coral, Solomon and Bismark Seas region
    — fisheries II-440
    — impacts of large urban areas II-439–40
    — industrial-scale impacts II-439
    — shipping and offshore accidents and impacts II-440–1
    — village-level impacts II-438
coastal and shallow water habitats, Puerto Rico I-582–4
coastal squeeze, causing habitat reduction III-353
coastal states, responsibility for pollution by seabed activities III-338
coastal terrestrial vegetation, Madagascar, varies II-119
coastal uses, Hawaiian Islands II-802–4
Coastal Warm Drift I-754
coastal waters
— Carolinas, flora and fauna of
    — benthic microalgae I-357–8
    — coastal benthic invertebrate communities I-360–2
    — estuarine and coastal finfish communities I-362–3
    — estuarine phytoplankton I-354–6
    — estuarine zooplankton I-359
    — macroalgae I-356–7
    — marine phytoplankton I-356
    — marine zooplankton I-360
    — offshore benthic communities I-362
    — seagrasses and other rooted submersed aquatic vegetation I-358–9
— Coral, Solomon and Bismark Seas region
    — impacts of land use II-435–6
    — subsistence and artisanal fisheries II-436–7
    — threats to sustainability II-437–8
— Norwegian, constituents of I-20
— West Guangdong coast, water quality II-555
— Xiamen region
    — bacterial pollution II-522–3
    — mainly in good condition II-521–2
    — Maluan Bay sediments a secondary pollution source II-524
    — oil pollution II-523
    — organic pollution, eutrophication and red tides II-522
    — sea dumping and disposal of solid waste II-523–4
    — threatened by rapid economic development II-522
coastal wave energy development
— current and future prospects III-315
    — Osprey 2000 III-315
— enclosed water column devices III-314, *III-315*
— location advantages and disadvantages III-314
— tapered channel concept III-314–15
coastal wetlands
— Argentina I-755–6
— Black Sea, changes in ecology of I-293
    — anthropogenic influences resulting in loss/degradation of I-293
— Côte d'Ivoire I-810–11
— Hueque, Venezuela, drainage and deforestation I-655
— loss of
    — northern New South Wales II-638
    — Yellow Sea II-493–4
— loss of, Southern California I-388
    — affecting seabirds I-398
— lost to landfill, Hawaiian Islands II-805
— mangrove destruction, the Bahamas I-428
— northern Gulf of Mexico, potential loss of I-446
— west Taiwan II-502–3
— *see also* khawrs; swamps; wetlands
coastal wilderness, Dutch Antilles I-601
Coastal Zone Management Act (USA) I-451–2

— National Estuarine Research Reserve System I-452
Coastal Zone Management Plan (CZMP), Sri Lanka II-185–6
— Special Area Management Plans II-186
coastal zone, Somali coast
— alterations of II-80
— geomorphic features II-68–72
    — Bajuni barrier islands II-69, *II-73*, *II-75*
    — braided channelized coast II-69–72
    — coastal reef terrace II-69, *II-70*
    — Merka Red Dune Complex II-68–9, *II-70*
    — rivers and alluvial plains II-68
— suffering degradation through lack of protective measures II-80
coastal zones I-4
— Argentina, main divisions I-751–2
— artisanal and non-industrial uses
    — El Salvador I-553
    — Madagascar II-125
    — Mozambique II-108–9
    — Taiwan II-509
— atmospheric deposition to III-205–7
    — atmospheric flow in, and atmosphere—surface exchange III-206–7
    — impacts in III-205–6
— Bay of Bengal
    — Calcutta and Howrah a significant detriment to II-148
    — mangroves II-149
— Belize I-503
    — aquaculture and fishing I-511
    — artisanal use of I-507
— and non-industrial use I-510–11
    — assessment of anthropogenic impacts I-507
    — effects of coastal erosion and landfill I-510
    — management of I-512
    — protected areas **I-513**
    — tourism I-511
— Brazilian tropical coast, for recreation and tourism I-728
— Cambodia traditional agriculture II-576
— Chile
    — flora and fauna I-707
    — industrial activity I-710–15
    — passage of rivers through I-704–5
    — southern marine habitats I-708, **I-708**
    — upwelling ecosystems and embayment ecosystems I-706
    — upwellings *I-705*
— Colombian Caribbean Coast, influence of Sierra Nevada de Santa Marta I-672
— community-based management, Hawaiian Islands II-809
— the Comoros, urban and industrial impacts II-251
— Côte d'Ivoire
    — cultural and historic sites I-818
    — mining of construction materials from I-816
— definitions of I-458
— east coast, Peninsular Malaysia
    — assessment of liquid and airborne pollution II-351
    — changes in catchment land use II-351
    — development of tourism II-351
    — relevant legislation and guidelines **II-356**
— eastern Australian region agriculture in II-636–8
— eastern Taiwan Strait II-501
— El Salvador I-547
    — coastline meso-tidal I-547
    — industrial uses I-554–5
— Guinea, pressure on natural resources I-801
— industrial effects I-511
— Integrated Management Plan for the Coastal zones of Brunei II-379
— legal definition in need of revision III-354
— Lesser Antilles, major features **I-632**
— Madagascar
    — cities II-125
    — industrial uses II-125

    — shipping and offshore accidents II-125
— Mozambique
    — industrial uses II-109
    — Mecúfi Coastal zones Management Projects II-111
    — a priority area II-110
    — Xai-Xai Sustainable Development Centre for the Coastal zones II-111
— need to understand physical systems III-352
— New Caledonia, potential effects of mining discharges II-731
— Nicaragua, Caribbean coast I-520
    — recreational and tourist destinations I-523
— Nicaragua, Pacific coast I-543
— Peru
    — arid semi-desert I-689
    — characteristics of I-689–90
    — climate influenced by Peruvian current I-690
    — 'wet desert' climate I-690
— pressure on and degradation of III-295
— shallow, human impact on III-7
— southeast India, domestic sewage a problem II-172
— southern Brazil
    — urbanization, tourism and industrialization I-734
    — very little is protected I-745
— Southern Gulf of Mexico, land classification I-475
— Sri Lanka
    — Coastal zones Management Plan (CZMP) II-185–6
    — pressures on II-181
— Taiwan
    — abused by public and private sector II-511
    — coastal land subsidence II-507
    — longshore currents II-506
— Tanzania II-85
— Tasmania
    — recreational use of II-655
    — urban development II-655
— tropical, sensors relevant for mapping III-284
— tropical Brazil, transgressive episodes I-722
— Venezuela, development pressure I-645
— Victoria, Australia
    — indigenous peoples II-666
    — white settlers II-666
coastal/marine structures
— effects on beach morphology I-258–9, I-276
— *see also* dikes and breakwaters
coastguards, for the Dutch Antilles I-609
coastline
— Gulf of Guinea, low-lying and swampy I-775
— Jamaica I-561
— Puerto Rico, shelf morphology I-581
— Venezuela
    — economic activities I-654–5
    — influential because accessible I-654
    — regions of I-645–6
— West Africa I-775
coastline modification
— Coral, Solomon and Bismark Seas region II-438
— French Polynesia II-821
coasts
— braided and channelized, southern Somalia II-69–72
— erosion, protection reduces sediment availability III-352
— North Sea, urban and artisanal use of I-54
Cobscook Bay, marine ecosystem bibliography I-316
coccolithophores, and algal bloom detection III-298
cockle harvesting/fisheries I-51–2, I-73
— disrupts the environment I-92
coconut tree replanting, Marshall Islands, benefits vs. impacts II-785
Cocos (Keeling) Atoll II-583
cod I-35, I-49, I-313
— Baltic Sea I-110
    — growth conditions less favourable I-128
    — concentrations of PCBs and DDT in **I-37**

"cold pool", connection to Georges Bank I-324
cold water dome, persistent, East China Sea II-508
cold-water plumes, Venezuelan coast I-646
collisional tectonics, Aegean area I-235
Colombia
— coral bleaching and mortality events III-54
— rehabilitation of key mangrove system I-463
Colombia, Caribbean Coast I-663–75
— coastal erosion I-672–3
— effects from urban and industrial activities I-673
— offshore systems I-669–71
— populations affecting the area I-671
— protective measures I-673
— rural factors I-671–2
— seasonality, currents, natural environmental variables I-665–7
— bimodal wet–dry seasonality I-665
— shallow water marine ecosystems and coastal habitats I-666–9
Colombia, Pacific Coast I-677–86
— coast is tectonically active I-682
— coastal erosion and landfill I-681–2
— effects from urban and industrial activities I-682–4
— major shallow water marine and coastal habitats I-679–80
— offshore systems I-680
— populations affecting the area I-680–1
— protective measures I-684–5
— Gene Bank of Fishing and Aquarian Resources I-684
— National Contingency Plan against Hydrocarbon Spills in Marine Waters I-685
— rural factors I-680–1
— seasonality, currents, natural environmental variables I-679
Colombian Current *I-678*
colonial powers, and the Malacca Strait II-337
colonisation, of North Sea still going on I-45
Commission de l'Océan Indien (COI) *see* Indian Ocean Commission
common dolphins, northern stock III-91
Common Fisheries Policy I-57
Common Market for Eastern and Southern Africa II-80
community-based coastal resources management, the Philippines II-418
Comores *see* Comoros Archipelago
Comoros Archipelago *II-114*, II-243–52, *II-244*
— effects from urban and industrial activities II-251
— major shallow water marine and coastal habitats II-245–6
— Anjouan II-246
— Grande Comore II-245
— Mayotte II-246
— Moheli II-246
— offshore systems II-246–7
— populations affecting the area II-247
— protective measures II-251–2
— regulations for environmental protection and management II-251
— watershed improvements II-252
— seasonality, currents, natural environmental variables II-245
— threats to the environment II-247–51
conch fishery I-591, I-635, I-637
— Belize I-509
— Jamaica I-572
conservation
— efforts in Belize I-512, I-514
— of whale stocks III-75–7
— Yellow Sea area, inhibiting factors II-495–6
conservation measures/policies
— Andaman, Nicobar and Lakshadweep Islands II-195–6
— for the Sundarbans II-157–8
Conservation Sensitive Management System (CSMS), analysis of coastal geomorphological sensitivity III-353
construction, and infilling of coastal mangroves I-637
consumption, and population III-396
contaminants
— chemical, safety limits in Norwegian fish and shellfish I-25
— chemical, secondary sources I-24
— entering the Southern California Bight
— in biota I-394
— DDT contamination I-390
— processes undergone I-390–1
— in sediments I-392–4
— in the water column I-391–2
— Gulf of Maine I-314–16
— input, transport and biological responses of, North Sea I-56
— levels in Greenland marine ecosystem I-5
— future trends I-14–15
— indirect evidence for I-9
— in the marine environment I-450
— Norway, within reach of tidal activity I-28
contamination I-158
— bacterial, Tagus estuary I-158
— biocide I-538
— chemical, from sewage, Arabian Gulf II-10
— Colombian Pacific coast, mainly transitory I-683
— marine, El Salvador I-555
— toxic, North Carolina estuaries I-360–1
— *see also* mercury contamination; microbiological contamination
continental collision II-19
continental seas, western Indonesia, uniqueness of II-383
continental shelf
— Norwegian I-23
— bottom fauna communities I-23
— bottom substrate I-23
— *see also* Argentinian continental shelf; South Brazil Bight
Continental Shelf Alternative (CSA) sites I-327
Convention on Biological Diversity I-608, III-341
Convention on the Conservation of Antarctic Marine Living Resources III-342
Convention for the Protection of the Black Sea against Pollution I-294
Convention on the Protection of the Marine Environment of the Baltic Sea Area (Helsinki Convention) I-101
Cook Inlet *I-374*, *I-375*
copepods I-814, I-831, II-54
— diapausing populations II-24
— Gulf of Alaska I-379–80
— North Carolina I-359
— Southern Bight I-48
copper (Cu) I-131
— Azorean amphipods I-213
— surficial sediments, Penobscot Bay *I-316*
copper levels, Island Copper Mine III-239
copper mining, Chile I-710
— lessons for the future, a case study I-713
— lessons for the future, case study, complaints about pollution and construction of new tailings lagoon I-713
— *see also* Island Copper Mine, Canada
copra, from Chagos Islands II-223, II-229
coprostanol, west Taiwan coast II-506, II-509
coquinas II-77
coral assemblages, Arabian Sea II-22
coral atolls
— American Samoa II-769
— Belize I-504–5
— Borneo II-370
— Chagos Archipelago II-223
— Lakshadweep Islands II-194
— Marshall Islands II-775
— the Seychelles II-236
— peripheral reefs II-236
— Western Australia II-693
— shelf edge
coral barrier, reef beaches, Cambodia II-574
coral bleaching I-419
— August 1998 I-421
— Belize reefs I-504
— related to water temperature I-409–10

coral carpets, Somali coast II-72
coral cays
— Borneo, Pulau Sangalaki II-371
— western Indonesia II-388–9
coral collection, Hawaiian Islands II-803
coral colonies, use of high-resolution airborne methods for status of III-288
coral communities
— Aruba, affected by oil pollution *I-596*, I-597
— British Virgin Islands I-618
    — effects of hurricanes I-618
— coastal reef terrace, Somalia II-69, *II-70*
— eastern Australian region II-633
— Gulf of Aden II-52
    — natural stressors II-52
— Gulf of Thailand II-304
— offshore islands, Taiwan Strait II-504
— southern Somali coast II-72
— West African coast I-780
— on Yemeni black basalt effusions II-51
coral degradation, tourist areas, Gulf of Thailand II-304
coral diseases
— global epidemic of III-226
— Hawaiian Islands II-807
coral harvesting, prohibited, the Bahamas I-430
coral mining II-315
— affecting reefs of Pulau Seribu II-397
— the Comoros II-248
    — uses of corals II-249
— contributes to reef and forest degradation, Tanzania II-91
— El Salvador I-554
— illegal, western Indonesia II-391
— Madagascar II-125
— the Maldives II-208
    — reefs show little sign of recovery II-208, *II-209*
    — sea-level rise effects exacerbated by II-215
— Mozambique II-108
— *see also* coral quarrying
coral mortality
— Colombian Caribbean Coast I-668
— due to Crown-of-Thorns starfish outbreaks II-616
coral pinnacles
— Bajuni Islands II-74
— Brazil I-723
coral quarrying, Gulf of Mannar II-165
coral reef areas
— no-take reserves III-386
— US Marine Protected Areas III-386
coral reef bleaching III-37
— 1998 event
    — Madagascar II-116
    — Sri Lanka II-180
— American Samoa II-712
— Chagos Archipelago
    — evidence of cover decline pre-1998 II-228, *II-228*
    — massive coral mortality after 1998 event II-228, *II-228*
— the Comoros
    — 1983 in Mayotte II-249
    — 1998 event, mass mortality II-249
— and coral death III-38
— Coral, Solomon and Bismark Seas region II-429
— Fiji II-711
— followed by coral disease III-224
— French Polynesia
    — 1991, 1994 and 1998 bleaching events II-822
    — problem of remoteness from coral recruitment II-822
— Great Barrier Reef II-617
— Gulf of Thailand II-304
— Hawaiian Islands II-807
— the Maldives II-202–3
    — 1998 event very extensive II-214

— consequences of on the socio-economic welfare of communities II-214
— Marshall Islands II-777
— and mortality, 1998 event III-43–57
    — Arabian Region III-47–8
    — Caribbean and Atlantic Ocean III-54–6
    — Central and Eastern Pacific Ocean III-53
    — Chagos Archipelago III-49
    — East Asia III-52
    — Indian Ocean III-48–9
    — interpretations and conclusions III-46–7
    — mechanisms of III-44
    — Pacific coast of the Americas III-53–4
    — Pacific Ocean, Northwest and Southwest III-52–3
    — Singapore, Thailand, Vietnam III-52
    — Southeast Asia III-49–52
— and mortality, Socotra II-52
— the Philippines II-408
— recent El Niño years, time too long for reef tolerances III-224
— several periods, Andaman Sea II-303
— the Seychelles
    — 1998 event II-236
    — massive mortalities II-237
— stress thought to be main reason with other factors II-137
— Vanuatu II-742
— western Indonesia II-389
coral reef communities
— east coast, Peninsular Malaysia II-348
— Moorea Island, French Polynesia II-819
coral reef ecosystems
— Andaman and Nicobar Islands II-191–2
— remote sensing of III-287–8
    — degrees of sophistications III-287
    — future challenges III-288
    — reef habitat maps III-287–8
    — representation of individual habitats III-287–8
    — use of colour aerial photography III-287
coral reef fish II-117, II-180
— The Maldives II-202
coral reef habitats, Sri Lanka II-180
coral reef microcosms, abundance in III-40
Coral Reef Monitoring Project, USEPA I-412
Coral Reef Rehabilitation and Management Project (COREMAP), western Indonesia II-393, II-399–400
coral reef species, Tanzania II-86–7
Coral Reef Symposia III-44
— late awareness of degradation III-37
coral reef zones
— Bermuda Islands I-227
    — rim and terrace reefs I-227–8
coral reefs II-693, III-33–42
— affected by hurricanes I-631
— American Samoa II-712, II-767–8
    — recovering from series of natural disturbances II-768
    — reef fish assemblage II-768
— Anguillan shelf I-618
— Australia II-582
    — vulnerable to eutrophication and sedimentation II-585
— the Bahamas I-421
    — near-shore health of I-429
— Bay of Bengal II-277
— biodiversity of III-34
    — based in calcium framework building III-35
— Borneo II-370–2
    — Berau Barrier reef system II-370
    — erosion of, Kota Kinabalu Bay and Tunku Abdul Rahman Park II-376
    — reef flats II-370–1
— Brazilian I-722, I-723–4
    — differ from well known coral reef models I-723
    — formed by coalescence of 'chapeiroes' I-723

— reef types I-723
— Cambodia II-574
— cf. rainforests **III-34**
— Chagos Islands II-223
　— biological patterns on the reef slopes II-227–8
　— changes over twenty years II-228–9
　— ecology of II-226–7
　— island ecology II-229
　— reef flats, algal ridges, spur and groove systems II-227
　— submerged banks and drowned atolls II-223, **II-224**
— Colombian Caribbean Coast I-668
　— best development in southwest I-668
　— general degradation I-668
— Colombian Pacific Coast I-680
— conservation, an international priority III-40–1
— damaged, Morrocoy National Park, Venezuela I-655
— decline in coral cover III-37
— degradation
　— early III-37
　— primary bases for III-40
— 'design' makes for vulnerability to changing environmental conditions I-413
— destructive fishing practices III-123
— Dutch Antilles
　— natural disasters I-607
　— and reefal algal beds I-599–600
— east coast, Peninsular Malaysia II-355
　— effects of creating Marine Parks II-355
— effects of sewage not well documented II-804
— endangered III-36–7, III-41
　— loss through destructive fishing methods III-37
　— lost through siltation and eutrophication III-37
— Fiji Islands II-755–7
　— reef provinces **II-756**
　— reef types II-756, **II-756**
　— studies on Suva reef II-756
— Florida Keys I-407–8
　— Carysfort Reef I-410–12
　— coral diseases I-410, *I-411*
　— degradation from agricultural and urban factors I-409
　— Florida Bay Hypothesis I-407
　— recruitment low I-410
　— stressed by environmental change I-408–9
— fossil, Red Sea coastline II-38
— French Polynesia II-817–19
　— Fangataufa atoll gastropod assemblage studies II-817
　— reef monitoring networks II-824
　— reef restoration schemes II-824
　— status of II-823
— Great Barrier Reef region II-616–17
　— Crown-of-Thorns Starfish outbreaks II-616–17
　— little evidence of long-term decline II-616
　— pressures and status II-616
— Gulf of Aden II-52
— Gulf of Mannar II-163–4
　— and the Gulf islands II-164
— Gulf of Mexico
　— impacts on some reefs **I-471**
　— principal characteristics **I-471**
— and hard bottoms, northern Gulf of Mexico I-442
　— decline of I-445
— Hawaiian Islands II-795, II-796–7
　— benthic reef life II-799–800
　— better management needed in northwestern islands II-810
　— Kane'ohe Bay, reef restoration II-796, II-809
　— monitoring of II-796
— indicators of oceanic health and global climate change I-413–14
— Jakarta Bay, once beautiful now almost destroyed II-397
— Jamaica I-562–4, I-566, *I-567*
　— changes in I-562
　— *Diadema* mass mortality I-563–4
　— differences in I-562
　— hurricane damage I-562, I-563
　— impacts of fishing I-568
　— impediments to recovery I-572
　— reef deterioration I-563
　— reef zonation I-562, I-562–3
　— some recovery I-563, I-564
— Kenya, result of predator overfishing III-128
— Lesser Antilles I-631
— Madagascar II-118
　— ancient II-118
— Malacca Strait II-336
— the Maldives II-202–3
　— need to keep reefs healthy II-218
　— protected from mining by tourism II-217
— and marine environments, pharmaceuticals from III-39
— Marshall Islands II-778–9
— Mauritius II-258
— natural perturbations III-37
— natural products, identification and extraction from III-40
— New Caledonia II-725–6
— Nicaragua, Caribbean coast I-520–2
　— *Diadema* mass mortality I-520
　— Miskito Coast Marine Reserve I-520, *I-521*
　— reef fish I-524, I-525
— Norwegian coast I-21–2
— Palk Bay II-166–7
— Papua New Guinea coastline II-429
— the Philippines II-408–9
　— effects of overfishing, sedimentation and destructive fishing II-415
　— primary productivity II-409
　— reef health II-408–9
— primary productivity III-35
— Puerto Rico I-582–4
　— fringing reefs I-582
　— patch reefs I-582
　— shelf reefs I-582, I-583–4
— Queensland Shelf II-619
— Quirimba Archipelago II-102
— Red Sea II-37, II-40–1
　— alignment of II-39
　— coral distributions II-41
　— effects of oil pollution II-43
　— fringing reefs II-40
— Reefs at Risk analysis III-44
— risk criteria classification evaluates potential risk from ports and harbours II-417–18
— role of nutrients in degradation III-38
　— sensitivity of to N, P and CO_2 III-38
— the Seychelles II-236
　— threatened II-236
— Singapore, smothered by siltation II-339
— social and economic value III-38, III-40–1
— Solomon Islands II-429–30
— Somali coast II-65
　— fringing reefs II-65
　— shelf and fringing II-72–4, II-80
— South Western Pacific Islands II-711–12
— Southern Gulf of Mexico I-469
　— and protected zones I-471–2, *I-472*
　— Vera Cruz reef system I-471
— Sri Lanka II-180
— Straits of Malacca II-314
— and submerged banks, Lakshadweep Islands II-194
　— degradation from siltation and sponge infestation II-194
— Taiwan II-503, II-504
— Taiwan Strait, increasingly threatened II-504
— Tanzania II-85–7
　— degraded sites II-86
　— restoration project, Dar es Salaam II-89

— Thailand
    — Andaman Sea II-302
    — Gulf of Thailand II-302–3
— threats pre-1998 III-44
— Torres Strait II-597
— true, Bar Al Hackman II-22
— Turks and Caicos Islands I-590
— Vanuatu II-741–2
    — condition/special features of coral communities **II-742**
    — status of II-742
— Venezuela I-650–1
    — coastline reefs, less diverse and under pressure I-650
    — Mochima Bay I-651
    — on offshore islands, pristine condition I-650
    — Turiamo Bay, diverse reef fauna I-650–1
— Vietnam II-565
— vulnerable to oil spills and their effects III-273–4
— Western Australia II-696
— western Indonesia II-388–9
— western Sumatra II-387
— *see also* coral atolls; fringing reefs; patch reefs; reefs
Coral Sea *II-426*, II-427, *II-612*
— Chesterfield Islands and Bellona reef II-711
— nutrient and sediment loads II-614
— western, circulation in II-614
Coral Sea Basin II-427
Coral Sea Coastal Current II-428, *II-594*
Coral Sea Island Territories, inclusion in Great Barrier Reef Marine Park? II-627
Coral Sea water II-581
Coral, Solomon and Bismark Seas Region II-425–46
— coastline change II-438
— human impacts on coastal seas II-438–41
— land and sea use factors impacting on coastal waters II-435–8
    — inability to manage stocks for sustainability II-437
— offshore systems II-431–3
— people, development and change II-433–5
    — inadequate information for establishing any form of baseline II-435
— provisions for the management and protection of coastal seas II-441–3
    — community-based management II-442–3
    — national administrative and legal arrangements II-441
    — protected species, habitats and areas II-441–2
    — regional cooperation II-442
— seasons, currents, seismicity, volcanicity and cyclonic storms II-428–9
— shallow water marine and coastal habitats II-429–31
coral-rubble mining I-604
coralligenous formations, Tyrrhenian Sea I-273–4
coralline coast, Mozambique *II-100*, II-101, II-104
corals
— Andaman and Nicobar Islands II-191–2
— Arabian Gulf II-7–8
— Brazilian, some endemics I-723
— Chagos Islands
    — most diverse site in the Indian Ocean II-225
    — soft corals II-227
    — a stepping stone for corals II-224–5
— collection for aquarium and shell trade II-438
— the Comoros, used in building II-249
— deep-sea, little studied III-379
— diversity of, Spratly Islands II-364
— French Polynesia II-819
— Great Barrier Reef, diversity and species assemblages II-616
— Gulf of Oman II-22
— hermatypic I-566, I-590, I-723, II-633
    — the Bahamas I-421
— Maldivian, zooxanthellate and azooxanthellate II-202
— Marshall Islands, great biodiversity II-778
— Papua New Guinea reefs II-431

— physiology of calcification III-35
    — calcification and nutrient uptake III-35–6
    — scleractinian corals III-35
— scleractinian
    — Andaman Sea II-302
    — Fiji II-711
    — Gulf of Mannar II-164
    — Lakshadweep Islands II-194
    — Madagascan, affinities of II-117
    — Palk Bay II-167
    — Vanuatu II-741
    — Venezuela I-650
    — western Indonesia II-389
— the Seychelles, growth affected by southeast trade winds II-236
— species diversity, Grand Récif of Toliara II-118
— Sri Lanka II-180
— stony II-796
    — east coast of Taiwan II-508
    — Fiji Islands II-757
— Taiwan II-504
— Turks and Caicos Islands I-589, I-590
— Vanuatu, similarities with Great Barrier Reef II-741
— varying growth conditions, Puerto Rico I-583
— Western Australia II-696
Coriolis effect III-311
Coriolis Force I-631, II-274, II-288, III-316
Coro, Gulf of *I-644*, I-646
corrales (stockyards), Spanish fishing technique I-176
'Corriente de Navidad' I-139
Corsica
— Lavezzi nature reserve I-280–1
— transboundary park in the Bocche di Bonifacio I-281
Costa Rica Dome Structure I-535
Costa Rican Current I-489, *I-678*, I-679
Côte d'Ivoire I-805–20
— coastal erosion and landfill I-816
— construction of Abidjan harbour, importance of I-815–16
— effects of urban and industrial activities I-817–18
— geological context and geographical limits I-807
— major shallow water marine and coastal habitats I-810–13
— offshore systems I-813–15
— population I-815–16
— protective measures I-818–19
    — International conventions, coastal and marine environment **I-818**
    — National Environmental Action Plan I-818
— rural factors I-816
— seasonality, currents, natural environmental variables I-807–10
— water inputs, from continent and ocean I-807
Cotentin Peninsula *I-66*
Cox's Bazaar sand beach II-289
crab fishery, and yields, Island Copper Mine III-244
crabs, pelagic II-54–5
Crepidula fornicata, altered benthic habitats I-47–8
Cretan Sea *I-234*, I-235
crocodiles I-505
Cromwell Current I-705
cross-correlation analyses III-217
Crown-of-Thorns starfish II-303
— a potential environmental problem II-214
Crown-of-Thorns starfish outbreaks II-711, II-712
— causes indeterminate II-617
— east coast, Peninsular Malaysia II-355
— French Polynesia II-823
— Great Barrier Reef II-616–17
— Hawaiian Islands II-807
— Marshall Islands II-777
cruise ship discharge, the Bahamas I-430
crustacean farming III-169, III-170, *III-170*
crustacean fishery
— Alaska, collapse of I-380

— KwaZulu-Natal coast II-137
crustaceans I-34, I-172, II-586
— Côte d'Ivoire I-815
— El Salvador I-550–1
— Faroes I-34, I-35
— fishing for I-73
— glacial relict I-125
— Gulf of Maine I-311–12
— PAHs in I-497
— Peru I-691
— Spanish north coast I-139, I-140
— Tagus estuary I-157
— Vietnam II-564
cryptomonads I-355
crystalline rock habitats, Sri Lanka II-180
cultivated land, decline in, Guangdong II-558
cultural convergence, Sundarbans II-152
cultural evolution, and cultural adaptation III-396
Curaçao I-596, I-597
— beach replenishment I-604
— coastal and marine habitats I-599
— coastal urbanization I-606
— fringing reefs I-599
— land-use planning I-608, I-609
— livestock decline I-601
— national parks I-609
— reef considered to be overfished I-602–3
— reefs damaged by ship groundings I-606
— sewage discharge I-606
— turtle grass I-601
Curaçao Dry Dock, contamination from I-605
curio trade, marine life collection for 251, II-91, II-108, II-125, II-212
— marine ornamentals trade II-393
— Marshall Islands II-782
Curonian Lagoon I-109
— fishery I-111
currents
— affecting the Bahamas I-418, I-418
— Arabian Sea, mirror seasonal wind direction II-20
— Belize, affected by prevailing winds I-503
— Cambodian Sea II-571, II-572
— Côte d'Ivoire coast II-807–8
— eastern Australian region II-631
— Great Australian Bight II-676
— Gulf of Aden II-50
— development of gyres II-50
— Gulf of Mexico I-437
— influencing Marsha, Islands II-775–6
— in the Malacca Straits II-311
— Mexican Pacific coast I-487
— offshore, Taiwan Strait II-501–2
— Sea of Okhotsk II-465–6
— South China Sea II-552
— surface water patterns II-366, II-366
— surface, Java Sea II-383–4
— through the Fijian Groups II-754
— Torres Strait II-596
— Xiamen coastal waters II-516
— residual currents in the Outer Harbour II-516
Currituck Sound I-352, I-360
customary fishing rights
— Fiji II-758, II-761
— more interest in marine protected areas now II-762
customary marine practices/law
— Coral, Solomon and Bismark Seas region II-434
— Marshall Islands II-779, II-786
Cuulong Project, Mekong Delta II-299
Cuvier's beaked whale III-92
Cuvumbi Island, Bajuni Islands, reef front and reef flat II-74, II-78
cyanide, and the live reef fish food trade II-373, II-392–3
cyanide fishing

— Indonesian Seas II-392–3
— the Philippines II-415
cyanobacteria I-743, I-821, III-3, III-221
— blooms III-298–9
— and marine mass mortalities III-218
— The Maldives II-204
cyanophytes II-10–11
— picoplanktonic, Neuse Estuary I-355
Cyclades I-234, I-235
cyclone shelters, Bangladesh II-288
cyclones I-439, II-596
— affecting Madagascar II-116, II-117
— Bay of Bengal II-148
— and storm surges II-288
— and coral cover II-616
— formation of, Coral, Solomon and Bismark Seas region II-429
— French Polynesia, related to abnormal El Niños II-815–16
— Gulf of Thailand II-301
— and high rainfall, Fiji II-753
— influence on Great Barrier Reef II-613
— the Maldives II-201
— Mascarene Plateau II-256
— Mexican Pacific coast I-488, I-493
— occasional
— Mozambique II-102
— Sri Lanka II-177
— the Philippines II-407
— South West Pacific Islands
— damage by associated waves II-710
— effects of II-709, II-710
— Vanuatu II-739–40
— damage by II-740
— west Seychelles, infrequent II-235–6
Cyprus I-254
cytochrome P450, from *Exxon Valdez* oil spill I-382
cytochrome P450-1A values, elevated, Prince William Sound III-274
cytochrome P450-aromatase systems, inhibited by TBT III-249

dabs I-49, I-70
Dahlak Islands II-36
Dall's porpoise III-90
— caught in Japanese salmon drift-net fishery III-379
— hunted in Japanese waters III-94
— possible endocrine disruption II-456–7, II-460
Dalmation coast, tourism I-277
dam building
— on major rivers, contributing to coastal erosion II-108
— protests against II-306
Damperien Province, Western Australia II-694
dams
— adverse effect on mangroves III-19
— and canals, effects of construction and maintenance of I-446–7
— effects of, Côte d'Ivoire I-810, I-816
— effects of, Gulf of Guinea main rivers I-789, I-792
— in estuaries, South Korea II-493
— hydroelectric power, reduce salmon populations I-128
— responsible for hydrological change, east coast, Peninsular Malaysia II-353
— Taiwan rivers, increasing coastal erosion II-506
— Tasmania, regulation of freshwater flow II-654
— a threat to freshwater fish II-697
Danish water, total nitrogen input to III-201
Danube River, increased loads of organic and inorganic pollutants I-303
Dardanelles I-234, I-243
data mining III-212
— and data models III-214–15, **III-215**, **III-216**
— for disturbance indicator types and pathogen toxin and disease combinations III-214
— and other research, to retrospectively derive new time series III-216

Davis Strait *I-6*, I-7
— cod abundance and temperature I-8
— increased Hg in upper sediments I-9
— trawler fishery for Greenland halibut I-8
Daymaniyat Islands *II-18*, II-22
— Daymaniyat Islands National Nature Reserve II-30
DDT II-545
— air and surface seawater, worldwide II-454, *II-457*
— in animals from the Greenland seas I-9, I-11
— in Asian developing region waters II-449
— contamination in Southern California I-390, I-392
    — total DDT of most concern I-393, **I-393**
    — well preserved in anoxic deep basin sediments I-393
— Faroes I-37–8, **I-37**
— Norwegian west coast, still an environmental problem I-25
— recent use, Vietnam II-565
— in river waters, El Salvador I-552
— as seed dressing I-635
— still reaching Baltic Sea via precipitation I-131
de la Mare, Dr William, on the New Management Procedure (IWC) III-78
debris, non-biodegradable, a danger to sea turtles III-67
decision making process, for coastal management, greater inclusiveness more sensible III-355
Declaration on the Protection of the Black Sea Environment I-294
deep water formation
— Greenland Sea I-7
— Tyrrhenian Sea I-272
deep water renewal, Norwegian fjords I-21
deep-sea smelt, Okhotsk Sea II-466, II-468
deepwater dumpsite 106 *I-322*, I-327
deforestation I-241, I-427–8
— Australian Rainforest II-585
— Belize I-508, I-510
— Borneo II-374–5
— catchment areas of Malacca Strait II-338
— causing soil erosion I-174–5
— in the Comoros II-247
— effects of, Andaman and Nicobar Islands II-193
— from mining, New Caledonia II-731
— Great Barrier Reef region catchments II-620–1
— Guangdong II-558
— Guinea I-801
— Gulf of Guinea I-786
— Jamaica I-568
    — reforestation I-572–3
— Kamchatka peninsula II-469
— and land conversion, cause of high sedimentation rates II-384
— Lesser Antilles I-635, **I-636**
— logging in Gulf of Papua watersheds II-604
— Mozambique II-109
— Nicaragua I-523, I-538
— the Philippines II-412
    — reduction in forest cover II-413
— round Sea of Japan II-475
— and soil erosion, Hawaiian Islands II-801–2
— in southeast India, affects coastal zone II-165
— southern Brazil
    — of Atlantic forest I-741
    — of mangroves I-740, I-741
    — for Rio-Santos road I-742
— Southern Gulf of Mexico
    — and erosion I-476
    — impact of I-479
— Sumatra and Kalimantan II-395
— Tanzania II-89–90
— through livestock grazing, Dutch Antilles I-601, *I-602*
— west coast Taiwan II-503
degradation
— southern Brazil I-740–3
    — main problems I-740

Delaware Bay *I-322*, I-329, *I-336*
Deltaic Sundarbans *II-146*, II-147
— brackish waters support phytoplankton, macrobenthic algae and zooplankton II-150
— effects of seasonal changes II-149
— Hugli estuary, carries industrial discharges from Haldia region II-155
— important morphotypes II-147
— major rural activities and environmental impact II-153–4
— monsoon period II-149
— physico-chemical characteristics II-149, **II-149**
— population extremely poor II-152
— post-monsoon periods II-149
— pre-monsoon period II-148–9
— rich in natural resources II-152
— a unique ecosystem II-156
denitrification I-106, III-262
— water-column, Arabian Sea II-24
Denmark
— Action Plan for Offshore Wind Farms in Danish Waters III-307
— acts on behalf of Faroes I-36
— monitoring and modelling of offshore wind energy technology III-308
    — Wind Atlas Analysis and Application Program (WA^{S}P)
— offshore wind energy production III-307–8
    — Tunø Knob installation III-308
    — Vindeby installation III-307–8
— pilot offshore wind energy projects III-306
Denmark Strait *I-6*, I-7, I-9
D'Entrecasteau Basin *II-426*
D'Entrecasteaux Channel *II-648*
— localised cooling II-651
deposition processes, from the atmosphere III-200–1
Derjugin's Basin *II-464*, II-465
desalination
— Arabian Gulf
    — by-products II-10
    — multistage flash evaporation plants II-10
    — and power plants II-10
— Oman II-28–9
— Red Sea, saline discharges from desalination plants II-43
desalination plants, Bay of Bengal, effects of saline discharge II-279
desert
— Namibia I-823
— Peru I-689–90
developed countries, marine reserves in III-386
developing countries
— a chance for coastal zone development with fewer problems III-357
— marine reserves in III-386
— prawn/shrimp fisheries, high utilisation of catch, poor utilisation of species III-143
development pressure
— English Channel coasts I-71
— Venezuelan coastal zone I-645
Dhofar *II-18*
— excessive cutting of firewood II-26
— limestone cliffs II-19, II-21
Diadema mass mortality I-410, I-421, I-503, I-520, I-607, I-632
— co-incidental with El Niño conditions III-223
— Jamaica I-563–4
diamonds, coastal deposits, South Africa II-142
diarrhetic shellfish poisoning (DSP) I-69, I-160
diatom blooms I-155
— Baltic Sea I-125
— Irish Sea I-89
— spring and autumn, English Channel I-67
— toxic, causing debilitating illness III-218
diatom sediment records, Portuguese shelf I-160
diatoms I-355, I-692, I-743, I-783, I-814, I-821, I-830–1, II-24, III-221
— Australia II-583

— Bay of Bengal II-275
— Campeche Sound I-474
— discriminated against III-259
— in northern Gulf of Mexico I-449
— West Guangdong coast II-553
dibutyltin (DBT) II-449
Diego Garcia, Chagos Archipelago *II-221*
— military development imported alien flora II-229
— occasional storms from cyclone fringes II-225
— only inhabited island II-223
— recreational fishing II-230
diffusional transfer 202
diffusiophoresis III-201
diffusive attenuation coefficient
— and water quality III-296–7
— K-maps III-296
dikes and breakwaters, construction interrupts sand drift I-176
dinoflagellate blooms I-827
— toxic, French Polynesia II-821
dinoflagellates I-692, I-783, I-814, I-831, II-320
— Australia II-583
— ichthyotoxic I-363–4
— increased number in the Arabian Gulf *II-12*, II-13
— toxic I-355, II-304, II-587, III-218, III-221
— benthic, produce tumour-promoting agents III-219
— *Pfiesteria piscicida* III-223
— winter-blooming I-355
dioxin
— discharge from old Norwegian magnesium plant I-25
— remobilization of I-25
discards I-146–7, **I-148**
— at sea, percentage of total catch III-377
— and by-catch III-123–5
— fate of III-127
— reasons for discarding III-137–8
disease
— an ecological opportunist III-225
— can devastate aquaculture ponds III-369
— chronic conditions within endangered populations, monitoring of III-227
— increase in extent and impact of III-225
— novel, possible introduction through aquaculture III-369
— reflect perturbations with ecosystems III-225
— transfer via ballast water III-355
— trophodynamically acquired, a global issue III-221
— water-related, Vanuatu II-745
disease disturbances III-225–6, **III-226**
diseases, water-borne, Marshall Islands II-781
Disko Bay, Greenland, shrimp fishing I-8
dispersants *see* oil dispersants
dissolved inorganic carbon (DIC), non-Redfield ratio depletion I-225–6
dissolved organic N (DON), varied origin III-199
dissolved oxygen I-240
— Campeche Sound I-469
— depletion of III-199
— and fish kills I-364
— Gulf of Aden II-51
— levels in the Cambodian Sea II-572–3
— variable, Canary Islands I-191, *I-192*
disturbance
— keystone-endangered and chronic cyclical III-227–8
— as a regular feature of an ecosystem III-227
disturbance categories/types III-217–28
— derivation III-217
— grouping of III-217
disturbance regimes III-227
— better understanding of the natural history of III-230
dive tourism
— artificial dive objects I-604
— Bonaire I-597

— Dutch Antilles I-604
— effects of on Borneo islands II-377–8
— Hawaiian Islands II-806
— SCUBA diving
— East Kalimantan II-377
— Sabah II-377
— Sri Lanka II-180
— Vanuatu II-742
diving and snorkelling, impacts from II-213
Djibouti *II-36*, II-49, II-55, *II-64*
— use of mangroves II-58
Dnestr River, pollutants carried I-303
Dogger Bank I-46, I-47
— plankton I-48
dogwhelks
— imposex I-25, I-36–7, I-55, I-75
— severe impact of TBT III-250–1
— southeast England, the Dumpton Syndrome III-252
Doldrums *see* Inter-Tropical Convergence Zone (ITCZ)
dolphin mortality
— by-catch is to some degree controllable III-139
— reduction in through changed purse-seining procedure III-141
— through tuna fishing III-142–3
dolphins I-90, I-834, II-24, II-205, II-368, III-90–1
— caught for shark bait II-57
— declining, Black Sea I-291
— Great Barrier Reef II-618–19
— Gulf of Guninea I-780
— illegal hunting in Black Sea III-94
— Java Sea II-363
— Marshall Islands II-780
— and mercury contamination III-95
— Patagonian coast I-756
— Sri Lanka II-180
— *see also* named varieties
dominance, pattern of altered by oil spills III-275
Doñana National Park, southern Spain *I-168*
— controversy over infrastructure improvements around I-173
— sand and marsh ecosystems I-171–2
Donax denticulatus, wash zone, dissipative beaches I-648
Dover *I-66*
Dover Straits *I-66*
— shipping through I-72
downwelling
— Ekman downwelling I-223
— nearshore, Alaskan Gulf I-376
dragnet fishery, southern Brazil, east coast, causing degradation I-740
drainage, a development issue, Arabian Gulf II-9
Drake Passage I-753
dredge and fill
— Gulf coastal plain, USA I-445
— navigation channels I-447
— Jamaica, damaging seagrasses I-569
— of mangrove swamps III-19
dredge fishery, Victoria Province, Australia II-668
dredge spoil, dumped I-77
dredging I-623, II-124, II-546, II-642, II-655
— Belize I-510
— in Chesapeake Bay I-348
— Curaçao I-602
— damaging to seagrass beds III-7, III-8–9
— destructive III-379
— detrimental effects of I-145
— and dumping, in the Comoros II-247–8
— and earth-moving activities, Marshall Islands II-782–3, *II-782*
— effects of, Tagus estuary I-161
— environmental impacts of, southern Spanish coast I-176–7
— and filling
— for construction purposes I-601–2
— Jamaica, damaging seagrasses I-569

— followed by reclamation, French Polynesia II-821
— and habitat destruction III-379–80
— harbour entrances and navigation channels, Buenos Aires I-762–3
— instream, degradation and erosion caused by I-444
— Kuwait II-9–10
— and land fill, Hawaiian Islands II-805
— and landfill, Red Sea II-43
— Malacca Strait II-339
— Maputo and Beira II-108
— northern Gulf of Mexico I-447–8
— now a necessity, Malacca Straits II-317
— to improve Gulf of Aden ports 58
drift netting, most popular, east coast, Peninsular Malaysia II-352
Driftnet Ban, United Nations III-140, III-145
drill cuttings, contaminated, discharge of III-363
drilling fluids, oil-based, toxicity of III-363
drilling waste management I-23–4
"drowned river system" *see* Chesapeake Bay
drowned river valleys, Tasmania II-653
Drupella snails, coral-eating II-696
dry deposition I-13
— measurement techniques III-203–5
  — field measurement of, current status III-205
  — use of surrogate surfaces III-204–5
  — wind tunnel experiments III-204
— modelling frameworks and algorithms III-202–3
  — current modelling uncertainties III-202–3
  — mathematical treatment of physical processes III-202
— of particles to water surfaces, processes and consequences III-197–209
  — atmospheric deposition to the coastal zone III-205–7
  — atmospheric particles III-199–200
  — deposition processes III-200–1
  — gas deposition and role of particles III-205
  — importance of in nutrient fluxes III-201
  — nutrient fluxes and aquatic cosystem responses III-198–9
dry deposition velocities III-200–1, III-202, III-204–5
— based on eddy correlation III-205
Dry Tortugas *I-406*
— corals little diseased I-410
— decline in pink shrimp landings I-447
— extreme low temperature, and death of corals I-410
— showing only small decline I-412
Dry Tortugas National Monument I-412
— redesignated as Dry Tortugas National Park I-413
Dry Tortugas National Park, coral bleaching (1998) III-56
duck plague virus III-225, III-227
dugongs II-54, II-90, II-123, II-336, II-431, II-574, II-587, II-601, II-615
— Bazaruto Islands II-104
— eastern Australian region II-637
— endangered in Gulf of Mannar II-165, II-166
— Great Barrier Reef
  — entanglement problems II-619
  — pressures and status II-619
— in the Mayotte lagoon II-250
— Red Sea II-42–3
— Sri Lanka II-180
— Vanuatu II-741
— western Indonesian seas II-401
dumping
— at sea III-338–9, III-342, III-365–6
  — of dredge spoils III-365
  — some prohibitions achieved III-365
— Baltic Proper, of chemical warfare agents I-117
— Irish Sea
  — blast furnace spoil I-94
  — munitions I-96
— New York Bight I-323, I-324
— North Sea
  — of industrial waste and sewage sludge finished I-53, I-55
  — of munitions I-52

— *see also* waste dumping
Dumping Convention **III-334**
Dumpton Syndrome III-252
Dungeness *I-66*, I-68
Durvillia habitat, exposed coasts, Tasmania II-652
Dutch Antilles I-595–614
— Aruba *I-596*, I-597
— coral bleaching and mortality events III-54
— definition and description I-597–8
— effects from urban and industrial activity I-502–7
  — artisanal and non-industrial uses I-602–4
  — cities I-606
  — cumulative impacts and ecological trends I-607
  — industrial use I-604–6
— environmental hindrances ordinances I-608
— environmental institutional capacity I-609
— Fishery Protection Law I-608
— government environmental protection and management funding, sparse I-612
— law enforcement is critical I-609
— Leeward Group *I-596*, I-597–8
— major shallow marine and coastal habitats I-598–601
— maritime legislation I-608
— National Parks Foundation of the Netherlands Antilles I-607–8
— overview of implementation of marine environmental policy and legislation I-609, **I-610–11**
— protective measures I-607–12
  — implementation and an evaluation I-609
  — policy development and legislation I-607–9
  — practical management measures I-609
  — prognoses and prospects I-609, I-612
— rural factors, coastal erosion, landfill and excavation I-60–2
— Windward Group *I-596*, I-598
dynamite fishing
— cessation of II-95
  — case study Mtwara, Tanzania II-87, II-95
— Gulf of Thailand II-304

earthquake activity
— the Philippines II-408
— Vanuatu II-744
— Xiamen region II-516–17
East African Coastal Current *II-64*
— important for larval dispersal and downwelling II-85
— influenced by the monsoons II-85
East Arabian Current II-50
— behaviour of II-20
East Australian Current II-581, *II-581*, II-614, II-619, II-631, *II-632*, II-649–50, II-664
— giving distinctive ecosystems around the Kent Group II-651
— varies with El Niño/Southern Oscillation cycle II-650
East China Sea *II-474*, *II-500*, II-508
East Greenland Current I-5, I-7
East India Coastal Current (EICC) II-274
East Madagascar Current *II-101*, II-245
East Pacific Plate I-487
East Sakhalin Current II-466
Easter Island I-708
eastern Atlantic flyway, migratory feeding, North Sea coasts I-49
Eastern North Atlantic Central waters I-153
Ebrié lagoon, Côte d'Ivoire I-811–12
— artificial Vridi canal I-812
— fish I-813
— freshwater inputs I-812
— hydrological zonation I-812
— physical framework I-811–12
— pollution in I-817
— sewage discharged to I-817
echinoderms I-35, I-239, III-242
Ecklonia radiata habitat II-652
ecological catastrophe, Aznalcóllar mining spillage disaster I-178

ecological problems, and their causes, west coast, Sea of Japan, Amursky Bay II-483–5
ecological resources
— exploitation of III-366–9
    — aquaculture III-368–9
    — marine capture fisheries III-366–8
ecological tariffs III-401
ecological tax reform III-401
economic development, and coastal management III-351
economic incentives, to achieve economic goals III-401
economic income, skewed III-396
ecosystem, defined I-705
ecosystem health information III-212
ecosystem instability and collapse, processes typically leading to III-227
ecosystem management, and the reserve concept III-377–8
ecosystem modelling I-50
ecosystems
— changes in structure and function following eutrophication III-263, **III-264**
— connectedness of I-478
— dynamics of III-259
— effects of fisheries on III-117–33
— measuring impact of by-catch on III-144
— refers to processes and functions III-259
— response to increased nutrient loads unpredictable III-366
— retrogression III-259–60
— vulnerable species targetted by opportunistic microorganisms III-227
ecosystems services, value of III-395
ecotourism I-4, II-89, II-95, III-376
— and Brazilian coral reefs I-723
— Fiji II-762
— Great Barrier Reef, Australia II-586
— Marshall Islands II-780
— Midway Atoll II-801
— a non-consumptive resource use II-141
— possible in the mangrove ecosystems of Pacific Colombia I-682–3
— potential for, Mayotte II-251
eddies I-46, I-630
— in Caribbean Basin I-580
— mesoscale, in cold water at Paria Peninsula I-646
— surface, Angola-Benguela front I-825
— SWODDIES I-138
— *see also* Sitka Eddy
eddy accumulation techniques, dry deposition measurement III-203–4
eddy correlation/covariance techniques, dry deposition measurement III-203
eddy currents, Canary Islands I-189
eddy diffusion, between deep and upper waters I-391
eddy systems I-85, I-223
eelgrass, Carolinas I-358
effluent disposal pipes
— KwaZulu-Natal II-142
— using Algulhas Current for dispersal, South Africa II-142
effluents
— from shallow-well injection, migration of I-409
— polluting Norwegian coastal waters I-24
Egypt *II-36*
— dive tourism II-44
— Red Sea coast, coastal pollution by oil II-43
Ekaterina Strait *II-464*, II-465
Ekman downwelling I-223
Ekman transport II-274
— drives coastal upwelling off Oman–Yemen coast II-19
El Niño
— affects Mexican Pacific coast I-486, I-488, I-489
— affects Nicaraguan Pacific coast I-534
— biological effects of, Peru I-692–3
— change to subtropical and tropical species I-693
— causes of I-702
— a complex, multi-level phenomenon I-703
— described I-679
— dramatic effects of I-679
— in El Salvador I-547
— *see also* Atlantic Niño; Benguela Niño years
El Niño events II-776
— affecting the California Current I-387
— affecting Gulf of Alaska mixed layer depth III-182
— biological effects of I-703–4
— bring drought, Coral, Solomon and Bismark Seas Region II-428–9
— and coral bleaching III-38
— extreme
    — and the 1997–98 coral bleaching event III-44–6
    — areas of bleaching III-46
    — impact on Indian and Southeast Asian winds III-44
    — interpretation and conclusions III-46
— impacts on Northeast Pacific Ocean III-183–4
    — effect of 1997–8 event III-183–4
— importance of I-702
— mechanism governing strength of I-702–3
— northern Chilean coast invaded by exotics I-707
— physical alterations due to I-703
— position of Subtropical Water Mass I-701
El Niño Southern Oscillation (ENSO)
— effects of I-223–4
— effects on climate and oceanography of Fiji Islands II-753–4
— effects on the Great Barrier Reef region II-613–14
— long-term variations I-702–4
— periodic influence in Australia II-582, II-632
El Salvador I-545–58
— coastal erosion and landfill I-552
— effects from urban and industrial activities I-553–6
— geography I-547
— major shallow water marine and coastal resources I-548–51
— offshore systems I-551
— populations affecting the area I-551
— protective measures I-556–7
    — Ley del Medio Ambiente (Environmental Law) I-556
    — protected areas I-556–7
— rural factors I-551–2
— seasonality, currents, natural environmental variables I-547–8
elasmobranchs, sensitive to ecological change, North Sea I-49
Electronic Chart Display and Information Service (ECDIS), Malacca Straits II-325
Elefsis Bay *I-234*, I-237
— decrease in invertebrate abundances I-239
— industrial pollution I-245
— metal pollution I-246
elephant seals I-756, I-757
Eleuthera *I-416*
EMECS (Environmental Management of Enclosed Coastal Seas) I-56
Emperor Seamounts
— fossil coral reefs II-800
— relationship to Hawaiian Islands II-794
enclosed shores, Hong Kong II-538
endangered species
— Abrolhos Bank–Cabo Frio, Southern Brazil I-736
— consequences of climatic irregularities III-224
— leatherback, green turtles and loggerheads III-62
— loss or imminent loss, Black Sea I-291, **I-292**
— Olive Ridley and Hemp's Ridley critically endangered III-63
— Philippines II-410
— sea turtles III-143
— vaquita III-90
endangered species protection, western Indonesia II-400–1
endemic species
— Coral, Solomon and Bismark Seas region II-431
— Hawaiian Islands II-797, II-800
— Marshall Islands II-779

endemism
— Australian Bight II-678–9
— islands of the southwest Indian Ocean II-251
— low, reefal communities, French Polynesia II-819
— southern African marine biota II-135
— southern Australia II-581
— Tanzanian coastal forests II-88
endocrine disrupters
— potential harm from III-86
— TBT III-249
energy
— from the oceans III-303–21
  — possibility of combined technologies III-311
— from seagrasses, detrital and direct grazing pathways III-3
engineering structures, causing beach erosion II-639
English Channel I-65–82
— anthropogenic impacts on I-74–7
— coastal and marine habitats I-68–71
— defined I-67
— formation of I-67
— lies at boundary of Boreal and Lusitanian biogeograpical regions I-68
— physical and biological environment I-67–8
— protective and remediation measures I-77–9
  — International Conventions and Agreements, and EC Directives **I-78**
— urban and rural populations I-71–4
Eniwetak Atoll, nuclear tests II-783
ENSO events
— 1997/1998, impacted mammals and seabirds in several regions III-225
— 1998 event II-236
— affected by climate change III-370
— effects of, Vanuatu II-740
— effects on seabirds III-110
— influence on New Caledonia II-725
— influencing sea surface temperatures II-384
— Marshall Islands II-776–7
  — coral bleaching II-777
  — Crown-of-Thorns starfish II-777
  — sea-level rise II-777
— the Philippines, manifestations of II-408
— South Western Pacific Islands, affect oceanography and climate II-710
— Teacapán—Agua Brava lagoon system I-494
entanglement
— of mammals and seabirds in floating synthetic debris III-125–6
— usually fatal for small cetaceans III-145
entanglement nets (jarife), Tanzania II-90
environmental assessment, for newer Paupuan mines II-439
environmental coastal protection, Chile I-715
environmental concerns
— Brazilian tropical coast I-727–8
— coral reefs, Hawaiian Islands II-807
— in the Maldives II-206–7, **II-206**
— mariculture, Hawaiian Islands II-803
— Marshall Islands II-780, **II-781**
  — related to cross-sectoral issues II-784–5
environmental conservation, restoration and improvement, Gulf of Mexico I-479
environmental criteria, concepts in II-202
environmental degradation
— Andaman and Nicobar Islands II-193
— Cabo Frio–Cabo de Santa Marta Grande, southern Brazil I-738
— China II-492
— coastal forested watershed, Cambodia II-575
— early, the Seychelles II-239
— east coast, Peninsular Malaysia
  — of coral reefs II-355
  — from tourism II-353–4
— Godavari-Krishna delta II-171–2

— Hong Kong II-543–6
— Lakshadweep Islands II-194
— Palk Bay–Madras coast II-169
— parts of North Spanish coast I-142
— the Philippines II-418
— potential, Tasmanian marine fish farms II-641
— Sabah coral reefs II-371, *II-372*
— southern Brazil I-740–3
— Taiwan's marine and coastal environment II-510
— Tasmanian estuaries II-654–5
— through land reclamation, West Guangdong II-557
— Western Australia, metropolitan areas II-697
environmental effects, of Australian fishing II-587
environmental impact assessment, Hawaiin Islands II-808
environmental impacts
— of alien species in Irish Sea I-89
— Chilean fishmeal industry I-713–14
— of coastal settlements, eastern Australian region II-640
— of coastal zone industries, Chile **I-711**
— and ecological trends, Dutch Antilles I-607
— from repetitive trawling II-624
— industrial fishing, Gulf of Guinea I-787–8
— of intensive fishing
  — on the Irish Sea environment I-92
  — on Irish Sea fisheries I-92
— of mining activities, New Caledonia II-731
— nickel mining, New Caledonia II-713
— of tourism II-212–13, **II-213**
environmental issues
— Australia II-586
  — marine environment **II-589**
— Mascarene Region II-265
— south-east Queensland II-641–2
environmental management
— poor, Vietnam II-566
— sound, expected of companies and organisations III-351
— Torres Strait and the Gulf of Papua II-607–8
environmental management strategy, Marshall Islands II-787
environmental matrices
— Gulf of Mexico
  — extensive cattle pasture I-479
  — offshore oil industry I-478–9
environmental opportunities and prospects, Marshall Islands II-787–8
environmental pollution, in the Bay of Bengal II-154–5
environmental problems
— Cambodia, no clear perception of II-576, II-577
— critical, Xiamen region II-521–4
  — bacterial pollution II-522–3
  — oil pollution II-523
  — organic pollution, eutrophication and red tides II-522
  — pollution from pesticides II-524
  — sea dumping and disposal of solid waste II-523–4
  — sediment deterioration and secondary pollution II-524
— Malacca Straits II-317–21
— Marshall Islands, related to population pressure II-780
— some east coast Indian cities II-278
environmental projects, French Polynesia II-824
environmental quality criteria, use in Norwegian fjords and coastal waters I-27
Environmental Risk Assessment, Malacca Strait II-320–1, *II-321*
— analysis of likelihood of adverse effects II-320–1
— retrospective analysis of decline in key habitats II-320, **II-321**
Environmental Sensitivity Index (ESI) III-269, **III-270**
environmental stress, and altered benthos, Southern California I-396
environmental variability, in fish stocks, southwestern Africa I-832
environments, deep, Red Sea II-39
EOS satellite, with colour sensor MODIS III-295
epi-continental seas I-45
epiphytes, seagrass, productivity of III-2
Equatorial Counter Current I-486, I-489, II-201

Equatorial Countercurrent *I-678*, I-679, I-701, *II-581*, II-614, II-619, II-724, II-767, II-775, II-780
Equatorial Current, and El Niño I-703
Equatorial Subsurface Water Mass, low dissolved oxygen content I-702
Equatorial Under Current *I-774*
equity and fairness, in fisheries management 157–8, III-155
— inter-generational equity III-157
— issue of relative deprivation and historical participation III-157
equity theory in fisheries management III-157–8
Eritrea *II-36*
EROD induction I-56
erosion
— of beaches through sand mining I-636
— due to hurricanes I-622
— Dutch Antilles I-601–2
— Nicaraguan Pacific coast I-538
erosion–accretion cycles, Hawaiian Islands II-804–5
erosion–deposition cycles
— Arabian Sea II-27
— Gulf of Aden II-58
— Inhaca Island II-105
ERTS-1 satellite, early recognition of cyanobacteria blooms III-299
ESA satellite ENVISAT, with ocean colour sensor MERIS III-295
Escheria coli
— Hong Kong bathing waters II-541
— Penang and Selangor II-319
Espichel, Cape *I-152*, I-153
Essential Fish Habitat (EFH) I-451
estuaries
— east coast, Peninsular Malaysia II-349
— eastern Australia II-634
— Gulf of Maine I-310, *I-311*
— New York Bight, altered condition I-327
— North Sea, importance of I-49
— northeastern Australia II-614
— northern Gulf of Mexico
— freshwater flushing rate a critical parameter I-447
— tidal and subtidal oyster reefs I-441–2
— South Africa II-135–6
— greatest threat will be lack of water II-136
— resource use II-136
— southwestern Africa I-829
— Taiwan Strait II-502, II-503
— Tasmania II-653
— degree of sedimentation II-654
— southern, tannin-stained II-652
estuarine crocodiles
— Great Barrier Reef II-618
— Solomon Islands, depleted II-431
estuarine ecosystem, Palk Bay–Madras coast II-169
estuarine habitats
— Channel coast of England I-68
— Xiamen region, Jiulongjiang estuary II-518
estuary wetland ecosystems, North China, affected by oil and agriculture II-494
EU
— Habitats Directive, omits to take natural change into account III-354–5
— MAST programme I-56
— nature conservation, a shared competency I-57
EU/EC
— Convention on International Trade in Endangered Species (CITES) I-608
— Migratory Species Treaty (Bonn Convention) I-608
EU/EC Directives **I-78**
— on Bathing Water Quality I-76, I-95
— concerning waste-water treatment I-145
— Conservation of Natural Habitats and of Wild Fauna and Flora I-52, I-57, I-77, I-96–7, I-162
— Conservation of Wild Birds I-52, I-57, I-77, I-162, I-214

— on the control of nitrates I-53
— for protection of European Seas **I-57**
— Shellfish Hygiene I-76
— on Shellfish Waters I-76
— to increase urban wastewater treatment, too many derogations I-55
Eucla Bioregion, Australian Bight II-678, II-679
euphotic layer, Côte d'Ivoire I-810
Euphrates, River II-3
Europe
— current and planned offshore wind farms III-309–11, *III-310*
— wind resources predicted III-308, *III-309*
European Environment Agency, arrangements for protecting North East Atlantic in the 1990s **I-57**
European Union, funding for off shore wind energy technology III-307
eustatic movements, tilting European landmass I-47
eutrophication I-137–8, I-174, I-278, II-542, II-770, II-771, III-366
— and algal blooms I-20
— areas of French channel coast contributing to I-74
— in Australian coastal waters II-585
— Baltic Proper I-103–8
— biological effects of I-109–10
— rivers bring pollutants I-103
— and supersaturation of oxygen I-104–5
— Baltic Sea I-130–1
— Bermuda I-228
— Black Sea I-290
— changing phytoplankton community composition I-290
— of coastal areas, western Indonesia II-396
— concepts for remediation III-263
— cultural, Western Australia II-700
— in fjords with restricted circulation I-24
— from aquaculture I-144
— from fish farming I-26
— Guanabara Bay, southern Brazil I-741
— Gulf of Guinea, coastal waters I-784, I-788–9
— Gulf of Thailand II-307
— Hawaiian Islands II-804
— Irish Sea, possible near English coast I-87
— in its broadest sense III-198–9
— lagoon, Vanuatu II-745
— leads to increase in N and P III-259
— local, Red Sea II-43
— of marine waters: effects on benthic microbial communities III-257–65
— coastal inlets of the southern Baltic Sea III-260–3
— definition and sources of pollution III-258
— effects of on marine communities III-259–60
— in most Carolinas estuaries I-363
— estuarine fish kills I-363–4
— measures for reduction I-366
— more stringent N and P reductions needed I-366
— North Korean artificial lakes II-493
— North Sea I-53
— northern Gulf of Mexico, growing problem I-443, I-450
— in Norwegian coastal waters, dependent on transboundary load I-24
— and nutrients, English Channel I-74
— Patagonia I-765
— the Philippines II-417
— potential for, Jamaica I-564
— in a range of aquatic environments III-205–6
— secondary effects I-108
— some North Carolina estuaries I-356, I-363
Euvoikos Gulf *I-234*, I-237, I-244
— affected by wastes I-246
evaporation
— in excess of precipitation
— Arabian Gulf II-3
— Red Sea II-39

— Gulf of California Coastal Province, Mexico I-487–8
evolutionary mechanisms, allowing toxic species to spread III-219
Exclusive Economic Zones (EEZs) I-2
— Andaman and Nicobar Islands II-191
— Azores Archipelago I-203
— the Bahamas I-422
— Cambodia II-571
— and fisheries management III-158
— Lakshadweep Islands II-194
— Mozambique II-105
— North Sea I-51
— the Philippines II-408
— South Africa II-137
— USA I-451
  — utilization of economic resources I-365–6
exotic species/introductions see alien/accidental/exotic/introduced organisms/species
exports, from Faroese fishery I-35–6
extinction—recolonization dynamics, Chilean coast I-707
extreme events, The Maldives II-203
Exuma Cays *I-416*, *I-419*
— patch reefs I-429–30
— shallow marine habitat I-421
Exuma Cays Land and Sea Park
— evidence for positive effect of protective measures I-431–2
— regulations I-431
Exuma Sound I-417, I-428
Exxon Valdez oil spill I-382, III-275
— biological effects at low levels of aromatic compounds III-364
— characteristics of III-272–3
— decrease in *F. gardneri* III-272
— oiled mussel beds III-271, III-274
— part of severity of effects on *Fucus* due to high-pressure-hot-washing III-275
Eyre Bioregion, Australian Bight II-678

faecal coliforms I-743
— areas of Gulf of Guinea coast I-789
— contaminates freshwater sources, Marshall Islands II-781
— Dar es Salaam and Zanzibar II-92, II-93
— Hong Kong bathing beaches II-541
— Johore Strait II-335
— Maputo Bay II-109
— Morrocoy National Park I-658
— Peninsular Malaysia west coast II-319, II-340
— Sepetiba Bay, southern Brazil I-741
— Suva region, Fiji II-760–1
  — Suva harbour II-719
— Sydney beaches II-642
— Xiamen coastal waters II-522–3
faecal contamination, Durban beaches II-142
faecal pollution, Guanabara Bay, southern Brazil I-741
Fal estuary *I-66*, I-74, I-75
— closure of shellfishery I-69
Falklands/Malvinas Current I-751
FAO
— catch statistics reveal "fishing down" of food chains III-368
— Code of Conduct for Responsible Fisheries III-138, III-341–2
— Code of Conduct for Responsible Fishing III-159
— data base to maintain records of high seas fishing vessels III-160
— international plan to reduce incidental catch of seabirds in longline fisheries III-146
Farasan Islands *II-36*, II-38, II-41
farming, 'modernization' of III-7
Faroe Islands I-31–41
— coasts and shallow waters I-33–5
— mariculture I-35
— oceanic climate I-33
— offshore resources I-35–8
— pilot whales hunted III-92
— protective measures and the future I-38–40

— typical upper and deeper layer water flows I-33, *I-34*
faroes I-505
'faros', Maldivian mini-atolls II-201
Federal and Islamic Republic of Comoros (FIRC) II-247
Federal Water Pollution Control Act Amendments (USA), Section 404 Program I-452–3
feral animals, impact of Australian Bight coast II-684
fertiliser pollution, from Mozambique's upstream neighbours II-109
fertiliser run-off, Madagascar, impacting coral reefs II-123, II-124
fertilizer consumption, Lesser Antilles I-635, **I-635**
Ficopomatus enigmatus, positive effects of I-48
Fiji Islands *II-706*, II-707, *II-708*, II-751–64
— agriculture and fisheries II-715–16
— biogeography II-753
— coastal erosion and landfill II-760
— coastal modifications II-717
— coral reefs II-711
  — degraded by pollution II-711
— effects of urban and industrial activities II-760–1
— island groups II-753
— major shallow water marine and coastal habitats II-754–8
— marine environment critical II-762
— offshore systems II-758
— overfishing of marine invertebrate species II-759
— population II-714, II-758
— protective measures II-761–2
  — private sanctuaries II-761
— rural factors II-758–60
— seasonality, currents, natural environmental variables II-753–4
— sensitive and endangered marine species II-757
Fiji plateau II-753
fin whales, depleted III-78
financial and technology flows, global, importance for environmental decision-making III-346
Findlater Jet *II-18*, II-19
finfish communities, estuarine and coastal, Carolinas coast I-362–3
finfish farming III-170
finfish fishery, Belize I-509
Finisterre Cape *I-136*, I-139
Finland, Gulf of *I-100*, I-122
fire, effects of, northern Gulf of Mexico I-449–50
fire cycle
— natural, important in maintaining coastal ecosystems I-449
  — changes in detrimental I-449
fish
— the Azores I-207
— Baltic Sea
  — butyltins in **I-115**
  — larval stages in *Fucus* belt I-125
  — mixture of freshwater and marine species I-128
— Belize lagoonal shelf I-506
— Cambodia II-575
— coastal, eastern Australian region II-634
— Côte d'Ivoire, Ebrié Lagoon I-813
— deep water, "boom and bust" cycles III-120
— El Salvador I-550
  — estuarine nursery habitats I-550
— in English Channel I-70
— fauna, Australian Bight II-679
— Great Barrier Reef II-617
— Gulf of Guinea
  — commercial species **I-784**
  — Guinean Trawling Survey I-783
  — populations I-783
— Gulf of Mannar II-164–5
— Lesser Antilles I-633
— life history traits (*r*- and *k*-selection) III-119
— Mauritius II-259
— nearshore, Hawaiian Islands, depleted II-803
— North Aegean Sea I-238
— North Sea I-49

— North and South Carolina coasts I-362–3
— offshore, Fiji Islands II-758
— organochlorine pollution in, Asian waters II-449–51
— possible extinction of groupers and Humphead wrasse II-373
— problem of size-selective fishing III-121–2
    — changes in size structure, community level III-121–2
— South Aegean Sea I-238
— Southern California, pollution effects on I-397
— southwestern Africa
    — mesopelagic I-832–3
    — surf zone I-830
— species diversity, Gulf of Mexico I-473
— species and groups, Vietnam II-564
— Tagus estuary, nursery grounds I-157
— target species, ecologist-market collision III-158
— vulnerable species, fishing of III-120
Fish Aggregating Devices (FADs) II-372
— New Caledonia coastal fishery II-730
— used in the Comoros II-249
fish barrages, intertidal, Madagascar II-126
fish communities
— natural factors for change I-398
— Southern California, impacts of anthropogenic pollution I-397–8
fish diseases, from pollution I-397
fish diversity
— Baltic, marine, decreases northwards I-109
— demersal, soft grounds, Spanish north coast I-141
fish farming
— Faroes I-35
— Norway I-26
    — and expanding industry I-28
    — mass mortalities in fish cages I-20
— potential for, east coast of India II-278
— round North Sea I-52
— southern Spain I-177
fish farms, marine, Yellow Sea II-494
fish mortality, Tanzania II-89
fish plant, Marshall Islands, benefits vs. impacts II-785
fish processing at sea, dumping of organic material III-127
fish recruitment I-92
fish stock assessments III-156
— and Fishers' insights III-157
— relationship between present and future stocks III-156
— validity of single-species models in multi-species ecosystems III-156–7
fish stocks
— Adriatic Sea I-277
— affected by commercial fisheries III-97
— Arabian Gulf, dwindling II-11
— assessments in the Maldives II-210
— Baltic Sea I-110–12
    — freshwater species I-111
— Bangladesh **II-292**
— Chesapeake Bay I-342–3
    — coastal ocean spawners I-343
— collapsing I-3, III-119–21, III-367
— commercial fisheries, Western Australia **II-698**
    — Fish Habitat Protection Areas II-699
— Coral, Solomon and Bismark Seas region
    — coastal, information on II-434–5
    — fisheries management measures II-442
— declining
    — English Channel I-76
    — Gulf of Maine and Georges Bank I-310
— demersal fish, Gulf of Aden II-54
— dolphin stocks III-42
— evaluation of global status of III-377
— Gulf of Guinea, significant changes in I-783
— Gulf of Maine and Georges Bank I-312
— Gulf of Thailand III-128
— increased off Alaskan coast I-376–7

— Jamaica I-564
    — reefs overfished I-568–9
— Lesser Antilles, shared assessment of I-633
— Malacca Strait II-337
— Malacca Straits II-314
    — declining II-320
— New Caledonia II-730
— New York Bight, overfished I-330
— North Sea, management by EU I-57
— northern Spanish coast *I-147*
    — over-exploited and depleted I-146
    — sardine stock critical I-146
— off coasts of Oman and Yemen II-27
— over-exploited III-367
    — Irish Sea I-91
— Queen Conch I-425
— Sea of Okhotsk, some decrease in II-470
— some declines, Victoria Province II-668
— southern Spain, over-exploited I-173
— Tanzania II-88
Fish Stocks Agreement III-334, III-340, III-341, III-343
— *see also* Straddling Stocks Agreement
fisheries III-366–8
— Adriatic Sea I-277, **I-278**
— Aegean Sea, overfished I-244
— American Samoa, changing local fishing patterns II-770
— Andaman Islands, poor development of II-192, **II-193**
— Andaman and Nicobar Islands, deep-sea, development of recommended II-196
— Anguilla
    — artisanal I-621
    — Fisheries Protection Ordinance and Turtle Ordinance I-624
— Arabian Gulf II-8
— Arabian Sea II-25
    — artisanal II-26–7
    — industrial II-29
— Argentina
    — commercial I-763
    — industrial I-763–4
    — offshore, impact on fauna I-757–8
    — semi-industrial and artisanal I-763–4
— Australia II-586–7
    — Australian Fishing Zone II-586
    — commercial II-584
    — decline through overfishing II-587
— the Azores I-209–10
    — demersal and pelagic *I-209*, I-210
    — little regulation I-214
— the Bahamas
    — commercial and artisanal I-422, I-424–5
    — other resources I-426–7
    — protective regulations I-430–1
    — use of poisonous substances prohibited I-430
— Baltic I-111–12
— Bangladesh II-292, **II-293**
    — resources II-294
— Bay of Bengal
    — artisanal and industrial II-293
    — brackish water II-278
    — marine II-277–8
— Belize
    — artisanal I-508–9, I-511
    — regulation of I-513
— British Virgin Islands
    — commercial trap fishing I-621
    — moratorium on large-scale foreign fishing I-624
— Cambodia II-574
    — foreign poachers II-576
    — freshwater II-575
    — inland, Mekong River II-575
    — inshore, depleted II-576

— multi-fishing practices II-576
— Canary Islands I-195–6
  — cetacean fishery I-195
  — tuna I-196
— Carolinas I-362–3
  — fisheries monitoring programs I-363
  — viable, stressed and overfished I-363
— Chesapeake Bay, decline in I-343, I-346
— Chile I-710–12
  — cash values I-712
  — new management tools I-463
— co-management in III-159
— Colombian Caribbean Coast I-669–71
  — artisanal I-671–2
  — industrial I-670–1
  — related to upwellings in northeast I-670
  — *Tarpon atlanticus*, depleted population I-669–70
— Colombian Pacific Coast
  — artisanal rights protected I-681
  — commercial landings I-684
  — industrial I-683
  — industrial fishing denounced I-681
  — in the mangrove forests I-682
  — protective measures I-684
— the Comoros, artisanal II-249
— Coral, Solomon and Bismark Seas region II-440
  — problems best solved by community-based initiatives II-442
  — subsistence and artisanal II-434, II-436–7
  — tuna fishery II-432, *II-433*
— Côte d'Ivoire I-816
  — changes in I-815–16
  — demersal I-815
  — pelagic I-815
— declining yields III-377
— destructive II-374, II-387, II-438, II-566
  — the Comoros II-249
  — Indonesian Seas II-392–3
— destructive fishing techniques II-377, II-503, III-37, III-122–3
  — the Philippines II-415
  — *see also* cyanide; dynamite fishing
— discards by ocean region **III-367**
— Dutch Antilles I-602–4
— east coast, Peninsular Malaysia II-357
  — artisanal II-352–3
  — main pelagic species II-352
— eastern Australian region
  — commercial II-640–1
  — commercial and recreational II-637
— Ebrié lagoon, Côte d'Ivoire I-813
— effects on ecosystems III-117–33
— El Salvador I-550, I-552–3
  — affected by oil spills I-556
  — artisanal I-553
— English Channel I-73–4
  — impacts arising from fishing I-76–7
— Faroes I-35–6
  — collapse then increase I-35
— Fiji
  — commercial catch data II-760
  — commercial/artisanal II-715–16
  — overfishing II-760
  — overfishing of marine invertebrates II-759
  — subsistence, impacts of II-758
— and the Fish Stocks Agreement III-341
— French Polynesia
  — lagoonal species II-819
  — subsistence and commercial II-819–20
— Godavari-Krishna delta II-170–1
— Great Australian Bight
  — commercial II-682–3, II-684
  — inshore commercial fishing II-682–3

— Great Barrier Reef region
  — commercial II-623
  — live food-fish export II-623
  — recreational II-621
— Greenland I-8
— Guangdong
  — closed fishing seasons II-560
  — outstripping sustainable capacity II-557–8
— Guinea
  — artisanal I-800, I-801
  — Fish Smoking Center I-800
  — industrial I-800–1
  — National Fishing Supervisions Center I-800
  — pelagic I-800
  — traditional I-800
— Gulf of Aden II-56–7
  — artisanal II-56
  — industrial, illegal and uncontrolled II-59
— Gulf of Guinea
  — artisanal I-787
  — illegal fishing I-787
  — industrial I-787–8
— Gulf of Maine and Georges Bank I-312–13
— Gulf of Mannar II-164–5
  — indiscriminate use of small nets II-165
— Gulf of Thailand II-305–6
  — conflict between subsistence and light-luring fishermen II-306
  — decline in the resource II-305
  — dominant demersal fish groups II-305
  — dominant pelagic fish groups II-306
  — mortalities of shellfish and finfish III-218–19
  — multi-gear, multi-species II-305
— Hawaiian Islands
  — commercial II-802–3
  — derelict fishing gear II-804
  — resistance to licensing/monitoring II-809–10
  — subsistence, artisanal and recreational II-802
— Hong Kong II-538
  — seriously affected by urban development II-543
— impact on seabirds III-111
— impact on small cetaceans III-97–8
— Indonesian II-391–4
  — artisanal II-391, II-394
  — commercial exploitation of offshore fisheries II-391
  — high fishing pressure on major stocks II-391
  — reef fisheries II-393
— industrial
  — effects of III-112
  — sequential overfishing III-120–1
— intensive fisheries management III-367
— Irish Sea I-91–2
  — artisanal I-91, I-93
  — comparison with North Sea I-92, **I-92**
— issues along the Black Sea Coast I-292, **I-293**
  — need for cooperative action I-292
— and its problems I-3
— Jamaica
  — management plans I-572
  — reef I-566, I-568
— Japan, drive fisheries, small cetaceans III-93
— Java Sea, pelagic catch II-391
— Lakshadweep Islands II-194
— large incidental capture of sea turtles III-66
— late development of, Somalia II-80
— Lesser Antilles **I-633**, I-636–7
  — artisanal methods can damage marine habitat I-635
  — ongoing assessment I-634
  — overfishing of nearshore waters I-633
— long-lived species replaced by shorter-lived species III-367
— Madagascar
  — artisanal II-121, II-122–3, II-125

— commercial II-121–2
— offshore, estimated potential 120, **II-120**
— Malacca Strait, mainly artisanal II-338
— Malacca Straits, capture fisheries and coastal mariculture II-314
— the Maldives II-210–12
    — coral reef fisheries II-211–12
    — lagoonal bait fisheries II-210–11
— management for multispecies, marine harvest refugia I-244–5
— Marshall Islands II-783–4
    — artisanal II-784
    — maximum sustainable yield not yet reached II-783–4
    — pelagic fisheries II-784
— Mascarene Plateau
    — artisanal II-261–2
    — banks fishery II-256
    — effects of artisanal fisheries II-262
    — potential for longline and deep-water fishery II-260
    — some foreign tuna fishery II-260
— may negatively affect survival of marine mammals and birds III-128–9
— Mediterranean, Israeli I-257
— Mexican Pacific coast I-489–91
    — artisanal I-490–1, *I-491*
    — benthic invertebrates I-489–90
    — foreign fleets I-490
— Montserrat I-621–2
— Mozambique II-105, II-107
    — artisanal II-107
— New Caledonia II-728
    — artisanal II-729–31
    — Beche-de-Mer and Trochus exploitation II-730
    — professional II-729–30
    — subsistence reef fisheries II-715, II-729–30
— New York Bight
    — "foreign fishing" I-330
    — Magnuson Act I-330
    — recreational I-323, I-329
— Nicaraguan Caribbean coast I-522, I-523–5
    — automatic jigging machines I-523
— Nicaraguan Pacific coast I-534
    — artisanal fishing I-537
    — big pelagic fish I-536
    — demersal fish I-536
    — economically important marine resources **I-536**
    — estimated biomass **I-536**
    — industrial I-539
— North Sea I-51–2
    — cockle fishery I-51–2
    — detritus from catch-processor ships III-127
    — shellfish I-51
— North Spanish coast
    — artisanal I-144
    — demersal I-138
    — Galicia I-142
    — negative impacts I-146
    — pelagic I-141
— northern Gulf of Mexico I-445
    — threatened by effects of navigation channels I-447–8
— Norwegian coast I-25–6
— Oman, artisanal II-21, II-26–7
— Palk Bay II-167
— Peru
    — artisanal I-693, *I-694*
    — demersal I-693
    — pelagic I-693
— the Philippines
    — coastal, affected by loss of mangroves II-414
    — intrusion of commercial fishers into municipal waters II-416
    — offshore II-411
— processing waste III-137
— purse-seine
    — by-catch from different methods III-124, *III-124*
    — dolphin mortality III-124
    — reduction in dolphin mortality III-141
    — tuna mortality, Eastern Tropical Pacific III-94
    — tuna, Pacific, incidental take III-379
— recreational I-323, I-329, I-362, I-397, I-430–1
— Red Sea II-42
    — artisanal II-43
    — reef fisheries II-43
— rejects and marketable catch III-137
— Sabah, reef, overfished II-372
— Samoa II-714
    — inshore II-716
— Sarawak II-372
— Sea of Okhotsk
    — commercial, ecosystem effects II-469–70
    — intensive II-470
— selective fishing III-136
    — may result in ecosystem imbalance III-139
    — Norway III-138
— the Seychelles
    — artisanal II-238
    — foreign vessels charged for fishing II-238
    — reef fisheries II-236
— small-scale, more important than large-scale commercial II-140
— South Africa
    — artisanal and subsistence II-139–41
    — commercial II-136–8
    — optimal utilisation of catch II-138
    — recreational angling II-138–9
    — target switching common II-137
— South Western Pacific Islands, Industrial II-718–19
— southern Brazil I-737
    — artisanal I-737, I-739
    — industrial I-739
— Southern California Bight
    — commercial I-396–7
    — recreational I-397
— Southern Gulf of Mexico I-475
    — artisanal I-475–6
    — species and production **I-476**
— southern Portugal
    — artisanal, Tagus and Sado estuaries I-161
    — coastal waters I-159
— southern Spain I-172
    — major economic activity, Gulf of Cadiz I-172
    — overexploited I-175
— southwestern Africa I-830, I-833
    — commercial trawling I-830, **I-830**
    — hake I-833–4
    — purse seiner catches *I-831*
    — secondary effects of fishing I-833
— Spratly Islands, pressure on recently increased II-364
— Sri Lanka
    — chank and sea cucumber II-183
    — coastal II-182
    — marine II-181–2
    — marine ornamental II-182–3
— Tagus and Sado estuaries I-159
— Taiwan Strait II-503
    — artisanal fishing II-506
    — "bull-ard" fishery II-504–5
    — coral reef II-504
    — decline in fishermen II-505
    — seasonal grey mullet fishery II-503–4
— Tanzania
    — artisanal II-90–1
    — commercial II-90
— target and non-target catches III-137
— Tasmania II-654
    — commercial II-655–6

— individual transferable quotas (ITQs) II-656
— offshore II-653
— Tonga, subsistence reef fisheries II-716
— Torres Strait and Gulf of Papua
    — commercial II-604
    — subsistence and artisanal II-601–3
— tropical seas, many low-value species in catches II-372
— Tyrrhenian Sea I-279
— Vanuatu
    — artisanal II-715
    — commercial II-744
    — community-based management schemes II-745–6, **II-746**
    — impacts of II-743–4
    — major area for development II-744
    — undeveloped II-744
— Venezuela I-653–4
    — worker conflicts, artisanal and trawl fisheries I-653
— Victoria Province
    — commercial II-667–8
    — specialised bay and inlet fishery II-667–8
— Western Australia, commercial II-697–9
— Xiamen region, capture fisheries II-519
— Yemen, artisanal II-56
— Yuzhnoprimorsky region II-480
— *see also* by-catch; discards; recreational fishing
fisheries agreements, regional III-340
fisheries law, global III-158–9
— international law, based on UNCLOS III-158–9
    — national approaches to III-158
— regional agreements III-158
fisheries management
— for artisanal and subsistence fisheries, South Africa II-140–1
— based on ecosystem principles III-377
— binational I-463
— and closed areas III-378
— designed for fish conservation III-157
— emergence of "epistemic community" in III-163
— emerging global fisheries laws III-158–9
— exploitation rate, implementation error in III-389
— Great Barrier Reef region, evaluation of II-624
— Hawaiian Islands II-803–4
    — need for restrictions II-810
— incentive structures III-155–6
— issues related to enforcement III-160–1
    — blacklisting III-160–1
    — flag state responsibility III-161
— issues related to legitimacy of III-155, III-157
    — substantive and process legitimacy III-155
— issues related to surveillance III-159–60
    — observer programs III-160
— limits to exploitation and target exploitation rates III-382
— Malacca Strait, little information to base sustainable use on II-341
— the Maldives, for single species fisheries II-211–12
— ocean dynamics and marine habitat changes becoming important III-185–6
— participation in III-159
— past, present and future, southwestern Africa I-836–7
— precautionary approach III-79
— social aspects of management issues III-161–2
    — restrictions on effort III-161
    — restrictions on technique III-161
— as a social science problem III-153–64
    — equity and fairness III-157–8
    — "governance" and "institutional embeddedness" III-155
    — scientific realism III-156–7
— Sri Lanka **II-186**
— Torres Strait and the Gulf of Papua II-605–7
Fisheries and Oceans Canada, view of passage of El Niño signal, Northeast Pacific III-184–5
fisheries regulations, Madagascar
— conservation of valuable species II-126

— management and policies II-126
— new local control law (GELOSE) II-126
fisheries resources, New Caledonia, not safe from certain threats II-731
fisheries science
— and Fishers' insights III-157
— a mandated science III-156
fisheries scientists I-2
Fisheries Sector Program (FSP), the Philippines II-418–19
fisheries—ecosystem interaction model III-118–19, *III-118*
Fishery Conservation and Management Act (USA) I-451
fishery systems, sources of uncertainty in analysis of III-389
fishing
— as a business III-155
— ecosystems effects of III-378–80
    — by-catch and incidental take III-379
    — debris and ghost fishing III-380
    — habitat destruction III-379–80
— illegal, "patterned deviant behaviour" III-155
— regulatory discards seen as perverse III-156
fishing discards *see* discards
fishing fleet, global, overlarge III-155
fishing fleets, influencing government over international laws III-158
fishing gear
— cause of seabird and marine mammal mortalities II-656
— derelict, dangerous nuisance II-804
— destructive II-27
— subsistence fishing II-436
fishing mortality
— directed, of target organisms III-118–22
    — density-dependent responses III-121
    — loss of genetic diversity III-122
    — overfishing III-119
    — population collapses III-119–21
    — population size III-118–19
    — size-selective fishing III-121–2
— indiscriminate III-122–6
    — by-catch and discard mortality III-123–5
    — caused by lost gear, ghost fishing III-125–6
    — from physical impacts and destructive practices III-122–3
— natural, indiscriminate changes in III-126–9
    — changes mediated by biological interactions: competition and predation III-127–9
    — changes mediated by dumped 'food subsidies' III-127
    — changes mediated by habitat degradation III-126
— underestimation of, and population depletion III-126
fishing practices
— concerns about direct and indirect effects of on marine ecosystems III-352
— destructive I-365, I-366, I-554
fishing pressure, above optimum, Irish Sea I-92
Fishing Reserves, Mascarene region II-265, **II-265**
fishing techniques
— controls on economically inefficient III-161
— potential for environmental impacts II-29
fishmeal, Chile
— effect on coastal zone I-713–14
— environmental impacts of I-713–14
— product of fishing industry I-710, I-712
fjord circulation
— creates environmentally significant gradients I-19
— two-layer flows I-19
fjord complexes, Greenland I-8
fjords
— as contaminant traps I-19
— described I-19
— Norwegian coast I-17, I-19
    — changes in shallow water communities I-22
    — deep, contain arctic bottom fauna I-20
    — deep water circulation I-19
    — main fluvial input at head I-19

— southern, stagnant bottom water I-19
— west coast, large and deep I-19
flash floods II-23
flatback turtles III-63–4
— Great Barrier Reef II-618
— restricted range III-64
— a vulnerable species III-64
Fleet lagoon *I-66*, I-68
Flinders Current *II-581*, *II-674*, II-676
Flinders Region, Victoria Province II-663
Flindersian Province
— southern Australia, marine fish fauna II-679
— Western Australia II-694
flood and coastal hazard maps, Hawaiian Islands II-805
flood risk, and rising sea level III-190–4
flood warning system, Firth of Clyde III-352
Flores Sea *II-382*
— atolls, reefs damaged by destructive fishing practices II-389
Florida
— marine reserves, key design principles III-390
— use of zoning III-389–90
Florida arm, Gulf Stream I-589
Florida Bay *I-406*, I-407, *I-436*
— effects of cyanobacteria III-218
— influx of freshwater, nutrients and sediments I-409
— seagrass die-off
— shown by SPOT XS imagery III-286
— a trend for coastal waters in the new Millennium? III-14–16
— western, importance of to Florida Keys reef tract III-16
Florida Current I-407
Florida Escarpment *I-436*
Florida Keys I-405–14, *I-436*
— agriculture and urban factors I-409
— area of sediment influx I-409
— coral bleaching and mortality events III-56
— coral vitality: long-term study of Carysfort reef I-410–12
— Florida Keys National Marine Sanctuary and Protection Act I-412
— global stresses I-409–10
— grim picture for the future of the reefs I-413
— industrial stresses I-409
— islands composed of calcium carbonate rock I-408
— localized ecological reef stress I-410
— major shallow water, marine and coastal habitats I-407–8
— populations affecting the area I-408
— protective measures I-412–13
— multi-use zoning concept I-413
— stress to the environment and reefs I-408–9
— USEPA coral reef monitoring project I-412
Florida Keys Coral Reef Disease Study I-412
Florida Keys National Marine Sanctuary I-408, I-409, I-452
Florida Keys Reef Ecosystem: Timeline **I-413**
Florida Loop Current I-437, I-439
— carries effects of Mississippi floods I-409, *I-410*
Florida Straits *I-416*, *I-436*, I-437, I-439
Flower Garden Banks *I-436*
— marine sanctuary I-442, I-452
— no real coral bleaching III-56
flushing time, North Sea I-46
fly ash disposal, China II-495
fog and cloud, Peruvian coast I-690
Fonseca, Gulf of *I-532*, I-537–8, *I-546*
— coastal diving birds I-549
— industrial shrimp farming I-538
— over-exploitation of resources I-537–8
Food and Agricultural Organisation *see* FAO
food chains
— energy transfer by zooplankton I-359
— marine, Greenland, heavy metals in I-9
food pollution, Asian developing countries and OCs in fish II-450
food webs, detrital III-3
forest clearance *see* deforestation

forestry
— Australia, environmental considerations II-584
— northern Spanish coastal area, causing soil loss I-142–3
forests
— Chesapeake Bay catchments I-345
— El Salvador I-547
— Gulf coastal plain, USA, conversion to farmland I-444
— Nicaragua I-519, I-520
— degradation of I-539
Forth Estuary Forum
— economic appraisal of the Forth Estuary III-351
— economic development a 'Flagship Project' III-351
Forth Kuril Strait *II-464*, II-465
Frailes Canyon I-487
Fram Strait *I-6*
Framework Convention on Climate Change III-340
France
— influence in French Polynesia II-823–4
— La Rance tidal energy project III-304, III-316
— nitrogen inputs to the English Channel and North Sea I-74
— offshore wind farm planned III-309
— prohibition of organotin-based paints on small boats III-251
franciscana (La Plata river dolphin) III-90
Fraser Island II-634
Freez Strait *II-464*, II-465
freezing, northern Gulf of Mexico, causes mass mortalities I-439
French Polynesia II-813–26
— effects of human activities II-820–3
— exploitation of resources II-819–20
— islands described II-815
— major shallow water marine and coastal habitats II-817–19
— offshore systems II-817
— population and politics II-815
— protective measures II-823–4
— Marine Area Management Plans II-824
— reef restoration schemes II-824
— seasonality, currents, natural environmental variables II-815–17
fresh water, lacking, Namibia I-835
freshet (spring flow), Chesapeake Bay I-340
freshwater
— a critical resource
— the Comoros II-251
— the Mascarene Region II-264–5
— dilution of, Hong Kong coastal waters II-537
— a future problem
— Marshall Islands II-780, II-781
— New Caledonia II-729
— reduced in estuaries, Victoria Province, Australia II-666–7
— resources scarce, Xiamen region II-518
freshwater coral kills, Hawaiian Islands II-807
freshwater flushing
— important for water bodies with poor circulation I-449–50
— rate a critical parameter in northern Gulf of Mexico estuaries I-447
freshwater influx
— affects seasonal patterns, Yellow Sea II-489
— Florida Bay I-409
— Gulf of Alaska I-378
— northern Gulf of Mexico I-437, I-443
— deprived of most of spring runoff I-448
— to Cambodian Sea II-571
freshwater lens, Belize continental shelf I-503
freshwater supplies, the Bahamas I-429
freshwater systems, contamination by nutrients and pesticides I-539
freshwater turtles, east coast, Peninsular Malaysia II-350
fringing reefs II-277
— Andaman and Nicobar Islands II-191, *II-192*
— Borneo II-370
— Brazil I-723
— Colombian Pacific Coast, Gorgona Island I-680
— Comoros Islands II-246
— many threats to II-248

— east coast, Peninsular Malaysia II-350
— Fiji Islands II-756
— French Polynesia II-817, *II-818*
— inner and outer, Tanzania II-85–6
— islands in Torres Strait II-597
— Madagascar II-118
— mainland Belize I-505–6
— Mauritius II-258
— the Philippines II-408
— Puerto Rico I-582
— San Andrés and Providencia Archipelago I-668
— the Seychelles II-236
— western Indonesia II-389
frontogenesis, subtropical I-223
Fucus I-70
— Baltic Sea, decreased through eutrophication I-130–1
— lead and zinc in, near major Greenland mines *I-12*
— response to oil spills III-273
Fucus serratus, scarcity of in Faroes I-33
Fucus vesiculosus
— community changes
— disappearances associated with fish/fisheries decline I-131
— southern Baltic I-110
— hard bottoms, Baltic Sea I-125
Fujeirah *II-18*, II-19
— construction of oily waste reception facilities II-29
Fuma Island, Bajuni Islands, sandy tail *II-75*, II-76
Fundy, Bay of *I-308*, I-311, *I-311*
— cetacean–gillnet interaction III-144–5
fur seals, Patagonian coast I-756

Galapagos Islands, coral bleaching and mortality events III-53
Galicia *I-136*, I-137
— dense coastal population I-142
Ganges River *II-146*
— mouths of *II-286*
gap-fraction analyses, used with remote sensing III-286
garbage disposal, poor, Sri Lanka II-183
garfish, Baltic Sea I-111
gases, solubility of in water, and removal by wet deposition III-200
Gdansk Basin I-101, I-106
Gdansk Deep I-101
— cyclic behaviour, phosphate and nitrate concentration I-108
Gdansk, Gulf of, anthropogenic heavy metals in sediments I-115
Gelidium sesquipedale beds
— Cantabrian coast I-140
— Portuguese coast I-159
Gene Bank of Fishing and Aquarian Resources, Colombia I-685
genetic diversity, loss of in exploited fish populations III-121, III-122
Geneva Convention, addresses military threats to the marine environment III-342
geological features, South Western Pacific Islands region II-712
geology, Gulf of Thailand II-299
Georges Bank
— closed areas III-386
— as larval source sites III-381, *III-382*
— and "cold pool" I-324
— nontidal surface circulation *I-324*
— primary production over I-310
Georges Bank (R. Backus and D. Bourne) I-309
Georges Basin I-309, *I-311*
Geosphere Biosphere Programme (IGBP), scientific definition of coastal zone III-295
German Bight
— cadmium budget I-54
— effects of oxygen depletion I-53
— organic pollutants in I-54
Ghana, mangroves I-779
ghost fishing III-380
— and fishing mortality III-125
giant clam culture, Marshall Islands II-781, II-782

giant clam fishery, Fiji, overexploited II-759
giant clam hatchery, South Sulawesi II-401
giant clams, endangered, the Philippines II-410
Gibraltar Strait *I-168*, I-169, I-172
gillnets
— and cetaceans III-144–5
— coastal
— give high small cetacean by-catch III-94–5
— and seabirds III-146
— drifting/driftnets, high incidental catches III-124–5
— and ghost fishing III-125
gillnetting
— by-catch from II-57
— a danger to dugongs and dolphins II-180
— and decline in Australian dugongs II-587
— prohibited in Gulf of Alaska I-383
— prohibited in Southern California I-397
— Western Australia II-697
Gippsland Lakes
— Victoria
— adjusting to changed environment II-667
— affected by high mercury levels II-669
glaciation, Antarctic I-706
glass-eels I-157
Global Climate Change (GCC) models, predict increase in extreme events III-47
global codification, learning-based approach to III-346–7
global commons, and sustainability III-360
Global Coral Reef Monitoring Network II-435
Global Ocean Observation System I-56
Global Plan of Action (GPA) III-344, III-345
global stresses, Florida Keys I-409–10
global warming
— an abdication of intergenerational responsibilities III-370
— potential effects on oceans III-98
— and range extension I-88
— response to climate change III-181–2
— and whale distribution III-86–7
Global Waste Survey III-339, III-345
— waste connectivity III-338–9
Glovers Atoll (Reef) *I-502*, I-504–5
— deep reef habitats I-506
GNP versus sustainable welfare III-395
GOA *see* Alaska, Gulf of
Godavari and Krishna deltaic coast II-170–3
— biodiversity II-170
— climate and coastal hydrography II-170
— environmental degradation II-171–2
— fish and fisheries II-170–1
— geological features II-170
— impacts of human activities on the ecosystem II-172
— mangrove ecosystem II-170
— suggestions and recommendations II-172–3
Golfe Normano-Breton, gradient in zooplankton type and biomass I-70
Gorgona Island National Park, Colombia I-681, I-684
Gotland Basin I-101, I-106
Gotland Deep *I-101*, I-102
gradient techniques, dry deposition measurement III-203
Grand Comore (Ngazidja/Ngazidia)
— establishment of a coelacanth park II-251
— small reefs and restricted seagrass beds II-245
— a volcanic island II-245
Grand Turk *I-588*, I-590
— tourists I-591
gravel habitats, sensitive to disturbance III-123
Great Australian Bight II-673–90
— coast of geological significance II-678
— coastal erosion and landfill II-684
— effects from urban and industrial activities II-684–6
— major shallow water marine and coastal habitats II-678–81

— need for greater research and conservation management II-686
— offshore systems II-681
— populations affecting the area II-682
— protective measures II-686–8
— Great Australian Bight Marine Park II-686–8
— research II-688
— rural factors II-682–4
— seasonality, currents, natural environmental variables II-675–7
— a single demersal biotone II-675
— Yalata Aboriginal Land Lease II-682
Great Australian Bight Marine Park II-686–8
— Benthic Protection Area II-687
— Commonwealth Marine Park II-687
— Marine Mammal Protection Area II-687
— potential pressures on marine conservation **II-688**
— Whale Sanctuary, Head of Bight II-686
Great Australian Bight Trawl Fishery II-682, II-683
Great Bahama Bank *I-416*, I-421
Great Barrier Reef II-595, *II-612*, III-41
— degradation of III-37
Great Barrier Reef Marine Park II-595
— aboriginal interests in II-620
— does not fisheries management II-624
— a model in large marine ecosystem conservation II-626
— protective measures II-624–7
— biodiversity conservation II-625–6
— evaluation of the management II-626–7
— fisheries management II-626
— human and financial resources II-624–5
— impact of tourism II-626
— management framework II-624
— management mechanisms II-625
— shipping and oil spills II-626
— water quality II-626
— tourism II-622
Great Barrier Reef Marine Park Act II-620
Great Chagos Bank *II-221*
— peat on Eagle Island II-229
— seismic tremors II-225–6
— world's largest atoll II-223
Great Corn Island, volcanic with reefs I-522
Great Lakes Fisheries Commission III-154
Great South Channel I-309, *I-311*
Great Whirl *II-18*, *II-48*, II-49, I-50, II-66
— high speed of currents II-54
Greater Antilles, coral bleaching and mortality events III-54, III-56
Greece
— protected areas I-250
— turtle protection not carried through III-355
green turtle nesting sites, Hawaiian Islands II-799
green turtles I-780, II-54, II-180, II-250, II-289, II-336, II-350, II-598, II-779, II-784
— an endangered species III-62
— Chagos Islands II-229
— death through exposure to okadaic acid III-219
— "edible turtle" III-62
— fibropapilloma incidence on, and turtle mortality III-227
— Fiji, endangered II-757
— Great Barrier Reef II-618
— growth slow III-62
— Hawaiian Islands, with fibropapillomatosis II-799
— The Maldives II-205
green urchins I-312
greenhouse gases
— atmospheric III-188
— commitments to reductions in III-304
greenhouse scenarios, for Australia II-585
greenhouse warming III-369
Greenland icecap, increase in Pb, Cd, Zn and Cu in snow and ice cores I-9
Greenland Sea *I-6*, I-7

Greenland seas I-5–16
— future threats for the marine environment I-14–15
— oceanography I-7–8
— bedrock shorelines I-8
— ocean surface circulation *I-7*
— pollutant sources I-11–14
— population and marine resources I-8–9
— present contaminant levels I-9–11
Greenland—Scotland Ridge, acts as partial barrier I-33
Grenada *I-644*
Grenadines, whaling I-636
grey seals I-49–50, III-95
— Cornwall and Scilly Isles I-70
— Irish Sea I-91
grey whales III-74, III-76
— census of III-79
— numbers of III-81
— for subsistence III-83
Grotius, freedom of the oceans concept (1609) I-2
groundings, shipping, South Western Pacific Islands II-718
groundwater
— the Azores I-207
— the Bahamas, vulnerable to contamination I-429
— pollution of, Borneo II-376
groundwater contamination, Nicaraguan Pacific coast I-539
groundwater extraction, Taiwan, causing subsidence and saline intrusion II-507, II-509
groundwater pollution
— Gulf of Cadiz I-174
— round Sea of Japan II-477
grouper fishery, the Maldives II-211–12
growth rates, seagrasses III-2
Guadalcanal Island II-427
— effects of earthquakes II-429
— small-scale aquaculture trial II-438
Guadalquivir River *I-168*
— estuary fishing I-175
— regulation of fishing I-183
Guadimar River *I-168*
— spillage from the Aznalcóllar mining disaster I-178
guano mining I-829
Guatemala, focus on an ecosystem framework for planning and management I-463–4
Guban coastal plain *II-64*
Guinea I-797–803
— continental shelf I-799
— fishing benefits from its EEZ I-801
— major shallow water marine and coastal habitats I-799–800
— main zones I-799
— offshore systems I-800–1
— populations affecting the area I-801–2
— protective measures I-802
— National Center of Supervision and Fishing Protection I-802
— party to London Convention I-802
— seasonality, currents, natural environmental variables I-799
Guinea Current *I-774*, I-778, I-781, I-807–8
Guinea, Gulf of I-773–96
— coastal erosion and landfill I-786–7
— defined region I-775
— geomorphic features of continental shelf I-775
— LME divided into subsystems I-775
— effects from urban and industrial activities I-787–93
— islands I-781
— major shallow water marine and coastal habitats I-779–81
— offshore systems I-781–4
— populations affecting the area I-784
— protective measures I-793–4
— Gulf of Guinea LME projects I-793–4
— marine and environmental legislation I-793
— protected areas I-793
— rural factors I-784–6

— seasonality, currents, natural environmental variables I-775–9
Guinea Under Current *I-774*, I-778, I-781
Guinea waters I-777
Gulf of Aqaba
— continuation of the Red Sea rift II-37
— sea-floor spreading II-37
Gulf Area Oil Companies Mutual Aid Organisation II-29
Gulf of California Coastal Province, Mexico I-487–8
— habitats I-487–8
Gulf Coast Fisheries Management Council I-525
Gulf Coast, USA, habitats of I-440
Gulf of Guinea LME projects I-793–4
Gulf of Maine Point Source Inventory I-316
Gulf of Mexico Aquatic Mortality Network III-228
— multi-jurisdictional monitoring III-229, III-231
Gulf of Mexico Fishery Management Council I-451
Gulf of Mexico states (USA), regulations addressing habitat degradation I-452
Gulf Stream I-189, I-204, *I-222*, I-223, I-227, I-356, I-362, I-419
— meanders, off the Carolinas coast I-353
— transient effects I-324
Gulf War
— oil spills II-11–13
— oil penetration aided by burrowing animals III-271
Gush Dan outfall, sewage sludge I-259, I-261
Guyana Current I-630, I-632
Gyrodinium aureolum blooms I-69, I-90

habitat changes, small, dramatic effects on whales III-87
habitat degradation III-130
— by fishing methods, effects of III-126
— and change, greatest effects on inshore small cetaceans III-96–7
— Coral, Solomon and Bismark Seas region, localised concerns II-437–8
— leads to decline in fish populations III-126
— Peruvian coast I-694–5
habitat destruction III-379–80
habitat diversity, northern Gulf of Mexico I-438
habitat infilling, North Sea coasts I-54
habitat loss
— American Samoa II-770
— coastal, through new building I-173
— east coast, Peninsular Malaysia II-353
habitat modification, can damage sea turtle populations III-66
habitats I-3
— restoration, Chesapeake Bay I-347
HABS *see* harmful algal blooms (HABs)
haddock I-35, I-49, I-313
Hadramout *II-18*
Haifa *I-254*
Haifa Bay I-255
— coastal zone outside I-261
— heavy metal monitoring I-262
— land-based pollution sources I-259–61
— nutrients I-259–60
— toxic metal and organic pollutants I-260–1
Hainan Current II-537, II-538
Hainan Island *II-550*, II-551
Hajar Mountains, Oman *II-18*, II-21
hake I-313
— Argentine Sea I-757
— in Peruvian fisheries I-693
— silver, contaminants in I-315
— southwestern Africa I-833, I-833–4
hake fishery, Argentina I-763–4
Halaniyat Islands *II-18*
— some coral cover II-22
halibut fisheries, individual quota system, US Pacific Northwest III-155
haline stratification
— Baie de Seine I-67

— off Lancashire–Cumbrian coasts I-86
haling, catch per unit whaling effort (CPUE) indices III-79
halocline, Black Sea, generation of I-288–9
halogenated hydrocarbons (HHCs) III-95
harbo(u)r porpoises I-50, III-90
— by-catch in the North Atlantic III-94
— English Channel I-70
— Irish Sea I-90
harbour seals I-49–50
harbours
— and marinas, accumulation of TBT in sediments III-250
— TBT contamination not reduced III-254
hard bottom habitats
— Norwegian coast I-21
— gradients of change I-22
— Southern California Bight I-388
— Tyrrhenian Sea I-273
hard bottoms I-442
hard substrate communities, Bahamas Archipelago **I-420**
hard-bottom outcrops, Carolinas coast I-353, I-362
hard-bottom/coral reef communities, Bahamas Archipelago **I-420**
hardgrounds, Puerto Rican shelf I-581–2
hardshores, the Azores I-205, *I-205*
— intertidal zonation pattern I-205, *I-206*
harmful algal blooms (HABs) III-98, III-213, III-298
— cause of mass lethal mortality disturbances III-223
— as indirect cause of morbidity and mortality III-218
— profit from wetland and mangrove destruction III-221–2
harp seals
— effect of commercial fisheries on III-97
— hunted in Greenland I-8
Hawai'i
— marine reserves III-386
— no-fishing 'kapu' zones III-386
Hawai'i Coastal Zone Management Plan II-808
Hawaiian Islands National Wildlife Refuge II-798
Hawaiian Islands (USA) II-791–812
— challenges for the new millennium II-809–10
— better management in the northwestern islands II-810
— community-based coastal area management II-809
— controls over fishing II-810
— preservation of natural tourism amenities II-810
— public educations II-809–10
— coastal uses II-802–4
— geological origin of II-793–4
— islands listed **II-794**
— major shallow-water marine and coastal habitats II-795–800
— offshore systems II-800–1
— populations affecting the area II-801
— protective measures II-807–9
— development plans II-808
— environmental impact assessment II-808
— environmental legislation II-807–8
— marine environmental restoration II-809
— marine protected areas II-808–9
— region defined II-793–4
— rural factors II-801–2
— seasonality, currents, natural environmental variables II-794
— urbanization factors II-804–7
Hawaiian Ocean Resources Management Plan II-808
hawksbill turtle I-780, II-205, II-229, II-250, II-350, II-779, II-784, II-799, III-62–3
— critically endangered III-62–3
— endangered, Fiji II-757
— Great Barrier Reef II-618
— harvested for tortoise shell III-62
hazardous substances, targets planned by North Sea states III-362–3
hazardous waste
— Marshall Islands, nuclear testing II-783
— shipments of II-320
HCHs II-545

— air and surface seawater, worldwide II-454, *II-456*
— in the English Channel I-75
heat flux, across Line-P, Northeast Pacific III-184
heavy metal contamination
— Dutch Antilles I-605
— riverine and coastal environment, western Indonesia II-396
— West Guangdong and Pearl River delta II-556, **II-556**
heavy metal mining I-738
— Australia II-639, II-641
heavy metal pollution I-279
— Amursky Bay, eastern Russia II-484
— Anzoategui State and Cariaco Gulf, Venezuela I-656
— from mining, Greenland I-11–12
— Johore Strait II-335
— Liverpool Bay I-95
— Malacca Straits II-318
— Morrocoy National Park I-657–8, **I-657**
— northern Spanish coast I-145, **I-146**
— Odiel saltmarshes I-170–1
— Pago Pago harbour, American Samoa II-712
— southern Brazil
    — Guanabara Bay I-741
    — and mangroves I-742
    — Sepetiba Bay I-741, I-741–2
— Tanzania II-92
— west Taiwan, and "green oysters" II-509–10
heavy metals
— affecting small cetaceans III-95
— Arabian Gulf
    — in sediments II-12
    — in sewage II-10
— Australia II-588
— Bay of Bengal
    — in cultured prawn tissue II-155
    — and estuaries II-155
— Bothnian Bay and Bothnian Sea I-131
— coastal and river waters, Gulf of Cadiz **I-180**
— dune mining of, KwaZulu-Natal II-142
— found in coral skeletons II-390
— from pesticides and fungicides, in shellfish I-493
— Great Barrier Reef region II-623
— in harbour muds, Belize City I-511
— in the Hoogly (Hugli) river II-278
— incorporated into estuarine trophic web I-179
— long-range airborne transport of I-115
— Norway I-24–5
— off west coast, Peninsular Malaysia II-341
— offshore, Guinea I-802, **I-802**
— one-hop contaminants I-13
— Patagonian coast I-765
    — localised I-761
— a potential problem, Gulf of Guinea I-792–3
— in sediments, Teacapán—Agua Brava lagoon system I-495, *I-495*
— in shrimps I-493, I-494
— in surficial sediments, Gulf of Thailand II-305, **II-305**
— Suva Harbour, Fiji II-719
Hector's dolphin III-91
— incidental mortality III-94
HEED (Health Ecological and Economic Dimensions) approach III-214
HELCOM Convention, on Baltic Marine Environmental Protection strategy I-132
Helsinki Conventions on the Protection of the Marine Environment of the Baltic Sea Area III-339
herpesvirus, ducks, geese and swans III-225
herring I-49, I-90, I-313
— Baltic Sea I-110, I-110–11, I-128
— Vistula Lagoon I-112
herring fishery, decline in Irish Sea I-91
hexachlorocyclohexane isomers *see* HCHs
Himalayas, and the Bay of Bengal II-287

Hiri Current II-619
Honduras, coral bleaching and mortality events III-54
Honduras, Gulf of *I-502*, I-503
Hong Kong II-535–47
— coastal erosion and landfill II-542–3
— coastline and coastal waters II-537
— effects from urban and industrial activities II-543–6
— Hong Kong Special Administrative Region II-537
— hydrography II-537
— Mai Po marshes II-539–41
— major shallow water marine and coastal habitats II-537–8
    — mixed subtropical fauna and flora II-537–8
— offshore systems II-538
— populations affecting the area II-538–42
— problems in Victoria Harbour II-542
— protective measures II-546
    — oil pollution control ordinances II-546
    — Water Control Zones II-546
— rural factors II-542
— Sewerage Master Plans II-546
Hoogly (Hugli) river, heavily polluted II-155, II-278–9
Hormuz, Straits of II-3, *II-18*, II-19
— corals in Musandam II-22
— important shipping lanes II-28
— inflowing water enriches the Gulf II-6
— water flow through II-4, *II-5*
horse mackerel I-141
— southwestern Africa I-832–3, *I-833*
horseshoe crabs
— captive breeding programmes, Sundarbans II-158–9
— potential source of bioactive substance II-159
hotspot sediments
— environmental implications of I-25
— Southern California Bight I-392
Houtman Abrolhos reefs, Western Australia II-696
Hudson Canyon *I-322*
Hudson Shelf Valley *I-322*, I-326, *I-326*
— up- and down-welling aids sediment transport I-323–4
Huelva *I-168*
— commercial port I-176
— urban-industrial development environmentally hazardous I-174
— waste dumped into the estuary I-179, **I-179**
Huizache y Caimanero lagoon system I-493–4
— fertilizers and agrochemicals from runoff found in shrimps I-494
— lagoon environmentally managed I-494
— marsh I-493–4
— shrimp culture I-494
human activities
— accelerating eutrophication, and its results III-198–9
— direct and indirect effects of seabirds III-111, III-114
— effects of, French Polynesia II-820–3
— increased spread rate of organisms and changed meaning of distance III-336–7
human breast milk
— contaminated with organochlorines
    — Hong Kong II-545
    — India and China II-452
Human Development Index, low, Papua New Guinea and the Solomon Islands III-434
human impacts, on coastal sediments III-352–3
human-assisted redistributions *see* alien/accidental/exotic/introduced organisms/species
Humber estuary I-54
— seasonal change between nutrient source and sink I-53
Humboldt Current *I-700*, I-701
— *see also* Peruvian Current
Humboldt Current ecosystem *see* south Pacific eastern margin ecosystem
humpback whales III-76
— Bermuda Islands I-226–7
— Hawaiian Islands II-800

— recovery of III-82
hurricane damage, and coral bleaching, Belize III-55
hurricanes I-340–1, I-408
— affecting Belize I-503–4
— Anguilla, British Virgin Islands and Montserrat, damage from I-622
— the Bahamas I-418–19
— cause ecological effects to Gulf of Mexico habitats I-439
— destructiveness of III-224
— dispersing anthropogenic rubbish in the marine habitat I-606
— Dutch Antilles, and oil slick movement I-605
— effects on Carolinas coast I-364
— El Salvador I-547
— Gulf of Mexico I-471, II-470
— intensity of increased I-631
— Jamaica I-561
 — damaging the coral reefs I-562, I-563, I-564
— most common path, Lesser Antilles *I-628*
— Nicaragua, Joan and Mitch **I-519**
— northeast Caribbean I-617
— Pacific Mexican coast I-487, I-488, I-489, I-493
— Puerto Rico I-578, I-579, I-580
— Turks and Caicos Islands I-589
— windward Dutch Antilles I-598
hydrocarbon pollution
— Australia II-588
— from tanker off Mozambique II-109
hydrocarbons
— enhanced values near refineries I-75
— surface marine sediments, Taiwan Strait II-510
hydroelectric power
— for heavy industry I-24
— regulation of river flow for I-19
hydrogen sulphide (H_2S) I-105–6, *I-105*, I-287
— in prawn ponds II-154
hydrography, Hong Kong II-537
hydrography and circulation, Gulf of Thailand
— forcing mechanisms in coastal seas II-300
— understanding of limited II-299–300
hydrologic cycle III-394
hydrology, Xiamen region, China II-515–16
hydrothermal vents, Guayamas basin I-487
hygroscopicity, of particles III-200
hyper-nutrification II-320
hypersaline conditions, south Texas and South Florida I-438
hypersalinisation, Magdalena Delta soils I-671
hyposalinity, Jiulongjiang estuary surface water II-518
hypoxia
— Black Sea I-201, I-303
— Florida Keys, near rivers and deltas I-408
— New York Bight I-325

ice cover
— Baltic Proper I-102
— in Chesapeake Bay I-340
— Greenland seas I-5
— winter, Sea of Okhotsk II-466
ice scouring, winter I-311
icefjords, Greenland I-8
— halibut fishing I-8
Iceland, ITQ system III-162
ICES *see* International Council for the Exploration of the Sea (ICES)
ICES
— Seabird Ecology working group III-112
— studies on ghost fishing III-125
— study group on bottom trawling III-126
ICZM *see* Integrated Coastal Zone Management (ICZM)
image data sources III-294–5
immigrants, behaviour of, Levantine basin I-257–8
IMO *see* International Maritime Organization (IMO)

imposex
— as an indicator of environmental contamination III-253
— causal link to TBT III-249
— dogwhelks I-25, I-36–7, I-55, I-56, I-158, III-250–1, *III-251*
— from tributyltin
 — Malacca Strait II-339
 — Singapore and Port Dickson II-319
— genetic aspects: the Dumpton Syndrome III-252
— occurrence in southwest England *III-251*
— whelks I-210–11
impoundments, lead to loss of wetland I-445–6, I-447
incentive structures, in fisheries management III-155–6
increased prevalence theory III-226
Index of Sustainable Economic Welfare (ISEW) III-395
India
— conservation policies for the Sundarbans II-157
— production of DDT and HCHs II-449
— Southeast II-161–73
 — Godavari and Krishna deltaic coast II-170–3
 — Gulf of Mannar II-163–6
 — Palk Bay II-166–8
 — Palk Bay-Madras Coast II-168–70
Indian Monsoon Current II-274
Indian NE Monsoon current *II-64*
Indian Ocean
— anticyclonic gyre in Mascarene region II-255
— biogeographic position of the Chagos Islands II-224–5
— Central, coral bleaching and mortality events III-48–9
— sever coral bleaching III-224
— Southern, coral bleaching and mortality events III-48
— western, oceanic platform II-235
Indian Ocean Commission (COI)
— joined by the Comoros II-247
— Madagascar a member II-127
— pilot ICZM operation, Mauritius and Reunion II-266
Indian Ocean Whale Sanctuary II-29, II-180
Indian—Australian plate margin, upthrust II-712
individual quota systems III-155, III-161
— best management tool for many fisheries III-162
individual transferable quotas (ITQs)
— ideal but complex and can cause problems III-162
— will lead to concentration of access rights III-162
Indo-Pacific beaked whale II-54
Indo-West Pacific Marine Province, Pan-Tethyan origin II-707
Indonesia
— coral bleaching and mortality events III-52
— Lembata Island, only subsistence traditional whaling operation III-84
— mangroves III-20
— marine conservation target II-401
— national management systems and legislations and the Malacca Straits II-321–3
— National Oil Spill Contingency Plan II-323
— national parks programme II-342
— northeastern, great species diversity II-389
Indonesia, Western, Continental Seas II-381–404
— land-based processes affecting Western Indonesian seas II-395–9
— major marine and coastal habitats II-385–90
— marine resource extraction II-390–5
— a microtidal region II-384
— oceanography II-383–5
— prospects for the future II-403
— protective measures II-399–403
 — endangered species protection II-400–1
 — Integrated Coastal Zone Management II-399–400
 — marine protected areas II-401, **II-402**
 — PROKASIH Program II-400
Indonesian Throughflow, Coral, Solomon and Bismark Seas Region II-428
Indus river dolphin III-90
industrial activities, Malacca Strait II-339

industrial centres, Russian sector, Sea of Japan basin **II-475**
industrial crowding, Calcutta II-155
industrial development
— coastal zone, Cambodia, no serious environmental pollution II-576–7
— Great Barrier Reef hinterland, loss of coastal habitat and water quality II-622
— limited, Coral, Solomon and Bismark Seas region II-439
— poor in Sundarbans II-153
— Tasmania II-654
— Victoria Province coast II-669
— Xiamen region II-518–19
industrial discharges/effluent
— Cabo Frio–Cabo de Santa Marta Grande, southern Brazil I-741
— Hawaiian Islands II-805–6
— Lesser Antilles I-638–9
— Madras, discharged into the estuaries II-169–70
— and marine pollution, Dutch Antilles I-605
— Tanzania II-92
industrial diversification II-263
industrial pollution
— east coast, Peninsular Malaysia II-351
— Fiji II-745
— Gulf of Guinea coast I-788–9
— Hong Kong II-542
— Malacca Straits II-317, **II-318**
— New Caledonia II-733
— Pago Pago harbour, American Samoa II-771
— Peru I-695
— Sarawak II-376
— southern Brazil, east coast I-740
— Sri Lanka II-184
industrial waste III-339
— dumping of, Tasmania II-654
— poorly handled, the Philippines II-417
— Tasmania
— disposal in coastal waters II-656
— dumping of II-654, II-656
industrialisation, at an early stage, Vietnam II-565
industry
— Argentine coast, Bahía Blanca, petrochemicals I-758
— around the North Sea I-52
— Brazilian tropical coast, effects of I-727–8
— causing pollution, Peruvian coast I-693, I-694–5, *I-694*
— development of round the Aegean I-241
— eastern Australian region, central New South Wales II-636
— El Salvador I-554–5
— English Channel coasts I-72
— expansion and diversification, Venezuelan coastal zone I-645
— growth of, Côte d'Ivoire I-815–16
— Guinea, uses old procedures I-802
— Gulf of Maine I-313
— development and change I-314
— heavy, Noumea and Suva II-718
— marine, Guangdong II-559, **II-559**
— Mexican Pacific Coast I-489
— causing environmental stress I-491
— Nicaraguan Caribbean coast I-525
— northern Spanish shoreline, effects of I-145–6
— Oman
— Muscat and Salalah II-25
— new developments, Sur and Sohar II-25, II-27
— and sources of employment, Gulf of Guinea countries **I-785**
— southern Brazil
— impinging on shallow water coastal and marine habitats I-736, I-737
— industrial complexes I-737–9
— Trinidad I-638
— Vanuatu II-745
infaunal biodiversity assessment, Island Copper Mine III-237
— large infaunal species III-242–3

— juveniles, seen infrequently III-242–3
— sampling design and procedures III-237–8, III-239
— similarity analyses of species and abundance data III-239
— tailings rate tolerable to infauna III-239
infections, in humans, from casual exposure to water III-219
informal settlements, South African coast, a pollution problem II-142–3
infrastructure
— causing coastal modification, Samoa II-717
— interrupting coastal sediments, Gulf of Guinea I-786–7
— leading coastal occupation, Brazilian tropical coast I-727
— the Maldives
— interference with natural erosion-deposition cycles II-208
— interfering with sand movement II-203
— Mascarene Islands, interferes with sand movement II-262
Inhaca and Portuguese Islands Reserves, Mozambique II-105
— slow soil recovery from slash and burn agriculture II-107
— turtle nesting beaches II-105
inorganic nutrients
— Neuse Estuary I-366
— Pamlico, Neuse and New River I-363–4, **I-363**
inshore habitats, Australian Bight II-678
— biogeographical regions II-678
Institute of Marine Affairs (IMA), Trinidad I-634
institutional mechanisms
— established under IMO and UNEP III-345
— importance of III-344–5
integrated coastal area management *see* Integrated Coastal Zone Management (ICZM)
integrated coastal management
— need for in Bangladesh II-295–6
— Xiamen region
— land—sea integration II-531–2
— movement towards II-528
Integrated Coastal Zone Management (ICZM) I-79
— and Agenda 21 III-350
— definitions I-458–9
— finance for III-356
— is it achievable? III-356–7
— must include economic development III-351
— needs to develop in Madagascar II-129
— Nicaragua I-541–2
— critical strategies for I-542
— MAIZCo I-526, I-528, I-534
— Nicaraguan Caribbean coast I-526–8
— the Philippines II-418
— a process not a solution III-350
— western Indonesia II-399–400
— CEPI projects II-399
integrated coastal zone management projects, Tanzania II-94
— constraints on development and implementation of II-96
— programmes attempting to put ICZM into practice II-94–5, **II-94**
— Tanga Coastal Zone Conservation and Development Programme II-94
Integrated Ecological Economic Modelling and Assessment (SCOPE project)
— basic framework III-398
— steps in III-399
integrated marine and coastal management (ICAM) III-341, III-344
integration, different kinds I-459
Inter-American Tropical Tuna Commission (IATTC) I-680
— Tuna-Dolphin Program, training skippers III-142
Inter-Tropical Convergence Zone (ITCZ) I-629, I-679, II-39, II-49, II-66, II-347, II-725, II-753, II-780
— Côte d'Ivoire I-808
— Gulf of Guinea I-776
— over Chagos Islands II-225
— and seasonal changes in the Caribbean sea surface waters I-646
Intermediate Antarctic Water Mass I-702
Intermediate Arctic Water I-679
internal boundary layers III-206–7
— concept III-206

International Commission for the South East Atlantic Fishery, and overfishing I-836
International Conferences on the Protection of the North Sea I-26
international conservation organisations, in the South Western Pacific Islands II-720
International Convention for the Conservation of Atlantic Tuna (ICCAT) III-158
— relies on trade restrictions III-159
International Convention on the Long-Range Transport of Air Pollutants I-132
International Convention for the Prevention of Pollution of the Sea by Oil (OILPOL) III-337
International Convention for the Regulation of Whaling III-340, III-341
International Coral Reef Initiative (ICRI), Regional Workshop for the Tropical Americas, Jamaica I-571
International Council for the Exploration of the Seas (ICES) I-8, I-50–1
— Geneva Convention III-76
— maximal/maximum sustainable catches/yields (MSY) III-75, III-81, III-118
International Geosphere-Biosphere Program(IGBP), GLOBEC program III-369–70, III-394
International Maritime Dangerous Goods Code III-343
International Maritime Organization (IMO)
— complete ban on TBT recommended I-75
— Great Barrier Reef a 'Particularly Sensitive Area' II-616
— Gulf of Aden a 'special area' II-59
— lack of oil reception facilities leads to pollution III-345
— Marine Environment Protection Committee I-640
— oil as a serious pollutant from shipping accidents III-336
— prevention of maritime pollution by non-traditional pollutants III-343
— see also MARPOL Convention
International Mussel Watch Programme/Project
— bivalves sampled, southern Brazil I-743
— Patagonian coast I-762
International North Sea Conferences I-57
International Ocean Institute Training Programmes III-328–9
International Pacific Halibut Commission III-158
international rivers, create problems for North Sea I-58–9
International Safety Management (ISM) Code III-343
International Seabed Authority, prevention of pollution from seabed mining III-338
International Whaling Commission (IWC) III-341
— aboriginal subsistence whaling management procedure (1982) III-83–4
  — a new procedure requested (1995) III-84
  — a persistent cause of bad feeling III-84
— beluga stocks III-90
— considered best body to govern small cetacean catches III-99
— Decades of Cetacean Research III-82
— effectiveness of III-99
— Indian Ocean Whale Sanctuary II-29
— limits finally to be set for all catches III-83
— mandate for the whaling industry III-77
— New Management Procedure III-78
— proposals for sustainable catch limits III-78
— Revised Management Procedure III-79, III-80
  — DNA fingerprinting to reveal sub-populations III-80
  — process error III-80
— and smaller cetaceans III-76
— Sodwana Declaration II-29
— UN requested ten year moratorium III-78
intertidal areas
— English south coast, important to migrating wildfowl I-71
— intertidal rocky communities, Spanish north coast I-140
  — coasts exposed to moderate wave action I-140
  — estuarine coasts I-140
— Long Island and New York, poor intertidal fauna I-326
— Southern Californian Bight, unique I-388
— Tyrrhenian Sea, 'trottoir' I-273
Intertropical Confluence Area see Inter-Tropical Convergence Zone (ITCZ)
intoxication events, Madagascan artisanal fishery II-123
invertebrate communities, coastal benthic, North and South Carolina I-360–2
invertebrate fisheries/gleaning
— Mozambique II-108–9
— South Africa II-139
  — catch per unit effort (CPUE) II-139
— Tanzania II-90
invertebrates
— coral reef, Marshall Islands II-778
— in the Deltaic Sundarbans II-150
— Great Barrier Reef II-617
— Lake Tyres, Victoria Province II-665
— land, Marshall Islands II-778
— marine, El Salvador I-550–1
— Spanish north coast I-139–40, I-140
  — Cantabrian shelf megabenthos I-141
Ionian Sea *I-268*
Iran *II-2*, II-21
— desalination plant at Bushehr II-10
— southern, Makran coast II-19
Iraq *II-2*
— devastated by Gulf War II-11
— projects possibly detrimental to the Gulf
  — drainage of marshes in South II-9
  — Third River (Main Outfall Drainage) *II-6*, II-9
Ireland, offshore wind farm being considered III-309
Irish Sea I-83–98
— coastal erosion and landfill I-93
— effects from urban and industrial activities I-93–6
— major benthic marine and coastal habitats I-87–9
  — range limit for some northern and southern species I-88
— natural environmental variables I-85–7
— offshore systems I-89–92
— populations affecting the area I-92–3
— protective measures I-96–7
— region defined I-85
— rural factors I-93
— topography and sediments I-85, *I-86*
Irish Sea Coast
— artisanal and non-industrial use I-93–4
— environmental impact of cities I-94–5
— industrial uses I-94
— ports and shipping I-95
Irminger Current I-7
Irminger Sea, redfish exploited I-8
Irrawaddy dolphin III-91–2
irrigation
— afalaj system, Arabia II-26
— and drainage, development poor, Somalia II-79
— Gulf coastal plain, USA, growing problem I-443–4
— and water shortage, southern Spain I-175
island communities, Great Barrier Reef region II-615
— high floral biodiversity II-615
Island Copper Mine, Canada, effect of mine tailings on the biodiversity of the seabed III-235–46
— discharge of tailings III-236
— no contamination problem from bioactivation of trace metals III-237
— recovering and unaffected stations III-240–1
'Island Domains', South Western Pacific II-707
islands
— and beaches
  — Marshall Islands II-777
  — The Maldives II-203
— offshore
  — and beaches, Hawaiian Islands II-797
  — and coral reefs, east coast, Peninsular Malaysia II-350

ISM Code *see* International Safety Management (ISM) Code
isostatic rebound I-375
Israel, coast of, and the Southeast Mediterranean I-253–65
— coastal erosion I-258–9
— effects of land-based pollution I-259–61
— natural characteristics I-255–6
— Nature Reserves and National Parks Authority I-262
— population I-258
— protective measures I-262–3
— — National Masterplan for the Mediterranean coast I-261
— shallow marine habitats, the Red Sea invaders I-256–8
ITCZ *see* Inter-Tropical Convergence Zone (ITCZ)
IUCN, and Project Tiger II-157
Ivittuut, Greenland *I-6*, I-12
Ivoirian Undercurrent I-808
Ivory Coast *see* Côte d'Ivoire
IWC *see* International Whaling Commission (IWC)
Izmir I-241
Izmir Bay *I-234*, I-236, I-237, I-244
— dinoflagellate red tides I-238
— eutrophication spreading I-244
— industrial development I-246–7
— wastes dumped with no treatment I-247

Jakarta Bay
— heavy metal loading II-396, **II-398**
— 'Where Have All the Reefs Gone'? the demise of Jakarta Bay and the final call for Pulau Seribu II-397–8
Jakarta Mandate on Marine and Coastal Biodiversity III-341
— integrated marine and coastal management (ICAM) III-341
Jamaica I-559–74
— coastal erosion and landfill I-569
— coral bleaching and mortality events III-55
— effects of urban and industrial activities I-569–70
— Environmental Protection Areas, Negril and Green Island watersheds I-571
— geography of I-561
— macroalgal mats smother old reefs III-223
— major shallow water marine and coastal habitats I-561–5
— Montego Bay and Negril Marine Parks I-571
— offshore systems I-565–6
— populations affecting the area I-567
— Portland Bight Fisheries Management Council I-572
— Portland Bight Sustainable Development Area I-572
— protective measures I-570–3
— — Council on Ocean and Coastal Zone Management I-570
— — Environmental Permit and Licensing system I-570
— — International Conventions participation I-570, **I-571**
— — National System of Protected Areas *I-571*
— — Natural Resources Conservation Authority I-570
— rural factors I-568–9
— seasonality, currents, natural environmental variables I-561
— stresses causing collapse of many reef species III-223
Jamaica coral reef action plan I-571
Jangxia Creek, China, tidal energy project III-316
Japan
— acquisition of whaling technology III-74
— against the Southern Ocean Whale Sanctuary III-85–6
— banned organotin-based paints completely III-251, III-253
— coral bleaching and mortality events III-52
— directed hunts of small cetaceans III-93–4
— mistaken killing of dolphins III-97
— resumed Antarctic whaling III-77
— still killing minke whales III-78
— whale meat
— — continued search for III-78
— — insatiable demand for III-78
Japan, Sea of II-473–86
— climate change II-483
— ecological problems and their causes II-483–5
— — Amursky Bay II-483–5
— — Rudnaya River Valley II-485
— landscapes and tourism II-482–3
— natural resources, species and protected areas II-480–1
— protected areas II-481–2
— state of marine, coastal and freshwater environment II-476–80
Java Sea *II-362*, II-363, II-363–4, *II-382*, II-552
— fish landings related to the monsoon II-372
— fishery production II-363
— fragile II-363
— monsoonal climate II-383–4
— pelagic resource base considered heavily exploited II-393
jellyfish I-832
jet drops and film droplets, cause enhancement to particle deposition III-202
Jiangsu Coastal Current II-489
Jinjira *see* St Martin's Island, Bangladesh
John Pennecamp Coral Reef State Park I-412
Johore Strait II-311, II-334–5
— East Strait II-334
— effect of causeway II-334
— increase in sewage and industrial waste discharges II-335
— low wave energy II-334
— poor water quality II-334, II-335
— seagrass beds II-334
— Sungei Buloh Nature Park II-335
— West Strait, problems II-334
Jordan Basin I-309, *I-311*
Juba-Lamu embayment *II-67*
Jutland Current *I-18*, I-23

Kakinada sand spit, Godavari and Krishna deltaic coast II-172, II-272
Kalimantan
— East, SCUBA diving II-377
— Northeastern *II-362*
— — coastal habitats II-365
— transmigration programme II-374
Kamchatka *II-464*
— "eastern channel", west coast II-466
Karimata Strait *II-382*
karst I-561
— formation I-417
Kattegatt *I-100*, *I-122*
kelp II-52
— Chile I-707
— Faroes, forests of *Laminaria hyperborea* I-33–5
— Galician coast I-140
— giant, extensive beds, Tasmanian waters II-652
— Gulf of Maine I-312
— Irish Sea I-89
— Norwegian coast I-21
— — harvesting causes public concern I-26
— — kelp forests denuded I-21
— response to oil spills III-273
— Southern California Bight I-388, I-395
kelp forests III-35
Kelvin wave dynamics III-184
Kelvin waves, and El Niño I-703
Kemp's ridley turtle III-63
— critically endangered III-63
Kent Group ecosystems, Tasmania II-651
Kerch Strait I-287, I-298
Kerguelen Islands II-583
Key Largo *I-406*, I-408
— reefs showing physical or biological stress I-410
Key Largo Formation I-408
— migration of effluent through I-409
Key Largo National Marine Sanctuary I-412
Key West *I-406*, I-408
khawrs
— brackish coastal wetlands II-23, *II-52*
— larger, may host mangroves II-53

— periodic opening to sea II-23, II-52–3
— roads may interfere with the process II-27
Kilinailau Trench *II-426*
Kishon River, Israel
— carrying nitrogen and phosphorus I-259
— heavy metal contamination in the estuary I-260–1
Kislaya, Bay of, tidal energy project III-316
Kizilirmak delta, Turkey, conservation area I-293
knowledge-based tools in coastal management III-355
Korea, disappearance of tidal marsh zone II-490
Korea Strait II-475
Korean Peninsula, Yellow Sea coast population II-491–2
Kosi Bay system, KwaZulu-Natal, estuarine resource use II-136
krill, Antarctic, food for seabirds III-108
Krishna-Godavari deltaic coast *II-162*, II-163
Kruzenshtern Strait *II-464*, II-465
Kuril Basin *II-464*, II-465
Kuril Islands *II-464*
Kuroshio Current II-489, II-516, II-537
— eastern Taiwan II-501, *II-502*, II-508
Kuwait *II-2*
— coral islands II-7, *II-8*
— coral reef fish fauna II-8
— devastated by Gulf War II-11
— dual-purpose power/desalination plants II-10
— hydrographic changes due to Third River project II-9
— landfill II-9–10
— mariculture II-8–9
— nutrients offshore II-6
— offshore circulation anomaly II-4
Kuwait Action Plan II-29
Kuwait Institute for Scientific Research, research related to pollution by petrochemical industries II-3–4
KwaZulu-Natal
— degraded state of estuaries II-135–6
— from management to co-management: the *Pomatomus saltatrix* fishery II-138
— importance of subsistence and artisanal fishermen II-140
— intertidal harvesting, Maputaland Marine Reserve II-140
— invertebrate fishery/gleaning II-139
— well managed II-139
— *see also* South Africa
Kyoto Agreement III-304
Kyoto Declaration, and Plan of Action III-139

La Hague, radionuclides transported from *I-14*, I-54–5, I-76
La Niña I-708, I-709
— strong
— associated coral bleaching III-46
— bleaching south of typhoon path III-47
La Perouse seamount II-257
Labrador Current *I-352*, I-356
Lac Badana National Park II-80
Laccadives *see* Lakshadweep Islands
lagoonal habitats, important in Mascarene Region II-256
lagoonal shelf, Belize I-506
lagoons
— east coast, Peninsular Malaysia II-349
— Marshall Islands, water residence time II-776
Laguna de Tacarigua, Venezuela I-652–3
— American crocodile I-652
— common birds and fish **I-652**
— eurihaline-mixohaline I-652
— planktonic productivity I-652
Laguna Madre *I-436*
lake waters, airborne spectrometer studies, and chlorophyll content III-297–8
Lakshadweep Islands II-20, *II-190*, II-194–5
— biodiversity II-194
— climate and coastal hydrography II-194

— conservation recommendations II-195
— environmental degradation II-194
— low numbers of seabirds recorded II-204
— population and tourism II-194
Lakshadweep (Laccadive)—Maldives—Chagos Ridge II-201, II-223
— formation of II-223
— spread of coral biodiversity along II-202
Laminaria, harvested for alginate I-70
Laminaria hyperborea forests I-33–4
land, an increasingly scarce resource, the Comoros II-251
land clearance, for agriculture, impact of silt, nutrients and contaminants I-508
land degradation
— Australia II-585
— Fiji II-758
land mosses, as biomonitors I-116
land reclamation
— for agriculture and spoil disposal I-144
— Coral, Solomon and Bismark Seas region II-438
— effects of, Western Port Bay, Victoria II-667
— Guangdong coast II-557
— and habitat loss, Singapore II-334
— large-scale, Hong Kong II-542–3, II-546, *II-546*
— Malé's artificial breakwater II-208–9
— pros and cons, the Maldives II-208
— small-scale, Tasmania II-655
— Taiwan II-505
— an earlier policy II-510
— effects of II-509
— for urban development, western Indonesia II-398, *II-399*
— west coast of Peninsular Malaysia II-339
— west coast Taiwan II-503
land subsidence, caused by artesian wells II-307
land tenure rights
— Marshall Islands II-786
— vs. governance system, Marshall Islands II-785
land use, Colombian Caribbean Coast I-672
land-slips, and associated sediment plumes, show a dynamic landscape II-429
land-use practices, environmental threat, the Comoros II-247
landfill
— American Samoa II-770
— Belize I-510
— biodiversity loss and monetary loss I-145
— Curaçao I-602
— Djibouti, impacting mangroves II-58
— due to urbanization, Arabian Gulf II-9–10
— and habitat loss, major concern, Cambodia II-576
— impacting some lowlands, Gulf of Mexico I-477, I-479
— Kota Kinabalu, Sabah II-375
— the Mascarenes II-263
— New Caledonia II-731
— North and South Korea II-493
— not common in Argentina I-762
— obvious form of estuarine alteration I-448
— a problem throughout the Gulf of Guinea I-787
— Sarawak II-375
Langstone Harbour *I-66*, I-174
Lanzarote, Los Jameos del Agua, endemic invertebrates I-195
Laperuz Strait *II-464*, II-465
Large Marine Ecosystems (LMEs) I-2
— Arabian Gulf II-1–16
— Arabian Sea *II-64*
— Australia *II-580*
— basis for assessing the health of III-230–1
— Bay of Bengal II-287
— Benguela Current LME I-823–37
— East African Marine Ecosystem II-85
— Gulf of Guinea I-773–96
— health impacted by morbidity, mortality and disease events III-218

— Lesser Antilles, Trinidad and Tobago I-627–41
— many becoming stressed III-218
— Mexico, northern Gulf of I-437
— Red Sea II-35–45
— Somali Coastal Current II-63–82
— South Western Pacific Islands Region II-707, II-709
— West Iberian large marine ecosystem I-137
Latin America, coastal management in I-457–66
— coastal zone defined I-458
— integrated coastal zone management defined I-458–9
— the Latin America and Caribbean focus I-460–3
— regional examples I-463–4
— some initiatives on coastal zone management **I-464**
— sustainable development defined I-459–60
Law of the Sea *see* UNCLOS (UN Convention on the Law of the Sea)
lead (Pb)
— Baja California I-495, **I-496**
— in Boston Harbor sediments I-315
— contamination by in Tagus estuary I-158
— decrease in Greenland snows I-14
— from petroleum combustion I-329
— in Greenland marine mammals I-9
— in the North Atlantic III-361
lead pollution
— Greenland I-12
— ships' paint I-37
lead and zinc mining, Chile I-710
leatherback turtles I-780, II-205, II-350, **II-355**, III-61–2
— considered endangered by IUCN III-62
Leeuwin Current II-581, *II-581*, *II-674*, II-676, II-677, II-694, *II-695*
— introduces an Indo-Pacific element II-676
— linked to population dynamics of West and South Australia's commercially important pelagic species II-676, II-679
Leeuwin Under Current *II-674*
Lesser Antilles
— coral bleaching and mortality events III-56
— defined I-629
— economies over reliant on coastal environment I-634
— southern, influence of Amazon and Orinoco rivers I-629–30
Lesser Antilles, Trinidad and Tobago I-627–41
— coastal erosion and landfill I-636
— effects from urban and industrial activities I-636–9
— major shallow water marine and coastal habitats I-631–2
— offshore systems I-632–4
— populations affecting the area I-633–5
— protective measures I-639–40
— rural factors I-635–6
— seasonality, currents, natural environmental variables I-629–31
Levantine Basin *I-254*, I-255
— deep chlorophyll maximum I-256
Levantine Intermediate Water I-255, I-256
— entering the Tyrrhenian Sea I-272
— silicates in I-271
Levantine Sea *I-234*, I-241
Levantine Surface Water I-255
life, beginning on Earth III-394
light availability, and seagrass growth I-358
light-stress-induced mortality, seagrasses III-15
Lighthouse Reef *I-502*, I-504–5
Ligurian Sea *I-268*
limestone, used for cement, Sri Lanka II-181
limestone mountains, Arabian coast II-19, II-21
limestone platforms, Red Sea, foundation for "Little Barrier Reef" II-40
limpets I-71
— harvesting of, the Azores I-206, I-210
lindane, Seine estuary I-75
liquefied natural gas (LNG), Indonesia II-390
Lisbon *I-152*
Lisbon embayment I-153
literacy rate, Gulf of Mannar II-165

litter
— conspicuous source of pollution, Fiji Islands II-761
— dumping of I-512
— and floating wastes, Borneo II-375–6
— and marine debris, western Indonesia II-398–9
— ocean and beach, Australia II-588, II-640, II-684, **II-685**
Little Andamans, reported deforestation II-193
Little Bahama Bank *I-416*, I-421
littoral communities, Faroes, response to wave exposure I-33
littoral transport
— Côte d'Ivoire I-808
— sand, Gulf of Guinea I-778
Littorina littorea, lacking in Faroes I-33
live reef fish food trade II-373, II-392–3
— Cambodia II-576
— Coral, Solomon and Bismark Seas region II-440
— Great Barrier Reef region II-623
Liverpool Bay *I-84*, *I-85*
— discharges to I-95, **I-95**
livestock grazing, and erosion I-601–2
livestock rearing
— changing pattern of, Oman and Yemen II-26
— Gulf of Aden coasts II-57–8
lobster fishery II-80
— Belize I-509
— Lesser Antilles I-635, I-637
— Madagascar, under local control II-126
— Nicaragua I-524
    — Pacific coast I-537
— Yemen II-59
lobster habitats, artificial I-425
local institutions, learning from III-400
loggerhead turtles II-205, II-574, III-63, III-355
— developmental migrations III-63
— Great Barrier Reef II-618
logging
— Alaska, leads to habitat degradation I-381, I-383
— Cambodia II-575
— effects of
    — Andaman and Nicobar Islands II-193, II-195
    — Coral, Solomon and Bismark Seas region II-435–6
— Gulf of Papua II-607
    — probably a threat to coastal and marine environments II-604
Lombok Strait *II-382*, II-383
— tanker traffic II-390
London Dumping Convention II-524, III-338, III-365
— 1996 Protocol III-339
— evolution of III-338–9
— greater prohibition of dumping III-339
— need for more comprehensive perspective on waste management III-339
— permitted amounts of dredged materials **III-365**
Long Bay *I-352*
Long Beach *I-322*
long-distance transport of pollutants
— atmospheric
    — Baltic Sea I-132
    — lead in the southern Baltic I-116, I-214
    — one-hop or multi-hop pathways I-13
— to the Sargasso Sea I-229
Long-Range Transboundary Air Pollution Convention III-340
longline fisheries
— incidental capture of sea turtles a problem for III-146
— Madagascar II-122
— and seabirds III-145–6
— some important by-catch III-125
longshore drift, Belize I-510
longshore sand transport, Israeli coast I-256
Looe Key National Marine Sanctuary I-412
Lophelia banks III-379
Lord Howe Island II-635, II-636, II-643

Louisade Archipelago *II-426*
Louisade Plateau *II-426*
Louisiana–Texas Slope and Plateau *I-436*
low energy environments
— biological effects aligned with oil persistence III-275–6
— oil dynamics III-271
lowland evergreen forest, Cambodia II-573–4
Lüderitz upwelling cell *I-822*, I-825
— eddies, filaments and superfilaments I-826
— separates Northern Benguela from Southern Benguela I-823
Lyme Bay *I-66*

Maamorilik, Greenland *I-6*
— sources of lead and zinc pollution identified I-12
Macao *II-550*
mackerel I-49, I-313
— Baltic Sea I-111
Macquarie Harbour, Tasmania, affected by mining pollution II-654
Macquarie Island II-583
macroalgae III-3
— the Bahamas I-429
— Baltic Proper I-109
— benthic I-70
    — Baltic, changes in I-109–10
— Carolinas coast
    — colonization limited I-357
    — offshore I-362
    — rich communities occur on rocky outcrops I-357
    — a transition zone I-356–7
— Colombian Caribbean Coast I-668–9
— communities, Gulf of Aden II-52
— eutrophication causing community changes I-130–1
— Great Barrier Reef region II-615
— increase in Belize reefs I-505
— marine I-490
— northern Patagonia I-755
— Norwegian coast I-21
— red, disappearance from Wadden Sea creeks I-48–9
— reef habitats, Tasmania II-652
— southern Arabia, luxuriant during the Southwest Monsoon *II-22*, II-23
— spring bloom, Gulf of Aden II-50, II-54
— Western Australia II-695, II-701
    — habitat alienation and fragmentation II-699
    — habitat loss II-699
— western Sumatra II-387
macroalgal blooms I-666
macroalgal mats I-53
macrobenthic assemblages/fauna
— intertidal sedimentary areas I-47
— North Sea I-48
macrobenthos II-277
— southern Yuzhnoprimorsky region, reduction in density of II-481
Macrocystis pyrifera, adjacent to *Lessonia* and *Phyllospora* habitats, Tasmania II-652
macrofauna
— intertidal zone, English Channel I-71
— North Sea, northern and southern species I-48
— Portuguese coastal waters I-159
    — some limits on life-span and size I-159
macromolluscs, Chagos Islands II-227–8
macrophytes
— free floating, Côte d'Ivoire reservoirs and rivers I-818
— growth restricted, the Bodden III-261
— replaced by nuisance algal species I-53
Madagascar II-101, II-113–31, *II-244*
— coastal erosion and landfill II-124–5
— effects from urban and industrial activities II-125
— islets and islands II-119
— major shallow water marine and coastal habitats II-117–19
— offshore systems II-120

— populations affecting the area II-120–3
— protective measures II-125–9
    — Environmental and other legislation II-126–7
— recommendations and prognosis II-129
— rural factors II-123–4
— seasonality, currents, natural environmental variables II-115–17
Madagascar Current *II-114*
Madeira Current I-204, *I-204*
maërl I-69–70
— commercial exploitation I-70
maërl vegetation I-69–70, I-140
Mafia Island, Tanzania *II-84*, II-85
— Multi User Marine Park II-86, II-94
    — unsustainable octopus harvesting II-90–1
Magdalena River
— discharges fertilize the Colombian Caribbean I-669
— mangroves in the delta I-671
    — rehabilitation project I-671
— sediment movement from I-667
Magdalena River Basin, Colombia I-665–6
Magellan Strait I-753, I-762
magellanic penguin, affected by oil spills I-765–6
Mai Po marshes, Hong Kong II-539–41
— Deep Bay area
    — conversion of mangroves to fishponds II-539
    — habitat loss a threat II-540
    — over-wintering site for cormorants II-539
    — shrimp ponds (Gei Wais) II-540
— mudflats II-539
— protective measures II-540
— a RAMSAR Site II-539
    — threats to II-540
— Site of Special Scientific Interest II-539
Maine, Gulf of and Georges Bank I-307–20, I-330
— boreal waters I-309
— circulation of the Gulf I-309, I-310, *I-311*
    — nontidal surface circulation I-324, *I-324*
— continual addition of new species I-312
— early settlement and development I-313, I-314
— effects from urban and industrial activities I-314–17
— forestland, development of I-313
— growth of environmental regulation I-313–14
— Gulf of Mine Habitat Workshop I-318
— hydrological regions I-310
— major shallow water marine and coastal habitats I-310–12
— natural environmental variables: currents, tides, waves and nutrients I-310
— offshore systems I-312
— population affecting the area I-312–14
— principal basins I-309
— urban areas I-313
— Working Group on Human Induced Biological Change I-318
— workshops and Proceedings I-309
    — primary research goals and tasks identified **I-318**
Maine, Gulf of, cetacean–gillnet interaction III-144–5
MAIZCo (ICZM), Nicaragua I-526, I-528, I-534
Makaronesy archipelagoes, physical data **I-187**
Makassar Strait *II-382*, II-383
— tanker traffic II-390
Malacca Strait *II-310*, *II-332*, *II-382*
— diluted by river discharges II-333
— including Singapore and Johore Straits II-331–44
    — coastal population II-337
    — general environmental setting II-333–4
    — impact of human activities II-337–40
    — protective measures and sustainable use II-341–2
    — water quality II-340–1
— main problems and issues II-341–2
— marine and coastal habitats II-335–7
— organotins II-449
— pre-European trading route II-337

— tanker traffic II-390
Malacca Straits II-309–29
— conclusions and recommendations II-325–7
— critical environmental problems II-317–21
— natural environmental conditions II-311–14
    — climatology and oceanography II-311–12
    — coastal and marine ecosystems II-312–14
    — geography II-311
    — populations affecting the area II-311
    — topography II-311
— ports, trade and navigation II-316–17
— protective measures II-321–5
    — coordination of the management of the Straits II-325
    — international legal regime governing the Straits II-324–5
    — national management systems and legislations II-321–4
— ratification by littoral states of international conventions **II-324**
— resource exploitation, utilization and conflicts II-314–17
— total net economic value, marine and coastal resources **II-326**, II-327
Malacca Straits Demonstration Project (MSDP) II-325
Malacca Straits Strategic Environment Management Plan needed II-327
malaria control I-476
Malaysia II-299
— approaches to prevent and control marine pollution II-323
— coral bleaching and mortality events III-51
— Kuala Selangor Nature Park II-342
— legislation on environmental protection II-323
— mangroves II-312
— Matang managed mangrove forest III-25
— Matang Mangrove Forest Reserve II-342
— move to fisheries management II-341
— National Oil Spill Contingency Plan II-323
— Peninsular, coastal plains and basins on west coast II-311
— regulation
    — of land-based pollution II-323
    — of toxic and hazardous wastes II-323
    — of vessel-related marine pollution II-323
— swine farming source of agricultural waste II-318–19
— water quality monitoring programme II-340–1, **II-341**
Malaysia, Peninsular, East Coast of II-345–59
— coastal erosion and land reclamation II-353
— effects from urban and industrial activities II-353–4
— impact on habitats and communities II-354–5
— major shallow-water marine and coastal habitats II-348–50
— populations affecting the area II-350–1
— protective measures II-355–7
    — 1985 Fisheries Act II-355–6
    — coastal erosion II-356–7
    — prognosis II-357
    — turtle hatcheries II-356
— rural factors II-351–3
— seasonality, currents, natural variables II-347–8
— the shallow seas II-350
Malaysia and the Philippines, coral bleaching and mortality events III-52–3
The Maldives II-199–219
— coastal erosion and landfill II-208–10
— effects from urban and industrial activities II-210–15
— major shallow water marine and coastal habitats II-202–5
— offshore systems II-205
— part of Lakshadweep (Laccadive)—Maldives—Chagos Ridge II-201
— populations affecting the area II-206–7
— protective measures II-215–18
    — carrying capacity, sustainability and future prospects II-217
    — environmental legislation and related measures **II-215**, II-216
    — environmental restoration II-217
    — initiation of Protected Area system II-216–17
    — multidisciplinary Environmental Impact Assessment (EIA) II-216
    — national and regional development plans II-216
— rural factors II-207

— seasonality, currents, natural environmental variables II-201–2
Malé Declaration II-216
Maluan Bay, Xiamen region
— chemical oxygen demand (COD) II-522
— ecosystem changes II-524–5
Malvinas/Falkland Islands *I-750*, I-751, *I-752*, I-753, I-766–7
— fishing industry I-763
— squid fishery I-763
mammals
— North Sea I-49–50
— as pests, Marshall Islands II-778
man
— and changes in marine mammals and seabirds I-9
— interference with river flows I-365
Man and Biosphere Programme (UNESCO), research project, Moorea and Takapoto reefs II-817
managed coastline retreat I-54, I-79, III-353
management I-3–4
management practices, customary, Madagascan marine and coastal resources II-126
management problems, Xiamen region II-526–8
— conflicting uses in marine waters II-526–7
— lack of knowledge and information II-527
— new management measures II-528–9
— transboundary problems II-527–8
management structures and policy, South Africa
— advisory bodies II-143
— legislation II-143
— principles II-143
Managua, Lake *I-532*, I-533
manatee grass I-601
manatees I-505, I-509, I-591, I-631
mandated science, fisheries management as III-156
mangrove communities, Borneo, mixed, extending up river valleys II-367
mangrove crab culture, sustainable community aquaculture II-375
mangrove deforestation II-438
— Mozambique II-109
— promotes coastal erosion II-108
— Torres Strait and Gulf of Papua II-605
mangrove destruction, West Guangdong II-559
mangrove ecosystem management, sustainable basis for needed, Malacca Strait II-341–2
mangrove ecosystems
— Godavari-Krishna delta II-170
    — impact of deforestation and prawn seed collection II-172
    — indiscriminate exploitation II-171
— interaction with other ecosystems III-18, III-24
— need to improve sustainable use III-29
— Palk Bay–Madras coast II-168–9, II-170
— rehabilitation of III-26–7
    — concern for the human factor III-28
    — criteria and practical considerations III-27
    — goals III-27
    — need for III-26–7
    — replanting programmes III-27
— species poor, but support biodiversity III-24
mangrove forests
— Borneo
    — distribution factors II-366–7
    — successional communities on accreting shores II-367
— Brazilian tropical coast I-722, I-727
— Colombian Pacific Coast I-679–80
    — exploitation of I-680, I-681
    — fisheries in I-682
    — uses of I-682–3
— Gulf of Papua II-597–8
    — lack low salinity species II-598
    — mangrove tree species, Fly River delta II-597
    — pristine II-597
— Indian Sundarbans II-147

— managed, Sundarbans the first II-148
— Palk Bay, areas cleared for salt pans II-167–8
— Palk Bay–Madras coast, reduced by human activity II-169
— the Philippines
    — loss of and loss of coastal productivity II-414
    — provide nursery grounds II-408
— Sundarbans
    — home of the Bengal tiger II-150
    — under serious threat II-151
    — zoned on the tidal flats II-149–50
— Thailand, use for shrimp farming III-175–6
— western Indonesia, reduced by clearing II-385, II-388
mangrove logging, Cambodia II-575
mangrove palm, Gulf of Guinea, a significant problem I-786
mangrove shrimp ponds III-21–2, III-25
— effects of acid release III-25–6
mangroves III-17–32
— Abrolhos Bank–Cabo Frio, Southern Brazil I-735, I-736
— aerial roots III-18, *III-19*
— American Samoa
    — limited occurrence II-768
    — loss of II-768
— Andaman and Nicobar Islands II-192
    — conservation activities initiated II-195–6
    — degraded sites, restoration of II-196
— Australia II-582
— the Bahamas I-421–2
    — typical zonation I-421–2
— Bangladesh II-290–1
    — land area changes and biodiversity II-290–1
    — *see also* Sundarbans
— Bay of Bengal II-277
    — depletion in Orissa II-279
— Belize I-506
    — some clearance I-510
— biodiversity and human communities III-27–8
— Borneo II-366–70
    — clearing for shrimp farms II-368–9
    — fauna II-368
    — non-conversion uses II-369
    — for wood chip industry II-369
— and braided channels *II-71*, II-76–7
    — low-energy intertidal environment II-76–7
— British Virgin Islands I-619
— Cabo Frio–Cabo de Santa Marta Grande, southern Brazil
    — degraded I-742–3
    — estuarine–lagoon complex I-736–7
    — lost to urban development I-743
    — seriously degraded I-736
— Central America, critical coastal habitat I-460
— clearance for aquaculture III-368
— coastal lagoons, Côte d'Ivoire I-818
— Colombian Caribbean Coast I-669
    — rehabilitation important I-669
— Coral, Solomon and Bismark Seas region II-430, II-438
    — species diversity II-431
— Côte d'Ivoire I-810–11
— degradation of, the Comoros II-248
— development limited, northwest Arabian Sea and Gulf of Oman II-23
— distribution *III-18*
    — patterns of III-20
— Dutch Antilles I-600–1
— east coast, Peninsular Malaysia II-348
    — destroyed by development, Pulau Redang II-353
    — estuaries, lagoons and mainland II-349
    — rate of destruction alarming II-354
— eastern Australian region II-632–3
    — loss of II-639
— ecological values III-22–4
    — importance to fish populations III-23

— mangrove litter III-23
— productivity III-22–3
— stabilisation of exposed land III-23–4
— El Salvador I-548, I-553
— Fiji Islands II-755
    — cleared for development II-760
— the future III-38–9
— Guinea, deforestation I-801
— Gulf of Aden II-53
— Gulf Coast, USA I-440
— Gulf of Guinea I-779–80
    — Ghana I-779
    — importance of I-779–80
    — over-exploited I-786
— Gulf of Mannar II-164
— Gulf of Thailand II-301
— Hawaiian Islands II-797, II-802
— Iran, Saudi Arabia and Bahrain II-7
— Jamaica
    — Black River Morass I-565
    — Negril Morass I-564
— killed by hypersalinity II-135
— Koh Kong Bay, Cambodia
    — may now be seriously degraded II-576
    — now cleared for shrimp farming II-577
    — pristine II-573
— Lesser Antilles I-631, I-632, I-637
— long term oil retention in sediments III-271
— Madagascar II-118–19
    — change in cover II-118–19, II-123
    — exploitation of II-125
    — harvesting regulations II-126
    — majority on the west coast II-118
— Magdalena River Delta I-671
— Malacca Strait II-335–6
    — conversion for aquaculture no longer valid II-342
— Malacca Straits II-312
    — a natural resource being lost II-314–15
— Malaysia, loss through land reclamation II-312
— The Maldives II-204
    — high species richness II-204
— Marshall Islands II-777, II-778, II-779
— Mauritius II-258
— Mayotte Island II-246
— Mexican Pacific coast I-488, I-492
— need for easier availability of existing knowledge III-28
— negative effects of oil spills III-273
— New Caledonia II-726
— Nicaraguan Caribbean coast I-522
— Nicaraguan Pacific coast I-535
    — contamination and degradation of the ecosystem I-534
    — protected areas I-535
    — reduction in, Gulf of Fonseca I-538
— northeastern Australia II-614–15
    — reclamation and draining threats II-615
— and other habitats, Tanzania II-87–8
    — mangrove harvesting II-89–90
    — restoration of, Dar es Salaam II-89
    — use of mangrove timber **II-88**
— patterns of use II-24–6
    — benefits from mangroves **I-25**
    — misuse through international agencies III-26
    — pressures for change III-24–5
    — recreation and ecotourism III-26
    — shrimp aquaculture III-21–2, III-25
— Peru I-691–2
— the Philippines II-409
    — source of fishery and forest products II-409
— present extent and loss III-19–22
    — areal statistics III-20, **III-20**
    — causes of loss **III-21**

— eastern and western groups III-19–20
— loss to shrimp farming III-21–2
— problems of human pressure III-18–19
— Puerto Rico I-582
    — clearance of, effects I-584
— the Quirimbas II-102–3
— red, Gulf of Mexico I-472–3
— Red Sea II-40
    — hard-bottom (reef) and soft-bottom mangals II-41
— rehabilitation of key system, Colombia I-463
— remote sensing of III-285–6
    — detection of change in resources III-285
    — future challenges III-286
    — mangrove leaf area index (LAI) III-285–6
— role in sediment stabilisation II-414
— Sinai Peninsula II-41, *II-41*
— Somalia II-71, *II-71*, II-77
— South Western Pacific Islands II-710
— spawning and nursery grounds II-335
— Sri Lanka II-179
    — damaged by shrimp aquaculture II-179
— Sumatra, zonation depending on tidal regime II-312
— Turks and Caicos Islands I-589–90
    — harvesting of mangroves I-592
— Vanuatu, species diversity II-740
— Venezuela
    — conversion to shrimp ponds I-655
    — mainly in Orinoco Delta and Paria Gulf I-651
— Victoria Province, Australia II-665
— Vietnam II-564, II-565
    — effects of destruction of II-566
— west coast Taiwan II-503
— Western Australia II-695
— western Indonesia II-385–8, **II-388**
— western Sumatra II-387
    — very productive II-388
— Xiamen region II-518
Mangueira Lagoon, southern Brazil *I-732*, I-737
Mannar, Gulf of II-163–6
— climate and coastal hydrography II-163
— fish and fisheries II-164–5
— human population and environmental degradation II-165
    — islands affected by new harbour at Tuticorin II-165
— impact of human activities on the ecosystem II-165–6
    — effects of industries and the power station II-165
— main marine ecosystems II-163–4
— National Marine Park and a Marine Biosphere Reserve II-163
Manning Shelf Bioregion, Australia II-634
Manus Basin *II-426*
Manus Trench *II-426*
Maracaibo Lake *I-644*, I-646
— ecosystem under extreme pressure I-656
— population round I-654
Maria Island, Tasmania
— changes in oceanographic climate recorded II-651
— Marine and Estuarine Protected Area II-658–9
mariculture I-52, I-69, I-319, I-362, II-314, II-339
— American Samoa II-770
— culturing of mangrove oysters I-569
— Gulf of Thailand II-306
— Hawaiian Islands, research and development II-803
— Kuwait II-8–9
— northern Gulf of Mexico I-448
— potential for, eastern Russia II-481, II-482
— Taiwan II-505
— Tanzania II-91–2
— Xiamen region II-519, **II-520**
marinas, sources of contaminants I-72
maninculture, Guangdong, main methods II-558
marine antifoulants III-247–56
— biocide-free 'non-stick' coatings, for the future? III-254

— effectiveness of regulations: measuring and monitoring TBT in the environment III-253–4
— leaching during normal operations III-364
— new self-polishing copolmer paints III-248
— organotin-based paints banned on smaller boats III-251
— TBT
    — ban on, finding safe alternatives III-254–5
    — environmental impacts of III-250–1
    — persistence of in the environment III-249–50
— TBT-based, legislative control of III-251, III-253
    — regulations have reduced contamination III-253
— *see also* tributyltin (TBT)
marine biota
— Azores I-203
— Chilean I-707
    — Peruvian Province and Magellanic Province I-707
— Southern California Bight, effects of anthropogenic activities I-394–9
marine birds
— Patagonian coast, Argentina I-756, I-757
    — breeding **I-756**
marine circulation
— Arabian Gulf II-4
— development of two-gyre system
    — Arabian Gulf II-4
    — Gulf of Aden II-50
    — Somali Indian Sea Coast II-66
marine climate data, accessibility of III-216
marine and coastal communities, Côte d'Ivoire, affected by seasonal factors I-809–10
marine coastal and estuarine ecosystems, trophic status categories III-258
marine and coastal habitats
— Baltic Sea I-124–8
    — biodiversity of I-124–5
    — hard-bottom communities I-125–6
    — pelagic communities I-126, I-128
    — soft-bottom communities I-126
— Chesapeake Bay I-341–3
— continental seas, western Indonesia II-385–90
— English channel coast I-68–9
    — loss of coastal habitats I-74
    — protective and remediation measures I-77–9
— French channel coast I-69
— Irish Sea
    — intertidal habitats I-87–9
    — sub-tidal habitats I-89
— Lesser Antilles, Trinidad and Tobago I-631–2
— Malacca Strait II-335–7
— offshore, English Channel I-69
— and offshore systems, Black Sea I-290–4
— Sargasso Sea and Bermuda Islands I-227–8
    — coral reef zones I-227–8
    — two inshore nutrient zones I-228
— shallow
    — Adriatic Sea I-272–3
    — Tyrrhenian Sea I-273–4
— shallow, Dutch Antilles I-598–601
    — coastal wilderness I-601
    — coral reefs/reefal algal beds I-599–600
    — mangroves I-600–1
    — salinãs I-601
    — seagrass beds I-601
— shallow, southern Spain I-170–2
    — Eastern sector I-172
    — Western sector I-170–2
    — Western sector described I-170–2
— shallow water
    — Australia II-582
    — the Azores I-205–9
    — the Bahamas I-419–22

— Bay of Bengal II-149–51
— Belize I-504–6
— Chagos Archipelago II-226–9
— Colombian Pacific Coast I-679–80
— the Comoros II-245–6
— Coral, Solomon and Bismark Seas region II-429–31
— Côte d'Ivoire I-810–13
— east coast, Peninsular Malaysia II-348–50
— eastern Australian region II-632–5
— El Salvador I-548–51
— Fiji Islands II-754–8
— French Polynesia II-817–19
— Great Australian Bight II-678–81
— Great Barrier Reef region II-614–19
— Guinea I-799–800
— Gulf of Aden II-51–4
— Gulf of Alaska I-375
— Gulf of Guinea I-779–81
— Gulf of Maine and Georges Bank I-310–12
— Hawaiian Islands II-795–800
— Hong Kong II-537–8
— Jamaica I-561–5
— Marshall Islands II-777–9
— Mascarene Region II-256–9
— Mozambique II-102–5
— New Caledonia II-725–6
— New York Bight I-325–7
— Nicaraguan Caribbean coast I-520–2
— Nicaraguan Pacific coast I-534–5
— North Sea I-47–50
— northern Gulf of Mexico I-440–2
— Oman 21–4
— the Philippines II-408–10
— Red Sea II-40–2
— Sea of Okhotsk II-467–8
— the Seychelles II-236–7
— Somali Indian Ocean coast II-72–7, II-78
— South Western Pacific Islands II-710–12
— southeast South American shelf marine ecosystem I-754–7
— southern Brazil I-735–7
— Southern Gulf of Mexico I-471–4
— southwestern Africa I-828–30
— Sri Lanka II-178–80
— Taiwan Strait II-502–3
— Tanzania II-85–8
— Tasmanian region II-651–3
— The Maldives II-202–5
— Torres Strait and Gulf of Papua II-597–8
— Turks and Caicos Islands I-589–91
— Vanuatu II-740–2
— Venezuela I-648–53
— Victoria Province, Australia II-664–6
— Western Australia II-695–6
— Xiamen region II-517–18
— Spanish north coast I-139–40
— intertidal rocky communities I-140
— soft-bottom communities I-139–40
— subtidal rocky communities I-140
— wetlands and marshes I-139
— Vietnam II-564–5
— littoral habitat II-564–5
marine and coastal protected areas, Madagascar II-127
marine communities
— changes in structure anddiversity III-127
— effects of eutrophication on III-259–60
— Gulf of Alaska, trophic shift in I-376–8
— ocean/climate variability I-375
— major increase in cod and ground fish I-375–6
— population changes of shrimp and forage fish I-375
marine conservation, Vanuatu, traditional and modern practices
  II-745–6, **II-746, II-747**

marine conservation areas, Borneo II-376–8
marine conservation and resource management, scope of III-376
marine ecosystem health
— concept III-213
— marine epidemiological model III-213
— tracking of HABs III-213
marine ecosystem health as an expression of morbidity, mortality
  and disease events III-211–34
— basis for assessing the health of large marine ecosystems III-230–1
— categories of disturbance III-217–28
  — anoxic/hypoxic disturbances III-219–21
  — biotoxins and exposure disturbances III-218–19, *III-220*, **III-220**
  — disease disturbances III-225–6
  — keystone-endangered and chronic cyclicaldisturbances
    III-227–8
  — mass lethal mortality disturbances III-223–4
  — new, novel occurrences and invasive disturbances III-226–7
  — physically forced (climate/oceanographic) disturbances
    III-224–5
  — trophic-magnification disturbances III-221–2
— data assimilation methods III-215–17
— disturbance type derivation III-217
— Gulf of Mexico Aquatic Mortality Network III-228
— HEED approach III-214
— network for developing standards and achieving consensus
  III-229–30
— survey methods III-214–15
marine ecosystems
— Canary Islands I-193–5
  — *Cymodocea–Caulerpa* communities I-195
  — deepwater species rise at night I-194
  — mesolittoral area I-194
  — pelagic system I-193–4
  — rocky and sandy seabeds I-195
  — sublittoral area I-194
  — supralittoral area I-194
— and coastal habitats, shallow water, Colombian Caribbean Coast
  I-666–9
— Gulf of Mannar II-163–4
— long-term concerns about direct and indirect effects of fishing
  practices III-352
— rocky strata I-194–5
— sustainability of human activities on III-359–73
  — exploitation of ecological resources III-366–9
  — global commons III-360
  — global trends III-369–71
  — introduction of hazardous substances and radioisotopes to the
    marine environment III-360–6
  — towards sustainability III-371
— western Sumatra III-386–7
  — disturbance and stress II-387
marine environment
— annual input of petroleum hydrocarbons **III-268**
— global legal instruments (at year 2000) III-331–48
  — Antarctic regime III-342
  — from present to future III-342–7
  — Law of the Sea III-332, III-336
  — marine organisms III-340–1
  — marine pollution III-336–40
  — protection from military activities III-342
  — "soft law" instruments III-339, III-344
  — taking stock III-332, **III-333–5**
— introduction of elevated nutrient loads III-366
— introduction of hazardous substances and radioisotopes to
  III-360–6
  — contminants from shipping III-364–5
  — dumping of wastes at sea III-365–6
  — land-based sources III-361–3
  — operational discharges from the offshore oil and gas sector
    III-363–4
— protection of from military activities III-342

— bilateral agreements III-342
— regional protection programmes III-363
marine environment strategies, Australia II-590
Marine Environmental Act, Faroes I-38
marine environmental agreements, global
— instruments and their institutional arrangement III-344–5
— move away from binding commitments leads to increased vulnerability III-345
marine environmental protection, legislative power of national governments I-57–9
Marine Environmental Protection Committee (MEPT: IMO), ban on TBT proposed III-365
marine environmental restoration, Hawaiian Islands II-809
marine environmental science, multinational training programmes in III-323–30
— CC:TRAIN III-327–8
— International Ocean Institute Training Programme III-328–9
— synthesis III-329–30
— TRAIN-COAST-SEA III-325–7
— UN training programmes, TRAIN-X strategy III-324–5
marine epidemiological information system, information flows III-229
Marine and Estuarine Protected Areas (MPAs), Tasmania II-658
marine fishery reserves I-526
marine habitats, smothered I-383
marine habitats, Brazilian tropical coast I-722–5
— bays I-725
— continental shelf I-722–3
— coral reefs 723–4
marine habitats, shallow
— the Bahamas I-421–2
— coral reefs I-421
— mangroves I-421–2
— seagrass I-421
— Israeli shelf, Red Sea invaders I-256–8
— Tagus and Sado coastal waters I-157–8
— marine biological resources I-157–8
— plankton I-157
— Tagus salt marshes I-157
marine harvesting, attempts to limit impacts of III-378
marine mammals
— abnormalities and perturbations due to toxic chemicals II-456
— accumulate toxins in blubber I-91
— Aegean Sea I-239
— Africa, southwestern I-834
— Alaska, food-limited I-377
— Australian Bight II-680
— Baltic Proper I-109
— Bazaruto Islands II-104–5
— Chesapeake Bay I-343
— and coastal mammals, El Salvador I-549
— contamination by and bioaccumulation of persistent organic organochlorines II-455–6
— Great Barrier Reef II-618–19
— hunted in Greenland I-8–9
— importance of polynyas to I-7
— Jamaica I-561–2
— little affected by radionuclides III-96
— loss of, Java Sea II-364
— Maldives II-205
— Marshall Islands II-779, II-780
— Patagonian coast, Argentina I-756–7
— breeding **I-756**
— the Philippines II-410
— Portuguese coastal waters I-159
— and reptiles, Hawaiian Islands II-800, **II-800**
— small, west coast, Sarawak II-368
— Southern California, effects of anthropogenic activities I-399
— Sri Lanka II-180
— susceptible to chemicals accumulation III-361
— threatened in Hong Kong waters II-542

marine monitoring, Southern California Bight, unable to assist environmental management I-400
Marine National Parks
— Andabar and Nicobar Islands **II-196**
— Madagascar II-127
— Mozambique
— Bazaruto, dugongs and turtles II-110
— Inhaca and Portuguese Islands II-110
— the Quirimbas may be next II-110
— the Seychelles II-239–40, **II-240**
— Curieuse and Sainte Anne II-240, II-241
marine nature reserves
— French Polynesia II-824
— Skomer Island *I-84*, I-96
— Strangford Lough *I-84*, I-96
marine organisms
— common, eastern coast of Taiwan II-508
— harmful effects of increased exposure to UV-B III-370
— legal instrument **III-334–5**
— migration circuit concept III-378
— protection of III-340–1
— TBT, accumulation of III-250
— TBT, toxic to III-250
— Yuzhnoprimorsky region
Marine Park of the North Sporades *I-234*, I-250
Marine Parks
— east coast, Peninsular Malaysia II-356, II-357
— Italy
— Adriatic coast I-280
— Tyrrhenian coast I-280
— Jamaica I-571
— Tanzania II-86, II-94
— further park proposed for Mnazi Bay, Mtwara II-94, II-95
— western Indonesia, zoning schemes II-401
marine plants, Aegean Sea I-239
marine pollution III-336–40
— atmospheric pollution III-340
— Australia II-588–9
— Bay of Bengal II-280
— by persistent organic organochlorines II-455–6
— dismantling of ships III-339
— dumping III-338–9
— from present to future III-3427
— learning-based approach to global codification III-346–7
— from wastes and sewage, Vanuatu II-747–8
— Hong Kong, land-based origin II-542
— Israeli legislation against I-261–2
— Java Sea II-398
— land-based III-339–40
— legal instruments **III-333–4**
— Malacca Strait II-339
— Malacca Straits, sea-based sources II-319–20
— oil and chemical spills II-319–20
— oily discharges II-319
— TBT II-319
— Mozambique II-109
— North Sea, changes in approaching control of I-59
— regulation of land-based sources, Sumatra II-322
— risk of, Lesser Antilles I-638
— sea-based, Indonesian legislation II-322
— seabed activities: peaceful exploration and exploitation III-336, III-338
— shipping III-336
— South China Sea II-354
— Sri Lanka, land-based sources II-184–5
— western Indonesia II-396–8
— Xiamen region
— management of II-532
— monitoring of II-531
marine pollution parameters
— West Guangdong coast II-554–5

— conventional water quality parameters II-555
— pollution sources and waste products II-554–5
— trace toxic organic contaminants in the Pearl River II-555–6
marine populations, dispersal patterns III-380–1
— early life dispersal distance III-381
— larval stage duration III-381
— potential dispersal ranges III-381
marine protected areas
— the Azores I-214, **I-215**
— — proposed **I-216**
— the Comoros II-251
— Djibouti II-55
— east coast of Sumatra **II-322**
— insufficient, Strait of Malacca II-342
— Irish Sea, slow process I-96
— a limited success rate, Tanzania II-93–4
— Sweden I-132
Marine Protected Areas (MPAs)
— American Samoa II-771, **II-771**
— Anguilla and British Virgin Islands I-625
— Australia II-590, II-686
— — Great Australian Bight Marine Park II-686–8
— Belize **I-513**, I-514
— Borneo, management objectives for II-377
— classification of III-377
— eastern Australian region II-643
— Fiji Islands, proposed **II-761**
— Hawaiian Islands **II-803**, II-808–9
— — need to extend coverage II-810
— healthy communities of endangered species at risk from severe storms III-224
— Lesser Antilles I-639–40, **I-639**
— Mexico I-497
— New Caledonia II-734, **II-734**
— Papua New Guinea II-441
— the Philippines II-419
— restrict fishing effort III-161
— South West Pacific Islands II-719–20
— Terminos Lagoon, Southern Gulf of Mexico I-472
— Turks and Caicos Islands I-592–3
— Vanuatu II-746–7
— Victoria Province, Australia II-670
— western Indonesia II-401, **II-402**, II-403
— — many only "paper parks" II-401
marine reptiles, Aegean Sea I-240
marine reserves
— assessing the effectiveness of III-384–9
— — regional summaries of reserve implementation III-386
— — within-reserve effects III-384–5
— demarcate boundaries
— — simply and reliably III-383–4
— — useful III-390
— design of III-380–4
— — dispersal III-380–1
— — edge effects III-383–4
— — minimum viable biomass III-382–3
— — number of III-384
— — optimum size III-383
— — reserve size and number III-381–2
— effectiveness depends on public acceptance, understanding and compliance III-390
— failures in resource management, reasons for III-378
— fine filter and coarse filter III-380
— framework for design of 'no-take' reserves and networks evolving III-376
— global distribution of *III-376*
— guidelines for the development of III-380
— Madagascar II-127
— Mayotte, Longogori Reserve (S passage) II-250
— need a robust approach to management III-377
— no-take zones III-376, III-377, III-378, III-386
— oceanographic setting crucial for egg and larval stages III-380–1
— placement in current/counter current and gyre systems III-381, *III-381*
— potential benefits of III-384
— potential to meet diverse management objectives III-376
— predicting the effects of III-386–9
— — dynamic pool models III-387, III-389
— — logistic models III-387
— — spatial harvesting models (yield per recruit) III-387, III-388
— and resource management III-375–92
— — closed areas and fisheries management III-378
— — dealing with uncertainty III-389
— — ecosystem effects of fishing III-378–80
— — ecosystem management and the reserve concept III-377–8
— social and economic considerations III-389–91
— — enforcement and compliance III-390
— — user participation in management process III-390–1
— — zoning III-389–90
— South Africa, Maputaland Marine Reserve II-140
marine resources
— Australia II-584
— exploitation of, Coral, Solomon and Bismark Seas region II-434
— Greenland
— — fishery I-8
— — hunting I-8–9
— Guangdong II-556
— utilised in the Bahamas **I-424**
marine scientists I-2
marine species
— and habitats, protection of, duplication or complementary **I-58**
— timing of reproduction and reproductive success II-51
— Xiamen coastal waters **II-517**
marine transgression, Kenya and Somalia II-67
Marine Turtle Specialist Group (IUCN) III-64–5
marine turtles, Fiji Islands II-757
marine zonation scheme, Xiamen region II-529–30, *II-530*, **II-530**
marine zoning III-389–90
— *see also* Marine Protected Areas (MPAs); marine reserves
Marmara Sea *I-234*, I-243
MARPOL II-324
— Gulf of Oman designated a 'Special Area' II-28, II-29
— Malacca Straits, possible designation as a 'Special Area' II-325
— special status for Wider Caribbean area I-640
MARPOL Convention I-38, I-55, I-147, I-248, I-262, I-451, II-658, III-343
— importance of III-337
— operational discharges regulated under III-364
marsh and estuarine systems, central Gulf of Mexico I-438
marsh vegetation, persistent oil effects III-276
Marshall Islands II-773–89
— changes in the country's sociopolitical status II-775
— coastal erosion and landfill II-782–3
— cultural and historical resources II-779
— degrading of Ebeye's natural resources II-783
— effects from urban and industrial activities II-783–4
— ENSOs II-776–7
— geographical gradients in physical features **II-776**
— key environmental and marine regulations **II-786**
— major shallow water marine and coastal habitats II-777–9
— need for greater level of protection II-785
— offshore systems II-779–80
— other environmental concerns II-784–5
— — community and individual conflicts linked to land tenure and governance II-785
— — conflicts from different user interests II-785
— — greater incorporation of environmental concerns in development planning II-785
— — limited understanding on cross-sectoral issues II-784
— overall assessment of environmental governance II-788
— populations affecting the area II-780
— protective measures II-786–8

— institutions II-786
— international legislation and regional programmes II-787
— land tenure and customary marine practices II-786
— national legislations II-786
— proposed environmental policies and strategies II-788
— responses to environmental issues: assessment and status II-787–8
— US Army, Kwajalein Atoll (USAKA) procedures and standards II-786–7
— rural factors II-781–2
— seasonality, currents, natural environmental variables II-775–6
— species, habitats and sites of conservational interest II-779
marshes
— coastal, Louisiana, managed by man I-448
— continental, Guadalquivir River I-170
— eastern Russia II-482
— *see also* coastal wetlands; saltmarsh; tidal marshes; wetlands
Martinique, effects of hurricanes I-631
Mascarene Anticyclone II-116
Mascarene Basin II-255
Mascarene Plateau II-255
— an obstacle to deep water flow II-255
— close to a tidal amphidrome II-256
— internal wave generation II-256
Mascarene Region II-116, II-253–68
— coastal erosion and landfill II-262
— effects from urban and industrial activities II-262–5
    — artisanal and non-industrial uses of the coast II-262–3
    — cities and sewage discharges II-264
    — freshwater II-264–5
    — light industry II-263
    — sand mining and lime production II-263
    — shipping, offshore accidents and impacts II-265
    — tourism II-264
— major shallow water marine and coastal habitats II-256–9
— offshore systems II-259–60
— populations affecting the area II-260
— protective measures II-265–6
— rural factors II-260–2
— seasonality, currents, natural environmental variables II-255–6
Mascarene Ridge *II-254*
Masirah Island *II-18*
— some coral cover II-22
mass lethal mortality disturbances III-223–4
— many reports coincide with climate extremes III-223–4
mass mortalities, cause ecosystem collapse and reorganization round a new stable state III-223
"maszoperie" I-110
Matang Mangrove Forest Reserve, Malaysia II-342
Mauritius *II-254*, II-255
— agriculture II-260–1
    — deforestation causes top soil loss II-260–1
— artisanal fishing II-261
— cities and sewage discharges II-264
— coral reefs II-258
    — soft corals II-258
    — spur and groove zone II-258
— eutrophication in the lagoons II-261
— Fishing Reserves II-265, **II-265**
— fishponds (barachois) II-262
— habitat diversity II-257–9
— mangroves II-258
— Nature Reserves II-265, **II-265**
— population II-260
— Round Island, free from introduces mammals and plans II-259
— sand mining and lime production, effects of II-263
— Southeast Trades drive most winds II-256
— stone-crushing plants, environmental effects of II-263
— sugar mills and textile plants II-263
— Terre Rouge Bird Sanctuary II-258
— tourism II-264

maximal/maximum sustainable catches/yields (MSY) III-75, III-81, III-118
Mayotte (Maore/Mahore) *II-244*
— coconut, bananas and ylang-ylang grown for export II-246
    — Iris Bank II-246
— forest more luxuriant now II-247
— fringing reef II-246
— introduction of sewage treatment II-251
— lagoon a series of hydrologic basins II-248–9
— land-based pollutants and the lagoon II-251
— Marine Reserve of Longogori II-250
— reef threatened by silting II-248
— remained with France II-247
— seagrass beds II-248
— sedimentation in the lagoon affect fishing II-249
Mediterranean
— development of marine aquaculture III-169
— offshore wind energy development slower III-311
— receives water from the Black Sea I-22
Mediterranean Action Plan (MAP) I-249
Mediterranean Climate, Great Australian Bight II-675
Mediterranean fauna
— poor in animal species I-240
— zooplankton I-240–1
Mediterranean Sea III-205
Mediterranean Water (MW) I-139, I-153, I-169, I-205
MEDPOL National Monitoring Programme (Greece) I-249
meiobenthos II-277
Mellish Plateau *II-426*
Menai Strait *I-84*, I-96
mercury contamination
— Colombian Pacific coast I-683
— Haifa Bay I-260, I-263
— Kalimantan rivers II-396
— southern Brazil, east coast I-740
mercury (Hg) I-5, I-95, I-116, I-793, II-390
— accumulation in the Everglades I-444
— in Azorean seabirds, fish and cephalods I-212
— Baja California I-495, **I-496**
— in cinnabar mine tailings, the Philippines II-416
— evidence of increasing atmospheric concentrations I-14
— in Faroese pilot whale meat I-31, I-36, I-39
— in Greenland seabirds and mammals I-9, *I-10*
— low levels in Arabian Gulf II-12
— a multi-hop contaminant I-13
— in seabirds of German North Sea coast I-49
— as seed dressing I-635
— in surficial sediments, Gulf of Thailand II-305, **II-305**
Merka Formation II-68
Merka Red Dune Complex II-68–9, *II-70*
— potential for a national glass industry II-79
MESA New York Bight Atlas Monograph Series I-323, I-325
Meso America *see* Central America
Messina Strait I-273, I-281–2
— animal communities I-281
    — Atlantic affinity species I-281
    — endemic species I-282
    — relict species I-281
— intense tidal currents I-281
Mestersvig, Greenland *I-6*
— pollution from lead—zinc mine I-12
metal biomagnification, absent in southern Baltic food chain I-116
metal pollution
— Baltic Sea I-114–16
    — atmospheric and riverine fluxes I-115–16
    — biota I-116
— Russian Far East II-476
    — of surface waters II-477
— *see also* heavy metal pollution
metal sequestration, Tagus salt marshes I-158

metals
— English Channel
— generally low I-74–5
— higher near estuaries and inshore I-75
— entering North Sea from mining and industry I-54
— in land-based pollution III-361
methane
— emissions from the Bodden III-262
— increased in the atmosphere III-188
methane gas hydrates, Blake Plateau I-367
methylmercury I-114, III-95
— in Faroese pilot whale meat I-36
metropolitan areas, located near estuaries, problems associated with climate change and sea level rise III-194
Metula oil spill, Chile III-276
— asphalt formation III-270
Mexican Basin, and Sigsbee Deep I-442
Mexican Pacific coast
— case studies I-492–7
— Altata-Ensenda del Pabellón I-493
— Huizache y Caimanero lagoon system I-493–4
— Navachiste—San Ignacio—Macapule bays I-492–3
— Teacapán—Agua Brava lagoon system I-494–5, I-497
Mexican Pacific coastline I-485.**I-486**
Mexico
— coastal zone of Campeche, analysis of environment and its problems I-464
— coasts have ecological and socioeconomic problems I-464
— and Gulf of Mexico, coral bleaching and mortality events III-56
— Pacific coast, coral bleaching and mortality events III-53–4
Mexico, Gulf of
— growth of hypoxic zone III-219
— problem of incidental mortality of juvenile red snappers III-143–4
Mexico, northern Gulf I-435–56
— coastal erosion and landfill I-445–8
— defined I-437–8
— eastern sector I-437
— effects from urban and industrial activities I-448–51
— interactions
— with North Atlantic Ocean I-437
— with waters and biota of the Caribbean I-437
— a large marine ecosystem I-437
— major shallow water marine and coastal habitats I-440–2
— offshore systems I-442–3
— populations affecting the area I-443
— rural factors I-443–5
— seasonality, currents, natural environmental variables I-438–40
— western and central sectors I-437
Mexico, Pacific coast I-483–9
— Californian Coastal Province I-486–7
— effects of urban development and industrial activities I-492–7
— case studies I-492–7
— North Pacific Ocean Province I-487–8
— Pacific Center Coastal Province I-488–9
— population I-489
— protective measures I-497–8
— included in international agreements I-497–8
— rural factors and fishing I-489–91
— Sea of Cortes Oceanic Province I-488
— Tropical South Pacific Ocean Province I-489
Mexico, Southern Gulf I-467–82
— coastal erosion and landfill I-476–7
— effects of human activities on natural processes I-478–9
— environmental framework I-469–71
— major shallow water marine and coastal habitats I-471–4
— offshore systems I-474–5
— populations affecting the area I-475–6
— protective measures I-479–80
— international agreements signed by Mexico **I-480**
— National Development Plan (1995–2000) I-479–80
— programs of sustainable regional development (Proders) I-480
— rural factors I-476
— urban and industrial activities I-477–8
micro-organics, sediment contamination by slight in the English Channel I-75
microalgae
— benthic I-357–8
— primary production I-357–8
— toxic I-356
microalgal blooms, increased due to due to seagrass die-off III-15
microbes
— oil degrading, in cyanobacterial mats II-12
— oxydation of organic material III-261
microbial diseases, and prawn mortality II-154
microbiological contamination
— English Channel
— impact on bathing water quality I-76
— impact on shellfish I-76
Micronesia, Marshall Islands a part of II-775
Micronesia, Federated States of, coral bleaching and mortality events III-52
microphytobenthos, North Sea I-48
Middle American Trench I-533
Middle Atlantic Bight I-323
— nontidal surface circulation *I-324*
Midway Atoll National Wildlife Refuge II-798
migration circuit concept, marine organisms III-378
migrations, seasonal, Okhotsk Sea II-466
military activities, protection from **III-335**
military usage
— French Polynesia II-822
— Hawaiian Islands II-804
— Marshall Islands II-783
Mindanao Current II-407
mine tailings
— copper mining, polluting Chilean coast I-713
— construction of large lagoon for tailings I-713
— effect of on the biodiversity of the seabed, Island Copper Mine, Canada III-235–46
— after mine closure sustainable ecological succession soon established III-245
— biodiversity in terms of higher taxa III-241–2
— crab fishery and yields III-244
— data set III-237–8
— habitat change since mine closure III-241
— large infaunal species III-242–3
— species evennness III-240–1
— species richness III-238–9
— tailings deposition levels affecting fauna III-239–40
— time- and cost-effectiveness of the benthos surveys III-243–4
— placement of for minimal and reversible environmental losses III-236
— risk of groundwater contamination I-710
mine tailings deposition, what is the tolerable rate? III-239–40
mineral extraction, areas of interest on Tyrrhenian sea floor I-279
mineral resources, Somalia II-79
mineral springs, west coast, Sea of Japan II-483
minerals
— coastal sands, Sri Lanka II-181
— Colombian Pacific coast I-683
— deep-sea deposits, Mascarene Basin II-260
— Godavari basin II-170
— Malacca Straits II-316
— Palk Bay–Madras coast II-168
— South Africa II-142
Minimata disease II-396, II-398
mining
— Alaska I-381
— Australia II-585
— Chile I-710
— lessons for the future, a case study I-713
— coastal, the Philippines II-415–16

— problems from cinnabar mines, Palawan II-415–16
— wastes and tailings serious threats to the marine environment II-415
— diamonds, Namibia I-835
— Great Barrier Reef hinterland II-622
— Greenland, and heavy metal pollution I-11–12
— Gulf of Papua II-607
— fate of sediments from II-603
— New Caledonia, and its effects II-731–2, *II-732*
— Nicaragua I-525–6, I-541
— Papua New Guinea II-439
— problem of mine discharges II-439
— tailings discharged direct to the sea II-439
— Peru I-695
— runoff from increasing sedimentation and turbidity I-444–5
— small-scale, the Philippines II-414–15
— discharge of untreated mine tailings II-414–15
— southern Spain
— Aznalcóllar disaster I-178
— problems of drainage from I-178
— Sumatra, surface and submarine II-395
— Tasmania, impact on coastal waters II-654
mining pollution I-710
minke whales III-78
— catch limits III-83
— counting of III-79–80
— number estimates III-82
— probably more than one biologically distinct population
Miskito Coast Marine Reserve, Nicaragua I-526
— corals I-520, *I-521*
— recommendations for **I-527**
Mississippi River *I-436*, I-443
— deltaic marshes I-440
— results of 1993 extreme flooding III-219
— silt and clay from I-437
Mnemiopsis, effects of introduction to the Black Sea I-290, I-304
Mogadishu Basin *II-67*
Moheli (Moili/Mwali), good soils II-246
molecular markers, used to identify contaminant sources I-392
mollusc aquaculture I-653
— Patagonia I-764
mollusc diversity, Sunda Shelf II-389
mollusc farming III-167, *III-168*
molluscan fauna, larger Norwegian fjords I-20–1
molluscs I-34, I-172, II-586
— carrying bacteria, Sri Lanka II-184
— culturing of II-278
— economically important, Colombian Pacific Coast I-684
— fishing for I-73
— lagoons, Gulf of Mexico I-473
— Nicaraguan Pacific coast I-534
— Spanish north coast I-139
— Tagus and Sado estuaries I-158
— Vietnam II-564
Monin—Obukhov theory 202
monitoring and enforcement, becoming easier on the high seas III-390
monitoring programmes, lacking, Yellow Sea II-496
monk seals
— Hawaiian Islands II-799, II-800
— Mediterranean Sea, decline in III-129
monsoons II-116, II-537
— affecting southeast India II-163, II-166
— affecting Tanzania II-85
— Bay of Bengal II-148
— causes high concentrations of heavy metals in Bay region II-155
— currents and gyres II-273–4, *II-274*
— Borneo II-365
— Cambodia II-571
— Coral, Solomon and Bismark Seas Region II-428
— east coast, Peninsular Malaysia II-347

— intermonsoon changeover period II-347
— winds, waves and currents II-347–8
— Gulf of Thailand, influence of II-300, II-301
— Indian system influences climate of northwestern Arabian Ocean II-19–20
— effects of Southwest Monsoon II-19
— influences in Malacca Straits II-311
— influencing northern Maldivian islands strongly II-201
— and the Intertropical Convergence Zone II-49
— and the Malacca Strait II-333
— the Philippines II-407
— South China Sea coast II-551
— and the Sri Lankan climate II-177
— Vietnam II-564
— winds affect northern and northwestern Madagascar II-117
— winter and summer, Arabian Gulf II-4
Mont Saint Michel Bay
— important bird location I-71
— marshes I-69
Montauk Point *I-322*, I-323
Montego Bay, Jamaica *I-560*
— reef restoration efforts I-546
Monterrey sardine I-488
Montserrat *I-616*, I-617, I-618, I-623
— climate I-617
— coral communities
— before volcanic activity I-618
— effects of volcanic activity I-618–19
— endemic birds I-620
— fishing I-621–2
— regulations not enforced I-624
— hurricane damage I-622
— loss of population due to volcanic activity I-620–1
— mangroves limited I-619
— seagrass beds I-618
— Sustainable Development Plan I-621
— volcanically active *I-616*, I-617
mooring system, British Virgin Islands I-624
moorings, the Bahamas I-431
morbilli virus *see* phocine distemper viruses
Morecambe Bay *I-84*
— shore communities I-88
Moreton Bay, Queensland
— environmental problems II-642
— high biodiversity *II-630*, II-634
— residential marinas II-639
Morondava, Madagascar, problems of land loss II-124–5
morphological abnormalities, indicator of system health III-225
Morrocoy National Park, Venezuela *I-644*
— much damage due to sedimentation I-655
— pollution estimates I-657–8
— subject to man-made disturbances I-655
Morrosquillo, Golfo de *I-664*, I-665, I-666
mother-of-pearl shell II-44
motor vehicles, emissions from I-345
Mozambique II-99–112
— coastal divisions II-101
— coastal erosion and landfill II-108
— effects from urban and industrial activities II-108–9
— high tidal range II-II-101
— major shallow water marine and coastal habitats II-102–5
— Bazaruto Islands II-104–5
— Inhaca and Portuguese Islands Reserves II-105
— Quirimba Archipelago II-102–4, II-245
— offshore systems II-105, II-107
— participation in International Conventions **II-105**
— population II-107
— protected by Madagascar II-101
— protective measures II-109–11
— Framework Environmental Law (1997) II-110
— rural factors II-107–8

— seasonality, currents, natural environmental variables II-101–2
— tropical humid to subhumid climate II-101
Mozambique Channel II-101, *II-114*, II-251
— carried high volume of crude oil traffic II-109
Mozambique Current II-101, *II-101*, *II-114*, II-245
Mozambique Gyre *II-101*, II-105
mucilaginous aggregates, effects of I-278
mud flats *see* soft shores
mud reefs I-583
muddy bottoms, Gulf of Papua II-597
mudflats
— and accreting mangroves, Perak and Selangor II-336
— extensive, Malacca Straits II-312
— Hong Kong II-539
Multiple Marine Ecological Disturbances (MMEDs) III-212
— episodic events and co-occurring anomalies III-214
— HEED approach III-214
— HEED database and GIS III-229–30
— indicators of decline in ecosystem health III-230
— observational reports, additional information for III-215–16
— pooling of co-occurring biological disturbance data, resulting evaluations III-216–17
— use of marine ecosystem health framework III-230–1, *III-230*
Multiple Marine Ecological Disturbances (MMEDs) program
— Health Ecological and Economic Dimensions (HEED) of III-215
— HEED system III-215, III-229
— scale in aggregation of anomaly indicators III-212
multiple-use management model, Great Barrier Reef Marine Park, criticism of II-616
Multivariate ENSO Index (MEI) I-703, *I-704*
munitions, dumping of in Beaufort's Dyke I-96
Murat Bioregion, Australian Bight II-678, II-679
Musandam *II-18*
Muscat *II-18*
— Qurm National Nature Reserve
    — at risk from urban development II-30
    — mangroves at risk II-28
mussel beds, protect sequestered oil from weathering III-271
mussel cultivation
— Albufeira lagoon I-156
— Galicia I-144
Mussel Watch Project I-452
— and contamination by organotins II-451, **II-453**
mussels I-34–5, I-51, I-290, III-167
— adapted to hypersaline conditions I-442
— Albufeira lagoon, purification needed I-158
— Island Copper Mine dock, bioaccumulation in III-244
— marine, Scope-for-Growth I-56
— metal contamination, Tagus estuary I-158
— response to oil spills III-273
— Venezuela I-649
Muthupet Lagoon, Palk Bay II-167
Mytilus edulis platensis I-754
— circalittoral banks, Buenos Aires Province I-755

N/P ratios
— Aegean Sea I-240
— Baltic Sea I-130
— trophic zone, Baltic Proper I-106
Namib Desert I-823
Namib Naukluft National Park I-835
Namibia
— claimed her EEZ at independence I-836
— diamonds I-835
— favourable factors for a sustainable fishery I-837
— the fishing industry I-830
— hake catches I-834
— Kudu gas field I-835
— Namib Naukluft National Park I-835
— other development and commercial activities I-837
— ports, Walvis Bay and Lüderitz I-835
— "sulphur eruptions" I-828
— sustainable harvesting of seals I-834
Nancowrie Biosphere Reserve, Andabar and Nicobar Islands II-196
Nares Strait I-5, *I-6*, I-13
narwhals III-76, III-90
Nassau Grouper fishery I-425
National Coastal Erosion Study, east coast, Peninsular Malaysia II-356
National Estuary Program (USA) I-452
National Marine Fisheries Service (NMFS; USA), publishes bycatch statistics III-159
National Marine, Protection, Research and Sanctuaries Act (USA) I-452
National Parks
— Colombian Pacific I-684
— Fiji Islands II-761–2
— Gulf of Cadiz I-182, **I-182**
— Lac Badana National Park, Somalia II-80
— Masoala National Park, Madagascar II-127
— Southern Gulf of Mexico, "Arrecife Alacranes" I-471
Natura 2000 network I-52, I-77
natural capital depletion tax III-401
natural events, hazards and uncertainty, Marshall Islands II-776
natural hazards, Xiamen region II-516–17
natural resource depletion I-634
natural resource extraction, Alaska I-381
natural resource regions, Russian East Coast II-480–1
— Severoprimorsky region II-480
— Yuzhnoprimorsky region II-480–1
Nature Reserves, Mascarene region II-265, **II-265**
Nauru Basin *II-426*
Navachiste—San Ignacio—Macapule bays, Mexican Pacific coast I-492–3
— dams I-492
— mangroves I-492
navigation channels, problems created by I-447–8
Nazareth Bank *II-254*, II-255
Nazca Plate I-701
nearshore habitats, Hong Kong II-538
needs, concept of, in Third World I-459–60
Negril Morass, Jamaica I-564
— hummocky swamp I-564
nehrungen *see* sand barriers, southern Baltic coast
nematodes I-48
Netherlands, offshore wind energy production in III-308
neurotoxic shellfish poisoning (NSP) III-221
Neuse River Estuary *I-352*, I-366, III-205
— picoplanktonic cyanophytes I-355
— productivity pulses and algal blooms I-354
— zooplankton abundance and planktonic trophic transfer I-359
New Amsterdam anticyclone *see* Mascarene Anticyclone
New Britain Trench II-427
New Caledonia *II-706*, II-707, *II-708*, II-723–36, II-729
— coastal erosion and landfill II-731
— coastal modifications II-717
— coral reefs II-711
— effects from urban and industrial activities II-731–4
— the lagoon II-715
— major shallow water marine and coastal habitats II-725–6
    — offshore systems II-726–7
— Marine Protected Areas (MPAs) II-719
— nickel mining II-729
    — case study II-713
— populations affecting the area II-713–14, II-727–9
    — activities affecting the sea II-728–9
— protective measures II-734–5
    — Marine Protected Areas (MPAs) II-734, **II-734**
    — reef-monitoring project II-734
    — species-specific regulations II-734
— rural factors II-729–31

— seasonality, currents, natural environmental variables II-725
— subsistence reef fisheries II-715
— ZoNéCo Programme II-727
New Guinea Basin *II-426*
New Guinea Coastal Undercurrent II-428
New Ireland Basin *II-426*
new, novel occurrences and invasive disturbances III-226–7, III-230, III-259
New Providence Island *I-416*
— coastal erosion I-428–9
— perturbations over the last fifty years I-430
New River Estuary *I-352*, I-366
New South Wales
— central, major habitats II-635
— fisheries management system, gives security within adaptive management III-400
— share system conceptual framework relevant to other fisheries III-400–1
New York Bight I-321–33
— Bight restoration plans I-330–1
— dump sites in I-323
— major physical, hydrographic and chemical factors I-323–5
— major shallow water marine and coastal habitats I-325–7
— "new ways forward" I-331
— New York harbor
    — anthropogenic impacts I-327
    — clean up efforts I-327
— offshore systems I-327
— opposing uses I-331
— populations and conditions affecting the Bight I-328–9
— resources at risk I-329–31
— some habitat improvement I-330
New York Metropolitan area, growth of landfill and dumping of wastes I-328
New Zealand, no-take areas III-386
New Zealand Fur Seal II-681
Newfoundland, collapse of Atlantic cod fishery III-120
NGO activities, Jamaica I-572
NGO participation, increased due to "soft law" instruments III-344
Nicaragua
— central highlands, drainage of I-533–4
— coasts *I-533*
— cotton cultivation, after-effects of I-538
— government in I-538
    — laws awaiting approval I-541
— major economic and political trends I-540
— Pacific volcanic chain I-533
Nicaragua, Caribbean coast I-517–42
— coastal resource management need modification of law I-527–8
— effects from urban and industrial activities I-525–6
— geography of I-519
— ICZM I-526–8
— major shallow water marine and coastal habitats I-520–2
— natural reserves I-526
— offshore systems I-522–3
— population I-523
    — eastern slopes of the central highlands I-523
— protective measures I-526
— rural factors I-523–5
— seasonality, currents, natural variables I-519–20
Nicaragua, Lake *I-532*, I-533
Nicaragua, Pacific coast I-531–42
— effects from urban and industrial activities I-539–41
— geography/geology of I-533–4
— major shallow water marine and coastal habitats I-534–5
— marine resources used by Honduras and Salvador I-538
— offshore systems I-535–7
— populations affecting the area I-537–8
— protective measures I-541–2
— rural factors I-538–9
— seasonality, currents geography/geology of I-534

Nicaraguan Center for Hydrobiological Research, estimate of fisheries biomass I-525
nickel mining
— New Caledonia II-713, II-729
    — destructiveness of II-713, II-731
    — environmental effects of processing still largely unknown II-733
Nicobar Islands *II-190*, *II-270*
Niger, river and delta I-776–7
— oil a source of conflict I-788
— onshore oil production in the Delta I-788
— sand movement in the delta I-778
Nigeria, mangroves III-20
Nile Delta, retreat of I-258
nitrate concentrations, Maria Island II-651
nitrate contamination/pollution I-174
nitrate enrichment, euphotic layer, Côte d'Ivoire I-810
nitrate salts, mining of Chile I-710
nitrates I-255–6
— increased load to Baltic Sea I-130
— and seagrass and other aquatic vegetation I-358
nitrification III-262
nitrogen
— anthropogenic
    — in Baltic Proper I-104, *I-104*
    — emissions to the atmosphere III-199
    — load to Chesapeake Bay I-345
— increased load to Baltic Sea I-130
— load to Chesapeake Bay I-344–5
— organic, and phosphorus at depth in Bay of Bengal II-274
nitrogen concentration, Orinoco River I-647
nitrogen cycle, investigations of the impact of eutrophication on III-262–3
nitrogen fixation, important in Baltic waters I-128
nitrogen flux, riverine, increase in III-366
nitrous oxide, increased in the atmosphere III-188
no-take reserves III-376, III-377, III-378, III-386
— for fisheries management, coral reefs III-386
NOAA
— environmental survey baseline investigation, DWD 106 I-327
— National Status and Trends Program I-452
noise pollution, effect on marine mammals III-97
nomadic jellyfish, an immigrant I-157
Nord-Pas de Calais coast I-67
Nordostrundingen *I-6*
Norfolk Island II-635, II-636, II-643
— catastrophic soil erosion II-638
Normandy *I-66*
"nortes" I-489
North Aegean Trough I-235
North Atlantic, Sverdrup transport in I-579
North Atlantic Central Water, South of Finisterre Cape I-139
North Atlantic Current I-5, *I-18*, I-20, I-23, I-204, *I-204*, I-223
North Atlantic Deep Water I-579
North Atlantic Oscillation I-223
— and exceptional North Sea conditions I-45
North Atlantic water, southern Spain I-169
North Brazilian Current I-630, I-722
— retroflection of causes eddies I-630
North Carolina
— demersal marine zooplankton I-360
— eutrophication in some estuaries I-356
— toxic contamination I-360–1
North Channel *I-84*, I-85
North East Monitoring Program I-315
North Equatorial Counter Current I-630, II-407, II-428
— Gulf of Guinea *I-774*, I-778
North Equatorial Current I-223, I-487.I-488, I-489, I-617–18, *I-628*, II-406, II-407
North Equatorial Drift *I-222*
North Equatorial Pacific Current II-775, II-779–80, II-795, II-797

North Inlet *I-352*, I-356, I-359
— salt marsh estuary I-353
North Korea, famine due to deforestation, soil erosion and natural disasters II-492
North Loyalty Basin *II-426*
North New Hebrides Trench *II-426*
North Pacific Current *I-374*, *III-180*
North Pacific Fisheries Management Council, observer program on all larger and some smaller fishing vessels III-147
North Pacific Groundfish Observer Program (NPGOP) III-160
North Pacific High I-378
North Pacific Oceanic Province, Mexico I-487-8
North Pacific Pressure Index (NPPI) I-376
North Pacific Subtropical Anticyclonic Gyre II-407, II-795
North Sea I-43-63
— atmospheric nitrogen deposition assessment III-201
— cessation of some polluting inputs I-56
— climate I-45
— coastal erosion and landfill I-53-4
— cyclonic circulation I-45-6
— disposal of dredged material in I-55
— duplication of responsibility for **I-58**
— effects of urban and industrial activities I-54-6
— eutrophication I-53
— general decline in populations of marine mammals and seabirds III-129
— groundfish assemblage, changes in III-128
— major shallow water marine and coastal habitats I-47-50
— partitioning of I-46
— populations affecting the North Sea I-50-3
— protective measures I-56-9
　— international arrangements affecting protection of **I-56**
　— protection at subregional level **I-58**
— reduction of nutrient input by bordering countries agreed III-220-1
— region defined I-45
— seabirds and fisheries in III-112
— seasonality and natural environmental variables I-45-7
North Sea management
— need to increase management plans I-59
— new system is required I-59
North Sea Task Force I-57
— Monitoring Master Plan I-75, I-77
Northeast Asia Regional Global Observing System (NEARGOOS) II-497
Northeast Atlantic, high by-catch mortality rates III-125
Northeast Pacific shelf, detritus from catch-processor ships III-127
Northeast Providence Channel *I-416*
Northern Adriatic I-269
— communities of I-172
— described I-269
— formation and circulation of water masses I-271
— primary production high in offshore systems I-174
— salinity I-271
northern Benguelan current region *see* Africa, southwestern
northern fur seals, enzyme induction by PCBs II-457, *II-460*
northern Gulf of Mexico shelf, pulses of shrimp, crabs and fish I-438
Northern New South Wales, major habitats II-634
Northwest Arabian Sea and Gulf of Oman II-17-33
— coastal erosion and landfill II-27-8
— georaphy and geology II-19
— major shallow water marine and coastal habitats II-21-4
— offshore systems II-24-5
— populations II-25
— protective measures II-29-31
— rural factors II-35-7
— seasonality, currents and natural environmental stresses II-19-21
— urban and industrial activities II-28-9
Northwest Atlantic Fisheries Organization (NAFO) I-8
Northwest Pacific Action Plan II-496
Northwest Providence Channel *I-416*

Northwestern African Upwelling I-190
Norway
— climate modifies fjord morphology I-19
— coastal wave energy development III-315
— data on contamination in organisms and bottom sediments I-27
— "no discards" policy forces selective fishing III-138
— use of antibiotics in fish cages III-369
— Whaling Act (1929) III-77
— whaling technology III-74-5
— wild fish may contain antibiotic residues III-369
Norwegian Atlantic Current *I-19*
Norwegian coast I-17-30
— anthropogenic influences I-24-6
— environmental setting I-19
— major shallow water marine and coastal habitats I-21-3
— monitoring programmes, environmental quality criteria and protective measures I-26-8
— natural environmental variables I-19-21
— north—south community gradient less than expected I-22-3
— offshore systems I-23-4
Norwegian Coastal Current *I-18*, I-20, I-23
Norwegian Sea, depletion of rorquals III-74
Norwegian Trench I-17
nuclear fuel reprocessing plant, La Hague I-72, I-76
nuclear power, India II-279
nuclear power plants/stations I-54, I-72
— southern Brazil I-738
Nuclear Test Ban Treaty III-342
nuclear waste dump sites, Sea of Japan II-478, *II-480*
nucleation III-200
Nullarbor National Park, Australian Bight II-682, II-684
Nullarbor Plain, Great Australian Bight region II-682
nursery areas, restriction of fishing in III-378
nutricline, Levantine basin I-256
nutrient burden, reduction of III-258-9
nutrient changes, drastic, signs of in the Bodden III-260-1
nutrient concentrations, reef and lagoonal water, French Polynesia II-816
nutrient discharges, to North Sea, changes in I-53
nutrient effects, of seagrasses III-2
nutrient enrichment
— of coastal waters of Florida reef tract I-409
— Cockburn Sound, Western Australia, effects of II-700-1
— from equatorial and open ocean upwelling I-535
— from sewage effluents, The Maldives II-204
— from upwelling, southwestern Africa I-825
— New York Bight I-324-5
— of northern Gulf of Mexico I-449
— problems caused by II-212
— a threat to coral reefs III-36
nutrient fluxes
— and aquatic ecosystem responses III-198-9
— importance of dry deposition processes III-201
nutrient loading, anthropogenic III-366
nutrient loads
— Chesapeake Bay I-344-5
　— reduction in I-346
— high
　— brought by the Mississippi I-437, I-438, I-443
　— introduction of to the marine environment III-366
— increased, North Carolina I-358
— Yellow Sea coastal waters II-495
nutrient ration, effects of alteration III-259
nutrient reduction, Norwegian obligation I-24, I-26
nutrient supply
— Northeast Pacific
　— effected by shallow mixed layers III-184
　— vulnerable to climatic change III-171
— reduced in Californian coastal wasters III-185
nutrients
— causing eutrophication, Hong Kong II-542

— from fertilisers, Australia II-584
— anthropogenic inputs, to the English Channel I-74
— Arabian Gulf II-5–6, **II-5**
— Argentine Sea I-757, *I-757*
— Baltic Proper, temporal and spatial variability in I-106–8
— calcification and uptake of III-35–6
— concentrations around Tasmania II-651
— concentrations of, Canary Islands I-191–2
— — and nitrite I-191–2, *I-193*
— from fish farming III-369
— from the land, oceans a sink for III-394
— Gulf of Alaska shelf, from deep water I-379, I-380
— high, nearshore, Hawaiian Islands II-795
— horizontal gradients, Tagus and Sado embayments I-155, *I-156*
— input into Haifa Bay I-259–60
— inputs during upwelling events and high algal growth, Côte d'Ivoire I-814
— Irish Sea
— — increasing I-87
— — sources of inputs I-93, **I-93**
— large increase of to Baltic Sea I-130
— low
— — in Australian waters II-582
— — off Central-eastern Australia II-632
— Red Sea II-39
— reduction of input to North Sea by bordering countries agreed III-220–1
— role of in coral reef degradation III-38
— and salinity, off Madagascar II-115–16
— sources of input into Gulf of Guinea I-779
— to Bay of Biscay from Cantabria I-143–4
— transferred from sea to land by birds, Chagos Islands II-229

ocean drift netting, banned II-719
ocean dumping II-322
ocean physics, and foraging seabirds, North Pacific III-113
ocean resources, intergenerational and interspatial effects of use of III-397
ocean species, common, affected by coastal pollution and ocean dumping I-330
Ocean Station Papa, Gulf of Alaska *III-180*
— deviation from normal salinity III-182, *III-182*
— mid-winter mixed layer depth trend III-182, *III-182*, III-185
ocean temperature zones, Australia II-581
ocean temperatures, effects of changes in III-182
Ocean Thermal Energy Conversion (OTEC) III-304
ocean thermal energy plants II-279
— India, impacts of II-279
oceanic islands, associated with Chile I-701
— habitats and faunas I-708
oceanic mixed layer, Gulf of Alaska III-182, III-185, III-814
oceanic productivity, may decrease with climate change III-369
oceanic swell, Chagos Islands, resistance of spur and groove system II-227
oceanographic conditions, as determinants of habitat boundaries, Lesser Antilles I-630
oceanography
— Australia II-583
— of the Bahamas I-419
— Case 1 and Case 2 waters III-294
— eastern Australian region II-631
— Fiji Islands II-754
— in Greece I-242
— Marshall Islands II-779–80
— Vanuatu II-740
oceanography/marine hydrology, Gulf of Guinea I-777–9
— littoral transport and marine sedimentology I-778
— ocean currents littoral transport and marine sedimentology I-778
— productivity and the seasonal cycle I-779
— sea water quality/structure I-777–8
— tides/waves I-778

— upwelling I-778–9
oceans
— biological divisions I-2
— biome scheme I-2
— common property and open access characteristics III-397
— ecological, economic and social importance of III-393–403
— — ecological importance III-394
— — economic importance of III-395
— — social importance III-396–7
— — sustainable governance III-397–402
— human impacts on III-360
— impediments to ecological or scientific divisions I-2–3
— low in nitrates and phosphate, Australia II-583
— missing areas, reasons for I-3
— political divisions I-2
— unique problems III-397
Oceans Policy, Australia II-590–1
OCs *see* organochlorines
octachlorostyrenes (OCSs) I-54
Oculina reefs, Florida
— experimental reserve III-390
— mostly destroyed by fishers III-379
Odiel River *I-168*
— estuary receives high metal load I-171, I-174
Odiel saltmarshes I-170–1
ODP oceanographic surveys, of the Somali Basin II-67
off shore wind energy III-304–11
— access issues III-306
— foundations, design concepts III-306
— problems of grid connection III-306
— prospects for the future III-308–11
— — new technology and research needs III-311
— review of current technology III-304–7
— status at the millennium III-3078
— suitable off shore areas constrained III-307
— wind farms offshore III-305–6
— — size of installation III-306
offshore habitat, Cambodia II-574–5
offshore petroleum installations, Norway, monitoring of bottom conditions I-27
Offshore Pollution Liability Agreement (northwest Europe) III-338
offshore systems
— Adriatic Sea I-274
— Aegean Sea I-240–1
— — deep-water fauna I-240–1
— American Samoa II-769
— Australia II-583
— — geomorphology II-583
— — oceanography II-583
— — offshore territories II-583
— the Azores I-209–10
— the Bahamas I-422–3
— — deep water channels and V-shaped canyons I-422
— Belize deep reef habitats I-506
— — pelagic waters I-506
— Borneo II-372–3
— Chagos Archipelago II-229–30
— Chesapeake Bay I-343–4
— Colombian Caribbean Coast, importance of fisheries I-669–71
— Colombian Pacific Coast I-680
— Comoros Archipelago II-246–7
— — Geyser and Zélée Bank II-246–7
— Coral, Solomon and Bismark Seas region II-431–3
— — the environment II-431–2
— — tuna fisheries II-432, *II-433*
— Côte d'Ivoire I-813–15
— — benthic assemblages I-814
— — phytoplankton I-814
— — zooplankton I-814–15
— eastern Australian region
— — Elizabeth and Middleton Reefs II-635

— Lord Howe Island II-635
— Norfolk Island II-635
— El Salvador I-551
— Fiji Islands II-758
— and fisheries resources, Bangladesh II-292–5
— French Polynesia II-817
— Great Australian Bight II-681
  — variable abundance of pelagic fish II-681
— Guinea I-800–1
  — fishing I-800
  — industrial fishing I-800–1
  — upwelling I-800
— Gulf of Aden II-54–5
— Gulf of Alaska I-380
— Gulf of Guinea I-781–4
  — interannual variability in upwelling I-781–2
  — upwellings I-781
— Gulf of Maine and Georges Bank I-312
— Gulf of Papua II-598–9
— Gulf of Thailand II-304–5
— Hawaiian Islands II-800–1
  — deep-sea cobalt manganese crusts II-800
— Hong Kong II-538
— Jamaica I-565–6
  — deep habitats not well known I-566
  — Discovery Bay I-566, *I-567*
  — Morant Cays I-565
  — Pedro Cays I-566
— Lesser Antilles I-632–3
  — low salinity lens of Amazon discharge, Tobago to Barbados I-633
— Madagascar II-120
— The Maldives II-205
— Marshall Islands II-779–80
  — oceanography II-779–80
— Mascarene Region II-259–60
— Mozambique II-105, II-107
— New Caledonia II-726–7
  — pelagic zone II-736–7
  — ZoNéCo programme II-727
— New York Bight I-327
— Nicaraguan Caribbean Coast I-522–3
— Nicaraguan Pacific coast I-535
  — oxygen-minimum layer I-535
— northeastern Australia II-619
— northern Gulf of Mexico I-442–3
  — neritic province I-442
  — oceanic province I-442
— Northwest Arabian Sea and Gulf of Oman II-24–5
— the Philippines II-410–11
  — offshore fisheries II-411
  — oil and gas II-411
  — productivity low II-410
  — upwelling and internal waves II-411
— Portuguese coastal waters I-159
— Red Sea II-42–3
— Sea of Okhotsk II-468
  — major and permanent zones of vertical intermixing II-468
  — mesopelagic layer II-468
— South Western Pacific Islands II-712–13
  — pelagic communities II-712–13
— southeast South American shelf marine ecosystem I-757–8
— Southern Gulf of Mexico I-474–5
  — fisheries I-475
  — upwelling and the Yucatan Current I-474–5
— southern Spain I-172
— southwestern Africa I-830–5
  — demersal zone I-833–5
  — environmental variability I-832
  — epipelagic zone I-830–2
  — fishing activity I-832
  — mesopelagic zone I-832–3
— Spanish North coast I-140–1
— Tanzania II-88
— Tasmania II-653
— Tyrrhenian Sea I-274–5
— Vanuatu II-743
— Victoria Province
  — pelagic system II-666
  — slope communities II-666
— West African continental shelf benthic communities I-782, **I-782**
  — fish populations I-783
  — pelagic variability fish populations
  — plankton productivity I-783
  — whale migrations I-783–4
— Western Australian region II-696
— Yellow Sea II-490–1
  — resident and migratory species II-491
  — spawning, nesting and nursery area II-490
offshore wave energy conversion systems III-313–14
— research now limited III-314
oil
— biodegradation increased with clay–oil flocculation III-271
— effects on near shore populations and communities III-271–5
  — effects on biological communities III-272–5
  — general effects III-271–2
— mutagenic effects of long-term exposure III-278
— penetration affected by viscosity and type III-269
— persistence of on shores III-268–71
  — beach wetting: adhesive properties of oil III-270
  — dynamics of in low-energy environments III-271
  — oil-contaminated sandy beaches III-270–1
  — permeability III-269–70
— stickiness of, possible effects on III-270
— total entering northern Gulf of Mexico I-450–1
— trapped in low energy areas I-450
oil contamination
— low levels Gulf of Aden beaches II-59
— southern Brazil I-740
oil deposits, Nicaragua I-520
oil dispersants I-556
— toxic I-76, I-450
— toxicity of III-276
oil exploration I-40
— in Greenland Seas I-14
— Nicaragua, Caribbean coast I-526
oil exploration and production
— Argentina I-766–7
— Gulf of Guinea I-792
— offshore, Australia II-588
— Peru I-696
— southern Brazil I-738
— Western Australia II-702
oil exports
— Colombian Caribbean Coast I-673
— Nigeria and Gabon I-792
oil and gas
— exploration and exploitation, Côte d'Ivoire I-817
— the Philippines II-411
— Trinidad I-638
oil and gas exploration
— drilling waste piles on Norwegian seabed I-17, I-23–4
— Irish Sea I-96
— Mozambique coast II-109
— North Carolina coast I-366–7
— North Sea I-51
  — exploitation increasing I-52–3
— and production, Gulf of Thailand II-304–5
— South Africa II-142
oil and gas fields
— Bass Strait II-669
— east coast, Peninsular Malaysia II-354

oil and gas industry
— Australia II-584
— Sakhalin Shelf II-470–1
oil and gas installations
— offshore
    — far-field effects possible III-364
    — marine pollution from III-360
    — operational discharges from III-363–4
    — other chemicals in use, possible effects of III-364
oil and gas potential, Somalia II-79
oil and gas production
— Adriatic Sea I-278
— Malacca Straits II-315–16, *II-315*
— North Sea, environmental impacts at all stages I-55–6
— northern Gulf of Mexico I-448
— offshore, pollution by III-338
— transport of crude oil and environmental threat I-450
— Yemen II-59
oil and gas resources, South China shelf II-557
oil industry, offshore, Gulf of Mexico I-478–9
oil installations, offshore, Brunei II-372
oil persistence III-278
— effects may change as oil weathers III-271
— from the Refinería Panama storage tank rupture III-273
oil pipelines II-603–4, II-605
— environmental effect, Gulf of Guinea I-792
— southern Brazil, effects of rupture I-742
oil pollution II-376
— Black Sea I-294
— chronic
    — Argentinian coast I-766
    — western Indonesia, from oil refineries and production facilities II-390
— Côte d'Ivoire beaches I-817
— Dutch Antilles I-604–5
— False Bay, South Africa II-142
— from shipping III-364
— Gulf of Guinea I-792
— Gulf of Oman
    — from routine tanker operations II-28
    — worsening II-29
— heavy near Chittagong and Chalna, Bangladesh II-294–5
— increasing, South China Sea II-354
— Jamaica I-570
— and loss of seabirds III-113
— marine ecosystem at Toamasina (Madagascar) threatened II-125
— of marine sediments I-638
— Mexican Pacific coast I-497
— northern Gulf of Mexico I-450–1
— not avoided by small cetaceans III-96
— Sakhalin Shelf II-471
— Saronikos Gulf I-245
— sensitivity of organisms to I-450
— southern Brazil, Sepetiba Bay I-741–2
— Sri Lanka II-185
— Straits of Hormuz II-28
— Sydney Harbour and Botany Bay II-641
— Tierra del Fuego I-762
— Tyrrhenian Sea coasts I-279
— Venezuela
    — eastern coastline I-655–6
    — western coastline I-656–9
— Xiamen coastal waters II-523
oil production
— Argentina I-761, I-763
— Chile I-712
— Guangdong II-557
— Gulf of Papua, pipelines for delivery of II-603–4
— Mexico I-476
— Niger Delta, onshore I-788
— northern Gulf of Mexico *I-443*
— Northern Sakhalin, causing serious concern II-477
— northwestern Arabian Sea II-28
— onshore, Wytch Farm, Dorset I-72
— western Indonesia II-390
    — effects of increase in II-390
oil refineries
— Aden II-59
— Aruba and Curaçao I-604–5
— Fawley I-72
— Malacca Straits *II-315*, II-316
— Mogadishu II-80
— South Korea II-495
— southern Brazil I-738
oil revenues, invested in infrastructure II-28
oil seeps
— Gulf of Alaska I-381
— resulting from salt tectonism, northern Gulf of Mexico I-442, I-451
oil slicks
— Bay of Bengal II-280
— from ballast water I-72
oil spill contingency planning, Coral, Solomon and Bismark Seas region II-441
oil spill response equipment, for the Malacca Straits II-325
oil spills I-512, I-638, II-354, II-417
— accidental, risk of I-14
— acute
    — lethal and sublethal effects III-272
    — sublethal and chronic effects from III-278
— Aegean Sea I-248
— affect sea turtles III-67
— Alaska, *Exxon Valdez* I-382
— amount spilled *III-268*
— Arabian Gulf II-3
    — and the Gulf War II-11–13
— Argentina I-765–7
    — "Metula" spill I-765
    — "San Jorge" spill I-765
— Australia II-587–8
— Bay of Cadiz I-181
— Black Sea, Nassia disaster I-294
— Chilean coast **I-712**
— differences in lead to different responses III-272–3
— direct causes of mortality III-272
— El Salvador I-556
— endangering western Indonesian coastlines II-388
— English Channel I-72, I-75–6
    — Amoco Cadiz, long term effects I-76
    — Torrey Canyon, major damage from oil dispersants I-76
— Galician coast I-146
— Great Australian Bight II-686
— Gulf of Aden II-59
— Gulf of Mexico I-473
— high potential for, western Indonesian seas II-390, **II-390**
— Hong Kong, mainly minor II-542
— and illegal discharges, North Sea I-55–6
— increasing, Marshall Islands II-783
— Malacca Strait II-339
— Malacca Straits II-319–20
    — serious impact on fragile ecosystems II-319
    — Standard Operation Procedures (SOP) II-322–3
— Maracaibo Lake I-656
— Mozambique Channel II-251
— Niger Delta I-788
— northern Gulf of Mexico I-450
— Norway, few I-26
— off Fujeirah II-28
— persistence on beaches III-168–71
— persistent oil effects and their causes III-275–6
— Peru I-696
— a risk for Mozambique II-109
— *Showa Maru* II-334

— small, Sri Lanka II-185
— South China Sea (western) II-559
— southern Brazil
    — Sao Sebastiao City I-742
    — Sepetiba Bay I-741–2
    — southeast coast, and degraded mangroves I-736
— Tasmania II-657
— timing of important III-269, III-270, III-272
— Tobago I-638
— Torres Strait, *Oceanic Grandeur* II-605
— treatment effects III-276–7
    — bioremediation III-277
    — dispersants III-276
    — injuries due to III-275, III-276
    — manual removal of oil III-276–7
    — problems with use of heavy machinery III-277
    — sand-blasting and high/low-pressure-water techniques III-277
— Xiamen coastal waters II-523
— Yellow Sea II-495
oil tanker traffic
— Malacca Strait II-339
— Malacca Straits II-316–17
— Sepetiba Bay, southern Brazil I-741–2
oil terminals, southern Brazil, Sao Sebastiao City I-739, I-742
oil-well fires II-11
okadaic acid III-40, III-219
Okhotsk, Sea of II-463–72
— currents II-465–6
— effects from urban and industrial activities II-469–71
— major shallow-water marine and coastal habitats II-467–8
— offshore systems II-468
— populations affecting the area II-468–9
— protective measures II-471
    — fishery regulations II-471
    — poaching and overfishing II-471
— rural factors II-469
— seasonality, currents, natural environmental variables II-466–7
    — seasonal changes in biota II-466
— winds drive winter water movement II-465
Okhotsk shelf, productive fish area II-467
Old Bahama Channel *I-416*, I-417
Olive Ridley turtle I-780, II-180, II-205, II-289, III-63
— captive breeding programmes, Sundarbans II-158
— critically endangered III-63
— distinctive nesting behaviour III-63
— threatened I-537
Oman *II-2, II-18*
— agriculture II-26
— construction of regional fishing harbours II-27
— coral communities
    — damaged by nets *II-30*
    — limiting factors II-22
— increase in beach tar II-28
— industrial diversification, new industrial development, Sohar and Sur II-25, II-27, II-28
— limestone mountains *II-18*
    — unique vegetation II-21
— network of conservation areas proposed II-30
    — national nature reserves II-30
— population II-25
— Ra's Al Hadd National Scenic Reserve II-30
— Ra's Al Junayz National Nature Reserve II-30
— seabirds III-110
— upwellings along coast affect local weather II-50
Oman, Gulf of *II-18*
— current flow *II-18*, II-20
— defined II-19
— fish biodiversity II-24
— low-energy environments II-21
— sea water temperature fluctuations II-21, II-22, *II-22*
— shallow water marine and coastal habitats II-21–4

— corals, reefs and macroalgae II-22–3
— seagrasses II-23–4
— turtles II-24
— *wadis*, *khawrs* and mangroves II-23
Ontong Java Plateau *II-426*
open ocean habitats I-224–7
— anguilla I-226
— biogeochemistry of the area round Bermuda I-224–6
— humpback whales I-226–7
— *Sargassum* community I-226
open sea banks I-47
Operation Raleigh I-590
optical remote sensing, governing processes involved III-295, *III-296*
orcas ("killer whale") III-76
— high levels of PCBs III-96
— widely distributed III-91
organic carbon, New York Bight, sources of I-324–5
organic matter/material
— Baltic Proper, sinks to soft bottoms I-128
— Baltic Sea I-112–14
— delivered to Baltic Proper **I-103**, I-104
organic pollutants
— Aegean Sea **I-248**
— Baltic Sea I-131
— Faroes I-37–8
— Nervión estuary, northern Spain I-145–6
— Sargasso Sea I-229
organic pollution
— Adriatic Sea I-270
— in the English Channel I-75
— Hong Kong coastal waters II-543
    — trace contaminants II-544
— Sea of Japan *II-479*
— West Guangdong II-559
    — Pearl River mouth II-555
— west Taiwan II-509
— Xiamen region II-522
organisms, sensitivity to oil III-272
organochlorine burdens
— Arctic beluga III-362
— slowing recovery of Baltic Sea seal populations III-361
organochlorine pollution
— fish from Asian waters II-449–51
— oceanwide II-454
organochlorine residues
— in Hoogly (Hugli) river sediments II-279
— in rivers and marine biota, Malaysia II-319, II-338, II-341
organochlorines
— in Australia's marine environment II-588
— Jakarta Bay II-398
— in Liverpool Bay I-95
— southern and western Baltic I-112, *I-113*
    — DDT in herring and perch I-112, *I-114*
— in USA dolphins III-96
— use of in China II-492
organochlorines, persistent
— Asian developing countries *II-448*, II-449, *II-453*
— contamination and bioaccumulation in marine mammals II-455–6
    — toxic effects II-456–8
— contamination in North and South Pacific II-454
— global fate II-455
— Hong Kong II-542
— major pollution sources now II-455
— river and estuarine sediments, Asian developing region II-449, *II-451*, *II-550*
— temporal trend of contamination II-458–60
    — in Antarctic minke whales II-458, *II-460*
— West Guangdong coast II-555–6
organophosphate compounds I-493
organotin pollution, in fish II-451, *II-454*

organotins
— in anti-fouling paints I-146
— Asian developing regions II-449
— in USA dolphins III-96
— *see also* tributyltin
Orinocco effects I-579
Orinoco Basin
— extent of and physiographic units I-646
— geochemical characteristics of rivers I-646–7
Orinoco River
— changed tidal effects after closure of Caño Manamo I-648
— delta *I-644*
    — functions like a wetland I-648
    — main zone I-648
— effects of fluctuation in discharge I-648
— influences of, Venezuelan Atlantic coast and Caribbean Sea I-646–8
Orinoco River Plume I-630
Ortegal Cape *I-136*, I-139
Oslo *I-18*
— contamination of harbour sediments I-27
Oslo and Paris Commissions (OSPARCOM) I-43, III-221
— monitoring contaminants in sediments, biota and waters I-77–8
— monitoring programmes (JMP and JAMP) I-27, I-77–8
OSPAR Convention I-26, I-57, **I-57**, I-147, I-162
— hazardous substance targets III-362–3
— stronger controls on off shore oil and gas III-362–3
Otway Region, Victoria Province II-663
over-exploitation
— of coastal fish resource, Sri Lanka II-182
— littoral fish, Canary Islands I-196
— Mexican artisanal shrimp fisheries I-490, I-491
— Northwest Pacific sardine fishery I-490
— of timber, Vanuatu II-743
— turtle fishery I-491
— Venezuelan coast zone I-645
over-harvesting, of renewable resources, Marshall Islands II-784
overcapacity, world's fishing fleets III-367, III-377
overfishing I-3, I-312, III-119, III-130, III-367
— Adriatic Sea I-278
— Carolinas I-366
— and catch decline, Mozambique II-107
— Colombia
    — Caribbean Coast I-671
    — Pacific Coast, shrimps I-684
— in the Comoros II-248
— Coral, Solomon and Bismark Seas region, boom-and-bust cycles II-437
— Malacca Straits II-320
— 'Malthusian overfishing' III-123
— of marine invertebrates, Fiji Islands II-759
— North Sea I-52
— northern Gulf of Mexico I-445
— Oman and Yemen II-27
    — with industrial methods II-29
— of predators III-128–9
— of salmon and seatrout, in the English Channel I-73–4
— South Western Pacific Islands II-716, II-719
— technique restrictions not effective against recruitment overfishing III-161
— threatens artisanal fishing, Belize I-509
— Vanuatu II-747
overgrazing
— effects of I-444
— Hawaiian Islands II-802
— Oman and Yemen II-26
overgrazing/overcropping eastern Australian region II-637
overharvesting, of fish resources, Palk Bay–Madras coast II-170
Owen Fracture Zone *II-64*, II-67
oxygen deficiency, Baltic Proper, affecting cod spawning I-110
oxygen depletion III-259
— between Mississippi and Sabine rivers I-443
— effects of, North Sea I-53
Oxygen Minimum Layer, Arabian Sea II-24
oyster banks, Chesapeake Bay I-341
oyster beds
— importance of, Carolinas coast I-362
— Texas, disappearing I-447
oyster culture, Venezuela I-653
oyster farms
— Korea II-494
— West Guangdong II-558–9
oyster fishery, small commercial, KwaZulu-Natal II-139
oyster habitat, Chesapeake Bay, destruction of III-126
oyster harvesting
— Chesapeake Bay I-341–2
— northern Gulf of Mexico I-442
oyster reefs, northern Gulf of Mexico I-441–2
oyster shell, for agricultural uses I-447
oysters III-167
— communities destroyed through burial I-447
— decline in production, Tagus estuary I-162
— dragnets and dredging damage seagrass beds III-8–9
— effects of TBT III-250
— El Salvador I-551
    — contamination in I-555
— farming of, Normandy and Brittany coasts I-73
— as indicators of environmental contamination III-253
— rock oysters II-52
— Texas, declines in linked to salinity perturbations I-439–40
— *see also* pearl oyster culture; pearl oyster fishery
ozone, tropospheric, Azores I-204
ozone depletion III-370–1
— a possible impact on small cetaceans III-98

Pacific basin, detection of mid-1970s regime shift III-224–5
Pacific Center Coastal Province, Mexico I-488
— habitats I-488
— important fishing area I-491
— varied marine and coastal fauna I-491
Pacific continental platform I-534
Pacific Deep Water I-679, I-702, II-583
Pacific Ocean
— Central and Eastern, coral bleaching and mortality events III-53
— equatorial current system I-679
— north, surface layer is fresh III-181
— Northwest, coral bleaching and mortality events III-52
— Southwest, coral bleaching and mortality events III-52–3
Pacific Ridley turtle, destroyed by landslides I-539
Pagassitikos Gulf *I-234*, I-236, I-237, I-244
— nutrient rich I-244
— some chlorinated biphenyls I-247–8
Pago Pago harbour, American Samoa
— dredging and filling causing reef loss II-717
— heavy metal pollution II-712, II-771
— industrial activities II-770–1
— loss of habitats II-770
PAH monitoring, Prince William Sound and Cook Inlet I-381
PAHs *see* polycyclic aromatic hydrocarbons (PAHs)
Palk Bay *II-162*, II-163, II-166–8
— climate and coastal hydrography II-166
— coastal ecosystems II-166–7
— fish and fisheries II-167
— geological features II-166
— human population and environmental degradation II-167
— Muthupet Lagoon II-167
— Vedaranyam wildlife sanctuary II-167
Palk Bay–Madras coast II-168–70
— biodiversity II-169
— climate and coastal hydrography II-168
— coastal ecosystems II-168–9
— environmental degradation II-169

— affected by decrease in freshwater II-169
— fish and fisheries biodiversity II-169
— geological features II-168
— impacts of human activities on the ecosystem II-169–70
Pamlico River Estuary *I-352*
— dinoflagellates I-355
— freshwater eelgrass I-359
— productivity pulses and algal blooms I-354
Pamlico Sound *I-352*, I-353
— environmental problems for bottom communities I-360
Panama Bight *I-678*
— seasonal upwelling cycle I-679
— tuna/anchovy/shrimp fishery I-680
Panama Current *I-678*
Panama, Gulf of *I-678*
Panamic Coastal Province, Mexico I-488–9
— habitats and communities I-489
"pantry" reserves, Marshall Islands II-786
Papua, Gulf of *II-426*, *II-594*
— agriculture II-600–1
— fisheries management II-606–7
— low level of development II-599
— water and sediment discharge to II-596, **II-596**
Papua New Guinea (PNG) II-427, **II-427**, II-595
— coral bleaching and mortality events III-53
— coral reefs in good conditions II-431
— effects of logging II-435
— impact of land use on coastal waters II-435–6
— level of social and economic development lo II-434
— mangroves II-430
— population and demography II-433–4, **II-434**
— research providing some information on marine biota II-435
— seagrass species diversity II-431
— sedimentary coast, backed by mountains II-427–8
Papuan Barrier Reef II-430, II-597
paralytic poisoning, and saxitoxin III-40
Paralytic Shellfish Poisoning (PSP) I-69, I-160, I-765, II-320, II-339, II-418, II-495, II-543, III-221
Paria, Gulf of *I-644*
— dry season cirulation of Orinoco waters I-648
Paria Peninsula, cold waters present I-646
Paris Convention, Protection of Marine Pollution from Land-based Sources III-339
particles
— hygroscopic growth of, and deposition velocities III-203
— removal by wet deposition III-200
Pas de Calais, zooplankton in waters off I-70
pastoral nomadism, Somalia II-77, II-80
Patagonian coast, Argentina I-751–2, *I-752*
— eutrophication I-765
— evolution of settlements I-758, I-761
— hydrocarbon concentrations I-761
— intensification of tidal currents I-754
— northern, characteristic sublittoral communities I-754–5
— unique environments I-756–7
— marine birds I-757
— marine mammals I-756–7
patch reefs I-505, II-7–8, II-65, II-191
— Glovers Atoll I-505
— lagoonal
— Bermuda I-227, I-228
— Mauritius II-258
— Madagascar II-118
— northern Kalimantan and eastern Sabah II-371
— Tanzania II-85–6
Patella, affected by oil dispersants I-76
Patos Lagoon, southern Brazil *I-732*, I-737
— fishes of I-740
PCBs I-54, I-279, I-392, I-555
— in animals from the Greenland seas I-9, I-11, I-14
— in Asian developing region waters II-449

— in Bergen fish and shellfish I-27
— in blubber II-455
— concentrations in Arctic species III-361, *III-362*
— coplanar PCBs considered more toxic II-457
— estimated loads in the global environment II-455, **II-455**
— in the Gulf of Maine I-317
— high levels in Irish Sea mammals I-95
— Hong Kong coastal waters II-544, **II-545**
— in marine sediments, Vietnam II-565–6
— marine transport of from European waters I-14, *I-14*
— pollution from, Thailand II-449
— reasons for persistence of III-362
— reduction in positively affecting Baltic seal populations I-131
— uniform distribution of in air and surface seawater II-454, *II-458*
— in whale blubber I-36
Peale's dolphins III-91, III-93
Pearl Cays complex, corals I-522
pearl oyster and Chank beds, Gulf of Mannar II-164
pearl oyster culture II-207
— French Polynesia, problems of II-820
— Marshall Islands II-781, II-782
— Solomon Islands II-438
pearl oyster fishery
— Arabian Gulf, in decline II-11
— Fiji, overexploited II-759
— French Polynesia II-820
pearl oysters II-598
Pearl River Delta *II-550*
— phytoplankton II-553
— tributaries II-551
Pearl River mouth
— irregular semi-diurnal tides dominate II-552
— and the Pearl River II-551
pearl shell fishery II-602
peat, Sri Lanka II-181
peat swamp forests, Malacca Strait II-336, II-342
peat swamps
— east Sumatra II-312
— Malacca Strait II-335
pelagic organisms
— Irish Sea I-90–1
— West Guangdong coast II-554
Pemba *II-84*, II-85
Penobscot Bay *I-308*, I-314
Pentland Firth, Scotland, considerable tidal currents III-318
peroxyacetylnitrate (PAN) I-205
Persian Gulf *see* Arabian Gulf
persistent organic compounds, contribute to seal decline I-50
persistent organic pollutants (POPs) I-5, I-39–40, I-381, III-360, III-361
— Asian developing regions II-447–62
— atmospheric transport to colder regions III-361
— decline very slow III-362
— Greenland I-9, I-11, *I-11*
— from Europe and Russia I-13
— may affect human health I-11
— toxicity risk to small cetaceans III-95–6
— transport and deposition in Europe I-56
Peru I-687–97
— biodiversity I-691–3
— Peruvian—Chilean province I-691
— Provincia Panameña I-691
— characteristics of the coast I-689–91
— coastal populations and the main sources of pollution I-693–6
— collapse of anchovy fishery III-119, III-129
— decline in guano birds III-111, III-129
— direct and indirect small cetacean catches III-93
— legislation on environmental protection I-696
— main populated and industrial areas I-695
— National Contingency Plan (oil spills) I-696
Peru (Humboldt) Current I-486, I-487, I-488, I-489, I-543

Peruvian Current
— high productivity I-692
— influences coastal climate I-690
— northward-flowing I-690
pesticide pollution
— and coastal ecosystems, the Seychelles II-239
— from Mozambique's upstream neighbours II-109
— Peru I-694–5
— Southern Gulf of Mexico I-476, I-479
— Xiamen coastal waters II-524
pesticides
— causing environmental stress, Altata-Ensenada del Pabellón I-493
— discharged to Black Sea I-301–2
— El Salvador
    — in oysters I-555
    — in river waters I-552
— golf courses, Hawaiian Islands II-806
— Gulf of Guinea I-784–5
— in land-based pollution III-361
— loss of from shipping III-364
— Malaysian waterways II-341
— organochlorine, Vietnam coastal waters II-565
— and other chemicals used in aquaculture III-369
— in shrimps I-493
— use and abuse, the Philippines II-412–13
— use round Sea of Japan II-475
— use of in Tanzania II-88–9
— used in Greece, concentration in the marine environment I-243–4
Peter the Great Gulf
— changes in bottom communities II-481
— radioactivity in II-470–80
Peter the Great Marine Reserve, endangered II-477
petrochemicals, southern Brazil I-739
petroleum, Venezuela's main export I-655
petroleum hydrocarbons, in Arabian Gulf sediments II-12
petroleum refineries *see* oil refineries
Pfisteria, ichthyotoxic I-363–4, I-366
Phaeocystis I-69, I-90, I-356
pharmaceuticals
— from marine environments and coral reefs III-39
— natural products and chemicals III-38–9
— natural products from coral reefs III-40
Philippine Tuna Research Project II-410
Philippines, The II-405–23, II-416–17
— the area and its natural environmental variables II-407–8
— coastal erosion and landfill II-413–14
— coral bleaching and mortality events III-51–2
— effects from urban and industrial activities II-414–18
— major shallow-water marine and coastal habitats II-408–10
— mangroves
    — clearance for aquaculture III-368
    — lost to fishponds III-21–2
— offshore systems II-410–11
— populations affecting the area II-411–12
— protective measures II-418–20
    — Coastal Environment Program (CEP) II-419
    — Coastal Resource Management Program II-419
    — community-based coastal resources management (CB-CRM) II-418
    — Fisheries Sector Program II-418–19
    — legalities, utilization, conservation and management of the coastal areas **II-420**
    — marine protected areas II-419
— rural factors II-412–13
— seimically active II-408
phocine distemper virus (PDV) (1988)
— killed seals on English south coast I-71
— reduced seal numbers, North Sea I-50
phocine distemper viruses III-225
phosphate I-255
— increased load to Baltic Sea I-130

— a limiting factor I-240
— release of, the Bodden III-262–3
— seawater, Vietnam II-565
phosphate enrichment, euphotic layer, Côte d'Ivoire I-810
phosphate mining, Makatea, Tuamotu Archipelago II-822
phosphoric acid manufacture I-72
phosphorite ore, Onslow Bay I-366
phosphorus
— from phosphate rock processing I-94
— increased load to Baltic Sea I-130
— as a limiting nutrient I-106
— load to Chesapeake Bay I-344–5
phosphorus level, low, Orinoco River I-647–8
photic layer, Panama Bight I-679
photoinhibition II-42
photosynthesis III-394
— anoxigenic III-263
phthalates, in coastal sediments, west Taiwan II-509
Phyllophera meadows, Black Sea, decrease in I-291, I-304
physical environmental anomalies, make entire populations vulnerable III-225
physically forced (climate/oceanographic) disturbances III-224–5
phytohydrographic associations, Arabian Gulf II-6–7, **II-7**
phytoplankton
— Carolinas
    — estuarine I-354–6
    — marine I-356
— Guinea I-800
— Gulf of Guinea I-779
— Gulf of Thailand II-304
— harmful effects of increased exposure to UV-B III-370
— key roles of III-198
— Malacca Strait II-333
— Malacca Straits II-312
— offshore, Côte d'Ivoire I-814
— species diversity, Vietnam II-564
— West Guangdong coast II-553
— wide range in UV-B sensitivity, effects of III-370–1
phytoplankton assemblages, Australia II-583
phytoplankton biomass
— and biodiversity, Pearl River mouth II-553–4
— coastal, Sunda Shelf II-384
— nutrient control of, Tuamoto archipelago II-817
phytoplankton blooms I-244, II-50, II-700
— Baltic Proper
    — changes in composition and dominance I-109
    — spring I-106
— Canary Islands I-192
— Irish Sea I-89
— Norway, spring and summer I-20
— Spanish north coast I-140, I-141
phytoplankton communities, Mascarene Region, nutrient-limited II-259
phytoplankton ecology, Bay of Bengal II-275, *II-276*
phytoplankton growth
— North Spanish coast I-138
— Tagus and Sado coastal waters I-155–6
— western English Channel I-67
phytoplankton processes, Spanish north coast, modified by oceanographic processes I-141
picoplankton, becoming dominant, the Bodden III-261
pilchard fishery
— Australian Bight II-683
    — may affect seabirds II-685
— southwestern Africa I-831
    — failure of I-832
— Victoria Province, Australia II-668
pilchards, killed by herpes virus II-587
pilot whales III-76, III-91
— hunted, Faroe Islands III-92
— long-finned I-31, I-36

Pinatubo, Mount, effects of eruptions II-408
pinnipeds, effects of decline in III-99
Pinus pinea forests, southern Spain I-170, I-181
Pitt Bank *II-222*
plaice I-49
— Baltic, stock decreased I-111
plankton
— abundance of, Argentine Sea I-757
— in the Black Sea I-290
— Côte d'Ivoire I-812–13
— eastern Australian region II-634
— North Sea I-48
— Peru, changed during El Niño I-692
— Tagus estuary I-157
— West Guangdong coast **II-554**
plankton assemblages, southeastern Taiwan II-508
plankton productivity, Gulf of Guinea I-783
planktonic communities, French Polynesian lagoons II-816–17
planktonic food web, model for, Takapoto atoll II-817
planktonic systems, Irish Sea I-89–90
Plantagenet Bank I-227
plate boundary, diffuse, through Chagos area II-226
platform reefs II-388
— the Seychelles II-236
plutonium (Pu) I-94
— fall out from nuclear tests I-117
PO_4 levels, affected by Tagus and Sado river discharges I-155, *I-157*
Po basin
— agricultural and industrial pollutants from I-275
— evolution of Po delta I-276
poaching, and marine reserves III-390
Poland
— discharged partly treated sewage to River Vistula I-115
— effect of increased standard of living I-112
— environmental contamination from mining I-114–15
— pollutants to Baltic Proper I-103–4
"poles of development", Chile I-710
politics, in fisheries management III-154
pollack I-313
pollock fishery
— Alaska III-147
— Bering Sea III-129
pollutant layers, elevated III-200
pollutants
— atmospheric transport of I-13
— baseline studies, Gulf of Alaska I-381, I-383
— brought by river to Black Sea I-287, I-298–9
— from Turkish Black Sea coast *I-200–301*, **I-300**
— Carolinas coast, sources of I-354
— carried to Baltic Proper by rivers I-103
— direct discharge into estuaries, Portugal I-162
— dispersion and deposition processes in the coastal zone III-206
— entering the Bay of Bengal II-278
— environmental, Baltic Sea I-112–17
— global redistribution of *III-361*
— higher sensitivity of Baltic populations I-129
— increase susceptibility to infection III-225
— and the internal boundary layer III-207
— Malacca Strait II-340
— marine current transport of I-13–14, *I-14*
— Palk Bay–Madras coast II-169
— and pathogens, waterborne, spread of in the Caribbean I-503
— reduction of phosphorus and nitrogen to North Sea I-24
— sea ice transport of I-14
— Southern California Bight
— largest reduction from publicly owned treatment works I-389–90
— multiple source discharges I-388–9, *I-389*
— reductions in I-389
— *see also* long-distance transport of pollutants
polluter pays principle II-323, II-327, III-339, III-401

— Jamaica I-570
pollution
— Aegean Sea I-245–9
— affecting eelgrass beds III-7–8
— affecting small cetaceans III-95–6
— airborne, Azores I-204
— atmospheric, from ships III-340
— Bahamas, effects on water quality and near-shore habitat I-429–30
— of beaches, Sri Lanka II-183
— Belize
— control through the Environmental Protection Act I-513
— from industry I-511
— urban I-512
— Black Sea, sources of I-287, I-297, **I-297**, I-298–9, **I-298**, *I-299*
— Cabo Frio–Cabo de Santa Marta Grande, southern Brazil I-738–9
— definition and sources of III-258
— degrading Gulf of Guinea coastal waters I-784
— effects of in the Baltic Sea I-130–1
— eutrophication I-130–1
— heavy metals I-131
— organic pollutants I-131
— pulp mill industry I-131
— effects of climate change and sea level rise III-195
— effects of land-based sources, Israel I-259–61
— Haifa Bay I-259–61
— entering the northern Gulf of Mexico I-449–50
— entering rivers and coastal seas, western Indonesia II-396
— estuarine and marine, South Africa II-142–3
— from fishponds, the Philippines II-414
— from ships, Coral, Solomon and Bismark Seas region II-440–1
— from shrimp farming and agrochemicals, Bangladesh II-294
— in the Great Barrier Reef region II-623–4
— Guinea I-801, I-802
— Gulf of Alaska, mainly from long-distance transport I-381
— Gulf of Mannar II-165
— impacting on seabirds I-49
— industrial
— and domestic, Peru I-694, I-695
— Tanzania II-92
— Kingston Harbour, Jamaica I-570
— land-based
— contributing to marine pollution III-339–40, III-340, III-360, III-361–3
— regulation of III-362–3
— slower treatment of III-339–40
— threat to coral reef biodiversity II-389
— land-based, Turkish Black Sea coast I-296–303
— from city sewerage systems I-300–1, *I-301*
— from rivers I-299–300
— monitoring I-291–302
— pesticides and PCBs I-301–2
— localised, Coral, Solomon and Bismark Seas region II-439
— Malacca Strait
— faecal coliform count II-340
— Indonesian side not systematically monitored II-341
— land-based II-340
— sea-based II-340
— Malacca Straits II-317–20
— agricultural waste II-318–19
— coliform contamination II-319
— land-based II-317–18, **II-318**
— main problem areas II-325–6
— sea-based sources II-319–20
— moderate in Canary Islands I-198
— Nicaraguan Pacific coast I-534
— organic I-835
— Papeete, Tahiti II-817
— point-source, Fiji II-760
— release to marine environment from point and non-point sources III-258
— Saronikos Gulf I-238

— and seabird deaths III-114
— secondary, from Maluan Bay sediments, Xiamen region II-524
— the Seychelles, from habitation and farms II-239
— Taiwan Strait, effects of tidal currents II-502
— through shipping operations and accidents I-146
— Venezuela I-654
— vessel-sourced, reduction of III-343
— Western Australia II-697
pollution abatement, some progress, North Sea countries I-52
pollution hotspots
— identified in Baltic I-132
— the Philippines II-418
pollution prevention programmes
— Yellow Sea
— economic problems a major impediment to II-496
— land-based, obstacles to II-495–6
pollution-contamination distinction difficult III-258
polybrominated diphenylethers (PBDEs) I-36
— presence in marine mammals III-362
polychaetes I-157, I-239
— Faroes I-34
— first to colonize mine tailings III-240
— Spanish North coast I-139, I-140
polychlorinated biphenyls *see* PCBs
polycyclic aromatic hydrocarbons (PAHs) I-56, I-392, I-555, II-417, III-361
— Arabian Gulf II-12
— attributed to oil seeps I-381
— in fjords I-24
— Great Barrier Reef region II-623
— Hong Kong coastal waters II-544, **II-545**
— posing a risk to ecosystems and seafood consumers II-545
— multi-hop contaminants I-13
— in Norwegian shellfish I-25
— Prince William Sound, delayed effects I-382
— in sediments, Baja California I-497
— southern Brazil
— Guanabara Bay I-741
— Santos Estuary I-743
— Taiwan Strait II-510
— *see also* PAH monitoring
polynyas
— Baffin Bay I-8
— Greenland Sea I-7
— Sea of Okhotsk II-466
Pomeranian Bay I-109, I-115
Poole Bay *I-66*
POPs *see* persistent organic pollutants
population
— Adriatic and Balkan coastlines I-275
— Aegean, ancient cultures, modern cities I-241–2
— Alaska I-381
— American Samoa II-769
— population growth rates II-769, *II-769*
— Andaman, Nicobar and Lakshadweep Islands II-193
— and increasing environmental degradation II-193
— Lakshadweeps II-194
— Anguilla, British Virgin Islands and Montserrat I-620–1
— Arabian Gulf II-9
— Argentinian coast I-758, I-761–2
— around the Irish Sea I-92–3
— around the North Sea, effects of I-50–3
— around the Sea of Japan II-475
— Australia II-586
— Aboriginal peoples II-583, II-682
— general community II-584
— indigenous communities II-583–4
— the Azores I-210
— the Bahamas I-423
— Baltic catchment I-50
— Bangladesh II-291–2, **II-292**

— Belize I-506–7
— Borneo II-373–4
— Brazil I-733
— Carolinas coast, growing I-354
— Chesapeake Bay area
— Europeans I-344
— growth and development I-344
— Native Americans I-344
— sprawl development I-345
— coastal
— Cambodia II-575
— surrounding the Yellow Sea II-491–2
— Vietnam II-563
— Colombian Pacific Coast I-680–1
— Comoros Archipelago II-247
— and consumption III-396
— Côte d'Ivoire I-815–16
— coastal cities I-815
— indigenous population I-815
— and demography
— Papua New Guinea II-433–4, **II-434**
— Solomon Islands II-433–4, **II-434**
— Dutch Antilles I-612
— leeward group I-597, I-598
— east coast, Peninsular Malaysia II-350–1, *II-351*
— urban growth rate II-351, **II-351**
— eastern Australian region II-635–6
— New South Wales II-636
— south-east Queensland II-636
— El Salvador I-551
— English Channel coasts I-71
— Faroes I-35
— Fiji II-758
— Florida Keys I-408
— Great Australian Bight region II-682
— European colonisation II-682
— Great Barrier Reef region II-619–20
— indigenous people II-619–20
— trends in II-620
— Greenland I-8
— growing beyond sustainable limits II-173
— Guinea coastal zone I-801–2
— Gulf of Aden states, rural and poor II-56, **II-56**
— Gulf of Guinea coast I-784
— country demographic information **I-785**
— Gulf of Maine I-314
— Hawaiian Islands II-801
— decline of native population II-801
— demographic patterns II-801
— tourism trends II-801
— Hong Kong II-538, II-541
— Huelva and Cadiz, southern Spain I-172
— human
— growth of III-396
— total impact of III-396
— and human settlements, Lesser Antilles I-634, **I-634**
— Iranian Gulf of Oman coast II-25
— Israel I-258
— Jamaica I-567
— development pressure I-561, I-567
— Madagascar II-120–1
— increase in major coastal towns II-121
— Marshall Islands
— coastal uses and environmental issues II-780
— demographic patterns II-780
— Mascarene Region
— cities and sewage discharges II-264
— coastal II-260
— Mexican Pacific coast I-489
— in medium and small communities I-492
— Mozambique, trend to urbanisation II-107

— New Caledonia
  — distribution of II-728
  — structure II-727–8
— of the New York Bight area I-328–9
— Nicaraguan Caribbean coast I-523
— Nicaraguan Pacific coast I-537–8
  — Estero Real, degraded natural resources I-537
  — Gulf of Fonseca I-537–8
  — population-related problems I-537
— northern coast of Spain I-142, *I-143*
— northern Gulf of Mexico I-443, I-448
— Norway I-17, I-24, I-28
— Oman II-25
— Palk Bay coast II-167
— Pearl River delta and West Guangdong II-556–8
— Peruvian coast I-693
  — cities I-695
— the Philippines II-411–12
  — coastal population depends on coastal fisheries II-414
  — growth rate II-412
  — rural–urban migration II-412
— Poland and Lithuania I-103–4
— and politics, French Polynesia II-815
— rates of change, Lesser Antilles I-634
— Red Sea coastline, mainly major ports and cities II-43
— regions round Sea of Okhotsk II-468–**II-469**
  — Sakhalinsky region, unique II-468
— round the Malacca Straits II-311
— rural, Cambodia II-575
— the Seychelles II-237–8
— Singapore II-337
— Somalia 77
  — migration to cities II-77, *II-79*
— South Western Pacific Islands II-713–14
  — population trends II-713
— southeast India, Gulf of Mannar II-165
— southern Black Sea coast, Turkey **I-296**, I-297
— southern Brazil
  — cities growing fast I-738
  — metropolitan areas I-737, I-738
  — Rio de Janeiro and Santos I-738
  — Santos Bay I-743
— Southern California I-388
— Southern Gulf of Mexico I-475
— Southern Portugal I-160–1
— Sri Lanka II-177, II-181
— of the Sundarbans II-153
  — scheduled castes/tribes II-153
— Tanzania II-85, II-88
— Tasmania II-653–4
— The Maldives II-206–7
  — demographic patterns II-206
— Torres Strait and Gulf of Papua II-599–600
— Turks and Caicos Islands I-591
— Tyrrhenian coasts I-275, I-279
— UAE II-25
— Vanuatu II-743
  — dual economic structure II-743
— Venezuela I-654–5
— Victoria, Australia II-666
  — indigenous people II-666
— Vietnam II-566
— west Taiwan coast II-505
— Western Australia II-696–7
— Yellow Sea, high densities inhibit conservation II-495
— Yemen II-25
population biomass, minimum viable in a marine reserve III-382–3
population density, Australia II-586
population growth rates, Oman and Yemen II-25
porpoises III-90
— *see also* Dall's porpoise ; harbo(u)r porpoises

port activities, Colombian Caribbean Coast I-673
port development
— Gulf of Guinea I-789
— Xiamen region II-519, II-521
  — deep harbours II-519
Portland, Maine *I-308*, I-313
Portsmouth Harbour *I-66*
— loss of salt marsh I-74
Portugal, Tagus and Sado estuaries I-151–65
— benthic vegetation I-159–60
— coastal erosion and landfill I-161
— the defined region I-153–4
— dredging I-161–2
— natural environmental variables and seasonality I-153–6
— offshore systems I-159
— populations affecting the area I-160–1
— protective measures I-162, **I-163**
— rural fishing I-161
— shallow marine habitats I-157–8
— upwelling effect on fisheries I-160
— urban and industrial effects I-162
Portuguese coastal waters I-153
Posidonia oceanica meadows I-274
power station effluent temperature, Hawaiian Islands II-805–6
power stations, conventional, North Sea coasts I-54
prawn culture
— Grand Bahama I-437
— the Sundarbans II-152–3, II-153–4
  — causing deterioration of coastal water bodies II-154
  — ecological crop loss II-154
prawn fishery
— Borneo, linked to mangroves II-368
— highest discard/catch ratios III-143
— large discard III-125
— northern New South Wales II-638
prawn trawling fishery
— Great Barrier Reef region II-623
  — effects on benthic communities II-623–4
— Torres Strait and Gulf of Papua II-604, II-606
Preah Sihanouk National Park, Cambodia II-571
precautionary action, principle of III-371
precautionary principle
— and equity III-157
— in Straddling Stocks Agreement III-158
precipitation
— Colombian Caribbean Coast I-665
— Gulf of Alaska, large freshwater flux I-378
— increasing III-188
— Norwegian coast I-19
primary production
— Adriatic Sea I-270
— Andaman and Nicobar Islands II-191
— Baltic Proper I-108
— Bay of Bengal coastal waters II-275, *II-276*
— by phytoplankton, usually N and P limited III-198
— Cambodian Sea II-573
— Campeche Sound I-474
— Canary Islands I-192, I-193
— Chagos Islands II-230
— coastal areas, Coral, Solomon and Bismark Seas region II-431–2
— coral reefs III-35
  — the Philippines II-409
— Gulf of Aden II-55
— Gulf of Cadiz I-172
— Gulf of Guinea I-779
— high, northern Gulf of Thailand II-302
— high, Sunda Shelf II-384–5
— important, Tasmania II-654
— inner New York Bight I-324, *I-325*
— Irish Sea, related to stratification of water masses and distribution of fronts I-89–90

— Izmir Bay I-244
— Malacca Strait II-334
— Malacca Straits II-312
— the Maldives II-202
— Norwegian coast, strong seasonality I-20
— and nutrient cycles, Gulf of Alaska I-379–80
   — nutrient source probably deep ocean I-379
— pelagic, Baltic Proper I-126
— percentage from the sea III-394
— Red Sea, low II-42
— stimulated by macronutrients III-259
— Tagus estuary I-157
— water column, Lesser Antilles I-633
— western English Channel I-67
primitive earth III-394
Prince William Sound *I-374*, I-375, I-380
— pollution of I-381
— southern, zooplankton community influenced by advection from the Alaskan Shelf I-380
— *see also* Exxon Valdez oil spill
principle components analysis, for grouping of disturbance types III-217
produced water
— containing mercury, Gulf of Thailand II-305
— effects of oil and chemicals in III-363
— from oil and gas production I-24, I-55
productivity
— Arabian Sea
   — and the Northeast Monsoon II-20
   — and the Southwest Monsoon II-19
— high, Gulf of Paria, Trinidad I-631
— and Nicaraguan Pacific coast fisheries I-536
— primary and secondary, of seagrasses III-2–3
— Somali Current LME II-66
PROKASIH (Clean Rivers Program), western Indonesia II-400
property rights, importance in regulation of the coast III-354
property rights institutions, sophisticated III-400
property rights regimes, and the oceans III-396–7
protected areas
— Borneo, mangroves in II-367–8
— Gulf of Guinea I-793
— Marshall Islands, resistance to II-788
— Turks and Caicos Islands **I-593**
Protected Natural Areas, Mexican Pacific I-497, **I-497**
protected species
— eastern Australian region II-643
— Madagascar II-17
— status and exploitation of, the Comoros II-249–50
   — coelacanths II-250
proton secretion III-35
— and nutrient uptake III-36
Providenciales, TCI
— pollution problem I-591
— surrounding waters, fishing and tourism I-591
— tourist destination I-591
Prymnesium parvum I-26
public health, Colombian Pacific coast I-680, **I-681**
Puerto Rico I-575–85
— coral bleaching and mortality events III-56
— geology I-577, *I-577*
— physical parameters I-577–81
— population development and land use:: effects from urban and industrial activities I-584
— shallow water and coastal habitats I-582–4
— shelf morphology and sediments I-581–2
— US National Estuarine Sanctuary I-584
Pulicat Lake, Palk Bay–Madras coast II-169, II-278
purse-seine fishery
— Madagascar II-122
— South Africa II-136–7
Puttalam Lagoon, Sri Lanka II-179

Puttalam Lagoon–Dutch Bay–Portugal Bay system, Sri Lanka, seagrass beds II-179
pyrite and evaporites, Laguna de Tacarigua I-652

Qatar *II-2*
— coral bleaching and mortality events III-47
queen conch I-590, I-592
Queen Conch fishery I-425
Queensland Plateau *II-426*, II-619
Queensland Trough *II-426*
Quirimba Archipelago, Mozambique
— fishing techniques II-107–8
— marine/coastal habitats II-102–4
— seagrass fishery, Montepuez Bay II-106
   — marema (basket traps), use of II-106
   — seagrass preferences II-102–4
— seagrass the most abundant habitat II-104
— source of productivity for South East Africa II-103–4
   — net primary productivity calculations II-103
   — unspoilt mixture of mangrove, seagrass and coral reefs II-103
quota systems, based on total allowable catch (TAC) III-161–2

radioactive discharges, Irish Sea I-94
radioactive oceanographic tracers, round the Azores I-204
radioactive pollution, Sea of Japan II-478–80
radioactive waste, stored in north Russia, cause for concern I-14–15
radionuclides
— ^{137}Cs, before and after Chernobyl I-116–17
— Black Sea, from Chernobyl disaster I-294
— from La Hague *I-14*, I-54–5, I-76, I-117
   — and Sellafield I-54–5, I-117, III-362
— in Greenland seas I-11
   — long-distance marine transport of I-13
— natural III-362
— not affecting small cetaceans III-96
— via Black Sea Water I-248–9
rainbow trout I-35
rainfall
— Coral, Solomon and Bismark Seas Region II-428
— Côte d'Ivoire I-808–9
— Dutch Antilles I-598
— Guinea I-799
— Gulf of Guinea I-776
   — Accra dry belt I-776
— Hawaiian Islands II-795
— Marshall Islands II-775
— Pearl River watershed II-551
— South West Pacific Islands II-709
— Torres Strait II-595
— Vanuatu II-739
— within the Fijian Group II-753
— *see also* precipitation
rainy seasons, Tanzania II-85
raised beaches II-349
Raleigh Bay *I-352*
RAMSAR Convention I-147, I-431, I-608, I-793, II-658, **III-334**, III-340
— TCI signed up to I-592
— Walvis Bay and Sandwich Bay wetlands, Namibia I-829
— Western Salt Ponds of Anegada (British Virgin Islands) accepted I-623
RAMSAR sites I-52, I-96, I-148
— Albufeira Lagoon I-162, *I-163*
— Inagua National Park, the Bahamas I-431
Ras al Hadd *II-18*, II-19
Ras Caseyr *II-64*
Ras Muhammed marine park *II-36*, II-44
Raso, Cape *I-152*, I-153
Ratak Submarine Ridge II-779
Recife de Fora Municipal Marine Park, Brazil I-728

recreational angling I-830, **I-830**, II-138–9
— conflict with commercial interests II-138–9
recreational boating, Dutch Antilles I-604
recreational fisheries, management of III-161
recreational fishing
— Australia II-587
— Great Barrier Reef region II-621
— — management of II-626
— Hawaiian Islands II-803
— New Caledonia II-729
— Tasmania II-655
— Victoria Province II-668
— Western Australia II-697
recreational industry, New York I-328
Red Sea II-35–45
— biogeographic position II-38
— coral bleaching and mortality events III-47
— endemic species II-38
— extent II-37
— geographical and historical setting II-37–8
— major shallow-water marine and coastal habitats II-40–2
— offshore systems II-42–3
— oil contamination throughout II-43
— population, urban and industrial activities II-43–4
— protective measures II-44
— receives continual supply of larvae for the Indian Ocean *II-36*, II-38
— seasonality, currents, natural environmental variables II-39–40
— turnover time II-40
Red Sea invaders, southeastern Mediterranean I-256–8
Red Sea rift II-37
Red Sea—Gulf of Aden water exchange II-39
red tides I-90, I-244, I-439, I-742, II-376, III-198, III-199, III-298
— Bay of Bengal II-275
— Benguela Current III-218
— and fish kills III-206
— Hong Kong II-542
— — impact on fish production II-543
— Izmir Bay I-238
— North Benguela region I-827
— Pearl River mouth II-553
— the Philippines II-418
— Rías Bajas I-141
— Straits of Malacca II-320
— Xiamen coastal waters II-522
— Yellow Sea II-495
Redfield ratios, P:N:Si, Arabian Gulf II-6
reed field, Liaohe Delta II-494
reef ecosystems, decline in abundance from poor fishing practices III-126
reef fish
— Marshall Islands II-778–9
— problem of size-selective fishing III-121
reef flats
— central Red Sea II-40, *II-40*
— Chagos Islands II-227
reef gleaning, Fiji Islands II-759
reef habitats, shallow, Tasmania II-652
reef and lagoonal water, French Polynesia II-816
reef systems, marine reserves, percentage adult population protected III-382
reefs
— artificial I-183
— Belize, buffered from urban pollution I-512
— deep reef habitats I-506
Refinería Panama storage tank rupture III-278
— long-term effects on physical structure of mangrove forest III-273, III-274
— mortality of subtidal corals III-273–4
Reflagging Agreement III-158, III-161
— requirements of III-160

regime shift, Gulf of Alaska shelf I-375
Regional Organisation for the Conservation of the Environment of the Red Sea and the Gulf of Aden (PERGS) II-59
Regional Organisation for the Protection of the Marine Environment (ROPME) II-3, II-29, II-44
Regional Seas Program III-340
— provision for protection against land-based pollution III-339
Relative Penis Size Index (RPSI) III-250
relative sea-level rise, and landform alteration, northern Gulf of Mexico I-446–7
relaxed eddy accumulation (REA) III-204
— application to particle measurement III-205
remineralisation processes, enhanced, the Bodden III-261
remittances, important in South Western Pacific Islands II-714
remote sensing
— applications of III-294
— Brazilian tropical coast *I-725*
— Coastal Zone Color Scanner information
— — Chagos Islands II-230
— — The Maldives II-202
— could aid seagrass research III-12
— in fisheries management III-160
— modern definition III-294
— provides large-scale synoptic data III-284
— satellite imaging of Chilean coast upwellings *I-705*, *I-706*
— and sea bottom types III-297
— of tropical coastal resource III-283–91
— — coral reef systems III-87–8
— — economic considerations III-288–9
— — mangroves III-285–6
— — sensors relevant to mapping tropical coastal zones III-284
— — tropical seagrass ecosystems III-286–7
— — types of data achievable III-284
— used to classify and quantify coastal marine habitats, Mauritius II-259
— *see also* AVHRR imagery
renewable energy, from the oceans III-304
reproductive disturbance
— in Baltic biota I-114
— in breeding colonies of seabirds, Southern California I-398
— from *Exxon Valdez* oil spill I-382
reproductive failure
— from organic chemicals III-95
— seabirds, and fish population collapse III-129
reptiles
— El Salvador I-549–50
— Great Barrier Reef II-618
reservoirs
— southern Spain, Guadiana and Guadalquivir basins I-175
— *see also* dams
residence time
— Baltic Sea water I-103
— fjord basin water I-21
residual currents, complex, Irish Sea I-85, *I-86*
resource depletion, the Philippines II-418
resource exploitation, French Polynesia II-819–20
resource management, USA potentially conflicting paradigms III-350
resource utilisation conflicts, Xiamen region **II-518**
resources
— non-renewable, South Africa II-142
— right of access to (South Africa) II-141
restingas
— Abrolhos Bank–Cabo Frio, Southern Brazil I-736
— Brazil I-727
— Santa Catarina State, southern Brazil I-737
retroflection eddies *see* eddies
Reunion *II-254*, II-255
— agriculture II-261
— artisanal fishing II-261–2
— ciguatera poisoning outbreaks II-261–2
— Fishing Reserves **II-265**, II-266

— steep volcanic surfaces support corals II-259
— tourism II-264
Revillagigedo Islands, Mexico I-487
— seasonal surges I-487
— some coral bleaching (1998) III-53–4
Rhine, River I-58–9
— Rhine Action Programme for rehabilitation of I-59
Rhine water, effects on southern North Sea I-46
Rhodos gyre I-236
Rías Bajas, Galicia I-137, I-138, I-141
Riau Archipelago, Sumatra II-311, II-337
— coral reefs II-314
— seagrass beds II-314, II-336
Riga, Gulf of *I-122*
right whales III-74, III-76, III-81
Rim Current, Black Sea I-287
Rincón region, Buenos Aires Province, significant biological activity I-755
ringed seals I-8, III-95
Rio de Janeiro metropolitan area
— industries and port I-738
— Jacarepaguá lagoon systems, receives industrial waste I-742
— pollution in I-741
— Guanabara Bay, sewage and industrial effluents I-741
Rio de la Plata basin system I-751
Rio de la Plata estuary *I-751*, I-759–61
— anthropogenic impacts I-760
— environmental gradients I-759
— formation of I-759
— human impact I-759–60
— impact of urban-industrial zone I-759
— nutrient discharge I-760
— pollutants I-760
— sectors of I-759
— state of knowledge I-760
— system characteristics I-759
Rio Declaration, on transgenerational responsibility III-360
river deltas, New Guinea coastline, support pristine mangrove forests II-597
river dolphins, India, DDT, PCBs and HCHs in II-452
river inflow
— Colombian Caribbean I-665
— Côte d'Ivoire I-807
— and run-off I-809
— Guinea I-799
— Gulf of Guinea I-776–7
river pollution, Russian Far East II-477
river run-off
— Bay of Bengal II-271, **II-272**, II-275
— lessened through irrigation withdrawals II-287
— and river impacts II-273
— Gulf of Thailand, freshwater discharge in addition to II-301
— in Malacca Strait II-333
river and wave material transport, east coast, Peninsular Malaysia II-582
rivers
— Chilean coast, various flow regimes I-704–5, **I-704**
— Gulf of Guinea coast, downstream effects of damming I-789, I-792
— influence on southern Spain I-169–70
— polluted, Venezuelan coastal zone I-645
— Southeast Asia, the most turbid II-306
Rocas atoll, Brazil I-723
— biological reserve I-724
rock lobster fishery II-56
— artisanal, Torres Strait II-602
— Australia II-586
— Gulf of Papua II-604
— South Africa II-137
— Tasmania II-655, II-656
— Victoria Province, Australia II-668
rock and surf angling II-138

rocky coastlines
— Irish Sea I-87–8
— communities of I-88
rocky reefs
— eastern Australian region II-634
— subtidal II-633
— El Salvador I-548–9
— Los Cóbanos the most extensive *I-546*, I-549, I-552
— habitats of, Australian Bight II-679
— Louisiana coast I-421
— Puerto Rico I-582–3
— Victoria Province, subtidal II-664
rocky and sandy seabeds, Canary Islands I-195
rocky shores
— Borneo II-666
— eastern Australian region II-633
— exposed, persistence of oil on III-276
— and headlands, east coast, Peninsular Malaysia II-349
— Malacca Strait II-336
— northern Argentina, macro and megafauna I-754
— Venezuela I-649–50
— algal zone I-649
— barnacle zone I-649
— *Littorina* zone I-649
— microhabitats I-649
— Victoria Province II-664
— western Sumatra II-387
Rodrigues *II-254*, II-255
— agriculture II-261
— artisanal fishing important II-262
— lagoons heavily silted II-259
— mangroves II-259
— population II-260
Rodrigues Bank *II-254*, II-255
Rompido Sand Cliffs *I-168*
Rudnaya River Valley, eastern Russia
— degradation/decay of ecosystems taking place II-485
— health of population requires improvement II-485
— polluted surface and groundwaters II-485
— a pre-crisis situation II-485
runoff
— agricultural, western Indonesia II-396
— annual, mainland Norway I-19–20
— and biota, Papua New Guinea coast II-597
— extensive, Tahiti and Moorea II-821
— increased by impervious surfaces I-445
— polluting Taiwan's coastal environment II-507, **II-510**
— silt-laden, increases turbidity I-446
— soil, Guangdong II-558
— urban, contains pollutants I-450
— *see also* soil runoff; stormwater runoff; surface runoff
Rupat Strait II-311
rural factors
— affecting the Aegean I-242–5
— affecting southern Spanish coastal zones I-174–6
rural land use
— impacts on western Indonesian seas II-395–6
— agriculture II-396
— deforestation II-395
— mining II-395–6
— *see also* land use
rural-urban migration I-567
— the Philippines II-412
— Tanzania II-92–3
Russians, whaling III-75

S:N ratio, Irish Sea I-87
Saba Bank *I-596*, I-598
— fishing by non-Saban fishermen I-603–4
— reefal areas I-600
Saba Island *I-596*, I-598, I-603

— coastal and marine habitats *I-600*
— sediments, due to erosion limiting factor for reef development I-601, *I-602*
Sabah *II-362*
— coastal habitats II-365
— illegal immigrants II-373
— population density II-373–4
— problem of Sipadan Island II-377
— reef destruction II-371, *II-372*
— SCUBA diving II-377
— Semporna Islands Park (proposed) II-377
    — current and potential threats to **II-378**
    — management plan for II-377
— Tunku Abdul Rahman Park, impacts on II-376
Sabellid worm colonies, Yemen II-51
sabkha
— described II-8
— Oman
    — at risk from rising sea levels II-28
    — Bar Al Hickman II-21
Saccostrea build-ups *II-76*, II-77
Sado estuary *I-152*, I-154
Saharan low pressure zone I-775
St Bees Head *I-84*
St Brandon Bank *II-254*, II-255
— lagoons with sandy floors II-257
— spur and groove regions with fish II-256–7
St Brandon Islands *II-254*, II-255
— artisanal fishing II-262
— seabirds and turtle nesting sites II-257
— temporary settlements II-260
St Brandon Sea *II-254*, II-257
St. Eustatius *I-596*, I-598, I-603
— coastal erosion I-602
— coastal and marine habitats *I-600*
— deep reef systems I-600
— manatee grass I-601
St. Helena Sound *I-352*, I-354
St. Lucia
— coral bleaching and mortality events III-56
— fuel-wood reforestation I-637
— Marine Islands Nature Reserve I-639
St Lucia system, KwaZulu-Natal, periodic hypersalinity in II-135
St. Maarten *I-586*, I-598, I-603
— coastal and marine habitats *I-600*
— decline in livestock I-601
— filling of saliñas and lagoons I-602
— manatee and turtle grasses I-601
— urbanization and tourist developments I-598
Saint Malo, Golfe de *I-66*
St Martin's Island, Bangladesh, focus for ecotourism II-289–90
saithe I-35, I-49
Sakhalinsky Bay *II-464*
Salalah *II-18*, II-19
saliñas, Dutch Antilles I-601
saline intrusion
— from over abstraction of groundwater II-26
— into aquifers, the Philippines II-414
— Marshall Islands II-781
— some Gulf of Aden coasts II-58
— Taiwan II-507
— to groundwater around Zanzibar II-93
salinisation
— Gulf of Mexico I-476
— of land II-584
salinity
— Aegean Sea I-236
— affecting Baltic biota I-123
— Arabian Gulf II-5
— Baltic Sea
    — and cod reproduction I-128
    — increase in I-111
— Bass Strait II-664
— in Bay of Bengal II-287
— Bien Dong Sea II-563, *II-563*
— Black Sea I-289
— Cambodian Sea II-571
— Campeche Sound I-469
— Canary Islands I-190–1, *I-191*, *I-192*
— changes at the deep salinity minimum, North Pacific III-182
— Chesapeake Bay I-339
— coastal waters, Xiamen region, China II-516
— Colombian Pacific I-679
— dry season, coastal zone of Gulf of Paria and Orinoco delta I-648
— El Salvador estuaries and open water I-547–8
— French Polynesia II-816
— Great Australian Bight II-677
— hypersalinity, Red Sea II-37–8
— impact on fisheries II-117
— Irish Sea I-86–7
— lagoon and open water, Vanuatu II-740
— lower near the Mississippi I-437
— Malacca Straits II-311–12
— Marshall Islands II-776
— near-bottom, Alaskan Gulf I-378–9
— negative anomaly, Alaskan coast III-183
— New York Bight I-323
— North Sea I-45, I-47
— ocean and lagoon, South Western Pacific Islands II-710
— off Guinea coast I-799
— Palk Bay II-166
— of Peruvian coastal waters I-691
— South China Sea II-348, II-552
— surface, Argentine Basin *I-753*, *I-753*
— surface water, Papua New Guinea coast II-596
— variation in estuaries, correlates with rainfall I-520
— variations, Côte d'Ivoire I-809
— Western Coral Sea II-614
— Yellow Sea II-490
salinity gradients
— Gulf of Suez II-39
— horizontal, Baltic Proper I-102
salinity stress I-364
salinity trend, North Pacific III-182
salinity–NO_3 relationship, northern Alaskan Gulf I-379
salmon
— Atlantic I-35, I-313
    — Baltic Sea I-110, I-111, I-128
    — farmed III-168
— Pacific
    — change in salmon survival rate in the open Pacific III-185
    — Okhotsk Sea II-467
    — possible biological effects of climatic change III-185–6
— sockeye, Alaska, accumulating PCBs and DDT I-381, I-383
salt diapirs I-442
salt extraction
— Gulf of Guinea I-786
— Southern Brazil I-738
salt flats, Lac Badana channel, Somalia II-77
salt pans
— Mauritius II-262
— Palk Bay, mangroves cleared for II-167–8
salt ponds
— Anguilla *I-619*, I-620
— British Virgin Islands I-620
— Cambodia II-574
— El Salvador I-553
salt tectonics, effects of I-442
salt wedge, Chesapeake's main tributaries I-337
salt-marsh plants, hybridisation of I-47
saltmarsh 24, I-49, I-358
— decreasing, Chesapeake Bay I-337

— eastern Australian region II-632
  — loss of II-639, II-640
— English Channel coasts I-68, I-69, *I-69*
— Gulf Coast, USA I-440
— Gulf of Maine I-311
— impact of dredging I-176
— lost in Cantabria and the Basque country I-144
— lost to infilling I-54
— Mar Chiquita coastal lagoon, Argentina I-755
— North Spanish coast I-139
— southern Spain I-170
  — Gulf of Cadiz I-177
— Sri Lanka II-180
— Tagus estuary I-158
— Tasmania II-653
— Victoria Province, Australia II-665
saltwater intrusion
— Belize I-512
— into coastal aquifers I-443, I-447
  — changes caused by I-448
Salvage Islands *I-186*, I-187
Salwa, Gulf of *II-2*
Samoa
— coastal modifications, serious impacts II-717
— coral reefs
  — cyclone physical destruction II-712
  — degraded II-712, II-714
— cyclone damage II-714
— population II-714
Samoa Group *II-706*, II-707, *II-709*
— agriculture and fisheries II-716
  — overfishing II-716
— coral reefs II-712
  — American Samoa II-712
  — Samoa II-712
— Marine Protected Areas (MPAs) II-720
— *see also* American Samoa
San Andrés and Providencia Archipelago
— coral reefs I-668
— strong wave energy I-666
— volcanic origin with reef developments I-667
San Jorge Gulf I-753
— crude oil production I-761
San Matias Gulf, fishing I-763–4
San Salvador Islands, mangroves stressed I-422
sand
— oolithic I-590
— white carbonate, Puerto Rico I-581
sand accumulation, Gulf of Mannar, Palk Strait and Palk Bay II-181
sand barriers, southern Baltic coast III-260
sand dunes I-170
— active, Brazilian tropical coast I-722
— Cabo Frio region, southern Brazil I-736
— embayments behind, nursery grounds I-326
— front edge recession I-429
— Great Australian Bight II-675
  — Yalata dunes II-678
— Mexican Pacific coast I-487
— removal of to enlarge beaches I-276
— scrub landscape, mammals in, Doñana National Park, southern Spain I-171–2
— with xerophytic grasses II-51
sand and gravel extraction
— east coast of England I-53
— El Salvador I-554
— English Channel I-74
  — impacts from I-77
— New York Harbor I-330
sand loss, from beaches I-176, I-177
sand mining I-604, I-786
— Anguilla I-622

— by suction, east coast, Peninsular Malaysia II-355
— causing beach erosion I-636
— the Comoros, forbidden II-248
— for construction, Israel I-258
— and coral mining, Marshall Islands II-782
— Côte d'Ivoire I-816
— and dredging, in the Bahamas I-428
— from beaches II-27
— Jakarta Bay II-391
— and lime production, Mauritius II-263
— Malacca Straits II-316
— the Maldives II-208
— the Mascarenes II-262
— offshore, Curaçao I-605
— Orissa, for heavy minerals II-279
— the Philippines II-415
— river, east coast, Peninsular Malaysia II-353
— Southern Brazil I-737, I-738
  — causing environmental degradation I-740, I-741
— Suva Reefs II-717
— Todos os Santos Bay, Brazil I-725
— Turks and Caicos Islands I-592
— Vanuatu II-744–5
sand spits/sand banks I-170, I-828–9
— east coast, Peninsular Malaysia II-349
— *see also* Kakinada sand spit, Godavari and Krishna deltaic coast
sand vegetation, Doñana National Park, southern Spain I-171
sand-eels I-49
sandstone/beach rock habitats, Sri Lanka II-180
sandy beaches
— Arabian Sea coast II-21
— the Azores I-207
— Borneo II-369–70
— Buenos Aires Province I-754
— Cabo Frio–Cabo de Santa Marta Grande, southern Brazil I-736
— east coast, Peninsular Malaysia II-349
— eastern Australian region II-633
— French channel coast I-69
— Gulf of Aden, deposits of ilmenite and rutile II-58
— Hawaiian Islands, colourful II-797
— high-energy, Yemeni and Somali coasts II-51
— loss of and beach nourishment I-176
— Malacca Strait II-337
— Marshall Islands II-777
— Mauritius II-258
— New York Bight I-325
— and oil spills III-269
— oil-contaminated III-270–1
— Oman II-21
— regeneration of causes serious impacts I-181
— replenished by wadis II-52
— southwestern Africa I-828
  — and rocky beaches I-829–30
— Taiwan II-502
— Venezuela I-648–9
  — dissipative beaches I-648–9
  — high energy beaches, zonation of I-649, *I-649*
— west coast of Malaysia II-311
— western Sumatra II-387
— Xiamen region, polluted II-519
— Yellow Sea II-490
— *see also* beach nourishment/replenishment
sandy bottom habitats I-548
sandy bottoms, southern Somali coast *II-70*, II-72, II-74
Sanganeb Atoll II-40
Santa Maria
— depauperate palagonitic tuff I-206–7
— limpet harvesting I-206
Santa Monica Bay *I-386*
Sarawak
— population II-374

— Pulau Bruit National Park II-368
— well managed coastal National Parks and Protected Areas II-376-7
sardine, horizontal migration of II-466-7
sardine fishery I-141, II-56
— beach seining II-26
— Northwest Pacific I-490
— Pacific I-396-7
— Peru I-693
— Portuguese I-160
— South Africa, collapse of II-137
— Venezuela I-653
sardinella fishery, southern Brazil I-739
Sardinia, 'Smeralda' Coast I-278, I-279
Sargasso Sea, defined I-223
Sargasso Sea and Bermuda I-221-31
— coastal erosion and landfill I-228-9
— effects from urban and industrial activities I-229
— major open ocean habitats I-224-7
— major shallow water marine and coastal habitats I-227-8
— populations I-228
— protective measures I-230
— seasonality, currents, natural environmental variables I-223-4
Sargassum community I-226
— displaced benthos I-226
Sargassum decurrens, distribution of, Australia II-695
Saronikos Gulf *I-234*, I-235-6, I-247
— industrial pollution I-245-6
— pollution in, affecting the plankton I-238
— seagrass beds and algae I-239
— water masses in I-237
satellite imagery, uses of III-287
satellite remote sensing III-394
— of the coastal ocean: water quality and algal blooms III-293-302
    — areas of application III-295
    — critical bottom depth III-297
    — semianalytical atmospheric radiation transfer models III-295
    — sensor measures upwelling radiance III-297
    — water quality parameters III-295-8
— coasts of III-294
— *see also* AVHRR imagery; remote sensing
satellite-based vessel monitoring systems III-160
Saudi Arabia *II-2*, *II-36*
— coral islands and patch reefs II-7-8
Saya de Malha Bank *II-254*, II-255
— sand, coral and green algae II-256
scallop fishery
— Argentina, collapse of I-764
— New Caledonia II-734
— Scotland I-90
scallop and Puelche oyster culture, Argentina I-764
scavengers, food subsidies from by-catch/discard/ and processing dumping III-127
scheduled castes/tribes
— Lakshadweep Islands II-194
— Sundarbans II-153
science and the seas III-394
scientific realism, in fisheries management III-155, III-156-7
Scilly Isles I-71
Scotian Shelf *I-311*
Scotland
— early attempts to determine effects of fishing on fish populations III-376
— legislation relating to the coastal and marine environment III-353
Scottish scallop fishery, Amnesic Shellfish Toxin contamination I-90
sea breezes, Borneo II-365
sea cliff retreat I-144, I-211
sea colour remote sensing III-294
— need for development III-298
— not as advanced as that for ocean colour III-300
Sea of Cortes Oceanic Province, Mexico I-488
— oceanic habitat I-488

sea cucumber fishery
— the Maldives II-212
— Sri Lanka II-183
sea cucumbers II-90, II-123, II-784
— exploitation of, Mozambique II-108
— Madagascar II-123, II-126
sea horses
— endangered, Palk Bay II-168
— Gulf of Mannar II-165
sea ice transport, of pollutants I-14
sea level
— anomalous rise, Alaskan coast, a propagating Kelvin wave III-84
— high seasonal oscillation, Bay of Bengal II-287
— history of, Brazilian tropical coast I-722
— rising III-188
— *see also* sea-level rise
Sea Moss, harvesting of I-635
sea otters, effected by *Exxon Valdez* oil spill III-274
sea salt extraction, southern Spain I-177
sea surface temperature (SST)
— Aegean Sea I-236
— American Samoa II-767, *II-767*
— Arabian Gulf and Straits of Hormuz II-4-5
— the Bahamas I-419
— Bay of Biscay I-14
— Black Sea I-289
— Brazilian tropical coast I-721
— Cambodian Sea II-572
— Canary Islands, spatial and temporal differences I-190*I-191*, *I-192*
— the Comoros II-245
— continental seas, western Indonesia II-384
— Dutch Antilles I-598
— east coast, Peninsular Malaysia II-348
— eastern Australian region II-631-2
— effect of Southwest Monsoon off Arabian coasts II-19
— French Polynesia II-816
— Great Australian Bight II-677
— Gulf of Guinea I-778
— Gulf of Thailand II-300
— higher, affecting tropical cyclones III-369
— increasing, central Gulf of Alaska III-182
— Indian Ocean and Southeast Asia (1998) *III-45*
    — 'hot spot migration III-44-6
— Irish Sea I-85-6
    — slight rise in I-86, *I-87*
— lagoon and open water, Vanuatu II-740
— long-term increase, Gulf of Guinea, linked to global warming? I-781
— Madagascar II-115
— Malacca Straits II-311
— Maldive Islands II-201
— New Caledonia II-725
— of the Nicaraguan Caribbean coast I-519-20
— off southern Angola *I-824*
— Somali coast II-66
— South China Sea II-407-8
— South China Sea coast II-552
— South Western Pacific Islands II-710
— Spanish north coast I-137
— Sri Lanka II-178
— sub-surface anomaly, Line-P, Northeast Pacific III-183-4
— Tasmania II-650-1
— varies with season, Yellow Sea II-490
— Venezuela
    — AVHRR images *I-647*
    — and upwellings I-646
— Victoria coastal waters II-664
— warmer, effects of III-195
— Western Coral Sea II-614
sea swell, North Spanish coast I-138
sea turtle conservation, Mozambique II-111

sea turtle fibropapilloma disease III-67
sea turtles 59–71, I-522, I-525, I-535, I-591, II-21, II-44, II-250, II-600, II-601, II-797
— Abrolhos National Park I-726
— Bahamas I-426
— breeding grounds, Chagos Islands II-229
— conservation priorities III-67–8
    — as flagship species for conservation III-68
    — international and regional levels III-68
    — national level III-67–8
— Coral, Solomon and Bismark Seas region II-431
— evolutionary history III-60
— Great Barrier Reef II-618
— Gulf of Guinea, status of I-780, **I-780**
— Gulf of Mexico I-473–4
— Hawaiian Islands II-799
— Java Sea II-364
— Lesser Antilles I-620, I-636
— life cycle III-60–1
    — delayed maturity III-61
    — pelagic migration drifting III-61
    — reproduction III-60–1
— living, biological status of 61–4
— and longline fisheries III-146
— Maldives II-205
— migrations, growth and population structure III-61
— modern research needs and tools III-64–5
    — information networks III-64–5
    — nesting beach studies III-64
    — use of PIT tags III-64
— Mozambique II-105
— nesting ground harvests difficult to control III-65
— the Philippines II-410
— protected by Indonesian Law II-400–1
— protected in El Salvador I-550
— Red Sea II-42
— Sri Lanka II-180
— Sunda Shelf II-400–1
— Tanzania II-87
— threats to species survival, historical and modern III-65–7
    — damage to nesting and foraging habitat III-66–7
    — direct harvest III-65
    — fisheries mortality III-65–6
— vulnerable to fisheries III-379
— Yucatan I-471
— *see also* named varieties
sea urchin harvest I-636–7
sea urchins
— effects on reefs, French Polynesia II-819, II-822
— Mauritius II-257–8
sea water temperatures, Turks and Caicos Islands I-589
sea waves, Brazilian tropical coast I-721
sea-bed activities, peaceful exploration and exploitation III-336, III-338
sea-bed current stress II-596
sea-bed exploration, the Philippines II-417
sea-bed and sediments, English Channel I-67, *I-68*, I-69
sea-floor spreading II-66
— Gulf of Aqaba II-37
sea-level rise
— adverse effect on coastal ecosystems III-98
— French Polynesia, effects of II-823
— Gulf of Guinea I-787
— impact on coastal habitats, northern Gulf of Mexico I-446, I-446–7
— Marshall Islands vulnerable to II-777
    — significant points II-777
— problem of Bay of, Bengal II-281, II-288, II-295
— puts reclaimed land at risk I-145
— South Western Pacific Islands, effects on II-716–17
— threat taken seriously in the Maldives II-214–15
— *see also* relative sea-level rise

sea-salt production, China II-494
sea-surface microlayer, enrichment of I-391–2
seabirds III-105–15
— Aegean Sea I-239–40
— American Samoa II-769
— Antarctic III-108
— "Arrecife Alacranes" national park, Yucatan I-471
— Australian Bight II-679–80
— the Azores
    — mercury in I-212
    — roosting/nesting sites I-207
    — some protection for I-214
— Chagos Islands, high diversity II-229
— El Salvador
    — nesting and migratory I-549
    — plentiful, "Colegio de las Aves" I-549
— English Channel, offshore, migrating and breeding species I-71
— evolution and history III-106–7
    — derivation III-106
    — Tertiary development III-106–7
— exploitation and conservation III-111–14
    — loss to growing populations III-111, III-114
    — ocean physics-foraging seabirds relationship, North Pacific III-113
    — seabirds and fisheries in the North Sea III-112
    — value to older cultures 111
— Faroese, concentrations of PCBs and DDT in **I-37**
— Fiji Islands II-758
    — threats to II-757
— food and fisheries III-111
— Great Australian Bight, threats to II-684–5
— Great Barrier Reef II-618, II-679–80
— habitat disturbance, Dutch Antilles I-604
— Hawaiian Islands II-798–9, **II-798**
    — threatened or endangered **II-799**
— and longline fisheries III-145–6
    — development from the Australian zone III-146
— of Madagascar II-120
    — breeding sites II-120
— the Maldives II-204–5
    — socio-economic importance of II-204
— Marshall Islands II-779
— modern III-107
    — main groups **III-106**
    — Miocene climax III-106
    — shorebirds/waders III-107
— movements and measurements III-107–11
    — breeding counts, "apparently occupied nest" III-109
    — coastal birds, fluctuating distribution III-109
    — complexities of breeding seasons III-109–11
    — distributions III-107
    — effects of ENSO events III-110
    — erratic productivity and mortality III-109
    — foraging ranges and feeding areas III-108
    — migrations III-107, III-110–11
    — seasonal fluctuations at sea III-109
— nesting and migratory, El Salvador I-549
— North Sea I-49
    — supported by fishing fleet discards I-51
— plentiful, El Salvador, "Colegio de las Aves" I-549
— reduced numbers of, Greenland I-8–9
— southern Baltic I-109
— Southern California
    — dramatic declines in I-398
    — effects of major impacts I-398
— and waders, Gulf of Aden coasts II-54
— *see also* fishing mortality; marine birds
seafood
— consumption, Torres Strait II-600, II-601
— health risks from, Southern California I-399–400
seagrass I-71, I-194, I-228, I-239, I-311, II-290

— Arabian Gulf II-7
— Australia II-582
    — major human-induced declines II-700, **II-700**
    — vulnerable to eutrophication and sedimentation II-585
— the Bahamas I-421
— Borneo II-370
— Carolinas I-361-2
    — and other rooted submersed aquatic vegetation I-358-9
    — remapping of I-366
— Coral, Solomon and Bismark Seas region II-430
— damaged by dredging and filling I-569
— distinguished from macroalgae, problem in remote sensing III-286
— east coast, Peninsular Malaysia II-348, II-349-50
    — threats to identified II-354-5, II-357
— eastern Australian region II-633
    — dieback, south-east Queensland II-637
— eastern sector, northern Gulf of Mexico I-438
— English Channel I-70
— epiphytes/epiphyte communities I-357-8
— Fiji Islands II-754-5
— Global Declines and Effects of Climate Change III-10-11
    — some losses documented III-10
— global status of III-1-16
    — declines and effects of climate change III-10-11
    — ecosystems services III-2-3
    — planning, management, policy, goals III-9, III-11
    — research priorities III-12
    — worldwide decline III-7-9
— Great Barrier Reef region II-615
— Guidelines for the Conservation and Restoration of in the USA and adjacent waters, a synopsis III-8
— Gulf of Aden II-53
— Gulf of Mannar, affected by trawling II-165
— Gulf of Thailand II-301
— Hawaiian Islands II-796, II-797
— high species diversity, Papua New Guinea and Torres Strait II-431
— impact on of oil and chemical spills I-450
— Jask and Char Bahar, Iran coast II-24
— Koh Kong Bay, Cambodia II-573
— lagoonal, Lakshadweep Islands II-194
— Malacca Strait II-336
— management possibilities, planning, policy and goals III-9, III-11
— Mar Chiquita coastal lagoon, Argentina I-755
— Marshall Islands II-779
— Mexican Pacific coast I-490
— North Sea, 1930s 'wasting disease' I-49
— Oman II-23
— Papua New Guinea coast II-598
— Red Sea II-42
— sediment accumulation and stabilization III-2
— Tasmania II-652
— The Maldives II-204
— Vanuatu II-740-1
— Victoria Province, Australia II-664-5
    — die-back, possibly due to catchment erosion II-667, II-668
    — epiphytic algae II-665
— wasting disease III-9, III-14
— Western Australia II-693, II-696
    — Kimberley coast II-694
— western Sumatra II-387
seagrass beds I-195, I-590
— Anguilla I-618, *I-619*
— Cambodia II-573, II-574
— Chesapeake Bay I-341
— Colombian Caribbean Coast I-668
— damaged by boat propellers III-9
— Dutch Antilles I-601
— El Salvador I-548
— fluctuations in salinity, water temperature and turbidity bad for I-441
— Johore Strait II-334
— in lagoons, Mauritius II-258
— Lesser Antilles, Trinidad and Tobago I-631
— little damaged by hurricanes III-2
— losses from low oxygen III-220
— Madagascar II-119
— Malacca Straits II-312, II-314
— New Caledonia II-726
— Nicaragua, Caribbean coast I-522
— northern Gulf of Mexico, high diversity of I-440-1, *I-441*
— nursery areas I-421, I-440-1
— nursery function III-3
— the Philippines II-410
    — conversion and utilization of II-413-14, **II-413**
— Quirimbas, increasing fishing pressure II-102
— role of, Coral, Solomon and Bismark Seas region II-430
— and sandy bottoms, southern Somali coast *II-70*, II-72, II-74, II-76
— the Seychelles II-236
— shelter function III-3
— South Western Pacific Islands II-711
— Sri Lanka II-179-80
— Tanzania II-8708
— Torres Strait II-598
    — seagrass abundance on reefs II-598
— Venezuela I-650
— West Florida Shelf I-438
— western Indonesia II-388
    — intertidal II-388
seagrass communities
— Australian Bight II-679
— Hong Kong, threatened II-541-2
seagrass dynamics, SPOT XS, Landsat TM and airborne methods all needed III-286-7
seagrass ecosystems
— affected by natural environmental impacts III-8
— divisions of animal community III-3
— nitrogen a rate-limiting factor III-3
— tropical, remote sensing of III-286-7
    — future challenges III-287
seagrass meadows *see* seagrass beds
seagulls, colonies dependent on groundfish trawling discards III-127
sealions I-756
seals I-834, II-797
— common I-109
    — French channel coast I-70
    — north east Irish coast I-91
— in the English Channel I-70-1
— Great Australian Bight II-680-1
— Northwest and Southwest Atlantic, effects of population recovery III-129
seamount fisheries, Emperor Seamounts II-802
seamounts
— the Comoros II-245
— fished and unfished, benthic biomass III-379
— Mascarene Region II-255, II-257
— offshore Tasmania II-653, II-659
— Vening Meinnesz Seamounts II-583
seasonality
— Belize I-503
— El Salvador I-547
— Jamaica I-561
— Nicaraguan Pacific coast I-534
— North Sea, temperature and salinity variability I-47
— Norwegian coastal ecosystem I-20, I-23
seawalls/shore structures, to protect the land, Hawaiian Islands II-805
seawater density, Gulf of Alaska III-181
seawater flooding, western Indonesia, carries contaminants and bacteria II-398
seaweed aquaculture, Fiji Islands II-754
seaweed farming III-167-8
— Tanzania II-91-2
seaweed harvesting, Norway I-26

seaweeds
— alien I-47
— Andaman and Nicobar Islands II-192
— brown and red, diversity decline, North Sea I-47
— Cadiz–Tarifa arc I-172
— cultivated, Yellow Sea II-494
— east coast of Taiwan II-508
— the Philippines, species diversity II-409
Secchi disk depth
— retrieved from satellite data III-296
— and water quality III-296
sediment accumulation, and stabilization, due to seagrasses III-2
sediment contamination, reduced in Gulf of Maine harbors I-317
sediment depletion, an increasing problem in coastal systems III-352
sediment drift I-47
sediment grain size, critical to permeability of oil III-269
sediment load
— high, western Indonesian rivers II-384, II-395
— south-east Queensland II-637
sediment mobility, from loss of aquatic vegetation III-259
sediment sinks, North Sea I-46
sediment supply
— Brazilian tropical coast I-722, I-728
— to Gulf of Guinea I-777
sediment transport
— alteration by coastal development, Western Australia II-699
— Bay of Bengal II-271–2, *II-272*
— into coastal zone from Orinoco River I-648
— patterns altered by harbour and reclamation engineering II-493
sedimentary basins, fault-controlled, cutting Somali coastline *I-67*, II-66
sedimentation
— affecting Andaman Sea coral reefs II-303
— between Rio and Santos I-742
— Borneo
    — a controlling factor in mangrove development II-367
    — high rates of, reducing coral cover and fish abundance II-366, *II-371*
    — a major concern II-375
— Brazilian tropical coast I-722
— can be serious in El Salvador I-552
— Chilka Lake II-279
— coastal, from deforestation, Australia II-666–7
— and contamination, near-shore environment, Belize I-508
— discharge into coastal waters, damage caused, Hawaiian Islands II-801–2
— excessive, Kuwait, and coral bleaching II-10
— Florida Keys, smothering coral I-408
— from logging, threat to Coral, Solomon and Bismark Seas region coral reefs II-436
— Great Barrier Reef region II-620–1
— Gulf of Mexico I-471
— high rates, western Indonesia II-384
— impact on Madagascan coral reefs II-123–4
— lagoonal, from mining, New Caledonia II-731
— Malacca Straits, a growing problem II-317
— Nicaraguan Pacific coast I-539
— Norwegian fjords I-21
— patterns in the Jiulongjiang estuary II-518
— Peruvian deltas I-696
— recent, Mexican coastal plain I-476
— reducing coral cover and fish abundance Borneo II-371, *II-371*
— Vanuatu, threatens some reefs II-742
— Western Australia, increased, carrying pollutants II-701
sedimentology, marine, Gulf of Guinea I-778
sediments
— accumulation of TBT in III-250
— Bay of Bengal II-273
— carbonates, West Florida shelf I-438
— containing agrochemicals, lagoons, French Polynesia II-821
— from rivers systems during the monsoon, Sri Lanka II-178
— Gulf of Aden, settle in deeper water II-53
— Gulf of Cadiz I-170, *I-171*
— high nutrient content I-493
— importance of availability of III-352
— influx to estuaries, El Salvador I-548
— Irish Sea bed I-89
— lagoonal, New Caledonia II-726
— marine
    — colonized by microorganisms III-261
    — trace metals in, Greenland I-9
— pulsed into Chesapeake Bay I-340–1
— shelf, Puerto Rico I-581–2
— size important in benthic community stratification II-665–6
— smothering coral polyps II-193
— Southern California Bight
    — contamination in I-392–4
    — quality assessed by toxicity studies I-396
— surface, Gulf of Maine, metal concentrations in I-315
— transport and deposition of III-352
— trapped by mangroves I-421
sediments loads, Fijian rivers II-758
sei whales III-82, III-82–3
— depletion of III-78
seine netting
— Montepuez Bay, Quirimba Island II-106
— Tanzania, destructive to reefs II-90
seismicity
— Coral, Solomon and Bismark Seas region II-429
— the Philippines II-408
— Vanuatu II-744
— Xiamen region II-516–17
— *see also* earthquake activity
seismology, sea level and island ages, Chagos Islands II-225–6
selenium (Se), detoxifying mercury in Greenland I-9
Sellafield, UK
— radioactive discharges to Irish Sea I-94
— radionuclides transported from I-11, I-13, *I-14*
sensors, satellite and airborne **III-284**
set-back lines, in coastal planning III-353
Setúbal embayment I-153
sewage
— affects coral reefs III-38
— in Bangladeshi rivers II-293
— Borneo
    — island treatment systems II-377–8
    — mainly reaches rivers untreated II-376
— causing problems in the Maldives II-212
— a disposal problem for small islands I-623
— domestic
    — and organic pollution, western Indonesia II-396
    — source of marine contamination, Coral, Solomon and Bismark Seas region II-440
— effluent discharged to deep injection wells I-429
— entering Aden Harbour II-58
— from Mexican towns, carried into Belize by currents I-512
— historically discharged contaminants from I-392
— impact of, Hawaiian Islands II-804, II-809
— inadequate facilities, Fiji Islands II-760–1
— Lesser Antilles I-638
— limited treatment, Gulf of Guinea I-784
— Madras, discharged into the estuaries II-169
— Malacca Strait, from the Sumatran coast II-340
— in Metro Manila II-416
— polluting Turkish Black Sea coast I-300–1, **I-301**
— Rio de la Plata estuary, effects of I-759
— seeps can lead to nutrient enrichment II-135–6
— and sludge discharges, Côte d'Ivoire I-817
— a threat to potable water supplies, Tanzania II-93
— Torres Strait and Gulf of Papua II-605
— treatment demanded in Canary Islands I-197
— untreated, problems of discharge to sea, Marshall Islands II-783

— and water use, the Bahamas I-429–30
sewage discharge
— direct to sea, east coast, Peninsular Malaysia II-351
— and high BOD, Malacca Straits II-317
— to groundwater lens, Bermuda I-228
— and treatment, Dutch Antilles I-606, I-608
— untreated
    — Gulf of Guinea I-789
    — to Irish Sea I-95
sewage disposal
— American Samoa II-770
— and deteriorating water quality, Jamaica I-567
— French Polynesia II-821–2
— into coastal waters, southeast India II-165
— Jamaica I-569
    — and deteriorating water quality I-567
— new scheme implemented, Hong Kong II-543
— a priority, New Caledonia II-733
— proper system needed for Lakshadweep Islands II-195
— Puerto Rico I-584
— Vanuatu, a concern II-745
sewage outfalls
— Southern California
    — contamination of sediments near I-393–4
    — improvement of condition of benthos near I-395–6
sewage pollution
— Cabo Frio–Cabo de Santa Marta Grande, southern Brazil I-741, I-742
— the Comoros II-251
— effects of, Gulf of Aden II-58
— impacting on coral reef building algae, Tanzania II-93
— Maputo Bay II-109
— Morrocoy National Park, Venezuela I-655
— of urban shorelines, Madagascar II-125
— Zanzibar, in nearshore waters II-93
sewage sludge
— disposed of in New York Bight I-327
    — ocean disposal ended I-330
— dumped at sea I-77
— dumping in Irish Sea I-96
sewage treatment
— Tasmania II-657
— Victoria Province, Australia II-669
    — facilities, Port Phillip Bay II-669
sewage treatment facilities
— lack of, Red Sea countries II-43
— Port Phillip Bay, Victoria Province, Australia II-669
sewage treatment plants
— poor design and maintenance I-638, **I-639**
— Tasmania, some improvement in II-657
sewage and waste management
— Australia II-586
— Great Barrier Reef Marine Park II-626
sewage wastes, reaching the Basque coast I-145
The Seychelles II-233–41
— coastal erosion and landfill II-239
— effects from urban and industrial activities II-239
— inner granitic islands II-235
    — Precambrian II-235
— major shallow water marine and coastal habitats II-236–7
— offshore systems II-237
— outer coralline islands and atolls II-235
    — high limestone islands II-235
— populations affecting the area II-237–8
— protective measures II-239–41
    — network of marine protected areas 239–40, **II-240**
    — ratification of international Conventions II-239, **II-240**
— rural factors II-238–9
— seasonality, currents, natural environmental variables II-235–6
— Seychelles National Land Use Plan II-238–9
Seychelles Bank II-235

shallow water ecosystems and biotic communities, Aegean Sea I-237–40
— benthic fauna I-238–9
— fish fauna I-238
— marine mammals I-239
— marine plants I-239
— marine reptiles I-240
— plankton I-237–8
— sea birds I-239–40
— zooplankton communities I-237
shallow water habitats
— Southern China coastal waters II-552–4
    — intertidal zone II-552–3
— Yellow Sea II-490
shark capture, Southern Gulf of Mexico I-475
shark fins, from by-catch, Marshall Island II-784
shark fishery
— Belize I-509
— El Salvador I-550, I-553–4
— Hawaiian Islands II-803
— Mozambique, linked to possible dugong population collapse II-104
— Nicaraguan Pacific coast I-539, I-540
— Yemeni II-57
sharks II-363
— as by-catch III-147
— Chagos Islands, drop in numbers II-228–9
— in the English Channel I-70
Sharm el Sheik *II-36*, II-44
Shatt al Arab *II-2*
— low oxygen saturation II-5
— nitrates, silicates and phosphates II-6
Shaumagin Islands *I-374*
Shebeli alluvial plain *II-67*, *II-70*
Shebeli River, Somalia *II-64*, II-68
shelf benthos
— eastern Australian region II-633–4
— Great Barrier Reef region II-615–16
shelf reefs, Puerto Rico I-582, I-583–4
shelf seas, vulnerable III-295
shelf-edge atolls, Western Australia II-696
Shelikof Strait *I-374*
— lower, interannual variation in copepod biomass I-380
Shelikov Bay *II-464*
shell deposits, Sri Lanka II-181
shellfish, spring mass mortalities, Taiwan II-506
shellfish culture plots, TBT in II-319
shellfisheries
— Albermarle—Pamlico Estuarine System I-360
— Canary Islands I-195
— Carolinas coast
    — barrier islands/coastal rivers region I-361–2
    — Northern Carolina, closed due to high bacterial counts I-364–5
— closure of I-76
— decline in Chesapeake Bay I-346
— Gulf of Cadiz I-175–6
— Irish Sea I-91
— New York Bight I-329
    — overharvested I-327
— northern Spanish coastal communities I-144
— San José Gulf I-764
— shell fish farming, Adriatic Sea I-277
shells, as a form of currency II-436–7
shifting/slash-and-burn cultivation II-88, II-435, II-601
— Gulf of Guinea countries I-786
— Lesser Antilles I-635
shingle areas, Channel coast of England I-68
shipping
— Australia II-587
— eastern Australian region II-641
— in the English Channel I-72–3

— Great Barrier Reef lagoon, inner and outer routes II-622–3
— Gulf of Aden II-55
— impacts of Bay of Bengal II-279–80
— impacts on the Gulf II-13
— intensive activity, South China Sea II-559
— Malacca Strait II-339
— Malacca Straits II-316–17, *II-316*
— and marine pollution III-336
    — inputs of contaminants from III-364–5
    — LOS legal framework III-333–4, III-336
— New Caledonia II-728
— New York Bight I-323
— offshore accidents and impacts
    — Lesser Antilles I-638
    — New Caledonia II-733–4
    — Tasmania II-657–8
— and offshore impacts
    — Belize I-512
    — El Salvador I-556
— potential source of marine pollution, Vietnam II-566
— and seaports, Nicaraguan Pacific coast I-540
— South Western Pacific Islands II-718
— Torres Strait II-605
— Victoria coast II-669
shipping accidents, potential for, the Maldives II-213
ships, dismantling of, potential pollutants III-339
shipwrecks, Madagascan coast II-125
shoalgrass, Carolinas I-358
shore birds
— coastal wetlands of Argentina I-755–6
— Colombian Pacific Coast I-680
— migratory
    — east coast, Peninsular Malaysia II-350
    — Hawaiian Islands II-798
shoreline erosion II-193
Shoreline Management Plans (SMPs) I-59
— England and Wales III-351–2
shorelines, effects of sea level rise III-188
shrimp aquaculture
— Colombian Caribbean Coast I-672
— developing at the expense of mangroves, around Beira II-109
— east coast of India
    — collapse of II-278
    — disease problem II-280
— El Salvador I-553
— environmental consequences cause great concern II-394
— extensive, Bangladesh II-292
— Guinea I-801
— Gulf of Mannar II-166
— Gulf of Thailand II-306
— Malacca Straits II-314
— mangrove clearance for, Rufiji Delta, Tanzania II-89, II-91
— Mexico I-492, I-493
    — extensive aquaculture I-494
— New Caledonia II-732, II-733
— a pollution source, Yellow Sea II-494
— Sri Lanka
    — damaging mangroves II-179
    — major pollution source II-184
— western Indonesia II-394
shrimp farming III-169, *III-170*, III-368
— Asia and Latin America III-368
— Belize I-511
— Cambodia II-576–7
— Colombian Pacific Coast I-683–4
— mangrove clearance for, Borneo II-368–9, *II-368*
— marine, Venezuela I-653–4
— Nicaraguan Caribbean coast I-526
— Nicaraguan Pacific coast I-535, I-538, I-539–40
shrimp fisheries II-27
— Arabian Gulf II-7, II-8, II-11

— Argentina I-763
— Brunei II-372
— Colombian Caribbean Coast I-670
— Côte d'Ivoire I-815
— deep Water, Fiji II-758
— Greenland I-8
— Guinea I-801
— Gulf of Guinea I-787
— Gulf of Thailand II-306
— highest discard/catch ratios III-143
— incidental capture of sea turtles III-66
— large discard III-125
— Madagascar
    — increasingly regulated II-126
    — more productive to the west II-117
    — substantial by-catch II-121
    — variable yields II-121
— Mexican Pacific coast I-490, I-493
    — artisanal I-490–1
— Morecambe Bay I-91
— Nicaraguan Caribbean coast I-524
— Nicaraguan Pacific coast
    — fishing gear not optimally designed I-539
    — penaeid and white shrimp I-534
— northwest Pacific Ocean, high discard rate III-137
— Pamlico Sound I-362–3
— Sofala Bank, Mozambique, changes in II-109
— South Carolina I-363
— southeast coast, Brazil I-739
— Southern Gulf of Mexico I-474
— and water temperature, Alaska I-377
shrimp nurseries I-488
shrimp ponds, western Indonesia II-388
shrimp trawling
— Gulf of Suez II-43
— Norway I-25
shrimp viral diseases III-226
shrimps, amphidromous and diadromous, El Salvador I-550–1
Si:N and Si:P ratios, decline in, northern Gulf of Mexico I-449
Siberian High I-376
— dominant over Black Sea in winter I-287
Siberian High Pressure Core II-465
Sicily, fishing ports I-279
Sierra Leone—Guinea Plateau, continental shelf I-775
Sierra Nevada de Santa Marta, Colombia I-672
Sikhote-Alin, eastern Russia, endemic species II-481
silicates
— Baltic Proper I-106
— indicators of eutrophication in Baltic Proper I-108
sills, associated with fjords I-19
Sinai Peninsula *II-36*
— mangroves II-41, *II-41*
Singapore *II-310*
— coastal modification II-317, II-339
— coral reefs, stressed II-314
— decline in fish catch II-320
— habitat loss severe II-334
— heavy metal pollution II-341
    — Keppel harbour II-318
— improvements in water quality II-341
— industrial centre II-339
— no regulation framework on the environment II-323–4
— Port, provides all major port services II-317
— Prevention of Pollution at Sea Act (1991) II-324
— regulation of land-based pollution II-324
— and sustainable development II-342
Singapore Strait *II-310*, II-311, II-333, II-334
— and South China Sea II-334
Singapore, Thailand, Vietnam, coral bleaching and mortality events III-52
SIORJI growth triangle II-337

Sites of Special Scientific Interest (SSSIs) I-96
Sitka Eddy I-375
Skagerrak *I-100*
Skomer Island *I-84*, I-96
Slovenia and Croatia, special nature reserves I-280
sludge dumping II-543
— New York Bight I-324
— Western Australia II-700
slumping, fjords I-21
small cetaceans III-89–103
— adressing the issues III-99–100
— by-catch in coastal gillnets III-94–5
— classification III-90–2
    — *odontoceti* (dolphins and porpoises) III-90
— and environmental change III-96–8
— human impacts III-92–8
    — direct and indirect catches III-92–5, III-99
    — pollution III-95–6
— populations in a fairly healthy state III-98
— vulnerable to bioaccumulation of toxins III-98
Small Island Developing States
— development problems II-265
— Federal and Islamic Republic of the Comoros II-247
— Sustainable Development of, Conference II-216
Snake Cays, fringing reefs I-505, *I-505*
social choice theory, conventional III-398
Socotra II-49, *II-49*, *II-64*
— artisanal fishery II-56
— coral communities II-52
— effective community management of traditional fisheries II-60
— no regulation on waste disposal II-58
— seabird and raptor nesting site II-54
— seagrass beds II-53
— terrestrial diversity II-49
— use of beach coral debris II-58
Socotra Archipelago II-49
— mangrove forests II-53
— to become a Biosphere Reserve? II-60
Socotra Eddy/Gyre *II-18*, *II-48*, II-50, II-66
sodium cyanide, fish asphyxiant III-123
soft bottom habitats, Malacca Strait II-337
soft corals II-742
soft shores
— eastern Australian region II-633
— Hong Kong II-538
soft-bottom benthos, Western Australia II-695
soft-bottom communities
— Baltic Sea I-126
— Spanish north coast I-139–40
— Tahiti II-819
— Tyrrhenian Sea I-274
soft-bottom habitats
— Southern California Bight I-388
— Taiwan Strait II-502–3
soft-bottom systems
— Carolinas coast I-353
    — diverse communities I-362
soft-sediment communities, Bahamas Archipelago **I-420**
soft-sediment habitats
— Tasmania II-652
— Victoria Province, Australia II-665–6
soil degradation, Lesser Antilles I-635
soil erosion I-142–3, I-174–5
— Australia II-585, II-637
— due to prawn culture II-153
— Fiji II-715, II-758
— Hawaiian Islands II-801–2
— high islands, French Polynesia II-821
— Jamaica I-568
— Mascarenes II-261, II-262
— northern end of Sumatra II-339

— a problem in China II-491
— round Sea of Japan II-475
— sediments raising river beds II-558
— though agriculture, the Philippines II-412
soil runoff, many causes, American Samoa II-770
solar radiation
— and heating of the Baltic Sea I-123–4
— Nicaragua **I-519**
— reduced by Gulf War smoke II-11
— southern Spain I-169
sole I-92, I-162
Solent *I-66*, I-72
solid waste disposal
— American Samoa, improvement in II-770
— inappropriate, Argentina I-765
— Marshall Islands *II-782*, II-783
— not understood, Cambodia II-576
solid wastes
— Côte d'Ivoire I-817
— dumped and burnt II-58, II-262
— dumping of
    — Bangladesh II-293–4
    — municipal, Dutch Antilles I-606–7
— poor disposal of
    — Fiji II-761
    — Vanuatu II-745
— a problem
    — Gulf of Guinea coast I-788–9
    — in the Maldives II-212
    — Metro Manila II-416
solitary waves, coastal erosion and habitat destruction issues III-355–6
Solomon Islands *II-426*, II-427, **II-427**
— concerns about pesticide pollution II-438
— coral reefs in good conditions II-431
— environmental assessment of the only mine II-439
— impact of land use on coastal waters II-435–6
— level of social and economic development low II-434
— mangroves II-430
— plantation agriculture II-435
— population and demography II-433–4, **II-434**
— positive effects of earthquake damage to reefs II-429
— rainforest logging II-435
— small ecosytems knowledge II-435
— tuna fishery II-423
    — "Tuna 2000" policy for sustainability II-432
Solomon Sea *II-426*
Solomon Strait *II-426*
Solway Firth *I-84*
Somali Current, seasonal reversal of II-50, *II-64*, II-66
Somali Natural Resources Management Programme II-60
Somali Plain *II-64*
Somalia II-49
— focus on peace and socio-economic considerations II-60
— Indian Ocean Coast of II-63–82
    — effects from human activities and protective measures II-80
    — Late Pleistocene to present-day event sequence II-72
    — major shallow water marine and coastal habitats 72–7, *II-78*
    — natural environmental parameters II-65–6
    — population and natural resources II-77–80
    — present day features of the coastal zone II-68–72
    — structural framework II-66–7
— location and extent II-65
— modern setting of the coast II-72
— poverty of II-56
— very low level of urbanisation and industrialisation II-58
Songo-Songo Archipelago, Tanzania, reef species richness differences II-86
Soudan Bank *II-254*, II-255
South Aegean volcano arc I-235
South Africa II-133–44

— the coastline II-135
— estuaries II-135–6
— management structures and policy II-143
— pollution and environmental quality II-142–3
— resources use II-136–42
South American Plate I-701
south Atlantic anticyclone I-823
South Atlantic Central Water
— Gulf of Guinea I-778
— South Brazil Bight I-735
South Atlantic Current I-824
South Atlantic high pressure cell I-721
South Atlantic subtropical gyre I-823
South Australia Current *II-674*
South Brazil Bight I-734–5
— bottom thermal front 734
— Inner Shelf, Middle Shelf and Outer shelf water bodies I-734–5
South Caicos, fishing centre I-591
South China Sea *II-310*, II-334, *II-346*, II-357, *II-362*, II-363, *II-382*, II-571
— coastal areas west of Hong Kong II-551
— fish landings related to the monsoon II-372
— marine pollution II-354
— monsoon generation II-365
— presence of thermocline and halocline II-348
— salinity II-348, II-552
— sea surface temperature (SST) II-407–8, II-552
— surface currents II-407
— surface water patterns II-366, *II-366*
— typhoon tracks II-365, *II-366*
— water quality **II-350**
South China Sea Current II-552
South China Sea warm current II-516
South East Anatolia Project(Turkey), may deprive Gulf of river flow and nutrients II-9, II-13
South East Asia, mangrove loss III-21
South East Asian—Great Barrier Reef Region, Indo-Pacific marine species diversity II-753
South East Queensland, major habitats II-634
South Equatorial Current I-598, *II-84*, II-101, *II-101*, II-254, II-255, II-428, *II-581*, II-614, II-619, *II-724*, II-767, II-775
— dominant round Madagascar *II-114*, II-115
— seed stock derives from further East II-117
— the Seychelles lie within II-236
South Florida Slope *I-436*
South Korea
— increase in agrochemical use II-493
— industrialization and urbanization II-492
South Pacific Central Water II-631
south Pacific eastern margin ecosystem I-705–7
— Antarctic ecosystem I-706
— central south Pacific gyre ecosystem
 — islands/archipelago ecosystems I-706
 — pelagic oceanic ecosystem I-706
— coastal ecosystem I-706
— sub-Antarctic ecosystem I-706
South Pacific Forum Fisheries Agency II-442, III-160
South Pacific Marine Region II-739
South Pacific Regional Environmental Program II-442, II-824
South Pacific subtropical gyre II-614
south Tasmanian bioregion, greater number of endemics II-652
South West Pacific Islands Region II-705–22
— aid projects short-lived II-720
— biogeography II-707–9
— coastal erosion and landfill II-716–18
— effects of urban and industrial activities II-718–19
— island groups *II-706*
— major shallow water marine and coastal habitats II-710–12
— new environmental initiatives at community level II-720
— offshore systems II-712–13
— protective measures II-719–20

— evaluation of marine environmental protection measures II-720
— marine protected areas II-719–20
— modern conservation practices II-719
— rural factors II-714–16
— seasonality, currents, natural environmental variables II-709–10
South Western Atlantic burrowing crab I-755
Southampton Water
— refinery discharges I-75
— vegetation loss I-74
Southeast Asia, coral bleaching and mortality events III-49–52
Southeast South American Shelf Marine Ecosystem I-749–71
— coastal erosion and landfill I-762–3
— effects from urban and industrial activities I-763–7
— human populations affecting the area I-758–62
— major shallow water marine and coastal habitats I-754–7
— offshore systems I-757–8
— protective measures I-767
— region defined I-751–3
— rural factors I-762
— seasonality, currents, natural environmental variables I-753–4
Southeast Trade Winds II-116, II-428, II-710, II-753
Southern Adriatic I-269
— beach-rocks habitat I-273
— central Oceanic community I-274
— surface circulation I-271
— water column divisions I-270–1
Southern Bluefin Tuna fishery, Australian Bight II-682, II-683
Southern Brazil Shelf I-735
Southern California I-385–404
— anthropogenic inputs
 — distribution and fate of I-390–4
 — and human contributions I-388–90
— biogeographic provinces and habitats I-387–8
— DDT contamination in I-390
— effects of anthropogenic activities on marine biota I-394–9
— geography and oceanography I-387
— human health concerns I-399–400
— monitoring and management of I-400
Southern California Bight *I-386*
— impaired water bodies I-392
— loss of wetlands affects fish nursery areas and migratory birds I-399
— variations in oceanic environment I-387
Southern China, Vietnam to Hong Kong II-549–60
— effects of urban and industrial activities II-558–9
— major shallow water biota II-552–4
— marine pollution parameters II-554–6
— populations affecting the area II-556–8
— protective measures II-559–60
 — closed fishing seasons II-560
 — local legislation II-559–60
— seasonality, currents, natural variables II-551–2
southern New South Wales, major habitats II-635
Southern Ocean Current *I-700*
Southern Oscillation I-702
Southern Pelagic Province of Australia II-675
southern right whale
— Australian Bight II-676, II-680, II-683
 — disturbance and threats to II-685
Southern Shark fishery, Australian Bight II-682, II-683
southeast Asian archipelago, a zone of megabiodiversity II-389
Southwest Monsoon drift-current II-552
Soya Current II-466
Spain, North Coast I-135–50
— coastal erosion and landfill I-144–5
— effects from urban and industrial activities I-145–7
— lies in Northeast Atlantic Shelf and Eastern Canaries Coastal provinces I-137
— major shallow water marine and coastal habitats I-139–40
— populations affecting the area I-142
— protected coastal areas *I-143*, I-148

— protective measures I-147–9
  — application of Coastal Law I-147–8
  — Law for the Conservation of Natural areas and the Wild Fauna and Flora I-148
  — preservation of marine habitats I-149
  — protected areas being degraded I-148
— region defined I-137–8
— rural factors I-142–4
— seasonality, currents and natural environmental variables I-138–9
Spain, North West, Basques originated whaling III-74
Spain, southern, Atlantic coast I-167–84
— climate I-169
— coastal erosion and landfill I-176–7
— effects from urban and industrial activities I-177–83
— major marine and coastal shallow water habitats I-170–2
— offshore systems I-172
— populations affecting the area I-172–4
— protection measures I-181–3
  — Guiding Plan for Use and Management of La Breña and Barbate saltmarshes I-182–3
  — law on Andalusian Territory Regulation I-182
  — programmes and actions affecting whole coast I-183
— rural factors I-174–6
— seasonality, currents and natural environmental variables I-169–70
Spartina anglica
— a hybrid I-47
— spread of I-68
spawning, herring and mackerel, North Sea I-49
spawning stock biomass (SSB)
— Irish Sea, decline in I-92
— percentage preservation III-382
— Southern Bluefin Tuna III-367
— to give maximum sustainable yield (MSS) III-157
Special Areas of Conservation (SACs) I-52, I-77
— marine I-96–7, *I-97*
Special Management Areas, Hawaiian Islands II-808–9
Special Protected Areas (SPAs) I-52, I-77, I-96, I-148, I-162
Specially Protected Natural Territories, eastern Russia II-481–2
— black fir—hardwood forest ecosystem II-482
— "Borisovskoye plato" natural preserve II-482
— "Kedrovaya pad" nature reserve II-482
— Lazovsky Reserve II-482
— Sikhote-Alin Biosphere Reserve II-481–2
species, diversity of functional groups buffers impacts of stressors III-227–8
species abundance, effects of oil spills on III-271
species abundance and distribution, effects of variation in climate, temperature and other factor I-67–8
species diversity
— and distribution, Faroes I-33
— high, northern Australia II-581
— macroalgae, high Carolinas coast I-357
species evenness, mine tailings, Island Copper Mine III-240–1
species extinction, coral reefs III-37
species richness
— of corals in the Chagos Islands II-224–5, II-227
— mine tailings
  — patterns emerging III-238–9
  — under impact and during recovery III-238
— Oman beaches II-21
— Red Sea II-38
species vulnerability, to oil spills III-271
sperm whales/whaling III-74, III-76
— Azores III-74
— numbers indeterminate III-81
spilled oil, persistence of on shores and its effects on biota III-267–81
— effects of oil on nearshore populations and communities III-271–5
— effects of treatments III-276–7
— influences of biota on oil persistence III-271
— oil persistence on shores III-268–71
— recovery of biota and the importance of persistent oil 1275–6

— synthesis III-277–8
  — common themes III-278
Spiny Lobster fishery I-424–5
— Hawaiian Islands II-803
— Nicaragua I-525
— rocky coasts, Arabian Sea II-27
— South Caicos I-591
  — vulnerability to recruitment overfishing I-592
— Turks and Caicos Islands I-592
sponge harvesting I-426, I-511
sponges II-633
— encrusting I-590
— Guinea I-800
— populations, Palk Bay II-167
sport fishing I-621, II-138, II-213, II-260, II-806
— a problem I-197
sprat, Baltic Sea I-110, I-128
Spratly Islands *II-362*, II-364
— disputed claims to II-364
— Layang Layang Island, SCUBA diving II-377
spur and groove structures/systems
— Chagos Island reefs II-227
— Mascarene Region II-256–7, II-258
— outer reef slopes, Tanzania II-85
squid I-90, II-137
squid fishery, Argentina I-763
^{90}Sr, levels of fallout decreased in Greenland seas I-11
Sri Lanka II-175–87, *II-270*
— coastal and marine shallow water habitats II-178–80
— coastal resources management II-185–6
— direct and indirect small cetacean catches III-93
— geographical setting II-177
— marine pollution II-184–5
— marine resource use and populations affecting the area II-181–3
— non-living resources II-180–1
— protected marine fish II-183
— regulatory route in coastal planning and management III-353
— rural and urban factors affecting the coastal environment II-183–4
— seasonality, currents, natural environmental variables II-177–8
  — climate and rainfall II-177
  — currents and tides II-178
— shipping activities and fishery harbours II-185
Sri Lanka Turtle Conservation Project II-180
SST *see* sea surface temperature
stagnation periods
— Baltic Proper I-106, I-108
  — and formation of hydrogen sulphide I-105–6, *I-105*
Standard Operation Procedures (SOP), for oil spill response, Malacca Straits II-322–3
State Offshore Island Seabird Sanctuaries (Hawaii) II-797
Steller sea lion, decline in III-98–9, III-99
Stono River Estuary *I-352*, I-354
storm surges I-631
— associated with typhoons III-190, *III-193–4*, *III-193*, *III-194*
— and estuarine flood risk, effects of climate change and sea level rise 193–4, III-190
— from cyclones, Fiji II-710
— from tsunami waves, Vanuatu II-744
— Mascarene Plateau II-256
— North Sea I-46
— return periods III-194, *III-195*
— Xiamen region II-517
storm waves, Victoria coast II-663–4
storm-water management, problems, Dutch Antilles I-601
stormwater runoff
— contributes nutrients and heavy metals to coastal waters, New South Wales II-642
— increase in I-445
— nutrients in II-586
— polluted
  — Tasmania II-657

— Victoria Province II-669
Straddling Stocks Agreement III-158
— important principles III-158–9
Straits of Malacca *see* Malacca Straits
strandflat I-19
Strangford Lough *I-84*, I-96
stratification
— of Caribbean waters I-579
— near-surface waters, Southern California Bight I-391
— north Pacific Ocean III-181
— North Sea I-46
— seasonal
— New York Bight I-323
— North Spanish coast I-139
— Tagus estuary I-153–4
stratospheric Quasi-Biennial Oscillation (QBO) I-224
stream channelization, impact of I-446
stresses
— causing coral reef deterioration I-563
— favours domination by smaller organisms III-227
— Florida Keys
— global I-409–10
— industrial I-409
— localized, mass mortality of *Diadema* I-410
— to the environment and reefs I-409
striped bass (rockfish) I-348
— juvenile habitat I-342–3
— spawning stock restored I-342
striped dolphin, hunted in Japanese waters III-91, III-94
sturgeon I-342
sub-Antarctic ecosystem I-706
Sub-Antarctic Water I-701–2, I-706, I-735, II-664
sub-cellular damage, in fish, from chlorinated hydrocarbons I-397
sub-Saharan drought, reduced flows of Gulf of Guinea rivers I-777
'sub-tropical underwater' I-666
Subarctic Boundary *III-180*, III-181
— represents abrupt change in stratification III-181
Subarctic Current *III-180*
submarine canyons *I-136*, I-137, I-170, I-667, I-701, I-799
— Australia II-583
— Bay of Bengal II-271
— and shelf valleys I-325–6, *I-326*
— Taiwan Strait II-501
— "Trou sans Fond" canyon, Côte d'Ivoire I-805, *I-806*, I-807, I-808
submarine mining, placer tin, causing high turbidity II-395
subsidence
— coastal
— northern Gulf of Mexico I-446
— Somali coast II-66, II-67
— Colombian Pacific coast, seismic I-682
— marginal Mesozoic, Somali coast II-66
— Sucre coast, Venezuela I-645
— through groundwater extraction, Taiwan's littoral zone II-507
subsidence phenomena, Po delta I-276
subsidies
— political attempt to create legitimacy III-155
— some necessary and justifiable III-155–6
"substituted industrialisation", Chile I-710
Subsurface Equatorial Water I-679
subtidal habitats
— Irish Sea
— hard substrates I-89
— soft substrates I-89
— rocky communities, Spanish north coast I-140
Subtropic Underwater I-579
Subtropical Convergence II-583, II-653
Subtropical Maximum Salinity Water II-255
subtropical mode water, formation of, Sargasso Sea I-223
Subtropical Surface Current *I-688*, I-690
subtropical underwater I-632
Subtropical Water Mass I-701

succession, pattern of altered by oil spills III-275
Sudan *II-36*
Suez Canal *I-254*, *II-36*
— and "Lessepian migration" II-38
— opened the Mediterranean to Red Sea migrants I-257–8
— *see also* Red Sea invaders
Suez, Gulf of *II-36*, II-37, II-39
— coastal pollution by oil II-43
Sulawesi Sea *II-362*, *II-382*
— coastal habitats II-365
sulphate, to the Sargasso I-229
Sulu Sea *II-362*
— circulation II-407
— coastal habitats II-365
— internal waves II-411
Sumatra
— accelerated natural erosion II-338
— central, alluvial coastal plain II-311
— east, peat swamps II-312
— maintenance of a mangrove buffer belt II-342
— mangroves II-312
— converted to aquaculture ponds II-314–15
— massive deforestation II-338
— North, traditional fisheries management II-341
— sources of industrial pollution II-317–18
— western, marine ecosystems of II-386–7
sunbelt development, northern New South Wales II-636
Sunda Shelf II-312, II-333
Sunda Shelf region II-383
— Asian floral and faunal realm II-389
— exposed during Pleistocene lowstands II-383
— river discharge on Indonesian portion of II-384, **II-385**
Sunda Strait *II-382*
Sundarban Biosphere Reserve II-147, II-156–7
— Project Tiger II-156, II-157
Sundarban ecosystem II-148, II-290
Sundarbans
— Bangladeshi *II-146*, II-147
— British occupation II-151
— cultural convergence II-152
— early settlers II-151
— a hostile environment for settlement II-151–2
— Indian *II-146*, II-147
— name derivation II-290
— natural protection by II-290
— a necessity for Bangladesh II-291
— salinity intrusion close to II-287
— spread of colonization II-152
— *see also* Deltaic Sundarbans
sunlight penetration, reduced by nuisance micro- and macro-algae III-219
superficial reefs, Brazil I-723
surf, high energy I-487,I-486
surf zone, southwestern Africa I-830
Surface Equatorial Current I-690
surface runoff
— and river flow, Texas I-437
— source of pollutants to the Southern Californian Bight I-390, **I-391**
— stormwater discharges, little regulation and control of I-390
surface temperature, mean global, rising III-188
surface waves, direction and velocity, Bay of Bengal II-148
surfaces, rainfall-impervious, effects on flood frequency/intensity I-341
surrogate surfaces, particle collection on III-204–5
surveillance, in fisheries management III-159–60
suspended particulate matter, Haifa Bay I-261
suspended sediments/solids
— estuarine waters, Perak and Johore II-338
— flowing into the Aegean I-242–3
— off southern Spanish coast I-175
sustainability
— adaptive management III-400

— in coastal management III-350, III-356
— Coral, Solomon and Bismark Seas region, threats to II-437–8
— of current and future practices III-360
— gauged against certain criteria III-360
— of human activities on marine ecosystems III-359–73
— requires precautionary approach to ecosystems management III-400
sustainable development I-465
— application to society III-350–1
— in coastal zones III-351, III-356
— definitions I-459–60
— elements of **I-459**
— Xiamen coastal waters, measures for II-531
sustainable energy, from the oceans III-304
Sustainable Fisheries Act (USA) I-451
sustainable governance, of the oceans III-397–402
— the deliberative process in governance III-397–8
— moving from public opinion to public judgement III-397
— use of "visions" III-397–8
— the deliberative process in governance two-tier social decision structure III-398, *III-398*
— integrated ecological–economic modelling III-398–9
— property rights regimes III-397
— conflict-resolution mechanisms III-399
— design principles III-399
— new III-399–401
— taxes and other economic incentives III-401–2
— ecological tax reform III-401
— shifting tax burden to ecological damage and consumption of non-renewables III-410–12
sustainable regional development, programs for, Mexico I-480
Suva Harbour, Fiji, pollution in II-719
Svalbard I-13, I-27
— fjords I-19
swamp forests
— Cambodia II-573
— Coral, Solomon and Bismark Seas region II-430–1
— Gulf of Mexico I-472
swamp lands, filled and canalised, no longer trap sediments II-136
swamps
— freshwater
— Coral, Solomon and Bismark Seas region II-430–1
— Gulf of Papua II-598
— Gulf of Mexico, Centla I-472
— Peru, 'Reserved Zone of Villa Swamps' I-692
Swatch of No Ground II-289
Sweden, offshore wind farm development III-309–10
swell waves I-190, I-618
— the Bahamas I-419
— Gulf of Aden II-50
— reaching Puerto Rico I-578
swine industry, Carolinas, a significant environmental threat I-366–7
synthetic drilling muds, impacting on marine benthic communities III-363
synthetic organic chemicals, in land-based pollution III-361–2
Szczecin Lagoon (Oder Haff) I-109
— fishery I-111
— metal pollution I-115

Tadjora, Golfe de II-49
— mangroves II-53
Tagus embayment, frontal boundary I-155
Tagus estuary *I-152*, I-153–4
— disposal of wastes in I-162, **I-162**
— effects of fertilizers and pesticides I-162
— flow patterns at the mouth I-154
— two distinct regions I-153
Tagus and Sado coastal waters I-155–6
Tagus and Sado coastal zone I-153
— protected areas I-162, *I-163*, **I-163**

Taiwan
— coral bleaching and mortality events III-50
— eastern II-508
Taiwan Current II-537, II-538
Taiwan Shoal *II-500*, II-501
Taiwan Strait II-499–512
— coastal erosion and landfill II-506–7
— effects from urban and industrial activities II-507–10
— major shallow water marine and coastal habitat II-502–3
— offshore systems II-503–5
— populations affecting the areas II-505
— protective measures II-510–11
— coastal zone management II-510–11
— instability of coastal policy II-510
— protection of coastal resources II-511
— rural factors II-506
— seasonality, currents, natural environmental variables II-501–2
Tangier Sound *I-336*
— decline of aquatic vegetation I-345, I-347
Tanker Safety and Pollution Prevention Conference III-337
Tanzania II-83–98
— coastal erosion II-93
— effects from urban and industrial activities II-92–3
— offshore systems II-88
— population II-88
— protected areas and integrated coastal management II-93–6
— rural factors II-88–92
— seasonality, currents, natural environmental variables II-85
— shallow water marine and coastal habitats II-85–8
— signatory to international conventions supporting ICZM **II-94**, II-95
— Southern, local community training and education II-95
Tanzanian Coastal Management Partnership II-95
tar
— on Aegean coasts I-248
— Bay of Bengal II-280
— contamination, Dutch Antilles beaches I-605
— on Israeli beaches I-262–3
— in the Sargasso Sea I-229
tar balls
— Belize I-512
— Gulf of Guinea beaches I-792
— Jamaica I-570
— Malacca Straits II-319
— Sri Lankan beaches II-185
Taranto, Gulf of *I-268*
Tasman Bay, destruction of coralline grounds III-126
Tasmania, catchment management policy II-655
Tasmanian Province II-649
Tasmanian region II-647–60
— coastal erosion and landfill II-655
— effects from urban and industrial activities II-655–8
— major shallow water marine and coastal habitats II-651–2
— offshore systems II-653
— populations affecting the area II-653–4
— protective measures II-658–9
— international conventions II-658
— Marine and Estuarine Protected Areas II-658–9
— state legislation II-658
— rural factors II-654–5
— seasonality, currents, natural variables II-649–51
Tasmanian Wilderness World Heritage Area II-653
Tatarsky Strait *II-464*, II-465, II-475
Taura Syndrome, spread by shrimp-eating birds? III-226
TBT *see* tributyltin (TBT)
TCI *see* Turks and Caicos Islands
Teacapán—Agua Brava lagoon system I-494–5, I-497
— adjacent agricultural area uses N and P fertilisers I-495
— estuarine system I-494
— mangroves I-494–5
^{99}Tc, increase in due in Greenland Seas from Sellafield I-11

technetium (Tc), conservative behaviour I-13
tectonic activity
— the Azores I-203, I-211
— extensional, Aegean Sea I-235
— northwestern Arabian Sea II-19, II-27
— off the Mexican Pacific coast I-487, I-488
Tehuantepec, Gulf of, fishery resources I-491
Tehuantepec winds I-488
'teleconnection', Pacific and Equatorial Atlantic, maybe? I-827
temperature, and the water column, Gulf of Alaska III-181
Terceira (Azores), destruction and creation of marshes I-207–9
Terminos Lagoon *I-468*
terraces, Somali coast II-72
terrestrial vegetation, Marshall Islands II-777–8
Texas Louisiana Shelf *I-436*
Thailand II-299
— butyltin compounds in sediments **II-452**
— mangrove loss, uncontrolled conversion to shrimp ponds III-21
— "shifting aquaculture" III-368
— shrimp farming III-175–6
    — move to more intensive methods III-176
    — regulations III-177, **III-177**
    — source of land for III-175–6
— species used in aquaculture III-171
Thailand, Gulf of II-297–308, *II-310*
— coastal erosion, land subsidence and sea-level rise II-307
— coastal habitats II-301–4
— defined II-299
— depletion through human intervention III-120
— effects from urban and industrial activities II-307
— geological description II-299
— offshore systems II-304–5
— physical oceanography II-299–300
— populations affecting the area II-306
    — rural factors II-306–7
— protective measures II-307
— ray and shark species reduced III-120
— seasonality, currents, natural environmental variables II-301
Thermaikos Gulf *I-234*, I-235, *I-237*, I-244
— increased phytoplankton abundance I-238
— organophosphorus pesticides present I-244
— river input I-244
— sewage pollution I-246
thermal pollution
— Curaçao I-605
— damaging to seagrass beds III-9
— discharges from desalination plants II-10, II-28
— Indonesian power plants II-390
thermal stratification
— around Bermuda I-225
— English Channel I-67
— summer
    — Irish Sea I-85
    — Spanish north coast I-140
thermal vents, deep sea, South Western Pacific Islands region II-712
thermal winds, Gulf of Aden II-50
thermohaline circulation, and climate change III-369–70, *III-370*
Thessaloniki *I-234*, I-244
— summer anoxia I-241
Third River (Main Outfall Drainage), Iraq *II-6*
— may seriously impact Gulf ecosystem II-9, II-13
— purpose of II-9
threshold hypothesis, for economic growth and welfare III-395
tidal amplitude
— Faroes I-33
— La Rance I-72
tidal barrages III-316–18
— construction will change the environment III-318
— double basin systems *318*, III-317
— electricity generation
    — ebb generation III-316–17, III-318

— flood generation III-317
— two-way generation III-317
— high costs of, but 21st century development likely III-319
— single basin schemes *317*, III-316
tidal bars *II-71*, II-76
tidal current generation III-318–19
— current and future prospects III-319
— SeaFlow Project III-319
— vertical axis and horizontal axis turbines III-318
    — problem of fixing III-319
tidal currents
— coastal waters, Xiamen region, China II-516
— Guinea coast I-799
— Puerto Rico I-580
    — Guayanilla Bay I-580
    — Mayagüez-Añasco Bay I-580
— strong
    — Irish Sea I-85
    — Yellow Sea II-490–1
— Taiwan Strait II-502
tidal energy III-304, III-315–19
— affects height of oil deposition III-269
— current and future prospects III-319
— harnessing the energy in tides III-316–19
— public perception of III-318
tidal flats
— clay–oil flocculation, reduces oil retention in fine sediments III-271
— vulnerable to oil spills III-271
— Xiamen region II-517–18
— Yellow Sea II-490
tidal marshes
— Cabo de Santa Marta Grande-Chui, southern Brazil I-737
— northwest Florida, impounded for mosquito control I-445–6
tidal power, electricity generation I-54
tidal range
— Baltic Sea I-124
— Galicia I-138-9
tidal regimes, effects of, Madagascar II-117
tidal residual, in homogeneous and stratified water III-188
tidal residual circulation, magnitude of III-188–90
tidal surges I-487, I-488, I-489
tidepools, with algae I-649
tides
— Chagos Islands II-225
    — water accumulates oxygen II-225
— Colombian Pacific Coast I-679
— and currents, affected by monsoons, Sri Lanka II-178
— diurnal and semi-diurnal, Puerto Rico I-580, I-581
— east coast, Peninsular Malaysia II-347
— extracting energy from III-315–16
    — Spring and Neap tides III-316
— harnessing the energy in III-316–19
    — tidal barrage methods III-316–18
    — tidal current generation III-318–19
— Indian Ocean II-66
— influences on III-316
— local tidal currents, Tanzania II-85
— Red Sea, annual, importance of II-39
— and tidal currents, effects of climate change and sea level rise III-188–90
Tierra del Fuego Island I-752–3, *I-753*
— whales I-756–7
Tierra del Fuego Province, coastal economy I-761–2
tiger reserves, India II-156, II-157
Tigris, River II-3
Tilapia, in aquaculture, Bahamas I-427
timber exploitation/harvesting
— Colombian Pacific Coast I-681
— problems from I-444
timber industry, Sakhalin, profitable but overharvesting II-469
TINRO Basin *II-464*, II-465

Tiran, Strait of *II-36*, II-37
titanium dioxide processing I-72
titanium—magnesium deposits, Yuzhnoprimorsky region II-480
Tivela mactroides, dissipative beaches I-648
Tobago *I-628*, *I-644*
Todos os Santos Bay, Brazil *I-720*
— mining of calcareous sand I-725
— Pinaunas Reef Environmental Protected Area I-728
Toliara region, Madagascar
— barrier reef II-116, II-118
    — signs of over-exploitation II-123
— changes to the Grand Récif, 1964–1996 II-123, II-124
— long swell II-116
— types of fishing and target species II-123
Tonga *II-706*, II-707, *II-709*
— agriculture and fisheries II-716
— at risk from sea-level rise II-716
— coastal modifications II-717
— coral reefs II-711–12
— Marine Protected Areas (MPAs) II-720
— population II-714
Tongatapu
— coral reefs II-711
    — degraded II-712
— Fanga'uta lagoon II-712
Tongue of the Ocean *I-416*, I-417, I-428
— V-shaped canyons I-422
Tordesillas, Treaty of (1494) I-2
Torres Strait *II-426*
— commercial use of marine resources II-600
— coral reefs II-597
— defined II-595
— environmental management II-608
— fisheries
    — artisanal II-602
    — management of II-606
    — problems of over-exploitation II-602
    — reef II-601
    — sedentary resources II-602
    — subsistence II-601
— high indigenous population II-600
— high seagrass species diversity II-431
— indigenous fishing rights protected II-584
— low level of development II-600
— population mainly on "Inner islands" II-599
— tidal circulation II-596
Torres Strait Baseline Study II-600
Torres Strait and the Gulf of Papua II-593–610
— coastal erosion and landfill II-603
— effects from urban and industrial activities II-603–5
— major shallow water marine and coastal habitats II-597–8
— offshore systems II-598–9
— populations affecting the area II-599–600
— protective measures II-605–8
    — environmental management II-607–8
    — fisheries management II-605–7
    — international agreements II-608
— rural factors II-600–3
— seasonality, currents and natural environmental variables II-595–7
Torres Strait Islands II-583
Torres Strait Treaty (Australia–Papua New Guinea) II-595, II-605–6
— Torres Strait Protected Zone (TSPZ) II-606
Torrey Canyon oil spill, increase in *F. vesiculosus* III-272, III-274, III-275
Total Allowable Catch (TAC)
— for exploited Baltic fish stocks I-112
— southwestern Africa I-836
— under CAP I-92
total suspended matter (TSM) III-297
tourism I-215, II-43–4
— Adriatic coasts I-277
— Alaska I-381, I-383

— Anguilla I-620, I-622
— Argentine coast I-758
— around the North Sea I-50, I-51
— Australia II-584
    — marine and coastal II-586
— awareness of, Côte d'Ivoire I-816
— the Bahamas, effects on the population I-423
— Belize I-511, I-514
    — demands on the coastal zone I-507
    — regulation of I-513
— Bermuda I-228
— British Virgin Islands I-620
— Canary Islands I-188–9, I-197
— coastal
    — development for unlikely to meet high standards, Mozambique II-108
    — Malacca Straits II-316, II-326
    — Xiamen region II-519, **II-520**, **II-521**
— and coastal pollution, Sri Lanka II-185
— Colombian Caribbean Coast I-673
— the Comoros II-251
— and coral reefs II-315
— coral reefs important for II-398
— east coast, Peninsular Malaysia II-353–4
— and economic value of coral reefs III-38, III-41
— English Channel coasts I-72
— Fiji Islands II-714, II-758
    — impacting on turtles II-757
— Florida Keys II-408
— French Polynesia, importance of II-823
— Godavari–Krishna delta II-172
— Great Australian Bight II-683–4
— Great Barrier Reef II-622
— Gulf of Maine and Georges Bank I-314
— Gulf of Mannar, affects coastal water quality II-166
— Hawaiian Islands II-797, II-806–7
    — boating, surfing and submarines II-807
    — development of Waikiki and shoreline erosion II-805
    — diving and snorkelling II-806
    — golf courses II-806
    — infrastructure II-806
    — sport fishing II-806
— impact of, Great Barrier Reef Marine Park II-626
— Irish Sea coast I-93
— Jamaica, demands on the environment I-567
— Lakshadweep Islands II-194
— and landscapes, west coast, Sea of Japan II-482–3
— Lesser Antilles I-637–8, **I-637**
— the Maldives
    — boating, fishing and other effects II-213
    — consideration of carrying capacity II-217
    — impacts from diving and snorkelling II-213
    — impacts from infrastructure II-212–13
    — many positive effects II-217
— Marshall Islands II-780
    — infrastructures and activities II-783
— Mauritius and Reunion, already well-developed II-264
— Mexican Pacific coast
    — development plans I-491
    — potential for I-491
— Namibia I-830
— New Caledonia, a developing sector II-732
— Nicaragua, Caribbean coast I-523
— North and South Carolina I-354
— Palk Bay area II-168
— potential, Nicaraguan Pacific coast I-535
— and recreation, Latin America I-462–3
— returning to Montserrat I-620
— round the Tyrrhenian Sea I-278–9
— the Seychelles II-238, II-240
    — Sainte Anne Marine National Park II-241

— small industry, Coral, Solomon and Bismark Seas region II-439
— southern Brazil I-737, I-738, I-739
    — destructive side of I-740
    — indiscriminate I-742
— Southern Gulf of Mexico, poorly developed I-475
— Southern Portugal I-161
    — pressure of, Gulf of Cadiz I-179, I-181
— southern Spain I-173, I-177
— southwestern Africa I-835
— Taiwan Strait coral reefs II-504
— Tanzania II-92
— Tonga II-714
— Turks and Caicos Islands I-591
— Tyrrhenian coast I-275
— upward trends in The Maldives II-206
— Venezuela I-654
tourism trends, Hawaiian Islands II-801
Townsville Trench *II-426*
toxic materials, accumulation in parts of Chesapeake Bay I-348
toxic residues, Peruvian coastal waters I-695
toxic waste
— deliberate dumping of II-59
— Indonesia, cradle-to-grave approach II-322
toxic waste trade, Nigeria I-788
toxicity
— in Argentine coast molluscs I-764–5
— of harmful algal blooms III-98
toxin bioaccumulation, from algal blooms III-199
toxins
— defensive III-39
— dinoflagellate III-40
trace elements
— atmospheric input into North Sea I-56
— Bay of Bengal II-274
trace metal pollution, the Philippines II-416
trace metals
— Faroes I-36–7
— Gulf of Maine I-314
— Hong Kong coastal waters II-544
— oysters and sediments, El Salvador I-555
— in sediments I-26
trade, Somalia II-80
trade agreement, affect fisheries management III-146
Trade Winds I-188, I-591, III-369
— Marshall Islands II-775
— Mexican Pacific coast I-488
— and the Northwestern African Upwelling I-190
— and the Puerto Rican wave climate I-577–8
— tropical coast of Brazil I-721
tragedy of the commons I-462
— and fishing management III-154
TRAIN-SEA-COAST III-325–7
— integrated management of coastal and marine areas III-325
— network III-326
    — capabilities of III-326
    — central unit responsibilities III-326
    — development units for specific training priorities III-326–7
    — range of training courses under development III-327
— UN/DOALOS (UN Division for Ocean Affairs and the Law of the Sea) III-325
    — programme of action III-325
TRAIN-X strategy (UN)
— main elements of III-324
— methodology III-325, III-329
— training networks III-324–5
— *see also* CC:TRAIN; TRAIN-SEA-COAST
trans-oceanic floating debris, even in Chagos Islands II-230
transboundary pollution
— affecting Mozambique II-109
— air pollution, Sea of Japan II-476
— oil spills in Malacca Straits II-325

— Xiamen region II-527–8
— Yellow Sea II-496
transboundary straddling stocks/species, the Philippines II-410
transgressive–regressive cycles, Holocene, Argentine coastlines I-762
Transkei coast
— artisanal and subsistence collectors II-140
— overexploitation of mussels II-141
transport, the Maldives, environmental concerns II-213–14
trap fisheries I-397
trawl fisheries
— eastern Bass Strait II-668
— restricted, Western Australia II-697
— southern Brazil I-739
— Venezuela I-653
trawl nets, by-catch excluders II-587
trawling
— affecting Black Sea biota I-290
— banned, by Indonesia II-320, II-338, II-341
— beam trawls, North Sea III-368
— bottom trawling
    — unselective III-124
    — very destructive III-379
— causing damage to the Norwegian continental shelf I-25
— destructive III-122–3
    — Gulf of Aden II-59
— detrimental effect on the English Channel I-76, I-77
— ecological impact of I-178
— and habitat destruction III-379–80
— North Sea I-51, *I-51*, I-52
— pelagic trawls more selective III-124
— restricted in southern Spain I-183
— Southern California Bight I-397
— trawls and dredges scour bottoms I-445
trawling exclusion zones, Gulf of Alaska I-383
treaties, traditional, substituted by global plans and programs of action III-344
Treaty of the Rio de la Plata and its Maritime Front I-463
tributyltin (TBT) I-72, I-95, II-12, II-339, II-449
— in Australia II-588, II-669, II-701
— in the Baltic I-114, I-131
— currently only from larger vessels III-364–5
— degradation of III-249–50
— and endocrine disrupter III-249
— environmental impacts of III-250–1
— high concentrations, Malacca Straits 319
— Hong Kong II-542
— and imposex I-25, I-36–7, I-55, I-75, I-210–11, III-364
— major antifoulant III-248
— a moiety III-248
— as part of free association paints III-248
— persistence of in the environment III-249–50
— in Prince William Sound I-381
— Suva Harbour, Fiji I-719, II-745
— Tagus and Sado estuaries I-158
Trinidad *I-628*
Trinidad and Tobago, estuarine conditions and turbid waters I-630
Triste, Gulf of *I-644*
— coastal retreat I-645
— contamination by oil, domestic and rural wastes I-645
— great industrial impact I-659
— high heavy metal levels I-659
Trobriand Trough *II-426*
trochus fishery
— Fiji, overexploited II-759
— New Caledonia II-730
— Torres Strait II-602
Trochus shell II-784, II-820
Tromelin seamount II-255, II-257, II-260
Trondheim *I-18*
— contaminated harbour I-27–8
trophic cascade model III-127

— cases of top-down controls in community structure III-128–9
trophic-magnification disturbances III-221–3
trophodynamic disturbances
— affect habitat supporting organisms III-221–2
— and habitat lost, significant portion of lost GDP III-222–3
Tropical Atlantic Central Water I-579
Tropical Cyclones I-223, I-224
— impact on the Sargasso Sea and Bermuda I-224
tropical habitats, Puerto Rico I-582–4
tropical lows, Puerto Rico I-578
Tropical South Pacific Oceanic Province, Mexico I-489
— oceanic habitat I-489
tropical storms I-488, I-489, I-617, I-631
— "Agnes" (1972), effects of I-341, I-346
— impacts on Chesapeake Bay I-340–1
— Xiamen region II-517
Tropical Surface Waters I-777
tropicalization, of Peruvian coastal waters I-692
troposphere, thermal stratification of, Canary Islands I-188
'trottoir', Tyrrhenian Sea I-273
"Trou sans Fond" canyon, Côte d'Ivoire I-805, *I-806*, I-807, I-808
trout, farming of III-168
Tsesis oil spill III-272
tsunamis II-744
— Colombian Pacific Coast I-679
— Coral, Solomon and Bismark Seas region II-429
— Fiji Islands II-753
Tubataha Reef *II-406*
tuna fisheries
— the Azores I-209
— Borneo II-373
— Canary Islands I-196
— Coral, Solomon and Bismark Seas region II-432, *II-433*
— French Polynesia II-819
— Gulf of Guinea I-787
— long line and purse-seine, Colombian Caribbean Coast I-670–1
— longline, Bay of Bengal II-277
— Madagascar II-121–2
— the Maldives II-210
— Marshall Islands II-784
— Mexico I-490
— Mozambique II-105, II-109
— Nicaraguan Pacific coast I-536
— the Philippines II-411
— recreational, Chagos Islands II-230
— South Western Pacific Islands II-718–19
— Southern Gulf of Mexico I-475
tuna–dolphin problem III-141–3
turbidity I-548
— affects kelp beds I-395
— changes in alter biological processes II-193
— coastal, Cameroon I-790
— Côte d'Ivoire I-809–10
— due to seagrass die-off III-14–15
— in fjords I-21
— harmful effects
— in estuaries I-365
— on habitats I-445
— high, inshore, northern Gulf of Mexico I-437
— Kuwait, caused by landfill II-9–10
— offshore, Guinea I-799
— a problem in Bermuda I-228–9
— Tagus estuary I-154
— of water, related to maximum depth of living reef II-397
— of water above seagrasses, Western Australia II-700
turbidity currents, responsible for Bahamian V-shaped canyons I-422
Turbo shell II-820
turbot, Baltic predator I-111
turbulence, thermally-induced or mechanically generated III-207
turbulent transport 202

Turkey
— Black Sea fishing industry I-292
— the southern Black Sea I-294–6
— characteristics and population of the coast I-295–6
— development of the coastal areas I-296
— land-based pollution I-296–303
— rural factors affecting the coast I-296
Turks Bank *I-588*
— reef areas I-590
Turks and Caicos Islands I-587–94
— artisanal and non-industrial uses of the coasts I-592
— coastal erosion and landfill I-592
— coastal habitat map III-289
— defined I-589
— major shallow water marine and coastal habitats I-589–91
— offshore systems I-591
— populations affecting the area I-591
— protective measures I-592–3
— regulations to protect local fisheries I-593
— rural areas I-591–2
— seasonality, currents, natural variables I-589
Turneffe Atoll (Island) *I-502*, I-504–5
turtle eggs
— collection now banned, east coast, Peninsular Malaysia II-355
— eaten, Socotra II-57
Turtle Excluder Devices (TEDs) II-127, III-66, III-141, III-143
turtle fisheries I-620
— Mexican Pacific coast I-491
turtle grass I-562, I-601, I-632
— die-off in Florida Bay, a trend for coastal waters in the new Millennium? III-14–16
turtle hatcheries, east coast, Peninsular Malaysia II-356
Turtle Island *II-406*
turtle nesting beaches II-24, II-54, II-350, II-377
— the Comoros II-250
— Mozambique II-105
— St Brandon Islands II-257
— St Martin's Island, Bangladesh II-289
turtle nesting sites I-550
— West Africa I-780
turtlegrass II-598
Tweed-Moreton Bioregion, Australia II-634
Twofold Region, Victoria Province II-663
typhoon shelters, Hong Kong, polluted *II-543*, II-544–5
typhoons II-483
— Cambodia II-571
— Marshall Islands II-776
— present and future effects III-190, III-193, *III-194*
— South China Sea II-552, *II-552*
— Xiamen region II-516, II-517, **II-517**
— Yellow Sea II-489, II-490
Tyrrhenian Sea *I-268*, I-271–2
— algal species of tropical origin I-282
— Atlantic Water in I-272
— Central-Southern, abyssal fauna I-274–5
— coastal erosion I-276–7
— coastline I-270
— deep waters I-272
— effects of urban and industrial activities I-278–9
— effects of winds I-272
— limits I-270
— nutrient availability I-271
— shipping activity intense I-275, I-279
— typical Mediterranean biocenoses I-273

UAE *see* United Arab Emirates
UK
— commitment to offshore wind energy III-310
— 'insensitive structures', potential for legal action III-354
— many laws relating to the coast III-353

— NERC models., southern North Sea I-50
— nutrient input into the English Channel I-74
UK government, White Paper, biodiversity issues in overseas territories I-623–4
UK Overseas Territories in the Northeast Caribbean: Anguilla, British Virgin Islands, Montserrat I-615–26
— Darwin Initiative funds I-623
— legislation and protective measures I-623–5
— prospects and prognoses I-625
UN Conference on Environment and Development (UNCED) III-343
— Agenda 21 III-158
— and the London Dumping Convention reviews III-338–9
— Preparatory Committee, process-oriented criteria III-346
UN Convention on Biodiversity I-147, I-162
UN Convention on the Law of the Sea (LOS: UNCLOS) *see* UNCLOS (UN Convention on the Law of the Sea)
UN Framework Convention on Climate Change *see* Climate Change Convention
UN/DOALOS (UN Division for Ocean Affairs and the Law of the Sea), and the TRAIN-SEA-COAST programme III-325–6
UNCHE III-343
UNCLOS (UN Convention on the Law of the Sea) I-56, I-147, III-158, III-332, **III-333**, III-336
— adopted by the Philippines II-408–10
— authorises littoral states to undertake enforcement measures II-324–5
— basis for international fisheries law, little detail on application III-158
— requirements of ships transitting Malacca Straits II-324
— *see also* Reflagging Agreement; Straddling Stocks Agreement
UNEP
— global conference on sewage III-345
— need to accurately assess program effectiveness III-346
— progress with global 'POPs' Convention III-363
— Regional Seas Convention for East Africa (Nairobi Convention) II-127
— Regional Seas Programme I-2, III-339, III-340
— — Kuwait Action Plan I-2
— — Mediterranean Action Plan (MAP) I-249
— — Red Sea Action Plan II-44
— — South Pacific Region I-2
Unguja Island II-85
unique environments, Patagonian coast, Argentina I-756–7
UNITAR (UN Institute for Training and Research)
— approach of I-327
— — country team approach I-327
— Climate Change Programme training packages III-328
— development of CC:TRAIN III-327
— regional partners III-327
United Arab Emirates *II-2*, *II-19*
— coral bleaching and mortality events III-47
— special protection proposed for Khawr Kalba II-31
— *see also* Fujeirah
United Joint Group of Experts on the Scientific Aspects of Marine Pollution (GESAMP), definition of pollution III-258
uplift, Venezuelan coast I-645
upper water column characteristics, the Maldives II-201–2
upwelling index, Alaskan coast I-378, I-380
upwellings II-614
— act as barrier to gene flow and marine organism distribution II-51
— Arabian Sea II-19, II-66
— — cool and nutrient rich II-19
— — open-ocean and coastal II-19
— — stimulate phytoplankton II-24
— Arabian Sea system II-50
— attractive to seabirds III-113
— Bay of Bengal II-274
— Benguela ecosystem II-135
— Cambodia II-571, II-572, II-574
— Chagos Islands II-230
— Colombian Caribbean Coast I-666
— causes special environmental conditions I-666
— recedes in rainy season I-666
— restrict coral formations I-668
— Colombian Pacific Coast I-679
— Côte d'Ivoire I-808, I-809
— equatorial and open ocean I-535
— Galician and Cantabrian coasts I-139, I-141
— Great Australian Bight, provide nutrients to surface waters II-676, II-677
— Guinea coast I-800
— Gulf of Aden II-50–1, II-54
— Gulf of Alaska I-376
— — from Ekman pumping I-380, I-381
— Gulf of Guinea I-781
— — central subsystem I-778–9
— — intensification of winter upwelling I-781
— — interannual variability in I-781–2
— intensity may reduce with global warming III-369
— Long Island and New Jersey coastlines I-324
— Madagascan coast II-115
— the Maldives II-202
— North Sea I-46
— Northern Benguela I-823
— and nutrient enrichment, Messina Strait I-281
— off Luzon, South China Sea II-411
— off Peru I-690
— — importance of in El Niño events I-691
— off West Greenland coast I-7
— Oman coast II-19, II-50, II-54
— Sea of Okhotsk II-468
— shelf-break area, Argentinian continental shelf I-754
— shelf-edge, Carolinas coast I-353
— Somali coast II-50, II-51
— southeast coast, southern Brazil I-733
— southern Baltic I-111
— Southern Gulf of Mexico I-475
— southern Portuguese coast I-153, I-155
— — effects on fisheries I-160
— southern Spanish coast I-172
— southwestern Africa
— — central Namibian region I-825–6
— — effects of remote forcing from the equatorial Atlantic I-825
— — interannual variability I-827
— — low-oxygenated bottom water, northern Namibia I-827
— — Lüderitz upwelling cell *I-822*, I-823, I-825, I-826–7
— — northern Namibian Region I-825
— — wind-driven I-824
— upwelling ecosystem, Chilean coast I-706
— Venezuelan coast I-646
— Vietnam II-564
— Yemeni coast II-54
uranium, in the Hoogly (Hugli) river II-278–9
urban development
— coastal, Dutch Antilles I-606
— Colombian Caribbean Coast I-673
— Côte d'Ivoire, encroaching on lagoons and estuaries I-817
— Hong Kong
— — pressure on local marine environment II-543
— — rapid habitat loss II-541–2
— poorly planned, impact of, Coral, Solomon and Bismark Seas region II-439–40
— Tasmania II-654
— — pressure on the marine environment II-657
— Torres Strait and Gulf of Papua II-605
— unplanned, southern Brazil I-740, I-742
— Victoria Province, Australia II-669, II-670
urban, environmental and health problems, the Maldives II-212
— sewage-related problems II-212
urban impacts
— Australia II-586
— Noumea, New Caledonia II-733

urban and industrial activities
— Gulf of Mexico I-477–8
  — artisanal and non-industrial uses I-477
  — industrial uses I-477
  — shipping and offshore accidents I-477–8
— Yellow Sea II-494–5
  — aquaculture and coastal industries II-494
  — oil and oil spills II-495
  — wastewater and solid waste discharges II-494–5
urban and industrial activities, effects of
— American Samoa
  — habitat loss II-770–1
  — industrial activities and impacts II-770–1
  — other urban activities and impacts II-770
— Argentine coast I-763–7
  — commercial fisheries I-763
— Bay of Bengal II-279–80
— British Virgin Islands I-623
— Cambodia II-576–7
— Colombian Caribbean Coast I-673
— Colombian Pacific Coast I-682–4
  — artisanal and non-industrial uses of the coast I-682–3
  — cities I-683
  — industrial uses of the coast I-683–4
— the Comoros II-251
— Côte d'Ivoire I-817–18
  — depletion and degradation of coastal habitats I-817–18
  — oil and gas exploration I-817
  — waste disposal I-817–18
— Dutch Antilles I-502–7
— east coast, Peninsular Malaysia II-353–4
— eastern Australian region II-639–42
  — cities II-640
  — coastal settlements II-640
  — commercial fisheries II-640–1
  — industrial uses II-640
  — mining and dredging II-641
  — ports and shipping II-641
  — regional issues II-641–2
— Fiji Islands II-760–1
— Great Australian Bight II-684–6
— Great Barrier Reef region II-622–4
— Gulf of Aden II-58–9
— Gulf of Guinea I-787–93
  — artisanal and non-artisanal coastal use I-787
  — cities I-788–9
  — dams I-789, I-792
  — industrial fishing I-787–8
  — onshore oil production I-788
  — shipping and offshore I-792–3
  — toxic waste trade I-788
— Gulf of Thailand II-307
— Hong Kong II-543–6
  — anthropogenic contamination widespread II-543
— Lesser Antilles, Trinidad and Tobago I-636–9
— Madagascar II-125
— the Maldives II-210–15
— Marshall Islands II-783–4
— the Mascarene Region II-262–5
— Mozambique II-108–9
— New Caledonia II-731–4
  — artisanal and industrial uses of the coast II-731–2
  — cities II-732–3
  — shipping and offshore accidents and impacts II-733–4
— the Philippines II-414–18
  — artisanal and non-industrial uses of the coast II-414–15
  — cities II-416–17
  — industrial uses of the coast II-415–16
— Sea of Okhotsk II-469–71
— the Seychelles II-239
— shipping and offshore accidents and impacts II-417–18
— pollution hot spots II-418
— South Western Pacific Islands II-718–19
— southwestern Africa I-835
— Taiwan Strait II-507–10
— Tanzania II-92–3
— Tasmania
  — artisanal and non-industrial coastal uses II-655
  — commercial usage of coastal resources II-655–6
— Torres Strait and Gulf of Papua II-603–5
  — artisanal and non-industrial coastal uses II-603
  — cities II-605
  — fisheries II-604–5
  — industrial uses II-603–4
  — shipping and offshore accidents, impacts II-605
— Vanuatu II-745
— Venezuela I-655–9
  — eastern coastline I-655–6
  — western coastline I-656–9
— Victorian Province, Australia II-667–9
  — commercial fisheries II-667–8
  — industrial uses of the coast II-669
  — recreational fisheries II-668
  — shipping and offshore accidents and impacts II-669
  — urban use (cities) II-669
— Western Guangdong coast II-558–9
  — artisanal and non-industrial uses II-558–9
  — cities II-559
  — industrial uses II-559
  — shipping and offshore accidents and impacts II-559
— western Indonesian II-396–9
urban, industrial and other activities, effects of, Australia II-586–9
urban pollution
— Borneo II-375
— Patagonia, and eutrophication I-765
urban sewage, a pollutant I-24
urban waste, a problem along entire Gulf of Guinea coast I-788–9
urbanisation
— adjoining an estuary, creates problems II-135–6
— of Carolinas coast I-365
— coastal and waste disposal, Tanzania II-92–3
— effects of, Hawaiian Islands II-804–7
— expansion allows debris to enter northern Gulf of Mexico I-450
— Florida Keys I-408
— Great Barrier Reef catchment II-622
— Pearl River delta II-554–5, II-558, II-559
— the Philippines
  — major industrial regions II-417
  — rapid, and informal settlements II-416
— rapid
  — east coast of India II-278
  — effects of I-449
— South Western Pacific Islands II-718
— Southern California I-388
— Vietnam, impact on the marine environment II-566
Uruguay, EcoPlata I-463
USA
— and adjacent waters, Guidelines for the Conservation and Restoration of seagrasses, a synopsis III-8
— Coastal Zone Management Act
  — CZM evaluation model III-347
  — independent evaluation of program effectiveness III-346–7
— eastern, atmospheric nitrogen pathway to watersheds III-201
— Endangered Species Act II-798
— marine reserves III-386
— property rights important in regulation of the coast III-354
— stringent environmental standards for uses of Kwajalein atoll II-786–7
user conflict
— conflict resolution over use of Aliwal Shoal, South Africa II-141, II-142
— from increased resource use II-142

Ushant *I-66*
USSR
— catch reports false, collusion with Japan III-77, III-86
— natives of eastern Siberia, need for whale meat III-83
UV radiation, increase in III-370

Vanuatu *II-706*, II-707, *II-708*, II-717, II-737–49
— agriculture II-714–15
— biogeography II-739
— coastal erosion and landfill II-744–5
— coral reefs II-711
   — cyclone damage II-711
— customary law and reef ownership II-741
— cyclone-prone II-710
— effects of urban and industrial activities II-745
— island groups II-739
— major shallow water marine and coastal habitats II-740–2
— Marine Protected Areas (MPAs) II-719
— offshore systems II-743
— population II-713
— population affecting the area II-743
— protective measures II-745–7
   — marine conservation II-745–6, **II-746**, **II-747**
   — Marine Protected Areas (MPAs) II-746–7
— rural factors II-743–4
— seagrasses II-711
— seasonality, currents, natural environmental variables II-739–40
— unsatisfied demand for fresh fish II-744
vaquita, Gulf of California
— endangered III-90
— incidental catch III-94
Vas Deferens Sequence Index (VDSI) III-250
Vedaranyam Wildlife Sanctuary, Palk Bay II-167
Venezuela I-643–61
— coastal erosion and landfill I-655
— the continental coastline I-645–6
— effects from urban and industrial activities I-655–9
— fisheries I-653–4
— major shallow water marine and coastal habitats I-648–53
— populations affecting the area I-654–5
Venezuela, Gulf of *I-644*, I-645, I-646
Venice, Gulf of *I-268*
— salinity I-271
Vening Meinesz Seamounts II-583
vermetid reefs *I-66*, I-255, I-256–7
vermin and insect infestations, Tanzanian coast II-88–9
vertebrates
— accumulation of TBT in III-250
— endangered, southern Brazil I-737
— Gulf of Aden II-54
Victoria Province, Australia II-661–71
— coastal erosion and landfill II-667
— effects from urban and industrial activities II-667–9
— effects of removal of native vegetation from catchments II-666–7
— Fisheries Management Plans II-670
— Gippsland Lakes, adjusting to changed environment II-667
— human populations affecting the area II-666
— major shallow water marine and coastal habitats II-664–6
— marine species, distribution patterns II-663
— offshore systems II-666
— problems in marine environment II-670
— protective measures II-670
   — endangered and vulnerable species II-670
— rural factors II-666–7
— seasonality, currents, natural variables II-663–4
Victorian Biodiversity Strategy II-670
Vietnam II-299
— coral bleaching and mortality events III-52
— mangroves lost to fishponds III-22, III-25
Vietnam and adjacent Bien Dong (South China Sea) II-561–8
— biodiversity II-564

— impacts from development II-566
— legislation II-566–7
   — Environmental Protection Law II-566–7
   — species in need of protection II-567, **II-567**
— marine and coastal habitats II-564–5
— physical parameters II-563–4
— regional setting II-546
— river systems and estuaries II-566
— water quality II-565–6
Virginian Sea *see* Middle Atlantic Bight
Vistula Lagoon I-109
— fishery I-111
volcanic islands, Mascarene region *see* Mauritius; Reunion; Rodrigues
volcanic pinnacle, off Tasmania II-583
volcanicity
— Coral, Solomon and Bismark Seas region II-429
— latent, American Samoa II-767
— the Philippines II-408
Volterra—Hjort formulation III-79
Vulnerability Index *see* Environmental Sensitivity Index (ESI)

Wadden Sea I-45, I-53
— black and white spots III-263
— decline in eelgrass beds III-7–8
— disappearance of red macroalgae I-48–9
— long-term changes from bottom trawling noted III-126
— suspended sediment I-46
— and *Zostera marina* III-6–7
   — decrease in area suitable for re-establishment III-7
   — large-scale decline III-6
wadi systems, show past erosional processes II-37
wadis
— Al Batinah II-27
— and development of khawrs II-23
— Gulf of Aden coast, flow from percolates into groundwater II-52
Wake Island II-775
— extinction of Wake Rail II-778
— war-time activities II-782
war-time effects, Marshall Islands II-782–3
warming events, El Niño, Northeast Pacific Ocean III-183–4, *III-183*
waste disposal I-634
— an environmental problem, Jamaica I-569
— Côte d'Ivoire I-817–18
— Faroes I-38–9
waste dumping
— English Channel I-77
   — regional, global and European regulation I-77, **I-78**
waste management
— difficulties of, Guinea I-801
— Vanuatu, by industry, minimal II-745
waste plastic pellets, dangers of, Arabian Gulf II-12
wastes
— disposal of, Bangladesh II-293–4
— domestic, dumpsite for, New Caledonia II-733
— dumping of
   — at sea III-365–6
   — Suva Harbour, Fiji II-719
   — in Xiamen coastal waters II-523–4
— Guangdong Province **II-555**
— Halong City, Vietnam II-566
— Turkish industry, input to Black Sea **I-298**, I-302
— urban and industrial, disposal of, Tasmania II-655
— *see also* chemical waste; hazardous waste; industrial waste; radioactive waste; solid waste; toxic waste; urban waste
wastewater
— direct discharge into coastal waters
   — Chinese Yellow sea II-494–5
   — West Guangdong coast II-555
— domestic and industrial, New York Bight I-323
— El Salvador

— domestic, untreated, discharged to marine environment I-555–6
— industrial discharges I-554–5
— entering Amursky Bay II-483
— Hong Kong, arrives in coastal waters II-538
— industrial, Chile I-712–13
— municipal, Chile I-713
　— final discharge to the sea I-713
　— submarine disposal solution I-715
— problems of discharge to sea, Marshall Islands II-783
— released to the Yellow Sea, Korea II-495
— reuse of, Oman and UAE II-29
— Tasmania, polluting II-657
— treatment seen as a priority, Mauritius II-263, II-264
— urban, causing contamination, southern Portugal I-179
— used for irrigation, Curaçao I-606
wastewater discharge
— major issue in Noumea, New Caledonia II-733
— Peru I-695
— to sea, Gulf of Aden II-58
Wastewater Management for Coastal Cities (C. Gunnerson and J. French) I-331
wastewater pollution, Arabian Gulf II-10
wastewater treatment
— Chesapeake Bay
　— biological nutrient removal I-347
　— improving I-347
— Norway I-27
water
— clean, needed for aquaculture III-170
— Colombian Pacific, from artisanal wells I-680
— potable, demand for, Gulf coast, USA I-448
— resources in Somalia II-79
water bodies and their circulations, around Puerto Rico I-579–81
water clarity/transparency
— Andaman Sea II-302
— decreasing, Baltic Sea I-130
water column habitats, Hawaiian Islands II-796
water column stability, Gulf of Alaska III-181
water masses
— Bien Dong Sea II-563–4, *II-564*
　— transformation and spreading of II-564
— Chilean coast I-701–2
— Gulf of Mexico *II-470*
water pollution
— local, Turkish Black Sea coast I-302
— Pearl River delta II-554–5
— the Philippines II-416–17
— Sea of Japan II-477
— sugar and rum industry, Jamaica I-569–70
— threatens Mai Po marshes RAMSAR site II-540
water quality
— American Samoa, a concern II-770
— and aquaculture III-172
— declining, Australia's inland waterways and lakes II-585
— degraded, areas of Victoria coast, Australia II-669
— Great Barrier Reef Marine Park II-626
— impaired
　— Adriatic coasts I-278
　— northern Gulf of Mexico I-449, I-449–50, I-450
— and industry, Nicaraguan Caribbean coast I-525
— Jakarta Bay and Pulau Seribu, influence of Java on II-397
— Jamaican beaches I-569
— Jiulongjiang estuary II-518
— Malacca Strait II-340–1
— marine, often exceeding standards, east coast, Peninsular Malaysia II-351–2, **II-351**, **II-352**
— New York Bight I-328–9
— off Carolinas coast I-353
— Papeete Harbour, Tahiti II-817
— parameters, optical sensing III-295–8

— bottom depth and reflectance III-297
— poor
　— central New South Wales metropolitan areas II-642
　— creeks and groundwater, some areas of French Polynesia II-822
　— northern New South Wales rivers II-637
— some improvement, Cubatao, southern Brazil I-743
— South China Sea **II-350**
— Southern Californian shoreline I-399
— Vietnam II-565
— west coast of Taiwan II-509–10
— West Guangdong coast II-555
— Western Australia II-699
　— declining II-700–1
water quality parameters, optical sensing
— chlorophyll pigment, coloured dissolved organic matter and total suspended matter III-297–8
　— retrieval of chlorophyll content from spectral radiance III-297
— secchi disk depth and diffusive attenuation coefficient III-296–7
water residence time, lagoonal waters, French Polynesia II-816
water scarcity, and dam building, side effects II-306
water surfaces, dry deposition to III-201–2
water temperature, Baltic Proper, seasonal variations I-102–3
water transit time, English Channel I-67
water weeds, invading mangrove canals, Côte d'Ivoire I-818
water-column nitrate inhibition I-358
water-leaving radiance, satellite data of, measurement of diffuse attenuation coefficient III-296
water-purification technology, Tyrrhenian coasts, Arno valley I-280
watershed management planning II-496
watershed management units, Jamaica I-565
wave climate
— deep wave heights I-578
— deep-water, off Fiji II-758
— Puerto Rico
　— generated by Trade Winds I-577–8
　— swell waves I-578
wave energy III-311–15
— current and future prospects III-315
— extraction of from waves III-311–12
— and persistence of oil III-269
— Puerto Rico I-578
— the resource III-313–15
wave energy programme, Chennai, India II-279
wave exposure, a key role in Belize atolls I-505
wave height, increase in I-46, I-68
wave intensity, and community structure I-22
wave power III-304
wave (swell) surges
— Marshall Islands II-776
— post-hurricane **I-519**
wave-power devices
— absorber mode III-313, *III-314*
— attenuator mode III-314
— design criteria III-313
— enclosed water column devices III-313
— flexible membrane devices III-313
— relative motion devices III-313
— tethered buoyant structures III-313
waves
— energy in is kinetic III-311, III-312
— from Antarctic winter storms, Hawaii II-795
— generation of at sea III-311–12, *III-312*
— Great Australian Bight, west coast swell environment II-675–6
— growth of through differential pressure distribution III-311, *III-312*
— Gulf of Guinea I-778
— interception of by an energy converter 312
— Levantine basin I-256
— modified by atolls, Belize I-503
— permanent surf, Côte d'Ivoire I-808
— the Philippines, generated by North Pacific storms II-407
— Significant Height III-311

— wind-induced, east coast, Peninsular Malaysia II-347
weather effects, the Comoros II-245
West African flyway, Gulf of Guinea a part of I-780
West African manatee I-780
West Bengal
— felling of mangrove forests II-151
— structure of Coastal Regulation Zone (CRZ) II-157–8, **II-158**
West Florida Shelf *I-436*
— seagrass beds I-438
West Greenland
— changes in marine climate I-7
— decline in marine mammal and seabird stocks I-8–9
— lower level of POPs than East Greenland I-13
West Greenland Current I-7
West Guangdong Province coast II-551
— intertidal habitats II-552–3
— phytoplankton and red tide organisms II-553–4
West Iberian large marine ecosystem I-137
West Kamchatka Current II-466
West Kamchatka Shelf, productive fish area II-467
West Wind cold water mass II-676
West Wind Drift Current (WWDC) I-701
Western Atlantic Ocean Experiment (WATOX) I-224, I-229
Western Australian region II-691–704
— coastal erosion, landfill and effects from urban and industrial activities II-700–2
— geomorphology of the coast II-693–4
— Kimberley coast, ria system II-694
— major shallow water marine and coastal habitats II-695–6
— offshore systems II-696
— populations affecting the area II-696–7
— protective measures II-702
    — Western Australian Marine Parks and Reserves Authority II-702
— rural factors II-697–700
— seasonality, currents, natural environmental variables II-694–5
Western Central South Pacific Current II-619
Western Indian Ocean Marine Scientists Association (WIOMSA) II-95–6
Western Somali Basin II-67
wet deposition I-13, I-56, I-223, III-200
wetland forest, Cambodia II-574
wetlands
— Bangladeshi coast II-289
— drained, Tasmania II-655
— forested, Gulf Coast, USA I-440
— Gulf of Guinea, some protected areas I-793
— Jamaica
    — Black River Morass I-565
    — Negril Morass I-564
— loss of
    — American Samoa II-770
    — northern Gulf of Mexico I-445–6, I-448
    — to agriculture, Peru I-695
    — to urban development *II-718*
— Mai Po marshes, Hong Kong II-539–41
— and marshes, North Spanish coast I-139
— miniature, unique, the Azores I-207, *I-208*, I-215
    — destruction and creation of I-207–9, I-211
— and protected areas, Peru I-691–2
    — National Sanctuary of the Mangroves of Tumbes I-691
    — National Sanctuary of Mejia Lagoons I-692
— reclamation of, Sri Lanka II-183–4
— tropical Brazil I-727
— Victoria Province, Australia II-666
Whale Research Programme under Special Permit (JARPA), Japan III-82
Whale Sanctuaries III-84–5
— entire Southern Ocean a sanctuary III-84–5
— further suggestions III-85
— in part of Antarctic Pacific sector III-77, III-84

— proposed for Indian Ocean III-78
whale watching II-251, III-85
— Great Australian Bight II-683
— south east Queensland II-636
whales I-343, I-591, I-834, II-24, II-205, II-582
— in the Azores I-213–14
— baleen (rorqual) whales III-76
— bottle-nosed III-76, III-383
— Colombia I-680
— Great Australian Bight II-680
— Great Barrier Reef II-618–19
— individual, tags and radio tags III-80
— Marshall Islands II-780
— Mexican Pacific coast I-487, I-488
— migrations of II-116
    — Gulf of Guinea I-783–4
— Mozambique waters II-105
— in the Southern California Bight I-399
— Sri Lanka II-180
— *see also* individual types
whales and whaling III-73–88
— assumption about whales as food competitors not sound III-86
— in a changing world III-86–7
— competition III-85–6
— conservation III-75–7
    — Mørch's memorandum III-75
    — quotas controversial III-77
— historical perspective III-74–5
    — expansion of whaling fleets III-75
    — factory ships III-75
    — modern whaling III-74
    — sale of whaling technology to Japan III-74
— how many? what is happening? III-80–3
— just lookin' III-85
— nations withdrawing from whaling III-78
— 'the orderly development of the whaling industry' III-77–8
    — installation of on-ship freezers III-77
— sanctuary III-84–5
— 'scientific' whaling III-78
— sharing resources III-84
— subsistence, and indigenous rights III-83–4
— visual counting and acoustic listening III-79–80
— whales in a changing world III-86–7
whaling I-3, I-36, I-636
— regulation of III-341
whiting I-49
Wider Caribbean I-460–2
— coastal crises I-461–2, **I-461**
— coastline activity **I-462**
— lack of port reception facilities for garbage III-345
— pollution along heavily urbanized/industrialized coasts I-461
widgeon-grass, Carolinas I-358
wildfowl and waders (shorebirds) III-107, III-110
Wildlife Management Areas
— Gulf of Papua II-608
— Papua New Guinea II-441
Wildlife Reserve, Kure Atoll, Hawaiian Islands II-798
Wilkinson Basin I-309, *I-311*
wind power III-304
wind shear III-305
wind speed, varies with height over different roughnesses III-305, *III-305*
wind stress, and residual flows in the English Channel I-67
wind turbines
— lifetimes of III-306
— modified to operate in sea conditions III-305
— new technology for installation in deeper water III-311
— power output of III-304–5, *III-305*
wind waves I-190
winds
— Aegean, summer Etesians I-236

— affecting the Bahama I-418, *I-418*
  — hurricanes I-418–19
— Arabian Gulf, Shamal and Kaus II-4
— Azores I-203–4
— Coral, Solomon and Bismark Seas Region II-428, *II-428*
— Côte d'Ivoire
  — monsoon I-808
  — Northeast Trade/Harmattan I-808
— Gulf of Aden, controlled by monsoons II-49–50
— Gulf of Alaska I-378
— Gulf of Guinea I-776
— Levantine basin I-256
— Mexican Pacific coast I-486, I-487
— monsoonal
  — Gulf of Thailand II-300
  — may cause damage, Palk Bay II-167
— Northern Benguela I-823
— northern Gulf of Mexico I-439
— northwest Arabian Sea II-19
— prevailing, Red Sea, and extreme sea breezes II-39
— Puerto Rico I-578–9
— *see also* hurricanes; Trade Winds; tropical storms
Winyah Bay *I-352*, I-353
women, trained to restore degraded mangrove forest, Dar es Salaam II-89
World Commission on Environment and Development I-459
World Conference on Fisheries Management and Development III-161
World Heritage Convention, TCI signatories to I-592
World Trade Organization (WTO), hindrance to environmental protection III-159

xerophytic shrub vegetation I-170
Xiamen region, China II-513–33
— clean up of Yuan Dang Lagoon II-531
— coastal waters of II-515
— critical environmental problems II-521–4
— ecosystem changes II-524–5
— geography of II-515
— local agencies with marine waters mandates II-526
— major shallow water marine and coastal habitats II-517–18
— protective measures II-525–32
  — a functional marine zonation scheme II-529–30
  — institutional arrangements for marine environment management II-526
  — land–sea integration II-531–2
  — laws and regulations II-525–6, II-529
  — major management problems II-526–8
  — Marine Management Co-ordination Committee (MMCC) II-528–9
  — marine pollution management II-532
  — monitoring, surveillance and emergency preparedness II-531
  — preparedness and response systems II-532
  — present limitations and recommendations II-531
— resource exploitation, utilisation and conflicts II-518–21
— seasonality, currents, natural environmental variables II-515–17
— a Special Economic Zone II-519
— towards integrated coastal management II-528
  — new management issues II-528–9
  — strategic management plan (SMP) and its implementation II-528

Yellow River, modern source of sediment to the Yellow Sea II-493
Yellow Sea *II-474*, II-487–98
— coastal erosion and landfill II-493–4
— main rivers entering II-489

— offshore systems II-490–1
— physical parameters and environmental variables II-489–90
— population II-491–2
— protection and conservation measures II-495–7
  — current conservation and marine protection measures II-496–7
  — factors inhibiting conservation II-495–6
— rural factors II-492–3
— shallow water habitats II-490
— shift away from demersal fish III-128
— urban and industrial activities II-494–5
Yellow Sea Large Marine Ecosystem (YSLME) programme II-496–7
Yellow Sea Warm Current II-489
Yellowfin fishery II-56
Yemen *II-18*, *II-36*, II-43, II-49
— beaches and mudflats important for seabirds and waders II-54
— dense and flourishing coastal algal community II-51
— fishing subsidies II-43
— limestone cliffs II-19
— new licensing round for oil and gas exploration II-59
— population II-25
— seasonal rainfall on southern mountains II-50

Zagros Mountains *II-18*, II-19
zährte, in southern Baltic rivers I-111
Zanzibar *II-84*, II-88
— problems of tourism expansion II-92
Zeehan Current II-649, *II-649*, II-650, *II-674*
Zhejiang—Fujian Coastal Current II-489, II-516
zinc pollution, Greenland I-12
zinc tolerance I-75
zinc (Zn) I-162
— Azorean amphipods I-213
zoogeographic provinces, South African shore II-135
zooplankton
— Antarctic I-707
— Bay of Bengal II-275–7
— Black Sea I-290–1
— Carolinas
  — estuarine I-359
  — marine I-360
— English Channel I-70
— Guinea upwellings, Senegal—Mauritanian, cold waters from I-800
— Gulf of Alaska I-379–80
— Gulf of Guinea I-779, I-783
— Irish Sea I-89–90
— Malacca Straits II-312
— the Maldives II-202
— North Spanish coast I-138, I-141
— offshore, Côte d'Ivoire I-814–15
— Peru, affected by El Niño I-692–3
— Southern Gulf of Mexico I-474
— southwestern Africa I-830, I-832
— Tagus estuary I-157
— Tyrrhenian Sea I-274
— Vietnam II-564
— West Guangdong coast II-554
zooplankton community, Deltaic Sundarbans II-150
Zostera I-70
— Amursky Bay, degradation of II-481
— *Z. marina*
  — propagation of from seed III-4–5
  — recovery of I-70
  — and the Wadden Sea III-6–7